HANDBOOK OF
BIOMEDICAL OPTICS

HANDBOOK OF
BIOMEDICAL OPTICS

EDITED BY

David A. Boas
Constantinos Pitris
Nimmi Ramanujam

CRC Press
Taylor & Francis Group
Boca Raton   London   New York

CRC Press is an imprint of the
Taylor & Francis Group, an **informa** business

CRC Press
Taylor & Francis Group
6000 Broken Sound Parkway NW, Suite 300
Boca Raton, FL 33487-2742

First issued in paperback 2020

© 2011 by Taylor & Francis Group, LLC
CRC Press is an imprint of Taylor & Francis Group, an Informa business

No claim to original U.S. Government works

ISBN-13: 978-0-367-57694-3 (pbk)
ISBN-13: 978-1-4200-9036-9 (hbk)

Library of Congress Cataloging-in-Publication Data

Handbook of biomedical optics / edited by David A. Boas, Constantinos Pitris, Nimmi Ramanujam.
    p. ; cm.
    Includes bibliographical references and index.
    ISBN 978-1-4200-9036-9 (hardcover : alk. paper)
    1. Diagnostic imaging--Digital techniques--Handbooks, manuals, etc. 2. Imaging systems in medicine--Handbooks, manuals, etc. I. Boas, David A. II. Pitris, Constantinos. III. Ramanujam, Nimmi.
    [DNLM: 1. Diagnostic Imaging--methods. 2. Laser Therapy, Low-Level--methods. 3. Optics and Photonics--methods. 4. Photochemistry--methods. WN 180]

RC78.7.D53.H353 2011
616.07'54--dc22
                                                                                    2010045399

Visit the Taylor & Francis Web site at
http://www.taylorandfrancis.com

and the CRC Press Web site at
http://www.crcpress.com

This book is dedicated to Professor Britton Chance (1913–2010), a pioneer in biophotonics and a highly respected leader since the inception of this rapidly developing field of the biomedical sciences. Through his passion for research and education, he enriched the lives of generations of scientists. His legacy will live on through the numerous scientists who trained under his mentorship. This book commemorates his relentless spirit of innovation and discovery.

Contents

Preface...xi

Editors...xiii

Contributors...xv

PART I   Background

1   Geometrical Optics 3
    Ting-Chung Poon

2   Diffraction Optics 11
    Colin Sheppard

3   Optics: Basic Physics 33
    Raghuveer Parthasarathy

4   Light Sources, Detectors, and Irradiation Guidelines 49
    Carlo Amadeo Alonzo, Malte C. Gather, Jeon Woong Kang, Giuliano Scarcelli, and Seok-Hyun Yun

5   Tissue Optical Properties...67
    Alexey N. Bashkatov, Elina A. Genina, and Valery V. Tuchin

PART II   Spectroscopy and Spectral Imaging

6   Reflectance Spectroscopy..103
    Sasha McGee, Jelena Mirkovic, and Michael Feld

7   Multi/Hyper-Spectral Imaging...131
    Costas Balas, Christos Pappas, and George Epitropou

8   Light Scattering Spectroscopy 165
    Le Qiu, Irving Itzkan, and Lev T. Perelman

9   Broadband Diffuse Optical Spectroscopic Imaging..181
    Bruce J. Tromberg, Albert E. Cerussi, So-Hyun Chung, Wendy Tanamai, and Amanda Durkin

10 Near-Infrared Diffuse Correlation Spectroscopy for Assessment of Tissue Blood Flow 195
    Guoqiang Yu, Turgut Durduran, Chao Zhou, Ran Cheng, and Arjun G. Yodh

11 Fluorescence Spectroscopy 217
    Darren Roblyer, Richard A. Schwarz, and Rebecca Rae Richards-Kortum

12 Raman, SERS, and FTIR Spectroscopy...233
    Andrew J. Berger

PART III   Tomographic Imaging

13 Optical Coherence Tomography: Introduction and Theory..255
Yu Chen, Evgenia Bousi, Constantinos Pitris, and James G. Fujimoto

14 Functional Optical Coherence Tomography in Preclinical Models.................................281
Melissa C. Skala, Yuankai K. Tao, Anjul M. Davis, and Joseph A. Izatt

15 Optical Coherence Tomography: Clinical Applications 303
Brian D. Goldberg, Melissa J. Suter, Guillermo J. Tearney, and Brett E. Bouma

16 Forward Models of Light Transport in Biological Tissue 319
Andreas H. Hielscher, Hyun Keol Kim, and Alexander D. Klose

17 Inverse Models of Light Transport 337
Simon R. Arridge, Martin Schweiger, and John C. Schotland

18 Laminar Optical Tomography 359
Sean A. Burgess and Elizabeth M. C. Hillman

19 Diffuse Optical Tomography Using CW and Frequency Domain Imaging Systems......373
Subhadra Srinivasan, Scott C. Davis, and Colin M. Carpenter

20 Diffuse Optical Tomography: Time Domain 395
Juliette Selb and Adam Gibson

21 Photoacoustic Tomography and Ultrasound-Modulated Optical Tomography...............419
Changhui Li, Chulhong Kim, and Lihong V. Wang

22 Optical and Opto-Acoustic Molecular Tomography..443
Vasilis Ntziachristos

PART IV   Microscopic Imaging

23 Assessing Microscopic Structural Features Using Fourier-Domain Low
Coherence Interferometry..463
Robert N. Graf, Francisco E. Robles, and Adam Wax

24 Phase Imaging Microscopy: Beyond Dark-Field, Phase Contrast, and Differential
Interference Contrast Microscopy..483
Chrysanthe Preza, Sharon V. King, Nicoleta M. Dragomir, and Carol J. Cogswell

25 Confocal Microscopy 517
William C. Warger, II, Charles A. DiMarzio, and Milind Rajadhyaksha

26 Fluorescence Microscopy with Structured Excitation Illumination 543
*Alexander Brunner, Gerrit Best, Paul Lemmer, Roman Amberger, Thomas Ach, Stefan Dithmar,
Rainer Heintzmann, and Christoph Cremer*

27 Nonlinear Optical Microscopy for Biology and Medicine...561
Daekeun Kim, Heejin Choi, Jae Won Cha, and Peter T. C. So

28 Fluorescence Lifetime Imaging: Microscopy, Endoscopy, and Tomography 589
*James McGinty, Clifford Talbot, Dylan Owen, David Grant, Sunil Kumar, Neil Galletly, Bebhinn Treanor,
Gordon Kennedy, Peter M. P. Lanigan, Ian Munro, Daniel S. Elson, Anthony Magee, Dan Davis, Gordon Stamp,
Mark Neil, Christopher Dunsby, and Paul M. W. French*

29 Application of Digital Holographic Microscopy in Biomedicine..617
Christian Depeursinge, Pierre Marquet, and Nicolas Pavillon

30 Polarized Light Imaging of Biological Tissues...649
Steven L. Jacques

PART V   Molecular Probe Development

31 Molecular Reporter Systems for Optical Imaging 673
Walter J. Akers and Samuel Achilefu

32 Nanoparticles for Targeted Therapeutics and Diagnostics 697
Timothy Larson, Kort Travis, Pratixa Joshi, and Konstantin Sokolov

33 Plasmonic Nanoprobes for Biomolecular Diagnostics of DNA Targets............................723
Tuan Vo-Dinh and Hsin-Neng Wang

PART VI   Phototherapy

34 Photodynamic Therapy...733
Jarod C. Finlay, Keith Cengel, Theresa M. Busch, and Timothy C. Zhu

35 Low Level Laser and Light Therapy...751
Ying-Ying Huang, Aaron C.-H. Chen, and Michael R. Hamblin

Index..771

Preface

There has been tremendous progress in medicine and biology over the past few decades. During this time, we have witnessed previously unimaginable advances in the understanding of human biology and physiology. One notable example is the human genome project, which has opened up new insights into disease diagnosis and therapy. Science and engineering have been instrumental in harnessing this new knowledge into clinically practical solutions. Within the field of medical imaging, the use of light plays an increasingly important role to reveal functional, structural, and molecular information nondestructively; also, interest in biomedical optics is growing rapidly. The use of light in medicine and biology offers great promise to deliver non- or minimally invasive diagnostics and targeted, customizable therapeutics more efficiently and safely. We are still far from the holy grail of "optical biopsies" and "optical therapeutics," but the strides we are making in biomedical optics foretell a future that may well deliver on the promise.

Biomedical optics is a broad discipline that covers the use of light in medicine and biology. It includes the use of light-based diagnostic methodologies as well as light-mediated therapeutics. Light can be used to both image microstructure and attain biochemical information such as the oxygenation of blood or the redox state of mitochondria. Optical imaging methodologies can cover from the microscopic to the macroscopic levels, with techniques including various forms of microscopy on the one hand and diffuse optical imaging on the other. In addition, diagnostically useful biochemical and functional information can be collected using spectroscopic techniques or molecular imaging. Laser surgery is now standard practice in certain medical disciplines, and photodynamic therapy offers treatment alternatives. More recently, tissue engineering and nanotechnology have entered the realm of biomedical optics, offering novel and potentially powerful possibilities both in the formation (e.g., with two-photon techniques) as well as the monitoring of tissue constructs.

The breadth and depth of the knowledge in biomedical optics has been expanding continuously and exponentially. We receive this progress with excitement; at the same time, it exemplifies the daunting amount of multidisciplinary information that is being amassed and the lack of a single source to serve as a reference and teaching tool for scientists in related fields. With this handbook, we aim to provide a comprehensive and detailed source of knowledge to serve as a research and teaching reference for both graduate and undergraduate students.

This handbook is organized into six parts: Background (Part I), Spectroscopy and Spectral Imaging (Part II), Tomographic Imaging (Part III), Microscopic Imaging (Part IV), Molecular Probe Development (Part V), and Phototherapy (Part VI). Part I provides introductory material on optics as well as the optical properties of tissue. Part II describes the various forms of spectroscopy and its application in medicine and biology, including methods that exploit intrinsic absorption and scattering contrast, dynamic contrast, as well as fluorescence and Raman contrast mechanisms. Tomographic imaging enables the formation of three-dimensional images despite the strong blurring of the highly scattered light within the tissue. Part III provides extensive coverage of tomography from the microscopic with optical coherence tomography to the macroscopic with diffuse optical tomography and photoacoustic tomography. Part IV discusses both conventional and cutting-edge technologies and their translation to biomedical applications in the basic sciences and clinical studies. Molecular imaging holds great promise for providing exquisite sensitivity to disease-specific markers. Its success is dependent on the development of exogenous agents that are detailed in Part V. Finally, Part VI provides examples of how light is being used to treat disease and injury.

We are indebted to the hard work of all of the authors to help create this comprehensive and cohesive handbook. We have learned a great deal about the process of compiling a handbook and thank Luna Han for all of her assistance and patience. We would also like to thank Christy Wanyo for her tremendous assistance at the end with handling the daunting tasks of formatting the chapters. Prof. Ramanujam would also like to thank Ms. Marlee Junker for managing many of the day to day tasks associated with compiling, reviewing, and handling portions of this handbook.

For MATLAB® and Simulink® product information, please contact:

The MathWorks, Inc.
3 Apple Hill Drive
Natick, MA, 01760-2098 USA
Tel: 508-647-7000
Fax: 508-647-7001
E-mail: info@mathworks.com
Web: www.mathworks.com

Editors

Dr. David A. Boas is an associate professor at the Harvard Medical School and associate physicist at Massachusetts General Hospital in Boston, Massachusetts. He received his bachelor's degree in physics from Rensselaer Polytechnic Institute, Troy, New York, in 1991 and his PhD in physics from the University of Pennsylvania, Philadelphia, Pennsylvania. His research interests include the following: photon migration in highly scattering media with an emphasis on diffuse optical tomography, clinical applications of diffuse optical tomography in brain and breast radiology, and fundamental studies of brain function and stroke using diffuse optical tomography and optical microscopy. Dr. Boas has been an associate editor of *Optics Express* and a guest editor of *Medical Physics* and the *Journal of Biomedical Optics*. He is a member of SPIE and the Optical Society of America (OSA) and has served as conference program chair for various OSA topical meetings.

Dr. Constantinos Pitris is an associate professor in the faculty of Electrical and Computer Engineering at the University of Cyprus, Nicosia, Cyprus. He completed his studies at the University of Texas at Austin (BS honors in electrical engineering, 1993, MS in electrical engineering, 1995), Massachusetts Institute of Technology (PhD in electrical and medical engineering, 2000), and Harvard Medical School (MD magna cum laude in medicine, 2002). He has worked as a research assistant at the University of Texas and Massachusetts Institute of Technology, and as a postdoctoral associate at the Wellman Laboratories of Photomedicine of the Massachusetts General Hospital and Harvard Medical School. His main research interests cover the areas of optics and biomedical imaging. The goal of this research is the introduction of new technologies in clinical applications for the improvement of diagnostic and therapeutic options. He is an active member of the OSA and a reviewer for *Optics Letters*, *Applied Optics*, and *Biomedical Optics*.

Dr. Nimmi Ramanujam is professor of biomedical engineering at Duke University. She received her PhD in biomedical engineering from the University of Texas, Austin, Texas, in 1995 and trained as an NIH postdoctoral fellow at the University of Pennsylvania from 1996 to 1999. Prior to her tenure at Duke, she was an assistant professor in the Department of Biophysics and Biochemistry at the University of Pennsylvania from 1999 to 2000 and in the Department of Biomedical Engineering at the University of Wisconsin, Madison, Wisconsin, from 2000 to 2005. Dr. Ramanujam's interests in the field of biophotonics are centered on research and technology development for applications to cancer. She is developing novel quantitative optical sensing and imaging tools for applications to breast, cervical, and head and neck cancers. She has been leading a multidisciplinary effort to translate these technologies into cancer patients. Dr. Ramanujam is a fellow of the Optical Society of America and is a member of the Department of Defense (DOD) Breast Cancer Research Program (BCRP) Integration Panel (IP). She has received several awards for her work in cancer research and technology development, including the Stansell Distinguished Research Award from the Pratt School of Engineering at Duke University, Era of Hope Scholar awards from the DOD, a Global Indus Technovator award from MIT, and the MIT TR100 innovator award.

Contributors

Thomas Ach
Department of Ophthalmology
University Hospital Heidelberg
Heidelberg, Germany

Samuel Achilefu
Department of Radiology
Washington University School of
Medicine
St. Louis, Missouri

Walter J. Akers
Department of Radiology
Washington University School of
Medicine
St. Louis, Missouri

Carlo Amadeo Alonzo
Wellman Center for Photomedicine
Massachusetts General Hospital
Boston, Massachusetts

Roman Amberger
Kirchhoff-Institute for Physics
Heidelberg University
Heidelberg, Germany

Simon R. Arridge
Department of Computer Science
University College London
London, United Kingdom

Costas Balas
Department of Electronic and Computer
Engineering
Technical University of Crete
Chania, Greece

Alexey N. Bashkatov
Institute of Optics and Biophotonics
Saratov State University
Saratov, Russia

Andrew J. Berger
The Institute of Optics
University of Rochester
Rochester, New York

Gerrit Best
Applied Optics and Information
Processing
Kirchhoff-Institute for Physics
Heidelberg University
Heidelberg, Germany

Brett E. Bouma
Wellman Center for Photomedicine
Massachusetts General Hospital
Boston, Massachusetts

Evgenia Bousi
Department of Electrical and Computer
Engineering
University of Cyprus
Nicosia, Cyprus

Alexander Brunner
Applied Optics and Information
Processing
Kirchhoff-Institute for Physics
Heidelberg University
Heidelberg, Germany

Sean A. Burgess
Department of Biomedical Engineering
Columbia University
New York, New York

Theresa M. Busch
Division of Medical Physics
University of Pennsylvania School of
Medicine
Philadelphia, Pennsylvania

Colin M. Carpenter
Radiation Oncology
Stanford University School of Medicine
Stanford, California

Keith Cengel
Division of Medical Physics
University of Pennsylvania School of
Medicine
Philadelphia, Pennsylvania

Albert E. Cerussi
Beckman Laser Institute and Medical
Clinic
University of California, Irvine
Irvine, California

Jae Won Cha
Department of Mechanical Engineering
Massachusetts Institute of Technology
Cambridge, Massachusetts

Aaron C.-H. Chen
Wellman Center for Photomedicine
Massachusetts General Hospital
and
Graduate Medical Sciences
Boston University School of Medicine
Boston, Massachusetts

Yu Chen
Fischell Department of Bioengineering
University of Maryland
College Park, Maryland

Ran Cheng
Center for Biomedical Engineering
University of Kentucky
Lexington, Kentucky

Heejin Choi
Department of Mechanical Engineering
Massachusetts Institute of Technology
Cambridge, Massachusetts

So-Hyun Chung
Beckman Laser Institute and Medical
    Clinic
University of California, Irvine
Irvine, California

Carol J. Cogswell
Department of Electrical, Computer, and
    Energy Engineering
University of Colorado
Boulder, Colorado

Christoph Cremer
Institute for Pharmacy and Molecular
    Biotechnology
Heidelberg University
Heidelberg, Germany

and

The Jackson Laboratory
Bar Harbor, Maine

and

Institute for Molecular Biophysics
University of Maine
Bar Harbor, Maine

Anjul M. Davis
Department of Biomedical Engineering
Duke University
Durham, North Carolina

Dan Davis
Department of Life Sciences
Imperial College London
London, United Kingdom

Scott C. Davis
Thayer School of Engineering
Dartmouth University
Hanover, New Hampshire

Christian Depeursinge
Advanced Photonics Laboratory
Federal Polytechnic School of Lausanne
Lausanne, Switzerland

Charles A. DiMarzio
Department of Electrical and Computer
    Engineering
Department of Mechanical and
    Industrial Engineering
Northeastern University
Boston, Massachusetts

Stefan Dithmar
Department of Ophthalmology
University Hospital Heidelberg
Heidelberg, Germany

Nicoleta M. Dragomir
Australian Institute of Physics
Victoria University
Melbourne, Victoria, Australia

Christopher Dunsby
Department of Physics
Imperial College London
London, United Kingdom

Turgut Durduran
Institute of Photonic Sciences
Barcelona, Spain

Amanda Durkin
Beckman Laser Institute and Medical
    Clinic
University of California, Irvine
Irvine, California

Daniel S. Elson
Department of Surgery and Cancer
Imperial College London
London, United Kingdom

George Epitropou
Department of Electronic and Computer
    Engineering
Technical University of Crete
Chania, Greece

Michael Feld
George R. Harrison Spectroscopy
    Laboratory
Massachusetts Institute of Technology
Cambridge, Massachusetts

Jarod C. Finlay
Division of Medical Physics
University of Pennsylvania School of
    Medicine
Philadelphia, Pennsylvania

Paul M.W. French
Wellman Center for Photomedicine
Massachusetts General Hospital
and
Department of Dermatology
Harvard Medical School
Boston, Massachusetts

James G. Fujimoto
Department of Electrical Engineering
    and Computer Science
Massachusetts Institute of Technology
Cambridge, Massachusetts

Neil Galletly
Department of Histopathology
Imperial College London
London, United Kingdom

Malte C. Gather
Wellman Center for Photomedicine
Massachusetts General Hospital
Boston, Massachusetts

Elina A. Genina
Institute of Optics and Biophotonics
Saratov State University
Saratov, Russia

Adam Gibson
Department of Medical Physics and
    Bioengineering
University College London
London, United Kingdom

Robert N. Graf
Department of Biomedical Engineering
Duke University
Durham, North Carolina

David Grant
Department of Physics
Imperial College London
London, United Kingdom

Michael R. Hamblin
Wellman Center for Photomedicine
Massachusetts General Hospital
and
Department of Dermatology
Harvard Medical School
Boston, Massachusetts
and
Harvard–MIT Division of Health
    Sciences and Technology
Cambridge, Massachusetts

Rainer Heintzmann
King's College London
London, United Kingdom
and
Institute of Photonic Technology
Jena, Germany
and
University of Jena
Jena, Germany

Andreas H. Hielscher
Departments of Biomedical Engineering,
    Radiology and Electrical Engineering
Columbia University
New York, New York

Elizabeth M.C. Hillman
Departments of Biomedical Engineering
    and Radiology
Columbia University
New York, New York

Ying-Ying Huang
Wellman Center for Photomedicine
Massachusetts General Hospital
and
Department of Dermatology
Harvard Medical School
Boston, Massachusetts
and
Aesthetic and Plastic Center of Guangxi
    Medical University
Nanning, People's Republic of China

Irving Itzkan
Beth Israel Deaconess Medical Center
Harvard University
Boston, Massachusetts

Joseph A. Izatt
Department of Biomedical Engineering
Duke University
Durham, North Carolina

Steven L. Jacques
Departments of Dermatology and
    Biomedical Engineering
Oregon Health & Science University
Portland, Oregon

Pratixa Joshi
Department of Biomedical Engineering
The University of Texas at Austin
Austin, Texas

Jeon Woong Kang
Wellman Center for Photomedicine
Massachusetts General Hospital
Boston, Massachusetts

Gordon Kennedy
Department of Physics
Imperial College London
London, United Kingdom

Chulhong Kim
Department of Biomedical Engineering
Washington University in St. Louis
St. Louis, Missouri

Daekeun Kim
Department of Biomedical Engineering
Dankook University
Yongin-si, Gyeonggi-do, Republic of Korea

Hyun Keol Kim
Department of Biomedical Engineering
Columbia University
New York, New York

Sharon V. King
Boulder Nonlinear Systems
Lafayette, Colorado

Alexander D. Klose
Department of Radiology
Columbia University
New York, New York

Sunil Kumar
Department of Physics
Imperial College London
London, United Kingdom

Peter M.P. Lanigan
Institute of Chemical Biology
Imperial College London
London, United Kingdom

Timothy Larson
Department of Biomedical Engineering
The University of Texas at Austin
Austin, Texas

Paul Lemmer
Kirchhoff-Institute for Physics
Heidelberg University
Heidelberg, Germany

Changhui Li
Department of Biomedical Engineering
Washington University in St. Louis
St. Louis, Missouri

Anthony Magee
Department of Life Sciences
Imperial College London
London, United Kingdom

Pierre Marquet
Neuropsychiatric Center
Lausanne University Hospital
Lausanne, Switzerland

Sasha McGee
George R. Harrison Spcetroscopy
    Laboratory
Massachusetts Institute of Technology
Cambridge, Massachusetts

James McGinty
Blackett Laboratory
Imperial College London
London, United Kingdom

Jelena Mirkovic
George R. Harrison Spectroscopy
    Laboratory
Massachusetts Institute of Technology
Cambridge, Massachusetts

Ian Munro
Department of Physics
Imperial College London
London, United Kingdom

Mark Neil
Department of Physics
Imperial College London
London, United Kingdom

Vasilis Ntziachristos
Institute for Biological and Medical
    Imaging
Technical University of Munich
and
Helmholtz Zentrum München
Munich, Germany

Dylan Owen
Department of Physics
Imperial College London
London, United Kingdom

Christos Pappas
Department of Electronic and Computer
    Engineering
Technical University of Crete
Chania, Greece

Raghuveer Parthasarathy
Department of Physics and Materials
    Science Institute
The University of Oregon
Eugene, Oregon

Nicolas Pavillon
Advanced Photonics Laboratory
Federal Polytechnic School of Lausanne
Lausanne, Switzerland

Lev T. Perelman
Beth Israel Deaconess Medical Center
Harvard University
Boston, Massachusetts

Constantinos Pitris
Department of Electrical and Computer
    Engineering
University of Cyprus
Nicosia, Cyprus

Ting-Chung Poon
Department of Electrical and Computer
    Engineering
Virginia Polytechnic Institute and State
    University
Blacksburg, Virginia

Chrysanthe Preza
Department of Electrical and Computer
    Engineering
The University of Memphis
Memphis, Tennessee

Le Qiu
Beth Israel Deaconess Medical Center
Harvard University
Boston, Massachusetts

Milind Rajadhyaksha
Dermatology Service
Memorial Sloan-Kettering Cancer Center
New York, New York

Rebecca Rae Richards-Kortum
Department of Bioengineering
Rice University
Houston, Texas

Francisco E. Robles
Medical Physics Department
Duke University
Durham, North Carolina

Darren Roblyer
The Beckman Laser Institute and
    Medical Clinic
University of California, Irvine
Irvine, California

Giuliano Scarcelli
Wellman Center for Photomedicine
Massachusetts General Hospital
Boston, Massachusetts

John C. Schotland
Department of Bioengineering
University of Pennsylvania
Philadelphia, Pennsylvania

Richard A. Schwarz
Department of Bioengineering
Rice University
Houston, Texas

Martin Schweiger
Department of Computer Science
University College London
London, United Kingdom

Juliette Selb
Martinos Center for Biomedical
    Imaging
Massachusetts General Hospital
Charlestown, Massachusetts

Colin Sheppard
Division of Bioengineering
National University of Singapore
Singapore

Melissa C. Skala
Department of Biomedical Engineering
Duke University
Durham, North Carolina

Peter T.C. So
Departments of Mechanical Engineering
    and Biological Engineering
Massachusetts Institute of Technology
Cambridge, Massachusetts

Konstantin Sokolov
Department of Biomedical Engineering
The University of Texas at Austin
Austin, Texas

and

Department of Imaging Physics
The University of Texas MD Anderson
    Cancer Center
Houston, Texas

Subhadra Srinivasan
Thayer School of Engineering
Dartmouth College
Hanover, New Hampshire

Gordon Stamp
Department of Histopathology
Imperial College London
London, United Kingdom

Melissa J. Suter
Wellman Center for Photomedicine
Massachusetts General Hospital
Boston, Massachusetts

Clifford Talbot
Department of Physics
Imperial College London
London, United Kingdom

Wendy Tanamai
Beckman Laser Institute and Medical
   Clinic
University of California, Irvine
Irvine, California

Yuankai K. Tao
Department of Biomedical Engineering
Duke University
Durham, North Carolina

Guillermo J. Tearney
Wellman Center for Photomedicine
Massachusetts General Hospital
Boston, Massachusetts

Kort Travis
Department of Physics
The University of Texas at Austin
Austin, Texas

Bebhinn Treanor
Department of Life Sciences
Imperial College London
London, United Kingdom

Bruce J. Tromberg
Beckman Laser Institute and Medical
   Clinic
University of California, Irvine
Irvine, California

Valery V. Tuchin
Institute of Optics and Biophotonics
Saratov State University
and
Institute of Precise Mechanics and
   Control
Russian Academy of Sciences
Saratov, Russia

Tuan Vo-Dinh
Departments of Biomedical Engineering
   and Chemistry
Duke University
Durham, North Carolina

Hsin-Neng Wang
Departments of Biomedical Engineering
   and Chemistry
Duke University
Durham, North Carolina

Lihong V. Wang
Department of Biomedical Engineering
Washington University in St. Louis
St. Louis, Missouri

William C. Warger, II
Wellman Center for Photomedicine
Massachusetts General Hospital
Boston, Massachusetts

Adam Wax
Department of Biomedical Engineering
   and Medical Physics Program
Duke University
Durham, North Carolina

Arjun G. Yodh
Department of Physics and Astronomy
University of Pennsylvania
Philadelphia, Pennsylvania

Guoqiang Yu
Center for Biomedical Engineering
University of Kentucky
Lexington, Kentucky

Seok-Hyun Yun
Wellman Center for Photomedicine
Massachusetts General Hospital
Boston, Massachusetts

Chao Zhou
Research Laboratory of Electronics
Massachusetts Institute of Technology
Cambridge, Massachusetts

Timothy C. Zhu
Division of Medical Physics
University of Pennsylvania School of
   Medicine
Philadelphia, Pennsylvania

I

Background

1 **Geometrical Optics** *Ting-Chung Poon* 3
Fermat's Principle • Matrix Method in Paraxial Optics • A Thin Converging Lens • Magnifying Lens • Compound Microscope • References

2 **Diffraction Optics** *Colin Sheppard* 11
Huygens' Principle • Fraunhofer and Fresnel Diffraction • Huygens' Diffraction Formula • Fraunhofer Diffraction • Fresnel Diffraction • Kirchhoff Diffraction Integral • Angular Spectrum of Plane Waves • Evanescent Waves • Diffraction by a Phase Screen • Thin Lens • Focus of a Lens • Circularly Symmetric Aperture • Effect of Defocus • Image Formation • Coherent Transfer Function • Spatial Filtering • Incoherent Imaging • References

3 **Optics: Basic Physics** *Raghuveer Parthasarathy* 33
Introduction • Electromagnetic Waves and Wave Motion • Diffraction • Refraction • Lenses • Reflection and Transmission (Fresnel's Equations) • Concluding Remarks • References

4 **Light Sources, Detectors, and Irradiation Guidelines** *Carlo Amadeo Alonzo, Malte C. Gather, Jeon Woong Kang, Giuliano Scarcelli, and Seok-Hyun Yun* 49
Introduction • Light Sources • Light Detectors • Irradiation Guidelines • References

5 **Tissue Optical Properties** *Alexey N. Bashkatov, Elina A. Genina, and Valery V. Tuchin* 67
Introduction • Basic Principles of Measurements of Tissue Optical Properties • Integrating Sphere Technique • Kubelka–Munk and Multi-Flux Approach • Inverse Adding-Doubling Method • Inverse Monte Carlo Method • Direct Measurement of the Scattering Phase Function • Optical Properties of Tissues • Summary • Acknowledgments • References

Geometrical Optics

1.1 Fermat's Principle 3
1.2 Matrix Method in Paraxial Optics 3
     Ray Transfer Matrix • Ray Tracing through Thin Converging Lens
1.3 A Thin Converging Lens 5
     Imaging • Numerical Aperture, Resolution, Depth of Focus, and Depth of Field
1.4 Magnifying Lens 8
1.5 Compound Microscope 9
References 10

Ting-Chung Poon
*Virginia Polytechnic Institute
and State University*

Geometrical optics is the study of light without diffraction or interference and is based on *Fermat's principle*. We treat light as particles of energy traveling through space. These particles follow trajectories that are called *rays*. Hence, geometrical optics is often called *ray optics*. Fermat's principle is a concise statement that contains all the physical laws, such as the *law of reflection* and the *law of refraction*, in geometrical optics (Poon and Kim 2006).

1.1 Fermat's Principle

Fermat's principle states that the path of a light ray follows is an extremum in comparison to nearby paths. The extremum may be a minimum, a maximum, or stationary with respect to variations in the ray path. However, the extremum is usually a minimum. For a simple example, as shown in Figure 1.1, the shortest distance (the minimum distance) between two points A and B is along a straight line (solid line) in a *homogeneous medium*, i.e., in a medium with a constant *refractive index*, instead of taking the nearby dotted line. Since the speed of light in a homogeneous medium is constant, the time it takes for the ray to traverse the solid line must be minimum. Hence Fermat's principle is often stated as a *principle of least time*. Under this context, the light ray would follow that path for which the time taken is minimum. For a more complicated example, we show the derivation of the well-known Snell's law of refraction. In Figure 1.2, θ_i and θ_t are the angles of incidence and transmission, respectively. The angles are measured from the normal NN′ to the interface MM′, which separated media 1 and 2, characterized by refractive indexes n_i and n_t, respectively. The total time taken to transit from point A to B is given by

$$t(z) = \frac{AO}{v_1} + \frac{OB}{v_2} = \frac{\sqrt{h_1^2 + z^2}}{v_1} + \frac{\sqrt{h_1^2 + (d-z)^2}}{v_2}, \quad (1.1)$$

where v_1 and v_2 are the light velocities in media 1 and 2, respectively. According to Fermat's principle, we are required to minimize the total time. In order to minimize $t(z)$, we set

$$\frac{dt(z)}{dz} = \frac{z}{v_1\sqrt{h_1^2 + z^2}} - \frac{d-z}{v_2\sqrt{h_1^2 + (d-z)^2}} = 0,$$

which gives

$$\frac{\sin\theta_i}{v_1} = \frac{\sin\theta_t}{v_2}.$$

Now, $v_1 = c/n_i$ and $v_2 = c/n_t$, where c is the speed of light particles in vacuum. Hence the above equation becomes

$$n_i \sin\theta_i = n_t \sin\theta_t, \quad (1.2)$$

which is called *Snell's law of refraction*.

1.2 Matrix Method in Paraxial Optics

1.2.1 Ray Transfer Matrix

We now consider how matrices may be used to describe ray propagation through optical systems comprising, for instance, a succession of lenses all centered on the same axis called the *optical axis*. We take the optical axis along the z-axis, which is the general direction in which the rays travel. We also consider those rays, called *paraxial rays*, that lie only in the x–z plane and that are close to the z-axis. To be precise, paraxial rays are rays with angles of incidence, reflection, and refraction at an interface, satisfying the small-angle approximation in that $\tan\theta \approx \sin\theta \approx \theta$ and $\cos\theta \approx 1$, where angle θ is measured in radians. *Paraxial*

FIGURE 1.1 In a homogeneous medium, light ray takes the shortest distance, a straight line (solid line), between two points.

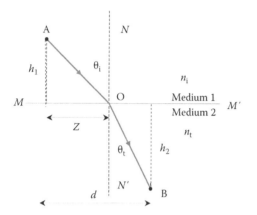

FIGURE 1.2 Law of refraction: incident (AO) and refracted (OB) rays.

optics deals with paraxial rays. Hence, in paraxial optics, Snell's law simplifies to

$$n_i \theta_i = n_t \theta_t. \tag{1.3}$$

We can now consider the propagation of a paraxial ray through an optical system shown in Figure 1.3, where a ray at a given x–z plane may be specified by its height x from the optical axis and by its angle θ or *slope*, which makes with the z-axis. The convention for the angle is anticlockwise positive measured from the z-axis. The height x of a point on a ray is taken positive if the point lies above the z-axis and negative if it is below the z-axis. The quantities (x, θ) represent the coordinates of the ray for a given z-plane. However, it is customary to replace the corresponding angle θ

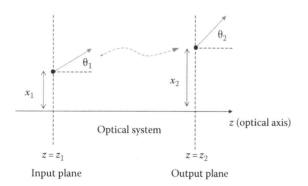

FIGURE 1.3 Input and output planes in an optical system.

by $v = n\theta$, where n is the refractive index at the z-constant plane. Therefore, as shown in Figure 1.3, the ray at $z = z_1$ passes through the input plane with input ray coordinates $(x_1, n_1\theta_1)$ or (x_1, v_1). After the ray has gone through the optical system, the output ray coordinates at $z = z_2$ are $(x_2, n_2\theta_2)$ or (x_2, v_2). In the matrix formalism of paraxial optics, we relate the input coordinates to the output coordinates by a 2×2 matrix as follows:

$$\begin{pmatrix} x_2 \\ v_2 \end{pmatrix} = \begin{pmatrix} A & B \\ C & D \end{pmatrix} \begin{pmatrix} x_1 \\ v_1 \end{pmatrix}. \tag{1.4}$$

The above ABCD matrix is called the *ray transfer matrix*, which can be made up of many matrices to account for the effects of ray passing through various optical elements such as lenses. To write out Equation 1.4, we have

$$x_2 = Ax_1 + Bv_1 \tag{1.5a}$$

and

$$v_2 = Cx_1 + Dv_1. \tag{1.5b}$$

As an example, let us formulate Snell's law in terms of the matrix formulism. From Equation 1.3, we can write it as

$$\begin{pmatrix} x_2 \\ v_2 \end{pmatrix} = \begin{pmatrix} 1 & 0 \\ 0 & 1 \end{pmatrix} \begin{pmatrix} x_1 \\ v_1 \end{pmatrix}, \tag{1.6}$$

where $v_2 = n_2\theta_t$ and $v_1 = n_1\theta_i$. Hence the ABCD matrix for Snell's law is $\begin{pmatrix} 1 & 0 \\ 0 & 1 \end{pmatrix}$, and Figure 1.4 summarizes the matrix for mulism for Snell's law. Note that the input and output planes in this case are the same plane at $z = z_1 = z_2$ and $x_1 = x_2$ as the heights of the input and output rays are the same. Other useful matrices are summarized in Figure 1.5. In Figure 1.5a, we have the *translation matrix* **T**, which describes the ray undergoing a translation of distance d in a homogenous medium characterized by n, and the matrix equation is given as follows:

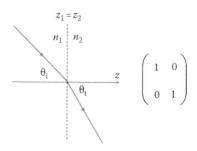

FIGURE 1.4 ABCD matrix for Snell's law.

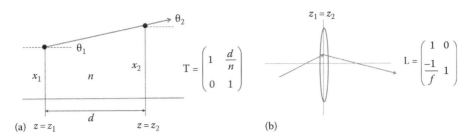

(a) $z = z_1$ $z = z_2$ (b)

FIGURE 1.5 (a) ABCD matrix for ray translation in homogeneous medium: translation matrix **T**. (b) ABCD matrix for ray diffraction by thin lens of focal length *f*: lens matrix *L*.

$$\begin{pmatrix} x_2 \\ v_2 \end{pmatrix} = \begin{pmatrix} 1 & \dfrac{d}{n} \\ 0 & 1 \end{pmatrix}\begin{pmatrix} x_1 \\ v_1 \end{pmatrix} = \mathbf{T}\begin{pmatrix} x_1 \\ v_1 \end{pmatrix}, \tag{1.7a}$$

where $v_2 = n\theta_2$ and $v_1 = n\theta_1$. Hence the translation matrix for ray propagating a distance *d* in a homogenous medium of *n* is

$$\mathbf{T} = \begin{pmatrix} 1 & \dfrac{d}{n} \\ 0 & 1 \end{pmatrix}. \tag{1.7b}$$

Note that when the ray is undergoing translation in a homogeneous medium, $\theta_1 = \theta_2$, and therefore $v_1 = v_2$. When a thin converging lens of focal length *f* is involved, the matrix equation is

$$\begin{pmatrix} x_2 \\ v_2 \end{pmatrix} = \begin{pmatrix} 1 & 0 \\ -\dfrac{1}{f} & 1 \end{pmatrix}\begin{pmatrix} x_1 \\ v_1 \end{pmatrix} = \mathbf{L}\begin{pmatrix} x_1 \\ v_1 \end{pmatrix}, \tag{1.8a}$$

where **L** is the *thin-lens matrix* given by

$$\mathbf{L} = \begin{pmatrix} 1 & 0 \\ -\dfrac{1}{f} & 1 \end{pmatrix} \tag{1.8b}$$

Note that by definition, a lens is thin when the thickness of it is assumed to be zero, and hence $x_1 = x_2$, i.e., the input and output planes have become the same plane or $z_1 = z_2$, as shown in Figure 1.5b.

1.2.2 Ray Tracing through Thin Converging Lens

1.2.2.1 Input Rays Traveling Parallel to the Optical Axis

From Equation 1.8a, we recognize that $x_1 = x_2$ as the heights of the input and output rays are the same for the thin lens as illustrated in Figure 1.5b. Now, according to Equation 1.8a, $v_2 = -(1/f)x_1 + v_1$. For $v_1 = 0$, i.e., the input rays are parallel to the

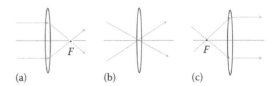

(a)   (b)   (c)

FIGURE 1.6 Ray tracing through a thin converging: (a) all parallel input rays converge to the back focal point *F*, (b) all input rays through the center of the lens pass undeviated, and (c) all input rays through the front focal point *F* give output rays parallel to the optics axis.

optical axis, $v_2 = -(1/f)x_1$. For positive x_1, $v_2 < 0$ as $f > 0$ for a converging lens. For negative x_1, $v_2 > 0$. Hence all input rays parallel to the optical axis converge behind the lens to the back focal point *F* (a distance of *f* away from the lens) of the lens as shown in Figure 1.6a. Note that for a thin lens, the front focal point is also a distance of *f* away from the lens.

1.2.2.2 Input Rays Traveling through the Center of the Lens

For input rays traveling through the center of the lens, their input ray coordinates are $(x_1, v_1) = (0, v_1)$. The output ray coordinates, according to Equation 1.8a, are $(x_2, v_2) = (0, v_1)$ because $v_2 = v_1$, Hence we see all rays traveling through the center of the lens will pass undeviated as shown in Figure 1.6b.

1.2.2.3 Input Rays Passing through the Front Focal Point of the Lens

For this case, the input ray coordinates are $(x_1, x_1/f)$, and, according to Equation 1.8a, the output ray coordinates are $(x_2 = x_1, 0)$, which states that all output rays will be parallel to the optical axis ($v_2 = 0$), as shown in Figure 1.6c.

1.3 A Thin Converging Lens

1.3.1 Imaging

In this section, we show an example of using the transfer matrices, namely, the translation matrix **T** and the lens matrix **L**, to analyze the problem of a single thin lens. Figure 1.7 shows an object OO′ located at a distance d_0 in front of a thin lens of focal length *f*. We first construct a *ray diagram* for the imaging system, and the knowledge obtained from the last section should help

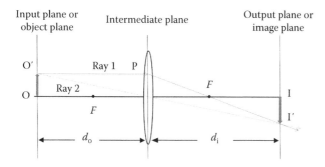

FIGURE 1.7 Imaging of a thin converging lens.

us to accomplish it. We send two rays from a point, say from O′, toward the lens. Ray 1 from O′ is incident parallel to the optical axis, and from part (a) from the last section, the input ray parallel to the optical axis converges behind the lens to the back focal point *F*. A second ray, i.e., ray 2 also from O′ may now be drawn through the center of the lens without bending—that is, the result from part (b) from the last section. The interception of the two rays on the other side of the lens will form an image point of O′. The image point of O′ is labeled at I′ in the diagram.

Now we investigate the imaging properties of the single thin lens using the matrix formalism. Let us consider the input plane, the immediate plane, and the output plane as shown in Figure 1.7. We also let (x_0, v_0), (x_p, v_p), and (x_i, v_i) represent the coordinates of the ray at O′, P (where P is on the immediate plane, and O′P defines the path where the parallel ray from O′ hitting the thin lens), and I′, respectively. We see that there are three matrices involved in the problem. From the input plane or the object plane to the immediate plane, it involves a translation matrix of distance d_0 (see Equation 1.7b, where we have assumed $n = 1$ for air):

$$\begin{pmatrix} x_p \\ v_p \end{pmatrix} = \begin{pmatrix} 1 & d_0 \\ 0 & 1 \end{pmatrix}\begin{pmatrix} x_0 \\ v_0 \end{pmatrix}, \qquad (1.9)$$

and then when the incident ray to the lens is diffracted from the intermediate plane, we have

$$\begin{pmatrix} x'_p \\ x'_p \end{pmatrix} = \begin{pmatrix} 1 & 0 \\ \frac{-1}{f} & 1 \end{pmatrix}\begin{pmatrix} x_p \\ v_p \end{pmatrix}, \qquad (1.10)$$

where (x_p, v_p) and (x'_p, v'_p) are the input ray and output ray coordinates due to the thin lens. Note that (x_p, v_p) and (x'_p, v'_p) are on the same plane—the intermediate plane as the lens is thin. Now finally, the ray exiting from the lens will translate for a distance d_i to reach the final output plane (or the image plane):

$$\begin{pmatrix} x_i \\ v_i \end{pmatrix} = \begin{pmatrix} 1 & d_i \\ 0 & 1 \end{pmatrix}\begin{pmatrix} x'_p \\ v'_p \end{pmatrix}. \qquad (1.11)$$

By substituting Equation 1.9 into Equation 1.10 and subsequently into Equation 1.11, we can relate the input coordinates (x_0, v_0) on the object plane to the output coordinates (x_i, v_i) on the image plane of the whole system as follows:

$$\begin{pmatrix} x_i \\ v_i \end{pmatrix} = \begin{pmatrix} 1 & d_i \\ 0 & 1 \end{pmatrix}\begin{pmatrix} 1 & 0 \\ \frac{-1}{f} & 1 \end{pmatrix}\begin{pmatrix} 1 & d_0 \\ 0 & 1 \end{pmatrix}\begin{pmatrix} x_0 \\ v_0 \end{pmatrix} = \mathbf{S}\begin{pmatrix} x_0 \\ v_0 \end{pmatrix}, \qquad (1.12a)$$

where

$$\mathbf{S} = \begin{pmatrix} 1 & d_i \\ 0 & 1 \end{pmatrix}\begin{pmatrix} 1 & 0 \\ \frac{-1}{f} & 1 \end{pmatrix}\begin{pmatrix} 1 & d_0 \\ 0 & 1 \end{pmatrix} = \mathbf{T_2 L T_1} \qquad (1.12b)$$

is called the *system matrix* of the entire imaging system. The overall system matrix **S** is expressed in terms of the product of three matrices $\mathbf{T_2 L T_1}$ written in order from right to left as the ray goes from left to right along the optical axis. Let **A** and **B** be the 2 × 2 matrices as follows:

$$\mathbf{A} = \begin{pmatrix} a & b \\ c & d \end{pmatrix} \quad \text{and} \quad \mathbf{B} = \begin{pmatrix} e & f \\ g & h \end{pmatrix}.$$

Then, the matrix product **AB** is

$$\mathbf{AB} = \begin{pmatrix} a & b \\ c & d \end{pmatrix}\begin{pmatrix} e & f \\ g & h \end{pmatrix} = \begin{pmatrix} ae + bg & af + bh \\ ce + dg & cf + dh \end{pmatrix}. \qquad (1.13)$$

According to the rule of matrix multiplication in Equation 1.13, Equation 1.12b can be simplified to

$$\mathbf{S} = \begin{pmatrix} 1 - d_i/f & d_0 + d_i - d_0 d_i/f \\ -1/f & 1 - d_i/f \end{pmatrix}. \qquad (1.14)$$

Hence Equation 1.12a becomes

$$\begin{pmatrix} x_i \\ v_i \end{pmatrix} = \begin{pmatrix} 1 - d_i/f & d_0 + d_i - d_0 d_i/f \\ -1/f & 1 - d_i/f \end{pmatrix}\begin{pmatrix} x_0 \\ v_0 \end{pmatrix} = \mathbf{S}\begin{pmatrix} x_0 \\ v_0 \end{pmatrix} = \begin{pmatrix} A & B \\ C & D \end{pmatrix}\begin{pmatrix} x_0 \\ v_0 \end{pmatrix}. \qquad (1.15)$$

To investigate the conditions for imaging, let us concentrate on the ABCD matrix of **S** in the above equation. For imaging, the B element of **S** must be zero, which leads to $x_i = Ax_0 + Bv_0 = Ax_0$. This means that all rays passing through the input plane at the same object point x_0 will pass through the same image point x_i in the output plane—a condition of imaging. In addition, $A = x_i/x_0$ is the *lateral magnification* of the imaging system. Now, in our case of thin-lens imaging, $B=0$ in Equation 1.15 leads to $d_0 + d_i - d_0 d_i/f = 0$, which gives the *thin-lens formula*:

$$\frac{1}{d_0} + \frac{1}{d_i} = \frac{1}{f}. \tag{1.16}$$

The sign convention is that the object distance d_0 is positive (negative) if the object is to the left (right) of the lens. If the image distance d_i is positive (negative), the image is to the right (left) of the lens, and it is real (virtual). In Figure 1.7, we have $d_0 > 0$, $d_i > 0$, and the image is therefore real, which means physically that light rays actually converge to the formed image.

Returning to Equation 1.15 with Equation 1.16, we have

$$\begin{pmatrix} x_i \\ v_i \end{pmatrix} = \begin{pmatrix} 1 - d_i/f & 0 \\ -1/f & 1 - d_i/f \end{pmatrix} \begin{pmatrix} x_0 \\ v_0 \end{pmatrix}, \tag{1.17}$$

which relates the input ray and output ray coordinates in the imaging system. The lateral magnification M of the imaging system, using Equations 1.16 and 1.17, is

$$M = A = \frac{x_i}{x_0} = \frac{1 - d_i}{f} = -\frac{d_i}{d_0}. \tag{1.18}$$

If $M > 0$, the image is erect and if $M < 0$, the image is inverted. As shown in Figure 1.7, we have inverted image as both d_i and d_0 are positive.

When imaging of a volume, we need to consider longitudinal magnification as well. *Longitudinal magnification* M_z is the ratio of an image displacement along the axial direction δd_i to the corresponding object displacement δd_0: $M_z = \delta d_i/\delta d_0$. Using Equation 1.16 and treating d_i and d_0 as variables, we can take the derivative of d_i with respect to d_0 to obtain

$$M_z = \frac{\delta d_i}{\delta d_0} = -M^2. \tag{1.19}$$

Equation 1.19 states that the longitudinal magnification is equal to the square of the lateral magnification. The minus sign in front of the equation means that the decrease in the distance of the object from the lens $|d_0|$ will result in the increase in the image distance $|d_i|$. The situation of a magnified volume is shown in Figure 1.8, where a cube volume (abcd plus the dimension into the paper) is imaged into a truncated pyramid with a–b imaged into a′–b′ and c–d imaged into c′–d′.

1.3.2 Numerical Aperture, Resolution, Depth of Focus, and Depth of Field

The *numerical aperture* NA of a lens is usually defined for an object or image located infinitely far away. It is a measure of

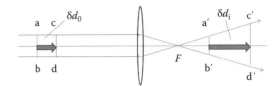

FIGURE 1.8 Longitudinal magnification: the image of a magnified volume is on the right side of the lens (side view).

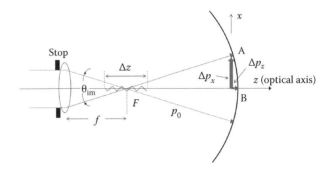

FIGURE 1.9 Uncertainty principle used to find resolution and depth of focus.

the light-gathering ability. Figure 1.9 shows an object located at infinity, which sends rays parallel to the lens. The angle θ_{im} used to define the NA on the image side is

$$NA_i = n_i \sin\left(\frac{\theta_{im}}{2}\right) \tag{1.20}$$

where n_i is refractive index in the image space. Let us now find the lateral resolution, Δx. Since we treat light as particles in geometrical optics, each particle can then be characterized by its momentum, p_0. Quantum mechanics relates the minimum uncertainty in a position of quantum, Δx, to the uncertainty of its momentum, Δp_x, according to the relationship

$$\Delta x \Delta p_x \geq h, \tag{1.21}$$

where

  h is *Planck's constant*

  Δp_x is the momentum difference between rays FB and FA along the x-direction, i.e., the transverse direction as shown in Figure 1.9

The momentum of the FB ray along the x-axis is zero, while the momentum of the FA ray along the x-axis is $p_0 \sin(\theta_{im}/2)$, where $p_0 = h/\lambda_0$ with λ_0 being the wavelength in the medium, i.e., in the image space. Hence Δp_x is $\Delta p_x = p_0 \sin(\theta_{im}/2)$. By substituting this into Equation 1.21, we have

$$\Delta x \geq \frac{h}{\Delta p_x} = \frac{h}{p_0 \sin(\theta_{im}/2)} = \frac{\lambda_0}{\sin(\theta_{im}/2)}. \tag{1.22}$$

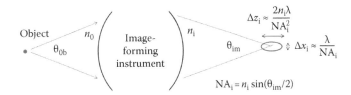

FIGURE 1.10 Image-forming instrument illustrating depth of focus and resolution in the image space.

Since the wavelength in the image space, λ_0 is equal to λ/n_i, where λ is the wavelength in air or in vacuum, Equation 1.22 becomes, using Equation 1.20,

$$\Delta x \geq \frac{\lambda}{n_i \sin(\theta_{im}/2)} = \frac{\lambda}{NA_i}. \qquad (1.23)$$

Similarly, we can calculate the *depth-of-focus*, Δz, using $\Delta z \Delta p_z \geq h$, where Δp_z is the momentum difference between rays FB and FA along the z-direction, as shown in Figure 1.9. Details of the derivation have been provided (Poon 2007), and we state the final result as follows:

$$\Delta z \approx \frac{2n_i\lambda}{NA_i^2}. \qquad (1.24)$$

To summarize the above results, Figure 1.10 shows an image-forming instrument with θ_{0b} and θ_{im} denoting the ray of maximum divergent angle and maximum convergent angle from the object side and the image side, respectively. If the resolutions in the object space are given by $\Delta x_0 \approx \lambda/NA_0$ and $\Delta z_0 \approx 2n_0\lambda/NA_0^2$, where $NA_0 = n_0 \sin(\theta_{0b}/2)$ is the NA and n_0 is the refractive index in the object space, and if the lateral magnification of the instrument is M, the resolutions on the image space are then $\Delta x_i \approx M\Delta x_0$ and $\Delta z_i \approx M^2\Delta z_0$. Take a 40×, $NA_0 \approx 0.6$ microscope objective as an example, we have $\Delta x_0 \approx 1\,\mu m$ for red light and Δz_0 in the object space is called *depth of field*, which is given by $\Delta z_0 \approx 2n_0\lambda/NA_0^2 \approx 3.3\,\mu m$ for $n_0 = 1$ in air. In the image space, the lateral resolution is $\Delta x_i \approx M\Delta x_0 \approx 40\,\mu m$, and the depth of focus is $\Delta z_i \approx M^2\Delta z_0 \approx 5.2\,mm$.

1.4 Magnifying Lens

The normal eye can form sharp images of objects as close to as about 250 mm away. The distance of 250 mm is known as the

near point (NP) or *least distance of distinct vision* of the eye. For the eye to view objects closer than the NP, a magnifying lens can be used. The magnifying lens is simply a converging lens. Figure 1.11a shows that the unaided eye forms an image of an object located at the NP, and Figure 1.11b shows that a magnifying lens forms a virtual image, where the object is well within the NP of the eye, and we assume that the eye is close to the lens. Therefore, the magnifying lens makes an erect and magnified virtual image of the object. From Equation 1.16, we can write

$$d_i = \frac{f}{d_0 - f} d_0, \qquad (1.25)$$

where

f is the focal length of the magnifying lens
d_i should be negative and greater in magnitude than d_0 as we recall $M = -d_i/d_0$

Note that, if $d_0 = f$, $d_i \to \infty$, the magnification M tends to be infinite. In practice, the paraxial approximation limits M to values about 10 for a single lens. In any case, the *magnifying power* or *angular magnification*, M_θ, is conventionally used for magnifying lenses or microscopes (Nelkon and Parker 1970):

$$M_\theta = \frac{\alpha}{\alpha_{NP}}, \qquad (1.26)$$

where α_{NP} and α have been defined in Figure 1.11a and b. Figure 1.11c shows the situation when $d_0 = f$, where the image of the object appears coming from infinity. The eye is most comfortable or relaxed (called *zero accommodation*) to see distant objects, whereas the eye must achieve *maximum accommodation* to see an image located at its NP (see Figure 1.11a and b when $d_i = -250\,mm$].

Now by definition $\alpha_{NP} = h_0/250$, where h_0 is the size of the object as shown in Figure 1.11a for an unaided eye. Also, $\alpha = h_0/d_0$ from Figure 1.11b. Hence according to Equation 1.26, we have

$$M_\theta = \frac{h_0/d_0}{h_0/250} = \frac{250}{d_0}, \qquad (1.27)$$

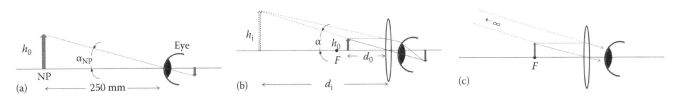

FIGURE 1.11 (a) Object at the NP of the eye, (b) object located within the focal length of the magnifying lens, and (c) object located at the front focal point F of the magnifying lens (eye relaxed as the eye receiving parallel rays).

where in Equation 1.27, d_0 is measured in millimeters. If the image is viewed at infinity, i.e., $d_0 = f$, the eye is relaxed, and we have

$$M_\theta = \frac{250}{f}, \tag{1.28a}$$

where the image is formed at infinity.

At the other extreme, if the image is viewed at the NP, we find $d_0 = f d_i / (d_i - f)$ from Equation 1.16. But $d_i = -250$, so we have $d_0 = -250f/(-250 - f)$. By substituting this quantity into Equation 1.27, we have

$$M_\theta = 1 + \frac{250}{f}, \tag{1.28b}$$

where the image formed at the near point of the eye.

This is the maximum value of the magnifying power that can be achieved with a single magnifying lens for the eye. In Equations 1.28a and b, $M_\theta > 0$ as $f > 0$ for a converging lens, which means that the image is always virtual and erect.

1.5 Compound Microscope

A compound microscope uses two converging lenses. The lens with short focal length f_0 facing the object is called the *objective*. The lens with focal length f_e in front of the eye is called the *eyepiece*. The objective makes a real, inverted, and magnified image of the object within the focal length of the eye-piece. The eye-piece then further magnifies the image from the objective by acting as a magnifying lens to eventually give a magnified, virtual image of the object. The situation is shown in Figure 1.12a. Let us now find the magnifying power of the compound microscope, M_θ^C, if the image is viewed at the NP of the eye. Again, using Equation 1.26, we have

$$M_\theta^C = \frac{\alpha_C}{\alpha_{NP}}, \tag{1.29}$$

where α_C is defined in Figure 1.12a. Let us further work on Equation 1.29. As usual, we assume that the eye is close to the eye-piece. The angle α_C subtended by the image of height h_2 is given by $\alpha_C = h_2/250$. With the unaided eye, the object of height h subtends an angle $\alpha_{NP} = h/250$. Therefore, we can rewrite Equation 1.29 as

$$M_\theta^C = \frac{\alpha_C}{\alpha_{NP}} = \frac{h_2/250}{h/250} = \frac{h_2}{h} = \frac{h_2/h_1}{h/h_1}. \tag{1.30}$$

The factor h_1/h is simply the lateral magnification of the objective, which is given by $M_0 = h_1/h = 1 - d_i/f_0$ (see Equation 1.18). As for h_2/h_1, we see that the angular magnification for the eye-piece by definition is $M_\theta^e = \alpha_C/\alpha_{NP} = (h_2/250)/(h_1/250) = h_2/h_1$, which is equal to $1 + (250/f_e)$ according to the results given by Equation 1.28b. Hence Equation 1.30 becomes

$$M_\theta^C = \frac{h_2/h_1}{h/h_1} = M_0 M_\theta^e = \left(1 - \frac{d_i}{f_0}\right)\left(1 + \frac{250}{f_e}\right) \tag{1.31}$$

for the image observed at the NP of the eye. Since $d_i > f_0$, M_θ^C is always negative (as $M_0 < 0$ and $M_\theta^e > 0$) to reflect that we observe a virtual and inverted final magnified image. From Equation 1.31, we recognize that the magnifying power of the compound microscope is conveniently expressed in terms of the product of the lateral magnification produced by the objective and the magnifying power of the eye-piece. Now if the final image is to be formed at infinity, the eye is unaccommodated or relaxed. In this case, we notice that the real image formed by the objective must be located at the focal point of the eyepiece. The situation is shown in Figure 1.12b. We, therefore, have

$$M_\theta^C = \frac{\alpha_\infty}{\alpha_{NP}} = \frac{h_1/f_e}{h/250} = \frac{h_1}{h}\frac{250}{f_e} = \left(1 - \frac{d_i}{f_0}\right)\frac{250}{f_e}. \tag{1.32}$$

Considering the objective only, the image distance must then be equal to $d_i = f_0 + L$, and the lateral magnification of the objective is subsequently given by

$$M_0 = 1 - \frac{d_i}{f_0} = \frac{f_0 - d_i}{f_0} = \frac{-L}{f_0}. \tag{1.33}$$

Incorporating this into Equation 1.32, the magnifying power of the compound microscope becomes

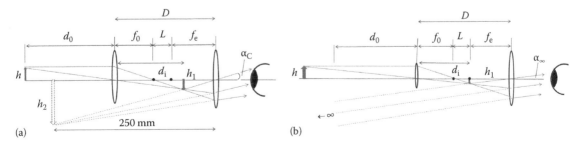

FIGURE 1.12 Compound microscope (a) image formed at the NP of the eye and (b) image formed at infinity (eye relaxed).

$$M_\theta = M_0 \frac{250}{f_e} = \frac{-L}{f_0} \frac{250}{f_e}. \qquad (1.34)$$

L is known as the *tube length* and in most US-made microscopes, the standard is that $L = 160$ mm. Therefore, the lateral magnification stated on the barrel of an objective usually assumes 160 mm tube length. Otherwise, L will be stated on the barrel. In practice, the tube length is large compared to either f_0 or f_e.

References

Nelkon, M. and Parker, P. 1970. *Advanced Level Physics*. London: Heinemann Educational Books Ltd.

Poon, T.-C. 2007. *Optical Scanning Holography with MATLAB*. New York: Springer.

Poon, T.-C. and Kim, T. 2006. *Engineering Optics with MATLAB*. Hackensack, NJ: World Scientific.

2

Diffraction Optics

2.1	Huygens' Principle	11
2.2	Fraunhofer and Fresnel Diffraction	11
2.3	Huygens' Diffraction Formula	12
2.4	Fraunhofer Diffraction	12
	Examples of Fraunhofer Diffraction—Single Slit • Rectangular Aperture • Circular Aperture • Annular Aperture	
2.5	Fresnel Diffraction	15
	Introduction • Circular Aperture • Rectangular Aperture • Single Slit • Half-Plane • Circular Obstruction	
2.6	Kirchhoff Diffraction Integral	18
2.7	Angular Spectrum of Plane Waves	19
2.8	Evanescent Waves	19
2.9	Diffraction by a Phase Screen	20
2.10	Thin Lens	21
2.11	Focus of a Lens	21
2.12	Circularly Symmetric Aperture	22
2.13	Effect of Defocus	22
2.14	Image Formation	23
	Special Cases	
2.15	Coherent Transfer Function	26
	A Grating Object • Square Wave Object	
2.16	Spatial Filtering	29
2.17	Incoherent Imaging	29
	Two-Point Object • Optical Transfer Function	
	References	31

Colin Sheppard
National University of Singapore

2.1 Huygens' Principle

According to Huygens, each point on a wave front serves as the source of a spherical secondary wavelet with the same frequency as the primary wave. The amplitude at any point is the superposition of these wavelets. Note that Huygens' principle considers diffraction as a summation of spherical waves, not as a summation of plane waves, as we will consider in Section 2.7. This theory gives a simple qualitative description of diffraction but needs to be adapted to give good agreement with more exact formulations that will be shown later in Section 2.6.

Consider the propagation of a plane wave. Each point in the wave front can be considered as a source of secondary waves (Figure 2.1). This describes the propagating wave correctly, but suggests the possibility that the wave can equally well propagate backward. This is one reason the model has to be modified to agree with exact theories.

2.2 Fraunhofer and Fresnel Diffraction

Consider an opaque screen illuminated with a plane wave. The light spreads as a result of diffraction. If the observation screen is far enough away from the aperture, the diffraction pattern does not change in structure, but merely changes in size, as the distance is further increased. This situation is called the Fraunhofer diffraction (Figure 2.2). Closer to the aperture the diffraction pattern does change with distance. This is called Fresnel diffraction. Calculation of Fresnel diffraction is based on an approximation, which eventually breaks down: closer to the aperture more advanced theories are required. We shall discuss these regions later. The Fraunhofer diffraction pattern is obtained at a very large distance from the aperture, but using a lens, an image of it can be formed at a finite distance. In this case, the regions either closer or further from the lens give Fresnel diffraction.

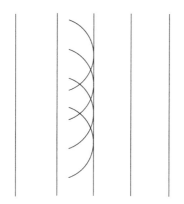

FIGURE 2.1   Huygens' principle for the propagation of a plane wave.

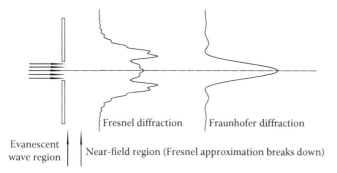

Fresnel diffraction   Fraunhofer diffraction

Evanescent
wave region   Near-field region (Fresnel approximation breaks down)

FIGURE 2.2   Regimes of diffraction.

2.3 Huygens' Diffraction Formula

According to Huygens' principle the amplitude at P is obtained by integrating the contribution from the points in the surface S. If the source and observation points are quite close to the axis, for unpolarized light, the field vectors can be represented by scalars because they are almost normal to the axis. We find that

$$U(P) = -\frac{i}{\lambda} \iint_S \frac{e^{ikr}}{r} U(Q) \, dS. \qquad (2.1)$$

This form of the equation is in a slightly different form compared with that of Hecht (1987), for example. $U(Q)$ gives the strength of the illumination at Q, e^{ikr}/r represents a spherical wave emanating from Q, and the factor $-i/\lambda$ results from the fact that Q is a driven dipole. The far-field of a dipole is $\pi/2$ out of phase with the forcing function, similar to the behavior of resonance.

This expression has been developed from a semiqualitative model: it is not strictly correct but gives a good prediction if, first, we are not too close to the aperture, and, second, if the aperture is large compared with the wavelength. In optics these two conditions are usually, but not always, true. The more rigorous Kirchhoff diffraction formula we will derive later, but usually this is not much of an improvement because it still does not take account of the fact that the illumination of the aperture is changed by the presence of the screen, and in addition vector effects are also neglected.

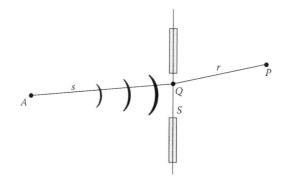

FIGURE 2.3   Geometry of diffraction.

If the aperture is illuminated with a spherical wave from a point a distance s away (Figure 2.3), then

$$U(Q) = A \frac{e^{iks}}{s},$$

and

$$U(P) = -A \frac{i}{\lambda} \iint_S \frac{e^{ik(r+s)}}{rs} \, dS. \qquad (2.2)$$

This is a form of the Huygens–Fresnel diffraction formula.

2.4 Fraunhofer Diffraction

In Equation 2.1, r does not change very much as Q varies over S. Although it is still necessary to take account of the changes in the exponent because this produces a multiplicative factor, in the denominator we can assume r is constant at a value r_0:

$$U_2(P) = -\frac{i}{\lambda r_0} \iint_S e^{ikr} U_1(Q) \, dS. \qquad (2.3)$$

Now, we have (Figure 2.4)

$$r^2 = (x_2 - x_1)^2 + (y_2 - y_1)^2 + z^2$$
$$= r_0^2 + (x_1^2 + y_1^2) - 2(x_1 x_2 + y_1 y_2). \qquad (2.4)$$

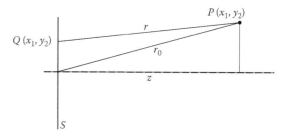

FIGURE 2.4   Geometry of the Fraunhofer diffraction.

Taking the square root of both sides, if $x_1, y_1 \ll r_0$, we can neglect the second term and expand the square root to give

$$r = r_0 - \frac{(x_1 x_2 + y_1 y_2)}{r_0}. \tag{2.5}$$

This is one form of the Fraunhofer approximation. Sometimes the expansion is made in terms of z, rather than r_0, but expansion in r_0 is valid for larger values of x_2, y_2. So for our case

$$U_2(P) = -\frac{i}{\lambda r_0} \exp(ikr_0) \iint_S U_1(x_1, y_1) \exp\left[-\frac{ik}{r_0}(x_1 x_2 + y_1 y_2)\right] dx_1 dy_1. \tag{2.6}$$

The condition $x_1, y_1 \ll r_0$ can be written

$$r_0 \gg \frac{k(x_1^2 + y_1^2)_{max}}{2}. \tag{2.7}$$

Introducing the Fresnel number N,

$$N = \frac{(x_1^2 + y_1^2)_{max}}{\lambda r_0}, \tag{2.8}$$

we have

$$N \ll \frac{1}{\pi}. \tag{2.9}$$

As an example, for $x_1^2 + y_1^2 = 1\,\text{mm}^2$, that is the width is $2\,\text{mm}$, then for the Fraunhofer condition to be valid, $r_0 \gg 6\,\text{m}$, which will not be the case in a laboratory experiment. But for $x_1 = 100\,\mu\text{m}$, $r_0 \gg 60\,\text{mm}$, which we can observe easily.

If we define the Fourier transform of $f(x)$ as

$$F[f(m)] = \int_{-\infty}^{+\infty} f(x) e^{-2\pi imx} dx, \tag{2.10}$$

Equation 2.6 is recognized as saying that

$$U_2(P) = \text{const.} \times F[U_1(x_1, y_1)]. \tag{2.11}$$

In these expressions $U(x_1, y_1)$ is taken as the value of the incident field in the aperture, and zero outside of the aperture. This is called the Kirchhoff boundary condition.

2.4.1 Examples of Fraunhofer Diffraction—Single Slit

For a long slit, length $2l$, width $2a$, in the plane $y_2 = 0$ illuminated uniformly

$$U_2(P) = -\frac{i}{\lambda r_0} \exp(ikr_0) \iint_S \exp\left(-\frac{2\pi i x_1 x_2}{\lambda r_0}\right) dx_1 dy_1$$

$$= -\frac{2\ell i}{\lambda r_0} \exp(ikr_0) \int_{-a}^{a} \exp\left(-\frac{2\pi i x_1 x_2}{\lambda r_0}\right) dx_1$$

$$= -\frac{2\ell i}{\lambda r_0} \exp(ikr_0) 2a \left[\frac{\sin(2\pi a x_2/\lambda r_0)}{2\pi a x_2/\lambda r_0}\right],$$

so

$$I = \frac{A^2}{\ell^2 r_0^2} \left[\frac{\sin(ka\sin\theta)}{ka\sin\theta}\right]^2, \tag{2.12}$$

where A is the area of the aperture.

2.4.2 Rectangular Aperture

Consider now a rectangular slit, sides $2a$ and $2b$, illuminated with a plane wave. The diffracted amplitude is

$$U(P) = -\frac{i}{\lambda r_0} \exp(ikr_0) \int_{-a}^{a} \exp\left(-\frac{2\pi i x_1 x_2}{\lambda r_0}\right) dx_1 \int_{-b}^{b} \exp\left(-\frac{2\pi i y_1 y_2}{\lambda r_0}\right) dy_1. \tag{2.13}$$

so that the intensity is

$$I = \left(\frac{A^2}{\lambda^2 r_0^2}\right)^2 \left[\frac{\sin(kax_2/r_0)}{(kax_2/r_2)}\right]^2 \left[\frac{\sin(kby_2/r_0)}{(kby_2/r_2)}\right]^2. \tag{2.14}$$

There are zeros along a series of perpendicular lines in the diffraction pattern (Figure 2.5).

2.4.3 Circular Aperture

Consider diffraction by a circular aperture, radius a. Introducing polar coordinates (Figure 2.6)

$$x_1 = R_1 \cos\phi, \quad x_2 = R_2 \cos\varphi,$$

$$y_1 = R_1 \sin\phi, \quad y_2 = R_2 \sin\varphi, \tag{2.15}$$

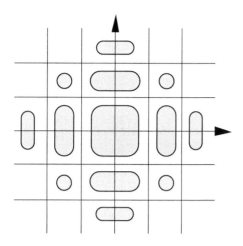

FIGURE 2.5 Fraunhofer diffraction by a rectangular aperture.

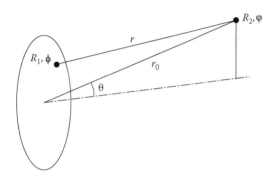

FIGURE 2.6 Diffraction by a circular aperture.

the diffracted field is

$$U_2(P) = -\frac{i}{\lambda r_0} \exp(ikr_0) \int_0^{2\pi} \int_0^a \exp\left[-\frac{ikR_1R_2\cos(\phi-\varphi)}{r_0}\right] R_1 dR_1 d\phi.$$

(2.16)

As the answer, by symmetry, is independent of φ, we can assume without loss of generality that $\varphi = 0$. The integral in ϕ is thus

$$\int_0^{2\pi} \exp\left(-\frac{ikR_1R_2\cos\phi}{r_0}\right) d\phi.$$

(2.17)

But we know that

$$J_0(z) = \frac{1}{2\pi} \int_0^{2\pi} \exp(iz\cos\phi) d\phi$$

(2.18)

as this is the definition of the Bessel function of the first kind of order zero, so

$$U_2(P) = -\frac{2\pi i}{\lambda r_0} \exp(ikr_0) \int_0^a J_0\left(\frac{kR_1R_2}{r_0}\right) R_1 dR_1.$$

(2.19)

This integral can be solved by use of the recurrence relationship (Abramowitz and Stegun 1965, Eq. 9.1.20)

$$\frac{1}{z}\frac{d}{dz}\left[zJ_1(z)\right] = J_0(z).$$

(2.20)

We have

$$\int J_0(z)z\,dz = zJ_1(z),$$

(2.21)

so the diffracted amplitude is

$$\begin{aligned}
U(R_2) &= -\frac{i}{\lambda r_0}\exp(ikr_0)\frac{r_0^2}{k^2R_2^2}\int_0^{kaR_0/r_0} J_0(\xi)\xi\,d\xi \\
&= -\frac{i}{\lambda R_2^2}\left[\xi J_1(\xi)\right]_0^{kaR_2 r_0} \\
&= \frac{i\pi a^2}{r_0\lambda}\exp(ikr_0)\left[\frac{2J_1(kaR_2/r_0)}{(kaR_2/r_0)}\right] \\
&= -\frac{iA}{\lambda r_0}\exp(ikr_0)\left[\frac{2J_1(ka\sin\theta)}{(ka\sin\theta)}\right],
\end{aligned}$$

(2.22)

where A is the area of the aperture, or

$$I(\theta) = \left(\frac{A}{\lambda r_0}\right)^2\left[\frac{2J_1(ka\sin\theta)}{(ka\sin\theta)}\right]^2.$$

(2.23)

This is illustrated in Figure 2.7. Note that we have written the equation in this form as $2J_1(\nu)/\nu \rightarrow 1$ for $\nu \rightarrow 0$.

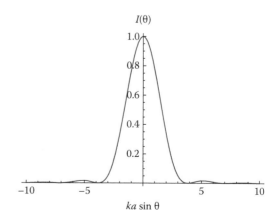

FIGURE 2.7 Fraunhofer diffraction by a circular aperture.

2.4.4 Annular Aperture

For an annular aperture (Figure 2.8), in Equation 2.6 we can put

$$U_1(R_1) = \mathrm{circ}\left(\frac{R_1}{a}\right) - \mathrm{circ}\left(\frac{R_1}{\varepsilon a}\right), \qquad (2.24)$$

where $\mathrm{circ}(x) = 1, x < 1; = 0, x > 1$. Therefore

$$U(R_2) = -\frac{i\pi a^2}{\lambda r_0}\exp(ikr_0)\left\{\left[\frac{2J_1(kaR_2/r_0)}{(kaR_2/r_0)}\right] - \varepsilon^2\left[\frac{2J_1(k\varepsilon aR_2/r_0)}{(k\varepsilon aR_2/r_0)}\right]\right\}. \qquad (2.25)$$

As the value of ε is increased (Figure 2.9), the central peak becomes narrower, but the side lobes become stronger. For the limiting case when $\varepsilon \to 1$,

$$U(R_2) = -\frac{i}{\lambda r_0}\exp(ikr_0)\int_0^\infty \delta(R_1 - a)J_0\left(\frac{2\pi R_1 R_2}{\lambda r_0}\right)2\pi R_1\,dR_1$$

$$= -\frac{i}{\lambda r_0}\exp(ikr_0)J_0\left(\frac{2\pi R_2 a}{\lambda r_0}\right). \qquad (2.26)$$

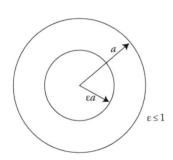

FIGURE 2.8   An annular aperture.

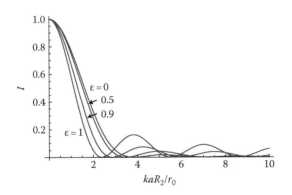

FIGURE 2.9   Normalized intensity in the Fraunhofer diffraction pattern of an annular aperture.

2.5 Fresnel Diffraction

2.5.1 Introduction

Knowing the field in a plane x_1, y_1 we can calculate the field in the plane x_2, y_2 using Equation 2.3.

Using the binomial theorem to expand the square root, Equation 2.5 now becomes (without neglecting the second term)

$$r = r_0 + \frac{x_1^2 + y_1^2}{2r_0} - \frac{x_1^2 x_1^2 + y_1 y_2}{r_0}. \qquad (2.27)$$

So

$$U_2(P) = -\frac{i}{\lambda r_0}\exp(ikr_0)\iint_S U_1(x_1, y_1)\exp\left[-\frac{ik}{r_0}(x_1 x_2 + y_1 y_2)\right]$$

$$\times \exp\left[\frac{ik}{2r_0}(x_1^2 + y_1^2)\right]dx_1\,dy_1. \qquad (2.28)$$

2.5.2 Circular Aperture

Consider a circular aperture illuminated with a plane wave. Using Equation 2.18 in Equation 2.28, we have

$$U_2(R_2, r_0) = -\frac{ik}{r_0}\exp(ikr_0)\int_0^a J_0\left(\frac{kR_1 R_2}{r_0}\right)\exp\left(\frac{ikR_1^2}{2r_0}\right)R_1\,dR_1. \qquad (2.29)$$

Unfortunately, this integral cannot be solved in terms of elementary functions, though it can be solved in terms of Lommel functions (Born and Wolf 1975). This approach is not particularly useful, and it is usually easier to solve it numerically.

But, along the axis it reduces to a simple form:

$$U_2(0, r_0) = -\frac{ik}{r_0}\exp(ikr_0)\int_0^a \exp\left(\frac{ikR_1^2}{2r_0}\right)R_1\,dR_1$$

$$= -\frac{ik}{r_0}\exp(ikr_0)\left[\exp\left(-\frac{ikR_1^2}{2r_0}\right)\right]_0^a$$

$$= -2i\exp(ikr_0)\exp\left(-\frac{ika^2}{4r_0}\right)\sin\left(\frac{ka^2}{4r_0}\right). \qquad (2.30)$$

or

$$I_2(0, r_0) = 4\sin^2\left(\frac{ka^2}{4r_0}\right). \qquad (2.31)$$

So we obtain a series of maxima and minima (zeros) in intensity along the axis (Figure 2.10). This can be explained in terms of

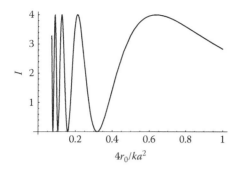

FIGURE 2.10 Intensity along the axis of a circular aperture according to the Fresnel diffraction theory.

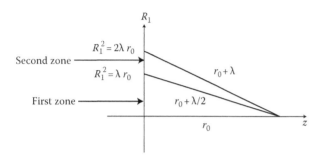

FIGURE 2.11 The Fresnel half-period zone construction.

Fresnel's zones in the following way. Contributions from successive zones (Figure 2.11) tend to cancel as the phases are different by 180°. If the number of zones is N, then $a^2 = n\lambda r$ and

$$N = \frac{a^2}{\lambda r}, \tag{2.32}$$

that is, it is equal to the Fresnel number, defined earlier (Equation 2.8).

Note that for a general axially symmetric U_1, along the axis, the amplitude is

$$U_2(0, r_0) = -\frac{ik}{r_0}\exp(ikr_0)\int_0^a U_1(R_1)\exp\left(\frac{ikR_1^2}{2r_0}\right)R_1\,dR_1. \tag{2.33}$$

Let $R_1^2/a^2 = t$, and put $U_1(R_1) = V_1(t)$. We have

$$2R_1\,dR_1 = a^2\,dt,$$

and thus

$$U_2(0, r_0) = -i\pi N\exp(ikr_0)\int_0^1 V_1(t)\exp(i\pi Nt)\,dt. \tag{2.34}$$

So the intensity is given by

$$I_2(N) = (\pi N)^2\left|FT\left[V_1(t)\right]\right|^2. \tag{2.35}$$

2.5.3 Rectangular Aperture

If a rectangular aperture, sides $2a$, $2b$, is illuminated by a plane wave, from Equation 2.28,

$$U_2(x_2, y_2) = -\frac{ie^{ikr_0}}{\lambda r_0}\int_{-a}^{+a}\exp\left(\frac{ikx_1^2}{2r_0}\right)\exp\left[-\frac{ik(x_1x_2)}{r_0}\right]dx_1$$

$$\times \int_{-b}^{+b}\exp\left(\frac{iky_1^2}{2r_0}\right)\exp\left[-\frac{ik(y_1y_2)}{r_0}\right]dy_1. \tag{2.36}$$

Now, consider the integral

$$I = \int_{-a}^{+a}\exp\left(\frac{ikx_1^2}{2r_0}\right)\exp\left[-\frac{ik(x_1x_2)}{r_0}\right]dx_1$$

$$= \int_{-a}^{+a}\exp\left[\frac{ik\left(x_1^2 - 2x_1x_2\right)}{2r_0}\right]dx_1. \tag{2.37}$$

By completing the square

$$I = \exp\left(-\frac{ikx_2^2}{2r_0}\right)\int_{-a}^{+a}\exp\left[\frac{ik(x_2 - x_1)^2}{2r_0}\right]dx_1$$

$$= \sqrt{\frac{r_0\lambda}{2}}\exp\left(-\frac{ikx_2^2}{2r_0}\right)\int_{-w_1}^{+w_1}\exp\left(\frac{i\pi w^2}{2}\right)dw, \tag{2.38}$$

where

$$w = \sqrt{\frac{2}{r_0\lambda}}(x_2 - x_1). \tag{2.39}$$

We now introduce the Fresnel integrals, defined as

$$C(w) = \int_0^w \cos\left(\frac{\pi w'^2}{2}\right)dw'.$$

$$S(w) = \int_0^w \sin\left(\frac{\pi w'^2}{2}\right)dw'.$$

$$F(w) = C + iS \tag{2.40}$$

So

$$I = 2\exp\left(-\frac{ikx_2^2}{2r_0}\right)\frac{F(w_2) - F(w_1)}{\sqrt{2/r_0\lambda}} \tag{2.41}$$

$$= 2\exp\left(-\frac{ikx_2^2}{2r_0}\right)\frac{F\left[\sqrt{\frac{2}{r_0\lambda}}(x_2 + a)\right] - F\left[\sqrt{\frac{2}{r_0\lambda}}(x_2 - a)\right]}{\sqrt{2/r_0\lambda}} \tag{2.42}$$

and thus

$$U_2(x_2, y_2) = -2i \exp(ikr_0) \exp\left(\frac{ikR_2^2}{2r_0}\right)\left[F(w_{x_2}) - F(w_{x_1})\right]$$

$$\times \left[F(w_{y_2}) - F(w_{y_1})\right]. \qquad (2.43)$$

The Fresnel integrals are tabulated. They have the following important properties:

$$\left.\begin{array}{l} C(w) = -C(-w) \\ S(w) = -S(-w) \end{array}\right\} \text{odd function,} \qquad (2.44)$$

$$C(0) = S(0) = 0, \quad F(0) = 0, \qquad (2.45)$$

$$C(\infty) = S(\infty) = \frac{1}{2}, \quad F(-\infty) = -\frac{1}{2}(1-i), \qquad (2.46)$$

$$C(-\infty) = S(-\infty) = -\frac{1}{2}, \quad F(-\infty) = -\frac{1}{2}(1-i). \qquad (2.47)$$

Let us consider first what happens as the aperture becomes very large. Then from Equation 2.43, at $x_2 = y_2 = 0$

$$U_2(0,0) = -\frac{i}{2}\exp(ikr_0)(1+i)^2 = \exp(ikr_0). \qquad (2.48)$$

So it reproduces exactly what is expected for a plane wave! This is remarkable, because the assumptions we have made such as the Fresnel approximation and that $r \approx r_0$ break down for points on the aperture that are far from the axis. It works because the contributions from these off-axis points are weak because they are far away. So now it is established that the method can be applied even for an infinitely large aperture, giving us some confidence to look at some other cases.

2.5.4 Single Slit

Here $b \to \infty$ and we can take $y_2 = 0$ without loss of generality, so that

$$U_2(x_2) = \frac{(1-i)}{2}\exp(ikr_0)\exp\left(-\frac{ikx_2^2}{2r_0}\right)$$

$$\times \left\{F\left[\sqrt{\frac{2}{r_0\lambda}}(x_2 + a)\right] - F\left[\sqrt{\frac{2}{r_0\lambda}}(x_2 - a)\right]\right\}. \qquad (2.49)$$

For slits we get intensity distributions such as those shown in Figure 2.12.

Along the axis, $x_2 = 0$, and

$$U_2(0) = -(1+i)\exp(ikr_0)F\left(\sqrt{\frac{2}{r_0\lambda}}a\right). \qquad (2.50)$$

The intensity along the axis (Figure 2.13) should be compared with that for a circular aperture. It should be noted that the minima are not zeros this time. The maxima correspond approximately to the case when the Fresnel number $N = a^2/\lambda\, r_0$ is

$$N = 2n + \frac{3}{4}, \qquad (2.51)$$

and the minima approximately to the case when

$$N = 2n + \frac{5}{4}, \qquad (2.52)$$

where n is zero or a positive integer.

2.5.5 Half-Plane

For diffraction by an edge, we can put in Equation 2.42

$$w_2 = \sqrt{\frac{2}{r_0\lambda}}x_2, \quad w_1 \to \infty, \qquad (2.53)$$

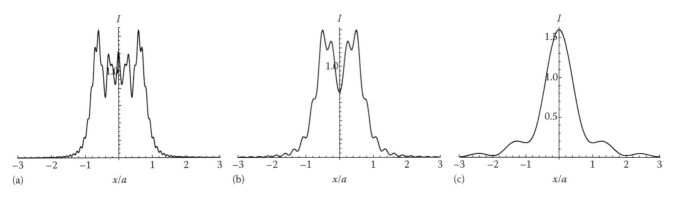

FIGURE 2.12 Fresnel diffraction by a slit: (a) $N = a^2/\lambda r_0 = 5$, (b) $N = 2$, and (c) $N = 0.5$.

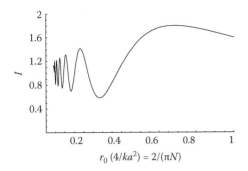

FIGURE 2.13 Intensity along the axis for a slit aperture according to Fresnel diffraction.

so that the intensity is

$$I = \frac{1}{2}\left\{\left[\frac{1}{2} - C\left(\sqrt{\frac{2}{r_0\lambda}}x_2\right)\right]^2 + \left[\frac{1}{2} - S\left(\sqrt{\frac{2}{r_0\lambda}}x_2\right)\right]^2\right\}. \quad (2.54)$$

Note that the diffraction pattern is the same at any distance, but of course it scales with distance (Figure 2.14). This is in contrast to the diffraction pattern for a slit, which changes with distance.

2.5.6 Circular Obstruction

Consider a circular obstruction, radius a. Then

$$U_2(0, r_0) = \frac{ik}{r_0}\exp(ikr_0)\int_a^\infty \exp\left(-\frac{ikR_1^2}{2r_0}\right)R_1\,dR_1$$

$$= \frac{ik}{r_0}\exp(ikr_0)\int_0^\infty \exp\left(-\frac{ikR_1^2}{2r_0}\right)R_1\,dR_1$$

$$-\frac{ik}{r_0}\exp(ikr_0)\int_0^a \exp\left(-\frac{ikR_1^2}{2r_0}\right)R_1\,dR_1. \quad (2.55)$$

The first term represents a plane wave and the second the diffracted field for a circular aperture, rather than an obstruction.

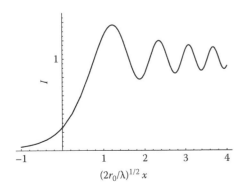

FIGURE 2.14 Fresnel diffraction by an edge.

This is a particular case of Babinet's principle, which states that the sum of the fields for two complementary screens is equal to the unobstructed disturbance.

The first term we evaluated in Equation 2.48: it gave just $\exp(ikr_0)$. So

$$U_2 = -\exp(ikr_0) + \left\{\exp(ikr_0)\left[\exp\left(\frac{ika^2}{2r_0}\right) + 1\right]\right\}$$

$$= \exp(ikr_0)\exp\left(\frac{ika^2}{2r_0}\right). \quad (2.56)$$

The intensity is thus constant along the axis (the approximations break down when we get too close to the obstruction). This is the Poisson (or Arago) spot. It can be regarded as being caused by light scattered from the edge of the disc. Comparing with the case of the circular aperture, Equation 2.30 represents interference between the edge-diffracted wave and the undiffracted wave ($R_1 = 0$). This is the principle of the boundary diffraction wave concept introduction by Young.

2.6 Kirchhoff Diffraction Integral

We now return to the problem of deriving the diffraction integral starting from the wave equation

$$\left(\nabla^2 + k^2\right)U = 0. \quad (2.57)$$

Consider a closed surface S with inward normal $\mathbf{n}$ (Figure 2.15). Then we can show using Green's theorem that at any point inside S

$$\iint_S \left(U\frac{\partial U'}{\partial n} - U'\frac{\partial U}{\partial n}\right)dS = 0, \quad (2.58)$$

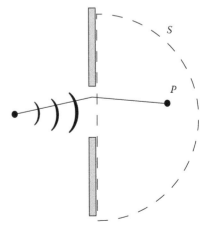

FIGURE 2.15 Diffraction according to Kirchhoff diffraction theory.

for any U' which also satisfies the wave equation. We can get different solutions by appropriate choice of the so-called Green function U' that satisfies the wave equation.

If we choose $U' = \exp(ikr)/r$, which satisfies the wave equation, and excluding the region around its singularity at $r = 0$, we obtain the Kirchhoff diffraction integral:

$$U_p = \frac{1}{4\pi} \iint \left[U \frac{\partial}{\partial n}\left(\frac{e^{ikr}}{r}\right) - \left(\frac{e^{ikr}}{r}\right)\frac{\partial U}{\partial n} \right] dS. \qquad (2.59)$$

It must be stressed that this expression is rigorously correct (in the scalar approximation).

Consider now diffraction by a planar screen with an aperture illuminated with a wave emanating from point S. We assume that the field in the aperture is the same as in the screen were absent. So in the aperture

$$U = \frac{U_0 e^{iks}}{s}, \quad \frac{\partial U}{\partial n} = \frac{U_0 e^{iks}}{s}\left(ik - \frac{1}{s}\right)\cos(n, s), \qquad (2.60)$$

where $\cos(n,s)$ is the cosine of the angle between the n and s directions, whereas on the screen

$$U = 0, \quad \frac{\partial U}{\partial n} = 0. \qquad (2.61)$$

These are called the Kirchhoff boundary conditions.

So, neglecting $1/s$ and $1/r$ in comparison to k,

$$U(P) = -\frac{iU_0}{\lambda} \iint\limits_{\text{Aperture}} \frac{e^{ik(r+s)}}{rs}\left[\frac{\cos(n, s) - \cos(n, r)}{2}\right] dS. \qquad (2.62)$$

Here we have assumed that the contribution from the spherical part of the surface S vanishes as its radius is made increasingly large. The factor in square brackets is called the obliquity factor. Apart from this, the expression is identical to Equation 2.22. Equation 2.62 is the Kirchhoff diffraction formula. Note that the surface of integration is arbitrary.

Although the Kirchhoff diffraction integral is exact, the assumed boundary conditions are not. So the Kirchhoff diffraction formula is also not exact. In particular it fails to reproduce the assumed field in the aperture. This is because the choice of U and $\partial U/\partial n$ in the aperture is inconsistent.

It can be shown that in fact, for the special case of integration over a planar region,

$$U_p = \frac{1}{2\pi} \iint U \frac{\partial}{\partial n}\left(\frac{e^{ikr}}{r}\right) dS \qquad (2.63)$$

and

$$U_p = -\frac{1}{2\pi} \iint \left(\frac{e^{ikr}}{r}\right)\frac{\partial U}{\partial n} dS \qquad (2.64)$$

These are called the Rayleigh–Sommerfeld I and II diffraction formulae. They have the advantage of being consistent as we need only assume U or $\partial U/\partial n$ in the aperture, but are not in practice any more accurate as we do not usually know either U or $\partial U/\partial n$ accurately. In fact, Equation 2.59 can be thought of as the average of Equations 2.63 and 2.64 and has been claimed to give a better prediction of the field at P.

2.7 Angular Spectrum of Plane Waves

Consider a single plane wave propagating at angle θ, for which

$$E = E_0 \exp(-ikx\sin\theta)\exp(-ikz\cos\theta). \qquad (2.65)$$

In the plane $z = 0$

$$E = E_0 \exp(-ikx\sin\theta),$$

which is a harmonic variation with spatial wavelength $\lambda/\sin\theta$. So if we alter θ, we alter the spatial wavelength in the x direction. According to Fourier synthesis, *any* field in the plane $z = 0$ can be represented by a sum of Fourier components:

$$E_y(x) = \int\limits_{-\infty}^{+\infty} G(\theta)\exp(-ikx\sin\theta)d(\sin\theta). \qquad (2.66)$$

$G(\theta)$ is the strength of a plane wave component traveling in direction θ. Here $G(\theta)$ is *complex* to account for the relative phase of the components. As a simple example, for the field

$$E_y(x) = A\cos\left(\frac{2\pi x}{\Lambda}\right)$$
$$= \frac{A}{2}\exp\left(\frac{2\pi i x}{\Lambda}\right) + \frac{A}{2}\exp\left(-\frac{2\pi i x}{\Lambda}\right), \qquad (2.67)$$

we have immediately $\sin\theta = \pm\lambda/\Lambda$, representing two plane waves. In this case, the diffraction pattern of the cosine grating in the far field is two bright spots.

2.8 Evanescent Waves

Note that

$$E(x) = \int\limits_{-\infty}^{+\infty} G(\theta)\exp(-ikx\sin\theta)d(\sin\theta) \qquad (2.68)$$

has limits $\pm\infty$. But $|\sin\theta| \leq 1$ for θ to be real. So $|\sin\theta| > 1$ corresponds to waves traveling at a complex angle. We have

$$k_x^2 + k_z^2 = k^2, \qquad (2.69)$$

so that

$$k_z^2 = k^2 - k_x^2. \tag{2.70}$$

If $k_x > k$, we can put

$$k_z = \pm i\sqrt{k_x^2 - k^2},$$

which is imaginary. Therefore the electric field is

$$E = E_0 \exp(-ik_x x)\exp\left[\pm\left(z\sqrt{k_x^2 - k^2}\right)\right], \tag{2.71}$$

representing a wave traveling in the x direction, but with an exponential decay in the z direction. Note that we take the positive root to give a physical solution for $z \geq 0$.

This is an *evanescent* wave. The integral integrates over all propagating *and* evanescent waves. In the far field, the evanescent waves have decayed, so they make no contribution: only the propagating waves remain.

2.9 Diffraction by a Phase Screen

Suppose we have a screen that has an amplitude transmittance $t(x,y)$. This is complex to account for amplitude and phase effects. Then if it is illuminated by a plane wave, amplitude U_0 the field immediately after the screen is $U_0 t$.

A thin lens can be thought of as such a screen. It consists of an amplitude term $P(x_1, y_1)$ that is called the pupil function of the lens, which is unity in the aperture and zero outside, and a phase term $e^{i\Phi(x,y)}$.

$$t(x_1, y_1) = P(x_1, y_1)e^{i\Phi(x,y)}, \tag{2.72}$$

or in polar coordinates

$$t(r_1, \theta_1) = P(r_1, \theta_1)e^{i\Phi(r_1, \theta_1)}. \tag{2.73}$$

In general, we can expand Φ as a power series in r and also as a series in $\cos n\theta$:

$$\Phi = \sum_{n,m} a_{mn}r^m \cos n\theta$$

$$= a_{00} + a_{20}r^2 + a_{40}r^4 + \cdots + a_{11}r\cos\theta$$

$$+ a_{21}r^2\cos\theta + \cdots + a_{22}r^2\cos 2\theta + \cdots, \tag{2.74}$$

or, collecting all the terms except the squared term,

$$\Phi = a_{20}r^2 + \Phi'(r, \theta). \tag{2.75}$$

So the amplitude transmittance of the lens can be written as

$$t(r_1, \theta_1) = P(r_1)\exp\left(ia_{20}r_1^2\right)e^{i\Phi'}. \tag{2.76}$$

After the lens, neglecting diffraction for the present,

$$U(r) = P(r)\exp\left(ia_{20}r^2\right)e^{i\Phi'}e^{ikz}$$

$$= P(r_1)e^{i\Phi'}\exp\left[ik\left(\frac{a_{20}r^2}{k} + z\right)\right]. \tag{2.77}$$

Neglecting Φ', which represents aberrations, the phase front through the origin is the paraboloid with equation (Figure 2.16)

$$\frac{a_{20}r^2}{k} + z = 0. \tag{2.78}$$

Consider a sphere, radius f, centered on $z = f$. Then

$$r^2 + (z - f)^2 = f^2,$$

and using the binomial theorem and assuming $r \ll f$

$$(z - f) = \sqrt{f^2 - r^2} = \pm f\left(1 - \frac{r^2}{2f}\right). \tag{2.79}$$

Taking the negative root and comparing Equations 2.78 and 2.79

$$\frac{a_{20}}{k} = -\frac{1}{2f}. \tag{2.80}$$

So the lens, including the aberration term, can be taken as

$$t(r_1, \theta_1) = P(r_1)\exp\left(-\frac{ikr_1^2}{2f}\right)e^{i\Phi'}, \tag{2.81}$$

where f is the focal length of the lens.

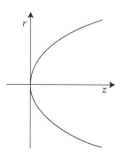

FIGURE 2.16 Phase front of a lens.

2.10 Thin Lens

For a thin lens, the field after the lens is (Goodman 1968)

$$U_2(x,y) = U_1(x,y)P(x,y)\exp\left[ik(n-1)\Delta(x,y)\right] \quad (2.82)$$

where

    Δ is the thickness of the glass
    n is its refractive index (Figure 2.17)

In the paraxial approximation, $x/R_1 \ll 1$ and higher powers can be neglected, so

$$\Delta(x,y) = \Delta_0 - \frac{x^2+y^2}{2}\left(\frac{1}{R_1} - \frac{1}{R_2}\right). \quad (2.83)$$

Thus,

$$U_2(x,y) = U_1(x,y)P(x,y)\exp\left[ik(n-1)\Delta_0\right]$$
$$\times \exp\left[-ik(n-1)\left(\frac{x^2+y^2}{2}\right)\left(\frac{1}{R_1} - \frac{1}{R_2}\right)\right]. \quad (2.84)$$

Putting

$$\frac{1}{f} = (n-1)\left(\frac{1}{R_1} - \frac{1}{R_2}\right), \quad (2.85)$$

we then have

$$U_2(x,y) = U_1(x,y)P(x,y)\exp\left[ik(n-1)\Delta_0\right]\exp\left[-\frac{ik}{2f}(x^2+y^2)\right]. \quad (2.86)$$

The term $\exp[ik(n-1)\Delta_0]$ is a constant phase term, which we neglect.

2.11 Focus of a Lens

We assume the lens is illuminated by a plane wave so that

$$U_2(x,y) = P(x,y)\exp\left[-\frac{ik(x^2+y^2)}{2f}\right]. \quad (2.87)$$

This represents a spherical wave convergent on the point F (Figure 2.18). But

$$U_3(x_3,y_3) = -\frac{ie^{ikf}}{\lambda f}\int_{-\infty}^{+\infty}\int_{-\infty}^{+\infty} U_2(x,y)$$
$$\times \exp\left\{\frac{ik}{2f}\left[(x_3-x)^2 + (y_3-y)^2\right]\right\}dx\,dy \quad (2.88)$$

as

$$r^2 = f^2 + (x_3-x)^2 + (y_3-y)^2,$$

so that if $(x_3 - x) \ll f$.

$$r \approx f + \frac{(x_3-x)^2}{2f} + \frac{(y_3-y)^2}{2f}.$$

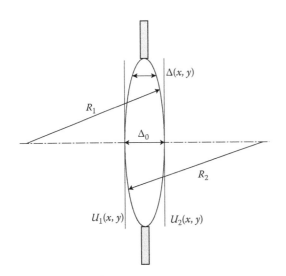

FIGURE 2.17 The thin lens.

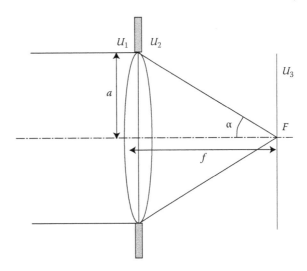

FIGURE 2.18 Focusing by a lens.

From now on we shall not write explicitly the limits $\pm\infty$ on the integrals. The field after the lens is

$$U_3(x_3, y_3) = -\frac{ie^{ikf}}{\lambda f} \iint P(x, y) \exp\left[-\frac{ik(x^2 + y^2)}{2f}\right] \exp\left[\frac{ik(x^2 + y^2)}{2f}\right]$$

$$\times \exp\left[\frac{ik(x_3^2 + y_3^2)}{2f}\right] \exp\left[-\frac{ik}{f}(xx_3 + yy_3)\right] dx\, dy$$

$$= -\frac{ie^{ikf}}{\lambda f} \exp\left[\frac{ik(x_3^2 + y_3^2)}{2f}\right] \iint P(x, y)$$

$$\times \exp\left[-\frac{ik}{f}(xx_3 + yy_3)\right] dx\, dy. \tag{2.89}$$

We now consider some important special cases for the pupil function.

2.12 Circularly Symmetric Aperture

Most real optical systems have circular symmetry, so we can put

$$P(x, y) = P(r)$$

and

$$U_3(r_3) = -\frac{ie^{ikf}}{\lambda f} \exp\left(\frac{i\pi r_3^2}{\lambda f}\right) \int_0^\infty P(r) J_0\left(\frac{2\pi r r_3}{\lambda f}\right) 2\pi r\, dr. \tag{2.90}$$

In particular, for a circular aperture of radius a

$$U_3(r_3) = -\frac{ie^{ikf}}{\lambda f} \exp\left(\frac{i\pi r_3^2}{\lambda f}\right) \pi a^2 \left[\frac{2J_1(2\pi r_3 a/\lambda f)}{2\pi r_3 a/\lambda f}\right]. \tag{2.91}$$

Let us define the *numerical aperture* of the lens:

$$NA = \sin\alpha = \frac{a}{f}. \tag{2.92}$$

We define a normalized optical coordinate

$$v = kr_3 \sin\alpha \approx \frac{2\pi r_3 a}{\lambda f}. \tag{2.93}$$

and also put

$$\rho = \frac{r}{a}. \tag{2.94}$$

Then in general

$$U_3(v) = -\frac{ie^{ikf}}{\lambda f} \exp\left(\frac{iv^2\lambda f}{4\pi a^2}\right) 2\pi a^2 \int_0^1 P(\rho) J_0(v\rho)\rho\, d\rho. \tag{2.95}$$

or introducing the Fresnel number

$$N = \frac{a^2}{\lambda f}, \tag{2.96}$$

and the field is

$$U_3(v) = -i\pi N e^{ikf} \exp\left(\frac{iv^2}{4\pi N}\right) \int_0^1 2P(\rho) J_0(v\rho)\rho\, d\rho. \tag{2.97}$$

For a plain circular aperture

$$U_3(v) = -i\pi N e^{ikf} \exp\left(\frac{iv^2}{4\pi N}\right) \left[\frac{2J_1(v)}{v}\right]. \tag{2.98}$$

This is called the amplitude point spread function or impulse response. Note that

$$N = \frac{a^2}{\lambda f} = \frac{a\sin\alpha}{\lambda} = \frac{f\sin^2\alpha}{\lambda},$$

that is, for a big lens (compared with the wavelength), N is very large and the exponential term is close to unity. The intensity is the modulus squared of the amplitude, giving the Airy disc:

$$I(v) = \pi^2 N^2 \left[\frac{2J_1(v)}{v}\right]^2. \tag{2.99}$$

The intensity variation is as shown in Figure 2.7. In Figure 2.9, the effect of a central obstruction is also shown. As the obstruction ratio ε increases, the point spread function becomes narrower, but with larger side-lobes.

2.13 Effect of Defocus

We now consider the field on a defocused plane, $z = f + \delta z$ (Figure 2.19). The field is

$$U_3(x_3, r_3) = -\frac{ie^{ikz}}{\lambda z} \iint P(x, y) \exp\left[-\frac{ik}{2f}(x^2 + y^2)\right]$$

$$\times \exp\left\{\frac{ik}{2z}\left[(x_3 - x)^2 + (y_3 - y)^2\right]\right\} dx\, dy$$

$$= -\frac{ie^{ikz}}{\lambda z} \exp\left[\frac{ik}{2z}(x_3^2 + y_3^2)\right] \iint P(x, y) \exp\left[-\frac{ik}{2f}(x^2 + y^2)\right]$$

$$\times \exp\left[\frac{ik}{2z}(x^2 + y^2)\right] \exp\left[-\frac{2\pi i}{\lambda z}(xx_3 + yy_3)\right] dx\, dy.$$

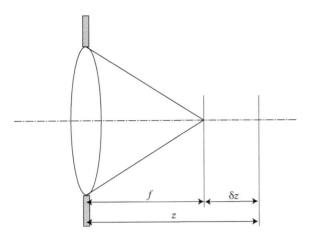

FIGURE 2.19 Geometry of defocus.

For a radially symmetric system

$$U_3(r_3) = -\frac{ie^{ikz}}{\lambda z}\exp\left(\frac{i\pi r_3^2}{\lambda z}\right)\int P(r)$$

$$\times \exp\left[-\frac{ikr^2}{2}\left(\frac{1}{f}-\frac{1}{z}\right)\right]J_0\left(\frac{2\pi rr_3}{\lambda z}\right)2\pi r\,dr. \quad (2.100)$$

For a circular aperture, again of radius a, put

$$v = \frac{2\pi a r_3}{\lambda z} = \frac{2\pi a r_3}{\lambda f} \quad (2.101)$$

$$u = \frac{2\pi a^2}{\lambda}\left(\frac{1}{f}-\frac{1}{z}\right) \approx \frac{2\pi a^2}{\lambda f^2}\delta z \quad (2.102)$$

if $\delta z \ll f$. So

$$U_3(v,u) = -i\pi Ne^{ikz}\exp\left(\frac{iv^2}{4\pi N}\right)\int_0^1 2J_0(v\rho)\exp\left(-\frac{iu\rho^2}{2}\right)\rho\,d\rho$$

$$(2.103)$$

Along the axis, $v=0$, the field is

$$U_3(v,u) = -i\pi Ne^{ikz}\exp\left(\frac{iv^2}{4\pi N}\right)\int_0^1 2\exp\left(-\frac{iu\rho^2}{2}\right)\rho\,d\rho$$

$$= -i\pi Ne^{ikz}\exp\left(\frac{iv^2}{4\pi N}\right)\left[-\frac{2}{iu}\exp\left(-\frac{iu\rho^2}{2}\right)\right]_0^1$$

$$= -i\pi Ne^{ikz}\exp\left(-\frac{iu}{4}\right)\left[\frac{\sin(u/4)}{u/4}\right]. \quad (2.104)$$

So the intensity along the axis is

$$I(0,u) = \pi^2 N^2\left[\frac{\sin(u/4)}{u/4}\right]^2, \quad (2.105)$$

The field at a general point can be calculated from Lommel functions or by numerical integration.

It is interesting to consider also the *annular* aperture $P(\rho) = \delta(\rho-1)$. Then from Equation 2.100

$$U_3(v,u) = -2i\pi Ne^{ikz}\exp\left(\frac{iv^2}{4\pi N}\right)\exp\left(-\frac{iu}{2}\right)J_0(v) \quad (2.106)$$

or

$$I(v,u) = 4\pi^2 N^2 J_0^2(v). \quad (2.107)$$

Note that the intensity does not change with u (within the range of validity of the equation). This represents a Bessel beam (a so-called diffraction-free beam), which is the subject of much research at present. Actually, it is very well known as a mode of free space in cylindrical coordinates, for example, of a circular waveguide. Power diffracts outward, but also *inward* from the large side lobes, to achieve a dynamic equilibrium.

2.14 Image Formation

Before the lens (Figure 2.20)

$$U_2(x_2,y_2) = -\frac{ie^{ikd_1}}{\lambda d_1}\iint U_1(x_1,y_1)$$

$$\times \exp\left\{\frac{ik}{2d_1}\left[(x_2-x_1)^2+(y_2-y_1)^2\right]\right\}dx_1dy_1.$$

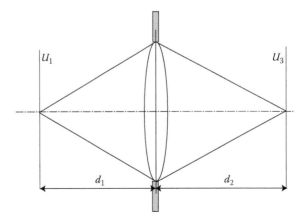

FIGURE 2.20 Imaging by a lens.

Multiplying by the pupil function, the amplitude after the lens is

$$
U_2(x_2, y_2) = -\frac{ie^{ikd_1}}{\lambda d_1} P(x_2, y_2) \exp\left[-\frac{ik}{2f}\left(x_2^2 + y_2^2\right)\right] \iint U_1(x_1, y_1)
$$

$$
\times \exp\left\{\frac{ik}{2d_1}\left[(x_2 - x_1)^2 + (y_2 - y_1)^2\right]\right\} dx_1 \, dy_1.
$$

So, finally the image amplitude is

$$
U_3(x_3, y_3) = -\frac{1}{\lambda^2 d_1 d_2} \exp\left[ik(d_1 + d_2)\right] \iiiint P(x_2, y_2) U_1(x_1, y_1)
$$

$$
\times \exp\left\{\frac{ik}{2d_1}\left[(x_2 - x_1)^2 + (y_2 - y_1)^2\right]\right\}
$$

$$
\times \exp\left\{\frac{ik}{2d_2}\left[(x_3 - x_2)^2 + (y_3 - y_2)^2\right]\right\}
$$

$$
\times \exp\left[-\frac{ik}{2f}\left(x_2^2 + y_2^2\right)\right] dx_1 \, dy_1 \, dx_2 \, dy_2.
$$

$$
= -\frac{1}{\lambda^2 d_1 d_2} \exp\left[ik(d_1 + d_2)\right] \iiiint P(x_2, y_2) U_1(x_1, y_1)
$$

$$
\times \exp\left[\frac{ik}{2d_1}\left(x_1^2 + y_1^2\right)\right] \exp\left[\frac{ik}{2d_2}\left(x_3^2 + y_3^2\right)\right]
$$

$$
\times \exp\left[\frac{ik}{2}\left(\frac{1}{d_1} + \frac{1}{d_2} - \frac{1}{f}\right)\left(x_2^2 + y_2^2\right)\right]
$$

$$
\times \exp\left\{-ik\left[x_2\left(\frac{x_1}{d_1} + \frac{x_3}{d_2}\right) + y_2\left(\frac{y_1}{d_1} + \frac{y_3}{d_2}\right)\right]\right\} dx_1 \, dy_1 \, dx_2 \, dy_2.
$$

$$
\tag{2.108}
$$

We now look at some special cases.

2.14.1 Special Cases

The first case that we consider is when the lens law is satisfied, $1/d_1 + 1/d_2 = 1/f$. Then, $d_2 = Md_1$, where M is magnification. The field is

$$
U_3(x_3, y_3) = -\frac{1}{\lambda^2 M d_1} \exp\left[ikd_1(1 + M)\right] \iiiint P(x_2, y_2) U_1(x_1, y_1)
$$

$$
\times \exp\left[\frac{ik}{2d_1}\left(x_1^2 + y_1^2\right)\right] \exp\left[\frac{ik}{2Md_1}\left(x_3^2 + y_3^2\right)\right]
$$

$$
\times \exp\left\{-\frac{ik}{d_1}\left[x_2\left(x_1 + \frac{x_3}{M}\right) + y_2\left(y_1 + \frac{y_3}{M}\right)\right]\right\} dx_1 \, dy_1 \, dx_2 \, dy_2.
$$

$$
\tag{2.109}
$$

Performing the integrals in terms of x_2, y_2, we have

$$
h(x, y) = \iint P(x_2, y_2) \exp\left[\frac{ik}{d_1}\left(x_2 x + y_2 y\right)\right] dx_2 \, dy_2 \tag{2.110}
$$

which is the point spread function, given by the Fourier transform of the pupil function. Then

$$
U_3(x_3, y_3) = -\frac{1}{\lambda^2 M d_1^2} \exp\left[ikd_1(1 + M)\right]
$$

$$
\times \exp\left[\frac{ik}{2Md_1}\left(x_3^2 + y_3^2\right)\right] \iint U_1(x_1, y_1)
$$

$$
\times \exp\left[\frac{ik}{d_1}\left(x_1^2 + y_1^2\right)\right] h\left(x_1 + \frac{x_3}{M}, y_1 + \frac{y_3}{M}\right) dx_1 \, dy_1.
$$

$$
\tag{2.111}
$$

Now, for good imaging, h falls off quickly, that is, $x_1 + x_3/M$ is small, or $x_1 \approx -x_3/M$, so that (Goodman 1968)

$$
U_3(x_3, y_3) = -\frac{1}{\lambda^2 M d_1^2} \exp\left[ikd_1(1 + M)\right]
$$

$$
\times \exp\left[-\frac{ik}{2Md_1}\left(x_3^2 + y_3^2\right)\left(1 + \frac{1}{M}\right)\right]
$$

$$
\times \iint U_1(x_1, y_1) h\left(x_1 + \frac{x_3}{M}, y_1 + \frac{y_3}{M}\right) dx_1 \, dy_1. \tag{2.112}
$$

The intensity can thus be written as

$$
I_3(x_3, y_3) = \frac{1}{\left(\lambda^2 M d_1^2\right)^2} \left|\left(U_1 \otimes h\right)\right|^2, \tag{2.113}
$$

where $\otimes$ represents the convolution operation. We can show that this is also valid with defocus: then h is then the defocused point spread function and

$$
u = \frac{2\pi a^2}{\lambda}\left[\frac{1}{f} - \left(\frac{1}{d_1} + \frac{1}{d_2}\right)\right]. \tag{2.114}
$$

So, for coherent imaging, for an object $t(x, y)$, the intensity in the image is

$$
I_3(x_3, y_3) = \frac{1}{\left(\lambda^2 M d_1^2\right)^2} \left|\left(t \otimes h\right)\right|^2. \tag{2.115}
$$

Imaging is *linear* in amplitude, and space invariant, that is the convolution means that each point of the object results in a distribution of *amplitude* in the image given by the *amplitude* point spread function. Finally, the *intensity* in the image is given by finding the modulus squared of the amplitude.

Note that Equation 2.112 shows that the image of a point (x_1, y_1) in the object occurs at a point $(x_1 + x_3/M = 0, y_1 + y_3/M = 0)$, the center of the point spread function h. That is, at $x_3 = -Mx_1$, $y_3 = -My_1$: thus the image is *inverted* and magnified by a factor M.

Next, we consider the defocused case, $1/d_1 + 1/d_2 \neq 1/f$. We take

$$\frac{1}{d_1} + \frac{1}{d_2} - \frac{1}{f} = \frac{1}{d_0}. \qquad (2.116)$$

From Equation 2.108 we can see that Equation 2.109 is still valid if we put $P_{\text{eff}}(x_2, y_2)$, given by

$$P_{\text{eff}}(x_2, y_2) = P(x_2, y_2) \exp\left[-\frac{ik}{2d_0}(x_2^2 + y_2^2) \right]. \qquad (2.117)$$

This effective pupil function is called the defocused pupil function. It is a complex quantity, given by multiplying the ordinary pupil function by a quadratic phase variation. Equation 2.118 is only true for small defocus, as the approximation in Equation 2.112 is otherwise not valid. This is because the point spread function becomes broader with defocus, that is, the *intensity* point spread function behaves as shown in Figure 2.21.

As the system is defocused, the peak intensity decreases, and the pattern spreads out. By conservation of energy the total energy in the pattern, $\iint |h(x,y)|^2 \, dx \, dy$, must be constant. The zeros in the pattern also disappear with defocus. Note that the *amplitude* in the point spread function is complex for the defocused case.

Next, we now look at an example when defocus is not small. We consider the special case when $d_1 = d_2 = f$. So,

$$\frac{1}{d_0} = \frac{1}{d_1} + \frac{1}{d_2} - \frac{1}{f} = \frac{1}{f}, \qquad (2.118)$$

and $M = 1$. To solve this case, we return to Equation 2.108. As it stands, the expression does not simplify much! The reason is that there is the Fresnel diffraction by the object, which is then truncated by the pupil (Figure 2.22). So it is quite a complicated problem. However, it can be solved if we consider the pupil to be

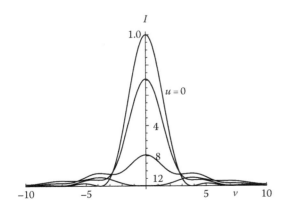

FIGURE 2.21 Defocused image of a point.

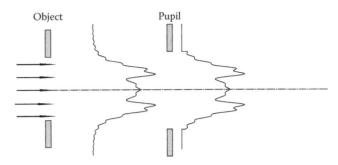

FIGURE 2.22 Truncation of the beam by the pupil.

very big, so there is negligible truncation. Then Equation 2.108 becomes

$$U_3(x_3, y_3) = -\frac{1}{\lambda^2 f^2} \exp(ikf) \iiiint U_1(x_1, y_1)$$

$$\times \exp\left[\frac{ik}{2f}(x_1^2 + y_1^2) \right] \exp\left[\frac{ik}{2f}(x_3^2 + y_3^2) \right]$$

$$\times \exp\left[-\frac{ik}{2f}(x_2^2 + y_2^2) \right]$$

$$\times \exp\left\{ -\frac{ik}{f}[x_2(x_1 + x_3) + y_2(y_1 + y_3)] \right\} dx_1 \, dy_1 \, dx_2 \, dy_2 \qquad (2.119)$$

So far, this is not much simpler! But we can do the integrals in x_2 and y_2 now. We have

$$\int_{-\infty}^{\infty} \int_{-\infty}^{\infty} \exp\left[-\frac{ik}{2f}(x_2^2 + y_2^2) \right] \exp\left\{ -\frac{ik}{f}[x_2(x_1 + x_3) + y_2(y_1 + y_3)] \right\} dx_2 \, dy_2. \qquad (2.120)$$

This is just the Fourier transform of a Gaussian (albeit of imaginary argument). You can get this from tables of Fourier transforms, or it can be evaluated using the properties of the Fresnel integrals. After putting $x = x_1 + x_3$, $y = y_1 + y_3$, it is

$$-\frac{2\pi f}{ik} \exp\left[-\frac{ik}{2f}(x^2 + y^2) \right]. \qquad (2.121)$$

The important features to notice are that it is independent of x_2, y_2, and when you put it back in Equation 2.119, the quadratic terms in x_1, x_3, y_1 and y_3, all cancel to give

$$U_3(x_3, y_3) = -\frac{ie^{2ikf}}{\lambda f} \iint U_1(x_1, y_1) \exp\left[-\frac{ik}{f}(x_1 x_3 + y_1 y_3) \right] dx_1 \, dy_1. \qquad (2.122)$$

Compare this with Equation 2.89. Again we have the 2D Fourier transform, but now the parabolic phase factor of Equation 2.89 is no longer present (Figure 2.23). This is a very important result, which is the basis of most Fourier optics systems. It allows us

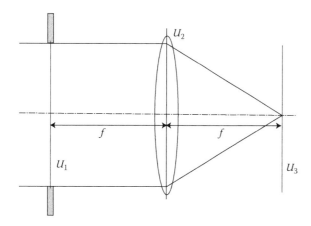

FIGURE 2.23   Imaging of an object in the front focal plane.

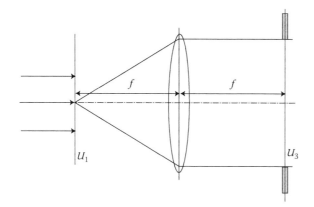

FIGURE 2.24   Fourier transformation by a lens.

to perform a 2D Fourier transform almost instantaneously (Figure 2.24), in the time for light to travel a distance 2*f*.

Note that if U_1 is a constant, assuming the pupil *P* is very big, U_3 is a delta function. If the pupil is finite, we get the point spread function. If U_1 is a circular function, a constant value within a circle, smaller than the pupil *P*, then the final amplitude is its Hankel function, that is the Airy disk $2J_1(v)/v$.

We have seen that if U_1 is a delta function, then U_3 is a constant. Two of these units may be coupled together, as in Figure 2.25. But the Fourier transform of the Fourier transform of $U_1(x)$ is $U_1(-x)$. So we just form an inverted, unity magnification image of U_1. This is called a *4f* optical system.

2.15 Coherent Transfer Function

We have looked at two imaging systems, shown in Figures 2.20 and 2.25. For the system in Figure 2.20, Equation 2.89 shows that the image amplitude is multiplied by a parabolic phase factor, but for the system in Figure 2.25 there is no phase factor, and the field U_3 in the output plane can be

$$U_3(x_3, y_3) = (U_1 \otimes h)(x_3, y_3). \tag{2.123}$$

In each case the amplitude point spread function *h* is given by the 2D Fourier transform of the pupil function.

Considering the system in Figure 2.25, we see that if the pupils are very big, then U_3 is a perfect image of U_1. If we think of U_1 as a grating, it produces various diffraction orders, and these are combined by the second lens to produce an image. However, the pupil *P* cuts off some of the diffraction orders and hence a perfect image is not formed in practice. *P* can therefore be thought of as having the effect of a coherent transfer function, a low pass filter. We resolve U_1 into gratings, and some of these orders get through the system. The strength of the spatial frequency components is multiplied by *P* to give their strength in the image. This is the principle of the Abbe theory of image formation in a microscope.

Mathematically, we introduce the Fourier transform of the object amplitude U_1, given by $\tilde{U}_1$

$$\tilde{U}_1(m, n) = \int_{-\infty}^{+\infty}\int_{-\infty}^{+\infty} U_1(x_1, y_1)\exp\left[-2\pi i(mx_1 + ny_1)\right]\mathrm{d}x_1\,\mathrm{d}y_1. \tag{2.124}$$

Inverting, we get

$$U_1(x_1, y_1) = \int_{-\infty}^{+\infty}\int_{-\infty}^{+\infty} \tilde{U}_1(m, n)\exp\left[2\pi i(mx_1 + ny_1)\right]\mathrm{d}m\,\mathrm{d}n. \tag{2.125}$$

But from Equation 2.123, and neglecting the multiplying constants we had previously

$$U_3(x_3, y_3) = \iint U_1(x_1, y_1)h\left(x_1 + \frac{x_3}{M}, y_1 + \frac{y_3}{M}\right)\mathrm{d}x_1\,\mathrm{d}y_1. \tag{2.126}$$

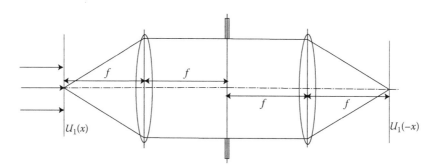

FIGURE 2.25   A *4f* system.

Substituting Equation 2.125 in Equation 2.136:

$$U_3(x_3, y_3) = \iiint \tilde{U}_1(m,n) h\left(x_1 + \frac{x_3}{M}, y_1 + \frac{y_3}{M}\right)$$
$$\times \exp\left[2\pi i (mx_1 + ny_1)\right] dm\, dn\, dx_1\, dy_1. \quad (2.127)$$

But we also have

$$h(x,y) = \iint P(x_2, y_2) \exp\left[-\frac{ik}{f_1}(x_2 x + y_2 y)\right] dx_2\, dy_2, \quad (2.128)$$

$$P(x_2, y_2) = \iint h(x,y) \exp\left[\frac{ik}{f_1}(x_2 x + y_2 y)\right] dx\, dy. \quad (2.129)$$

So, doing the integrals in x_1, y_1 in Equation 2.127, and putting $x = x_1 + x_3/M$ and so on,

$$\iint h_1(x,y) \exp\left\{2\pi i \left[m\left(x - \frac{x_3}{M}\right) + n\left(y - \frac{y_3}{M}\right)\right]\right\} dx\, dy$$

$$= P(m\lambda f_1, n\lambda f_1) \exp\left[-2\pi i \left(\frac{mx_3}{M} + \frac{ny_3}{M}\right)\right].$$

So Equation 2.127 can be written as

$$U_3(x_3, y_3) = \iint \tilde{U}_1(m,n) P(m\lambda f_1, n\lambda f_1)$$
$$\times \exp\left[-2\pi i \left(\frac{mx_3}{M} + \frac{ny_3}{M}\right)\right] dm\, dn. \quad (2.130)$$

So if pupil function is as shown in Figure 2.26a, the coherent transfer function is as shown in Figure 2.26b. For an object that is only a function of x, $n = 0$ and the corresponding coherent transfer function is given by a section through the two-dimensional transfer function (Figure 2.27). Note that in general

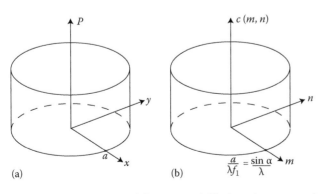

FIGURE 2.26 (a) The pupil function and (b) the coherent transfer function.

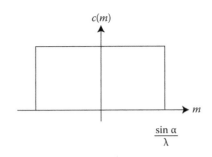

FIGURE 2.27 The coherent transfer function for imaging of a line structure.

there will be both positive and negative spatial frequencies. The imaging system behaves as a low-pass filter, that is, it transmits spatial frequencies less than $(\sin \alpha)/\lambda$. Note the sudden cutoff, corresponding to the edge of the pupil, as compared with transfer functions of electrical systems, which roll off smoothly.

2.15.1 A Grating Object

Consider as an example an amplitude transmittance that is a cosinusoidal variation on a constant background (Figure 2.28a):

$$U_1 = 1 + a\cos(2\pi \nu x). \quad (2.131)$$

The period of grating is $\Lambda = 1/\nu$. The spectrum of the object is the Fourier transform of U_1. As

$$U_1 = 1 + \frac{a}{2} e^{2\pi i \nu x} + \frac{a}{2} e^{-2\pi i \nu x}$$

we have for the object spectrum (Figure 2.28b)

$$T(m) = \tilde{U}_1 = \left[\delta(m) + \frac{a}{2}\delta(m-\nu) + \frac{a}{2}\delta(m-\nu)\right]\delta(n). \quad (2.132)$$

So, in this case the image amplitude is

$$U_3(x_3, y_3) = \iint \left[\delta(m) + \frac{a}{2}\delta(m-\nu) + \frac{a}{2}\delta(m+\nu)\right] c(m,n)\delta(n)$$

$$\times \exp\left[-\frac{2\pi i}{M}(mx_3 + ny_3)\right] dm\, dn$$

$$= c(0) + \frac{a}{2} c(\nu) \exp\left(-\frac{2\pi i \nu x_3}{M}\right) + \frac{a}{2} c(-\nu) \exp\left(\frac{2\pi i \nu x_3}{M}\right).$$

For a circular pupil, c is even; so $c(-\nu) = c(+\nu)$, and we obtain for the image amplitude

$$U_3(x_3, y_3) = c(0) + a c(\nu) \cos\left(\frac{2\pi \nu x_3}{M}\right).$$

In this case, if $\nu < a/\lambda f_1$ and $c(\nu) = 1$, the image is the same as the object but magnified by M.

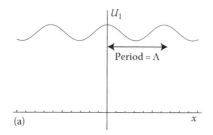 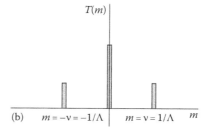

FIGURE 2.28 (a) A grating object and (b) its spatial frequency content.

The image *intensity* is thus

$$I_3\left(x_3, y_3\right) = \left| c(0) + ac(v)\cos\left(\frac{2\pi v x_3}{M}\right) \right|^2.$$

For the moment, take *a* as *real*, and also *c* as real; then

$$I_3(x_3, y_3) = c^2(0) + 2c(0)c(v)a\cos\left(\frac{2\pi v x_3}{M}\right) + a^2 c^2(v)\cos^2\left(\frac{2\pi v x_3}{M}\right). \tag{2.133}$$

Using the identity $\cos(2\theta) = \cos^2\theta - 1$, we obtain

$$I_3\left(x_3, y_3\right) = \left[c^2(0) + \frac{a^2}{2}c^2(v) \right] + 2c(0)c(v)a\cos\left(\frac{2\pi v x_3}{M}\right)$$
$$+ \frac{a^2}{2}c^2(v)\cos\left(\frac{4\pi v x_3}{M}\right). \tag{2.134}$$

Note that a second harmonic term is introduced by the squaring operation, but this can be neglected if *a* is small.

Now let us consider what happens if *a* and *c* are complex, and |*a*| is small. Now Equation 2.133 becomes

$$I_3\left(x_3, y_3\right) = \left| c(0) \right|^2 + 2c(0)\operatorname{Re}\left[ac(v) \right]\cos\left(\frac{2\pi v x_3}{M}\right). \tag{2.135}$$

If we consider an object whose *thickness l* varies in a cosinusoidal fashion

$$l = l_0 + l_1 \cos(2\pi v x),$$

Then if its refractive index is *n*, the phase change on passing through it is *nl*, and so the amplitude on the far side if it is illuminated with a plane wave is

$$t = \exp\left\{ik\left[l_0 + l_1 \cos(2\pi v x) \right]\right\}$$
$$= \exp(ikl_0)\exp\left[ikl_1 \cos(2\pi v x) \right]. \tag{2.136}$$

The first part of this is just a constant phase term and hence can be ignored. We then expand the second part into its Fourier components. Actually, this is identical to frequency modulation (FM) in communication theory, and the strengths of the various harmonics are given by Bessel functions. Let us just look at the much simpler case when $kl_1 \ll 1$ however. Then

$$t = 1 + ikl_1 \cos(2\pi v x). \tag{2.137}$$

We can see that this is the same as Equation 2.131 with *a* given by an imaginary quantity. Thus a phase object is represented by imaginary *a*. We see from Equation 2.135, that if *c*(v) is *real* then there is no image, and we just see a constant intensity. Note that this is not in general true for a strong phase object where terms of strength a^2 cannot be ignored.

Previously, we introduced the concept of the defocused pupil function. From Equations 2.100 or 2.117 we have

$$P_{\text{eff}}\left(\rho\right) = P(\rho)\exp\left(-\frac{iu}{2}\rho^2\right), \tag{2.138}$$

so that now we introduce the defocused transfer function *c*(*m*,*u*), as shown in Figure 2.29. For *u* = 0, that is the focused case, *c*(*m*) is purely real. As *u* is increased, the imaginary part increases in strength. For *u* > π, the real part starts to go negative, which is

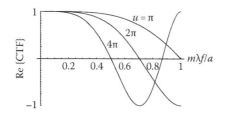 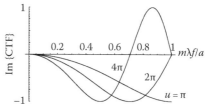

FIGURE 2.29 Defocused coherent transfer function.

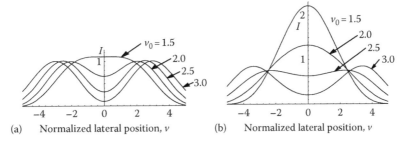

FIGURE 2.30   Images of a two-point object: (a) incoherent imaging and (b) coherent imaging.

not good for imaging of amplitude information. The imaginary part images phase information, but this also starts to go negative for $u > 2\pi$. So we only get good imaging of phase information if $u \leq 2\pi$. If u is negative the imaginary part of $c(m, u)$ also becomes positive, so contrast is reversed. For a phase object, there is no contrast at the focus, and positive or negative contrast on either side of focus. Defocusing of the microscope was often used to image weak phase structures, as for example in biological samples, before more modern methods of phase imaging were invented.

2.15.2 Square Wave Object

Consider a square wave object, with transmittance as shown in Figure 2.30. Then

$$U_1(x) = \frac{1}{2} + \frac{2}{\pi}\left[\cos(2\pi vx) - \frac{1}{3}\cos(6\pi vx) + \frac{1}{5}\cos(10\pi vx) - \cdots\right].$$
(2.139)

The harmonics are transmitted up to the cutoff frequency. By making $1/v$ very large, we can calculate the image of a single edge. Interestingly, although the number of terms then is very large we still get an overshoot as in the image of a square wave object. The image of an edge can be expressed in terms of Fresnel integrals. The fact that the wiggles do not disappear for a large number of terms is a consequence of the Gibbs phenomenon.

2.16 Spatial Filtering

The principle of spatial filtering is to alter the strength in the image of the Fourier components of the object by putting in an appropriate mask: For example, suppose we change the phase everywhere except at $m = 0$ by $\pi/2$. Then the transfer function is

$$c(v) = \begin{cases} -i & v \neq 0 \\ 1 & v = 0 \end{cases}.$$
(2.140)

Then from Equation 2.143

$$I_3 = 1 + 2\,\mathrm{Re}\left(-ai\right)\cos\left(\frac{2\pi vx_3}{M}\right)$$

$$= 1 + 2\,\mathrm{Im}(a)\cos\left(\frac{2\pi vx_3}{M}\right).$$
(2.141)

So again we have managed to image the phase information. This method is called Zernike phase contrast, for the invention of which Zernike received the Nobel Prize. In practice, it is easier to change the phase just of the direct beam $m = 0$, rather than vice versa, but the result is exactly the same. Note that changing the phase by $-\pi/2$ rather than by $+\pi/2$ reverses the contrast. Also, if we make the filter

$$c(v) = \begin{cases} -bi, & v \neq 0, \\ 1, & v = 0, \end{cases}$$
(2.142)

we then have

$$I_3 = 1 + 2b\,\mathrm{Im}(a)\cos\left(\frac{2\pi vx_3}{M}\right).$$
(2.143)

If $b > 1$, we also enhance the contrast by a factor b, making very weak phase objects visible.

2.17 Incoherent Imaging

Most of what we have said so far is applicable to *coherent* optical systems. This requires that the point spread function of the imaging system is small in extent compared with the lateral spatial coherence of the illumination. The opposite condition, where the point spread function is large compared with the spatial coherence, results in *incoherent* imaging. Incoherent imaging is in everyday life a much more common phenomenon. For example photography, or just seeing, is normally an incoherent process. Fluorescence also results in incoherent image formation. The in-between case, where the point spread function is about the size of the spatial coherence of the illumination, is *partially coherent*. This case is much more complicated, and arises for example in the theory of microscope imaging. In incoherent imaging, because there is no coherent interference between neighboring points, we have to sum the intensities for these points, rather than the complex amplitudes. We return to the geometry of Section 2.14.

Each point in the object results in a diffraction blur in the image. To obtain the total image we add the contributions from the *intensities* of the individual points.

For *coherent* imaging, we had in Equation 2.108

$$I_3(x_3, y_3) = \frac{1}{\left(\lambda^2 M d_1^2\right)^2} |t \otimes h|^2,$$

but now for *incoherent* imaging, we have

$$I_3(x_3, y_3) = \frac{1}{\left(\lambda^2 M d_1^2\right)^2} \left(|t|^2 \otimes |h|^2\right). \tag{2.144}$$

That is we must convolve the object *intensity* $|t|^2$ with the *intensity* point spread function $|h|^2$. Note that the term point spread function can refer to either the amplitude or the intensity point spread function in the literature.

We can derive Equation 2.144 properly by considering a single point in the object U_1 at (x,y). Its image from Equation 2.111 is

$$U_3(x_3, y_3) = \frac{1}{\lambda^2 M d_1^2} \exp\left[ikd_1(1+M)\right] \exp\left[\frac{ik}{2Md_1}\left(x_3^2 + y_3^2\right)\right]$$

$$\times \iint \delta(x_1 - x)\delta(y_1 - y)$$

$$\times \exp\left[\frac{ik}{d_1}\left(x_1^2 + y_1^2\right)\right] h\left(x_1 + \frac{x_3}{M}, y_1 + \frac{y_3}{M}\right) dx_1 dy_1$$

$$= \frac{1}{\lambda^2 M d_1^2} \exp\left[ikd_1(1+M)\right] \exp\left[\frac{ik}{2Md_1}\left(x_3^2 + y_3^2\right)\right]$$

$$\times \exp\left[\frac{ik}{d_1}\left(x^2 + y^2\right)\right] h\left(x + \frac{x_3}{M}, y + \frac{y_3}{M}\right).$$

So the intensity in the image of a single point is simply

$$I_3(x_3, y_3) = \frac{1}{\left(\lambda^2 M d_1^2\right)^2} \left| h\left(x + \frac{x_3}{M}, y + \frac{y_3}{M}\right) \right|^2. \tag{2.145}$$

Finally, adding up for the points of the object

$$I_3(x_3, y_3) = \frac{1}{\left(\lambda^2 M d_1^2\right)^2} \iint \left| h\left(x + \frac{x_3}{M}, y + \frac{y_3}{M}\right) \right|^2 |t(x,y)|^2 \, dx \, dy \tag{2.146}$$

which represents Equation 2.144.

2.17.1 Two-Point Object

One of the most important theoretical objects is two bright points in a dark background (e.g., two pinholes in an opaque screen, or two stars). The image is then, for different normalized separations $2v_0$, as shown in Figure 2.31a. For $v_0 = 2.5$ the points are well resolved. For $v_0 = 1.5$ they are not: they just almost look like one point. It is traditional to say that the points are just resolved when the maximum of one point spread function is placed over the zero of the other. This is called the Rayleigh criterion, and occurs when $v_0 = 1.92$. It is found that for a circular pupil aperture, the intensity in the middle is then 0.735 times the intensity at the points themselves. Note that the Rayleigh criterion applies for incoherent imaging. In general, though, we may introduce the *generalized Rayleigh criterion*, which states that the points are just resolved if the intensity at the center is 0.735 times that at the points. The intensity at the points need not be the same as the intensity at the maximum: some published papers have got this wrong! By putting in the values for the Bessel function we obtain, for the incoherent case

$$\left(\Delta x\right)_{\min} = 0.61 \frac{\lambda}{\sin\alpha}. \tag{2.147}$$

This is often written as $1.22 f \lambda/D$ in terms of the diameter D of the pupil, rather than the radius. A typical value for a high power microscope is about $0.5\,\mu m$. For the coherent case, similar plots are shown in Figure 2.31b. We find

$$\left(\Delta x\right)_{\min} = 0.82 \frac{\lambda}{\sin\alpha}, \tag{2.148}$$

that is, the resolution is not as good.

2.17.2 Optical Transfer Function

Equation 2.146 is linear in intensity. This means we can introduce a transfer function, usually called the optical transfer function (OTF). Note that this operates on *intensities*, rather than the amplitudes for the coherent transfer function described earlier. So they are not strictly comparable.

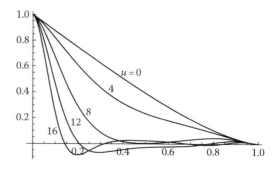

FIGURE 2.31 Defocused optical transfer function for a circular pupil.

We introduce the object *intensity* spectrum, given by the Fourier transform of its intensity, which is in contrast with Equation 2.124,

$$V_1(m,n) = \int\limits_{-\infty}^{+\infty} \int\limits_{-\infty}^{+\infty} |U_1(x_1,y_1)|^2 \exp\left[-2\pi i(mx_1 + ny_1)\right] dx_1 \, dy_1.$$

(2.149)

Then we can show by exactly the same method as in Section 2.16 that

$$I(x_3,y_3) = \iint V_1(m,n)C(m,n)\exp\left[-2\pi i\left(\frac{mx_3}{M} + \frac{ny_3}{M}\right)\right]dm\,dn$$

(2.150)

where the OTF, $C(m,n)$, is given by the Fourier transform of the intensity point spread function $|h|^2$. As for the coherent case we had $P(m\lambda f, n\lambda f) = F(h)$, where $F(.)$ represents the Fourier transform, now we have

$$C(m,n) = F(|h|^2)$$

$$= P(m\lambda f, n\lambda f) \otimes P^*(m\lambda f, n\lambda f),$$ (2.151)

This 2D convolution represents the area of overlap of the two pupils. We can show that the area of overlap for two circles is

$$\frac{2}{\pi}\left[\cos^{-1}(\tilde{m}) - \tilde{m}\sqrt{1-\tilde{m}^2}\right],$$

(2.152)

which is shown in Figure 2.30. The cutoff frequency is twice that for a coherent system.

If the system is defocused, we must integrate over the area of overlap taking into account the phase of the pupil. This cannot be done analytically, but can be expressed in terms of a single integral (Hopkins 1955). The result is also shown in Figure 2.31. The response drops off with defocus, that is, imaging of these particular spatial frequency components is worse. It is the mid-spatial frequencies that are most strongly affected, resulting in poorer imaging. Note that the OTF must always be purely real. An important feature is that the OTF can go negative with defocus. This means that some spatial frequency components have their contrast reversed. This results in optical artifacts, which means that you can see something that is not really in the object.

References

Abramowitz, M. and Stegun, I. A. 1965. *Handbook of Mathematical Functions*. New York: Dover.

Born, M. and Wolf, E. 1975. *Principles of Optics*. Oxford: Pergamon Press.

Goodman, J. W. 1968. *Introduction to Fourier Optics*. New York: McGraw-Hill.

Hecht, E. 1987. *Optics*. Reading, MA: Addison-Wesley.

Hopkins, H. H. 1955. The frequency response of a defocused optical system. *Proc. R. Soc. Lond. Ser. A* 231: 91–103.

Optics: Basic Physics

3.1 Introduction 33
3.2 Electromagnetic Waves and Wave Motion 33
Light Is an Electromagnetic Wave • Waves—Notation • Superposition • Further Properties of Electromagnetic Waves • The Electromagnetic Spectrum • Coherence
3.3 Diffraction..36
Two-Slit Interference • *N*-Slit Interference • The Diffraction Grating as a Monochromator • Single-Slit Interference • Diffraction and Resolution
3.4 Refraction...41
Snell's Law • Fermat's Principle • Total Internal Reflection
3.5 Lenses..42
A Spherical Interface • A Spherical Interface—The Paraxial Regime • Focal Points • Real and Virtual Images • Concave Lenses • Thin Lenses • Magnification • Resolution
3.6 Reflection and Transmission (Fresnel's Equations) 46
Case I: $\vec{E}$ Perpendicular to the Plane of Incidence • Case II: $\vec{E}$ Parallel to the Plane of Incidence • Brewster's Angle
3.7 Concluding Remarks...47
References...48

Raghuveer Parthasarathy
University of Oregon

3.1 Introduction

Optics is both a very old and a very contemporary field of research. Mirrors made millennia ago as well as the advanced imaging methods that decorate recent years' lists of scientific breakthroughs (Betzig et al. 2006, Bates et al. 2007, Hell 2007, Abbott 2009) can all be understood with the same physical framework. We will explore the basic physics of optics in this chapter, intended to serve as a review of elementary principles, or as an introduction for readers new to optics. Our treatment is necessarily brief and minimal—the reader interested in further elaboration should consult a textbook devoted to optics, such as Hecht (2002) or Born and Wolf (1997).

3.2 Electromagnetic Waves and Wave Motion

3.2.1 Light Is an Electromagnetic Wave

Insightful experiments by Hans Christian Ørsted, Michael Faraday, and others in the first half of the nineteenth century revealed the principle of electromagnetic induction: a changing magnetic field gives rise to an electric field, and, conversely, a changing electric field creates a magnetic field. Later, James Clerk Maxwell synthesized these and other observations into a set of succinct mathematical expressions, known as Maxwell's equations, which encapsulate the core of classical electromagnetism.

It follows simply from these that electric ($\vec{E}$) and magnetic ($\vec{B}$) fields in vacuum can be connected by the relations

$$\nabla^2 \vec{E} = \int \varepsilon_0 \mu_0 \frac{\partial^2 \vec{E}}{\partial t^2},$$

$$\nabla^2 \vec{B} = \varepsilon_0 \mu_0 \frac{\partial^2 \vec{B}}{\partial t^2},$$

where

ε_0 is the permittivity of free space (a constant)

μ_0 is the permeability of free space (another constant)

t is the time

Both of these expressions have form of a wave equation

$$\nabla^2 \psi = \frac{1}{v^2} \frac{\partial^2 \psi}{\partial t^2},$$

which admits solutions of the form $\psi(\vec{r}, t) = f(\vec{r} - \vec{v}t)$, i.e., waves traveling through space ($\vec{r}$) and time (t) with velocity $\vec{v}$. Maxwell therefore realized that electric and magnetic fields can propagate as traveling waves, with a speed that is a simple function of electrostatic and magnetostatic constants: $c = (\varepsilon_0 \mu_0)^{-1/2}$. Inserting values for ε_0 and μ_0 yields $c = 3.0 \times 10^8$ m/s, in striking correspondence to the speed of light, which had been measured with few-percent accuracy by the mid-nineteenth century. Especially following the experiments of Heinrich Rudolph Hertz, in which electromagnetic waves were generated and detected, it became clear that light is an electromagnetic wave and that visible light is but one part of a broader electromagnetic spectrum.

3.2.2 Waves—Notation

As discussed above, electric and magnetic fields in space obey wave equations. To define the terms and symbols related to wave motion, we will first consider a one-dimensional wave equation, the simplest solution to which is a sinusoidal traveling wave of *amplitude A*, *wavenumber k*, and *angular frequency* ω:

$$\psi(x,t) = A\cos(kx - \omega t - \delta) = \text{Re}\left\{A\exp\left[j(kx - \omega t - \delta)\right]\right\}, \quad \text{where}$$

$j = \sqrt{-1}$ and δ is a phase offset. (We generally will not bother explicitly writing that the real part of the complex exponential is to be considered.) The *wavelength* is given by $\lambda = 2\pi/k$, and the *frequency* by $f = \omega/2\pi$; if we consider a particular position in space, ψ oscillates with *period* $T = f^{-1}$. The wave speed is related to the other variables by $v = \omega k^{-1} = \lambda f$, as the reader may wish to illustrate by drawing the wave for various values of t. The argument of the oscillatory function is often referred to as the phase: $\phi(x,t) = kx - \omega t - \delta$. Considering a particular moment in time, the phase advances by 2π over a distance is given by λ; over an arbitrary distance Δx, the phase shift is $\Delta\phi = 2\pi\Delta x\lambda^{-1}$.

For the one-dimensional traveling wave noted above, each point in space corresponds to a particular phase. In two- or three-dimensions, more complex structures arise. It is useful to consider points of equal phase, which we will refer to as *wavefronts*.

Plane waves. A simple and very useful construction is the *plane wave*. Let us illustrate this for a two-dimensional wave (Figure 3.1), in which we can plot the value of ψ along the third dimension.

Note that ψ only varies along one spatial dimension (in this case, x). Contours of equal phase (i.e., wavefronts) are lines in the xy plane. As the wave travels, for the example shown in Figure 3.1, it moves in the $\hat{x}$-direction—i.e., parallel to a wavevector, $\vec{k}$,

that is perpendicular to these lines of constant phase and parallel to $\hat{x}$. We can write $\psi(x,y) = A\exp[j(kx - \omega - \delta)]$, or $\psi(\vec{r}) = A\exp[j(\vec{k}\cdot\vec{r} - \omega t - \delta)]$, where $\vec{r}$ is a vector in the xy plane—note that the dot product selects the x-component of $\vec{r}$.

For a three-dimensional plane wave, positions of constant phase (i.e., wavefronts) form a set of parallel planes. This is a good description of many sorts of light beams. Furthermore, any three-dimensional wave can be expressed as a combination of plane waves by Fourier analysis. The three-dimensional plane wave is described by $\psi(\vec{r}) = A\exp[j(\vec{k}\cdot\vec{r} - \omega t - \delta)]$, where $\vec{r}$ is any vector in three-dimensional space. We will show this explicitly: consider a position vector $\vec{r} = x\hat{x} + y\hat{y} + z\hat{z}$, where ^ indicates a unit vector, and some particular vector $\vec{r}_0$. Their difference:

$$\vec{r} - \vec{r}_0 = (x - x_0)\hat{x} + (y - y_0)\hat{y} + (z - z_0)\hat{z}.$$

Consider the set of points $\{\vec{r}\}$ described by $(\vec{r} - \vec{r}_0)\cdot\vec{k} = 0$. As $\vec{r}$ varies, this sweeps out a plane perpendicular to $\vec{k}$. Expanding this: $(\vec{r} - \vec{r}_0)\cdot\vec{k} = k_x(x - x_0) + k_y(y - y_0) + k_z(z - z_0) = 0$, or $k_x x + k_y y + k_z z = a$, where $a = k_x x_0 + k_y y_0 + k_z z_0$ is a constant. Therefore, the equation of a plane perpendicular to $\vec{k}$ is $\vec{k}\cdot\vec{r} = \text{constant} = a$. The set of planes over which $\psi(\vec{r})$ (at $t = 0$) varies sinusoidally is $\psi(\vec{r}) = A\cos(\vec{k}\cdot\vec{r})$ or $\psi(\vec{r}) = A\exp(j\vec{k}\cdot\vec{r})$. This function is periodic if $\vec{k}\cdot\vec{r}$ changes by 2π, i.e., $|\vec{k}|\lambda = 2\pi$, or $k = |\vec{k}| = 2\pi/\lambda$, as expected. The traveling plane wave is described by $\psi(\vec{r}) = A\exp[j(\vec{k}\cdot\vec{r} - \omega t - \delta)]$. To reiterate, the wavefronts of a three-dimensional plane wave are planes. Typically, we will only draw wavefronts that are separated in phase by $\Delta\phi = 2\pi$, which are therefore spatially separated by distance λ. The wavevector $\vec{k}$ points perpendicular to these planes. Often, one describes the wave by a *ray* that points along $\vec{k}$.

Spherical waves. A point-source of light emits *spherical waves*—the wavefronts are concentric spheres that travel away from the point. The wave function is

$$\psi(\vec{r},t) = \frac{A}{r}\exp\left[j(kr - \omega t - \delta)\right],$$

where

A is a constant

r is the distance from the source

The amplitude decreases with r, for reasons that will become clear shortly.

Cylindrical waves. A line-source of light, for example, a slit, emits *cylindrical waves*—the wavefronts are concentric cylinders that travel away from the line. The wave function is

$$\psi(\vec{r},t) = \frac{A}{\sqrt{r}}\exp\left[j(kr - \omega t - \delta)\right],$$

where

A is a constant

r is the distance from the line

Again, the amplitude decreases with r.

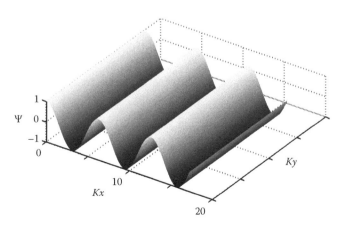

FIGURE 3.1 A two-dimensional plane wave: $\psi(x,y) = \cos(kx - \omega t)$, plotted at time $t = 0$.

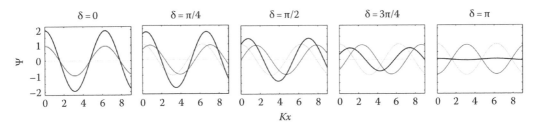

FIGURE 3.2 The sum $\psi = \psi_1 + \psi_2$ (black) of the waves $\psi_1 = 1.0 \cos(kx)$ (light gray) and $\psi_2 = 0.95 \cos(kx - \delta)$ (medium gray), plotted for various values of δ.

3.2.3 Superposition

The wave equation is linear in ψ; therefore, its solutions obey the principle of *superposition*: If ψ_1 and ψ_2 each satisfy the wave equation, then $\psi = \psi_1 + \psi_2$ is also a solution. The relative phase difference between ψ_1 and ψ_2 is important in determining their *interference*:

Figure 3.2 shows an illustration of the superposition of two sine waves. I have plotted $\psi_1 = 1.0 \cos(kx)$, $\psi_2 = 0.95 \cos(kx - \delta)$, and $\psi = \psi_1 + \psi_2$ for various values of δ. (I have chosen slightly different amplitudes for these two waves to make the illustrations clearer.)

Note that a phase difference $\delta = 0$ leads to *constructive interference*, and a phase difference $\delta = \pi$ leads to *destructive interference*.

3.2.4 Further Properties of Electromagnetic Waves

3.2.4.1 Electromagnetic Waves in Matter

We noted above that electric and magnetic fields can propagate in free space as waves, with speed $c = 3.0 \times 10^8$ m/s. In a transparent material of index of refraction n (related to the polarizability of the material), fields also propagate as waves, but more slowly, with speed $v = c/n$. For air at 20°C and atmospheric pressure, $n = 1.0003$. For water at 20°C, $n = 1.33$. For typical glass, $n = 1.46$. The *frequency* of the wave is unchanged from its value in vacuum—the rate of oscillation of the atoms excited by the electric field is constant. The *wavelength* of the light is different from its value in vacuum and obeys the general relation encountered earlier: $v = \lambda f$. Therefore, waves in matter are shorter than in free space: $\lambda = v/f = c/nf$. The wavelength in matter $\lambda = \lambda_0/n$, where λ_0 is the free space wavelength, and so the phase shift corresponding to a change in position Δx along the wave is $\Delta\phi = 2\pi(\Delta x/\lambda) = 2\pi n(\Delta x/\lambda_0)$.

Generally, when one states a wavelength for light, it is the *free space wavelength*, λ_0, that is being referred to—we say that orange light has a wavelength of ≈ 600 nm, even though when it enters your eye ($n = 1.3$), its wavelength shortens to ≈ 450 nm.

3.2.4.2 Polarization

Another consequence of electrodynamics is that the electric and magnetic field vectors at any point are perpendicular to one another and to the wave's propagation direction (Figure 3.3). The

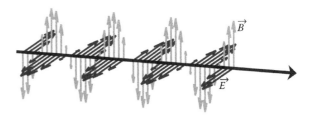

FIGURE 3.3 Electric (dark gray) and magnetic (light gray) fields of a plane-polarized electromagnetic wave. The black arrow indicates the propagation direction, perpendicular to $\vec{E}$ and $\vec{B}$.

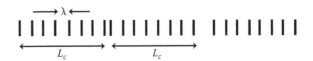

FIGURE 3.4 The coherence length, L_C, describes the spatial extent over which wavefronts (planes that differ by a phase shift of 2π) are separated by integer multiples of the wavelength. Over distances larger than $\approx L_C$, the coherence of the wave with itself—the ability to translate by an integer number of wavelengths and "match up"—is lost.

magnitudes of the field amplitudes are related by $|\vec{E}| = v|\vec{B}|$, where v is the speed. The direction of $\vec{E}$ specifies the *polarization* of the wave. If this direction is constant, as in Figure 3.4, we say that the wave is *linearly polarized* (or *plane polarized*). In Figure 3.4, for example, note that $\vec{E}$ is always parallel to the x-axis (in other words, $\vec{E} = E(z,t)\hat{x}$). Waves do not have to be plane polarized and can do a variety of interesting things. If the direction of $\vec{E}$ rotates as the wave propagates, then we have *circular* or elliptical polarization. (We will not go into the difference between the two—the reader can explore this as well as other constructions, such as radial polarization.)

3.2.4.3 Energy and Momentum

Electromagnetic waves carry energy and momentum. The power per unit area crossing a surface is $\vec{S} = c^2 \epsilon_0 \vec{E} \times \vec{B}$, known as the Poynting vector. Note that it points along the propagation direction (i.e., parallel to $\vec{k}$), not surprisingly. Because $|\vec{E}| = v|\vec{B}|$, $\vec{S}$ is proportional to $|\vec{E}|^2$.

The *intensity* (or *irradiance*), *I*, of the wave is the average energy carried per unit area per unit time, i.e., the power per unit area. It is the intensity, not the electric field directly, that we "see" as brightness. "Average" means that we consider the average power over a period. (Note that the intensity is a *number*, not a vector.) Since $\vec{S}$ is proportional to $\left|\vec{E}\right|^2$, the intensity of an electromagnetic wave is proportional to $\left|\vec{E}\right|^2$ as well. This is in general true for vibrations and waves: the energy of a wave is proportional to the square of its amplitude, and therefore

$$I \propto \left|\vec{E}\right|^2 = \vec{E} \cdot \vec{E}^*.$$

The principle of conservation of energy and the proportionality of $\vec{S}$ and *I* is on $\left|\vec{E}\right|^2$ explain the decaying amplitude of the spherical and cylindrical waves discussed in Section 3.2.2. For a spherical wave, integrating the power carried by the wave over a shell of radius *r* surrounding the source must give a result that is independent of *r*—all the power must cross the shell, regardless of the size of the shell. The shell area scales as r^2 and so $\vec{S}$ must scale as r^{-2} for the product to be independent of *r*, from which we conclude that the amplitude scales as $\sqrt{r^{-2}} = r^{-1}$ for a three-dimensional spherical wave.

We often can ignore the constant of proportionality, being concerned with relative intensities. However, for completeness: $I = (1/2)\varepsilon_0 c E_0^2$ in vacuum, where E_0 is the electric field amplitude. In matter, $I = (1/2)\varepsilon_0 v E_0^2$, where *v* is the speed of the wave and ε is the permittivity of the medium. The ability of light to carry energy and momentum has become especially important in recent years with the development of optical trapping techniques, in which light itself is used to grab, pull, push, and twist microscopic objects.

Though it travels as a wave, light carries energy in discrete, quantized packets. This realization, primarily by Max Planck and Albert Einstein in the early twentieth century, marked the birth of quantum mechanics. The energy of a *photon*, the quantized "unit" of light, is proportional to its frequency and hence inversely proportional to its wavelength. More precisely, the photon energy $E = hf = hc/\lambda$, where $h = 6.626 \times 10^{-34}$ m^2 kg/s is Planck's constant. Photons of lower wavelength have more energy. This explains, for example, why the emission of light from fluorescent molecules necessarily occurs at higher wavelengths than does absorption: a photon is absorbed, and some energy is converted into nonradiative (e.g., vibrational) modes, leaving a smaller quantum of energy for emission.

3.2.5 The Electromagnetic Spectrum

The range of wavelengths of electromagnetic waves that are relevant to science and technology is enormous, ranging from very high-energy gamma rays spouting from astrophysical sources ($\lambda \approx 10^{-13}$ m) to x-rays used to probe molecular structure ($\lambda \approx 10^{-10}$ m) to microwaves ($\lambda \approx 10^{-2}$ m) to radio waves ($\lambda > 1$ m). "Visible light" spans the tiny range of wavelengths from about 400 (violet/blue) to 800 nm (red), yet its correspondence to the energetics of electronic transitions in molecules and, relatedly, our ability to see it, makes it an immensely useful part of the electromagnetic spectrum.

3.2.6 Coherence

We have been considering waves as ideal sinusoidal forms that oscillate at a unique frequency and extend infinitely through space. For such a perfectly *coherent* wave, the wavefronts are always separated by distance λ, and knowing the phase at one point specifies it at all points. For real waves, this is not exactly the case. The wavefronts of light from a real, imperfect wave, are separated by λ if we consider some finite span of approximate size L_C, but if we look at larger lengths, the phase relations appear "randomized"—see Figure 3.4.

This lack of perfect coherence arises from the emission of any real source not being perfectly monochromatic, but rather consisting of a range of output frequencies, Δf. (This is due to factors such as the finite linewidth of electronic transitions and the thermal velocities of atoms and molecules.) Roughly, $L_C = c/n \, \Delta f$. Furthermore, the light from extended sources such as an incandescent light bulb or the sun is emitted by many independent sources throughout the object, and each emitted wave has a random phase relative to any other. Such sources are referred to as *incoherent* light sources. The length L_C referred to above is called the *coherence length*—it is about 10 µm (around 20 λ) for a light bulb. (There is a more precise way to define the coherence length that we will not go into here.)

A *laser* is a *coherent* light source—all the waves emitted by the device have the same phase. Moreover, L_C is typically around 1 m ($>10^6 \, \lambda$) and can even be kilometers in length—a good approximation to our ideal infinite wave. This remarkable property of lasers contributes to their tremendous utility, as will be evident later in this book.

3.3 Diffraction

For centuries, debate raged over whether light is a wave or a particle—an interesting history that we will not go into. Whether or not it is important to consider the wave-nature of light in describing its propagation, rather than simply imagining rays of light that travel in simple geometric paths, depends on the spatial scale of the phenomena being considered. For features that are not large compared to the wavelength (λ), for example, visible light passing through micron-sized slits or kilometer-sized radio waves detected by an array of dishes, the wave nature of light is inescapable. Light's interference with itself determines its intensity profile, and diffraction—this interference being induced by barriers or obstacles—is paramount. This regime in which the wave nature of light is important is called *physical optics*. The regime in which the system size is much greater than the wavelength of light, and hence wave properties are relatively unimportant, is called *geometric optics* or *ray optics*.

Diffraction is a general property of waves, and the phenomena we will explore in this section also apply to water waves, sound waves, etc.

3.3.1 Two-Slit Interference

Consider a plane wave incident on a barrier with two slits, separated by a distance D (Figure 3.5). (Imagine the slits themselves to have negligible width—we will return to this later.) Each slit acts as a point-source for waves, which continue propagating to the right in the figure. Far to the right is a screen. We want to know the *intensity*, I, of the light hitting the screen as a function of θ, the angle relative to a line perpendicular to the barrier (see Figure 3.5).

The electric field of the incident wave is

$$\vec{E} = \vec{E}_0 \exp\left[j(kx - \omega t) \right],$$

with $k = 2\pi/\lambda$, as usual (see Section 3.2). We could add any phase offset to this—it does not matter, as we will see shortly. We are concerned with the light hitting a far-off screen, at angle θ. If the screen were close by, a ray would have to leave slit #1 at some angle θ_1 and slit #2 at some angle θ_2, where θ_1 and θ_2 may be different, to both reach the screen at angle θ. However, as the screen moves farther and farther away, both θ_1 and θ_2 approach θ—try drawing this if it's not evident. So, to consider $I(\theta)$, we need to consider rays leaving each slit at angle θ. Let us define our coordinates so that the barrier is at $x = 0$.

The two rays that travel at angle θ are indicated in Figure 3.6; their fields are

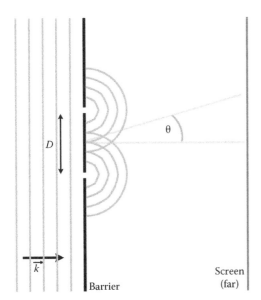

FIGURE 3.5 Two-slit interference: A plane wave is incident from the left on two slits of negligible width separated by distance D. Each slit acts like a point source for waves continuing to the right; the two resulting waves interfere with one another. This interference is manifested in the pattern of light intensity observed on a distant screen, and is a function of the wavelength, D, and the angle θ.

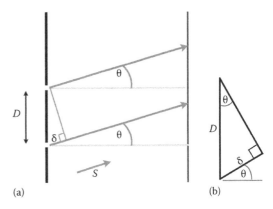

FIGURE 3.6 The geometry of light propagation for two-slit interference. (a) For the angle θ illustrated, light traveling from the lower slit (slit #2) travels a greater distance than light from slit #1. The extra path length is denoted δ and is the reason for a phase difference between the two waves. (b) A "zoomed in" view of the geometry relating D, δ, and θ.

$$\vec{E}_1 = \vec{E}_0 \exp\left[j(ks - \omega t) \right],$$

$$\vec{E}_2 = \vec{E}_0 \exp\left[j\left(k(s + \delta) - \omega t\right) \right],$$

where we have defined s as the coordinate in the "tilted" θ-direction, and we have indicated the extra distance that ray 2 has to travel by δ. Note that $\vec{E}_1(s = 0)$ and $\vec{E}_2(s = -\delta)$ have the same phase, as they should, since they come from the same incident wave.

Graphically, we can see that if δ is an integer multiple of λ, the two waves will add constructively (Figure 3.7). If δ is a half-integer multiple of λ, the two waves will add destructively and give zero light intensity.

Let us examine this mathematically. The superposition of the two electric fields:

$$\vec{E} = \vec{E}_1 + \vec{E}_2 = \vec{E}_0 \exp\left[j(ks - \omega t) \right]\left\{1 + \exp\left[jk\delta \right]\right\}.$$

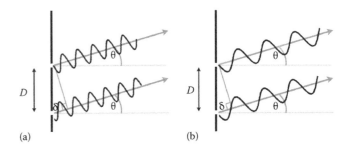

FIGURE 3.7 (a) If the extra path length, δ, between the two paths is an integer multiple of the wavelength, the two waves will constructively interfere, leading to high intensity at the screen. (b) If the extra path length δ between the two paths is a half-integer multiple of the wavelength, the two waves will destructively interfere, leading to zero intensity at the screen—note that when wave #1 is "up," wave #2 is "down" and vice versa.

From geometry, $\delta = D \sin \theta$ (Figure 3.6b), so

$$\vec{E} = \vec{E}_0 \exp\left[j(ks - \omega t) \right]\left\{ 1 + \exp\left[jkD \sin \theta \right] \right\},$$

$$\vec{E} = \vec{E}_0 \exp\left[j(ks - \omega t) \right]\left\{ 1 + \exp\left[2\pi jD \sin \theta / \lambda \right] \right\}.$$

The intensity (see Section 3.2.4.3) is given by $I \propto \left| \vec{E} \right|^2 = \vec{E} \cdot \vec{E}^*$. Therefore,

$$I \propto \left| \vec{E}_0 \right|^2 (1)\left\{ 1 + \exp\left[2\pi jD \sin \frac{\theta}{\lambda} \right] \right\}\left\{ 1 + \exp\left[-2\pi jD \sin \frac{\theta}{\lambda} \right] \right\},$$

$$I \propto \left| \vec{E}_0 \right|^2 \left[2 + 2\cos\left(\frac{2\pi}{\lambda} D \sin \theta \right) \right],$$

making use of the Euler relation $\cos(x) = (1/2)(\exp(jx) + \exp(-jx))$. Via the identity $2[1 + \cos(2x)] = \cos^2(x)$, the intensity becomes

$$I \propto 4\left| \vec{E}_0 \right|^2 \cos^2\left(\pi \frac{D \sin \theta}{\lambda} \right)$$

Note that *without* interference—just considering the incident plane wave, for example, $I \propto \left| \vec{E}_0 \right|^2$, with the same constant of proportionality (*c*'s etc.)—we will define this intensity as I_0. Therefore,

$$I = 4I_0 \cos^2\left(\pi \frac{D \sin \theta}{\lambda} \right).$$

As we saw graphically, if $D \sin \theta = m\lambda$, where m is an integer, the $\cos^2$ factor is maximal, and we have *constructive* interference. If $D \sin \theta = (m/2)\lambda$, where m is an odd integer (i.e., $m/2 = 1/2, 3/2, 5/2, \ldots$), the $\cos^2$ factor is zero, and we have *destructive* interference. The intensity pattern, we see on the screen, therefore, is *not uniform* but rather has a sequence of maxima and minima. This is plotted in Figure 3.8.

Note that the maximal value of the intensity is four times that of a single wave. If interference "did not exist," we would have light from the two slits combining to simply give twice the single-wave intensity. With interference, we have bright peaks with four times the intensity and dark minima with zero intensity.

3.3.2 N-Slit Interference

Now consider N slits, *each* separated by distance D (drawn in Figure 3.9a for $N = 5$).

Building on our $N = 2$ analysis above, we can write the total electric field as

$$\vec{E} = \vec{E}_0 \exp\left[j(ks - \omega t) \right]\left\{ 1 + \exp\left[j\alpha \right] + \exp\left[j2\alpha \right] \right. $$
$$\left. + \exp\left[j3\alpha \right] + \cdots + \exp\left[j(N-1)\alpha \right] \right\},$$

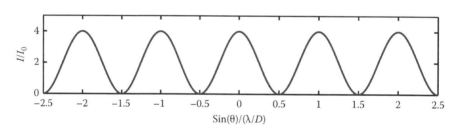

FIGURE 3.8 The two-slit intensity function: $I = 4I_0 \cos^2 (\pi D \sin \theta \, \lambda^{-1})$.

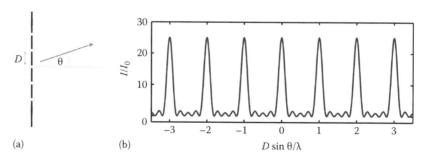

FIGURE 3.9 *N*-slit interference. (a) Geometry. Each slit is of negligible width and is separated from its neighbor by distance D. In the example drawn, $N = 5$. (b) The *N*-slit intensity function, $I = I_0 \dfrac{\sin^2 (N\pi D \sin \theta / \lambda)}{\sin^2 (\pi D \sin \theta / \lambda)}$, plotted for $N = 5$. Note that there are infinitely many large maxima located at angles for which $\sin(\theta) = m\lambda/D$, where m is any integer. Between each pair of these peaks are $N - 2$ smaller local maxima and $N - 1$ zeros. In this example, the zeros are located at $\sin(\theta) = \lambda/5D, 2\lambda/5D, 3\lambda/5D, 4\lambda/5D$.

where, for convenience, we have defined $\alpha \equiv (2\pi/\lambda)\, D \sin \theta$. Note that this is

$$\vec{E} = \vec{E}_0 \exp\left[j(ks - \omega t) \right] \left\{ 1 + \left(\exp\left[j\alpha \right] \right) + \left(\exp\left[j\alpha \right] \right)^2 \right.$$
$$\left. + \left(\exp\left[j\alpha \right] \right)^3 + \cdots + \left(\exp\left[j\alpha \right] \right)^{(N-1)} \right\}.$$

The terms in the braces form a finite geometric series, since each term is equal to the preceding one times $e^{j\alpha}$. Therefore,

$$\vec{E} = \vec{E}_0 \exp\left[j(ks - \omega t) \right] \left(\frac{1 - \exp(jN\alpha)}{1 - \exp(j\alpha)} \right).$$

We can simplify the expression in the parentheses by factoring out exponentials from the numerator and denominator:

$$\frac{1 - \exp(jN\alpha)}{1 - \exp(j\alpha)} = \frac{\exp(-jN\alpha/2)\left[\exp(jN\alpha/2) - \exp(-jN\alpha/2) \right]}{\exp(-j\alpha/2)\left[\exp(j\alpha/2) - \exp(-j\alpha/2) \right]}$$
$$= \exp\left(\frac{-j(N-1)\alpha}{2} \right) \frac{\sin(N\alpha/2)}{\sin(\alpha/2)},$$

using the Euler relation, $\sin(x) = (1/2j)(\exp(jx) - \exp(-jx))$.

Therefore, $\vec{E} = \vec{E}_0 \exp\left[j(ks - \omega t) \right] \exp\left(-\frac{j(N-1)\alpha}{2} \right) \frac{\sin(N\alpha/2)}{\sin(\alpha/2)}$.

The intensity $I \propto \left| \vec{E} \right|^2$:

$$I = I_0 \frac{\sin^2(N\alpha/2)}{\sin^2(\alpha/2)}.$$

Explicitly writing the α's:

$$I = I_0 \frac{\sin^2(N\pi D \sin\theta/\lambda)}{\sin^2(\pi D \sin\theta/\lambda)}.$$

This is plotted in Figure 3.9b for $N = 5$.

Maxima and minima. We see that the *numerator* of $I(\theta)$ is zero when $N\pi D \sin\theta/\lambda = m\pi$, i.e., $D \sin\theta/\lambda = m/N$, where m is an integer—but note that *both* numerator and denominator are zero if m is an integer multiple of N. We see that the *denominator* is zero when $\pi D \sin\theta/\lambda = m'\pi$, i.e., $D \sin\theta/\lambda = m'$, where m' is an integer—in this case, however, the numerator must also be zero, since $N\pi D \sin\theta/\lambda = Nm'\pi$ and N is an integer. If both the numerator and denominator are zero $I \to I_0 N^2$. A more detailed summary of the locations of maxima and minima is a useful exercise for the reader.

As illustrated in the plot of $I(\theta)$ for $N = 5$ slits (Figure 3.9), there are large maxima separated in angle by $\sin\theta = \lambda/D$.

The form of $I(\theta)$ reveals that this *angular spacing* between the peaks is independent of the number of slits. The *angular width* of the large peaks is approximately $\Delta \sin\theta \approx \lambda/ND$—half the distance in angle to the first local minimum—which gets sharper as we increase the number of slits. This is a very useful feature, as we will see shortly.

3.3.3 The Diffraction Grating as a Monochromator

Suppose we have a telescope that collects light from a star, and we want to measure the star's spectrum—i.e., the intensity as a function of wavelength, $I(\lambda)$. How can we do this? Our detector (like most good detectors, at least over some range of wavelengths) simply measures intensity, regardless of the wavelength of the light hitting it.

We can pass the light though an N-slit grating, or, equivalently, reflect it off a surface with N mirrors—a *diffraction grating*. How does this help? Light of wavelength λ_1 is deflected to angle λ_1/D. By this, we mean that the maximal intensity peak for light of this "color" is at the angle given by $\sin\theta_1 = \lambda_1/D$, and integer multiples, as in Section 3.3.2; typically, the angles involved are small, so $\sin\theta \approx \theta$. Light of wavelength λ_2 is deflected to angle λ_2/D, etc. Moving our detector to various positions on the screen and measuring the intensity as a function of *angle* on the screen reveals the intensity as a function of wavelength! (In other words $I(\lambda_1) = I(\theta_1)$, $I(\lambda_2) = I(\theta_2)$, etc.).

The *sharper* the diffraction peaks (high N), the *finer* the resolution in λ—see the end of the preceding section. The discovery (within the past ≈ 10 years) of planets outside our solar system—one of the most remarkable discoveries of recent history—used the approach outlined above to measure tiny shifts in stellar spectra due to the influence of the orbiting planets. The typical N of the diffraction gratings was around 100,000!

3.3.4 Single-Slit Interference

In our initial discussion of two-slit interference, we neglected the finite width of the diffraction grating. This finite width is important—just as waves from each slit interfere with one another, waves traversing various paths through a *single* slit will interfere with one another, and lead to diffraction. Fortunately, it is easy to analyze single-slit interference—it is simply the limit of the N-slit case discussed in Section 3.2.2 as $N \to \infty$, $D \to 0$, and the product $ND \to a$, where a is the width of the slit. The reader can verify that

$$I(\theta) = I_0 \left(\frac{\sin\beta}{\beta} \right)^2,$$

where $\beta = \pi a \sin\theta \lambda^{-1}$, as plotted in Figure 3.10.

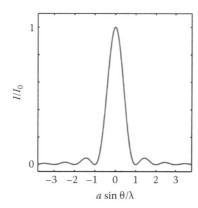

FIGURE 3.10 The intensity function of a single slit of width *a*. Note that the angular width of the peak is approximately λ/*a*.

3.3.5 Diffraction and Resolution

This single-slit diffraction pattern is exceptionally important. Any optical element—the pupil of your eye, a telescope mirror, a microscope lens, etc—is an aperture, and the $I(\theta)$ above describes how light travels through it. Why?

We have been considering light leaving an aperture, i.e., being "transmitted," and reaching a screen, where it is "received." But look carefully at Figure 3.5, 3.7, or 3.9—our wave interference scenarios. Our analysis did not invoke at all the *direction* the waves were traveling, only the path length difference between various paths. So we would get the same interference effects if light were *transmitted* from a point source at angle θ on the screen, passed through aperture(s), and were detected at the left.

Consider light from a point source (e.g., a star) located at the far-off "screen." We observe the point source by detecting the intensity passing through a single-slit aperture of width *a* (e.g., a telescope lens plus an intensity detector). We tilt the barrier containing our slit (e.g., our telescope), so that the angular position of the star of interest is $\theta_1 = 0$; this angle gives the maximum of the single slit intensity function, and we happily detect light from the star. We tilt the telescope; at the new $\theta_2 = 0$ there is no star; we see no light. We tilt further; at this third $\theta_3 = 0$ there is light again. "Aha!," we say, "We have seen two stars!"

Now suppose there were two stars very close to one another in angular position—let us say the difference in sin θ is just 0.1λ/*a*. (We typically deal with small angles, by the way, so sin θ ≈ θ.) Since the width of our interference function is ≈ λ/*a*, no matter how precisely we point at one star, we will be detecting a sizeable fraction of the intensity of the other—*there is no way we can tell that we are looking at two stars rather than only one!*

The *angular limit of resolution*, often just referred to as the *resolution*, of our single-slit aperture—the minimum angular separation that two objects must have in order to be able to distinguish them—is $\theta_{res} \approx \lambda/a$, where *a* is the aperture size. (It is an "approximately equals" sign, because there are slightly different ways of defining criteria for distinguishability that will not concern us here; most commonly, one uses the "Rayleigh criterion" $\theta_{res} = 1.22\,\lambda/a$.) Note that smaller θ_{res} means that we can

more finely distinguish objects—we can "see" better—and that this can be achieved by increasing the size of our aperture. This is why one builds big telescopes. (Big telescopes have another, unrelated, advantage: they collect more light.)

This issue of diffraction sets the fundamental limit on the performance of telescopes, microscopes, and other optical devices. Regarding microscopy, and all the diverse applications of it described, for example, in this book, the above angular description of resolution together with expressions for the focusing ability of lenses (Section 3.5 and Chapter 1) set a spatial limit for optical resolution. Roughly, objects separated by distance less than $\Delta x \approx \lambda$ cannot be resolved as separate entities. We will revisit the diffraction limit on resolution in Section 3.5.8.

The diffraction of light as it passes through an aperture maps the emission of an ideal pointlike source onto the observed pattern of intensity. For a one-dimensional slit, this mapping is given by the profile shown in Figure 3.10. In microscopy, one is interested in the analogous pattern caused by diffraction through the two-dimensional, typically circular, objective lens. One refers to the resulting intensity profile of a point source as the point spread function (PSF). In other words, the image of a point source (e.g., a fluorescent molecule) will not look like a point, but will look like the PSF. For an ideal aberration-free circular lens of radius *a* and focal length *f* (defined in Section 3.5), the PSF is given (Gu 1999) by

$$I(v) = I_0 \left[\frac{2J_1(v)}{v} \right]^2,$$

where

$$v = \frac{2\pi}{\lambda}\frac{a}{f}r, \ r \text{ is the radial coordinate in the imaging plane}$$

I_0 is the central intensity
J_1 is a Bessel function of the first kind

This function is illustrated in Figure 3.11; note that the width of the intensity peak is roughly λ/2.

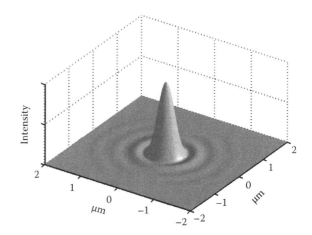

FIGURE 3.11 The PSF for an ideal circular aperture, plotted for $a/f = 0.7$ and $\lambda = 0.6\,\mu m$.

3.4 Refraction

When light travels from one medium to another, it may change direction. This phenomenon—familiar whenever we see the "bent" shape of a straw poking out of a glass of water—is known as *refraction*. (The light may also change its intensity at a boundary between media, which we will discuss in Section 3.6.) Refraction, like diffraction, is inherently a consequence of the wave nature of light, and our analysis below applies also to waves in water, sound waves, etc.

3.4.1 Snell's Law

The basic setup for issues of refraction is shown in Figure 3.12a: a ray of light crosses the boundary between two media, with indices of refraction n_1 and n_2, making angles θ_1 and θ_2 with respect to the normal in each medium, respectively. The question is: *How are θ_1 and θ_2 related?* The answer is crucial to the propagation of light and to the design of lenses and other optical elements.

The answer, as we will show, is that light obeys *Snell's law*:

$$n_1 \sin \theta_1 = n_2 \sin \theta_2.$$

There are several ways to derive this result. We could directly examine the wave equations for electromagnetic fields and look for solutions consistent with the presence of a boundary between two media, but this would be both painful and unilluminating. There are, fortunately, simpler ways of thinking about light propagation.

3.4.2 Fermat's Principle

3.4.2.1 Minimal Time Paths and Snell's Law

A general principle describing wave motion was put forth by Pierre de Fermat in the seventeenth century. It is sometimes stated as "light travels from one point to another along the path that takes the minimal amount of time." This is not quite correct—we will fix it in a few paragraphs—but it is a good place to start. We will also return to justifying Fermat's principle shortly. First, let us use it to derive Snell's law.

Imagine you are on a beach, and someone in the ocean is drowning. You rush out to help, which requires both running on land and swimming in the water. You can run faster than you can swim. What path should you take? With a bit of thought, you will realize that a straight line between you and the drowning person is not the best idea—rather, you should reduce the length of the swim to minimize the overall time to your target. How much should you run and how much should you swim?

The same dilemma is encountered by our light beam, traveling from position A in a medium of index of refraction n_1 (your position on the beach, in the above analogy) to position B in a medium of n_2 (the swimmer's position, in the water) in Figure 3.12b. The speed of light in medium 1 is $v_1 = c/n_1$ and in medium 2 is $v_2 = c/n_2$. Within each medium, the light travels in a straight line—itself a consequence of Fermat's principle, as you can convince yourself. There are many possible paths between A and B, as illustrated in the figure, that we can label based on the position x at which they cross the interface. One of these—let us call it the one that goes through position x_0—minimizes the total travel time. What is this path? (What is x_0?)

The travel time in medium 1, t_1, is the distance traveled in medium 1 divided by the speed in medium 1: $t_1 = (n_1/c)\sqrt{y_1^2 + x^2}$; similarly, the travel time in medium 2 is $t_2 = (n_2/c)\sqrt{y_2^2 + (L-x)^2}$. The total travel time is $t = t_1 = t_2$. To find the minimal time, we determine the x for which $dt/dx = 0$; call this x_0:

$$\frac{dt}{dx} = \frac{n_1}{c}\frac{x}{\sqrt{y_1^2 + x^2}} - \frac{n_2}{c}\frac{(L-x)}{\sqrt{y_2^2 + (L-x)^2}},$$

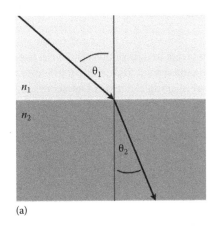
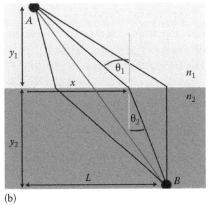

(a)   (b)

FIGURE 3.12 Refraction. (a) The path taken by light traveling between two media bends at the interface; the relation between the angles θ_1 and θ_2 depends on the indices of refraction of the two materials, and is given by Snell's law. (b) Light traveling from point A in medium 1 to point B in medium 2 can follow infinitely many possible paths, four of which are illustrated here. Fermat's principle states that the actual path the light takes is that for which the total travel time is minimal (actually extremal, as discussed in the text), from which Snell's law follows.

$$\left.\frac{dt}{dx}\right|_{x=x_0} = 0 \rightarrow n_1 \frac{x_0}{\sqrt{y_1^2 + x_0^2}} = n_2 \frac{(L - x_0)}{\sqrt{y_2^2 + (L - x_0)^2}}.$$

(To show that x_0 is a minimum, we should also examine the second derivative of t, but as we will see later, it does not actually matter if x_0 is the site of a minimum or a maximum. Furthermore, we can intuit from the form of $t(x)$ that this extremum is, in fact, a minimum.)

Note from geometry that

$$\sin\theta_1 = \frac{x_0}{\sqrt{y_1^2 + x_0^2}},$$

$$\sin\theta_2 = \frac{(L - x_0)}{\sqrt{y_2^2 + (L - x_0)^2}}.$$

Therefore, the above condition becomes

$$n_1 \sin\theta_1 = n_2 \sin\theta_2.$$

We have shown that when the above condition is met, the travel time for light propagation is minimized. This is Snell's law.

We can also use Fermat's principle to derive Snell's *law of reflection*, which states that the *reflected* ray makes the same angle with the interface as the incident ray.

3.4.2.2 Explaining Fermat's Principle

Now let us explain Fermat's principle. Suppose light travels along many paths, all of which interfere with one another. Paths for which the phase difference is near zero will constructively interfere. Consider the minimal time path. As we saw in our derivation of Snell's law above, this is the path for which $\sum n_i d_i$ is minimal, where the sum runs over all the segments being considered (two segments in the above example), and d_i is the length of segment i. From Section 3.2, minimizing $\sum n_i d_i$ is the equivalent to minimizing the *phase* traversed by the wave along the path. Therefore, the minimal time path is also the path of minimal phase and is also the path of minimal $\sum n_i d_i$—all these statements are equivalent. This sum $\sum n_i d_i$ is more properly written as an integral and is called the optical path length (OPL): $\text{OPL} = \int_A^B n(x)\, dx$.

Why should the path of minimal OPL be the path light takes? Let us call this path P. By construction, $\left.\frac{d(\text{OPL})}{d''s''}\right|_P$, is zero, where s indicates any variable that characterizes the paths. Therefore, nearby paths are similar in phase and so constructively interfere. Consider a path for which $\frac{d(\text{OPL})}{d''s''}$ is *not* zero—moving to a slightly different path, the OPL can change appreciably, perhaps

higher in one direction, lower in another, etc., and so we would not expect constructive interference.

You may be thinking: the minimal OPL path is not the only one for which we can guarantee constructive interference. What about the *maximal* OPL path? This too provides constructive interference. And so the proper formulation of Fermat's principle is that *light travels along paths of extremal optical path length*. Typically, these are minimal OPL paths; but, in certain geometries, they can be maximal OPL paths as well.

3.4.3 Total Internal Reflection

Let us look more carefully at Snell's law. What if $n_1 > n_2$, and θ_1 is large, so that $(n_1/n_2)\sin\theta_1 > 1$? What θ_2 can satisfy Snell's law: $n_1 \sin\theta_1 = n_2 \sin\theta_2$? None. This means that *there can be no wave transmitted to medium 2*; the light from medium 1 is totally reflected at the interface, a condition known as *total internal reflection*.

Fiber optics, which underpin much of modern communication, work because of total internal reflection. Consider a glass fiber ($n = 1.5$) surrounded by air ($n = 1.0$). We want the light to travel along the fiber and not leak out into the air. This is automatically enforced by total internal reflection due to the higher index of refraction of the glass and the large incident angles of light traveling along the fiber (as long as the fiber is not severely bent). In a fiber optic cable, light can propagate for kilometers with losses of a fraction of a percent!

A careful treatment of electromagnetic fields would show that the light intensity in medium 2 is not exactly zero. Rather, an "evanescent wave" decays to zero over a short distance, comparable to λ, in medium 2. This attribute finds applications in biophysical imaging. In total internal reflection fluorescence microscopy, excitation of fluorescent probes by the evanescent wave allows discrimination of molecules near interface despite the presence of large concentrations of probes in the bulk (Axelrod et al. 1984, Axelrod 2001, Groves et al. 2009).

3.5 Lenses

We often wish to collect and reshape electromagnetic wavefronts to create images of objects. *Lenses* are powerful tools for achieving these goals and are obviously very useful, forming the essential imaging elements of telescopes, microscopes, cameras, your eyes, and many other devices. The "ideal" shape of a lens surface is generally some non-spherical conic section (hyperbola, parabola, etc.), but, in practice, spherical lenses are typically used, since they are much easier to make than aspheric (non-spherical) lenses. Typically, one uses spherical lenses and then corrects for their aberrations (nonideal behavior), e.g., by using combinations of lenses. We will briefly explore lenses.

3.5.1 A Spherical Interface

Consider a point source emitting spherical waves from point S, in a medium of index of refraction n_1 (see Figure 3.13). Can

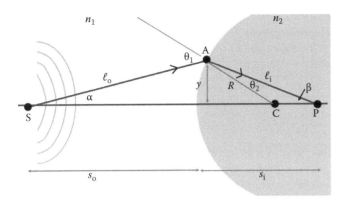

FIGURE 3.13 A spherical interface. C, S, A, and P refer to particular points—the center of the spherical interface, the object point, the point at which the ray drawn hits the interface, and the image point, respectively. Italicized letters refer to distances. Greek letters refer to angles—note that $\alpha = \angle ASC$ and $\beta = \angle CPA$.

we construct a spherical interface of radius R that focuses the emitted light to point P, regardless of where it hits the interface? What should R be? Point P is embedded in a medium of index of refraction n_2; we are considering the shape of the interface between media 1 and 2. Consider $n_2 > n_1$, so that the rays from S will be refracted "inwards."

In Figure 3.13, Point C is the center of the sphere of radius R. The distance between the "object" point, S, and the interface is s_o, and the distance between the "image" point, P, and the interface is s_i. The angles that the incident and reflected rays make with respect to the normal to the interface are θ_1 and θ_2. As usual θ_1 and θ_2 are related by Snell's law: $n_1 \sin \theta_1 = n_2 \sin \theta_2$. We can relate θ_2 to β via the law of sines: $(\sin \beta / R) = (\sin \theta_2 / (s_i - R))$. Relating θ_1 to α is not quite as transparent; first note that $\angle SAC = \pi - \theta_1$, so $\sin(\angle SAC) = \sin(\pi - \theta_1) = \sin \theta_1$, and then apply the law of sines to $\triangle SAC$ to get $(\sin \alpha / R) = (\sin \theta_1 / (R + s_o))$. Inserting all this into Snell's law:

$$n_1 \frac{R + s_o}{R} \sin \alpha = n_2 \frac{s_i - R}{R} \sin \beta,$$

i.e., $n_1 (R + s_o) \sin \alpha = n_2 (s_i - R) \sin \beta.$

More geometry: $\sin \alpha = \dfrac{y}{l_o} = \dfrac{y}{\sqrt{s_o^2 + y^2}}$ and $\sin \beta = \dfrac{y}{l_i} = \dfrac{y}{\sqrt{s_i^2 + y^2}}.$

From which $n_1 (R + s_o) s_i \sqrt{1 + \left(\dfrac{y}{s_i}\right)^2} = n_2 (s_i - R) s_o \sqrt{1 + \left(\dfrac{y}{s_o}\right)^2}$

We have derived a relation that must hold for focusing at P to occur. In other words, we know what R we need—the R that satisfies the above expression. Unfortunately, it depends on y, the position at which our ray hits the interface! Therefore, different rays will not focus to the same image spot.

3.5.2 A Spherical Interface—The Paraxial Regime

What we have shown is that a truly spherical interface will not serve as an ideal lens. There is a way out of this, however, which is to limit ourselves to the paraxial regime, meaning that we consider only light that is nearly parallel with the optical axis, *SP*. In other words, we consider small α and β. Therefore, y/s_o and y/s_i are small, allowing us to neglect them in the above equation: $n_1 (R + s_o) s_i \approx n_2 (s_i - R) s_o$, from which $(n_1/s_o) + (n_2/s_i) = (n_2 - n_1)/R$. A simple, useful relation! (By the way, we could also have derived this directly from Fermat's principle, by determining the R for which SAP is an extremal path for any A.)

Should we be bothered by limiting ourselves to the paraxial case? Yes and no. In practice one *does* try to design optical systems such that beams are close to the center of spherical lens elements or, equivalently, to have one's image and object distances be large compared to the size of the lens. If one does this, the above relation works very well. In practice, one works in the paraxial regime and applies additional corrections if necessary. We will continue limiting ourselves to the paraxial regime.

3.5.3 Focal Points

If R, n_1, and n_2 are fixed, decreasing s_o means that s_i increases (and vice versa), from the above boxed relation. Let us increase s_i until $s_i \to \infty$; in other words, parallel rays emerge from the interface; what is s_o? From above: $(n_1/s_o) + (n_2/\infty) = (n_2 - n_1)/R$, therefore $s_o = (n_1/(n_2 - n_1))R$—an object at this distance focuses "to infinity." We will call this distance the *object focal length*, $f_o \equiv (n_1/(n_2 - n_1))R$. The spherical waves from the point source turn into plane waves.

The same holds if we do not consider a "semi-infinite" medium on the right, but rather a finite lens with a spherical surface at the left and a flat surface at the right—a *planoconvex lens* (see Figure 3.14). Note that since the right edge is flat, all rays are normal to it, and there is no "bending" of the rays due to refraction.

We can of course consider the opposite situation, in which plane waves (parallel rays from $s_o = \infty$) are focused to an image at some s_i. This particular s_i is denoted f_i, the *image focal length*.

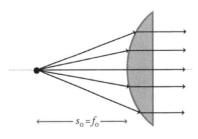

FIGURE 3.14 A planoconvex lens. Light emanating from a source located at the object focal length is focused to an image distance of infinity (i.e., the rays become parallel).

3.5.4 Real and Virtual Images

Solving the lens equation above for s_i, we have

$$s_i = n_2 \left(\frac{n_2 - n_1}{R} - \frac{n_1}{s_o} \right)^{-1} = \frac{n_2}{n_1} \left(\frac{1}{f_o} - \frac{1}{s_o} \right)^{-1}.$$

If $s_o > f_o$, then $s_i > 0$, and point P is to the right of the interface. The rays from S converge at P. To an observer at the right, *it looks as if light is emanating from point* P. We have what is called a *real image* at P (see Figure 3.15a). If, for example, we put a power meter at P, we detect a high degree of power due to the focused light.

If $s_o < f_o$, then $s_i < 0$, and point P is to the left of the interface. The rays do not actually hit point P, but they *appear* to an observer at the right as if they are emanating from P (see Figure 3.15b). We have what is called a *virtual image* at P. If, for example, we put a power meter at P, we *do not* detect a high-intensity focused spot, since there is no "spot" there.

3.5.5 Concave Lenses

The same analysis works for concave lenses, but we treat R as *negative* ($R < 0$). Since

$$\frac{n_1}{s_o} + \frac{n_2}{s_i} = \frac{n_2 - n_1}{R}, \text{ if } n_2 > n_1 \text{ then } s_i < 0 \text{—we have a virtual image.}$$

3.5.6 Thin Lenses

Let us glue one lens of radius of curvature R_1 onto another of R_2 (see Figure 3.16). We will consider thin lenses and so neglect the

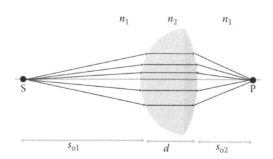

(a)   (b)

FIGURE 3.15 (a) Real and (b) virtual images. (a) Light emanates from P. (b) Light looks to an observer like it is emanating from point P located to the left of the interface.

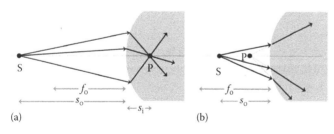

FIGURE 3.16 Focusing light with a thin lens (imagine d is small).

lens thickness d (i.e., we are assuming d is smaller than other lengths involved).

The object and image lengths for "lens 1" (the left half of the lens) are related by

$$\frac{n_1}{s_{o1}} + \frac{n_2}{s_{i1}} = \frac{n_2 - n_1}{R_1}.$$

The image of lens 1 provides the "object" for lens 2. Therefore, $s_{o2} = -s_{i1} + d \approx -s_{i1}$, where the negative sign arises, because, as defined above, a positive image length and a positive object length lie in opposite directions. Considering lens 2:

$$\frac{n_2}{-s_{i1}} + \frac{n_1}{s_{i2}} = \frac{n_1 - n_2}{R_2},$$

where we keep track of which index of refraction is which.

We need to adopt a consistent set of sign conventions for the radii. As noted above, a convex "left" lens has $R > 0$, and a concave "left" lens has $R < 0$. For the right side lens, these are switched. Returning to our thin lens, adding the two expressions above:

$$\frac{n_1}{s_{o1}} + \frac{n_1}{s_{i2}} = (n_2 - n_1) \left(\frac{1}{R_1} - \frac{1}{R_2} \right).$$

For a thin lens in air, $n_1 \approx 1$; $n_2 = n_{lens}$, giving us the *thin lens equation*, or *Lensmaker's formula*:

$$\frac{1}{s_o} + \frac{1}{s_i} = (n_{lens} - 1) \left(\frac{1}{R_1} - \frac{1}{R_2} \right).$$

The *focal length*, f, is given either by s_o or $s_i \to \infty$ (it does not matter which):

$$\frac{1}{f} = (n_{lens} - 1) \left(\frac{1}{R_1} - \frac{1}{R_2} \right).$$

We can then write the thin lens equation as $(1/s_o) + (1/s_i) = 1/f$, also known as the Gaussian Lens Formula. This is one the most important relations for the design of optical systems.

For example: Consider parallel rays incident on the flat side of a glass ($n = 1.5$) planoconvex lens with a radius of curvature of 50 mm. Where will these rays be focused to? Answer: $R_1 = \infty$, $R_2 = -50$ mm. $1/f = (1.5 - 1)(1/50$ mm$)$, so $f = 100$ mm, $s_i = 100$ mm. The rays will focus to a point 100 mm beyond the curved side of the lens.

3.5.7 Magnification

Lenses magnify objects. The magnification can be >1 or <1. See Figure 3.17 depicting a thin lens, which is magnifying an extended object (i.e., not a point source)—in this case, a pear.

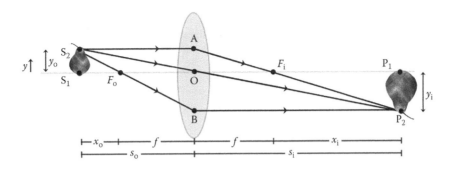

FIGURE 3.17 Magnification of an image by a lens. Rays emanating from point S_1, on the optical axis, are focused to point P_1 (not drawn). Rays from S_2 are focused to P_2.

Points F_o and F_i are each a distance f, the focal length, from the lens. Consider light emanating from the top of the pear. The ray that goes through F_o will emerge from the lens parallel to the axis (think about why this is). The ray the leaves the pear parallel to the axis will go through F_i. The ray that goes through the center of the lens will be undeflected in the thin lens limit (Hecht 2002).

The *magnification*, M_T, is defined to be the height of the image relative to the height of the object—i.e., $M_T \equiv y_i/y_o$. Triangle S_1S_2O is similar to triangle P_1P_2O, so $y_o/s_o = -y_i/s_i$, so $M_T \equiv -s_i/s_o$; the negative sign shows that the image is *inverted*.

Triangle AOF_i is similar to triangle $P_1P_2F_i$, so $y_o/f = |y_i|/(s_i-f)$.

Triangle BOF_o is similar to triangle $S_1S_2F_o$, so $y_o/(s_o-f) = |y_i|/f$. Combining these, $(f/(s_o-f)) = ((s_i-f)/f)$. Note that $s_i-f = x_i$ (see figure). Using the last similar triangle relation again, $(f/(s_o-f)) = (-y_i/y_o)$. And so, $M_T = (y_i/y_o) = -(x_i/f)$. We could also have written: $M_T = -(f/x_o)$.

As the object distance x_o is lowered, the magnification increases. The reader may think about what happens if $x_o < 0$, i.e., the object is closer than the focal point. Drawing rays, convince yourself that the lens cannot form an image of the object.

Another important aspect of lenses that follows from the diagrams above, and for which it is useful to develop an intuition: Parallel rays that are also parallel to the optical axis are focused to the focal point on the optical axis. Parallel rays that are *tilted* with respect to the optical axis are also focused to a point at distance f, but *off* the optical axis; such points define the *focal plane*. The reader may wish to draw such rays.

3.5.8 Resolution

When considering single-slit diffraction in Section 3.3, we realized that the angular resolution of a device is given by $\theta_{min} \approx \lambda/a$, where λ is the wavelength of light and a is the diameter of the imaging aperture. Two objects must have an angular separation of at least θ_{min} if they are to be resolved as separate objects. Using lenses to magnify objects, this angular resolution criterion still holds. Moreover, the fact that the object distance cannot be closer than the focal length turns

our resolution relation into a distance criterion. We will briefly sketch this:

Consider two objects separated in position by Δy at a distance s from a lens (see Figure 3.18). For the two to be resolvable, we need $\theta > \lambda/a$, where $\theta \approx \Delta y/s$. Therefore, we need $\Delta y > s\lambda/a$. Since $s > f$, $f = (n_1/(n_2-n_1))R$, and $R > a$, we can write $s > (n_1/(n_2-n_1))a$. Combining the two inequalities: $\Delta y > (n_1/(n_2-n_1))\lambda$. The numerical factor $n_1/(n_2-n_1) \approx 1$ in our rough treatment. Therefore, our minimum resolvable spatial separation is $\Delta y_{min} \approx \lambda$. We cannot resolve objects smaller than (approximately) the wavelength of light.

More precise statements of optical resolvability can be constructed. Typically, one invokes the Abbe criterion that the minimal $\Delta y = \lambda/(2n\sin\theta)$, where n is the index of refraction of the medium and θ is the maximum angle over which light is collected. The value of $\sin \theta$ is bounded by 1, so at best $\Delta y = \lambda/(2n)$. For $\lambda = 400$ nm (blue) light in water ($n = 1.3$), the theoretical resolution limit is about 150 nm. (In practice, any aberrations or imperfections further reduce the resolution.) Hence, the wavelengths of visible light set a limit of roughly a few hundred nanometers as the minimal size of resolvable structures—smaller, for example, than cells but far larger than the characteristic sizes of proteins or small molecules.

It is important to keep in mind that the issues of resolution discussed above govern the discrimination of two (or more) objects. If one knows that only a *single* point source contributes to an image, giving an intensity profile like that of Figure 3.11, for example, the center of this profile can be determined to arbitrarily high precision (in practice, a few nanometers typically). A few of the many applications of this principle are illustrated in (Crocker

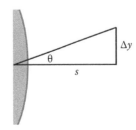

FIGURE 3.18 Schematic, for considering spatial resolution.

and Grier 1996, Weihs et al. 2006, Crocker and Hoffman 2007, Roichman et al. 2008, Kong and Parthasarathy 2009).

Despite the fundamental nature of the diffraction-limited resolution, the past decade or so has seen the birth of several very clever techniques for surmounting it, using interferometry, nonlinear optical processes, or single-molecule imaging (e.g., Betzig et al. 2006, Bates et al. 2007, Hell 2007, Abbott 2009) to yield optical information at scales an order of magnitude smaller than what was traditionally thought possible.

3.6 Reflection and Transmission (Fresnel's Equations)

The law of reflection ($\theta_r = \theta_i$, where r and i refer to reflected and incident rays; Figure 3.19a) and Snell's law ($n_i \sin \theta_i = n_t \sin \theta_t$, where t refers to the transmitted ray) give the *directions* of reflected and transmitted rays at boundaries. What are the *amplitudes* of the electromagnetic waves? In other words, *how much* light is reflected and transmitted? Similar questions arise when considering other sorts of waves hitting boundaries—for example, waves on strings, incident at an interface between two media with different propagation speeds. In all these situations, transmission and reflection are analyzed by considering the boundary conditions imposed by the junction.

To consider the general case of a plane electromagnetic wave hitting a surface at some angle θ_i (with respect to the normal), we will have to separately consider the components with electric field perpendicular and parallel to the *plane of incidence*. (The incident, reflected, and transmitted rays all lie in and define the plane of incidence, "POI," which also includes the normal to the surface.)

Recall from Section 3.2.4.2 some properties of electromagnetic waves:

- The electric and magnetic field vectors of an electromagnetic wave are perpendicular to each other.
- $\vec{E} \times \vec{B}$ points along $\vec{k}$, the wavevector, i.e., along the direction of propagation.
- The field amplitudes are related by $|\vec{E}| = v|\vec{B}|$, where $v = c/n$ is the wave speed.

The boundary conditions that govern electric and magnetic fields at the interface between media are

i. The tangential (i.e., parallel to the interface) components of the electric field, $\vec{E}$ are continuous across the boundary.
ii. The tangential (i.e., parallel to the interface) components of $\vec{B}/\mu$, where μ is the magnetic permeability of the medium, are continuous across the boundary.

Let us consider the two cases.

3.6.1 Case I: $\vec{E}$ Perpendicular to the Plane of Incidence

Note that a circle with a dot in it indicates a vector that points out of the page towards you (see Figure 3.19b). The electric field vectors are completely tangential to the interface. The magnetic field vectors are not. Applying boundary condition (i) to the amplitudes (E_0) of the electric fields,

$$E_{0i} + E_{0r} = E_{0t}$$

Applying boundary condition (ii) to the amplitudes (B_0) of the magnetic fields, $-\dfrac{B_{0i}}{\mu_i}\cos\theta_i + \dfrac{B_{0r}}{\mu_i}\cos\theta_r = -\dfrac{B_{0t}}{\mu_t}\cos\theta_t$ (see Figure 3.19 to understand the signs).

Using $B_0 = E_0/v$ (from above), $v_i = v_r$ (since they are in the same media), $\theta_i = \theta_r$ (law of reflection), and $v_i = c/n_i$, we can write the above relation as

$$\frac{n_i}{\mu_i}\left(E_{0i} - E_{0r}\right)\cos\theta_i = \frac{n_t}{\mu_t}E_{0t}\cos\theta_t.$$

Combining this with the boundary condition (i) equation above, substituting to eliminate E_{0t}, we can solve for the ratio of the reflected wave amplitude to the incident wave amplitude:

$$\left(\frac{E_{0r}}{E_{0i}}\right)_{\perp} = \frac{n_i\mu_i^{-1}\cos\theta_i - n_t\mu_t^{-1}\cos\theta_t}{n_i\mu_i^{-1}\cos\theta_i + n_t\mu_i^{-1}\cos\theta_t}.$$

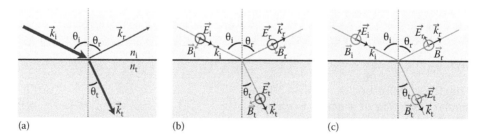

(a)   (b)   (c)

FIGURE 3.19 Reflection and refraction at an interface. The incident wave (wavevector $\vec{k}_i$) is reflected (wavevector $\vec{k}_r$) and transmitted (wavevector $\vec{k}_t$). Both the angles of the reflected and transmitted waves and their amplitudes are determined by the dielectric properties of the materials that comprise the interface. (a) Electric and magnetic field vectors for light polarized with $\vec{E}$ perpendicular to the plane of incidence. (b) Electric and magnetic field vectors for light polarized with $\vec{E}$ parallel to the plane of incidence.

Similarly solving instead for the ratio of the transmitted wave amplitude to the incident wave amplitude,

$$\left(\frac{E_{0t}}{E_{0i}}\right)_{\perp} = \frac{2n_i\mu_i^{-1}\cos\theta_i}{n_i\mu_i^{-1}\cos\theta_i + n_t\mu_t^{-1}\cos\theta_t}.$$

Typically, one deals with nonmagnetic materials: $\mu \approx \mu_0$, the permeability of free space. The above equations simplify, yielding two of the four *Fresnel equations*, for the amplitude reflection coefficient, $r_{\perp}$ and the amplitude transmission coefficient, $t_{\perp}$.

$$r_{\perp} = \left(\frac{E_{0r}}{E_{0i}}\right)_{\perp} = \frac{n_i\cos\theta_i - n_t\cos\theta_t}{n_i\cos\theta_i + n_t\cos\theta_t},$$

$$t_{\perp} = \left(\frac{E_{0t}}{E_{0i}}\right)_{\perp} = \frac{2n_i\cos\theta_i}{n_i\cos\theta_i + n_t\cos\theta_t}.$$

3.6.2 Case II: $\vec{E}$ Parallel to the Plane of Incidence

Applying the boundary conditions to this geometry leads to (see Figure 3.19c):

$$\left(\frac{E_{0r}}{E_{0i}}\right)_{\parallel} = \frac{n_t\mu_t^{-1}\cos\theta_i - n_i\mu_i^{-1}\cos\theta_t}{n_i\mu_i^{-1}\cos\theta_i + n_t\mu_t^{-1}\cos\theta_t},$$

and

$$\left(\frac{E_{0t}}{E_{0i}}\right)_{\perp} = \frac{2n_i\mu_i^{-1}\cos\theta_i}{n_i\mu_i^{-1}\cos\theta_t + n_t\mu_t^{-1}\cos\theta_i}.$$

For typical nonmagnetic media, we get the other two *Fresnel equations*:

$$r_{\parallel} = \left(\frac{E_{0r}}{E_{0i}}\right)_{\parallel} = \frac{n_t\cos\theta_i - n_i\cos\theta_t}{n_i\cos\theta_t + n_t\cos\theta_i},$$

$$t_{\parallel} = \left(\frac{E_{0t}}{E_{0i}}\right)_{\parallel} = \frac{2n_i\cos\theta_i}{n_i\cos\theta_t + n_t\cos\theta_i}.$$

3.6.3 Brewster's Angle

Let us plot $r_{\perp}$ and $r_{\parallel}$ as a function of θ_i for light incident from air ($n_i = 1$) to water ($n_t = 1.33$)—see Figure 3.20. We notice something very interesting: a particular θ_i for which the reflection coefficient is *zero* for light with its electric field parallel the plane of incidence. There is no such angle for the perpendicular polarization.

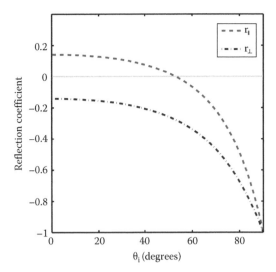

FIGURE 3.20 Fresnel coefficients $r_{\perp}$ and $r_{\parallel}$ for light incident from air to water, as a function of incidence angle.

The reader can work through the algebra and show that if $n_t > n_i$, $r_{\parallel} = 0$ at one particular incident angle, θ_i. This angle is called *Brewster's angle*, θ_p, and is given by $\tan\theta_p = n_t/n_i$. All the parallel-polarized light is transmitted. What about the perpendicular polarization? One can show that there is *no* angle that gives $r_{\perp} = 0$. Therefore, shining randomly polarized light incident at the Brewster angle, the reflected light is completely polarized with its electric field perpendicular to the plane of incidence. This property, together with a desk lamp, a sink, and a large bowl, once saved your author from misfortune, as he found himself with a much-needed linear polarizer whose transmission axis was unlabeled. The reader can test his or her mystery-solving skills by figuring out how he determined the polarizer's axis and thereby saved the day.

The polarization-dependence of reflection at interfaces also underpins Brewster angle microscopy, in which an interface is imaged with parallel-polarized light incident at the Brewster angle of the two media. The presence of interfacial molecules distinct from those of the two media—for example, lipids organized at an air–water interface—alters the local index of refraction, leading to a nonzero reflection coefficient. The intensity of the reflected light therefore provides a sensitive measure of interfacial molecular organization.

3.7 Concluding Remarks

In the preceding pages, we have explored the basic elements of optics. All of these topics can be explored much further, uncovering still more depth and beauty than we have been able to sketch, and also illuminating applications of great importance to science and technology. As will be evident throughout this book, these two aspects of optics—its formal elegance and its practical utility—are intertwined. Demands from fields as diverse as biological imaging, astronomy, and telecommunications drive the search for deeper insights into the behavior of light. Conversely, explorations of the intricacies of electromagnetic wave propagation have yielded, and will undoubtedly continue to yield, remarkable new tools.

References

Abbott, A. 2009. Microscopic marvels: The glorious resolution. *Nature* 459: 638–639.

Axelrod, D. 2001. Total internal reflection fluorescence microscopy in cell biology. *Traffic* 2: 764–774.

Axelrod, D., Burghardt, T. P., and Thompson, N. L. 1984. Total internal-reflection fluorescence. *Annu. Rev. Biophys. Bioeng.* 13: 247–268.

Bates, M., Huang, B., Dempsey, G. T., and Zhuang, X. 2007. Multicolor super-resolution imaging with photo-switchable fluorescent probes. *Science* 317: 1749–1753.

Betzig, E., Patterson, G. H., Sougrat, R. et al. 2006. Imaging intracellular fluorescent proteins at nanometer resolution. *Science* 313: 1642–1645.

Born, M. and Wolf, E. 1997. *Principles of Optics. Electromagnetic Theory of Propagation, Interference and Diffraction of Light*, 6th edn. Cambridge, U.K.: Cambridge University Press.

Crocker, J. C. and Grier, D. G. 1996. Methods of digital video microscopy for colloidal studies. *J. Coll. Interf. Sci.* 179: 298–310.

Crocker, J. C. and Hoffman, B. D. 2007. Multiple-particle tracking and two-point microrheology in cells. *Methods Cell. Biol.* 83: 141–178.

Groves, J. T., Parthasarathy, R., and Forstner, M. B. 2009. Fluorescence imaging of membrane dynamics. *Annu. Rev. Biomed. Eng.* 10: 311–338.

Gu, M. 1999. *Advanced Optical Imaging Theory*. Berlin, Germany: Springer.

Hecht, E. 2002. *Optics*, 4th edn. San Francisco, CA: Pearson Addison Wesley.

Hell, S. W. 2007. Far-field optical nanoscopy. *Science* 316: 1153–1158.

Kong, Y. and Parthasarathy, R. 2009. Modulation of attractive colloidal interactions by lipid membrane functionalization. *Soft Matter* 5: 2027–2029.

Roichman, Y., Sun, B., Roichman, Y., Amato-Grill, J., and Grier, D. G. 2008. Optical forces arising from phase gradients. *Phys. Rev. Lett.* 100: 013602–013604.

Weihs, D., Mason, T. G., and Teitell, M. A. 2006. Biomicrorheology: A frontier in microrheology. *Biophys. J.* 91: 4296–305.

4

Light Sources, Detectors, and Irradiation Guidelines

Carlo Amadeo Alonzo
Massachusetts General Hospital

Malte C. Gather
Massachusetts General Hospital

Jeon Woong Kang
Massachusetts General Hospital

Giuliano Scarcelli
Massachusetts General Hospital

Seok-Hyun Yun
Massachusetts General Hospital

4.1 Introduction 49
4.2 Light Sources 49
    Introduction • Nonlaser Sources • Laser Sources
4.3 Light Detectors...55
    Introduction • Photodiodes • Photomultiplier Tubes • Arrayed Detectors
4.4 Irradiation Guidelines 61
    Introduction • Radiation Effects at the Tissue Level • Radiation Effects at the
    Cellular Level • Safety Practices
References.. 64

4.1 Introduction

Biomedical optics is an interdisciplinary study that demands proficiency across a broad range of topics, stretching from fundamental concepts in biology and biochemistry to technical instrumentation in optics. The latter can sometimes be quite intimidating itself, particularly when one considers the wide diversity of devices associated with modern techniques and applications in biomedical optics, as seen in later chapters of this book. It is easier to make sense of various instrumentation schemes when guided by a basic understanding of the properties and operating principles of devices involved in producing and detecting light. Light sources and light detectors represent the beginning and end, respectively, of any optical system used to study biological structures or mechanisms. Different types of these devices provide access to the different properties of light that, in turn, reveal different aspects of the biological cells and tissues under study.

Section 4.2 briefly describes some nonlaser light sources: high-pressure arc lamps, low-pressure vapor lamps, incandescent lamps, and light-emitting diodes (LEDs). Greater emphasis is placed on laser light sources as these play a more significant role in state-of-art biomedical optics techniques. Some basic properties that make laser radiation particularly useful in biomedical optics applications are discussed. The fundamental components of a laser are explained in the terms of their function during laser operation. As much progress in biomedical optics has been enabled by nanosecond- to femtosecond-duration laser pulses, some basic concepts of pulsed laser operation are also introduced. Finally, four types of lasers—gas, solid-state, dye, and

semiconductor—are described and differentiated. Some specific examples of each type are also enumerated in the context of relevant biomedical applications.

Section 4.3 offers a complementary discussion of devices used to measure the intensity and distribution of light. Photodiodes, photomultiplier tubes (PMTs), and arrayed detectors such as charge coupled devices (CCDs) lie at the heart of all schemes to measure properties of light, even in more complex instruments, such as spectrometers and streak cameras. Each of these types of detectors is described in terms of their respective operating principles, unique characteristics, and typical applications.

Section 4.4 describes how exposure to optical radiation can trigger adverse effects to the cells and tissue. Depending on optical frequency, power, and duration of exposure, these range from photochemical, thermal, and thermoacoustic damage in tissues, to photobleaching, photodamage, and phototoxicity in cells. Such phenomena are discussed in order to provide a guide to safe and effective use of optical radiation in biomedicine.

4.2 Light Sources

4.2.1 Introduction

Light is an electromagnetic wave. Although the region of major concern in optics extends from the infrared, across the visible, to the ultraviolet, it is only a small portion of the vast electromagnetic spectrum, as shown in Figure 4.1. While all electromagnetic waves are of the same fundamental physical phenomenon, the particular time, length, and energy scale of each spectral

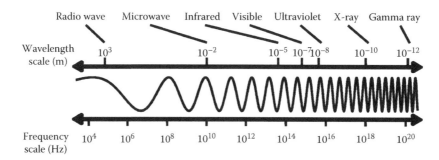

FIGURE 4.1 Electromagnetic spectrum. The upper scale presents a representative wavelength scale for each class of electromagnetic (EM) wave. A corresponding frequency scale is drawn below for reference. It should be noted that there are no specific boundaries between classes of EM waves. The most well-defined region would be for visible light, usually considered between 400 and 750 nm in wavelength. Otherwise, there is no particular wavelength that separates infrared from microwave radiation for example.

band lead to some specific properties that differentiate "light" from the shorter-wavelength x-rays and gamma rays, as well as the longer microwaves and radio waves, when propagation through and interaction with biological tissues are considered.

4.2.1.1 X-Rays and Gamma Rays

The application of x-rays and gamma rays in medicine is associated with the field of radiology and nuclear medicine. Short wavelengths ($<10^{-9}$ m) make these rays less prone to diffraction, yielding a ballistic trajectory through biological tissues—they behave more like streams of particles rather than propagating waves. The corresponding high oscillation frequencies ($>10^{17}$ Hz) also make them deeply penetrating, with minimal absorption and scattering in less dense materials. In fact, the propensity of x-rays to pass through soft tissues, but with different transmission coefficients, is what makes them so useful for diagnostic imaging. X-ray projection and computed tomography (CT) can visualize human anatomy using the tissue-dependent transmittance as contrast. However, when high-frequency radiation does interact with tissues, it is strongly ionizing—stripping away electrons from the constituent atoms and molecules. For gamma rays in particular, this destructive capacity in tissues is utilized in applications such as antimicrobial sterilization and the treatment of cancerous tumors. Gamma rays, at very low radiation level, are used in positron emission tomography (PET) and single-photon emission computer tomography (SPECT), widely used clinical imaging modalities.

4.2.1.2 Microwaves and Radiowaves

Microwaves and radio waves lie at the other end of the electromagnetic spectrum. These low frequency waves also interact weakly with biological tissues leading to deep penetration. Preparation of tissue with a strong magnetic field allows radio waves to yield images with high contrast between different tissue types through magnetic resonance imaging (MRI). Although they are nonionizing, radio frequency and microwave radiation can induce localized heating in tissues, and are thus useful for some surgical interventions.

4.2.1.3 Infrared, Visible, and Ultraviolet Light

Light, from infrared to visible and ultraviolet, occupies a window of particular relevance in biomedical applications.

Electromagnetic waves at these optical frequencies are easily collected into beams, directed, and focused using mirrors and lenses. Light interacts in a number of ways with biological specimens and can yield a wealth of information. Absorption and scattering limit the penetration of light in tissues but provide good contrast to reveal detailed structures down to the cellular and subcellular level. Specialized optical components can also be used to visualize changes in the properties of light, such as phase and polarization, as it is transmitted through a sample. Photons of ultraviolet light, as well as the shorter wavelengths of visible light, carry just the right energy to induce the phenomena of fluorescence. Particular molecules, whether introduced (exogenous) or intrinsic (endogenous) to the tissue, convert short wavelength light to longer wavelengths of a particular signature, simultaneously revealing both their presence and identity.

In general, light is produced by the energy-level transitions of atomic electrons. As electrons leap down from more energetic to less energetic states, the energy difference is emitted as photons of light. Very large energy gaps correspond to high-frequency ultraviolet light. Smaller electronic transitions yield visible light. Low-frequency infrared light is produced by even closer transitions, or by changes in molecular vibrational and rotational states.

4.2.1.4 The Sun as a Light Source

The sun is the most universally accessible source of light. The heat of the sun generates a spectral pattern that approximately follows blackbody radiation at a temperature of about 5800 K (see Figure 4.2). Peak emission is in the region from 400 to 750 nm, corresponding to the human visual range.

Sunlight was a convenient and ubiquitous source of illumination during the development of the microscope—the earliest technology to harness light in studying biological tissues. The sun provided bright light and its paraxial rays were easy to collect and focus. However, obvious limitations such as inconsistent light quality and availability motivated the development of more controllable light sources. Over the past century, light sources and biomedical applications have progressed in parallel. Technological innovations in light sources have spurred further developments in biomedical applications, while ever more specialized applications have inspired more sophisticated new sources.

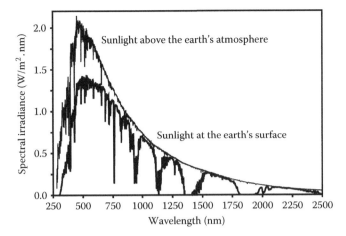

FIGURE 4.2 Spectral profile of sunlight above the Earth's atmosphere and at the Earth's surface. The difference between the two profiles is due to reflection and absorption of light within the atmosphere, primarily by molecules of water, oxygen, and carbon dioxide. (Based on data from American Society for Testing and Materials (ASTM) Terrestrial Reference Spectra for Photovoltaic Performance Evaluation, ASTM G-173-03.)

4.2.1.5 Man-Made Light Sources

Gas discharge and incandescent lamp sources were initially introduced as proxies for the incoherent radiation provided by the sun, but providing more consistent control and availability of light. These devices remain efficient work horses for many applications, particularly in microscopy and spectroscopy. However, it was the introduction of laser light sources that revolutionized the field of biomedical optics. Novel properties of coherent laser light opened new avenues of interrogation and interaction with biological tissues. As laser systems continue to advance not only in capability, but also in ease of use, optical techniques become more important and more pervasive in the practice of medicine. Man-made light sources may be categorized into two classes: nonlasers and lasers, as described in the following sections.

4.2.2 Nonlaser Sources

4.2.2.1 High-Pressure Arc Lamps

High-pressure arc lamps generate high intensity quasi-continuum spectra ranging from the ultraviolet to near-infrared. A gas (typically xenon, mercury, or a mixture of these) is filled into a quartz envelope with two tungsten electrodes. An applied voltage ionizes the gas and creates a bright arc between the electrodes. The electrical current driving these gas discharge lamps need to be regulated otherwise they would quickly burn out. The output power of commercially available high-pressure arc lamps ranges from a few watts to several kilowatts.

The xenon arc lamp is widely used as a steady state spectroscopy source because of its smooth spectral profile between 250 and 700 nm. Other arc lamps, such as the mercury-filled variant, feature prominent spectral lines more suitable for applications

needing single wavelength excitation. Being an isotropic source, the emission from arc lamps is usually omnidirectional. Parabolic and ellipsoidal reflectors may be used to collect the generated light and focus it to a sample or specimen.

4.2.2.2 Low-Pressure Vapor Lamps

Low-pressure vapor lamps operate via gas discharge similar to high-pressure arc lamps, but can be operated with less complex power supplies. They are favored as high-intensity sources of stable spectral lines as each elemental gas generates a known set of characteristic lines. For example, a low-pressure mercury lamp has a dominant wavelength peak at 253.7 nm as well as a prominent triplet at 365.0/365.5/366.3 nm. Because of such characteristic discrete lines, these lamps are used in simple filter type spectrometers and also serve as calibration light sources.

4.2.2.3 Incandescent Lamps

An incandescent lamp generates light by heating a metal filament, typically tungsten. This provides a continuous blackbody radiation spectrum that is useful for, among other things, intensity calibration of various types of detectors including spectrometers. Incandescent lamps are inexpensive and very simple to operate with no external regulating requirements. They are very convenient sources for broadband illumination at infrared to visible wavelengths. However, ultraviolet output from these lamps is usually very low.

4.2.2.4 Light-Emitting Diodes

LEDs are solid-state, semiconductor-based light sources. Light is produced via electroluminescence when positive and negative charge carriers recombine at a semiconductor junction. Their intrinsic emission is close to monochromatic and wavelength is determined by the characteristic energy bandgap of the semiconductor or semiconductor alloy used. The range of available wavelengths from LEDs stretches from the near-infrared (~1050 nm) to the ultraviolet (UV) (~250 nm) and continues to be expanded. The addition of phosphors and integration of multiple emitters in single devices have also made broadband and white-light LEDs possible.

Since the output of LEDs can be modulated at high frequencies (>100 MHz), they are popular light sources in applications requiring fast response. Additional advantages are their compact size and the low level of heat generated during operation, which render them well suited for use in tight spaces and potentially even for in vivo use. Although LEDs are very energy efficient, they are still unable to match the high output powers of incandescent lamps and arc lamps.

4.2.3 Laser Sources

4.2.3.1 Properties of Lasers

Low-coherence lamps and LEDs are useful for many applications, particularly when low cost and simple operation are the primary concern. However, many modern techniques in biomedical optics are practical only when applied with laser sources due to several important characteristics of coherent laser radiation.

4.2.3.1.1 Directionality

Directionality is the most immediately evident property of laser radiation. While light from incoherent lamp sources tends to spread out in all directions, light from a laser propagates as a beam along a particular direction. Laser beams can travel long distances with little loss of intensity and can be focused into very small beam spots, limited ultimately by diffraction. Tight focusing is essential for high resolution imaging, such as in confocal microscopy. Conservation of power along the beam also means that very high light intensities can be achieved at the focal spot. This property is key to achieving practical signal-to-noise ratios in many biomedical imaging applications.

4.2.3.1.2 Monochromaticity

Monochromaticity is another reason lasers appear much brighter than typical lamp sources for the same total output power. In general, lasers carry power over a very narrow spectral span, as opposed to broadband sources where power is spread over a wider range of wavelengths. This spectral concentration is very useful for techniques that require selective but efficient optical excitation. Laser hair removal is a specific example. Although "monochromatic" implies a single discrete frequency for laser oscillation, in reality all lasers posses a finite spectral bandwidth. Some lasers even operate with multiple modes, that is, multiple regularly spaced spectral lines. A laser may also be tunable such that the emission can be shifted across a range of wavelengths.

4.2.3.1.3 Coherence

Coherence refers to a consistent phase relationship between distant points on a light wave. In practice, coherence is determined by observing the interference patterns of overlapping light waves. Coherent light produces sharp contrast between dark and bright fringes. Fringes produced by partially coherent light appear washed out with poor contrast. Completely incoherent light does not produce any interference pattern at all.

We differentiate between spatial coherence and temporal coherence. Spatial coherence compares light from two points spatially separated across a single wavefront of light. The maximum separation for which interference can still be observed is called the spatial coherence length. Temporal coherence compares two sequential points along a wave train. Coherence time is defined as the maximum time interval over which a light wave can still interfere with a previous segment of itself. Since light is a traveling wave in both space and time, coherence time is often measured as an equivalent temporal coherence length.

Lasers can have spatial and temporal coherence lengths reaching several meters. In contrast, light from an incandescent lamp has coherence lengths in the order of just a few micrometers. Coherent laser sources enable the practical application of interferometry and holography that rely on the diffraction and interference of light. Coherence also leads to the formation of randomly distributed bright and dark interference spots, that is, speckle, when laser light illuminates an extended area. Speckle is often considered undesirable noise in imaging applications, but may sometimes be utilized to reveal information about surfaces.

4.2.3.1.4 Short Pulse Duration

Lasers may operate in either continuous wave (CW) or pulsed modes. A CW laser operates with uniform output power over time, while a pulsed laser emits light in a sequence of short bursts. Pulsed operation can be advantageous as it concentrates energy into a short window of time. This allows very high levels of instantaneous power even while maintaining modest average power. Ultrashort-pulsed lasers with pulse durations in the order of femtoseconds (10^{-15} s) have made high-resolution multiphoton microscopy techniques practical. High-energy pulsed lasers also enable very precise surgical procedures through very controlled and localized ablation of tissue.

4.2.3.2 Fundamental Laser Components

Lasers seem to come in a boundless variety of shapes, sizes, and materials—from microchip-sized semiconductor lasers, to fiber lasers tens of meters long, or even amplified laser systems that can fill a large room. However, regardless of the specific implementation, there are three basic components that can be identified in any operational laser—the gain medium, excitation or pump mechanism, and resonant cavity. These elements are diagrammed schematically in Figure 4.3.

4.2.3.2.1 Gain Medium

Lasers are often identified by specifying the gain media, for example, a helium–neon (HeNe) laser, or a neodymium:yttrium aluminum garnet (Nd:YAG) laser. The gain medium, alternatively referred to as the active medium, is where light amplification takes place. The material may be a solid, liquid, or gas. The electronic energy-level structure of a material determines its emission wavelengths and bandwidths; thus the gain medium partly determines the central wavelength and wavelength span of laser emission. Gas media, in particular, have narrow gain bandwidths. They amplify light over a very narrow spectral range ($<10^{-3}$ nm) and are practically single-wavelength sources.

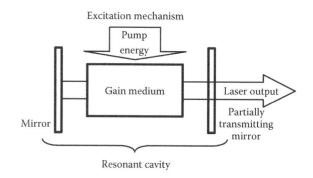

FIGURE 4.3 Fundamental laser components. The gain medium, excitation mechanism, and resonant cavity are the three basic components that can be identified in any operational laser.

Solid and liquid materials have broad gain bandwidths and can support lasing action over much wider intervals. Broad bandwidths enable the design of tunable lasers and ultrashort-pulsed lasers.

4.2.3.2.2 Excitation Mechanism

Before a gain medium can amplify light, atoms in the material must be suitably excited. An excitation mechanism is needed to pump energy into the gain medium. This energy is stored by atoms in their excited state, and released via either spontaneous or stimulated emission as the atoms relax back to their original ground state (see Figure 4.4). Excitation may be through electrical pumping, such as when injecting an electrical current into a semiconductor or gaseous gain medium. Alternatively, a laser may use an optical pumping scheme where the gain medium absorbs light, either from a flash lamp or another laser, and converts it to a different wavelength for emission.

4.2.3.2.3 Resonant Cavity

Spontaneous emission from an excited gain medium does not posses the coherent properties of laser radiation. A resonant cavity is necessary to provide optical feedback for stimulated emission to take place. Reintroducing previously emitted light knocks down excited atoms back to the ground state. These stimulated transitions emit coherent radiation without absorbing the optical feedback. Thus, the incident light is amplified, particularly if the excitation mechanism has induced population inversion in the gain medium, that is, there are more atoms

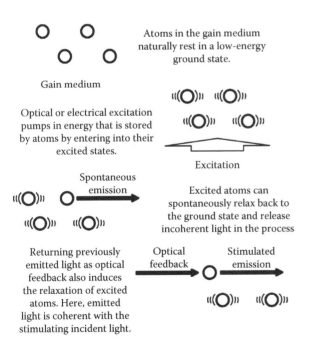

FIGURE 4.4 Fundamental laser operation. Optical or electrical excitation provides energy that is stored by atoms in the gain medium. This energy is released as incoherent light through spontaneous emission, or as coherent light through stimulated emission in the presence of optical feedback.

in the excited state than the ground state. A resonant cavity is typically constructed by placing the gain medium between two mirrors, forming a Fabry–Perot resonator. The resonator may also be as simple as the cleaved ends of a semiconductor gain medium. Other configurations for optical feedback include ring resonators and distributed feedback via scattering.

Ultimately, the resonant modes of the cavity, in conjunction with the gain profile of the active medium, determine the wavelengths at which a laser operates. This is referred to as the longitudinal modes of a laser. The resonator also determines the cross-sectional profile, or transverse modes, of the laser beam. Most often, the cavity is designed to produce a singular circular lobe of the fundamental Gaussian transverse mode.

Some laser designs may also feature other components for expanded functionality. For example, modulators can be used to induce pulsed laser operation. Spectral filters may be inserted to control the specific output wavelength in a tunable laser. However, the three fundamental components: gain medium, excitation mechanism, and resonant cavity remain the defining parts of a laser, and will always be present in any laser system.

4.2.3.3 Pulsed Laser Operation

Many biomedical applications require coherent light to be delivered as a series of laser pulses. Depending on the specific application and corresponding power density requirements, pulse durations may reach as short as several femtoseconds (10^{-15} s). Pulses longer than a few microseconds (10^{-6} s) are easily produced by directly modulating the CW output of a laser via mechanical, electro-optic, or acousto-optic shutter mechanisms. Shorter pulse durations, in the order of nanoseconds (10^{-9} s) or less, require more specialized laser designs. Two important methods for generating short laser pulses are Q-switching and modelocking.

4.2.3.3.1 Q-Switching

Q-switched lasers produce high-energy pulses that range from picoseconds to nanoseconds in duration. In preparation for pulse formation, optical feedback is prevented in the laser cavity by an absorbing or scattering component. The resonant cavity is then said to be in a low-Q state. Without optical feedback, energy builds up in the gain medium as there is no stimulated emission to deplete it. When the gain has reached its peak level (i.e., saturation), a high-Q state is restored by removing the source of loss. Stimulated emission then builds up exponentially while the stored energy in the gain medium is rapidly depleted. The result is an optical pulse that carries the maximum amount of energy that can be delivered by the active medium. Pulse energy is limited by the excited state lifetime of the gain medium. Longer lifetimes allow more energy to build up. Pulse duration is limited by cavity lifetime which measures the time it takes to extract energy from the resonant cavity. Q-switched lasers typically achieve nanosecond pulsewidths. In order to maximize pulse energy, repetition rates of Q-switched lasers are generally kept slow such that the gain has sufficient time to reach saturation.

4.2.3.3.2 Modelocking

Modelocking can produce laser pulses from several picoseconds to less than 10 fs in width. Pulse formation in modelocked lasers can be interpreted in terms of interference between multiple longitudinal modes. The greater the number of modes involved in the interference, the shorter the resulting pulses. Very broad gain bandwidth is thus necessary to achieve the narrowest pulses. The necessary interference can only occur if the different longitudinal laser modes maintain a fixed phase relationship between themselves. This is not an inherent behavior in lasers, but can be induced by creating modulation sidebands that link, that is, lock, adjacent modes to each other. The modulation frequency is determined by the round-trip travel time of pulses circulating in the cavity, typically in the order of tens of megahertz.

4.2.3.4 Types of Lasers

Lasers can be broadly categorized according to the type of gain media they utilize. The most generally recognized types of lasers are gas, dye, solid-state, and semiconductor. Below, some basic properties representative of each type are described. These are followed by some specific examples of laser systems with particular relevance in biomedical applications.

4.2.3.4.1 Gas Lasers

Gas lasers principally use electric current discharge as an excitation mechanism. Gas molecules receive energy via collisions with accelerating electrons. Often, a mixture of gases is used in the gain medium to improve the efficiency of energy transfer. However, emissions originate from just a single species. Different gas lasers cover a wide range of laser wavelengths from the infrared, across the visible, and into the ultraviolet. Very narrow spectral emission profiles are an important distinguishing property gas lasers have over other laser types. The sparse density of the gaseous medium minimizes perturbations to the internal energy levels of the active molecules. In general, gas lasers are also quite robust against thermal damage even when operating at high power levels, thanks to large volumes of gain media and gas flow systems that allow heat to dissipate quickly.

The CO_2 gas laser was one of the earliest lasers to be developed, but it remains relevant even today. CO_2 lasers typically operate between at 9.6 and 10.6 μm by accessing vibrational transitions of gaseous CO_2 molecules. High efficiency and powerful output levels—reaching over 100 kW in CW operation and 10 kJ per pulse in pulsed operation—are important advantages of this type of laser. The strong absorption by water at these wavelengths make the CO_2 laser particularly suited for application in laser surgery and other photothermal therapies such as skin resurfacing.

The HeNe laser has long served as a convenient source of red (632.8 nm) CW-laser radiation at low powers (1–100 mW). Although today it has been replaced in most applications by less expensive semiconductor lasers operating at similar wavelengths, this lasers still maintains one key advantage—long

coherence length. HeNe lasers remain important light sources for holography and interferometry applications.

Excimer lasers are a class of gas lasers that use noble gas or noble gas halide complexes as gain media. The term excimer is a contraction of "excited dimer" that describes the excited state of the gain medium. Some specific examples of this type include: xenon fluoride (351 nm), xenon chloride (308 nm), krypton fluoride (248 nm), and argon fluoride (193 nm). Pulsed operation at ultraviolet wavelengths is characteristic of these lasers. Typical pulse durations are a few tens of nanoseconds with pulse energy falling between 0.2 and 1.0 J. The combination of short pulses, high energy, and UV wavelengths make excimer lasers useful for highly precise surgical procedures, such as sculpting the surface of the cornea to correct refractive errors in vision. The short UV pulses can remove very small amounts of tissue in a highly controlled fashion without any damage to surrounding areas. However, this approach is only suitable for exposed surfaces or those accessible with an endoscope as UV light does not penetrate very far into tissue.

4.2.3.4.2 Solid-State Lasers

Solid-state lasers refer specifically to lasers where the gain media are metal-ion dopants in crystalline or glass host materials. The dopants may be rare-earth ions such as neodymium (Nd) and erbium (Er), or transition-metal ions like titanium (Ti) and chromium (Cr). These lasers are optically pumped with either flashlamps or other laser devices. Laser emission typically centers about the near-infrared region, although some solid-state lasers may emit at slightly longer infrared wavelengths (~2 μm). Gain bandwidth is generally broad as the intrinsic electric field of the host material perturbs the electronic energy levels of the dopant ions. Thus, solid-state gain media are useful for building tunable lasers and pulsed lasers utilizing either Q-switching or modelocking mechanisms. Depending on the specific host material and bulk configuration used, solid-state lasers can also be designed to deliver high output power levels.

Erbium- and neodymium-based lasers are popular sources for near-infrared wavelengths close to 1 μm. Laser emission results from electronic transitions in the dopant ions, but energy level perturbations by the host material significantly affect the emission wavelength. For example, an Nd:YAG ($Y_3Al_5O_{12}$) laser will typically operate at 1.064 μm, while an Nd:glass laser operates at 1.054 μm. The Nd:YAG is a popular platform for producing Q-switched pulses because of its long excited state lifetime, as well as the good mechanical stability and thermal conductivity of the YAG crystal that allow it to withstand high energy pulses. Such lasers have become very useful for photodisruption of tissues in laser surgery. Erbium shows an even wider host-dependent variation of lasing wavelength. Er:YAG output is at 2.94 μm, while Er:glass emission centers at 1.54 μm. The longer wavelength of the Er:YAG laser makes it very useful for phototherapy on the surface of the skin where it is quickly absorbed by water molecules.

Solid-state lasers doped with titanium and chromium also operate in the near-infrared, but are distinguished by very broad gain bandwidths, ranging from 100 to 400 nm. The Ti:sapphire (Ti:Al$_2$O$_3$) is perhaps the most widely used among these types of lasers today. It may be configured as a CW laser tunable from 700 to 980 nm, or as a modelocked laser capable of producing femtosecond pulses. The modelocked Ti:sapphire laser has become the workhorse for multiphoton microscopy as its wavelength range is suitable for two-photon excitation of many standard fluorescent dyes and fluorescent proteins— green fluorescent protein (GFP) stands out among these. Cr^{4+}:Fosterite(Cr^{4+}:Mg$_2$SiO$_4$) is another significant solid-state laser of this type. It operates around 1.25 μm and provides deeper tissue penetration compared to the Ti:sapphire. These broadband solid-state lasers are also important light sources for optical coherence tomography.

4.2.3.4.3 Dye Lasers

Dye lasers have gain media that consist of organic dyes dissolved in a liquid such as ethyl or methyl alcohol, glycerol, or water. These dyes exhibit wide emission bands across the visible region, and sometimes stretching into near-infrared wavelengths. Similar to solid-state lasers, optical pumping is also achieved with flashlamps or other laser sources. Tunable laser operation across the visible wavelengths is an important feature of dye lasers, and it is broad bandwidth that enables pulsed operation via either Q-switching or modelocking.

The complex and potentially hazardous handling of liquid dyes and solvents has tended to limit general interest in dye lasers for applications beyond spectroscopic studies in research laboratories. However, the ability of dye lasers to produce high energy pulses at wavelengths strongly absorbed by hemoglobin has been proven useful in the treatment of vascular lesions and the removal of scars and tattoos from the skin.

4.2.3.4.4 Semiconductor Lasers

Semiconductor lasers, or laser diodes, differ significantly from the solid-state lasers described above in terms of the mechanisms for excitation and emission. Semiconductor gain media are pumped via direct injection of electrical current. Electrons and holes are delivered from opposite sides of a semiconductor junction. As in LEDs, the recombination of these negative and positive charge carriers results in the emission of light at an optical frequency proportional to the semiconductor band gap. Given sufficient current density and optical feedback, stimulated emission and laser action take place. Emission bandwidths are typically broad, with center wavelengths in the red to near-infrared region. More recently developed wide bandgap semiconductor lasers with blue-violet wavelengths have also become commercially important.

Simple operation, physically compact and robust devices, and low cost compared to other lasers are among the advantages that make semiconductor lasers very attractive sources for applications requiring low to moderate power (i.e., from milliwatts to several watts). Significant biomedical applications of semiconductor lasers range from laser hair removal to the treatment of macular degeneration and retinopathy in the eye.

4.3 Light Detectors

4.3.1 Introduction

The vast variety of light-sources and techniques to image, investigate, or manipulate biomedical samples, calls for light detectors tailored to specific requirements. The result is an equally abundant multitude of light detection schemes. But at the heart of these schemes, there actually lies just a handful of different detector device types.

The basic functionality of most light detectors is the same: They make use of interactions between radiation and matter to convert the light intensity into a proportional electrical signal. More sophisticated optical detection systems allow measurements of various other properties such as the spatial, spectral and temporal profile, polarization, and phase of the incident light. However, such systems still contain simple intensity detectors as their basic building blocks. The most common among these are photodiodes and PMTs, both of which are single channel detectors that only measure light at a single point. In order to simultaneously measure a spatial distribution of light intensity, an arrayed detector is necessary. CCDs are the most common example of these, although complementary metal-oxide-semiconductor (CMOS) detectors and photodiode or PMT arrays are gradually becoming more important.

4.3.2 Photodiodes

4.3.2.1 Principle of Operation

Photodiodes are solid-state photodetectors that operate on the basis of the internal photoelectric effect. The material providing the sensitivity to incident light is a semiconductor. In pure semiconductors, the electrons fill up all available energy states in the valence band, which is one of the two relevant manifolds of energetic states in semiconductors. The other relevant band, the conductance band, is separated from the valence band by the bandgap and is higher in energy. At room temperature and in the absence of any light, the lack of any unoccupied states in the valence band means that the semiconductor cannot support a net movement of charges. It then has a very low (ideally zero) conductivity. Thus, little or no current is measured if one applies an electric field across the material. However, incident photons with energies larger than the bandgap can excite electrons from the valence to the conduction band where they are free to move. This results in an increase of conductivity, which is (over a certain range) proportional to the number of incident photons. Consequently, the application of an electric field now results in a measurable current that is proportional to the number of the incident photons.

Most photodiodes used today are based on p–i–n structures. Here, a layer of an intrinsic semiconductor, that is, a semiconductor free of any impurities, is sandwiched between a p- and an n-doped region. In these doped regions a defined number of mobile holes (p-doped) or electrons (n-doped) are created by adding a well-controlled amount of impurities. The p–i–n

configuration induces a permanent electric field in the intrinsic region of the structure and thus facilitates generation of a photo-induced current, even if no external electric field is present. In this short-circuit or photovoltaic mode, *p–i–n* photodiodes feature low noise levels (due to the absence of significant shot-noise, see below) but have a limited response time. Application of an external voltage (reverse-bias mode) improves the response time at the cost of increased noise (also see APDs below).

The bandgap of the semiconductor on which a photodiode is based defines the wavelength range over which the device is sensitive. Photons with energies below the bandgap will not excite electrons into the conductance band and thus will not generate a signal. Independent of its energy, one photon only excites a single electron in a conventional photodiode. This reduces the response (relative to the incident power) for short wavelengths and usually defines the lower end of the spectral response curve of a photodiode.

4.3.2.2 Characteristics and Applications

Photodiodes are possibly the most widely used photodetector today. Among them, silicon-based devices are the most common. The prevalence of silicon is due to the extreme maturity of the silicon technology, which is the result of some 50 years of development. In particular, silicon devices are relatively low-cost, offer linear response, are very rugged, and can be easily integrated with other optical and electronic components. Silicon photodiodes can be used to detect light in the near UV, the entire visible, and some part of the near infrared (NIR) region of the spectrum. At the maximum of the spectral response curve, which is located at around 800 nm, *p–i–n* silicon photodiodes can have a quantum efficiency approaching unity, that is, nearly every incident photon causes excitation and subsequent extraction of an electron.

Silicon photodiodes feature a sharp drop-off in spectral response at wavelengths longer than 1.1 μm. Thus, alternative materials are required to access the full NIR region (0.75–1.6 μm). This spectral region is important for many biomedical optical techniques since most biological samples show relatively low absorption at these wavelengths. The band structure of germanium renders photodiodes based on this material sensitive to light with wavelengths up to 1.8 μm. Good response in the NIR can also be obtained in hetero-junction structures where the *p*- and the *n*-type part of the diode consists of different materials, for example, heterojunction photodiodes based on the semiconductor alloys InGaAsP and InP. Such devices achieve a quantum efficiency approaching 75%. Despite the significant progress in heterojunction technology over recent years, the use of different materials in one detector still considerably increases the cost of manufacturing.

The inherent spectral response of photodiodes is relatively broad. Various coatings and filters can be used to block certain spectral regions, thereby defining a narrow range of wavelengths to which the detector is sensitive. In addition, anti-reflection coatings are routinely employed to improve the sensitivity of photodiodes and various other photodetectors.

In reverse-bias configuration, *p–i–n* type photodiodes can achieve response times on the order of a few tens of picoseconds rendering them well suited for the detection of fast processes provided the absolute intensity of the incident light is sufficiently high to obtain reliable measurements. The ultrafast detection of small optical signals requires more sophisticated metrology.

An ideal photodiode generates an electrical signal that is simply proportional to the optical input. However, even in the absence of light, a photodiode will generate a finite electrical output, referred to as dark current. In addition, both the dark current and the real signal will have a random component, known as noise. The total noise level is a superposition of different contributions. The two most important to be aware of are the shot-noise and the thermal noise. The presence of shot-noise is a direct result of the photon nature of light. Each incident photon generates a tiny signal pulse. Even for a constant optical signal the incident photon stream will thus have random fluctuations that translate into electrical noise at the output of the detector. In this context, the term shot-limited noise level is sometimes used to indicate that the noise of the output signal is at or close to this fundamental limit. Thermal noise results from random motion of charges within the photodiode and can be reduced by cooling of the detector. To quantify the overall noise level for a photodiode or any other detector or detection system, one usually quotes the noise equivalent power (NEP). The NEP is defined as the radiant power of an incident signal that results in a signal-to-noise ratio of unity at the detector output. The NEP is measured for a sinusoidally modulated input signal and varies with the frequency of the input signal.

It is important to note that the properties of nearly all semiconductor-based photodetectors significantly depend on temperature. We have already seen that the noise level increases with temperature and that cooling of the photodiode might be necessary to measure small signals. However, even for fairly large signals an active temperature control might be required to ensure repeatable measurements if the photodiode is used in an environment with significant temperature variation. In addition, external mechanical stress may considerably increase the noise level and electromagnetic interference (EMI) can constitute a serious source of noise that calls for measures to protect the photodiode, for example, by shielding or adequate packaging.

4.3.2.3 Avalanche Photodiodes

Photodiodes that are designed for operation at a high reverse bias (usually 50–300 V) are known as avalanche photodiodes (APDs). The reverse bias generates a high electric field inside the photodiode that accelerates the photo-generated charges. Collision of these accelerated (primary) charges with the crystal lattice leads to ionization, which in turn generates secondary electron-hole pairs. In a chain reaction (avalanche process), these secondary charges are then also accelerated and thus generate additional charge by further ionization processes. Depending on the reverse bias, avalanche ionization amplifies the photocurrent by 50 to several 100 times. One usually refers to this amplification as the gain or multiplication factor of the APD.

If the reverse-bias is too high (i.e. above the breakdown voltage), the arrival of just a single photon triggers a continuing avalanche process (Geiger mode). While this is useful for single-photon counting experiments and allows measuring the arrival of a single photon with high time resolution ($\leq$50 ps), the reset required to stop the avalanche process results in a dead time of approximately 50 ns. One can therefore not use the Geiger mode when dealing with large optical signals or considerable amounts of background light.

Both, the breakdown voltage and the gain are strongly temperature dependent. In addition, if the APD is operated close to the breakdown voltage, the generated photocurrent itself can lead to a significant reduction of the electric field in the avalanche region of the device. This creates an unwanted situation in which the photocurrent is not proportional to the input signal.

Due to high gain factors, APDs are particularly useful when measuring small optical signals that one might not be able to detect with conventional photodiodes. However, the increased sensitivity brings in some limitations: (1) operation close to the breakdown voltage results in nonlinear response; (2) the active area of APDs is very limited in size since defects in the crystal structure and strain must be avoided to achieve the high breakdown voltages required for an efficient avalanche process; (3) the noise level is often higher than in conventional photodiodes since amplification noise constitutes an additional source of noise; and (4) the readout circuit is generally more complex due to the need for a high operating voltage.

4.3.3 Photomultiplier Tubes

4.3.3.1 Principle of Operation

In contrast to photodiodes, photomultiplier tubes (PMTs) employ the external photoelectric effect. As illustrated in Figure 4.5, incident photons hit a photocathode and if their energy is sufficiently high, extract electrons from the material by ionization. The critical energy for this photoemission process depends on the photocathode material: Photoemission only takes place if the energy of the incident photon is above the work function W of the photocathode or in other words if the wavelength of the light is shorter than hc/W, where h is Planck's constant and c is the speed of light. However, even if the photons provide sufficient energy, electron-trapping in the bulk of the photocathode my still prevent their extraction. Therefore, the quantum efficiency of photoemission, that is, the ratio of extracted electrons over incident photons, is well below unity in real devices. The extracted electrons are focused by a pair of electrodes and then accelerated toward a secondary electrode, or *dynode*, which is kept at an attractive electric potential. Upon impact, each incident primary electron can extract several secondary electrons by ionization processes. Using again a drop in electric potential, the secondary electrons are collected by a third electrode, the *anode*. The flow of electrons from photocathode to anode gives rise to a photocurrent that forms the output signal. The components of a PMT are housed in a vacuum tube to allow electrons to travel over macroscopic distances once they have been extracted from the cathode.

4.3.3.2 Characteristics and Applications

The PMT may be regarded as the vacuum analogue of the APD. However, the possibility to add multiple amplification stages by integrating several dynodes (usually between 6 and 14) allows substantially higher amplification factors. Depending on the voltage drop between the electrodes and the number of dynodes, a gain between 10^5 and 10^8 can be achieved. The quantum efficiency of the primary photoemission process, however, is considerably lower than in semiconductor photodiodes. Traditionally, metals and metal alloys—especially alloys of alkali metals—have been used as photocathodes. Due to partial reflection of the incident light and trapping of electrons, these electrodes typically have quantum efficiencies around 10%. Today, metal photocathodes have been complemented by semiconductor alloy cathodes with lower reflectivity and quantum efficiency up to 30%.

As with photodiodes, the spectral response of PMTs is strongly dependent on the photocathode material used, but is also influenced by the transmission profile of the vacuum tube's glass window. PMTs can cover the UV range (down to about 120 nm) and the entire visible part of the spectrum. The sensitivity of most PMTs drops sharply at wavelengths above 1 μm, which limits their application in the IR. Exceptions are devices using p-n-junction photocathodes based on InP/InGaAs that provide reasonable sensitivity up to 1.6 μm.

Again in analogy to the situation for APDs, the design of the peripheral circuitry required to operate the PMT and to measure the output signal is not trivial. The overall voltage drop between photocathode and anode usually ranges between 500 and 3000 V and a set of voltage-dividing resistors is required to adjust the voltage gradient between the individual elements of the PMT. Special care must be taken to avoid introduction of additional noise from the peripheral circuitry. For instance, improper selection of the grounding scheme will result in a considerable

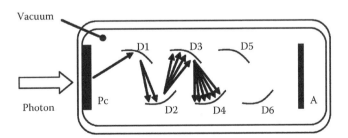

FIGURE 4.5 Schematic illustration of a photomultiplier tube. Incident photons extract electrons from the photocathode (PC). The electrons are accelerated toward the first dynode (D1). At each dynode (D1 to D6) the incident electrons stimulate emission of additional secondary electrons. This avalanche process results in the buildup of a large current pulse that is collected at the anode (A).

increase of the noise level and can even cause permanent damage of the PMT. Depending on the measurement scheme, PMTs can be used to either measure DC or AC signals or can be operated in photon-counting mode.

Assuming an appropriate driving circuit, the largest source of noise in PMTs is thermal electron extraction from the photocathode. This effect is most pronounced for materials with a low work function since the chance of electron extraction by thermal fluctuations is relatively high in this case. Vice versa, photocathodes with high work function and sensitivity in the UV or blue end of the visible spectrum are virtually immune to thermal noise. Cooling of the photocathode, and in certain cases also of the first dynode, with a Peltier element generally reduces thermal noise.

Additional potential sources of noise include external electric or magnetic fields that will distort the electron trajectories in the PMT. The strong signal amplification in PMTs and the fact that PMTs are usually housed in evacuated glass or quartz tubes also means that they are extremely susceptible to stray light. Therefore, appropriate housing and shielding are particularly important for PMTs. Often the outer surface of PMTs is coated with black conductive paint, also referred to as hydroxyapatite HA coating. "HA coating" is a term commonly used by manufacturers to refer to black conductive paint applied to the outer surface of the PMT glass envelope in order to reduce noise. It is not composed of hydroxyapatite.

Depending on the dynode geometry the *signal rise time* of PMTs varies between 1 and 20 ns. The limiting factor is the spread in electron transit time that results from the fact that electrons follow different trajectories on their way from the photocathode to the anode. Certain dynode designs can limit this effect. However, this usually comes at the cost an overall reduction in sensitivity or leads to reduced uniformity across the active area of the photocathode.

4.3.3.3 Microchannel Plates

Microchannel plates (MCPs) are a refinement of the PMT concept and offer improvements mainly in terms of response time, compactness, reduced power consumption, and lower sensitivity to magnetic fields. Instead of using a series of discrete dynodes, MCPs are based on a disc-shaped array of capillaries with diameters in the 10-μm range and lengths on the order of 1 mm. The inner walls of the capillaries have a high Ohmic resistance coating. If a voltage is applied to both ends of the capillary, a gradient in electric potential is generated between one end and the other. Primary electrons extracted from the photocathode of the MCP are absorbed by the capillary array and passed toward the opposite end. Due to the high aspect ratio of the capillaries, these electrons impinge on the capillary walls multiple times and thus generate an avalanche of secondary electrons that then exits the capillary at the opposite end.

MCPs can achieve a temporal resolution (signal *rise time*) on the order of 100 ps since all electron trajectories inside the microchannel array are nearly parallel and thus have almost equal lengths. In addition, the absolute path lengths are much shorter than in conventional PMTs. Therefore, all electrons have a similar transit time, which is a principal requirement for fast response.

4.3.4 Arrayed Detectors

The photodetectors discussed so far are single element detectors. However, biomedical optics often requires the simultaneous measurement of multiple spatially separated optical signals, for example, to create an image of a specimen or to measure the spectral distribution of an optical signal. If a large detector array is required, as it is the case for image generation, detection and read-out should be integrated on a single chip. Separate wiring of discrete detectors would be impractical in terms of cost and device size.

4.3.4.1 Charge Coupled Device—Principle of Operation

A very common detector array scheme based on a photodiode-like detection principle is the CCD. Instead of using the *p–i–n* diodes described before, CCD sensors consist of an array of metal–oxide–semiconductor (MOS) capacitors as photosensitive elements. These structures generate and store an amount of charge that is proportional to the number of photons collected since the detector was last read-out, which means that CCD chips are integrating detectors. During the measurement, all photocapacitors record the incident optical signal simultaneously. The subsequent read-out of the charge stored in each capacitor or pixel, however, is a sequential process. One common read-out scheme, which is to embed charge transfer channels under each row of the CCD array, is illustrated in Figure 4.6. These channels consist of a series of capacitors, referred to as wells, each of which is capable of storing the charge generated by one of the pixels during exposure. By manipulating the voltage applied to these wells, the generated charge can be transferred along the channel and is eventually collected at the end of the channel by an amplifier. The spatial origin of each charge package can be inferred from the sequence of arrival, which ultimately allows the reconstruction of the image. To avoid intermixing of the charge packages accumulated in neighboring pixels, at least three wells are required for every pixel.

4.3.4.2 Characteristics and Applications of CCDs

Most CCD chips are based on silicon as the active and photosensitive material. Therefore, they achieve good sensitivity in the visible and near IR (≤1 μm) range of the spectrum with quantum efficiencies between 20% and 40%. Conventional front-illuminated CCDs have limited sensitivity in the UV due to significant absorption of UV light in the top electrodes of the CCD chip. Good UV sensitivity can either be achieved by down-conversion of the UV light to visible wavelengths using phosphors or preferably by using a back-illumination scheme. In this case, the wafer on which the CCD is fabricated is back-thinned so that the light can be passed to the photodetectors on the chip through the backside of the substrate. This scheme greatly reduces absorption losses and yields good sensitivity over the UV, visible, and near IR range with quantum efficiency between 50% and 90% across the entire spectral range. CCD chips can also be used for direct detection of x-rays. In this case, the generation of multiple

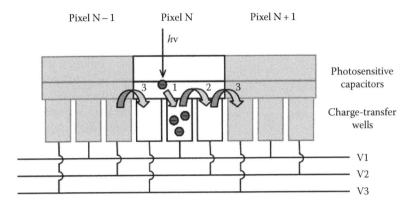

Pixel N − 1   Pixel N   Pixel N + 1

FIGURE 4.6 Illustration of the charge transfer mechanism used to read out the photo-generated charge from a CCD detector. Incident photons (*h*v) charge the photosensitive capacitor. Application of a voltage to line V1 transfers the charges to one of the three wells connected to each pixel. The charge is transported to the adjacent well when line V2 is switched on and V1 is switched off. The charge is then transferred to the first well of the neighboring pixel by activation of line V3. Repetitive voltage cycles applied to V1, V2, and V3 thus transport the charge generated in each pixel along the channel without intermixing charge from different pixels.

electron-hole pairs per incident photon even enables determining the energy of the incident photons.

Like conventional photodiodes, the noise generated in a CCD chip can be reduced by active cooling. Sensitive CCD chips are often equipped with integrated thermoelectric (TE) coolers to ensure a homogenous and constant temperature distribution of the chip.

The dynamic range of any detector is defined as the ratio between the largest input signal that can be measured without saturating the detector and the smallest signal that can be distinguished from noise. The dynamic range is of particular importance for image detectors as a low dynamic range will reduce the contrast of recorded images. For CCDs, the saturation limit is given by the capacitance of the wells storing the photo-generated charge, whereas the noise limit at low light intensities is usually dominated by read-out noise—at least if the device is TE-cooled. Simple CCDs typically have an 8 bit dynamic range, that is, the saturation level is 256 times larger than the noise level. More sophisticated devices can reach dynamic ranges of up to 16 bit or 65,000. The dynamic range must not be confused with the bit-depth, which refers to the number of digitization steps provided by the analogue-to-digital converter (ADC) on the chip. Ideally, the bit-depth should be equal or slightly larger than the dynamic range. However, CCDs with bit-depths that greatly exceed the actual dynamic range are frequently offered, in the hope that the large number of digitization steps gives the (false) impression of a large dynamic range.

CCD image sensors were developed in the 1970s, and have since then been tremendously improved with regard to their dynamic range, sensitivity, production cost, speed, and noise level. Today, they have replaced many other imaging sensors, including the photographic film, which was for a long time superior to electronic image detection in terms of dynamic range, sensitivity, and cost. In addition, electronic imaging schemes such as CCD imaging have enabled new kinds of data collection, in particular by allowing continuous data acquisition and convenient image analysis. It is worth noting that compared to other technological breakthroughs mediated by the semiconductor technology, this is a relatively recent advance and that the widespread use of high quality CCD image sensors in biomedical labs has only begun some 15 years ago.

4.3.4.3 CCD Spectrometers

Measurements of the spectral distribution of an optical signal can often provide very useful information about the composition or state of a specimen. Traditional techniques include absorption and fluorescence spectroscopy, but Raman-spectroscopy and various time-resolved and nonlinear spectroscopy techniques have complemented these methods and can often provide additional or more precise information. Similar to the situation with CCD image sensors, traditional spectroscopic measurement techniques such as scanning spectrometers were to a large extent replaced by CCD-based spectrometers over the past 10–15 years. In addition, CCD-based systems have enabled various measurements that have not been possible before, as their fast and parallel data acquisition allows the measurement of rapidly changing spectral profiles.

Spectrometers come in various configurations and refinements. A typical configuration is a CCD-based fiber spectrometer that consists of an entrance slit, a diffraction grating, and a linear CCD detector (see Figure 4.7). The light enters through the entrance slit, which restricts the spatial distribution of the incoming light to a narrow line. The light is then collimated and passed onto the diffraction grating, which is the main functional component of the spectrometer. The grating reflects light of different wavelengths under different angles, which enables spatial separation of different spectral components in the input signal. The diffracted light is subsequently focused onto the CCD detector, which consists of a single line of sensitive elements. The position of the focus on the CCD detector depends on the diffraction angle and thus on the wavelength of the incident light. The spectrometer is calibrated using a light

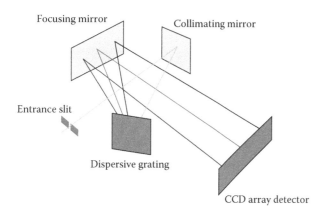

FIGURE 4.7 Schematic illustration of a crossed Czerny-Turner CCD spectrometer. Light enters through the slit, is collimated by a first mirror, and then divided into its spectral components by a dispersive grating. The focusing mirror creates an image of the entrance slit on the CCD detector. Since the position of the image depends on the wavelength of the light, the CCD signal yields the spectrum of the incident light.

source with a set of spectrally narrow emission lines at known wavelengths.

In addition to the previously discussed general performance parameters of photodetectors, spectrometers can be characterized by their spectral range and spectral resolution. The first refers to the wavelength range that can be unambiguously analyzed with a spectrometer; the latter describes the minimal wavelength difference of two spectral lines that can be resolved. It is technologically challenging to combine a large spectral range and a high spectral resolution in one device. In addition, improving the spectral resolution usually reduces the sensitivity of the spectrometer. The optimal trade-off between these conflicting performance parameters therefore depends very much on the intended application. For instance, Raman spectroscopy requires excellent spectral resolution and high sensitivity but only a limited spectral range. By contrast, a low spectral resolution is often acceptable for fluorescence spectroscopy since the fluorescence spectra of many dyes used in biomedical optics are spectrally broad.

4.3.4.4 Complementary Metal–Oxide–Semiconductor Detectors

CMOS is an alternative technology to integrate light measurement and read-out for a large array of detectors on a single chip. Instead of using charge collecting wells and a charge transfer process to transfer the information about the light intensity, each pixel of a CMOS sensor comprises a separate circuit to process and digitize the signal before transferring the data to the output interface of the chip. This scheme adds complexity to the design of the chip and sacrifices uniformity of detection and fill-factor (i.e., the fraction of the chip surface that is light sensitive). In the past, these issues prevented fabrication of CMOS-based image sensors with acceptable performance. Advances in the CMOS process technology, however,

have resulted in a rediscovery of the CMOS sensor concept over recent years. The CMOS concept is very attractive since the required process technology is widely used in other fields of the electronics industry, which offers the potential of lower production cost. In addition, the use of standard semiconductor technology also enables integration of image processing functions on the chip, which reduces the complexity of off-chip circuitry. Finally, CMOS sensors usually consume less power than comparable CCD chips. While many of these features were originally thought to render CMOS sensors particularly suitable for consumer applications, they may also turn out useful for biomedical optics, especially in high-throughput experiments where cost is often a limiting factor or in situations where highly integrated electronics ("camera on a chip") can enable new types of measurements. Some examples are endoscopes containing ultraminiaturized cameras and data acquisition units, which are often also single use.

4.3.4.5 APD and PMT Arrays

Both the CCD and the CMOS sensor can be regarded as an arrayed version of conventional photodiodes combined with an efficient read-out scheme. Other sensor types, in particular APDs and PMTs, are also available as arrays. However, the number of detectors per array is significantly lower in these cases due to technical limitations. While CCDs and CMOS chips with pixel counts in excess of 10^7 are commercially available, APD and PMT arrays are usually limited to well below 100 detectors. Therefore, they cannot be directly used for imaging. However, they are very useful to simultaneously collect weak optical signals at different wavelengths, as required, for example, in scanning imaging techniques such as confocal or two-photon microscopy.

4.3.4.6 Intensified CCD Cameras

Intensified CCD cameras can be used to perform direct imaging of low-level optical signals. These devices are a combination of a MCP–PMT and a sensitive CCD chip. The optical signal is incident on the front surface of the MCP and the primary electrons generated at the photocathode are amplified as they pass through the capillaries of the MCP. Instead of collecting the photocurrent across the MCP, the electrons are directed toward a phosphor screen at the output facet of the MCP to reconstruct the spatial distribution of the input signal. Due to the large gain of the MCP (usually $\geq 10^5$), the image on the phosphor screen is orders of magnitude brighter and can therefore be recorded with the CCD chip. The spectral response of intensified CCD cameras is determined by the response of the photocathode. Different cameras with a range of photocathodes covering the UV, visible, and NIR are available today.

A prominent variant of the intensified CCD camera is the streak camera that can measure the temporal profile of fast optical signals with a subpicosecond time resolution. Since streak cameras can measure the temporal profile of many channels simultaneously, they can be combined with a spectrograph to perform ultrafast spectroscopic measurements. As shown

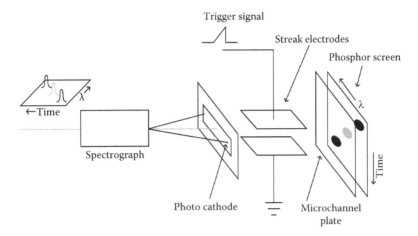

FIGURE 4.8 Schematic of a streak camera setup for ultrafast spectroscopic measurements. The optical signal is passed through a spectrograph in order to split up the different spectral components in *x*-axis. The photo cathode then converts the optical signal into electrons. The streak electrodes deflect the electrons along the *y*-axis by different amounts depending on their time of arrival. The electron signal is then amplified by the microchannel plate and converted back into optical information by a phosphor screen that is coupled to a sensitive CCD chip.

in Figure 4.8, a voltage ramp, which is synchronized with the optical signal, is applied to a pair of additional electrodes that is mounted in between the photocathode and the actual MCP. Depending on their arrival time at the detector, the primary electrons generated at the photocathode are exposed to a different electric field. Therefore, they are deflected by different distances, impinge on the MCP at different positions, and are thus laterally resolved on the CCD chip.

4.4 Irradiation Guidelines

4.4.1 Introduction

The common intent of all techniques described in this book is to use the interaction of light with tissue or cells to cure and diagnose diseases or to understand basic biological mechanisms. However, whenever biological tissue or living cells are exposed to optical radiation, adverse processes or responses can be triggered depending on the nature and state of cells and tissue, and most importantly depending on wavelength, power, and duration of optical exposure. In this section, we will deal with such phenomena to provide a guide to safe and effective use of optical radiation in biomedicine. In the process, we will see that often times, damage mechanisms have been turned into useful technologies to gather unique information or devise novel therapy treatments.

4.4.2 Radiation Effects at the Tissue Level

In this section, we will briefly explain the harmful phenomena that may occur when light is delivered to tissue. Such physical mechanisms are the basis for establishing damage thresholds for light applications. For an exact calculation of damage thresholds in the specific circumstances of a particular experimental situation, the American National Standards Institute

(ANSI) publishes exhaustive guidelines for the safe use of lasers and other light sources both in clinical and in research settings (ANSI 2000).

4.4.2.1 Photochemical Damage

At low power densities, on the order of $1 W/cm^2$, prolonged exposure (>1 s) to lasers or other light sources can cause unwanted chemical reactions in tissue that may produce damage identified as photochemical. If a tissue component is excited by high-energy visible radiation, the energy released upon de-excitation may split a bond in another molecule producing reactive oxygen species, the most common of which are singlet oxygen, hydrogen peroxide, hydroxyl radicals, and other free radicals. Reactive oxygen species are very aggressive and very dangerous to tissue especially because they can attack and break cell membranes.

In the retina, this type of damage is particularly interesting since the retina is devoted to visual perception and transduction of photons in this region, yet it is vulnerable to blue light because of photooxidation of photoreceptors and lipofuscin pigments in the retinal pigment epithelium (Wu et al. 2006, Delori et al. 2007). The damage threshold associated to this mechanism is independent of exposure duration time, although for very long exposure times (>1 day) one has to take into account incipient repair mechanisms.

4.4.2.2 UV-Induced Risks

Ultraviolet radiation covers the spectral range of 180–400 nm, the most harmful region of the optical spectrum because each photon carries sufficient energy (>3 eV) to induce molecular transitions inside a cell. UV-induced risks are more significant than any other radiation given the ubiquitous presence of UV in sunlight. A significant decrease in cell viability of all types has been reported after UV irradiation, even moderate, with evidence suggesting that DNA damage is the primary cause of

UV-induced cell death (Regan and Parrish 1982, deGruijl 1997). UV radiation does not penetrate the body beyond the skin; therefore skin and ocular tissue are the two most important targets of UV-induced damage.

As far as skin is concerned, the only useful interaction between skin and UV radiation is the synthesis of vitamin D. Apart from that, acute responses of the skin to UV radiation usually start with inflammation, that is, redness and swelling, followed by repair at the cellular level and by morphologic changes, for example, melanization, epidermal thickening, aimed at protecting the skin from further damage. In the long term, exposure to UV radiation has been found to induce chronic effects such as premature aging of the skin, wrinkling, and increased skin fragility. In addition, UV radiation can cause mutations in tumor suppressor genes controlling the cell cycle in epidermal cells, and thus massive exposure to UV radiation leads to increased probability of developing skin cancer and suppression of immunity against infections.

With regard to the eye, acute effects of UV radiation include keratoconjuncitivitis, a painful inflammation of the cornea also known as snow blindness given its higher occurrence in mountain environments where solar radiation is reflected at any angle by the snow surface. Long-term exposure to UV has also been linked to cataract formation (Taylor et al. 1988) since ultraviolet radiation can favor photooxidation of proteins of the crystalline lens (Davies and Truscott 2001).

4.4.2.3 Thermal Damage

The primary cause of thermal damage in tissue is usually absorption of light by the tissue and subsequent conversion into thermal energy. The temperature increase of tissue induces protein denaturation, hence loss of mechanical strength, and eventually hydrolysis and gelatinization. In the visible region, the main tissue components responsible for light absorption in tissue are melanin and hemoglobin; water absorption, instead, starts to become significant in the NIR region. Depending on the duration of exposure, two thermal damage regimes can be distinguished. When light is shined for more than $20\,\mu s$, heat can diffuse out of the exposed area; in this situation, tissue damage shows a quasilinear dependence of illumination power versus exposure duration. When light is delivered in short bursts ($<20\,\mu s$), we operate in the so-called thermal confinement regime where heat is localized to the exposed area; in this case, the illumination power to induce damage is independent of the global exposure time (Delori et al. 2007).

Thermal damage to tissue can be harmful in many circumstances, but it is also extensively used for therapeutic and aesthetic purposes. For example, selective photothermolysis (Anderson and Parrish 1983) cleverly uses the hugely different absorption coefficients of various components of skin tissue to selectively target strong absorbers via thermal damage (e.g. blood vessel coagulation, laser hair removal); fractional photothermolysis (Manstein et al. 2004), on the other hand, uses the ability of skin to repair thermally damaged areas for minimally invasive skin resurfacing.

4.4.2.4 Thermoacoustic Damage

When short laser pulses, less than 1 ns, are delivered to tissue, damage to the retina or other tissues containing strong light absorbers is caused by mechanisms different but connected to heating. The most important effect induced by short powerful irradiation is to create significant thermal transients. Rapid heating can lead to the generation of thermoelastic stress as the heated volume cools down. Thermoelastic stress is most prominent when the laser pulse duration is shorter than the characteristic time of stress wave propagation across the heated volume (acoustic confinement regime). In this case, internal stress is confined to the heated region and is maximal.

When acoustic confinement regime is not achieved, the heated volume undergoes significant thermal expansion that generates thermoelastic stress propagating outside the heated area. Other phenomena that may occur as a consequence of short pulse laser irradiation are surface vaporization, bubble formation and explosion, nonlinear absorption, self-focusing and laser-induced breakdown. Due to the high spatiotemporal confinement of optical radiation, the energy required to cause damage with short and ultrashort pulses via thermoacoustic damage is much less than for longer pulses or continuous radiation.

Damage induced via this mechanism has found significant application in medicine. For example, laser tissue ablation, has been largely used for many years as a surgical tool in dermatology for skin rejuvenation, in ophthalmology for intraocular surgery, and in orthopedics (Vogel and Venugopalan 2003).

4.4.3 Radiation Effects at the Cellular Level

Fluorescence-related optical technology account for the majority of optical methods used to investigate cell behavior. Here, we will review processes that can hinder fluorescence microscopy either because of induced changes on the fluorophores or because of alteration of normal cell functions; we will also notice that some of these "adverse" effects have been turned into effective technologies to gather additional information on cell properties.

4.4.3.1 Photobleaching

Photobleaching is probably the most common reason for fluorescent signals not being as strong as expected. Bleaching generally defines a class of phenomena that cause fluorescent probes to fade permanently. Fluorescence works by exciting a fluorophore from the ground state to an excited state (singlet state). The light emitted from the de-excitation to the ground state is captured by a microscope for fluorescent imaging or spectroscopy. In principle, a fluorescent probe can cycle this process of excitation-emission for an unlimited number of times. However, in practice, no more than 40,000 cycles usually occur before permanent beaching occurs (Lichtman and Conchello 2005). The most common mechanism by which a fluorophore bleaches is its excitation to a long-lived triplet state rather than the desired singled state. While in the triplet state, fluorophores may interact with other molecules to produce irreversible modifications. In this respect, molecular oxygen can play an important role in bleaching. When fluorescent molecules react with oxygen, fluorescence is permanently inhibited and highly reactive singlet oxygen species are liberated that can be toxic to living cells.

Photobleaching can be reduced by limiting the duration and/ or power of illumination. A variety of anti-bleaching molecules have also been created for fixed samples that usually contain a mixture of water–glycerin and a chemical that reduces the production or the lifetime of singlet oxygen. In multiphoton microscopy, the photobleaching rate in the focal plane scales nonlinearly with fluorescent signal generated, thus representing a bigger problem (Hopt and Neher 2001). However, while standard microscopy bleaches samples uniformly in all illuminated areas, multiphoton microscopy only induces bleaching at the focal plane because multiphoton fluorescent probes have negligible one-photon absorption. Therefore, when 3D images with many axial sections are needed, multiphoton fluorescence is advantageous in terms of photobleaching.

In certain situations, photobleaching can be used to gain unique information about samples. A specific technique called fluorescent recovery after photobleaching (FRAP) has been recently developed using this effect (Lippincott-Schwartz et al. 2001). In FRAP, a region of the sample is intentionally bleached with massive light exposure, then, the fluorescence emission intensity is monitored to detect its recovery due to unbleached fluorescent molecules diffusing to the target region. This is a unique way to measure the mobility of fluorescently labeled species in small regions of cells or tissue.

4.4.3.2 Photodamage

The term photodamage is generally used to indicate a physical damage to the sample such as membrane permeability changes or DNA strand breaks. As described in the tissue level section, the most prevalent phenomenon causing photodamage is the heating caused by light absorption from fluorophores or cell themselves. Heating can be the direct cause of cell damage or can be the cause of other effects such as thermoelastic stresses or bubble formation. Absorption, then, is the most significant parameter to take into account when considering photodamage. With respect to photodamage, multiphoton microscopy offers a definite advantage over standard confocal microscopy, fluorescence or reflectance. In single-photon microscopy techniques, continuous visible illumination of a large area is used to get information on a thin slice of the sample; on the other hand, in two-photon microscopy, short laser pulses with low average power (mW) usually in the NIR region are used. Due to the negligible absorption in the NIR by typical fluorophore and cell types and the square-power dependence of the effect, two-photon effects only occur in a subfemtoliter region of the sample, and photodamage (although locally increased with respect to single photon transition) is reduced overall (So et al. 2000).

Two-photon microscopy is usually performed using 10–100 fs laser pulses in the NIR region of the spectrum (700–1100 nm) with average powers between 0.1 and 10 mW. In this regime, while average power adverse effects are nearly nonexistent, the relevant damage mechanisms are due to high peak intensities (up to 10^{12} W/cm^2) delivered by highly confined radiation both in space and time. At these levels of peak intensity, impaired cell division, uncontrolled cell growth, and cell fragmentation have been reported. The damage thresholds have been shown to be inversely proportional to pulse duration and proportional to the square of the mean power, suggesting a two-photon absorption phenomenon as the primary effect of the damage (Konig et al. 1999).

Cellular and tissue photodamage has been recently used to devise highly localized surgical techniques (Vogel et al. 2005). Femtosecond laser nanosurgery uses higher levels of optical intensity with respect to cellular and tissue damage thresholds. The nonlinear absorption induced by the high spatiotemporal confinement of optical radiation inside the cellular medium can cause electron ionization within the focal volume, and the resulting free-electron plasma induces intracellular optical breakdown and physical disruption of the material. As a result, selected tissue ablation with submicron resolution provides a precise nanoscalpel for novel biological studies such as single cell DNA transfection, or selective disruption of individual subcellular components.

4.4.3.3 Phototoxicity

Phototoxicity refers to the generation of harmful chemical species, most notably reactive oxygen species (ROS), as a result of laser illumination of fluorescently labeled or unlabeled cells. ROS can react with fluorophores, leading to bleaching; or with oxidizable subcellular components such as proteins, nucleic acids, and lipids leading to cell cycle arrest or cell death. As with the case of photobleaching, the production of ROS mainly depends on the fluorophore chemical properties and on the amount of total energy of radiation that is delivered.

Phototoxicity can be minimized by reducing illumination power; by deoxygenating the sample environment (in fixed specimens) and by using antioxidative agents (mostly in fixed specimens). Also in this case, two-photon microscopy reduces the overall phototoxicity by confining the adverse effects to the small focal region. However, in multiphoton microscopy, phototoxicity rapidly grows with incident illumination power: a dependence on the square of average power has been found at low excitation powers, but a faster dependence has been reported for higher excitation powers.

Based on the principle of phototoxicity, a successful technique known as photodynamic therapy was developed (Dougherty et al. 1998). Photodynamic therapy aims at the selective destruction of a specific cell type by binding cells with a photosensitizer (fluorophore) and illuminating them with light. The photosensitizer is first excited to its singlet state, and then goes into the longer-lived tripled state that interacts with molecular oxygen to create highly aggressive ROS. ROS, in turn, react with the nearby cells to induce cell damage and cell death. Photodynamic therapy has been widely studied and widely used for many applications including skin diseases, macular degeneration, and cancer, where it may provide localized and specific, thus less invasive, treatment of cancerous cells.

4.4.4 Safety Practices

4.4.4.1 Laboratory Safety

Whenever light is used to treat patients or for research purposes, the most vulnerable and most exposed organs of human body are the skin and the eyes. When entering a laboratory where lasers are used, a series of common sense rules are largely agreed upon and should be followed. First, one has to be aware of what radiation sources are present and their basic characteristics, that is, wavelength and power. This is particularly important when invisible laser radiation is in use.

Once the basic specification of the lasers are known, protective eye goggles can be chosen that block (absorb) the corresponding radiation that poses any danger. Useful protective eyewear is labeled with two parameters: (1) the range of wavelength they absorb and (2) the amount of light they absorb. This information is given as an optical density (OD) rating. OD is a logarithmic measure of the absorption; for example, OD 3 means that the incident light is attenuated by a factor of 1000.

Optical experiments should be designed such that stray laser light does not leak out of the optical table to avoid unwanted exposure to bystanders or distracted users. For the same reason, it is particularly dangerous to wear shiny watches or jewelry that might reflect light in random direction outside the working environment.

Alignment and modification of optical instruments that utilize lasers should be performed at minimal power. Particular care should to be taken to avoid placing the eyes in the same plane as the optical radiation path. A distraction may easily cause direct exposure of the eyes to the laser light. Similarly, no parts of the skin should be left uncovered during operation of a laser system as optical radiation may be harmful under short- and long-term exposure.

4.4.4.2 Dealing with Skin Exposure

While we should try to avoid light exposure, particularly in the laboratory environment, in some circumstances light exposure is inevitable. The most frequent of these situations is the exposure to sunlight. The sun is the most common and probably most dangerous source of UV radiation. Given the serious tissue damage that UV light can induce, a series of sunscreen products have been developed. Topical sunscreens are creamy solutions made of organic and inorganic chemical that can be applied to the skin and that absorb, reflect, and scatter light so as to avoid sunburns. Oral sunscreens are systemic photoprotection agents, such as beta-carotene, that act as quenchers of free radicals thus reducing adverse effects of UV radiation on skin cells (Regan and Parrish 1982).

Laser treatments, such as photothermolysis or laser ablation procedures, necessarily involve direct application of laser light to skin. Safety concerns in these treatments include the possibility of diffuse thermal damage as heat may diffuse to tissues surrounding the treated area. Such damage can be prevented through cooling mechanisms such as air blowing, cryogen sprays, and gels. These are usually applied to the skin before and after laser procedures.

References

Anderson, R. R. and Parrish, J. A. 1983. Selective photothermolysis—Precise microsurgery by selective absorption of pulsed radiation. *Science* 220: 524–527.

ANSI. 2000. *American National Standard for Safe Use of Lasers (ANSI 136.1-2000)*. Orlando, FL: The Laser Institute of America.

Davies, M. J. and Truscott, R. J. W. 2001. Photo-oxidation of proteins and its role in cataractogenesis. *J. Photochem. Photobiol. B* 63: 114–125.

deGruijl, F. R. 1997. Health effects from solar UV radiation. *Radiat. Protect. Dosim.* 72: 177–196.

Delori, F. C., Webb, R. H., and Sliney, D. H. 2007. Maximum permissible exposures for ocular safety (ANSI 2000), with emphasis on ophthalmic devices. *J. Opt. Soc. Am. A* 24: 1250–1265.

Dougherty, T. J., Gomer, C. J., Henderson, B. W. et al. 1998. Photodynamic therapy. *J. Natl. Cancer Inst.* 90: 889–905.

Hopt, A. and Neher, E. 2001. Highly nonlinear photodamage in two-photon fluorescence microscopy. *Biophys. J.* 80: 2029–2036.

Konig, K., Becker, T. W., Fischer, P., Riemann, I., and Halbhuber, K. J. 1999. Pulse-length dependence of cellular response to intense near-infrared laser pulses in multiphoton microscopes. *Opt. Lett.* 24: 113–115.

Lichtman, J. W. and Conchello, J. A. 2005. Fluorescence microscopy. *Nat. Methods* 2: 910–919.

Lippincott-Schwartz, J., Snapp, E., and Kenworthy, A. 2001. Studying protein dynamics in living cells. *Nat. Rev. Mol. Cell Biol.* 2: 444–456.

Manstein, D., Herron, G. S., Sink, R. K., Tanner, H., and Anderson, R. R. 2004. Fractional photothermolysis: a new concept for cutaneous remodeling using microscopic patterns of thermal injury. *Lasers Surg. Med.* 34: 426–438.

Regan, J. D. and Parrish, J. A., eds. 1982. *The Science of Photomedicine*. New York: Plenum Press.

Smith, F. G., King, T. A., and Wilkins, D. 2007. *Optics and Photonics: An Introduction*. West Sussex, U.K.: John Wiley & Sons.

So, P. T. C., Dong, C. Y., Masters, B. R., and Berland, K. M. 2000. Two-photon excitation fluorescence microscopy. *Ann. Rev. Biomed. Eng.* 2: 399–429.

Taylor, H. R., West, S. K., Rosenthal, F. S. et al. 1988. Effect of ultraviolet-radiation on cataract formation. *New Engl. J. Med.* 319: 1429–1433.

Vogel, A., Noack, J., Huttman, G., and Paltauf, G. 2005. Mechanisms of femtosecond laser nanosurgery of cells and tissues. *Appl. Phys. B* 81: 1015–1047.

Vogel, A. and Venugopalan, V. 2003. Mechanisms of pulsed laser ablation of biological tissues. *Chem. Rev.* 103: 577–644.

Wu, J. M., Seregard, S., and Algvere, P. V. 2006. Photochemical damage of the retina. *Surv. Ophthalmol.* 51: 461–481.

Bibliography

Hecht, J. 1992. *The Laser Guidebook.* New York: McGraw-Hill.

Johnson, M. 2003. *Photodetection and Measurement: Maximizing Performance in Optical Systems.* New York: McGraw-Hill Professional.

Niemz, M. H. 2007. *Laser–Tissue Interactions: Fundamentals and Applications.* Berlin, Germany: Springer-Verlag.

Rieke, G. 2002. *Detection of Light: From the Ultraviolet to the Submillimeter.* Cambridge, U.K.: Cambridge University Press.

Siegman, A. E. 1986. *Lasers.* Mill Valley, CA: University Science Books.

Svelto, O. 1998. *Principles of Lasers.* Trans. David C. Hanna. New York: Plenum Press.

Vo-Dinh, T., ed. 2003. *Biomedical Photonics Handbook.* Boca Raton, FL: CRC Press.

Waynant, R. W. and Ediger, M. N. 2000. *Electro-Optics Handbook.* New York: McGraw-Hill.

5

Tissue Optical Properties

5.1	Introduction 67
5.2	Basic Principles of Measurements of Tissue Optical Properties.....................67
5.3	Integrating Sphere Technique..69
5.4	Kubelka–Munk and Multi-Flux Approach...70
5.5	Inverse Adding-Doubling Method 71
5.6	Inverse Monte Carlo Method..72
5.7	Direct Measurement of the Scattering Phase Function.................................74
5.8	Optical Properties of Tissues 75
	Skin and Subcutaneous Tissue Optical Properties • Ocular Tissue Optical Properties • Head/Brain Tissue Optical Properties • Epithelial/Mucous Tissue Optical Properties • Breast Tissue Optical Properties • Cartilage • Liver • Muscle • Aorta • Lung Tissue • Myocardium
5.9	Summary...95
	Acknowledgments...95
	References...95

Alexey N. Bashkatov
Saratov State University

Elina A. Genina
Saratov State University

Valery V. Tuchin
*Saratov State University and
Russian Academy of Sciences*

5.1 Introduction

Recent technological advancements in the photonics industry have spurred real progress toward the development of clinical functional imaging and surgical and therapeutic systems. The development of the optical methods in modern medicine in the areas of diagnostics, surgery, and therapy has stimulated the investigation of optical properties of human tissues, since the efficacy of optical probing of the tissues depends on the photon propagation and fluence rate distribution within irradiated tissues. Examples of diagnostic use are the following: the monitoring of blood oxygenation and tissue metabolism, detection of stomach malignancies, and recently suggested various techniques for optical imaging.

Therapeutic usage mostly includes applications in laser surgery and photodynamic therapy. For these applications, the knowledge of tissue optical properties is of great importance for the interpretation and quantification of the diagnostic data, and for the prediction of light distribution and absorbed energy for therapeutic and surgical use. Numerous investigations related to the determination of tissue optical properties are available in literature; however, the optical properties of many tissues have not been studied in a wide wavelength range.

In this chapter we present an overview of tissue optical properties measured in a wide range of wavelength using the integrating sphere spectroscopy technique.

5.2 Basic Principles of Measurements of Tissue Optical Properties

Methods for determining the optical parameters of tissues can be divided into two large groups, direct and indirect methods (Mueller and Sliney 1989, Cheong et al. 1990, Duck 1990, Welch and van Gemert 1992, Niemz 1996, Tuchin 1997, 2002, 2007, Vo-Dinh 2003). Direct methods include those based on some fundamental concepts and rules such as the Bouguer–Beer–Lambert law, the single-scattering phase function for thin samples, or the effective light penetration depth for slabs. The parameters measured are the collimated light transmission T_c and the scattering indicatrix $I(\theta)$ (angular dependence of the scattered light intensity, W/cm² sr) for thin samples or the fluence rate distribution inside a slab. The normalized scattering indicatrix is equal to the scattering phase function $I(\theta)/I(0) \equiv p(\theta)$, 1/sr. These methods are advantageous in that they use very simple analytic expressions for data processing. Their disadvantages are related to the necessity to strictly fulfill experimental conditions dictated by the selected model (single scattering in thin samples, exclusion of the effects of light polarization, and refraction at cuvette edges, etc.); in the case of slabs with multiple scattering, the recording detector (usually a fiber light guide with an isotropically scattering ball at the tip end) must be placed far from both the light source and the medium boundaries).

Indirect methods obtain the solution of the inverse scattering problem using a theoretical model of light propagation in a medium. They are in turn divided into iterative and noniterative models. The former use equations in which the optical properties are defined through parameters directly related to the quantities being evaluated. The latter are based on the two-flux Kubelka–Munk model and multi-flux models (Kubelka and Munk 1931, Yoon et al. 1987, Cheong et al. 1990, LaMuraglia et al. 1990, Ishimaru 1997, Ebert et al. 1998, Farrar et al. 1999, Phylips-Invernizzi et al. 2001, Ragain and Johntson 2001, Yang et al. 2002, 2004, 2007, Wei et al. 2003, Yang and Kruse 2004, Yang and Miklavcic 2005, Donner and Jensen 2006, Tuchin 2007, Hebert and Becker 2008, Kokhanovsky and Hopkinson 2008). In indirect iterative methods, the optical properties are implicitly defined through measured parameters. Quantities determining the optical properties of a scattering medium are enumerated until the estimated and measured values for reflectance and transmittance coincide with the desired accuracy. These methods are cumbersome, but the optical models currently in use may be even more complicated than those underlying noniterative methods (examples include the diffusion theory (Farrell et al. 1992, Hayakawa et al. 2004, Bargo et al. 2005, Dimofte et al. 2005, Zhang et al. 2005, Comsa et al. 2006), inverse adding-doubling (IAD) (Pickering et al. 1993a,b, Prahl et al. 1993, Qu et al. 1994, de Vries et al. 1999, Sardar et al. 2001, 2004, 2005, 2007, Bashkatov et al. 2004, 2005a,b, 2006b, 2009, 2010, Chen et al. 2005, Wei et al. 2005, Chandra et al. 2006, Gebhart et al. 2006, Zhu et al. 2007), and inverse Monte Carlo (IMC) (Marchesini et al. 1992, Graaff et al. 1993b, van der Zee 1993, Hammer et al. 1995, Hourdakis and Perris 1995, Wang et al. 1995, Yaroslavsky et al. 1996a, 2002b, Simpson et al. 1998, Ripley et al. 1999, Roggan et al. 1999a,b, Dam et al. 2000, Hayakawa et al. 2001, Bashkatov et al. 2006a, 2007, Friebel et al. 2006, 2009, Palmer and Ramanujam 2006a,b, Salomatina et al. 2006, Meinke et al. 2007a,b) methods).

The optical parameters of tissue samples (μ_a, μ_s, and g) could be measured by different methods. Here μ_a is the absorption coefficient, μ_s is the scattering coefficient, and g is the anisotropy factor of scattering. The single- or double-integrating sphere method combined with collimated transmittance measurements is most often used for in vitro studies of tissues (see Figures 5.1 and 5.2). This approach implies either sequential or simultaneous determination of three parameters: collimated transmittance $T_c = I_d/I_0$ (I_d is the intensity of transmitted light measured using a distant photodetector with a small aperture, W/cm², and I_0 is the intensity of incident radiation, W/cm²), total transmittance $T_t = T_c + T_d$ (T_d being diffuse transmittance), and diffuse reflectance R_d. The optical parameters of the tissue are deduced from these measurements using different theoretical expressions or numerical methods (two-flux and multi-flux models, the IMC or IAD methods) relating μ_a, μ_s, and g to the parameters being investigated.

Any three measurements from the following five are sufficient for the evaluation of all three optical parameters (Cheong et al. 1990):

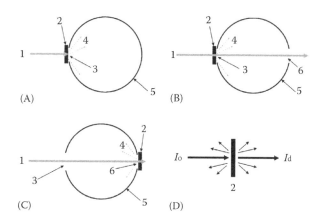

FIGURE 5.1 Measurement of tissue optical properties using an integrating sphere. The surface of the integrating sphere is coated with BaSO₄, MgO, or Spectralon which have nearly 100% diffuse remittance over the entire optical spectrum. (A) Total transmittance mode, (B) diffuse transmittance mode, (C) diffuse reflectance mode, (D) collimated transmittance mode. 1, the incident beam; 2, the tissue sample; 3, the entrance port; 4, the transmitted (or diffuse reflected) radiation; 5, the integrating sphere; 6, the exit port.

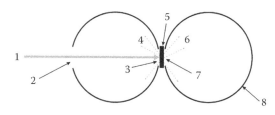

FIGURE 5.2 The double-integrating sphere setup. 1, the incident beam; 2,7, the entrance port; 3, the exit port; 4, the diffuse reflected radiation; 5, the tissue sample; 6, the transmitted radiation; 8, the integrating sphere.

1. Total (or diffuse) transmittance for collimated or diffuse radiation
2. Total (or diffuse) reflectance for collimated or diffuse radiation
3. Absorption by a sample placed inside an integrating sphere
4. Collimated transmittance (of unscattered light)
5. Angular distribution of radiation scattered by the sample

Iterative methods normally take into account discrepancies between refractive indices at sample boundaries as well as the multilayer nature of the sample. The following factors are responsible for the errors in the estimated values of optical coefficients and need to be borne in mind in a comparative analysis of optical parameters obtained in different experiments (Cheong et al. 1990):

- The physiological conditions of tissues (the degree of hydration, homogeneity, species-specific variability, frozen/thawed or fixed/unfixed state, in vitro/in vivo measurements, smooth/rough surface)
- The geometry of irradiation

- The matching/mismatching interface refractive indices
- The angular resolution of photodetectors
- The separation of radiation experiencing forward scattering from unscattered radiation
- The theory used to solve the inverse problem

To analyze light propagation under multiple scattering conditions, it is assumed that absorbing and scattering centers are uniformly distributed across the tissue. UV-A, visible, or near-infrared (NIR) radiation is normally subject to anisotropic scattering characterized by a clearly apparent direction of photons undergoing single scattering, which may be due to the presence of large cellular organelles (mitochondria, lysosomes, Golgi apparatus, etc.) (Mueller and Sliney 1989, Cheong et al. 1990, Duck 1990, Welch and van Gemert 1992, Niemz 1996, Tuchin 1997, 2002, 2007, Vo-Dinh 2003).

When the scattering medium is illuminated by unpolarized light and/or only the intensity of multiply scattered light needs to be computed, a sufficiently strict mathematical description of continuous wave (CW) light propagation in a medium is possible in the framework of the scalar stationary radiation transfer theory (RTT) (Mueller and Sliney 1989, Cheong et al. 1990, Duck 1990, Welch and van Gemert 1992, Niemz 1996, Tuchin 1997, 2002, 2007, Vo-Dinh 2003). This theory is valid for an ensemble of scatterers located far from one another and has been successfully used to work out some practical aspects of tissue optics. The main stationary equation of RTT for monochromatic light has the form

$$\frac{\partial I(\vec{r}, \vec{s})}{\partial s} = -\mu_t I(\vec{r}, \vec{s}) + \frac{\mu_s}{4\pi} \int_{4\pi} I(\vec{r}, \vec{s}) p(\vec{s}, \vec{s}') d\Omega' \quad (5.1)$$

where

$I(\vec{r}, \vec{s})$ is the radiance (or specific intensity) – average power flux density at point $\vec{r}$ in the given direction $\vec{s}$, W/cm^2 sr

$p(\vec{s}, \vec{s}')$ is the scattering phase function, 1/sr

$d\Omega'$ is the unit solid angle about the direction $\vec{s}'$, sr

$\mu_t = \mu_a + \mu_s$ is the total attenuation coefficient

It is assumed that there are no radiation sources inside the medium.

The scalar approximation of the radiative transfer equation (RTE) gives poor accuracy when the size of the scattering particles is much smaller than the wavelength, but provides acceptable results for particles comparable to and larger than the wavelength (Chandrasekhar 1960, Mishchenko et al. 2002).

The phase function $p(\vec{s}, \vec{s}')$ describes the scattering properties of the medium and is in fact the probability density function for scattering in the direction $\vec{s}'$ of a photon traveling in the direction $\vec{s}$; in other words, it characterizes an elementary scattering act. If scattering is symmetric relative to the direction of the incident wave, then the phase function depends only on the scattering angle θ (angle between directions $\vec{s}$ and $\vec{s}'$), that is, $p(\vec{s}, \vec{s}') = p(\theta)$. The assumption of random distribution of scatterers in a medium (i.e., the absence of spatial correlation in the tissue structure) leads to normalization: $\int_0^\pi p(\theta) 2\pi \sin\theta \, d\theta = 1$. In practice, the phase function is usually well approximated with the aid of the following postulated Henyey–Greenstein function (Henyey and Greenstein 1941, Yoon et al. 1987, Mueller and Sliney 1989, Cheong et al. 1990, Duck 1990, Welch and van Gemert 1992, Marchesini et al. 1992, Graaff et al. 1993b, Pickering et al. 1993a,b, Prahl et al. 1993, Qu et al. 1994, Hammer et al. 1995, Hourdakis and Perris 1995, Wang et al. 1995, Chan et al. 1996a,b, Nemati et al. 1996, Niemz 1996, Yaroslavsky et al. 1996a, 2002b, Tuchin 1997, 2002, 2007, Simpson et al. 1998, de Vries et al. 1999, Ripley et al. 1999, Roggan et al. 1999a,b, Dam et al. 2000, Hayakawa et al. 2001, Sardar et al. 2001, 2004, 2005, 2007, Vo-Dinh 2003, Bashkatov et al. 2004, 2005a,b, 2006a,b, 2007, 2009, Chen et al. 2005, Wei et al. 2005, Chandra et al. 2006, Gebhart et al. 2006, Palmer and Ramanujam 2006a,b, Salomatina et al. 2006, Zhu et al. 2007):

$$p(\theta) = \frac{1}{4\pi} \frac{1 - g^2}{\left(1 + g^2 - 2g\cos\theta\right)^{3/2}} \quad (5.2)$$

where g is the scattering anisotropy parameter (mean cosine of the scattering angle θ)

$$g \equiv \langle\cos\theta\rangle = \int_0^\pi p(\theta)\cos\theta \cdot 2\pi\sin\theta \, d\theta$$

The value of g varies in the range from -1 to 1; $g = 0$ corresponds to isotropic (Rayleigh) scattering, $g = 1$ to total forward scattering (Mie scattering at large particles), and $g = -1$ to total backward scattering (Mueller and Sliney 1989, Cheong et al. 1990, Duck 1990, Welch and van Gemert 1992, Niemz 1996, Tuchin 1997, 2002, 2007, Ishimaru 1997, Mishchenko et al. 2002, Vo-Dinh 2003).

5.3 Integrating Sphere Technique

One of the indirect methods to determine optical properties of tissues in vitro is the integrating sphere technique (Spitzer and Ten Bosch 1975, Cheong et al. 1990, Marchesini et al. 1992, Graaff et al. 1993b, Pickering et al. 1993a,b, Prahl et al. 1993, van der Zee et al. 1993, Qu et al. 1994, Hammer et al. 1995, Hourdakis and Perris 1995, Chan et al. 1996a,b, Nemati et al. 1996, Yaroslavsky et al. 1996b, 2002b, Ebert et al. 1998, Simpson et al. 1998, de Vries et al. 1999, Farrar et al. 1999, Ripley et al. 1999, Roggan et al. 1999a,b, Dam et al. 2000, Hayakawa et al. 2001, Sardar et al. 2001, 2004, 2005, 2007, Bashkatov et al. 2004, 2005a,b, 2006b, 2007, 2009, Dimofte et al. 2005, Wei et al. 2005, Friebel et al. 2006, 2009, Gebhart et al. 2006, Palmer and Ramanujam 2006b, Salomatina et al. 2006, Meinke et al. 2007a,b, Tuchin 2007, Zhu et al. 2007). Diffuse reflectance R_d, total T_t and/or diffuse transmittance T_d, and collimated transmittance T_c are measured. In general, the absorption and scattering coefficients and anisotropy factor can be obtained from these data using an inverse method based on the radiative transfer theory. When the scattering phase function $p(\theta)$ is available from goniophotometry, g can be readily calculated. In this case, for the determination of μ_a and μ_s it is sufficient to measure R_d and T_t only. Sometimes in experiments with tissue and blood samples, a double-integrating sphere configuration is preferable, since in this case both reflectance and transmittance can be measured simultaneously and less degradation of the sample is expected during measurements (see Figure 5.2). Nevertheless, in the case of a double-integrating sphere arrangement of the experiment in addition to the single-integrating sphere corrections of measured signals, multiple exchange of light between the spheres should be accounted for (Pickering et al. 1993b, Yaroslavsky et al. 2002a).

Some tissues (e.g., melanin containing) and blood have high total attenuation coefficients in the visible and NIR spectral range. Therefore, the collimated transmittance measurement for such samples (e.g., the undiluted blood layer with a moderate thickness ≈0.1 mm) (Roggan et al. 1999a) is a technically difficult task. To solve this problem, a powerful light source combined with a sensitive detector must be used (Nilsson et al. 1997). Alternatively, it is possible to collect the collimated light together with some forward-scattered light using the third integrating sphere (Yaroslavsky et al. 1999a). In this case, the collimated transmittance is separated from the scattered flux on the stage of the data processing using, for example, a Monte Carlo (MC) technique (Yaroslavsky et al. 1996a) or a small angle approximation (Yaroslavsky et al. 1998). Another approach was used in papers by Bashkatov et al. (2007), Friebel et al. (2006), Hayakawa et al. (2001), and Meinke et al. (2007a,b). In these studies the diffuse reflectance, total and diffuse transmittance have been measured (see Figure 5.1), and IMC algorithm, taking into account geometry of the measurement, has been used for treatment of the experimental data.

5.4 Kubelka–Munk and Multi-Flux Approach

To separate the light beam attenuation due to absorption from the loss due to scattering, the one-dimensional, two-flux Kubelka–Munk (KM) model can be used as the simplest approach to solve the problem. This approach has been widely used to determine the absorption and scattering coefficients of biological tissues, provided the scattering is significantly dominant over the absorption (Kubelka and Munk 1931, Spitzer and Ten Bosch 1975, Anderson and Parrish 1981, Yoon et al. 1987, van Gemert et al. 1989, Cheong et al. 1990, LaMuraglia et al. 1990, Ishimaru 1997, Ebert et al. 1998, Farrar et al. 1999, Phylips-Invernizzi et al. 2001, Ragain and Johntson 2001, Yang et al. 2002, 2004, 2007, Wei et al. 2003, Yang and Kruse 2004, Yang and Miklavcic 2005, Donner and Jensen 2006, Tuchin 2007, Hebert and Becker 2008, Kokhanovsky and Hopkinson 2008). The KM model assumes that light incident on a slab of tissue because of interaction with the scattering media can be modeled by two fluxes, counterpropagating in the tissue slab. The optical flux, which propagates in the same direction as the incident

flux, is decreased by absorption and scattering processes and is also increased by backscattering of the counterpropagating flux in the same direction. Changes in counterpropagating flux are determined in an analogous manner. The fraction of each flux lost by absorption per unit path length is denoted as K, while the fraction lost due to scattering is called S. The main assumptions of the KM model: K and S parameters are assumed to be uniform throughout the tissue slab; all light fluxes are diffuse; and the amount of light lost from the edges of the sample during reflectance measurements is negligible. Basic KM model does not account for reflections at boundaries at which index of refraction mismatches exist.

Following the KM model and diffusion approximation of the radiative transfer equation, the KM parameters were expressed in terms of light transport theory: the absorption and scattering coefficients and scattering anisotropy factor (van Gemert et al. 1989, Cheong et al. 1990). Thus, when scattering significantly prevails on absorption, a simple experimental method using modified KM model expressions can be successfully employed as

$$S = \frac{1}{bd} = \ln\left[\frac{1 - R_d(a - b)}{T_t}\right];$$

$$K = S(a - 1); \quad a = \frac{1 - T_t^2 + R_d^2}{2R_d}; \quad b = \sqrt{a^2 - 1};$$

$$K = 2\mu_a; \quad S = \frac{3}{4}\mu_s(1 - g) - \frac{1}{4}\mu_a;$$

$$\mu_t = \mu_a + \mu_s; \quad \mu_s' = \mu_s(1 - g) > \mu_a, \tag{5.3}$$

where μ_t is determined basing on Bouguer–Beer–Lambert law ($T_c = \exp(-\mu_t l)$) from measured values of collimated transmittance T_c, where l is tissue sample thickness. Thus, all three parameters (μ_a, μ_s, g) can be found from the experimental data for total transmittance T_t, diffuse reflectance R_d, and collimated transmittance T_c of the sample.

Often, such simple methods as the KM model (Prahl et al. 1993, Qu et al. 1994, Hammer et al. 1995, Chan et al. 1996a,b, Nemati et al. 1996, Roggan et al. 1999a,b, Sardar et al. 2001, 2004, 2005, 2007, Bashkatov et al. 2004, 2005a,b, 2006a,b, 2007, 2009, Wei et al. 2005, Friebel et al. 2006, 2009, Gebhart et al. 2006, Meinke et al. 2007a,b, Zhu et al. 2007) or diffusion approximation (Yaroslavsky et al. 1996a, 2002b, Hayakawa et al. 2004, Bargo et al. 2005, Salomatina et al. 2006) are used as the first step of the inverse algorithm for estimation of the optical properties of tissues and blood. The estimated values of the optical properties are then used to calculate the reflected and transmitted signals, employing one of the more sophisticated models of light propagation in tissue or blood. At the next step, the calculated values are compared with the measured ones. If the required accuracy is not achieved, the current optical properties are altered using one of the optimization algorithms. The procedures of altering the optical properties and calculating the reflected and transmitted signals are repeated until the calculated values match the measured values with the required accuracy.

5.5 Inverse Adding-Doubling Method

The IAD method provides a tool for the rapid and accurate solution of inverse scattering problem (Pickering et al. 1993a,b, Prahl et al. 1993, Qu et al. 1994, Chan et al. 1996a,b, Nemati et al. 1996, de Vries et al. 1999, Sardar et al. 2001, 2004, 2005, 2007, Troy and Thennadil 2001, Bashkatov et al. 2004, 2005a,b, 2006b, 2009, Chen et al. 2005, Wei et al. 2005, Chandra et al. 2006, Gebhart et al. 2006, Zhu et al. 2007). It is based on the general method for the solution of the transport equation for plane-parallel layers suggested by van de Hulst (van de Hulst 1980) and introduced to tissue optics by Prahl (1988, 2010). An important advantage of the IAD method when applied to tissue optics is the possibility of rapidly obtaining iterative solutions with the aid of up-to-date microcomputers; moreover, it is flexible enough to take into account anisotropy of scattering and the internal reflection from the sample boundaries. The method includes the following steps:

1. The choice of optical parameters to be measured
2. Counting reflections and transmissions
3. Comparison of calculated and measured reflectance and transmittance
4. Repetition of the procedure until the estimated and measured values coincide with the desired accuracy

In principle, the method allows for any intended accuracy to be achieved for all the parameters being measured, provided the necessary computer time is available. An error of 3% or less is considered acceptable (Prahl et al. 1993). Also, the method may be used to directly correct experimental findings obtained with the aid of integrating spheres. The term "doubling" in the name of the method means that the reflection and transmission estimates for a layer at certain ingoing and outgoing light angles may be used to calculate both the transmittance and reflectance for a layer twice as thick by means of superimposing one upon the other and summing the contributions of each layer to the total reflectance and transmittance. Reflection and transmission in a layer having an arbitrary thickness are calculated in consecutive order, first for the thin layer with the same optical characteristics (single scattering), then by consecutive doubling of the thickness, for any selected layer. The term "adding" indicates that the doubling procedure may be extended to heterogeneous layers for modeling multilayer tissues or taking into account internal reflections related to abrupt change in refractive index (Prahl et al. 1993).

The adding-doubling technique is a numerical method for solving the one-dimensional transport equation in slab geometry (Prahl 1988, 2010, Prahl et al. 1993). It can be used for media with an arbitrary phase function and arbitrary angular distribution of the spatially uniform incident radiation. Thus, finite beam size and side losses of light cannot be taken into account. The method is based on the observation that for an arbitrary incident radiance angular distribution $I_{in}(\eta_c)$, where η_c is the cosine of the polar angle, the angular distribution of the reflected radiance (normalized to an incident diffuse flux) is given by Prahl 1988, 2010, Prahl et al. 1993, and Yaroslavsky et al. 2002a:

$$I_{ref}(\eta_c) = \int_0^1 I_{in}(\eta_c')R(\eta_c', \eta_c)2\eta_c'\, d\eta_c', \tag{5.4}$$

where $R(\eta_c', \eta_c)$ is the reflection redistribution function determined by the optical properties of the slab.

The distribution of the transmitted radiance can be expressed in a similar manner, with obvious substitution of the transmission redistribution function $T(\eta_c', \eta_c)$. If M quadrature points are selected to span over the interval $(0, 1)$, the respective matrices can approximate the reflection and transmission redistribution functions:

$$R\left(\eta_{ci}', \eta_{cj}\right) \to R_{ij}; \quad T\left(\eta_{ci}', \eta_{cj}\right) \to T_{ij}. \tag{5.5}$$

These matrices are referred to as the reflection and transmission operators, respectively. If a slab with boundaries indexed as 0 and 2 is comprised of two layers, (01) and (12), with an internal interface 1 between the layers, the reflection and transmission operators for the whole slab (02) can be expressed as

$$\mathbf{T}^{02} = \mathbf{T}^{12}(\mathbf{E} - \mathbf{R}^{10}\mathbf{R}^{12})^{-1}\mathbf{T}^{01},$$

$$\mathbf{R}^{20} = \mathbf{T}^{12}(\mathbf{E} - \mathbf{R}^{10}\mathbf{R}^{12})^{-1}\mathbf{R}^{10}\mathbf{T}^{21} + \mathbf{R}^{21},$$

$$\mathbf{T}^{20} = \mathbf{T}^{10}(\mathbf{E} - \mathbf{R}^{12}\mathbf{R}^{10})^{-1}\mathbf{T}^{21},$$

$$\mathbf{R}^{02} = \mathbf{T}^{10}(\mathbf{E} - \mathbf{R}^{12}\mathbf{R}^{10})^{-1}\mathbf{R}^{12}\mathbf{T}^{01} + \mathbf{R}^{01}, \tag{5.6}$$

where $\mathbf{E}$ is the identity matrix defined in this case as

$$E_{ij} = \frac{1}{2\eta_{ci}w_i}\delta_{ij}, \tag{5.7}$$

where

w_i is the weight assigned to the ith quadrature point
δ_{ij} is a Kroneker delta symbol, $\delta_{ij} = 1$ if $i = j$, and $\delta_{ij} = 0$ if $i \neq j$

The definition of the matrix multiplication also slightly differs from the standard. Specifically,

$$(\mathbf{AB})_{ik} \equiv \sum_{j=1}^{M} A_{ij} 2\eta_{cj}w_j B_{jk}. \tag{5.8}$$

Equations 5.6 allow one to calculate the reflection and transmission operators of a slab when those of the comprising layers are known. The idea of method is to start with a thin layer for which the RTE can be simplified and solved with relative ease, producing the reflection and transmission operators for the thin layer, then to proceed by doubling the thickness of the layer until the thickness of the whole slab is reached. Several techniques exist for layer initialization. The single-scattering equations for reflection and transmission for the Henyey–Greenstein function are given by van de Hulst (1980) and Prahl (1988). The refractive index mismatch can be taken into account by adding effective boundary layers of zero thickness and having the reflection and transmission operators determined by Fresnel's formulas. The total transmittance and reflectance of the slab are obtained by straightforward integration of Equation 5.4. Different methods of performing the integration and the IAD program provided by Prahl (1988, 2010) allows one to obtain the absorption and the scattering coefficients from the measured diffuse reflectance R_d and diffuse transmittance T_d of the tissue slab. This program is the numerical solution to the steady-state RTE (see Equation 5.1) realizing an iterative process, which estimates the reflectance and transmittance from a set of optical parameters until the calculated reflectance and transmittance match the measured values. Values for the anisotropy factor g and the refractive index n must be provided to the program as input parameters.

It was shown that using only four quadrature points, the IAD method provides optical parameters that are accurate to within 2%–3% (Prahl et al. 1993), as was mentioned earlier; higher accuracy, however, can be obtained by using more quadrature points, but it would require increased computation time. Another valuable feature of the IAD method is its validity for the study of samples with comparable absorption and scattering coefficients (Prahl et al. 1993), since other methods based on only diffusion approximation are inadequate. Furthermore, since both anisotropic phase function and Fresnel reflection at boundaries are accurately approximated, the IAD technique is well suited to optical measurements for biological tissues and blood held between two glass slides. The adding-doubling method provides accurate results in cases when the side losses are not significant, but it is less flexible than the MC technique.

The IAD method has been successfully applied to determine optical parameters of blood (Sardar and Levy 1998); human and animal dermis (Chan et al. 1996b, Troy and Thennadil 2001, Bashkatov et al. 2005b); brain tissues (Gebhart et al. 2006); bronchial tissue (Qu et al. 1994); ocular tissues such as retina (Sardar et al. 2004, 2005), choroids, sclera, conjunctiva, and ciliary body (Chan et al. 1996b, Nemati et al. 1996, Bashkatov et al. 2009, 2010); mucous tissue (Bashkatov et al. 2004); subcutaneous tissue (Bashkatov et al. 2005a,b); cranial bone (Bashkatov et al. 2006b); aorta (Chan et al. 1996b); and other soft tissues in the wide range of the wavelengths (Cheong et al. 1990, Tuchin 2007).

5.6 Inverse Monte Carlo Method

Both the real geometry of the experiment and the tissue structure may be complicated. Therefore, inverse Monte Carlo (IMC) method should be used if reliable estimates are to be obtained. A number of algorithms to use the IMC method are available now in the literature (Firbank et al. 1993, Graaff et al. 1993b, van der Zee 1993a, Roggan et al. 1994, 1995, 1999a,b, Bevilacqua et al. 1995, 1999, Hammer et al. 1995, Hourdakis and Perris 1995, Kienle and Patterson 1996, Yaroslavsky et al. 1996a, 2002b, Laufer et al. 1998, Simpson et al. 1998, Ripley et al. 1999, Dam

et al. 2000, Alexandrakis et al. 2001, Du et al. 2001, Hayakawa et al. 2001, Preece and Claridge 2002, Khalil et al. 2003, Pfefer et al. 2003, Nishidate et al. 2003, 2004, Thueler et al. 2003, Kumar et al. 2004, Ugryumova et al. 2004, Bashkatov et al. 2006a, 2007, Friebel et al. 2006, 2009, Palmer and Ramanujam 2006a,b, Salomatina et al. 2006, Zhu et al. 2006, Kotova et al. 2007, Meinke et al. 2007a,b, Reif et al. 2007, Seo et al. 2007, Tuchin 2007, Prerana et al. 2008). Many researchers use the MC simulation algorithm and program provided by Jacques and Wang (Jacques and Wang 1995, Wang et al. 1995). Among the firstly designed IMC algorithms, a similar algorithm for determining all three optical parameters of the tissue (μ_a, μ_s, and g) based on the in vitro evaluation of the total transmittance, diffuse reflectance, and collimated transmittance using a spectrophotometer with integrating spheres can be mentioned (Marchesini et al. 1992, Firbank et al. 1993, Graaff et al. 1993b, van der Zee et al. 1993a,b, Roggan et al. 1994, 1995, 1999a,b, Hammer et al. 1995, Hourdakis and Perris 1995, Yaroslavsky et al. 1996a, 2002b, Laufer et al. 1998, Simpson et al. 1998, Ripley et al. 1999, Dam et al. 2000, Du et al. 2001, Hayakawa et al. 2001, Ugryumova et al. 2004, Nishidate et al. 2004, Friebel et al. 2006, 2009, Salomatina et al. 2006, Bashkatov et al. 2007, Meinke et al. 2007a,b). The initial approximation (to speed up the procedure) was achieved with the help of the Kubelka–Munk theory, specifically its four-flux variant (Roggan et al. 1994, 1995, 1999a,b, Hammer et al. 1995, Yaroslavsky et al. 1996b, 2002b, Dam et al. 2000, Friebel et al. 2006, 2009, Salomatina et al. 2006, Bashkatov et al. 2007, Meinke et al. 2007a,b). The algorithms take into consideration the sideways loss of photons, which becomes essential in sufficiently thick samples. Similar results were obtained using the condensed IMC method (Firbank et al. 1993, Graaff et al. 1993b, van der Zee 1993a,b, Hourdakis and Perris 1995, Ripley et al. 1999, Ugryumova et al. 2004).

The MC technique is employed as a method to solve the forward problem in the inverse algorithm for the determination of the optical properties of tissues and blood. The MC method is based on the formalism of the RTT, where the absorption coefficient is defined as a probability of a photon to be absorbed per unit length, and the scattering coefficient is defined as the probability of a photon to be scattered per unit length. Using these probabilities, a random sampling of photon trajectories is generated.

The basic algorithm for the generation of photon trajectories can be shortly described as follows (Jacques and Wang 1995, Wang et al. 1995, Yaroslavsky et al. 1996a, 2002a). A photon described by three spatial coordinates and two angles (x, y, z, θ, ϕ) is assigned its weight $W = W_0$ and placed in its initial position, depending on the source characteristics. The step size s of the photon is determined as $s = -\ln(\xi)/\mu_t$, where ξ is a random number between [0, 1]. The direction of the photon's next movement is determined by the scattering phase function substituted as the probability density distribution. Several approximations for the scattering phase function of tissue and blood have been used in MC simulations. These include the two empirical phase functions widely used to approximate the scattering phase function of tissue and blood, Henyey–Greenstein phase function (HGPF) (see Equation 5.2), the Gegenbauer kernel phase function (GKPF) (Reynolds and McCormick 1980), and theoretical Mie phase function (Bohren and Huffman 1983). The HGPF has one parameter g that may be represented as the infinite series of Legendre polynomials $P_n^1(\cos\theta)$

$$p_{HG}(\theta) = \frac{1}{4\pi} \sum_{n=0}^{\infty} (2n+1) f_n P_n^1(\cos\theta), \quad (5.9)$$

where $f_n = g^n$ is the nth order moment of the phase function.

The GKPF has two variable parameters, α and g,

$$p_{GK}(\theta) = K[1 + g^2 - 2g\cos(\theta)]^{-(\alpha+1)}, \quad (5.10)$$

where

$$K = \alpha g \pi^{-1}(1-g^2)^{2\alpha}[(1+g)^{2\alpha} - (1-g)^{2\alpha}]^{-1}, \quad \alpha > -1/2, \; |g| \le 1$$

The GKPF is a generalization of the HGPF and can be reduced to HGPF by setting $\alpha = 0.5$. The GKPF may be represented as the infinite series of Gegenbauer polynomials, C_n^α (Reynolds and McCormick 1980, Barber and Hill 1990):

$$p_{GK}(\theta) = \frac{2K}{(1-g^2)} \sum_{n=0}^{\infty} \left(1 + \frac{n}{\alpha}\right) C_n^\alpha[\cos(\theta)]g^n. \quad (5.11)$$

The HGPF and GKPF are widely employed in radiative transport calculations for the description of the single-scattering process in whole blood (Nilsson et al. 1997, Sardar and Levy 1998, Roggan et al. 1999a, Yaroslavsky et al. 2002a, Friebel et al. 2006, 2009, Meinke et al. 2007a,b) and tissues (Farrell et al. 1992, Graaff et al. 1993b, Pickering et al. 1993b, van der Zee 1993a, Qu et al. 1994, Bevilacqua et al. 1995, Hammer et al. 1995, Hourdakis and Perris 1995, Chan et al. 1996a,b, Nemati et al. 1996, Laufer et al. 1998, Simpson et al. 1998, Ripley et al. 1999, Roggan et al. 1999b, Dam et al. 2000, Du et al. 2001, Sardar et al. 2001, 2004, 2005, 2007, Troy and Thennadil 2001, Preece and Claridge 2002, Yaroslavsky et al. 2002a, Nishidate et al. 2003, 2004, Pfefer et al. 2003, Thueler et al. 2003, Bashkatov et al. 2004, 2005a,b, 2006a,b, 2007, 2009, Kumar et al. 2004, Ugryumova et al. 2004, Wei et al. 2005, Chandra et al. 2006, Gebhart et al. 2006, Salomatina et al. 2006, Seo et al. 2007) because of their mathematical simplicity. However, it is clear that the HGPF and GKPF cannot be used for accurate calculations of the angular light distribution scattered by a single erythrocyte. For some calculations, theoretical Mie phase function may be useful (Bohren and Huffman 1983):

$$p(\theta) = \frac{1}{k^2 r^2}\left(|S_1|^2 + |S_2|^2\right), \quad (5.12)$$

where S_1 and S_2 are functions of the polar scattering angle and can be obtained from the Mie theory as

$$S_1(\theta) = \sum_{n=1}^{\infty} \frac{2n+1}{n(n+1)}\{a_n\pi_n(\cos\theta) + b_n\tau_n(\cos\theta)\},$$

$$S_2(\theta) = \sum_{n=1}^{\infty} \frac{2n+1}{n(n+1)}\{b_n\pi_n(\cos\theta) + a_n\tau_n(\cos\theta)\}. \quad (5.13)$$

The parameters π_n and τ_n represent

$$\pi_n(\cos\theta) = \frac{1}{\sin\theta}P_n^1(\cos\theta),$$

$$\tau_n(\cos\theta) = \frac{d}{d\theta}P_n^1(\cos\theta), \quad (5.14)$$

where $P_n^1(\cos\theta)$ is the associated Legendre polynomial. The following recursive relationships are used to calculate π_n and τ_n:

$$\pi_n = \frac{2n-1}{n-1}\pi_{n-1}\cos\theta - \frac{n}{n-1}\pi_{n-2},$$

$$\tau_n = n\pi_n\cos\theta - (n+1)\pi_{n-1}, \quad (5.15)$$

and the initial values are

$$\begin{cases} \pi_1 = 1, \pi_2 = \cos\theta, \\ \tau_1 = \cos\theta, \tau_2 = 3\cos 2\theta. \end{cases} \quad (5.16)$$

The coefficients a_n and b_n are defined as

$$a_n = \frac{\psi_n(\alpha)\psi_n'(m\alpha) - m\psi_n(m\alpha)\psi_n'(\alpha)}{\zeta_n(\alpha)\psi_n'(m\alpha) - m\psi_n(m\alpha)\zeta_n'(\alpha)}$$

$$b_n = \frac{m\psi_n(\alpha)\psi_n'(m\alpha) - \psi_n(m\alpha)\psi_n'(\alpha)}{m\zeta_n(\alpha)\psi_n'(m\alpha) - \psi_n(m\alpha)\zeta_n'(\alpha)} \quad (5.17)$$

$$m = n_p/n_0 \quad \alpha = 2\pi a n_0/\lambda_0$$

where

a is the radius of spherical particles
λ_0 is the light wavelength in vacuum
$\psi_n, \zeta_n, \psi_n', \zeta_n'$ are the Riccati–Bessel functions of the first and second kind
n_0 is the refractive index of the ground material
n_p is the refractive index of scattering particles

For the HGPF, the random scattering angle θ_{rnd}^{HG} is given by

$$\theta_{rnd}^{HG} = \arccos\left\{\frac{1}{2g}\left[1 + g^2 - \left(\frac{1-g^2}{1-g+2g\xi}\right)^2\right]\right\}. \quad (5.18)$$

For the GKPF, the random scattering angle θ_{rnd}^{GK} is determined as (Yaroslavsky et al. 1999b)

$$\theta_{rnd}^{GK} = \arccos\left[(1 + g^2 - 1/\sqrt[\alpha]{\zeta_{rnd}})/2g\right], \qquad (5.19)$$

where $\zeta_{rnd} = 2\alpha g \xi/K + (1 + g)^{-2\alpha}$, and α and K are defined in Equation 5.10.

If experimental scattering phase function is known for the discrete set of scattering angles θ_i, $f(\theta) = f(\theta_i)$, it can be determined in the total angular range using the spline-interpolation technique. Then, the value of the function $F_n = \int_0^{\theta_n} f(\theta)d\theta$ can be calculated numerically for any value of θ_n. It is easy to see that F_n is a nondecreasing function that is mapping the interval (0, 1). Therefore, when random value γ is sampled, θ_{rnd}^{exp} is determined by setting $F_n = \xi$.

The Mie phase function can be tabulated and treated in the same way as the experimental phase function (Yaroslavsky et al. 1999b). In most cases azimuthal symmetry is assumed. This leads to $p(\phi) = 1/2\pi$ and, consequently, $\phi_{rnd} = 2\pi\xi$. At each step, the photon loses part of its weight due to absorption: $W = W(1 - \Lambda)$, where $\Lambda = \mu_s/\mu_t$ is the albedo of the medium.

When the photon reaches the boundary, part of its weight is transmitted according to the Fresnel equations. The amount transmitted through the boundary is added to the reflectance or transmittance. Since the refraction angle is determined by the Snell's law, the angular distribution of the outgoing light can be calculated. The photon with the remaining part of the weight is specularly reflected and continues its random walk.

When the photon's weight becomes lower than a predetermined minimal value, the photon can be terminated using "Russian roulette" procedure (Keijzer et al. 1989, Yaroslavsky and Tuchin 1992, Jacques and Wang 1995, Wang et al. 1995, Yaroslavsky et al. 1996a, 2002a). This procedure saves time, since it does not make sense to continue the random walk of the photon, which will not essentially contribute to the measured signal. On the other hand, it ensures that the energy balance is maintained throughout the simulation process.

The MC method has several advantages over the other methods because it may take into account mismatched medium–glass and glass–air interfaces, losses of light at the edges of the sample, any phase function of the medium, and the finite size and arbitrary angular distribution of the incident beam. If the collimated transmittance is measured, then the contribution of scattered light into the measured collimated signal can be accounted for (Wang et al. 1995, Yaroslavsky et al. 1996a, Tuchin 2007). The only disadvantage of this method is the long time needed to ensure good statistical convergence, since it is a statistical approach. The standard deviation of a quantity (diffuse reflectance, transmittance, etc.) approximated by MC technique decreases proportionally to $1/\sqrt{N}$, where N is the total number of launched photons.

It is worth noticing that stable operation of the algorithm was maintained by generation of from 10^5 to 5×10^5 photons

per iteration. Two to five iterations were usually necessary to estimate the optical parameters with approximately 2% accuracy. The computer time required can be reduced not only by the condensed IMC method (Graaff et al. 1993b) but also by means of graphical solution of the inverse problem (Khalil et al. 2003, Pfefer et al. 2003, Thueler et al. 2003) or by means of producing a lookup table (van der Zee 1993a,b, Firbank et al. 1993, Hourdakis and Perris 1995, Kienle and Patterson 1996, Ripley et al. 1999, Ugryumova et al. 2004, Kotova et al. 2007) following a preliminary MC simulation. In the last case, a linear or spline interpolation (Press et al. 1992) between the data points can be used to improve the accuracy of the selection process.

In general, in vivo μ_a and μ_s' values for human skin proved to be significantly smaller than those obtained in vitro (about 10 and 2 times, respectively) (Graaff et al. 1993b, Laufer et al. 1998, Simpson et al. 1998, Doornbos et al. 1999, Ripley et al. 1999, Tuchin 2007, Salomatina and Yaroslavsky 2008). For μ_a, the discrepancy may be attributed to the low sensitivity of the double-integrating sphere, and goniometric techniques have been applied for in vitro measurements at weak absorption combined with strong scattering ($\mu_a \ll \mu_s$) and the sample preparation methods. For μ_s', the discrepancy may be related to the strong dependence of the method on variations in the relative refractive index of scatterers and the ground medium of the tissue m, $\mu_s' \sim (m - 1)^2$, which can be quite different for living and sampled tissue (Graaff et al. 1992, Khalil et al. 2003). The ex vivo measurements using the integrating sphere technique with corresponding IMC models, and very carefully prepared human tissue samples allow for accurate evaluation of μ_a and μ_s' which are very close to in vivo measurements (Graaff et al. 1993a, Laufer et al. 1998, Simpson et al. 1998, Doornbos et al. 1999, Ripley et al. 1999, Roggan et al. 1999b, Salomatina and Yaroslavsky 2008).

5.7 Direct Measurement of the Scattering Phase Function

Direct measurement of the scattering phase function $p(\theta)$ is important for the choice of an adequate model for the tissue being examined (van der Zee 1993a, Tuchin 1997, 2007, Yaroslavsky et al. 1999b, 2002a, Binzoni et al. 2006). The scattering phase function is usually determined from goniophotometric measurements in relatively thin tissue samples (Hardy et al. 1956, Bruls and van der Leun 1984, Fine et al. 1985, Jacques et al. 1987, Yoon et al. 1987, Steinke and Shepherd 1988, Marchesini et al. 1989, Duck 1990, Firbank et al. 1993, van der Zee 1993, van der Zee et al. 1993, Yaroslavskaya et al. 1994, Fried et al. 1995, Treweek and Barbenel 1996, Sardar and Levy 1998, Zijp and Ten Bosch 1998, Choukeife and L'Huillier 1999, Drezek et al. 1999, Yaroslavsky et al. 1999b, Hammer et al. 2001, Sardar et al. 2001, 2004, 2005, Darling et al. 2006, Forster et al. 2006, Turcu et al. 2006, Tuchin 2007, Pop and Neamtu 2008). The measured scattering indicatrix can be approximated either by the HGPF (Jacques et al. 1987, Yoon et al. 1987, Marchesini et al. 1989, Yaroslavskaya et al. 1994, Drezek et al. 1999, Yaroslavsky et al. 1999a, Hammer et al. 2001, Sardar et al. 2001, 2004, 2005) (see Equation 5.2) or by a set of HGPFs, with each function characterizing the type of scatterers and specific contribution to the indicatrix (see Equation 5.9) (Forster et al. 2006, Tuchin 2007). In the limiting case of a two-component model of a medium containing large and small (compared with the wavelength) scatterers, the indicatrix is represented in the form of anisotropic and isotropic components (Jacques et al. 1987, Yoon et al. 1987, Marchesini et al. 1989, Firbank et al. 1993, Graaff et al. 1993a, Fried et al. 1995, Zijp and Ten Bosch 1998, Bashkatov et al. 2005a, 2007, Darling et al. 2006). Other approximating functions are equally useful, e.g., those obtained from the Rayleigh–Gans approximation (Graaff et al. 1993a), ensuing from the Mie theory (Steinke and Shepherd 1988, Bevilacqua et al. 1995, 1999, Drezek et al. 1999), or a two-parameter GKPF (HGPF is a special simpler case of this phase function, see Equation 5.10) (Yaroslavsky et al. 1996b, 1999b, Hammer et al. 2001). Some of these types of approximations were used to find the dependence of the scattering anisotropy factor g for dermis and epidermis on the wavelengths in the range 300–1300 nm, which proved to coincide fairly well with the empirical formula (van Gemert et al. 1989),

$$g_e \sim g_d \sim 0.62 + \lambda \times 0.29 \times 10^{-3}, \qquad (5.20)$$

on the assumption of a 10% contribution of isotropic scattering (at wavelength 633 nm) (Jacques et al. 1987). The wavelength λ is given in nanometers. Another form of the spectral dependence of anisotropy factor can be presented as

$$g(\lambda) = A + B\left(1 - \exp\left(-(\lambda - C)/D\right)\right). \qquad (5.21)$$

The experimental values of anisotropy factor g for many types of human and animal tissues are presented in Tables 5.1 through 5.11 and approximated using Equation 5.21.

It should be noted that the correct prediction of light transport in tissues depends on the exact form of the phase function used for calculations (van der Zee 1993, Yaroslavsky et al. 1999b, 2002a, Thueler et al. 2003, Friebel et al. 2006, Tuchin 2007). Simulations performed with different forms of $p(\theta)$ (HGPF, Mie, and GKPF) with the same value of $<\cos(\theta)>$ result in the collection of significantly different fractions of the incident photons, particularly when small numerical-aperture delivery and collection fibers (small source-detection separation) are employed (Yaroslavsky et al. 1996b, 1999b, 2002a, Bevilacqua et al. 1999).

Moreover, for media with high anisotropy factors, precise measurements of the scattering phase function in the total angle range from 0° to 180° is a difficult technical task, demanding an extremely large dynamic range of measuring equipment. Most of the scattered radiation lies in the range from 0° to 30°, counting from the direction of the incident beam. In addition, measurements at angles close to 90° are strongly affected by scattering of higher orders, even for the samples of moderate optical thickness (Khlebtsov et al. 2002).

5.8 Optical Properties of Tissues

Above-discussed methods and techniques were successfully applied for the estimation of optical properties of a wide number of tissues. Measurements done in vitro and ex vivo by different research groups are summarized in Tables 5.1 through 5.11. Evidently, many types of animal and human tissues may have very close optical properties, but some specificity is expected. Early published data on optical properties of both human and animal tissues are presented in the following papers: Ebert et al. (1998), Farrar et al. (1999), Wei et al. (2003, 2005), Bargo et al. (2005), Zhang et al. (2005), Gebhart et al. (2006), Qu et al. (1994), Sardar and Levy (1998), Sardar et al. (2001, 2004, 2005, 2007), Bashkatov (2002), Bashkatov et al. (2004, 2005a,b, 2006b, 2007, 2010), Friebel et al. (2006, 2009), Hammer et al. (1995), Marchesini et al. (1992), Meinke et al. (2007a,b), Ripley et al. (1999), Roggan et al. (1999a,b), Salomatina et al. (2006), Salomatina and Yaroslavsky (2008), Simpson et al. (1998), van der Zee et al. (1993b), Yaroslavsky et al. (1996b, 2002b), Chan et al. (1996a,b), Nemati et al. (1996), Spitzer and Ten Bosch (1975), Troy and Thennadil (2001), Prahl (1988), Laufer et al. (1998), Du et al. (2001), Firbank et al. (1993), Ugryumova et al. (2004), Graaff et al. (1993a), Peters et al. (1990), Ghosh et al. (2001), Schmitt and Kumar (1998), Wang (2000), Cilesiz and Welch (1993), Lin et al. (1996), Genina et al. (2005), Vargas et al. (1999), Jacques (1996), Beek et al. (1997), Germer et al. (1998), Ma et al. (2005), Maitland et al. (1993), Nilsson et al. (1995), Parsa et al. (1989), Patwardhan et al. (2005), Ritz et al. (2001), Schwarzmaier et al. (1998), Youn et al. (2000), and partially summarized in review papers, book chapters and books Cheong et al. (1990), Duck (1990), Tuchin (2002, 2007), Vo-Dinh (2003), Mobley and Vo-Dinh (2003), Roggan et al. (1995), Cheong (1995).

Data presented in Tables 5.1 through 5.11 well reflect the situation in the field of tissue optical parameters measurements. It is clearly seen that the major attention was paid to skin and underlying tissues and head/brain optical properties investigations because of great importance and perspectives of optical tomography of subcutaneous tumors and optical monitoring and treatment of mental diseases. Optical properties of female breast are also well studied due perspectives to mammography. Nevertheless, in general, not many data for optical transport parameters are available in the literature. Moreover, these data are dependent on the tissue preparation technique, sample storage procedure, applied measuring method and inverse problem-solving algorithm, measuring instrumentation noise, and systematic errors.

5.8.1 Skin and Subcutaneous Tissue Optical Properties

Skin presents a complex heterogeneous medium where blood and pigment content are spatially distributed variably in depth. Skin consists of three main visible layers from surface: epidermis (100–150 μm thick, the blood-free layer), dermis (1–4 mm thick,

TABLE 5.1   Optical Properties of Human and Animal Skin and Subcutaneous Tissues Measured In Vitro and Ex Vivo

Tissue	λ, nm	μ_a, cm^{-1}	μ_s, cm^{-1}	g	μ_s', cm^{-1}	Remarks
Caucasian skin ($n = 21$)	400	3.76(0.35)	—	—	71.8(9.4)	IS, IAD; whole skin, 1–6 mm; postmortem; <24 hr after death; stored at 20°C in saline; measurements at room temperature; in the spectral range 400–2000 nm: $\mu_s' = 1.1 \times 10^{12} \lambda^{-4} + 73.7 \lambda^{-0.22}$, [$\lambda$] in nm; Bashkatov et al. (2005a)
	500	1.19(0.16)	—	—	32.5(4.2)	
	600	0.69(0.13)	—	—	21.8(3.0)	
	700	0.48(0.11)	—	—	16.7(2.3)	
	800	0.43(0.11)	—	—	14.0(1.9)	
	900	0.33(0.02)	—	—	15.7(2.1)	
	1000	0.27(0.03)	—	—	16.8(2.8)	
	1100	0.16(0.04)	—	—	17.1(2.7)	
	1200	0.54(0.04)	—	—	16.7(2.9)	
	1300	0.41(0.07)	—	—	14.7(2.6)	
	1400	1.64(0.31)	—	—	14.3(3.7)	
	1500	1.69(0.35)	—	—	14.4(3.8)	
	1600	1.19(0.22)	—	—	14.2(3.4)	
	1700	1.55(0.28)	—	—	14.7(3.5)	
	1800	1.44(0.22)	—	—	13.4(2.9)	
	1900	2.14(0.28)	—	—	12.2(3.1)	
	2000	1.74(0.29)	—	—	12.0(2.9)	
Caucasian skin ($n = 3$)	400	13.48	—	—	34.28	IS, IAD; whole skin, female skin; in the spectral range 400–1800 nm: $\mu_s' = 2.85 \times 10^7 \lambda^{-2.311} + 209.311 \lambda^{-0.518}$, [$\lambda$] in nm; data from graphs (Chan et al. 1996b)
	500	6.19	—	—	25.05	
	600	3.77	—	—	18.67	
	700	2.41	—	—	14.82	
	800	1.94	—	—	12.42	
	900	1.76	—	—	10.57	
	1000	1.55	—	—	9.23	
	1100	1.33	—	—	7.91	
	1200	1.76	—	—	7.11	
	1300	1.76	—	—	6.60	
	1400	10.29	—	—	6.21	
	1500	16.21	—	—	5.47	
	1600	5.44	—	—	5.87	
	1700	4.11	—	—	5.59	
	1800	6.05	—	—	5.68	
Caucasian skin ($n = 22$)	1000	0.98	—	—	12.58	DIS, IAD; whole skin, 22 samples taken from 14 subjects; measured within 24 h of excision; data from graphs (Troy and Thennadil 2001); in the spectral range 1000–1250 nm: $\mu_s' = 7.59 \times 10^7 \lambda^{-2.503} + 165.86 \lambda^{-0.402}$, [$\lambda$] in nm
	1100	0.98	—	—	11.77	
	1200	1.87	—	—	11.08	
	1300	1.77	—	—	10.69	
	1400	7.94	—	—	11.39	
	1500	13.1	—	—	11.38	
	1600	5.20	—	—	10.10	
	1700	4.85	—	—	9.96	
	1800	6.50	—	—	9.96	
	1900	13.0	—	—	10.63	
Rabbit skin (epidermis + dermis)	630	0.94(0.13)	213(21)	0.812(0.017)	40(2.2)	DIS, IAD, Beek et al. (1997)
	632.8	0.33(0.02)	306(12)	0.898(0.007)	31.6(2.2)	
	790	0.70(0.07)	321(8)	0.94(0.003)	18.4(0.5)	
Piglet skin (epidermis + dermis)	632.8	1.0(0.1)	492(17)	0.953(0.001)	22.7(0.8)	DIS, IAD, Beek et al. (1997)
	790	2.4(0.2)	409(14)	0.952(0.001)	19.3(0.6)	
	850	1.6(0.1)	403(20)	0.962(0.005)	14.3(1.5)	
Stratum corneum	350	25.92	500	0.902	48.99	Data from graphs (Patwardhan et al. 2005) (with references to papers by Bruls and van der Leun 1984, Jacques et al. 1987, van Gemert et al. 1989); in the spectral range 400–700 nm: $g = 0.918 + 0.304(1 - \exp(-(\lambda - 507.4)/2404))$
	400	17.28	500	0.903	48.44	
	450	11.63	500	0.910	45.24	
	500	10.47	500	0.916	41.93	
	550	9.83	500	0.923	38.69	
	600	8.67	500	0.930	35.02	
	650	8.21	500	0.936	32.21	
	700	8.15	500	0.942	28.93	

TABLE 5.1 (continued) Optical Properties of Human and Animal Skin and Subcutaneous Tissues Measured In Vitro and Ex Vivo

Tissue	λ, nm	μ_a, cm^{-1}	μ_s, cm^{-1}	g	μ_s', cm^{-1}	Remarks
Epidermis (lightly pigmented/medium pigmented/highly pigmented)	350	9.99/30.16/69.8	210.4	0.702	—	Data from graphs (Patwardhan et al. 2005) (with references to papers by Bruls and van der Leun 1984, Jacques et al. 1987, van Gemert et al. 1989); in the spectral range 350–700 nm: $g = 0.745 + 0.546(1 - \exp(-(\lambda - 500)/1806))$; $\mu_s = 1.752 \times 10^8\, \lambda^{-2.33} + 134.67\, \lambda^{-0.494}$, $[\lambda]$ in nm
	400	6.77/20.2/46.67	156.3	0.712	—	
	450	4.41/13.5/31.52	121.6	0.728	—	
	500	2.58/9.77/21.82	93.01	0.745	—	
	550	1.63/6.85/16.13	74.70	0.759	—	
	600	1.47/5.22/12.35	63.76	0.774	—	
	650	1.2/3.68/9.15	55.48	0.787	—	
	700	1.06/3.07/7.11	54.66	0.804	—	
Human epidermis ($n = 10$)	400	26.36	—	—	31.73	IS, 1D diffusion approximation, 10 samples from Caucasian skin; $\mu_s' = 1.175 \times 10^3\, \lambda^{-0.6}$, $[\lambda]$ in nm; data from graphs (Marchesini et al. 1992)
	450	13.84	—	—	30.11	
	500	7.79	—	—	28.27	
	550	5.73	—	—	26.82	
	600	3.01	—	—	25.29	
	650	1.58	—	—	24.14	
	700	0.89	—	—	22.97	
	750	0.49	—	—	22.09	
	800	0.35	—	—	21.27	
Human epidermis ($n = 7$)	400	12.96(1.44)	—	—	106.2(11)	IS, IMC, in the spectral range 370–1400 nm: $\mu_s' = 1.08 \times 10^8\, \lambda^{-2.364} + 135.71\, \lambda^{-0.267}$, $[\lambda]$ in nm; Salomatina et al. (2006)
	500	7.07(0.66)	—	—	70.6(7)	
	600	3.08(0.76)	—	—	51.4(5.0)	
	700	2.58(0.77)	—	—	42.7(4.1)	
	800	1.71(0.59)	—	—	36.8(3.6)	
	900	0.80(0.45)	—	—	33.6(3.5)	
	1000	0.45(0.28)	—	—	30.6(3.4)	
	1100	0.17(0.14)	—	—	29.2(3.2)	
	1200	0.71(0.44)	—	—	26.5(3.1)	
	1300	0.71(0.42)	—	—	25.7(3.1)	
	1400	15.53(2.5)	—	—	27.5(3.6)	
	1500	23.69(3.5)	—	—	28.3(4.2)	
	1600	7.49(1.64)	—	—	23.0(3.3)	
Human dermis	350	20.74	212.7	0.715	—	Data from graphs (Patwardhan et al. 2005) (with references to papers by Bruls and van der Leun 1984, Jacques et al. 1987, van Gemert et al. 1989); in the spectral range 350–700 nm: $g = 0.715 + 3.8 \times 10^{-4}(1 - \exp(-(\lambda - 542)/1129))$; $\mu_s = 1.752 \times 10^8\, \lambda^{-2.33} + 134.67\, \lambda^{-0.494}$, $[\lambda]$ in nm
	400	13.82	159.9	0.715	—	
	450	9.31	124.1	0.715	—	
	500	8.37	92.24	0.715	—	
	550	7.86	77.22	0.715	—	
	600	6.94	63.09	0.715	—	
	650	6.57	55.98	0.715	—	
	700	6.52	53.62	0.715	—	
Human dermis ($n = 8$)	400	9.13(1.18)	—	—	76.8(11)	IS, IMC, in the spectral range 370–1400 nm: $\mu_s' = 1.19 \times 10^8\, \lambda^{-2.427} + 71.476\, \lambda^{-0.258}$, $[\lambda]$ in nm; Salomatina et al. (2006)
	500	3.36(0.43)	—	—	46.2(4.6)	
	600	1.72(0.24)	—	—	32.2(2.9)	
	700	1.53(0.25)	—	—	26.4(2.5)	
	800	1.22(0.21)	—	—	22.5(2.3)	
	900	0.83(0.17)	—	—	20.1(2.3)	
	1000	0.79(0.18)	—	—	18.6(2.2)	
	1100	0.46(0.17)	—	—	17.6(2.1)	
	1200	1.33(0.22)	—	—	16.6(2.0)	
	1300	1.19(0.24)	—	—	16.2(2.0)	
	1400	11.7(1.14)	—	—	18.6(1.9)	
	1500	17.5(1.48)	—	—	19.9(2.0)	
	1600	6.63(0.57)	—	—	15.9(1.8)	

(continued)

TABLE 5.1 (continued) Optical Properties of Human and Animal Skin and Subcutaneous Tissues Measured In Vitro and Ex Vivo

Tissue	λ, nm	μ_a, cm^{-1}	μ_s, cm^{-1}	g	μ_s', cm^{-1}	Remarks
Caucasian dermis ($n = 12$)	633	0.32	—	—	26.99	IS, IMC; samples from abdominal and breast tissue obtained from plastic surgery or *postmortem* examinations; data from graphs (Simpson et al. 1998); $\mu_s' = 1.66 \times 10^5 \, \lambda^{-1.356}$, [$\lambda$] in nm
	700	0.12	—	—	23.02	
	750	0.09	—	—	20.62	
	800	0.02	—	—	18.80	
	850	0.01	—	—	17.41	
	900	0.03	—	—	16.18	
	950	0.22	—	—	15.10	
	1000	0.39	—	—	14.68	
Negroid dermis ($n = 5$)	633	2.45	—	—	32.29	IS, IMC; samples from abdominal and breast tissue obtained from plastic surgery or *postmortem* examinations; data from graphs (Simpson et al. 1998); $\mu_s' = 3.33 \times 10^5 \, \lambda^{-1.438}$, [$\lambda$] in nm
	700	1.51	—	—	27.24	
	750	1.12	—	—	24.02	
	800	0.80	—	—	21.26	
	850	0.61	—	—	19.70	
	900	0.46	—	—	18.55	
	950	0.49	—	—	17.67	
	1000	0.49	—	—	16.83	
Caucasian dermis	450	5.13	134.9	0.054	—	IS, IAD; data from graph of reference (Prahl 1988); in the spectral range 450–800 nm: $\mu_s = 2.97 \times 10^5 \, \lambda^{-1.257}$, [$\lambda$] in nm; in the spectral range 500–800 nm: $g = 0.334 + 0.217(1 - \exp(-(\lambda - 567)/90.76))$, [$\lambda$] in nm
	500	3.45	119.9	0.120	—	
	550	2.28	108.1	0.288	—	
	600	1.81	97.38	0.410	—	
	650	1.44	87.89	0.461	—	
	700	1.16	78.48	0.500	—	
	750	1.03	72.29	0.519	—	
	800	0.88	65.89	0.531	—	
Piglet skin dermis	632.8	0.89(0.1)	289(7)	0.926(0.002)	21.1(0.4)	DIS, IAD, Beek et al. (1997)
	790	1.8(0.2)	254(5)	0.945(0.001)	13.9(0.3)	
	850	0.33(0.03)	285(5)	0.968(0.001)	9(0.2)	
Porcine skin dermis ($n = 44$)	900	0.06	282.6	0.904	—	IS, IMC; postmortem time 2 ~ 10 h; data from graphs (Du et al. 2001); in the spectral range 1000–1300 nm: $\mu_s = 440.2 \, \lambda^{-0.072}$, [$\lambda$] in nm
	1000	0.12	270.4	0.904	—	
	1100	0.17	267.2	0.904	—	
	1200	1.74	263.9	0.903	—	
	1300	1.04	262.8	0.903	—	
	1400	9.11	246.2	0.872	—	
	1500	7.32	259.6	0.873	—	
Porcine skin dermis ($n = 40$)	325	5.6	220	0.38	—	IS, IMC; data from graphs (Ma et al. 2005); in the spectral range: $g = 0.653 + 0.219(1 - \exp(-(\lambda - 530.2)/242.8))$; $\mu_s = 3.286 \times 10^8 \, \lambda^{-2.487} + 80.454 \, \lambda^{-0.215}$, [$\lambda$] in nm
	442	1.9	89	0.36	—	
	532	1.4	69	0.64	—	
	633	0.7	58	0.72	—	
	850	1.6	90	0.88	—	
	1064	3.1	26	0.86	—	
	1310	6.2	40	0.87	—	
	1557	10.4	21	0.82	—	
Subdermis (primarily globular fat cells) ($n = 12$)	633	0.12	—	—	12.58	IS, IMC; samples from abdominal and breast tissue obtained from plastic surgery or *postmortem* examinations; data from graphs (Simpson et al. 1998); $\mu_s' = 139.24 \, \lambda^{-0.373}$, [$\lambda$] in nm
	700	0.09	—	—	12.10	
	750	0.09	—	—	11.75	
	800	0.08	—	—	11.40	
	850	0.09	—	—	11.17	
	900	0.12	—	—	10.95	
	950	0.15	—	—	10.81	
	1000	0.12	—	—	10.71	

TABLE 5.1 (continued)   Optical Properties of Human and Animal Skin and Subcutaneous Tissues Measured In Vitro and Ex Vivo

Tissue	λ, nm	μ_a, cm^{-1}	μ_s, cm^{-1}	g	μ_s', cm^{-1}	Remarks
Human subcutaneous adipose tissue ($n=6$)	400	2.26(0.24)	—	—	13.4(2.8)	IS, IAD; tissue slabs, 1–3 mm; <6 h after surgery; stored at 20°C in saline; measurements at room temperature; in the spectral range 600–1500 nm: $\mu_s' = 1.05 \times 10^3\, \lambda^{-0.68}$, [λ] in nm; Bashkatov et al. (2005a)
	500	1.49(0.06)	—	—	13.8(4.0)	
	600	1.18(0.02)	—	—	13.4(4.7)	
	700	1.11(0.05)	—	—	12.2(4.4)	
	800	1.07(0.11)	—	—	11.6(4.6)	
	900	1.07(0.07)	—	—	10.0(3.4)	
	1000	1.06(0.06)	—	—	9.39(3.3)	
	1100	1.01(0.05)	—	—	8.74(3.3)	
	1200	1.06(0.07)	—	—	7.91(3.2)	
	1300	0.89(0.07)	—	—	7.81(3.2)	
	1400	1.08(0.03)	—	—	7.51(3.3)	
	1500	1.05(0.02)	—	—	7.36(3.4)	
	1600	0.89(0.04)	—	—	7.16(3.2)	
	1700	1.26(0.07)	—	—	7.53(3.3)	
	1800	1.21(0.01)	—	—	7.5(3.48)	
	1900	1.62(0.06)	—	—	8.72(4.2)	
	2000	1.43(0.09)	—	—	8.24(4.0)	
Human subcutaneous adipose tissue ($n=10$)	400	15.98(3.2)	—	—	49.5(6.5)	IS, IMC, in the spectral range 370–1300 nm: $\mu_s' = 1.08 \times 10^8\, \lambda^{-2.525} + 157.494\, \lambda^{-0.345}$, [λ] in nm; Salomatina et al. (2006)
	500	5.50(0.69)	—	—	35.4(4.5)	
	600	1.89(0.40)	—	—	27.0(3.2)	
	700	1.27(0.24)	—	—	23.0(2.5)	
	800	1.08(0.23)	—	—	20.2(2.1)	
	900	0.95(0.22)	—	—	18.5(1.8)	
	1000	0.89(0.25)	—	—	17.4(1.7)	
	1100	0.74(0.22)	—	—	16.6(1.5)	
	1200	1.65(0.30)	—	—	16.1(1.5)	
	1300	1.05(0.27)	—	—	15.8(1.4)	
	1400	6.27(0.88)	—	—	16.8(1.6)	
	1500	8.52(1.46)	—	—	17.6(1.8)	
	1600	3.60(0.61)	—	—	15.7(1.6)	
Rat subcutaneous adipose tissue ($n=10$)	400	2.25(1.34)	—	—	19.8(6.3)	IS, IAD; tissue slabs, 1–2 mm; <6 h after surgery; stored at 20°C in saline; measurements at room temperature; in the spectral range 600–1400 nm: $\mu_s' = 25.51\, \lambda^{-0.12}$, [λ] in nm; Bashkatov et al. (2005b)
	500	0.64(0.34)	—	—	14.3(4.1)	
	600	0.64(0.33)	—	—	12.2(3.5)	
	700	0.75(0.36)	—	—	11.4(3.2)	
	800	1.05(0.47)	—	—	11.0(3.1)	
	900	1.25(0.55)	—	—	10.8(3.0)	
	1000	1.43(0.61)	—	—	10.9(3.0)	
	1100	1.43(0.61)	—	—	10.6(2.8)	
	1200	2.07(0.99)	—	—	11.2(2.9)	
	1300	1.43(0.64)	—	—	10.5(2.8)	
	1400	2.29(1.20)	—	—	10.7(3.0)	
	1500	2.03(1.07)	—	—	10.3(3.0)	
	1600	1.40(0.72)	—	—	9.33(2.7)	
	1700	3.04(1.69)	—	—	11.6(3.4)	
	1800	2.67(1.43)	—	—	10.8(3.2)	
	1900	4.55(2.65)	—	—	13.8(4.5)	
	2000	3.99(2.28)	—	—	12.7(4.2)	
	2100	2.76(1.53)	—	—	11.3(3.6)	
	2200	2.65(1.48)	—	—	12.2(3.6)	
	2300	6.92(3.67)	—	—	22.7(6.0)	
	2400	6.54(3.52)	—	—	24.0(6.1)	
	2500	5.58(3.04)	—	—	23.9(6.4)	

Note: rms values are given in parentheses. IS, single integrating sphere; DIS, double integrating sphere.

TABLE 5.2 Optical Properties of Ocular Tissues Measured In Vitro and Ex Vivo

Tissue	λ, nm	μ_a, cm^{-1}	μ_s, cm^{-1}	g	μ_s', cm^{-1}	Remarks
Human sclera ($n = 10$)	400	4.25(0.25)	—	—	78.63(9.35)	IS, IAD; stored at 20°C in saline; measurements at room temperature; in the spectral range 370–1800 nm: $\mu_s' = 2.411 \times 10^5 \, \lambda^{-1.325}$, [$\lambda$] in nm; Bashkatov et al. (2010)
	500	1.84(0.21)	—	—	61.49(5.56)	
	600	0.85(0.21)	—	—	51.95(4.94)	
	700	0.59(0.21)	—	—	43.62(4.91)	
	800	0.48(0.21)	—	—	36.68(4.92)	
	900	0.52(0.21)	—	—	31.47(4.51)	
	1000	0.77(0.25)	—	—	26.30(3.78)	
	1100	0.57(0.22)	—	—	21.57(3.12)	
	1200	1.56(0.34)	—	—	19.09(2.89)	
	1300	1.58(0.33)	—	—	16.47(2.44)	
	1400	9.50(1.19)	—	—	21.23(3.39)	
	1500	14.1(1.91)	—	—	23.26(4.19)	
	1600	6.05(1.13)	—	—	15.23(2.32)	
	1700	5.33(1.02)	—	—	13.39(2.00)	
	1800	6.94(1.27)	—	—	13.84(2.07)	
	1900	28.8(3.98)	—	—	28.85(6.11)	
	2000	38.4(6.25)	—	—	36.07(8.54)	
	2100	20.1(2.56)	—	—	22.32(4.18)	
	2200	15.97(1.89)	—	—	18.50(2.97)	
	2300	19.4(2.23)	—	—	21.56(3.14)	
	2400	28.7(3.46)	—	—	28.57(5.07)	
	2500	49.16(5.72)	—	—	39.31(8.88)	
Porcine sclera	413	7.28	—	—	146.5	DIS, IMC, data from graphs (Hammer et al. 1995); in the spectral range: $\mu_s' = 5.5 \times 10^5 \, \lambda^{-1.37}$, [$\lambda$] in nm
	430	6.73	—	—	136.7	
	450	5.95	—	—	125.9	
	488	5.40	—	—	115.0	
	514	4.85	—	—	108.5	
	540	4.63	—	—	99.85	
	559	4.19	—	—	91.16	
	586	3.86	—	—	88.99	
	633	2.43	—	—	81.40	
	700	1.76	—	—	69.46	
	780	0.44	—	—	60.78	
	1064	0.22	—	—	42.33	
Porcine sclera ($n = 4$)	650	4.25	—	—	42	IS, IAD, data from graphs (Chan et al. 1996b); in the spectral range: $\mu_s' = 51.65 - 0.023 \, \lambda + 1.45 \times 10^6 \, \lambda^{-2}$, [$\lambda$] in nm
	700	3.75	—	—	42	
	750	3	—	—	41	
	800	2.5	—	—	40.5	
	850	2	—	—	40.5	
	900	1.75	—	—	39.6	
	950	2	—	—	38.7	
	1000	1.75	—	—	37.8	
	1050	1.5	—	—	35.1	
	1100	1.25	—	—	32.4	
	1150	2.25	—	—	30.6	
	1200	2.25	—	—	28.8	
	1250	2	—	—	26.1	
	1300	2.5	—	—	23.4	
	1350	3.75	—	—	21.6	
	1400	15	—	—	20.7	
	1425	20	—	—	19.8	
	1450	21.5	—	—	18.9	

TABLE 5.2 (continued)   Optical Properties of Ocular Tissues Measured In Vitro and Ex Vivo

Tissue	λ, nm	μ_a, cm^{-1}	μ_s, cm^{-1}	g	μ_s', cm^{-1}	Remarks
	1500	12.5	—	—	18	
	1550	7.5	—	—	18	
	1600	5.75	—	—	15.3	
	1650	5	—	—	14.4	
	1700	5.5	—	—	13.5	
	1750	7.5	—	—	12.6	
	1800	8	—	—	10.8	
Rabbit sclera ($n=8$)	500	3.9(1.2)	—	—	85.96(29.08)	IS, IAD; in the spectral range: $\mu_s' = 1.001 \times 10^8 \lambda^{-2.234}$, [$\lambda$] in nm; Nemati et al. (1996)
	550	3.7(1.1)	—	—	75.54(25.16)	
	700	2.2(0.6)	—	—	50.40(15.39)	
	850	1.6(0.5)	—	—	29.83(7.50)	
	1050	1.4(0.6)	—	—	16.28(3.20)	
Rabbit conjunctiva ($n=3$)	500	1.4(0.8)	—	—	19.4(9.78)	IS, IAD; in the spectral range: $\mu_s' = 5.28 \times 10^7 \lambda^{-2.478} + 129.974 \lambda^{-0.437}$, [$\lambda$] in nm; Nemati et al. (1996)
	550	1.8(0.9)	—	—	17.07(7.25)	
	700	0.6(0.3)	—	—	11.55(4.74)	
	850	0.7(0.4)	—	—	9.54(3.32)	
	1050	0.5(0.4)	—	—	8.41(3.06)	
Rabbit ciliary body ($n=8$)	500	52.8(28.6)	—	—	65.75(38.84)	IS, IAD; in the spectral range: $\mu_s' = 889.37 \lambda^{-0.417}$, [$\lambda$] in nm; Nemati et al. (1996)
	550	51.2(27.7)	—	—	64.18(36.97)	
	700	40.8(20.7)	—	—	59.66(29.32)	
	850	23.9(11.0)	—	—	51.18(22.41)	
	1050	8.2(2.5)	—	—	49.40(19.62)	
Bovine retina	413	11.63	—	—	14.34	DIS, IMC, data from graphs (Hammer et al. 1995); in the spectral range: $\mu_s' = 6.99 \times 10^7 \lambda^{-2.689} + 97.656 \lambda^{-0.426}$, [$\lambda$] in nm
	430	10.72	—	—	13.64	
	450	6.97	—	—	12.48	
	488	3.82	—	—	11.06	
	514	3.90	—	—	10.49	
	540	4.76	—	—	9.91	
	559	4.66	—	—	9.35	
	586	3.80	—	—	8.64	
	633	1.58	—	—	7.75	
	700	1.12	—	—	7.29	
	780	0.82	—	—	6.91	
	1064	0.65	—	—	5.59	
Porcine retinal pigment epithelium	413	1216.6	—	—	192.4	DIS, IMC, data from graphs (Hammer et al. 1995)
	430	1197.0	—	—	192.1	
	450	1168.0	—	—	198.5	
	488	1111.6	—	—	197.8	
	514	1117.7	—	—	180.3	
	540	1062.6	—	—	186.5	
	559	1027.0	—	—	184.5	
	586	970.97	—	—	187.6	
	633	880.67	—	—	185.8	
	700	678.86	—	—	228.1	
	780	454.77	—	—	258.3	
	1064	75.306	—	—	258.2	
Porcine blood-free choroid	413	239.7	—	—	93.5	DIS, IMC, data from graphs (Hammer et al. 1995)
	430	243.2	—	—	90.0	
	450	236.4	—	—	85.0	
	488	218.5	—	—	80.2	
	514	215.8	—	—	79.6	
	540	207.3	—	—	78.8	
	559	205.4	—	—	77.8	

(continued)

TABLE 5.2 (continued)   Optical Properties of Ocular Tissues Measured In Vitro and Ex Vivo

Tissue	λ, nm	μ_a, cm^{-1}	μ_s, cm^{-1}	g	μ_s', cm^{-1}	Remarks
	586	195.6	—	—	76.9	
	633	171.1	—	—	77.0	
	700	131.3	—	—	79.6	
	780	87.79	—	—	79.9	
	1064	5.491	—	—	69.2	
Human	400	9.62	—	—	21.61	IS, Kubelka–Munk model, the K–M coefficients were calculated and
trabecular	450	7.57	—	—	17.49	converted to linear transport coefficients; data from graph (Farrar et al.
meshwork	500	5.28	—	—	14.61	1999); in the spectral range: $\mu_s' = 7.8 \times 10^7 \lambda^{-2.631} + 117.4\, \lambda^{-0.417}$, [$\lambda$] in nm
($n = 10$)	550	3.61	—	—	12.71	
	600	2.29	—	—	11.47	
	650	1.32	—	—	10.86	
	700	0.70	—	—	10.34	
	750	0.32	—	—	9.89	
	800	0.16	—	—	9.56	

Note: rms values are given in parentheses.

vascularized layer), and subcutaneous fat (from 1 to 6 mm thick, in dependence from the body site).

The randomly inhomogeneous distribution of blood and various chromophores and pigments in skin produces variations of average optical properties of skin layers. Nonetheless, it is possible to define the regions in the skin where the gradient of skin cells' structure, chromophores, or blood amounts changing with a depth equals roughly zero. This allows subdividing these layers into sublayers regarding the physiological nature, physical and optical properties of their cells, and pigments content. The epidermis can be subdivided into the two sublayers: nonliving and living epidermis. Nonliving epidermis or stratum corneum (~20 μm thick) consists of only dead squamous cells, which are highly keratinized with a high lipid and protein content, and has a relatively low water content. Living epidermis (~100 μm thick) contains most of the skin pigmentation, mainly melanin, which is produced in the melanocytes.

Dermis is a vascularized layer and the main absorbers in the visible spectral range are the blood hemoglobin, carotene, and bilirubin. In the IR spectral range, absorption properties of skin dermis are defined by absorption of water. The scattering properties of the dermis are mainly defined by the fibrous structure of the tissue, where collagen fibrils are packed in collagen bundles and have lamellae structure. The light scatters on both single fibrils and scattering centers, which are formed by the interlacement of the collagen fibrils and bundles. To sum up, the average scattering properties of the skin are defined by the scattering properties of the reticular dermis because of relatively big thickness of the layer (up to 4 mm) and comparable scattering coefficients of the epidermis and the reticular dermis. Absorption of hemoglobin and water of skin dermis and lipids of skin epidermis define absorption properties of whole skin.

The subcutaneous adipose tissue is formed by aggregation of fat cells (adipocytes) containing stored fat (lipids) in the form of a number of small droplets for lean or normal humans and a few or even single big drop in each cell for obesity humans; and the lipids of mostly presented by triglycerides. The diameters of the adipocytes are in the ranges from 15 to 250 μm and their mean diameter is varied from 50 to 120 μm. In the spaces between the cells there are blood capillaries (arterial and venous plexus), nerves, and reticular fibrils connecting each cell and providing metabolic activity of fat tissue. Absorption of the human adipose tissue is defined by the absorption of hemoglobin, lipids, and water. The main scatterers of adipose tissue are spherical droplets of lipids, which are uniformly distributed within adipocytes.

The skin and subcutaneous tissue optical properties have been measured with integrating sphere technique in the visible and NIR spectral ranges (Prahl 1988, van Gemert et al. 1989, Marchesini et al. 1992, Chan et al. 1996b, Beek et al. 1997, Simpson et al. 1998, Du et al. 2001, Troy and Thennadil 2001, Bashkatov et al. 2005a,b, Ma et al. 2005, Patwardhan et al. 2005, Salomatina et al. 2006) and the in vitro and ex vivo results are presented and summarized in Table 5.1.

5.8.2 Ocular Tissue Optical Properties

A large number of studies in past years have supported the use of transscleral cyclophotocoagulation (TSCPC) as a treatment for advanced glaucoma. The procedure involves the delivery of light, usually with a contact probe, through the superficial layers of conjunctiva and sclera to the ciliary body. In spite of the large number of studies that have been reported on the use of TSCPC, the choice of lasers has been based largely on a clinical comparison of results with various lasers. Knowledge of the ocular tissue optical properties allows one to develop new and optimize already existing laser procedures of glaucoma treatment.

Besides the TSCPC procedures, laser trabeculoplasty as a method of surgical intervention, is successfully used in the treatment of primary open angle glaucoma for increasing the outflow of aqueous humor. For optimization of the procedure, the knowledge of human trabecular meshwork structure is very important. In addition to the procedures of glaucoma treatment,

TABLE 5.3 Optical Properties of Head/Brain Tissues Measured In Vitro and Ex Vivo

Tissue	λ, nm	μ_a, cm^{-1}	μ_s, cm^{-1}	g	μ_s', cm^{-1}	Remarks
Human white matter ($n=19$)	400	16.47	—	—	85.29	IS, IAD; data from graphs (Gebhart et al. 2006); in the spectral range: $\mu_s' = 2.67 \times 10^7 \lambda^{-2.188} + 399.6\,\lambda^{-0.396}$, [$\lambda$] in nm
	500	3.38	—	—	67.58	
	600	2.12	—	—	56.35	
	700	1.42	—	—	47.70	
	800	1.35	—	—	41.64	
	900	1.41	—	—	37.42	
	1000	1.76	—	—	32.56	
	1100	1.57	—	—	29.33	
	1200	2.81	—	—	26.63	
	1300	2.32	—	—	24.35	
Human gray matter ($n=25$)	400	20.33	—	—	27.33	IS, IAD; data from graphs (Gebhart et al. 2006); in the spectral range: $\mu_s' = 9.21 \times 10^7 \lambda^{-2.564} + 99.4\,\lambda^{-0.473}$, [$\lambda$] in nm
	500	4.08	—	—	15.30	
	600	2.61	—	—	11.05	
	700	1.41	—	—	9.18	
	800	1.07	—	—	8.00	
	900	1.05	—	—	7.23	
	1000	1.23	—	—	5.21	
	1100	1.12	—	—	5.28	
	1200	1.97	—	—	4.69	
	1300	1.75	—	—	4.33	
Tumor (glioma) ($n=39$)	400	21.5	—	—	38.07	IS, IAD; data from graphs (Gebhart et al. 2006); in the spectral range: $\mu_s' = 2.25 \times 10^7 \lambda^{-2.279} + 266.6\,\lambda^{-0.495}$, [$\lambda$] in nm
	500	4.09	—	—	28.13	
	600	2.35	—	—	22.74	
	700	1.42	—	—	18.34	
	800	1.35	—	—	15.58	
	900	1.40	—	—	14.47	
	1000	1.89	—	—	11.28	
	1100	1.70	—	—	10.00	
	1200	2.86	—	—	8.95	
	1300	2.62	—	—	8.88	
Pig cranial bone ($n=24$)	650	0.37	350.1	0.923	27.25	IS, IMC; data from graphs (Firbank et al. 1993); in the spectral range: $\mu_s = 1.82 \times 10^5 \lambda^{-0.965}$; $\mu_s' = 1.357 \times 10^6 \lambda^{-1.675}$; $g = 0.724 + 0.216(1 - \exp(-(\lambda - 398.1)/97.8))$, [$\lambda$] in nm
	700	0.24	325.6	0.930	23.13	
	750	0.25	307.1	0.934	20.45	
	800	0.25	287.1	0.936	18.53	
	850	0.26	271.7	0.938	17.04	
	900	0.35	257.1	0.941	15.57	
	950	0.47	242.1	0.945	13.36	
Cortical bone (taken from the leg of horse)	550	11.2	—	—	162.8	IS, IMC, in the spectral range: $\mu_s' = 7.216 \times 10^5 \lambda^{-1.424} + 38.86\,\lambda^{-0.094}$, [$\lambda$] in nm; Ugryumova et al. (2004)
	600	8.11	—	—	150.6	
	650	6.90	—	—	142.5	
	700	5.71	—	—	135.2	
	750	4.63	—	—	128.7	
	800	4.10	—	—	125.6	
	850	3.80	—	—	122.3	
	900	3.90	—	—	119.8	
	950	3.83	—	—	107.9	
Human cranial bone ($n=10$)	800	0.11(0.02)	—	—	19.48(1.52)	IS, IAD; in the spectral range: $\mu_s' = 1533.02 \lambda^{-0.65}$, [$\lambda$] in nm; Bashkatov et al. (2006b)
	900	0.15(0.02)	—	—	18.03(1.19)	
	1000	0.22(0.03)	—	—	17.10(0.91)	
	1100	0.15(0.02)	—	—	16.20(0.80)	
	1200	0.65(0.06)	—	—	16.38(0.84)	
	1300	0.49(0.05)	—	—	14.92(0.78)	
	1400	1.37(0.13)	—	—	16.10(1.10)	

(continued)

TABLE 5.3 (continued)   Optical Properties of Head/Brain Tissues Measured In Vitro and Ex Vivo

Tissue	λ, nm	μ_a, cm^{-1}	μ_s, cm^{-1}	g	μ_s', cm^{-1}	Remarks
	1500	3.13(0.26)	—	—	15.96(1.37)	
	1600	2.47(0.40)	—	—	15.84(3.05)	
	1700	2.77(0.46)	—	—	16.12(3.72)	
	1800	2.97(0.62)	—	—	15.42(3.98)	
	1900	4.39(1.33)	—	—	11.37(2.76)	
	2000	4.47(1.18)	—	—	11.48(2.01)	
Human white matter (native samples) ($n = 7$)	400	2.80	413.4	0.756	—	IS, IMC; data from graphs (Yaroslavsky et al. 2002b); in the spectral range 700–1100: $\mu_s = 1.67 \times 10^6 \lambda^{-1.375} + 702.8 \lambda^{-0.192}$, in the spectral range 360–1100: $g = 0.8 + 0.099(1 - \exp(-(\lambda - 484.7)/216.18))$, [λ] in nm
	500	0.97	422.8	0.807	—	
	600	0.80	408.3	0.838	—	
	700	0.75	395.2	0.862	—	
	800	0.87	369.3	0.875	—	
	900	1.01	335.7	0.883	—	
	1000	1.16	309.6	0.887	—	
	1100	1.01	297.0	0.893	—	
Human white matter (coagulated) ($n = 7$)	400	9.08	532.8	0.827	—	IS, IMC; data from graphs (Yaroslavsky et al. 2002b); in the spectral range 600–1100: $\mu_s = 1.92 \times 10^6 \lambda^{-1.434} + 846.94 \lambda^{-0.168}$, in the spectral range 360–1100: $g = 0.859 + 0.082(1 - \exp(-(\lambda - 468.2)/200.3))$, [λ] in nm
	500	3.30	498.5	0.870	—	
	600	2.05	479.5	0.899	—	
	700	1.61	443.3	0.916	—	
	800	1.66	412.2	0.926	—	
	900	1.96	380.7	0.932	—	
	1000	2.17	357.8	0.937	—	
	1100	2.72	344.2	0.935	—	
Human gray matter (native samples) ($n = 7$)	400	2.30	124.6	0.865	—	IS, IMC; data from graphs (Yaroslavsky et al. 2002b); in the spectral range: $\mu_s = 2.08 \times 10^4 \lambda^{-0.847}$; $g = 0.883 + 0.019(1 - \exp(-(\lambda - 482.8)/105.6))$, [λ] in nm
	500	0.47	106.5	0.884	—	
	600	0.22	92.80	0.894	—	
	700	0.16	81.11	0.899	—	
	800	0.20	75.67	0.899	—	
	900	0.32	65.16	0.901	—	
	1000	0.49	58.89	0.905	—	
	1100	0.48	53.04	0.906	—	
Human gray matter (coagulated) ($n = 7$)	400	8.12	297.7	0.785	—	IS, IMC; data from graphs (Yaroslavsky et al. 2002b); in the spectral range: $g = 0.833 + 0.046(1 - \exp(-(\lambda - 459.4)/90.9))$, [λ] in nm
	500	2.01	330.9	0.852	—	
	600	0.80	342.0	0.870	—	
	700	0.87	332.3	0.883	—	
	800	0.96	261.5	0.875	—	
	900	1.10	226.2	0.872	—	
	1000	1.55	201.6	0.879	—	
	1100	1.65	192.2	0.879	—	
Human cerebellum (native samples) ($n = 7$)	400	4.60	279.1	0.802	—	IS, IMC; data from graphs (Yaroslavsky et al. 2002b); in the spectral range: $g = 0.836 + 0.067(1 - \exp(-(\lambda - 459)/160.5))$, [λ] in nm
	500	1.30	284.6	0.850	—	
	600	0.79	274.1	0.872	—	
	700	0.58	267.1	0.890	—	
	800	0.58	248.5	0.896	—	
	900	0.64	226.6	0.899	—	
	1000	0.73	210.9	0.901	—	
	1100	0.67	204.2	0.900	—	
Human cerebellum (coagulated) ($n = 7$)	400	19.5	575.4	0.599	—	IS, IMC; data from graphs (Yaroslavsky et al. 2002b); in the spectral range: $g = 0.743 + 0.184(1 - \exp(-(\lambda - 519.7)/217))$, [λ] in nm
	500	4.75	474.7	0.744	—	
	600	2.79	466.3	0.789	—	
	700	1.50	488.6	0.851	—	
	800	1.01	459.9	0.877	—	
	900	0.99	455.6	0.894	—	
	1000	1.04	427.8	0.906	—	
	1100	0.87	418.9	0.913	—	

TABLE 5.3 (continued)   Optical Properties of Head/Brain Tissues Measured In Vitro and Ex Vivo

Tissue	λ, nm	μ_a, cm^{-1}	μ_s, cm^{-1}	g	μ_s', cm^{-1}	Remarks
Human pons (native samples) ($n=7$)	400	3.29	162.3	0.897	—	IS, IMC; data from graphs (Yaroslavsky et al. 2002b); in the spectral range: $\mu_s = 4.332 \times 10^4 \lambda^{-0.934}$; $g = 0.908 + 0.012(1 - \exp(-(\lambda - 484.8)/153.7))$, [$\lambda$] in nm
	500	0.83	129.0	0.911	—	
	600	0.57	108.2	0.914	—	
	700	0.48	94.01	0.916	—	
	800	0.67	83.57	0.917	—	
	900	0.77	75.61	0.919	—	
	1000	0.99	68.93	0.920	—	
	1100	0.96	64.29	0.921	—	
Human pons (coagulated) ($n=7$)	400	15.9	728.3	0.842	—	IS, IMC; data from graphs (Yaroslavsky et al. 2002b); in the spectral range: $\mu_s = 9.779 \times 10^5 \lambda^{-1.2}$; $g = 0.86 + 0.027(1 - \exp(-(\lambda - 425.1)/56.4))$, [$\lambda$] in nm
	500	8.77	637.6	0.884	—	
	600	7.64	524.0	0.892	—	
	700	7.08	407.0	0.889	—	
	800	6.48	331.8	0.887	—	
	900	6.01	278.9	0.883	—	
	1000	5.90	237.1	0.881	—	
	1100	5.96	218.3	0.883	—	
Human thalamus (native samples) ($n=7$)	400	3.19	154.1	0.847	—	IS, IMC; data from graphs (Yaroslavsky et al. 2002b); in the spectral range 500–1100 nm: $\mu_s = 1.793 \times 10^3 \lambda^{-0.363}$; in the spectral range 360–1100 nm: $g = 0.865 + 0.04(1 - \exp(-(\lambda - 486.9)/239.9))$, [$\lambda$] in nm
	500	1.01	187.0	0.867	—	
	600	0.66	175.4	0.882	—	
	700	0.50	165.5	0.888	—	
	800	0.58	158.7	0.893	—	
	900	0.68	153.1	0.897	—	
	1000	0.87	145.1	0.899	—	
	1100	0.80	143.7	0.906	—	
Human thalamus (coagulated) ($n=7$)	400	14.9	434.0	0.831	—	IS, IMC; data from graphs (Yaroslavsky et al. 2002b); in the spectral range 500–1100 nm: $\mu_s = 5.578 \times 10^4 \lambda^{-0.792}$; in the spectral range 360–1100 nm: $g = 0.864 + 0.068(1 - \exp(-(\lambda - 431.3)/115.6))$, [$\lambda$] in nm
	500	3.90	403.1	0.903	—	
	600	1.59	354.2	0.916	—	
	700	1.23	316.1	0.921	—	
	800	1.08	279.8	0.928	—	
	900	1.06	253.5	0.932	—	
	1000	1.24	236.0	0.932	—	
	1100	1.32	222.4	0.935	—	
Brain tumor (Meningioma) ($n=6$)	400	3.75	196.2	0.872	—	IS, IMC; data from graphs (Yaroslavsky et al. 2002b); in the spectral range 500–1100 nm: $\mu_s = 1.69 \times 10^4 \lambda^{-0.718}$; in the spectral range 360–1100 nm: $g = 0.889 + 0.07(1 - \exp(-(\lambda - 418.6)/78.6))$, [$\lambda$] in nm
	500	1.07	183.4	0.932	—	
	600	0.67	175.0	0.953	—	
	700	0.30	154.9	0.956	—	
	800	0.23	138.9	0.959	—	
	900	0.22	125.9	0.958	—	
	1000	0.37	116.8	0.956	—	
	1100	0.64	115.2	0.965	—	
Brain tumor (Astrocytoma WHO grade II) ($n=4$)	400	16.0	200.3	0.889	—	IS, IMC; data from graphs (Yaroslavsky et al. 2002b); in the spectral range: $\mu_s = 9.254 \times 10^4 \lambda^{-1.025}$; $g = 0.903 + 0.06(1 - \exp(-(\lambda - 410.5)/33.7))$, [$\lambda$] in nm
	500	2.03	155.4	0.958	—	
	600	1.19	130.4	0.962	—	
	700	0.41	112.0	0.960	—	
	800	0.50	96.94	0.967	—	
	900	0.32	86.62	0.963	—	
	1000	0.44	78.81	0.961	—	
	1100	0.45	72.22	0.968	—	
Human dura mater ($n=10$)	400	3.08	—	—	22.35	IS, IAD, in the spectral range: $\mu_s' = 2.887 \times 10^4 \lambda^{-1.164}$, [$\lambda$] in nm; Genina et al. (2005)
	450	1.51	—	—	22.89	
	500	1.09	—	—	21.60	
	550	1.10	—	—	18.48	
	600	0.80	—	—	17.11	
	650	0.70	—	—	15.51	
	700	0.74	—	—	13.99	

Note: rms values are given in parentheses.

TABLE 5.4 Optical Properties of Epithelial/Mucous Tissues Measured In Vitro and Ex Vivo

Tissue	λ, nm	μ_a, cm^{-1}	μ_s, cm^{-1}	g	μ_s', cm^{-1}	Remarks
Normal human colon mucosa/ submucosa ($n = 13$)	476.5	2.32(0.09)	214(5.35)	0.885(0.019)	—	IS, IAD; data from graphs (Wei et al. 2005)
	488.0	3.27(0.13)	228(5.69)	0.891(0.021)	—	
	496.5	2.58(0.10)	212(5.27)	0.897(0.024)	—	
	514.5	3.12(0.12)	216(5.38)	0.902(0.026)	—	
	532.0	3.33(0.14)	208(5.16)	0.908(0.029)	—	
Adenomatous human colon mucosa/submucosa ($n = 13$)	476.5	5.27(0.21)	233(5.72)	0.897(0.023)	—	IS, IAD; data from graphs (Wei et al. 2005)
	488.0	5.34(0.22)	238(5.84)	0.903(0.027)	—	
	496.5	4.87(0.19)	228(5.67)	0.907(0.028)	—	
	514.5	4.37(0.17)	231(5.69)	0.917(0.033)	—	
	532.0	5.16(0.20)	223(5.63)	0.913(0.031)	—	
Normal human colon muscle layer/ chorion ($n = 13$)	476.5	1.31(0.05)	221(5.61)	0.923(0.037)	—	IS, IAD; data from graphs (Wei et al. 2005)
	488.0	1.73(0.07)	215(5.33)	0.932(0.044)	—	
	496.5	1.27(0.05)	200(5.08)	0.927(0.041)	—	
	514.5	1.14(0.04)	189(5.03)	0.933(0.045)	—	
	532.0	1.53(0.06)	193(5.05)	0.941(0.048)	—	
Adenomatous human colon muscle layer/ chorion ($n = 13$)	476.5	3.17(0.12)	233(5.71)	0.927(0.042)	—	IS, IAD; data from graphs (Wei et al. 2005)
	488.0	3.51(0.14)	223(5.62)	0.935(0.046)	—	
	496.5	2.90(0.11)	216(5.36)	0.933(0.044)	—	
	514.5	2.57(0.09)	198(5.07)	0.936(0.047)	—	
	532.0	2.75(0.10)	208(5.14)	0.945(0.049)	—	
Human maxillary sinuses mucous membrane ($n = 10$)	400	4.89(0.92)	—	—	36.01(6.41)	IS, IAD; stored at 20°C in saline; measurements at room temperature; in the spectral range 400—2000 nm: $\mu_s' = 443742.6\ \lambda^{-1.62}$, [$\lambda$] in nm; Bashkatov et al. (2004)
	500	1.13(0.18)	—	—	17.69(2.84)	
	600	0.45(0.23)	—	—	13.81(2.43)	
	700	0.16(0.24)	—	—	11.53(2.02)	
	800	0.13(0.16)	—	—	9.79(1.68)	
	900	0.12(0.09)	—	—	7.62(0.92)	
	1000	0.27(0.21)	—	—	6.14(0.74)	
	1100	0.16(0.14)	—	—	5.19(0.58)	
	1200	0.57(0.31)	—	—	4.43(0.43)	
	1300	0.67(0.35)	—	—	3.90(0.38)	
	1400	4.84(1.79)	—	—	5.07(0.71)	
	1500	6.06(2.38)	—	—	4.95(1.21)	
	1600	2.83(1.01)	—	—	3.13(0.55)	
	1700	2.26(0.79)	—	—	2.84(0.51)	
	1800	3.04(1.15)	—	—	3.04(0.57)	
	1900	9.23(2.69)	—	—	7.01(3.57)	
	2000	9.30(2.28)	—	—	6.26(3.56)	
Human stomach wall mucosa ($n = 15$)	400	13.4(2.09)	53.0(3.23)	0.037(0.107)	51.06(6.85)	IS, IMC; stored at 20°C in saline; measurements at room temperature; in the spectral range 400–2000 nm: $\mu_s' = 1.027 \times 10^{12}\ \lambda^{-4} + 164.3\ \lambda^{-0.446}$; $g = 0.498 + 0.319(1 - \exp(-(\lambda - 533.7)/138.7))$, [$\lambda$] in nm; Bashkatov et al. (2007)
	500	2.07(0.25)	44.1(2.47)	0.439(0.108)	24.77(2.24)	
	600	1.37(0.22)	46.72(1.69)	0.622(0.109)	17.65(1.54)	
	700	0.75(0.15)	48.02(1.19)	0.714(0.086)	13.72(1.09)	
	800	0.78(0.17)	46.99(0.89)	0.761(0.108)	11.22(0.81)	
	900	0.92(0.20)	44.65(0.83)	0.784(0.108)	9.64(0.76)	
	1000	1.18(0.23)	42.17(0.74)	0.793(0.108)	8.73(0.67)	
	1100	1.11(0.24)	42.66(0.66)	0.817(0.108)	7.80(0.60)	
	1200	1.76(0.28)	40.83(0.59)	0.819(0.109)	7.38(0.53)	
	1300	1.76(0.27)	39.99(0.51)	0.833(0.109)	6.67(0.47)	
	1400	8.70(1.02)	25.72(0.97)	0.637(0.109)	9.34(0.88)	
	1500	11.9(1.85)	23.45(1.56)	0.574(0.097)	10.0(1.42)	
	1600	5.0(0.57)	31.25(0.56)	0.790(0.105)	6.56(0.51)	
	1700	4.74(0.56)	34.11(0.47)	0.824(0.105)	5.99(0.43)	
	1800	6.05(0.68)	31.15(0.61)	0.794(0.105)	6.43(0.55)	
	1900	19.5(2.78)	17.34(2.62)	0.294(0.104)	12.25(2.38)	
	2000	20.9(3.53)	26.28(3.80)	0.563(0.105)	11.48(3.45)	

TABLE 5.4 (continued)   Optical Properties of Epithelial/Mucous Tissues Measured In Vitro and Ex Vivo

Tissue	λ, nm	μ_a, cm^{-1}	μ_s, cm^{-1}	g	μ_s', cm^{-1}	Remarks
Human uterus (Myometrium) ($n=6$)	610	0.45	—	—	13.30	IS, IMC; in the spectral range: $\mu_s' = 9.597 \times 10^5$ $\lambda^{-1.731}$, [λ] in nm; data from graphs (Ripley et al. 1999)
	700	0.19	—	—	11.47	
	800	0.10	—	—	9.08	
	900	0.10	—	—	7.32	
	1000	0.38	—	—	6.09	
Human uterus (Leiomyoma (fibroid) ($n=6$)	610	0.15	—	—	10.99	IS, IMC; in the spectral range: $\mu_s' = 1.029 \times 10^6$ $\lambda^{-1.783}$, [λ] in nm; data from graphs (Ripley et al. 1999)
	700	0.07	—	—	8.73	
	800	0.03	—	—	6.75	
	900	0.03	—	—	5.47	
	1000	0.32	—	—	4.73	
Gallbladder tissue ($n=6$)	350	37.62	—	—	29.34	IS, diffusion approximation; in the spectral range 350–1850 nm: $\mu_s' = 1.761 \times 10^8$ $\lambda^{-2.692} + 5.95\,\lambda^{-0.061}$, [$\lambda$] in nm; data from graphs (Maitland et al. 1993)
	450	23.17	—	—	16.66	
	550	13.63	—	—	10.76	
	650	5.16	—	—	8.19	
	750	4.84	—	—	6.67	
	850	4.43	—	—	5.92	
	950	4.53	—	—	5.44	
	1050	4.50	—	—	4.75	
	1150	5.27	—	—	4.50	
	1250	5.29	—	—	4.39	
	1350	6.95	—	—	4.56	
	1450	24.43	—	—	5.92	
	1550	11.36	—	—	4.62	
	1650	8.07	—	—	4.15	
	1750	9.24	—	—	4.45	
	1850	14.03	—	—	5.25	
	1950	78.15	—	—	10.94	
	2050	36.11	—	—	6.38	
	2150	20.20	—	—	4.99	
	2250	20.35	—	—	4.33	
	2350	30.86	—	—	5.80	
	2450	49.57	—	—	10.76	
Gallbladder bile ($n=5$)	350	32.98	—	—	16.39	IS, diffusion approximation; data from graphs (Maitland et al. 1993)
	450	75.81	—	—	26.01	
	550	2.58	—	—	2.48	
	650	1.81	—	—	1.74	
	750	0.53	—	—	1.67	
	850	0.39	—	—	1.91	
	950	0.41	—	—	1.97	
	1050	0.92	—	—	1.74	
	1150	1.89	—	—	1.97	
	1250	2.30	—	—	2.21	
	1350	3.99	—	—	2.94	
	1450	25.13	—	—	9.01	
	1550	9.98	—	—	3.98	
	1650	5.82	—	—	3.05	
	1750	7.54	—	—	3.44	
	1850	13.18	—	—	5.52	
	1950	71.63	—	—	20.77	
	2050	38.09	—	—	9.16	
	2150	19.31	—	—	5.84	
	2250	19.48	—	—	4.45	
	2350	31.84	—	—	9.18	
	2450	53.18	—	—	15.22	

Note: rms values are given in parentheses.

TABLE 5.5 Optical Properties of Breast Tissue Measured In Vitro and Ex Vivo

Tissue	λ, nm	μ_a, cm^{-1}	μ_s, cm^{-1}	g	μ'_s, cm^{-1}	Remarks
Human breast (normal glandular tissue) ($n=3$)	500	3.42	461.8	0.947	—	IS, IMC; in the spectral range: $g = 0.737 + 0.229(1 - \exp(-(\lambda - 282.7)/93.5))$, [$\lambda$] in nm; data from graphs (Peters et al. 1990)
	600	0.92	431.5	0.959	—	
	700	0.48	409.1	0.965	—	
	800	0.55	332.7	0.965	—	
	900	0.67	275.1	0.965	—	
	1000	0.90	213.7	0.957	—	
	1100	0.82	200.2	0.961	—	
Human breast (normal adipose tissue) ($n=7$)	500	2.73	313.8	0.971	—	IS, IMC; in the spectral range: $g = 0.741 + 0.236(1 - \exp(-(\lambda + 23.3)/148))$, [$\lambda$] in nm; data from graphs (Peters et al. 1990)
	600	0.99	294.1	0.972	—	
	700	0.81	306.3	0.9749	—	
	800	0.82	313.8	0.976	—	
	900	0.84	306.1	0.976	—	
	1000	0.90	306.2	0.976	—	
	1100	1.14	332.2	0.977	—	
Human breast (fibrocystic tissue) ($n=8$)	500	2.28	879.2	0.980	—	IS, IMC; in the spectral range: $g = 0.749 + 0.234(1 - \exp(-(\lambda + 15.1)/177.8))$, [$\lambda$] in nm; data from graphs (Peters et al. 1990)
	600	0.47	623.5	0.977	—	
	700	0.24	568.0	0.978	—	
	800	0.28	548.8	0.981	—	
	900	0.39	536.5	0.981	—	
	1000	0.63	485.6	0.982	—	
	1100	0.84	465.9	0.983	—	
Human breast (fibroadenoma) ($n=6$)	500	4.02	447.7	0.970	—	IS, IMC; in the spectral range: $g = 0.749 + 0.235(1 - \exp(-(\lambda - 255.9)/89.4))$, [$\lambda$] in nm; data from graphs (Peters et al. 1990)
	600	1.76	492.6	0.979	—	
	700	0.53	438.4	0.982	—	
	800	0.34	384.0	0.983	—	
	900	0.79	327.3	0.983	—	
	1000	1.57	269.1	0.982	—	
	1100	1.48	209.2	0.979	—	
Human breast (ductal carcinoma) ($n=9$)	500	2.60	426.4	0.954	—	IS, IMC; in the spectral range: $g = 0.727 + 0.236(1 - \exp(-(\lambda)/156.5))$, [$\lambda$] in nm; data from graphs (Peters et al. 1990)
	600	1.61	337.5	0.958	—	
	700	0.44	277.4	0.961	—	
	800	0.34	233.0	0.962	—	
	900	0.45	181.2	0.957	—	
	1000	0.64	143.7	0.950	—	
	1100	0.52	123.0	0.946	—	

Note: rms values are given in parentheses.

ophthalmologic inspection of the eye fundus is very important as a tool for diagnosis and therapy.

Thus, a more fundamental knowledge of the optical properties of the ocular tissues is necessary to obtain correct applications and interpretations. The optical properties of ocular tissues have been measured with integrating sphere technique in the visible and NIR spectral ranges (Hammer et al. 1995, Chan et al. 1996b, Nemati et al. 1996, Farrar et al. 1999, Bashkatov et al. 2010) and are presented and summarized in Table 5.2.

5.8.3 Head/Brain Tissue Optical Properties

The most detailed in vitro investigations of normal and coagulated brain tissues (gray matter, white matter, cerebellum, pons, and thalamus), as well as of native tumor tissues (astrocytoma WHO grade II and meningioma), using single-integrating-sphere spectral measurements in the spectral range from 360 to 1100 nm and IMC algorithm for data processing are described by Yaroslavsky et al. (2002b) (see Table 5.3). As it follows from Table 5.3, all brain tissues under study shared qualitatively similar dependencies of the optical properties on the wavelength. The scattering coefficient decreased and the anisotropy factor increased with the wavelength, which can be explained by the lowering of the contribution of Rayleigh scattering and growing of the contribution of Mie scattering with the wavelength. The wavelength-dependent absorption coefficient behavior of all brain tissues resembled a mixture of oxy- and deoxy-hemoglobin absorption spectra. This means that in spite of careful preparation of the samples, it was not possible to remove all blood residuals from the tissue sections.

TABLE 5.6 Optical Properties of Cartilage Measured In Vitro and Ex Vivo

Tissue	λ, nm	μ_a, cm^{-1}	μ_s, cm^{-1}	g	μ_s', cm^{-1}	Remarks
Cartilage (rabbit)	632.8	0.33(0.05)	214(0.2)	0.909(0.005)	19.4(1.1)	DIS, IAD, Beek et al. (1997)
Equine articular cartilage ($n = 18$)	350	3.28	—	—	53.90	IS, Kubelka–Munk model, the K–M coefficients were calculated and converted to linear transport coefficients; data from graph (Ebert et al. 1998); in the spectral range 400–850: $\mu_s' = 5.448 \times 10^7 \lambda^{-2.333}$, [$\lambda$] in nm
	400	1.35	—	—	47.09	
	450	0.79	—	—	35.35	
	500	0.60	—	—	27.70	
	550	0.53	—	—	22.26	
	600	0.53	—	—	17.94	
	650	0.55	—	—	14.74	
	700	0.58	—	—	12.39	
	750	0.64	—	—	10.62	
	800	0.67	—	—	9.22	
	850	0.71	—	—	7.32	
Nasal septal cartilage (porcine) ($n = 25$)	400	0.50	—	—	18.14	IS, IAD, data from graphs (Youn et al. 2000); in the spectral range 400–1150: $\mu_s' = 1.29 \times 10^8 \lambda^{-2.638} + 6.72 \lambda^{-0.308}$, [$\lambda$] in nm
	500	0.26	—	—	10.79	
	600	0.22	—	—	6.88	
	700	0.20	—	—	4.71	
	800	0.18	—	—	3.60	
	900	0.23	—	—	2.88	
	1000	0.39	—	—	2.51	
	1100	0.24	—	—	2.21	
	1200	0.74	—	—	2.32	
	1300	0.65	—	—	2.27	
	1400	2.31	—	—	5.51	
Porcine aural cartilage ($n = 2$)	400	1.03	—	—	60.6	IS, IAD, data from graphs (Youn et al. 2000); in the spectral range 400–1300: $\mu_s' = 1.624 \times 10^7 \lambda^{-2.249} + 16.04 \lambda^{-0.088}$, [$\lambda$] in nm
	500	0.28	—	—	39.5	
	600	0.20	—	—	28.8	
	700	0.19	—	—	23.0	
	800	0.27	—	—	19.5	
	900	0.42	—	—	17.2	
	1000	0.72	—	—	15.5	
	1100	0.55	—	—	13.9	
	1200	1.60	—	—	13.5	
	1300	1.57	—	—	12.6	
Mouse ears ($n = 8$)	400	7.54(1.71)	64.72(6.80)	0.289(0.110)	—	IS, IMC; postmortem time 5 ~ 10 min; in the spectral range 400–1300 nm: $\mu_s' = 1.224 \times 10^8 \lambda^{-2.508} + 28.84 \lambda^{0.03}$; $g = 0.405 + 0.232(1 - \exp(-(\lambda - 554.7)/340))$, [$\lambda$] in nm; Salomatina and Yaroslavsky (2008)
	500	2.41(0.43)	54.37(0.86)	0.341(0.044)	—	
	600	1.95(0.30)	49.88(1.13)	0.451(0.039)	—	
	700	0.97(0.09)	45.90(1.66)	0.485(0.049)	—	
	800	0.61(0.06)	40.99(1.42)	0.511(0.047)	—	
	900	0.49(0.07)	39.92(1.56)	0.546(0.045)	—	
	1000	0.45(0.10)	38.66(1.16)	0.573(0.051)	—	
	1100	0.26(0.09)	37.64(1.21)	0.592(0.044)	—	
	1200	0.85(0.17)	38.21(1.31)	0.607(0.038)	—	
	1300	0.72(0.20)	37.56(1.22)	0.612(0.035)	—	
	1400	5.68(1.28)	44.10(1.14)	0.582(0.047)	—	
	1500	8.22(1.93)	46.28(1.65)	0.549(0.058)	—	
	1600	3.45(0.81)	40.80(1.09)	0.625(0.032)	—	
Mouse ears ($n = 10$)	400	7.23(1.14)	83.93(9.55)	0.327(0.048)	—	IS, IMC; postmortem time 72 h; in the spectral range 400–1300 nm: $\mu_s' = 1.293 \times 10^8 \lambda^{-2.478} + 33.7 \lambda^{0.048}$; $g = 0.454 + 0.186(1 - \exp(-(\lambda - 523)/249.2))$, [$\lambda$] in nm; Salomatina and Yaroslavsky (2008)
	500	2.33(0.37)	70.37(3.83)	0.417(0.037)	—	
	600	1.69(0.23)	64.58(3.13)	0.521(0.028)	—	
	700	1.07(0.13)	60.61(2.78)	0.558(0.027)	—	
	800	0.80(0.11)	54.14(2.48)	0.569(0.023)	—	
	900	0.65(0.11)	52.67(2.56)	0.591(0.021)	—	
	1000	0.67(0.14)	50.58(2.09)	0.591(0.023)	—	

(continued)

TABLE 5.6 (continued)   Optical Properties of Cartilage Measured In Vitro and Ex Vivo

Tissue	λ, nm	μ_a, cm^{-1}	μ_s, cm^{-1}	g	μ_s', cm^{-1}	Remarks
	1100	0.52(0.14)	50.03(2.32)	0.620(0.020)	—	
	1200	0.98(0.17)	50.03(2.35)	0.638(0.020)	—	
	1300	0.90(0.17)	49.52(2.35)	0.651(0.020)	—	
	1400	4.69(0.43)	55.74(2.91)	0.663(0.018)	—	
	1500	7.26(0.74)	57.20(3.09)	0.645(0.018)	—	
	1600	3.23(0.39)	52.45(2.26)	0.679(0.015)	—	
Mouse ears	400	10.87(2.93)	90.93(10.6)	0.401(0.038)	—	IS, IMC; postmortem time 73 h; in the spectral
(frozen-thawed)	500	2.44(0.51)	69.17(3.72)	0.453(0.025)	—	range 400–1300 nm: $\mu_s' = 1.523 \times 10^8$
($n = 10$)	600	1.44(0.24)	63.24(3.10)	0.546(0.021)	—	$\lambda^{-2.487} + 29.12\,\lambda^{0.062}$; $g = 0.466 + 0.189$
	700	0.94(0.18)	59.19(2.84)	0.576(0.016)	—	$(1 - \exp(-(\lambda - 486)/292.2))$, [$\lambda$] in nm;
	800	0.65(0.17)	52.90(2.80)	0.580(0.015)	—	Salomatina and Yaroslavsky (2008)
	900	0.52(0.16)	51.50(2.58)	0.605(0.011)	—	
	1000	0.52(0.16)	48.98(2.49)	0.600(0.009)	—	
	1100	0.37(0.15)	48.64(2.50)	0.631(0.008)	—	
	1200	0.81(0.19)	48.51(2.48)	0.649(0.008)	—	
	1300	0.74(0.18)	47.92(2.46)	0.658(0.009)	—	
	1400	4.51(0.45)	54.03(2.74)	0.672(0.010)	—	
	1500	7.09(0.68)	55.20(2.91)	0.649(0.012)	—	
	1600	3.05(0.36)	50.62(2.94)	0.680(0.012)	—	

Note: rms values are given in parentheses.

At the same time, the differences in the spectral characteristics of the brain tissues have been observed. For example, the total attenuation coefficients ($\mu_t = \mu_a + \mu_s$) of white matter are substantially higher than those of gray matter. The two brain stem tissues (pons and thalamus) also have different optical properties. The tumors are generally macroscopically less homogeneous than any normal tissues; thus, their scattering coefficients and anisotropy factors are slightly higher than those of normal gray matter.

After coagulation, the values of absorption and scattering coefficients increased for all tissues. The extent of this increase, however, is different for each tissue type, and is characterized by factor from 2 to 5. It was shown (Yaroslavsky et al. 2002b) that a significant increase of both interaction coefficients is a result of substantial structure changes, caused mostly by tissue shrinkage and condensation, as well as collagen swelling and homogenization of the vessel walls. Tissue shrinkage caused by loosing water at coagulation makes tissue more dense, which leads to increase of both scattering and absorption coefficients in the spectral range where water absorption is weak (up to 1100–1300 nm). The refractive index microscopic redistribution of a tissue due to cellular and fiber proteins' denaturation and homogenization at thermal action also may have a strong inclusion in alteration of scattering and absorption properties. The similar increase of both absorption (by a factor of 2–10) and scattering (by a factor of 2–4) coefficients in the wavelength range from 500 to 1100 nm was found for coagulated human blood.

The reduced scattering coefficient of skull bone is considerably less than that of brain white matter and is comparable with that of gray matter, cerebellum, and brain stem tissues (pons and thalamus). It is also comparable with scalp tissue values. At

coagulation of soft brain tissues, their reduced scattering coefficient may considerably exceed that of skull bone for all tissues presented in Table 5.3. This would imply that in NIR spectroscopy on the adult head, the effect of light scattering by the skull is of the same order of magnitude as that of surrounding scalp tissue and brain. A possible reason for this is the high values of scattering anisotropy factor g due to the specific structure of bone. For example, the cortical bone consists of an underlying matrix of collagen fibers, around which calcium-bearing hydroxyapatite crystals are deposited. These crystals are the major scatterers of bone (Ugryumova et al. 2004, Bashkatov et al. 2006b); they are big sized and have a high refraction power, and therefore may be responsible for the high values of g.

The head/brain tissues' optical properties have been measured with integrating sphere technique in the visible and NIR spectral ranges (Firbank et al. 1993, Yaroslavsky et al. 2002b, Ugryumova et al. 2004, Genina et al. 2005, Bashkatov et al. 2006b, Gebhart et al. 2006) and summarized in Table 5.3.

5.8.4 Epithelial/Mucous Tissue Optical Properties

The investigation of the epithelial/mucous optical properties is necessary for light dosimetry at photodynamic therapy of bladder, colon, esophagus, stomach, etc. The treatment of purulent maxillary sinusitis is an important problem in modern rhinology despite the wide application of surgical and pharmaceutical methods. One of the new methods of treatment of this disease is photodynamic therapy of the mucous membrane of the maxillary sinus. The epithelial/mucous tissues' optical properties have been measured with integrating sphere technique in the visible

TABLE 5.7 Optical Properties of Liver Tissues Measured In Vitro and Ex Vivo

Tissue	λ, nm	μ_a, cm^{-1}	μ_s, cm^{-1}	g	μ_s', cm^{-1}	Remarks
Rat liver ($n=9$)	500	7.19	106.5	0.812	—	IS, IMC, data from graphs (Nilsson et al.
	550	11.41	111.5	0.814	—	1995); in the spectral range 550–800:
	600	5.27	103.4	0.875	—	$\mu_s = 4.446 \times 10^3\ \lambda^{-0.588}$; $g = 0.814 + 0.108$
	650	2.20	98.0	0.910	—	$(1 - \exp(-(\lambda - 555.6)/50.3))$, $[\lambda]$ in nm
	700	1.34	92.7	0.915	—	
	750	1.13	92.1	0.918	—	
	800	0.99	88.9	0.919	—	
Rat liver ($n=11$)	400	62.68	242.32	0.895	—	IS, seven-flux model with δ-Eddington
	500	12.70	167.34	0.933	—	phase function, data from graphs
	600	10.56	150.12	0.946	—	(Parsa et al. 1989); in the spectral range
	700	5.45	120.25	0.950	—	400–1800: $\mu_s = 1.046 \times 10^6\ \lambda^{-1.4}$, $[\lambda]$
	800	5.55	95.16	0.945	—	in nm
	900	5.74	64.45	0.916	—	
	1000	6.18	67.18	0.932	—	
	1100	5.67	58.98	0.925	—	
	1200	6.31	50.83	0.917	—	
	1300	5.90	46.75	0.911	—	
	1400	18.70	45.98	0.899	—	
	1500	20.56	38.36	0.907	—	
	1600	9.26	34.92	0.902	—	
	1700	8.86	31.02	0.891	—	
	1800	11.07	22.62	0.817	—	
	1900	52.51	86.16	0.826	—	
	2000	52.45	42.24	0.742	—	
	2100	27.46	26.02	0.813	—	
	2200	23.54	16.29	0.769	—	
Porcine liver (native tissue) ($n=15$)	400	44.26	123.2	0.767	—	DIS, IMC, data from graphs (Ritz et al.
	500	9.88	93.65	0.880	—	2001); in the spectral range 400–2400:
	600	6.56	75.96	0.903	—	$\mu_s = 6.847 \times 10^4\ \lambda^{-1.059}$; in the spectral
	700	1.44	64.53	0.923	—	range 400–1300: $g = 0.725 + 0.208$
	800	0.80	56.62	0.929	—	$(1 - \exp(-(\lambda - 363.4)/111.9))$, $[\lambda]$ in nm
	900	0.61	51.95	0.932	—	
	1000	0.48	47.52	0.934	—	
	1100	0.21	43.12	0.933	—	
	1200	0.99	37.66	0.930	—	
	1300	1.10	33.63	0.931	—	
	1400	10.02	32.21	0.915	—	
	1500	17.54	29.09	0.881	—	
	1600	7.78	25.86	0.928	—	
	1700	6.76	24.33	0.930	—	
	1800	8.40	23.84	0.911	—	
	1900	51.56	24.68	0.795	—	
	2000	68.84	20.96	0.735	—	
	2100	29.46	19.86	0.779	—	
	2200	21.44	19.90	0.815	—	
	2300	28.11	18.68	0.757	—	
	2400	41.48	21.00	0.703	—	
Porcine liver (coagulated) ($n=15$)	400	91.83	711.29	0.760	—	DIS, IMC, data from graphs (Ritz et al.
	500	15.04	640.29	0.838	—	2001); in the spectral range 400–1800:
	600	7.52	503.6	0.861	—	$\mu_s = 1.442 \times 10^5\ \lambda^{-0.882}$; in the spectral
	700	2.38	441.8	0.885	—	range 400–1800: $g = 0.699 + 0.226$
	800	1.01	393.6	0.899	—	$(1 - \exp(-(\lambda - 316.5)/220))$, $[\lambda]$ in nm
	900	0.54	355.3	0.908	—	

(continued)

TABLE 5.7 (continued) Optical Properties of Liver Tissues Measured In Vitro and Ex Vivo

Tissue	λ, nm	μ_a, cm^{-1}	μ_s, cm^{-1}	g	μ_s', cm^{-1}	Remarks
	1000	0.28	326.9	0.913	—	
	1100	0.14	300.4	0.919	—	
	1200	0.52	272.7	0.922	—	
	1300	0.50	252.5	0.923	—	
	1400	12.93	237.9	0.919	—	
	1500	17.88	223.3	0.919	—	
	1600	8.02	207.6	0.928	—	
	1700	7.89	198.6	0.930	—	
	1800	8.97	189.7	0.924	—	
	1900	60.58	192.5	0.879	—	
	2000	59.32	182.2	0.865	—	
	2100	29.82	167.1	0.897	—	
	2200	30.03	164.1	0.904	—	
	2300	34.41	162.3	0.899	—	
	2400	50.44	165.5	0.889	—	
Liver (rabbit)	632.8	11.3(5.2)	190(41)	0.934(0.023)	8.9(3.9)	DIS, IAD, Beek et al. (1997)
Liver (goat)	632.8	12.3(9.0)	491(72)	0.980(0.011)	8.7(4.6)	DIS, IAD, Beek et al. (1997)
Liver (rat)	632.8	3.8(0.2)	289(10)	0.952(0.004)	13.0(1.0)	DIS, IAD, Beek et al. (1997)
Liver (rat)	1064	2.0(0.3)	151(6)	0.948(0.005)	7.9(0.7)	DIS, IAD, Beek et al. (1997)
Human liver tissue (normal) ($n = 10$)	850	1.0(0.2)	204(36)	0.955(0.01)	—	DIS, IMC, Germer et al. (1998)
	980	0.8(0.1)	182(33)	0.955(0.01)	—	
	1064	0.5(0.1)	169(33)	0.952(0.01)	—	
Human liver tissue (coagulated) ($n = 10$)	850	0.7(0.2)	236(47)	0.887(0.02)	—	DIS, IMC, Germer et al. (1998)
	980	0.5(0.1)	210(27)	0.896(0.02)	—	
	1064	0.2(0.1)	200(26.8)	0.904(0.01)	—	

Note: rms values are given in parentheses.

TABLE 5.8 Optical Properties of Muscle Tissues Measured In Vitro and Ex Vivo

Tissue	λ, nm	μ_a, cm^{-1}	μ_s, cm^{-1}	g	μ_s', cm^{-1}	Remarks
Muscle ($n = 1$)	633	1.23	—	—	8.94	IS, IMC; samples from abdominal and breast tissue obtained from plastic surgery or *postmortem* examinations; data from graphs (Simpson et al. 1998); $\mu_s' = 7.67 \times 10^3 \, \lambda^{-1.045}$, [$\lambda$] in nm
	700	0.48	—	—	8.18	
	750	0.41	—	—	7.71	
	800	0.28	—	—	7.04	
	850	0.30	—	—	6.67	
	900	0.32	—	—	6.21	
	950	0.46	—	—	5.90	
	1000	0.51	—	—	5.73	
Muscle	630	1.4(0.2)	110(5)	0.846(0.009)	16.5(0.7)	Rabbit; DIS, IAD, Beek et al. (1997)
	632.8	0.74(0.06)	140(6)	0.968(0.002)	4.4(0.3)	
	790	2.3(0.2)	157(11)	0.95(0.005)	6.8(0.7)	
Muscle	630	1.2(0.1)	239(16)	0.732(0.013)	62.1(2)	Piglet; DIS, IAD, Beek et al. (1997)
	632.8	0.59(0.01)	179(12)	0.858(0.012)	24.7(0.7)	
Muscle ($n = 9$)	500	1.17	89.2	0.903	—	Rat; IS, IMC, data from graphs (Nilsson et al. 1995); in the spectral range: $\mu_s = 2.39 \times 10^7 \, \lambda^{-2.215} + 376.94 \, \lambda^{-0.274}$; $g = 0.883 + 0.051 \, (1 - \exp(-(\lambda - 469.3)/84.11))$, [$\lambda$] in nm
	550	1.66	88.2	0.909	—	
	600	0.95	83.3	0.926	—	
	650	0.56	79.0	0.930	—	
	700	0.52	73.56	0.930	—	
	750	0.52	71.3	0.931	—	
	800	0.54	66.7	0.930	—	

Note: rms values are given in parentheses.

TABLE 5.9 Optical Properties of Aorta Tissues Measured In Vitro and Ex Vivo

Tissue	λ, nm	μ_a, cm^{-1}	μ_s, cm^{-1}	g	μ_s', cm^{-1}	Remarks
Bovine aorta	400	2.93	—	—	29.61	IS, IAD; in the spectral range 500–1350 nm: $\mu_s' = 883.96\,\lambda^{-0.527}$,
($n=1$)	500	1.15	—	—	34.37	[λ] in nm; data from graphs (Chan et al. 1996b)
	600	0.61	—	—	30.99	
	700	0.41	—	—	27.73	
	800	0.41	—	—	25.31	
	900	0.57	—	—	23.53	
	1000	0.98	—	—	23.36	
	1100	0.70	—	—	22.50	
	1200	1.79	—	—	21.30	
	1300	1.70	—	—	20.40	
	1400	6.82	—	—	15.82	
	1500	8.29	—	—	12.78	
	1600	5.38	—	—	17.01	
	1700	4.80	—	—	17.93	
	1800	5.94	—	—	16.62	
Human aorta	300	53.65	—	—	71.26	IS, diffusion approximation with δ-Eddington phase function;
($n=9$)	400	10.72	—	—	49.06	in the spectral range 300–1800 nm: $\mu_s' = 2.78 \times 10^5\,\lambda^{-1.443}$, [$\lambda$]
	500	4.82	—	—	33.41	in nm; data from graphs (Cilesiz and Welch 1993)
	600	2.95	—	—	26.56	
	700	2.90	—	—	21.95	
	800	2.99	—	—	18.56	
	900	2.99	—	—	16.27	
	1000	3.37	—	—	13.89	
	1100	3.02	—	—	12.03	
	1200	4.51	—	—	10.56	
	1300	4.59	—	—	9.47	
	1400	26.44	—	—	7.49	
	1500	25.63	—	—	6.04	
	1600	10.83	—	—	7.10	
	1700	10.04	—	—	6.65	
	1800	12.96	—	—	6.24	

Note: rms values are given in parentheses.

TABLE 5.10 Optical Properties of Lung Tissues Measured In Vitro and Ex Vivo

Tissue	λ, nm	μ_a, cm^{-1}	μ_s, cm^{-1}	g	μ_s', cm^{-1}	Remarks
Lung epithelium	400	3.41	355.8	0.938	—	IS, IAD; data from graphs (Qu et al. 1994); in
($n=9$)	500	2.03	286.2	0.945	—	the spectral range: $\mu_s = 1.219 \times 10^6\,\lambda^{-1.347}$;
	600	1.29	211.8	0.954	—	$g = 0.946 + 0.047(1 - \exp(-(\lambda - 500.9)/556.3))$,
	700	1.10	189.5	0.967	—	[λ] in nm
Lung submucosa	400	38.8	263.1	0.911	—	IS, IAD; data from graphs (Qu et al. 1994); in
($n=15$)	500	4.03	241.0	0.923	—	the spectral range: $\mu_s = 6.036 \times 10^3\,\lambda^{-0.52}$;
	600	2.21	212.4	0.935	—	$g = 0.922 + 0.084(1 - \exp(-(\lambda - 488.2)/599.4))$,
	700	1.49	205.0	0.946	—	[λ] in nm
Lung cartilage	400	15.1	300.9	0.902	—	IS, IAD; data from graphs (Qu et al. 1994); in
($n=12$)	500	2.53	275.1	0.929	—	the spectral range: $\mu_s = 3.524 \times 10^3\,\lambda^{-0.409}$;
	600	1.13	255.0	0.948	—	$g = 0.929 + 0.072(1 - \exp(-(\lambda - 499.7)/309.1))$,
	700	0.87	246.3	0.965	—	[λ] in nm
Lung (rabbit)	632.8	2.8(0.2)	330(21)	0.904(0.012)	30.8(3.2)	DIS, IAD, Beek et al. (1997)
Lung (piglet)	632.8	2.0(0.1)	301(22)	0.933(0.003)	19.7(1.4)	DIS, IAD, Beek et al. (1997)
Lung (dog)	632.8	3.2(0.7)	230(5)	0.935(0.017)	15.4(4.3)	DIS, IAD, Beek et al. (1997)
Lung (piglet)	790	2.4(0.3)	263(18)	0.926(0.004)	20.0(1.7)	DIS, IAD, Beek et al. (1997)
Lung (piglet)	850	0.76(0.07)	278(21)	0.957(0.002)	10.9(0.7)	DIS, IAD, Beek et al. (1997)

Note: rms values are given in parentheses.

TABLE 5.11   Optical Properties of Myocardium Tissues Measured In Vitro and Ex Vivo

Tissue	λ, nm	μ_a, cm^{-1}	μ_s, cm^{-1}	g	μ_s', cm^{-1}	Remarks
Myocardium (dog)	630	2.0(0.2)	159(6)	0.854(0.015)	23.0(1.4)	DIS, IAD, Beek et al. (1997)
Myocardium (dog)	632.8	2.1(0.1)	191(16)	0.940(0.004)	11.3(1.2)	DIS, IAD, Beek et al. (1997)
Myocardium (dog)	790	0.98(0.2)	164(10)	0.943(0.004)	6.0(1.6)	DIS, IAD, Beek et al. (1997)
Bovine myocardium (native tissue) ($n=7$)	1000	2.40	86.65	0.937	—	DIS, IMC; data from graphs (Schwarzmaier et al. 1998); in the spectral range 1050–1450 nm: $\mu_s = 9.213 \times 10^3\,\lambda^{-0.673}$, [$\lambda$] in nm
	1100	2.32	82.92	0.939	—	
	1200	3.23	77.93	0.942	—	
	1300	3.19	73.38	0.938	—	
	1400	7.25	69.70	0.915	—	
	1500	10.17	68.64	0.876	—	
Bovine myocardium (laser coagulated tissue) ($n=7$)	1000	4.64	349.9	0.910	—	DIS, IMC; data from graphs (Schwarzmaier et al. 1998); in the spectral range 1050–1450 nm: $\mu_s = 3.906 \times 10^4\,\lambda^{-0.676}$, [$\lambda$] in nm
	1100	4.64	343.8	0.923	—	
	1200	5.84	324.0	0.922	—	
	1300	5.74	303.5	0.924	—	
	1400	11.04	294.5	0.915	—	
	1500	16.24	261.0	0.875	—	
Bovine myocardium (thermocoagulated tissue) ($n=7$)	1000	5.69	405.0	0.915	—	DIS, IMC; data from graphs (Schwarzmaier et al. 1998); in the spectral range 1050–1450 nm: $\mu_s = 5.827 \times 10^3\,\lambda^{-0.379}$, [$\lambda$] in nm
	1100	6.12	411.0	0.931	—	
	1200	8.37	394.5	0.934	—	
	1300	8.42	383.7	0.938	—	
	1400	13.90	373.0	0.937	—	
	1500	19.94	351.5	0.904	—	

Note: rms values are given in parentheses.

and NIR spectral ranges (Maitland et al. 1993, Ripley et al. 1999, Bashkatov et al. 2004, 2007, Wei et al. 2005) and summarized in Table 5.4.

5.8.5 Breast Tissue Optical Properties

Studies of optical breast imaging have usually focused on the clinical evaluation of specific source-detector systems. Clinically, optical imaging has been shown to be useful in distinguishing cystic from solid lesions and is particularly valuable in the diagnosis of hematoma. It has been able to detect some cancers which were not demonstrated by mammography and may, therefore, be a useful complementary procedure. However, optical method is still an experimental technique. Although it has the advantage of being a risk-free method of evaluating breast disease, current procedures are limited in their ability to detect small, deep lesions. The optimization of optical imaging requires a better understanding of the basic optical properties of breast tissues and the imaging process. The breast tissue optical properties have been measured with integrating sphere technique in the visible and NIR spectral ranges (Peters et al. 1990) and summarized in Table 5.5.

5.8.6 Cartilage

Knowledge of the optical and thermal properties of cartilage may aid for diagnostics and laser dosimetry determination in laser-assisted cartilage reshaping studies. Notwithstanding the

importance of the optical properties of cartilage, limited studies have been reported providing absorption and scattering coefficient data. Ebert et al. (1998) investigated optical properties of equine articular cartilage in the 300–850 nm spectral range. Although most cartilage in the body is composed of similar chemical structures such as water, collagen, and proteoglycans, these constituents are present in different proportions. Because of the varying composition, the optical properties of nasal septal cartilage are distinct from those of articular cartilage. Knowledge of optical properties of cartilage may be important for the development of noninvasive optical diagnostics to minimize nonspecific thermal damage in laser-assisted cartilage reshaping procedures. The tissue optical properties have been measured with integrating sphere technique in the visible and NIR spectral ranges (Beek et al. 1997, Ebert et al. 1998, Youn et al. 2000, Salomatina and Yaroslavsky 2008) and summarized in Table 5.6.

5.8.7 Liver

The liver is the most common manifestation site of distant metastases from colorectal carcinomas. In ~25% of the patients, liver metastases are concomitantly detected already at the time of primary diagnosis. Another 50% develop metachronous hepatic metastases, the liver frequently being the only site of metastasis. However, only a maximum of 30% of patients can be considered for surgical resection, which is currently the only established standard procedure for the treatment of liver metastases. For this reason, it is necessary to standardize other treatment concepts such as laser-induced thermotherapy (LITT). Precise knowledge about the spatial distribution of induced thermal tissue damage and the temperature distribution in the specific target organ as well as its dependency on the selected application parameters is of decisive importance for the safe application of LITT. Ideally, it should already be possible to plan the required application parameters in advance so that treatment can be precisely adjusted to the individual findings. The calculation of laser-induced thermal tissue reactions is a complex task in which the computation of laser light distribution in scattering and absorbing media is of primary importance. This requires knowledge about the optical parameters (absorption, scattering, anisotropy) of the target tissue, which may considerably differ depending on the tissue structure. Especially in LITT of liver metastases, it is necessary to determine these parameters not only in the healthy liver but also in metastatic tissue.

The tissue optical properties have been measured with integrating sphere technique in the visible and NIR spectral ranges (Parsa et al. 1989, Nilsson et al. 1995, Beek et al. 1997, Germer et al. 1998, Ritz et al. 2001) and summarized in Table 5.7.

5.8.8 Muscle

The optical properties of muscle have been measured with integrating sphere technique in the visible and NIR spectral ranges (Nilsson et al. 1995, Beek et al. 1997, Simpson et al. 1998) and summarized in Table 5.8.

5.8.9 Aorta

Aorta is a turbid tissue composed of interwoven elastin and collagen fibers, arranged in a trilayer structure of intima, media, and adventitia. Its appearance ranges from opaque white (porcine) to a pinkish-white in cadaveric samples. The tissue optical properties have been measured with integrating sphere technique in the visible and NIR spectral ranges (Cilesiz and Welch 1993, Chan et al. 1996b) and summarized in Table 5.9.

5.8.10 Lung Tissue

The lung tissue optical properties have been measured with integrating sphere technique in the visible and NIR spectral ranges (Qu et al. 1994, Beek et al. 1997) and summarized in Table 5.10.

5.8.11 Myocardium

The myocardium tissue optical properties have been measured with integrating sphere technique in the visible and NIR spectral ranges (Beek et al. 1997, Schwarzmaier et al. 1998) and summarized in Table 5.11.

5.9 Summary

We believe that this overview of tissue optical properties will give to users a possibility to predict optical properties of tissues under their interest and evaluate light distribution in the organ under examination or treatment. Authors tried to collect as complete as possible data on tissue optical properties and presented some of these data in the form of approximation formulas as a function of the wavelength to be easy to use these data. In spite of this and availability of other reviews (Cheong et al. 1990, Duck 1990, Cheong 1995, Müller and Roggan 1995, Roggan et al. 1995, Mobley and Vo-Dinh 2003, Tuchin 2007, 2009, Altshuler and Tuchin 2009), evidently, the data collection and measurements should be continued in order to have more complete and precise information about different tissues in norm and pathology, to recognize age-related, disease-related, and treatment-related changes of optical properties.

Laser photodynamic therapy and laser-induced interstitial thermal therapy (LITT) of deep tumors are the most promising techniques among the least invasive therapies of cancer. In this case, besides the knowledge of the optical properties of tumor tissue and the surrounding substances, the knowledge of the blood content and its optical properties is essential for therapy planning and for exact dosimetry. Data on blood optical properties can be found in the following papers, book chapters, and books: Tuchin (2007, 2009), Friebel et al. (2006, 2009), Meinke et al. (2007a,b), Roggan et al. (1994, 1995, 1999a), Yaroslavsky et al. (1996b, 1999b, 2002a), Hammer et al. (2001), Turcu et al. (2006), Mobley and Vo-Dinh (2003).

Acknowledgments

The research described in this chapter has been made possible by grants: 224014 Photonics4life-FP7-ICT-2007-2; Grant RUB1-2932-SR-08 CRDF; RF Ministry of Science and Education 2.1.1/4989 and 2.2.1.1/2950, Project 1.4.09 of Federal Agency of Education of RF; RFBR-08-02-92224-NNSF_a (RF-P.R. China); RFBR-09-02-90487 Ukr_a; RFBR-10-02-90039-Bel_a; the Special Program of RF "Scientific and Pedagogical Personnel of Innovative Russia," Governmental contracts 02.740.11.0484 and 02.740.11.0770.

References

Alexandrakis, G., Busch, D. R., Faris, G. W. et al. 2001. Determination of the optical properties of two-layer turbid media by use of a frequency-domain hybrid Monte Carlo diffusion model. *Appl. Opt.* 40: 3810–3821.

Altshuler, G. B. and Tuchin, V. V. 2009. Physics behind the light-based technology: Skin and hair follicle interactions with light. In *Cosmetic Applications of Laser & Light-Based Systems*, ed. G. Ahluwalia. Norwich, NY: William Andrew, Inc., pp. 49–109.

Anderson, R. R. and Parrish, J. A. 1981. The optics of human skin. *J. Invest. Dermatol.* 77: 13–19.

Barber, P. W. and Hill, S. C. 1990. *Light Scattering by Particles: Computational Methods*. Singapore: World Scientific.

Bargo, P. R., Prahl, S. A., Goodell, T. T. et al. 2005. In vivo determination of optical properties of normal and tumor tissue with while light reflectance and an empirical light transport model during endoscopy. *J. Biomed. Opt.* 10: 034018.

Bashkatov, A. N. 2002. Control of tissue optical properties by means of osmotically active immersion liquids, PhD thesis, Saratov State University, Saratov, Russia.

Bashkatov, A. N., Genina, E. A., Kochubey, V. I. et al. 2004. Optical properties of mucous membrane in the spectral range 350–2000 nm. *Opt. Spectrosc.* 97: 978–983.

Bashkatov, A. N., Genina, E. A., Kochubey, V. I. et al. 2005a. Optical properties of human skin, subcutaneous and mucous tissues in the wavelength range from 400 to 2000 nm. *J. Phys. D Appl. Phys.* 38: 2543–2555.

Bashkatov, A. N., Genina, E. A., Kochubey, V. I. et al. 2005b. Optical properties of the subcutaneous adipose tissue in the spectral range 400–2500 nm. *Opt. Spectrosc.* 99: 836–842.

Bashkatov, A. N., Genina, E. A., Kochubey, V. I. et al. 2006a. Estimate of the melanin content in human hairs by the inverse Monte-Carlo method using a system for digital image analysis. *Quantum Electron.* 36: 1111–1118.

Bashkatov, A. N., Genina, E. A., Kochubey, V. I. et al. 2006b. Optical properties of human cranial bone in the spectral range from 800 to 2000 nm. *Proc SPIE* 6163: 616310.

Bashkatov, A. N., Genina, E. A., Kochubey, V. I. et al. 2007. Optical properties of human stomach mucosa in the spectral range from 400 to 2000 nm: Prognosis for gastroenterology. *Med. Laser Appl.* 22: 95–104.

Bashkatov, A. N., Genina, E. A., Kochubey, V. I. et al. 2009. Optical clearing of human eye sclera. *Proc SPIE* 7163: 71631R.

Bashkatov, A. N., Genina, E. A., Kochubey, V. I. et al. 2010. Optical properties of human eye sclera in the spectral range 370–2500 nm. *Opt. Spectrosc.* 109: 1282–1290.

Beek, J. F., Blokland, P., Posthumus, P. et al. 1997. In vitro double-integrating-sphere optical properties of tissues between 630 and 1064 nm. *Phys. Med. Biol.* 42: 2255–2261.

Bevilacqua, F., Marquet, P., Depeursinge, C. et al. 1995. Determination of reduced scattering and absorption coefficients by a single charge-coupled-device array measurement, Part II: Measurements on biological tissues. *Opt. Eng.* 34: 2064–2069.

Bevilacqua, F., Piguet, D., Marquet, P. et al. 1999. In vivo local determination of tissue optical properties: applications to human brain. *Appl. Opt.* 38: 4939–4950.

Binzoni, T., Leung, T. S., Gandjbakhche, A. H. et al. 2006. The use of the Henyey–Greenstein phase function in Monte Carlo simulations in biomedical optics. *Phys. Med. Biol.* 51: N313–N322.

Bohren, C. F. and Huffman, D. R. 1983. *Absorption and Scattering of Light by Small Particles*. New York: Wiley.

Bruls, W. A. G. and van der Leun, J. C. 1984. Forward scattering properties of human epidermal layers. *Photochem. Photobiol.* 40: 231–242.

Chan, E., Menovsky, T., and Welch, A. J. 1996a. Effects of cryogenic grinding on soft-tissue optical properties. *Appl. Opt.* 35: 4526–4532.

Chan, E. K., Sorg, B., Protsenko, D. et al. 1996b. Effects of compression on soft tissue optical properties. *IEEE J. Sel. Top. Quantum Electron.* 2: 943–950.

Chandra, M., Vishwanath, K., Fichter, G. D. et al. 2006. Quantitative molecular sensing in biological tissues: An approach to non-invasive optical characterization. *Opt. Express* 14: 6157–6171.

Chandrasekhar, C. 1960. *Radiative Transfer*. Toronto, Ontario, Canada: Dover.

Chen, Y.-C., Ferracane, J. L., and Prahl, S. A. 2005. A pilot study of a simple photon migration model for predicting depth of cure in dental composite. *Dental Mater.* 21: 1075–1086.

Cheong, W. F. 1995. Summary of optical properties. In *Optical-Thermal Response of Laser-Irradiated Tissue*, eds. A. J. Welch and M. J. C. van Gemert. New York: Plenum Press, pp. 274–303.

Cheong, W.-F., Prahl, S. A., and Welch, A. J. 1990. A review of the optical properties of biological tissues. *IEEE J. Quantum Electron.* 26: 2166–2185.

Choukeife, J. E. and L'Huillier, J. P. 1999. Measurements of scattering effects within tissue-like media at two wavelengths of 632.8 nm and 680 nm. *Lasers Med. Sci.* 14: 286–296.

Cilesiz, I. F. and Welch, A. J. 1993. Light dosimetry: Effects of dehydration and thermal damage on the optical properties of the human aorta. *Appl. Opt.* 32: 477–487.

Comsa, D. C., Farrell, T. J., and Patterson, M. S. 2006. Quantification of bioluminescence images of point source objects using diffusion theory models. *Phys. Med. Biol.* 51: 3733–3746.

Dam, J. S., Dalgaard, T., Fabricius, P. E. et al. 2000. Multiple polynomial regression method for determination of biomedical optical properties from integrating sphere measurements. *Appl. Opt.* 39: 1202–1209.

Darling, C. L., Huynh, G. D., and Fried, D. 2006. Light scattering properties of natural and artificially demineralized dental enamel at 1310 nm. *J. Biomed. Opt.* 11: 034023.

de Vries, G., Beek, J. F., Lucassen, G. W. et al. 1999. The effect of light losses in double integrating spheres on optical properties estimation. *IEEE J. Sel. Top. Quant. Electron.* 5: 944–947.

Dimofte, A., Finlay, J. C., and Zhu, T. C. 2005. A method for determination of the absorption and scattering properties interstitially in turbid media. *Phys. Med. Biol.* 50: 2291–2311.

Donner, C. and Jensen, H. W. 2006. Rapid simulation of steady-state spatially resolved reflectance and transmittance profiles of multilayered turbid materials. *J. Opt. Soc. Am. A* 23: 1382–1390.

Doornbos, R. M. P., Lang, R., Aalders, M. C. et al. 1999. The determination of in vivo human tissue optical properties and absolute chromophore concentrations using spatially resolved steady-state diffuse reflectance spectroscopy. *Phys. Med. Biol.* 44: 967–981.

Drezek, R., Dunn, A., and Richards-Kortum, R. 1999. Light scattering from cells: Finite-difference time-domain simulations and goniometric measurements. *Appl. Opt.* 38: 3651–3661.

Du, Y., Hu, X. H., Cariveau, M. et al. 2001. Optical properties of porcine skin dermis between 900 nm and 1500 nm. *Phys. Med. Biol.* 46: 167–181.

Duck, F. A. 1990. *Physical Properties of Tissue: A Comprehensive Reference Book.* London, U.K.: Academic Press.

Ebert, D. W., Roberts, C., Farrar, S. K. et al. 1998. Articular cartilage optical properties in the spectral range 300–850 nm. *J. Biomed. Opt.* 3: 326–333.

Farrar, S. K., Roberts, C., Johnston, W. M. et al. 1999. Optical properties of human trabecular meshwork in the visible and near-infrared region. *Lasers Surg. Med.* 25: 348–362.

Farrell, T. J., Patterson, M. S., and Wilson, B. C. 1992. A diffusion theory model of spatially resolved, steady-state diffuse reflectance for the noninvasive determination of tissue optical properties in vivo. *Med. Phys.* 19: 879–888.

Fine, I., Loewinger, E., Weinreb, A. et al. 1985. Optical properties of the sclera. *Phys. Med. Biol.* 30: 565–571.

Firbank, M., Hiraoka, M., Essenpreis, M. et al. 1993. Measurement of the optical properties of the skull in the wavelength range 650–950 nm. *Phys. Med. Biol.* 38: 503–510.

Forster, F. K., Kienle, A., Michels, R. et al. 2006. Phase function measurements on nonspherical scatterers using a two-axis goniometer. *J. Biomed. Opt.* 11: 024018.

Friebel, M., Helfmann, J., Netz, U. et al. 2009. Influence of oxygen saturation on the optical scattering properties of human red blood cells in the spectral range 250 to 2000 nm. *J. Biomed. Opt.* 14: 034001.

Friebel, M., Roggan, A., Muller, G. et al. 2006. Determination of optical properties of human blood in the spectral range 250 to 1100 nm using Monte Carlo simulations with hematocrit-dependent effective scattering phase functions. *J. Biomed. Opt.* 11: 034021.

Fried, D., Glena, R. E., Featherstone, J. D. B. et al. 1995. Nature of light scattering in dental enamel and dentin at visible and near-infrared wavelengths. *Appl. Opt.* 34: 1278–1285.

Gebhart, S. C., Lin, W.-C., and Mahadevan-Jansen, A. 2006. In vitro determination of normal and neoplastic human brain tissue optical properties using inverse adding-doubling. *Phys. Med. Biol.* 51: 2011–2027.

Genina, E. A., Bashkatov, A. N., Kochubey, V. I. et al. 2005. Optical clearing of human dura mater. *Opt. Spectrosc.* 98: 470–476.

Germer, C.-T., Roggan, A., Ritz, J. P. et al. 1998. Optical properties of native and coagulated human liver tissue and liver metastases in the near infrared range. *Lasers Surg. Med.* 23: 194–203.

Ghosh, N., Mohanty, S. K., Majumder, S. K. et al. 2001. Measurement of optical transport properties of normal and malignant human breast tissue. *Appl. Opt.* 40: 176–184.

Graaff, R., Aarnoudse, J. G., Zijp, J. R. et al. 1992. Reduced light-scattering properties for mixtures of spherical particles: A simple approximation derived from Mie calculations. *Appl. Opt.* 31: 1370–1376.

Graaff, R., Dassel, A. C. M., Koelink, M. H. et al. 1993a. Optical properties of human dermis in vitro and in vivo. *Appl. Opt.* 32: 435–447.

Graaff, R., Koelink, M. H., de Mul, F. F. M. et al. 1993b. Condensed Monte Carlo simulations for the description of light transport. *Appl. Opt.* 32: 426–434.

Hammer, M., Roggan, A., Schweitzer, D. et al. 1995. Optical properties of ocular fundus tissues—An in vitro study using the double-integrating-sphere technique and inverse Monte Carlo simulation. *Phys. Med. Biol.* 40: 963–978.

Hammer, M., Yaroslavsky, A. N., and Schweitzer, D. 2001. A scattering phase function for blood with physiological haematoctit. *Phys. Med. Biol.* 46: N65–N69.

Hardy, J. D., Hammel, H. T., and Murgatroyd, D. 1956. Spectral transmittance and reflectance of excised human skin. *J. Appl. Physiol.* 9: 257–264.

Hayakawa, C. K., Hill, B. Y., You, J. S. et al. 2004. Use of the δ-P1 approximation for recovery of optical absorption, scattering, and asymmetry coefficients in turbid media. *Appl. Opt.* 43: 4677–4688.

Hayakawa, C. K., Spanier, J., Bevilacqua, F. et al. 2001. Perturbation Monte Carlo methods to solve inverse photon migration problems in heterogeneous tissues. *Opt. Lett.* 26: 1335–1337.

Hebert, M. and Becker, J.-M. 2008. Correspondence between continuous and discrete two-flux models for reflectance and transmittance of diffusing layers. *J. Opt. A Pure Appl. Opt.* 10: 035006.

Henyey, L. G. and Greenstein, J. L. 1941. Diffuse radiation in the galaxy. *Astrophys. J.* 93: 70–83.

Hourdakis, C. J. and Perris, A. 1995. A Monte Carlo estimation of tissue optical properties for use in laser dosimetry. *Phys. Med. Biol.* 40: 351–364.

Ishimaru, A. 1997. *Wave Propagation and Scattering in Random Media.* New York: IEEE Press.

Jacques, S. L. 1996. Origins of tissue optical properties in the UVA, visible and NIR regions. In *Advances in Optical Imaging and Photon Migration*, eds. R. R. Alfano and J. G. Fujimoto. Washington, DC: OSA TOPS 2: Optical Society of America, pp. 364–371.

Jacques, S. L., Alter, C. A., and Prahl, S. A. 1987. Angular dependence of the He-Ne laser light scattering by human dermis. *Lasers Life Sci.* 1: 309–333.

Jacques, S. L. and Wang, L. 1995. Monte Carlo modeling of light transport in tissue. In *Optical–Thermal Response of Laser-Irradiated Tissue*, eds. A. J. Welch, M. J. C. van Gemert. New York: Plenum Press, pp. 73–100.

Keijzer, M., Jacques, S. L., Prahl, S. A. et al. 1989. Light distribution in artery tissue: Monte Carlo simulation for finite-diameter laser beams. *Lasers Surg. Med.* 9: 148–154.

Khalil, O., Yeh, S.-J, Lowery, M. G. et al. 2003. Temperature modulation of the visible and near infrared absorption and scattering coefficients of human skin. *J. Biomed. Opt.* 8: 191–205.

Khlebtsov, N. G., Maksimova, I. L., Tuchin, V. V. et al. 2002. Introduction to light scattering by biological objects. In *Handbook of Optical Biomedical Diagnostics*, ed. PM107:V. V. Tuchin. Bellingham, WA: SPIE Press, pp. 31–167, Chap. 1.

Kienle, A. and Patterson, M. S. 1996. Determination of the optical properties of turbid media from a single Monte Carlo simulation. *Phys. Med. Biol.* 41: 2221–2227.

Kokhanovsky, A. and Hopkinson, I. 2008. Some analytical approximations to radiative transfer theory and their application for the analysis of reflectance data. *J. Opt. A Pure Appl. Opt.* 10: 035001.

Kotova, S. P., Mayorov, I. V., and Mayorova, A. M. 2007. Application of neutral networks for determining optical parameters of strongly scattering media from the intensity profile of backscattered radiation. *Quantum Electron.* 37: 22–26.

Kubelka, P. and Munk, F. 1931. Ein Beitrag zur Optik der Farbanstriche. *Z. Tech. Phys.* 12: 593–601.

Kumar, D., Srinivasan, R., and Singh, M. 2004. Optical characterization of mammalian tissues by laser reflectometry and Monte Carlo simulation. *Med. Eng. Phys.* 26: 363–369.

LaMuraglia, G. M., Prince, M. R., Nishioka, N. S. et al. 1990. Optical properties of human arterial thrombus, vascular grafts, and sutures: Implications for selective laser thrombus ablation. *IEEE J. Quantum Electron.* 26: 2200–2206.

Laufer, J., Simpson, C. R., Kohl, M. et al. 1998. Effect of temperature on the optical properties of ex vivo human dermis and subdermis. *Phys. Med. Biol.* 43: 2479–2489.

Lin, W.-C., Motamedi, M., and Welch, A. J. 1996. Dynamics of tissue optics during laser heating of turbid media. *Appl. Opt.* 35: 3413–3420.

Ma, X., Lu, J. Q., Din, H. G. et al. 2005. Bulk optical parameters of porcine skin dermis at eight wavelengths from 325 to 1557 nm. *Opt. Lett.* 30: 412–414.

Maitland, D. J., Walsh, J. T. Jr., and Prystowsky, J. B. 1993. Optical properties of human gallbladder tissue and bile. *Appl. Opt.* 32: 586–591.

Marchesini, R., Bertoni, A., Andreola, S. et al. 1989. Extinction and absorption coefficients and scattering phase functions of human tissues in vitro. *Appl. Opt.* 28: 2318–2324.

Marchesini, R., Clemente, C., Pignoli, E. et al. 1992. Optical properties of in vitro epidermis and their possible relationship with optical properties of in vivo skin. *J. Photochem. Photobiol. B* 16: 127–140.

Meinke, M., Muller, G., Helfmann, J. et al. 2007a. Empirical model functions to calculate hematocrit-dependent optical properties of human blood. *Appl. Opt.* 46: 1742–1753.

Meinke, M., Muller, G., Helfmann, J. et al. 2007b. Optical properties of platelets and blood plasma and their influence on the optical behavior of whole blood in the visible to near infrared wavelength range. *J. Biomed. Opt.* 12: 014024.

Mishchenko, M. I., Travis, L. D., and Lacis, A. A. 2002. *Scattering, Absorption, and Emission of Light by Small Particles.* Cambridge, U.K.: Cambridge University Press.

Mobley, J. and Vo-Dinh, T. 2003. Optical properties of tissues. In *Biomedical Photonics Handbook*, ed. T. Vo-Dinh. Boca Raton, FL: CRC Press, pp. 2-1–2-75.

Mueller, G. J. and Sliney, D. H. (eds.) 1989. *Dosimetry of Laser Radiation in Medicine and Biology*, vol. IS5. Bellingham, WA: SPIE Press.

Müller, G. and Roggan, A. (eds.) 1995. *Laser–Induced Interstitial Thermotherapy*, vol. PM25. Bellingham, WA: SPIE Press.

Nemati, B., Rylander III, H. G., and Welch, A. J. 1996. Optical properties of conjunctiva, sclera, and the ciliary body and their consequences for transscleral cyclophotocoagulation. *Appl. Opt.* 35: 3321–3327.

Niemz, H. 1996. *Laser–Tissue Interactions. Fundamentals and Applications.* Berlin, Germany: Springer Verlag.

Nilsson, A. M. K., Berg, R., and Andersson-Engels, S. 1995. Measurements of the optical properties of tissue in conjunction with photodynamic therapy. *Appl. Opt.* 34: 4609–4619.

Nilsson, A. M. K., Lucassen, G. W., Verkruysse, W. et al. 1997. Changes in optical properties of human whole blood in vitro due to slow heating. *Photochem. Photobiol.* 65: 366–373.

Nishidate, I., Aizu, Y., and Mishina, H. 2003. Estimation of absorbing components in a local layer embedded in the turbid media on the basis of visible to near-infrared (VIS-NIR) reflectance spectra. *Opt. Rev.* 10: 427–435.

Nishidate, I., Aizu, Y., and Mishina, H. 2004. Estimation of melanin and hemoglobin in skin tissue using multiple regression analysis aided by Monte Carlo simulation. *J. Biomed. Opt.* 9: 700–710.

Palmer, G. M. and Ramanujam, N. 2006a. Monte Carlo-based inverse model for calculating tissue optical properties. Part I: Theory and validation on synthetic phantoms. *Appl. Opt.* 45: 1062–1071.

Palmer, G. M. and Ramanujam, N. 2006b. Monte Carlo-based inverse model for calculating tissue optical properties. Part II: Application to breast cancer diagnosis. *Appl. Opt.* 45: 1072–1078.

Parsa, P., Jacques, S. L., and Nishioka, N. S. 1989. Optical properties of rat liver between 350–2200 nm. *Appl. Opt.* 28: 2325–2330.

Patwardhan, S. V., Dhawan, A. P., and Relue, P. A. 2005. Monte Carlo simulation of light–tissue interaction: Three-dimensional simulation for trans-illumination-based imaging of skin lesions. *IEEE Trans. Biomed. Eng.* 52: 1227–1236.

Peters, V. G., Wyman, D. R., Patterson, M. S. et al. 1990. Optical properties of normal and diseased human breast tissues in the visible and near infrared. *Phys. Med. Biol.* 35: 1317–1334.

Pfefer, T. J., Matchette, L. S., Bennett, C. L. et al. 2003. Reflectance-based determination of optical properties in highly attenuating tissue. *J. Biomed. Opt.* 8: 206–215.

Phylips-Invernizzi, B., Dupont, D., and Caze, C. 2001. Bibliographical review for reflectance of diffusing media. *Opt. Eng.* 40: 1082–1092.

Pickering, J. W., Bosman, S., Posthumus, P. et al. 1993a. Changes in the optical properties (at 632.8 nm) of slowly heated myocardium. *Appl. Opt.* 32: 367–371.

Pickering, J. W., Prahl, S. A., van Wieringen, N. et al. 1993b. Double-integrating-sphere system for measuring the optical properties of tissue. *Appl. Opt.* 32: 399–410.

Pop, C. V. L. and Neamtu, S. 2008. Aggregation of red blood cells in suspension: Study by light-scattering technique at small angles. *J. Biomed. Opt.* 13: 041308.

Prahl, S. A. 1988. Light transport in tissue, PhD Thesis, University of Texas at Austin, Austin, TX.

Prahl, S. A. 2010. Inverse adding-doubling for optical properties measurements. http://omlc.ogi.edu/software/iad/index.html

Prahl, S. A., van Gemert, M. J. C., and Welch, A. J. 1993. Determining the optical properties of turbid media by using the adding-doubling method. *Appl. Opt.* 32: 559–568.

Preece, S. J. and Claridge, E. 2002. Monte Carlo modelling of the spectral reflectance of the human eye. *Phys. Med. Biol.* 47: 2863–2877.

Prerana, Shenoy, M. R., and Pal, B. P. 2008. Method to determine the optical properties of turbid media. *Appl. Opt.* 47: 3216–3220.

Press, W. H., Tuekolsky, S. A., Vettering, W. T. et al. 1992. *Numerical Recipes in C: The Art of Scientific Computing.* Cambridge, U.K.: Cambridge University Press.

Qu, J., MacAulay, C., Lam, S. et al. 1994. Optical properties of normal and carcinomatous bronchial tissue. *Appl. Opt.* 33: 7397–7405.

Ragain, J. C. and Johnston, W. M. 2001. Accuracy of Kubelka–Munk reflectance theory applied to human dentin and enamel. *J. Dent. Res.* 80: 449–452.

Reif, R., A'Amar, O., and Bigio, I. J. 2007. Analytical model of light reflectance for extraction of the optical properties in small volumes of turbid media. *Appl. Opt.* 46: 7317–7328.

Reynolds, L. O. and McCormick, N. J. 1980. Approximate two-parameter phase function for light scattering. *J. Opt. Soc. Am.* 70: 1206–1212.

Ripley, P. M., Laufer, J. G., Gordon, A. D. et al. 1999. Near-infrared optical properties of ex vivo human uterus determined by the Monte Carlo inversion technique. *Phys. Med. Biol.* 44: 2451–2462.

Ritz, J.-P., Roggan, A., Isbert, C. et al. 2001. Optical properties of native and coagulated porcine liver tissue between 400 and 2400 nm. *Lasers Surg. Med.* 29: 205–212.

Roggan, A., Dörschel, K., Minet, O. et al. 1995. The optical properties of biological tissue in the near infrared wavelength range—Review and measurements. In *Laser–Induced Interstitial Thermotherapy, PM25,* eds. G. Müller and A. Roggan. Bellingham, WA: SPIE Press, pp. 10–44.

Roggan, A., Friebel, M., Dorschel, K. et al. 1999a. Optical properties of circulating human blood in the wavelength range 400–2500 nm. *J. Biomed. Opt.* 4: 36–46.

Roggan, A., Minet, O., Schroder, C. et al. 1994. The determination of optical tissue properties with double integrating sphere technique and Monte Carlo simulations. *Proc SPIE* 2100: 42–56.

Roggan, A., Schadel, D., Netz, U. et al. 1999b. The effect of preparation technique on the optical parameters of biological tissue. *Appl. Phys. B Lasers Opt.* 69: 445–453.

Salomatina, E., Jiang, B., Novak, J. et al. 2006. Optical properties of normal and cancerous human skin in the visible and near-infrared spectral range. *J. Biomed. Opt.* 11: 064026.

Salomatina, E. and Yaroslavsky, A. N. 2008. Evaluation of the in vivo and ex vivo optical properties in a mouse ear model. *Phys. Med. Biol.* 53: 2797–2807.

Sardar, D. K. and Levy, L. B. 1998. Optical properties of whole blood. *Lasers Med. Sci.* 13: 106–111.

Sardar, D. K., Mayo, M. L., and Glickman, R. D. 2001. Optical characterization of melanin. *J. Biomed. Opt.* 6: 404–411.

Sardar, D. K., Salinas, F. S., Perez, J. J. et al. 2004. Optical characterization of bovine retinal tissues. *J. Biomed. Opt.* 9: 624–631.

Sardar, D. K., Swanland, G.-Y., Yow, R. M. et al. 2007. Optical properties of ocular tissues in the near infrared region. *Lasers Med. Sci.* 22: 46–52.

Sardar, D. K., Yow, R. M., Tsin, A. T. C. et al. 2005. Optical scattering, absorption, and polarization of healthy and neovascularized human retinal tissues. *J. Biomed. Opt.* 10: 051501.

Schmitt, J. M. and Kumar, G. 1998. Optical scattering properties of soft tissue: A discrete particle model. *Appl. Opt.* 37: 2788–2797.

Schwarzmaier, H.-J., Yaroslavsky, A. N., Terenji, A. et al. 1998. Changes in the optical properties of laser coagulated and thermally coagulated bovine myocardium. *Proc SPIE* 3254: 361–365.

Seo, I., You, J. S., Hayakawa, C. K. et al. 2007. Perturbation and differential Monte Carlo methods for measurement of optical properties in a layered epithelial tissue model. *J. Biomed. Opt.* 12: 014030.

Simpson, C. R., Kohl, M., Essenpreis, M. et al. 1998. Near-infrared optical properties of ex vivo human skin and subcutaneous tissues measured using the Monte Carlo inversion technique. *Phys. Med. Biol.* 43: 2465–2478.

Spitzer, D. and Ten Bosch, J. J. 1975. The absorption and scattering of light in bovine and human dental enamel. *Calc. Tiss. Res.* 17: 129–137.

Steinke, J. M. and Shepherd, A. P. 1988. Comparison of Mie theory and the light scattering of red blood cells. *Appl. Opt.* 27: 4027–4033.

Thueler, F., Charvet, I., Bevilacqua, F. et al. 2003. In vivo endoscopic tissue diagnostics based on spectroscopic absorption, scattering, and phase function properties. *J. Biomed. Opt.* 8: 495–503.

Treweek, S. P. and Barbenel, J. C. 1996. Direct measurement of the optical properties of human breast skin. *Med. Biol. Eng. Comput.* 34: 285–289.

Troy, T. L. and Thennadil, S. N. 2001. Optical properties of human skin in the near infrared wavelength range of 1000 to 2200 nm. *J. Biomed. Opt.* 6: 167–176.

Tuchin, V. V. 1997. Light scattering study of tissues. *Phys. Usp.* 40: 495–515.

Tuchin, V. V. (ed.) 2002. *Handbook of Optical Biomedical Diagnostics PM107.* Bellingham, WA: SPIE Press.

Tuchin, V. V. 2007. *Tissue Optics: Light Scattering Methods and Instruments for Medical Diagnosis.* Bellingham, WA: SPIE Press.

Tuchin, V. V. 2009. Optical spectroscopy of biological materials. In *Encyclopedia of Applied Spectroscopy*, ed. D. L. Andrews. Weinheim: Wiley-VCH Verlag GmbH & Co. KGaA, pp. 555–626.

Turcu, I., Pop, C. V. L., and Neamtu, S. 2006. High-resolution angle-resolved measurements of light scattered at small angles by red blood cells in suspension. *Appl. Opt.* 45: 1964–1971.

Ugryumova, N., Matcher, S. J., and Attenburrow, D. P. 2004. Measurement of bone mineral density via light scattering. *Phys. Med. Biol.* 49: 469–483.

van de Hulst, H. C., 1980. *Multiple Light Scattering: Tables, Formulas and Applications.* New York: Academic Press.

van der Zee, P. 1993. Methods for measuring the optical properties of tissue samples in the visible and near infrared wavelength range. In *Medical Optical Tomography: Functional Imaging and Monitoring*, eds. G. Mueller, B. Chance, R. Alfano et al., Vol. 11. Bellingham, WA: SPIE Press, Institute Series, pp. 166–192.

van der Zee, P., Essenpreis, M., and Delpy, D. T. 1993. Optical properties of brain tissue. *Proc SPIE* 1888: 454–465.

van Gemert, M. J. C., Jacques, S. L., Sterenborg, H. J. C. M. et al. 1989. Skin optics. *IEEE Trans. Biomed. Eng.* 36: 1146–1154.

Vargas, G., Chan, E. K., Barton, J. K. et al. 1999. Use of an agent to reduce scattering in skin. *Lasers Surg. Med.* 24: 133–141.

Vo-Dinh, T. (ed.) 2003. *Biomedical Photonics Handbook.* Boca Raton, FL: CRC Press.

Wang, R. K. 2000. Modelling optical properties of soft tissue by fractal distribution of scatterers. *J. Modern Opt.* 47: 103–120.

Wang, L., Jacques, S. L., and Zheng, L. 1995. MCML—Monte Carlo modeling of light transport in multi-layered tissues. *Comput. Methods Prog. Biomed.* 47: 131–146.

Wei, H.-J., Xing, D., Lu, J.-J. et al. 2005. Determination of optical properties of normal and adenomatous human colon tissues in vitro using integrating sphere techniques. *World J. Gastroenterol.* 11: 2413–2419.

Wei, H.-J., Xing, D., Wu, G.-Y. et al. 2003. Optical properties of human normal small intestine tissue determined by Kubelka–Munk method in vitro. *World J. Gastroenterol.* 9: 2068–2072.

Welch, A. J. and van Gemert, M. C. J. (eds.) 1992. *Tissue Optics.* New York: Academic Press.

Yang, Y., Celmer, E. J., Koutcher, J. A. et al. 2002. DNA and protein changes caused by disease in human breast tissues probed by the Kubelka–Munk spectral function. *Photochem. Photobiol.* 75: 627–632.

Yang, L. and Kruse, B. 2004. Revised Kubelka–Munk theory. I. Theory and application. *J. Opt. Soc. Am. A* 21: 1933–1941.

Yang, L., Kruse, B., and Miklavcic, S. J. 2004. Revised Kubelka–Munk theory. II. Unified framework for homogeneous and inhomogeneous optical media. *J. Opt. Soc. Am. A* 21: 1942–1952.

Yang, L. and Miklavcic, S. J. 2005. Revised Kubelka–Munk theory. III. A general theory of light propagation in scattering and absorptive media. *J. Opt. Soc. Am. A* 22: 1866–1873.

Yang, L., Miklavcic, S. J., and Kruse, B. 2007. Qualifying the arguments used in the derivation of the revised Kubelka–Munk theory: Reply. *J. Opt. Soc. Am. A* 24: 557–560.

Yaroslavsky, A. N., Priezzhev, A. V., Rodriguez, J. et al. 2002a. Optics of blood. In *Handbook of Optical Biomedical Diagnostics*, vol. PM107, ed. V. V. Tuchin. Bellingham, WA: SPIE Press, pp. 169–216, Chap. 2.

Yaroslavsky, A. N., Schulze, P. C., Yaroslavsky, I. V. et al. 2002b. Optical properties of selected native and coagulated human brain tissues in vitro in the visible and near infrared spectral range. *Phys. Med. Biol.* 47: 2059–2073.

Yaroslavsky, I. V. and Tuchin, V. V. 1992. Light propagation in multilayer scattering media. Modeling by the Monte Carlo method. *Opt. Spectrosc.* 72: 505–509.

Yaroslavskaya, A. N., Utz, S. R., Tatarintsev, S. N. et al. 1994. Angular scattering properties of human epidermal layers. *Proc SPIE* 2100: 38–41.

Yaroslavsky, A. N., Vervoorts, A., Priezzhev, A. V. et al. 1999a. Can tumor cell suspension serve as an optical model of tumor tissue in situ? *Proc SPIE* 3565: 165–173.

Yaroslavsky, I. V., Yaroslavsky, A. N., Goldbach, T. et al. 1996a. Inverse hybrid technique for determining the optical properties of turbid media from integrating-sphere measurements. *Appl. Opt.* 35: 6797–6809.

Yaroslavsky, A. N. Yaroslavsky, I. V., Goldbach, T. et al. 1996b. Optical properties of blood in the near-infrared spectral range. *Proc SPIE* 2678: 314–324.

Yaroslavsky, A. N., Yaroslavsky, I. V., Goldbach, T. et al. 1999b. Influence of the scattering phase function approximation on the optical properties of blood determined from the integrating sphere measurements. *J. Biomed. Opt.* 4: 47–53.

Yaroslavsky, A. N., Yaroslavsky, I. V., and Schwarzmaier, H.-J. 1998. Small-angle approximation to determine radiance distribution of a finite beam propagating through turbid medium. *Proc SPIE* 3195: 110–120.

Yoon, G., Welch, A. J., Motamedi, M. et al. 1987. Development and application of three-dimensional light distribution model for laser irradiated tissue. *IEEE J. Quantum Electron.* 23: 1721–1733.

Youn, J.-I., Telenkov, S. A., Kim, E. et al. 2000. Optical and thermal properties of nasal septal cartilage. *Lasers Surg. Med.* 27: 119–128.

Zhang, R., Verkruysse, W., Choi, B. et al. 2005. Determination of human skin optical properties from spectrophotometric measurements based on optimization by genetic algorithms. *J. Biomed. Opt.* 10: 024030.

Zhu, D., Lu, W., Zeng, S. et al. 2007. Effect of losses of sample between two integrating spheres on optical properties estimation. *J. Biomed. Opt.* 12: 064004.

Zhu, C., Palmer, G. M., Breslin, T. M. et al. 2006. Diagnosis of breast cancer using diffuse reflectance spectroscopy: Comparison of a Monte Carlo versus partial least squares analysis based feature extraction technique. *Lasers Surg. Med.* 38: 714–724.

Zijp, J. R. and ten Bosch, J. J. 1998. Optical properties of bovine muscle tissue in vitro; A comparison of methods. *Phys. Med. Biol.* 43: 3065–3081.

II

Spectroscopy and Spectral Imaging

6 **Reflectance Spectroscopy** *Sasha McGee, Jelena Mirkovic, and Michael Feld* 103
Introduction • Background • Current Biomedical Applications of Reflectance Spectroscopy • Critical
Discussion • Summary • Future Perspective • References

7 **Multi/Hyper-Spectral Imaging** *Costas Balas, Christos Pappas, and George Epitropou* 131
Introduction • Color vs. Spectral Imaging • SI Camera Hardware Configurations and Calibration • Scanning SI
Systems Based on Electronically Tunable Filters • Single Exposure or Instantaneous Spectral Imagers • Spectral Data
Classification, Unmixing, and Visualization • Conclusion and Future Perspectives • References

8 **Light Scattering Spectroscopy** *Le Qiu, Irving Itzkan, and Lev T. Perelman* 165
Introduction • Basic Principles of Light Absorption and Scattering • Light Scattering Spectroscopy • Confocal Light
Absorption and Scattering Spectroscopic Microscopy • Conclusion • Acknowledgment • References

9 **Broadband Diffuse Optical Spectroscopic Imaging** *Bruce J. Tromberg, Albert E. Cerussi, So-Hyun Chung,
Wendy Tanamai, and Amanda Durkin* 181
Introduction • Background • Presentation of State of the Art • Critical Discussion • Summary • Future
Perspective • Acknowledgments • References

10 **Near-Infrared Diffuse Correlation Spectroscopy for Assessment of Tissue Blood Flow** *Guoqiang Yu,
Turgut Durduran, Chao Zhou, Ran Cheng, and Arjun G. Yodh* 195
Introduction • Near-Infrared DCS Technology Development • Fundamentals of Diffuse
Correlation Spectroscopy • DCS Technology • Validation Work • In Vivo Applications of
DCS • Summary • Acknowledgments • Grant Acknowledgments • References

11 **Fluorescence Spectroscopy** *Darren Roblyer, Richard A. Schwarz, and Rebecca Rae Richards-Kortum* 217
Introduction • Cancer and Endogenous Fluorescence • Fluorophore Localization • Instrumentation • Empirical
Analysis Methods • Model-Based Analysis Methods • Imaging Spectroscopy • Discussion • References

12 **Raman, SERS, and FTIR Spectroscopy** *Andrew J. Berger* 233
Introduction • Definitions and Fundamentals • Comparison to Other Spectroscopies • Equipment for Acquiring
Raman Spectra • Mathematical Processing of Vibrational Spectral Data • State of the Art: Raman Biomedical
Applications • Enhanced Raman Scattering • Mid-IR Spectroscopy • References

6

Reflectance Spectroscopy

Sasha McGee
Massachusetts Institute of Technology

Jelena Mirkovic
Massachusetts Institute of Technology

Michael Feld
Massachusetts Institute of Technology

6.1 Introduction 103
6.2 Background...103
  Elastic Scattering • Absorption • Instrumentation • Modeling and Data Analysis
6.3 Current Biomedical Applications of Reflectance Spectroscopy......................................111
  Tissue Diagnostics • Tissue Therapeutics
6.4 Critical Discussion 123
  Discriminating Disease in the Context of Normal Tissue Variations • Developing
  Clinically Relevant Spectroscopic Tools • Simplification of the Instrumentation and
  Analysis Procedures • Progressing toward Large, Prospective Multisite Studies
6.5 Summary.. 125
6.6 Future Perspective 125
References.. 125

6.1 Introduction

In this chapter, we present an overview of reflectance spectroscopy (RS), a technique which examines light that has undergone a combination of elastic scattering and absorption. The reflectance signal measured from epithelial tissue is determined by the structural and biochemical properties of the tissue; therefore, analyzing changes in optical properties can be used to noninvasively probe the tissue microenvironment. The first part of this chapter describes the physics, instrumentation, and modeling of the scattering and absorption processes. In the latter part of the chapter, we discuss the use of this technique in two areas: tissue diagnostics and tissue therapeutics.

6.2 Background

6.2.1 Elastic Scattering

Elastic scattering occurs when the path of a light ray in a medium is redirected as a result of inhomogeneities in the refractive index (RI), and subsequently reemitted without a change in the incident energy. The extent to which light is scattered is a function of the density of particles, the sizes of the particles relative to the wavelength of the incident light, and the ratio of the refractive indices (RIs) of the particles relative to that of the surrounding medium. Upon entering a medium, light can be reemitted after undergoing only a single scattering event or after multiple scattering events. Multiple scattering within the tissue is responsible for the turbid appearance of biological tissue. If the incident light undergoes multiple scattering events in the medium to the extent that it becomes randomized in direction, this is known as diffuse reflectance. Thus, the diffuse reflectance signal generally contains information about scattering and absorbing components deeper within the tissue as compared to singly scattered light. Typically, the measured reflectance signal represents light that has sampled a variety of sampling depths within the tissue, and is therefore an average measure of the properties over a certain volume of tissue. Significant scatterers present in tissue include collagen, keratin, nuclei, and mitochondria; however, other smaller subcellular components (e.g., lysosomes, membranes) also scatter light (Saidi et al. 1995, Mourant et al. 1998, 2007 Collier et al. 2005).

6.2.2 Absorption

In the process of absorption, the incident light excites electrons in molecules and atoms to higher energy levels by converting the light's energy into internal energy. The energy from the incident light is subsequently converted into heat as the electrons return to the ground energy level. The energy levels exist as discrete levels and can only be excited by photons with a frequency that corresponds to the difference between the initial and final energy level. The structure of the molecule determines which incident frequencies will be absorbed. When tissue is irradiated, the process of absorption attenuates the light which returns to the surface to be collected.

Table 6.1 lists the major tissue absorbers in the ultraviolet (UV) and visible regions of the electromagnetic (EM) spectrum (Richards-Kortum and SevickMuraca. 1996, Du et al. 1998, Prahl 2008, Vo-Dinh 2003). Nucleic acids and amino acids are the major absorbers in the UV region of the spectrum. Of all the amino acids, tryptophan exhibits the most intense absorption.

TABLE 6.1 Summary of the Major Endogenous Tissue Absorbers in the Ultraviolet and Visible Regions of the Spectrum and Their Absorption Maxima or Absorption Range

Absorber	Absorption Peak(s)/Range [nm]
Nucleic acids: DNA, RNA	258
Amino acid: Tyrosine	275
Amino acid: Tryptophan	280
Amino acid: Phenylalanine	260
Oxyhemoglobin	415, 542, 576
Deoxyhemoglobin	433, 556
β-carotene	<300 nm, ~450
Melanin	400–700

Light is readily absorbed by tissue in the UV region; therefore, it can only penetrate the first few cell layers. Hemoglobin is the principal tissue absorber in the visible region of the spectrum and strongly absorbs blue light. The removal of the blue component from the incident light results in its characteristic red color. Melanin and β-carotene are additional absorbers in the visible region. In the wavelength region from 600 to 1000 nm, known as the "therapeutic window," scattering predominates over absorption, and light can penetrate as deep as ~10 cm before being collected (Richards-Kortum and SevickMuraca 1996). The extinction spectra of several tissue absorbers in the UV and visible regions of the spectrum are shown in Figure 6.1. Water is another major tissue absorber (not shown), particularly in the near-infrared (NIR) region of the spectrum, and has an absorption peak at 970 nm (Prahl 2009).

6.2.3 Instrumentation

A typical setup for measuring reflectance spectra (Figure 6.2) consists of a broadband light source (e.g., xenon or tungsten lamp), a flexible optical fiber probe to convey the excitation light to the tissue and collect the reemitted light, a spectrograph to disperse the collected light, and a charge-coupled device (CCD) to detect the measured signal. A computer is used to control data acquisition and store the measured signal.

In addition to the relatively inexpensive and simple instrumentation requirements, another advantage of RS is that it can be performed very rapidly. Although the optical properties of the tissue are independent of the instrumentation, the measured scattering and absorption signal are highly dependent on the instrumentation. Therefore, by including additional components or introducing alterations to this basic setup, the device can be optimized for a specific application.

6.2.3.1 Point-Probe Instruments

An optical fiber probe is typically used to deliver the excitation light and collect the returned light when RS is used in biomedical applications. As a result, the measured signal is greatly impacted by the properties of the illumination and collection

fibers, including the numerical aperture (NA), diameter of the fibers, angle, and spacing (source–detector separation) between the illumination and collection fibers (Bargo et al. 2003a,b, 2005, Utzinger and Richards-Kortum 2003, Arifler et al. 2005). A number of probe designs have been explored in order to optimize the collection of light returning from a specific depth within the tissue or differentiate the signal returned from various layers (Amelink et al. 2003, Hattery et al. 2004, Nieman et al. 2004, Skala et al. 2004, Arifler et al. 2005, Liu and Ramanujam 2006, Wang et al. 2007, Nieman et al. 2009). For epithelial cancers, the reflectance signal measured from the epithelium (the most superficial layer of the tissue) is frequently targeted since these cancers originate within this layer as opposed to the deeper stroma. By isolating light emitted from the structures/layers of interest (e.g., epithelial layer, vasculature) within the tissue from other extraneous signals, the sensitivity of the spectral signal to disease-related changes may be enhanced. Knowledge of the origin of the scattering and absorption signal also facilitates accurate modeling and interpretation of the reflectance data. We briefly highlight a few approaches investigators have developed to optimize the probe design.

Figure 6.3 shows a diagram depicting the sampling depth of the collection fiber as a function of source–detector separation (distance between the illumination and collection fibers).

As the distance between the illumination fiber and collection fibers increases, the measured reflectance predominantly reflects light rays that have sampled deeper layers in the medium, whereas for small source–detector separations, the light is redirected into the collection fiber after only a few scattering events and thus only the more superficial layers of the medium are probed. Because of the localization of blood vessels in the deeper stromal layer, the sensitivity of the reflectance measurements to hemoglobin absorption increases with increased source–detector separations.

Schwarz et al. have developed a ball lens-coupled optical fiber probe that enables depth-sensitive measurements of epithelial tissue by varying the source–detector distances (Schwarz et al. 2005). A clinical system was developed that employed a multifiber probe design consisting of channels with different depth response characteristics ("shallow," "medium," and "deep") to selectively target the epithelium or stroma (Schwarz et al. 2008). Monte-Carlo modeling and measurements in water were used to estimate the probe depth response. The shallow channel primarily measures light that has sampled the first 300–500 μm (mostly epithelium). The medium channel collects light that has sampled a range of depths (up to 1 mm) that includes the epithelium and stroma, the proportion of which depends on the thickness of the epithelium. The deep channel largely measures the properties of the stroma. Zhu et al. developed an optical fiber probe consisting of 19 illumination fibers surrounded by concentric rings of collection fibers at three different source–detector distances to sample light returning from different depths and potentially enhance their discrimination of malignant and nonmalignant tissues (Zhu et al. 2005). Figure 6.4 shows a schematic of the probe design.

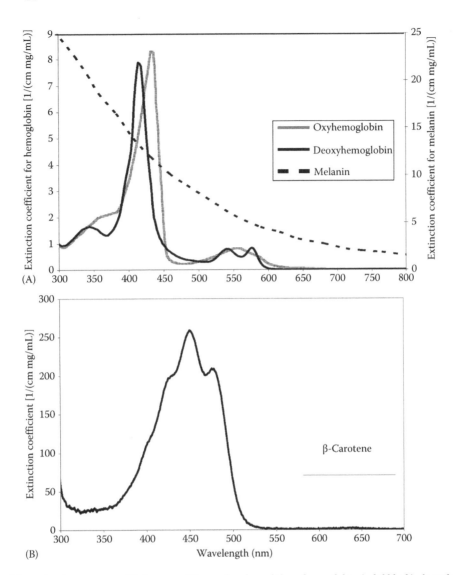

FIGURE 6.1 A plot of the extinction spectra, ε(λ), for several tissue absorbers: (A) oxyhemoglobin (solid black), deoxyhemoglobin (shaded gray line), and melanin (dashed black line); (B) β-carotene.

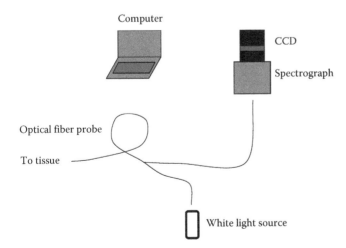

FIGURE 6.2 A schematic of a typical setup for acquiring reflectance spectra.

The distance between the center of the illumination core and that of the three rings of collection fibers is 735, 980, and 1225 μm, respectively. The system measures spectra at all three source–detector separations simultaneously. From their measurements, they could evaluate the reflectance spectra recorded at each of the three source–detector separations, as well as the diffuse reflectance intensity integrated over the entire spectrum as a function of the source–detector separation. To optimize the optical fiber probe for tissue diagnostic applications, Amelink et al. developed a fiber-optic technique that selectively interrogates light returning from the most superficial layers of the tissue known as differential path-length spectroscopy (Amelink et al. 2004a,b). The setup consists of two fibers: a larger diameter dual delivery and collection (dc) fiber and a second smaller diameter optical fiber, which exclusively collects the returned light. The fibers are in close contact such that the distance between the two fibers is minimized. The differential

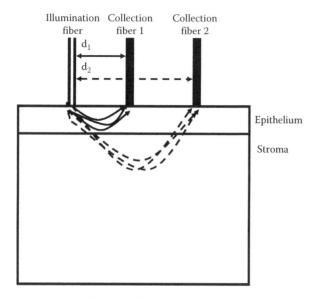

FIGURE 6.3 A diagram illustrating the impact of the distance between the illumination and collection fibers (source–detector separation) on the sampling depth of the measured reflectance signal.

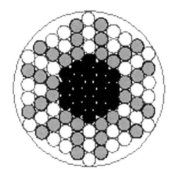

FIGURE 6.4 The common end of the fiber-optic probe that comes in contact with the tissue (the black circles correspond to the illumination fibers, the gray circles correspond to the collections fibers that form the three collection rings, and the white circles correspond to dead fibers for bundle packing. (From Zhu, C. et al., *J. Biomed. Optics*, 10(2), 024032. With permission.)

reflectance signal is obtained by subtracting the signal collected from the fiber used solely for collection from the dc fiber. The light collected by the collection fiber has traveled on average longer distances through the medium (more multiple-scattering) as compared to the light collected in the dc fiber (largely backscattering). Because of the small separation between the fibers, the contribution of multiply scattered light in the dc fiber can be approximated by that sampled by the collection fiber, thus the differential reflectance signal mostly represents light that has been singly scattered close to the probe tip.

Nieman et al. developed oblique polarized RS, a technique in which the optical fiber probe includes polarizers and an angled illumination–collection design (Nieman et al. 2004, 2008). The three optical fibers are positioned to form a single line. Polarizing film was placed at the tip of the central illumination fiber and the two identical collection fibers on either side. The polarization state of the collection fibers are set either to a parallel or perpendicular polarization state compared to the illumination polarization state. This design permits the collection of the parallel and perpendicular signal, which can be processed to yield information about scatterer size (e.g., epithelial nuclei) and size distribution. The two collection fibers are additionally angled at 37° with respect to the delivery fiber. This geometry maximizes the overlap between the acceptance angle of the collection fiber with that of the illumination fiber, thus enhancing the collection of light from the superficial tissue layers. In order to collect diffuse reflectance spectra from the colon wall during colonoscopies, Wang et al. developed an optical fiber probe with a side-viewing geometry in which the fiber tip was cut to 45° and polished to bend the illumination light or returned light at 90° (Wang et al. 2009). Figure 6.5 shows a photograph and schematic of the probe. The optical fiber probe consists of a single illumination fiber and four collection fibers arranged side-by-side spaced at four different source–detector separation distances (0.6, 1.5, 2.5, and 4 mm). The diameter of the probe is small enough such that the probe can be advanced through the accessory channel of a colonoscope and placed in contact with the colon tissue during measurements. The use of angled delivery and/or collection fibers is common in many studies in order to minimize the collection of specularly reflected light (Tunnell et al. 2003, Amelink et al. 2004a, van Veen et al. 2005).

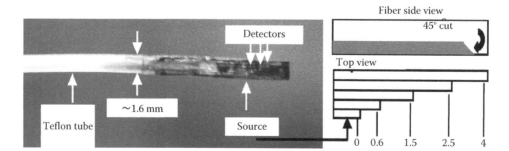

FIGURE 6.5 A photograph and schematic diagram of the endoscopic fiber-optic probe. (From Wang, H.W. et al., *Opt. Express*, 17(4): 2805. With permission.)

6.2.3.2 Imaging Instruments

The point-probe-based devices described earlier provide detailed information about a small area of tissue, but are not suitable for evaluating a large surface area. Several investigators have developed reflectance imaging instruments to enable wide-area tissue surveillance (Zeng et al. 2004, Schomacker et al. 2006, Gebhart et al. 2007, Lo et al. 2009). Imaging offers a distinct advantage over probe systems in that measurements can be performed in a noncontact manner, thus eliminating errors introduced into the reflectance measurements due to variations in probe pressure and angle during contact (Reif et al. 2008, Ti and Lin 2008).

Schomacker et al. developed an optical detection system (ODS) to evaluate cervical tissue, which performs reflectance imaging in combination with fluorescence (Schomacker et al. 2006). The ODS system consists of three major components: (1) a console to house the light sources, coupling optics, electronics, and spectrometer, (2) an optical probe head that operates on an articulating arm and supports the cables, and (3) a single-use sheath to cover the optical probe. The entire system is computer-controlled. Following a measurement, the system displays a video image of the cervix with areas of suspected high-grade disease indicated using a false-color overlay. Spectroscopic measurements are acquired using a noncontact illumination probe. Two xenon flash lamps serve as the light sources for the reflectance measurements and flood illuminate a broad (3.2 cm) area of tissue. A complete optical scan of the cervix covers a 25 mm diameter circular area and probes 499 sites along the tissue surface in 12 s. The system also includes a color video camera to visualize the cervix and capture images during measurements

to track motion and to focus the probe. The system was designed to identify high-grade lesions ≥2 mm in size.

Yu et al. developed a noncontact, quantitative spectroscopic imaging (QSI) system (Yu et al. 2008). The instrument images a "virtual" probe (~1 mm diameter spot size) at the tissue surface that is then raster scanned to interrogate a wide tissue area (~4 cm^2). The small area of tissue sampled by the virtual probe is similar to that of the contact probe used in previous studies (Georgakoudi et al. 2002, Muller et al. 2003, McGee et al. 2008, Mirkovic et al. 2009). Because the virtual probe has a fixed illumination-collection geometry and defined spot size, it provides a scale parameter with which to extract $\mu'_s(\lambda)$ and $\mu_a(\lambda)$ for each of the areas sampled by the virtual probe. This method also allowed the investigators to directly transfer the data analysis procedures and quantitative parameters extracted in the point-probe studies to the imaging studies.

Figure 6.6a shows a schematic depicting the components of the optical head of the QSI system. During measurements, the physician maneuvers the optical head on an articulating arm. The bulkier components, including the computer, spectrograph, CCD, and white light source are housed in a mobile cart. By modeling the measured reflectance spectra, the investigators extracted quantitative parameter maps. Figure 6.6b shows an example of a map of the A parameter, which is related to the magnitude of scattering.

A comprehensive discussion of multi-/hyper-spectral imaging devices that have been developed to measure tissue optical properties is presented in Chapters 7 through 9; therefore, in the subsequent sections, we focus our review primarily on studies conducted using point-probe devices.

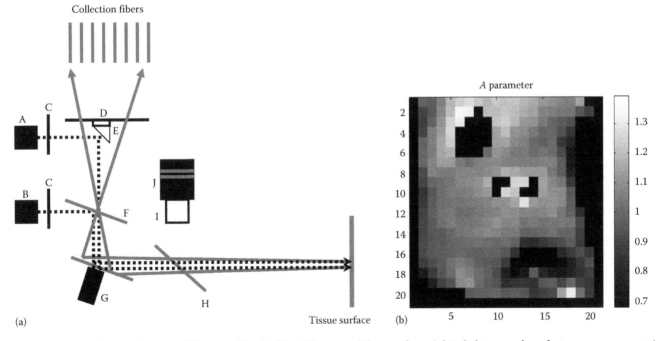

FIGURE 6.6 (a) Schematic diagram of the optical head of the QSI system. (A) xenon lamp (white light source for reflectance measurements), (B) nitrogen laser (source for fluorescence measurements), (C) lens, (D) aperture, (E) rod mirror, (F) dichroic mirror, (G) 2D scanning mirror, (H) 90T/10R beamsplitter, (I) zoom lens, (J) color video CCD. (b) Example of a map for the *A* parameter (related to the magnitude of the scattering) based on a measurement of cervical tissue. The 21 × 21 pixels correspond to the 2.1 cm × 2.1 cm QSI scan area.

6.2.4 Modeling and Data Analysis

6.2.4.1 Introduction

In this section we briefly discuss the models that are commonly used to describe the propagation of light in tissue and the measured reflectance signal. The models are described in more detail in Chapter 16. It is important to note that the selection of a model should take into account both the scattering and absorption properties of the tissue under study, as well as the design of the instrument used to measure the reflectance (e.g., optical fiber probe geometry). Furthermore, many of these models make simplifying assumptions, such as neglecting the multilayer tissue architecture, which may be acceptable for some but not all tissue types. A major challenge in evaluating reflectance spectra is the reliable extraction of both the scattering and absorption coefficients independently given the coupling of these processes in determining the intensity of the reemitted light.

6.2.4.2 Tissue Scattering and Absorption

There are two important wavelength-dependent optical parameters used to characterize the scattering and absorption properties of a medium, respectively: the scattering coefficient $\mu_s(\lambda)$ and absorption coefficient $\mu_a(\lambda)$. The scattering of light in tissue is not isotropic (an equal probability of scattering in all directions), but strongly forward-directed. Therefore, the scattering in tissue is most often described using the reduced scattering coefficient, $\mu_s'(\lambda)$. Together, the scattering and absorption coefficients determine the intensity of light reemitted at the surface of the medium. The parameters are expressed in units of inverse length (e.g., mm^{-1} or cm^{-1}). The inverse of these coefficients represent the average distance (step-size) traveled by a photon before encountering a scatterer or absorber, respectively. With an increase in the scattering in a medium, there is a corresponding increase in the scattering coefficient; similarly, with an increase in absorption in a medium, there is a corresponding increase in the absorption coefficient.

6.2.4.3 Scattering Coefficient

In 1908, Gustav Mie developed an exact solution to Maxwell's equations to describe the scattered intensity of a plane EM wave by a homogeneous sphere of uniform RI. The scattered intensity was determined by two quantities: the size parameter (*x*) and the relative RI (*m*). The size parameter is a dimensionless quantity that is calculated as follows:

$$x = \frac{2\pi a \cdot n_{\mathrm{med}}}{\lambda},\qquad(6.1)$$

where

    a is the radius of the sphere
    n_{med} is the RI of the surrounding medium
    λ is the wavelength of the light in the medium

The relative RI, *m*, is defined as n_s/n_{med}, where n_s is the RI of the sphere. From the Mie calculations, one can obtain the scattering efficiency, $Q_s(\lambda, a)$ and then calculate $\mu_s(\lambda)$ as follows:

$$\mu_s(\lambda) = Q_s(\lambda, n_s, a) \cdot \rho(a) \cdot \pi a^2,\qquad(6.2)$$

where ρ is the number density of scatterers. Using Mie theory to model the measured $\mu_s'(\lambda)$, one can derive information about the physical properties of the scatters, including the size of the scatterers (*a*) and the density of scatterers (ρ). As shown by Equation 6.2, the scatterers can be considered to be uniform in size or have a specific size distribution. Using the anisotropy parameter, $g(\lambda)$, which is also derived from Mie calculations, $\mu_s'(\lambda)$ can be calculated as follows:

$$\mu_s'(\lambda) = \mu_s(\lambda) \cdot [1 - g(\lambda)],\quad [\mathrm{mm}^{-1}].\qquad(6.3)$$

The anisotropy parameter is a measure of the amount of forward direction retained after a single scattering event and is equal to the average cosine of the scattering angle.

For a wide range of values of the size parameter relevant to tissue, the decrease in μ_s' as a function of wavelength has been shown to be well described by a power law (Graaff et al. 1992, Mourant et al. 1997). A single or double power law expression, such as those shown in Equations 6.4 and 6.5, have been widely used to describe $\mu_s'(\lambda)$ (Mourant et al. 1998, Amelink et al. 2004a,b Bargo et al. 2005, Hidovic-Rowe and Claridge 2005, Wang et al. 2005, Mirkovic et al. 2009).

$$\mu_s'(\lambda) = A \cdot \lambda^{-B}\qquad(6.4)$$

$$\mu_s'(\lambda) = A \cdot \left(\frac{\lambda}{\lambda_o}\right)^{-B} + C \cdot \left(\frac{\lambda}{\lambda_o}\right)^{-4}.\qquad(6.5)$$

Two or more parameters (*A*, *B*, and/or *C* parameters) can be extracted to describe the tissue scattering properties by modeling $\mu_s'(\lambda)$ using these equations. The values of the exponents are related to the size of the scatterers (Mourant et al. 1997). The second power law in Equation 6.5 (the wavelength term raised to the negative fourth power) is consistent with the Rayleigh scattering regime, in which the scatterers are significantly smaller than the wavelength of the incident light (small values of *x*). The first power law of in Equation 6.5 describes Mie scattering regime in which the size of the scatterers (Characterized by the B parameter) is similar to the wavelength of the incident light. The coefficients in the power law are related to the magnitude of scattering by larger Mie scatterers and smaller Rayleigh scatterers (A and C parameters, respectively).

6.2.4.4 Absorption Coefficient

The Beer–Lambert law describes the intensity that emerges after passing through an absorbing solution, *I*, as follows:

$$I(\lambda) = I_o(\lambda)e^{-\varepsilon(\lambda)\cdot c \cdot L},\qquad(6.6)$$

where

- I_o is the intensity of the light launched into a solution containing the absorber
- ε is the extinction coefficient of the absorber
- c is the concentration of the absorber
- L is the optical path length, the distance the light travels through the absorbing solution

Because the quantity I/I_o is the transmittance, T, and absorption, A, can be defined as the negative natural logarithm of T, $-\ln(T)$, Equation 6.6 can be rearranged to calculate A as follows:

$$A(\lambda) = \varepsilon(\lambda) \cdot c \cdot L \qquad (6.7)$$

where

- ε is the extinction coefficient of the absorber
- c is the concentration of the absorber
- L is the optical path length, the distance the light travels through the absorbing solution

The absorption parameter, $\mu_a(\lambda)$, can be calculated after rearranging Equation 6.8 to form the following equation:

$$\mu_a(\lambda) = \varepsilon(\lambda) \cdot c, \qquad (6.8)$$

and the absorption coefficient for hemoglobin, $\mu_a^{Hb}(\lambda)$, can be calculated as follows:

$$\mu_a^{Hb}(\lambda) = cHb_T[\alpha\varepsilon_{HbO_2}(\lambda) + (1-\alpha)\varepsilon_{Hb}(\lambda)]. \qquad (6.9)$$

The total concentration of hemoglobin, cHb_T, represents the sum of the concentrations of oxyhemoglobin ($cHbO_2$) and deoxyhemoglobin (cHb). The hemoglobin oxygen saturation, α, is defined as $cHbO_2/(cHbO_2 + cHb)$. The extinction coefficients for oxyhemoglobin and deoxyhemoglobin are given by $\varepsilon_{HbO2}(\lambda)$ and $\varepsilon_{Hb}(\lambda)$, respectively.

Rather than considering hemoglobin to be homogenously distributed throughout the tissue, a number of models have been developed which account for the alterations in the lineshape and intensity of the hemoglobin absorption spectrum as a result of the absorber being confined to discrete compartments (e.g., red blood cells or finite-sized blood vessels) (Liu et al. 1995, Svaasand et al. 1995, Verkruysse et al. 1997, van Veen et al. 2002, Finlay and Foster 2004). This phenomenon is known as pigment packaging. In a model developed by Svaasand et al., the absorption coefficient of hemoglobin is represented by the product of the absorption coefficient of hemoglobin in whole blood, μ_a^{blood}, a wavelength-dependent correction factor that alters the shape of the absorption spectrum, K, and the volume fraction of blood sampled, v (Svaasand et al. 1995, van Veen et al. 2002). The absorption coefficient for hemoglobin is calculated as follows:

$$\mu_a^{Hb}(\lambda) = K(\lambda) \cdot v \cdot \mu_a^{blood}(\lambda)[mm^{-1}]. \qquad (6.10)$$

The correction factor K is calculated by the following equation:

$$K(\lambda) = \left\{ \frac{1 - \exp[-2 \times \mu_a^{blood}(\lambda) \cdot R]}{2 \cdot R \cdot \mu_a^{blood}(\lambda)} \right\}, \qquad (6.11)$$

where R is the effective vessel radius in units of millimeters and the absorption coefficient of blood, $\mu_a^{blood}(\lambda)$ is in units of mm^{-1}. The volume fraction, v, is described by the following ratio:

$$v = \frac{cHb_T}{150}, \qquad (6.12)$$

where cHb_T is the extracted total concentration of hemoglobin in units of mg/mL. The concentration of hemoglobin in whole blood has assigned a value of 150 mg/mL (Amelink et al. 2004b). The equation for μ_a^{blood} is given by

$$\mu_a^{blood}(\lambda) = 0.1 \cdot \ln(10)$$
$$\times 150 \left[\alpha \times \varepsilon_{HbO_2}(\lambda) + (1-\alpha) \cdot \varepsilon_{Hb}(\lambda) \right], \left[mm^{-1} \right] \qquad (6.13)$$

where

- ε is the extinction coefficient in units of $1/[cm\cdot(mg/mL)]$
- ln is the natural logarithm

This model assumes that there is a single-sized vessel distributed uniformly throughout the tissue. The dramatic effect of the correction factor, K, on the lineshape and intensity of the absorption coefficient of hemoglobin is shown in Figure 6.7 (Prahl 1998).

Figure 6.7 shows synthetic reflectance curves generated with the model of Zonios et al. using optical parameters consistent

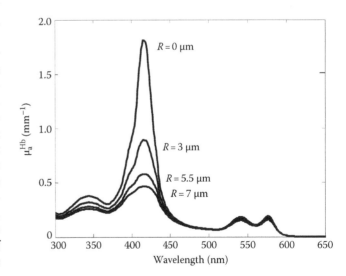

FIGURE 6.7 A plot of the absorption coefficient of hemoglobin (μ_a^{Hb}) as a function of wavelength for a vessel radius (R) of 0 (no pigment packaging), 3, 5.5, and 7 μm, respectively.

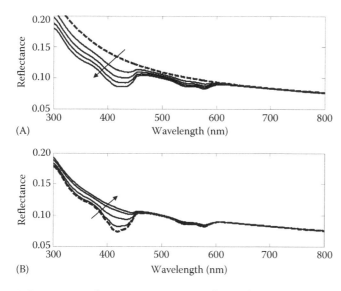

(A)
Wavelength (nm)

(B)
Wavelength (nm)

FIGURE 6.8 Reflectance spectra generated using the model of Zonios et al. (1999) using absorption and scattering parameters consistent with those obtained from oral cavity tissue. In each plot, the dashed curve represents the curve for the minimum value of the parameter being varied and the arrow indicates the direction of increasing value. (A) Hemoglobin concentration was varied between 0 and 2 mg/mL (vessel radius held constant). (B) Vessel radii were varied between 2.5 and 20 μm (hemoglobin concentration held constant). *(From Zonios, G. et al., Appl. Opt., 38, 6628, 1999.)*

with oral cavity tissue to calculate $\mu_s'(\lambda)$ and $\mu_a(\lambda)$ (Zonios et al. 1999). Figure 6.8A shows the effect of varying cHb_T while the vessel radius, R, is held constant. The plot shows that the depth of the 420 and 540/580 nm hemoglobin peaks increases with increasing hemoglobin concentration. Figure 6.8B shows the effect of varying the vessel radius, while cHb_T is held constant. In this case, the intensity of the 420 nm absorption peak decreases relative to that of the 540/580 nm peaks, while the latter peaks remain essentially unchanged.

Accounting for vessel packaging has been shown to improve the accuracy of modeled fits to the hemoglobin absorption features in tissue reflectance spectra.

The absorption coefficient for a tissue can be calculated by summing the absorption coefficients from each individual absorber, if one assumes the absorbers are homogeneously distributed. Additional tissue absorbers can be modeled based on Equation 6.8. The concentrations of the absorbers extracted from the reflectance spectra represent the average concentration over the tissue volume probed.

6.2.4.5 Radiative Transport Equation

The radiative transport equation (RTE) describes the transport of energy through a medium containing particles, and has been applied to a variety of scattering phenomena (Ishimaru 1978, Cheong et al. 1990). We briefly mention this subject here as the RTE is presented in detail in Chapter 16. In this description of light propagation, light is treated as a collection of photons and the field properties are neglected. The RTE can only be solved

numerically, and requires further simplifications (e.g., such as diffusion approximation) in order to obtain a closed-form solution.

6.2.4.6 Diffusion Theory

Several models describing the propagation of light in tissue have been developed based on the diffusion approximation to the RTE (Farrell et al. 1992, Zonios et al. 1999, Wang et al. 2005). The diffusion approximation assumes that light transport is dominated by scattering (i.e., absorption is minimal), such that photons are scattered uniformly in all directions and engage in what can be considered a random walk through the medium. One of the requirements for the diffusion approximation is that scattering of light is isotropic—which is not the case for tissue. Therefore, to satisfy the diffusion approximation, $\mu_s'(\lambda)$ is used to describe the diffusion of photons rather than $\mu_s(\lambda)$. By applying the correction to the scattering coefficient using $g(\lambda)$, scattering can be treated in a manner equivalent to the isotropic case. In addition to this requirement, there must be a sufficient number of scattering events before a photon is absorbed in order for the diffusion approximation to be valid. Both the optical properties of the sample and the source–detector separation of the probe determine whether the assumptions of the diffusion approximation are satisfied. An approximate rule to ensure this criterion is met is that $\mu_s'(\lambda)/\mu_a$ should exceed 10 (Jacques and Pogue 2008). For probe-based devices, the diffusion approximation is accurate for modeling the reemitted light for large source–detector separations—more than one or two transport mean free paths (e.g., several millimeters), where the mean free path is defined as $1/[\mu_s'(\lambda)+\mu_a]$ (Jacques and Pogue 2008). In some applications, when models based on the steady-state diffusion approximation assume a semi-infinite homogeneous medium and are applied to reflectance data measured from layered tissue, this can lead to fitting errors and considerable cross talk between the extracted $\mu_s'(\lambda)$ and $\mu_a(\lambda)$ (Farrell et al. 1998).

Models based on higher-order solutions to the transport equation, such as the P_3 approximation (Dickey et al. 2001, Hull and Foster 2001), have been developed to model tissue reflectance (Wang et al. 2009). This algebraic solution expresses the radiance in a truncated series of Legendre polynomials (Dickey et al. 2001). This technique for approximating the solution to the transport equation is more generally known as the P_n approximation, where n is the order of the Legendre polynomial used in the calculation. The P_3 approximation is more appropriate than diffusion theory for highly absorbing tissues, for small source–detector separations and near a boundary of mismatched optical properties (Jacques and Pogue 2008). When the expansion is terminated after the first term (P_1), the result is the steady-state diffusion equation described earlier.

6.2.4.7 Monte Carlo Simulations

Monte Carlo simulations use a computer to generate an exact solution to light transport in tissue (Jacques 1996). In this technique, a large number of photons (e.g., 5,000,000) are propagated into a medium characterized by a specified set of optical properties, including the thickness of the medium, RI (ambient and within the medium), $\mu_s(\lambda)$, $\mu_a(\lambda)$, scattering phase function, and $g(\lambda)$ (Wang et al. 1995). Upon interaction with a scatterer or absorber, light is either redirected or attenuated, respectively. The reflectance, transmittance, and absorption are calculated by recording the trajectories of the photons launched into the medium. A number of additional parameters can be incorporated into the simulation to ensure that the output accurately models the actual signal measured by an instrument. Some examples include specifying the properties of the illumination beam and incorporating multiple layers, each with distinct optical properties. Monte Carlo simulations can be time consuming to perform because they use an iterative algorithm and must track a large number of photons. The optical properties of the tissue under study must be known *a priori* or estimated for this technique. Unlike the diffusion approximation, this method is accurate for all source–detector separations.

Several studies have used this technique to model tissue reflectance spectra (Thueler et al. 2003, Palmer and Ramanujam 2006, Bender et al. 2009, Chang et al. 2009). In the Monte Carlo-based inverse model developed by Palmer et al., the free parameters in the model included the absorber concentrations and the scatterer size and density (Palmer and Ramanujam 2006). The absorption coefficient was calculated based on the absorber concentrations and the extinction coefficients of the absorbers (fixed values). The reduced scattering coefficient was calculated using Mie theory based on the scatterer size and density and using a fixed value for the RI mismatch. After applying an initial set of guess values for the free parameters, the modeled reflectance was determined based upon a lookup table of diffuse reflectance values for a wide range of values for μ_a and μ_s. This procedure greatly minimized the time required to fit the spectra. To reduce the time required to generate the lookup table, a single Monte Carlo simulation was performed, and a scaling approach was used to generate the reflectance resulting from different sets of optical properties. Cubic splines were used to interpolate between table values. The sum of the squares of the differences between the modeled and measured reflectance spectra was minimized using the Guess–Newton nonlinear least-squares algorithm to iteratively vary the free parameters. The global minimum was established after using several random sets of initial values for the free parameters.

6.2.4.8 Additional Methods

In addition to model-based approaches, principal components analysis (PCA) is another technique commonly used to reduce the dimensionality of reflectance data without a significant loss of information (de Veld et al. 2005, Bard et al. 2006). PCA is a statistical technique that linearly transforms the original spectral data into a set of eigenvectors (principal components), each of which accounts for a certain percentage of the total variance in the full data set. In addition, the eigenvalues (scores) for each sample is calculated. The scores describe the contribution of each principal component to the measured spectrum for a given sample. Usually, only the first few principal components are needed to describe the majority of the variance in the full

data set. The scores for these principal components are used for sample classification.

6.3 Current Biomedical Applications of Reflectance Spectroscopy

In this section, we describe how the information extracted from the reflectance signal is utilized for biomedical applications. We specifically focus on the performance of these techniques in the areas of tissue diagnostics and therapeutics.

6.3.1 Tissue Diagnostics

Greater than 85% of all cancers originate in the epithelium. The accessibility of the epithelium to examination by light means that optical spectroscopy (OS) may be used to detect epithelial cancers. Many of the tissue components affected by the disease process also interact with light through processes such as elastic scattering and absorption. Some of the alterations in tissue morphology associated with disease progression that can affect the scattering signal include hyperplasia, nuclear crowding, degradation of collagen in the extracellular matrix by matrix metalloproteinases (MMPs), and increased nuclear/cytoplasmic ratio (Tosios et al. 1998, Cotran et al. 1999, Jordan et al. 2004, Zigrino et al. 2005, Arifler et al. 2007). With disease progression, hemoglobin absorption features may be affected by several changes, including angiogenesis and tissue hypoxia (Wu 1996, Macluskey et al. 2000, Semenza 2002). Therefore, the impact of disease should lead to changes in the patterns in the light reflected from the tissue. The actual spectroscopic changes observed with carcinogenesis will depend on a combination of these changes, as well as factors such as the tissue under study and the methods and instrumentation used to probe the tissue.

6.3.1.1 Cervix

RS has been extensively investigated in studies of the cervix to determine whether spectroscopy can serve as an adjunct to colposcopy (Nordstrom et al. 2001, Georgakoudi et al. 2002, Mirabal et al. 2002, Huh et al. 2004, Chang et al. 2005, Marin et al. 2005, Mourant et al. 2007, 2009, Weber et al. 2008, Mirkovic et al. 2009). Colposcopy is the current clinical standard for cervical dysplasia diagnosis and is performed in women who demonstrate an abnormal result based on screening by a Papanicolaou (Pap) smear test. Colposcopy involves the application of acetic acid as a contrast agent in the identification of abnormal tissue, visual inspection, and biopsy of at-risk tissue (Ferris 2004). The tissue specimens are then evaluated by histopathology. During histopathological evaluation, the specimens are classified as non-dysplastic, cervical intraepithelial neoplasia 1 (CIN 1) (a low-grade lesion), CIN 2 or CIN 3 (high-grade lesions), cancer, or indefinite for dysplasia. In adult women, CIN 2 is used as the threshold for lesions that require treatment; thus, the set of lesions requiring treatment (CIN 2, 3 or cancer) is often referred to as CIN 2+ (Wright et al. 2007).

There are several limitations in the use of colposcopy to evaluate cervical tissue. The reported sensitivity and specificity for colposcopy varies widely in the literature but overall demonstrates its limited accuracy (Mitchell et al. 1998, Solomon et al. 2003, Gage et al. 2006). In expert hands, colposcopy was reported to be quite sensitive (96%) but not very specific (48%) for detection of cervical dysplasia (Mitchell et al. 1998). In other studies, colposcopy has shown a poor sensitivity (Solomon and Grp 2003, Gage et al. 2006). Biopsy of suspicious tissue is invasive and specimen processing and diagnosis are time consuming, requiring hours to days before a diagnosis can be rendered. Finally, diagnosis of dysplasia can be unreliable because of interobserver disagreement among pathologists (Devet et al. 1990). In contrast, RS can be developed to provide diagnostic information noninvasively, in real-time, and in an objective and quantitative manner. As an objective, real-time diagnostic tool, spectroscopy could potentially reduce the number of unnecessary biopsies during colposcopy and poor reproducibility in the histopathological diagnosis that is subsequently rendered. The reflectance studies in the discussion to follow demonstrate the potential of spectroscopy to enable targeted biopsy during colposcopic examinations and thus improve the accuracy of disease detection. Table 6.2 summarizes the results of point-probe studies in the cervix.

Using a probe that incorporated fibers with four different source–detector separations, Mirabal et al. collected *in vivo* reflectance spectra from 324 sites in 161 patients during colposcopy (Mirabal et al. 2002). Algorithms were developed according to the following scheme: (1) selection of the source–detector separations to analyze, (2) data reduction using PCA, and (3) feature selection and classification using Mahalanobis distances with cross-validation. The data set included colposcopically normal squamous tissue ($n = 227$), normal columnar samples that were colposcopically abnormal ($n = 18$), human papilloma virus (HPV) infection ($n = 52$), CIN 1 ($n = 9$), CIN 2 ($n = 3$), and CIN 3/Carcinoma-in-situ (CIS) ($n = 15$). Normal columnar tissue that was colposcopically abnormal could be distinguished from HPV/CIN 1 with a sensitivity of 75% and specificity of 89%. Normal columnar tissue that was colposcopically abnormal could be distinguished from CIN2,3 and CIS with a sensitivity of 72% and specificity of 83%. The investigators found that the diagnostic performance was high when using only a single source–detector separation, and did not increase appreciably when data from additional separations were included.

Mourant et al. evaluated the potential of RS to detect precancerous and cancerous lesions of the cervix using a probe designed to collect unpolarized (1 fiber), horizontally polarized (2 fibers: collection fibers 1 and 4), and vertically polarized light (1 fiber: collection fiber 3) returning from the tissue (Mourant et al. 2009). From the tissue spectra, they extracted four absorption parameters (cHb_T, α, blood vessel diameter, water concentration), three scattering parameters related to the unpolarized light (amplitude and exponent of a power law and slope), and two ratios calculated from the polarized light signals. Figure 6.9

TABLE 6.2 Summary of Results from Point-Probe Reflectance Studies in the Cervix

Study	Technique	Num. Patients (Num. Samples)	Distinction	Sensitivity [%]	Specificity [%]
Mirabal et al. (2002)	Ref.	161 (324)	CNS vs. colposcopically abnormal (biopsied) NC	89	77
			CNS vs. HPV/CIN1	72	56
			CNS vs. CIN2$^+$	72	81
			NC vs. HPV/CIN1	75	89
			NC vs. CIN2$^+$	72	83
			CNS, SQM, MSE vs. LSIL/HSIL	62	82
Mourant et al. (2009)	Ref.	151 (362)	Non-HSIL (CNS, biopsied normal, cervicitis, LSIL) vs. HSIL/Cancer	77	62
			Non-HSIL (biopsied normal, cervicitis, LSIL) vs. HSIL/cancer	77	44
			Non-HSIL (CNS, biopsied normal, cervicitis) vs. HSIL/Cancer	77	68
			Non-HSIL (biopsied normal, cervicitis) vs. HSIL/cancer	79	47
Mourant et al. (2007)	Ref.	29 (88)	Normal, cervicitis, LSIL vs. CIN2/CIN3 (HSIL)	100	80
Georgakoudi et al. (2002)	Ref.	44 (84)	MSE, SQM (both colposcopically abnormal) vs. LSIL/HSIL	69	57
			CNS, SQM, MSE vs. LSIL/HSIL	62	82

Notes: CNS, colposcopically normal squamous; NC, normal columnar; HPV, human papilloma virus; CIN, cervical intraepithelial neoplasia; MSE, mature squamous epithelium; SQM, squamous metaplasia; LSIL, low-grade squamous intraepithelial lesions; HSIL, high-grade squamous intraepithelial lesions.

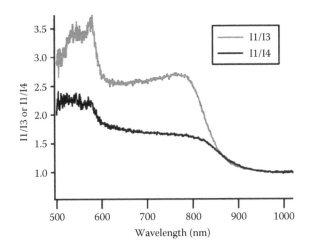

FIGURE 6.9 Representative I1/I3 and I1/I4 spectra. (From Mourant, J.R. et al., *Appl. Optics*, 48(10): D26. With permission.)

shows examples of the wavelength-dependent ratios of the measured polarized light signals. I1/I3 represents the ratio of the intensity of light measured by collection fiber 1 (I1) to that measured by collection fiber 3 (I3). I1/I3 increases for more strongly scattering tissue. I1/I4 represents the ratio of I1 to the intensity of light measured by collected by fiber 4 (I4). This ratio increases as the average size of the scatterer decreases. The average value for a narrow band of wavelengths was calculated for each of the two ratio spectra and used as diagnostic parameters.

In vivo spectra were collected from colposcopically abnormal sites and normal sites in 151 patients. The data set included unbiopsied, colposcopically normal sites ($n=181$), and biopsied sites, which were classified as normal ($n=36$), cervicitis ($n=44$), low-grade squamous intraepithelial lesions (LSIL; $n=43$), high-grade squamous intraepithelial lesions (HSIL; $n=56$), and cancer ($n=2$). A classification scheme was developed based on the values of I1/I4, slope, I1/I3, and cHb_T. After determining the optimal thresholds for these parameters, they were used to cast a positive or negative vote toward the probability of disease, and the sensitivity and specificity of the diagnostic algorithms was determined based on fivefold cross-validation. The investigators found that a number of parameters needed to be corrected for several confounding factors, including the particular doctor taking the probe measurements, differences among probes, and patient age, menopausal or menstrual cycle status. For colposcopically abnormal tissue, a sensitivity of $77.1\% \pm 4.5\%$ and specificity of $43.8\% \pm 3.3\%$ was obtained when distinguishing HSIL/cancer from LSIL/non-dysplastic samples, and a sensitivity of $78.9\% \pm 4.4\%$ and specificity of $47.3\% \pm 3.7\%$ was obtained when distinguishing HSIL/cancer from non-dysplastic samples. Values of I1/I3, I1/I4, and slope were found to increase for HSIL, indicating increased scattering and a decrease in the scatterer size.

Georgakoudi et al. measured diffuse reflectance spectra (in addition to fluorescence spectra) from patients undergoing colposcopy (Georgakoudi et al. 2002). Data were collected from

44 patients and analyzed using a physical model, which provided quantitative parameters relating to tissue morphology and biochemistry. The data set included the following biopsied sites: benign/mature squamous epithelium (MSE; $n=5$), squamous metaplasia (SQM; $n=16$), LSIL ($n=2$), and HSIL ($n=11$). Spectroscopic data were also collected from colposcopically normal squamous (CNS) ectocervical tissue sites that were not biopsied ($n=50$). Diagnostic models were developed using logistic regression and the performance evaluated using leave-one-out cross-validation. Diffuse reflectance spectra were analyzed using a diffusion-based model. For the discrimination of biopsied sites, a sensitivity of 69% and sensitivity of 57% was obtained for the distinction of MSE/SQM from SIL. A sensitivity of 62% and a specificity of 82% was obtained when CNS sites were combined with MSE/SQM sites and compared to the SIL category. For both distinctions, analysis using fluorescence alone yielded considerably higher specificities (a 10% increase), but equivalent or slightly lower sensitivities compared to reflectance alone. Chang et al. also evaluated the diagnostic performance of reflectance and fluorescence for the discrimination of cervical lesions (Chang et al. 2005). They found that fluorescence alone yielded superior performance values as compared to reflectance alone, and that the combination of reflectance and fluorescence provided only a modest improvement in diagnostic performance.

A few studies have also examined the potential of a spectroscopic-based tool to serve as an adjunct to Pap smear and improve the sensitivity of the detection of high-grade disease. The TruScreen [Polarprobe] (Polartechnics Limited, Sydney, Australia) is a point-probe device that uses a combination of electrical and reflectance signals to classify cervical tissue in real time (Coppleson et al. 1994, Singer et al. 2003).

Significant progress has been made in combining spectroscopy with imaging for the detection of cervical dysplasia (Ferris et al. 2001, Milbourne et al. 2005, Alvarez et al. 2007, DeSantis et al. 2007, Park et al. 2008). Milbourne et al. and Park et al. conducted three color reflectance and fluorescence spectroscopic imaging and observed that dysplastic cervical tissue emits light at a slightly different wavelength than normal cervical tissue when excited with UV or blue excitation light. A study by Alvarez et al. describes results from a large multicenter, two-arm randomized trial comparing the effectiveness of colposcopy alone to colposcopy in combination with an optical detection system (ODS) method based on white light tissue reflectance, fluorescence and cervical video imaging. The results of the study showed that the use of ODS in conjunction with colposcopy increased the true positive rate for detection of CIN 2,3 in women with atypical squamous cell (ASC) and LSIL referral cytology by 26.5%. The use of the ODS did not improve detection of CIN 2,3 in women with HSIL referral cytology.

A number of studies employing model-based techniques to extract spectral parameters relating to tissue physiology have evaluated the changes in the optical properties with increasing disease severity. In a study by Chang et al., the investigators observed that the mean $\mu_s'(\lambda)$ from 450 to 600 nm was significantly lower in CIN as compared to normal samples (Chang et al. 2009). The cHb$_T$ was significantly higher in CIN 2$^+$ samples as compared to normal/CIN 1, as well as for CIN compared to normal samples; however, there was no significant difference in α. The trend in cHb$_T$ was largely due to an increase in cHbO$_2$. The trends in cHb$_T$ and μ$_s'$(λ) were preserved but were no longer statistically significant when colposcopically normal (non-biopsied) samples were excluded and only colposcopically abnormal, biopsy-confirmed normal samples were considered. The parameter trends in most studies indicate an increase in parameters related to hemoglobin concentration and a decrease in hemoglobin oxygen saturation and parameters related to the magnitude of the scattering (Georgakoudi et al. 2002, Weber et al. 2008, Mirkovic et al. 2009). However, studies discriminating HSIL from non-HSILs among clinically suspicious sites report no significant change in cHb$_T$ between the two groups (Georgakoudi et al. 2002, Mourant et al. 2007).

6.3.1.2 Breast

There are several potential applications of a reflectance-based tool for evaluating breast tissue that have been investigated including the following: (1) the assessment of surgical tumor margins, (2) intraoperative evaluation of sentinel lymph nodes, and (3) to serve as an adjunct to core-needle biopsy (CNB) procedures.

During a lumpectomy surgery, a procedure in which the breast tumor is removed, it is difficult to predict tumor-free (clear) margins in the tissue surrounding the lesion. Analysis of frozen sections can provide a relatively rapid means of diagnosing tissue surrounding the tumor; however, an experienced pathologist is needed to interpret the findings because of the artifacts resulting from the freezing process and the poor quality of the specimens as compared to traditional permanent sections (Cendán et al. 2005). Because disease may still be missed, postoperative pathological examination of permanent sections is used to establish the final diagnosis. A number of patients (20%–40%) will have to undergo a re-excision procedure because tumor cells are identified at or near the margins (Pleijhuis et al. 2009). Spectroscopy may be able to provide an objective, accurate intraoperative assessment of the margins for invisible disease.

Determining the presence or absence of metastatic disease in the lymph nodes is critical for determining the best treatment for breast cancer. Sentinel lymph nodes, any lymph node with a direct lymphatic connection to the tumor, are among the first to be invaded during the spread of breast cancers. Currently, lymph nodes are evaluated intraoperatively by performing histological evaluations of frozen sections. As previously mentioned, this method has a low sensitivity and the results are highly dependent on the skill level of the pathologist (Van Diest et al. 1999). By providing a diagnosis in real time, spectroscopy may be able to eliminate the need for a second procedure and the associated costs.

The current standard screening procedure for breast malignancy is mammography. If an abnormality is identified, a diagnostic biopsy is performed (e.g., by procedures such as CNB and fine needle aspiration) followed by histopathological evaluation

to establish a diagnosis. Approximately 95% of the abnormalities observed during screening mammograms are nonmalignant, and the results are highly dependent on the skill of the radiologist (Fletcher and Elmore 2003). Furthermore, histopathological evaluation requires time for sample processing and is subjective. By providing a diagnosis in real-time and enabling targeted biopsy during core-needle procedures, RS can potentially reduce the number of unnecessary biopsies, ease patient anxiety, and eliminate the requirement for a second visit to obtain a diagnosis.

Majumder et al. performed a comparative study of several spectroscopic techniques (fluorescence, reflectance, Raman spectroscopy) for the classification of breast lesions (Majumder et al. 2008). Spectra were acquired from freshly frozen tissue samples obtained during reduction mammoplasty, uninvolved areas from radical mastectomy procedures, or surgically removed breast lesions. A total of 293 unique tissue sites were examined from 74 tissue samples. The classification algorithm consisted of two steps: (1) extraction of diagnostic features from the spectra using nonlinear maximum representation and discrimination feature (MRDF) and (2) development of a probabilistic scheme of classification based on sparse multinomial logistic regression (SMLR) for classifying the nonlinear features into four tissue categories (normal, fibroadenoma (FA), ductal carcinoma *in situ* (DCIS), and invasive ductal carcinoma (IDC)). Table 6.3 summarizes the results from this study for reflectance alone and in combination with fluorescence spectroscopy (FS). The diagnostic algorithm yielded an overall classification accuracy of 72% (211 out of 293) based on leave-one-out cross-validation. The investigators note that RS misclassified 45% of IDC samples as normal, 10% of normal samples as IDC, and 50% of DCIS samples as IDC. For many tissue categories, the misclassified samples displayed unique morphological features, which were not present for the majority of the correctly classified samples.

Volynskaya et al. evaluated 202 reflectance spectra from 104 freshly excised breast surgical specimens in 17 patients (Volynskaya et al. 2008). The data set included samples classified by histopathology as normal ($n=31$), fibrocystic change (FCC; $n=55$), FA ($n=9$), and infiltrating ductal carcinoma (IFDC; $n=9$). Figure 6.10 shows representative reflectance spectra and modeled fits for each histopathology category.

Using a physical model based on the diffusion approximation, they modeled μ$_s'$(λ) with a power law and extracted two scattering parameters (*A* and *B* parameters) related to the magnitude and size of the scatterers, respectively. The extracted absorption parameters included the relative concentration of oxyhemoglobin and β-carotene, respectively. A stepwise diagnostic algorithm was developed to distinguish the four histopathology categories based on a combination of reflectance and fluorescence parameters. The investigators found that the concentration of β-carotene contributed to the discrimination of normal specimens from all other categories (first step) and the *A* parameter contributed to the discrimination of FA from FCC and infiltrating ductal carcinoma (second step). For the final step of the algorithm, the discrimination of IFDC from FCC, the concentration of oxyhemoglobin was found to be diagnostic. Figure 6.11 shows

TABLE 6.3 Summary of Results from Multistep Diagnostic Algorithms to Discriminate among Several Breast Histopathology Categories

	Histopathology Categories	Num. Tissue Samples (Num. Spectra)	Optical Technique	Classification Accuracy for Each Histopathology Category				
				A [%]	B [%]	C [%]	D [%]	Overall Accuracy [%]
Majumder et al. (2008)	Normal (A)	32 (134)	Ref.	86	86	28	51	72
	FA (B)	11 (55)						
	DCIS (**C**)	6 (18)	Ref. and Fl. combined	86	98	89	72	84
	IDC (D)	25 (86)						
Volynskaya et al. (2008)	Normal (A)	31 (60)	Ref.	81	89	85	100	81
	FA (B)	9 (18)						
	FCC (C)	55 (110)	Ref. and Fl. combined	84	100	93	100	91
	IFDC (D)	9 (14)						

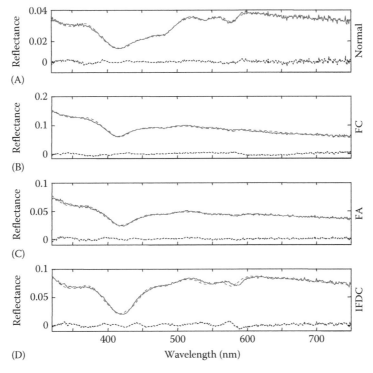

(A)

(B)

(C)

(D)

FIGURE 6.10 Representative spectra of diffuse reflectance spectra for all pathologies: (A) normal tissue, (B) fibrocystic change (FCC) lesion, (C) fibroadenoma (FA) lesion, and (D) infiltrating ductal carcinoma (IFDC) lesions. Reflectance data are shown in solid grey, the modeled fit is shown in the dashed gray line, and the residual is shown by the black dash-dotted line. (From Volynskaya, Z. et al., *J. Biomed. Optics*, 13(2), 024012. With permission.)

boxplots for the *A* parameter (related to the magnitude of scattering) and β-carotene for each histopathology category.

When the performance of the stepwise diagnostic algorithm was evaluated using only reflectance parameters, a sensitivity, specificity, positive predictive value (PPV), negative predictive value (NPV), and total efficiency of 100%, 100%, 100%, 100%, and 81% (85/104), respectively, was obtained for the discrimination of benign from malignant lesions. The PPV represents the probability that a patient has disease (IFDC) among all people who test positive for disease by spectroscopy; the NPV represents the probability of the absence of disease among all people who test negative for disease by spectroscopy.

The parameters extracted from the reflectance spectra were important for accurately discriminating benign from malignant lesions, while those extracted from the fluorescence spectra played a major role in the subclassification of normal and benign

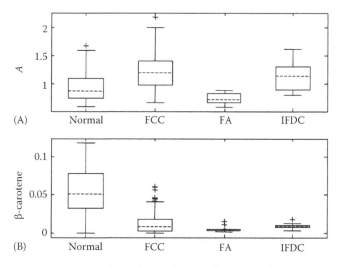

FIGURE 6.11 Boxplots of normal tissue, fibrocystic change (FCC), fibroadenoma (FA), and infiltrating ductal carcinoma (IFDC) for diagnostically relevant parameters. (A) A scattering parameter and (B) β-carotene. (From Volynskaya, Z. et al., *J. Biomed. Optics*, 13(2), 024012. With permission.)

lesions. The performance results for this multistep algorithm are shown in Table 6.3 along with those of the previous study by Majumder et al. Both studies demonstrate that the combination of the RS and FS improves overall accuracy, largely due to improvements in the subclassification of normal and benign lesions.

Table 6.4 summarizes the results for several point-probe reflectance studies in the breast for the discrimination of malignant tissue from nonmalignant tissue. Results for the combination of fluorescence and RS are shown if a separate analysis of RS was also performed. A few of these studies are summarized in further detail in the text to follow.

Zhu et al. used a point-probe device to measure reflectance spectra from freshly excised breast tissue and discriminate malignant (*n* = 54 samples) from nonmalignant (normal/benign fibrous/normal adipose tissue; *n* = 70 samples) tissue (Zhu et al. 2008b). Data were collected from patients with breast cancer or patients undergoing breast reduction procedures. A scalable

Monte Carlo inverse model was used to extract cHb_T, α, concentration of β-carotene and mean $\mu_s'(\lambda)$ from the spectra.

A linear support vector machine (SVM) classifier was used to discriminate malignant from nonmalignant breast tissue samples and the sensitivity and specificity were calculated using both a holdout validation and leave-one-out cross-validation method. An average sensitivity of 81.7% and specificity of 88.6% was obtained for the discrimination of malignant from nonmalignant samples based on the holdout method (83.3% and 87.1%, respectively, based on leave-one-out cross-validation).

Johnson et al. investigated the potential of RS to evaluate excised lymph nodes in patients with breast cancer undergoing either sentinel node biopsy or axillary node clearance (Johnson et al. 2004). The investigators used a point-probe device to measure 782 reflectance spectra from 139 excised nodes (~2–20 spectra were collected per node). The lymph nodes were classified as either normal (394 spectra from 86 nodes) or metastatic (388 spectra from 53 nodes). Because a one-to-one correlation between each site in the node interrogated by spectroscopy and pathology was not possible since it would require microdissection of the node, metastatic nodes were further classified as demonstrating partial, minute areas (micrometastases), complete replacement by cancer or cancer detectable only by immunohistochemistry (IHC). PCA was performed on the spectra followed by linear discriminant analysis (LDA) to discriminate normal from metastatic nodes. Sensitivities and specificities were assessed by leave-one-out cross-validation or multiple training-set/test-set combinations. A per-spectrum analysis was performed, as well as a per-node analysis, in which each node was treated as a single entity. In the per-node analysis, if any single spectrum from a particular node was classified as metastatic by LDA, the entire node was regarded as metastatic. Only the normal nodes and those with total metastatic replacement were used in the training sets for both the per-node and per-spectrum analyses since histology was only available on a per-node basis. However, partial-, micro-, and IHC nodes were used in the test sets for per-node analysis. A sensitivity of 84% and specificity of 91% was demonstrated in the discrimination of normal nodes from completely metastatic nodes (per-spectrum analysis). An

TABLE 6.4 Results from Point-Probe Reflectance Studies in the Breast for the Discrimination of Malignant Tissue from Nonmalignant Tissue

Study	Num. Patients (Num. Samples)	*In Vivo*	Technique	Sensitivity [%]	Specificity [%]	PPV [%]	NPV [%]
Volynskaya et al. (2008)	17 (104)	No	Ref.	100	100	100	100
			Ref. and Fl. combined	100	100	100	100
Zhu et al. (2008a)	73 (124)	No	Ref.	82	89		
			Ref. and Fl. combined	84	89		
Bigio et al. (2000)	24 (72)	Yes	Ref.	69	85	40–75	80–96
Palmer et al. (2003)	32 (56)	No	Ref.	30	78		
Palmer et al. (2006)	23 (41)	No	Ref.	82	92		

Note: Studies that report results for the combination of reflectance and fluorescence spectroscopy are shown if separate results for reflectance are also reported.

average sensitivity of 75% and specificity of 89% was obtained for the discrimination of normal nodes from totally metastatic nodes (per-node analysis), although the detection of partial metastases was less consistent with sensitivities ranging between 19% and 86%. These figures compared well with the touch imprint cytology (TIC) technique currently used at the study hospital for sentinel node assessment (sensitivity of 75% and specificity of 99%), although TIC requires the presence of an experienced cytologist. The investigators note that assessing the nodes more thoroughly (e.g., increased sampling) may further improve performance values. Bigio et al. also performed a study in which they acquired *ex vivo* spectra from the sentinel lymph nodes of 21 patients (Bigio et al. 2000). Using an artificial neural network (ANN) classification method, they obtained a sensitivity of 58% and specificity of 93% for distinguishing malignant specimens from normal specimens. These values were derived by randomly selecting 80% of the data for training and the remaining 20% for testing, and averaging the statistics from five separate replications of testing. Hierarchical cluster analysis (HCA) was also tested and yielded a sensitivity of 91% and specificity of 77% using leave-one-out cross-validation. The extent of metastasis in the node was not accounted for in this preliminary analysis.

Only a few studies in the breast have been performed *in vivo* (Bigio et al. 2000, van Veen et al. 2005, Brown et al. 2009). Bigio et al. evaluated the potential of RS to identify disease during transdermal needle measurements and for assessing the status of tumor margins during surgery (Bigio et al. 2000). Two different probe designs were employed for these two applications. For the transdermal needle measurements, a smaller probe diameter ($\leq$1 mm) was used so that it could fit in the CNBs. For the second application, the handheld probe had a sheath approximately 5 mm in diameter; however, the fiber dimensions and separation were identical to those used in the first design. Data were collected from 24 patients and included 13 cancer sites and 59 normal sites. ANN and HCA were used to classify the sites as normal or cancer. A sensitivity of 69% and a specificity of 85% were obtained for the distinction of cancer from normal tissue using ANN, and a sensitivity of 67% and a specificity of 79% were obtained using HCA. van Veen et al. used differential path-length spectroscopy in combination with CNB to measure and compare the optical properties in a small volume of healthy and malignant breast tissue (van Veen et al. 2005). Brown et al. used RS to quantitatively assess the hemoglobin concentration and oxygen saturation of malignant and nonmalignant breast tissue during CNB procedures (Brown et al. 2009). The status of the estrogen, progesterone, epidermal growth factor, and HER2/neu receptors of malignant samples were also evaluated and correlated with the hemoglobin parameters. HER2/neu-amplified breast cancers exhibited significantly higher cHb_T and a significantly higher α.

Many studies in the breast combine RS with other spectroscopic modalities, in particular FS (Palmer et al. 2003, Breslin et al. 2004, Zhu et al. 2005, 2008b, Volynskaya et al. 2008). Zhu et al. and Volynskaya et al. found that the combination of reflectance and fluorescence parameters yielded sensitivities and specificities similar to those obtained based on reflectance alone when discriminating between malignant from nonmalignant tissue (Volynskaya et al. 2008, Zhu et al. 2008a). Palmer et al. found fluorescence produced superior results in the distinction of malignant from nonmalignant tissue and no improvement with the combination of information extracted from the reflectance and fluorescence spectra (Palmer et al. 2003). In contrast to both of these findings, Breslin et al. found that the combination of reflectance and fluorescence improved diagnostic accuracy for this distinction compared to reflectance alone (Breslin et al. 2004).

Both hemoglobin and β-carotene are the major absorbers found in breast tissue. Volynskaya et al. observed that malignant tissue demonstrated a higher concentration of hemoglobin as compared to benign lesions (Volynskaya et al. 2008); however, other studies have found no statistically significant difference (Palmer et al. 2006, Zhu et al. 2008b, Brown et al. 2009) in the concentration of hemoglobin for these two groups. The values of α are lower in malignant tissue as compared to nonmalignant tissue (van Veen et al. 2005, Palmer et al. 2006, Zhu et al. 2008b, Brown et al. 2009). β-carotene is a lipid-soluble molecule and therefore found in abundance in breast tissue, which is largely composed of adipose tissue. The concentration of this absorber has been shown to be lower in malignant tissue as compared to nonmalignant breast tissue (Ghosh et al. 2001, Breslin et al. 2004, Palmer et al. 2006, Volynskaya et al. 2008, Zhu et al. 2008b). These findings are consistent with the fact that normal breast tissue is mostly composed of fat, while there is an increase in fibrous tissue in breast lesions (Cotran et al. 1999). An *in vivo* study by van Veen et al., however, compared normal and malignant tissue and did not observe a statistically significant decrease in β-carotene in malignant tissue (van Veen et al. 2005). Studies have also shown a significant increase in the mean $\mu_s'(\lambda)$ in malignant tissue as compared to nonmalignant tissues (Ghosh et al. 2001, Palmer et al. 2006, Zhu et al. 2008b).

6.3.1.3 Oral Cavity

The current gold standard for the diagnosis of oral cancer is biopsy followed by histopathological assessment. Suspicious lesions are identified by the physician based on visible changes in the tissue mucosa as observed under standard white light illumination. There are a number of limitations in the detection scheme. First, the invasiveness of the biopsy procedure limits the number of tissue areas that can be evaluated. Significant underdiagnosis has been noted with biopsy, particularly when only a single biopsy is performed and the lesion is nonhomogeneous (Pentenero et al. 2003, Lee et al. 2007). Second, the selection of the biopsy site is dependent on the experience of the physician. The selection of the tissue area to be biopsied is important because the pathology for this specimen is assumed to be representative of the extent of disease in the suspicious lesion as a whole, and this often determines whether treatment is indicated (Pentenero et al. 2003). Finally, the accuracy of histopathological classification is limited due to significant inter- and intra-observer disagreement (vanderWaal et al. 1997, Fischer et al. 2004, 2005).

Leukoplakia ("white plaque") is the most commonly observed suspicious lesion found in the oral cavity (Cawson and Odell 2002). In a study of 3256 biopsied leukoplakias, only 20% demonstrated dysplasia, CIS, or infiltrating squamous cell carcinoma (SCC); therefore, a number of unnecessary biopsies are performed (Waldron and Shafer 1975). The major goal of reflectance studies in the oral cavity has been to develop a tool to noninvasively and objectively discriminate diseased from nondiseased oral lesions. Table 6.5 shows a summary of the results from several point-probe reflectance studies in the oral cavity. As shown in the table, some studies perform analyses on a single anatomic site (e.g., buccalmucosa) or a specific set of anatomic sites, in addition to the combination of all anatomic sites. In the following discussion, we present a few examples of studies that have been performed in the oral cavity.

Sharwani et al. collected 25 *in vivo* reflectance spectra from 25 patients with leukoplakia (Sharwani et al. 2006). Data were classified as normal ($n=4$), benign ($n=10$), dysplastic ($n=10$), and CIS ($n=1$). Using LDA followed by leave-one-out cross-validation, they obtained a sensitivity and specificity of 72.7% and 75%, respectively, for distinguishing normal from dysplastic sites.

Muller et al. measured reflectance spectra *in vivo* from 15 patients with lesions and 8 healthy volunteers (Muller et al. 2003). The data set from patients included normal ($n=16$),

benign ($n=6$), dysplastic ($n=19$), and SCC ($n=12$) samples. A quantitative, physical model based on the diffusion approximation was used to extract absorption and scattering parameters. Differences in the scattering signal were noted between keratinized (gingiva, palate, and dorsal surface of the tongue) and nonkeratinized (all other sites) anatomic sites in healthy volunteers; therefore, separate diagnostic algorithms were developed for each group in addition to those developed for the combination of all anatomic sites. The diagnostic parameters extracted by modeling the reflectance spectra based on a linear fit to $\mu_s'(\lambda)$ included the scattering slope and intercept. The sensitivity and specificity were evaluated using leave-one-out cross-validation. For the combination of all sites, normal tissue could be distinguished from dysplasia/cancer with a sensitivity and specificity of 74%, and dysplasia could be distinguished from cancer with a sensitivity of 67% and specificity of 37%.

Schwarz et al. used a depth-sensitive optical fiber probe to measure reflectance (and fluorescence) spectra from patients with oral lesions and healthy subjects (Schwarz et al. 2009). All data were collected *in vivo* and included several anatomic sites. The final data set used in the analysis consisted of 281 measurements from 60 patients and 271 measurements from 64 healthy subjects without lesions. The probe measures reemitted light predominantly from the shallow, medium, and deep tissue layers via three source–detector distances. Figure 6.12 shows examples of reflectance spectra for the four histopathology categories collected from the three channels.

TABLE 6.5 Summary of the Results from Point-Probe Reflectance Studies in the Oral Cavity

Study	Num. Patients (Num. Spectra)	Num. Healthy Subjects (Num. Spectra)	Distinction	Sensitivity [%]	Specificity [%]	AUC
Sharwani et al. (2006)	25 (25)		Normal vs. dysplasia	73	75	
Muller et al. (2003)	15 (53)	8 (38)	Normal vs. dysplasia/cancer	74	74	
			Normal vs. dysplasia/cancer: nonkeratinized sites	88	68	
			Normal vs. dysplasia/cancer: keratinized sites	64	81	
			Dysplasia vs. cancer	67	37	
			Dysplasia vs. cancer: nonkeratinized sites	25	67	
			Dysplasia vs. cancer: keratinized sites	45	67	
de Veld et al. (2005)	155 (172)	70 (581)	Normal vs. cancer	82	88	0.93
			Normal vs. benign/dysplasia/cancer	89	80	0.90
			Benign vs. dysplasia/cancer	69	77	0.77
Nieman et al. (2008)	27 (57)		Normal vs. moderate dysplasia/severe Dysplasia/cancer	90	86	0.89
			Benign vs. moderate dysplasia/severe Dysplasia/cancer	100	85	0.91
			Mild dysplasia vs. moderate dysplasia/severe Dysplasia/cancer	80	83	0.87
Mallia et al. (2008[a])	29 (48)	35	Normal vs. hyperplasia	70	63	
	13 (21)		Normal vs. hyperplasia: buccal mucosa	97	86	
	29 (48)		Hyperplasia vs. dysplasia	100	80	
	13 (21)		Hyperplasia vs. dysplasia: buccal mucosa	100	86	
	29 (48)		Dysplasia vs. cancer	91	100	
	13 (21)		Dysplasia vs. cancer: buccal mucosa	96	100	

Note: Area under the receiver–operator characteristic curve (AUC).

[a] No validation technique was used in this analysis.

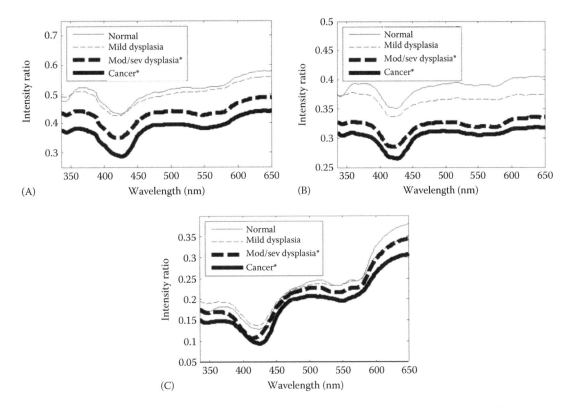

FIGURE 6.12 Average spectra of nonkeratinized tissue by diagnosis, illustrating differences in data obtained at different depths. Reflectance spectra with white light illumination. (A) Shallow, (B) medium, and (C) deep probe channels, respectively. An asterisk (*) next to a diagnostic category indicates that differences in the mean intensities of normal tissue and that diagnostic category were statistically significant (two-tailed Student t test, $P < .05/6$, correcting for 6 comparisons per panel). (From Schwarz, R.A. et al., *Cancer*, 115, 1669. With permission.)

Separate diagnostic algorithms were developed for keratinized sites (gingiva and hard palate) and nonkeratinized sites (buccal mucosa, tongue, floor of the mouth, and the lip) using LDA, and a receiver operator characteristic (ROC) curve was used to determine the optimal sensitivity and specificity. For nonkeratinized sites, reflectance intensities or intensity ratios from the medium and shallow channels (which measure the optical properties mostly from the epithelium and shallow stromal regions, respectively) were diagnostic for the discrimination of normal/mildly dysplastic samples from samples demonstrating moderate/severe dysplasia or cancer. The diagnostic performance of models developed for keratinized sites were not as accurate as those developed for nonkeratinized sites. Separate performance values were not reported for RS alone.

A study by de Veld et al. examined diffuse reflectance and fluorescence spectra from oral tissue from 172 oral lesions and 70 healthy volunteers (de Veld et al. 2005). They analyzed the spectra using PCA, after applying various normalization methods. Multiple classification algorithms were also tested for distinguishing various subsets of the data. They found that malignant lesions could successfully be distinguished from healthy tissue (sensitivity of 82% and specificity of 88%), and that the best separation for distinguishing benign lesions from dysplastic/malignant lesions yielded a sensitivity, specificity, and area under the ROC curve (AUC) was 69%, 77% and 0.77, respectively.

Combining reflectance and fluorescence only slightly improved diagnostic values.

Nieman et al. analyzed 57 reflectance spectra from 27 patients using oblique polarized RS (Nieman et al. 2008). The data set included 22 spectra from clinically normal tissue and 35 spectra from clinically abnormal tissue (13 benign, 12 moderately dysplastic, 10 severely dysplastic) from the tongue, buccal mucosa, floor of the mouth, gingiva, and soft palate. From the measured signal, they analyzed the parallel, perpendicular, sum of the parallel and perpendicular components (diffuse reflectance), difference between parallel and perpendicularly polarized light (depolarization ratio), and ratio of parallel to perpendicular components. Two features were extracted from each of these five spectral types, and nuclear size was also extracted from the polarization ratio (using Mie theory). Thus, for each classification, diagnostic parameters were identified from among these 11 features and combined to maximize the diagnostic accuracy. Classification models were generated by LDA, and leave-one-out cross-validation was used to train and test the models. The performance values for various models were compared based on the AUC. The LDA model developed using the best combination of diagnostic features yielded an AUC of 0.89 for the distinction of clinically normal sites from clinically abnormal sites, which were classified as moderate dysplasia, severe dysplasia, and cancer (MD/SD/C). An AUC of 0.91 was obtained for the distinction of benign lesions from MD/SD/C, and 0.87 for the distinction of mildly dysplastic lesions from MD/SD/C.

In examining the parameter trends with disease, McGee et al. found that dysplastic/malignant lesions were associated with increased scattering by smaller particles (Rayleigh scattering), higher cHb_T, and decreased scattering amplitude and β-carotene as compared to benign lesions (McGee et al. 2009). Mueller et al. observed a decrease in the slope of $\mu_s'(\lambda)$ with disease progression and found that cHb_T was not correlated with disease (Muller et al. 2003). In comparing normal tissue to malignant tissue, Amelink et al. noted a statistically significant decrease in α and scattering amplitude, and increase in blood volume fraction and scattering by smaller particles with disease (Amelink et al. 2008).

6.3.1.4 Lungs

White light bronchoscopy is a commonly used technique for the definitive diagnosis and staging of lung cancer. However, many premalignant lesions in the lung, including dysplastic lesions and CIS, are not detectable by this technique since there are no visible changes in the mucosa and because of the small size of these lesions (Hirsch et al. 2001, Wistuba and Gazdar 2006). The goal of many spectroscopic studies in the lungs has been to improve the detection of these premalignant lesions.

Bard et al. evaluated the diagnostic performance of diffuse RS, FS, and differential path-length spectroscopy for the classification of malignancies of the bronchial tree (Bard et al. 2006). Spectroscopic measurements were acquired from endobronchial lesions in 107 patients. PCA followed by LDA was performed to distinguish "low-grade" lesions (54 normal samples, 67 metaplastic samples, and 7 mildly dysplastic lesions) from "high-grade" lesions (6 severely dysplastic/CIS samples and 57 invasive carcinoma samples). The classification accuracy of diffuse RS as measured by the sensitivity, specificity, and AUC were 86%, 81%, and 0.90 ± 0.02, respectively. Analysis of the signal derived from differential path-length spectroscopy yielded a sensitivity, specificity, and AUC of 0.81%, 88%, and 89 ± 0.03, respectively. The accuracy of a diagnostic model based on the fluorescence signal acquired by the same system yielded poorer sensitivity and AUC values, 73% and 0.82 ± 0.03, respectively, but a similar specificity, 82%. Differential path-length spectroscopy was also used to specifically evaluate tissue hypoxia in endobronchial lesions as an important prognostic indicator of disease severity and survival (Bard et al. 2005). The investigators observed that tumors were characterized by a significantly lower α and Mie scattering amplitude, and a higher blood volume fraction compared to normal or metaplastic/mildly dysplastic bronchial tissue.

Aerts et al. evaluated the differential path-length (and fluorescence) signal from 17 endoscopically abnormal tissue sites in nine patients during bronchoscopy (Aerts et al. 2007). In addition to standard histopathology, quantification of HIF1A expression levels (a marker for hypoxia) was performed on the biopsied tissue samples and graded as negative/weakly or positive/strongly staining. The extracted spectroscopic parameters included α, local blood volume fraction, apparent average vessel diameter, and the *B* parameter (related to the size of the Mie scattering particles). Samples exhibiting severe dysplasia/malignancy ($n=6$) demonstrated a statistically significant decrease in α as compared to nonmalignant samples ($n=11$). A significant difference was also observed in the spectroscopically determined values for α between HIF1A positive and negative expression biopsies.

Fawzy et al. performed an *in vivo* imaging study of lung tissue using an integrated reflectance/fluorescence endoscopic imaging and spectroscopy system (Zeng et al. 2004, Fawzy et al. 2006). For the imaging measurements, the bronchial tissue was illuminated by the endoscope with a broad beam (~2 cm beam diameter) and the reemitted signal was collected from the tissue surface in a noncontact manner by the imaging fiber bundle of the endoscope. For the spectroscopy measurements, an optical fiber collected the reflectance signal or fluorescence signal from a spot at the center of the image. The spectroscopic measurements were also performed in a noncontact manner by using an intermediate image plane, thus eliminating the inconvenience of passing the optical fibers through the instrument channel of the endoscope. If a suspicious lesion was visualized based on the imaging view, spectroscopic measurements were then performed at the site. A set of 100 spectra were collected from 22 patients and classified as normal ($n=21$), hyperplasia (benign; $n=29$), mild dysplasia ($n=3$), or malignant ($n=50$). Because of the limited number of samples, data were reclassified as follows: (1) malignant (for moderate dysplasia or more severe disease) or (2) normal/benign (sites classified as mildly dysplastic or demonstrating less significant disease). An additional biopsy and corresponding spectral measurement was acquired from each patient from either a normal or benign appearing tissue site in order to assess the performance of the spectroscopic diagnosis relatively independent of the performance of the imaging diagnosis. Five histopathology-confirmed malignant lesions were found by these random biopsies. The investigators developed a light transport model to fit the reflectance signal based on the general diffusion approximation and a two-layer architecture in order to obtain $\mu_s(\lambda)$, $g(\lambda)$, and $\mu_a(\lambda)$. The tissue scatterer volume fraction and size distribution (in each layer), microvascular blood volume fraction and α were then derived using a numerical inversion fitting algorithm based on a Newton-type iteration scheme. Discriminant function analysis (DFA) was used to identify diagnostic parameters and develop models for tissue classification. Leave-one-out cross-validation was used to evaluate the sensitivity and specificity. Malignant tissue was characterized by a significant increase in blood volume fraction and significant decrease in α and mucosal layer scattering compared to normal/benign tissue. A moderately significant decrease in the size distribution parameter of the mucosa was also observed, which indicates an increase in the average size of the scatterers. Three significant parameters for the discrimination between malignant and nonmalignant tissue were identified by DFA: blood volume fraction, α and the scattering volume fraction in the top layer. Using these parameters, a sensitivity of 83% and specificity of 81% was obtained for the distinction of malignant lesions from normal/benign tissue. The sensitivity and specificity based on the physician's subjective identification of disease using imaging (combined reflectance and fluorescence) for the same patient population were 87% and 43%, respectively. As observed in the investigations previously described, malignant lesions demonstrated a decrease in α and scattering amplitude and an increase in blood volume fraction as compared to non-malignant lesions.

6.3.1.5 Esophagus and Gastrointestinal Tract

Endoscopic surveillance with four-quadrant biopsies along the length of the esophagus is the standard procedure for the management of patients with Barrett's esophagus (BE). Detection of high-grade dysplasia (HGD) is critical as the 5 year cancer risk exceeds 30% in individuals with HGD and surgical intervention is required (Bresalier 2009). Some limitations in the surveillance for dysplasia include the inability to sample large areas of at-risk esophagus by biopsy and interobserver variation in grading the degree of dysplasia. Several investigators have used RS to detect dysplasia and cancer in patients with BE.

Georgakoudi et al. used reflectance, fluorescence, and light-scattering spectroscopy to evaluate tissue classified as low-grade dysplasia (LGD) and HGD in 16 patients with known BE undergoing standard endoscopic surveillance (Georgakoudi et al. 2001). Reflectance spectra were analyzed using a model based on the diffusion approximation. Their analysis showed that the amplitude of the extracted $\mu_s'(\lambda)$ decreases with the development of dysplasia. Values of the slope extracted from linear fits to $\mu_s'(\lambda)$ also decreased with the development of dysplasia. Using logistic regression and leave-one-out cross-validation, the sensitivity and specificity for differentiating HGD from LGD/non-dysplastic Barrett's tissue (NDB) were 86% and 100%, respectively. The sensitivity and specificity for differentiating LGD/HGD from NDB were 79% and 88%, respectively. The results based on RS yielded a slightly lower overall diagnostic accuracy as compared to fluorescence and light-scattering spectroscopy. The combination of all three techniques resulted in the best overall sensitivity and specificity for distinguishing HGD from non-HGD and dysplastic from non-dysplastic epithelium.

In a study by Lovat et al., *in vivo* reflectance spectra were collected from 81 patients by passing the optical fiber probe through the biopsy channel of the endoscope (Lovat et al. 2006). The data set included 129 non-dysplastic, 5 LGD, 35 HGD, and 10 cancerous sites. HGD and cancerous sites were further classified as having extensive disease (>50% of the surface epithelium) or focal disease. PCA was performed for data reduction. LDA was used to generate the diagnostic algorithms. The distinction of extensive HGD/cancer from non-dysplastic/LGD sites yielded a sensitivity of 92% and specificity of 60% based on leave-one-out cross-validation (87% and 59%, respectively, based on the block validation method). The distinction of focal HGD/cancer sites from non-dysplastic/LGD sites yielded a sensitivity of 85% and specificity of 60% (80% and 59%, respectively, based on the block validation method).

HGD/cancer could be distinguished from moderate/severe inflammation (56 sites) with a sensitivity of 79% and specificity of 79% (77% and 76%, respectively, based on the block validation method).

Amelink et al. used narrow band imaging (NBI) and differential path-length spectroscopy to evaluate α *in vivo* from 7 normal, 10 LGD, 7 HGD, and 4 cancerous esophageal tissue samples (Amelink et al. 2009). NBI involves using filters to restrict the illumination light to a few narrow bands of light centered around the peaks of maximum absorption by hemoglobin (Gono et al. 2004). The measured signal is limited to the superficial vasculature because blue and green light have a relatively shallow penetration depth. They sought to determine whether they could noninvasively detect hypoxia, which is known to increase in the Barrett's metaplasia-dysplasia-adenocarcinoma sequence (Griffiths et al. 2007). The optical fiber probe was passed through the working channel of the endoscope and measurements were collected from 15 patients. The results showed that there was no significant difference in α among the four histopathology categories. The authors hypothesize that these findings may be due to the fact that the cancerous lesions in the study were early stage superficial adenocarcinomas. Additionally, cellular hypoxia may not always be reflected by low oxygen saturations in the microvasculature due to a number of factors related to the tissue microenvironment. No significant differences were found between histopathology categories in the scattering amplitude or size parameters, and the vessel diameters.

Colonoscopy is an endoscopic procedure used to detect high-risk lesions and cancer in the colon and rectum. Prevention of colorectal cancer largely depends on the detection of adenomatous polyps (premalignant lesions) and adenocarcinomas. During this procedure, biopsy of suspicious tissue sites and/or definitive treatment of polyps by polypectomy can be performed. Polyps that are large (≥10 mm), or demonstrate high-grade dysplasia or significant villous components by histology are considered to be advanced adenomas (Levin et al. 2008). There are several limitations to colonoscopy. The accuracy of this procedure is dependent on the skill of the operator and this technique may miss 6%–12% of large adenomas and 5% of cancers (Levin et al. 2008). Spectroscopy may improve the detection of premalignant lesions and diseased sites and enable disease identification without the need for biopsy.

Dhar et al. measured *in vivo* spectra from 138 colonic sites from 45 patients in order to evaluate the potential of RS to differentiate lesions of the colon (Dhar et al. 2006). The data set included spectra from 290 normal samples and 193 pathological samples (hyperplastic polyps, adenomatous polyps, chronic colitis, and cancer). All spectra were intensity corrected at 650 nm and then analyzed by PCA. LDA was used to classify the lesions and evaluate the sensitivity and specificity. To improve the accuracy of their results, 60% of the data was used to train the statistical diagnostic algorithm and 40% used as the test data set. The investigators obtained a sensitivity of 80% and specificity of 75% for distinguishing cancer from adenomatous polyps and a sensitivity of 85% and specificity of 88% for distinguishing dysplastic polyps from chronic colitis.

Zonios et al. evaluated reflectance spectra from adenomatous colon polyps using a physical model based on the diffusion approximation (Zonios et al. 1999). Data were collected *in vivo* from adenomatous polyps and corresponding sites in the surrounding normal mucosa in 13 patients undergoing routine colonoscopy. From the diffuse reflectance spectra, they extracted the cHb_T, α, scatterer density, and scatterer size. Adenomatous polyps were characterized by an increased cHb_T, a lower scatterer density, and a larger scatterer size. There was no trend in α with disease.

Wang et al. used diffuse RS to noninvasively measure cHb_T and α in colon tissue as indicators of angiogenesis and hypoxia, respectively (Wang et al. 2009). The goal of this pilot study was to further understand their role in disease progress and prognosis. *In vivo* spectra were acquired during colonoscopies from normal tissue (2 patients), polyps (17 patients), and cancer (8 patients) by advancing the probe through the accessory channel of the colonoscope. The probe employed a side-viewing geometry and consisted of a single illumination fiber and four collection fibers spaced at different distances (see Figure 6.5). Using a hybrid P_1 and P_3 diffuse reflectance model, they extracted cHb_T and α. They found that cHb_T increased and α decreased with colon carcinogenesis. They also noted considerable inter- and intra-patient heterogeneity in total cHb_T and α and that these quantities did not correlate with polyp size. In contrast to the findings of Zonios et al., adenomatous polyps exhibited a statistically significant decrease in α and no difference in cHb_T compared to surrounding normal tissue.

6.3.1.6 Additional Tissue Sites

RS has also been used to evaluate brain tumors (Lin et al. 2001, Gebhart et al. 2007, Gonga et al. 2008), ovarian cancer (Utzinger et al. 2001), bladder cancer (Mourant et al. 1995, Koenig et al. 1998), and skin cancer (Borisova et al. 2008, Rajaram et al. 2008, Drakaki et al. 2009).

6.3.2 Tissue Therapeutics

One application of RS that has been extensively evaluated is noninvasive monitoring of hemoglobin concentration and α for dosimetry during photodynamic therapy (PDT) (Solonenko et al. 2002, Korbelik et al. 2003, Woodhams et al. 2004, Amelink et al. 2005, Thompson et al. 2005, Wang et al. 2005, Cottrell et al. 2006, Johansson et al. 2007, Kruijt et al. 2008, Larsen et al. 2008, Lee et al. 2008). PDT is described in greater detail in Chapter 34; therefore, we just briefly discuss this application in this section. The cytotoxic effects of PDT are dependent on the availability of oxygen and the formation of reactive oxygen species. Therefore, continuous monitoring of tissue oxygenation during or after treatment based on α may be one means of predicting treatment efficacy (Woodhams et al. 2007). The hemoglobin oxygenation saturation within tumors has also been shown to have prognostic value (Wang et al. 2004). To improve the use of

RS for dosimetry during PDT, accurate modeling of the measured reflectance is critical. If the fitted optical properties are to be used to calculate the fluence rate, accounting for the layered architecture of tissue may be important as one layer algorithms may result in significant inaccuracies in the calculation (Farrell et al. 1998). Although measurement of tissue oxygenation based on hemoglobin oxygen saturation is an indirect method, there is a close association between the two values, and in contrast to other direct measurement methods, spectroscopy is noninvasive and relatively simple to implement in the clinic (Woodhams et al. 2007).

The evaluation of tissue absorption properties by RS has also been used to evaluate the efficacy of radiation and chemotherapy (Jakubowski et al. 2004, Sunar et al. 2006, Cerussi et al. 2007). Tissue oxygen concentration has been shown to be important in the response of human tumors to radiation therapy, chemotherapy, and other treatment modalities (Evans and Koch 2003). Sunar et al. used spectroscopy to monitor the response of eight patients with head and neck tumors to treatment with a combination of chemotherapy and radiation therapy (chemoradiation therapy) (Sunar et al. 2006). RS and diffuse correlation spectroscopy were performed before (baseline) and during chemoradiation therapy. From the former technique, they extracted the concentration of $cHbO_2$ and cHb, and calculated cHb_T and α; from the latter, they extracted information related to tumor blood flow. Diffuse correlation spectroscopy is discussed in greater detail in Chapter 10. Reflectance spectra were acquired using a four-channel frequency domain instrument. Optical fiber bundles were placed in contact with the tissue (*in vivo*) and used to deliver and measure the light returning from the tissue at four different source–detector separations. Measurements were performed weekly from baseline (week 0) for a period of 4 weeks. The preliminary data from this feasibility study demonstrated statistically significant changes in cHb_T and α in the first 2 weeks of treatment. These results demonstrated that it may be possible to assess the early response of patients to therapy using spectroscopy. Cerussi et al. investigated whether diffuse OS measurements obtained before and after the first week of a 3 month adriamycin/cytoxan neoadjuvant chemotherapy regimen could be used to predict the final postsurgical response in 11 patients with breast cancer (Cerussi et al. 2007). For each patient, spectra were collected *in vivo* using a probe-based instrument from malignant breast tissue and from the same regions on the contralateral normal breast tissue. Subjects were classified as responders if there was a 50% change in the maximum-tumor axis in final pathology dimensions relative to the initial maximum axis dimension; the remaining subjects were classified as nonresponders. The investigators extracted $cHbO_2$, cHb, the concentration of water, and bulk lipid content from the NIR spectra. At baseline, the only statistically significant difference between responders ($n=6$) and nonresponders ($n=5$) was in the ratio of $cHbO_2$ in tumor tissue to that of normal tissue. The best single predictor of therapeutic response after 1 week of treatment was the cHb of the tumor tissue; responders demonstrated a larger decline than nonresponders.

6.4 Critical Discussion

In this section, we discuss four major areas in which studies evaluating the application of RS to tissue diagnostics can be improved. We concentrate most of our discussion points on those tissues in which the most extensive research has been performed, the cervix, breast, and oral cavity.

6.4.1 Discriminating Disease in the Context of Normal Tissue Variations

The normal variations of biological tissue must be carefully assessed and taken into account when evaluating the performance of spectroscopic diagnostic algorithms. If these intrinsic tissue variations are not considered, differences in the optical properties between diseased and non-disease tissue cannot be definitively attributed to the disease process, as they may in fact reflect the influence of the tissue microenvironment. Several studies have shown that normal tissue variations are important sources of spectroscopic contrast and thus these variations can confound the evaluation of spectroscopic contrast due to disease. In cervical tissue, variations have been observed in the reflectance signal between normal ectocervical (squamous tissue), transformation zone (mixture of squamous and columnar tissue), and endocervical tissue (columnar tissue) (Georgakoudi et al. 2002, Freeberg et al. 2007b, Mirkovic et al. 2009). Age and menopausal status are other factors that can produce changes in the microanatomy, and as a result, changes in the spectral properties (Arifler et al. 2006, Freeberg et al. 2007b, Mourant et al. 2009). Large variations in the optical properties have been noted in breast tissue as a result of the presence of various tissue types, including fibrous, glandular, and adipose tissue (van Veen et al. 2004, Palmer et al. 2006, Zhu et al. 2008b). The composition of breast tissue also changes with age and can vary with body mass index (i.e., lipid content) (Cerussi et al. 2001, Taroni et al. 2003). In oral tissue, many investigators have noted differences in the reflectance properties of various anatomic sites (e.g., buccal mucosa, gingiva) in healthy volunteers (Muller et al. 2003, Mallia et al. 2008, McGee et al. 2008). The differences in optical properties have been attributed to the varying histological architecture of the anatomic sites in terms of keratinization, the density and size of the underlying blood vessels, epithelial thickness, and the composition of the stroma (e.g., bone, muscle, or fat) (McGee et al. 2008, Schwarz et al. 2008). For example, the buccal mucosa is a non-keratinized site with a thick epithelium, while the hard palate is keratinized and has a comparably thinner epithelium. Therefore, the optical properties of these two sites may differ significantly. Figure 6.13 shows a binary scatter plot of cHb_T versus the A scattering parameter (related to the magnitude of Mie scattering) for the buccal mucosa and hard palate based on data measured in healthy volunteers.

There is a clear distinction in the spectroscopy parameters for these sites as they can be discriminated with a sensitivity of 89% and specificity of 95%. Developing diagnostic algorithms specific to a single anatomic site or for specific groups of anatomic sites

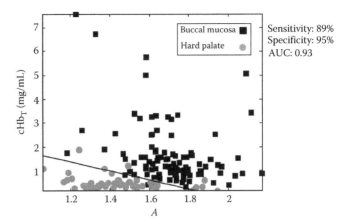

FIGURE 6.13 A binary scatter plot of the total hemoglobin concentration (cHb_T) versus the A scattering parameter for the buccal mucosa (black squares) and hard palate (gray circles). A model was developed using logistic regression to separate the two sites and the decision line and performance results are shown.

with similar spectroscopic properties has been found to significantly improve diagnostic performance compared to algorithms developed for all sites combined (Muller et al. 2003, Mallia et al. 2008, McGee et al. 2009). Thus, improved discrimination between diseased and non-diseased tissue may be an important additional benefit of minimizing the confounding effects of normal tissue variations.

6.4.2 Developing Clinically Relevant Spectroscopic Tools

Spectroscopic tools should provide information that is useful for improving clinical decision-making, in particular, identifying disease and distinguishing lesions, which require intervention from those that do not. Therefore, diagnostic algorithms must be designed to address clinically relevant distinctions. To reliably access the potential of RS to serve as an adjunct to colposcopy, it is important to distinguish diseased from non-diseased tissue among colposcopically abnormal sites. Specifically, the distinction of CIN2+ from low-grade lesions within the transformation zone is vital. The vast majority of diseased lesions in the cervix are located within the transformation zone, and normal ectocervix and endocervix are easily identified by colposcopic examination (Blaustein and Kurman 2002). Historically, many clinical studies have examined less relevant distinctions, such as the discrimination of diseased samples (colposcopically abnormal) either from colposcopically normal samples or a combination of colposcopically normal tissue and colposcopically abnormal nondysplastic/low-grade disease tissue samples (Nordstrom et al. 2001, Mirabal et al. 2002, Chang et al. 2009). Several studies have shown that the inclusion of colposcopically normal sites in the non-diseased category leads to higher diagnostic performance values (Mourant et al. 2007, 2009). Furthermore, as the percentage of normal samples increases in the non-diseased category, there is a corresponding increase in the magnitude of the performance values, thus artificially enhancing the apparent diagnostic accuracy (Mirkovic et al. 2009). Combining colposcopically normal samples (typically squamous ectocervical tissue) with non-dysplastic or LGD samples (mostly in the transformation zone) is not helpful from a clinical standpoint and furthermore ignores the intrinsic differences between these two tissues. By focusing on the clinically relevant distinction, however, the impact of these normal tissue variations can be minimized. Distinguishing diseased from non-diseased tissue among colposcopically abnormal samples is a more challenging distinction as compared to when colposcopically normal sites are included in the non-diseased group (Chang et al. 2005, Mirkovic et al. 2009, Mourant et al. 2009). For the distinction of HSIL/cancer from LSIL/non-dysplastic sites, Mourant et al. observed a decrease in specificity from 62.2% to 43.8% in the training set, when colposcopically normal samples were excluded and only colposcopically abnormal samples were distinguished. The investigators noted than some of the diagnostic parameters identified in the distinction of HSIL from non-HSIL were no longer significant when colposcopically normal samples were excluded. A study by Mirkovic et al. noted similar findings and that the spectral differences observed between clinically normal squamous and HSIL sites were largely due to differences between squamous mucosa and the mixed squamous/columnar tissue of the transformation zone, and not due to a disease (Mirkovic et al. 2009).

The major clinical challenge in the assessment of surgical margins surrounding excised breast lesions is the distinction of diseased from non-diseased sites for areas of tissue that appear normal by gross inspection. Therefore, spectra from grossly normal tissue, which is classified as diseased based on histopathology, must be acquired and distinguished from spectra from grossly normal, histopathology-confirmed normal tissue. Reflectance studies in the breast have typically evaluated *ex vivo* tissue. To evaluate the true performance accuracy of this technique in diagnostic procedures, such as CNB, there is a need to transition to *in vivo* studies. While a significant portion of the work performed in the breast has focused on developing models for distinguishing normal/benign from malignant tissue, it is also important to discriminate subgroups within these categories. Benign lesions, such as those demonstrating fibrocystic change, do not increase a woman's risk for developing breast cancer. However, proliferative benign lesions such as fibroadenomas must be identified because they are associated with a modest increase in the risk for developing breast cancer, and therefore require intervention (Cotran et al. 1999).

In the oral cavity, de Veld et al. found that RS could reliably distinguish healthy (clinically normal) oral tissue samples from clinically abnormal malignant samples (not a clinically relevant distinction). In contrast, distinguishing among clinically abnormal lesions versus dysplastic/malignant lesions) led to considerably weaker performance values (de Veld et al. 2005). Other studies have also demonstrated this finding (Muller et al. 2003, McGee et al. 2009). In the discrimination of lesion categories, clinically normal oral tissue samples are often grouped with benign lesions. This is an inappropriate grouping given that

clinically normal tissue can be easily distinguished from the visible changes in the tissue mucosa associated with lesions, and also because it can lead to misleading results when evaluating diagnostic algorithms. Just as for cervical tissue, combining clinically normal oral tissue samples with clinically abnormal benign lesions for distinction from clinically abnormal diseased samples can increase the apparent diagnostic accuracy, and the magnitude of the increase is dependent on the proportion of healthy samples (McGee et al. 2009). McGee et al. also noted that while several diagnostic parameters were identified for the distinction of healthy mucosa from lesions (all types), very few were diagnostic in the more challenging distinction, discriminating benign from dysplastic/malignant lesions.

6.4.3 Simplification of the Instrumentation and Analysis Procedures

Many of the clinical studies involve instrumentation with complex probe designs, prolonged times for data acquisition, and intricate data analysis procedures (e.g., when multiple spectroscopic techniques are used) in order to extract the maximum amount of information with which to detect disease-related contrast and construct diagnostic algorithms. However, the results of many studies demonstrate that certain elements of the instrumentation or analysis may be unnecessary and a more simplified approach may be sufficient (Mirabal et al. 2002, de Veld et al. 2005, Zhu et al. 2005, Sunar et al. 2006, Freeberg et al. 2007a, Schwarz et al. 2009). Only those aspects of the instrumentation and analysis required to extract parameters or spectral features necessary for discriminating the histopathology categories of interest (clinically relevant distinctions) need to be retained. This may also help reduce the instrumentation costs and size, and data acquisition time.

6.4.4 Progressing toward Large, Prospective Multisite Studies

There is a need to transition from smaller pilot studies focused on validating instrumentation, analysis techniques, and developing diagnostic algorithms to larger scale clinical studies in which the instrumentation and diagnostic algorithms are prospectively tested on a large number of patients at multiple study sites. Considerable progress in this area has been made in the cervix. The diagnostic effectiveness of RS (either alone or in combination with other spectroscopic techniques), in both contact probe and imaging modes, has been tested in various stages of clinical studies from pilot to phase III clinical trials (Nordstrom et al. 2001, Georgakoudi et al. 2002, Mirabal et al. 2002, Huh et al. 2004, Chang et al. 2005, Marin et al. 2005, Alvarez et al. 2007, Mourant et al. 2007). In order to prove the clinical validity of RS in tissue diagnostics in other tissue types, studies need to be performed in large cohorts of patients to determine whether the performance meets or exceeds that of standard procedures given the significant biological variability among patients and within lesion categories.

6.5 Summary

RS examines the interplay of elastic scattering and absorption. Measurements can be performed in a fraction of a second and with relatively simple instrumentation that can easily be adapted for a number of biomedical applications. The major goal of RS in tissue diagnostics has been primarily to enable targeted biopsies of diseased tissue or provide a minimally/noninvasive alternative to biopsy and histopathological evaluation. The potential of this technique to diagnose disease has been demonstrated in a wide range of tissue types, both alone or in combination with other spectroscopic modalities. Major areas in which reflectance studies focusing on tissue diagnostics can be improved were discussed including the following: (1) discriminating disease in the context of normal tissue variations, (2) developing clinically relevant spectroscopic tools, (3) simplification of the instrumentation and analysis methods, and (4) progressing toward large, prospective multisite studies. Advances in these areas will significantly strengthen the translation of these research instruments into effective clinical tools.

6.6 Future Perspective

Future RS studies focused on tissue diagnostics should involve large, prospective studies using simplified instrumentation and analysis methods that have been optimized to address clinically relevant distinctions. Diagnostic algorithms should be robust enough to distinguish disease contrast from other sources of normal tissue contrast and accurate even in the context of large biological variability among patients and within lesion categories.

Advancing the potential of reflectance-based spectroscopic tools to provide real-time or rapid diagnoses is also an important goal. Developments in these areas will enable this technique to significantly impact the lives of patients.

References

Aerts, J., Amelink, A., van der Leest, C. et al. 2007. HIF1a expression in bronchial biopsies correlates with tumor microvascular saturation determined using optical spectroscopy. *Lung Cancer* 57: 317–321.

Alvarez, R. D., Wright, T. C., and Optical Detection Group. 2007. Effective cervical neoplasia detection with a novel optical detection system: A randomized trial. *Gynecol. Oncol.* 104: 281–289.

Amelink, A., Bard, M. P. L., Burgers, S. A. et al. 2003. Single-scattering spectroscopy for the endoscopic analysis of particle size in superficial layers of turbid media. *Appl. Opt.* 42: 4095–4101.

Amelink, A., Haringsma, J., and Sterenborg, H. 2009. Noninvasive measurement of oxygen saturation of the microvascular blood in Barrett's dysplasia by use of optical spectroscopy. *Gastroint. Endosc.* 70: 1–6.

Amelink, A., van den Heuvel, A. P., de Wolf, W. J. et al. 2005. Monitoring PDT by means of superficial reflectance spectroscopy. *J. Photochem. Photobiol. B Biol.* 79: 243–251.

Amelink, A., Kaspers, O. P., Sterenborg, H. et al. 2008. Noninvasive measurement of the morphology and physiology of oral mucosa by use of optical spectroscopy. *Oral Oncol.* 44: 65–71.

Amelink, A. and Sterenborg, H. J. C. M. 2004a. Measurement of the local optical properties of turbid media by differential path-length spectroscopy. *Appl. Opt.* 43: 3048–3054.

Amelink, A., Sterenborg, H. J. C. M., Bard, M. P. L. et al. 2004b. In vivo measurement of the local optical properties of tissue by use of differential path-length spectroscopy. *Opt. Lett.* 29: 1087–1089.

Arifler, D., MacAulay, C., Follen, M. et al. 2006. Spatially resolved reflectance spectroscopy for diagnosis of cervical precancer: Monte Carlo modeling and comparison to clinical measurements. *J. Biomed. Opt.* 11: 16.

Arifler, D., Pavlova, I., Gillenwater, A. et al. 2007. Light scattering from collagen fiber networks: Micro-optical properties of normal and neoplastic stroma. *Biophys. J.* 92: 3260–3274.

Arifler, D., Schwarz, R. A., Chang, S. K. et al. 2005. Reflectance spectroscopy for diagnosis of epithelial precancer: Model-based analysis of fiber-optic probe designs to resolve spectral information from epithelium and stroma. *Appl. Opt.* 44: 4291–4305.

Bard, M. P. L., Amelink, A., Hegt, V. N. et al. 2005. Measurement of hypoxia-related parameters in bronchial mucosa by use of optical spectroscopy. *Am. J. Respir. Crit. Care Med.* 171: 1178–1184.

Bard, M. P. L., Amelink, A., Skurichina, M. et al. 2006. Optical spectroscopy for the classification of malignant lesions of the bronchial tree. *Chest* 129: 995–1001.

Bargo, P. R., Prahl, S. A., Goodell, T. T. et al. 2005. In vivo determination of optical properties of normal and tumor tissue with white light reflectance and an empirical light transport model during endoscopy. *J. Biomed. Opt.* 10: 034018.

Bargo, P. R., Prahl, S. A., and Jacques, S. L. 2003a. Collection efficiency of a single optical fiber in turbid media. *Appl. Opt.* 42: 3187–3197.

Bargo, P. R., Prahl, S. A., and Jacques, S. L. 2003b. Optical properties effects upon the collection efficiency of optical fibers in different probe configurations. *IEEE J. Sel. Top. Quant. Electron.* 9: 314–321.

Bender, J. E., Vishwanath, K., Moore, L. K. et al. 2009. A robust Monte Carlo model for the extraction of biological absorption and scattering in vivo. *IEEE Trans. Biomed. Eng.* 56: 960–968.

Bigio, I. J., Bown, S. G., Briggs, G. et al. 2000. Diagnosis of breast cancer using elastic-scattering spectroscopy: Preliminary clinical results. *J. Biomed. Opt.* 5: 221–228.

Blaustein, A. and Kurman, R. J. 2002. *Blaustein's Pathology of the Female Genital Tract*. New York: Springer.

Borisova, E., Troyanova, P., Pavlova, P. et al. 2008. Diagnostics of pigmented skin tumors based on laser-induced autofluorescence and diffuse reflectance spectroscopy. *Quantum Electron.* 38: 597–605.

Bresalier, R. S. 2009. Barrett's esophagus and esophageal adenocarcinoma. *Annu. Rev. Med.* 60: 221–231.

Breslin, T. M., Xu, F. S., Palmer, G. M. et al. 2004. Autofluorescence and diffuse reflectance properties of malignant and benign breast tissues. *Ann. Surg. Oncol.* 11: 65–70.

Brown, J. Q., Wilke, L. G., Geradts, J. et al. 2009. Quantitative optical spectroscopy: A robust tool for direct measurement of breast cancer vascular oxygenation and total hemoglobin content in vivo. *Cancer Res.* 69: 2919–2926.

Cawson, R. A. and Odell, E. W. 2002. *Essentials of Oral Pathology and Oral Medicine.* New York: Churchill Livingstone.

Cendán, J. C., Coco, D., and Copeland III, E. M. 2005. Accuracy of intraoperative frozen-section analysis of breast cancer lumpectomy-bed margins. *J. Am. Coll. Surg.* 201: 194–198.

Cerussi, A. E., Berger, A. J., Bevilacqua, F. et al. 2001. Sources of absorption and scattering contrast for near-infrared optical mammography. *Acad. Radiol.* 8: 211–218.

Cerussi, A., Hsiang, D., Shah, N. et al. 2007. Predicting response to breast cancer neoadjuvant chemotherapy using diffuse optical spectroscopy. *Proc. Natl. Acad. Sci.* 104: 4014–4019.

Chang, V. T. C., Cartwright, P. S., Bean, S. M. et al. 2009. Quantitative physiology of the precancerous cervix in vivo through optical spectroscopy. *Neoplasia* 11: 325–332.

Chang, S. K., Mirabal, Y. N., Atkinson, E. N. et al. 2005. Combined reflectance and fluorescence spectroscopy for in vivo detection of cervical pre-cancer. *J. Biomed. Opt.* 10: 024031.

Cheong, W. F., Prahl, S. A., and Welch, A. J. 1990. A review of the optical-properties of biological tissues. *IEEE J. Quantum Electron.* 26: 2166–2185.

Collier, T., Follen, M., Malpica, A. et al. 2005. Sources of scattering in cervical tissue: Determination of the scattering coefficient by confocal microscopy. *Appl. Opt.* 44: 2072–2081.

Coppleson, M., Reid, B. L., Skladnev, V. N. et al. 1994. An electronic approach to the detection of pre-cancer and cancer of the uterine cervix: A preliminary evaluation of polarprobe. *Int. J. Gynecol. Cancer* 4: 79–83.

Cotran, R. S., Kumar, V., Collins, T. et al. 1999. *Robbins Pathologic Basis of Disease.* Philadelphia, PA: Saunders.

Cottrell, W. J., Oseroff, A. R., and Foster, T. H. 2006. Portable instrument that integrates irradiation with fluorescence and reflectance spectroscopies during clinical photodynamic therapy of cutaneous disease. *Rev. Sci. Instrum.* 77: 8.

de Veld, D. C. G., Skurichina, M., Wities, M. J. H. et al. 2005. Autofluorescence and diffuse reflectance spectroscopy for oral oncology. *Lasers Surg. Med.* 36: 356–364.

DeSantis, T., Chakhtoura, N., Twiggs, L. et al. 2007. Spectroscopic imaging as a triage test for cervical disease: A prospective multicenter clinical trial. *J. Low Genit. Tract. Dis.* 11: 18–24.

Devet, H. C. W., Knipschild, P. G., Schouten, H. J. A. et al. 1990. Interobserver variation in histopathological grading of cervical dysplasia. *J. Clin. Epidemiol.* 43: 1395–1398.

Dhar, A., Johnson, K. S., Novelli, M. R. et al. 2006. Elastic scattering spectroscopy for the diagnosis of colonic lesions: Initial results of a novel optical biopsy technique. *Gastroint. Endosc.* 63: 257–261.

Dickey, D. J., Moore, R. B., Rayner, D. C. et al. 2001. Light dosimetry using the P3 approximation. *Phys. Med. Biol.* 46: 2359–2370.

Drakaki, E., Kaselouris, E., Makropoulou, M. et al. 2009. Laser-induced fluorescence and reflectance spectroscopy for the discrimination of basal cell carcinoma from the surrounding normal skin tissue. *Skin Pharmacol. Physiol.* 22: 158–165.

Du, H., Fuh, R. C. A., Li, J. Z. et al. 1998. PhotochemCAD: A computer-aided design and research tool in photochemistry. *Photochem. Photobiol.* 68: 141–142.

Evans, S. M. and Koch, C. J. 2003. Prognostic significance of tumor oxygenation in humans. *Cancer Lett.* 195: 1–16.

Farrell, T. J., Patterson, M. S., and Essenpreis, M. 1998. Influence of layered tissue architecture on estimates of tissue optical properties obtained from spatially resolved diffuse reflectometry. *Appl. Opt.* 37: 1958–1972.

Farrell, T. J., Patterson, M. S., and Wilson, B. 1992. A diffusion-theory model of spatially resolved, steady-state diffuse reflectance for the noninvasive determination of tissue optical-properties in vivo. *Med. Phys.* 19: 879–888.

Fawzy, Y. S., Petek, M., Tercelj, M. et al. 2006. In vivo assessment and evaluation of lung tissue morphologic and physiological changes from non-contact endoscopic reflectance spectroscopy for improving lung cancer detection. *J. Biomed. Opt.* 11: 12.

Ferris, D. G. 2004. *Modern Colposcopy: Textbook and Atlas.* Dubuque, IA: Kendall/Hunt.

Ferris, D. G., Lawhead, R. A., Dickman, E. D. et al. 2001. Multimodal hyperspectral imaging for the noninvasive diagnosis of cervical neoplasia. *J. Low Genit. Tract. Dis.* 5: 65–72.

Finlay, J. C. and Foster, T. H. 2004. Effect of pigment packaging on diffuse reflectance spectroscopy of samples containing red blood cells. *Opt. Lett.* 29: 965–967.

Fischer, D. J., Epstein, J. B., Morton, T. H. et al. 2004. Interobserver reliability in the histopathologic diagnosis of oral premalignant and malignant lesions. *J. Oral Pathol. Med.* 33: 65–70.

Fischer, D. J., Epstein, J. B., Morton, T. H. et al. 2005. Reliability of histologic diagnosis of clinically normal intraoral tissue adjacent to clinically suspicious lesions in former upper aerodigestive tract cancer patients. *Oral Oncol.* 41: 489–496.

Fletcher, S. W. and Elmore, J. G. 2003. Mammographic screening for breast cancer. *N. Engl. J. Med.* 348: 1672–1680.

Freeberg, J. A., Benedet, J. L., MacAulay, C. et al. 2007a. The performance of fluorescence and reflectance spectroscopy for the in vivo diagnosis of cervical neoplasia; point probe versus multispectral approaches. *Gynecol. Oncol.* 107: S248–S255.

Freeberg, J. A., Serachitopol, D. M., McKinnon, N. et al. 2007b. Fluorescence and reflectance device variability throughout the progression of a phase II clinical trial to detect and screen for cervical neoplasia using a fiber optic probe. *J. Biomed. Opt.* 12: 034015.

Gage, J. C., Hanson, V. W., Abbey, K. et al. 2006. Number of cervical biopsies and sensitivity of colposcopy. *Obstet. Gynecol.* 108: 264–272.

Gebhart, S. C., Thompson, R. C., and Mahadevan-Jansen, A. 2007. Liquid-crystal tunable filter spectral imaging for brain tumor demarcation. *Appl. Opt.* 46: 1896–1910.

Georgakoudi, I., Jacobson, B. C., Van Dam, J. et al. 2001. Fluorescence, reflectance, and light-scattering spectroscopy for evaluating dysplasia in patients with Barrett's esophagus. *Gastroenterology* 120: 1620–1629.

Georgakoudi, I., Sheets, E. E., Muller, M. G. et al. 2002. Tri-modal spectroscopy for the detection and characterization of cervical precancers *in vivo*. *Am. J. Obstet. Gynecol.* 186: 374–382.

Ghosh, N., Mohanty, S. K., Majumder, S. K. et al. 2001. Measurement of optical transport properties of normal and malignant human breast tissue. *Appl. Opt.* 40: 176–184.

Gonga, J., Yi, J., Turzhitsky, V. M. et al. 2008. Characterization of malignant brain tumor using elastic light scattering spectroscopy. *Dis. Markers* 25: 303–312.

Gono, K., Obi, T., Yamaguchi, M. et al. 2004. Appearance of enhanced tissue features in narrow-band endoscopic imaging. *J. Biomed. Opt.* 9: 568–577.

Graaff, R., Aarnoudse, J. G., Zijp, J. R. et al. 1992. Reduced light-scattering properties for mixtures of spherical-particles—A simple approximation derived from Mie calculations. *Appl. Opt.* 31: 1370–1376.

Griffiths, E. A., Pritchard, S. A., McGrath, S. M. et al. 2007. Increasing expression of hypoxia-inducible proteins in the Barrett's metaplasia-dysplasia-adenocarcinoma sequence. *Br. J. Cancer* 96: 1377–1383.

Hattery, D., Hattery, B., Chernomordik, V. et al. 2004. Differential oblique angle spectroscopy of the oral epithelium. *J. Biomed. Opt.* 9: 951–960.

Hidovic-Rowe, D. and Claridge, E. 2005. Modelling and validation of spectral reflectance for the colon. *Phys. Med. Biol.* 50: 1071–1093.

Hirsch, F. R., Prindiville, S. A., Miller, Y. et al. 2001. Fluorescence versus white-light bronchoscopy for detection of preneoplastic lesions: a randomized study. *J. Natl. Cancer Inst.* 93: 1385–1391.

Huh, W. K., Cestero, R. M., Garcia, F. A. et al. 2004. Optical detection of high-grade cervical intraepithelial neoplasia in vivo: Results of a 604-patient study. *Am. J. Obstet. Gynecol.* 190: 1249–1257.

Hull, E. L. and Foster, T. H. 2001. Steady-state reflectance spectroscopy in the P-3 approximation. *J. Opt. Soc. Am. A Opt. Image Sci. Vis.* 18: 584–599.

Ishimaru, A. 1978. *Wave Propagation and Scattering in Random Media*. New York: Academic Press.

Jacques, S. L. 1996. Modelling light propagation in tissue. In *Biomedical Optical Instrumentation and Laser-Assisted Biotechnology, NATO ASI Series. Series E, Applied Sciences No. 325*, ed. A. M. Verga Scheggi, vol. xxviii. Boston, MA: Kluwer Academic, 407pp.

Jacques, S. L. and Pogue, B. W. 2008. Tutorial on diffuse light transport. *J. Biomed. Opt.* 13: 19.

Jakubowski, D. B., Cerussi, A. E., Bevilacqua, F. et al. 2004. Monitoring neoadjuvant chemotherapy in breast cancer using quantitative diffuse optical spectroscopy: A case study. *J. Biomed. Opt.* 9: 230–238.

Johansson, A., Svensson, J., Bendsoe, N. et al. 2007. Fluorescence and absorption assessment of a lipid mTHPC formulation following topical application in a non-melanotic skin tumor model. *J. Biomed. Opt.* 12: 9.

Johnson, K. S., Chicken, D. W., Pickard, D. C. O. et al. 2004. Elastic scattering spectroscopy for intraoperative determination of sentinel lymph node status in the breast. *J. Biomed. Opt.* 9: 1122–1128.

Jordan, R. C. K., Macabeo-Ong, M., Shiboski, C. H. et al. 2004. Overexpression of matrix metalloproteinase-1 and-9 mRNA is associated with progression of oral dysplasia to cancer. *Clin. Cancer Res.* 10: 6460–6465.

Koenig, F., Larne, R., Enquist, H. et al. 1998. Spectroscopic measurement of diffuse reflectance for enhanced detection of bladder carcinoma. *Urology* 51: 342–345.

Korbelik, M., Sun, J., and Zeng, H. 2003. Ischaemia-reperfusion injury in photodynamic therapy-treated mouse tumours. *Br. J. Cancer* 88: 760–766.

Kruijt, B., de Bruijn, H. S., van der Ploeg-van den Heuvel, A. et al. 2008. Monitoring ALA-induced PpIX photodynamic therapy in the rat esophagus using fluorescence and reflectance spectroscopy. *Photochem. Photobiol.* 84: 1515–1527.

Larsen, E. L. P., Randeberg, L. L., Gederaas, O. A. et al. 2008. Monitoring of hexyl 5-aminolevulinate-induced photodynamic therapy in rat bladder cancer by optical spectroscopy. *J. Biomed. Opt.* 13: 9.

Lee, T. K., Baron, E. D., and Foster, T. H. 2008. Monitoring Pc 4 photodynamic therapy in clinical trials of cutaneous T-cell lymphoma using noninvasive spectroscopy. *J. Biomed. Opt.* 13: 3.

Lee, J. J., Hung, H. C., Cheng, S. J. et al. 2007. Factors associated with underdiagnosis from incisional biopsy of oral leukoplakic lesions. *Oral Surg. Oral Med. Oral Pathol. Oral Radiol. Endodontol.* 104: 217–225.

Levin, B., Lieberman, D. A., McFarland, B. et al. 2008. Screening and surveillance for the early detection of colorectal cancer and adenomatous polyps, 2008: A joint guideline from the American Cancer Society, the US Multi-Society Task Force on Colorectal Cancer, and the American College of Radiology. *Gastroenterology* 134: 1570–1595.

Lin, W. C., Toms, S. A., Johnson, M. et al. 2001. In vivo brain tumor demarcation using optical spectroscopy. *Photochem. Photobiol.* 73: 396–402.

Liu, H., Chance, B., Hielscher, A. H. et al. 1995. Influence of blood-vessels on the measurement of hemoglobin oxygenation as determined by time-resolved reflectance spectroscopy. *Med. Phys.* 22: 1209–1217.

Liu, Q. and Ramanujam, N. 2006. Sequential estimation of optical properties of a two-layered epithelial tissue model from depth-resolved ultraviolet-visible diffuse reflectance spectra. *Appl. Opt.* 45: 4776–4790.

Lo, J. Y., Yu, B., Fu, H. L. et al. 2009. A strategy for quantitative spectral imaging of tissue absorption and scattering using light emitting diodes and photodiodes. *Opt. Express* 17: 1372–1384.

Lovat, L. B., Johnson, K., Mackenzie, G. D. et al. 2006. Elastic scattering spectroscopy accurately detects high grade dysplasia and cancer in Barrett's oesophagus. *Gut* 55: 1078–1083.

Macluskey, M., Chandrachud, L. M., Pazouki, S. et al. 2000. Apoptosis, proliferation, and angiogenesis in oral tissues. Possible relevance to tumour progression. *J. Pathol.* 191: 368–375.

Majumder, S. K., Keller, M. D., Boulos, F. I. et al. 2008. Comparison of autofluorescence, diffuse reflectance, and Raman spectroscopy for breast tissue discrimination. *J. Biomed. Opt.* 13:

Mallia, R. J., Thomas, S. S., Mathews, A. et al. 2008. Oxygenated hemoglobin diffuse reflectance ratio for *in vivo* detection of oral pre-cancer. *J. Biomed. Opt.* 13: 041306.

Marin, N. M., Milbourne, A., Rhodes, H. et al. 2005. Diffuse reflectance patterns in cervical spectroscopy. *Gynecol. Oncol.* 99: S116–S120.

McGee, S. A., Mardirossian, V., Elackattu, A. et al. 2009. Anatomy-based algorithms for detecting oral cancer using reflectance and fluorescence spectroscopy. *Ann. Otol. Rhinol. Laryngol.* 118: 817–826.

McGee, S. A., Mirkovic, J., Mardirossian, V. et al. 2008. Model-based spectroscopic analysis of the oral cavity: Impact of anatomy. *J. Biomed. Opt.* 13: 064034.

Milbourne, A., Park, S. Y., Benedet, J. L. et al. 2005. Results of a pilot study of multispectral digital colposcopy for the in vivo detection of cervical intraepithelial neoplasia. *Gynecol. Oncol.* 99: S67–S75.

Mirabal, Y. N., Chang, S. K., Atkinson, E. N. et al. 2002. Reflectance spectroscopy for in vivo detection of cervical precancer. *J. Biomed. Opt.* 7: 587–594.

Mirkovic, J., Lau, C., McGee, S. et al. 2009. Effect of anatomy on spectroscopic detection of cervical dysplasia *J. Biomed. Opt.* 14: 044021.

Mitchell, M. F., Schottenfeld, D., Tortolero-Luna, G. et al. 1998. Colposcopy for the diagnosis of squamous intraepithelial lesions: A meta-analysis. *Obstet. Gynecol.* 91: 626–631.

Mourant, J. R., Bigio, I., Boyer, J. et al. 1995. Spectroscopic diagnosis of bladder cancer with elastic scattering spectroscopy. *Lasers Surg. Med.* 17: 350–357.

Mourant, J. R., Bocklage, T. J., Powers, T. M. et al. 2007. In vivo light scattering measurements for detection of precancerous conditions of the cervix. *Gynecol. Oncol.* 105: 439–445.

Mourant, J. R., Freyer, J. P., Hielscher, A. H. et al. 1998. Mechanisms of light scattering from biological cells relevant to noninvasive optical-tissue diagnostics. *Appl. Opt.* 37: 3586–3593.

Mourant, J. R., Fuselier, T., Boyer, J. et al. 1997. Predictions and measurements of scattering and absorption over broad wavelength ranges in tissue phantoms. *Appl. Opt.* 36: 949–957.

Mourant, J. R., Powers, T. M., Bocklage, T. J. et al. 2009. In vivo light scattering for the detection of cancerous and precancerous lesions of the cervix. *Appl. Opt.* 48: D26–D35.

Muller, M. G., Valdez, T. A., Georgakoudi, I. et al. 2003. Spectroscopic detection and evaluation of morphologic and biochemical changes in early human oral carcinoma. *Cancer* 97: 1681–1692.

Nieman, L. T., Jakovljevic, M., and Sokolov, K. 2009. Compact beveled fiber optic probe design for enhanced depth discrimination in epithelial tissues. *Opt. Express* 17: 2780–2796.

Nieman, L. T., Kan, C. W., Gillenwater, A. et al. 2008. Probing local tissue changes in the oral cavity for early detection of cancer using oblique polarized reflectance spectroscopy: A pilot clinical trial. *J. Biomed. Opt.* 13:

Nieman, L., Myakov, A., Aaron, J. et al. 2004. Optical sectioning using a fiber probe with an angled illumination-collection geometry: Evaluation in engineered tissue phantoms. *Appl. Opt.* 43: 1308–1319.

Nordstrom, R. J., Burke, L., Niloff, J. M. et al. 2001. Identification of cervical intraepithelial neoplasia (CIN) using UV-excited fluorescence and diffuse-reflectance tissue spectroscopy. *Lasers Surg. Med.* 29: 118–127.

Palmer, G. M. and Ramanujam, N. 2006. Monte Carlo-based inverse model for calculating tissue optical properties. Part I: Theory and validation on synthetic phantoms. *Appl. Opt.* 45: 1062–1071.

Palmer, G. M., Zhu, C. F., Breslin, T. M. et al. 2003. Comparison of multiexcitation fluorescence and diffuse reflectance spectroscopy for the diagnosis of breast cancer (March 2003). *IEEE Trans. Biomed. Eng.* 50: 1233–1242.

Palmer, G. M., Zhu, C. F., Breslin, T. M. et al. 2006. Monte Carlo-based inverse model for calculating tissue optical properties. Part II: Application to breast cancer diagnosis. *Appl. Opt.* 45: 1072–1078.

Park, S. Y., Follen, M., Milbourne, A. et al. 2008. Automated image analysis of digital colposcopy for the detection of cervical neoplasia. *J. Biomed. Opt.* 13: 014029.

Pentenero, M., Carrozzo, M., Pagano, M. et al. 2003. Oral mucosal dysplastic lesions and early squamous cell carcinomas: Underdiagnosis from incisional biopsy. *Oral Dis.* 9: 68–72.

Pleijhuis, R. G., Graafland, M., de Vries, J. et al. 2009. Obtaining adequate surgical margins in breast-conserving therapy for patients with early-stage breast cancer: Current modalities and future directions. *Ann. Surg. Oncol.* 16: 2717–2730.

Prahl, S. A. 2008. Tabulated molar extinction coefficient for hemoglobin in water, Oregon Medical Laser Center website, http://omlc.ogi.edu/spectra/hemoglobin/summary.html (accessed March 8, 2008).

Prahl, S. 2009. Optical absorption of water, Oregon Medical Laser Center website, http://omlc.ogi.edu/spectra/water/index.html (accessed September 23, 2009).

Rajaram, N., Nguyen, T. H., and Tunnell, J. W. 2008. Lookup table-based inverse model for determining optical properties of turbid media. *J. Biomed. Opt.* 13: 3.

Reif, R., Amorosino, M. S., Calabro, K. W. et al. 2008. Analysis of changes in reflectance measurements on biological tissues subjected to different probe pressures. *J. Biomed. Opt.* 13: 3.

Richards-Kortum, R. and SevickMuraca, E. 1996. Quantitative optical spectroscopy for tissue diagnosis. *Annu. Rev. Phys. Chem.* 47: 555–606.

Saidi, I. S., Jacques, S. L., and Tittel, F. K. 1995. Mie and Rayleigh modeling of visible-light scattering in neonatal skin. *Appl. Opt.* 34: 7410–7418.

Schomacker, K. T., Meese, T. M., Jiang, C. S. et al. 2006. Novel optical detection system for in vivo identification and localization of cervical intraepithelial neoplasia. *J. Biomed. Opt.* 11: 034009.

Schwarz, R. A., Arifler, D., Chang, S. K. et al. 2005. Ball lens coupled fiber-optic probe for depth-resolved spectroscopy of epithelial tissue. *Opt. Lett.* 30: 1159–1161.

Schwarz, R. A., Gao, W., Daye, D. et al. 2008. Autofluorescence and diffuse reflectance spectroscopy of oral epithelial tissue using a depth-sensitive fiber-optic probe. *Appl. Opt.* 47: 825–834.

Schwarz, R. A., Gao, W., Weber, C. R. et al. 2009. Noninvasive evaluation of oral lesions using depth-sensitive optical spectroscopy. *Cancer* 115: 1669–1679.

Semenza, G. L. 2002. HIF-1 and tumor progression: pathophysiology and therapeutics. *Trends Mol. Med.* 8: S62–S67.

Sharwani, A., Jerjes, W., Salih, V. et al. 2006. Assessment of oral premalignancy using elastic scattering spectroscopy. *Oral Oncol.* 42: 343–349.

Singer, A., Coppleson, M., Canfell, K. et al. 2003. A real time optoelectronic device as an adjunct to the Pap smear for cervical screening: A multicenter evaluation. *Int. J. Gynecol. Cancer* 13: 804–811.

Skala, M. C., Palmer, G. M., Zhu, C. F. et al. 2004. Investigation of fiber-optic probe designs for optical spectroscopic diagnosis of epithelial pre-cancers. *Lasers Surg. Med.* 34: 25–38.

Solomon, D. and Grp, A. 2003. Results of a randomized trial on the management of cytology interpretations of atypical squamous cells of undetermined significance. *Am. J. Obstet. Gynecol.* 188: 1383–1392.

Solomon, D., Schiffman, M., Tarone, R. et al. 2003. A randomized trial on the management of low-grade squamous intraepithelial lesion cytology interpretations. *Am. J. Obstet. Gynecol.* 188: 1393–1400.

Solonenko, M., Cheung, R., Busch, T. M. et al. 2002. In vivo reflectance measurement of optical properties, blood oxygenation and motexafin lutetium uptake in canine large bowels, kidneys and prostates. *Phys. Med. Biol.* 47: 857–873.

Sunar, U., Quon, H., Durduran, T. et al. 2006. Noninvasive diffuse optical measurement of blood flow and blood oxygenation for monitoring radiation therapy in patients with head and neck tumors: a pilot study. *J. Biomed. Opt.* 11: 064021–13.

Svaasand, L. O., Fiskerstrand, E. J., Kopstad, G. et al. 1995. Therapeutic response during pulsed laser treatment of portwine stains: Dependence on vessel diameter and depth in dermis. *Lasers Med. Sci.* 10: 235–243.

Taroni, P., Pifferi, A., Torricelli, A. et al. 2003. In vivo absorption and scattering spectroscopy of biological tissues. *Photochem. Photobiol. Sci.* 2: 124–129.

Thompson, M. S., Johansson, A., Johansson, T. et al. 2005. Clinical system for interstitial photodynamic therapy with combined on-line dosimetry measurements. *Appl. Opt.* 44: 4023–4031.

Thueler, P., Charvet, I., Bevilacqua, F. et al. 2003. In vivo endoscopic tissue diagnostics based on spectroscopic absorption, scattering, and phase function properties. *J. Biomed. Opt.* 8: 495–503.

Ti, Y. L. and Lin, W. C. 2008. Effects of probe contact pressure on in vivo optical spectroscopy. *Opt. Express* 16: 4250–4262.

Tosios, K. I., Kapranos, N., and Papanicolaou, S. I. 1998. Loss of basement membrane components laminin and type IV collagen parallels the progression of oral epithelial neoplasia. *Histopathology* 33: 261–268.

Tunnell, J. W., Desjardins, A. E., Galindo, L. et al. 2003. Instrumentation for multi-modal spectroscopic diagnosis of epithelial dysplasia. *Technol. Cancer Res. Treat.* 2: 505–514.

Utzinger, U., Brewer, M., Silva, E. et al. 2001. Reflectance spectroscopy for in vivo characterization of ovarian tissue. *Lasers Surg. Med.* 28: 56–66.

Utzinger, U. and Richards-Kortum, R. 2003. Fiber optic probes for biomedical optical spectroscopy. *J. Biomed. Opt.* 8: 121–147.

Van Diest, P. J., Torrenga, H., Borgstein, P. J. et al. 1999. Reliability of intraoperativefrozen section and imprint cytological investigation of sentinel lymph nodes in breast cancer. *Histopathology* 35: 14–18.

van Veen, R. L. P., Amelink, A., Menke-Pluymers, M. et al. 2005. Optical biopsy of breast tissue using differential path-length spectroscopy. *Phys. Med. Biol.* 50: 2573–2581.

van Veen, R. L. P., Sterenborg, H., Marinelli, A. et al. 2004. Intraoperatively assessed optical properties of malignant and healthy breast tissue used to determine the optimum wavelength of contrast for optical mammography. *J. Biomed. Opt.* 9: 1129–1136.

van Veen, R. L. P., Verkruysse, W., and Sterenborg, H. J. C. M. 2002. Diffuse-reflectance spectroscopy from 500 to 1060 nm by correction for inhomogeneously distributed absorbers. *Opt. Lett.* 27: 246–248.

vanderWaal, I., Schepman, K. P., vanderMeij, E. H. et al. 1997. Oral leukoplakia: A clinicopathological review. *Oral Oncol.* 33: 291–301.

Verkruysse, W., Lucassen, G. W., deBoer, J. F. et al. 1997. Modelling light distributions of homogeneous versus discrete absorbers in light irradiated turbid media. *Phys. Med. Biol.* 42: 51–65.

Vo-Dinh, T. 2003. *Biomedical Photonics Handbook*. Boca Raton, FL: CRC Press.

Volynskaya, Z., Haka, A. S., Bechtel, K. L. et al. 2008. Diagnosing breast cancer using diffuse reflectance spectroscopy and intrinsic fluorescence spectroscopy. *J. Biomed. Opt.* 13: 024012.

Waldron, C. A. and Shafer, W. G. 1975. Leukoplakia revisited—Clinicopathologic study 3256 oral leukoplakias. *Cancer* 36: 1386–1392.

Wang, L. H., Jacques, S. L., and Zheng, L. Q. 1995. MCML—Monte Carlo modeling of light transport in multilayered tissues. *Comput. Methods Prog. Biomed.* 47: 131–146.

Wang, H. W., Jiang, J. K., Lin, C. H. et al. 2009. Diffuse reflectance spectroscopy detects increased hemoglobin concentration and decreased oxygenation during colon carcinogenesis from normal to malignant tumors. *Opt. Express* 17: 2805–2817.

Wang, A., Nammalavar, V., and Drezek, R. 2007. Experimental evaluation of angularly variable fiber geometry for targeting depth-resolved reflectance from layered epithelial tissue phantoms. *J. Biomed. Opt.* 12: 044011.

Wang, H. W., Putt, M. E., Emanuele, M. J. et al. 2004. Treatment-induced changes in tumor oxygenation predict photodynamic therapy outcome. *Cancer Res.* 64: 7553–7561.

Wang, H. W., Zhu, T. C., Putt, M. E. et al. 2005. Broadband reflectance measurements of light penetration, blood oxygenation, hemoglobin concentration, and drug concentration in human intraperitoneal tissues before and after photodynamic therapy. *J. Biomed. Opt.* 10: 13.

Weber, C. R., Schwarz, R. A., Atkinson, E. N. et al. 2008. Model-based analysis of reflectance and fluorescence spectra for in vivo detection of cervical dysplasia and cancer. *J. Biomed. Opt.* 13: 10.

Wistuba, II and Gazdar, A. F. 2006. Lung cancer preneoplasia. *Annu. Rev. Pathol. Mech. Dis.* 1: 331–348.

Woodhams, J. H., Kunz, L., Bown, S. G. et al. 2004. Correlation of real-time haemoglobin oxygen saturation monitoring during photodynamic therapy with microvascular effects and tissue necrosis in normal rat liver. *Br. J. Cancer* 91: 788–794.

Woodhams, J. H., MacRobert, A. J., and Bown, S. G. 2007. The role of oxygen monitoring during photodynamic therapy and its potential for treatment dosimetry. *Photochem. Photobiol. Sci.* 6: 1246–1256.

Wright, T. C., Massad, L. S., Dunton, C. J. et al. 2007. 2006 consensus guidelines for the management of women with cervical intraepithelial neoplasia or adenocarcinoma in situ. *Am. J. Obstet. Gynecol.* 197: 340–345.

Wu, J. X. 1996. Apoptosis and angiogenesis: Two promising tumor markers in breast cancer. *Anticancer Res.* 16: 2233–2239.

Yu, C. C., Lau, C., O'Donoghue, G. et al. 2008. Quantitative spectroscopic imaging for non-invasive early cancer detection. *Opt. Express* 16: 16227–16239.

Zeng, H. S., Petek, M., Zorman, M. T. et al. 2004. Integrated endoscopy system for simultaneous imaging and spectroscopy for early lung cancer detection. *Opt. Lett.* 29: 587–589.

Zhu, C. F., Breslin, T. M., Harter, J. et al. 2008b. Model based and empirical spectral analysis for the diagnosis of breast cancer. *Opt. Express* 16: 14961–14978.

Zhu, C. F., Palmer, G. M., Breslin, T. M. et al. 2005. Use of a multiseparation fiber optic probe for the optical diagnosis of breast cancer. *J. Biomed. Opt.* 10(2): 024032.

Zhu, C. F., Palmer, G. M., Breslin, T. M. et al. 2008b. Diagnosis of breast cancer using fluorescence and diffuse reflectance spectroscopy: A Monte-Carlo-model-based approach. *J. Biomed. Opt.* 13: 15.

Zigrino, P., Loffek, S., and Mauch, C. 2005. Tumor-stroma interactions: Their role in the control of tumor cell invasion. *Biochimie* 87: 321–328.

Zonios, G., Perelman, L. T., Backman, V. M. et al. 1999. Diffuse reflectance spectroscopy of human adenomatous colon polyps in vivo. *Appl. Opt.* 38: 6628–6637.

7

Multi/Hyper-Spectral Imaging

7.1	Introduction	131
7.2	Color vs. Spectral Imaging	132
7.3	SI Camera Hardware Configurations and Calibration	133
7.4	Scanning SI Systems Based on Electronically Tunable Filters	135

Overview • SI Systems Based on Electromechanically Tunable Filters • SI Systems Based on Electronically Tunable Filters

7.5	Single Exposure or Instantaneous Spectral Imagers	140

Overview • SE–SI Systems Based on Image Replication, Filtering, and Projection Optics • SE–SI Based on Coded Image Capturing • SE–SI Systems Based on Surface Plasmons and Nanodevices

7.6	Spectral Data Classification, Unmixing, and Visualization	148

Overview • Preprocessing for Reducing the Dimensionality of the Data • Spectral Classification and Unmixing Methods • Unsupervised Clustering Methods: The k-Means Algorithm • Supervised Classification Methods Based on Distance Similarity Measures • Supervised Classification Methods Based on Nongeometric Similarity Measures • Spectral Unmixing Algorithms • Classification Performance Measures • Summary and Basic Taxonomy of Spectral Cube Data Processing Processes

7.7	Conclusion and Future Perspectives	159
	References	159

Costas Balas
Technical University of Crete

Christos Pappas
Technical University of Crete

George Epitropou
Technical University of Crete

7.1 Introduction

Important new developments in the field of biomedical optical imaging (OI) allow for unprecedented visualization of the tissue microstructure and enable quantitative mapping of disease-specific endogenous and exogenous substances (Balas 2009). Spectral imaging (SI) is one of the most promising OI modalities, belonging to this general field, and it will be reviewed in more detail in this chapter.

SI combines spectroscopy with imaging. A spectral imager provides spectral information at each pixel of a two-dimensional (2D) detector array. The SI systems acquire a three-dimensional (3D) data set of spectral and spatial information, known as *spectral cube*. The spectral cube can be considered as a stack of images, each of them acquired at a different wavelength. Combined spatial and spectral information offers great potential for the nondestructive/invasive investigation of a variety of studied samples.

Spectroscopy finds applications in analytical chemistry since a long time. Different spectroscopy types and modalities exist, depending on the optical property that it is intended to be measured, namely, absorption, spontaneous emission (fluorescence, phosphorescence), scattering (Rayleigh elastic, Raman inelastic) spectroscopy, etc. As the light travels into the sample, photons are experiencing absorption, which may result in fluorescence emission and multiple scattering due to the local variation of the index of refraction. Spectrometers measure the intensity of the light emerging from the sample as a function of the wavelength. The collected light passes through a light-dispersing element (grating), which spatially splits the light wavelengths onto the surface of an optical sensor array, interfaced with a computer for recording and processing the spectrum. Sample illumination can be provided by either a broadband (e.g., white light) or a narrowband light source. In the first case, the measured spectra provide information for the absorption and scattering characteristics of the tissue. In the second case, the measured spectra probe the fluorescence characteristics of the sample. Particularly, in steady-state fluorescence spectroscopy, a narrowband light source is used for fluorescence excitation, such as lasers, LEDs, or filtered light sources, emitting typically in the blue-ultraviolet band. A sensitive optical sensor is used for collecting the emission spectra.

The collected emission spectra can provide diagnostic information for the compositional status of the sample. This makes spectroscopy an indispensible tool for nondestructive analysis and for the development of novel, noninvasive diagnostic approaches. In biomedical sciences particularly, the diagnostic potential of tissue spectroscopy is based on the assumption that the absorption, fluorescence, and scattering characteristics of the tissue change during the progress of the disease.

Over the last 20 years, spectroscopy has been extensively investigated as a tool for identifying various pathologic conditions on the basis of their spectral signatures. It has been demonstrated that spectroscopy can successfully probe intrinsic or extrinsic chromophores and fluorophores, the concentration of which changes during the development of the disease. In its conventional configuration, spectroscopy uses single-point probes that cannot easily sample large areas or small areas at high spatial resolution (SR). It is obvious that this configuration is clearly suboptimal when solid and highly heterogeneous materials, such as the biological tissues, are examined. In these cases, the collected spectrum is the result of the integration of the light emitted from a great number of area points. This has the effect of mixing together signals originating from both pathologic and healthy areas, which makes the spectral signature-based identification problematic. Looking at the same problem from another perspective, point spectroscopies are considered as inefficient in cases where the mapping of some characteristic, spectrally identifiable property, is of the utmost importance.

Spectroscopy probes optical signals with high spectral resolution but with poor SR. The vastly improved computational power together with the recent technological developments in tunable optical filter and imaging sensor technologies have become the catalysts for merging together imaging and spectroscopy. Both areas—imaging and spectroscopy—continue to be impacted by technological innovations that enable faster acquisition of superior-quality data. SI has the unique feature of combining the advantages of both imaging and spectroscopy (high spatial and spectral resolution) in a single instrument. In SI, light intensity is recorded as a function of both wavelength and location. In the image domain, the data set includes a full image at each individual wavelength. In the spectroscopy domain, a fully resolved spectrum at each individual pixel can be recorded. These devices can measure the spectral content of light energy at every point in an image. Multiple images of the same scene at different wavelengths are acquired for obtaining the spectra. As an example, an SI device integrating an imaging sensor with 1000×1000 pixels provides 1 million individual spectra. A spectrum containing 100 data points results from an equal number of spectral images. Assuming that the intensity in each pixel is sampled at 8 bits, then the size of the resulting spectral cube equals to 100 Mbytes. Due to the huge size of the collected data sets, SI data processing, analysis, and storage require fast computers and huge mass memory devices. Several mathematical approaches are used for spectral classification and image segmentation on the basis of the acquired spectral characteristics. The spectra are classified using spectral similarity measures, and the resulting different spectral classes are recognized as color-coded image clusters. SI can be easily adapted to a variety of OI instruments such as camera lenses, telescopes, microscopes, endoscopes, etc. For this reason, applications of SI span from planet and earth inspection (remote sensing) to internal medicine and molecular biology. Particularly, in clinical diagnosis, spectral mapping can provide information for the topography, the severity, and the margins of the lesion for diagnosis and for guiding treatment.

SI is a wide, rapidly evolving interdisciplinary field that can hardly fit in a single chapter. For this reason, an extended reference list has been included. Readers willing to deepen into some particular subjects should use these references in order to expand their knowledge in the field.

7.2 Color vs. Spectral Imaging

Photons encountering the pixels of an imaging sensor create electrons in pixel cells (photoelectric effect); thereby, the number of photons is proportional to the number of electrons. The photon's wavelength information, however, is not "transferred" to the electrons. Hence, unfiltered imaging chips are color blind. Color or SI devices employ optical filters placed in front of the imaging chip. Color imagers use either Si charge-coupled devices (CCD) or C-MOS sensors, which are sensitive in the visible and in the near- infrared (NIR) part of the spectrum (400–1000 nm). A band-pass filter is used for rejecting the NIR band (700–1000 nm). In 3-chip configurations, three photon channels are created with the aid of a trichroic prism assembly, which directs the appropriate wavelength ranges of light to their respective sensors. Camera electronics combine the red, green, and blue (R, G, B) imaging channels composing a high-quality color image, which is delivered to external devices through an analog or digital interface. An alternative, cheaper, and more popular color camera configuration employs a single chip, where the color filters are spread, similar to a mosaic, across all pixels of the sensor. Due to the fact that each pixel "sees" only one primary color, three pixels are required to record the color of the corresponding area of the object. This reduces significantly the SR of the imager. This unwanted effect is partially compensated with a method called "spatial color interpolation" carried out by the camera electronics. The interpolation algorithm estimates the two missing primary color values for a certain pixel by analyzing the values of its adjacent pixels. In practice, even the most excellent color space interpolation methods cause a low-pass effect. Thus, single chip cameras yield images that are more blurred than those of 3-chip or of monochrome cameras. This is especially evident in cases of subtle, fiber-shaped image structures. Color cameras emulate the human vision for color reproduction and are real-time devices since they record three spectral bands simultaneously at very high frame rates. Human vision-emulating color imaging devices usually describe color with three parameters (RGB values), which are easy to interpret since they model familiar color perception processes. They share, however, the limitations of human color vision. Color cameras and human color vision allocate the incoming light to three color coordinates, thus missing significant spectral information. Due to this fact, objects emitting or remitting light with completely different spectral components can have precisely the same RGB coordinates, a phenomenon known as metamerism. The direct impact

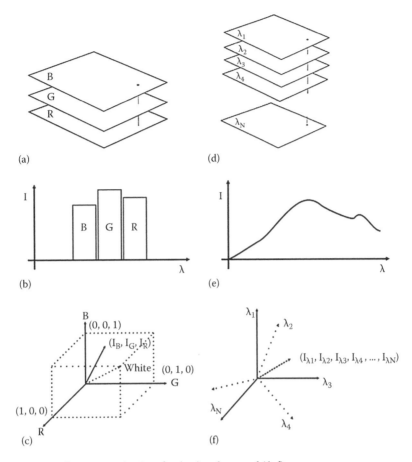

FIGURE 7.1 Image data acquisition and representation in color (a–c) and spectral (d–f) cameras.

of the metamerism is the inability of the color imaging systems to distinguish between materials having the same color appearance but different chemical composition. This sets serious limitations to their analytical power and consequently to their diagnostic capabilities.

Unlike images taken with standard color (RGB) cameras, SI information is not discernible to the human eye. In SI, a series of images is acquired at many wavelengths, producing a spectral cube. Each pixel in the spectral cube, therefore, represents the spectrum of the scene at that point. The nature of imagery data is typically multidimensional, spanning spatial and spectral dimensions (x, y, λ).

Figure 7.1 illustrates the data sets that are generated by color and SI systems for comparison purposes. A color camera captures typically three images corresponding to the band-pass characteristics of the RGB primary color filters (Figure 7.1a). Color image pixels miss significant spectral information as it is integrated into three, broad spectral bands (Figure 7.1b). The color of a pixel can be represented as vector in a three-dimensional "color space" having the RGB values as coordinates. SI systems collect a stack of pictures, where each image is acquired at a narrow spectral band and all together compose the spectral cube (Figure 7.1d). A complete spectrum can be calculated for every image pixel (Figure 7.1e), which can

be otherwise represented as a vector in a "multidimensional spectral space" (Figure 7.1f).

7.3 SI Camera Hardware Configurations and Calibration

SI camera systems consist of a monochrome sensor, an electronically controlled spatial or spectral scanning mechanism, imaging optics, and a computer platform for storage, display analysis, and processing of imaging data. Control electronics synchronize the scanning and the data capturing processes, so that a set of spectral images are collected as members of the spectral cube.

The number of the spectral bands that an SI system is capable of acquiring determines the distinction between multispectral imaging (MSI) and hyperspectral imaging (HSI). MSI devices typically acquire 5–20 spectral bands, while HSI systems acquire up to a few hundreds of spectral bands. Ultraspectral imaging (USI) devices are currently under development with the capacity of acquiring thousands of very narrow spectral bands.

SI systems use monochrome sensors or sensor arrays, which can capture only two of the three spectral dimensions of the spectral cube at a time. To capture the third dimension, spatial or spectral scanning is required. Depending on the method employed for building the spectral cube, SI devices are classified as follows: (a) whiskbroom SI devices, where a linear sensor array is used to collect the spectrum (λ dimension) from a single point at a time; the other two spatial coordinates are collected with (*x*, *y*) spatial scanning; (b) pushbroom SI devices in which a 2D sensor array is used, the one dimension of which captures the first spatial (*x*) coordinate and the other the spectral coordinate in each camera frame; the second spatial coordinate (*y*) is captured with line (slit) scanning; (c) staring SI devices, where a 2D sensor array is coupled with an imaging monochromator, which is tuned to scan the spectral domain and in each scanning step, a full spectral image frame is recorded. A tunable light source is also suitable for staring SI, but in that case, the possible contribution of the broadband ambient light should be eliminated. Whiskbroom and pushbroom imagers utilizing spatial scanning for building the spectral cube do not provide live display of spectral images, since they are calculated from the spectra after the completion of the spatial scanning of the corresponding area. Staring imagers, on the other hand, are based on the tuning of the imaging wavelength and the spectra are calculated from the spectral cube composed by the spectral images that are captured in time sequence. Compared to the other approaches, staring imagers have the advantage of displaying live spectral images, which is essential for aiming and focusing. Spatial scanning SI instruments have been originally developed long time ago for remote sensing applications (RSA). These instruments are typically carried by a platform, such as satellites, airplanes, etc., and have been extensively used mainly for Earth resources monitoring. In these applications, the movement of the platform comprises the spatial scanning mechanism, and the spectrally scanned lines are stitched together to construct spectral images of the scene. The spatial scanning SI concept has been transferred to biomedical imaging and especially to laser microscopy. A typical microscope operating on the basis of the spatial scanning principle is the confocal microscope. Here, the sample is spatially scanned with a laser and the resulting fluorescence emission spectrum is collected for each point by a detector after passing through the confocal aperture of the microscope. Staring SI devices have also been used for RSA. However, they are more suitable for civilian applications and more specifically for the nondestructive analysis of man-made or natural targets. In biomedical imaging applications particularly, staring SI imagers are more suitable when wide field, broadband illumination is used. Several *in vivo* applications use such light sources and require real-time imaging at selected spectral bands, such as in endoscopy where inspection of live images is essential. For such applications, staring SI imagers comprise the technology of choice.

However, the considerable spatial or spectral scanning time restricts, in general, our ability to obtain meaningful spectral cubes only in static imaging conditions. In other words, correct spectral measurements are feasible only in cases where the target remains stationary and/or spectrally unaltered during the instrument's spatial/spectral scanning time. In endoscopic applications, for example, it is practically impossible to avoid organ or endoscopic tip motions during this time. The same applies in retina imaging where saccadic eye movements are always present. Another example referring to both in vivo and in vitro imaging is the case where the spectral or color characteristics of the target change as a result of biomarker-tissue interaction. SI enables the selective imaging of different biomarkers even though their spectra are broad and are overlapping. However, it is very difficult to monitor the uptake kinetics separately for each biomarker because color changes are in several cases occurring at shorter times than the typical scanning time of SI devices. The solution to this problem is staring imaging systems operating with electronically tunable filters (ETF) without moving parts. The tuning time of these filters is very short (millisecond, microsecond regime), enabling the capturing of a number of preselected spectral images in less than a second time. In that case however, the spectral information is reduced to the MSI level. For moving targets and for rapid time-varying events, other techniques allow for the construction of sensors that can capture the three (*x*, *y*, λ) dimensions in a single camera exposure. But again, this is achieved at the cost of the spectral information since only a number of bands typical for an MSI system can be recorded with the current level of development of these technologies.

Selecting the SI camera optimal configuration and components requires a "systems" approach. The intended application determines the SI system's spectral range and resolution. Si CCD detectors can be used to cover the spectral range ultraviolet (UV)-visible and NIR up to 1 μm. InGaAs detectors are suitable for the up to 1.7 μm NIR range. For longer infrared wavelengths, HgCdTe or InSb cameras must be used. Ideally, the wavelength range of the monochromators should match at least a significant part of the spectral range within which the selected detector is sensitive. Narrowband imaging and monochromator optics reduce the overall light throughput of an SI system. Moreover, the light throughput of the monochromator depends on the wavelength. Furthermore, the quantum efficiency (QE) of the detector also changes with the wavelength. Particularly, the QE of a Si CCD and the ETF's throughput are both pure in the UV-blue range of the spectrum. This problem can be compensated by either using a light source with strong emission in this wavelength region or by using a back-illuminated CCD, which has much better QE in this band or by increasing the exposure time of the sensor, if the application permits, or combinations. In general, fast optics can improve the overall light throughput of the system and should be used in any case. In some applications, increasing the output power of a light source is not an option because it could overheat the sample or provoke photochemical reactions, such as photobleaching of fluorophores. Acquisition of meaningful spectral data requires the compensation for the wavelength dependence of the detector's QE, monochromator's throughput, and illuminating source's emission. SI system's calibration is very essential in order to achieve "device-independent" spectral measurements. Calibration can be performed with the aid of a reference sample displaying a known or a flat spectrum over the entire operating wavelength range. A calibration curve or a lookup table can be obtained by comparing the measured with the actual spectral characteristics of the calibration sample with that measured by the SI system spectra. Image brightness can be corrected on the basis of the calibration data, after spectral image acquisition. The calibration curve or the lookup table can also be integrated into the system's software for controlling the detector's exposure time during image acquisition, in all tuning steps of the filter. By changing the detector's exposure or gain settings, the wavelength dependence of the SI system's response is compensated and the spectral images that are acquired and captured are calibrated.

In the following two paragraphs, the most important scanning (paragraph 4) and single exposure (SE) (Section 7.5) SI systems will be outlined, together with some indicative applications in biomedical sciences.

7.4 Scanning SI Systems Based on Electronically Tunable Filters

7.4.1 Overview

Spectral scanning SI systems can be categorized on the basis of the technological approaches employed for tuning the imaging wavelength. Two general classes of devices can be defined depending on whether their imaging monochromator contains or not moving parts. In the first case, mechanical parts, onto which filters or mirrors, etc. are mounted, are spatially translated for selecting the imaging wavelength. We can therefore name this category as *electromechanical tunable filters* (EMTF). The second category refers to systems based on nonmoving optical modules whose spectral transmission can be electronically controlled through the application of voltage or acoustic signal, etc. For this reason, the members of this class of instruments are known as ETF.

An alternative categorization could be done depending on whether the SI system provides or not live (real time) display of spectral images. A simple and rather trivial EMTF is a set of discrete band-pass filters, which are swapping in front of the sensor. The filters can be mounted on a rotating disk or on a translating stage. *Linearly variable filters* (LVF), with transmission centre wavelength varying linearly along their surface, comprise an alternative to discret filters option but they are more suitable as monochromators of the illuminating source. In all these cases, the motion of the filter arrangements is driven by a stepper motor, and in most cases, it is synchronized with the successive image capturing.

A more sophisticated setup belonging to the EMTF class of imagers is based on the interferometric Fourier transformed approach. Here, the controllable translation of the optical components determines the wavelengths at which constructive interference occurs. *Fourier transform imaging* (FTI) systems acquire image interferograms from which the spectral cube is produced with post-acquisition processing. Due to this fact, they do not provide live SI and, for this reason, are more suitable for in vitro applications of stationary targets.

ETF-based systems without moving parts are more elegant solutions. Their spectral transmission can be electronically controlled through the application of voltage or acoustic signal, etc.

By mounting an ETF in front of a monochrome camera, one can acquire a stack of images at contiguous spectral bands. A tunable filter, similarly to a filter wheel, allows for the selection of one band at a time and the display live spectral images of the target. The continuous tuning of wavelengths provides the capability for finer spectral sampling. Also, unlike filter wheels, ETFs permit rapid and random switching between spectral bands. ETFs are sophisticated spectral scanners, which are inherently robust and compact and can be tuned much faster than a typical filter wheel. Well-established EFTs with numerous applications in biomedical imaging are the liquid crystal tunable filters (LCTF) and the acousto-optic tunable filter (AOTF). The recently developed liquid crystal tunable Fabry–Perot (LCFP) filters belong also to the ETF family. LCFPs provide very narrowband imaging but they cannot be tuned over a wide spectral range. Due to their particular characteristics, they have been proven to be very useful, mainly in astrophysical applications, while their biomedical applications have not been developed yet.

7.4.2 SI Systems Based on Electromechanically Tunable Filters

7.4.2.1 Discrete Filters/Spatially Variable Filters

The simplest implementation of a spectral scanning imager is a rotating disk with a number of band-pass filters located in its periphery, optically coupled with an imaging sensor. The rotation of the disk can be driven by a stepper motor so that the filters are swapping in front of the sensor. It is obvious that the size of the disk will be proportional to the number of the filters and consequently to the number of the spectral images. Device size restrictions limit the number of discrete filters so that only MSI configurations are allowable. When a limited number of *a priori* selected wavelengths is needed, this solution can be very efficient in terms of light throughput, cost, and ease of implementation. On the other hand, a wide selection of bandwidths is not readily available and the user/designer has to rely on costly custom-made glass filters. Furthermore, the number and the specifications of the filters referring to both central wavelength and bandwidth, is crucial and sometimes can only be decided after dense spectral data have been collected and analyzed. This may require the use of multiple filter wheels, each with different filter numbers and bandwidths. Lastly, electromechanical filter wheels have the following inherent drawbacks: (a) slow band switching, (b) small number of filters, (c) only sequential access to spectral bands, (d) cumbersome design, (e) limited versatility, and (f) image misalignment effects.

As it has been discussed earlier in this chapter, SI devices can also be built upon a tunable light source. Such light sources may be tunable lasers or broadband light sources, coupled with light-dispersing elements (grating) or band-pass filters. Band-pass filtering can be achieved with a set of discrete filters or an LVF. LVFs are band-pass filters, the transmission centre wavelength of which varies linearly along their surface. LVF-based tunable light source designs are simpler, offering also good throughput. The full width half maximum (FWHM) for a certain wavelength is typically 1%–1.5% of the peak wavelength. Circular variable filters (CVF) comprise an alternative option having transmission that varies along the rotation angle of the filter. Typically, tunable light sources based on either gratings, or LVFs, CVFs employ rectangular-to-circular fiber optic bundles, with the rectangular end being attached to the source's exit slit. The circular end is attached to other optical instruments or illuminates directly the target. It should be noted that by decreasing the slit's width, the FWHM can decrease down to the filter's specification, but at the same time, light throughput decreases.

A variety of SI systems based on either sensor or light source filtering, using the techniques described in this subparagraph, have been extensively used as research instruments in a variety of *in vivo* and *in vitro* applications, with very good results. In a number of applications, SI devices of this type have been converted to commercial devices. Indicative applications include bronchoscopy for lung cancer detection (Palcic et al. 1991), gastrointestinal endoscopy for assisting cancer diagnosis (Gono 2008), colposcopy for cervical cancer detection (Alvarez and Wright 2007), histology for assessing estrogen and progesterone receptors in breast cancer (Papadakis et al. 2003), and hematology for discriminating lymphocytes from lymphoblasts in acute lymphoblastic leukemia (Katzilakis et al. 2004) etc.

7.4.2.2 Fourier Transform Imaging Spectrometer

The *Fourier transform imaging spectrometer* (FTIS) can be categorized to the EMTF scanners. Compared to the other members of this family, it differs in that scanning refers to optical path difference domain. An FTIS is essentially an interferometric Fourier spectroscopy device combined with an imaging sensor array. In FTIS, an interferogram is captured for each pixel as a function of the optical path difference and then, as implied by the name, a Fourier transform is applied for transforming the recorded interferogram into spectral information. Several devices have

been developed for research applications, each one of them based, in one way or another, on a Michelson interferometer. In a typical Michelson interferometer, a beam splitter divides the incoming light into two equal parts, which are directed toward two orthogonal paths, each one terminated by a mirror. One arm has fixed length, while the other arm's length can be tuned to achieve a variable delay time. The beams are reflected from the mirrors back to the beam splitter and finally are recombined onto the detectors surface. Depending upon the delay time, the two beams may interfere constructively or destructively. The interference pattern is recorded by the detector, as a function of the optical path, and from the resulting amount of data (the cube of the interferogram), the spectral cube is calculated with the aid of a fast Fourier transformation.

Conventional FTISs have moving parts, making them vulnerable to misalignments or mechanical failures. Despite of that, they have been proved to perform satisfactory in demanding and harsh environments (Persky 1995). Alternative configurations of the same principle have been proposed for improving the robustness of the design, such as the ones based on a Sagnac interferometer (Malik et al. 1996) or on a birefringence interferometer (Harvey and Fletcher-Holmes 2004). A Sagnac interferometer is constituted of a beam splitter and reflective optics and has a single optical path, which is traversed in opposite directions by the split components of the initial beam (Figure 7.2a). The optical path difference is induced by changing the angle of incidence of the initial beam with respect to the beam splitter. The interference pattern produced by the recombined beams is measured at different angles of incidence and the resulting data are Fourier transformed into spectral information. Fourier transform interferometers (FTIR) based on Sagnac cavities have been launched to the market and have found application in scientific research (Yoneya et al. 2002, Ito et al. 2008). The second approach refers to the birefringence interferometer proposed

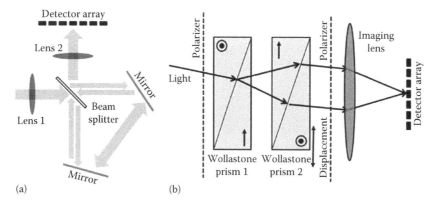

FIGURE 7.2 (a) Optical layout of a common-path Sagnac interferometer. The incoming light is collected by collimating optics (lens 1) and directed toward a beam splitter, which divides the light beam in two parts. The split beams are reflected by the mirrors, traverse the cavity of the interferometer in opposite directions and are finally collected by the lens 2 and recombined onto the surface of the detector; (b) the Harvey Fletcher-Holmes interferometer with two Wollastone prisms. The incoming light is resolved by a linear polarizer placed at 45° with respect to the optical axis of the Wollastone prisms. The extraordinary and ordinary ray components are resolved by the prism1 and exit the prism 2 propagating in collinear directions. The two rays are finally recombined on the surface of the detector array through an imaging lens. The optical axis of each segment of the Wollastone prisms is denoted by arrow or circle.

by Harvey and Fletcher-Holmes (2004). In that case, the setup consisted of two Wollastone prisms, and the optical path variation is achieved by performing a transverse displacement of the center of the second prism with respect to the center of the first prism (Figure 7.2b). With this configuration, the demand for scanning accuracy is dramatically reduced (over two orders of magnitude) in comparison to a typical Michelson configuration. Indicatively, tens of nanometers accuracy is required for a Michelson-type interferometer. It should be noted, however, that in the configuration proposed by Harvey and Fletcher-Holmes, the throughput is decreased due to the use of polarizers and birefringent materials. Despite setup differences, the inteferometric devices share working principles and problems. If the object moves, data acquisition errors will occur. Apart from object movements, these devices are sensitive to external vibrations, which randomly alter the difference of the optical paths (phase noise). Sophisticated mechanisms for achieving precise control over the moving parts are often employed to compensate for motion artefacts, but this increases the complexity. In any case, FTIR exhibit important advantages over other scanning techniques as, for example, high throughput, high signal-to-noise ratio (SNR), high spectral resolution and finally, high spectral range limited only by the spectral range of the detector, the beam splitting, and the reflective elements used. Last, it should be mentioned that some electrically or heat tunable Fourier transform interferometers have been reported in the literature with their range of applications being limited thus far (Shi et al. 2007, Boer et al. 2004).

FTIS have found many applications thus far, as, for example, in cytology where the ability of the technique to reveal spectral information in living cells correlated to enzymatic reactions has been demonstrated (Malik et al. 1996). The technique has also been proved efficient in discerning the differences in the fluorescence spectral characteristics of chromosome-specific staining probes, enabling the classification of human chromosomes (Schrock et al. 1996). In clinical studies, FTIS adapted to a fundus camera, has been used in SI and analysis of the retina. In particular, Yoneya et al. (2002) used an FTIS system based on a Sagnac interferometer (ASI Co) for evaluating the level of the ischemia in retina while in a more recent study, Ito et al. (2008) used the same device for measuring the oxygen saturation level in the retina. In both studies, the tissue was scanned in the spectral range 480–600 nm for about 6 s, using a broadband light source. The resolution of the final image was 284 × 224 pixels.

7.4.3 SI Systems Based on Electronically Tunable Filters

7.4.3.1 Liquid Crystal Tunable Filters

A liquid crystal tunable filter (LCTF) is based on a Lyot filter, invented by Bernard Lyot in 1920. The Lyot filter employs polarizing interferometry to yield a narrowband filter suitable for recording monochromatic images. It consists of N birefringent plates (retarders), each positioned between parallel polarizers

oriented such that the polarizer axis forms an angle of 45° to the optic axis.

As it is known, a birefringent crystal, such as calcite or boron nitride, exhibits two different indices of refraction. Because of that, linearly polarized light propagating through the crystal is resolved into orthogonally polarized components, called "ordinary" and "extraordinary" rays, traveling at different speeds. This introduces a mutual optical path difference ($d\Delta n$) between these two components, where $\Delta n = n_o - n_e$ is the birefringence, n_o and n_e are the ordinary and extraordinary RIs, and d the thickness of the crystal. The associated mutual phase retardation is then $\Gamma = 2\pi d\Delta n/\lambda$, where λ is the wavelength. The 45° angle between the polarizer and optic axes indicates that equal ordinary and extraordinary components are transmitted. At the crystal's exit, these two components are superimposed and their combined polarization depends on their mutual phase retardation. Phase retardation can turn the polarization plan of the light transmitted through the crystal so that it can in general form a nonzero angle with the axis of the output polarizer (analyzer). The projected to the analyzer's axis light intensity component will be finally transmitted from the polarizer–retarder–analyzer stage. The transmitted intensity or the transfer function of the stage is given by the following formula:

$$T = \cos^2\left(\frac{\Gamma}{2}\right) = \cos^2\left(\frac{\pi d\Delta n}{\lambda}\right) \qquad (7.1)$$

As seen in Equation 7.1, the light transmittance of the stage is a function of d, Δn, and λ. Assuming a birefringent crystal with a given d and Δn, and broadband light entering the stage, a certain set of wavelengths will be transmitted, while the remaining wavelengths will be rejected as not fulfilling the transmission criteria. In order to reduce the bandwidth and the number of the transmitted spectral bands, several cascaded polarizer–retarder–analyzer stages are assembled in series, with the output of one being the input of the other (Figure 7.3).

The thickness of the retarder varies from stage to stage and more specifically, the ratio between the thicknesses of consecutive

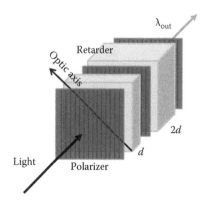

FIGURE 7.3 A two-stage Lyot filter assembly composed of polarizers and retarding elements, where d is the thickness of the first retarder and λ_{out} is the transmitted wavelength.

retarders (waveplates) is a factor of two. Particularly, if the thickness of the thinnest plate is *d*, then the thickness of the *j*th plate is $2^{j-1}d$. The spectral transmission function of an *N*-waveplate Lyot filter is the product of the transmission functions of all constituent waveplate/polarizer assembly and is given by

$$T = \prod_{j=1}^{N} \cos^2\left(2^{j-2}\Gamma\right) \qquad (7.2)$$

Figure 7.4a illustrates the transmittance (*T*) as a function of the wavelength for a Lyot filter comprised of two stages. The transmittance of the first stage is $\cos^2(\pi d \Delta n/\lambda)$ and of the second is $\cos^2(2\pi d \Delta n/\lambda)$. Their product is illustrated in Figure 7.4b. Figure 7.4c depicts the composite transmittance for a stack of five stages with retarder thickness *d*, 2*d*, 4*d*, 8*d*, 16*d*. It is clearly seen that the wavelength selectivity and spectral bandwidth can be decreased by increasing the number of retarders *N*; however, adding more Lyot stages means adding retarders with thicknesses of *d*, 2*d*, 4*d*, 8*d*, or more, which makes the tuning speed very slow and increases the attenuation due to the additional interfaces and polarizers.

Apart from the optimum number of stages determining the spectral discrimination performance of the Lyot filter, it can be proved that the thickness of the first stage determines the free spectral range and the thickness of the final stage determines the spectral resolving power (Brady 2009).

What has been described so far is a multistage Lyot filter that, with a given set of alternating birefringent retarders and analyzers, behaves as a band-pass filter, transmitting certain spectral band(s). Tunability of the transmitted bands is achieved by modifying the Lyot filter for enabling the external control of the birefringence and hence the control of phase retardation (Equation 7.2). The Lyot filter becomes LCTF when thin liquid crystal layers, acting as electronically controlled phase retarders, are added to each stage side-by-side with the fixed retarders.

Two transparent electrodes are placed on either side of the liquid crystal waveplate. The liquid crystal cell contains nematic crystals that are aligned with their long axis nearly perpendicular to the light path. The inner surfaces of the liquid crystal cell are prepared in such a way that the molecules have a preferred orientation parallel to the surface. However, when a voltage is applied across the electrodes, an electric field parallel to the light path generates a torque that twists the liquid crystals into the direction of the electric field. The molecules are then aligning with the applied electric field, thus decreasing the retardance through the liquid crystal waveplate (Slawson et al. 1999). The electronically adjustable retardance enables the tuning of the transmitted wavelength. The LCTF's spectral transmittance dependents on the phase shift, which inherently depends on the total retardance by both the fixed and the liquid crystal waveplate adjustable retarders. LCTF is a wonderful device offering versatile and relatively fast wavelength selection, without much image distortion or shift. Switching speed is limited by relaxation time of the liquid crystals, being of the order of ~50 ms. Special devices can be designed for fast switching (~5 ms) through a short sequence of wavelengths. LCTFs can cover both visible and infrared spectral bands but with different module assemblies, each spanning approximately one octave of wavelength (e.g., 400–750 nm). The minimum output bandwidth is 5 nm increasing with the wavelength up to 30 nm or more. The major problem of LCTFs is their poor light throughput since half of the light corresponding to one polarization state is rejected by the input polarizer, and peak transmission of the other half probably is less than 40%. Poor filter transmittance will require long sample and sensor exposure times, especially in low-light applications. Long light exposure

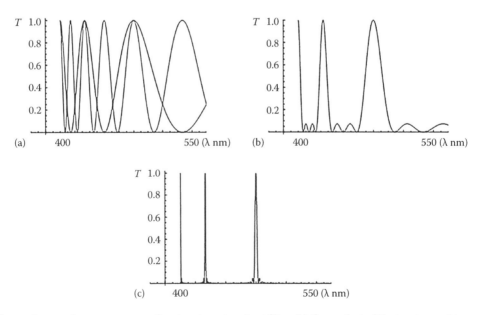

FIGURE 7.4 (a) Spectral transmittances corresponding to a two-stage Lyot filter; (b) the product of the two transmittances; (c) the composite transmittance for a stack of five stages.

of the sample would result in unwanted effects such as fluorophore photobleaching. Long camera integration times would in practice cancel the fast switching ability of a spectral imager integrating an LCTF. Insufficient out-of-band rejection is a second problem of LCTFs requiring extra optics for blocking light leaks. Finally, due to the thermal instability of the liquid crystals, maximum operating temperature of LCTFs is low, ranging between 10° and 40°, which should be taken into account when the filter is indented to be used for filtering light sources.

LCTF's science and technology evolves very rapidly because of its proven applicability in a variety of biomedical imaging applications. Recently, a number of more efficient, compact, and less expensive liquid crystal-based tunable filter configurations have been developed, departing from the Lyot's filter concept. Amongst the most interesting implementations, are the Holographic polymer-dispersed liquid crystals (H-PDLC) (Woltman et al. 2007, Bowley et al. 2001, Qi et al. 2004, Fox et al. 2007) and the vertically aligned deformed helix ferroelectric liquid crystals (VA-DHFLC) (Woltman et al. 2007, McMurdy et al. 2006). H-PDLCs and VA-DHFLCs form photonic crystals with a tunable photonic band gap determining a range of wavelengths, which are reflected by the crystal, according to the coupled mode theory. In each case, the optical anisotropy and the responsivity of liquid crystals to electric field or heating is exploited for altering or deactivating the wavelength selectivity of the photonic crystal. The spectral resolution of both the H-PDLC and the VA-DHFLC is in the range of 15–20 nm. Another very exciting progress in liquid crystal tunable filter science and technology is the LCFP. Fabry–Perot cavities are optical resonators consisting of two planar, partially transparent, reflective surfaces, separated by an intermediate medium, which can be air or some kind of dielectric material. A Fabry–Perot filter is penetrable by wavelengths for which a standing wave condition is satisfied inside the cavity. In a LCFP, a liquid crystal is used as the medium. By applying potential to the crystal, the refractive index (RI) of the medium changes and therefore the transmission modes supported by the cavity are affected. This enables the control of the transmitted wavelength. Using this configuration, switching times of the order of few milliseconds can be achieved. The minimum bandwidth in each tuning step is very narrow (0.05 nm), but the tuning range is not larger than 100 nm. A succession of cavities may be used for broadening the range of frequencies but such a configuration is inefficient in terms of throughput, being at the same time more expensive.

LCTFs have been proved as an indispensable tool in biomedical OI in wide range of *in vivo* and *in vitro* applications. Attached to an endoscope, LCTFs were proved efficient in detecting small differences in fluorescent properties of malignant versus non-malignant mice tissues (Martin et al. 2006). Integrated with a colposcope, the LCTF was used to identify the optimum spectral bands for monitoring the uptake kinetics of biomarkers for improving diagnostic accuracy of cervical neoplasia (Balas 2001). LCTFs have also been used in hemodynamic imaging studies, in mapping of oxygen transfer and oxygen saturation levels in tumors (Sorg et al. 2005, Skala et al. 2009). In this study, the variation of the oxy- and deoxyhemoglobin optical signatures in the red/NIR range of wavelengths is investigated and the information obtained is used for understanding of tumor growth and proliferation. LCTFs have been used for studying retinal oxygen saturation in retina (Hirohara et al. 2007). Moreover, combination of LCTFs with photostable, bright, narrow bandwidth, quantum dots-based fluorescent probes has been used for *in vivo* fluorescence imaging of small animals (Gao et al. 2004) and in human tissues (Tholouli et al. 2006). Last, LCTFs have been used in two-photon fluorescence microscopy (Lansford et al. 2001), for discriminating mixtures of florescent proteins.

7.4.3.2 Acousto-Optic Tunable Filters

An acousto-optic tunable filter (AOTF) (Tran 2003, Stratis et al. 2001, Bei et al. 2004) consists of an acousto-optic crystal, a birefringent material whose optical properties can be altered by applying acoustic frequencies to the crystal. Crystals that are commonly used in the construction of AOTFs are tellurium dioxide (TeO_2), lithium niobate ($LiNbO_3$), and calcium molybdate ($CaMoO_4$), among others. Operation of the AOTF relies on the interaction of light and acoustic waves traveling through a birefringent crystal. The density of the atoms making up the crystal is modulated by the acoustic or pressure wave. This results in modulation of the RI, which is directly affected by the structure of the atoms. Therefore, a volume grating is formed and moves with the speed of sound inside the crystal.

An AOTF works in so-called noncollinear configuration when the acoustic and optical waves propagate at quite different angles through the crystal. A piezoelectric transducer, bonded to the one side of the crystal, emits acoustic waves, usually at radio frequencies (50–200 MHz). As these acoustic waves propagate through the crystal, they cause the crystal lattice to be alternately compressed and relaxed. The resultant density changes produce periodic RI variations via the elasto-optic effect. This periodic perturbation acts like a transmission diffraction grating, diffracting a narrow band of spectral frequencies at a time. The undiffracted wavelengths exit the crystal at the same angle as the incident light beam (zero-order beam). The diffracted wavelength is determined by the momentum matching condition for a transmission diffraction grating, described as

$$K_d = K_i \pm K_g \tag{7.3}$$

where k_d, k_i, and k_g are the wave vectors of the diffracted beam, the incident beam, and the acoustic wave, respectively. The corresponding magnitudes of the vectors are $k_d = 2\pi n_d/\lambda_0$, $k_i = 2\pi n_i/\lambda_0$, and $k_g = 2\pi f/V$, where f is the acoustic frequency, V is the acoustic speed in the crystal, λ_0 is the center wavelength, and n_i, n_d the indices of the refracted and diffracted beams, respectively. Equation 7.3 describes a phonon–photon interaction in which photons of appropriate wavelength interact with phonons—the "packets of energy" transferred through the acoustic waves.

Figure 7.5 illustrates the principle of operation of a noncollinear AOTF. When unpolarized white light is incident on an AOTF, light of specific wavelength will be diffracted in two

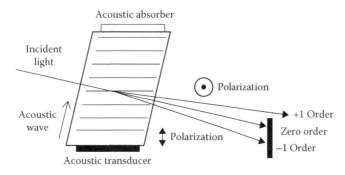

FIGURE 7.5 Schematic representation of a noncollinear AOTF.

directions with orthogonal polarizations. One has Doppler upshifted (+ diffracted beam) and the other has Doppler downshifted (− diffracted beam) optical frequency for the applied radio frequency (RF). When the zero-order beam is used as output, the AOTF functions as a notch filter; if the ±1-order beam is used, the AOTF functions as a band-pass filter. The unwanted beams are rejected with a beam stop. Changing the frequency of the acoustic wave has the effect of changing the grating spacing, thus adjusting the wavelength of the diffracted beam. The wavelength of the diffracted beam is selectable at a bandwidth less than 2 nm by electronically varying the RF. When multiple RF signals are applied, multiple wavelengths are simultaneously diffracted in the same direction. This is an advantage over LCTFs, which generate only a single bandpass at a time. One should however keep in mind that AOTF devices require the incident light to be collimated.

The center wavelength is depended on the applied acoustic frequency, according to the following equation:

$$\lambda_0 = \frac{V \Delta n}{f} \sqrt{\left(\sin^4 \theta_i + \sin^2 2\theta_i\right)} \qquad (7.4)$$

where

$\Delta n = |n_e - n_0|$ is the difference in the RIs between the ordinary and the extraordinary ray

θ_i is the incident angle

The spectral operation range of an AOTF is quite extended and depends on the material used for the fabrication of the filter. For example, for TeO_2, the operation range is 350–4500 nm while for quartz crystal, the range is extended to the UV range (120–6500 nm). The bandwidth of the transmitted wavelength ranges from about 1 nm in the UV range (using quartz crystal) to 15–25 nm in the 1.5–2.5 μm IR range (using the TeO_2 crystal). Transmission efficiency is high, reaching the value of 98% when the incoming beam is polarized. Tuning speeds are often less than 50 μs, determined by the transit time of the acoustic wave inside the crystal.

The fact that the diffracted rays do not exit the acousto-optic crystal at a unique direction, as determined by Equation 7.3, but over range of diffraction angles, causes the unwanted effects of

image distortion and blurring. These effects restricted the maximum resolution to less than 1 μm and this was for a period of time the main reason that the AOTFs were not widely used for imaging applications. The first attempt to correct this deficiency is dated back to 1996 when Wachman et al. (1996) investigated the origins of AOTF image blurring. By employing image processing techniques, the resolution increased at 0.35 μm. Since then, several other methods have been applied for improving image quality involving optical correction (Wachman et al., 1997, Suhre and Gupta 2005) or digital image postprocessing (Suhre and Villa 1998, Frances et al. 2006).

AOTF-based spectral imagers have been used in several biomedical OI applications (Rajwa et al. 2005, Treado et al. 1992). Martin et al. (2006) demonstrated a dual-modality hyperspectral imager based on an AOTF, which was used for distinguishing normal from malignant mouse tissues. Wachman et al. (1997) presented an AOTF-SI adapted to a microscope and recorded spectral images of oxyhemoglobin and deoxyhemoglobin in the brain of a living mouse. A dispersive prism was added in the experimental setup to compensate for blurring effects. AOTFs have also been used in conjunction with fluorescence microscopes for multispectral image cytometry applications (Rajwa et al. 2005), with confocal microscopes (Song and Gweon 2008) and with endoscopes (Vo-Dinh et al. 2004).

7.5 Single Exposure or Instantaneous Spectral Imagers

7.5.1 Overview

As discussed in the previous paragraphs, tunable filter-based SI systems record spectral images in a time-sequential manner and obtain the spectra from *post hoc* assembly of the time-sequential data. A major disadvantage of systems based on scanning is that they are fundamentally unsuitable for recording phenomena that are changing on a timescale that is shorter than the duration required for recording the spectral cube. The scanning/acquisition time will typically be tens of seconds or even several minutes, depending on the spectral resolution and SR required. This imposes severe limitations in the application of scanning SI to dynamically changing processes. This is because moving or transient scenes can cause accurate coregistration of spectral images, required for calculating the spectra, to be highly problematic. A typical example relates with the need for simultaneous monitoring the uptake kinetics of multiple biomarkers, administered to the same biological sample and exhibiting different excitation/emission characteristics. Another example is the irregular mutual translation between the scene to be spectrally imaged and the imager as occurs in SI of the eye, in endoscopy, in flow cytometry (FCM), etc. There are numerous other biomedical and non-biomedical applications requiring SI and analysis of transient or moving scenes such as combustion studies, imaging from aerial platforms, where scanning time needs to be very short or, in the best case, zero. This need motivated the development of "single shot" or "single exposure" or "instantaneous" or "simultaneous" spectral imagers capable of recording the information required for constructing the spectral cube without spectral or spatial scanning. SE–SI systems exhibit several distinct advantages over scanning devices, including fast acquisition of accurately registered images, high device robustness and reliability, low cost, automation, simplicity of operation, etc. On the other hand, however, current technological limitations set a trade-off between the spatial and the spectral resolution that can be captured simultaneously with SE devices. The current level of development of SE systems allows for the simultaneous acquisition of a relatively small number of spectral images, and in some designs, the SR is inversely proportional to the number of spectral bands. Scanning devices are preferable for imaging stable or invariant scenes and in general in applications where scanning time is not critical, since they display superior spatial and spectral resolution as compared to the ones provided by SE–SI imagers. However, a great number of applications require the acquisition of only a small number of predetermined spectral bands and therefore all these applications can be more effectively addressed with SE–SI systems. Owing to its distinct advantages (low cost, robustness, automation, etc.), SE–SI concept is becoming the basis for the development of several biomedical imaging devices suitable for direct clinical implementation. For this reason, the attractiveness of this concept is exponentially increasing over the last years and research efforts are targeting the elimination of the associated spectral and spatial trade-offs, toward the development of advanced medical devices useful in a variety of applications. This paragraph starts with the basic description of some simple designs enabling SE-SI, progresses with more sophisticated approaches, and ends with the most recent developments in the field.

7.5.2 SE-SI Systems Based on Image Replication, Filtering, and Projection Optics

7.5.2.1 Dichroic/Polychroic Optics

Color cameras can be considered as the simplest form of SE-SI devices. As is has been described in the previous paragraph, their output is a color image composed by standard primary color images. The spectral transmission characteristics of the primary color filters, deposited onto the sensor's area in mosaic arrangement are broad and standard, not providing flexibility in selecting the imaging spectral bands. One reasonable approach to overcome this restriction would be the design of a mosaic arrangement composed by more than three filters, each having the size of a single pixel and desired light transmission characteristics. By overlaying such a multi-filter mask onto a black and white sensor, the spatial and spectral information can be captured simultaneously and the corresponding set of spectral images can be reconstructed. It is obvious that by increasing the number of distinct filters in the mask, the spectral information increases but at the cost of the SR. Although this could be acceptable in several applications, this solution is rather

impractical because it is very difficult to tailor and construct application-specific mosaic multi-filter masks with pixel-size elements. A more practical approach is to generate a number of filtered image replicas and project each of them onto different subareas of the imaging detector, or, alternatively, to project each image to a different sensor array. In multisensor configurations, the SR is not sacrificed to the benefit of the spectral information, but the system's complexity and cost increase. In the simplest optical designs, image replicas can be generated by employing either light-splitting elements or a microlens arrays placed in the imaging ray path. Referring to image splitting-based single aperture (monocular) systems, an optical copy of the incident light is generated using a beam splitter, such as a half-silvered mirror. In that case, two image copies are generated each at half intensity. These images can be filtered with, for example, band-pass interference filters and then be recorded by sensor array(s). However, all the out-of-band light is rejected, which results in significant light loses. Dichroic mirrors/filters provide a more efficient alternative by enabling almost lossless splitting and filtering with the same element. Although for two- and three-sensor configurations ad-hoc designs are often effective, the design problem becomes challenging as the number of sensors (and of spectral bands) increases. In these cases, the design can take the form of optical splitting trees, where the optical path topology takes the shape of a tree because of recursive beam splitting. However, multiple beam splitting and filtering elements occupy typically a significant volume that cannot fit in the space limited by the focal length of the objective lens. For this reason, secondary (relay) lenses are often required to re-project the spited images onto the corresponding imaging sensors. As it has been discussed earlier, multiple sensor designs have the advantage of not sacrificing the spatial information to the benefit of the spectral information. On the other hand, as the number of the required spectral bands increases, the number of sensors and consequently the system's volume and complexity increase accordingly. Due to the recent availability of (very) high-resolution imaging sensors, more robust systems can be developed using a single detector. In that case, splitting and image projection optics are designed so that the sensor's area is tiled with spectrally resolved and spatially separated replicas of the image formed by the objective lens. The SR of the images captured simultaneously is the result of the division of the sensor's SR by the number of the projected images. One typical representative of the single sensor SE-SI concept is the "QV2" (Quad-View) module (Photometrics, The United States), which can project up to four, spatially separated spectral images onto a single CCD. This module is adaptable to almost any OI apparatus (microscope, fundus camera, etc.), converting them to simultaneous multichannel imaging devices. Applications in light microscopy include multicolor fluorescence, quantitative radiometric imaging, fluorescence resonance energy transfer (FRET) imaging, fluorescence *in situ* hybridization imaging, etc. The same device, adapted to a fundus camera, has been used for assessing hemoglobin oxygen saturation in retinal vessels, *in vivo* (Hardarson et al. 2006).

A second device belonging to the same category of SE–SI systems is the "ImageStream" (Amnis Corporation, The United States), which captures images at six spectral bands simultaneously. Spectral decomposition is achieved with the aid of a set of dichroic mirrors placed in a fanned configuration in the back side of an objective lens. The first mirror reflects a certain spectral band and the remaining wavelengths are transmitted and reflected by the second mirror and so on. The mirrors form a small angle with each other, which causes the reflected spectral images by each mirror to be projected at different lateral positions onto the same CCD sensor. ImageStream is an integrated SI system containing spectral decomposition elements, illumination sources, and objective optics, and it is optimized for use in FCM. Amongst the interesting applications of this system is the discrimination between live, necrotic, and apoptotic cells for studying cell death procedures (George et al. 2004).

SE–SI systems based on dichroic or trichroic beam splitting elements are suitable for applications requiring the simultaneous acquisition of (typically) 2–6 spectral bands. When attempting to increase the number of spectral bands, the design becomes bulky, composed by multiple light-splitting elements. The direct consequence of this will be significant light losses, while alignment of multiple images is becoming a technically challenging task. Apart from the limitation associated with the number of spectral bands, special care should be given for the minimization of ghost image artifacts, produced by second-order light reflections by the mirrors and of image distortion effects. Nevertheless, most current applications require a small number of spectral bands to be captured simultaneously. In these cases, SE systems based on dichroic mirrors/filters comprise often a cost-effective solution since (a) they are polarization independent, (b) they display high light gathering efficiency, (c) they can be easily reconfigured to adapt to different applications, and (d) imaging data require very little post-capturing processing, which enables not only the real-time image capturing but also the real-time display of the captured spectral images. The latter is very essential in several *in vivo* applications, where the areas of interest are selected on the basis of the feedback provided by the displayed spectral images. Real-time display is also essential for image focusing and for achieving fast examination.

The need for devices capable of acquiring a greater number of spectral bands continues to drive the developments in the field. Several alternative devices have been proposed offering improved spectral information, although in some cases, one or more of the aforementioned advantages are sacrificed.

7.5.2.2 SE–SI Using Multiple Retarder/ Wollastone Prism Stages

The snapshot SI device integrating a set of Wollaston prisms in combination with retarders (Harvey and Fletcher-Holmes 2003, Harvey et al. 2005) adopts a similar configuration with the Lyot filter (Section 7.4.3.1). The difference is that the polarizers of the Lyot filter have being replaced by Wollaston prisms (see Figure 7.3).

The Wollaston prism is a polarizing beam splitter consisted of two right triangle prisms with perpendicular optical axes,

cemented together. At the interface, the ordinary ray becomes extraordinary and vice versa, which causes the diverging of the two polarized beams from the direction of the propagation of the incident light, with the divergence angle being determined by the wedge angle of the prisms. Thus, this assembly has the effect of both image splitting and spectral separation. The spectrally separated images are projected on different, not overlapping areas on the same CCD. For reconstructing the spectral cube, registration algorithms are applied for spectral image registration and for compensating image distortion effects caused by the optical components. The IRIS device implementing this principle of operation has been used for fluorophore separation in FRET imaging and retinal imaging (Alabboud et al. 2007, Harvey et al. 2003, Harvey et al. 2005). In retinal imaging particularly, configurations enabling the demultiplexing of eight narrow spectral bands were used for assessing blood oxygenation in retina. Blood oxygenation is of particular interest for the clinical diagnosis of retinal diseases, such as diabetes, glaucoma, and age-related macular degeneration.

7.5.2.3 SE–SI Using Microlens Arrays

An alternative snapshot SI design is based on multi-aperture systems in combination with band-pass filters (Mathews 2008 and Shogenji et al. 2004) for projecting several spectral bands onto the same sensor. In multi-aperture imaging systems, an array of lenslets is placed right in front of a large-format CCD. Spectral separation is achieved by accommodating band-pass filters in front of each microlens. With this simple setup, several filtered image replicas are projected in different areas of the imaging sensor and recorded simultaneously. Amongst the system's drawbacks is the inefficient use of light and the image distortion, both caused by the lens array. Distortions can be compensated by using image correction algorithms and/or by extracting the central, undistorted part of the images. Despite these limitations, multi-aperture configurations enable the capturing of spatio-spectral information at video rates. Multi-aperture systems have been used for measuring and mapping of blood oxygenation in human retina (Ramella-Roman and Mathews 2007). Typical setups integrate arrays with 6–18 lenslets, which can capture an equal number of spectral images using large-format detector arrays.

7.5.3 SE–SI Based on Coded Image Capturing

7.5.3.1 Computed Tomography

More sophisticated SE-SI approaches involve capturing of a single multiplexed coded image, which is subsequently uncoded for reconstructing the spatio-spectral information. The computed tomography imaging spectrometer (CTIS) (Descour and Dereniak 1995, Volin et al. 1998, Okamoto and Yamaguchi 1991) belongs to the category of devices that are capturing coded images. It is based on a 2D grating disperser, which is used to project spectral and spatial information onto a large-format focal plane array. The holographic disperser comprises a set of holographic gratings. The specific type of the diffractive optical element used is called a computer-generated hologram (CGH). It operates at multiple orders, creating an array of images on the sensor. The CGH diffracts the incident light at different directions as a function of the wavelength. Multiple diffraction angles corresponding to diffraction orders are possible for a certain wavelength. Objective optics project the image onto the disperser and the resulting dispersed images, corresponding to a set of diffraction orders, including the zero-order undispersed image, are projected onto the detector. Therefore, each image is not simply composed of single wavelengths; spatial and spectral information from each object pixel is multiplexed over the entire array. Each dispersed image may be interpreted as a projection of the 3D spectral cube onto the plane of the detector from a certain projection angle. The zero-order image is an undispersed broadband image, which, as it needs no processing, is displayed in real time and can be used for aiming and focusing. The image cube in wavelength space is then reconstructed from a single image. The models and algorithms employed in CTIS imaging for the reconstruction of the spectral cube are similar to that of positron-emission tomography, hence the name "computed tomography" imaging spectrometer. Fourier analysis is employed for the reconstruction of the spectral images from the captured raw data. Due to the fact that the full range of the projection angles cannot be measured, the system is undersampled in the Fourier domain (the so-called missing cone problem), meaning that SR and field of view are traded off with regard to spectral information. The constraints imposed by the undersampling are essential in case of objects with little spatial contrast and sharp spectral transitions. CTIS can operate at video rates, over wavelengths ranging from 450 to 750 nm (Ford et al. 2001). The image pixel size varies, depending on the number of the resolved spectral bands. For example, 30 spectral bands can be separated with 83×83 pixels SR (Ford et al. 2001). The technique has been tested in proof-of-concept experiments (Volin et al. 1998) and has been used in a series of applications, such as in resolving the signal responses of molecules whose fluorescence spectrum shifts in response to changes in cell pH (Ford et al. 2001), in spectral unmixing of fluorophores (Volin et al. 1998, Descour et al. 2003), and in pH measurement using combination of fluorescent probes (Descour et al. 2003). The diagnostic potential of CTIS in ophthalmology has been also investigated. In a study conducted by Johnson et al. (2007), spectral cubes of different parts of the human retina were acquired in the spectral range 450–700 nm, at acquisition times of the order of 3 ms, and oxygen saturation data were obtained.

7.5.3.2 Coded Apertures

A *pinhole camera*, also known as *camera obscura*, is a simple OI device in the shape of a closed box with a small hole in one side. Light from a scene passes through this small hole and projects an inverted image on the opposite side of the box. The earliest description of this type of imaging device dates from the fifth century BC. Pinhole cameras were used to observe the solar eclipse, sun movement (solography), and to examine the laws of light projection. During the twentieth century, electronic of film

pinhole cameras were extensively used for surveillance applications and in x-ray and gamma-ray photography, since these radiations cannot be focused by an ordinary camera lens. Pinhole cameras have a couple of advantages over lenses—they have infinite depth of field, and they do not suffer from chromatic aberration. For achieving good image quality, the hole should be thin and narrow with typical diameters of 0.20–0.35 mm. The image resolution is inversely proportional with the pinhole's size. An extremely small hole, however, can produce significant diffraction effects and a less clear image. Due to their small apertures, pinholes let very little light to pass through and therefore the light sensitivity of the camera is very poor. A possible way to overpass this problem would be the use of multiple small holes, but this results in a confusing montage of overlapping images. Pinhole cameras share the same limitation regarding the low light throughput with slit spectrometers where the narrower the slit, the greater the spectral resolution.

In 1968, R.H. Dicke at Princeton University's (Dicke 1968) described a multiple pinhole camera for astronomical applications, which provided a means of obtaining good (high) spatial or angular resolution with a large aperture. It has been shown that the replacement of the entrance slit with a *coded aperture* can increase the throughput and the SNR by at least one order of magnitude without sacrificing spectral resolution at all. A coded aperture is a more complicated pattern than a single hole or slit, composed of opaque and transparent features, with the overall transmissive area being much greater. Coded apertures are masks with hole patterns based on Hadarmard or S-matrices. A Hadamard matrix is a square matrix whose rows are mutually orthogonal and the elements in the matrix are either +1 or −1. In several cases, it is more convenient to use Sylvester matrices (simplex or S-matrices), which are derived from Hadamard matrices, with elements 0 and 1. The rows of an S-matrix are not orthogonal themselves but in decoding processes can be converted to include 1 and −1 elements. In an S-matrix encoded mask, 0 and 1 correspond to opaque and transparent areas, respectively. Reflective masks comprise an alternative configuration using light more efficiently, but this makes the design of the system more complicated.

Golay (Golay 1949), first proposed the replacement of the input slit of a conventional dispersion spectrometer with a coded aperture for encountering the low throughput problem. The basic operation of a coded aperture spectrometer is described below using a rather simplistic demonstration of the principle of operation (Mende et al. 1993). The light that passes the coded mask (replacing the slit of the spectrometer) is dispersed by a prism or a grating and finally it is collected by a detector array. The image is a convolution of the coded-mask pattern and the spectral profile of the light illuminating the coded mask. In general, the sequence of clear (1) and opaque (2) slits of the mask could be chosen arbitrary. For simplicity, we assume a seven element S-matrix mask of coding pattern 1110100. We also assume that the input light has a spectral distribution described by seven spectral intensities, $\lambda_1, \lambda_2, \lambda_3, \lambda_4, \lambda_5, \lambda_6, \lambda_7$, that need to be measured by the spectrometer. Light from the mask elements is displaced by the spectrometer linearly with the wavelength and each particular detector element (pixel), records overlapped information for almost all λ intensities. It is therefore obvious that the unprocessed output readings of the detector cannot provide the λ intensities separately. For this reason, unmixing of the spectral information is required on the basis of algorithms described below. Figure 7.6a illustrates the basic light detection scheme. Spectral intensity λ_1 is diffracted most to the left and the corresponding (modulated) intensity is deposited to detector elements D_1–D_7. Accordingly, λ_2 intensity is recorded by detector elements D_2–D_8 and the rest of the wavelengths are shifted to the right in a similar fashion. By adding the content of columns D_8–D_{13} to the elements of columns D_1–D_6, we end up to a cyclic S-matrix, which is given below (Figure 7.6b):

In a cyclic S-matrix, successive rows are generated by cycling the first row through all its cyclic permutations. Cyclic difference sets have interesting and unique properties that allow the calculation of the λ intensities separately. Calculation of the λ_4 intensity, for example, involves the calculation of the sum of all detector elements, while assigning appropriate signs. The sign minus is assigned to 0 and the plus to 1. In the case of λ_4, the sums are produced according to the signs associated with the fourth row $(+--++\,+-)$. Sums are then produced on the basis of the fourth row rule for all elements containing λ_4. In the fourth row, we obtain $4\lambda_4$ and, interestingly, all other rows not containing λ_4 become zero. This result needs to be normalized (divided) by $(N+1)/2$, where N is the order of the S-matrix. In current case, $N=7$ (seven elements in the mask) and $(7+1)/2=4$ clear apertures. This procedure can be repeated for all rows, using the corresponding sign rule each time, for calculating all the other λ intensities. Thus, by recognizing that the signal in the detector elements is equivalent to the column sum of the

matrix, we can recover the individual spectral intensities using a linearized deconvolution technique. If another coding scheme is used, which does not have the orthogonality property, the inverse operation will be more complex. Generally speaking, if the recorded intensity matrix is m and the single wavelength intensity matrix is X then,

$$m = FX \quad \text{or} \quad X = F^{-1}m \tag{7.5}$$

where F is the Hadamard or the S matrix.

Coded aperture single point spectroscopy has been recently advanced to single-shot spectral imagers capable of providing more than 30 spectral images in a single camera exposure (zero scanning time). A typical representative of this class of instruments is the coded aperture snapshot spectral imagers (CASSI) (Wagadarikar et al. 2007, 2009, Gehm et al. 2007, Fernandez et al. 2009). The optical layout of CASSI system is shown in Figure 7.7a.

The code used is a random 256×248 element binary pattern developed lithographically on a quartz substrate (Figure 7.7b). This limits the SR of the reconstructed spectral images to an equal (256×248) or smaller number of pixels. The dispersive element is a double-Amici prism, a complex prism composed of segments of three separate prisms, for eliminating anamorphic distortion and for enabling the linear configuration of the optical setup. The mask modulates the rays, which are then dispersed horizontally across the CCD, with the amount of the dispersion depending on the wavelength. The CCD records spatial and spectral information in a mixed fashion. It is therefore essential to reconstruct the 3D spatio-spectral data cube from the 2D data recorded by the CCD. This is a very challenging task because the number of voxels in the spatio-spectral data cube to be reconstructed exceeds the number of pixels in the CCD measurements. The system is therefore underdetermined, hence having infinitely many solution candidates. The solution to this problem has been provided by the outstanding recent development in the emerging field of *compressive sensing* (Candes and Tao 2005, Candes 2006, Candes and Wakin 2008, Tsaig and Donoho 2006). The basic idea of this theory is that when the signal of interest is very sparse (i.e., zero-valued at most locations) or compressible, relatively few incoherent observations are necessary to reconstruct the most significant nonzero signal components.

According to Shannon's theorem (also called the Shannon–Nyquist sampling theorem), the resolution of an image is proportional to the number of measurements. However, the compressed-sensing concept states something that is completely different: the achievable resolution is controlled primarily by the information content of the image. Particularly, it is well known that in standard compression methods, images can be represented by a sum of wavelets. The (few) large coefficients are identifying wavelets that make a significant contribution to the image. The small coefficients are discarded in compression and the image is reconstructed from only the remaining wavelets. Generally speaking, images have a sparse representation in some

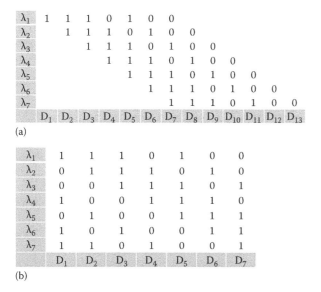

(a)

(b)

FIGURE 7.6 (a) Light detection scheme of a coded-aperture spectrometer, where λ_{1-7} are the wavelength intensities; (b) the resulting cyclic S-matrix.

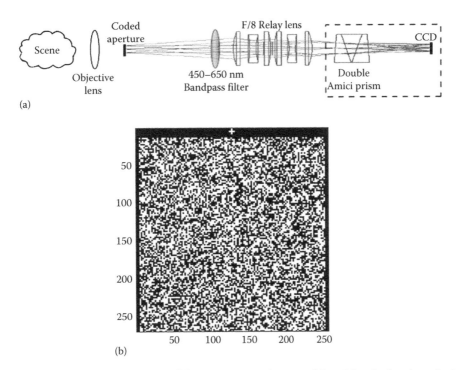

FIGURE 7.7 The schematic (a) and the aperture code (b) of the CASSI imager. (Reprinted from Wagadarikar, A. et al., *Opt. Express*, 17(8), 6368, 2009. With permission.)

basis, such as a wavelet basis and in that case, a variety of non-linear reconstruction methods can be applied, which can dramatically outperform linear reconstructions. Nonlinear image reconstruction based upon sparse representations of images has received widespread attention recently with the advent of "compressive sensing." This emerging theory indicates that, subject to a *restricted isometry property* (Candes and Tao 2005) condition on the observation matrix (CCD), the reconstruction problem has unique solution, and very high-dimensional vectors (3D spatio-spectral data) can be recovered with astonishing accuracy from a much smaller dimensional observation (2D information captured by the sensor). This becomes possible by applying random sampling to the sparse data or more precisely, to the imaging data that have "sparse" representation in some basis, meaning that only a few nonzero basis coefficients contain the vast majority of the image information. Detailed analysis of this theory and the relevant formulations lies outside the scope of this chapter. For more information on this exciting new theory, the reader should consult relevant references sited in this document and elsewhere.

The net effect of CASSI device integrating this nonlinear reconstruction method is the calculation of 32 spectral images in the visible spectral range with 10 nm spectral resolution, from the coded and overlapped spatio-spectral data, captured by the CCD sensor. Impressively, all these images are captured with a single camera exposure and multiple spectral cubes can be captured at video rates (30 frames per second). It should be noted, however, that capturing the information at video rates does mean that the spectral images are available and can be displayed at the same rate. In fact, processing of each frame requires 45 s

(with SunFire X4100 M2 workstation) (Wagadarikar et al. 2009). Nevertheless, it is expected that special-purpose hardware and multi-core processors shorten remarkably the processing time, making feasible for the reconstruction process to be real time, matching the coded image capturing video rate.

CASSI system is of low cost and offers a unique means for capturing both spatial and spectral information in dynamically changing objects in light microscopy. Recently (Fernandez et al. 2009), it has been used in steady-state fluorescence microscopy to discriminate between fluorescent beads with peak spectral emissions separated by 10–20 nm in the spectral range 500–700 nm. The application of CASSI to fluorescence microscopy has shown comparative results with a multispectral confocal microscope.

Very recently, an alternative to CASSI optical design is based on *phase-coded aperture* (Chi and George 2009). The authors describe a system based on a phase-coded aperture and have shown, theoretically, that it can be used for SE–SI imaging. The design is much simpler that one of CASSI, including just a phase screen and an imaging detector. Image reconstruction algorithms are different to those used in CASSI, employing correlation functions. The aperture pattern of a uniformly redundant array, used in x-ray imaging, is changed to a bandlimited pattern and used as the point spread function (PSF) of the optical subsystem. The intermediate image is coded by this complicated PSF, and a correlation type of processing is used to decode the image and recover the object.

The basic setup of the phase-coded aperture SE-SI is illustrated in Figure 7.8. It is based on the controlling of the thickness variations of the phase plate at the aperture plane. Generally, if the phase delay functions for two wavelengths are different,

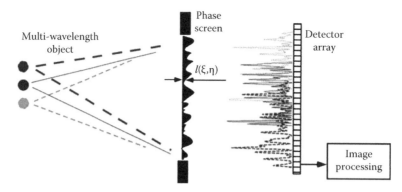

FIGURE 7.8 The schematic illustration of the phase-coded aperture snapshot imager, where $l(\xi,n)$ is the phase plate thickness. (Reprinted from Chi, W. and George, N., *Opt. Commun.*, 282, 2110, 2009. With permission.)

then their diffraction patterns would be uncorrelated for the two wavelengths. Experimental validation of this alternative concept is still pending. Subject to the successful implementation of this idea to working experimental prototypes, this approach could have a great number of applications in both color and SI regimes.

7.5.4 SE-SI Systems Based on Surface Plasmons and Nanodevices

Surface plasmons are charge density waves formed on the surface of materials with free electrons (metals). These electromagnetic (EM) waves are confined on conductor's surface and are propagating along the metal/dielectric (or metal/air) interface. Surface plasmons can be excited by both electrons and photons. The result of the coupling of plasmons with photons is known as *polariton* and the phenomenon as surface *plasma polaritons* (SPP). In that case, the electric field of the optical radiation excites free electron oscillations, and it is possible to transform photons into plasmons, use their unique properties to manipulate the original light, and reconvert the plasmons into photons. Such a unique property of SPP is the so-called *extraordinary transmission* of optical radiation through subwavelength holes, discovered by (Ebbesen et al. 1998). According to the Bethe theory of small apertures, it was believed that optical transmission through subwavelength apertures must be negligible. However, in the recent decade, some experiments have shown that an extraordinary optical transmission is possible if there is a periodic array of holes or slits in a metallic film. A working knowledge of the effect is provided next, without detailing its theoretical basis (Battula and Chena 2006).

It is well known that the dielectric response, $\varepsilon(\omega)$, of a metal is governed mainly by its free electron plasma, and for frequencies smaller than the plasma frequency, the real part of $\varepsilon(\omega)$ is negative. Negative ε, indicates that SSP modes can be supported in the metal–dielectric interface. The dispersion relation of these modes for a flat interface does not cross the light cone and so SPP cannot be excited by a direct incident plane wave. However, if the metal–dielectric interface is modulated periodically, the SPP bands fold allowing the external radiation to excite the SPP modes. Therefore, when the frequency ω and the in-plane wave

vector $\mathbf{k}_\parallel$ of the light scattered by the periodic structure (grating) vector match the surface mode (or SPP mode) of the structure, the transmittance and reflectance are modified. In other words, the enhanced transmission in metallic subwavelength apertures has been considered to be due to the coupling of the incident light via diffraction to the SPP modes of the metallic structure. SPP resonance modes on lower and upper surfaces of the gratings are excited when one of the diffracted waves is coupled to the SPP modes. However, there are other kinds of EM resonances, such as the cavity modes or waveguide resonances, which may also play an important role in enhanced transmission. It was reported that in order to have high transmission through metallic gratings, the waveguide mode resonance in the channel is the only effective way, even at the SPP resonance mode excitation wavelength. Experimental assessment of extraordinary transmission has shown that the fraction of exposed area divided by the fraction of the hole's area, is ≥ 2, which implies that twice as much light is transmitted as the illuminated hole's area together. The standard aperture theory would predict a transmission efficiency of 0.01 (Martín-Moreno et al. 2001).

The advent of nanotechnology has greatly accelerated the study of the science of periodic structures and tiny holes in metal films. In recent years, the interest in SPP is fuelled by the emergence of photonic materials, metamaterials, and nanophotonics. As shown in recent experiments, SPPs can be manipulated using waveguides and resonators. SPPs hold promise for application in sensing, imaging, photovoltaics, telecommunications, and optoelectronic circuit integration due to their ability to concentrate and guide EM energy at the nanoscale. Ebbesen and coworkers (Laux et al. 2008) have very recently exploited this plasmon-boosted hole transmission for SI. By focused ion-beam milling, they fabricated several overlapping circular groove areas in a silver film, each with a different period and centered on a different hole. The periodic grooves act like an antenna for the incoming light by converting it to SPs and enhancing transmission through the aperture. Interestingly, the dispersion of the plasmons on either side of the hole array dictates at which wavelength an optical transmission peak occurs, which provides a means for separating the spectral content of the incoming light. The transmission peak wavelength of the aperture can be

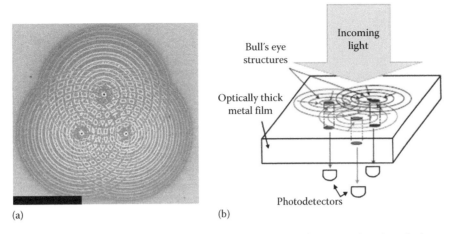

FIGURE 7.9 (a) Periodic structures adjusted to give rise to resonances at three different wavelengths, which are selectively transmitted through the holes; (b) the spatial filtering and detection process for the incoming white light through three overlapping structures. (Reprinted from Laux, E. et al., *Nat. Photonics*, 2,161, 2008. With permission.)

tuned by controlling the groove periodicity. The transmission spectrum is also modulated by the other structural parameters, such as groove depth, width, aperture shape, and size. In order to achieve spectral separation, a number of overlapping light-collection structures are milled on an optically thick Ag film (Figure 7.9a). The periods have been adjusted to give rise to resonances at three different wavelengths, which are transmitted through the holes. By optimizing the groove depth, it became possible to design a device with up to 70% overlap between the groove areas while preserving the spectral separation efficiency.

With photodetectors placed underneath the apertures, a miniature spectrometer is obtained capable of recording the spectrum of light incident on the overlap area. The different colors are separated as they couple to different gratings and are redirected toward three distinct photodetectors through apertures in the metal film (Figure 7.9b). It is obvious that the periodic structure of the system can be configured to provide several spectral bands separately. An array of such spectrometers may be placed in the focal plane of an imaging lens, resulting in a compact SI system.

The plasmon-assisted light harvesting has some distinct advantages over several other simultaneous SI approaches. With the standard Bayer filter approach (see Section 7.2), for example, light wavelengths that do not match the light transmission characteristics of the RGB filters are rejected, never reaching the detector's surface. The direct consequence of this is resolution and throughput losses. In contrast, plasmonic photon sorters act as an antenna, harvesting in principle all of the light from an area around the hole and the spectral information is sorted through plasmon resonances. Furthermore, it is important to note that the antenna effect of the plasmonic structure permits reduction in the underlying photodetector size, with potential gains in speed and chip dimensions. However SPP-based spectral imagers face several challenges that need to be addressed in order to be used effectively in practice (van Hulst 2008). First, transmission efficiency is a critical issue. Despite plasmonic effects, the absolute transmission is a few percent at best, which is below the efficiency of conventional filter devices. Second, the number of spectral bands will probably be limited by overlap, crosstalk, and losses to not many more than three. The SPP devices could be probably proved more efficient in the NIR spectral range, where surface plasmon resonances become sharper with higher quality factors, enabling higher spectral resolution.

Very recently, Diest et al. (2009) reported a method for filtering white light into individual colors using metal–insulator–metal (MIM) resonators. MIM waveguides are plasmonic resonators that couple incident light into EM modes, propagating along a metal–dielectric interface. Lithium niobate and silicon nitride are typical insulators used in MIMs coated on one side with gold and with silver on the other side. MIM waveguides have received increasing attention because the top and bottom metal layers in an MIM waveguide can serve as both cladding layers and electrodes and so they can work as both passive and active devices.

Figure 7.10a illustrates the MIM resonator geometry. This structure has input and output slits milled into the top and bottom cladding layers, respectively. The separation between the input and output slits is labeled "d_{sep}," and the depth to which the output slit is etched into the lithium niobate is labeled "d_{output}." By varying the RI and the thickness of the dielectric layer within MIM resonators, as well as the d_{sep} and d_{output} characteristics, the spectral transmittance can be tuned over a wide range. Light enters the resonant through a slit at the top layer of the cladding and the output color is received through a slit at the bottom layer. Color filtering is achieved due to interference between the optical modes supported inside the cavity of the resonator. In addition, the electro-optic effect of lithium niobate (RI control though external voltage) allows the selected output color to be shifted across the red, green, and blue regions of the visible spectrum. MIMs can be easily integrated into a high-density array of MIM resonator color filters. The typical dimension of such a pixel would be ~3 to −5 μm, which is much smaller than the filters of a color camera. Hexagonal pixels could be tiled together

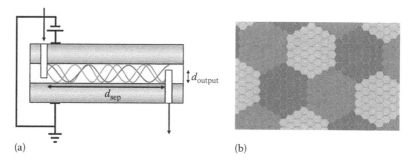

(a)   (b)

FIGURE 7.10   (a) The schematic diagram of a MIM waveguide with the insulator sandwiched between metal coatings acting as both grading and electrodes; (b) hexagonal packing of multiple MIMs forming pixels with different spectral transmissions. (Reprinted from Diest, K. et al., *Nano Lett.*, 9(7), 2579, 2009. With permission.)

to form "superpixels," as shown in Figure 7.10b. A tunable pixel array overlaid onto an imaging sensor may enable the development of imaging devices providing simultaneous color and/or spectral image capturing and freedom in selecting and controlling the imaging wavelength.

7.6 Spectral Data Classification, Unmixing, and Visualization

7.6.1 Overview

As it has been discussed in the previous paragraphs of this chapter, spectral imagery of either macroscopic or microscopic origin is usually depicted in a spectral cube, a registered set of images, featuring one spectral and two spatial dimensions (pixel coordinates). SI data are transforming the RGB color space to a high-dimensional representation of intensity vectors corresponding to different wavelengths. The number of the spectral dimensions is proportional to the spectral resolution of the spectral camera and, in principle, the higher the spectral resolution, the greater the diagnostic information that it is contained to the acquired spectra. The number of spectra that are acquired by a spectral camera is proportional to the SR of its image sensor and, therefore, the higher the SR, the greater the topological information that can be obtained for the localization of a target feature. Generally speaking, hardware improvements toward increasing the system's spectral and spatial resolutions impact positively on its accuracy in identifying and mapping compositional features of diagnostic importance. It should be noted, however, that the spectral resolution and SR requirements depend heavily on both sample's complexity and operating wavelength range (e.g., ultraviolet-visible-infrared spectroscopy). Solid tissues are typically characterized by a high spatial heterogeneity, which sets high SR requirements in case that sample surface details need to be depicted. The spectral resolution requirements depend largely on the operating spectroscopy type. For example, mid-infrared spectra contain fine features, such as sharp peaks and deeps, while visible and NIR spectral profiles are typically smooth and broad. Higher spectral resolution is required for acquiring spectra containing fine features, but these spectra are much less

informative for the composition of the sample, as compared to the smooth spectra. Smooth and broad spectra are, in most cases, the result of the superimposition of several spectra, corresponding to a number of co-localized compositional features. These spectra can be captured at a much lesser resolution, without significant loss of information. However, sophisticated spectral data processing and analysis methods are required for unmixing, extracting, and classifying the diagnostic information that might be contained in the smooth spectra, produced by visible and NIR absorption, scattering, and fluorescence spectroscopy. These spectroscopies, although not very informative for compositional analysis, are the ones most commonly used in biomedical applications, mainly due to their lower equipment cost and to their adaptability to common OI instruments (microscopes, endoscopes, etc.). The extraction of pathology-specific diagnostic information from these spectra is a complex process involving pattern recognition, image processing, statistical analysis, and many others (Schott 2007, Landgrebe 2003). These methods, applied to the spectral cube data set, are aiming at addressing the challenge to identify the link between the spectral content of the acquired images and status of the examined sample.

For the purpose of introducing the basic ideas of these processes, we consider the spectral cube of a tissue sample with two types of conditions present: benign and malignant. A first course of action for determining the pixels belonging in each of these two categories or classes is the implementation of a simple classification scheme. In Figure 7.11a and b, the spectra of two pixels are considered as the most representative for benign and malignant tissue. The coordinates of these two "spectral signatures are $\mathbf{y} = (y_1, y_2, \ldots, y_N)$ and $\mathbf{x} = (x_1, x_2, \ldots, x_N)$ for the benign and malignant tissue, respectively, where y_1, x_1 are the measured intensities at λ_1 wavelength, y_2, x_2 are the measured intensities at λ_2 wavelength, etc. By observing the plots, the most informative wavelengths for aiding a classification task are found in the spectral range $[\lambda, \lambda'']$, where the two spectra differ considerably. The comparison of the spectral response of each pixel $\mathbf{p} = (p_1, p_2, \ldots, p_N)$ at wavelength λ', establishes a simple rule of classification:

$$\text{if} \left| p_{\lambda'} - y_{\lambda'} \right| < \left| p_{\lambda'} - x_{\lambda'} \right| \quad \text{then} \quad p \rightarrow \text{"benign"} \quad (7.6)$$

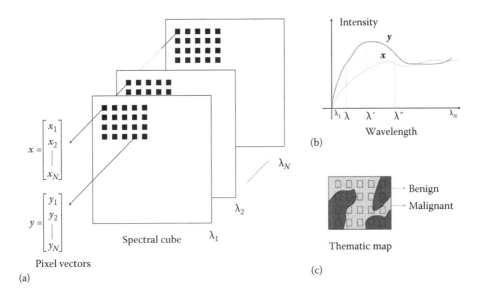

FIGURE 7.11 (a) A spectral cube comprises a set of images acquired at *N* different wavelengths; (b) spectral signatures are a collection of pixels representing categories of materials with different structural and/or chemical composition; (c) a thematic map is the outcome of a classification or unmixing process and it is constructed by annotating different artificial colors or gray shades to pixel clusters belonging to different classes.

In practical applications, where more classes are involved, using simply an intensity value corresponding to a single wavelength is not enough. More effective approaches, which will be discussed in the next paragraph, use the entire set of wavelengths within the range $[\lambda, \lambda'']$. Moreover, the assumption of a class represented by only one spectrum captures neither interclass variations nor potential variability due to instrument's noise. Usually, a "training set" of spectra for each class is collected, with the size of the set depending on the complexity of the material under investigation. Assembling an optimal training set depends on how completely and precisely a spectra collection of different categories is indented to be modeled. A classification process usually involves assigning each pixel to a predefined class. Classes should be exhaustive, meaning that they should include not only the detection targets, but also all the other distinct categories present in a scene. For instance, while a classifier would decide for the presence of malignant or benign pixel areas, other components, such as normal tissue, should be included in the list of classes. Generally speaking, a sufficient number of different categories reflecting the complexity of the material should be defined in order for all the pixels to be classified correctly. Another important issue, when collecting spectra for training, is that classes must be separable in terms of available spectral features. Low separability may be due to the high similarity of spectral signatures corresponding to different classes, which would comprise the foundation for pure results, regardless of the efficiency of the classifier. Separability criteria for SI data include empirical measures such as transformed divergence or Bhattacharrya distance (Swain 1978).

The discussion so far was based on the assumption that only one class can be assigned to an area corresponding to a single pixel. In reality, however, this area is often composed by several substances mixed together and, therefore, the spectrum from that pixel curries convoluted information for all these

components. Strictly speaking, information for the components of the mixtures is missed when classifying whole pixels. The recognition that pixels of interest are frequently a combination of numerous disparate components has introduced a need to quantitatively decompose, or "unmix," these mixtures. The need for subpixel analysis is addressed with spectral unmixing methods, where a mixed pixel is decomposed into a collection of constituent spectra, or endmembers, and a set of corresponding fractions, or abundances, that indicate the proportion of each endmember present in the pixel. When the unmixed spectra are classified, the color coding of the corresponding thematic maps may express the spatial distribution of the abundances of endmember classes.

A major advantage of acquiring 2D spectral information from a series of complete images—as opposed to acquisition of point information—is that both spectral and spatial information can be depicted and displayed comprehensively with the aid of the so-called thematic maps. A thematic map may be the outcome of a classification or unmixing process and it is constructed by annotating different artificial colors or gray shades to pixel clusters belonging to different classes. This facilitates the direct visualization of the various image clusters corresponding to the identified classes. In Figure 7.11c, two different gray shades are used for generating a thematic map representation, depicting the spatial distribution of the two (benign, malignant) classes over the surface of the sample. When more than two classes are present, it is more convenient to represent them using a pseudocolor scale, with each color indicating the presence of an area within the sample, having same or similar spectral characteristics. The diagnostic value of these thematic maps depends on whether the identified classes can be effectively assigned to different compositional/pathological conditions. In order to establish such a correlation, the spectral profiles corresponding to selected tissue sites are analyzed together with the findings of reference methods, for example, biopsy samples, taken from the same sites. If the findings of the reference method can be depicted in the spectral characteristics, then the thematic map would in fact be equivalent with a "pathology map" indicating a successful outcome of the entire process. SI methods with validated thematic maps depicting successfully the pathological status of the tissue have great clinical potential since they would enable the *in vivo* detection and grading of a pathology, provide map-based guiding of surgical treatment, facilitate objective follow-up, etc.

Last, but not least, adopting the most appropriate classification performance evaluation strategy comprises a key aspect throughout the entire process. Performance evaluation metrics is very essential for both optimizing the algorithms employed and for assessing objectively, if possible, the end result. Referring specifically to biological tissues, verification of algorithm results comprises a scientific challenge itself. The primary reason has been the lack of ground truth or, in other words, the detailed knowledge of the components comprising each pixel in a scene.

7.6.2 Preprocessing for Reducing the Dimensionality of the Data

In most spectral cube data sets, not all of the measured intensities at every wavelength are important for understanding the underlying phenomena of interest. As processing the entire data set may be computationally intensive and time consuming, transforming the image set into a reduced dimensionality structure is often of essential importance. This preprocessing step simplifies also the calculations performed by the classifier, thus increasing its efficiency, while in parallel making the process relatively immune to the so-called curse of dimensionality issue (Bellman 1961).

As it is expected, spectral bands at which the spectra are very similar can be omitted at no or minimum cost in the classification accuracy. It is therefore possible to isolate the most informative wavelengths by examining the spectra, as described in the example of Figure 7.11b. In more demanding applications, however, sophisticated methods are employed, which can perform the data dimensionality reduction in an automated manner. Indicatively, principal components analysis (PCA) (Rodarmel and Shan 2002) is the best, in the mean-square error sense, linear dimension reduction technique. PCA reexpress the original spectral bands by combining their orthogonal linear components with the largest variance. It computes an orthonormal basis for a series of images, transforming the spectral cube to a "noiseless" structure with lower dimension. The best low-dimensional basis is derived from the eigenvectors of the covariance matrix. These eigenvectors correspond to the largest eigenvalues, known also as "principal components." In PCA, the data set is organized as an $m \times n$ matrix X, where m is the number of spectral bands and n is the number of pixels. The Covariance matrix $\sum_{m \times m}$ is computed by subtracting the sample mean (μ) of each row of matrix X from each particular row element as follows:

$$\sum_{m \times m} = \frac{1}{n-1}(X - \mu)(X - \mu)^{\mathrm{T}} \tag{7.7}$$

Note that since $(X-\mu)(X-\mu)^{\mathrm{T}}$ is a symmetric matrix, it can be diagonalized by its orthonormal eigenvectors. Thus, $XX^{\mathrm{T}} = U\Lambda U^{\mathrm{T}}$, where $\Lambda = \mathrm{diag}[\lambda_1 \cdot \lambda_2 \cdots \lambda_m]$ is the diagonal matrix of the ordered eigenvalues $\lambda_1 < \cdot \lambda_2 < \cdots \lambda_m$ and U is an $m \times m$ orthogonal matrix containing the eigenvectors. Using the eigenvalues to form a new basis, any pixel vector **p** can be expressed as a linear combination of the eigenvectors:

$$
\mathbf{p} = \begin{bmatrix} p_1 \\ p_2 \\ \vdots \\ p_m \end{bmatrix} \xrightarrow{\text{PCA}} \begin{bmatrix} p_{pc1} \\ p_{pc2} \\ \vdots \\ p_{pcm} \end{bmatrix}
$$

$$
= \begin{bmatrix} U_{11}p_1 & + & U_{12}p_2 & + & \cdots & + & U_{1m}p_m \\ U_{21}p_1 & + & U_{22}p_2 & + & \cdots & + & U_{2m}p_m \\ & \vdots & & & & & \vdots \\ U_{m1}p_1 & + & U_{m2}p_2 & + & \cdots & + & U_{mm}p_m \end{bmatrix}
$$

$$
= \begin{bmatrix} \sum_{i=1}^{m} U_{1i}p_i \\ \sum_{i=1}^{m} U_{2i}p_i \\ \vdots \\ \sum_{i=1}^{m} U_{mi}p_i \end{bmatrix} \tag{7.8}
$$

Considering the variance among the ordered eigenvalues, large values are encountered in the first k eigenvalues and then decline sharply. The procedure of defining the appropriate principal components (parameter k) is associated with a trade-off between dimensionality reduction and loss of information. Interpretation of PCs is generally neither easy nor straightforward, because while possessing orthogonality and uncorrelatedness, they derive from mathematical definitions without holding direct physical meanings. However, in a PCA transformed spectral cube, each projection can still be considered as a grayscale image, after normalizing to a 0–255 range.

Figure 7.12 illustrates the image of a histology sample stained for estrogen (ER) and progesterone (PR) receptors with fast red (FR) and with hematoxylin, which stains blue/purple the non-ER/PgR reactive cell nuclei.

Figure 7.12a is the color microscopy image of a subarea of the sample. Figure 7.12b through d correspond to the first, second, and third principal component, respectively. These principal components have been derived from a spectral cube composed of 22 spectral bands in the range 420–700 nm. It is clearly seen that only the first two components are informative and that the dark areas in images (b) and (c), and, therefore, the 22 grayscale spectral images can be reduced to a grayscale representation of the first two principal components.

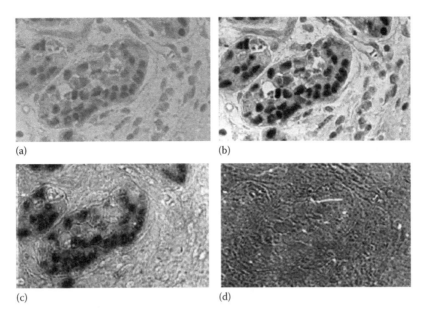

FIGURE 7.12 The color microscopy image of a histology sample (a), and the first three principal components (images b, c, and d) corresponding to the same image area, calculated from a spectral cube composed of 22 spectral bands in the visible part of the spectrum. Only the first two principal components (b, c) are informative.

Several PCA variants, such as independent component analysis (ICA), selective PCA, Kernel-PCA (Yao and Tian 2003, Gu et al. 2006, Bayliss et al. 1997), or wavelet transform (Mallet et al. 1997, Pu and Gong 2004), have been successfully used for spectral cube data processing or as a preprocessing step for subsequent classification. Additionally, Filter (De Backer et al. 2005) and wrapper methods (Bazi and Melgani 2006) as well as combinations (Mao 2004, Tan et al. 2008, Yu et al. 2002) have been employed to select the best subset of spectral bands for optimal classification.

7.6.3 Spectral Classification and Unmixing Methods

After extracting suitable features for representing different categories, the next step is to assign every pixel to one of these categories according to its level of resemblance (or similarity) with the aid of a classification method. In general, selection of the classification model is highly application dependent, requiring extensive research for identifying the most suitable algorithm, along with the optimization of its parameters. Image classification can be performed using either supervised or unsupervised techniques, depending on the human involvement in the configuration of the classification model. If the classifier utilizes parameters of a hypothesized distribution for the training set, then classification is considered a parametric procedure. Commonly used parametric classifiers employ statistical parameters of mean and covariance requiring a Gaussian distribution, which doesn't always hold for small training sets.

Assuming that no previous knowledge exists about the exact location of representative for the class spectra (spectral signatures), unsupervised classification provides an overview of the different groups (clusters) present in the data set. In complex biomedical samples, for example, it is not always possible to distinguish unique spectra and collect a training set of separable classes. Unsupervised classification techniques usually require specifying only a small number of parameters concerning the number and characteristics of the clusters. Several clustering algorithms exist for detecting the tendencies of pixels to group, iterating to a (local) minimum of an average distance from each pixel to the nearest cluster means.

7.6.4 Unsupervised Clustering Methods: The *k*-Means Algorithm

k-Means (MacQueen 1967) is a well-known unsupervised learning algorithm, which is widely used for data classification. The algorithm partitions a set of observations into separate groupings, such that an observation in a given group is more similar to another observation in the same group than to another observation in a different group. The canonical form of *k*-means defines the similarity measure using the Euclidean metric: For $y, z \in \mathcal{R}^d$, x, is more similar to y than z, if and only if

$$\|x - y\| \le \|x - z\| \tag{7.9}$$

Applied to SI data, the output of the *k*-means algorithm is pixel clusters having similar spectral profiles. In other words, it classifies a given set of pixels (N) into k, a priory defined partitions (clusters), where $k < N$, based on their spectral attributes and on the assumption that these attributes form a vector space. The procedure starts by arbitrary defining k centroids, one for each cluster. The next step is to associate each pixel to the nearest centroid by calculating the Euclidean distance (see Section 7.6.5.3 for more details). A preliminary clustering is completed when this has been done for all pixels. Next, k new centroids are calculated as the center of mass of the observations within the clusters resulting from the previous step. This is followed by a new binding between the same data set points and the nearest, new centroid. Putting this procedure into a loop, the k centroids change their location in each iteration step and after a number of repetitions, they do not move any more. This is because the k-means algorithm aims at minimizing *an objective function*, in this case a squared error function, given by the following formula:

$$J = \sum_{J=1}^{k} \sum_{I=1}^{N} \left\| x_i^{(j)} - c_j \right\|^2 \qquad (7.10)$$

where $\left\| x_i^{(j)} - c_j \right\|^2$ is a chosen distance measure between a data point $x_i^{(j)}$ and the cluster center c_j.

The objective function is an indicator of the distance of the N data points from their respective cluster centers and its convergence to its minimum value indicates that the minimum total intra-cluster variance has been reached. When convergence of J to a minimum has been achieved, the k-means algorithm outputs the coordinates of the centers of each of the k clusters and the membership index for each pixel. The minimum value of J comprises also a measure of cluster scatter.

A drawback of the k-means algorithm is that the number of clusters k is an input parameter. An inappropriate choice of k may yield poor results, while selecting the optimal number of clusters is a trial and error procedure. The indentified clusters in a spectral cube are often visualized with the aid of a thematic map, where different pseudocolors represent pixel clusters having similar spectral characteristics. This comprises a spectral mapping tool that enables the comprehensive representation of numerous spectra by grouping them in a small set of clusters. Grouping the spectral data to a set of "spectral families" can facilitate several diagnostic tasks since these tendencies may represent different pathologic conditions. On the other hand, unsupervised clustering can comprise a preprocessing step for supervised algorithms if more detailed and accurate analysis is required. Figure 7.13 illustrates an example of the application of the k-means algorithm to the same with Figure 7.12 data set.

Figure 7.13b illustrates a thematic map, calculated with k-means unsupervised clustering. The various pseudocolors represent the spatial distribution of the classes, a priory defined. These classes correspond to dark red (positive for receptors), blue/purple (negative for receptors), purple (nonspecific staining), and slide background (no staining). As it can be seen, the algorithm separates the classes decently, although positively and negatively stained areas are underestimated due to the fact that the algorithm cannot discriminate effectively specific from unspecific staining due to spectral similarity.

A number of alternative to k-means algorithms, such as Projection Pursuit (Ifarraguerri and Chang 2000), Hidden Marcov Chain models (Mercier et al. 2003), Expectation Maximization variations, etc., belong to the list of the unsupervised classification algorithms that have found numerous applications in a variety of remote sensing and nondestructive analysis fields.

7.6.5 Supervised Classification Methods Based on Distance Similarity Measures

7.6.5.1 Spectral Angle Mapper

The spectral angle mapper (SAM) (Kruse et al. 1993) is a physically based spectral classifier that determines the spectral similarity between the measured and the reference spectra. The spectra are treated as vectors in a space with dimensionality equal to the number of bands and the angle that is formed between these vectors is used as a metric of the spectral similarity (Figure 7.14). Smaller angles represent closer matches to the reference spectrum. SAM has also been used as a feature selection method for selecting an optimal subset of spectral bands (Chen et al. 2008, Keshava 2004).

The angle (θ) between pixel vectors as a discrimination measure is given by the following formula:

$$\text{SAM}(x,y) = \cos^{-1}\left(\frac{(x,y)}{\|x\|\|y\|}\right) = \cos^{-1}\left(\frac{\sum_{i=1}^{M} x_i y_i}{\sqrt{\sum_{i=1}^{M} x_i^2} \sqrt{\sum_{i=1}^{M} y_i^2}}\right) \qquad (7.11)$$

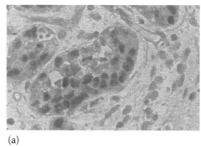
(a)

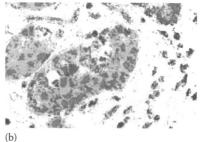
(b)

FIGURE 7.13 (See color insert.) (a) The color microscopy image of a histology sample and (b) k-means classification of the same image area based on a spectral cube data set composed of 22 bands in the visible part of the spectrum using the k-means classifier.

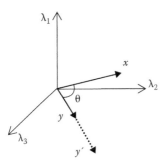

FIGURE 7.14 Vector representation of **x, y** pixels in a 3D orthonormal space corresponding to three different wavelengths. SAM essentially calculates the angle θ between them.

where **x, y** are two M-dimensional spectra in comparison.

This technique, when used on calibrated reference data, is relative intensive to illumination and albedo effects. This is illustrated in Figure 7.14, where an increase in the illumination has the effect of increasing the magnitude of the vector $y'(\|y'\| > \|y\|)$, but it has no effect to the angle formed between y'(dashed line) and x, so that SAM(x, y) = SAM(x, y').

Figure 7.15 illustrates a the application of SAM algorithm, trained with sets of spectral signatures corresponding to different classes, to the same data set that has been used in Figures 7.2 and 7.3.

By comparing Figures 7.13b and 7.15b, it can be concluded that SAM performs much better than *k*-means in separating the classes. This is evidenced by fact that the areas of the positively and negatively stained areas overlap substantially with the corresponding morphological features of the cells in the color microscopic image.

7.6.5.2 Spectral Correlation Mapper

Another minimum distance classifier is spectral correlation mapper (SCM) or angle (SCA) (Carvalho and Meneses 2000). SAM cannot distinguish between negative and positive correlations because only the absolute value is considered. SCM has been generated as an improvement on the SAM. The main difference is that SCM standardizes the data, centralizing itself in the mean of *x* and *y*.

SCM calculates a statistical measure of independence known as Pearson correlation coefficient. In probability theory and statistics, correlation indicates the strength and direction of a linear relationship between two random variables. SCM similarity metric is calculated using the following formula:

$$\text{SCM}(x, y) = \frac{\sum_i (x_i - \bar{x})(y_i - \bar{y})}{\sqrt{\sum_i (x_i - \bar{x})^2} \sqrt{\sum_i (y_i - \bar{y})^2}} \quad (7.12)$$

where x, y, $\bar{x}, \bar{y}$ are two M-dimensional spectra and sample means, respectively.

While in SAM the angle θ presents a variation anywhere between 0° and 90° with the positive and the negative correlations having (falsely) an equal value, SCM values range between +1 for positive correlation and –1 for negative correlation. Apart from that, SCM has been proved to be more efficient in eliminating shading effects. Unlike the SAM, the SCM measure has no direct physical meaning associated with spectral cube data.

7.6.5.3 Euclidian Distance Measure

The Euclidian distance (ED) algorithm calculates the distance between the spectral signatures of two pixel vectors *x* and *y* in the *n*-dimensional spectral feature according to the simple formula:

$$\text{ED}(x, y) = \|x - y\| = \sqrt{\sum_{i=1}^{n} (x_i - y_i)^2} \quad (7.13)$$

If both *x* and *y* are normalized to unity, the relationship between ED(x, y) and SAM(x, y) can be established as follows:

$$\text{ED}(x, y) = 2\sqrt{1 - \cos(\text{SAM}(x, y))} = 2\sin\left(\frac{\text{SAM}(x, y)}{2}\right) \quad (7.14)$$

When SAM(**x**, **y**) is small, $2\sin\left(\frac{\text{SAM}(x, y)}{2}\right) \approx \text{SAM}(x, y)$, in which case SAM(**x**, **y**) is nearly the same as ED(**x**, **y**). Figure 7.16 shows the geometric interpretation of ED in relation with SAM.

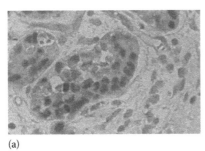

(a)

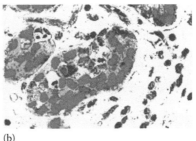

(b)

FIGURE 7.15 (See color insert.) (a) The color microscopy image of a histology sample and (b) classification of the same image area using the SAM algorithm and a training set of spectra, representative to the classes present in the sample.

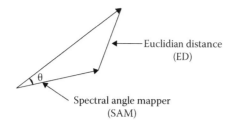

FIGURE 7.16 The geometric relationship between ED and SAM.

The main difference of the Euclidean distance measure as compared to SAM is that the ED takes into account the brightness difference between the two vectors, whereas SAM and SCM are invariant to with brightness. This implies that the ED algorithm may be more suitable for class discrimination in cases where calibrated spectra differ mainly in intensity characteristics.

7.6.5.4 Spectral Information Divergence

Spectral information divergence (SID) is based on the concept of divergence in information theory (Chang 2000). The SID algorithm is formulated as follows:

$$\text{SID}(\mathbf{x},\mathbf{y}) = D(\mathbf{x}\|\mathbf{y}) + D(\mathbf{y}\|\mathbf{x}) \tag{7.15}$$

where,

$$D(\mathbf{x}\|\mathbf{y}) = \sum_{i=1}^{M} p_i \log\left(\frac{p_i}{q_i}\right) \tag{7.16}$$

$$D(y\|x) = \sum_{i=1}^{M} q_i \log\left(\frac{q_i}{p_i}\right) \tag{7.17}$$

The $p_j = x_j / \sum_{i=1}^{M} x_i$, $q_j = y_j / \sum_{i=1}^{M} y_i$, $p = \{p_m\}_{m=1}^{M}$, $q = \{q_m\}_{m=1}^{M}$ are probability vectors resulting from M-dimensional pixel vectors $\mathbf{x}$, $\mathbf{y}$, respectively. SID can be used to measure the spectral similarity between two pixels x and y by making use of relative entropy to account for the spectral information provided in each pixel. SID can be used to measure the spectral similarity between two pixels x and y by utilizing the relative entropy to account for the spectral information contained in each pixel.

An essential issue related with the use of classifiers based on distance metrics is that verification is often needed on whether the pixel spectra that have been used for the training of the classifier are truly representative of the classes. For these classifiers, each pixel vector should be tested against all of the members of each class, while the class membership is decided according to the lowest value of the classifier's distance metric. In the case of ED, for instance, for a training set **T** of S samples, finding the ED between pixel **p** and every member $\mathbf{t} = (t_1, t_2, \ldots, t_N)$ of **T**, is required, so that

$$\text{if } j = \arg\min_{\substack{1 \le j \le S \\ t \in T}} \sqrt{\sum_{i=\lambda}^{\lambda''} (p_i - t_{i_j})^2} \tag{7.18}$$

Then, p belongs to class (j)

7.6.6 Supervised Classification Methods Based on Nongeometric Similarity Measures

The simplest probabilistic classifiers are based on Bayes theorem or on maximum likelihood estimations. Assuming that the probability density function for each class is normal (Gaussian), the (multivariate) maximum likelihood classifier generates estimates for the covariance matrix and the mean of each class. These two parameters represent pixel variability within a particular class as well as correlations between wavelengths. The covariance matrix and mean estimations can be easily calculated by the sample mean (μ) and sample covariance (Σ) from the training set. A discriminant function g employing these estimations to determine class membership is as follows:

$$g_i(\mathbf{x}) = -\frac{1}{2}(\mathbf{x} - \mu_i)^T \sum_{i}^{-1} (\mathbf{x} - \mu_i) - \frac{1}{2}\ln\left|\sum_{i}\right| + \ln P(\omega_i) \tag{7.19}$$

where

i = class index

$\mathbf{x}$ = M-dimensional pixel vector

$P(\omega_i)$ = probability of class ω_i (can be assumed the same for all classes)

Σ_i, μ_i = covariance matrix and mean vector of class ω_i

More advanced choices for hyperspectral classification include neural networks (Bishop 2005, Civco 1993, Villmann et al. 2003), kernel methods (Camps-Valls and Bruzzone 2005), and, especially, support vector machines (SVMs) (Vapnik 2000, Melgani and Bruzzone 2004, Pal and Mather 2004), decision trees (Friedl and Brodley 1997, Ham et al. 2005), expert systems (Buchanan 1986, Chiou 1985), linear (constrained) discriminant analysis (Du and Chang 2001), etc. Referring specifically to nonparametric methods, such as SVMs, very good performance can be achieved, even though a small number of training samples is available. This is done by maximizing the margin between the classification boundary and the nearest data point of each class. A large separation margin can improve classification accuracy for unseen data. To visualize this concept, two types of pixels are plotted in Figure 7.17 for two wavelengths as black and white symbols.

Consider x as a pixel vector; $y = \{+1, -1\}$ denotes the binary class label. Pixels in the training set are separable by the hyperplane $w^t x + b = 0$, if there exist a vector w and a scalar b such that

$$\begin{cases} w^t x + b \ge 1, & \text{if } y = +1 \\ w^t x + b < -1, & \text{if } y = -1 \end{cases} \rightarrow y(w^t x + b) - 1 \ge 0 \tag{7.20}$$

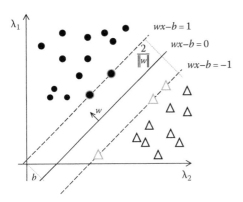

FIGURE 7.17 Maximum-margin hyperplane and margins for an SVM trained with samples from two classes. Pixels on the margin are called the support vectors.

Choosing the maximum margin ($2/\|\boldsymbol{w}\|$) as a "separation guide" instead of using directly the hyperplane $\boldsymbol{w}^t \boldsymbol{x} + b = 0$, can be justified by the fact that it produces the fittest boundary. Between the black and white support vectors, multiple separation planes for the two classes can be found. By "biasing" the boundary to directions parallel to the support vectors (pixels on the separation margin, Figure 7.17), it is ensured that even if the location of this hyperplane is not identified with high accuracy, the margins are informative on the least chance that causes a misclassification. To maximize the margin, $\|\mathbf{w}\|$ needs to be minimized under the constraint $y_j(\boldsymbol{w}^t \boldsymbol{x}_j + b) - 1 \geq 0$ (j indices indicate class membership). The Lagrangian $L(\boldsymbol{w}, b, a) = (1/2)\|\boldsymbol{w}\|^2 - \sum_{j=1}^{l} a_j \left[y_j \left(\boldsymbol{w}^t \boldsymbol{x}_j + b \right) - 1 \right]$ (α_j are the Lagrange multipliers), formulates the minimization of $\|\mathbf{w}\|$ as a linear constrained optimization problem, whose solutions, in turn, are used in the discriminant function D to classify an unknown pixel $\boldsymbol{p}$:

$$D(\boldsymbol{p}) = \text{sgn} \left(\sum_j y_j a_j \langle \boldsymbol{p} \boldsymbol{x}_j \rangle + b \right) \quad (7.21)$$

$D(\boldsymbol{p})$ indicates that the decision function depends on the inner product between the pixels and a (small) set of support vectors $\boldsymbol{x}_j$. In the case of classes not linearly separable, a nonlinear transform function Φ maps vectors $\boldsymbol{p}$, $\boldsymbol{x}_j$ to a higher dimensional feature space, where it is more likely to be separable by a linear decision boundary. The nonlinear transform functions $\Phi(\boldsymbol{p})$ and $\Phi(\boldsymbol{x}_j)$ need not be specified explicitly as a Kernel function $k(\boldsymbol{p}, \boldsymbol{x}_j) = \Phi(\boldsymbol{p})\Phi(\boldsymbol{x}_j)$ is applied directly in the inner product of D. Using the kernel, the discriminant function can be rewritten as

$$D(\boldsymbol{p}) = \text{sgn} \left(\sum_j y_j a_j K(\boldsymbol{p}, \boldsymbol{x}_j) + b \right) \quad (7.22)$$

For example, assuming a Gaussian Kernel, the discriminant function becomes

$$D(\boldsymbol{p}) = \text{sgn} \left(\sum_j y_j a_j e^{-\frac{|p - x_j|^2}{2\sigma^2}} + b \right) \quad (7.23)$$

Choosing the correct kernel (polynomial, sigmoid, Gaussian radial basis, etc.) and associated parameters is a nontrivial task and research is ongoing for optimizing the kernel design. Several, specially tailored for the SI domain versions of SVM algorithm exist, for example, anomaly detection for hyperspectral imagery using support vectors (Banerjee et al. 2006) or kernels induced from spectral angle metrics (Mercier and Lennon 2006). As an example in SI biomedical applications, SVMs have been used for discriminating benign and malignant tissue cells from human colon (Rajpoot et al. 2004). Spectral cubes of $1024 \times 1024 \times 20$ dimensions, between 450 and 640 nm, acquired from archival H&E stained microarray tissue sections. Following a feature extraction procedure from four ICA components based on k-means segmentation, SVM achieved 87% classification accuracy (89% sensitivity and 85% specificity).

A series of other more sophisticated approaches have been developed and used mainly in remote sensing, which would also be useful in biomedical sciences. Indicatively, knowledge-based and contextual approaches are exploiting the spatial features of groups of pixels together with the spectral information for achieving better classification accuracy. Such spatial features include texture and/or other morphological features (Coburn and Robert 2004, Wei et al. 2008, Casasent and Chen 2004).

7.6.7 Spectral Unmixing Algorithms

An image area assigned by a single pixel usually contains a lot of different materials. These materials are mixed together and the pixel spectrum recorded by an SI sensor is a combination of the spectral characteristics of the various individual materials. To get more information from a single pixel, the proportions of these materials can be approximated using a spectral mixing model. Using such a model, the mixed pixel can be reconstructed from known spectra in the image or the mixed pixel can be divided into components. Spectral unmixing (Keshava 2003, Parra et al. 2000, Plaza et al. 2002) is the procedure by which the measured spectrum corresponding to a linearly or nonlinearly mixed pixel is decomposed into a collection of constituent spectra, or endmembers, and a set of corresponding fractions, known as abundances, which indicate the proportion of each endmember present in the pixel. Endmembers are the spectral signatures that are assumed to belong only to one class corresponding to a pure material. If the total surface area is considered to be divided proportionally according to the fractional abundances of the endmembers, then the reflected radiation will convey the characteristics of the associated media with the same proportions. In this sense, there exists a linear relationship between the fractional abundance of the substances comprising the area being imaged and the spectra in the emitted radiation. The linear mixing model (LMM) (Keshava 2003) is expressed as

$$p = \sum_{i=1}^{n} x_i a_i + w = Ax + w \qquad (7.24)$$

where

$A \in \mathbb{R}_+^{m \times n}$ is the endmember matrix
a_i is the spectral signature of the *i*th endmember
$x \in \mathbb{R}_+^n$ is the vector of the abundances
$p \in \mathbb{R}^m$ is the measured pixel spectrum
w is the noise vector
n is the number of endmembers

A is considered positive in order to have physical meaning; in addition, the abundance vector needs to satisfy $x \geq 0$ and $\sum_n x_i \leq 1$ or $\sum_n x_i = 1$ in order for the abundance estimation to have a physical meaning as well.

A nonlinear mixing model is needed in case the components of interest are mixed on spatial scales smaller than the path length of photons in the mixture. In that case, light typically interacts with more than one component as it is multiply scattered. It has been proved that if an LMM is used in a nonlinear problem, then this will result in a significant error in the calculation of the abundances (Keshava 2003).

It has been observed that only a limited number of endmembers (typically <10) can be practically identified with spectral unmixing methods. This number increases with the number of the acquisition bands and with the relative difference between the spectra of the components under analysis (Papadakis et al. 2003).

The typical processes followed for solving the unmixing problem are: (a) reduction of the data set's dimensionality using, for example, PCA, (b) determination of the endmembers using empirical or algorithmic methods, and (c) inversion, where the abundances are calculated for each pixel and usually the result has the form of thematic map(s). Endmember determination includes both interactive and automated methods. Interactive methods are based on the selection of image endmembers with the maximum abundance by either using algorithms or field knowledge. Often, spectral libraries or reference endmembers are used to achieve a better link to the ground truth. Automated methods estimate endmembers without requiring specific assumptions on the probabilistic densities of the data. Referring, for instance, to nonparametric methods, the *k*-means centroids serve as estimates of endmembers. As an extension of classification algorithms that assign the class label of the nearest centroid to pixels, abundance estimates that implicitly observe the nonnegativity and full additivity conditions are derived from the relative proximity of a pixel to each centroid (Keshava 2003).

Generally speaking, unmixing algorithms do not estimate the endmember and the abundances simultaneously. Ignoring noise and assuming that pure pixels are available, searching for an expression of *p* as a product of endmembers and abundances is

$$\hat{x} = \text{argmin}_x \|p - Ax\|^2 \qquad (7.25)$$

Least square techniques are commonly used for minimizing this square error, while as further assumptions are imposed, such as $x \geq 0$ and $\sum_{i=1}^{n} x_i = 1$, more advanced methods are employed. For example, the nonnegative least square method (NNLS), introduced first by Lawson and Hanson (Lawson and Hanson 1995), was used to find nonnegative solutions to a linear system. Essentially, in the NNLS algorithm, every time a negative solution is found for x_i, the *i*th column is removed from the set of equations and x_i is set to zero, reestimating the solution. The process is repeated until all solutions are positive. While NNLS is an iterative algorithm that only satisfies the positive constraint, several least squares variants have been adopted for hyperspectral data. In biomedical SI microscopy, spectral unmixing is mainly based on least square techniques. An example is illustrated in Figure 7.18 (Mansfield et al. 2008), where Figure 7.18a depicts the RGB representation of a spectral image from ductal breast carcinoma section stained forER with DAB, PR, with FR and nuclei with hematoxylin. The unmixed images from this sample can be seen in the grayscale images B-D of Figure 7.18. In these images, pixels with larger abundances appear darker. However, in the case of multiple stained pixels, pseudocolored images 8E and F, may be more useful as pseoudocolors facilitate the visualization of different abundancies. In Figure 7.18e, a composite image maps the hematoxylin in blue, the ER in red, and the PR in green. Figure 7.18f shows only the ER images (in red) and PR images (in green), without hematoxylin. Pixels stained with both ER and PR appear yellow (red combined with green).

Although endmembers are usually derived with aid of reference samples, co-localized clinical stains can exhibit considerable variability of their spectra. Moreover, most algorithms for automatic endmember extraction operate under the assumption that pure pixels exist in the scene, but this assumption is not always valid. Despite these facts, commercially available software packages for microscopy applications implement often unsupervised decomposition algorithms for the purpose of improving automation, although this may lead to erroneous results due to the lack of reference endmember set.

7.6.8 Classification Performance Measures

The performance of the various classifiers is traditionally assessed quantitatively with the aid of the so-called confusion matrix, which contains information about actual and predicted classifications. As it is well understood, classifier might perform well for a class that accounts for a large proportion of the data creating a (positive) bias in overall accuracy, although low accuracy might be observed in classes consisted by a small set of data.

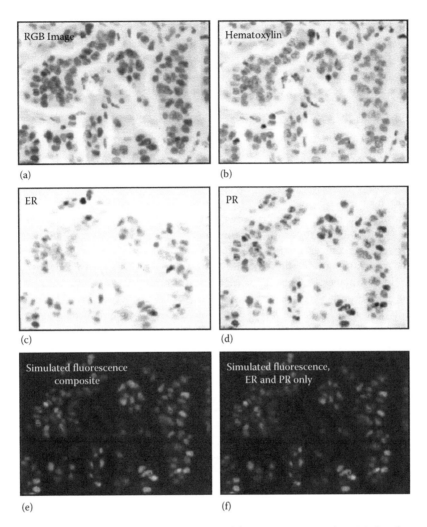

FIGURE 7.18 (See color insert.) Invasive ductal carcinoma section stained for estrogen receptor (ER, DAB) and progesterone receptor (PR, Fast Red), and counterstained with hematoxylin. (a) RGB representation of data set; (b–d) unmixed images corresponding to hematoxylin, ER, and PR, respectively; (e) simulated fluorescence composite with hematoxylin in blue, ER in red, and PR in green; (f) simulated fluorescence composite showing only ER (red) and PR (green). (Reprinted from Mansfield, J.R. et al., *Curr. Protoc. Mol. Biol.*, 84, 1, 2008.)

For the purpose of avoiding such a bias, individual class accuracies are usually considered. Individual class accuracy is obtained by dividing the number of correctly classified pixels in a class, by the total number of pixels classified in that class. This parameter is called producer's accuracy or sensitivity. A misclassification error is not only an omission from the correct class but also a commission into another class. This is expressed by the user's accuracy or specificity, which is obtained by dividing the total number of correctly classified pixels in a class, by the total number of pixels of that class committed to other classes. A long list of alternatives to the sensitivity and specificity percentages measures are used for the same purpose, including receiver operating characteristic curves, correlation, (relative) entropy, mutual information (Baldi et al. 2000), kappa statistics (Sokal 1989), and *t*-tests (Dietterich 1998, Zimmerman 1997).

The aforementioned measures are useful for assessing performance, but estimating the abilities of a classification model to generalize for unknown samples not used in the training set is equally important. Particularly, while a classifier is possible to maximize the classification accuracy for a specific training set by selecting the best available algorithm, true performance for unknown samples that have not been used for training can be significantly inferior, due to *overfitting*. A classification model is considered trained if it is able to predict the correct output for a pixel collection of known class. However, if the training set is particularly small, the model may adjust to the very specific pattern of the training data and thus not being able to generalize for types of samples not presented during training. Other forms of overfitting occur when selecting the parameters of a complex learning algorithm (kernel parameters for SVM classifiers) solely based on the training set performance.

Cross validation (Kohavi 1995) is a common methodology used to estimate the generalization error by perturbing the training, validation and test data sets. The key idea is to utilize various ways of distributing the samples (resampling of the same data set) at the expense of higher computation costs, due to repeated classification. Typical distribution strategies involve (a) *random subsampling*, where the total available pixels are split randomly in a fixed number of samples for training set and validation sets, (b) *k-fold Cross Validation (CV)*, where the classification procedure is performed k times, that is, the total available pixels are split into k folds, containing equal number of pixels. For each experiment, $k-1$ folds are used for training and the remaining fold for validating. The procedure is repeated until all possible combinations of different folds for training/validation sets have been applied, while the overall performance can be calculated by averaging the classification error rate of each experiment, (c) *Leave one out Cross Validation (LOOCV)*, where the classification procedure is performed as many times as the number of available samples. For example, in a data set of N samples, for each experiment, $N-1$ pixels are used for training and the remaining pixel for validation. The main advantage of k-Fold CV is that all the pixels in a data set are eventually used for both training and validating (each pixel is used for validation exactly once), whereas LOOCV cannot be replicated. A 10-fold CV is routinely used although for large data sets, even threefold CV can be quite accurate.

7.6.9 Summary and Basic Taxonomy of Spectral Cube Data Processing Processes

The spectral cube data processing processes are summarized in the block diagram of Figure 7.19. This schematic is indicative since processing is specialized depending upon the intended application. The analysis of the spectral cube data set begins with the preprocessing step. The basic task of this step is data dimensionality reduction and/or the creation of a list of classes that is suitably exhaustive and that includes the classes of user interest. If necessary, it may also include data correction tasks such as image registration. Image registration is required to compensate for target motions or deformations during the acquisition of the spectral cube or for relative displacements of spectral images eventually caused by the optical filters.

Next, the analysis may follow the route of spectral unmixing in case that the intention is to recover subpixel compositional information or the route of pixel level information for performing classification of whole pixels. Classification may be supervised or unsupervised and the corresponding algorithms may also be applied to the unmixed spectra. Supervised classification and spectral unmixing algorithms require a suitable training data set. Using the various spectral similarity measurement algorithms that have been outlined earlier in this paragraph, the similarity between a reference signature and each pixel signature in the image is calculated and the pixel or the subpixel component is placed into the specified class and represented as a specific color in a thematic map. All the algorithms are tested against a validation data set for preliminary evaluation and optimization before applied to the test sample and if needed, the procedure is iterated until achieving satisfactory performance. The output of all these processes is one or more than one thematic maps. Spectral unmixing processes produce typically one map per unmixed component, which may also be combined together to form a single map. Finally, the test data set is examined and the performance of the algorithms is evaluated using a reference method as golden standard. Reference

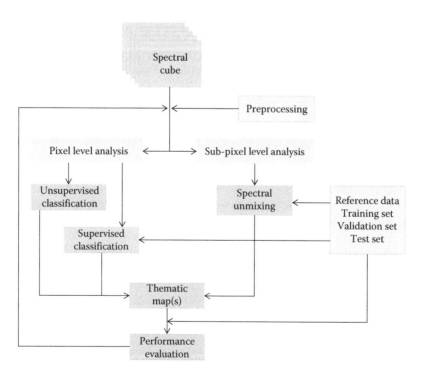

FIGURE 7.19 Basic taxonomy of spectral cube data processing processes.

methods may be the results from the analysis of the subject material using, for example, destructive chemical analysis, biopsies expert's knowledge, etc. This establishes the performance characteristics of the system, which determines its clinical value.

7.7 Conclusion and Future Perspectives

The demanding fields of optical diagnosis, remote sensing, and nondestructive analysis, together with the recent technological improvements in computer, optical filter, and imaging sensor technologies have historically become the catalysts for merging together imaging and spectroscopy. During the last three decades, SI technology, implemented on satellite and aeroplane-based platforms, has found an enormous range of applications in many subjects, such as mineralogical mapping of the earth's surface, agricultural and environment control, and defense.

In biomedical applications, SI has shown great potential but it still remains largely unexplored. The so-far results from laboratory experiments and clinical tests suggest strongly that SI can remarkably advance biomedical research and improve clinical practices in detecting and characterizing tissue pathologies. Translational research is currently underway, aiming at transferring laboratory setups and experimental results to the clinical practice. Today, the dramatically increased computational power will facilitate this process, since it enables the rapid acquisition and processing of the massive amount of data generated by SI systems. SI can provide quantitative maps of intrinsic tissue components based on their spectral signatures, which has great clinical value. We expect that optical signal amplification methods using the newly developed biomarkers will substantially enhance diagnostic information and will accelerate the clinical adoption of SI. On the other hand, SI has the potential to become an indispensable tool in optical molecular imaging for the development and detection of novel biomarkers. Combined with advanced spectral unmixing and deconvolution algorithms, SI comprises the technology of choice for separating multiple biomarkers with overlapping spectra, used to stain different targets within the same tissue sample. This will contribute substantially to the conversion of current empirical approaches into a modern medical practice, centered on the principles of molecular biology.

The major scientific and technological challenge is to develop new solutions that will enable not only the acquisition of the spectral cube in video rates but also the calculation and display of the diagnostic thematic maps in real time. Efforts are necessary in this direction, since there is a clear need for novel fast, high throughput, and affordable tunable filter technologies. We believe that further technological advances will lead to the development of new architectures for detector systems and new optical methods for real-time spectral discrimination. The continuing advances in photonic crystal and metamaterial will eventually enable the integration of multiple filters directly on focal plane arrays.

References

Alabboud, I., Muyo, G., Gorman, A. et al. 2007. New spectral imaging techniques for blood oximetry in the retina. *Proc. SPIE* 6631: 66310L.

Alvarez, R. D. and Wright, T. C. Jr. 2007. Increased detection of high-grade cervical intraepithelial neoplasia utilizing an optical detection system as an adjunct to colposcopy. *Gynecol. Oncol.* 106: 23–28.

Balas, C. 2001. A novel optical imaging method for the early detection, quantitative grating, and mapping of cancerous and precancerous lesions of cervix. *IEEE Trans. Biomed. Eng.* 48(1): 96–104.

Balas, C. 2009. Review of biomedical optical imaging—a powerful, non-invasive, non-ionizing technology for improving *in vivo* diagnosis. *Meas. Sci. Technol.* 20: 104020.

Baldi, P., Brunak, S., Chauvin, Y., Andersen, C.A.F., and Nielsen H. 2000. Assessing the accuracy of prediction algorithms for classification: An overview. *Bioinformatics* 16(5): 412–424.

Banerjee, A., Burlina, P., and Diehl, C. 2006. A support vector method for anomaly detection in hyperspectral imagery. *IEEE Trans. Geosci. Remote Sensing* 44(8): 2282–2291.

Battula, A. and Chena, S. C. 2006. Extraordinary transmission in a narrow energy band for metallic gratings with converging-diverging channels. *Appl. Phys. Lett.* 89: 131113.

Bayliss, J., Gualtieri, J., and Cromp, R. 1997. Analyzing hyperspectral data with independent component analysis. *Proc. SPIE* 3240: 133–143.

Bazi, Y. and Melgani, F. 2006. Toward an optimal SVM classification system for hyperspectral remote sensing images. *IEEE Trans. Geosci. Remote Sensing* 44(11): 3374–3385.

Bei, L., Dennis, G. I., Miller, H. M., Thomas, Spaine, W., and Carnahan, J. W. 2004. Acousto-optic tunable filters: Fundamentals and applications as applied to chemical analysis techniques. *Progr. Quantum Electron.* 28: 67–87.

Bellman, R. 1961. *Adaptive Control Processes: A Guided Tour.* Princeton, NJ: Princeton University Press.

Bishop, C. 2005. *Neural Networks for Pattern Recognition.* Oxford, U.K.: Oxford University Press.

Boer, G., Ruffieux, P., Scharf, T., Seitz, P., and Dandliker, R. 2004. Compact liquid-crystal-polymer Fourier-transform spectrometer. *Appl. Opt.* 43(11): 2201–2208.

Bowley, C. C., Kossyrev, P. A., and Crawford, G. P. 2001. Variable-wavelength switchable Bragg gratings formed in polymer-dispersed liquid crystal. *Appl. Phys. Lett.* 79(1): 9–11.

Brady, D. J. 2009. *Optical Imaging and Spectroscopy.* Danvers, MA: John Wiley & Sons, Inc.

Buchanan, B. 1986. Expert systems: Working systems and the research literature. *Expert Syst.* 3(1): 32–50.

Camps-Valls, G. and Bruzzone, L. 2005. Kernel-based methods for hyperspectral image classification. *IEEE Trans. Geosci. Remote Sensing* 43(6): 1351–1362.

Candes, E. J. 2006. Compressive sampling. *Proceedings of the International Congress of Mathematicians*, Madrid, Spain, 2006. European Mathematical Society, Madrid, Spain.

Candes, E. J. and Tao, T. 2005. Decoding by linear programming. *IEEE Trans. Inform. Theory* 15(12): 4203–4215.

Candes, E. J. and Wakin, M. 2008. An introduction to compressive sampling. *IEEE Signal Process. Mag.* 25(2): 21–30.

Carvalho, O. and Meneses, P. 2000. Spectral Correlation Mapper (SCM): An improving spectral angle mapper. *Ninth JPL Airborne Earth Science Workshop*, Pasadena, CA, 2000, JPL Publication 00-18: 65–74.

Casasent, D. and Chen, X. W. 2004. Feature reduction and morphological processing for hyperspectral image data. *Appl. Opt.* 43: 227–236.

Chang, C. 2000. An information-theoretic approach to spectral variability, similarity, and discrimination for hyperspectral image analysis. *IEEE Trans. Inform. Theory* 46(5): 1927–1932.

Chen, J., Wang, R., and Wang, C. 2008. A multiresolution spectral angle-based hyperspectral classification method. *Int. J. Remote Sensing* 29(11): 3159–3169.

Chi, W. and George, N. 2009. Phase-coded aperture for optical imaging. *Opt. Commun.* 282: 2110–2117.

Chiou, S. 1985. NASA image-based geological expert system development project for hyperspectral image analysis. *Appl. Opt.* 24(14): 2085–2091.

Civco, D. 1993. Artificial neural networks for land-cover classification and mapping. *Int. J. Geographical Inform. Sci.* 7(2): 173–186.

Coburn, C. A. and Roberts, A. C. B. 2004. A multiscale texture analysis procedure for improved forest stand classification. *Int. J. Remote Sensing* 25: 4287–4308.

De Backer, S., Kempeneers, P., Debruyn, W., and Scheunders, P. 2005. A band selection technique for spectral classification. *IEEE Geosci. Remote Sensing Lett.* 2(3): 319–323.

Descour, M. and Dereniak, E. 1995. Computed-tomography imaging spectrometer: Experimental calibration and reconstruction results. *Appl. Opt.* 34(22): 4817–4826.

Descour, M. R., Tkaczyk, T. S., Ford, B. K., Lynch, R. M., Locke, A., and Dereniak, E. 2003. *The Computed Tomography Imaging Spectrometer*. Hoboken, NJ: Wiley, 460–461.

Dicke, R. H. 1968. Scatter-hole cameras for x-rays and gamma rays. *Astrophys. J.* 153, L101–L106.

Diest, K., Dionne, J. A., Spain, M., Atwater, H. A., and Watson, T. J. 2009. Tunable color filters based on metal-insulator-metal resonators. *Nano Lett.* 9(7): 2579–2583.

Dietterich, T. 1998. Approximate statistical tests for comparing supervised classification learning algorithms. *Neural Comput.* 10(7): 1895–1923.

Du, Q. and Chang, C. 2001. A linear constrained distance-based discriminant analysis for hyperspectral image classification. *Pattern Recogn.* 34(2): 361–373.

Ebbesen, T. W., Lezec, H. J., Ghaemi, H. F., Thio, T., and Wolff, P. A. 1998. Extraordinary optical transmission through sub-wavelength hole arrays. *Nature* 391: 667.

Fernandez, C. A., Wagadarikar, A., Brady, D. J., McCain, S. C., and Oliver, T. 2009. Fluorescence microscopy with a coded aperture snapshot spectral imager. *Proc. SPIE* 7184: 71840Z-1–11.

Ford, B. K., Volin, C. E., Murphy, S. M., Lynch, R. M., and Descour M. R. 2001. Computed tomography based spectral imaging fro fluorescence microscopy. *Biophys. J.* 80: 986.

Fox, A. E, Rai, K., and Fontecchio, A. K. 2007. Holographically formed polymer dispersed liquid crystal films for transmission mode spectrometer applications. *Appl. Opt.* 46(25): 6277–6282.

Frances, J. V., Maravilla, J. C., Mari, J. M. et al. 2006. Configurable-bandwidth imaging spectrometer based on an acousto-optic tunable filter. *Rev. Sci. Instrum.* 77: 073108-1–073108-10.

Friedl, M. and Brodley, C. 1997. Decision tree classification of land cover from remotely sensed data. *Remote Sensing Environ.* 61(3): 399–409.

Gao, X., Cui, Y., Levenson, R. M. et al. 2004. In vivo cancer targeting and imaging with semiconductor quantum dots. *Nat. Biotechnol.* 22(8): 969–976.

Gehm, M. E., John, R., Brady, D. J., Willett, R. M., and Schulz, T. J. 2007. Single-shot compressive spectral imaging with a dual-disperser architecture. *Opt. Express* 15(21): 14013–14027.

George, C. T., Basiji, D. A., Hall, B. E. et al. 2004. Distinguishing modes of cell death using the ImageStream multispectral imaging flow cytometer. *Cytometry* 59A: 237–245.

Golay, M. J. E. 1949. Multislit spectroscopy. *J. Opt. Soc. Am.* 39: 437–444.

Gono, K. 2008. Multifunctional endoscopic imaging system for support of early cancer diagnosis. *IEEE J. Sel. Top. Quantum Electron.* 14: 62–69.

Gu, Y., Liu, Y., and Zhang, Y. 2006. A selective kernel PCA algorithm for anomaly detection in hyperspectral imagery. *Proc. IEEE Int. Conf. Acoust. Speech Signal Process.* 2: 14–19.

Ham, J., Chen, Y., Crawford, M. M., and Ghosh, J. 2005. Investigation of the random forest framework for classification of hyperspectral data. *IEEE Trans. Geosci. Remote Sensing* 43(3): 492–501.

Hardarson, S. H., Harris, A., Karlsson A. R. et al. 2006. Automatic retinal oximetry. *Investig. Ophthalmol. Vis. Sci.* 47: 5011–5016.

Harvey, A. R. and Fletcher-Holmes, D. W. 2003. High-throughput snapshot spectral imaging in two dimensions. *Proc. SPIE* 4959: 46–54.

Harvey, A. R. and Fletcher-Holmes, D. W. 2004. Birefringent Fourier-transform imaging spectrometer. *Opt. Express* 12(22): 5368–5374.

Harvey, A. R., Fletcher-Holmes, D. W., Gorman, A., Altenbach, K., Arlt, J., and Read, N. D. 2005. Spectral imaging in a snapshot. *Proc. SPIE* 5694: 110–119.

Hirohara, Y., Okawa, Y., Mihashi, T. et al. 2007. Validity of retinal oxygen saturation analysis: Hyperspectral imaging in visible wavelength with fundus camera and liquid crystal wavelength tunable filter. *Opt. Rev.* 14(3): 151–158.

Ifarraguerri, A. and Chang, C. 2000. Unsupervised hyperspectral image analysis with projection pursuit. *IEEE Trans. Geosci. Remote Sensing* 38(6): 2529–2538.

Ito, M., Murayama, K., Deguchi, T. et al. 2008. Oxygen saturation levels in the juxta-papillary retina in eyes with glaucoma. *Experimental Eye Res.* 86(3): 512–518.

Johnson, W. R., Wilson, D. W., Fink, W., Humayun, M., and Bearman, G. 2007. Snapshot hyperspectral imaging in ophthalmology. *J. Biomed. Opt.* 12: 014036-1–014036-7.

Katzilakis, N., Stiakaki, E., Papadakis, A. et al. 2004. Spectral characteristics of acute lymphoblastic leukemia in childhood. *Leuk. Res.* 28: 1159–1164.

Keshava, N. 2003. A survey of spectral unmixing algorithms. *Lincoln Lab. J.* 14(1): 55–78.

Keshava, N. 2004. Distance metrics and band selection in hyperspectral processing with applications to material identification and spectral libraries. *IEEE Trans. Geosci. Remote Sensing* 42(7): 1552–1565.

Kohavi, R. 1995. A study of cross-validation and bootstrap for accuracy estimation and model selection. *Proceedings of the International Joint Conference On Artificial Intelligence (IJCAI-95)*, Stanford University, Stanford, CA, 1995, pp. 1137–1143.

Kruse, F. A., Lefkoff, A. B., Boardman, J. W., Heidebrecht, J. W., Shapiro, K. B., Barloon, A. T., and Goetz, P. J. 1993. The spectral image processing system (SIPS)—Interactive visualization and analysis of imaging spectrometer data. *Remote Sensing Environ.* 44: 145–163.

Landgrebe, D. A. 2003. *Signal Theory Methods in Multispectral Remote Sensing.* Hoboken, NJ: John Wiley & Sons.

Lansford, R., Bearman, G., and Fraser, S. E. 2001. Resolution of multiple green fluorescent protein color variants and dyes using two-photon microscopy and imaging spectroscopy. *J. Biomed. Opt.* 6(3): 311–318.

Laux, E., Genet, C., Skauli, T., and Ebbesen, T. W. 2008. Plasmonic photon sorters for spectral and polarimetric imaging. *Nat. Photonics* 2: 161–164.

Lawson, C. and Hanson, R. 1995. *Solving Least Squares Problems.* Philadelphia, PA: Society for Industrial Mathematics.

MacQueen, J. B. 1967. Some methods for classification and analysis of multivariate observations. *Proceedings of 5th Berkeley Symposium on Mathematical Statistics and Probability*, University of California Press, Berkeley, CA, 1967, Vol. 1, pp. 281–297.

Malik, Z., Cabibi, D., Buckwald, R. A., Talmi, A., Garini, Y., and Lipson, S. G. 1996. Fourier transform multiplex spectroscopy for the quantitative cytology. *J. Microsc.* 182: 133–140.

Mallet, Y., Coomans, D., Kautsky, J., and Vel De, O. 1997. Classification using adaptive wavelets for feature extraction. *IEEE Trans. Pattern Anal. Mach. Intell.* 19(10): 1058–1066.

Mansfield, J. R., Hoyt, C., and Levenson, R. M. 2008. Visualization of microscopy-based spectral imaging data from multi-label tissue sections. *Curr. Protoc. Mol. Biol.* 84: 1–14.

Mao, K. 2004. Feature subset selection for support vector machines through discriminative function pruning analysis. *IEEE Trans. Syst. Man Cybern.* 34(1): 60–67.

Martin, E., Wabuyele, M. B., Chen, K. et al. 2006. Development of an advanced hyperspectral imaging (HSI) system with applications for cancer detection. *Ann. Biomed. Eng.* 34(6): 1061–1068.

Martin, M. E., Wabuyele, M., Panjehpour, M. et al. 2006. An AOTF-based dual-modality hyperspectral imaging system (DMHSI) capable of simultaneous fluorescence and reflectance imaging. *Med. Eng. Phys.* 28: 149–155.

Martín-Moreno, L., Garcial-Vidal, F. J., Lezec, H. J., Pellerin, K. M., Thio, T., and Pendry, J. B. 2001. Theory of extraordinary optical transmission through subwavelength hole arrays. *Phys. Rev. Lett.* 86: 1114–1117.

Mathews, S. A. 2008. Design and fabrication of a low-cost, multispectral imaging system. *Appl. Opt.* 47(28): 71–76.

McMurdy, J. W., Crawford, G. P., and Jay, G. D. 2006. Monolithic microspectrometer using tunable ferroelectric liquid crystals. *Appl. Phys. Lett.* 89: 081105.

Melgani, F. and Bruzzone, L. 2004. Classification of hyperspectral remote sensing images with support vector machines. *IEEE Trans. Geosci. Remote Sensing* 42(8): 1778–1790.

Mende, S. B., Claflin, E. S., Rairden, R. L., and Swenson, G. R. 1993. Hadamard spectroscopy with a two-dimensional detecting array. *Appl. Opt.* 32(34): 7095–7105.

Mercier, G., Derrode, S., and Lennon, M. 2003. Hyperspectral image segmentation with Markov chain model. *Proceedings of the IEEE IGARSS'03*, Toulouse, France, 2003, Vol. 6.

Mercier, G. and Lennon, M. 2003. Support vector machines for hyperspectral image classification with spectral-based kernels. *IEEE Geosci. Remote Sensing Symp.* 1: 288–290.

Okamoto, T. and Yamaguchi, I. 1991. Simultaneous acquisition of spectral image information. *Opt. Lett.* 16: 1277.

Pal, M. and Mather, P. 2004. Assessment of the effectiveness of support vector machines for hyperspectral data. *Future Generation Comput. Syst.* 20(7): 1215–1225.

Palcic, B., Lam, S., Hung, J., and MacAulay, C. 1991. Detection and localization of early lung cancer by imaging techniques. *Chest* 99: 742–743.

Papadakis, A., Stathopoulos, E., Delides, G., Berberides, K., Nikiforidis, G., and Balas, C. 2003. A novel spectral microscope system: Application in quantitative pathology. *IEEE Trans. Biomed. Eng.* 50 (2): 207–217.

Parra, L., Spence, C., Sajda, P., Ziehe, A., and Muller, K. R. 2000. Unmixing hyperspectral data. *Adv. Neural Infor. Process. Syst.* 12: 942–948.

Persky, M. J. 1995. A review of spaceborn infrared Fourier transform spectrometers for remote sensing. *Rev. Sci. Instrum.* 66(10): 4763–4797.

Plaza, A., Martinez, P., Perez, R., and Plaza J. 2002. Spatial/spectral endmember extraction by multidimensional morphological operations. *IEEE Trans. Geosci. Remote Sensing* 40(9): 2025–2041.

Pu, R. and Gong, P. 2004. Wavelet transform applied to EO-1 hyperspectral data for forest LAI and crown closure mapping. *Remote Sensing Environ.* 91(2): 212–224.

Qi, J., Li, L., De Sarkar, M., and Crawford, G. P. 2004. Nonlocal photopolymerization effect in the formation of reflective holographic polymer-dispersed liquid crystals. *J. Appl. Phys.* 96(5): 2443–2450.

Rajpoot, K., Rajpoot, N., and Turner, M. 2004. Hyperspectral colon tissue cell classification. *Proceedings of the SPIE Medical Imaging (MI'04)*, San Diego, CA, 2004.

Rajwa, B., Ahmed, W., Venkatapathi, M. et al. 2005. AOTF-based system for image cytometry. *Proceedings of the SPIE—Spectral Imaging: Instrumentation, Applications, and Analysis III*, San Jose, CA, 2005, Vol. 5694.

Ramella-Roman, J. C. and Mathews, S. A. 2007. Spectroscopic measurement of oxygen saturation in the retina. *IEEE J. Sel. Top. Quantum Electron.* 13: 1697–1703.

Rodarmel, C. and Shan, J. 2002. Principal component analysis for hyperspectral image classification. *Surveying Land Infor. Syst.* 62(2): 115–122.

Schott, J. 2007. *Remote Sensing: The Image Chain Approach.* New York: Oxford University Press.

Schrock, E., du Manoir, C., Veldman, T. et al. 1996. Multicolor spectral karyotyping of human chromosomes. *Science* 26: 494.

Shi, Z., Boyd, R. W., Camacho, R. M., Vudyasetu, P. K., and Howell, J. C. 2007. Slow-light Fourier transform interferometer. *Phys. Rev. Lett.* 99: 240801–240804.

Shogenji, R., Yamada, K., Miyatake, S., and Tanida, J. 2004. Multispectral imaging using compact compound optics. *Opt. Express* 12(8): 1643–1655.

Skala, M. C., Fontanella, A., Hendargo, H., et al. 2009. Combined hyperspectral domain optical coherence tomography microscope for noninvasive hemodynamic imaging. *Opt. Express* 34(3): 289–291.

Slawson, R. W., Ninkov, Z., and Horch, E. P. 1999. Hyperspectral imaging: Wide-area spectrophotometry using a liquid crystal tunable filter. *Publ. Astron. Soc. Pac.* 111: 621–626.

Sokal, R. 1989. Nonparametric statistics for the behavioral sciences. *Q. Rev. Biol.* 64(2): 242.

Song, I. C. and Gweon, D.G. 2008. A spectral detector using the dispersion of an acousto-optic tunable filter for confocal spectral imaging microscopy. *Meas. Sci. Technol.* 19: 085504 (7pp).

Sorg, B. S., Moeller, B. J., Donovan, O., Cao, Y., and Dewhirst, M. W. 2005. Hyperspectral imaging of hemoglobin saturation in tumor microvasculature and tumor hypoxia development. *J. Biomed. Opt.* 10: 044004.

Stratis, D. N., Eland, K. L., Carter, J. C., Tomlinson, S. J., and Angel, M. 2001. Comparison of acousto-optic and liquid crystal tunable filters for laser-induced breakdown spectroscopy. *Appl. Spectrosc.* 55(8): 999–1004.

Suhre, D. R. and Gupta, N. 2005. Acousto-optic tunable filter side-lobe analysis and reduction with telecentric confocal optics. *Appl. Opt.* 44(27): 5797–5801.

Suhre, D. R. and Villa, E. 1998. Imaging spectroradiometer for the 8–12-μm region with a 3-cm^{-1} passband acousto-optic tunable filter. *Appl. Opt.* 37(12): 2340–2345.

Swain, P. 1978. *Fundamentals of Pattern Recognition in Remote Sensing. Remote Sensing: The Quantitative Approach.* New York: McGraw-Hill International Book Co., pp. 136–187.

Tan, F., Fu, X., Zhang, Y., and Bourgeois, A. G. 2008. A genetic algorithm-based method for feature subset selection. *Soft Comput. A Fusion Foundations, Methodol. Appl.* 12(2): 111–120.

Tholouli, E., Hoyland, J. A., and Di Vizio, D. 2006. Imaging of multiple mRNA targets using quantum dot based in situ hybridization and spectral deconvolution in clinical biopsies. *Biochem. Biophys. Res. Commun.* 348: 628–636.

Tran, C. D. 2003. Infrared multispectral imaging: Principles and instrumentation. *Appl. Spectrosc. Rev.* 38(2): 133–153.

Treado, P. J., Levin, I. W., and Lewis, E. N. 1992. High-fidelity Raman imaging spectrometry—a rapid method using an acoustooptic tunable filter. *Appl. Spectrosc.* 46: 1211–1216.

Tsaig, Y. and Donoho, D. L. 2006. Extensions of compressed sensing. *Signal Process.* 86: 549–571.

van Hulst, N. F. 2008. Plasmonics: Sorting colours. *Nat. Photonics* 2: 139–140.

Vapnik, V. 2000. *The Nature of Statistical Learning Theory.* New York: Springer Verlag.

Villmann T., Mernyi, E., and Hammer, B. 2003. Neural maps in remote sensing image analysis. *Neural Netw.* 16(3–4): 389–403.

Vo-Dinh, T., Stokes, D. L., Wabuyele, M. B. et al. 2004. A hyperspectral imaging system for in vivo optical diagnostics. *IEEE Eng. Med. Biol. Mag.* 23(5): 40–49.

Volin, C. E., Ford, B. K., Descour, M. R., Wilson, D. W., Maker P. M., and Bearman G. H. 1998. High speed spectral imager for imaging transient fluorescence phenomena. *Appl. Opt.* 37: 8112.

Wachman, E. S., Niu, W., and Farkas, D. L. 1996. Imaging acousto-optic tunable filter with 0.35-micrometer spatial resolution. *Appl. Opt.* 35: 5220–5226.

Wachman, E. S., Niu, W., and Farkas, D. L. 1997. AOTF microscope for imaging with increased speed and spectral versatility. *Biophys. J.* 73: 1215–1222.

Wagadarikar, A. A., John, R., Willett, R., Brady, D. J. 2007. Single disperser design for compressive, single-snapshot spectral imaging. *Proc. SPIE* 6714: 67140A-1–67140A-9.

Wagadarikar, A. A., Pitsianis, N. P., Sun, X., and Brady, D. J. 2009. Video rate spectral imaging using a coded aperture snapshot spectral imager. *Opt. Express* 17(8): 6368–6388.

Wei, S., Li, J., Chen Y. et al. 2008. Textural and local spatial statistics for the object-oriented classification of urban areas using high resolution imagery. *Int. J. Remote Sensing* 29: 3105–3117.

Woltman, S. J., Jay, G. D., and Crawford, G. P. 2007. Liquid-crystal materials find a new order in biomedical applications. *Nat. Mater.* 6: 929–938.

Yao, H. and Tian, L. 2003. A genetic-algorithm-based selective principal component analysis (GA-SPCA) method for high-dimensional data feature extraction. *IEEE Trans. Geosci. Remote Sensing* 41(6): 1469–1478.

Yoneya, S., Saito, T., Nishiyama, Y. et al. 2002. Retinal oxygen saturation levels in patients with central retinal vein occlusion. *Ophthalmology* 109(8): 1521–1526.

Yu, S., De Backer, S., and Scheunders, P. 2002. Genetic feature selection combined with composite fuzzy nearest neighbor classifiers for hyperspectral satellite imagery. *Pattern Recogn. Lett.* 23(1–3): 183–190.

Zimmerman, D. 1997. Teacher's corner: A note on interpretation of the paired-samples *t* test. *J. Educ. Behav. Stat.* 22(3): 349.

8

Light Scattering Spectroscopy

8.1 Introduction 165
8.2 Basic Principles of Light Absorption and Scattering 166
   Absorption Mechanisms • Scattering Mechanisms • Polarization
8.3 Light Scattering Spectroscopy 167
   Light Scattering from Cells and Subcellular Structures • Analytical Approximations for
   Light Scattering from Cells and Subcellular Organelles • Principles of Light Scattering
   Spectroscopy • Polarized Light Scattering Spectroscopy • Biomedical Applications of
   Light Scattering Spectroscopy • Imaging of Early Cancer and Precancerous Lesions with
   Endoscopic Polarized Light Scattering Spectroscopy
8.4 Confocal Light Absorption and Scattering Spectroscopic Microscopy.........174
   Principles of CLASS Microscopy • Applications of CLASS Microscopy in Cell
   Biology • Spectroscopy of Single Nanoparticles
8.5 Conclusion 178
Acknowledgment..179
References..179

Le Qiu
Harvard University

Irving Itzkan
Harvard University

Lev T. Perelman
Harvard University

8.1 Introduction

Optical spectroscopic techniques have shown promising results in the detection of diseases on cellular scale. They do not require tissue removal, can be performed in vivo, rapidly interrogate large tissue surfaces, and permit the diagnosis to be made in real time. While fluorescence and Raman spectroscopy are effective in revealing the molecular properties of tissue, the technique called light scattering spectroscopy (LSS) is capable of characterizing the structural properties of tissue on a cellular and subcellular scale. LSS connects the spectroscopic properties of light elastically scattered by small particles to their size, refractive index, and shape. Light scattering in biological tissues originates from tissue inhomogeneities such as cellular organelles, extracellular matrix, and blood vessels. This often translates into unique angular, polarization, and spectroscopic features of scattered light emerging from tissue, and therefore information about the tissues macroscopic and microscopic structure can be obtained from the characteristics of the scattered light.

Bigio and Mourant (1997) and Mourant et al. (1995) demonstrated that the spectroscopic features of elastically scattered light could be used to detect transitional carcinoma of the urinary bladder and adenoma and adenocarcinoma of the colon and rectum with good accuracy. Over the last decade, LSS has been used to characterize changes in tissue on the cellular scale and to detect dysplastic changes in epithelial tissues. In 1998, Perelman et al. (1998) observed the characteristic LSS spectral behavior in light backscattered from human intestinal cells. Later, these studies were extended to additional organs by Backman et al. (2000) and Wallace et al. (2000).

Several techniques have been used to separate the multiple scattered light in LSS measurements from the single-scattered light component and improve the extraction of the single-scattered light component, which contains the information about the epithelial cells. The diffuse reflectance spectroscopy technique described in Zonios et al. (1999) and Georgakoudi et al. (2001), which models the multiple scattered diffuse background light, can be quite useful since the multiple scattered diffuse background light provides valuable information about the biochemical and morphological organization of the submucosa and the degree of angiogenesis. At the same time, experimentally based techniques for diffuse background removal have the advantage of being less sensitive to tissue variability. In order to remove multiple scattered light, Backman et al. (1999) and Sokolov et al. (1999) used LSS in combination with polarized light, sometimes called polarized LSS. In the coherence-gating method developed by Wax et al. (2005), multiple scattered light is removed by employing angle-resolved low-coherence interferometry, which combines the LSS methodology of obtaining nuclear sizes from Mie theory with the ideas of optical coherence tomography, capable of depth sensing and removing multiple scattered light. Another technique targeting the removal of multiple scattered light is azimuthal LSS described in Yu et al. (2006), where the LSS signal is calculated by subtracting reflectance signals measured at two azimuthal angles, 0 and 90 degrees. Recently, by combining LSS with confocal microscopy, Itzkan et al. (2007)

and Fang et al. (2007) developed confocal light absorption and scattering spectroscopic (CLASS) microscopy, a new technique that can also remove multiple scattered light and achieve depth sensing. In addition, the multispectral nature of LSS enables CLASS to measure internal cell structures much smaller than the diffraction limit without damaging the cell or requiring contrast agents common to optical microscopy, which may affect cell function.

8.2 Basic Principles of Light Absorption and Scattering

8.2.1 Absorption Mechanisms

The mechanism of absorption depends on the specific molecule and the energy of the incident photon. The molecule is excited to a higher energy state as is governed by the selection rules. The transition probability can be calculated using various quantum mechanical approximations, such as perturbation theory. It can be converted into an absorption cross section, which is useful for macroscopic calculations. In a biological sample, it is often a particular group within a molecule called chromophore, which is responsible for the absorption. Usually, electrons in a covalent or ionic bond undergo a transition to a higher energy level requiring energy from the ultraviolet (UV) (200–400 nm) to the visible range (400–750 nm). (Radiation below 200 nm is called the vacuum UV, because oxygen in air is a strong absorber in this region.) These absorptions can be detected spectroscopically. Examples of biochemical chromophores include (1) peptide bonds in amino acids, which have a strong absorption band in the far UV; (2) purine and pyrimidine bases and their derivatives in nucleic acids (DNA, RNA, and NADH), which absorb energy in the UV (250–350 nm); (3) highly conjugated systems such as porphyrin in red blood cells have a strong absorption in the UV and visible regions; (4) absorption in metal complexes is usually explained as d–d transitions that cover the whole visible and some of the near infrared range; and (5) charge transfer in Heme proteins (Campbell and Dwek 1984).

When light propagates through a homogeneous absorbing medium, it can be described using the Beer–Lambert–Bouger law or simply Beer's law, which states that the intensity of light, I, traversing the medium is

$$I = I_0 e^{-\mu_a \cdot l} \tag{8.1}$$

where

I_0 is the intensity of the incoming light

l is the distance that the light propagates in the medium

μ_a is the absorption coefficient, which depend on the properties of the medium

The absorption coefficient is proportional to the sum of the molar concentrations of various chromophores present in the medium c_i multiplied by their molar extinction coefficients ε_i:

$$\mu_a = \sum_i c_i \varepsilon_i \tag{8.2}$$

The exponential factor $\mu_a l$ is also often referred to as the optical density.

Absorption is one of the most basic and longest used sources of contrast in biomedical optics. However, tissue usually exhibits relatively weak and spectrally broad absorptions. The same is true for live cells, which do not exhibit prominent chromophores. The only prominent common absorber in live tissue is the hemoglobin present in red blood cells. Lack of prominent chromophores in tissue is the major limitation of absorption when used as the sole source of contrast in biomedical optics.

8.2.2 Scattering Mechanisms

If in addition to absorption, the medium also scatters light, the situation becomes significantly more complex. Light scattering processes can be divided into elastic and inelastic scattering. In case of elastic scattering, photon energy (and thus wavelength) is preserved. The electrons in the scattering medium are excited to a virtual state and return to the ground state. This type of molecular scattering is known as Rayleigh scattering and is a common scattering process in the visible and near infrared spectral regions.

There are a variety of inelastic processes, where the energy (and wavelength) of the photon is not conserved. A frequently used technique employs fluorescence scattering. Fluorescence is a three-step process, and emission occurs at wavelengths longer than the excitation wavelength due to the photon-transferring part of its energy to heat the medium via molecular rotations and vibrations. The timescale for excitation is femtoseconds, and the relaxation to heat happens on a picosecond to nanosecond timescale. Fluorescent scattering is currently one of the most popular approaches for imaging cells and tissues. However, since native fluorescence of cells and subcellular structures is often very weak or nonexistent, and limited mainly to NADH and FAD, fluorescence scattering is primarily used in combination with exogenous fluorescence labels. Since delivery of those labels is not always trivial and the labels can also affect cell function in tissue, there is a need for techniques that use native sources of contrast in tissue and cells.

Phosphorescence scattering is similar to fluorescence scattering, except that the excited electron transits to a metastable state by an intersystem crossing process. Because the transition from this metastable state to ground state is forbidden, electrons at the metastable level stay populated for a long time until the thermal energy raises the electrons to a state where relaxation is permitted. The timescale varies from microseconds to seconds.

Raman scattering is another interesting example of inelastic scattering, which is now used in biomedical optics. In case of nonresonant Raman scattering, a molecule is excited to a virtual state and has a small but nonzero probability to relax to a different vibrational level in the ground state. The emission photon for this process has either less energy (Stokes line) or more energy (anti-Stokes line) than the excitation photon. The probability for such a process, i.e., the Raman scattering cross section, is usually quite small, many orders of magnitude smaller than elastic or even fluorescence scattering cross sections in the majority of biological media (Campbell and Dwek 1984). On the other hand, Raman scattering is quite specific and exhibits very narrow spectral lines, making it a good candidate for various biomedical optics applications.

8.2.3 Polarization

Description of light propagation using the intensity of light, I, as in Equation 8.1, is a simplification. In reality, light should be described as a vector wave. According to Maxwell's equations, the electric vector is perpendicular to the direction of propagation of the wave. Thus, in the simplest case of a plane wave, the electric vector $\mathbf{E}$ can be written as a sum of two components:

$$\mathbf{E} = E_\parallel \mathbf{e}_\parallel + E_\perp \mathbf{e}_\perp \tag{8.3}$$

where unit vectors $\mathbf{e}_\parallel$ and $\mathbf{e}_\perp$ are perpendicular to each other and to the direction of propagation $\mathbf{e}_\perp \times \mathbf{e}_\parallel$ while $E_\parallel$ and $E_\perp$ are complex oscillating functions. For a plane wave, the choice of unit vectors $\mathbf{e}_\parallel$ and $\mathbf{e}_\perp$ is arbitrary. However, in the case of light scattering, we can project both the incident and scattered fields on the axes parallel and perpendicular to the scattering plane formed by the incident and outgoing wave vector. In that case, we can choose $\mathbf{e}_\parallel$ to be along the parallel axis and $\mathbf{e}_\perp$ to be along the perpendicular axis.

If the field we are describing is scattered by a particle, the distance to the point in space where the scattering is observed r is always much larger than the particle's characteristic size and wavelength λ. The components of the scattering field $E_{sca}^\parallel$ and $E_{sca}^\perp$ can then be expressed using a scattering matrix $\mathbf{S}$ as

$$\begin{bmatrix} E_{sca}^\parallel \\ E_{sca}^\perp \end{bmatrix} = \frac{e^{i(kr-\omega t)}}{ikr} \mathbf{S} \begin{bmatrix} E_{inc}^\parallel \\ E_{inc}^\perp \end{bmatrix} \tag{8.4}$$

where $E_{inc}^\parallel$ and $E_{inc}^\perp$ are components of the incident field. The component $E_{sca}^\parallel$ then describes light with parallel polarization and component $E_{sca}^\perp$ light with perpendicular polarization. In turn, the scattering matrix

$$\mathbf{S} = \begin{bmatrix} S_2 & S_3 \\ S_4 & S_1 \end{bmatrix}$$

depends on the directions of incident and scattered waves as well as the scatterer's size, shape, composition, and orientation (van de Hulst 1957).

Since many subcellular organelles have an approximately spherical shape, the expression for the scattering matrix $\mathbf{S}$ for a sphere is important when describing scattering of polarized light by cells and their internal structures. In case of plane wave scattering, this expression is provided by Mie theory. It predicts that if the incident light is completely collimated, light scattered by a spherical scatterer directly backward will be polarized parallel to the incident light polarization. However, in case of nonzero incident and collection, solid angles both polarizations will be present in the scattered light. The following useful approximate expression for the difference of intensities of parallel $I_\parallel$ and perpendicular $I_\perp$ polarized light scattered by a spherical scatterer is provided in Backman et al. (1999)

$$I_\parallel - I_\perp = \frac{I_0}{k^2 r^2} \int\limits_{\pi}^{\pi - \sqrt{\Delta\Omega/\pi}} \mathrm{Re}\left[S_1^\star(\vartheta) S_2(\vartheta) \right] \sin\vartheta \, d\vartheta \tag{8.5}$$

where

I_0 is the incident intensity of the linearly polarized light
$\Delta\Omega$ is the solid angle of the delivery and collection optics

8.3 Light Scattering Spectroscopy

8.3.1 Light Scattering from Cells and Subcellular Structures

Though there are hundreds of human cell types, the subcellular compartments in different cells are rather similar and are limited in number (Perelman and Backman 2002). Every cell is bounded by a membrane, a phospholipid bilayer approximately 10 nm in thickness. Two major cell compartments are the nucleus, which is approximately spherical with a diameter of 7–10 μm and the surrounding cytoplasm. The cytoplasm contains various organelles and inclusions. Mitochondria are oblong with a large dimension of 1–5 μm and a transverse diameter between 0.2 and 0.8 μm. Other smaller organelles include lysosomes, which are 250–800 nm in size and of various shapes and peroxisomes, which are 200 nm–1.0 μm spheroidal bodies of lower densities than the lysosomes. Peroxisomes are more abundant in metabolically active cells such as hepatocytes where they are counted in the hundreds.

Most cell organelles and inclusions are themselves complex objects with spatially varying refractive indices (Beuthan et al. 1996). Many organelles such as mitochondria, lysosomes, and nuclei possess an average refractive index substantially different from that of their surrounding. Therefore, an accurate model acknowledges subcellular compartments of various sizes with a refractive index differing from that of the surrounding.

Sizes, shapes, and refractive indices of major cellular and subcellular structures are presented in Figure 8.1. In this figure, we also show the relevant approximations that can be used to describe light scattering from these objects.

Studies of light scattering by cells have a long history. Brunstin and Mullaney (1974) initiated a series of experiments connecting the internal structure of living cells to the scattering pattern by measuring forward and near-forward scattering in cell suspensions using a rigorous quantitative approach. They

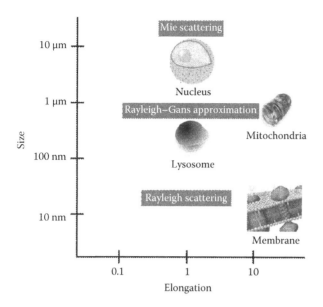

FIGURE 8.1 Hierarchy of dimensional scales inside of a cell. Hierarchy of scales and optical properties of subcellular structures along with the relevant approximations that can be used to describe light scattering from those objects. The refractive index of cytoplasm is approximately in the 1.36–1.37 and organelles in 1.41–1.44 range. (Adapted from Perelman, L.T. and Backman, V., Light scattering spectroscopy of epithelial tissues: Principles and applications, in: *Handbook on Optical Biomedical Diagnostics*, V.V. Tuchin (ed.), SPIE Press, Bellingham, WA, 2002, pp. 675–724; Beuthan, J. et al., *Phys. Med. Biol.*, 41, 369, 1996; Sloot, P.M.A. et al., *Cytometry*, 9, 636, 1988.)

used cells of several types including Chinese hamster's oocytes (CHO) and HeLa cells. They compared the resulting angular distribution of the scattered light for CHO cells with the predictions of Mie theory and found very good agreement between theory and experiment. This was achieved by approximating a cell as a denser sphere imbedded into a larger less dense sphere. The sizes of these spheres corresponded to the average sizes of the cell nuclei and cells, respectively. Particles large compared to a wavelength produce a scatter field that peaks in the forward and near-backward directions in contrast to smaller particles, which scatter light more uniformly. Despite nonhomogeneity and the lack of a perfectly spherical shape of cells and their nuclei, experimental results agreed well with Mie theory, which deals with uniform spheres. These results were supported by Sloot et al. (1988) in experiments with white blood cells (leukocytes), who found that light scattering by the leukocytes in the near-forward direction can be explained if each cell is approximated by two concentric spheres, one being the cell itself and the other being the nucleus, and Hammer et al. (1998), who showed that near-forward scattering of light by red blood cells can be accurately described using the van de Hulst approximation to Mie theory, which is derived for large particles of spherical shapes rather than the actual concave–convex disks that are red blood cells.

Studies of the angular dependence of cellular light scattering by Mourant et al. (1998) showed that the cell structures responsible for light scattering can be correlated with the angle of scattering. When a cell is suspended in a buffer solution of lower refractive index, the cell itself is responsible for small angle scattering. This result was used in flow cytometry to estimate cell sizes (Watson 1991). At slightly larger angles, the nucleus is primarily responsible for scattering, and it is the major scatterer in forward directions when the cell is a part of a contiguous layer.

Smaller organelles, cell inclusions, suborganelles, and subnuclear inhomogeneities are responsible for scattering at larger angles. Scattering may originate from organelles themselves or their internal components. Angular dependence may reveal whether the scattering originates from the objects of regular or irregular shape, spherical or elongated, and inhomogeneous or uniform. In some cases, large angle scattering can be attributed to a specific predominant organelle. Research conducted by Beauvoit et al. (1994) provided strong evidence that mitochondria are responsible for approximately 50% of light scattering from hepatocytes.

Components of organelles can also scatter light. Finite-difference time-domain (FDTD) simulations provide a means to study spectral and angular features of light scattered by arbitrary particles of complex shape and density. Using FDTD simulations, Drezek et al. (1999) investigated the influence of cell morphology on the scattering pattern and demonstrated that as the spatial frequency of refractive index variation increases, the scattering intensity increases at large angles.

8.3.2 Analytical Approximations for Light Scattering from Cells and Subcellular Organelles

To find the matrix elements of the scattering matrix **S** for cells and subcellular organelles, one needs to solve Maxwell's wave equations with proper boundary conditions for electric and magnetic fields for those structures. However, exact solutions of the Maxwell's equations for majority of even simple shapes are not known. As a matter of fact, there are just a few cases when the analytical solution to the wave equation has been found. In 1907, Mie (1908) obtained the solution for the scattering of a plane wave by a uniform sphere, which we mentioned in Section 8.2.3. In this solution, the only two nonzero elements of the scattering matrix S_1 and S_2 are expressed as infinite series of Bessel functions of two parameters, ka and kma, where k the wave number, a the diameter of the sphere, and m the relative refractive index of the sphere. Other examples of particles for which the scattering problem has been solved analytically are cylinders, coated spheres, uniform and coated spheroids, strips, and planes (Bohren and Huffman 1983). Even for these "simple" cases, the elements of the scattering matrix can be expressed only as infinite series, which are often poorly converging.

Difficulties with finding exact solutions of the wave equations have led to the development of approximate methods of solving the scattering problem. One class of approximations was originally found by Rayleigh in 1871 (Rayleigh 1871) and is known as Rayleigh scattering. Rayleigh scattering describes light scattering by particles small comparing to the wavelength and is a very important approximation for biomedical optics since a great variety of the structures from which cells organelles are built, such as the tubules of the endoplasmic reticulum, or the cisternae of Golgi apparatus, etc., fall into this category. In the Rayleigh limit, the scattering cross section scales as the six power of the particle's linear dimension and varies inversely with $\lambda 4$.

For larger particles with sizes comparable to the wavelength, the Rayleigh approximation fails, and one can use another solution called Rayleigh–Gans approximation (van de Hulst 1957). It is applicable if the relative refractive index of the particle is close to unity and the phase shift across the particle is small. Since the refractive index of most subcellular organelles ranges from 1.38 to 1.42 (Sloot et al. 1988, Beauvoit et al. 1994, Beuthan et al. 1996), and the refractive index of the cytoplasm varies from 1.34 to 1.36, both conditions of the Rayleigh–Gans approximation are satisfied for the majority of small organelles.

If a particle is relatively homogenous, then the Rayleigh–Gans approximation gives

$$\begin{pmatrix} S_2 & S_3 \\ S_4 & S_1 \end{pmatrix} = \frac{ik^3(m-1)V}{2\pi} R(\theta, \phi) \begin{pmatrix} \cos\theta & 0 \\ 0 & 1 \end{pmatrix}, \quad (8.6)$$

where m is the relative refractive index averaged over the volume of the particle, and the function $R(\theta,\phi) = 1/V \int e^{i\delta} dV$ is the so-called form factor. From (8.6), it is easy to see that the total intensity of light scattered by a small organelle increases with the increase in its refractive index as $(m-1)^2$ and as the six power of its size. However, the angular distribution of the scattered light differs from that of Rayleigh scattering. For $\theta = 0$, the form factor equals unity. In other directions $|R| < 1$, the scattering has a maximum in forward direction.

Unfortunately, none of the above-mentioned approximations can be applied to the cell nucleus whose size is significantly larger than that of a wavelength. An approximate theory of light scattering by large particles was first proposed by van de Hulst in 1957 (van de Hulst 1957) and originally formulated for spherical particles. However, it can be extended to large particles of arbitrary shape. Although the van de Hulst theory does not provide universal rules for finding the scattering matrix for all scattering angles even in the case of a homogenous sphere, it can be used to obtain scattering amplitudes in the near-forward direction and the scattering cross section:

$$\sigma_s \approx 2\pi a^2 \left\{ 1 - \frac{\sin(2x(m-1))}{x(m-1)} + \left(\frac{\sin(x(m-1))}{x(m-1)} \right)^2 \right\}, \quad (8.7)$$

where $x = ka$ is called the size parameter.

It shows that large spheres give rise to a very different type of scattering than the small particles considered above. Both the intensity of the forward scattering and the scattering cross section are not monotonic functions of wavelength. Rather, they exhibit oscillations with the wavelength; the frequency of these oscillations is proportional to $x(m-1)$, so it increases with the sphere size and refractive index.

8.3.3 Principles of Light Scattering Spectroscopy

The strong dependence of the scattering cross section (Equation 8.7) on size and refractive index of the scatterer, such as the cell nucleus, as well as on the wavelength suggests that it should be possible to design a spectroscopic technique, which can differentiate cellular tissues by the sizes of the nuclei. Indeed, the hollow organs of the body are lined with a thin, highly cellular surface layer of epithelial tissue, which is supported by underlying, relatively acellular connective tissue. There are four main types of epithelial tissue: squamous, cuboidal, columnar, and transitional, which can be found in different organs of the human body. Depending on type, the epithelium consists either of a single layer of cells or several cellular layers. Here, to make the treatment of the problem more apparent, we consider epithelial layers consisting of a single well-organized layer of cells, such as simple columnar epithelium or simple squamous epithelium. For example, in healthy columnar epithelial tissues, the epithelial cells often have an en-face diameter of 10–20 μm and height of 25 μm. In dysplastic epithelium, the cells proliferate, and their nuclei enlarge and appear darker (hyperchromatic) when stained (Cotran et al. 1994).

LSS can be used to measure these changes. The details of the method have been published by Perelman et al. (1998) and will only be briefly summarized here. Consider a beam of light incident on an epithelial layer of tissue. A portion of this light is backscattered from the epithelial nuclei, while the remainder is transmitted to deeper tissue layers, where it undergoes multiple scattering and becomes randomized before returning to the surface.

Epithelial nuclei can be treated as spheroidal Mie scatters with a refractive index, which is higher than that of the surrounding cytoplasm (Sloot et al. 1988, Beuthan et al. 1996, Perelman and Backman 2002). Normal nuclei have a characteristic size of 4–7 μm. In contrast, the size of dysplastic nuclei varies widely and can be as large as 20 μm, occupying almost the entire cell volume. In the visible range, where the wavelength is much smaller than the size of the nuclei, the van de Hulst approximation (Equation 8.7) can be used to describe the elastic scattering cross section of the nuclei. Equation 8.7 reveals a component of the scattering cross section, which varies periodically with inverse wavelength. This, in turn, gives rise to a periodic component in tissue reflectance. Since the frequency of this variation (in the inverse wavelength space) is proportional to the particle size, the nuclear size distribution can be obtained from that periodic component.

However, single scattering events cannot be measured directly in biological tissue. Because of multiple scattering, information about tissue scatterers is randomized as light propagates into the tissue, typically over one effective scattering length (0.5–1 mm, depending on the wavelength). Nevertheless, the light in the thin layer at the tissue surface is not completely randomized. In this thin region, the details of the elastic scattering process are preserved. The total signal reflected from tissue can be divided into two parts: single backscattering from the uppermost tissue structures such as cell nuclei and a background of diffusely scattered light. To analyze the single scattering component of the reflected light, the diffusive background must be removed. This can be achieved either by modeling using diffuse reflectance spectroscopy (Zonios et al. 1999, Georgakoudi et al. 2001) or by other techniques such as polarization background subtraction (Backman et al. 1999, Sokolov et al. 1999, Jacques 2000), coherence gating method (Wax et al. 2005), or azimuthal LSS (Yu et al. 2006).

There are several techniques that can be employed to obtain the nuclear size distribution from the remaining single scattering component of the backreflected light, which can be called the LSS spectrum. A good approximation for the nuclear size distribution can be obtained from the Fourier transform of the periodic component as described (Perelman et al. 1998). A more advanced technique introduced by Fang et al. in 2003 (Fang et al. 2003) and described here is based on linear least squares with a non-negativity constraints algorithm.

The experimentally measured reflectance spectrum consists of a large diffusive background plus the component of forward scattered and backscattered light from the nuclei in the epithelial layer. For a thin slab of epithelial tissue containing nuclei with size distribution $N(\delta)$ (number of nuclei per unit area (mm²) and per unit interval of nuclear diameter (μm)), the approximate solution of the transport equation for the backscattered component is a linear combination of the backscattering spectra of the nuclei of different sizes:

$$S(\lambda) = \int_0^\infty I(\lambda, \delta) N(\delta) d\delta + \varepsilon(\lambda) \qquad (8.8)$$

where

$I(\lambda, \delta)$ is the LSS spectrum of a single scatterer with diameter δ
$\varepsilon(\lambda)$ is the experimental noise

It is convenient to write this in a matrix form $\hat{S} = \hat{I} \cdot \hat{N} + \hat{E}$, where $\hat{S}$ is the experimental spectrum measured at discrete wavelength points, $\hat{N}$ is a discreet nuclear size distribution, $\hat{I}$ is the LSS spectrum of a single scatterer with diameter δ, and $\hat{E}$ is the experimental noise. Since the LSS spectrum $\hat{I}$ is a highly singular matrix and a certain amount of noise is present in the experimental spectrum $\hat{S}$, it is not feasible to calculate the size distribution $\hat{N}$ by directly inverting the matrix $\hat{I}$. Instead, we can multiply both sides of the equation $\hat{S} = \hat{I} \cdot \hat{N} + \hat{E}$ by the transpose matrix $\hat{I}^T$ and introduce the matrix $\hat{C} = \hat{I}^T \cdot \hat{I}$. We can now compute matrix **C** eigenvalues $\alpha_1, \alpha_2, \ldots$ and sequence them from large to small. This can be done because $\hat{C}$ is a square symmetric matrix. Then, we use the linear least squares with non-negativity

constraints algorithm (Craig and Brown 1986) to solve the set of equations

$$\hat{I}^T \hat{S} - (\hat{C} + \alpha_k \hat{H})\hat{N} \to \min$$
$$\hat{N} \ge 0 \qquad (8.9)$$

where

$\alpha_k \hat{H} \hat{N}$ is the regularization term
matrix H represents the second derivative of the spectrum

The use of the non-negativity constraint and the regularization procedure is critical to find the correct distribution $\hat{N}$. By using this algorithm, we can accurately reconstruct the nuclear size distribution.

8.3.4 Polarized Light Scattering Spectroscopy

Here, we discuss in greater details one of the techniques used to remove the multiple scattered light in LSS measurements and improve the extraction of the single-scattered light component. This technique is sometimes called polarized LSS or PLSS.

In the PLSS technique, the tissue is illuminated with a polarized light. The light backscattered from the superficial epithelial layer retains its polarization, i.e., it is polarized parallel to the incoming light. The light backscattered from the deeper tissues becomes depolarized and contains equal amounts of parallel and perpendicular polarizations. By subtracting the two, one can cancel out the contribution of the deeper tissues, and the resulting signal is proportional only to the signal from the superficial epithelial layer, which contains information about early precancerous changes. The residual of the parallel and perpendicular components $I_{PLSS}(\lambda) = I_{II}(\lambda) - I_\perp(\lambda)$ can be related to the properties of scatterers in the superficial epithelial layer using relation (Equation 8.7) and then processed using the algorithm described in the previous section.

To verify this, Backman et al. (1999) employed an instrument that delivers collimated polarized light to the tissue and separates the two orthogonal polarizations of backscattered light. In this system, light from a broadband source is collimated and then refocused at a small solid angle onto the sample, using lenses and an aperture. Studies have shown that the unpolarized component of the reflected light can be canceled by subtracting the perpendicular spectral component from the parallel component allowing the single scattering signal to be extracted. These residual spectra were fitted to the model based on Mie theory.

Backman et al. (1999) performed experiments with monolayers of normal intestinal epithelial cells and T84 intestinal malignant cells placed above a thick layer of gel containing blood and BaSO₄, placed underneath to simulate underlying tissue. For normal intestinal epithelial cells, the best fit was obtained using $d = 5.0\,\mu m$, $\Delta d = 0.5\,\mu m$, and $n = 1.035$. For T84 intestinal malignant cells, the corresponding values were $d = 9.8\,\mu m$, $\Delta d = 1.5\,\mu m$, and $n = 1.04$. To check these results, the distribution

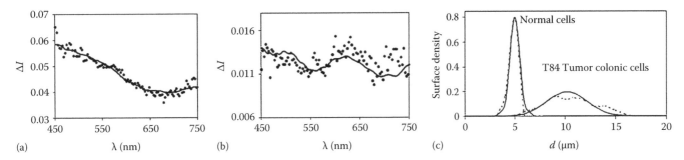

FIGURE 8.2 Spectra of polarized components of backscattered light. Spectrum of (a) normal intestinal cells, (b) T84 intestinal malignant cells, and (c) corresponding nuclear size distributions. In each case, the solid line is the distribution extracted from the data, and the dashed line is the distribution measured using light microscopy. (From Backman, V. et al., *IEEE J. Sel. Top. Quantum Electron.*, 5, 1019, 1999. With permission.)

of the average size of the cell nuclei was measured by morphometry on identical cell preparations that were processed in parallel for light microscopy. The nuclear sizes and their standard deviations were found to be in very good agreement with the parameters extracted from Mie theory (Figure 8.2). In addition, a larger value of n was obtained for T84 intestinal malignant cells, which is in agreement with the hyperchromaticity of cancer cell nuclei observed in conventional histopathology of stained tissue sections.

8.3.5 Biomedical Applications of Light Scattering Spectroscopy

The ability of LSS to diagnose dysplasia and carcinoma in situ (CIS) was tested in in vivo human studies in four different organs and in three different types of epithelia: columnar epithelium of the colon and Barrett's esophagus (BE), transitional epithelium of the urinary bladder, and stratified squamous epithelium of the oral cavity (Backman et al. 2000, Wallace et al. 2000). All clinical studies were performed during routine endoscopic screening or surveillance procedures. In all of the studies, an optical fiber probe delivered white light from a xenon arc lamp to the tissue surface and collected the returned light. The probe tip was brought into gentle contact with the tissue to be studied. Immediately after the measurement, a biopsy was taken from the same tissue site. The biopsied tissue was prepared and examined histologically by an experienced pathologist in the conventional manner. The spectrum of the reflected light was analyzed and the nuclear size distribution determined. The majority of distributions of dysplastic cell nuclei extended to larger size.

These size distributions were then used to obtain the percentage of nuclei larger than 10 micron and the total number of nuclei per unit area (population density). As noted above, these parameters quantitatively characterize the degree of nuclear enlargement and crowding, respectively.

Figure 8.3 displays these LSS parameters in binary plots to show the degree of correlation with histological diagnoses. In all four organs, there is a clear distinction between dysplastic and nondysplastic epithelia. Both dysplasia and CIS have a higher percentage of enlarged nuclei and, on average, a higher population density, which can be used as the basis for spectroscopic tissue diagnosis.

These results show the promise of LSS as a real-time, minimally invasive clinical tool for accurately and reliably classifying invisible dysplasia. Although the presented data sets are limited in size, the effectiveness of LSS in diagnosing early cancerous lesions is again clearly demonstrated, and this suggests the general applicability of the technique.

8.3.6 Imaging of Early Cancer and Precancerous Lesions with Endoscopic Polarized Light Scattering Spectroscopy

LSS-based detection of dysplasia in BE was successfully demonstrated using a simple proof-of-principle single-point instrument (Jacques 2000, Wax et al. 2005). This instrument was capable of collecting data at randomly selected sites, which were then biopsied. The data was processed off-line, and a comparison with biopsy results was made at a later time. However, to fully realize its potential, the LSS technique needed to be extended to an endoscope-compatible clinical high-resolution imaging modality in order to probe and analyze large areas of the epithelial surfaces of various internal organs in real time. This will enable the physician to take confirming biopsies at suspicious sites and minimize the number of biopsies taken at nondysplastic sites.

A recently developed (Qiu et al. 2009, 2010) clinical endoscopic polarized scanning spectroscopy (EPSS) instrument that is compatible with existing endoscopes represents such an extension (Figure 8.4a). It scans large areas of the esophagus chosen by the physician and has the software and algorithms necessary to obtain quantitative, objective data about tissue structure and composition, which can be translated into diagnostic information in real time. Biopsies of suspicious sites can be taken and biopsies of nondysplastic sites avoided.

The instrument uses the polarization technique to extract diagnostic information about dysplasia in the epithelial layer; however, the EPSS instrument also sums the two polarizations to permit the use of diffuse reflectance spectroscopy, which can also provide information about early stages of adenocarcinoma (Georgakoudi et al. 2001).

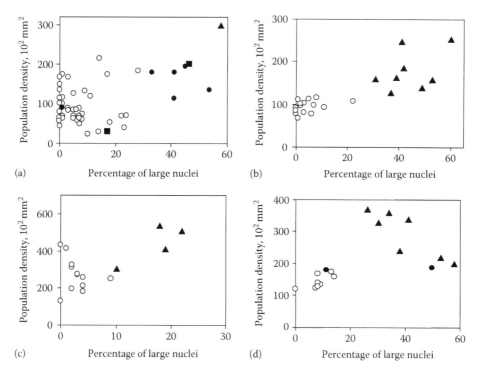

FIGURE 8.3 Dysplasia/CIS classifications for four types of tissue obtained clinically with LSS compared with histologic diagnosis. In each case, the ordinate indicates the percentage of enlarged nuclei, and the abscissa indicates the population density of the nuclei, which parameterizes nuclear crowding. (a) BE: nondysplastic Barrett's mucosa (O), indefinite for dysplasia (■), low-grade dysplasia (●), high-grade dysplasia (▲); (b) colon: normal colonic mucosa (O), adenomatous polyp (▲); (c) urinary bladder: benign bladder mucosa (O), transitional cell CIS (▲); (d) oral cavity: normal (O), low-grade dysplasia (●), squamous cell CIS (▲). (Adapted from Backman, V. et al., *Nature*, 406, 35, 2000.)

The EPSS instrument is a significant advance over the single-point fiber-optic instrument in that (1) it scans the esophagus and has the software and algorithms necessary to obtain quantitative, objective data about tissue structure, and composition, which can be translated into diagnostic information and guide biopsy in real time; (2) it employs collimated illumination and collection optics, which enables the instrument to generate maps of epithelial tissue not affected by the distance between the probe tip and the mucosal surface, making it dramatically less sensitive to peristaltic motion; (3) it incorporates both the polarization technique for removing the unwanted background in the LSS signal from the epithelium and diffuse reflectance spectroscopy to provide information about the underlying submucosa; (4) it integrates the data analysis software with the instrument in order to provide the physician with real-time diagnostic information; and (5) it combines LSS information with diffuse reflectance spectroscopy information measured by the same instrument, thereby improving the diagnostic assessment capability.

A block diagram of the EPSS instrument is shown in Figure 8.4b. The instrument uses commercially available gastroscopes and video processors. A standard PC is adapted to control the system. Commercially available spectrometers are also employed.

Qiu et al. (2010) performed clinical measurements using the EPSS instrument during a routine endoscopic procedure at the Interventional Endoscopy Center at Beth Israel Deaconess Medical Center for a patient with suspected dysplasia in BE. Figure 8.5 is a view through the endoscope camera of the probe and the "flying spot" during this procedure. Spectroscopic data collected during the clinical procedures confirm that the polarization technique is very effective in removing the unwanted background signals. For example, data from a random spatial location presented in Figure 8.6a show that the perpendicular polarization spectral component exhibits standard diffuse reflectance features originating in the deeper tissue layers, with the hemoglobin absorption bands clearly observable in the 540–580 nm region. At the same time, the parallel polarization spectral component, in addition to diffuse features, exhibits a very clear oscillatory structure, characteristic of the diagnostically important nuclear scattering originating in the uppermost epithelial layer.

Another important observation is that spectra collected from multiple spatial locations inside the BE are insensitive to the effects of motion. During a procedure, it is difficult to maintain a fixed distance between the optical probe head and the esophageal surface, due to peristaltic motion and other factors. Therefore, an important feature of the EPSS instrument is its ability to collect spectra of epithelial tissue that are not affected by the orientation and the distance between the distal tip of the probe and the mucosal surface. This is achieved with collimated illumination and collection optics. By comparing parallel polarization spectra collected at multiple locations inside a BE during

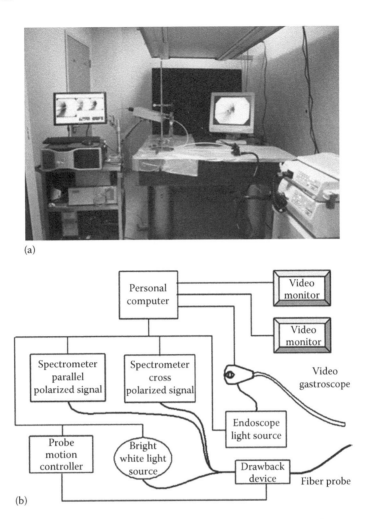

(a)

(b)

FIGURE 8.4 Clinical EPSS instrument. (a) EPSS instrument performing measurements in ex vivo bovine esophagus. During the procedure, the fiberoptic probe is inserted into the gastroscope's instrument channel; (b) instrument system block diagram. (Adapted from Qiu, L. et al., *IEEE Proc. Eng. Med. Biol. Soc.*, 1997, 2009.)

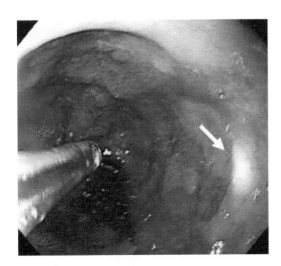

FIGURE 8.5 Photograph of the EPSS probe performing scanning of BE during a clinical procedure obtained via the endoscope video channel. The "flying spot" is clearly seen on the esophagus wall to the right. (Adapted from Qiu, L. et al., *IEEE Proc. Eng. Med. Biol. Soc.*, 1997, 2009.)

a standard clinical procedure (see the 10 spectra in Figure 8.6b), one finds that although the amplitudes of the spectra differ from location to location, the spectral shape is practically unchanged, and, what is even more significant, the oscillatory structure containing the diagnostic information is intact (Qiu et al. 2010). Thus, the issue of peristaltic motion appears to be addressed in the EPSS instrument.

The backscattered spectrum at each individual spatial location was extracted by subtracting perpendicular from parallel polarized reflectance spectra. The backscattered spectra were then normalized to remove amplitude variations due to peristalsis. The mean of the normalized spectra was calculated. The difference from the mean for each site was calculated, squared, and summed over all spectral points. A site was considered likely to be dysplastic if this parameter differed by more than 10% of the summed mean square. No data points were used for the calibration of this simple diagnostic rule. This analysis is straightforward and can be done in real time. By extracting the nuclear size distributions from the backscattered spectra for each individual spatial location, Qiu et al. (2010) found that this simple rule is

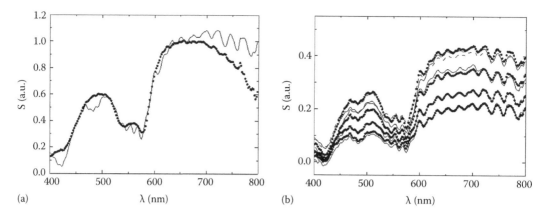

FIGURE 8.6 (a) Parallel (solid line) and perpendicular (dotted line) polarization spectra collected with the EPSS instrument from a single spatial location in a patient with BE. (b) Parallel polarization spectra from 10 different locations in the same patient. (Adapted from Qiu, L. et al., *IEEE Proc. Eng. Med. Biol. Soc.*, 1997, 2009.)

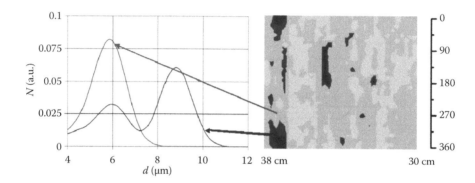

FIGURE 8.7 Nuclear size distributions for one HGD site and one nondysplastic site in BE. Black and Dark regions of the map indicate areas suspicious for dysplasia based on nuclear size distributions extracted from the backscattering spectra for each individual spatial location. Nondysplastic BE sites had nuclear size distributions centered about 5–6 μm diameter while sites marked as suspicious for dysplasia have nuclear size distributions with a main peak centered from 9 to 15 μm. The arrows indicate the specific locations on the esophageal surface from which the size distributions were extracted from the PLSS data. (Adapted from Qiu, L. et al., *Nat. Med.*, 16, 603, 2010.)

approximately equivalent to a contribution of more than 25% from enlarged nuclei greater than 10 microns in diameter.

The result for one of the first patients imaged with the EPSS instrument is presented in the form of pseudo-color maps in Figure 8.7. Black and Dark gray regions of the map indicate areas suspicious for dysplasia based on nuclear size distributions extracted from the backscattered spectra for each individual spatial location. Nondysplastic BE sites had nuclear size distributions centered around 5–6 μm diameter while sites marked as suspicious for dysplasia have nuclear size distributions with a main peak centered from 9 to 15 μm. The arrows indicate the specific locations on the esophageal surface from which the size distributions are extracted from the PLSS data.

Pathologic examination of tissue biopsies taken in a pattern prescribed by the present standard-of-care for a recent patient revealed no dysplasia, and the patient was dismissed. However, an EPSS scan taken at the same time indicated several probable sites of focal dysplasia, which were located in regions where biopsies had not been taken. The patient was recalled, and biopsies were taken at the suspicious sites indicated by EPSS. Pathology

confirmed high-grade dysplasia at these locations. This patient will now be given appropriate treatment. Standard-of-care procedures, even when diligently performed by the most highly skilled and experienced gastroenterologists, can miss focal dysplasia because these procedures biopsy only a very small fraction of esophageal tissue, blindly, and by necessity according to prescribed protocol. The capability of EPSS to examine the entire esophageal epithelium millimeter-by-millimeter enables detection of dysplastic cells and guidance of confirmative biopsy, greatly increasing the probability of early detection and treatment and, in all likelihood, of saving lives.

8.4 Confocal Light Absorption and Scattering Spectroscopic Microscopy

8.4.1 Principles of CLASS Microscopy

Another application of LSS, very different from what we have discussed above, is to detect sources of highly specific native contrast within internal cell structures using optical microscopy.

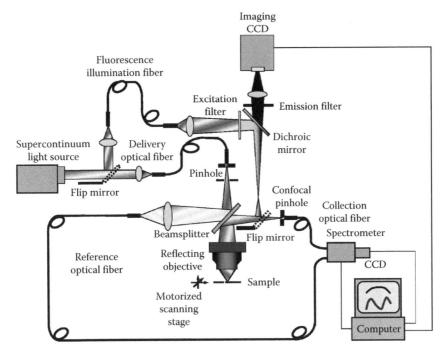

FIGURE 8.8 Schematic of the prototype CLASS/fluorescence microscope. (Adapted from Itzkan, I. et al., *Proc. Natl. Acad. Sci. USA*, 104, 17255, 2007.)

Recently, we developed confocal light absorption and scattering spectroscopic (CLASS) microscopy (Fang et al. 2007, Itzkan et al. 2007), which combines LSS with the principles of confocal microscopy. The multispectral nature of LSS enables it to measure internal cell structures much smaller than the diffraction limit without damaging the cell or requiring exogenous markers, which could affect cell function. CLASS microscopy approaches the accuracy of electron microscopy but is nondestructive and does not require the contrast agents common to traditional optical microscopy. In Section 8.4, we discuss the basic physical principles of CLASS microscopy and its applications to such diverse areas as cell biology, drug discovery, prenatal diagnosis, and cellular and tissue imaging with nanoparticulate markers.

A schematic of the CLASS microscope is shown in Figure 8.8. Broadband illumination is provided from either a Xe arc lamp, for measurements performed on extracted organelles in suspension, or a supercontinuum laser (Fianium SC-450-2), for measurements performed on organelles in living cells. Both sources use an optical fiber to deliver light to the sample. To insure that CLASS microscopy detects organelles inside living cells and correctly identifies them, Itzkan et al. (2007) augmented the CLASS instrument with a wide field fluorescence microscopy arm, which shares a major part of the CLASS optical train.

Depth sectioning characteristics of a CLASS microscope can be determined by translating a mirror located near the focal point and aligned normal to the optical axis of the objective using five wavelengths spanning the principal spectral range of the instrument (Figure 8.9). The half-width of the detected signal is approximately 2 μm, which is close to the theoretical value for the 30 μm pinhole and 36× objective used (Wilson and

Carlini 1987). In addition, the shapes of all five spectra, shown in Figure 8.9, are almost identical (500, 550, 600, 650, and 700 nm), which demonstrates the excellent chromatic characteristics of the instrument. Small maxima and minima on either side of the main peak are due to diffraction from the pinhole. The slight asymmetry is due to spherical aberration in the reflective objective (Scalettar et al. 1996).

To establish the imaging capabilities of the CLASS microscope, several tests were performed using suspensions of carboxylate-modified Invitrogen microspheres, which exhibit red fluorescence emission at a wavelength of 605 nm with excitation at 580 nm. The microspheres were effectively constrained

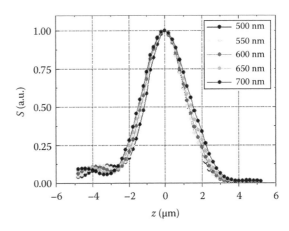

FIGURE 8.9 Depth sectioning of CLASS microscope along vertical axis. (Adapted from Itzkan, I. et al., *Proc. Natl. Acad. Sci. USA*, 104, 17255, 2007.)

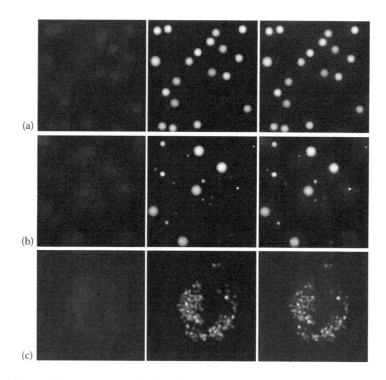

FIGURE 8.10   (a) Fluorescence image of the suspensions of carboxylate-modified 1.9 μm diameter microspheres exhibiting red fluorescence (left side), the image reconstructed from the CLASS data (middle), and the overlay of the images (right side). (b) Image of the mixture of three sizes of fluorescent beads with sizes 0.5, 1.1, and 1.9 μm mixed in a ratio of 4:2:1 (left side), the image reconstructed from the CLASS data (middle) and the overlay of the images (right side). (c) Image of live 16HBE14o-human bronchial epithelial cells with lysosomes stained with lysosome-specific fluorescence dye (left side), the image reconstructed from the CLASS data (middle), and the overlay of the images (right side). (Adapted from Itzkan, I. et al., *Proc. Natl. Acad. Sci. USA*, 104, 17255, 2007.)

to a single layer geometry by two thin microscope slides coated with a refractive index matching optical gel. Figure 8.10a shows (from left to right) the fluorescence image of a layer of 1.9 μm diameter microspheres, the image reconstructed from the CLASS data, and the overlay of the images. Figure 8.10b shows the images obtained from a mixture of three different sizes of fluorescent beads with sizes 0.5, 1.1, and 1.9 μm mixed in a ratio of 4:2:1. Note the misleading size information evident in the conventional fluorescence image. A 0.5 μm microsphere that is either close to the focal plane of the fluorescence microscope or carries a high load of fluorescent label produces a spot, which is significantly larger than the microsphere's actual size. The CLASS image (middle panel of Figure 8.10b), on the other hand, does not make this error and correctly reconstructs the real size of the small microsphere located on the right side near the center. We can also see that prior fluorescence labeling does not affect the size determination of these objects with CLASS measurements.

8.4.2 Applications of CLASS Microscopy in Cell Biology

To confirm the ability of CLASS to detect and identify specific organelles in a live cell, simultaneous CLASS and fluorescence imaging of live 16HBE14o-human bronchial epithelial cells,

with the lysosomes stained with a lysosome-specific fluorescent dye, was performed using the combined CLASS/Fluorescence Instrument (Itzkan et al. 2007). The fluorescence image of the bronchial epithelial cell, the CLASS reconstructed image of the lysosomes, and the overlay of two images are provided in Figure 8.10c.

The overall agreement is very good; however, as expected, there is not always a precise, one-to-one correspondence between organelles appearing in the CLASS image and the fluorescence image. This is because the CLASS image comes from a single, well-defined confocal image plane within the cell, while the fluorescence image comes from several focal "planes" within the cell, throughout the thicker depth of field produced by the conventional fluorescence microscope. Thus, in the fluorescence image, one observes the superposition of several focal "planes" and therefore additional organelles above and below those in the single, well-defined confocal image plane of the CLASS microscope.

Another interesting experiment checked the ability of the CLASS microscopy to do time sequencing on a single cell. The cell was incubated with DHA, a substance which induces apoptosis, for 21 h. The time indicated in each image is the time elapsed after the cell was removed from the incubator. In Figure 8.11, the nucleus, which appears as the large blue organelle, has its actual shape and density reconstructed from

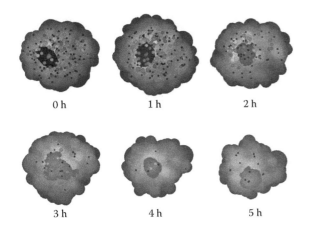

0 h   1 h   2 h

3 h   4 h   5 h

FIGURE 8.11 The time sequence of CLASS microscope reconstructed images of a single cell. The cell was treated with DHA and incubated for 21 h. The time indicated in each image is the time elapsed after the cell was removed from the incubator. (Adapted from Itzkan, I. et al., *Proc. Natl. Acad. Sci. USA*, 104, 17255, 2007.)

the CLASS spectra obtained using point-by-point scanning. The remaining individual organelles reconstructed from the CLASS spectra are represented simply as spheroids whose size, elongation, and darkness indicate different organelles. The small red spheres are peroxisomes, and the intermediate size green spheres are lysosomes. Organelles with sizes in the 1000–1300 nm range are mitochondria and are shown as large yellow spheroids. By the third hour, the shape of the nucleus has changed dramatically, and the nuclear density, indicated by darkness depth, has decreased with time. The organelles have almost completely vanished by 4 h.

8.4.3 Spectroscopy of Single Nanoparticles

Recently, significant attention has been directed toward the applications of metal nanoparticles, such as gold nanorods, to medical problems, primarily as extremely bright molecular marker labels for fluorescence, absorption, or scattering imaging of living tissue (Durr et al. 2007). Nanoparticles with sizes small compared to the wavelength of light made from metals with a specific complex index of refraction, such as gold or silver, have absorption and scattering resonance lines in the visible part of the spectrum. These lines are due to in-phase oscillation of free electrons and are called surface plasmon resonances.

However, samples containing a large number of gold nanorods usually exhibit relatively broad spectral lines. This observed linewidth did not agree with theoretical calculations, which predict significantly narrower absorption and scattering lines. As we have shown in Qiu et al. (2007), the spectral peak of nanorods is dependent on their aspect ratio, and the linewidth discrepancy is explained by the inhomogeneous line broadening caused by the contribution of nanorods with assorted aspect ratios.

This broadened linewidth limits the use of nanorods with uncontrolled aspect ratios as effective molecular labels, since it would be rather difficult to image several types of nanorod

markers simultaneously. However, this suggests that nanorod-based molecular markers selected for a prescribed aspect ratio and, to a lesser degree, size distribution should provide spectral lines sufficiently narrow for effective biomedical imaging.

We performed (Qiu et al. 2007) optical transmission measurements of gold nanorod spectra in aqueous solutions using a standard transmission arrangement for extinction measurements described in Bohren and Huffman (1983). Concentrations of the solutions were chosen to be close to 1010 nanoparticles per milliliter of the solvent to eliminate optical interference. The measured longitudinal plasmon mode of the nanorods is presented as a dotted curve on Figure 8.12. It shows that multiple nanorods in aqueous solution have width at half maximum of approximately 90 nm. This line is significantly wider than the line one would get from either T-matrix calculations or the dipole approximation. The solid line in Figure 8.12 shows the plasmon spectral line calculated using the T-matrix for nanorods with a length and width of 48.9 and 16.4 nm, respectively. These are the mean values of the sizes of the multiple nanorods in the aqueous solution. The theoretical line is also centered at 700 nm but has width of approximately 30 nm. The ensemble spectrum is three times broader than the single particle spectrum.

The CLASS microscope described above is at present uniquely capable of performing single nanoparticle measurements. To determine experimentally that individual gold nanorods indeed exhibit narrow spectral lines, single gold nanorods were selected and their scattering spectra measured using the CLASS microscope.

The nanorods were synthesized in a two-step procedure adapted from Jana et al. (2001). Sizes of 404 nanorods were measured from TEM images. The average length and width of the nanorods and standard deviations were found to be 48.9 ± 5.0 nm

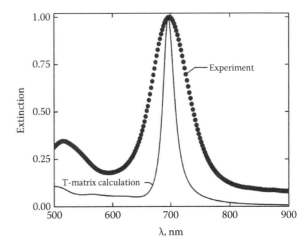

FIGURE 8.12 Optical properties of an ensemble of gold nanorods. Normalized extinction of the sample of gold nanorods in aqueous solution. Dots, experiment; dashed line, T-matrix calculation for a single size nanorod with length and width of 48.9 and 16.4 nm, respectively. (Adapted from Qiu, L. et al., *Appl. Phys. Lett.*, 93, 153106, 2008. With permission.)

and 16.4 ± 2.1 nm, respectively, with an average aspect ratio of 3.0 ± 0.4.

The concentration of gold nanorods used was approximately one nanorod per $100 \mu m^3$ of solvent. The dimensions of the confocal volume have a weak wavelength dependence and were measured to be $0.5 \mu m$ in the lateral direction and $2 \mu m$ in the longitudinal direction at 700 nm, which gives a probability of about 0.5% to find a particle in the confocal volume. Thus, it is unlikely that more than one nanorod is present in the confocal volume. To locate individual nanorods, a 64 by 64 raster scan was performed with $0.5 \mu m$ steps. The integrated spectral signal was monitored from 650 to 750 nm, and, when a sudden jump in the magnitude of a signal was observed, it was clear that a nanorod is present in the confocal volume. Then, a complete spectrum for this particle was collected.

Scattering spectra from nine individual gold nanorods, all of which had a linewidth of approximately 30 nm, were measured (Qiu et al. 2008). The experimental data from one of these spectra are shown in Figure 8.13. These measurements were compared with numerical calculations, which use the complex refractive index of gold (Johnson and Christy 1972), and various values of the phenomenological A-parameter correction (Kreibig and Vollmer 1995) were used to account for finite size and interface effects. The curve for $A = 0.13$ is the best fit for measurements made on eight different nanorods. This agrees very well with an A-parameter calculated using a quantum mechanical jellium model. Thus, using the CLASS microscope, we have detected the plasmon scattering spectra of single gold nanorods (Qiu et al. 2007, 2008). From these measurements, one can draw the conclusion that single gold nanorods exhibit a scattering line significantly narrower than the lines routinely observed in experiments that involve multiple nanorods. Narrow, easily tunable

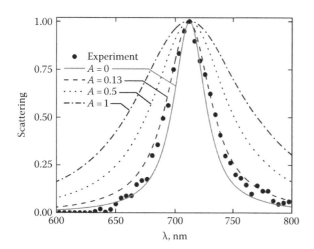

FIGURE 8.13 Normalized scattering spectrum for a single gold nanorod. Dots, CLASS measurements. Other lines are T-matrix calculations for a nanorod with an aspect ratio of 3.25 and a diameter of 16.2 nm and various A values. Solid line is for the natural linewidth, $A = 0$. Also included are lines for $A = 0.5$ and $A = 1$. The curve for $A = 0.13$ is the best fit for measurements made on eight different nanorods. (Adapted from Qiu, L. et al., *Appl. Phys. Lett.*, 93, 153106, 2008. With permission.)

spectra would allow several biochemical species to be imaged simultaneously with molecular markers, which employ gold nanorods of several different controlled aspect ratios as labels. These markers could be used for cellular microscopic imaging where even a single nanorod could be detected. Minimizing the number of nanoparticles should reduce possible damage to a living cell. For optical imaging of tumors, multiple gold nanorods with a narrow aspect ratio distribution might be used. A possible technique for obtaining a narrow aspect ratio distribution might employ devices already developed for cell sorting. These would use the position of the narrow plasmon spectral line for particle discrimination.

8.5 Conclusion

One of the major hurdles any medical technology has to overcome on its way to the clinic is convincing practicing physicians and regulators that it can deliver on its promises. Often, emerging optical diagnostic techniques promised that they would be capable of providing optically based diagnoses of early cancer and precancer, but the implementation has been elusive. Other successful "seeing through the human body" techniques such as magnetic resonance imaging and spectroscopy, x-ray computed tomography, and ultrasound do not promise direct diagnostic information. What they promise and what they deliver very successfully are images from which the physician has to draw appropriate conclusions. The same is true for optical technologies, which have already made their way to clinic, such as a pulse oximeter or an optical bilirubin monitor, which provide the physician with important information but not with a conclusive diagnosis.

In this regard, the light-scattering spectroscopic-based techniques discussed here, which assess epithelial cell morphology in vivo, provide information for performing in vivo real-time pathology and are a step closer to providing the diagnosis itself. Thus, the technology based on LSS recently developed for the detection of early cancer could be of great clinical importance. Such technology can probe and analyze the epithelial surfaces in real time, targeting cell morphology on micron and submicron scales. At the same time, especially because tissue removal is not required, LSS-based technology can be used to examine extended tissue surfaces and thereby provide better guidance for biopsy placement. Detection with LSS can lead to early treatment and prevent the development of cancer in various organs. The clinical light-scattering spectroscopic imaging instruments can thus be used in practical medicine during routine videoscopic/endoscopic procedures in various organs to detect early precancerous changes in vivo in real time. The patient referred to in Section 8.3.6 whose high-grade dysplasia was caught by the EPSS instrument and which would have been missed otherwise is a forerunner of the possibility of saving many lives using LSS.

Another advantage of LSS-based technology is its ability to track changes in morphology in cells. Unlike biopsy, the assessed cells remain intact and can be monitored for future changes. This may enable researchers to perform progressive studies and discover the conditions, which lead to the development of cancer, and other diseases, which involve pathological changes in tissue structure.

Acknowledgment

This study was supported by National Institutes of Health grants EB003472 and RR017361 and National Science Foundation grant BES0116833.

References

Backman, V., Gurjar, R., Badizadegan, K. R. et al. 1999. Polarized light scattering spectroscopy for quantitative measurement of epithelial cellular structures in situ. *IEEE J. Sel. Top. Quantum Electron.* 5: 1019–1027.

Backman, V., Wallace, M. B., Perelman, L. T. et al. 2000. Detection of preinvasive cancer cells. *Nature* 406: 35–36.

Beauvoit, B., Kitai, T., and Chance, B. 1994. Contribution of the mitochondrial compartment to the optical-properties of the rat-liver—A theoretical and practical approach. *Biophys. J.* 67: 2501–2510.

Beuthan, J., Minet, O., Helfmann, J. et al. 1996. The spatial variation of the refractive index in biological cells. *Phys. Med. Biol.* 41: 369–382.

Bigio, I. J. and Mourant, J. R. 1997. Ultraviolet and visible spectroscopies for tissue diagnostics: Fluorescence spectroscopy and elastic-scattering spectroscopy. *Phys. Med. Biol.* 42: 803–814.

Bohren, C. F. and Huffman, D. R. 1983. *Absorption and Scattering of Light by Small Particles.* New York: Wiley.

Brunstin, A. and Mullaney, P. F. 1974. Differential light-scattering from spherical mammalian-cells. *Biophys. J.* 14: 439–453.

Campbell, I. D. and Dwek, R. A. 1984. *Biological Spectroscopy.* Menlo Park, CA: Benjamin/Cummings Publishing Company.

Cotran, R. S., Robbins, S. L., and Kumar, V. 1994. *Robbins Pathological Basis of Disease.* Philadelphia, PA: W. B. Saunders Company.

Craig, I. J. D. and Brown, J. C. 1986. *Inverse Problems in Astronomy: A Guide to Inversion Strategies for Remotely Sensed Data.* London: A. Hilger.

Drezek, R., Dunn, A., and Richards-Kortum, R. 1999. Light scattering from cells: Finite-difference time-domain simulations and goniometric measurements. *Appl. Opt.* 38: 3651–3661.

Durr, N. J., Larson, T., Smith, D. K. et al. 2007. Two-photon luminescence imaging of cancer cells using molecularly targeted gold nanorods. *Nano Lett.* 7: 941–945.

Fang, H., Ollero, M., Vitkin, E. et al. 2003. Noninvasive sizing of subcellular organelles with light scattering spectroscopy. *IEEE J. Sel. Top. Quantum Electron.* 9: 267–276.

Fang, H., Qiu, L., Vitkin, E. et al. 2007. Confocal light absorption and scattering spectroscopic microscopy. *Appl. Opt.* 46: 1760–1769.

Georgakoudi, I., Jacobson, B. C., Van Dam, J. et al. 2001. Fluorescence, reflectance and light scattering spectroscopies for evaluating dysplasia in patients with Barrett's esophagus. *Gastroentorolgy* 120: 1620–1629.

Hammer, M., Schweitzer, D., Michel, B. et al. 1998. Single scattering by red blood cells. *Appl. Opt.* 37: 7410–7418.

Itzkan, I., Qiu, L., Fang, H. et al. 2007. Confocal light absorption and scattering spectroscopic microscopy monitors organelles in live cells with no exogenous labels. *Proc. Natl. Acad. Sci. USA* 104: 17255–17260.

Jacques, S. L., Roman, J. R. and Lee, K. 2000. Imaging superficial tissues with polarized light. *Las. Surg. Med.* 26: 119–129.

Jana, N. R., Gearheart, L., and Murphy, C. J. 2001. Wet chemical synthesis of high aspect ratio cylindrical gold nanorods. *J. Phys. Chem. B* 105: 4065–4067.

Johnson, P. B. and Christy, R. W. 1972. Optical-constants of noblemetals. *Phys. Rev. B* 6: 4370–4379.

Kreibig, U. and Vollmer, M. 1995. *Optical Properties of Metal Clusters.* Berlin, Germany/New York: Springer-Verlag.

Mie, G. 1908. Beiträge zur optik trüber medien, speziell kolloidaler metallösungen. *Ann. Phys.* 330: 377–445.

Mourant, J. R., Bigio, I. J., Boyer, J. et al. 1995. Spectroscopic diagnosis of bladder cancer with elastic light scattering. *Lasers Surg. Med.* 17: 350–357.

Mourant, J. R., Freyer, J. P., Hielscher, A. H. et al. 1998. Mechanisms of light scattering from biological cells relevant to noninvasive optical-tissue diagnostics. *Appl. Opt.* 37: 3586–3593.

Perelman, L. T. and Backman, V. 2002. Light scattering spectroscopy of epithelial tissues: Principles and applications. In: *Handbook on Optical Biomedical Diagnostics*, ed. V. V. Tuchin. Bellingham, WA: SPIE Press, pp. 675–724.

Perelman, L. T., Backman, V., Wallace, M. et al. 1998. Observation of periodic fine structure in reflectance from biological tissue: A new technique for measuring nuclear size distribution. *Phys. Rev. Lett.* 80: 627–630.

Qiu, L., Chuttani, R., Zhang, S. et al. 2009. Diagnostic imaging of esophageal epithelium with clinical polarized LSS endoscopic scanning instrument. *IEEE Proc. Eng. Med. Biol. Soc.* 1997–2000.

Qiu, L., Larson, T. A., Smith, D. K. et al. 2007. Single gold nanorod detection using confocal light absorption and scattering spectroscopy. *IEEE J. Sel. Top. Quantum Electron.* 13: 1730–1738.

Qiu, L., Larson, T. A., Smith, D. et al. 2008. Observation of plasmon line broadening in single gold nanorods. *Appl. Phys. Lett.* 93: 153106.

Qiu, L., Pleskow, D. K., Chuttani, R. et al. 2010. Multispectral scanning during endoscopy guides biopsy of dysplasia in Barrett's esophagus. *Nat. Med.* 16: 603–606.

Rayleigh (Strutt, J. W.) 1871. On the light from the sky, its polarization and color. *Phil. Mag.* 41: 107–120, 274–279.

Scalettar, B. A., Swedlow, J. R., Sedat, J. W., and Agard, D. A. 1996. Dispersion, aberration and deconvolution in multi-wavelength fluorescence images. *J. Microsc.* 182: 50–60.

Sloot, P. M. A., Hoekstra, A. G., and Figdor, C. G. 1988. Osmotic response of lymphocytes measured by means of forward light-scattering—Theoretical considerations. *Cytometry* 9: 636–641.

Sokolov, K., Drezek, R., Gossage, K., and Richards-Kortum, R. 1999. Reflectance spectroscopy with polarized light: Is it sensitive to cellular and nuclear morphology. *Opt. Exp.* 5: 302–317.

van de Hulst, H. C. 1957. *Light Scattering by Small Particles*. New York: Wiley.

Wallace, M., Perelman, L. T., Backman, V. et al. 2000. Endoscopic detection of dysplasia in patients with Barrett's esophagus using light scattering spectroscopy. *Gastroenterology* 119: 677–682.

Watson, J. V. 1991. *Introduction to Flow Cytometry*. Cambridge, MA: Cambridge University Press.

Wax, A., Pyhtila, J. W., Graf, R. N. et al. 2005. Prospective grading of neoplastic change in rat esophagus epithelium using angle-resolved low-coherence interferometry. *J. Biomed. Opt.* 10: 051604-1–051604–10.

Wilson, T. and Carlini, A. R. 1987. Size of the detector in confocal imaging systems. *Opt. Lett.* 12: 227–229.

Yu, C. C., Lau, C., Tunnell, J. W. et al. 2006. Assessing epithelial cell nuclear morphology by using azimuthal light scattering spectroscopy. *Opt. Lett.* 31: 3119–3121.

Zonios, G., Perelman, L. T., Backman, V. et al. 1999. Diffuse reflectance spectroscopy of human adenomatous colon polyps in vivo. *Appl. Opt.* 38: 6628–6637.

9

Broadband Diffuse Optical Spectroscopic Imaging

Bruce J. Tromberg
University of California, Irvine

Albert E. Cerussi
University of California, Irvine

So-Hyun Chung
University of California, Irvine

Wendy Tanamai
University of California, Irvine

Amanda Durkin
University of California, Irvine

9.1 Introduction 181
9.2 Background...181
9.3 Presentation of State of the Art 182
DOS Instrumentation and Theoretical Framework • Calculation of Tissue Chromophore Concentration • Calculation of Tissue Bound Water Content • Formation of Diffuse Optical Spectroscopic Images
9.4 Critical Discussion 185
Breast Cancer Detection • Monitoring the Effects of Chemotherapy in Breast Tumors
9.5 Summary...189
9.6 Future Perspective 189
Development of Advanced Spectroscopic Analysis Tools • Co-Registration with Conventional Radiologic Methods • Improve Spatial Localization • Utilization of Molecular-Targeted Exogenous Contrast
Acknowledgments...190
References...190

9.1 Introduction

Diffuse optical spectroscopy (DOS) employs model-based photon-migration methods to quantitatively separate light absorption spectra from scattering spectra in multiple scattering thick tissues. Diffuse optical spectroscopic imaging (DOSI) extends this technology to an imaging format in order to map spectra from discrete regions of tissue. Bandwidth is achieved by combining time-independent spectroscopy with frequency-domain photon migration (FDPM), reducing dependence on the availability of specific diode lasers while expanding spectral content to hundreds of near-infrared (NIR) wavelengths (~600–1000 nm). This combined approach improves the accuracy of absorption and scattering spectra calculations and allows simultaneous characterization of absorption contributions from several endogenous and exogenous chromophores (Bevilacqua et al. 2000).

In this chapter, we review the development of DOSI instrumentation as well as techniques for *forming quantitative image maps of the tissue* concentration of oxy-hemoglobin (ctO$_2$Hb), deoxy-hemoglobin (ctHHb), water (ctH$_2$O), and bulk lipid. Special emphasis is placed on the unique functional information that can be obtained from high-resolution DOSI spectra. This includes hemoglobin parameters related to microvascular function, such as tissue oxygen delivery, oxygen consumption, and blood volume, as well as water macromolecular-binding state determined from the shape and position of the ~975 nm NIR water absorption feature. Although conventional NIR spectroscopy (NIRS) has been widely used in brain, muscle, and breast characterization, this chapter highlights breast cancer imaging in human subjects where the majority of our DOSI studies have been applied.

9.2 Background

Optical methods provide a single scalable platform that can be used to image structure and function in tissues ranging from cells to organs. Imaging performance, including contrast, resolution, and depth, can be controlled by manipulating spatial, spectral, and temporal properties of the light source. Specific interaction mechanisms between light and tissue on a microscopic scale determine the biologic origins of contrast and ultimately limit the performance of macroscopic diffuse optical imaging.

At depths from approximately 1–2 mm and greater, multiple scattering dominates NIR light propagation in tissue. Under these conditions, optical-phase relationships become randomized, and coherent waves are not easily detectable. In this "diffusion regime," light transport can be modeled as a diffusive process where photons behave as stochastic particles that move in proportion to a gradient, much like the bulk transport of molecules or heat. Quantitative tissue measurements can be obtained by separating light absorption from scattering using spatially or temporally modulated photon-migration technologies. The underlying physical principle of these methods is based on the fact that light absorption events, which are a consequence of molecular interactions, are less likely to occur than light scattering. This is due to the fact that typical tissue NIR absorption lengths (i.e., mean-free paths) are ~10 cm, while scattering lengths are ~20 μm.

We first introduced broadband DOS as an alternative to conventional time-independent (continuous-wave, CW) NIRS by combining frequency-domain and broadband CW methods. Our goal was to quantitatively separate bulk absorption and reduced scattering *spectra* in centimeter-thick tissues rather than restrict tissue optical property measurements to a few discrete wavelengths (Bevilacqua et al. 2000, Jakubowski et al. 2009). The combination of broadband DOS with spatial scanning allows one to perform DOSI. DOSI can be used to form quantitative image maps of local tissue concentrations of ctO_2Hb, $ctHHb$, ctH_2O, and bulk lipid. DOSI also maps the wavelength-dependence of the reduced tissue scattering parameter $[\mu_s'(\lambda)]$, which is related to intracellular structures such as mitochondria (Beauvoit and Chance 1998) and nuclei (Mourant et al. 2000) and extracellular components such as collagen (Cerussi et al. 2002, Pogue et al. 2004, Spinelli et al. 2004).

Although images are formed, DOSI is inherently a low-resolution technique because of multiple light scattering. Diffuse photons probe a large sample volume, providing macroscopically averaged tissue absorption and scattering properties at depths up to several centimeters. Consequently, the resolution of diffuse optical imaging methods is on the order of a few transport scattering lengths (~5 mm to 1 cm; Gandjbakhche et al. 1994, Boas et al. 1997, Pogue et al. 2006). However, the potential limitations of DOSI in localizing small lesions are not important in characterizing tissue function. This is due to the fact that diffuse optical methods are inherently sensitive to endogenous biochemical composition and tissue pathological state (Cerussi et al. 2007).

For example, DOS measurements of hemoglobin reflect the status of tissue microvasculature (Liu et al. 1995) and may be a direct measure of angiogenesis (Tromberg et al. 2000, Pogue et al. 2001, Zhu et al. 2005). Changes in tumor hemoglobin concentration have been measured for chemotherapy agents known to possess antiangiogenic capabilities (Cerussi et al. 2007). Tissue concentrations of water and lipids are substantially different in breast cancer versus normal tissue (Taroni et al. 2004, Cerussi et al. 2006), and we have shown that DOS can reveal unique cancer-specific absorption spectral signatures not found in normal breast (Kukreti et al. 2007). Changes in breast tissue water and lipids have been associated with long-term alterations during neoadjuvant chemotherapy (NAC) as seen with both DOS (Jakubowski et al. 2004) and magnetic resonance imaging (MRI; Jagannathan et al. 1998). DOS-measured tumor water concentration and water-binding state scale with the Nottingham Bloom–Richardson histopathology score and may be proportional to tissue cellularity and extracellular matrix (ECM) composition (Cerussi et al. 2006, Chung et al. 2008, Chung 2009). Finally, the extent of water macromolecular binding determined from DOS spectral content is similar to MRI measures of water

apparent diffusion coefficient, and water spectral variations can be used to measure deep tissue temperature (Chung et al. 2008, Chung 2009).

A practical advantage of DOSI is that it can be administered frequently at the bedside in settings such a doctor's office or medical center. DOSI is relatively simple to perform and interpret compared to technologies such as mammography, MRI, and positron emission tomography (PET). Because of its size and portability, DOSI is a low-barrier-to-access technology, potentially creating new opportunities for patients to receive personalized medical treatment, and for physicians to gain new insight into disease progression and therapy.

9.3 Presentation of State of the Art

9.3.1 DOS Instrumentation and Theoretical Framework

DOS combines multifrequency FDPM and CW NIRS technologies. Detailed descriptions are reported elsewhere (Bevilacqua et al. 2000, Cerussi et al. 2006, Tromberg et al. 2008, Jakubowski et al. 2009), but a summary of the main features is provided here and in Figure 9.1.

The FDPM component employs diode lasers at several (e.g., six) wavelengths (typically 658, 682, 785, 810, 830, and 850 nm). A steady-state current source is mixed with RF power provided by a network analyzer (Agilent, 8753E) in a bias network. Each source is intensity-modulated by the ~20 dBm swept RF output at frequencies that range between 50 and 500 MHz. The phase shift and amplitude drop of the modulated wave are measured in the test material (e.g., tissue) compared to a standard reference material of known optical properties (e.g., silicone tissue phantom), using a high-bandwidth avalanche photodiode (APD) detector (module C5658 with S-6045-03 silicon APD, Hamamatsu, Japan) that transmits signals to the network analyzer. The average optical power of the diode lasers is about 10–30 mW at the measurement site.

For the CW spectroscopy component, a high-intensity tungsten–halogen source (Micropak HL2000-HP, Ocean Optics, Dunedin, Florida) and an autocalibrated spectrometer (B&W Tek, Newark, DE, model 611) are employed to obtain spectra with absolute wavelength accuracy from 650 to 1010 nm. More than 1000 wavelengths are obtained in this range with ~0.4 nm resolution. The instrument response for CW spectroscopy is calibrated using a spectraflect-coated reflectance standard.

In our current configuration, a handpiece contains a fiber bundle coupled to each diode laser and a fiber bundle coupled to the broadband light source (Figure 9.1). The source fibers are placed approximately 3 cm from the APD detector that measures the FDPM signal component, while a 1 mm core diameter plastic-clad silica fiber is used to collect and transmit broadband signals to the spectrometer. Source and detector fibers are arranged in a low-angle crossing pattern in order to optimize field of view overlap. The entire handpiece is placed in contact with the tissue for both FDPM and CW signals. Data are acquired by moving

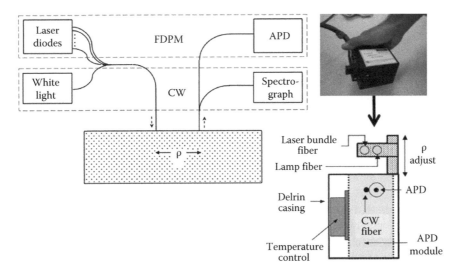

FIGURE 9.1 *Left:* Schematic of DOSI instrument illustrating combination of FDPM and CW subsystems. *Right:* DOSI handpiece with APD detector for FDPM measurements and fiber coupled to spectrometer for CW measurements. Source-detector separation, ρ, can be varied.

the handpiece to different tissue locations. A minimum of three FDPM wavelengths is sampled by serially applying a predetermined DC bias above lasing threshold and sweeping through RF. This process is repeated for each diode laser and interwoven with CW spectra acquisition. Modulated RF signals are detected by the APD, while steady-state light is directed by an optical fiber to the spectrometer. The source-detector separation, ρ, can range from 18 to 40 mm, with typical breast measurements fixed at 28 mm. Total acquisition time is approximately 3–6 s per location.

DOS employs computational models for quantifying absorption and scattering spectra from tissue. In order to calculate μ_s' and μ_a, the nonlinear frequency-dependent FDPM phase and amplitude data are fit to a P1 diffusion approximation with semi-infinite boundary conditions (Haskell et al. 1994, Fishkin et al. 1996). Theoretical reflectance, *R, is calculated using the FDPM-determined* μ_s' and μ_a values, and the measured CW reflectance spectrum, R_0, is calibrated using *R*. The FDPM-generated wavelength dependence of scattering, $\mu_s'(\lambda)$, is fit to a scattering power law, $A\lambda^{-b}$, based on Mie theory with a wide distribution of scatter centers (Graaff et al. 1992, Mourant et al. 1997). This allows us to obtain a continuous scattering spectrum and the structural contrast parameters *A* (scatter prefactor) and *b* (scatter power). The calibrated R_0 is deconvolved with $\mu_s'(\lambda)$ in order to obtain a continuous μ_a spectrum (Jakubowski et al. 2009).

9.3.2 Calculation of Tissue Chromophore Concentration

Measurements of absorption coefficients at each wavelength, combined with the wavelength-dependent molar extinction (ε) values of each chromophore, enable us to calculate tissue chromophore concentration (*C*) using Equation 9.1:

$$\vec{\mu}_a(\lambda) = 2.303 \left[\varepsilon(\lambda)\right]\vec{C} \qquad (9.1)$$

where
   the square brackets [] represent a matrix
   the upper arrows indicate a vector
   λ is the wavelength

We assume that the measured absorption spectra are linear combinations of the component chromophore spectra, and we solve the equation by adjusting μ_a and *C* into the appropriate vectors, ε into a matrix, and solving the matrix equation. No *a priori* knowledge about chromophore concentration is assumed (other than standard non-negative constraints).

These calculations quantify the tissue concentration (μM) of oxyhemoglobin (ctO$_2$Hb), deoxyhemoglobin (ctHHb), total hemoglobin (ctTHb = ctO$_2$Hb + ctHHb), and % oxygen saturation (stO$_2$ = 100 × ctO$_2$Hb/ctTHb) using published hemoglobin extinction coefficients (Zijlistra et al. 2000). Percent ctH$_2$O and bulk lipid (ctLipid) are also determined from tissue absorption spectra using pure water and lipid spectra at 100% concentration, which corresponds to 55.6 M and 0.9 g mL^{-1}, respectively (Cubeddu et al. 2000, Cerussi et al. 2002, 2007). Water extinction coefficients are obtained by measuring distilled water in a cuvette using a spectrophotometer (Beckman DU 650) at various temperatures in order to account for possible temperature effects (Chung 2009). Lipid extinction coefficient values are obtained from mammalian fat spectra (van Veen et al. 2005).

9.3.3 Calculation of Tissue Bound Water Content

In the presence of water binding to macromolecules such as proteins, the water absorption peak at 975 nm undergoes both broadening and red shifting (Pauling 1960, Pimentel and McClelland 1960, Bellamy 1968, Eisenberg and Kauzmann 1969, Chung et al. 2008). These spectral changes appear as a consequence of variations in the relative contributions of harmonic overtones from fundamental O–H vibrations at 3.05 and 2.87 μm (Eisenberg and Kauzmann 1969).

Tissue absorption spectra are postprocessed to measure bound water state using a three-step process (Figure 9.2a–c). First, a tissue water spectrum (μ_a, tissue water) is obtained by eliminating the contributions of other chromophores by subtracting the extinction coefficient spectra of oxy- and deoxy-hemoglobin and bulk lipid multiplied by their concentrations from the overall tissue absorption spectrum. Second, the residual between the tissue (μ_a, tissue water) and pure water spectra (μ_a, pure water) is calculated by subtracting the pure water spectrum (at physiological temperature, 36°C for breast tissues (Jeffrey et al. 1999)) from a normalized tissue water spectrum in the wavelength range of 935–998 nm. Finally, in order to represent the residual using an index, the absolute values of the differences are combined and divided by the number of points in the sum to form a bound water index (BWI):

$$BWI = \frac{\sum_i \left| \frac{\mu_{a,\text{tissue water}}(\lambda_i)}{ctH_2O} - \mu_{a,\text{pure water}}(\lambda_i) \right|}{N} \times 1000 \quad (9.2)$$

where

λ_i is the ith wavelength (935 nm ≤ λ_i ≤ 998 nm)
ctH_2O is the water concentration described above
N is the number of wavelength points in the sum

Note that, above 935 nm, the tissue water absorption spectrum is linearly shifted to higher wavelengths relative to the pure water spectrum (Chung et al. 2008).

BWI was validated using bound water phantoms that were fabricated with varying amounts of gelatin. The DOSI-measured BWI index was compared to gold standard measurements based on magnetic resonance spectroscopy (MRS) and conductivity. Our results in a series of gelatin phantoms confirm that BWI signals are proportional to bound water state (Chung et al. 2008), Because temperature impacts the pure water spectrum, we calculated BWI over the entire expected physiological range of ~34°C to 38°C. These simulations show that under these conditions, BWI varies by a small amount; up to 4% (Chung et al. 2008). As a result, we have selected 36°C (Jeffrey et al. 1999) for all in vivo BWI calculations.

9.3.4 Formation of Diffuse Optical Spectroscopic Images

Data in the clinical setting are acquired by placing the hand-held probe on the tissue and scanning over the desired area (Figure 9.3). In the case of breast measurements, a large scan area is used to cover regions of the breast both with lesion and surrounding normal tissues. The nipple is taken to be the origin of a 2-D grid with 10 mm spacing. Each point measurement on the grid forms an element of the spectroscopic image and 2-D nearest-neighbor cubic splines are used to round out the discrete shapes. All images are scaled using the maximum and minimum values. A four-stage color bar is used with RGB values of (0, 0, 0), (0, 0, 255), (0, 255, 255), and (255, 255, 255) to display all maps. Figure 9.3 shows an example of a deoxy-hemoglobin image from a 31-year-old premenopausal subject with a 37 × 18 × 20 mm fibroadenoma (Tanamai et al. 2009).

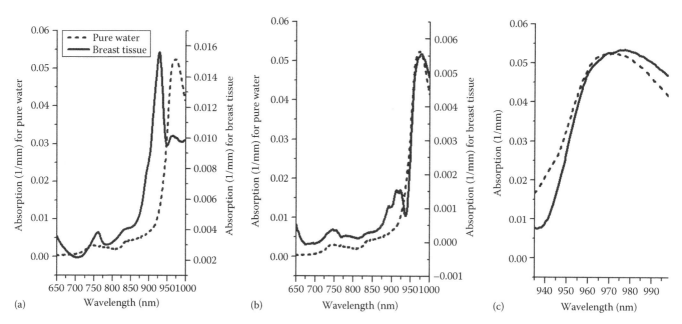

FIGURE 9.2 (a) In vivo tissue absorption spectrum (solid line) from normal breast tissue. (b) Tissue water spectrum after subtracting other tissue components' spectra (solid line). (c) Normalized tissue water spectrum at 935–998 nm (solid line). The pure water spectrum at 36°C is shown in each panel (a, b, and c, dashed lines) for comparison. Tissue bound water effect is clearly seen in (c). (Adapted from Chung, S.H. et al., *Phys. Med. Biol.*, 53, 6713, 2008. With permission.)

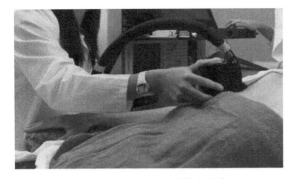

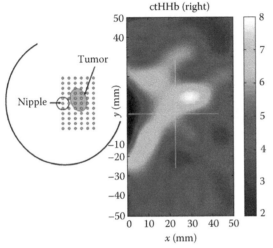

FIGURE 9.3 DOSI handpiece is scanned in a standard grid pattern on patient. A deoxy-hemoglobin image is shown from a 31-year-old fibroadenoma patient (From Tanamai, W. et al., *J. Biomed. Opt.*, 14, 014024, 2009.)

9.4 Critical Discussion

9.4.1 Breast Cancer Detection

Controversial guidelines recently issued by the U.S. Preventative Services Task Force (USPSTF) conclude that the risk of false positives and complications from biopsies and similar invasive procedures is too high for mammography to be used routinely in pre- and perimenopausal women, i.e., up to age 50 (USPSTF 2009). This is due to the fact that the sensitivity of mammography is poor: approximately 60% for younger women and women with dense breast tissue. Although the risk of breast cancer is low in this population, women in their forties account for at least a quarter of breast cancer diagnoses and up to 17% of breast cancer deaths each year (USPSTF 2009). Additional complications arise due to the fact that in premenopausal women, mammographic density and false negative rates are greater during the luteal versus follicular phase of the menstrual cycle (White et al. 1998). Similarly, the use of hormone replacement therapy in postmenopausal women is known to increase mammographic density and has been shown to impede the effectiveness of mammographic screening (Laya et al. 1996, Litherland et al. 1999). It is estimated that, up to 20% of all breast cancers, roughly, 40,000 cases per year in the United States are not discovered by screening mammography (National Cancer Institute).

The intense debate surrounding mammography recommendations underscores the need for new technologies that can detect breast cancer and distinguish between malignant and benign tumors in younger women and women with mammographically dense tissue. Because DOSI measures several functional contrast elements related to cellular metabolism, angiogenesis, and ECM, optical methods have potential for use in this population.

In order to assess the performance of DOSI and evaluate optical contrast elements that are important to breast cancer, we have recently characterized large, palpable tumors in a population of 58 human subjects (Cerussi et al. 2006). Figure 9.4 shows an in vivo DOSI functional map of a $1.3 \times 0.9 \times 1.0$-cm infiltrating ducal carcinoma tumor in a 37-year-old subject. The subject provided informed written consent as per the guidelines of an institution-approved protocol (UCI no. 02-2306). The DOSI map was generated by moving the handheld probe over a 4×7 cm grid in 1-cm steps. The origin (0, 0) represents the nipple, and nongrid points were interpolated using 2D nearest neighbors. We have mapped the tissue optical index (TOI), which is a composite contrast function defined as $TOI = (ctHHb \times ctH_2O)/ctLipid$, which has proven useful in discriminating between normal and stage II/III malignant breast lesions (Cerussi et al. 2006). Three specific regions are identified within the map: a normal region far from the lesion site (C), a region directly over the lesion site (B), and a second suspicious location near the primary lesion site (A). Broadband absorption and scattering spectra are shown from each region to the right.

Regions (B) and (A) show statistically verified alterations in tissue physiology consistent with malignancy (Cerussi et al. 2006). Below 850 nm, increased hemoglobin absorption (both oxy and deoxy) is observed in the tumor tissues relative to normal tissues (Franceschini et al. 1997, Tromberg et al. 2000, Pogue et al. 2001, Taroni et al. 2004, Chance et al. 2005). Near 980 nm, tumors display greater absorption due to increased water (i.e., OH bonds) as well as spectral shifts suggestive of changes in the water-binding state (Cerussi et al. 2006, Chung et al. 2008). The distinctive peak in the normal spectrum at 930 nm is due to high bulk lipid concentration, which is lower in tumor tissues. The scattering amplitude (i.e., prefactor, A) and the wavelength dependence of scattering (i.e., scatter power, b) differ between tumor and normal tissues, suggesting that malignancy changes the density and size of tissue-scattering centers (Mourant et al. 1997, Nilsson et al. 1998, Carpenter et al. 2007).

In addition to hemoglobin contrast, we have recently shown that spectral alterations in the water and lipid "signature regions" can be used to distinguish between tumor and normal tissues as well as malignant and benign tumors (Kukreti et al. 2007). This is likely due to differences in the biochemical disposition of these components in malignant tumors, such as the association of water with macromolecules and changes in lipid metabolism. In order to develop a more detailed understanding of these phenomena, we have examined bound water states in 18 breast cancer patients (Chung et al. 2008).

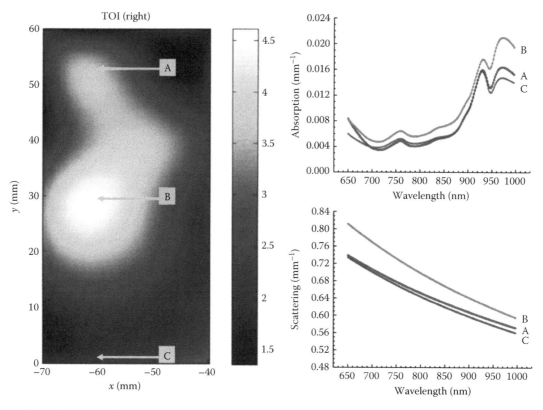

FIGURE 9.4 *Left:* TOI (TOI = ctHHb × ctH$_2$O)/ctLipid) map of 13-mm diameter infiltrating ducal carcinoma tumor in a 37-year-old subject. *Right:* NIR absorption and reduced scattering spectra obtained by DOSI shown for specific regions of the image.

Figure 9.5 compares BWI values from malignant (BWI = 1.96 ± 0.3) and normal tissues (BWI = 2.77 ± 0.47) measured in 18 subjects. Malignant tumors had significantly lower BWI than normal tissues ($p < 0.0001$), suggesting that tumors appear to have more free than bound water. The relationship between bulk water concentration and BWI of normal and malignant tissues was also studied. In general, BWI is inversely proportional to tissue water content. In normal tissues, the water content range is relatively narrow (10%–28%), and the BWI spans between 2.3 and 4.3. However, in cancer tissues, the bulk water content values are high and widely spread (22%–73%) while the BWI is clustered between 1.4 and 2.4.

In normal well-differentiated tissues, we expect a relatively narrow range of BWI and water content values due to the

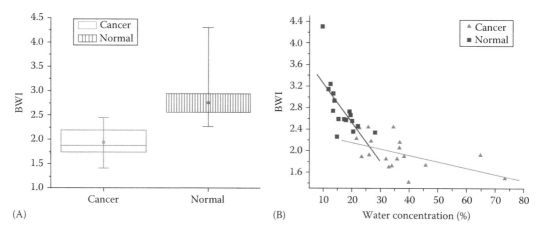

FIGURE 9.5 (A) Box plots of BWI of malignant (1.96 ± 0.3) and normal breast tissues (2.77 ± 0.47) for 18 subjects ($p < 0.0001$). (B) BWI versus bulk water concentration of normal (blue squares) and malignant (red triangles) breast tissue. Both bulk water concentration and BWI values were acquired from the same spatial locations. (Adapted from Chung, S.H. et al., *Phys. Med. Biol.*, 53, 6713, 2008. With permission.)

absence of significant disorder and the expectation that normal tissues have an intrinsic maximum water content and binding capacity. With the appearance of cancer and loss of differentiation, water content increases dramatically, and the tissue structural diversity increases. This is evident from the broad water content range and reduced BWI values seen in malignant tumors.

9.4.2 Monitoring the Effects of Chemotherapy in Breast Tumors

NAC (i.e., presurgical) is increasingly prescribed for patients with locally advanced or inoperable breast cancers in order to reduce primary tumor size, minimize metastatic impact, and improve breast tissue conservation during surgery. Locally advanced breast cancer represents 5%–20% of all newly diagnosed breast cancers in the United States, with a higher incidence in medically underserved areas (Mankoff et al. 1999). Patient prognosis depends on both initial extent of disease and tumor pathological response to NAC. Complete pathological response is an important clinical endpoint that strongly correlates with patient survival (Fisher et al. 1998, 2002). Conventional clinical assessments such as physical exam, ultrasound, and mammography have been shown to be inadequate for predicting pathological response (Mankoff et al. 1999, Yeh et al. 2005, Tardivon et al. 2006). Thus, there is considerable interest in developing new quantitative imaging methods to monitor and predict chemotherapy response, both prior to and as early as possible during the course of treatment.

A recent study evaluating palpation, mammography, ultrasound, and MRI showed 19%, 26%, 35%, and 71% agreement, respectively, with final pathological response (Yeh et al. 2005).

Anatomical changes in tumor presentation are not reliable predictors of final pathological state (Feldman et al. 1986, Helvie et al. 1996, Vinnicombe et al. 1996). Functional measurements of tumors from contrast-enhanced MRI (Warren et al. 2004, Hylton 2006), MRS (Chenevert et al. 2000, Meisamy et al. 2004), and PET (Mankoff et al. 2003, Kim et al. 2004, McDermott et al. 2006) have shown substantial improvement over conventional anatomic imaging. However, these techniques are difficult for advanced stage cancer patients (lengthy scan times, exogenous contrast), particularly if frequent measurements are desired.

We reported the first use of DOSI to track tumor response to NAC in a 54-year-old human subject with a $2.3 \times 2.5\,cm$ mass (Jakubowski et al. 2004). Several DOSI measurements were performed during a single stage, three-cycle, 68-day adriamycin/cytoxan (AC) treatment. ctTHb and tumor ctH_2O dropped by 56% and 67%, respectively. The drop in ctTHb was consistent with diminished vessel density, which is a signature of NAC response (Makris et al. 1999). Tissue bulk lipids increased by 28%, suggesting normal tissue reformation. Tumor to control water/lipid ratio dropped 4.4-fold, which was in excellent agreement with previous findings from MRS that showed similar changes (Jagannathan et al. 1998). Subsequent studies have supported these findings by direct comparison of optical imaging with MRI (Choe et al. 2005, Shah et al. 2005) and ultrasound (Zhu et al. 2005).

The sensitivity of this approach is illustrated in Figure 9.6, which shows TOI (TOI = ctHHb × ctH_2O)/lipid) images acquired prior to NAC and after the conclusion of a 20-week regimen (combination anthracyclines and herceptin). This 45-year-old patient started with a 30 mm infiltrating ductal carcinoma and treatment resulted in a complete pathological response. At the conclusion of all chemotherapy, the absorption spectrum of the

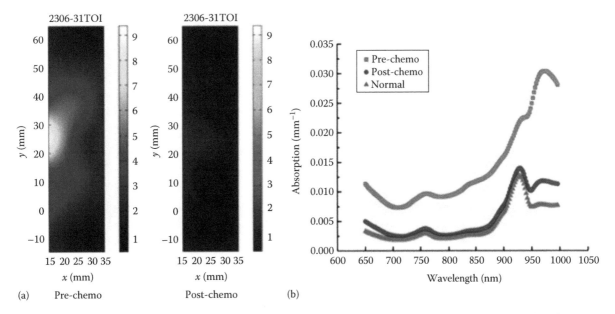

FIGURE 9.6 (a) DOSI measurements of TOI (TOI = ctHHb × ctH_2O)/lipid) in 45-year-old patient with 30-mm invasive ductal carcinoma prior to NAC and at the conclusion of therapy. (b) Changes in tumor spectra show return to normal tissue properties.

lesion was comparable to that of contralateral normal tissues (Figure 9.6b).

In an 11-patient pilot study, early DOSI measurements were able to predict final tumor pathological state (Cerussi et al. 2005, 2007). DOSI was performed within 1 week prior to chemotherapy (1.8 ± 4.5 days) and approximately 1 week after (6.5 ± 1.4 days) the start of chemotherapy. Treatment for all 11 subjects consisted of three to four cycles of doxorubicin/cyclophosphamide (A/C); nine of these patients also received an additional three to four cycles of Taxanes. Each chemotherapy cycle lasted 3 weeks. DOSI-measured tumor concentrations of ctHHb, ctO$_2$Hb, and ctH$_2$O dropped $27\% \pm 15\%$, $33\% \pm 7\%$, and $11\% \pm 15\%$, respectively, within 1 week of the first treatment for pathology-confirmed responders ($N=6$), while nonresponders ($N=5$) and normal side controls showed no significant changes (Table 9.1). All values were calculated as the ratio of the measurements, so that "no change" = unity (1). Unlike ctHHb, both groups dropped in ctO$_2$Hb: nonresponders

by 18% and responders by 33%. Changes in ctO$_2$Hb reflect changes in both local and systemic oxygen delivery, a consequence of chemotherapy-induced anemia (Jakubowski et al. 2004). These data support the idea that DOSI-measured response may provide early biochemical clues to predict final pathological outcome.

A particularly powerful feature of DOSI is the ability to monitor chemotherapy response on a frequent, even daily, basis (Cerussi et al. 2010). Figure 9.7 shows a series of DOSI measurements obtained from a 51-year-old female with an initial 60×27 mm lesion in the left upper outer quadrant of the breast. Standard biopsy prior to treatment identified the lesion as infiltrating ductal carcinoma. The lesion size after all chemotherapy (see below) was 4 mm. Residual tumor appeared on MRI and DOSI and was classified as a partial pathological response.

Three days after the initial A/C treatment, we observed a large decrease in tumor ctHHb (Figure 9.7). Tumor ctHHb levels remained lower 4 days post-treatment. However, ctHHb increased over the course of the next few days. Interestingly, we note that although there was an initial ctHHb decrease, after approximately 6 days ctHHb levels returned to near baseline levels. In the previously described 11-patient A/C study, we found that in cases where ctHHb returns to baseline within, on average, 6 days, the end pathological response was poorer than for those who had significantly decreased levels on day 6 (Cerussi et al. 2007).

The individual variability in response trajectories highlights the fact that, for the best functional view of the tissue, multiple time points may be required. The ctHHb ratio

TABLE 9.1 Tumor Physiological Changes from NAC Measured by DOSI

	ctHHb	ctO$_2$Hb	ctH$_2$O (Rel)	Lipid	SP
RESP ($N=6$)	0.73 ± 0.17	0.67 ± 0.06	0.80 ± 0.08	1.30 ± 0.3	0.88 ± 0.20
NON ($N=5$)	1.02 ± 0.05	0.82 ± 0.10	0.96 ± 0.03	1.11 ± 0.14	0.97 ± 0.20
Z	0.008^a	0.03^a	0.008^a	0.41	0.93

Source: From Cerussi, A. et al. *Proc. Natl. Acad. Sci. USA*, 104, 4014, 2007.
[a] Indicates significant result.

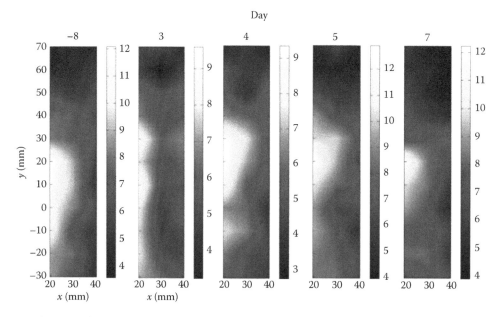

FIGURE 9.7 DOSI serial images of ctHHb in 51-year-old patient with 60×27 mm invasive ductal carcinoma taken prior to treatment (-8 days) and on days 3, 4, 5, and 7 after the A/C chemotherapy. Note each image scaled independently. The maximum ctHHb value of the tumor dropped significantly after the treatment (days 3 and 4) but quickly returned to pretreatment levels (days 5 and 7). (Adapted from Cerussi, A. et al., *Acad. Radiol.*, 17(8), 1031, 2010.)

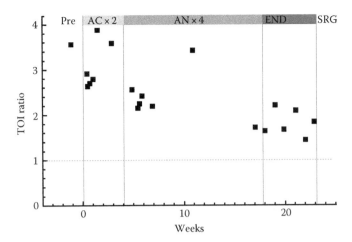

FIGURE 9.8 TOI (TOI = ctHHb × ctH$_2$O)/lipid) ratio (tumor/normal) over the entire course of treatment. PRE = before therapy; AC × 2 = two cycles of A/C chemotherapy; AN × 4 = four cycles of combination Avastin chemotherapy; END = after all chemotherapy; SRG = before final surgery. (Adapted from Cerussi, A. et al., *Acad. Radiol.*, 17(8), 1031, 2010.)

(tumor-to-normal) experienced only a modest decrease (1.52–1.3). Thus, there remained a significant amount of ctHHb at the conclusion of two cycles of A/C, which in our experience is still a strong indicator of the presence of disease (Cerussi et al. 2007). The TOI ratio (tumor-to-normal) had also dropped significantly (about 3.5–2.5), nevertheless a discernable lesion signature remained.

Figure 9.8 displays the long-term TOI measurements in this patient for multiple chemotherapy stages (expressed as a ratio of tumor-to-normal). The shaded bars on the top refer to each stage of the therapy: before all treatment (PRE), during two cycles of A/C (AC × 2), during four cycles of a combination cocktail, which included the antiangiogenic agent Avastin (Bevacizumab) (AN × 4), after all chemotherapy (END), and before final surgery (SRG). At baseline, the TOI ratio was 3.5, which is comparable to the average malignant tumor contrast values found in our last population study (Cerussi et al. 2006). At the conclusion of the A/C stage, the TOI ratio dropped significantly, but had not reached normal levels. The Avastin stage demonstrated further decreases in tumor area, as evidenced by the drop in TOI ratio to about 1.5 just prior to surgery. Interestingly, this above-normal presurgical value was confirmed by postsurgical observation: infiltrating ductal carcinoma with a 4-mm residual invasive component, a partial pathological response.

9.5 Summary

DOSI is a low-resolution functional imaging technique that provides macroscopically averaged tissue absorption and scattering properties at depths up to several centimeters. Potential limitations of DOSI in localizing small (e.g., millimeter) lesions

may not be important in certain breast cancer applications where conventional radiologic methods perform poorly. These include detecting early stage tumors (e.g., 5–10 mm) in younger women and high-risk subjects with mammographically dense breasts, and characterizing NAC response in large, palpable stages II–IV tumors. This is due to the fact that DOSI is highly sensitive to vascular and cellular physiology and metabolism as well as endogenous biochemical composition. These quantitative "functional optical biomarkers" are easy to interpret and provide objective measures to distinguish between malignant and benign tumors and predict therapeutic outcome.

The capability of DOSI to map broadband, high-resolution absorption, and scattering spectra in thick tissues is an important feature that can potentially reveal new contrast mechanisms beyond conventional NIRS oximetry. For example, the extent of bound water, as measured by the spectrally derived BWI, is a promising approach for improving breast cancer characterization. The BWI is significantly lower in malignant tumors, indicating that the water associated with breast cancers is more like free water. The origin of BWI signals is likely due to increased vascular permeability, edema, and necrosis and could be a surrogate marker of high-interstitial fluid pressures and alterations in malignant tumor ECM that promote water entrapment and swelling (Chung et al. 2008).

An important practical advantage of DOSI is that it can be used frequently at the bedside in unconventional settings such a doctor's office or medical center. DOSI is relatively simple to perform and interpret compared to technologies such as mammography, MRI, and PET. Because of its size, portability, and low cost, DOSI is a low-barrier-to-access technology, potentially creating new opportunities for patients to receive personalized treatment and for physicians to gain insight into mechanisms of cancer appearance and response to therapy.

9.6 Future Perspective

Although DOSI parameters lack the specificity of conventional gene or protein-based biomarkers, DOSI measures endogenous biochemical composition and tumor pathological state (Cerussi et al. 2007). Because DOSI is a rich source of spatial–spectral information, there are many strategies for further improvements in contrast, sensitivity and specificity. Broadly, these approaches include (1) development of advanced spectroscopic analysis tools, (2) co-registration with conventional radiologic methods, (3) improving tumor spatial localization, and (4) combining with molecular-targeted exogenous contrast agents.

9.6.1 Development of Advanced Spectroscopic Analysis Tools

Tumor absorption spectra communicate rich information regarding the tissue physiological state. We commonly utilize linear methods for quantifying biochemical components in order to characterize tumors and assess chemotherapy response. We recently demonstrated that by examining spectral shifts using a differential spectroscopy method, we were able to gain new insight into tissue molecular disposition. This significantly improved our ability to detect tumors (Kukreti et al. 2007) and distinguish between malignant and benign lesions (Kukreti et al. 2010). Thus, a major future direction is to continue to advance sophisticated differential spectroscopy approaches for cancer detection and therapeutic monitoring. These spectroscopic "biomarkers" can potentially provide information on tumor growth rate, hypoxia, angiogenic phenotype, and the status of tumor stromal and epithelial elements. This work may improve clinical performance for problems such as breast cancer detection, differential diagnosis, and predicting chemotherapy outcome. In addition, these studies could help elucidate mechanisms and possible targets of therapies and contrast agents.

9.6.2 Co-Registration with Conventional Radiologic Methods

Image fusion is potentially a powerful approach for visualizing the combination multidimensional structural and functional data. Several groups have explored the integration of DOSI with anatomic imaging methods such as MRI, ultrasound, and mammography (Azar and Intes 2008). A significant amount of work remains to advance these combination modalities into a single interactive image that can provide scientists and clinicians with more information than either approach alone. However, diffuse optical methods provide a relatively low cost and simple addition to conventional imaging platforms. By providing a convenient approach for simultaneously visualizing structural and functional information, multimodality DOSI is expected to improve clinical insight and decision making.

9.6.3 Improve Spatial Localization

The use of anatomic imaging methods in conjunction with diffuse optics can significantly improve accuracy in localizing and characterizing subsurface tumors. This is typically accomplished by using radiologic data as prior knowledge in image reconstruction. An important issue for future study involves determining the origin of DOSI contrast with respect to other methods. Sources of optical, ultrasound, x-ray, and MRI signals are different, and this can impact the accuracy of image reconstruction. We have shown that *measured* DOSI breast cancer contrast typically exceeds *predicted* values if the spatial extent of tumors is limited to dimensions determined by ultrasound (Holboke et al. 2000, Li et al. 2008). This implies that the spatial extent of optical contrast may extend beyond anatomic features. Opportunities for improved spatial localization include methods that parameterize the tumor, constraining it to a realistic physiological description based on the expected spatial extent of optical contrast. We have recently described an approach based on a Gaussian representation that effectively serves as prior knowledge. In practical terms, this can reduce the dependence on a large number of source-detector views, making the concept of spectrally rich "sparse tomography" a realistic possibility (Liu et al. 2010).

9.6.4 Utilization of Molecular-Targeted Exogenous Contrast

There is considerable activity in the optical imaging community to develop probe strategies that rely on specific ligand–receptor interactions as well as conventional blood volume localization (Licha 2002, Frangioni 2008). Although exogenous contrast agents can potentially significantly improve spatial localization and enhance functional information content, simultaneous advances in both instrumentation and probes are required. DOSI is a spectrally and temporally broadband technology that can provide quantitative data from single or multiple endogenous and exogenous contrast elements simultaneously. Thus, it is a convenient platform to simultaneously explore relationships between intrinsic biochemical composition and specific molecular pathways. In addition, DOSI broadband tissue optical properties can be used to inform quantitative analysis methods for characterizing exogenous molecular probes that require knowledge of tissue absorption and scattering at both excitation and emission wavelengths.

Acknowledgments

This work was supported by the National Institutes of Health under grants P41-RR01192 (Laser Microbeam and Medical Program; LAMMP), U54-CA105480 (Network for Translational Research in Optical Imaging; NTROI), U54CA136400, and NCI-2P30CA62203 (University of California, Irvine Cancer Center Support Grant). Additional support came from the California Breast Cancer Research Program. BLI programmatic support from the Beckman Foundation and the Air Force Research Laboratory, under agreement number FA9550-04-1-0101, is acknowledged. The authors thank Montana Compton and Drs. Rita Mehta, David Hsiang, and John Butler for their assistance as well as the patients who generously volunteered their time for these studies.

References

Azar, F. S. and Intes, X. 2008. *Translational Multimodal Optical Imaging.* Boston, MA: Artech House.

Beauvoit, B. and Chance, B. 1998. Time-resolved spectroscopy of mitochondria, cells and tissues under normal and pathological conditions. *Mol. Cell Biochem.* 184: 445–455.

Bellamy, L. J. 1968. *Advances in Infrared Group Frequencies.* London, U.K.: Chapman and Hall.

Bevilacqua, F., Berger, A. J., Cerussi, A. E. et al. 2000. Broadband absorption spectroscopy in turbid media by combined frequency-domain and steady-state methods. *Appl. Opt.* 39: 6498–6507.

Boas, D. A., O'Leary, M. A., Chance, B. et al. 1997. Detection and characterization of optical inhomogeneities with diffuse photon density waves: A signal-to-noise analysis. *Appl. Opt.* 36: 75–92.

Carpenter, C. M., Pogue, B. W., Jiang, S. et al. 2007. Image-guided optical spectroscopy provides molecular-specific information in vivo: MRI-guided spectroscopy of breast cancer hemoglobin, water, and scatterer size. *Opt. Lett.* 32: 933–935.

Cerussi, A., Hsiang, D., Shah, N. et al. 2007. Predicting response to breast cancer neoadjuvant chemotherapy using diffuse optical spectroscopy. *Proc. Natl. Acad. Sci. U. S. A.* 104: 4014–4019.

Cerussi, A., Shah, N., Hsiang, D. et al. 2006. In vivo absorption, scattering, and physiologic properties of 58 malignant breast tumors determined by broadband diffuse optical spectroscopy. *J. Biomed. Opt.* 11: 044005.

Cerussi, A., Shah, N., Hsiang, D. et al. 2005. Can diffuse optical spectroscopy predict the final pathological response of neoadjuvant chemotherapy?: A retrospective pilot study. In: *Proceedings of the San Antonio Breast Cancer Research Symposium*, San Antonio, TX, 2005.

Cerussi, A., Tanamai, V., Mehta, R. et al. 2010. Frequent optical imaging during breast cancer neoadjuvant chemotherapy reveals dynamic tumor physiology in an individual patient. *Acad. Radiol.* 17(8):1031–1039.

Cerussi, A. E., Jakubowski, D., Shah, N. et al. 2002. Spectroscopy enhances the information content of optical mammography. *J. Biomed. Opt.* 7: 60–71.

Chance, B., Nioka, S., Zhang, J. et al. 2005. Breast cancer detection based on incremental biochemical and physiological properties of breast cancers: A six-year, two-site study. *Acad. Radiol.* 12: 925–933.

Chenevert, T. L., Stegman, L. D., Taylor, J. M. et al. 2000. Diffusion magnetic resonance imaging: An early surrogate marker of therapeutic efficacy in brain tumors. *J. Natl. Cancer Inst.* 92: 2029–2036.

Choe, R., Corlu, A., Lee, K. et al. 2005. Diffuse optical tomography of breast cancer during neoadjuvant chemotherapy: A case study with comparison to MRI. *Med. Phys.* 32: 1128–1139.

Chung, S. H. 2009. Characterization of water molecular state in in-vivo thick tissues using diffuse optical spectroscopic imaging. Ph.D. dissertation, University of California, Irvine, CA.

Chung, S. H., Cerussi, A. E., Klifa, C. et al. 2008. In vivo water state measurements in breast cancer using broadband diffuse optical spectroscopy. *Phys. Med. Biol.* 53: 6713–6727.

Cubeddu, R., D'Andrea, C., Pifferi, A. et al. 2000. Effects of the menstrual cycle on the red and near-infrared optical properties of the human breast. *Photochem. Photobiol.* 72: 383–391.

Eisenberg, D. and Kauzmann, W. 1969. *The Structure and Properties of Water.* Oxford, U.K.: Clarendon Press.

Feldman, L. D., Hortobagyi, G. N., Buzdar, A. U. et al. 1986. Pathological assessment of response to induction chemotherapy in breast cancer. *Cancer Res.* 46: 2578–2581.

Fisher, B., Bryant, J., Wolmark, N. et al. 1998. Effect of preoperative chemotherapy on the outcome of women with operable breast cancer. *J. Clin. Oncol.* 16: 2672–2685.

Fisher, E. R., Wang, J., Bryant, J. et al. 2002. Pathobiology of preoperative chemotherapy: Findings from the National Surgical Adjuvant Breast and Bowel (NSABP) protocol B-18. *Cancer* 95: 681–695.

Fishkin, J. B., Fantini, S., VandeVen, M. J. et al. 1996. Gigahertz photon density waves in a turbid medium: Theory and experiments. *Phys. Rev. E* 53: 2307–2319.

Franceschini, M. A., Moesta, K. T., Fantini, S. et al. 1997. Frequency-domain techniques enhance optical mammography: Initial clinical results. *Proc. Natl. Acad. Sci. U. S. A.* 94: 6468–6473.

Frangioni, J. V. 2008. New technologies for human cancer imaging. *J. Clin. Oncol.* 26: 4012–4021.

Gandjbakhche, A. H., Nossal, R., and Bonner, R. F. 1994. Resolution limits for optical transillumination of abnormalities deeply embedded in tissues. *Med. Phys.* 21: 185–191.

Graaff, R., Aarnoudse, J. G., Zijp, J. R. et al. 1992. Reduced light-scattering properties for mixtures of spherical-particles — a simple approximation derived from Mie calculations. *Appl. Opt.* 31: 1370–1376.

Haskell, R. C., Svaasand, L. O., Tsong-Tseh, T. et al. 1994. Boundary conditions for the diffusion equation in radiative transfer. *J. Opt. Soc. Am. A* 11: 2727–2741.

Helvie, M. A., Joynt, L. K., Cody, R. L. et al. 1996. Locally advanced breast carcinoma: Accuracy of mammography versus clinical examination in the prediction of residual disease after chemotherapy. *Radiology* 198: 327–332.

Holboke, M. J., Tromberg, B. J., Li, X. et al. 2000. Three-dimensional diffuse optical mammography with ultrasound localization in a human subject. *J. Biomed. Opt.* 5: 237–247.

Hylton, N. 2006. MR imaging for assessment of breast cancer response to neoadjuvant chemotherapy. *Magn. Reson. Imaging Clin. North Am.* 14: 383–389, vii.

Jagannathan, N. R., Singh, M., Govindaraju, V. et al. 1998. Volume localized in vivo proton MR spectroscopy of breast carcinoma: Variation of water-fat ratio in patients receiving chemotherapy. *NMR Biomed.* 11: 414–422.

Jakubowski, D., Bevilacqua, F., Merritt, S. et al. 2009. Quantitative absorption and scattering spectra in thick tissues using broadband diffuse optical spectroscopy. In: *Biomedical Optical Imaging*, eds. J. Fujimoto and D. Farkas. New York: Oxford University Press, pp. 330–354.

Jakubowski, D. B., Cerussi, A. E., Bevilacqua, F. et al. 2004. Monitoring neoadjuvant chemotherapy in breast cancer using quantitative diffuse optical spectroscopy: A case study. *J. Biomed. Opt.* 9: 230–238.

Jeffrey, S. S., Birdwell, R. L., Ikeda, D. M. et al. 1999. Radiofrequency ablation of breast cancer: First report of an emerging technology. *Arch. Surg.* 134: 1064–1068.

Kim, S. J., Kim, S. K., Lee, E. S. et al. 2004. Predictive value of [18F] FDG PET for pathological response of breast cancer to neoadjuvant chemotherapy. *Ann. Oncol.* 15: 1352–1357.

Kukreti, S., Cerussi, A., Tromberg, B. et al. 2007. Intrinsic tumor biomarkers revealed by novel double-differential spectroscopic analysis of near-infrared spectra. *J. Biomed. Opt.* 12: 020509.

Kukreti, S., Cerussi, A., Tanamai, W. et al. 2010. Characterization of metabolic differences between benign and malignant tumors: High-spectral-resolution diffuse optical spectroscopy, *Radiology* 254(1): 277–284.

Laya, M. B., Larson, E. B., Taplin, S. H. et al. 1996. Effect of estrogen replacement therapy on the specificity and sensitivity of screening mammography. *J. Natl. Cancer Inst.* 88: 643–649.

Li, A., Liu, J., Tanamai, W. et al. 2008. Assessing the spatial extent of breast tumor intrinsic optical contrast using ultrasound and diffuse optical spectroscopy. *J. Biomed. Opt.* 13: 030504.

Licha, K. 2002. Contrast agents for optical imaging. In: *Top. Curr. Chem.* 222: 1–29.

Litherland, J. C., Stallard, S., Hole, D. et al. 1999. The effect of hormone replacement therapy on the sensitivity of screening mammograms. *Clin. Radiol.* 54: 285–288.

Liu, H., Chance, B., Hielscher, A. H. et al. 1995. Influence of blood vessels on the measurement of hemoglobin oxygenation as determined by time-resolved reflectance spectroscopy. *Med. Phys.* 22: 1209–1217.

Liu, J., Li, A., Cerussi, A. E. et al. 2010. Parametric diffuse optical imaging in reflectance geometry. *IEEE J. Select. Top. Quantum Electron.* 16(3): 555–564.

Makris, A., Powles, T. J., Kakolyris, S. et al. 1999. Reduction in angiogenesis after neoadjuvant chemoendocrine therapy in patients with operable breast carcinoma. *Cancer* 85: 1996–2000.

Mankoff, D. A., Dunnwald, L. K., Gralow, J. R. et al. 1999. Monitoring the response of patients with locally advanced breast carcinoma to neoadjuvant chemotherapy using [technetium 99 m]-sestamibi scintimammography. *Cancer* 85: 2410–2423.

Mankoff, D. A., Dunnwald, L. K., Gralow, J. R. et al. 2003. Changes in blood flow and metabolism in locally advanced breast cancer treated with neoadjuvant chemotherapy. *J. Nucl. Med.* 44: 1806–1814.

McDermott, G. M., Welch, A., Staff, R. T. et al. 2006. Monitoring primary breast cancer throughout chemotherapy using FDG-PET. *Breast Cancer Res. Treat* 102: 75–84.

Meisamy, S., Bolan, P. J., Baker, E. H. et al. 2004. Neoadjuvant chemotherapy of locally advanced breast cancer: Predicting response with in vivo (1)H MR spectroscopy—A pilot study at 4 T. *Radiology* 233: 424–431.

Mourant, J. R., Canpolat, M., Brocker, C. et al. 2000. Light scattering from cells: The contribution of the nucleus and the effects of proliferative status. *J. Biomed. Opt.* 5: 131–137.

Mourant, J. R., Fuselier, T., Boyer, J. et al. 1997. Predictions and measurements of scattering and absorption over broad wavelength ranges in tissue phantoms. *Appl. Opt.* 36: 949–957.

National Cancer Institute, Cancer topics. http://www.cancer.gov/cancertopics/factsheet/Detection/mammograms

Nilsson, A. M. K., Sturesson, C., Liu, D. L. et al. 1998. Changes in spectral shape of tissue optical properties in conjunction with laser-induced thermotherapy. *Appl. Opt.* 37: 1256–1267.

Pauling, L. 1960. *The Nature of the Chemical Bond.* Ithaca, NY: Cornell University Press.

Pimentel, G. C. and McClellan, A. L. 1960. *The Hydrogen Bond.* San Francisco, CA: W. H. Freeman and Company.

Pogue, B. W., Davis, S. C., Song, X. et al. 2006. Image analysis methods for diffuse optical tomography. *J. Biomed. Opt.* 11: 33001.

Pogue, B. W., Jiang, S., Dehghani, H. et al. 2004. Characterization of hemoglobin, water, and NIR scattering in breast tissue: Analysis of intersubject variability and menstrual cycle changes. *J. Biomed. Opt.* 9: 541–552.

Pogue, B. W., Poplack, S. P., McBride, T. O. et al. 2001. Quantitative hemoglobin tomography with diffuse near-infrared spectroscopy: Pilot results in the breast. *Radiology* 218: 261–266.

Shah, N., Gibbs, J., Wolverton, D. et al. 2005. Combined diffuse optical spectroscopy and contrast-enhanced magnetic resonance imaging for monitoring breast cancer neoadjuvant chemotherapy: A case study. *J. Biomed. Opt.* 10: 51503.

Spinelli, L., Torricelli, A., Pifferi, A. et al. 2004. Bulk optical properties and tissue components in the female breast from multiwavelength time-resolved optical mammography. *J. Biomed. Opt.* 9: 1137–1142.

Tanamai, W., Chen, C., Siavoshi, S. et al. 2009. Diffuse optical spectroscopy measurements of healing in breast tissue after core biopsy: Case study. *J. Biomed. Opt.* 14: 014024.

Tardivon, A. A., Ollivier, L., El Khoury, C. et al. 2006. Monitoring therapeutic efficacy in breast carcinomas. *Eur. Radiol.* 16: 2549–2558.

Taroni, P., Danesini, G., Torricelli, A. et al. 2004. Clinical trial of time-resolved scanning optical mammography at 4 wavelengths between 683 and 975 nm. *J. Biomed. Opt.* 9: 464–473.

Tromberg, B. J., Pogue, B. W., Paulsen, K. D. et al. 2008. Assessing the future of diffuse optical imaging technologies for breast cancer management. *Med. Phys.* 35: 2443–2451.

Tromberg, B. J., Shah, N., Lanning, R. et al. 2000. Non-invasive in vivo characterization of breast tumors using photon migration spectroscopy. *Neoplasia* 2: 26–40.

USPSTF 2009. Screening for breast cancer: U.S. Preventive Services Task Force recommendation statement. *Ann. Intern. Med.* 151: 716–726, W-236.

van Veen, R. L., Sterenborg, H. J., Pifferi, A. et al. 2005. Determination of visible near-IR absorption coefficients of mammalian fat using time- and spatially resolved diffuse reflectance and transmission spectroscopy. *J. Biomed. Opt.* 10: 054004.

Vinnicombe, S. J., MacVicar, A. D., Guy, R. L. et al. 1996. Primary breast cancer: Mammographic changes after neoadjuvant chemotherapy, with pathologic correlation. *Radiology* 198: 333–340.

Warren, R. M., Bobrow, L. G., Earl, H. M. et al. 2004. Can breast MRI help in the management of women with breast cancer treated by neoadjuvant chemotherapy? *Br. J. Cancer* 90: 1349–1360.

White, E., Velentgas, P., Mandelson, M. T. et al. 1998. Variation in mammographic breast density by time in menstrual cycle among women aged 40–49 years. *J. Natl. Cancer Inst.* 90: 906–910.

Yeh, E., Slanetz, P., Kopans, D. B. et al. 2005. Prospective comparison of mammography, sonography, and MRI in patients undergoing neoadjuvant chemotherapy for palpable breast cancer. *AJR Am. J. Roentgenol.* 184: 868–877.

Zhu, Q., Kurtzma, S. H., Hegde, P. et al. 2005. Utilizing optical tomography with ultrasound localization to image heterogeneous hemoglobin distribution in large breast cancers. *Neoplasia* 7: 263–270.

Zijlistra, W. G., Buursma, A., and Assendelft, O. W. V. 2000. *Visible and Near Infrared Absorption Spectra of Human and Animal Haemoglobin: Determination and Application*. Zeist, the Netherlands: VSP Publications.

Near-Infrared Diffuse Correlation Spectroscopy for Assessment of Tissue Blood Flow

Guoqiang Yu
University of Kentucky

Turgut Durduran
Institute of Photonic Sciences

Chao Zhou
Massachusetts Institute of Technology

Ran Cheng
University of Kentucky

Arjun G. Yodh
University of Pennsylvania

10.1 Introduction 195
10.2 Near-Infrared DCS Technology Development 196
10.3 Fundamentals of Diffuse Correlation Spectroscopy.......................................197
  Single Scattering • Multiple Scattering Limit (DWS) • Correlation Diffusion Equation (DCS)
10.4 DCS Technology 200
  DCS System • Fiber-Optic Probes
10.5 Validation Work..203
  DCS versus Doppler Ultrasound in Premature Infant Brain • DCS versus ASL-MRI in Human Muscle
10.6 In Vivo Applications of DCS 204
  Cancer Therapy Monitoring • Cerebral Physiology and Disease • Skeletal Muscle Hemodynamics
10.7 Summary...211
Acknowledgments..212
Grant Acknowledgments..212
References...212

10.1 Introduction

Microvascular blood flow (BF) delivers nutrients such as oxygen (O_2) to tissue and removes metabolic by-products from tissue. Normal microvascular BF is thus critical for tissue function. Abnormal BF is associated with conditions such as cardiovascular disease, stroke, head trauma, peripheral arterial disease (PAD), and cancer (Yu et al. 2005a, b, 2006, Durduran et al. 2009b, Zhou et al. 2009). Therefore, measurement of BF holds potential to provide useful information for diagnosis of tissue disease and for monitoring therapeutic effects.

The ideal BF measurement should provide quantitative information about macro- and microvasculature with millisecond temporal resolution. The measurements should be carried out continuously, noninvasively, and without risk to subjects. Furthermore, ideal measurements would not be limited to the tissue surface, i.e., it is desirable to probe BF in deep tissues. Unfortunately, no such ideal modality exists. We thus begin with a brief review of the "imperfect" technologies currently utilized in the clinic.

A variety of noninvasive methods are employed for the measurement of BF and blood cell velocity (Wintermark et al. 2005). Ultrasound, for example, penetrates through the skin, making transcutaneous flow measurements practical. When a cellular target recedes from the fixed sound source, the frequency of the reflected sound wave is lowered because of the Doppler effect (Hoskins 1990). For small changes, the fractional change in sound wave frequency equals the fractional change in cell velocity. Doppler ultrasound can image large vessels in three dimensions with relatively high spatial (~mm) and temporal (~ms) resolution. Unfortunately, this technique is limited to large vessels; extension to the microvasculature requires exogenous contrast agents such as microbubbles (Sehgal et al. 2000, Yu et al. 2005b, Sunar et al. 2007).

Other imaging techniques have been developed to evaluate tissue hemodynamics at the level of the microvasculature (Wintermark et al. 2005), including positron emission tomography (PET) (Baron 1999), single photon emission computed tomography (SPECT) (Mahagne et al. 2004), xenon-enhanced computed tomography (XeCT) (Latchaw et al. 2003), dynamic perfusion computed tomography (PCT) (Wintermark et al. 2000), dynamic susceptibility contrast MRI (DSC-MRI) (Kidwell et al. 2003), and arterial spin labeling magnetic resonance imaging (ASL-MRI) (Detre et al. 1992, Williams et al. 1992). These techniques use endogenous (ASL-MRI) and exogenous tracers (PET, SPECT, XeCT, PCT, and DSC-MRI), and their temporal and spatial resolutions vary. For example, the data acquisition time for ASL-MRI, PCT, and DSC-MRI is in the range of seconds to minutes, approximately 10 times faster than the other techniques. The spatial resolution for PCT, DSC-MRI, and ASL-MRI can be as small as 2 mm and is typically 4–6 mm.

All of these medical diagnostics, however, have limitations that preclude their routine use in the clinic. PET, SPECT, and xenon-CT require exposure to ionizing radiation, and PET and SPECT require arterial blood sampling for quantification of BF. The MRI methods cannot be used in patients with pacemakers, metal implants, or claustrophobia. Furthermore, most (if not all) of these imaging methods are costly and employ major nonportable instrumentation, requiring patient transport. The methods are largely incompatible with serial measurements and so tend to be used only once in the patient's hospitalization, typically in association with a specific research protocol.

Other surface-sensitive imaging techniques for the measurement of microvascular flow include scanning laser Doppler (Liu et al. 1997), laser speckle imaging (Dunn et al. 2001, Durduran et al. 2004a), and Doppler optical coherence tomography (DOCT) (Chen et al. 1998). These methods are used primarily for noninvasive monitoring of BF in tissues located within a few hundred microns below the tissue surface.

It should be apparent, from the discussion above, that a major unfilled niche remains as per the quest for an "ideal" monitor: "bedside" measurement of BF in deep tissues. To this end, near-infrared (NIR) diffuse optical technologies (Gopinath et al. 1993, Hielscher et al. 1993, Chance 1998, Chance et al. 1998, Fantini et al. 1999, Durduran et al. 2002, Wolf et al. 2003, Yu et al. 2003, Choe et al. 2009, Li et al. 2009) provide a fast and portable alternative to the more costly medical diagnostics (e.g., MRI or CT). A well-known spectral window exists in the NIR (700–900 nm), wherein tissue absorption is relatively low, so that light can penetrate into deep/thick volumes of tissue (up to several centimeters). Traditional near-infrared spectroscopy (NIRS) has long been used for continuous measurements of hemoglobin concentration, blood oxygen saturation, and, indirectly, for BF assessment using exogenous tracers (e.g., indocyanine green dye; Kuebler 2008). The present chapter is concerned with a qualitatively different technique for the measurement of BF. The technique was originally introduced for the study of complex fluids and was dubbed diffusing-wave spectroscopy (DWS) (Maret and Wolf 1987, Pine et al. 1988, Stephen 1988, Maret 1997); more recently, diffuse correlation spectroscopy (DCS) (Boas et al. 1995, Boas and Yodh 1997), a differential formulation of DWS, has been developed and vigorously applied to probe BF in deep tissue vasculature. The aforementioned photon-correlation methods share the advantages of NIRS (i.e., deep/thick tissue penetration), but provide a more direct and robust measure of BF (Boas et al. 1995, Boas 1996, Boas and Yodh 1997, Cheung et al. 2001, Durduran 2004, Durduran et al. 2004b, Li et al. 2005, Yu et al. 2005a,b, 2006, Zhou et al. 2007, Durduran et al. 2009b, Shang et al. 2009).

DCS offers several attractive new features for BF measurement. Among these features are noninvasiveness (i.e., no ionizing radiation, no contrast agents), high-temporal resolution (up to 100 Hz) (Dietsche et al. 2007), and relatively large penetration depth (up to several centimeters) (Durduran et al. 2004b, Li et al. 2005). On the other hand, DCS has relatively poor spatial resolution, about 0.5 mm near the surface and degrades with depth. For measurement in adult humans, DCS holds potential to provide information complementary to that available from imaging techniques that measure tissue morphology, such as MRI, or tissue function, such as PET. Perhaps, most importantly, DCS can be easily deployed at the bedside in the clinic (Yu et al. 2006, Durduran et al. 2009b) and, therefore, can be utilized for continuous monitoring.

In this chapter, we sketch the historical development and physical basis of DCS. We then describe DCS instrumentation and provide examples of its validation. Finally, we provide some in vivo application examples. This chapter is not intended to discuss every detail of every problem; rather, it is intended to provide a flavor for the method and a snap-shot of recent progress.

10.2 Near-Infrared DCS Technology Development

Near-infrared diffuse optical spectroscopies naturally separate into "static" and "dynamic" regimes. Here, we use the terms "static" and "dynamic" to distinguish methods that probe the *motions* of scatterers. NIRS is a "static" method; it primarily measures the relatively slow variations in tissue absorption and scattering. The most common applications of NIRS focus on the measurement of oxy- and deoxy-hemoglobin concentration. These concentrations are then utilized to derive microvascular total hemoglobin (THC) concentration and blood oxygen saturation. Very fast NIRS methods (Fantini et al. 1999, Wolf et al. 2003) have been employed to measure rapid (~100 Hz) changes in tissue scattering, but we nevertheless still refer to these methods as "static," since they probe changes in the "amount of scattering" rather than scatterer motion. "Dynamic" methods, on the other hand, directly measure the motions of scatterers (Tanaka et al. 1974, Stern 1975, Feke and Riva 1978, Maret and Wolf 1987, Pine et al. 1988, Stephen 1988, Tong et al. 1988, Boas et al. 1995, Boas and Yodh 1997). In the case of tissues, the critical moving scatterers are red blood cells (RBCs). The dynamic or correlation methods achieve this goal, typically using coherent sources and monitoring temporal statistics (or frequency-domain analogs of temporal statistics) of the speckle fluctuations of the scattered light. In most of these dynamic experiments, the electric field temporal autocorrelation function or its Fourier transform is measured. The detected signal is related to the motion of the RBCs, and BF can be derived using a model for photon propagation through tissues.

A well-known and notable methodology for measuring BF is "laser Doppler flowmetry" (Bonner and Nossal 1981). In this case, pairs of very closely separated (<1 mm) sources and detectors are used to detect single-scattered light from tissue samples. In laser Doppler flowmetry, the frequency-broadening or frequency-shift of the detected speckle fluctuations are fit to a model within the single-scattering approximation. The single-scattering approximation simplifies the experimental analysis, but it also limits the reliability and amount of information that can be extracted from real tissue samples. A related technique is "laser speckle flowmetry" or "laser speckle contrast analysis" (LASCA) (Briers 2001). LASCA utilizes the spatial blurring of speckles during a CCD exposure time to obtain large-field-of-view images of tissue motions in a single shot. Both methods are mostly limited to superficial tissues (~1 mm), although recently laser Doppler flowmetry with larger source-detector separations was utilized to extend its reach to ~1 cm (Binzoni et al. 2004).

Studies of deep tissues required the development of multiple-scattering models of photon propagation. Various extensions of photon correlation spectroscopy from single- to highly scattering systems were made in the 1980s (Bonner and Nossal 1981, Valkov and Romanov 1986, Bonner et al. 1987, Maret and Wolf 1987, Pine et al. 1988). The technique-dubbed DWS was developed to study optically dense complex fluids that multiply scattered light. DWS was a brilliant insight that had transformative effects on the soft matter field. From the biomedical optics perspective, a key advance was the development of an understanding of the analogy between correlation transport (Ackerson et al. 1992) and photon transport (in the early-mid 1990s). A theory based on the diffusion equation for temporal correlation transport was first introduced by Boas and coworkers (Boas et al. 1995, Boas and Yodh 1997); they called the technique DCS in order to avoid confusion with terminology associated with diffuse absorption spectroscopy. Applications in biological tissues ensued rapidly thereafter because of the connection thus established between traditional diffuse optics (NIRS) and the diffuse models for transport of electric field temporal autocorrelation functions through turbid media. The DCS technique was thus built on a rigorous mathematical description, which clearly showed how motional fluctuations were impressed upon the temporal correlations of diffuse light fields propagating in tissue. One advantage of the correlation diffusion equation theory over previous theories was the ease with which predictions could be made for turbid media with spatially varying dynamics and optical properties using numerical and analytic tools similar to those of NIRS (see for example, Boas et al. 1995, Boas and Yodh 1997, and Heckmeier et al. 1997).

Instrumentation and application of DCS for in vivo measurements quickly followed this early research (Gisler et al. 1995, 1997, Cheung et al. 2001, Culver et al. 2003a, Durduran et al. 2004b, 2009b, 2005, Li et al. 2005, 2008, Yu et al. 2005a,b, 2006, 2007, Zhou et al. 2006, 2007, 2009, Dietsche et al. 2007, Sunar et al. 2007, Gagnon et al. 2008, 2009, Buckley et al. 2009, Varma et al. 2009, Kim et al. 2010, Roche-Labarbe et al. 2010, Shang et al. 2010, Mesquita et al. 2010, Zirak et al. 2010, Carp et al. 2010, Belau et al. 2010). We next describe the theoretical basis of DCS.

10.3 Fundamentals of Diffuse Correlation Spectroscopy

DCS is an extension of single-scattering dynamic light scattering (DLS) (Berne and Pecora 1990, Chu 1991, Brown 1993) (or quasi-elastic light scattering, QELS) to the multiple scattering limit. DLS has been widely used to study particle suspension properties such as particle size and shape. However, DLS is only applicable to optically thin samples. DCS is a multiple-scattering technique that extends the methodology of DLS to the study of optically thick samples. In this section, we start with a description of single-scattering theory, and then we extend the discussion to the multiple scattering regime. Detailed reviews of the theoretical development for DCS can also be found in the references (Boas et al. 1995, Boas 1996, Boas and Yodh 1997, Durduran 2004, 2010b, Zhou 2007).

10.3.1 Single Scattering

In a single-scattering laboratory sample, photons are usually scattered once (or not at all) before they leave the sample (see Figure 10.1a). A pointlike photon detector is placed at an angle θ relative to the input beam propagation direction. If the scatterers are particle-like objects that move, then the total electric field will vary in time, and intensity fluctuations are observed. The

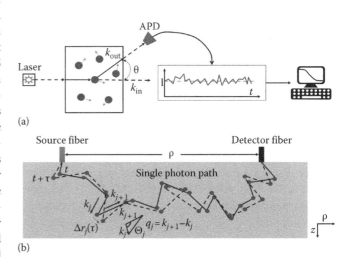

(a)

(b)

FIGURE 10.1 (a) Illustration schematic of a single-scattering dynamic light-scattering experiment. (b) Schematic of multiple-scattering setup showing a typical photon path through the turbid media. $\mathbf{k}_j$ and $\mathbf{k}_{j+1}$ are the wave vectors associated with the photon (or electric field) before and after the jth scattering event, respectively. $\mathbf{q}_j = \mathbf{k}_{j+1} - \mathbf{k}_j$ is the scattering wave vector or momentum transfer, and θ_j is the scattering angle of the jth scattering event. The solid line indicates the photon path at time t, while the dotted line represents the photon path at time $t+\tau$. During the delay time τ, the jth scatterer is displaced by $\Delta\mathbf{r}_j(\tau)$. Also indicated is the source-detector separation on the sample surface, ρ, and the direction z is defined normal to the sample surface. (From Zhou, C., In-vivo optical imaging and spectroscopy of cerebral hemodynamics, PhD dissertation, University of Pennsylvania, Philadelphia, PA, 2007.)

fluctuations of the electric field and intensity carry information about the dynamic properties of the medium, i.e., the motion of the particles. The normalized temporal autocorrelation function of the scattered electric field ($\mathbf{E}(t)$) is

$$g_1(\tau) = \frac{\langle \mathbf{E}(t) \cdot \mathbf{E}^*(t+\tau) \rangle}{\langle \mathbf{E}(t) \cdot \mathbf{E}^*(t) \rangle} = e^{i\omega\tau} e^{-\frac{1}{6}q^2 \langle \Delta r^2(\tau) \rangle}. \quad (10.1)$$

Here, τ is the autocorrelation delay time, ω is the angular frequency of the input light field, $q^2 = 2k_0^2(1-\cos\theta)$ is the square of the scattering wavevector, $k_0 = 2\pi n/\lambda$ is the wavevector magnitude of the incident light field, n is the index of refraction of the medium, λ is the wavelength of the light field, and $\langle \Delta r^2(\tau) \rangle$ is the mean square displacement of the scatterers in the medium, which directly characterizes the particle movement. In most experiments, the normalized temporal intensity autocorrelation function, $g_2(\tau) = \langle I(t)I(t+\tau) \rangle / \langle I(t) \rangle^2$, is calculated from the intensity fluctuations of the scattered light. $g_2(\tau)$ is related to the temporal field autocorrelation function, $g_1(\tau)$, through the Siegert relationship (Rice 1954):

$$g_2(\tau) = 1 + \beta |g_1(\tau)|^2, \quad (10.2)$$

where β depends on the detection optics and is inversely proportional to the number of detected speckles or modes. The β value can be determined experimentally for each measurement from the intercept of the intensity autocorrelation function as the delay time τ approaches zero.

10.3.2 Multiple Scattering Limit (DWS)

Multiple scattering effects must be included in applications involving most biological tissues. In this case, each scattering event from a moving scatterer contributes to the accumulation of the phase shift and therefore the decay of the correlation function. If we assume that the field from individual photon paths (see Figure 10.1b) is uncorrelated, the total temporal field autocorrelation function can be expressed as the weighted sum of the field autocorrelation function from each photon path. Furthermore, if we assume a homogeneous medium and further assume that each scattering event is independent and that the scatterer displacements are uncorrelated, the field autocorrelation function from a single path can be written as

$$g_1(\tau)_{\text{onepath}} = e^{i\omega\tau} e^{-\frac{1}{3}k_0^2 Y \langle \Delta r^2(\tau) \rangle}, \quad (10.3)$$

where $Y = N \cdot (1 - \langle \cos\theta \rangle_N)$ and $\langle \cos\theta \rangle_N$ are the average value of cosine over all the N scattering events along the path. When N is large, the average value approaches the ensemble average, $\langle \cos\theta \rangle$, which is usually denoted by the so-called anisotropy factor (g) of the medium. The reduced photon-scattering length or random-walk step length, $l_s^* = 1/\mu_s'$, where μ_s' is the reduced scattering

coefficient of the medium. The photon-scattering length, $l_s = 1/\mu_s$, where μ_s is the scattering coefficient of the medium. Thus,

$$l_s^* = \frac{1}{\mu_s'} = \frac{1}{\mu_s(1-g)} = \frac{l_s}{1 - \langle \cos\theta \rangle} \quad \text{(Wolf et al. 1988). Let } s \text{ repre-}$$

sents the total pathlength associated with a particular photon path. Then, the number of scattering events associated with this same path is $N = s/l_s$, and $Y = s/l_s^*$ equals the total number of photon *random-walk* steps associated with the photon path.

The final detected field autocorrelation function contains the contributions of all photon paths. If we use $P(Y)$ to represent the probability distribution for photon paths with a number of random walk steps, Y, then the total electric field autocorrelation function can be computed by incoherently integrating the contributions from each photon path (Maret and Wolf 1987, Middleton and Fisher 1991), i.e.,

$$g_1(\tau) = e^{i\omega\tau} \int_0^\infty P(Y) e^{-\frac{1}{3}k_0^2 Y \langle \Delta r^2(\tau) \rangle} dY. \quad (10.4)$$

In a highly scattering medium, Equation 10.4 can be equivalently expressed as an integral over all possible pathlengths using the pathlength distribution (Maret and Wolf 1987, Pine et al. 1988, MacKintosh and John 1989), $P(s)$, i.e.,

$$g_1(\tau) = e^{i\omega\tau} \int_0^\infty P(s) e^{-\frac{s}{3l_s^*} k_0^2 \langle \Delta r^2(\tau) \rangle} ds. \quad (10.5)$$

Equation 10.5 is the primary result from DWS for a homogeneous turbid scattering medium composed of moving particle-like scatterers. Note, the DWS correlation function is typically measured at some points inside the sample or on its surface, and $P(s)$ depends implicitly on both measurement location and source position. The distribution of $P(Y)$ can be readily obtained from Monte Carlo simulation (Zhou 2007). Alternatively, derivation of $P(s)$ can be achieved experimentally, e.g., from a time-resolved spectroscopy measurement (Jacques 1989, Patterson et al. 1989, Yodh et al. 1990, Benaron and Stevenson 1993). Analytical solutions for $P(s)$ can also be obtained for simple geometries, such as infinite, semi-infinite, and slab, by solving the photon diffusion equation with appropriate boundary conditions (see Chapter 16).

10.3.3 Correlation Diffusion Equation (DCS)

Later, Boas et al. (Boas et al. 1995, Boas 1996, Boas and Yodh 1997) derived a correlation diffusion equation from correlation transport theory (Ackerson and Pusey 1988, Ackerson et al. 1992). The correlation diffusion equation aptly described the propagation of the unnormalized electric field temporal autocorrelation function in turbid media. This differential equation approach is particularly attractive for investigation of heterogeneous media (Boas et al. 1995, Boas and Yodh 1997, Heckmeier et al. 1997) and provides a natural framework for tomographic reconstruction of tissue dynamics (Boas and Yodh 1997, Culver et al. 2003a, Zhou et al. 2006). Rather than reproduce the earlier derivations, here, we simply remind the reader about the diffusion equation for photon fluence rate, and then we write out the analogous result for photon-electric field correlation. The rigorous step-by-step derivation of the correlation diffusion equation can be found elsewhere (Boas et al. 1995, Boas 1996, Boas and Yodh 1997).

The well-known photon diffusion equation, which describes the photon propagation in tissue, is indicated below. In highly scattering media, such as tissue, the photon fluence rate, $\Phi(\mathbf{r},t)$ [W·cm^{-2}], obeys the time-dependent diffusion equation:

$$\nabla \cdot (D\nabla\Phi(\mathbf{r},t)) - v\mu_a\Phi(\mathbf{r},t) + vS(\mathbf{r},t) = \frac{\partial\Phi(\mathbf{r},t)}{\partial t}, \qquad (10.6)$$

where

  r is the position vector
  t [s] is time
  v [cm·s^{-1}] is the speed of light in the medium

μ_a [cm^{-1}] is the medium absorption coefficient, and $D \approx v/3\mu_s'$ is the photon diffusion coefficient in the medium, where μ_s' [cm^{-1}] is the reduced scattering coefficient of the medium. $S(\mathbf{r},t)$ [Watt·cm^{-3}] is the isotropic source term.

It turns out that under essentially the same set of approximations, the *unnormalized* temporal field autocorrelation function, $G_1(\mathbf{r},\tau) = \langle \mathbf{E}(\mathbf{r},t) \cdot \mathbf{E}^*(\mathbf{r},t+\tau) \rangle$, obeys a formally similar diffusion equation, i.e.,

$$\nabla \cdot (D\nabla G_1(\mathbf{r},\tau)) - (v\mu_a + \tfrac{1}{3}v\mu_s'k_0^2\langle\Delta r^2(\tau)\rangle)G_1(\mathbf{r},\tau)$$
$$= -vS_0e^{i\omega\tau}\delta^3(\mathbf{r} - \mathbf{r_s}). \qquad (10.7)$$

Here, the source term is continuous wave (CW), and scatterer movement (i.e., $\langle\Delta r^2(\tau)\rangle$) combines with photon absorption to give an *effective* "absorption" term for the attenuation of unnormalized electric field temporal autocorrelation function as it travels through the medium. The formal similarity of Equations 10.6 and 10.7 suggests that their solutions will also be formally similar. In semi-infinite homogeneous media (see Figure 10.2), for example, the solution to Equation 10.7 can be obtained as (Boas 1996)

$$G_1(\rho,\tau) = \frac{vS_0e^{i\omega\tau}}{4\pi D}\left(\frac{e^{-K(\tau)r_1}}{r_1} - \frac{e^{-K(\tau)r_2}}{r_2}\right). \qquad (10.8)$$

Here, ρ is the distance between the source and detector fiber, $r_1 = \sqrt{\rho^2 + z_0^2}$, $r_2 = \sqrt{\rho^2 + (z_0 + 2z_b)^2}$, $z_0 = 1/\mu_s'$, $z_b = \frac{2}{3\mu_s'}\frac{1+R_{\text{eff}}}{1-R_{\text{eff}}}$, $R_{\text{eff}} = -1.44n^{-2} + 0.71n^{-1} + 0.668 + 0.064n$, is the effective reflection coefficient determined by the ratio of the refraction indices

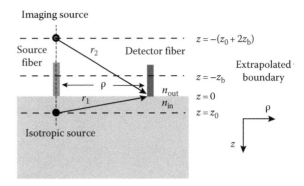

FIGURE 10.2 Illustration of the semi-infinite geometry. The collimated source is usually approximated as an isotropic source located at $z = z_0 = 1/\mu_s'$ into the medium. The boundary condition requirement leads to a signal size of zero (i.e., for fluence rate in the case of a NIR diffuse reflectance measurement or for the temporal autocorrelation function in the case of the DCS measurement) at $z = -z_b = -\frac{2}{3\mu_s'}\frac{1+R_{\text{eff}}}{1-R_{\text{eff}}}$, which is generally called the extrapolated zero-boundary condition (Haskell et al. 1994). For the semi-infinite homogeneous geometry, the extrapolated zero boundary condition can be satisfied by considering a negative isotropic imaging source located at $z = -(z_0 + 2z_b)$. (From Zhou, C., In-vivo optical imaging and spectroscopy of cerebral hemodynamics, PhD dissertation, University of Pennsylvania, Philadelphia, PA, 2007.)

inside and outside the medium ($n = n_{\text{in}}/n_{\text{out}}$, see Figure 10.2), and $K^2(\tau) = 3\mu_a\mu_s' + \mu_s'^2k_0^2\langle\Delta r^2(\tau)\rangle$.

One other modification of the correlation diffusion equation is required for biological tissues. Generally, biological tissues contain *static* (or very slow moving) scatterers (e.g., organelles and mitochondria) and *moving* scatterers (e.g., RBCs). The scattering events from the static objects in tissue do not contribute significantly to the phase shift and correlation function temporal decay in Equation 10.7. To account for this effect, we introduce a unitless factor, α, which represents the fraction of light-scattering events from "moving" scatterers. Formally, the factor α is included as a prefix to the *effective* "absorption" term in Equation 10.7 (i.e., $\frac{1}{3}v\mu_s'k_0^2\alpha\langle\Delta r^2(\tau)\rangle$), and so it arises as well as in the definition of $K^2(\tau)$ (e.g., $K^2(\tau) = 3\mu_a\mu_s' + \mu_s'^2k_0^2\alpha\langle\Delta r^2(\tau)\rangle$). We thus see that the decay of the temporal field correlation function depends on tissue optical properties, μ_a, μ_s', the mean-square-displacement of the moving scatterers, $\langle\Delta r^2(\tau)\rangle$, and the factor α, which accounts for the presence of static scatterers.

For the case of random ballistic flow, $\langle\Delta r^2(\tau)\rangle = V^2\tau^2$, where V^2 is the second moment of the cell velocity distribution. For the case of diffusive motion, $\langle\Delta r^2(\tau)\rangle = 6D_b\tau$, where D_b is the *effective* Brownian diffusion coefficient of the tissue scatterers and is distinct from the well-known thermal Brownian diffusion coefficient due originally to Einstein (1905). RBCs pass through capillaries in single file and experience shear flow in larger vessels; the RBCs also experience tumbling motions in addition to translation. Intuitively, the random ballistic flow model might be considered the best model with which to fit DCS data. In

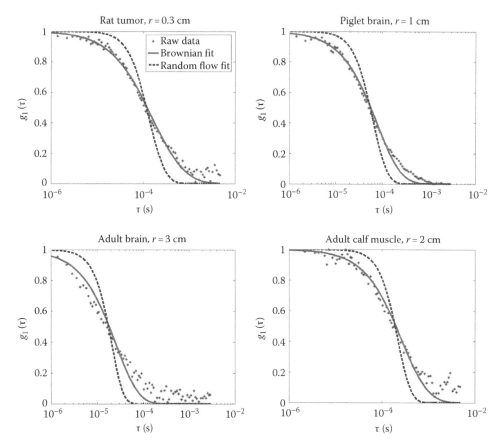

FIGURE 10.3 DCS data (i.e., the normalized electric field autocorrelation function) from a rat tumor, piglet brain, human calf muscle, and adult human brain. The source–detector separations are also indicated on each plot. Dots are experimental data. The dashed line is a fit with $\langle \Delta r^2 \rangle \sim \tau^2$ (random flow), and the solid line is a fit with $\langle \Delta r^2 \rangle \sim \tau$ (Brownian motion). Note, the fitting accuracy depends on the delay time (τ); at the longer delay times, the data tend to deviate further from the fits in the measurements. This effect arises mainly because later delay times in the autocorrelation function are due to the photons that probe the more superficial tissues; the presence of the skull and or the skin, with its relatively minimal vasculature, therefore alters the long-delay time portion of the autocorrelation functions. (From Zhou, C., In-vivo optical imaging and spectroscopy of cerebral hemodynamics, PhD dissertation, University of Pennsylvania, Philadelphia, PA, 2007.)

practice, however, we have observed that the diffusion model, i.e., $\langle \Delta r^2(\tau) \rangle = 6 D_b \tau$, fits the autocorrelation curves rather well over a broad range of tissue types (see Figure 10.3), ranging from rat brain (Cheung et al. 2001, Zhou et al. 2006) and mouse tumor (Menon et al. 2003, Yu et al. 2005b), to piglet brain (Zhou et al. 2009), adult human skeletal muscle (Yu et al. 2005a, Shang et al. 2009), adult human tumors (Sunar et al. 2006, Zhou et al. 2007), premature brain (Buckley et al. 2009, Roche-Labarbe et al. 2010), and adult brain (Durduran et al. 2004b, Li et al. 2005, 2008, Durduran et al. 2009b). The D_b value obtained from such experiments is generally several orders of magnitude larger than the thermal (i.e., Einstein) Brownian diffusion coefficient of RBCs in plasma. The reason for the Brownian-motion-like correlation curves is still not apparent and more investigation is needed to sort through these issues. Nevertheless, an empirical approach has been adopted by researchers in biomedical optics applying DCS for the measurement of tissue BF.

Although the unit of αD_b (cm²/s) is different from the traditional blood perfusion unit [mL/min/100 g], changes in this flow-index (αD_b) have been found to correlate quite well with other BF measurement modalities (see Section 10.5). Therefore, αD_b is used as the BF index (BFI = αD_b) and relative BF, rBF = BFI/BFI$_0$, where BFI$_0$ is the measurement at the baseline, is used to indicate relative BF changes for DCS throughout this chapter.

10.4 DCS Technology

In this section, we briefly review the system design and implementation of DCS technology. Detailed descriptions of specific systems can be found in the references.

10.4.1 DCS System

Figure 10.4a shows the diagram of a typical DCS system. A long coherence length laser is required as the light source for DCS. Output from the laser can be delivered to the tissue through a multimode source fiber. Single-mode (or few-mode) fibers should be used to collect photons from a single (or a few) speckle(s). Fast photon-counting avalanche photodiodes (APDs) (typically from Perkin Elmer, CA) are generally used as detectors. A multi-tau

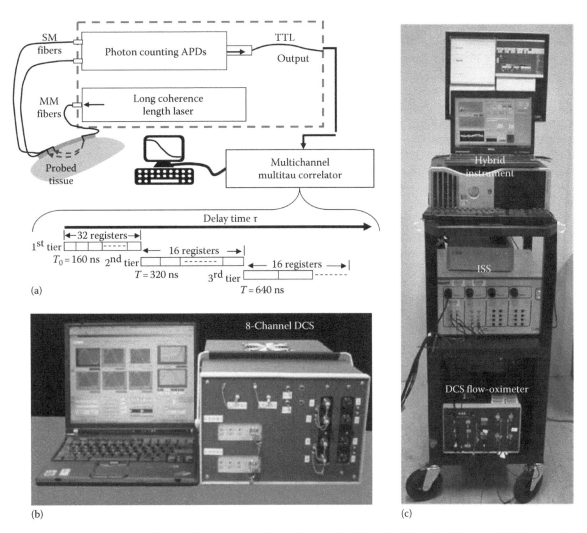

FIGURE 10.4 (a) Schematic diagram of a typical DCS system. (b) A photograph of the eight-channel DCS system. (c) Photograph of a hybrid diffuse optical system consisting of a DCS flow oximeter and a frequency-domain tissue oximeter (From Imagent, ISS Inc.).

correlator board (Correlator.com, NJ) takes the TTL output from the APDs and calculates normalized temporal intensity autocorrelation functions of the detected signal. Since the correlator board is at the heart of the DCS measurements, we will briefly describe its structure and the algorithm used for calculating the correlation curves.

The correlator board utilizes a "multi-tau" scheme for the calculation of the autocorrelation functions. Figure 10.4a shows the structure of a typical multi-tau correlator design. In this case, the first 32 registers (the first tier) of the correlator have a bin width $T_0 = 160$ ns. After that, the bin width doubles every 16 registers. For example, registers 33–48 (the second tier) have a bin width $T = 320$ ns, and registers 49–64 (the third tier) have a bin width $T = 640$ ns. When a measurement starts, a digital counter reports the number of TTL pulses (photon counts) detected within each binning time (160 ns) of the first register. The value in each register is passed to its right as a new value is coming in from the left. Note that a register with a bin width $T = 160$ ns is updated every 160 ns, while the register with a 320 ns bin width

updates every 320 ns, etc. The temporal intensity autocorrelation functions are calculated before each shift. For example, the unnormalized intensity autocorrelation coefficient for the ith register, $G_2(\tau_i)$, is calculated as

$$G_2(\tau_i) = \langle n_i \cdot n_0 \rangle. \tag{10.9}$$

Here

    n_i indicates the photon count in the ith register
    n_0 is the photon count at zero delay time ($\tau = 0$) with the *same bin width* as the ith register
    τ_i is the delay time between n_i and n_0

The autocorrelation function is continuously averaged over the entire acquisition time t. Note that n_0 is the photon count in the first register when n_i is in the first tier, while n_0 is the summation of the photon count in the first *four* registers when n_i is in the third tier ($T/T_0 = 4$), etc. The delay time τ_i is calculated as the summation of all the bin widths on the left of the *i*th register. Using the multi-tau scheme, a delay time span of many orders of magnitude (i.e., from hundred of nanoseconds to minutes) can be achieved with only a few hundred register channels, and the computation load is greatly reduced compared to a linear autocorrelator.

The DCS technologies have been implemented in tissues by several research groups (Cheung et al. 2001, Culver et al. 2003a, Durduran et al. 2004b, 2005, Li et al. 2005, Yu et al. 2005a,b, Roche-Labarbe et al. 2010, Shang et al. 2009), yielding BF information in animal models and in human subjects/patients. To fully utilize the BF information provided by DCS, hybrid systems combining DCS and NIRS have been demonstrated to provide more comprehensive information for the calculation of tissue BF, blood oxygenation, and oxygenation metabolism (see an example in Figure 10.4c) (Cheung et al. 2001, Culver et al. 2003a, Durduran 2004, Yu et al. 2005a, Roche-Labarbe et al. 2010). In addition, due to the recent development of novel solid-state laser technologies, the DCS system can be made very compact (Zhou et al. 2007, Shang et al. 2009). Figure 10.4b shows an eight-channel DCS instrument, which only measures 18 cm × 28 cm × 33 cm. Very recently, Shang et al. (2009) demonstrated a dual wavelength DCS flow oximeter, which permits tissue BF and oxygenation to be measured simultaneously by a single compact system (bottom of Figure 10.4c). The device is truly portable and is suitable for bedside and *en route* monitoring of tissue hemodynamics.

10.4.2 Fiber-Optic Probes

In an analogous fashion to NIRS, DCS also enables use of a large variety of probes. The most basic probe employs one or more source-fibers (multimode) and one or more detector fibers (single- or few-mode). A key point in DCS is the use of single- or few-mode fibers, which limit the detector fiber diameter to tens of micrometers. We have shown in the past (Zhou et al. 2006) that enlarging the fiber diameter to cover multiple speckles increases the signal intensity but also increases the noise in a proportional fashion, and therefore the signal-to-noise ratio is not necessarily improved. In our laboratories, we routinely employ single-mode

fibers to detect light from a single speckle, and we often bundle 2–8 fibers nearby and detect the collected photons with individual, independent detectors (see Figure 10.4a).

In Figure 10.5, we show three example probes. The first probe (see Figure 10.5a) is a typical probe used in brain and muscle studies on humans (Durduran et al. 2004b, Yu et al. 2005a). Straight, 90 degree bent or side-firing fibers are utilized. This probe can readily be made MRI compatible (Yu et al. 2005a, Durduran et al. 2010a). The second probe (see Figure 10.5b) is a typical "noncontact" probe, where an array of source-detector fibers is placed together at the focal plane of a "camera." In many experiments, we utilize an old SLR camera; this camera holds a 1:1 imaging lens array and provides a light tight box. In this arrangement, cross-polarizers are often utilized between source and detector fibers to reduce the detrimental effects of surface reflections and single-scattered light. This approach improves the "β" value (see Section 10.3) and while we lose photons, we gain in reduced noise (Zhou et al. 2006). Finally, the third type of probe (see Figure 10.5c) shows a surgical device, wherein side-firing fibers were embedded in a catheter, which could be inserted into tissues (Yu et al. 2006) or sutured onto the tissue surface. Practically all ideas from experiences in NIRS can be adapted for DCS use, and the instruments are hybridized by adding extra source and/or detector fibers.

Here, we also highlight a probe utilized by Gisler and coworkers that employ up to 32 few-mode fibers to detect light from multiple speckles simultaneously (Dietsche et al. 2007). Their goal was to maximize the number of detected photons per fiber by utilizing few-mode fibers, which collect data from several speckles and also at the same time acquire many correlation functions in parallel. With this approach, they have reduced the integration time down to 6.5 ms and are able to resolve changes in BF due to arterial pulsation in an analogous fashion to pulse-oximetry. Figure 10.6a and b shows the arrangement of the fibers and collection tip (Dietsche et al. 2007). As shown in Figure 10.6c, they were able to obtain the pulsation dynamics (inset) and some initial insights into the dependence of the shape of the field autocorrelation functions on the pulsatility of the vasculature. This experimental approach may turn out to be useful for improving our understanding of the physical basis of

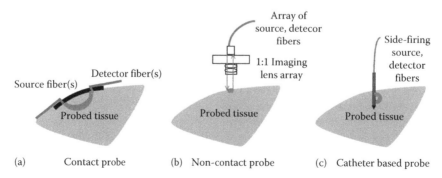

FIGURE 10.5 Three schematic examples of DCS tissue probes: (a) contact probe, (b) noncontact probe, and (c) side-illumination catheter-based probe.

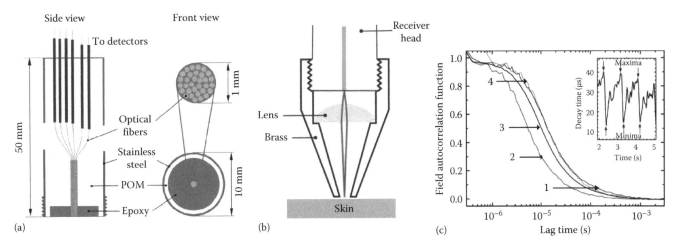

FIGURE 10.6 (a) Side and top views of the multifiber receiver head, (b) receiver head with attached collection optics, and (c) the electric field autocorrelation functions $\left\langle g_j^{(1)}(\tau)\right\rangle_N$ *averaged over N fibers* as a function of lag time τ at the diastolic maxima (1) and systolic minima (2) of mean decay time (τ_d) measured at the fingertip. The number of receiver channels is $N=23$. The integration time per field autocorrelation function is 26 ms. Data are averages over 11 field autocorrelation functions measured at the maxima and minima of the τ_d curve (top and bottom, respectively, in inset). (3) $\left\langle g_j^{(1)}(\tau)\right\rangle_N$ averaged over 10 s, which mimics the "standard" DCS measurement, where the data are averaged across the heart-beat cycle. For easier comparison with the diastolic data, the systolic field autocorrelation function was shifted in time (4). Note that the shape of the curves are slightly different in a complex manner, possibly indicating different amounts of photon penetration at maxima and minima, differing optical properties and subtle changes in physiology. This is an indication that with further improved understanding of the physical basis of these curves, more information about the underlying physiology may be accessible. (From Dietsche, G. et al., *Appl. Opt.*, 46, 8506, 2007. With permission.)

the detailed shapes of the measured autocorrelation functions. Furthermore, this unique probe may become more popular in the future with increased parallelization of DCS detection and corresponding autocorrelator electronics.

10.5 Validation Work

The biomedical application of DCS for deep tissue measurements of BF is a relatively new development. Therefore, a substantial amount of technique validation work has been carried out, and, arguably, more work is needed to establish its medical uses. This validation research is needed even though the basis of DCS is DWS, a method that has been widely utilized in condensed matter research (Maret 1997). However, DCS application to in vivo measurement of deep tissue BF was (and continues to be) novel and demanded careful examination in this context. During the last decade, DCS measurements of BF variation in various tissues/organs have been compared to other standards, including power Doppler ultrasound in murine tumors (Menon et al. 2003, Yu et al. 2005b), laser Doppler in rat brain (Durduran 2004), xenon-CT in traumatic brain (Kim et al. 2010), Doppler ultrasound in premature infant brain (Buckley et al. 2009, Roche-Labarbe et al. 2010), fluorescent microsphere measurement of cerebral BF in piglet brain (Durduran et al. 2008, Zhou et al. 2009), ASL-MRI in human brain and muscle (Durduran et al. 2004b, Yu et al. 2007), and to reports in the literature (Cheung et al. 2001, Culver et al. 2003a, c, Durduran 2004, Durduran et al. 2005, Dietsche et al. 2007, Li et al. 2008). This validation research has progressed hand-in-hand with

numerical, theoretical studies (Boas et al. 1995, Durduran 2004, Jaillon et al. 2006, Zhou et al. 2006, Zhou 2007, Gagnon et al. 2008, Varma et al. 2009), and with studies of tissue simulating phantoms (Boas et al. 2002, Cheung et al. 2003, Durduran 2004, Zhou 2007), wherein the medium's viscosity and/or the flow speed of scatterers were varied (Boas 1996, Cheung et al. 2001, Durduran 2004). Overall, these validation studies have shown that DCS measurements of BF variations are in good agreement with theoretical expectation, computer simulation, and other biomedical measurement techniques. Some of these validation studies are described below.

10.5.1 DCS versus Doppler Ultrasound in Premature Infant Brain

Two recent papers (Buckley et al. 2009, Roche-Labarbe et al. 2010) have utilized DCS in prematurely born infants and have compared DCS microvascular BF findings to transcranial Doppler ultrasound measurements of large artery BF velocity. In work by Buckley et al. (2009), the authors monitored very low birth weight, very premature infants during a 12° postural elevation. DCS was used to measure *microvascular* cerebral BF (CBF) and transcranial Doppler ultrasound (TCD) to measure *macrovascular* BF velocity in the middle cerebral artery (Buckley et al. 2009, Roche-Labarbe et al. 2010). Population-averaged DCS and TCD data yielded no significant hemodynamic response to this postural change ($p > 0.05$), indicating overall agreement between the two modalities in response to this mild challenge. More interestingly, *absolute* DCS data

(αD_b) correlated significantly with peak systolic, end diastolic, and mean velocities measured by TCD ($p = 0.036, 0.036$, and 0.047). Roche-Labarbe et al. (2010) have also reported a similar finding ($p = 0.04$) comparing *absolute* DCS data to mean velocities measured by TCD. Overall, these two studies demonstrate that DCS has a strong potential for use in premature infants and agrees well with the established TCD measures, despite the fact that the two techniques measure related but different quantities (i.e., microvascular local CBF vs. large artery, global CBF velocity). The studies also suggest that with further understanding of the physical basis of the photon-RBC interactions at the microscopic level, and/or with improved calibration, DCS could be used to measure absolute CBF.

10.5.2 DCS versus ASL-MRI in Human Muscle

The DCS BF measurement was also validated against flow measurements by arterial-spin-labeled perfusion MRI (ASL-MRI) using human calf-muscle and brain (Durduran 2004, Durduran et al. 2004b, 2007, 2009a, Yu et al. 2007). For example, a contact optical probe (see Figure 10.5a) was placed on the calf-muscles of seven healthy subjects for concurrent measurements with ASL-MRI (Yu et al. 2007). The calf (with the optical probe) was then placed into the MRI knee coil (see Figure 10.7a). The optical probe in the MRI room was connected to the DCS instrument in the control room by long optical fibers through a port in a magnetic-field-shielded wall. After a period of baseline, a large leg cuff on the thigh was inflated rapidly to occlude BF to the lower leg for 5 min. The BF indices measured by DCS were compared to the absolute ASL-MRI flow around the peak of the hyperemia after cuff release (note: one limitation of this measurement was the reliability (actually, lack thereof) of ASL-MRI at low-baseline flow levels (<10 mL/100 g/min [Petersen et al. 2006]), which, in turn, prevented accurate calculations of relative flow changes from MRI. A good correlation ($R^2 > 0.6$) was observed with both the relative (see Figure 10.7b) and absolute (see Figure 10.7c) flow indices from DCS and absolute ASL-MRI flow. These observations suggest that with further systematic calibration along with improved modeling, it may be feasible to estimate absolute BF values.

10.6 In Vivo Applications of DCS

As mentioned above, the utility of DCS technology for monitoring tissue BF has been demonstrated in tumors (Menon et al. 2003, Wang et al. 2004, Durduran et al. 2005, Yu et al. 2005b, 2006, Sunar et al. 2006), brain (Culver et al. 2003a, c, Durduran 2004, Zhou et al. 2006), and skeletal muscles (Yu et al. 2005a, 2007). The early stages of many of these studies focused on BF in animal models (e.g., murine tumors (Yu et al. 2005b), rat and piglet brain (Cheung et al. 2001, Culver et al. 2003a, Zhou et al. 2006, 2009), and pig limb muscles (Xing et al. 2007). More recently, the DCS technique has been a key component in a variety of clinical studies (e.g., human cancers of prostate, breast and head and neck, cerebral functional activities, cerebral stroke, traumatic brain injury (TBI), and skeletal muscle physiology). In these preclinical and clinical investigations, DCS was used for quantification of tissue hemodynamic status, for diagnosis of disease, and for continuous monitoring and evaluation of therapeutic effects. The optical techniques were often validated as part of these measurements, e.g., by comparison to other diagnostic modalities (see Section 10.5). In total, the research demonstrated the utility of the DCS method and the range of its capabilities. Selected applications in cancer, brain, and muscle are given below.

10.6.1 Cancer Therapy Monitoring

DCS has been utilized to monitor tumor contrast in breast cancer (Durduran et al. 2005, Zhou et al. 2007), to monitor early physiological changes in breast cancer in response to chemotherapy (Zhou et al. 2007), to monitor the effects of chemoradiation therapy on head-and-neck cancer (Sunar et al. 2006), to monitor response to photodynamic therapy (PDT) in prostate cancer (Du et al. 2006, Yu et al. 2006), and to assess the efficacy of cancer therapy in murine tumor models (Menon et al. 2003, Yu et al. 2005b, Song et al. 2007, Sunar et al. 2007, Busch et al. 2009, Cerniglia et al. 2009). Measurement and assessment of tissue/tumor hemodynamic changes during cancer treatment is particularly attractive for cancer therapies

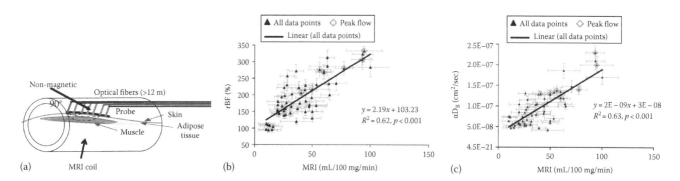

FIGURE 10.7 (a) Configuration of concurrent optical-MRI measurement. (b) Correlation between peak MRI flow and relative peak DCS flow. (c) Correlation of peak MRI flow and peak absolute DCS flow-index αD_b. Error bars were estimated from measurement repeatability. (From Yu, G. et al., *Opt. Express*, 15, 1064, 2007. With permission.)

that require tissue oxygen for treatment efficacy. For example, PDT requires tissue oxygen because the process creates singlet oxygen, which, in turn, kills tumor cells, and damages local vasculature (Yu et al. 2005b). PDT is well known to be less efficacious in patients with hypoxic tumors (Busch et al. 2000, Leach et al. 2002). Factors that modulate tissue oxygen include BF, blood oxygenation, and oxygen metabolism, i.e., factors that these new measurement tools can probe. Furthermore, cancer therapy can alter tumor hemodynamic/metabolic status, which further impacts treatment outcome. It is thus anticipated that functional assessment of tumor hemodynamic status during cancer therapy may provide information useful for early prediction of long-term treatment outcomes, thus enabling clinicians to optimize and individualize treatment. Tumor hemodynamics, however, are not routinely measured during cancer therapy due, in part, to a paucity of appropriate technologies.

In this subsection, we first describe a preclinical example that demonstrates the use of DCS for real-time monitoring of BF responses to PDT in murine tumors. A strong correlation between tumor hemodynamic changes during treatment and long-term treatment efficacy was found in this study, indicative of the clinical dosimetry potential of DCS for prediction of cancer therapy efficacy. A clinical translational study using DCS for monitoring and evaluation of prostate cancer therapy in humans is then described.

10.6.1.1 Prediction of Photodynamic Therapy in Murine Tumors

PDT requires administration of a photosensitizer that localizes in tumor tissue and is subsequently activated by exposure to optical radiation (Dougherty et al. 1998). The photoexcited photosensitizer initiates a cascade of chemical reactions, involving highly reactive oxygen intermediates that can cause necrosis and apoptosis of cells. Many studies suggest that PDT-mediated vascular damage significantly affects tissue oxygen

supply (BF) and thus contributes to long-term tumor response to therapy.

Monitoring of tumor hemodynamic responses during PDT, however, has proven difficult due to interference between measurement and treatment. In this study, a noncontact probe with source and detector fibers on the back image-plane of a camera was employed to avoid blocking the treatment light (see Figure 10.8a) (Yu et al. 2005b). Source-detector separations ranged from 1 to 4 mm, permitting light to penetrate to depths of ~0.5–2 mm below tumor surface. An optical filter mounted in front of the camera lens attenuated treatment light. DCS and the noncontact probe were employed to monitor the BF of murine tumors ($n = 15$) during light illumination in Photofrin-mediated PDT. Measurements were also made at specific time points after treatment (Yu et al. 2005b).

Figure 10.8b shows relative changes of tumor BF (rBF) during PDT. Within minutes of the start of PDT, rBF rapidly increased, followed by a decline, and subsequent peaks and declines with variable kinetics. The experiments discovered that the slope (flow-reduction rate) and duration (interval time, data not shown) over which rBF decreased following the initial PDT-induced increase was highly associated with treatment durability (see Figure 10.8c); here, treatment durability was defined as the time of tumor growth to a volume of 400 mm³ (pretreatment tumor volume was ~100 mm³). These findings were consistent with the hypothesis that treatment efficacy is a function of tumor oxygenation during PDT; under oxygen-limited conditions (e.g., such as might arise with rapidly declining BF), treatment efficacy was abrogated. After PDT, all animals showed decreases in rBF at 3 and 6 h, and rBF, at these time points, was also predictive of tumor response (data not shown) (Yu et al. 2005b).

Thus, these data demonstrate that DCS-measured changes in tumor rBF during and after Photofrin-PDT are predictive of treatment efficacy. The data further suggest that in situ BF monitoring during therapy may be very useful for real-time adjustment and optimization of PDT in humans.

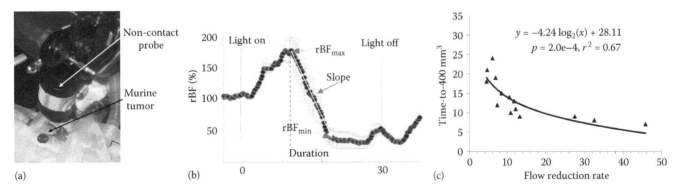

FIGURE 10.8 (a) Photograph of the noncontact DCS preclinical model probe. (b) Representative murine tumor BF responses to PDT. Plot shows rBF versus time (before, during, and just after PDT). Points with error bars represent the average ± SD. rBF$_{max}$ and rBF$_{min}$ depict the maximum and minimum, respectively, of the first peak in BF. Notice the substantial fluctuations in relative BF during PDT. Flow-reduction rate is defined by the slope of the decrease in BF after its initial PDT-induced increase. (c) Correlation between treatment durability/efficacy (i.e., tumor time-to-400 mm³) and flow-reduction rate (slope). The results indicate that tumor rBF during Photofrin-PDT is predictive of treatment efficacy. (From Yu, G. et al., *Clin. Cancer Res.*, 11, 3543, 2005b.)

10.6.1.2 Real-Time *In Situ* Monitoring of Human Prostate PDT

Armed with promising results from the murine models above, Yu et al. (2006) proceeded to adapt the DCS system for use in a Phase I clinical trial of interstitial human prostate PDT. A thin side-illumination fiber-optic probe (see Figure 10.9a) containing source and detector fibers was constructed with multiple source-detector separations (0.5–1.5 cm) (Yu et al. 2006). The fiber-optic probe was placed inside an 18-gauge catheter that had already been inserted into the patient's prostate gland. Five patients with locally recurrent prostate cancer in the Phase I trial of motexafin lutetium (MLu)-mediated PDT were measured using DCS and the side-illumination probe. The prostate was illuminated sequentially in several quadrants (Q1 – Q4) until the entire gland was treated.

Measured BF variation showed a similar trend in each individual. Figure 10.9b and c shows typical responses in BF over the course of PDT in two prostatic tumors. As was the case for murine tumors, a sharp decrease in prostate BF was observed [−41 ± 12% (*n* = 5)], suggesting that MLu-mediated PDT has an antivascular effect. The slope (flow-reduction rate) during PDT showed large interprostate heterogeneities; 15 %/min in Figure 10.9b versus 10 %/min in Figure 10.9c measured from two prostates during PDT. On average (*n* = 5), the flow-reduction rate from the five subjects was 12 ± 5 (%/min).

Although, in this case, the study did not attempt to correlate clinical outcome with DCS-measured flow response during PDT, clearly PDT-induced flow responses hold potential for prediction of treatment outcomes in humans, as shown in marine tumor models. The present study took a step in this direction.

10.6.2 Cerebral Physiology and Disease

Noninvasive CBF measurements provide physiological insight critical for both preclinical models and in clinical applications.

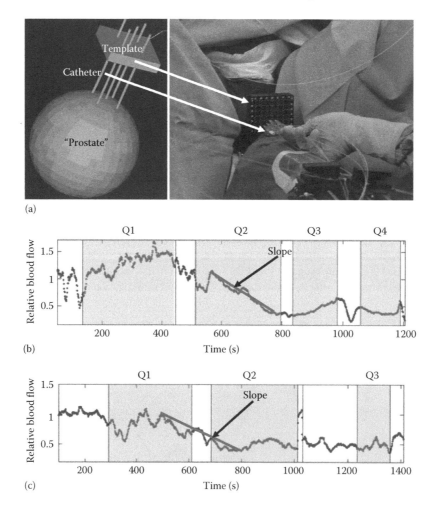

FIGURE 10.9 (a) Custom-made template for guidance of placing catheters for prostate PDT. The 732-nm treatment light was administered through the cylindrical diffusing fibers inside the catheters. Our small side-illumination probe was placed in the center of the prostate before PDT through one of the catheters. The catheter remained in place throughout PDT. (b) and (c) Tumor BF responses during PDT as a function of time measured from two prostates. Four quadrants of the prostate (Q1 – Q4) were illumination sequentially until the entire gland was treated. The illumination periods are presented as shaded areas. Flow-reduction rate is defined by the slope of the decrease in BF. (From Yu, G. et al., *Photochem. Photobiol.*, 82, 1279, 2006. With permission.)

DCS was first utilized in rat brain models in a monitoring capacity (Cheung et al. 2001) and for demonstration of 3D DCS tomography (Culver et al. 2003a,b, Zhou et al. 2006). It has also been utilized in neonatal piglet models of head trauma, illustrating potential for continuous long-term "bedside" monitoring (Zhou et al. 2009). In 2004, its use in human brain (through intact skull) was first demonstrated (Durduran 2004), and subsequently, the techniques have been applied for functional studies of CBF in healthy adults (Durduran et al. 2004b, Li et al. 2005, 2008). Noteworthy comparative studies demonstrated that hemodynamic responses to external functional stimuli of the sensorimotor cortex were in line with ASL-MRI and fMRI studies. In the clinical settings, DCS use has been reported in premature infants (Buckley et al. 2009, Roche-Labarbe et al. 2010), in neonates with congenital heart defects (Durduran et al. 2010a), in adults with acute ischemic stroke (Durduran et al. 2009b), and in TBI patients (Kim et al. 2010). In this subsection, we describe several preclinical and clinical examples of the use of DCS for CBF measurement.

10.6.2.1 3D rCBF Tomography of Cortical Spreading Depression in Rat Brain

Zhou et al. (2006) demonstrated the feasibility of in vivo 3D tomographic reconstruction of relative cerebral blood flow (rCBF) using DCS to probe a rat brain cortical spreading depression (CSD) model (see Figure 10.10a). CSD is a wave of excitation and depolarization of neuronal cells that spreads radially with a speed of 2–5 mm/min over the cerebral cortex (Leao 1944). CSD is accompanied by robust and localized (on the cortex) BF changes (Nielsen 2000, Ayata 2004). Thus, CSD is a good model for testing the feasibility of 3D diffuse optical tomography of BF.

Figure 10.10b shows reconstructed rCBF images localized at the cortex layer of the rat brain (~1 mm deep) during CSD. In this case, CSD was induced by placing a 1-mm^3 filter paper soaked in 2 mol/L potassium chloride (KCl) onto the dura. Images ((a), from left to right, from top to bottom) are shown roughly every 20 s from immediately before KCl was applied ($t \approx 26$ s) until the end of the second CSD peak. Notice that a strong increase in BF appears from the top of the image and proceeds to the bottom of the image. After the first peak, a sustained decrease in BF is observed (~3 min), which covers most of the image area. Three regions of interest (ROI) were selected, and rCBF changes therein are plotted in Figure 10.10b. The propagation of the CSD waves can be clearly identified from the delay between each curve. Figure 10.10c shows the dependence of maximal rCBF changes on depth using the data from the second ROI-2 in Figure 10.10a. The maximal change occurs at 1 mm (i.e., just below the skull), corresponding to the cortex surface. The peak spreads ~0.5 mm above and below the cortical surface as expected from the "resolution" broadening of the diffuse photons. No significant change is observed at the surface ($z = 0$ mm) nor in the deep region ($z = 3$ mm). Clearly, three-dimensional tomographic in vivo relative BF information is revealed. A movie demonstrating rCBF changes at different brain layers during CSD accompanies the original publication in *Optics Express* (Zhou et al. 2006).

10.6.2.2 Cerebral Cortical Blood Flow Responses to Functional Activity

The first reported use of DCS in human brain probed local CBF with DCS in motor cortex during sensorimotor stimuli (Durduran et al. 2004b). Durduran et al. (2004b) and later Gisler and coworkers (Li et al. 2005) reported measurements of cortical BF to finger-tapping stimulation. Importantly, Durduran et al. employed a hybrid optical instrument that combined DCS with NIRS to measure CBF as well as the concentrations of oxygenated hemoglobin (HbO$_2$), deoxygenated hemoglobin (Hb), and THC. They then combined this information in a model for cerebral metabolic rate of oxygen consumption (CMRO$_2$) to derive the variations of CMRO$_2$ during sensorimotor activation by an *all-optical* method. CMRO$_2$, in particular, is of great interest to the neuroscientists. The population-averaged results exhibited a robust change, which correlated with the activation. Mean changes observed were 39 ± 10% for rCBF, 12.5 ± 2.8 μM for HbO$_2$, −3.8 ± 0.8 μM for Hb, 8.3 ± 2.3 μM for THC, and 10.1% ± 4.4% for rCMRO$_2$. The CBF changes measured were well within the literature values (Roland et al. 1980, Colebatch et al. 1991, Seitz and Roland 1992, Ye et al. 1999, Kastrup et al. 2002). NIRS results were harder to cross-validate, since similar data are not available from other modalities. However, the NIRS results were in qualitative agreement with BOLD–fMRI (Kastrup et al. 2002, Mehagnoul-Schipper et al. 2002). Most interestingly, the increase in CMRO$_2$ is also within the range of values from hybrid MRI measurements (Hoge et al. 1999, Kastrup et al. 2002). Similar results have been recently reported in visual stimuli as well (Li et al. 2008).

10.6.2.3 CMRO$_2$ Estimates by NIRS-Only and NIRS-DCS in Premature Infants

In Section 10.5.1, experiments by Roche-Labarbe et al. (2010) were described; these measurements utilized DCS in premature born infants and validated its use against TCD (ultrasound) measurements. In the same report, the authors also demonstrate that the combination of NIRS and DCS to derive CMRO$_2$ could be more robust than NIRS-only models that they had utilized previously. They obtained measurements at multiple positions on the head and on a weekly basis for the first 6 weeks of life as shown in Figure 10.11. Interestingly, the THC concentration and the blood oxygen-saturation decrease as the premature born infant matures over time. This is in contrast to increasing CBF measured by DCS. If THC concentration is converted to cerebral blood volume (CBV) utilizing a simple model, there is no trend in CBV over time. Furthermore, if CBV is then utilized to derive CMRO$_2$ as is done in NIRS-only approaches, then both CBV and CMRO$_2$ appear unaltered over time. On the other hand, if CMRO$_2$ is derived using NIRS–DCS hybrid data, a linear increase with time is observed. This is in qualitative agreement with physiological expectations, which dictate that blood oxygen-saturation should decrease while

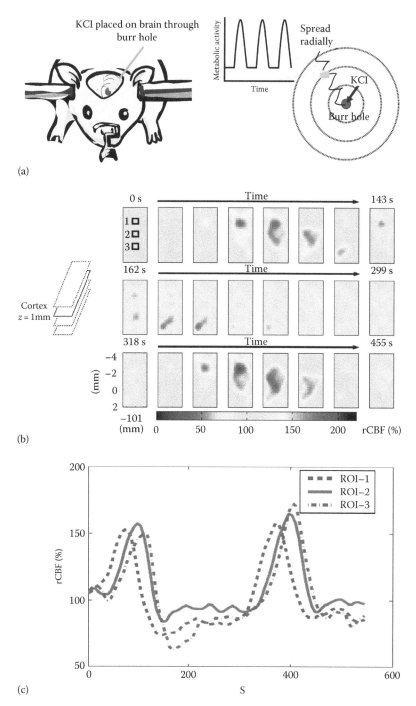

FIGURE 10.10 (See color insert.) CSD in rat brain. (a) A rat was fixed on a stereotaxic frame with the scalp retracted and the skull intact. CSD was induced by placing KCl solution on the rat brain through a small hole drilled through the skull. Periodic activations and deactivations of the neurons then spread out radially on the cortex. (b) rCBF changes on the cortex of the rat brain (~1 mm deep) as a function of time during CSD. rCBF images from the cortex (from left to right, from top to bottom) are shown roughly every 20 s starting immediately before KCl was applied until the end of the second CSD peak. A strong increase in BF appears from the top and proceeds to the bottom of the image. After the peak, there is a sustained decrease in BF, which covers most of the image area. The amount of rCBF changes is reflected in the image as a spectrum of color, with deep blue indicates a decrease in rCBF and dark red indicates an increase in rCBF. (c) Temporal rCBF curves from three ROI indicated in the 0 s image in (b), demonstrating the propagation of the CSD waves. (From Zhou, C. et al., *Opt. Express*, 14, 1125, 2006. With permission.)

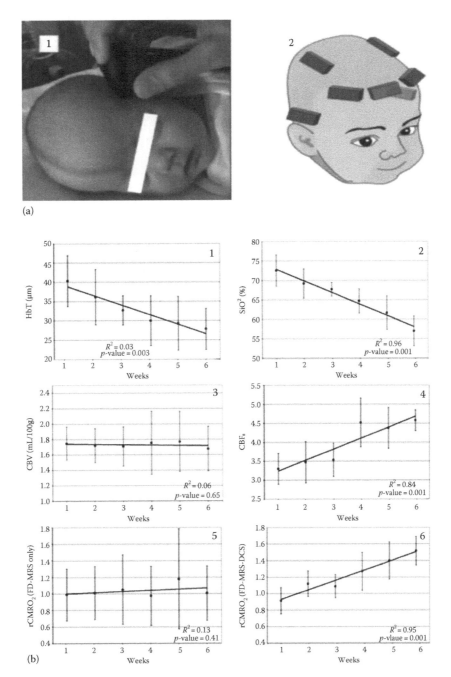

FIGURE 10.11 (a1) Picture schematic of a subject during measurement. (a2) Locations of recording on the subject's head. (b) Averages ± 95% confidence intervals as a function of age. (b1) THC concentration (HbT). (b2) Microvascular blood StO_2. (b3) CBV. (b4) Cerebral BF index (CBFix). (b5) $rCMRO_2$ estimated from FD-NIRS only. (b6) $rCMRO_2$ calculated by combining FD-NIRS and DCS. Bold lines are linear regressions; R^2 and P-value of the regressions are in the bottom right corner of each graph. HbT and StO_2 decrease while CBF increases as the premature born infant matures over time. When HbT is converted to CBV, the trends vanish and therefore if NIRS alone is used to derive $CMRO_2$, there are no trends in $CMRO_2$. However, if $CMRO_2$ is derived using NIRS–DCS hybrid data, a linear increase with time is observed. (From Roche-Labarbe, N. et al., *Hum. Brain Mapp.*, 31(3), 341, 2010. With permission.)

CBF is increasing; that is, oxygen utilization as well as oxygen metabolism should be increasing with the infant's age. The authors argue that these findings demonstrate the robustness of the NIRS-DCS combined model (notice large error bars in subfigure (5)), which might be expected since this model relies on fewer approximations than the NIRS-only model (Boas et al. 2003). Taken together with other recent reports (Durduran et al. 2010a) of DCS-NIRS use in neonates to estimate $CMRO_2$, current research demonstrates the feasibility and the importance of concurrent DCS and NIRS; the two complementary methodologies permit measurement of microvascular hemodynamics in a previously inaccessible way.

10.6.2.4 Cerebrovascular Hemodynamics in Adult Neurointensive Care Units

Working closely with neurologists, optical scientists have identified an unfilled niche application for DCS–NIRS hybrid approach as a bedside monitor in neurointensive care units for adults. To this end, we have studied the responses of a cohort of acute ischemic stroke patients (Durduran et al. 2009b) and TBI patients (Kim et al. 2010). In TBI patients, we have validated DCS against portable xenon-CT (Kim et al. 2008) and demonstrated good agreement with invasive measurements of intracranial pressure, cerebral perfusion pressure, and the partial pressure of oxygen during postural changes and hyperoxia. Here, we highlight our findings in acute ischemic stroke patients.

In the acute ischemic stroke population, we have induced mild orthostatic stress by changes in head-of-bed (HOB) positioning as shown in Figure 10.12a, with probes placed on the forehead near the frontal poles. CBF and hemoglobin concentrations were measured sequentially for 5 min at each HOB positions: 30, 15, 0, −5, and 0 degrees and normalized to their values at 30 degrees. In Figure 10.12b (left), the infarcted hemisphere (peri-infarct) shows a very large CBF increase in response to lowering of HOB position, whereas the opposite hemisphere (contrainfarct) shows minimal changes that are similar to those observed on healthy people. While this was expected and observed in most (~75%) of the people ($n = 17$), others have shown a "paradoxical" response

(see Figure 10.12b, right) where CBF decreased in response to lowering of the HOB position. Larger changes in peri-infarct hemisphere are presumably due to damaged cerebral autoregulation and are observed in both types of responses. The existence of the paradoxical response is an indicator for the potential usefulness of a bedside monitor for individualized stroke care. In control populations (Durduran et al. 2009b, Edlow et al. 2010), we have shown that both hemispheres behave in an identical fashion with postural challenge, as expected. Among other things, the example illustrates that diffuse optical instrumentation can be deployed at the neurointensive unit to directly monitor injured tissues.

10.6.3 Skeletal Muscle Hemodynamics

Characterization of the oxygen supply and metabolism with optical methods in skeletal muscles has important implications in exercise medicine, and for treatment, screening, and understanding of diseases such as the PAD. In addition, these types of measurements hold potential to improve fundamental understanding of muscle function within the context of the cardiovascular disease (Cheatle et al. 1991, Wallace et al. 1999).

In a pilot study (Yu et al. 2005a), issues of light penetration and flow sensitivity were addressed by experimentally investigating tissue layer responses during prolonged cuff occlusion.

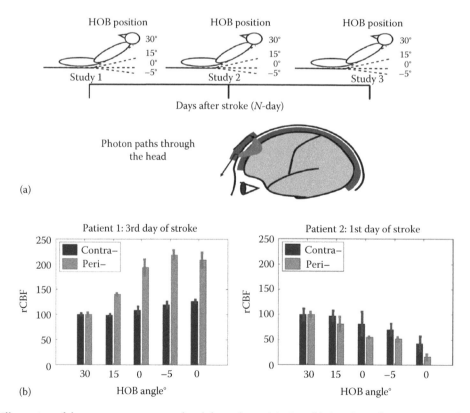

FIGURE 10.12 (a) Illustration of the measurement protocol and the probe positioning. (b) CBF data taken over 25 min from two representative subjects. A representative plot (left) was observed in about 75% of the subjects ($n = 17$); the infarcted hemisphere (peri-infarct) shows a very large CBF increase in response to lowering of HOB position. Others have shown a "paradoxical" response (right) where CBF decreased in response to lowering of the HOB position. (From Durduran, T. et al., *Opt. Express*, 17, 3884, 2009b. With permission.)

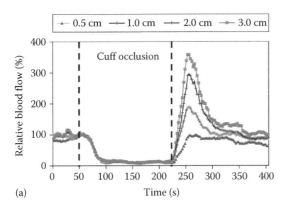
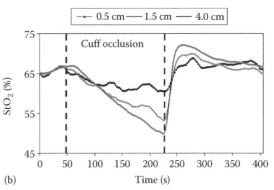

FIGURE 10.13 Representative curves of relative BF (a) and tissue blood oxygen saturation (StO$_2$) (b) as function of time during arterial cuff occlusion. Data are shown from different source-detector pairs measured on a healthy leg. The source-detector separations in the figure are 0.5, 1, 2, and 3 cm for DCS flow measurement and 0.5, 1.5, and 4.0 cm for NIRS oxygenation measurement, respectively. Vertical lines indicate the beginning and end of the occlusion period. Stronger reactive hyperemia after the release of occlusion and deoxygenation during occlusion were derived from source-detector pairs with large separations, i.e., 2 and 3 cm for DCS and 1.5 and 4 cm for NIRS, respectively. (From Yu, G. et al., *J. Biomed. Opt.*, 10, 024027, 2005a. With permission.)

A contact fiber-optical probe with source-detector separations of 0.5, 1, 2, and 3 cm was employed for this study (see Figure 10.5a). Based on diffusion theory, the light penetration depth depends on tissue optical properties and source-detector separation (typically, penetration is approximately one-third to one-half of the source-detector separation on the tissue surface).

Ten healthy subjects and one patient with PAD were measured. A 3 min cuff occlusion protocol was used to investigate the different layer responses in skeletal muscles in order to estimate light penetration depth and to validate results in the ischemic states. A skinfold caliper was used to mechanically measure the thickness of the upper layers (skin and adipose) above muscle. The thickness of the upper layers (skin and adipose tissues) above leg flexors and wrist flexors was 5.5 ± 0.4 mm and 2.8 ± 0.6 mm, respectively (n = 10). Therefore, the optical signals detected from the large separations (≥2 cm) are mainly from the deep muscle tissues.

Figure 10.13 shows the typical responses of rBF and blood O$_2$ saturation (StO$_2$) during leg arterial occlusion from the different source-detector pairs on a healthy individual. Hemodynamic responses derived from source-detector pairs with large separations (≥2 cm) were significantly stronger than those from the shortest source-detector separations (0.5 cm), consistent with the larger responses expected for muscle tissue compared to top skin and adipose tissues.

Table 10.1 lists the hemodynamic responses in cuff occlusions from 10 healthy volunteers and one PAD patient. For the healthy volunteers, cuff occlusion of the leg flexor and arm flexor muscles produced a similar response. The hemodynamic response trends in the PAD patient (data not shown) were similar to those of the healthy volunteers, and different responses were not found in the arm muscles of healthy controls compared to the patient. However, in the PAD patient leg muscle, the relative magnitude of reactive hyperemia was ~1/2× of the controls, and the

recovery half-time of both StO$_2$ and rBF after occlusion was ~3× those of the controls.

Further investigations will test the capability of diffuse optical techniques for screening and diagnosis of PAD. In a different vein, the community has begun to consider DCS measurement during exercise, but these measurements can have motional artifacts; better understanding and characterization of these motional artifacts is needed and may require analysis of the methodologies at a fundamental level.

10.7 Summary

In this chapter, we have outlined the development and application of DCS. The technique has now been adopted by several research groups around the world for measurement of BF in deep tissues. In a relatively short time, we have witnessed the translation of DCS from its theoretical conception to its application in intensive care units at the hospital. These diffuse optical methods are flexible, portable, and rapid and can be combined with other modalities such as NIRS or tomography, MRI, PET, and CT. Currently, its use is limited to measurements of relative BF, but promising results are indicative of potential for absolute measurements in the future.

Open questions about the technique still remain and will be interesting to consider. For example, the interaction of photons with moving red-blood cells in the complex environment of the tissue microvasculature is only partially understood (see Section 10.3.3). Thus, empirical approaches to defining BF indices have been adopted. While these empirical indices have withstood extensive cross-validation (see Section 10.5), it remains desirable to generate a more fundamental understanding of the origins of this BF index. We expect improved physical understanding, along with many more clinical applications, to emerge rapidly as the technique is adopted by others.

TABLE 10.1 Responses in Cuff Occlusions from 10 Healthy Volunteers and 1 PAD Patient

Parameters	Subjects	T_m (s)	Max Δ	T_{50} (s)	OS (Δ)
Leg Occlusion					
StO$_2$ (%)	Healthy	177.1 ± 20.7	-16.4 ± 4.4	33.7 ± 26.0	3.8 ± 1.7
	PAD	180.0	-15.0	96.0*	3.0
THC (μM)	Healthy	88.1 ± 81.9	-1.8 ± 5.9	16.2 ± 18.3	2.8 ± 3.1
	PAD	25.0	-10.0	36.0	5.0
rBF (%)	Healthy	51.0 ± 11.5	-90.0 ± 2.4	25.6 ± 14.5	311.4 ± 90.8
	PAD	60.0	-93.0	90.0*	165.0*
Arm Occlusion					
StO$_2$ (%)	Healthy	174.7 ± 15.3	-25.1 ± 8.2	19.4 ± 15.2	11.4 ± 5.0
	PAD	180.0	-23.0	23.0	10.0
THC (μM)	Healthy	46.6 ± 61.2	-1.4 ± 6.4	13.6 ± 7.3	8.6 ± 5.0
	PAD	111.0	-16.0	20.0	22.0
rBF (%)	Healthy	14.0 ± 7.4	-90.3 ± 3.8	11.3 ± 6.1	445.1 ± 194.1
	PAD	11.0	-92.0	12.0	450.0

Source: Yu, G. et al., *J. Biomed. Opt.*, 10, 024027, 2005a, Table 1. With permission.

Time to reach maximal change (T_m), maximal change (Max Δ), recovery half-time (T_{50}), and hyperemic overshoot (OS) are shown for StO$_2$ (%), THC (μM), and rBF (%). The 100% is assigned for baseline BF. Means $\pm$ SD are reported.

* Substantially different, healthy volunteers versus the PAD patient.

Acknowledgments

We gratefully acknowledge collaborations and discussions with numerous scientists at the University of Pennsylvania and in the biomedical optics community in general. Our work would not have been possible without the support and guidance of our clinical and physiological collaborators. Much of the research described in this review was facilitated by collaborations over many years with many colleagues, including David Boas, Erin Buckley, Theresa Busch, Britton Chance, Cecil Cheung, Joseph Culver, Regine Choe, John Detre, Lixin Dong, Douglas Fraker, Thomas Floyd, Joel Greenberg, Steven Hahn, Andrew Kofke, Meeri Kim, Daniel Licht, Gwen Lech, Emilie Mohler, Susan Margulies, Goro Nishimura, Harry Quon, Mark Rosen, Mitchell Schnall, Ulas Sunar, Bruce Tromberg, Jionjiong Wang, Xiaoman Xing, and Leonid Zubkov. One of us (AGY) particularly acknowledges the collaboration and camaraderie of Britton Chance, who for two-decades provided us with his institution encouragement, and willing ear.

Grant Acknowledgments

We gratefully acknowledge support from

- **NIH grants**: CA-149274 (G. Yu), HL-083225, HL-57835 (A. G. Yodh), NS-60653 (A. G. Yodh), RR-02305 (A. G. Yodh and R. Reddy), EB-07610 (T. Durduran), Ns-45839 (A. G. Yodh), CA-126187 (A. G. Yodh)
- **AHA grants:** BGIA 2350015 (G. Yu), BGIA 0665446U (G. Yu)
- **DOD award:** W81XWH-04-1-0006 (G. Yu)
- **Thrasher Research Fund**: NR 0016 (T. Durduran)
- **Fundacio Cellex Barcelona** (T. Durduran)

References

Ackerson, B. J., Dougherty, R. L., Reguigui, N. M., and Nobbman, U. 1992. Correlation transfer: Application of radiative transfer solution methods to photon correlation problems. *J. Thermophys. Heat Trans.* 6: 577–588.

Ackerson, B. J. and Pusey, P. N. 1988. Shear-induced order in suspensions of hard-spheres. *Phys. Rev. Lett.* 61: 1033–1036.

Ayata, C., Shin, H. K., Salomone, S. et al. 2004. Pronounced hypoperfusion during spreading depression in mouse cortex. *J. Cereb. Blood Flow Metab.* 24(10): 1172–1182.

Baron, J. C. 1999. Mapping the ischaemic penumbra with PET: Implications for acute stroke treatment. *Cerebrovasc. Dis.* 9: 193–201.

Belau, M., Ninck, M., Hering, G. et al. 2010. Noninvasive observation of skeletal muscle contraction using near-infrared time-resolved reflectance and diffusing-wave spectroscopy. *J. Biomed. Opt.* 15: 057007.

Benaron, D. A. and Stevenson, D. K. 1993. Optical time-of-flight and absorbance imaging of biologic media. *Science* 259: 1463–1466.

Berne, B. J. and Pecora, R. 1990. *Dynamic Light Scattering with Applications to Chemistry, Biology, and Physics.* Malabar, FL: Krieger.

Binzoni, T., Leung, T. S., Seghier, M. L., and Delpy, D. T. 2004. Translational and Brownian motion in laser-Doppler flowmetry of large tissue volumes. *Phys. Med. Biol.* 49: 5445–5458.

Boas, D. 1996. Diffuse photon probes of structural and dynamical properties of turbid media: Theory and biomedical applications. PhD dissertation, University of Pennsylvania, Philadelphia, PA.

Boas, D. A., Campbell, L. E., and Yodh, A. G. 1995. Scattering and imaging with diffusing temporal field correlations. *Phys. Rev. Lett.* 75: 1855–1858.

Boas, D. A., Culver, J. P., Stott, J. J., and Dunn, A. K. 2002. Three dimensional Monte Carlo code for photon migration through complex heterogeneous media including the adult human head. *Opt. Express* 10: 159–170.

Boas, D. A., Strangman, G., Culver, J. P. et al. 2003. Can the cerebral metabolic rate of oxygen be estimated with near-infrared spectroscopy? *Phys. Med. Biol.* 48: 2405–2418.

Boas, D. A. and Yodh, A. G. 1997. Spatially varying dynamical properties of turbid media probed with diffusing temporal light correlation. *J. Opt. Soc. Am. A Opt. Image Sci. Vis.* 14: 192–215.

Bonner, R. and Nossal, R. 1981. Model for laser Doppler measurements of blood flow in tissue. *Appl. Opt.* 20: 2097–2107.

Bonner, R. F., Nossal, R., Havlin, S., and Weiss, G. H. 1987. Model for photon migration in turbid biological media. *J. Opt. Soc. Am. A* 4: 423–432.

Briers, J. D. 2001. Laser Doppler, speckle and related techniques for blood perfusion mapping and imaging. *Physiol. Meas.* 22: R35–R66.

Brown, W. 1993. *Dynamic Light Scattering: The Method and Some Applications.* New York: Clarendon Press.

Buckley, E. M., Cook, N. M., Durduran, T. et al. 2009. Cerebral hemodynamics in preterm infants during positional intervention measured with diffuse correlation spectroscopy and transcranial Doppler ultrasound. *Opt. Express* 17: 12571–12581.

Busch, T. M., Hahn, S. M., Evans, S. M., and Koch, C. J. 2000. Depletion of tumor oxygenation during photodynamic therapy: Detection by the hypoxia marker EF3 [2-(2-nitro-imidazol-1[H]-yl)-*N*-(3,3,3-trifluoropropyl)acetamide]. *Cancer Res.* 60: 2636–2642.

Busch, T. M., Xing, X., Yu, G. et al. 2009. Fluence rate-dependent intratumor heterogeneity in physiologic and cytotoxic responses to Photofrin photodynamic therapy. *Photochem. Photobiol. Sci.* 8: 1683–1693.

Carp, S. A., Dai, G. P., Boad, D. A. et al. 2010. Validation of diffuse correlation spectroscopy measurements of rodent cerebral blood flow with simultaneous arterial spin labeling MRI; towards MRI-optical continuous cerebral metabolic monitoring. *Biomed. Opt. Express* 1(2): 553–565.

Cerniglia, G. J., Pore, N., Tsai, J. H. et al. 2009. Epidermal growth factor receptor inhibition modulates the microenvironment by vascular normalization to improve chemotherapy and radiotherapy efficacy. *PLoS One* 4: e6539.

Chance, B. 1998. Near-infrared images using continuous, phase-modulated, and pulsed light with quantitation of blood and blood oxygenation. *Adv. Opt. Biopsy Opt. Mammography Ann. NY Acad. Sci.* 838: 19–45.

Chance, B., Anday, E., Nioka, S. et al. 1998. A novel method for fast imaging of brain function, non-invasively, with light. *Opt. Express* 2: 411–423.

Cheatle, T. R., Potter, L. A., Cope, M. et al. 1991. Near-infrared spectroscopy in peripheral vascular disease. *Br. J. Surg.* 78: 405–408.

Chen, Z., Milner, T. E., Wang, X., Srinivas, S., and Nelson, J. S. 1998. Optical Doppler tomography: Imaging in vivo blood flow dynamics following pharmacological intervention and photodynamic therapy. *Photochem. Photobiol.* 67: 56–60.

Cheung, C., Culver, J. P., Takahashi, K., Greenberg, J. H., and Yodh, A. G. 2001. In vivo cerebrovascular measurement combining diffuse near-infrared absorption and correlation spectroscopies. *Phys. Med. Biol.* 46: 2053–2065.

Cheung, R., Solonenko, M., Busch, T. M. et al. 2003. Correlation of in vivo photosensitizer fluorescence and photodynamic-therapy-induced depth of necrosis in a murine tumor model. *J. Biomed. Opt.* 8: 248–252.

Choe, R., Konecky, S. D., Corlu, A. et al. 2009. Differentiation of benign and malignant breast tumors by in-vivo three-dimensional parallel-plate diffuse optical tomography. *J. Biomed. Opt.* 14: 024020.

Chu, B. (ed.) (1991) *Laser Light Scattering: Basic Principles and Practice.* New York: Academic Press.

Colebatch, J. G., Deiber, M. P., Passingham, R. E., Friston, K. J., and Frackowiak, R. S. J. 1991. Regional cerebral blood-flow during voluntary arm and hand movements in human subjects. *J. Neurophysiol.* 65: 1392–1401.

Culver, J. P., Durduran, T., Furuya, T. et al. 2003a. Diffuse optical tomography of cerebral blood flow, oxygenation, and metabolism in rat during focal ischemia. *J. Cereb. Blood Flow Metab.* 23: 911–924.

Culver, J. P., Siegel, A. M., Stott, J. J., and Boas, D. A. 2003b. Volumetric diffuse optical tomography of brain activity. *Opt. Lett.* 28: 2061–2063.

Culver J. P., Durduran T., Cheung C., Furuya D., Greenberg J. H., and Yodh A. G. 2003c. Diffuse optical measurement of hemoglobin and cerebral blood flow in rat brain during hypercapnia, hypoxia and cardiac arrest. *Adv Exp Med Biol.* 510: 293–297.

Detre, J. A., Leigh, J. S., Williams, D. S., and Koretsky, A. P. 1992. Perfusion imaging. *Magn. Reson. Med.* 23: 37–45.

Dietsche, G., Ninck, M., Ortolf, C. et al. 2007. Fiber-based multispeckle detection for time-resolved diffusing-wave spectroscopy: Characterization and application to blood flow detection in deep tissue. *Appl. Opt.* 46: 8506–8514.

Dougherty, T. J., Gomer, C. J., Henderson, B. W., Jori, G., Kessel, D., Korbelik, M., Moan, J., and Peng, Q. 1998. Photodynamic therapy. *J. Nat. Cancer Inst.* 90: 889–905.

Du, K. L., Mick, R., Busch, T. et al. 2006. Preliminary results of interstitial motexafin lutetium-mediated PDT for prostate cancer. *Lasers Surg. Med.* 38: 427–434.

Dunn, A. K., Bolay, T., Moskowitz, M. A., and Boas, D. A. 2001. Dynamic imaging of cerebral blood flow using laser speckle. *J. Cereb. Blood Flow Metab.* 21: 195–201.

Durduran, T. 2004. Non-invasive measurements of tissue hemo-dynamics with hybrid diffuse optical methods. PhD thesis, University of Pennsylvania, Philadelphia, PA.

Durduran, T., Burnett, M. G., Yu, G. et al. 2004a. Spatiotemporal quantification of cerebral blood flow during functional activation in rat somatosensory cortex using laser-speckle flow-metry. *J. Cereb. Blood Flow Metab.* 24: 518–525.

Durduran, T., Choe, R., Culver, J. P. et al. 2002. Bulk optical properties of healthy female breast tissue. *Phys. Med. Biol.* 47: 2847–2861.

Durduran, T., Choe, R., Yu, G. et al. 2005. Diffuse optical measurement of blood flow in breast tumors. *Opt. Lett.* 30: 2915–2917.

Durduran, T., Kim, M. N., Buckley, E. M. et al. 2009a. Validation of diffuse correlation spectroscopy for measurement of cerebral blood flow across spatial scales and against multiple modalities. *SPIE Photonics West.* San Jose, CA, 2009, SPIE.

Durduran, T., Kim, M. N., Buckley, E. M. et al. 2008. Diffuse optical monitoring of cerebral oxygen metabolism at the bedside in cerebrovascular disorders. *OSA: Annual Meeting, Frontiers in Optics.* Rochester, NY, 2008, OSA.

Durduran, T., Yu, G., Burnett, M. G. et al. 2004b. Diffuse optical measurement of blood flow, blood oxygenation, and metabolism in a human brain during sensorimotor cortex activation. *Opt. Lett.* 29: 1766–1768.

Durduran, T., Zhou, C., Buckley, E. M. et al. 2010a. Optical measurement of cerebral hemodynamics and oxygen metabolism in neonates with congenital heart defects. *J. Biomed. Opt.* 15(3): 037004.

Durduran, T., Choe, R., Baker, W. B. et al. 2010b. Diffuse optics for tissue monitoring and tomography. *Rep. Prog. Phys.* 73(2010): 076701.

Durduran, T., Zhou, C., Edlow, B. L. et al. 2009b. Transcranial optical monitoring of cerebrovascular hemodynamics in acute stroke patients. *Opt. Express* 17: 3884–3902.

Durduran, T., Zhou, C., Yu, G. et al. 2007. Preoperative measurement of CO_2 reactivity and cerebral autoregulation in neonates with severe congenital heart defects. *SPIE Photonics West.* San Jose, CA, 2007, SPIE.

Edlow, B. L., Kim, M. N., Durduran, T. et al. 2010. The effect of healthy aging on cerebral hemodynamic responses to posture change. *Physiol. Meas.,* 31(4):477–495.

Einstein, A. 1905. On the motion of small particles suspended in liquids at rest required by the molecular-kinetic theory of heat. *Ann. Phys.* 17: 549–560.

Fantini, S., Hueber, D., Franceschini, M. A. et al. 1999. Non-invasive optical monitoring of the newborn piglet brain using continuous-wave and frequency-domain spectroscopy. *Phys. Med. Biol.* 44: 1543–1563.

Feke, G. T. and Riva, C. E. 1978. Laser Doppler measurements of blood velocity in human retinal-vessels. *J. Opt. Soc. Am.* 68: 526–531.

Gagnon, L., Desjardins, M., Bherer, L., and Lesage, F. 2009. Double layer estimation of flow changes using diffuse correlation spectroscopy. *Optical Tomography and Spectroscopy of Tissue VIII.* San Jose, CA, 2009, Proc. SPIE.

Gagnon, L., Desjardins, M., Jehanne-Lacasse, J., Bherer, L., and Lesage, F. 2008. Investigation of diffuse correlation spectroscopy in multi-layered media including the human head. *Opt. Express* 16: 15514–15530.

Gisler, T., Ruger, H., Egelhaaf, S. U. et al. 1995. Mode-selective dynamic light scattering: theory versus experimental realization. *Appl. Opt.* 34: 3546–3553.

Gisler, T., Rueger, H., Engelhaaf, S. U. et al. 1997. Mode-selective dynamic light scattering: theory versus experimental realization. *Appl. Opt. (Reprint)* 34: 3546–3553.

Gopinath, S. P., Robertson, C. S., Grossman, R. G., and Chance, B. 1993. Near-infrared spectroscopic localization of intracranial hematomas. *J. Neurosurg.* 79: 43–47.

Haskell, R. C., Svaasand, L. O., Tsay, T. et al. 1994. Boundary conditions for the diffusion equation in radiative transfer. *J. Opt. Soc. Am. A* 11: 2727–2741.

Heckmeier, M., Skipetrov, S. E., Maret, G., and Maynard, R. 1997. Imaging of dynamic heterogeneities in multiple-scattering media. *J. Opt. Soc. Am.* 14: 185–191.

Hielscher, A., Tittel, F. K., and Jacques, S. L. 1993. Non-invasive monitoring of blood oxygenation by phase resolved transmission spectroscopy. In: *Photon Migration and Imaging in Random Media and Tissues,* eds. B. Chance and R. R. Alfano. Los Angeles, CA: SPIE, pp. 275–288.

Hoge, R. D., Atkinson, J., Gill, B. et al. 1999. Linear coupling between cerebral blood flow and oxygen consumption in activated human cortex. *Proc. Natl. Acad. Sci. U. S. A.* 96: 9403–9408.

Hoskins, P. R. 1990. Measurement of arterial blood-flow by Doppler ultrasound. *Clin. Phys. Physiol. Meas.* 11: 1–26.

Jacques, S. L. 1989. Time-resolved reflectance spectroscopy in turbid tissues. *IEEE Trans. Biomed. Eng.* 36: 1155–1161.

Jaillon, F., Skipetrov, S. E., Li, J. et al. 2006. Diffusing-wave spectroscopy from head-like tissue phantoms: Influence of a non-scattering layer. *Opt. Express* 14: 10181–10194.

Kastrup, A., Kruger, G., Neumann-Haefelin, T., Glover, G. H., and Moseley, M. E. 2002. Changes of cerebral blood flow, oxygenation, and oxidative metabolism during graded motor activation. *Neuroimage* 15: 74–82.

Kidwell, C. S., Alger, J. R., and Saver, J. L. 2003. Beyond mismatch: Evolving paradigms in imaging the ischemic penumbra with multimodal magnetic resonance imaging. *Stroke* 34: 2729–2735.

Kim, M. N., Durduran, T., Frangos, S. et al. 2008. Validation of diffuse correlation spectroscopy against xenon CTCBF in humans after traumatic brain injury or subarachnoid hemorrhage. *Neurocritical Care Society Annual Meeting.* Miami, FL, 2008.

Kim, M. N., Durduran, T., Frangos, S. et al. 2010. Noninvasive measurement of cerebral blood flow and blood oxygenation using near-infrared and diffuse correlation spectroscopies in critically brain-injured adults. *Neurocrit. Care* 12(2): 173–180.

Kuebler, W. M. 2008. How NIR is the future in blood flow monitoring? *J. Appl. Physiol.* 104: 905–906.

Latchaw, R. E., Yonas, H., Hunter, G. J. et al. 2003. Guidelines and recommendations for perfusion imaging in cerebral ischemia: A scientific statement for healthcare professionals by the writing group on perfusion imaging, from the Council on Cardiovascular Radiology of the American Heart Association. *Stroke* 34: 1084–1104.

Leach, R. M., Hill, H. S., Snetkov, V. A., and Ward, J. P. T. 2002. Hypoxia, energy state and pulmonary vasomotor tone. *Respir. Physiol. Neurobiol.* 132: 55–67.

Leao, A. A. P. 1944. Spreading depression of activity in the cerebral cortex. *J. Neurophysiol.* 7: 359–390.

Li, J., Dietsche, G., Iftime, D. et al. 2005. Noninvasive detection of functional brain activity with near-infrared diffusing-wave spectroscopy. *J. Biomed. Opt.* 10: 044002-1–044002-12.

Li, J., Ninck, M., Koban, L. et al. 2008. Transient functional blood flow change in the human brain measured noninvasively by diffusing-wave spectroscopy. *Opt. Lett.* 33: 2233–2235.

Li, Z., Krishnaswamy, V., Davis, S. C. et al. 2009. Video-rate near infrared tomography to image pulsatile absorption properties in thick tissue. *Opt. Express* 17: 12043–12056.

Liu, D. L., Svanberg, K., Wang, I., Andersson-Engels, S., and Svanberg, S. 1997. Laser Doppler perfusion imaging: New technique for determination of perfusion and reperfusion of splanchnic organs and tumor tissue. *Lasers Surg. Med.* 20: 473–479.

MacKintosh, F. C. and John, S. 1989. Diffusing-wave spectroscopy and multiple scattering of light in correlated random media. *Phys. Rev. B* 40: 2382–2406.

Mahagne, M. H., David, O., Darcourt, J. et al. 2004. Voxel-based mapping of cortical ischemic damage using Tc 99m L,L-ethyl cysteinate dimer SPECT in acute stroke. *J Neuroimaging* 14: 23–32.

Maret, G. 1997. Diffusing-wave spectroscopy. *Curr. Opin. Colloid Interface Sci.* 2: 251–257.

Maret, G. and Wolf, P. E. 1987. Multiple light scattering from disordered media. The effect of brownian motion of scatterers. *Z. Phys. B* 65: 409–413.

Mehagnoul-Schipper, D. J., Van Der Kallen, B. F. W., Colier, W. et al. 2002. Simultaneous measurements of cerebral oxygenation changes during brain activation by near-infrared spectroscopy and functional magnetic resonance imaging in healthy young and elderly subjects. *Hum. Brain Mapp.* 16: 14–23.

Menon, C., Polin, G. M., Prabakaran, I. et al. 2003. An integrated approach to measuring tumor oxygen status using human melanoma xenografts as a model. *Cancer Res.* 63: 7232–7240.

Mesquita, R. C., Skuli, N., Kim, M. N. et al. 2010. Hemodynamic and metabolic diffuse optical monitoring in a mouse model of hindlimb ischemia. *Biomed. Opt. Express* 1(4): 1173–1187.

Middleton, A. A. and Fisher, D. S. 1991. Discrete scatterers and autocorrelations of multiply scattered light. *Phys. Rev. B* 43: 5934–5938.

Nielsen, A. N., Fabricius, M., and Lauritzen, M. 2000. Scanning laser-doppler flowmetry of rat cerebral circulation during cortical spreading depression. *J. Vasc. Res.* 37(6): 513–522.

Patterson, M. S., Chance, B., and Wilson, B. C. 1989. Time resolved reflectance and transmittance for the non-invasive measurement of tissue optical properties. *Appl. Opt.* 28: 2331–2336.

Petersen, E. T., Zimine, I., Ho, Y.C.L. & Golay, X. 2006. Noninvasive measurement of perfusion: A critical review of arterial spin labelling techniques. *Br. J. Radiol.* 79: 688–701.

Pine, D. J., Weitz, D. A., Chaikin, P. M., and Herbolzheimer, E. 1988. Diffusing-wave spectroscopy. *Phys. Rev. Lett.* 60: 1134–1137.

Rice, S. O. 1954. Mathematical analysis of random noise. In: *Noise and Stochastic Processes*, ed. N. Wax. New York: Dover, p. 133.

Roche-Labarbe, N., Carp, S. A., Surova, A. et al. 2010 Noninvasive optical measures of CBV, StO2, CBF Index, and rCMRO2 in human premature neonates' brains in the first six weeks of life. *Hum. Brain Mapp.* 31(3): 341–352.

Roland, P. E., Larsen, B., Lassen, N. A., and Skinhoj, E. 1980. Supplementary motor area and other cortical areas in organization of voluntary movements in man. *J. Neurophysiol.* 43: 118–136.

Sehgal, C. M., Arger, P. H., Rowling, S. E. et al. 2000. Quantitative vascularity of breast masses by Doppler imaging: Regional variations and diagnostic implications. *J. Ultrasound Med.* 19: 427–40; quiz 441–442.

Seitz, R. J. and Roland, P. E. 1992. Learning of sequential finger movements in man—a combined kinematic and positron emission tomography (PET) study. *Eur. J. Neurosci.* 4: 154–165.

Shang, Y., Symons, T. B., Durduran, T. et al. 2010. Effects of muscle fiber motion on diffuse correlation spectroscopy blood flow measurements during exercise. *Biomed. Opt. Express* 1(2): 500–511.

Shang, Y., Zhao, Y., Cheng, R. et al. 2009. Portable optical tissue flow oximeter based on diffuse correlation spectroscopy. *Opt. Lett.* 34: 3556–3558.

Song, L., Li, H., Sunar, U. et al. 2007. Naphthalocyanine-reconstituted LDL nanoparticles for in vivo cancer imaging and treatment. *Int. J. Nanomed.* 2: 767–774.

Stephen, M. J. 1988. Temporal fluctuations in wave propagation in random media. *Phys. Rev. B* 37: 1–5.

Stern, M. D. 1975. In vivo evaluation of microcirculation by coherent light scattering. *Nature* 254: 56–58.

Sunar, U., Makonnen, S., Zhou, C. et al. 2007. Hemodynamic responses to antivascular therapy and ionizing radiation assessed by diffuse optical spectroscopies. *Opt. Express* 15: 15507–15516.

Sunar, U., Quon, H., Durduran, T. et al. 2006. Noninvasive diffuse optical measurement of blood flow and blood oxygenation for monitoring radiation therapy in patients with head and neck tumors: A pilot study. *J. Biomed. Opt.* 11(6): 064021–064021.

Tanaka, T., Riva, C., and Ben-Sira, I. 1974. Blood velocity measurements in human retinal vessels. *Science* 186: 830–831.

Tong, P., Goldburg, W. I., Chan, C. K., and Sirivat, A. 1988. Turbulent transition by photon-correlation spectroscopy. *Phys. Rev. A* 37: 2125–2133.

Valkov, A. Y. and Romanov, V. P. 1986. Characteristics of propagation and scattering of light in nematic liquid crystals. *Sov. Phys. JETP* 63: 737–743.

Varma, H. M., Nandakumaran, A. K., and Vasu, R. M. 2009. Study of turbid media with light: Recovery of mechanical and optical properties from boundary measurement of intensity autocorrelation of light. *J. Opt. Soc. Am. A Opt. Image Sci. Vis.* 26: 1472–1483.

Wallace, D. J., Michener, B., Choudhury, D. et al. 1999. Summary of the results of a 95 subject human clinical trial for the diagnosis of peripheral vascular disease using a near infrared frequency domain hemoglobin spectrometer. *Proc. SPIE*: 300–316.

Wang, H. W., Putt, M. E., Emanuele, M. J. et al. 2004. Treatment-induced changes in tumor oxygenation predict photodynamic therapy outcome. *Cancer Res.* 64: 7553–7561.

Williams, D. S., Detre, J. A., Leigh, J. S., and Koretsky, A. P. 1992. Magnetic resonance imaging of perfusion using spin inversion of arterial water. *PNAS* 89: 212–216.

Wintermark, M., Maeder, P., Verdun, F. R. et al. 2000. Using 80 kVp versus 120 kVp in perfusion CT measurement of regional cerebral blood flow. *Am. J. Neuroradiol.* 21: 1881–1884.

Wintermark, M., Sesay, M., Barbier, E. et al. 2005. Comparative overview of brain perfusion imaging techniques. *Stroke* 36: 83–99.

Wolf, M., Franceschini, M. A., Paunescu, L. A. et al. 2003. Absolute frequency-domain pulse oximetry of the brain: Methodology and measurements. *Adv. Exp. Med. Biol.* 530: 61–73.

Wolf, R. L., Maret, G., Akkermans, E., and Maynard, R. 1988. Optical coherent backscattering by random media: An experimental study. *J. Phys. France* 49: 60–75.

Xing, X., Mohler, E. R., Zhou, C. et al. 2007. Hemodynamic changes in diabetic pig muscle. *SVMB 18th Annual Meeting*, Baltimore, MA, 2007.

Ye, F. Q., Yang, Y. H., Duyn, J. et al. 1999. Quantitation of regional cerebral blood flow increases during motor activation: A multislice, steady-state, arterial spin tagging study. *Magn. Reson. Med.* 42: 404–407.

Yodh, A. G., Kaplan, P. D., and Pine, D. J. 1990. Pulsed diffusing-wave spectroscopy: High resolution through nonlinear optical gating. *Phys. Rev. B* 42: 4744–4747.

Yu, G., Durduran, T., Furuya, D., Greenberg, J. H., and Yodh, A. G. 2003. Frequency-domain multiplexing system for in vivo diffuse light measurements of rapid cerebral hemodynamics. *Appl. Opt.* 42: 2931–2939.

Yu, G., Durduran, T., Lech, G. et al. 2005a. Time-dependent blood flow and oxygenation in human skeletal muscles measured with noninvasive near-infrared diffuse optical spectroscopies. *J. Biomed. Opt.* 10: 024027.

Yu, G., Durduran, T., Zhou, C. et al. 2005b. Noninvasive monitoring of murine tumor blood flow during and after photodynamic therapy provides early assessment of therapeutic efficacy. *Clin. Cancer Res.* 11: 3543–3552.

Yu, G., Durduran, T., Zhou, C. et al. 2006. Real-time in situ monitoring of human prostate photodynamic therapy with diffuse light. *Photochem. Photobiol.* 82: 1279–1284.

Yu, G., Floyd, T., Durduran, T. et al. 2007. Validation of diffuse correlation spectroscopy for muscle blood flow with concurrent arterial spin labeled perfusion MRI. *Opt. Express* 15: 1064–1075.

Zhou, C. 2007. In-vivo optical imaging and spectroscopy of cerebral hemodynamics. PhD dissertation, University of Pennsylvania, Philadephia, PA.

Zhou, C., Choe, R., Shah, N. et al. 2007. Diffuse optical monitoring of blood flow and oxygenation in human breast cancer during early stages of neoadjuvant chemotherapy. *J. Biomed. Opt.* 12: 051903.

Zhou, C., Eucker, S., Durduran, T. et al. 2009. Diffuse optical monitoring of hemodynamic changes in Piglet brain with closed head injury. *J. Biomed. Opt.* 14: 034015.

Zhou, C., Yu, G., Furuya, D. et al. 2006. Diffuse optical correlation tomography of cerebral blood flow during cortical spreading depression in rat brain. *Opt. Express* 14: 1125–1144.

Zirak, P., Delgado-Mederos, R., Martí-Fàbregas, J. et al. 2010. Effects of acetazolamide on the micro- and macro-vascular cerebral hemodynamics: A diffuse optical and transcranial doppler ultrasound study. *Biomed. Opt. Express* 1(5): 1443–1459.

11

Fluorescence Spectroscopy

Darren Roblyer
University of California, Irvine

Richard A. Schwarz
Rice University

Rebecca Rae
Richards-Kortum
Rice University

11.1 Introduction 217
11.2 Cancer and Endogenous Fluorescence...218
11.3 Fluorophore Localization 219
11.4 Instrumentation..220
11.5 Empirical Analysis Methods...222
    Oral Cavity • Uterine Cervix • Brain • Bronchus • Other Sites
11.6 Model-Based Analysis Methods..224
    Probabilistic Photon Migration • Diffusion Theory • Combined Analytical
    Methods • Monte Carlo
11.7 Imaging Spectroscopy...228
11.8 Discussion...228
References...228

11.1 Introduction

Fluorescence-based techniques have become important tools in biomedical research and clinical medicine. A wide range of common biomedical applications are based on the measurement of fluorescence, including fluorescence microscopy, DNA sequencing, flow cytometry, and fluorescence-based immunoassays (Lakowicz 2006). The use of fluorescence spectroscopy (FS) for biomedical diagnostics has been the subject of extensive research and has included efforts to characterize cardiac tissue, analyze blood, and aid in screening and early detection of cancer. FS is of particular interest as a noninvasive in vivo diagnostic technique, which may provide an alternative to invasive biopsy.

In this chapter, we will frame our discussion of FS in the context of early detection of cancer, highlighting examples from organ systems including the cervix, oral cavity, breast, brain, skin, esophagus, and bronchus. Preclinical and clinical studies have demonstrated the potential value of FS to improve early cancer diagnosis and have demonstrated important advantages over current clinical tools. Currently, early detection of oral and skin cancers relies on visual examination of tissue at risk with white light illumination. For less accessible mucosal surfaces (e.g., bladder, colon, esophagus), visual examination is performed through an endoscope. Because the sensitivity and specificity of visual examination can be low (Lingen et al. 2008), areas of tissue that appear suspicious for neoplasia require a biopsy and histopathologic analysis to confirm the presence of precancer or cancer. Many at-risk patients do not have access to the significant medical infrastructure and expertise required for screening and early detection. As an alternative, FS provides a minimally invasive tool to assess cancer-related changes in tissue morphology and biochemistry. Furthermore, recent advances in quantitative analysis of FS signals can reduce subjectivity and provide a diagnostic result immediately at the point-of-care.

Achieving the potential of FS for improved early detection of cancer relies on (1) a thorough understanding of the tissue chromophores that contribute to FS signal, (2) instrumentation designs optimized to utilize the most diagnostically relevant excitation and collection wavelengths and that target the most diagnostically relevant tissue depths, and (3) empirical or model-based methods to analyze the tissue fluorescence signal and to discriminate among normal, neoplastic, and benign tissue. These three points are made more challenging in the face of confounding factors that may affect FS signals such as variations in patient age, tissue type, menopausal status, and many others.

This chapter provides an overview of current research to develop and optimize the use of FS for early detection of cancer with a focus on probe and instrumentation design. We begin with a discussion of endogenous fluorophores and their role in cancer. We then describe studies that have identified tissue fluorophores in specific organ sites and their location in epithelial tissue. We then describe instrumentation for tissue FS, including recent advances in probe design. Finally, we describe the history and current status of both empirical and model-based approaches to analyze tissue FS for the early detection of cancer.

11.2 Cancer and Endogenous Fluorescence

The epithelium is the outermost layer of tissue of the internal and external surfaces of the human body. Connective tissue underlying the epithelium, called the extracellular matrix or stroma, provides structural support to tissue and contains vasculature. The epithelium is separated from the stroma by the basement membrane. Normal proliferation of epithelial cells, such as skin cells or the cells composing the lining of the digestive tract, begins near the basement membrane. As these cells mature, they migrate toward the tissue surface where they may be sloughed off and replaced by new cells.

The vast majority (>85%) of human cancers develop in epithelial tissues. Malignant transformation most commonly begins near the basement membrane as a result of genetic disregulation and abnormal cellular signaling of proliferating cells. A front of these abnormally proliferating cells progresses upward through the epithelium. The transformation is accompanied by atypical cellular architecture, uncontrolled cell division, and abnormal cellular metabolism. When confined to the epithelium, these changes are referred to as dysplasia. The thickness of the epithelium may be altered by dysplastic progression. Invasive carcinoma is characterized by the infiltration of this front downward through the basement membrane, a situation that allows for metastases through the surrounding blood vessels. The transformation from normal tissue to cancer may occur over a widely varying amount of time from several months to years. The accessibility of the epithelium for screening and treatment differs among cancers. For example, oral cancers present in the relatively accessible squamous epithelium of the oral mucosal lining while breast cancers most commonly present in the cuboidal or squamous epithelium of milk ducts, often several centimeters from the surface of the breast.

Biological tissues contain native fluorophores that may be excited by the introduction of external electromagnetic energy (Alfano et al. 1984, Lakowicz 2006). It has long been known that changes in endogenous fluorescence may accompany malignant progression. In 1924, Policard used a Wood's lamp to view red fluorescence from necrotic rat sarcomas (Policard 1924). More recently, native fluorophores related to cellular metabolism and tissue structure in the epithelium have been used as biomarkers to detect neoplastic transformation. Some commonly identified fluorophores associated with neoplastic progression are indicated in Table 11.1. Mitochondrial cofactors involved in cellular metabolism such as reduced nicotinamide adenine dinucleotide (NADH) and flavin adenine dinucleotide (FAD) have been identified as important fluorophores originating from the cytoplasm of cells in epithelial tissues such as oral mucosa, breast, esophagus, and cervix (Richards-Kortum and Sevick-Muraca 1996, Drezek et al. 2001b, Georgakoudi et al. 2002a, Pavlova et al. 2003, 2008b). These cofactors have peak emission in the blue-green portion of the visible spectrum with the excitation–emission peaks of 340–460 nm for NADH and 460–520 nm for FAD (Brookner et al. 2000, Heintzelman et al. 2000a, Pitts et al. 2001, Sokolov et al. 2002). It is hypothesized that levels of NADH and FAD change during neoplastic progression due to increased mitochondrial metabolic activity (Pavlova et al. 2003). The ratio of FAD to NADH is commonly called the redox ratio and is often used to assess the metabolic activity of tissue. Britton Chance and colleagues were the first to use fluorescence to assess cellular redox and current methods rely heavily on their important work during the previous 50 years (Chance et al. 1962a,b, 1979). The redox ratio provides an indication of the oxidation-reduction state of a cell or tissue and a decrease in the redox ratio is generally indicative of increased metabolic activity (Chance et al. 1962a,b).

Structural proteins such as collagen and elastin fibers, their cross-links, and keratin have been identified as important sources of autofluorescence in the connective tissue of the breast, cervix, esophagus, and oral cavity (Richards-Kortum and Sevick-Muraca 1996). The emission spectra of these proteins are blue-green having a peak near 400 nm (Sokolov et al. 2002). It is believed that the autofluorescence signal from collagen and elastin cross-links decreases during neoplastic progression in several organ sites (Drezek et al. 2001b, Georgakoudi et al. 2002a, Pavlova et al. 2003, 2008b).

TABLE 11.1 Commonly Identified Fluorophores in FS Cancer Studies

Fluorophore	Excitation (λ)	Emission (λ)	Tissue	References
NADH	290–370	340–460	Oral, cervical, breast, esophageal	Pavlova et al. (2003, 2008), Drezek et al. (2001a), Palmer et al. (2003), Georgakoudi et al. (2002a)
FAD	430–460	460–520	Oral, cervical, breast	Pavlova et al. (2003), Palmer et al. (2003)
Collagen and elastin	300–370	400, 440	Oral, cervical, breast, esophageal, skin, bronchus	Pavlova et al. (2003), Palmer et al. (2003), Richards-Kortum and Sevick-Muraca (1996), Georgakoudi et al. (2002a), Brancaleon et al. (2001), Uehlinger et al. (2009)
Keratin	340	400	Cervical	Chang et al. (2006)
Tryptophan	280	340	Cervical, breast, skin	Ingrams et al. (1997), Palmer et al. (2003), Richards-Kortum and Sevick-Muraca, (1996), Brancaleon et al. (2001)
Tyrosine	270	300	Cervical	Pavlova et al. (2003), Richards-Kortum and Sevick-Muraca (1996)
Retinol/vitamin A	360	480	Breast	Zhu et al. (2008a,b)
Protoporphyrin IX	410	635	Oral	Ingrams et al. (1997), Betz et al. (2002), Inaguma and Hashimoto (1999)

Note: Approximate excitation and emission peaks are shown as well as human tissue locations where the fluorophores have been identified.

Other sources of fluorescence in epithelial tissue include the aromatic amino acids tryptophan, tryrosine, and phenylalanine. These chromophores have peak emission in the UV and near UV (Richards-Kortum and Sevick-Muraca 1996). In the breast, tryptophan has been identified as a major contributor with excitation-emission maxima 280–340 nm (Palmer et al. 2003), and retinol or vitamin A has also been identified as a possible contributing fluorophore (Zhu et al. 2008a). Porphyrins, which have peak emissions near 600 nm, have been identified in breast, oral tissue, skin, and others (Inaguma and Hashimoto, 1999).

11.3 Fluorophore Localization

Endogenous fluorophore concentrations vary depending on tissue type, tissue layer or depth, subject age, and other factors (Brookner et al. 2003, Cox et al. 2003). The excitation and emission characteristics of various individual fluorophores have been measured using laboratory spectrofluorometers; however, it is important to note that these characteristics are dependent on their local optical environment and may change in the context of other scattering and absorption components, additional fluorophores, and whether fluorophores are bound to proteins (Richards-Kortum and Sevick-Muraca 1996, Skala et al. 2007).

Microscopy studies using frozen and fresh human tissue slices have been performed in order to elucidate the spatial distribution of major fluorophores in the layered epithelium of some organ sites. Lohmann et al. used fluorescence microscopy of cryosectioned, unstained tissue slices from human cervical tissue with confirmed pathologic diagnoses of normal, dysplasia, and invasive carcinoma (Lohmann et al. 1989). They observed a decrease in stromal fluorescence during malignant progression. To study changes in tissue fluorescence in live epithelial cells, Drezek et al. obtained tissue from fresh biopsies and observed statistically significant increases in epithelial fluorescence at 380 nm excitation and an accompanying decrease in redox ratio (indicating increased metabolic activity) in dysplastic compared to normal tissue. They also observed a decrease in stromal fluorescence at 380 and 460 nm excitation. Pavlova et al. colocalized the cytoplasmic fluorescence in the epithelium of cervical tissue slices with metabolically active mitochondria using a mitochondrial vital stain and laser scanning fluorescence confocal microscopy. This study supported the hypothesis that epithelial fluorescence is dominated by mitochondrial NADH and FAD (Drezek et al. 2001a, Pavlova et al. 2003).

In fresh oral tissue, Pavlova et al. also observed an increase in epithelial fluorescence with UV excitation and a decrease in stromal fluorescence in dysplastic and cancerous tissue at UV and 488 nm excitation compared to normal tissue (Pavlova et al. 2008b). These observations are demonstrated in Figure 11.1 which shows fluorescence confocal microscopy images and corresponding histology images of normal and dysplastic tissue. The stromal fluorescence shows a marked decrease in Figure 11.1b compared to Figure 11.1a. Additionally, they showed that benign inflammation showed decreased epithelial fluorescence compared to normal and that fluorescence patterns in general varied by anatomic site, a finding confirmed by several other groups (Ingrams et al. 1997, de Veld et al. 2003).

FS signal is affected by the binding states of fluorophores and the presence of quenchers. Skala et al. used multiphoton microscopy to elucidate the spatial location of NADH and FAD fluorescence intensity and lifetimes in a hamster cheek pouch cancer model. The relative contribution of protein-bound and -unbound NADH and FAD was determined by their respective fluorescence lifetimes. They found that the contribution of protein bound NADH and FAD decreased in precancers

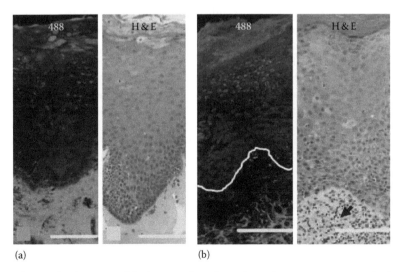

(a)   (b)

FIGURE 11.1 (See color insert.) Confocal images of fresh tissue slices from the oral cavity at 488 nm excitation with corresponding H & E stained pathology section. (a) Normal oral tissue. (b) Tissue with mild dysplasia. Note that in the fluorescence images there is a decrease in intensity in the stroma, located at the bottom of the images, in the mild dysplasia compared to the normal tissue. The basement membrane is illustrated with a white line in B. (Adapted from Pavlova, I. et al., *Clin. Caner Res.*, 14(8), 2396, 2008. With permission.)

compared to normal tissue and speculated that this was caused by a metabolic shift from oxidative phosphorylation to glycolysis. Additionally, they discovered high spatial variability of the redox ratio (FAD/NADH) in precancerous tissue compared to normal (Skala et al. 2007).

An increase in epithelial fluorescence due to NADH and FAD and a decrease in stromal fluorescence due to the breakdown of collagen cross-links during malignant progression has been implicated in many organ sites in addition to the ones mentioned above, including the pancreas (Malavika et al. 2007), bronchus (Zellweger et al. 2001), and others (Richards-Kortum and Sevick-Muraca 1996). Instrumentation design and model-based methods have utilized this common observation to improve the diagnostic ability of FS and to help better understand the biological basis of the measured FS signal.

11.4 Instrumentation

Devices used for FS in the lab and in the clinic generally include a light source, components for monochromatic or narrowband excitation, a probe that delivers the excitation light to and collects light from the interrogated specimen, components for spectrally resolved measurements of emission (e.g., a spectrograph), and a detector. Both broadband and laser sources are commonly used for excitation. When a broadband source is used, monochromatic or narrowband excitation is achieved using either a monochromator or excitation filters. Arc plasma lamps are commonly used to provide a broad range of excitation wavelengths. Xenon arc lamps have a generally uniform output from 300 to 750 nm, whereas mercury and metal halide lamps have intensity peaks at 313, 334, 365, 406, 435, 546, and 578 nm. These sharp intensity peaks can be beneficial in FS when attempting to excite specific fluorophores. A spectrograph is used to disperse the emission signal into chromatic components, which can be measured with a CCD array or linear detector array, or scanned with a single detector such as an avalanche photodiode (APD).

Many of the clinical devices used for cancer applications combine diffuse reflectance measurements with FS. As described in detail later in this chapter, measuring both reflectance and fluorescence spectra can aid in analyzing data using analytical models designed to separate the effects of scattering and absorption. These reflectance measurements are typically made using the same probe coupled with a broadband source. Typically, measurements of broadband reflectance and FS are performed in serial.

Figure 11.2 shows the instrumentation setup for two different clinical devices. Figure 11.2a is a system designed for brain tumor demarcation (Lin et al. 2000). The system is capable of collecting both reflectance and fluorescence signal. A broadband halogen lamp is used for reflectance illumination and a nitrogen laser with 337 nm excitation is used for fluorescence. A fiber-optic probe consisting of seven fibers is used to illuminate and collect light from the tissue. The six fibers surrounding the center fiber are angled to affect the illumination/collection overlap region in measured tissue. Figure 11.2b shows a device designed

for bronchial cancer detection (Zellweger et al. 2001). Excitation light is provided by a xenon lamp and a spectrograph is used to provide 10 nm FWHM bands of light between 350 and 480 nm. A quartz fiber bundle consisting of four excitation fibers and three emission fibers is used to illuminate and collect light from the tissue. The fiber bundle is designed to be placed down the biopsy channel of a standard flexible bronchoscope. An emission spectrograph and cooled CCD detector are used to measure emission.

The majority of FS devices for clinical use utilize a point probe to interrogate a small (several mm) area of tissue. Probe designs vary according to the type of tissue measured. For tissues which are more readily accessible such as the skin, oral cavity, and cervix, probes are often designed so that they may be held in the operator's hand and placed on the tissue surface during measurements. Probes may also be designed to be placed in the accessory channel of an endoscope to examine the mucosal surface of hollow organs, such as the bronchus, esophagus, lung, or colon (Tunnell et al. 2003).

One or more optical fibers incorporated into the probe are typically used to deliver and collect light from the tissue. The number and spatial location of the illumination and collection fibers at the distal end of the probe may be strategically placed in order to maximize collection efficiency or to interrogate specific tissue volumes or depths. Fiber diameter, probe–tissue spacing, and optics placed between the fibers and tissue are important considerations for probe design when considering depth resolution and sensitivity (Utzinger et al. 2003).

Depth-sensitivity is of particular interest for applications in detection of early cancer and its precursors. During the progression from normal tissue to dysplasia to micro-invasive cancer, changes occur in the fluorescence of both the epithelium, which typically accounts for the first few hundred microns of superficial tissue, and the deeper connective tissue or stroma (Lohmann et al. 1989, Pavlova et al. 2003). In the oral cavity, it has been suggested that the autofluorescence from the epithelium may provide information that can help discriminate dysplasia from benign inflammation (Drezek et al. 2001a, Pavlova et al. 2008b). However, it has been shown that the majority of fluorescence signals collected using many traditional probe designs is dominated by photons arising in the deeper stromal layers (Pavlova et al. 2008a). Several design strategies have been attempted to separate signal from the epithelial and stromal layers of interrogated tissue.

Probe configurations using a single illumination/collection fiber have been shown to be sensitive to singly scattered photons from the upper epithelial layers (Canpolat and Mourant 2001, Amelink et al. 2003). However, specular reflections and lower signal to noise due to autofluorescence from impurities in the fiber core are limiting factors for single fibers (Utzinger et al. 2003). Probe designs using multiple and separate collection and illumination fibers can reduce specular reflection and autofluorescence. Different source detector separations can adjust the overlap of illumination and collection cones allowing sensitivity to different tissue depths (Zhu et al. 2003, Skala et al. 2004).

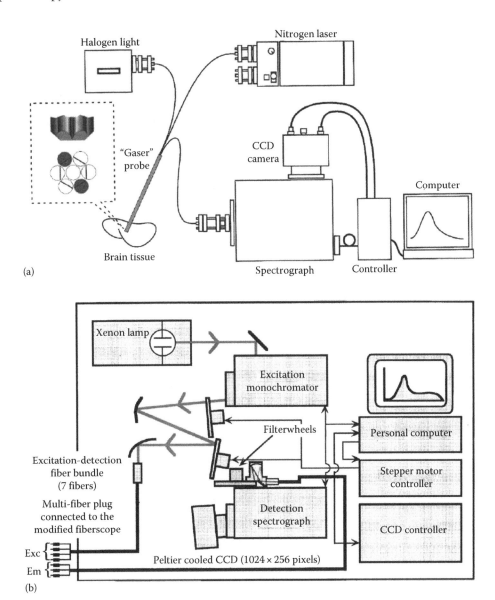

FIGURE 11.2 Two fluorescence spectroscopy systems for cancer detection. (a) This system is designed for brain tumor demarcation. Fluorescence and reflectance signal are collected. A 7-fiber probe with several angled fibers is used for illumination and collection. (Adapted from Lin, W.C. et al., *J. Biomed. Opt.*, 5(2), 214, 2000. With permission.) (b) This system is designed for detecting bronchial cancer. The fiber bundle is designed to be inserted into the biopsy channel of a standard bronchoscope. (Adapted from Zellweger, M. et al., *J. Biomed. Opt.*, 6(1), 41, 2001. With permission.)

Tilted or angled fibers have been used to reduce the penetration depth of the excitation light by altering the angle of incidence compared to the surface of the tissue (Pfefer et al. 2003, 2005, Utzinger et al. 2003, Liu and Ramanujam 2004, Skala et al. 2004). Figure 11.3a illustrates five possible probe geometries that may be used. The depth and location of the illumination/collection overlap region is dependent on whether a single fiber is used (a and b) or, if separate fibers are used, the angle of those fibers relative to the tissue surface and each other (d–f.) (Pfefer et al. 2005). Monte Carlo simulations, phantom experiments, and in vivo experiments have been conducted to estimate the depth of penetration (Pfefer et al. 2005). Figure 11.3b and c show results from a Monte Carlo simulation of two probe designs with the same illumination–collection fiber separation but with different

incidence and collection angles (Skala et al. 2004, Pfefer et al. 2005). In Figure 11.3b, the illumination and collection fibers are at a 0° angle in reference to a normal from the simulated tissue surface. In Figure 11.3c, the fibers are tilted 45° to the normal and are directed toward each other. The scale in the figure represents the number of collected photons collected from locations in the tissue. In this example, the tilted configuration collects many more photons from a superficial tissue depth compared to the normal incidence case.

Optical elements at the distal probe tip can help achieve better depth sensitivity as well. Figure 11.4 shows a probe with a ball lens configuration (Schwarz et al. 2008); this probe has four channels, one for "shallow," one for "medium," and two for "deep" measurements. By placing the illumination and

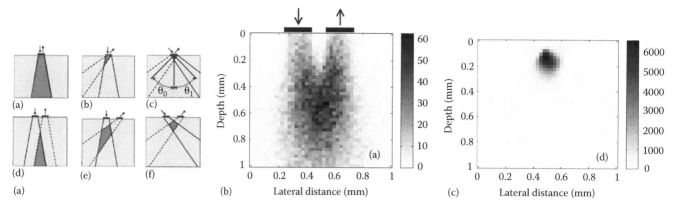

FIGURE 11.3 (a) Several possible probe designs are illustrated with corresponding illumination-collection overlap regions. (b and c) Monte Carlo simulation results from two different probe geometries. (b) Results from a geometry with illumination and collection fibers at normal incidence to the tissue. (c) Geometry with the probes at a 45° angle pointed toward each other. The scale shows the number of photons collected. (Adapted from Pfefer, T.J. et al., *J Biomed. Opt.*, 10(4), 44016, 2005. With permission.)

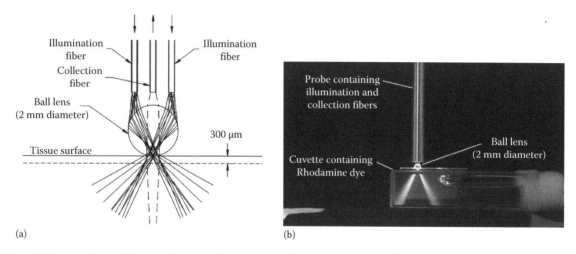

FIGURE 11.4 (a) Probe with a ball lens configuration. Illumination and collection fibers are shown. (b) The same probe illuminating a cuvette with Rhodamine dye. The illumination rays converge at a very superficial depth. (Adapted from Schwarz, R.A. et al., *Appl. Opt.*, 47(6), 825, 2008. With permission.)

collection fibers in different spatial locations relative to a 2 mm diameter sapphire ball lens, excitation light is refracted at different incidence angles compared to the tissue surface, which allows for depth sensitivity in measurements. A view of the probe and ball lens in "shallow" mode is shown in Figure 11.4a. A picture of the probe using the "shallow" fibers illuminating a cuvette with Rhodamine dye is shown in Figure 11.4b demonstrating the highly divergent source rays.

Experiments using water and tissue phantoms have confirmed that the "shallow" channel provides a significant increase in sensitivity to the upper layers of tissue. Measurements from water show that fluorescence intensity drops to 10% of its peak value by 550 μm from the probe tip (Schwarz et al. 2008). Monte Carlo simulations have indicated between 88% and 93% of the measured fluorescence signal originates in the epithelium and upper stroma of oral tissue (depending on the epithelial thickness) (Arifler et al. 2005).

The experimental devices discussed above employ a wide range of probe designs, excitation and collection wavelengths, and collection geometries. Analysis of data is necessary to identify clinically and diagnostically relevant features collected from these devices. Empirical analysis methods use features extracted from the raw data to discriminate disease for nondiseased cases while model-based analysis attempt to extract specific chromophore concentrations by fitting measurements to a mathematical model, which can then be used for diagnostic classification.

11.5 Empirical Analysis Methods

Empirical analysis uses relatively straightforward features or aspects of measured fluorescence spectra to discriminate diseased from nondiseased tissue. Examples of these features may be the location or intensity of spectral peaks or the ratio of two spectral peaks. The majority of human studies using

FS have utilized empirical analysis methods to classify spectra and we present here a review of work done in several organ sites. Empirical methods are limited in that they do not separate absorption and scattering effects in tissue and therefore may be prone to potential confounding factors such as hemoglobin absorption or differences in tissue geometry. Section 11.6 describes model-based methods that have attempted to address this limitation. Despite this concern, initial studies using empirical analysis have achieved highly promising results.

The use of FS and empirical data analysis methods dates back to at least 1984, when Alfano et al. measured fluorescence spectra of normal and cancerous rat kidney and prostate and cancerous rat and mouse bladder (Alfano et al. 1984). Using the same device, this group performed measurements on human breast and lung tissue (Alfano et al. 1987). Figure 11.5 shows example spectra from their 1987 study, illustrating a primary peak at 514 and 517 nm for normal and cancerous breast tissue, respectively. Alfano et al. used the location of this peak as a feature to discriminate cancerous from normal tissue.

Since these early studies, a plethora of preclinical and clinical FS studies have been conducted for cancer diagnosis in the oral cavity (Schwarz et al. 2008), cervix (Ramanujam et al. 1996b), brain (Lin et al. 2000, 2001), skin (Brancaleon et al. 2001), breast (Zhu et al. 2008a), bronchus (Zellweger et al. 2001), and other organ sites. Many groups have collected data at multiple excitation wavelengths, resulting in a large body of multivariate data. Diagnostically relevant features are often not immediately apparent. Subtle changes contained in the spectral data may be related to biochemical changes occurring during malignant progression. In order to analyze the high information density produced from FS, data reduction methods are commonly used to select particular features of measured spectra, which have diagnostic

value. Often, multivariate feature selection algorithms are used to empirically identify spectral parameters, which correlate with diagnosis. Examples of these features are intensities at specific wavelengths, peak intensity wavelengths, or ratios of intensities. Care must be taken when using multiple features for diagnostic classification, however; when using a small data set, multiple features may discriminate the desired classes well, but this ability to classify may not generalize well to new data.

In many cases, diagnostic algorithms developed using empirical data analysis methods have yielded high diagnostic sensitivity and specificity. We review this work beginning with the oral cavity, the uterine cervix, brain, and several other organ sites.

11.5.1 Oral Cavity

The oral cavity is easily accessible making it an ideal location for FS measurements. The majority of oral cancers originate from the epithelial layers of the oral mucosa (American Cancer Society 2006). Tissue structure and morphology differ substantially in the distinct anatomical regions of the oral cavity. For example, the hard palate and gingiva have a thick keratin layer, which is largely absent in the buccal mucosa, floor of mouth, and lateral and ventral tongue. These changes lead to spectroscopic differences (Ingrams et al. 1997, de Veld et al. 2003). Significant work toward understanding the effects of excitation wavelength (Gillenwater et al. 1998, Heintzelman et al. 2000b) and benign conditions such as fibrosis (Tsai et al. 2003) on diagnosis has been conducted. Despite these and other potential confounding factors, much progress has been made toward using FS as a diagnostic tool for early detection of oral dysplasia and cancer (de Veld et al. 2005).

Several groups have been able to discriminate normal or low grade oral precancers from high grade precancers using FS in the UV-VIS together with empirically derived analysis algorithms. Principal component analysis has been frequently used to reduce the high information density of measurements (de Veld et al. 2003, 2004, Chu et al. 2006). Following data reduction, a variety of classification algorithms have been used, including linear discriminant classifiers (Schwarz et al. 2009), regression (Muller et al. 2003, Chu et al. 2006, Majumder et al. 2006), neural networks (Wang et al. 2003, de Veld et al. 2004), and support vector machines (Majumder et al. 2005). Classification accuracy has generally compared favorably to direct visual inspection, the current standard of care for screening (Downer et al. 2004, de Veld et al. 2005).

Recently, Schwarz et al. reported a sensitivity of 82% and specificity of 87% in an independent validation set from the classification of normal and mild dysplasia from moderate dysplasia, severe dysplasia, and carcinoma using FS and diffuse reflectance spectroscopy (Schwarz et al. 2009). The classification was done on nonkeratinized anatomical regions because of the significant difference in spectral characteristics between keratinized and nonkeratinized sites (Ingrams et al. 1997). A depth-resolved fiber-optic probe was utilized to collect signal and a xenon arc lamp with multiple narrowband filters was used for excitation.

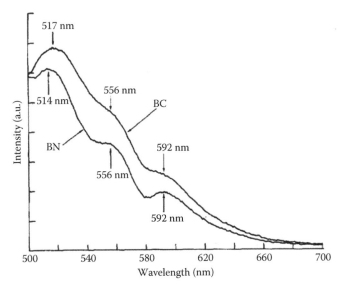

FIGURE 11.5 FS measurements from normal (BN) and cancerous (BC) human breast tissue. The tissue was excited at 488 nm. (Adapted from Alfano, R. et al., *IEEE J. Quantum Electron.*, 23(10), 1806, 1987. With permission, © 1987 IEEE.)

For a summary of classification results from oral cancer FS studies in the literature we refer the reader to Schwarz et al. and specifically to Figure 3 in this article (Schwarz et al. 2009). It should be noted that to date, FS studies have been conducted on small patient populations (<100) and typically in tertiary care centers.

11.5.2 Uterine Cervix

The standard screening test for cervical cancer and its precursors is the Papanicolaou smear, which suffers from low specificity and long wait times for diagnosis. An abnormal Pap smear is followed by colposcopy, in which a low power microscope is used to visualize light reflected from the cervix. Upon application of acetic acid, suspicious areas appear white. Reflectance images with green illumination highlight changes in vasculature associated with cervical cancer and its precursors. Although colposcopy has a high sensitivity, it has poor specificity and suspicious lesions require confirmatory biopsy. FS has been studied in the cervix as an alternative method for detection of intraepithelial lesions.

Pilot clinical studies utilizing FS alone and in combination with reflectance spectroscopy have yielded encouraging results (Ramanujam et al. 1996a,b, Chang et al. 2002, 2005, Georgakoudi et al. 2002b). Chang et al. were able to achieve a sensitivity of 83% and a specificity of 80% for discriminating normal squamous tissue from high-grade lesions and a sensitivity of 72% and specificity of 78% for discriminating normal columnar tissue from high grade lesions using a point fiber-optic probe with xenon lamp illumination and excitation between 330 and 480 nm (Chang et al. 2005).

Promising pilot studies in the cervix have led to investigations into confounding factors affecting FS diagnosis. Menstrual cycle status, menopausal status, age, and tissue type (squamous versus columnar versus transitional) have been found to significantly affect FS measurements (Georgakoudi et al. 2002b, Brookner et al. 2003, Cox et al. 2003). It was found that particular care must be taken when comparing FS measurements from different tissue types within the cervix. The cervix is composed of both columnar and squamous epithelium separated by a transformation zone. Precancerous lesions typically arise in the squamous region of the squamo-columnar junction and FS signal changes significantly with both disease state and tissue type (Ramanujam et al. 1996a). An investigation of probe pressure during measurements found that this factor does not significantly affect results (Nath et al. 2004).

11.5.3 Brain

In the brain, FS has been used to discriminate normal brain tissue from primary and secondary brain tumors including glioblastomas and astocytomas. This discrimination is important during tumor resection in order to maintain neurological function and minimize risk for reoccurrence. Lin et al. were able to discriminate normal tissue from tumor with a sensitivity of 97%

and specificity of 96% in one study and were able to discriminate infiltrating tumor margins from normal brain with a sensitivity of 100% and specificity of 76% in another study (Lin et al. 2000, 2001). The authors used a nitrogen laser (337 nm) for excitation and used a point fiber-optic probe to collect signal. A fluorescence emission peak at 460 nm was identified and used for discrimination, but the origin of this peak is unclear.

11.5.4 Bronchus

Using both excised specimens and in vivo measurements through bronchoscopes, it has been demonstrated that FS may be able to discriminate preneoplastic and neoplastic tissue from normal tissue. By measuring excised, formalin-fixed specimens, Gebrecht et al. hypothesized that the differences in autofluorescence signal from healthy and preneoplastic tissue were due to both epithelial thickening and blood absorption below the epithelial layer (Gabrecht et al. 2007). Ueherlinger et al. used time resolved spectroscopy in vivo and determined that fluorescence signal in the bronchus was likely due to a single dominant fluorophore, probably elastin (Uehlinger et al. 2009).

11.5.5 Other Sites

It has been observed from in vivo FS measurements of nonmelanoma skin cancer patients that an increase in tryptophan fluorescence and a decrease in collagen fluorescence likely accompanies malignant progression (Brancaleon et al. 2001). In the pancreas, FS measurements from excised human tissue showed spectral differences between normal and adenocarcinoma (Malavika et al. 2007). The authors attribute an increase in fluorescence intensity at 400 nm in adenocarcinoma to an increase in collagen signal, which is consistent with the increased fibrosis associated with disease. FS may also be useful for cancer diagnosis in the ovary (Brewer et al. 2001), larynx (Arens et al. 2006), and colon (Anandasabapathy 2008).

11.6 Model-Based Analysis Methods

While empirical analysis may be able to discriminate diseased tissue in many cases, there may not be a clear link between the diagnostically useful features and the underlying biological processes responsible for those features. A measured fluorescence spectrum may be analyzed to determine the spectral contributions of the constituent components that produced the fluorescence signal, yielding concentrations of individual fluorophores, which can provide insight into biology. However, tissue is a turbid medium in which fluorescence is affected by multiple factors such as the concentration of fluorophores, their localization, scattering and absorption by other chromophores, protein binding of fluorophores, and the excitation/collection geometry of the probe. These effects can distort the spectral shape and intensity of measurements. For example, FS measured from two different tissue samples with the same fluorophore composition and concentrations but with different blood volume will display significantly different FS spectra because of the highly absorbing properties of hemoglobin. Without correction of the FS measurements obtained from samples with both scattering and absorption components, accurate determination of fluorophore concentrations cannot be achieved.

Jobsis et al. and Mayevsky and Chance were among the first to suggest taking into account the absorbing properties of hemoglobin when analyzing fluorescence from tissue in the 1970s (Jobsis et al. 1971, Mayevsky and Chance 1974, 1982). Since then, many methods, both experimental and model based, have been utilized in order to extract intrinsic fluorescence from measured fluorescence of turbid media. Following extraction from measured signal, intrinsic fluorescence spectra can be fit as a linear combination of the emission spectra of the constituent fluorophores contained in the sample.

Experimental approaches have been utilized to reduce the effect of scattering and absorption on measured fluorescence spectra, but have the limitations of requiring specific conditions and may only provide intrinsic fluorescence measurements at a subset of emission wavelengths. Liu et al. and Anidjar et al. reduced the effects of hemoglobin absorption, a major absorber in tissue, on the measured fluorescence signal by looking at the ratio of fluorescence spectra at two emission wavelengths where hemoglobin absorption is very similar (Liu et al. 1992, Maurice et al. 1996). This ratio, taken near 350 and 440 nm for both studies, is largely independent of hemoglobin absorption. These studies suffered from the limitation of only being able to correct for these effects at the two specific emission wavelengths. Weersink et al. collected fluorescence spectra at one source–detector separation and a diffuse reflectance spectra at another and was able to use these measurements together to cancel out a majority of the scattering and absorption contributions to the fluorescence spectra (Weersink et al. 2001). Biswal et al. used polarized fluorescence and elastic scattering measurements to accomplish a similar result (Biswal et al. 2003).

Alternatively, mathematical models have been used to extract intrinsic fluorescence. Several classes of models have been proposed over the last 30 years including analytical models based on probabilistic photon migration (Wu et al. 1993a, Zhang et al. 2000, Muller et al. 2001), diffusion theory (Hyde et al. 2001, Nair et al. 2002, Stasic et al. 2003), and combinations of different analytical methods (Chang et al. 2004, 2006). Analytical models have the advantage of providing relatively simple closed form solutions. These methods have the disadvantage of often requiring specific boundary conditions and are only applicable if specific criteria concerning the optical properties of the measured media and probe geometry are met.

Monte Carlo methods have also been utilized (Ramanujam et al. 1994, Qu et al. 1995, Zonios et al. 1996, Welch et al. 1997, Drezek et al. 2001b, Palmer and Ramanujam 2008, Pavlova et al. 2008a, 2009, Zhu et al. 2008a,b). This approach has the advantage of being adaptable to arbitrary probe geometries, complex tissue geometries, and a wide range of tissue optical properties. It has the disadvantage of requiring a relatively high degree of

computational power, as the paths of many photons must be computed and recorded.

We describe here several common model-based methods for the analysis of fluorescence from tissue.

11.6.1 Probabilistic Photon Migration

Probabilistic photon migration is a model-based approach that utilizes distribution functions to describe photon interactions and paths in a turbid media (Bonner et al. 1987, Nossal et al. 1989). This modeling method treats photons as particles that experience discrete absorption, scattering, or fluorescence interactions. This method differs from the Monte Carlo method, which will be described in Section 11.6.4, in that the individual paths of each photon are not simulated and recorded. One of the significant advantages of this model is that it can provide relatively simple analytical expressions to separate scattering, absorption, and anisotropy effects. An early application in biomedical optics was the modeling of diffuse reflectance spectroscopy measurements in tissue (Wu et al. 1993b). Using this model, the optical properties, μ_s and μ_a, could be readily calculated from reflectance measurements. Wu et al. expanded this model to describe fluorescence in the visible wavelength range where absorption from hemoglobin and water are not significant (>500 nm) (Wu et al. 1993a). This model was later expanded to include a broad range of emission wavelengths, 370–700 nm (Zhang et al. 2000, Muller et al. 2001). By measuring both the diffuse reflectance and fluorescence from tissue, the intrinsic fluorescence can be extracted using this photon migration model. The resulting intrinsic fluorescence spectra can then be fit with a linear combination of emission spectra from fluorophores present in the measured media in order to obtain individual fluorophore concentrations (Georgakoudi et al. 2002b). This process can provide insight into the biochemical changes occurring during malignant progression.

Probabilistic photon migration for the extraction of intrinsic fluorescence is limited in that it relies on the assumption that fluorescence and reflectance have similar photon paths (Muller et al. 2001). This assumption may not hold when absorption is significantly higher than scattering ($\mu_a > 3\mu_s'$). The model also may not hold when fluorophores in the measured media contribute to a high degree to scattering and absorption such as with high quantum yield fluorophores (i.e. Rhodamine dye and tryptophan). An additional limitation is that proposed models to date have only considered single layer, homogenous media. This assumption may neglect the important differential influence of tissue layers in the development of dysplasia and cancer in epithelial tissues.

Müller et al. experimentally tested this model on minced and bulk tissue specimens from the oral cavity (Drezek et al. 2001b, Muller et al. 2001). Figure 11.6 shows both measured and extracted intrinsic fluorescence excitation emission matrices (EEMs) from in vitro oral buccal mucosa. In the measured EEM in Figure 11.6a, the valley seen near 420 nm emission is attributed to hemoglobin. In Figure 11.6b, the extracted intrinsic fluorescence does not show this feature demonstrating the

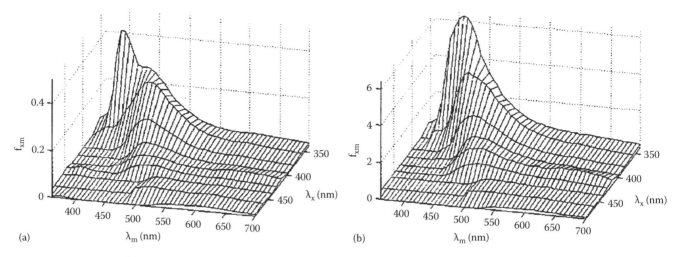

FIGURE 11.6 Measured and intrinsic fluorescence EEMs from minced oral tissue. The intrinsic fluorescence EEM has the effects of hemoglobin absorption removed. (Adapted from Muller, M.G. et al., *Appl. Opt.*, 40(25), 4633, 2001. With permission.)

ability of the model to remove the affects of strong absorption on measured fluorescence.

Georgakoudi et al. used this model to extract intrinsic fluorescence from in vivo and resected cervical and esophageal tissue (Georgakoudi et al. 2002a). Measurements were taken using a FS probe passed through the biopsy channel of an endoscope for esophageal measurements. The intrinsic fluorescence spectra from both tissue types were well represented as a linear combination of NADH and collagen fluorescence. Measurements taken before and after ligation of esophageal varices, or swollen blood vessels, showed a decrease in fluorescence intensity and a red shift in peak wavelength of emission spectra. The authors attributed these changes to the induced reduction in metabolism caused by asphyxiation and the subsequent reduction in NADH fluorescence. Good discrimination between normal sites and dysplastic/high grade sites was achieved by using the relative contributions of collagen and NADH fluorescence for both the esophageal and cervical tissue as shown in Figure 11.7.

A similar analysis was conducted in oral tissue where it was discovered that the collagen and NADH contributions could help to discriminate normal from dysplastic and cancerous tissue, and to a lesser extent, dysplastic from cancerous tissue (Muller et al. 2003). When FS was combined with diffuse reflectance and light scattering spectroscopy, a sensitivity of 96% and specificity of 96% were achieved for discrimination of normal and abnormal oral tissue. The method has also been applied to breast cancer with highly favorable results (Volynskaya et al. 2008).

11.6.2 Diffusion Theory

In media where scattering dominates absorption and light propagates with only a weak directional flux, light propagation may be described using the radiative transport equation or, after simplifying assumptions, the diffusion equation. These criteria are generally satisfied in human tissue in the VIS-IR at locations far from the source.

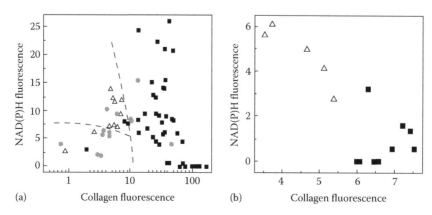

FIGURE 11.7 Scatter plots showing the NADH and collagen fluorescence extracted from intrinsic fluorescence measurements of cervical (a) and esophageal (b) tissue. (a) Black squares show 43 colposcopically normal ectocervical sites, gray circles show 12 squamous metaplastic sites, and triangles show 10 high grade lesions. (b) Black squares show 10 nondysplastic sites and triangles show 5 high-grade dysplastic sites. (Adapted from Georgakoudi, I. et al., *Cancer Res.*, 62(3), 982, 2002. With permission.)

Analytical solutions to the diffusion equation have been used to describe fluorescent light propagation in tissue such as the breast (Hyde et al. 2001, Nair et al. 2002). Nair et al. and Hyde et al. used both reflectance and fluorescence measurements to extract intrinsic fluorescence from tissue. Nair et al. analyzed intrinsic fluorescence spectra at radial distances from the source while Hyde et al. analyzed fluorophore concentration versus depth in a layered tissue. It is important to note that these boundary value problems are highly dependent on illumination/ collection geometry.

11.6.3 Combined Analytical Methods

In epithelial tissues, both the epithelium and the stroma contribute to measured fluorescence spectra. It has been suggested that it may be necessary to take into account the differential biochemical and morphological changes occurring in each layer in order to elucidate diagnostic information. Two-layer models based on diffusion theory have been demonstrated, but the relatively weak scattering of the thin epithelial layer prevents an accurate description of light propagation in this layer with diffusion theory (Hyde et al. 2001). In order to address this area, models using a combination of analytical methods have been proposed.

Chang et al. proposed a two-layer model for the UV-VIS region, with optical properties matching the epithelium in the upper layer, and stroma in the bottom layer (Chang et al. 2004). Light propagation in the epithelium was modeled using the Beer–Lambert law and was justified by the observation that albedo is relatively small in the epithelium and the thickness of the epithelium is comparable to the mean free path, reducing the effects of scattering. Diffusion theory was used to model light propagation in the thicker and more highly scattering stroma. Results from the model were compared with Monte Carlo simulations with good agreement. The model was used to analyze in vivo fluorescence measurements from 292 patients with cervical intraepithelial dysplasia (Chang et al. 2006). NADH, FAD, keratin, and three collagen components were considered as the contributing fluorophores in the model and concentrations were determined from the measured data. It was found that there was an increase in epithelial fluorescence from FAD and a decrease from keratin in measurements from dysplastic tissue compared to normal. Stromal fluorescence from collagen also decreased. These results are consistent with previous studies (Drezek et al. 2001a, Pavlova et al. 2003).

11.6.4 Monte Carlo

Monte Carlo modeling of light propagation provides a robust simulation tool that is highly adaptable to different tissue organizations (including layered structures), probe geometry, and is valid for arbitrary optical properties and wavelengths. Monte Carlo is ubiquitously used as a gold standard for other analytical models. A disadvantage of the Monte Carlo method is that it is computationally intensive, since the paths of millions or more

photons must be calculated and recorded in order to achieve statistical significance. As a result, Monte Carlo techniques are limited in their ability to provide intrinsic spectra in near real time, which may be desired in a clinical setting. Emerging scaled Monte Carlo models that utilize lookup tables are a potential solution to this problem. The Monte Carlo method is a powerful tool for forward models that help to explain the biological basis of observed measurements in the cervix, breast, oral cavity, and other organ sites (Drezek et al. 2001b, Palmer and Ramanujam 2008, Pavlova et al. 2008a).

Palmer et al. recently developed a Monte Carlo model to extract intrinsic fluorescence and applied it to FS measurements of the breast (Palmer and Ramanujam 2008, Zhu et al. 2008a). Their model assumed a single layer with homogenous optical properties. A reflectance Monte Carlo model was first used to obtain scattering and absorption coefficients from measured broadband diffuse reflectance measurements. These were then used as inputs to their fluorescence Monte Carlo model. A simulation was used to model emission photons and reciprocity was then used to describe excitation photons. This eliminated the need for two separate simulations, one for excitation and one for emission. The model was scalable so as to be able to account for any optical properties. This model was applied to FS measurements of excised breast tissue from patients with carcinoma and fibrous or benign conditions. They assumed NADH, retinol, and collagen to be the contributing fluorophores and by using the extracted fluorophore results they were able to classify malignant tissue from nonmalignant tissue with a sensitivity of 82.6% and a specificity of 75.7% in a validation set. The majority of the misclassified samples were either fibrous or benign tissue. It was suggested that information provided from fluorescence and reflectance data may be complementary although classification performance was not improved in this study when using both modalities.

Drezek et al. developed a two-layer Monte Carlo model to describe fluorescence in the cervix. The upper layer represented the epithelium and the second layer the stroma. NADH and collagen were assigned as the fluorophores in the top and bottom layer, respectively. Confocal fluorescence images of resected oral tissue were used to obtain layer and fluorescence properties of normal, normal with inflammation, and neoplastic samples. This method provided more certainty in the Monte Carlo input parameters and helped explain observed spectral differences. The model matched measured FS data well and suggested that NADH fluorescence increases and collagen fluorescence decreases during malignant progression. The model was sensitive to epithelial thickness, which is known to decrease with age and increase with dysplasia. It was also noted that the majority of fluorescence signal originated in the stroma (70%–80% for dysplastic tissue).

As described in Section 11.4, there have been several strategies for depth selectivity of FS measurements to take advantage of the differential signal between the epithelium and stroma. In particular, probes that can measure a greater percentage of epithelial fluorescence than described in the previous study may be able to diagnose tissue with greater accuracy and help reduce false positives from inflammation (Drezek et al. 2001b, Pavlova et al. 2008b). Pavlova et al. recently developed a five-layer Monte Carlo model to for describing fluorescence in oral tissue measured with a ball-lens depth sensitive probe (Pavlova et al. 2009). The contributing fluorophores used in the model were keratin, FAD, NADH, and collagen. Spectra predicted from the Monte Carlo were then compared to clinically obtained FS measurements from the same sites. It was observed that in nonkeratinized tissue, when using the depth sensitive probe, fluorescence intensity decreased and peak wavelength red shifted during malignant progression. This result was reproduced by the Monte Carlo simulations and was attributed to a decrease in fluorescence from the stromal layers and an increase in the epithelium. It was noted that 90% of the detected signal originated from the epithelium using the depth selective probe.

11.7 Imaging Spectroscopy

Point probe measurements are inherently limited in that they are only able to interrogate a small region of tissue with each measurement. In order to address this issue, several groups have looked to hyperspectral imaging. These techniques provide spectral information from a larger field of view by either scanning tissue spatially or spectrally. The end result is a data cube in which a spectrum is available at many spatial locations.

Martin et al. have used a liquid crystal tunable filter (LCTF) to obtain hyperspectral data for cancer diagnostics (Martin et al. 2006). Gebhart et al. used a LCTF hyperspectral system to discriminate normal human brain tissue from tumors (Gebhart et al. 2007). Yu et al. demonstrated a scanning spectral imager that combines hyperspectral imaging with a photon migration model to obtain images of fluorophore concentrations (Yu et al. 2008).

The high information density yielded from spectroscopic imaging systems may complicate analysis. Multispectral fluorescence imaging approaches, which use only a few excitation and collection wavelengths, offers a simplification of the hybridization of FS and imaging. As noted in the *direct fluorescence spectroscopy* section, many studies that initially collected spectrally dense data ultimately used only a small subset of specific features from the spectra, such as intensities at chosen wavelengths and ratios of emissions, in order to classify the data. If these features can be realized using multispectral imaging then there is an added advantage of obtaining wide field measurements instead of point measurements. Lam et al. utilized the ratio of red and green emission band in the lung to discriminate cancer (Lam et al. 1998). Roblyer et al. utilized red-to-green ratios at a specific excitation wavelength in order to discriminate dysplasia and cancer in the oral cavity (Roblyer et al. 2009).

11.8 Discussion

The preclinical and clinical studies discussed above have shown that FS offers promise to improve early cancer diagnosis in a variety of organ sites. Multiple studies have revealed a commonality in precancerous lesions in various organ sites: an increase in epithelial fluorescence coupled with a decrease in fluorescence from the underlying stroma compared to normal tissue. New probe designs and analytical tools are being developed to differentially target changes in optical properties in epithelial and stromal tissue. While these approaches show promise to improve diagnostic specificity, significant improvements are still required prior to large scale adoption of FS in a clinical setting. In particular, an improvement must be made in the understanding of confounding factors associated with benign conditions.

Recent advances in computational modeling have added important information to the field, helping to uncover and explain the biological basis of observed FS signals. The identification of the important fluorophores present in particular tissues has been of particular interest. Model-based analysis has revealed that in the breast, for example, NADH, collagen, and retinol are main contributors in the UV-VIS range (Zhu et al. 2008a) while in the cervix, NADH, FAD, keratin, and collagen are the dominant fluorophores (Chang et al. 2006, Pavlova et al. 2008a). Scalable Monte Carlo models may hold the potential for near real-time extraction of optical tissue properties, answering some of the limitations of analytical solutions such as multilayered geometry and flexibility in modeling probe design. One remaining drawback of point probe FS measurements is the small sampling size. Recent advances in hyperspectral instrumentation may allow for FS diagnostics using a larger field of view.

FS occupies an important segment of biomedical optics and of the larger field of medical diagnostics. Instrumentation and analysis techniques have developed to the point of being able to quantify the biochemical composition of human tissues. This is not possible using the most common structural medical imaging techniques such as computed tomography and magnetic resonance imaging. Further study, larger clinical trials, and commercialization of FS technology hold the promise of making FS a standard clinical tool for medical diagnostics.

References

Alfano, R., Tang, G., Pradhan, A. et al. 1987. Fluorescence spectra from cancerous and normal human breast and lung tissues. *IEEE J. Quantum Electron.* 23(10): 1806–1811.

Alfano, R., Tata, D., Cordero, J. J. et al. 1984. Laser induced fluorescence spectroscopy from native cancerous and normal tissue. *IEEE J. Quantum Electron.* 20(12): 1507–1511.

Amelink, A., Bard, M. P. L., Burgers, S. A. et al. 2003. Single-scattering spectroscopy for the endoscopic analysis of particle size in superficial layers of turbid media. *Appl. Opt.* 42(19): 4095–4101.

American Cancer Society. 2006. *Oral Cavity and Oropharyngeal Cancer.* American Cancer Society.

Anandasabapathy, S. 2008. Endoscopic imaging: Emerging optical techniques for the detection of colorectal neoplasia. *Curr. Opin. Gastroenterol.* 24(1): 64–69.

Arens, C., Reussner, D., Neubacher, H. et al. 2006. Spectrometric measurement in laryngeal cancer. *Eur. Arch. Otorhinolaryngol.* 263(11): 1001–1007.

Arifler, D., Schwarz, R. A., Chang, S. K. et al. 2005. Reflectance spectroscopy for diagnosis of epithelial precancer: Model-based analysis of fiber-optic probe designs to resolve spectral information from epithelium and stroma. *Appl. Opt.* 44(20): 4291–4305.

Biswal, N. C., Gupta, S., Ghosh, N. et al. 2003. Recovery of turbidity free fluorescence from measured fluorescence: An experimental approach. *Opt. Express* 11(24): 3320–3331.

Bonner, R. F., Nossal, R., Havlin, S. et al. 1987. Model for photon migration in turbid biological media. *J. Opt. Soc. Am. A* 4(3): 423–432.

Brancaleon, L., Durkin, A. J., Tu, J. H. et al. 2001. In vivo fluorescence spectroscopy of nonmelanoma skin cancer. *Photochem. Photobiol.* 73(2): 178–183.

Brewer, M., Utzinger, U., Silva, E. et al. 2001. Fluorescence spectroscopy for in vivo characterization of ovarian tissue. *Lasers Surg. Med.* 29(2): 128–135.

Brookner, C. K., Follen, M., Boiko, I. et al. 2000. Autofluorescence patterns in short-term cultures of normal cervical tissue. *Photochem. Photobiol.* 71(6): 730–736.

Brookner, C., Utzinger, U., Follen, M. et al. 2003. Effects of biographical variables on cervical fluorescence emission spectra. *J. Biomed. Opt.* 8(3): 479–483.

Canpolat, M. and Mourant, J. R. 2001. Particle size analysis of turbid media with a single optical fiber in contact with the medium to deliver and detect white light. *Appl. Opt.* 40(22): 3792–3799.

Chance, B., Cohen, P., Jobsis, F. et al. 1962a. Intracellular oxidation-reduction states in vivo. *Science* 137: 499–508.

Chance, B., Cohen, P., Jobsis, F. et al. 1962b. Localized fluorometry of oxidation-reduction states of intracellular pyridine nucleotide in brain and kidney cortex of the anesthetized rat. *Science* 136(3513): 325.

Chance, B., Schoener, B., Oshino, R. et al. 1979. Oxidation-reduction ratio studies of mitochondria in freeze-trapped samples. NADH and flavoprotein fluorescence signals. *J. Biol. Chem.* 254(11): 4764–4771.

Chang, S. K., Arifler, D., Drezek, R. et al. 2004. Analytical model to describe fluorescence spectra of normal and preneoplastic epithelial tissue: Comparison with Monte Carlo simulations and clinical measurements. *J. Biomed. Opt.* 9(3): 511–522.

Chang, S. K., Follen, M., Malpica, A. et al. 2002. Optimal excitation wavelengths for discrimination of cervical neoplasia. *IEEE Trans. Biomed. Eng.* 49(10): 1102–1111.

Chang, S. K., Marin, N., Follen, M. et al. 2006. Model-based analysis of clinical fluorescence spectroscopy for in vivo detection of cervical intraepithelial dysplasia. *J. Biomed. Opt.* 11(2): 024008.

Chang, S. K., Mirabal, Y. N., Atkinson, E. N. et al. 2005. Combined reflectance and fluorescence spectroscopy for in vivo detection of cervical pre-cancer. *J. Biomed. Opt.* 10(2): 024031.

Chu, S. C., Hsiao, T. C. R., Lin, J. K. et al. 2006. Comparison of the performance of linear multivariate analysis methods for normal and dyplasia tissues differentiation using autofluorescence spectroscopy. *IEEE Trans. Biomed. Eng.* 53(11): 2265–2273.

Cox, D. D., Chang, S. K., Dawood, M. Y. et al. 2003. Detecting the signal of the menstrual cycle in fluorescence spectroscopy of the cervix. *Appl. Spectrosc.* 57(1): 67–72.

de Veld, D. C., Skurichina, M., Witjes, M. J. H. et al. 2003. Autofluorescence characteristics of healthy oral mucosa at different anatomical sites. *Lasers Surg. Med.* 32(5): 367–376.

de Veld, D. C., Skurichina, M., Witjes, M. J. H. et al. 2004. Clinical study for classification of benign, dysplastic, and malignant oral lesions using autofluorescence spectroscopy. *J. Biomed. Opt.* 9(5): 940–950.

de Veld, D. C., Witjes, M. J., Sterenborg, H. J. et al. 2005. The status of in vivo autofluorescence spectroscopy and imaging for oral oncology. *Oral Oncol.* 41(2): 117–131.

Downer, M. C., Moles, D. R., Palmer, S. et al. 2004. A systematic review of test performance in screening for oral cancer and precancer. *Oral Oncol.* 40(3): 264–273.

Drezek, R., Brookner, C., Pavlova, I. et al. 2001a. Autofluorescence microscopy of fresh cervical-tissue sections reveals alterations in tissue biochemistry with dysplasia. *Photochem. Photobiol.* 73(6): 636–641.

Drezek, R., Sokolov, K., Utzinger, J. et al. 2001b. Understanding the contributions of NADH and collagen to cervical tissue fluorescence spectra: Modeling, measurements, and implications. *J. Biomed. Opt.* 6(4): 385–396.

Gabrecht, T., Andrejevic-Blant, S., Wagneires, G. 2007. Blue-violet excited autofluorescence spectroscopy and imaging of normal and cancerous human bronchial tissue after formalin fixation. *Photochem. Photobiol.* 83(2): 450–458.

Gebhart, S. C., Thompson, R. C., and Mahadevan-Jansen, A. 2007. Liquid-crystal tunable filter spectral imaging for brain tumor demarcation. *Appl. Opt.* 46(10): 1896–1910.

Georgakoudi, I., Jacobson, B. C., Muller, M. G. et al. 2002a. NAD(P)H and collagen as in vivo quantitative fluorescent biomarkers of epithelial precancerous changes. *Cancer Res.* 62(3): 682–687.

Georgakoudi, I., Sheets, E. E., Muller, M. G. et al. 2002b. Trimodal spectroscopy for the detection and characterization of cervical precancers in vivo. *Am. J. Obstet. Gynecol.* 186(3): 374–382.

Gillenwater, A., Jacob, R., Ganeshappa, R. et al. 1998. Noninvasive diagnosis of oral neoplasia based on fluorescence spectroscopy and native tissue autofluorescence. *Arch. Otolaryngol. Head Neck Surg.* 124(11): 1251–1258.

Heintzelman, D. L., Lotan, R., and Richards-Kortum, R. R. 2000a. Characterization of the autofluorescence of polymorphonuclear leukocytes, mononuclear leukocytes and cervical epithelial cancer cells for improved spectroscopic discrimination of inflammation from dysplasia. *Photochem. Photobiol.* 71(3): 327–332.

Heintzelman, D. L., Utzinger, U., Fuchs, H. et al. 2000b. Optimal excitation wavelengths for in vivo detection of oral neoplasia using fluorescence spectroscopy. *Photochem. Photobiol.* 72(1): 103–113.

Hyde, D. E., Farrell, T. J., Pattersson, M. S. et al. 2001. A diffusion theory model of spatially resolved fluorescence from depth-dependent fluorophore concentrations. *Phys. Med. Biol.* 46(2): 369–383.

Inaguma, M. and Hashimoto, K. 1999. Porphyrin-like fluorescence in oral cancer: In vivo fluorescence spectral characterization of lesions by use of a near-ultraviolet excited autofluorescence diagnosis system and separation of fluorescent extracts by capillary electrophoresis. *Cancer* 86(11): 2201–2211.

Ingrams, D. R., Dhingra, J. K., Roy, K. et al. 1997. Autofluorescence characteristics of oral mucosa. *Head Neck* 19(1): 27–32.

Jobsis, F. F., Oconnor, M., Vitale, A. et al. 1971. Intracellular redox changes in functioning cerebral cortex. 1. Metabolic effects of epileptiform activity. *J. Neurophysiol.* 34(5): 735–749.

Lakowicz, J. R. 2006. *Principles of Fluorescence Spectroscopy.* Berlin, Germany: Springer.

Lam, S., Kennedy, T., Unger, M. et al. 1998. Localization of bronchial intraepithelial neoplastic lesions by fluorescence bronchoscopy. *Chest* 113(2): 696–702.

Lin, W. C., Toms, S. A., Johnson, M. et al. 2001. In vivo brain tumor demarcation using optical spectroscopy. *Photochem. Photobiol.* 73(4): 396–402.

Lin, W. C., Toms, S. A., Motamedi, M. et al. 2000. Brain tumor demarcation using optical spectroscopy; an in vitro study. *J. Biomed. Opt.* 5(2): 214–220.

Lingen, M. W., Kalmar, J. R., Karrison, T. et al. 2008. Critical evaluation of diagnostic aids for the detection of oral cancer. *Oral. Oncol.* 44(1): 10–22.

Liu, C. H., Das, B. B., Sha Glassman, W. L. et al. 1992. Raman, fluorescence, and time-resolved light-scattering as optical diagnostic-techniques to separate diseased and normal biomedical media. *J. Photochem. Photobiol. B* 16(2): 187–209.

Liu, Q. and Ramanujam, N. 2004. Experimental proof of the feasibility of using an angled fiber-optic probe for depth-sensitive fluorescence spectroscopy of turbid media. *Opt. Lett.* 29(17): 2034–2036.

Lohmann, W., Mussmann, J., Lohman, C. et al. 1989. Native fluorescence of the cervix uteri as a marker for dysplasia and invasive carcinoma. *Eur. J. Obstet. Gynecol. Reprod. Biol.* 31(3): 249–253.

Majumder, S. K., Ghosh, N., Gupta, P. K. 2005. Support vector machine for optical diagnosis of cancer. *J. Biomed. Opt.* 10(2): 024034.

Majumder, S. K., Gupta, A., Gupta, S. et al. 2006. Multi-class classification algorithm for optical diagnosis of oral cancer. *J. Photochem. Photobiol. B* 85(2): 109–117.

Malavika, C., James, S., Heidt, D. et al. 2007. Probing pancreatic disease using tissue optical spectroscopy. *J. Biomed. Opt.* 12(6): 060501.

Martin, M. E., Wabuyele, M. B., Chen, K. et al. 2006. Development of an advanced hyperspectral imaging (HSI) system with applications for cancer detection. *Ann. Biomed. Eng.* 34(6): 1061–1068.

Maurice, A., Oliver, C., Avrillier, S. et al. 1996. Ultraviolet laser-induced autofluorescence distinction between malignant and normal urothelial cells and tissues. *J. Biomed. Opt.* 1(3): 335–341.

Mayevsky, A. and Chance, B. 1974. Repetitive patterns of metabolic changes during cortical spreading depression of awake rat. *Brain Res.* 65(3): 529–533.

Mayevsky, A. and Chance, B. 1982. Intracellular oxidation-reduction state measured in situ by a multichannel fiber-optic surface fluorometer. *Science* 217(4559): 537–540.

Muller, M. G., Georgakoudi, I., Zhang, Q. et al. 2001. Intrinsic fluorescence spectroscopy in turbid media: Disentangling effects of scattering and absorption. *Appl. Opt.* 40(25): 4633–4646.

Muller, M. G., Valdez, T. A., Georgakoudi, I. et al. 2003. Spectroscopic detection and evaluation of morphologic and biochemical changes in early human oral carcinoma. *Cancer* 97(7): 1681–1692.

Nair, M. S., Ghosh, N., Raju, N. S. et al. 2002. Determination of optical parameters of human breast tissue from spatially resolved fluorescence: a diffusion theory model. *Appl. Opt.* 41(19): 4024–4035.

Nath, A., Rivoire, K., Chang, S. et al. 2004. Effect of probe pressure on cervical fluorescence spectroscopy measurements. *J. Biomed. Opt.* 9(3): 523–533.

Nossal, R., Bonner, R. F., Weiss, G. H. et al. 1989. Influence of pathlength on remote optical sensing of properties of biological tissue. *Appl. Opt.* 28(12): 2238–2244.

Palmer, G. M., Keely, P. J., Bereslin, T. M. et al. 2003. Autofluorescence spectroscopy of normal and malignant human breast cell lines. *Photochem. Photobiol.* 78(5): 462–469.

Palmer, G. M. and Ramanujam, N. 2008. Monte-Carlo-based model for the extraction of intrinsic fluorescence from turbid media. *J. Biomed. Opt.* 13(2): 024017.

Pavlova, I., Sokolov, K., Drezek, R. et al. 2003. Microanatomical and biochemical origins of normal and precancerous cervical autofluorescence using laser-scanning fluorescence confocal microscopy. *Photochem. Photobiol.* 77(5): 550–555.

Pavlova, I., Weber, C. R., Schwarz, R. A. et al. 2008a. Monte Carlo model to describe depth selective fluorescence spectra of epithelial tissue: Applications for diagnosis of oral precancer. *J. Biomed. Opt.* 13(6): 024017.

Pavlova, I., Weber, C. R., Schwarz, R. A. et al. 2009. Fluorescence spectroscopy of oral tissue: Monte Carlo modeling with site-specific tissue properties. *J. Biomed. Opt.* 14(1): 014009.

Pavlova, I., Williams, M., El-Naggar, A. et al. 2008b. Understanding the biological basis of autofluorescence imaging for oral cancer detection: High-resolution fluorescence microscopy in viable tissue. *Clin. Cancer Res.* 14(8): 2396–2404.

Pfefer, T. J., Agrawal, A., Drezek, R. A. et al. 2005. Oblique-incidence illumination and collection for depth-selective fluorescence spectroscopy. *J. Biomed. Opt.* 10(4): 44016.

Pfefer, T. J., Matchette, L. S., Ross, A. M. et al. 2003. Selective detection of fluorophore layers in turbid media: The role of fiber-optic probe design. *Opt. Lett.* 28(2): 120–122.

Pitts, J. D., Sloboda, R. D., Dragnev, K. H. et al. 2001. Autofluorescence characteristics of immortalized and carcinogen-transformed human bronchial epithelial cells. *J. Biomed. Opt.* 6(1): 31–40.

Policard, A. 1924. Etude sur les aspects offerts par des tumeurs expérimentales examinées a la lumière de Wood. *C. R. Soc. Biol. (Paris)* 91: 1423–1424.

Qu, J. N., Macaulay, C., Lam, S. et al. 1995. Laser-induced fluorescence spectroscopy at endoscopy—Tissue optics, Monte-Carlo modeling, and in-vivo measurements. *Opt. Eng.* 34(11): 3334–3343.

Ramanujam, N., Mitchell, M. F., Mahadevan-Jansen, A. et al. 1994. In vivo diagnosis of cervical intraepithelial neoplasia using 337-nm-excited laser-induced fluorescence. *Proc. Natl. Acad. Sci. USA* 91(21): 10193–10197.

Ramanujam, N., Mitchell, M. F., Mahadevan-Jansen, A. et al. 1996a. Cervical precancer detection using a multivariate statistical algorithm based on laser-induced fluorescence spectra at multiple excitation wavelengths. *Photochem. Photobiol.* 64(4): 720–735.

Ramanujam, N., Mitchell, M. F., Mahadevan-Jansen, A. et al. 1996b. Spectroscopic diagnosis of cervical intraepithelial neoplasia (CIN) in vivo using laser-induced fluorescence spectra at multiple excitation wavelengths. *Lasers Surg. Med.* 19(1): 63–74.

Richards-Kortum, R. and Sevick-Muraca, E. 1996. Quantitative optical spectroscopy for tissue diagnosis. *Annu. Rev. Phys. Chem.* 47: 555–606.

Roblyer, D., Kurachi, C., Stepanek, V. et al. 2009. Objective detection and delineation of oral neoplasia using autofluorescence imaging. *Cancer Prev. Res.* 2(5): 423–431.

Schwarz, R. A., Gao, W., Daye, D. et al. 2008. Autofluorescence and diffuse reflectance spectroscopy of oral epithelial tissue using a depth-sensitive fiber-optic probe. *Appl. Opt.* 47(6): 825–834.

Schwarz, R. A., Gao, W., Weber, C. et al. 2009. Noninvasive evaluation of oral lesions using depth-sensitive optical spectroscopy. *Cancer* 115(8): 1669–1679.

Skala, M. C., Palmer, G. M., Zhu, C. et al. 2004. Investigation of fiber-optic probe designs for optical spectroscopic diagnosis of epithelial pre-cancers. *Lasers Surg. Med.* 34(1): 25–38.

Skala, M. C., Riching, K. M., Gendron-Fitzpatrick, A. et al. 2007. In vivo multiphoton microscopy of NADH and FAD redox states, fluorescence lifetimes, and cellular morphology in precancerous epithelia. *Proc. Natl. Acad. Sci. USA* 104(49): 19494–19499.

Sokolov, K., Galvan, J., Myakov, A. et al. 2002. Realistic three-dimensional epithelial tissue phantoms for biomedical optics. *J. Biomed. Opt.* 7(1): 148–156.

Stasic, D., Farrell, T. J., Patterson, M. S. 2003. The use of spatially resolved fluorescence and reflectance to determine interface depth in layered fluorophore distributions. *Phys. Med. Biol.* 48(21): 3459–3474.

Tsai, T., Chen, H. M., Wang, C. Y. et al. 2003. In vivo autofluorescence spectroscopy of oral premalignant and malignant lesions: Distortion of fluorescence intensity by submucous fibrosis. *Lasers Surg. Med.* 33(1): 40–47.

Tunnell, J. W., Desjardins, A. E., Galindo, L. et al. 2003. Instrumentation for multi-modal spectroscopic diagnosis of epithelial dysplasia. *Technol. Cancer Res. Treat.* 2(6): 505–514.

Uehlinger, P., Gabrecht, T., Glanzman, T. et al. 2009. In vivo time-resolved spectroscopy of the human bronchial early cancer autofluorescence. *J. Biomed. Opt.* 14(2): 024011.

Utzinger, U. and Richards-Kortum, R. R. 2003. Fiber optic probes for biomedical optical spectroscopy. *J. Biomed. Opt.* 8(1): 121–147.

Volynskaya, Z., Haka, A. S., Bechtel, K. L. et al. 2008. Diagnosing breast cancer using diffuse reflectance spectroscopy and intrinsic fluorescence spectroscopy. *J. Biomed. Opt.* 13(2): 024012.

Wang, C. Y., Tsai, T., Chen, H. M. et al. 2003. PLS-ANN based classification model for oral submucous fibrosis and oral carcinogenesis. *Lasers Surg. Med.* 32(4): 318–326.

Weersink, R., Patterson, M. S., Diamond, K. et al. 2001. Noninvasive measurement of fluorophore concentration in turbid media with a simple fluorescence/reflectance ratio technique. *Appl. Opt.* 40(34): 6389–6395.

Welch, A. J., Gardner, C., Richards-Kortum, R. et al. 1997. Propagation of fluorescent light. *Lasers Surg. Med.* 21(2): 166–178.

Wu, J., Feld, M. S., and Rava, R. P. 1993a. Analytical model for extracting intrinsic fluorescence in turbid media. *Appl. Opt.* 32(19): 3585–3595.

Wu, J., Partovi, F., and Rava, R. P. 1993b. Diffuse reflectance from turbid media—An analytical model of photon migration. *Appl. Opt.* 32(7): 1115–1121.

Yu, C. C., Lau, C., O'Donoghue, G. et al. 2008. Quantitative spectroscopic imaging for non-invasive early cancer detection. *Opt. Express* 16(20): 16227–16239.

Zellweger, M., Grosjean, P., Goujon, D. et al. 2001. In vivo autofluorescence spectroscopy of human bronchial tissue to optimize the detection and imaging of early cancers. *J. Biomed. Opt.* 6(1): 41–51.

Zhang, Q. G., Muller, M. G., Wu, J. et al. 2000. Turbidity-free fluorescence spectroscopy of biological tissue. *Opt. Lett.* 25(19): 1451–1453.

Zhu, C. F., Breslin, T. M., Harter, J. et al. 2008a. Model based and empirical spectral analysis for the diagnosis of breast cancer. *Opt. Express* 16(19): 14961–14978.

Zhu, C. F., Liu, Q., and Ramanujam, N. 2003. Effect of fiber optic probe geometry on depth-resolved fluorescence measurements from epithelial tissues: A Monte Carlo simulation. *J. Biomed. Opt.* 8(2): 237–247.

Zhu, C., Palmer, G. M., Breslin, T. M. et al. 2008b. Diagnosis of breast cancer using fluorescence and diffuse reflectance spectroscopy: A Monte-Carlo-model-based approach. *J. Biomed. Opt.* 13(3): 034015.

Zonios, G. I., Cothren, R. M., Arendt, J. T. et al. 1996. Morphological model of human colon tissue fluorescence. *IEEE Trans. Biomed. Eng.* 43(2): 113–122.

12

Raman, SERS, and FTIR Spectroscopy

12.1 Introduction 233
12.2 Definitions and Fundamentals...233
    Wave Number Units • Reading a Spectrum • Energetics of Raman
    Scattering • Initial States • Classical Model of the Raman Shift
12.3 Comparison to Other Spectroscopies.. 237
    Fluorescence • Near-Infrared Absorption • Vibrational Absorption: Infrared Spectroscopy
12.4 Equipment for Acquiring Raman Spectra 239
    Lasers • Detectors • Sampling Geometries
12.5 Mathematical Processing of Vibrational Spectral Data 242
    Quantification • Classification
12.6 State of the Art: Raman Biomedical Applications......................... 243
    Soft Tissue • Hard Tissue • Biofluids • Microbes • Single Cells • Multimodal Systems
12.7 Enhanced Raman Scattering 246
    Resonance Raman Applications • Coherently Stimulated Raman • SERS
12.8 Mid-IR Spectroscopy... 248
References.. 249

Andrew J. Berger
University of Rochester

12.1 Introduction

This chapter is designed to introduce techniques that probe the vibrational properties of molecules' chemical bonds: Raman scattering and mid-infrared (MIR) absorption. The greatest strength of these techniques is that they are intrinsically highly chemically specific, compared to other spectroscopies such as fluorescence or visible/near-infrared (NIR) absorption.

The two techniques, Raman and MIR absorption, are complementary. While they provide similar types of information, much is different at the levels of instrumentation and application. In this chapter, the Raman technique will be presented first, with occasional acknowledgment of important links to MIR absorption. At the end, particular additional aspects of MIR absorption will be discussed.

Looking broadly, Raman scattering and MIR absorption are key technologies in several biomedical research areas. They are excellent at detecting slight differences between types of cells or tissues. This makes them good for subtle classification tasks, either in point sampling (e.g., bacterial identification) or in image analysis (e.g., tissue histology). Often, vibrational spectroscopy can detect differences between specimens when other, more widespread optical methods cannot. Vibrational spectroscopy is also well-suited for measuring biochemical concentrations, particularly in situations where the target chemical exists in the presence of other, higher-concentration interferents. In both cases (classification and quantification), vibrational spectroscopy accomplishes this on the strength of the unusually rich spectrum of information it provides.

Vibrational spectroscopy also has significant limitations. Most immediately, the Raman effect is very weak, precluding rapid data acquisition, particularly of images; and MIR absorption has very shallow penetration depths, eliminating its straightforward application on bulk samples.

12.2 Definitions and Fundamentals

Vibrational spectroscopy probes the frequencies at which molecular atoms resonantly vibrate. In the Raman scattering process, the target molecule either gains or loses a quantum of vibrational energy, resulting in a scattered photon whose wavelength is different from the incident one. In the MIR absorption process, the molecule always gains the quantum, and no light is emitted. In order to understand these processes better, one must start by analyzing molecular vibrations themselves.

Just like idealized balls on springs, atoms connected by electronic bonds vibrate at normal-mode frequencies determined by masses and spring constants. To get a sense of the typical frequencies involved, consider two atoms engaged in a linear stretching vibration. The potential energy U of a typical bond is ~5 eV, and the internuclear stretching Δr required to go from equilibrium to dissociation is about 0.5 Å. Approximating the potential as harmonic, we have $U = (1/2)k\Delta r$; solving for the spring constant k gives a value of 6×10^2 J/m². If the two atoms are carbons with masses of 12 amu, then the reduced mass μ of the oscillator is 6 amu, and the frequency of the simple harmonic motion will be $\nu = (1/2\pi)\sqrt{k/\mu} = 4.0 \times 10^{13}$ Hz. Most biomedical vibrations occur at frequencies in this regime, from about 1.2 to 5.5×10^{13} Hz, called the "fingerprint region" because the sequence of vibrational resonances for different molecules is so distinctive in this range. Since hydrogen is so much lighter than all other atoms present in biomolecules, simple vibrations involving a hydrogen (such as a C–H stretch) occur in a distinct regime of higher frequencies, in the neighborhood of 9×10^{13} Hz.

12.2.1 Wave Number Units

Vibrational spectra are typically plotted not versus the vibrational frequency, but rather versus the energy of vibration. For a harmonic molecular oscillation at frequency ν, the spacing ΔE of allowed energy levels is $\Delta E = h\nu$, where h is Planck's constant (6.626×10^{-34} J · s). Real internuclear potentials are approximately harmonic near equilibrium, so the lowest few vibrational states are appropriately understood via this model. This ΔE is, therefore, the energy required to promote the molecule from its ground vibrational state to its first excited state. Converting the fingerprint regime of frequencies to energies, the values in SI units of Joules run from 8.0 to 36×10^{-21}, or in eV units from 0.049 to 0.220.

The conventional energy-related unit for vibrational spectroscopy is the *wave number*. It is usually indicated by the variable $\tilde{\nu}$ (i.e., the Greek letter ν with a tilde), and is related to the energy by the following formula:

$$\tilde{\nu} = \frac{\Delta E}{hc}, \qquad (12.1)$$

where

ΔE is the energy of the excited vibrational mode above the ground state
c is the speed of light in vacuum (3.00×10^8 m/s)

The product hc has the value 2.0×10^{-25} J · m. Equating expressions for ΔE, the wave number is related to the vibrational frequency by $\tilde{\nu} = \nu/c$.

As Equation 12.1 shows, wave numbers do not have dimensions of energy, but rather energy divided by the fundamental constants h and c. Since the energy of a photon can be written as $E_{photon} = hc/\lambda$, where λ is the wavelength, wave numbers have dimensions of *inverse wavelength*. This inverse length has a straightforward physical interpretation in absorption mode, where the entire energy of the photon has been consumed. In that case, $\Delta E = E_{photon}$, i.e., the entire energy of the photon has been taken up by the molecule, and the wave number of the transition is simply the inverse of the absorbed photon's wavelength. In the case of Raman scattering, where the photon gives only *some* of its energy to the molecule, the wave number of the transition is the *difference* between two inverse lengths, those of the incident and scattered photons. The Raman shift is given by

$$\tilde{\nu} = \frac{1}{\lambda_i} - \frac{1}{\lambda_s},$$

and its units are sometimes called *relative* wave numbers to emphasize the shift.

In wave numbers, the fingerprint region is approximately 400–1800 cm⁻¹. Inverse centimeters are the conventional way to present wave numbers in vibrational spectroscopy, as the values can be reported as 3- or 4-digit whole numbers with the accuracy of measurement typically lying in the range of 1–20 on this scale.

12.2.2 Reading a Spectrum

Figure 12.1 shows the vibrational Raman spectrum of a clump of bacteria (genus *Streptococcus*), with the *x*-axis in wave numbers. The increasing values on the wave number axis correspond to increasingly large quanta of energy transferred from the photon to a molecular oscillation. The region on the left is the fingerprint region, ranging as noted earlier from approximately 400 to 1800 cm⁻¹. As indicated in the figure, the observed bands can be often assigned to the motion of particular molecular oscillators; the assignments shown here are taken from a publication by Puppels (1999). As noted above, there also exists a separate higher-energy realm of vibrational transitions, corresponding to resonances involving hydrogen. In biomedical Raman spectroscopy, this regime is simply called the *high wave number region*. This regime offers practical advantages in certain situations to be discussed in Section 12.4.3, but is exploited far less often than the fingerprint regime.

12.2.3 Energetics of Raman Scattering

A diagram of electronic energy levels and vibrational sublevels, typically called a *Jablonski diagram*, is helpful in discussing Raman scattering and other related optical processes. As shown in Figure 12.2, a Jablonski diagram sketches the potential energy of a diatomic bond as a function of nuclear separation. This particular diagram concentrates on just two curves: the lower for the ground electronic state of the bond, and the upper for the first electronic excited state. In each case, a local energy minimum exists at some internuclear separation; near that position, the potential is essentially quadratic, and the first several vibrational sublevels are nearly those of a simple harmonic oscillator. The upward and downward arrows on this Jablonski diagram describe a Raman scattering transition: the molecule is lifted to an intermediate state by the energy of the incident photon and then is lowered to a final state by losing the energy

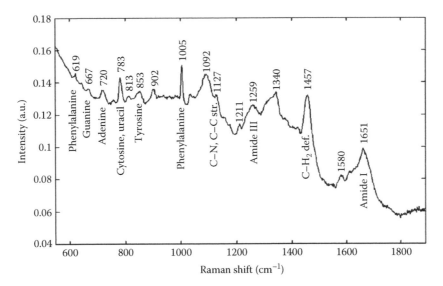

FIGURE 12.1 Typical Raman spectrum of a bacterium (genus *Streptococcus*); band assignments from Puppels (1999).

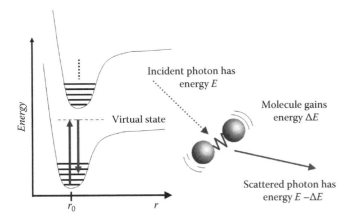

FIGURE 12.2 Jablonski energy level diagram for depicting Raman scattering. Horizontal lines represent vibrational energy levels, while the two large curves depict electronic energy versus internuclear separation coordinate *r*. The upward and downward arrows indicate the energy of the incident and scattered photons, as shown in the cartoon on the right.

of the scattered photon. The schematic cartoon on the right of the figure indicates the incident and scattered photon paths corresponding to the up/down arrows.

In Figure 12.3, the energy level spacings from the previous figure are isolated, without regard for the nuclear coordinate. This makes it easier to discuss the differences between different types of transitions. On the left, the fluorescence process is depicted, consisting of an absorption, a non-radiative relaxation, and an emission back to the ground electronic manifold. The molecule starts in an energy state labeled $|i\rangle$ and finishes in a state labeled $|f\rangle$. The emitted photon has less energy than the incident photon for two reasons: (1) non-radiative relaxation of the upper electronic state to its lowest vibrational level, and (2) radiative relaxation down to a state $|f\rangle$ that is (in this case) more energetic than $|i\rangle$.

The second section of Figure 12.3 depicts the Raman scattering process. Note that the molecule begins and ends the process in the same states $|i\rangle$ and $|f\rangle$, respectively, as the fluorescent example. In standard ("nonresonant") Raman scattering, however, the incident photon's energy places the molecule in an energetically *forbidden* state. This is indicated by the dashed line, and is commonly called a "virtual" state. Since the state is not energetically allowed, it can only persist for the time permitted by the Heisenberg uncertainty principle, $\Delta t \sim \hbar/\Delta E$. As this is on the femtosecond scale, energy-changing interactions with the environment are essentially impossible before the virtual state must radiatively decay. Quantum-mechanically, by far the greatest probability amplitude is for the state to decay directly back to the initial state $|i\rangle$, resulting in the emission of a photon with the same wavelength as the incident one (hence *elastic scattering*). There is some small probability amplitude, however, that the molecule will return to a different vibrational energy level $|f\rangle$ (inelastic or Raman scattering). Since no energy is exchanged with the environment, $\Delta E_{\text{molecule}} = E_{\text{(photon in)}} - E_{\text{(photon out)}}$. This is not true of fluorescence, where there is additional energy loss due to non-radiative processes.

12.2.4 Initial States

According to the energy conservation rule just noted, the emitted photon can either lose energy (called a Stokes or redshift) or gain energy (anti-Stokes or blueshift), depending upon the initial and final states of the molecular oscillation. The anti-Stokes process is shown third in Figure 12.3, with the energies of $|i\rangle$ and $|f\rangle$ reversed. Under normal conditions, however, biomolecular oscillators are overwhelmingly likely to be in their lowest vibrational states. Recall again that the typical range of biomedical vibrational energies in the fingerprint region is from 400 to 1800 cm^{-1}. These energies are large compared to the room- or

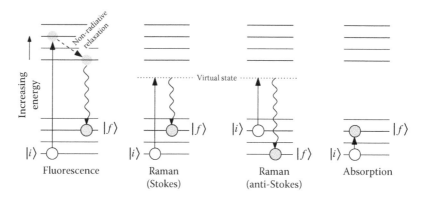

FIGURE 12.3 Simplified energy level diagrams, based upon Figure 12.2, illustrate the differences between fluorescence, Raman, and absorption processes.

body temperature thermal fluctuation energy kT (where k is Boltzmann's constant). In wave numbers, the thermal energy is $\tilde{v}_{thermal} = kT/hc$; at $T = 310\,K$ (37°C), this is equal to $214\,cm^{-1}$.

Due to the size of kT, the resting equilibrium ratio N_1/N_0 of molecular oscillators in the first excited state and the ground state, which is prescribed by the Boltzmann distribution law as

$$\frac{N_1}{N_0} = \exp\left(\frac{-\Delta E}{kT}\right) = \exp\left(\frac{-\tilde{v}_{transition}}{\tilde{v}_{thermal}}\right),$$

is about 0.009, or 0.9%, for a representative vibrational band of $\tilde{v}_{transition} = 1000\,cm^{-1}$, in the middle of the fingerprint region. As a result, photons are almost 100 times more likely to encounter molecules in this oscillation's ground state, from which no anti-Stokes scattering is possible. As a result, the Stokes Raman band will be correspondingly about 100 times stronger than the anti-Stokes band. (Simultaneous recording of Stokes and anti-Stokes scattering intensities provides, in fact, a noninvasive measurement of sample temperature.) Even for $\tilde{v}_{transition} = 400\,cm^{-1}$, the upper-state population percentage is only 15%. The relative weakness of anti-Stokes scattering in the fingerprint regime limits its utility in routine biomedical Raman experiments, where even the Stokes signal typically requires seconds to collect. When spontaneous anti-Stokes signals are strong enough to detect, however, they offer the advantage of being blueshifted from the laser line and, therefore, isolated from fluorescence, which solely occurs at redshifts. Also, if the equilibrium of states is actively manipulated by optical pumping, the anti-Stokes signal can be greatly enhanced; this is exploited in a technique called coherent anti-Stokes Raman scattering (CARS), as will be discussed below.

The last portion of Figure 12.3 depicts a transition from $|i\rangle$ up to $|f\rangle$ mediated by absorption. The molecule starts and ends in the same states as in the fluorescence and Stokes-Raman cases, but in this instance, there is no emitted photon. The incident photon exactly matches the energy of the transition, so there is nothing left over to emit.

12.2.5 Classical Model of the Raman Shift

A full derivation of the Raman scattering process is beyond the scope of this chapter (see, e.g., the chapter on Raman spectroscopy from Demtroder (1998)and its subsequent citations), but a simple classical model provides useful insight into key parameters. "Scattering," at the single-molecule level, means that an incident light field shakes a molecule's electrons and induces a varying dipole moment p. This dipole antenna then broadcasts its own radiation at whatever frequency(ies) it is oscillating. If nuclear motion of the molecule is ignored, then the induced dipole simply oscillates at the driving frequency, producing light of unchanged wavelength. But because the molecule is executing nuclear motion at a different frequency Ω, the two frequencies combine into a more complicated dipole oscillation that includes terms at frequencies ω, $\omega + \Omega$, and $\omega - \Omega$ (more discussion to follow below). The first term corresponds to elastic scattering (unchanged wavelength), while the latter two are Raman-shifted by an amount $\pm\Omega$ that characterizes the vibrational motion. For such dipole-like scattering, the cross-section scales generally as λ^{-4} for both elastic and Raman scattering, meaning that Raman scattering is almost five times stronger at $532\,nm$ excitation than at $785\,nm$.

Nuclear motion (i.e., vibration) influences the radiated frequencies because it can affect the polarizability of the electron distribution. Polarizability α is defined as the proportionality between applied electric field E and induced dipole moment p, i.e., it is proportional to the applied electric field E according to

$$p = \alpha E. \tag{12.2}$$

If nuclear motion along a coordinate Q (where $Q = 0$ is the equilibrium position) changes the polarizability, then to first order in a Taylor expansion

$$\alpha \approx \alpha_0 + \left.\frac{d\alpha}{dQ}\right|_{Q=0} Q + \cdots. \tag{12.3}$$

If the electric field oscillates at optical frequency ω and the coordinate Q oscillates with maximum displacement Q_0 at vibrational frequency Ω, then Equation 12.2 becomes

$$p(t) = \left[\alpha_0 + \frac{d\alpha}{dQ}(Q_0 \cos\Omega t)\right]E_o \cos\omega t$$

$$= \alpha_0 E_o \cos\omega t + \frac{d\alpha}{dQ}Q_0 E_o \cos\Omega t \cos\omega t$$

$$= \alpha_0 E_o \cos\omega t + \frac{1}{2}\frac{d\alpha}{dQ}Q_0 E_o \left\{\cos[\underbrace{(\omega-\Omega)}_{S}t] + \cos[\underbrace{(\omega+\Omega)}_{AS}t]\right\}.$$

$$(12.4)$$

The first term on the right of Equation 12.4 is at the optical excitation frequency, leading to elastic scattering. The induced dipole moment of the vibrating molecule, however, contains Stokes and anti-Stokes terms, labeled in Equation 12.4 by S and AS, due to the multiplication of the sinusoidal ω and Ω terms. The prefactor of $d\alpha/dQ$ captures another important point, that the nuclear motion must alter the polarizability for a mode to be Raman active. Under some symmetry conditions, vibrational motion may not alter the polarizability, in which case, there is no Raman scattering. Elastic scattering will still take place, however, since it only requires α_0 itself to be nonzero.

12.3 Comparison to Other Spectroscopies

12.3.1 Fluorescence

The most direct comparison with Raman is made with fluorescence, since both phenomena occur simultaneously when a laser excites a biological sample. The most notable shortcoming of Raman scattering relative to fluorescence (as well as most other biomedical spectroscopy techniques) is that it is weaker. In situations that are favorable for exciting fluorescence, such as blue or near-UV excitation of intact tissue specimens, Raman contributions are typically many orders of magnitude weaker, low enough to be undetectable for typical fluorescence acquisition times. Even when conditions are more favorable for Raman, such as when studying isolated cells (which fluoresce less) or when using NIR excitation light (which is less strongly absorbed and, therefore, generates less fluorescence), even the strongest Raman bands in the fingerprint region are rarely more than 50% as strong as a broad emissive background that is usually attributed to some residual amount of fluorescence. It is worth emphasizing, however, that fluorescence spectroscopy or imaging must be performed using the limited range of wavelengths that are absorbed by the target fluorophore, whereas Raman scattering can be elicited using any excitation wavelength more energetic than the targeted vibrational transition.

As a consequence of its weakness, spontaneous biomedical Raman scattering does not lend itself to video-rate imaging or even to point monitoring of sub-second changes. In almost all biomedical applications, Raman spectral bands are simply too weak to be useful when gathered for significantly less than a second. This is a major downside compared to fluorescence, which can routinely provide video-rate tissue imaging simply by placing a Stokes-shifted band-pass filter ($\Delta\lambda = 10-30$ nm) in front of the collection optics, either in widefield mode or using a raster-scanned spot to achieve confocal sectioning. Only when a Raman transition is enhanced by several orders of magnitude can signals reach strengths compatible with video-rate pixel dwell times. Such enhancements can be achieved by coherent, multiphoton emission techniques such as CARS; resonance excitation, where the laser wavelength is close to an absorption band of a molecular species; and surface-enhanced Raman scattering, where a local field enhancement is provided by a metallic nanostructure. All of these approaches will be discussed briefly later in this chapter.

Although Raman spectroscopy loses out to fluorescence in terms of sensitivity, its greatest advantage over fluorescence is that it is far more *specific*. Raman bands are typically tens of wave numbers wide, which translates to only a couple of nanometers. This is much sharper than fluorescence peaks, which can span 100 or more nanometers (c.f. the broad background in Figure 12.1). As such, over a fixed spectral detection range, Raman scattering can pack more information than fluorescence; in other words, the spectra are more distinctive, just as the phrase "fingerprint region" implies. As a consequence, Raman spectroscopy is well-suited to the particular challenge of decomposing a measured spectrum into contributions from several known chromophores. Compared to fluorescence, the inversion process is more robust using Raman because the degree of spectral overlap among the chromophores' signatures is less severe. In addition, nearly all biological molecules give off Raman scattering in nearly equal strengths proportional to concentration, whereas fluorescence spectra are typically dominated by a few major fluorophores (structural proteins collagen and elastin, coenzymes NADH and FAD+, aromatic amino acids, and porphyrins), rendering most other biological molecules undetectable without exogenous labeling. Also, NADH and FAD+ fluorescence levels (among others) are oxygenation-dependent and therefore change substantially between in vivo and ex vivo conditions, whereas few Raman signatures are substantially altered.

Another key difference between Raman and fluorescence is in how the information is encoded in the emitted light. In fluorescence, since the excited molecule relaxes to its upper electronic level's ground state before radiatively decaying, memory of the excitation wavelength is essentially lost, and the conserved quantity is the emitted spectrum's shape versus *wavelength*. In Raman, by contrast, the energy shift is always measured relative to the incident wavelength, and the conserved quantity is the emitted spectrum's shape versus *relative wave number*. These points underscore the fact that fluorescence must be excited at a wavelength that is sufficiently absorbed, whereas Raman scattering can be performed at any excitation wavelength. Tuning the excitation wavelength will shift all of the Raman peaks, while fluorescence features remain at fixed positions. This phenomenon can be exploited to discriminate between the two types of emission; taking the difference between two such spectra suppresses the fluorescence and highlights Raman-related features. This technique is called shifted-excitation Raman difference spectroscopy (Shreve et al. 1992).

12.3.2 Near-Infrared Absorption

As just noted, the tunability of Raman excitation means that it can be performed at whatever wavelength is most convenient, as long as the photon energy exceeds the vibrational energy. Since tissue specimens readily autofluoresce under visible and near-UV excitation, Raman spectroscopy of such samples is usually performed in the NIR (700–1100 nm), where endogenous absorption and, therefore, fluorescence are much weaker (although, as noted above, the scattering is several times weaker than when visible excitation is used). This spectral regime is also employed for NIR reflectance spectroscopy, as discussed in Chapter 6. Unlike fluorescence or Raman, NIR absorption spectroscopy requires a broadband source in order to obtain broadband information. Raman's relative merits compared to NIR absorption are similar to the Raman/fluorescence comparison. NIR absorption bands are much stronger than Raman bands, but the features (due to vibrational overtones and combination bands) are much broader. The principal NIR absorbers are oxy- and deoxyhemoglobin, cytochrome, water, and lipid; it is difficult to study other chemicals using NIR absorption, whereas Raman spectroscopy can detect many less abundant components.

12.3.3 Vibrational Absorption: Infrared Spectroscopy

The vibrational modes of biomolecules can also be probed using absorption rather than scattering. As noted earlier, if an incident photon has a frequency corresponding to a molecular vibration, it can be absorbed entirely rather than scattered inelastically, with all of its energy going to promote the oscillator to a higher energy level.

Unlike Raman scattering, which can be generated using any sufficiently energetic excitation wavelength, vibrational absorption is confined to a fixed spectral range. As noted above, the vibrational fingerprint regime for biomolecules spans from about 400 to 1800 cm^{-1}. In wavelength, that corresponds to the range 5.6–25 μm, which lies in the MIR. Vibrational absorption spectroscopy is equivalently known as MIR spectroscopy, or by convention simply infrared (IR) spectroscopy. Figure 12.4 emphasizes via an energy level diagram that an upward transition of ΔE for a molecular oscillator can be accomplished at only one wavelength by absorption, but multiple ways (three depicted) by scattering.

MIR absorption spectra resemble Raman spectra in that both are plotted versus wave number and the major features

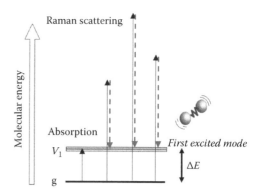

FIGURE 12.4 Comparison of energy level changes for vibrational absorption and scattering. While both processes leave the molecule in an excited vibrational state, the total energy of the absorbed photon must be ΔE, while the total energy of the scattered photon is a variable that depends upon the incident photon's energy.

correspond to molecular vibrational motion. The quantum–mechanical selection rules are different, however; for the simplest molecular oscillations involving a few atoms, a strong Raman-active vibration will be absent in IR, and vice versa. Raman and IR are sometimes called "complementary" techniques for this reason, because they fill in gaps in each other's vibrational information. In clinically oriented biomedical spectroscopy, it is extremely uncommon for both methods to be used on the same sample, because the modalities of operation are so different and the goal is rarely to gain fundamental understanding of the molecular dynamics. In addition, because biological molecules are more complex and exist in heterogeneous environments, the strict selections rules for some oscillations are weakly violated, leading to bands that appear in both modes.

The different wavelength regimes for Raman and MIR absorption lead to some profound biomedical consequences. Water happens to be a strong absorber in the MIR, which limits penetration to only a few hundreds of microns for that modality. Since biological specimens naturally contain water, MIR absorption spectroscopy of tissue can only be performed on regions a few cell layers thick, either in transmission through slices or in attenuated total internal reflection from bulk specimens (to be discussed below). Due to the deleterious effects of water, MIR transmission spectroscopy is performed on dried samples whenever possible. In contrast, Raman spectroscopy is tunable and, as noted, is often performed using NIR excitation. In addition to reducing fluorescence relative to the visible, NIR light is weakly absorbed by water, enabling Raman spectroscopy to sense millimeters deep into samples that include living tissue. At the same time, because NIR Raman utilizes shorter wavelengths than MIR absorption (0.7–1.1 microns versus 5–25 microns), diffraction-limited focused spot sizes are an order of magnitude smaller. This difference in focusing power enables Raman microscopy to interrogate subcellular regions 1 micron in feature size or less, while MIR absorption cannot resolve better than at the scale of entire cells.

A final distinction between Raman and MIR absorption is that Raman scattering can be stimulated coherently. This permits nonlinear increases in signal strength that can be dramatic when sub-nanosecond pulses of high peak power are used. MIR, being an absorptive technique, does not have an analogous coherently driven avenue available.

12.4 Equipment for Acquiring Raman Spectra

Ordinary Stokes-shifted Raman spectroscopy is performed using a setup conceptually similar to fluorescence spectroscopy, as shown in Figure 12.5. In both methods, excitation light impinges upon a sample, emitted light is collected over some solid angle, and a spectrometer records some portion of the emitted spectrum. It remains essential to block as much of the unshifted light as possible before it reaches the spectrograph, as otherwise it will drown out the desired signal even in the most well-designed of instruments. Just as in fluorescence, the blocking is accomplished by placing filters with nonoverlapping transmission spectra in the excitation and collection paths. If the collection path overlaps the excitation path (epi-detection), as is frequently the case, then either one of these two filters must be used in both reflection and transmission mode, or else an additional dichroic filter specifically for that purpose is used. Present state-of-the-art filters of all these types routinely provide >90% transmission in their desired passbands while rejecting four or more orders of magnitude in spectral regions only a few hundred wave numbers away.

12.4.1 Lasers

All practical biomedical Raman spectroscopy is performed using lasers, because no other sources supply enough photons within the necessarily narrow bandwidth of ~10 cm^{-1} that prevents undue broadening of the peaks. Due to the aforementioned weakness of spontaneous Raman scattering, more powerful lasers are required than are commonly employed for fluorescence. Depending upon the sample and the application,

tens or even hundreds of mW may be needed in order to obtain sufficient signal quality on the scale of seconds or a few minutes.

Fixed-wavelength diode or solid-state lasers have become by far the most common choice for biomedical Raman spectroscopy. Suitable lasers in the visible (for cell work, where fluorescence is low) and NIR (for all biomedical purposes) providing sufficient power have gotten small enough, down to a few centimeters on a side, to incorporate into cart-mounted or even handheld devices. All diode-based lasers for Raman spectroscopy incorporate some sort of linewidth-narrowing element, such as an external cavity and grating, to keep the bandwidth within the few cm^{-1} range. The cost of such lasers is currently more than 25 times less than a tunable Ti:sapphire laser. Broad tunability is rarely used in Raman biomedical applications except in exploratory work because high-performance bandpass, dichroic, and notch filters only function over a few nm without requiring realignment or replacement. The most commonly reported excitation wavelengths for the biomedical Raman spectroscopy discussed in this chapter are 785 and 830 nm, two NIR wavelengths that happen (currently) to be standards for the laser diode industry.

12.4.2 Detectors

At present, essentially all Raman spectroscopy, biomedical or otherwise, uses charge-coupled device (CCD) array detectors to record spectra. Spectrographs using ruled or holographic gratings, typically 600 or 1200 lines/mm, disperse images of a spot or slit onto the CCD array, which records all the wavelength information in parallel. Crucially, the CCD elements are thermoelectrically or liquid-nitrogen cooled to −50°C or lower to reduce thermal noise, which decreases by about an order of magnitude for each 10 degrees of cooling. At such temperatures, CCD pixels provide shot-noise-limited detection for most Raman spectra obtained with exposure times of a second or more. Only in cases of unusually low fluorescence background will the thermal or readout counts be significant sources of noise.

The bandgap of silicon detectors limits quantum efficiency past 1100 nm, which therefore makes 850 nm about the outer limit of what excitation wavelength can be used. Quantum

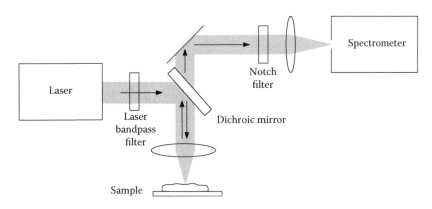

FIGURE 12.5 Essential components for gathering a Raman spectrum.

efficiencies currently can exceed 50% past 900 nm for some detector types, and improvements are ongoing as the demand for scientific-grade NIR CCDs continues to increase. The highest such quantum efficiencies are currently obtained using back-thinned, deep-depletion CCDs. One concern for applications with low Raman-to-fluorescence ratios is that back-thinned CCDs will always exhibit some amount of etaloning, a spectral rippling effect caused by interferences between reflections from the front and back surfaces of the CCD chip. Etaloning modulates the effective quantum efficiency (QE) of the pixels, and unfortunately, the irregular period often mimics the width of typical Raman peaks. If the fluorescence level is high, then even a 1% or less variation in the detector's QE will impart a rippling to the fluorescence that will be comparable to the amplitude of the Raman bands. For certain demanding applications, then, back-thinned CCDs should therefore be avoided, as they only provide a factor of ~2 advantage in overall QE.

12.4.3 Sampling Geometries

All of the typical geometries employed in fluorescence spectroscopy of ex vivo samples are also available for Raman spectroscopy. Figure 12.5 illustrates common-path epi-illumination, where the emitted light retraces the illumination path until separated by a dichroic beam splitter. By using a common focusing/collecting lens, this approach ensures that the two arms of the optical system stare at the same location, even if the focal point is displaced or aberrated by the sample and/or its holder (such as the curved surface of a vial containing a liquid sample). Such built-in collection efficiency is particularly important for Raman spectroscopy, due to its low sensitivity.

Non-common-path geometries are also common. Oblique-incidence illumination is often used for turbid bulk tissue specimens such as biopsies. As illustrated in Figure 12.6a, the laser enters outside the numerical aperture of the collection lens. In this case, the illumination and collection paths are non-coaxial and relatively easy to set up. An alternative uses a small hole in, or folding mirror in front of, a collection lens; the latter is indicated in Figure 12.6b. This makes the illumination and collection coaxial, but the solid angles do not overlap. In both cases, the nonoverlapping geometry eliminates the need for a dichroic mirror and eliminates specular reflection from surfaces normal to the collection axis. On the downside, the system becomes more sensitive to misalignment because any optical distortions affect the delivery and collection differently.

12.4.3.1 Microscopic Volumes

For high-resolution (few-micron scale or less) interrogation of a sample, the collecting objective obscures too much solid angle to permit a separate delivery beam, so the common-path geometry is necessary. Confocal sectioning (as discussed in Chapter 25) reduces the signal from bulk targets but provides the highest spatial resolution, down to submicron scales. To make the common-path geometry of Figure 12.5 confocal, a spatial filter should be placed at the entrance plane of the spectrograph or a conjugate plane, slightly exceeding the magnified image of the focused laser spot. This is often accomplished by placing a multimode optical fiber at the final lens's focus instead, and then placing the other end of the fiber at the spectrograph's entrance plane. The fiber itself then serves both as the spatial filter (with the rejected light never even arriving at the spectrograph) and as an easy-to-align relay of the filtered light to the spectrograph.

12.4.3.2 Imaging

Although spontaneous Raman scattering is too weak for real-time image acquisition, some microscopic imaging applications are still pursued. As in fluorescence imaging, there are two alternative methods of acquiring Raman images. One, using the confocal approach just described, is to raster-scan the sample or the excitation beam in two dimensions, either $x–y$ or $x–z$. This intrinsically provides full Raman spectra at each location (i.e., hyperspectral imaging), but the spatial locations are scanned serially in time. Since each spectrum usually requires at least a few seconds, building up even a 100-micron-square image with few-micron resolution usually requires many hours.

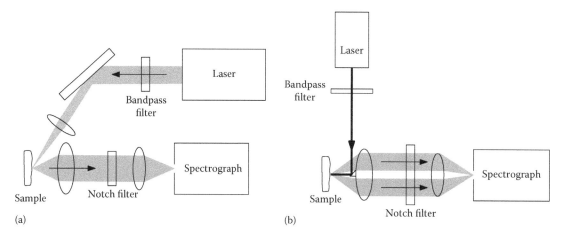

FIGURE 12.6 Nonoverlapping excitation/collection schemes. (a) Oblique incidence. (b) Folding mirror.

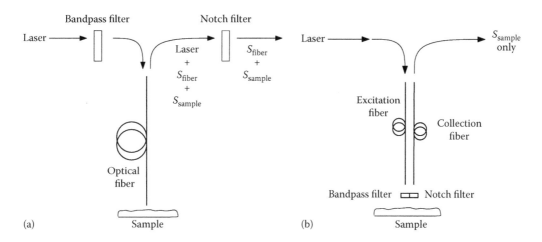

FIGURE 12.7 Fiber probe geometries. (a) In a conventional fluorescence probe, the filters are placed at the proximal (operator) end of the probe. While this prevents the laser light from reaching the spectrograph, significant Stokes-shifted signal S_{fiber} generated by the fiber material gets through along with the desired S_{sample}. (b) By using separate excitation/collection fibers and placing the filters at the probe's distal (sample) end, the material in the fibers no longer corrupts the spectrum, and only S_{sample} reaches the spectrograph.

The other alternative is to acquire a sequence of wide-field images, each recorded over a different spectral passband using a variable filter. Instruments that can sweep a narrow (~10 cm^{-1}) passband include liquid crystal and acousto-optic tunable filters (LCTFs and AOTFs). For dedicated applications, this permits the user to acquire information only in spectral regions that are deemed necessary; this can save significant time. The wide-field approach, however, is intrinsically both en face and non-confocal, and therefore best suited for thin specimens (e.g., tissue slices) where optical sectioning is not crucial.

12.4.3.3 Fiber-Connected Probes

Like many other optical methods discussed in earlier chapters, Raman spectroscopy can be performed through flexible, optical fiber-based probes, either at the surface of the body or endoscopically. A problem unique to Raman spectroscopy, however, is that silica-based optical fibers emit significant fluorescence signal when excited in the NIR (which, as noted above, is the desirable excitation regime for avoiding tissue fluorescence). Figure 12.7a illustrates the geometry of a single-fiber probe, in which the complementary filters are located at the proximal (operator's) end. Laser light traversing the probe in both directions gives rise to Stokes-shifted light S_{fiber} that reaches the spectrograph (both paths contribute equally at the spectrograph, aside from slight wavelength-dependent variations in sample and fiber optical properties). Even the small lengths of fiber (~1–10 m) required to make a useful probe give rise to enough fluorescence that the associated shot noise strongly degrades the Raman spectral quality.

Due to this effect, single-fiber probes presently cannot be used to both deliver and collect light in the simple manner of reflectance and fluorescence spectroscopy or optical coherence tomography. In particular, this rules out the simplest of all endoscopic fiber-probe designs, namely a single fiber (perhaps with some terminating microlens and/or steering prism)

placed directly against or immersed within biological material as in Figure 12.7a. No other multimode fiber core material has yet been found that provides a sufficiently low-noise background in the fingerprint regime under NIR excitation.

Instead, Raman-compatible probes must place the complementary filters distal to the fiber conduit, as shown in Figure 12.7b. Now, although Stokes-shifted signal is still generated in the excitation fiber, it is blocked from entering the sample, and no Stokes-shifted signal is created in the collection fiber because the notch filter prevents laser light from entering. Many geometries have been employed, both common-path and nonoverlapping (Shim et al. 1996, Utzinger and Richards-Kortum 2003, Motz et al. 2004).

The challenge in using distal filters for endoscopic probes is that the distal tip of the probe needs to be small. Obtaining and aligning filters with dimensions all smaller than 2 mm is non-routine, and has stifled commercial development and deployment of Raman endoscopic probes to date. Individual research groups continue to develop narrow-bore probes with such filters, however (Motz et al. 2004, Chau et al. 2008). For applications that do not demand such narrow-bore probe tips, the distal end of the probe can contain bulkier (~ 10 cm) free-space optics for filtering, retaining the advantages of a fiber-connected, hand-held device that can be moved freely relative to the rest of the instrument (Lieber and Mahadevan-Jansen 2007).

One way around the fiber-signal problem is to acquire spectra in the high-wave number regime rather than the fingerprint regime (c.f. Section 12.2). Under NIR excitation, signals from silica fiber are negligible in the high-wave number regime, which has allowed single-fiber probes to be built in the simple manner common in reflectance and fluorescence. While the high-wave number regime has more highly overlapping peaks, they are strong, and probes using this approach have recently been used to discriminate biopsies of basal cell carcinoma from non-involved skin (Koljenovic et al. 2005) and to determine lipid concentrations in human coronary artery samples (Nazemi and Brennan 2009).

12.5 Mathematical Processing of Vibrational Spectral Data

Biomedical research using vibrational spectroscopy sometimes requires nothing more than integrating the area under one or more peaks and drawing conclusions directly from those values. As Figure 12.1 shows, the peaks usually have direct chemical interpretations. In the best situations, the peak areas are interpreted in terms of chemical levels, and these levels in turn correlate with some property of the sample (e.g., degree of tissue malignancy), which is desirable to measure in a noninvasive optical manner rather than by some more invasive reference technique.

More often, however, the correlation between the vibrational spectral data and the desired sample properties is not obvious. The signature of the desired property can be spread out over multiple peaks, which in turn are overlapped by bands from other, unrelated species. In such cases, more sophisticated methods can be invoked to "see through" the multivariate nature of the data and identify key aspects that may be useful.

These techniques fall into two major categories: quantification of continuous quantities (analogous to taking the area under a peak), and classification into discrete categories (analogous to making a diagnosis based upon that peak's area). Although these multivariate techniques are by no means unique to vibrational spectroscopy, within biomedical optics, they are applied more commonly in this field than elsewhere, and as such are more central to an appreciation of the state of the art.

12.5.1 Quantification

As noted in the introductory section, biomedical vibrational spectroscopy finds its niche in situations where specificity is important. One such application is the quantitative assessment of multiple biochemicals' concentrations in a single sample. It does not need to be argued here that the medical field is full of scenarios in which multiple biochemicals' concentrations hold great significance. As just one example, blood, the most thoroughly analyzed human specimen, contains dozens of simple chemicals whose levels are routinely assessed.

As noted already (p. 237), the distinctive patterns of sharp peaks (i.e., fingerprints) in Raman spectra make it relatively easy (compared to fluorescence or visible/NIR absorption) to identify the contribution from one chemical species, even when others are contributing in the same spectral region. (All of the mathematical comments about Raman peaks in this section are equally applicable to MIR absorption bands.) Quantification of some target chemical's concentration c_t is actually made more straightforward by the fact that Raman scattering is such a weak effect. Because even the total amount of Raman scattering from a biological sample barely depletes the excitation light,

the concentration of one species has essentially no effect on the Raman signal measured from another. By the same logic, doubling the concentration of the target chemical simply doubles its Raman signal.

For the purposes of quantifying c_t, then, a measured Raman spectrum (denoted by vector s) can be considered to be a contribution $c_t \mathbf{R}_t$ from the target chemical (normalized Raman lineshape $\mathbf{R}_t$) and a superposition of contributions $c_i \mathbf{R}_i$ from N other Raman scattering species, with all contributions linearly superposing as

$$s = c_t \mathbf{R}_t + \sum_{i=1}^{N} c_i \mathbf{R}_i \qquad (12.5)$$

This doubly linear model (linear superposition of spectra, linearity between spectral response and concentration) is often invoked in spectroscopy but is particularly appropriate for describing spontaneous Raman scattering due to the weakness of the effect. The concentrations and associated lineshapes can correspond strictly to chemicals (such as glucose), or to any other set of building blocks that together comprise the sample (such as tissue layers or cell types).

12.5.1.1 Multivariate Calibration Techniques

With multiple spectral contributions due to different chemicals, Raman spectra lend themselves to multivariate data analysis. It is beyond the goals of this handbook to provide a detailed introduction to this topic (see Geladi and Kowalski (1986) and Haaland and Thomas (1988)), but some important points about calibration and prediction must be noted. In some cases, it is possible to acquire a complete set of all the contributing Raman lineshapes, $\mathbf{R}_t$ and $\mathbf{R}_{1..N}$. Measuring these "basis spectra" provides the most straightforward and explicit calibration possible. To estimate the concentration of the target chemical in a new sample, one simply measures the sample's spectrum s, models it using the basis spectra (typically with a least-squares criterion for the residual), and reports the coefficient c_t from the fit. This approach is called ordinary least squares (OLS). The second row of Table 12.1 summarizes these points. Note that in OLS there is nothing special about the target chemical; the concentrations of the other N constituents are predicted as well. A related method, called classical least squares, operates in a similarly complete way, requiring a set of representative "training" samples in which the concentration of each Raman scattering species is known by some reference method.

In most scenarios, however, the full set of basis spectra may be difficult to obtain, and only one or a few target chemicals are of interest. In such situations, it is possible to perform a so-called implicit calibration that still relies upon the model of Equation 12.5. Instead of measuring basis spectra or complete concentration lists, one instead records spectra from J calibration samples of varying composition (e.g., blood specimens from multiple donors) and uses a reference method to measure

TABLE 12.1 Summary of Inputs and Outputs for Various Multivariate Methods of Concentration Prediction

Calibration Input	Output for Future Samples	Technique Name(s)
Pure spectra of all $N+1$ constituents $\mathbf{R}_t, \mathbf{R}_{1\text{-}N}$	$c_t, c_{1\text{-}N}$	Ordinary least squares
Concentration of all $N+1$ constituents for J $\{c_t, c_{1\text{-}N}\}J$ samples	$c_t, c_{1\text{-}N}$	Classical least squares
Target concentration only for J samples $\{c_t\}_J$	c_t	Partial least squares
Target spectrum, target chemical $\mathbf{R}_t, \{c_t\}_J$ concentration for J samples	c_t	Hybrid linear analysis, constrained regularization

only c_t in each. Somewhat surprisingly, this is sufficient to predict the concentration c_t in future specimens robustly, as long as enough samples are scanned (J must exceed N, and should in practice be at least three times greater). In the absence of noise sources and other nonidealities, implicit calibration methods are just as accurate as OLS. Implicit calibration is in fact usually more robust, because it naturally characterizes real-world sources of additional spectral variation, such as baseline drift and residual fluorescence, just as it would a genuine Raman component.

The workhorse for implicit calibration in biomedical Raman spectroscopy is a technique called partial least squares (PLS). As noted in Table 12.1, this method requires nothing but the concentrations c_t from the calibration set (in addition to the spectra of those samples). Even the spectrum $\mathbf{R}_t$ of the purified target chemical is not needed. Often, however, $\mathbf{R}_t$ is also known. In such cases, this information can be folded into the calibration in various ways (Berger et al. 1998, Shih et al. 2007), slightly augmenting the performance of PLS.

12.5.2 Classification

Raman spectroscopy's specificity also makes it a good tool for biomedical classification tasks. Here, the goal is to assign spectra to discrete groups, rather than to extract values of a continuous variable such as glucose concentration. The discrete groups might, for instance, be different microbial species or different histological tissue classes.

There are two broad classes of classification: unsupervised and supervised. Unsupervised classification places spectra into groups based solely upon the spectral features in the data. One can imagine a bunch of bullets shot at a blank wall; unless the bullets land in a perfectly even two-dimensional distribution, there will be various clusters formed. The same is true of spectra, which define points in a space with one dimension per recorded wavelength. Unsupervised classification methods use various metrics to identify the clusters and to quantify how distinct they are. The user can then check whether the clustering correlates with some known property of the samples.

A popular method is called hierarchical cluster analysis (HCA) (Ward 1963). In essence, going back to the points-in-space view, HCA inflates a small balloon at each point. As they inflate, some of them start to touch, defining a linkage. Larger and larger groups get linked together as the inflation continues. A plot of linkages versus balloon radius generates a family-tree-like diagram that emphasizes groupings. The spectrum of an "unknown" sample can then be classified into a group based upon where it falls in this space.

Supervised classification forces the groups to correspond to predetermined categories, rather than being dictated solely by the spectra data. In the points-in-space view, this corresponds to coloring some of the points red, others blue, and finding the plane that separates red from blue with the highest success rate. A popular linear method for choosing this plane is linear discriminant analysis (LDA) (Sharma 1996), which operates by minimizing within-group variance while maximizing between-group variance. Linear models, however, are poorly suited for "yes/no" classification tasks; a better choice is a model that is forced to predict a number between 0 and 1. An example of such a method is logistic regression (Sharma 1996), which linearly weights the input variables and sends the result to a Fermi–Dirac function, which transitions smoothly from 0 to 1 at its limits. The weight coefficients are adjusted to optimize the assignment of near-0's and near-1's to the red and blue training samples, respectively.

12.6 State of the Art: Raman Biomedical Applications

This section describes some of the most significant and recent areas of progress in Raman biomedical spectroscopy. The intent here is to provide a sense of what kinds of questions Raman spectroscopy can help to address. For comprehensive surveys of particular Raman applications, including lists of the many organs that have been investigated in vivo, several excellent review articles are available, including Hanlon et al. (2000), Krafft and Sergo (2006), and Nijssen et al. (2009). The last of these three review articles focuses particularly on oncological applications.

12.6.1 Soft Tissue

Like diffuse reflectance and fluorescence, Raman spectroscopy has been used to study a range of soft tissues, mainly ex vivo but increasingly on live animals and humans. As noted above, a signature virtue of Raman spectroscopy is the ability to perform robust spectral "unmixing" that provides information equivalent or analogous to a biochemical assay. In a paradigmatic example, it has been shown repeatedly that Raman spectroscopy of ex vivo human artery specimens can provide information similar to what is given by traditional biochemical assays (e.g., see Römer et al. (1998)). The Raman approach performs OLS fits to artery spectra, using a basis set of spectra from purified chemical constituents such as triglycerides, protein, and calcification minerals. The relative weights assigned by the spectral fitting approach correlate well with the relative percentages determined from assay. As such, Raman spectroscopy provides a nondestructive, simple, and potentially endoscopic method of chemically characterizing artery tissue. Clinically, this has relevance for scanning the inner lumen of blood vessels to find regions that are in danger of rupturing and releasing plaques that can cause heart attack and stroke. Several groups have shown that fiber-optic probes can acquire artery spectra with good signal-to-noise, permitting quantitative OLS analysis, in as little as 1 s (Motz et al. 2004, Nazemi and Brennan 2009).

Similar explicit modeling has been done for other soft tissues, such as breast. By modeling the spectra from ex vivo breast tissue samples in terms of the fundamental chemical and morphological constituents (such as fat and collagen), cancerous tissues can be distinguished from normal and benign specimens with greater than 90% sensitivity and specificity (Shafer-Peltier et al. 2002, Haka et al. 2005). Logistic regression was used to convert the extracted morphological and chemical parameters into disease classifications.

It is also possible to proceed directly from spectra to disease classification, without the intermediate step of chemical or cellular interpretation. For example, Crow et al. have used this approach to show that Raman spectra can accurately classify cells from different prostatic adenocarcinoma cell lines (Crow et al. 2005), and Kanter et al. have classified multiple types of cervical precancers (Kanter et al. 2009).

12.6.2 Hard Tissue

Hard tissues, specifically bones, are particularly well-suited for Raman spectroscopy because they generate much more Raman signal than an equivalent volume of soft tissue. This is partly because of their overall higher density, and partly because of their mineral content. Bone typically generates strong scattering that produces high signal-to-noise spectra in a few seconds even at long working distances for the collection optics. A typical bone spectrum is shown in Figure 12.8, along with band assignments taken from Carden and Morris (2000).

Some of the Raman bands in bone spectra are exquisitely sensitive to the local structure of the bone, and therefore to its mechanical properties. Raman spectroscopy, therefore, offers a nice modality for characterizing bone structure in studies of fracture, wound healing, arthritis, and osteoporosis (McCreadie et al. 2006).

Because bone signatures are so intense, they can be detected through many millimeters of overlying soft tissue. By offsetting the excitation and collection locations on the surface of the soft tissue, one preferentially detects a more deeply penetrating photon population, just as in elastically scattered diffuse photon migration experiments (see Chapter 11). In an animal model, by combining measurements at different source–detector separations, Morris et al. have been able to compute automatically the bone layer spectrum and have shown good agreement with a spectrum of that same bone after surgically exposing it

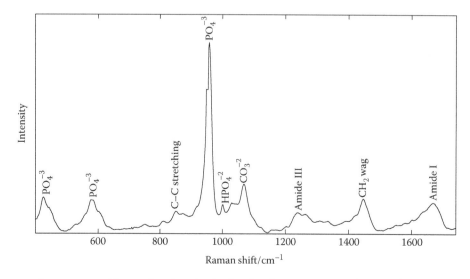

FIGURE 12.8   Representative spectrum of bone, taken from bovine jaw. Band assignments are from Carden and Morris (2000).

(Schulmerich et al. 2007). The same group has multiplexed the source and detector locations and shown that tomographic Raman-based reconstructions of bone location in canine tibia are possible (Schulmerich et al. 2008).

12.6.3 Biofluids

As noted earlier, Raman spectroscopy is particularly well suited to the challenge of quantifying a target chemical's concentration in the presence of strongly contributing and overlapping signatures. This ability maps nicely onto the challenge of analyzing the chemical content of blood samples, which are routinely analyzed in hospitals, medical offices, and home settings for the concentrations of multiple chemicals. Glucose and various cholesterol-related species are only the most popularly known targets. Since the blood samples are already removed from the body, ex vivo measurements in cuvettes or vials are straightforward.

Unlike the OLS-based analysis of artery and breast tissue mentioned above, blood analysis routinely employs implicit models such as PLS. Implicit modeling suits the challenge well, as it is often possible to obtain reference concentrations for one or more targets of interest (especially glucose or cholesterol) but arduous to obtain a complete set of spectra that describe all contributions to a blood spectrum. Typical experiments require on the scale of 50–200 separate blood spectra (from separate subjects) in order to produce robust PLS-based predictions of target species in subsequent samples.

Hospital analyzers usually operate upon blood plasma (blood with cells removed), and plasma is a convenient target because it is optically clear. The paradigmatic experiment is to acquire a spectrum of each sample along with a full analyzer report, and then search one by one for targets that exhibit some predictability under PLS calibration. PLS-based studies on plasma have clearly demonstrated the ability to obtain accuracy comparable to that of hospital analyzers for about 15 target species, including total protein, serum albumin, and glucose (Berger et al. 1999, Qu et al. 1999, Qi and Berger 2007). A concentration accuracy limit on the scale of 1 mg/dL appears to exist for most chemicals (Rohleder et al. 2005), which perhaps could be nudged downward with higher signal-to-noise ratios (SNRs) (i.e., longer integration times, more efficient collection, or higher laser flux), but is ultimately tied to the degree of spectral overlap between the species' Raman signatures and the number of datapoints per spectrum. At present, at least, this concentration limit precludes the detection of many clinically relevant targets that exist at lower concentrations.

Whole blood is also a desirable target, both for simpler (less sample preparation) ex vivo analysis and especially for potential in vivo applications. The above results for blood serum have been mostly reproduced for ex vivo whole blood samples (Enejder et al. 2002). There have been attempts to correlate glucose concentrations with transcutaneous measurements, but the published work remains preliminary (Chaiken et al. 2005, Enejder et al. 2005).

Ex vivo urine analysis can be performed, and the concentration of at least one key analyte, creatinine, has been quantified with clinical accuracy via Raman spectroscopy on the same timescale as blood measurements, namely 20–300 s (Qi and Berger 2007). There has been little action in this area because urine concentrations have variable amounts of dilution due to water, and clinically useful analysis requires many samples to be acquired over a 24 h cycle.

12.6.4 Microbes

Vibrational spectroscopy, in both Raman and IR absorption modes, has shown powerful ability to discriminate and identify microbial samples at the species and strain (subspecies) level. The spectra can be acquired from smaller cultures than are required for traditional analysis, which means that fewer hours of growing time are needed. A major motivation is the ability to identify particular pathogens, e.g., in hospital settings, where faster recognition can lead to faster clinical responses such as quarantining of particular patients or areas.

Several studies have been performed using up to several dozen different categories, in both Raman and MIR absorption modes, for instance to identify *Acenetobacter* strains (Maquelin et al. 2006) and mycobacteria (Rebuffo-Scheer et al. 2007, Buijtels et al. 2008). In all cases, unsupervised clustering analysis (HCA) has been sufficient. When large sets of spectra are clustered into a family-tree dendrogram, the groupings wind up according to species and subspecies. The fundamental chemical reasons for the successful groupings have not, to date, been firmly established.

12.6.5 Single Cells

Raman spectroscopy has the ability to analyze individual, living cells, and even subcellular domains without the need for exogenous dyes or other labeling agents. This is accomplished by placing the cells under a high-power microscope objective and using confocal excitation and collection optics to limit the interrogation volume to a few microns or less in all dimensions. If the beam is focused to subcellular dimensions, the optical gradients can be powerful enough to trap smaller cells in three dimensions, providing trapping and Raman excitation with a single beam.

Single-cell studies find particular relevance in immunology, where there is interest in cancer (leukemia) (Chan et al. 2008) and in T-cell activation (Mannie et al. 2005). In the latter case, Raman spectroscopy offers the ability to watch a single cell over time. Raman spectroscopy has also been used extensively to identify single bacteria, as has been reviewed recently by Harz et al. (2009).

The ability to analyze single cells suggests future applications in cell sorting, in the manner of flow cytometry. The advantage of not having to use labels would mean that all sorting could be "negative," i.e., resulting in populations that could potentially be reintroduced without alterations back into a living system.

Intrinsic Raman sorting based on spontaneous Raman scattering will never, however, touch the speed of traditional flow cytometry due to the much longer time needed to acquire a signal from a cell (at least several seconds).

12.6.6 Multimodal Systems

Raman spectroscopy involves many components, such as lasers, filters, and microscope objectives, which make it compatible with other techniques in biomedical optics. Confocal laser-scanning reflectance microscopes have been developed with a separate channel for Raman interrogation. This only requires a dichroic filter to combine the reflectance and Raman lasers (in principle, the same laser could be used for both, but the power required for Raman is greater than would be desired for routine reflectance microscopy). The beam combining can be done either downstream (Caspers et al. 2001) or upstream (Dickensheets et al. 2000) of the beam scanning optics. The former provides a static Raman interrogation at the center of the confocal field of view; the latter enables the Raman sampling to occur at any point in the field, provided that the scanning optics (e.g., a micro-electromechanical, two-axis mirror) can be pointed stably for enough time.

Optical coherence tomography (Patil et al. 2008) and angularly resolved elastic scattering (Smith and Berger 2008) are other laser-based diagnostic methods that have been combined with Raman.

12.7 Enhanced Raman Scattering

Since ordinary, spontaneous Raman scattering is so weak, it is difficult to acquire a useful spectrum from most biomedical targets in much less than a second. Exacting applications such as discriminating between microbial species or examining single cells require significantly more time, often a minute or more. These timescales limit the scope of what can be accomplished. In particular, real-time imaging and high-throughput specimen testing cannot be pursued at this time. Also, as mentioned in the section on biofluids, signals from analytes at concentrations below about 1 mg/dL are challenging to measure in practical amounts of time.

As noted earlier, there are several techniques that can enhance the strength of Raman scattering, either of one band or of all bands. Three important methods are briefly discussed below.

12.7.1 Resonance Raman Applications

Resonance Raman scattering occurs when the incident photon's energy comes close to an allowed absorption by the molecule. Due to this reduced ΔE, the virtual state is energetically allowed to exist for longer, and the likelihood of an inelastic scattering interaction increases greatly, scaling inversely with ΔE. Depending upon how close the incident photon comes to the absorption, the increase in scattering is called either resonant or pre-resonant enhancement. The signal can increase by a few orders of magnitude.

For analyzing biological specimens, resonance enhancement provides the useful ability to increase the signal of one target while leaving other targets alone. UV resonance Raman spectroscopy exploits this to selectively enhance DNA and protein bands at different excitation wavelengths (Boustany et al. 2000). In passing, it is worth noting that the fluorescence from excitation well into the UV (<300 nm) is strongly redshifted, due to the initial non-radiative relaxation to the lowest level of the first excited electronic state. This places the Raman scattering in a different spectral range from the fluorescence, and represents a distinct, though rarely used, approach to avoiding contamination from fluorescence.

A rare but useful visible-wavelength resonance Raman application on intact tissues is the enhancement of carotenoid signals, enabling their detection above the fluorescence background in both the retina and in skin (Hata et al. 2000, Ermakov et al. 2001). As Figure 12.9 shows, 488 nm excitation of both forearm and oral cavity tissue exhibits noticeable Raman peaks (upper traces) that are pronounced upon subtraction of the broad fluorescent background (lower traces). Non-enhanced Raman bands are completely undetectable due to the shot noise form the fluorescence in this wavelength regime.

As for the NIR, the flip side of its being a relatively fluorescence-free (and therefore absorption-free) window is that there are no species that have significant enough absorption bands there to provide measurable resonance enhancement.

As noted above in Section 12.7.1, resonance enhancement of bands requires particular wavelengths of excitation (usually in the ultraviolet) that may be incompatible with other task requirements, such as keeping the sample alive or performing

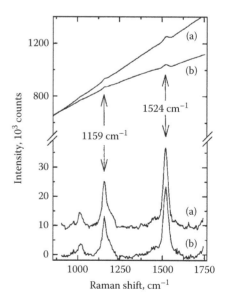

FIGURE 12.9 Resonance Raman spectra of (a) human forearm skin and (b) oral cavity tissue, both recorded in vivo with 488 nm laser excitation that resonantly enhances contributions from carotenoids. Upper traces show the original data; lower traces, after mathematical subtraction of the fluorescent background. (Reprinted from Ermakov, I. V. et al., *Opt. Lett.*, 26, 1179, 2001. With permission.)

the measurement through an overlying layer. If nothing else, resonance enhancement, by its nature of exploiting an absorption band, always increases absorption and therefore heating. In addition, only one species is enhanced at a time, in general. These drawbacks prevent resonance Raman from providing a general-purpose approach to increasing Raman signal strength.

12.7.2 Coherently Stimulated Raman

There are, however, two major ways of enhancing the strength of Raman scattering, that are compatible with a wide range of excitation wavelengths. One is to drive the emission process coherently, rather than relying upon spontaneous, individual scattering events. The energy-level diagrams in Figure 12.10 illustrate two ways in which a multiphoton sequence can stimulate Raman emission. In CARS (Zumbusch et al. 1999), a combination of two photons at frequency ω_P (for "pump") and one at ω_S (for "Stokes") drives ground state oscillators to a virtual level from which anti-Stokes Raman scattering is emitted at frequency ω_{AS}. The process only happens when the energy difference between the driving photon frequencies, $\hbar(\omega_P - \omega_S)$, matches the energy spacing between the oscillator's ground state and an excited vibrational level; hence the signal is strong only for specific tunings of the Stokes beam relative to the pump beam. The CARS process, therefore, generates strong signal corresponding to a particular vibrational mode, and enjoys the anti-Stokes advantage of spectrally avoiding fluorescence.

Since the driving process in CARS uses three photons rather than one, the emitted signal is nonlinearly proportional to the input intensity, scaling as $I_P^2 I_S$. For that reason, biological CARS is always performed using picosecond or femtosecond pulses and tightly focused spots in order to obtain high peak

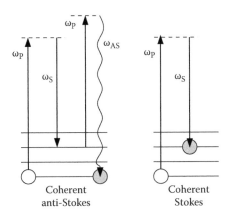

FIGURE 12.10 Coherent processes for generating stimulated anti-Stokes and Stokes Raman scattering. Both cases require two phase-locked beams (pump and Stokes) with photons whose energy difference $\hbar(\omega_P - \omega_S)$ equals the spacing between the ground state and a vibrationally excited level, as shown. In the diagram on the left, three photons prepare a virtual state from which anti-Stokes scattering occurs at a new frequency, ω_{AS}; in the right, the pump beam prepares the virtual state and the Stokes beam stimulates Stokes scattering at its own frequency of ω_S.

intensities. Creating and managing these laser pulses adds a significant extra technical and financial burden to the experiment, although instrumentation is rapidly becoming more turnkey. Also, the anti-Stokes advantage of avoiding fluorescence can be offset by the presence of non-resonant emission at the anti-Stokes wavelength due to energy-level pathways other than the one sketched in Figure 12.10.

In biomedical applications to date, the CARS process has mostly been used for video-rate imaging at a single spectral band, usually the CH_2 stretch at $2845\,cm^{-1}$, a strong band prominent in lipid-rich structures such as sebaceous glands and adipocytes (Evans et al. 2005). In such cases, one uses picosecond pulses in order to obtain the narrow bandwidth (tens of wave numbers or less) needed to isolate a particular vibrational band; sub-picosecond pulses intrinsically have too large a wavelength spread. Multiwavelength CARS spectroscopy is also possible, either by scanning the wavelength of one of the picosecond-pulsed beams or by substituting a sub-picosecond pulse with broader wavelength content (e.g., supercontinuum generation) for either the pump or Stokes beam. In this case, one stimulates emission of a spectrum rather than a single band, but each individual band is much less intensely stimulated because the power is spread over a larger spectral range.

Most recently, stimulated Stokes Raman scattering has also been applied to cellular imaging (Freudiger et al. 2008); the right-hand side of Figure 12.10 shows the process, which similarly involves photons of two wavelengths with a tuned energy difference. The detection process is fundamentally different, in that the stimulated scattering is now at the same wavelength at the Stokes beam rather than at a new frequency. Detection therefore entails detection of small changes in the intensity of a beam. It is worth noting that, at biomedical concentration levels, the signal strength of a stimulated spectrum is not necessarily greater than that of spontaneous Raman scattering for the same average excitation power (Cui et al. 2009).

12.7.3 SERS

The other technique for enhancing Raman signal is to create regions of increased field strength using metallic structures with nanoscale features that act as lightning rods. The structures can be random, such as roughened metal surfaces or colloids in suspension, or they can be arrayed in an orderly fashion, such as can be achieved using lithography or self-assembly techniques. Light interacting with these structures gets concentrated into enhanced fields that persist on a scale of nanometers away from the sharp points of the structures themselves. Molecules that reside within a few nanometers of the metal therefore experience this enhanced field and undergo what is called surface-enhanced Raman scattering (SERS), with orders of magnitude reported up to the high teens.

SERS is an excellent technique for detecting ultralow concentrations of chemicals that cannot be sensed using ordinary Raman. There are two approaches. Most fundamentally, one can enhance Raman scattering from the target chemical itself. For

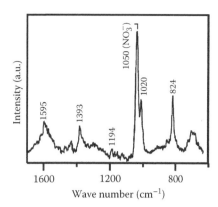

FIGURE 12.11 SERS spectrum of *Bacillus subtilis* spores (a research surrogate for anthrax) at 2.1×10^{-14} M concentration, acquired using enhancement from silver film over nanospheres (AgFON) substrates (600 nm spheres with 200 nm silver coating). Excitation parameters: 50 mW of 785 nm laser light, acquisition time, 1 min. (Reprinted from Zhang, X. et al., *J. Am. Chem. Soc.*, 127(12), 4488, 2005. With permission.)

instance, detection of calcium dipicolinate from *Bacillus subtilis* spores (surrogates for anthrax) was accomplished at the 10^{-13} M level using SERS substrates consisting of silver film deposited on nanospheres (Zhang et al. 2005), whereas in spontaneous Raman, it is unusual to detect a molecule's presence below the micromolar level. Figure 12.11 shows a representative spectrum from this experiment, exhibiting peaks that could never been observed at such low concentrations using spontaneous, unenhanced Raman scattering.

Depending upon the target molecule, however, it can be nontrivial to achieve consistent binding within the few-nanometer range of strong enhancement. Another problem is that molecules other than the target can be as likely or more likely to occupy the enhancement locations. The second SERS approach, therefore, is to conjugate an exogenous SERS "beacon" to a molecularly specific antibody that binds to the target of interest. The beacon consists of a chosen metal/molecule pairing that gives off a strong SERS signal, without the need to make the target chemical itself bind to the metal. The sensitivity is thus conferred by the efficient and stable SERS beacon, while the specificity is provided not by the Raman spectral bands themselves, but by the conjugated antibody. This second approach is conceptually identical to molecularly specific labeling with fluorophores, but the relative narrowness of Raman peaks allows for greater multiplexing of different beacons within the same spectral range (Qian et al. 2008).

While SERS unquestionably extends the detection range of Raman spectroscopy, some compromises are worth noting. Unlike all the other manifestations of Raman spectroscopy presented above, SERS departs from the paradigm of using only photons to achieve its results. The use of metallic structures, and possibly exogenous SERS beacons, raises biocompatibility issues that are absent in other applications. The intensified fields also lead to heating, which limits the amount of light that can be

used. Another compromise is that the enhancement is limited to only a few nanometers separation from the metal. As with all exogenous contrast agents, this makes it more challenging to detect intracellular material, because now the SERS substrate as well as the photons must penetrate the cellular membrane. Yet another hurdle is that since SERS requires binding of the target, either to the metal or to an antibody, an extended-use sensor needs to permit reversible binding, which can be challenging to ensure. Also, while in principle SERS can be performed at any excitation wavelength, the metallic nanostructures will provide strongest enhancement at particular wavelengths that excite plasmon resonances, so the technique requires a careful matching of excitation wavelength and metallic enhancer. Lastly, and least obviously, spontaneous Raman can be a reasonable competitor to SERS in sensitivity in situations where there is an abundance of molecules, i.e., many more than can be enhanced by SERS due to lack of binding sites. In essence, while the sensitivity per molecule is enormously greater for SERS, its effective sampling volume can be minute compared to the spontaneous Raman volume of interrogation.

12.8 Mid-IR Spectroscopy

This chapter has focused primarily on Raman spectroscopy, but as noted earlier, the complementary technique of MIR absorption provides similar molecularly specific information. Figure 12.12 shows, as an example, the MIR absorption spectra of breast tissue constituents, in both the fingerprint and high-wave number regimes. When the regions of higher absorption are plotted as positive-going peaks, as shown here, the spectrum strongly resembles a Raman spectrum in both the widths and the general locations of the peaks.

The technique has recently experienced a surge due to improved MIR detector technology. This has enabled construction of dense two-dimensional array detectors, leading to wide-field hyperspectral imaging capabilities analogous to the tunable-filter-based Raman imaging mentioned earlier. Since transmissive MIR spectroscopy is typically performed upon samples that have been sliced thin and dried to reduce water absorption, the non-confocal nature of the wide-field imaging is irrelevant.

An alternative geometry for MIR spectroscopy places biological samples upon a high refractive-index crystal and sends light through the crystal at high bouncing angles. Total internal reflection occurs between the crystal and the sample, with an evanescent wave penetrating into the sample to a depth comparable to a wavelength, namely a few microns. There is thus an effective pathlength of about twice this much, about 10 microns, at each such bounce, of which there can be several for a few-inch-long crystal. This technique, called attenuated total internal reflectance (ATR), permits bulk samples to be analyzed without slicing, but still only samples a thickness of about 10 microns. Recently, flexible optical fibers made of silver halide-enabled ATR to be performed using a handpiece, where an exposed loop of the fiber material was simply placed in contact with tissue

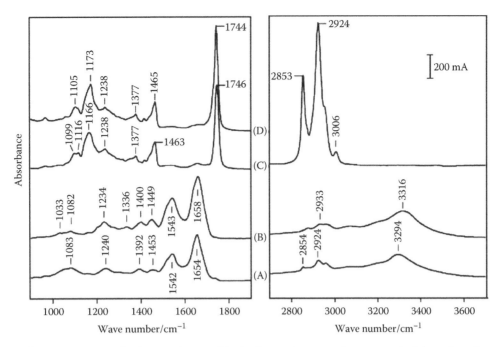

FIGURE 12.12 MIR absorption spectra of major constituents of benign breast tissue constituents, acquired through a microscope from ~30 μm regions of tissue slices. (A) epithelium of fibroadenoma; (B) connective tissue; (C) adipose tissue; (D) milk secretion in ducts. (Reprinted from Fabian, H. et al., *Biochim. Biophys. Acta Biomembranes*, 1758(7), 874, 2006. With permission.)

specimens. Mackanos et al. used such a probe to acquire spectra of fresh colonic biopsies, with the only sample preparation being to blow cool air over the mucosal surface for a few minutes to reduce water content (and, therefore, increase MIR penetration). Normal, dysplastic, and hyperplastic regions were distinguished with greater than 90% success across a range of metrics (Mackanos et al. 2009).

Rather than employing tunable filters, MIR hyperspectral images are obtained interferometrically, with a scanning reference mirror varying the pathlength difference between sample and reference beams. Interferograms are acquired in parallel for each *xy* location in the image. Because these signals versus mirror position are converted to intensities versus wavelength via a Fourier transform, the technique is commonly called "FT-IR," without needing to specify that the wavelength regime is the MIR in particular. The interferometric nature of the setup permits trade-offs between acquisition time and spectral resolution and/or range.

A key application space for FT-IR is the analysis of unfixed, unstained tissue slices as an adjunct or possible replacement for traditional histologically prepared slides. Because it exploits an intrinsically stronger optical interaction, FT-IR imaging requires an order of magnitude less time than Raman spectroscopy to acquire a comparable data cube in space and wave number, although due to the longer wavelength, the resolution is not as high. Recent progress in this field has been reviewed by Srinivasan and Bhargava (2007). For example, FT-IR imaging microscopy has been used to classify cancerous sections of tissue from breast (Fabian et al. 2006) and prostate (Bhargava 2007).

References

Berger, A. J., Koo, T.-W., Itzkan, I., and Feld, M. S. 1998. An enhanced algorithm for linear multivariate calibration. *Anal. Chem.* 70: 623–627.

Berger, A. J., Koo, T.-W., Itzkan, I., Horowitz, G., and Feld, M. S. 1999. Multicomponent blood analysis by near-infrared Raman spectroscopy. *Appl. Opt.* 38: 2916–2926.

Bhargava, R. 2007. Towards a practical Fourier transform infrared chemical imaging protocol for cancer histopathology. *Anal. Bioanal. Chem.* 389: 1155–1169.

Boustany, N., Manoharan, R., Dasari, R., and Feld, M. 2000. Ultraviolet resonance Raman spectroscopy of bulk and microscopic human colon tissue. *Appl. Spectrosc.* 54: 24–30.

Buijtels, P. C. A. M., Willemse-Erix, H. F. M., Petit, P. L. C. et al. 2008. Rapid identification of mycobacteria by Raman spectroscopy. *J. Clin. Microbiol.* 46: 961–965.

Carden, A. and Morris, M. D. 2000. Application of vibrational spectroscopy to the study of mineralized tissues (review). *J. Biomed. Opt.* 5: 259–268.

Caspers, P. J., Lucassen, G. W., Carter, E. A., Bruining, H. A., and Puppels, G. J. 2001. In vivo confocal Raman microspectroscopy of the skin: Noninvasive determination of molecular concentration profiles. *J. Investig. Dermatol.* 116: 434–442.

Chaiken, J., Finney, W., Knudson, P. E. et al. 2005. Effect of hemoglobin concentration variation on the accuracy and precision of glucose analysis using tissue modulated, noninvasive, in vivo Raman spectroscopy of human blood: A small clinical study. *J. Biomed. Opt.* 10: 031111.

Chan, J. W., Taylor, D. S., Lane, S. M., Zwerdling, T., Tuscano, J., and Huser, T. 2008. Nondestructive identification of individual leukemia cells by laser trapping Raman spectroscopy. *Anal. Chem.* 80: 2180–2187.

Chau, A. H., Motz, J. T., Gardecki, J. A., Waxman, S., Bouma, B. E., and Tearney, G. J. 2008. Fingerprint and high-wavenumber Raman spectroscopy in a human-swine coronary xenograft in vivo. *J. Biomed. Opt.* 13: 040501.

Crow, P., Barrass, B., Kendall, C. et al. 2005. The use of Raman spectroscopy to differentiate between different prostatic adenocarcinoma cell lines. *Br. J. Cancer* 92: 2166–2170.

Cui, M., Bachler, B. R., and Ogilvie, J. P. 2009. Comparing coherent and spontaneous Raman scattering under biological imaging conditions. *Opt. Lett.* 34: 773–775.

Demtroder, W. 1998. *Laser Spectroscopy.* Berlin, Germany: Springer-Verlag, pp. 489–493.

Dickensheets, D. L., Wynn-Williams, D. D., Edwards, H. G. M., Schoen, C., Crowder, C., and Newton, E. M. 2000. A novel miniature confocal microscope/Raman spectrometer system for biomolecular analysis on future mars missions after Antarctic trials. *J. Raman Spectrosc.* 31: 633–635.

Enejder, A. M. K., Koo, T.-W., Oh, J. et al. 2002. Blood analysis by Raman spectroscopy. *Opt. Lett.* 27: 2004–2007.

Enejder, A. M. K., Scecina, T. G., Oh, J. et al. 2005. Raman spectroscopy for noninvasive glucose measurements. *J. Biomed. Opt.* 10: 031114.

Ermakov, I. V., Ermakova, M. R., McClane, R. W., and Gellermann, W. 2001. Resonance Raman detection of carotenoid antioxidants in living human tissues. *Opt. Lett.* 26: 1179–1181.

Evans, C. L., Potma, E. O., Puoris'haag, M., Ct, D., Lin, C. P., and Xie, X. S. 2005. Chemical imaging of tissue *in vivo* with video-rate coherent anti-Stokes Raman scattering microscopy. *Proc. Nat. Acad. Sci.* 102: 16807–16812.

Fabian, H., Thi, N. A. N., Eiden, M., Lasch, P., Schmitt, J., and Naumann, D. 2006. Diagnosing benign and malignant lesions in breast tissue sections by using IR-microspectroscopy. *Biochim. Biophys. Acta Biomembranes* 1758(7): 874–882.

Freudiger, C. W., Min, W., Saar, B. G. et al. 2008. Label-free biomedical imaging with high sensitivity by stimulated Raman scattering microscopy. *Science* 322: 1857–1861.

Geladi, P. and Kowalski, B. R. 1986. Partial least-squares regression: A tutorial. *Anal. Chim. Acta* 185: 1–17.

Haaland, D. M. and Thomas, E. V. 1988. Partial least-squares methods for spectral analysis: Relation to other quantitative calibration methods and the extraction of quantitative information. *Anal. Chem.* 60: 1193–1202.

Haka, A. S., Shafer-Peltier, K. E., Fitzmaurice, M., Crowe, J., Dasari, R. R., and Feld, M. S. 2005. Diagnosing breast cancer by using Raman spectroscopy. *Proc. Natl. Acad. Sci.* 102: 12371–12376.

Hanlon, E. B., Manoharan, R., Koo, T.-W. et al. 2000. Prospects for *in vivo* Raman spectroscopy. *Phys. Med. Biol.* 45: R1–R59.

Harz, A., Roesch, P., and Popp, J. 2009. Vibrational spectroscopy—A powerful tool for the rapid identification of microbial cells at the single-cell level. *Cytometry A* 75A: 104–113.

Hata, T. R., Scholz, T. A., Ermakov, I. V. et al. 2000. Non-invasive Raman spectroscopic detection of carotenoids in human skin. *J. Investig. Dermatol.* 115: 441–447.

Kanter, E. M., Majumder, S., Vargis, E. et al. 2009. Multiclass discrimination of cervical precancers using Raman spectroscopy. *J. Raman Spectrosc.* 40: 205–211.

Koljenovic, S., Schut, T., Wolthuis, R. et al. 2005. Tissue characterization using high wave number Raman spectroscopy. *J. Biomed. Opt.* 10: 031116.

Krafft, C. and Sergo, V. 2006. Biomedical applications of Raman and infrared spectroscopy to diagnose tissues. *Spectroscopy* 20: 195–218.

Lieber, C. A. and Mahadevan-Jansen, A. 2007. Development of a handheld Raman microspectrometer for clinical dermatologic applications. *Opt. Express* 15: 11874–11882.

Mackanos, M. A., Hargrove, J., Wolters, R. et al. 2009. Use of an endoscope-compatible probe to detect colonic dysplasia with Fourier transform infrared spectroscopy. *J. Biomed. Opt.* 14: 044006.

Mannie, M. D., McConnell, T. J., Xie, C., and Li, Y.-Q. 2005. Activation-dependent phases of T cells distinguished by use of optical tweezers and near infrared Raman spectroscopy. *J. Immunol. Methods* 297: 53–60.

Maquelin, K., Dijkshoorn, L., van der Reijden, T. J., and Puppels, G. J. 2006. Rapid epidemiological analysis of *Acinetobacter* strains by Raman spectroscopy. *J. Microbiol. Methods* 64: 126–131.

McCreadie, B. R., Morris, M. D., Ching Chen, T. et al. 2006. Bone tissue compositional differences in women with and without osteoporotic fracture. *Bone* 39: 1190–1195.

Motz, J. T., Hunter, M., Galindo, L. H. et al. 2004. Optical fiber probe for biomedical Raman spectroscopy. *Appl. Opt.* 43: 542–554.

Nazemi, J. H. and Brennan III J. F. 2009. Lipid concentrations in human coronary artery determined with high wavenumber Raman shifted light. *J. Biomed. Opt.* 14: 034009.

Nijssen, A., Koljenovic, S., Schut, T. C. B., Caspers, P. J., and Puppels, G. J. 2009. Towards oncological application of Raman spectroscopy. *J. Biophotonics* 2: 29–36.

Patil, C. A., Bosschaart, N., Keller, M. D., van Leeuwen, T. G., and Mahadevan-Jansen, A. 2008. Combined Raman spectroscopy and optical coherence tomography device for tissue characterization. *Opt. Lett.* 33: 1135–1137.

Puppels, G. J. 1999. *Fluorescent and Luminescent Probes for Biological Activity*, Chap. 29, 2nd edn. New York: Academic Press, pp. 377–406.

Qi, D. and Berger, A. J. 2007. Chemical concentration measurement in blood serum and urine samples using liquid-core optical fiber Raman spectroscopy. *Appl. Opt.* 46: 1726–1734.

Qian, X., Peng, X.-H., Ansari, D. O. et al. 2008. *In vivo* tumor targeting and spectroscopic detection with surface-enhanced Raman nanoparticle tags. *Nat. Biotechnol.* 26: 83–90.

Qu, J. Y., Wilson, B. C., and Suria, D. 1999. Concentration measurement of multiple analytes in human sera by near-infrared laser Raman spectroscopy. *Appl. Opt.* 38: 5491–5498.

Rebuffo-Scheer, C. A., Kirschner, C., Staemmler, M., and Naumann, D. 2007. Rapid species and strain differentiation of non-tuberculous mycobacteria by Fourier-transform infrared micro spectroscopy. *J. Microbiol. Methods* 68: 282–290.

Rohleder, D., Kocherscheidt, G., Gerber, K., Köhler, W., Möcks, J., and Petrich, W. 2005. Comparison of mid-infrared and Raman spectroscopy in the quantitative analysis of serum. *J. Biomed. Opt.* 10: 031108-1–031108-10.

Römer, T. J., Brennan III J. F., Fitzmaurice, M. et al. 1998. Histopathology of human coronary atherosclerosis by quantifying its chemical composition with Raman spectroscopy. *Circulation* 97: 878–885.

Schulmerich, M. V., Cole, J. H., Dooley, K. A. et al. 2008. Noninvasive Raman tomographic imaging of canine bone tissue. *J. Biomed. Opt.* 13: 020506.

Schulmerich, M. V., Dooley, K. A., Vanasse, T. M., Goldstein, S. A., and Morris, M. D. 2007. Subsurface and transcutaneous Raman spectroscopy and mapping using concentric illumination rings and collection with a circular fiber-optic array. *Appl. Spectrosc.* 61: 671–678.

Shafer-Peltier, K. E., Haka, A. S., Fitzmaurice, M. et al. 2002. Raman microspectroscopic model of human breast tissue: Implications for breast cancer diagnosis *in vivo. J. Raman Spectrosc.* 33: 552–563.

Sharma, S. 1996. *Applied Multivariate Techniques.* New York: Wiley.

Shih, W.-C., Bechtel, K. L., and Feld, M. S. 2007. Constrained regularization: Hybrid method for multivariate calibration. *Anal. Chem.* 79: 234–239.

Shim, M. G., Wilson, B. C., Marple, E., and Wach, M. 1999. Study of fiber-optic probes for *in vivo* medical Raman spectroscopy. *Appl. Spectrosc.* 53: 619–627.

Shreve, A. P., Cherepy, N. J., and Mathies, R. A. 1992. Effective rejection of fluorescence interference using a shifted excitation difference technique. *Appl. Spectrosc.* 46: 707–711.

Smith, Z. J. and Berger, A. J. 2008. Integrated Raman- and angular-scattering microscopy. *Opt. Lett.* 33: 714–716.

Srinivasan, G. and Bhargava, R. 2007. Fourier transform-infrared spectroscopic imaging: The emerging evolution from a microscopy tool to a cancer imaging modality. *Spectroscopy* 22: 30–43.

Utzinger, U. and Richards-Kortum, R. 2003. Fiber optic probes for biomedical spectroscopy. *J. Biomed. Opt.* 8: 121–147.

Ward Jr. J. H. 1963. Hierarchical grouping to optimize an objective function. *J. Am. Stat. Assoc.* 58: 236–244.

Zhang, X., Young, M. A., Lyandres, O., and Duyne, R. P. V. 2005. Rapid detection of an anthrax biomarker by surface-enhanced Raman spectroscopy. *J. Am. Chem. Soc.* 127(12): 4484–4489.

Zumbusch, A., Holtom, G. R., and Xie, X. S. 1999. Three-dimensional vibrational imaging by coherent anti-Stokes Raman scattering. *Phys. Rev. Lett.* 82: 4142–4145.

III

Tomographic Imaging

13 **Optical Coherence Tomography: Introduction and Theory**   *Yu Chen, Evgenia Bousie, Constantinos Pitris, and James G. Fujimoto* 255
Introduction • Development of OCT: Historical Perspective • OCT Theory • Delivery Devices • Clinical Applications of OCT • Commercial OCT Systems • Conclusions • Acknowledgments • References

14 **Functional Optical Coherence Tomography in Preclinical Models**   *Melissa C. Skala, Yuankai K. Tao, Anjul M. Davis, and Joseph A. Izatt* 281
Introduction • Hemodynamic Imaging • Nanoparticle-Based Contrast • Discussion • References

15 **Optical Coherence Tomography: Clinical Applications**   *Brian D. Goldberg, Melissa J. Suter, Guillermo J. Tearney, and Brett E. Bouma* 303
Introduction • Ophthalmology • Cardiology • Gastroenterology • Other Clinical Applications • Summary • References

16 **Forward Models of Light Transport in Biological Tissue**   *Andreas H. Hielscher, Hyun Keol Kim, and Alexander D. Klose*..319
Introduction • RTE-Based Models of Light Transport in Biological Tissue • Numerical Methods • Examples in Biomedical Research • Summary • Acknowledgments • References

17 **Inverse Models of Light Transport**   *Simon R. Arridge, Martin Schweiger, and John C. Schotland* 337
Introduction • Definition of Problems in Optical Tomography • Methods • Examples • Summary • References

18 **Laminar Optical Tomography**   *Sean A. Burgess and Elizabeth M. C. Hillman* 359
Introduction • Background • 3D Image Reconstruction • Optical Property and Basic Geometry Determination • Monte Carlo Modeling for LOT • Presentation of State of the Art • Critical Discussion • Summary • Future Perspectives • References

19 **Diffuse Optical Tomography Using CW and Frequency Domain Imaging Systems**   *Subhadra Srinivasan, Scott C. Davis, and Colin M. Carpenter*..373
Introduction • Theory and Instrumentation • In Vivo Diffuse Optical Tomography • Challenges and Critical Discussion • References

20 **Diffuse Optical Tomography: Time Domain**   *Juliette Selb and Adam Gibson* 395
Introduction • Background • Time-Domain Instrumentation • Time-Domain Data Analysis • Reconstructing Optical Properties from Time-Domain Data • In Vivo Applications • Critical Discussion • References

21 **Photoacoustic Tomography and Ultrasound-Modulated Optical Tomography**   *Changhui Li, Chulhong Kim, and Lihong V. Wang*..419
Introduction to Photoacoustic Tomography • PAT Modalities • PA Spectroscopy and Functional Imaging • PA Contrast Agents and Molecular Imaging • Comparison of PAT with Other Imaging Modalities • Summary of PAT • Introduction to Ultrasound-Modulated Optical Tomography • Detection Techniques in UOT • The Use of a Powerful Long Coherent Pulsed Laser in UOT • Summary and Discussion on UOT • References

22 **Optical and Opto-Acoustic Molecular Tomography**   *Vasilis Ntziachristos*..443
Introduction • Reporter Technologies • Small Animal Opto-Acoustic Molecular Tomography • Small Animal Hybrid Optical Tomography • Discussion • Acknowledgments • Glossary • References

Optical Coherence Tomography: Introduction and Theory

13.1 Introduction 255
13.2 Development of OCT: Historical Perspective 255
    Optical Ranging of Biological Tissues • Time-Domain OCT • Fourier-Domain OCT
13.3 OCT Theory..258
    Low Coherence Interferometry • OCT Signal from Multiple Scatterers • Time Domain Optical Coherence Tomography • Spatial Resolution in Optical Coherence Tomography • Spectral Radar: Optical Coherence Tomography in the Fourier Domain • Wavelength Tuning (Swept Source) OCT
13.4 Delivery Devices 267
13.5 Clinical Applications of OCT 268
    Ophthalmology • Cardiology • Oncology • Other Applications • OCT in Biology
13.6 Commercial OCT Systems...271
13.7 Conclusions..271
Acknowledgments...272
References...272

Yu Chen
University of Maryland

Evgenia Bousi
University of Cyprus

Constantinos Pitris
University of Cyprus

James G. Fujimoto
Massachusetts Institute of Technology

13.1 Introduction

Optical coherence tomography (OCT) is an emerging medical imaging technology that enables micron scale, cross-sectional, and 3D imaging of microstructure of biological tissues in situ and in real time (Huang et al. 1991a, Fujimoto et al. 2000, Fujimoto 2003). OCT can function as a type of "optical biopsy": imaging tissue microstructure with 1–10 μm resolutions and 1–2 mm penetration depths, approaching those of standard excisional biopsy and histopathology, but without the need to remove and process tissue specimens (Fujimoto et al. 1995, Brezinski et al. 1996b, Tearney et al. 1997b). OCT is analogous to ultrasound B mode imaging, except that imaging is performed by measuring the echo time delay and intensity of back-reflected or backscattered light rather than sound. An optical beam is scanned across the tissue and echoes of backscattered light are measured as a function of axial range (depth) and transverse position (see Figure 13.1). Two-dimensional cross-sectional OCT images of tissue are constructed by juxtaposing a series of axial measurements of backscattered light at different transverse positions. The resulting data set is a 2D array that represents the optical backscattering within a cross-sectional slice of the tissue. Three-dimensional imaging can also be performed by stacking the 2D cross-sectional images at different transverse positions.

OCT imaging has a number of features that make it attractive for a broad range of applications. OCT can perform imaging with resolutions approaching that of conventional histopathology, but imaging is possible in situ and in real time. OCT can be performed fiber-optically using devices such as handheld probes, endoscopes, catheters, laparoscopes, and needles, which enable noninvasive or minimally invasive internal body imaging. OCT can be performed in real time, allowing guidance of excisional biopsy or interventional procedures. 3D-OCT data sets provide comprehensive, volumetric information on architectural morphology. Cross-sectional images with arbitrary orientations as well as projection and rendered views can be generated. OCT data is in digital form, facilitating quantitative image processing techniques as well as electronic storage and transmission.

13.2 Development of OCT: Historical Perspective

13.2.1 Optical Ranging of Biological Tissues

The development of OCT originates from the techniques for precise measurement of the dimensions of the structures within the tissue ("see through the tissue"), a process similar to object ranging using sonar or radar. The dimensions of structures can be determined by measuring the echo delay time it takes for a short pulse of sound or electromagnetic wave to be back-reflected from different structures at various distances. Ultrasound utilizes this principle to generate depth-resolved measurement of biological

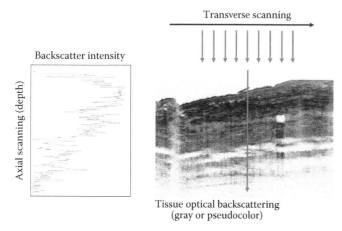

FIGURE 13.1 Principle of OCT. OCT measures the backscattering intensity as a function of depth. Cross-sectional images are generated by scanning a beam across the tissue. A gray scale or false color image can then be displayed.

tissues (A-mode). The speed of sound in tissue is approximately 1500 m/s, therefore, the measurement of distances with a resolution of 100 μm (typical ultrasound resolution) corresponds to a temporal resolution in the range of 100 ns, which is well within the limits of fast electronic detection.

However, ultrasound imaging resolution is not sufficient to reveal fine biological structures. Optical methods have the advantage of higher resolution due to the shorter wavelength of light compared to sound. Accordingly, the echo delay times associated with light are extremely short. For example, the measurement of a structure with a resolution on the 10 μm scale (typical OCT resolution) will correspond to a temporal resolution on the range of 10 fs, given that the speed of light (3×10^8 m/s in air) is much faster than sound. Direct electronic detection is not possible on this time scale. Therefore, several techniques have been developed to measure the ultrafast optical echoes.

The most direct approach to measure the ultrafast optical echoes is to use high-speed optical gating first proposed by Duguay et al. around 40 years ago (Duguay and Hansen 1969, 1971, Duguay 1971, Duguay and Mattick 1971, Bruckner 1978). Duguay demonstrated an ultrafast optical Kerr shutter to photograph light in flight. The Kerr shutter can achieve picosecond or femtosecond resolution by using an intense ultrashort light pulse rather than electrical pulse to induce the birefringence (Kerr effect) in an optical medium between two crossed polarizers. In optical scattering medium such as biological tissues, the high-speed shutter could be used to gate out unwanted scattered light and detect light echoes from internal structures. This technology could noninvasively image internal biological structures.

The major disadvantage of the Kerr shutter is that it requires high intensity laser pulses to induce the Kerr effect. Alternatively, the high-speed gating can be achieved by nonlinear cross-correlation (Duguay and Hansen 1968) or optical heterodyning detection (Kompfner and Park 1976, Park et al. 1981). In nonlinear cross-correlation, the specimen is illuminated with short pulses, and the backscattered light is parametric converted (sum

frequency generation or second harmonic generation) with a reference pulse in a nonlinear optical crystal. The reference pulse is delayed by a variable time τ from the illuminating pulse, and the nonlinear process creates a high-speed optical gate. The response function $S(\tau)$ can be expressed as

$$S(\tau) \sim \int_{-\infty}^{\infty} I_s(t) I_r(t-\tau) \mathrm{d}t \qquad (13.1)$$

where

$I_s(t)$ is the signal
$I_r(t)$ is the reference pulse

The temporal resolution in nonlinear optical gating is determined by the pulse duration, and the sensitivity is determined by the conversion efficiency of the nonlinear process. From Equation 13.1, the second harmonic signal is proportional to the product of I_s and I_r, weak signals backscattered from the sample can be amplified by using a strong reference intensity. Detection sensitivity of 10^7 can be achieved. In other words, the backscattered signal as small as 10^{-7} of the incident pulse energy can be detected. Optical time-of-flight ranging measurements were first demonstrated in biological tissues by Fujimoto et al. to measure the thickness of corneal, stratum corneum, and epidermis (Fujimoto et al. 1986). Nonlinear cross-correlation does not require pulses of as high intensity as the Kerr shutter, but still requires the use of short pulses.

Interferometric detection overcomes many of the limitations of nonlinear gating techniques, and can measure the echo time delay of backscattered light with high sensitivity and dynamic range. This technique is analogous to coherent optical detection in optical communications. OCT is based on low coherence interferometry, which has been previously applied in photonic devices to perform optical ranging (Takada et al. 1987, Youngquist et al. 1987, Gilgen et al. 1989). Figure 13.2 shows a schematic of how low coherence interferometry works.

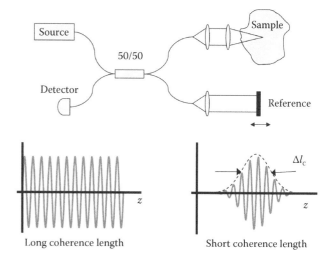

FIGURE 13.2 Schematic of low coherence interferometry.

Measurements are performed using a Michelson interferometer with a low coherence length light source. One arm of the interferometer illuminates the light on the tissue and collects the backscattered light (typically referred to as "sample arm"). A second arm of the interferometer has a reference path delay that is scanned as a function of time (typically referred to as "reference arm"). Optical interference between the light from the sample and reference arms occurs only when the optical delays match to within the coherence length of the light source. Low coherence interferometry enables 1D measurement of the magnitude of backscattered light from internal tissue microstructures versus the echo delay time (and hence the depth) with high time accuracy. The first biological application of low coherence interferometry was in ophthalmology for the measurement of eye length (Fercher et al. 1988). A dual beam interferometer was used to perform the first in vivo measurement of axial eye length (Hitzenberger 1991), and the high-resolution measurements of corneal thickness in vivo were demonstrated using low coherence interferometry (Huang et al. 1991, Hitzenberger 1992, Swanson et al. 1992). Low coherence interferometry has been applied to the precise biometric measurements in the highly scattering biological tissues (Clivaz et al. 1992, Schmitt et al. 1993). The successful demonstration of promising biological applications using low coherence interferometry soon triggered the development of OCT, which is basically imaging using low coherence interferometry. Two major forms of OCT has been developed independently, namely, time-domain OCT (TD OCT) and Fourier-domain OCT (FD OCT).

13.2.2 Time-Domain OCT

OCT was first demonstrated in 1991 for cross-sectional imaging of retina and coronary artery (Huang et al. 1991a). The first in vivo OCT imaging of the human retina were performed in 1993 (Fercher et al. 1993, Swanson et al. 1993). Since then, OCT has rapidly developed as a noninvasive biomedical imaging modality that enables cross-sectional visualization of tissue microstructures in vivo (Fujimoto et al. 1995, Schmitt 1999, Fercher et al. 2003, Fujimoto 2003). The rapid development of OCT increasingly impact clinical medicine by translating OCT technology from bench to various clinical settings including ophthalmology (Izatt et al. 1994b, Hee et al. 1995a,b, 1998, Puliafito et al. 1996, Gaudric et al. 1999, Antcliff et al. 2000, 2001, Jumper et al. 2000, Apostolopoulos et al. 2002, Sato et al. 2003, Schuman et al. 2004, Wollstein et al. 2005, Ko et al. 2006), cardiology (Brezinski et al. 1996a,b, 1997, Fujimoto et al. 1999, Jang et al. 2001, 2002, 2005, Tearney et al. 2003), gastroenterology (Izatt et al. 1996, Sergeev et al. 1997, Tearney et al. 1997c, Kobayashi et al. 1998, Bouma et al. 2000, Jäckle et al. 2000, Li et al. 2000b, Sivak et al. 2000, Poneros et al. 2001, 2002, Zuccaro et al. 2001, Shen et al. 2004, Isenberg et al. 2005, Yang et al. 2005a, Evans et al. 2006, Chen et al. 2007), dermatology (Welzel et al. 1997, 1998, 2004), dentistry (Otis et al. 2000a,b, 2003, de Melo et al. 2005, Freitas et al. 2006), urology (Tearney et al. 1997d, D'Amico et al. 2000, Zagaynova et al. 2002, Manyak et al. 2005),

gynecology (Boppart et al. 1999, Pitris et al. 1999, Brewer et al. 2004), among others.

The majority of the OCT systems reported between 1991 and 2003 are based on the time-domain detection. In TD OCT, the optical depth ranging is achieved by physically scanning the reference arm mirror. The backscattered light coming from the internal tissue structures with optical pathlength different from the distance selected by the reference mirror (either closer or further) will be rejected by the low coherence gating. TD OCT can achieve >100 dB detection sensitivity and up to several kHz axial scan speed (Tearney et al. 1997a, Rollins et al. 1998, Herz et al. 2004a), which enables real time imaging of tissue at a frame rate on the order of 1–10 frames per second.

The growing interests in OCT applications foster rapid technical development. Important imaging parameters such as axial (depth) resolution, transverse resolution, imaging speed, detection sensitivity, have been improved over time. Axial resolution has been significantly improved during this period to better visualize the internal tissue microstructure. Standard OCT achieves axial image resolutions of 10–20 μm using the superluminescent diodes (SLDs) (Huang et al. 1991a, Swanson et al. 1993, Hee et al. 1995b). Using Kerr-lens mode-locked Ti:Sapphire or Cr:Forsterite laser, axial resolution around 5 μm can be achieved (Bouma et al. 1995, 1996, Boppart et al. 1998a). Ultrahigh axial resolutions approaching 1–2 μm has been demonstrated (Drexler et al. 1999, 2001) and continues to be an active research topic (Hartl et al. 2001, Bizheva et al. 2003, Bourquin et al. 2003, Wang et al. 2003, Drexler 2004). Transverse imaging resolution can be improved by using high numerical aperture (NA) objectives to focus the light to a tighter spot. Typical OCT imaging systems use 10–20 μm transverse resolution to obtain sufficient depth of field. Optical coherence microscopy (OCM), which combines confocal microscopy with coherence detection, achieves cellular level transverse resolution (Izatt et al. 1994a). OCM has been demonstrated to image human tissues ex vivo (Izatt et al. 1996) and in vivo (Aguirre et al. 2003). The improvement of image resolution greatly enhances the capability of OCT (and OCM) to visualize fine tissue structures.

Another active research area for OCT technology is the development of imaging probe devices to enable the internal body imaging. Fiber-based catheter imaging was first demonstrated in 1996 (Tearney et al. 1996a,b), and in vivo endoscopic imaging has been shown in 1997 (Tearney et al. 1997b). This design was using the side-viewing configuration in order to visualize the structures in the lumen wall. Forward imaging devices such as handheld probes and laparoscopes were also demonstrated (Boppart et al. 1997a). Needle-based imaging device (~0.4 mm in diameter) has been developed to enable solid organ imaging (Li et al. 2000a). Nowadays, various OCT imaging probes have been developed for different clinical applications (Sergeev et al. 1997, Pan et al. 2001, 2003, Xie et al. 2003, 2005, 2006, Zara et al. 2003, Herz et al. 2004b, Jain et al. 2004, Liu et al. 2004, Tran et al. 2004, Tumlinson et al. 2004, Yang et al. 2005b, Yeow et al. 2005, Bohringer et al. 2006, Jung et al. 2006, Wu et al. 2006, Zara and Patterson 2006, Aguirre et al. 2007, Han et al. 2008).

13.2.3 Fourier-Domain OCT

In 2003, several research groups independently demonstrated that Fourier-domain detection enables 10–100 fold improvements in detection sensitivity and speed over the time-domain configuration (Choma et al. 2003, de Boer et al. 2003, Leitgeb et al. 2003). These advances not only greatly improve the performance of OCT, but enables 3D-OCT imaging in vivo. 3D-OCT promises to be a powerful advancement because 3D comprehensive volumetric data will enable new visualization and processing techniques such as the generation of cross-sectional images with arbitrary orientation, the generation of projection views similar to en face microscopy images, improved quantitative measurements of morphology, improved image processing techniques to reduce speckle and enhance contrast, and virtual manipulation of tissue geometry for the visualization of morphology.

Fourier-domain detection can achieve very high detection sensitivities by measuring the interference signals in the Fourier or frequency domain (Fercher et al. 1995, Choma et al. 2003, de Boer et al. 2003, Leitgeb et al. 2003). In Fourier-domain detection, the reference mirror position is fixed, and echoes of light are obtained by Fourier transforming the interference spectrum. These techniques are somewhat analogous to Fourier transform spectroscopy and have a significant sensitivity and speed advantage compared to previous TD OCT because they measure the optical echo signals from different depths along the entire A-scan simultaneously rather than sequentially.

Fourier-domain OCT can be performed using two complementary techniques, known as spectral/Fourier domain OCT and swept source/Fourier domain OCT (SS-OCT, also known as Optical Frequency Domain Imaging, OFDI). Spectral/Fourier domain detection uses a spectrometer and a high-speed line scan camera to measure the interference spectrum in parallel. Although the basic principle of spectral/Fourier domain OCT has been known since 1995 (Fercher et al. 1995), limitations in CCD technology and the lack of recognition of performance advantages have delayed the use of this technology for nearly a decade. The first demonstration of in vivo retinal imaging using spectral/Fourier domain OCT was reported in 2002 (Wojtkowski et al. 2002), and high-speed imaging using line-scan CCDs was demonstrated in 2003 (de Boer et al. 2003). Spectral/Fourier domain OCT was rapidly developed since then. Spectral/Fourier domain OCT can perform high-speed imaging at 15–75 kHz axial scan rates using line-scan CCD cameras (Yun et al. 2003a, Cense et al. 2004, Nassif et al. 2004, Wojtkowski et al. 2004, Zhang et al. 2006, Makita et al. 2008), and 70–300 kHz axial scan rates using CMOS cameras (Potsaid et al. 2008). Spectral/Fourier domain OCT had a powerful impact on OCT imaging because it enables ultrahigh resolutions as well as 3D-OCT imaging (Wojtkowski et al. 2005, Srinivasan et al. 2006, Ruggeri et al. 2007, Vizzeri et al. 2009).

In contrast, swept source/Fourier domain OCT uses a frequency-swept laser light source and a photodetector to measure the interference spectrum (Chinn et al. 1997, Golubovic et al. 1997, Yun et al. 2003, Oh et al. 2006). Swept source/Fourier domain OCT technology has the advantage that it can perform imaging at longer wavelengths of 1000 and 1300 nm. Imaging at these wavelengths is important because of reduces optical scattering and improves image penetration depths (Brezinski et al. 1996b). Swept source OCT was first demonstrated in 1997, but performance was limited by available laser technologies (Chinn et al. 1997, Golubovic et al. 1997). Recent advances in frequency swept lasers enable high-speed imaging. OCT imaging with 19 kHz axial scan rates was demonstrated in 2003 (Yun et al. 2003b), and 115 kHz axial scan rates were achieved in 2005 (Oh et al. 2005). Using advanced laser technology, ultrahigh speed SS-OCT imaging has been recently demonstrated at record imaging speed of 370 kHz axial scan rates, ~100 times faster than previous TD OCT technology (Huber et al. 2006, Adler 2007b). Swept source OCT enables 3D OCT imaging of highly scattering tissues and has a great impact in disease diagnosis (Yun et al. 2006, Adler et al. 2007a, Vakoc et al. 2007a,b, Tsai et al. 2008, Adler et al. 2009).

13.3 OCT Theory

13.3.1 Low Coherence Interferometry

When two light beams combine, their fields are added resulting in a new wave pattern. This phenomenon is known as interference. Low coherence interferometry (LCI), which is the basis of OCT, is the interference between a reference and a sample beam limited by the short coherence length of the light source used. This, so-called, "coherence gating" results in high-resolution localization of the origin of the interfering beams and can be used for ranging or, in 2D, for imaging of complex scattering samples.

A simplified schematic of a low coherence Michelson interferometer is shown in Figure 13.3. In the interferometer, the beam from the light source is directed onto a beam splitter that divides the beam into a reference and a sample, that is, a measurement, beam. If the sample is a perfectly reflecting mirror and the polarization effects of light are ignored, then the back-reflection

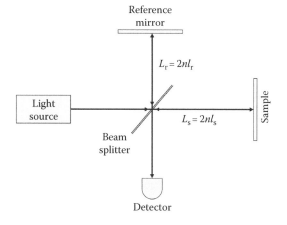

FIGURE 13.3 Michelson interferometer used for low coherence interferometry.

of the beam from the sample and the reference mirror can be represented by the complex functions

$$E_s\left(t - \frac{L_s}{c}\right)$$

and

$$E_r\left(t - \frac{L_r}{c}\right)$$

respectively. L_s and L_r are the corresponding optical path lengths of the arms of the interferometer and c is the speed of light. Thus, the resulting light intensity, I_d, at the photodetector is

$$I_d(\tau) = \langle [E_s(t) + E_r(t+\tau)][E_s(t) + E_r(t+\tau)]^* \rangle, \quad (13.2)$$

or

$$I_d = \langle E_s(t)E_s^*(t) + E_s(t)E_r^*(t+\tau) + E_s^*(t)E_r(t+\tau) + E_r^*(t+\tau)E_r(t+\tau) \rangle$$

where the angular brackets denote the time average over the integration time at the detector and the asterisk symbol indicates the complex conjugate operation. The time delay, τ, corresponding to the round-trip optical path length difference between the two arms is

$$\tau = \frac{\Delta L}{c} = \frac{L_s - L_r}{c} = \frac{2n(\ell_s - \ell_r)}{c}$$

where

n is the refractive index of the medium
ℓ_s and ℓ_r are the geometric lengths of the arms

Since the intensities of the sample and reference beams can be represented as

$$I_s = \langle E_s(t)E_s^*(t) \rangle$$

and

$$I_r = \langle E_r(t+\tau)E_r^*(t+\tau) \rangle$$

and

$$\langle E_s(t)E_r^*(t+\tau) + E_s^*(t)E_r(t+\tau) \rangle = 2\,\text{Re}\{\langle E_s^*(t)E_r(t+\tau) \rangle\}$$

the intensity at the photodetector becomes

$$I_d(\tau) = I_s + I_r + 2\,\text{Re}\{\langle E_s^*(t)E_r(t+\tau) \rangle\}$$

or

$$I_d(\tau) = I_s + I_r + 2\sqrt{I_s I_r}\,\text{Re}\{V_{mc}(\tau)\}, \quad (13.3)$$

where

$$V_{mc}(\tau) = \frac{\langle E_s(t)E_r^*(t+\tau) \rangle}{\sqrt{I_s I_r}}, \quad (13.4)$$

The first two components of I_d, I_s and I_r, are the backscattered intensities by the sample and the reference arm, respectively, and the third term is the interference signal. The interference signal carries the information about the structure of the sample, and depends on the optical path delay between the reference and sample arm. The normalized mutual coherence function $V_{mc}(\tau)$ in Equation 13.4 is a measure of the degree to which the temporal and spatial characteristics of E_s and E_r match.

If the complex spectral components are ignored, the phase difference, with respect to the optical path delay, can be expressed as $\varphi = 2\pi f_0\tau$, where f_0 is the center frequency of the light source. If E_s and E_r originate from a single wave front, spatial coherence can be neglected and the complex mutual coherence function reduces to self-coherence, if a mirror is placed in sample arm (Pan et al. 1995, Schmitt 1999, Wang and Tuchin 2004). Equation 13.3 can be rewritten into the form

$$I_d(\tau) = I_s + I_r + 2\sqrt{I_s I_r}\,\text{Re}\{V_{tc}(\tau)\}, \quad (13.5)$$

where

$$\text{Re}\{V_{tc}(\tau)\} = \frac{\langle E(t)E^*(t+\tau) \rangle}{\sqrt{I_s I_r}}.$$

If V_{tc} is expressed as

$$V_{tc}(t) = A(t)e^{i2\pi f_0\tau}$$

then

$$\text{Re}\{V_{tc}(t)\} = A(t)\cos(2\pi f_0\tau) = |V_{tc}|\cos(2\pi f_0\tau)$$

which finally results in

$$I_d(\tau) = I_s + I_r + 2\sqrt{I_s I_r}\,|V_{tc}(\tau)|\cos(2\pi f_0\tau). \quad (13.6)$$

The Wiener–Khinchin theorem (also known as the Wiener–Khinchin–Einstein theorem or the Khinchin–Kolmogorov theorem) states that the power spectral density of a wide-sense-stationary random process is the Fourier transform of the corresponding autocorrelation function, that is,

$$S_{xx}(f) = \int_{-\infty}^{\infty} r_{xx}(\tau)e^{-j2\pi f \tau}\mathrm{d}\tau$$

where

$$r_{xx}(\tau) = \mathrm{E}[x(t)x^*(t-\tau)]$$

is the autocorrelation function. According to the Wiener–Khinchin theorem, the temporal coherence function V_{tc} is actually the Fourier transform of the power spectral density $S(f)$ of the light source, which is fully characterized by its shape, its spectral bandwidth, and its center wavelength (Gilgen et al. 1989, Schmitt 1999, Wang 1999, Akcay et al. 2002, Wang and Tuchin 2004):

$$V_{tc}(\tau) = \Im\{S(k)\} = \int_{0}^{\infty} S(f)\exp(-j2\pi f \tau)\mathrm{d}f. \qquad (13.7)$$

This relationship reveals that the shape and width of the emission spectrum of the light source are important variables since they influence the resolution of the low coherence interferometer.

Equation 13.6 can be rewritten so that I_d is presented as a function of ΔL based on

$$2\pi f_0 \tau = 2\pi \frac{c}{\lambda_0}\tau = \frac{2\pi\Delta L}{\lambda_0} = k_0 \Delta L$$

$$k_0 = \frac{2\pi}{\lambda_0}, \quad f_0 = \frac{c}{\lambda_0}, \quad c\tau = \Delta L.$$

resulting in

$$I_d(\Delta L) = I_s + I_r + 2\sqrt{I_s I_r}\left|\Im\{S(k)\}\right|\cos(k_0 \Delta L) \qquad (13.8)$$

where $k_0 = 2\pi/\lambda_0$ is the average wave number and the relation $\lambda_0 = c/f_0$ is used to transform from the time domain to the path domain (Pan et al. 1995, Schmitt 1999).

So far, including Equation 13.8, it was assumed that the sample is a perfectly reflecting mirror that induces a time delay but does not affect the amplitude and coherence of the sample beam. In reality, the light reflected from the sample is composed of single or least scattered light and diffuse or multiply scattered light. In contrast to diffuse backscattered light, the light that undergoes only single scattering or very little scattering maintains its coherence and contributes to the LCI signal. Taking into account the scattering path inside the media, the total round-trip path length L_s of the sample arm is

$$L_s = L_{so} + L_s' \qquad (13.9)$$

where

L_{so} is the round-trip path length to the sample surface
L_s' is the total round-trip path length inside the sample to the scatterer site

13.3.2 OCT Signal from Multiple Scatterers

Any given sample can be expressed as a series of scatterers, that is,

$$L_s' = \sum n_s \ell_i \qquad (13.10)$$

where

n_s is the refractive index of each layer of the medium
ℓ_i the scattering path of light inside the medium

In such a case, the intensity given in Equation 13.2 can be written in the form

$$I_d(\tau) = \left\langle \left[\int_{L_{so}}^{\infty} E_s'(t,L_s)\mathrm{d}L_s + E_r(t+\tau)\right]\left[\int_{L_{so}}^{\infty} E_s'(t,L_s)\mathrm{d}L_s + E_r(t+\tau)\right]^* \right\rangle \qquad (13.11)$$

where $E_s'(t,L_s)$ is the path length-resolved field density. Equation 13.11 yields

$$I_d(L_s, L_r) = I_s + I_r + 2\mathrm{Re}\int_{L_{so}}^{+\infty}\left\langle E_s'(t,L_s)\mathrm{d}L_s * E_r(t+\tau)\right\rangle$$

Since

$$E_s'(t,L_s) = E_s(t) * R(L_s) = E_r(t) * R(L_s)$$

then

$$I_d(L_s, L_r) = I_s + I_r + 2\sqrt{I_s I_r}\int_{-\infty}^{+\infty}\sqrt{R(L_s)}\,\mathrm{Re}\{V_{tc}(\Delta L)\}\mathrm{d}L \qquad (13.12)$$

or

$$I_d(L_s, L_r) = I_s + I_r + 2\sqrt{I_s I_r}\int_{-\infty}^{+\infty}\sqrt{R(L_s)}\left|V_{tc}(\Delta L)\right|\cos(k_0\Delta L)\mathrm{d}L \qquad (13.13)$$

where $R(L_s)$ is the normalized path length-resolved diffuse reflectance, that is, the normalized derivative of the intensity depth distribution of the sample wave, representing the fraction of power reflected from the layer located at position L_s within the object.

The signal-carrying interference term, that is, the interference modulation, in Equation 13.12 can also be expressed as a convolution

$$I_{signal}(L_s, L_r) = 2\sqrt{I_s I_r}\left[\sqrt{R(L_s)} \otimes \text{Re}\{V_{tc}(\Delta L)\}\right]$$

$$\text{Re}\{V_{tc}(\Delta L)\} = \frac{\langle E(L_s)^* E^*(L_s - L_r)\rangle}{\sqrt{I_s I_r}} \quad (13.14)$$

where

Re$\{V_{tc}(\Delta L)\}$ is the coherence function, that is, the interferometric response in the ideal case of a mirror in both arms
$\otimes$ denotes the convolution operation

The function Re$\{V_{tc}(\Delta L)\}$ is also called the point spread function (PSF) of the system, since it defines the resolution of OCT, and is a function of the properties of the light source (Pan et al. 1995, 1996, Kulkarni et al. 1997, Schmitt 1999, Yung et al. 1999).

In order to examine the spectral equivalent of Equation 13.14, let r_{ii} describe the source autocorrelation, that is, $r_{ii}(\Delta L) = \langle E_{source}(L_s)E^*_{source}(L_s + L_r)\rangle$.

The cross-correlation function for an arbitrary specimen in the sample arm of the interferometer is

$$r_{is}(\Delta L) = \langle E_s(L_s)E^*_s(L_s - L_r)\rangle$$

and is equal to

$$r_{is}(\Delta L) = \sqrt{R(L_s)} \otimes [r_{ii}(\Delta L)].$$

Using the correlation terms, Equation 13.14 can be rewritten as

$$I_d(L_s, L_r) = I_s + I_r + 2I_s I_r [r_{is}(\Delta L)]. \quad (13.15)$$

Finally, the corresponding spectral relations can be obtained using the Wiener-Khinchin theorem. First, the power spectrum of the light source is obtained as the Fourier transform of its autocorrelation, that is,

$$S(k) = \Im\{(r_{ii}(\Delta L)\} \quad (13.16)$$

Furthermore, the cross spectral density function of two waves is obtained as the Fourier transform of the cross-correlation function as

$$S_{is}(k) = \Im\{r_{is}(\Delta L)\} \quad (13.17)$$

The following spectral interference law is obtained from Equation 13.15:

$$I(k) = S_i(k) + S_s(k) + 2\,\text{Re}\{S_{is}(k)\} \quad (13.18)$$

13.3.3 Time Domain Optical Coherence Tomography

The term *tomography* is used whenever 2D data is derived from a 3D object to construct a slice image of the object's internal structure. In OCT, multiple parallel LCI scans are performed to generate the 2D image. A typical measurement system consists of a Michelson interferometer illuminated by a low temporal coherence light source, such as a superluminescent diode or a broad bandwidth laser. In OCT, the object to be measured is placed in one arm of the interferometer. A measurement beam emitted by the light source is reflected or backscattered from the object with different delay times, depending on the various optical properties of different layers within the object. A longitudinal profile of reflectivity versus depth is obtained by translating the reference mirror, or by other means changing the pathlength of the reference arm, and synchronously recording the magnitude of the intensity of the resulting interference fringes. A fringe signal is evident at the detector only when the optical path difference in the interferometer is less than the coherence length of the light source. Locating the maximum fringe visibility position allows one to determine the location of internal structures of the object with a resolution in the micrometer scale (Hee et al. 1995a, Masters 1999, Vabre et al. 2000). A simple TD OCT system with the interference signal and the A-Scan envelope is shown in Figure 13.4.

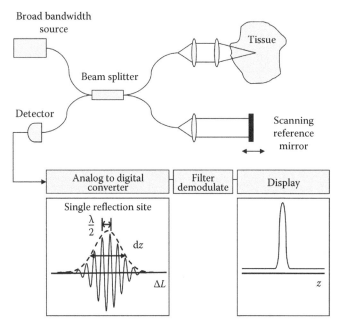

FIGURE 13.4 Typical fiber-optic implementation of time-domain OCT system, interferogram, and the A-Scan envelope.

The modulation of the optical time delay between the arms, for example, by translating the reference mirror by a constant speed, shifts the interference signal to the corresponding Doppler frequency defined by

$$f_d = \frac{2v}{\lambda_0},$$

where

v is the speed of the moving mirror
λ_0 is the center wavelength of the light source

Shifting to the Doppler frequency facilitates the removal of the dc background components and other low-frequency noise during demodulation (Schmitt 1999). To extract the signal carrying component, the detection circuit usually contains three main components: (1) a transimpedance amplifier, (2) a band-pass filter centered at f_d, and (3) an amplitude demodulator to extract the envelope of the interferometric signal (Hee 2002).

13.3.4 Spatial Resolution in Optical Coherence Tomography

The spatial resolution of an OCT system is usually different in the axial and transverse directions with the mechanisms that govern each being independent. The axial resolution depends on the characteristics of the light source, whereas the transverse resolution is determined by the optics of the imaging device.

The axial resolution in OCT is determined by the coherence length of the light source implying that high axial resolution can be achieved independently of the beam-focusing conditions. The coherence length is the spatial width of the field autocorrelation produced by the interferometer. As shown before, the envelope of the field autocorrelation is the Fourier transform of the power spectrum. Thus, the width of the autocorrelation function, or the axial resolution, is inversely proportional to the width of the power spectrum. Therefore, for a source with a Gaussian spectral distribution, the axial resolution, Δz, is

$$\Delta z = \frac{2\ln 2 \lambda_0^2}{\pi \Delta \lambda} \tag{13.19}$$

where $\Delta \lambda$ is the full width at half maximum (FWHM) wavelength range of the light source (Brezinski and Fujimoto 1999). Equation 13.19 is valid only in a vacuum. In the usual case where the sample is dispersive, the refractive index of the sample material depends on the wavelength and affects the system's resolution. This effect is particularly significant when broadband sources are used. Distances measured with LCI are optical distances. The geometric equivalent is derived by dividing the optical distance by the group refractive index n_g of the media.

In real materials, not only is the refractive index n a function of wavelength, but also the group index n_g:

$$n_g = n - \lambda \frac{dn}{d\lambda} \tag{13.20}$$

This dispersion effect, when different in the two arms of the interferometer, results in a broadening of the interferogram and an increase in coherence length which, in turn, translates into a deterioration in resolution. In a dispersive medium, the coherence length Δz_m can be calculated by

$$\Delta z_m = \sqrt{\Delta z^2 + \left(\frac{dn_g}{d\lambda} d_g \Delta \lambda \right)^2} \tag{13.21}$$

where d_g is the geometric depth in the medium (Drexler et al. 1998, Hitzenberger et al. 1999).

The transverse resolution of an OCT system is determined by the optics of the imaging device and does not depend on the axial resolution. The selection of optics is a tradeoff between the transverse resolution and the imaging depth range, that is, the depth of focus. The transverse resolution Δx can be approximated by

$$\Delta x \simeq \frac{4\lambda}{\pi} \left(\frac{f_1}{d}\right) = \frac{4\lambda}{\pi \text{NA}} \tag{13.22}$$

where

f_1 is the focal length of the focusing lens
d is the light beam diameter on the lens aperture
NA is the numerical aperture of the lens

High transverse resolution can be obtained by using a large NA and focusing the beam to a small spot size. The transverse resolution is also related to the Rayleigh range, z_r, which is a measure of the depth of focus, that is,

$$z_r = \frac{\pi \Delta x^2}{4\lambda} \tag{13.23}$$

The Rayleigh range is the distance from the focal plane to the point where the light beam diameter has increased by a factor of $\sqrt{2}$. Given the above relationships, it is obvious that improving the transverse resolution produces a decrease in the depth of focus, a significant tradeoff in the design of OCT imaging devices.

13.3.5 Spectral Radar: Optical Coherence Tomography in the Fourier Domain

The basis of Fourier Domain OCT (FD OCT) is Equation 13.17. In contrast to TD OCT, depth information in FD OCT is provided by an inverse Fourier transform of the spectrum of the backscattered light (Born and Wolf 1999). The amplitude of the spectrum of the backscattered sample light amplitude, $I(k)$, is obtained using a spectrometer. The inverse Fourier transform of the recorded spectral intensity yields the same signal as obtained by standard low coherence interferometry and provides a back-reflection profile as a function of depth.

A simplified optical setup of an FD OCT interferometer is shown in Figure 13.5. The broadband source used is similar to TD OCT. In contrast to time-domain techniques, the time-consuming mechanical OCT depth scan is replaced by a spectral measurement. FD OCT measures the signal in the Fourier domain and, by Fourier transformation, delivers the scattering profile in the spatial domain. The interference spectrum $I(k)$ for a single scatterer at a distance z_1 from the reference plane is a cosine function multiplied by the spectrum of the source, $S(k)$. The Fourier transform provides the location of the peak at that frequency that corresponds to the scatterer location. The backscattering profile of a complicated, real, sample is the sum of many such signals.

Analytically, the signal in the sample arm, backscattered from different depths within the sample, is combined with the signal in the reference arm to produce an interferogram. The total interference signal is given by the spectral intensity distribution of the light source times the square of the sum of the backreflected reference and sample signals:

$$I(k) = S(k)\left| a_R \exp(j2kr) + \int_0^\infty a(z)\exp(j2k(r+n(z).z)\mathrm{d}z \right|^2 \quad (13.24)$$

where
k is the wavenumber $k = 2\pi/\lambda$
r is the path length in the reference arm
$r + z$ is the path length in the object arm
z is the path length in the object arm, measured from the reference plane

z_0 is the offset distance between reference plane and object surface
n is the refractive index ($n = 1$ for $z < z_0$ and varying depending on the sample for longitudinal positions in the object $z > z_0$)
a_R is the reflection coefficient of the reference
$a(z)$ is the backscattering coefficient of the object signal, $a(z)$ is zero for $z < z_0$
$S(k)$ is the spectral intensity distribution of the light source

Assuming $a_R = 1$, the interference signal $I(k)$ can be written as

$$I(k) = S(k)\left| 1 + \int_0^\infty a(z)\exp(j2knz)\mathrm{d}z \right|^2 \quad (13.25)$$

Expanding the complex exponential results in

$$I(k) = S(k)\left| 1 + \int_0^\infty a(z)\cos(2knz)\mathrm{d}z + j\int_0^\infty a(z)\sin(2knz)\mathrm{d}z \right|^2$$

The absolute value of a complex three-term can be evaluated by multiplying with the complex conjugate term

$$I(k) = S(k)\left(\left(1 + \int_0^\infty a(z)\cos(2knz)\mathrm{d}z \right) + j\int_0^\infty a(z)\sin(2knz)\mathrm{d}z \right)$$
$$\times \left(\left(1 + \int_0^\infty a(z)\cos(2knz)\mathrm{d}z \right) - j\int_0^\infty a(z)\sin(2knz)\mathrm{d}z \right)$$

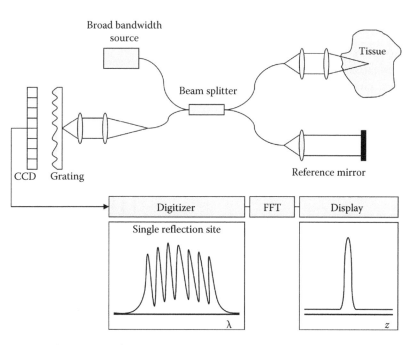

FIGURE 13.5 Typical fiber-optic implementation of Fourier domain OCT, interferogram, and backreflection profile.

The result is

$$I(k) = S(k)\left(\left(1 + \int_0^\infty a(z)\cos(2knz)\mathrm{d}z\right)^2 - \left(j\int_0^\infty a(z)\sin(2knz)\mathrm{d}z\right)^2\right)$$

$$I(k) = S(k)\left[1 + 2\int_0^\infty a(z)\cos(2knz)\mathrm{d}z\right.$$

$$\left. + \frac{\left(\int_0^\infty a(z)\cos(2knz)\mathrm{d}z\right)^2 - \left(j\int_0^\infty a(z)\sin(2knz)\mathrm{d}z\right)^2}{\left|\int_0^\infty a(z)\exp(j2kz)\mathrm{d}z\right|^2}\right]$$

Substituting

$$\left|\int_0^\infty a(z)\exp(j2kz)\mathrm{d}z\right|^2 = \left(\int_0^\infty a(z)\exp(-j2kz)\mathrm{d}z\right)\left(\int_0^\infty a(z')\exp(j2kz')\mathrm{d}z'\right)$$

$$= \int_0^\infty\int_0^\infty a(z)a(z')\exp(-j2kz)\exp(j2kz')\mathrm{d}z\mathrm{d}z'$$

in the above equation results in an interference signal of the form

$$I(k) = S(k)\left(1 + 2\int_0^\infty a(z)\cos(2knz)\mathrm{d}z + \int_0^\infty\int_0^\infty a(z)a(z')\mathrm{e}^{-j2kn(z-z')}\mathrm{d}z\mathrm{d}z'\right)$$

(13.26)

Therefore, the interference signal is a sum of three terms. The first is a DC term. The second term encodes the depth information of the object. The backreflection intensity is incorporated in $a(z)$, while the corresponding optical pathlength difference is found in the argument of the cosine term. The third term describes the mutual interference of all elementary waves. Assuming that $a(z)$ is symmetrical with respect to z and also $a(z) = 0$ for all $z < z_0$, suggests that $a(z)$ can be replaced by the symmetrical expansion $\hat{a}(z) = a(z) + a(-z)$ and, therefore,

$$I(k) = S(k)\left(1 + \int_0^\infty a(z)\cos(2knz)\mathrm{d}z + \int_0^\infty a(z)\cos(2knz)\mathrm{d}z\right.$$

$$+ \frac{1}{2}\int_0^\infty\frac{1}{2}\int_0^\infty a(z)a(z')\mathrm{e}^{-j2kn(z-z')}\mathrm{d}z\mathrm{d}z'$$

$$\left. + \frac{1}{2}\int_0^\infty\frac{1}{2}\int_0^\infty a(z)a(z')\mathrm{e}^{-j2kn(z-z')}\mathrm{d}z\mathrm{d}z'\right)$$

$$I(k) = S(k)\left(1 + \int_0^\infty a(z)\cos(2knz)\mathrm{d}z + \int_{-\infty}^0 a(-z)\cos(2knz)\mathrm{d}z\right.$$

$$+ \frac{1}{2}\int_0^\infty\frac{1}{2}\int_0^\infty a(z)a(z')\mathrm{e}^{-j2kn(z-z')}\mathrm{d}z\mathrm{d}z'$$

$$\left. + \frac{1}{2}\int_{-\infty}^0\frac{1}{2}\int_{-\infty}^0 a(-z)a(-z')\mathrm{e}^{-j2kn(z-z')}\mathrm{d}z\mathrm{d}z'\right)$$

$$I(k) = S(k)\left(1 + \int_{-\infty}^\infty \hat{a}(z)\cos(2knz)\mathrm{d}z\right.$$

$$\left. + \frac{1}{4}\int_{-\infty}^\infty\int_{-\infty}^\infty \hat{a}(z)\hat{a}(z')\mathrm{e}^{-j2kn(z-z')}\mathrm{d}z\mathrm{d}z'\right)$$

Expanding the second term in complex exponentials and sum using, again, the symmetrical properties of $a(z)$ results in

$$\int_{-\infty}^\infty \hat{a}(z)\cos(2knz)\mathrm{d}z = \int_{-\infty}^\infty \hat{a}(z)\left(\frac{\mathrm{e}^{j2knz} + \mathrm{e}^{-j2knz}}{2}\right)\mathrm{d}z$$

$$= \frac{1}{2}\int_{-\infty}^\infty \hat{a}(z)\mathrm{e}^{j2knz}\mathrm{d}z + \frac{1}{2}\int_{-\infty}^\infty \hat{a}(z)\mathrm{e}^{-j2knz}\mathrm{d}z$$

$$= \frac{1}{2}\int_{-\infty}^\infty \hat{a}(-z)\mathrm{e}^{-j2knz}\mathrm{d}z + \frac{1}{2}\int_{-\infty}^\infty \hat{a}(z)\mathrm{e}^{-j2knz}\mathrm{d}z$$

$$= \int_{-\infty}^\infty \hat{a}(z)\mathrm{e}^{-j2knz}\mathrm{d}z$$

Since the amplitude is real, the third term can be re-formed as a function of the autocorrelation term

$$AC[\hat{a}(z)] = \int_{-\infty}^\infty \hat{a}(z)\,\hat{a}^*(z_0 + z')\mathrm{d}z$$

resulting in

$$\frac{1}{4}\int_{-\infty}^\infty\int_{-\infty}^\infty \hat{a}(z)\hat{a}(z')\mathrm{e}^{-j2kn(z-z')}\mathrm{d}z\mathrm{d}z'$$

$$= \frac{1}{4}\int_{-\infty}^\infty\int_{-\infty}^\infty \hat{a}(z)\hat{a}(z+z'')\mathrm{e}^{-j2kn(z)}\mathrm{e}^{j2kn(z_0+z'')}\mathrm{d}z\mathrm{d}z''$$

$$= \frac{1}{4}\int_{-\infty}^\infty \mathrm{e}^{-j2kn(z)}\mathrm{d}z\underbrace{\int_{-\infty}^\infty \hat{a}(z)\mathrm{e}^{j2kn(z_0)}\hat{a}(z+z'')\mathrm{e}^{j2kn(z'')}\mathrm{d}z''}_{AC[\hat{a}(z)]}$$

$$= \frac{1}{4}\int_{-\infty}^\infty AC[\hat{a}(z)]\mathrm{e}^{-j2kn(z)}\mathrm{d}z$$

where $z' = z_0 + z''$ (with z_0 an offset constant) and $dz' = dz''$. Incorporating the above results, the intensity of the spectral interferometric signal becomes

$$I(k) = S(k)\left(1 + \int_{-\infty}^{\infty} \hat{a}(z)e^{-j2knz}dz + \frac{1}{4}\int_{-\infty}^{\infty} AC[\hat{a}(z)]e^{-j2kn(z)}dz\right)$$

The second and the third terms are actually Fourier transforms of the scattering amplitude $\hat{a}(z)$ and of the autocorrelation term $AC[\hat{a}(z)]$ over the variable z. By using the Fourier transform scaling property, the interferogram can be rewritten as

$$I(k) = S(k)\left(1 + \int_{-\infty}^{\infty} \hat{a}(z)e^{-jknz}\overset{\text{time scale}}{\overset{(2)}{\frown}} dz + \frac{1}{4}\int_{-\infty}^{\infty} AC[\hat{a}(z)]e^{-jknz}\overset{\text{time scale}}{\overset{(2)}{\frown}} dz\right)$$

$$I(k) = S(k)\left(1 + \frac{1}{2}\mathfrak{I}_z\{\hat{a}(z)\} + \frac{1}{8}\mathfrak{I}_z\{AC[\hat{a}(z)]\}\right)$$

Performing the inverse Fourier transformation generates the following relationship

$$F^{-1}\{I(k)\} = F^{-1}\{S(k)\} \otimes \left(\left[\delta(z)\right] + \frac{1}{2}\hat{a}(z) + \frac{1}{8}AC[\hat{a}(z)]\right)$$

$$= A \otimes (B + C + D) \tag{13.27}$$

where $\otimes$ denotes convolution. The signal C is the symmetrical scattering amplitude $\hat{a}(z)$ and therefore the strength of the scattering versus the depth of the sample.

The first convolution $A \otimes B$ is the Fourier transformation of the source spectrum (correlogram) located around $z = 0$. It can be separated from the information-carrying signal by shortening the optical pathlength in the reference arm, compared to the sample arm, thus increasing the frequency separation between the terms (Figure 13.6). $A \otimes D$ designates the

autocorrelation terms, which describe the mutual interference of all scattered elementary waves. In strongly scattering media, such as skin, the influence of D can be neglected because the autocorrelation term is much weaker than the signal term, which is weighted by the strong reference amplitude. Moreover, these terms are also located around $z = 0$. Therefore, the auto-correlation term can also be separated from the object signal $a(z)$ with a small offset z_0 (Figure 13.6). However, if the object exhibits high backscattering from large depths, there is still the possibility for the $A \otimes D$ term to interfere with the object signal. High backscattering deep within tissue can lead to overlap of the terms that can, in turn, result in artifacts in the data. A solution is to perform, at each position, a second measurement with no reference signal and subtract that signal from $I(k)$. Also notice in Equation 13.27 that the signal C is convoluted with the correlogram of the source. To achieve high-resolution measurements, a light source with a sufficiently broad and smooth spectrum, without noise and ripple, must be selected so that the resulting convoluted peaks from individual scatterers will be sufficiently narrow.

13.3.5.1 Axial Range Limitations

The measurable axial range of SD OCT is limited by the resolution of spectrometer. This is in sharp contrast with TD OCT where the axial range is defined by the mechanical scanning range of the optical delay-line. A large difference between the object and reference optical paths results in a high-frequency interferogram in the spectral domain. According to the sampling theorem, the sampling frequency of the spectrometer has to be twice as large as the highest occurring frequency in the spectrum. The interference spectrum $I(k)$ for a mirror at a distance z_1 from the reference plane contains the cosine function $a(z)\cos(2knz_1)$ multiplied by $S(k)$. For $z = z_{max}$, the period of the cosine fringes becomes

$$dk = \frac{\pi}{nz_{max}}.$$

Therefore, the spectrometer must resolve at least $dk/2$. Since

$$k = \frac{2\pi}{\lambda} \Rightarrow dk = \frac{2\pi(d\lambda)}{\lambda^2}$$

then the maximum resolvable depth is

$$Z_{max} = \frac{\lambda_0^2}{4n(d\lambda)}. \tag{13.28}$$

where $d\lambda$ denotes the wavelength sampling interval defined by the detector separation of a linear detector array.

One of the drawbacks of frequency domain detection is a signal decay, or roll-off, with depth. The major contributor to this roll-off is the spectrometer resolution. The finite number of pixels in the CCD acts as a low-pass filter that rejects the

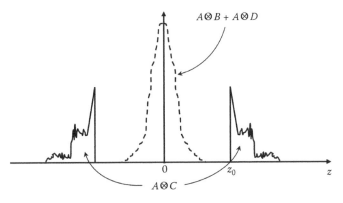

FIGURE 13.6 Amplitude of the results of the Fourier transform of the spectral interferogram recorded with Fourier-domain OCT.

higher frequencies associated with scattering from the deeper layers of the sample. This corresponds to a window function in the spatial domain that reduces the intensity of the signal as a function of depth even if the reflectivity remains constant. In addition, other components of the OCT system, such as the grating and focusing optics in particular, can contribute to the roll-off. Aberrations and diffraction, which degrade the wavelength resolution, result in signal decay. An additional limitation of Fourier-domain imaging stems from the fact that since the DFT of a real function is always an even function; each signal peak in the positive frequency range will have a mirror image in the negative frequency range. Any signal that exceeds the maximum ranging depth, as defined by the CCD resolution, will appear as an aliased signal.

13.3.5.2 Sensitivity Advantage

The main advantage of FD OCT over conventional TD OCT is higher sensitivity that also results in increased imaging speed. In TD OCT, the light backreflected from a series of scatterers is measured sequentially. Light that is scattered back from each scatterer contributes to the interference signal only if the distance between the reference and the scatterer plane is less than the coherence length. Therefore, only a small fraction of the light that is scattered back during the entire scanning time is utilized. In contrast, FD OCT detects the backscattered light from the entire depth range in parallel (not directly, but with frequency-domain detection). Such a detection scheme results in longer exposure time per pixel and, consequently, higher sensitivity, significantly superior to that of TD OCT.

In TD OCT, the excess photon noise will dominate photon shot noise for the conditions of high-speed scanning (Podoleanu and Jackson 1999). Thus, balanced detection is required for shot-noise-limited detection. The shot-noise-limited sensitivity of TD OCT is defined as

$$S_{\mathrm{TD}}^{\mathrm{shot}} = \frac{\eta}{h\nu_0 BW}\varepsilon P_{\mathrm{s}} \qquad (13.29)$$

where

η is the efficiency of the detector
$h\nu_0$ is the energy of the incoming photons
BW is the detection bandwidth

In addition, $\varepsilon P_{\mathrm{s}} = R_{\mathrm{s}}P_0$, P_{s} is the power of light illuminating the sample and ε is a parameter of the interferometer that is defined as the efficiency of light transmitted from the probing arm to the detection arm. R_{s} is the sample reflectivity and P_0 the power of light source.

FD OCT is characterized by parallel detection in the frequency domain. The integration time is relatively long, thus, shot noise dominates the intensity noise (Rollins et al. 1998, Podoleanu and Jackson 1999, Choma et al. 2003, de Boer et al. 2003, Leitgeb et al. 2003, 2004). The sensitivity approaches the photon shot noise limit even with high-speed imaging and without balanced

detection (Leitgeb et al. 2004). The shot-noise-limited sensitivity of SD-OCT is given by

$$S_{\mathrm{FD}}^{\mathrm{shot}} = \frac{n\tau}{h\nu_0}\varepsilon P_{\mathrm{s}} \qquad (13.30)$$

where τ is the integration time.

To compare the sensitivity of TD and FD OCT, assume that the same light source and interferometer configuration are used and that the quantum efficiencies of the detectors are the same, that is, P_0, ν_0, ε, and η are equal. If the spectrum of the light source has a Gaussian shape, the optimal BW can be expressed as (Rollins et al. 1998)

$$BW = \frac{2\Delta\lambda_{\mathrm{FWHM}}\nu_{\mathrm{g}}}{\lambda_0^2} \qquad (13.31)$$

where

$\Delta\lambda_{\mathrm{FWHM}}$ is the bandwidth of the light source
λ_0 is the center wavelength
ν_{g} denotes the group velocity induced by an optical delay-line

When the depth is scanned with a constant ν_{g} in a TD OCT system and the observation time of one axial scan and the ranging depth are the same for both TD and FD OCT, the relationship between the detection bandwidth and the integration time τ is given by

$$BW = \frac{2\Delta\lambda_{\mathrm{FWHM}}nZ_{\mathrm{max}}}{\lambda_0^2\tau} \qquad (13.32)$$

where Z_{max} is the ranging depth and n is the refractive index of tissue. Substituting the ranging depth for FD OCT from Equation 13.28, the relationship between the shot-noise limited sensitivities of FD OCT and TD OCT is expressed as

$$S_{\mathrm{FD}}^{\mathrm{shot}} = \frac{2\Delta\lambda_{\mathrm{FWHM}}}{d\lambda}S_{\mathrm{TD}}^{\mathrm{shot}} \qquad (13.33)$$

In Equation 13.33, the sensitivity gain is proportional to the ratio of the ranging depth to the axial resolution

$$\frac{\Delta\lambda_{\mathrm{FWHM}}}{d\lambda} \approx \frac{Z_{\mathrm{RD}}}{dz}.$$

Thus, the sensitivity gain of FD OCT over TD OCT becomes even more apparent as the ranging depth increases and the axial resolution decreases.

13.3.6 Wavelength Tuning (Swept Source) OCT

In Wavelength Tuning or Swept Source OCT (SS OCT) the wavelength-dependent intensity data are not recorded simultaneously

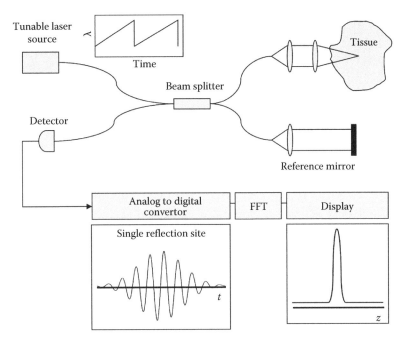

FIGURE 13.7 Typical fiber-optic implementation of Swept Source OCT, interferogram, and backreflection profile.

by using a broadband light source and a spectrometer. Instead, they are recorded sequentially, with a single photodetector, at the detector side, while tuning the wavelength of the light source. If the wavelength λ of the tunable laser in Figure 13.7 is constant, the intensity at the photodetector can be calculated as (Fujimoto 2002):

$$I = I_s + I_r + 2\sqrt{I_s I_r}\cos(2\pi\Delta\Phi) \tag{13.34}$$

where I_s and I_r are the light intensities reflected from the sample and reference, respectively. $\Delta\Phi$ is the phase difference between the two beams:

$$\Delta\Phi = 2\frac{L}{\lambda} = 2L\frac{k}{2\pi} \tag{13.35}$$

where k is the wavenumber corresponding to λ. If the wavenumber is changed, the phase difference changes accordingly. This causes the intensity at the photodetector to oscillate with a frequency

$$f = \frac{d\Delta\Phi}{dt} = \frac{d\Delta\Phi}{dk}\frac{dk}{dt} = \frac{L}{\pi}\frac{dk}{dt} \tag{13.36}$$

Hence, the frequency is directly proportional to the tuning rate of the wavenumber dk/dt and to the path difference, L. If dk/dt is constant, L can be obtained by a Fourier transform of the time-dependent intensity signal recorded by the photodetector during tuning. Fourier transforming the time-depended beat signal yields the sample depth structure, that is, the magnitude of the beat signal defines the amplitude reflectance and the beat

frequency defines the depth position of light scattering sites in the sample.

As in the case of FD OCT, the main advantage of this technique, compared to standard OCT techniques, is that the reference arm length is fixed and no moving parts are required. This significantly increases the speed of scanning. The use of a single photodetector provides the added advantage of simple elimination of the unwanted dc intensity terms by high-pass filtering of the photodetector signal or heterodyne detection. This enhances the usable dynamic range of the detection system considerably. Compared to FD OCT, SS OCT offers similar high-speed data acquisition but without the drawbacks of the spectral limitations of the CCD camera. The price currently to be paid is that the light sources are expensive and so far only available for a limited range of wavelengths. However, it is expected that these limitations will be overcome in the near future.

13.4 Delivery Devices

The success of OCT in clinical applications will depend, on a large part, on the design and availability of delivery mechanisms that will allow seamless integration with current and new diagnostic modalities. Such integration will lead to the introduction of OCT in mainstream diagnostic applications and allow clinicians to use the enhanced imaging capabilities of this technique to benefit the patients without the need of extensive re-training or significant increase in procedure time. Such interplay between imaging technologies and clinical diagnostics will both improve outcome and reduce the cost of therapy.

Applications such as imaging during open surgery, laparoscopic imaging, and imaging of the skin or oral cavity require

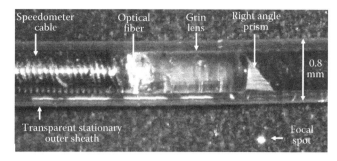

FIGURE 13.8 Fiber-optic catheter/endoscope.

forward imaging probes. Such devices can have a variety of shapes and sizes, sharing the main characteristic of imaging in the forward direction. Scanning can be achieved with one of a number of different methods. Approaches include piezoelectric bending, toroidal scanning, or galvanometric mirror scanning, and, more recently, micromechanical machine systems (MEMS) (Boppart et al. 1997a, Yaqoob et al. 2006).

Intraluminal and intravascular imaging requires small, fiber-optic catheters (Tearney et al. 1996). In most cases, the catheter/endoscope consists of a hollow, speedometer-type cable carrying a single-mode optical fiber (Figure 13.8). The beam from the distal end of the fiber is focused by a gradient-index (GRIN) lens and is directed perpendicular to the catheter axis by a microprism or micromirror. The distal optics are encased in a transparent housing. The beam is scanned either circumferentially (by rotation) or linearly (by translation) of the cable inside a transparent, static housing. These catheters/endoscopes can have a diameter of <1 mm, which is small enough to allow imaging in a human coronary artery or access through the ports of standard endoscopes. The diameter can also be further reduced. In fact, OCT imaging can be performed even through a needle-based device, small enough to be introduced anywhere in a living organ, with minimal disruption to the tissue or structure, even in solid masses (Li et al. 2000a).

13.5 Clinical Applications of OCT

Several features of OCT point to its potential of becoming a very powerful imaging technology for the diagnosis of a wide range of pathologies:

1. OCT can image with axial resolutions of 1–10 μm, one to two orders of magnitude higher than competing technologies such as conventional ultrasound. This resolution approaches that of histopathology, allowing architectural morphology as well as cellular features to be resolved. Unlike ultrasound, imaging can be performed directly through air without requiring direct contact with the tissue or a transducing medium.

2. Imaging can be performed in situ, without the need to excise a specimen. This enables imaging of structures in which biopsy would be hazardous or impossible. It also allows better tissue or organ area coverage, reducing the sampling errors associated with excisional biopsy.

3. Imaging can be performed in real time, without the need to process a specimen as in conventional biopsy and histopathology. This allows pathology to be monitored on screen and stored in high resolution. Real-time imaging can enable real-time diagnosis, and coupling this information with surgery, can enable surgical guidance.

4. OCT is fiber-optically based, therefore can be interfaced to a wide range of instruments including catheters, endoscopes, laparoscopes, and surgical probes. This enables access to a wide range of internal organ systems.

5. Finally, OCT is compact and portable, which is an important feature for a clinically viable device.

There are three general application scenarios that are envisioned for OCT in the diagnosis of disease. First, OCT can enable guiding standard excisional biopsy to reduce sampling errors and false negative results. This can improve the accuracy of biopsy as well as reduce the number of biopsies that are taken, resulting in better prognosis as well as significant cost savings. Second, after more extensive clinical studies have been performed, it may be possible to use OCT to directly diagnose or grade disease. This application will be more challenging since it implies making a diagnosis on the basis of OCT rather than conventional pathology. Applications include situations where OCT might be used to grade early neoplastic changes or determine the depth of neoplastic invasion. Third, there may be scenarios where diagnosis and treatment are performed in real time based on OCT imaging. This would require the OCT diagnostic information to be directly and immediately coupled to treatment decisions. The integration of diagnosis and treatment could reduce the number of patient visits, yielding a significant reduction in health care costs and improve patient compliance. Each of these scenarios requires a different level of OCT performance not only in its ability to image tissue pathology, but to achieve the required level of sensitivity and specificity for each clinical situation.

The most developed clinical OCT applications are those focusing on ophthalmic, cardiovascular, and gastrointestinal imaging. These will be covered in greater detail in the next chapters of the Handbook. In this section, a brief overview of the major contributions of OCT in medicine and biology will be presented.

13.5.1 Ophthalmology

The relatively transparent nature and accessibility of the human eye made it an ideal target for early OCT systems. Ophthalmic applications of OCT have since expanded rapidly, significantly contributing to the establishment of the technology as an invaluable diagnostic tool in the areas of retinal diseases and glaucoma (Hee et al. 1995a,b, Puliafito et al. 1995, Dogra et al. 2004, Costa et al. 2006). The technique was commercialized by Carl Zeiss Meditec, with a suite of companies following in recent years. The transition from bench to bedside has been so successful that OCT is now considered superior to the current standards of care for the evaluation of a number of different retinal conditions (Alexander and Choate 2004). Despite the commercialization of ophthalmic technology, OCT also remains a very dynamic research tool, contributing to ocular disease and therapy research and to the understanding of previously unknown clinical features and microscopic manifestations.

13.5.2 Cardiology

Over the past decade, the application of OCT in the filed of cardiology has been extensively investigated. Initially, OCT was applied to the examination of coronary artery structure and the evaluation of atherosclerotic plague morphology and stenting complications. Subsequently, cellular, mechanical, and molecular analysis was performed including the estimation of macrophage load (Jang et al. 2001, Bouma et al. 2003, Tearney et al. 2003, Chau et al. 2004). The application of OCT to cardiology was greatly enhanced by technological developments such as rotational catheter-based probes, very high imaging speed systems, and functional OCT modalities. Currently, imaging and validation studies are being performed in vivo. OCT has been found to be superior to the current state of the art, that is, intravascular ultrasound (IVUS), since it provides significantly higher resolution, by as much as an order of magnitude (Fujimoto et al. 1999, Jang et al. 2002).

13.5.3 Oncology

From very early in the development of OCT, cancer imaging has been an area of intense research interest. Imaging has been performed in a wide range of malignancies including gastrointestinal, respiratory, and reproductive tract, skin, breast, bladder, brain, ear, nose, and throat cancers. OCT has been used to evaluate the larynx, and has been shown to effectively quantify the thickness of the epithelium and evaluate the integrity of the basement membrane. It was also used to visualize the structure of the lamina propria (Armstrong et al. 2006b). In addition, preliminary studies have been conducted to evaluate the application of OCT to the oral cavity, oropharynx, vocal folds, and nasal mucosa (Armstrong et al. 2006b). Numerous studies have been devoted to the investigation of OCT imaging in the gastrointestinal tract (Bouma et al. 2000, Li et al. 2000b, Pitris et al. 2000, Sivak et al. 2000, Poneros et al. 2001, Pfau et al. 2003, Shen et al. 2004, Evans et al. 2006, Chen et al. 2007). Figure 13.9 shows a representative endoscopic OCT image of the esophagus with the corresponding histology. The feasibility of using OCT to identify dysplasia includes a recent blinded clinical trial that showed an accuracy of 78% in the detection of dysplasia in patients with Barrett's esophagus (Isenberg et al. 2005), and a sensitivity of 83% and a specificity of 75% for detecting high-grade dysplasia and intramucosal carcinoma using a numeric scoring system based on the surface maturation and glandular architecture (Evans et al. 2006).

In addition to imaging neoplasia inside tubular systems, such as the gastrointestinal tract, OCT also holds promise for the detection of cancers in solid organs such as breast (Boppart et al. 2004, Hsiung et al. 2007). Figure 13.10 shows an example of OCT imaging of the human breast. Using quantitative signal analyses including slope, standard deviation, and spatial frequency, high sensitive tissue classification has been shown with normal and tumor tissues (Zysk and Boppart 2006, Goldberg et al. 2008). Recent studies have reported that OCT can characterize breast tissues (including tumor) with 80%–100% sensitivity and 78%–88% specificity with overall accuracy >90% (Goldberg et al. 2008, Iftimia et al. 2009, Mujat et al. 2009, Nguyen et al. 2009, Zysk et al. 2009). With improvements in clinician training, feature definition, and system resolution, OCT can become a strong candidate for future clinical adoption for image-guided interventions for breast cancer, including guiding breast biopsy and identifying surgical margins during open surgeries.

13.5.4 Other Applications

In addition to ophthalmology, cardiology, and gastroenterology, which are the most developed fields in OCT, novel applications are constantly being explored. Important new OCT developments include quantitative imaging and 3D mapping of the upper airways, musculoskeletal imaging with standard and polarization sensitive OCT, and evaluation of dental structures (Otis et al. 2000a,b, de Melo et al. 2005, Li et al. 2005, Armstrong

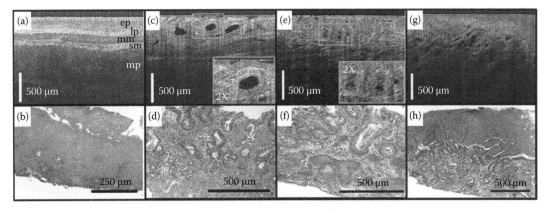

FIGURE 13.9 In vivo ultrahigh resolution endoscopic OCT image (top row) and the corresponding histology (bottom row). a–b: normal esophagus; c–d: Barrett's esophagus; e–f: High-grade dysplasia; g–h: Adenocarcinoma. (Adapted from Chen, Y. et al., *Endoscopy*, 39, 599, 2007. With permission.)

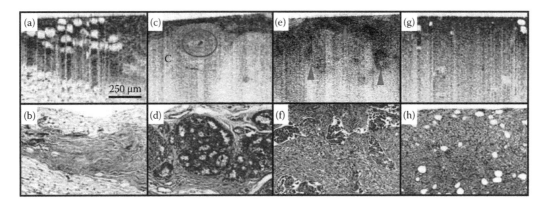

FIGURE 13.10 Ultrahigh resolution OCT image (top row) and the corresponding histology (Hematoxylin-eosin stain, bottom row) of human breast tissues ex vivo. (a) OCT image of normal fibroadipose tissue. (b) Histologic specimen corresponding to OCT image. (c) OCT image of DCIS lesions in lobules. (d) Histologic specimen corresponding to OCT image. (e) OCT image of infiltrating ductal carcinoma. (f) Histologic specimen corresponding to OCT image. (g) OCT image of a solid variant infiltrative lobular carcinoma. (h) Histologic specimen corresponding to OCT image. (Adapted from Hsiung, P.L. et al., *Radiology*, 244, 865, 2007. With permission.)

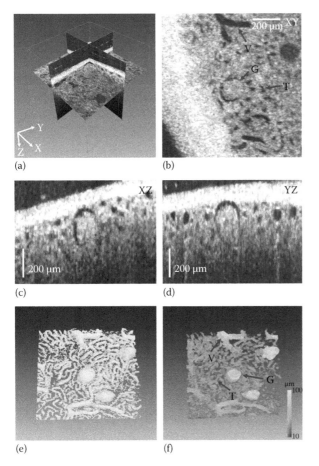

FIGURE 13.11 OCT imaging of human kidney ex vivo. (a) 3D cut-through view of the human kidney showing a glomerulus (G) surrounded by tubules (T) and vessels (V). (b–d): Representative OCT images in the XY, XZ, and YZ planes. **(See color insert.)** (e) 3D volumetric image of the segmented glomeruli and tubular network. (f) Automatically quantified luminal and interstitial dimensions and color-coded on the structural image. (Adapted from Li, Q. et al., *Opt. Express*, 17, 16000, 2009. With permission.)

et al. 2006a). In dentistry, the detection of early decay not visible on standard x-ray, has been shown to have great promise and is being actively commercialized. In addition, applications in areas such as neurology (Boppart et al. 1998, Jafri et al. 2005), gynecology, for the diagnosis of ovarian cancer (Boppart et al. 1999, Hariri et al. 2009), and urology, for the diagnosis of prostate cancer (D'Amico et al. 2000) and kidney disease (Li et al. 2009, Onozato et al. 2010) are advancing rapidly (Figure 13.11). Further investigation is needed to evaluate the role of OCT-guided procedures in a variety of clinical areas. Future clinical trials will prove the role of OCT in providing point-of-care screening or diagnosis and positively affecting patient outcome.

13.5.5 OCT in Biology

Imaging embryonic morphology is important to understand the process of development and differentiation as well as elucidating aspects of genetic expression, regulation, and control. The technologies currently used to provide structural information about microscopic specimens are elaborate, complex, and expensive (Morton et al. 1990). Not even confocal microscopy is practical since imaging depths are limited to less than 0.5 mm in scattering tissues (Jester et al. 1991). Currently, morphological changes, occurring during development, are studied by histological sections of sacrificed embryos and thus only reveal developmental events at discrete time points. OCT can provide the means to in vivo image morphology in situ and continuously in the same subject.

One of the most well-explored areas, in the application of OCT in developmental biology, is embryonic cardiac imaging. Studies have shown that Xenopus-laevis hearts can be readily imaged. Estimates of ejection fraction were performed by extrapolating the M-scan image data to a volume (assuming an ellipsoid-shaped ventricle) (Boppart et al. 1997b). Embryonic chick hearts in 2D (in vivo) and 3D (fixed) were also imaged (Yelbuz et al. 2002). 4D images of embryonic chick hearts were obtained by pacing excised hearts and gating the image

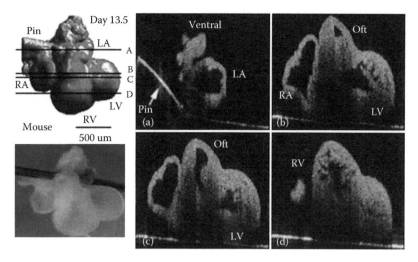

FIGURE 13.12 3D reconstruction of a 13.5 dpc embryonic mouse heart, ventral view (top left panel). A digital image of the same mouse heart is below the rendered image. The scale bar applies to all of the images. Panels a–d are samples of transverse 2D OCT "slices" through the heart at levels represented by the labeled lines in the reconstruction. (From Jenkins, M.W. et al., *Opt. Express*, 14, 736, 2006, With permission.)

acquisition to the cardiac cycle with one representative 3D volume, shown in Figure 13.12 (Jenkins et al. 2006). In addition, fixed embryonic murine hearts and paced excised embryonic murine hearts were imaged using gated OCT (Luo et al. 2006). In both studying genetically altered adult murine skeletal muscle (Pasquesi et al. 2006) and genetically altered hearts, compared to wild-type, OCT promises to provide new insights into developmental biology.

13.6 Commercial OCT Systems

The first commercial OCT devices were developed by Humphrey Systems (now a part Carl Zeiss Meditec, Inc.) for retinal imaging. The first device was released in 1996 and gained Federal Drug Administration (FDA) approval in 2002. Since then, there have been several iterations of the device, with the Stratus OCT™ systems selling more than 6000 units (Carl Zeiss Meditec 2006). In recent years, several new OCT systems are being offered commercially. In addition to Carl Zeiss Meditec Inc, systems are being developed and sold by LightLab Imaging, Imalux Corporation,, ISIS Optronics GmbH, OCT Medical Imaging, Inc, Michelson Diagnostics, Ltd, Novacam Technologies, Inc, and Lantis Laser Inc. High-speed retinal imaging is the focus of several, recently established, companies, including OptoVue, Inc, Topcon Corporation, Optol Technology, Heidelberg Engineering, and Opthalmic Technologies, Inc. Today, it is estimated that more than 37,000 OCT scans are performed daily in the United States (Walz 2006). In addition to the clinical systems, a number of research-oriented devices are commercially available from Thorlabs, Inc. and Bioptigen, Inc. Individual OCT components can be obtained from companies such as Femtolasers Produktions, Nippon Telegraph and Telephone Corporation, Thorlabs, Inc, and MenloSystems GmbH.

Most recently available, commercial systems operate in the Fourier domain, which offers significant advantages both in speed and sensitivity. The axial resolution of most commercial devices is in the range of 7–12 μm. They are mainly targeted for ophthalmic applications, for both the anterior segment and retinal imaging, most are combined with a slit-lamp, and there are even devices that combine OCT and fluorescent angiography. There are also some commercially available nonophthalmic systems. LightLab Imaging, for example, specializes in cardiovascular and endoscopic, catheter-base, OCT imaging. Bioptigen offers a system with advanced visualization software and compatible with a variety of scanners. Thorlabs has recently introduced OCT systems with bench-top and hand-held imaging probes. Michelson Diagnostics offers a system with a novel quadruple-beam focusing for extended depth of focus without the usual resolution reduction.

13.7 Conclusions

OCT can perform a type of optical biopsy, that is, imaging of tissue morphology at the micron-scale, in situ, and in real time. Image information is available immediately without the need for excision and histologic processing of a specimen. The development of high-resolution and high-speed OCT technology, as well as OCT compatible catheter/endoscopes and other delivery devices, represent enabling steps for many future OCT imaging clinical applications. More research remains to be done and numerous clinical studies must be performed to determine in which clinical situations OCT can play a role. However, the unique capabilities of OCT imaging suggest that it has the potential to significantly impact the diagnosis and clinical management of many diseases and improve patient prognosis. In addition, OCT can be a powerful research tool for applications in many areas, including developmental biology, where nondestructive, serial imaging, on the same samples can offer new insights and significant cost and time advantages.

Acknowledgments

We would like to acknowledge the contributions from colleagues at the Massachusetts Institute of Technology (Drs. Aaron Aguirre, Paul Herz, and Pei-Lin Hsiung), LightLab Imaging (Dr. Joseph Schmitt), Harvard Medical School (Drs. Hiroshi Mashimo and Qin Huang), University of Maryland (Chao-Wei Chen and Qian Li), Georgetown University (Drs. Peter Andrews and Maristela Onozato). This research was sponsored in part by the National Science Foundation BES-0522845 and ECS-0501478 (JGF); the National Institutes of Health R01-CA75289-11 and R01-EY11289-20 (JGF); the Air Force Office of Scientific Research FA9550-040-1-0011 and FA9550-040-1-0046 (JGF); and the Prevent Cancer Foundation (YC), the Nano-Biotechnology Award of the State of Maryland (YC), the Minta Martin Foundation (YC), the General Research Board (GRB) Award and Seed Grant Program of the University of Maryland (YC).

References

Adler, D. C., Chen, Y., Huber, R., Schmitt, J., Connolly, J., and Fujimoto, J. G. 2007a. Three-dimensional endomicroscopy using optical coherence tomography. *Nat. Photonics* 1: 709–716.

Adler, D. C., Huber, R., and Fujimoto, J. G. 2007b. Phase-sensitive optical coherence tomography at up to 370,000 lines per second using buffered Fourier domain mode-locked lasers. *Opt. Lett.* 32: 626–628.

Adler, D. C., Zhou, C., Tsai, T. H., Schmitt, J., Huang, Q., Mashimo, H., and Fujimoto, J. G. 2009. Three-dimensional endomicroscopy of the human colon using optical coherence tomography. *Opt. Express* 17: 784–796.

Aguirre, A. D., Herz, P. R., Chen, Y., Fujimoto, J. G., Piyawattanametha, W., Fan, L., and Wu, M. C. 2007. Two-axis MEMS scanning catheter for ultrahigh resolution three-dimensional and en face imaging. *Opt. Express* 15: 2445–2453.

Aguirre, A. D., Hsiung, P., Ko, T. H., Hartl, I., and Fujimoto, J. G. 2003. High-resolution optical coherence microscopy for high-speed, in vivo cellular imaging. *Opt. Lett.* 28: 2064–2066.

Akcay, C., Parrein, P., and Rolland, J. P. 2002. Estimation of longitudinal resolution in optical coherence imaging. *Appl. Opt.* 41: 5256–5262.

Alexander, L. and Choate, W. 2004. Optical coherence tomography: Rewriting the standard of care in diagnosis, management and interventional assessment. *Rev. Optometry* 141: 1CE–8CE.

Antcliff, R. J., Spalton, D. J., Stanford, M. R., Graham, E. M., ffytche, T. J. and Marshall, J. 2001. Intravitreal triamcinolone for uveitic cystoid macular edema: An optical coherence tomography study. *Ophthalmology* 108: 765–772.

Antcliff, R. J., Stanford, M. R., Chauhan, D. S., Graham, E. M., Spalton, D. J., Shilling, J. S., Ffytche, T. J. and Marshall, J. 2000. Comparison between optical coherence tomography and fundus fluorescein angiography for the detection of cystoid macular edema in patients with uveitis. *Ophthalmology* 107: 593–599.

Apostolopoulos, M. N., Koutsandrea, C. N., Moschos, M. N., Alonistiotis, D. A., Papaspyrou, A. E., Mallias, J. A., Kyriaki, T. E., Theodossiadis, P. G., and Theodossiadis, G. P. 2002. Evaluation of successful macular hole surgery by optical coherence tomography and multifocal electroretinography. *Am. J. Ophthalmol.* 134: 667–674.

Armstrong, J. J., Leigh, M. S., Sampson, D. D., Walsh, J. H., Hillman, D. R., and Eastwood, P. R. 2006a. Quantitative upper airway imaging with anatomic optical coherence tomography. *Am. J. Respir. Crit. Care Med.* 173: 226–233.

Armstrong, W. B., Ridgway, J. M., Vokes, D. E., Guo, S., Perez, J., Jackson, R. P., Gu, M., Su, J., Crumley, R. L., Shibuya, T. Y., Mahmood, U., Chen, Z. and Wong, B. J. 2006b. Optical coherence tomography of laryngeal cancer. *Laryngoscope* 116: 1107–1113.

Bizheva, K., Povazay, B., Hermann, B., Sattmann, H., Drexler, W., Mei, M., Holzwarth, R., Hoelzenbein, T., Wacheck, V. and Pehamberger, H. 2003. Compact, broad-bandwidth fiber laser for sub-2-microm axial resolution optical coherence tomography in the 1300-nm wavelength region. *Opt. Lett.* 28: 707–709.

Bohringer, H. J., Lankenau, E., Rohde, V., Huttmann, G., and Giese, A. 2006. Optical coherence tomography for experimental neuroendoscopy. *Minim. Invasive Neurosurg.* 49: 269–275.

Boppart, S. A., Bouma, B. E., Pitris, C., Southern, J. F., Brezinski, M. E., and Fujimoto, J. G. 1998a. In vivo cellular optical coherence tomography imaging. *Nat. Med.* 4: 861–865.

Boppart, S. A., Bouma, B. E., Pitris, C., Tearney, G. J., Fujimoto, J. G., and Brezinski, M. E. 1997a. Forward-imaging instruments for optical coherence tomography. *Opt. Lett.* 22: 1618–1620.

Boppart, S. A., Brezinski, M. E., Pitris, C., and Fujimoto, J. G. 1998b. Optical coherence tomography for neurosurgical imaging of human intracortical melanoma. *Neurosurgery* 43: 834–841.

Boppart, S. A., Goodman, A., Libus, J., Pitris, C., Jesser, C. A., Brezinski, M. E., and Fujimoto, J. G. 1999. High resolution imaging of endometriosis and ovarian carcinoma with optical coherence tomography: Feasibility for laparoscopic-based imaging. *Br. J. Obstet. Gynaecol.* 106: 1071–1077.

Boppart, S. A., Luo, W., Marks, D. L., and Singletary, K. W. 2004. Optical coherence tomography: Feasibility for basic research and image-guided surgery of breast cancer. *Breast Cancer Res. Treat.* 84: 85–97.

Boppart, S. A., Tearney, G. J., Bouma, B. E., Southern, J. F., Brezinski, M. E., and Fujimoto, J. G. 1997b. Noninvasive assessment of the developing Xenopus cardiovascular system using optical coherence tomography. *Proc. Natl. Acad. Sci. U. S. A.* 94: 4256–4261.

Born, M. and Wolf, E. 1999. *Principles of Optics*. Cambridge, U.K.: Cambridge University Press.

Bouma, B., Tearney, G. J., Boppart, S. A., Hee, M. R., Brezinski, M. E., and Fujimoto, J. G. 1995. High-resolution optical coherence tomographic imaging using a mode-locked Ti:Al$_2$O$_3$ laser source. *Opt. Lett.* 20: 1486–1488.

Bouma, B. E., Tearney, G. J., Bilinsky, I. P., Golubovic, B., and Fujimoto, J. G. 1996. Self-phase-modulated Kerr-lens mode-locked Cr:forsterite laser source for optical coherence tomography. *Opt. Lett.* 21: 1839–1841.

Bouma, B. E., Tearney, G. J., Compton, C. C., and Nishioka, N. S. 2000. High-resolution imaging of the human esophagus and stomach in vivo using optical coherence tomography. *Gastrointest. Endosc.* 51: 467–474.

Bouma, B. E., Tearney, G. J., Yabushita, H., Shishkov, M., Kauffman, C. R., DeJoseph Gauthier, D., MacNeill, B. D., Houser, S. L., Aretz, H. T., Halpern, E. F., and Jang, I. K. 2003. Evaluation of intracoronary stenting by intravascular optical coherence tomography. *Heart* 89: 317–320.

Bourquin, S., Aguirre, A. D., Hartl, I., Hsiung, P., Ko, T. H., Fujimoto, J. G., Birks, T. A., Wadsworth, W. J., Bunting, U., and Kopf, D. 2003. Ultrahigh resolution real time OCT imaging using a compact femtosecond Nd: Glass laser and nonlinear fiber. *Opt. Express* 11: 3290–3297.

Brewer, M. A., Utzinger, U., Barton, J. K., Hoying, J. B., Kirkpatrick, N. D., Brands, W. R., Davis, J. R., Hunt, K., Stevens, S. J., and Gmitro, A. F. 2004. Imaging of the ovary. *Technol. Cancer Res. Treat.* 3: 617–627.

Brezinski, M. E. and Fujimoto, J. G. 1999. Optical coherence tomography: High-resolution imaging in nontransparent tissue. *IEEE J. Sel. Top. Quantum Electron.* 5: 1185–1192.

Brezinski, M. E., Tearney, G. J., Bouma, B. E., Boppart, S. A., Hee, M. R., Swanson, E. A., Southern, J. F., and Fujimoto, J. G. 1996a. Imaging of coronary artery microstructure (in vitro) with optical coherence tomography. *Am. J. Cardiol.* 77: 92–93.

Brezinski, M. E., Tearney, G. J., Bouma, B. E., Izatt, J. A., Hee, M. R., Swanson, E. A., Southern, J. F., and Fujimoto, J. G. 1996b. Optical coherence tomography for optical biopsy. Properties and demonstration of vascular pathology. *Circulation* 93: 1206–1213.

Brezinski, M. E., Tearney, G. J., Weissman, N. J., Boppart, S. A., Bouma, B. E., Hee, M. R., Weyman, A. E., Swanson, E. A., Southern, J. F., and Fujimoto, J. G. 1997. Assessing atherosclerotic plaque morphology: Comparison of optical coherence tomography and high frequency intravascular ultrasound. *Heart* 77: 397–403.

Bruckner, A. P. 1978. Picosecond light-scattering measurements of cataract microstructure. *Appl. Opt.* 17: 3177–3183.

Carl Zeiss Meditec. 2006. Carl Zeiss Meditec announces installation of 6,000th Stratus OCT. Dublin, CA: Carl Zeiss Meditec, Inc.

Cense, B., Nassif, N., Chen, T. C., Pierce, M. C., Yun, S., Park, B. H., Bouma, B., Tearney, G., and de Boer, J. F. 2004. Ultrahigh-resolution high-speed retinal imaging using spectral-domain optical coherence tomography. *Opt. Express* 12: 2435–2447.

Chau, A. H., Chan, R. C., Shishkov, M., MacNeill, B., Iftimia, N., Tearney, G. J., Kamm, R. D., Bouma, B. E., and Kaazempur-Mofrad, M. R. 2004. Mechanical analysis of atherosclerotic plaques based on optical coherence tomography. *Ann. Biomed. Eng.* 32: 1494–1503.

Chen, Y., Aguirre, A. D., Hsiung, P. L., Desai, S., Herz, P. R., Pedrosa, M., Huang, Q., Figueiredo, M., Huang, S. W., Koski, A., Schmitt, J. M., Fujimoto, J. G., and Mashimo, H. 2007. Ultrahigh resolution optical coherence tomography of Barrett's esophagus: Preliminary descriptive clinical study correlating images with histology. *Endoscopy* 39: 599–605.

Chinn, S. R., Swanson, E. A., and Fujimoto, J. G. 1997. Optical coherence tomography using a frequency-tunable optical source. *Opt. Lett.* 22: 340–342.

Choma, M. A., Sarunic, M. V., Yang, C. H., and Izatt, J. A. 2003. Sensitivity advantage of swept source and Fourier domain optical coherence tomography. *Opt. Express* 11: 2183–2189.

Clivaz, X., Marquisweible, F., Salathe, R. P., Novak, R. P., and Gilgen, H. H. 1992. High-resolution reflectometry in biological tissues. *Opt. Lett.* 17: 4–6.

Costa, R. A., Skaf, M., Melo, L. A. Jr., Calucci, D., Cardillo, J. A., Castro, J. C., Huang, D., and Wojtkowski, M. 2006. Retinal assessment using optical coherence tomography. *Prog. Retin. Eye Res.* 25: 325–353.

D'Amico, A. V., Weinstein, M., Li, X., Richie, J. P., and Fujimoto, J. 2000. Optical coherence tomography as a method for identifying benign and malignant microscopic structures in the prostate gland. *Urology* 55: 783–787.

de Boer, J. F., Cense, B., Park, B. H., Pierce, M. C., Tearney, G. J., and Bouma, B. E. 2003. Improved signal-to-noise ratio in spectral-domain compared with time-domain optical coherence tomography. *Opt. Lett.* 28: 2067–2069.

de Melo, L. S., de Araujo, R. E., Freitas, A. Z., Zezell, D., Vieira, N. D., Girkin, J., Hall, A., Carvalho, M. T., and Gomes, A. S. 2005. Evaluation of enamel dental restoration interface by optical coherence tomography. *J. Biomed. Opt.* 10: 064027.

Dogra, M. R., Gupta, A., and Gupta, V. 2004. *Atlas of Optical Coherence Tomography of Macular Diseases.* New York: Taylor & Francis.

Drexler, W. 2004. Ultrahigh-resolution optical coherence tomography. *J. Biomed. Opt.* 9: 47–74.

Drexler, W., Hitzenberger, C. K., Baumgartner, A., Findl, O., Sattmann, H., and Fercher, A. F. 1998. Investigation of dispersion effects in ocular media by multiple wavelength partial coherence interferometry. *Exp. Eye Res.* 66: 25–33.

Drexler, W., Morgner, U., Ghanta, R. K., Kärtner, F. X., Schuman, J. S., and Fujimoto, J. G. 2001. Ultrahigh-resolution ophthalmic optical coherence tomography. *Nat. Med.* 7: 502–507.

Drexler, W., Morgner, U., Kartner, F. X., Pitris, C., Boppart, S. A., Li, X. D., Ippen, E. P., and Fujimoto, J. G. 1999. In vivo ultrahigh-resolution optical coherence tomography. *Opt. Lett.* 24: 1221–1223.

Duguay, M. A. 1971. Light photographed in flight. *Am. Sci.* 59: 551–556.

Duguay, M. A. and Hansen, J. W. 1968. Optical sampling of subnanosecond light pulses. *Appl. Phys. Lett.* 13: 178–180.

Duguay, M. A. and Hansen, J. W. 1969. An ultrafast light gate. *Appl. Phys. Lett.* 15: 192.

Duguay, M. A. and Hansen, J. W. 1971. Ultrahigh-speed photography of picosecond light pulses. *IEEE J. Quantum Electron.* Qe 7: 37–39.

Duguay, M. A. and Mattick, A. T. 1971. Ultrahigh speed photography of picosecond light pulses and echoes. *Appl. Opt.* 10: 2162–2170.

Evans, J. A., Poneros, J. M., Bouma, B. E., Bressner, J., Halpern, E. F., Shishkov, M., Lauwers, G. Y., Mino-Kenudson, M., Nishioka, N. S., and Tearney, G. J. 2006. Optical coherence tomography to identify intramucosal carcinoma and high-grade dysplasia in Barrett's esophagus. *Clin. Gastroenterol. Hepatol.* 4: 38–43.

Fercher, A. F., Drexler, W., Hitzenberger, C. K., and Lasser, T. 2003. Optical coherence tomography-principles and applications. *Rep. Prog. Phys.* 66: 239–303.

Fercher, A. F., Hitzenberger, C. K., Drexler, W., Kamp, G., and Sattmann, H. 1993. In vivo optical coherence tomography. *Am. J. Ophthalmol.* 116: 113–114.

Fercher, A. F., Hitzenberger, C. K., Kamp, G., and Elzaiat, S. Y. 1995. Measurement of intraocular distances by backscattering spectral interferometry. *Opt. Commun.* 117: 43–48.

Fercher, A. F., Mengedoht, K., and Werner, W. 1988. Eye-length measurement by interferometry with partially coherent light. *Opt. Lett.* 13: 1867–1869.

Freitas, A. Z., Zezell, D. M., Vieira, N. D., Ribeiro, A. C., and Gomes, A. S. L. 2006. Imaging carious human dental tissue with optical coherence tomography. *J. Appl. Phys.* 99: 024906.

Fujimoto, J. G. 2002. Optical coherence tomography: Introduction (Chapter 1) In *Handbook of Optical Coherence Tomography*, eds. B. E. Bouma and G. J. Tearney. New York: Marcel Dekker, Inc.

Fujimoto, J. G. 2003. Optical coherence tomography for ultrahigh resolution in vivo imaging. *Nat. Biotechnol.* 21: 1361–1367.

Fujimoto, J. G., Boppart, S. A., Tearney, G. J., Bouma, B. E., Pitris, C., and Brezinski, M. E. 1999. High resolution in vivo intra-arterial imaging with optical coherence tomography. *Heart* 82: 128–133.

Fujimoto, J. G., Brezinski, M. E., Tearney, G. J., Boppart, S. A., Bouma, B., Hee, M. R., Southern, J. F., and Swanson, E. A. 1995. Optical biopsy and imaging using optical coherence tomography. *Nat. Med.* 1: 970–972.

Fujimoto, J. G., De Silvestri, S., Ippen, E. P., Puliafito, C. A., Margolis, R., and Oseroff, A. 1986. Femtosecond optical ranging in biological systems. *Opt. Lett.* 11: 150–153.

Fujimoto, J. G., Pitris, C., Boppart, S. A., and Brezinski, M. E. 2000. Optical coherence tomography: An emerging technology for biomedical imaging and optical biopsy. *Neoplasia* 2: 9–25.

Gaudric, A., Haouchine, B., Massin, P., Paques, M., Blain, P., and Erginay, A. 1999. Macular hole formation: New data provided by optical coherence tomography. *Arch. Ophthalmol.* 117: 744–751.

Gilgen, H. H., Novak, R. P., Salathe, R. P., Hodel, W. and Beaud, P. 1989. Submillimeter optical reflectometry. *IEEE J. Lightw. Technol.* 7: 1225–1233.

Goldberg, B. D., Iftimia, N. V., Bressner, J. E., Pitman, M. B., Halpern, E., Bouma, B. E., and Tearney, G. J. 2008. Automated algorithm for differentiation of human breast tissue using low coherence interferometry for fine needle aspiration biopsy guidance. *J. Biomed. Opt.* 13: 014014.

Golubovic, B., Bouma, B. E., Tearney, G. J., and Fujimoto, J. G. 1997. Optical frequency-domain reflectometry using rapid wavelength tuning of a Cr4+:forsterite laser. *Opt. Lett.* 22: 1704–1706.

Han, S., Sarunic, M. V., Wu, J., Humayun, M., and Yang, C. H. 2008. Handheld forward-imaging needle endoscope for ophthalmic optical coherence tomography inspection. *J. Biomed. Opt.* 13: 020505.

Hariri, L. P., Bonnema, G. T., Schmidt, K., Winkler, A. M., Korde, V., Hatch, K. D., Davis, J. R., Brewer, M. A., and Barton, J. K. 2009. Laparoscopic optical coherence tomography imaging of human ovarian cancer. *Gynecol. Oncol.* 114: 188–194.

Hartl, I., Li, X. D., Chudoba, C., Hganta, R. K., Ko, T. H., Fujimoto, J. G., Ranka, J. K., and Windeler, R. S. 2001. Ultrahigh-resolution optical coherence tomography using continuum generation in an air-silica microstructure optical fiber. *Opt. Lett.* 26: 608–610.

Hee, M. R. 2002. Optical coherence tomography: Theory, Chapter 2. In: Bouma, B.E and Tearney, G.J. (Eds.), *Handbook of Optical Coherence Tomography*, New York: Marcel Dekker, Inc.

Hee, M. R., Izatt, J. A., Swanson, E. A., Huang, D., Schuman, J. S., Lin, C. P., Puliafito, C. A., and Fujimoto, J. G. 1995a. Optical coherence tomography for ophthalmic imaging: New technique delivers micron-scale resolution. *IEEE Eng. Med. Biol. Mag.* 14: 67–76.

Hee, M. R., Izatt, J. A., Swanson, E. A., Huang, D., Schuman, J. S., Lin, C. P., Puliafito, C. A., and Fujimoto, J. G. 1995b. Optical coherence tomography of the human retina. *Arch. Ophthalmol.* 113: 325–332.

Hee, M. R., Puliafito, C. A., Duker, J. S., Reichel, E., Coker, J. G., Wilkins, J. R., Schuman, J. S., Swanson, E. A., and Fujimoto, J. G. 1998. Topography of diabetic macular edema with optical coherence tomography. *Ophthalmology* 105: 360–370.

Hee, M. R., Puliafito, C. A., Wong, C., Duker, J. S., Reichel, E., Rutledge, B., Schuman, J. S., Swanson, E. A., and Fujimoto, J. G. 1995c. Quantitative assessment of macular edema with optical coherence tomography. *Arch. Ophthalmol.* 113: 1019–1029.

Herz, P. R., Chen, Y., Aguirre, A. D., Fujimoto, J. G., Mashimo, H., Schmitt, J., Koski, A., Goodnow, J., and Petersen, C. 2004a. Ultrahigh resolution optical biopsy with endoscopic optical coherence tomography. *Opt. Express* 12: 3532–3542.

Herz, P. R., Chen, Y., Aguirre, A. D., Schneider, K., Hsiung, P., Fujimoto, J. G., Madden, K., Schmitt, J., Goodnow, J., and Petersen, C. 2004b. Micromotor endoscope catheter for in vivo, ultrahigh-resolution optical coherence tomography. *Opt. Lett.* 29: 2261–2263.

Hitzenberger, C. K. 1991. Optical measurement of the axial eye length by laser Doppler interferometry. *Investig. Ophthalmol. Vis. Sci.* 32: 616–624.

Hitzenberger, C. K. 1992. Measurement of corneal thickness by low-coherence interferometry. *Appl. Opt.* 31: 6637–6642.

Hitzenberger, C. K., Baumgartner, A., Drexler, W., and Fercher, A. F. 1999. Dispersion effects in partial coherence interferometry: Implications for intraocular ranging. *J. Biomed. Opt.* 4: 144–151.

Hsiung, P. L., Phatak, D. R., Chen, Y., Aguirre, A. D., Fujimoto, J. G., and Connolly, J. L. 2007. Benign and malignant lesion in the human breast depicted with ultrahigh resolution and dimensional optical coherence tomography. *Radiology* 244: 865–874.

Huang, D., Swanson, E. A., Lin, C. P., Schuman, J. S., Stinson, W. G., Chang, W., Hee, M. R., Flotte, T., Gregory, K., Puliafito, C. A., and Fujimoto, J. G. 1991a. Optical coherence tomography. *Science* 254: 1178–1181.

Huang, D., Wang, J., Lin, C. P., Puliafito, C. A., and Fujimoto, J. G. 1991b. Micron-resolution ranging of cornea and anterior chamber by optical reflectometry. *Lasers Surg. Med.* 11: 419–425.

Huber, R., Wojtkowski, M., and Fujimoto, J. G. 2006. Fourier Domain Mode Locking (FDML): A new laser operating regime and applications for optical coherence tomography. *Opt. Express* 14: 3225–3237.

Iftimia, N. V., Mujat, M., Ustun, T., Ferguson, R. D., Danthu, V., and Hammer, D. X. 2009. Spectral-domain low coherence interferometry/optical coherence tomography system for fine needle breast biopsy guidance. *Rev. Sci. Instrum.* 80: 024302.

Isenberg, G., Sivak, M. V., Chak, A., Wong, R. C. K., Willis, J. E., Wolf, B., Rowland, D. Y., Das, A., and Rollins, A. 2005. Accuracy of endoscopic optical coherence tomography in the detection of dysplasia in Barrett's esophagus: A prospective, double-blinded study. *Gastrointest. Endosc.* 62: 825–831.

Izatt, J. A., Hee, M. R., Owen, G. M., Swanson, E. A., and Fujimoto, J. G. 1994a. Optical coherence microscopy in scattering media. *Opt. Lett.* 19: 590–592.

Izatt, J. A., Hee, M. R., Swanson, E. A., Lin, C. P., Huang, D., Schuman, J. S., Puliafito, C. A., and Fujimoto, J. G. 1994b. Micrometer-scale resolution imaging of the anterior eye in vivo with optical coherence tomography. *Arch. Ophthalmol.* 112: 1584–1589.

Izatt, J. A., Kulkarni, M. D., Wang, H.-W., Kobayashi, K., and Sivak, M. V. Jr. 1996. Optical coherence tomography and microscopy in gastrointestinal tissues. *IEEE J. Sel. Top. Quantum Electron.* 2: 1017–1028.

Jäckle, S., Gladkova, N., Feldchtein, F., Terentieva, A., Brand, B., Gelikonov, G., Gelikonov, V., Sergeev, A., Fritscher-Ravens, A., Freund, J., Seitz, U., Soehendra, S., and Schrödern, N. 2000. In vivo endoscopic optical coherence tomography of the human gastrointestinal tract—Toward optical biopsy. *Endoscopy* 32: 743–749.

Jafri, M. S., Farhang, S., Tang, R. S., Desai, N., Fishman, P. S., Rohwer, R. G., Tang, C. M., and Schmitt, J. M. 2005. Optical coherence tomography in the diagnosis and treatment of neurological disorders. *J. Biomed. Opt.* 10: 051603.

Jain, A., Kopa, A., Pan, Y. T., Fedder, G. K., and Xie, H. K. 2004. A two-axis electrothermal micromirror for endoscopic optical coherence tomography. *IEEE J. Sel. Top. Quantum Electron.* 10: 636–642.

Jang, I. K., Bouma, B. E., Kang, D. H., Park, S. J., Park, S. W., Seung, K. B., Choi, K. B., Shishkov, M., Schlendorf, K., Pomerantsev, E., Houser, S. L., Aretz, H. T., and Tearney, G. J. 2002. Visualization of coronary atherosclerotic plaques in patients using optical coherence tomography: Comparison with intravascular ultrasound. *J. Am. Coll. Cardiol.* 39: 604–609.

Jang, I. K., Tearney, G., and Bouma, B. 2001. Visualization of tissue prolapse between coronary stent struts by optical coherence tomography: Comparison with intravascular ultrasound. *Circulation* 104: 2754.

Jang, I. K., Tearney, G. J., MacNeill, B., Takano, M., Moselewski, F., Iftima, N., Shishkov, M., Houser, S., Aretz, H. T., Halpern, E. F., and Bouma, B. E. 2005. In vivo characterization of coronary atherosclerotic plaque by use of optical coherence tomography. *Circulation* 111: 1551–1555.

Jenkins, M. W., Rothenberg, F., Roy, D., Nikolski, V. P., Hu, Z., Watanabe, M., Wilson, D. L., Efimov, I. R., and Rollins, A. M. 2006. 4D embryonic cardiography using gated optical coherence tomography. *Opt. Express* 14: 736–748.

Jester, J. V., Andrews, P. M., Petroll, W. M., Lemp, M. A., and Cavanagh, H. D. 1991. In vivo, real-time confocal imaging. *J. Electron. Microsc. Tech.* 18: 50–60.

Jumper, J. M., Gallemore, R. P., McCuen, B. W. II, and Toth, C. A. 2000. Features of macular hole closure in the early postoperative period using optical coherence tomography. *Retina* 20: 232–237.

Jung, W., McCormick, D. T., Zhang, J., Wang, L., Tien, N. C., and Chen, Z. P. 2006. Three-dimensional endoscopic optical coherence tomography by use of a two-axis micro-electromechanical scanning mirror. *Appl. Phys. Lett.* 88: 163901.

Ko, T. H., Witkin, A. J., Fujimoto, J. G., Chan, A., Rogers, A. H., Baumal, C. R., Schuman, J. S., Drexler, W., Reichel, E., and Duker, J. S. 2006. Ultrahigh-resolution optical coherence tomography of surgically closed macular holes. *Arch. Ophthalmol.* 124: 827–836.

Kobayashi, K., Izatt, J. A., Kulkarni, M. D., Willis. J., and Sivak, M. V. Jr. 1998. High-resolution cross-sectional imaging of the gastrointestinal tract using optical coherence tomography: Preliminary results. *Gastrointest. Endosc.* 47: 515–523.

Kompfner, R. and Park, H. 1976. High-resolution heterodyne coincidence detection of optical pulse streams. *Int. J. Electron.* 41: 317–323.

Kulkarni, M. D., Thomas, C. W., and Izatt, J. A. 1997. Image enhancement in optical coherence tomography using deconvolution. *Electron. Lett.* 33: 1365–1367.

Leitgeb, R., Hitzenberger, C. K., and Fercher, A. F. 2003. Performance of Fourier domain vs. time domain optical coherence tomography. *Opt. Express* 11: 889–894.

Leitgeb, R. A., Drexler, W., Unterhuber, A., Hermann, B., Bajraszewski, T., Le, T., Stingl, A., and Fercher, A. F. 2004. Ultrahigh resolution Fourier domain optical coherence tomography. *Opt. Express* 12: 2156–2165.

Li, Q., Onozato, M. L., Andrews, P. M., Chen, C. W., Paek, A., Naphas, R., Yuan, S., Jiang, J., Cable, A., and Chen, Y. 2009. Automated quantification of microstructural dimensions of the human kidney using optical coherence tomography (OCT). *Opt. Express* 17: 16000–16016.

Li, X. D., Chudoba, C., Ko, T., Pitris, C., and Fujimoto, J. G. 2000a. Imaging needle for optical coherence tomography. *Opt. Lett.* 25: 1520–1522.

Li, X. D., Martin, S., Pitris, C., Ghanta, R., Stamper, D. L., Harman, M., Fujimoto, J. G., and Brezinski, M. E. 2005. High-resolution optical coherence tomographic imaging of osteoarthritic cartilage during open knee surgery. *Arthritis Res. Ther.* 7: R318–R323.

Li, X. D., Boppart, S. A., Van Dam, J., Mashimo, H., Mutinga, M., Drexler, W., Klein, M., Pitris, C., Krinsky, M. L., Brezinski, M. E., and Fujimoto, J. G. 2000b. Optical coherence tomography: Advanced technology for the endoscopic imaging of Barrett's esophagus. *Endoscopy* 32: 921–930.

Liu, X. M., Cobb, M. J., Chen, Y. C., Kimmey, M. B., and Li, X. D. 2004. Rapid-scanning forward-imaging miniature endoscope for real-time optical coherence tomography. *Opt. Lett.* 29: 1763–1765.

Luo, W., Marks, D. L., Ralston, T. S., and Boppart, S. A. 2006. Three-dimensional optical coherence tomography of the embryonic murine cardiovascular system. *J. Biomed. Opt.* 11: 021014.

Makita, S., Fabritius, T., and Yasuno, Y. 2008. Full-range, high-speed, high-resolution 1-mu m spectral-domain optical coherence tomography using BM-scan for volumetric imaging of the human posterior eye. *Opt. Express* 16: 8406–8420.

Manyak, M. J., Gladkova, N. D., Makari, J. H., Schwartz, A. M., Zagaynova, E. V., Zolfaghari, L., Zara, J. M., Iksanov, R., and Feldchtein, F. I. 2005. Evaluation of superficial bladder transitional-cell carcinoma by optical coherence tomography. *J. Endourol.* 19: 570–574.

Masters, B. R. 1999. Early development of optical low-coherence reflectometry and some recent biomedical applications. *J. Biomed. Opt.* 4: 236–247.

Morton, E. J., Webb, S., Bateman, J. E., Clarke, L. J., and Shelton, C. G. 1990. Three-dimensional x-ray microtomography for medical and biological applications. *Phys. Med. Biol.* 35: 805–820.

Mujat, M., Ferguson, R. D., Hammer, D. X., Gittins, C., and Iftimia, N. 2009. Automated algorithm for breast tissue differentiation in optical coherence tomography. *J. Biomed. Opt.* 14: 034040.

Nassif, N., Cense, B., Park, B. H., Yun, S. H., Chen, T. C., Bouma, B. E., Tearney, G. J., and de Boer, J. F. 2004. In vivo human retinal imaging by ultrahigh-speed spectral domain optical coherence tomography. *Opt. Lett.* 29: 480–482.

Nguyen, F. T., Zysk, A. M., Chaney, E. J., Kotynek, J. G., Oliphant, U. J., Bellafiore, F. J., Rowland, K. M., Johnson, P. A., and Boppart, S. A. 2009. Intraoperative evaluation of breast tumor margins with optical coherence tomography. *Cancer Res.* 69: 8790–8796.

Oh, W. Y., Yun, S. H., Tearney, G. J., and Bouma, B. E. 2005. 115 kHz tuning repetition rate ultrahigh-speed wavelength-swept semiconductor laser. *Opt. Lett.* 30: 3159–3161.

Oh, W. Y., Yun, S. H., Vakoc, B. J., Tearney, G. J. and Bouma, B. E. 2006. Ultrahigh-speed optical frequency domain imaging and application to laser ablation monitoring. *Appl. Phys. Lett.* 88: 103902.

Onozato, M. L., Andrews, P. M., Li, Q., Jiang, J., Cable, A., and Chen, Y. 2010. Optical coherence tomography of human kidney. *J. Urol.* 183: 2090–2094.

Otis, L. L., al-Sadhan, R. I., Meiers, J., and Redford-Badwal, D. 2003. Identification of occlusal sealants using optical coherence tomography. *J. Clin. Dentist.* 14: 7–10.

Otis, L. L., Colston, B. W. Jr., Everett, M. J., and Nathel, H. 2000a. Dental optical coherence tomography: A comparison of two in vitro systems. *Dento Maxillofac. Radiol.* 29: 85–89.

Otis, L. L., Everett, M. J., Sathyam, U. S., and Colston, B. W. Jr. 2000b. Optical coherence tomography: A new imaging technology for dentistry. *J. Am. Dental Assoc.* 131: 511–514.

Pan, Y., Birngruber, R., Rosperich, J., and Engelhardt, R. 1995. Low-coherence optical tomography in turbid tissue: Theoretical analysis. *Appl. Opt.* 34: 6564–6574.

Pan, Y., Lankenou, E., Welzel, J., Birngruber, R., and Engelhardt, R. 1996. Optical coherence-gated imaging of biological tissues. *IEEE J. Sel. Top. Quantum Electron.* 2: 1029–1034.

Pan, Y., Li, Z., Xie, T., and Chu, C. R. 2003. Hand-held arthroscopic optical coherence tomography for in vivo high-resolution imaging of articular cartilage. *J. Biomed. Opt.* 8: 648–654.

Pan, Y., Xie, H., and Fedder, G. K. 2001. Endoscopic optical coherence tomography based on a microelectromechanical mirror. *Opt. Lett.* 26: 1966–1968.

Park, H., Chodorow, M., and Kompfner, R. 1981. High-resolution optical ranging system. *Appl. Opt.* 20: 2389–2394.

Pasquesi, J. J., Schlachter, S. C., Boppart, M. D., Chaney, E., Kaufman, S. J., and Boppart, S. A. 2006. In vivo detection of exercised-induced ultrastructural changes in genetically-altered murine skeletal muscle using polarization-sensitive optical coherence tomography. *Opt. Express* 14: 1547–1556.

Pfau, P. R., Sivak, M. V., Chak, A., Kinnard, M., Wong, R. C. K., Isenberg, G. A., Izatt, J. A., Rollins, A., and Westphal, V. 2003. Criteria for the diagnosis of dysplasia by endoscopic optical coherence tomography. *Gastrointest. Endosc.* 58: 196–202.

Pitris, C., Goodman, A., Boppart, S. A., Libus, J. J., Fujimoto, J. G., and Brezinski, M. E. 1999. High-resolution imaging of gynecologic neoplasms using optical coherence tomography. *Obstet. Gynecol.* 93: 135–139.

Pitris, C., Jesser, C., Boppart, S. A., Stamper, D., Brezinski, M. E., and Fujimoto, J. G. 2000. Feasibility of optical coherence tomography for high-resolution imaging of human gastrointestinal tract malignancies. *J. Gastroenterol.* 35: 87–92.

Podoleanu, A. G. and Jackson, D. A. 1999. Noise analysis of a combined optical coherence tomograph and a confocal scanning ophthalmoscope. *Appl. Opt.* 38: 2116–2127.

Poneros, J. M., Brand, S., Bouma, B. E., Tearney, G. J., Compton, C. C., and Nishioka, N. S. 2001. Diagnosis of specialized intestinal metaplasia by optical coherence tomography. *Gastroenterology* 120: 7–12.

Poneros, J. M., Tearney, G. J., Shiskov, M., Kelsey, P. B., Lauwers, G. Y., Nishioka, N. S., and Bouma, B. E. 2002. Optical coherence tomography of the biliary tree during ERCP. *Gastrointest. Endosc.* 55: 84–88.

Potsaid, B., Gorczynska, I., Srinivasan, V. J., Chen, Y. L., Jiang, J., Cable, A., and Fujimoto, J. G. 2008. Ultrahigh speed spectral/Fourier domain OCT ophthalmic imaging at 70,000 to 312,500 axial scans per second. *Opt. Express* 16: 15149–15169.

Puliafito, C. A., Hee, M. R., Schuman, J. S., and Fujimoto, J. G. 1995. *Optical Coherence Tomography of Ocular Diseases.* Thorofare, NJ: Slack, Inc.

Puliafito, C. A., Hee, M. R., Schuman, J. S., and Fujimoto, J. G. 1996. *Optical Coherence Tomography of Ocular Diseases.* Thorofare, NJ: Slack Inc.

Rollins, A. M., Kulkarni, M. D., Yazdanfar, S., Ung-arunyawee, R., and Izatt, J. A. 1998. In vivo video rate optical coherence tomography. *Opt. Express* 3(6): 219–229.

Ruggeri, M., Webbe, H., Jiao, S. L., Gregori, G., Jockovich, M. E., Hackam, A., Duan, Y. L., and Puliafito, C. A. 2007. In vivo three-dimensional high-resolution imaging of rodent retina with spectral-domain optical coherence tomography. *Investig. Ophthalmol. Vis. Sci.* 48: 1808–1814.

Sato, H., Kawasaki, R., and Yamashita, H. 2003. Observation of idiopathic full-thickness macular hole closure in early postoperative period as evaluated by optical coherence tomography. *Am. J. Ophthalmol.* 136: 185–187.

Schmitt, J. M. 1999. Optical coherence tomography (OCT): A review. *IEEE J. Sel. Top. Quantum Electron.* 5: 1205–1215.

Schmitt, J. M., Knuttel, A., and Bonner, R. F. 1993. Measurement of optical-properties of biological tissues by low-coherence reflectometry. *Appl. Opt.* 32: 6032–6042.

Schuman, J. S., Puliafito, C. A., and Fujimoto, J. G. 2004. *Optical Coherence Tomography of Ocular Diseases*, 2nd edn. Thorofare, NJ: Slack Inc.

Sergeev, A. M., Gelikonov, V. M., Gelikonov, G. V., Feldchtein, F. I., Kuranov, R. V., Gladkova, N. D., Shakhova, N. M., Suopova, L. B., Shakhov, A. V., Kuznetzova, I. A., Denisenko, A. N., Pochinko, V. V., Chumakov, Y. P., and Streltzova, O. S. 1997. In vivo endoscopic OCT imaging of precancer and cancer states of human mucosa. *Opt. Express* 1: 432–440.

Shen, B., Zuccaro, G. Jr., Gramlich, T. L., Gladkova, N., Trolli, P., Kareta, M., Delaney, C. P., Connor, J. T., Lashner, B. A., Bevins, C. L., Feldchtein, F., Remzi, F. H., Bambrick, M. L.,

and Fazio, V. W. 2004. In vivo colonoscopic optical coherence tomography for transmural inflammation in inflammatory bowel disease. *Clin. Gastroenterol. Hepatol.* 2: 1080–1087.

Sivak, M. V., Jr., Kobayashi, K., Izatt, J. A., Rollins, A. M., Ung-Runyawee, R., Chak, A., Wong, R. C., Isenberg, G. A., and Willis, J. 2000. High-resolution endoscopic imaging of the GI tract using optical coherence tomography. *Gastrointest. Endosc.* 51: 474–479.

Srinivasan, V. J., Wojtkowski, M., Witkin, A. J., Duker, J. S., Ko, T. H., Carvalho, M., Schuman, J. S., Kowalczyk, A., and Fujimoto, J. G. 2006. High-definition and 3-dimensional imaging of macular pathologies with high-speed ultrahigh-resolution optical coherence tomography. *Ophthalmology* 113: 2054–2065.

Swanson, E. A., Huang, D., Hee, M. R., Fujimoto, J. G., Lin, C. P., and Puliafito, C. A. 1992. High-speed optical coherence domain reflectometry. *Opt. Lett.* 17: 151–153.

Swanson, E. A., Izatt, J. A., Hee, M. R., Huang, D., Lin, C. P., Schuman, J. S., Puliafito, C. A., and Fujimoto, J. G. 1993. In vivo retinal imaging by optical coherence tomography. *Opt. Lett.* 18: 1864–1866.

Takada, K., Yokohama, I., Chida, K., and Noda, J. 1987. New measurement system for fault location in optical waveguide devices based on an interferometric technique. *Appl. Opt.* 26: 1603–1608.

Tearney, G. J., Boppart, S. A., Bouma, B. E., Brezinski, M. E., Weissman, N. J., Southern, J. F., and Fujimoto, J. G. 1996a. Scanning single-mode fiber optic catheter-endoscope for optical coherence tomography. *Opt. Lett.* 21: 543–545.

Tearney, G. J., Bouma, B. E., and Fujimoto, J. G. 1997a. High-speed phase- and group-delay scanning with a grating-based phase control delay line. *Opt. Lett.* 22: 1811–1813.

Tearney, G. J., Brezinski, M. E., Boppart, S. A., Bouma, B. E., Weissman, N., Southern, J. F., Swanson, E. A., and Fujimoto, J. G. 1996b. Catheter-based optical imaging of a human coronary artery. *Circulation* 94: 3013.

Tearney, G. J., Brezinski, M. E., Bouma, B. E., Boppart, S. A., Pitvis, C., Southern, J. F., and Fujimoto, J. G. 1997b. In vivo endoscopic optical biopsy with optical coherence tomography. *Science* 276: 2037–2039.

Tearney, G. J., Brezinski, M. E., Southern, J. F., Bouma, B. E., Boppart, S. A., and Fujimoto, J. G. 1997c. Optical biopsy in human gastrointestinal tissue using optical coherence tomography. *Am. J. Gastroenterol.* 92: 1800–1804.

Tearney, G. J., Brezinski, M. E., Southern, J. F., Bouma, B. E., Boppart, S. A., and Fujimoto, J. G. 1997d. Optical biopsy in human urologic tissue using optical coherence tomography. *J. Urol.* 157: 1915–1919.

Tearney, G. J., Yabushita, H., Houser, S. L., Aretz, H. T., Jang, I. K., Schlendorf, K. H., Kauffman, C. R., Shishkov, M., Halpern, E. F., and Bouma, B. E. 2003. Quantification of macrophage content in atherosclerotic plaques by optical coherence tomography. *Circulation* 107: 113–119.

Tran, P. H., Mukai, D. S., Brenner, M., and Chen, Z. P. 2004. In vivo endoscopic optical coherence tomography by use of a rotational microelectromechanical system probe. *Opt. Lett.* 29: 1236–1238.

Tsai, M. T., Lee, H. C., Lu, C. W., Wang, Y. M., Lee, C. K., Yang, C. C., and Chiang, C. P. 2008. Delineation of an oral cancer lesion with swept-source optical coherence tomography. *J. Biomed. Opt.* 13: 044012.

Tumlinson, A. R., Hariri, L. P., Utzinger, U., and Barton, J. K. 2004. Miniature endoscope for simultaneous optical coherence tomography and laser-induced fluorescence measurement. *Appl. Opt.* 43: 113–121.

Vabre, L., Dubois, A., Beaurepaire, E., and Boccara, C. 2000. Optical coherence microscopy for high resolution biological imaging. *Proc. SPIE* 4160: 24–30.

Vakoc, B. J., Shishko, M., Yun, S. H., Oh, W. Y., Suter, M. J., Desjardins, A. E., Evans, J. A., Nishioka, N. S., Tearney, G. J., and Bouma, B. E. 2007a. Comprehensive esophageal microscopy by using optical frequency-domain imaging (with video). *Gastrointest. Endosc.* 65: 898–905.

Vakoc, B. J., Tearney, G. J., and Bouma, B. E. 2007b. Real-time microscopic visualization of tissue response to laser thermal therapy. *J. Biomed. Opt.* 12: 020501–020503.

Vizzeri, G., Balasubramanian, M., Bowd, C., Weinreb, R. N., Medeiros, F. A., and Zangwill, L. M. 2009. Spectral domain-optical coherence tomography to detect localized retinal nerve fiber layer defects in glaucomatous eyes. *Opt. Express* 17: 4004–4018.

Walz, M. 2006. Hot technologies for 2007: OCT: Imaging of the future. *R&D Mag.* 6.

Wang, R. K. 1999. Resolution improved optical coherence-gated tomography for imaging through biological tissues. *J. Mod. Opt.* 46: 1905–1912.

Wang, R. K. and Tuchin, V. V. 2004. Optical coherence tomography—light scattering and imaging enhancement. In: *Biomedical Diagnostics, Environmental and Material Science*, Volume 2, Part IV (Optical Coherence Tomography), Chapter 13, ed. V. V. Tuchin. Boston, MA: Kluwer Academic Publishers.

Wang, Y., Zhao, Y., Nelson, J. S., Chen, Z., and Windeler, R. S. 2003. Ultrahigh-resolution optical coherence tomography by broadband continuum generation from a photonic crystal fiber. *Opt. Lett.* 28: 182–184.

Welzel, J., Lankenau, E., Birngruber, R., and Engelhardt, R. 1997. Optical coherence tomography of the human skin. *J. Am. Acad. Dermatol.* 37: 958–963.

Welzel, J., Lankenau, E., Birngruber, R., and Engelhardt, R. 1998. Optical coherence tomography of the skin. *Curr. Probl. Dermatol.* 26: 27–37.

Welzel, J., Reinhardt, C., Lankenau, E., Winter, C., and Wolff, H. H. 2004. Changes in function and morphology of normal human skin: Evaluation using optical coherence tomography. *Br. J. Dermatol.* 150: 220–225.

Wojtkowski, M., Leitgeb, R., Kowalczyk, A., Bajraszewski, T., and Fercher, A. F. 2002. In vivo human retinal imaging by Fourier domain optical coherence tomography. *J. Biomed. Opt.* 7: 457–463.

Wojtkowski, M., Srinivasan, V., Fujimoto, J. G., Ko, T., Schuman, J. S., Kowalczyk, A., and Duker, J. S. 2005. Three-dimensional retinal imaging with high-speed ultrahigh-resolution optical coherence tomography. *Ophthalmology* 112: 1734–1746.

Wojtkowski, M., Srinivasan, V. J., Ko, T. H., Fujimoto, J. G., Kowalevicz, A., and Duker, J. S. 2004. Ultrahigh resolution, high speed, Fourier domain optical coherence tomography and methods for dispersion compensation. *Opt. Express* 12: 2404–2422.

Wollstein, G., Paunescu, L. A., Ko, T. H., Fujimoto, J. G., Kowalevicz, A., Hartl, I., Beaton, S., Ishikawa, H., Mattox, C., Singh, O., Duker, J., Drexler, W., and Schuman, J. S. 2005. Ultrahigh-resolution optical coherence tomography in glaucoma. *Ophthalmology* 112: 229–237.

Wu, J. G., Conry, M., Gu, C. H., Wang, F., Yaqoob, Z., and Yang, C. H. 2006. Paired-angle-rotation scanning optical coherence tomography forward-imaging probe. *Opt. Lett.* 31: 1265–1267.

Xie, H., Pan, Y., and Fedder, G. K. 2003. Endoscopic optical coherence tomographic imaging with a CMOS-MEMS micromirror. *Sensors and Actuators A (Physical)* A103: 237–241.

Xie, T., Mukai, D., Guo, S., Brenner, M., and Chen, Z. 2005. Fiber-optic-bundle-based optical coherence tomography. *Opt. Lett.* 30: 1803–1805.

Xie, T. Q., Guo, S. G., Chen, Z. P., Mukai, D., and Brenner, M. 2006. GRIN lens rod based probe for endoscopic spectral domain optical coherence tomography with fast dynamic focus tracking. *Opt. Express* 14: 3238–3246.

Yang, V. X. D., Tang, S. J., Gordon, M. L., Qi, B., Gardiner, G., Cirocco, M., Kortan, P., Haber, G. B., Kandel, G., Vitkin, I. A., Wilson, B. C., and Marcon, N. E. 2005a. Endoscopic Doppler optical coherence tomography in the human GI tract: Initial experience. *Gastrointest. Endosc.* 61: 879–890.

Yang, V. X. D., Mao, Y. X., Munce, N., Standish, B., Kucharczyk, W., Marcon, N. E., Wilson, B. C., and Vitkin, I. A. 2005b. Interstitial Doppler optical coherence tomography. *Opt. Lett.* 30: 1791–1793.

Yaqoob, Z., Wu, J., McDowell, E. J., Heng, X., and Yang, C. 2006. Methods and application areas of endoscopic optical coherence tomography. *J. Biomed. Opt.* 11: 063001.

Yelbuz, T. M., Choma, M. A., Thrane, L., Kirby, M. L., and Izatt, J. A. 2002. Optical coherence tomography: A new high-resolution imaging technology to study cardiac development in chick embryos. *Circulation* 106: 2771–2774.

Yeow, J. T. W., Yang, V. X. D., Chahwan, A., Gordon, M. L., Qi, B., Vitkin, I. A., Wilson, B. C., and Goldenberg, A. A. 2005. Micromachined 2-D scanner for 3-D optical coherence tomography. *Sensors Actuators A: Phys.* 117: 331–340.

Youngquist, R., Carr, S., and Davies, D.1987. Optical coherence-domain reflectometry: A new optical evaluation technique. *Opt. Lett.* 12: 158–160.

Yun, S. H., Tearney, G. J., Bouma, B. E., Park, B. H., and de Boer, J. F. 2003a. High-speed spectral-domain optical coherence tomography at 1.3 mu m wavelength. *Opt. Express* 11: 3598–3604.

Yun, S. H., Tearney, G. J., de Boer, J. F., Iftimia, N., and Bouma, B. E. 2003b. High-speed optical frequency-domain imaging. *Opt. Express* 11: 2953–2963.

Yun, S. H., Tearney, G. J., Vakoc, B. J., Shishkov, M., Oh, W. Y., Desjardins, A. E., Suter, M. J., Chan, R. C., Evans, J. A., Jang, I. K., Nishioka, N. S., de Boer, J. F., and Bouma, B. E. 2006. Comprehensive volumetric optical microscopy in vivo. *Nat. Med.* 12: 1429–1433.

Yung, K. M., Lee, S. L., and Schmitt, J. M. 1999. Phase-domain processing of optical coherence tomography images. *J. Biomed. Opt.* 4: 125–136.

Zagaynova, E. V., Streltsova, O. S., Gladkova, N. D., Snopova, L. B., Gelikonov, G. V., Feldchtein, F. I., and Morozov, A. N. 2002. In vivo optical coherence tomography feasibility for bladder disease. *J. Urol.* 167: 1492–1496.

Zara, J. M. and Patterson, P. E. 2006. Polyimide amplified piezoelectric scanning mirror for spectral domain optical coherence tomography. *Appl. Phys. Lett.* 89: 263901.

Zara, J. M., Yazdanfar, S., Rao, K. D., Izatt, J. A., and Smith, S. W. 2003. Electrostatic micromachine scanning mirror for optical coherence tomography. *Opt. Lett.* 28: 628–630.

Zhang, Y., Cense, B., Rha, J., Jonnal, R. S., Gao, W., Zawadzki, R. J., Werner, J. S., Jones, S., Olivier, S., and Miller, D. T. 2006. High-speed volumetric imaging of cone photoreceptors with adaptive optics spectral-domain optical coherence tomography. *Opt. Express* 14: 4380–4394.

Zuccaro, G., Gladkova, N., Vargo, J., Feldchtein, F., Zagaynova, E., Conwell, D., Falk, G., Goldblum, J., Dumot, J., Ponsky, J., Gelikonov, G., Davros, B., Donchenko, E. and Richter, J.2001. Optical coherence tomography of the esophagus and proximal stomach in health and disease. *Am. J. Gastroenterol.* 96: 2633–2639.

Zysk, A. M. and Boppart, S. A. 2006. Computational methods for analysis of human breast tumor tissue in optical coherence tomography images. *J Biomed. Opt.* 11: 054015.

Zysk, A. M., Nguyen, F. T., Chaney, E. J., Kotynek, J. G., Oliphant, U. J., Bellafiore, F. J., Johnson, P. A., Rowland, K. M., and Boppart, S. A. 2009. Clinical feasibility of microscopically-guided breast needle biopsy using a fiber-optic probe with computer-aided detection. *Technol. Cancer Res. Treat.* 8: 315–321.

Functional Optical Coherence Tomography in Preclinical Models

Melissa C. Skala
Duke University

Yuankai K. Tao
Duke University

Anjul M. Davis
Duke University

Joseph A. Izatt
Duke University

14.1 Introduction 281
14.2 Hemodynamic Imaging.. 282
    Doppler OCT • Volumetric Doppler OCT of Bidirectional Flow • OCT of Functional
    Brain Activation • Speckle Variance OCT • Multifunctional Imaging with Doppler OCT
14.3 Nanoparticle-Based Contrast 289
    Backscattering-Based Nanoparticle Contrast • Photothermal OCT • Backscattering
    Albedo • Magnetomotive OCT • Improved Delivery for In Vivo Imaging
14.4 Discussion 297
References... 298

14.1 Introduction

Functional imaging is a powerful tool for investigating biological signaling, disease processes, and potential therapies in both in vivo and in vitro systems. Microscopy, including confocal and multiphoton microscopy, has been the standard for high-resolution functional imaging in live cells and tissues. However, these microscopy techniques suffer from relatively shallow imaging depths. Magnetic resonance imaging (MRI) and PET have been the standard for functional imaging deep within the body, with the caveat of relatively poor resolution. Optical coherence tomography (OCT) fills a niche between high-resolution microscopy techniques and whole-body imaging techniques with relatively good resolution ($\sim$1–10 μm) and penetration depths ($\sim$1–2 mm) in tissue. OCT thus provides structural information noninvasively in vivo at a resolution similar to that of histology. Combined functional and structural imaging in this regime could become a powerful tool for small animal studies.

One of the strengths of OCT is its hemodynamic imaging capabilities. Microvessel morphology (Mariampillai et al. 2008, Tao et al. 2008, Skala et al. 2009, Wang and An 2009), blood flow velocities (Izatt et al. 1997, Yazdanfar et al. 1997, Iftimia et al. 2008, Davis et al. 2008, 2009, Larina et al. 2009, Skala et al. 2009), hematocrit (Xu et al. 2008, Faber and van Leeuwen 2009), and changes in total blood content (Aguirre et al. 2006, Chen et al. 2009) can be detected in three dimensions using the phase and/or magnitude provided by an OCT interferogram. OCT is attractive for small animal imaging studies of hemodynamics because contrast agents are not required, and thus longitudinal measurements of disease processes are simplified. This combination of multiparameter hemodynamic imaging on the microvessel level (vessel network structure, blood velocity and vessel flow profiles, hematocrit, etc.), with intrinsic contrast imaging and high-resolution structural information provides OCT with an attractive niche for studies of blood supply in animal models. This information can be applied to problems in a wide variety of disciplines including developmental biology, cancer, and neuroscience.

Molecular imaging (MI) is an important piece of any functional imaging platform. However, OCT is intrinsically insensitive to incoherent scattering processes such as fluorescence and spontaneous Raman scattering, which are central to optical MI, because OCT depends on coherent detection of scattered light. Thus, creative means are necessary to achieve molecular contrast in OCT. Previous studies have achieved molecular contrast using dyes (Vinegoni et al. 2004, Xu et al. 2004, Yang et al. 2004b, Applegate and Izatt 2006), proteins (Yang et al. 2004a), and intrinsic tissue molecules (Faber et al. 2003, Applegate and Izatt 2006), or functional contrast using magnetomotive methods (Oldenburg et al. 2005, 2008, Oh et al. 2007, 2008). Recently, there has been significant interest in the development of plasmon-resonant nanoparticles for molecular contrast in OCT. Plasmon-resonant nanoparticles are small ($\sim$40–200 nm in diameter) metallic particles (usually silver or gold) that scatter light with high efficiency (Homola et al. 1999). Nanoparticles respond strongly to light because the conduction electrons in the metal undergo a collective resonance called a surface plasmon resonance. Metal nanoparticles are attractive contrast agents for OCT because these particles can be tuned to resonate in the near-infrared (NIR) region where light penetrates deepest in tissue, and the absorption and scattering properties of the particles can also be tuned based on the particle's size, shape, and material composition.

As the hemodynamic and MI capabilities of OCT mature, they can be combined to provide a multifunctional picture of drug delivery and disease processes in animal models. These images would offer a comprehensive set of endpoints that allow for quantitative evaluation of potential therapies. In the future, these functional OCT systems could become standard imaging tools for the biological and medical sciences.

14.2 Hemodynamic Imaging

Functional extensions of OCT allow for noninvasive imaging of blood flow, vessel morphology, and changes in total blood content in living systems. Doppler OCT is a method for quantitative cross-sectional imaging of microvessel blood flow rates in vivo (Izatt et al. 1997, Yazdanfar et al. 1997, 2000, 2003, Skala et al. 2009) and is based on phase changes induced by flowing erythrocytes in blood vessels. Other methods, including single-pass volumetric bidirectional flow imaging (SPFI) (Tao et al. 2008), OCT fractional changes (Aguirre et al. 2006, Chen et al. 2009), and speckle variance OCT (Mariampillai et al. 2008), detect blood vessels or changes in blood content using the magnitude of the OCT interferogram. The combination of OCT with other optical imaging (OI) techniques, such as absorption-based spectral imaging of hemoglobin oxygen saturation, can provide a complete picture of hemodynamics in vivo (Skala et al. 2009).

14.2.1 Doppler OCT

Doppler OCT provides quantitative information on microvessel flow rates in vivo (Izatt et al. 1997, Yazdanfar et al. 1997, 2000, 2003, Bower et al. 2007, Skala et al. 2009). Doppler shifts arise from motion in the sample (such as flowing erythrocytes in blood vessels), and Doppler OCT detects this motion through phase-sensitive detection of interference between the probe and reference beam. Phase changes are calculated from multiple A-scans collected at the same position in the volume, and phase changes can be related to flow velocity profiles within vessels by measuring the angle between the blood flow and the incident beam in the three-dimensional (3D) volume.

In Doppler OCT, the Doppler frequency shift (f_D) is defined as

$$f_D = \frac{\Delta\varphi}{2\pi T} \tag{14.1}$$

where

    $\Delta\varphi$ is the average phase shift over the repeated A-scans
    T is the integration time of the camera

The minimum Doppler frequency is given by

$$f_{min} = \frac{\phi_{noise}}{2\pi T} \tag{14.2}$$

where ϕ_{noise} represents the standard deviation of the phase in the absence of flow in the sample. To measure ϕ_{noise}, a stationary mirror is placed in the sample arm and the standard deviation of the phase is recorded at the surface of the mirror over multiple A-scans. The phase noise is also affected by speckle noise and transverse motion (Park et al. 2003). The maximum frequency that can be unambiguously detected is defined by

$$f_{max} = \frac{\phi_{wrap}}{2\pi T} \tag{14.3}$$

where ϕ_{wrap} is π radians. For accumulated phase changes larger than $\pm\pi$ over the acquisition time T, the measured frequency wraps by 2π, similar to other phase imaging techniques. This phase can be unwrapped with techniques from phase microscopy (Hendargo et al. 2009).

From the Doppler frequency data, flow velocity can be determined by

$$v = \frac{\Delta\phi \cdot \lambda_0}{4\pi \cdot T \cdot n \cdot \cos(\alpha)} = \frac{f_D \cdot \lambda_0}{2 \cdot n \cdot \cos(\alpha)} \tag{14.4}$$

where

    λ_0 is the source center wavelength
    n is the refractive index (RI) of the medium
    α is the angle of incidence of the probe beam relative to the moving scatterer

Accurate quantification of flow velocity cannot be made using a single B-mode image (Figure 14.1a), unless the vessel is oriented so that the direction of flow lies in the plane of the OCT beam scan (equivalent to the y–x plane in Figure 14.1b). To measure the angle of flow in independently oriented vessels, volume images must be acquired around each region of interest. Volume renderings allow

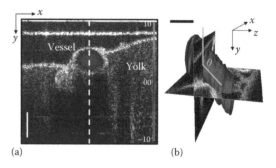

(a)　　　　　　　　　　　　　　　　(b)

FIGURE 14.1 (See color insert.) Doppler OCT measurement and volume rendering of chicken embryo heart vessel. Doppler OCT image (blue) superimposed on an OCT intensity image of a vessel cross section (a). Three-dimensional volume rendering of the same vessel (b). The volume rendering is used to measure the angle of blood flow (green), relative to the OCT scanning beam. Two orthogonal OCT planes are displayed, and the x–y plane corresponds to the image shown in (a). Scale bar is 200 μm. (From Davis, A. et al., *Anat. Rec. (Hoboken)*, 292, 311, 2009. With permission.)

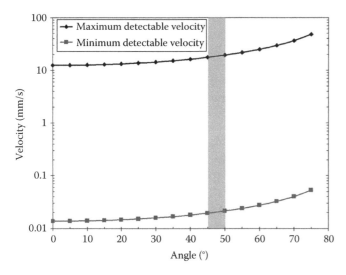

FIGURE 14.2 Flow velocity range of the Davis et al. system (19 kHz 1310 nm) calculated from Equation 14.4. The minimum and maximum (no phase wrapping) detectable flow velocity depends on the angle of flow relative to the OCT scanning beam (Equation 14.4). For the chick embryo imaging studies, the flow angles of the vessels were between 45° and 50°. (From Davis, A. et al., *Anat. Rec. (Hoboken)*, 292, 311, 2009. With permission.)

for a direct measure of the angle of the center of the vessel lumen relative to the OCT scanning beam (Figure 14.1b).

When flow rates are faster than the readout time of the CCD camera, the measured signal becomes phase-wrapped and velocity is not uniquely extractable from the phase. To address this artifact, Doppler measurements can be low-pass filtered to reduce the phase noise, and two-dimensional (2D) phase unwrapping algorithms can be applied (Ghiglia and Pritt 1998). Figure 14.2 contains a plot of the minimum and maximum velocities that can be resolved with the Davis et al. system for flow angles between 0° and 75°, calculated using Equation 14.4

and the features of their system (1310 nm center wavelength, 84 nm bandwidth, 53 µs camera readout time, RI 1.33) (Davis et al. 2009). In the Davis et al. study, the vessels were usually oriented between 45° and 50° relative to the OCT beam, which corresponds to a maximum velocity range between 13 and 14.5 mm/s and a minimum detectable velocity of ~0.02 mm/s.

To test the accuracy of velocity and flow measurements using Doppler OCT, Davis et al. measured the flow of 10% Intralipid through a 1.2 mm inner diameter (*D*) glass capillary tube at five different flow rates (0.25, 0.5, 1.0, 1.25, and 1.5 mL/min) using a Harvard Apparatus syringe pump. The cross-sectional velocity profiles from five images were averaged for each flow rate (Figure 14.3a). The peak velocity (V_{max}) was determined from the peak of a second-order polynomial fit to each averaged profile (Figure 14.3a). Figure 14.3b shows the measured Doppler OCT flow [$Q = \pi*(D/2)^2*V_{max}/2$] and the calibrated flow from the syringe pump, which are in good agreement (*x* = *y* line, *R*-squared = 0.98).

Next, Davis et al. correlated blood flow dynamics with heart development in the chick embryo using Doppler OCT (Davis et al. 2008). Structural volumes and Doppler velocity profiles from the outflow tract of the chick embryo heart were acquired at multiple stages of cardiac development (10 µm spot size on the sample). Figure 14.4 contains a volumetric reconstruction, surface rendering, and Doppler measurements from one stage of heart development. The volume rendering (Figure 14.4a) shows that the heart tube has begun to bend in the ventricle region, indicating the initial phase of the looping process. Measurements of velocity versus time and depth versus time were simultaneously acquired along the dashed yellow line through the center of the outflow tract (Figure 14.4a, inset), and allow for correlation of blood flow with expansion and dilation of the outflow tract. A volume reconstruction of the heart is also provided in the bottom of Figure 14.4a. At contraction, there is nearly zero net blood flow (Figure 14.4b, blue dashed line) due to blood constriction from the closing of the outflow tract. Peak blood flow velocity

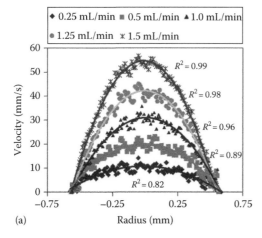

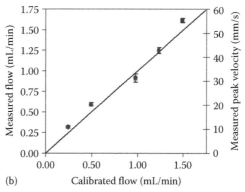

FIGURE 14.3 Validation of Doppler flow measurements. Doppler measurements were acquired from 10% Intralipid flowed through a 1.2 mm inner diameter capillary tube at rates of 0.25, 0.5, 1.0, 1.25, and 1.5 mL/min (a). Averaged Doppler velocity profiles (dots) and corresponding quadratic fit (solid line) are shown with each flow rate along with the *R*-squared values for each fit. Comparison of Doppler OCT measured and syringe-pump calibrated flow rates (b) indicate good agreement (*x* = *y* line, *R*-squared = 0.98). (From Davis, A. et al., *Anat. Rec. (Hoboken)*, 292, 311, 2009. With permission.)

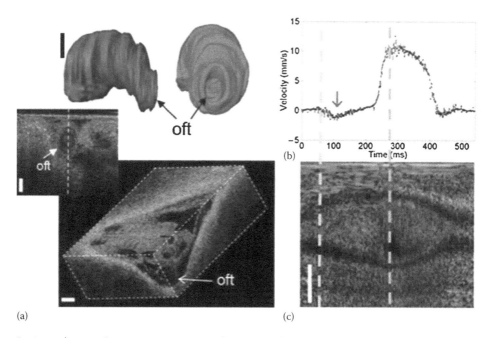

FIGURE 14.4 (See color insert.) Doppler OCT measurements of one stage of chick embryo heart development (Hamburger–Hamilton stage 11). Surface rendering and volume reconstruction of the heart tube (a). Inset shows Doppler OCT images of blood flow out of the outflow tract (out of the plane). Doppler velocity measurements were acquired along the yellow dashed line. Blood flow velocity dynamics (b). M-mode images (c) where the vertical axis is depth and the horizontal is time. These M-mode images are acquired simultaneously with the velocity measurement, enabling correlation of heart tube contractions. Doppler velocity measurements were taken along the green dotted line in (c). oft, outflow tract, scale bar = 250 µm. (From Davis, A.M. et al., *J. Opt. Soc. Am. A: Opt. Image Sci. Vis.*, 25, 3134, 2008. With permission.)

at this stage of development was 11 mm/s, in agreement with previous measurements using laser Doppler velocimetry (Hu and Clark 1989), microparticle image velocimetry (Vennemann et al. 2006), and pulsed Doppler ultrasound (Butcher et al. 2007, McQuinn et al. 2007) techniques. M-mode images (Figure 14.4c) are acquired simultaneously with the velocity measurement, enabling correlation of heart tube contractions. Measurements such as these can be used to determine the mechanisms of blood flow in models of early heart development, thus elucidating the genetic markers for abnormal heart development.

14.2.2 Volumetric Doppler OCT of Bidirectional Flow

Alternative processing techniques based on the magnitude of the OCT interferogram can also be used to detect and map blood vessels in animal models. One recent technique uses a modified Hilbert transform algorithm to separate moving and nonmoving scatterers within a depth (Tao et al. 2008). The resulting images map the components of moving scatterers flowing into and out of the imaging axis onto opposite image half-planes, enabling volumetric bidirectional flow mapping.

The SPFI-SDOCT processing steps are outlined in Figure 14.5. Previous work (Yasuno et al. 2006, Baumann et al. 2007, Wang 2007, Wang et al. 2007) has shown that a constant phase shift imposed between A-scans along a single lateral scan direction creates a carrier frequency that can isolate real and complex conjugate peaks to opposite spatial frequency spaces. Windowing out

the conjugate peaks, inverse transforming back to k-space (Hilbert transform), and then applying conventional spectral Fourier transforms on the data allows for separation of moving (flow) and nonmoving (structure) scatterers to opposite image half-planes. Imaging of bidirectional flow requires applying positive and negative carrier frequencies and processing the two associated datasets separately. SPFI-SDOCT recognizes that without the use of carrier frequencies, the spatial frequencies of moving and nonmoving scatterers do not overlap at spatial frequencies above the nonmoving scatterer bandwidth (Figure 14.5b). An analytic signal for the spectral interferogram can be obtained by applying a Heaviside function ($H[u - f_T]$), frequency-shifted outside of the structural bandwidth (Figure 14.5c), and then inverse Fourier transforming the result (Figure 14.5d). A threshold frequency (f_T) defines the minimum detectable velocity in SPFI-SDOCT and is related to the spatial correlation of sequential A-scans. Therefore, lateral oversampling of A-scans allows for a reduced threshold frequency and thus slower minimum detectable velocities. All spatial frequencies above f_T can be detected given that their associated accumulated phases are above the system phase-noise floor. Since the resulting complex interferometric signal is a sum of positive moving scatterers and the conjugate of negative moving scatterers, application of conventional spectral domain optical coherence tomography (SDOCT) processing yields a flow image where flow in opposite directions (into and out of the A-scan axis) is imaged to opposite sides of DC (Figure 14.5e).

SPFI-SDOCT was demonstrated in a mouse window chamber model implanted with a 4T1 metastatic mouse mammary

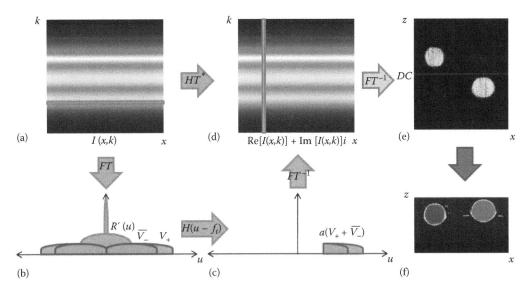

FIGURE 14.5 (See color insert.) Flowchart of SPFI-SDOCT processing. A lateral Fourier transform of the B-scan (a) yields the spatial frequencies of the sample centered at DC (structural data, yellow) and spatial frequencies of moving scatterers shifted by their respective Doppler frequencies (positive velocities in red, negative velocities in blue) (b). Applying a frequency-shifted Heaviside step function (c) and then applying an inverse Fourier transform of the resulting spatial frequencies recreates the analytic interferometric signal (modified Hilbert transform) (d). A spectral inverse Fourier transform of the analytic interferometric signal maps the depth-resolved reflectivities of scatterers moving in opposite directions (into or out of the A-scan axis) on opposite image half-planes (e), which can then be overlaid for vessel identification (f). (From Tao, Y.K. et al., *Opt. Express*, 16, 12350, 2008. With permission.)

adenocarcinoma. The animals were imaged using a spectral domain OCT system centered at 860 nm (100 nm bandwidth) with a 9 µm spot size on the sample. A 6 mm × 5 mm volume mosaic was created by acquiring 30 1 mm × 1 mm volumes imaged with 1800 A-scans/frame and 100 frames/volume. Each SPFI-SDOCT reconstructed frame was separated into two halves and combined to create bidirectional flow maps with grayscale intensities corresponding to the reflectivity of scatterers moving into or out of the A-scan axis. Vessels supplying the surrounding muscle of the window chamber (Figure 14.6) are clearly visible in the mouse model. SPFI-SDOCT imaged vessels are tubular, as expected, and range from 15 to 230 µm in diameter. The bright regions on the top and bottom of the image denote regions of intensity saturation from high reference reflections. Since these mouse models were imaged in a common-path configuration, where the reflection from one side of the window was used as the reference reflector, these saturation artifacts could not be mitigated. The small, tortuous vessel networks surrounding the central veins are indicative of tumor vasculature. These initial studies demonstrate that SPFI-SDOCT allows for bidirectional volumetric flow imaging without acquiring multiple A-scans at each lateral position. Recent extensions of this technique also allow for velocity-resolved maps of vessel networks in the human eye using a moving spatial frequency window (Tao et al. 2009).

14.2.3 OCT of Functional Brain Activation

Changes in blood content can also be determined from changes in reflectivity of OCT magnitude images (Chen et al. 2009). Chen et al. measured OCT magnitude images (time-domain

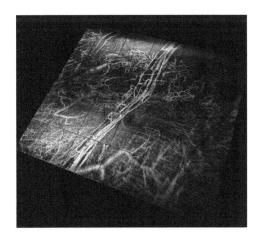

FIGURE 14.6 Two-dimensional projection of bidirectional volumetric flow in a mouse window chamber tumor model. The 6 × 5 mm volume is comprised of 30 1 × 1 mm volumes sampled with a 1 ms integration time with 2048 (depth) × 1800 × 100 pixels. Vessel sizes range from 15 to 230 µm.

system centered at 1060 nm with a 40 nm bandwidth and an axial/lateral resolution of 13/36 µm) from the rat somatosensory cortex through a thinned skull during forepaw electrical stimulation. The ratio of post-stimulus scans over pre-stimulus baseline images was computed for each individual trial to provide an OCT fractional change map. The cross-sectional OCT fractional change maps were compared with en-face optical intrinsic signal imaging (OISI) images from the same animals at 570 nm illumination (isosbestic point for oxygenated and deoxygenated hemoglobin). OISI provides 2D, depth-integrated activation maps of brain activity by imaging the optical reflectivity of the brain onto a CCD camera and calculating the ratio of post-stimulus scans over pre-stimulus images for each individual trial (identical to the OCT calculations). OISI provides an image of the change in total hemoglobin concentration, which is proportional to total blood volume if the hematocrit is assumed constant.

Figure 14.7 shows the co-registered OISI and OCT images in the region of functional activation. The functional signal was computed as a ratio of the reflectance at each time point to the mean reflectance in the pre-stimulus period and

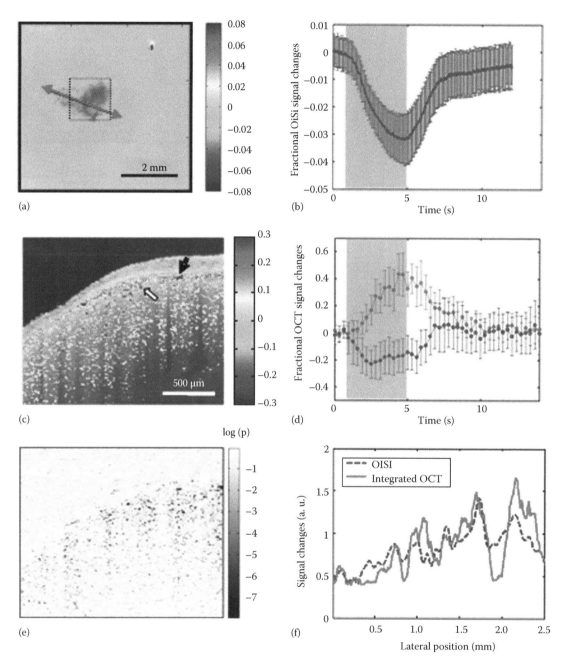

FIGURE 14.7 (See color insert.) Representative results of OISI and OCT imaging of functional activation. En face functional activation map from OISI with co-registered OCT scan indicated by the red arrow (a). The averaged time course for OISI over the region denoted by dashed box in (a) with stimulation period shaded grey (b). Functional OCT image superimposed with the structural image (c). In order to simultaneously display these two images, we threshold the functional OCT signals to those with responses >5% (or <0.5%). Representative functional OCT time courses for regions of interest specified by solid arrow (positive changes) and hollow arrow (negative changes) (d). Map of p-values showing the statistical significance of the fractional OCT changes from the baseline (e). The lateral line profile of OISI signal changes and the depth-integrated OCT signal changes showing the spatial correspondence of OCT and OISI signals (f). The OISI and OCT signal changes are scaled to overlap in order to provide a visual comparison. (From Chen, Y. et al., *J. Neurosci. Methods*, 178, 162, 2009. With permission.)

is, therefore, representative of a percent signal change from the baseline. Figure 14.7a shows a representative ratio OISI image from the time window around the peak of maximal activation ($t = 4$–6 s). The region of functional activation corresponding to forepaw stimulation is clearly delineated (blue region). A decrease in the reflectance at 570 nm corresponds to increased concentration of hemoglobin. Figure 14.7b shows the time-course of the averaged fractional reflectivity changes in the region denoted by the dashed box in Figure 14.7a. The decreased signal intensity during the stimulation period arises from increased blood absorption during the stimulus (due to increased blood volume).

Figure 14.7c shows the functional OCT image from the time window around the peak of maximal activation ($t = 4$–6 s). The OCT scan location is shown over the region of activation indicated by the red arrow in Figure 14.7a. Similar to the OISI fractional change map, the OCT fractional change map was computed by normalizing all time points to the baseline pre-stimulus period. The OCT functional map reveals a localized area of activation with both positive and negative changes in the cortex. The temporal sequences of activation for the representative regions of interest in Figure 14.7c are plotted in Figure 14.7d. The OCT functional signal time courses reveal clear increases and decreases that deviate from baseline, reach a peak near the cessation of the stimulus, and then gradually return to baseline. The time course of OCT signal changes correlate well with the co-registered OISI time course as shown in Figure 14.7b. The observed changes in the OCT functional signal are statistically different from baseline as shown in Figure 14.7e. In addition, the lateral line profiles of OISI signal changes and the depth-integrated OCT signal changes are plotted in Figure 14.7f, showing the spatial correspondence of OCT and OISI signals.

This study demonstrates that OCT fractional change maps correlate well with that of the OISI signals, and provide depth-resolved, layer-specific dynamics of the functional activation patterns of the rat somatosensory cortex (Chen et al. 2009). Thus, OCT may serve as a promising technology to provide complementary information to OISI for functional neuroimaging.

14.2.4 Speckle Variance OCT

A final OCT-based method for mapping blood vessels in animal models relies on the speckle pattern in OCT magnitude images (Mariampillai et al. 2008). The texture of OCT images usually has a speckled appearance due to coherent interference (Fercher 2008). In the absence of flow, this speckled appearance remains constant from frame to frame. Blood moving through the frame causes variation in the speckle pattern that can be used to identify blood vessels. Speckle variance techniques based on structural image intensity have previously been used to image vasculature with high-frequency ultrasound (Cheung et al. 2007). This concept has recently been applied to OCT to image microvessels in the mouse dorsal skin fold window chamber using a Fourier domain mode locking (FDML) swept-source (SS) OCT system (1310 nm center wavelength, 112 nm sweeping range, 15 μm spot size on the sample) (Mariampillai et al. 2008). In this study, interframe speckle variance images (SV_{ijk}) of the structural OCT intensity (I_{ijk}) were calculated across $N = 3$ B-mode images as follows:

$$SV_{ijk} = \frac{1}{N} \sum_{i=1}^{N} \left(I_{ijk} - I_{\text{mean}} \right)^2 \tag{14.5}$$

where

j and k are lateral and depth indices of the B-mode images
i denotes the B-mode slice index with I_{mean} as the average over the same set of pixels

A flow phantom study was performed to evaluate the relationship between speckle variance, flow velocity, and the Doppler angle. Figure 14.8 shows Doppler and speckle variance images

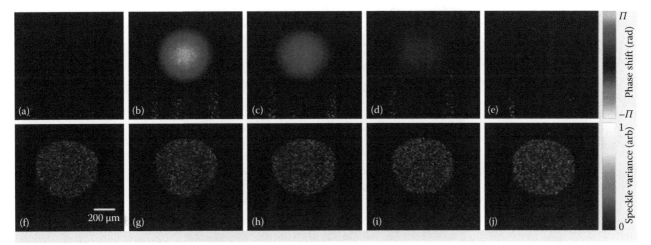

FIGURE 14.8 (See color insert.) Doppler phase shift measured for stationary Intralipid within a 600 μm tube at an 80° Doppler angle (a). Doppler phase shift measured from Intralipid moving through the tube at a rate of 12 mL/h for Doppler angles of 75°, 80°, 85°, 89.5°, respectively (b–e). Corresponding normalized (to fixed value for all images) speckle variance images (f–j). (From Mariampillai, A. et al., *Opt. Lett.*, 33, 1530, 2008. With permission.)

of a 600 µm inner diameter polymer tube filled with 0.5% Intralipid driven by a syringe pump. The tube was embedded in Agarose gel mixed with Intralipid such that the fluid and gel had similar scattering properties. SS-OCT images were acquired at four different Doppler angles (75°, 80°, 85°, and 89.5°) with the pump off or on (24 mm/s peak velocity, assuming laminar flow). Doppler images were processed using phase-based color Doppler signal processing techniques (White et al. 2003) with an ensemble length of 16 A-scans. For speckle variance imaging, four adjacent A-scans were averaged together to improve the signal-to-noise ratio of the structural image (Yang et al. 2003, Barton and Stromski 2005), leading to a decrease in the speckle variance noise floor. Figure 14.8 demonstrates that the speckle variance has little dependence on the Doppler angle or flow velocity and can even distinguish the Intralipid fluid from the surrounding gel at a zero bulk flow rate.

Speckle variance imaging was directly compared with intravital fluorescence confocal microscopy using a dorsal skin fold window chamber model in athymic nude mice (Figure 14.9). Fluorescence confocal imaging (LSM 510 Meta NLO, Zeiss, excitation 488 nm, collection 530 ± 15 nm bandpass) was performed using 5 mg/kg of 500 kD fluorescein labeled dextran via tail vein injection. A 5× (NA = 0.25) objective was used for imaging a 1.8 mm × 1.8 mm area over a 200 µm depth (Figure 14.9c). Three-dimensional (1000 × 2000 × 512 pixels) speckle variance OCT was performed in the same region (Figure 14.9d) with a total volumetric imaging time of approximately 1 min. The smallest vessels detectable by speckle variance imaging were ~25 µm in diameter. The capillary bed, consisting of microvessels ~5 µm in diameter as detected by fluorescence confocal microscopy, was below the lateral resolution of the OCT system (15 µm).

The main advantage of the speckle variance processing technique compared with conventional Doppler OCT is that it can be implemented in real time to provide angle-independent microvascular information with little additional computational complexity. Speckle variance is based on intrinsic contrast, so this technique may also be advantageous compared to fluorescence microscopy, especially in the presence of neovasculature that is subject to fluorescent marker extravasation. Similar to phase-variance-based techniques (Fingler et al. 2007), speckle variance detection also suffers from multiple scattering-induced artifacts leading to artificial speckle variance contrast beneath the blood vessels (i.e., shadowing effects). Another key disadvantage is the effect of interframe bulk tissue motion, which can dominate the speckle variance. However, higher-frame-rate (>1000 fps) FDML OCT systems are currently under development, and these systems may diminish tissue motion artifacts. Although faster systems may allow for velocity-resolved speckle variance imaging, this technique is currently limited to vessel morphology mapping.

14.2.5 Multifunctional Imaging with Doppler OCT

Hemodynamic imaging with coherence-based methods can be particularly powerful when combined with other OI modalities. For example, the combination of speckle variance OCT of microvessel structure, Doppler OCT of blood flow rates and absorption-based imaging of blood oxygen saturation could provide valuable information on oxygen supply in tissues. A recent study has combined speckle variance OCT, Doppler OCT, and hyperspectral imaging to investigate oxygen supply in the mouse dorsal skin fold window chamber tumor model (Skala et al. 2009).

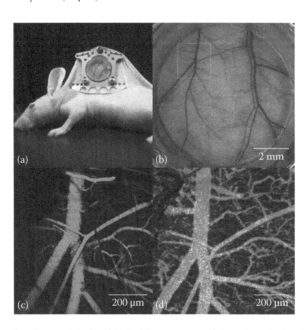

FIGURE 14.9 Dorsal skinfold window chamber model (a). White light microscopy of the entire window (b). The white box represents the approximate location of confocal and OCT imaging. Maximum intensity projection image of a fluorescence confocal z stack obtained using 500 kD fluorescein labeled dextran (1.8 × 1.8 mm area) (c). Speckle variance OCT en face projection image of vasculature without the use of any external contrast agents (1.8 × 1.8 mm area) (d). (From Mariampillai, A. et al., *Opt. Lett.*, 33, 1530, 2008. With permission.)

Hyperspectral (hemoglobin oxygen saturation) and spectral domain OCT (SDOCT) (blood flow and vascular morphology) arms were connected through two separate baseports in an inverted microscope (Carl Zeiss Axiovert 200). Two-dimensional hyperspectral images were collected with transillumination, and detection was achieved with a liquid-crystal tunable filter placed in front of a DVC 1412 CCD camera (DVC Company) (Sorg et al. 2005). Images were collected every 10 nm between 500 and 610 nm, and hemoglobin saturation images were calculated by applying an extension of the Beer–Lambert law to the wavelength-dependent absorption at each pixel (assuming oxygenated and deoxygenated hemoglobin as the only absorbers) (Sorg et al. 2005). The system, software, and analysis techniques have previously been validated on liquid phantoms with an accuracy of approximately 1% and in vivo (Sorg et al. 2005). The common-path spectral domain OCT arm consists of a Ti:sapphire laser source (Femtolasers) centered at 790 nm with a 90 nm FWHM. OCT images were collected in the common-path geometry, such that the reference reflection originates from the surface of the window chamber coverslip. This geometry minimizes the effects of bulk motion (Vakhtin et al. 2003). The interferogram is detected using a custom-made spectrometer with a line-scan CCD camera (Atmel, Aviiva). Three-dimensional blood flow images were collected with Doppler OCT, which measures phase changes due to flowing erythrocytes. Phase changes are calculated from multiple A-scans collected at the same position in the volume, and can be related to flow velocity profiles by measuring the angle of incidence in the 3D volume. Three-dimensional morphology was collected with speckle variance OCT, which detects the presence of blood vessels due to changes in the speckle pattern of OCT magnitude images (Mariampillai et al. 2008). Speckle variance is advantageous for morphology mapping because it is independent of the angle between the blood flow and the incident beam. However, it is currently not capable of accurately quantifying blood velocity or flow direction.

Hyperspectral, Doppler OCT, and speckle variance OCT images were collected from 4T1 tumors implanted in dorsal skin fold window chambers in nude mice. The transmission image of the window chamber morphology (Figure 14.10a) and hyperspectral image of the percent hemoglobin oxygenation saturation in the vessels (Figure 14.10c) were taken with a 2.5× objective (NA = 0.12). Speckle variance OCT (Figure 14.10b) provides 3D vessel morphology, and Doppler OCT (Figure 14.10d) provides 3D vessel flow velocities and flow direction. Speckle variance and Doppler OCT images were collected over a 1×1 mm area with 250×125 pixels, and 1024 pixels in the depth dimension. OCT images were collected with a 4× objective (NA = 0.1) with a 1 ms integration time for each A line. Eight repeated B scans were collected for speckle variance OCT, registered using "stackreg" for ImageJ, and 10 repeated A-scans were collected for Doppler OCT. The Doppler flow profiles in Figures 14.10e and f were fit to a second-order polynomial and corrected for the angle of incidence to provide velocity in mm/s. Maximum blood velocity is determined from the peak of this fit, vessel diameter is determined from the zero crossings of the fit, and the shear

rate on the vessel wall is calculated from the derivative of the fit (dv_z/dr), assuming a Newtonian fluid (van Leeuwen et al. 1999). Co-registration of speckle variance OCT, hyperspectral, and Doppler OCT images allow for vessel morphology, percent hemoglobin oxygen saturation, blood velocity and direction of flow, and shear rate on the vessel wall to be determined at any point within the window chamber.

The relationship between the hemoglobin oxygen saturation and the maximum velocity, flow rate, and shear rate for three vessels measured every 6 h for 24 h in the same tumor is shown in Figure 14.11. The flow rate [$Q = \pi^*(D/2)^2 * V_{max}/2$ (mm^3/s)] was calculated from the vessel diameter (D) and maximum vessel velocity (V_{max}). For all three vessels of interest (VOIs) and five time points grouped together, there was a strong correlation between velocity and hemoglobin saturation (Figure 14.11d, $r = 0.78$, $p = 0.0006$, $N = 15$), and between flow rate and hemoglobin saturation (Figure 14.11e, $r = 0.89$, $p = 9 \times 10^{-6}$, $N = 15$), and a moderate correlation between shear rate on the vessel wall and hemoglobin saturation (Figure 14.11f, $r = 0.56$, $p = 0.03$, $N = 15$). The extracted maximum velocities and vessel diameters (Dewhirst et al. 1990), shear rate (Tsuzuki et al. 2000), and hemoglobin saturation (Sorg et al. 2005, 2008, Dedeugd et al. 2009) values reported here are in agreement with previous studies.

The combination of hyperspectral microscopy and OCT creates a tool that can dynamically and noninvasively monitor vessel structure and function. These combined measurements will allow for unprecedented insight into the relationship between vessel structure, hemoglobin saturation, blood flow, and hypoxic episodes in tissues. This hemodynamic imaging approach has broad applicability in the study of diseases that are affected by oxygen supply, such as cerebral hypoxia (stroke), cancer, cardiac ischemia, and Alzheimer's disease. Moreover, these processes can be studied with multifunctional OI in vivo and do not require contrast-enhancing dyes.

14.3 Nanoparticle-Based Contrast

The use of nanoparticles as contrast agents for molecularly targeted OCT is attractive for several reasons. Unlike organic fluorophores, metal particles are not subject to photobleaching or cytotoxicity. The wide range of shapes (such as nanorods, nanocubes, nanoshells, etc.) and sizes of nanoparticles allows for the plasmon resonance of these particles to be tuned throughout the visible and NIR wavelength regions. This tunability allows the user to select the optimal imaging wavelength for their application, and also allows for multiple molecules to be differentiated based on multiple target resonant peaks. The added benefits of coherent detection allows for high-resolution molecular images to be collected at deeper depths than traditional microscopy techniques, and also allows for additional function measurements provided by OCT such as hemodynamic and collagen content imaging (Applegate et al. 2004). In the future, molecular OCT techniques could augment standard fluorescence microscopy as a method for deep-tissue, depth-resolved MI with relatively high resolution and target sensitivity.

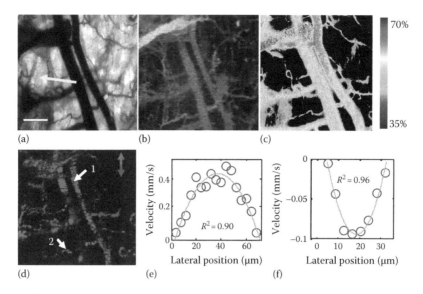

FIGURE 14.10 (See color insert.) Multidimensional functional imaging of 4T1 tumor vasculature in the dorsal skin fold mouse window chamber. Transmission image of the window chamber, with a portion of the tumor visible (arrow, a). Scale bar is 200 μm. Speckle variance OCT image of vessel morphology shown as an en face average intensity projection over 1 mm depth (b). Hyperspectral image of percent hemoglobin oxygen saturation, thresholded for the *R*-squared value of the linear fit and total absorption value at each pixel (background pixels are black) (c). Doppler OCT image of vessel blood flow direction shown as an en face maximum intensity projection over depth, with red vessels (□□□1) flowing toward the top of the image and blue vessels (2) flowing toward the bottom of the image (d). The flow profiles at points 1 (e) and 2 (f) were fit to a second-order polynomial and corrected for the angle of incidence to provide velocity in millimeters per second. (From Skala, M.C. et al., *Opt. Lett.*, 34, 289, 2009. With permission.)

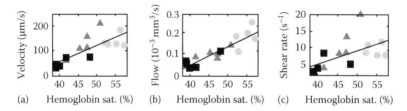

FIGURE 14.11 Scatter plots of maximum velocity versus hemoglobin saturation (a), flow (vessel cross-sectional area times 1/2 maximum velocity) versus hemoglobin saturation (b) and shear rate versus hemoglobin saturation (c) for three vessels of interest (VOI) measured every 6 h for 24 h in the same tumor. When all VOI (● VOI 1; ▲ VOI 2; ■ VOI 3) and time points were grouped together, there was a strong correlation between maximum velocity and hemoglobin saturation ($r = 0.78$, $p = 0.0006$, $N = 15$), as well as between flow and hemoglobin saturation ($r = 0.89$, $p = 9 \times 10^{-6}$, $N = 15$), and a moderate correlation between shear rate on the vessel wall and hemoglobin saturation ($r = 0.56$, $p = 0.03$, $N = 15$). (From Skala, M.C. et al., *Opt. Lett.*, 34, 289, 2009. With permission.)

Gold nanoparticles currently under development for molecular contrast OCT include gold nanoshells (Agrawal et al. 2006, Gobin et al. 2007, Adler et al. 2008), gold nanocages (Cang et al. 2005, Chen et al. 2005, Skrabalak et al. 2007), gold nanospheres (Skala et al. 2008), and gold nanorods (Oldenburg et al. 2006, Troutman et al. 2007). The source of contrast for these small metal particles can be based on their scattering (Agrawal et al. 2006, Gobin et al. 2007, Troutman et al. 2007) or absorption (Cang et al. 2005, Chen et al. 2005, Adler et al. 2008, Skala et al. 2008) properties, or on a combination of their scattering and absorption properties (Oldenburg et al. 2006). In addition, iron oxide nanoparticles, which are FDA approved as contrast agents for MRI, have been developed as contrast agents in OCT using magnetomotive OCT (Oldenburg et al. 2005, 2008, Oh et al. 2007, 2008).

14.3.1 Backscattering-Based Nanoparticle Contrast

Gold nanoshells are an attractive contrast agent for OCT because they resonate in the NIR, where OCT typically operates, and they have a high backscattering coefficient (Barton et al. 2004, Loo et al. 2004, Agrawal et al. 2006, Gobin et al. 2007). Nanoshells consist of a dielectric core nanoparticle surrounded by a metal shell with tunable plasmon resonances. Nanoshells offer the ability to manipulate both the resonant wavelength and the relative scattering and absorption efficiencies of the particle through the size and composition of each layer of the nanoshell (Oldenburg et al. 1998, 1999).

A recent study characterized the contrast enhancement of gold nanoshells in OCT images acquired from water and turbid tissue simulating phantoms at 1310 nm (time-domain system with 50 nm bandwidth with a 15 µm spot size on the sample) (Agrawal et al. 2006). The effect of nanoshell concentration, core diameter, and shell thickness on signal enhancement was characterized. Experimental results indicated trends that were consistent with predicted optical properties, including a monotonic increase in signal intensity and attenuation with increasing shell and core size. Threshold concentrations for a 2 dB OCT signal intensity gain were determined for several nanoshell geometries. Increases in the core diameter over the range of 126–291 nm and shell thickness over the range of 15–25 nm produced significant increases in OCT signal intensity at 1310 nm (Figure 14.12). For the most highly backscattering nanoshells tested (291 nm core diameter, 25 nm shell thickness), a concentration of 10^9 nanoshells/mL was needed to produce a 2 dB intensity gain.

This threshold concentration of 10^9 nanoshells/mL can be compared to the concentrations that may be achieved in tumors in vivo (Agrawal et al. 2006). First, consider the case of nanoshells labeled to target a tumor. A tumor marker such as the epidermal growth factor receptor can be found at levels of 10^6 per cell on cancerous cells (versus 10^3–10^5 per cell on normal cells) (Carpenter 1987, Kwok and Sutherland 1991). For the case of cells with 10 µm diameter and nanoshells with 300 nm diameter, up to 4000 nanoshells could be accommodated on the cell surface, and at least 100 receptors would be available to bind to each antibody-conjugated nanoshell. If only 10% of each cell's surface were covered with nanoshells (400 nanoshells/cell), the nanoshell concentration would be 6.4×10^{11} nanoshells/mL, two orders of magnitude above the threshold concentration. Quantitative experimental studies such as these will allow for methodical optimization of OCT diagnostics based on nanoparticle molecular contrast (Zagaynova et al. 2008, Kirillin et al. 2009).

Gold nanoshells can be used for contrast-enhanced OCT imaging based on their backscattering properties, and also for photothermal therapy due to their absorption properties. The attributes of this combined imaging and therapeutic platform have been demonstrated in mouse models of cancer in vivo (Gobin et al. 2007). PEG-modified nanoshells (119 ± 11 nm diameter core, ~12 nm shell thickness) were injected intravenously (tail vein injection 20 h prior to OCT imaging, 150 µL nanoshells in PBS at 1.5×10^{11} nanoshells/mL) in tumor-bearing mice and allowed to passively accumulate in the tumor tissue due to the leakiness of the tumor vasculature. The significant accumulation of particles within the tumor tissue dramatically increased the NIR scattering within the tumor, enhancing the OCT contrast. Histological examination of tumors using silver staining confirmed the presence of nanoshells throughout the tumors. Additionally, neutron activation analysis verified nanoshells present in a representative tumor at 12.5 ppm, which is equivalent to approximately 3 million nanoshells per gram of tumor tissue, compared to 0 ppm for tumors of mice injected with just phosphate-buffered saline (PBS).

OCT images were collected for nanoshell-injected and control (PBS-injected) mice and analyzed to assess the increase in contrast provided by the nanoshells in tumors compared to normal tissue. This study used a commercially available OCT imaging system (Niris Imaging System, Imalux). The axial and transverse resolutions were measured to be approximately 10 and 15 µm, respectively. Figure 14.13 shows OCT imaging results obtained after PBS injection (Figure 14.13a and c), and after nanoshell injections (Figure 14.13b and d). The strata of the skin and

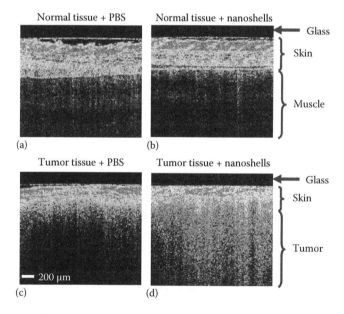

FIGURE 14.13 Representative OCT images from normal skin and muscle tissue areas of mice systemically injected with nanoshells (a) or with PBS (b). Representative OCT images from tumors of mice systemically injected with nanoshells (c) or with PBS (d). Analysis of all images shows a significant increase in contrast intensity after nanoshell injection in the tumors of mice treated with nanoshells while no increase in intensity is observed in the normal tissue. The glass of the probe is 200 µm thick and shows as a dark non-scattering layer. (From Gobin, A.M. et al., *Nano Lett.*, 7, 1929, 2007. With permission.)

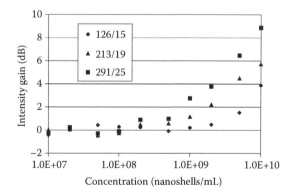

FIGURE 14.12 Measured OCT intensity gain (at 1310 nm) as a function of nanoshell type and concentration. Data are shown for three different nanoshell types (legend, core diameter/shell thickness) added to a turbid tissue phantom ($\mu_s = 100$ cm^{-1}, $g = 0.9$). Maximum standard deviation ($n = 3$) was 0.8 dB. Error bars are not shown for clarity. (From Agrawal, A. et al., *J. Biomed. Opt.*, 11, 041121, 2006. With permission.)

underlying muscle can be seen in the OCT image of the normal tissue (Figure 14.13a and b). There is no enhancement in layers of normal tissue in mice treated with nanoshells compared with the PBS-injected mouse because nanoshells do not extravasate appreciably into normal tissues.

The enhanced brightness of the image in Figure 14.13d relative to Figure 14.13c indicates that gold nanoshells can provide substantial contrast in OCT imaging. Figure 14.13d also shows that the borders of tumors (on the left) can more easily be discerned in the OCT images of nanoshell-treated mice compared to those mice that had only PBS injection (Figure 14.13c). Figure 14.14 shows the quantification of the image intensity of OCT images of normal tissue ($n = 3$) and tumor tissue ($n = 6$) with PBS injections and nanoshell injections. Images were analyzed to first quantify the contrast and then analyzed using a Student's t-test of the two populations of images from PBS-treated and nanoshell-treated mice. The data show a significant increase in the optical contrast of tumor compared to normal tissue when nanoshells are used, $p < 0.00002$. No statistical difference is observed in the intensity of the optical contrast of images of normal tissue whether nanoshells are used or PBS.

Next, tumor-bearing mice receiving either nanoshell or PBS injections were randomly assigned to control (untreated), sham (PBS + laser), and treatment (nanoshell + laser) groups. Following OCT imaging, animals in the treatment and sham groups had their tumors irradiated with a NIR laser. In vivo irradiation was accomplished using an Integrated Fiber Array Packet, FAP-I System, with a wavelength of 808 nm (Coherent) at a power density of 4 W/cm² and a spot size of 5 mm diameter for 3 min. Twelve days after treatment, tumors on all but two (out of six) nanoshell-treated mice had completely regressed. Kaplan–Meier statistical analysis shows a median survival of 14 days for the PBS + laser group and 10 days for the untreated control group. By day 21, the survival of the nanoshell + laser group was significantly greater than either the control or sham groups ($p < 0.001$), and this continued for the duration of the study (7 weeks).

These studies demonstrate that nanoparticle-enhanced OCT imaging is feasible in vivo, that significant differences exist between the nanoparticle-based OCT contrast of tumors and normal tissues, and that integration of diagnostic and therapeutic technologies is possible with a single nanoparticle (Gobin et al. 2007).

Finally, a study by Troutman et al. investigated the backscattering-based contrast of gold nanorods for OCT imaging in tissue phantoms (Troutman et al. 2007). A time-domain OCT system (890 nm center wavelength, 150 nm bandwidth, and 6/14 µm axial/lateral resolution) was used to image nanorods in polyacrylamide-based phantoms with tissue-like scattering properties ($\mu_s = 30$ cm⁻¹; $g = 0.89$). A void within the phantom was then filled with nanorods that resonate within the spectral bandwidth of the OCT source (912 nm peak resonance), with nanorods that resonate outside of the OCT source bandwidth (750 nm peak resonance), or with water alone. The OCT magnitude measured from gold nanorods was dependent on whether the plasmon resonance peak overlapped with the OCT source bandwidth, confirming the resonant character of enhancement (Figure 14.15). Gold nanorods (1.25 g gold/L water) with a plasmon resonance peak that overlaps with the OCT source (Figure 14.15c) yielded a signal-to-background ratio (SBR) of 4.5 dB, relative to the tissue phantom. Nanorods with a resonance at 750 nm and water alone yielded a 1.9 dB and a –1.3 dB SBR, respectively. One key feature of gold nanorods is their sharp optical resonance, with a full width at half-maximum as little as 80 nm and typically around 120 nm (Troutman et al. 2007). Thus, spectral multiplexing with gold nanorods may be possible, such that contributions from multiple nanorod configurations can be resolved in OCT images. Another possibility is differential OCT imaging, in which two images at shifted OCT center wavelengths are collected so that both images of the tissue (typically containing no resonant structures) remain similar, whereas the contrast agent is visible in one image but not the other. In comparison, other nanostructures such as nanoshells have much broader scattering bands (Gobin et al. 2007).

These scattering-based experiments demonstrate that OCT contrast enhancement using plasmonic gold nanoparticles can be achieved with conventional OCT systems and image processing software. As nanoparticles of different sizes, shapes, and materials are developed, these scattering-based plasmonic OCT contrast enhancement techniques will continue to improve.

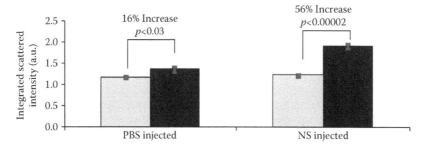

FIGURE 14.14 Quantification of OCT images shows a significant increase in intensity of images of tumors from mice with systemic nanoshell injection. Normal muscle tissue is shown in the gray bar while tumor tissue is shown in the black bar. Values are average ± standard error. (From Gobin, A.M. et al., *Nano Lett.*, 7, 1929, 2007. With permission.)

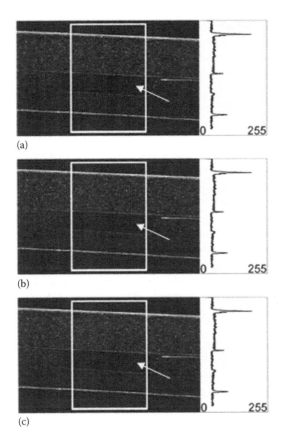

FIGURE 14.15 OCT images of different preparations loaded into the center cell (arrow, a) of polyacrylamide tissue phantoms: (a) water, (b) nanorods with resonance at 750 nm, (c) nanorods with resonance at 912 nm. Intensity profiles of the boxed regions are shown at the right and represent average signal intensity (0–255 range, horizontal axis) versus depth (vertical axis). Image dimensions are 3.2 by 2 mm. (From Troutman, T.S. et al., *Opt. Lett.*, 32, 1438, 2007. With permission.)

14.3.2 Photothermal OCT

One caveat of scattering-based detection of nanoparticle targets for molecular OCT is that these methods suffer from limited contrast over a highly scattering tissue background. Photothermal imaging provides one potential method for increasing the molecular contrast of OCT over a highly scattering background, and has been the focus of recent work (Yablon et al. 1999, Telenkov et al. 2004, Kim et al. 2006, Nakata et al. 2006, Adler et al. 2008, Skala et al. 2008). In photothermal imaging, strong optical absorption of a small metal particle at its plasmon resonance results in a change in temperature around the particle (photothermal effect). This temperature change leads to a variation in the local index of refraction that can be optically detected with an amplitude-modulated heating beam that spatially overlaps with the focus position of the sample arm of an interferometer. Previous work has shown that photothermal interference contrast images of gold nanoparticles from a modified DIC microscope are insensitive to a highly scattering background (Boyer et al. 2002). Recent work has applied this concept to OCT with the added benefits of depth-resolution and increased imaging depth (Skala et al. 2008). This photothermal OCT system has a measured sensitivity of 14 parts per million (ppm, weight/weight) to 60 nm diameter gold nanospheres in tissue-like phantoms, and has been used to measure epidermal growth factor receptor expression from live cells in monolayer and in 3D tissue constructs (Skala et al. 2008).

The photothermal OCT system shown in Figure 14.16 consists of a diode pumped, solid-state frequency doubled Nd:YAG as the heating source (Coherent, Verdi) and a common-path spectral domain (SDOCT) system. The SDOCT system operates at a 20 kHz A-scan rate, with a super luminescent diode light source (Superlum) centered at 840 nm with full-width half-maximum bandwidth of 52 nm (6 μm axial resolution, 20 μm imaging spot size on the sample). The interferogram is detected using a custom-made spectrometer with a 1024 pixel line-scan CCD camera (Atmel, Aviiva). The focused spot of the pump beam overlaps with the focused spot of the imaging beam (20 μm pump beam spot size on the sample) and is amplitude-modulated with a chopper (Thorlabs).

Analytical expressions (derived from the equation of heat conduction) for the photothermal signal from gold nanoparticles detected interferometrically in a modified DIC microscope have previously been published (Boyer et al. 2002). The peak-to-peak phase shift near the heated gold nanoparticle is given by

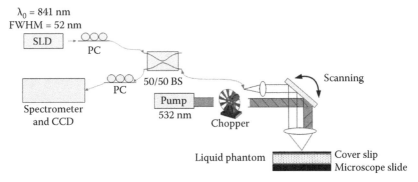

FIGURE 14.16 Schematic of the photothermal OCT setup. PC, polarization controller; SLD, super luminescent diode; BS, beam splitter. (From Skala, M.C. et al., *Nano Lett.*, 8, 3461, 2008. With permission.)

$$\Delta\phi_{PP} = 0.64 \left[\frac{\partial n}{\partial T} \frac{1}{\kappa} \right] \left[\frac{I(\vec{r}_0)}{\lambda} \right] N\sigma \qquad (14.6)$$

where

$\partial n / \partial T$ is the variation of the RI per unit of temperature for the medium ($\sim 10^{-4}$ K^{-1} for water at room temperature (Tilton and Taylor 1938))

κ is the thermal conductivity of the medium (0.19 W/K/m for water)

$I(\vec{r}_0)$ is the intensity of the heating beam at the nanoparticle position

λ is the imaging wavelength

N is the number of nanoparticles in the focal spot

σ is the absorption cross section of the nanoparticle at the heating wavelength (1.1×10^{-14} m^2 for 60 nm diameter gold nanospheres in water)

All calculations and experiments are at $I = 8$ kW/cm^2 and $\omega = 25$ Hz unless otherwise stated, and at these settings, $\Delta\phi_{PP} \sim 4 \times 10^{-4}$ radian for one nanosphere. Analytical expressions for temperature variations due to photothermal imaging were also derived by Cognet et al. (2003). For our system, the theoretical maximum instantaneous temperature change on the nanosphere surface is 24 K (12 K average temperature change), and this temperature change falls off exponentially with distance from the heat source (5 K 100 nm from the sphere surface).

The sensitivity of the photothermal OCT system was tested with liquid phantoms. Phantoms consisted of 1.55 µm diameter polystyrene spheres added to water/nanosphere solution to give $\mu_s = 100$ cm^{-1} ($g = 0.92$). The liquid phantom was sandwiched between a coverslip and microscope slide (~ 120 µm liquid depth). The pump beam was amplitude modulated at 25 Hz. The characteristic phase oscillations can be seen in a phantom with 84 ppm (Figure 14.17a). The photothermal signal (Figure 14.17b) was defined as the height of the 25 Hz peak in the Fourier transformed phase at the microscope slide surface, minus the background (mean of the Fourier transformed phase from 27 to 50 Hz). A total of 10 measurements were collected at each nanoparticle concentration ranging from 0 to 84 ppm. A linear relationship was found between the photothermal signal and the nanoparticle concentration, as expected (Figure 14.17c). The empirically measured sensitivity of the Skala et al. photothermal system at 8 kW/cm^2 pump power was 14 ppm (0 vs. 14 ppm, $p = 0.006$). The photothermal signal increases linearly with increasing pump power, as expected (Figure 14.17d). Pump beam intensities do not exceed the maximum permissible exposure levels in skin specified in the ANSI Standard for the Safe Use of Lasers (American National Standards Institute 2007).

Molecular targeting was tested in cells that overexpress the epidermal growth factor receptor (EGFR) (MDA-MB-468) and minimally express EGFR (MDA-MB-435). Additional controls

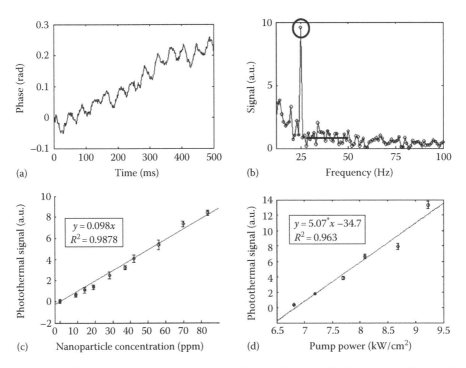

FIGURE 14.17 Phase of the tissue-like phantom (polystyrene spheres with µs = 100 cm^{-1}) with 84 ppm nanospheres and 25 Hz pump frequency (a). The definition of the photothermal signal (b) in the Fourier transformed phase is the peak at 25 Hz minus the background (27–50 Hz). The photothermal signal is linearly dependent on the nanoparticle concentration (c) in the scattering phantom, and the empirically measured sensitivity of the system is 14 ppm (0 vs. 14 ppm, $p = 0.006$). The photothermal signal is linearly dependent on the pump laser power (d). All error bars are standard error of $N = 10$ measurements. (From Skala, M.C. et al., *Nano Lett.*, 8, 3461, 2008. With permission.)

were performed with the same labeling protocol in a previous study (Curry et al. 2008). Dark-field microscopy was used as an independent method to confirm cell labeling (Curry et al. 2008). Trypan blue exclusion confirmed the viability of the cells after each experiment.

Three-dimensional cell constructs containing MDA-MB-468 (EGFR+) cells or MDA-MB-435 (EGFR-) cells were grown to demonstrate molecular photothermal imaging in thick samples with topical application of antibody-conjugated nanospheres (Figure 14.18). In each of the experimental and two control groups, $n = 17$ images were collected from two constructs (for a total of 51 images from six constructs). Nonparametric Wilcoxon rank sum tests were performed to determine whether the photothermal signals of experimental (EGFR+/nanosphere+) and control (EGFR–/nanosphere+ and EGFR+/nanosphere–) samples were significantly different, with a Bonferroni correction for the $n = 2$ comparisons. Images of the EGFR+/nanosphere+ construct had significantly higher photothermal signals ($p < 0.0001$ for EGFR+/nanosphere+ vs. EGFR–/nanosphere+ and EGFR+/nanosphere+ vs. EGFR+/nanosphere–) than the two controls (Figure 14.18d). There was no significant difference between the two controls ($p > 0.05$). These results verify MI of a targeted molecule (EGFR) using photothermal OCT with a topically applied contrast agent (antibody-conjugated nanospheres).

This study demonstrated a molecular OCT imaging technique capable of detecting a target with 14 ppm sensitivity (Skala et al. 2008). Molecular targeting was confirmed with EGFR-specific imaging in cell monolayer and in 3D tissue constructs. Photothermal OCT provides one example of how modified OCT imaging technology and image processing algorithms can potentially improve OCT contrast based on plasmonic gold nanoparticles.

14.3.3 Backscattering Albedo

As shown above, plasmon-resonant gold nanorods can serve as a scattering-based contrast agent for OCT (Troutman et al. 2007). However, novel signal processing algorithms can be used to further enhance the OCT contrast provided by gold nanorods in a tissue-like environment (Oldenburg et al. 2006). The mechanism of contrast investigated by Oldenburg et al. is based on the backscattering albedo (a'), defined as the ratio of the backscattering (μ_b) to extinction (μ_t) coefficient. Low backscattering albedo gold nanorods (14×44 nm; $\lambda_{max} = 780$ nm) within a high backscattering albedo tissue phantom were readily detected at 82 ppm (w/w) in a regime where extinction alone could not discriminate nanorods. The estimated threshold of detection was 30 ppm. Images of gold nanorods in tissue phantoms consisting of 2% Intralipid confirmed that the addition of nanorods to tissue decreases a' while increasing μ_t. Higher backscattering albedo silica spheres were investigated as a comparison, to demonstrate the relative merits of scattering- versus absorption-dominated contrast.

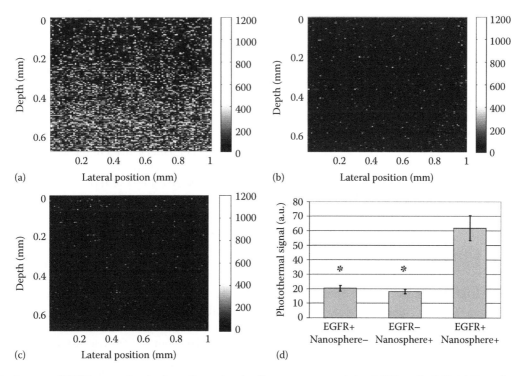

FIGURE 14.18 Images of EGFR expression in three-dimensional cell constructs containing EGFR+ cells (MDA-MB-468) with and without antibody-conjugated nanospheres (a and c, respectively) and EGFR– cells (MDA-MB-435) with antibody-conjugated nanospheres (b). There was a significant increase in the photothermal signal from EGFR overexpressing cell constructs labeled with antibody-conjugated nanospheres (d) compared to the two controls (EGFR+/Nanosphere– and EGFR–/Nanosphere+). $N = 17$ images for each group (*, $p < 0.0001$). Pump power 8.5 kW/cm². (From Skala, M.C. et al., *Nano Lett.*, 8, 3461, 2008. With permission.)

Gold nanorods (with dominant absorption) were found to provide higher contrast in tissue-like phantoms than the silica spheres (with dominant scattering).

14.3.4 Magnetomotive OCT

A final class of nanoparticle-based contrast for OCT is centered not on optical contrast mechanisms (as is the case with plasmonic gold nanoparticles), but on the tiny displacements of magnetic nanoparticles induced by an external magnetic field. Magnetic nanoparticles are attractive contrast agents for OCT (Oldenburg et al. 2005, 2008, Oh et al. 2007, 2008) because they are already FDA approved as contrast agents for MRI (Anker and Kopelman 2003). Magnetic nanoparticles composed of biocompatible iron oxides exhibit magnetic susceptibilities that are typically $>10^5$ times larger than that of human tissue, including red blood cells (Schenck 2005). Magnetic nanoparticles can be externally controlled with a magnetic field after they have been introduced into the tissue, and thus offer the possibility of lock-in detection to small magnetic nanoparticle displacements induced by the external magnetic field. Phase-sensitive OCT is particularly attractive for this detection scheme because it is sensitive to nanometer-scale displacements.

Oldenburg et al. recently demonstrated magnetomotive OCT with a sensitivity of 27 µg/g (~2 nM) to iron oxide magnetic nanoparticles (~20 nm diameter) in tissue phantoms (Oldenburg et al. 2008). This sensitivity was achieved with phase-resolved OCT in tissue phantoms exposed to ~800 G magnetic fields modulated at 56 and 100 Hz to mechanically actuate iron oxide magnetic nanoparticles embedded in the tissue phantoms. Paramagnetic vibrations were filtered from diamagnetic vibrations in these studies using a mechanical phase lag filter equation.

Oldenburg et al. also used magnetomotive OCT to monitor magnetic nanoparticle diffusion into an excised rat tumor over a 2 h time period (Figure 14.19) (Oldenburg et al. 2008). Ex vivo rat mammary tumors were immersed in a saline solution with

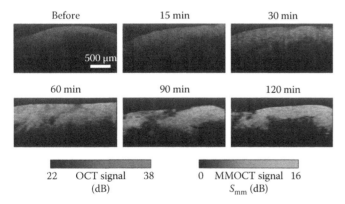

FIGURE 14.19 (See color insert.) Representative magnetomotive OCT images of iron oxide nanoparticle diffusion in tumors versus time. Red and green display the structural (OCT) and magnetomotive (MMOCT) image channels, respectively, as indicated by the colored scale bars. (From Oldenburg, A.L. et al., *Opt. Express*, 16, 11525, 2008. With permission.)

4 mg/g COOH-magnetic nanoparticles for 15 min at room temperature, rinsed in pure saline for 1 min, and then imaged using magnetomotive OCT. The magnetomotive OCT images exhibit a useful dynamic range of approximately 16 dB. There appears to be structure in the magnetomotive OCT image correlated with the structure in the OCT image, which may be due to liquid voids. While there is some transverse variation in the magnetomotive OCT signals, such as in the 15 and 60 min images, there is no depth-dependent gradient that one might expect to observe during a diffusion process. This may be because the diffusion length at ≥15 min is much longer than the imaging depths of ~600 µm achieved in this tumor. However, further studies are needed to fully characterize the depth-dependent magnetomotive OCT signal. These and other studies (Oldenburg et al. 2005, 2008, Oh et al. 2007, 2008) demonstrate that magnetomotive OCT is a promising method for molecular contrast in OCT imaging, and may aid in the development of magnetic nanoparticles for biomedical applications such as hyperthermic therapy.

14.3.5 Improved Delivery for In Vivo Imaging

One challenge for translating nanoparticle-based OCT imaging in vivo is efficient delivery and distribution of nanoparticles in the tissue of interest (Kim et al. 2009). Topical delivery of nanoparticles is advantageous for OCT because the imaging depth of OCT is only ~2 mm in tissue. Topical delivery can avoid adverse systemic effects while delivering a large quantity of contrast agent to the epithelial layers where many types of oral, skin, and gastrointestinal cancers develop. The penetration of nanoparticles to subsurface epithelial layers is greatly hampered by various biological barriers such as the stratum corneum (the 10–15 µm thick outermost nonliving tissue layer) (Figure 14.20a) (Coulman et al. 2006). A recent study (Kim et al. 2009) hypothesized that this obstacle to nanoparticle delivery for OCT imaging could be overcome by generating micropassages through the stratum corneum using microneedles, followed by enhanced distribution of nanoparticles in the underlying epithelial layers with ultrasonic forces (Figure 14.20b and c) (Larina et al. 2005). This multimodal delivery of antibody-conjugated gold nanoparticles (71 nm diameter gold spheres conjugated to anti-EGFR) resulted in enhanced contrast of in vivo OCT images of oral dysplasia in a hamster model.

The hamster cheek pouch model, which mimics the development of human oral cancer by chronic treatment with DMBA (9, 10-dimethyl-1,2-benzanthracene), was used to test the efficacy of multimodal nanoparticle delivery (Kim et al. 2009). CR3 roller microneedles (MTS dermaroller, clinical resolution laboratory) with holes of 70 µm diameter and 300 µm depth were rolled on both DMBA-treated and DMBA-untreated hamster cheek pouches three times at three different angles (0°, 45°, and 90°). 200 µL of the anti-EGFR conjugated PEGylated gold nanoparticle solution (1.78×10^{10} particles/mL) was topically applied to the hamster's cheek pouch for 10 min by dropping it into a well on the tissue surface. After nanoparticle application, 0.3 W/cm² of 1 MHz ultrasonic force was applied to the

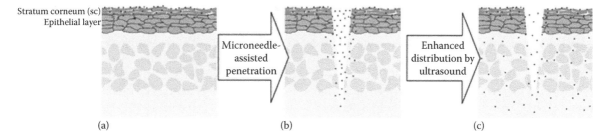

Stratum corneum (sc)
Epithelial layer

Microneedle-assisted penetration

Enhanced distribution by ultrasound

(a)   (b)   (c)

FIGURE 14.20 Schematic of a potential delivery strategy for in vivo topical application of gold nanoparticles. The stratum corneum and underlying epithelial layers are a biological barrier to gold nanoparticle transport (a). Penetration of gold nanoparticles into the subsurface epithelial layers could be enhanced through a passage generated by a microneedle (b). Enhanced gold nanoparticle distribution could be achieved in tissue by ultrasonic force (c). (From Kim, C.S. et al., *J. Biomed. Opt.*, 14, 034008, 2009. With permission.)

cheek pouch for 1 min using a Dynatron 125 ultrasonicator (Dynatronics Corporation).

Kim et al. found that the microneedle treated, ultrasound-enhanced tissue topically exposed to gold nanoparticles (Figure 14.21b) shows higher OCT contrast than tissue only topically exposed to gold nanoparticles (Figure 14.21a). The mean intensity of normal and DMBA-treated tissue using multimodal delivery of gold nanoparticles increased by 149% and 177%, respectively, compared to simple topical delivery of gold nanoparticles.

The duration of gold nanoparticle-induced OCT signal enhancement was also investigated by Kim et al. (2009). Normal hamster cheek pouches were imaged over a period of 40 days after initial administration of gold nanoparticles using microneedles and ultrasound. No noticeable changes were observed for 2 days. OCT signals in epithelial and connective tissue regions were significantly diminished after a week, indicating that the gold nanoparticles were cleared from these deeper tissue layers. However, enhanced signals from the stratum corneum persisted over the entire 40 days, perhaps due to poor active transport in this tissue layer. Histological analysis also found that no inflammation or other adverse effects were observed in the hamster cheek pouch over a period of 40 days, indicating the potential biocompatibility of the gold nanoparticles. However, no selective antibody-mediated binding of the gold nanoparticles was observed in these studies. This may be due to the enhanced permeation and retention effects on relatively large particles in tissue. Due to their size (71 nm), the gold nanoparticles may have been immobilized in the tissue with a low EGFR-encountering frequency. Further studies investigating methods for improved

penetration and distribution of topically applied nanoparticles will greatly enhance the effectiveness of nanoparticle-based contrast for in vivo OCT imaging.

14.4 Discussion

As the functional extensions of OCT continue to develop, the impact of this technology can be seen in preclinical studies outside of the OCT development community. Doppler OCT has been used to test the efficacy of photodynamic tumor therapies in vivo (Collins et al. 2008, Standish et al. 2008), magnetomotive OCT has been used to detect macrophages in atherosclerotic plaque (Oh et al. 2008), and nanoshell-enhanced OCT images have been used to monitor photothermal therapy in tumors in vivo (Loo et al. 2004, Gobin et al. 2007). With the development of more sophisticated functional OCT technologies, multimodal OCT systems that combine structural, hemodynamic, and MI may become routine for the development of new drugs in preclinical models of cancer, stroke, Alzheimer's, developmental defects, etc.

OCT metrics for microvascular network structure, blood flow, hematocrit, blood oxygenation, and total blood content are still in the early stages of development and, thus, require robust validation. These preclinical validation studies could include comparisons with established imaging techniques, such as high-frequency ultrasound, micro-PET, MRI, and micro-CT, or with histology and immunohistochemistry. Multimodal studies could also compare OCT and other OI techniques such as multiphoton microscopy, diffuse reflectance spectroscopy, and photon migration. These validation studies would not only define the capabilities of OCT for hemodynamic imaging, but could also establish a niche for hemodynamic OCT within the medical and OI communities.

Molecular contrast OCT is within an even earlier stage of development than hemodynamic OCT. However, significant progress has been made to establish functional contrast mechanisms for intrinsic tissue molecules, extrinsic dyes, and most recently for nanoparticles. The success of nanoparticle-based OCT imaging will depend on the delivery, specific uptake, and clearance of nanoparticles in small animals. OCT is a surface-imaging technique, so the simplest and most practical route of

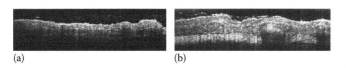

(a)   (b)

FIGURE 14.21 In vivo OCT images of DMBA-treated hamster cheek pouches exposed to gold nanoparticles (a) and microneedle treated, exposed to gold nanoparticles and ultrasound (b). Images are approximately 4 mm in length and 1 mm in depth. (From Kim, C.S. et al., *J. Biomed. Opt.*, 14, 034008, 2009. With permission.)

administration is topical delivery of nanoparticles. However, there is great interest in systemic delivery of nanoparticles in the biomaterials community and these studies could be used to guide intravenous injection of nanoparticle contrast agents for OCT. For example, De Jong et al. (2008) performed a kinetic study to evaluate the effects of gold nanosphere size (10, 50, 100, and 250 nm diameter) on the in vivo distribution of nanoparticles in rats. The authors showed that the tissue distribution of gold nanoparticles (24 h after intravenous injection) is size dependent with the smallest 10 nm nanoparticles possessing the most widespread organ distribution in rats. The number of particles found in the liver was the highest, followed by blood, spleen, kidney, lungs, brain, reproductive organs, thymus, and heart, whereas the 50, 100, and 250 nm particles were almost solely distributed to the liver, spleen, and blood. Further in vivo experiments (Terentyuk et al. 2009) coupled with in vitro toxicity studies (Lewinski et al. 2008) will begin to elucidate the biocompatibility of nanoparticles, and the best route for administering nanoparticles as contrast agents for OCT in small animals.

References

Adler, D. C., Huang, S.-W., Huber, R., and Fujimoto, J. G. 2008. Photothermal detection of gold nanoparticlesusing phase-sensitive optical coherencetomography. *Opt. Express* 16: 4376–4393.

Agrawal, A., Huang, S., Wei Haw Lin, A. et al. 2006. Quantitative evaluation of optical coherence tomography signal enhancement with gold nanoshells. *J. Biomed. Opt.* 11: 041121.

Aguirre, A. D., Chen, Y., Fujimoto, J. G., et al. 2006. Depth-resolved imaging of functional activation in the rat cerebral cortex using optical coherence tomography. *Opt. Lett.* 31: 3459–3461.

American National Standards Institute, I. 2007. *ANSI Z136.1-2007 American National Standard for Safe Use of Lasers.* Orlando, FL: Laser Institute of America.

Anker, J. N. and Kopelman, R. 2003. Magnetically modulated optical nanoprobes. *J. Appl. Phys.* 93: 6698–6700.

Applegate, B. E. and Izatt, J. A. 2006. Molecular imaging of endogenous and exogenous chromophores using ground state recovery pump-probe optical coherence tomography. *Opt. Express* 14: 9142–9155.

Applegate, B. E., Yang, C., Rollins, A. M., and Izatt, J. A. 2004. Polarization-resolved second-harmonic-generation optical coherence tomography in collagen. *Opt. Lett.* 29: 2252–2254.

Barton, J. and Stromski, S. 2005. Flow measurement without phase information in optical coherence tomography images. *Opt. Express* 13: 5234–5239.

Barton, J. K., Halas, N. J., West, J. L., and Drezek, R. A. 2004. Nanoshells as an optical coherence tomography contrast agent. *Proc. SPIE* 5316: 99

Baumann, B., Pircher, M., Götzinger, E., and Hitzenberger, C. 2007. Full range complex spectral domain optical coherence tomography without additional phase shifters. *Opt. Express* 15: 13375–13387.

Bower, B. A., Zhao, M., Zawadzki, R. J., and Izatt, J. A. 2007. Real-time spectral domain Doppler optical coherence tomography and investigation of human retinal vessel autoregulation. *J. Biomed. Opt.* 12: 041214.

Boyer, D., Tamarat, P., Maali, A., Lounis, B., and Orrit, M. 2002. Photothermal imaging of nanometer-sized metal particles among scatterers. *Science* 297: 1160–1163.

Butcher, J. T., McQuinn, T. C., Sedmera, D., Turner, D., and Markwald, R. R. 2007. Transitions in early embryonic atrioventricular valvular function correspond with changes in cushion biomechanics that are predictable by tissue composition. *Circ. Res.* 100: 1503–1511.

Cang, H., Sun, T., Li, Z. Y. et al. 2005. Gold nanocages as contrast agents for spectroscopic optical coherence tomography. *Opt. Lett.* 30: 3048–3050.

Carpenter, G. 1987. Receptors for epidermal growth factor and other polypeptide mitogens. *Annu. Rev. Biochem.* 56: 881–914.

Chen, J., Saeki, F., Wiley, B. J. et al. 2005. Gold nanocages: Bioconjugation and their potential use as optical imaging contrast agents. *Nano Lett.* 5: 473–477.

Chen, Y., Aguirre, A. D., Ruvinskaya, L. et al. 2009. Optical coherence tomography (OCT) reveals depth-resolved dynamics during functional brain activation. *J. Neurosci. Methods* 178: 162–173.

Cheung, A. M., Brown, A. S., Cucevic, V. et al. 2007. Detecting vascular changes in tumour xenografts using micro-ultrasound and micro-ct following treatment with VEGFR-2 blocking antibodies. *Ultrasound Med. Biol.* 33: 1259–1268.

Cognet, L., Tardin, C., Boyer, D. et al. 2003. Single metallic nanoparticle imaging for protein detection in cells. *Proc. Natl. Acad. Sci. USA* 100: 11350–11355.

Collins, H. A., Khurana, M., Moriyama, E. H. et al. 2008. Blood-vessel closure using photosensitizers engineered for two-photon excitation. *Nat. Photonics* 2: 420–424.

Coulman, S. A., Barrow, D., Anstey, A. et al. 2006. Minimally invasive cutaneous delivery of macromolecules and plasmid DNA via microneedles. *Curr. Drug Deliv.* 3: 65–75.

Curry, A. C., Crow, M., and Wax, A. 2008. Molecular imaging of epidermal growth factor receptor in live cells with refractive index sensitivity using dark-field microspectroscopy and immunotargeted nanoparticles. *J. Biomed. Opt.* 13: 014022.

Davis, A., Izatt, J., and Rothenberg, F. 2009. Quantitative measurement of blood flow dynamics in embryonic vasculature using spectral Doppler velocimetry. *Anat. Rec. (Hoboken)* 292: 311–319.

Davis, A. M., Rothenberg, F. G., Shepherd, N., and Izatt, J. A. 2008. In vivo spectral domain optical coherence tomography volumetric imaging and spectral Doppler velocimetry of early stage embryonic chicken heart development. *J. Opt. Soc. Am. A: Opt. Image Sci. Vis.* 25: 3134–3143.

De Jong, W. H., Hagens, W. I., Krystek, P. et al. 2008. Particle size-dependent organ distribution of gold nanoparticles after intravenous administration. *Biomaterials* 29: 1912–1919.

Dedeugd, C., Wankhede, M., and Sorg, B. S. 2009. Multimodal optical imaging of microvessel network convective oxygen transport dynamics. *Appl. Opt.* 48: D187–197.

Dewhirst, M. W., Oliver, R., Tso, C. Y. et al. 1990. Heterogeneity in tumor microvascular response to radiation. *Int. J. Radiat. Oncol. Biol. Phys.* 18: 559–568.

Faber, D. J., Mik, E. G., Aalders, M. C. G., and van Leeuwen, T. G. 2003. Light absorption of (oxy-)hemoglobin assessed by spectroscopic optical coherence tomography. *Opt. Lett.* 28: 1436–1438.

Faber, D. J. and van Leeuwen, T. G. 2009. Are quantitative attenuation measurements of blood by optical coherence tomography feasible? *Opt. Lett.* 34: 1435–1437.

Fercher, A. F. 2008. Inverse scattering, dispersion, and speckle in optical coherence tomography. In: Drexler, W. and Fujimoto, J. G. (eds.). *Optical Coherence Tomography*. Berlin: Springer, pp. 119–146.

Fingler, J., Schwartz, D., Yang, C., and Fraser, S. E. 2007. Mobility and transverse flow visualization using phase variance contrast with spectral domain optical coherence tomography. *Opt. Express* 15: 12636–12653.

Ghiglia, D. C. and Pritt, M. D. 1998. *Two-Dimensional Phase Unwrapping: Theory, Algorithms, and Software*. New York: Wiley.

Gobin, A. M., Lee, M. H., Halas, N. J. et al. 2007. Near-infrared resonant nanoshells for combined optical imaging and photothermal cancer therapy. *Nano Lett.* 7: 1929–1934.

Hendargo, H. C., Zhao, M., Shepherd, N., and Izatt, J. A. 2009. Synthetic wavelength based phase unwrapping in spectral domain optical coherence tomography. *Opt. Express* 17: 5039–5051.

Homola, J., Yee, S. S., and Gauglitz, G. 1999. Surface plasmon resonance sensors: Review. *Sens. Actuators B* 54: 3–15.

Hu, N. and Clark, E. B. 1989. Hemodynamics of the stage 12 to stage 29 chick embryo. *Circ.Res.* 65: 1665–1670.

Iftimia, N. V., Hammer, D. X., Ferguson, R. D. et al. 2008. Dual-beam Fourier domain optical Doppler tomography of zebrafish. *Opt. Express* 16: 13624–13636.

Izatt, J. A., Kulkarni, M. D., Yazdanfar, S., Barton, J. K., and Welch, A. J. 1997. In vivo bidirectional color Doppler flow imaging of picoliter blood volumes using optical coherence tomography. *Opt. Lett.* 22: 1439–1441.

Kim, C. S., Wilder-Smith, P., Ahn, Y. C. et al. 2009. Enhanced detection of early-stage oral cancer in vivo by optical coherence tomography using multimodal delivery of gold nanoparticles. *J. Biomed. Opt.* 14: 034008.

Kim, J., Oh, J., and Milner, T. E. 2006. Measurement of optical path length change following pulsed laser irradiation using differential phase optical coherence tomography. *J. Biomed. Opt.* 11: 041122.

Kirillin, M., Shirmanova, M., Sirotkina, M. et al. 2009. Contrasting properties of gold nanoshells and titanium dioxide nanoparticles for optical coherence tomography imaging of skin: Monte Carlo simulations and in vivo study. *J. Biomed. Opt.* 14: 021017.

Kwok, T. T. and Sutherland, R. M. 1991. Differences in EGF related radiosensitisation of human squamous carcinoma cells with high and low numbers of EGF receptors. *Br. J. Cancer* 64: 251–254.

Larina, I. V., Evers, B. M., Ashitkov, T. V. et al. 2005. Enhancement of drug delivery in tumors by using interaction of nanoparticles with ultrasound radiation. *Technol. Cancer Res. Treat.* 4: 217–226.

Larina, I. V., Ivers, S., Syed, S., Dickinson, M. E., and Larin, K. V. 2009. Hemodynamic measurements from individual blood cells in early mammalian embryos with Doppler swept source OCT. *Opt. Lett.* 34: 986–988.

Lewinski, N., Colvin, V., and Drezek, R. 2008. Cytotoxicity of nanoparticles. *Small* 4: 26–49.

Loo, C., Lin, A., Hirsch, L. et al. 2004. Nanoshell-enabled photonics-based imaging and therapy of cancer. *Technol. Cancer Res. Treat.* 3: 33–40.

Mariampillai, A., Standish, B. A., Moriyama, E. H. et al. 2008. Speckle variance detection of microvasculature using swept-source optical coherence tomography. *Opt. Lett.* 33: 1530–1532.

McQuinn, T. C., Bratoeva, M., Dealmeida, A. et al. 2007. High-frequency ultrasonographic imaging of avian cardiovascular development. *Dev. Dyn.* 236: 3503–3513.

Nakata, T., Yoshimura, K., and Ninomiya, T. 2006. Real-time photodisplacement microscope for high-sensitivity simultaneous surface and subsurface inspection. *Appl. Opt.* 45: 2643–2655.

Oh, J., Feldman, M. D., Kim, J. et al. 2007. Magneto-motive detection of tissue-based macrophages by differential phase optical coherence tomography. *Lasers Surg. Med.* 39: 266–272.

Oh, J., Feldman, M. D., Kim, J. et al. 2008. Detection of macrophages in atherosclerotic tissue using magnetic nanoparticles and differential phase optical coherence tomography. *J. Biomed. Opt.* 13: 054006.

Oldenburg, A. L., Crecea, V., Rinne, S. A., and Boppart, S. A. 2008. Phase-resolved magnetomotive OCT for imaging nanomolar concentrations of magnetic nanoparticles in tissues. *Opt. Express* 16: 11525–11539.

Oldenburg, A. L., Gunther, J. R., and Boppart, S. A. 2005. Imaging magnetically labeled cells with magnetomotive optical coherence tomography. *Opt. Lett.* 30: 747–749.

Oldenburg, A. L., Hansen, M. N., Zweifel, D. A., Wei, A., and Boppart, S. A. 2006. Plasmon-resonant gold nanorods as low backscattering albedo contrast agents for optical coherence tomography. *Opt. Express* 14: 6724–6738.

Oldenburg, S. J., Averitt, R. D., Westcott, S. L., and Halas, N. J. 1998. Nanoengineering of optical resonances. *Chem. Phys. Lett.* 288: 243–247.

Oldenburg, S. J., Jackson, J. B., Westcott, S. L., and Halas, N. J. 1999. Infrared extinction properties of gold nanoshells. *Appl. Phys. Lett.* 75: 2897–2899.

Park, B., Pierce, M., Cense, B., and de Boer, J. 2003. Real-time multi-functional optical coherence tomography. *Opt. Express* 11: 782–793.

Schenck, J. F. 2005. Physical interactions of static magnetic fields with living tissues. *Prog. Biophys. Mol. Biol.* 87: 185–204.

Skala, M. C., Crow, M. J., Wax, A., and Izatt, J. A. 2008. Photothermal optical coherence tomography of epidermal growth factor receptor in live cells using immunotargeted gold nanospheres. *Nano Lett.* 8: 3461–3467.

Skala, M. C., Fontanella, A., Hendargo, H., Dewhirst, M. W., and Izatt, J. A. 2009. Combined hyperspectral and spectral domain optical coherence tomography microscope for non-invasive hemodynamic imaging. *Opt. Lett.* 34: 289–291.

Skrabalak, S. E., Au, L., Lu, X., Li, X., and Xia, Y. 2007. Gold nanocages for cancer detection and treatment. *Nanomedicine* 2: 657–668.

Sorg, B. S., Hardee, M. E., Agarwal, N., Moeller, B. J., and Dewhirst, M. W. 2008. Spectral imaging facilitates visualization and measurements of unstable and abnormal microvascular oxygen transport in tumors. *J. Biomed. Opt.* 13: 014026.

Sorg, B. S., Moeller, B. J., Donovan, O., Cao, Y., and Dewhirst, M. W. 2005. Hyperspectral imaging of hemoglobin saturation in tumor microvasculature and tumor hypoxia development. *J. Biomed. Opt.* 10: 44004.

Standish, B. A., Lee, K. K., Jin, X. et al. 2008. Interstitial Doppler optical coherence tomography as a local tumor necrosis predictor in photodynamic therapy of prostatic carcinoma: An in vivo study. *Cancer Res.* 68: 9987–9995.

Tao, Y. K., Davis, A. M., and Izatt, J. A. 2008. Single-pass volumetric bidirectional blood flow imaging spectral domain optical coherence tomography using a modified Hilbert transform. *Opt. Express* 16: 12350–12361.

Tao, Y. K., Kennedy, K. M., and Izatt, J. A. 2009. Velocity-resolved 3D retinal microvessel imaging using single-pass flow imaging spectral domain optical coherence tomography. *Opt. Express* 17: 4177–4188.

Telenkov, S. A., Dave, D. P., Sethuraman, S., Akkin, T., and Milner, T. E. 2004. Differential phase optical coherence probe for depth-resolved detection of photothermal response in tissue. *Phys. Med. Biol.* 49: 111–119.

Terentyuk, G. S., Maslyakova, G. N., Suleymanova, L. V. et al. 2009. Circulation and distribution of gold nanoparticles and induced alterations of tissue morphology at intravenous particle delivery. *J. Biophotonics* 2: 292–302.

Tilton, L. W. and Taylor, J. K. 1938. Refractive index and dispersion of distilled water for visible radiation, at temperatures 0 to 60°C. *J. Res. Natl. Bur. Stand.* 20: 419–425.

Troutman, T. S., Barton, J. K., and Romanowski, M. 2007. Optical coherence tomography with plasmon resonant nanorods of gold. *Opt. Lett.* 32: 1438–1440.

Tsuzuki, Y., Fukumura, D., Oosthuyse, B. et al. 2000. Vascular endothelial growth factor (VEGF) modulation by targeting hypoxia-inducible factor-1alpha → hypoxia response element → VEGF cascade differentially regulates vascular response and growth rate in tumors. *Cancer Res.* 60: 6248–6252.

Vakhtin, A. B., Kane, D. J., Wood, W. R., and Peterson, K. A. 2003. Common-path interferometer for frequency-domain optical coherence tomography. *Appl. Opt.* 42: 6953–6958.

van Leeuwen, T. G., Kulkarni, M. D., Yazdanfar, S., Rollins, A. M., and Izatt, J. A. 1999. High-flow-velocity and shear-rate imaging by use of color Doppler optical coherence tomography. *Opt. Lett.* 24: 1584–1586.

Vennemann, P., Kiger, K. T., Lindken, R. et al. 2006. In vivo micro particle image velocimetry measurements of blood-plasma in the embryonic avian heart. *J. Biomech.* 39: 1191–1200.

Vinegoni, C., Bredfeldt, J., Marks, D., and Boppart, S. 2004. Nonlinear optical contrast enhancement for optical coherence tomography. *Opt. Express* 12: 331–341.

Wang, R. 2007. In vivo full range complex Fourier domain optical coherence tomography. *Appl. Phys. Lett.* 90: 054103.

Wang, R., Jacques, S., Ma, Z. et al. 2007. Three dimensional optical angiography. *Opt. Express* 15: 4083–4097.

Wang, R. K. and An, L. 2009. Doppler optical micro-angiography for volumetric imaging of vascular perfusion in vivo. *Opt. Express* 17: 8926–8940.

White, B., Pierce, M., Nassif, N. et al. 2003. In vivo dynamic human retinal blood flow imaging using ultra-high-speed spectral domain optical coherence tomography. *Opt. Express* 11: 3490–3497.

Xu, C., Ye, J., Marks, D. L., and Boppart, S. A. 2004. Near-infrared dyes as contrast-enhancing agents for spectroscopic opticalcoherence tomography. *Opt. Lett.* 29: 1647–1649.

Xu, X., Yu, L., and Chen, Z. 2008. Effect of erythrocyte aggregation on hematocrit measurement using spectral-domain optical coherence tomography. *IEEE Trans. Biomed. Eng.* 55: 2753–2758.

Yablon, A. D., Nishioka, N. S., Mikic, B. B., and Venugopalan, V. 1999. Measurement of tissue absorption coefficients by use of interferometric photothermal spectroscopy. *Appl. Opt.* 38: 1259–1272.

Yang, C., Choma, M. A., Lamb, L. E., Simon, J. D., and Izatt, J. A. 2004a. Protein-based molecular contrast optical coherence tomography with phytochrome as the contrast agent. *Opt. Lett.* 29: 1396–1398.

Yang, C., McGuckin, L. E., Simon, J. D. et al. 2004b. Spectral triangulation molecular contrast optical coherence tomography with indocyanine green as the contrast agent. *Opt. Lett.* 29: 2016–2018.

Yang, V., Gordon, M., Qi, B. et al. 2003. High speed, wide velocity dynamic range Doppler optical coherence tomography (Part I): System design, signal processing, and performance. *Opt. Express* 11: 794–809.

Yasuno, Y., Makita, S., Endo, T. et al. 2006. Simultaneous BM-mode scanning method for real-time full-range Fourier domain optical coherence tomography. *Appl. Opt.* 45: 1861–1865.

Yazdanfar, S., Kulkarni, M. D., and Izatt, J. A. 1997. High resolution imaging of in vivo cardiac dynamics using color Doppler optical coherence tomography. *Opt. Exp.* 1: 424–431.

Yazdanfar, S., Rollins, A. M., and Izatt, J. A. 2000. Imaging and velocimetry of the human retinal circulation with color Doppler optical coherence tomography. *Opt. Lett.* 25: 1448–1450.

Yazdanfar, S., Rollins, A. M., and Izatt, J. A. 2003. In vivo imaging of human retinal flow dynamics by color Doppler optical coherence tomography. *Arch. Ophthalmol.* 121: 235–239.

Zagaynova, E. V., Shirmanova, M. V., Kirillin, M. Y. et al. 2008. Contrasting properties of gold nanoparticles for optical coherence tomography: Phantom, in vivo studies and Monte Carlo simulation. *Phys. Med. Biol.* 53: 4995–5009.

Optical Coherence Tomography: Clinical Applications

Brian D. Goldberg
Axsun Technologies

Melissa J. Suter
Massachusetts General Hospital

Guillermo J. Tearney
Massachusetts General Hospital

Brett E. Bouma
Massachusetts General Hospital

15.1 Introduction 303
15.2 Ophthalmology 303
15.3 Cardiology 304
    OCT Imaging of Atherosclerotic Plaque • OCT Imaging of Coronary Stents • Noncoronary
    Cardiac Applications
15.4 Gastroenterology 308
    Esophagus • Colon • Duodenum • Pancreatobiliary Ductal System
15.5 Other Clinical Applications 314
    Breast Cancer • Pulmonology
15.6 Summary...315
References..315

15.1 Introduction

The first in vivo clinical application of optical coherence tomography (OCT) was in the early 1990s in the field of ophthalmology (Fercher et al. 1993, Swanson et al. 1993). Since that time, OCT has been applied to a wide range of clinical fields, including gastroenterology, cardiology, oncology, pulmonology, dermatology, dentistry, and experimental biology. The axial, or depth, resolution of OCT is on the order of a few microns, similar to that of a standard histology slice. Thus, OCT is often referred to as an "optical biopsy" and images resemble histologic sections obtained by standard biopsy or staining techniques. The micron-scale resolution places OCT in between higher resolution optical techniques such as confocal microscopy and lower resolution nonoptical techniques such as ultrasound, MRI, and CT.

This chapter focuses on the clinical applications of OCT. A brief treatment of OCT in ophthalmology will be presented followed by discussions of the two largest emerging applications of OCT—in cardiology for studying vulnerable plaques in coronary arteries and in gastroenterology for Barrett's esophagus (BE) detection. In addition to both preclinical and clinical imaging work in these fields, the development of various endoscopic and catheter-based probes will be discussed.

15.2 Ophthalmology

The use of OCT in clinical ophthalmology is by far the most advanced and has become a critical tool for studying and diagnosing both anterior and posterior eye disease. OCT is particularly well suited for imaging the eye because it is noncontact, noninvasive, and can achieve high resolution of the entire retinal field in a short amount of time. Typically, ophthalmic OCT systems use near-infrared light sources in the 800 nm range. This wavelength is chosen to minimize the effects of water absorption in the vitreous humor. Other wavelengths such as 930 and 1060 nm are becoming increasingly more common and provide a tradeoff between imaging depth and axial resolution. Commercially available OCT systems designed for ophthalmic applications have axial resolutions ranging from 3 to 5 μm and A-line rates in the 20–50,000 A-scans/s range (Gabriele et al. 2008).

In the posterior eye, OCT has been used to identify all the major surface and subsurface retinal features in a highly reproducible fashion. Figure 15.1 shows a comparison between retinal images using a swept source at 1050 nm and a spectral domain (SD)-OCT system at 840 nm (Lee et al. 2006). OCT retinal imaging has been used as a tool to diagnose and investigate macular hole formation (Michalewska et al. 2009), macular edema, which is correlated with diabetic retinopathy (Hannouche and de Avila 2009), and age-related macular degeneration (de Bruin et al. 2008). In addition, measurements of the retinal nerve fiber layer (RNFL) (Cense et al. 2004, Anderson et al. 2005), a key indicator of glaucomatous damage have become an important tool for detecting and monitoring glaucoma, a leading cause of blindness worldwide.

In the anterior eye, also referred to as the anterior segment, OCT has been used to monitor corneal thickness and shape during descemet stripping and endothelial keratoplasty (DSEK) (Tarnawska and Wylegala 2010) and laser-assisted in situ keratomileusis (LASIK) (Kucumen et al. 2009) surgeries. Anterior segment OCT is also used in biometry to measure intraocular lens power (Konstantopoulos et al. 2007, Ramos et al. 2009) as

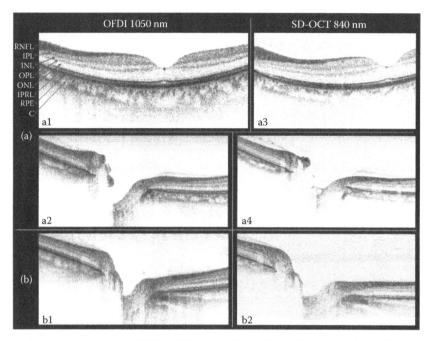

FIGURE 15.1 Comparison of two imaging systems (OFDI at 1050 nm and SD-OCT at 840 nm). a1 and a2: OFDI images at fovea and optic nerve head, respectively, from volunteer A, 36-year-old Asian male. a3 and a4: SD-OCT images from the same volunteer at similar tissue locations. b1 and b2: OFDI and SD-OCT images, respectively, obtained from volunteer B, 41-year-old Caucasian male. OFDI images exhibit considerably deeper penetration in tissue than SD-OCT images in all the data sets. The OFDI image (a1) shows the anatomical layered structure: RNFL; retinal nerve fiber layer, IPL; inner plexiform layer, INL; inner nuclear layer, OPL; outer plexiform layer, ONL; outer nuclear layer, IPRL; interface between the inner and outer segments of the photoreceptor layer, RPE; retinal pigmented epithelium, and C; choriocapillaris and choroid. (From Lee, E.C. et al., *Opt. Express*, 14(10), 4403, 2006. With permission.)

well as to measure ocular angle, which is a key indicator of intra-ocular pressure (Nolan 2008). Figure 15.2 shows a cross-sectional image of the anterior segment taken with a high-speed complementary metal oxide semiconductor (CMOS) SD-OCT system at 135,000 A-scans/s (Grulkowski et al. 2009). A 3D rendering of the anterior eye is shown in Figure 15.3 taken with a 1300 nm source from Axsun Technologies at 50,000 A-scans/s (Wojtkowski 2010).

For a more detailed discussion of ophthalmic OCT applications, please refer to numerous review articles (Costa et al. 2006, van Velthoven et al. 2007, Simpson et al. 2008, Ramos et al. 2009, Sakata et al. 2009) as well as books (Schuman et al. 2004, Steinert and Huang 2008) on the subject.

15.3 Cardiology

15.3.1 OCT Imaging of Atherosclerotic Plaque

In the field of cardiology, OCT has emerged as a powerful tool for studying disease within coronary arteries. The main goal of intracoronary imaging is to detect, study, and treat coronary disease that may lead to both acute coronary events such as myocardial infarctions (AMI) as well as more chronic disease such as ischemia and angina caused by vessel occlusion. The ability to image within a coronary vessel with high speed and high resolution enables the clinician and scientist a new tool for studying these dynamics. Currently, commercial systems

for intracoronary OCT are available from Lightlab Imaging (Westford, Massachusetts) based on both time-domain (TD)-OCT (M2x) and swept source (SS)-OCT approaches (C7-XR™). The C7-XR currently has both CE mark and FDA approval. Other companies are currently developing systems based on SS-OCT technologies (Volcano Corporation, Terumo Medical) for this emerging clinical application.

The leading hypothesis for the cause of acute events is rupture, thrombus formation, and vessel occlusion of a vulnerable atherosclerotic plaque. There are many types of atherosclerotic plaque including thin and thick-capped fibroatheromas as well as fibrocalcific, lipid-rich, and necrotic core lesions. A detailed treatment of different plaque types is outside the scope of this book. However, it should be noted that plaque type, morphology, and inflammation levels are all thought to play a role in determining rupture vulnerability. The majority of vulnerable plaques are thought to be thin-cap fibroatheromas (TCFAs) and described as having a thin fibrous cap (<65 μm), large lipid pool underlying the thin cap, and increased macrophage activity (Libby 2001).

Intracoronary OCT is most often compared with intravascular ultrasound (IVUS). In general, OCT has a roughly 10× improvement in axial (~15–20 μm vs. 100–200 μm) and transverse resolution (~20–40 μm vs. 200–300 μm) compared with IVUS. However, IVUS has greater imaging depth due to limited penetration of the OCT infrared light (1–2.4 mm vs. 10 mm) (Prati et al. 2010). Therefore, IVUS does not have the spatial resolution

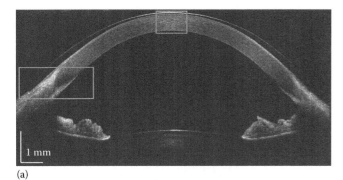

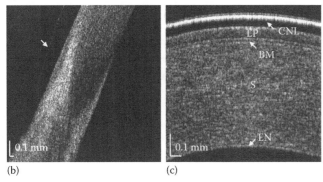

(a)

(b)   (c)

FIGURE 15.2 High resolution and high density imaging of the anterior segment. (a) SOCT cross-sectional image of the anterior segment of myopic eye with a contact lens, single line exposure time 70 μs, repetition rate: ~14,000 A-scans/s, 15,000×4096 pixels, 16 mm. The red rectangles indicate location of the magnified regions shown in panels b and c. (b) Magnification of the limbal region with the contact lens (indicated by the arrow). (c) Magnification of the apex region of the cornea. CNL, contact lens; EP, epithelium; BM, Bowman's membrane; S, stroma; EN, endothelium. (From Grulkowski, I. et al., *Opt. Express*, 17(6), 4842, 2009. With permission.)

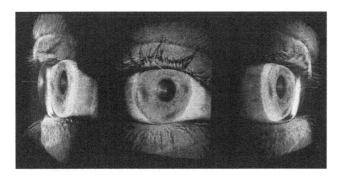

FIGURE 15.3 Large scale OCT imaging of the anterior segment of the human eye in vivo. Volume rendering from data acquired by a table top swept source OCT instrument using 1,300 nm tunable laser (Axsun Technologies) with 50,000 A-scans/s. (From Wojtkowski, M., *Appl. Opt.*, 49(16), D30, 2010. With permission.)

for identifying TCFAs and several studies have shown that OCT is superior to IVUS for identifying various plaque types.

Early studies of coronary morphology with OCT were all done ex vivo and with benchtop scanning mirrors not suitable for in vivo clinical applications. The first catheter-based sample

TABLE 15.1 Signal Characteristics of Plaque and Coronary Tissue with OCT

Tissue Type	
Fibrous plaque	Homogeneous, signal-rich region
Lipid pool	Homogeneous, signal-poor region with diffuse borders
Calcification	Heterogeneous region of signal-rich and signal-poor areas with sharp edges
Macrophages	Increased heterogeneity fibrous cap
Red thrombus	High signal attenuation within vessel lumen
White thrombus	Low signal attenuation within vessel lumen

probe for intracoronary OCT was developed in 1996 (Tearney et al. 1996) and followed by in vivo OCT aorta images in rabbit (Fujimoto et al. 1999) and then coronary imaging in man (Jang et al. 2002). One challenge in intracoronary OCT is that blood must be cleared from the image field due to significant scattering of the infrared light by red blood cells. Current intracoronary OCT techniques employ continuous flushing of an optically translucent contrast agent (e.g., Visipaque™) at 1–4 mL/s for 2–3 s that flows down the vessel and displaces the blood. As the contrast agent flows down the vessel, the OCT catheter is pulled back and blood-free images can be obtained.

Several studies have demonstrated OCT's ability to image plaque microstructure and described qualitative image classification schemes (Jang et al. 2002, Yabushita et al. 2002, Kume et al. 2006). Table 15.1 summarizes the image characteristics for various plaque and coronary tissue types imaged with OCT and representative OCT images for fibrous, fibro-calcific, and lipid-rich plaque types are shown in Figures 15.4 through 15.6. The sensitivity and specificity for identifying fibrous, fibrocalcific, and lipid-rich plaques was also measured (Yabushita et al. 2002, Kume et al. 2006) Sensitivities ranged from 71% to 79% for fibrous, 95% to 96% for fibro-calcific, and 85% to 94% for lipid-rich plaque types. Specificities were consistently high for all plaque types and were in the 88%–99% range.

The detection of macrophage content within atheroscleoritc plaque has also been described both qualitatively and quantitatively both ex vivo (Tearney et al. 2002) and in vivo (MacNeill et al. 2004). This is important because macrophage content is thought to play a role in plaque vulnerability. It has been shown that macrophage density is higher in patients presenting with ST-elevation myocardial infarction (STEMI) and acute coronary syndrome (ACS) compared with stable angina pectoris (SAP). In addition, macrophage density is higher in ruptured plaques compared with adjacent nonruptured sites (MacNeill et al. 2004).

Chronic total occlusions (CTOs), as seen on angiograms, are defined as completely blocked vessels that are older than 12 weeks. CTOs present a particular challenge for percutaneous coronary intervention (PCI) because it is often challenging to pass a guide wire through the occlusion in order to perform angioplasty. OCT has demonstrated the ability to differentiate red and white thrombus (Kume et al. 2006). As listed in Table 15.1, red thrombus is characterized by high signal attenuation as a result of the high blood content within the thrombus.

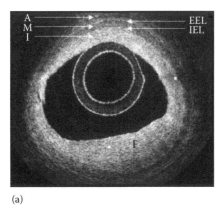
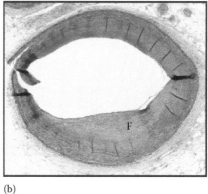

(a)   (b)

FIGURE 15.4 (a) OCT image of a fibrous coronary plaque showing a homogeneous, signal-rich interior (F). An area of intimal hyperplasia is seen opposite fibrous lesion, demonstrating intima (I, with intimal hyperplasia), internal elastic lamina (IEL), media (M), external elastic lamina (EEL), and adventitia (a). (b) Corresponding histology (Movat's pentachrome; magnification ×40). Tick marks, 1 mm. (From Yabushita, H. et al., *Circulation*, 106(13), 1640, 2002. With permission.)

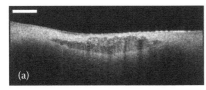
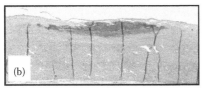

FIGURE 15.5 (a) OCT image of a fibrocalcific aortic plaque showing a sharply delineated region with a signal-poor interior. (b) Corresponding histology (H&E; magnification ×40). Bar = 500 μm. (From Yabushita, H. et al., *Circulation*, 106(13), 1640, 2002. With permission.)

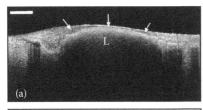
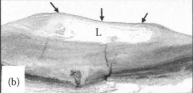

FIGURE 15.6 (a) OCT image of a lipid-rich carotid plaque showing a signal-poor lipid pool (L) with poorly delineated borders beneath a thin homogeneous band, corresponding to fibrous cap (arrows). (b) Corresponding histology (Movat's pentachrome; magnification ×20). Bar = 500 μm. (From Yabushita, H. et al., *Circulation*, 106(13), 1640, 2002. With permission.)

Conversely, white thrombus, which has relatively low red blood cell number has a low signal attenuation. This finding has led investigators to develop OCT as a tool for guidance during CTO-PCI (Nigel et al. 2007, 2010, Regar et al. 2007). Figure 15.7 shows example OCT images of red and white thrombus.

Using these descriptions as a baseline, researchers have used OCT as a tool for identifying the culprit lesion during AMI. In one study (Kubo et al. 2007), culprit lesions were assessed by OCT, IVUS, and coronary angioscopy (CAS). OCT was far superior for identifying plaque rupture (73% OCT, 40% IVUS, 47% CAS) and fibrous cap erosion (OCT 23%, IVUS 0%, CAS 3%) and was equal with CAS in indentifying thrombus (OCT and CAS 100%, IVUS 33%). In the future, it is expected that OCT will enable identification of vulnerable plaques prior to rupture and allow the clinician and researcher to study the evolution of these plaques in response to different treatment options.

The qualitative image analysis and interpretation in the above studies leaves open the possibility for intraobserver variability. In addition, qualitative only image analysis requires advanced training for the physician or end user. It is anticipated that advanced image and signal processing techniques will play a critical role in enhancing the usability of OCT within a clinical environment. Several groups have targeted the optical attenuation coefficient as a means for automatically identifying plaque type within OCT images (Levitz et al. 2004, van der Meer et al. 2005, Freek et al. 2006, Chenyang et al. 2008, van Soest et al. 2010). Figure 15.8 shows a representative example in which an advanced necrotic core lies underneath a calcification. This approach is similar to virtual histology IVUS (VH®-IVUS), which has proven to be a valuable tool in enhancing IVUS's imaging capabilities by automatically detecting vulnerable plaques (Tanaka et al. 2006, Wu et al. 2009).

Other automated and semiautomated image analysis may also prove beneficial in a clinical environment. These include metrics such as lumen area, maximum and minimum lumen diameter, percentage of luminal stenosis, and length of plaque burden.

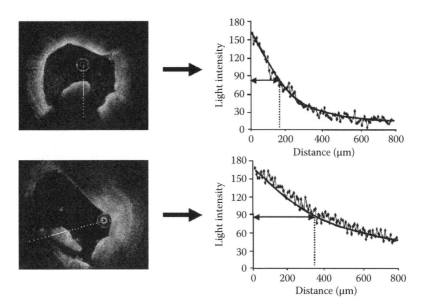

FIGURE 15.7 Representative analysis of intensity attenuation of red and white thrombi. Intensity attenuation of red thrombus (top) was greater than that of white thrombus (bottom) in the region of interest (dashed line). The half-width of signal intensity attenuation in white thrombus was greater than that in red thrombus (arrows). (From Kume, T. et al., *Am. J. Cardiol.*, 97(12), 1713, 2006. With permission.)

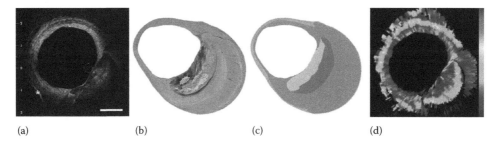

(a)   (b)   (c)   (d)

FIGURE 15.8 (See color insert.) (a) OCT image of a coronary atheroclerotic lesion ex vivo: * indicates the needle used for marking the imaged site, the length of the white scale bar is 1 mm; (b) corresponding histology, H&E stain; (c) cartoon histology, overlayed on the original histology slide, indicating an advanced necrotic core red behind a calcification gray, and a slight fibrotic green circumferential intimal thickening; and (d) OCT-derived attenuation coefficient μ_t, plotted on a continuous linear color scale from 0 to 15 mm^{-1}. The area corresponding to the necrotic core exhibits a higher attenuation coefficient (8–10 mm^{-1}) than the calcification next to it or the surrounding fibrous tissue (2–3 mm^{-1}). (From van Soest, G. et al., *J. Biomed. Opt.*, 15(1), 011105, 2010. With permission.)

In addition, fast reconstruction algorithms, 3D rendering, and image segmentation will allow the clinician to quickly interpret the OCT images.

15.3.2 OCT Imaging of Coronary Stents

As of 2003, approximately 86% of all patients undergoing a PCI received a stent (CDC 2003). Of those, nearly one-third were drug eluting stents (DES) and two-thirds were nondrug eluting or bare metal stents (BMS). Stents provide a mechanical mechanism to maintain sufficient vessel lumen size in the setting of advanced CAD and flow limiting stenotic vessels. While it has been shown that stent placement is beneficial for reducing angina and short-term exercise tolerance, several studies have shown no significant benefit for patients in terms of death, myocardial infarction (MI), or other long-term cardiac events. There are several theories regarding the failure of stents including inadequate stent placement, insufficient stent expansion, and restenosis. DES were thought to limit restenosis rates by eluting drugs to reduce inflammation and thrombus formation. However, DES increase the rate of so-called late stent thrombosis because they may limit protective reendothelialization over the stent struts.

Intracoronary OCT is particularly well suited as a tool for guiding acute stent placement, as well as studying the long-term effects of stent placement, vessel healing, and resulting failure rates for both BM and DE stents. As of early 2010, there were 28 either completed or ongoing clinical trials involving OCT and stents.

Figure 15.9 show in vivo OCT images of stent struts (Tearney et al. 2008). Stent struts can be seen as high reflectivity signals followed by a characteristic shadow that obscures the underlying tissue. The high resolution of OCT enables observation and measurement of morphologic features relating to stent placement. In

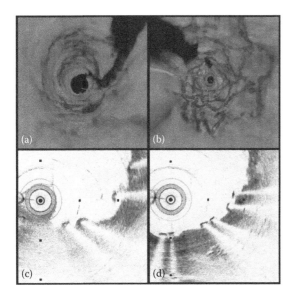

FIGURE 15.9 (See color insert.) (a) Fly-through view (distal-proximal) of the BMS shows covered struts underneath the surface of the artery wall, as well as some struts that appear near the luminal surface. (b) Fly-through view (distal-proximal) of the DES demonstrates uncovered struts. (c and d) OFDI cross-sectional images of the BMS and DES, respectively. The OFDI appearance of the struts is different for the 2 stents. Tick marks, 1 mm. (From Tearney, G.J. et al., *J. Am. Coll. Cardiol. Imaging*, 1(6), 752, 2008. With permission.)

a small study of 42 stents placed in 39 patients, incomplete stent apposition was found in 17% of cases (Bouma et al. 2003), but only 7% when detected with IVUS. The effect of stent malapposition is not fully understood, but it is thought that it could lead to subacute thrombosis and delay the healing process. OCT may play a future role in guiding stent implantation. Preimplantation OCT may be useful for determining the appropriate stent size and type. In addition, postimplantation OCT can check for stent apposition and level of stent expansion, and automated algorithms are being developed to detect and characterize stent placement with OCT (Unal et al. 2010).

Post-stent follow up is also an important area where OCT can serve as a valuable research and clinical imaging tool. Figure 15.10 shows an example of varying degrees of restenosis along a single 10-year old DES (Cypher®). Several clinical studies are ongoing using OCT to study neointimal growth following stent placement. Studies have compared BM and DE stents (Xie et al. 2008) as well as compared different DES (Chunyu et al. 2010). In order to address the problem of late stent thrombosis, investigators and stent companies are developing biodegradable stents and OCT is also being used to study their effectiveness (Barlis et al. 2010). It is anticipated that OCT will play a key role in the development and assessment of new DES and enable a better understanding of the mechanisms behind late-stent thrombosis and how to prevent it.

15.3.3 Noncoronary Cardiac Applications

OCT is also being applied to noncoronary cardiac applications. A straightforward extension of intracoronary OCT is catheter-based imaging in the peripheral arteries to assess atherosclerosis, vessel size, and to guide balloon and stent placement. Other noncoronary applications include guidance for radio frequency ablation procedures (Fleming et al. 2009), quantifying the flow of lipid through carotid arteries (Ghosn et al. 2009), and in developmental biology for studying heart development (Yelbuz et al. 2002, Luo et al. 2006, Jenkins et al. 2007).

15.4 Gastroenterology*

While OCT imaging has the potential of being applied to essentially any luminal organ within the considerable extents of the gastrointestinal tract, clinical trials have been predominantly focused on imaging the esophagus, and to a smaller extent the colon, duodenum, and pancreas and biliary ducts. These clinical studies are discussed in detail below.

15.4.1 Esophagus

The vast majority of OCT imaging in the gastrointestinal tract has occurred in the esophagus primarily for the purpose of screening for, and the management of, patients with BE. BE is a condition in which healthy squamous epithelium is replaced with specialized intestinal metaplasia (SIM), placing individuals at a higher risk for developing esophageal adenocarcinoma. The development of esophageal adenocarcinoma is thought to involve a stepwise progression from gastroesophageal reflux disease, through SIM, low-grade dysplasia (LGD), high-grade dysplasia (HGD), and finally adenocarcinoma (Nandurkar et al. 1999). Current guidelines for screening of patients with reflux disease for SIM, and surveillance of patients with SIM for dysplasia, recommend regular endoscopy procedures accompanied with the collection of tissue specimens, through forceps biopsy, for histological evaluation (Incarbone et al. 2002). There is much debate about the efficacy of this approach for screening and surveillance due to the large sampling error associated with the systematic but unguided biopsy acquisition, the low diagnostic yield, the high cost associated with the endoscopy procedure, and the knowledge that only a relatively small percentage of individuals will ultimately develop neoplastic changes.

In vivo OCT images of the human esophagus were first published in 1997 by Sergeev et al. using a transverse galvanometric scanning catheter (Sergeev et al. 1997). This initial proof of principle pilot study was soon followed by larger in vivo studies by other groups (Sergeev et al. 1997, Bouma et al. 2000, Jackle et al. 2000a,b Sivak et al. 2000). Early in vivo OCT studies of the upper gastrointestinal tract implied that OCT may be a powerful diagnostic tool for the discrimination of different forms of esophageal pathology; however it was not until 2001 that the first study was conducted to develop and prospectively test

---

* The following section on clinical OCT within the GI tract was adapted directly from a book chapter on Clinical advances in optical imaging written by Suter et al. and reprinted with permission of John Wiley & Sons, Inc. (Suter et al. 2010).

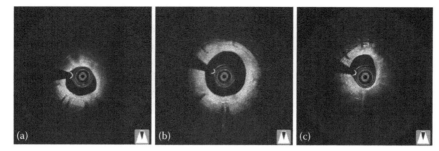

FIGURE 15.10   Three transverse sections showing differing degrees of in-stent restenosis with a single stent. The distal (a) and proximal (c) ends of the stent are highly stenosed, whereas the middle (b) portion of the stent remains open. (Images courtesy of Volcano Corporation.)

possible diagnostic criteria to support this supposition (Poneros et al. 2001). For OCT to be a clinically viable diagnostic device, identification of SIM at the squamocolumnar junction (SCJ) for screening, and identification and grading of dysplasia in patients with SIM for surveillance is necessary.

15.4.1.1 Normal Esophageal Squamous Mucosa and Gastric Cardia

A number of investigators have reported appearance of normal esophageal squamous mucosa and gastric cardia on the OCT (Bouma et al. 2000, Brand et al. 2000, Jackle et al. 2000a, Li et al. 2000). OCT characteristics of the healthy esophagus reveal a horizontal layered morphology of the squamous epithelium. Figure 15.11a shows an example of normal squamous mucosa with clear delineation of five layers of the mucosa including the epithelium,

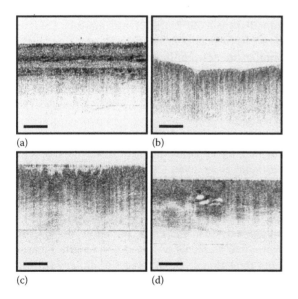

FIGURE 15.11   (a) OCT image of normal squamous epithelium (bar = 0.5 mm) shows a 5-layered appearance (epithelium, lamina propria, muscularis mucosa, submucosa, and muscularis propria). (b) OCT image of gastric mucosa with "pit and crypt" architecture. (c) OCT image of specialized intestinal metaplasia without dysplasia with irregular mucosal surface and absence of a layered or pit and crypt architecture. (d) OCT image of specialized intestinal metaplasia without dysplasia with submucosal glands.

lamina propria, muscularis mucosa, submucosa, and the muscularis propria. The presence of vessels and glands in the mucosa may also be observed. The OCT image penetration is greater in the squamous mucosa than in other gastrointestinal (GI) tissues, and without significant compression, is approximately 2 mm.

The gastroesophageal junction (GEJ) is the transition region of the GI tract where the esophagus meets the stomach. OCT images of the gastric cardia, the mucosa at the GEJ, show a regular vertical banding architecture with a "pit and crypt" appearance, a high superficial surface reflectivity, decreased image penetration (<1 mm) and broad regular glandular architecture. The morphology of the squamous mucosa (Figure 15.11a) and gastric cardia (Figure 15.11b) observed in the OCT images are easily identifiable and clearly distinct.

15.4.1.2 Identification of Specialized Intestinal Metaplasia

OCT images of SIM in the human esophagus have been characterized in a number of OCT studies. A lack of the horizontal layered structure of the squamous mucosa, the lack of the vertical pit and crypt architecture of cardia, heterogeneous scattering, irregular surface topology, and glands in the epithelium with layered architecture are all OCT indicators of SIM (Poneros et al. 2001, Evans et al. 2007) (Figure 15.11c and d). Poneros et al. developed diagnostic criteria for the detection of SIM by collecting OCT/biopsy correlated pairs.(Poneros et al. 2001) The investigators performed in vivo OCT imaging using a linear scanning catheter in 121 patients. In addition to the OCT imaging beam, a visible guiding beam was directed through the imaging catheter to highlight the OCT image location within the esophagus. Following OCT imaging the catheter was withdrawn from the endoscope and a forceps biopsy specimen of the target location was obtained. A total of 288 biopsy-correlated OCT images were obtained. Seventy-five image pairs were used to develop the diagnostic criteria. The criteria were subsequently retrospectively applied to the entire training set of 166 image pairs resulting in a sensitivity of 100% and a specificity of 93%. When the same diagnostic criteria were applied to an independent validation set comprised of 122 OCT/biopsy correlated pairs, the criteria were found to be 97% sensitive and 92% specific for SIM with or without dysplasia (Poneros et al. 2001). Though this study made significant advancements toward

TABLE 15.2 OCT Diagnostic Criteria Developed and Prospectively Tested to Diagnose SIM and to Grade Dysplasia

Diagnosis	OCT Finding
Normal squamous mucosa	1. Layered architecture
Normal cardia[a]	1. Vertical pit and crypt architecture
	2. Highly reflective surface
	3. Broad, regular glandular architecture
	4. Poor image penetration
SIM[b]	1. Lack of layered or vertical pit and crypt architecture
	2. Heterogeneous scattering
	3. Irregular surface
	4. Glands in epithelium with layered architecture
HGD/IMC[c]	1. Increased surface/subsurface reflectivity (score 0–2)
	2. Irregular gland/duct architecture (score 0–2)

Source: Poneros, J.M. et al., *Gastroenterology*, 120(1), 7, 2001; Evans, J.A. et al., *Clin. Gastroenterol. Hepatol.*, 4(1), 38, 2006; Evans, J.A. et al., *Gastrointest. Endosc.*, 65, 50, 2007.

[a] See algorithm in Ref. 12.
[b] 2 of first 3 criteria or 4th criterion indicate SIM.
[c] Total score of ≥2 indicates HGD/IMC.

confirmation that OCT may be a powerful diagnostic tool for the detection of SIM when compared to other upper GI tissues, a number of false-positives, secondary to inflamed cardia, were noted. For this reason the same group conducted a second study in 2007 focused specifically on the identification of SIM at the SCJ (Evans et al. 2007). A similar study design was defined with the acquisition of 196 OCT/biopsy pairs, this time limited to the SCJ region. With refined diagnostic criteria, the investigators reported a sensitivity of 85% and specificity of 95% when applied retrospectively to a training set of 40 images, and a sensitivity of 81% and a specificity ranging from 57% to 66% when applied prospectively to a validation set of 156 images (Evans et al. 2007). A summary of the diagnostic criteria used in the above two studies can be viewed in Table 15.2.

15.4.1.3 Identification of Dysplasia

Identification and diagnosis of dysplasia in SIM is essential for effective surveillance of Barrett's patients. Two groups have investigated the diagnostic potential of OCT to detect dysplasia in the esophagus (Isenberg et al. 2005, Evans et al. 2006). In 2005, Isenberg et al. obtained OCT/biopsy pairs with a circumferential scanning catheter held at a fixed 30-degree angle with respect to the esophageal wall (Isenberg et al. 2005). A total of 314 usable OCT/biopsy pairs were obtained from 33 patients. The endoscopist performing the procedure rendered a diagnosis of each of the OCT images immediately following the procedure, and therefore with a priori knowledge of the patient history and prior diagnoses. The pathologist interpreting the biopsies however, was blinded to the OCT findings. If the endoscopist determined that the OCT image contained dysplasia, the image was further categorized as LGD, HGD, or intramucosal carcinoma (IMC). The authors state that two criteria were used to determine

a diagnosis of dysplasia (1) reduced light scattering and (2) loss of tissue architecture that they deduced from an earlier pilot study (Das et al. 2001). The reported sensitivity and specificity for the detection of dysplasia were 68% and 82%, respectively, and if the analysis was limited to the detection of HGD, the sensitivity dropped to 50% with a corresponding specificity of 72% (Isenberg et al. 2005).

Evans et al. performed a similar study using the same linear scanning probe and approach published in their studies aimed at the detection of SIM (Poneros et al. 2001, Evans et al. 2006, 2007). The study was designed to develop OCT diagnostic criteria for the identification of HGD and IMC in SIM. The investigators assessed a total of 177 OCT/biopsy correlated images from 55 patients undergoing surveillance endoscopy for BE. OCT diagnostic criteria were based on histopathologic characteristics, specifically a scoring of surface maturation and gland architecture (Table 15.2.) A blinded investigator prospectively analyzed all of the OCT images and assigned a score and corresponding diagnosis. A score of greater than or equal to 2 was considered positive for IMC/HGD. The sensitivity of distinguishing IMC/HGD from all other tissue types was determined to be 83.3% with a corresponding specificity of 75% (Evans et al. 2006). Figure 15.12 shows OCT images of different forms of dysplasia highlighting features used as diagnostic criteria. A summary of the three studies conducted by this group is presented in Table 15.3.

15.4.1.4 Second Generation OCT Imaging of the Esophagus

Studies have suggested that OCT imaging of the esophagus can detect and diagnose esophageal pathology relevant to screening and surveillance for BE. Until recently however, OCT imaging of the esophagus has been restricted to imaging small areas, in a similar point sampling fashion to random biopsy. With <1% of the involved tissue generally being assessed, traditional OCT imaging of BE suffers from the same sampling error as the standard surveillance biopsy approach, limiting the clinical utility of the technology. Recently, a pilot clinical study was published where large area comprehensive optical frequency domain imaging (OFDI) of the entire distal esophagus was performed (Suter et al. 2008). A balloon-based imaging catheter was used to dilate the esophagus and center the optical imaging core within the lumen. Combined circumferential and longitudinal scanning techniques were implemented to produce spiral cross-sectional images of the entire distal esophagus in longitudinal segments up to 6 cm in length. Figures 15.13 and 15.14 show sample OFDI images of BE, and BE with dysplasia, acquired with the balloon-based imaging catheter. The endoscopy images in Figure 15.13a and Figure 15.14a depict an irregular transition zone (z-line) between the squamous mucosa (pale pink in color) and the glandular tissue of the gastric cardia (dark pink in appearance), which is consistent with the presence of SIM. Endoscopy alone however, is generally insufficient to detect further changes such as the presence of dysplasia, which was successfully detected with OFDI in Figure 15.14e. Although this study has demonstrated that large area

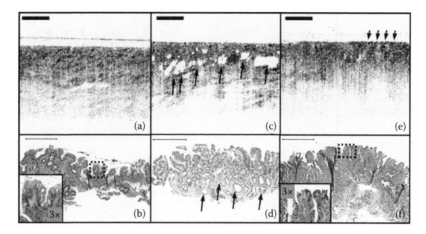

FIGURE 15.12 OCT images of SIM with and without IMC/HGD. (a) OCT image of SIM without dysplasia demonstrates glandular architecture with a relatively low reflectivity. (b) Corresponding histology to (a) with inset demonstrates a low nuclear to cytoplasm ratio in the superficial epithelium. (c) OCT image of IMC/HGD enables visualization of large and irregular glands (arrows). (d) Irregular, dilated glands are also seen in the corresponding histology to (c) (arrows). (e) OCT image of IMC/HGD shows a disorganized architecture and increased surface reflectivity (arrows). (f) Corresponding histology for (e) demonstrates abnormal glandular architecture and an increased superficial nuclear-to-cytoplasm ration (inset). Histology hematoxylin-eosin; original magnification, 40×. Scale bars, 0.5 mm.

TABLE 15.3 Summary of OCT Clinical Studies in Barrett's Esophagus

Test	#Patients	# Biopsies	Sensitivity	Specificity
SIM vs. other GI tract tissues	121	288	97%	92%
SIM vs. cardia at SCJ	113	196	81%	57%–66%
IMC/HGD vs. Indeterminate-grade dysplasia/LGD/SIM without dysplasia	55	242	83%	75%

Source: Poneros, J.M. et al., *Gastroenterology*, 120(1), 7, 2001; Evans, J.A. et al., *Clin. Gastroenterol. Hepatol.*, 4(1), 38, 2006; Evans, J.A. et al. *Gastrointest. Endosc.*, 65, 50, 2007.

high-resolution OFDI imaging of the entire distal esophagus is possible, to date the investigators could not perform direct correspondence of the OFDI data with histopathologic analysis of biopsy specimens, the gold standard. Two solutions are suggested to address this limitation. The first involves reimaging the esophagus following biopsy and coregistering the pre- and post-biopsy OFDI datasets to determine the biopsy locations. The second approach involves esophageal marking immediately following OFDI imaging with the balloon catheter still in place, which would provide the foundation for a guided biopsy platform. Direct registration of the OFDI datasets and histopathology would enable a study to compare the accuracy of the

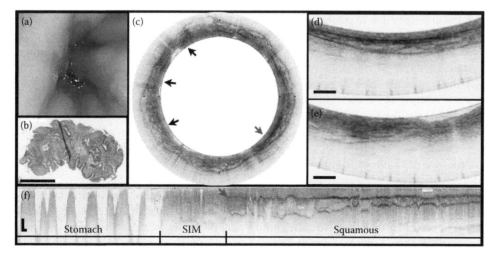

FIGURE 15.13 Barrett's esophagus (a) Videoendoscopic image demonstrates an irregular SCJ. (b) Histopathologic image of a biopsy specimen obtained from the SCJ demonstrates SIM without dysplasia (H&E, orig. mag. ×2). (c) Cross-sectional OFDI image reveals both the normal layered appearance of squamous mucosa (*lower right arrow*, expanded in d) and tissue that satisfies the OCT criteria for SIM (*left arrows*, expanded in e). (f) Longitudinal section across the gastroesophageal junction shows the transition from squamous mucosa to SIM to cardia. The length of the BE segment is 7 mm in this OFDI reconstruction. Scale bars (b, d, e, and f) and tick marks (c) represent 1 mm.

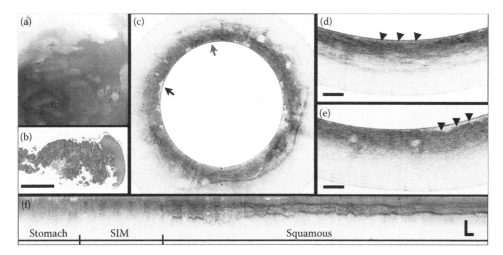

FIGURE 15.14 Barrett's esophagus with dysplasia. (a) Videoendoscope image reveals a patchy mucosa consistent with SIM. (b) Histopathologic image of the biopsy taken from the SCJ demonstrates intestinal metaplasia and low-grade dysplasia. (c) Cross-sectional OFDI image, demonstrating regions consistent with SIM without dysplasia (top arrow) and SIM with HGD (left arrow). (d) Expanded view of C, taken from the region denoted by the top arrow in C, demonstrating good surface maturation (arrowheads), which is consistent with SIM without dysplasia. (e) Expanded view of C, taken from the region denoted by the left arrow in C, demonstrating features consistent with HGD, including poor surface maturation (surface arrowheads) and the presence of dilated glands (deep arrowheads) in the mucosa. (f) A longitudinal slice highlights the transition from gastric cardia, through a 9 mm segment of SIM, and finally into squamous mucosa. Scale bars (b, d, e, and f) and tick marks (c) represent 1 mm.

previously developed OCT diagnostic criteria when applied to the balloon-based OFDI images, a critical step in translating this technology to clinical practice. If successful, the capability to noninvasively obtain microscopic image data over large epithelial surface areas may aid in early detection, diagnosis, and intervention, resulting in a consequent reduction in morbidity and mortality associated with BE.

15.4.2 Colon

OCT imaging of the colon has been suggested to have potential clinical utility for the differentiation of colonic polyps, the detection and staging of early carcinoma of the colon (Jackle et al. 2000a, Sivak et al. 2000), and for the evaluation of inflammation in Crohn's disease and ulcerative colitis (Zhu et al. 2005). The first in vivo OCT imaging of the colon demonstrated that OCT provides sufficient resolution and contrast to identify the mucosa including colonic crypts, the muscularis mucosa, and the submucosa of the colon (Jackle et al. 2000a, Sivak et al. 2000). As is the case in the esophagus, OCT images of carcinoma of the colon reveal a loss of the layered architecture found in healthy mucosa (Jackle et al. 2000a). Future studies are required to further investigate the diagnostic potential of OCT for the detection of carcinoma of the colon.

Differentiation of Crohn's disease from ulcerative colitis, both inflammatory bowl diseases, is important as treatment options may vary greatly. The correct diagnosis is often difficult, even by histopathology, as the depth of endoscopic biopsy is insufficient to determine transmural inflammation, which is an indicator of Crohn's disease. In 2004, Shen et al. demonstrated the diagnostic potential of OCT to differentiate Crohn's disease from

ulcerative colitis in an ex vivo imaging study of 585 histology correlated OCT images obtained from 48 patients (Zhu et al. 2005). The sensitivity for the detection of transmural disease by OCT was found to be 86% sensitive and 91% specific. The same investigators have also performed an in vivo OCT study using the diagnostic criteria previously developed in the ex vivo study (Zhu et al. 2005). A total of 601 OCT images were obtained from 70 patients with prior clinical diagnosis of either Crohn's disease or ulcerative colitis. The in vivo sensitivity and specificity for the diagnosis of transmural inflammation in Crohn's disease was found to be 90.0% and 83.3%, respectively (Zhu et al. 2005). These studies demonstrate that OCT can accurately detect a disruption in the normal layered structure of the colon wall indicative of transmural inflammation, a hallmark feature of Crohn's disease.

15.4.3 Duodenum

Investigating the duodenal mucosa with OCT may have a clinical utility in the early detection of celiac disease in the pediatric population, and to a lesser extent in adults. OCT imaging of the human duodenum has been previously demonstrated in vivo in a number of pilot studies (Sivak et al. 2000, Testoni et al. 2007). In 2000, Sivak et al. used a circumferential scanning catheter to obtain images from the healthy gastrointestinal tissues including 12 duodenum sites (Sivak et al. 2000). The acquired OCT images allow differentiation of individual villi, however, when the imaging probe was in contact with the duodenum, differentiation of individual villi was more ambiguous. In 2007, Masci et al. published the first OCT images of the duodenum of individuals with celiac disease (Testoni et al. 2007). The duodenum of

18 patients with, and 22 individuals without suspicion of celiac disease were imaged with OCT and the imaged locations were subsequently biopsied for histopathologic analysis. Blinded interpretation of the villi morphology in the OCT images and in the histology slides was identical in all samples (Testoni et al. 2007). The promising results from this pilot study indicate that OCT imaging of the duodenum may aid in the detection and diagnosis of celiac disease. In addition, comprehensive large area OCT imaging of the duodenum may reduce the large sampling error associated with the detection of celiac disease where the distribution of mucosal lesions is often sporadic.

15.4.4 Pancreatobiliary Ductal System

Diagnosis of cholangiocarcinoma, an adenocarcinoma of the bile ducts, and adenocarcinomas associated with the pancreatic ducts, typically occur in the advanced stages leaving patients with a poor prognosis. When suspected, an endoscopic retrograde cholangiopancreatography (ERCP) imaging procedure may be performed to investigate the biliary and pancreatic ducts and to provide access for diagnostic tests such as forceps biopsy or brush cytology. Unfortunately, the diagnostic yields of forceps biopsy and brush cytology are low, are difficult to obtain, and may be associated with a high risk of complications. For this reason, high resolution imaging of the pancreatic and hepatobiliary ductal system may prove beneficial.

The first clinical OCT demonstration in the common bile duct was performed in 2001 (Seitz et al. 2001). Seitz et al. utilized both a 2.8 mm diameter forward viewing catheter in three patients, and a longitudinal scanning catheter in one patient (Seitz et al. 2001). The OCT imaging probe was delivered through the working channel of a standard duodenoscope during an ERCP procedure. OCT images of the biliary duct wall revealed a layered architecture as seen in histology. In addition, the presence of glands and vessels in the stroma were identified (Seitz et al. 2001). This pilot study both demonstrated the feasibility of OCT for investigating the human bile ducts and provided preliminary results suggesting that OCT may afford images with sufficient detail to detect biliary pathology.

In 2002, Poneros et al. performed in vivo OCT imaging of the extra and intra hepatic bile duct using a 2.6 mm circumferential scanning catheter (Poneros et al. 2002). The goal of this pilot study was to demonstrate the feasibility of intraductal OCT of the biliary tract as a diagnostic tool. In vivo images were obtained of the bile ducts in five patients. The architectural features identified in the OCT images acquired in vivo were compared to OCT/histology image pairs obtained from five cadaver specimens. Figure 15.15 shows OCT images obtained in vivo of the extrahepatic biliary epithelium. Based on the layered architecture including the epithelium, a highly scattering submucoa, and the serosa, the OCT image was classified as normal. Figure 15.16 shows an example of an OCT image of a presumed cholangiocarcinoma obtained in vivo. Histologically, diagnosis of cholangiocarcinoma may be made by the presence of villiform papillary morphology. These features were observed in the OCT

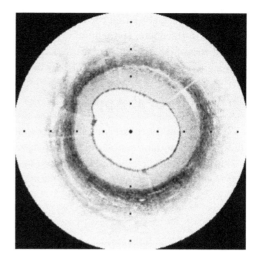

FIGURE 15.15 Circularized OCT image of the extrahepatic biliary epithelium obtained in vivo demonstrating different layers (bar = 0.5 mm).

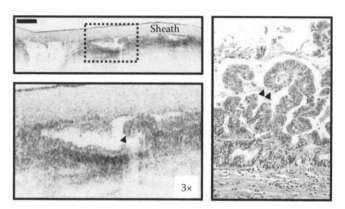

FIGURE 15.16 (a) OCT image of cholangiocarcinoma obtained in vivo (Scale bar = 0.5 mm). (b) Expanded view of (a) taken from the region denoted by the box, showing papillary fronds. (c) Photomicrography of a reference histologic section of papillary cholangiocarcinoma showing papillary fronds similar to those seen by OCT. (H&E, orig. mag. ×40).

images of Figure 15.16 (arrow). Although images obtained during this study indicate that OCT imaging of the biliary tree does provide images with sufficient resolution and contrast for the evaluation of biliary strictures, to date no further rigorous clinical studies have been performed.

Testoni et al. published the first demonstration of in vivo OCT of the pancreatic duct in 2007 (Testoni et al. 2007). The investigators imaged the main pancreatic ducts of 12 patients with main pancreatic duct strictures using a 1.2 mm diameter circumferential scanning catheter during an ERCP procedure (Testoni et al. 2007). A layered architecture was observed in all cases of normal pancreatic ductal epithelium, and a loss of the layered architecture was observed in neoplastic lesions (Testoni et al. 2007). This pilot study demonstrates that OCT is a feasible tool for the investigation of pancreatic strictures during an ERCP procedure and may aid in the diagnosis of neoplasia. Further clinical studies are needed to determine the diagnostic utility of OCT imaging of the pancreatic duct, however, given that the vast majority of pancreatic cancers arise in the ductal epithelium, screening of patients at an increased risk of developing pancreatic malignancies with an optical imaging technology such as OCT may be warranted.

15.5 Other Clinical Applications

There are a variety of additional and growing clinical fields using OCT including oncology, pulmonology, dermatology, dentistry, developmental biology, neurology, and in musculoskeletal imaging. A brief treatment of a small number of these examples is included below.

15.5.1 Breast Cancer

OCT has shown promise in breast cancer for biopsy guidance and margin detection during tumor resection (Boppart 2004). There is a strong difference in the characteristic OCT signature from adipose and fibroglandular tissue, from which most breast tumors arise. Adipose tissue is characterized by multiple high reflectivity peaks arising from the border between the cell membrane and large lipid core, whereas fibroglandular tissue contains scattering centers that are much closer together and produce a more homogenous looking OCT signal. Analysis of the scattering coefficient and spatial frequency content of the OCT signal has enabled investigators to developed quantitative metrics for automatically differentiating these tissue types (Zysk and Boppart 2006, Goldberg et al. 2009, Mujat et al. 2009). It has been more challenging to differentiate normal fibroglandular tissue from tumor tissues, although it is generally accepted that

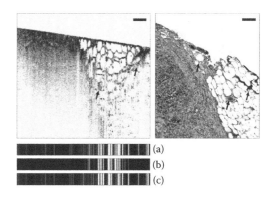

FIGURE 15.17 OCT image of a vertical tumor margin assessed along all vertical scans using three detection techniques (left) and the corresponding H&E stained histology (right). The correspondence between the OCT image and histology is very strong. Highly correlated image features are indicated with arrows. The bottom three boxes show the results of (a) combined analysis, (b) Fourier-domain classification, and (c) periodicity analysis, for each scan line within the image. Black, white, and gray regions represent tumor, adipose, and stroma classifications, respectively. Scale bars are 200 μm. (From Zysk, A.M. and Boppart, S.A., *J. Biomed. Opt.*, 11(5), 054015, 2006. With permission.)

tumor tissue is more dense that normal fibroglandular tissue, which manifests as increased scattering and attenuation of the OCT signal.

Figure 15.17 shows an example of tumor margin identification by OCT (Zysk and Boppart 2006). In another study (Nguyen et al. 2009), a 100% sensitivity and 92% specificity for positive tumor margins was reported indicating OCT may be a valuable

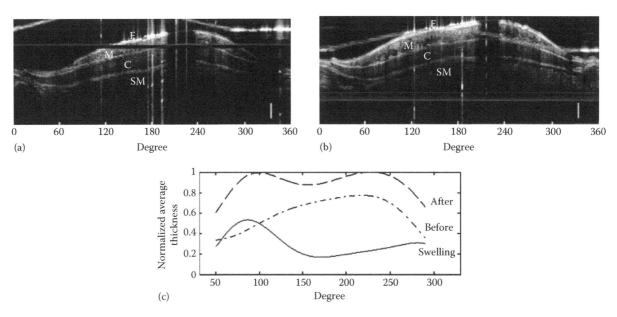

FIGURE 15.18 (a) Baseline and (b) 2-h postexposure circumferential images of rabbit trachea with mucosa layer outlined; (c) normalized average mucosa thickness and swelling distribution along tracheal circumference, calculated based on six characteristic circumferential B-scan images: E, epithelium; M, mucosa; C, cartilage ring; SM, submucosa; scale bar, 500 μm. (From Jiechen, Y. et al., *J. Biomed. Opt.*, 14(6), 060503, 2009. With permission.)

intraoperative tool for margin detection. OCT has also been used for imaging sentinel lymph nodes (Luo 2005). In addition, OCT has been applied in needle-based settings for guiding both core- and fine-needle aspiration biopsies (Zysk et al. 2007, 2009, Goldberg et al. 2009).

15.5.2 Pulmonology

Similar to coronary and esophageal OCT, OCT imaging in the airway is particularly attractive because small diameter optical fiber probes can be made that enable high resolution of the upper and lower airways. Quantitative size measurements in the upper airway have been described for studying airway collapse during sleep apnea (Armstrong et al. 2003) as well as a way to correlate changes in lung function testing (Coxson et al. 2008). OCT is also being investigated for studying smoke-induced airway injury (Jiechen et al. 2009) and Figure 15.18 shows an example of mucosal swelling following smoke inhalation.

15.6 Summary

OCT imaging has been applied to a variety of clinical applications of which ophthalmology, cardiology, and gastroenterology are the most developed. Future development will come from all portions of the OCT research and commercial arena. The research community will continue to define new imaging applications for OCT; the clinical community will help to define the clinical utility of OCT for various medical indications; and the commercial community will develop robust and reliable OCT products suitable for use in the clinical environment. It is expected that OCT development in these different spheres will overlap greatly in the areas of signal and image processing, source and detector development, and catheter/probe development. Combined, these efforts should help transition from a purely research imaging tool, to a fully developed and widely utilized clinical imaging modality.

References

Anderson, S. C., Wagner, P.S., Doyle, C. et al. 2005. Long-term variability of the retinal nerve fiber layer thickness (RNFL) in glaucoma measured by optical coherence tomography (OCT). *Investig. Ophthalmol. Vis. Sci.* 46: 637.

Armstrong, J. J., Leigh, M. S., Walton, I. D. et al. 2003. In vivo size and shape measurement of the human upper airway using endoscopic long-range optical coherence tomography. *Opt. Express* 11(15): 1817–1826.

Barlis, P., Regar E., Serruys P. W. et al. 2010. An optical coherence tomography study of a biodegradable vs. durable polymer-coated limus-eluting stent: A LEADERS trial sub-study. *Eur. Heart J.* 31(2): 165–176.

Boppart, S. A., Luo, W., Marks, D. L., and Singletary, K. W. 2004. Optical coherence tomography: feasibility for basic research and image-guided surgery of breast cancer. *Breast Cancer Res. Treat.* 84: 85–87.

Bouma, B. E., Tearney, G. J., Compton, C. C., and Nishioka, N. S. 2000. High-resolution imaging of the human esophagus and stomach in vivo using optical coherence tomography. *Gastrointest. Endosc.* 51: 467–474.

Bouma, B. E., Tearney, G. J., Yabushita, H. et al. 2003. Evaluation of intracoronary stenting by intravascular optical coherence tomography. *Heart* 89(3): 317–320.

Brand, S., Poneros, J. M., Bouma, B. E. et al. 2000. Optical coherence tomography in the gastrointestinal tract. *Endoscopy* 32(10): 796–803.

CDC 2003. National Hospital Discharge Survey: 2003 *Annual Summary with Detailed Diagnosis and Procedure Data.*

Cense, B., Chen, T. C., Park, B. H., Pierce, M. C., and de Boer, J. F. 2004. Thickness and birefringence of healthy retinal nerve fiber layer tissue measured with polarization-sensitive optical coherence tomography. *Investig. Ophthalmol. Vis. Sci.* 45(8): 2606–2612.

Chenyang, X., Joseph, M. S., Stephane, G. C., and Renu, V. 2008. Characterization of atherosclerosis plaques by measuring both backscattering and attenuation coefficients in optical coherence tomography. *J. Biomed. Opt.* 13(3): 034003.

Chunyu, F., Kim, J. S., Lee, J. M. et al. 2010. Different vascular healing patterns with various drug-eluting stents in primary percutaneous coronary intervention for ST-segment elevation myocardial infarction: Optical coherence tomographic findings. *Am. J. Cardiol.* 105(7): 972–976.

Costa, R. A., Skaf, M., Melo, L. A. Jr. et al. 2006. Retinal assessment using optical coherence tomography. *Prog. Retin. Eye Res.* 25(3): 325–353.

Coxson, H. O., Quiney, B., Sin, D. D. et al. 2008. Airway wall thickness assessed using computed tomography and optical coherence tomography. *Am. J. Respir. Crit. Care Med.* 177(11): 1201–1206.

Das, A., Sivak, M. V. Jr., Chak, A. et al. 2001. High-resolution endoscopic imaging of the GI tract: A comparative study of optical coherence tomography versus high-frequency catheter probe EUS. *Gastrointest. Endosc.* 54(2): 219–224.

de Bruin, D. M., Burnes, D. L., Loewenstein, J. et al. 2008. In-vivo three-dimensional imaging of neovascular age related macular degeneration using optical frequency domain imaging at 1050 nm. *Investig. Ophthalmol. Vis. Sci.* 49(10): 4545–4552.

Evans, J. A., Poneros, J. M., Bouma, B. E. et al. 2006. Optical coherence tomography to identify intramucosal carcinoma and high-grade dysplasia in Barrett's esophagus. *Clin. Gastroenterol. Hepatol.* 4(1): 38–43.

Evans, J. A., Bouma, B. E., Bressner, J. et al. 2007. Identifying intestinal metaplasia at the squamocolumnar junction using optical coherence tomography. *Gastrointest. Endosc.* 65: 50–56.

Fercher, A. F., Hitzenberger, C. K., Drexler, W., Kamp, G., and Sattmann, H. 1993. In vivo optical coherence tomography. *Am. J. Ophthalmol.* 116: 113–115.

Fleming, C. P., Quan, K. J., Wang, H., Amit, G., and Rollins, A. M. 2009. In vitro characterization of cardiac radiofrequency ablation lesions using optical coherence tomography. *Opt. Express* 18(3): 3079–3092.

Freek, J. V. D. M., Dirk, J. F., Inci, C., Martin, J. C. V. G., and Ton, G. V. L. 2006. Temperature-dependent optical properties of individual vascular wall components measured by optical coherence tomography. *J. Biomed. Opt.* 11(4): 041120.

Fujimoto, J. G., Boppart, S.A., Tearney, G. J. et al. 1999. High resolution in vivo intra-arterial imaging with optical coherence tomography. *Heart* 82(2): 128–133.

Gabriele, M. L., Wollstein, G., and Schuman, J. S. 2008. Spectral domain optical coherence tomograph: The potential role of this technology in glaucoma. *Glaucoma Today* 6(2): 35–39.

Ghosn, M. G., Leba, M., Vijayananda, A. et al. 2009. Effect of temperature on permeation of low-density lipoprotein particles through human carotid artery tissues. *J. Biophotonics* 2(10): 573–580.

Goldberg, B. D., Vakoc, B. J., Oh, W. Y. et al. 2009. Performance of reduced bit-depth acquisition for optical frequency domain imaging. *Opt. Express* 17(19): 16957–16968.

Grulkowski, I., Gora, M., Szkulmowski, M. et al. 2009. Anterior segment imaging with spectral OCT system using a high-speed CMOS camera. *Opt. Express* 17(6): 4842–4858.

Hannouche, R. Z. and de Avila, M. P. 2009. Retinal thickness measurement and evaluation of natural history of the diabetic macular edema through optical coherence tomography. *Arq. Bras. Oftalmol.* 72(4): 433–438.

Incarbone, R., Bonavina, L., Saino, G., Bona, D., and Peracchia, A. 2002. Outcome of esophageal adenocarcinoma detected during endoscopic biopsy surveillance for Barrett's esophagus. *Surg. Endosc.* 16: 263–266.

Isenberg, G., Sivak, M. V., Jr., Chak, A. et al. 2005. Accuracy of endoscopic optical coherence tomography in the detection of dysplasia in Barrett's esophagus: A prospective, double-blinded study. *Gastrointest. Endosc.* 62(6): 825–831.

Jackle, S., Gladkova, N., Feldchtein, F. et al. 2000a. In vivo endoscopic optical coherence tomography of the human gastrointestinal tract—Toward optical biopsy. *Endoscopy* 32(10): 743–749.

Jackle, S., Gladkova, N., Feldchtein, F. et al. 2000b. In vivo endoscopic optical coherence tomography of esophagitis, Barrett's esophagus, and adenocarcinoma of the esophagus. *Endoscopy* 32(10): 750–755.

Jang, I.-K., Bouma, B. E., Kang, D. H. et al. 2002. Visualization of coronary atherosclerotic plaques in patients using optical coherence tomography: Comparison with intravascular ultrasound. *J. Am. Coll. Cardiol.* 39(4): 604–609.

Jenkins, M. W., Chughtai, O. Q., Basavanhally, A. N., Watanabe, M., and Rollins, A. M. 2007. In vivo gated 4D imaging of the embryonic heart using optical coherence tomography. *J. Biomed. Opt.* 12(3): 030505.

Jiechen, Y., Liu, G., Zhang, J. et al. 2009. In vivo early detection of smoke-induced airway injury using three-dimensional swept-source optical coherence tomography. *J. Biomed. Opt.* 14(6): 060503.

Konstantopoulos, A., Hossain, P., and Anderson, D. F. 2007. Recent advances in ophthalmic anterior segment imaging: A new era for ophthalmic diagnosis? *Br. J. Ophthalmol.* 91(4): 551–557.

Kubo, T., Imanishi, T., Takarada, S. et al. 2007. Assessment of culprit lesion morphology in acute myocardial infarction—Ability of optical coherence tomography compared with intravascular ultrasound and coronary angioscopy. *J. Am. Coll. Cardiol.* 50(10): 933–939.

Kucumen, R., Dinc, U., Yenerel, N., Gorgun, E., and Alimgil, M. 2009. Immediate evaluation of the flaps created by femtosecond laser using anterior segment optical coherence tomography. *Ophthal. Surg. Lasers Imaging* 40(3): 251–254.

Kume, T., Akasaka, T., Kawamoto, T. et al. 2006. Assessment of coronary arterial thrombus by optical coherence tomography. *Am. J. Cardiol.* 97(12): 1713–1717.

Lee, E. C., de Boer, J. F., Mujat, M., Lim, H., and Yun, S. H. 2006. In vivo optical frequency domain imaging of human retina and choroid. *Opt. Express* 14(10): 4403–4411.

Levitz, D., Thrane, L, Frosz, M. et al. 2004. Determination of optical scattering properties of highly-scattering media in optical coherence tomography images. *Opt. Express* 12(2): 249–259.

Li, X. D., Boppart, S. A., Van Dam, J. et al. 2000. Optical coherence tomography: Advanced technology for the endoscopic imaging of Barrett's esophagus. *Endoscopy* 32: 921–930.

Libby, P. 2001. Current concepts of the pathogenesis of the acute coronary syndromes. *Circulation* 104(3): 365–372.

Luo, W., Nguyen, F.T., Zysk, A. M., Ralson, T. S., Brocken brough, J. Marks, D. L., Oldenburg, A. L., and Boppart, S. A. 2005. Optical biopsy of lymph node morphology using optical colurence tomography. *Technol. Cancer Res. Treat.* 4(5): 539–547.

Luo, W., Marks, D. L., Ralston, T. S., and Boppart, S. A. 2006. Three-dimensional optical coherence tomography of the embryonic murine cardiovascular system. *J. Biomed. Opt.* 11(2): 021014.

MacNeill, B. D., Jang, I. K., Bouma, B. E. et al. 2004. Focal and multi-focal plaque macrophage distributions in patients with acute and stable presentations of coronary artery disease. *J. Am. Coll. Cardiol.* 44(5): 972–979.

Michalewska, Z., Michalewski, J., Sikorski, B. L. et al. 2009. A study of macular hole formation by serial spectral optical coherence tomography. *Clin. Exp. Ophthalmol.* 37(4): 373–383.

Mujat, M., Ferguson, R. D., Hammer, D. X., Gittins, C., and Iftimia, N. 2009. Automated algorithm for breast tissue differentiation in optical coherence tomography. *J. Biomed. Opt.* 14(3): 034040.

Nandurkar, S., Talley, N. J., Martin, C. J. et al. 1999. Gastroesophageal reflux and adenocarcinoma of the esophagus. *N. Engl. J. Med.* 341(7): 536–538.

Nguyen, F. T., Zysk, A. M., Chaney, E. J. et al. 2009. Intraoperative evaluation of breast tumor margins with optical coherence tomography. *Cancer Res.* 69(22): 8790–8796.

Nigel, R. M., Wright, G. A., Mariampillai, A. et al. 2010. Doppler optical coherence tomography for interventional cardiovascular guidance: In vivo feasibility and forward-viewing probe flow phantom demonstration. *J. Biomed. Opt.* 15(1): 011103.

Nigel, R. M., Yang, V. X., Standish, B. A. et al. 2007. Ex vivo imaging of chronic total occlusions using forward-looking optical coherence tomography. *Lasers Surg. Med.* 39(1): 28–35.

Nolan, W. 2008. Anterior segment imaging: Ultrasound biomicroscopy and anterior segment optical coherence tomography. *Curr. Opin. Ophthalmol.* 19(2): 115–121.

Poneros, J. M., Brand, S., Bouma, B. E. et al. 2001. Diagnosis of specialized intestinal metaplasia by optical coherence tomography. *Gastroenterology* 120(1): 7–12.

Poneros, J. M., Tearney, G. J., Shiskov, M. et al. 2002. Optical coherence tomography of the biliary tree during ERCP. *Gastrointest. Endosc.* 55(1): 84–88.

Prati, F., Regar, E., Mintz, G. et al. 2010. Expert review document on methodology, terminology, and clinical applications of optical coherence tomography: Physical principles, methodology of image acquisition, and clinical application for assessment of coronary arteries and atherosclerosis. *Eur. Heart J.* 31(4): 401–415.

Ramos, J. L. B., Yan, L., and Huang, D. 2009. Clinical and research applications of anterior segment optical coherence tomography a review. *Clin. Exp. Ophthalmol.* 37(1): 81–89.

Regar, E., Leeuwen, T. G. V., and Serruys, P. W. J. 2007. *Optical Coherence Tomography in Cardiovascular Research.* Abingdon, U.K.: Informa Healthcare.

Sakata, L. M., DeLeon-Ortega, J., Sakata, V., and Girkin, C. A. 2009. Optical coherence tomography of the retina and optic nerve—a review. *Clin. Exp. Ophthalmol.* 37(1): 90–99.

Schuman, J. S., Puliafito, C. A., and Fujimoto, J. G. 2004. *Optical Coherence Tomography of Ocular Diseases.* Thorofare, NJ: Slack Incorporated.

Seitz, U., Freund, S., Jaeckle, F. et al. 2001. First in vivo optical coherence tomography in the human bile duct. *Endoscopy* 33(12): 1018–1021.

Sergeev, A. M., Gelikonov, V., Gelikonov, G. et al. 1997. In vivo endoscopic OCT imaging of precancer and cancer states of human mucosa. *Opt. Express* 1: 432–440.

Simpson, T., Optom, D., Fonn, D., Optom, D., and Optom, M. 2008. Optical coherence tomography of the anterior segment. *Ocul. Surf.* 6(3): 117–127.

Sivak, M. V., Jr., Kobayashi, K., and Izatt, J. A. 2000. High-resolution endoscopic imaging of the GI tract using optical coherence tomography. *Gastrointest. Endosc.* 51: 474–479.

Steinert, R. and Huang, D. 2008. *Anterior Segment Optical Coherence Tomography.* Thorofare, NJ: Slack Incorporated.

Suter, M. J., Bouma, B. E., and Tearney, G. J. 2010. High-resolution optical coherence tomography imaging in gastroenterology. *Advances in Optical Imaging for Clinical Medicine.* Hoboken, NJ: Wiley & Sons.

Suter, M. J., Vakoc, B.J., Yachimski, P.S. et al. 2008. Comprehensive microscopy of the esophagus in human patients with optical frequency domain imaging. *Gastrointest. Endosc.* 68(4): 745–753.

Swanson, E. A., Izatt, J. A., Hee, M. R. et al. 1993. In vivo retinal imaging by optical coherence tomography. *Opt. Lett.* 18(21): 1864–1866.

Tanaka, K. et al. 2006. High risk fibroatheroma lesions are remote from the minimal lumen area site: A virtual histology IVUS analysis from the PROSPECT study. *Am. J. Cardiol.* 98(8A): 94M.

Tarnawska, D. and Wylegala, E. 2010. Monitoring cornea and graft morphometric dynamics after descemet stripping and endothelial keratoplasty with anterior segment optical coherence tomography. *Cornea* 29(3): 272–277.

Tearney, G. J., Brezinski, M. E., Boppart, S. A. et al. 1996. Catheter-based optical imaging of a human coronary artery. *Circulation* 94(11): 3013.

Tearney, G. J., Waxman, S., Shishkov, M. et al. 2008. Three-dimensional coronary artery microscopy by intracoronary optical frequency domain imaging. *J. Am. Coll. Cardiol. Imaging* 1(6): 752–761.

Tearney, G. J., Yabushita, H., Houser, S. L. et al. 2002. Optical coherence tomography is capable of quantifying macrophage content in atherosclerotic plaque fibrous caps. *Circulation* 106(19): 3238.

Testoni, P. A., Mariani, A., Mangiavillano, B. et al. 2007. Intraductal optical coherence tomography for investigating main pancreatic duct structures. *Am. J. Gastroenterol.* 102(2): 269–274.

Unal, G., Gurmeric, S., and Carlier, S. 2010. Stent implant follow-up in intravascular optical coherence tomography images. *Int. J. Cardiovasc. Imaging (formerly Cardiac Imaging)* 26(7): 809–816.

van der Meer, F. J., Faber, D. J., Baraznji Sassoon, D. M. et al. 2005. Localized measurement of optical attenuation coefficients of atherosclerotic plaque constituents by quantitative optical coherence tomography. *IEEE Trans. Med. Imaging* 24(10): 1369–1376.

van Soest, G., Goderie, T., Regar, E. et al. 2010. Atherosclerotic tissue characterization in vivo by optical coherence tomography attenuation imaging. *J. Biomed. Opt.* 15(1): 011105.

van Velthoven, M. E. J., Faber, D. J., Verbraak, F. D., van Leeuwen, T. G., and de Smet, M. D. 2007. Recent developments in optical coherence tomography for imaging the retina. *Prog. Retin. Eye Res.* 26(1): 57–77.

Wojtkowski, M. 2010. High-speed optical coherence tomography: Basics and applications. *Appl. Opt.* 49(16): D30–D61.

Wu, X. F. et al. 2009. Virtual histology intravascular ultrasound analysis of attenuated plaques detected by greyscale intravascular ultrasound in patients with acute coronary syndromes: A prospect substudy. *J. Am. Coll. Cardiol.* 53(10): A99.

Xie, Y., Takano, M., Murakami, D. et al. 2008. Comparison of neointimal coverage by optical coherence tomography of a sirolimus-eluting stent versus a bare-metal stent three months after implantation. *Am. J. Cardiol.* 102(1): 27–31.

Yabushita, H., Bouma, B. E., Houser, S. L. et al. 2002. Characterization of human atherosclerosis by optical coherence tomography. *Circulation* 106(13): 1640–1645.

Yelbuz, T. M., Choma, M. A., Thrane, L., Kirby, M. L., and Izatt, J. A. 2002. Optical coherence tomography—A new high-resolution imaging technology to study cardiac development in chick embryos. *Circulation* 106(22): 2771–2774.

Zhu, C., Palmer, G. M., Breslin, T. M., Xu, F., and Ramanujam, N. 2005. Use of a multiseparation fiber optic probe for the optical diagnosis of breast cancer. *J. Biomed. Opt.* 10(2): 024032.

Zysk, A. M., Adie, S. G., Armstrong, J. J. et al. 2007. Needle-based refractive index measurement using low-coherence interferometry. *Opt. Lett.* 32(4): 385–387.

Zysk, A. M. and Boppart, S. A. 2006. Computational methods for analysis of human breast tumor tissue in optical coherence tomography images. *J. Biomed. Opt.* 11(5): 054015.

Zysk, A. M., Nguyen, F. T., Chaney, E. J. et al. 2009. Clinical feasibility of microscopically-guided breast needle biopsy using a fiber-optic probe with computer-aided detection. *Technol. Cancer Res. Treat.* 8(5): 315–321.

Forward Models of Light Transport in Biological Tissue

16.1 Introduction 319
16.2 RTE-Based Models of Light Transport in Biological Tissue..............................321
16.3 Numerical Methods.. 322
  Introduction • Finite-Difference Methods for Regular Geometries • Finite-Difference
  Methods for Irregular Geometries • Delta-Eddington Method • Finite-Volume
  Discrete-Ordinates Method • Finite-Element Even-Parity Method
16.4 Examples in Biomedical Research 327
  Diffusion versus Transport Solution • Influence of Collimated Light Sources
  (Delta-Eddington Approach) • SNR Analysis Using the Frequency-Domain RTE • Light
  Transmission through Finger Joints Using the Frequency-Domain RTE • Modeling
  Fluorescence with FD Blocking-Off Method
16.5 Summary..332
Acknowledgments...333
References..333

Andreas H. Hielscher
Columbia University

Hyun Keol Kim
Columbia University

Alexander D. Klose
Columbia University

16.1 Introduction

Over the last two decades, considerable advances have been made toward modeling light propagation in tissue. These advances have been mainly driven by the development of optical tomographic imaging techniques, which rely on accurate modeling of light-tissue interactions in biomedical tissue. These techniques employ near-infrared (NIR) light in the wavelength range of 600–900 nm, to probe biological tissue and image physiological functions (Gibson et al. 2005, Nissilä et al. 2005, Gibson and Dehghani 2009). The propagation of light in the NIR is governed by the spatial distribution of the tissue's scattering and absorption coefficients, given by μ_s and μ_a, respectively. The scattering coefficient takes on values in the range of $\mu_s = 20, \ldots,$ 200 cm^{-1} (Cheong et al. 1990, Cheong 1995, Collier et al. 2005). Also, often used to describe optical properties of tissue is the reduced scattering coefficient, $\mu_s' = (1 - g)\mu_s$, which is a scaled scattering coefficient that takes into account the mean scattering cosine g (anisotropy factor) of a single scattering event. Typical values of g in tissue are between 0.8 and 0.98. Difference in the refractive index (RI) between intracellular and extracellular fluids, and various subcellular components, such as mitochondria or nuclei, as well as varying tissue densities give rise to differences in scattering coefficient and g-factors (Mourant et al. 1998, 2000). A multitude of chromophores inside tissue cause absorption at various wavelengths. The absorption coefficient covers a wide range of values from $\mu_a = 0.01$ to 0.5 cm^{-1}, for NIR (Utzinger

et al. 2001, Culver et al. 2003, Drezek et al. 2003, Srinivasan et al. 2003). Endogenous chromophores are, for example, hemoglobin, cytochrome, flavins, and porphyrins. Differences in chromophore content and concentration lead to different absorption coefficients $\mu_a(r)$.

In optical tomographic imaging, also often referred to as diffuse optical tomography (DOT), one seeks to provide the spatial distribution of these optical properties and related physiological parameters by probing the tissue with light and measuring the transmitted light distribution on the tissue surface. A variety of different instruments have been developed that perform highly sensitive measurements of transmitted light intensities on various body parts (Schmidt et al. 2000, McBride et al. 2001, Schmitz et al. 2002, Thompson and Sevick-Muraca 2003, Patwardhan et al. 2005, Gulsen et al. 2006, Joseph et al. 2006, Li et al. 2006, Lasker et al. 2007). The data are input to so-called model-based iterative image reconstruction schemes, which typically consist of two parts: (1) a forward model for light propagation and (2) an inverse model. The forward model predicts the detector readings on the tissue boundary given a distribution of optical properties inside the medium. The inverse model determines the optical parameters inside the tissue, given a set of detector readings and detector predictions on the boundary of the tissue. The forward and inverse models are iteratively employed within an optimization method. The predicted detector readings of the forward model on the tissue surface are subsequently compared to measured detector readings by defining an objective function, for example, a χ^2-error norm. The objective function is minimized with an optimization method for nonlinear functions, such as conjugate gradient or quasi-Newton methods. The optical parameters at the minimum of the error function are considered as the solution of the inverse problem (Hielscher et al. 1999, Klose and Hielscher 1999). Inverse models of light transport are treated in detail in the next chapter. Here we focus on forward models of light transport.

The light propagation in biological tissue is most accurately described by the radiative transfer equation (RTE). The RTE is typically solved with some numerical methods since no analytical solutions of the RTE exist for nonuniform media with complex geometries. These numerical methods are either low-order approximations to the RTE, such as the diffusion approximation, or high-order approximations to the RTE, for example, discrete ordinates (S_n) or spherical harmonics (P_N) methods (Carlson and Lathrop 1968). The diffusion approximation is widely employed for DOT and has been successfully applied in tissues where $\mu'_s \gg \mu_a$ (diffusion regime), for example, imaging of the female breast or imaging of muscular tissue (Tromberg et al. 2008). However, the diffusion approximation has limited applications when applied to tissue with small geometries, tissue with high absorption, or tissue containing void-like areas (Hielscher et al. 1998). In that case, high-order approximations of the RTE are required. In all cases, the spatial discretization of the tissue domain can be performed with either finite-difference (FD), finite-volume (FV), or finite-element (FE) techniques. Chapters 9, 10, and 17 through 22 present insights into theory and applications of the diffusion approximation. Here, we continue with an in-depth discussion of modeling light-tissue interactions with the RTE.

That radiative energy transport in scattering media can be described by the RTE, which has been known for over 50 years (Chandrasekhar 1960, Case and Zweifel 1967, Duderstadt and Martin 1979, Ishimaru 1996). Initially, it was used to model processes and phenomena in nuclear physics, astrophysics, atmospheric physics, and oceanography. Numerical solution methods and applications that considered the special case of biomedical optics only started to emerge in the 1990s. Aronson et al. (1991) was the first to suggest that the RTE could be used for biomedical imaging, and Hielscher et al. (1998) were the first to perform extensive numerical studies assessing the validity of the RTE and the diffusion equation in biological tissue, especially when large absorption coefficients and void-like domains are present. Hielscher et al. used a time-independent diffusion-accelerated FD-discrete ordinate (S_n) method for calculating the light distributions inside a human brain model. Later, Aydin et al. (2004, 2005) developed an FE-spherical harmonics (P_N) method for studying light propagation in void-like domains. Khan and Thomas (2005) employed a P_N approximation to the RTE for studying the impact of the RI in photon propagation in tissue.

The time-independent RTE has also been applied as forward model within DOT codes. Klose et al. (2002) were the first to use an FD-S_n technique based on the steady-state (SS) RTE for reconstructing the scattering and absorption properties from experimental data obtained with tissue phantoms containing voids (Klose et al. 2002). This code was later used to reconstruct spatial maps of scattering and absorption coefficients of arthritic finger joints (Hielscher et al. 2004, Scheel et al. 2005). The joint cavity contains a low scattering fluid that constitutes a nondiffusive regime for NIR light. Hence, the small finger geometry and the void-like joint cavity require an RTE-based image reconstruction method. Subsequently, Cai et al. (2003) utilized an analytic solution to the time-independent RTE for reconstructing the optical properties in numerical tissue models. Arridge and Dorn (2006) and Tarvainen et al. (2005a,b, 2006) developed a hybrid method consisting of a diffusion model and a transport model that increased the speed of the image reconstruction process, by limiting the use of the RTE to areas in the medium, where the diffusion approximation cannot be applied.

The RTE has also been applied to DOT in time-dependent cases. Dorn (1998) introduced a generalized algebraic reconstruction technique in conjunction with an FD-S_n method based on the time-dependent RTE. Using numerical data, he showed that the code was able to reconstruct scattering and absorption coefficients. Similarly, Boulanger and Charette (2005a,b) focused on problems in which short-light pulses are injected into a tissue and developed a discrete-ordinate code that employs a so-called piecewise parabolic method, which is particularly suited for the treatment of short-pulse propagation in turbid media. Ren et al. (2004, 2006) developed an FV method based on the RTE in the frequency domain, where the source term is sinusoidally modulated in the 100–1000 MHz range. Using numerical tissue models, they showed that the absorption and scattering coefficients can be better separated using frequency-domain techniques, as compared to the time-independent case. Abdoulaev and Hielscher (2003) developed the first RTE-based FE code for biomedical application, which will be discussed in more detail later. Abdoulaev et al. (2005) also introduced a so-called PDE-constrained optimization technique in the frequency domain for reconstructing optical properties, which was later refined by Kim and Hielscher (2009). This approach leads to considerable shorter convergence times for RTE-based codes. Most recently, a simplified spherical harmonics (SP_N) method has been proposed by Klose et al. (Klose and Larsen 2006). This approach also leads to considerable speed increase in forward and inverse models, while maintaining much of the accuracy of an unapproximated RTE solver. This is achieved by reducing the number of equations of the P_N technique leading to a transport approximation for highly absorbing media based on coupled diffusion equations.

Advances have also been made in image reconstruction of fluorescent and bioluminescence light sources in tissue for optical molecular tomography. Klose et al. were the first to develop an FD-S_n method for optical molecular tomography that recovers the fluorescent probe concentration distribution (Klose and Hielscher 2003a, Klose et al. 2005, Klose 2009a,b). Kim and Moscoso (2006) established an analytical approach for recovering the fluorescent probe distribution in scattering tissue. Klose was also the first to develop RTE-based codes that can model bioluminescence effects (Klose 2007, 2009a,b). Most recently, Lu et al. (2009) employed the SP_N methods to analyze first experimental bioluminescence data. Molecular imaging codes are currently used in preclinical studies involving small animals. For example, in fluorescence molecular tomography (FMT), biochemical fluorescent markers are injected into a cancer-bearing mouse, and upon excitation with an external light source, the fluorescent marker will emit light. The fluorescent light is detected with a CCD camera on the tissue surface. Subsequently, a spatial map of the fluorescent marker concentration is reconstructed that, for example, allows to monitor the growth of a tumor or the effectiveness of new anticancer drugs.

In the following sections, we will provide more details concerning the major aspects of numerically implementing RTE-based forward codes, and show some applications that illustrate the performance of these codes.

16.2 RTE-Based Models of Light Transport in Biological Tissue

The fundamental quantity in tissue optics is the radiance or angular flux of photons, $\Psi(r,\Omega)$, with units of $W\ cm^{-2}\ sr^{-1}$, at the spatial position r and unit direction Ω. In general, the RTE can be derived as a balance equation for the angular flux Ψ. In the time-dependent case, this balance equation is given by

$$\frac{\partial \Psi(r,\Omega,t)}{c\partial t} + \Omega \cdot \nabla \Psi(r,\Omega,t) + (\mu_a + \mu_s)\Psi(r,\Omega,t)$$

$$= \mu_s \int_{4\pi} p(\Omega,\Omega')\Psi(r,\Omega',t)d\Omega' + Q(r,\Omega,t) \qquad (16.1)$$

The RTE is an integro-differential equation for the radiance Ψ as a function of the direction Ω and location r for time point t. The radiance is given in units of photons $cm^{-2}\ sr^{-1}$. The source term $Q(r, \Omega, t)$ represents the power injected into a solid angle centered on Ω in a unit volume at r inside of the scattering medium. In the special case that the sources amplitude is sinusoidally modulated, we get the FD RTE, given by

$$\Omega \cdot \nabla \Psi(r,\Omega,\omega) + \left(\mu_a + \mu_s + \frac{i\omega}{c}\right)\Psi(r,\Omega,\omega)$$

$$= \mu_s \int_{4\pi} p(\Omega,\Omega')\Psi(r,\Omega',\omega)d\Omega' + Q(r,\Omega,\omega) \qquad (16.2)$$

When the source term is time-independent, this simplifies to the SS equation:

$$\Omega \cdot \nabla \Psi(r,\Omega) + (\mu_a + \mu_s)\Psi(r,\Omega)$$

$$= \mu_s \int_{4\pi} p(\Omega,\Omega')\Psi(r,\Omega')d\Omega' + Q(r,\Omega). \qquad (16.3)$$

Other quantities besides the radiance Ψ that are included in the RTE are the scattering coefficient μ_s, the absorption coefficient μ_a, both given in units of cm^{-1}, and the scattering phase function $p(\Omega,\Omega')$ with units of sr^{-1}, which all describe the optical properties of the medium. The scattering phase function $p(\Omega,\Omega')$ is a probability function that describes the anisotropic scattering behavior of photons in biological tissue. The angle θ encloses the two directions formed by Ω and Ω' in the interval $[0,\pi]$ with $\cos\theta = (\Omega \cdot \Omega')$ and its anisotropy factor $g = \langle\cos\theta\rangle$ (where $\langle\ldots\rangle$ denotes the expectation value). A commonly used scattering phase function for biomedical applications is the Henyey–Greenstein (HG) phase function

$$p_{HG} = \frac{1-g^2}{4\pi(1+g^2-2g\cos\theta)^{3/2}}. \qquad (16.4)$$

At the tissue boundary, light can also be partially reflected back into the tissue. The amount of light that is reflected back along direction Ω is given by the boundary condition

$$\Psi(r,\Omega) = R(\Omega' \cdot n)\Psi(r,\Omega') + S(r,\Omega) \quad \text{for } \Omega \cdot n < 0. \qquad (16.5)$$

and the reflectivity R,

$$R = \frac{\sin^2(\beta-\alpha)}{\sin^2(\beta+\alpha)} + \frac{\tan^2(\beta-\alpha)}{\tan^2(\beta+\alpha)}. \qquad (16.6)$$

The two quantities α and β, are angles between the surface normal and the directions Ω' of outward-directed light and Ω'' of refracted light. Exterior light sources $S(r,\Omega)$, such as laser diodes for probing the tissue, constitute additional boundary sources besides the partially reflected light. The fluence $\Phi(r)$ is obtained by integrating the radiance $\Psi(r,\Omega)$ over all directions Ω:

$$\Phi = \int_{4\pi} \Psi(r,\Omega)d\Omega. \qquad (16.7)$$

If one wants to model fluorescence effects, an additional source term, besides the boundary source, has to be included into the RTE that takes light-emitting fluorophores inside the tissue into account. This second term describes the propagation of fluorescent light after fluorophores have been excited with an external light source at excitation wavelength λ^x. Therefore, we obtain the following RTE at the emission wavelength λ^m:

$$\Omega \cdot \nabla \Psi(r,\Omega) + (\mu_a + \mu_s)\Psi(r,\Omega)$$

$$= \frac{1}{4\pi}\mu_a^{x\to m}\eta\Phi^x + \mu_s \int_{4\pi} p(\Omega,\Omega')\Psi(r,\Omega')d\Omega'. \qquad (16.8)$$

In the time-independent case, the excitation field Φ^x is calculated with Equations 16.3 and 16.5. The energy of the excitation field at

λ^x is partially absorbed by the fluorophore having an absorption coefficient $\mu_a^{x \to m}$. The fluorophore absorption coefficient is determined by the fluorophore concentration, c, inside tissue and the extinction coefficient, ε, at the excitation wavelength λ^x. The term η represents the quantum yield of the fluorophore, which determines the amount of light that will be reemitted at λ^m for a given amount of excitation light.

Another application of the RTE in tissue optics is bioluminescence tomography (BLT). This imaging approach does not have boundary sources, but only interior sources. Therefore, BLT is similar to FMT, except that no externally imposed excitation field Φ^x is given. Bioluminescent sources utilize biochemical energy for the optical excitation of probes. The RTE for bioluminescence light originating from light-emitting sources $Q(r)$ inside tissue is given by

$$\Omega \cdot \nabla \Psi(r, \Omega) + (\mu_a + \mu_s) \Psi(r, \Omega)$$
$$= Q(r) + \mu_s \int_{4\pi} p(\Omega, \Omega') \Psi(r, \Omega') d\Omega'. \quad (16.9)$$

Equations 16.1 through 16.9 provide a complete set of RTE-based models for simulating light propagation in biomedical tissue. In the following, we will provide detailed examples on how to find numerical solutions to these equations.

16.3 Numerical Methods

16.3.1 Introduction

While only a few analytical solutions for the RTE exist, there are many ways to solve the RTE numerically. The first step in any methods is the discretization of the spatial and angular domains. For spatial discretization, one can distinguish between three commonly employed approaches, which are FD, FV, and FE methods. For the angular discretization, the spherical harmonics (P_N) and discrete ordinates (S_n) methods are most often used. Whatever combination of spatial and angular discretization methods are applied, they all lead to a particular system of algebraic equations. These algebraic systems of equations are typically solved with some iterative methods.

Iterative solvers are mainly based on so-called transport sweeps along the direction of light propagation through the medium, which are often referred to as the source-iteration (SI) technique. The Gauss-Seidel (GS) method and successive-over-relaxation approach belong to this category (see, e.g., Klose et al. 2002, Boulanger and Charette 2005a,b, Kim and Charette 2007). These methods are stable and easy to implement and are widely used in structured meshes, but can be quite slow at convergence when highly scattering media are considered. Another approach to solving the discretized radiative transfer equations is to use gradient-search techniques based on the Krylov subspace method. The generalized minimum residual (GMRES) (Saad 1986) and biconjugate gradients with stabilization (BI-CGSTAB)

(van den Vorst 1992) belong in this class, and are mainly applied to large sparse linear systems resulting from unstructured mesh discretization. In most cases, these gradient-search techniques are a factor of 10–30 faster than conventional SI techniques. Preconditioning techniques (see, e.g., Patton and Holloway 2002) may be employed to further accelerate Krylov-subspace methods.

In the following sections, a more detailed discussion of the discretization procedures is presented together with different iterative methods. The main focus is here on finite-difference discrete-ordinates (FD-DO), the finite-volume discrete-ordinates (FV-DO), and the finite-element discrete-ordinates (FE-DO) methods, since these are most widely used in tissue optics.

16.3.2 Finite-Difference Methods for Regular Geometries

When one applies the FD-S_n methods, the direction Ω is replaced with a set of discrete ordinates Ω_k. The total number, K, of ordinates Ω_k is given by $K = N(N+2)$ and N being the number of direction cosines of the S_n method. The integral in the RTE is approximated with a quadrature rule:

$$\int_{4\pi} p(\Omega, \Omega') \Psi(r, \Omega') d\Omega' \cong \sum_{k'=1}^{K} w_{k'} p_{kk'} \Psi_{k'}(r) \quad (16.10)$$

where w_k are weights determined by full level symmetry of the ordinates. Thus, the angular discretization yields a set of K coupled differential equations for the radiance $\Psi_k(r)$ in the directions Ω_k:

$$\Omega_k \cdot \nabla \Psi_k(r) + (\mu_a + \mu_s) \Psi_k(r) = \mu_s \sum_{k'=1}^{K} w_{k'} p_{kk'} \Psi_{k'}(r) \quad (16.11)$$

Next, the continuous spatial variable r has to be discretized on a structured grid with grid points at $r = (x_m, y_n, z_l)$ and (m, n, l) being grid point indices of $[(1...M), (1...N), (1....L)]$. The grid spacing between adjacent grid points is given by Δx, Δy, and Δz.

The spatial derivatives in Equations 16.1 through 16.3 can be substituted with first-order FD approximations, known as step method, or with second-order approximations, given by the diamond method. The *step method* depends on the direction Ω_k of the angular-dependent radiance $\Psi_k(r)$. Thus, the set of all angular directions Ω_k are subdivided into eight octants and we get eight different FD formulas for the radiance $\Psi_k(r)$ depending on the directions. The octants are defined by the direction cosines $(\Omega_x, \Omega_y, \Omega_z)$:

$$\text{1. Octant: } \Omega_x < 0, \Omega_y < 0, \Omega_z < 0 \quad (16.12a)$$

$$\text{2. Octant: } \Omega_x > 0, \Omega_y < 0, \Omega_z < 0 \quad (16.12b)$$

3. Octant: $\Omega_x > 0, \Omega_y > 0, \Omega_z < 0$ (16.12c)

4. Octant: $\Omega_x < 0, \Omega_y > 0, \Omega_z < 0$ (16.12d)

5. Octant: $\Omega_x < 0, \Omega_y < 0, \Omega_z > 0$ (16.12e)

6. Octant: $\Omega_x > 0, \Omega_y < 0, \Omega_z > 0$ (16.12f)

7. Octant: $\Omega_x > 0, \Omega_y > 0, \Omega_z > 0$ (16.12g)

8. Octant: $\Omega_x < 0, \Omega_y > 0, \Omega_z > 0$ (16.12h)

Therefore, we obtain eight different difference terms. For example, the 1. Octant yields

$$\frac{\partial \Psi}{\partial x} \cong \frac{\Psi(\Omega_k)_{i+1jl} - \Psi(\Omega_k)_{ijl}}{\Delta x}, \quad \frac{\partial \Psi}{\partial y} \cong \frac{\Psi(\Omega_k)_{ij+1l} - \Psi(\Omega_k)_{ijl}}{\Delta y},$$

$$\frac{\partial \Psi}{\partial z} \cong \frac{\Psi(\Omega_k)_{ijl+1} - \Psi(\Omega_k)_{ijl}}{\Delta z},$$

(16.13)

and the RTE for the radiance $\Psi(\Omega_k)$ at grid point (ijl) can now be written as

$$\Omega_x \frac{\Psi_{i+1jl} - \Psi_{ijl}}{\Delta x} + \Omega_y \frac{\Psi_{ij+1l} - \Psi_{ijl}}{\Delta y} + \Omega_z \frac{\Psi_{ijl+1} - \Psi_{ijl}}{\Delta z}$$

$$+ (\mu_s + \mu_a)_{ijl} \Psi_{ijl} = \frac{Q_{ijl}}{4\pi} + (\mu_s)_{ijl} \sum_{k'} w_{k'} p_{kk'} \Psi_{ijlk'} \quad (16.14)$$

Alternatively a so-called diamond-differencing scheme can be employed. This scheme is a second-order FD approach that, unlike the step method, does not require fine structured grids for attaining sufficient numerical accuracy. The diamond-differencing equations are constructed by means of auxiliary grid points and difference equations. The grid point (ijl) is at the center of a grid cell with six surfaces. Each surface has a centered cell-surface grid point $(i + 1/2jl, i - 1/2jl, ij + 1/2l, ij - 1/2l, ijl + 1/2, ijl - 1/2)$, which are connected to the cell-centered grid point (ijl). The radiance $\Psi(\Omega_k)$ at grid point (ijl) is given by the radiances at the cell-surface points:

$$\Psi_{ijlk} = \frac{1}{2}(\Psi_{i+1/2jlk} + \Psi_{i-1/2jlk}); \quad \Psi_{ijlk} = \frac{1}{2}(\Psi_{ij+1/2lk} + \Psi_{ij-1/2lk});$$

$$\Psi_{ijl} = \frac{1}{2}(\Psi_{ijl+1/2k} - \Psi_{ijl-1/2k})$$

(16.15)

and the RTE is given by the relationships of the cell-surface grid points:

$$\Omega_x \frac{\Psi_{i+1/2jl} - \Psi_{i-1/2jl}}{\Delta x} + \Omega_y \frac{\Psi_{ij+1/2l} - \Psi_{ij-1/2l}}{\Delta y}$$

$$+ \Omega_z \frac{\Psi_{ijl+1/2} - \Psi_{ijl-1/2}}{\Delta z} + (\mu_s + \mu_a)_{ijl} \Psi_{ijl}$$

$$= \frac{Q_{ijl}}{4\pi} + (\mu_s)_{ijl} \sum_{k'} w_{k'} p_{kk'} \Psi_{ijlk'} \quad (16.16)$$

Depending on the octant, the cell-surface grid points are expressed by the cell-centered points, and we obtain, for example, for the 7. Octant with $\Omega_x > 0, \Omega_y > 0, \Omega_z > 0$ the following:

$$\Psi_{i+\frac{1}{2}jl} = 2\Psi_{ijl} - \Psi_{i-\frac{1}{2}jl} \quad (16.17a)$$

$$\Psi_{ij+\frac{1}{2}l} = 2\Psi_{ijl} - \Psi_{ij-\frac{1}{2}l} \quad (16.17b)$$

$$\Psi_{ijl+\frac{1}{2}} = 2\Psi_{ijl} - \Psi_{ijl-\frac{1}{2}} \quad (16.17c)$$

Equations 16.17a–c are used to eliminate the number of unknowns in Equation 16.16.

Both, the step method and the diamond method, yield an algebraic system of linear equations for the unknown radiance Ψ:

$$A\Psi = B\Psi + Q/4\pi \quad (16.18)$$

where

    A depicts the discretized streaming and collision operator
    B is the discretized integral operator
    Q constitutes the source term

This algebraic system of equations can either be solved with an SI technique for the SS case or with an Euler-differencing (ED) method for the time-dependent case.

The SI technique sweeps through the grid along the directions Ω_k. For example, for the direction Ω_k with cosines $\Omega_x > 0, \Omega_y > 0, \Omega_z > 0$, we obtained the following explicit iteration rule for the iteration step $h + 1$, given step h:

$$\Psi_{ijl}^{h+1} = \frac{Q_{ijl} + [\mu_s]_{ijl} \sum_{k'=1}^{K} w_{k'} p_{kk'} \Psi_{ijlk'}^{h} + \frac{\Omega_{xk}}{\Delta x} \Psi_{i-1jlk}^{h+1} + \frac{\Omega_{yk}}{\Delta y} \Psi_{ij-1lk}^{h+1} + \frac{\Omega_{zk}}{\Delta z} \Psi_{ijl-1k}^{h+1}}{\frac{\Omega_{xk}}{\Delta x} + \frac{\Omega_{yk}}{\Delta y} + \frac{\Omega_{zk}}{\Delta z} + [\mu_t]_{ijl}}$$

(16.19)

The difference terms in Equation 16.19 will differ for ordinates of the remaining seven octants. The time-dependent RTE can be solved with an Euler-backward (EB) differencing scheme for the time derivative in Equation 16.1 (Dorn 1998). Employing the

EB step method for the FD discretization yields the following discretized RTE (Figures 16.1 and 16.2):

$$\frac{\Psi_{ijlk}^{t+1} - \Psi_{ijlk}^{t}}{c\Delta t} + \Omega_{xk}\frac{\Psi_{ijlk}^{t+1} - \Psi_{i-1jlk}^{t+1}}{\Delta x} + \Omega_{yk}\frac{\Psi_{ijlk}^{t+1} - \Psi_{ij-1lk}^{t+1}}{\Delta y}$$

$$+ \Omega_{zk}\frac{\Psi_{ijlk}^{t+1} - \Psi_{ijl-1k}^{t+1}}{\Delta z} + [\mu_s + \mu_a]_{ijl}\Psi_{ijlk}^{t+1}$$

$$= Q_{ijlk}^{t+1} + [\mu_s]_{ijl}\sum_{k'} w_{k'}p_{kk'}\Psi_{ijlk'}^{t} \qquad (16.20)$$

16.3.3 Finite-Difference Methods for Irregular Geometries

Extra steps have to be taken if the FD methods are to be employed in a domain with non-rectangular surfaces. Such irregular geometries can be treated with the FD methods on unstructured as well as on structured grids (Thompson et al. 1985, 1999, Farrashkhalvat and Miles 2003). Structured grids are characterized by regular connectivity, that is, the points of the grid can be indexed and the neighbors of each point can be calculated rather than be looked up. Therefore, they offer less computational overhead than unstructured grids and are the preferred choice when simple irregular geometries are considered.

FD methods for structured grids applied to irregular geometries are, for example, the body-fitted-grids methods, the blocked-off region method, and the adaptive mesh refinement methods (Chai et al. 1994, Lang and Walter 1995, Jessee et al. 1998). Body-fitted-grids methods describe the boundary most

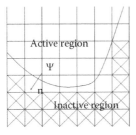

FIGURE 16.1 Finite difference grid with physical domain, active region, and blocked-off region (crosses). The solution is only sought in the active region.

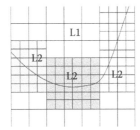

FIGURE 16.2 Adaptive mesh refinement with two different grid levels L1 and L2. Grid level L2 is refined in the vicinity of boundary.

accurately, but often become intractable and fail to perform if the boundaries are too complicated. Therefore, the other two methods are more widely applied. When using a *Blocking-off method* approach, a physical domain is modeled using a so-called nominal domain that is divided into two regions (Klose and Hielscher 2006). The "active" region is that part of the physical domain where the solution is sought. The "inactive" region lies outside the physical boundary. The blocked-off region method will sustain a high numerical accuracy if the mesh size is sufficiently fine. The disadvantages of this method are long computation times and large memory requirement. This can be overcome by combining the blocked-off region method with an adaptive mesh refinement technique.

The adaptive-mesh-refinement method is less computational expensive because it uses different levels of grid refinements. The adaptive mesh refinement method uses a structured grid that is clipped at the boundary with local grid refinement near sharp features on the boundary. This will reduce the truncation error at the boundary. The adaptive mesh refinement method solves the RTE on a composite grid L with different spatial refinement levels L_1, L_2, L_3,...,L_x. Grid L_1 is the very coarse grid, and grid L_2 is a refined grid where the spatial difference between adjacent grid points of L_1 is cut into half. For a given ordinate direction, the standard solution algorithm sweeps the spatial grid on the coarse level L_1 from upstream to downstream according to the requirements of the step difference method. Next, the conditions at the coarse–fine level interface are interpolated and applied to the boundary of the fine grid level L_2. The RTE is solved on that spatial subdomain with grid level L_2. At the fine–coarse level, interface radiances leaving the fine grid level L_2 are then averaged and applied to the boundary of the coarse grid level L_1. Using the modified radiance distribution on grid level L_1, a new coarse grid solution is sought again by sweeping through the grid. At this time, the solution on L_1 is influenced by the solution of the fine grid level L_2. When a radiance ray passes an interface from the coarse grid level L_1 to the finer grid L_2, the interface radiance is interpolated with a piecewise constant operator, whereas when a ray passes from the fine to the coarse grid level, the radiance is averaged using an area averaged operator. Iteration between the levels continues until the composite solution is converged. When a ray passes from a coarse to fine grid, the interface radiance is interpolated, and when a ray passes from a fine to a coarse grid, the interface radiance is averaged.

16.3.4 Delta-Eddington Method

Scattering in biological media is typically strongly forward directed. That means that during each individual scattering event, light is not changing its direction very much ($g > 0.9$). In this case, large sets of discrete ordinates are needed to accurately describe light propagation in tissue, which increases the computational burden considerable. To alleviate this problem, the Delta-Eddington (DE) methods can be applied (Joseph et al. 1976, Klose and Hielscher 2003b). This method is derived from a Lengendre polynomial expansion of the HG scattering function, which is given by

$$p_{HG}(\cos\theta) = \frac{1}{4\pi}\sum_{n=0}^{N}(2n+1)b_nP_n(\cos\theta). \quad (16.21)$$

Here, $P_n(\cos\theta)$ are the Legendre polynomials, and the coefficients b_n are given by

$$b_n = 2\pi\int_{-1}^{1}p_{HG}(\cos\theta)P_n(\cos\theta)\mathrm{d}\cos\theta = g^n. \quad (16.22)$$

Many Legendre polynomials are needed when describing highly anisotropically scattering media with $g > 0.9$ and many discrete ordinates are required when solving the RTE with an S_n method. The DE method reduces the number of discrete ordinates and Legendre polynomials by transforming the phase function into two separate terms: The first part is a fraction f of scattered photons and is described by a δ-function and the remaining fraction $(1-f)$ out of scattered photons is described by a modified HG function p'_{HG}

$$p_{HG}(\cos\theta) \cong p_{\delta-HG}(\cos\theta) = (1-f)p'_{HG}(\cos\theta) + \frac{1}{2\pi}f\delta(1-\cos\theta) \quad (16.23)$$

The function p'_{HG} is expanded into a series of Legendre polynomials:

$$p'_{HG}(\cos\theta) = \frac{1}{4\pi}\sum_{n=0}^{N}(2n+1)b'_nP_n(\cos\theta) \quad (16.24)$$

and substituted into Equation 16.23. The new coefficients of Legendre polynomial expansion are obtained with

$$b'_n = \frac{b_n - f}{1-f}. \quad (16.25)$$

With N being the number of terms (typically $N = 5$) after the series expansion in Equation 16.24 is terminated. Hence, the DE approach yields a new RTE

$$\Omega \cdot \nabla\Psi(r,\Omega) + (\mu_a + (1-f)\mu_s)\Psi(r,\Omega)$$
$$= (1-f)\mu_s\int_{4\pi}p'_{HG}(\Omega,\Omega')\Psi(r,\Omega')\mathrm{d}\Omega', \quad (16.26)$$

and smaller sets of discrete ordinates can be employed.

16.3.5 Finite-Volume Discrete-Ordinates Method

16.3.5.1 Introduction

As an alternative strategy to FD methods that employ structured grids, complicated tissue geometries can also be handled by unstructured meshes, typically triangulations in 2D and tetrahedral elements in 3D, within an FV framework. Compared to the FD methods, FV methods allows for the energy conservative formulation based on the integration of the radiative transfer equation (RTE) over a discretized control volume, which ensures the conservation of the radiative energy within each control volume.

To implement an FV method on an unstructured mesh, we can apply the cell-centered or vertex (or node)-centered methods for the spatial differencing. Both methods are widely used and, for example, have recently been implemented for FD optical tomography (see, e.g., Ren et al. 2004, 2006, Kim and Hielscher 2009). The node-centered FV method takes advantage of beneficial properties of both the FE and FV methods, thus combining the conservation properties of the FV formulation and the geometric flexibility of the FE approach. The node-centered scheme is more accurate in the flux calculation at boundary grids when the grids are highly skewed, but it produces a slightly larger local truncation error in comparison to the cell-centered scheme on the same grid. Another advantage of the node-centered scheme is that it allows for a direct positioning of sources and detectors in DOT while the cell-centered scheme requires special treatment. A more detailed comparison of both methods can be found in Meese (1998).

Figure 16.3 shows the construction of the control volumes based on the node-centered scheme in the 2D case. The triangulation is first done for the computation domain and polygonal control volumes are constructed around each node of the grid by connecting the centroid of each triangular element to the midpoints of the corresponding sides. All the unknowns are stored on the grid nodes.

As with the structured mesh, the node-centered FV-DO formulation is obtained by integrating the RTE in the FD over the control volume with a divergence theorem (Kim et al. 2008, Kim and Hielscher 2009) as

$$\sum_{j=1}^{N_f}(\vec{n}_j\cdot\Omega^m)\Psi_j^m dA_j + \left(\mu_a + \mu_s + \frac{i\omega}{c}\right)\Psi_N^m\Delta V_N$$
$$= \sum_{m'=1}^{M}p^{m'm}\Psi_N^{m'}w^{m'}\Delta V_N + q_N^m\Delta V_N \quad (16.27)$$

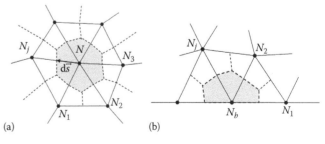

FIGURE 16.3 Typical triangular elements and node-centered control volumes in 2D unstructured meshes: (a) internal node and (b) boundary node.

where

> N_f and M are the number of surfaces surrounding the node N and the number of discrete ordinates, respectively
>
> $\vec{n}_j$ and Ψ_j^m denote the surface normal vector and the radiation intensity defined on the jth surface

Also, the surface intensity Ψ_j^m is related to the nodal intensity Ψ_N^m by the appropriate spatial differencing scheme that can be applied to unstructured meshes.

16.3.5.2 Spatial Differencing Schemes

The *step-differencing scheme* simply replaces the surface intensity Ψ_j^m by the value at the upstream node, which can be expressed as the following general form:

$$\Psi_j^m = \max(n_j \cdot \Omega^m dA_j, 0)\Psi_N^m - \max(-n_j \cdot \Omega^m dA_j, 0)\Psi_{N_j}^m \quad (16.28)$$

where N_j denotes the neighbor node of the main node N with the jth surface as a common face between the node N and its neighbor node N_j. Then, the final discretized form of (16.27) is obtained as

$$\left[\left(\mu_a + \mu_s + \frac{i\omega}{c} \right) \Delta V_N + \sum_{j=1}^{N_f} \max(n_j \cdot \Omega^m dA_j, 0) \right] \Psi_N^m$$

$$= \sum_{j=1}^{N_f} \max(-n_j \Omega^m dA_j, 0) \Psi_{N_j}^m$$

$$+ \mu_s \sum_{m'=1}^{M} p^{m'm} \Psi_N^{m'} w^{m'} \Delta V_N + q_N^m \Delta V_N, \quad (16.29)$$

which is numerically stable and easy to implement but is only first-order accurate.

To achieve high-order accuracy, least-square or interpolation schemes can be employed. However, high-order schemes used without a first-order component can produce physically unrealistic solutions such as negative values or overshoots or undershoots; therefore the high-order schemes are often used in combination with a first-order step scheme.

In unstructured meshes, for example, the surface intensity Ψ_j^m can be obtained as

$$\Psi_j^m = \Psi_{up}^m + \alpha \nabla \Psi_{r,up}^m \cdot d\vec{s} \quad (16.30)$$

where

> $\nabla \Psi_{r,up}^m$ is the reconstructed gradient at the upstream control volume
>
> $d\vec{s}$ is the vector from the center of the upstream control volume to the center of the surface (see Figure 16.3a) (Venkatakrishnan, 1993)

The parameter α is a limiter (varying from 0 to 1) used to limit oscillations due to numerical instability. In practice, the first term Ψ_{up}^m is treated implicitly and the second term with $\nabla \Psi_{r,up}^m$ is computed explicitly from the previous iteration. A detailed comparison study on the use of different limiter functions can be found in Coelho (2002).

The major drawback of this limiter-based approach is, however, that the scheme does not fully achieve second-order accuracy; it only takes somewhere between first-order and second-order accuracy. To overcome this problem, the linear discontinuous (LD) scheme has been proposed by Morel et al. (1996), which can lead to a significant improvement in both numerical stability and accuracy. The LD scheme is known to be third-order accurate if a slant path of light through a control volume is not greater than three mean free paths. An example of using the LD scheme for optical tomography can be found in Rasmussen et al. (2006).

16.3.5.3 Solution of Resulting Linear Equations

After discretization of the RTE for all nodes, Equation 16.20 can be written into the matrix-vector form as follows:

$$A\Psi = b \quad (16.31)$$

where each line of the matrix A contains the coefficients of the discretized equation at node number N and direction m. The vector Ψ denotes the radiation intensity vector and the vector b is the vector of implementing both the discretized boundary condition and the source term. The large, sparse matrix A contains complex-valued entries resulting from the FD equation of radiative transfer. As a result, the complex-valued sparse, linear system of equations given by (16.24) is iteratively solved for intensity into a discrete-ordinate direction by using any complex- or real-value versions of gradient-search techniques such as Bi-CGSTAB and GMRES solvers (Ren et al. 2004, Gu et al. 2007, Kim et al. 2008, Kim and Hielscher 2009), which fall into the class of Krylov subspace methods.

The basic idea of Krylov subspace methods is as follows: Let Ψ^0 be an initial guess of radiance Ψ and $r_0 = b - A\Psi^0$ be the initial residual, then the L-dimensional Krylov subspace can be constructed as

$$K_L(A, r_0) = \text{span}\{r_0, Ar_0, \ldots, A^L r_0\}. \quad (16.32)$$

Here, the initial residual r_0 is the source that drives the correction to the initial guess Ψ^0. GMRES algorithms seek an optimal solution of the next iterate, given by

$$\Psi = \Psi^0 + \sum_{l=0}^{L-1} a_l A^l r_0 \quad (16.33)$$

that minimizes the residual $\|A\Psi - b\|$. More details can be found in Patton and Holloway (2002), including suitable preconditioning techniques for GMRES methods.

A brief comparison of the GMRES method and the SI methods, in terms of their computation times and accuracy, is presented here. It is well known that the convergence of the SI method is restricted by the scattering ratio ($\varpi = \mu s/\mu t$) regardless of the cell width (Reed 1971). Hence, for the problem $\varpi \ll 1$ (transport limit), the SI method converges rapidly. For the case $\varpi \sim 1$ (diffuse limit), the convergence is quite slow due to the strong angular coupling between intensities, which may restrict the efficiency of the SI method. On the other hand, the Krylov subspace methods, such as the GMRES method (Saad 1986), shows fast convergence, especially when applied to highly scattering media typically encountered in biological tissues. To be specific, the Krylov subspace-based GMRES method first moves the whole scattering integral over to the left-hand side of the equation, and constructs the large, sparse, nonsymmetric matrix represented by A. In this form, the coupling of the angular radiances for each spatial node and each discrete ordinates direction is explicitly displayed. Hence, the resulting sparse matrix A contains all of the streaming, absorption, and scattering information, and then the sparse linear equation of the form $A\Psi = b$ can be solved in an efficient manner by using the Krylov subspace-based GMRES method (Oliveria 1995, Patton 1996). Thus, the GMRES method provides fast convergence for the cases of highly scattering media such as biological tissues where the efficiency of the SI method is degenerated. With suitable preconditioning (Patton and Holloway 2002), the GMRES solver allows for the additional speedup for the solution of such sparse nonsymmetric matrix problems. Generally, the GMRES solver is computationally more efficient than the SI method, but it should be noted that its memory requirement and implementation complexity are higher.

16.3.6 Finite-Element Even-Parity Method

Besides FD and FV methods, FE methods can be employed (Abdoulaev and Hielscher 2003). Here, we will demonstrate the derivation of the even-parity equations for the general case of interior sources. The special case $Q = 0$ pertains to transillumination tomography with boundary sources S. The RTE (16.1) as part of an even-parity approach with interior sources Q can be written for all ordinates $-\Omega$ as follows:

$$-\Omega \cdot \nabla \Psi(r,-\Omega) + (\mu_a + \mu_s)\Psi(r,-\Omega)$$
$$= Q + \mu_s \int_{4\pi} p(-\Omega,\Omega')\Psi(r,\Omega')d\Omega'. \quad (16.34)$$

Adding and subtracting Equations 16.34 and 16.1 with interior sources yields a system of two coupled equations:

$$\Omega \cdot \nabla \Psi^-(r,\Omega) + (\mu_a + \mu_s)\Psi^+(r,\Omega)$$
$$= Q + \mu_s \int_{4\pi} p^+(\Omega,\Omega')\Psi^+(r,\Omega')d\Omega' \quad (16.35)$$

$$\Omega \cdot \nabla \Psi^+(r,\Omega) + (\mu_a + \mu_s)\Psi^-(r,\Omega)$$
$$= Q + \mu_s \int_{4\pi} p^-(\Omega,\Omega')\Psi^-(r,\Omega')d\Omega',$$

where the superscripts "−" and "+" denote even and odd components of a function (such as Ψ or p) with respect to the ordinates Ω. For example, in the case of Ψ, we have

$$\Psi^\pm(r,\Omega) = \frac{\Psi(r,\Omega) \pm \Psi(r,-\Omega)}{2}. \quad (16.36)$$

One can rewrite Equations 16.35 in the following operator form:

$$\Omega \cdot \nabla \Psi^\mp(r,\Omega) + H_\pm \Psi^\pm(r,\Omega) = Q \quad (16.37)$$

where the operators H_- and H_+ are defined as

$$H_\pm f(r,\Omega) = (\mu + \mu_s)f(r,\Omega) - \mu_s \int_{4\pi} p^\pm(\Omega \cdot \Omega')f(r,\Omega')d\Omega'. \quad (16.38)$$

(here, $f(r,\Omega)$ is an arbitrary function that depends on r and Ω, e.g., Ψ^- or Ψ^+).

After exclusion of Ψ^- from Equation 16.37, one arrives at

$$-\Omega \cdot \nabla \left\{ H_-^{-1} \left[\Omega \cdot \nabla \Psi^+(r,\Omega) \right] \right\} + H_\pm \Psi^\pm(r,\Omega)$$
$$= Q^+(r,\Omega) - \Omega \cdot \nabla \left\{ H_-^{-1} Q^-(r,\Omega) \right\}. \quad (16.39)$$

Equation 16.39 is solved by inverting the operator H_- explicitly. In case of isotropic scattering, H_- can easily be inverted since we have the simplified relation $H_-^{-1} f = (\mu_a + \mu_s)^{-1} f$. If the source Q is also isotropic, then we arrive at

$$-\Omega \cdot \nabla \left([\mu_a + \mu_s]^{-1} \Omega \cdot \nabla \Psi^+ \right) + (\mu_a + \mu_s)\Psi^+(r,\Omega)$$
$$= Q(r) + \frac{\mu_s}{4\pi} \int_{4\pi} \Psi^+(r,\Omega')d\Omega'. \quad (16.40)$$

The integral in Equation 16.40 can be approximated by a discrete ordinates quadrature formula. The resulting set of algebraic equations is then solved with a Galerkin FE method.

16.4 Examples in Biomedical Research

In the following, we will present some examples of how the numerical methods introduced in Section 16.3 can be used in biomedical research. The results include comparison of transport and diffusion simulations, the influence of the scattering phase function of light propagation, and an analysis of the signal-to-noise ratio (SNR) in experimental data and its effects on optimal source-modulation frequencies to be used in small animal imaging. Solutions of the inverse problems are given in the next chapter by Simon Arridge.

16.4.1 Diffusion versus Transport Solution

Numerical simulations can be used to determine what model of light transport is most appropriate. For example, many studies have been performed to determine in what cases the diffusion approximation to the RTE is accurate and in what case the DA is not sufficient. Figure 16.4 shows an example for a medium in which the DA is insufficient. The figure shows results of numerical calculation using an FD-S_n transport code and diffusion code for the case of a 4×4 cm^2 medium that contains a 2.8 cm inner-diameter void-like ring (thickness = 0.1 cm), which is low in absorption and scattering $\mu_a = 0.01$ cm, $\mu'_s = 0.1$, $g = 0.8$). The optical properties of the background medium were chosen to be $\mu_a = 0.35$, $\mu'_s = 11.6$, $g = 0.8$. The simulations were performed on a spatial 240×240 spatial grid with $dx = dy = 0.0167$ cm, and 32 ordinates. It can be seen that the diffusion solution differs substantially from the transport solution. In the diffusion case, it appears that light is channeled along the void-like region, while the transport solution shows a much weaker influence of the void ring.

This behavior was confirmed in experimental studies in which we prepared tissue phantoms that were composed of clear epoxy resin into which silicon-dioxide (SiO$_2$) monospheres and ink were mixed to generate optical properties as used in the simulations. The phantoms were continuously illuminated with NIR light from a laser diode at (wavelength $\lambda = 678$ nm). As the

phantoms were three-dimensional but the calculations were two-dimensional, we had to provide a z-axis-independent fluence. This was achieved by illuminating the phantom with an extended line source along the z-axis (Figure 16.5 left). The line source was realized by a laser diode with a light-emission angle of 20° along the z-axis, while along the x-axis, the beam was collimated. In this way, the source consisted of a 6.0 cm long line with a width of 0.1 cm on the surface of the phantom. The source power applied to an area of 0.01 cm^2 was 0.2 mW. Measurements were taken with the line source positioned at different locations along the x-axis. To measure the fluence, we used an avalanche photodiode. The detector was placed in the x–y plane of the phantom at the midpoint of the z-axis ($z = 7.0$ cm), and could be translated around the phantom along the x-axis and y-axis (Figure 16.5 left). The detection area at the boundary of the phantom was limited by a pinhole, which had a diameter of 0.1 cm. Two lenses projected the detection area onto the APD chip, whose diameter was 0.3 cm. An iris was placed in between the lenses, in order to adjust the aperture angle to 45°.

Figure 16.5 right shows the results for a light source at position A and measurements performed along the y-axis. Clearly

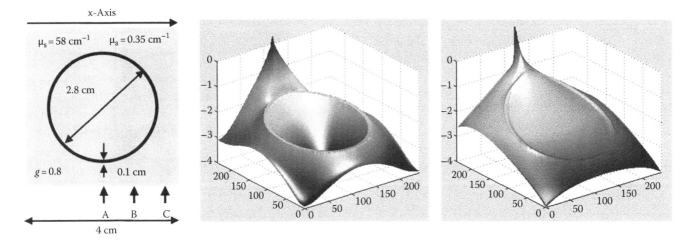

FIGURE 16.4 Comparison of the fluence distributions as calculated by using an RTE-based code (right) and a DA-based code (middle) for a highly scattering medium ($\mu_s = 11.6$ cm^{-1}, $g = 0.8$, $\mu_a = 0.35$) that contains a void-like ring ($\mu'_s = 0.1$, $g = 0.8$, $\mu_a = 0.01$ cm) (left). The fluences are plotted on a logarithmic scale in normalized arbitrary units.

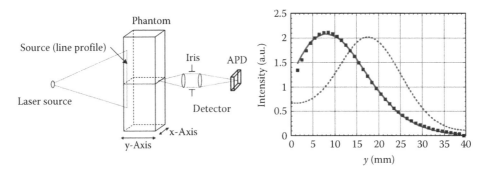

FIGURE 16.5 Comparison of experimental data (black squares) and numerical simulations using codes that employ the transport theory (solid line) and diffusion theory (dotted line), respectively. (From Klose, A.D. et al., *J. Quant. Spectrosc. Radiat. Transf.*, 72(5), 691, 2002. With permission.)

visible is the good agreement between measurements and transport simulations. The diffusion-theory-based simulation differs significantly from measurements and transport simulations. More details and additional results can be found in Klose et al. (2002) and Hielscher et al. (1998).

16.4.2 Influence of Collimated Light Sources (Delta-Eddington Approach)

Light sources used in medical application are typically collimated (e.g., a laser beam that hits the surface of a tissue). The importance of modeling this light source accurately is illustrated in the following example. To deal with the collimated light source and a medium that is strongly forward scattering ($g = 0.95$), we employ here the DE approach that was discussed in Section 16.3.4. Figure 16.6 shows the ratio of two calculations, one in which an isotropic source was employed and a second one in which a collimate beam was used. The medium was simulated on a 41×41 grid, with $dx = dy = 0.025$ cm. As optical properties, we chose $\mu_a = 0.2$ cm^{-1}, $\mu_s' = 10$ cm^{-1} in this case. The calculations were carried out with a S6 method (48 discrete ordinates). Figure 16.6 shows the ratio of the fluence calculations with collimated beam divided by the results obtained with an isotropic source. As can be seen, a significantly larger fluence is observed for the collimated source. This underscores the necessity to use an accurate model for the source. More details and other examples can be found in Klose and Hielscher (2003b).

16.4.3 SNR Analysis Using the Frequency-Domain RTE

16.4.3.1 Introduction

Time-dependent data are typically considered superior to time-independent data, since it carries more information. For example, in the frequency domain, in addition to total amplitude that SS measurements provide, phase information is provided. However, at small source modulation frequencies, the phase can be relatively small, and may be difficult to measure. An example is shown in Figure 16.7. Here, we used an FV-S$_n$ frequency domain RTE code (Ren et al. 2004) to calculate the phase shift at $\omega = 100$ and $\omega = 600$ MHz in a numerical model of a mouse. One can see that at $\omega = 100$ MHz, only phase shifts up to ~4° occur. With reported phase measurement errors between 0.1° and 1° (Chance et al. 1998, Ramanujam et al. 1998, Nissilä et al. 2002, Godavarty et al. 2003), a 100 MHz system is of limited use for small animal imaging. At $\omega = 600$ MHz, phase shifts up to ~25° occur. However, while the phase shift increases with increasing frequency, the noise in the phase measurements increases too. As we show next, RTE codes can be used to show that this leads to some optimal frequency at which the SNR is largest.

16.4.3.2 SNR Analysis

To perform the SNR analysis, we consider a shot-noise model introduced by Toronov et al. (2003) and modified by Gu et al. (2007), which is given by

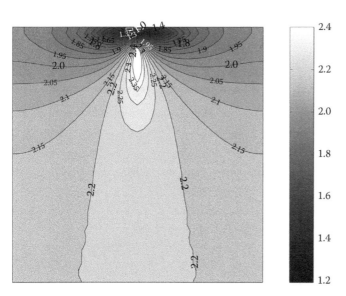

FIGURE 16.6 Ratio of fluence calculated for a collimated and a diffuse source that is located in the center of the medium boundary. (From Klose, A.D. and Hielscher, A.H., Modeling photon propagation in anisotropically scattering media with the equation of radiative transfer, in *Optical Tomography and Spectroscopy of Tissue V*, Chance, B. et al. (eds), Proceedings of the SPIE—The International Society for Optical Engineering, Vol. 4955, pp. 624–633, 2003b. With permission.)

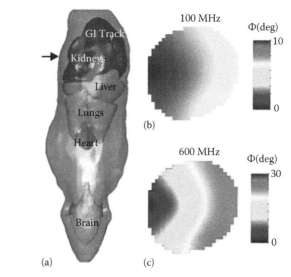

FIGURE 16.7 (See color insert.) Phase shift at 100 MHz (b) and 600 MHz (c) in an axial cross section through a mouse model (a). The position of the cross section is indicated by the black arrow in (a). The calculations were done with a FV transport-theory code (Ren et al. 2004). The optical properties (μ_a, μ_s') in units of cm^{-1} were chosen to be as follows: brain (0.3, 15), GI track (0.27, 12.5), heart (0.1, 5), kidneys (0.67, 6), liver (0.33, 7.5), and lung (0.81, 20). The anisotropy factor was assumed constant $g = 0.9$. (Adapted from Ren, K. et al., *Opt. Lett.*, 29(6), 578, 2004.)

$$\mathrm{SNR_{AC}} = \frac{\langle \mathrm{AC} \rangle}{\sigma_{\mathrm{AC}}} \propto \frac{\langle \mathrm{AC} \rangle}{\sqrt{\mathrm{DC}}}; \quad \mathrm{SNR_\Phi} = \frac{\langle \Phi \rangle}{\sigma_\Phi} \propto \langle \Phi \rangle \mathrm{SNR_{AC}}, \quad (16.41)$$

Here, σ_{AC} and σ_Φ are the standard deviations of the amplitude and the phase shift, respectively, and $\langle \cdot \rangle$ represent the ensemble average of the corresponding quantity. σ_{AC} and σ_Φ are proportional to $\sqrt{\mathrm{DC}}$ and $\sqrt{\mathrm{DC}}/\mathrm{AC}$, respectively, that is, the phase noise increases with the frequency.

To demonstrate the effects of increasing the source-modulation frequency, we consider a $3 \times 3\,\mathrm{cm}^2$ phantom with a homogenous distribution of optical properties in the medium. The synthetic data are generated by solving the FD transport equation on an unstructured grid with the FV-Sn method described Section 16.3.5. We corrupt the data by adding Gaussian random noise according to the noise model given in Equation 16.41. We have obtained the SNR values at specified locations (see Figure 16.8c) from running forward simulations in the range of $0.05\,\mathrm{cm}^{-1} \leq \mu_a \leq 0.5\,\mathrm{cm}^{-1}$ for the absorption coefficient and in the range of $5.0\,\mathrm{cm}^{-1} \leq \mu_s' \leq 20\,\mathrm{cm}^{-1}$ for the reduced scattering coefficient.

The SNR values for the amplitude and the phase shift are shown in Figures 16.8 and 16.9. Note that all the values are normalized to the SNR values at 100 MHz since the proportionality constants in Equation 16.41 are not explicitly known. Comparing Figure 16.8a and b, we observed that keeping μ_a constant and increasing μ_s', the frequency dependence of the SNR becomes more pronounced. On the other hand, when we keep μ_s' constant and increase μ_a from 0.05 to 0.5 cm^{-1}, the frequency dependence of the amplitude SNR becomes less pronounced (see Figure 16.8a and c). These results indicate that the change in μ_a

has a stronger impact on the amplitude signal than the change in μ_s. That is why the curves on Figure 16.8c and d look somewhat frequency-independent.

For the frequency dependence of the phase SNR value, it is obvious from Equation 16.41 that the phase SNR equals the amplitude SNR multiplied by a phase shift. As the source modulation frequency increases, the amplitude SNR decreases, while the phase shift increases. Therefore, there will be a certain frequency for which the phase SNR takes on a maximum value. Indeed, Figure 16.9a and b shows that there exist a modulation frequency between 400 and 800 MHz for which the phase SNR is largest at all positions except for locations 1 and 2. We also found that the position of this maximum moves to higher frequencies when μ_a is increased (Figure 16.9a vs. 16.9c) and μ_s' is decreased (Figure 16.9b vs. 16.9a). Another finding is that the SNR of the phase shift improves with frequency regardless of μ_s' when μ_a is increased (Figure 16.9c and 16.9d). More details on the SNR results can be found in literature (Kim et al. 2008).

16.4.4 Light Transmission through Finger Joints Using the Frequency-Domain RTE

Light transmission through finger joints can be used to obtain information on the physiological state of the joint, such as evaluating the inflammatory stage in rheumatoid arthritis (RA). To this end, we consider the finger phantom of well-characterized optical properties as introduced by Netz et al. (2008). As shown in Figure 16.10a, the capsule is highly absorbing, weakly scattering relative to the skin-tissue complex and a bone has inhomogeneity in absorption only and its scattering coefficient is

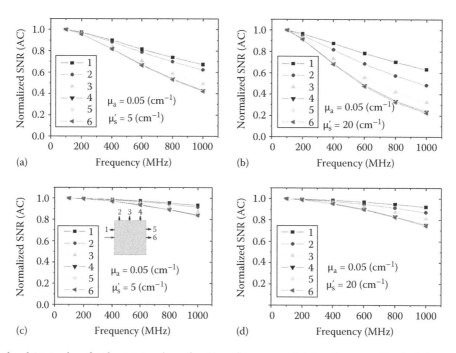

FIGURE 16.8 Normalized SNR values for the AC signal as a function of source-modulation frequency for six different source–detector pairs. (From Kim, H.K. et al., *Opt. Exp.*, 16(22), 18082, 2008. With permission.)

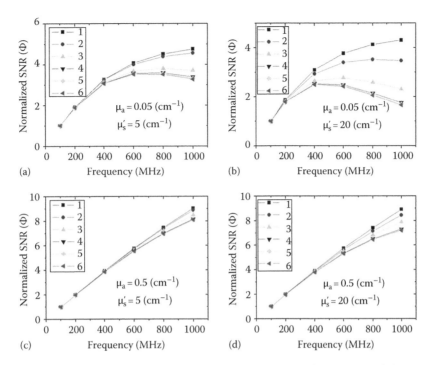

FIGURE 16.9 Normalized SNR of phase shifts at various frequencies. (From Kim, H.K. et al., *Opt. Exp.*, 16(22), 18082, 2008. With permission.)

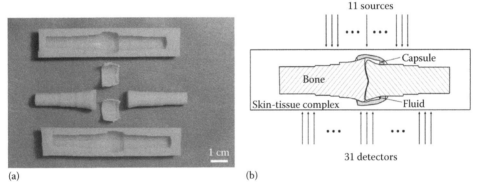

FIGURE 16.10 Structure of a finger phantom: (a) layout of major parts and (b) schematic of source–detector configuration. (From Kim, H. K. et al., *Opt. Exp.*, 16(22), 18082, 2008. With permission.)

identical to that of the skin-tissue medium. The synovial fluid in the joint gap can be optically almost transparent for the healthy finger; hence its absorption and scattering coefficients are as small as ~0.04 cm⁻¹. An onset of a disease accompanies the change of optical properties in the affected area, that is, it typically increases both μ_a and μ'_s in the capsule and the fluid. Therefore, we examined the two finger phantoms: one for the RA case and one for the no-RA case.

For simulation of light transmission through the finger phantom, we employed the FD RTE as a light propagation model in finger joints since the DA is of limited accuracy in this small-tissue fluid-filled case. As shown in Figure 16.10b, we illuminated the phantom with a laser source scanning the 11 locations, and evaluated the transmitted photon density wave data at the 31 positions. Figure 16.11 shows the predictions

of amplitude and phase delay obtained with 600 MHz, for the no-RA and the RA cases, respectively. As shown in Figure 16.11a, for the no-RA phantom, the amplitude gradually increases toward the joint area and shows a rapid increase at the joint center by the existence of the clear fluid (Figure 16.10). On the other hand, the RA phantom shows only a small increase at joint center, which is due to the increase of both μ_a and μ'_s not only in the joint fluid, but also in the capsule. As for the phase delay, the RA phantom yields larger phase delay by the increase of scattering in joint fluid, whereas the no-RA phantom shows smaller phase shift (Figure 16.11b).

Moreover, we measured the amplitudes and phase shifts at 600 MHz for the same phantom. As before, we extracted the 31 detector readings from the CCD image and compared the difference in amplitude and phase shift obtained for the RA and

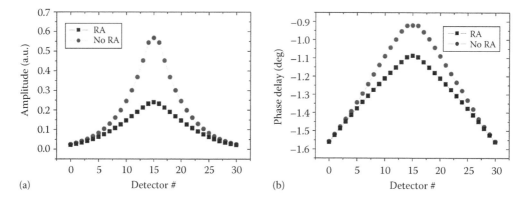

FIGURE 16.11 Calculated profiles of amplitude and phase shift obtained along the line of measurement at 600 MHz. The laser source is located at the joint center.

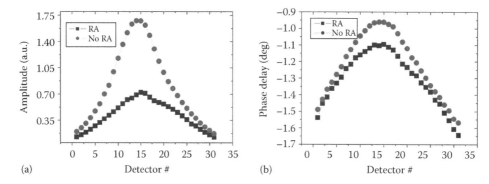

FIGURE 16.12 Experimentally measured profiles of amplitude and phase shift obtained along the line of measurement at 600 MHz. The laser source is located at the joint center. (From Kim, H.K. et al., *Opt. Exp.*, 16(22), 18082, 2008. With permission.)

no-RA phantoms. As shown in Figure 16.12, measurements are very similar to predictions both with respect to the amplitude and phase shift patterns, for the no-RA and RA phantoms.

16.4.5 Modeling Fluorescence with FD Blocking-Off Method

In the final example, we show how fluorescence effects in small animals can be modeled using an FD discrete ordinates method for curved geometries using a structured grid in conjunction with a blocking-off region method described in Section 16.3.3. A numerical mouse model was generated from a complete magnetic resonance imaging (MRI) data set consisting of 91 slices. The MRI slices were segmented and merged to obtain a 3D numerical mouse model. Overall, this model consisted of 81,000 voxels or grid points. Each voxel inside the mouse volume was assigned the optical tissue parameters $\mu_s' = 5\,\mathrm{cm}^{-1}$ and $\mu_a = 0.4\,\mathrm{cm}^{-1}$. We added a fluorescent source of size $0.2 \times 0.2 \times 0.2\,\mathrm{cm}^3$ to the numerical model. The mouse model was illuminated with a point source and the fluorescence light was detected at 119 detector points surrounding the numerical model opposite to the source point. Figure 16.13 (left) shows the fluence distribution on the tissue surface for the excitation source. Figure 16.13 (right) shows the light distribution on the

tissue surface, which originates from the fluorescent source inside the mouse.

16.5 Summary

We have presented various approaches to numerically model light propagation in highly scattering biological tissue. We specifically focused on algorithms that employ the equation of radiative transfer to model light propagation in this medium. By not relying on the diffusion approximation, these algorithms have the potential to provide more accurate solutions in cases where the diffusion approximation fails, such as in media with small geometries or media containing void-like areas. We presented implementations of such algorithms with FD, FV, and FE schemes, and discussed their numerical implementation on structured as well as unstructured grids. To illustrate the performance of these codes, we showed examples that focus on the differences between transport-theory- and diffusion-theory-based algorithms, the importance of accurately modeling the light source, and a signal-to-noise analysis for FD data. Furthermore, simulations were compared to experimental data in studies involving finger joints and propagation of fluorescence light in an animal model was discussed.

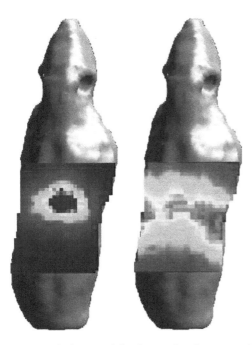

FIGURE 16.13 Calculations of the fluence distribution on the tissue surface of a numerical mouse model. The left images shows the excitation field originating from a point source on the surface of the animal. The right image shows the result fluorescence field. (From Klose, A.D. and Hielscher, A. H., Fluorescence molecular tomography of small animals using the radiative transfer equation for curved geometries, in *Proceedings of the OSA Biomedical Topical Meeting*, Fort Lauderdale, FL, 2006b. With permission.)

Acknowledgments

This work was supported in part by the National Institute of Arthritis and Musculoskeletal and Skin Diseases (NIAMS-grant 2R01 AR46255), the National Institute of Biomedical Imaging and Bioengineering (NIBIB-grants 5R01-EB001900 and 1R21EB011772-01A1), the National Cancer Institute (NCI-grants 5R33-CA91807, 4R33CA118666, U54CA126513), and the National Center for Research Resources (NCRR-grants UL1RR024156 and 1R44RR030701-01).

References

Abdoulaev, G. S. and Hielscher, A. H. 2003. Three-dimensional optical tomography with the equation of radiative transfer. *J. Electron. Imaging* 12(4): 594–601.

Abdoulaev, G. S., Ren, K., Hielseher, A. H. 2005. Optical tomography as a PDE-constrained optimization problem. *Inverse Prob.* 21(5): 1507–1530.

Aronson, R., Barbour, R. L., Lubowsky, J., and Graber, H. 1991. Application of transport theory to infra-red medical imaging. In: *Modern Mathematical Methods in Transport Theory*. Basel: Birkhauser, pp. 64–67.

Arridge, S. R. and Dorn, O. 2006. Reconstruction of subdomain boundaries of piecewise constant coefficients of the radiative transfer equation from optical tomography data. *Inverse Prob.* 22(6): 2175–2196.

Aydin, E. D., de Oliveira, C. R. E., and Goddard, A. J. H. 2004. A finite element-spherical harmonics radiation transport model for photon migration in turbid media. *J. Quant. Spectrosc. Radiat.* 84(3): 247–260.

Aydin, E. D., Katsimichas, S., and de Oliveira, C. R. E. 2005. Time-dependent diffusion and transport calculations using a finite-element-spherical harmonics method. *J. Quant. Spectrosc. Radiat.* 95(3): 349–363.

Boulanger, J. and Charette, A. 2005a. Numerical developments for short-pulsed near infra-red laser spectroscopy. Part I: Direct treatment. *J. Quant. Spectrosc. Radiat.* 91(2): 189–209.

Boulanger, J. and Charette, A. 2005b. Numerical developments for short-pulsed near infra-red laser spectroscopy. Part II: Inverse treatment. *J. Quant. Spectrosc. Radiat.* 91(3): 297–318.

Cai, W., Xu, M., and Alfano, R. R. 2003. Three-dimensional radiative transfer tomography for turbid media. *IEEE J. Sel. Top. Quant.* 9(2): 189–198.

Carlson, B. G. and Lathrop, K. D. 1968. Transport theory—The method of discrete ordinates. In: *Computing Methods in Reactor Physics*, ed. H. Greenspan. New York: Gordon & Breach, pp. 166–266.

Case, K. M. and Zweifel, P. F. 1967. *Linear Transport Theory*. Reading, MA: Addison-Wesley.

Chai, J., Lee, H. S., and Patankar, S. V. 1994. Treatment of irregular geometries using a Cartesian coordinates finite-volume radiation heat transfer procedure. *Num. Heat Transf. Part B* 26: 225–235.

Chance, B., Cope, M., Gratton, E., Ramanujam, N., and Tromberg, B. 1998. Phase measurement of light and absorption and scatter in human tissue. *Rev. Sci. Instrum.* 69: 3457–3481.

Chandrasekhar, S. 1960. *Radiative Transfer*. New York: Dover Publications Inc.

Cheong, W. F. 1995. Summary of optical properties. In: *Optical-Thermal Response of Laser-Irradiated Tissue*, ed. A. J. Welch, Chap. 8. New York: Plenum Press.

Cheong, W. F., Prahl, S. A., and Welch, A. J. 1990. A review of the optical properties of biological tissue. *IEEE J. Quantum Electron.* 26: 2166–2185.

Coelho, P. J. 2002. Bounded skew high-order resolution schemes for the discrete ordinates method. *J. Comput. Phys.* 175: 412–437.

Collier, T., Follen, M., Malpica, A., and Richards-Kortum, R. 2005. Sources of scattering in cervical tissue: determination of the scattering coefficient by confocal microscopy. *Appl. Opt.* 44(11): 2072–2081.

Culver, J. P., Durduran, T., Furuya, D., Cheung, C., Greenberg, J. H., and Yodh, A. G. 2003. Diffuse optical tomography of cerebral blood flow, oxygenation, and metabolism in rat during focal ischemia. *J. Cereb. Blood Flow Metab.* 23: 911–924.

Dorn, O. 1998. A transport-backtransport method for optical tomography. *Inverse Prob.* 14: 1107–1130.

Drezek, R. A., Richards-Kortum, R., Brewer, M. A., Feld, M. S., Pitris, C., Ferenczy, A., Faupel, M. L., and Follen, M. 2003. Optical imaging of the cervix. *Cancer* 98(9 Suppl): 2015–2027.

Duderstadt, J. J. and Martin, W. R. 1979. *Transport Theory*. New York: John Wiley & Sons.

Farrashkhalvat, M. and Miles, J. P. 2003. *Basic Structured Grid Generation*. Oxford, U.K.: Butterworth & Heinemann.

Gibson, A. and Dehghani, H. 2009. Diffuse optical imaging. *Phil. Trans. R. Soc. A.* 367: 3055–3072.

Gibson, A. P., Hebden, J. C., and Arridge, S. R. 2005. Recent advances in diffuse optical imaging. *Phys. Med. Biol.* 50(4): R01–R43.

Godavarty, A., Eppstein, M. J., Zhang, C. Y., Theru, S., Thompson, A. B., Gurfinkel, M., and Sevick-Muraca, E. M. 2003. Fluorescence-enhanced optical imaging in large tissue volumes using a gain-modulated ICCD camera. *Phys. Med. Biol.* 48(12): 1701–1720.

Gu, X., Ren, K., and Hielscher, A. H. 2007. Frequency-domain sensitivity analysis for small imaging domains using the equation of radiative transfer. *Appl. Opt.* 46: 1624–1632.

Gulsen, G., Xiong, B., Birgul, O., and Nalcioglu, O. 2006. Design and implementation of a multifrequency near-infrared diffuse optical tomography system. *J. Biomed. Opt.* 11(1): 014020.

Hielscher, A. H., Alcouffe, R. E., and Barbour, R. L. 1998. Comparison of finite-difference transport and diffusion calculations for photon migration in homogeneous and heterogeneous tissues. *Phys. Med. Biol.* 43: 1285–1302.

Hielscher, A. H., Klose, A. D., and Hanson, K. M. 1999. Gradient-based iterative image reconstruction scheme for time-resolved optical tomography. *IEEE Trans. Med. Imag.* 18(3): 262–271.

Hielscher, A. H., Klose, A. D., Scheel, A. K., Moa-Anderson, B., Backhaus, M., Netz, U., and Beuthan, J. 2004. Sagittal laser optical tomography for imaging of rheumatoid finger joints. *Phys. Med. Biol.* 49(7): 1147–1163.

Ishimaru, A. 1996. *Wave Propagation and Scattering in Random Media*. Piscataway, NJ: IEEE Press.

Jessee, J. P., Fiveland, W. A., Howell, L. H., Colella, P., and Pember, R. B. 1998. An adaptive mesh refinement algorithm for the radiative transport equation. *J. Comp. Phys.* 139: 380–398.

Joseph, D. K, Huppert, T. J., Franceschini, M. A., Boas, D. A. 2006. Diffuse optical tomography system to image brain activation with improved spatial resolution and validation with functional magnetic resonance imaging. *Appl. Opt.* 45(31): 8142–8151.

Joseph, J. H., Wiscombe, W. J., and Weinman, J. A. 1976. The Delta-Eddington approximation for radiative transfer. *J. Atmos. Sci.* 33: 2452–2459.

Khan, T. and Thomas, A. 2005. Comparison of P-N or spherical harmonics approximation for scattering media with spatially varying and spatially constant refractive indices. *Opt. Commun.* 255(1–3): 130–166.

Kim, H. K. and Charette, A. 2007. A sensitivity function-based conjugate gradient method for optical tomography based on the frequency-domain equation of radiative transfer. *J. Quant. Spectrosc. Radiat. Transf.* 104: 24–39.

Kim, H. K. and Hielscher, A. H. 2009. A PDE-constrained SQP algorithm for optical tomography based on the frequency-domain equation of radiative transfer. *Inverse Prob.* 25:015010 (20pp).

Kim, A. D. and Moscoso, M. 2006. Radiative transport theory for optical molecular imaging. *Inverse Prob.* 22(1): 23–42.

Kim, H. K., Netz, U. J., Beuthan, J., and Hielscher, A. H. 2008. Optimal source-modulation frequencies for transport-theory-based optical tomography of small-tissue volumes. *Opt. Exp.* 16(22):18082–18101.

Klose, A. D. 2007. Transport-theory-based stochastic image reconstruction of bioluminescent sources. *J. Opt. Soc. Am. A.* 24(6): 1601–1608.

Klose, A. D. 2009a. Radiative transfer of luminescence light in biological tissue. In: *Light Scattering Reviews*, Vol. 4, ed. A. A. Kokhanovsky. Berlin, Germany: Springer, pp. 293–345.

Klose, A. D. 2009b. Hyperspectral excitation-resolved fluorescence tomography of quantum dots. *Opt. Lett.* 34(16): 2477–2479.

Klose, A. D. and Hielscher, A. H. 1999. Iterative reconstruction scheme for optical tomography based on the equation of radiative transfer. *Med. Phys.* 26(8): 1698–1707.

Klose, A. D. and Hielscher, A. H. 2003a. Fluorescence tomography with simulated data based on the equation of radiative transfer. *Opt. Lett.* 28(12): 1019–1021.

Klose, A. D. and Hielscher, A. H. 2003b. Modeling photon propagation in anisotropically scattering media with the equation of radiative transfer. In: *Optical Tomography and Spectroscopy of Tissue V*, ed. B. Chance, R. R. Alfano, B. J. Tromberg, M. Tamura, and E. M. Sevick-Muraca. Proceedings of the SPIE—The International Society for Optical Engineering, Vol. 4955, pp. 624–633.

Klose, A. D. and Hielscher, A. H. 2006. Fluorescence molecular tomography of small animals using the radiative transfer equation for curved geometries. In: *Proceedings of the OSA Biomedical Topical Meeting*, Fort Lauderdale, FL.

Klose, A. D. and Larsen, E. W. 2006. Light transport in biological tissue based on the simplified spherical harmonics equations. *J. Comput. Phys.* 220: 441–470.

Klose, A. D., Netz, U., Beuthan, J., and Hielscher A. H. 2002. Optical tomography using the time-independent equation of radiative transfer. Part I: Forward model. *J. Quant. Spectrosc. Radiat. Transf.* 72(5): 691–713.

Klose, A. D., Ntzachristos, V., and Hielscher, A. H. 2005. The inverse source problem based on the radiative transfer equation in molecular optical imaging. *J. Comput. Phys.* 202: 323–345.

Lang, J. and Walter, A. 1995. An adaptive discontinuous finite element method for the transport equation. *J. Comp. Phys.* 117: 28–34.

Lasker, J. M., Masciotti, J., Schoenecker, M., Schmitz, C., and Hielscher, A. H. 2007. Digital-signal-processor-based dynamic optical tomography imaging system. *Rev. Sci. Instrum.* 78(8): 083706.

Li, C., Zhao, H., Anderson, B., and Jiang, H. 2006. Multispectral breast imaging using a ten-wavelength, 64×64 source/detector channels silicon photodiode-based diffuse optical tomography system. *Med. Phys.* 33(3): 627–636.

Lu, Y., Machado, H. B., Douraghy, A., Stout, D., Herschman, H., and Chatziioannou, A. F. 2009. Experimental bioluminescence tomography with fully parallel radiative-transfer-based reconstruction framework. *Opt. Exp.* 17(19): 16681–16695.

McBride, T. O., Pogue, B. W., Jiang, S., Osterberg, U. L., and Paulsen, K. D. 2001. A parallel-detection frequency-domain near-infrared tomography system for hemoglobin imaging of the breast in vivo. *Rev. Sci. Instrum.* 72(3): 1817–1824.

Meese, E. A. 1998. Finite volume methods for the incompressible Navier–Stokes equations on Unstructured grids. Ph.D. thesis, Norwegian University of Science and Technology, Trondheim, Norway.

Morel, J. E., Wareing, T. A., and Smith, K. 1996. A linear-discontinuous spatial differencing scheme for Sn radiative transfer calculations. *J. Comput. Phys.* 128: 445–462.

Mourant, J. R., Freyer, J. P., Hielscher, A. H., Eick, A. A., Shen, D., and Johnson, T. M. 1998. Mechanisms of light scattering from biological cells relevant to noninvasive optical-tissue diagnostics. *Appl. Opt.* 37(16): 3586–3593.

Mourant, J. R., Canpolat, M., Brocker, C. et al. 2000. Light scattering from cells: the contribution of the nucleus and the effects of proliferative status. *J. Biomed. Opt.* 5(2): 131–137.

Netz, U. J., Beuthan, J., and Hielscher, A. H. 2008. Multipixel system for gigahertz frequency-domain optical imaging of finger joints. *Rev. Sci. Instrum.* 79: 034301.

Nissilä, I., Kotilahti, K., Fallström, K., and Katila, T. 2002. Instrumentation for the accurate measurement of phase and amplitude in optical tomography. *Rev. Sci. Instrum.* 73(9): 3306–3312.

Nissilä, I., Noponen, T., Heino, J., Kajava, T., and Katila, T. 2005. Diffuse optical imaging. In: *Advances in Electromagnetic Fields in Living Systems*, ed. J. C. Lin. New York: Springer, pp. 77–129.

Oliveria, S. 1995. Multigrid and Krylov subspace methods for transport equations: absorption case. In: *Proceedings of the 7th Copper Mountain Conference on Multigrid Methods*, Copper Mountain, CO.

Patton, B. W. 1996. Application of Krylov subspace iterative techniques to the numerical solution of the neutron transport equation. PhD thesis, University of Michigan, Dearborn, MI.

Patton, B. W. and Holloway, J. P. 2002. Application of preconditioned GMRES to the numerical solution of the neutron transport equation. *Ann. Nucl. Energy* 29: 109–136.

Patwardhan, S., Bloch, S., Achilefu, S., and Culver, J. 2005. Time-dependent whole-body fluorescence tomography of probe bio-distributions in mice. *Opt. Exp.* 13(7): 2564–2577.

Ramanujam, N., Du, C., Ma, H. Y., and Chance, B. 1998. Sources of phase noise in homodyne and heterodyne phase modulation devices used for tissue oximetry studies. *Rev. Sci. Instrum.* 69(8): 3042–3054.

Rasmussen, J. C., Joshi, A., Pan, T., Wareing, T., McGhee, J., and Sevick-Muraca, E. M. 2006. Radiative transport in fluorescence-enhanced frequency domain photon migration. *Med. Phys.* 33(12): 4685–4700.

Reed, W. 1971. New difference schemes for the neutron transport equation. *Nucl. Sci. Eng.* 46: 31–39.

Ren, K., Abdoulaev, G. S, Bal, G., and Hielscher, A. H. 2004. Algorithm for solving the equation of radiative transfer in the frequency domain. *Opt. Lett.* 29(6): 578–580.

Ren, K., Bal, G., and Hielscher, A. H. 2006. Frequency domain optical tomography based on the equation of radiative transfer. *SIAM J. Sci. Comput.* 28(4): 1463–1489.

Saad, Y. 1986. GMRES: A generalized minimum residual algorithm for solving nonsymmetric linear systems. *SIAM J. Sci. Stat. Comput.* 7: 856–869.

Scheel, A. K., Backhaus, M., Klose, A. D. et al. 2005. Comparison of sagittal laser optical tomography with ultrasound and clinical examination for diagnosis of synovitis in PIP joints. *Ann. Rheum. Dis.* 64: 239–245.

Schmidt, F. E. W., Fry, M. E., Hillman, E. M. C., Hebden, J. C., and Delpy, D. T. 2000. A 32-channel time-resolved instrument for medical optical tomography. *Rev. Sci. Instrum.* 71(1): 256–265.

Schmitz, C. H., Löcker, M., Lasker, J. M., Hielscher, A. H., and Barbour, R. L. 2002. Instrumentation for fast functional optical tomography. *Rev. Sci. Instrum.* 73(2): 429–439.

Srinivasan, S., Pogue, B. W., Jiang, S. et al. 2003. Interpreting hemoglobin and water concentration, oxygen saturation, and scattering measured by near-infrared tomography of normal breast in vivo. *Proc. Natl. Acad. Sci. U.S.A.* 100(21): 12349–12354.

Tarvainen, T., Vauhkonen, M., and Kolehmainen, V. 2005a. Hybrid radiative-transfer-diffusion model for optical tomography. *Appl. Opt.* 44(6): 876–886.

Tarvainen, T., Vauhkonen, M., and Kolehmainen, V. 2005b. Coupled radiative transfer equation and diffusion approximation model for photon migration in turbid medium with low-scattering and non-scattering regions. *Phys. Med. Biol.* 50(20): 4913–4930.

Tarvainen, T., Vauhkonen, M., and Kolehmainen, V. 2006. Finite element model for the coupled radiative transfer equation and diffusion approximation. *Int. J. Numer. Meth. Eng.* 65(3): 383–405.

Thompson, A. B. and Sevick-Muraca, E. M. 2003. Near-infrared fluorescence contrast-enhanced imaging with intensified charge-coupled device homodyne detection: measurement precision and accuracy. *J. Biomed. Opt.* 8: 111–120.

Thompson, J. F., Soni, B. K., and Weatherill, N. P. 1999. *Handbook of Grid Generation*. Boca Raton: CRC Press.

Thompson, J. F., Warsi, Z. U. A., and Wayne Mastin, C. 1985. *Numerical Grid Generation*. New York: North-Holland.

Toronov, V., D'Amico, E., Hueber, D., Gratton, E., Barbieri, B., and Webb, A. 2003. Optimization of the signal-to-noise ratio of frequency-domain instrument for near-infrared spectro-imaging of the human brain. *Opt. Exp.* 11: 2117–729.

Tromberg, B. J., Pogue, B. W., Paulsen, K. D., Yodh, A. G., Boas, D. A., and Cerussi, A. E. 2008. Assessing the future of diffuse optical imaging technologies for breast cancer management. *Med. Phys.* 35(6): 2443–2451

Utzinger, U., Brewer, M., Silva, E., Gershenson, D., Blast, Jr., R. C., Follen, M., and Richards-Kortum, R. 2001. Reflectance spectroscopy for in vivo characterization of ovarian tissue. *Lasers Surg. Med.* 28(1): 56–66.

van den Vorst, H. A. 1992. BI-CGSTAB: A fast and smoothly converging variant of BI-CG for the solution of non-symmetric linear systems. *SIAM J. Sci. Stat. Comput.* 13: 631–644.

Venkatakrishnan, V. 1993. On the accuracy of limiters and convergence to steady state solutions. *AIAA* 93: 0880.

17

Inverse Models of Light Transport

Simon R. Arridge
University College London

Martin Schweiger
University College London

John C. Schotland
University of Pennsylvania

17.1 Introduction 337
17.2 Definition of Problems in Optical Tomography 337
  Forward Mappings and Their Derivatives • Physical Models • Optical Tomography
  Modalities
17.3 Methods 341
  Forward Models • Direct and Semi-Analytic Inverse Methods • Numerical Inversion • Use
  of Priors
17.4 Examples 351
  Comparison of Reconstruction Methods • Direct Reconstruction Method • Structured Light
  Results
17.5 Summary 355
References 355

17.1 Introduction

The inverse problem in optical tomography covers the topic of image reconstruction in particular, or, slightly more generally, the recovery from measured data of properties of a medium, which may include shapes, volumes, dynamic features, and others. In a typical experiment, a highly scattering medium is illuminated by a narrowly collimated beam and the light that propagates through the medium is collected by an array of detectors. There are many variants of this basic scenario. For instance, the source may be pulsed or time-harmonic, coherent or incoherent, and the illumination may be spatially structured or multispectral. Likewise, the detector may be time- or frequency-resolved, polarization or phase sensitive, located in the near- or far-field, and so on. The mathematical formulations of the forward problems considered in the previous chapter are dictated primarily by spatial and temporal *scale*. In this chapter, we consider only the spatial so-called macroscale on which light is considered as an intensity rather than an amplitude and is essentially incoherent, and for which timescales are in the nanosecond range. Some more general aspects are considered in the review article by Arridge and Schotland (2009).

17.2 Definition of Problems in Optical Tomography

There are several ways in which the inverse problem may be stated.

- We may distinguish between forward models based on diffusion or on radiative transport. Radiative transport

spans the range from micro- to macroscopic scales whereas diffusion is essentially only a macroscopic model.

- Two different approaches have commonly been used: methods based on scattering theory and methods based on optimization. The former explicitly invokes the concept of a *Green's function* whereas the latter usually only involves this concept implicitly.

- The inverse problem may be considered linear or nonlinear. Nonlinearity is a familiar problem for optimization-based approaches. Scattering theory can deal with nonlinearity but in an essentially different way.

17.2.1 Forward Mappings and Their Derivatives

We consider a domain Ω with boundary $\partial\Omega$. Experiments are performed by applying inwardly directed photon currents $J^-(r_s), r_s \in \partial\Omega$ and measuring outgoing photon currents $J^+(r_m), r_m \in \partial\Omega$. In the "noncontact" application, these functions are applied and/or measured by free-space propagation from $\partial\Omega$ to an exterior surface Σ_{ext}. The propagation of light inside the domain is governed by a differential operator $\mathcal{L}(x)$ parameterized by functions $x(r), r \in \Omega$. For a particular applied source current J_i^-, the measureable current is found by solving

$$\mathcal{L}(x)U_i = 0 \tag{17.1}$$

$$\mathcal{B}^- U_i = J_i^- \tag{17.2}$$

$$J_i^+ = \mathcal{B}^+ U_i \tag{17.3}$$

where B^-, B^+ are the boundary conditions obeyed by $\mathcal{L}$. In the case where J_i^- is a δ-function located at $\mathbf{r}_s$, the solution U_i defines the *Green's function* $G(\mathbf{r}, \mathbf{r}_s)$, and we define the linear Green's operator

$$U_i = \mathcal{G}(x)J_i^- = \int_{\partial\Omega} G(\mathbf{r},\mathbf{r}_s)J_i^-(\mathbf{r}_s)\,\mathrm{d}\mathbf{r}_s, \qquad (17.4)$$

The definitions (17.1)–(17.3) lead to the definition of the forward mapping

$$J_i^+ = F_i(x). \qquad (17.5)$$

Measured data are obtained as the result of a measurement operator $\mathcal{M}$, which, when combined with the forward map, defines a forward operator

$$\mathcal{F} = \mathcal{M}F. \qquad (17.6)$$

The key tool for the study of the inverse problem is the linearization of the forward map F. Consider a perturbation in the parameters $x(\mathbf{r}) \rightarrow x(\mathbf{r}) + \eta(\mathbf{r})$. The Fréchet derivative of the forward mapping is the linear mapping defined by

$$F_i'(x)\eta = F_i(x+\eta) - F_i(x) + o(\|\eta\|^2). \qquad (17.7)$$

The Fréchet derivative is a linear integral operator

$$F_i'\eta = \mathcal{G}'[J_i^- \otimes \eta] = \int_{\partial\Omega}\int_\Omega G'(\mathbf{r}_m,\mathbf{r},\mathbf{r}_s)\eta(\mathbf{r})J^-(\mathbf{r}_s)\,\mathrm{d}\mathbf{r}\,\mathrm{d}\mathbf{r}_s \qquad (17.8)$$

The operator $\mathcal{G}'$ has kernel $G'(\xi_m,\xi,\xi_s)$ whose form depends on the particular problem considered. We may define in general terms a potential operator

$$\mathcal{V}(\eta) = \mathcal{L}(x+\eta) - \mathcal{L}(x), \qquad (17.9)$$

which leads to

$$\mathcal{G}' = -\mathcal{G}\mathcal{V}\mathcal{G}, \quad \text{and} \quad \mathcal{G}'^* = -\mathcal{G}^*\mathcal{V}\mathcal{G}. \qquad (17.10)$$

where $\mathcal{G}^*$ is the adjoint Green's operator.

The second Fréchet derivative is given by

$$F_i''(x)(\eta_1,\eta_2) = F_i(x+\eta_1+\eta_2) - F_i'(x)(\eta_1+\eta_2) - F_i(x) + o(\|\eta\|^3), \qquad (17.11)$$

with corresponding second derivative Green's operator $\mathcal{G}''$ and kernel G'' given by

$$\mathcal{G}'' = \mathcal{G}\mathcal{V}\mathcal{G}\mathcal{V}\mathcal{G}, \qquad (17.12)$$

and in general, we can define the nonlinear mapping F as

$$F_i(x+\eta)$$
$$= [\mathcal{G}(x) - \mathcal{G}(x)\mathcal{V}(\eta)\mathcal{G}(x) + \cdots + (-1)^n\mathcal{G}(x)(\mathcal{V}(\eta)\mathcal{G}(x))^n + \cdots]J_i^- \qquad (17.13)$$

which is the Born series.

17.2.2 Physical Models

17.2.2.1 Radiative Transport

In radiative transport theory, the propagation of light through a material medium is formulated in terms of a conservation law that accounts for gains and losses of photons due to scattering and absorption (Case and Zweifel 1967, Ishimaru 1978). The fundamental quantity of interest is the specific intensity $I(\mathbf{r},\hat{\mathbf{s}})$, defined as the intensity at the position $\mathbf{r}$ in the direction $\hat{\mathbf{s}}$. The specific intensity obeys the radiative transport equation (RTE):

$$\hat{\mathbf{s}}\cdot\nabla I + (\mu_a + \mu_s)I = \mu_s\int p(\hat{\mathbf{s}},\hat{\mathbf{s}}')I(\mathbf{r},\hat{\mathbf{s}}')\,\mathrm{d}\hat{\mathbf{s}}', \quad \mathbf{r}\in\Omega, \qquad (17.14)$$

which we have written in its time-independent form. Here, μ_a and μ_s are the absorption and scattering coefficients and p is the phase function. The specific intensity also satisfies the half-range boundary condition

$$I(\mathbf{r},\hat{\mathbf{s}}) = J^-(\mathbf{r},\hat{\mathbf{s}}), \quad \hat{\mathbf{s}}\cdot\hat{\mathbf{v}}<0, \quad \mathbf{r}\in\partial\Omega, \qquad (17.15)$$

where

$\hat{\mathbf{v}}$ is the outward unit normal to $\partial\Omega$
J^- is the incident specific intensity at the boundary

The above choice of boundary condition guarantees the uniqueness of solutions to the RTE (Case and Zweifel 1967). The phase function p is symmetric with respect to interchange of its arguments and obeys the normalization condition

$$\int p(\hat{\mathbf{s}},\hat{\mathbf{s}}')\mathrm{d}\hat{\mathbf{s}}' = 1, \qquad (17.16)$$

for all $\hat{\mathbf{s}}$. We will often assume that $p(\hat{\mathbf{s}}, \hat{\mathbf{s}}')$ depends only upon the angle between $\hat{\mathbf{s}}$ and $\hat{\mathbf{s}}''$, which holds for scattering by spherically symmetric particles. Note that the choice $p = 1/(4\pi)$ corresponds to isotropic scattering.

The total power P passing through a surface Σ is related to the specific intensity by

$$P = \int_\Sigma \mathrm{d}\mathbf{r}\int_{s^2}\mathrm{d}\hat{\mathbf{s}}\,I(\mathbf{r},\hat{\mathbf{s}})\hat{\mathbf{s}}\cdot\hat{\mathbf{v}}. \qquad (17.17)$$

The energy density Φ is obtained by integrating out the angular dependence of the specific intensity:

$$\Phi(\boldsymbol{r}) = \frac{1}{c} \int I(\boldsymbol{r}, \hat{\boldsymbol{s}}) \mathrm{d}\hat{\boldsymbol{s}}. \tag{17.18}$$

We note that the RTE allows for the addition of intensities. As a result, it cannot explain certain wavelike phenomena.

17.2.2.2 Diffuse Light

The diffusion approximation (DA) to the RTE is widely used in applications. It is valid in the regime where the scattering length $l_s = 1/\mu_s$ is small compared to the distance of propagation. The standard approach to the DA is through the P_N approximation, in which the angular dependence of the specific intensity is expanded in spherical harmonics. The DA is obtained if the expansion is truncated at first order. The DA may also be derived using asymptotic methods (Larsen and Keller 1974). The DA holds when the scattering coefficient is large, the absorption coefficient is small, the point of observation is far from the boundary of the medium, and the timescale is sufficiently long. Both the P_N expansion and the asymptotic method result in the diffusion equation for the energy density Φ:

$$\frac{\partial}{\partial t} \Phi(\boldsymbol{r}, t) = \nabla \cdot [D(\boldsymbol{r}) \nabla \Phi(\boldsymbol{r}, t)] - c\mu_a(\boldsymbol{r}) u(\boldsymbol{r}, t), \tag{17.19}$$

but the expression for the diffusion coefficient is different. For the asymptotic method, it is given by

$$D = \frac{1}{3} c\ell^\star, \quad \ell^\star = \frac{1}{(1-g)\mu_t}, \tag{17.20}$$

where

$\ell^\star$ is known as the transport mean free path $\mu_t = \mu_a + \mu_s$ is the attenuation coefficient, and where the anisotropy g is given by

$$g = \int \hat{\boldsymbol{s}} \cdot \hat{\boldsymbol{s}}' p(\hat{\boldsymbol{s}} \cdot \hat{\boldsymbol{s}}') \mathrm{d}\hat{\boldsymbol{s}}', \tag{17.21}$$

with $-1 < g < 1$. Note that $g = 0$ corresponds to isotropic scattering and $g = 1$ to extreme forward scattering. The expression for $\ell^\star$ obtained from the P_N method is

$$\ell^\star = \frac{1}{(1-g)\mu_s + \mu_a}. \tag{17.22}$$

To get the theory for frequency domain measurements, we assume $e^{i\omega t}$ time-dependence with modulation frequency ω. The energy density obeys the equation

$$-\nabla \cdot [D(\boldsymbol{r}) \nabla \Phi(\boldsymbol{r})] + (c\mu_a(\boldsymbol{r}) + i\omega) \Phi(\boldsymbol{r}) = 0, \tag{17.23}$$

The solution to (17.23) obeys the Lippmann–Schwinger equation

$$\Phi = \Phi_0 - \mathcal{G}\mathcal{V}\Phi_0, \quad \Phi_0 = \mathcal{G}J^-, \tag{17.24}$$

where $\mathcal{G}$ is the Green's function for a homogeneous medium with absorption $\bar{\mu}_a$ and diffusion coefficient D_0. We have also made use of the specific form of the generalized potential operator introduced in (17.9), which for the diffusion equation is given by

$$\mathcal{V} = c\delta\mu_a - \nabla \cdot (\delta D \nabla), \tag{17.25}$$

where $\delta\mu_a = \mu_a - \bar{\mu}_a$ and $\delta D = D - \bar{D}$, with $\bar{\mu}_a$ and $\bar{D}$ constant. The unperturbed Green's function $G(\boldsymbol{r}, \boldsymbol{r}')$ satisfies

$$(\nabla^2 - k^2) G(\boldsymbol{r}, \boldsymbol{r}') = -\frac{1}{D_0} \delta(\boldsymbol{r} - \boldsymbol{r}'), \tag{17.26}$$

where the diffuse wave number k is given by

$$k^2 = \frac{c\bar{\mu}_a + i\omega}{\bar{D}}. \tag{17.27}$$

We note here that the fundamental solution to the diffusion equation is given by

$$G(\boldsymbol{r}, \boldsymbol{r}') = \frac{1}{4\pi D} \frac{e^{-k|\boldsymbol{r}-\boldsymbol{r}'|}}{|\boldsymbol{r} - \boldsymbol{r}'|}. \tag{17.28}$$

By iterating (17.24) beginning with $\Phi = \Phi_0$, we obtain (compare to (17.13))

$$\Phi = \Phi_0 - \mathcal{G}\mathcal{V}\Phi_0 + \mathcal{G}\mathcal{V}\mathcal{G}\mathcal{V}\Phi_0 + \cdots, \tag{17.29}$$

which is the analog of the Born series for diffuse waves.

17.2.3 Optical Tomography Modalities

17.2.3.1 Diffuse Optical Tomography (DOT)

The nonlinear inverse problem in DOT is to recover functions μ_a, D from measurements y. The data may be collected either in the time domain or in the frequency domain, either at one or several modulation frequencies ω (Hebden et al. 1997). If the former is used, it is typical to Fourier transform the data and to develop the analysis in the frequency domain. At the zero frequency $\omega = 0$ ("DC" measurements), the data simply represent a total photon count without any phase information, and the recovery of both μ_a and D suffers from nonuniqueness (Arridge and Lionheart 1998), but see Harrach (2009). In this case, a simpler problem is commonly defined in which one parameter (usually scattering) is assumed known and the forward mapping is restricted to recovery of the second function.

For the DOT problem, the Fréchet derivative (17.7) has a kernel based on the density functions for absorption and diffusion

$$\begin{pmatrix} \rho_{\mu_a}(\boldsymbol{r};\omega) \\ \rho_D(\boldsymbol{r};\omega) \end{pmatrix} = \begin{pmatrix} \Phi^\star(\boldsymbol{r};\omega)\Phi(\boldsymbol{r};\omega) \\ \nabla\Phi^\star(\boldsymbol{r};\omega)\cdot\nabla\Phi(\boldsymbol{r};\omega) \end{pmatrix} \qquad (17.30)$$

$$\begin{pmatrix} \rho_{\mu_a}(\boldsymbol{r},t) \\ \rho_D(\boldsymbol{r},t) \end{pmatrix} = \begin{pmatrix} \displaystyle\int_0^t \Phi^\star(\boldsymbol{r},-t')\Phi(\boldsymbol{r},t-t')\mathrm{d}t' \\ \displaystyle\int_0^t \nabla\Phi^\star(\boldsymbol{r},t')\cdot\nabla\Phi(\boldsymbol{r},t-t')\mathrm{d}t' \end{pmatrix} \qquad (17.31)$$

where we note that the adjoint problem leads to a backward-time equation.

Although DOT is a nonlinear inverse problem, a linearized version is often considered, which we refer to as *difference diffuse optical tomography* (DDOT). Here we take differences in measurements $\delta y = y_2 - y_1$ and formulate a linear mapping. Clearly the relevant linear mapping is given by the linear Fréchet derivative operator, evaluated at $x_1 = (\mu_{a1}, D_1)$. As in nonlinear DOT, the simplified case of absorption only imaging is frequently considered. The success of this approach depends on how sensitive the reconstruction is to the correct choice of linearization point x_1. This typically involves careful calibration and consideration of the modeling errors.

17.2.3.2 Fluorescence Diffuse Optical Tomography (FDOT)

In fluorescence optical tomography, sources are introduced at an *excitation* wavelength λ^e giving rise to an excitation field Φ^e. Fluorescence is regarded as a function $h(\boldsymbol{r})$ which governs the absorption of radiation at wavelength λ^e and (partial) reemission as a source at the longer wavelength (lower energy) λ^f. Measurements are taken at λ^f and λ^e. Using the frequency domain formulation, we consider two coupled PDEs

$$-\nabla\cdot D^e(\boldsymbol{r})\nabla\Phi^e(\boldsymbol{r};\omega) + (c\mu_a^e(\boldsymbol{r}) + i\omega)\Phi^e(\boldsymbol{r};\omega) = 0 \qquad (17.32)$$

$$-\nabla\cdot D^f(\boldsymbol{r})\nabla\Phi^{e\to f}(\boldsymbol{r};\omega) + (c\mu_a^f(\boldsymbol{r}) + i\omega)\Phi^{e\to f}(\boldsymbol{r};\omega) = h(\boldsymbol{r};\omega)\Phi^e(\boldsymbol{r};\omega) \qquad (17.33)$$

The quantity h is a product of the concentration of fluorescent material and its conversion efficiency. Since reemission is a Poisson process, it is characterized by a lifetime τ. In the frequency domain, this leads to a complex valued parameter

$$h(\boldsymbol{r},\omega) = h_0(\boldsymbol{r})\frac{1}{1 + i\omega\tau(\boldsymbol{r})}. \qquad (17.34)$$

The simplest problem considered is to recover h from measurements $\{y^e, y^f, y^{e\to f}\}$. Consider the frequency-domain version. The

kernel of the Fréchet derivative (17.7) in this case is the density function with respect to h given by

$$\rho_h(\boldsymbol{r};\omega) = \Phi^{e\to f^\star}(\boldsymbol{r};\omega)\Phi^e(\boldsymbol{r};\omega), \qquad (17.35)$$

giving a complex-valued reconstruction from which h_0, τ can be recovered using (17.34). Clearly, for DC measurements, only the fluorescence h_0 can be recovered, not the lifetime; *fluorescence lifetime imaging tomography* (FLIM tomography) is a term used to emphasize that both h_0 and τ are being recovered. Time domain measurements may of course be used to solve this problem. If the data are Fourier transformed into the frequency domain, the techniques for image reconstruction are unchanged. It may also be expressed directly in the time domain. In this case, the kernel is given by a double convolution

$$\rho_h(\boldsymbol{r};t) = \int_0^t\int_0^{t'} \Phi^{f^\star}(\boldsymbol{r};-t'')\Phi^e(\boldsymbol{r};t-t')h(\boldsymbol{r},t''-t')\mathrm{d}t'\mathrm{d}t'' \qquad (17.36)$$

where

$$h(\boldsymbol{r},t''-t') = h_0(\boldsymbol{r})e^{-\frac{|t''-t'|}{\tau(\boldsymbol{r})}} \qquad (17.37)$$

gives explicitly the exponential decay of the Poisson process. In practice, the half-life does not have a discrete value but a distribution, and the linear integral equation with (17.36) as its kernel has to also be convolved with the finite time spread function of the measurement system (Soloviev et al. 2007).

17.2.3.3 Multispectral Diffuse Optical Tomography

Whereas each of the above problems could be considered at a range of spectral samples and the spectral variation of the recovered solutions determined, the idea in multispectral DOT is to reformulate the problem into the recovery of a set of images of known chromophores with well-characterized spectral dependence:

$$\mu_a(\lambda_j) = \sum_i \epsilon_i(\lambda_j)c_i \to \mu_a(\lambda) = \boldsymbol{\epsilon}\mathbf{c} \qquad (17.38)$$

where ε is a known matrix. Similarly a spectral dependence of scattering can be written as

$$\mu_s'(\lambda) = a\lambda^{-b} \qquad (17.39)$$

We note that the above model for the wavelength dependence of μ_s' corresponds to Rayleigh scattering when $b = 4$. In general, subdominant power-law corrections may be necessary to accurately represent the scattering behavior of tissue.

The use of the spectral constraints (17.38)–(17.39) allows for a reparameterization from a set of uncoupled inverse problems (one for each spectral sample) into a problem in recovery of the parameters $\{a, b, c_k; k = 1 \ldots K\}$. As well as reducing the inverse problem dimension, this approach can ameliorate some of the aspects of nonuniqueness between absorption and scattering in the case of DC measurements because the necessary conditions for functions to be in the problem null-space are not possible under the spectral constraint conditions (Corlu et al. 2003, 2005).

Similarly to MSDOT in *multispectral fluorescence DOT* (MSFDOT), the fluorescence may be considered a linear combination of fluorophores emitting radiation in a known spectral pattern

$$h(\lambda_j) = \sum_i \epsilon_i(\lambda_j)p_i \rightarrow h(\boldsymbol{\lambda}) = \boldsymbol{\epsilon}^f \mathbf{p}. \tag{17.40}$$

In this case, a linear forward operator may be constructed, as in Zacharopoulos et al. (2009).

17.2.3.4 Structured Illumination Methods

In the above description, it is assumed that the light sources are localized, which allows for the use of expressions for Green's functions resulting from a point source. However, the requirement from an inverse problems point of view is simply that the incoming radiation can be used to construct a basis spanning functions on the object boundary. It is natural therefore to consider distributed light sources that globally illuminate the input surface.

Spatially modulated light has been widely used in biomedical optics, ranging from microscopy (Neil et al. 1997) to diffuse optics (Cuccia et al. 2005). In diffuse optics, a pattern of structured light is typically projected on the sample (Cuccia et al. 2005, Joshi et al. 2006b, Bassi et al. 2008) and images of reflectance (or transmittance) are acquired.

17.3 Methods

17.3.1 Forward Models

The inversion methods described in the sequel are dependent on the accuracy and efficiency with which Green's functions can be computed. The analytic forms of such functions for the DA are known for several problems and geometries. The format for the RTE is generally limited to one-dimensional and other special cases

17.3.1.1 Volume Discretization Methods

In the finite element method (FEM), the volume Ω is discretized to a mesh

$$\Omega \rightarrow \{\mathbb{T}_\Omega, \mathbb{N}_\Omega, \mathbb{U}_\Omega\} \tag{17.41}$$

where

$\mathbb{T}_\Omega$ is the set of P elements Δ_e. $e = 1, \ldots P$
$\mathbb{N}_\Omega$ is the set of N vertices $\boldsymbol{N}_p$; $p = 1, \ldots, N$
$\mathbb{U}_\Omega$ is the set of N locally supported basis functions $u_p(\boldsymbol{r})$; $p = 1, \ldots, N$

For the diffusion equation, writing

$$\Phi(\boldsymbol{r}) \simeq \Phi^h(\boldsymbol{r}) = \sum_p \Phi_p u_p(\boldsymbol{r}), \tag{17.42}$$

results in a discrete system

$$\mathsf{K}(x)\boldsymbol{\Phi} = \boldsymbol{q}, \tag{17.43}$$

where $\mathsf{K}(x)$ has matrix elements

$$K_{lm} = \int_\Omega (D(\boldsymbol{r})\nabla u_l(\boldsymbol{r}) \cdot \nabla u_m(\boldsymbol{r}) + (c\mu_a(\boldsymbol{r}) + i\omega)u_l(\boldsymbol{r}))\mathrm{d}\boldsymbol{r}$$

$$+ \frac{1}{2\zeta} \int_{\partial\Omega} u_l(\boldsymbol{r})u_m(\boldsymbol{r})\mathrm{d}\boldsymbol{r}, \tag{17.44}$$

and $\boldsymbol{q}$ represents the discretization of the boundary conditions. Making use of a basis expansion allows the representation of $\mathsf{K}(x)$ as

$$\mathsf{K} = \mathsf{S} + i\omega\mathsf{B} + \sum_k (D_k \mathsf{K}_k^D + c\mu_k \mathsf{K}_k^\mu), \tag{17.45}$$

where

B is the mass matrix
S is the matrix of surface integrals with elements given by the last term on the right in (17.45)
$\mathsf{K}_k^D \equiv \partial\mathsf{K}/\partial D_k$, $\mathsf{K}_k^\mu \equiv \partial\mathsf{K}/\partial_{c\mu k}$ are given by

$$\mathsf{K}_{k,lm}^D = \int_\Omega b_k(\boldsymbol{r})\nabla u_l(\boldsymbol{r}) \cdot \nabla u_m(\boldsymbol{r})\mathrm{d}\boldsymbol{r} \tag{17.46}$$

$$\mathsf{K}_{k,lm}^\mu = \int_\Omega b_k(\boldsymbol{r})u_l(\boldsymbol{r})u_m(\boldsymbol{r})\mathrm{d}\boldsymbol{r}. \tag{17.47}$$

In RTE FEM, a basis is also defined for the angular directions

$$S^{n-1} \rightarrow \mathbb{V}_{S^{n-1}} = \{\mathbb{T}_{S^{n-1}}, \mathbb{N}_{S^{n-1}}, \mathbb{U}_{S^{n-1}}\}. \tag{17.48}$$

Different schemes result by choosing different bases on the unit circle S^1 or sphere S^2

1. The discrete ordinate method chooses a discrete set: $\mathbb{U}_{S^n} = \{\delta(\hat{\boldsymbol{s}} - \hat{\boldsymbol{s}}_p)\}$ which are chosen to give exact quadrature points for a spherical harmonic expansion of the angular variable. The radiance is thus represented explicitly in a set of ray directions (Abdoulaev and Hielscher 2003).

2. The P_N method chooses the spherical harmonics (rotated into real functions) directly: $\mathbb{U}_{S^n} = \{\tilde{Y}_p(\hat{\boldsymbol{s}})\}$. This allows explicitly the representation as diffusion in the lowest

order with higher terms representing higher order effects (Aydin et al. 2002).

3. The local basis function method chooses a set of locally supported basis functions on $\mathbb{T}_{S^{n-1}}$. This allows essentially the same machinery of sparse matrix manipulation as for the spatial variables (Richling et al. 2001, Tarvainen et al. 2005b, 2006).

4. The wavelet basis chooses a hierarchical set of wavelets on the sphere to allow for variable and adaptive angular discretization (*Buchan et al. 2005*).

The RTE has been utilized as the forward model for optical tomography reconstruction in a few studies. In most of these papers, the forward solution of the RTE has been based either on the finite difference method (FDM) or the finite volume method (FVM) and the discrete ordinate formulation of the RTE—see Dorn (1998), Klose and Hielscher (1999), Klose and Hielscher (2002), Abdoulaev et al. (2005), and Ren et al. (2006). In the FEM solution of the RTE in low-scattering or nonscattering regions, the ray-effect may become visible (Lathrop 1968, 1971). In Tarvainen et al. (2005a), the FEM model of Tarvainen et al. (2005b) and Tarvainen et al. (2006) was augmented with the streamline diffusion modification to stabilize the forward solution in this case. The streamline diffusion modification has been found to stabilize numerical solutions of the RTE in situations in which standard techniques produce oscillating results (Kanschat 1998, Richling et al. 2001, Tarvainen et al. 2005a). Due to the heavy computational and memory requirements of meshes, adaptive techniques have been used within an FVM approach (Soloviev 2006, Soloviev and Krasnosselskaia 2006), as well as the FEM scheme (Joshi et al. 2006a, Kwon et al. 2006, Guven et al. 2007a,b, Lee et al. 2007). In order to overcome the difficulty of 3D meshing, meshless methods can be used (Qin et al. 2008).

Finally, we mention briefly Monte Carlo (MC) methods which have a long pedigree in optical tomography and are an efficient and general model especially relevant for mesoscopic problems (Prahl et al. 1989, Wang et al. 1995, Boas et al. 2002, Heiskala et al. 2005). Use of the MC model within image reconstruction is usually limited to the construction of a linear perturbation model for reconstructing the difference in optical properties from changes in the data (Heiskala 2009, Heiskala et al. 2009) with Steinbrink et al. (2001) describing the use of MC to recover optical absorption changes in a layered model.

17.3.1.2 Boundary Discretization Methods

Rather than meshing a volume, we may consider the domain as the union of a number of subdomains

$$\Omega = \cup_\ell \Omega_\ell, \quad \ell = 1 \ldots L, \tag{17.49}$$

together with constant within subdomain optical parameters $\{x_\ell\}$ and a set of interfaces between domains

$$\Sigma_j = \partial\Omega_{\ell,\ell'}, \quad j = 1 \ldots J. \tag{17.50}$$

We consider the diffusion approximation and introduce the following notation:

$$U_i := \Phi_{i-1}\big|_{\Sigma_i} = \Phi_i\big|_{\Sigma_i}, \tag{17.51}$$

$$J_i := D_{i-1}\frac{\partial\Phi_{i-1}}{\partial\nu_{i-1}}\bigg|_{\Sigma_i} = -D_i\frac{\partial\Phi_i}{\partial\nu_i}\bigg|_{\Sigma_i} \equiv D_i\frac{\partial\Phi_{i-1}}{\partial\nu}\bigg|_{\Sigma_i} = -D_i\frac{\partial\Phi_i}{\partial\nu}\bigg|_{\Sigma_i}. \tag{17.52}$$

For simplicity, we develop the discussion using only one region. For a more general implementation, we refer to Sikora et al. (2006). For homogeneous region Ω, Green's second theorem provides the following:

$$\Phi(\boldsymbol{r}) + \int_{\partial\Omega}\left(\frac{\partial G(\boldsymbol{r},\boldsymbol{r}'_m)}{\partial\nu} + \frac{G(\boldsymbol{r},\boldsymbol{r}'_m)}{2\zeta D}\right)U(\boldsymbol{r}'_m)\,\mathrm{d}\boldsymbol{r}'_m = Q(\boldsymbol{r}) \tag{17.53}$$

where we used the Robin boundary condition to eliminate J and where

$$Q(\boldsymbol{r}) = \int_\Omega G(\boldsymbol{r},\boldsymbol{r}')q(\boldsymbol{r}')\,\mathrm{d}\boldsymbol{r}'. \tag{17.54}$$

Taking the limit as $\boldsymbol{r} \in \Omega \to \partial\Omega$ results in a boundary integral equation (BIE) in the unknown function U.

In contrast to (17.41), only the boundary $\partial\Omega$ is discretized to a mesh

$$\mathbb{V} = \{\mathbb{T}, \mathbb{N}, \mathbb{U}\} \tag{17.55}$$

where

$\mathbb{T}$ is the set of P surface elements Δ_e; $e = 1, \ldots, P$
$\mathbb{N}$ is the set of N vertices $\boldsymbol{N}_p$; $p = 1, \ldots, N$
$\mathbb{U}$ is the set of N locally supported basis functions $u_p(\boldsymbol{r})$; $p = 1, \ldots, N$

Then the function U is represented as

$$U(\boldsymbol{r}) \approx \sum_{p=1}^{N} U_p u_p(\boldsymbol{r}) \tag{17.56}$$

Using the representation (17.56) in the BIE system (17.53) followed by sampling at the nodal points, we obtain a discrete system for the collocation Boundary Element Method (Becker 1992, Beer 2001, Aliabadi 2002, Sikora and Arridge 2002, Sikora et al. 2004) as a linear matrix equation

$$\left(\varrho\mathsf{I} + \mathsf{A} + \frac{1}{2\zeta}\mathsf{B}\right)U = Q \tag{17.57}$$

where

  U is the vectors of coefficients U_p in (17.56)

  ϱ is a factor accounting for boundary curvature ($\varrho = \frac{1}{2}$ for smooth boundaries)

The matrices A, B in (17.57) are dense and are constructed by integrating the product of shape functions and either Green's functions (single layer potentials) or their normal derivatives (double layer potentials), which is straightforward except where the surface element contains a singularity, when appropriate numerical techniques are required (Sikora et al. 2006).

After calculation of the boundary functions U, it can be used to construct internal fields through the application of (17.53), and thence to the construction of a linearized problem, such as in Section 17.3.2. In particular, for a single homogeneous region of arbitrary shape, we would write for an internal point $r \in \Omega$ and surface points $r_p \in \partial\Omega$

$$\Phi(r) = G(r,r_s) - \sum_{p=1}^{N}\left(\frac{\partial G(r,r_p)}{\partial \mathbf{v}} + \frac{G(r,r_p)}{2\zeta D}\right)U_p. \quad (17.58)$$

Another approximation is obtained if we write (17.57) in the form

$$(\mathsf{I} - \mathsf{G})U = \mathbf{Q}, \quad (17.59)$$

together with a Neumann series approximation

$$U = (\mathsf{I} + \mathsf{G} + \mathsf{G}^2 + \cdots)\mathbf{Q}. \quad (17.60)$$

Truncation of this series at the *n*th power in G constitutes the *n*th-order Kirchhoff approximation (Ripoll et al. 2001, Ripoll and Ntziachristos 2006)

$$\Phi(r) = G(r,r_s) + \sum_{p=1}^{N}\left(\frac{\partial G(r,r_p)}{\partial \mathbf{v}} + \frac{G(r,r_p)}{2\zeta D}\right)G(r_p,r_s)$$

$$+ \sum_{p=1}^{N}\sum_{p'=1}^{N}\left(\frac{\partial G(r,r_p)}{\partial \mathbf{v}} + \frac{G(r,r_p)}{2\zeta D}\right)$$

$$\times \left(\frac{\partial G(r_p,r_{p'})}{\partial \mathbf{v}} + \frac{G(r_p,r_{p'})}{2\zeta D}\right)G(r_{p'},r_s)$$

$$\vdots \qquad \qquad (17.61)$$

The Kirchhoff approximations are thus seen to be of the same form as the Born series, with the perturbations being taken as single and double layer potential operators on the boundary.

17.3.2 Direct and Semi-Analytic Inverse Methods

17.3.2.1 Direct Methods

By direct inversion we mean the use of inversion formulas and associated fast image reconstruction algorithms. In optical tomography, such formulas exist for particular experimental geometries, including those with planar, cylindrical, and spherical boundaries (Schotland and Markel 2001, Markel and Schotland 2001a,b, 2002a,b,c). The essential idea in such approaches is to exploit the symmetry of the problem so that the linear forward operator admits a block-decomposition. In this section, we restrict ourselves to the case of the infinite slab geometry.

Let Σ_0 be the plane $z = 0$ and Σ_L the plane $z = L$, and let $\Omega = \{(x,y,z)|0 < z < L\}$ be the domain of an infinite slab. Let vectors $\rho_s \in \Sigma_0$, $\rho_m \in \Sigma_L$ parameterize the incident and outgoing radiation on their respective planes.

We write the linearized problem in DOT as

$$\delta y(\rho_m,\rho_s) = \int_{\Omega} K_{\mu_a}(\rho_m,\rho_s,r)\delta\mu_a(r) + K_D(\rho_m,\rho_s,r)\delta D(r)\mathrm{d}^n r \quad (17.62)$$

where

  $\delta\mu_a$ is a small change in μ_a

  δD is a small change in D

In general, the kernels K_{μ_a}, K_D are nonstationary, but in the particular case of translational invariance of Ω, we have

$$K_{\mu_a}(\rho_m,\rho_s,r) = G*(|\rho_m - \rho|,z)G(|\rho_s - \rho|,z) \quad (17.63)$$

$$K_D(\rho_m,\rho_s,r) = \nabla G*(|\rho_m - \rho|,z) \cdot \nabla G(|\rho_s - \rho|,z) \quad (17.64)$$

where

  $r = (\rho,z)$

  $G(\rho,z)$ is the Greens function depending only on the distance between source and observation point

Consider the Fourier transform of δy [$k_s, k_m \in \mathbb{R}^2$]:

$$\delta\hat{y}(k_m,k_s) = \int_{\mathbb{R}^2} e^{-ik_s \cdot \rho_s} \int_{\mathbb{R}^2} e^{-ik_m \cdot \rho_m}$$

$$\times \left[\int_{\Omega} K_{\mu_a}(\rho_m,\rho_s,r)\delta\mu_a(r) + K_D(\rho_m,\rho_s,r)\delta D(r)\mathrm{d}^2\rho\,\mathrm{d}z\mathrm{d}^2\rho_m\mathrm{d}^2\rho_s\right] \quad (17.65)$$

Applying the substitution of variables

$$\mathbf{v}_s = \rho_s - \rho \Rightarrow \rho_s = \mathbf{v}_s + \rho$$

$$\mathbf{v}_m = \rho_m - \rho \Rightarrow \rho_m = \mathbf{v}_m + \rho$$

Then (using also (17.63) and (17.64)) (17.65) becomes

$$\delta\hat{y}\,(\boldsymbol{k}_{\mathrm{m}},\boldsymbol{k}_{\mathrm{s}}) = \int_{\mathbb{R}^2} e^{-i\boldsymbol{k}_{\mathrm{s}}\cdot\boldsymbol{\upsilon}_s} \int_{\mathbb{R}^2} e^{-i\boldsymbol{k}_{\mathrm{m}}\cdot\boldsymbol{\upsilon}_m} \int_{\mathbb{R}^2} e^{-i(\boldsymbol{k}_{\mathrm{m}}+\boldsymbol{k}_{\mathrm{s}})\cdot\boldsymbol{\rho}}$$

$$\left[\int_0^L G^*(\upsilon_m,z)G(\upsilon_p,z)\delta\mu_{\mathrm{a}}(r) \right.$$

$$\left. + \nabla G^*(\upsilon_m,z)\cdot\nabla G(\upsilon_p,z)\delta D(r)\mathrm{d}z \right] \mathrm{d}^2\boldsymbol{\rho}\,\mathrm{d}^2\boldsymbol{\rho}_m\mathrm{d}^2\boldsymbol{\rho}_s. \tag{17.66}$$

Rearranging, we have

$$\delta\hat{y}(\boldsymbol{k}_{\mathrm{m}},\boldsymbol{k}_{\mathrm{s}}) = \int_0^L \left[\hat{G}^*(\boldsymbol{k}_{\mathrm{m}},z)\hat{G}(\boldsymbol{k}_{\mathrm{s}},z)\delta\hat{\mu}_{\mathrm{a}}(\boldsymbol{k}_{\mathrm{s}}+\boldsymbol{k}_{\mathrm{m}},z) \right.$$

$$\left. +(\boldsymbol{k}_{\mathrm{m}}\cdot\boldsymbol{k}_{\mathrm{s}})\frac{\partial\hat{G}^*(\boldsymbol{k}_{\mathrm{m}},z)}{\partial z}\frac{\partial\hat{G}(\boldsymbol{k}_{\mathrm{s}},z)}{\partial z}\delta\hat{D}(\boldsymbol{k}_{\mathrm{s}}+\boldsymbol{k}_{\mathrm{m}},z) \right] \mathrm{d}z \tag{17.67}$$

where $\hat{G}^*(\boldsymbol{k},z), \hat{G}(\boldsymbol{k},z), \widehat{\delta\mu}_{\mathrm{a}}(\boldsymbol{k},z), \widehat{\delta D}(\boldsymbol{k},z)$ are the two-dimensional transverse Fourier transforms of the functions G^*, G, $\delta\mu_{\mathrm{a}}$, δD, respectively. We can get an infinite number of other equations by adding an arbitrary vector υ to $\boldsymbol{k}_{\mathrm{m}}$ and subtracting the same from $\boldsymbol{k}_{\mathrm{s}}$

$$\delta\hat{y}(\boldsymbol{k}_{\mathrm{m}}+\boldsymbol{\upsilon},\boldsymbol{k}_{\mathrm{s}}-\boldsymbol{\upsilon}) = \int_0^L \left[\hat{G}^*(\boldsymbol{k}_{\mathrm{m}}+\boldsymbol{\upsilon},z)\hat{G}(\boldsymbol{k}_{\mathrm{s}}-\boldsymbol{\upsilon},z)\delta\hat{\mu}_{\mathrm{a}}(\boldsymbol{k}_{\mathrm{s}}+\boldsymbol{k}_{\mathrm{m}},z) \right.$$

$$+(\boldsymbol{k}_{\mathrm{m}}+\boldsymbol{\upsilon})\cdot(\boldsymbol{k}_{\mathrm{s}}-\boldsymbol{\upsilon})\frac{\partial\hat{G}^*(\boldsymbol{k}_{\mathrm{m}}+\boldsymbol{\upsilon},z)}{\partial z}$$

$$\left. \times\frac{\partial\hat{G}^*(\boldsymbol{k}_{\mathrm{s}}-\boldsymbol{\upsilon},z)}{\partial z}\delta\hat{D}(\boldsymbol{k}_{\mathrm{s}}+\boldsymbol{k}_{\mathrm{m}},z) \right] \mathrm{d}z \tag{17.68}$$

The surface $(\boldsymbol{k}_{\mathrm{m}}+\boldsymbol{\upsilon}, \boldsymbol{k}_{\mathrm{s}}-\boldsymbol{\upsilon})$ is a plane in $\mathbb{R}^4$, and (17.68) represents an operator from a line in solution space to a plane in data space (see Figure 17.1) The main result, therefore, is that we have decoupled the original equation (17.62) into a set of operator equations

$$\delta\hat{y}_{\boldsymbol{k}_{\mathrm{m}},\boldsymbol{k}_{\mathrm{s}}}(\boldsymbol{\upsilon}) = \mathcal{K}_{\boldsymbol{k}_{\mathrm{m}},\boldsymbol{k}_{\mathrm{s}}} \left[\begin{pmatrix} \delta\hat{\mu}_{\mathrm{a}\boldsymbol{k}_{\mathrm{m}},\boldsymbol{k}_{\mathrm{s}}}(z) \\ \delta\hat{D}_{\boldsymbol{k}_{\mathrm{m}},\boldsymbol{k}_{\mathrm{s}}}(z) \end{pmatrix} \right] \tag{17.69}$$

Solution of these for selected $\{\boldsymbol{k}_{\mathrm{m}}, \boldsymbol{k}_{\mathrm{s}}\}$ allows building a representation of the transverse Fourier transform of $(\delta\mu_{\mathrm{a}}, \delta D)$, from which the spatial functions can be recovered by (2D) inverse Fourier transform.

The kernels in (17.63) and (17.64) and the Fourier transform version (17.68) require the Green's functions of the

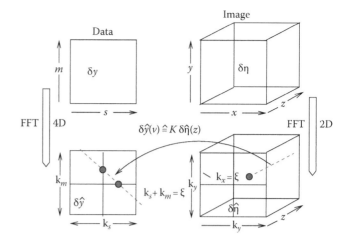

FIGURE 17.1 A three-dimensional illustration of the Fourier method. A 2D Fourier transform along x, y maps lines in $\delta\hat{\eta}$ onto planes in $\delta\hat{y}$.

(homogeneous) diffusion equation for a plane wave source, which we obtain from

$$\left[-D\nabla^2 + \mu_{\mathrm{a}} + \frac{i\omega}{c} \right] G_{\boldsymbol{k}_{\mathrm{s}}}(\rho_{\mathrm{s}},z) = e^{-i\boldsymbol{k}_{\mathrm{s}}\cdot\rho_{\mathrm{s}}}. \tag{17.70}$$

Transverse Fourier transformation gives

$$\left[-D\left(\frac{\partial}{\partial z} - k_{\mathrm{s}}^2\right) + \mu_{\mathrm{a}} + \frac{i\omega n}{c} \right] \hat{G}(\boldsymbol{k}_{\mathrm{s}},z) = \delta(z), \tag{17.71}$$

from which

$$\hat{G}(\boldsymbol{k}_{\mathrm{s}},z) = \exp\left[-\left(\frac{\mu_{\mathrm{a}} + Dk_{\mathrm{s}}^2 + \frac{i\omega}{c}}{D} \right)^{1/2} z \right]. \tag{17.72}$$

Other forms obeying the boundary conditions on Σ_0, Σ_L can also be derived.

The direct method therefore consists of inverting the linear operators (17.69) separately to build lines in the 2D Fourier transform $\delta\hat{\eta}$, followed by a final inverse Fourier transform that is computationally fast. The algorithm has recently been tested in noncontact optical tomography experiments. Quantitative reconstructions of complex phantoms with millimeter-scale features located centimeters within a highly-scattering medium have been reported (Wang et al. 2005, Konecky et al. 2008) (see Section 17.4.2). A nonlinear extension of the direct method has recently been developed (Markel et al. 2003, Moskow and Schotland 2008). In this method, a combination of forward operators with the direct inversion method is used to correct higher order terms in the Born series.

17.3.2.2 Semi-Analytic Methods

These methods describe the approach wherein a linear problem is constructed by defining a fixed grid (not necessarily regular) and sampling the forward and adjoint Green's functions at the grid points, for each source–detector pair in the data

$$
\begin{pmatrix}
\delta y(r_{d,1}, r_{s,1}) \\
\delta y(r_{d,2}, r_{s,1}) \\
\vdots \\
\delta y(r_{d,i}, r_{s,j}) \\
\vdots
\end{pmatrix}
=
\begin{pmatrix}
G^\star(r_1, r_{d,1})G(r_1, r_{s,1}) & G^\star(r_2, r_{d,1})G(r_2, r_{s,1}) & \cdots \\
G^\star(r_1, r_{d,2})G(r_1, r_{s,1}) & G^\star(r_2, r_{d,2})G(r_2, r_{s,1}) & \cdots \\
\vdots & \vdots & \ddots \\
G^\star(r_1, r_{d,i})G(r_1, r_{s,j}) & G^\star(r_2, r_{d,i})G(r_2, r_{s,j}) & \cdots \\
\vdots & \vdots & \ddots
\end{pmatrix}
$$
$$
\times
\begin{pmatrix}
\delta\mu_a(r_1) \\
\delta\mu_a(r_2) \\
\vdots
\end{pmatrix}
\tag{17.73}
$$

with a similar expression for the diffusion kernel utilizing

$$
\begin{pmatrix}
\vdots \\
\delta y(r_{d,i}, r_{s,j}) \\
\vdots
\end{pmatrix}
=
\begin{pmatrix}
\vdots & & \vdots & \ddots \\
\cdots & \nabla G^\star(r_k, r_{d,i}) \cdot \nabla G(r_k, r_{s,j}) & \cdots \\
\vdots & & \vdots & \ddots
\end{pmatrix}
$$
$$
\times
\begin{pmatrix}
\vdots \\
\delta D(r_k) \\
\vdots
\end{pmatrix}.
\tag{17.74}
$$

The matrices in (17.73)–(17.74) are sometimes referred to as *weight matrices*, or *Jacobians*, the latter term emphasizing their connection to the linearization of a forward mapping.

The matrices are essentially a discretization of the first term in the Born series as given in (17.29) with the initial field Φ_0 itself given by the Green's function $G(r_k, r_{s,j})$. It is also widely adopted to use a matrix based on the *Rytov* series (O'Leary et al. 1995) which leads to the forms

$$
\frac{G^\star(r_k, r_{d,i})G(r_k, r_{s,j})}{G(r_{d,k}, r_{s,j})}, \quad \frac{\nabla G^\star(r_k, r_{d,i}) \cdot \nabla G(r_k, r_{s,j})}{G(r_{d,k}, r_{s,j})}
\tag{17.75}
$$

The Rytov series is straightforwardly derived by considering the linearization of log $\mathcal{F}$ in place of $\mathcal{F}$. More generally, within a Bayesian framework, the forward model calls for a "whitening" matrix based on the covariance of the data. We return to this point in Section 17.3.4.

The form of the Green's function used in these models can be any of the forms from Section 17.3.1. Because no symmetry properties are being exploited, the matrices involved can be very large. Inversion schemes for such approaches are discussed in the next section.

17.3.3 Numerical Inversion

The starting point for numerical inversion methods is the definition of a variational form whose minimum represents the solution. The most commonly used is a regularized weighted least squares functional

$$
\mathcal{E}(y, \mathcal{F}(x)) = \frac{1}{2}\|y - \mathcal{F}(x)\|^2_{\Gamma_e^{-1}} + \frac{1}{2}\|x\|^2_{\Gamma_x^{-1}}
\tag{17.76}
$$

From the Bayesian point of view, (17.76) represents the negative log of the posterior probability density function

$$
\pi(x|y) = \exp(-\mathcal{E}(y, \mathcal{F}(x))) = \pi(y|x)\pi(x)
\tag{17.77}
$$

where $\pi(y|x)$ is the Gaussian (normal) probability density function of the distribution of the noise in the data with mean zero and covariance Γ_e

$$
\pi(y|x) = \pi_{\text{noise}}(y - \mathcal{F}(x)) = \mathcal{N}(0, \Gamma_e)
\tag{17.78}
$$

and $\pi(x)$ is the Gaussian (normal) prior probability density function of the distribution of x with mean zero and covariance Γ_x. Thus minimization of (17.76) represents the *maximum a posteriori* (MAP) estimate of (17.77).

The forms in (17.76) and (17.77) can be generalized in two main ways. First, the prior may be assumed not normal but of the form

$$
\pi(x) = \exp(-\Psi(x))
\tag{17.79}
$$

Second, the noise need not be Gaussian distributed but could follow an alternative probability model such as Poisson. This consideration leads to alternative *data fit functionals* of which the commonest is the cross-entropy or KL divergence

$$
KL(y, \mathcal{F}(x)) = \int_Y \left(y \log\left[\frac{y}{\mathcal{F}(x)}\right] - y + \mathcal{F}(x) \right) dy
\tag{17.80}
$$

From an optimization point of view, the minimization of (17.76) is known as the regularized output least squares solution and can be achieved with classical methods if $\varepsilon(y, \mathcal{F}(x))$ is convex. Nonconvexity does not normally arise from the forward mappings but may arise if the prior is nonconvex. An alternative formulation of the problem which has its origins in control theory is known as the "all-at-once" or "PDE-constrained method." In this formulation, the solution of the forward problem is treated as a constraint

$$
\text{minimize:} \quad \|y - \mathcal{M}U\|^2_{\Gamma_e^{-1}} + \alpha\Psi(x)
$$
$$
\text{subject to:} \quad (17.1) - (17.3)
\tag{17.81}
$$

Several of the problems in optical tomography are linear. In addition, it is frequently the case that a linearized problem is considered even when the forward problem is nonlinear. Suppose we assume a known linearization point x_0 and we consider the true solution to be a perturbation from this point $x = x_0 + \eta$, then we consider the minimization

$$\eta_* = \arg\min_\eta \left[\frac{1}{2} \left\| y - \mathcal{F}(x_0) - \mathcal{F}'(x_0)\eta \right\|_{\Gamma_e^{-1}}^2 + \alpha \frac{1}{2} \|\eta\|_{\Gamma_x^{-1}}^2 \right] \quad (17.82)$$

$$= \arg\min_\eta \left[\frac{1}{2} \left\| \delta y - \mathcal{F}'(x_0)\eta \right\|_{\Gamma_e^{-1}}^2 + \alpha \frac{1}{2} \|\eta\|_{\Gamma_x^{-1}}^2 \right] \quad (17.83)$$

where $\delta y = y - \mathcal{F}(x_0)$ is the change in measurement assumed to be linearly related to η. In the next section, we consider several generic linear solvers as applied in optical tomography.

17.3.3.1 Linear Methods

Consider the solution of a linear discrete problem (such as (17.73) or (17.74))

$$\delta y = A\eta \quad \delta y \in \mathbb{R}^M, \quad \eta \in \mathbb{R}^N, \quad A \in \mathbb{R}^M \times \mathbb{R}^N. \quad (17.84)$$

The weighted least-squares term to be minimized is

$$\eta_* = \arg\min_\eta \left[\frac{1}{2} \|\delta y - A\eta\|_{\Gamma_e^{-1}}^2 + \alpha \frac{1}{2} \|\eta\|_{\Gamma_x^{-1}}^2 \right]$$

$$= \arg\min_\eta \left[\frac{1}{2} (\delta y - A\eta)^T \Gamma_e^{-1} (\delta y - A\eta) + \alpha \frac{1}{2} \eta^T \Gamma_x^{-1} \eta \right] \quad (17.85)$$

We define factorizations

$$L_e^T L_e = \Gamma_e^{-1} \qquad L_x^T L_x = \Gamma_x^{-1} \quad (17.86)$$

and the canonical (dimensionless) variables

$$\tilde{y} = L_e \delta y \qquad \tilde{A} = L_e A L_x^{-1} \qquad \tilde{x} = L_x \eta \quad (17.87)$$

Then the solution to (17.85) is given by

$$\tilde{x}_* = \arg\min_{\tilde{x}} \left[\tilde{\varepsilon}_\alpha(\tilde{y}, \tilde{A}\tilde{x}) := \frac{1}{2} \|\tilde{y} - \tilde{A}\tilde{x}\|^2 + \alpha \frac{1}{2} \|\tilde{x}\|^2 \right] \quad (17.88)$$

In purely algebraic terms, the transformation $A \to \tilde{A}$ given in (17.87) is a conditioning step, with L_e, L_x^{-1} the left and right preconditioners, respectively (Calvetti and Somersalo 2007). We now consider some linear inversion methods for (17.88). A Newton solution for an over determined system ($M > N$) is written as

$$\tilde{x}_* = (\tilde{A}^T \tilde{A} + \alpha I)^{-1} \tilde{A}^T \tilde{y} \quad (17.89)$$

$$L_x \eta_* = \left(L_x^{-T} A^T \Gamma_e^{-1} A L_x^{-1} + \alpha I \right)^{-1} L_x^{-T} A^T \Gamma_e^{-1} \delta y$$

$$\eta_* = \left(A^T \Gamma_e^{-1} A + \alpha \Gamma_x^{-1} \right)^{-1} A^T \Gamma_e^{-1} \delta y \quad (17.90)$$

A Newton solution for an underdetermined system ($M < N$) is written as

$$\tilde{x}_* = \tilde{A}^T (\tilde{A}\tilde{A}^T + \alpha I)^{-1} \tilde{y} \quad (17.91)$$

$$L_x \eta_* = L_x^{-T} A^T L_e^T (L_e A \Gamma_x A^T L_e^T + \alpha I)^{-1} L_e \delta y$$

$$\eta_* = \Gamma_x A^T (A \Gamma_x A^T + \alpha \Gamma_e)^{-1} \delta y \quad (17.92)$$

The steepest descent method is an iterative reconstruction scheme. Beginning with an arbitrary initial estimate $\tilde{x}^0$ (usually all zeros), we iterate solutions using

$$\tilde{x}_{k+1} = \tilde{x}_k + \tau_k \tilde{s}_k, \quad (17.93)$$

where the steepest direction for (17.88) is given by

$$\tilde{s}_k = \tilde{A}^T (\tilde{y} - \tilde{A}\tilde{x}_k) - \alpha \tilde{x}_k = \tilde{s}_{k-1} - \tau_{k-1} \tilde{A}^T \tilde{A} s_{k-1}, \quad (17.94)$$

and

$$\tau_k = \frac{\|\tilde{s}_k\|^2}{\|\tilde{A}\tilde{s}_k\|^2 + \alpha \|s_k\|^2}, \quad (17.95)$$

is the step length in direction $\tilde{s}^{(k)}$ that minimizes the one-dimensional error function

$$\tilde{\varepsilon}(\tau) := \tilde{\varepsilon}_\alpha(\tilde{y}, \tilde{A}(\tilde{x} + \tau\tilde{s}_k)). \quad (17.96)$$

The Landweber method replaces (17.93) with

$$\tilde{x}_{k+1} = \tilde{x}_k + \tau \tilde{A}^T (\tilde{y} - \tilde{A}\tilde{x}_k), \quad (17.97)$$

where τ is a relaxation parameter. Rather than take the exact step that would move the solution to the minimum of the error function in the direction $\tilde{s}^{(k)}$, the Landweber method takes a fixed step. Thus, its convergence is slower than the steepest descent method.

Considering the form of (17.97), we have

$$\eta^{(k+1)} = \eta^{(k)} + \tau \Gamma_x A^T \Gamma_e^{-1} (\delta y - A\eta^{(k)}). \quad (17.98)$$

Krylov methods are the method of choice for large-scale linear ill-posed problems due to their ability to reach a solution in a subspace of relatively low dimension. Deriving the Krylov sequences for the problems (17.89) or (17.91), both give rise to the same set

$$\tilde{\mathbb{K}}_\alpha = \left\{ \tilde{A}^T \tilde{y}, \tilde{A}^T \tilde{A} \tilde{A}^T \tilde{y} + \alpha \tilde{A}^T \tilde{y}, \ldots, \sum_{j'=0}^{j} \alpha^{j'} \left(\tilde{A}^T \tilde{A} \right)^{j-j'} \tilde{A}^T \tilde{y} \ldots \right\}$$

$$=: \{ \tilde{\mathfrak{v}}_\alpha^{(0)}, \tilde{\mathfrak{v}}_\alpha^{(1)}, \ldots, \tilde{\mathfrak{v}}_\alpha^{(j)} \ldots \} \qquad (17.99)$$

This basis set is clearly a linear combination of the basis for the unregularized Krylov space of dimension J given by

$$\tilde{\mathbb{K}} \equiv \{ \tilde{\mathfrak{v}}^{(j)} \} := \left\{ \left(\tilde{A}^T \tilde{A} \right)^j \tilde{A}^T \tilde{y}, \quad j = 0 \ldots J-1 \right\}, \quad (17.100)$$

which implies that the Krylov spaces spanned by (17.99) and (17.100) are the same. Finally, we may construct the Krylov space for the original parameters:

$$\mathbb{K}_\alpha = \begin{cases} \mathfrak{v}_\alpha^{(0)} = \Gamma_x A^T \Gamma_e^{-1} \delta y \\ \mathfrak{v}_\alpha^{(1)} = (\Gamma_x A^T \Gamma_e^{-1} A + \alpha) \mathfrak{v}_\alpha^{(0)} \\ \vdots \\ \mathfrak{v}_\alpha^{(j)} = (\Gamma_x A^T \Gamma_e^{-1} A + \alpha) \mathfrak{v}_\alpha^{(j-1)} \end{cases} \qquad (17.101)$$

In the conjugate gradient method, a second sequence of $\tilde{A}$-conjugate directions $\{\tilde{p}_1, \tilde{p}_2, \ldots\}$ is constructed and (17.94) becomes

$$\tilde{x}_{k+1} = \tilde{x}_k + \tau_k \tilde{p}_k. \qquad (17.102)$$

The direction $\tilde{p}_k$ is given by

$$\tilde{p}_{k+1} = \tilde{s}_{k+1} + \beta_k \tilde{p}_k \qquad (17.103)$$

where the Gram–Schmidt constants β_k are given by one of the methods

$$\beta_k = \begin{cases} \text{Fletcher-Reeves:} \dfrac{\left\| \tilde{s}_{k+1} \right\|^2}{\left\| \tilde{s}_k \right\|^2} \\[3mm] \text{Polak-Ribiere:} \dfrac{\langle \tilde{s}_{k+1}, \tilde{s}_{k+1} - \tilde{s}_k \rangle}{\langle \tilde{p}_{k+1}, \tilde{s}_{k+1} - \tilde{s}_k \rangle} \end{cases} \qquad (17.104)$$

and the step length (c.f. (17.95)) is given by

$$\tau_k = \frac{\left\| s_k \right\|^2}{\left\| \hat{A} \tilde{p}_k \right\|^2 + \alpha \left\| \tilde{p}_k \right\|^2} \qquad (17.105)$$

We note that given a Krylov vector (an image in parameter space), the next vector is formed by the steps

forward project → filter in data space

$$\rightarrow \text{back project} \rightarrow \text{filter in image space} \qquad (17.106)$$

Thus, the sequence can be formed without building the matrix A and constitutes a *matrix free* approach.

Let us define $a_i^R \in \mathbb{R}^N$ as the ith row of A and $a_j^C \in \mathbb{R}^M$ as the jth column. Then we can define a number of purely algebraic conditioning matrices

$$R_1 = \text{diag}\left[\left\{ \left\| a_i^R \right\|_1 ; i = 1 \ldots M \right\} \right]$$

$$C_1 = \text{diag}\left[\left\{ \left\| a_j^C \right\|_1 ; j = 1 \ldots N \right\} \right] \qquad (17.107)$$

$$R_2^2 = \text{diag}\left[\left\{ \left\| a_j^R \right\|_2^2 ; i = 1 \ldots M \right\} \right] = \text{diag}\left[AA^T \right]$$

$$C_2^2 = \text{diag}\left[\left\{ \left\| a_j^C \right\|_2^2 ; j = 1 \ldots N \right\} \right] = \text{diag}\left[A^T A \right] \qquad (17.108)$$

where the notation $\| \cdot \|_p := (\Sigma .^p)^{1/p}$ defines the p-norm of a vector.

For the special case where A contains purely nonnegative elements the forms R_1, C_1 can be constructed as row and columns sums:

$$R_1 = A\mathbf{1}, \quad C_1 = A^T \mathbf{1}. \qquad (17.109)$$

The choice

$$\Gamma_e \leftarrow R_1 \quad \Gamma_x^{-1} \leftarrow C_1$$

leads to the *simultaneous algebraic reconstruction technique* (SART)

$$\eta^{(k+1)} = \eta^{(k)} + \tau C_1^{-1} A^T R_1^{-1} (\delta y - A \eta^{(k)}). \qquad (17.110)$$

This is suitable for emission tomography and also for fluorescence optical tomography, where the sought for parameters are positive and the matrix A represents a probability of a photon emitted at voxel i arriving at detector j.

The choice

$$\Gamma_e \leftarrow R_2^2 \quad \Gamma_x \leftarrow I$$

leads to the *simultaneous iterative reconstruction technique* (SIRT)

$$\eta^{(k+1)} = \eta^{(k)} + \tau A^T R_2^{-2} (\delta y - A \eta^{(k)}). \qquad (17.111)$$

A particular preconditioner that is used in conjugate gradient solvers is given by Golub and van Loan (1983)

$$\Gamma_x^{-1} = M = \text{diag}\left[A^T \Gamma_e^{-1} A\right], \qquad (17.112)$$

which seeks to equalize the curvature of the objective function along the coordinate axes ("sphering").

The SIRT and SART algorithms derived from the Landweber method can also be thought of as derived from the Algebraic reconstruction technique (ART) or Kaczmarz method (Kaczmarz 1937). This is a row-action method that computes an update to the solution by processing one row of the linearized system at a time. The inner loop generates an update from row i by

$$\eta^{(k+1)} = \eta^{(k)} + \varpi_k \frac{\delta y_i - \langle a_i^R, \eta^{(k)} \rangle}{\|a_i^R\|^2} a_i^R, \qquad (17.113)$$

where ϖ_k is a relaxation parameter. The loop over rows of A is repeated n times. The rows i are processed in randomized order, which has been shown to improve the convergence rate of the method (Strohmer and Vershynin 2006).

We mention without derivation two multiplicative algorithms: the *multiplicative ART* (MART) scheme

$$x_{k+1} = x_k \odot \exp\left[A^T \log\left(\frac{\delta y}{A x_k}\right)\right]. \qquad (17.114)$$

and the *maximum likelihood expectation maximization* (MLEM) algorithm

$$x_{k+1} = \frac{x_k}{A^T \mathbf{1}} \odot \left[A^T \left(\frac{y}{\mathcal{F}(x_k)}\right)\right] \qquad (17.115)$$

In the MLEM case, the explicit regularization leads to the MAP-EM algorithm

$$x_{k+1} = \frac{x_k}{A^T \mathbf{1} + \alpha \Psi'(x_k)} \odot \left[A^T \left(\frac{y}{\mathcal{F}(x_k)}\right)\right]. \qquad (17.116)$$

Alternatively, a two-step method reconstruction followed by filtering can be employed

$$x_{k+\frac{1}{2}} = \frac{x_k}{A^T \mathbf{1}} \odot \left[A^T \left(\frac{y}{\mathcal{F}(x_k)}\right)\right], \qquad (17.117)$$

$$x_{k+1} = \mathcal{P} x_{k+\frac{1}{2}}, \qquad (17.118)$$

where $\mathcal{P}$ is a projection such as positivity, scale-space filtering or frequency band pass.

17.3.3.2 Nonlinear Methods

The Gauss–Newton method can be considered as the iterative minimization of the quadratic Taylor-series approximation to $\varepsilon(y, \mathcal{F}(x))$

$$\varepsilon(\delta y, \mathcal{F}(x_k + \eta)) \simeq \frac{1}{2} \|y - \mathcal{F}(x_k) - \mathcal{F}'(x_k)\eta\|^2_{\Gamma_e^{-1}}$$
$$+ \alpha(\Psi(x_k) + 2\langle \eta, \mathcal{W} x_k \rangle + \langle \eta, \mathcal{W} \eta \rangle) \qquad (17.119)$$

where $\mathcal{W}(x_k)$ represents the mapping induced by the linearization of the functional Ψ. Minimization of (17.119) is given by the solution to

$$(A^T \Gamma_e^{-1} A + \alpha \mathcal{W})\eta = A^T \Gamma_e^{-1}(y - \mathcal{F}(x_k)) - \mathcal{W} x_k, \qquad (17.120)$$

$$H\eta = -g \qquad (17.121)$$

with Hessian $H = (A^T \Gamma_e^{-1} A + \alpha \mathcal{W})$ and gradient $g = \mathcal{W} x_k - A^T \Gamma_e^{-1}(y - \mathcal{F}(x_k))$. From the form of the Hessian, we can see it is guaranteed to be symmetric nonnegative definite provided that Ψ is convex, and therefore η is guaranteed to be in a descent direction for ε. Solution of (17.121) can be carried out with any of the methods in Section 17.3.3. In particular, the *Gauss–Newton–Krylov* method uses a Krylov solver for (17.121). For a badly ill-posed problem, the size of the requisite Krylov space can be quite small and may be constructed entirely through forward and adjoint solutions and image and data filtering operations as in (17.106), and therefore also as a matrix-free approach.

Since the approximation in (17.119) is only locally quadratic, the update given by solving (17.121) may (1) not be optimal or may (2) not be a descent step (if the Hessian is not symmetric positive definite). There are two strategies for resolving these problems. The *damped Gauss–Newton* method is a globalization strategy that addresses the first problem. In this approach, the update direction is used in a one-dimensional line search to find an update step τ_k that minimizes ε along this direction:

$$\tau_k = \arg\min_\tau \varepsilon(y, \mathcal{F}(x_k + \tau \eta)) \qquad (17.122)$$

Note that the full nonlinear mapping $\mathcal{F}$ is used in (17.122), not its linearization. The *Levenberg–Marquardt* method can also address the second problem. In this approach, a control parameter γ is used to modify the Hessian

$$H \to H + \gamma I.$$

When γ is large, the update η tends toward the steepest descent direction with increasingly shorter steps. When γ is small, the update tends toward the Newton direction. The idea in the algorithm is to decrease γ whenever the preceding step reduced ε

but to increase it and re-solve (17.119) if the preceding step increased ε. The role of γ also serves to modify the eigenspectrum of H and force it to be symmetric positive definite. The role of column normalization of A is often crucial in the use of the Levenberg–Marquardt algorithm since it can "sphere" the level sets of the local quadratic approximation. Typically, the Levenberg–Marquardt method is required for highly nonlinear problems, which is not the case in optical tomography. A comparison of Levenberg–Marquardt and damped Gauss–Newton methods can be found in Schweiger et al. (2005). The Gauss–Newton method was used for the inverse RTE problem in optical tomography in Tarvainen et al. (2008).

The nonlinear conjugate gradient method (NCG) is effectively the same as linear CG with line search as in (17.122) replacing the calculation of the step length given in (17.105). Rather than present it in the canonical variables, we can present it in the normal variables (Fletcher–Reeves version)

$$\text{Set } \boldsymbol{g}_0 = \alpha \Psi'(\boldsymbol{x}_0) - \mathcal{F}'^* \Gamma_e^{-1}(\boldsymbol{y} - \mathcal{F}(\boldsymbol{x}_0))$$

$$\boldsymbol{p}_0 = -\Gamma_x \boldsymbol{g}_0$$

for $k = 1, \ldots \cdot$ **do**

$$\tau_k = \arg \min_\tau \varepsilon(\boldsymbol{x}_{k-1} + \tau \boldsymbol{p}_{k-1})$$

$$\boldsymbol{x}_k = \boldsymbol{x}_{k-1} + \tau_k \boldsymbol{p}_{k-1}$$

$$\boldsymbol{g}_k = \alpha \Psi'(\boldsymbol{x}_k) - \mathcal{F}'^* \Gamma_e^{-1}(\boldsymbol{y} - \mathcal{F}(\boldsymbol{x}_k))$$

$$\beta_k = \frac{\langle \boldsymbol{g}_k, \Gamma_x \boldsymbol{g}_k \rangle}{\langle \boldsymbol{g}_{k-1}, \Gamma_x \boldsymbol{g}_{k-1} \rangle}$$

$$\boldsymbol{p}_k = -\Gamma_x \boldsymbol{g}_k + \beta_k \boldsymbol{p}_{k-1}$$

end for

Note that $\tilde{\boldsymbol{s}}_k = -\Gamma_x \boldsymbol{g}_k$ represents the application of the preconditioner $\boldsymbol{M}^{-1} = \Gamma_x$ to the gradient. If we assume $\Psi'(\boldsymbol{x}) \equiv \mathcal{W}\boldsymbol{x} = \mathsf{L}_x^\mathsf{T}\mathsf{L}_x\boldsymbol{x} = \Gamma_x^{-1}\boldsymbol{x}$ then this conditioned gradient will be

$$\tilde{\tilde{\boldsymbol{s}}}_k = \Gamma_x \mathcal{F}'^* \Gamma_e^{-1}(\boldsymbol{y} - \mathcal{F}(\boldsymbol{x}_k)) - \alpha \boldsymbol{x}_k$$

Note the similarity to the general scheme in (17.106). Also note that the variables $\tilde{\boldsymbol{s}}$ are no longer dimensionless. They have the reciprocal dimensions of the original parameters themselves.

Nonlinear conjugate gradients are usually restarted after a certain number of iterations. As for any descent method, NCG may be combined with projection onto convex sets for enforcement of constraints such as lower and upper bounds, but at the expense of loss of its conjugacy properties. Nonlinear CG was used in optical tomography for the diffusion-based problem in Arridge and Schweiger (1998) and for the RTE-based problem in Kim and Charette (2007).

The BFGS (Broyden–Fletcher–Goldfarb–Shanno) algorithm is a quasi-Newton approach that builds up estimates $\tilde{\mathsf{H}}^{-1}$ of the inverse of the Hessian matrix H (Liu and Nocedal 1989, Nocedal and Wright 1999). The update rule

$$\boldsymbol{x}_{k+1} = \boldsymbol{x}_k - \lambda_k \tilde{\mathsf{H}}_k^{-1} \boldsymbol{g}_k \qquad (17.123)$$

is employed, where $\tilde{\mathsf{H}}_k$ is updated at each iteration by the formula

$$\tilde{\mathsf{H}}_{k+1}^{-1} = \mathsf{V}_k^\mathsf{T} \tilde{\mathsf{H}}_k^{-1} \mathsf{V}_k + \rho_k \boldsymbol{d}_k \boldsymbol{d}_k^\mathsf{T}, \qquad (17.124)$$

$$\rho_k = \frac{1}{\langle z_k, \boldsymbol{d}_k \rangle}, \qquad (17.125)$$

$$\mathsf{V}_k = \mathsf{I} - \rho_k z_k \boldsymbol{d}_k^\mathsf{T}, \qquad (17.126)$$

$$\boldsymbol{d}_k = \boldsymbol{x}_{k+1} - \boldsymbol{x}_k, \qquad (17.127)$$

$$z_k = \boldsymbol{g}_{k+1} - \boldsymbol{g}_k \qquad (17.128)$$

In the limited memory version of the algorithm (L-BFGS), the approximate matrices $\tilde{\mathsf{H}}_k^{-1}$ are not stored explicitly but described implicitly by a limited number of pairs of vectors $\{\boldsymbol{d}_i, z_i\}$, where in each iteration, a new vector pair is added and the oldest pair is discarded.

The nonlinear Kaczmarz method is widely used in nonlinear tomography (Natterer and Wübbeling 2001). It is applicable for problems such as DOT, which use multiple sources. Using one source at a time, the subset of the data from this source is back-projected and added to the solution

$$\boldsymbol{x}_{k+1} = \boldsymbol{x}_k + \mathcal{C}\mathcal{F}'^*_{i(k)}\Gamma_e^{-1}(\boldsymbol{y}_{i(k)} - \mathcal{F}_{i(k)}(\boldsymbol{x}_k)) \qquad (17.129)$$

where the index $i(k)$ refers to the subset of the data accessed on the kth iteration of the algorithm. The operator $\mathcal{C}$ is a simple operator playing the role of a right preconditioner. A natural choice would be

$$\mathcal{C} = \left(\mathcal{F}_{i(k)}\mathcal{F}'^*_{i(k)} + \alpha\mathsf{I} \right)^{-1} \qquad (17.130)$$

but this may be hard to compute. Applications in optical tomography can be found in (Dorn 1998, Gonzalez-Rodriguez and Kim 2009).

A relatively simple approach to generating step directions is to update one pixel at a time. It is equivalent to taking a descent direction $\boldsymbol{s}$, projecting to one unit axis, and iterating through the dimensions of the solution space. This is effectively a Gibbs sampling procedure for the posterior distribution within the Bayesian framework. Writing $\hat{\boldsymbol{e}}_k$ for a unit vector with 1 in the kth entry and 0 otherwise, the ICD method considers

$$\tau_k = \arg \min_k \left[\varepsilon(\boldsymbol{x}_k + \tau \hat{\boldsymbol{e}}_k) \right] \qquad (17.131)$$

$$\boldsymbol{x}_{k+1} = \boldsymbol{x}_k + \tau_k \hat{\boldsymbol{e}}_k \qquad (17.132)$$

Taking a local linearization around x_k the one-dimensional objective functional to be minimized is

$$\varepsilon\left(x_k + \tau\hat{e}_k\right) = \frac{1}{2}\left\|y - \mathcal{F}(x_k)\right\|^2_{\Gamma_e^{-1}} - \tau\left\langle y - \mathcal{F}(x_k), \Gamma_e^{-1}a_k^C\right\rangle$$
$$+ \frac{1}{2}\tau^2\left\|a_k^C\right\|^2_{\Gamma_e^{-1}} + \alpha\Psi(x_k + \tau\hat{e}_k) \qquad (17.133)$$

If the prior is Gaussian, then (17.133) is a weighted least-squares problem with minimum

$$\tau_k \frac{\left\langle y - \mathcal{F}(x_k), \Gamma_e^{-1}a_k^C\right\rangle - \left\langle\hat{e}_k, \Gamma_x^{-1}x_k\right\rangle}{\left\|a_k^C\right\|^2_{\Gamma_e^{-1}} + \left\langle\hat{e}_k, \Gamma_x^{-1}\hat{e}_k\right\rangle} \qquad (17.134)$$

but it is relatively simple also for a non-Gaussian prior since evaluation of the prior usually only involves neighboring pixels. Projection onto constraint sets is also efficient since once the constraints are imposed for one pixel it is not revisited. However, the evaluation of the likelihood is only efficient for an explicit matrix method and not for the matrix-free approach.

The ICD method was used in optical tomography by Bouman and Webb (Ye et al. 1999, Milstein et al. 2003). Acceleration was achieved by a multiresolution strategy in Ye et al. (2001).

The PDE constrained method considers the approach presented in (17.82). To implement the method, we consider Lagrangian (dual) fields Z and define an objective function

$$\mathcal{J}_i(x, U_i, Z_i) = \left\|y_i - \mathcal{M}U_i\right\|^2_{\Gamma_e^{-1}} + \alpha\Psi(x) + \left\langle Z_i, (\mathcal{L}(x)U_i - q_i)\right\rangle_\Omega \qquad (17.135)$$

where q_i represents an equivalent source for the boundary condition (17.2). When considering all sources, the full Lagrangian is

$$\mathcal{J}(x, U, Z) = \sum_i \mathcal{J}_i(x, U_i, Z_i) \qquad (17.136)$$

The minimum of (17.135) occurs where the first variation

$$\mathcal{J}_{i,x} = \alpha\Psi'(x) + \left\langle Z_i, \mathcal{L}_x U_i\right\rangle_\Omega \qquad (17.137)$$

$$\mathcal{J}_{i,U} = \mathcal{L}^*Z_i - \mathcal{M}^*\Gamma_e^{-1}(y_i - \mathcal{M}U_i) \qquad (17.138)$$

$$\mathcal{J}_{i,z} = \mathcal{L}U_i - q_i \qquad (17.139)$$

becomes zero. We recognize (17.138) as the equation for the backprojected field of the residual difference between the data y_i and the measurement of the direct field U_i. We

recognize $\mathcal{L}_x = V(\eta)$ as the potential operator introduced in (17.9). Therefore, if (17.138) and (17.139) are both equated to zero, (17.137) is the gradient of the negative log posterior of a MAP estimation scheme. If instead $\{-\mathcal{J}_{i,x}, -\mathcal{J}_{i,U}, -\mathcal{J}_{i,Z}\}$ is used as an update direction for $\{x, U, Z\}$, then we may solve for all variables simultaneously without requiring convergence for any one variable until termination.

Taking the second variation of $\mathcal{J}$ results to a Newton system

$$\begin{pmatrix} \alpha\Psi''(x) & \mathcal{L}_x^\star Z_i & \mathcal{L}_x U_i \\ \mathcal{L}_x^\star Z_i & \mathcal{M}^\star\Gamma_e^{-1}\mathcal{M} & \mathcal{L} \\ \mathcal{L}_x U_i & \mathcal{L} & 0 \end{pmatrix}\begin{pmatrix} x^\delta \\ U_i^\delta \\ Z_i^\delta \end{pmatrix}$$
$$= -\begin{pmatrix} \alpha\Psi'(x) + \left\langle Z_i, \mathcal{L}_x U_i\right\rangle_\Omega \\ \mathcal{L}^\star Z_i - \mathcal{M}^\star\Gamma_e^{-1}(y - \mathcal{M}U_i) \\ \mathcal{L}U_i - q \end{pmatrix} \qquad (17.140)$$

In many applications, the system in (17.140) is simplified by taking the Schur complement of the complete Hessian and by using the Gauss–Newton or other quasi-Newton approximations for the solution scheme (Abdoulaev et al. 2005, Bangerth and Joshi 2008, Kim and Hielscher 2009). For the time-domain problem, even with such Hessian reduction techniques, Newton methods are infeasible but first-order descent schemes can still be used (Soloviev et al. 2008).

17.3.4 Use of Priors

So far, we did not discuss the forms of the data covariances Γ_e or the prior $\Psi(x)$. The format of noise will usually be predicted on physical grounds. It is often taken to be zero-mean Gaussian noise with a possibly nonwhite covariance. When considering photon-counting measurements, the implication is that the noise should be Poisson with variance $\sigma_j^2 = y_j$. Assuming sufficient signal to approach the central limit theorem would lead to an equivalent Gaussian model of additive noise with

$$\Gamma_e = \text{diag}[y] \quad \Leftrightarrow \quad \mathsf{L}_e = \text{diag}\left[\frac{1}{y^{1/2}}\right] \qquad (17.141)$$

A more commonly used model assumes that equal numbers of photons are collected at each detector leading to a constant relative error and the formal covariance structure

$$\Gamma_e = \text{diag}[y^2] \quad \Leftrightarrow \quad \mathsf{L}_e = \text{diag}\left[\frac{1}{y}\right] \qquad (17.142)$$

which also corresponds to the error model implicit in using the Rytov series in place of the Born series for the linearized model. Within this photon-counting paradigm, there is no correlation between errors on different detectors. However, when considering the actual errors between measured and modeled data, the correlation is far from being negligible, and their mean is far from being zero. This discrepancy can be understood by formally considering the modeling error as a random variable (Kaipio and Somersalo 2005):

$$\pi(y^{\text{meas}} - \mathcal{F}_h(x_{\text{true}})) \sim \mathcal{N}(\bar{\epsilon}, \Gamma_e + \Gamma_\epsilon) \qquad (17.143)$$

where

$\mathcal{F}_h$ is an approximate model with the accuracy of the computational method employed

$\bar{\epsilon}$ is the model bias representing the discrepancy between this model and the real Physics

Γ_e is the model error covariance

In practical applications of optical tomography, an estimate of the bias is made by measuring a reference problem and comparing it to a reference model

$$\bar{\epsilon} = y^{\text{ref}} - \mathcal{F}_h(x_{\text{ref}}) \qquad (17.144)$$

which leads to a corrected model

$$\mathcal{F}_{\text{corrected}}(x) = \mathcal{F}_h(x) + y^{\text{ref}} - \mathcal{F}_h(x_{\text{ref}}) \qquad (17.145)$$

Clearly, the corrected model and the measured data agree exactly for the value $x = x_{\text{ref}}$, and the assumption is that they will also agree at nearby values. This is questionable.

A more principled approach is the *approximation error method* (Kaipio and Somersalo 2005). In this approach, the statistical properties of the modeling error are estimated by sampling over a plausible distribution of solutions and comparing the modeled data from each sample with "measured" data from this sample. In practice, the measured data could be also be modeled but with a much more accurate technique. In Arridge et al. (2006), this technique was shown to result in reconstructed images using a relatively inaccurate forward model that were of almost equal quality to those using a more accurate forward model; the increase in computational efficiency was an order of magnitude.

Choice of regularization Ψ is critical and difficult to justify unequivocally. Assuming a form

$$\Psi(x) = \int_\Omega \psi(|\nabla x|)) \mathrm{d}r, \qquad (17.146)$$

where ψ is an image to image mapping leads to the linearization

$$\Gamma_x^{-1} = \mathcal{W}(x) = \nabla^{\mathrm{T}} k(r) \nabla \qquad (17.147)$$

where the *diffusivity* is given by

$$k := \frac{\psi'(|\nabla x|)}{|\nabla x|} \qquad (17.148)$$

Note that the matrix Γ_x is a covariance whose entries represent a correlation between pixels. On the other hand, Γ_x^{-1} is sparse, representing the local relationship between neighbors. We could therefore specify Γ_x in a number of ways:

1. From a database of representative images $\{x_k\}$ so that $\Gamma_x = E[(x - \bar{x})(x - \bar{x})^{\mathrm{T}}]$
2. By specifying a Markov random field $\Gamma_x^{-1} \equiv \mathsf{W}$
3. By specifying a PDE $\Gamma_x^{-1} \equiv \nabla^{\mathrm{T}} k \nabla$

Methods such as first-order Tikhonov (Phillips-Twomey), total variation (TV), or generalized Gaussian Markov random field methods (GGMRF) (Bouman and Sauer 1993) make statistical assumptions about the distribution of edges. They pose assumptions about the regularity (e.g., smoothness) of the solution, the regularity being measured in terms of some norm of the solution and its differentials (Douiri et al. 2005). In addition, the "lumpy background" prior provides an effective method of estimating information content of different imaging systems (Pineda et al. 2006).

Image processing methods for computer vision offer a large set of techniques for de-noising and segmentation based on anisotropic diffusion processes (Geman and Geman 1984, Rudin et al. 1992, Acar and Vogel 1994, Vogel and Oman 1996). These techniques formulate directly a PDE for image flow rather than as the Euler equation of a variational form and can be considered more general. More specific prior information can be incorporated if we assume that some approximate knowledge of the objects such as shape topology and intra-region parameter regions is available (Kolehmainen et al. 1999).

17.4 Examples

17.4.1 Comparison of Reconstruction Methods

The reconstruction methods described in this chapter were compared for a cylindrical test case, from which data were generated by using the FEM forward solver.

The object consisted of a cylinder with radius 25 mm and height 50 mm, with background optical parameters of absorption $\mu_a = 0.01$ mm^{-1}, scattering $\mu'_s = 1$ mm^{-1}, and refractive index $n = 1.4$. Various spherical and ellipsoidal inclusions with different contrast in absorption and scattering were embedded in the cylinder. The layout of the insertions is shown in Figure 17.2, where absorption inclusions are shown in light gray and scattering inclusions in dark gray.

Sources and detector positions were arranged in five rings at elevations −20, −10, 0, +10, and +20 mm around the cylinder mantle. Each ring contained 16 sources and 16 detectors at equal angular spacing, giving a total of 80 sources and 80 detectors. The simulated measurements were calculated for all

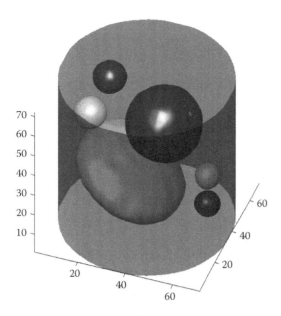

FIGURE 17.2 Layout of target object: absorption contrast inclusions are shown in light gray, scattering contrast inclusions in dark gray.

$80 \times 80 = 6400$ source–detector combinations. Measurements consisted of the logarithmic amplitude and phase shift at each detector position, given an amplitude-modulated input signal of 100 MHz.

The forward data were generated with a finite element model, using an unstructured mesh consisting of 83,142 nodes and 444,278 four-noded tetrahedral elements, using piecewise linear basis functions.

For the reconstructions, a coarser mesh consisting of 27,084 nodes and 141,702 four-noded tetrahedral elements was used to compute the forward data. The reconstruction was performed on a regular grid of trilinear basis functions of dimension $32 \times 32 \times 32$.

The reconstruction results from different inversion methods and using simulated data that were contaminated with 1% Gaussian-distributed additive random noise are shown in Figure 17.3.

In each image, the left two columns are the target images in the original mesh for the forward data generation and mapped into the mesh used for the inverse solvers. Columns 3–8 show reconstruction results obtained with Gauss–Newton using GMRES linear solver, Gauss–Newton using BiCGSTAB linear solver, nonlinear conjugate gradient solver, limited-memory BFGS solver using 20 basis vector to approximate the inverse Hessian, random-order ART solver, and Block-ART solver.

All reconstructions used a first-order Tikhonov regularization scheme with hyperparameter $\tau = 10^{-5}$. The stopping criteria were set as follows:

$$\varepsilon_n / \varepsilon_0 < \sigma_a \quad \text{(convergence)}$$

$$\varepsilon_{n-1} - \varepsilon_n < \sigma_b \quad \text{(stagnation)}$$

with objective function ε_n at iteration n, and the tolerance limits set to $\sigma_a = \sigma_b = 10^{-8}$.

It can be seen that all reconstruction schemes are able to reconstruct the larger absorption and scattering inclusions with good quantitation, but generally underestimate the contrast of small objects. Sharp boundary features appear blurred in the reconstruction, which is in part also due to the smoothing prior. The scattering parameter distribution generally appears to be reconstructed with a higher spatial resolution than the absorption parameter. The differences in reconstruction results between the different methods are small, except for BlockART and ART, which exhibit lower resolution and less distinct features in the reconstructed images.

To compare the performance of the different reconstruction algorithms, the convergence of the objective function as a function of iteration number and of computer runtime is shown in Figure 17.4. It can be seen that all reconstruction algorithms converge to similar levels of the objective function at convergence. However, the convergence rate with respect to iteration count and runtime differs significantly between the different methods. The Gauss–Newton methods converge fastest of the tested methods both with iteration count and with runtime. Note that the GMRES and BiCGSTAB versions show nearly identical convergence behavior and are therefore not plotted separately. The conjugate gradient solver converges significantly slower, while the L-BFGS solver shows a similar convergence rate to conjugate gradients with respect to iteration count, but achieves a similar performance to the Gauss–Newton solver with respect to runtime, owing to its lower computational cost per iteration. The slowest convergence is shown by the ART solver, while Block-ART is comparable in performance to the conjugate gradient solver.

17.4.2 Direct Reconstruction Method

We show here the results of reconstruction of structures in a slab geometry from CCD measurements and a scanning source. The slab had a depth of 6 cm and a background optical properties $\mu_a = 0.005 \, \text{mm}^{-1}$ $\mu_s' = 0.75 \, \text{mm}^{-1}$, the source was scanned on a 35×35 grid at 4 mm spacing, and the cameras recorded 512×512 data for each source position, leading to a total data set of order 10^8 source–detector pairs Figure 17.5 is a reconstruction of bar targets 7–9 mm thick placed at a depth of 1 and 3 cm in the slab. Figure 17.6 is a reconstruction of complex structures (Wang et al. 2005, Konecky et al. 2008). The images were reconstructed in approximately 1 min of CPU time on a 1.5 GHz computer (Wang et al. 2005).

17.4.3 Structured Light Results

A linear reconstruction scheme was developed directly in the Fourier domain (Bassi et al. 2009), using 12 spatial frequencies ($\mathbf{k}$ from -0.4 to $0.4 \, \text{rad/mm}$). CCD images $A_{\mathbf{k}} \exp(i\theta)$, considered as complex numbers, were each Fourier Transformed and

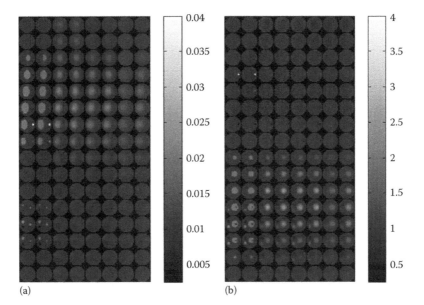

FIGURE 17.3 Cross sections through target and reconstruction images. (a) Sub-figure absorption, (b) sub-figure scattering. Columns within sub-figures, from left to right: (i) Target distribution for forward data generation, (ii) target distribution mapped into reconstruction mesh, (iii) reconstruction with Gauss–Newton solver using GMRES linear solver, (iv) reconstruction with Gauss–Newton solver using BiCGSTAB linear solver, (v) reconstruction with nonlinear conjugate gradient solver, (vi) reconstruction with limited-memory BFGS solver, (vii) reconstruction with ART solver, and (viii) reconstruction with Block-ART solver.

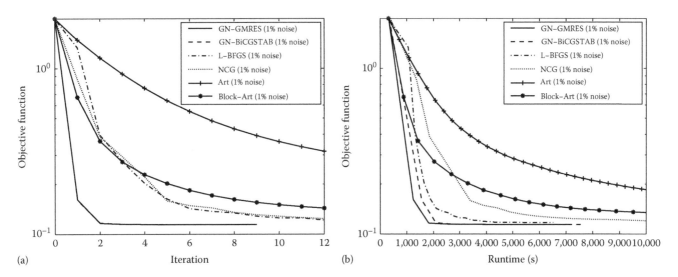

FIGURE 17.4 Convergence of objective function as a function of iteration number (a) and runtime (b) for different inverse solvers. Simulated data with 1% random Gaussian data noise.

sampled at the same 12 spatial frequencies to give 144 data samples δy. The set of spatial frequencies were used as source profiles to generate forward fields Φ_k and as measurement profiles to generate adjoint fields Φ_k^*, from which a Jacobian J with 144 rows and 327,680 columns was constructed. Partitioning into real and imaginary parts to give data and Jacobian with 288 rows, the solution was obtained by solving

$$\eta = A^T(AA^T + \lambda I)^{-1}\delta y$$

The method was firstly tested with data simulated using finite elements. The reconstructed objects are presented in Figure 17.7. Then, the algorithm was applied to experimental data. The considered phantom ($64 \times 64 \times 10$ mm) had background optical parameters of: $\mu_a = 0.012$ mm^{-1}, $\mu_s' = 0.81$ mm^{-1}. At the center of the phantom, a 3 mm diameter cylindrical inclusion was positioned, with eight times absorption than the background. Reconstruction time took approximately 1 s on a Quad Core 2.4 GHz and 2 GB RAM. The results are

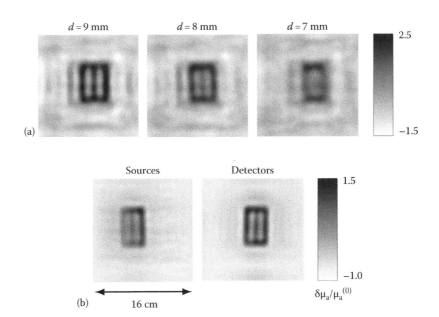

FIGURE 17.5   Reconstructions of bar targets. (From Konecky, S. et al., *Opt. Express*, 16, 5048, 2008. With permission.)

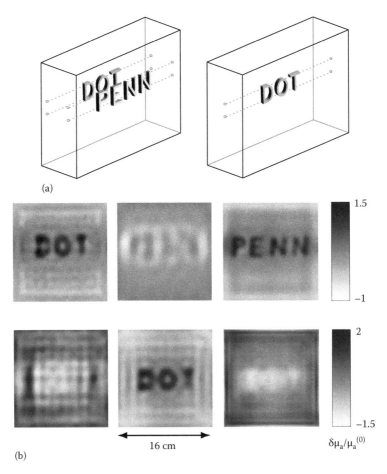

FIGURE 17.6   Reconstructions of a complex phantom. (From Konecky, S. et al., *Opt. Express*, 16, 5048, 2008. With permission.)

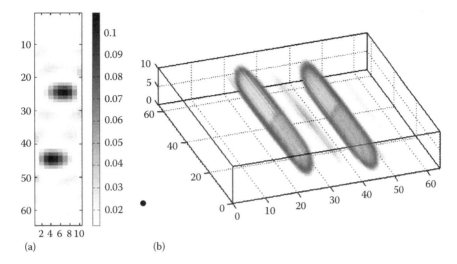

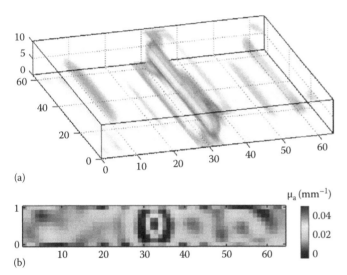

FIGURE 17.7 Reconstructed absorption coefficient with simulated data: profile at the central plane of the phantom (a) and 3D reconstruction (b).

FIGURE 17.8 Reconstructed absorption coefficient with experimental data: profile at the central plane of the phantom (b) and 3D reconstruction (a).

shown in Figure 17.8. The resulting absorption coefficient is underestimated and artifacts are present around the inclusion, which may be partly due to the fact that the data have not been calibrated in amplitude. These results demonstrate that by acquiring a small number of spatial frequencies and by using a reconstruction algorithm working in the Fourier domain, objects can be accurately and quickly localized in three dimensions inside scattering media.

17.5 Summary

In this chapter, we summarized methods for optical tomography reconstruction using both scattering theory and optimization methods. Lack of space precludes the discussion of further topics of interest which include shape-based methods, refractive index variation, anisotropy, and mutlimodality imaging. For further details, we refer to Arridge and Schotland (2009).

References

Abdoulaev, G. S. and Hielscher, A. H. 2003. Three-dimensional optical tomography with the equation of radiative transfer. *J. Electron. Imag.* 12(4): 594–601.

Abdoulaev, G. S., Ren, K., and Hielscher, A. H. 2005. Optical tomography as a PDE-constrained optimisation problem. *Inverse Probl.* 21: 1507–1530.

Acar, R. and Vogel, C. R. 1994. Analysis of bounded variation penalty methods for ill-posed problems. *Inverse Probl.* 10: 1217–1229.

Buchan, A. G., Pain, C., Eaton, M., Smedley-Stevenson, R., and Goddard, A. 2005. Linear and quadratic octahedral wavelets on the sphere for angular discretisations of the Boltzmann transport equation. *Ann. Nucl. Energy* 32(11): 1224–1273.

Aliabadi, M. H. 2002. *The Boundary Element Method*. Chichester, U.K.: Wiley.

Arridge, S. R., Kaipio, J. P., Kolehmainen, V. et al. 2006. Approximation errors and model reduction with an application in optical diffusion tomography. *Inverse Probl.* 22(1): 175–196.

Arridge, S. R. and Lionheart, W. R. B. 1998. Non-uniqueness in diffusion-based optical tomography. *Opt. Lett.* 23: 882–884.

Arridge, S. R. and Schotland, J. 2009. Optical tomography: Forward and inverse problems. *Inverse Probl.* 25(12): 123010 (59 pp).

Arridge, S. R. and Schweiger, M. 1998. A gradient-based optimisation scheme for optical tomography. *Opt. Express* 2(6): 213–226.

Aydin, E. D., de Oliveira, C., and Goddard, A. J. H. 2002. A comparison between transport and diffusion calculations using a finite element-spherical harmonics radiation transport method. *Med. Phys.* 2(9): 2013–2023.

Bangerth, W. and Joshi, A. 2008. Adaptive finite element methods for the solution of inverse problems in optical tomography. *Inverse Probl.* 24(3): 034011 (22 pp).

Bassi, A., D'Andrea, C., Valentini, G., Cubeddu, R., and Arridge, S. 2008. Temporal propagation of spatial information in turbid media. *Opt. Lett.* 33(23): 2339–2341.

Bassi, A., D'Andrea, C., Valentini, G., Cubeddu, R., and Arridge, S. 2009. Detection of inhomogeneities in diffusive media using spatially modulated light. *Opt. Lett.* 34: 2156–2158.

Becker, A. A. 1992. *The Boundary Element Method in Engineering.* Maidenhead, U.K.: McGraw-Hill.

Beer, G. 2001. *Programming the Boundary Element Method. An Introduction for Engineers.* Chichester, U.K.: John Wiley & Sons.

Boas, D. A., Culver, J. P., Stott, J. J., and Dunn, A. K. 2002. Three dimensional Monte Carlo code for photon migration through complex heterogenous media including the adult human head. *Opt. Express* 10: 159–170.

Bouman, C. A. and Sauer, K. 1993. A generalised Gaussian image model for edge-preserving MAP estimation. *IEEE Trans. Image Process.* 2(3): 296–310.

Calvetti, D. and Somersalo, E. 2007. *Introduction to Bayesian Scientific Computing.* New York: Springer.

Case, M. C. and Zweifel, P. F. 1967. *Linear Transport Theory.* New York: Addison-Wesley.

Corlu, A., Choe, R., Durduran, T. et al. 2005. Diffuse optical tomography with spectral constraints and wavelength optimisation. *Appl. Opt.* 44(11): 2082–2093.

Corlu, A., Durduran, T., Choe, R. et al. 2003. Uniqueness and wavelength optimization in continuous-wave multispectral diffuse optical tomography. *Opt. Lett.* 28: 23.

Cuccia, D. J., Bevilacqua, F., Durkin, A. J., and Tromberg, B. J. 2005. Modulated imaging: quantitative analysis and tomography of turbid media in the spatial-frequency domain. *Opt. Lett.* 30: 1354.

Dorn, O. 1998. A transport-backtransport method for optical tomography. *Inverse Probl.* 14(5): 1107–1130.

Douiri, A., Schweiger, M., Riley, J., and Arridge, S. R. 2005. Local diffusion regularization method for optical tomography reconstruction by using robust statistics. *Opt. Lett.* 30: 2439–2441.

Geman, S. and Geman, D. 1984. Stochastic relaxation, Gibbs distributions, and the Bayesian restoration of images. *IEEE Trans. Pattern Anal.* 6(6): 721–741.

Golub, G. H. and van Loan, C. F. 1983. *Matrix Computations.* Baltimore, MD: John Hopkins University.

Gonzalez-Rodriguez, P. and Kim, A. D. 2009. Comparison of light scattering models for diffuse optical tomography. *Opt. Express* 17: 8756–8774.

Guven, M., Yazici, B., Kwon, K., Giladi, E., and Intes, X. 2007a. Effect of discretization error and adaptive mesh generation in diffuse optical absorption imaging: I. *Inverse Probl.* 23: 1115–1133.

Guven, M., Yazici, B., Kwon, K., Giladi, E., and Intes, X. 2007b. Effect of discretization error and adaptive mesh generation in diffuse optical absorption imaging: II. *Inverse Probl.* 23: 1135–1150.

Harrach, B. 2009. On uniqueness in diffuse optical tomography. *Inverse Probl.* 25(5): 055010.

Hebden, J. C., Arridge, S. R., and Delpy, D. T. 1997. Optical imaging in medicine: I. Experimental techniques. *Phys. Med. Biol.* 42: 825–840.

Heiskala, J. 2009. Accurate modeling of tissue properties in diffuse optical imaging of the human brain. PhD thesis, Helsinki University of Technology, Espoo, Finland.

Heiskala, J., Hiltunen, P., and Nissila, I. 2009. Significance of background optical properties, time-resolved information and optode arrangement in diffuse optical imaging of term neonates. *Phys. Med. Biol.* 54(3): 535–554.

Heiskala, J., Nissilä, I., Neuvonen, T., Järvenpää, S., and Somersalo, E. 2005. Modeling anisotropic light propagation in a realistic model of the human head. *Appl. Opt.* 44(11): 2049–2057.

Ishimaru, A. 1978. *Wave Propagation and Scattering in Random Media*, vol. 1. New York: Academic.

Joshi, A., Bangerth, W., Hwan, K., Rasmussen, J. C., and Sevick-Muraca, E. M. 2006a. Fully adaptive fem based fluorescence tomography from time-dependant measurements with area illumination and detection. *Med. Phys.* 33: 1299–1310.

Joshi, A., Bangerth, W., and Sevick-Muraca, E. 2006b. Non-contact fluorescence optical tomography with scanning patterned illumination. *Opt. Express* 14: 6516–6534.

Kaczmarz, S. 1937. Angenäherte auflösung von systemen linearer gleichungen. *Bull. Internat. Acad. Polon. Sci. Lett. A* 35: 335–357.

Kaipio, J. and Somersalo, E. 2005. *Statistical and Computational Inverse Problems.* New York: Springer.

Kanschat, G. 1998. A robust finite element discretization for radiative transfer problems with scattering. *East-West J. Numer. Math.* 6(4): 265–272.

Kim, H. K. and Charette, A. 2007. A sensitivity function-based conjugate gradient method for optical tomography with the frequency-domain equation of radiative transfer. *J. Quant. Spectrosc. Radiat. Transf.* 104: 24–39.

Kim, H. K. and Hielscher, A. H. 2009. A PDE-constrained SQP algorithm for optical tomography based on the frequency-domain equation of radiative transfer. *Inverse Probl.* 25(1): 015010.

Klose, A. D. and Hielscher, A. H. 1999. Iterative reconstruction scheme for optical tomography based on the equation of radiative transfer. *Med. Phys.* 26: 1698–1707.

Klose, A. D. and Hielscher, A. H. 2002. Optical tomography using the time-independent equation of radiative transfer - Part 2: Inverse model. *J. Quant. Spectrosc. Radiat. Transf.* 72: 715–732.

Kolehmainen, V., Arridge, S. R., Lionheart, W. R. B., Vauhkonen, M., and Kaipio, J. P. 1999. Recovery of region boundaries of piecewise constant coefficients of an elliptic PDE from boundary data. *Inverse Probl.* 15: 1375–1391.

Konecky, S., Panasyuk, G. Y., Lee, K., Markel, V., Yodh, A. G., and Schotland, J. C. 2008. Imaging complex structures with diffuse light. *Opt. Express* 16: 5048–5060.

Kwon, K., Yazici, B., and Guven, M. 2006. Two-level domain decomposition methods for diffuse optical tomography. *Inverse Probl.* 22: 1533–1559.

Larsen, E. and Keller, J. 1974. Asymptotic solution of neutron-transport problems for small mean free paths. *J. Math. Phys.* 15: 75–81.

Lathrop, K. D. 1968. Ray effects in discrete ordinates equations. *Nucl. Sci. Eng.* 32: 357–369.

Lathrop, K. D. 1971. Remedies for ray effects. *Nucl. Sci. Eng.* 45: 255–268.

Lee, J., Joshi, A., and Sevick-Muraca, E. 2007. Fully adaptive finite element based tomography using tetrahedral dual-meshing for fluorescence enhanced optical imaging in tissue. *Opt. Express* 15: 6955–6975.

Liu, D. C. and Nocedal, J. 1989. On the limited memory BFGS method for large scale optimization. *Math. Prog. B* 45(3): 503–528.

Markel, V., O'Sullivan, J., and Schotland, J. 2003. Inverse problem in optical diffusion tomography. IV. Nonlinear inversion formulas. *J. Opt. Soc. Am. A* 20: 903–912.

Markel, V. and Schotland, J. C. 2001a. Inverse problem in optical diffusion tomography. I. Fourier-Laplace inversion formulas. *J. Opt. Soc. Am. A* 18: 1336–1347.

Markel, V. and Schotland, J. C. 2001b. Inverse scattering for the diffusion equation with general boundary conditions. *Phys. Rev. E [Rapid Commun.]* 64: 035601.

Markel, V. and Schotland, J. C. 2002a. Effects of sampling and limited data in optical tomography. *App. Phys. Lett.* 81: 1180–1182.

Markel, V. and Schotland, J. C. 2002b. Inverse problem in optical diffusion tomography. II. Role of boundary conditions. *J. Opt. Soc. Am. A* 19: 558–566.

Markel, V. and Schotland, J. C. 2002c. Scanning paraxial optical tomography. *Opt. Lett.* 27: 1123–1125.

Milstein, A. B., Oh, S., Webb, K. J. et al. 2003. Fluorescence optical diffusion tomography. *Appl. Opt.* 42(16): 3081–3094.

Moskow, S. and Schotland, J. 2008. Convergence and stability of the inverse scattering series for diffuse waves. *Inverse Probl.* 24: 065005 (16 pp).

Natterer, F. and Wübbeling, F. 2001. *Mathematical Methods in Image Reconstruction.* Philadelphia, PA: SIAM.

Neil, M. A. A., Juskaitis, R., and Wilson, T. 1997. Method of obtaining optical sectioning by using structured light in a conventional microscope. *Opt. Lett.* 22: 19057.

Nocedal, J. and Wright, S. J. 1999. *Numerical Optimization.* New York: Springer Verlag.

O'Leary, M. A., Boas, D. A., Chance, B., and Yodh, A. G. 1995. Experimental images of heterogeneous turbid media by frequency-domain diffusing-photon tomography. *Opt. Lett.* 20: 426–428.

Pineda, A. R., Schweiger, M., Arridge, S., and Barrett, H. H. 2006. Information content of data types in time-domain optical tomography. *J. Opt. Soc. Am. A* 12: 2989–2996.

Prahl, S. A., Keijzer, M., Jacques, S. L., and Welch, A. J. 1989. A Monte Carlo model of light propagation in tissue. In:

Dosimetry of Laser Radiation in Medicine and Biology, vol. 5, eds. G. J. Müller and D. H. Sliney. Bellingham, WA: SPIE IS, pp. 102–111.

Qin, C., Tian, J., Yang, X. et al. 2008. Galerkin-based meshless methods for photon transport in the biological tissue. *Opt. Express* 16(25): 20317–20333.

Ren, K., Bal, G., and Hielscher, A. 2006. Frequency domain optical tomography based on the equation of radiative transfer. *SIAM J. Sci. Comput.* 28(4): 1463–1489.

Richling, S., Meinköhn, E., Kryzhevoi, N., and Kanschat, G. 2001. Radiative transfer with finite elements I Basic method and tests. *Astron. Astrophys.* 380(2): 776–788.

Ripoll, J. and Ntziachristos, V. 2006. From finite to infinite volumes: Removal of boundaries in diffuse wave imaging. *Phys. Rev. Lett.* 96(17): 173903 (4 pp).

Ripoll, J., Ntziachristos, V., and Nieto-Vesperinas, M. 2001. The Kirchhoff approximation for diffusive waves. *Phys. Rev. E.* 64: 1–8.

Rudin, L. I., Osher, S., and Fatemi, E. 1992. Nonlinear total variation based noise removal algorithm. *Physica D* 60: 259–268.

Schotland, J. C. and Markel, V. 2001. Inverse scattering with diffusing waves. *J. Opt. Soc. Am. A* 18: 2767–2777.

Schweiger, M., Arridge, S. R., and Nissilä, I. 2005. Gauss-Newton method for image reconstruction in diffuse optical tomography. *Phys. Med. Biol.* 50: 2365–2386.

Sikora, J. and Arridge, S. 2002. Some numerical aspects of 3D BEM application to optical tomography. In: *IV International Workshop Computational Problems of Electrical Engineering*, Zakopane, Poland, pp. 59–62.

Sikora, J., Riley, J., Arridge, S., Zacharopoulos, A., and Ripoll, J. 2004. Light propagation in diffusive media with non-scattering regions using 3D BEM. In: Wilde, S. C. (ed.), *Proceedings of Third International Conference on Boundary Integral Methods: Theory and Applications*, University of Reading, Reading, U.K.

Sikora, J., Zacharopoulos, A., Douiri, A. et al. 2006. Diffuse photon propagation in multilayered geometries. *Phys. Med. Biol.* 51: 497–516.

Soloviev, V. 2006. Mesh adaptation technique for Fourier-domain fluorescence lifetime imaging. *Med. Phys.* 33: 4176–4183.

Soloviev, V., D'Andrea, C., Brambilla, M. et al. 2008. Adjoint time domain method for fluorescent imaging in turbid media. *Appl. Opt.* 47: 2303–2311.

Soloviev, V. and Krasnosselskaia, L. 2006. Dynamically adaptive mesh refinement technique for image reconstruction in optical tomography. *Appl. Opt.* 45: 2828–2837.

Soloviev, V., Tahir, K., McGinty, J. et al. 2007. Fluorescence lifetime imaging by using time gated data acquisition. *Appl. Opt.* 46: 7384–7391.

Steinbrink, J., Wabnitz, H., Obrig, H., Villringer, A., and Rinneberg, H. 2001. Determining changes in NIR absorption using a layered model of the human head. *Phys. Med. Biol.* 46: 879–896.

Strohmer, T. and Vershynin, R. 2006. A randomized solver for linear systems with exponential convergence. *Approximation,*

Randomization, and Combinatorial Optimization. Algorithms and Techniques, Lecture Notes in Computer Science, 4110: 499–507.

Tarvainen, T., Vauhkonen, M., and Arridge, S. R. 2008. Image reconstruction in optical tomography using the finite element solution of the frequency domain radiative transfer equation. *J. Qaunt. Spect. Rad. Transf.* 109: 2767–2278.

Tarvainen, T., Vauhkonen, M., Kolehmainen, V., Arridge, S. R., and Kaipio, J. P. 2005a. Coupled radiative transfer equation and diffusion approximation model for photon migration in turbid medium with low-scattering and non-scattering regions. *Phys. Med. Biol.,* 50: 4913–4930.

Tarvainen, T., Vauhkonen, M., Kolehmainen, V., and Kaipio, J. 2006. Finite element model for the coupled radiative transfer equation and diffusion approximation. *Int. J. Numer. Meth. Eng.* 65(3): 383–405.

Tarvainen, T., Vauhkonen, M., Kolehmainen, V., and Kaipio, J. P. 2005b. A hybrid radiative transfer - diffusion model for optical tomography. *Appl. Opt.,* 44(6): 876–886.

Vogel, C. R. and Oman, M. E. 1996. Iterative methods for total variation denoising. *SIAM J. Sci. Comp.,* 17: 227–238.

Wang, L., Jacques, S. L., and Zheng, L. 1995. MCML-Monte Carlo modeling of light transport in multi-layered tissues. *Comput. Methods Programs Biomed.* 47: 131–146.

Wang, Z., Panasyuk, G., Markel, V., and Schotland, J. 2005. Experimental demonstration of an analytic method for image reconstruction in optical diffusion tomography with large data sets. *Opt. Lett.* 30: 3338–3340.

Ye, J. C., Bouman, C. A., Webb, K. J., and Millane, R. P. 2001. Nonlinear multigrid algorithms for Bayesian optical diffusion tomography. *IEEE Trans. Image Process.* 10(5): 909–922.

Ye, J. C., Webb, K. J., Bouman, C. A., and Millane, R. P. 1999. Optical diffusion tomography by iterative coordinate-descent optimization in a Bayesian framework. *J. Opt. Soc. Am. A* 16(10): 2400–2412.

Zacharopoulos, A. D., Svenmarker, P., Axelsson, J., Schweiger, M., Arridge, S. R., and Andersson-Engels, S. 2009. A matrix-free algorithm for multiple wavelength fluorescence tomography. *Opt. Express* 17: 3042–3051.

18

Laminar Optical Tomography

18.1 Introduction 359
18.2 Background..359
    Optical Contrast in the Body • Basic Principles of LOT •
    LOT Instrumentation • Modeling • Uses of Modeling for LOT
18.3 3D Image Reconstruction... 362
18.4 Optical Property and Basic Geometry Determination........................ 364
    Types of Modeling
18.5 Monte Carlo Modeling for LOT 365
18.6 Presentation of State of the Art 366
    Fluorescence LOT • Simultaneous Multiwavelength LOT • Combined OCT and
    Line-Scan LOT
18.7 Critical Discussion 369
18.8 Summary...370
18.9 Future Perspectives...370
References..371

Sean A. Burgess
Columbia University

Elizabeth M. C. Hillman
Columbia University

18.1 Introduction

In microscopy, light is minimally scattered allowing high-resolution imaging but with shallow penetration depths (a few hundred micrometers). While this is suitable for imaging thin tissue sections ex vivo, it can be limiting for in vivo imaging of intact tissues. In contrast to microscopy, diffuse optical tomography (DOT) is a technique more similar to x-ray computed tomography (CT), which utilizes light to image large tissue volumes such as the breast or brain (Arridge 1999). DOT relies on a model of light scattering within the tissue to compensate for the stochastic nature of light propagation in turbid media. The uncertainty of the photon paths manifests as blurring and uncertainty in the reconstructed images, such that resolutions between 5 and 10 mm are typical. In between these two extremes lie a wide range of mesoscopic optical imaging techniques that aim to harness the value of optical contrast while accommodating the effects of scattering and maximizing resolution and penetration depth. Several technologies for mesoscopic imaging have been developed in recent years. In this chapter, we examine laminar optical tomography (LOT), a mesoscopic imaging technique that measures multiply scattered light providing absorption and fluorescence contrast. The measurements collected by LOT comprise a dense tomography-like data set covering the spatial regime between that covered by microscopic and diffuse techniques. LOT was first implemented in 2004 (Hillman et al. 2004) where its use was demonstrated by imaging rat cortex through a thinned skull (Hillman et al. 2007b). Since then,

several advances have been made to the technique incorporating fluorescence (Hillman et al. 2007a, Yuan et al. 2009a) and multispectral imaging (Burgess et al. 2008) and imaging various tissues in vivo.

18.2 Background

18.2.1 Optical Contrast in the Body

A practical problem associated with optical imaging is the strong attenuation of light in biological tissues. This effect prevents whole body examinations, as is readily achievable using x-ray and magnetic resonance imaging. Optical techniques become more practical as the tissue thickness decreases, since attenuation is a function of the distance that the light travels in the tissue. Light attenuation in tissue is the result of two optical phenomena: absorption and scattering. During absorption, the energy of a photon is taken up by matter thereby reducing the intensity of the transmitted light. This effect is wavelength dependent, yielding absorption spectra that are characteristic of the constituents of the media. In principle, the measurement of the absorption spectrum of tissue allows for the identification of its individual absorbers. The higher the concentration of an absorber, the stronger its relative contribution to the overall absorption properties of the tissue. There are many chromophores in the body that can absorb visible light, with the three major chromophores being oxygenated hemoglobin (HbO$_2$), deoxygenated hemoglobin (HbR), and melanin. Figure 18.1

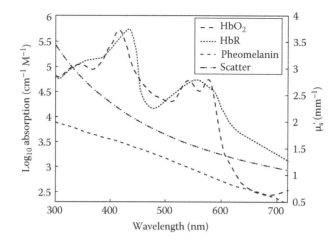

FIGURE 18.1 The three major absorbers of light in tissue are oxy- and deoxy-hemoglobin (HbO$_2$ and HbR) and melanin (shown on log scale, from (Prahl, Prahl)). Differences in absorption at a given wavelength can be exploited using optical techniques to generate contrast. Reduced scattering coefficient (linear right-hand axis) is estimated as $s' = A - b$ where $A = 1.14 \times 10^{-8}$ and $b = 1.3$ for tissue (Corlu et al. 2005) is also shown. (Based on data from Corlu, A. et al., *Appl. Opt.*, 44(11), 2082, 2005.)

shows the absorption spectra for these chromophores. Oxy- and deoxy-hemoglobin found in blood are highly absorbing in the visible wavelengths while melanin found in the epidermis of skin is responsible for absorbing harmful ultraviolet light. Optical absorption measurements can have important diagnostic applications such as the blood oxygenation measures provided by pulse oximetry (Severinghaus and Astrup 1986). As we will see later in this chapter, LOT is sensitive to absorption contrast in the body.

Upon absorbing a photon, some molecules emit a lower-energy (longer wavelength) photon resulting in another source of contrast: fluorescence. Like absorption, fluorescence is wavelength dependent. Fluorophores have characteristic excitation/emission spectra referring to the wavelengths of the absorbed light and wavelengths of the resulting emitted light. Many substances within the body, when excited with the proper wavelength, will fluoresce including elastin, keratin, tryptophan, nicotinamide adenine dinucleotide (NADH), and flavin adenine dinucleotide (FAD) (Ramanujam 2000). In addition to these endogenous sources of fluorescence contrast, exogenous optical contrast agents can also be employed and are continually being developed (Miyawaki et al. 2003, Tsien 1998, Zhang et al. 2002). Optical dyes are capable of enhancing contrast or providing contrast to molecules and structures that would not otherwise be visualized.

Scattering of light can also provide contrast in living tissues since light scattering is a function of the size, density, and shape of the particles causing the scattering as well as the wavelength of light (see Figure 18.1). When light undergoes scattering events, photons are redirected such that the intensity of the incident light is reduced. Photons are scattered most strongly by structures of the same size as the optical wavelength with large refractive index mismatches making cell nuclei and mitochondria the main scatterers in tissue (Mourant et al. 1998). Scattering also represents the most significant obstacle for in vivo optical imaging,

as scattering leads to the loss of focus of light at depth within a tissue, and uncertainty in optical paths. While LOT does not explicitly utilize scattering contrast for imaging, as we will see in the next section, it does exploit tissue scattering to acquire depth-sensitive measurements. With an understanding of the contrast yielding light interactions in tissue, we now shift our attention to how LOT exploits these phenomena to generate images.

18.2.2 Basic Principles of LOT

LOT measurements rely on scanning a point source of light, and detecting light that is emitted from the tissue close to each source position (from 0 to 3 mm away). The position to which light is delivered is referred to as the source and the position from which light is collected from is considered the position of a detector. The wider the distance between the source position and the detector position, the deeper on average the detected light has traveled (see Figure 18.2). As the source position is scanned across the surface of the sample, for each position several detector measurements are made. The resulting tomography-like data

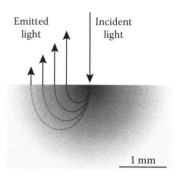

FIGURE 18.2 Upon entering the sample, light is multiply scattered. Light emitting from further distance from the source position will, on average, have traveled more deeply into the sample.

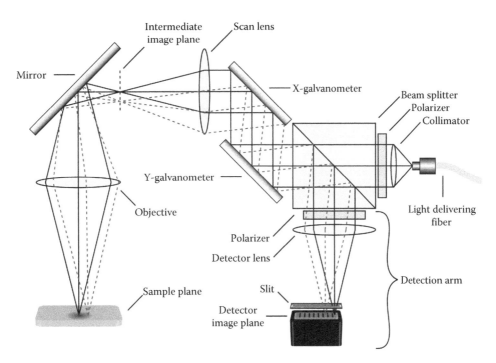

FIGURE 18.3 Design of an LOT system: The incident light is scanned using galvanometer mirrors, which also de-scan the returning light. The photodetector array is positioned in a plane conjugate to the sample plane. Different elements in the array detect light emerging from the sample at different distances from the scanning spot.

set contains information about the depth-resolved properties of the tissue. Whereas other microscopy techniques, such as confocal microscopy and optical coherence tomography (OCT), reject scattered light, LOT relies on scattering to collect off-axis backscattered light measurements.

Since LOT does not reject scattered light, there is no explicit "scattering limit" as encountered in confocal microscopy. Imaging to tissue depths beyond those attainable with microscopy can feasibly be achieved if wider source-detector (S-D) separations are used, and lower image resolution is acceptable. In the limit of wider S-D separations, the measurements that LOT acquires are equivalent to reflectance-geometry DOT measurements (Arridge 1999, Zeff et al. 2007). The two techniques differ in that with the narrower S-D separations used in LOT, the detected light has been scattered and attenuated less, and can therefore yield higher resolution images with better signal to noise. In addition, LOT instrumentation differs from DOT, which typically uses optical fibers positioned in a grid pattern on the tissue's surface (Franceschini et al. 2000, Schmitz et al. 2005, Zeff 2007). LOT uses noncontact laser scanning, allowing rapid-frame-rate, multispectral imaging with an adaptable field of view and measurement density. The basic design of LOT instrumentation is described next.

18.2.3 LOT Instrumentation

LOT instrumentation uses a measurement geometry that is similar to that used in laser scanning confocal microscopy. A focused laser beam is scanned over the sample plane. The galvanometer

mirrors used to scan the beam also act to de-scan the backscatter emitted light. The de-scanned light is detected by a linear array photomultiplier tube (PMT) positioned in a conjugate image plane. The basic layout for absorption contrast measurements is shown in Figure 18.3.

The incident light is shown as solid lines originating from an optical fiber. The optical fiber transmits light from the laser to the instrumentation optics. The light is collimated and passed through a beam splitter before being reflected by a set of galvanometer mirrors. These mirrors are electronically controlled to raster scan the beam in *x* and *y* directions. The beam then passes through a scan lens, which converts the angular deviation of the collimated light into a lateral translation of the scanning spot. The raster scanning spot at the intermediate image plane of the scan lens can then be imaged onto the sample plane using an objective lens arrangement. On a scattering sample, the source spot will appear to have a bright center, with a glow of light around it corresponding to the light that has entered and scattered deeper and more laterally within the sample before emerging from the surface.

Backscattered light originating from the center of the spot on the sample will travel back through the optical system with the same path as the incident light. However, upon reaching the beam splitter, the returning light is reflected toward the detection arm of the system. The detection arm focuses the returning light to the detector plane (conjugate to the sample plane) where a slit and photodetector array are positioned to measure the intensity of the returning light. The use of a slit is analogous to the pinhole used in confocal microscopy. However, rather than rejecting off-axis light, LOT measures this light with the detector array. As the galvanometer mirrors deflect the incident light, causing it to pan back and forth over the tissue, the light traces the same path back through the system, being de-scanned by the galvanometer mirrors such that the focus of the scanning spot will always be imaged back to the center of the detection plane. In essence, the direction change that the galvanometers cause in the forward direction is undone when the light reflects off them on the return path.

Off-axis backscattered light (Figure 18.3, dashed line) follows the path of light emerging from the sample at a position laterally displaced from the focus of the incident beam. This light also travels back through the system and is de-scanned by the galvanometer mirrors. As the light reflects off the beam splitter, it focuses to a point adjacent to the focal point of light from the center of the scanning beam. Effectively, the system creates an image of the scanning spot at the detection plane, consisting of both the spot's bright center and its more diffuse surround. As with the confocal light, for all scanning source positions, the light emerging from the tissue at some distance away from the spot's focus will be imaged onto a detector that is similarly displaced from the center of the scanning spot in the detector plane. It is this scattered surround of light that provides information about the deeper layers of the tissue. Each detector in the linear array therefore separates light emerging at specific distance from the focus.

Careful lens selection allows LOT to be configured for a range of S-D separation increments and fields of view. The system illustrated in Figure 18.3 is drawn with 1× magnification. The scan lens and detector lens have the same focal lengths, and the objective lens is a single lens creating 1× magnification. In this configuration, the physical separation between the detector channels on the detector array will equal the absolute distance between the source and detector positions at the tissue. If the magnification of the objective lens is changed, both the field of view, and the relative separations between the source and detector positions will vary. If only the focal length of the detector lens is changed, the separation spacing between the source and detectors can be changed independently of the field of view. A 4× magnification on the detection side of the system maps a 1 mm pitch between detector array elements to a 250 μm separation at the tissue surface. Changing the focal length of the scan lens will change both the S-D separations and the field of view for a given angular deviation of the galvanometer mirrors.

Other important considerations for the system are the use of polarization and the method of separating the returning backscattered light from the incident light. A polarization-maintaining fiber is used to deliver horizontally p-polarized laser light to the system. This light is passed through a polarizer before reaching the sample plane, ensuring that only p-polarized light reaches the tissue. As the light scatters within the tissue, it gradually loses its polarization, such that emerging light contains both s and p-polarized light. Specular reflections occurring at the surface of the tissue, or from lenses in the system, maintain their p-polarization and can be detrimental to image quality.

By placing a second polarizer in the detection arm, rotated 90° relative to the first polarizer, returning p-polarized light is rejected. This effectively blocks specular reflections, while also preferentially selecting photons that have scattered more deeply into this tissue. This technique of light detection comes at the expense of the amount of available detected light. However, the use of a polarizing beam splitter cube can significantly improve light efficiency, since almost 100% of the p-polarized laser light will be transmitted to the sample, and 50% of returning light (100% of s-polarized light) will be reflected toward the detector plane, such that the polarizers in Figure 18.3 are required only to "clean up" the already polarized light. If a 50:50 nonpolarizing beam splitter were used instead, 50% of incoming light would be lost, as would 50% of the returning light. The detection arm polarizer would discard an additional 50% of the returning light if p-polarized reflections needed to be blocked.

18.2.4 Modeling

A critical aspect of LOT is mathematical modeling of light propagation. The fact that light emerges from the same surface of the tissue as it enters is because it has scattered in a random path. In a homogenous scattering medium, light detected at further distances from the point where it entered will have traveled, on average, more deeply into the medium than light that emerges close to the point of incidence. This can be predicted using mathematical models that simulate the way that light is likely to travel in the tissue, assuming that the scattering and absorbing properties of the medium, as well as the illumination and detection geometry are known (Dunn and Boas 2000, Wang et al. 1995, Arridge 1999). More complex geometries and heterogeneous tissues can also be investigated, although the accuracy of the model will depend on the amount of a priori information available.

18.2.5 Uses of Modeling for LOT

Modeling can be used to analyze LOT data in a variety of ways, from prediction of bulk optical properties to 3D image reconstruction as described in the following.

18.3 3D Image Reconstruction

Conventional tomographic image reconstruction creates a cross-sectional image from a series of projection measurements through an object. In x-ray CT, several hundred projections are acquired at all angles through the object (Kak and Slaney 1987). Because of light scattering, it is possible to acquire a set of projections through an object in reflectance geometry when using optical techniques, as illustrated in Figure 18.2. A series of overlapping projections that sample across the surface, with different S-D separations probing different depths, can therefore be used to reconstruct a 3D image of objects within the scattering tissue in an analogous way to x-ray CT.

In both x-ray CT and LOT, knowledge of the sensitivity distribution of each of the projection measurements is required.

In x-ray CT, this is relatively straightforward since detected x-rays can be restricted to those that have passed through the body in straight lines at a known series of angles. However, with LOT, the sensitivity of each measurement path is a function of the optical scattering and absorption properties of the tissue being probed. In addition to estimates of the background properties of the tissue, it is necessary to model the likely paths that detected light has taken through the tissue. To date, Monte Carlo modeling methods have been used to simulate these so-called sensitivity functions $J_{s,d}(r)$ that correspond to maps of the likely change in measurement $\Delta M_{s,d}$ between a source and detector position s,d that will result from a unit change in absorption $\Delta\mu_a(r)$ at position r:

$$J_{s,d}(r) = \frac{\partial M_{s,d}}{\partial\mu_a(r)} \qquad (18.1)$$

which can be simplified to

$$\Delta M_{s,d} = J_{s,d}(r)\Delta\mu_a(r) \qquad (18.2)$$

in cases where the change in absorption $\Delta\mu_a(r)$ is small. Assuming that $J_{s,d}(r)$ accurately represents the physical measurements from LOT, Equation 18.2 can be inverted using a range of regularization approaches to produce images of $\Delta\mu_a(r)$ (Arridge 1999). Compared to x-ray CT methods, optical tomographic image

reconstructions are less able to completely describe the structure of an object owing to the inherent uncertainty in the paths that photons have traveled. This uncertainty results in poorer quantitation and resolution than for x-ray imaging, in addition to spatially dependent resolution.

In its simplest form, LOT image reconstruction can assume that the tissue being imaged is homogenous, its optical properties well known, and that the image to be reconstructed is of a small change in absorption $\Delta\mu_a(r)$ such that measurements $\Delta M_{s,d}$ correspond to the difference between measurements before and after that change. In this case, a fairly simple Monte Carlo–based forward model can be used to generate $J_{s,d}(r)$, and Tikhonov regularization or other iterative solvers can be used to invert Equation 18.2. Images reconstructed in this way are shown in Figure 18.4 and in the following publications (Hillman et al. 2004, 2006, 2007a,b, Yuan et al. 2009b).

More advanced image reconstruction approaches could include using nonlinear image reconstruction schemes that attempt to accommodate background optical property heterogeneity by remodeling light propagation based on changing estimates of the baseline (Arridge 1999, Hielscher et al. 1999, Hillman et al. 2000, 2001). Iterative nonlinear reconstructions have been developed for DOT; however, implementation is greatly facilitated by the availability of analytic solutions to the diffusion approximation, which enable rapid finite element and finite difference-based forward modeling (Arridge et al. 1992). As described in the following, the diffusion approximation is not

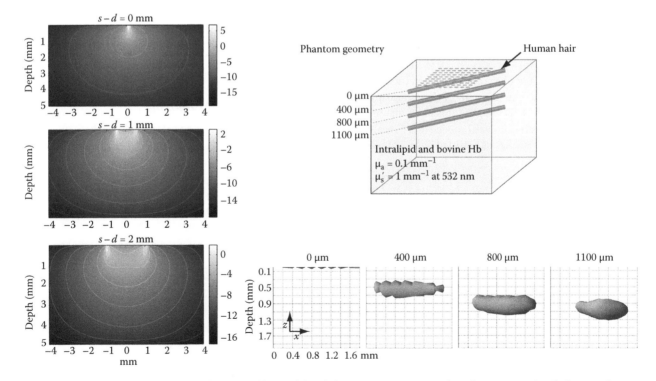

FIGURE 18.4 Left: sensitivity functions ($J_{s,d}$) generated by modeling light propagation in a sample with $\mu_a = 0.1\,\text{mm}^{-1}$ and $\mu'_s = 1\,\text{mm}^{-1}$ at 532 nm. LOT measurements ($\Delta M_{s,d}$) were made on a phantom (top right) consisting of human hair incrementally positioned to different depths in a background medium. Tikhonov regularization was used to generate a 3D reconstruction of $\Delta\mu_a(r)$ from $\Delta M_{s,d}$ and $J_{s,d}$ (bottom right). (Adapted from Hillman, E.M.C. et al., *Opt. Lett.*, 29, 1650, 2004; Hillman, E.M.C. et al., *Proc. SPIE (Medical Imaging)*, 6143, 61431M, 2006.)

valid for LOT data, and Monte Carlo modeling is typically too slow for implementation within iterative nonlinear reconstructions. Furthermore, to date, LOT has only been implemented for continuous wave (CW) measurements, which in general cannot accurately distinguish between the effects of absorption and scattering (Arridge and Lionheart 1998). This makes it unfeasible, at present, to develop an "absolute" image reconstruction scheme that can predict and account for both scattering and absorbing baseline heterogeneities in tissue using LOT.

18.4 Optical Property and Basic Geometry Determination

An alternative approach to a full 3D image reconstruction is to utilize models of light propagation to perform simplified analysis of raw LOT data. This approach is valuable in situations where a full 3D image reconstruction is not required, but simple parameters relating to a tissue's optical properties or structure are sought. For example, Figure 18.5 shows a phantom with three embedded absorbing objects intended to mimic melanotic skin lesions at different depths below the surface. In skin cancer screening, determining the so-called Breslow thickness or invasion depth of a tumor is a critical prognostic factor that can determine how treatment should proceed. LOT data were acquired by scanning the focused laser beam over a 20 mm square area in a 200×200 position grid. Emerging light was detected at seven different distances from the beam's focus (between 0 and 1.75 mm), generating seven 2D "raw data" images. The measurements at every point in these images can be modeled. If the same *x–y* pixel from each of these images is selected, it is possible to plot the intensity of remitted light as a function of S-D separation, and the shape of this curve is characteristic of both the absorbing properties of the object, its depth, and its thickness (see Figure 18.5). Rather than using all of these measurements to attempt to reconstruct a 3D image of this phantom, instead a series of Monte Carlo–based forward models were created that predicted the relative intensities of remitted light that would be

expected at each S-D separation for a wide range of different object absorption coefficients and depths. Using least-squares fitting, it was possible to match the raw data plots extracted from the LOT raw data images to the predicted values in the simulated lookup table. This approach provided correct prediction of absorber depth and prediction of the absorption coefficient to within 20% accuracy.

While CW measurements have previously been demonstrated to be insufficient for DOT image reconstruction of both absorbing and scattering properties (Arridge and Lionheart 1998), LOT does hold the potential to allow characterization of the optical properties of certain simple superficial tissues. Standard double integrating sphere measurements rely on the complementarity of optical measurements made in transmission and reflectance geometries, whereby increase absorption will attenuate both measurements, but increased scattering will increase reflectance and decrease transmittance (Patterson et al. 1991). As demonstrated by Bevilacqua et al. (1999), measurements made in LOT-like geometries capture elements of both reflectance and transmittance, since very short S-D separations will see increases in light detected in more highly scattering media, whereas more distant detectors are more similar to transmittance measurements and would decrease. As such, it is feasible that if sufficient multi-distance measurements are made on a medium with homogenous optical properties, that the relative intensities at each separation could uniquely describe both the absorbing and scattering properties of the tissue (Cuccia et al. 2005). The caveat to this approach is that as the geometry of the object becomes more complex and unknown, the proper fitting between a forward model and measured data becomes highly challenging and could result in significant misinterpretation of results.

18.4.1 Types of Modeling

LOT imaging spans length scales from less than 100 μm to several millimeters, and to date imaging has been predominantly

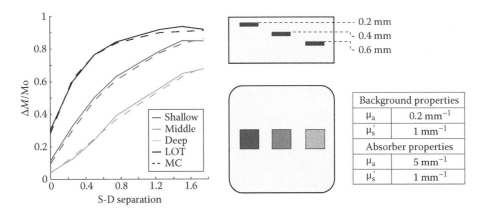

FIGURE 18.5 Simplified analysis of LOT data. A phantom with absorbers embedded at different depths was imaged with LOT. The resulting LOT data was compared to data simulated by a forward model (left) to deduce the depth of the absorbers and predict the absorption coefficient of the absorber to within 20%. (Adapted from Burgess, S.A. et al., *Opt. Lett.*, 33, 2710, 2008.)

performed using visible light wavelengths between 470 and 650 nm. The typical absorption coefficients (μ_a) of tissue within this wavelength range are between 0.1 and 0.5 mm^{-1}, and reduced scattering coefficients (μ_s') are between 0.5 and 2 mm^{-1} (corresponding to anisotropy g between 0.75 and 0.9 and scattering coefficients μ_s between 5 and 20 mm^{-1}). Light propagation in scattering tissue can be modeled by the radiative transport equation (RTE). However, solving this equation has proved challenging until very recently, making it difficult to use effectively for routine image reconstruction. A simplification of the RTE is the diffusion approximation, which neglects information about the directionality of scattering by scaling the scattering coefficient according to the anisotropy of scatter to create the reduced scattering coefficient such that $\mu_s' = \mu_s(1-g)$, and assuming that all scattering events are isotropic. The diffusion approximation imposes limitations on the applicability of the model, including the requirement that μ_s' is much greater than μ_a and that the distances being considered are much greater than $1/\mu_s'\mu_s'$. Neither of these conditions are true for LOT and therefore the diffusion approximation to the RTE is not a suitable model for this application. Instead LOT must utilize a model that can account for the directionality of scattering of light on small length scales, leaving the option to either utilize the complete RTE for modeling, or to employ Monte Carlo modeling. While analytic solutions to the RTE are now becoming available, they have yet to be applied to LOT image reconstructions (Gonzalez-Rodriguez et al. 2007, Gonzalez-Rodriguez and Kim 2009). As such, approaches to Monte Carlo modeling are most well developed for LOT, as described in the following.

18.5 Monte Carlo Modeling for LOT

Monte Carlo modeling has been widely used for optical imaging and spectroscopy applications (Wang et al. 1995). Monte Carlo modeling involves setting up a geometry with a particular source configuration and optical properties, and then launching individual photons into the medium and allowing them to propagate until they interact with a scatterer. The distance traveled between scattering events is randomly selected from a range of lengths based on the density of scatterers given by μ_s. Upon scattering, the new direction of propagation is again randomly assigned, but weighted by the scattering anistotropy of the medium g. The photon is then allowed to continue until it meets another scatterer, and then another until it exits the tissue. The effects of absorption can be applied to the model if the pathlength traveled by the photon is logged at each point within the volume. While in actuality, a photon can either exist or not, rather than annihilating photons based on the likelihood of absorption having occurred, it is typical to reduce the "weight" of the photon to a number less than 1 as absorbing events occur. This allows the net probability of the photon existing to be determined with fewer launched photons.

The main drawback of Monte Carlo modeling of photon migration is the extensive computational burden. Variations to the model, including the use of photon weight reduction rather than termination as described earlier, can be used to improve the modeling speed. The development of faster computer processors and utilization of parallel computing with graphics processors (Alerstam et al. 2008) are also making Monte Carlo more feasible.

Modeling can be used to predict LOT measurements $M_{s,d}$ by determining which photons leave the object's surface within areas and angular distributions corresponding to each LOT detector. By creating a grid within the volume and recording the pathlength $x_{n,N}$ of detected photons N visiting particular regions r_n of this grid, it is possible to infer the changes in measurement that would result from differences in absorption $\mu_a(r_n)$ within those regions using Beer's law:

$$M_{s,d} = \sum_N e^{-\sum_n \mu_a(r_n)x_{n,N}} \tag{18.3}$$

This kind of modeling does not rely on any linearization assumptions, and can be used to generate "lookup table" data as described earlier.

It is also possible to directly generate $J_{s,d}(r)$ using Monte Carlo modeling by effectively distributing the weight of every detected photon into the grid elements along its path (scaled by the pathlength of the photon's visit within each grid element). The resultant distribution is equal to $J_{s,d}(r)$ under the assumption that changes in absorption $\Delta\mu_a$ are small. However, this approach can only be used for absorption imaging (not fluorescence) and can be slow and noisy for larger S-D separations since all areas of a 3D grid must be frequently visited for smooth distributions.

An aspect that is important to consider for both of the preceding cases is that the majority of Monte Carlo models implemented to date reduce data to a grid whose elements are larger than the scattering length $1/\mu_s'$ of the tissue (approximately 1 mm). Typically, these models only consider a photon to have interacted with a voxel if a scattering event occurred within it. For LOT, we are interested in absorption events that occur over length scales of around 50 μm. In conventional models, unless a photon's scattering event happened to occur within this region, the influence of absorption within this region would not be accounted for if the photon had simply passed through the region without scattering. A move accurate model for LOT needs to account for the behavior of each photon as it passes through every grid element regardless of where its scattering events occur.

Another method to calculate $J_{s,d}(r)$ with much improved computation time and signal to noise is to exploit reciprocity. Reciprocity assumes that a photon traveling from a source position to a particular position in a medium has the same probability of existing as a photon originating from that point within the medium getting back to the source position. It is therefore possible to infer the probability of a photon entering tissue at a source position, visiting a particular location, and then exiting at the detector position by taking the product of the probability of the photon reaching the visited position when incident at the source position, and the probability of it reaching the same point when incident at the detector position. If the geometry has symmetry, only a single forward model of light propagation within the medium is required to derive $J_{s,d}(r)$ for any S-D separation.

Reciprocity is routinely exploited for modeling using the diffusion approximation (Arridge 1999), and in Monte Carlo models that consider volumes with centimeter dimensions (Boas and Dale 2005). However, for LOT, the directionality of the photon is of critical importance. In general, LOT light is injected into tissue with a relatively low NA (e.g., 0.05). This means that the majority of the photons are initially forward facing into the tissue. Similarly, photons that are to be detected must emerge from the tissue within a narrow cone, and so therefore must leave the tissue with a suitable scattering trajectory. As a result, the angularly independent probability of a photon reaching a particular position is not sufficient information; the direction of propagation of that photon is also required. In order to infer the probability that a photon will visit a particular position, the dot product of the relative angles of photons reaching that position from the source position, and from the detector position must be calculated. If properly implemented, this method can improve speed, but requires additional memory to store both the spatial and angular information for each photon (Dunn and Boas 2000, Iranmahboob and Hillman 2008).

The analysis presented earlier has been specific to modeling the effects of absorption changes. Fluorescence modeling requires different considerations. This is because when a photon is absorbed by a fluorophore, it does not retain its directionality and is reemitted isotropically (in any direction) as a photon of lower energy. Therefore, the direction that the photon is traveling in when it interacts with the fluorophore is irrelevant, and the direction at which the returning photon is launched will be random. However, it is still necessary to account for the directionality of the paths taken by incident and detected photons (Hyatt et al. 2008). Fluorescence Monte Carlo models utilizing reciprocity reported to date have not accounted for the directionality of incident and detected light and are therefore unsuitable for LOT modeling (Swartling et al. 2003). An additional aspect that must be accounted for is the need to consider the differing optical properties of the background medium at the incident (excitation) light wavelength, and the detected (emission) wavelength of light from the fluorophore. This means that two different models are required in addition to estimates of the absorption and scattering properties of the medium are required at both wavelengths. Overall, these considerations mean that sensitivity functions $J_{s,d}(r)$ for fluorescence LOT measurements will be different to those derived for absorption changes, and could even have differing depth sensitivities, particularly if fluorescence is present in the background (Hillman et al. 2007a, Hillman and Burgess 2009).

In summary, modeling is a critical aspect of LOT, which can allow both reconstruction of 3D images, and quantification of optical properties and geometrical parameters of measured tissues. Special considerations are required for LOT modeling to account for the much smaller length scales of LOT measurements compared to DOT image reconstruction.

18.6 Presentation of State of the Art

18.6.1 Fluorescence LOT

The basic layout of an LOT system shown in Figure 18.3 is for absorption contrast measurements. If fluorescence measurements are instead desired, the detection arm can be modified by replacing the system's beam splitter with a dichroic, and placing a long-pass cleanup filter in front of the detector. The detector arm configuration used to image the fluorescence of voltage-sensitive dye Di-4-ANEPPS in cardiac tissue in (Hillman et al. 2007a) is shown in Figure 18.6.

The system shown utilizes an excitation wavelength of 532 nm and reflects the returning fluorescence light with a 540 nm short-pass dichroic toward the detector image plane. A linear fiber bundle positioned in the detector image plane was used to couple the returning light into a series of avalanche photodiode detectors, as an alternative to using a linear PMT array. The detectors are simultaneously sampled for each scanning source position. A longpass filter included in the detection arm removes any residual excitation light.

Figure 18.7 shows fluorescence measurements obtained with the LOT system described earlier. A phantom, consisting of a fluorescent rod lowered into an absorbing and scattering background, was constructed. A Monte Carlo–based model of light propagation within the phantom was used to generate measurement sensitivity functions. The raw data are shown for the fluorescent rod positioned at depths of 0, 800, and 1200 μm. With the rod positioned at the surface of the phantom, a bright line appears in the raw data for the narrow S-D separations. When the scanning spot is directly over the fluorescent rod, the detectors closest to the source are exposed to an increase in fluorescence light. As the S-D separation increases, a second bright line appears due to the increase in fluorescence light detected when the source has passed the rod but the detector is positioned over the rod. While the raw data images provide depth-sensitive measurements, true depth sections require 3D image reconstruction as shown in Figure 18.7, bottom. Image reconstruction was performed utilizing Tikhonov regularization.

The growing number of fluorescent dyes available today makes fluorescent LOT appealing for functional imaging of tissues. Fluorescence LOT was used to measure electrical propagations in the heart using the voltage-sensitive dye Di-4-ANEPPS in (Hillman et al. 2007a). Measurements were acquired from the right ventricle of a Langendorf perfused rat heart. Due to the rapidly beating heart, the LOT system scan was modified to perform a rapid line scans triggered by successive heart beats. While this technique requires the heart beat to be repetitive and repeatable, the effective frame rate was 667 frames per second. The fluorescence LOT measurements that were obtained were able to detect the direction of propagation of an electric

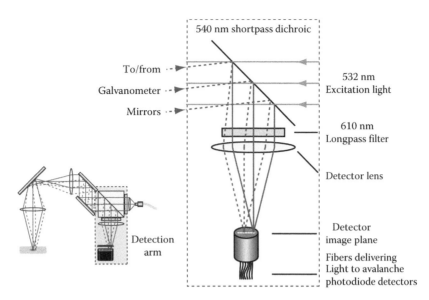

FIGURE 18.6 The detection arm of an LOT system can be modified to measure fluorescence through use of a dichroic beam splitter, to pass excitation light and reflect emitted fluorescence light, and a longpass filter to remove residual excitation light.

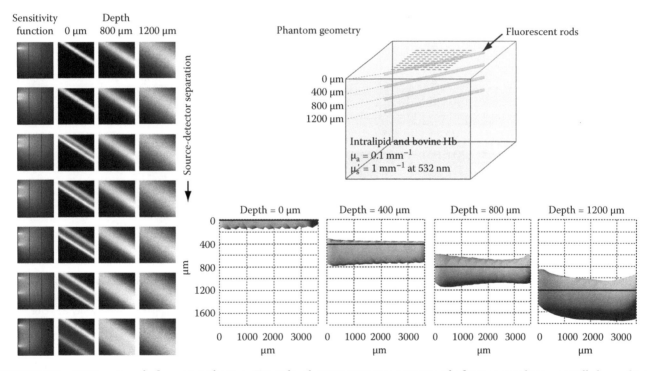

FIGURE 18.7 LOT imaging of a fluorescent phantom. Top right: phantom geometry consisting of a fluorescent rod incrementally lowered into an absorbing and scattering background media. Left: sensitivity functions for each S-D separation and raw LOT data acquired for each rod depth. A 3D reconstruction of the rods from the LOT data set (right, bottom) shows poorer resolution at deeper depths.

wave within the cardiac wall. This study highlights some of the advantages of LOT. The noncontact measurement geometry and rapid scanning point source allowed for imaging a moving object. While this system used an alternate scan configuration to achieve higher frame rates, more recent LOT systems, using a raster scanning pattern, are capable of imaging at frame rates up to 100 frames per second.

18.6.2 Simultaneous Multiwavelength LOT

To exploit the absorption spectrum of tissues, measurements at multiple wavelengths are required. While the use of shutters to sequentially illuminate the sample permits such measures, the acquisition frame rate is hindered. Furthermore, since images at each wavelength are not acquired simultaneously, the motion of

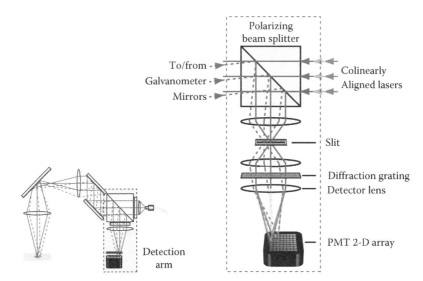

FIGURE 18.8 Simultaneous measurements at multiple wavelengths are achieved by including a dispersive element and 2D PMT array in the detection arm. Detector separations increase across rows of the detector, while the wavelength of detected light changes down the columns of the detector.

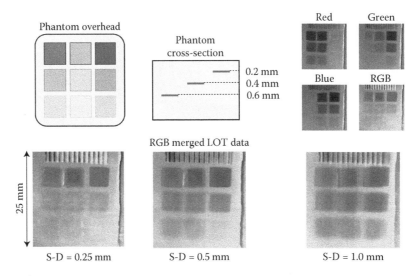

FIGURE 18.9 (See color insert.) Simultaneous multiwavelength phantom imaging. Top left: layout of the phantom showing different colored absorbers across rows and deeper depths down each column. Top right: raw LOT data simultaneously collected for each illumination wavelength. Bottom: red-green-blue merge of the raw LOT data for 3 S-D separations. Deeper absorbers become more apparent in the wider detector separations. (Adapted from Burgess, S.A. et al., *Opt. Lett.*, 33, 2710, 2008.)

a living sample, or fast dynamic changes, can impair pixel-by-pixel spectral analysis. Figure 18.8 shows how the detection arm of an LOT system can be modified to simultaneously acquire images at multiple wavelengths in parallel. Collinearly aligned lasers simultaneously illuminate the sample. The returning backscattered multiwavelength light passes through a dispersive element in the detection arm such as a prism or a diffraction grating, which physically separates the light into its wavelength components. A 2D PMT array replaces the linear array, simultaneously acquiring each of the illumination wavelengths along one dimension, and different S-D separations along the other dimension.

Figure 18.9 shows raw LOT data acquired from a simultaneous multiwavelength system. The phantom consisted of red, green, and blue absorbers embedded at different depths within a scattering background medium. The raw data from each illumination wavelength (Figure 18.9, grayscale images) were merged to form photograph-like images of the color phantom. The deeper absorbers become more apparent in the wider S-D separation channels demonstrating the increasing depth sensitivity of the measurements.

A simultaneous multiwavelength LOT system with fluorescence was implemented in (Burgess et al. 2008) and is shown in Figure 18.10. Two detection arms are included in the system for

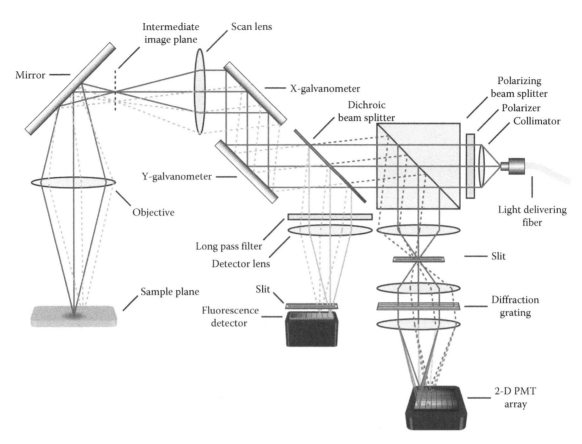

FIGURE 18.10 (See color insert.) Simultaneous multiwavelength and fluorescence LOT system layout. Lasers with a wavelength of 488, 532, and 638 nm are delivered together into the system via a polarization-maintaining optical fiber. Careful selection of the dichroic beam splitter allows returning absorption contrast light to be passed while fluorescence light is reflected. A transmission diffraction grating and an 8 × 8 2D PMT array in the absorption detection arm permits simultaneous measurements of absorption contrast at multiple wavelengths.

both fluorescence and absorption contrast measurements. The fluorescence section includes a dichroic beam splitter, passing 488, 535, and 638 nm light and reflecting all other visible wavelengths. The system is capable of acquiring a 200 × 200 pixel (source position) image at 23 frames per second and a 45 × 45 pixel image at 100 frames per second.

18.6.3 Combined OCT and Line-Scan LOT

LOT can be combined with other optical imaging techniques as presented in Yuan et al. (2009b), where combined OCT and line-scan LOT were demonstrated. While OCT can produce high-resolution images to depths of up to 2 mm in tissue, OCT image contrast is primarily from back-scattering differences within the tissue, fluorescence cannot be measured and measurements are generally insensitive to tissue absorption. The advantage of a multimodality approach is therefore that OCT can provide higher resolution structural images, which can be co-registered with images of fluorescence generated using LOT, providing both morphological and functional information.

The fluorescence and multiwavelength LOT systems described thus far have utilized a point-based raster scanning data acquisition approach. Line-scanning LOT is an alternative approach

in which an illumination line is projected onto the sample and a 2D detector array is used to acquire the multiple S-D separations for the entire line of source positions. This setup uses a cylindrical lens to generate a laser line and a single galvanometer mirror to scan the line over the sample plane. As before, the galvanometer also acts to de-scan the line onto a 2D detector array with the detector array elements adjacent to the de-scanned line measuring the off-axis backscattered light. Figure 18.11 shows co-registered OCT and fluorescence LOT images of a Cy5.5 dye-filled capillary tube suspended in a scattering medium. The left image shows an overlap of the OCT surface rendering with a 3D reconstruction of the fluorescent LOT data. The white arrows in the OCT images (center column) indicate the position of the tube and correlate with the location of the tube in the fluorescence LOT images (right column). The combined OCT–LOT system imaged the phantom at 0.1 Hz, limited by the OCT scan speed of the system.

18.7 Critical Discussion

At a first glance, LOT appears most similar to DOT owing to its detection of scattered light. It is therefore useful to identify key differences between the two. LOT has many advantages

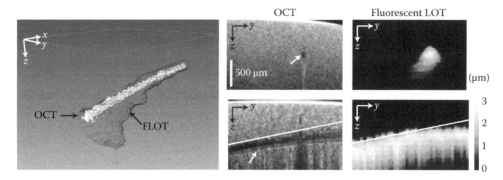

FIGURE 18.11 Co-registered OCT and fluorescence LOT. The combined system was demonstrated by imaging a phantom consisting of a capillary tube filled Cy5.5 dye embedded in a background absorbing and scattering media. Left: co-registered 3D reconstruction of OCT and fluorescence LOT data. Middle: OCT image with white arrow indicating the position of the embedded tube. Right: slice along the same planes of the reconstructed fluorescence LOT data. (From Yuan, S. et al., *Opt. Lett.*, 34, 1615, 2009b. With permission.)

compared to DOT-like imaging geometries where multiple source and detector fibers are placed in contact with, or imaged onto the tissue (Culver et al. 2003a,b, Schmitz et al. 2005). For LOT, the number of detector elements is determined by the desired number of S-D separations, and only one light source is needed (for a given wavelength). Fiber-based systems are typically arranged in an array pattern, with the source position varied by either using a fiber switch to route source light into each fiber, or by using multiple light sources that can be switched on and off, each attached to its own source fiber. Each detector fiber is generally attached to its own detector. Detector fibers close to an illuminated source fiber will see high light intensities, but the same detector will see much lower intensities when the source is positioned at another point in the grid. As a result, a fiber-based grid system consisting of 8 sources and 8 detectors can acquire 64 measurements at a single wavelength; however, each detector's dynamic range must be sufficient to detect 8 different signal amplitudes corresponding to each separation. In LOT, the number of source positions can be increased by quickly moving the laser beam to additional positions. Further, each detector in LOT is positioned to detect light for only one given S-D separation; therefore, detectors at wider S-D separations can be configured to have higher gain than the closer detectors. As the source position changes, the relative positions of these detectors do not change, so they do not require the same dynamic range as that of detectors in fiber-based systems.

Another important distinction is the difference in the light detected by narrow S-D separations in LOT compared to the much wider separations used in DOT. In DOT, large objects such as the breast or infant head are imaged, and circumferential measurements are commonly acquired. Diffusion-approximation-based models are typically used for image reconstruction with resolution scales being 0.5–1 cm for a 10–15 cm object. Surface artifacts are common, owing to the generally sparse distribution of sources and detectors, and resolution is typically poorer at the center of the object. In LOT, measurement densities are higher and image reconstructions are attempted within much smaller volumes of tissue (several millimeters). Since fewer scattering

events happen between a source and detector that are 1–2 mm apart (rather than cm's for DOT) there is less uncertainty in LOT, and image resolution can be much higher (100–200 μm). However, owing to the reflectance geometry of LOT measurements, deeper structures are probed much more poorly than superficial ones. As a result, reconstructed image resolution and accuracy in LOT will degrade with increasing depth.

The combined use of OCT and LOT highlights another important feature of LOT: the ability to provide functional contrast. LOT is sensitive to hemoglobin concentration and oxygenation. Not all optical techniques, including OCT, can directly sense optical absorption and some have limited imaging depths due to scattering. LOT can provide functional imaging beyond the scattering limits of microscopy, making it a valuable tool that can offer insight into biological processes.

18.8 Summary

In summary, we have described the fundamental light interactions in tissue that LOT relies upon to generate depth-sensitive measurements. The details and design of a basic LOT system through more advanced configurations, including fluorescence, multiwavelength and multimodality LOT systems have been described. LOT data sets are comprised of depth-sensitive measurements requiring analysis and image reconstruction techniques that have been discussed. Examples of recent applications of LOT were provided, and key LOT features were compared to other existing techniques. LOT is an important new technique for in vivo optical imaging, providing a unique set of advantages over existing methods including noncontact, high-speed, quantitative, depth-resolved imaging of both absorption and fluorescence contrast.

18.9 Future Perspectives

LOT has seen exciting advances in recent years incorporating fluorescence and absorption contrast measurements in the same instrument, and combining LOT with other optical imaging modalities. The sensitivity of LOT to contrast in superficial regions of tissues has led to its use in research settings, including imaging of exposed rodent cortex and cardiac tissue. Dermal imaging is a strong candidate for clinical LOT imaging. LOT is sensitive to depths into the dermis and could provide valuable depth-sensitive measures of HbO_2 and HbR hemoglobin. Miniaturization of LOT instrumentation to allow endoscopic imaging could provide additional opportunities for LOT imaging.

References

Alerstam, E., Svensson, T., and Andersson-Engels, S. 2008. Parallel computing with graphics processing units for high-speed Monte Carlo simulation of photon migration. *J. Biomed. Opt.* 13: 060504.

Arridge, S. R. 1999. Optical tomography in medical imaging. *Inverse Probl.* 15: 41–93.

Arridge, S. R., Cope, M., and Delpy, D. T. 1992. The theoretical basis for the determination of optical pathlengths in tissue: Temporal and frequency analysis. *Phys. Med. Biol.* 37: 1531–1559.

Arridge, S. R. and Lionheart, W. R. B. 1998. Nonuniqueness in diffusion-based optical tomography. *Opt. Lett.* 23: 882–884.

Bevilacqua, F., Piguet, D., Marguett, P. et al. 1999. In vivo local determination of tissue optical properties applications to human brain. *Appl. Opt.* 22: 4939–4950.

Boas, D. A. and Dale, A. M. 2005. Simulation study of magnetic resonance imaging-guided cortically constrained diffuse optical tomography of human brain function. *Appl. Opt.* 44: 1957–1968.

Burgess, S. A., Bouchard, M. B., Yuan, B., and Hillman, E. M. 2008. Simultaneous multiwavelength laminar optical tomography. *Opt. Lett.* 33: 2710–2712.

Corlu, A., Choe, R., Durduran, T., Lee, K., Schweiger, M., Arridge, S. R., Hillman, E.M.C., and Yodh, A.G. 2005. Diffuse optical tomography with spectral constraints and wavelength optimization. *Appl. Opt.* 44(11): 2082–2093.

Cuccia, D. J., Bevilacqua, F., Durkin, A. J., and Tromberg, B. J. 2005. Modulated imaging: Quantitative analysis and tomography of turbid media in the spatial-frequency domain. *Opt. Lett.* 30: 1354–1356.

Culver, J. P., Durduran, T., Furuya, D. et al. 2003a. Diffuse optical tomography of cerebral blood flow, oxygenation, and metabolism in rat during focal ischemia. *J. Cereb. Blood Flow Metab.* 23: 911–924.

Culver, J. P., Siegel, A. M., Stott, J. J., and Boas, D. A. 2003b. Volumetric diffuse optical tomography of brain activity. *Opt. Lett.* 28: 2061–2063.

Dunn, A. K. and Boas, D. A. 2000. Transport-based image reconstruction in turbid media with small source–detector separations. *Opt. Lett.* 25: 1777–1779.

Franceschini, M. A., Toronov, V., Filiaci, M. E., Gratton, E., and Fantini, S. 2000. On-line optical imaging of the human brain with 160-ms temporal resolution. *Opt. Express* 6: 49–57.

Gonzalez-Rodriguez, P. and Kim, A. D. 2009. Reflectance optical tomography in epithelial tissues. *Inverse Probl.* 25: 015001.

Gonzalez-Rodriguez, P., Kim, A. D., and Moscoso, M. 2007. Reconstructing a thin absorbing obstacle in a half space of tissue. *J. Opt. Soc. Am. A* 24: 3456–3466.

Hielscher, A. H., Klose, A. D., and Hanson, K. M. 1999. Gradient-based iterative image reconstruction scheme for time-resolved optical tomography. *IEEE Trans. Med. Imaging* 18: 262–271.

Hillman, E. M. C., Bernus, O., Pease, E., Bouchard, M. B., and Pertsov, A. 2007a. Depth-resolved optical imaging of transmural electrical propagation in perfused heart. *Opt. Express* 15: 17827–17841.

Hillman, E. M. C., Boas, D. A., Dale, A. M., and Dunn, A. K. 2004. Laminar optical tomography: Demonstration of millimeter-scale depth-resolved imaging in turbid media. *Opt. Lett.* 29: 1650–1652.

Hillman, E. M. C. and Burgess, S. A. 2009. Sub-millimeter resolution 3D optical imaging of living tissue using laminar optical tomography. *Laser Photonics Rev.* 3: 159–179.

Hillman, E. M. C., Dehghani, H., Hebden, J. C. et al. 2001. Differential imaging in heterogeneous media: Limitations of linearization assumptions in optical tomography. *Proc. SPIE* 4250: 327–338.

Hillman, E. M. C., Devor, A., Bouchard, M. et al. 2007b. Depth-resolved optical imaging and microscopy of vascular compartment dynamics during somatosensory stimulation. *Neuroimage* 35: 89–104.

Hillman, E. M. C., Devor, A., Dunn, A. K., and Boas, D. A. 2006. Laminar optical tomography: High-resolution 3D functional imaging of superficial tissues. *Proc. SPIE (Medical Imaging)* 6143: 61431M.

Hillman, E. M. C., Hebden, J. C., Schmidt, F. E. W. et al. 2000. Calibration techniques and datatype extraction for time-resolved optical tomography. *Rev. Sci. Instrum.* 71: 3415–3427.

Hyatt, C. J., Zemlin, C. W., Smith, R. M. et al. 2008. Reconstructing subsurface electrical wave orientation from cardiac epifluorescence recordings: Monte Carlo versus diffusion approximation. *Opt. Express* 16: 13758–13772.

Iranmahboob, A. K. and Hillman, E. M. C. 2008. Diffusion vs. Monte Carlo for Image reconstruction in mesoscopic volumes. In: *Biomedical Topical Meetings, OSA Technical Digest, Optical Society of America*, Washington DC, 16–19 March, 2008.

Kak, A. C. and Slaney, M. 1987. *Principles of Computerized Tomographic Imaging.* New York: IEEE Press.

Miyawaki, A., Sawano, A., and Kogure, T. 2003. Lighting up cells: Labelling proteins with fluorophores. *Nat. Cell Biol.* Suppl: S1–S7.

Mourant, J. R., Freyer, J. P., Hielscher, A. H. et al. 1998. Mechanisms of light scattering from biological cells relevant to noninvasive optical-tissue diagnostics. *Appl. Opt.* 37: 3586–3593.

Patterson, M. S., Wilson, B. C., and Wyman, D. R. 1991. The propagation of optical radiation in tissue II. Optical properties of tissues and resulting fluence distributions. *Lasers Med. Sci.* 6: 379–390.

Ramanujam, N. 2000. Fluorescence spectroscopy in vivo. In: *Encyclopedia of Analytical Chemistry*, 1st edn., ed. R. A. Meyers, Chichester, U.K.: John Wiley & Sons, pp. 20–56.

Schmitz, C. H., Graber, H. L., Pei, Y. et al. 2005. Dynamic studies of small animals with a four-color DOT imager. *Rev. Sci. Instrum.* 76: 094302.

Severinghaus, J. W. and Astrup, P. B. 1986. History of blood gas analysis. VI. Oximetry. *J. Clin. Monit.* 2: 270–288.

Swartling, J., Pifferi, A., Enejder, A. M. K., and Andersson-Engels, S. 2003. Accelerated Monte Carlo models to simulate fluorescence spectra from layered tissues. *J. Opt. Soc. Am. A* 20: 714–727.

Tsien, R. Y. 1998. The green fluorescent protein. *Annu. Rev. Biochem.* 67: 509–544.

Wang, L.-H., Jacques, S. L., and Zheng, L.-Q. 1995. MCML—Monte Carlo modeling of photon transport in multi-layered tissues. *Comput. Methods Program. Biomed.* 47: 131–146.

Yuan, B., Burgess, S. A., Bouchard, M. B. et al. 2009a. A system for high-resolution depth-resolved optical imaging of fluorescence and absorption contrast. *Rev. Sci. Instrum.* 80: 043706–043711.

Yuan, S., Li, Q., Jiang, J., Cable, A., and Chen, Y. 2009b. Three-dimensional coregistered optical coherence tomography and line-scanning fluorescence laminar optical tomography. *Opt. Lett.* 34: 1615–1617.

Zeff, B. W., White, B. R., Dehghani, H., Schlaggar, B. L., and Culver, J. P. 2007. Retinotopic mapping of adult human visual cortex with high-density diffuse optical tomography. *Proc. Natl. Acad. Sci.* 104:12169–12174.

Zhang, J., Campbell, R. E., Ting, A. Y., and Tsien, R. Y. 2002. Creating new fluorescent probes for cell biology. *Nat. Rev. Mol. Cell Biol.* 3: 906–918.

Diffuse Optical Tomography Using CW and Frequency Domain Imaging Systems

Subhadra Srinivasan
Dartmouth College

Scott C. Davis
Dartmouth College

Colin M. Carpenter
Stanford University

19.1 Introduction 373
19.2 Theory and Instrumentation 373
　　　Theory of Modeling Light Propagation and Image Reconstruction for FD Systems • Extension to Continuous Wave Measurements • Spectral Fitting • Multispectral Image Reconstruction • Effect of Wavelengths • Multimodality Imaging
19.3 In Vivo Diffuse Optical Tomography.. 382
　　　Optical Imaging of the Breast • Optical Imaging of the Brain
19.4 Challenges and Critical Discussion 387
References... 389

19.1 Introduction

In this chapter, we discuss the principles of diffuse optical tomography (DOT) imaging using continuous wave (CW) and frequency domain (FD) measurement systems. Section 19.2 describes the instruments and imaging configurations commonly used for acquiring CW and FD data. The numerical approaches to transforming the measured data into meaningful images in tissue are also discussed in Section 19.2, and the importance of resolving the spectral content of the reemitted light to extract physiologically relevant parameters is described. These measurements can be used to reduce inaccuracies in the final image caused by coupling errors, but also facilitate the direct recovery of oxyhemoglobin (HBO) and deoxyhemoglobin (Hb) concentration, water and lipid content, effective scatterer size, and density (using Mie theory) in the image. In addition to spectral content, cutting-edge systems are being developed to couple DOT systems with non-optically based medical imaging devices, such as MRI and ultrasound. Several methods to couple the data sets from these hybrid imaging systems are described.

The most prominent clinical applications of DOT systems are detailed in Section 19.3. These include imaging of the breast, brain, finger joints, and prostate. Finally, we examine some of the remaining challenges in the field of DOT imaging including acquisition speed, imaging sensitivity, computational load, especially for volumetric imaging in three dimensions, and the critical question of whether DOT can provide adequate image contrast to impact clinical practice.

The reader is referred elsewhere in this work for a detailed discussion on fluorescence molecular tomography.

19.2 Theory and Instrumentation

FD imaging systems for DOT utilize intensity-modulated light sources at high frequencies (a few hundred megahertz). This allows the measurement of amplitude and phase shift of scattered intensity signal relative to the incident signal, providing the basis for FD tomography. FD systems are simple and robust with a higher signal-to-noise ratio (SNR) and have been utilized for clinical use because of the speed and reliability. Fishkin (Fishkin et al. 1991) and Patterson (Patterson et al. 1991) presented some of the early work in demonstrating FD systems. By using these measurements along with a model for diffuse light propagation (Arridge et al. 1993), one can reconstruct images of absorption and reduced scattering coefficients of tissue in vivo. By repeating this process at multiple wavelengths, recovery of functional images is possible, since the main absorbers (called chromophores) in tissue are HbO and Hb. Jobsis (1977) first showed in 1977 that using two wavelengths appropriately placed, noninvasive monitoring of HbO and Hb changes was possible. Following this, researchers have used multiwavelength interrogation of tissue to recover concentrations of the primary chromophores such as HbO and Hb, water, lipids, and used it to monitor changes in physiology of the breast (Pogue et al. 2001, Shah et al. 2001, Durduran et al. 2002), brain (Gratton et al. 1997, Hueber et al. 2001), finger-joints (Hielscher et al. 2004), etc.

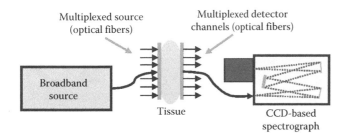

FIGURE 19.1 This schematic of a multispectral CW DOT system illustrates the general configuration of DOT imaging. CW systems are the simplest to implement and can be assembled using a wide array of light sources and detectors.

The measurement techniques employed for diffuse optical imaging can be organized into three categories, namely, CW, which covers time-independent measurements, FD, and time domain (TD). From an experimental standpoint, CW systems are the simplest to implement since source and detector instruments do not require rapid time responses. This reduces cost pressures on instrumentation and allows broadband detection with charged coupled device array (CCD)-based spectrometers. Figure 19.1 is a diagram of a simple spectrally resolved imaging system operating in CW mode. In this example, a light source is sequentially coupled into a series of optical fibers in contact with the tissue. For each illuminated source position, measurements of the light transmitted through the tissue are recorded at all detector positions around the imaging volume. This may be accomplished with sequential coupling of the detector to the detection optical fibers or by parallel detection if multiple detectors are available. The light source in this system may be composed of a broadband lamp, or more likely, a bank of LEDs or laser diodes (LDs). The CCD-based spectrograph detection system measures the entire near-infrared (NIR) spectrum of the transmitted light for each source-detector measurement, facilitating the quantification of tissue chromophores based on spectral signatures. In general, TD and FD systems are not able to provide such highly resolved spectral information with a single measurement; however, additional information about photon pathlength revealed by TD and FD measurements dramatically improve the imaging capability of the system.

The basic principle of FD imaging is illustrated in Figure 19.2. The intensity of a light source is modulated prior to entering the

tissue. In most cases, the modulation frequency is in hundreds of megahertz and is accomplished by modulating the intensity of solid-state emitters such as LDs. As the sinusoidal intensity wave propagates through the tissue, the extended pathlength caused by photon scattering delays the phase of the modulated signal. After exiting the tissue, the sinusoid has been attenuated and shifted in phase as compared to the incident light. The measurable changes in amplitude and phase contain information about the absorbing and scattering properties of the interrogated tissue volume and when properly calibrated are used in the imaging algorithms to recover these optical properties.

An example of an FD DOT system operating in heterodyning mode is schematically depicted in Figure 19.3. In this example, two RF generators are used, one operating at 100 MHz and the other at 100.001 MHz. Ninety-nine to one splitters are used to extract and mix a portion of the signal from each RF generator to produce a 1 kHz reference channel. The 100 MHz signal is coupled to a DC signal through a bias-T, the output of which powers the LD, now modulated at 100 MHz. This is the signal that interrogates the tissue volume. On the detection side, the output signal from the PMT is mixed with the 100.001 MHz signal and thus becomes the product of the 100.001 MHz signal and the phase-shifted 100 MHz signal that has been transmitted through the tissue. This data channel is then amplified and low pass filtered to remove the high-frequency component. Both reference and data channels are acquired with a data acquisition (DAQ) board and processed using lock-in detection at 1 kHz to extract the AC amplitude and phase shift between data and reference.

The example in Figure 19.3 is a simplified version of a typical FD DOT system operating in heterodyned acquisition mode. In practice, an array of laser sources at different wavelengths may be used sequentially and detectors may be fully parallelized to speed acquisition. Parallel illumination may also be incorporated by expanding the array of RF modulators and introducing different phase offsets or modulation frequencies for each laser channel. These sources may be decoded using lock-in detection, a strategy that facilitates fast DOT image acquisition to measure dynamic changes in tissue physiology.

The advantage of heterodyne detection is that lock-in detection is performed on a low-frequency signal, which improves the accuracy in detecting the phase. However, heterodyning

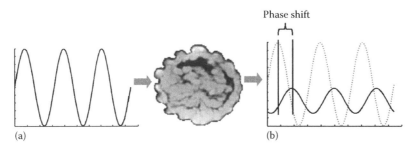

FIGURE 19.2 In an FD DOT imaging system, a modulated light source illuminates the tissue surface (a). After propagating through the tissue, the intensity wave has been attenuated and shifted in phase (b, solid line) as compared to the illumination signal (dotted line).

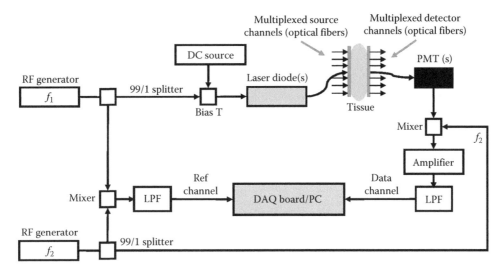

FIGURE 19.3 A schematic of a heterodyned FD DOT system. Heterodyning requires two signal generators (RF generator) operating at f_1 and f_2 to compare the down-converted data signal against a similarly down-converted reference channel. In this figure, LPF refers to low-pass filtering to remove the high frequency component of mixed RF signals.

techniques require an additional reference oscillator that needs to be phase locked to the laser modulator. Alternatively, the homodyne technique uses the light source frequency, f_1, as the reference channel. This technique has been successfully used for DOT imaging in the FD (Culver et al. 2003, Thompson and Sevick-Muraca 2003, Netz et al. 2008). Because the same oscillator is used for the laser driver and the reference, lower-cost frequency generators may be used. Thus, heterodyning schemes offer more accurate signal detection, but are more expensive due to the requirement of an additional, stable oscillator.

A wide variety of light sources may be selected depending on geometry, data type, and detection capabilities. Filament or gas lamps (tungsten and xenon), laser systems (including gas, crystal, and diode lasers), LEDs, or tunable laser systems such as Ti:Sapphire lasers may be used for CW imaging. While broadband lamps are inexpensive and flexible in terms of wavelength selection, the isotropic emission of lamps limits the ability to focus sufficient power on the tissue surface for imaging through relatively thick tissue volumes. They are also not suitable for time-dependent measurements that require rapid modulation of the source intensity.

FD systems typically make use of LEDs and LDs modulated in the 100s of MHz. Diode light sources are available only for discrete or relatively narrow wavelength bands and thus multiple diode devices are required for spectrally resolved imaging throughout the NIR wavelength range. Tunable lasers provide flexibility in terms of wavelength but are often very expensive and require special environmental controls and physical stabilization not required for LDs. Available optical detectors sensitive enough for deep tissue imaging include CCDs, intensified CCDs, electron multiplied CCDs (EMCCD), avalanche photodiodes (APDs), and photomultiplier tubes (PMTs). PMTs are by far the most sensitive instruments available, provide a wide dynamic range, and are commonly used for FD imaging.

However, PMTs may be prohibitively expensive, are impractical for large, high-resolution arrays, are often nonlinear in output voltage, and have nonuniform sensitivity. For example, a popular PMT used for optical imaging is the Hamamatsu R6357, which is used for its high gain capabilities; however, it suffers from poor sensitivity in NIR wavelengths above 850 nm. Less-sensitive APDs cost less, are sufficiently fast for FD imaging, and have flat sensitivity across the wavelengths of interest for NIR imaging. When designing DOT imaging systems, these trade-offs must be considered careful in the context of the imaging geometry and biological information to be investigated.

CCD arrays are sensitive across the NIR range, while their large, high-resolution imaging arrays facilitate parallel detection of detector positions on the tissue surface or spectrally resolved measurements when coupled to a spectrograph. Their relatively long integration times limit their use to CW mode unless modified with an intensifier or electron multiplier. Several researchers have developed FD DOT imaging systems using intensified CCDs (Netz et al. 2008, Thompson and Sevick-Muraca 2003, Lee et al. 2007).

Data measured from the boundary of the tissue is used to recover tomographic images of optical properties or chromophores inside the tissue volume. In the next section, we describe the principles behind this model-based image recovery process using CW and FD measurements.

19.2.1 Theory of Modeling Light Propagation and Image Reconstruction for FD Systems

Here we discuss the use of the diffusion approximation for modeling light propagation in tissue. The diffusion equation can be used to model light propagation in highly scattering tissues and under the assumption that the angular distribution of the light intensity is almost uniformly isotropic (Ishimaru 1978, Patterson et al. 1990a), and is given by

$$-\nabla \cdot \kappa(r) \nabla \Phi(r, \omega) + \left(\mu_a(r) + \frac{i\omega}{c} \right) \Phi(r, \omega) = q_0(r, \omega) \quad (19.1)$$

where

$\Phi(r, \omega)$ is the isotropic fluence at modulation frequency ω and position r
$\kappa(r)$ is the diffusion coefficient
$\mu_a(r)$ is the absorption coefficient
c is the speed of light in the medium
$q_0(r, \omega)$ is the isotropic source

The diffusion coefficient can be written as

$$\kappa(r) = \frac{1}{3(\mu_a(r) + \mu_s'(r))}, \quad (19.2)$$

where μ_s' is the reduced scattering coefficient.

The diffusion equation can be solved numerically using finite element discretization (Arridge et al. 1993, Paulsen and Jiang1995). This represents the forward problem where the fluence distribution $\Phi(r, \omega)$ is obtained in the domain as a function F of *known* optical properties $\mu(r)$ represented by

$$\Phi = F(\mu), \quad (19.3)$$

where $\mu = [\mu_a, \kappa]$

In the image reconstruction procedure, the inverse problem is solved where given measurements of $\Phi(r, \omega)$ at the boundary, the distribution of *unknown* optical properties in the domain is recovered. This is obtained by solving the inverse:

$$\mu = F^{-1}(\Phi) \quad (19.4)$$

The solution of the inverse problem typically involves minimization of an appropriate error functional. This can be represented by a least squares norm of the difference between the measured data and the model data, where data refers to measurements of intensity and phase at the boundary:

$$\chi^2 = \frac{1}{2} \sum_{j=1}^{M} \left((\Phi_{meas} - F(\mu))^2 \right) \quad (19.5)$$

where

Φ_{meas} is the measurements at the boundary
$F(\mu)$ is the calculated fluence measurements($= \Phi_{calc}$) using the forward model

This can be solved in a direct single step (linear reconstruction) or iteratively using gradient-based approaches. In the iterative procedure, an initial estimate for the optical properties is used to calculate the model data, and the error functional is minimized, by solving an update equation giving a new estimate for the optical properties. Calculating the derivative of Equation 19.5, and setting the derivative to zero for minimization leads to the update equation as follows:

$$\frac{\partial \chi^2}{\partial \mu} = J^{T}(\Phi_{meas} - F(\mu)) = 0, \quad (19.6)$$

where

$$J = \frac{\partial F(\mu)}{\partial \mu} = \frac{\partial \Phi_{calc}}{\partial \mu} \quad (19.7)$$

The Jacobian J represents the sensitivity matrix that provides the perturbation of each amplitude and phase measurement to change in optical properties in the imaging domain. The adjoint method given by Arridge and Schweiger (1995) provides an elegant approach to calculate the Jacobian using the forward model. The Jacobian has the general structure

$$J = [J_{\mu a}, J_{\kappa}] \quad (19.8)$$

where $J_{\mu a}$ and J_{κ} are the sensitivity matrices for absorption and diffusion coefficients, respectively.

Using Taylor's expansion, one can obtain the solution iteratively, starting at an initial estimate around a point μ_{i-1} to obtain an updated solution μ_i:

$$F(\mu_i) = F(\mu_{i-1}) + J(\mu_i - \mu_{i-1}) + \cdots \quad (19.9)$$

Substituting in Equation 19.6 and using $\Delta\mu = \mu_i - \mu_{i-1}$, we get

$$J^T \left(\Phi_{meas} - F(\mu_{i-1}) - J\Delta\mu \right) = 0 \quad (19.10)$$

Using

$$\Delta\Phi = \Phi_{meas} - F(\mu_{i-1}) \quad (19.11)$$

$$J^T \Delta\Phi = J^T J \Delta\mu$$

$$\Rightarrow \Delta\mu = (J^T J)^{-1} J^T \Delta\Phi \quad (19.12)$$

where the data difference is the least squares norm of the difference between the measured and calculated data at wavelength λ for all measurements M.

$$\Delta\Phi = \left\{\begin{array}{c} \Phi_{meas}^1 - \Phi_{calc}^1 \\ \Phi_{meas}^2 - \Phi_{calc}^2 \\ . \\ . \\ \Phi_{meas}^M - \Phi_{calc}^M \end{array}\right\}_\lambda \qquad (19.13)$$

and

$$\mu = \left\{\begin{array}{c} \mu_a \\ \kappa \end{array}\right\} \qquad (19.14)$$

This inverse problem is ill-posed and in order to solve this problem successfully, two aspects have to be kept in mind: (1) initial estimate of the properties and (2) the regularization method. The derivation based on Taylor's series assumes a close initial point around which the function is approximated; and for this reason, a reasonable close initial estimate is important. Typically, this initial estimate can be obtained by using a homogeneous fitting procedure where the analytical solution to an infinite medium is used to derive homogeneous initial values that are used in combination with the Newton–Raphson method for fitting the slope of logarithm of intensity and phase with distance (Pogue et al. 2000). Use of the slopes makes this fitting procedure robust to noise, and this has been found to yield good starting values for image recovery (McBride et al. 2003).

The second aspect is regularization. Since the matrix to be inverted (represented by the Hessian = J^TJ) is ill-conditioned, this requires regularization to control the update of the properties. This regularization can be based on the Levenberg–Marquardt (Marquardt 1963) method in which case, Equation 19.12 is modified to

$$\Delta\mu = (J^TJ + \alpha I)^{-1} J^T \Delta\Phi \qquad (19.15)$$

where α is the regularization parameter that is decreased monotonically with iterations (Marquardt 1963). L-curve methods can be employed to find an optimum α by plotting the norm of the solution against the residual data difference for all values of α (Hansen 1998). The choice of α can also be empirical based on experimental data. Alternatively, the equation for Tikhnov regularization can be obtained by minimization of an altered error functional that takes into account the a priori knowledge regarding the solution and data (Hansen 1998). The iterative image reconstruction procedure can be depicted by the flow chart in Figure 19.4.

19.2.2 Extension to Continuous Wave Measurements

In the case of CW measurements, only intensity data is available (i.e., there is no phase information) and it is difficult to

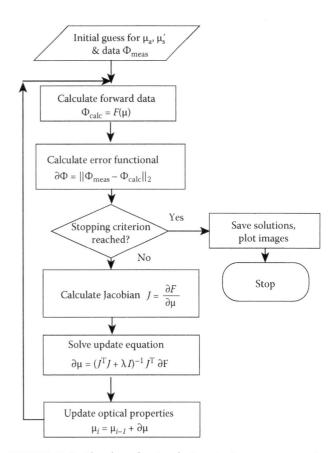

FIGURE 19.4 Flowchart showing the iterative image reconstruction procedure to obtain images of optical properties from boundary FD measurements of intensity and phase shift.

separate the effects of absorption from the effects of scattering. Hence, in this case, some assumptions regarding the scattering properties of the medium are generally used to reconstruct images of the absorption properties. This limits the quantitative accuracy of the results, which depend on the scattering coefficients given for the reconstruction. However, it can still be used to reconstruct changes in absorption successfully if the underlying scattering properties do not change significantly and in applications such as monitoring pulsatile dynamics (Li et al. 2009).

19.2.3 Spectral Fitting

The underlying physiological mechanisms responsible for absorption are the chromophores such as HbO, Hb, and water. The spectral behavior of some of these chromophores is shown in Figure 19.5 using data from Boulnois (1986) and Hale and Querry (1973). By reconstructing optical properties at multiple wavelengths, the concentrations of these chromophores and scattering parameters can be obtained by spectral fitting. The main chromophores available depend on the tissue type and wavelengths. For example, in the breast, the primary chromophores are HbO, Hb, water, and lipids. The concentrations of these chromophores can be obtained by decomposing

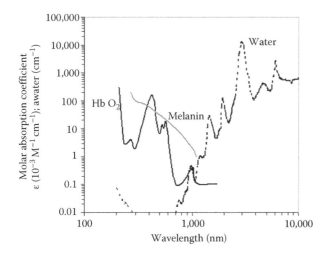

FIGURE 19.5   Molar absorption spectra of HbO, melanin, and water. (Data obtained from Boulnois, J.L., *Lasers Med. Sci.*, 1, 47, 1986; Hale, G.M., and Querry, M.R., *Appl. Opt.*, 12(3), 555, 1973.)

the reconstructed absorption coefficients at multiwavelengths using

$$\mu_a(\lambda) = \sum_{i=1}^{nc} \varepsilon_i(\lambda)c_i \qquad (19.16)$$

where

> nc is the number of chromophores contributing to the absorption
> $\varepsilon_i(\lambda)$ is the molar absorption coefficient at wavelength λ
> c is the concentration of the chromophores

Written in matrix form as

$$\mu_a = [\varepsilon]c \qquad (19.17)$$

Similarly, spectral fitting of scattering parameters can also yield wavelength-independent parameters. Scattering in tissue in the NIR wavelength regime occurs due to cells and their organelles, and refractive index mismatch between organelles and cell membranes. This can be modeled by Mie theory (Steinke and Shepherd 1988), which assumes homogeneous dielectric spheres. Studies have shown that a simpler approximation to Mie theory can be used to successfully model scattering spectra, and is given by (Mourant et al. 1997, van Staveren et al. 1991):

$$\mu_s'(\lambda) = A\lambda^{-b} \qquad (19.18)$$

This approximation fits the scattering coefficients at multiwavelengths to a power relationship, whose slope is governed by scatter power (b); and offset by scatter amplitude (A). Scatter amplitude has been found to depend on the density of the scattering particles whereas scatter power typically depends on the diameter

of the scattering particles. Together these two parameters reflect changes in structural composition of tissue and its constituents such as collagen.

19.2.4 Multispectral Image Reconstruction

The direct multispectral image reconstruction (MSIR) takes the spectral signatures of the primary chromophores in tissue, and the scattering behavior with wavelength, as constraints directly into the minimization procedure. In this reconstruction, instead of recovering optical properties, one can directly estimate the concentrations of chromophores and scatter parameters by using multiwavelength measurements simultaneously. This type of reconstruction is more robust to noise and cross-talk between parameters as the spectral signatures are embedded into the recovery (Srinivasan et al. 2004, Li et al. 2004, Corlu et al. 2005). The minimization functional can be written as

$$\chi^2 = \sum_{j=1}^{Mk} (\Phi_{meas} - \Phi_{calc})^2 \qquad (19.19)$$

This functional is now over all measurements M at each single wavelength, and over all wavelengths k. This can be minimized in a similar fashion as before to give the update equation

$$\Delta c = \left(J_{sp}^T J_{sp} + \alpha_{sp} I\right)^{-1} J_{sp}^T \Delta\Phi_{sp} \qquad (19.20)$$

where J_{sp} is the spectral Jacobian representing the perturbation in the measurements due to a change in the chromophore concentrations or scatter parameters, and the data difference $\Delta\Phi_{sp}$ is for multiwavelength measurements simultaneously given by

$$\Delta\Phi_{sp} = \begin{Bmatrix} \Delta\Phi_{\lambda 1} \\ \Delta\Phi_{\lambda 2} \\ \vdots \\ \Delta\Phi_{\lambda k} \end{Bmatrix}^2 \qquad (19.21)$$

where

$$\Delta\Phi_{\lambda} = \Phi_{meas,\lambda} - \Phi_{calc,\lambda}. \qquad (19.22)$$

$\Delta c = c_i - c_{i-1}$ and c is defined by

$$c = \begin{Bmatrix} c_1 \\ c_2 \\ \vdots \\ c_n \\ A \\ b \end{Bmatrix} \qquad (19.23)$$

The spectral Jacobian is given by

$$
J_{sp} = \begin{bmatrix}
J_{c1,\lambda1} & J_{c2,\lambda1} & \cdots & J_{cn,\lambda1} & J_{A,\lambda1} & J_{b,\lambda1} \\
J_{c1,\lambda2} & J_{c2,\lambda2} & \cdots & J_{cn,\lambda2} & J_{A,\lambda2} & J_{b,\lambda2} \\
\vdots & \vdots & & \vdots & \vdots & \vdots \\
J_{c1,\lambda k} & J_{c2,\lambda k} & \cdots & J_{cn,\lambda k} & J_{A,\lambda k} & J_{b,\lambda k}
\end{bmatrix} \qquad (19.24)
$$

where

$J_{cn,\lambda k}$ is the sensitivity matrix for the nth chromophore at the kth wavelength

$J_{A,\lambda k}$ and $J_{b,\lambda k}$ are the sensitivity matrices for scatter parameters A and b, respectively, at the kth wavelength

The relationships between J_{sp} in Equation 19.24 and J in Equation 19.15 can be derived as follows:

$$
J_{c,\lambda} = \frac{\partial \Phi}{\partial c}\bigg|_{\lambda} = \frac{\partial \Phi}{\partial \mu_a} \frac{\partial \mu_a}{\partial c}\bigg|_{\lambda} \qquad (19.25)
$$

for each chromophore in the model. Using Equation 19.17 we get $\partial \mu_a = \varepsilon \partial c$ so that, substituting for $\partial \mu_a / \partial c$ in Equation 19.25,

$$
J_{c,\lambda} = \frac{\partial \Phi}{\partial c}\bigg|_{\lambda} = \frac{\partial \Phi}{\partial \mu_a}\varepsilon\bigg|_{\lambda} = \left(J_{\mu a,\lambda}\right) \otimes \left(\varepsilon_{\lambda}^{c1,c2..cn}\right) \qquad (19.26)
$$

where $\otimes$ refers to the Kronecker tensor product. Similarly

$$
J_{A,\lambda} = \frac{\partial \Phi}{\partial A}\bigg|_{\lambda} = \frac{\partial \Phi}{\partial \kappa} \frac{\partial \kappa}{\partial A}\bigg|_{\lambda} \qquad (19.27)
$$

$$
\frac{\partial \kappa}{\partial A} = \left(\frac{\partial \kappa}{\partial \mu_s'}\right)\left(\frac{\partial \mu_s'}{\partial A}\right) \qquad (19.28)
$$

using Equation 19.2

$$
\frac{\partial \kappa}{\partial \mu_s'} = \frac{1}{3}\left(\frac{-1}{(\mu_a + \mu_s')^2}\right) = \frac{1}{3}\left(-9\kappa^2\right) = -3\kappa^2, \qquad (19.29)
$$

and using Equation 19.18

$$
\frac{\partial \mu_s'}{\partial A} = \lambda^{-b} \qquad (19.30)
$$

Equation 19.27 now becomes

$$
J_{A,\lambda} = \frac{\partial \Phi}{\partial A}\bigg|_{\lambda} = \frac{\partial \Phi}{\partial \kappa} \frac{\partial \kappa}{\partial A}\bigg|_{\lambda} = J_{\kappa}\left(-3\kappa^2\right)\left(\lambda^{-b}\right)\bigg|_{\lambda} \qquad (19.31)
$$

Similarly, for the scatter power, the relationship can be derived as (Srinivasan et al. 2004)

$$
J_{b,\lambda} = \frac{\partial \Phi}{\partial b}\bigg|_{\lambda} = \frac{\partial \Phi}{\partial \kappa} \frac{\partial \kappa}{\partial b}\bigg|_{\lambda} = J_{\kappa}\left(-3\kappa^2\right)\left(\mu_s'\right)\left(-\ln \lambda\right)\bigg|_{\lambda} \qquad (19.32)
$$

The iterative procedure used for the MSIR is shown in Figure 19.6. Results from this method, in comparison to the conventional approach of reconstructing first for optical properties, are shown in Figure 19.7, using experimental FD measurements on a gelatin phantom made with blood (Brooksby et al. 2005b).

An extension to the MSIR is the spectral derivative image reconstruction (SDIR), which provides inherent insensitivity to geometry and coupling errors (Xu et al. 2005a). This is based on the rationale that errors due to geometry and coupling are largely wavelength independent. Hence, by using difference at two adequate wavelengths, these errors are canceled, while still

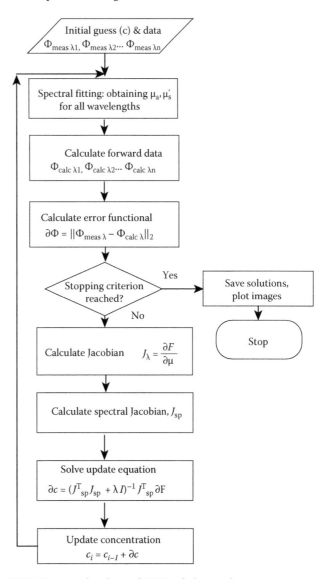

FIGURE 19.6 Flowchart of MSIR of chromophore concentrations and scattering.

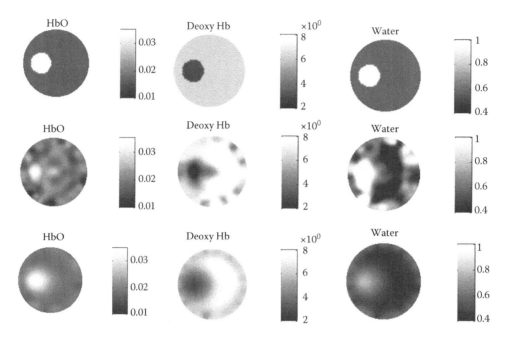

FIGURE 19.7 Top row shows expected values for total hemoglobin, oxygen saturation, water, scatter amplitude, and scatter power. Middle row shows the results using conventional approach of spectral fitting after recovery of optical properties and bottom row shows results from MSIR.

retaining the information regarding the heterogeneity of the tissue. In this method, the data difference is given by

$$\Delta\Phi_{sp} = \left\{\begin{array}{c} \Delta\Phi_{\lambda 1} - \Delta\Phi_{\lambda 2} \\ \Delta\Phi_{\lambda 2} - \Delta\Phi_{\lambda 3} \\ \vdots \\ \Delta\Phi_{\lambda(k-1)} - \Delta\Phi_{\lambda k} \end{array}\right\}^{2} \quad\quad (19.33)$$

and the Jacobian is modified as

$$J_{sp} = \left[\begin{array}{cccc} J_{c1,\lambda 1} - J_{c1,\lambda 2} & J_{c2,\lambda 1} - J_{c2,\lambda 2} & \cdots \\ J_{c1,\lambda 2} - J_{c1,\lambda 3} & J_{c2,\lambda 2} - J_{c2,\lambda 3} & \cdots \\ \vdots & \vdots & \\ J_{c1,\lambda(k-1)} - J_{c1,\lambda k} & J_{c2,\lambda(k-1)} - J_{c2,\lambda k} & \cdots \\[1em] J_{cn,\lambda 1} - J_{cn,\lambda 2} & J_{A,\lambda 1} - J_{A,\lambda 2} & J_{b,\lambda 1} - J_{b,\lambda 2} \\ J_{cn,\lambda 2} - J_{cn,\lambda 3} & J_{A,\lambda 2} - J_{A,\lambda 3} & J_{b,\lambda 2} - J_{b,\lambda 3} \\ \vdots & \vdots & \vdots \\ J_{cn,\lambda(k-1)} - J_{cn,\lambda k} & J_{A,\lambda(k-1)} - J_{A,\lambda k} & J_{b,\lambda(k-1)} - J_{b,\lambda k} \end{array}\right]$$

$$(19.34)$$

The regularization parameter may be different for MSIR from the conventional reconstruction for optical properties; and can be found through the L-curve method or empirically. Scattering may need higher regularization than chromophore concentration images, due to higher sensitivity to noise. The reconstruction of scatter amplitude and scatter power can also lead to images of effective scatterer size and density using Mie theory (Wang et al. 2005). Extension to CW assumes that prior information regarding scatter power is available in which case, the chromophore

concentrations and scatter amplitude A can be reconstructed provided measurements are made with a sufficient number of wavelengths. In the CW scenario, additional optimization needs to be carried out to find optimal wavelengths that minimize cross-talk between the parameters. A detailed discussion of this technique can be found in Corlu et al. (2005).

19.2.5 Effect of Wavelengths

The effects of (1) number of wavelengths used for the spectral reconstruction and (2) the use of longer wavelengths at which water features are stronger are shown via results from a simulation study. Results from a simulation using six wavelengths in the range 660–850 nm versus 14 wavelengths in the range 660–960 nm are shown in Figure 19.8 using FD-simulated measurements at 100 MHz, with 1% random Gaussian noise in the amplitude and phase data and using MSIR. The cross-talk in water from HbO is clearly reduced by the addition of the longer wavelengths.

19.2.6 Multimodality Imaging

19.2.6.1 Inclusion of Complimentary Image Data

Incorporating image data from alternate imaging modalities may be used to improve spatial and quantitative recovery of tissue properties. A great majority of studies investigating multimodality systems have combined a high-spatial resolution modality with optics to improve its spatial limitations, similar to the benefits that computed tomography (CT) provides positron emission tomography (PET) in a PET/CT system. To date, x-ray mammography, x-ray computed tomography, magnetic resonance imaging (MRI), and ultrasound (US) have been used to provide more accurate boundaries and locations of optical

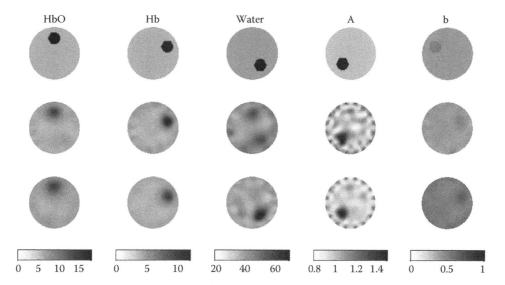

FIGURE 19.8 Top row shows the expected images for HbO, Hb, water, A, and b for the test phantom used to generate FD measurements. Middle row shows the results using six of the wavelengths with MSIR and bottom row shows the images obtained by the use of 14 wavelengths uniformly distributed in the range 660–960 nm, which includes the six wavelengths used for middle row images and MSIR.

fibers. Additional complimentary information may be derived from these modalities, including perfusion, cellular density, or biochemical content. Similarly, magnetoecephalography (MEG) may be combined with optics to provide additional information about neuronal activation (Ou et al. 2009).

The identification of tissue boundaries with structural modalities has been used to provide accurate structure for numerical modeling. Dehghani et al. (2004) demonstrated reduced artifacts and improved imaging accuracy when the air/tissue boundary was modeled with an anatomically correct mesh

shape determined with a surface camera and a mechanical deformation model rather than a conical mesh. A more accurate approach used the outer surface defined by x-ray mammography (Li et al. 2003, Fang et al. 2009), MRI (Ntziachristos et al. 2002, Carpenter et al. 2008), or US (Zhu et al. 2003) to define the numerical boundary. The structural modality can also be used to determine fiber locations with respect to this boundary with high accuracy, using radiological fiducial markers and the appropriate "locator" scans (Zhang et al. 2005b, Brooksby et al. 2006). These images are demonstrated in Figure 19.9.

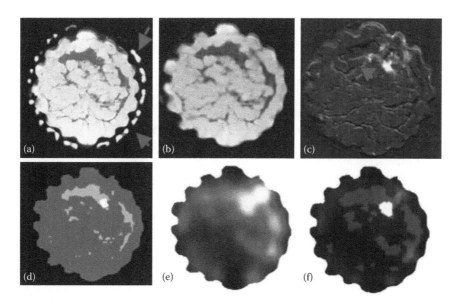

FIGURE 19.9 (a–d) Structural imaging information that may be incorporated into the optical imaging reconstruction includes (a) locating the optical fibers via the placement of radiological fiducial markers, (b) determining air–tissue boundaries to aid the light propagation model, and (b, c) separating tissue types (in this case, into adipose, fibroglandular, and suspect lesion). This information may be (d) segmented into a binary mask to define external and internal tissue boundaries. The end result, (e, f), are optical images (total hemoglobin in this case) that may either (e) loosely or (f) rigidly incorporate the structural modality.

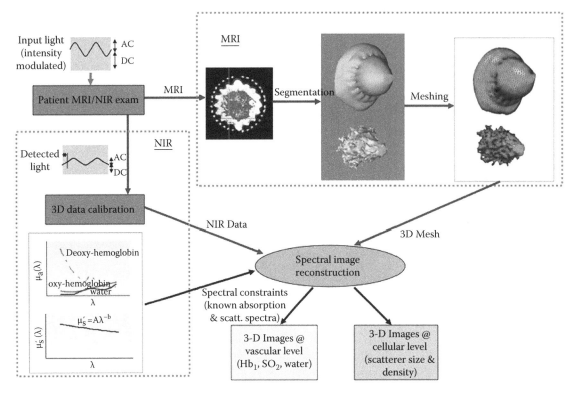

FIGURE 19.10 (See color insert.) Schematic of processes involved in multimodality MRI-optical imaging with specific application to breast tissue characterization for diagnosis and treatment monitoring.

Interior boundaries defined by structural modalities have also been used to dramatically improve imaging accuracy (Brooksby et al. 2005) by influencing the distribution of optical contrast. Both x-ray mammography images (Li et al. 2003, Fang et al. 2009) and MR images (Ntziachristos et al. 2002, Brooksby et al. 2004) have been used to aid in the optical reconstruction. MR imaging appears to be optimal for defining interior boundaries because of its high soft tissue contrast. In addition, its high sensitivity in detecting abnormal lesions facilitates using the MR completely for spatial details, while using the optical imaging to obtain molecular signatures of tissue types defined by MR (Carpenter et al. 2008). These techniques are sensitive not only to the choice in interior boundaries, but also to the segmentation error that may result in co-registration of the interior boundaries (Boverman et al. 2005, Carpenter et al. 2007a). Images that have been used to define the boundaries using MRI are shown in Figure 19.9.

Various approaches have been implemented to adjust the influence of the boundaries defined by structural modalities on the optical images. Several approaches have implemented spatial filters to influence optical contrast (Intes et al. 2004, Guven et al. 2005, Brooksby et al. 2005, Yalavarthy et al. 2007), which include parameters to adjust the influence of boundaries. Other approaches have used the boundary information as an edge constraint to force optical contrast to lie within certain tissue types, such as the cortex in functional brain imaging (Boas and Dale 2005, Schweiger and Arridge 1999), or the adipose, fibroglandular, and suspect lesion regions in breast imaging (Barbour et al.

1995, Brooksby et al. 2003, Carpenter et al. 2008). An example of using different approaches to incorporate those boundaries is shown in Figure 19.9e and f. The optimal approach for incorporating the internal tissue boundaries depends on the strength of the correlation between the modalities.

The methodology to integrate optical and structural multimodality imaging is described by the flowchart in Figure 19.10. These individual steps of segmentation, meshing, and visualization may be specific to applications and need to be optimized in the future for enabling truly multimodal imaging to be performed in a regular manner.

19.3 In Vivo Diffuse Optical Tomography

19.3.1 Optical Imaging of the Breast

19.3.1.1 Historical Developments

Many of the major developments in diffuse optical imaging have been advanced through imaging of breast cancer. Breast imaging offers opportunities for optics because of the need for more effective, lower-cost, nonionizing alternatives to x-ray and MR mammography. In addition, accurate numerical models of light propagation in the breast, and the ability to detect light transmitted through the breast, improve the accuracy and utility of optical mammography.

Early breast imaging systems measured the attenuation of light intensity in projections, similar to x-ray mammography.

This technique, called transilliumination light scanning, or diaphanography, used projections of CW light from a lamp to determine areas of high absorption. The pioneering work by Cutler et al. (Cutler 1929) was followed by Bartrum and Crow (1984) and others (Alveryd et al. 1990, Gisvold et al. 1986) who used more modernized equipment to guide visible and NIR light to the breast with fiber optics and detect breast projections with video equipment. These studies were suspended after reporting lower sensitivity and higher numbers of false positives compared to x-ray mammography (Alveryd et al. 1990), with poor spatial resolution, even with sophisticated CT image reconstruction techniques (Jackson et al. 1987). Later attempts used laser light sources to localize the incident light source, with limited improvements (Jarry et al. 1984).

Noting that the tissue scatter properties needed to be quantified to more accurately determine tissue light absorption properties, TD and FD techniques were introduced to separate absorption from scatter (Patterson et al. 1989, 1991). These techniques quantify the time delay of photons caused by light scatter by either recording the temporal spread of photons resulting from an ultrafast laser pulse (Chance et al. 1988), or the phase shift of frequency-locked laser light (Gratton et al. 1993). Using FD techniques, empirical or model-based approaches have been shown to be useful to determine tissue properties.

The empirical approach, termed the method of *N*-images, or the edge-correction method, accounts for intensity attenuation due to tissue thickness and photon loss through the edge of the detected volume by utilizing the nearly linear dependency of phase on source-detector distance (Fantini et al. 1996). To account for photon losses, the ac intensity measurement is amplified by a correction factor, which includes the phase (related to source-detector distance) and a reference measurement. This method produces spatial maps of corrected intensity (*N*) that have smaller dynamic range and are more sensitive to inhomogeneities. This method was shown in several case studies to offer improved images of breast cancer (Franceschini et al. 1997).

Alternatively, the model-based approach fits measurements of amplitude and phase to a light propagation model (Arridge and Schweiger 1997). In the breast, the lossy diffusion equation is universally used to model light transport, because of the high scattering albedo (Patterson et al. 1990). One drawback of this approach is that it requires the solving of a nonlinear inverse problem to determine tissue properties. Despite the inherent complexity, this method is by far the most popular approach in current optical breast imaging (Poplack et al. 2007, Choe et al. 2009, Jiang et al. 2001, Culver et al. 2003, Fang et al. 2009).

The advantage of the model-based approach is that it quantifies the absorption and scattering of tissue. These images may be used to determine blood content, oxygen saturation, water, and lipid content (Corlu et al. 2005, Li et al. 2004, Srinivasan et al. 2004), and cellular features related to tissue scatter (Wang et al. 2005, Li et al. 2008). These contrasts are promising for cancer detection, because cancer is characterized by many of these physiological properties, including high vascularity and reduced oxygenation (Vaupel et al. 1989). Additionally, tumors may have localized regions of edema, which may be determined in water content images.

19.3.1.2 Clinical Viability

Clinical studies investigating physiological properties with optical imaging have fallen into two categories: (1) the identification of healthy tissue properties (Durduran et al. 2002, Srinivasan et al. 2003, Brooksby et al. 2006, Pogue et al. 2004, Poplack et al. 2004) and (2) the performance of identifying diseased breast tissue such as invasive ductal carcinomas from benign lesions (Pogue et al. 2001, Choe et al. 2009, Chance et al. 2005, Zhu et al. 2005, Ntziachristos et al. 2002, Poplack et al. 2007), and benign cysts (van de Ven et al. 2009, Gu et al. 2004). Although geometries, photodetectors, and techniques vary between the different studies, clinical outcomes appear consistent (Leff et al. 2008).

Clinical studies consistently show high contrast in total hemoglobin between malignant tumors and both healthy and benign lesions. Total hemoglobin contrast ratios of 1.16:1 (Choe et al. 2009), 1.35:1 (Poplack et al. 2007), 1.5:1 (Grosenick et al. 2005), 1.36:1 (Intes 2005), and 5.8:1 (Spinelli et al. 2005) between malignant lesions and the background tissue have been reported for various systems around the world. Although tumors are known to have local regions of hypoxia, disease separation based on tissue oxygenation has shown inconclusive trends in the literature. Several larger studies (Grosenick et al. 2005, Choe et al. 2005, Spinelli et al. 2005) have shown no significant separation between malignant tissue oxygenation and the background. This is probably due to the poor spatial resolution of optical imaging combined with the fact that it measures hemoglobin oxygen saturation rather than tissue oxygenation.

Tissue scattering may be a promising contrast in breast imaging because it is sensitive to cellular size and packing density (Wang et al. 2005), and breast tumors are known to have highly dense vasculature (Bigio et al. 2000). The scattering properties of the breast reported in the literature have been inconsistent, however, probably due to the difficulty in obtaining sufficient stability in phase measurements. However, more recent studies have shown promise in characterizing the scattering properties of malignant tissue (Choe et al. 2009, Wang et al. 2006, Carpenter et al. 2007b, Li et al. 2008).

19.3.1.3 Additional Disease Contrast

An advantage of optical imaging over other clinical modalities such as PET, CT, and MRI is the vast number of contrast agents, owing to developments in cellular tagging in microscopy. Contrast agents in general offer potentially better disease-specific contrast, because they may be designed to specifically bind to diseased tissue. Furthermore, contrast washout dynamics may yield more information about function and pathology of tissue (Alacam et al. 2008). In breast imaging, studies have focused exclusively on indocyanine green (ICG), because of the limited number of approved optical contrast agents. Two pioneering studies demonstrated the promise in ICG imaging of the breast. Ntziachristos et al. (2000) noted an increase in ICG uptake in regions, which were identified in MRI to be tumors. Intes et al. (2003) identified slower contrast kinetics in malignant tumors than healthy tissue, although another study did not show such differentiation (Rinneberg et al. 2008). More discussion of the utilization of fluorescent agents in diffuse optical imaging can be found in Chapters 22 and 31.

Dynamic breast compression has also been proposed as a means to improve contrast between diseased and normal tissue (Jiang et al. 2003). Several studies have identified the decrease in total hemoglobin (Jiang et al. 2009, Carp et al. 2008) and water (Jiang et al. 2003) under compression in the healthy breast. Although the physiology of tumors under compression has not been investigated thoroughly, it is believed that the higher interstitial pressures in the diseased tissue (Carp et al. 2006) and the stiffer tissue matrix (Van Houten et al. 2003) may prevent the reduction in total hemoglobin and water, and thus increase contrast under compression (Boverman et al. 2007).

19.3.1.4 Future Improvement in Imaging Accuracy

Limitations in CW and FD optical imaging hardware have spurred new hardware developments to improve quantification and image accuracy. Systems with broadband wavelengths across the NIR range have improved tissue chromophore separation and quantification (Li et al. 2006). Combined CW and FD systems that use frequency-modulated light in the PMT optical sensitivity range and CW light for longer wavelengths have been shown to dramatically improve water estimates (Wang et al. 2008). The use of multiple frequency data has resulted in a reduction in image artifacts (Intes and Chance 2005). Video-rate DAQ methods have enabled the imaging of hemodynamics or contrast kinetics in real time (Piao et al. 2005), albeit in small imaging domains.

Multimodality methods are a logical step in improving quantification and imaging accuracy. Optical imaging has been successfully co-registered with x-ray (Zhang et al. 2005a, Fang et al. 2009), MRI (Ntziachristos et al. 2000, Brooksby et al. 2004, Carpenter et al. 2007b), and PET (Konecky et al. 2008), and select studies have shown improved tissue property recovery (Brooksby et al. 2005, Yalavarthy et al. 2007). These imaging systems may offer an ideal breast imaging modality by combining high detection rates with high functional information for lesion characterization.

19.3.2 Optical Imaging of the Brain

Functional brain imaging is a technique that measures the neurophysical response to physical, visual, or aural stimulations. During functional activation, the metabolic demands of the tissue increase. The tissue oxygenation in the brain is strictly controlled; therefore, this activation is followed by a local increase in blood flow. Functional brain images aim to determine the spatial location and intensity of the blood flow, blood volume, and blood oxygenation response to brain activation. Knowledge of functional activation may be used to investigate the extent of neurological diseases, such as Alzheimer's disease (Greicius et al. 2004), epilepsy (Detre et al. 1998), stroke (Cramer et al. 1997), and others.

Currently, the established tool to measure hemodynamic changes in the brain is functional MRI (fMRI), which measures changes in Hb using a method called blood oxygen level-dependent (BOLD) MRI. Optical imaging's role in functional brain imaging is promising for two reasons. First, optical imaging measures both HbO and Hb. Therefore, it has the ability to decouple physiological changes due to blood volume and oxygenation, a primary limitation in fMRI. Additionally, optical imaging may be low cost and portable. This enables the imaging of subjects that may not be suitable for MRI, such as infants or patients undergoing critical care (Hillman 2007).

Most optical brain images in the literature are topographical maps of the surface of the cortex. These images consist of multiple spectroscopic measurements that are interpolated to form the hemodynamic topography of the brain (Villringer and Chance 1997, Miki et al. 2005, Franceschini et al. 2007, Perrey 2008). The disadvantage of this approach is that depth sensitivity is completely dependent upon the source detector distance, and volumetric images are not computed. DOT, on the other hand, allows volumetric probing of functional activation (Boas et al. 2004a,b).

DOT is inherently more challenging in the brain than in the breast. This is due to the higher temporal resolution needed to observe functional signals, the larger volumes in the brain, and the difficulty in modeling light propagation due to the presence of the skull and the cerebral spinal fluid. Most DOT studies in the brain have used time-resolved photon measurements, and have enabled images of oxygen saturation changes due to functional stimulation in healthy adults (Gibson et al. 2006) and ill neonates (Benaron et al. 2000). Since TD techniques may be prohibitively expensive and impractical in clinical settings, CW systems have been designed to image functional brain activation. These systems use frequency-encoded wavelengths and temporally encoded source locations with APD-based detection hardware to achieve high temporal resolution images (Culver et al. 2006, Joseph et al. 2006). Using this technique, Bluestone et al. (2001) recovered images of tissue hemodynamics during the Valsalva maneuver. Zeff et al. (2007) imaged a spatially varying visual stimulus and found similar results to those reported by fMRI and PET.

Recent efforts in brain activation optical imaging have concentrated on multimodality methods. These approaches have been encouraged by the use of fMRI's higher spatial resolution to aid the spatial accuracy of optical imaging (Boas and Dale 2005). To better understand the complimentary imaging data, several studies have been performed to correlate BOLD and optically measured hemodynamics. These studies have demonstrated positive correlations between the BOLD signal and HbO in the rat brain during whisker stimulation (Kennerley et al. 2005), forepaw stimulation (Siegel et al. 2003), hypoxic respiratory challenge (Chen et al. 2003, Xu et al. 2005), and in the human brain during visual (Toronov et al. 2007) and motor stimulation (Sassaroli et al. 2006, Huppert et al. 2006). Both

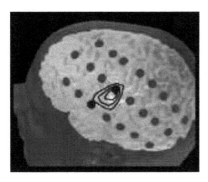

FIGURE 19.11 Activation maps for four subjects in Ou et al. (2009). Neural activation estimated from MEG measurements at 35 ms after the onset of the first stimulus in the stimulus train is depicted in yellow. The contour plot corresponds to 90%, 75%, and 50% thresholds of maximum signal decrease in HbR estimated from the DOI measurements in the interval 3–5 s after stimulation onset. Three rows of DOI source–detector pairs are shown as red and blue circles. The blue circles indicate the source–detector pair that detected the strongest vascular response. (Courtesy of Ou, W. et al., *Neuroimage*, 46(3), 624, 2009. With permission.)

positive correlation (Huppert et al. 2006, Siegel et al. 2003) and negative correlation (Toronov et al. 2007, Kennerley et al. 2005, Chen et al. 2003, Seiyama et al. 2004) between BOLD and Hb have been reported. By providing accurate and quantitative hemodynamic information through the combination of diffuse optics and fMRI, a superior instrument to noninvasively monitor brain function seems achievable.

A more advantageous multimodality approach may result from the addition of MEG to optical imaging. Adding MEG provides information about electrical response in the cortex, which may compliment the vascular response measured by optical imaging. Ou et al. (2009) demonstrated a spatial correlation between the two modalities during median-nerve electrical stimulation. An example of this spatial correlation in a human volunteer is shown in Figure 19.11. This study also showed a significant correlation between the late neural components (>30 ms) and both HbO and Hb. These types of complimentary experiments have great potential to understand the basic mechanisms of neural activity in the brain.

19.3.2.1 Diffuse Optical Imaging and Characterization of Rheumatoid Arthritis in Finger Joints

Rheumatoid arthritis (RA) is a progressive inflammatory disease of the joints, which afflicts 1.6 million people in the United States (Helmick et al. 2008). In the early stages, RA is characterized by inflammation of the synovial fluid and synovial membrane, also known as synovitis. As the disease progresses, the inflamed tissue invades the joint cavity, recruiting new blood vessel growth as it proliferates, a process known as neovascularization. The progression results in the destruction of cartilage and the erosion of the bone structure. A complete loss of function of the joint is probable in severe cases. The standard diagnosis of RA is completed with a combination of clinical examinations, lab workups and x-ray imaging (Rindfleisch et al. 2005). The latter can show detailed bone structure to reveal bone erosion later in the progression of RA; however, it is insensitive to changes in the synovial fluid and synovial membranes (synovitis), which are indications of the early stages of disease. Alternative imaging approaches capable of characterizing the soft-tissue and fluid phases in the joint more directly would help detect RA earlier, quantify the extent of the disease, and monitor the disease state over time.

In the mid-1990s, researchers reported changes in the optical properties of the synovial fluid in patients of RA (Beuthan et al. 1996, Prapavat et al. 1998). It was shown that as disease progresses, the synovial fluid loses optical transparency and becomes cloudy and more highly attenuating. The spectral changes of the optical properties were documented in Beuthan et al. (1996) and generally show an increase in both μ_a and μ_s with the onset of disease. These findings inspired further research activity using transillumination NIR light to characterize the property changes and diagnose the disease state. Scheel et al. (2002) and Schwaighofer et al. (2003) demonstrated that NIR transillumination is sensitive to these changes and can achieve sensitivity and specificity rates of over 80% when detecting changes from initial baseline readings. Meanwhile, researchers developing optical tomographic techniques to image the internal distribution of optical properties in tissue volumes, such as the breast and brain, began exploring the feasibility of applying model-based imaging techniques to the finger joint (Klose et al. 1998). It is thought that imaging the internal distribution of the optical properties will produce more accurate values for the optical properties, better characterize the extent of disease, and eliminate the need to measure changes against baseline measurements.

Xu et al. (2001) published the first DOT images of finger joints in a healthy volunteer using a PMT-based CW imaging system at 785 nm. The patient interface consisted of a single plane of optical fibers surrounding the finger joint in a coronal geometry. Shortly after this paper appeared, a CW imaging system based on silicon photodiodes and configured in the sagittal imaging geometry was introduced for tomographic finger joint imaging (Beuthan et al. 2002, Netz et al. 2001). Data from this system were processed using the radiative transport theory rather than the diffusion approximation for the recovery of optical properties in the finger joints, a more accurate, though computationally expensive, modeling approach. Three-dimensional DOT imaging of finger joints was first demonstrated by Xu et al. (2002) and detailed analyses of a small number of cases in human finger joint imaging were published by Hielscher et al. (2004) and Zhang and Jiang 2005 in the following years. While limited in scope and statistical power, these studies showed that optical imaging could detect difference between diseased and healthy joints and lead to a larger clinical trial.

The most extensive clinical trial of optical imaging of finger joints was completed by Scheel et al. (2005) in 2004, though it should be noted that the reconstruction algorithm used in this study was based upon the radiative transport equation (RTE) rather than the diffusion approximation. Still, it is a noteworthy paper in this discussion since it uses similar acquisition procedures and relies on changes in the same optical properties to produce disease-to-normal tissue contrast. RTE-based optical tomograms of affected and unaffected finger joints in 13 RA patients were compared with clinical diagnosis and ultrasound imaging of the finger joints. Receiver operator characteristic curves for the optical data produced relatively modest values for sensitivity and specificity, in the 70% range. While the ability to discriminate healthy and normal was encouraging, it was clear that more advanced approaches would be required to improve the diagnostic performance of the new modality. Following this study, finger joint optical imaging systems became more sophisticated with the introduction of rapid multispectral DAQ, FD measurements, and spatial-prior implementations.

Prior to the publication of a 2007 paper by Lasker et al. (2007), optical tomography of finger joints had been performed using single-wavelength systems capable only of extracting the properties μ_a and μ'_s. While changes in optical properties may reveal underlying changes that are related to physiologically relevant parameters, the spectral content of optical data can extract physiologically meaningful data directly in the form of hemoglobin concentration and water content. In the Lasker study, a two-wavelength CW system was used to investigate RA-induced angiogenesis in the synovium by imaging HbO and Hb concentrations. The system was also designed to image at approximately 2.4 Hz and thus was used to measure dynamic changes caused by venous and arterial occlusion. In this preliminary study, the patient sample size was quite small, precluding conclusions about diagnostic power of the technique; however, indications of angiogenesis were observed in finger joints with clinically diagnosed RA based on hemoglobin content.

To date, all systems that have been used in vivo have used CW DAQ with silicon photodiodes, PMTs or CCD cameras. In 2008, Netz et al. (2008) introduced a FD finger joint imaging system composed of an intensified CCD detection unit. The system operates at frequencies as high as 1 GHz and is able to provide valuable phase shift information to more accurately extract tissue scattering properties. This represents the only FD system dedicated to finger joint imaging, though in vivo data have not been published using this system as of the publication date of this book.

The trend toward coupling high-resolution structural imaging with diffuse tomography systems has recently been extended to finger joint imaging by Yuan et al. (2008). In a 2008 paper, they describe a hybrid x-ray tomosynthesis/DOT imaging system dedicated to finger joint imaging. The optical component consists of four planes of optical probes that circle the finger for 3D DAQ and a CW CCD detector. While the system operates using only a single wavelength (as published to date), the innovation for the finger joint imaging field is the introduction of spatially guided image recovery routines for the optical properties, which improves quantitative and qualitative accuracy. This system has been tested with only a few patients to date, but is a promising step toward more accurate optical images.

While the most recently introduced finger joint imaging systems incorporate some of the cutting-edge technologies found in other parts of the DOT community, in general, the advances have not kept pace with the capabilities of the DOT breast and brain imaging systems. The improvements in imaging performance using FD data, as well as spectral and spatial prior implementations have been well documented for breast and brain imaging over the past several years. The most recent iterations of finger joint imaging systems are migrating toward these more capable technologies and thus offer more promising platforms for RA diagnosis.

19.3.2.2 Prostate Imaging

Most DOT imaging systems have been developed as noninvasive imaging platforms. Thus, the patient interfaces are usually designed to surround and transilluminate the organ of interest, such as in breast and finger joint imaging, or sample a portion of the organ from the patient's exterior in reflectance mode, such as in brain imaging. This limits the application of the technique to a modest number of organs accessible from the exterior of the patient. Researchers are now investigating imaging organs from within the body using surgical implantation of optical fibers or endoscopic imaging probes adapted with optical source-detector arrangements. In particular, interest in imaging the prostate originated from the need for researchers developing photodynamic therapy (PDT) techniques to treat prostate cancer to determine the dose delivered to the normal and diseased tissues. Zhu (Zhu et al. 2005) and other researchers (Ali et al. 2004, Arnfield et al. 1993, Svensson et al. 2007) have conducted extensive studies to determine the optical properties of the prostate CW and TD probe systems. In Zhu's work, optical probes implanted directly into the prostate gland through the rectal cavity during the PDT surgical procedure were used to interrogate the tissue properties using CW measurements before and after treatment. A natural extension of this work is to acquire full tomographic data sets for imaging recovery and numerical simulations (Zhou and Zhu 2006) and preliminary clinical imaging studies (Zhou et al. 2007) have been published demonstrating the feasibility of this approach.

The contributions of the dosimetry studies led some researchers to develop systems capable of diagnosing and characterize prostatic lesions based on the endogenous optical contrast (Ali et al. 2004, Jiang et al. 2008). DOT imaging using endoscopy introduces challenges associated with working in restricted volumes. Unlike breast and brain imaging systems that can have fairly large patient interface arrays compared to the organ under investigation, the size of the probe head on a DOT endoscopic device is severely restricted. The use of endoscopic probes generally precludes transillumination of the tissue and therefore must operate in the reflectance geometry. The problem of variable depth sensitivity, an issue even for transillumination DOT imaging systems, is magnified in this geometry. Furthermore, positioning of the probe to interrogate the tissue of interest is also more difficult in an endoscopic configuration and thus video rate imaging may be necessary to help locate the probe head with respect to the diseased tissue.

The most advanced prostate endoscopy system to date was introduced in a series of papers published by Xu et al. (2008) and Jiang et al. (2008). This system combines a novel optical imaging array with a standard clinical transrectal ultrasound probe. Light is transmitted to and from the tissue through optical fibers that are coupled to micro-optics prisms and grin lenses on the probe head. These devices redirect light between the fibers and the tissue in both source and detection channels. The entire unit fits on a clinical transrectal ultrasound transducer for simultaneous ultrasound and optical imaging of the tissue. For near-video-rate imaging, the single laser source is spectrally encoded to allow all source positions to be illuminated simultaneously. Thus, each source position transmits light at a given wavelength that is slightly different from all other source positions. The detection channel is based on a CCD coupled to a spectrograph, which detects all source-detector pairs simultaneously and decodes the measured signals using the spectral information. This approach allows the system to acquire images at approximately 10 Hz. The structural information provided by the transrectal ultrasound images is used as spatial prior information in the optical reconstructions to reduce depth-dependent sensitivity and improve overall imaging accuracy. To date, this unique system has been used in canine cadavers and ex vivo tissue samples.

19.4 Challenges and Critical Discussion

Though DOT has demonstrated sensitivity to a variety of important physiological parameters that are unquantifiable with conventional medical imaging modalities, the nature of light propagation through tissue puts relatively tight limits on what regions of the body may be imaging with DOT. NIR light is detectable through a maximum depth of approximately 10 cm in fatty and fibro-glandular tissues. Highly absorbing tissue with high concentrations of hemoglobin, such as muscle tissue, reduce the tissue penetration even further. These limits of light penetration make measurements through the whole body infeasible. Additionally, since DOT is a model-based imaging method, it is critical that the diffusion model accurately describes the photon propagation through tissue, precluding DOT imaging over short distances (approximately less than 1 cm) or through regions in the body with low scattering properties, such as the airways, unless proper photon migration models are implemented (see Chapter 18 for an extension of DOT to short distances). Imaging geometry is a critical design consideration for interrogating the tissue volume of interest. While most DOT systems have focused on relatively accessible organs such as the breast, and brain, migrating the technology to endoscopic imaging of the body's interior requires more innovative design.

There are many trade-offs in selecting the appropriate photo-detection hardware for optical imaging, including speed, sensitivity, cost, dynamic range, and ease-of-use. Generally, because of the low light intensities involved, photodetectors with the highest gain, such as APDs and PMTs, are used because they offer the highest photon sensitivity for imaging tissue at depth. These may be used over a large dynamic range, and may even be heavily biased to enable single photon counting. These detectors can achieve an adequate SNR for up to 10 cm of tissue, which is common in breast imaging. However, these detectors require the selection of voltage gains to acquire signal with an optimal SNR, which can be time consuming. If the experimental conditions allow detector calibration, these gains may be predetermined before data collection to improve acquisition speed. The nonuniformity in spectral sensitivity is an additional challenge to the selection of the appropriate detector. PMTs are far less uniform than solid-state photodetectors because of the use of a photocathode, and may even preclude adequate SNR from all wavelengths of interest. Additionally, PMTs are expensive compared to solid-state detectors, due to the manufacturing difficulties in constructing low-noise electron tubes, and can be easily damaged during exposure to high light levels. In general, the optimal choice of detector depends on the tissue spectrum, the desired temporal resolution, and the budget.

Innovations in efficient numerical modeling and reconstruction algorithms, and the use of parallelized computing resources are effective methods to address the difficulties in processing large, information-rich data sets produced by increasingly sophisticated experimental systems. The Intel MKL library offers an attractive option through PARDISO, a parallel direct solver, which can be used to solve the forward problem in a multiprocessing environment.

The Moore–Penrose inverse using

$$\Delta \mu = J^{\mathrm{T}}(JJ^{\mathrm{T}} + \alpha I)^{-1}\Delta\Phi \qquad (19.35)$$

provides an efficient inversion for large 3D problems when J is under-determined, that is, the number of unknowns is much larger than measurements available (Arridge et al. 1997). Using this inverse formulation, the size of the Hessian is now a function of measurement size rather than mesh size.

The other challenge in 3D is with respect to MSIR. Figure 19.12 shows a schematic of the change in Jacobian size going

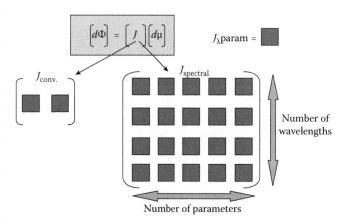

FIGURE 19.12 Change in size of Jacobian from conventional reconstruction to MSIR.

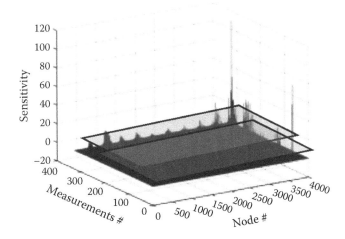

FIGURE 19.13 Plot of the magnitude of the Jacobian for a rectangular imaging domain Threshold values can be set (see planes in the figure) below which the sensitivity of the measurements is very low.

19.14). The results between the two algorithms varied by less than 2%.

Ultimately, advancing the technology beyond laboratory research requires extensive clinical evaluation of the imaging systems, starting with the assessment of the imaging limits. These metrics can be determined using the standards of the conventional medical imaging community. One approach that has been applied to DOT imaging is contrast-detail analysis (Davis et al. 2005, Pogue et al. 2000, Song et al. 2004), which examines the trade-off between the minimum detectable size and contrast of an abnormality. These studies are typically completed using tissue-simulating phantoms containing objects of various diameters and contrasts, and thus true values of the object parameters are known. Images of the phantoms are used to determine the detection limits for the range of object sizes and contrasts imaged. These detectable limits are determined either by human readers or by parameters such as contrast-to-noise ratio calculated from the image. Typically, smaller objects require higher contrasts to be detected while larger objects may be detected with less contrast.

While controlled studies such as contrast-detail analysis provide an effective approach for analyzing the device in the context of the contrast mechanisms expected in vivo, the question of whether the technology is adequately sensitive to biologically relevant perturbation in the tissue arising from disease is of critical importance. Not only must these changes by detectable by the imaging system, but they must be sufficiently prevalent in either the general patient population or a subset of that population to facilitate reliable diagnosis. Disease heterogeneity throughout the population can erode the diagnostic performance of the imaging technology, reducing sensitivity and specificity rates. DOT imaging of cancer relies on the perturbations in hemoglobin concentration caused by neovasculature, metabolic activity and oxygen consumption, water content, as well as the size

from conventional image reconstruction to MSIR. This requires large memory capabilities and methods to reduce and optimize this requirement are needed for successful volumetric imaging.

Several approaches have been used to counter this. One such method is based on storing the Jacobian as a sparse matrix. This is accomplished by setting a threshold on the Jacobian values, below which the elements of the matrix are set to zero. Figure 19.13 shows a plot of the Jacobian, and from the figure it is obvious that a large portion of the Jacobian has low sensitivity values, which do not contribute much information regarding the domain to the reconstruction. By setting a threshold, a majority of these values will become zero so that the size of the Hessian (J^TJ) requires less memory. In the example shown, the sparse Hessian required 75% of the memory required by the full matrix when stored as JJ^T and only 12% of the when stored as J^TJ (Figure

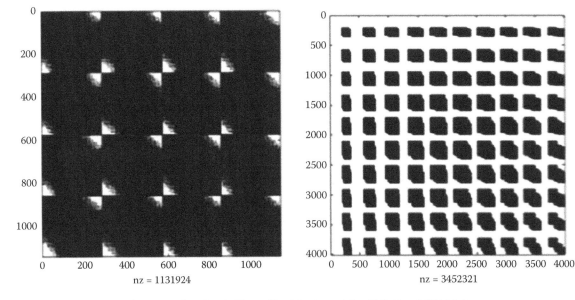

FIGURE 19.14 Sparsity patterns of Hessian when thresholding of Jacobian is used: for JJ^T (left) and J^TJ (right).

and density of organelles and cellular membranes caused by the disease. While not all of these contrast mechanisms have demonstrated diagnostic value, it may be possible to enhance sensitivity to these perturbations or exploit additional contrast mechanisms by using more sophisticated imaging devices capable of rapidly measuring changes due to a stimulus, such as tissue compression, or detecting and imaging exogenously administered optical contrast agents. The latter is addressed more fully in Chapters 22 and 31.

References

Alacam, B., Yazici, B., Intes, X., Nioka, S., and Chance, B. 2008. Pharmacokinetic-rate images of indocyanine green for breast tumors using near-infrared optical methods. *Phys. Med. Biol.* 53(4): 837–859.

Ali, J. H., Wang, W. B., Zevallos, M., and Alfano, R. R. 2004. Near infrared spectroscopy and imaging to probe differences in water content in normal and cancer human prostate tissues. *Technol. Cancer Res. Treat.* 3(5): 491–497.

Alveryd, A., Andersson, I., Aspegren, K., Balldin, G., Bjurstam, N., Edstrom, G., Fagerberg, G., Glas, U., Jarlman, O., and Larsson, S. A. 1990. Lightscanning versus mammography for the detection of breast cancer in screening and clinical practice. A Swedish multicenter study. *Cancer* 65(8): 1671–1677.

Arnfield, M. R., Chapman, J. D., Tulip, J., Fenning, M. C., and McPhee, M. S. 1993. Optical properties of experimental prostate tumors in vivo. *Photochem. Photobiol.* 57(2): 306–311.

Arridge, S. R. and Schweiger, M. 1995. Photon-measurement density functions. Part 2: Finite-element-method calculations. *Appl. Opt.* 34: 8026–8037.

Arridge, S. R. and Schweiger, M. 1997. Image reconstruction in optical tomography. *Philos. Trans. R. Soc. Lond. B* 352: 717–726.

Arridge, S. R., Schweiger, M., Hiraoka, M., and Delpy, D. T. 1993. A finite-element approach for modeling photon transport in tissue. *Med. Phys.* 20(2): 299–309.

Barbour, R. L., Graber, H. L., Chang, J., Barbour, S. S., Koo, P. C., and Aronson, R. 1995. MRI-guided optical tomography: Prospects and computation for a new imaging method. *IEEE Comp. Sci. Eng.* 2: 63–77.

Bartrum, R. J. and Crow, H. C. 1984. Transillumination light scanning to diagnose breast cancer: A feasibility study. *Am. J. Roentg.* 142: 409–414.

Benaron, D. A., Hintz, S. R., Villringer, A., Boas, D., Kleinschmidt, A., Frahm, J., Hirth, C., Obrig, H., van Houten, J. C., Kermit, E. L., Cheong, W. F., and Stevenson, D. K. 2000. Noninvasive functional imaging of human brain using light. *J. Cereb. Blood Flow Metab.* 20(3): 469–477.

Beuthan, J., Netz, U., Minet, O., Klose, A. D., Hielscher, A. H., Scheel, A., Henniger, J., and Muller, G. 2002. Light scattering study of rheumatoid arthritis. *Quantum Electron.* 32: 945–952.

Beuthan, J., Prapavat, V., Naber, R. D., Minet, O., and Mueller, G. J. 1996. In: *Proceedings SPIE 2676, Advanced Bioimaging Techniques and Devices*, San Jose, CA, 1996, pp. 43–53.

Bigio, I. R., Bown, S. G., Briggs, G., Kelley, C., Lakhani, S., Pickard, D., Ripley, P. M., Rose, I. G., and Sanders, C. 2000. Diagnosis of breast cancer using elastic-scattering spectroscopy: preliminary clinical results. *J. Biomed. Opt.* 5(2): 221–228.

Bluestone, A., Abdoulaev, G., Schmitz, C., Barbour, R., and Hielscher, A. 2001. Three-dimensional optical tomography of hemodynamics in the human head. *Opt. Express* 9(6): 272–286.

Boas, D. and Dale, A. 2005. Simulation study of magnetic resonance imaging-guided cortically constrained diffuse optical tomography of the human brain function. *Appl. Opt.* 44(10): 1957–1968.

Boas, D. A., Chen, K., Grebert, D., and Francheschini, M. A. 2004a. Improving the diffuse optical imaging spatial resolution of the cerebral hemodynamic response to brain activation in humans. *Opt. Lett.* 29(13): 1506–1508.

Boas, D. A., Dale, A. M., and Franceschini, M. A. 2004b. Diffuse optical imaging of brain activation: Approaches to optimizing image sensitivity, resolution, and accuracy. *Neuroimage* 23(Suppl 1): S275–S288.

Boulnois, J. L. 1986. Photophysical processes in recent medical laser developments: A review. *Lasers Med. Sci.* 1: 47–66.

Boverman, G., Fang, Q., Carp, S. A., Miller, E. L., Brooks, D. H., Selb, J., Moore, R. H., Kopans, D. B., and Boas, D. A. 2007. Spatio-temporal imaging of the hemoglobin in the compressed breast with diffuse optical tomography. *Phys. Med. Biol.* 52(12): 3619–3641.

Boverman, G., Miller, E., Li, A., Zhang, Q., Chaves, T., Brooks, D., and Boas, D. 2005. Quantitative spectroscopic diffuse optical tomography of the breast guided by imperfect a priori structural information. *Phys. Med. Biol.* 50: 3941–3956.

Brooksby, B., Jiang, S., Dehghani, H., Pogue, B. W., Paulsen, K. D., Kogel, C., Doyley, M., Weaver, J. B., and Poplack, S. P. 2004. Magnetic resonance-guided near-infrared tomography of the breast. *Rev. Sci. Instrum.* 75(12): 5262–5270.

Brooksby, B., Jiang, S., Dehghani, H., Pogue, B. W., Paulsen, K. D., Weaver, J. B., Kogel, C., and Poplack, S. P. 2005a. Combining near infrared tomography and magnetic resonance imaging to study in vivo breast tissue: Implementation of a Laplacian-type regularization to incorporate MR structure. *J. Biomed. Opt.* 10(5): 050504-1–050504-10.

Brooksby, B., Pogue, B. W., Jiang, S., Dehghani, H., Srinivasan, S., Kogel, C., Tosteson, T., Weaver, J. B., Poplack, S. P., and Paulsen, K. D. 2006. Imaging breast adipose and fibroglandular tissue molecular signatures using hybrid MRI-guided near-infrared spectral tomography. *Proc. Natl. Acad. Sci. U. S. A.* 103: 8828–8833.

Brooksby, B., Srinivasan, S., Jiang, S., Dehghani, H., Pogue, B. W., Paulsen, K. D., Weaver, J., Kogel, C., and Poplack, S. P. 2005b. Spectral-prior information improves Near-Infrared diffuse tomography more than spatial-prior. *Opt. Lett.* 30(15): 1968–1970.

Brooksby, B. A., Dehghani, H., Pogue, B. W., and Paulsen, K. D. 2003. Near-infrared tomography breast image reconstruction with a priori structural information from MRI: Algorithm development for reconstructing heterogeneities. *IEEE JSTQE* 9(2): 199–209.

Carp, S. A., Kauffman, T., Fang, Q., Rafferty, E., Moore, R., Kopans, D., and Boas, D. 2006. Compression-induced changes in the physiological state of the breast as observed through frequency domain photon migration measurements. *J. Biomed. Opt.* 11(6): 064016.

Carp, S. A., Selb, J., Fang, Q., Moore, R., Kopans, D. B., Rafferty, E., and Boas, D. A. 2008. Dynamic functional and mechanical response of breast tissue to compression. *Opt. Express* 16(20): 16064–16078.

Carpenter, C., Srinivasan, S., Pogue, B., and Paulsen, K. 2008. Methodology development for three-dimensional MR-guided near infrared spectroscopy of breast tumors. *Opt. Express* 16(22): 17903–17914.

Carpenter, C. M., Srinivasan, S., Pogue, B. W., Dehghani, H., and Paulsen, K. D. 2007a. A Comparison of Edge Constrained Optical Reconstruction Methods Incorporating Spectral and MR- Derived Spatial Information. In *Proceedings of SPIE*, 64310K. San Jose, CA, 2007.

Carpenter, C. M., Pogue, B. W., Jiang, S. J., Dehghani, H., Wang, X., Paulsen, K. D., Wells, W. A., Forero, J., Kogel, C., Weaver, J. B., Poplack, S. P., and Kaufman, P. A. 2007b. Image-guided spectroscopy provides molecular specific information in vivo: MRI-guided spectroscopy of breast cancer hemoglobin, water, and scatterer size. *Opt. Lett.* 32(8): 933–935.

Chance, B., Leigh, J. S., Miyake, H., Smith, D. S., Nioka, S., Greenfeld, R., Finander, M., Kaufmann, K., Levy, W., Young, M., Cohen, P., Yoshioka, H., and Boretsky, R. 1988. Comparison of time-resolved and -unresolved measurements of deoxyhemoglobin in brain. *Proc. Natl. Acad. Sci. U. S. A.* 85: 4971–4975.

Chance, B., Nioka, S., Zhang, J., Conant, E. F., Hwang, E., Briest, S., Orel, S. G., D., S. M., and Czerniecki, B. J. 2005. Breast cancer detection based on incremental biochemical and physiological properties of breast cancers: A six-year, two-site study. *Acad. Radiol.* 12(8): 925–933.

Chen, Y., Tailor, D., Intes, X., and Chance, B. 2003. Correlation between near-infrared spectroscopy and magnetic resonance imaging of rat brain oxygenation modulation. *Phys. Med. Biol.* 48: 417–427.

Choe, R., Corlu, A., Lee, K., Duduran, T., Konecky, S., Grosicha-Kopyra, M., Arridge, S., Czerniecki, B., Fraker, D., DeMichele, A., Chance, B., Rosen, M., and Yodh, A. 2005. Diffuse optical tomography of breast cancer during neoadjuvant chemotherapy: A case study with comparison to MRI. *Med. Phys.* 32(4): 1128–1139.

Choe, R., Konecky, S. D., Corlu, A., Lee, K., Durduran, T., Busch, D. R., Czerniecki, B., Tchou, J., Fraker, D. L., DeMichele, A., Chance, B., Putt, E., Schnall, M. D., Rosen, M. A., and Yodh, A. G. 2009. Differentiation of benign and malignant breast lesions by in-vivo three-dimensional diffuse optical tomography. *Cancer Res.* 69(2): 102S.

Corlu, A., Choe, R., Durduran, T., Lee, K., Schweiger, M., Arridge, S. R., Hillman, E. M., and Yodh, A. G. 2005. Diffuse optical tomography with spectral constraints and wavelength optimization. *Appl. Opt.* 44(11): 2082–2093.

Cramer, S. C., Nelles, G., Benson, R. R., Kaplan, J. D., Parker, R. A., Kwong, K. K., Kennedy, D. N., Finklestein, S. P., and Rosen, B. R. 1997. A functional MRI study of subjects recovered from hemiparetic stroke. *Stroke* 28(12): 2518–2527.

Culver, J. P., Choe, R., Holboke, M. J., Zubkov, L., Durduran, T., Slemp, A., Ntziachristos, V., Chance, B., and Yodh, A. G. 2003. Three-dimensional diffuse optical tomography in the parallel plane transmission geometry: Evaluation of a hybrid frequency domain/continuous wave clinical system for breast imaging. *Med. Phys.* 30(2): 235–247.

Culver, J. P., Schlaggar, B. L., Dehghani, H., and Zeff, B. W. 2006. In: *Life Science Systems and Applications Workshop*, Bethesda, MD, 2006, pp. 1–2.

Cutler, M. 1929. Transillumination as an aid in the diagnosis of breast lesions. *Surg. Gyn. Obst.* 48: 721–729.

Davis, S. C., Pogue, B. W., Dehghani, H., and Paulsen, K. D. 2005. Contrast-detail analysis characterizes diffuse optical fluorescence tomography image reconstruction. *J. Biomed. Opt.* 10(5): 050501-1–050501-3.

Dehghani, H., Doyley, M., pogue, B. W., Jiang, S., Geng, J., and Paulsen, K. D. 2004. Breast deformation modeling for image reconstruction in near-infrared tomography. *Phys. Med. Biol.* 49: 1131–1145.

Detre, J. A., Maccotta, L., King, D., Alsop, D. C., Glosser, G., D'Esposito, M., Zarahn, E., Aguirre, G. K., and French, J. A. 1998. Functional MRI lateralization of memory in temporal lobe epilepsy. *Neurology* 50(4): 926–932.

Durduran, T., Choe, R., Culver, J. P., Zubkov, L., Holboke, M. J., Giammarco, J., Chance, B., and Yodh, A. G. 2002. Bulk optical properties of healthy female breast tissue. *Phys. Med. Biol.* 47(16): 2847–2861.

Fang, Q. Q., Carp, S. A., Selb, J., Boverman, G., Zhang, Q., Kopans, D. B., Moore, R. H., Miller, E. L., Brooks, D. H., and Boas, D. A. 2009. Combined optical imaging and mammography of the healthy breast: Optical contrast derived from breast structure and compression. *IEEE Trans. Med. Imaging* 28(1): 30–42.

Fantini, S., Franceschini, M. A., Gaida, G., Gratton, E., Jess, H., Mantulin, W. W., Moesta, K. T., Schlag, P. M., and Kaschke, M. 1996. Frequency-domain optical mammography: Edge effect corrections. *Med. Phys.* 23(1): 149–157.

Fishkin, J., Gratton, E., van de Ven, M. J., and Mantulin, W. W. 1991. Diffusion of intensity modulated near infrared light in turbid media. *Proc. SPIE* 1431: 122–135.

Franceschini, M. A., Moesta, K. T., Fantini, S., Gaida, G., Gratton, E., Jess, H., Mantulin, W. W., Seeber, M., Schlag, P. M., and Kaschke, M. 1997. Frequency-domain techniques enhance optical mammography: Initial clinical results. *Proc. Natl. Acad. Sci. U. S. A.* 94(12): 6468–6473.

Franceschini, M. A., Thaker, S., Themelis, G., Krishnamoorthy, K. K., Bortfeld, H., Diamond, S. G., Boas, D. A., Arvin, K., and Grant, P. E. 2007. Assessment of infant brain development with frequency-domain near-infrared spectroscopy. *Pediatr. Res.* 61(5 Pt 1): 546–551.

Gibson, A. P., Austin, T., Everdell, N. L., Schweiger, M., Arridge, S. R., Meek, J. H., Wyatt, J. S., Delpy, D. T., and Hebden, J. C. 2006. Three-dimensional whole-head optical tomography of passive motor evoked responses in the neonate. *Neuroimage* 30(2): 521–528.

Gisvold, J. J., Brown, L. R., Swee, R. G., Raygor, D. J., Dickerson, N., and Ranfranz, M. K. 1986. Comparison of mammography and transillumination light scanning in the detection of breast lesions. *AJR Am. J. Roentgenol.* 147(1): 191–194.

Gratton, E., Fantini, S., Franceschini, M. A., Gratton, G., and Fabiani, M. 1997. Measurements of scattering and absorption changes in muscle and brain. *Philos. Trans. R. Soc. Lond. Ser. B: Biol. Sci.* 352(1354): 727–735.

Gratton, E., Mantulin, W. W., van de Ven, M. J., Fishkin, J. B., Maris, M. B., and Chance, B. 1993. A novel approach to laser tomography. *Bioimaging* 1: 40–46.

Greicius, M. D., Srivastava, G., Reiss, A. L., and Menon, V. 2004. Default-mode network activity distinguishes Alzheimer's disease from healthy aging: Evidence from functional MRI. *Proc. Natl. Acad. Sci. U. S. A.* 101(13): 4637–4642.

Grosenick, D., Moesta, K. T., Moller, M., Mucke, J., Wabnitz, H., Gebauer, B., Stroszczynski, C., Wassermann, B., Schlag, P. M., and Rinneberg, H. 2005. Time-domain scanning optical mammography: I. Recording and assessment of mammograms of 154 patients. *Phys. Med. Biol.* 50(11): 2429–2449.

Gu, X., Zhang, Q., Bartlett, M., Schutz, L., Fajardo, L. L., and Jiang, H. 2004. Differentiation of cysts from solid tumors in the breast with diffuse optical tomography. *Acad. Radiol.* 11(1): 53–60.

Guven, M., Yazici, B., Intes, X., and Chance, B. 2005. Diffuse optical tomography with a priori anatomical information. *Phys. Med. Biol.* 12: 2837–2858.

Hale, G. M. and Querry, M. R. 1973. Optical constants of water in the 200-nm to 200-um wavelength region. *Appl. Opt.* 12(3): 555–563.

Hansen, P. C.. 1998. *Rank-Deficient and Discrete Ill-Posed Problems: Numerical Aspects of Linear Inversion*. Philadelphia, PA: SIAM.

Helmick, C. G., Felson, D. T., Lawrence, R. C., Gabriel, S., Hirsch, R., Kwoh, C. K., Liang, M. H., Kremers, H. M., Mayes, M. D., Merkel, P. A., Pillemer, S. R., Reveille, J. D., Stone, J. H., and National Arthritis Data Workgroup. 2008. Estimates of the prevalence of arthritis and other rheumatic conditions in the United States. Part I. *Arthritis Rheum.* 58(1): 15–25.

Hielscher, A. H., Klose, A. D., Scheel, A. K., Moa-Anderson, B., Backhaus, M., Netz, U., and Beuthan, J. 2004. Sagittal laser optical tomography for imaging of rheumatoid finger joints. *Phys. Med. Biol.* 49(7): 1147–1163.

Hillman, E. M. C. 2007. Optical brain imaging in vivo: Techniques and applications from animal to man. *J. Biomed. Opt.* 15(2): 051402.

Hueber, D. M., Franceschini, M. A., Ma, H. Y., Zhang, Q., Ballesteros, J. R., Fantini, S., Wallace, D., Ntziachristos, V., and Chance, B. 2001. Non-invasive and quantitative near-infrared haemoglobin spectrometry in the piglet brain during hypoxic stress, using a frequency-domain multidistance instrument. *Phys. Med. Biol.* 46(1): 41–62.

Huppert, T., Hoge, R., Dale, A., Franceschini, M., and Boas, D. 2006. Quantitative spatial comparison of diffuse optical imaging with blood oxygen level-dependent and arterial spin labeling-based functional magnetic resonance imaging. *J. Biomed. Opt.* 11(6): 064018.

Intes, X. 2005. Time-domain optical mammography SoftScan initial results. *Acad. Radiol.* 12(8): 934–947.

Intes, X. and Chance, B. 2005. Multi-frequency diffuse optical tomography. *J. Mod. Opt.* 52(15): 2139–2159.

Intes, X., Maloux, C., Guven, M., Yazici, T., and Chance, B. 2004. Diffuse optical tomography with physiological and spatial a priori constraints. *Phys. Med. Biol.* 49: N155–N163.

Intes, X., Ripoll, J., Chen, Y., Nioka, S., Yodh, A. G., and Chance, B. 2003. In vivo continuous-wave optical breast imaging enhanced with Indocyanine Green. *Med. Phys.* 30(6): 1039–1047.

Ishimaru, A. 1978. *Wave Propagation and Scattering in Random Media*. New York: Academic Press.

Jackson, P. C., Stevens, P. H., Smith, J. H., Kear, D., Key, H., and Wells, P. N. T. 1987. The development of a system for transillumination computed tomography. *Br. J. Radiol.* 60: 375–380.

Jarry, G., Ghesquiere, S., Maarek, J. M., Fraysse, F., Debray, S., Bui-Mong-Hung, and Laurent, D. 1984. Imaging mammalian tissues and organs using laser collimated transillumination. *J. Biomed. Eng.* 6(1): 70–74.

Jiang, H., Xu, Y., Iftimia, N., Eggert, J., Klove, K., Baron, L., and Fajardo, L. 2001. Three-dimensional optical tomographic imaging of breast in a human subject. *IEEE Trans. Med. Imaging* 20(12): 1334–1340.

Jiang, S., Pogue, B., Laughney, A., Kogel, C., and Paulsen, K. 2009. Measurement of pressure-displacement kinetics of hemoglobin in normal breast tissue with near-infrared spectral imaging. *Appl. Opt.* 48(10): D130–D136.

Jiang, S., Pogue, B. W., Paulsen, K. D., Kogel, C., and Poplack, S. P. 2003. In vivo near-infrared spectral detection of pressure-induced changes in breast tissue. *Opt. Lett.* 28(14): 1212–1214.

Jiang, Z., Piao, D., Xu, G., Ritchey, J. W., Holyoak, G. R., Bartels, K. E., Bunting, C. F., Slobodov, G., and Krasinski, J. S. 2008. Trans-rectal ultrasound-coupled near-infrared optical tomography of the prostate, part II: Experimental demonstration. *Opt. Express* 16(22): 17505–17520.

Jobsis, F. F. 1977. Non-invasive, infra-red monitoring of cerebral and myocardial oxygen sufficiency and circulatory parameters. *Science* 198: 1264.

Joseph, D. K., Huppert, T. J., Franceschini, M. A., and Boas, D. A. 2006. Diffuse optical tomography system to image brain activation with improved spatial resolution and validation with functional magnetic resonance imaging. *Appl. Opt.* 45(31): 8142–8151.

Kennerley, A., Beswick, J., Martindale, J., Johnston, D., Papadakis, N., and Mayhew, J. 2005. Concurrent fMRI and optical measures for the investigation of the hemodynamic response function. *Magn. Reson. Med.* 54: 354–365.

Klose, A. D., Hielscher, A. H., Hanson, K. M., and Beuthan, J. 1998. Two- and three-dimensional optical tomography of finger joints for diagnostics of rheumatoid arthritis. In *Proceedings of the SPIE Vol. 3566, Photon Propagation in Tissues IV*, San Jose, CA, 1998, pp. 151–160.

Konecky, S. D., Choe, R., Corlu, A., Lee, K., Wiener, R., Srinivas, S. M., Saffer, J. R., Freifelder, R., Karp, J. S., Hajjioui, N., Azar, F., and Yodh, A. G. 2008. Comparison of diffuse optical tomography of human breast with whole-body and breast-only positron emission tomography. *Med. Phys.* 35(2): 446–455.

Lasker, J. M., Fong, C. J., Ginat, D. T., Dwyer, E., and Hielscher, A. H. 2007. Dynamic optical imaging of vascular and metabolic reactivity in rheumatoid joints. *J. Biomed. Opt.* 12(5): 052001.

Lee, K., Konecky, S. D., Choe, R., Ban, A., Corlu, A., Durduran, T., and Yodh, A. G. 2007. In: *Diffuse Optical Imaging of Tissue*, Munich, Germany, 2007.

Leff, D. R., Warren, O. J., Enfield, L. C., Gibson, A., Athanasiou, T., Patten, D. K., Hebden, J., Yang, G. Z., and Darzi, A. 2008. Diffuse optical imaging of the healthy and diseased breast: a systematic review. *Breast Cancer Res. Treat.* 108(1): 9–22.

Li, A., Miller, E. L., Kilmer, M. E., Brukilaccio, T. J., Chaves, T., Stott, J., Zhang, Q., Wu, T., Choriton, M., Moore, R. H., Kopans, D. B., and Boas, D. A. 2003. Tomographic optical breast imaging guided by three-dimensional mammography. *Appl. Opt.* 42(25): 5181–5190.

Li, A., Zhang, Q., Culver, J., Miller, E., and Boas, D. 2004. Reconstructing chromosphere concentration images directly by continuous-wave diffuse optical tomography. *Opt. Lett.* 29(3): 256–258.

Li, C., Grobmyer, S. R., Massol, N., Liang, X., Zhang, Q., Chen, L., Fajardo, L. L., and Jiang, H. 2008. Noninvasive in vivo tomographic optical imaging of cellular morphology in the breast: Possible convergence of microscopic pathology and macroscopic radiology. *Med. Phys.* 35(6): 2493–2501.

Li, C., Zhao, H., Anderson, B., and Jiang, H. 2006. Multispectral breast imaging using a ten-wavelength, 64 × 64 source/ detector channels silicon photodiode-based diffuse optical tomography system. *Med. Phys.* 33(3): 627–636.

Li, Z., Krishnaswamy, V., Davis, S. C., Srinivasan, S., Paulsen, K. D., and Pogue, B. W. 2009. Video-rate near infrared tomography to image pulsatile absorption properties in thick tissue. *Opt. Express* 17(14): 12043–12056.

Marquardt, D. W. 1963. An algorithm for least-squares estimation of nonlinear parameters. *J. Soc. Ind. Appl. Math.* 11(2): 431–441.

McBride, T. O., Pogue, B. W., Osterberg, U. L., and Paulsen, K. D. 2003. Strategies for absolute calibration of near infrared tomographic tissue imaging. *Adv. Exp. Med. Biol.* 530: 85–99.

Miki, A., Nakajima, T., Takagi, M., Usui, T., Abe, H., Liu, C. J., and Liu, G. T. 2005. Near-infrared spectroscopy of the visual cortex in unilateral optic neuritis. *Am. J. Ophthalmol.* 139(2): 353–356.

Mourant, J. R., Fuselier, T., Boyer, J., Johnson, T. M., and Bigio, I. J. 1997. Predictions and measurements of scattering and absorption over broad wavelength ranges in tissue phantoms. *Appl. Opt.* 36(4): 949.

Netz, U., Beuthan, J., Cappius, H. J., Koch, H. C., Klose, A. D., and Hielscher, A. H. 2001. Imaging of rheumatoid arthritis in finger joints by sagittal optical tomography. *Med. Laser Appl.* 16: 306–310.

Netz, U. J., Beuthan, J., and Hielscher, A. H. 2008. Multipixel system for gigahertz frequency-domain optical imaging of finger joints. *Rev. Sci. Instrum.* 79(3): 034301.

Ntziachristos, V., Yodh, A. G., Schnall, M., and Chance, B. 2000. Concurrent MRI and diffuse optical tomography of breast after indocyanine green enhancement. *Proc. Natl. Acad. Sci. U. .S. A.* 97(6): 2767–2772.

Ntziachristos, V., Yodh, A. G., Schnall, M. D., and Chance, B. 2002. MRI-guided diffuse optical spectroscopy of malignant and benign breast lesions. *Neoplasia* 4(4): 347–354.

Ou, W., Nissilä, I., Radhakrishnan, H., Boas, D. A., Hämäläinen, M. S., and Franceschini, M. A. 2009. Study of neurovascular coupling in humans via simultaneous magnetoencephalography and diffuse optical imaging acquisition. *Neuroimage* 46(3): 624–632.

Patterson, M. S., Chance, B., and Wilson, B. C. 1989. Time resolved reflectance and transmittance for the noninvasive measurement of tissue optical-properties. *Appl. Opt.* 28(12): 2331–2336.

Patterson, M. S., Moulton, J. D., Wilson, B. C., Berndt, K. W., and Lakowicz, J. R. 1991. Frequency-domain reflectance for the determination of the scattering and absorption properties of tissue. *Appl. Opt.* 30(31): 4474–4476.

Patterson, M. S., Wilson, B. C., and Wyman, D. R. 1990a. The propagation of optical radiation in tissue I. Models of radiation transport and their application. *Lasers Med. Sci.* 6: 155–168.

Patterson, M. S., Wilson, B. C., and Wyman, D. R. 1990b. The propagation of optical radiation in tissue II. Optical properties of tissues and resulting fluence distributions. *Lasers Med. Sci.* 6: 379–390.

Paulsen, K. D. and Jiang, H. 1995. Spatially varying optical property reconstruction using a finite element diffusion equation approximation. *Med. Phys.* 22(6): 691–701.

Perrey, S. 2008. Non-invasive NIR spectroscopy of human brain function during exercise. *Methods* 45(4): 289–299.

Piao, D., Jiang, S., Srinivasan, S., Dehghani, H., and Pogue, B. W. 2005. Video-rate near-infrared optical tomography using spectrally-encoded parallel light delivery. *Opt. Lett.* 30(19): 2593–2595.

Pogue, B. W., Jiang, S., Dehghani, H., Kogel, C., Soho, S., Srinivasan, S., Song, X., Tosteson, T. D., Poplack, S. P., and Paulsen, K. D. 2004. Characterization of hemoglobin, water, and NIR scattering in breast tissue: Analysis of intersubject variability and menstrual cycle changes. *J. Biomed. Opt.* 9(3): 541–552.

Pogue, B. W., Paulsen, K. D., Abele, C., and Kaufman, H. 2000. Calibration of near-infrared frequency-domain tissue spectroscopy for absolute absorption coefficient quantitation in neonatal head-simulating phantoms. *J. Biomed. Opt.* 5(2): 185–193.

Pogue, B. W., Poplack, S. P., McBride, T. O., Wells, W. A., Osterman, S. K., Osterberg, U. L., and Paulsen, K. D. 2001. Quantitative hemoglobin tomography with diffuse near-infrared spectroscopy: Pilot results in the breast. *Radiology* 218(1): 261–266.

Pogue, B. W., Willscher, C., McBride, T. O., Osterberg, U. L., and Paulsen, K. D. 2000. Contrast-detail analysis for detection and characterization with near-infrared diffuse tomography. *Med. Phys.* 27(12): 2693–2700.

Poplack, S. P., Paulsen, K. D., Hartov, A., Meaney, P. M., Pogue, B. W., Tosteson, T. D., Grove, M. R., Soho, S. K., and Wells, W. A. 2004. Electromagnetic breast imaging: Average tissue property values in women with negative clinical findings. *Radiology* 231(2): 571–580.

Poplack, S. P., Tosteson, T. D., Wells, W. A., Pogue, B. W., Meaney, P. M., Hartov, A., Kogel, C. A., Soho, S. K., Gibson, J. J., and Paulsen, K. D. 2007. Electromagnetic breast imaging: Results of a pilot study in women with abnormal mammograms. *Radiology* 243(2): 350–359.

Prapavat, V., Luhmann, T., Krause, A., Backhaus, J., Beuthan, J., and Mueller, G. J. 1998. Evaluation of early rheumatic disorders in PIP joints using a cw-transillumination method: first clinical results. In: *Proceedings of the SPIE*, Vol. 3196 (Optical and Imaging Techniques for Biomonitoring III), San Remo, Italy, 1998.

Rindfleisch, J. A. and Muller, D. 2005. Diagnosis and management of rheumatoid arthritis. *Am. Fam. Physician* 72: 1037–1047.

Rinneberg, H., Grosenick, D., Moesta, K. T., Wabnitz, H., Mucke, J., Wubbeler, G., Macdonald, R., and Schlag, P. M. 2008. Detection and characterization of breast tumours by time-domain scanning optical mammography. *Opto-Electron. Rev.* 16(2): 147–162.

Sassaroli, A., Frederick, B., Tong, Y., Renshaw, P., and Fantini, S. 2006. Spatially weighted BOLD signal for comparison of functional magnetic resonance imaging and near-infrared imaging of the brain. *NeuroImage* 33: 505–514.

Scheel, A. K., Backhaus, M., Klose, A. D., Moa-Anderson, B., Netz, U., Hermann, K. G., Beuthan, J., Muller, G., Burmester, G. R., and Hielscher, A. H. 2005. First clinical evaluation of sagittal laser optical tomography for detection of synovitis in arthritic finger joints. *Ann. Rheum. Dis.* 64: 239–245.

Scheel, A. K., Krause, A., Rheinbaben, I. M., Metzger, G., Rost, H., Tresp, V., Mayer, P., Reuss-Borst, M., and Müller, G. A. 2002. Assessment of proximal finger joint inflammation in patients with rheumatoid arthritis, using a novel laser-based imaging technique. *Arthritis Rheum.* 46(5): 1177–1184.

Schwaighofer, A., Tresp, V., Mayer, P., Krause, A., Beuthan, J., Rost, H., Metzger, G., Müller, G. A., and Scheel, A. K. 2003. Classification of rheumatoid joint inflammation based on laser imaging. *IEEE Trans. Biomed. Eng.* 50(3): 375–382.

Schweiger, M. and Arridge, S. R. 1999. Optical tomographic reconstruction in a complex head model using a priori region boundary information. *Phys. Med. Biol.* 44(11): 2703–2721.

Seiyama, A., Seki, J., Tanabe, H. C., Sase, I., Takatsuki, A., Miyauchi, S., Eda, H., Hayashi, S., Imaruoka, T., Iwakura, T., and Yanagida, T. 2004. Circulatory basis of fMRI signals: Relationship between changes in the hemodynamic parameters and BOLD signal intensity. *Neuroimage* 21(4): 1204–1214.

Shah, N., Cerussi, A., Eker, C., Espinoza, J., Butler, J., Fishkin, J., Hornung, R., and Tromberg, B. 2001. Noninvasive functional optical spectroscopy of human breast tissue. *Proc. Natl. Acad. Sci. U. S. A.* 98(8): 4420–4425.

Siegel, A., Culver, J., Mandeville, J., and Boas, D. 2003. Temporal comparison of functional brain imaging with diffuse optical tomography and fMRI during rat forepaw stimulations. *Phys. Med. Biol.* 48: 1391–1403.

Song, X., Pogue, B. W., Jiang, S., Doyley, M. M., Dehghani, H., Tosteson, T. D., and Paulsen, K. D. 2004. Automated region detection based on the contrast-to-noise ratio in near-infrared tomography. *Appl. Opt.* 43(5): 1053–1062.

Spinelli, L., Torricelli, A., Pifferi, A., Taroni, P., Danesini, G., and Cubeddu, R. 2005. Characterization of female breast lesions from multi-wavelength time-resolved optical mammography. *Phys. Med. Biol.* 50(11): 2489–2502.

Srinivasan, S., Pogue, B. W., Jiang, S., Dehghani, H., Kogel, C., Soho, S., Chambers, J. G., Tosteson, T. D., Poplack, S. P., and Paulsen, K. D. 2003. Interpreting hemoglobin and water concentration, oxygen saturation, and scattering measured by near-infrared tomography of normal breast in vivo. *Proc. Nat. Acad. Sci. U. S. A.* 100(21): 12349–12354.

Srinivasan, S., Pogue, B. W., Jiang, S., Dehghani, H., and Paulsen, K. D. 2004. Spectrally constrained chromophore and scattering NIR tomography improves quantification and robustness of reconstruction. *Appl. Opt.* 44(10): 1858–1869.

Steinke, J. M. and Shepherd, A. P. 1988. Comparison of Mie theory and the light scattering of red blood cells. *Appl. Opt.* 27: 4027–4033.

Svensson, T., Andersson-Engels, S., Einarsdóttir, M., and Svanberg, K. 2007. In vivo optical characterization of human prostate tissue using near-infrared time-resolved spectroscopy. *J. Biomed. Opt.* 12(1): 014022.

Thompson, A. B. and Sevick-Muraca, E. M. 2003. Near-infrared fluorescence contrast-enhanced imaging with intensified charge-coupled device homodyne detection: Measurement precision and accuracy. *J. Biomed. Opt.* 8(1): 111–120.

Toronov, V., Zhang, X., and Webb, A. 2007. A spatial and temporal comparison of hemodynamic signals measured using optical and functional magnetic resonance imaging during activation in the human primary visual cortex. *Neuroimage* 34(3): 1136–1148.

van de Ven, S., Elias, S., Wiethoff, A., van der Voort, M., Leproux, A., Nielsen, T., Brendel, B., Bakker, L., van der Mark, M., Mali, W., and Luijten, P. 2009. Diffuse optical tomography of the breast: initial validation in benign cysts. *Mol. Imaging Biol.* 11(2): 64–70.

Van Houten, E. E. W., Doyley, M. M., Kennedy, F. E., Weaver, J. B., and Paulsen, K. D. 2003. Initial in vivo experience with steady-state subzone-based MR elastography of the human breast. *J. Magn. Reson. Imaging* 17(1): 72–85.

Vaupel, P., Kallinowski, F., and Okunieff, P. 1989. Blood flow, oxygen and nutrient supply, and metabolic microenvironment of human tumors: A review. *Cancer Res.* 49: 6449–6465.

Villringer, A. and Chance, B. 1997. Non-invasive optical spectroscopy and imaging of human brain function. *Trends Neurosci.* 20(10): 435–442.

van Staveren, H. J., Moes, C. J. M., van Marle, J., Prahl, S. A., and van Gemert, M. J. C. 1991. Light scattering in intralipid-10% in the wavelength range of 400–1100 nm. *Appl. Opt.* 30(31): 4507–4514.

Wang, J., Davis, S. C., Srinivasan, S., Jiang, S. D., Pogue, B. W., and Paulsen, K. D. 2008. Spectral tomography with diffuse near-infrared light: Inclusion of broadband frequency domain spectral data. *J. Biomed. Opt.* 13(4): 041305.

Wang, X., Pogue, B. W., Jiang, S., Dehghani, H., Song, X., Srinivasan, S., Brooksby, B. A., Paulsen, K. D., Kogel, C., Poplack, A. P., and Wells, W. A. 2006. Image Reconstruction of effective Mie scattering parameters of breast tissue in vivo with near-infrared tomography. *J. Biomed. Opt.* 11(4): 041106.

Wang, X., Pogue, B. W., Jiang, S., Song, X., Paulsen, K. D., Kogel, C., Poplack, S. P., and Wells, W. A. 2005. Approximation of Mie scattering parameters from near-infrared tomography of health breast tissue in vivo. *J. Biomed. Opt.* 10(5): 051704-1–051704-10.

Xu, G., Piao, D., Musgrove, C. H., Bunting, C. F., and Dehghani, H. 2008. Trans-rectal ultrasound-coupled near-infrared optical tomography of the prostate, part I: Simulation. *Opt. Express* 16(22): 17484–17504.

Xu, H., Pogue, B., Springett, R., and Dehghani, H. 2005a. Spectral derivative based image reconstruction provides inherent insensitivity to coupling and geometric errors. *Opt. Lett.* 30(21): 2912–2914.

Xu, H., Springett, R., Dehghani, H., Pogue, B. W., Paulsen, K. D., and Dunn, J. F. 2005b. Magnetic-resonance-imaging—coupled broadband near-infrared tomography system for small animal brain studies. *Appl. Opt.* 44(10): 2177–2188.

Xu, Y., Iftimia, N., Jiang, H., Key, L., and Bolster, M. 2001. Imaging of in vitro and in vivo bones and joints with continuous-wave diffuse optical tomography. *Opt. Express* 8(7): 447–451.

Xu, Y., Iftimia, N., Jiang, H., Key, L. L., and Bolster, M. B. 2002. Three-dimensional diffuse optical tomography of bones and joints. *J. Biomed. Opt.* 7(1): 88–92.

Yalavarthy, P. K., Pogue, B. W., Dehghani, H., Carpenter, C. M., Jiang, S. J., and Paulsen, K. D. 2007. Structural information within regularization matrices improves near infrared diffuse optical tomography. *Opt. Express* 15(13): 8043–8058.

Yuan, Z., Zhang, Q., Sobel, E. S., and Jiang, H. 2008. Tomographic x-ray-guided three-dimensional diffuse optical tomography of osteoarthritis in the finger joints. *J. Biomed. Opt.* 13(4): 044006.

Zeff, B. W., White, B. R., Dehghani, H., Schlaggar, B. L., and Culver, J. P. 2007. Retinotopic mapping of adult human visual cortex with high-density diffuse optical tomography. *Proc. Natl. Acad. Sci. U. S. A.* 104(29): 12169–12174.

Zhang, Q., Brukilacchio, T., Li, A., Stott, J., Chaves, T., Wu, T., Chorlton, M., Rafferty, E., Moore, R., Kopans, D., and Boas, D. 2005a. Coregistered tomographic x-ray and optical breast imaging: initial results. *J. Biomed. Opt.* 10(2): 024033.

Zhang, Q. and Jiang, H. 2005. Three-dimensional diffuse optical imaging of hand joints: System description and phantom studies. *Opt. Lasers Eng.* 43: 1237–1251.

Zhang, X., Toronov, V., and Webb, A. 2005b. Simultaneous integrated diffuse optical tomography and functional magnetic resonance imaging of the human brain. *Opt. Express* 13(14): 5513–5521.

Zhou, X. and Zhu, T. C. 2006. Image reconstruction of continuous wave diffuse optical tomography (DOT) of Human Prostate. In *Proceedings of the COMSOL Users Conference*, Boston, MA, COMSOL.

Zhou, X., Zhu, T. C., Finlay, J. C., Li, J., Dimofte, A., and Hahn, S. M. 2007. Two-dimensional/three-dimensional hybrid interstitial diffuse optical tomography of human prostate during photodynamic therapy: phantom and clinical results. In: *Proceedings of SPIE 6434, Optical Tomography and Spectroscopy of Tissue VII*, San Jose, CA, 2007.

Zhu, Q., Chen, N., and Kurtzman, S. H. 2003. Imaging tumor angiogenesis by use of combined near-infrared diffusive light and ultrasound. *Opt. Lett.* 28(5): 337–339.

Zhu, Q., Cronin, E., Currier, A., Vine, H., Huang, M., Chen, N., and Xu, C. 2005. Benign versus malignant breast masses: Optical differentiation with US-guided optical imaging reconstruction. *Radiology* 237(1): 57–66.

Zhu, T. C., Finlay, J. C., and Hahn, S. M. 2005. Determination of the distribution of light, optical properties, drug concentration, and tissue oxygenation in-vivo in human prostate during motexafin lutetium-mediated photodynamic therapy. *J. Photochem. Photobiol. B* 79(3): 231–241.

Diffuse Optical Tomography: Time Domain

20.1 Introduction 395
20.2 Background..395
History • Comparison with Frequency-Domain Measurements
20.3 Time-Domain Instrumentation 398
Overview • TCSPC Systems • Time-Gated Systems
20.4 Time-Domain Data Analysis..403
The Temporal Point Spread Function • Modeling the TPSF • Characterizing the TPSF Using
Datatypes
20.5 Reconstructing Optical Properties from Time-Domain Data 406
Absolute Measurement of Baseline Optical Properties • Relative Measurement of Changes in
Optical Properties • Tomographic Imaging
20.6 In Vivo Applications 408
Breast Characterization and Imaging • Baseline Optical Properties of the Brain • Functional
Imaging with Depth Sensitivity • Tomographic Brain Imaging
20.7 Critical Discussion 413
References..414

Juliette Selb
Massachusetts General Hospital

Adam Gibson
University College London

20.1 Introduction

When a pulse of light enters a diffusive medium such as tissue, it undergoes a number of changes. Most obviously, its intensity is attenuated due to the scattering and absorption properties of tissue. The measurement of this attenuation forms the basis of continuous wave (CW) imaging, as described in Chapter 19. However, the pulse is also delayed and broadened. If this broadening is also measured, we can obtain additional information about the tissue's optical properties. This concept forms the basis of time-domain imaging, which is in some senses analogous to frequency-domain imaging also described in Chapter 19. Formally, frequency-domain data are the Fourier transform of time-domain data, expressed at a particular modulation frequency. Time-domain data are therefore somewhat more general than the equivalent frequency-domain data, as they include information from all modulation frequencies. They also contain information about higher moments of the time distribution of the detected light, and can reveal information about measurement errors such as poor contact and unintentional reflections or room light. Furthermore, measuring time-domain data by recording the flight times of individual photons affords the opportunity of maximizing sensitivity, thereby allowing measurements across larger volumes of tissue than other methods.

In this chapter, we review the various methods by which time-domain data may be acquired, and discuss some of the unique theoretical features of time-domain imaging that result from the extra information inherent in data acquired using this method. We then describe some of the applications of time-domain imaging, including the determination of optical properties, both by local measurements and by imaging, and we look at the major clinical uses of the method such as optical mammography and imaging the brain.

20.2 Background

20.2.1 History

When a short pulse of light (a few picoseconds in length) is transmitted through a highly scattering medium such as tissue, the resulting temporal distribution of photons is known as the temporal point spread function (TPSF). When measured across a few centimeters of tissue, this curve (an example is shown in Figure 20.1) typically reaches a peak after a few nanoseconds, equivalent to 5–10 times the direct distance between source and detector, indicating the high degree of scatter inherent in diffuse optical measurements. The shape of this distribution carries information about the tissue's absorption and scatter coefficients (Hebden and Delpy 1994).

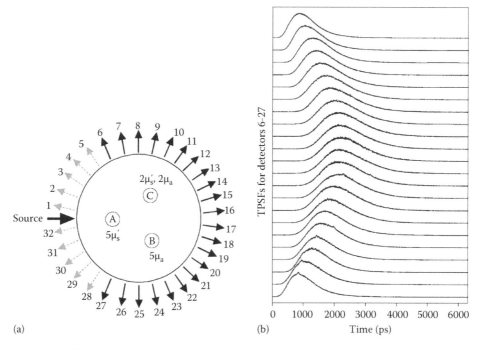

FIGURE 20.1   TPSFs (b) recorded by the 22 active detectors opposite source no. 1 (a). Each TPSF consists of up to ~10^6 photons, and the curves are normalized to their respective maximum values. (From Schmidt, F.E.W., Development of a time-resolved optical tomography system for neonatal brain imaging. PhD thesis, UCL, 1999.)

Some of the earliest attempts at obtaining time-domain data relied on time gating to isolate those photons that emerge earliest and that are assumed to have traveled the shortest distance through tissue. These early-arrival photons are scattered less than late-arrival photons, so their path may be modeled as a straight line, allowing more straightforward analysis, and image reconstruction based on the Radon transform, as used in computed tomography (CT) (Kalender 2006). However, the number of unscattered photons that can be measured through relevant thicknesses of tissue is vanishingly small, and this approach was soon abandoned for imaging large volumes of tissue. Recently, however, early-arrival photons have been successfully used to image fluorescence in small animals, where the effective tissue thickness is much smaller (Niedre et al. 2008).

More sophisticated attempts to measure the entire TPSF were reviewed by Hebden et al. (1997) who discussed the use of streak cameras and time-to-amplitude converter (TAC) systems. Both measure the time elapsed since a trigger signal, the streak camera by causing a time-varying deflection of the light (or photoelectrons) across a spatially resolved detector, and a TAC by measuring the voltage across a capacitor that integrates with time between a trigger and a stop signal. The streak camera offers excellent temporal resolution (~100 fs) but its small light collection area limits sensitivity. Hebden et al. concluded in 1997 that TAC-based systems provide the best option for time-domain optical systems and many groups since have followed this advice.

Pulsed laser diodes now provide sources that are portable, low cost, and flexible, and advances in TAC-based systems have meant that multi-detector time-domain systems are now competitive with frequency-domain systems in terms of cost. The latest TAC-based systems are often based on time-correlated single-photon counting (TCSPC) hardware, which gives very flexible measurements of photon arrival times with high count rates and good temporal resolution (Becker 2005). These systems are generally used in conjunction with a photon-counting photomultiplier tube (PMT) detector.

An alternative approach is to use time-gating techniques. These methods rely on the TPSF being smooth with few high temporal frequency components. The detector (often a charged-couple device (CCD) camera) is made to be sensitive for a brief period, typically 500 ps, following a controllable delay. As the delay is stepped, recording for a few hundred meters at each stage, the TPSF can be built up. This approach provides fast sampling of the TPSF from many different detectors in parallel. It has been used in optical mammography to distinguish between late-arrival photons, which mainly depend on absorption, and early-arrival photons, which are affected by both absorption and scatter (Grosenick et al. 2003). Selb et al. (2004, 2006) used a similar approach to obtain depth-resolved measurements from the brain by distinguishing between early-arrival photons, which probe superficial layers, and later-arrival photons, which are more likely to have passed through deeper tissues.

20.2.2 Comparison with Frequency-Domain Measurements

Most optical imaging studies to date have used intensity-only measurements, despite Arridge and Lionheart (1998) showing that in the absence of prior information, such data are unable to distinguish between absorption and scatter. Additional information about the dispersion of the light can be obtained using frequency-domain or time-domain systems. Frequency-domain systems (reviewed in Chapter 19) are fast and relatively inexpensive compared to time-domain systems and so are frequently used for imaging across relatively small thicknesses of tissue (up to about 6 cm), and have been applied to imaging the compressed breast and cortical mapping.

Some of the advantages of time-domain measurements compared to other methods are listed in the following:

- Time-domain measurements can be made with photon-counting PMT detectors and high-power pulsed laser sources, leading to the highest possible sensitivity. Such TCSPC systems are the only way of measuring light across large volumes of tissue across which only a few photons per second may be measured.
- The TPSF contains temporal frequency components up to several GHz, so a time-domain measurement includes information from a wide frequency range compared to frequency-domain measurements that are obtained at a single frequency. Having said this, the TPSF is smooth, so it is unclear what additional useful information this provides.
- Phase-wrapping can occur in frequency-domain systems if the change in phase is greater than 2π. This can be avoided by reducing the modulation frequency at the expense of reducing the full-scale deflection of the measurement and reducing the maximum frequency that can be used to probe the medium.
- The TSPF can be processed to give datatypes other than intensity and mean photon flight time (equivalent to intensity and phase of a frequency-domain system). Schweiger and Arridge (1999) showed that some other datatypes can improve the discrimination between absorption and scatter. However, because higher-order datatypes such as variance and skew are more sensitive to the small number of late-arriving photons, they are more sensitive to noise and have not been successfully applied to experimental imaging.
- Perhaps the most underrated advantage of time-domain imaging is its ability to identify poor data. Experimental data, and especially clinical data, can be contaminated with unwanted light, which is still correlated with the measurement, and so is difficult to exclude using lock-in amplifiers or time-gating. Such noise may come from instrumentation error or more commonly from light that has leaked around the object being imaged. This is a particular problem when the geometry of the object being examined is complex or the optical connectors are not held robustly against the surface of the object. For example, Figure 20.2 was taken from data acquired during 3D optical tomography of the brain using an adjustable helmet. Poor connection between the optode and the head can be seen to manifest itself as light appearing before the main

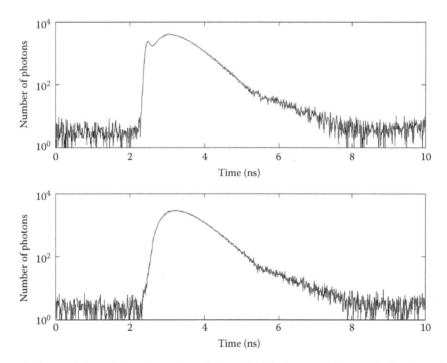

FIGURE 20.2 TPSFs acquired on an infant. The top image shows data acquired during stimulation, the bottom image is the corresponding reference data. The pre-peak visible on the top image is due to light escaping from the head and will overwhelm any changes in the TPSF due to the stimulation. Effects such as these can be eliminated by examination of time-domain data. (From Gibson, A.P. et al., *Neuroimage*, 30, 521, 2006. With permission.)

part of the TPSF (Gibson et al. 2006). Such artifacts would lead to significant errors in the calculation of intensity and mean photon flight time, and would lead to artifacts in the image. It is hard to imagine how such artifacts could be identified without access to the time-domain data.

20.3 Time-Domain Instrumentation

20.3.1 Overview

Time-domain instrumentation can be generally divided into two groups: TCSPC systems and gated systems. TCSPC systems measure the arrival times of individual photons. They have high sensitivity and good temporal resolution but are relatively slow and expensive. Gated systems work by measuring light within a short time window. The TPSF is built up by measuring at different time delays. This approach is less expensive than TCSPC systems, and in particular, the cost and acquisition time do not scale with the number of channels, allowing many more detectors.

20.3.2 TCSPC Systems

20.3.2.1 Principle of Operation

TCSPC systems rely on measuring the arrival times of individual photons. This requires a detection system that can identify the arrival of a photon, distinguish it from background events, measure its arrival time, and then store the arrival time in a convenient way. Furthermore, the probability of detecting a photon in a single counting period must be much less than one. If this condition holds, then we can discount the possibility of two photons arriving in a period, and therefore we do not need to compensate for any errors that might occur due to multiple photons arriving. For a thorough overview of TCSPC methods and applications, see Becker (2005).

First-generation TCSPC systems such as University College London (UCL)'s original time-domain optical tomography system MONSTIR (Schmidt 1999, Schmidt et al. 2000) use a PMT as a detector to produce an electronic pulse when a photon is detected. This pulse may vary in height and shape so a constant fraction discriminator (CFD; Ortec Model 935, EG&G, USA) is used to create a logic pulse of fixed size and shape. The CFD also eliminates low-amplitude pulses that are likely to be thermal in origin. The photon arrival times are measured using a picosecond time analyzer (PTA; Ortec Model 9208, EG&G, USA) that compares the arrival times to two sinusoidal reference waveforms at different frequencies arranged as an "electronic vernier." The measured pulse starts the timing and a (delayed) reference pulse synchronized to the laser stops it. This operation, where the arrival of a photon begins the timing measurement and the reference pulse ends it, is the opposite of what might be conventionally expected. It is called "reversed start-stop" operation and is used because the timing circuit is active only when a photon is measured rather than every time a reference pulse is generated, improving efficiency. The PTA has its own on-board memory that stores all the measured photon arrival times as a histogram.

An alternative way to measure the arrival time is to use a TAC. When a pulse is detected, the TAC begins to charge a capacitor. The reference pulse stops the charging, so the voltage across the TAC is proportional to the time elapsed between the two pulses. This signal is then amplified and fed to an analog-to-digital converter (ADC), which in turn feeds a memory bank that assembles the histogram. The PTA is more expensive than TAC but also more accurate.

This first-generation system is accurate and sensitive, but bulky and relatively slow. It is also *one-dimensional* in that it measures the arrival time of a single train of photons. Second-generation TCSPC devices (Becker 2005, Becker et al. 2005) combine the CFDs and PTAs (or equivalently the CFDs, TACs, and ADCs) into a single printed circuit board, which is mounted within a PC. This leads to an increase in speed of up to a factor of 100 and a similar scale reduction in size. The size reduction is particularly beneficial for a clinical system, which needs to work in a restricted space on a hospital ward.

A further advantage of the new technique is that it can be described as *multidimensional*. This means that when a photon is detected, its arrival time is not merely streamed to memory, but it can be sent to a specific location in the memory that can be controlled by the user. In this way, a single TCSPC unit can record from multiple detectors and multiple sources. This could be used to distinguish between different wavelengths (of the source or the detector), different source or detector locations, or the *x–y* coordinates of a scanning system. This requires an extra signal from a router that, for example, may include a signal showing which detector measured a particular photon, and then ensure that the data from that photon is routed to the appropriate memory bank.

Such TCSPC modules have been incorporated into a second-generation MONSTIR system (Becker et al. 2005, Jennions 2008). The CFDs and PTAs (which took up about half the total volume of the instrument) have been replaced by four printed circuit boards (SPC-144, Becker & Hickl GmbH, Berlin) in an industrial PC. A full image data set now takes about 3 min to acquire instead of 12 min previously. This is largely because the new TCSPC cards have a dual-memory architecture, meaning that data can be downloaded to the computer from one set of memory while data are being acquired by the other. A further advantage is the reduced warm-up time. The PTA in MONSTIR was designed to be continually on and "run hot." Clearly this is not possible if the unit is to be moved to the hospital, and we found that it needed to be left on for about 12 h to stabilize. The current PC-based TCSPC electronics take less than 2 h to warm up.

20.3.2.2 Light Sources

The light sources for time-domain optical imaging are determined both by the requirements of TCSPC and of optical imaging. The wavelengths are generally around 800 nm, and almost always between 700 and 1000 nm. The selection of the optimal wavelength is complex and may not yet be fully understood (Gibson et al. 2005). The power must conform to relevant national and international safety limits. Time-domain imaging demands that the light be pulsed.

There are two approaches. Most commonly, pulsed laser diodes are used. These are sufficiently low in cost that every source fiber can have its own source laser, meaning that multiplexing is not necessary. They are available in a number of wavelengths with narrow bandwidths and are easy to control electronically. They are widely used in time-domain optical imaging systems (Ntziachristos et al. 1998, Grosenick et al. 2005a, Taroni et al. 2005b). However, the power available for very short pulses in the near infrared (NIR) is currently limited to a few mW, although it is rising rapidly.

Other types of laser can provide much greater power, but may need realigning or calibrating after movement, which is unacceptable in a portable clinical system. A compromise was reached in MONSTIR with a fiber-laser (IMRA Inc., USA). This is a custom-built laser containing two coiled optical fibers that act as the lasing cavities pumped by laser diodes. This is portable and can provide 10–30 mW of power in an 80 MHz train of 3 ps pulses at alternating wavelengths.

20.3.2.3 Detectors

In optical tomography, we need to record TPSFs from low-intensity signals with high accuracy. The duration of the TPSF is of the order of a few ns, so to resolve this, the impulse response function (IRF) of the detector must be a few hundred ps or less. More importantly, the temporal stability must be better than the changes in the size of the signal being measured. The change in mean photon flight time between photons measured across the brain under different conditions may be only a few ps, so achieving adequate stability can be challenging. Moreover, we are measuring low-intensity diffuse light so in order to collect as much light as possible, we need a detector with a large collection area. We therefore need stable, sensitive detectors, which are efficient in the NIR and with a fast IRF.

At the moment, the only detectors that meet these requirements are PMTs (see www.hamamatsu.com). The most promising alternative is likely to be single-photon avalanche photodiodes (SPADs), but these are currently available with small (<1 mm^2) collection areas, so their sensitivity is currently inadequate.

20.3.2.4 TCSPC Systems

Two general approaches have been taken to TCSPC-based diffuse optical tomography: either having all channels fully parallel or by sharing channels by scanning the source, detector, or both.

The UCL optical tomography system, MONSTIR, discussed in detail earlier, and by Schmidt et al. (2000) and Jennions (2008), takes the former approach. Even though there is only one laser, its output passes through a 32-way switch so the light is sent into one of 32 parallel source channels. The source fibers and detector fiber bundles are combined into coaxial fibers, each of which can supply or collect light. This leads to an elegant calibration procedure (Hebden et al. 2003) and halves the number of connections that must be made to the head or breast. One fiber is used to deliver light while the remaining 31 collect light that has traveled across the body. Each detector fiber bundle transmits light to a microchannel-plate photomultiplier tube (MCP-PMT; R411OU-05MOD, Hamamatsu, Japan), which has a shorter transit-time spread (TTS) compared to a standard PMT, leading to a faster IRF and improved temporal stability. Thirty-two separate variable optical attenuators reduce the intensity of the more intense sources, to prevent saturating the detectors and reduce the dynamic range of the measured light.

The latter approach was taken by groups in Berlin (Grosenick et al. 1999, 2005a) and Milan (Pifferi et al. 2003, Taroni et al. 2005b), as well as by the Montreal-based company Advanced Research Technologies Inc. (Intes 2005). The Berlin and Milan systems both use a single laser diode source at each wavelength of interest, which is multiplexed into a single pulse train traveling along a source fiber. Data may be collected using a scanning fiber bundle placed opposite the source, or by a small number of fiber bundles that are offset from the source to give off-axis data. The source fiber and detector fiber bundle(s) are scanned across the area of interest, which in both these cases is the gently compressed breast. Data are collected at 1 mm intervals. Both systems use PMTs, which supply a signal that is processed by a TCSPC system in a PC. The ART system is based on a Ti:Sapphire laser, which is tuned sequentially at four discrete wavelengths. On the detection side, the light is collected by five multimode fibers in an X geometry facing a streak camera. The emission and detection fibers are scanned simultaneously over the region of examination.

Using a modern PC-based TCSPC system and scanning a single source and detector across the gently compressed breast so that light levels are relatively high leads to a fast and efficient time-domain system, which is competitive with frequency-domain systems in terms of cost and speed, whilst providing the additional data inherent in the time-domain.

20.3.3 Time-Gated Systems

20.3.3.1 Principle of Operation

While time-domain devices using TCSPC detection are the most broadly employed, an alternative approach, based on the temporally gated measurement of the TPSF, has been developed by a few teams (D'Andrea et al. 2003, Selb et al. 2005a, 2006, Turner et al. 2005, Pifferi et al. 2008). These instruments also employ a pulsed source and time-resolved detection of the broadened and attenuated TPSF. However, rather than building a histogram of the arrival times of the photons, a temporally gated detector integrates the light intensity within a narrow delay window, acting as a very fast shutter, which can be temporally scanned relative to the laser pulse (Figure 20.3). This method has stemmed from earlier gating technologies that focused on the detection of early ballistic and snake photons in transmission geometry (Feng et al. 1993, 1994, Hee et al. 1993, Kalpaxis et al. 1993, Wang et al. 1993), and reconstructed images with algorithms similar to those developed for CT (Kalender 2006). However, as previously described, the number of ballistic photons becomes

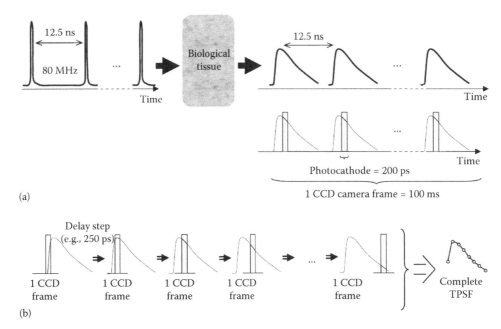

FIGURE 20.3 Principle of time-gated detection. (a) After propagation through the scattering medium, the injected pulses of light exit attenuated and broadened. A fast detector (e.g., photocathode) synchronized with the pulse emission detects photons through a narrow delay window. An integrating device (e.g., CCD camera) integrates light over several thousands of successive pulses. (b) To record the complete TPSF, the delay of the narrow temporal window is scanned over the TPSF, each CCD array corresponding to one particular delay.

vanishingly small after a few centimeters of propagation in biological tissue. Rather than exclusively detecting early photons, time-gated instruments now record photons arriving within different delay windows, either in transmission or in reflectance geometries. The window is delayed temporally step by step relative to the laser pulse emission, so that the complete TPSF can be recorded (Figure 20.3). Because of the low intensity contained in individual light pulses, the light from each delay window needs to be integrated over many cycles.

20.3.3.2 Light Sources

Gated systems are based on pulsed sources of light with similar requirements as those employed with TCSPC detection. However, the gated detectors typically require more photons for a reasonable signal-to-noise ratio (SNR). In particular, gated systems can measure multiple photons per laser cycle, unlike systems using TCSPC technology. Pulsed sources with higher power are therefore generally used in gated measurements.

D'Andrea et al. (2003) developed the first gated system, based on a mode-locked Argon laser (CR-18, Coherent) emitting at 514 nm with a repetition rate of 80 MHz, a pulse width of 120 ps, and source power under 5 mW. The Time-Domain Imager (TDI) of the Photon Migration Laboratory at MGH (Selb et al. 2005a, 2006) uses a pulsed Ti:Sapphire laser (MaiTai, Spectra Physics), with a repetition rate of 80 MHz. This source provides very narrow temporal pulses, high power, and wavelength flexibility. To respect skin illumination safety norms, the laser power of 1 W is attenuated with crossed polarizers to typically 15 mW when a single illuminating fiber is used, or around 100 mW when the

source is time-multiplexed to several fibers. The pulses are about 150 fs in duration, which is 3 orders of magnitude better than necessary, considering that the IRF is limited by the detector response width, on the order of a few hundreds of picoseconds. The source wavelength can be tuned between 750 and 850 nm, but more recent laser models permit a broader spectrum with higher power. Pulsed laser diodes as those typically used in TCSPC systems currently do not have sufficient power to provide a competitive alternative to solid laser sources. However, the technological advances in this field are fast, and such sources could become available within a few years.

20.3.3.3 Gated Detectors

Because the measured TPSF results from the convolution of the tissue TPSF by the detector IRF, the ideal temporal gate should be very narrow, and have steep rising and falling edges. Furthermore, for fast recording of the whole TPSF, the delay between pulse emission and detection should be rapidly adjustable. Typically the gating operation is achieved with a light-amplifying device in which gain can be modulated on and off extremely rapidly by a controlling voltage.

The solution first developed by D'Andrea et al. (2003), and subsequently adopted by Selb et al. (2005a) and Turner et al. (2005), is an intensified charge-coupled device (ICCD) camera, which consists of an image intensifier (the gating device) and a CCD camera (the integrating device). The image intensifier is a time-gated photocathode married to a high spatial resolution microchannel plate (MCP) and a phosphor layer. The principle of the image intensifier is the following (Figure 20.4): a photon

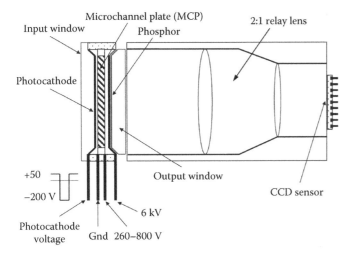

FIGURE 20.4 Schematic of an ICCD camera. (From Brukilacchio, T.J., A diffuse optical tomography system combined with x-ray mammography for improved breast cancer detection, PhD thesis, Electrical Engineering, Tufts University, 2003.)

incident on the photocathode is converted into a primary photoelectron. When the photocathode is "on" (gate open), the primary photoelectrons are accelerated by a high potential of several hundred volts toward the input face of the MCP. When no potential is applied, the photocathode response is effectively turned off, providing the gating effect. A fraction of the primary photoelectrons actually enters the MCP active area, where they each strike the wall of one of the many channels in the MCP. This causes the release of secondary emitted electrons, which in turn strike the wall to generate further electrons. Secondary photoelectrons are thereby produced and multiplied with a gain, depending on the high voltage applied to the MCP, similarly to the amplification process in a PMT. Finally, a large potential accelerates the exiting electron cloud toward the phosphor layer, where they produce green photons. The image intensifier is coupled to a CCD camera where the green photons are imaged. The CCD converts the green photons into photoelectrons and stores them in wells for the specified integration time. A more detailed description of the whole process, as well as a thorough characterization of the sources of noise at every stage of the detection, can be found in Brukilacchio (2003). The gate width provided by these devices and reported by D'Andrea et al. (2003) and Selb et al. (2005a) is on the order of a few hundred picoseconds (200–500 ps), while the integration times used on the CCD are on the order of a few tens of milliseconds (50–200 ms). The CCD therefore integrates light over typically a million cycles of the pulsed source. The ICCD detection is generally implemented in two different ways: either the delay gate is temporally scanned relative to the pulsed source, permitting a step-by-step recording of the whole TPSF, or the gate stays adjusted at a particular delay, and the evolution of intensity within this fixed gate is recorded (D'Andrea et al. 2003, Selb et al. 2006).

Recently, a novel experimental approach suggested by Spinelli et al. (2006) required a somewhat different gated detection. The authors first showed theoretically that a "null" source-detector separation in reflectance geometry can provide higher contrast and spatial resolution in the reconstructed images than the usual source-detector distance of a few centimeters. The experimental implementation of this idea, however, presents technological challenges. At very short source-detection spacing (a few millimeters), early photons largely dominate the signal, and the useful late photons are several orders of magnitude less numerous than those of the peak of the TPSF. Detectors with dynamic range high enough to measure the whole TPSF are not available. The technological solution proposed by Pifferi et al. (2008) is to "gate out" early photons. This approach is different from the previously described ICCD-based gating method, in the fact that it still makes use of the TCSPC technology. However, it is based on a single-photon avalanche diode (SPAD) operated in time-gated mode, which provides the gating capacity. The SPAD can be switched on and off at a fast rate, higher than 50 MHz, and with fast rise and fall times (few hundreds of picoseconds) due to the relatively low bias voltage required and the low capacity loading.

20.3.3.4 Time-Gated Devices

D'Andrea et al. (2003) developed the first time-domain system based on an ICCD camera, which worked in transmission geometry. The system was based on an argon laser (514 nm) delivering pulses on one side of the sample. The output side of the sample was imaged on the ICCD camera through a macro lens. The intensifier tube (HRI, Kentech, Didcot, UK) of the ICCD was synchronized with the 80 MHz laser pulses through a trigger unit and a delay generator. A 300 ps gate width was typically used. The light intensifier was coupled to a 12 bit CCD camera (PCO GmbH, Göttingen, Germany), where light was integrated over 100 ms frames. A binning of 4 × 4 pixels was used on the CCD camera to obtain a spatial resolution equivalent to that of the image intensifier. This setup allowed the recording of the whole TPSF with a 50 ps delay step in about 30 s. Importantly, the temporal response of the whole field of view (typically 6 × 6 cm²) was recorded *in parallel* on all pixels of the CCD camera. The team pursued two types of data analysis on the data sets: first, the TPSF at each pixel was fitted to the theoretical response from a slab obtained from the random walk theory, yielding the local optical properties of the medium; then the whole frame of the CCD was visualized within a specific delay window, clearly revealing the shadow of any inclusion. The authors suggested a combination of both analyses for breast screening. The second modality can be applied for a first fast inspection of a large area, followed by more accurate characterization of regions of interest where a suspected lesion has been detected. The system was tested on homogeneous phantoms with optical properties similar to those of the human breast, as well as on phantoms containing a small inclusion (1 cm in diameter) with varied absorption and scattering properties. However, the gated technology was later abandoned by the Milan team and replaced with TCSPC technology (Taroni et al. 2004a).

Selb et al. (2005a, 2006) translated the ICCD technology to brain imaging using the TDI system described above. For this application, the image intensifier is not used directly to image the organ, but in combination with optical fibers directing the light from diverse locations on the head onto the ICCD array. An objective lens images the tips of the fibers onto the photocathode of the image intensifier. In this configuration, the CCD camera does not act as an actual imager device, but as a multi detector, where each fiber is detected on a limited number of pixels. A technological innovation was also implemented to permit the detection of the light coming from different delay gates simultaneously. Each detector position consists of a bundle of seven fibers of different lengths (by 10 cm increments), all imaged in parallel on the CCD (Figure 20.5). Each fiber thus introduces a different optical delay in the response (by 500 ps increment). In that way the seven fibers record the same TPSF, but with a relative delay depending on their length. The evolution of intensity can therefore be recorded simultaneously at multiple delays in the same CCD frame. In the current configuration of the time-domain device, 175 fibers are imaged simultaneously on the detector, corresponding to 25 detectors positions at 7 delays each (Selb et al. 2006). During each CCD frame, the laser beam is scanned through 8 source fiber positions, staying approximately 10 ms on each. The eight source locations are selected so that they can be illuminated during the same camera frame while minimizing cross-talk from second nearest sources. This system is currently limited to a single wavelength at a time, which prevents spectroscopic applications. Measurements at 800 nm, the isobestic point of oxy- and deoxy-hemoglobin absorption spectra, are sensitive to blood volume only (total hemoglobin), without contamination by oxygenation variations. Measurements at other wavelengths will present mixed contributions from variations in blood volume and blood oxygenation, without the ability to distinguish between them. The use of two wavelengths would require a second laser, which is a costly part of the equipment. A solution such as the fiber-laser developed at UCL for the MONSTIR device would be a viable alternative, as well as newly commercially available high-power pulsed laser diodes. This is one of the reasons why the TDI system is currently restricted to proof of principle type of experiments, and is inadequate for clinical applications in its current design.

Turner et al. (2005) have developed a system based on the same gating technology, but for the collection of early photons in transmission only. Because of the limitation in the number of photons after several centimeters of tissue, the applications of their technology are limited to small animal imaging (Niedre et al. 2006).

Finally, the system based on a single-photon avalanche diode developed by the group in Milan uses a single source and detector at short separation (Pifferi et al. 2008). The laser source is a

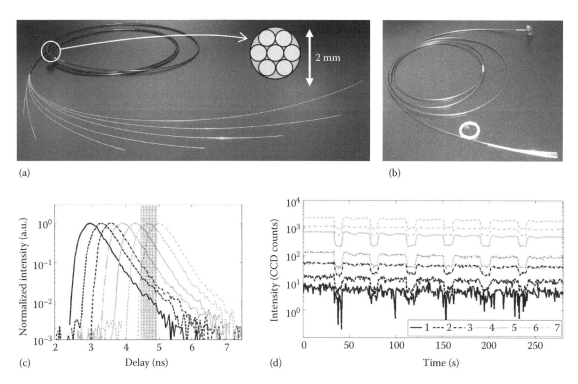

(a)   (b)   (c) Delay (ns)   (d) Time (s)

FIGURE 20.5 Simultaneous detection in multiple temporal gates through optical delay of the TPSF in fibers of different lengths, implemented at the Photon Migration Imaging Laboratory, Boston. (a–b) Each detector position consists of a bundle of seven fibers of different lengths by increment of 10 cm. (c) Example of recorded TPSF: the seven fibers record the same information, delayed by 500 ps. (d) By fixing the delay to an appropriate value, the changes in intensity are recorded at seven sample delays over the whole TPSF. (Adapted from Selb, J. et al., *J. Biomed. Opt.*, 11, 044008, 2006. With permission.)

pulsed laser diode (672 or 758 nm). The system operates in retrodiffusion mode, and was tested on phantoms as well as on a healthy volunteer to test its capacity at detecting brain activation.

20.4 Time-Domain Data Analysis

20.4.1 The Temporal Point Spread Function

The TPSF fully characterizes light transport across the object of interest. It contains information about the optical absorption, scatter and the refractive index, as well as noise and artifacts. If scatter increases, the TPSF is broadened as photons on average will travel further. If absorption increases, the TPSF is narrower as an increase in absorption preferentially removes photons that have traveled further. Simple visual inspection of the TPSF can therefore give a good impression of the optical properties of a medium. More specifically, the absorption coefficient is given by the gradient of the log of the TPSF at longer flight times (Chance et al. 1988, Patterson et al. 1989).

The measured TPSF may include light from other sources in addition to that which has traveled across the object of interest. This is illustrated in Figure 20.6 (Hillman et al. 2000). The measured TPSF is a convolution of IRFs from the laser source, the source and detector optics, and the TPSF due to the medium. There is also additive noise. These confounding signals must be extracted before the TPSF can be claimed to represent the medium under examination. The laser pulse can be characterized experimentally. The pulse width may be a few ps (pulses from the fiber laser used in the UCL system are about 3–5 ps in duration) and is therefore negligible compared to the overall width of the TPSF, which is likely to be of the order of a few ns. Pulses from laser diodes or LEDs, however, may be longer and might need to be formally measured and deconvolved from the measured TPSF. The source and detector IRFs depend on the optical fibers (their length affects the delay and dispersion affects the width of the TPSF), coupling between different optical components and between the optics and the object being examined, and reflections within the optics. Reflections manifest themselves as a delayed peak such as the one visible in the detector IRF in Figure 20.6.

The source and detector IRFs are generally removed by some calibration procedure. This may involve measurement on a test phantom of known optical properties (Selb et al. 2006), or taking specific measurements designed to extract the sources and detector IRFs explicitly (Hillman et al. 2000, Hebden et al. 2003), and then deconvolving the source and detector IRFs from the measured TPSF. However, deconvolution can introduce further errors as some of the photon counts are low and dominated by noise, particularly at long photon flight times. Hillman et al. (2000) pointed out that convolution of the time-domain TPSFs is equivalent to addition of the mean photon flight time and multiplication of the intensities. Similarly, convolution is equivalent to addition in the Fourier domain. So a faster and potentially more robust calibration procedure may be obtained by preprocessing the measured TPSF to give the intensity and mean photon flight time and then adding and subtracting the calibration data appropriately.

Alternatively, if we are interested in imaging the difference between two states (say reference and data) during which the source and detector IRFs remain constant, it may not be necessary to perform a formal calibration. Instead, directly deconvolving the reference TPSF from the data TPSF (or equivalently finding the difference in mean photon flight time or ratio of intensities) may be sufficient.

A further source of error, which has been examined in some depth, is the effect of coupling losses and delays between the source (or detector) fibers and the object being imaged. This cannot be characterized using a phantom as the coupling will vary between the phantom and the object under examination. Hair, for example, can lead to losses of more than half of the signal. In the time-domain, measurements of mean photon flight time are generally robust to surface effects, so one approach would be to reconstruct using mean flight time only, but this leads to an even less well-posed, underdetermined, and nonunique inverse problem. As before, imaging the difference between two states during which the coupling remains constant can be an effective solution. However, more sophisticated solutions have been proposed (Schmitz et al. 2000, Boas et al. 2001, Oh et al. 2002, Vilhunen et al. 2004, Tarvainen et al. 2005), which solve for the coupling coefficients, either by assuming symmetry (and therefore identifying

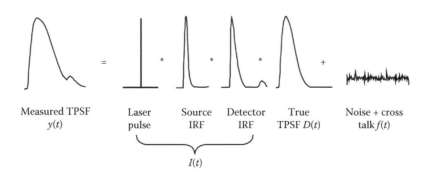

Measured TPSF $y(t)$ Laser pulse   Source IRF   Detector IRF   True TPSF $D(t)$ Noise + cross talk $f(t)$

$I(t)$

FIGURE 20.6 Measured TPSFs are a convolution of the true object TPSF, the source IRF, detector IRF, and the laser pulse, plus background noise and cross-talk. (From Hillman, E.M.C., Experimental and theoretical investigations of near infrared tomographic imaging methods and clinical applications, PhD thesis, University of London, 2002.)

source–detector pairs, which should be identical apart from the coupling) or by treating the coupling coefficients as additional unknowns in the image reconstruction.

The final contribution to the TPSF is noise. Noise can manifest itself in a number of ways. Most obviously, there is photon counting noise, which follows Poisson statistics. For a large number of photons, this approximates to Gaussian and so the noise is equal to the square root of the number of photons. As a rule of thumb, around 10^4–10^5 photons must be measured if data derived from the TPSF can be assumed to be reliable. This limits the maximum separation between source and detector from which useful data can be obtained. Photon counting noise is temporally uncorrelated—it is not synchronous with the source pulse. Other uncorrelated noise can occur, such as background light from the room, dark counts arising from the detector electronics, and drift and jitter in the electronics. These noise sources lead to raised background level in the TPSF. They can sometimes be reduced by changes to the instrumentation (cooling the detectors, changing the CFD thresholds, allowing more time for the electronics to warm up and settle, or even turning down the room lights).

Correlated noise, on the other hand, may arise from light from the source leaking around the object being imaged. This can be seen as a pre-peak in Figure 20.2, and is a particular problem when an adjustable helmet is used for 3D optical tomography of the brain (Gibson et al. 2006). Time-domain measurements allow such artifacts to be identified in the data and be either corrected or eliminated. Other sources of correlated noise include cross-talk in the detectors and reflections in the optics, which may be reduced by appropriate calibration procedures.

Time-domain measurements allow these different sources of noise to be identified, and also provide some information as to the possible sources of noise. This is particularly useful when testing a new patient interface, when light leaks or reflections could be significant but are hard to identify using other systems. A final source of error occurs when there is a failure of a source or detector. Time-domain measurements allow such a problem to be immediately identified and may also provide some diagnostic information as to their source.

20.4.2 Modeling the TPSF

Experimentally, the TPSF is the measured response to a very short impulse of light incident on the tissue of interest. From a theoretical point of view, the measured response can be modeled by convolving the input function with the appropriate Green's function. The Green's function is the solution to a source, which is a delta function in space and time, and can be found by solving the diffusion equation for the particular geometry under examination. Green's function solutions have been presented for simple homogeneous objects (Arridge et al. 1992) and some more complex geometries (Martelli et al. 2002, Shendeleva 2004).

Analytical calculation of the TPSF can be used to analyze data, such as by fitting the Green function to the measured TPSF (Liebert et al. 2003b). Green's functions can also be used

to provide forward data for image reconstruction, particularly where the geometry can be simplified to a slab (such as in compressed-breast optical mammography) or infinite half-space (when imaging a small region of the scalp).

20.4.3 Characterizing the TPSF Using Datatypes

A single TPSF contains all available information about photon flight times. It may be divided into typically between 5 and 2000 bins, depending on whether the system is based on time-gating or TCSPC. The full TPSF is therefore very computationally intensive to process as well as being highly redundant (Arridge and Schweiger 1995). These two factors mean that it is usual to preprocess the TPSF to generate parameters that characterize it more efficiently. These parameters are known as *datatypes*.

The most intuitive and most commonly used datatypes are the first two temporal moments (integrated intensity and mean photon flight time). These are calculated formally as the Mellin transform of the TPSF, $\Gamma(t)$, using Equation 20.1 (Arridge and Schweiger 1995, Hebden et al. 1998)

$$M_n\big(\Gamma(t)\big) = E^{-1} \int_0^\infty t^n \Gamma(t)\,\mathrm{d}t \tag{20.1}$$

where
    E is the integrated intensity
    n is the order of the moment

The intensity and mean photon flight time are calculated by setting $n = 0$ and 1, respectively. The factor E^{-1} is not applied when calculating the integrated intensity, but for higher-order moments, it normalizes the datatype so it is independent of the overall light intensity. This decreases the sensitivity of normalized datatypes to coupling and other surface and instrumentation effects.

The central moments (variance, skew, and kurtosis) are calculated similarly from Equation 20.2

$$C_n\big(\Gamma(t)\big) = E^{-1} \int_0^\infty \big(t - \langle t \rangle\big)^n \Gamma(t)\,\mathrm{d}t \tag{20.2}$$

where $\langle t \rangle$ is the mean photon flight time (equal to M_1). In practice, the higher-order moments are increasingly sensitive to late-arriving photons (Hebden et al. 1998). There are relatively fewer of these, so these datatypes rapidly become dominated by noise. Intensity and mean photon flight time have a number of advantages:

- They have an intuitive, physical meaning. This is particularly helpful when looking at the raw data. If we know that the mean photon flight time is, say, 3 ns, then we can identify outliers that are much bigger or smaller than this.

Moreover, if there is an error of, say, 100 ps, then we can immediately say that this is equivalent to a light distance of about 3 cm. This makes the error meaningful and can help to identify the source of the error.

- They can be converted to the frequency-domain equivalent by a Fourier transform. This allows a frequency-domain system to be directly compared to a time-domain system or, more importantly, allows time-domain data to be analyzed and reconstructed using a reconstruction algorithm developed for the frequency-domain.
- Calibrating the full TPSF can require ill-posed deconvolution operations. If moments of the data and calibration measurements are calculated, the calibration reduces to simple addition and subtraction (Hillman et al. 2000).
- Perhaps most importantly, certain datatypes can be directly calculated using the finite element method (FEM). This approach was introduced by Arridge and Schweiger (1995). They took the Mellin transform of the Green's function and noticed that it could be simplified to a form that allowed it to be calculated directly using FEM. The calculation is performed repeatedly for each n, so higher-order moments take longer to calculate than lower-order moments. In a 2D simulation, Arridge and Schweiger (1995) showed that calculating intensity and M_1 (the mean photon flight time) was up to 400 times faster than calculating the full TPSF, once the FEM system had been set up.

Even though intensity and meantime do have certain advantages as datatypes, other parameters have also been examined. The Laplace transform may have particular advantages as it effectively weights the TPSF with an exponential filter, reducing the effect of longer photon flight times (Schweiger and Arridge 1997). This reduces noise, as well as emphasizing the contribution of photons, which may have taken a more direct path. It therefore has some similarities with time-gating techniques. The Laplace transform of the TPSF, $L[\Gamma(t), s]$, is then normalized to the total intensity [Equation 20.3]

$$L\left[\Gamma(t),s\right]=E^{-1}\int_0^\infty e^{-st}\Gamma(t)\mathrm{d}t \qquad (20.3)$$

where s determines the rate of exponential decay and must be determined by the user. As with the Mellin transform, the Laplace transform of the TPSF can be calculated directly using FEM.

The Laplace transform is sensitive to noise at short photon flight times, so a further class of datatypes was proposed: the Mellin–Laplace transform (Hebden et al. 1998), which provides a temporal filter that has zero weight at $t=0$ and $t=\infty$ by combining the Mellin and Laplace transforms [Equation 20.4]:

$$ML_n\left[\Gamma(t),s\right]=E^{-1}\int_0^\infty t^n e^{-st}\Gamma(t)\mathrm{d}t. \qquad (20.4)$$

The datatypes discussed so far all have the advantage that they can be calculated directly from the FEM, leading to computationally efficient reconstructions. They are also differentiable and so can be incorporated into gradient-based iterative image reconstruction methods (Arridge 1999). However, other datatypes such as the time to peak intensity or the log of the slope of the TPSF have been considered (Chance et al. 1988, Patterson et al. 1989). An alternative approach is to use fitted datatypes. The absorption and scatter coefficients may be derived by fitting a Green's function to the TPSF (Arridge et al. 1992), which can then be used in a backprojection-type algorithm (Benaron et al. 1994).

Other approaches include time-gating, for example, Grosenick et al. (1999), Selb et al. (2006), Taroni et al. (2004b), and taking the Fourier transform of the TPSF to give the equivalent frequency-domain datatypes of intensity and phase.

Multiple datatypes can be incorporated into image reconstruction by appending additional data to the measurement vector ΔM and by appending the appropriate Jacobian J_μ^M. It is possible to solve for different parameters $\Delta\mu_a$ using the same method (Equation 20.5).

$$\begin{bmatrix}\Delta M_1\\\Delta M_2\\\vdots\\\Delta M_n\end{bmatrix}=\begin{bmatrix}J_{\mu_1}^{M_1}&J_{\mu_2}^{M_1}&\cdots&J_{\mu_m}^{M_1}\\J_{\mu_1}^{M_2}&J_{\mu_2}^{M_2}&\cdots&J_{\mu_m}^{M_2}\\\vdots&\vdots&\ddots&\vdots\\J_{\mu_1}^{M_n}&J_{\mu_2}^{M_n}&\cdots&J_{\mu_m}^{M_n}\end{bmatrix}\begin{bmatrix}\Delta\mu_1\\\Delta\mu_2\\\vdots\\\Delta\mu_m\end{bmatrix} \qquad (20.5)$$

Clearly, more measurements types ΔM_n can lead to a better posed inverse problem by increasing the amount of information. However, this must be done effectively. Arridge and Lionheart (1998) showed that reconstructions that use intensity data only are unable to uniquely distinguish between absorption and scatter in the absence of prior information. Despite this, it is common to reconstruct for changes in absorption from changes in intensity, particularly in optical topography, and some groups report absolute images from intensity data only. Other datatypes have been used far less frequently, although Hebden et al. (1999) used mean flight time and variance to reconstruct images from a tissue-equivalent phantom. It is reasonable to expect that choosing appropriate datatypes may assist in distinguishing between absorption and scatter. A range of studies have looked theoretically at the optimal combinations of datatypes, mainly from two different points of view: maximizing the information content and minimizing the effect of noise.

Schweiger and Arridge (1999) studied combinations of Mellin–Laplace datatypes and their ability to simultaneously reconstruct absorption and scatter. They plotted the objective function as a function of absorption and scatter. An ideal objective function would have a single, well-defined minimum, which can be approached equally from all directions. Meantime and variance were close to this ideal. Intensity showed an elongated valley of the form $\mu_s\mu_s=\mathrm{const}$, confirming the inability of intensity measurements to distinguish between absorption and scatter. Combinations of datatypes provided closer approximation to the ideal objective function map. The optimum single datatype was $L(s = 0.001)$, and the optimum combination of two datatypes was C_3 (skew) and $L(s = 0.001)$. In general, central moments appeared to be more sensitive to absorption but relatively insensitive to scatter.

Hebden et al. (1998) examined a different criterion: the sensitivity of different datatypes to noise. Analytical expressions for the standard deviation of the datatypes were derived and used to calculate the uncertainty as a function of number of photons in a simulated TPSF. They found that pairs of datatypes where one of the pairs was total intensity generally performed well, probably because the elongated valley described above tended to sharpen the objective function of the datatype it was paired with. The best combination of datatypes was found to be intensity and mean photon flight time. This pair of datatypes has since been widely and successfully used in time-domain image reconstruction (Arridge et al. 2000, Hebden et al. 2002).

Finally, one may decide that the advantages of using datatypes are insufficient to justify the resulting loss of information. Gao et al. (2002) reconstructed images of tissue-equivalent phantoms using the full TPSF. Image reconstruction took 12 times longer than using meantime and variance, but contrast, spatial resolution, and quantitation all appeared to be improved compared to images reconstructed using datatypes. This work was extended by Zhao et al. (2007).

20.5 Reconstructing Optical Properties from Time-Domain Data

20.5.1 Absolute Measurement of Baseline Optical Properties

By fitting the experimental TPSF to a model of light propagation through the tissue, one can estimate the bulk optical properties of the tissue (absorption and scattering) at one or multiple wavelengths. From spectroscopic measurements, the concentrations of oxy- and deoxy-hemoglobin can then be extracted. This approach was first suggested and implemented by Patterson et al. (1989) who derived the analytical solution of the diffusion equation in a homogeneous semi-infinite medium or a slab, measured either in transmission or reflectance. A detailed description of the theory can be found in Part II of this book. They validated the method by fitting Monte-Carlo simulations to the analytical expression, and applied the method to the measurement of phantoms with known optical properties. Liu et al. (1993) suggested an alternative empirical approach, where a small biological sample of unknown optical properties is placed in a large size host medium whose controlled optical properties can be adjusted until they match those of the inclusion. This technique is however extremely fastidious and cannot be used in vivo. The theoretical modeling of Patterson et al. (1989) was subsequently improved by Haskell et al. (1994) who introduced extrapolated boundary conditions. The method has

been validated on phantoms (Ntziachristos and Chance 2001, Swartling et al. 2003, Pifferi et al. 2005) and applied to a number of different tissues types, including muscle, breast, and head. Ntziachristos and Chance (2001) studied the experimental errors of time-domain measurement of baseline optical properties, and demonstrated common systematic quantification errors of 10%. They showed that higher accuracy is obtained on media with high scattering and low absorption. Modeling accuracy is important when reconstructing for absolute optical properties: Sassaroli et al. (2001) studied the effect of the medium curvature and showed typical errors of about 10% on the optical properties when a cylindrical or spherical geometry was erroneously assumed to be a semi-infinite medium geometry. Swartling et al. (2005) used a supercontinuum light pulse providing a very broad spectrum of illumination wavelength. By fitting for scattering and absorption at all wavelengths, they retrieve the hemoglobin concentration with more confidence (Bassi et al. 2004, 2006, 2007). Rather than fitting the whole TPSF, Liebert et al. (2003b) have presented a new method based on fitting simply the mean time of flight and the variance of the TPSF, enabling faster recovery of the optical properties. In all these approaches, careful characterization of the system IRF is crucial to retrieve accurately the optical properties (Liebert et al. 2003a).

Clearly, the application of these approaches to in vivo measurements is limited, as human organs are generally far from homogeneous. Measurements based on a homogeneous model therefore only retrieve some local average value of the optical properties. One way to implement this simple data analysis to heterogeneous structures is to scan the source and detector over the tissue. At each position, the measurements are fitted to the analytical solution of the diffusion equation for a homogeneous slab. An image is then reconstructed simply by interpolating the local optical properties. This allows for variations in space even though the adopted model is homogeneous and therefore much simpler than an actual 3D reconstruction. The technique has been applied in optical mammography by the teams in Berlin and Milan, who use a single source-detector pair in transmission scanned over the whole breast (Grosenick et al. 2005a, Taroni et al. 2005b). Another approach consists of directly producing images with some features of the TPSF, without any actual reconstruction of the optical properties. For instance, Grosenick et al. (2005a) have done a series of point by point measurements in transmission on the breast, and have compared the contrast and sensitivity of different datatypes.

Alternatively, more elaborate modeling of the tissue can be implemented to account for the complex structures of in vivo tissue. As first approximations, analytical solutions to the diffusion equation for simple geometries have been developed, in particular, for a homogeneous slab with a single spherical inclusion (Grosenick et al. 2007), and for a two-layer medium (Kienle et al. 1998b, Martelli et al. 2003a). In the case of brain or muscle, the main heterogeneity arises from the layered structure of the organ. The brain is covered by multiple layers of different tissues: scalp, skull (the bone itself consisting of different layers), and cerebrospinal fluid (CSF). To a lesser extent, a muscle is hidden beneath layers of skin and adipose tissue. Measurements of these tissues based on the homogeneous medium approximation will result in a mixing between the optical properties of superficial and deep tissues. On the adult head in particular, Comelli et al. (2007) have shown with Monte-Carlo simulation that such measurements mostly reflect the optical properties of the skin and scalp for typical adult geometry. To overcome this issue, Kienle et al. (1998b) solved the diffusion equation for a two-layer medium in the steady state, frequency domain and time domain. They compared a 2D spatial Fourier transform approach against Monte-Carlo simulations (Kienle et al. 1998a,b), and applied the method in vivo to the characterization of optical properties of muscle (Kienle and Glanzmann 1999). Martelli et al. (2003a) suggested an alternative procedure, where the analytical solution of the two-layer diffusion equation is obtained with the eigenfunction method (Martelli et al. 2003b). They similarly validated the method on Monte-Carlo simulations (Martelli et al. 2003a) and phantom measurements (Martelli et al. 2004) and applied it to in vivo characterization of muscle (Martelli et al. 2004). The implementation on head measurements is rendered more complicated because of thicker superficial layers. So far, only one study by Gagnon et al. (2008) reports measurements on the head with the discrimination between superficial layers and brain with a time-domain system. Martelli et al. (2007) went one step further in the modeling and developed the analytical solution for a three-layer model, representing for instance skin/scalp/brain. However, the method was only validated with Monte-Carlo simulations, and has not been applied to in vivo measurements.

For breast imaging, some authors have modeled the breast as a homogenous medium with a lesion as a single spherical inclusion (Grosenick et al. 2005b), and derived the analytical solution in the time domain for this particular geometry (Grosenick et al. 2007).

Finally, an approach intermediate between the ones described so far and full 3D tomography consists of introducing a more complex and realistic tissue structure obtained through an anatomical imaging modality, and segmenting the measured organ into a number of tissue types. For the breast, these tissue types are typically adipose tissue, glandular tissue, and tumor or other breast lesion. For brain imaging, the actual structure of a subject head can be segmented into scalp, skull, CSF, and gray and white matter. Instead of reconstructing a full 3D image of the organ, the inverse problem consists of retrieving the absorption and scattering of each homogeneous tissue type, which greatly reduces the number of unknown and the ill-posedness of the problem. Barnett et al. (2003) suggested using the actual segmented head geometry, and applying a Bayesian reconstruction framework. For this type of inverse problem, however, analytical solutions to the diffusion equation are not available, and numerical (finite element, boundary element, finite difference) or statistical (Monte-Carlo, random walk) modeling of light propagation in the medium is required.

20.5.2 Relative Measurement of Changes in Optical Properties

Time-domain technology can also be used to measure variations in optical properties over time. Usually the scattering coefficient is assumed constant, and changes in the TPSF are considered to be arising from small changes in absorption only.

Ntziachristos et al. (1999) published the first study of functional imaging using time-domain technology. They showed that time-domain data enable characterization of absorption changes with higher accuracy than CW measurements even if the baseline optical properties are approximated very roughly.

The most valuable feature of time-domain functional imaging over other modalities is the depth information contained in the data. The main idea behind depth-resolved imaging with time-domain data is the fact that photon time of flight contains information about the average depth to which they traveled. Steinbrink et al. (2001) suggested using different time delays to retrieve information about different depths. Liebert et al. (2004) recognized that higher moments are more sensitive to deeper tissues, as they give more weight to late photons. In practice, three moments of the TPSF are sufficient for depth discrimination. The integrated intensity is simply the equivalent of the CW signal and it is highly sensitive to superficial layers. The mean time of flight has a sensitivity profile that goes deeper, while the variance has even deeper sensitivity (Figure 20.7). Similarly, Selb et al. (2005a) used the gated intensity at different delays to probe different depths of light penetration. While all delay gates bear a similar contribution from the superficial layers, the contribution from deep layers, that is, the cortical layers, increases with longer delays. One advantage of using the moments of the TPSF is that the deconvolution of the IRF simplifies into a subtraction, hence it cancels out when measuring changes in these moments (Liebert et al. 2004). Whether gated intensity or moments of the TPSF are used, both techniques depend only slightly on the tissue baseline optical properties. Furthermore, time-domain technology enables the prior estimation of the baseline properties, as described above.

20.5.3 Tomographic Imaging

Time-domain measurements when combined with photon-counting detectors are ideally suited to tomographic imaging through large volumes of tissue due to their high sensitivity. A photon-counting PMT is the most sensitive type of detector and is the most suitable method for tomographic imaging across the head or breast. A single 800 nm pulse from a skin-safe 50 mW laser pulsing at 80 MHz contains ~10^{20} photons. Source losses (10^{-1}) are dominated by coupling light into the fiber and the source switch, and detector losses (10^{-3}) by the limited numerical aperture of the detector fiber and the quantum efficiency of the photomultiplier (Schmidt 1999). Further losses occur when coupling the light into and out of the object under examination. The coupling coefficient can be calculated (Schmitz et al. 2000,

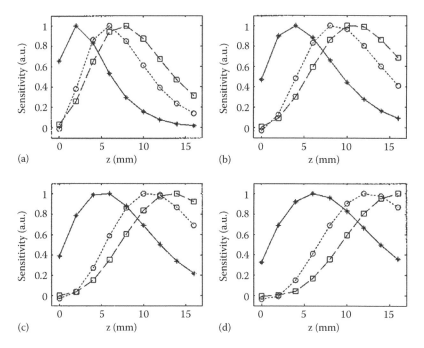

FIGURE 20.7 Depth sensitivity profile of (*) integrated intensity, (O) mean time of flight, and (□) width of the TPSF for a source-detector separation of (a) 14 mm, (b) 26 mm, (c) 38 mm, and (d) 50 mm. Each curve is normalized to its maximum value. The results were obtained by Monte-Carlo simulations in a 10-layer medium with identical background optical properties of $\mu_a = 0.1\,\text{cm}^{-1}$ and $\mu_s' = 10\,\text{cm}^{-1}$. (From Liebert, A. et al., *Appl. Opt.*, 43, 3037, 2004. With permission.)

Boas et al. 2001, Tarvainen et al. 2005, Vilhunen et al. 2004), but values are rarely quoted, possibly because they are very variable. Anecdotally, the coupling coefficients are of the order of 10^{-1} but can be more if the skin is dark or hair is present. This gives, very roughly, a total loss of the order of 10^{-6}, meaning that 10^{14} photons may be available per pulse. A common rule of thumb is that NIR light traveling through tissue loses an order of magnitude in intensity for each centimeter traversed, so this suggests that if we measure across 12 cm of tissue, we should record 100 photons per second. Experience using the MONSTIR time-domain imaging system at UCL (Schmidt et al. 2000) suggests that adequate data may be recorded across a term baby's head (average diameter 12 cm; (Austin et al. 2006)) or across a 16 cm diameter hemisphere used for breast imaging (Enfield et al. 2007). This indicates the lower absorption and scatter seen in the breast compared to the brain.

However, this sensitivity comes at a price, namely speed. Two things limit speed: the signal-to-noise ratio (SNR) and dead-time in the detectors. When measuring only tens or hundreds of photons per second across a large distance, it is necessary to count for a long period in order to obtain a temporal point-spread function with an adequate SNR. There may be a trade-off between acquisition speed and SNR; data have been acquired across the head or breast by counting for 10–15 s from each source. When measuring across smaller distances, where the photon flux is higher, detector dead-time may become significant and limits the number of photons, which can be counted to $\sim10^5\,\text{s}^{-1}$.

The reconstruction of full 3D tomographic images requires sophisticated modeling and nonlinear image reconstruction routines, which are explained in detail in Chapters 16 and 17.

20.6 In Vivo Applications

20.6.1 Breast Characterization and Imaging

Optical mammography is a major area of application of diffuse NIR imaging. Breast cancer is among the leading causes of death in women (Jemal et al. 2008), but early detection enabling appropriate therapy can reduce mortality (Tabar et al. 2004). X-ray mammography is the clinical standard for screening of breast cancer (Fletcher and Elmore 2003), but it is still impeded by a number of drawbacks: ionizing radiation limits its frequency of utilization; insufficient specificity leads to subsequent diagnostic core biopsies; poor sensitivity in dense breasts; and debatable benefit in young women. NIR spectroscopy or tomography has been suggested as an alternative or complementary technique because of its high sensitivity to blood content and oxygenation. Breast tumors exhibit elevated vascularization (Rice and Quinn 2002), and so should be detectable using diffuse optical imaging. Furthermore, the typical dimension of the breast makes it adequate for transmission imaging. Basically, two geometries have been used: breast in compression between two plates (similar to traditional x-ray mammography but with lighter compression), or rings of optical connectors surrounding the breast. Optical mammography has been investigated as a stand-alone breast screening modality for tumor detection; in combination with more traditional imaging techniques (x-ray mammography, MRI) for the characterization of detected breast lesions; and for the monitoring of response to therapy. A lot of initial work has also been dedicated to the characterization of the healthy breast and the variation in its optical properties with a number of demographic factors (age, body mass index, menstrual cycle, and breast density). A recent article by Leff et al. (2008) presents a thorough review of all clinically oriented studies using optical mammography before 2006. Much of the work done to date has used frequency-domain or continuous wave systems (see Schmitz et al. 2005, Carpenter et al. 2007, Fang et al. 2009, and elsewhere in this book), although Leff identified 14 studies based on time-domain technology, stemming from five different centers.

The first time-domain instrument developed for breast imaging was described by Suzuki et al. (1994) who used a single wavelength (753 nm) and a single source-detector separation of 3 or 4 cm in reflectance geometry. The TPSF measured at a single position was fitted to the analytical solution of the diffusion equation to extract the local absorption and scattering coefficients of the breast. The system was applied to 30 healthy women. The authors showed a high negative correlation of absorption with age, which they suggested was due to the greater content of fibroglandular tissue in the mammographically dense breasts of young women. Both absorption and scattering correlated with body mass index.

Ntziachristos et al. developed a multichannel TCSPC imaging system that could be used either independently or in combination with an MRI scanner (Ntziachristos et al. 1998) and imaged 10 subjects with suspicious breast lesions (Ntziachristos et al. 2002). The breast was gently compressed between two plates, which hold 24 sources (three time-multiplexed wavelengths: 690, 780, and 830 nm) and eight detectors. From these measurements they deduced both the average optical properties of the breast based on a homogeneous slab model, and the optical properties of the tumor and the background based on tissue segmentation obtained from MRI scans. They consistently observed higher total hemoglobin content and lower oxygen saturation in tumors but the results were more diverse for benign lesions: fibroadenomas were mildly hypoxic with increased vascular content; other benign lesions also showed increased vascularization with raised oxygenation; and cysts had low absorption.

The largest collection of clinical data with time-domain optical mammography was obtained jointly by groups in Milan and Berlin under a European Union funded consortium known as OPTIMAMM (Hebden and Rinneberg 2005). Both teams developed multi-wavelength time-domain systems based on a single source-detector pair in transmission, spatially scanned in two dimensions over the breast and in total imaged over 300 women with suspected breast lesions. A series of articles by Grosenick et al. present the evolution of the system and data analysis (Grosenick et al. 1997, 1999), as well as the clinical results on a population of over 150 women (Grosenick et al. 1999, 2003, 2004, 2005a,b). Out of 102, 92 confirmed cancerous lesions

were detected optically, due to their increased total hemoglobin concentration. Some examples are presented in Figure 20.8. Generally, the datatype that provided the highest contrast and contrast-to-noise ratios was the inverse of the total photon counts in selected time windows. However, late time windows yielded images much less affected by edge effects than those obtained with early photons or all photons. Cancerous lesions presented higher absorption coefficients than those of surrounding healthy breast tissue by a factor of 2–4, while scattering coefficients only presented an increase of 20% on average in tumors. Blood oxygen saturation was found to be a poor discriminator for tumors and healthy breast tissue.

The group in Milan has developed a very similar system, with a special emphasis on spectral sampling (Spinelli et al. 2004, 2005, Taroni et al. 2004a,c, 2005a,b). By adding wavelengths beyond 900 nm (683, 785, 913, and 975 nm), they can better characterize lipid content in the breast. Over the course of 1 year (2002), 101 patients with malignant and benign lesions were analyzed retrospectively. Projected images were reconstructed using the scattering coefficient derived from a homogeneous model, as well as late gated intensity, more sensitive to absorption contrasts. More than 80% of malignant lesions were detected, usually due to their strong blood absorption at short wavelengths. Cysts offered a similar detection rate, with a contrast originating from their low scattering. Fibroadenomas presented low absorption at 913 nm and high at 975 nm, but were difficult to distinguish from other structures. Both the Berlin and Milan groups reported sensitivity and specificity of 80%–85%.

The optical group at UCL has developed a time-domain imager, which images the uncompressed breast suspended in a hemispherical cup filled with a highly scattering fluid, and the sources and detectors are coupled around the cup. First 2D (Yates et al. 2005a,b) and later 3D images of absorption and images were reconstructed using the finite element method based package TOAST (time-resolved optical absorption and scatter tomography) developed at UCL (Arridge et al. 2000). Thirty-eight patients with a range of benign and malignant conditions were scanned with the 3D device (Enfield et al. 2007). The diseased breast and, when possible, the healthy breast were both scanned, with each scan taking a little more than 10 min. In common with other studies, the authors reported high contrast in tumors due to their hypervascularization, and in cysts because of their low scattering properties, but low detection rates on fibroadenomas. The sensitivity was 86% and the specificity was 67% (Enfield et al. 2008).

The Canadian company ART (Advanced Research Technologies Inc.) has developed a time-domain breast imaging device marketed under the name SoftScan, and approved for sale in Canada and Europe. Initial results on 49 volunteers, both healthy and presenting a breast lesion, were consistent with other published studies, showing a reduction in total hemoglobin concentration with body mass index and higher total hemoglobin content in breast lesion compared to healthy surrounding tissues (Intes 2005). However, unlike the majority of other studies, they also

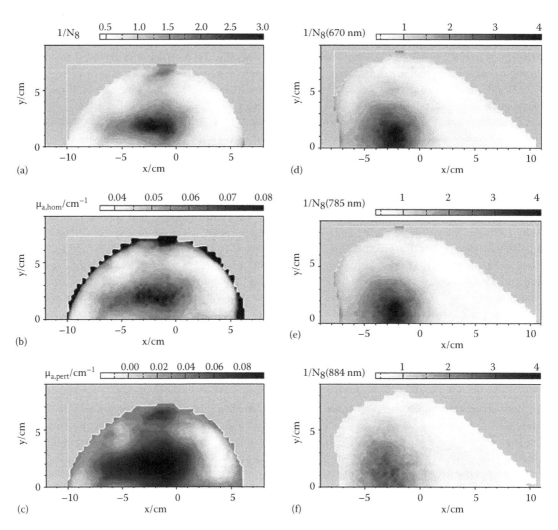

FIGURE 20.8 Optical mammograms obtained with a scanned single source and detector on-axis time-domain system in transmission geometry. The figure presents images obtained from different datatypes extracted from the TPSFs. Left: Patient with invasive lobular carcinoma. (a) Inverse photon counts in a late window, (b) absorption coefficients retrieved from a homogeneous model at each position, (c) absorption coefficients from a perturbation model with a single spherical inclusion. Right: Patient with invasive ductal carcinoma. The images display inverse photon counts at various wavelengths: (d) 670 nm, (e) 785 nm, and (f) 884 nm. (From Grosenick, D. et al., *Phys. Med. Biol.*, 50, 2429, 2005a. With permission.)

showed significant difference in blood volume between benign and malignant lesions.

The general conclusions of these optical mammography studies can be summarized as follows: all teams report detection rates of 80%–85% for cancerous lesions when optical imaging is used as a stand-alone modality. The contrast of tumors originates mainly from their increased vascularization, and sometimes lowered oxygen saturation, though hypoxia appears to be a poor discriminator in general. Similar detection rates are reported for cysts, which offer a strong scattering contrast. On the other hand, fibroadenomas are much more problematic, and are reported as having optical properties either similar to the background or to malignant lesions. They are therefore missed or not distinguished from other structures in about 50% of the cases, leading to both low sensitivity and specificity for these benign lesions. In summary, a major limitation of optical mammography pointed out by all studies is the difficulty to distinguish between benign

lesions and tumors or background, because of similar optical properties, severely limiting the specificity of the modality in a stand-alone configuration. Also, all studies have mentioned the difficulty of detecting lesions close to the chest wall, due to simple geometry considerations. Current improvements of the instrumentation are oriented toward increased spectral and spatial resolution, in order to better characterize the different types of lesions. In particular, the Berlin team is investigating off-axis measurements (source and detector not aligned) to improve the angular sampling of the tissue (Grosenick et al. 2005a), while the Milan group has extended the range of wavelengths in order to maximize sensitivity to fat (Taroni et al. 2005b).

It is becoming increasingly apparent that optical mammography lacks the spatial resolution and specificity necessary to be successfully promoted as an alternative to x-ray mammography for screening. However, particularly when used together with other imaging modalities, its lack of ionizing radiation and functional information may provide an important role in niche applications such as imaging young women or women with denser breasts, distinguishing between malignant and benign lesions, and assessing the success of neoadjuvant chemotherapy (Tromberg et al. 2008).

20.6.2 Baseline Optical Properties of the Brain

The quantitative measurement of the baseline optical properties of the brain is of interest for longitudinal and cross-sectional studies. They allow comparison of cerebral oxygenation and blood volume between subjects, as well as the monitoring of their evolution within a subject over days or weeks, opening the door to a number of diagnostic and treatment monitoring applications, for instance in the context of neurocritical care.

One growing application of optical imaging is neonatal intensive care monitoring. Perinatal brain development is extremely fast, but difficult to evaluate in this fragile population. Optical characterization of brain development and health in infants is a very promising alternative to invasive PET and SPECT and restrictive MRI modalities. The low cost and noninvasiveness of optical technology make it an ideal candidate for routine clinical monitoring, and the small sizes of neonates' heads facilitate the accurate characterization of cerebral blood volume (CBV) and oxygenation.

Both frequency-domain (Franceschini et al. 2007) and time-domain (Ijichi et al. 2005) measurements have been used to measure baseline cerebral oxygenation and blood volume. This alleviates the need for a tracer—an invasive injection of indocyanine green (ICG), or an oxyhemoglobin bolus induced by a change in arterial saturation—which is necessary if CBV is to be measured using CW devices (Wyatt et al. 1990, Brun and Greisen 1994, Leung et al. 2004). In particular, Ijichi et al. (2005) used time-resolved spectroscopy to measure scattering, CBV, and oxygenation in 22 neonates with gestational age ranging from 30 to 42 weeks. They showed a significant negative linear relationship between postconceptional age and tissue oxygenation, and a significant positive relationship between postconceptional age and CBV.

In adults, the measurement of brain optical properties is rendered more challenging by the thickness of the superficial layers of skin, skull, and CSF. Ohmae et al. (2006) performed time-domain recordings on the foreheads of five healthy volunteers during acetazolamide injections, simultaneously with PET ^{15}O-CO measurements. They used three wavelengths (761, 791, and 836 nm) and four source-detector separations from 2 to 5 cm. Data at each distance were fitted independently to the analytical solution of the diffusion equation for semi-infinite homogeneous medium using a nonlinear least square fit of the whole TPSF. Comparing absolute and relative change in CBV, as inferred by PET and NIRS, they observed a good correlation at all distances, though higher at 4 and 5 cm separation. From their measurements, they also estimated the contribution of cerebral tissue to the signal to be below 50% when the optode spacing was 2 cm, increasing to 50%–75% at 4 and 5 cm, with increasing sensitivity at longer wavelengths. Their study perfectly illustrates the challenges of baseline measurements in the adult. Ohmae et al. (2007) further applied the technique to monitoring cerebral oxygenation during coronary artery bypass surgery in 23 patients (see Figure 20.9).

Comelli et al. (2007) performed time-resolved broadband (700–1000 nm) spectroscopic characterization on the forehead of five healthy volunteers with a probe consisting of one source and five detectors with separations from 2 to 6 cm. They

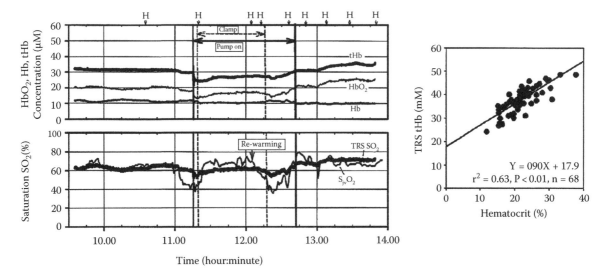

FIGURE 20.9 (a) Optical monitoring of cerebral hemoglobin concentrations and blood oxygenation in one patient during cardiopulmonary bypass surgery with a time-resolved device (Hamamatsu TRS-10). The time domain modality enables characterization of absolute values of oxy- (HbO$_2$), deoxy- (Hb), and total (tHb) hemoglobin concentrations as well as oxygen saturation (SO$_2$). The jugular venous oxygen saturation (SjvO$_2$) is monitored simultaneously. The trends of SO$_2$ and SjvO$_2$ match during the surgery, even though SjvO$_2$ present higher variations than SO$_2$, which could be partially explained by different spatial sensitivity. (b) In nine patients undergoing the same procedure, arterial blood hematocrit and TRS-derived total hemoglobin concentration show high correlation. (From Ohmae, E. et al., *J. Biomed. Opt.*, 12, 062112, 2007. With permission.)

retrieved the optical properties of the medium with a nonlinear least square fit of the whole TPSF to the analytical solution of the diffusion equation for a semi-infinite homogeneous medium. Their study shows high variation between subjects, but consistent results within each subject. They also performed Monte-Carlo simulations on a four-layer model to test the accuracy of their approach, and showed that the retrieved values are closer to the optical properties of superficial than deep tissues, even at long distances. Surprisingly, increasing the source-detector separation did not yield better results, in disagreement with the study by Ohmae et al. Gagnon et al. (2008) applied for the first time the two-layer model developed by Kienle et al. (1998a,b) to adult head measurements. They performed simultaneous measurement and fitting at four optode spacings from 1 to 3 cm and four NIR wavelengths on the forehead of five subjects. The thickness of the first layer was obtained from MRI anatomical scans of each subject. In agreement with Comelli et al. (2007), the study shows large inter-subject variability. Monte-Carlo simulations confirm the advantage of the two-layer model over 1 cm or 3 cm fitting alone. Their simulations also show that a 10% error in the thickness estimation will introduce 10%–15% error in the retrieved absorption of the brain (Gagnon et al. 2008).

20.6.3 Functional Imaging with Depth Sensitivity

By selecting different time windows or datatypes from the TPSF, it is possible to isolate later-arriving photons that are more likely to have penetrated deeper tissues, such as the brain. This approach has been investigated as a method for detecting the hemodynamic response to brain activation with better depth selectivity than is available from CW measurements. Most studies of this nature have been "proof of principle" type of experiments on healthy subjects (Steinbrink et al. 2001, Liebert et al. 2004, Torricelli et al. 2005, Montcel et al. 2006, Selb et al. 2006,

Kacprzak et al. 2007), and the number of physiological and clinical applications is still extremely limited (Liebert et al. 2005, Ohmae et al. 2007, Mackert et al. 2008).

Steinbrink et al. (2001) reported the in vivo implementation of their method to obtain depth resolution from time-domain relative measurements on one healthy volunteer during a finger-tapping task, and a Valsalva maneuver. They showed they were able to distinguish between cortical activation and a systemic response, as shown in Figure 20.10. The similar method suggested by Liebert et al. (2004) and based on moments of the TPSF has been applied in vivo to single or multiple source-detector pairs (Kacprzak et al. 2007). The technique was also applied in combination with magnetoencephalography (MEG) to study neurovascular coupling in healthy subjects (Mackert et al. 2005, 2008, Sander et al. 2007a,b).

The same approach was also applied to cerebral blood flow estimation in combination with ICG bolus injection (Liebert et al. 2004), which strongly increases the absorption contrast. The depth-resolved analysis showed a clear distinction between the dynamic behavior of the bolus passage in the superficial and deep layers. The deep layers presented a rapid arrival and transit time of the bolus, while the superficial layers showed a delayed initial rise in absorption, followed by a slower washout period. These dynamics are consistent with intra- and extra-cerebral compartment behaviors, as observed in MRI studies using a gadolinium tracer for blood flow estimation. The group applied the method to assess cerebral perfusion in two stroke patients (Liebert et al. 2005). In particular, they successfully observed, in one patient with acute middle cerebral artery ischemia, a significant difference in bolus transit time between both hemispheres, which subsequently disappeared after 1 day.

Selb et al. (2005a) applied their time-domain gated device to the study of a healthy volunteer during finger-tapping, and successfully distinguished the superficial systemic response from cortical activation, which could not be achieved by CW

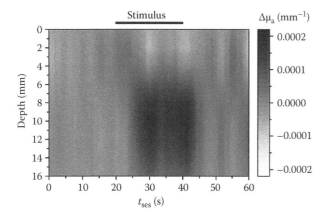
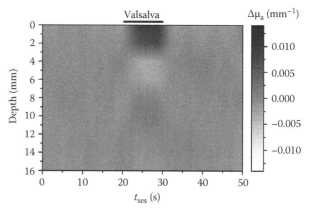

FIGURE 20.10 Depth resolved absorption changes at 805 nm during a finger-tapping task (a), and a Valsalva maneuver (b). The depth information was recovered from the changes in intensity at all delays by inversion of the matrix of temporal mean partial pathlengths in each layer. For the motor stimulation, the absorption change takes place in deep layers of the head, in agreement with the expected cortical location of hemoglobin changes. For the Valsalva maneuver, the absorption changes appear more superficial, suggesting extracerebral origin. (From Steinbrink, J. et al., *Phys. Med. Biol.*, 46, 879, 2001. With permission.)

measurements with a single source-detector separation. They minimized superficial contamination by using the first delay gate, assumed to reflect only superficial changes, to normalize all other gates, which bear contribution both from superficial and cortical layers (Selb et al. 2005b). The method can be seen as a simplified version of the depth reconstruction method, where only two layers are considered, and the sensitivity to the first layer is identical at all time delays. This increased the signal-to-background ratio during cerebral activation, where the background is the physiological contamination from superficial layers. With their extended imaging system enabling bi-hemisphere measurements, they showed localized activation on the contralateral hemisphere during a motor task, with the increased signal at later gate suggesting deep (i.e., cortical) origin (Selb et al. 2006). A smaller ipsilateral activation was also visible.

Similar results were reported by Montcel et al. (2005, 2006) using a TCSPC-based system. The authors measured a subject during a finger-tapping task and a Valsalva maneuver. Rather than studying moments of the TPSF or gated intensity, they studied the evolution of intensity at all delays along the TPSF. They used finite element modeling of the forward problem to retrieve the depth of the response to each task.

20.6.4 Tomographic Brain Imaging

The high sensitivity of optical tomography, together with its noninvasiveness, portability, and relative robustness to movement, makes it the only medical imaging technique that is suited for functional imaging of the neonatal brain at the bedside. Hypoxic-ischemic brain injury can affect the brain at and around birth, and optical imaging may provide a clinically acceptable way of diagnosing and monitoring this important and poorly understood range of conditions.

The first time-domain imaging of the brain was carried out by the group of Benaron at Stanford University (USA). It took up to 6 h to acquire a full image data set across the neonatal head (Hintz et al. 1999, Benaron et al. 2000). Tomographic adult imaging required four 6 h sessions, partly because the average light power of the system was only 0.1 mW. A headband held 34 sources and detectors onto the scalp in a ring arrangement and TPSFs were acquired (Hintz et al. 1998). The absorption and scatter coefficients were derived from the TPSF and used in a linear backprojection algorithm for image reconstruction (Benaron et al. 1994). In their first study, eight images were acquired from six babies, four of whom had intraventricular hemorrhage (IVH). Six of the eight images were deemed to accurately represent the bleed as seen on ultrasound, CT, or MRI (Hintz et al. 1999). The images were exceptional, particularly given the relatively simple instrumentation and image reconstruction, and suggested that time-resolved optical imaging may be able to offer an important role in brain-centered neonatal intensive care.

This work led to the development of a time-resolved imaging system (MONSTIR) and nonlinear image reconstruction software (TOAST) at UCL, built primarily for imaging the neonatal brain (Arridge et al. 2000, Schmidt et al. 2000). This work has been reviewed in detail elsewhere (Austin et al. 2006, Austin 2007, Hebden and Austin 2007). Briefly, a foam-lined adjustable helmet is used to couple sources and detectors onto the head. When the study is complete, the positions of the connectors are measured using a 3D digitizing arm (Microscribe 3D, Immersion Co, USA) and used to define the surface of a 3D finite element mesh for image reconstruction (Gibson et al. 2003). The system has been used to image brain pathology such as IVH (Hebden et al. 2002, Gibson et al. 2005) and changes in brain oxygenation due to changing ventilator settings (Hebden et al. 2004). It has also been used to image evoked functional changes by activating each of 12 sources in turn (Figure 20.11). A full image study (a reference data set and a data set during activation) took about 20 min to acquire (Gibson et al. 2006).

It is still unclear how useful optical tomography of the neonatal brain will ultimately be in the clinical setting. In principle, it can provide clinically relevant information about the distribution of blood and oxygen, which cannot be obtained in other ways. This could potentially provide an indicator of hypoxic-ischemic brain injury, cerebral maturity (including myelination), and prognosis, as well as allowing therapy to be monitored. However, the relatively low spatial resolution and uncertainties about quantification restrict its use at the moment, as does the clinical practicality of the technique. Although imaging can be performed at the bedside, a large instrument and team of researchers are required at the bedside for a number of hours, which complicates access in a busy hospital ward. The use of anatomical prior information in the image reconstruction is intended to address the first concern and a smaller, faster, second-generation imaging system the second.

20.7 Critical Discussion

Time-domain optical imaging has certain advantages over CW and frequency-domain measurements. By acquiring the full TPSF, we obtain more data about the optical properties of the object being examined than by any other diffuse imaging method. The TPSF can be processed to provide a range of datatypes and to identify poor data. Moreover, photon counting with TCSPC systems provides the maximum possible sensitivity across large volumes of tissue. Because of these advantages, we may ask why time-domain systems are not more widespread. They do have a number of disadvantages: they are more complex and therefore expensive (although new sources and detectors are reducing the cost); the time taken to acquire data may be prohibitively long for some applications (particularly with TSCPC); and to use them to their best advantage, complex modeling and sophisticated nonlinear image reconstruction are required. And in any case, the additional information available in the full TPSF compared to amplitude and phase measurements provided by frequency-domain systems may be limited.

However, for certain applications, time-domain systems do provide real advantages over other systems. These advantages can be reduced to two main areas where time-domain systems are arguably the "gold standard" diffuse optical imaging systems: imaging large volumes and reconstructing absolute optical properties.

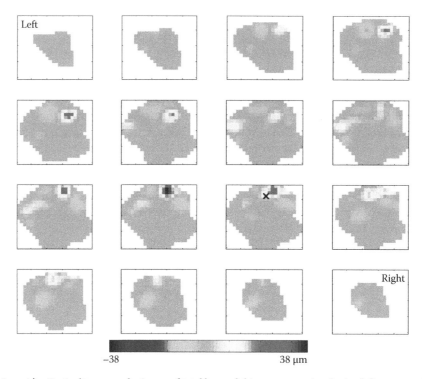

−38   38 μm

FIGURE 20.11 (See color insert.) Optical tomography image of total hemoglobin concentration during left motor activity in the neonatal brain, showing an increase near the estimated position of the motor cortex (denoted by a cross). This figure shows 16 sagittal slices from the left (top right) to the right (bottom left) of the head. (From Gibson, A.P. et al., *Neuroimage*, 30, 521, 2006. With permission.)

Only time-domain systems have been shown to have the sensitivity required to image the entire neonatal brain. Optical images have been reconstructed from infants showing intraventricular hemorrhage and functional activation, providing the only medical imaging modality suitable for functional imaging of this patient group at the bedside. Similarly, the breast is well suited for time-domain imaging. While more images have been acquired using CW and frequency-domain systems, time-domain systems have an important role in measuring the absolute optical properties of the breast as well as in more general applications.

The extra information inherent in time-domain measurements means they are well suited to determining the baseline or average optical properties of tissue. This has been used in a number of applications from single channel measurements of tissue assumed to be homogeneous, to multichannel measurements of inhomogeneous or layered tissue, taking advantage of the depth resolution available from time-gated measurements.

In summary, time-domain measurements have a major role in diffuse optical imaging, which can be expected to increase as sources and detectors improve in quality and reduce in cost.

References

Arridge, S. R. 1999. Optical tomography in medical imaging. *Inverse Probl.* 15: R41-R93.

Arridge, S. R., Cope, M., and Delpy, D. T. 1992. The theoretical basis for the determination of optical pathlengths in tissue: Temporal and frequency analysis. *Phys. Med. Biol.* 37: 1531–1559.

Arridge, S. R., Hebden, J. C., Schweiger, M. et al. 2000. A method for 3D time-resolved optical tomography. *Int. J. Imag. Syst. Technol.* 11: 2–11.

Arridge, S. R. and Lionheart, W. R. B. 1998. Nonuniqueness in diffusion-based optical tomography. *Opt. Lett.* 23: 882–884.

Arridge, S. R. and Schweiger, M. 1995. Direct calculation of the moments of the distribution of photon time-of-flight in tissue with a finite-element method. *Appl. Opt.* 34: 2683–2687.

Austin, T. 2007. Optical imaging of the neonatal brain. *Arch. Dis. Child. Fetal Neonatal Ed.* 92: 238–241.

Austin, T., Gibson, A. P., Branco, G. et al. 2006. Three dimensional optical imaging of blood volume and oxygenation in the neonatal brain. *Neuroimage* 31: 1426–1433.

Barnett, A. H., Culver, J. P., Sorensen, A. G. et al. 2003. Robust inference of baseline optical properties of the human head with three-dimensional segmentation from magnetic resonance imaging. *Appl. Opt.* 42: 3095–3108.

Bassi, A., Spinelli, L., D'Andrea, C. et al. 2006. Feasibility of white-light time-resolved optical mammography. *J. Biomed. Opt.* 11: 054035.

Bassi, A., Swartling, J., D'Andrea, C. et al. 2004. Time-resolved spectrophotometer for turbid media based on supercontinuum generation in a photonic crystal fiber. *Opt. Lett.* 29: 2405–2407.

Bassi, A., Farina, A., D'Andrea, C. et al. 2007. Portable, large-bandwidth time-resolved system for diffuse optical spectroscopy. *Opt. Exp.* 15: 14482–14487.

Becker, W. 2005. *Advanced Time-Correlated Single Photon Counting Techniques.* Berlin, Germany: Springer.

Becker, W., Bergmann, A., Gibson, A. et al. 2005. Multi-dimensional time-correlated single photon counting applied to diffuse optical tomography. *Proc. SPIE* 5693.

Benaron, D. A., Ho, D. C., Spilman, S. D. et al. 1994. Non-recursive linear algorithms for optical imaging in diffusive media. *Adv. Exp. Med. Biol.* 361: 215–222.

Benaron, D. A., Hintz, S. R., Villringer, A. et al. 2000. Noninvasive functional imaging of human brain using light. *J. Cereb. Blood Flow Metab.* 20: 469–477.

Boas, D. A., Gaudette, T., and Arridge, S. R. 2001. Simultaneous imaging and optode calibration with diffuse optical tomography. *Opt. Exp.* 8: 263–270.

Brukilacchio, T. J. 2003. A diffuse optical tomography system combined with x-ray mammography for improved breast cancer detection. PhD thesis, Electrical Engineering, Tufts University.

Brun, N. C. and Greisen, G. 1994. Cerebrovascular responses to carbon dioxide as detected by near-infrared spectrophotometry: Comparison of three different measures. *Pediatr. Res.* 36: 20–24.

Carpenter, C. M., Pogue, B. W., Jiang, S. et al. 2007. Image-guided optical spectroscopy provides molecular-specific information in vivo: MRI-guided spectroscopy of breast cancer hemoglobin, water, and scatterer size. *Opt. Lett.* 32: 933–935.

Chance, B., Leigh, J. S., Miyake, H. et al. 1988. Comparison of time-resolved and unresolved measurements of deoxyhemoglobin in brain. *Proc. Natl. Acad. Sci. U.S.A.* 85: 4971–4975.

Comelli, D., Bassi, A., Pifferi, A. et al. 2007. In vivo time-resolved reflectance spectroscopy of the human forehead. *Appl. Opt.* 46: 1717–1725.

D'Andrea, C., Comelli, D., Pifferi, A. et al. 2003. Time-resolved optical imaging through turbid media using a fast data acquisition system based on a gated CCD camera. *J. Phys. D-Appl. Phys.* 36: 1675–1681.

Enfield, L. C., Gibson, A. P., Everdell, N. L. et al. 2007. Three-dimensional time-resolved optical mammography of the uncompressed breast. *Appl. Opt.* 46: 3628–3638.

Enfield, L. C., Gibson, A. P., Everdell, N. L. et al. 2008. Sensitivity and specificity of 3D optical mammography. In: *OSA Biomedical Topical Meetings* St. Petersburg, Florida.

Fang, Q., Carp, S. A., Selb, J. et al. 2009. Combined optical imaging and mammography of the healthy breast: Optical contrast derived from breast structure and compression. *IEEE Trans. Med. Imag.* 28: 30–42.

Feng, L., Yoo, K. M., and Alfano, R. R. 1993. Ultrafast laser-pulse transmission and imaging through biological tissues. *Appl. Opt.* 32: 554–558.

Feng, L., Yoo, K. M., and Alfano, R. R. 1994. Transmitted photon intensity through biological tissues within various time windows. *Opt. Lett.* 19: 740–742.

Fletcher, S. W. and Elmore, J. G. 2003. Mammographic screening for breast cancer. *New Engl. J. Med.* 348: 1672–1680.

Franceschini, M. A., Thaker, S., Themelis, G. et al. 2007. Assessment of infant brain development with frequency-domain near-infrared spectroscopy. *Pediatr. Res.* 61: 546–551.

Gagnon, L., Gauthier, C., Hoge, R. D. et al. 2008. Double-layer estimation of intra- and extracerebral hemoglobin concentration with a time-resolved system. *J. Biomed. Opt.* 13: 054019.

Gao, F., Zhao, H. J., and Yamada, Y. 2002. Improvement of image quality in diffuse optical tomography by use of full time-resolved data. *Appl. Opt.* 41: 778–791.

Gibson, A. P., Riley, J., Schweiger, M. et al. 2003. A method for generating patient-specific finite element meshes for head modelling. *Phys. Med. Biol.* 48: 481–495.

Gibson, A. P., Hebden, J. C. and Arridge, S. R. 2005. Recent advances in diffuse optical imaging. *Phys. Med. Biol.* 50: R1-R43.

Gibson, A. P., Austin, T., Everdell, N. L. et al. 2006. Three-dimensional whole-head optical passive motor evoked responses in the tomography of neonate. *Neuroimage* 30: 521–528.

Grosenick, D., Wabnitz, H., and Rinneberg, H. 1997. Time-resolved imaging of solid phantoms for optical mammography. *Appl. Opt.* 36: 221–231.

Grosenick, D., Wabnitz, H., Rinneberg, H. H. et al. 1999. Development of a time-domain optical mammograph and first in vivo applications. *Appl. Opt.* 38: 2927–2943.

Grosenick, D., Moesta, K. T., Wabnitz, H. et al. 2003. Time-domain optical mammography: Initial clinical results on detection and characterization of breast tumors. *Appl. Opt.* 42: 3170–3186.

Grosenick, D., Wabnitz, H., Moesta, K. T. et al. 2004. Concentration and oxygen saturation of haemoglobin of 50 breast tumours determined by time-domain optical mammography. *Phys. Med. Biol.* 49: 1165–1181.

Grosenick, D., Moesta, K. T., Moller, M. et al. 2005a. Time-domain scanning optical mammography: I. Recording and assessment of mammograms of 154 patients. *Phys. Med. Biol.* 50: 2429–2449.

Grosenick, D., Wabnitz, H., Moesta, K. T. et al. 2005b. Time-domain scanning optical mammography: II. Optical properties and tissue parameters of 87 carcinomas. *Phys. Med. Biol.* 50: 2451–2468.

Grosenick, D., Kummrow, A., Macdonald, R. et al. 2007. Evaluation of higher-order time-domain perturbation theory of photon diffusion on breast-equivalent phantoms and optical mammograms. *Phys. Rev. E* 76: 061908.

Haskell, R. C., Svaasand, L. O., Tsay, T. T. et al. 1994. Boundary conditions for the diffusion equation in radiative transfer. *J. Opt. Soc. Am. A—Opt. Image Sci. Vis.* 11: 2727–2741.

Hebden, J. C. and Delpy, D. T. 1994. Enhanced time resolved imaging using a diffusion model of photon transport. *Opt. Lett.* 19: 311–313.

Hebden, J. C., Arridge, S. R., and Delpy, D. T. 1997. Optical imaging in medicine: I. Experimental techniques. *Phys. Med. Biol.* 42: 825–840.

Hebden, J. C., Arridge, S. R., and Schweiger, M. 1998. Investigation of alternative data types for time resolved optical tomography. In: *OSA Technical Digest, Biomedical Topical Meetings*, OSA, Washington, DC, 21.

Hebden, J. C., Schmidt, F. E., Fry, M. E. et al. 1999. Simultaneous reconstruction of absorption and scattering images by multichannel measurement of purely temporal data. *Opt. Lett.* 24: 534–536.

Hebden, J. C., Gibson, A., Yusof, R. M. et al. 2002. Three-dimensional optical tomography of the premature infant brain. *Phys. Med. Biol.* 47: 4155–4166.

Hebden, J. C., Gonzalez, F. M., Gibson, A. et al. 2003. Assessment of an in situ temporal calibration method for time-resolved optical tomography. *J. Biomed. Opt.* 8: 87–92.

Hebden, J. C., Gibson, A., Austin, T. et al. 2004. Imaging changes in blood volume and oxygenation in the newborn infant brain using three-dimensional optical tomography. *Phys. Med. Biol.* 49: 1117–1130.

Hebden, J. C. and Rinneberg, H. 2005. Optical mammography: Imaging and characterization of breast lesions by pulsed near-infrared laser light (OPTIMAMM). *Phys. Med. Biol.* 50 (Editorial).

Hebden, J. C. and Austin, T. 2007. Optical tomography of the neonatal brain. *Eur. Radiol.* 17: 2926–2933.

Hee, M. R., Izatt, J. A., Swanson, E. A. et al. 1993. Femtosecond transillumination tomography in thick tissues. *Opt. Lett.* 18: 1107–1109.

Hillman, E. M. C. 2002. Experimental and theoretical investigations of near infrared tomographic imaging methods and clinical applications. PhD thesis, University of London.

Hillman, E. M. C., Hebden, J. C., Schmidt, F. E. W. et al. 2000. Calibration techniques and datatype extraction for time-resolved optical tomography. *Rev. Sci. Instrum.* 71: 3415–3427.

Hintz, S. R., Benaron, D. A., Houten, J. P. V. et al. 1998. Stationary headband for clinical time-of-flight optical imaging at the bedside. *Photochem. Photobiol.* 68: 361–369.

Hintz, S. R., Cheong, W. F., Van Houten, J. P. et al. 1999. Bedside imaging of intracranial hemorrhage in the neonate using light: Comparison with ultrasound, computed tomography, and magnetic resonance imaging. *Pediatr. Res.* 45: 54–59.

Ijichi, S., Kusaka, T., Isobe, K. et al. 2005. Quantification of cerebral hemoglobin as a function of oxygenation using near-infrared time-resolved spectroscopy in a piglet model of hypoxia. *J. Biomed. Opt.* 10: 024026.

Intes, X. 2005. Time-domain optical mammography SoftScan: Initial results. *Acad. Radiol.* 12: 934–947.

Jemal, A., Siegel, R., Ward, E. et al. 2008. Cancer statistics, 2008. *CA Cancer J. Clin.* 58: 71–96.

Jennions, D. 2008. Time-resolved optical tomography instrumentation for fast 3D functional imaging. PhD thesis, University of London.

Kacprzak, M., Liebert, A., Sawosz, P. et al. 2007. Time-resolved optical imager for assessment of cerebral oxygenation. *J. Biomed. Opt.* 12: 034019.

Kalender, W. A. 2006. X-ray computed tomography. *Phys. Med. Biol.* 51: R29-R43.

Kalpaxis, L. L., Wang, L. M., Galland, P. et al. 1993. Three-dimensional temporal image reconstruction of an object hidden in highly scattering media by time-gated optical tomography. *Opt. Lett.* 18: 1691–1693.

Kienle, A., Glanzmann, T., Wagnieres, G. et al. 1998a. Investigation of two-layered turbid media with time-resolved reflectance. *Appl. Opt.* 37: 6852–6862.

Kienle, A., Patterson, M. S., Dognitz, N. et al. 1998b. Noninvasive determination of the optical properties of two-layered turbid media. *Appl. Opt.* 37: 779–791.

Kienle, A. and Glanzmann, T. 1999. In vivo determination of the optical properties of muscle with time-resolved reflectance using a layered model. *Phys. Med. Biol.* 44: 2689–2702.

Leff, D. R., Warren, O. J., Enfield, L. C. et al. 2008. Diffuse optical imaging of the healthy and diseased breast: A systematic review. *Breast Cancer Res. Treat.* 108: 9–22.

Leung, T. S., Aladangady, N., Elwell, C. E. et al. 2004. A new method for the measurement of cerebral blood volume and total circulating blood volume using near infrared spatially resolved spectroscopy and indocyanine green: Application and validation in neonates. *Pediatr. Res.* 55: 134–141.

Liebert, A., Wabnitz, H., Grosenick, D. et al. 2003a. Fiber dispersion in time domain measurements compromising the accuracy of determination of optical properties of strongly scattering media. *J. Biomed. Opt.* 8: 512–516.

Liebert, A., Wabnitz, H., Grosenick, D. et al. 2003b. Evaluation of optical properties of highly scattering media by moments of distributions of times of flight of photons. *Appl. Opt.* 42: 5785–5792.

Liebert, A., Wabnitz, H., Steinbrink, J. et al. 2004. Time-resolved multidistance near-infrared spectroscopy of the adult head: Intracerebral and extracerebral absorption changes from moments of distribution of times of flight of photons. *Appl. Opt.* 43: 3037–3047.

Liebert, A., Wabnitz, H., Steinbrink, J. et al. 2005. Bed-side assessment of cerebral perfusion in stroke patients based on optical monitoring of a dye bolus by time-resolved diffuse reflectance. *Neuroimage* 24: 426–435.

Liu, H., Miwa, M., Beauvoit, B. et al. 1993. Characterization of absorption and scattering properties of small-volume biological samples using time-resolved spectroscopy. *Anal. Biochem.* 213: 378–385.

Mackert, B. M., Leistner, S., Sander, T. et al. 2005. Combined DC-Magnetoencephalography and time-resolved near-infrared spectroscopy: An approach to characterise neurovascular coupling non-invasively in the human brain. *J. Neurol. Sci.* 238: S260-S261.

Mackert, B. M., Leistner, S., Sander, T. et al. 2008. Dynamics of cortical neurovascular coupling analyzed by simultaneous DC-magnetoencephalography and time-resolved near-infrared spectroscopy. *Neuroimage* 39: 979–986.

Martelli, F., Sassaroli, A., Yamada, Y. et al. 2002. Analytical approximate solutions of the time-domain diffusion equation in layered slabs. *J. Opt. Soc. Am. A Opt. Image Sci. Vis.* 19: 71–80.

Martelli, F., Del Bianco, S., and Zaccanti, G. 2003a. Procedure for retrieving the optical properties of a two-layered medium from time-resolved reflectance measurements. *Opt. Lett.* 28: 1236–1238.

Martelli, F., Sassaroli, A., Del Bianco, S. et al. 2003b. Solution of the time-dependent diffusion equation for layered diffusive media by the eigenfunction method. *Phys. Rev. E* 67: 056623.

Martelli, F., Del Bianco, S., Zaccanti, G. et al. 2004. Phantom validation and in vivo application of an inversion procedure for retrieving the optical properties of diffusive layered media from time-resolved reflectance measurements. *Opt. Lett.* 29: 2037–2039.

Martelli, F., Sassaroli, A., Del Bianco, S. et al. 2007. Solution of the time-dependent diffusion equation for a three-layer medium: Application to study photon migration through a simplified adult head model. *Phys. Med. Biol.* 52: 2827–2843.

Montcel, B., Chabrier, R., and Poulet, P. 2005. Detection of cortical activation with time-resolved diffuse optical methods. *Appl. Opt.* 44: 1942–1947.

Montcel, B., Chabrier, R., and Poulet, P. 2006. Time-resolved absorption and hemoglobin concentration difference maps: A method to retrieve depth-related information on cerebral hemodynamics. *Opt. Exp.* 14: 12271–12287.

Niedre, M. J., Turner, G. M., and Ntziachristos, V. 2006. Time-resolved imaging of optical coefficients through murine chest cavities. *J. Biomed. Opt.* 11: 064017.

Niedre, M. J., De Kleine, R. H., Aikawa, E. et al. 2008. Early photon tomography allows fluorescence detection of lung carcinomas and disease progression in mice in vivo. *Proc. Natl. Acad. Sci. U.S.A.* 105: 19126–19131.

Ntziachristos, V., Ma, X. H., and Chance, B. 1998. Time-correlated single photon counting imager for simultaneous magnetic resonance and near-infrared mammography. *Rev. Sci. Instrum.* 69: 4221–4233.

Ntziachristos, V., Ma, X. H., Yodh, A. G. et al. 1999. Multichannel photon counting instrument for spatially resolved near infrared spectroscopy. *Rev. Sci. Instrum.* 70: 193–201.

Ntziachristos, V. and Chance, B. 2001. Accuracy limits in the determination of absolute optical properties using time-resolved NIR spectroscopy. *Med. Phys.* 28: 1115–1124.

Ntziachristos, V., Yodh, A. G., Schnall, M. D. et al. 2002. MRI-guided diffuse optical spectroscopy of malignant and benign breast lesions. *Neoplasia* 4: 347–354.

Oh, S., Milstein, A. B., Millane, R. P. et al. 2002. Source-detector calibration in three-dimensional Bayesian optical diffusion tomography. *J. Opt. Soc. Am. A Opt. Image Sci. Vis.* 19: 1983–1993.

Ohmaea, E., Ouchib, Y., Odaa, M. et al. 2006. Cerebral hemodynamics evaluation by near-infrared time-resolved spectroscopy: Correlation with simultaneous positron emission tomography measurements. *NeuroImage* 29 (3): 697–705.

Ohmae, E., Oda, M., Suzuki, T. et al. 2007. Clinical evaluation of time-resolved spectroscopy by measuring cerebral hemodynamics during cardiopulmonary bypass surgery. *J. Biomed. Opt.* 12: 062112.

Patterson, M. S., Chance, B., and Wilson, B. C. 1989. Time resolved reflectance and transmittance for the noninvasive measurement of tissue optical-properties. *Appl. Opt.* 28: 2331–2336.

Pifferi, A., Taroni, P., Torricelli, A. et al. 2003. Four-wavelength time-resolved optical mammography in the 680–980-nm range. *Opt. Lett.* 28: 1138–1140.

Pifferi, A., Torricelli, A., Bassi, A. et al. 2005. Performance assessment of photon migration instruments: The MEDPHOT protocol. *Appl. Opt.* 44: 2104–2114.

Pifferi, A., Torricelli, A., Spinelli, L. et al. 2008. Time-resolved diffuse reflectance using small source-detector separation and fast single-photon gating. *Phys. Rev. Lett.* 100: 138101

Rice, A. and Quinn, C. M. 2002. Angiogenesis, thrombospondin, and ductal carcinoma in situ of the breast. *J. Clin. Pathol.* 55: 569–574.

Sander, T. H., Liebert, A., Burghoff, M. et al. 2007a. Cross-correlation analysis of the correspondence between magnetoencephalographic and near-infrared cortical signals. *Meth. Inform. Med.* 46: 164–168.

Sander, T. H., Liebert, A., Mackert, B. M. et al. 2007b. DC-magnetoencephalography and time-resolved near-infrared spectroscopy combined to study neuronal and vascular brain responses. *Physiol. Meas.* 28: 651–664.

Sassaroli, A., Martelli, F., Zaccanti, G. et al. 2001. Performance of fitting procedures in curved geometry for retrieval of the optical properties of tissue from time-resolved measurements. *Appl. Opt.* 40: 185–197.

Schmidt, F. E. W. 1999. Development of a time-resolved optical tomography system for neonatal brain imaging. PhD thesis, UCL.

Schmidt, F. E. W., Fry, M. E., Hillman, E. M. C. et al. 2000. A 32-channel time-resolved instrument for medical optical tomography. *Rev. Sci. Instrum.* 71: 256–265.

Schmitz, C. H., Graber, H. L., Luo, H. et al. 2000. Instrumentation and calibration protocol for imaging dynamic features in dense-scattering media by optical tomography. *Appl. Opt.* 39: 6466–6486.

Schmitz, C. H., Klemer, D. P., Hardin, R. et al. 2005. Design and implementation of dynamic near-infrared optical tomographic imaging instrumentation for simultaneous dual-breast measurements. *Appl. Opt.* 44: 2140–2153.

Schweiger, M. and Arridge, S. R. 1997. Direct calculation with a finite-element method of the Laplace transform of the distribution of photon flight time in tissue. *Appl. Opt.* 36: 9042–9049.

Schweiger, M. and Arridge, S. R. 1999. Application of temporal filters to time-resolved data in optical tomography. *Phys. Med. Biol.* 44: 1699–1717.

Selb, J., Stott, J. J., Franceschini, M. A. et al. 2004. Improvement of depth sensitivity to cerebral hemodynamics with a time domain system. In: *OSA Biomedical Topical Meetings*, Miami FC3.

Selb, J., Franceschini, M. A., Sorensen, A. G. et al. 2005a. Improved sensitivity to cerebral hemodynamics during brain activation with a time-gated optical system: Analytical model and experimental validation. *J. Biomed. Opt.* 10: 011013.

Selb, J., Hillman, E. M. C., Joseph, D. et al. 2005b. Discrimination between superficial and cerebral signals during functional brain imaging with a time-gated system. In: *European Conference on Biomedical Optics*, Munich, Germany.

Selb, J., Joseph, D. K., and Boas, D. A. 2006. Time-gated optical system for depth-resolved functional brain imaging. *J. Biomed. Opt.* 11: 044008.

Shendeleva, M. 2004. Green functions for diffuse light in a medium comprising two turbid half-spaces. *Appl. Opt.* 43: 5334–5342.

Spinelli, L., Martelli, F., Del Bianco, S. et al. 2006. Absorption and scattering perturbations in homogeneous and layered diffusive media probed by time-resolved reflectance at null source-detector separation. *Phys. Rev. E* 74: 021919.

Spinelli, L., Torricelli, A., Pifferi, A. et al. 2004. Bulk optical properties and tissue components in the female breast from multiwavelength time-resolved optical mammography. *J. Biomed. Opt.* 9: 1137–1142.

Spinelli, L., Torricelli, A., Pifferi, A. et al. 2005. Characterization of female breast lesions from multi-wavelength time-resolved optical mammography. *Phys. Med. Biol.* 50: 2489–2502.

Steinbrink, J., Wabnitz, H., Obrig, H. et al. 2001. Determining changes in NIR absorption using a layered model of the human head. *Phys. Med. Biol.* 46: 879–896.

Suzuki, K., Yamashita, Y., Ohta, K. et al. 1994. Quantitative measurement of optical parameters in the breast using time-resolved spectroscopy. Phantom and preliminary in vivo results. *Invest. Radiol.* 29: 410–414.

Swartling, J., Dam, J. S., and Andersson-Engels, S. 2003. Comparison of spatially and temporally resolved diffuse-reflectance measurement systems for determination of biomedical optical properties. *Appl. Opt.* 42: 4612–4620.

Swartling, J., Bassi, A., D'Andrea, C. et al. 2005. Dynamic time-resolved diffuse spectroscopy based on supercontinuum light pulses. *Appl. Opt.* 44: 4684–4692.

Tabar, L., Vitak, B., Yen, M. F. A. et al. 2004. Number needed to screen: Lives saved over 20 years of follow-up in mammographic screening. *J. Med. Screen.* 11: 126–129.

Taroni, P., Danesini, G., Torricelli, A. et al. 2004a. Clinical trial of time-resolved scanning optical mammography at 4 wavelengths between 683 and 975 nm. *J. Biomed. Opt.* 9: 464–473.

Taroni, P., Pallaro, L., Pifferi, A. et al. 2004b. Multi-wavelength time-resolved optical mammography. In: *OSA Biomedical Topical Meetings*, Miami ThB3.

Taroni, P., Pifferi, A., Torricelli, A. et al. 2004c. Time-resolved optical spectroscopy and imaging of breast. *Opto-Electron. Rev.* 12: 249–253.

Taroni, P., Spinelli, L., Torricelli, A. et al. 2005a. Multi-wavelength time domain optical mammography. *Technol. Can. Res. Treat.* 4: 527–537.

Taroni, P., Torricelli, A., Spinelli, L. et al. 2005b. Time-resolved optical mammography between 637 and 985 nm: Clinical study on the detection and identification of breast lesions. *Phys. Med. Biol.* 50: 2469–2488.

Tarvainen, T., Kohlemainen, V., Vauhkonen, M. et al. 2005. Computational calibration method for optical tomography. *Appl. Opt.* 44: 1879–1888.

Torricelli, A., Pifferi, A., Spinelli, L. et al. 2005. Time-resolved reflectance at null source-detector separation: Improving contrast and resolution in diffuse optical imaging. *Phys. Rev. Lett.* 95: 078101.

Tromberg, B. J., Pogue, B. W., Paulsen, K. D. et al. 2008. Assessing the future of diffuse optical imaging technologies for breast cancer management. *Med. Phys.* 35: 2443–2451.

Turner, G. M., Zacharakis, G., Soubret, A. et al. 2005. Complete-angle projection diffuse optical tomography by use of early photons. *Opt. Lett.* 30: 409–411.

Vilhunen, T., Kohlemainen, V., Vauhkonen, M. et al. 2004. Computational calibration method for optical tomography. In: *OSA Biomedical Topical Meetings*, Miami ThD3.

Wang, L., Ho, P. P., Liang, X. et al. 1993. Kerr-Fourier imaging of hidden objects in thick turbid media. *Opt. Lett.* 18: 241–243.

Wyatt, J. S., Cope, M., Delpy, D. T. et al. 1990. Quantitation of cerebral blood volume in human infants by near-infrared spectroscopy. *J. Appl. Physiol.* 68: 1086–1091.

Yates, T. D., Hebden, J. C., Gibson, A. P. et al. 2005a. Time-resolved optical mammography using a liquid coupled interface. *J. Biomed. Opt.* 10: 054011.

Yates, T. D., Hebden, J. C., Gibson, A. P. et al. 2005b. Optical tomography of the breast using a multi-channel time-resolved imager. *Phys. Med. Biol.* 50: 2503–2517.

Zhao, H., Gao, F., Tanikawa, Y. et al. 2007. Time-resolved diffuse optical tomography and its application to in vitro and in vivo imaging. *J. Biomed. Opt.* 12: 062107.

Photoacoustic Tomography and Ultrasound-Modulated Optical Tomography

21.1 Introduction to Photoacoustic Tomography 419
 Background • PA Signal Generation in Tissue • Wave Equations • Radiation Safety
21.2 PAT Modalities 422
 PA Computed Tomography • Acoustic-Resolution and Optical-Resolution PA
 Microscopy • PAT-Based on an Acoustic Lens System
21.3 PA Spectroscopy and Functional Imaging 426
21.4 PA Contrast Agents and Molecular Imaging 426
 PA Contrast Agents • PA Molecular Imaging
21.5 Comparison of PAT with Other Imaging Modalities 428
 Imaging Contrast • Imaging Depth and Spatial Resolution • Imaging Temporal
 Resolution • Safety
21.6 Summary of PAT 429
21.7 Introduction to Ultrasound-Modulated Optical Tomography 429
 Motivation • Mechanisms of UOT
21.8 Detection Techniques in UOT 431
 Photorefractive Interferometer–Based Detection • Confocal Fabry–Perot Interferometer–
 Based Detection • Spectral-Hole-Burning-Based Detection
21.9 The Use of a Powerful Long Coherent Pulsed Laser in UOT 437
21.10 Summary and Discussion on UOT 439
 References 439

Changhui Li
Washington University in St. Louis

Chulhong Kim
Washington University in St. Louis

Lihong V. Wang
Washington University in St. Louis

21.1 Introduction to Photoacoustic Tomography

21.1.1 Background

Photoacoustic (PA) tomography (PAT), also referred to as optoacoustic tomography, is an emerging biomedical imaging method based on the PA effect (Xu and Wang 2006, Li and Wang 2009, Oraevsky and Karabutov 2003). The PA effect was reported by Alexander Graham Bell in 1880 (Bell 1880), and refers to the generation of acoustic waves by substances being illuminated by electromagnetic (EM) waves of varying power. The PA effect results from the variational thermal expansion within the material as it absorbs the EM radiative energy. Since the ability of EM absorption reflects a material's characteristics, much physical and chemical information can be revealed by studying PA signals. Historically, the PA effect was of little research interest for a long time, but the invention of laser, an adequate PA illumination source, changed matters drastically. Now, various detection methods based on PA effect have been widely implemented in physics, chemistry, biology, engineering, and medicine (Rosencwaig 1980, Gusev et al. 1993).

PA applications in biomedicine began in the 1970s (Maugh 1975); however, rapid development in both theoretical and technological PAT began in the 1990s (Kruger 1994, Karabutov et al. 1996, Oraevsky et al. 1997, Wang et al. 1999). By using optical, microwave, or radio-frequency (RF) EM waves as the illumination sources, PAT detects PA signals generated by absorbers in tissue and retrieves the spatial distribution of the absorbers, as shown in Figure 21.1. PAT combines sensitive EM absorption contrast and acoustic detection, and it is noninvasive and nonionizing. Compared with pure high-resolution optical imaging methods, such as optical coherence tomography (OCT) and confocal microscopy, PAT can image deeper into tissue because multiple scattered photons are utilized. EM absorption properties are highly sensitive to physiological conditions, so PAT is suitable for functional imaging. In addition, by using bio-conjugated PA contrast agents, PAT

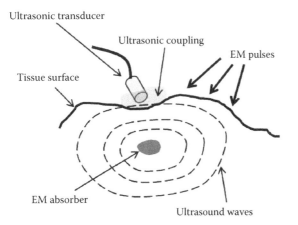

FIGURE 21.1 Illustration of the photoacoustic mechanism. An EM absorber in tissue absorbs the radiation energy of EM pulses, and then ultrasound waves are generated and propagated. An ultrasonic transducer is placed outside the tissue surface to detect the stress waves. An ultrasound coupling medium, such as water or gel, is applied between the tissue surface and the ultrasonic transducer.

can provide molecular imaging. PAT also inherits the scalability of ultrasound—imaging from several micrometer-size red blood cells (RBC) to centimeter-size breast tumors, at depths ranging from hundreds of micrometers to several centimeters (Wang 2009).

21.1.2 PA Signal Generation in Tissue

As mentioned, the PA signal in tissue is generated by the variational thermal expansion processes after illumination energy is absorbed and transformed into heat. Currently, EM sources used in PAT include short EM pulses and intensity-modulated continuous-wave (CW) EM waves. However, EM pulses dominate most embodiments of PAT because they presently generate higher signal-to-noise ratio (SNR) (Maslov and Wang 2008), and the information about the distance between the EM absorber

and the ultrasonic detector is contained in the time-resolved signal. Thus, unless otherwise noted, the EM source for PAT in this chapter always refers to short EM pulses, either laser or RF wave pulses.

Interactions between EM waves and tissues include scattering and absorption. All of these effects have been incorporated in optical detection: for instance, confocal optical microscopy depends on light scattering by tissue, and PAT relies on heat deposition by EM illumination. Although the absorbed EM energy can also convert into chemical energy, or reemit as fluorescence, in PAT, the heating effect dominates the absorption interaction in tissue. It is because PAT uses EM radiation power within the safety limit, guaranteeing no chemical reactions, and the quantum yield for fluorescence in natural tissue is small within the primary EM spectrum used by PAT.

Many endogenous EM absorbers are in tissues, the three primary ones of which are blood, melanosomes, and water. Blood behaves as a strong light-absorbing medium because it contains hemoglobin. The hemoglobin molecule has two conformations, oxygenated hemoglobin (HbO_2), and deoxygenated hemoglobin (Hb), which have markedly different light-absorbing characteristics. Figure 21.2a plots the molar extinction coefficients of HbO_2 and Hb. The concentrations of the two hemoglobin molecules, denoted by $[HbO_2]$ and $[Hb]$, depend on the type of blood and local physiological conditions, such as the metabolic rate. One important physiological parameter is the ratio of $[HbO_2]$ to the total hemoglobin concentration, which is called hemoglobin oxygen saturation (SO_2) and defined by

$$SO_2 = \frac{[HbO_2]}{[HbO_2]+[Hb]}. \qquad (21.1)$$

SO_2 is closely related to the local metabolic rate, for instance, abnormally low SO_2 is generally a characteristic sign for cancerous tissue. Since blood generally contains a mixture of HbO_2 and Hb, both of them contribute to blood optical absorption.

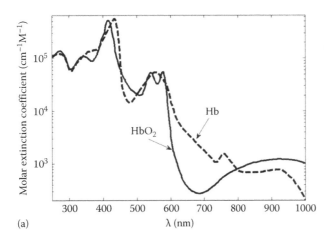

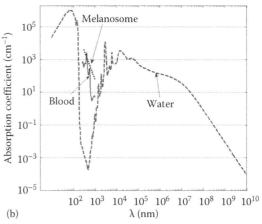

FIGURE 21.2 (a) Molar extinction coefficients of oxygenated and deoxygenated hemoglobin. (b) Absorption coefficients of melanosomes, blood (at $SO_2 = 80\%$ and total hemoglobin concentration of 2.33×10^{-3} mol/L), and water. (Calculated from http://omlc.ogi.edu/spectra)

Mathematically, the blood optical absorption coefficient μ_a at position $\mathbf{r}$ and wavelength λ is calculated by

$$\mu_a(\lambda,\mathbf{r}) = 2.303 \times \left[\varepsilon_{Hb}(\lambda)[Hb](\mathbf{r}) + \varepsilon_{HbO_2}(\lambda)[HbO_2](\mathbf{r}) \right] \quad (21.2)$$

where

λ is the wavelength

ε_{Hb} and ε_{HbO_2} are Hb and HbO$_2$ molar extinction coefficients (cm^{-1} M^{-1}), respectively

Due to the spectral dependence of absorption, PAT is suitable for spectroscopic imaging, which will be discussed in detail later. Melanosomes contain large amounts of melanin, a strong light-absorbing molecule, and are another endogenous EM absorber. Water can also become important in EM absorption in the near-infrared (NIR) spectrum and RF region. Figure 21.2b presents their absorption coefficients.

In addition to endogenous absorbers as mentioned, PAT can also target exogenous absorbers, including PA contrast agents, artificial implants, and foreign materials from trauma.

21.1.3 Wave Equations

In applications of PAT, the generated acoustic pressure in tissue is generally so weak that the nonlinear acoustic mechanism can be safely ignored. Moreover, the radiation pulse width, τ, is generally short enough that thermal diffusion can be neglected as well. Mathematically, this is called thermal confinement condition, that is,

$$\tau \ll \tau_{th} = \frac{d_c^2}{D_T}, \quad (21.3)$$

where

τ_{th} is the thermal confinement threshold

d_c is the characteristic dimension size (targeted spatial resolution)

D_T is thermal diffusivity (~0.14 mm^2/s for soft tissue)

For example, a 5-μm resolution corresponds to $\tau_{th} \sim 179\,\mu$s. The typical pulse widths for laser and RF wave used in PAT are 10 ns and 0.5 μs, respectively.

Under thermal confinement condition, the PA pressure p is an acoustically homogenous, nonviscous, and nondispersive medium with constant acoustic velocity v_s is described by (Morse and Ingard 1986)

$$\nabla^2 p(\mathbf{r},t) - \frac{1}{v_s^2}\frac{\partial^2}{\partial t^2} p(\mathbf{r},t) = -\frac{\beta}{C_p}\frac{\partial}{\partial t} H(\mathbf{r},t), \quad (21.4)$$

where

$H(\mathbf{r},t)$ is a heating function referring to the thermal energy converted at spatial position $\mathbf{r}$ and time t by the EM radiation per unit volume per unit time

β is the isobaric volume expansion coefficient in K^{-1}

C_p is the isobaric specific heat in J/K·kg

Depending on the illumination type, laser or RF wave, the heating function can be explicitly expressed as (Li et al. 2008a)

$$H(\mathbf{r},t) = \begin{cases} \mu_a(\mathbf{r})\Phi(\mathbf{r},t) & \text{(for optical illumination)} \\ \sigma(\mathbf{r})\langle E^2(\mathbf{r},t)\rangle & \text{(for RF wave illumination)} \end{cases}, \quad (21.5)$$

where

Φ is the optical fluence rate in W/m^2

σ is the electrical conductivity in S/m

E is the electrical field

$\langle\cdot\rangle$ represents short-time averaging

A more strict confinement condition, called acoustic stress confinement, is satisfied in many PA cases. In acoustic stress confinement, τ is so short that the volume expansion of the object within τ is negligible compared with the targeted spatial resolution, which is mathematically described as

$$\tau \ll \tau_{st} = \frac{d_c}{v_s}, \quad (21.6)$$

where τ_{st} is the acoustic stress confinement threshold. Under both thermal and acoustic stress confinement conditions, the heating process can be approximately treated as a Dirac delta heating, that is, $H(\mathbf{r},t) \approx A(\mathbf{r})\,\delta(t)$, and $A(\mathbf{r})$ is the total energy converted into heat at location $\mathbf{r}$ during one illumination pulse. An initial pressure, p_0, is built up just after the EM illumination:

$$p_0(\mathbf{r}) = \Gamma A(\mathbf{r}). \quad (21.7)$$

Here Γ is a dimensionless parameter called the Grueneisen parameter, which is defined as

$$\Gamma = \frac{v_s^2 \beta}{C_p}. \quad (21.8)$$

The Grueneisen parameter is not a constant: it depends on several parameters, such as the temperature.

Under thermal confinement condition, the solution of Equation 21.4 in an infinite homogenous medium is

$$\begin{aligned} p(\mathbf{r},t) &= \int dt \int d\mathbf{r}' G(\mathbf{r},t;\mathbf{r}',t') \frac{\beta}{C_p} \frac{\partial H(\mathbf{r}',t')}{\partial t'} \\ &= \frac{\beta}{4\pi C_p} \int d\mathbf{r}' \frac{1}{|\mathbf{r}-\mathbf{r}'|} \frac{\partial H\left(\mathbf{r}',t-|\mathbf{r}-\mathbf{r}'|/v_s\right)}{\partial t} \\ &= \frac{\beta}{4\pi C_p} \frac{\partial}{\partial t} \int d\mathbf{r}' \frac{1}{|\mathbf{r}-\mathbf{r}'|} H\left(\mathbf{r}',t-\frac{|\mathbf{r}-\mathbf{r}'|}{v_s}\right). \end{aligned} \quad (21.9)$$

If the illumination pulse also satisfies stress confinement, the solution can be further simplified to

$$p(\mathbf{r}, t) = \frac{\beta}{4\pi C_p} \frac{\partial}{\partial t} \int d\mathbf{r}' \frac{1}{|\mathbf{r} - \mathbf{r}'|} A(\mathbf{r}') \delta\left(t - \frac{|\mathbf{r} - \mathbf{r}'|}{v_s}\right)$$

$$= \frac{v_s^2}{4\pi} \frac{\partial}{\partial t} \frac{1}{v_s t} \iint_{|\mathbf{r} - \mathbf{r}'| = v_s t} p_0(\mathbf{r}') ds' \qquad (21.10)$$

Its frequency-domain counterpart is

$$\tilde{p}(\mathbf{r}, k) = -ik \iiint_{V'} p_0(\mathbf{r}') \tilde{G}_k^{(out)}(\mathbf{r}, \mathbf{r}') d^3 r', \tilde{G}_k^{(out)}(\mathbf{r}, \mathbf{r}') = \frac{e^{ik|\mathbf{r} - \mathbf{r}'|}}{4\pi|\mathbf{r} - \mathbf{r}'|}, \qquad (21.11)$$

where $k = \omega/v_s$, $\tilde{p}(\mathbf{r}, k) = \int_{-\infty}^{+\infty} p(\mathbf{r}, \bar{t}) e^{ik\bar{t}} d\bar{t}$, and $\bar{t} = v_s t$.

Thus, the PA pressure at $\mathbf{r}$ and time t comes from initial PA sources over a spherical shell with a radius of $v_s t$. Since the solution for an illumination source with a finite pulse width can be calculated through the temporal convolution of the pulse profile with the solution of the delta pulse source (Xu and Wang 2006), the illumination is always assumed to be a delta EM pulse in the following discussion, unless otherwise noted.

21.1.4 Radiation Safety

Most PAT laser sources are in the visible to NIR spectral range, whereas TAT sources are in the hundreds of MHz to several GHz frequency range. Although PAT uses nonionizing photons having much less energy than ionizing x-ray photons, the EM radiation, either laser or RF waves, could still damage tissue in case of high illumination energy or power density. To guarantee safe PA applications in humans, the radiation strength of the EM illumination must be controlled and is usually below the safety limits set by the American National Standard Institute (ANSI) (ANSI 2000), the Institute of Electrical and Electronics Engineers (IEEE) (IEEE 1999), and the Federal Drug Administration (FDA) (Department of Health and Human Services 2004).

RF heating is measured as the specific absorption rate (SAR) in units of W/kg. According to the IEEE standards, at frequencies between 100 kHz and 6 GHz in a controlled environment, that is, where the person is aware of the potential for exposure, the maximum SAR is less than 0.4 W/kg as averaged over the whole body, and the spatial peak SAR is below 8.0 W/kg as averaged over any 1 g of tissue. The SARs are averaged over any 6 min interval. The FDA standards are more relaxed than the IEEE counterparts.

Laser safety in PAT depends on the wavelength, pulse duration, exposure duration, and exposure aperture. The ANSI laser safety standards provide the safety limits. In the spectral region

of 400–700 nm, the maximum permissible exposure (MPE) on the skin surface from any single laser pulse must not exceed 20 mJ/cm². In the 700–1050 nm region, the MPE increases with the wavelength λ in nm, as given by $20 \times 10^2(\lambda - 700)/1000$ mJ/cm². In the 1050–1400 nm region, the MPE increases to 100 mJ/cm². In addition, if the same area on the skin is exposed to multiple laser pulses, the safety limit also depends on the illumination time. For instance, if the exposure period is more than 10 s, the mean irradiance should not exceed 200 mJ/cm² in the 400–700 nm region, $200 \times 10^2(\lambda - 700)/1000$ mJ/cm² in the 700–1050 nm region, and 1000 mJ/cm² in the 1050–1400 region. More details are given in the ANSI standards.

21.2 PAT Modalities

Over the past decade, many innovative forms of PAT have been developed, and some of them have been successfully applied to in vivo imaging. Based on the imaging method, most of the current PAT methods can be categorized into three groups, or hybrids between different groups, as follows.

- PA computed tomography (PACT) relies on image reconstruction algorithms to obtain images from the detected PA signals.
- PA microscopy (PAM) detects PA signals by using a positively focused ultrasonic transducer that suppresses the PA signals coming from outside the acoustic focal zone. The lateral resolution is determined by either optical or ultrasonic focusing, and each data acquisition gives a 1D image.
- PAT using an acoustic lens system is analogous to an optical imaging system, except the optical lens is replaced with an acoustic lens. Acoustic pressures are measured on the imaging plane.

Each imaging type has its advantages and applications. In the following, we give brief introductions to each modality, as well as typical examples.

21.2.1 PA Computed Tomography

PACT is often simply called PAT. If the illumination source is RF waves, PAT is also called thermoacoustic tomography (TAT). PACT is widely studied due to its flexibility and convenience. Various types of acoustic detectors have been explored, including single-element transducers, acoustic arrays, integrated transducers (line or large plane), and virtual transducers. The basic procedure of PACT is to first obtain PA signals at multiple locations over the tissue, then process the acquired data by using image reconstruction algorithms to form the final image.

The simplest PACT setup is to scan a single-element ultrasonic transducer over the tissue surface. Owing to its simplicity and high sensitivity, this setup has been widely implemented (Wang et al. 2003, Xiang et al. 2007, Zhang et al. 2008b). A typical setup for scanning a single transducer, developed by Wang et al. (2003), is shown in Figure 21.3, where a single ultrasonic transducer scans the target over a horizontal circle. By using an

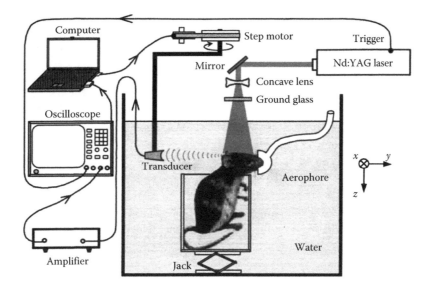

FIGURE 21.3 PAT setup for noninvasive transdermal and transcranial imaging of the rat brain in vivo, with the skin and skull intact. (From Wang, X.D. et al., *Nat. Biotechnol.*, 21, 803, 2003. With permission.)

ultrasound array and the parallel data acquisition imaging speed can be markedly increased.

Instead of using transducers made of piezoelectric materials, optical methods have also been used to detect PA signals (Paltauf and SchmidtKloiber 1997, Beard et al. 1999, Paltauf et al. 2007b, Niederhauser et al. 2004). Compared with most available commercial ultrasonic transducers, optical methods have wide, uniform sensitivity. In addition, some optical detection methods can provide better sensitivity than piezoelectric transducers at high frequencies (Beard et al. 1999, Zhang et al. 2008a). An example is the imaging system based on a planar Fabry–Perot (FP) film sensing interferometer (FPI) (Zhang et al. 2008a), shown in Figure 21.4.

In addition to using transducers that are much smaller than the region of interest, integrated ultrasonic transducers of large planar shape (Haltmeier et al. 2004) and line shape (Paltauf et al. 2007b) have also been studied. Mathematically, under certain

detection geometries, both large planar and line detectors can use the inverse Radon transformation to achieve exact image reconstruction.

As mentioned, PACT could be speeded up by using ultrasonic transducer arrays. Without major modifications, several commercially available ultrasound arrays have been implemented for PAT (Kruger et al. 2003, Yang et al. 2007a, Witte et al. 2008). To construct a compact system, the commercial ultrasound array can also be provided with optical fibers for light delivery, forming a handheld PAT system (Niederhauser et al. 2005). However, the pure ultrasound imaging uses both ultrasound emission and detection, PAT only requires the ultrasound detection. Moreover, the weak PA signal generally has a wide spectrum. Thus, in most PA applications, the desired ultrasonic detector has a high sensitivity and wide band detection. Therefore, specially designed ultrasound arrays are very valuable for PAT. Up to now, several custom designed arrays have been developed, including the high-frequency array (Zemp et al. 2007, 2008), which has 48 elements with a center frequency at 30 MHz; the 512-element 5 MHz full-ring array (Gamelin et al. 2008), with each element cylindrically focused; and an arc array (Oraevsky et al. 2007). Figure 21.5a presents the breast imaging system, called laser optoacoustic imaging system (LOIS), which uses an arch array system with 64 elements. Figure 21.5b is a PAT image acquired from this system, showing a tumor (Ermilov et al. 2009).

21.2.1.1 Reconstruction Algorithms

Both the imaging speed and image quality of PAT rely on the reconstruction algorithms. We present a brief introduction to reconstruction algorithms, and details are available in the literature (Xu and Wang 2006, Kuchment and Kunyansky 2008).

Mathematically, the reconstruction of the initial PA absorption distribution in an acoustically homogeneous, nondispersive, and nonviscous medium belongs to the inverse spherical Radon

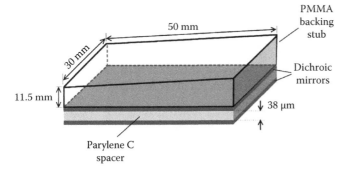

FIGURE 21.4 Schematic of an FP sensor head. The sensing structure comprises a 38 μm polymer (Parylene C) film spacer sandwiched between two dichroic mirrors, forming an FPI. The latter overlays a PMMA backing stub that is wedged to eliminate parasitic interference between light reflected from its upper surface and the FPI. (From Zhang, E. et al., *Appl. Opt.*, 47, 561, 2008a. With permission.)

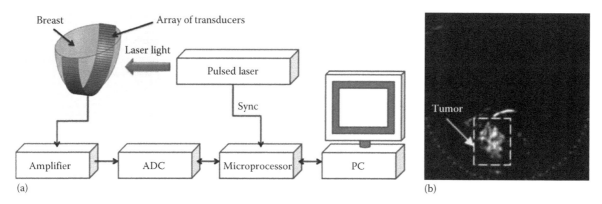

FIGURE 21.5   (a) Schematic diagram of LOIS-64. (b) Mediolateral photoacoustic images. High contrast of the object in the photoacoustic image implies the advanced angiogenesis indicative of a malignant tumor. (From Ermilov, S.A. et al., *J. Biomed. Opt.*, 14, 024007, 2009. With permission.)

transformation, which is discussed in (Xu and Wang 2006, Patch and Scherzer 2007, Anastasio et al. 2007). Analytical exact reconstruction algorithms have been developed for specific detection geometries, including spherical surfaces (Xu and Wang 2002, Finch et al. 2004, Kunyansky 2007a), cylindrical surfaces (Xu et al. 2002b), and planar surfaces (Xu et al. 2002a). A universal reconstruction formulation for these three geometries was also provided by Xu and Wang (2005, 2007), whose time-domain formulation is expressed as

$$p_0(\mathbf{r}) = \frac{1}{\Omega_0} \int_s d\Omega \left[2p(\mathbf{r}_d, t) - 2t \frac{\partial p(\mathbf{r}_d, t)}{\partial t} \right]_{t=|\mathbf{r}-\mathbf{r}_d|/v_s} , \quad (21.12)$$

where $d\Omega = dS / |\mathbf{r} - \mathbf{r}_d|^2 \cdot \left[n_d^s \cdot (\mathbf{r} - \mathbf{r}_d) / |\mathbf{r} - \mathbf{r}_d| \right]$ is the infinitesimal solid angle at detection position $\mathbf{r}_d$ with respect to the reconstruction point $\mathbf{r}$, and $\Omega_0 = \int d\Omega$. For spherical and cylindrical detection geometries, $\Omega_0 = 4\pi$; for a planar detection geometry, $\Omega_0 = 2\pi$. An exact series solution for the inversion of the spherical Radon transformation for any close detection surfaces that have explicit Dirichlet Laplacian eigenfunctions is presented in Kunyansky (2007b). This series solution extends the exact analytical solutions available for more detection surfaces, including cubes, ellipsoids, etc. Up to now, there have been no exact closed-form analytical solutions valid for arbitrary detection surfaces except those mentioned. Nonetheless, an exact numerical reconstruction algorithm over arbitrary closed surfaces has been developed (Burgholzer et al. 2007) based on the time-reversal concept (Xu and Wang 2004). Using integrated large detectors of planar and line shapes can take advantage of the existing inverse Radon transformation to reconstruct the PA image, as discussed in (Burgholzer et al. 2005, Paltauf et al. 2007a,b). However, exact reconstructions are valid only for limited scanning geometries: the large planar detector scans tangentially over a sphere that encloses the object; and the line detector scans perpendicularly around a circle or along an infinitely long line.

In practice, most PAT detection surfaces are both finite and non-closed. Using exact reconstruction algorithms in these

limited-view surfaces can cause image distortion and generate artifacts (Xu et al. 2004). Moreover, real ultrasonic detectors have finite sizes, limited bandwidths, and also limited detection angles, and these factors also adversely affect the quality of PAT images (Xu and Wang 2003).

Besides the mentioned exact reconstruction algorithms, many approximate reconstruction algorithms have also been developed, including the delay and sum algorithm (also called synthetic aperture) and several filtered back-projection reconstruction algorithms (Kruger et al. 1999, Xu and Wang 2005, Burgholzer et al. 2007).

Numerical iterative reconstruction algorithms are also used in PAT. Although iterative reconstructions may consume more computation time, they can use fewer PA detections (Paltauf et al. 2002). Moreover, real biological tissues are generally acoustically heterogeneous, and directly using the mentioned exact or approximate reconstruction algorithms degenerates the resolution and causes image artifacts. For instance, severe image distortions occur in the reconstructed images of the cerebral cortex of a monkey head (Yang and Wang 2008), where the PA pressure waves have to pass through a thick skull. Numerical iterative reconstruction algorithms have the potential to provide more accurate image reconstruction for PA sources in heterogeneous media. In the next section, we discuss PAT modalities that do not depend on reconstruction algorithms.

21.2.2 Acoustic-Resolution and Optical-Resolution PA Microscopy

PA microscopy (PAM) detects signals with a positively focused ultrasonic transducer, and focusing ultrasound in the tissue is much easier than focusing light at depths. According to the principle of reciprocity, a focused ultrasonic detector detects PA waves coming primarily from the focal zone, suppressing signals from outside. In addition, due to the relatively slow speed of ultrasound, PAM can also provide depth information from the time-resolved signal. Unlike PACT, PAM scans over the tissue surface, and at each scanning position it provides a 1D image without image reconstruction. B-scan and 3D images are obtained by

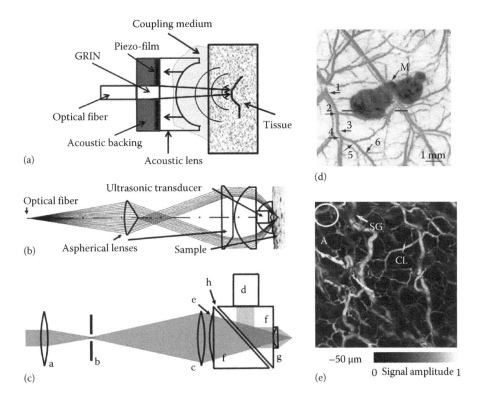

FIGURE 21.6 (a) Diagram of a bright-field confocal photoacoustic microscope in the backward detection mode. [Author's note: GRIN is gradient index optical condenser] (b) Schematic of the photoacoustic sensor of a dark-field reflection-mode photoacoustic microscope. (From Maslov, K. et al., *Opt. Lett.*, 30, 625, 2005. With permission.) (c) Schematic of the OR-PAM system: a, condenser lens; b, pinhole; c, microscope objective; d, ultrasonic transducer; e, correcting lens; f, isosceles prism; g, acoustic lens; h, silicon oil. (From Maslov, K. et al., *Opt. Lett.*, 33, 929, 2008. With permission.) (d) A composite of the two maximum amplitude projection (MAP) images projected along the z-axis, where an MAP image is formed by projecting the maximum photoacoustic amplitudes along a direction to its orthogonal plane. Here, blood vessels are pseudocolored red in the 584 nm image and the melanoma is pseudo-colored brown in the 764 nm image. As many as six orders of vessel branching can be observed in the image as indicated by numbers 1–6. (From Zhang, H.F. et al., *Nat. Biotechnol.*, 24, 848, 2006. With permission.) (e) In vivo OR-PAM image. (From Maslov, K. et al., *Opt. Lett.*, 33, 929, 2008. With permission.)

1D and 2D scanning, respectively. PAM does not exclude optical focusing, which can still be used in imaging very small targets within the ballistic or quasi-ballistic regimes. When using optical focusing, PAM is also referred to as optical-resolution PAM (OR-PAM) since its lateral resolution is determined by optical focusing. Consequently, PAM, using pure acoustic focusing, can be called acoustic-resolution PAM (AR-PAM).

PAM can be made both in transmission mode (the illumination and detection system are on the opposite sides of the tissue) and reflection mode (the illumination and detection systems are on the same side of the tissue). However, reflection mode detection is more convenient for in vivo tissue imaging. In reflection mode AR-PAM, bright field illumination can suffer from strong acoustic reverberations due to strong absorption near the surface, where considerable melanin may be present. The strong PA signal generated by these strong absorbers can reverberate in the detection system, such as the acoustic lens, and thus overshadow later arriving weak PA signals. Maslov et al. introduced dark-field confocal PAM (Maslov et al. 2005), a design that mitigates the problem. Dark-field PAM can achieve 15 μm axial resolution, 45 μm lateral resolution, and 3 mm imaging depth at

50 MHz center ultrasonic frequency, providing a powerful imaging tool for subcutaneous tissues. Moreover, OR-PAM can currently provide a 5 μm lateral resolution, with an imaging depth up to 700 μm (Maslov et al. 2008). Typical setups for PAM and several acquired images are shown in Figure 21.6. Figure 21.6a is an early PAM design, which uses the bright field illumination (Oraevsky and Karabutov 2003); Figure 21.6b is the design for dark-field PAM (Maslov et al. 2005); and Figure 21.6c is the design for OR-PAM (Maslov et al. 2008). Since PAT is scalable, another version of PAM (Song and Wang 2007), using lower frequency transducers, was developed to image deeper at reduced resolution of several hundreds of micrometers. The corresponding imaging depth achieved can reach several centimeters, useful for imaging internal organs.

21.2.3 PAT-Based on an Acoustic Lens System

Analogous to optical imaging system using an optical lens, PAT can also be achieved by using an acoustic lens to detect the projected PA signal at the image plane. However, unlike optical imaging, PA detection takes place after a time period due to the slow acoustic speed compared with that of light. There are several advantages to PAT using the acoustic lens: (a) no image reconstruction is needed, (b) the transducer array can be put at the imaging plane to achieve real-time detection, and (c) the acoustic lens can project the initial PA pressure distribution from the optically turbid medium to an optically clear medium, where optical detection can be applied. An example of a PAT based on an acoustic lens system is shown in Figure 21.7, where an acoustical 4f lens system is used. The image quality of PAT by using an acoustic lens system is inherently limited by the aperture and aberrations of the lens. Thus far, however, only a few acoustic lens systems have been built for PAT, and the imaging targets have been limited in phantom objects (Niederhauser et al. 2004, Chen et al. 2006, 2007, Zhang et al. 2007a).

The classification of three types of PAT imaging modalities in this chapter is by no means exclusive. There are various other designs of acoustic detectors, as well as various detection methods.

In addition to short EM pulses as the illumination sources, PAT can also be achieved in the frequency domain. Telenkov and Mandelis first presented a frequency-domain version of PAM (Telenkov and Mandelis 2006). They modulated the laser power by linearly chirping the frequency. The axial resolution of the frequency-domain system is determined by the bandwidth. The lateral resolution is given by the ultrasonic focusing. Maslov and Wang demonstrated another method, modulating the laser power at a fixed frequency (Maslov and Wang 2008). This latter system, without the frequency chirping, increased the SNR by using a narrowband transducer, which has a resonance frequency the same as the modulation frequency. However, no axial resolution is provided for a planar feature. The advantages of using intensity-modulated CW lasers include inexpensiveness, more stable illumination sources, and narrowband detection. More studies in this direction are expected.

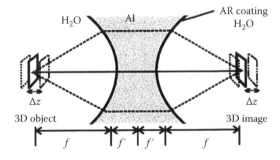

FIGURE 21.7 The acoustic lens system is arranged in a 4f configuration. This guarantees axial and lateral unit magnification of the imaged transient pressure distribution. A displacement of the object plane by Δz results in a displacement of the focused image plane by the same value Δz. The low speed of sound compared to the speed of light preserves axial resolution by the propagation time for transient acoustic imaging. [AR, antireflection]. (From Niederhauser, J.J. et al., *Appl. Phys. Lett.*, 85, 846, 2004. With permission.)

21.3 PA Spectroscopy and Functional Imaging

Since optical or RF absorption in tissue is highly sensitive to biological activities, such as metabolic rate, PAT is suitable for functional imaging. Moreover, as a hybrid method, PAT inherits the spectral imaging ability of optical imaging, and spectroscopic PAT can reveal more functional information about the tissue. Although there are many potential absorbers, most PA functional imaging research targets the circulation system, such as blood vessels.

Blood flow is affected by many normal and abnormal biological conditions, including neural activities, local metabolic rate, and blood vessel resistance. PAT provides a unique noninvasive way to study hemodynamics in vivo. By using a monochromatic illumination source, which is usually the isosbestic wavelength for Hb and HbO_2 absorption, PAT can monitor blood volume changes due to mechanical and chemical stimulation or vessel occlusion (Wang et al. 2003, Yang et al. 2007b).

Owing to the spectrally dependent absorption by blood, spectroscopic PAT is particularly useful for functional imaging of SO_2. SO_2 is closely related to the local metabolic rate of oxygen, and blood vessels in tumors generally have abnormally lower SO_2 values than blood vessels in surrounding healthy tissues. Thus, imaging SO_2 distribution helps to detect tumors. Since light absorption by blood depends on both [Hb] and [HbO_2], detecting SO_2 requires multiple wavelengths. The mathematics of spectroscopic PAT for calculating SO_2 is described in (Zhang et al. 2007b). Since light scattering and absorption in tissue depend on the wavelength, calculating SO_2 requires information about local fluence. Several research efforts have mapped SO_2 by using PA spectroscopic tomography, including PA computed tomography (PACT) (Wang et al. 2006) and PAM (Zhang et al. 2006). Figure 21.8 shows the SO_2 mapping by using PAM. Moreover, OR-PAM can provide local SO_2 at the capillary level.

Except for the optical-resolution PAM, where the illumination field is within the ballistic regime, quantitative SO_2 mapping generally requires knowledge of the relative values of the local fluence at multiple wavelengths. However, due to the complexity in the optical properties in the real tissue, it is challenging to acquire such information accurately. In practice, the invasive method was used (Zhang et al. 2006), and numerical iterative methods to solve the coupled propagation equations of the ultrasound and light are also explored (Cox et al. 2009, Laufer et al. 2007, Paltauf et al. 2002, Yin et al. 2007).

21.4 PA Contrast Agents and Molecular Imaging

21.4.1 PA Contrast Agents

Although tissue has some high endogenous EM absorbers, such as the blood and melanin, not all tissues are good targets for PAT; lymphatic fluid, for example, has a low optical absorption. As in x-ray-based computed tomography (CT), positron emission

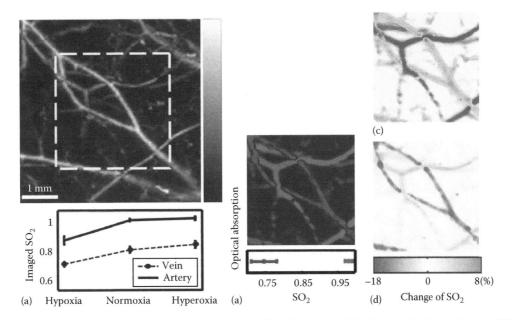

FIGURE 21.8 PAM imaging of variations in SO_2 in single subcutaneous blood vessels in a 200 g Sprague Dawley rat in vivo. (a) Structural image reflecting the total hemoglobin concentration acquired at the 584 nm optical wavelength under hyperoxia. (b) Static SO_2 image within the marked region in panel A under normoxia, where arteries and veins are pseudocolored red and blue, respectively, based on the imaged SO_2 values. (c) Image of the SO_2 changes from normoxia to hypoxia (hypoxia value-normoxia value). (d) Image of SO_2 changes from normoxia to hyperoxia (hyperoxia value-normoxia value). (e) Typical imaged values of SO_2 in venous and arterial bloods under all three physiological states, where different trends of variation are observed. (From Zhang, H.F. et al., *Appl. Phys. Lett.*, 90, 053901, 2007b. With permission.)

tomography (PET), and magnetic resonance imaging (MRI), PA contrast agents are also widely used to enhance the image quality and specificity. Primary PA contrast agents include dyes and nanoparticles (Ku and Wang 2005, Wang et al. 2004b, Yang et al. 2007c, Eghtedari et al. 2007). Among these, the contrast agents having high absorption in the NIR spectrum are especially desirable since NIR light can penetrate deeper into tissue.

Contrast agents made of dyes can increase PA image qualities of the circulation system. A typical dye is indocyanine green (ICG), a dye approved by FDA. ICG has high absorption in the NIR spectral region. Enhancement of PA signals from the blood has been demonstrated by using ICG in the NIR spectrum (Wang et al. 2004a). Unlike blood, the lymph fluid is generally transparent, making it difficult to image without contrast agents. Recently, methylene blue was used as a contrast agent to detect sentinel lymph nodes (Song et al. 2008). The advantage of using dyes is that many dyes are nontoxic. However, dyes, without binding to other chemicals, often suffer from a short circulation time, limiting their applications in long-term PA monitoring.

Compared with dyes, nanoparticles possess a high and tunable absorption spectrum and a longer circulation time. The unique advantage of nanoparticles is that their absorption peak is tunable by changing the shape and size of the particle. Various nanoparticles have been used for PAT, including carbon nanotubes (De La Zerda et al. 2008, Pramanik et al. 2009), gold nanorods (Li et al. 2005, Agarwal et al. 2007) and nanocages (Song et al. 2009), and nanoshells (Wang et al. 2004b). Figure 21.9 demonstrates that the injection of nanoshells can markedly increase the SNR of the PAT image. Even without

bio-conjugation, small nanoparticles (60–400 nm diameters) tend to accumulate in tumor sites due to the enhanced vascular permeability and retention caused by the tumor (Maeda et al. 2003). However, potential toxicity limits nanoparticles' in vivo applications. So far, no nanoparticle has been approved by the FDA for in vivo human application.

Most available PA contrast agents are for absorbing laser wavelengths, especially in the NIR spectral range. However, RF contrast agents are also desirable due to the superior penetration depth of RF in the body. There are only a few PA contrast agent for the RF region (Nie et al. 2007, Pramanik et al. 2009).

21.4.2 PA Molecular Imaging

Since EM absorption is a characteristic property of molecules, PAT has the potential to do molecular imaging. As mentioned, measuring SO_2 is an example of PAT used for imaging changes in molecules. However, PA molecular imaging generally refers to using PA contrast agents targeting certain molecules or molecular activities. The PA molecular imaging is done via two major mechanisms: (a) Certain contrast agents change their optical or RF absorption properties, triggered by changes at the molecular level (Li et al. 2007a). (b) Contrast agents with specific absorption spectra are bio-conjugated with certain proteins, such as antibodies, targeting specific molecules (Li et al. 2007b, Copland et al. 2004, Agarwal et al. 2007). Figure 21.10 is an example of PA molecular imaging by using single-wall carbon nanotubes (SWNT), with and without bio-conjugated cyclic Arg-Gly-Asp (RGD) peptides.

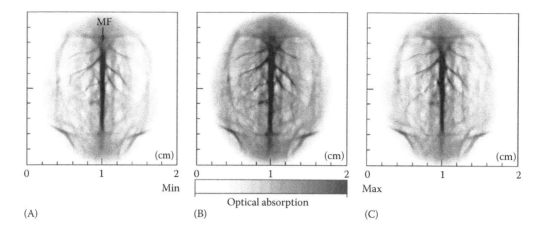

FIGURE 21.9 Noninvasive PAT of a rat brain in vivo employing the nanoshell contrast agent and NIR light at a wavelength of 800 nm. (A) Photoacoustic image acquired before the administrations of nanoshells. *MF*: median fissure. (B) Photoacoustic image obtained ~20 min after the third administration of nanoshells. (C) Photoacoustic image acquired ~350 min after the third administration of nanoshells. (From Wang, Y.W. et al., *Nano Lett.*, 4, 1689, 2004b. With permission.)

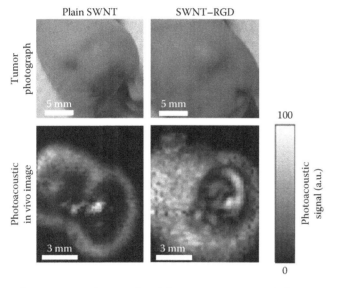

FIGURE 21.10 Photographs of the tumors in mice and the corresponding photoacoustic subtraction images (green) shown as horizontal slices through the tumors. Mice injected with plain single-walled carbon nanotubes (left-hand column) showed low photoacoustic signals compared with mice injected with SWNT–RGD (right-hand column). The tumors are in the same orientation in all images. (From De La Zerda, A. et al., *Nat. Nanotechnol.*, 3, 557, 2008. With permission.)

21.5 Comparison of PAT with Other Imaging Modalities

Up to now, the mechanism and various implementations of PAT have been discussed. A hybrid of optics and ultrasound, PAT inherits many merits of optical imaging, such as its sensitive optical contrast and spectroscopic imaging capability. The differences between PAT and pure optical imaging are also prominent. Pure optical imaging methods provide high spatial resolution only within the "soft limit," which is ~1 mm deep in optically

scattering tissue. Otherwise, the spatial resolution decreases significantly as the imaging depth increases. PAT can not only achieve optical resolution within the soft limit, but also break through the soft limit with ultrasonic resolution. Moreover, PAT is much more sensitive to optical absorption in tissue than optical microscopy, confocal microscopy, and OCT. In what follows, PAT is briefly compared with several commercialized noninvasive imaging methods, including ultrasound, x-ray computed tomography (CT), and magnetic resonance imaging (MRI).

21.5.1 Imaging Contrast

The imaging contrast of PAT comes from the difference in EM absorption by various tissues. Contrast in ultrasound, MRI, and CT comes primarily from differences in acoustic scattering, proton density, and electron density, respectively. In addition to endogenous contrast agents, PAT, like other imaging modalities, can also use exogenous contrast agents, such as nanoparticles. Exogenous contrast agents enable PAT to provide molecular imaging.

21.5.2 Imaging Depth and Spatial Resolution

The imaging depth and resolution of PAT are scalable, as they are with ultrasound. The ratio of the imaging depth to the spatial resolution can be greater than 100. For instance, PAT provides spatial resolutions in the micrometer to submillimeter ranges, with corresponding imaging depths from hundreds of micrometers to several centimeters. However, both PAT and ultrasound suffer from the presence of tissue cavities, such as in the lung, which can block ultrasonic waves. By comparison, CT and MRI can image through the whole body with uniform high resolution <1 mm.

21.5.3 Imaging Temporal Resolution

Using ultrasonic arrays, PAT can provide real-time imaging. The upper limit of the temporal resolution of PAT is similar to that of Ultrasound, which is on the order of microseconds. Multiple source-detectors CT can also provide real-time imaging with a temporal resolution of milliseconds. However, MRI has less temporal resolution as a result of the prolonged relaxation period, which can be on the order of hundreds of milliseconds to seconds.

21.5.4 Safety

Like ultrasound and MRI, PAT uses nonionizing illumination sources, including laser and RF waves. In clinical applications, the illumination intensity of PAT can be set to within accepted safety standards. CT, however, uses ionizing radiation, which limits its in vivo applications.

In addition, the cost of a commercialized PAT instrument is estimated to be comparable to that of an ultrasound system, which is much cheaper than either CT or MRI. Moreover, compact PAT can serve as a bedside medical imaging modality. In conclusion, PAT would be a great complement to many current imaging modalities.

21.6 Summary of PAT

PAT is a hybrid innovative biomedical imaging modality, combining EM absorption contrast and acoustic detection. The PA signal can be generated using either pulsed or intensity-modulated CW EM sources. Endogenous biological absorbers include blood, melanin, and water.

Three major PAT detecting methods are introduced in this chapter. The resolution of PAT is scalable, ranging from sub-millimeters to several micrometers. For PACT, both exact and approximate reconstruction algorithms have been developed. For PAM, the lateral resolution can be determined by either acoustical focusing or optical focusing. PAT can provide not only structural imaging, but also functional imaging by using spectroscopic imaging.

PA contrast agents have been extensively studied, including dyes and nanoparticles. Contrast agents can improve the sensitivity and specificity of PA detection. In order to detect deeper into scattering tissue, contrast agents with high absorption in the NIR region are desirable. Moreover, contrast agents enable PA molecular imaging.

Overall, PA techniques for biomedical imaging are maturing. They have been widely used to study both animal and human tissues. Recently, more research has focused on clinical applications. Commercialized PA systems are expected to be available in the near future, and wide clinical PA applications are foreseen.

21.7 Introduction to Ultrasound-Modulated Optical Tomography

21.7.1 Motivation

Ultrasound-modulated optical tomography (UOT), first proposed and demonstrated in the early 1990s (Marks et al. 1993,

Wang et al. 1995), is a novel bio-imaging technique that offers strong optical contrast with high ultrasonic spatial resolution. The principle of this technique is that a high-coherence laser source illuminates a light-scattering medium, which is insonified with ultrasound. Light passing through the ultrasonic focal volume is acoustically tagged so that each optical speckle spot at a detector surface has a time-varying modulation component due to the ultrasonic interaction. By measuring the degree of modulation at each scanned ultrasonic focal volume, images representative of the optical properties as well as acoustic properties of the tissue are formed. Since any light (scattered or unscattered) modulated by acousto-optic interactions contributes to UOT signals, the imaging depth can be extended to the quasi-diffusive or diffusive regimes, beyond one optical transport mean free path (~1 mm). In addition, within the reach of diffusive photons, the image resolution and the maximum imaging depth are scalable with the ultrasonic frequency.

Compared to diffuse optical tomography (DOT) (Zeff et al. 2007), a pure optical imaging technique, due to ultrasonic modulation of light, UOT provides higher spatial resolution while offering the same optical contrast (absorption and scattering) as DOT. Compared to PA imaging (Wang et al. 2003, Zhang et al. 2006), another hybrid imaging modality with strong optical contrast and high ultrasonic resolution, UOT is capable of supplying optical scattering (Kothapalli et al. 2007) and mechanical properties (Bossy et al. 2007) as well as optical absorption properties, whereas PA imaging is sensitive only to optical absorption. Although UOT images (UOT signals are proportional to local optical fluence) are not as sharp as PA images (PA signals are proportional to the product of the local absorption coefficient and local optical fluence), there is no rejection of unresolvable background signals in UOT images (which is also true for DOT). Moreover, because the image contrast of UOT is based on optical properties, this technique is potentially able to supply morphological and functional (total hemoglobin concentration and oxygen saturation of hemoglobin) information of biological tissues (Kim and Wang 2007).

A major challenge in UOT is the low signal-to-noise ratio (SNR) of the measurements resulting from low modulation efficiency and uncorrelated phases between speckle grains. A number of detection techniques have been explored to detect low UOT signals effectively, including parallel detection (Leveque et al. 1999) and speckle contrast detection (Li et al. 2002) with CCD cameras, a Fabry–Perot-interferometry-based technique (Sakadzic and Wang 2004), a photorefractive interferometry–based technique (Murray et al. 2004), and a spectral-hole-burning-based technique. (Li et al. 2008c) Furthermore, to overcome the low SNR in UOT, rather than exploring new detection techniques, intense ultrasound bursts have been applied instead of continuous-wave or pulsed ultrasound (Kim et al. 2006). By reducing the duty cycle on the ultrasonic transducer, it is possible to use much higher pressures, thus improving the ultrasound-modulated light signal level. Additionally, during timescales on the order of milliseconds, intense acoustic bursts can induce large localized tissue displacement through acoustic radiation force.

21.7.2 Mechanisms of UOT (Leutz and Maret 1995, Wang 2001a,b, Sakadzic and Wang 2006)

Three mechanisms have been identified for ultrasonic modulation in an optically scattering medium. The first mechanism is the incoherent intensity modulation of light due to variations of the optical properties of a medium caused by ultrasound. The medium is compressed and rarified depending on the time and location, while being insonified. Due to the disturbance of density, the optical properties (absorption coefficient, scattering coefficient, and refractive index) vary, so the detected light contains intensity-modulation information. However, ultrasonic modulation of incoherent light has been too weak to be detected experimentally.

The second mechanism is based on changes in the optical phase due to ultrasound-induced displacements of optical scatterers. The displacements of scatterers, assumed to follow ultrasonic amplitudes, modulate the physical path lengths of light passing through the ultrasonic field. Multiply scattered light accumulates modulated physical path lengths along its path. Consequently, the intensity of the speckles formed by the multiply scattered light fluctuates with the ultrasonic wave.

The third mechanism is based on changes in the optical phase due to ultrasonic modulation of the optical refractive index of the medium. As a result of ultrasonic modulation of the index of refraction, the physical path lengths of light passing through the ultrasonic field are modulated. As in the second mechanism, multiply scattered light accumulates modulated phases along its path, and the modulated phases cause intensity fluctuations of the speckles formed. Unlike the first mechanism, both the second and the third mechanisms require the use of coherent light.

In the analytical model (Wang 2001b), a plane wave of ultrasound uniformly insonifies an optically homogenous scattering medium. The analytical model is valid under the following two assumptions: (1) the optical mean free path is much longer than the optical wavelength (weak scattering) and (2) the change in the optical path length due to ultrasonic interaction is much less than the optical wavelength (weak modulation). Because of the weak-scattering approximation, the accumulative correlations between photons traveling different paths are negligible in comparison with the correlations from the photons traveling through the same paths. The autocorrelation function, $G_1(\tau)$, of the scalar electric field, $E(t)$, of the scattered light is as follows:

$$G_1(\tau) = \int_0^\infty p(s) < E_s *(t+\tau)E_s(t) > ds, \quad (21.13)$$

where

$<>$ denotes time and ensemble averaging

E_s denotes the electric field of the scattered light of a path length s

$p(s)$ is the probability density function of a path length s

The effect of Brownian motion on the autocorrelation function is ignored here. A coherent optical beam (plane wave) illuminates a slab normally (thickness: L), and the transmitted light is detected by a point detector. The solution of $p(s)$ is derived from diffusion theory with a zero-boundary condition. The autocorrelation function, $G_1(\tau)$, can be obtained as follows:

$$G_1(\tau) = \frac{(L/l)\sinh\left(\left\{\varepsilon[1 - \cos(\omega_a\tau)]\right\}^{1/2}\right)}{\sinh\left((L/l)\left\{\varepsilon[1 - \cos(\omega_a\tau)]\right\}^{1/2}\right)}, \quad (21.14)$$

$$\varepsilon = 6(\delta_n + \delta_d)(n_0 k_0 A)^2, \quad (21.15)$$

$$\delta_n = (\alpha_{n1} + \alpha_{n2})\eta^2, \quad (21.16)$$

$$\alpha_{n1} = \frac{k_a l \tan^{-1}(k_a l)}{2}, \quad (21.17)$$

$$\alpha_{n2} = \frac{\alpha_{n1}}{\left[(k_a l)/\tan^{-1}(k_a l) - 1\right]}, \quad (21.18)$$

$$\delta_d = \frac{1}{6}, \quad (21.19)$$

where

ω_a is the acoustic angular frequency

n_o is the background refractive index

k_o is the optical wave vector *in vacuo*

A is the acoustic amplitude, proportional to the acoustic pressure

k_a is the acoustic wave vector

l is the optical transport mean free path

η is the elasto-optical coefficient, related to the adiabatic piezo-optical coefficient of the material $\partial n/\partial p$, the density ρ, and the acoustic velocity v_a: $\eta = (\partial n/\partial p)\rho v_a^2$

δ_n and δ_d are related to the average contributions to ultrasonic modulation from changes in refractive index and particle displacements per free path (or per scattering event), respectively. δ_n increases with $k_a l$ because a greater phase modulation is accumulated from a longer optical mean path. By contrast, δ_d stays constant at 1/6, independent of k_a and l. As a result, the ratio between δ_n and δ_d increases with $k_a l$. Using the Wiener–Khinchin theorem and the Fourier transformation, the power spectral density of the ultrasound-modulated speckle can be derived as follows:

$$S(\omega) = \int_{-\infty}^{\infty} G_1(\tau)\exp(i\omega\tau)d\tau, \quad (21.20)$$

where ω is the relative frequency to the original optical frequency of the unmodulated light (ω_o) because $G_1(\tau)$ implicitly contains $\exp(-i\omega_o\tau)$. In other words, when ω is equal to 0 in $S(\omega)$, the frequency is the original optical frequency ω_o. Since $G_1(\tau)$ is an even periodic function in terms of τ, the spectral intensity at frequency $n\omega_a$ can be calculated by

$$I_n = \frac{1}{T_a} \int_0^{T_a} \cos(n\omega_a\tau) G_1(\tau) d\tau, \qquad (21.21)$$

where

T_a is the acoustic period
n is an integer

Figure 21.11 shows the power spectrum of a UOT signal observed experimentally using a spectral-hole burning detection technique with a 1 MHz ultrasound transducer. The one-sided modulation depth (M), defined as the ratio of the fundamental frequency I_1 to the unmodulated intensity I_0, often is of experimental interest.

Under the weak-modulation approximation (i.e., (L/l) $\varepsilon^{1/2} = \ll 1$), Equation 21.14 can be simplified to

$$G_1(\tau) = 1 - \frac{1}{6}\left(\frac{L}{l}\right)^2 \varepsilon\left[1 - \cos(\omega_a\tau)\right]. \qquad (21.22)$$

Thus, the modulation depth can be obtained as follows:

$$M = \frac{1}{12}\left(\frac{L}{l}\right)^2 \varepsilon \propto A^2. \qquad (21.23)$$

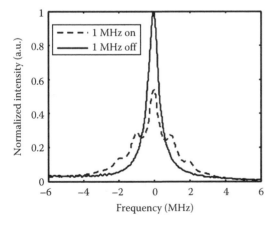

FIGURE 21.11 Experimental power spectrum of a UOT signal measured using a spectral-hole-burning-based detection technique with a 1 MHz ultrasound transducer. (From Li, Y. et al., *Opt. Express*, 16, 14862, 2008b. With permission.)

The relationship between M and the acoustic amplitude A is quadratic, and was observed experimentally using a Fabry–Perot interferometer. (Leutz and Maret 1995)

21.8 Detection Techniques in UOT

21.8.1 Photorefractive Interferometer–Based Detection (Murray et al. 2004, Sui et al. 2005)

In this detection system, a local oscillator (LO) that is wave front matched to the diffusive scattered light (signal beam) from the sample is derived through two-wave mixing (TWM). The signal beam and the LO interfere at the photodetector, where the ultrasonic phase modulation in the signal beam is converted to an intensity modulation. This wave front matched LO allows for the detection of phase modulation over the entire collected speckle field, giving a large *etendue* (acceptance solid angle × detection area). In addition, the photorefractive crystal (PRC) is adaptive and compensates for low-frequency shifts in the speckle patterns, avoiding speckle shifts on longer timescales than the response time of the crystal.

The experimental setup is shown in Figure 21.12. The light (a frequency-doubled Nd:YAG laser source, 532 nm) is first sent to a beam splitter (signal beam: reference beam = 25: 1). The reference beam is directed to the PRC, a Bismuth silicon oxide (BSO) crystal (5 mm × 5 mm × 7 mm along the X, Z, and Y axes, respectively). The signal beam illuminates a tissue mimicking phantom. The transmitted diffusive signal beam from the phantom is collected by a lens and directed into the PRC. Therefore, both the signal and reference beams interfere in the crystal. A 4 KHz, 10 KV/cm peak-to-peak AC field is applied to the crystal to enhance the grating strength and the TWM gain. After the crystal, both the signal and diffracted reference beams are detected by an avalanche photodiode (APD). The UOT signal from the APD is amplified, low-pass-filtered at 500 KHz, and displayed on an oscilloscope. The center frequency of the ultrasonic transducer is 1.1 MHz. The focal zone of the transducer is ~9 mm in length and 1.5 mm in width. A peak pressure of ~1 MPa is applied. The transducer is mounted on a 3D-automated translation stage controlled by a computer. The pulse repetition rate is typically 100 Hz. An optically scattering tissue phantom (4 cm × 2.7 cm × 4 cm along the X, Y, and Z axes, respectively) is made of gelatin and 400 nm diameter polystyrene microspheres. The reduced scattering coefficient of the phantom is 2 cm^{-1}. The optically absorptive target (5 mm × 8 mm × 5 mm along X, Y, and Z axes, respectively) is made of India ink. The absorption coefficient of the target is 3 cm^{-1}.

Figure 21.13a shows three typical A-lines (one-dimensional images along the ultrasound propagation direction, the Z-axis.) corresponding to Figure 21.13b. The middle peak signal in Figure 21.13a (2) corresponds to the embedded target. By scanning the ultrasound transducer along the X-axis, the B-scan (the depth-resolved two-dimensional image) is formed, representing the embedded target in Figure 21.13c.

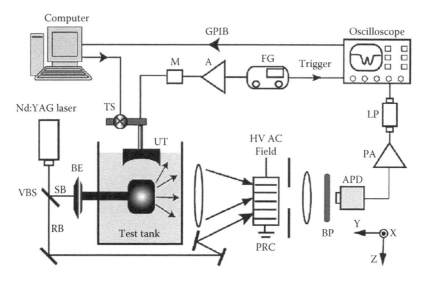

FIGURE 21.12 Experimental setup of the PRC-based detection technique. FG, function generator; A, power amplifier; M, impedance matching box; TS, translation stage; UT, ultrasound transducer; VBS, variable beam splitter; R, reference beam; SB, signal beam; BE, beam expander; BP, optical bandpass filter; APD, avalanche photodiode; PA, preamplifier; LP, low-pass filter. (From Sui, L. et al., *Appl. Opt.*, 44, 4041, 2005. With permission.)

21.8.2 Confocal Fabry–Perot Interferometer–Based Detection (Sakadzic and Wang 2004, Kothapalli and Wang 2008, 2009)

UOT can be implemented with a long-cavity scanning confocal Fabry–Perot interferometer (CFPI). The CFPI has a greater *etendue* than most CCD cameras and provides parallel speckle processing. In addition, the CFPI can detect the propagation of high-frequency ultrasound pulses in real time while tolerating speckle decorrelation.

The experimental setup is shown in Figure 21.14. Samples are gently pressed through a slit along the Z-axis to create a semicylindrical bump. The ultrasound and illuminating light are focused to the same spot orthogonally inside the sample (Figure 21.14b). Diffusive transmitted light is collected by a single multimode fiber (600 μm core diameter). This geometry is the semi-reflection mode while minimizing the detection of unmodulated light from the shallow surface. Further, due to orthogonal interactions between light and ultrasound, the modulation efficiency can be maximized. A 15 MHz-focused ultrasound transducer is driven by a pulser-echo amplifier. The other (30, 50, and 75 MHz) transducers are driven by the square bipolar pulses with periods of 34, 20, and 15 ns, respectively. Ultrasound peak pressures at the focal spot are applied up to 4 MPa for all transducers. The laser light (a frequency-doubled Nd:YAG laser source, 532 nm) is focused onto a spot of ~100 μm diameter in optically scattering medium. The sample is placed on a three-axes (X1, Y1, and Z1) translational stage. The collected light is coupled into the CFPI, operated in a transmission mode (50 cm cavity length, 0.1 *sr* mm² *etendue*, and ~20 finesse). A small portion of light from the beam splitter is used for a cavity tuning procedure. First the cavity is chirped through one free spectral range to find the position of the central frequency

of the original incident light. Then one CFPI mirror is repositioned by a calibrated amount so that the cavity is tuned to the frequency of one sideband of the ultrasound-modulated light (15–75 MHz greater than the original incident light frequency). An APD is used as a detector. By measuring ultrasound-modulated light intensity during the time of ultrasound flight, A-lines are acquired. In each operational cycle, first the resonant frequency of the CFPI is tuned and then data from 4000 ultrasound pulses are acquired in 1 s. By scanning the sample along the Z direction and acquiring each corresponding A-line, B-scan images can be obtained.

Figure 21.15 shows typical A-line (Figure 21.15c) and B-scan (Figure 21.15b) images, obtained using the 75 MHz transducer, of an optical absorptive patterned target (Figure 21.15a) embedded 2 mm below the optical scattering medium. The target has dimensions of about 400 μm × 100 μm × 500 μm along the X, Y, and Z axes, respectively. The B-scan images from Figure 21.15b agree well with the photograph in Figure 21.15a. Figure 21.15c represents A-lines along the X-axis for the three cuts in Figure 21.15b. These profiles show that axial widths of 75, 30, and 15 μm are resolved with 70%, 55%, and 30% contrast, respectively.

To investigate the resolutions of various ultrasound transducers (15, 30, 50, and 75 MHz central frequencies), the edge spread functions obtained from the B-scan images of an optically absorptive target are compared. The target, buried inside the chicken breast tissue, is 100 μm thick (along the X and Y axes) and 1 mm long (along the Z-axis). A photograph of this absorptive target is shown in Figure 21.16a. Figure 21.16b–e show B-scan images of Figure 21.16a obtained with 15, 30, 50, and 75 MHz central frequency of ultrasonic transducers at approximate depths of 2.5, 2, 1.5, and 1.2 mm below the chicken breast tissue, respectively. In pure ultrasound imaging, the

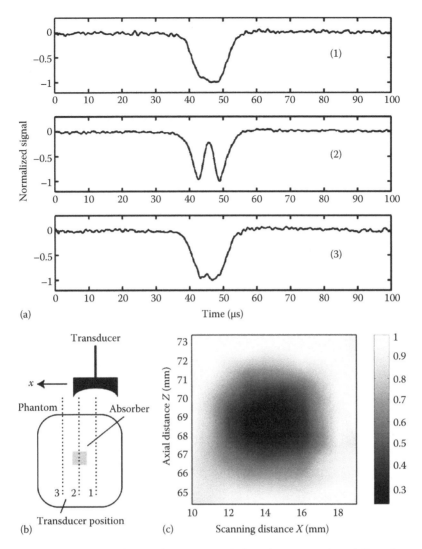

FIGURE 21.13 (a) Typical A-lines detected when the transducer is scanned along the *X*-axis at three different locations corresponding to positions 1, 2, and 3 in (b). (c) Two-dimensional B-scan image (*X–Z* plane) obtained by scanning the transducer. (From Sui, L. et al., *Appl. Opt.*, 44, 4041, 2005. With permission.)

lateral resolution (λ*F/D*) is related to the ultrasonic focal width, where λ is the wavelength of the ultrasound, *F* is the focal length of the transducer, and *D* is the element diameter of the transducer. The axial resolution depends on the length of the ultrasound pulse.

Table 21.1 shows the experimental measurements of focal width and focal length in water for all four transducers, and the results are compared with the lateral and axial resolutions of UOT images estimated from Figure 21.16b–e.

Figure 21.17a and b show typical axial and lateral edge spread functions of the absorbing object, respectively, obtained with all four transducers. The spatial resolution in UOT is defined as the one-way distance between the 25% and 75% points (as marked by dashed arrows in Figure 21.17) of the respective edge spread functions. For the transducer with 15 MHz central frequency, a lateral resolution of 140 μm and an axial resolution of 74 μm at a depth of 3 mm inside the chicken breast tissue are measured.

Similarly, 30, 50, and 75 MHz central frequency transducers provided axial resolutions of 60 μm (at a depth of 2.5 mm), 48 μm (depth 2.3 mm), and 31 μm (depth 1.8 mm), respectively, inside the chicken breast tissue. Their lateral resolutions were 70, 55, and 38 μm, respectively.

21.8.3 Spectral-Hole-Burning-Based Detection (Li et al. 2008b,c)

UOT can be implemented with spectral-hole-burning (SHB)-based detection as a narrowband spectral filter. This technique is able to filter one sideband of the UOT signal while suppressing the strong unmodulated light (DC) and the other sideband like Fabry–Perot interferometry. This detection method has several advantages: First, it provides a large *etendue*. In principle, the acceptance angle of the absorption-based quantum filters can be nearly 2π, yielding an *etendue* of 1131 *sr* mm² for a

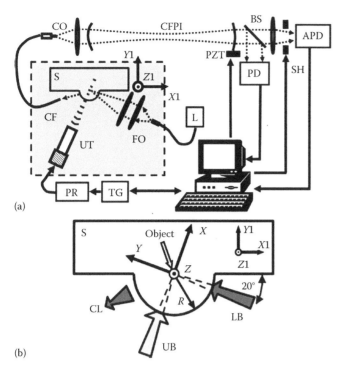

(a)

(b)

FIGURE 21.14 (a) Experimental setup of the CFPI-based detection technique. L, laser; TG, trigger generator; PR, pulser–receiver; UT, ultrasonic transducer; FO, focusing optics; CF, collecting fiber; S, sample; CO, coupling optics; PZT, piezoelectric transducer; BS, beam splitter; SH, shutter; PD, photodetector. (b) Top view of the sample (S): UB, ultrasound beam; LB, incident light beam; CL, collected light; R, radius of curvature. (From Sakadzic, S. and Wang, L.H.V., *Opt. Lett.*, 29, 2770, 2004. With permission.)

10×9 mm crystal aperture. The second advantage is that parallel speckle processing is possible. The third advantage is that this system is insensitive to speckle decorrelation because this UOT system detects transmitted ultrasound-modulated light through a frequency-dependent absorptive filter constructed

from an SHB crystal. Therefore, it can potentially provide in vivo imaging.

The SHB medium is a rare-earth ion–doped, inhomogeneously broadened optical absorber at room temperature. When cryogenically cooled (4.2 K), it has a sub-MHz homogeneous linewidth $\Delta\Gamma_H$ and an inhomogeneous bandwidth $\Delta\Gamma_I$ of GHz (Figure 21.18). When a narrowband light at frequency f_L illuminates a cryogenically cooled SHB crystal, the resonant ions are excited from the ground state to the excited state, resulting in an absorption coefficient $\alpha(f)$ change at the frequency. Sufficiently intense illumination excites the resonant ions until two states are equally populated, and as a consequence, a spectral hole is engraved at the frequency f_L with linewidth of $\Delta\Gamma_H$. For an optically dense crystal ($\alpha(f)L_c \gg 1$, where L_c is the crystal thickness), the optical absorption coefficient is significantly reduced at f_L, while the absorption coefficients at other frequencies remain. By matching the ultrasonic frequency with f_L, the ultrasound-modulated light can pass through the crystal without absorption, and be detected by a detector. The strong suppression of unmodulated light provides a high signal-to-noise ratio.

The experimental setup of the SHB UOT system is shown in Figure 21.19. The SHB crystal is a $10 \times 9 \times 1.5$ mm^3 2%@ Tm^{3+}:YAG crystal working at 793 nm. The optical absorption length is measured as $\alpha(f)L = 4$, yielding a DC suppression of 18 dB (Beer's law $I_{DC}^{out} = I_{DC}^{in}\exp(-\alpha L)$). The laser beam (a continuous Ti:Sapphire laser, 793.38 nm, the laser power of 2 W) is first modulated at 70 MHz with an acousto-optic modulator (AOM), labeled as AOMRef in Figure 21.19, generating 980 mW with 3.3 ms long optical pulses. These pulses are used as the pump beam and expanded to cover the whole crystal surface. The crystal is cryogenically cooled to ~4.2 K. The pump beam engraves transparent spectral holes across the crystal ($\alpha(70 \text{ MHz})L = 0.9$, and $\alpha(f \neq 70 \text{ MHz})L = 4$, corresponding to a 14 dB transmission improvement at 70 MHz) with a lifetime of about 10 ms. Once the spectral hole is generated, the AOMRef

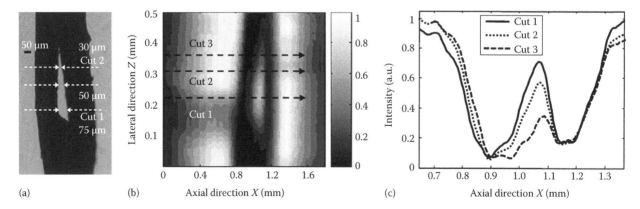

(a)   (b)   Axial direction X (mm)   (c)   Axial direction X (mm)

FIGURE 21.15 Imaging an optically absorptive patterned target located 2.0 mm inside the tissue mimicking phantom, using 75 MHz transducer. (a) Photograph of the target marked with two different sizes of gap, cut 1 = 75 µm and cut 2 = 30 µm (b) B-scan UOT image of (a) marked with the respective cuts and cut 3 = 15 µm. (c) Three 1D axial profiles, from the intensity data in (b), corresponding to three different cuts shown in (b). (From Kothapalli, S.R. and Wang, L.H.V., *J. Biomed. Opt.*, 13(5), 054046, 2008. With permission.)

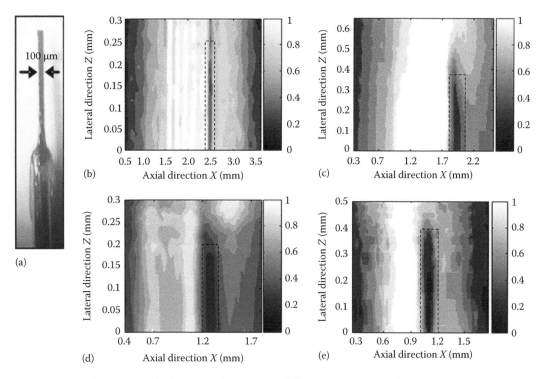

FIGURE 21.16 UOT images of a 100 μm wide absorptive object, using all four transducers. (a) Photograph of the object. B-scan UOT images of (a) inside the chicken breast tissue sample obtained with, (b) the 15 MHz transducer at 2.5 mm deep, (c) 30 MHz transducer at 2 mm deep, (d) 50 MHz transducer at 1.5 mm deep, and (e) 75 MHz transducer at 1 mm deep. In all of the above B-scan images, the object is enclosed with dotted rectangular box. (From Kothapalli, S.R. and Wang, L.H.V., *J. Biomed. Opt.*, 13(5), 054046, 2008. With permission.)

TABLE 21.1 Summary of Ultrasonic Parameters, Measured Axial and Lateral Resolutions, and Imaging Depth in UOT Experiments for All Four Transducers

Frequency (MHz)	Band Width (%)	Focal Width (μm)	Focal Depth (mm)	Resolution Using Edge-Spread Function		Imaging Depth (mm)
				Axial (μm)	Lateral (μm)	
15	100	90	4.7	74	140	3.0
30	80	55	5.5	60	70	2.5
50	80	33	5.5	48	55	2.3
75	85	25	3.0	31	38	1.8

Source: Kothapalli, S.R. and Wang, L.H.V., *J. Biomed. Opt.*, 13(5), 054046, 2008. With permission.

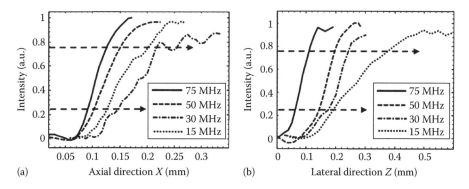

FIGURE 21.17 Measurement of the axial and lateral resolutions of UOT images shown in Figure 23.16, obtained with all four transducers. (a) Axial edge spread functions and (b) lateral edge spread functions of UOT signals for the respective high-frequency ultrasound transducers. (From Kothapalli, S.R. and Wang, L.H.V., *J. Biomed. Opt.* 13(5), 054046, 2008. With permission.)

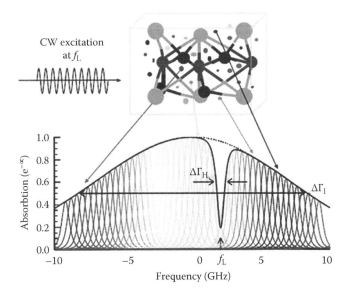

FIGURE 21.18 Normalized spectral absorption profile of an SHB crystal with a spectral hole burned at f_L. $\Delta\Gamma_I$: inhomogeneous bandwidth. $\Delta\Gamma_H$: homogeneous linewidth. (From Li, Y. et al., *Opt. Express*, 16, 14862, 2008b. With permission.)

is turned off and another AOM (AOMSam) is turned on with a driven frequency at 75 MHz. The modulated light illuminates a tissue mimicking phantom. Two cycles of ultrasound pulses with a peak pressure of 4.3 MPa are simultaneously applied into the phantom through a spherically focused transducer (central frequency, 5 MHz; focal length, 16.2 mm; focal spot size, 0.5 mm). The transmitted diffusive light has three major frequency components: two ultrasound-modulated light at 70 and 80 MHz, and strong unmodulated light at 75 MHz. Since the spectral hole is generated at 70 MHz, the ultrasound-modulated light at 70 MHz can transmit through the crystal and be detected by a Si detector, and the other two components are significantly suppressed.

Figure 21.20 shows B-scan images of an optical absorptive patterned object embedded inside 1 cm thick optically scattering medium. The target is clearly imaged with only 16 times averaging in Figure 21.20b. As shown in Figure 21.20c, a fair image can be obtained with only 4 times averaging, so the acquisition time per A-line can potentially be 80 µs. In addition, a thicker crystal or a crystal with higher doping can further reject the unmodulated light.

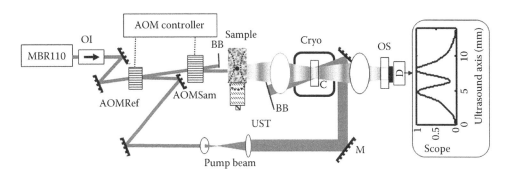

FIGURE 21.19 Experimental setup of the SHB-based detection technique. OI, optical isolator; AOM, acousto-optic modulator; UST, ultrasound transducer; Cryo, cryostat; C, crystal; BB, beam block; M, mirror; OS, optical shutter; D, detector. (From Li, Y.Z. et al., *Appl. Phys. Lett.*, 93(1), 11111, 2008c. With permission.)

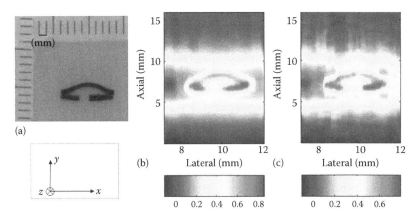

FIGURE 21.20 B-scan UOT images. (a) Photograph of the optical absorber buried inside the tissue phantom. (b) B-scan UOT image with 16 times averaging. (c) B-scan UOT image with 4 times averaging. (From Li, Y.Z. et al., *Appl. Phys. Lett.*, 93(1), 11111, 2008c. With permission.)

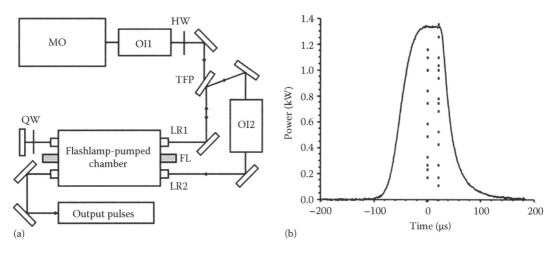

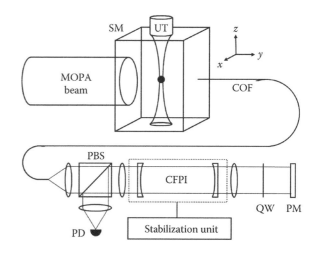

FIGURE 21.21 (a) Schematic of the Nd:YAG master oscillator power amplifier (MOPA). MO: continuous-wave master oscillator, OI1-2: optical isolators, LR1-2: Nd:YAG laser rods, FL: flashlamp, HW: half-wave plate, QW: quarter-wave plate, TFP: thin-film polarizer. Other components are plane dielectric mirrors. (b) Typical pulse intensity profile at the exit of the MOPA. The useful part of the pulse is enclosed between the dotted lines. (From Rousseau, G. et al., Ultrasound-modulated optical imaging using a confocal Fabry-Perot interferometer and a powerful long pulse laser, *Proceedings of SPIE, Photons Plus Ultrasound: Imaging and Sensing 2009*, 7177, 71771E 1–9, 2009. With permission.)

21.9 The Use of a Powerful Long Coherent Pulsed Laser in UOT (Rousseau et al. 2008, 2009)

A direct way to enhance ultrasound-modulated light is to increase the incident optical power. However, to keep the maximum permeable exposure (MPE) within the ANSI safety limit in terms of average power, the increase in laser power forces to reduce the laser exposure time. By using a train of laser pulses with a high peak power and an appropriate duty cycle, the amount of light can be significantly increased only during the time of flight of ultrasound in the medium. For a given average optical power, the use of a low-duty cycle laser pulse train has a benefit in SNR, and ultimately shot-noise-limited detection will be possible. By amplifying a single-frequency continuous-wave laser beam with a flash lamp–pumped gain-switched amplifier, a powerful long coherent pulsed laser can be implemented. Figure 21.21a shows the schematic of the master oscillator power amplifier (MOPA). Figure 21.21b presents a typical pulse intensity profile from the MOPA. More than 1 kW can be attained at a repetition rate of 25 Hz. The pulse duration is adjustable and typically is set at 90 μs to attain the maximum power. A 20-μs long part of the pulse (between the dotted line in Figure 21.21b) is typically used for UOT experiments, and the duration would be appropriate for the time of flight of ultrasound in biological tissues.

In the first implementation of the CFPI-based detection scheme (mentioned in Section 21.8.2), only high-frequency ultrasound can be used to reject the unmodulated light. With the single-pass configuration in CFPI, the detection of faint spectral lines close to a strong one is limited by the technically attainable values of the finesse and resonator length. Double-pass configuration can resolve this difficulty. Figure 21.22 shows the experimental setup of a single-frequency pulse laser and a double-pass

CFPI. The laser pulses (1064 nm) emitted by a Nd:YAG master MOPA illuminate the optically scattering medium (SM) while being insonified with a 5 MHz focused ultrasonic transducer (19 mm diameter, 45 mm focal length, 10 cycles). The transmitted diffusive light from the SM is collected by a multimode collecting optical fiber (COF: 1 mm core diameter, numerical aperture: 0.36) and sent toward the double-pass CFPI. The light exiting from the COF is collimated by a lens and polarized by a polarizing

FIGURE 21.22 Experimental setup of the double-pass CFPI-based detection technique. MOPA, master oscillator power amplifier; SM, scattering medium; UT, ultrasonic transducer; COF, collecting optical fiber; PBS, polarizing beam splitter; CFPI, confocal Fabry–Perot interferometer; QW, quarter-wave plate; PM, plane mirror; PD, photodetector. (From Rousseau, G. et al., Ultrasound-modulated optical imaging using a confocal Fabry-Perot interferometer and a powerful long pulse laser, *Proceedings of SPIE, Photons Plus Ultrasound: Imaging and Sensing 2009*, 7177, 71771E 1–9, 2009. With permission.)

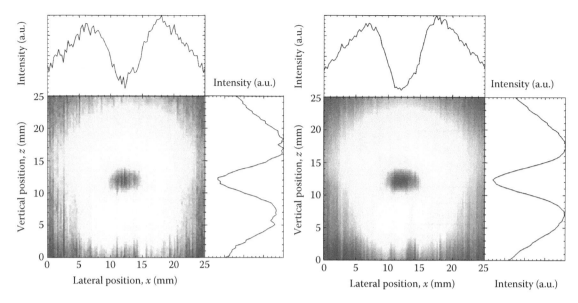

FIGURE 21.23 B-scan UOT images of an optically absorbing cube embedded in a 30 mm thick scattering medium. The vertical A-line (curves on right) and the horizontal profile (upper curves) passing through the optically absorbing object are also shown. The image on the left was obtained without any single averaging, whereas the right image was obtained with 16 times averaging. (From Rousseau, G. et al., Ultrasound-modulated optical imaging using a confocal Fabry-Perot interferometer and a powerful long pulse laser, *Proceedings of SPIE, Photons Plus Ultrasound: Imaging and Sensing 2009,* 7177, 71771E 1–9, 2009. With permission.)

beam splitter. The light beam is then coupled in the 1 m long CFPI (free spectral range, 75 MHz; finesse, 28). For double-pass configuration, the transmitted light from the CFPI is reflected back to CFPI. A quarter-wave plate rotates the polarization state of the reflected beam by 90°. After the second pass, the filtered light is detected by an InGaAs-PIN photodiode. The *etendue* of the CFPI (0.38 sr mm^2) matches that of the COF (0.33 sr mm^2). In the double-pass configuration, the full width at half maximum of the spectral filter is improved to 1.8 MHz. The suppression of the unmodulated light is 22 dB when the transmission peak of CFPI is positioned at 5 MHz, and 33 dB at 10 MHz.

A liquid optical phantom (60 mm × 30 mm × 90 mm along the *X*, *Y*, and *Z* axes, respectively) based on sunflower oil (optical absorbers) and titanium dioxide particles (optical scatterers) is used. An optically absorptive cube of polyvinyl alcohol cryogel (PVA-C) is used as a target. The maximum pressure of ~1.9 MPa is applied into the medium, so the mechanical index (MI) at this frequency is 0.85, a value below the safety limit of 1.9. The scattering medium is typically illuminated on a 30 mm diameter surface with 25 mJ pulses (energy in the useful part of the pulse: 20 μs duration and 1.3 kW peak power) at a repetition rate of 25 Hz. Consequently, the average irradiance is ~90 mW/cm^2, which is below the ANSI limit (1 W/cm^2 at 1064 nm). Figure 21.23 shows B-scan UOT images of an optical absorptive and acoustically transparent cube. It can be seen from Figure 21.23 (left) that the absorber is clearly imaged without any averaging (single pulse per A-line). The image in Figure 21.23 (right) is obtained by averaging 16 A-lines. The SNR is clearly acceptable without averaging.

Figure 21.24 shows a B-scan UOT image of a 60 mm thick phantom containing an optically absorptive cube. Each A-line

is averaged 256 times. The optically absorptive cube is clearly seen. The incident laser power and acoustic pressure can be increased without exceeding both optical and acoustic safety limits. Consequently, this approach can potentially provide in vivo imaging at deep tissues.

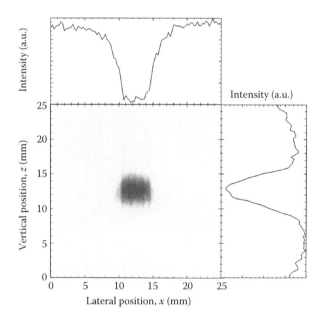

FIGURE 21.24 B-scan UOT images of an optically absorbing cube embedded in a 60 mm thick scattering medium. (From Rousseau, G. et al., Ultrasound-modulated optical imaging using a confocal Fabry-Perot interferometer and a powerful long pulse laser, *Proceedings of SPIE, Photons Plus Ultrasound: Imaging and Sensing 2009,* 7177, 71771E 1–9, 2009. With permission.)

21.10 Summary and Discussion on UOT

UOT is a novel hybrid bio-photonic imaging technique that can supply strong optical contrast and high ultrasonic resolution. By varying the ultrasonic parameters, the imaging depth and the spatial resolution are scalable, within the reach of diffusive photons. The major challenge in UOT is the low SNR due to uncorrelated speckle grains and low modulation efficiency. To overcome this problem, many detection techniques have been developed including PRC-based detection, CFPI-based detection, and SHB-based detection. Although the PRC-based detection has a larger *etendue*, its applicability is still limited by the response time, $\tau_{PR} \approx 100$ ms, of a bismuth silicon oxide crystal, since the speckle decorrelation time is less than 1 ms in vivo. CFPI- and SHB-based techniques have been employed as real-time frequency filters of one sideband of the UOT signals, while suppressing the unmodulated light and the other modulation sideband. Both detection techniques are immune to speckle decorrelation and are able to process parallel speckles. However, CFPI-based detection has relatively small *etendue* compared to PRC- and SHB-based detections. By employing a powerful long coherence laser, SNR in UOT can be improved significantly, and further, the maximum imaging depth and spatial resolution can also be improved. In addition, the laser can be adapted easily to the other detection methods such as SHB-based detection.

References

Agarwal, A., Huang, S. W., O'Donnell, M., Day, K. C., Day, M., Kotov, N., and Ashkenazi, S. 2007. Targeted gold nanorod contrast agent for prostate cancer detection by photoacoustic imaging. *J. Appl. Phys.* 102: 064701.

Anastasio, M. A., Zhang, J., Modgil, D., and la Riviere, P. J. 2007. Application of inverse source concepts to photoacoustic tomography. *Inverse Probl.* 23: S21–S35.

ANSI. 2000. *American National Standard for Safe Use of Lasers*, ed. A. N. S. In Institute, Z136.1-2000. New York: ANSI Standard.

Beard, P. C., Perennes, F., and Mills, T. N. 1999. Transduction mechanisms of the Fabry-Perot polymer film sensing concept for wideband ultrasound detection. *IEEE Trans. Ultrason. Ferroelectr. Freq. Control* 46: 1575–1582.

Bell, A. G. 1880. On the production of sound by light. *Am. J. Sci.* 20: 305.

Bossy, E., Funke, A. R., Daoudi, K., Boccara, A. C., Tanter, M., and Fink, M. 2007. Transient optoelastography in optically diffusive media. *Appl. Phys. Lett.* 90: 174111.

Burgholzer, P., Hofer, C., Paltauf, G., Haltmeier, M., and Scherzer, O. 2005. Thermoacoustic tomography with integrating area and line detectors. *IEEE Trans. Ultrason. Ferroelectr. Freq. Control* 52: 1577–1583.

Burgholzer, P., Matt, G. J., Haltmeier, M., and Paltauf, G. 2007. Exact and approximative imaging methods for photoacoustic tomography using an arbitrary detection surface. *Phys. Rev. E* 75: 046706.

Chen, Z. X., Tang, Z. L., and Wan, W. 2007. Photoacoustic tomography imaging based on a 4f acoustic lens imaging system. *Opt. Express* 15: 4966–4976.

Chen, Z. X., Tang, Z. L., Wan, W., and He, Y. H. 2006. Photoacoustic tomography imaging based on an acoustic lens imaging system. *Acta Phys. Sin.* 55: 4365–4370.

Copland, J. A., Eghtedari, M., Popov, V. L., Kotov, N., Mamedova, N., Motamedi, M., and Oraevsky, A. A. 2004. Bioconjugated gold nanoparticles as a molecular based contrast agent: Implications for imaging of deep tumors using optoacoustic tomography. *Mol. Imaging Biol.* 6: 341–349.

Cox, B. T., Arridge, S. R., and Beard, P. C. 2009. Estimating chromophore distributions from multiwavelength photoacoustic images. *J. Opt. Soc. Am. A* 26: 443–455.

De La Zerda, A., Zavaleta, C., Keren, S., Vaithilingam, S., Bodapati, S., Liu, Z., Levi, J., Smith, B. R., Ma, T. J., Oralkan, O., Cheng, Z., Chen, X. Y., Dai, H. J., Khuri-Yakub, B. T., and Gambhir, S. S. 2008. Carbon nanotubes as photoacoustic molecular imaging agents in living mice. *Nat. Nanotechnol.* 3: 557–562.

Department of Health and Human Services. 2004. Food and Drug Administration Modernization Act of 1997; Modifications to the list of recognized standards, recognition list number: 011, *Federal Register* 69 (191), 59240–59250.

Eghtedari, M., Oraevsky, A., Copland, J. A., Kotov, N. A., Conjusteau, A., and Motamedi, M. 2007. High sensitivity of in vivo detection of gold nanorods using a laser optoacoustic imaging system. *Nano Lett.* 7: 1914–1918.

Ermilov, S. A., Khamapirad, T., Conjusteau, A., Leonard, M. H., Lacewell, R., Mehta, K., Miller, T., and Oraevsky, A. A. 2009. Laser optoacoustic imaging system for detection of breast cancer. *J. Biomed. Opt.* 14: 024007.

Finch, D., Patch, S. K., and Rakesh 2004. Determining a function from its mean values over a family of spheres. *SIAM J. Math. Anal.* 35: 1213–1240.

Gamelin, J., Aguirre, A., Maurudis, A., Huang, F., Castillo, D., Wang, L. V., and Zhu, Q. 2008. Curved array photoacoustic tomographic system for small animal imaging. *J. Biomed. Opt.* 13: 024007.

Gusev, V. E., Karabutov, A. A., and Hendzel, K. 1993. *Laser Optoacoustics*. New York: AIP Press.

Haltmeier, M., Scherzer, O., Burgholzer, P., and Paltauf, G. 2004. Thermoacoustic computed tomography with large planar receivers. *Inverse Probl.* 20: 1663–1673.

IEEE. 1999. *IEEE Standard for Safety Levels with Respect to Human Exposure to Radio Frequency Electromagnetic Fields 3 kHz to 300 GHz*, ed., The Institute of Electrical and Electronics Engineers, Inc., New York.

Karabutov, A. A., Podymova, N. B., and Letokhov, V. S. 1996. Time-resolved laser optoacoustic tomography of inhomogeneous media. *Appl. Phys. B: Lasers Opt.* 63: 545–563.

Kim, C., Zemp, R. J., and Wang, L. H. V. 2006. Intense acoustic bursts as a signal-enhancement mechanism in ultrasound-modulated optical tomography. *Opt. Lett.* 31: 2423–2425.

Kim, C. H. and Wang, L. V. 2007. Multi-optical-wavelength ultrasound-modulated optical tomography: A phantom study. *Opt. Lett.* 32: 2285–2287.

Kothapalli, S. R., Sakadzic, S., Kim, C., and Wang, L. V. 2007. Imaging optically scattering objects with ultrasound-modulated optical tomography. *Opt. Lett.* 32: 2351–2353.

Kothapalli, S. R. and Wang, L. H. V. 2009. Ex vivo blood vessel imaging using ultrasound-modulated optical microscopy. *J. Biomed. Opt.* 14(1): 014015.

Kothapalli, S. R. and Wang, L. V. H. 2008. Ultrasound-modulated optical microscopy. *J. Biomed. Opt.* 13(5): 054046.

Kruger, R. A. 1994. Photoacoustic ultrasound. *Med. Phys.* 21: 127–131.

Kruger, R. A., Kiser, W. L., Reinecke, D. R., and Kruger, G. A. 2003. Thermoacoustic computed tomography using a conventional linear transducer array. *Med. Phys.* 30: 856–860.

Kruger, R. A., Reinecke, D. R., and Kruger, G. A. 1999. Thermoacoustic computed tomography-technical considerations. *Med. Phys.* 26: 1832–1837.

Ku, G. and Wang, L. H. V. 2005. Deeply penetrating photoacoustic tomography in biological tissues enhanced with an optical contrast agent. *Opt. Lett.* 30: 507–509.

Kuchment, P. and Kunyansky, L. 2008. A survey in mathematics for industry: Mathematics of thermoacoustic tomography. *Eur. J. Appl. Math.* 19: 191–224.

Kunyansky, L. A. 2007a. Explicit inversion formulae for the spherical mean Radon transform. *Inverse Probl.* 23: 373–383.

Kunyansky, L. A. 2007b. A series solution and a fast algorithm for the inversion of the spherical mean Radon transform. *Inverse Probl.* 23: S11–S20.

Laufer, J., Delpy, D., Elwell, C., and Beard, P. 2007. Quantitative spatially resolved measurement of tissue chromophore concentrations using photoacoustic spectroscopy: Application to the measurement of blood oxygenation and haemoglobin concentration. *Phys. Med. Biol.* 52: 141–168.

Leutz, W. and Maret, G. 1995. Ultrasonic modulation of multiply scattered-light. *Physica B* 204: 14–19.

Leveque, S., Boccara, A. C., Lebec, M., and Saint-Jalmes, H. 1999. Ultrasonic tagging of photon paths in scattering media: parallel speckle modulation processing. *Opt. Lett.* 24: 181–183.

Li, C. and Wang, L., V. 2009. Photoacoustic tomography and sensing in biomedicine. *Phys. Med. Biol.* 54: R59.

Li, C. H., Pramanik, M., Ku, G., and Wang, L. V. 2008a. Image distortion in thermoacoustic tomography caused by microwave diffraction. *Phys. Rev. E* 77: 031923.

Li, J., Ku, G., and Wang, L. H. V. 2002. Ultrasound-modulated optical tomography of biological tissue by use of contrast of laser speckles. *Appl. Opt.* 41: 6030–6035.

Li, L., Zemp, R. J., Lungu, G., Stoica, G., and Wang, L. H. V. 2007a. Photoacoustic imaging of lacZ gene expression in vivo. *J. Biomed. Opt.* 12: 020504.

Li, P. C., Huang, S. W., Wei, C. W., Chiou, Y. C., Chen, C. D., and Wang, C. R. C. 2005. Photoacoustic flow measurements by use of laser-induced shape transitions of gold nanorods. *Opt. Lett.* 30: 3341–3343.

Li, P. C., Wei, C. W., Liao, C. K., Chen, C. D., Pao, K. C., Wang, C. R. C., Wu, Y. N., and Shieh, D. B. 2007b. Photoacoustic imaging of multiple targets using gold nanorods. *IEEE Trans. Ultrason. Ferroelectr. Freq. Control* 54: 1642–1647.

Li, Y., Hemmer, P., Kim, C., Zhang, H., and Wang, L. V. 2008b. Detection of ultrasound-modulated diffuse photons using spectral-hole burning. *Opt. Express* 16: 14862–14874.

Li, Y. Z., Zhang, H. L., Kim, C. H., Wagner, K. H., Hemmer, P., and Wang, L. V. 2008c. Pulsed ultrasound-modulated optical tomography using spectral-hole burning as a narrowband spectral filter. *Appl. Phys. Lett.* 93(1): 11111.

Maeda, H., Fang, J., Inutsuka, T., and Kitamoto, Y. 2003. Vascular permeability enhancement in solid tumor: Various factors, mechanisms involved and its implications. *Int. Immunopharmacol.* 3: 319–328.

Marks, F. A., Tomlinson, H. W., and Brooksby, G. W. 1993. A comprehensive approach to breast cancer detection using light: photon localization by ultrasound modulation and tissue characterization by spectral discrimination. *Proceedings of SPIE, Photon Migration and Imaging in Random Media and Tissues*, Vol. 1888, pp. 500–510.

Maslov, K., Stoica, G., and Wang, L. V. H. 2005. In vivo dark-field reflection-mode photoacoustic microscopy. *Opt. Lett.* 30: 625–627.

Maslov, K. and Wang, L. V. 2008. Photoacoustic imaging of biological tissue with intensity-modulated continuous-wave laser. *J. Biomed. Opt.* 13: 024006.

Maslov, K., Zhang, H. F., Hu, S., and Wang, L. V. 2008. Optical-resolution photoacoustic microscopy for in vivo imaging of single capillaries. *Opt. Lett.* 33: 929–931.

Maugh II, T. H. 1975. Photoacoustic spectroscopy: New uses for an old technique. *Science* 188: 38–39.

Morse, P. M. and Ingard, K. U. 1986. *Theoretical Acoustics.* Princeton, NJ: Princeton University Press.

Murray, T. W., Sui, L., Maguluri, G., Roy, R. A., Nieva, A., Blonigen, F., and Dimarzio, C. A. 2004. Detection of ultrasound-modulated photons in diffuse media using the photorefractive effect. *Opt. Lett.* 29: 2509–2511.

Nie, L. M., Xing, D., Yang, D. W., and Zeng, L. M. 2007. Microwave-induced thermoacoustic imaging enhanced with a microwave contrast agent. *IEEE/ICME International Conference on Complex Medical Engineering, 2007 (CME 2007)*, Beijing, China, 2007.

Niederhauser, J. J., Jaeger, M., and Frenz, M. 2004. Real-time three-dimensional optoacoustic imaging using an acoustic lens system. *Appl. Phys. Lett.* 85: 846–848.

Niederhauser, J. J., Jaeger, M., Lemor, R., Weber, P., and Frenz, M. 2005. Combined ultrasound and optoacoustic system for real-time high-contrast vascular imaging in vivo. *IEEE Trans. Med. Imaging* 24: 436–440.

Oraevsky, A. and Karabutov, A. 2003. Optoacoustic tomography. In: *Biomedical Photonics Handbook*, ed., T. Vo-Dinh Boca Raton, FL: CRC Press.

Oraevsky, A. A., Ermilov, S. A., Conjusteau, A., Miller, T., Gharieb, R. R., Lacewell, R., Mehta, K., Radulescu, E. G., Herzog,

D., Thompson, S., Stein, A., Mccorvey, M., Otto, P., and Khamapirad, T. 2007. Initial clinical evaluation of laser optoacoustic imaging system for diagnostic imaging of breast cancer. *Breast Cancer Res. Treat.* 106: S47–S47.

Oraevsky, A. A., Jacques, S. L., and Tittel, F. K. 1997. Measurement of tissue optical properties by time-resolved detection of laser-induced transient stress. *Appl. Opt.* 36: 402–415.

Paltauf, G., Nuster, R., Haltmeier, M., and Burgholzer, P. 2007a. Experimental evaluation of reconstruction algorithms for limited view photoacoustic tomography with line detectors. *Inverse Probl.* 23: S81–S94.

Paltauf, G., Nuster, R., Haltmeier, M., and Burgholzer, P. 2007b. Photoacoustic tomography using a Mach-Zehnder interferometer as an acoustic line detector. *Appl. Opt.* 46: 3352–3358.

Paltauf, G. and Schmidtkloiber, H. 1997. Measurement of laser-induced acoustic waves with a calibrated optical transducer. *J. Appl. Phys.* 82: 1525–1531.

Paltauf, G., Viator, J. A., Prahl, S. A., and Jacques, S. L. 2002. Iterative reconstruction algorithm for optoacoustic imaging. *J. Acoust. Soc. Am.* 112: 1536–1544.

Patch, S. K. and Scherzer, O. 2007. Photo- and thermo-acoustic Imaging. *Inverse Probl.* 23: S1–S10.

Pramanik, M., Swierczewska, M., Green, D., Sitharaman, B., and Wang, L. H. V. 2009. Single-walled carbon nanotubes as a multimodal—thermoacoustic and photoacoustic—contrast agent. *J. Biomed. Opt.* 14: 034018.

Rosencwaig, A. 1980. *Photoacoustics and Photoacoustic Spectroscopy.* New York: Wiley.

Rousseau, G., Blouin, A., and Monchalin, J. P. 2008. Ultrasound-modulated optical imaging using a powerful long pulse laser. *Opt. Express* 16: 12577–12590.

Rousseau, G., Blouin, A., and Monchalin, J. P. 2009. Ultrasound-modulated optical imaging using a confocal Fabry-Perot interferometer and a powerful long pulse laser. *Proceedings of SPIE, Photons Plus Ultrasound: Imaging and Sensing 2009,* 7177, 71771E 1–9.

Sakadzic, S. and Wang, L. H. V. 2004. High-resolution ultrasound-modulated optical tomography in biological tissues. *Opt. Lett.* 29: 2770–2772.

Sakadzic, S. and Wang, L. H. V. 2006. Correlation transfer and diffusion of ultrasound-modulated multiply scattered light. *Phys. Rev. Lett.* 96(16): 163902.

Song, K. H., Kim, C. H., Cobley, C. M., Xia, Y. N., and Wang, L. V. 2009. Near-infrared gold nanocages as a new class of tracers for photoacoustic sentinel lymph node mapping on a rat model. *Nano Lett.* 9: 183–188.

Song, K. H., Stein, E. W., Margenthaler, J. A., and Wang, L. V. 2008. Noninvasive photoacoustic identification of sentinel lymph nodes containing methylene blue in vivo in a rat model. *J. Biomed. Opt.* 13: 054033–054036.

Song, K. H. and Wang, L. V. 2007. Deep reflection-mode photoacoustic imaging of biological tissue. *J. Biomed. Opt.* 12: 060503.

Sui, L., Roy, R. A., Dimarzio, C. A., and Murray, T. W. 2005. Imaging in diffuse media with pulsed-ultrasound-modulated light and the photorefractive effect. *Appl. Opt.* 44: 4041–4048.

Telenkov, S. A. and Mandelis, A. 2006. Fourier-domain biophotoacoustic subsurface depth selective amplitude and phase imaging of turbid phantoms and biological tissue. *J. Biomed. Opt.* 11: 044006.

Wang, L. H., Jacques, S. L., and Zhao, X. M. 1995. Continuous-wave ultrasonic modulation of scattered laser-light to image objects in turbid media. *Opt. Lett.* 20: 629–631.

Wang, L. H. V. 2001a. Mechanisms of ultrasonic modulation of multiply scattered coherent light: a Monte Carlo model. *Opt. Lett.* 26: 1191–1193.

Wang, L. H. V. 2001b. Mechanisms of ultrasonic modulation of multiply scattered coherent light: An analytic model. *Phys. Rev. Lett.* 87(4): 043903.

Wang, L. H. V. 2003. Ultrasound-mediated biophotonic imaging: A review of acousto-optical tomography and photo-acoustic tomography. *Dis. Markers* 19: 123–138.

Wang, L. H. V., Zhao, X. M., Sun, H. T., and Ku, G. 1999. Microwave-induced acoustic imaging of biological tissues. *Rev. Sci. Instrum.* 70: 3744–3748.

Wang, L. H. V. 2009. Multiscale photoacoustic microscopy and computed tomography. *Nat. Photonics* 3: 503–509.

Wang, X. D., Ku, G., Wegiel, M. A., Bornhop, D. J., Stoica, G., and Wang, L. H. V. 2004a. Noninvasive photoacoustic angiography of animal brains in vivo with near-infrared light and an optical contrast agent. *Opt. Lett.* 29: 730–732.

Wang, X. D., Pang, Y. J., Ku, G., Xie, X. Y., Stoica, G., and Wang, L. H. V. 2003. Noninvasive laser-induced photoacoustic tomography for structural and functional in vivo imaging of the brain. *Nat. Biotechnol.* 21: 803–806.

Wang, X. D., Xie, X. Y., Ku, G. N., and Wang, L. H. V. 2006. Noninvasive imaging of hemoglobin concentration and oxygenation in the rat brain using high-resolution photoacoustic tomography. *J. Biomed. Opt.* 11: 024015.

Wang, Y. W., Xie, X. Y., Wang, X. D., Ku, G., Gill, K. L., O'Neal, D. P., Stoica, G., and Wang, L. V. 2004b. Photoacoustic tomography of a nanoshell contrast agent in the in vivo rat brain. *Nano Lett.* 4: 1689–1692.

Witte, R. S., Kim, K., Agarwal, A., Fan, W., Kopelman, R., Kotov, N., Kipke, D., and O'Donnell, M. 2008. Enhanced photoacoustic neuroimaging with gold nanorods and PEBBLEs. *Proceedings of the SPIE—The International Society for Optical Engineering,* San Jose, CA, 2008.

Xiang, L. Z., Xing, D., Gu, H. M., Yang, D. W., Yang, S. H., Zeng, L. M., and Chen, W. R. 2007. Real-time optoacoustic monitoring of vascular damage during photodynamic therapy treatment of tumor. *J. Biomed. Opt.* 12.

Xu, M. and Wang, L. V. 2002. Time-domain reconstruction for thermoacoustic tomography in a spherical geometry. *IEEE Trans. Med. Imaging* 21: 814–822.

Xu, M. and Wang, L. V. 2006. Photoacoustic imaging in biomedicine. *Rev. Sci. Instrum.* 77: 41101–41122.

Xu, M. H. and Wang, L. H. V. 2005. Universal back-projection algorithm for photoacoustic computed tomography. *Phys. Rev. E* 71: 016706.

Xu, M. H. and Wang, L. V. 2003. Analytic explanation of spatial resolution related to bandwidth and detector aperture size in thermoacoustic or photoacoustic reconstruction. *Phys. Rev. E* 67: 056605.

Xu, M. H. and Wang, L. V. 2007. Universal back-projection algorithm for photoacoustic computed tomography (vol 71, art no 016706, 2005). *Phys. Rev. E* 75: 059903.

Xu, Y., Feng, D., and Wang, L. V. 2002a. Exact frequency-domain reconstruction for thermoacoustic tomography. I. Planar geometry. *IEEE Trans. Med. Imaging* 21: 823–828.

Xu, Y. and Wang, L. H. V. 2004. Time reversal and its application to tomography with diffracting sources. *Phys. Rev. Lett.* 92: 033902.

Xu, Y., Wang, L. V., Ambartsoumian, G., and Kuchment, P. 2004. Reconstructions in limited-view thermoacoustic tomography. *Med. Phys.* 31: 724–733.

Xu, Y., Xu, M., and Wang, L. V. 2002b. Exact frequency-domain reconstruction for thermoacoustic tomography. II. Cylindrical geometry. *IEEE Trans. Med. Imaging* 21: 829–833.

Yang, D. W., Xing, D., Yang, S. H., and Xiang, L. Z. 2007a. Fast full-view photoacoustic imaging by combined scanning with a linear transducer array. *Opt. Express* 15: 15566–15575.

Yang, S., Xing, D., Zhou, Q., Xiang, L., and Lao, Y. 2007b. Functional imaging of cerebrovascular activities in small animals using high-resolution photoacoustic tomography. *Med. Phys.* 34: 3294–3301.

Yang, X. M., Skrabalak, S. E., Li, Z. Y., Xia, Y. N., and Wang, L. H. V. 2007c. Photoacoustic tomography of a rat cerebral cortex in vivo with au nanocages as an optical contrast agent. *Nano Lett.* 7: 3798–3802.

Yang, X. M. and Wang, L. V. 2008. Monkey brain cortex imaging by photoacoustic tomography. *J. Biomed. Opt.* 13: 044009.

Yin, L., Wang, Q., Zhang, Q. Z., and Jiang, H. B. 2007. Tomographic imaging of absolute optical absorption coefficient in turbid media using combined photoacoustic and diffusing light measurements. *Opt. Lett.* 32: 2556–2558.

Zeff, B. W., White, B. R., Dehghani, H., Schlaggar, B. L., and Culver, J. P. 2007. Retinotopic mapping of adult human visual cortex with high-density diffuse optical tomography. *Proc. Natl. Acad. Sci. U. S. A.* 104: 12169–74.

Zemp, R. J., Bitton, R., Li, M. L., Shung, K. K., Stoica, G., and Wang, L. V. 2007. Photoacoustic imaging of the microvasculature with a high-frequency ultrasound array transducer. *J. Biomed. Opt.* 12: 010501.

Zemp, R. J., Song, L. A., Bitton, R., Shung, K. K., and Wang, L. H. V. 2008. Realtime photoacoustic microscopy in vivo with a 30-MHz ultrasound array transducer. *Opt. Express* 16: 7915–7928.

Zhang, E., Laufer, J., and Beard, P. 2008a. Backward-mode multiwavelength photoacoustic scanner using a planar Fabry-Perot polymer film ultrasound sensor for high-resolution three-dimensional imaging of biological tissues. *Appl. Opt.* 47: 561–577.

Zhang, H., Tang, Z., He, Y., and Guo, L. 2007a. Two dimensional photoacoustic imaging based on an acoustic lens and the peak-hold technology. *Rev. Sci. Instrum.* 78: 064902–4.

Zhang, H. F., Maslov, K., Sivaramakrishnan, M., Stoica, G., and Wang, L. H. V. 2007b. Imaging of hemoglobin oxygen saturation variations in single vessels in vivo using photoacoustic microscopy. *Appl. Phys. Lett.* 90: 053901.

Zhang, H. F., Maslov, K., Stoica, G., and Wang, L. H. V. 2006. Functional photoacoustic microscopy for high-resolution and noninvasive in vivo imaging. *Nat. Biotechnol.* 24: 848–851.

Zhang, Q. Z., Liu, Z., Carney, P. R., Yuan, Z., Chen, H. X., Roper, S. N., and Jiang, H. B. 2008b. Non-invasive imaging of epileptic seizures in vivo using photoacoustic tomography. *Phys. Med. Biol.* 53: 1921–1931.

22

Optical and Opto-Acoustic Molecular Tomography

22.1 Introduction 443
22.2 Reporter Technologies 444
    Direct Fluorescence Imaging • Indirect Fluorescence Imaging • Other Photonic Reporter Technologies
22.3 Small Animal Opto-Acoustic Molecular Tomography 446
    Multispectral Opto-Acoustic Tomography • MSOT Small Animal Imaging of Fluorochromes • MSOT Technology Components
22.4 Small Animal Hybrid Optical Tomography 451
    Calculation of the Forward Problem • Inversion • Use of Priors • Hybrid Fluorescence Molecular Tomography
22.5 Discussion 454
Acknowledgments 455
Glossary 455
References 456

Vasilis Ntziachristos
Technical University of Munich
and
Helmholtz Zentrum München

22.1 Introduction

This chapter summarizes emerging methods for small animal optical molecular imaging and it is based on an adaptation of the reviews of the subject in Deliolanis et al. (2011) and Ntziachristos and Razansky (2010). In particular, it presents current progress in high-performance fluorescence visualization in vivo through entire animal bodies and tissues. This is a field of the imaging sciences that is only currently maturing to robust methodologies for accurate, high-resolution photonic tissue imaging due to the use of hybrid approaches, such as the combination of anatomical and molecular imaging approaches or the use of opto-acoustics. The performance currently achieved with these methods has never been seen before and it is expected to revolutionize biological discovery and select clinical applications.

Fluorescence is a highly beneficial modality for biomedical imaging as it imparts versatile contrast of cellular and subcellular function and structure. As a result, it has been overwhelmingly utilized in the biomedical laboratory for performing in vitro assays, immunohistochemistry, or the superficial visualization of cells in vivo. The compelling advantages of fluorescence have more recently driven the development of powerful classes of fluorescent tags that can stain functional and molecular processes in vivo. The most widely acknowledged technology is naturally the 2008 Nobel-prize-awarded fluorescent protein (FP), which offers perhaps the most versatile tool for biological imaging (Giepmans et al. 2006). FPs attain the ability to tag cellular motility and subcellular process, from gene expression and signaling pathways to protein function and interactions, merging optimally with post-genomic "-omics" investigations and interrogating biology at the systems level. Promising new developments include the introduction of truly near-infrared (NIR)-shifted FPs, with excitation and emission spectra in the >650 nm (Shu et al. 2009). Such performance opens exciting possibilities for whole body animal imaging as it allows high-sensitivity imaging through several centimeters of tissue, due to the low photon attenuation by tissue in the 650–950 nm range, that is, the NIR spectral region. In parallel, a plethora of extrinsically administered probes are being developed, also operating in the NIR region (Tsien 2005, Weissleder and Pittet 2008). Fluorescent probes are agents that can probe tissue constituents and their function by staining in vivo certain classes of cells, receptors, proteases, and other moieties of cellular or subcellular activity. During the last decade, a large number of experimental and commercially available fluorescent agents and probes are increasingly being offered, from nonspecific fluorescent dyes and various FPs to targeted or activatable photoproteins and fluorogenic-substrate-sensitive fluorochromes to enable a highly potent field for biological imaging. So far, these contrast mechanisms were proven efficient in a number of small-animal applications but many of these agents attain strong potential for clinical translation as well. In addition, voltage sensitive dyes, fluorescence resonance energy transfer approaches, and lifetime measurements further allow the sensing of ions, protein–protein interactions, or the effects of the biochemical environment on the fluorochrome (Homma et al. 2009). Using fluorescence, therefore, previously invisible processes associated with tissue and disease growth and treatment can be sensed and visualized in real time and longitudinally. Naturally, fluorescence is increasingly used in basic biological discovery, drug discovery, and even considered for clinical studies of cancer, inflammation, neurodegenerative disease, cardiovascular disease, to name a few examples.

In vivo imaging of cell monolayers is typically achieved using conventional fluorescence microscopy. In the past two decades, significant progress has been achieved by using confocal and multiphoton microscopy to image deeper in tissues (Denk et al. 1990, Webb 1999). By offering technology that can account for scattering, confocal microscopy can reach depths of 100–200 microns when imaging in vivo whereas two- and multiphoton microscopy can penetrate some 500 microns. While these depths have yielded unparalleled insights into in vivo cellular function, these depths do not allow the sampling of entire structures—for example, entire tumors or visualize organs and events that are not superficial. Optical projection tomography (OPT) (Sharpe et al. 2002) and selective plane illumination microscopy (SPIM) (Huisken et al. 2004), developed recently as alternative volumetric imaging methods to confocal and nonlinear microscopy, are also strongly limited by tissue scattering. Therefore, while they are ideally suited to imaging mostly transparent samples, offering practical and efficient implementations, they cannot penetrate as deep as two-photon microscopy when it comes to highly scattering tissue imaging in vivo.

By utilizing fluorochromes that operate in the NIR, detection through several centimeters in tissues can be achieved. This is because of the low light absorption by tissue in the far-red and NIR regions (650–900 nm). In this spectral region, imaging becomes complicated largely due to tissue scattering, that is, changes in the direction of photon travel upon interaction with cellular interfaces and organelles. This leads to photon diffusion that reduces significantly the resolution and the ability to quantify properties of the tissue. In response, reliable imaging methods in the NIR utilize technology that account for tissue scattering and tissue absorption in order to improve the imaging performance in terms of image fidelity, resolution, and quantification. These technologies are generally based on the use of tomographic approaches, which can resolve in three dimensions the fluorochrome bio-distribution (Ntziachristos et al. 2005). Original systems utilized optical measurements only (Ntziachristos et al. 2002) to improve upon simplistic photographic approaches developed for superficial imaging, but yielded poor imaging performance and quantification. More recently, significant improvements in imaging performance comes from hybrid approaches

such as opto-acoustic methods (Wang et al. 2003, Razansky and Ntziachristos 2009) or the combination of fluorescence tomography with high-resolution methods such as x-ray computed tomography (CT) or magnetic resonance imaging (MRI) (Davis et al. 2008, Fang et al. 2009, Hyde et al. 2009, Schulz and Ntziachristos 2010). These methods rise well above early phase attempts on optical imaging and come with performance that renders them a robust and high-performing approach that is poised to shift the paradigm in biological imaging and select clinical regimes. As such, these methods offer the macroscopic extension of the microscope and are emerging as important tools for the biological laboratory for characterizing the biodistribution of fluorescent agents and probes or probing the effects of chemicals and environmental factors in vivo in entire organisms. It is also expected that these methods will transcend the boundaries.

In the following, we discuss techniques to impart contrast in small animals using fluorescence. We then describe currently utilized methods to macroscopically resolve this expression in small animals. In particular, we discuss multispectral opto-acoustic tomography, a technique developed to visualize fluorochromes distributed within several millimeters to centimeters of tissue with resolutions in the range of 20–100 microns. This is an emerging technology that senses photo-absorbing agents based on the emission of ultrasonic waves generated in response to a local transient temperature rise due to light absorption by these agents. To sense fluorochromes over other tissue chromophores with high specificity, the use of spectral approaches is necessary that can differentiate the unique spectra of the various fluorochromes of interest. In addition, we discuss hybrid fluorescence molecular tomography (FMT) approaches that can couple optical measurements to non-optical high-resolution modalities, such as x-ray CT or MRI, and not only improve the information content of the non-optical modality but by use of image priors based on the non-optical modality yield a more accurate optical imaging method.

22.2 Reporter Technologies

This section was more analytically presented in Ntziachristos (2006). Reporter technologies refers to mechanisms that generate non-endogenous (extrinsic) contrast in tissues, in order to tag with specificity otherwise invisible tissue and cellular processes and moieties, such as receptors, enzymes, signaling molecules and pathways, cellular motility, and traffic. Fluorescence has been the most common technology to impart contrast, but other methodologies are briefly reviewed. The following paragraphs briefly review the two major fluorescence reporter strategies that have been developed for in vivo imaging, that is, direct and indirect methods.

22.2.1 Direct Fluorescence Imaging

Direct imaging is associated with the administration of a fluorescent probe (i.e., a fluorescent molecule or nanoparticle) that targets a cellular element. Typically, the probes are *active probes*, that is, fluorochromes that are attached to an affinity ligand specific for a certain target. This paradigm is similar to probe design practices seen in nuclear imaging, except that a fluorochrome is used in the place of the isotope. Examples of affinity ligands include monoclonal antibodies and antibody fragments (Folli et al. 1994, Neri et al. 1997, Ramjiawan et al. 2000), modified or synthetic peptides (Licha et al. 2001, Ke et al. 2003, Chen et al. 2004, Achilefu et al. 2005), and other labeled small molecules (Zaheer et al. 2001, Moon et al. 2003). A characteristic of active probes is that they continuously fluoresce, even if they are not bound to the intended target. Other forms of *activatable probes* have also been developed. These probes carry quenched fluorochromes (Weissleder et al. 1999, Funovics et al. 2003) so that they do not emit fluorescence in their base state (Tung 2004). However, upon interaction with a target, they undergo a conformational change that unquenches the fluorochromes, which can then emit light upon excitation. Examples for identifying a series of proteases have been reported for in vivo imaging (Bremer et al. 2001, Funovics et al. 2003, Wunder et al. 2004) and many other activation examples exists. In contrast to active probes, activatable probes minimize background signals because they are essentially dark in the absence of the target and can improve contrast and detection sensitivity.

Fluorescence probes target specific cellular and subcellular events, and this ability differentiates them from nonspecific dyes, such as indocyanine green (ICG), which reveals generic functional characteristics such as vascular volume and permeability. By utilizing fluorochromes in the NIR spectral region, imaging can be performed through several centimeters of tissue.

22.2.2 Indirect Fluorescence Imaging

Indirect imaging is a strategy that evolved from corresponding in vitro reporting assays and is well suited to study gene expression and gene regulation. The most common practice is the introduction of a transgene (called reporter gene) in the cell. The transgene encodes for an FP, which acts as an intrinsically produced reporter probe. Transcription of the gene leads to the production of the FP, which can then be detected with optical imaging methods (Tsien 1998). Therefore, gene expression and regulation are imaged indirectly by visualizing and quantifying the presence of FPs in tissues. Cells can be stably transfected to express FP and report on their position for cell trafficking studies, or the transgene can be placed under promoters of interest for studying regulation. In addition, using the FP encoding gene to tag a gene of interest offers a platform for visualizing virtually every protein in vivo. This approach yields a chimeric protein that maintains the functionality of the original protein but is tagged with the FP so it can be visualized in vivo. It is also possible to transcribe and separately translate the protein of interest and the FP under control of the same promoter using a transgene containing an internal ribosomal entry site (IRES) between the genes encoding for the FP and the gene of interest (Mohrs et al. 2001). Therefore, the protein of interest remains intact while the FP still reports on gene transcription. Several different FP approaches have been developed to allow interrogation of protein–protein interactions through the utilization of fluorescence energy transfer (FRET) techniques (Chamberlain and Hahn 2000, van Roessel and Brand 2002), although these techniques have been primarily associated with microscopy and not macroscopy.

In recent years there has been a constant growth of different FP's, following the original development of the enhanced mutants of the green fluorescent protein (GFP) (Tsien 2005). The development of red-shifted FPs (Matz et al. 1999, Zhang et al. 2002, Wang et al. 2004a) has been more recently followed by proteins that emit in the NIR and are also appropriate for deep-tissue imaging (Shu et al. 2009).

22.2.3 Other Photonic Reporter Technologies

The following segment is adapted from Ntziachristos and Razansky (2010). In addition to fluorescence, other chromophoric agents can be used for reporting on tissue function using optical and opto-acoustic techniques. Besides intrinsic molecules such as the strong absorbers oxy- and deoxy-hemoglobin, of special interest are compounds having high molar extinction (absorption) coefficient within the NIR spectral band. Several dedicated agents have been considered for enhancing contrast, especially when utilizing opto-acoustic techniques. Due to high plasmonic absorption in the NIR and visible spectra, gold-based agents have been extensively explored in recent years and were shown to increase opto-acoustic signals ex vivo and within living subjects. Rayavarupu et al. used gold nanoparticles conjugated with monoclonal antibody specific to HER2 overexpressing SKBR3 for possible detection of breast carcinoma cells (Rayavarupu et al. 2007). Gold nanorods were detected with high sensitivity in vivo (Eghtedari et al. 2007) and later conjugated with Etanercept and used for noninvasive monitoring of anti-TNF drug delivery (Chamberland 2008). Gold nanoshells (Wang et al. 2004b) and nanocages (Yang et al. 2007) were used to enhance contrast in rat cerebral cortex in vivo. Other nanoparticulate agents were shown to enhance contrast in opto-acoustic imaging studies. A recent longitudinal study has demonstrated that single-walled carbon nanotubes (SWNT) conjugated with cyclic Arg-Gly-Asp (RGD) peptides can be used as a contrast agent for photoacoustic imaging of tumors (De la Zerda et al. 2008). Intravenous administration of these targeted nanotubes into mice bearing tumors showed eight times greater photoacoustic signal in the tumor in comparison to mice injected with nontargeted nanotubes. Shishkov et al. suggested using conjugated quantum dots as multimodal contrast agents for integrated fluorescent, photothermal, and photoacoustic detection and imaging (Shashkov et al. 2008). Activatable chromogenic-assays have also been considered for imparting tissue contrast using LacZ gene encoding for the X-gal chromogenic substrate (Li et al. 2007).

22.3 Small Animal Opto-Acoustic Molecular Tomography

Fluorescence visualization at depths appropriate for animal imaging is complicated by tissue scattering, similarly to diffuse optical imaging described in Chapters 16 through 20. Recent technical developments utilize hybrid approaches, in particular opto-acoustics and image priors, to overcome the effects of scattering and yield high-resolution quantitative images as described in the following.

22.3.1 Multispectral Opto-Acoustic Tomography

Opto-acoustic imaging is based on the generation of acoustic waves following the absorption of light pulses of ultrashort duration. While the opto-acoustic phenomenon has been known for more than a century (Bell 1880), its utilization for biomedical applications, such as spectroscopy or imaging was only considered in the past few decades, further intensifying in the past years (Rosencwaig 1973, Oraevsky et al. 1994). By combining now commercially available pulsed laser technology in the nanosecond range and sensitive acoustic detectors it is now possible to generate opto-acoustic responses of blood vessels in tissues (Wang et al. 2003, Siphanto et al. 2004) or fluorochromes distributed within tissues (Razansky et al. 2007).

Hemoglobin trapped in subsurface blood vessels offers a localized absorber with strong opto-acoustic contrast and can be detected easily in single-wavelength planar images (Wang et al. 2003, Siphanto et al. 2004). Conversely, volumetric imaging of fluorochromes or other photo-absorbing agents typically need to be resolved on top of the background absorption, due to intrinsic tissue chromophores such as hemoglobin, melanin, lipids, etc. Multispectral opto-acoustic tomography (MSOT) was developed to identify spectral signatures of reporter chromophores in tissues and relies on the spectral identification of common fluorochromes or other chromophores on top of the background tissue absorption, due, for example, to hemoglobin. In the sample experiment shown in Figure 22.1, which explains the principle of operation, tissue is illuminated with light pulses of duration in the 1–100 ns. Pulses of different wavelengths are used, in a time-shared fashion, whereas the wavelengths are selected in such a way so that they can sample a spectral transient in fluorochrome absorption as, for example, shown in Figure 22.1c. In response to the fast absorption of light pulses, photo-absorbing molecules in tissue that can absorb the photon energy undergo a thermo-elastic expansion that emits mechanical waves at ultrasonic frequencies. These waves can then be detected by acoustic detectors placed in proximity to the illuminated tissue. Using appropriate mathematical methods, images of the absorbed energy can then be reconstructed in analogy to the formation of ultrasound images or x-ray CT images. The amplitude of the generated broadband ultrasound waves depends on the optical absorption properties. The spatial resolution of the method is therefore solely determined by the diffraction limit of ultrasound waves or the available bandwidth of the ultrasonic detector.

While images can be generated at each light wavelength separately, multiwavelength illumination and spectral processing are necessary for identifying the unique spectral signatures of fluorochromes within other absorbing tissue chromophores, most notably hemoglobin, water, and lipids, especially in the NIR. Molecules with absorption spectra that are different than the ones of background tissue are best suited for MSOT imaging. It is possible to also resolve fluorochromes based on image subtraction between images obtained with the fluorochrome present and images obtained prior to fluorochrome administration. However, the use of spectral processing is particularly necessary when baseline measurements are not possible, for example, when long bio-distribution times are required for probe accumulation and/or targeting or activation.

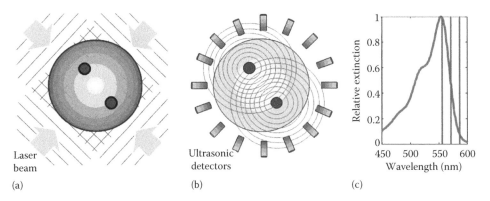

FIGURE 22.1 Principle of MSOT operation. (a) Pulsed light of time-shared multiple wavelengths illuminates the tissue of interest and establishes transient photon fields in tissue. (b) In response to the fast absorption transients by tissue elements, acoustic responses are generated via the thermo-acoustic phenomenon, which are then detected with acoustic detectors. By modeling photon and acoustic propagation in tissues and using inversion (tomographic) methods images can then be generated and spectrally unmixed to yield the bio-distribution of reporter molecules and tissue biomarkers. (c) Light of three different wavelengths (vertical lines) are selected to target the absorption transient of the fluorochrome as selected for spectral differentiation; here shown with the absorption spectrum of the fluorescent protein DsRed2. (From Ntziachristos, V., *Am. Chem. Soc.* 2795, 2009.)

The simplest form of spectral processing for resolving fluorochromes is the subtraction between images obtained at two adjacent wavelengths under the assumption that tissue exhibits similar attenuation in these wavelengths. Therefore, subtraction can effectively cancel out the effects of the background optical properties on the images. This concept has been shown to work in optically homogenous phantoms (Kruger et al. 2003) but its use with tissues in vivo is problematic since tissues are generally optically heterogeneous and there are often measurable attenuation changes between the wavelengths employed that are not accurately cancelled out with simple subtraction techniques. More generally, multiple wavelengths can be used and processed in order to accurately reveal the fluorochromes spectral signature. For accurately operating in vivo and in a volumetric fashion, MSOT needs to account for light propagation in heterogeneous tissues, accounting for the effects of intrinsic tissue optical properties (absorption and scattering) on the photon distribution. This is because each pixel in the raw opto-acoustic image is the product of tissue absorption at that pixel times the photon energy deposited at that pixel. While superficial opto-acoustic imaging can assume a piecewise homogenous distribution of light in tissues in the first hundred microns to millimeters and generate meaningful images, images generated by volumetric tissue opto-acoustic imaging do not accurately reflect tissue absorption but the combined effect of tissue absorption and light distribution in tissue. Unprocessed single wavelength images may offer details that can be identified as certain tissue structures but they are not quantitatively accurate unless they are adequately processed. This is particularly important for fluorochrome identification, which is based on the processing of multiple single-wavelength images and requires true absorption bio-distribution maps to yield accurate spectral identification. Therefore a significant aspect of MSOT technology is the use of appropriate methods that decompose the effects of photon distribution from absorption spatial changes in order to yield accurate and quantitative images of fluorochromes (or other photo-absorbing agents) in tissues.

In summary, there are two basic components required for opto-acoustic imaging of fluorochrome bio-distribution in tissues:

1. Multispectral separation is necessary in order to distinguish with high sensitivity the fluorescence spectral signature over the nonspecific background absorption. For accomplishing this operation it is important that the photo-absorbing agent has a distinct spectrum that is comparable to or larger than the background absorption spectrum. FPs and organic fluorochromes, for example, attain this characteristic by offering spectrally narrow absorption spectra that also can offer multispectral identification of several spectrally separated fluorochromes.

2. Accounting for the effects of light propagation in tissues in order to decompose true tissue and fluorochrome absorption from energy deposition in tissue.

22.3.2 MSOT Small Animal Imaging of Fluorochromes

The MSOT ability to detect reporter molecules from tissues has been showcased by visualizing fluorochromes and FPs in mice, fish, and other biologically relevant organisms. Figure 22.2 depicts results from Razansky et al. (2007) resolving a common organic fluorochrome (AlexaFluor 750™) injected in a mouse leg. Figure 22.2a–c depicts cross-sectional opto-acoustic tomographic reconstructions acquired at 750, 770, and 790 nm. Figure 22.2e shows an MSOT image (in color) generated by spectral processing of Figure 22.2a–c superimposed on Figure 22.2b (grayscale). Figure 22.2f depicts a corresponding ultrasonic image (25 MHz transducer) whereas Figure 22.2g shows a planar epi-fluorescence image of the dissected tissue that confirms the fluorochrome location, in comparison to the MSOT image in Figure 22.2e.

Detection herein is achieved without the need of baseline measurements obtained before the administration of the probe. As discussed previously, MSOT operates optimally by selecting fluorochromes (or other chromophores) with a steep drop in their absorption spectrum or otherwise a distinct spectral signature over tissue background absorption, as shown in Figure 22.2d. When utilizing fluorescent dyes, the emphasis is on low quantum-yield fluorochromes, which are particularly useful for photoacoustic signal excitation. The absorption spectrum of AF750 that was employed drops significantly in the spectral window 750–790 nm, compared to the smooth absorption variation of the spectra of common tissue chromophores in the NIR. Therefore, intrinsic tissue contrast can be readily suppressed with a multispectral approach, yielding highly sensitive imaging of fluorochrome distribution in tissue obtained by spectral matching of photoacoustic images acquired at several different adjacent wavelengths.

High-resolution MSOT imaging of FP, beyond the depths accessible by modern optical microscopy, was recently demonstrated (Razansky and Ntziachristos 2009) and opens exciting possibilities for the wide utilization of FP MSOT imaging. Figure 22.3 depicts results from whole-body visualization of deep-seated optical reporter molecules in mature diffuse organisms with high (mesoscopic) resolution (currently on the order of 38 microns), while simultaneously providing the necessary reference anatomical images. The images were three-dimensionally (3D) acquired in vivo through the brain of an adult (6 months old) mCherry-expressing transgenic zebrafish with a cross-section diameter of around 6 mm. The results demonstrate the ability to reveal many high-resolution morphological features, as evident from Figure 22.3a and b, supported by the corresponding histology (Figure 22.3c). Moreover, multispectral reconstructions accurately resolve FP expression in the brain of an intact living animal (Figure 22.3d), in high congruence with the corresponding epi-fluorescence images of the dissected brain (Figure 22.3e).

A particular strength of opto-acoustic imaging is the ability to simultaneously deliver anatomical, functional, and molecular

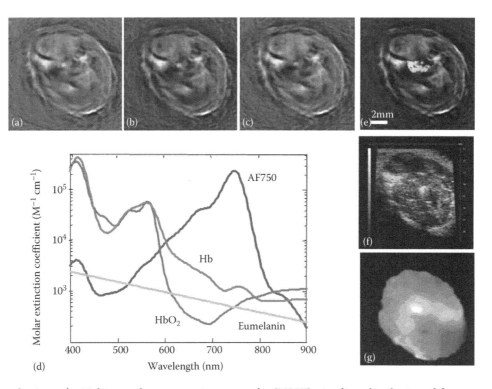

FIGURE 22.2 (See color insert.) Multispectral opto-acoustic tomography (MSOT) visualizes distribution of fluorescent molecular probe (AlexaFluor 750™) in a mouse leg[13]. (a), (b), and (c) are cross-sectional opto-acoustic tomographic reconstructions acquired at 750, 770, and 790 nm, respectively. (d) Absorption as a function of wavelength for AF750 fluorescent probe as compared to some intrinsic tissue chromophores (not in scale). Three wavelengths are used to spectrally resolve the probe location. (e) Spectrally resolved MSOT image that incorporates measurements at all the three wavelengths (in color), superimposed onto a single-wavelength anatomical image. (f) Corresponding ultrasonic image, acquired approximately at the same imaging plane, using 25 MHz high-resolution ultrasound system. (g) Planar epi-fluorescence image of dissected tissue confirms the fluorochrome location. (From Ntziachristos, V., *Am. Chem. Soc.* 2795, 2009.)

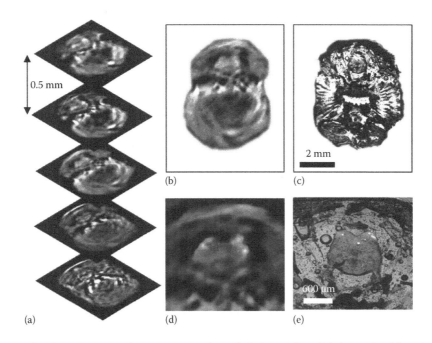

FIGURE 22.3 (See color insert.) Three-dimensional in vivo imaging through the brain of an adult (6 months old) mCherry-expressing transgenic zebrafish. (a) Five transverse opto-acoustic imaging slices through the brain area of living zebrafish taken at 585 nm. Examples of imaged slice and its corresponding histological section are shown in (b) and (c), respectively. (d) MSOT image of the brain (zoom-in) with mCherry expression shown in color. (e) Corresponding fluorescent histology made through excised brain. (From Razansky, D. et al., *Nat. Photonics*, 3, 412, 2009.)

contrast from tissues using a single modality, a characteristic typically achieved with multimodality imaging (Cherry 2006). Opto-acoustic technology scales well with different tissue sizes, especially when employing NIR light. Indeed, its range of operation begins at depths where two-photon microscopy ends (~0.5–1 mm) achieving resolutions in the sub 30 micron range and can visualize specimen of several centimeters in size (diameter) with resolutions ranging in the 100 micron. The limits of penetration of the technique depend on the ability to deposit sufficient photon energy in tissues, a property that eventually depends on the tissue optical properties that absorb and scatter light, reducing its strength as a function of propagation distance. Typically small animal imaging dimensions are possible (2–3 cm) but larger organs, such as the human breast have also been visualized (Stein et al. 2006). Finally, opto-acoustics is an inherently fast imaging technology therefore it holds great potential for real-time imaging of dynamic processes, such as pharmacokinetics, in living organisms,

22.3.3 MSOT Technology Components

In the following section, we briefly review the fundamental mathematics and technology aspects required for MSOT image formation and quantification for the interested reader. The vast majority of biomedical opto-acoustic imaging applications employ intense pulsed laser sources, even though other methods of opto-acoustic signal generation exist, for example, using modulated continuous wave sources (Maslov and Wang 2008). After the acoustic wave is created by light absorption, its magnitude is proportional to the local light intensity, optical absorption coefficient, and thermoelastic properties of the imaged tissue. The induced acoustic wave spectrum is mainly dependent upon the spatial frequency of the optical absorption variations and duration of the light pulse. For laser pulse durations in the nanosecond range, a biologically relevant spectrum of opto-acoustically induced signals is usually of ultra-wideband nature with useful information contained in the ultrasonic spectrum between several hundreds of kilohertz and several tenths of megahertz.

For imaging or sensing purposes, ultrasonic detectors are placed in the vicinity of the imaged object, as shown in Figure 22.1. Similarly to ultrasonic imaging, opto-acoustics is a time-resolved method, that is, time of arrival of the pressure wave directly indicates the distance to the opto-acoustic source in the imaged object. If the detector is placed in position $\vec{r}'$, it will sense an integrated pressure wave, namely,

$$p(\vec{r}',t) = \frac{\beta}{4\pi C} \int_V \frac{\partial \psi(\vec{r},t')}{\partial t'} \frac{d^3\vec{r}}{|\vec{r}-\vec{r}'|}\Bigg|_{t'=t-|\vec{r}-\vec{r}'|/v_s}, \quad (22.1)$$

where the spatiotemporal distribution of the absorbed light energy $\psi(\vec{r}, t')$ is measured in [W/cm³]. β is the isobaric thermal expansion coefficient, which describes the relative increase of the tissue volume per temperature increase under constant pressure (isobaric condition), v_s is the speed of sound in tissue and

C_p is the specific heat capacity at constant pressure, describing the energy required by a unit tissue mass to raise its temperature by a temperature unit. The typical parameters for biological tissue are $\beta = 3 \times 10^{-4}$ [°C⁻¹]; $v_s = 15 \times 10^{-4}$ [cm s⁻¹], and $C_p = 4.186$ [J g⁻¹ °C⁻¹]. Equation 22.1 can be interpreted such that, for each time point, the detected pressure variation has been created by integration over all the opto-acoustic sources located on a spherical shell of radius $|\vec{r} - \vec{r}'|$ surrounding the detector point $\vec{r}'$. The shape of the opto-acoustic signal created by an absorber depends on its size and optical absorption properties (Razansky and Ntziachristos 2009).

The ultimate goal of opto-acoustic tomographic imaging is the reconstruction of the underlining optical absorption contrast from a set of measured ultrasonic pressures $p(\vec{r}, t)$. Optical absorption can then be directly related to concentrations of intrinsic tissue chromophores and other biomarkers of interest. Often, focused detectors are used in order to reduce the dimension of the image reconstruction problem. For instance, in the case of a spherically focused acoustic detector, the detected waveforms directly represent a distribution of optical absorbers along the focus line, the so-called A-scan. To obtain 2D or 3D reconstruction, the detector is simply scanned along the object or, alternatively, beamforming with phased arrays is employed. Likewise, cylindrically focused detectors reduce the reconstruction problem into two dimensions, in which case tomographic inversion is necessary. Back-projection algorithms have so far been widely used for image reconstruction in opto-acoustic tomography. These algorithms are based on approximate closed-form inversion formulas expressed in two or three dimensions and are analogues to the Radon transform. Back-projection formulas exist for several detection geometries and are implemented either in the spatiotemporal domain (Xu and Wang 2005) or in the Fourier domain (Köstli and Board 2003).

By using multiwavelength illumination and subsequent spectral matching, for example, by incorporating a priori known or measured spectra of the biomarker(s) of interest, the fluorochrome/biomarker bio-distribution is resolved. To achieve this, it is assumed that every pixel in a single-wavelength opto-acoustic image represents a combined contribution of the photo-absorber(s) of interest with known molar extinction spectrum α_b and unknown concentration c_b and other background chromophores with known spectra α_m and unknown concentrations c_m ($m = 1, …, M$). This can be written in a form of a linear equation system, that is,

$$\mu_a(\lambda_n) = \alpha_b(\lambda_n)c_b + \sum_{m=1}^M \alpha_m(\lambda_n)c_m, \quad n = 1,…,N. \quad (22.2)$$

Using the measured absorption values and known spectra, the bio-distribution of the biomarker(s)/fluorochrome(s) of interest and other chromophores can be subsequently reconstructed from the above linear equations on a per-pixel basis with a linear regression method. In order for this approach to work accurately, methods are required that impart quantification.

The retrieval of images that reflect true absorption values in tissue, independently of the particulars of photon propagation, is therefore an essential aspect of MSOT, as described in the following section.

Although current opto-acoustic imaging is largely based on explicit (analytical closed-form) back-projection algorithms that are generally convenient and fast, they are not exact and may lead to the appearance of substantial artifacts in the reconstructed images. A common problem is the suppression of slowly varying image components and the accentuation of fast changes in the image (small details), which is usually also accompanied by negative optical-absorption values that otherwise have no physical interpretation. In addition, back-projection algorithms are based on an ideal description of the acoustic wave propagation and detection as well as on specific detection geometries; therefore, they cannot be easily generalized into more realistic opto-acoustic illumination-detection models that incorporate configuration and instrumentation-dependent factors.

Accurate evaluation of biomarker distribution, disease state, and treatment efficacy demand good quantification abilities. It has been shown that back-projection-related artifacts can be avoided by use of so-called model-based inverse methods (Paltauf et al. 2002). In contrast to back-projection algorithms, model-based schemes solve the forward problem numerically and iteratively. Ideally, this approach can yield artifact-free quantified reconstructions. However, the computational complexity involved with model-based schemes has so far severely limited their achievable resolution. Two contributing factors to the high complexity are the low efficiency/accuracy of the numerical forward solution and the need for a high number of iterations. Recently, a novel semi-analytical model-based inversion scheme for quantitative opto-acoustic image reconstruction was suggested (Rosenthal et al. 2009), where the presented semi-analytical solution is exact for piecewise planar acoustic-source functions, which significantly improves the accuracy and computational speed. The method eliminates image artifacts associated with the approximated back-projection formulations, that is, no negative absorption values are produced and the reconstructed image corresponds to the true light attenuation and energy deposition within the object.

Another quantification challenge arises from the fact that opto-acoustic signals do not directly convey information about the underlying optical absorption coefficient (μ_a) but rather on the local energy absorption ψ in tissue. The latter is simply the raw images that represent the product between μ_a and the local light fluence U. In Cox et al. (2006), a simple iterative reconstruction algorithm is suggested for planar geometry that can possibly resolve this imaging problem. In this algorithm, the optical absorption coefficient, obtained at each iteration, is used to calculate the fluence in the succeeding iteration. Theoretical simulations have shown that under ideal conditions, the algorithm converges and accurately recovers the absorption coefficient. In order to minimize the sensitivity of the algorithm to noise or to modeling errors, a regularizing term was added to the equation

to ensure convergence, albeit at the expense of reducing overall accuracy. In the recent work of Jetzfellner et al. (2009), the performance of an iterative opto-acoustic inversion scheme was verified with experimental data from phantoms resembling small animal imaging configurations. The influence of iterations and optical properties on the resulting inversions in terms of image spatial accuracy and quantification was considered. The findings indicate that the implementation of iterative inversion opto-acoustic schemes can potentially improve image quality in opto-acoustic tomographic imaging but their implementation needs to consider certain experimental factors that relate to the ability of accurately calculating the photon fluence distribution in the imaged object. In particular, knowledge of the absorption and reduced scattering coefficient is not always straightforward. While the absorption coefficient can be directly linked to the opto-acoustic signals, determination of the scattering coefficient is problematic as no accurate method exists to volumetrically determine its distribution, especially when it is spatially varying within tissue. Even with exact knowledge of the optical properties, however, it was found that the iterative method did not converge but was rather influenced by inconsistencies between experiment and theory, including noise and artifacts. Therefore, for best performance in realistic imaging scenarios, the iterative algorithm should preferably be used with some approximate or a priori information on the imaged object, including a well-defined convergence criterion and a good estimate of the background optical properties.

A more advanced method for light attenuation correction relies on the general properties of the optical fluence, rather than specific light-propagation model. The method sparsely decomposes the opto-acoustic image into two components: a slowly varying global component attributed to the diffusive photon fluence in the medium and a localized high spatial frequency component representing variations of the absorption coefficient (Rosenthal et al. 2009). This decomposition is based on the assumption that, owing to light diffusion, the photon fluence exhibits a slowly varying spatial dependence in contrast to fast spatial variations of the absorption coefficient, more typically associated with variations in structures that have well-defined boundaries that introduce high spatial frequency components. The decomposition in these two components relies on sparse representation, a relatively new and fast emerging area in the field of signal and image processing, which assumes that natural images can be approximated by a sum of a small number of elementary functions. In order to exploit this approximation, one needs to find a suitable set of these functions, often referred to as a library, and perform the decomposition numerically. Sparse representation has played an important role in the development of denoising and compression algorithms as well as in the field of compressive sensing. For opto-acoustic image representation, a simple library was suggested that is composed of two bases, one that sparsely describes the absorption coefficient and one sparsely describing the light fluence. The sparse representation of the opto-acoustic image in the library directly yields both the absorption coefficient and the light fluence.

22.4 Small Animal Hybrid Optical Tomography

Optical tomography aims at reconstructing the internal distribution of fluorochromes or chromophores in tissues based on light measurements collected at the tissue boundary in three dimensions. The principle of operation resembles that of x-ray CT, in that tissue is illuminated at different points or projections and the collected light is used in combination with a mathematical formulation that describes photon propagation in tissues. One point of distinction of optical tomography compared to tomographic methods based on high energy rays is that photons in the NIR or visible are highly scattered by tissue. While optoacoustics rely on the high resolution achieved by the acoustic detection, pure optical detection can benefit from the use of another, high-resolution modality that can be used for improving the optical imaging performance, in particular when data of the optical and the high-resolution modality are acquired simultaneously, or under identical geometrical conditions.

Principles of tomographic optical imaging and the methodologies used for the formulation of the theoretical model have been reviewed (Arridge 1999, Gibson et al. 2005). Most of the mathematical problems used in tomography attempt to model the photon propagation in tissues as a diffusive process, that is, utilize solutions of the diffusion equation. Since a higher number of spatial frequencies can be sampled with a point source, most solutions are obtained for a point source and point detector and form the basis for the forward problem by describing the expected response of each detector for an assumed medium. In the following, we outline some key equations used in the tomographic problem following the analytical approach. Their purpose is to give a flavor to the reader of the basic steps toward formulating an optical tomography problem, which are by no means exhaustive or descriptive of a wealth of methods and approaches that have been described in the literature. For the purposes, however, of this review, a tomographic problem generally assumes a distribution $O(r)$ of an optical property; let us assume here for simplicity the absorption coefficient around an average homogenous optical property value. Then an integral equation can be reached that relates the measured field ϕ_{sc} to this optical property variation and the field ϕ established in the medium due to the source at r_s (Kak and Slaney 1988) (Chapter 16), that is,

$$\phi_{sc}(r, r_s, \omega) = \int g(r, r', \omega)O(r')\phi(r', r_s, \omega)dr', \quad (22.3)$$

where $g(r, r')$ is the Green's function solution of the diffusion equation for a single delta function at r'. In practice, this function indicates the attenuation of the photon field when it propagates from position r' to position r where the detector is placed. The field $\phi(r', r_s)$ describes the photon distribution inside the tissue and it is generally a function of $\phi_{sc}(r', r_s)$ since this photon distribution depends on $O(r)$. The modulation frequency of the

photon beam is denoted by ω. For light of constant intensity, $\omega = 0$. To solve (22.3) in an analytical manner, a linearization is performed using an approximation such as the Born or the Rytov approximation (Kak and Slaney 1988, Oleary et al. 1995). In both approximations, the quantity $\phi(r', r_s)$ is essentially assumed equal to the photon field $\phi_0(r', r_s)$ that is established by the source at r_s in a geometrically similar but optically homogenous medium with the average optical properties those of the tissue investigated. Equation 22.3 can then be written as

$$\phi_{sc}(r, r_s, \omega) = \int g(r, r', \omega)O(r')\phi_0(r', r_s, \omega)dr', \quad (22.4)$$

written herein in the Born approximation sense (a similar solution is reached for the Rytov approximation case). Equation 22.4 describes a general derivation of a linear solution from the diffusion equation assuming a small absorption heterogeneous distribution (absorption perturbation). Interestingly, solutions reached for absorption heterogeneity or solutions for a fluorescence distribution reach very similar expressions (Li et al. 1996, Oleary et al. 1996). For fluorescence, which is the major focus of this review, a linear dependence of the fluorescence strength emitted from a volume element and the term $O(r')\phi_0(r', r_s)$ exists (Oleary et al. 1996). Then, the solution of a coupled set of two equations (Patterson and Pogue 1994, Hutchinson et al. 1996, Li et al. 1996), one describing the photon propagation at the excitation wavelength λ_{ex} and one at the emission wavelength λ_{em} takes the simple form (Oleary et al. 1996, Chang et al. 1997):

$$\phi_{fl}(r, r_s, \omega) = \int g^{\lambda_{em}}(r, r', \omega)O^f(r', \omega)\phi_0^{\lambda_{ex}}(r', r_s, \omega)dr', \quad (22.5)$$

which is virtually identical to (22.4) except two notable changes: (1) the Green functions solution is now calculated for the emission wavelength and (2) the distribution of fluorescence O^f is complex and depends on the modulation frequency as well as on the quantum yield γ the extinction coefficient of the dye ε the fluorochrome concentration [F] and the fluorochrome lifetime τ, that is,

$$O^f(r', \omega) \sim \frac{\gamma \varepsilon F(r')}{1 - i\omega\tau}, \quad (22.6)$$

Equations 22.4 and 22.5 can then be converted to a linear equation by discretization of the volume of interest into a number N of volume elements (voxels)

$$\phi_{sc}(r, r_s) = \sum_{n=1}^{N} W(r, r_n, r_s)O(r_n), \quad (22.7)$$

where $W(r, r_n, r_s)$ represents a "weight" that associates the effect of the optical property $O(r_n)$ at position r_n to a measurement at r due to a source at r_s and is found in this case as the product $g(r, r_n)\phi_0(r_n, r_s)$. Given M measurements, a system of linear equations is then obtained resulting in a matrix equation

$$y = Wx, \qquad (22.8)$$

where

    W is the weight matrix

    x represents the distribution $O(r_n)$ of optical property in each of the N voxels

    y is the corresponding measurement vector

22.4.1 Calculation of the Forward Problem

In practice, the established photon field ϕ_0 and the Green's function solution g in (22.4) and (22.5) are calculated for bounded media using analytical (Oleary et al. 1996, Chang et al. 1997, Ntziachristos and Weissleder 2001) or numerical solutions (Jiang et al. 1996, Paithankar et al. 1997, Jiang 1998, Hielscher and Bartel 2001, Roy and Sevick-Muraca 2001, Dehghani et al. 2003, Joshi et al. 2004, Schweiger et al. 2005). In addition, it is possible to reach solutions of (22.3), which are a more accurate description of the forward field, by iteratively solving (22.4) and updating ϕ_0 in each iterative step. Such solutions are generally implemented using numerical inversion schemes (Arridge and Hebden 1997). Furthermore, a number of experimentally validated methods have been proposed to independently account for changes in propagation between excitation and emission wavelengths (Milstein et al. 2003) and offer elegant ways to handle experimental uncertainties and noise in the data as well as prior information on image specifics (Eppstein et al. 2001, 2002, Milstein et al. 2003).

A particular approach that improves accuracy and overall image quality is the use of differential measurements. Fluorescence offers a significant advantage in that respect: fluorescence measurements can be always referenced to emission measurements obtained under identical experimental characteristics by use of different filter sets. One such approach that has facilitated the first in vivo demonstration of fluorescence tomography (Ntziachristos et al. 2002) is the use of the normalized Born approximation (Ntziachristos and Weissleder 2001). Under this method (22.4) is divided by a measurement at the emission wavelength and the problem inverted is

$$\frac{\phi_{fl}(r, r_s, \omega)}{\phi_{sc}(r, r_s, \omega)} = \frac{\Theta}{\phi_0^{\lambda_{ex}}(r', r_s, \omega)} \int g^{\lambda_{em}}(r, r', \omega) O^f(r', \omega) \phi_0^{\lambda_{ex}}(r', r_s, \omega) dr', \qquad (22.9)$$

where Θ is a constant that accounts for gain factors, mainly the changes in attenuation of the two filter sets employed. In essence, this scheme constructs a composite measurement vector of ratios. Inversion solves again for the same quantity that (22.5) did, that is, O^f. This approach has been shown to offer several experimental and reconstruction advantages (Ntziachristos and Weissleder 2001, Roy et al. 2003), that is, reduced sensitivity

to theoretical inaccuracies, unequal gain factors between different sources and detectors and high robustness in imaging even at highly optically heterogeneous backgrounds (Soubret et al. 2005) as will be also discussed in the following.

While the use of the diffusion equation as a forward model is appropriate for a variety of optical tomography schemes of tissues, there are cases where more accurate forward models are required, especially when void (nondiffusive) regions intersect photon propagation trajectories or when geometries using short source-detector separations are considered. For such regimes, solutions of the radiative transport equation (Chang et al. 1997, Klose and Hielscher 2003a) or diffusive solutions merged with radiosity principles have been proposed (Dehghani et al. 2000). These algorithms generally come with high computational cost and the need to select additional regularization parameters. However, experimental verification has been performed in most of these methods, some also at in vivo imaging (Klose et al. 2005).

22.4.2 Inversion

Inversion of (22.8) results in a so-called discrete ill-posed problem. This means in the general sense that the problem needs to be treated with a regularization process for efficient inversion. The literature is rich in methods for solving such systems and solutions and in these cases, it is noted that the more the knowledge of the model and the solution prior to inversion, the more optimal the reconstruction is (Hansen 1998). Popular inversion methods include the use of singular value decomposition methods, also in the form of direct inversion formulas (Schotland and Markel 2001), algebraic reconstruction techniques (Kak and Slaney 1988), Krylov subspace methods (Arridge 1999), Newton-based methods (Roy and Sevick-Muraca 2001, Klose and Hielscher 2003b), hybrid methods (Schweiger et al. 2005), and, more generally, efficient minimization methods that can be applied to large ill-posed inverse problems. To overcome the ill-posed nature of the inverse problem, different regularization methods can be used. Amongst the most popular is the Tikhonov regularization, which offers to minimize a linear combination of the residual $\|Wx - y\|_2^2$ and the weighted norm of the solution, that is,

$$\min\left(\|Wx - y\|_2^2 + a\|x\|_2^2\right), \qquad (22.10)$$

where a is a parameter associated with the amount of regularization (smoothness) of the solution. Equation 22.10 can be more generally written as a minimization of a function C:

$$C(\vec{x}) = \|\vec{y} - W\vec{x}\|^2 + Q(\vec{x}). \qquad (22.11)$$

The function $Q(\vec{x})$ is a penalty function, which is used in the stand-alone reconstructions as a regularization function. A typical implementation of $Q(\vec{x})$ then is given under the Tikhonov regularization shown in (22.10), where $Q(\vec{x}) = \alpha\|\vec{x}\|_2^2$, although different forms of functions can be constructed for minimization (Gibson et al. 2005). Minimization of (22.10) or (22.11) can be performed using conjugate gradient methods.

22.4.3 Use of Priors

While the above steps illustrate methods for stand-alone fluorescence tomography, significant improvements in the imaging performance come from the utilization of high-resolution image data that are co-registered with the optical measurements and can be used for improving the accuracy of the inversion by reducing the ill-posed nature of the inverse problem and for allowing for the construction of a more accurate forward problem—in analogy to the attenuation correction practices used in hybrid x-ray CT—positron emission tomography (PET) approaches.

We will refer now to hybrid fluorescence tomography and x-ray CT systems (Hyde et al. 2009) for simplicity, but the concepts described herein equally apply to other co-registered imaging modalities as well. Implementation of a priori information requires segmentation of x-ray images and identifying different organs and structures (e.g., the lung, heart, bone, mixed tissue, and tumors). Image segmentation can be based on simple intensity criteria but more accurate segmentation methods can yield improved performance in separating different tissue structures.

Then, utilization of prior information can occur at two levels. The first can utilize the segmented information to write a more accurate forward problem, using, for example, a finite element method (FEM) numerical solution. In this case, different structures need to be assigned with optical properties that can be usually retrieved from lookup tables as average tissue properties for the segmented tissues. Using (22.9) the sensitivity of the solution to optical properties is reduced (in particular absorption properties) and imaging accuracy can be retained, even at approximate estimations of the tissue optical property.

The second implementation level of image priors can occur at the inversion level. There are several methodologies that can include prior information (Brooksby et al. 2006, Hyde et al. 2009). A common approach is to modify the Tikhonov-type regularization term in (22.9) or (22.10), for example, by assuming

$$Q(\vec{x}) = \lambda \left\| L\vec{x} \right\|_2^2. \tag{22.12}$$

where L is a diagonal matrix. Each element on the diagonal corresponds to a certain pixel/voxel in the corresponding image. The values in each element are determined in accordance with certain assumptions of the tissue type that the voxel represented belongs to. For example, certain assumptions that can be implemented are those of zero fluorescence presence in certain voxels, if such voxels are known, or an assumption of a constant value across entire regions. The solution of the above minimization can also be computed using maximum likelihood approaches based on the statistics of the data collected (Hyde et al. 2007).

22.4.4 Hybrid Fluorescence Molecular Tomography

FMT has evolved as a tomographic method combining the theoretical mainframe of (22.3) through (22.12) with advanced instrumentation in order to overcome many of the limitations of planar imaging (reflectance and transillumination) and yield a robust and quantitative modality for fluorescent reporters in vivo. Original tomographic systems and methods were based on the use of fibers to couple light to and from tissue and the use of matching fluids to improve fiber coupling or simplify the boundary conditions used in the forward problem. However, such approaches imposed experimental difficulties due to the use of matching fluids and compromised the quality and size of the data collected.

Central to the new generation of systems developed for fluorescence tomography is the utilization of noncontact measurements, that is, measurements where the sources and detectors utilized do not come in physical contact with the tissue (Graves et al. 2003, Patwardhan et al. 2005). Besides the experimental simplicity that such design entails compared to fiber-based systems, it is further essential for collecting large information content and high-quality data sets, especially when using direct coupling with CCD cameras (Ripoll and Ntziachristos 2004). These methods typically also utilize surface measurements (e.g., using photogrammetry) to obtain the tissue surface and combine this information with the appropriate theoretical models in order to obtain accurate description of the forward model of photon propagation in diffuse media and air (Ripoll et al. 2003, Schultz et al. 2003). It has been experimentally shown that these methods can provide accurate reconstructions (Schultz et al. 2003, 2004) from phantoms and animals.

Original FMT feasibility studies resolved proteases in animal brains using a circular geometry and fiber-based systems (Ntziachristos et al. 2002). Newer generation prototypes based on noncontact techniques have allowed superior imaging quality demonstrating sub-resolution imaging capacity (Graves et al. 2003) and sensitivity that reaches well below a pico-mole of fluorescent dye (value reported for the Cy5.5. dye excited at 672 nm). Similarly, newer systems based on flying spot illumination technology confirmed these sensitivity findings and have reported further advances such as rapid whole body imaging (Patwardhan et al. 2005). In addition, the ability to tomographically image at the visible (Zacharakis et al. 2005) or to offer complete projection tomography (Meyer et al. 2005, Turner et al. 2005) has been showcased. Such advanced setups have been used for imaging probe distribution (Patwardhan et al. 2005), angiogenesis (Montet et al. 2005), proteases (Graves et al. 2003), or the effects of chemotherapy on tumors (Ntziachristos et al. 2004).

These implementations further allowed the utilization of free-space 360° illumination, a geometry that combines well with other gantry-based systems, such as x-ray CT, as shown in Figure 22.4. The utilization of hybrid FMT and x-ray CT systems, using image priors was showcased in an APP23 tg mouse model that develops neurodegeneration with a phenotype that resembles

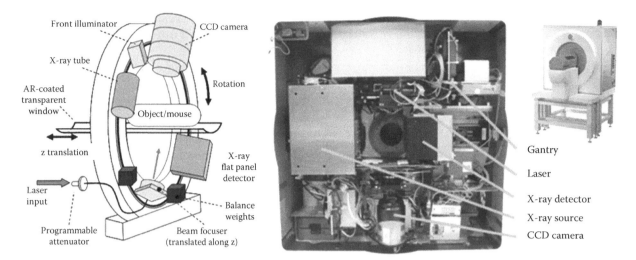

FIGURE 22.4 Novel hybrid FMT-XCT system developed at the Helmholtz Zentrum München, the German Research Center for Environmental Health and the Technische Universität München by adding an FMT system into a GE Explore Locus x-ray CT system.

human Alzheimer's disease (Hyde et al. 2009), using an NIR oxazine dye previously showcased to in vivo target amyloid-beta deposits (Hintersteiner et al. 2005). The following excerpt has been adapted from Hyde et al. (2009). FMT animal imaging was performed using a recently developed state-of-the-art FMT system employing a free-space geometry and full angular data collection in transmission mode. This system allows for use of arbitrary source-laser and detector point patterns on the surface of the animal. The collection of photon propagation data from multiple angles increases the information content of the signal and allows for improvements in spatial resolution and quantification. Localization of the animal surface was done using multiple silhouette images and a previously described volume carving algorithm (Lasser et al. 2008). X-ray CT was performed on a Gamma Medica scanner. The data were analyzed and segmented to differentiate brain and cortical regions, bone, and soft tissue. An affine transform was employed to co-register the surface computed by the FMT system with an isosurface from the CT image. Forward photon propagation calculations were performed using the FEM solutions to the diffusion approximation and were computed within a head geometry constructed from the CT structural information. The sensitivity functions resulting from this computation were used to construct a first-order Born model of light propagation (Ntziachristos and Weissleder 2001). The inverse problem, determining an unknown fluorescence distribution from measured data, was implemented by preferentially regularizing voxels based on an adaptive two-step method that assigns regularization values based on information contained in the dual wavelength measurements acquired. The first step assumes a subdivision of the three tissue types segmented in the classical sense, that is, to brain, brain cortex, and remaining tissues. All non-brain tissues were treated as a single type for simplicity and to prevent introduction of artifacts into the solutions. This improved the numerical properties of the problem and resulted in significantly more stable results. The non-cortical brain was divided into left and right hemispheres,

while the cortical region was broken into dorsal, left, and right lateral fragments. This information was incorporated into an approximate maximum likelihood solution (Hyde et al. 2007) as an additive penalty term, with relative weights based on the solution of a low-dimensional parameterized problem, which assumed a piecewise constant basis defined by the subdivisions. Essentially this step accurately determines the strength of fluorescence in each of the subdivisions by inverting a low parametric and easily solved problem. The determination of the piecewise subdivision-specific fluorescence strengths was employed to scale corresponding regularization parameters. The importance of this step is that the prior information on fluorescence strength is implemented based on the information contained in the collected data, that is, not by user-depended selection. Figure 22.5 depicts pertinent results from this study. Images constructed using the structurally guided multimodal algorithm (Figure 22.5f) consistently showed a marked improvement in quality and significant decrease in artifacts compared to those regularized using standard Tikhonov regularization with the identity matrix (Figure 22.5e) as compared to ex vivo imaging.

22.5 Discussion

New optical tomography approaches offer revolutionary performance characteristics compared to stand-alone implementations using stand-alone optical measurements and not only open up exciting possibilities for molecular imaging investigations in small animals but also possibly select clinical applications. MSOT offers high resolution and sensitivity, very versatile molecular contrast, portability, cost effectiveness, and the use of nonionizing radiation. While some existing modalities attain some of these features, *none combines all of them in one package*, as in MSOT. As such, MSOT is expected to become a method of choice in small animal imaging research. When considering clinical applications, MSOT has several niche focus points due to the high sensitivity, resolution, and portability, and it can shift

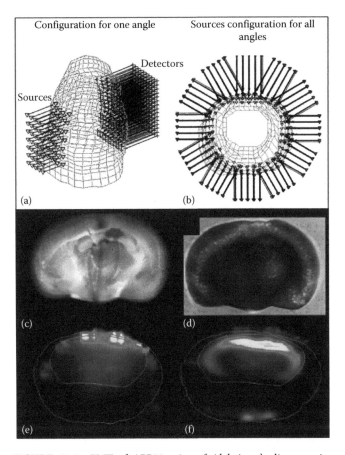

Configuration for one angle | Sources configuration for all angles

Detectors

Sources

(a) (b)

FIGURE 22.5 FMT of APP23 mice of Alzheimer's disease using image priors. (a,b) Visual representation of geometrical aspects. Photon measurements occurring in the absence of fluids require the capture of the surface, as shown in (a), obtained from the mouse head. The blue lines represent photon trajectories for the sources and the detectors in free space. (b) Axial representation in two dimensions of different illumination projections established after repositioning a scanning laser beam along different angles (projections). (c) White light and (d) fluorescence images of a brain slice obtained ex vivo from an APP23 mouse for confirmation of the in vivo data. The contrast in the fluorescence image represents amyloid-β distribution, stained with an NIR oxazine dye administered intravenously before animal euthanasia. (e) Stand-alone FMT reconstructions and (f) FMT reconstructions using structural image priors obtained from correlative x-ray CT images (not shown), showcasing improvements in image fidelity as compared to (e).

the paradigm of healthcare offering a safe point-of-care imaging modality for highly disseminated imaging. Foreseen application areas are all the endoscopic areas, where compared to similar photographic/color imaging it can impart superior quantification and offer depth imaging as well. In addition to breast imaging, joint imaging and vascular imaging either intravascular or noninvasive are all within the performance capacity of the method. Correspondingly, MSOT is expected to enter focused segments of the therapeutic efficacy and possibly the diagnostic segments, especially in areas not well served by current imaging modalities. The only MSOT limitation is depth penetration (i.e., 4–5 cm depth, more in low absorbing structures such as breast

tissue). Therefore, its application segment does not compete with the MRI and PET markets, but defines new operational areas, similar to those of ultrasound, but with different contrast and imaging performance characteristics, which enable specific imaging of tissue biomarkers, not available by ultrasound imaging. Overall therefore, MSOT is seen as highly complementary to established radiology and nuclear imaging modalities but also conventional opto-acoustic methods developed for tissue absorption imaging.

Hybrid (multimodal) optical tomography methods also offer an exciting alternative to nuclear imaging methods. The use of priors can significantly improve imaging performance, and the use of free-space, non-fluid approaches allow the integration of FMT with established anatomical modalities such as x-ray CT or MRI. With the compelling advantages of optical imaging, such as highly diverse contrast mechanisms and easy and safe usability and with imaging performance characteristics that have come not of age and exhibit high fidelity, sensitivity and resolution photonic imaging is bound to play a major role in biomedical research and drug discovery applications. This is because it brings a new standard of performance in small animal imaging and it can lead to significant niche clinical applications as well, especially in regimes that optical imaging is already an accepted modality, such as endoscopic application, but also new deeper resolving clinical areas. As such it can play a vital role in the drug discovery process, from accelerating the decision of potential candidates during in vivo screening applications and toxicology animal studies to playing an increasingly important role through Phase 0–III clinical trials by offering a method that can yield quantitative markers of treatment while it is not limited by application repetition due to cost or use of ionizing radiation. Therefore, it is expected that photonic imaging will define several new application areas and it will become the method of choice in small animal and select clinical imaging applications.

Acknowledgments

We are thankful to the IBMI scientists for assistance and discussions in association with imaging performance metrics reported herein. VN acknowledges support from an ERC Advanced Investigator Award and FP7 European Commission support under grant FMT-XCT.

Glossary

CT Computed tomography
FEM Finite element method
FP Fluorescent protein
GFP Green fluorescent protein
MRI Magnetic resonance imaging
MSOT Multispectral opto-acoustic tomography
NIR Near-infrared
RFP Red fluorescent protein
SNR Signal-to-noise ratio

References

Achilefu, S., Bloch, S., Markiewicz, M. A., Zhong, T., Ye, Y., Dorshow, R. B., Chance, B., and Liang, K. 2005, May 31. Synergistic effects of light-emitting probes and peptides for targeting and monitoring integrin expression. *Proc. Natl. Acad. Sci. USA* 102: 7976–7981.

Arridge, S. R. 1999. Optical tomography in medical imaging. *Inverse Problems* 15: R41–R93.

Arridge, S. R. and Hebden, J. C. 1997, May. Optical imaging in medicine. 2. Modelling and reconstruction. *Phys. Med. Biol.* 42: 841–853.

Bell, A. G. 1880. Upon the production of sound by radiant energy. *Am. J. Sci.* 20: 305–324.

Bremer, C., Tung, C. H., and Weissleder, R. 2001, June. In vivo molecular target assessment of matrix metalloproteinase inhibition. *Nat. Med.* 7: 743–748.

Brooksby, B., Pogue, B. W., Jiang, S., Dehghani, H., Srinivasan, S., Kogel, C., Tosteson, T. D., Weaver, J., Poplack, S. P., and Paulsen, K. D. 2006, June 6. Imaging breast adipose and fibroglandular tissue molecular signatures by using hybrid MRI-guided near-infrared spectral tomography. *Proc. Natl. Acad. Sci. USA* 103: 8828–8833.

Chamberlain, C. and Hahn, K. M. 2000, October. Watching proteins in the wild: Fluorescence methods to study protein dynamics in living cells. *Traffic* 1: 755–762.

Chamberland, D. L., Agarwal, A., Kotov, N., Fowlkes, J. B., Carson, P. L., and Wang, X. 2008. Photoacoustic tomography of joints aided by an Etanercept-conjugated gold nanoparticle contrast agent—An ex vivo preliminary rat study. *Nanotechnology* 19: 095101.

Chang, J. W., Graber, H. L., and Barbour, R. L. 1997. Luminescence optical tomography of dense scattering media. *J. Opt. Soc. Am. A Opt. Image Sci. Vis.* 14: 288–299.

Chen, X., Conti, P. S., and Moats, R. A. 2004, November 1. In vivo near-infrared fluorescence imaging of integrin alphavbeta3 in brain tumor xenografts. *Cancer Res.* 64: 8009–8014.

Cherry, S. R. 2006. Multimodality in vivo imaging systems: Twice the power or double the trouble? *Annu. Rev. Biomed. Eng.* 8: 35–62.

Cox, B. T., Arridge, S. R., Kostli, K. P., and Beard, P. C. 2006, March 10. Two-dimensional quantitative photoacoustic image reconstruction of absorption distributions in scattering media by use of a simple iterative method. *Appl. Opt.* 45: 1866–1875.

Davis, S. C., Pogue, B. W., Springett, R., Leussler, C., Mazurkewitz, P., Tuttle, S. B., Gibbs-Strauss, S. L., Jiang, S. S., Dehghani, H., and Paulsen, K. D. 2008, June. Magnetic resonance-coupled fluorescence tomography scanner for molecular imaging of tissue. *Rev. Sci. Instrum.* 79: 064302.

De la Zerda, A., Zavaleta, C., Keren, S., Vaithilingam, S., Bodapati, S., Liu, Z., Levi, J., Smith, B. R., Ma, T. J., Oralkan, O., Cheng, Z., Chen, X., Dai, H., Khuri-Yakub, B. T., and Gambhir, S. S. 2008, September. Carbon nanotubes as photoacoustic molecular imaging agents in living mice. *Nat. Nanotechnol.* 3: 557–562.

Dehghani, H., Arridge, S. R., Schweiger, M., and Delpy, D. T. 2000. Optical tomography in the presence of void regions. *J. Opt. Soc. Am. A Opt. Image Sci. Vis.* 17: 1659–1670.

Dehghani, H., Pogue, B. W., Jiang, S. D., Brooksby, B., and Paulsen, K. D. 2003, July, 1. Three-dimensional optical tomography: Resolution in small-object imaging. *Appl. Opt.* 42: 3117–3128.

Deliolanis, N. C., Wurdinger, T., Pike, L., Tannous, B. A., Breakefield, X. O., Weissleder, R., and Ntziachristos, V. 2011. In vivo tomographic imaging of red-shifted fluorescent proteins. *Biomed. Opt. Exp.*, 2: 887–900.

Denk, W., Strickler, J. H., and Webb, W. W. 1990. Two-photon laser scanning fluorescence microscopy. *Science* 248: 73–76.

Eghtedari, M., Oraevsky, A., Copland, J. A., Kotov, N. A., Conjusteau, A., and Motamedi, M. 2007, July. High sensitivity of in vivo detection of gold nanorods using a laser optoacoustic imaging system. *Nano Lett.* 7: 1914–1918.

Eppstein, M. J., Dougherty, D. E., Hawrysz, D. J., and Sevick-Muraca, E. M. 2001. Three-dimensional Bayesian optical image reconstruction with domain decomposition. *IEEE Trans. Med. Imaging* 20: 147–163.

Eppstein, M. J., Hawrysz, D. J., Godavarty, A., and Sevick-Muraca, E. M. 2002, July 23. Three-dimensional, Bayesian image reconstruction from sparse and noisy data sets: Near-infrared fluorescence tomography. *Proc. Natl. Acad. Sci. USA* 99: 9619–9624.

Fang, Q., Carp, S. A., Selb, J., Boverman, G., Zhang, Q., Kopans, D. B., Moore, R. H., Miller, E. L., Brooks, D. H., and Boas, D. A. 2009, January. Combined optical imaging and mammography of the healthy breast: Optical contrast derived from breast structure and compression. *IEEE Trans. Med. Imaging* 28: 30–42.

Folli, S., Westermann, P., Braichotte, D., Pelegrin, A., Wagnieres, G., van den Bergh, H., and Mach, J. P. 1994, May 15. Antibody-indocyanin conjugates for immunophotodetection of human squamous cell carcinoma in nude mice. *Cancer Res.* 54: 2643–2649.

Funovics, M., Weissleder, R., and Tung, C. H. 2003, November. Protease sensors for bioimaging. *Anal. Bioanal. Chem.* 377: 956–963.

Gibson, A. P., Hebden, J. C., and Arridge, S. R. 2005, February 21. Recent advances in diffuse optical imaging. *Phys. Med. Biol.* 50: R1–R43.

Giepmans, B. N. G., Adams, S. R., Ellisman, M. H., and Tsien, R. Y. 2006, April. Review—The fluorescent toolbox for assessing protein location and function. *Science* 312: 217–224.

Graves, E., Ripoll, J., Weissleder, R., and Ntziachristos, V. 2003. A sub-millimeter resolution fluorescence molecular imaging system for small animal imaging. *Med. Phys.* 30: 901–911.

Hansen, P. C. 1998. *Rank-Deficient and Discrete Ill-Posed Problems*. Philadelphia, PA: SIAM.

Hielscher, A. H. and Bartel, S. 2001. Use of penalty terms in gradient-based iterative reconstruction schemes for optical tomography. *J. Biomed. Opt.* 6: 183–192.

Hintersteiner, M., Enz, A., Frey, P., Jaton, A. L., Kinzy, W., Kneuer, R., Neumann, U., Rudin, M., Staufenbiel, M., Stoeckli, M., Wiederhold, K. H., and Gremlich, H. U. 2005, May. In vivo detection of amyloid-beta deposits by near-infrared imaging using an oxazine-derivative probe. *Nat. Biotechnol.* 23: 577–583.

Homma, R., Baker, B. J., Jin, L., Garaschuk, O., Konnerth, A., Cohen, L. B., Bleau, C. X., Canepari, M., Djurisic, M., and Zecevic, D. 2009. Wide-field and two-photon imaging of brain activity with voltage- and calcium-sensitive dyes. *Methods Mol. Biol.* 489: 43–79.

Huisken, J., Swoger, J., Del Bene, F., Wittbrodt, J., and Stelzer, E. H. K. 2004, August 13. Optical sectioning deep inside live embryos by selective plane illumination microscopy. *Science* 305: 1007–1009.

Hutchinson, C. L., Troy, T. L., and SevickMuraca, E. M. 1996, May 1. Fluorescence-lifetime determination in tissues or other scattering media from measurement of excitation and emission kinetics. *Appl. Opt.* 35: 2325–2332.

Hyde, D., de Kleine, R., MacLaurin, S., Miller, E., Brooks, D., Krucker, T., and Ntziachristos, V. 2009. Hybrid FMT-CT imaging of amyloid-beta plaques in a murine Alzheimer's disease model. *Neuroimage* 44: 1304–1311.

Hyde, D., Miller, E., Brooks, D. H., and Ntziachristos, V. 2007, July. A statistical approach to inverting the born ratio. *IEEE Trans. Med. Imaging* 26: 893–905.

Jetzfellner, T., Razansky, D., Rosenthal, A., Schulz, R., Englmeier, K.-H., and Ntziachristos, V. 2009. Performance of iterative optoacoustic tomography with experimental data. *Appl. Phys. Lett.* 95: 013703.

Jiang, H. B. 1998, August 1. Frequency-domain fluorescent diffusion tomography: A finite-element-based algorithm and simulations. *Appl. Opt.* 37: 5337–5343.

Jiang, H. B., Paulsen, K. D., Osterberg, U. L., Pogue, B. W., and Patterson, M. S. 1996, February. Optical image reconstruction using frequency-domain data: Simulations and experiments. *J. Opt. Soc. Am. A Opt. Image Sci. Vis.* 13: 253–266.

Joshi, A., Bangerth, W., and Sevick-Muraca, E. M. 2004, November 1. Adaptive finite element based tomography for fluorescence optical imaging in tissue. *Opt. Express* 12: 5402–5417.

Kak, A. and Slaney, M. 1998. *Principles of Computerized Tomographic Imaging.* New York: IEEE Press.

Ke, S., Wen, X., Gurfinkel, M., Charnsangavej, C., Wallace, S., Sevick-Muraca, E. M., and Li, C. 2003, November 15. Near-infrared optical imaging of epidermal growth factor receptor in breast cancer xenografts. *Cancer Res.* 63: 7870–7875.

Klose, A. D. and Hielscher, A. H. 2003a, June 15. Fluorescence tomography with simulated data based on the equation of radiative transfer. *Opt. Lett.* 28: 1019–1021.

Klose, A. D. and Hielscher, A. H. 2003b, April. Quasi-Newton methods in optical tomograpic image reconstruction. *Inverse Problems* 19: 387–409.

Klose, A. D., Ntziachristos, V., and Hielscher, A. H. 2005, January 1. The inverse source problem based on the radiative transfer equation in optical molecular imaging. *J. Comput. Phys.* 202: 323–345.

Köstli, K. P. and Board, P. C. 2003. Two dimensional photoacoustic imaging by use of Fourier transform image reconstruction and a detector with anisotropic response. *Appl. Opt.* 42(10): 1899–1908.

Kruger, R. A., Kiser, W. L., Reinecke, D. R., Kruger, G. A., and Miller, K. D. 2003, April. Thermoacoustic molecular imaging of small animals. *Mol Imaging* 2: 113–123.

Lasser, T., Soubret, A., Ripoll, J., and Ntziachristos, V. 2008, February. Surface reconstruction for free-space 360 degrees fluorescence molecular tomography and the effects of animal motion. *IEEE Trans. Med. Imaging* 27: 188–194.

Li, L., Zemp, R. J., Lungu, G., Stoica, G., and Wang, L. V. 2007, March–April. Photoacoustic imaging of lacZ gene expression in vivo. *J. Biomed. Opt.* 12: 020504.

Li, X. D., Oleary, M. A., Boas, D. A., Chance, B., and Yodh, A. G. 1996. Fluorescent diffuse photon: Density waves in homogeneous and heterogeneous turbid media: Analytic solutions and applications. *Appl. Opt.* 35: 3746–3758.

Licha, K., Hessenius, C., Becker, A., Henklein, P., Bauer, M., Wisniewski, S., Wiedenmann, B., and Semmler, W. 2001, January–February. Synthesis, characterization, and biological properties of cyanine-labeled somatostatin analogues as receptor-targeted fluorescent probes. *Bioconjug. Chem.* 12: 44–50.

Maslov, K. and Wang, L. V. 2008, March–April. Photoacoustic imaging of biological tissue with intensity-modulated continuous-wave laser. *J. Biomed. Opt.* 13: 024006.

Matz, M. V., Fradkov, A. F., Labas, Y. A., Savitsky, A. P., Zaraisky, A. G., Markelov, M. L., and Lukyanov, S. A. 1999, October. Fluorescent proteins from nonbioluminescent Anthozoa species. *Nat. Biotechnol.* 17: 969–973.

Meyer, H., Garofalakis, A., Zacharakis, G., Economou, E., Mamalaki, C., Kioussis, D., Ntziachristos, V., and Ripoll, J. 2005. A multi-projection non-contact fluorescence tomography setup for imaging arbitrary geometries. In: *Optical Tomography and Spectroscopy of Tissue VI*, Proceedings of SPIE Vol. 5693, pp. 246–254.

Milstein, A. B., Oh, S., Webb, K. J., Bouman, C. A., Zhang, Q., Boas, D. A., and Millane, R. P. 2003. Fluorescence optical diffusion tomography. *Appl. Opt.* 42: 3081–3094.

Mohrs, M., Shinkai, K., Mohrs, K., and Locksley, R. M. 2001, August. Analysis of type 2 immunity in vivo with a bicistronic IL-4 reporter. *Immunity* 15: 303–311.

Montet, X., Ntziachristos, V., Grimm, J., and Weissleder, R. 2005, July 15. Tomographic fluorescence mapping of tumor targets. *Cancer Res.* 65: 6330–6336.

Moon, W. K., Lin, Y., O'Loughlin, T., Tang, Y., Kim, D. E., Weissleder, R., and Tung, C. H., 2003, May–June. Enhanced tumor detection using a folate receptor-targeted near-infrared fluorochrome conjugate. *Bioconjug. Chem.* 14: 539–545.

Neri, D., Carnemolla, B., Nissim, A., Leprini, A., Querze, G., Balza, E., Pini, A., Tarli, L., Halin, C., Neri, P., Zardi, L., and Winter, G. 1997, November. Targeting by affinity-matured recombinant antibody fragments of an angiogenesis associated fibronectin isoform. *Nat. Biotechnol.* 15: 1271–1275.

Ntziachristos, V. 2006. Fluorescence molecular imaging. *Annu. Rev. Biomed. Eng.* 8: 1–33.

Ntziachristos, V. and Razansky, D. 2010. Fluorescence imaging by means of multi-spectral opto-acoustic tomography (MSOT). *ACR Chem. Rev.* 140(5): 2783–2794.

Ntziachristos, V., Ripoll, J., Wang, L. V., and Weissleder, R. 2005, March. Looking and listening to light: The evolution of whole-body photonic imaging. *Nat. Biotechnol.* 23: 313–320.

Ntziachristos, V., Schellenberger, E. A., Ripoll, J., Yessayan, D., Graves, E., Bogdanov, A., Josephson, L., and Weissleder, R. 2004, August 17. Visualization of antitumor treatment by means of fluorescence molecular tomography with an annexin V-Cy5.5 conjugate. *Proc. Natl. Acad. Sci. USA* 101: 12294–12299.

Ntziachristos, V., Tung, C., Bremer, C., and Weissleder, R. 2002. Fluorescence-mediated tomography resolves protease activity in vivo. *Nat. Med.* 8: 757–760.

Ntziachristos, V. and Weissleder, R. 2001. Experimental three-dimensional fluorescence reconstruction of diffuse media using a normalized Born approximation. *Opt. Lett.* 26: 893–895.

Oleary, M. A., Boas, D. A., Chance, B., and Yodh, A. G. 1995. Experimental images of heterogeneous turbid media by frequency-domain diffusing-photon tomography. *Opt. Lett.* 20: 426–428.

Oleary, M. A., Boas, D. A., Li, X. D., Chance, B., and Yodh, A. G. 1996. Fluorescence lifetime imaging in turbid media. *Opt. Lett.* 21: 158–160.

Oraevsky, A. A., Jacques, S. L., Esenaliev, R. O., and Tittel, F. K. 1994. Laser based optoacoustic imaging in biological tissues. *Proc. SPIE* 2134A: 122–128.

Paithankar, D. Y., Chen, A. U., Pogue, B. W., Patterson, M. S., and SevickMuraca, E. M. 1997. Imaging of fluorescent yield and lifetime from multiply scattered light reemitted from random media. *Appl. Opt.* 36: 2260–2272.

Paltauf, G., Viator, J., Prahl, S. A., and Jacques, S. L. 2002. Iterative reconstruction algorithm for optoacoustic imaging. *J. Acoust. Soc. Am.* 112: 1536–1544.

Patterson, M. S. and Pogue, B. W. 1994, April 1. Mathematical-model for time-resolved and frequency-domain fluorescence spectroscopy in biological tissue. *Appl. Opt.* 33: 1963–1974.

Patwardhan, S. V., Bloch, S. R., Achilefu, S., and Culver, J. P. 2005, April 4. Time-dependent whole-body fluorescence tomography of probe bio-distributions in mice. *Opt. Express* 13: 2564–2577.

Ramjiawan, B., Maiti, P., Aftanas, A., Kaplan, H., Fast, D., Mantsch, H. H., and Jackson, M. 2000, September 1. Noninvasive localization of tumors by immunofluorescence imaging using a single chain Fv fragment of a human monoclonal antibody with broad cancer specificity. *Cancer* 89: 1134–1144.

Rayavarapu, R. G., Petersen, W., Ungureanu, C., Post, J. N., van Leeuwen, T. G., and Manohar, S. 2007. Synthesis and bio-conjugation of gold nanoparticles as potential molecular probes for light-based imaging techniques. *Int. J. Biomed. Imaging*, 2007: 29817.

Razansky, D., Vinegoni, C., Distel, M., Ma R., Perrimon, N., Koster, R. W., and Ntziachristos, V. 2009. Multispectral optoacoustic tomography of deep-seated flourescent proteins in vivo. *Nat. Photon.* 3: 412–417.

Razansky, D. and Ntziachristos, V. 2009. Imaging of mesoscopic targets using selective-plane optoacoustic tomography. *Phys. Med. Biol.* 54(9): 2769–2777.

Razansky, D., Vinegoni, C., and Ntziachristos, V. 2007, October. Multispectral photoacoustic imaging of fluorochromes in small animals. *Opt. Lett.* 32: 2891–2893.

Ripoll, J. and Ntziachristos, V. 2004, December 20. Imaging scattering media from a distance: Theory and applications of noncontact optical tomography. *Mod. Phys. Lett. B* 18: 1403–1431.

Ripoll, J., Schultz, R., and Ntziachristos, V. 2003. Free-space propagation of diffuse light: Theory and experiments. *Phys. Rev. Lett.* 91: 103901-1–103901-4.

Rosencwaig, A. 1973, August 17. Photoacoustic spectroscopy of biological materials. *Science* 181: 657–658.

Rosenthal, A., Razansky, D., Ntziachristos V. 2009. Quantitative optoacoustic tomography using sparse signal representation. *IEEE Trans. Med. Imag*ing 28(12): 1997–2006.

Roy, R., Godavarty, A., and Sevick-Muraca, E. M. 2003. Fluorescence-enhanced optical tomography using referenced measurements of heterogeneous media. *IEEE Trans. Med. Imaging* 22: 824–836.

Roy, R. and Sevick-Muraca, E. M. 2001, May 1. Three-dimensional unconstrained and constrained image-reconstruction techniques applied to fluorescence, frequency-domain photon migration. *Appl. Opt.* 40: 2206–2215.

Schotland, J. C. and Markel, V. A. 2001. Inverse scattering with diffusing waves. *J. Opt. Soc. Am. A Opt. Image Sci. Vis.* 18: 2767–2777.

Schultz, R., Ripoll, J., and Ntziachristos, V. 2003. Experimental fluorescence tomography of arbitrarily shaped diffuse objects using non-contact measurements. *Opt. Lett.* 28: 1701–1703.

Schultz, R., Ripoll, J., and Ntziachristos, V. 2004. Fluorescence tomography of tissues with non-contact measurements. *IEEE Med. Imaging* 23: 492–500.

Schulz, R. and Ntziachristos, V. 2010. Hybrid x-ray CT—Fluorescence molecular tomography scanner. *IEEE Trans. Med. Imaging.* 29(2): 465–473.

Schweiger, M., Arridge, S. R., and Nissila, I. 2005, May 21. Gauss–Newton method for image reconstruction in diffuse optical tomography. *Phys. Med. Biol.* 50: 2365–2386.

Sharpe, J., Ahlgren, U., Perry, P., Hill, B., Ross, A., Hecksher-Sorensen, J., Baldock, R., and Davidson, D. 2002, April 19. Optical projection tomography as a tool for 3D microscopy and gene expression studies. *Science* 296: 541–545.

Shashkov, E. V., Everts, M., Galanzha, E. I., and Zharov, V. P. 2008, November. Quantum dots as multimodal photoacoustic and photothermal contrast agents. *Nano Lett.*, 8: 3953–3958.

Shu, X., Royant, A., Lin, M., Aguilera, T., Lev-Ram, V., Steinbach, P., and Tsien, R. 2009. Mammalian expression of infrared fluorescent proteins engineered from a bacterial phytochrome. *Science* 8: 804–807.

Siphanto, R. I., Kolkman, R. G. M., Huisjes, A., Pilatou, M. C., de Mul, F. F. M., Steenbergen, W., and van Adrichem, L. N. A. 2004. Imaging of small vessels using photoacoustics: An in vivo study. *Lasers Surg. Med.* 35: 354–362.

Soubret, A., Ripoll, J., and Ntziachristos, V. 2005. Accuracy of fluorescent tomography in the presence of heterogeneities: Study of the normalized Born ratio. *IEEE Trans. Med. Imaging* 24: 1377–1386.

Stein, A., Otto, P., McCorvey, B. M., Khamapirad, T., and Oraevsky, A. 2006. Opto-acoustic tomography system for diagnostic imaging of breast cancer. *Breast Cancer Res. Treat.* 100: S168–S168.

Tsien, R. Y. 1998. The green fluorescent protein. *Annu. Rev. Biochem.* 67: 509–544.

Tsien, R. Y. 2005, February 7. Building and breeding molecules to spy on cells and tumors. *Febs Lett.* 579: 927–932.

Tung, C. H. 2004. Fluorescent peptide probes for in vivo diagnostic imaging. *Biopolymers* 76: 391–403.

Turner, G. M., Zacharakis, G., Soubret, A., Ripoll, J., and Ntziachristos, V. 2005, February 15. Complete-angle projection diffuse optical tomography by use of early photons. *Opt. Lett.* 30: 409–411.

van Roessel, P. and Brand, A. H. 2002, January. Imaging into the future: Visualizing gene expression and protein interactions with fluorescent proteins. *Nat. Cell Biol.* 4: E15–E20.

Wang, L., Jackson, W. C., Steinbach, P. A., and Tsien, R. Y. 2004a, November 30. Evolution of new nonantibody proteins via iterative somatic hypermutation. *Proc. Natl. Acad. Sci. USA* 101: 16745–16749.

Wang, X., Pang, Y., Ku, G., Xie, X., Stoica, G., and Wang, L. V. 2003, July. Noninvasive laser-induced photoacoustic tomography for structural and functional in vivo imaging of the brain. *Nat. Biotechnol.* 21: 803–806.

Wang, Y., Xie, X., Wang, X., Ku, G., Gill, K. L., O'Neal, D. P., Stoica, G., Wang, L. V. 2004b. Photoacoustic tomography of a nanoshell contrast agent in the in vivo rat brain. *Nano Lett.* 4: 1689–1692.

Webb, R. H. 1999. Theoretical basis of confocal microscopy. *Methods Enzymol.* 307: 3–20.

Weissleder, R. and Pittet, M. 2008. Imaging in the era of molecular oncology. *Nature* 452: 580–589.

Weissleder, R., Tung, C. H., Mahmood, U., and Bogdanov, A. Jr. 1999, April. In vivo imaging of tumors with protease-activated near-infrared fluorescent probes. *Nat. Biotechnol.* 17: 375–378.

Wunder, A., Tung, C. H., Muller-Ladner, U., Weissleder, R., and Mahmood, U. 2004, August. In vivo imaging of protease activity in arthritis: A novel approach for monitoring treatment response. *Arthritis Rheum.* 50: 2459–2465.

Xu, M. and Wang, L. V. 2005, January. Universal back-projection algorithm for photoacoustic computed tomography. *Phys. Rev. E Stat. Nonlin. Soft Matter Phys.* 71: 016706.

Yang, X., Skrabalak, S. E., Li, Z. Y., Xia, Y., and Wang, L. V. 2007, December. Photoacoustic tomography of a rat cerebral cortex in vivo with au nanocages as an optical contrast agent. *Nano Lett.* 7: 3798–3802.

Zacharakis, G., Ripoll, J., Weissleder, R., and Ntziachristos, V. 2005, July. Fluorescent protein tomography scanner for small animal imaging. *IEEE Trans. Med. Imaging* 24: 878–885.

Zaheer, A., Lenkinski, R. E., Mahmood, A., Jones, A. G., Cantley, L. C., and Frangioni, J. V. 2001, December. In vivo near-infrared fluorescence imaging of osteoblastic activity. *Nat. Biotechnol.* 19: 1148–1154.

Zhang, J., Campbell, R. E., Ting, A. Y., and Tsien, R. Y. 2002, December. Creating new fluorescent probes for cell biology. *Nat. Rev. Mol. Cell Biol.* 3: 906–918.

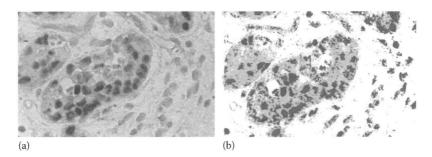

(a) (b)

FIGURE 7.13 (a) The color microscopy image of a histology sample and (b) *k*-means classification of the same image area based on a spectral cube data set composed of 22 bands in the visible part of the spectrum using the *k*-means classifier.

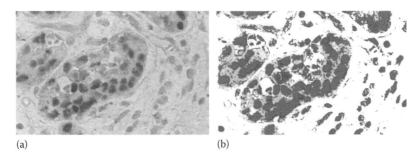

(a) (b)

FIGURE 7.15 (a) The color microscopy image of a histology sample and (b) classification of the same image area using the SAM algorithm and a training set of spectra, representative to the classes present in the sample.

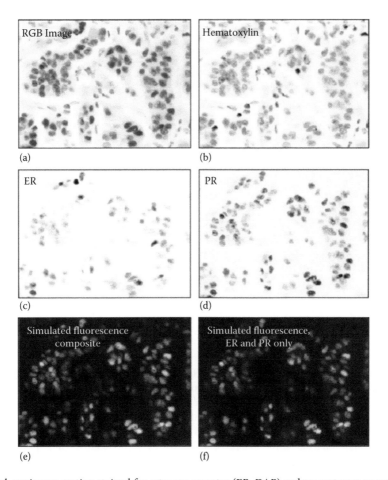

RGB Image Hematoxylin

(a) (b)

ER PR

(c) (d)

Simulated fluorescence composite Simulated fluorescence, ER and PR only

(e) (f)

FIGURE 7.18 Invasive ductal carcinoma section stained for estrogen receptor (ER, DAB) and progesterone receptor (PR, Fast Red), and counterstained with hematoxylin. (a) RGB representation of data set; (b–d) unmixed images corresponding to hematoxylin, ER, and PR, respectively; (e) simulated fluorescence composite with hematoxylin in blue, ER in red, and PR in green; (f) simulated fluorescence composite showing only ER (red) and PR (green). (Reprinted from Mansfield, J.R. et al., *Curr. Protoc. Mol. Biol.* 84, 1–14, 2008.)

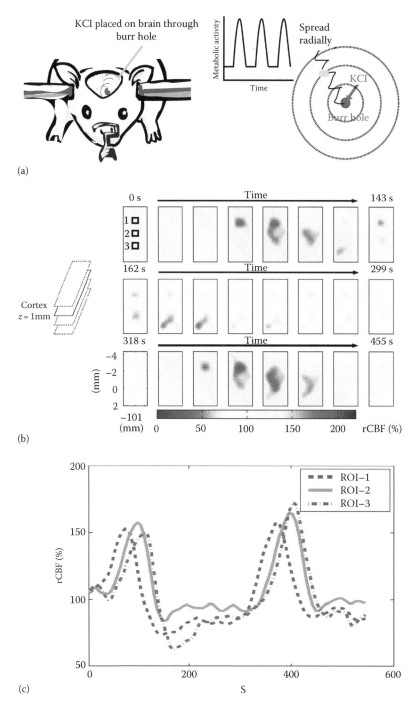

FIGURE 10.10 CSD in rat brain. (a) A rat was fixed on a stereotaxic frame with the scalp retracted and the skull intact. CSD was induced by placing KCl solution on the rat brain through a small hole drilled through the skull. Periodic activations and deactivations of the neurons then spread out radially on the cortex. (b) rCBF changes on the cortex of the rat brain (~1 mm deep) as a function of time during CSD. rCBF images from the cortex (from left to right, from top to bottom) are shown roughly every 20 s starting immediately before KCl was applied until the end of the second CSD peak. A strong increase in BF appears from the top and proceeds to the bottom of the image. After the peak, there is a sustained decrease in BF, which covers most of the image area. The amount of rCBF changes is reflected in the image as a spectrum of color, with deep blue indicates a decrease in rCBF and dark red indicates an increase in rCBF. (c) Temporal rCBF curves from three ROI indicated in the 0 s image in (b), demonstrating the propagation of the CSD waves. (From Zhou, C. et al., *Opt. Express*, 14, 1125, 2006. With permission.)

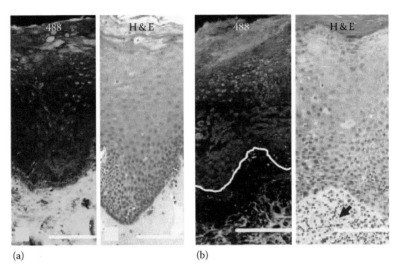

FIGURE 11.1 Confocal images of fresh tissue slices from the oral cavity at 488 nm excitation with corresponding H & E stained pathology section. (a) Normal oral tissue. (b) Tissue with mild dysplasia. Note that in the fluorescence images there is a decrease in intensity in the stroma, located at the bottom of the images, in the mild dysplasia compared to the normal tissue. The basement membrane is illustrated with a white line in B. (Adapted from Pavlova, I. et al., *Clin. Caner Res.*, 14(8), 2396, 2008. With permission.)

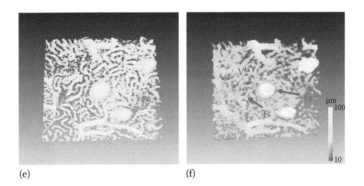

FIGURE 13.11 OCT imaging of human kidney ex vivo. (e) 3D volumetric image of the segmented glomeruli and tubular network. (f) Automatically quantified luminal and interstitial dimensions and color-coded on the structural image. (Adapted from Li, Q. et al., *Opt. Express*, 17, 16000, 2009. With permission.)

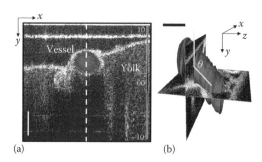

FIGURE 14.1 Doppler OCT measurement and volume rendering of chicken embryo heart vessel. Doppler OCT image (blue) superimposed on an OCT intensity image of a vessel cross section (a). Three-dimensional volume rendering of the same vessel (b). The volume rendering is used to measure the angle of blood flow (green), relative to the OCT scanning beam. Two orthogonal OCT planes are displayed, and the x–y plane corresponds to the image shown in (a). Scale bar is 200 μm. (From Davis, A. et al., *Anat. Rec. (Hoboken)*, 292, 311, 2009. With permission.)

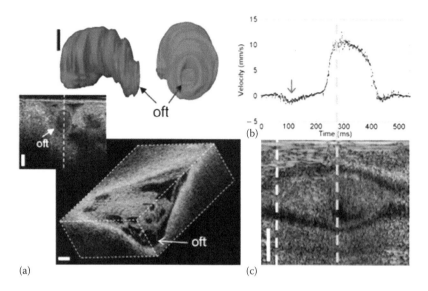

FIGURE 14.4 Doppler OCT measurements of one stage of chick embryo heart development (Hamburger–Hamilton stage 11). Surface rendering and volume reconstruction of the heart tube (a). Inset shows Doppler OCT images of blood flow out of the outflow tract (out of the plane). Doppler velocity measurements were acquired along the yellow dashed line. Blood flow velocity dynamics (b). M-mode images (c) where the vertical axis is depth and the horizontal is time. These M-mode images are acquired simultaneously with the velocity measurement, enabling correlation of heart tube contractions. Doppler velocity measurements were taken along the green dotted line in (c). oft, outflow tract, scale bar = 250 μm. (From Davis, A.M. et al., *J. Opt. Soc. Am. A: Opt. Image Sci. Vis.*, 25, 3134, 2008. With permission.)

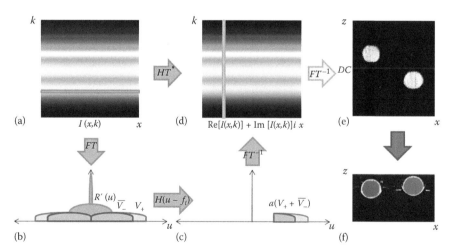

FIGURE 14.5 Flowchart of SPFI-SDOCT processing. A lateral Fourier transform of the B-scan (a) yields the spatial frequencies of the sample centered at DC (structural data, yellow) and spatial frequencies of moving scatterers shifted by their respective Doppler frequencies (positive velocities in red, negative velocities in blue) (b). Applying a frequency-shifted Heaviside step function (c) and then applying an inverse Fourier transform of the resulting spatial frequencies recreates the analytic interferometric signal (modified Hilbert transform) (d). A spectral inverse Fourier transform of the analytic interferometric signal maps the depth-resolved reflectivities of scatterers moving in opposite directions (into or out of the A-scan axis) on opposite image half-planes (e), which can then be overlaid for vessel identification (f). (From Tao, Y.K. et al., *Opt. Express*, 16, 12350, 2008. With permission.)

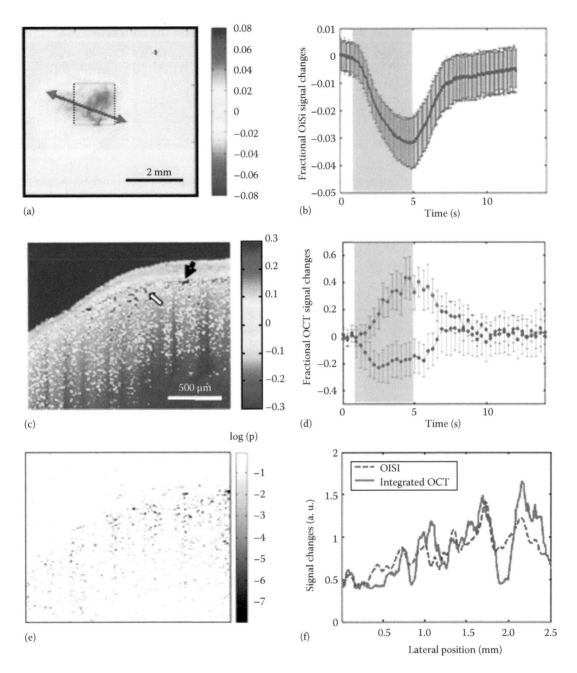

FIGURE 14.7 Representative results of OISI and OCT imaging of functional activation. En face functional activation map from OISI with co-registered OCT scan indicated by the red arrow (a). The averaged time course for OISI over the region denoted by dashed box in (a) with stimulation period shaded grey (b). Functional OCT image superimposed with the structural image (c). In order to simultaneously display these two images, we threshold the functional OCT signals to those with responses >5% (or <0.5%). Representative functional OCT time courses for regions of interest specified by solid arrow (positive changes) and hollow arrow (negative changes) (d). Map of *p*-values showing the statistical significance of the fractional OCT changes from the baseline (e). The lateral line profile of OISI signal changes and the depth-integrated OCT signal changes showing the spatial correspondence of OCT and OISI signals (f). The OISI and OCT signal changes are scaled to overlap in order to provide a visual comparison. (From Chen, Y. et al., *J. Neurosci. Methods*, 178, 162, 2009. With permission.)

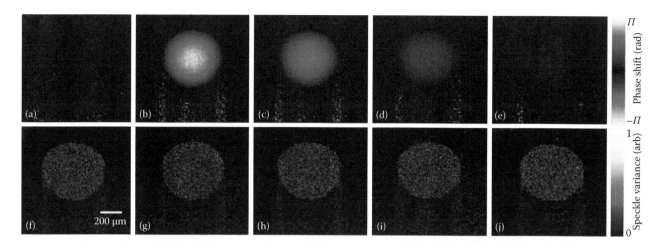

FIGURE 14.8 Doppler phase shift measured for stationary Intralipid within a 600 μm tube at an 80° Doppler angle (a). Doppler phase shift measured from Intralipid moving through the tube at a rate of 12 mL/h for Doppler angles of 75°, 80°, 85°, 89.5°, respectively (b–e). Corresponding normalized (to fixed value for all images) speckle variance images (f–j). (From Mariampillai, A. et al., *Opt. Lett.*, 33, 1530, 2008. With permission.)

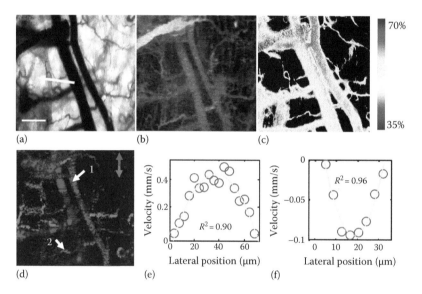

FIGURE 14.10 Multidimensional functional imaging of 4T1 tumor vasculature in the dorsal skin fold mouse window chamber. Transmission image of the window chamber, with a portion of the tumor visible (arrow, a). Scale bar is 200 μm. Speckle variance OCT image of vessel morphology shown as an en face average intensity projection over 1 mm depth (b). Hyperspectral image of percent hemoglobin oxygen saturation, thresholded for the R-squared value of the linear fit and total absorption value at each pixel (background pixels are black) (c). Doppler OCT image of vessel blood flow direction shown as an en face maximum intensity projection over depth, with red vessels (◻◦◻1) flowing toward the top of the image and blue vessels (2) flowing toward the bottom of the image (d). The flow profiles at points 1 (e) and 2 (f) were fit to a second-order polynomial and corrected for the angle of incidence to provide velocity in millimeters per second. (From Skala, M.C. et al., *Opt. Lett.*, 34, 289, 2009. With permission.)

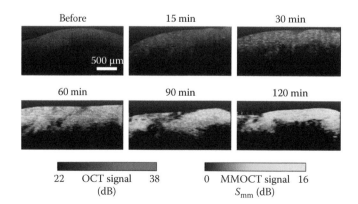

FIGURE 14.19 Representative magnetomotive OCT images of iron oxide nanoparticle diffusion in tumors versus time. Red and green display the structural (OCT) and magnetomotive (MMOCT) image channels, respectively, as indicated by the colored scale bars. (From Oldenburg, A.L. et al., *Opt. Express*, 16, 11525, 2008. With permission.)

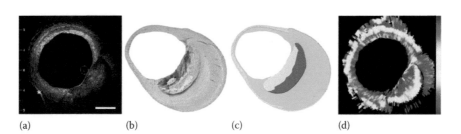

(a)   (b)   (c)   (d)

FIGURE 15.8   (a) OCT image of a coronary atheroclerotic lesion ex vivo: * indicates the needle used for marking the imaged site, the length of the white scale bar is 1 mm; (b) corresponding histology, H&E stain; (c) cartoon histology, overlayed on the original histology slide, indicating an advanced necrotic core red behind a calcification gray, and a slight fibrotic green circumferential intimal thickening; and (d) OCT-derived attenuation coefficient μ_t, plotted on a continuous linear color scale from 0 to 15 mm^{-1}. The area corresponding to the necrotic core exhibits a higher attenuation coefficient (8–10 mm^{-1}) than the calcification next to it or the surrounding fibrous tissue (2–3 mm^{-1}). (From van Soest, G. et al., *J. Biomed. Opt.*, 15(1), 011105, 2010. With permission.)

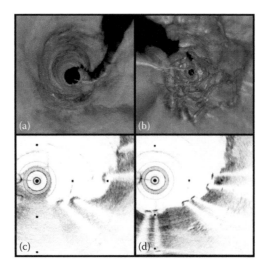

FIGURE 15.9   (a) Fly-through view (distal-proximal) of the BMS shows covered struts underneath the surface of the artery wall, as well as some struts that appear near the luminal surface. (b) Fly-through view (distal-proximal) of the DES demonstrates uncovered struts. (c and d) OFDI cross-sectional images of the BMS and DES, respectively. The OFDI appearance of the struts is different for the 2 stents. Tick marks, 1 mm. (From Tearney, G.J. et al., *J. Am. Coll. Cardiol. Imaging*, 1(6), 752, 2008. With permission.)

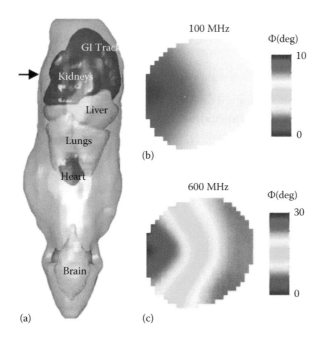

FIGURE 16.7   Phase shift at 100 MHz (b) and 600 MHz (c) in an axial cross section through a mouse model (a). The position of the cross section is indicated by the black arrow in (a). The calculations were done with a FV transport-theory code (Ren et al., 2004). The optical properties (μ_a, μ_s') in units of cm^{-1} were chosen to be as follows: brain (0.3, 15), GI track (0.27, 12.5), heart (0.1, 5), kidneys (0.67, 6), liver (0.33, 7.5), and lung (0.81, 20). The anisotropy factor was assumed constant g = 0.9. (Adapted from Ren, K. et al., *Opt. Lett.*, 29(6), 578, 2004.)

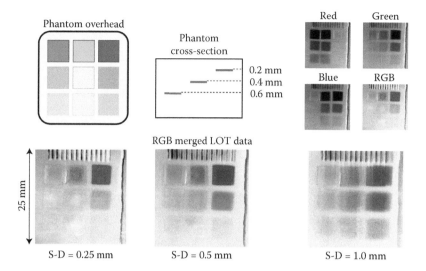

FIGURE 18.9 Simultaneous multiwavelength phantom imaging. Top left: layout of the phantom showing different colored absorbers across rows and deeper depths down each column. Top right: raw LOT data simultaneously collected for each illumination wavelength. Bottom: red-green-blue merge of the raw LOT data for 3 S-D separations. Deeper absorbers become more apparent in the wider detector separations. (Adapted from Burgess, S. A. et al., *Opt. Lett.*, 33, 2710, 2008.)

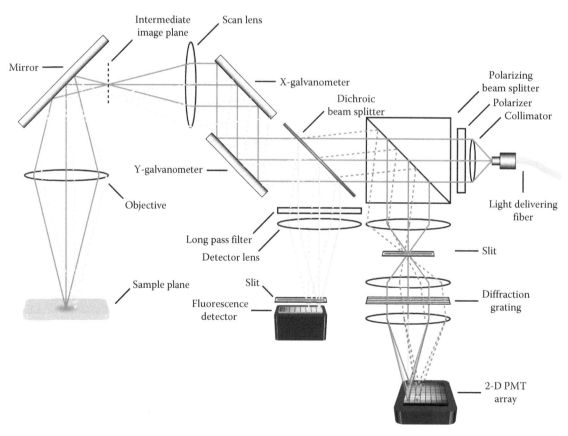

FIGURE 18.10 Simultaneous multiwavelength and fluorescence LOT system layout. Lasers with a wavelength of 488, 532, and 638 nm are delivered together into the system via a polarization-maintaining optical fiber. Careful selection of the dichroic beam splitter allows returning absorption contrast light to be passed while fluorescence light is reflected. A transmission diffraction grating and an 8 × 8 2D PMT array in the absorption detection arm permits simultaneous measurements of absorption contrast at multiple wavelengths.

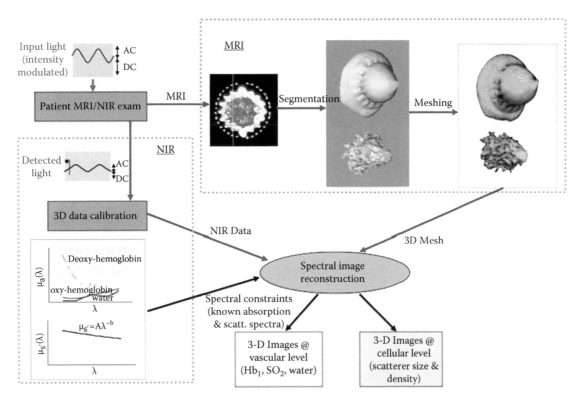

FIGURE 19.10 Schematic of processes involved in multimodality MRI-optical imaging with specific application to breast tissue characterization for diagnosis and treatment monitoring.

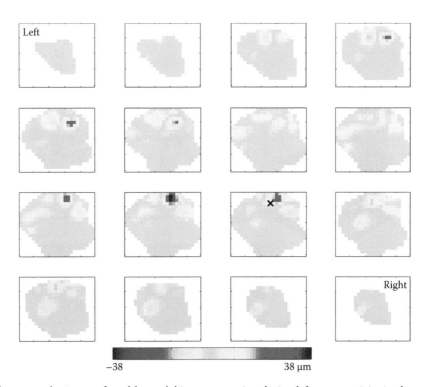

FIGURE 20.11 Optical tomography image of total hemoglobin concentration during left motor activity in the neonatal brain, showing an increase near the estimated position of the motor cortex (denoted by a cross). This figure shows 16 sagittal slices from the left (top right) to the right (bottom left) of the head. (From Gibson, A.P. et al., *Neuroimage*, 30, 521, 2006. With permission.)

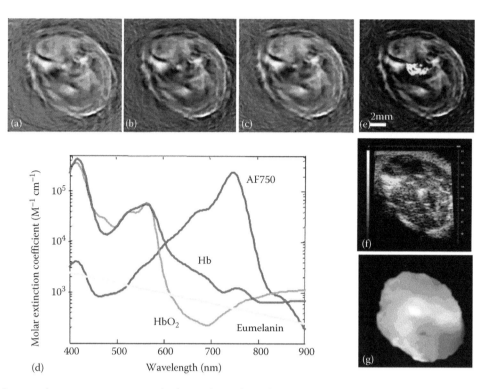

FIGURE 22.2 Multispectral opto-acoustic tomography (MSOT) visualizes distribution of fluorescent molecular probe (AlexaFluor 750™) in a mouse leg[13]. (a), (b), and (c) are cross-sectional opto-acoustic tomographic reconstructions acquired at 750, 770, and 790 nm, respectively. (d) Absorption as a function of wavelength for AF750 fluorescent probe as compared to some intrinsic tissue chromophores (not in scale). Three wavelengths are used to spectrally resolve the probe location. (e) Spectrally resolved MSOT image that incorporates measurements at all the three wavelengths (in color), superimposed onto a single-wavelength anatomical image. (f) Corresponding ultrasonic image, acquired approximately at the same imaging plane, using 25 MHz high-resolution ultrasound system. (g) Planar epi-fluorescence image of dissected tissue confirms the fluorochrome location. (From Ntziachristos, V., *Am. Chem. Soc.* 2795, 2009.)

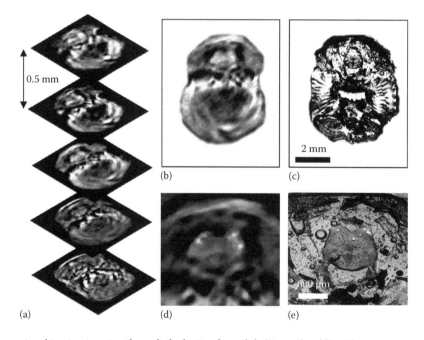

FIGURE 22.3 Three-dimensional in vivo imaging through the brain of an adult (6 months old) mCherry-expressing transgenic zebrafish. (a) Five transverse opto-acoustic imaging slices through the brain area of living zebrafish taken at 585 nm. Examples of imaged slice and its corresponding histological section are shown in (b) and (c), respectively. (d) MSOT image of the brain (zoom-in) with mCherry expression shown in color. (e) Corresponding fluorescent histology made through excised brain. (From Razansky, D. et al., *Nat. Photonics*, 3, 412, 2009.)

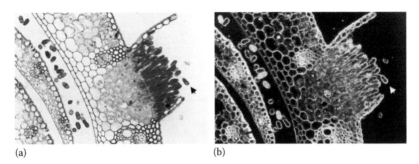

(a)   (b)

FIGURE 24.9 Bright-field (a) versus dark-field (b) images obtained from a stained object. Dark-field images enhance highly diffractive features, such as cell wall boundaries, that are comprised of strong refractive index variations between the feature and its surround. By comparison, bright-field image contrast is produced by variations in absorption of colored dyes or natural pigments that often fill the entire cell, making the cell wall or any internal structures less apparent. Arrows point to a spore that exemplifies the difference between these two microscope configurations. In the dark-field image, the cell wall is more clearly delineated but also shows some slight broadening due to flare, which may cause some loss of fine detail. Subject is a stained section of a fungal spore body of the wheat stem rust, *Puccinia graminis*. Images recorded using a color sensor and white light illumination. (Magnification approx. 100×).

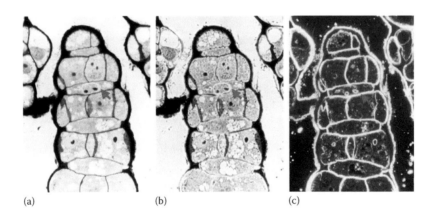

(a)   (b)   (c)

FIGURE 24.12 Bright-field (a) versus phase contrast (b) versus dark-field (c) images of a green alga, *Chara*, stained with hematoxylin-PAS. Although phase contrast is generally less useful for imaging stained objects, it can increase contrast and visibility of cell components, such as the granularity of the cell cytoplasm seen here. As expected, the dark-field image clearly shows cell wall structures better than the other modes, but often does not image other features as readily. For example, the red arrows point out a subcellular structure that appears as a bright oval ring in the bright-field and phase contrast images but is not visible in the dark-field result. Images recorded using a color sensor and white light illumination.

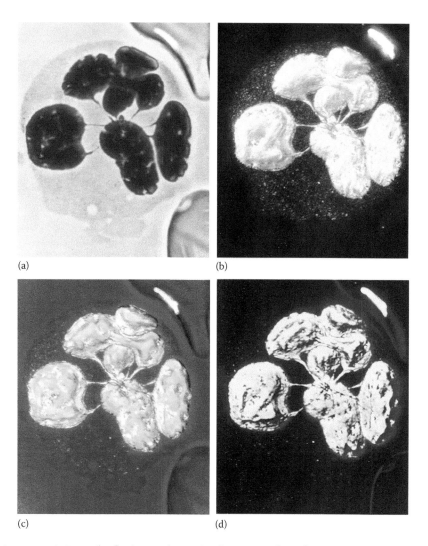

FIGURE 24.13 Combining transmission and reflection modes to visualize more subject features. (a) transmitted phase contrast (white light illumination), (b) reflected DIC (white light illumination), (c) combined transmitted phase contrast (with red filter in the transmitted white light illumination path) and reflected DIC, and (d) combined transmitted dark field (with red filter in illumination path) and reflected DIC. Subject is a stained white blood cell (neutrophil) from the salamander *Oedipina*. Color of the nucleus is due to the absorption and reflection properties of the dye, which appears magenta in transmitted phase contrast and green in reflected DIC. The red background and highlights in (c) and (d) are due to a red filter in the transmission illumination path to provide contrast with the green nucleus from the reflected light DIC imaging mode. Note how the combined mode (c) preserves the advantages of both the transmitted phase contrast (visibility of the granular cytoplasm and the transparent regions of the nuclear lobes) and the reflection DIC (visibility of the central convergence of strands from each nuclear lobe). In (d), the transmitted dark-field image combined with reflection DIC shows even more fine structure in the nuclear lobes. Images recorded using a color sensor and white light illumination. (Magnification 1500×).

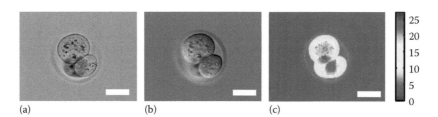

(a)　　　　　(b)　　　　　(c)

FIGURE 25.14 Multimodal images of live 3-cell mouse embryo in (a) brightfield, (b) DIC, (c) optical quadrature microscopy (From Warger, W.C. II and DiMarzio, C.A., *Opt. Express* 17: 2400, 2009).

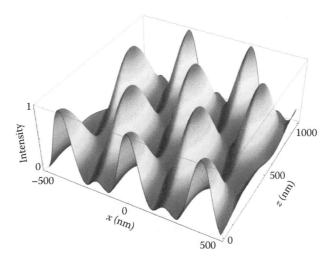

FIGURE 26.7 Intensity distribution for a three-beam interference ($\alpha = 60°$, $\lambda_0 = 488\,\text{nm}$, $n = 1.45$, $\text{NA} = 1.4$, $E_0 = E_{-1} = E_{+1}$). In comparison to a two-beam interference, the intensity distribution is additionally modulated along the z-axis. (The intensity is in arbitrary units.)

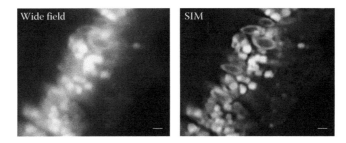

FIGURE 26.15 Two-color image (green: 488 nm excitation; red: 568 nm excitation) of lipofuscin accretions in human retinal pigment epithelium. In addition to the lateral resolution enhancement, the improved suppression of out-of-focus light of SIM is beneficial. Scale bars are 1 μm.

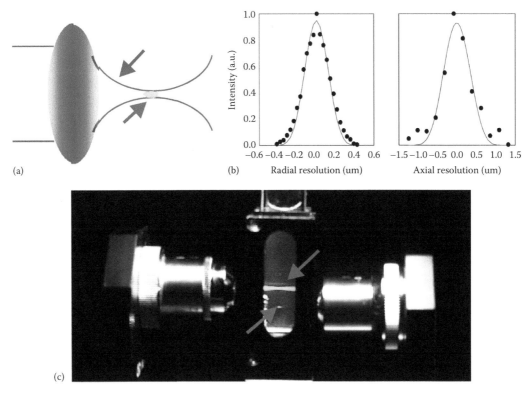

FIGURE 27.3 Point spread function of two-photon fluorescence microscope. (a) In traditional one-photon excitation, fluorescence is generated throughout the double inverted cones (blue arrow). Two-photon excitation generates fluorescence only at the focal point (red arrow). (b) The submicron point spread function (PSF) of two-photon excitation at 960 nm. The full widths at half maximum are 0.3 μm radially and 0.9 μm axially. (c) An experimental visualization of the small excitation volume of two-photon fluorescence. One- and two-photon excitation light are focused by two objectives (equal numerical aperture) onto a fluorescein solution. Fluorescence is generated all along the path in the one-photon excitation case (blue arrow) whereas fluorescence is generated only in a 3D confined focal spot for two-photon excitation (red arrow). The reduced excitation volume is thought to lead to less photodamage.

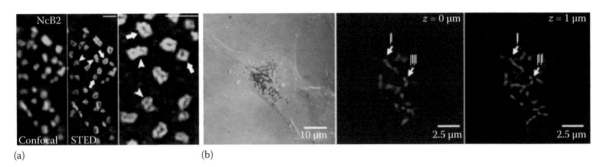

(a)   (b)

FIGURE 27.6 (a) *Drosophila* coiled-coil domain protein Bruchpilot imaged using confocal (left) and STED microscopy (middle and right). Scale bars represent 1 μm. (Adapted from Kittel, R.J. et al., *Science*, 312, 1051, 2006.) (b) Mitochondria network imaged using wide field two-photon imaging with PALM. DIC and epi-fluorescence view (left) and two-photon wide field imaging combined with PALM views at two depths. (Adapted from Vaziri, A. et al., *Proc. Natl. Acad. Sci. U. S. A.*, 105: 20221, 2008.)

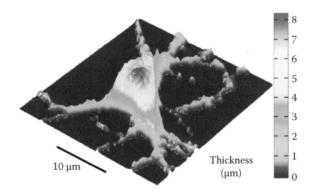

FIGURE 29.9 Absolute phase image, obtained for a living neuron in culture, represented in 3D. Provided that the refractive index of the cytoplasm can be determined, this image is a very good representation of the topology of the cell surface.

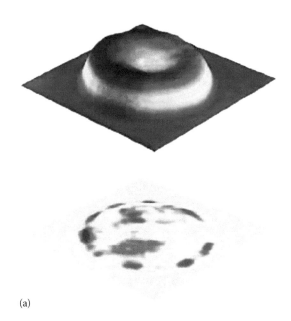

(a)

FIGURE 29.10 (a) Phase image of a living RBC on the surface of a cover slip. The lower part of the image shows the dynamical behavior of the living RBC: cell membrane fluctuations with amplitude of around 20 nm in a spectrum reaching a few tens of hertz. (From Kühn, J. et al., *Opt. Lett.*, 34, 653, 2009.)

FIGURE 29.13 Tomographic image of a testate amoebae *Hyalosphenia papilio*: tomographic image taken in the middle of the specimen, the refractive index is coded in color from around 1.4 to 1.6. The tomography has been obtained by rotating the specimen by 180°, one slice of the specimen is less than 3 µm. (From Pavillon, N. et al., Optical tomography by digital holographic microscopy. In: *Novel Optical Instrumentation for Biomedical Applications IV*, SPIE, 2009a.)

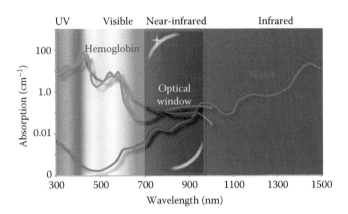

FIGURE 31.2 Wavelength-dependent light absorption in biological tissues due to dominant chromophores: oxy- and deoxy-hemoglobin in the visible and water in the infrared regions, respectively. Absorption of light is 1–2 orders of magnitude less in the near-infrared region, resulting in relative transparency termed the "Optical Window."

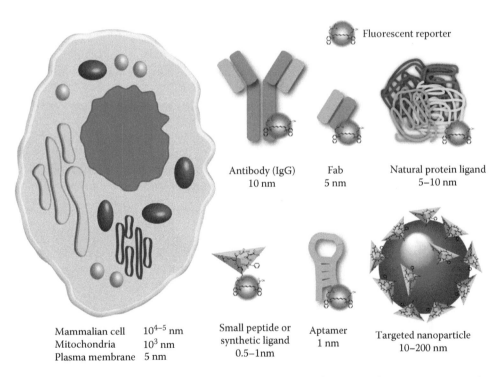

FIGURE 31.6 Representative fluorescent molecular imaging agents demonstrating the variety of targeting agents used in optical molecular imaging. The approximate sizes of a mammalian cell and organelles are given for comparison (right).

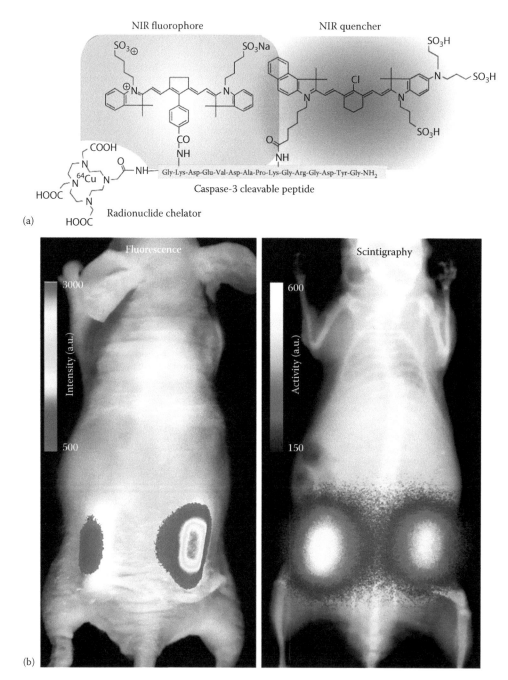

NIR fluorophore

NIR quencher

Caspase-3 cleavable peptide

Gly-Lys-Asp-Glu-Val-Asp-Ala-Pro-Lys-Gly-Arg-Gly-Asp-Tyr-Gly-NH$_2$

(a)

Radionuclide chelator

Fluorescence

Scintigraphy

Intensity (a.u.)

Activity (a.u.)

(b)

FIGURE 31.8 (a) Multimodal nuclear and activatable fluorescent probe for caspase-3 activity for detecting apoptosis. Radiolabeling with Cu[64] via metal chelator (blue) enables PET imaging. Cleavage of peptide sequence by Caspase-3 separates the NIR quencher (gray) from the NIR fluorophore (orange), enhancing fluorescence yield. (b) Fluorescence (left) and scintigraphic (right) imaging of a mouse with subcutaneously implanted tubes containing equal concentrations of **A** in solution with either BSA (left side) or caspase-3 enzyme (right side) 2 h after implantation. The scintigraphy signal was used to quantitatively measure the probe concentration while the fluorescence signal indicated the enzyme activity with approximately sixfold increase in intensity.

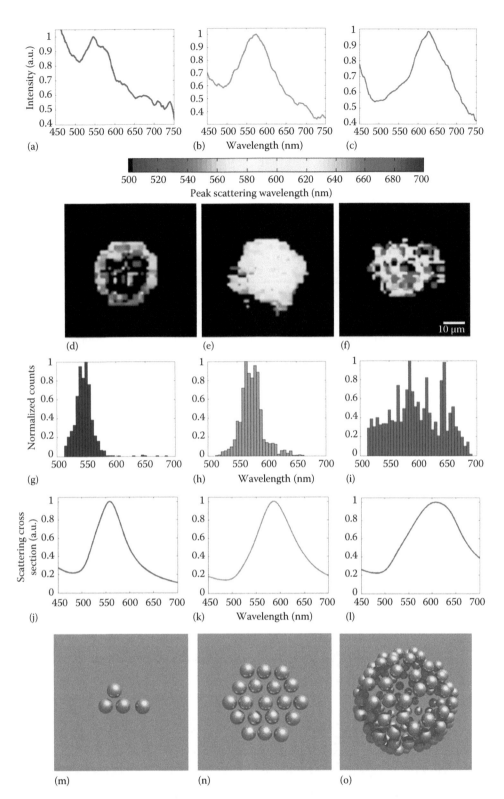

FIGURE 32.1 Quantitative relationship between regulatory states of EGFR and scattering from EGFR-bound gold nanoparticles. Hyperspectral darkfield microscopy of cells labeled at 4°C (left column, a, d, g), 25°C (middle column, b, e, h), and 37°C (right column, c, f, i). As temperature increases, EGFR complexes with nanoparticles undergo a gradual transition from small clusters to a more 3D volume-filling morphology that is associated with the trafficking of EGF receptors from cell membrane to early and, finally, to late endosomes. Representative single-pixel spectra from a cell image display scattering peaks at ca. 546 (a), 576 (b), and 601 nm (c). Cell images in (d–f) are color-coded according to the scattering peak position at each pixel in the field of view. Distributions of the peak scattering wavelengths indicated in (d–f) are shown in (g–i). Electrodynamic simulations of scattering from nanoparticle aggregates are shown in (j–l) for the following three typical structures: a 4-particle chain-like structure (j, m), a 19-particle disklike structure (k, n), and a 130-particle volume-like structure (l, o). The total scattering cross section is plotted for each structure alongside a rendering of the corresponding detailed particle arrangement (m–o). Distinct redshifting and broadening of the scattering spectra is due to both the effect of the increased number of contributing particles and the effect of the transition from a 2D to a more 3D volume-filling morphology. (From Aaron, J. et al., *Nano Lett.*, 9, 3612, 2009. With permission.)

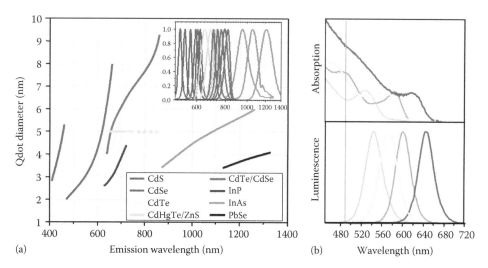

FIGURE 32.3 (a) Emission maxima and sizes of quantum dots of different composition. The insert shows representative emission spectra for the many types of quantum dots. (b) Absorption and the corresponding luminescence spectra for different types of QDs. (From Michalet, X. et al., *Science*, 307, 538, 2005. With permission.)

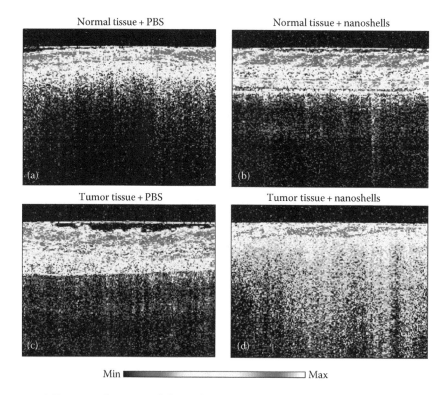

FIGURE 32.9 Representative OCT images from normal skin and muscle tissue area of mice injected systemically with either PBS (a) or gold nanoshells (b). (c and d) show representative OCT images from tumor xenografts following systemic injection of either PBS (c) or gold nanoshells (d). Scale bar is 200 microns. (From Gobin, A.M. et al., *Nano Lett.*, 7, 1929, 2007. With permission.)

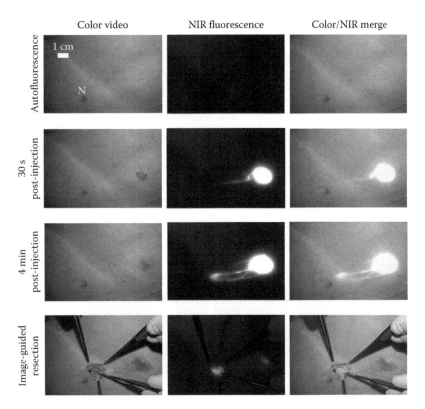

FIGURE 32.11 Fluorescence-guided surgical resectioning of sentinel lymph node using QD contrast agents. The blue color is trypan blue, the current clinical standard for detecting sentinel lymph nodes. Quantum dots were observed to offer much greater contrast for detecting sentinel lymph nodes in combination with a fluorescence imaging system. (From Kim, S. et al., *Nat. Biotechnol.*, 22, 93, 2004. With permission.)

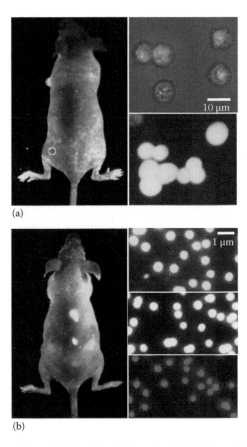

FIGURE 32.12 Sensitivity and multicolor capability of QD imaging in live animals. (a) Comparison between QD-tagged (right flank) and GFP-transfected cancer cells (left flank) shows the difficulty of detecting implanted GFP-expressing cancer cells in vivo while QD-tagged cells are easily observed. (b) Simultaneous in vivo imaging of multicolor QD-encoded microbeads demonstrates the ability to image multiple colors of quantum dots simultaneously. (From Gao, X. et al., *Nat. Biotechnol.*, 22, 969, 2004. With permission.)

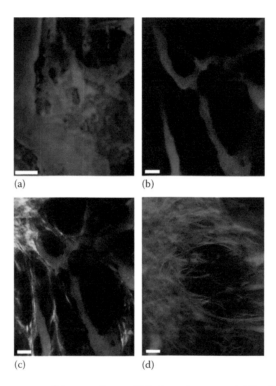

FIGURE 32.14 In vivo imaging of a mouse xenograft tumor using multiphoton microscopy. (a) Depth projection of 20 images taken at 5 micron intervals demonstrates extravasation of fluorescent dextran causing blurring of tumor blood vessels. (b) Depth projection of 15 images at 5 micron intervals with QD470 contrast agent present in bloodstream. (c) Simultaneous imaging of GFP and QD470 used to visualize both perivascular cells and blood vessels, taken in the same region as (b). (d) Imaging of endogenous tissue second-harmonic generation in blue and QD660 contrast agents to visualize the blood vessels in red. (From Stroh, M. et al., *Nat. Med.*, 11, 678, 2005. With permission.)

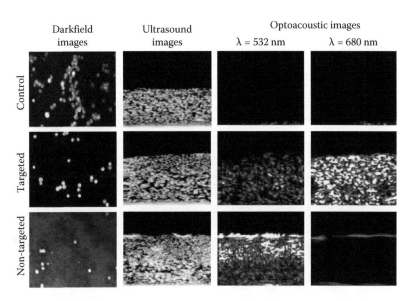

FIGURE 32.15 Dark field, ultrasound, and optoacoustic images of tissue phantoms taken at 532 and 680 nm excitation wavelength. The control phantom contained cells and collagen, the nontargeted phantom contained cells and PEGylated spherical gold nanoparticles of 40 nm diameter, and the targeted tissue phantom contained cells specifically labeled with anti-EGFR gold nanoparticles, as seen in the dark field images. Photoacoustic imaging shows an increase in signal associated with receptor-mediated nanoparticle aggregation as seen by the difference in signal at 680 nm excitation between the PEGylated and EGFR-targeted nanoparticle tissue phantoms. (From Mallidi, S. et al., *Opt. Express*, 15, 6583, 2007.)

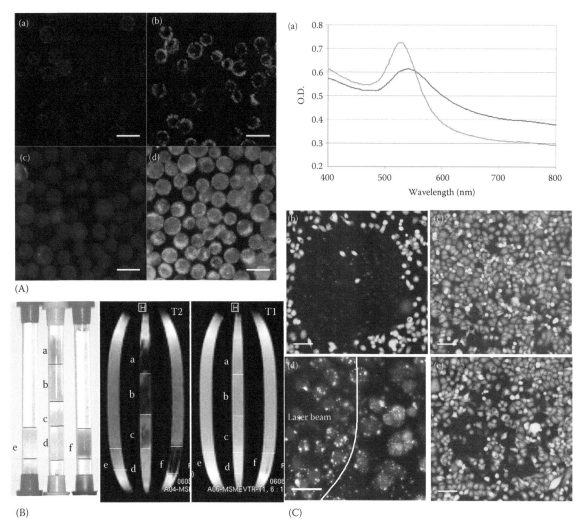

FIGURE 32.18 Hybrid plasmonic magnetic nanoparticles (AuFeOx) with both optical and magnetic properties allow a combination of optical and magnetic resonance imaging and photothermal therapy. (A) Confocal (top row) and dark field (bottom row) reflectance microscopy demonstrate clear difference in signal between EGFR(+) (left column) and EGFR(– cells (right column) after incubation with EGFR-targeted AuFeOx nanoparticles. (B) MRI images of tissue phantoms containing different concentrations of cells labeled with gold nanoparticles. Negative T2 contrast increases as the concentration of cells labeled with AuFeOx nanoparticles increases from (d) to (a) in the middle MRI tube. (C) (a) Absorbance spectra of unlabeled cells in the presence of PEGylated hybrid nanoparticles (green) and of labeled cells (red). The total amount of nanoparticles is the same in both cell suspensions. Fluorescence images of cells labeled with anti-EGFR hybrid nanoparticles (b) and cells pre-exposed to PEG-coated particles (c) after one 7 ns, 400 mJ/cm² laser pulse at 700 nm. Green fluorescence is due to calcein AM live stain that indicates cellular survival. (d) Dark field reflectance image of the irradiated labeled cells at the boundary of the laser beam spot shows that the cells do not detach after laser treatment. (e) Fluorescence image of cells pre-exposed to PEG-coated hybrid nanoparticles after 600 hundred 7 ns, 400 mJ/cm² pulses at 700 nm. The scale bars are ca. 100 μm in fluorescence images and ca. 50 μm in the dark field image. (From Larson, T.A. et al., *Nanotechnology*, 18, 325101, 2007. With permission.)

IV

Microscopic Imaging

23 Assessing Microscopic Structural Features Using Fourier-Domain Low Coherence Interferometry
Robert N. Graf, Francisco E. Robles, and Adam Wax 463
Introduction • Background • Theoretical Basis of fLCI • Imaging Systems: Design and Validation • Animal Tissue
Studies • Conclusion • Acknowledgments • References

**24 Phase Imaging Microscopy: Beyond Dark-Field, Phase Contrast, and Differential Interference Contrast
Microscopy** *Chrysanthe Preza, Sharon V. King, Nicoleta M. Dragomir, and Carol J. Cogswell* 483
Introduction • Phase Imaging Microscopy: Past and Present • Recent Advancements for Phase Imaging Microscopy:
Development and New Directions • Conclusion • References

25 Confocal Microscopy *William C. Warger, II, Charles A. DiMarzio, and Milind Rajadhyaksha* 517
Introduction to Confocal Imaging • Confocal Imaging
Parameters • Instrumentation • Summary • Acknowledgments • References

26 Fluorescence Microscopy with Structured Excitation Illumination *Alexander Brunner, Gerrit Best,
Paul Lemmer, Roman Amberger, Thomas Ach, Stefan Dithmar, Rainer Heintzmann, and Christoph Cremer* 543
Introduction • Abbe's Resolution Definition • Focusing SEI Methods • Wide-Field SEI Methods • The Conceptual
Basis of the SIM Method • Discussion and Outlook • Acknowledgments • References

27 Nonlinear Optical Microscopy for Biology and Medicine *Daekeun Kim, Heejin Choi, Jae Won Cha, and
Peter T. C. So* 561
Introduction • Principles and Methodology • Frontiers of Nonlinear Optical
Microscopy • Conclusion • Acknowledgments • References

28 Fluorescence Lifetime Imaging: Microscopy, Endoscopy, and Tomography *James McGinty, Clifford Talbot,
Dylan Owen, David Grant, Sunil Kumar, Neil Galletly, Bebhinn Treanor, Gordon Kennedy, Peter M. P. Lanigan,
Ian Munro, Daniel S. Elson, Anthony Magee, Dan Davis, Gordon Stamp, Mark Neil, Christopher Dunsby, and
Paul M. W. French* 589
Introduction • FLIM Techniques and Technology • Applications of FLIM Microscopy • FLIM Endoscopy • FLIM
Tomography • Outlook • Acknowledgments • References

29 Application of Digital Holographic Microscopy in Biomedicine *Christian Depeursinge, Pierre Marquet,
and Nicolas Pavillon* 617
Introduction • The Concept of Coherence • Holography • Realization of Optical Setups for Holographic
Microscopy • Space-Bandwidth Adaptation • "Electronic Focusing" and Extended Depth of Field • DHM
Instrument • Further Developments • Conclusions and Outlook • Acknowledgments • References

30 Polarized Light Imaging of Biological Tissues *Steven L. Jacques* 649
Introduction • Toolkit • Depolarization Coefficients: μ_{LP} and μ_{CP} [cm^{-1}] • Experimental
Measurements of Depolarization in Tissues • A Polarized Light Camera for Clinical Use • Future
Directions • Acknowledgments • Glossary • References

23

Assessing Microscopic Structural Features Using Fourier-Domain Low Coherence Interferometry

23.1 Introduction 463
23.2 Background.. 464
23.3 Theoretical Basis of fLCI 466
Time-Frequency Distribution • Dual Window Method • Simulations: Properties
of the DW Method • Spatial Coherence in White Light Interferometry
23.4 Imaging Systems: Design and Validation 470
Common Path fLCI System • Imaging fLCI System • Dual Window Processing Method
Validation
23.5 Animal Tissue Studies 477
23.6 Conclusion..479
Acknowledgments...481
References..481

Robert N. Graf
Duke University

Francisco E. Robles
Duke University

Adam Wax
Duke University

23.1 Introduction

Cancers typically develop slowly over time, beginning with just a few abnormal cells that grow and proliferate. The majority of malignancies develop through precancerous states characterized by varying levels of architectural and cytologic abnormality (Kumar et al. 2005). Detecting these structural and chemical changes in tissues at the earliest possible stages can greatly reduce rates of mortality and morbidity.

The current "gold standard" for detecting cancer of the epithelial tissue is the histopathologic analysis of biopsy samples. Biopsy samples are removed from the patient before being fixed, stained, and examined by a pathologist for morphological abnormalities. Although this procedure is the standard practice for cancer diagnosis, there are several drawbacks to its current form, including limited interobserver agreement, latency between biopsy and diagnosis, and limited coverage of at-risk tissues. In recent years, a large amount of research has focused on developing optical detection techniques that overcome these limitations of the traditional biopsy.

One particular method for optical biopsy, based on analysis of elastically scattered light, has shown much promise for detecting early cancerous and precancerous changes. This approach obtains information about the structures with which the light interacts. In the case of application to biology and medicine, elastic scattering spectroscopy (ESS) has been used to investigate the cellular morphology of in vivo and ex vivo tissue samples in order to identify alterations in structure due to disease.

Perelman et al. used ESS to probe cellular morphology by analyzing periodic fine structures in the detected scattering signals (Perelman et al. 1998). Later, Backman et al. measured the intensity of backscattered light with sensitivity to angle, polarization, and wavelength in order to determine the size and distribution of in vitro cell nuclei (Backman et al. 2001). This study was significant as it drew attention to nuclear morphology as a strategic target for tissue diagnostics. Not only is the scattering intensity from the nucleus large due to the index of refraction difference between the nuclear membrane and the surrounding cytoplasm, but enlargement of the nuclear diameter is also a key indicator of precancerous growth (Kumar et al. 2005). Several subsequent studies also focused on probing the morphology of cell nuclei and other intercellular organelles (Mourant et al. 2002; Wax et al. 2002; Wilson et al. 2005) and clinical applications of ESS (Bigio et al. 2000; Dhar et al. 2006; Lovat et al. 2006).

These advancements have paved the way for a new elastic light scattering technique known as Fourier domain low coherence interferometry (fLCI). fLCI seeks to analyze depth-resolved spectroscopic information in order to recover nuclear morphology from specific subsurface tissue layers. Spectral analysis of these tissue layers aims to detect enlargement of cell nuclei to identify the earliest stages of precancerous transformation. This biomarker, either alone or in conjunction with other information derived from the light scattering signal, can provide the quantitative information necessary to distinguish between normal and dysplastic epithelial tissue with high sensitivity and specificity.

23.2 Background

The fLCI technique combines the sensitivity of light scattering methods to changes in structure with the depth resolution of optical coherence tomography (OCT). Depth resolution is obtained by using a low coherence interferometry scheme. Light from a low coherence source is split into sample and reference beams, which are incident upon an experimental sample and a reference mirror, respectively. The reflected beams are recombined and the interference signal is detected as a function of wavelength, converted to wave number, and Fourier transformed to obtain a depth-resolved reflection profile. In OCT, a focused beam is scanned across a sample to generate a tomographic image. In fLCI, structural information about scatterers within the sample is obtained by examining the wavelength dependence of elastically scattered light.

A schematic of the first fLCI system is shown in Figure 23.1. In this approach, broadband light from a white light source is input to a Michelson interferometer geometry. Light from the source is coupled into a fiber and recollimated before being incident on a beamsplitter (BS), which splits the light into a reference arm, incident on a mirror, and a sample arm, incident on the experimental sample. Light scattered by the sample is mixed with the reference field at BS, and the combined field is focused into a fiber and detected with a spectrometer.

fLCI generates depth resolution by low coherence interferometry in the frequency domain as in Fourier domain (FD) OCT. The equation for the intensity of the signal detected by the spectrometer is

$$I = E_S^2 + E_S^2 + 2 \cdot \mathrm{Re}\left[E_S E_R^\star\right]\cos\left(2 \cdot \Delta z \cdot k + \phi\right). \quad (23.1)$$

It should be noted that the interference information, contained in the third term of Equation 23.1, is modulated by a cosine, which oscillates as a function of k, the wave number. As a result, the interference spectrum detected by the fLCI system will feature oscillations with frequencies dependent on Δz or the optical path length difference between the reference and sample arms of the system. The axial spatial cross correlation function between the sample and reference fields (i.e., a depth scan or A-scan) can be obtained by Fourier transforming the interference term. The resulting cross correlation function will display peaks in intensity corresponding to the locations of individual reflectors in the experimental sample. fLCI utilizes this method to selectively analyze specific subsurface layers in probed samples.

In order to obtain depth-resolved spectroscopic information, fLCI requires more than simply depth resolution. The data must be processed to simultaneously obtain depth resolution and spectral resolution. In order to generate depth-resolved spectroscopic information from data acquired in a single domain, fLCI typically employs a short-time Fourier transform, as is done in spectroscopic OCT (Oldenburg et al. 2007). With this approach, a Gaussian window is applied to the interference term before Fourier transforming to produce a depth scan centered about a particular center wave number. By shifting of the center of the Gaussian window and repeating the process, a data set with both depth and spectral resolution can be generated. It should be noted, however, that with this approach, increases in depth resolution result in degradation of spectral resolution and vice versa. The implications of this relationship will be discussed further below.

Upon producing depth-resolved spectra, fLCI seeks to determine scatterer structure by analyzing wave number dependent oscillations originating from specific depths of interest. More specifically, fLCI seeks to distinguish between normal and dysplastic epithelial tissue by detecting enlargement in the cell nucleus that occurs at the earliest stages of precancerous development. Figure 23.2a shows an illustration representing the scattering events that take place at both the front and back surface of each nucleus where an index of refraction change is present. The reflections from the front and back surfaces of the nuclei will interfere with one another, producing constructive or destructive interference, depending on the wavelength of the incident light. If the incident field is sufficiently broadband, and if the intensity of the scattered signal can be resolved as a function of wave number, the detected spectrum will exhibit oscillations between constructive and destructive interference as shown in Figure 23.2b. The frequency of this oscillation is directly dependent on the diameter of the scatterer with larger particles resulting in a higher frequency of oscillation and smaller particles resulting in a lower frequency of oscillation. fLCI seeks to detect these wave number dependent oscillations in diagnostically relevant tissue layers in order to measure nuclear enlargement associated with a developing precancer.

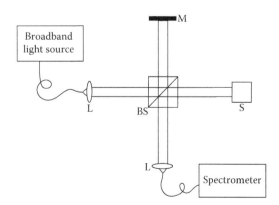

FIGURE 23.1 Schematic of the original fLCI system. L, lenses; BS, beamsplitter; M, reference mirror; S, sample.

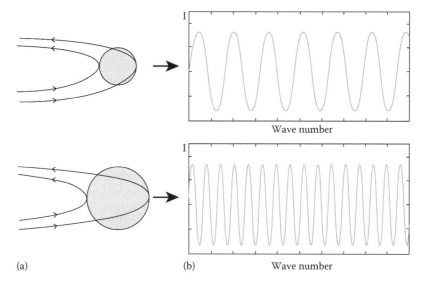

FIGURE 23.2 (a) Cell nuclei with incident and scattered fields indicated. (b) Interference spectra with wave number dependent oscillations caused by interference between front and back surface reflections.

Wax et al. first demonstrated the ability of fLCI to obtain depth-resolved spectra for determining scatterer sizes from elastic light scattering (Wax et al. 2003). This study reported the ability to isolate the scattering from polystyrene microspheres dried on the back surface of a cover glass. Fourier transform analysis of the spectral periodicity induced by the microspheres yielded a measured mean diameter of 1.65 μm ± 0.33 μm which compared well to the NIST certified 1.55 μm diameter. The results of the proof of principle experiment showed the ability of fLCI to size particles with subwavelength accuracy, actually surpassing the imaging resolution of the white light source.

Following this first initial demonstration, additional efforts sought to use spectroscopic OCT to provide functional information based on scattering properties. A 2004 study by Adler et al. (Adler et al. 2004) employed an autocorrelation metric to generate imaging contrast in SOCT images. This approach focused on using the autocorrelation of the depth-resolved spectrum as a contrast mechanism rather than looking for shifts in the center wavelength of the localized spectrum, producing an improvement in imaging contrast of biological specimens. This study did not quantitatively analyze spectral information to assess scatterer structure but instead employed a statistical approach to account for the effects of multiple scatterers in the focus of the incident imaging beam.

Further studies on using spectroscopic OCT were undertaken by Xu et al. (Xu et al. 2005a, 2006). In the first of these studies (Xu et al. 2005a), the effects of scatterer size on the light scattering spectrum were analyzed using ideal spherical particles. This study showed that the scattering spectra were sensitive to the size of an individual particle but that when using a sharply focused beam, the spectral signature shifted significantly. In addition, the effects of multiple scatterers within the focal region were found to further confound inversion of the spectral data to assess scatterer structure. Further work from this group

combined the use of spectroscopic OCT with multiphoton microscopy (Xu et al. 2006) to generate contrast using scatterer structure. In this work, the first peak in the Fourier transformed spectral data was used as a metric to provide imaging contrast. In studies of rat tissues and fibroblasts, this approach was shown to produce discrimination between tissue types and, significantly, was able to isolate the scattering contribution from the cell nucleus. This work also pointed toward a new direction for avoiding the trade-off between spectral and depth information by relying on the confocal effect of a tightly focused beam to maintain depth resolution in spectral data.

As an alternative to using spectral data to assess scattering properties using low coherence interferometry, Dyer et al. (Dyer et al. 2006) proposed an analysis method based on phase dispersion, that is, the variation of phase with frequency. This study analyzed scattering by phantoms with spherical scatterers and obtained qualitative agreement with Mie theory but also showed that the use of a focused beam complicated analysis of the scattering data. A follow on study by this group (Dennis et al. 2008) used phantoms with more elaborate structures and demonstrated quantitative sizing of scatterers and functional imaging based on raster scanning.

The studies described here have used both Fourier transform methods and Mie theory-based analysis to assess structure based on spectral data. The remainder of this chapter will focus on the use of the former approach; however, the use of theoretical treatments to infer scattering features has been widely applied in biomedical optics. For a more thorough overview, the reader is directed to the volume, *Biomedical Applications of Light Scattering* (Wax and Backman 2009), which contains detailed information on the subject, including a detailed presentation of light scattering formalisms, theoretical analyses, and experimental applications to analysis of biological and biomedical problems.

The following section focuses on the extraction of depth-resolved spectral information from interferometric signals.

23.3 Theoretical Basis of fLCI

23.3.1 Time-Frequency Distribution

The development of the fLCI technique requires an in-depth analysis of the temporal and spatial coherence of the detected signal. The signal processing literature points to a particularly relevant approach based on examining time-frequency distributions (TFDs). These distributions are of particular interest because the depth-resolved spectroscopic data obtained in fLCI can be viewed as an adapted form of a TFD with depth analogous to time and wave number analogous to frequency. In order to obtain information about the interferogram signal (Equation 23.1) in both the time and frequency domains simultaneously, the signal must be processed using one of two main approaches. Linear operations can be applied to the signal to yield a linear TFD or higher order functions can be calculated such as the bilinear TFDs which comprise the Cohen class of functions (Cohen 1989).

In order to generate a TFD from data acquired in a single domain, a linear operation such as the STFT is applied. Here a window is generated mathematically with a finite width and center and applied to the acquired signal. In the case of a time domain (TD) signal, I_T, a temporal window W of width T and center t_o can be applied,

$$I_T(k,t_o) = \int I_T(t) W(t,T,t_o) e^{ikt} dt, \qquad (23.2)$$

to yield the spectrum of light associated with the signal at time t_o. By executing this operation at several values of t_o successively, the TFD of the signal is obtained. Although this approach is commonly used in spectroscopic OCT (SOCT), it has the drawback that improved knowledge of the frequency k comes with reduced knowledge of the temporal distribution. Alternately, for data acquired in the frequency domain, a spectral window can be applied to obtain the TFD with a similar trade-off between time and frequency resolutions.

In bilinear signal representations, higher-order correlations are analyzed to obtain the joint TFD. As an example, one bilinear representation is the Wigner distribution, which is defined as

$$W(k,t) = \int \left\langle E\left(k+\frac{q}{2}\right) E^*\left(k-\frac{q}{2}\right)\right\rangle e^{-iqt} dq \qquad (23.3)$$

for an electrical field E, where $\langle \cdots \rangle$ denotes a statistical average and k is the wave vector which is related to the frequency as $k = \omega/c$. For a multicomponent signal, such as that found in the interferogram, the Wigner distribution of the total electric field, $E_T = E_R + E_S$, can be written as

$$W_T(k,t) = \int \left\langle E_T\left(k+\frac{q}{2}\right) E_T^*\left(k-\frac{q}{2}\right)\right\rangle e^{-iqt} dq = W_R + W_S + W_{Cross} \qquad (23.4)$$

where

W_R and W_S denote the individual Wigner distributions for the reference and sample fields, respectively

W_{Cross} gives the Wigner function for the cross terms

Although the cross terms are often regarded as undesired artifacts, they contain useful information about the temporal coherence of the sample field which is useful for coherence gated measurements.

Analysis of OCT signals by Graf and Wax (2007) showed that the bilinear representation of the interferogram signal, obtained from the Fourier transform (indicated by ~) of the square of the interference term:

$$\left|\tilde{I}_{int}(k)\right|^2 = \left|\Gamma(z)\right|^2$$

$$= \int E_R(k) E_S^*(k) \exp(ikz) dk \int E_R^*(k') E_S(k') \exp(-ik'z) dk' \qquad (23.5)$$

is given by the overlap of the Wigner distribution of the sample field with that of the reference field:

$$\left|\Gamma(z)\right|^2 = (2\pi)^2 \iint W_S(k,z') W_R(k,z'+z) dk\, dz' \qquad (23.6)$$

Similarly, the TFD for an SOCT signal can be written as the overlap of the sample Wigner distribution with an effective Wigner distribution for the window function:

$$S(k_w,z) = 2\pi \left|E_R(k_w)\right|^2 \int W_S(\bar{k},z') W_w(\bar{k}-k_w, z+z') d\bar{k}\, dz'. \qquad (23.7)$$

In Equation 23.7, the SOCT signal has been processed to yield the temporal (depth) profile of the sample field at the specific frequency (wavelength) given by the center of the window k_w. By systematically varying the center frequency of the window, the TFD of the SOCT signal is generated. A more detailed discussion of the use of TFDs to analyze SOCT signals is presented in Graf and Wax (2007), with an emphasis on the temporal coherence induced in the field due to structures in the sample and how that information can be accessed and used to assess structural features from fLCI depth-resolved spectroscopy measurements.

23.3.2 Dual Window Method

The dual window (DW) method for processing SOCT signals was introduced in Robles et al. (2009) as a method to avoid the limitations of the STFT. The DW technique allows the reconstruction of the Wigner TFD of an SOCT signal using two orthogonal windows which independently determine spectral and temporal resolution, avoiding the time-frequency resolution trade-off that limits current SOCT signal processing.

The DW method is based on calculating two separate STFTs and then combining the results. The first STFT uses a broad spectral Gaussian window to obtain high temporal/depth resolution while the second STFT uses a narrow spectral window to generate high spectroscopic resolution. The two resulting TFDs are then multiplied together to obtain a single TFD with simultaneously high spectral and temporal resolutions.

Consider the TFDs resulting from two STFTs, S_1 and S_2, generated by a narrow spectral window and a wide spectral window, respectively. Assuming that the reference field is slowly varying over the frequencies of interest, the processed signal is given by

$$\mathrm{DW}(k,z) = S_1(k,z) \cdot S_2^*(k,z)$$

$$= \iint 4\mathrm{Re}\left(E_S^*(k_1)E_S(k_2) \cdot \cos(k_1 \cdot d)\cos(k_2 \cdot d)\right)$$

$$\times e^{-(k_1-k)^2/2a^2} \cdot e^{-(k_1-k)^2/2b^2} \cdot e^{-i(k_1-k_2)z}\mathrm{d}k_1\,\mathrm{d}k_2 \qquad (23.8)$$

where a and b are independent parameters that set the widths of the windows, and $b \gg a$.

Upon some manipulation, the DW signal simplifies to

$$\mathrm{DW}(k,z) = 4b\sqrt{\pi}\iint W_S(\Omega,\zeta) \cdot e^{-2(\Omega-k)^2/b^2}e^{-2(d+\zeta+z)^2 a^2}$$

$$\times \cos(2\Omega \cdot d) \cdot \mathrm{d}\Omega\,\mathrm{d}\zeta. \qquad (23.9)$$

Equation 23.9 shows that the DW method is equivalent to probing the Wigner TFD of the sample field with two orthogonal Gaussian windows, one with a standard deviation of $b/2$ in the spectral dimension and another with a standard deviation of $1/(2 \cdot a)$ in the spatial/temporal dimension. Furthermore, a and b can independently tune the spectral and spatial/temporal resolutions, respectively, thus avoiding the trade-off that hinders the STFT. Equation 23.9 also shows that the processed signal is modulated by an oscillation that depends on the constant path difference, d, between the sample and reference arms. This phenomenon is also observed in the cross terms of the Wigner TFD, which have been identified to contain valuable information about phase differences that can be used to characterize the structure of a sample (Graf and Wax 2007).

Another interesting result is obtained from Equation 23.9 in the limit where a approaches zero and b is much larger than the bandwidth of the source, Δk. In these limits, the window with standard deviation $a \to 0$ approaches the delta function, while the second window with standard deviation $b \gg \Delta k$, becomes a constant across the spectrum. If we write the interferometric signal as $F(k) = 2\mathrm{Re}(E_R E_S \cdot \cos(k \cdot d))$, with $f(z) \Leftrightarrow F(k)$ a Fourier transform pair, we obtain

$$\mathrm{DW}(k,z)\big|_{a\to 0, b \gg \Delta k} = S_1(k,z)\big|_{a\to 0} S_2(k,z)_{b \gg \Delta k} = \frac{1}{\sqrt{2\pi}}f(z)F(k)e^{-ik\cdot z}.$$

$$(23.10)$$

Equation 23.10 shows that in this limit, the DW signal is equivalent to the Kirkwood & Rihaczek TFD, and if the real part is taken, it is equal to the Margenau & Hill (MH) TFD. Either of these two distributions can be simply transformed to produce any of the Cohen's class functions, such as the Wigner TFD (Cohen 1989).

23.3.3 Simulations: Properties of the DW Method

To illustrate the power of the DW method, two different simulations are presented. In the first, a signal consisting of two optical fields separated in depth and center wave number is simulated. The total sample field is given by $E_S = E_1 + E_2$, where $E_1 = E_0\exp(-z^2)\exp(i \cdot k_1 \cdot z)$, $E_2 = E_0\exp(-(z-z_0)^2)\exp(i \cdot k_2 \cdot z)$, and $k_1 > k_2$. The Wigner distribution of the total sample field is given by

$$W(k,z) = \frac{1}{2\pi}\int E_S^*\left(z - \frac{\zeta}{2}\right)E_S\left(z + \frac{\zeta}{2}\right)e^{ik\zeta}\mathrm{d}\zeta, \qquad (23.11)$$

and the MH distribution of the total sample field is given by

$$\mathrm{MH}(k,z) = \mathrm{Re}\frac{1}{\sqrt{2\pi}}\bar{E}_s(k)E_s(z)e^{-ikz}, \qquad (23.12)$$

where $\bar{E}_S(k) \Leftrightarrow E_S(z)$ is a Fourier transform pair. Figure 23.3 illustrates the resulting TFDs.

The ideal TFD, shown in Figure 23.3a, contains two well-separated pulses with Gaussian shapes in both the temporal and spectral dimensions but can only be constructed with prior knowledge of each of the complex fields. Figure 23.3b through d show different TFDs that can be generated from this single mixed field. The Wigner distribution (Figure 23.3b) shows the two Gaussian pulses and a modulated cross term, which can reveal temporal coherence information (Graf and Wax 2007), but is more often viewed as an undesirable artifact since it presents nonzero values at times/depths and frequencies that do not exist in the field. The MH distribution (Figure 23.3c) contains the two pulses comprising the signal field and two artifact pulses known as "reflections in time" (Cohen 1989), which also present nonzero intensities at times and frequencies that should contain no signal. The TFD generated using the DW method (Figure 23.3d) is computed using the product of two STFTs, processed with wide and narrow spectral windows, respectively. Here, the cross terms seen in the Wigner and MH distributions are eliminated as a result of using two orthogonal windows.

The second simulation models an SOCT signal from a Michelson interferometer with an experimental sample containing two distinct reflecting surfaces. The first sample surface

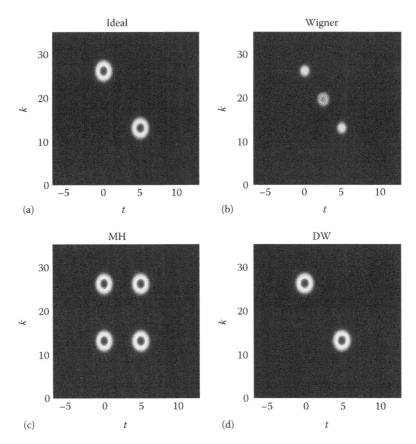

FIGURE 23.3 (a) Ideal TFD with E_1 centered at $z_0 = 5$ and $k_1 = 13$ and E_2 centered at $z_0 = 0$ and $k_2 = 26$. (b) Wigner TFD. (c) MH TFD. (d) Dual window method. (From Robles, F. et al., *Opt. Express*, 17(8), 6799, 2009.)

reflects the entire Gaussian spectrum of the source while the second sample surface absorbs the high-frequency portion (upper half) of the source spectrum. This simulation is analogous to the absorption phantom experiment discussed in Section 23.4.3 and illustrated in Figure 23.14. In the scenario of this simulation, that is, an SOCT system, neither the Wigner nor the MH distributions can be constructed because the detected signal is the intensity of the field and therefore the phase information is lost. Thus, TFDs are reconstructed via the STFT and the DW method.

Figure 23.4a through c shows the ideal TFD of the simulated signal and the TFDs generated by the STFT using narrow and wide spectral windows, respectively. In each case, the effects of the time-frequency resolution trade-off are obvious. The TFD generated with the wide spectral window suffers from degraded temporal resolution while the TFD generated with the narrow spectral window suffers from degraded spectral resolution. As Xu et al. showed, the STFT window can be optimized for specific applications, but regardless of the window size, a resolution trade-off must be made (Xu et al. 2005b). Figure 23.4d shows the TFD generated using the DW method, which computes the product of the TFDs shown in Figure 23.4b and c. Figure 23.4e shows the time marginals computed from Figure 23.4b through d to demonstrate that the DW method resolves the two sample surfaces with comparable resolution to the ideal

case, but the narrow spectral window STFT does not. Figure 23.4f shows the spectral profile of the rear surface reflection in Figure 23.4b through d illustrating that the DW method maintains higher spectral fidelity than the wide spectral window STFT. Note that the DW method is able to accurately portray the sample absorption, while the wide spectral window STFT does not. The DW frequency profile also reveals the same spectral modulation that is seen in the narrow window STFT and that is characteristic of the Wigner TFD. This modulation results from cross correlations between field components that overlap in time.

Graf and Wax have shown previously that temporal coherence information from cross terms in the Wigner TFD can be used to gain structural knowledge of samples via the SOCT signal (Graf and Wax 2007). TFDs obtained using the DW method exhibit the same phenomenon observed in the cross terms of the Wigner TFD, that is, a cosine term whose frequency depends on the constant path difference, d, between signal components. These oscillations can provide valuable information about phase differences.

Figure 23.5b shows the frequency profile from the front reflecting surface of the sample in the above simulation (Figure 23.4). This frequency spectrum is taken from depth 3 (dashed line) of the TFD generated by the DW method (Figure 23.5a). The observed spectral modulation, termed local oscillations, can

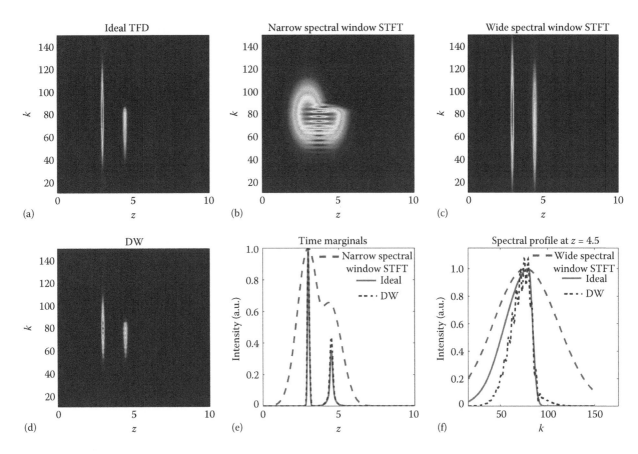

FIGURE 23.4 (a) Ideal TFD with simulated source bandwidth of $\Delta k = 35$ length^{-1} units. (b) Narrow spectral window STFT with standard deviation = 2 length^{-1} units. (c) Wide spectral window STFT with standard deviation = 45 length^{-1} units. (d) DW method using the two windows used in (b) and (c). (e) Time marginals (depth profile) of (a), (b), and (d). (f) Spectral profile at $t = 4.5$ in (a), (c), and (d). (From Robles, F. et al., *Opt. Express*, 17(8), 6799, 2009.)

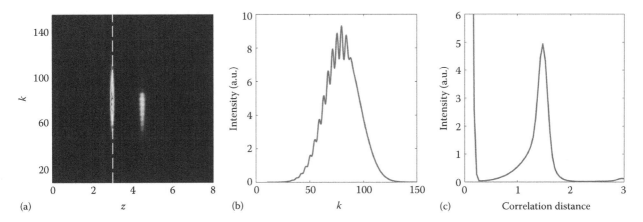

FIGURE 23.5 (a) TFD of simulation 2 generated by the DW processing method. (b) Spectral profile corresponding to the dashed line in (a). (c) Correlation plot with peak corresponding to sample spacing distance of 1.5 units. (From Robles, F. et al., *Opt. Express*, 17(8), 6799, 2009.)

reveal structural information about the simulated experimental sample. Fourier transforming the spectrum from Figure 23.5b generates a correlation plot (Figure 23.5c) with a clear peak at 1.5 units, in good agreement with the spacing of the surfaces in the simulated sample, thus providing additional information about its structure.

23.3.4 Spatial Coherence in White Light Interferometry

The utility of the Wigner distribution for analyzing optical signals is not confined to investigations of temporal coherence; but can also be used to analyze the role that *spatial* coherence plays in the combination of OCT signals. While the vast majority of current OCT systems employ a spatially coherent light source such as a superluminescent diode (SLD), as an alternative, white light sources offer the advantage of a significantly larger spectral bandwidth centered at a lower wavelength, enabling superior depth resolution. White light sources can also provide access to biologically useful spectral windows that are not readily available using SLDs, but they also present unique challenges due to the extremely low spatial coherence of the source light.

OCT with a white light source has been previously demonstrated in both the time and frequency domains (Fercher et al. 2000; Grajciar et al. 2005), but prior to our analysis, OCT imaging of biological samples using a white light source has only been demonstrated using TD OCT. By examining the Wigner TFDs of multicomponent signals, it can be seen that signals combine differently in FDOCT compared to TDOCT, leading to specific design criteria for white light FDOCT imaging of biological samples.

As shown above (Equation 23.5), the Wigner distribution of a two component electric field can be represented as the sum of the Wigner distributions of the sample field (E_S) and reference field (E_R), along with that of the cross terms representing the interference between them. For a sample field of n components, the Wigner function for the total field consists of the sum of the Wigner distribution of the reference field, n sample Wigner distributions (combined into W_S), and the Wigner distributions of the n cross terms between the reference field and each component of the signal field:

$$W_T = W_R + W_S + \sum_n W_{cross}^n. \tag{23.13}$$

In the frequency domain, our detected signal is the frequency marginal of the Wigner distribution (Graf et al. 2008):

$$S_{FD}(\omega) = \int dt \sum_n \cos(T_n(\omega + \omega_0))$$

$$\times \exp\left(\frac{-a^2}{2}(\omega + \omega_0)^2\right) \exp\left(\frac{-2}{a^2}\left(t - \frac{T_n}{2}\right)^2\right)$$

$$= a\sqrt{\frac{\pi}{2}} \exp\left(\frac{-a^2}{2}(\omega + \omega_0)^2\right) \sum_n \cos(T_n(\omega + \omega_0)) \tag{23.14}$$

where

a is the temporal coherence length of the source

ω_0 is the center frequency

T_n is the time delay of the nth component of the signal field relative to the reference field

The result is a sum of truncated cosine terms. Because these terms add coherently, phase differences between terms will reduce the fringe visibility. In OCT, reduced fringe visibility equates to a degradation of depth scan signal strength.

Alternatively, in the TD, the detected signal is a function of the time delay or path-length difference between the sample and reference fields. Each point of the detected signal is the time-averaged time marginal of the Wigner distribution. Here the power spectrum of the signal is taken by including the signal from both the in-phase (cos) and out-of-phase (sin) quadratures, as is commonly done in TDOCT:

$$S_{TD}(T_n) = \iint dt\, d\omega \sum_n \left(\cos(T_n(\omega + \omega_0))\right.$$

$$\left. + i\sin(T_n(\omega + \omega_0))\right) \exp\left(\frac{-a^2}{2}(\omega + \omega_0)^2\right) \exp\left(\frac{-2}{a^2}\left(t - \frac{T_n}{2}\right)^2\right)$$

$$= \sqrt{\frac{\pi}{2}} \sum_n \exp\left(\frac{-T_n^2}{2a^2}\right). \tag{23.15}$$

The resulting expression is a sum of Gaussian functions of T_n, the time delay. Each Gaussian pulse represents the power contained in the interference signal due to the nth field component. Unlike the coherent sum of energy signals in the frequency domain (Equation 23.14), the power signals in the TD sum *incoherently* and are less vulnerable to phase differences between field components.

The differences in signal combination result in increased difficulty in performing OCT with a white light source in the frequency domain. In FDOCT, the resolution of the detected signal by coherence mode is critical in minimizing the destructive effects that phase differences between field components can have on the coherent sum of energy signals. For many FDOCT systems, this is not a great concern as the spatial coherence length of light from a SLD can be an order of magnitude larger than that from a thermal source. However, for white light OCT, the resolution of the detected signal by mode is absolutely necessary. This effect is the exact opposite of that seen with TDOCT which combines signals incoherently and, as a result, interference efficiency is improved as more modes are combined.

23.4 Imaging Systems: Design and Validation

In this section, the instrumentation of two fLCI optical systems is presented. First, the design and development of the common path fLCI system based on a modified Michelson interferometer is discussed. The common path geometry provides design simplicity and ease of alignment but its effectiveness is limited to thin experimental samples. This drawback led to the development of the parallel fLCI system, which spatially resolves the optical signal from source to detector in order to meet the design criteria identified above enabling imaging of thick experimental samples, up to approximately 1 mm.

23.4.1 Common Path fLCI System

In the common path fLCI scheme (Graf and Wax 2005), shown in Figure 23.6, white light from an Xe arc lamp source (250W, Newport Oriel, Stratford, CT) is coupled into a multimode fiber (200 μm core diameter). The output of the fiber is collimated by an achromatic lens (L1, $f_1 = 10$ mm) to produce a pencil beam approximately 5 mm in diameter. The white light beam is transmitted to the sample by the BS, 50/50. In this scheme, the reference field is generated by placing the experimental sample on a glass coverslip and using the reflection from the front surface of the cover glass. This field is combined with light backscattered by the experimental sample and directed by the BS to a second achromatic lens (L2, $f_2 = 10$ mm) which collects the light and couples to a second multimode fiber (200 μm core diameter). The output of the fiber is coincident with the input slit of a high resolution (0.1 nm) spectrometer (InSpectrum, Princeton Instruments, Trenton, NJ).

To demonstrate the applicability of the fLCI technique to biological structures, specifically to accurately measure nuclear size,

a series of experiments which probed the nuclear morphology of in vitro T84 epithelial cell monolayers was conducted (Graf and Wax 2005). A typical spectrum of light backscattered by the sample is shown in Figure 23.7a, presented as a function of wave number, over the range of 10–12 μm⁻¹ (525–625 nm). The data show a prominent oscillation which arises from the interference between the reflection from front of the cover glass (reference field) and that from the rear surface (signal field). This study was completed before the development of the DW processing method, and thus the STFT was used to obtain depth-resolved spectroscopic information. Figure 23.7b shows a contour plot for the processed data resulting from this transform. The signal localized at the depth of 450 μm, corresponding to the roundtrip optical path length through the cover glass, gives the interference between the front and back surface reflections.

The scattering efficiency of the cells (Figure 23.8a) is examined by taking the ratio of the field scattered by the cells to that obtained from a blank coverslip. The size of the cell nuclei can be determined by Fourier transforming this ratio and analyzing the correlation information which is obtained (Figure 23.8b). The correlation function shows a sharp peak at a round trip path of 19.15 μm which is equal to $2 \cdot n \cdot d$, where n is the index of refraction of the cell nuclei (1.395) (Tuchin 2000) and d is the nuclear diameter. This analysis yields a nuclear diameter of 6.86 ± 1.37 μm, with the uncertainty given by the pixel size of this correlation function.

To gauge the accuracy of the fLCI measurements, the nuclear diameters of T84 cells were also measured using fluorescence and confocal microscopy. A typical fluorescence microscopy image of T84 nuclei can be seen in Figure 23.9a. Confocal Z-stacks were also acquired to assess the nuclear dimensions in the X–Z plane (Figure 23.9b).

The nuclear size determined using fLCI yielded consistent results across measurements of all 11 samples examined in this study. The fLCI measurements generated a mean longitudinal diameter of 6.87 ± 0.81 μm. Image analysis of confocal microscopy Z-stack images yielded a longitudinal diameter of 6.8 ± 1.1 μm and a transverse diameter of 10.5 ± 1.9 μm. Fluorescence microscopy was also used to investigate the transverse profile of the cell nuclei, yielding a mean transverse diameter of 10.4 ± 1.4 μm. The

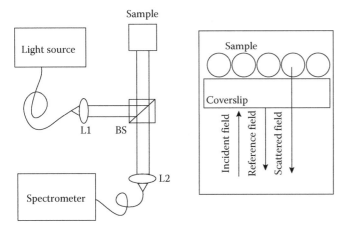

FIGURE 23.6 Schematic of the common-path fLCI system. Light source—250 W Xe arc lamp, L1 and L2—lenses, and BS—beamsplitter. Inset: Sample arm geometry. Reflection from front surface of coverslip is used as reference field with which light backscattered by the sample can interfere. (From Graf, R.N. and Wax, A., *Opt. Express* 13(12), 4693, 2005.)

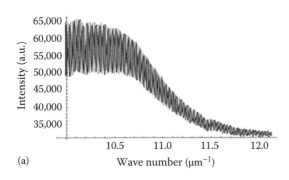 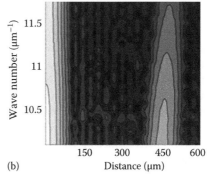

FIGURE 23.7 (a) Typical spectrum of light scattered by the in vitro cell sample. (b) Contour plot showing the depth resolved spectral data for the T84 cell sample. (From Graf, R.N. and Wax, A., *Opt. Express* 13(12), 4693, 2005.)

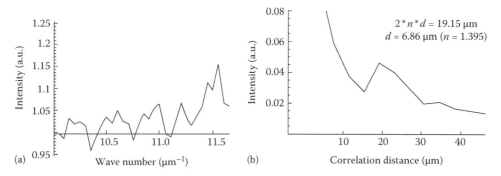

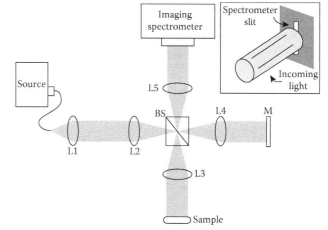

FIGURE 23.8 (a) Spectrum showing the scattering efficiency of the T84 cell nuclei. (b) Correlation function obtained by Fourier transforming ratioed data shown in (a). (From Graf, R.N. and Wax, A., *Opt. Express* 13(12), 4693, 2005.)

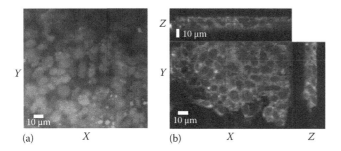

FIGURE 23.9 (a) Fluorescence microscopy image of T84 cell nuclei (Hoechst 33342 stain). (b) Confocal Z-stack image of T84 cell nuclei (LDS-751 stain) with all three axes of resolution visible. (From Graf, R.N. and Wax, A., *Opt. Express* 13(12), 4693, 2005.)

standard deviation of the fLCI measurements is lower than that of the microscopy measurements due to the high number of cells probed with each fLCI measurement compared to that examined using image analysis. Because the fLCI system is designed to only detect light directly backscattered within a narrow angular range, fLCI necessarily probes only the longitudinal dimension of the cells. Therefore, it is most appropriate to compare the fLCI measurements with the longitudinal confocal microscopy measurements, which reveals that fLCI measures the longitudinal diameter of epithelial cell nuclei with high precision and accuracy.

23.4.2 Imaging fLCI System

The second generation imaging fLCI system (Graf et al. 2008), shown in Figure 23.10, is based on a modified Michelson interferometer geometry, the 4f interferometer (Wax et al. 2001). White light from a Xenon arc lamp (150W, Newport Oriel, Stratford, CT) is coupled into a multimode fiber (200 μm core diameter) before being collimated by an achromatic lens (L1, $f_1 = 10$ mm). The resulting pencil beam spans 5 mm in diameter and enters the 4f interferometer formed by lenses L2, L3, L4, and L5 ($f_{2-5} = 10$ cm) along with the BS. The light is separated by BS into a reference arm, incident upon a reference mirror, and a sample arm, incident on the experimental sample. Light scattered by the sample is recombined with the reference signal by

FIGURE 23.10 Schematic of the imaging fLCI system. Light source—250 W Xe arc lamp, L1 through L5—lenses, BS—beamsplitter, and M—reference mirror. Inset: Incoming light incident on the spectrometer slit. Slit allows only a small slice of incoming light to enter the imaging spectrometer. (From Graf, R.N. et al., *Opt. Lett.*, 33(12), 1285, 2008.)

BS and reimaged onto the detection plane by the 4f imaging system formed by lenses L3 and L5.

The 4f interferometer uses two 4f imaging systems to spatially resolve light from the source to the detector. A schematic of each 4f imaging system is shown in Figure 23.11a. For clarity, the BS has been removed and the light paths have been unfolded. Light scattered by the experimental sample is reimaged onto the detection plane by the 4f system formed by lenses L3 and L5. Light reflected by the reference mirror is reimaged onto the detection plane by the 4f system formed by lenses L4 and L5.

The detection plane of the imaging system coincides with the entrance slit of an imaging spectrometer (Shamrock 303i, Andor Technology, South Windsor, CT) which spatially resolves 255 detection channels. The spectrometer utilizes a 1200 lines/mm diffraction grating with a blaze wavelength of 500 nm. Detection is accomplished with a front-illuminated CCD camera (iDus DV420, Andor Technology, South Windsor, CT) with a 1024 by 255 pixel array. The camera uses thermoelectric cooling to operate at −50°C resulting in a negligible dark current. The

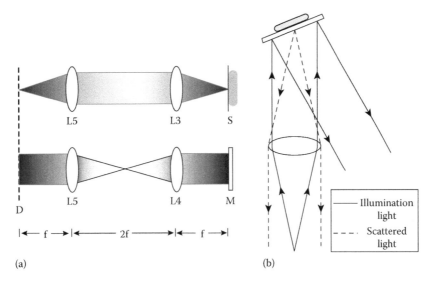

FIGURE 23.11 (a) Unfolded imaging optics of the 4f imaging interferometer showing 4f systems in sample arm (top) and reference arm (bottom). L3, L4, and L5—lenses, S—experimental sample, M—reference mirror, and D—detection plane. (b) Sample arm operating in scatter-mode in which specular reflection is angled away from collection optics.

imaging spectrometer optics, along with the combination of the 1200 lines/mm grating and the 1024 pixel CCD array, limits the detected spectrum to the 500–625 nm range. The high resolution of this grating is necessary to achieve the depth range desired in the coherence images. Other spectral regions can be accessed by tuning the center wavelength of the span or employing a multichannel spectrometer for simultaneous measurements. Data from the spectrometer are downloaded in real time to a laptop PC via the USB 2.0 interface, and spectrometer control and data acquisition is controlled using custom LabVIEW (National Instruments, Austin, TX) software.

The imaging fLCI system is capable of ultrahigh depth resolution parallel frequency domain OCT (pfdOCT) imaging. By spatially resolving the optical signal from the source to the detector, the system is capable of generating depth scans from 255 lateral spatial points in the experimental sample. The FDOCT capabilities are referred to as "parallel" because B-mode images are produced in a single acquisition, without the need for scanning the beam spatially or scanning a reference mirror.

The ultra-broad bandwidth and low center wavelength of the thermal light source yield a theoretical axial resolution of 1.03 μm, which is comparable to today's best ultrahigh resolution TD and FDOCT systems. The experimental depth resolution of the system was determined using a mirror for a sample to be 1.22 μm FWHM (Graf et al. 2008). The lateral resolution of the system is 26 μm, equal to the width of the detector pixels. This is poorer than many current TD and FDOCT systems, but it is achieved in a single shot without the need for beam scanning. The SNR of the presented image was determined by a separate experiment to be 89 dB, slightly worse than the SNR of typical coherent-source TD and FDOCT systems.

Unlike most TD or FDOCT systems, the sample arm of the presented system was designed to operate in "scatter-mode." Figure 23.11b illustrates the scatter-mode imaging schematic.

In this geometry, the light incident on the sample is collimated rather than focused, and the sample is tilted such that specularly reflected light is not directed back to the BS. Instead, only light backscattered by the sample is collected by lens L3 and mixed with the reference field at BS. This adjustment of sample arm geometry allows for the analysis of scattering media by avoiding saturation of the detector pixels by the large surface reflections. These reflections can be orders of magnitude greater than the signal originating from deeper tissue layers.

As discussed in Section 23.3.4, in order to obtain OCT images in the frequency domain with a thermal source, it is necessary to limit the number of modes that contribute to each detected signal. Here, each mode is considered to span an area given by the square of the transverse coherence length. The contributing modes are limited by spatially resolving the detected signal by coherence area at the detection plane using the 4f imaging system and imaging spectrometer.

To validate the imaging capabilities of the fLCI system, an experiment demonstrating the signal degradation caused by mixing uncorrelated spatial modes was conducted. Data from a scattering standard sample were acquired consisting of spectra from 255 adjacent spatial points in the sample. These interference spectra exhibited random phase changes between the adjacent spatial points, which provide a good basis for illustrating coherent and incoherent summation of the signal components.

Figure 23.12a shows the raw interference spectra collected from the scattering sample by the imaging spectrometer. The random phases of the interference fringes occur as a result of the random location of scatterers within the sample. Figure 23.12b shows depth scans generated by taking the magnitude of the Fourier transformed wave number spectra, as is typically done in FDOCT. The signals have been summed across the image in different ways. In the case of the solid curve, the raw spectra

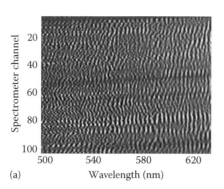
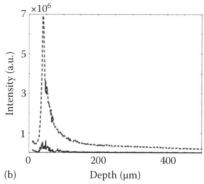

(a) Wavelength (nm)   (b) Depth (µm)

FIGURE 23.12 (a) Experimental spectra obtained from a scattering standard sample. (b) Depth scans generated from raw spectra in part (a). Depth information is degraded by summing spectra before Fourier transforming (solid) but preserved by summing channels after Fourier transforming (dashed). (From Graf, R.N. et al., *Opt. Lett.*, 33(12), 1285, 2008.)

from the different spatial points were summed together *before* the Fourier transform was performed. This operation is analogous to an FDOCT detection scheme, which sums many modes at the detection plane. As a result of the coherent sum of the energy signals, the phase differences between the signal components reduce interference efficiency and therefore degrade depth information.

Alternatively, the dashed curve in Figure 23.12b shows the depth scan generated by summing the information from different spatial points *after* the individual wave number spectra had been individually Fourier transformed. This is analogous to an incoherent sum where the power from each spatial coherence channel is combined. By utilizing the imaging spectrometer to maintain the spatial resolution of the sample field in the detected signal, the individual energy signals maintain their interference efficiency.

As a demonstration of the imaging abilities of this system, Figure 23.13 shows a scatter-mode image of the hamster cheek pouch epithelium. Ex vivo cheek pouch tissue was placed on a cover glass and moistened with phosphate buffered solution (PBS) before imaging by the fLCI system. The presented image contains the average of 25 successive scans in order to improve the signal to noise ratio. Several tissue layers are indicated by arrows including the epithelium (E),

mucosa (M), and submucosa (S). Because the cheek pouch was not stretched during imaging, the tissue also contained folds (F) at points where the sample did not adhere to the cover glass. The included scale bars correspond to 50 µm in the depth (vertical) dimension and 125 µm in the lateral (horizontal) dimension. The successful imaging of a thick sample using the fLCI imaging system verifies the need to resolve the interferogram by spatial coherence modes when using thermal light.

23.4.3 Dual Window Processing Method Validation

To validate the ability of the DW processing method to generate TFDs with simultaneously high spectral and temporal resolution, several experiments were completed using an absorption phantom as an experimental sample (Robles et al. 2009). All data were collected using the imaging fLCI system presented above. The absorption phantom consisted of a glass wedge filled with an absorbing dye as shown in Figure 23.14a. Figure 23.14b shows a pfdOCT scan of the absorption phantom with the two inner glass surfaces clearly visible. Note that the signal from the rear surface is significantly attenuated at the thicker end of the wedge due to considerable signal absorption from the greater volume of absorbing dye present. Because the experimental system operates in the visible wavelength band, a visible absorbing dye consisting of a red food-coloring gel and water solution was used. Figure 23.14c shows the transmission spectrum of the absorbing dye, which shows strong absorption in the high wave number range of the detected spectrum. From the structure of the phantom, signals returning from the front surface of the phantom are expected to mirror the source spectrum, while signals reflected by the back surface of the phantom are expected to exhibit spectra with attenuation of the higher wave numbers, due to absorption by the dye.

The interferometric data from the channel corresponding to the position of the dashed line in Figure 23.14b were processed with four different methods to yield the four TFDs shown in

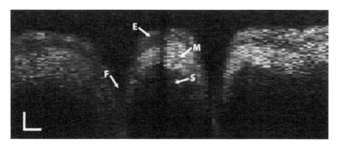

FIGURE 23.13 Parallel FDOCT image of a hamster cheek pouch. E, squamous epithelium; M, mucosa; S, subucosa; F, tissue fold. Scale bars correspond to 50 µm vertically and 125 µm horizontally. (From Graf, R.N. et al., *Opt. Lett.*, 33(12), 1285, 2008.)

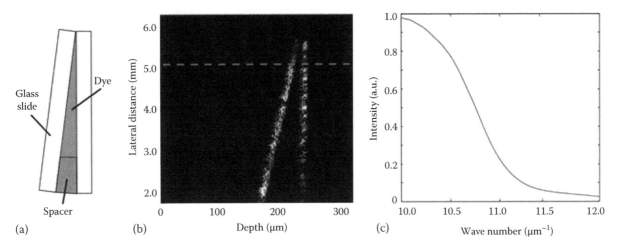

(a)   (b)   (c)

FIGURE 23.14 (a) Illustration of absorption phantom. (b) pfdOCT image of absorption phantom. (c) Transmission spectrum of absorbing dye used in absorption phantom. (From Robles, F. et al., *Opt. Express*, 17(8), 6799, 2009.)

Figure 23.15. Figure 23.15a was generated using the STFT processing method with a narrow spectral window of $0.0405\,\mu m^{-1}$. The resulting TFD has excellent spectral resolution, showing a spectrum which emulates the source spectrum at the depth corresponding to the front surface of the phantom. The sharp spectral cutoff at high wave numbers, characteristic of the dye absorption, is evident at deeper depths. However, the narrow spectral window used to generate this TFD yields very poor temporal resolution, resulting in an inability to resolve the two surfaces of the phantom. Figure 23.15b was also processed using the STFT method, but in this case a wide spectral window of $0.665\,\mu m^{-1}$ was used. The resulting TFD has excellent temporal resolution, clearly resolving the two surfaces of the phantom. However, the spectral resolution of the resulting TFD is too poor to resolve the spectral modulation expected for the rear surface spectrum. Figure 23.15c shows the TFD generated using

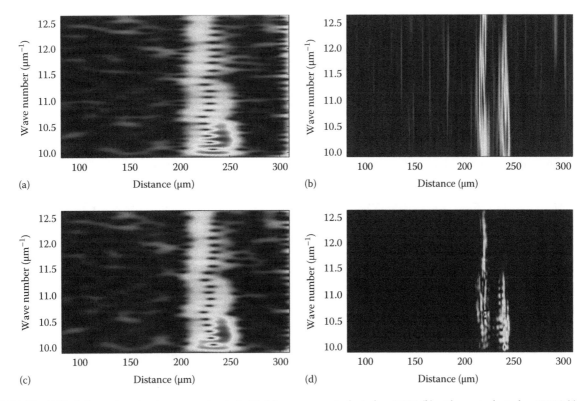

(a)   (b)

(c)   (d)

FIGURE 23.15 TFD of absorption phantom reconstructed with (a) narrow spectral window STFT, (b) wide spectral window STFT, (c) moderate spectral window STFT, and (d) double window method. (From Robles, F. et al., *Opt. Express*, 17(8), 6799, 2009.)

the STFT method with a window of moderate spectral width, 0.048 μm⁻¹. As expected, the spectral and temporal resolutions of the resulting TFD fall between those of Figure 23.15a and b, illustrating the temporal–spectral resolution trade-off associated with the STFT processing method. While the spectral characteristics of the absorbing dye are apparent in this TFD, the two phantom surfaces still cannot be resolved.

The TFD in Figure 23.15d was generated using the DW method. By processing the raw data with both a narrow and a wide spectral window, the TFD simultaneously achieves high spectral and temporal resolution. The front surface of the phantom exhibits a spectrum which emulates the source spectrum while the rear surface spectrum clearly reveals a spectral cutoff at high wave numbers due to the absorbing dye through which the signal field has passed. Additionally, the front and back surfaces of the phantom are clearly resolved in depth.

The utility of the DW processing method is further demonstrated by examining spectral cross-sections and time marginals of the generated TFDs. Figure 23.16a displays spectral profiles from depths corresponding to the absorption phantom's rear surface in the TFDs of Figure 23.15c and d. For reference, the absorbing dye transmission spectrum is displayed as well. Figure 23.16b shows spectral cross-sections from depths corresponding to the phantom's front surface, along with the source spectrum for reference. The time marginals of each TFD are displayed in Figure 23.16c along with the corresponding A-scan from Figure 23.14b. It is evident that the TFD generated by the DW method maintains the ability to resolve the two peaks of the absorption

phantom, while the TFD generated by the STFT method does not. Because the DW method produces a bilinear TFD, the noise floor of the resulting time marginal is lower than that of the time marginal produced by the STFT.

In addition to limiting the resolution trade-off associated with the STFT, the DW method also achieves an increase in the spectral fidelity of generated TFDs. The normalized spectra from Figure 23.16a and b are plotted in Figure 23.17 with the high-frequency modulation removed by a low-pass filter. By separating the low-frequency content from the high-frequency local oscillations, the fidelity with which each processing method recreates the ideal spectrum can be assessed. Chi-squared values for each processing method were calculated to assess goodness-of-fit. Table 23.1 summarizes the chi-squared values. For both the rear surface spectra in Figure 23.17a and the front surface spectra in Figure 23.17b, the chi-squared values indicate that the DW method recreates the ideal signal with greater spectral fidelity. In addition, the goodness-of-fit for the square of the STFT is calculated to account for the fact that the DW method produces a bilinear distribution. The DW method is also seen to produce superior spectral fidelity than the STFT squared.

As with the simulated SOCT signals in Section 23.3.3, the local oscillations seen in the TFD obtained from probing the absorption phantom (Figure 23.15) can also be analyzed to gain structural information about the experimental sample. Figure 23.18b shows the spectral profile from the front surface of the absorption phantom indicated by the dashed line in Figure 23.18a (same as Figure 23.15d). Fourier transforming this spectrum produces a

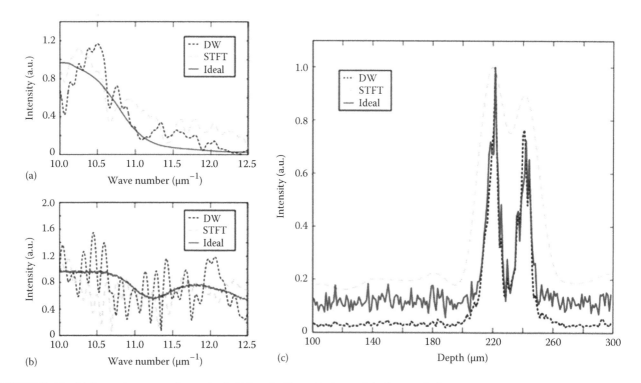

FIGURE 23.16 (a) Spectral cross-section at depth 245 μm in Figure 23.15c and d, along with dye transmission spectrum. (b) Spectral cross-section at depth 220 μm in Figure 15c and d, along with source spectrum. (c) Time marginals from Figure 23.15c and d, along with corresponding A-scan. (From Robles, F. et al., *Opt. Express*, 17(8), 6799, 2009.)

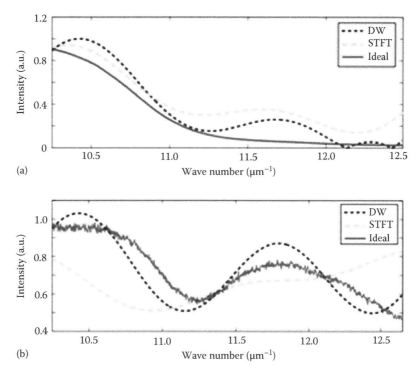

FIGURE 23.17 (a) Spectral profiles of Figure 23.16a with high-frequency modulations removed. (b) Spectral profiles of Figure 23.16b with high-frequency modulations removed. (From Robles, F. et al., *Opt. Express*, 17(18), 6799, 2009.)

TABLE 23.1 Chi-Squared Calculations

	DW	STFT	STFT²
Rear surface spectrum	0.0980	0.1329	0.1245
Front surface spectrum	0.0248	0.0317	0.0305

correlation plot as shown in Figure 23.18c with a clear correlation peak corresponding to a physical distance of $20.60 \pm 0.57\,\mu m$. This measurement represents the spacing between the phantom surfaces and is in excellent agreement with the spacing measured in the OCT image of the phantom, $20.60 \pm 5.97\,\mu m$. Here, the measurement uncertainty is larger than the $1.22\,\mu m$ depth resolution due to the fact that the glass surface was slightly abraded to increase the signal, producing a broader range of path lengths.

By comparing the performance of the DW and STFT processing methods in analyzing SOCT signals from an absorption phantom, it has been shown that the DW method recovers TFDs with superior fidelity while simultaneously maintaining high spectral and temporal resolution. In addition, this analysis shows that the local oscillations, contained in the TFDs generated by the DW method, contain valuable information about the structure of experimental samples. Unfortunately, processing methods such as the STFT and CWT are limited by an inherent trade-off between spectroscopic and depth resolution. This time-frequency trade-off greatly reduces the utility of the analysis by degrading either the depth or spectral resolution to the point

that important features cannot be accurately reconstructed. By avoiding this trade-off, the DW processing method will enable new directions in SOCT and depth-resolved spectroscopy.

23.5 Animal Tissue Studies

To demonstrate the ability of the fLCI system to distinguish between normal and dysplastic epithelial tissue through light scattering measurements of tissue microstructure, an animal tissue study was conducted using the hamster cheek pouch carcinogenesis model (Graf et al. 2009). For the animal studies, all experimental protocols were approved by the Institutional Animal Care and Use Committees of Duke University and North Carolina Central University and in accordance with the National Institutes of Health (NIH). Briefly, the left cheek pouch of each of 21 animals was topically treated with $100\,\mu L$ of 0.5% 7,12-dimethylbenz[*a*]anthracene (DMBA) (Sigma Chemical Company, St. Louis, MO) in mineral oil with a paintbrush three times per week for 6 weeks. The right cheek pouch was left untreated and served as the control group. At 24 weeks, after the initial treatment of DMBA, the hamsters were euthanized by CO_2 asphyxiation, and the entire left and right cheek pouches were excised and cut into two pieces. Each sample was laid flat between two cover glasses, moistened with PBS, and immediately scanned by the fLCI system. Following optical scanning, scanned areas were marked with India ink and processed for histopathological analysis. Further details of the animal model can be found in Graf et al. (2009).

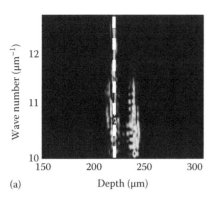
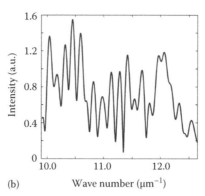
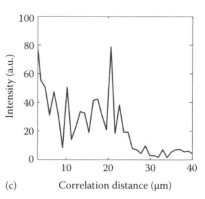

(a) Depth (µm)   (b) Wave number (µm⁻¹)   (c) Correlation distance (µm)

FIGURE 23.18 (a) Absorption phantom TFD generated with the DW method. (b) Spectrum corresponding to the dashed red line in (a). (c) Correlation plot with peak corresponding to phantom spacing distance of $20.60 \pm 0.57\,\mu m$, in good agreement with the OCT thickness measurement. (From Robles, F. et al., *Opt. Express*, 17(8), 6799, 2009.)

The raw data acquired by the pfdOCT system consisted of 120 spectra, each of which originates from adjacent 26 µm diameter spatial points on the experimental sample. The raw data, along with the plots of three such spectra, are shown in Figure 23.19a. The diameter of the signal beam was shaped to illuminate only 120 of the 255 spectral channels of the imaging spectrometer to preserve the signal to noise ratio of the measurements.

To analyze spectra from specific tissue layers, the spectrum detected by each channel of the imaging spectrometer was processed using the DW processing method, described above. A custom MATLAB® program was used to process the data with both a narrow spectral window of 0.0405 µm⁻¹ FWHM and a wide spectral window of 0.665 µm⁻¹ FWHM. The depth-resolved spectra generated by each window were multiplied together to produce a plot with simultaneously high spectral and depth resolution. The resulting 120 depth-resolved spectroscopic plots were summed together to improve the signal-to-noise ratio, producing a single depth-resolved spectroscopic plot for each tissue sample as shown in Figure 23.19b.

In neoplastic transformation, nuclear morphology changes are first observed in the basal layer of the epithelial tissue. In hamster buccal pouch tissue, the basal layer lies approximately 30–50 µm beneath the surface for normal tissue, and approximately 50–150 µm beneath the surface for dysplastic tissue. Because examination of the basal layer offers the earliest opportunity for detecting developing dysplasia, it is the target tissue layer for the fLCI technique and for this study. In order to target the basal layer of the epithelium, the raw experimental data were first processed to yield a parallel FDOCT image by a line-by-line Fourier transform. These "B-mode" images were summed across the transverse axis to generate single depth plots (A-scan) like those presented in Figure 23.20. Several important histological features can be identified in the depth scans and co-registered with the corresponding histopathology images. Figure 23.20 indicates the location of a keratinized layer (green arrow), the basal layer of the epithelium (red arrow), and the underlying lamina propria (blue arrow) in the micrographs of fixed and stained histological sections from untreated and treated tissue

samples. Scattering peaks corresponding to the same tissue layers were identified in each depth scan. To correlate the distances in the histology images with distances in the depth scans, the index of refraction of the tissue was taken into account.

An average refractive index for the tissue of $n = 1.38$ was used to convert depth scan distances to optical path lengths (Zysk et al. 2007a,b). Variation of the refractive index within the tissue is a potential limitation of the current method and is discussed further below. For each sample, a 15 µm depth segment corresponding to the location of the basal layer was selected from the depth scan and used to guide analysis of the depth-resolved spectroscopic plot, as shown in Figure 23.21a. The spectra from the depth identified with the basal layer in each A-scan were averaged to generate a single spectrum for light scattered by the basal layer. As shown in Figure 23.21b, a power law curve of the form $y = b \cdot x^{\alpha}$ was initially fit to each spectrum, modeling the spectral dependence resulting from the fractal structure of cellular organelles (Schmitt and Kumar 1996; Tuchin 2000; Wax et al. 2002). The residual of each spectrum was calculated by subtracting the power law curve from the experimental spectrum to produce a normalized spectrum which isolates the oscillatory features as shown in Figure 23.21c.

The normalized spectra showed clear oscillations resulting from interference produced by scattering from the front and back surfaces of basal cell nuclei. Each normalized spectrum was Fourier transformed to generate a correlation plot similar to that shown in Figure 23.21d, which shows a clear peak corresponding to the dominant frequency in the normalized spectrum. Peak detection was carried out by an automated, custom MATLAB program. The script first high-pass filtered the spectrum with a cutoff of 4 cycles to remove low frequency content which was not removed by the power law fit. The location of the peak in the correlation plot was then automatically detected by the MATLAB script and related to scatterer diameter with the simple equation $d = $ correlation distance$/(2n)$, where n is the refractive index and d is the diameter of the cell nuclei. A nuclear index of refraction of $n = 1.395$ was assumed (Graf and Wax 2005).

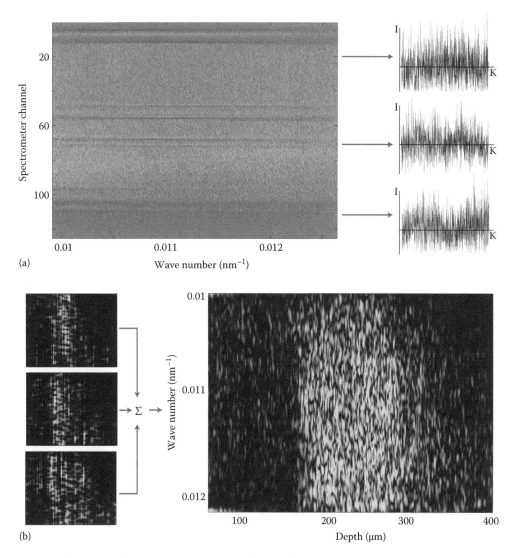

FIGURE 23.19 (a) Raw data from animal experiments with spectra from 3 of 120 spectrometer channels shown. (b) Three depth-resolved spectroscopic plots produced by DW processing the data in part (a). Summing the plots from all channels produces the final TFD as shown. (From Graf, R.N. et al., *J. Biomed. Opt.*, 14(6), 064030-8, 2009.)

The results of the fLCI/DW animal study are summarized in Figure 23.22. The 16 untreated tissue samples had a mean basal layer nuclear diameter of 4.28 μm with a standard deviation of 0.69 μm. The 21 scanned treated tissue samples had a mean basal layer nuclear diameter of 9.50 μm with a standard deviation of 2.08 μm. A statistical *t*-test revealed a *p*-value of less than 0.0001, indicating an extremely statistically significant difference between the basal layer nuclear diameters of the two populations. The presented decision line results in excellent separation, correctly categorizing 21 of 21 treated samples (100% sensitivity) and 16 of 16 untreated samples (100% specificity).

23.6 Conclusion

This chapter has presented the development of the fLCI approach for depth-resolved spectroscopy and illustrated its potential application to detecting precancerous tissues using structural measurements. Although the results of the above animal study were extremely promising, current fLCI methods have several limitations which will guide the future development of the technology. First, recent studies (Yi et al. 2009; Robles and Wax 2010) have shown that it is possible to measure localized spectral features in controlled phantoms such as absorption (Yi et al. 2009) and scatterer diameter (Robles and Wax 2010). The controlled measurements in these studies have shown that artifacts such as correlated scatterers can influence analysis results, an aspect that must be thoroughly characterized to enable meaningful clinical application of the approach. Additionally, more complete scattering models for animal tissue should be developed and employed. Finally, since the optical path length measurements are dependent on refractive index, the fLCI technique must account for this in determining the location of specific

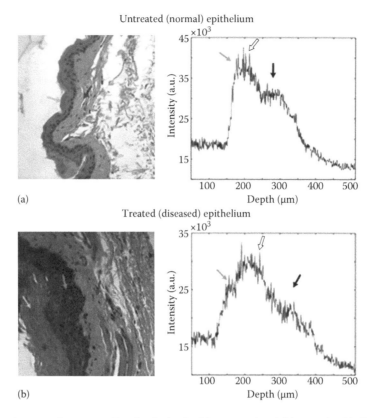

FIGURE 23.20 Histopathology image and corresponding depth plot for (a) untreated and (b) treated epithelium. Arrows indicate keratinized layer (light gray), basal layer (open arrow), and lamina propria (dark gray). (From Graf, R.N. et al., *J. Biomed. Opt.*, 14(6), 064030-8, 2009.)

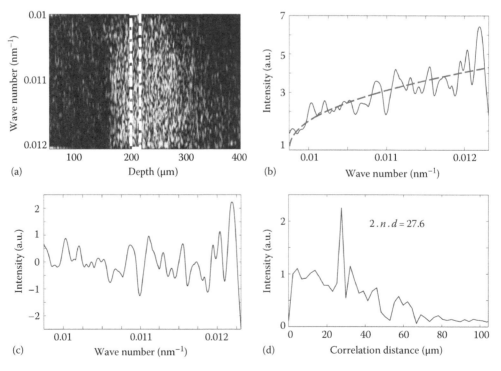

FIGURE 23.21 (a) Depth-resolved spectroscopic plot with basal layer indicated by dashed box. (b) Spectrum from basal tissue layer along with power law fit. (c) Residual spectrum from basal tissue layer. (d) Correlation plot generated by Fourier transforming spectrum in (c). Peak correlation distance can be related directly to scatterer size. (From Graf, R.N. et al., *J. Biomed. Opt.*, 14(6), 064030-8, 2009.)

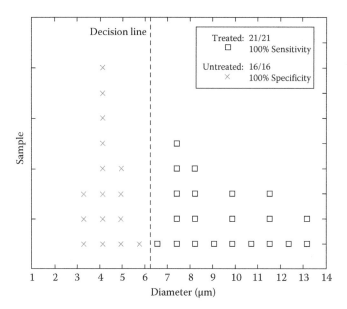

FIGURE 23.22 Nuclear diameter measurements for each sample of the complete animal trial. Decision line results in 100% sensitivity and 100% specificity. (From Graf, R.N. et al., *J. Biomed. Opt.*, 14(6), 064030-8, 2009.)

tissue layers. A deeper understanding of the dynamics of refractive index in dysplastic animal tissue must be developed.

The fLCI method's nuclear sizing algorithm, while effective in the animal study, is likely overly simplistic and should be further developed. Understanding how correlations between neighboring cellular structures and correlations between tissue layers contribute to generated correlation plots will facilitate the development of an advanced scattering model. Additionally, an analysis assessing the effectiveness of using Mie theory for fLCI scatterer size determination should also be undertaken. The measurement of fractal dimension and spectral slope as metrics for distinguishing between normal and diseased tissue should also be explored. The completion of these aims as well as the development of more advanced experimental schemes, compatible with in vivo studies, will move the fLCI approach closer toward the goal of becoming a clinically viable early cancer detection technique.

Acknowledgments

This research has been supported by grants from the National Institutes of Health (NCI R21-CA120128) and the National Science Foundation (BES 03-48204).

References

Adler, D., Ko, T., Herz, P., and Fujimoto, J. 2004. Optical coherence tomography contrast enhancement using spectroscopic analysis with spectral autocorrelation. *Opt. Express* 12(22): 5487–5501.

Backman, V., Gopal, V., Kalashnikov, M. et al. 2001. Measuring cellular structure at submicrometer scale with light scattering spectroscopy. *IEEE J. Sel. Topics Quantum Electron.* 7(6): 887–893.

Bigio, I. J., Bown, S. G., Briggs, G. et al. 2000. Diagnosis of breast cancer using elastic-scattering spectroscopy: Preliminary clinical results. *J. Biomed. Opt.* 5(2): 221–228.

Cohen, L. 1989. Time frequency-distributions—A review. *Proc. IEEE* 77(7): 941–981.

Dennis, T., Dyer, S. D., Dienstfrey, A., Singh, G., and Rice, P. 2008. Analyzing quantitative light scattering spectra of phantoms measured with optical coherence tomography. *J. Biomed. Opt.* 13(2): 024004.

Dhar, A., Johnson, K. S., Novelli, M. R. et al. 2006. Elastic scattering spectroscopy for the diagnosis of colonic lesions: Initial results of a novel optical biopsy technique. *Gastrointest. Endosc.* 63(2): 257–261.

Dyer, S. D., Dennis, T., Street, L. K., Etzel, S. M., Germer, T. A., and Dienstfrey, A. 2006. Spectroscopic phase-dispersion optical coherence tomography measurements of scattering phantoms. *Opt. Express* 14(18): 8138–8153.

Fercher, A. F., Hitzenberger, C. K., Sticker, M. et al. 2000. A thermal light source technique for optical coherence tomography. *Opt. Commun.* 185(1–3): 57–64.

Graf, R. N., Brown, W. J., and Wax, A. 2008. Parallel frequency-domain optical coherence tomography scatter-mode imaging of the hamster cheek pouch using a thermal light source. *Opt. Lett.* 33(12): 1285–1287.

Graf, R. N., Robles, F. E., Chen, X., and Wax, A. 2009. Detecting precancerous lesions in the hamster cheek pouch using spectroscopic white-light optical coherence tomography to assess nuclear morphology via spectral oscillations. *J. Biomed. Opt.* 14(6): 064030-8.

Graf, R. N. and Wax, A. 2005. Nuclear morphology measurements using Fourier domain low coherence interferometry. *Opt. Express* 13(12): 4693–4698.

Graf, R. N. and Wax, A. 2007. Temporal coherence and time-frequency distributions in spectroscopic optical coherence tomography. *J. Opt. Soc. Am. A: Opt. Image Sci. Vis.* 24(8): 2186–2195.

Grajciar, B., Pircher, M., Fercher, A., and Leitgeb, R. 2005. Parallel Fourier domain optical coherence tomography for in vivo measurement of the human eye. *Opt. Express* 13(4): 1131–1137.

Kumar, V., Abbas, A. K., Fausto, N., and Aster, J. 2005. *Robbins and Cotran pathologic basis of disease.* Philadelphia, Elsevier/Saunders.

Lovat, L. B., Johnson, K., Mackenzie, G. D. et al. 2006. Elastic scattering spectroscopy accurately detects high grade dysplasia and cancer in Barrett's oesophagus. *Gut* 55(8): 1078–1083.

Mourant, J. R., Johnson, T. M., Carpenter, S., Guerra, A., Aida, T., and Freyer, J. P. 2002. Polarized angular dependent spectroscopy of epithelial cells and epithelial cell nuclei to determine the size scale of scattering structures. *J. Biomed. Opt.* 7(3): 378–387.

Oldenburg, A. L., Chenyang, X., and Boppart, S. A. 2007. Spectroscopic optical coherence tomography and microscopy. *IEEE J. Sel. Topics Quantum Electron.* 13(6): 1629–1640.

Perelman, L. T., Backman, V., Wallace, M. et al. 1998. Observation of periodic fine structure in reflectance from biological tissue: A new technique for measuring nuclear size distribution. *Phys. Rev. Lett.* 80(3): 627–630.

Robles, F., Graf, R. N., and Wax, A. 2009. Dual window method for processing spectroscopic optical coherence tomography signals with simultaneously high spectral and temporal resolution. *Opt. Express* 17(8): 6799–6812.

Robles, F. E. and Wax, A. 2010. Measuring morphological features using light-scattering spectroscopy and Fourier-domain low-coherence interferometry. *Opt. Lett.* 35(3): 360–362.

Schmitt, J. M. and Kumar, G. 1996. Turbulent nature of refractive-index variations in biological tissue. *Opt. Lett.* 21(16): 1310–1312.

Tuchin, V. 2000. *Tissue Optics: Light Scattering Methods and Instruments for Medical Diagnosis.* Bellingham, WA: SPIE.

Wax, A. and Backman, V. (eds.). 2009. *Biomedical Applications of Light Scattering.* New York: McGraw-Hill.

Wax, A., Yang, C. H., Backman, V. et al. 2002. Cellular organization and substructure measured using angle-resolved low-coherence interferometry. *Biophys. J.* 82(4): 2256–2264.

Wax, A., Yang, C. H., Dasari, R. R., and Feld, M. S. 2001. Measurement of angular distributions by use of low-coherence interferometry for light-scattering spectroscopy. *Opt. Lett.* 26(6): 322–324.

Wax, A., Yang, C. H., and Izatt, J. A. 2003. Fourier-domain low-coherence interferometry for light-scattering spectroscopy. *Opt. Lett.* 28(14): 1230–1232.

Wilson, J. D., Bigelow, C. E., Calkins, D. J., and Foster, T. H. 2005. Light scattering from intact cells reports oxidative-stress-induced mitochondrial swelling. *Biophys. J.* 88(4): 2929–2938.

Xu, C. Y., Carney, P. S., and Boppart, S. 2005a. Wavelength-dependent scattering in spectroscopic optical coherence tomography. *Opt. Express* 13(14): 5450–5462.

Xu, C. Y., Kamalabadi, F., and Boppart, S. A. 2005b. Comparative performance analysis of time-frequency distributions for spectroscopic optical coherence tomography. *Appl. Opt.* 44(10): 1813–1822.

Xu, C. Y., Vinegoni, C., Ralston, T. S., Luo, W., Tan, W., and Boppart, S. A. 2006. Spectroscopic spectral-domain optical coherence microscopy. *Opt. Lett.* 31(8): 1079–1081.

Yi, J., Gong, J., and Li, X. 2009. Analyzing absorption and scattering spectra of micro-scale structures with spectroscopic optical coherence tomography. *Opt. Express* 17(15): 13157–13167.

Zysk, A. M., Adie, S. G., Armstrond, J. J. et al. 2007a. Needle-based refractive index measurement using low-coherence interferometry. *Opt. Lett.* 32(4): 385–387.

Zysk, A. M., Marks, D. L., Liu, D. Y., and Boppart, S. A. 2007b. Needle-based reflection refractometry of scattering samples using coherence-gated detection. *Opt. Express* 15(8): 4787–4794.

Phase Imaging Microscopy: Beyond Dark-Field, Phase Contrast, and Differential Interference Contrast Microscopy

Chrysanthe Preza
The University of Memphis

Sharon V. King
Boulder Nonlinear Systems

Nicoleta M. Dragomir
Victoria University

Carol J. Cogswell
University of Colorado

24.1 Introduction 483
24.2 Phase Imaging Microscopy: Past and Present.......................... 484
 Historical Perspective • Imaging System Configuration, Basic Operation, and
 Applications • Comparison of the Three Imaging Techniques: Advantages and
 Disadvantages • Examples of System Enhancements • Toward Quantitative Imaging:
 Modeling of Image Formation
24.3 Recent Advancements for Phase Imaging Microscopy: Development and New
 Directions..501
 Quantitative 2D Phase Imaging • Three-Dimensional Phase Imaging
24.4 Conclusion 510
References..510

24.1 Introduction

Biomedical applications aimed toward understanding live-cell dynamics have rekindled an interest in optical microscope modes that may provide additional information beyond the widely used fluorescence modalities that include confocal, structured illumination microscopy, multiphoton, and STED imaging. The three microscope modes discussed in this chapter—dark-field, phase contrast, and differential interference contrast (DIC) microscopy—are techniques that readily address the challenges of live-cell imaging. Their ability to image cell structure dynamics without having to introduce any chemical probes or dyes can be of great advantage in ensuring that cell function is not being altered. In addition, the ability to observe living cells over time, without introducing the potentially damaging effects of high-intensity light sources, offered by these techniques is often desirable for many biological investigations.

All three of these microscopy modes have been available for over 50 years and their development was aimed at the need to visualize transparent biological structures that do not absorb enough of the illuminating light to be seen with standard bright-field microscopy. Instead, these three modes produce varying image intensities (contrast) through manipulation of the many additional ways (beyond absorption) that light interacts with specimen structures. These include the well-known optical properties of refraction, diffraction, and scattering that occur whenever a specimen structure has a refractive index that varies from its surrounding medium. In general, the variations in refractive index and thickness among biological structures impart a phase shift to the light that transmits through the structure. Phase contrast and DIC microscopes are particularly designed to convert phase shifts into visible contrast using interference techniques. Dark-field microscopy, on the other hand, is more directed to imaging features that scatter and refract the illuminating light. The traditional design principles of all three of these modes, including a critical evaluation of their imaging advantages and potential drawbacks, are presented in Sections 24.2.1 through 24.2.3. Section 24.2.4 reviews some examples of system enhancements for the three microscopy modes.

Perhaps most significant to this discussion is that recent technological advancements have been made in the fields of optics and digital signal processing that are bringing these modalities back into the forefront of biomedical imaging. Specifically, the ongoing drive toward developing microscopy modes that integrate innovative optics and computational methods capable of detecting and imaging molecules at the nanoscale has sparked an interest in developing more precise methods for observing and measuring small variations in cells and subcellular components. In addition, more accurate image formation models to predict the intensity observed in measured microscopy images, acquired with the three traditional phase microscopy modes discussed in this chapter, have been developed over the last 15 years. These models, discussed in Section 24.2.5, have facilitated the development of specialized computational methods for the extraction of quantitative information that relates to fundamental properties of the specimens under examination. All of these efforts have provided motivation for the recent development of two-dimensional (2D) and three-dimensional (3D) quantitative phase imaging microscopy techniques discussed in Section 24.3.

24.2 Phase Imaging Microscopy: Past and Present

Many biological specimens absorb little or no light and are not visible when in focus under an ordinary bright-field light microscope configuration. Such specimens do have the property of changing (retarding or advancing) the phase of light that passes through them due to spatial variations in their refractive index and/or thickness in the direction of wave propagation. The phase of the light governs its interference and diffraction as it propagates through a microscope. Since the light phase variations relate to the specimen's fundamental properties, transparent specimens can be visualized with specialized microscopes that convert optical path phase variations into image intensity variations (contrast). Specimens that alter the phase of light that passes through them are called phase specimens and they can be characterized by a phase function defined as

$$\phi(\mathbf{x}) = \frac{2\pi}{\lambda} \mathrm{OPL}(\mathbf{x}) \qquad (24.1)$$

where

$\mathbf{x} = (x, y)$ are spatial coordinates
λ is the average wavelength of the illuminating light

$$\mathrm{OPL}(\mathbf{x}) = \int_{z_2(\mathbf{x})}^{z_1(\mathbf{x})} n(x, y, z)\mathrm{d}z$$

is the optical path-length (OPL) function of the specimen, which depends on the integral of the refractive index distribution of the specimen $n(x, y, z)$ over the thickness of the specimen that could vary spatially and is thus modeled by the difference $z_1(x) - z_2(x)$ along the z (optical) axis. A difference in OPL between two paths is called the optical path difference and it plays an important role in the amount of contrast generated in the image. The primary determining factors of OPL difference are the refractive index difference between the specimen and its surrounding medium, and the geometrical distance the light wavefront travels between two points on the optical path. Typical OPL differences for individual cells in culture are about 0.125 μm, or about a quarter wavelength of green light ($\lambda = 0.5$ μm), while for subcellular structures the OPL difference is much smaller. This observation has influenced the choice of optical elements used to convert a conventional microscope to one that has the ability to image

phase changes; for example, the phase plate in a phase contrast microscope discussed in Section 24.2.2.2.

The amount of contrast achieved in a microscope image depends on the type of illumination used, the interaction of the illumination with the specimen, the specimen morphology (i.e., refractive index and thickness variations), and, finally, how the waves leaving the specimen are treated (Inoue 1986). These facts, studied and proven with experiments and theory over the years, have led to the development of different microscope configurations capable of contrast generation that enables phase imaging.

In dark-field microscopy, contrast is generated in the image because the illumination light is altered. In phase contrast and DIC microscopy the illumination is also altered. However, in these modes, a complementary optical accessory (e.g., filter or prism) that manipulates the light after it has interacted with the specimen is used in combination with a conditioner that alters the light before it strikes the specimen. This provides more desirable imaging features (e.g., higher resolution, reduced artifacts due to dust or out-of-focus objects) than possible with dark-field microscopy.

24.2.1 Historical Perspective

For many years, dark-field microscopy was the only contrast generating mode for the observation of transparent specimens such as unstained live cells (Loveland 1981, Pluta 1988, 1989b, Molecular Expressions 2009b). Fritz Zernike's Nobel Prize winning phase contrast microscope, introduced in 1935, allowed the observation of such objects by converting phase changes in the imaging light to intensity changes visualized in an image through interferometry for the first time (Zernike 1942a). The most comprehensive and detailed description of phase contrast in its traditional form is found in references (Zernike 1942a,b, Pluta 1989a). Following the same principle of converting phase changes to intensity changes through interferometry, Francis Smith introduced the DIC microscope in 1955. This technique was further refined by Georges Nomarski to allow less aberrated high-resolution phase contrast imaging through Nomarski DIC microscopy (Nomarski 1955). A detailed description of DIC microscopy in its traditional form is found in references (Allen et al. 1969, Pluta 1989b). It is worth noting that useful interactive tutorials (Molecular Expressions 2009a, Nikon Miscoscopy U 2009) and videos (Centonze 2008) highlighting the use and operation of these microscope modes are freely available on the internet.

Given the many references that exist on the description and operation of these microscope modes (Inoue 1986, Pluta 1989b, Murphy 2001, Cox 2007, Molecular Expressions 2009a), the following paragraphs (Section 24.2.2) briefly summarize their basic system configuration and operation. Although some of these microscopy modes are available in both reflected- and transmitted-light configurations, the discussion in this section is focused primarily on transmitted-light systems. A comparison of the advantages and disadvantages of dark-field, phase contrast, and DIC imaging is presented in Section 24.2.3. Section 24.2.4 presents some examples of efforts to improve microscope system performance with electro-optical system enhancements. Section 24.2 concludes with a theoretical framework that enables image modeling in a transmitted-light microscope under partially coherent illumination suitable for the understanding and prediction of images formed with phase imaging systems (Section 24.2.5).

24.2.2 Imaging System Configuration, Basic Operation, and Applications

24.2.2.1 Dark-Field Microscopy

Dark-field microscopy, also known as dark-ground microscopy, renders a transparent specimen clearly visible against a dark background by imaging only the light that is scattered by the specimen while avoiding direct (non-diffracted) light. This is necessary because direct illumination light can be much brighter than the light scattered by the specimen and can dominate in the image plane making the transparent specimen invisible. A useful analogy is to consider how the faint light from stars is only visible during the night (with better visibility on a dark night) but remains invisible during the day when the bright sunlight dominates. An example dark-field image is shown in Figure 24.1b. Additional examples of dark-field images can be found at (Molecular Expressions 2009c).

Dark-field microscopy can be accomplished on a conventional light microscope at a low cost. A special condenser is used that provides symmetrical oblique illumination to the specimen from all sides (Figure 24.2). The oblique light gets scattered as it passes through the specimen due to optical discontinuities (such as cell membranes, organelles, nuclei). In this configuration of cardioid condenser/objective lens pair, the scattered light enters the objective lens while the direct or unscattered light remains an oblique hollow light cone that misses the objective because of its steep angle. If transmitted dark-field microscopy is considered, the numerical aperture (NA) of the condenser has to be about 15% higher than the NA of the objective. This is contrary to all other contrast methods, where the NA of the objective has a higher or equal NA than the condenser. A high NA objective would allow some of the direct illumination to enter the objective destroying the dark-field contrast. In the case of reflected dark-field microscopy, the central light stop is commonly located within the cube of the reflected light attachment such that the objective delivers the light to the specimen through the outer ring. In this case, the scattered light from the specimen is then collected in the central part of the objective as image-forming light rays.

The cardioid condenser lens is one of the most popular designs as it is an aberration-free system with high NA (1.2–1.4). Because the maximum NA for the objective lens that may be accepted by the cardioid lens condenser, when oil is used, is 0.9–1.0, high-NA objective lenses utilized in dark-field microscopy have an iris diaphragm that permits the necessary reduction of the NA.

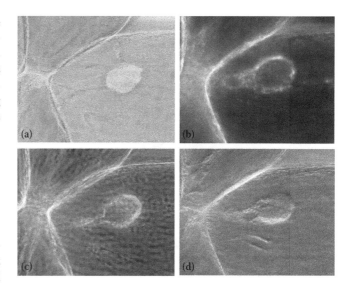

FIGURE 24.1 Comparison of four microscope modes for imaging a living stamen hair cell that is actively exhibiting cytoplasmic streaming. (a) *bright field* shows the nucleus, cell walls, and some of the naturally pigmented cytoplasm due to partial absorption of the illuminating beam; (b) *dark field* shows only the light scattering features such as those with large refractive index compared to their surroundings, for example, cell walls and streaming particles; (c) *phase contrast* shows similar light scattering features as in dark field but also displays blurred striated structures from out-of-focus cell surfaces that make the streaming particles difficult to discern; and (d) *DIC* shows the nucleus and some of the in-focus streaming particles, appearing in the characteristic pseudo 3D (bas-relief), and with no out-of-focus artifacts. Note that some cytoplasmic channels appear to have changed position within the four images due to their being acquired at different times. Images recorded using a black and white CCD camera and white light illumination. Objective: 40×, NA 1.25 (bright field and phase contrast); 40×, NA 0.75 (dark field and DIC), which accounts for some loss in lateral resolution.

A detailed discussion on dark-field condensers is readily available (Pluta 1989b, Murphy 2001, Molecular Expressions 2009b).

The direct or unscattered light, rejected by dark-field illumination, forms the background light in bright-field microscopy. Thus in terms of Fourier analysis, the direct light corresponds to the zero-order in the diffraction pattern formed at the back focal plane (Fourier plane) of the objective (i.e., the central spatial frequency of the optical Fourier transform). Because dark-field illumination removes this zero-order, it is equivalent to a spatial filtering technique without the use of a filter in the back focal plane. It follows that, alternatively, dark-field imaging can be accomplished by spatial filtering in the Fourier plane. This technique is called central dark-field microscopy (Pluta 1989b). In central dark-field microscopy, an opaque mask is placed in the back focal plane of the objective lens and thus it is implemented in the same way as phase contrast microscopy discussed in Section 24.2.2.2.

The main limitations of dark-field microscopy are that it is susceptible to artifacts due to scattering of dust particles and the imaging NA is limited to 1.0. A detailed procedure on how to

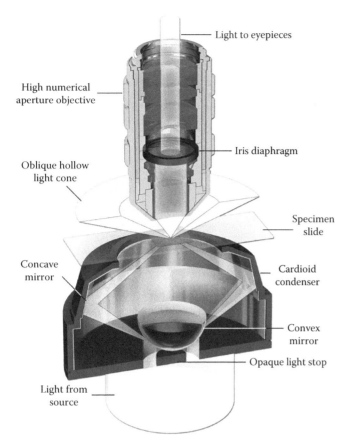

High numerical
aperture objective

Light to eyepieces

Oblique hollow
light cone

Iris diaphragm

Specimen
slide

Concave
mirror

Cardioid
condenser

Convex
mirror

Opaque light stop

Light from
source

FIGURE 24.2 High NA cardioid dark-field condenser. The specimen slide is illuminated with an oblique hollow light cone, such that unscattered light (light that does not interact with the specimen) passes outside the acceptance angle (NA) of the imaging objective lens. This is accomplished by partially blocking the source with an annular opaque light stop. Thus, only light scattered by the specimen into the imaging objective lens is observed in the field of view and the background is therefore otherwise dark. Dark-field illumination makes transparent specimens visible, even when the specimen does not absorb light and scatters it only weakly. This cardioid condenser configuration uses a convex spherical mirror and a concave cardioid mirror to project an oblique hollow cone of light onto the specimen at a very high angle. In lower NA dark-field condenser configurations, the opaque light stop is located in the front focal plane of a standard condenser lens assembly (no mirrors are necessary). (From Molecular Expressions, Darkfield illumination, 2009b, http://micro.magnet.fsu.edu/primer/techniques/darkfield.html. With permission.)

configure and align the dark-field microscope can be found at (Molecular Expressions 2009d). Solutions to numerous common problems associated with dark-field microscopy such as insufficient illumination and condenser misalignment can be found at Molecular Expressions (2009e).

Conventional dark-field microscopy is well established in practice with applications ranging from industrial inspection of both transparent and opaque objects, to marine biology for the observation of water organisms. Other examples include visualization of impurities and defects in transparent objects (e.g., glass, ceramics, and plastics) and surface micro-defects in opaque

objects (Pluta 1989b). The development of phase contrast microscopy reduced interest in the use of dark-field microscopy in biology and biomedicine for some years. However, recent studies in the characterization of single molecule interactions (Dietrich et al. 2007), assessment of cell viability (Wei et al. 2008), and real-time visualization of giant liposome transformation processes (Honda et al. 2006) using dark-field microscopy, as well as an example study of the use of nanoparticles to create contrast in dark-field microscopy (Bogatyrev et al. 2006), show a renewed interest. In addition, new optical methods employing dark-field illumination such as dark-field digital holographic microscopy (Dubois and Grosfils 2008), 3D dark-field photoacoustic microscopy (Song et al. 2008), and confocal dark-field microscopy (Schmelzeisen et al. 2008) have recently been reported exploiting further the benefits of this type of illumination.

24.2.2.2 Phase Contrast Microscopy

Although there are several different optical configurations for imaging phase variations, most of them are based on the Zernike phase-contrast approach discussed in Zernike (1942a,b), Pluta (1989b), Molecular Expressions (2009a), Nikon Miscoscopy (U 2009) and briefly summarized here. In a Zernike phase-contrast microscope (Figure 24.3) two additional optical components (the condenser annulus and the phase plate) are introduced in the light path of a conventional light microscope to ensure that the light waves diffracted by the phase object will interfere (either destructively or constructively) with the undiffracted illuminating wave. This forms a visible image in which intensity variations are due to variations in the specimen's phase function (Equation 24.1). The contrast generation principle requires that the interfering waves are sufficiently different in phase while their amplitudes are comparable. This condition is met by phase-shifting and attenuating the light waves that do not get diffracted by the specimen.

In the phase contrast microscope, the illuminating light is focused by the collector lens on the condenser annulus (placed in front of the condenser). The condenser annulus consists of an opaque circular plate with a transparent ring whose size allows the light to enter the objective lens (unlike in the case of the light stop used in dark-field microscopy discussed earlier). This illumination modification allows the specimen to be illuminated by defocused light waves emanating from the transparent annular ring (Figure 24.4). Light waves that do not interact with the specimen (undeviated light) are refocused into a ring-shaped region on the back focal plane of the objective, while the specimen diffracted light will appear everywhere in the back focal plane. This ring-shaped region contains partially absorbing material and a phase plate (both of which are components of a phase objective lens) that are complementary in shape to the condenser annulus. The absorbing material reduces the intensity of the undeviated light so it more closely matches the diffracted signal intensities from the specimen, while the phase plate alters the light's phase by 90°, either advancing or retarding depending on whether the phase plate has a reduced or increased opacity in the ring-shaped region (Figure 24.5). A phase of 90° (or $\pi/2$ rad) corresponds to an OPL equal to $\lambda/4$ (see Equation 24.1). Therefore, this type of

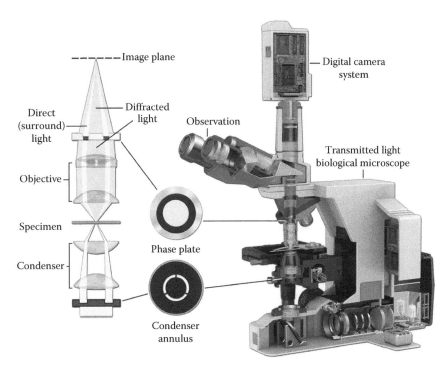

FIGURE 24.3 Phase contrast microscope configuration. The phase plate and condenser annulus, the key optical components that create phase contrast, are shown in relation to their location in the optical train (left) and the upright microscope (right). The condenser annulus, a transparent annular ring, is positioned in the front focal plane (aperture) of the condenser so that the specimen can be illuminated by parallel light wavefronts emanating from the ring. This oblique hollow light cone illumination ensures that unscattered light passes through the annular phase plate (ring) located in the back focal plane of the imaging objective. The quarter wave retardance of the phase plate results in an amplitude reduction and a phase change of the unscattered light (labeled as surround light) prior to its interference with the light weakly scattered by the specimen (labeled as diffracted light). The intensity at the image plane, observed by the eyepiece or detected by the camera, is due to the interference between the retarded unscattered light and the diffracted light. (From Nikon Microscopy U., Introduction to phase contrast microscopy, 2009, http://www.microscopyu.com/articles/dic/desenarmontdicintro.html. With permission.)

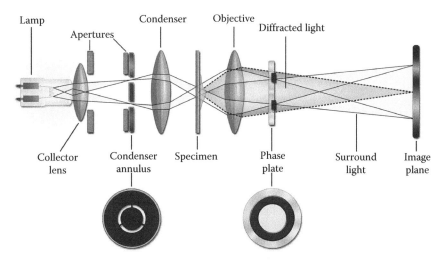

FIGURE 24.4 Phase contrast microscope optical train. Detail of the optical train further illustrates the separation of the diffracted and undiffracted light. Partially coherent illumination is shaped by the condenser annulus into a hollow cone and focused onto the specimen. Undiffracted light is imaged onto the annular phase plate (the conjugate plane) where it is retarded by a quarter wavelength. This configuration maximizes the differences in the diffracted and surround light resulting in larger intensity contrast in the interference pattern observed at the image plane than that observed with bright-field microscopy (i.e., without the phase contrast optical accessories). However, as diffracted light passes through the objective lens, a small portion inevitably also passes through the annular phase plate. Thus, part of the diffracted light gets retarded as well, which contributes to the formation of the halo artifact observed in phase contrast images. (From Nikon Microscopy U., Introduction to phase contrast microscopy, 2009, http://www.microscopyu.com/articles/dic/desenarmontdicintro.html. With permission.)

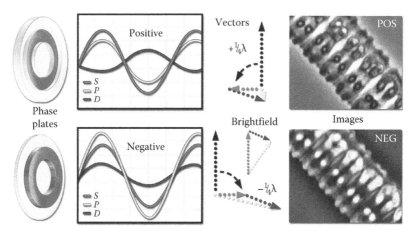

FIGURE 24.5 Positive and negative phase contrast systems. Image intensity may increase or decrease (due to an increase in specimen OPL) depending on the design of the phase plate in the objective. The ring-shaped region of the plate may either advance (top row images) or retard (bottom row images) the surround or undiffracted wave (S) with respect to the diffracted wave (D). This creates a resultant combined wave (P) with amplitude either lower due to destructive wave interference (positive phase contrast) or higher due to constructive wave interference (negative phase contrast) than the amplitude of the S wave, respectively. The object appears relatively darker than the background with positive phase contrast (right top image labeled POS) while with negative phase contrast the object appears bright against a darker background (right bottom image labeled NEG). Positive phase contrast optics is the traditional form produced by most manufacturers. Vector diagrams illustrating the relationship between individual wavefronts in bright-field and phase contrast microscopy are shown between the graphs and the images. In these diagrams, the vector length represents the amplitude of each wave, while the angle of rotation of the S vector relative to a fixed reference (vector diagram for bright field) indicates the change in the phase of the *S* wave by a quarter-wavelength ($\lambda/4$), which is shown as 90° counterclockwise rotation in positive phase contrast. In bright-field microscopy, the specimen remains invisible on a bright background because the *S* and *P* waves have similar amplitudes resulting in lack of contrast generation in the image. (From Nikon Microscopy U., Introduction to phase contrast microscopy, 2009, http://www.microscopyu.com/articles/dic/desenarmontdicintro.html. With permission.)

phase plate is referred to as a *quarter wavelength* plate. Since light waves are filtered by the phase plate in the back focal plane—the Fourier plane of the objective lens—phase contrast imaging is also known as a spatial filtering approach (Goodman 1996).

Destructive or constructive interference between the diffracted and undiffracted waves occurs at the image plane depending on the net phase difference between the waves. This net phase difference depends on the amount the diffracted light has been advanced or retarded by the object's phase and the effect of the phase plate. Destructive interference results in what is called *positive phase contrast* while constructive interference results in *negative phase contrast,* yielding rather different image intensities (Figure 24.5). Interpretation of phase contrast images can be difficult because of unexpected contrast reversal due to phase differences that are integer multiples of 180° or due to system artifacts (halo and shading off) discussed in the following. As illustrated in Figure 24.4, some of the light diffracted by the specimen inevitably passes through the annular phase plate, acquiring the same phase change as the undeviated light. Typically, the part of the light diffracted by the specimen that gets retarded by the phase plate corresponds to low spatial frequencies because of the location of the phase-altering ring (Murphy 2001). This results in the lack of interference between this low spatial frequency light and the undeviated light producing a localized reversal in the observed contrast, which is visualized as a halo around object boundaries. The halo artifact appears brighter in intensity than the background

in positive phase contrast images and darker in negative phase images (Figure 24.5). An additional artifact in phase contrast images is the shading-off effect in which uniform OPL in the specimen is not visualized as uniform intensity in the image. For more details about this artifact see (Pluta 1989b, Murphy 2001, Molecular Expressions 2009a). The halo and shading-off artifacts depend on the geometrical and optical properties of the phase plate and the specimen under observation (for a summary of results see Pluta (1989b)). Due to these artifacts, the intensity observed in phase contrast images is not linearly related to the specimen OPL (or phase) function. An example of a phase contrast image is shown in Figure 24.1c.

Valuable information about the correct use and alignment of phase contrast microscopy is readily available (Pluta 1989b, Molecular Expressions 2009a). Phase contrast microscopy is widely used in many applications to study living cells (Gundlach 1993), tissues (Peltroche-Llacsahuanga et al. 2000, Mantilla et al. 2001, Pot et al. 2009), microorganisms (Charles et al. 2000, Donders 2007, Carneiro et al. 2009, Molecular Expressions 2009a), cells in culture (Jung et al. 2009) and in the diagnosis of tumor cells (Berman and Crump 2008, Wu et al. 2009).

24.2.2.3 Differential Interference Contrast Microscopy

A large amount of information about Nomarski's DIC microscope as a qualitative technique, how to use DIC, as well as a large volume of information about the Nomarski prism design

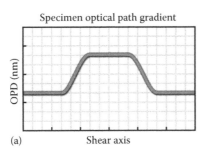
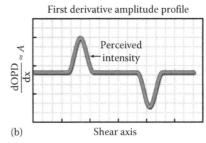

Specimen optical path gradient

First derivative amplitude profile

(a) Shear axis

(b) Shear axis

FIGURE 24.6 Origin of contrast in DIC. For slowly varying phase objects, the DIC image intensity is a qualitative approximation of the gradient (or derivative) of the phase delay in a single direction. The relationship of DIC intensity to the phase of the specimen is shown by comparison of (a) an illustration of the optical path delay (OPD) of a specimen and (b) the first derivative of the specimen phase profile. Contrast in the DIC image is created by the interference of two beams with differential phase delay. The total phase delay is the sum of both the differential delay introduced by the specimen and the delay of the prism or prism bias. Positive and negative differential phase delay relative to the prism bias creates the shadow-cast or relief appearance of DIC. Note that the sign of the phase gradient in DIC is dependent on the configuration of the optical train and that positive intensity does not necessarily correspond to an increase in slope. Image contrast for specific objects can be optimized by varying the prism bias (Pluta 1989b). (From Nikon Microscopy U., Introduction to phase contrast microscopy, 2009, http://www.microscopyu.com/articles/dic/desenarmontdicintro.html. With permission.)

is readily available in the literature (Allen et al. 1969, Lang 1969, Pluta 1989b). Although other DIC configurations exist currently, the configuration based on Nomarski's DIC is used almost exclusively (which is referred to here as DIC microscopy). In DIC microscopy, the image is formed from the difference between the amplitudes of two images that have a lateral differential displacement of a few tenths of a micrometer and are phase-shifted relative to each other. The differential nature of DIC microscopy translates into less restrictive phase wrapping conditions as compared with other interferometric phase imaging techniques. DIC imaging is based on the principle that gradients of the specimen's OPL function, along a certain direction (shear axis), are converted into intensity differences (Figure 24.6) that can be visualized as improved contrast in the DIC image. Thus, in DIC microscopy, the observed image has a shadow-cast appearance. An example of a DIC image is shown in Figure 24.1d.

The key optical components of a DIC microscope are a polarizer, an analyzer, and two Wollaston prisms arranged as shown in Figure 24.7. These optical components have certain properties that are fundamental to the operation of the DIC microscope. First, a polarizer (or an analyzer) that is an optical element with a single plane of vibration* has the property of letting light pass through it only in its plane of vibration. Second, a Wollaston prism, which consists of two prisms (wedges) made of birefringent, uniaxial[†] material cemented together, is used as a beam splitter.

When light emanating from a source passes through a polarizer, it becomes plane or linearly polarized. In the DIC microscope,

the orientation of the polarizer is set so that its direction of light vibration is at 45° with respect to the optical axis of the bottom wedge of the Wollaston prism, and at 90° with respect to the analyzer's direction of vibration. Thus, a ray emerging from the polarizer and entering the compensator Wollaston prism splits into two orthogonal components (Figure 24.8). The lateral separation distance between the two ray components, called the *shear*, is of the order of the resolution limit of the objective lens. The two wave components then pass through the condenser into different parts of the specimen, where each is either phase advanced or retarded depending on the refractive index and thickness of the specimen it travels through. The components are then collected and focused by the objective lens into the sliding Wollaston prism, which is oriented opposite to the first prism. When the second prism is perfectly aligned with the first one, as in Figure 24.8, the phase shift introduced to the two wave components by the first prism is canceled by the second prism, and thus any remaining phase difference is due only to the specimen.

By sliding the second prism along the direction of shear (perpendicular to the optical axis of the microscope) an additional uniform phase difference between the two components, called the *bias retardation*, is introduced. At the interface between the two halves of the sliding prism, the waves are combined to form a single beam (i.e., the shear is removed), but they are still plane-polarized with perpendicular vibration directions. In order for the two wave components to interfere and produce the desired image, their vibration planes must be forced to coincide. This is achieved by passing the recombined wave through the polarizing analyzer. The interference of the two components results in constructive and destructive superposition of the waves, which gives rise to dark and bright areas in the final DIC image. Thus, a DIC microscope converts phase differences to intensity variations. Specimens with birefringence properties are not suitable for DIC imaging. Birefringence introduces different phase delays to the orthogonally polarized beams of DIC microscopy, effectively causing local variations in the bias and disrupting the contrast mechanism of DIC imaging.

---

* Light propagates though space as an electromagnetic wave which is made up from a combination of electric vectors and magnetic vectors. In polarized light, the electric vector vibrates in the plane of vibration along the direction denoted by the Poynting vector.

† An uniaxial crystal has only a single optical axis, which indicates the direction in which the crystal behaves like an isotropic, that is, not a birefringent, material. A birefringent material has the property of exhibiting two different refractive indices with different directions of orientation to waves that enter it with different polarizations.

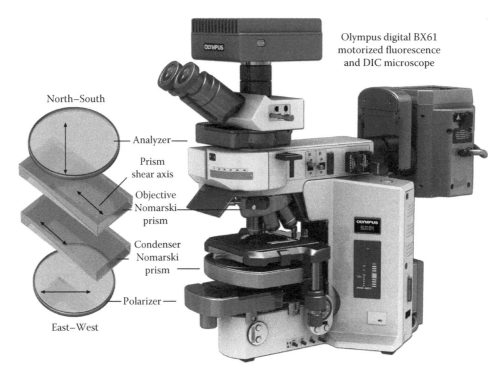

FIGURE 24.7 DIC microscope configuration. The key components that create contrast in DIC microscopy are the polarizer, the condenser and objective Nomarski prisms, and the analyzer. These four components (at left) are shown in relation to their location in the upright microscope (right). Double-ended arrows indicate the orientation of the polarizer and analyzer, as well as the shear direction of the Nomarski prisms. The analyzer is oriented orthogonal to the polarizer. The shear direction of the Nomarski prism is set at 45° with respect to the polarizer thereby orienting the prism structure such that the incident power is split evenly between the sheared orthogonal polarizations. (From Molecular Expressions, Differential interference contrast, 2009, http://micro.magnet.fsu.edu/primer/techniques/dic/dicintro.html. With permission.)

DIC microscopy is widely used in many applications to study living cells (Tsunoda et al. 2008, Lu et al. 2009, Szekely et al. 2009) and thick multicellular organisms (Gundlach 1993) over time (Danuser et al. 2000, Hamahashi et al. 2005). Another common use of DIC microscopy is the visualization of human embryos (Newmark et al. 2007) and sperm (Flaherty et al. 1995, Ito et al. 2009) in fertility clinics. DIC microscopy is also often used in combination with other microscopy modes because it provides complementary information about the underlying specimen (Farkas et al. 1993, Choi et al. 2008).

24.2.3 Comparison of the Three Imaging Techniques: Advantages and Disadvantages

Understanding the basic optical principles and implementation procedures for the three conventional phase imaging microscope modes presented in this chapter still leaves the following questions unanswered:

1. Which modes are best for imaging a particular biological specimen?
2. How does an observer evaluate whether certain image structures correspond to real object features or are instead an artifact of the microscope optical imaging system?

Attempting to answer these questions forms the basis for this section.

An illustration of the same sample imaged with the three microscopy imaging modes (dark field, phase contrast, and DIC) as compared to bright-field microscopy is shown in Figure 24.1 to provide an initial overview of the visual appearance of each image type before beginning a more detailed discussion of the advantages and disadvantages of each. Figure 24.1 shows a comparison of images of a live stamen hair cell from the spiderwort plant (*Tradescantia*) that contains some natural pigment as well as several transparent structures. Figure 24.1a shows contrast between some features due to differential absorption of the pigment, while the dark-field image (Figure 24.1b) enhances scattering and diffractive structures that appear more transparent in the bright-field mode. The phase contrast image (Figure 24.1c) and the DIC image (Figure 24.1d) provide contrast due to optical path differences (object phase) that are converted to intensity variations through optical interference.

24.2.3.1 Dark-Field Microscopy Compared to Bright-Field Microscopy

Dark-field microscopy is best for visualizing diffracting and scattering object features such as cell walls, organelles, and any other inclusions where refractive index variations occur between features and their surround. It is also useful for "detecting"

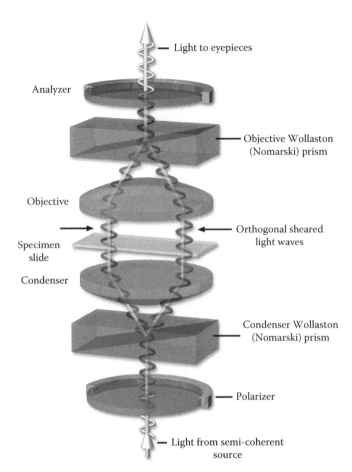

- Light to eyepieces
- Analyzer
- Objective Wollaston (Nomarski) prism
- Objective
- Orthogonal sheared light waves
- Specimen slide
- Condenser
- Condenser Wollaston (Nomarski) prism
- Polarizer
- Light from semi-coherent source

FIGURE 24.8 DIC microscope schematic. Light passing through the first polarizer is sheared into two laterally separated beams with orthogonal polarization by the condenser Nomarski prism before passing through the condenser lens and specimen. With the condenser adjusted correctly for Kohler illumination, each point of the specimen is illuminated by all points of the light source. The wavefronts of the two sheared beams are delayed differentially as they pass through different areas of the object. The second prism is designed to refractively recombine the two beams and can be shifted laterally in the beam path to introduce a phase difference bias between the orthogonally polarized wavefronts. This bias creates the shadow-cast or relief appearance of DIC. The orientation of the analyzer causes the orthogonal polarizations to destructively interfere at the image plane. (From Molecular Expressions, Differential interference contrast, 2009, http://micro.magnet.fsu.edu/ primer/techniques/dic/dicoverview.html. With permission.)

sub-resolution-sized objects including dielectric scatterers, metal particles (such as immuno-gold labels) and nanostructures. This is because even a small amount of light scattered by a tiny object can be detected whenever the remainder of the image field is dark.

Some advantages of dark-field microscopy as compared to bright-field microscopy are shown in Figure 24.9, with images from a stained section of wheat rust (fungus). The contrast producing mechanism for these two modes is distinctly different. In the bright-field image (Figure 24.9a), partially absorbing pigments linked to cell material make features appear darker

than their transparent surround. In the dark-field image (Figure 24.9b), the contrast producing mechanism is from diffraction (and refraction) at cell boundaries where there are refractive index variations between cell structures and their surround. This mechanism can often uncover additional information about cell makeup (indicated by arrows) that would be overlooked using just bright field. Another characteristic of dark-field microscopy is its ability to detect sub-resolution-sized particles that appear in this image as tiny bright points against the dark background and are otherwise absent in the bright-field image.

Some disadvantages of dark-field microscopy are its tendency to produce image flare artifacts around highly diffracting objects. These may make feature boundaries appear to extend over a slightly greater area than in reality. An example of this type of artifact can be seen when comparing some of the brightest cell walls in the dark-field image (Figure 24.9b) with corresponding features that appear thinner in the bright-field image (Figure 24.9a).

In some instances, the ability to detect tiny scattering objects as bright points can become a disadvantage rather than an advantage. For example, if a microscope preparation contains dirt specks, air bubbles, or other contaminants, these will be imaged as bright scatterers along with the features of interest and can produce a confusing result.

24.2.3.2 Phase Contrast Microscopy Compared to DIC Microscopy

Both of these microscope types utilize changes in OPL (phase variations) to produce image contrast by interference of light. However, the resulting images can be visually quite different, and each has advantages as well as potential loss of information or artifacts. Figure 24.10 shows a comparison of phase contrast and DIC microscopy when used to image a highly transparent biological structure (cattle mite). Several points of difference are apparent and need further clarification before the observer can determine the answers to the questions of which mode is best to use and how to interpret the resulting image features.

A phase contrast image (Figure 24.10a) usually has both brighter and darker features (from constructive and destructive interference) against a gray background, which can help delineate subtle feature differences not possible with dark-field microscopy (as seen when comparing Figure 24.1b and c). These bright and dark regions in phase contrast microscopy result from OPL differences through the specimen, with regions of greater optical path delay (e.g., higher refractive index) generally appearing dark and regions of weaker phase delay appearing bright. Also apparent in Figure 24.10a are "halo" artifacts, typical of phase contrast images, which appear as very bright outlines along diffractive feature boundaries and, as a result, can mask fine detail.

By comparison, the DIC image (Figure 24.10b) shows brighter and darker contrast at feature boundaries in response to variations in OPL (phase gradients) through the object along the shear direction of the DIC system. Compared to the phase contrast image, Figure 24.10b shows sharper feature boundaries (no halo

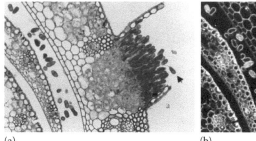
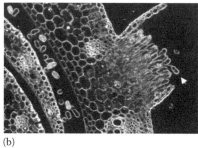

(a)   (b)

FIGURE 24.9 (See color insert.)   Bright-field (a) versus dark-field (b) images obtained from a stained object. Dark-field images enhance highly diffractive features, such as cell wall boundaries, that are comprised of strong refractive index variations between the feature and its surround. By comparison, bright-field image contrast is produced by variations in absorption of colored dyes or natural pigments that often fill the entire cell, making the cell wall or any internal structures less apparent. Arrows point to a spore that exemplifies the difference between these two microscope configurations. In the dark-field image, the cell wall is more clearly delineated but also shows some slight broadening due to flare, which may cause some loss of fine detail. Subject is a stained section of a fungal spore body of the wheat stem rust, *Puccinia graminis*. Images recorded using a color sensor and white light illumination. (Magnification approx. 100×.)

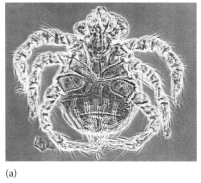
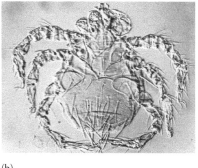

(a)   (b)

FIGURE 24.10   Phase contrast versus DIC microscopy images obtained from an unstained transparent object (cattle mite). Contrast in the phase image (a) is due to OPL variations through the mite specimen where greater optical densities appear darker (and lower densities appear brighter) relative to the gray background. Very bright "halo" artifacts are also visible around the highly diffractive edges. By comparison, the DIC image (b) has better edge delineation (no halo artifacts) due to its contrast being produced by specimen-induced phase gradients along the direction of shear (oriented here at 45° to the horizontal image axis). In DIC microscopy, the bright and dark regions correlate to the phase gradient directionality along feature boundaries and produce a characteristic pseudo-3D appearance that can aid the observer in interpreting feature characteristics. Subject is an unstained mite, *Eutrombicula sp.*, found in cattle. Images recorded using a black and white sensor and white light illumination. (Magnification approx. 100×.)

artifacts) and also illustrates the pseudo 3D (bas-relief) appearance that can help an observer differentiate specific features. However, DIC images also have limitations that must be considered when analyzing object features. One problem is that DIC microscopy does not work well with birefringent object features (such as plant cell walls) due to their impact on light polarization angles, which are critical for the proper working of the DIC prisms. Another limitation is that the bright and dark image intensities cannot be used as a quantitative measure of optical path through the specimen (e.g., as a measure of refractive index or thickness) due to nonlinearities in the image formation process. These will be discussed further in Section 24.2.5.

A final drawback of DIC microscopy is that object phase gradients are imaged with greatest contrast when they vary along the direction of the shear of the DIC prisms, whereas object phase variations orthogonal to the DIC shear direction can appear virtually invisible in the image. Figure 24.11 shows two DIC images of a live plant stamen hair cell whose cell surface contains long, parallel striations. Figure 24.11a shows the cell oriented in such a way that these surface striation features are perpendicular to the shear direction and hence are visible as alternating highlight and shadow structures related to the phase gradients produced by their edges. Figure 24.11b shows the same cell rotated by 90° on the microscope stage. Now the long axes of the striations are parallel to the DIC shear direction with the result that they have become virtually invisible. A rotating stage is often used with DIC microscopy to facilitate specimen visualization in different directions. However, specimen rotation is not always possible when imaging live cells in specialized chambers and controlled environments.

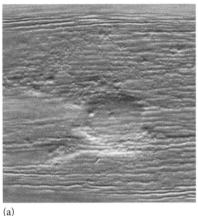 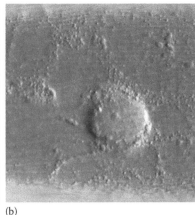

(a)   (b)

FIGURE 24.11 Effects of shear direction on DIC microscope images. Two images of the same live stamen hair cell with orthogonal shear directions are shown. The images were acquired by rotating the sample by 90° on the microscope stage. In image (a), the shear direction is vertical and thus enhances the edges (phase gradients) of the cell surface striations that appear here as horizontal lines. In image (b), these striations are much less visible because they are now parallel to the shear direction that has been rotated to be horizontal. The large circular body near the center of the two images is the cell nucleus. Images recorded using a black and white CCD camera and white light illumination. Objective: 40×, NA 0.75.

A final important distinction between phase contrast and DIC modalities is their response to features that fall outside the depth of field of the microscope objective (i.e., their response to optical defocus aberrations). Phase contrast images of thick objects are particularly prone to defocus artifacts in which out-of-focus regions are visible as blurred features superimposed on those that are in focus. In addition, defocused features can exhibit a contrast reversal (inversion of bright and dark image regions) that further complicates interpretation of image contents. In contrast, DIC microscopy is much less apt to show image defocus artifacts because out-of-focus features tend to produce low contrast and hence are indistinguishable from the gray background. This distinction between phase contrast and DIC microscopy defocus response is illustrated in Figure 24.10. In the phase contrast image (Figure 24.10a), the entire body of the mite appears visually complicated by the presence of defocused light (blurred image region) and dark features. This difficulty is much less severe in the DIC image (Figure 24.10b).

The choice of whether to use phase contrast or DIC microscopy requires careful consideration. Phase contrast microscopy can produce excellent results, especially for thin biological objects where defocus artifacts are minimized. New apodized phase contrast objectives are also available to reduce the halo artifacts common to more traditional systems. In addition, phase contrast microscopy can be used for imaging birefringent specimen features or objects mounted in birefringent plastic culture dishes, which produce problems when using DIC microscopy as explained earlier. The contrast producing mechanisms of the two systems are generally different and DIC becomes preferred for many complex biological objects for the following reasons. Returning to Figure 24.10a and b, most observers would agree that the bristles on the mite body are more clearly delineated in the DIC image than in the phase contrast image, although with the caveat that linear features may be affected by the shear direction of the DIC system. This preference for DIC microscopy

is primarily because features in phase contrast images can appear as a complex mixture—brighter or darker than the background—whereas the same features in DIC exhibit a gradient of bright or dark (bas-relief) that is more easily interpreted as being from the edge of a specific feature. DIC microscopy has better edge delineation (no halo artifacts) even when compared to the apodized phase contrast techniques described in Section 24.2.4.1. Finally, for imaging thick objects, DIC microscopy is generally preferred due to its ability to minimize the contrast from defocused features that can produce confusing artifacts in phase contrast imaging.

24.2.3.3 Phase Contrast Microscopy Compared to Dark-Field and Bright-Field Microscopy

Phase contrast microscopy can also provide advantages in some biological applications where stained or pigmented cells are imaged (although this is more typically considered true for DIC imaging). As long as the observer is aware of potential imaging artifacts in phase contrast microscopy (such as described in Figure 24.10a), more information about certain specimen features may be obtained. Figure 24.12 shows an example of a stained alga preparation, comparing bright-field, phase contrast, and dark-field imaging modes. In this example, the phase contrast image (Figure 24.12b) is similar to bright field (Figure 24.12a) but provides more contrast for viewing the fine structure of the cell interiors. Because the preparation is a thin section, the phase contrast image has minimal artifacts. As expected, the dark-field image (Figure 24.12c) clearly delineates the structure of the highly diffracting cell walls and shows tiny bright specks from scattering particles (in the surrounding medium) that may be smaller than the microscope resolution limit. On careful inspection, certain features may become more visible in one mode as compared to the others. For example, the red arrow points to a white ringlike structure that is visible in the phase contrast and bright-field images but is virtually invisible

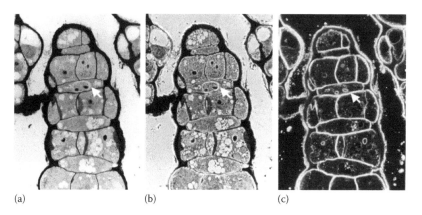

(a)   (b)   (c)

FIGURE 24.12 (See color insert.)   Bright-field (a) versus phase contrast (b) versus dark-field (c) images of a green alga, *Chara*, stained with hematoxylin-PAS. Although phase contrast is generally less useful for imaging stained objects, it can increase contrast and visibility of cell components, such as the granularity of the cell cytoplasm seen here. As expected, the dark-field image clearly shows cell wall structures better than the other modes, but often does not image other features as readily. For example, the red arrows point out a subcellular structure that appears as a bright oval ring in the bright-field and phase contrast images but is not visible in the dark-field result. Images recorded using a color sensor and white light illumination.

in the dark-field image. It is also interesting to note that because the contrast of this ringlike structure is increased in the phase contrast image, but invisible in the dark-field image, it can be deduced from the two contrast mechanisms that it is a weakly diffracting object that nevertheless has a strong phase delay relative to its background.

24.2.3.4 Combining Microscopy Modes

A frequently overlooked feature of microscopy is that some imaging modes can be used in combination to provide added information about specimen features. Two of the modes discussed in this chapter—dark field and DIC—can be used in a reflection microscope configuration as well as in transmitted light. It is common practice for biologists to use a transmitted-light mode such as phase contrast or DIC in combination with fluorescence to help locate a probe within the larger cell volume. However, it is also possible to combine just the modes described in this chapter to good effect. In Figure 24.13, some examples of the advantages of using combined reflection and transmission modes are presented. In this example, a phase contrast image of a newt white blood cell (neutrophil) from a transmitted light configuration (Figure 24.13a) is shown in comparison to a reflected light DIC image (Figure 24.13b).* The lobed nucleus has been stained but the surrounding cytoplasm is mostly transparent. Here, the reflection DIC image clearly shows the central convergence of strands from each nuclear lobe that, by comparison,

cannot be as clearly visualized in the phase contrast image. However, when using these two microscope modes in combination (Figure 24.13c), more information about the structure of the nuclear lobes and surrounding cytoplasm can be gained (see Figure 24.13 caption for details). Figure 24.13d is also included to illustrate that combining reflection DIC with transmitted dark-field imaging modes may also yield complementary useful information about specimen features.

The comparisons discussed in this section have illustrated the following conclusions: When imaging a particular preparation for the first time, it is always beneficial to create images using more than one optical mode of the microscope to ensure that observed features relate to true features of the object and are not an artifact of the optical imaging system. In addition, some microstructures may have such low contrast in some microscope modes as to be overlooked by the observer unless a complementary imaging mode is also utilized for comparison. Finally, only through using some of the recently developed quantitative imaging techniques (described elsewhere in this chapter) can a true representation of the fundamental object parameters be obtained.

Further comparisons of these modes in additional biological applications can be found in the references provided at the end of this chapter (Inoue 1986, Pluta 1989b, Murphy 2001, Cox 2007, Molecular Expressions 2009a, Nikon Miscoscopy U 2009). In addition, a study comparing bright-field, dark-field, and DIC microscopy can be found in Li et al. (2007) while a comparison of phase contrast and DIC microscopy is presented in Gundlach (1993).

24.2.4 Examples of System Enhancements

The basic configurations of these three phase imaging modes continue to be used relatively unchanged since their original development. Yet, there is ongoing exploration of innovative

---

* The reflection DIC mode requires a reflecting surface with phase gradient variations. In reflection DIC, there is still a mid-tone gray (DC offset) value (arbitrarily set by the user when adjusting the exposure) but it occurs only where there is light reflecting from the specimen. Otherwise, where there is nothing in the sample to reflect the light, there is no DC gray value and that part of the image appears dark. In regions where light is reflected, OPL variations (due to thickness variations in the sample) produce the bright and dark bas-relief effect (relative to the DC offset) just as in transmitted DIC.

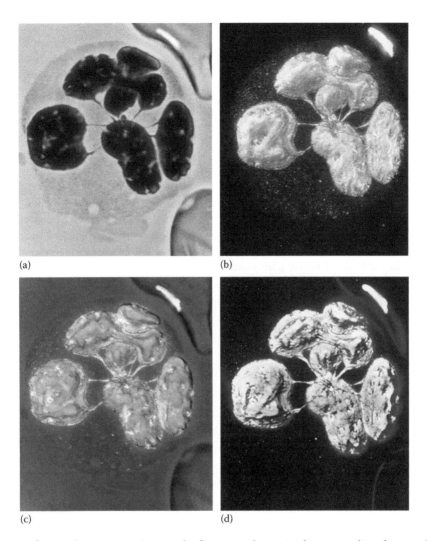

FIGURE 24.13 (See color insert.) Combining transmission and reflection modes to visualize more subject features. (a) transmitted phase contrast (white light illumination), (b) reflected DIC (white light illumination), (c) combined transmitted phase contrast (with red filter in the transmitted white light illumination path) and reflected DIC, and (d) combined transmitted dark field (with red filter in illumination path) and reflected DIC. Subject is a stained white blood cell (neutrophil) from the salamander *Oedipina*. Color of the nucleus is due to the absorption and reflection properties of the dye, which appears magenta in transmitted phase contrast and green in reflected DIC. The red background and highlights in (c) and (d) are due to a red filter in the transmission illumination path to provide contrast with the green nucleus from the reflected light DIC imaging mode. Note how the combined mode (c) preserves the advantages of both the transmitted phase contrast (visibility of the granular cytoplasm and the transparent regions of the nuclear lobes) and the reflection DIC (visibility of the central convergence of strands from each nuclear lobe). In (d), the transmitted dark-field image combined with reflection DIC shows even more fine structure in the nuclear lobes. Images recorded using a color sensor and white light illumination. (Magnification 1500×.)

methods to improve image quality by increasing resolution and contrast. These improvements employ new technology as it emerges and are developed based on an understanding of the method by which phase contrast is created. Two recent improvements to phase contrast imaging, discussed in the following, involve changes to the optical train of the microscope designed to reduce artifacts and improve contrast. Although video-enhanced DIC is not a new system enhancement, it is still used, and it is an early example of a system that integrates electronics and computations with optics to enhance imaging. Some recent enhancements to DIC microscopy are also discussed.

24.2.4.1 Apodized Phase Contrast Microscopy

Phase contrast can be improved and halo artifacts reduced by replacing the conventional phase plate found in the back focal plane of the phase contrast objective with an apodized phase plate consisting of a quarter wave phase shift ring with a 25% transmittance, and a pair of adjacent rings, which have 50% transmittance (Otaki 2000). Assuming the illuminating wavefront is a plane wave and the phase introduced by the object is small, wave theory predicts that apodization of the direct light increases the phase contrast by increasing the relative proportion of the signal from diffracted light. In initial experiment, images of human liver cells show a reduction in halo artifacts in the apodized phase contrast as compared to conventional phase contrast. The size of the rings used in the acquisition of these images was computed using a theoretical approximation of the object diffraction angle. The approximation assumes the object is spherical with a $10\,\mu m$ diameter and that the angle of diffraction is defined by the first minimum of the 2D Fraunhofer diffraction pattern of a circular aperture. Variants of this original design are now commercially available and provide a $2\times$ improvement in contrast as the result of halo reduction (Nikon brochure 2007).

24.2.4.2 Phase Contrast Microscopy with High Numerical Aperture

Phase contrast artifacts can also be reduced, and spatial resolution improved, by replacing the conventional illumination ring with a randomly distributed set of point apertures and the Zernike phase ring with a matched set of point-like phase shifters. Image resolution is optimized by allowing light from the full range of the NA to be collected. Implementation of this method, with an actively changing illumination (and filtering) pattern from a programmable spatial light modulator phase hologram, can be used to find the optimum phase pattern (Maurer et al. 2008).

24.2.4.3 Video-Enhanced DIC Microscopy

The field of video-enhanced DIC has extended the application of electronic signal processing techniques to DIC imaging of real-time dynamic processes. It was first introduced by Allen et al. (1981) as AVEC-DIC. Using this technique, a background signal can be continuously subtracted in real time to gain contrast using digital signal processing hardware rather than postprocessing in software. High-contrast images obtained with rapid time resolution are used for precision tracking and to quantify dynamic live-cell processes. Reference to many successful studies with important contributions to biology can be found in Salmon and Tran (2007). As in DIC microscopy, video-enhanced phase contrast microscopy is a successful technique in improving image contrast (Inoue 1986).

24.2.4.4 Recent DIC Microscopy Enhancements

A recently patented DIC microscope design (Shribak 2005) allows the shear to be rotated in two orthogonal directions of shear optically without mechanical rotation (and without rotating the specimen) through a combination of waveplates and electro-optic liquid crystal retarders (Shribak and Inoue 2006, Shribak et al. 2008).

The well-known microscope manufacturer, Carl Zeiss, Inc., introduced as a new product another non-confocal approach, the reflection-only C-DIC microscope (Danz and Gretscher 2004). The C-DIC design allows the direction of shear to be rotated, although not without simultaneously changing the bias.

A Senarmont compensator DIC microscope configuration (King et al. 2008, Molecular Expressions 2009a) is often used in applications that require precise control of the phase bias. For example, phase bias control is essential for phase-shifting DIC, discussed in Section 24.3. Experimental and simulated results have shown that phase-shifting, which is primarily useful as an image preprocessing step in quantitative phase imaging, also increases image contrast (King et al. 2008).

24.2.5 Toward Quantitative Imaging: Modeling of Image Formation

A *qualitative* description of image formation with dark-field, phase contrast, and DIC microscopy has been presented in Section 24.2.2 where system configuration and operation for each mode was reviewed. The first step toward *quantitative* imaging approaches, discussed in Section 24.3, is imaging system modeling. Intensity in microscope images can be quantified with theoretical modeling of the image formation and detection process in a microscope system. Frequency transfer modeling via Fourier analysis provides insight about the fundamental properties of the microscope system. In addition, prediction of microscope image intensity facilitates a more accurate interpretation of the images.

Over the years, several models have been developed based on different approaches and assumptions. All models are approximations and their accuracy depends on the assumptions upon which they are based. Assumptions are often made to trade off computational complexity and accuracy. In this section, a theoretical framework is provided that enables modeling image formation with phase imaging microscopes under partially coherent illumination. In addition, this section briefly reviews some of the key contributions made toward modeling microscope images from phase objects. The discussion here assumes familiarity with terms used in scalar diffraction theory (Goodman 1996, Born and Wolf 1999) and statistical optics (Goodman 1984).

Models developed assuming partial coherent imaging (Goodman 1984) are most accurate because they have the ability to model full-field imaging and capture the resolution possible by the use of condensers with a high NA. Coherent imaging assumes that the condenser's aperture is closed down to a pinhole, which is a very limiting case and never done in practice, especially in DIC microscopy. The increase in the computational complexity for the evaluation of partially coherent models is significant and thus many studies have focused on either a coherent imaging model or even a geometric optics model.

24.2.5.1 General Modeling Approach for Image Formation

A general model for image formation in partially coherent light for transmitted-light optics can be derived based on a fundamental approach of propagating the complex amplitude of the illuminating wave field under Köhler illumination through a thin specimen and the optical system, while accounting for the diffraction effects during light propagation. The approach presented here is based on a wave optics approximation using a scalar representation to model diffraction effects in the system (Goodman 1996) and follows the approach adapted by Preza

1998 and Preza et al. (1996, 1999) for DIC modeling. In general, a vectorial representation is more accurate as it allows modeling electromagnetic scattering through arbitrary objects (Richards and Wolf 1959), thus, taking into account additional effects due to high NA apertures and changes in the polarization state of the light as it propagates through the system—an effect occurring in DIC microscopy, for example, see Munro and Torok (2005).

Although specimens are three-dimensional, they are often assumed to be "thin." The assumption that the object is "thin" is valid in the paraxial approximation* and when the thickness of the object is sufficiently small so that the incident illumination exits the object at approximately the same transverse location as it enters (Goodman 1984). The following sections first discuss a 2D imaging model based on this assumption and then extend this model to three dimensions.

24.2.5.1.1 Two-Dimensional Imaging Model

In the 2D imaging model discussed in this section, paraxial illumination is assumed as well as a planar semitransparent object that can be modeled by a 2D complex transmission function that is independent of the angle of the paraxial illumination (i.e., the angle between the normal to the illuminating wave and the optical axis). The specimen's complex transmittance function $f(\mathbf{x}_o) = |f(\mathbf{x}_o)| e^{j\phi(\mathbf{x}_o)}$, where $j = \sqrt{-1}$ and $\phi(\mathbf{x}_o)$ is given by Equation 24.1, not only affects the phase of the illuminating plane wave, but also may attenuate the amplitude of the plane wave due to its magnitude $|f(\mathbf{x}_o)|$, which is typically less than one for an absorptive specimen. For a totally transparent specimen: $|f(\mathbf{x}_o)| = 1$.

The optical system's amplitude point spread function (PSF), $h(\mathbf{x})$, plays a key role in modeling image formation as a superposition integral of amplitudes shown as the inner integral of Equation 24.2:

$$i(\mathbf{x}) = \int_{D_c} \alpha(\xi) \left| \int_{-\infty}^{\infty} h_c(\xi;\mathbf{x}_o) f(\mathbf{x}_o) h(\mathbf{x} - \mathbf{x}_o) d\mathbf{x}_o \right|^2 d\xi \quad (24.2)$$

where

- $\mathbf{x} = (x, y)$ is a point in the image plane (i.e., in the plane where the image is formed at a distance z along the optical axis)
- $\mathbf{x}_o = (x_o, y_o)$ is a point in the object plane (i.e., where the specimen slide is positioned)
- $\xi = (u, v)$ is a point in the front focal plane of the condenser (see Figure 24.4).

The function $\alpha(\xi)$ in Equation 24.2 defines the intensity of the illumination over a region D_c in the condenser's aperture and it is zero outside this region. The function $\alpha(\xi)$ takes into account the fact that some points in the condenser's aperture

give rise to oblique plane waves that illuminate the specimen, while the remaining points are excluded either because the aperture is partially closed or because the light is partially blocked by an optical component (e.g., by a condenser annulus in the case of phase contrast microscopy or an opaque light stop in the case of dark-field microscopy). The complex amplitude of the oblique illuminating plane waves is defined by the function

$$h_c(\xi;\mathbf{x}_o) = \frac{1}{j\lambda f_{con}} e^{j2\pi(x_o\xi + y_o\eta)/\lambda f_{con}}$$

in Equation 24.2, where f_{con} is the focal length of the condenser lens (Goodman 1996, Preza et al. 1996) and λ is the mean wavelength[†] of the illuminating wave.

The actual form of the PSF $h(\mathbf{x})$ is based on the amplitude PSF for bright-field microscopy $k(\mathbf{x})$, defined as the Fourier transform (or diffraction pattern) of the system's exit pupil function (Goodman 1996), which is then modified mathematically to account for the optical accessories present in a specific phase imaging system. For example, the PSF for DIC imaging has been shown to be given by

$$h_{DIC}(\mathbf{x}) = (1 - R)e^{-j\Delta\theta}k(x - \Delta x, y) - Re^{j\Delta\theta}k(x + \Delta x, y)$$

where

- $2\Delta\theta$ is the DIC bias introduced by the sliding objective Wollaston prism
- $2\Delta x$ is the shear distance between the two illuminating waves due to the condenser Wollaston prism (indicated to be along the x-axis here without loss of generality), described in Section 24.2.2.3
- R is the amplitude ratio, that is, the amplitude of one wave field divided by the sum of amplitudes of the sheared wave fields and is often assumed to be equal to 0.5 (Preza et al. 1999)

In phase contrast microscopy, the PSF is computed to account for the effect of the ring-shaped phase plate (shown in Figures 24.4 and 24.5), which can be modeled by a complex-valued function $r(\mathbf{u}) = |r(\mathbf{u})| e^{j\phi_r(\mathbf{u})}$, where $\mathbf{u} = (u, v)$ is a point in the back focal plane of the objective lens. In the ring-shaped region, the magnitude $|r(\mathbf{u})| = a$ is less than one and the phase $\phi_r(\mathbf{u}) = \pm\pi/2$, while both are zero outside this region. The function $r(\mathbf{u})$ multiplies the pupil function of the microscope resulting in a new modified annular exit pupil function, which can be expressed as the difference of two concentric circular functions with different diameters multiplied by $ae^{\pm j\pi/2} = \pm ja$. Thus, the PSF for

---

* The paraxial approximation assumes that the illumination is a plane wave (or set of plane waves) incident on the object at small angles (Saleh and Teich 1991).

[†] For quantitative imaging, a narrowband illumination filter is used to convert the white broad-spectrum light source to a quasi-monochromatic (almost monochromatic) light source that can be modeled with a mean wavelength corresponding to the full width at half maximum wavelength of the filter spectral response.

phase contrast microscopy can be computed from the Fourier transform of this modified exit pupil function that results in

$$h_{PC}(x,y) = \pm ja[k(x,y) - k_{ic}(x,y)]$$

where it is assumed that the outer diameter of the phase ring coincides with that of the system's exit pupil and that the Fourier transform of the outer circle is equal to the amplitude PSF for bright-field microscopy. $k_{ic}(x, y)$ is defined as the Fourier transform of the inner circle of the ring. As discussed in Section 24.2.2.2, the illumination is also modified by the condenser annulus that consists of an opaque circular plate with a transparent ring. The effect of the illuminating ring is accounted for in Equation 24.2 by the functions $\alpha(\xi)$ and $h_c(\xi;\mathbf{x}_o)$, which define the region and obliquity of illumination, respectively. $\alpha(\xi)$ is a ring-shaped function equal to 1 inside the ring and zero outside.

For dark-field microscopy, the PSF is identical to the amplitude PSF for bright-field microscopy $k(\mathbf{x})$, because the system modification is in the condenser only. This configuration is sufficiently modeled via the functions $\alpha(\xi)$ and $h_c(\xi;\mathbf{x}_o)$ that define the region and obliquity of the illumination, respectively.

24.2.5.1.2 Three-Dimensional Imaging Model

Several efforts have been focused on modeling 3D phase imaging to gain a better understanding of the underlying imaging systems, to validate acquired images, and to facilitate the determination of a better specimen representation from detected microscopy images using computational methods (see discussion of computational methods in Section 24.3). Rigorous modeling of 3D imaging requires modeling the physical phenomenon of scattering, which is a challenging task, see Gong et al. (2008) and references therein, resulting in computationally intensive models and approaches for the solution of the inverse problem.

Approximations such as those of Born (Born and Wolf 1999) and Rytov (Sierra et al. 2009b) are often employed in modeling 3D imaging. The Born approximation is valid when the scattered field is weaker than the incident field, meaning that a linear relation between the object function and the phase delay can be employed in solving the wave equation. By contrast, the Rytov approximation introduces a condition such that the scattered field is considered independent of the specimen size and is more sensitive to the phase gradient. This condition states that the phase change must be linearly proportional to the index variation in the sample over the length of a wavelength. Thus, the validity conditions for the Rytov approximation are slightly different and somewhat less restrictive than the Born approximation. Both approximations simplify the modeling process and facilitate the extension of 2D models to 3D imaging using the principle of superposition (Nemoto 1988, Preza et al. 1999).

The extension of the 2D model in Equation 24.2 to three dimensions based on the first Born approximation can be formulated using a 3D PSF and a linear superposition in all three dimensions. This approach is possible because multiple refraction through the specimen can be neglected as a consequence of the first Born approximation. An easy-to-follow mathematical

derivation demonstrating this consequence can be found in (Sierra et al. 2009b). The 3D PSF of the microscope is obtained from the defocused PSF for transmitted-light optics, $k(x, y, \Delta z)$ defined as the 2D Fourier transform of the complex-valued, generalized pupil function (Goodman 1996)

$$p(x, y, \Delta z) = p(x, y)\exp[j2\pi w(x, y, \Delta z)/\lambda],$$

where

$p(x, y)$ is the real-valued exit pupil function of the system
$w(x, y, \Delta z)$ is an effective path-length error
$\Delta z = z_f - z_o$ is the amount of defect in the focus that is due to a displacement of the plane z_f at which the microscope is focused along the z (optical) axis (Preza et al. 1999).

Detailed derivations of $k(x, y, \Delta z)$ can be found in Born and Wolf (1999), Gibson and Lanni (1989), Hopkins (1955). The intensity in a 3D image acquired at the image plane z that is conjugate to the plane z_f at which the microscope is focused can be expressed by

$$i(\mathbf{x},z) = \int_{D_c} \alpha(\xi) \left| \int_{-\infty}^{\infty}\int_{-\infty}^{\infty} h_c(\xi;\mathbf{x}_o)f(\mathbf{x}_o,z_o)h(\mathbf{x}-\mathbf{x}_o,z-z_o)d\mathbf{x}_o dz_o \right|^2 d\xi$$

$$(24.3)$$

where the illumination functions $\alpha(\xi)$ and $h_c(\xi;\mathbf{x}_o)$ are the same as in Equation 24.2. The phase of the object's 3D complex transmittance function, $f(\mathbf{x}_o, z_o) = |f(\mathbf{x}_o, z_o)|e^{j\phi(\mathbf{x}_o, z_o)}$, is no longer modeled as a projection over the thickness of the object as in Equation 24.1, but rather at each object depth (at plane z_o) it is defined over a differential thickness, dz_o, that is,

$$\phi(\mathbf{x}_o, z_o) = 2\pi/\lambda \, n(\mathbf{x}_o, z_o)dz_o.$$

3D transfer functions for microscope imaging, defined in the back focal plane of the objective, have also been presented based on the Born approximation (Streibl 1985, Nemoto 1988, Sheppard and Mao 1989, Sheppard et al. 1994, Preza et al. 1996). For a recent review of relevant 3D modeling efforts based on these approximations see Sierra et al. (2009b), Trattner et al. (2009) and references therein. Although several of these models have worked well for thin objects, they are not accurate enough for objects that are optically thick or strongly heterogeneous because of significant scattering effects when the light interacts with such objects (as discussed in earlier chapters of this handbook). The reason for this is due to the fact that the Born and Rytov approximations place strong constraints on the maximum size of the object and the refractive index contrast distribution (Sierra et al. 2009b, Trattner et al. 2009). The validity of the Born approximation in the context of DIC microscopy has been rigorously examined in a recent study (Trattner et al. 2009). This study suggests a new validity criterion that is less restrictive than the known theoretical bound, justifying the use of the Born approximation to model larger specimens and larger differences in the refractive index between the specimen and the medium that surrounds it than currently permitted by the known theoretical bound.

Recently, a different 3D phase imaging model for a pure phase object based on a product-of-convolutions (POC) approach has been presented in Sierra et al. (2009b). In the POC approach, each slice of the object's function, defined at discrete depths within the object and with a certain thickness, Δz_o, is represented by an image in the image plane. These images are computed as the 2D convolution of each object slice with the defocused 2D PSF (defined earlier) at the corresponding depth. An object function in the image plane is computed as the product of these slice-images. The final image in the image plane is computed by multiplying the object function in the image plane with the unperturbed incident wavefield, which is propagated to the image plane via convolution with the in-focus PSF. Thus, the phase change introduced along the optical axis as the light propagates through the object is accounted for via a multiplicative effect rather than treating each object slice's contribution additively as in approaches based on the Born (e.g., as in Equation 24.3) and the Rytov approximations.

A comparison of 3D phase imaging models based on the Born approximation, the Rytov approximation, and the POC approach has been presented recently in Sierra et al. (2009b). In this study, the models based on the Born and Rytov approximations were also based on a first-order Taylor series approximation for the exponential term of the object's function, which is valid only for phase functions with very small values. This approximation was not made in the model based on the POC approach because it is not necessary (nor in Equation 24.3). The study used the three models to predict 2D phase images for a variety of thick objects for transmitted phase microscopy. Comparisons with both simulated and experimental images of common test objects show that the Rytov approximation is less restrictive than the Born approximation, and that the POC approach provides a more consistent representation of phase images for thick objects than models based on the first Born and the Rytov approximations.

24.2.5.2 Phase Contrast Microscopy Imaging Models

A mathematical description of the intensity in a phase contrast image of a 2D pure phase object with small phase variations has been derived based on the spatial filtering properties of a phase dot (not an annular phase plate) placed along the optical axis in the back focal plane (Pluta 1989b, Born and Wolf 1999). Based on this simplified configuration, the resulting model relates image intensity linearly with the 2D phase function of the object provided that the phase values are small and nonlinear terms can be ignored:

$$i(\mathbf{x}) \approx 2 \, |C|^2 \, [a^2 \pm 2a\phi(\mathbf{x})]$$

where

C is a constant

a is the magnitude of the phase plate and the sign in front of the phase function depends on the sign in the exponent of the phase plate defined in Section 24.2.5.1.1 (recall $\phi_r(\mathbf{u}) = \pm\pi/2$).

Although this simplification does not model the actual phase contrast microscope, it provides useful insights. Thus, the approximate image intensity in a phase contrast image varies linearly with the phase of the object only when the object's phase is small (i.e., only for a "weak" phase object).

Recently, a new expression for the intensity in a phase contrast microscopy image has been developed for an electron microscope (Beleggia 2008). The study extends the expression for the intensity in a phase contrast image derived by Born and Wolf (1999) for periodic phase objects in that it is not limited to objects that have "weak" phase and may be applicable to light microscopes.

3D phase imaging with conventional phase contrast microscopy has also been studied, for a weak phase object only, because of the nonlinearity of the system (Takahashi and Nemoto 1988).* Additionally, a 3D phase-transfer function was developed based on the Born approximation for a modified phase contrast system that does not include the phase plate (Noda et al. 1990a). In this case, the principles of a linear shift-invariant system are applicable, thus the characterization of the modified phase contrast microscope by a phase-transfer function is possible.

24.2.5.3 Dark-Field Microscopy Imaging Model

As discussed in Section 24.2.2.1, dark-field imaging is equivalent to a spatial filtering technique and therefore can also be implemented by placing an opaque mask in the back focal plane of the objective lens in the same way as phase contrast microscopy. Thus, an expression for the dark-field microscopy image intensity can be obtained by removing the effects of the phase plate (i.e., by setting the magnitude of the phase plate $a = 0$) in the general expression for the intensity in a phase contrast microscopy image derived by Born and Wolf (1999) yielding

$$i(\mathbf{x}) = 2 \, |C|^2 \, [1 - \cos\phi(\mathbf{x})]$$

where C is a constant. This clearly shows that the observed intensity is not linearly related to the specimen's phase and thus not indicative of the specimen's OPL distribution.

24.2.5.4 DIC Microscopy Imaging Models

Several models of the DIC microscope have been developed to date. The first model for traditional DIC microscopy is based on geometric optics and it has appeared in several publications where many useful results were provided particularly on modeling a Wollaston prism (Lessor et al. 1979, Hartman et al. 1980, Pluta 1989b, Born and Wolf 1999). The geometric optics model approximates the DIC image intensity as a linear function of the object's phase gradient along the direction of shear. This approximation provides great insight into the basic principle of DIC imaging, but it ignores the blurring effects of the DIC PSF. The first model to compute the image of a point source, which relates to the DIC PSF, was presented in Galbraith (1982). The

---

* This study has only been published in Japanese.

inclusion of diffraction effects due to the PSF and the partially coherent illumination in DIC imaging models has demonstrated that the intensity in a DIC image is a nonlinear function of the object's phase gradient along the direction of shear and that blurring effects due to the PSF are significant (Preza et al. 1999). In addition, these DIC models based on diffraction, facilitated early demonstrations of the quantitative capabilities of DIC microscopy and increased interest in the potential of quantitative images (see discussion in Section 24.3).

A paraxial coherent imaging model for DIC was first developed using Fourier optics analysis with the intention of understanding the signal processing characteristics of a DIC microscope (Holmes and Levy 1987). The model resulted in the characterization of the 2D DIC frequency transfer function but the study considered only weak phase objects and also assumed that the objective lens does not truncate the spectrum of the illumination field. Extensions of the 2D frequency transfer theory to a partially coherent imaging model, which is more accurate for DIC imaging, introduced the concept of a "phase gradient transfer function" for weakly scattering objects in a scanning DIC system (Cogswell and Sheppard 1992) and the transmission cross-coefficient for a conventional DIC microscope (Preza et al. 1999).

A recent study explored, for the first time, the effects of illumination coherence and DIC bias on the frequency transfer characteristics of three DIC systems with different transmitted illumination configurations (Mehta and Sheppard 2008). In this study, the approach used to model the beam shearing and DIC bias effect is different than the one used in Preza et al. (1999) resulting in a different but not necessarily a more accurate model for the Nomarski DIC configuration.

2D partially coherent transfer functions developed in Mehta and Sheppard (2008) show that when the condenser's aperture is equal to 40% of the objective's aperture, the different DIC system configurations have similar imaging characteristics. The trend between degree of coherence and image contrast predicted by this transfer function study captures the effect observed in images acquired experimentally with two of the configurations. Finally, a 3D DIC transfer function developed assuming a weakly scattering object has offered some insight into the properties of 3D DIC imaging (Kou and Sheppard 2007a).

2D and 3D general models were also developed using a theoretical spatial domain description of light propagation in a DIC microscope for quasi-monochromatic paraxial partially coherent imaging (Preza et al. 1996, 1998, 1999). Effects of the DIC bias on the PSF and transfer function were also demonstrated. Model evaluation, accomplished by comparing experimental images to simulated images using a common test object, demonstrated that both the 2D and 3D model predictions are useful approximations but that improvements may be achieved with new models in which the first Born approximation is not employed in the derivation of the 3D model (Preza et al. 1999). In addition, as part of this model development, the connection between simpler or other existing models was also derived by systematically applying simplifying assumptions to the general

model. For example, under the geometric optics assumption, blurring effects due to diffraction are ignored and thus an ideal DIC PSF can be modeled as (Preza et al. 1996)

$$h_{\text{DIC-ideal}}(\mathbf{x}) = 0.5e^{-j\Delta\theta}\delta(x - \Delta x, y) - 0.5e^{j\Delta\theta}\delta(x + \Delta x, y),$$

where $\delta(x, y)$ is the 2D Dirac delta function (Goodman 1996). With this PSF and a pure phase object, that is, $f(\mathbf{x}_{\text{o}}) = e^{j\phi(\mathbf{x}_{\text{o}})}$, Equation 24.2 reduces to

$$i(x, y) = a_1 \sin^2\left\{0.5[\phi(x - \Delta x, y) - \phi(x + \Delta x, y) + \Delta\theta\right\},$$

where a_1 is a constant. A further assumption that can be made is that the DIC shear, $2\Delta x$, is very small (differential) relative to the size of the specimen, and thus the phase difference in the equation above can be written approximately as a phase gradient resulting in the expression

$$i(x, y) \approx a_1 \sin^2\left[\Delta x \frac{\partial\phi(x, y)}{\partial x} + \Delta\theta\right],$$

which approximates the 2D DIC image intensity as a linear function of the object's phase gradient along the direction of shear. This predicts the geometric optics expression for the DIC image intensity.

Different mathematical approaches have also been used. For example, an approach based on a generalized ray tracing method and the law of energy conservation to simulate DIC image intensity also led to a 3D DIC model (Kagalwala et al. 1999). In general, ray tracing, based on ray optics theory, can facilitate calculation of light refraction or reflection while propagating through a complex optical system such as the DIC microscope; however, it cannot describe phenomena such as interference and diffraction, which require wave optics theory. In addition, approaches based on Jones matrices have been used to represent a DIC geometric imaging model, an approximate DIC modulation transfer function in the weak phase object regime (Aoki et al. 2006, Ishiwata et al. 2006) and to compare traditional DIC imaging to Senarmont DIC imaging (King 2008). The latter clearly demonstrates the mathematical connection between DIC and Senarmont DIC imaging and that the two approaches do not always yield the same images (King 2008). The advantage of the Jones matrices approach is that it allows specification of the phase of each optical element's (such as the Wollaston prism, polarizer, and analyzer) in terms of a common reference.

A vectorial model for DIC imaging under coherent illumination and a high NA objective lens was also developed based on the bright-field imaging model for a coherent confocal scanning reflection mode (Munro and Torok 2005). The model was used to rigorously simulate the effect of the shear and bias on DIC images. A vectorial approach captures polarization effects essential for DIC imaging and allows more accurate modeling of these effects created by a high NA lens.

24.3 Recent Advancements for Phase Imaging Microscopy: Development and New Directions

The dark-field, phase contrast, and DIC modes of microscopy (described in Section 24.2), are primarily qualitative phase imaging techniques. Studies of images from these modes reveal qualities of objects such as shape and relationships to other objects in the field of view. By comparison, recent research has led to the development of *quantitative* phase imaging techniques that produce images with measurable attributes of the object, such as its geometry and OPL distribution. These quantitative phase imaging approaches have sparked much interest recently and have been shown to provide alternative and complementary information to fluorescence microscopy imaging.

In live-cell imaging, particularly of very small scale dynamics or features, the ability to observe certain structures without tagging can be beneficial. For example, observation of multiple cell types can require acquisition of image series at multiple fluorescent wavelengths. Furthermore, fluorophores that bind to proteins of interest cannot always be assumed not to interfere with the dynamics of the system under study.

The emerging technologies described in this section are possible as the result of technological advances in detectors and diffractive optical elements (such as spatial light modulators), computers, and advances in modeling and algorithmic development suitable for partially coherent, nonlinear microscopic imaging. Digital detection, for example, allows the integration of computational methods into optical systems. The combination of optical and computational methods described in the following is a significant change from the traditional methods already discussed in this chapter in which phase contrast, created only through manipulation of the optical field, was simply recorded. It is this new approach to phase microscopy that has led to the significant development of quantitative phase imaging techniques. A quantitative phase image, in which intensity is directly proportional to phase, is shown in relationship to traditional DIC and fluorescence images in Figure 24.14.

As mentioned in Section 24.2.5, the various image formation models that have been developed for phase imaging have inspired the development of computational methods for the extraction of quantitative phase information from images acquired with phase imaging microscopes. As in the case with model development, there is a trade-off between computational complexity and accuracy in the computed quantitative phase information. Models approximate the image formation and detection process and become the abstraction on which computational methods are developed. The achieved accuracy depends on the accuracy of the model upon which the methodology is based. For iterative computational methods, accuracy depends also on the number of iterations and the convergence properties of the method. In Section 24.3.1, techniques focused on 2D phase imaging are reviewed and their limitations are discussed. 3D phase imaging is discussed in Section 24.3.2.

24.3.1 Quantitative 2D Phase Imaging

While there are several methods used to create quantitative phase images, multiple DIC-based phase imaging methods have been under development because of the widespread use of DIC, a mode available on most light microscopes. Table 24.1 summarizes the quantitative DIC-based phase imaging methods reviewed in this section. As mentioned earlier, DIC became popular because it can be applied using high NA objectives and produces high-contrast, full-field images with very few phase artifacts. For quantitative phase imaging, these attributes are again advantageous because DIC-based methods can be used to accurately quantify weak phase properties of submicron-scale features. Images of a septum wall forming across a fission yeast

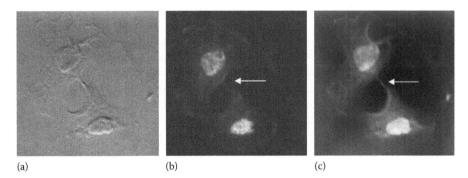

(a)   (b)   (c)

FIGURE 24.14 Comparison of a traditional DIC image (a) of a BPAE cell in the mitosis stage of the cell cycle, a DAPI fluorescence image (b) of the cell, and a quantitative phase image (c) computed from DIC images of the same cell all imaged with 40×/0.75 NA. The quantitative phase image was computed from experimental DIC images at 63×, 1.4 NA, using phase-shifting and spiral phase integration methods (SPI). Arrows on the quantitative phase image (c) and on the DAPI fluorescence image (b) highlight the correspondence between the two images, as well as the possible remnant of a mitotic spindle. Note the lateral resolution of this dense dot is excellent in the quantitative phase image (c) as compared to both the traditional DIC (a) and fluorescence (b) images. Phase changes that are hardly recognizable above the background in the traditional DIC image (a) are seen with much higher contrast in the quantitative phase image (c). (From King, S.V., Quantitative phase information from differential interference contrast microscopy, PhD dissertation, University of Colorado, Boulder, CO, 2008.)

TABLE 24.1 Quantitative DIC-Based Phase Imaging Methods Reviewed in Section 24.3.1.1

Year	Method	Main Advantages	Main Disadvantages
1980	Computation from infinite plane wave interference (Hartman et al. 1980)	Reported ± 0.2° accuracy	Limited to surface topography measurement from Nomarski reflection DIC
1997	Nonlinear iterative phase estimation from rotationally diverse images (Preza et al. 1997b)	Based on rigorous diffraction model and estimates phase only	Iterative computation Assumes no object absorption
2003	Nonlinear iterative phase estimation from generalized ray tracing model and conservation of energy (Kagalwala and Kanade 2003)	Based on model of diffraction effects	Assumes no object absorption
2004	Phase-shifting and spiral phase integration (Arnison et al. 2004)	Non-iterative, non-scanning	Paraxial imaging model, requires calibration
2004	Phase-shifting and Abel transform integration (Liu et al. 2004)	Non-iterative, non-scanning	Limited to rotationally symmetric objects
2004	Iterative phase estimate from phase-shifted DIC images (Axelrod et al. 2004)	Based on rigorous diffraction model	Uses bright-field image for amplitude estimation
2006	Phase estimation from retardation modulated DIC and an approximate DIC OTF (Ishiwata et al. 2006)	Takes into account defocus aberrations	Assumes a weak phase object
2007	Nonlinear iterative complex function estimation from rotationally diverse images (Preza and O'Sullivan 2007)	Based on rigorous diffraction model and estimates both phase and amplitude	Iterative computation

cell in Figure 24.15 illustrate the lateral resolution of quantitative DIC-based phase imaging as compared to Zernike phase contrast. The differential nature of DIC can also be a computational advantage because it imposes less restrictive phase wrapping conditions as compared with other interferometric phase imaging techniques.

The most significant disadvantage of DIC-based phase imaging methods is that they are not applicable to birefringent materials and they require rotation of the object with respect to the microscope. In some cases, modest modifications to the DIC microscope are required to facilitate data acquisition suitable for the technique. For example, this could include a stage that allows calibrated specimen rotation, a Wollaston prism that allows calibration of the DIC bias, or usage of a Senarmont configuration.

24.3.1.1 Computational Differential Interference Contrast Microscopy

The inclusion of diffraction in the DIC imaging model and demonstrations of the quantitative capabilities of computational DIC microscopy have increased interest in the potential of DIC-based quantitative phase imaging, primarily in fields that would benefit significantly from detection and measurement of small phase variations. This section reviews techniques focused on 2D phase imaging that have been developed based on DIC microscopy. The use of DIC microscopy in quantitative phase imaging has captured the interest of engineers and biologists since 1980, when the slope angles of large metal surfaces were determined with very high accuracy, from reflected-DIC images using simple equations describing the interference of plane waves (Hartman et al. 1980). Current DIC-based quantitative phase imaging methods primarily follow two basic approaches: (1) a two-step method that involves first extracting the phase gradient information from DIC images using a preprocessing phase-shifting method followed by integration to yield the desired specimen

phase and (2) direct estimation of the specimen phase from DIC images using some type of an iterative model-based algorithm. Development of the first approach was motivated by the approximate image model (Section 24.2.5.4) that relates DIC image intensity to the object's phase gradient and it was set in motion by post-processing of DIC images using pseudo-inverse filtering and integration, presented in van Munster et al. (1997) and Kam (1998), respectively. The development of the second type of approach was inspired by the work of Holmes and Levy, who formulated a description of DIC using Fourier domain analysis and first discussed the feasibility, in simulation, of phase restoration using an iterative modified Gerchberg–Saxton algorithm with constraints on amplitude, pass band, nonnegativity and finite extent, and a paraxial coherent imaging model (Holmes and Levy 1987, Holmes 1988). What follows is a review of methods based on these two approaches.

The first demonstration that the application of phase shifting to DIC images isolates the object's phase gradient information from the object's amplitude information (Cogswell et al. 1997) suggested the potential for quantitative phase measurement from integration. Images in Figure 24.16 illustrate the results of phase-shifting DIC images and the phase contrast produced by integration. Phase-shifting is a method derived from a geometric imaging model to compute object gradient phase information from four DIC images of the same field of view acquired with accurately known phase bias (Cogswell et al. 1997). The phase bias between the two orthogonally polarized beams can be accurately controlled using a calibrated Nomarski prism or a Senarmont compensator (Shimada et al. 1990, Hariharan 1993). Currently, all methods of integration of phase-shifted DIC (PS-DIC) images rely on a phase-shifting algorithm.

Two successfully demonstrated methods of integrating PS-DIC images are the spiral phase integration (SPI) method (Arnison et al. 2004) and integration via the Abel transform

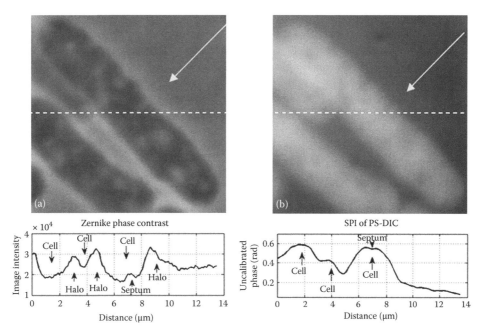

FIGURE 24.15 DIC-based quantitative phase imaging produces high-resolution results with few artifacts. Images shown are a Zernike phase contrast image (a) (objective: 63×, 1.25 NA) compared with a quantitative phase image (b) of the formation of a septum wall in a fixed preparation of fission yeast cells. The quantitative phase image was computed from experimental DIC images (objective: 63×, 1.4 NA.) using phase-shifting and a spiral phase integration method described in Section 24.3.1.1. Horizontal profiles through both images are plotted along the dashed white line and are shown below each image. The quantitative phase result exhibits the same cell structure and fine detail as the Zernike phase contrast image, in reverse contrast, but without the "halos" around the outer edge of the cells. Consequently, the septum wall, indicated with a white arrow, is easier to identify as a cell structure in the quantitative phase result (b) than in the phase contrast image (a). Images recorded using a black and white CCD camera and a 515–565 nm band-pass filtered illumination. (From King, S.V., Quantitative phase information from differential interference contrast microscopy, PhD dissertation, University of Colorado, Boulder, CO, 2008.)

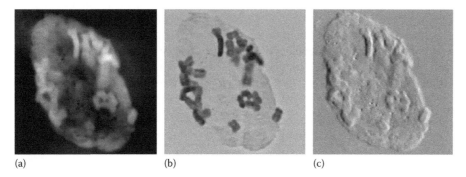

FIGURE 24.16 Phase contrast is computed through the integration of phase gradient information extracted from DIC images of partially absorbing samples using a phase-shifting process. The quantitative phase result (a) shows contrast from OPL rather than from the dye stain shown in the phase-shifted amplitude image (b). The phase-shifted amplitude image contains the absorption information from the dye that is separated from the phase-shifted DIC (PS-DIC) image (c). Phase shifting applied to two sets of experimental DIC images, with orthogonal shear directions, results in two PS-DIC images, one of which is shown in (c). The SPI result is obtained by integrating the two PS-DIC images that are a better representation of the object's phase gradient than the DIC images that include both amplitude and phase gradient information. The specimen shown is a squash preparation of a species of the orchid genus *Paphiopedilum* (common name, slipper orchid) arrested in metaphase stage of mitosis and stained with feulgen. Objective: 100×, 1.3 NA. (From King, S.V., Quantitative phase information from differential interference contrast microscopy, PhD dissertation, University of Colorado, Boulder, CO, 2008.)

(Liu et al. 2004). Although the Abel transform method assumes that the object has at least one direction of rotational symmetry, it has been demonstrated to measure refractive index profiles of optical fiber. The SPI method uses as input two PS-DIC images that have orthogonal shear directions. SPI is applicable regardless of object symmetry and it has been demonstrated to measure refractive index of polystyrene beads embedded in optical cement (King et al. 2007, 2008).

The benefit of using two DIC images with orthogonal shear directions for the quantitative computation of object phase was first demonstrated with a rotational-diversity iterative method that uses multiple DIC images with diverse shear directions as an input (Preza et al. 1997a,b, 1998). Such rotationally diverse DIC images can be acquired by rotating the specimen and equivalently by shear rotation. Simulated and experimental studies (Preza et al. 1998, Preza 2000) with this method demonstrated that when more than two rotationally diverse images are used, the accuracy in the result improves. However, the main improvement is achieved when at least two DIC images with orthogonal shear directions are used instead of one. The rotational-diversity phase estimation method is based on a partially coherent imaging model (Equation 24.2) and the assumption that the object does not absorb (i.e., it is an unstained transparent object modeled by a phase-only function with magnitude equal to 1).

The development of iterative model-based phase estimation algorithms has incorporated more rigorous imaging models, and image estimation frameworks that can incorporate regularization approaches, that have the potential to improve the accuracy in the computed phase as presented in (Preza et al. 2006, Preza and O'Sullivan 2009). Other iterative phase estimation methods, based on different modeling approaches (also discussed in Section 24.2.5.4), have proven successful but they also do not account for object absorption (Kagalwala and Kanade 2003, Aoki et al. 2006, Ishiwata et al. 2006).

The first iterative quantitative phase estimation method accounting for absorptive objects was developed for reflection DIC microscopy and it incorporates the use of a bright-field image to determine the object absorption (Axelrod et al. 2004). This work also demonstrates the broad phase wrapping limits of DIC microscopy and uses an atomic force microscope (AFM) to overcome phase wrapping at a very sharp edge.

Recently, the development of an alternating minimization iterative algorithm for the simultaneous computation of object amplitude and phase information, from two DIC images with orthogonal shear directions (Preza and O'Sullivan 2007, 2009, 2010, O'Sullivan and Preza 2008) has overcome limitations of iterative methods that do not account for absorptive objects. This approach has extended quantitative phase imaging to quantitative imaging of stained or partially absorbing objects using DIC microscopy.

Experimental testing and further development of quantitative phase imaging methods based on more accurate DIC models continues to further advance the capability of DIC-based quantitative phase imaging.

24.3.1.2 Other Types of Quantitative Phase Imaging Microscopy/Methods

Alternative approaches to quantitative phase microscopy that do not rely on DIC microscopy are summarized in Table 24.2. Unlike DIC-based quantitative phase imaging, all of these techniques, described briefly in the following, are applicable to birefringent objects because they do not rely on manipulation of polarization states to control phase.

24.3.1.2.1 Optical Coherence Tomography

Optical coherence tomography (OCT), pioneered in the late 1980s and early 1990s, measures reflectivity at a known depth within a scattering medium by using a Michelson interferometer illuminated with a low coherence (broad-spectrum) source (Fercher and Hitzenberger 2002). OCT employs the same technique with the addition of a high NA objective lens in the sample arm. Quantitative measurement of sample *phase* properties with nanometer scale axial resolution using OCT was first reported in 1995 (Tearney et al. 1995). This configuration is a scanning technique in which either the source or a mirror in the reference arm is scanned to build up a 2D image (laterally or along the *z*-axis, respectively). The depth of the reflection plane in the sample arm is selected by the length of the reference arm. Constructive interference from the reflection plane is confined to the coherence length of the source. By scanning in both directions, a 3D image

TABLE 24.2 Quantitative Non-DIC-Based Phase Imaging Methods Reviewed in Section 24.3.1.2

Year	Method	Main Advantages	Main Disadvantages
1995	Optical coherence microscopy or tomography (Tearney et al. 1995)	Nanometer scale axial resolution	Low lateral resolution
1999	Digital holographic microscopy (Cuche et al. 1999)	Non-scanning 3D reconstruction	Coherent noise, phase unwrapping required
2000	Phase-dispersion microscopy (Yang et al. 2000)	Potential to separate thickness from refractive index	Phase unwrapping required, difficult alignment
2002	Quantitative phase amplitude microscopy (Barone-Nugent et al. 2002)	Computation applied to bright-field microscopy	Iterative, low NA, very dependent on starting point
2004	Fast Fourier microscopy (Popescu et al. 2004)	Easy implementation on available microscopes	Halo-like artifacts contrast, geometric principles
2005	Hilbert phase microscopy (Ikeda et al. 2005)	Non-iterative, non-scanning	Requires phase unwrapping
2006	Spiral phase contrast microscopy (Bernet et al. 2006)	Non-iterative, non-scanning	Reduced spatial resolution
2006	Scanning transmission microscopy with a position-sensitive detector (Ayres and McLeod 2006)	3D depth section of phase information without a pinhole	Slow due to scanning
2007	Quantitative structured illumination phase imaging (Pavani et al. 2007)	Useful for thick objects	For homogenous objects only
2008	Jones phase microscopy (Wang et al. 2008)	Non-iterative, non-scanning	Requires phase unwrapping

can be compiled (for a discussion of 3D phase imaging, see Section 24.3.2). Spectral domain optical coherence tomography (SD-OCT) (Fercher et al. 1995) has also proven to be a successful quantitative phase imaging method (Joo et al. 2005). In this configuration, depth scans are acquired, without the movement of a reference mirror, by detection of the spectrum of the interferogram. Clinical applications of OCT include retinal imaging for diagnosis of retinal disease and guiding of cardiac catheterization for treatment of arteriosclerosis (Drexler et al. 2001, Chia et al. 2008). However, phase contrast OCT is restricted in lateral resolution by the Abbe diffraction limit, as well as limited by a low signal-to-noise ratio (S/N) and modulus π phase wrapping common to all traditional interferometric techniques.

24.3.1.2.2 Quantitative Phase Amplitude Microscopy

Quantitative phase amplitude (QPA) *microscopy* (Barone-Nugent et al. 2002) refers to a computational *non-interferometric* phase imaging technique (unlike most other quantitative phase imaging methods that work under the auspices of interferometric setups) that has found success in the commercial microscope market (IATIA Vision Sciences). This method uses an iterative phase estimation technique that takes as input one in-focus and two out-of-focus images acquired with a bright-field transmission microscope and is based on a mathematical model of wave propagation to estimate the phase of the wave field in the in-focus plane (Barone-Nugent et al. 2002). However, this iterative method is susceptible to obtaining different solutions given different initial guesses.

QPA microscopy is especially convenient for manufacturing of fiber-optic wave guides, for example (Dragomir et al. 2006), and applications of cell biology because it requires only a bright-field transmission microscope with no additional optics (Dragomir et al. 2007).

24.3.1.2.3 Digital Holographic Phase Microscopy

First introduced in 1999 (Cuche et al. 1999), *digital holographic phase microscopy* records a hologram of an object on a CCD camera and reconstructs the object digitally by simulating Fresnel propagation of a wavefront through the digital hologram. As in the case of optical coherence microscopy, digital holographic microscopy has been successfully used in biological applications with 2°–4° accuracy (Marquet et al. 2005, Dubois et al. 2006, Tishko et al. 2008). Despite its success, the method is limited by modulus π phase wrapping and suffers a loss in resolution due to coherent noise from dust, scratches, and stray reflections. Continuous monitoring of a reference value in the background medium is used to compensate for mechanical and thermal instabilities.

24.3.1.2.4 Phase-Dispersion Microscopy

Harmonic phase-dispersion microscopy (PDM) is an imaging technique in which contrast is provided by differences in refractive index at two harmonically related wavelengths (Ahn et al. 2005). Results shown from this method appear to have high contrast but are very noisy despite the use of heterodyne detection techniques.

24.3.1.2.5 Fast Fourier Phase Microscopy

In *fast Fourier microscopy* (Lue et al. 2007), the output of a bright-field transmission microscope is reflected off a programmable phase modulator, which is used to phase shift only the diffracted field while leaving the DC field component unmodulated. The two fields interfere at a CCD detector and the phase of the object is computed from the phase-shifted images. This method is based on the fundamental ideas of the Zernike phase contrast microscope (Zernike 1942a). The phase of the object is obtained directly, and does not require integration or calibration. Fast Fourier microscopy is limited in speed only by the liquid crystal response time of the programmable phase modulator. However, a pixilated phase modulator can introduce phase errors. The method's potential for the real-time study of dynamic biological processes has been demonstrated (Popescu et al. 2004). As in the case of digital holography, it is limited by modulus π phase wrapping. Additionally, the spatial coherence of the broad band illumination must be maintained by coupling into a single-mode fiber.

24.3.1.2.6 Hilbert Phase Microscopy

In *Hilbert phase microscopy*, sample and reference beams with identical magnification and optical resolution interfere in a Mach–Zender interferometer configuration with the reference arm at a slight tilt (Ikeda et al. 2005). The frequency of the interference fringes is adjusted so that it matches the cutoff frequency of the imaging system. The interference term is then isolated with a high-pass filter and combined with its Hilbert transform to form a complex analytical signal. The phase of the object is computed from the phase of the complex-valued analytical signal (after a phase unwrapping step).

24.3.1.2.7 Spiral Phase Contrast Microscopy

In a method related to both SPI of PS-DIC and the concept of fast Fourier microscopy described earlier, *spiral phase contrast microscopy* uses a liquid crystal spatial light modulator in the Fourier plane of an imaging system to generate a spiral phase wavefront. This digitally programmed wavefront has been shown to create the same type of images as a DIC microscope (Bernet et al. 2006). Reconstruction of quantitative phase images, through inverse filtering, of epithelial and red blood cells were demonstrated with this method.

24.3.1.2.8 Scanning Transmission Microscopy with a Position-Sensitive Detector

A more recent non-interferometric phase imaging technique is *scanning transmission microscopy* with a position-sensitive detector (Ayres and McLeod 2006). A position-sensitive detector outputs the position of the illumination spot as well as its intensity. The unique attribute of scanning transmission microscopy with this technology is that the detector itself creates an optical sectioning characteristic similar to that of confocal imaging in a transmissive imaging system. Thus far, it has proven to successfully image the phase of weakly diffracting objects.

24.3.1.2.9 Quantitative Structured Illumination Phase Imaging

Quantitative structured illumination phase imaging (QSIP) (Pavani et al. 2008) is easily implemented in any existing bright-field transmission microscope since it requires only an amplitude pattern mask in the illumination path and a post processing algorithm. Phase is calculated from deformations, caused by the object's phase variations, to the known amplitude pattern used to illuminate the object. In its current form, however, QSIP is applicable only to homogeneous objects.

24.3.1.2.10 Jones Phase Microscopy

Jones phase microscopy (JPM) is a computational method of extracting amplitude and phase from two interference microscope images using Jones matrices. The setup is the same as in Hilbert phase microscopy (discussed in Section 24.3.1.2.6), but it also includes polarized illumination in the two arms of the interferometer and an analyzer placed before the CCD camera. Four interference images are collected: two with the sample arm polarization at +45° with respect to the polarization of the reference arm (with the analyzer at first parallel, then perpendicular to the sample arm polarization) and two with the sample arm at −45° (with the analyzer parallel and then perpendicular to the sample arm polarization). These four images plus a JPM measurement with no sample, only taken once, give all the variables necessary to solve a set of four equations for the four components of a Jones matrix representing the object.

As seen in Tables 24.1 and 24.2, the list of quantitative phase microscopy methods encompasses techniques developed over the past 20 years. For some of the methods, particularly for optical coherence microscopy, but also for digital holographic microscopy and quantitative phase amplitude microscopy, the techniques have generated new, growing fields of study. A large volume of literature exists describing the basics as well as variations and improvements on the original techniques. The number of reported research activities indicates an increasing interest in quantitative phase microscopy as well as the need for multiple approaches to suit a wide variety of applications. For example, optical coherence microscopy is an excellent noninvasive tool for a clinical biologist diagnosing a patient, but it is not the best solution for a cell biologist studying the development of live cells in culture who can acquire images in transmission mode and thereby take advantage of phase imaging methods that produce results with higher contrast.

24.3.1.3 Current Limitations

The preceding paragraphs have pointed out the advantages and disadvantages of specific quantitative phase imaging methods. More generally however, these methods enable quantitative investigation of thickness and refractive index (or density) properties, which are not otherwise easily accessible, such as small spatial variations, variations embedded within a transparent medium, and weakly absorbing variations not marked by invasive contrast agents. The very combination of digital detection and post-processing methods that make quantitative phase imaging possible can, however, be a limitation. Methods based on time-consuming iterative algorithms or point scanning acquisition methods limit the ability to conduct time-resolved studies. Other methods are also hampered by slow data acquisition and/or a relatively small field of view. These are engineering challenges that are likely to be overcome in the future.

24.3.2 Three-Dimensional Phase Imaging

In Section 24.2, it was explained that phase contrast microscopy allows the study of morphology of an unstained object. Unfortunately, phase contrast images are two-dimensional projections through the refractive index (RI) and so fail to provide an adequate idea of the microstructure in three dimensions. The study of many transparent biological specimens and their organization is an intrinsically three-dimensional endeavor (Charrière et al., 2006a, b, Choi et al. 2007, Fauver et al., 2005, Heise et al., 2005, Vishnyakov and Levin 1998). Many methodologies delivering 3D specimen information have also been developed for fluorescence microscopy, such as confocal and multiphoton microscopy discussed in earlier chapters.

Several phase-based 3D imaging methods capable of resolving cellular and subcellular structures of biological cells have been developed since the early 1990s. Computational approaches for improving the 3D information from phase contrast microscopy images examined different techniques for nonabsorptive specimens, focusing on modeling the characterization of microscope images, leading to better understanding of the underlying imaging system. Although improvements in the 3D phase recovery have evolved over the years from qualitative information of the RI for thin (Noda et al. 1990b, Vishnyakov and Levin 1998) or thick specimens (Noda et al. 1992) to quantitative information with high accuracy in the reconstructed RI (Charrière et al. 2006a, Choi et al. 2007, Dragomir et al. 2008b, Lue et al. 2008b, Sung et al. 2009), many questions about how the 3D structures affect the phase images of thick specimens are still under debate (Gong et al. 2008, Sierra et al. 2009a).

Since refractive index serves as an important intrinsic contrast agent in visualizing transparent specimens, quantitative observation of transparent phase objects, and subsequently their 3D RI distribution, are subjects of long-standing interests in the imaging community (Choi et al. 2007, Cogswell et al. 1997, Kemper et al. 2007, Dragomir et al. 2008a,b, Rappaz et al. 2009, Sung et al. 2009). Additionally, the refractive indices of live cells can only be observed with phase contrast microscopy (Ross 1976) or DIC microscopy (Stephens and Allan 2003). Noninvasive measurement of local optical properties within various living biological cells is often desired and in the last two decades efforts have been made to develop modern technologies able to deliver such 3D information. Furthermore, reconstructing the 3D RI spatial distribution using tomographic techniques (Shaw 1990) can complement the images obtained using other contrast methods, for example, staining of the specimen (Cogswell et al. 1996).

In the early work of Noda et al. (1990) the specimen has been considered to be an optically thin and weak phase object, that is, the Born approximation (Goodman 1984) is satisfied (see discussion in Section 24.2.5.1.2). In this case, the phase contrast microscope can be characterized by a phase-transfer function, and a linear shift-invariant system principle applies. Using this approach, qualitative 3D phase distributions of tobacco cells were obtained in a transmission microscope with annular illumination under oblique illumination and normal Köhler conditions (Noda et al. 1990b).

Since phase contrast images are 2D projections along the optical axis (z) of the 3D phase function, often the phase microscopy methods developed to date are able to provide only the average refractive index. A successful strategy to determine 3D RI is based on measurement of projections in multiple directions as is done in x-ray tomography in which the projection of absorption is measured. This can be performed through rotation of the illumination (Noda et al. 1992, Vishnyakov and Levin 1998, Lauer 2002, Choi et al. 2007) or rotation of the object (Barty et al. 2000, Charrière et al. 2006a,b, Dragomir et al. 2008a,b, Kobayashi et al. 2006b). Figure 24.17 illustrates various sample mounting geometries for tomographic imaging. When enough projections are taken and interpolated, the spatial content becomes a volume that can be reconstructed by employing standard principles of computed tomography (Kak and Slaney 1988). The majority of tomographic phase reconstruction approaches developed to date work under the assumption that the phase of the transmitted field may be considered as a line integral of the RI along the

propagation direction. The 3D phase reconstruction is based on a two-step approach. First, the phase function is found, generally in the form of Radon projections. Second, a filtered back-projection algorithm based on the inverse Radon transform is applied to computationally reconstruct the 3D RI. A more detailed description of tomographic methods has been described in other works (Ramachandran and Lakshminarayanan 1971, Wedberg et al. 1995, Bronnikov 2002) and elsewhere in this volume.

24.3.2.1 Review of State-of-the-Art 3D Phase Imaging Methods

The following paragraphs describe the most recent developments in 3D phase tomography with an emphasis in quantitative 3D RI determination. The advantage and disadvantages of these methods are summarized in Table 24.3.

24.3.2.1.1 Rotational Oblique Illumination Computed Tomography

The examination of thick objects in conjunction with a tomographic approach was first suggested by Noda et al. (1992) and is based on a conventional transmission laser microscope that has been modified to suit the requirements of rotational oblique illumination (Noda et al. 1992). This was achieved by rotating the point source about the optical axis and collecting the corresponding projection images. These projection images are first mapped to a 3D image and then inverse filtered to produce the final result. The reconstructed 3D RI, based on the first Born approximation formulation, was discussed qualitatively.

24.3.2.1.2 Beam-Tilting Phase-Shifting Interferometry

Another beam-tilting tomographic approach is based on phase-shifting interferometry (Vishnyakov and Levin 1998). Here the oblique illumination is combined with a Linnik interference microscope under coherent illumination for studying the 3D spatial distribution of a thin (<10 μm) phase object (Vishnyakov and Levin 1998). A tomogram is reconstructed following a two-step approach: first the phase is calculated then the 3D RI is computed via an iterative algorithm (Herschberg). The number of projections is limited by the NA of the objective lenses.

24.3.2.1.3 Quantitative Phase Amplitude Tomography

Quantitative phase amplitude tomography involves rotation of the sample under constant beam illumination. Phase reconstruction is performed within the weak phase object approximation using a phase retrieval algorithm (Paganin and Nugent 1998) implemented for a shift-invariant linear imaging system (Barty et al. 2000). Applied to translationally invariant specimens such as optical fibers (Dragomir et al. 2008b), the method has high accuracy in the reconstructed RI, 1.74×10^{-4} (Goh et al. 2007), Table 24.3. Three-dimensional RI is computationally obtained from transverse phase images and a combination of a filtered back-projection algorithm (Kak and Slaney 1999) applied in a slice-by-slice implementation followed by a ramp (Ramlak) filter to produce quantitative results of the refractive

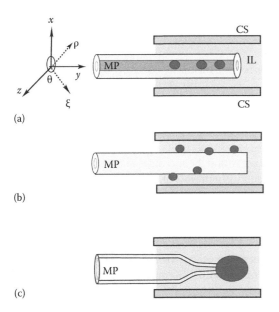

FIGURE 24.17 Sample mounting geometries for tomographic phase contrast imaging: MP micropipette, CS coverslip, IL immersion liquid. (a) Specimen embedded inside MP of known internal diameter, (b) specimen mounted onto thin MP of known external diameter, and (c) specimen clamped at the end of MP by patch-clamp technique. Geometries (a) and (b) are suitable for small specimens (<30 μm diameter), while (c) is suitable for larger specimens (>30 μm diameter). In general, a micromotor is used to rotate the MP over large angles eucentrically, to a precision better than 2°.

TABLE 24.3 Comparison of 3D Phase Imaging Methods Reviewed in Section 24.3.3

Year	Method	Advantages	Disadvantages
1992	Rotational oblique illumination computed tomography (Noda et al. 1992)	Good optical sectioning capability	*Qualitative*: Incomplete reconstruction of the lateral frequency due to the mismatch between the angle of the oblique illumination and the critical illumination angle, given by the NA of the objective lens
1998	Beam-tilting phase-shifting interferometry combined with Linnik interference microscope (Vishnyakov and Levin 1998)	Has potential for dynamic 3D RI Specimen can be observed under physiological conditions	*Qualitative*: The effect of coherent noise is present in the reconstructed RI Angular projections limited by the NA f the imaging lens
2000	Quantitative phase amplitude tomography involves rotation of the sample under constant beam illumination (Section 24.3.2.2.3)	Method suitable for translationally invariant specimens such as optical fibers Improved RI accuracy down to 0.001 is possible	Viscosity of the suspension liquid needs to be high enough to avoid object movement during image acquisition
2002	Optical diffraction RI tomography uses beam rotation to form tomographic images (Section 24.3.2.2.4)	Suitable for absorptive samples Diffraction-limited resolution in the horizontal direction	Highly dependent on the RI match between the sample and the suspension liquid Reduction in spatial resolution along the optical axis compared to the lateral directions due to the missing cone in the frequency domain *Quantitative*: RI information not provided
2006	DHM tomography rotation of the specimen in a micropipette relative to a fixed beam illumination (Charrière et al. 2006a) DHM tomography specimen rotated via a patch-clamp technique (Charrière et al. 2006b)	RI accuracy ~0.01 Suitable for biological specimens Micrometer spatial resolution in all three dimensions Method suitable for live specimens (>30 μm external diameter) RI better than 0.01	Index matching to the cylindrical surface of the capillary and the buffer medium is necessary Incompatibility between buffer medium RI and conditions given by the microscope objective Spatial resolution of 3 μm in the reconstructed volume Reconstruction algorithm suffers from 2π phase jumps
2006	Optical tweezers–based rotation phase tomography (Section 24.3.2.2.6)	Imaging a live small (~20 μm diameter) cell in buffer solution	Prone to sample drift artifacts RI accuracy and spatial resolution not specified
2007	Mach–Zehnder-interferometer-based tomographic phase microscopy	Applicable to suspended or substrate-attached cells and multicellular organisms	3D reconstruction suffers from missing projections due to limited illuminations angles (<60°) caused by the NA of the imaging system
	Tilting the illumination wave wile the specimen is kept fixed	Does not require sample perturbation of immersion of the specimen in special media	RI accuracy dependent on the diffraction constraints adopted
	Born approximation (Choi et al. 2007) Rytov approximation (Sung et al. 2009)	Submicrometer spatial resolution in the reconstructed tomogram	Difficult for thick and larger specimens, but mosaic tomogram is suggested
2008	Synthetic aperture tomographic phase microscopy sTPM (Lue et al. 2008b)	RI accuracy of 0.001 with axial resolution of 1 μm can be obtained in the 3D reconstruction of moving cells	The dynamic range of the detecting system limits the angular range of the synthesized plane waves

index (Bronnikov 2002). When applied to isolated, small partly absorbing specimens, either embedded in a capillary tube or glued onto it (Figure 24.17a and b), limitations due to the weak phase object approximation are evident (Dragomir et al. 2008a).

24.3.2.1.4 Optical Diffraction RI Tomography

The theoretical basis of optical diffraction tomography (ODT) was developed in the early 1970s (Wolf 1969, Dändliker and Weiss 1970). ODT records the complex wavefront diffracted by a transparent or semitransparent object under varying illumination angles (Lauer 2002). The Born approximation framework is adopted to make the relation linear between the complex RI of the object and the electric field. Lauer (2002) used phase-shifting holography to record the phase and amplitude of the scattered wave. The resulting 3D RI information is limited by the NA of the imaging lens, that is, by the angle between the direction of the optical axis and the direction of the plane wave that passes through the objective lens. That is to say, the illumination beams

that are not within the aperture limit of the imaging lens cannot be used resulting in the loss of corresponding frequencies. Thus, due to lack of low-frequency representation, the axial spatial resolution is considerably lower compared to the horizontal resolution.

24.3.2.1.5 Digital Holographic Microscopic Tomography

As discussed in other parts of this volume, digital holographic microscopy (DHM) (Kemper and Von Bally 2008) exploits the narrow spectrum of the laser light by using its long coherence length to produce clear interference phenomena. Various experimental setups have been made in modern DHM systems to extract phase, as well as 3D information, from different types of phase objects. The 3D RI distribution of a semitransparent object, a pollen grain, was reconstructed from back-projecting OPL values collected with DHM on a series of projections taken at various incidence angles. This system used the specimen embedded in a capillary micropipette of known internal diameter (Figure 24.17a) (Charrière et al. 2006a). The micropipette walls introduce aberrations that lead to loss of RI accuracy. To overcome these deficiencies, a patch-clamp technique was later proposed and investigated (Charrière et al. 2006b). Using this clamping technique (Figure 24.17c), these authors demonstrated the value of transmission off-axis DHM tomography for examining a living specimen (*Hyalosphenia papilio*) (Charrière et al. 2006b). The authors capitalized on the fact that the phase errors due to the capillary material are eliminated since the light path is going only through the specimen. Despite this advancement, the RI accuracy remained the same as in previous setups, that is 0.01, and the spatial resolution achieved in the reconstructed volume was only around $3\,\mu m$. Some claims are made that a whole volume of 3D information can be obtained using one shot DHM performed at a specified wavelength (Rappaz et al. 2009). This, however, may actually be only a topographic representation as a result of the OPL difference observed, since quantitative phase information derived from DHM can give a 2½D, not a true 3D imaging (Kou and Sheppard 2007b).

24.3.2.1.6 Optical Tweezers–Based Rotation Phase Tomography

Given the need to precisely characterize at cellular level the biodynamics of cells without resorting to fluorescent labeling, optical tweezers (Kobayashi et al. 2006a) have been used in conjunction with a conventional phase-contrast microscope (Kobayashi et al. 2006b). This unique approach harnesses the light pressure in the optical axis to apply a rotating torque to a floating small cell (~20 µm diameter) making possible various rotations in the range of 0 to 360°. The experimental setup consists of a filtered white light source necessary for the phase-contrast imaging and a green laser (532.6 nm) necessary for trapping and rotating the cell (Kobayashi et al. 2006a). The reconstruction algorithm is based on the first Born approximation. Reconstructed cross sections of RI have been qualitatively described.

24.3.2.1.7 Mach–Zehnder-Interferometric-Based Tomographic Phase Microscopy

The tomographic phase microscope introduced by Choi et al. (2007) is based on a Mach–Zehnder heterodyne interferometer, which enables phase imaging from time-independent interference patterns induced by the frequency shifting of a reference beam relative to a sample beam. The 3D RI reconstruction algorithm assumes that the scattered field is weaker than the incident field, that is, the phase of the transmitted field is interpreted as a line integral of the refractive index along the propagation direction. Thus, use of the filtered back-projection algorithm based on the inverse Radon transform is justified. The rotating beam geometry considers angular images ranging within ±60° enabling fast acquisition and avoiding the perturbation of the live specimen during investigation.

RI changes in the reconstructed volume of the order of 0.0285 with a submicrometer spatial resolution (0.5 µm in the transverse and 0.75 µm in the axial direction) have been obtained. Tomograms of thin and thick specimens require different computation time. For larger biological structures (>30 µm thickness), sample-induced optical aberrations are eliminated and

the range of tomographic imaging is extended by considering tomograms centered at different depths within the sample. The final reconstruction is a mosaic tomogram covering the entire sample formed by combining the in-focus slices series. The obvious limitations of this method come from the multiple executions of the inverse Radon transform.

More recently, these limitations have been overcome by taking into account diffraction while considering the Rytov approximation (Sung et al. 2009) rather than the Born approximation in solving for the wave equation as defined by Wolf's scalar theory (Wolf 1969). In this context, the 3D RI of a cell is produced without the requirement for propagation of the reconstruction algorithm. The implementation of this approximation is performed in the Fourier diffraction theory by introducing the complex phase into the wave equation according to Devaney (1981). The experimental implementation is based on angle-beam illumination to avoid specimen perturbation during acquisition. The experimental setup is similar to previous work but with an increased number of acquisition angles (±70°) using high NA (1.4) lenses in both the illumination and the acquisition process. However, the final reconstruction suffers from artifacts due to the missing angles. This is accounted for and corrected by considering prior knowledge of the object function via an iterative constraint algorithm (Tam and Perezmendez 1981). The spatial resolution achieved experimentally in the reconstructed volume is close to diffraction resolution (0.35 µm in the transverse and 0.7 µm in the axial direction), partly due to the weak contrast exhibited by the specimen and partly due to the limited sensitivity of the instrument at high angles of scattering.

24.3.2.1.8 Synthetic Aperture Tomographic Phase Microscopy

A different approach to account for the missing angles and to improve the axial resolution, referred to as synthetic tomographic phase microscopy (sTPM) takes into account the Huygens principle (Lue et al. 2008b). According to this principle, any directional plane wave can be created from a set of position-dependent focused beams. The phase and the amplitude can be measured while translating the sample across the beam, which makes it possible to synthesize plane waves from any incident direction. The idea behind the synthetic aperture was first proposed in ultrasound imaging (Nahamoo et al. 1984). In sTPM, a similar principle to confocal microscopy makes use of a focused Gaussian beam that can be approximated to a good extent as a spherical wave (Lue et al. 2008a). This is transformed into a line beam by a cylindrical lens, allowing single directional scanning in a transverse plane to generate plane waves. Furthermore, this allows a separation of the axes in the Fourier planes before and after the cylindrical lens in the focused beam axis and the plane wave axis, respectively. It is asserted that this separation also reduces the reconstruction problem to 2D instead of 3D. The phase and amplitude of the transmitted field are recorded interferometrically (Born and Wolf 1999) based on a heterodyne Mach–Zehnder interferometer (Fang-Yen et al. 2007). The final 3D reconstruction is based on a conventional filtered back-projection algorithm that computes the RI by applying an inverse Radon transform (Kak and Slaney 1988). The final axial resolution in the reconstructed volume, of about 1 µm, is good given the flow condition in which the acquisition is performed. Similarly the accuracy in the refractive index is governed by the first Born approximation condition as in previous tomographic phase microscopic techniques.

24.3.2.2 Concluding Remarks

Since the number of objects being investigated with phase imaging–based tomography is rapidly growing, further exciting 3D views due to the RI contrast of intact cells, organelles, or multicellular organisms can be expected in the future. Furthermore, the future of 3D RI lies not only on expanding the base of application but also on decreasing the image acquisition time as well as capturing and visualizing dynamic 3D biological events. A main drawback for tomography is that it is not really suitable for live-cell imaging in which any specimen perturbation such as rotation must be avoided.

24.4 Conclusion

Although the initial goal of dark-field, phase contrast, and DIC systems was to provide methods for making transparent biomedical specimen features visible, recent developments in optics, detectors, and digital image processing have greatly improved the precision and measurement capabilities of modern-day versions of these modes as well as facilitated the development of different types of microscope configurations capable of quantitative phase imaging. Their improved sensitivity for detecting fine detail, while minimizing the potential for damage from intense light sources, has opened new ways for investigation of molecular dynamics within living cells that are often not possible with fluorescence microscope techniques alone.

This chapter has focused on phase imaging microscopy accomplished with three conventional optical microscopy techniques (dark field, phase contrast, and DIC) and a variety of more recently developed methodologies that integrate new optical configurations and computational methods. The information discussed in this chapter demonstrates how the unique properties of these phase imaging methodologies can provide specimen feature information that is different from, and often complementary to, fluorescence microscopy. In addition, the chapter summarizes the evolution of phase microscopy from a qualitative imaging capability to a quantitative imaging technique, now accomplished by several 2D and 3D imaging methodologies that provide a wealth of new quantitative information.

References

Ahn, A., Yang, C., Wax, A. et al. 2005. Harmonic phase-dispersion microscope with a Mach–Zehnder interferometer. *Appl. Opt.* 44: 1188–1190.

Allen, R. D., Allen, N. S., and Travis, J. L. 1981. Video-enhanced contrast differential interference contrast (AVEC-DIC) microscopy: A new method capable of analyzing microtubule-related motility in the reticulopodial network of allogromia laticollaris. *Cell Motility* 1: 291–302.

Allen, R. D., David, G. B., and Nomarski, G. 1969. The Zeiss-Nomarski differential interference equipment for transmitted-light microscopy. *Zeitschrift fur Wissenschaftliche Mikroskopie und Mikroskopische Technik* 69: 193–221.

Aoki, G., Itoh, M., Yasuno, Y., and Yatagai, T. 2006. Three-dimensional measurement of microorganism by retardation modulated differential interference contrast microscope. *Proc. SPIE* 6090: 0A-1–8.

Arnison, M. R., Larkin, K. G., Sheppard, C. J. R., Smith, N. I., and Cogswell, C. J. 2004. Linear phase imaging using differential interference contrast microscopy. *J. Micro.* 214: 7–12.

Axelrod, N., Radko, A. A. L., and Yosef, N. B. 2004. Topographic profiling and refractive-index analysis by use of differential interference contrast with bright-field intensity and atomic force imaging. *Appl. Opt.* 43: 2272–2284.

Ayres, M. R. and McLeod, R. R. 2006. Scanning transmission microscopy using a position-sensitive detector. *Appl. Opt.* 45: 8410–8418.

Barone-Nugent, E. D., Barty, A., and Nugent, K. A. 2002. Quantitative phase-amplitude microscopy I: Optical microscopy. *J. Micro.* 206: 194–203.

Barty, A., Nugent, K. A., Roberts, A., and Paganin, D. 2000. Quantitative phase tomography. *Opt. Commun.* 175: 329–336.

Beleggia, M. 2008. A formula for the image intensity of phase objects in Zernike mode. *Ultramicroscopy* 108: 953–958.

Berman, D. W. and Crump, K. S. 2008. A meta-analysis of asbestos-related cancer risk that addresses fiber size and mineral type. *Crit. Rev. Toxicol.* 38: 49–73.

Bernet, S., Jesacher, A., Fürhapter, S., Maurer, C., and Ritsch-Marte, M. 2006. Quantitative imaging of complex samples by spiral phase contrast microscopy. *Opt. Express* 14: 3792–3805.

Bogatyrev, V. A., Dykman, L. A., Alekseeva, A. V. et al. 2006. Observation of time-dependent single-particle light scattering from gold nanorods and nanospheres by using unpolarized dark-field microscopy—art. no. 616401. In: *Proceedings of the SPIE*. eds. Zimnyakov, D. A. and Khlebtsov, N. G., pp. 16401–16401.

Born, M. and Wolf, E. 1999. *Principles of Optics*. Cambridge: Cambridge University Press.

Bronnikov, A. V. 2002. Theory of quantitative phase-contrast computed tomography. *J. Opt. Soc. Am.* 19: 472–480.

Carneiro, S., Amaral, A. L., Veloso, A. C. et al. 2009. Assessment of physiological conditions in *E. coli* fermentations by epifluorescent microscopy and image analysis. *Biotechol. Prog.* 25: 882–891.

Centonze, F. V. 2008. Phase contrast and differential interference contrast (DIC) microscopy. *J. Vis. Exp.* 18. Retrieved from http://www.jove.com/index/Details.stp?ID=844 on 8/26/2009.

Charles, C. H., Vincent, J. W., Borycheski, L. et al. 2000. Effect of an essential oil-containing dentifrice on dental plaque microbial composition. *Am. J. Dentist.* 13: 26C–30C.

Charrière, F., Marian, A., Montfort, F. et al. 2006a. Cell refractive index tomography by digital holographic microscopy. *Opt. Lett.* 31: 178–180.

Charrière, F., Pavillon, N., Colomb, T., and Depeursinge, C. 2006b. Living specimen tomography by digital holographic microscopy: Morphometry of testate amoeba. *Opt. Express* 14: 7005–7013.

Chia, S., Raffel, O. C., Takano, M. et al. 2008. Association of statin therapy with reduced coronary plaque rupture: An optical coherence tomography study. *Coron. Art. Dis.* 19: 237–242.

Choi, W., Fang-Yen, C., Badizadegan, K. et al. 2007. Tomographic phase microscopy. *Nat. Methods* 4: 717–719.

Choi, C., Margraves, C., English, A., and Kihm, K. 2008. Multicontrast microscopy technique to dynamically fingerprint live-cell focal contacts during exposure and replacement of a cytotoxic medium. *J. Biomed. Opt.* 13: 054069.

Cogswell, C. J., Larkin, K. G., and Klemm, H. U. 1996. Fluorescence microtomography and multi-angle image acquisition and 3D digital image reconstruction. *Proc. SPIE* 2655: 109–115.

Cogswell, C. J. and Sheppard, C. J. R. 1992. Confocal differential interference contrast (DIC) microscopy: Including a theoretical analysis of conventional and confocal DIC imaging. *J. Microsc. Oxf.* 165: 81–101.

Cogswell, C. J., Smith, N. L., Larkin, K. G., and Hariharan, P. 1997. Quantitative DIC microscopy using a geometric phase shifter. *Proc. SPIE* 2984: 72–81.

Cox, G. 2007. *Optical Imaging Techniques in Cell Biology*. Boca Raton, FL: CRC Press.

Cuche, E., Bevilacqua, F., and Depeursinge, C. 1999. Digital holography for quantitative phase-contrast imaging. *Opt. Lett.* 24: 291–293.

Dändliker, R. and Weiss, K. 1970. Reconstruction of three-dimensional refractive index from scattered waves. *Opt. Commun.* 1: 323–328.

Danuser, G., Tran, P. T., and Salmon, E. D. 2000. Tracking differential interference contrast diffraction line images with nanometre sensitivity. *J. Microsc. Oxf.* 198: 34–53.

Danz, R. and Gretscher, P. 2004. C-DIC: A new microscopy method for rotational study of phase structures in incident light arrangement. *Thin Solid Films* 462–463: 257–262.

Devaney, A. J. 1981. Inverse-scattering theory within Rytov approximation. *Opt. Lett.* 6: 374–376.

Dietrich, H. R. C., Vermolen, B. J., Rieger, B., Young, I. T., and Garini, Y. 2007. A new optical method for characterizing single molecule interactions based on dark field microscopy—art. no. 644403. In: *Ultrasensitive and Single-Molecule Detection Technologies II: Proceedings of the SPIE*. eds. Enderlein, J., and Gryczynski, Z. K., pp. 44403–44403.

Donders, G. G. G. 2007. Definition and classification of abnormal vaginal flora. *Best Pract. Res. Clin. Obstet. Gynaecol.* 21: 355–373.

Dragomir, N. M., Baxter, G. W., and Roberts, A. 2006. Phase-sensitive imaging techniques applied to optical fibre characterization. *IEEE Proc. Optoelectron.* 153: 217–221.

Dragomir, N. M., Goh, X. M., Curl, C., Delbridge, L. M. D., and Roberts, A. 2007. Quantitative polarized phase microscopy for birefringence imaging of unstained live specimens. *Opt. Express* 15: 17690–17698.

Dragomir, N. M., Goh, X. M., and Roberts, A. 2008a. Three-dimensional quantitative phase imaging: Current and future perspectives. In: *Proc SPIE Three-Dimensional and Multidimensional Microscopy: Image Acquisition and Processing XV*. Concello, J. A., ed. San Jose, CA: SPIE, p. 686106.1–686106.11.

Dragomir, N. M., Goh, X. M., and Roberts, A. 2008b. Three-dimensional refractive index reconstruction with quantitative phase tomography. *Microsc. Res. Tech.* 71: 5–10.

Drexler, W., Morgner, U., Ghanta, R. K. et al. 2001. Ultrahigh-resolution ophthalmic optical coherence tomography. *Nat. Med.* 7: 502–507.

Dubois, F. and Grosfils, P. 2008. Dark-field digital holographic microscopy to investigate objects that are nanosized or smaller than the optical resolution. *Opt. Lett.* 33: 2605–2607.

Dubois, F., Yourassowsky, C., Monnom, O. et al. 2006. Digital holographic microscopy for the three-dimensional dynamic analysis of in vitro cancer cell migration. *J. Biomed. Opt.* 11: 054032-1-5.

Fang-Yen, C., Oh, S., Park, Y. et al. 2007. Imaging voltage-dependent cell motions with heterodyne Mach–Zehnder phase microscopy. *Opt. Lett.* 32: 1572.

Farkas, D. L., Baxter, G., DeBiasio, R. L. et al. 1993. Multimode light microscopy and the dynamics of molecules, cells and tissues. *Annu. Rev. Physiol.* 55: 785–817.

Fauver, M., Seibel, E. J., Rahn, J.R., et al. 2005. Three-dimensional imaging of single isolated cell nuclei using optical projection tomography. *Opt. Express* 13: 4210–4223.

Fercher, A. F. and Hitzenberger, C. K. 2002. *Optical Coherence Tomography*. North-Holland, Amsterdam: Elsevier Science.

Fercher, A. F., Hitzenberger, C. K., Kamp, G., and El-Zaiat, S. Y. 1995. Measurement of intraocular distances by backscattering spectral interferometry. *Opt. Commun.* 117: 43–48.

Flaherty, S. P., Payne, D., Swann, N. J., and Matthews, C. D. 1995. Assessment of fertilization failure and abnormal fertilization after intracytoplasmic sperm injection (ICSI). *Reprod. Fert. Dev.* 7: 197–210.

Galbraith, W. 1982. The image of a point of light in differential interference contrast microscopy: Computer simulation. *Microscopica ACTA* 85: 233–254.

Gibson, S. F. and Lanni, F. 1989. Diffraction by a circular aperture as a model for three-dimensional optical microscopy. *J. Opt. Soc. Am. Opt. Image Sci. Vis.* 6: 1357–1367.

Goh, X. M., Dragomir, N. M., Jamieson, D. N., Roberts, A., and Belton, D. X. 2007. Optical tomographic reconstruction of ion beam induced refractive index changes in silica. *Appl. Phy. Lett.* 91: 181102.

Gong, W., Si, K., and Sheppard, C. J. R. 2008. Modeling phase functions in biological tissue. *Opt. Lett.* 33: 1599–1601.

Goodman, J. 1996. *Introduction to Fourier Optics*. New York: McGraw-Hill.

Goodman, J. W. 1984. *Statistical Optics*. New York: Wiley & Sons.

Gundlach, H. 1993. Phase contrast and differential interference contrast instrumentation and applications in cell, developmental, and marine biology. *Opt. Eng.* 32: 3223–3228.

Hamahashi, S., Onami, S., and Kitano, H. 2005. Detection of nuclei in 4D Nomarski DIC microscope images of early *Caenorhabditis elegans* embryos using local image entropy and object tracking. *BMC Bioinformatics* 5: 125.

Hariharan, P. 1993. The Senarmont compensator: An early application of geometric phase. *J. Modern Opt.* 40: 2061–2064.

Hartman, J. S., Gordon, R. L., and Lessor, D. L. 1980. Quantitative surface topography determination by Nomarski reflection microscopy. II Microscope modification, calibration and planar sample experiments. *Appl. Opt.*19: 2998–3009.

Heise, B., Sonnleitner, A., Heise, B. et al. 2005. DIC image reconstruction on large cell scans. *Microsc. Res. & Techn.* 66(6):312–320.

Holmes, T. J. 1988. Signal-processing characteristics of differential-interference-contrast microscopy. 2: Noise considerations. *Appl. Opt.* 27: 1302–1309.

Holmes, T. J. and Levy, W. J. 1987. Signal-processing characteristics of differential-interference-contrast microscopy. *Appl. Opt.* 26: 3929–3939.

Honda, M., Tanaka-Takiguchi, Y., Inaba, T., Hotani, H., and Takiguchi, K. 2006. Direct observation of transformation process of giant liposome induced by reconstructed cytoskeletons. In: *IEEE International Symposium on Micro-NanoMechatronics and Human Science*, pp. 282–287.

Hopkins, H. H. 1955. The frequency response of a defocused optical system. *Proc. R. Soc. Lond. Ser. Math. Phys. Sci.* 231: 91–103.

IATIA Vision Sciences. (retrieved www.iatia.com.au).

Ikeda, T., Popescu, G., Dasari, R. R., and Feld, M. S. 2005. Hilbert phase microscopy for investigating fast dynamics in transparent systems. *Opt. Lett.* 30: 1165.

Inoue, S. 1986. *Video Microscopy.* New York: Plenum Press, Inc.

Ishiwata, H., Itoh, M., and Yatagai, T. 2006. A new method of three-dimensional measurement by differential interference contrast microscope. *Opt. Commun.* 260: 117–126.

Ito, C., Akutsu, H., Yao, R. et al. 2009. Oocyte activation ability correlates with head flatness and presence of perinuclear theca substance in human and mouse sperm. *Hum. Reprod.* 24: 2588–2595.

Joo, C., Akkin, T., Cenxe, B., Park, B. H., and De Boer, J. F. 2005. Spectral-domain optical coherence phase microscopy for quantitative phase-contrast imaging. *Opt. Lett.* 30: 2131–2133.

Jung, J. E., Moon, J. Y., Ghil, S. H., and Yoo, B. S. 2009. 2,3,7,8-Tetrachlorodibenzo-p-dioxin (TCDD) inhibits neurite outgrowth in differentiating human SH-SY5Y neuroblastoma cells. *Toxicol. Lett.* 188: 153–156.

Kagalwala, F. and Kanade, T. 2003. Reconstructing specimens using DIC microscope images. *IEEE Trans. Syst. Man Cybernet. B* 33: 728–737.

Kagalwala, F., Lanni, F., and Kanade, T. 1999. Simulating DIC microscope images: From physical principles to a computational model. In: *IEEE Workshop on Photonic Modeling for Computer Vision Graph*, 48–55.

Kak, A. C. and Slaney, M. 1988. *Principles of Computerized Tomographic Imaging.* New York: Institute of Electrical and Electronics Engineers.

Kak, A. C. and Slaney, M. 1999. *Principles of Computerized Tomographic Imaging.* New York: IEEE Press Book.

Kam, Z. 1998. Microscopic differential interference contrast image processing by line integration (LID) and deconvolution. *Bioimaging* 6: 166–176.

Kemper, B. and Von Bally, G. 2008. Digital holographic microscopy for live cell applications and technical inspection. *Appl. Opt.* 47: A52–61.

Kemper, B., Kosmeier, S., Langehanenberg, P. et al. 2007. Integral refractive index determination of living suspension cells by multifocus digital holographic phase contrast microscopy. *J. Biomed. Opt.* 12: 054009.

King, S. V. 2008. Quantitative phase information from differential interference contrast microscopy. PhD dissertation, University of Colorado, Boulder, CO. Available at http://gradworks.uni.com/33/37/3337213.html.

King, S. V., Libertun, A. R., Piestun, R., Cogswell, C. J., and Preza, C. 2008. Quantitative phase microscopy through differential interference imaging. *J. Biomed. Opt.* 13: 024020.

King, S. V., Libertun, A. R., Preza, C., and Cogswell, C. J. 2007. Calibration of a phase-shifting DIC microscope for quantitative phase imaging. *Proc. SPIE* 6443: 64430M.

Kobayashi, H., Ishimaru, I., Hyodo, R. et al. 2006a. A precise method for rotating single cells. *Appl. Phy. Lett.* 88: 131103.

Kobayashi, H., Ishimaru, I., Yasokawa, T. et al. 2006b. Three-dimensional phase-contrast imaging of single floating cells. *Appl. Phy. Lett.* 89: 241117.

Kou, S. S. and Sheppard, C. J. R. 2007a. Comparison of three dimensional transfer function analysis of alternative phase imaging methods. *Proc. SPIE* 6443: 0Q-1-6.

Kou, S. S. and Sheppard, C. J. R. 2007b. Imaging in digital holographic microscopy. *Opt. Express* 15: 13640–13648.

Lang, W. 1969. Nomarski differential interference contrast microscopy II. Formation of the interference image. *Zeiss Inf.* 17: 12–16.

Lauer, V. 2002. New approach to optical diffraction tomography yielding a vector equation of diffraction tomography and a novel tomographic microscope. *J. Microsc.* 205: 165–176.

Lessor, D. L., Hartman, J. S., and Gordon, R. L. 1979. Quantitative surface topography determination by Nomarski reflection microscopy. I. Theory. *J. Opt. Soc. Am.* 69: 357–366.

Li, H. W., McCloskey, M., He, Y., and Yeung, E. S. 2007. Real-time dynamics of label-free single mast cell granules revealed by differential interference contrast microscopy. *Anal. Bioanal. Chem.* 387: 63–69.

Liu, Z., Dong, Chen, Q. et al. 2004. Nondestructive measurement of an optical fiber refractive-index profile by a transmitted-light differential interference contact microscope. *Appl. Opt.* 43: 1485–1492.

Loveland, R. P. 1981. *Photomicrography: A Comprehensive Treatise.* Malabar, FL: Robert E. Krieger Publishing Company.

Lu, N., Yu, X., He, X., and Zhou, Z. 2009. Detecting apoptotic cells and monitoring their clearance in the nematode *Caenorhabditis elegans. Methods Mol. Biol.* 559: 357–370.

Lue, N., Choi, W., Popescu, G. et al. 2007. Quantitative phase imaging of live cells using fast Fourier phase microscopy. *Appl. Opt.* 46: 1836–1842.

Lue, N., Choi, W., Popescu, G. et al. 2008a. Confocal diffraction phase microscopy of live cells. *Opt. Lett.* 33: 2074–2076.

Lue, N., Choi, W., Popescu, G. et al. 2008b. Synthetic aperture tomographic phase microscopy for 3D imaging of live cells in translational motion. *Opt. Express* 16: 16240–16246.

Mantilla, G. S., Danser, M. M., Sipos, P. M. et al. 2001. Tongue coating and salivary bacterial counts in healthy/gingivitis subjects and periodontitis patients. *J. Clin. Periodontol.* 28: 970–978.

Marquet, P., Rappaz, B., Magistretti, P. J. et al. 2005. Digital holographic microscopy: A noninvasive contrast imaging technique allowing quantitative visualization of living cells with subwavelength axial accuracy. *Opt. Lett.* 30: 468–470.

Maurer, C., Jesacher, A., Bernet, S., and Marte, M. R. 2008. Phase contrast microscopy with full numerical aperture illumination. *Opt. Express* 16: 19821–19829.

Mehta, S. B. and Sheppard, C. J. 2008. Partially coherent image formation in differential interference contrast (DIC) microscope. *Opt. Express* 16: 19462–19479.

Molecular Expressions. 2009a. Retrieved from http://micro.magnet.fsu.edu/micro/about.html, accessed on 7/13/2009.

Molecular Expressions. 2009b. Darkfield illumination. Retrieved from http://micro.magnet.fsu.edu/primer/techniques/darkfieldindex.html, accessed on 7/13/2009.

Molecular Expressions. 2009c. Gallery of darkfield images. Retrieved from http://micro.magnet.fsu.edu/primer/techniques/darkfieldgallery.html, accessed on 7/13/2009.

Molecular Expressions. 2009d. Set up of darkfield microscopy. Retrieved from http://micro.magnet.fsu.edu/primer/techniques/darkfieldsetup.html, accessed on 7/13/2009.

Molecular Expressions. 2009e. Troubleshooting darkfield microscopy. Retrieved from http://micro.magnet.fsu.edu/primer/techniques/darkfieldtrouble.html, accessed on 7/13/2009.

Munro, P. R. T. and Torok, P. 2005. Vectorial, high numerical aperture study of Nomarski's differential interference contrast microscope. *Opt. Express* 13: 6833–6847.

Murphy, D. B. 2001. *Fundamentals of Light Microscopy and Electronic Imaging.* New York: Wiley-Liss.

Nahamoo, D., Pan, S. X., and Kak, A. C. 1984. Synthetic aperture diffraction tomography and its interpolation-free computer implementation. *Sonics Ultrasound* 31: 218–229.

Nemoto, I. 1988. Three-dimensional imaging in microscopy as an extension of the theory of two-dimensional imaging. *J. Opt. Soc. Am. Opt. Image Sci. Vis.* 5: 1848–1851.

Newmark, J. A., Warger, W. C. II, Chang, C. et al. 2007. Determination of the number of cells in preimplantation embryos by using noninvasive optical quadrature microscopy in conjunction with differential interference contrast microscopy. *Microsc. Microanal.* 13: 118–127.

Nikon brochure. 2007. Objectives for biological microscopes. Retrieved from http://www.nikoninstruments.com/images/stories/PDFs/CFI60_Optics_Brochure.pdf, accessed on 7/13/2009.

Nikon Microscopy U. 2009. The source for microscopy education. Retrieved from http://www.microscopyu.com/, accessed on 7/13/2009.

Noda, T., Kawata, S., and Minami, S. 1990. Three-dimensional phase contrast imaging by an annular illumination microscope. *Appl. Opt.* 29: 3810–3815.

Noda, T., Kawata, S., and Minami, S. 1992. Three-dimensional phase-contrast imaging by a computed-tomography microscope. *Appl. Opt.* 31: 670–674.

Nomarski, G. 1955. Microinterf'erom`etre differential 'a ondes polaris'ees. *J. Phys. Rad.* 16: 9s.

O'Sullivan, J. A. and Preza, C. 2008. Alternating minimization algorithm for quantitative differential interference contrast microscopy. *Proc. SPIE* 6814: 6814.

Otaki, T. 2000. Artifact halo reduction in phase contrast microscopy using apodization. *Opt. Rev.* 7: 119–122.

Paganin, D. and Nugent, K. A. 1998. Noninterferometric phase imaging with partially coherent light. *Phy. Rev. Lett.* 80: 2586–2589.

Pavani, S. R. P., Libertun, A. R., and Cogswell, C. J. 2007. Structured-illumination quantitative phase microscopy. In: *Adaptive Optics: Analysis and Methods/Computational Optical Sensing and Imaging/Information Photonics/Signal Recovery and Synthesis Topical Meetings.* Washington, DC: Optical Society of America, pp. CMB4.

Pavani, S. R. P., Libertun, A. R., King, S. V., and Cogswell, C. J. 2008. Quantitative structured-illumination phase microscopy. *Appl. Opt.* 47: 15–24.

Peltroche-Llacsahuanga, H., Reichhart, E., Schmitt, W., Lütticken, R., and Haase, G. 2000. Investigation of infectious organisms causing pericoronitis of the mandibular third molar. *J. Oral Maxillofac. Surg.* 58: 611–666.

Pluta, M. 1988. *Advanced Light Microscopy: Principles and Basic Properties.* Warszawa: PWN—Polish Scientific Publishers.

Pluta, M. 1989a. *Advanced Light Microscopy: Specialized Methods,* pp. 146–197. Amsterdam, the Netherlands: Elsevier.

Pluta, M. 1989b. *Advanced Light Microscopy: Specialized Methods.* Amsterdam, the Netherlands: Elsevier.

Popescu, G., DeFlores, L. P., Vaughan, J. C. et al. 2004. Fourier phase microscopy for investigation of biological structures and dynamics. *Opt. Lett.* 29: 2503–2505.

Pot, S. A., Chandler, H., Colitz, C. M. et al. 2009 (article in press). Selenium functionalized intraocular lenses inhibit posterior capsule opacification in an ex vivo canine lens capsular bag assay. *Exp. Eye Res.*

Preza, C. 1998. Phase estimation using rotational diversity for differential interference contrast microscopy. Sever Institute of Technology, St. Louis, MO: Washington University.

Preza, C. 2000. Rotational diversity phase estimation from differential interference contrast microscopy images. *J. Opt. Soc. Am. Opt. Image Sci. Vis.* 17: 415–424.

Preza, C., King, S. V., and Cogswell, C. J. 2006. Algorithms for extracting true phase from rotationally-diverse and phase-shifted DIC images. In: *Proceedings of the SPIE: 3D and Multidimensional Microscopy: Image Acquisition and Processing XIV,* eds. Conchello, J. A, Cogswell, C. J., Wilson, T., p. 60900E.

Preza, C. and O'Sullivan, J. A. 2007. Quantitative determination of specimen properties using computational differential-interference contrast (DIC) microscopy. In: *Proceedings of the SPIE*, eds. Wilson, T. and Periasam, A., p. 66300E.

Preza, C. and O'Sullivan, J. A. 2009. Quantitative phase and amplitude imaging using differential-interference contrast (DIC) microscopy. In: *Proceedings of the SPIE*, eds. Bouman, C. A., Miiller, E. L., and Pollak, I., p. 724604.

Preza, C. and O'Sullivan, J. A. 2010. Implementation and evaluation of a penalized alternating minimization algorithm for computational DIC microscopy. In: *Proceedings of the SPIE*. eds. Bouman, C. A., Pollak, I., and Wolfe, P. J., pp. 75330E–75330E-11.

Preza, C., Snyder, D. L., and Conchello, J.-A. 1996. Imaging models for three-dimensional transmitted-light DIC microscopy. In: *Proceedings of the IS&T/SPIE Symposium on Electronic Imaging, Science and Technology*, eds. Cogswell, C. J., Kino, G. S., and Wilson, T., pp. 245–257.

Preza, C., Snyder, D. L., and Conchello, J. A. 1997a. Image reconstruction for three-dimensional transmitted-light DIC microscopy. In: *Proceedings of the SPIE*, eds. Cogswell, C. J., Conchello, J.-A., and Wilson, T., pp. 220–231.

Preza, C., Snyder, D. L., and Conchello, J.-A. 1999. Theoretical development and experimental evaluation of imaging models for differential interference contrast microscopy. *J. Opt. Soc. Am. Opt. Image Sci. Vis.* 16: 2185–2199.

Preza, C., Snyder, D. L., Rosenberger, F. U., Markham, J., and Conchello, J. A. 1997b. Phase estimation from transmitted-light DIC images using rotational diversity. In: *Proceedings of the SPIE*. pp. 97–107.

Preza, C., van Munster, E. B., Aten, J. A., Snyder, D. L., and Rosenberger, F. U. 1998. Determination of direction independent optical path length distribution of cells using rotational diversity transmitted light differential interference contrast (DIC) images. *Proc. SPIE* 3261A: 60–70.

Ramachandran, G. N. and Lakshminarayanan, L. V. 1971. Three-dimensional reconstruction from radiographs and electron micrographs: Application of convolution instead of fourier transforms. *Proc. Natl. Acad. Sci. US.* 68(9): 2236–2240.

Rappaz, B., Barbul, A., Hoffmann, A. et al. 2009. Spatial analysis of erythrocyte membrane fluctuations by digital holographic microscopy. *Blood Cells Mol. Dis.* 42: 228–232.

Richards, B. and Wolf, E. 1959. Electromagnetic diffraction in optical systems II. Structure of the image field in an aplanatic system. *Proc. R. Soc. Lond. Ser. Math. Phy. Sci.* 253: 358–379.

Ross, K. 1976. *Phase Contrast and Interference Microscopy for Cell Biologists*. London, U.K.: Edward Arnold Publishers.

Saleh, B. E. A. and Teich, M. C. 1991. *Fundamentals of Photonics*, John Wiley & Sons, Inc., New York.

Salmon, E. D. and Tran, P. 2007. High-resolution video-enhanced differential interference contrast light microscopy. In: *Methods in Cell Biology, 81, Digital Microscopy*, eds., Sluder, G. and Wolf, D. E., 3rd Edition, pp. 335–364, San Diego: Academic Press of Elsevier.

Schmelzeisen, M., Austermann, J., and Kreiter, M. 2008. Plasmon mediated confocal dark-field microscopy. *Opt. Express* 16: 17826–17841.

Shaw, P. J. 1990. Three-dimensional optical microscopy using tilted views. *J. Microsc.* 158: 165–172.

Sheppard, C. J., Gu, M., Kawata, Y., and Kawata, S. 1994. Three-dimensional transfer functions for high-aperture systems. *J. Opt. Soc. Am. Opt. Image Sci. Vis.* 11: 593–598.

Sheppard, C. J. R. and Mao, X. Q. 1989. Three-dimensional imaging in a microscope. *J. Opt. Soc. Am. Opt. Image Sci. Vis.* 6: 1260–1269.

Shimada, W., Sato, T., and Yatagai, T. 1990. Optical surface microtopography using phase shifting Normaski microscope. *Proc. SPIE* 1332: 525–529.

Shribak, M. 2005. Orientation-independent differential interference contrast microscopy technique and device. U.S. Patent Application. 0152030 U.S.A.

Shribak, M. and Inoue, S. 2006. Orientation-independent differential interference contrast microscopy. *Appl. Opt.* 45: 460–469.

Shribak, M. and LaFountain, J. 2008. Orientation-independent differential interference contrast microscopy and its combination with an orientation-independent polarization system. *J. Biomed. Opt.* 13: 014011.

Sierra, H., DiMarzio, C. A., and Brooks, D. H. 2009a. Modeling phase microscopy of transparent three-dimensional objects: A product-of convolutions approach. *J. Opt. Soc. Am.* 26: 1268–1276.

Sierra, H., DiMarzio, C. A., and Brooks, D. H. 2009b. Modelling phase microscopy of transparent three-dimensional objects: A product-of convolutions approach. *J. Opt. Soc. Am.* 26: 1268–1276.

Song, L., Maslov, K., Bitton, R., Shung, K. K., and Wang, L. V. 2008. Fast 3-D dark-field reflection-mode photoacoustic microscopy in vivo with a 30-MHz ultrasound linear array. *J. Biomed. Opt.* 13: 054028.

Stephens, D. J. and Allan, V. J. 2003. Light microscopy techniques for live cell imaging. *Science*: 82–86.

Streibl, N. 1985. Three-dimensional imaging by a microscope. *J. Opt. Soc. Am. Opt. Image Sci. Vis.* 2: 121–127.

Sung, Y., Choi, W., Fang-Yen, C. et al. 2009. Optical diffraction tomography for high resolution live cell imaging. *Opt. Express* 17: 266–277.

Szekely, D., Yau, T., and Kuchel, P. 2009. Human erythrocyte flickering: Temperature, ATP concentration, water transport, and cell aging, plus a computer simulation. *Eur. Biophys. J.* 38: 923–939.

Takahashi, A. and Nemoto, I. 1988. Inverse problem in microscopy. *Shingakugihou MBE88-58* (in Japanese).

Tam, K. C. and Perezmendez, V. 1981. Tomographical imaging with limited-angle input. *J. Opt. Soc. Am.* 71: 582–592.

Tearney, G. J., Brezinski, M. E., Southern, J. F., and Bouma, B. E. 1995. Determination of the refractive index of highly scattering human tissue by optical coherence tomography. *Opt. Lett.* 20: 2258–2260.

Tishko, T. V., Titar, V. P., Tishko, D. N., and Nosov, K. V. 2008. Digital holographic interference microscopy in the study of the 3D morphology and functionality of human blood erythrocytes. *Laser Phy.* 18: 486–490.

Trattner, S., Feigin, M., Greenspan, H., and Sochen, N. 2009. Validity criterion for the Born approximation convergence in microscopy imaging. *J. Opt. Soc. Am. Opt. Image Sci. Vis.* 26: 1147–1156.

Tsunoda, M., Isailovic, D., and Yeung, E. S. 2008. Real-time three-dimensional imaging of cell division by differential interference contrast microscopy. *J. Microsc. Oxf.* 232: 207–211.

van Munster, E. B., van Vliet, L. J., and Aten, J. A. 1997. Reconstruction of optical pathlength distributions from images obtained by a wide-field differential interference contrast microscope. *J. Microsc. Oxf.* 188: 149–157.

Vishnyakov, G. N. and Levin, G. G. 1998. Optical microtomography of phase objects. *Opt. Spectrosc.* 85: 73–77.

Wang, Z., Millet, L. J., Gillette, M. U., and Popescu, G. 2008. Jones phase microscopy of transparent and anisotropic sample. *Opt. Lett.* 33: 1270–1272.

Wedberg, T. C., Stamnes, J. J. and Singer, W. 1995. Comparison of the filtered bacpropagation and the filtered bacprojection algorithms for quantitative tomography. *Appl. Opt.* 34 (28): 6575–6581.

Wei, N., Flaschel, E., Friehs, K., and Nattkemper, T. W. 2008. A machine vision system for automated non-invasive assessment of cell viability via dark field microscopy, wavelet feature selection and classification. *BMC Bioinformatics* 9: 449.

Wolf, E. 1969. Three-dimensional structure determination of semi-transparent object from holographic data. *Opt. Commun.* 1: 153–156.

Wu, X. X., Kakehi, Y., Jin, X. H., Inui, M., and Sugimoto, M. 2009. Induction of apoptosis in human renal cell carcinoma cells by vitamin E succinate in caspase-independent manner. *Urology* 72: 193–199.

Yang, C., Wax, A., Georgakoudi, I. et al. 2000. Interferometric phase-dispersion microscopy. *Opt. Lett.* 25: 1526–1528.

Zernike, F. 1942a. Phase contrast, a new method for the microscopic observation of phase objects. Part I. *Physica* IX: 686–693.

Zernike, F. 1942b. Phase contrast, a new method for the microscopic observation of transparent objects. Part II. *Physica* IX: 974–986.

25

Confocal Microscopy

25.1 Introduction to Confocal Imaging 517
    Early Developments of Biomedical Confocal Imaging • Confocal Imaging
    Concept • Scanning
25.2 Confocal Imaging Parameters..522
    Resolution • Field of View • Beam Walk • Signal-to-Noise Ratio • Imaging in
    Turbid Tissue
25.3 Instrumentation...530
    Summary of Recent Innovations • Multimodal Microscopy • Line Scanning • Spectral
    Encoding • Coherence Reflectance Confocal Microscopy
25.4 Summary...538
Acknowledgments...538
References...538

William C. Warger, II
Massachusetts General Hospital

Charles A. DiMarzio
Northeastern University

Milind Rajadhyaksha
*Memorial Sloan-Kettering
Cancer Center*

25.1 Introduction to Confocal Imaging

A confocal imaging system creates a high-contrast image of a single plane within an optically transparent or translucent material by blocking most of the out-of-focus light that degrades the quality of the image with conventional microscopy techniques. Reducing out-of-focus light provides the ability to image individual planes (optical sections) within the sample without having to disturb the natural morphology by freezing and cutting the sample into thin slices, as is the current procedure for histology. A confocal microscope uses a point source of light, such as a focused laser beam, to illuminate a spot within the sample. Scattered or fluorescent light from the focal spot images through a point aperture positioned at the image plane of the objective, such that only the light originating from the focal spot passes to the detector to provide the signal for one pixel. The light source, illuminated spot, and detector aperture are in optically conjugate focal planes leading to the name "confocal." A two-dimensional image is created by scanning the focused spot within an area of the sample.

Confocal microscopy has been used extensively in biology and medicine for imaging human and animal tissues in the body (in vivo) and after biopsy (ex vivo) to view morphological, cellular, and nuclear detail with improved resolution. A confocal system generally provides reflectance or fluorescence contrast. In a confocal reflectance system, the signal is proportional to the backscatter of the illumination light provided by native variations in the refractive indices of organelles and microstructures within the tissue (Dunn et al. 1996). In a confocal fluorescence system, the signal is proportional to the excitation of exogenous fluorophores that are administered to label specific microstructures, or endogenous fluorophores from the

microstructures themselves (autofluorescence) (Rajadhyaksha and Gonzalez 2003).

Figure 25.1 illustrates the significance of rejecting out-of-focus light with epi-fluorescence, confocal fluorescence, confocal reflectance, and brightfield images of 1 µm fluorescent beads immersed within a thick gel. The highlighted beads in the focal plane of the fluorescence modalities (left two columns) show strong localized signals from the fluorophore bound to the surface of the beads. However, the signal from each bead blurs over an increasing area in the out-of-focus epi-fluorescence images such that the total power remains constant over the image with a decreased peak irradiance. The effect is more apparent with the signal from the out-of-focus beads that contribute to the strong "background" signal that reduces the overall contrast of the image. The signal from out-of-focus beads in the confocal images also blur over a larger area, but the peak irradiance decreases faster than the area increases. The total power from each bead then decreases with defocus (optical sectioning) thereby reducing the contribution of the out-of-focus beads and improving the contrast and resolution. The right two columns provide images with contrast related to the scattering of the illumination light from the refractive-index difference between the bead and the gel. The confocal reflectance images in the third column provide high contrast, similar to the confocal fluorescence images. However, the beads in the brightfield image are much more difficult to differentiate because the signals from the beads add coherently. The field at each pixel is then the superposition of the unscattered field and the scattered fields resulting in a high background and poor contrast and resolution.

Within this chapter, we provide a conceptual overview of confocal imaging within Section 25.1, a brief discussion of the

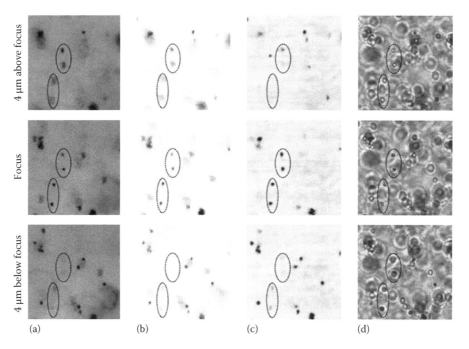

4 μm above focus | Focus | 4 μm below focus

(a)   (b)   (c)   (d)

FIGURE 25.1 Multimodal images of fluorescent beads with (a) epi-fluorescence, (b) confocal fluorescence, (c) confocal reflectance, and (d) brightfield to demonstrate optical sectioning provided by confocal microscopes. Three successive images shown in each column separated in axial direction by 4 μm with dashed ellipses to highlight same two beads in each image. Colorscale of epi-fluorescence and confocal images have been inverted to match colorscale of brightfield transmission image. Each image was collected with 20x, 0.75 NA objective on Keck 3D Fusion Microscope with 0.5 μm pixel size and 100 × 100 pixels area of interest. This equipment is discussed in Warger et al. 2007a.

underlying theory and characterization in Section 25.2, and a summary of technological developments and recent advancements in Section 25.3. The reader interested in more detail will benefit from several excellent books (Wilson 1990, Gu 1996, Sheppard and Shotton 1997, Diaspro 2002, Pawley 2006) and the references included herein.

25.1.1 Early Developments of Biomedical Confocal Imaging

The confocal imaging microscope was first conceptualized and patented by Marvin Minsky at Harvard University in 1955 (Minsky 1955, 1988). A nonimaging confocal system was developed by Hiroto Naora at the University of Tokyo in 1951 (Naora 1951), but Minsky has been credited as the founder of the microscopy technique. Petran developed the first confocal microscope for use within a biological application in 1967, which used the sun as a light source to image excised dorsal root ganglia of frogs and observe the brains of anesthetized salamanders for extended time periods up to 5 h (Egger and Petran 1967, Petran and Hadravsky 1970). Brakenhoff, Sheppard, and Wilson developed much of the theory underlying confocal imaging during the 1970s (Sheppard and Choudhury 1977, Sheppard and Wilson 1978, 1979, Brakenhoff et al. 1979). However, it was not until the 1980s that confocal imaging found wide acceptance in the biology community when the first commercial device was made available that incorporated laser illumination within a confocal fluorescence imaging system and used a computer for

data collection and display (Amos and White 2003). Since the mid-1980s, confocal imaging systems have continued to become faster, smaller, and more reliable, with subsequent development of diverse applications within the biomedical community.

25.1.2 Confocal Imaging Concept

In a conventional transmission microscope, a light source illuminates the entire field of view for full-field imaging. A schematic for such a system using a non-infinity-corrected objective lens, which provides a fixed distance between the objective and the image plane (tube or focal length), is shown in Figure 25.2. An infinity-corrected objective images the object at infinity and is used in conjunction with a second lens (tube lens) to create an intermediate image plane. A point object located at the object plane of the objective (within the sample) will image to a diffraction-limited spot in the image plane of the objective (-), where a two-dimensional detector is mounted. Identical point objects positioned along the optic axis and in out-of-focus transverse planes before or after the object plane will image to planes after or before the image plane (--). Assuming a diffraction-limited system, and a uniform, nonabsorbing sample, the irradiance from each of the point objects will be the same at the image plane, but the light from the out-of-focus point objects will spread across a larger area. The spreading of the light in transverse planes through the focal region of a lens is shown in Figure 25.3. A point object from the object plane of the objective will appear in focus, but the spreading of light from the out-of-focus

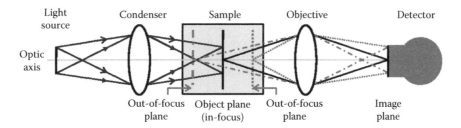

FIGURE 25.2 Full-field imaging system. Light source illuminates two-dimensional area with condenser lens, and objective lens images each point within two-dimensional field of view (-) onto two-dimensional detector array. Light from point objects positioned in out-of-focus planes (--) spreads over detector area and degrades contrast of in-focus image.

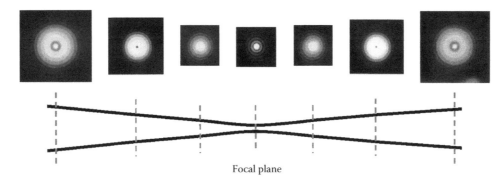

FIGURE 25.3 Intensity distribution of light within transverse planes through focus of a lens. (Adapted from Cagnet, M., *Atlas of Optical Phenomena*, Prentice Hall, Englewood Cliffs, NJ, 1962, p. 23.)

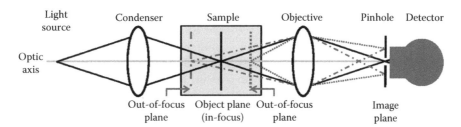

FIGURE 25.4 Confocal imaging system. Condenser lens illuminates focused spot within sample with point source of light. Objective lens focuses light from a point at object plane to spot at the image plane (-), while out-of-focus points (--) are in focus in planes before and after the image plane. Pinhole blocks most of the out-of-focus light that is spread across image plane and passes light from illuminated spot in object plane through pinhole to detector.

planes will contribute to blur that inhibits the ability to discriminate one transverse plane in the sample from another. The thickness of the section that appears in focus is referred to as the depth of field of the imaging system and will be discussed in Section 25.2.1.3. It is important to note that the distribution of the light at the focal plane in Figure 25.3 corresponds to a two-dimensional Airy pattern, and the finite diameter between the first zeros from center (Airy disk) is dependent on the diffraction limit of the optics, as will be discussed in Section 25.2.1.

In a transmission confocal microscope shown in Figure 25.4, the condenser lens focuses the light source to a diffraction-limited spot within the sample (creating a single-point imaging system) and a pinhole is mounted in the image plane. The light source is then optically conjugate to the focused spot within the sample *via* the condenser lens, because the light source is in the

object plane and the focused spot is in the image plane (conjugate image planes), and the focused spot is optically conjugate to the pinhole *via* the objective lens. Light from point objects along the optic axis passes through the objective as in the case for full-field imaging, but the pinhole blocks most of the light from the out-of-focus objects that spread across a larger area in the image plane, and allows the light from the point object at the object plane to pass through to the detector. In theory, the size of the pinhole (detector aperture) is matched to the size of the illuminated spot, thereby blocking most of the light that does not originate from the object plane.

Confocal microscopes generally utilize a reflection or epi-illumination imaging configuration, as shown in Figure 25.5, where the objective lens illuminates and collects the backscattered or fluorescent signal within the first 100–500 μm of the

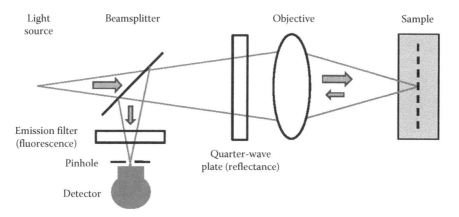

Light   Beamsplitter   Objective   Sample
source

Emission filter
(fluorescence)

Pinhole

Detector

Quarter-wave
plate (reflectance)

FIGURE 25.5 Reflection or epi-illumination confocal imaging system. Objective lens focuses illumination light to a focused spot within sample, and collects resultant backscattered or emitted fluorescent light. Dichroic BS separates illumination wavelengths from emission wavelengths in confocal fluorescence system, and PBS and quarter-wave plate separate illumination light and backscattered light in confocal reflectance system.

sample surface depending on the numerical aperture (NA) of the objective, the wavelength of the illumination light, and the turbidity of the sample (Smithpeter et al. 1998b, Gareau et al. 2008b). A confocal fluorescence microscope uses a dichroic beamsplitter (BS), which is designed to reflect a range of wavelengths while transmitting others, to separate the excitation wavelengths that illuminate the sample (illumination path) and the emission (fluorescence) wavelengths that are emitted from the sample (detection path). A confocal reflectance microscope typically uses polarization schemes to separate the illumination and detection paths and also increase the throughput of the system. The illumination light transmits through a polarizing beamsplitter (PBS) with *P*-polarization and becomes circularly polarized after transmitting through a quarter-wave plate with its optic axes oriented at 45° with respect to the laser polarization. The circularly polarized light illuminates a spot within the sample and the backscattered light from the object plane passes back through the objective and quarter-wave plate. On the return path, the light has passed through the quarter-wave plate a second time converting the incident *P*-polarized light into *S*-polarized light that will reflect from the PBS, provided the returning light retains its state of polarization as it interacts with the sample, and then focus through the pinhole to the detector. The detected light is then primarily single-scattered (ballistic) light that retains circular polarization before passing back through the quarter-wave plate and a small fraction of the multiply scattered light that returns randomly polarized.

25.1.3 Scanning

The benefit of confocal imaging over conventional full-field imaging techniques is the significant reduction of the out-of-focus signal that otherwise, under conventional (non-confocal) conditions, decreases the contrast and degrades the resolution of an image. However, confocal imaging is inherently slower than full-field imaging because each pixel is typically acquired separately, while all of the pixels are captured simultaneously

in a full-field imaging system. The time required to capture one confocal image (acquisition time) is primarily limited by the speed at which the illumination point scans across each pixel within the field of view and the acquisition speed of the electronics.

Some confocal microscopes use a fixed optical path and physically translate the sample incrementally. This technique provides diffraction-limited imaging throughout the field of view within a very stable, compact system, but is limited to objects that can be translated on a microscope stage. In addition, potential vibrations induced by sudden starts and stops of the stage limit the scanning speed because the sample may move between the collection of individual pixels. This limitation reduces the ability to scan the sample quickly and generally produces long acquisition times on the order of seconds for a field of view comparable to a conventional microscope.

To increase the versatility of confocal imaging, microscope developers have used beam-steering techniques to keep the sample stationary while the illumination light scans to each pixel location within the sample. The fundamental concept behind most scanning techniques is the relation of tilt and translation between the back-focal (pupil) and front-focal planes of a lens. A collimated beam traveling along the optic axis will focus to a spot at the intersection of the optic axis and the focal plane of the lens, as shown in Figure 25.6. Applying a tilt of angle β to the collimated beam, such that the beam is still centered on the optic axis at the pupil plane of the lens, results in a translation $L/2$ of the focused spot from the optic axis in the focal plane, where $L = 2f \tan \beta / M$ is the total distance scanned in one dimension of the field of view for a system with magnification M. Each pixel within the field of view can then be illuminated by adjusting the angle of the collimated beam in the pupil plane of the lens.

Traditionally, the pupil plane was always positioned at the shoulder of the objective housing where the threads meet the barrel. Recently, however, objective manufacturers are embedding the pupil within the objective housing, making it less accessible. Unless an on-axis beam-scanning mechanism is

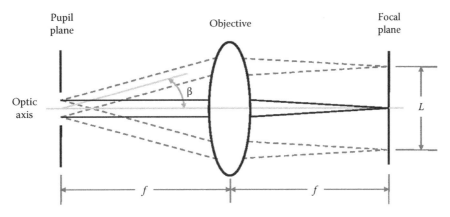

FIGURE 25.6 Concept of optical scanning. Collimated beam parallel to the optic axis focuses to spot on optic axis at focal length *f* of the lens. Collimated beam tilted at angle β and centered on optic axis in the pupil plane of the lens focuses to spot *L/2* from optic axis at focal length of the lens.

incorporated inside the objective housing, the pupil must be relayed to a conjugate plane outside of the objective housing in order to pivot the beam in the pupil while taking advantage of the full-pupil diameter.

Most confocal microscopes have incorporated a raster scan where the focused spot incrementally scans across the field of view in one direction (fast scan) while scanning one pixel in the orthogonal direction (slow scan) to sample a rectangular field of view. One technique to create a raster scan uses two single-axis pivoting mirrors (either a polygon and a galvanometric (galvo) mirror or a pair of galvanometric mirrors), where one scans the "horizontal" direction and the second scans the "vertical" direction. However, this technique requires two relay telescopes for both mirrors to pivot exactly in a pupil plane. The optical layout for a confocal reflectance microscope that utilizes two galvanometric scanners is shown in Figure 25.7. Collimated illumination light exits the laser and passes through the PBS. The first scanner reflects the beam through a deviation angle $\pm\beta_1$ to sample

points along a one-dimensional scan line. Each point within the scan line travels through the first relay telescope between the two scanners such that the pivot plane of the first scanner (P_1) and the pivot plane of the second scanner (P_2) are optically conjugate. The second scanner then reflects each point along the scan line through a second deviation angle $\pm\beta_2$ in the orthogonal direction to sample points throughout a two-dimensional scan area. Each point within the scan area travels through the second relay telescope between the second scanner and the quarter-wave plate such that the pivot plane of the second scanner (P_2) and the pupil plane of the objective (P_3) are optically conjugate. This optical configuration ensures that the beam will always be centered on the optic axis at the pupil plane of the objective and have a deviation angle dependent on the combined contributions of both scanners (through the magnification of the relay telescope). Each point fills the back pupil plane of the objective lens to use the full numerical aperture (NA) and illuminates a single point within the sample at the object plane O_1.

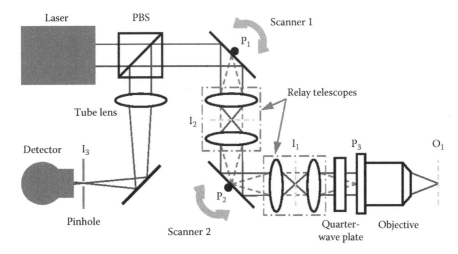

FIGURE 25.7 Optical layout for confocal reflectance microscope that uses two scanners to scan focused spot in object plane O_1. Two relay telescopes produce two optically conjugate pupil planes P_1 and P_2, such that the two one-dimensional scanners can scan the beam in the inaccessible pupil plane of the objective.

The backscattered light collected at the front-focal plane and within the NA of the infinity-corrected objective lens returns collimated and passes back through the quarter-wave plate and relay telescope. After the backscattered light reflects from Scanner 2, the light is descanned from a two-dimensional scan area to a one-dimensional scan line. In turn, the scan line is descanned by Scanner 1 to a single stationary point. The descanned light reflects from the PBS and the tube lens focuses the light through a stationary pinhole to a point detector.

The scanners and relay telescopes add size and weight to the overall instrument and generally result in a large, bulky device. Alternative techniques have been developed to reduce the size and increase the speed of confocal scanning systems, as will be discussed in Section 25.3. It is important to note that some full-field imaging techniques, such as structured illumination (Neil et al. 1997, Gustafson 2000) and deconvolution (McNally et al. 1999, Wallace et al. 2001), provide optical sectioning capabilities. However, these techniques are more complicated than conventional full-field microscopes.

25.2 Confocal Imaging Parameters

25.2.1 Resolution

The theory underlying the resolution of a confocal microscope becomes complex when all of the various parameters are included, such as the shape and illumination of the pupil, NA of the objective, and shape and size of the aperture in front of the detector. Here we provide a brief overview of the mathematics for the most common point-scanning confocal system that is circularly symmetric and uses a circular pinhole aperture.

25.2.1.1 Conventional Full-Field Microscopes

The performance of any imaging instrument is generally described in terms of the system's point spread function (PSF). The PSF of a lens corresponds to the shape of the focused spot, which includes geometric optics and diffraction effects, and provides a measurement of the instrument's ability to image a point object.

Mathematically, the PSF of a lens provides the amplitude distribution of the field in the planes near the geometric focal point:

$$h(u,v) = U(u,v) = -i \frac{2\pi}{\lambda} \frac{NA^2}{n} A e^{i\frac{n^2}{NA^2}u} \int_0^1 J_0(v\rho) e^{-i\frac{1}{2}u\rho^2} \rho \, d\rho,$$

$$(25.1)$$

assuming the lens has uniform, plane-wave illumination in the pupil plane and follows the paraxial approximation, and where J_0 is the zeroth-order Bessel function, i is $\sqrt{-1}$, n is the refractive index of the immersion medium for the lens, and A and ρ are the amplitude of the field and the radial coordinate centered on the optic axis in the pupil plane, respectively (Born and Wolf 1999). From a linear systems perspective (Gaskill 1978), the PSF is often considered the impulse response of the imaging system for a point object, which is generally described by the variable h,

so the variables $U(u,v)$ and $h(u,v)$ have been used interchangeably in the literature, depending on whether the author is describing the PSF of an imaging system $h(u,v)$, or the electromagnetic field within the focal region of a lens $U(u,v)$. The optical coordinates v and u near the focal plane of the lens relate to the transverse (lateral) direction from the optic axis (r) and the axial direction parallel to the optic axis (z) by

$$v = \frac{2\pi}{\lambda} NA r, \quad u = \frac{2\pi}{\lambda} \frac{NA^2}{n} z. \qquad (25.2)$$

The optical coordinates in Equation 25.2 are provided in terms of NA to stay consistent with the resolution equations that are typically provided in terms of NA. However, the optical units are often expressed in terms of the half angle of illumination θ or the radius of the pupil a and focal length of the lens f from the relation $a/f \approx \sin \theta$, which results from the paraxial approximation ($\theta \approx \sin \theta \approx \tan \theta$). The product of the half angle of illumination and the refractive index of the immersion medium after the lens n corrects for refraction and provides the numerical aperture NA $\approx n \sin \theta$. The lens does not follow the paraxial approximation for high NA, and the effect will be discussed in Section 25.2.1.6. It is important to emphasize that any physical distance can be converted into optical units by substitution into Equation 25.2.

Equation 25.1 is also considered the coherent point spread function because it includes the magnitude and phase information. Because the amplitude in Equation 25.1 cannot be seen or measured directly, the PSF is often described by the incoherent point spread function

$$I(u,v) = |h(u,v)|^2 = I_0 \left| \int_0^1 J_0(v\rho) e^{-i\frac{1}{2}u\rho^2} \rho \, d\rho \right|^2, \qquad (25.3)$$

where

$$I_0 = \left(\frac{2\pi}{\lambda} \frac{NA^2}{n} A \right)^2 \qquad (25.4)$$

is the irradiance at the geometric focal point $u = v = 0$.

In a conventional microscope, the condenser lens illuminates the entire field of view, and the resolution is primarily based on the objective's ability to resolve point objects. The optics within the system are generally assumed to be diffraction limited with the objective lens as the limiting resolving element, so the PSF of the microscope is simply the PSF of the objective lens in Equation 25.3. Assuming uniform plane-wave illumination in the pupil plane and substituting the relation $\int_0^x x' J_0(x') dx' = x J_1(x)$, the irradiance distribution of the focused spot along the lateral and axial axes normalized by I_0 reduces to

$$I_{conv}(0,v) = \left[\frac{2J_1(v)}{v} \right]^2 \qquad (25.5)$$

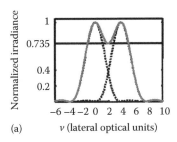
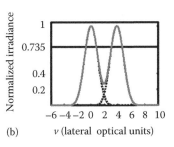
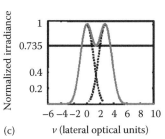

(a) v (lateral optical units) (b) v (lateral optical units) (c) v (lateral optical units)

FIGURE 25.8 (a) Two Airy patterns (--) representing image (-) of two point objects displaced by minimum resolvable distance ($v = 3.83$) using uniformly illuminated circular aperture in conventional microscope. (b) Two Airy patterns for same two point objects within confocal microscope separated by minimum resolvable distance in conventional microscope. (c) Two Airy patterns from two points sources in confocal microscope separated by minimum resolvable distance ($v = 2.77$) in confocal microscope.

$$I_{conv}(u,0) = \left[\frac{\sin(u/4)}{u/4}\right]^2, \qquad (25.6)$$

respectively, where $I_{conv}(0,v)$ is a one-dimensional Airy pattern. The Rayleigh criterion for resolution states that two point objects of equal irradiance are resolvable if the points are separated by the radius of the Airy disk, where the diameter of the Airy disk is defined as the distance between the first zeros of the Airy function. Thus, according to the Rayleigh criterion, the lateral resolution of a conventional microscope is determined when $J_1(v) = 0$, or $v = (2\pi/\lambda)\,\mathrm{NA}\,r = 3.83$, which reduces to the well-known

$$\Delta x_{conv} = 0.61\frac{\lambda}{\mathrm{NA}} \qquad (25.7)$$

for an evenly illuminated circular aperture. Following the same procedure in the axial direction provides the axial resolution

$$\Delta z_{conv} = 2\frac{n\lambda}{\mathrm{NA}^2}. \qquad (25.8)$$

It is important to emphasize that the axial resolution should not be confused with optical sectioning. Optical sectioning is dependent on the integrated irradiance across each transverse plane through the focused spot (Sheppard and Wilson 1978), and is a measure of the sample thickness that contributes to an image. As will be discussed in Section 25.2.1.3, a conventional microscope does not have optical sectioning capabilities because every transverse plane contributes to the final image. To distinguish axial resolution from optical sectioning, it may be helpful to think of the axial resolution in Equation 25.8 as the smallest distance two point objects can be separated and still be resolved along the axial direction, and optical sectioning as the ability to distinguish one layer of such objects from the layers above and below, where a single layer (optical section) may contain many point objects separated by an axial distance equal to the axial resolution.

Resolution may also be interpreted according to the saddle-to-peak irradiance ratio, where the two point objects are resolvable

if the dip in the summed irradiance from the two Airy functions reaches a critical value (Born and Wolf 1999). Figure 25.8a shows the plot for two Airy functions from Equation 25.5 separated by the Rayleigh criterion, and shows that the two points will be resolved if the saddle-to-peak irradiance ratio is 0.735 or the total signal drops 26.5% from the peak irradiance.

25.2.1.2 Conventional Confocal Microscopes

In a confocal microscope, the condenser lens focuses the illumination light to a diffraction-limited spot according to Equation 25.3, and the objective lens images the resulting scattered or fluorescent light from the illumination spot to a diffraction-limited spot at the pinhole. The PSF of a confocal microscope is then the PSF of the condenser lens multiplied by the PSF of the objective lens. In typical epi-illumination confocal systems shown in Figure 25.5, the objective lens acts as both the condenser and imaging lenses and the normalized irradiance distribution of the focused spot along the lateral and axial axes reduces to

$$I_{confocal}(v) = \left[\frac{2J_1(v_{ex})}{v_{ex}}\right]^2\left[\frac{2J_1(v_{em})}{v_{em}}\right]^2 = \left[\frac{2J_1(v)}{v}\right]^4 \qquad (25.9)$$

$$I_{confocal}(u) = \left[\frac{\sin(u_{ex}/4)}{u_{ex}/4}\right]^2\left[\frac{\sin(u_{em}/4)}{u_{em}/4}\right]^2 = \left[\frac{\sin(u/4)}{u/4}\right]^4,$$

$$(25.10)$$

where the subscripts ex and em correspond to the optical units for the excitation and emission wavelengths in a confocal fluorescence system, respectively, and the optical units without subscripts correspond to a confocal reflectance system where the same wavelength is used for illumination and detection. Strictly following the Rayleigh criterion, the resolution of a confocal microscope would be considered the same as a conventional microscope because the first minima of Equations 25.9 and 25.10 are the same as the first minima of Equations 25.7 and 25.8. However, the plot in Figure 25.8b shows that two point objects separated by the Rayleigh criterion in a confocal microscope have a much larger saddle-to-peak irradiance ratio and will be much easier to distinguish than two points in a conventional microscope. Figure 25.8c shows that by following the 26.5% dip in irradiance, the two points will be resolved in a confocal reflectance system if they are separated by at least $\nu = (2\pi/\lambda)\text{NA}$ $r = 2.77$, which results in a lateral resolution

$$\Delta x_{\text{confocal}} = 0.44\frac{\lambda}{\text{NA}}, \qquad (25.11)$$

assuming uniform plane-wave illumination in the pupil of the objective and a delta function for the pinhole. Following the same procedure in the axial direction provides the axial resolution:

$$\Delta z_{\text{confocal}} = 1.52\frac{n\lambda}{\text{NA}^2}. \qquad (25.12)$$

25.2.1.3 Optical Sectioning

Optical sectioning describes the ability of a microscope to provide fine details within a small axial thickness of the sample without degradation by out-of-focus light. Another interpretation of optical sectioning is a measure of the sample thickness in the axial direction that contributes to the resultant image. Mathematically (Sheppard and Wilson 1978), optical sectioning

is dependent on the sum of the irradiance distribution of the focused spot over each lateral plane along the axial direction expressed by

$$I_{\text{int}}(u) = \int_0^\infty I(u,v)v\,\mathrm{d}v. \qquad (25.13)$$

The optical section thickness can be measured by plotting the integrated irradiance *versus* the axial coordinates through the focused spot. A contour map of the irradiance distribution through the focused spot of a conventional microscope described by Equation 25.3 is shown in Figure 25.9a, where the dashed lines correspond to the geometric focus of the light described by ray tracing. A plot of the normalized integral of each plane over the field of view in Figure 25.9c shows the contribution of each plane to the resultant image. Theoretically, the plot in Figure 25.9c should be a straight line according to conservation of energy and Parseval's theorem, but the limit on the integral was bounded by the field of view in this example and not infinity as described in Equation 25.13. In practice, the integral will be bounded by the size of the two-dimensional detector used to image the point object. The straight line result of Equation 25.13 means all planes will contribute equally to the image, and for this reason a conventional microscope does not have optical sectioning capabilities.

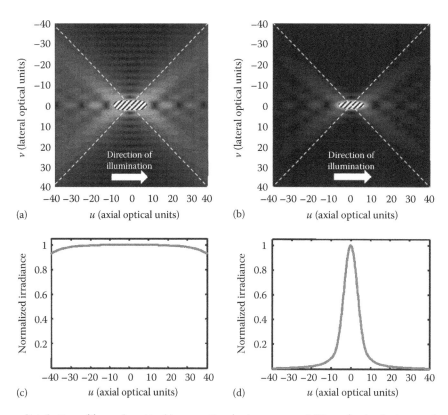

FIGURE 25.9 Irradiance distribution of focused spot in (a) conventional microscope and (b) confocal reflectance microscope (calculated from Born and Wolf 1999). Dashed line (--) corresponds to geometric focal point found by ray tracing, and shaded contours (//) highlight $1/e^2$ volume. Contour lines plotted for normalized irradiances: 0.9, 0.7, 0.5, 0.3, 0.2, 0.135, 0.05, 0.03, 0.02, 0.015, 0.01, 0.005, 0.002, 0.001, 0.0005, 0.0002, 0.0001, 0.00005, and 0.00002. Plot of contribution of each plane to final image from Equation 25.13 that shows (c) conventional microscope does not possess optical sectioning capabilities, while (d) confocal microscope can perform optical sectioning.

A contour map of the irradiance distribution through the focused spot of a confocal reflectance microscope with a pinhole size below the resolution limit, described by the square of Equation 25.3, is shown in Figure 25.9b. A confocal fluorescence system will have a slightly broader result because a longer wavelength will be substituted into the optical units for the fluorescence emission in the detection path (Jonkman and Stelzer 2002). The contour plot in Figure 25.9b shows that the light contributing to the image is from a much smaller area as compared to a conventional microscope. The significant drop in the plot of the normalized integral of each plane over the field of view shown in Figure 25.9d confirms that only the light from a finite thickness of the sample will contribute to the image. There are many techniques used to calculate the optical section thickness from the plot in Figure 25.9d, as will be discussed in Section 25.2.1.7. Taking the measurement of the full width at half maximum (FWHM) of the plot provides a diffraction-limited optical-section thickness of

$$\Delta z_{\text{FWHM}} = 1.36 \frac{n\lambda}{\text{NA}^2} \qquad (25.14)$$

for a confocal reflectance system *using uniform, plane-wave illumination in the pupil of the objective and a delta function for the pinhole*. The result of using a finite-sized pinhole will be described in Section 25.2.1.5.

In a conventional microscope, the depth of field is defined by the ability to discriminate between the in-focus and out-of-focus planes of the sample. A large depth of field corresponds to a larger section of the sample appearing in focus in the final image and *vice versa* for a small or shallow depth of field. The depth of field is increased if the pixel size is greater than the resolution limit. One definition of the depth of field (Inoue and Spring 1997) is

$$d_z = \frac{n\lambda}{\text{NA}^2} + \frac{n}{\text{NA}} d_{\text{pixel}}, \qquad (25.15)$$

where d_{pixel} is the width of a pixel in the object plane found by dividing the physical pixel size by the total magnification of the system. In a confocal microscope, d_{pixel} corresponds to the diameter of the pinhole in the object plane and the depth of field is synonymous with the optical section. The desired depth of field of an imaging system is dependent on the application. The shallow depth of field of a confocal microscope provides fine details within a small section of the sample, while the large depth of field of a conventional microscope provides information about the entire sample, but in less detail.

25.2.1.4 Pupil Apodization

The solution for Equation 25.3 that derives resolution Equations 25.7, 25.8, 25.11, and 25.12 assumes uniform, plane-wave illumination in the pupil of the objective. However, the light sources used in most confocal microscopes contain a Gaussian distribution. Sheppard and Gu have derived the solution for Equation

25.3 with a Gaussian pupil function, and have shown that the changes in the PSF are small assuming the beam fills the pupil of the objective and the pinhole is approximately equal to the size of the focused spot (Sheppard and Gu 1994). When using an infinity-corrected objective lens, we can assume a Gaussian distribution of the collimated beam illuminating the pupil of the objective:

$$I(\rho) = I_0 e^{-2\rho^2 / (h\sigma)^2}, \qquad (25.16)$$

where

I_0 is the irradiance of the beam along the optic axis

h is the fill factor found by dividing the diameter of the beam by the diameter of the pupil

σ is the radius of the collimated beam, assuming the measured beam diameter corresponds to the $1/e^2$ irradiance points.

When the beam under-fills the objective ($h < 1$), less of the NA will be used, thereby reducing the focusing power of the lens and broadening the PSF (worse resolution and optical sectioning). When the beam over-fills the objective ($h > 1$), the beam approaches a plane wave in the pupil and the resolution approaches the ideal case, but the laser power that does not pass through the pupil will be lost. Matching the spot diameter to the pupil diameter ($h = 1$) only loses 14% of the illumination light with a slight broadening of the PSF (Webb 1996).

25.2.1.5 Pinhole Diameter

The confocal resolution Equations 25.11 through 25.13 were calculated assuming the pinhole could be represented as a delta function. As the pinhole diameter increases, more out-of-focus light will pass to the detector resulting in a larger optical section. Wilson and Carlini showed this mathematically (Wilson and Carlini 1987) by imaging the pinhole within the focal region to determine the PSF. An image of an object can be interpreted as the convolution of the object with the PSF of the imaging system. By reciprocity, the same is true from the object to the pinhole, where the convolution of a pinhole function $D(v)$ with the PSF of the optical system provides the PSF for the detection path at the object. A pinhole function is a binary function that describes the shape of the pinhole. Including the PSF of the detection path in Equation 25.13

$$I(u) = \int_0^{v_p} I(u,v) \big[I(u,v) * D(v) \big] v \, dv, \qquad (25.17)$$

provides the optical section equation for a finite pinhole, where $*$ is a convolution and v_p is the physical radius of the pinhole in optical units divided by the total magnification of the system.

The effect of a finite pinhole diameter on the lateral resolution can be determined by substituting the modified PSF for the detection path in Equation 25.9, measuring the FWHM of the Airy function, and dividing by 2π to determine the

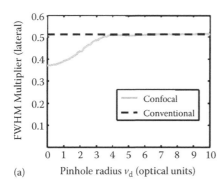
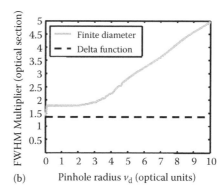

FIGURE 25.10 Effect of pinhole radius on resolution. (a) Lateral resolution FWHM multiplier approaches conventional microscope and (b) optical sectioning FWHM multiplier increases linearly with pinhole diameter ≥3 optical units. (Plots recalculated from Wilson, T. and Carlini, A.R., *Opt. Lett.*, 12, 227, 1987.)

multiplicative factor for λ/NA. A plot of the multiplicative factors for pinhole radii between a delta function ($v_d = 0$) and 10 optical units in Figure 25.10a shows that ideal lateral resolution can be obtained with a pinhole radius ≤1 optical unit, but the lateral resolution approaches that of a conventional microscope as the pinhole radius approaches 4 optical units. The effect of pinhole diameter on the optical sectioning can be determined by plotting the normalized irradiance from the focal plane in Equation 25.17, where an increased pinhole diameter passes more out-of-focus light and creates a larger optical section thickness. The FWHM of the optical sectioning plot divided by 2π provides the multiplicative factor for the FWHM of the optical section thickness in Figure 25.10b. The plots in Figure 25.10a and b show that using a pinhole radius less than 3 optical units provides a slight improvement to the lateral resolution, but the optical sectioning will remain unchanged. As a result, a pinhole radius ≥3 optical units is generally used to collect more light from the sample, and no improvement in lateral resolution is realized with a confocal microscope. It is important to note that the pinhole diameter can also be expressed in terms of Airy units (AU) that are equivalent to the radius of the Airy disk (1 AU = 3.83 optical units).

25.2.1.6 High-NA Objectives

All of the theory presented thus far has made use of the paraxial approximation that allows the radius of the objective pupil divided by the focal length to be approximated by the sine of the half angle of illumination (θ). Strictly speaking, the paraxial approximation is only valid for NA ≤ $1/\sqrt{2}$, but the theory can be extended for use with θ ≤ 60° (NA ≤ 0.87n) (Sheppard and Matthews 1987) after substitution of the axial coordinate

$$u = \frac{8\pi}{\lambda} n z \sin^2 \frac{\theta}{2}, \tag{25.18}$$

and ensuring the refractive index of the sample is greater than the refractive index of the immersion medium n (Wilson and Juskaitis 1995). High-NA objective lenses (NA > 0.87n) require

a more in-depth mathematical treatment that accounts for the vectorial nature of the electromagnetic field and the amplitude variation over the exit pupil (Sheppard and Matthews 1987, Wilson and Juskaitis 1995, Torok and Wilson 1997).

25.2.1.7 Measuring Resolution

A standard resolution metric is necessary to compare the various confocal microscopes that have been developed by different groups around the world. In this section, we provide a brief description of FWHM, $1/e^2$, and 10%–90% edge width (90/10 points), describe how to determine the appropriate multiplicative factors when calculating diffraction-limited resolution, and discuss the most widely accepted techniques to measure resolution.

The irradiance distribution of a collimated laser beam generally corresponds to a Gaussian distribution described by Equation 25.16 and plotted in Figure 25.11a. The FWHM is the distance between the points where the irradiance is equal to 50% of the maximum value, and the $1/e^2$ diameter, where $e \approx 2.718$ is the distance between the points of the Gaussian distribution where the irradiance is approximately equal to 13.5% of the maximum value. The $1/e^2$ diameter is also called the spot size and corresponds to the measurable diameter of the beam. The 90/10 points correspond to the distance between the irradiance points that are at 90% and 10% of the maximum normalized edge-spread function plotted in Figure 25.11b. It is important to note that the experimentally measured edge-spread function will contain noise in the tails of the function and the location of the 90/10 points must be approximated. The edge-spread function derives from a focused spot scanning across an intersection from a highly transmissive surface (minimum signal) to a highly reflective surface (maximum signal) in a confocal reflectance system, and can be described mathematically by the integral of the two-dimensional distribution, which can be interpreted in closed form for a Gaussian. For computational purposes, the discrete form of the edge-spread function can be represented by the sum of the two-dimensional Gaussian in one direction and the cumulative sum in the orthogonal direction.

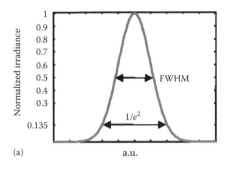 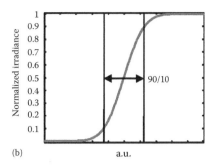

(a)   a.u.   (b)   a.u.

FIGURE 25.11 (a) FWHM and $1/e^2$ points within Gaussian reside at 0.5 and 0.135 of normalized irradiance, respectively. (b) 90/10 points of edge-spread function reside at 90% and 10% of normalized integral and cumulative sum of two-dimensional Gaussian. It is important to note that the plots represent ideal case and experimental measurements will contain fluctuations that result from noise within system.

Fourier optics theory states that the expression for the diffracted light that has propagated from the pupil plane to the front-focal plane of a lens reduces to the Fourier transform of the irradiance distribution within the pupil (Goodman 2004). Taking the Fourier transform of a circ function, which corresponds to a plane wave apodized by a circular pupil, results in an Airy function that matches the ideal result of Equation 25.5, where the product of the pixel size and the number of pixels corresponding to the radius of the Airy disk r_{plane} is equivalent to $0.61\lambda/NA$ in Equation 25.7. As a first approximation, we can assume the irradiance distribution within the pupil is equivalent to the Gaussian distribution of the illumination beam, and the Fourier transform provides a Gaussian spot at the focus of the lens. We can then convert between the lateral resolution measurements with the relations 90/10 points = 1.09* FWHM = 0.64* $1/e^2$ diameter = 1.28* $1/e^2$ radius (Rajadhyaksha et al. 1999a), and compare the experimental result to the diffraction-limited lateral resolution in Equation 25.11 that assumes plane-wave illumination in the pupil and a pinhole equivalent to a delta function.

To calculate a more accurate diffraction-limited lateral resolution for an experimental system, we must also account for the filling of the pupil. A Gaussian distribution with the appropriate fill factor can be apodized by a circ function according to Section 25.2.1.4, and then Fourier-transformed to provide the resultant focal spot. The number of pixels can then be measured for any of the three resolution points r_{gauss}, and the new multiplier m for the lateral resolution can be determined by equating the ratios $r_{gauss}/m = r_{plane}/0.61$. This results in the diffraction-limited lateral resolution

$$\Delta x = \frac{0.61 r_{gauss}}{r_{plane}} \frac{\lambda}{NA}, \qquad (25.19)$$

for a confocal system with Gaussian illumination in the pupil and a pinhole radius ≥3 optical units.

The lateral resolution has been measured for most microscopes using standard targets, such as Ronchi rulings or the 1951 USAF resolution target. While these targets provide a measure of the uniformity of the image throughout the field of view, which is a very important measurement in addition to resolution, their use to evaluate resolution can be considered subjective because the user determines the smallest resolvable set of line pairs by eye. In addition, these targets are generally only made of reflective materials and cannot be used for a fluorescence imaging system. The lateral resolution of a confocal fluorescence microscope is generally found by imaging fluorescent beads with a diameter less than the resolution of the imaging system, such that the bead will act as a delta function and image to an Airy function based on the microscope's PSF. Measuring the distance between the maximum value and the first null of the resultant Airy function provides the limiting lateral resolution of the system, which can be compared to the diffraction-limited case. However, multiple measures must be completed for beads displaced throughout the field of view to measure the resolution off-axis where aberrations produce the greatest effect. Because a point target does not scatter enough light, the lateral resolution of a confocal reflectance system has generally been measured from the edge-spread function of the system. Acquiring an image of a chrome-on-glass edge and plotting the detected power distribution for pixels normal to the line will provide an edge-spread function similar to Figure 25.11b. The lateral resolution can then be reported as the average distance between the 90/10 points for multiple edge-spread functions across the field of view.

The optical sectioning of a confocal fluorescence or reflectance system has been measured by translating a thin layer of fluorescent dye or a mirror through the focus of the objective, plotting the normalized detector power *versus* axial distance from focus, and measuring the FWHM. However, it is important to note that the FWHM of the reflection from a planar object, such as a mirror, does not directly correlate to the FWHM used to derive Equation 25.14 and to the plot in Figure 25.10c. When the mirror is some distance u in front of $(-u)$ or behind $(+u)$ the focal point of the objective as in Figure 25.12 (Sheppard and Wilson 1978), the light collected by the objective will mimic the reflection from a virtual point $2u$ in front of or behind the focal plane, respectively. Thus, $2u$ must be substituted in for u in Equations 25.13 and 25.17 resulting in twice the multiplicative factor in Equation 25.14 and in the plot in Figure 25.10c.

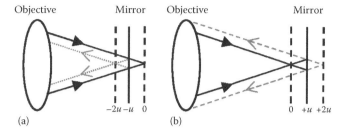

FIGURE 25.12 Measuring optical section with mirror translating through focus. Objective focuses light to focal plane at 0, and collects rays that reflect from planar object positioned an axial distance u (a) in front of or (b) behind the focal plane that appear to originate from a point a distance $2u$ from focus.

25.2.2 Field of View

The field of view for a single confocal image is dependent on the scanning method incorporated within the device. The main advantage of a stage-scanning system is that the field of view is only limited by the total translation available on the stage. A beam-steering system that uses a scanner to point to each pixel within the field of view is limited by the magnification of the system and the acceptance scan angle of the objective lens (typically ±3°). The total magnification of the system is not simply the magnification of the objective, and must be calculated from the relation

$$M = \frac{f_{\text{tube}}}{f_{\text{obj}}} \left(\frac{f_{1,1}}{f_{1,2}} \frac{f_{2,1}}{f_{2,2}} \cdots \frac{f_{n,1}}{f_{n,2}} \right), \qquad (25.20)$$

where

f_{tube} is the focal length of the tube lens before the detector
f_{obj} is the focal length of the objective
$f_{n,1}$ and $f_{n,2}$ are the focal lengths of the first and second lenses of relay telescope n in the order of the illumination path, respectively.

The typical f_{tube} values for various infinity-corrected objective manufacturers are 200 mm for Leica and Nikon, 180 mm for Olympus, and 164.5 mm for Zeiss, and the value is printed on the objective barrel for a non-infinity-corrected objective lens. The f_{tube} for the objective can also be calculated from the sine condition $f_{\text{tube}} = MD_{\text{pupil}}/(2\,\text{NA})$, where D_{pupil} is the diameter of the pupil in mm. One technique used to image samples that are larger than the typical 0.25–1 mm² field of view provided by beam-scanning systems is to create a mosaic of the sample by stitching individual images together in software. However, this technique still requires translation of the sample with respect to the imaging system (Rajadhyaksha et al. 2001, Patel et al. 2007, Gareau et al. 2008, 2009b,c) or of the imaging probe with respect to the sample (Vercauteren et al. 2006, Becker et al. 2007). However, these techniques currently only provide the final mosaic after each of the individual images (tiles) are acquired and corrected for registration and aberrations.

Another important aspect of the field of view is the pixel size because the advantage of a large field of view can be negated by a pixel size larger than the imaged features. Since a confocal image is generally collected in a plane parallel to the image surface (*en face*), the pixel size, limited by the sampling of the scanner electronics, ideally should be smaller than the lateral resolution by an amount sufficient to satisfy the Nyquist sampling theorem. As described in the previous sections, any object smaller than the lateral resolution of the instrument will image to a spot equivalent to the PSF. To properly sample the field of view, the Nyquist criterion states that there must be two pixels within the diameter of the Airy disk in every direction. Since the scanning elements generally translate the beam continuously through the deviation angle and are not stepped, the analog signal from the detector must be sampled every time the focused spot translates a distance equivalent to the Airy disk radius (Nyquist rate). Oversampling the image (sampling faster than the Nyquist rate) will provide excess information and require increased bandwidth and larger file sizes, while undersampling the image (sampling slower than the Nyquist rate) will produce a pixilated image that contains aliasing and appears to have a lateral resolution worse than the optics provide.

25.2.3 Beam Walk

As discussed in Section 25.1.3, a scanning element pivots the beam about the optic axis in a pupil plane that is conjugate to the pupil plane of the objective to scan the beam without added aberration. When the scanning element creates a pivot point that is misaligned by some distance along the optic axis from the conjugate pupil plane, the resulting pivot point will also be misaligned with the pupil plane of the objective by a proportional distance. This will scan (walk) the beam around the pupil plane of the objective, resulting in clipping of the beam by the pupil aperture. Clipping of the beam contributes to vignetting, where a drop in intensity and a broadening of the focused spot leads to uneven illumination and inconsistent lateral resolution throughout the field of view, and can occur at any point along the optical path if the microscope housing clips a part of the beam. Some systems have reduced the effects of beam walk by overfilling the pupil with a beam diameter much larger than the pupil diameter such that the displacement of the beam will produce a small change in the irradiance distribution. However, this compensation is not recommended for low signal applications where the throughput of the illumination light is critical. It is important to note that a scanning element can be positioned within the depth of focus of the first lens of the relay telescope and still be considered within the pupil plane. The depth of focus is defined as $\pm 2\pi d_{\text{spot}}^2/\lambda$, where d_{spot} is the diameter of the focused spot calculated by the $1/e^2$ points for the PSF of the lens (Saleh and Teich 1991).

25.2.4 Signal-to-Noise Ratio

The resolution measurements described in the previous sections have also assumed a large signal-to-noise ratio (SNR) such that the noise within the system was negligible. As the SNR decreases, the noise begins to corrupt the signal with random fluctuations that will eventually result in the inability to resolve the 26.5% dip in irradiance between two Airy functions. A system is quantum-noise limited when the only significant noise is shot (photon) noise. Shot noise results from the random arrival time of photons and is proportional to the square root of the number of photons detected (N) during the acquisition of one pixel (pixel dwell or integration time):

$$\sqrt{N} = \sqrt{\frac{\eta P_d}{h\nu} t_{pixel}},$$
(25.21)

where

η is the quantum efficiency of the detector (electrons/photon)

P_d is the power of the light that passes through the pinhole (W)

h is Planck's constant (6.6×10^{-34} W s^2)

ν is the frequency of the detected light (1/s)

t_{pixel} is the pixel dwell time (s).

When the system is quantum-noise limited, the number of detected photons is $N = \eta P_d t_{pixel}/h\nu$ and the SNR can be expressed as N divided by the square root of N, or simply the square root of N in Equation 25.21.

It is important to note that the relation of P_d to the output power of the light source (P_i) can be determined empirically for a confocal fluorescence or reflectance configuration by measuring the output power of the laser, the power incident upon the sample, and the resulting power that passes through the pinhole, or mathematically for a confocal reflectance system:

$$P_d = \frac{\iint_{-\infty}^{\infty} P_i h_t(u,v) h_r(u,v) \sigma_{sc} N_v \left[\left(1 - \sqrt{1-NA^2}\right)/4\pi \right] dA\,du}{\int_{-\infty}^{\infty} h_t\,dA \int_{-\infty}^{\infty} h_r\,dA}$$
(25.22)

where

$h_t(u,v)$ and $h_r(u,v)$ are the PSFs for the transmit (illumination) and receive (detection) paths

σ_{sc} is the scattering cross section

N_v is the number of scatterers per unit volume

and the expression within the brackets describes the fraction of scattered light that passes back through the NA of the objective. The expression for P_d in a confocal fluorescence configuration replaces the scattering cross section with the absorption cross section, and also takes into account the quantum yield of the fluorophore, saturation effects, the disparity between the excitation and emission wavelengths, and photobleaching (Sheppard et al. 2006). Substituting P_D into Equation 25.21 allows the fastest frame rate (shortest pixel dwell time) to be calculated that will provide an acceptable SNR for the application.

25.2.5 Imaging in Turbid Tissue

The descriptions of resolution and the PSF to this point have assumed imaging of an isotropic, non-scattering, homogeneous medium. However, biological samples contain numerous scattering components (Dunn et al. 1996b) and are often imaged through various refractive-index mismatches, such as the transitions between the immersion medium for the objective, the coverslip, and the immersion medium for the sample. Imaging through refractive-index mismatches produces spherical aberration, degrades the optical sectioning response by (1) decreasing the maximum response, (2) shifting the maximum location from the expected focus of the objective, (3) broadening the FWHM, and (4) increasing the background (increase in the side lobes and minima signal) (Torok et al. 1997, Sheppard 2000), and ultimately limits the penetration depth (Smithpeter et al. 1998b). There have been numerous analytical models presented to help understand the propagation of light through turbid media, including three primary models: (1) Mie scattering analysis, which calculates the scattered field for homogeneous spherical objects (Rajadhyaksha et al. 2004); (2) Monte Carlo models, which utilize a random walk algorithm to simulate the propagation of multiply scattered photons in bulk tissue (Schmitt et al. 1994, Bigelow and Foster 2006); and (3) finite-difference time-domain (FDTD) computations, which provide the complete solution to Maxwell's equations for an inhomogeneous object of arbitrary shape and size (Dunn and Richards-Kortum 1996, Smithpeter et al. 1998a, Drezek et al. 1999, Brock et al. 2006, Simon and DiMarzio 2007).

Analytical modeling and experimental measurement of resolution and PSF are typically reported in the objective's immersion medium or tissue phantoms such as Intralipid. The measurements in immersion medium provide an important metric for the comparison amongst the various system configurations, but the measures provide the resolution for the best case scenario and are not representative of the imaging capabilities within tissue. Phantoms provide heterogeneity and accurately model the scattering effects, but they do not model the aberrations inherent in tissue. Aberrations occur due to the heterogeneous and layered structure that results in "random phase plate" behavior of tissue, where each tissue layer acts like a phase plate that induces a random phase to the illumination and detection paths, and can strongly influence confocal imaging performance. Such effects have been accounted for in recent analytical models (Simon and DiMarzio, 2007) and experimental work in skin (Dwyer et al. 2006, 2007, Gareau et al. 2009a). By placing a thin slice of tissue on the test target, the effects of both scattering and aberrations become evident under the actual conditions of tissue and provide a representative measure of the expected resolution within the tissue of interest. Figure 25.13 shows an example of the nominal line spread function and edge spread function for a line-scanning confocal microscope, and the subsequent degradation through full-thickness human epidermis. Depending on the design of the microscope, the FWHM optical sectioning may degrade by two to five times and the lateral

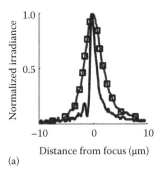 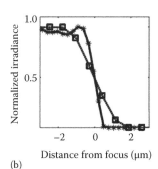 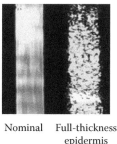

(a)   (b)   (c)

FIGURE 25.13 Effect of turbid tissue on (a) axial line spread function and (b) edge-spread function in line-scanning confocal microscope. Nominal functions (-) degrade through full-thickness human epidermis (□) due to tissue-induced scattering and aberrations. Full-thickness epidermis is typically 50–150 μm of tissue, consisting of 5–15 layers of cells. (c) Image of test grating target under nominal conditions and through full-thickness epidermis from which edge-spread functions are determined. (Courtesy of Daniel Gareau, Dermatology Service, Memorial Sloan-Kettering Cancer Center, New York.)

resolution by two times when imaging deeper in the dermis, through full-thickness epidermis in human skin. The amount of degradation with depth will, of course, depend on the properties of the specific tissue of interest.

25.3 Instrumentation

Confocal microscopes have been a staple within the biomedical community over the past two decades, yet research has continued to improve the instrumentation and integrate the technique into clinical applications. Within this section we discuss a sampling of the recent innovations to reduce the size of the instrumentation and enhance the imaging technique beyond the optical sectioning capabilities. For the sake of brevity, we provide a brief summary to introduce the broad range of techniques and point the interested reader to the pertinent references, followed by a more in-depth discussion on multimodal microscopy, line scanning, spectral encoding, and coherent reflectance.

25.3.1 Summary of Recent Innovations

25.3.1.1 Increased Acquisition Rate

Real-time (≥5 fps) and video-rate (30 fps) acquisition is possible with point-scanning configurations (Rajadhyaksha et al. 1999a), but the maximum acquisition rate is limited by the minimum pixel dwell time that provides an acceptable SNR. The simultaneous illumination and detection of multiple pixels with line- or disk-scanning techniques allows increased acquisition speed to image cellular dynamics without decreasing the pixel dwell time (Vermot et al. 2008). Line-scanning (slit-scanning) systems use anamorphic optics, such as a cylindrical lens, to transform the circularly symmetric Gaussian spot into an elliptical spot that illuminates the sample with a line, thereby simultaneously sampling an entire row of pixels within the image with a linear complementary metal oxide semiconductor (CMOS) or charged coupled device (CCD) array (Im et al. 2005, Wolleschensky et al. 2006, Dwyer et al. 2006, 2007) and

providing the ability to increase the acquisition speed in excess of 100 Hz for a 512 × 512 pixels field of view at the focal plane (Im et al. 2005, Wolleschensky et al. 2006). Disk-scanning (spinning-disk) systems use a spiral pattern of pinholes on a Nipkow disk to illuminate hundreds or thousands of pixels simultaneously, which are then acquired with a two-dimensional array detector such as a cooled CCD camera (Toomre and Pawley 2006) to provide an acquisition rate up to 1000 Hz (Tanaami et al. 2002).

25.3.1.2 Miniaturization

Optical imaging techniques provide images with high-resolution on the order of micrometers, but the limited penetration depth of ballistic imaging techniques requires endoscopic delivery with optical fiber to image within a patient. One approach is to separate the scanning instrumentation from a miniature objective lens with a fiber bundle containing tens of thousands of single-mode optical fibers tightly packed together, where each fiber corresponds to a pixel within the image (Gmitro and Aziz 1993, Rouse et al. 2004, Carlson et al. 2005, Makhlouf et al. 2008). The two ends of the fiber bundle may also be immersed in index-matching oil (Liang et al. 2001) or the illumination may be pulsed with a time-gated detector (Nakao et al. 2008) to reduce the specular reflection from the bundle tips in a confocal reflectance geometry. A second approach is to miniaturize the scanning mechanism such that it can be included within the endoscope. These devices physically translate the illumination tip of the single-mode optical fiber (Dabbs and Glass 1992, Seibel and Smithwick 2002) or incorporate a micro-electromechanical systems (MEMS) scanner at the distal end of the endoscope to scan the fiber output before a miniature objective lens (Dickensheets and Kino 1996, Maitland et al. 2006, Shin et al. 2007, Liu et al. 2007, Ra et al. 2007, 2008, Kumar et al. 2008, Poland et al. 2009). A third approach uses a grating to spread the wavelengths of a broadband light source spatially along a line such that each pixel along a row within the image corresponds to a different wavelength, leading to the name spectrally encoded confocal microscopy (SECM) (Tearney et al. 1998).

25.3.1.3 In-Plane Scanning

One reason for the large overall size of the scanning instrumentation is the prevalent use of reflective elements that fold the optical path, such as spinning polygons and galvanometric mirrors. One technique to reduce the size is to use a transmissive scanning device such as a dual-wedge (Risley prism) scanner where two identical prisms, mounted with the flat sides facing each other, rotate about the optic axis to scan the illumination spot within a spiral or rosette pattern (Warger and DiMarzio 2007b). The close proximity of the prisms leads to the possibility of positioning the prisms directly within the pupil of a custom-designed objective lens to eliminate the need for a relay telescope. The circular symmetry of the imaging system can also be corrected for the circularly symmetric aberrations inherent within optical devices. Another technique to reduce the size of the scanning system is to incorporate scanning within the light source (Poher et al. 2007, 2008). This device uses a micro-structured InGaN LED consisting of 120 elements, 17 μm wide and 3600 μm long with 34 μm center-to-center spacing to illuminate individual lines on the sample such that the electronics act as the scanning instrumentation to produce a confocal system without any moving parts.

25.3.1.4 Axial Scanning

The ability to image subsequent optical sections along the axial direction with diffraction-limited resolution and without limiting the acquisition speed is a scanning limitation that hinders all microscopy techniques. Endoscopic and handheld confocal instruments typically rely on mechanical devices to translate the objective, which limits the acquisition speed between successive optical sections and adds complexity and size to the overall device. A technique shown to overcome this limitation uses two objective lenses and two tube lenses to create an intermediate three-dimensional image that is free from aberrations within ±35 μm of the focus that can then be imaged with a translating mirror and a third tube lens (Botcherby et al. 2007, 2008). The technique has also been expanded from a point-scanning configuration to a line-scanning system, where the acquisition rate of the volumetric imaging was limited by the frame rate of the detector (Botcherby et al. 2009).

25.3.1.5 Improved Optical Sectioning/ Increased Working Distance

The optical configurations discussed thus far have included collinear illumination and detection paths between the BS and the focal spot. Intersecting the illumination and detection paths at the focus of a single objective in divided-aperture microscopy or at the foci of two separate objectives in theta microscopy can improve the optical sectioning by reducing the overlap of the focal volumes (angular gating) (Lindek and Stelzer 1996) or provide the same optical sectioning with an extended working distance using lower NA objectives (Webb and Rogomentich 1999). Another technique to improve optical sectioning is to incorporate coherence gating with confocal spatial filtering by interfering a portion of the unperturbed light source with the backscattered light from the sample. The total axial response then becomes the product of the confocal PSF along the axial direction and the axial coherence gate. With the addition of coherence gating, the NA of the objective can be reduced with the optical sectioning maintained to provide a longer working distance and greater penetration depth (Wang et al. 2006).

25.3.2 Multimodal Microscopy

25.3.2.1 Keck 3D Fusion Microscope

Until recent years, most microscopy techniques were developed into separate instruments such that multimodal microscopy was limited to the combination of images from separate microscopes with image processing techniques. The Keck 3D Fusion Microscope (Warger et al. 2007a) was then developed to ensure spatial registration of the images while maintaining physiologically correct structure and function during imaging with brightfield, differential interference contrast (DIC), epi-fluorescence, optical quadrature microscopy (OQM), confocal reflectance, confocal fluorescence, two-photon fluorescence, and more recently second-harmonic generation (SHG) and line-scanning fluorescence recovery after photobleaching (FRAP) on a single Nikon TE2000 microscope base. The microscope allows the user to place a sample on the microscope stage and acquire any combination of images without having to disturb the sample. An example of the images acquired of a live 3-cell mouse embryo stained with Hoechst and MitoTracker Green dyes is shown in Figure 25.14.

Brightfield images are acquired with a white light source, such as a halogen or xenon lamp, that illuminates the entire field of view. The contrast in a brightfield image is related to the absorption, scattering, refraction, and diffraction of the illumination light, which is extremely low in a transparent or translucent object providing a flat appearance for the embryo in Figure 25.14a. DIC microscopy provides contrast related to the one-dimensional gradient of the object's phase (Murphy 2001), thereby producing a shading effect along the edges of the embryo in Figure 25.14b. DIC uses the same optical path as brightfield, but includes two polarizers and two Wollaston prisms. As discussed previously, fluorescence microscopy images the distribution of individual molecules tagged to specific organelles by collecting only the fluorescent wavelengths that are emitted after excitation with a particular band of wavelengths (Murphy 2001). An epi-fluorescence system uses the objective to illuminate the sample with the excitation wavelengths and collect the fluorescent wavelengths from the excited fluorophores within the sample. Figure 25.14d shows an epi-fluorescence image of Hoechst 33342 dye (Invitrogen Corporation, Carlsbad, California) that binds the fluorophores to the DNA within the nucleus of each cell, and Figure 25.14e shows an epi-fluorescence image of MitoTracker Green dye (Invitrogen Corporation, Carlsbad, California) that binds the fluorophores to the mitochondria throughout the cells. These images were acquired with a mercury lamp for illumination and ultraviolet

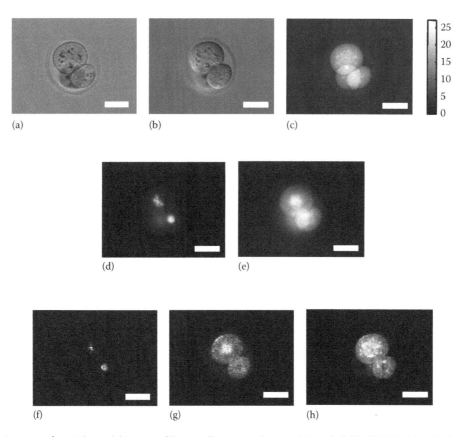

FIGURE 25.14 (See color insert.) Multimodal images of live 3-cell mouse embryo in (a) brightfield, (b) DIC, (c) optical quadrature microscopy (technique is discussed in Warger and Di Marzio, 2009), (d) epi-fluorescence with Hoechst dye, (e) epi-fluorescence with MitoTracker Green dye, (f) two-photon fluorescence with Hoechst dye, (g) confocal fluorescence with MitoTracker Green dye, and (h) confocal reflectance. Hoechst dye binds to DNA within nuclei and MitoTracker Green dye binds to mitochondria. Note optical sectioning capabilities of two-photon fluorescence, confocal fluorescence, and confocal reflectance only provide signal within focal plane and signal from third out-of-focus cell does not contribute to image. Scale bar is 50 μm. Embryos prepared by Judy Newmark and Carol Warner, Northeastern University. (technique discussed in Newmark et al., 2007.)

excitation (360 nm ± 20 nm excitation filter, 400 nm long-pass dichroic, and 460 nm ± 25 nm) and blue excitation (480 ± 15 nm excitation filter, 505 nm long-pass dichroic, and 535 nm ± 20 nm emission filter) filter cubes for the Hoechst and MitoTracker dyes, respectively. The brightfield, DIC, and epi-fluorescence images were acquired with an air-cooled CCD camera (Diagnostic Instruments, Inc., Sterling Heights, Michigan).

Optical quadrature microscopy (OQM) is an interferometric imaging technique that reconstructs the change in magnitude and phase of the coherent light that traverses through an optically transparent object (Warger and DiMarzio 2009). Assuming a non-scattering sample with a single uniform refractive index, the change in phase α is related to the thickness h and refractive index n_s of the object by $\alpha = (2\pi/\lambda)^* (n_s - n_0)h$, where λ is the wavelength of the laser and n_0 is the refractive index of the immersion medium. The phase image in Figure 25.14c can then be generalized as values proportional to the thickness of the embryo, where the large phase values correspond to cell overlap. OQM uses a linearly polarized 633 nm helium-neon laser (31-2082-000, Coherent) within a Mach–Zehnder interferometer,

where the signal path travels along the optical axis of the microscope through the camera port and mixes with the reference path outside of the microscope housing. The signal path is polarized at 45° to provide equal amounts of the two orthogonal polarization states, and the reference path is circularly polarized to provide a 90° phase shift between the two orthogonal polarization states. The reference and signal paths recombine at a nonpolarizing 50/50 BS and a PBS after each output of the 50/50 separates the two orthogonal polarization states that are acquired with 4 synchronized CCD cameras (XC-75, Sony Electronics Inc.). A reconstruction algorithm uses images from the four CCD cameras to reconstruct the magnitude and phase changes in the signal path induced by the sample.

As discussed throughout this chapter, confocal microscopy uses a pinhole in the image plane to block most of the out-of-focus light that spreads across the image plane. In a confocal fluorescence system, the signal is proportional to the excitation of the same fluorophores imaged in an epi-fluorescence system, but only the fluorophores within an optical section contribute to the image. This can be seen by comparing Figure 25.14e and g where the MitoTracker dye from the third out-of-focus cell degrades the quality of the epi-fluorescence image but does not contribute to the confocal fluorescence image. The image in Figure 25.14g was acquired with a line-tunable Argon laser (35LAP431-208, Melles Griot) at 488 nm, 505 nm longpass dichroic, 535 ± 15 nm emission filter, 300 μm pinhole, and a photomultiplier tube (PMT) (HC124-02, Hamamatsu). In a confocal reflectance system, the signal is proportional to the backscatter of the illumination light provided by native variations in the refractive index of organelles and microstructures within the sample (Dunn et al. 1996). The image in Figure 25.14h visualizes all of the refractive-index discontinuities within an optical section of the embryo that correspond to the MitoTracker seen in the confocal fluorescence image in addition to other organelles that do not bind to the dye. The image in Figure 25.14h was acquired with a titanium-sapphire laser (Tsunami, Spectra-Physics) in continuous-wave mode at 920 nm, 150 μm pinhole, and an avalanche photodiode (APD) (Model C5460, Hamamatsu).

Two-photon microscopy (Göppert 1929, Sheppard and Kompfner 1978, Denk et al. 1990) images the same fluorescence signal as epi-fluorescence and confocal fluorescence, but the illumination light uses two photons at twice the wavelength (half the frequency) to excite the fluorophore. One photon contains the energy $E = h\nu = hc/\lambda$, where h is Planck's constant, ν is the frequency of the light, c is the speed of light, and λ is the wavelength of the photon. In epi-fluorescence and confocal fluorescence, a single photon excites an electron to an energy state $h\nu$, but in two-photon, an electron is excited to the same energy state when two photons at half the frequency $(2h\nu/2)$ or twice the wavelength are absorbed simultaneously. The probability of two photons being absorbed simultaneously is proportional to the square of the light intensity (Göppert 1929, Göppert-Mayer 1931, Masters and So 2008). This leads to a PSF that is equivalent to a confocal reflectance PSF with optical units computed using the two-photon excitation wavelength. Since the fluorescent light that results from two-photon illumination originates from the focal spot alone, all of the fluorescent light can be collected without descanning to a pinhole. In addition to the optical sectioning capabilities, the advantage of two-photon imaging is the use of red to near-infrared wavelengths to excite the UV and blue excitation dyes, because UV and blue light is highly absorbed in tissue compared to red and infrared light. Two-photon microscopy generally provides improved optical sectioning and deeper penetration compared to confocal fluorescence due to the redshifted illumination wavelength, but the technique requires an expensive, pulsed laser source. Figure 25.14f shows that two-photon images of Hoechst-stained nuclei provide much improved image quality of the individual nuclei compared to the epi-fluorescence image in Figure 25.14e. The image in Figure 25.14f was acquired with a titanium-sapphire laser mode-locked at 920 ± 10 nm and a 675 nm long-pass hot mirror for excitation, and a 505 nm long-pass dichroic and 535 ± 20 nm emission filter for detection with a PMT (HC124-02, Hamamatsu).

The combination of imaging modalities on one microscope stage is very convenient to observe the various contrasts available from each mode, but the real power of the Keck 3D Fusion Microscope is the pixel-to-pixel registration for static samples among the image modalities (Tsai et al. 2008). Image fusion allows the contrast and measures provided by the separate images to be combined to form new measures that would not have been possible from using any of the image techniques individually (Warger et al. 2008).

25.3.2.2 Handheld Confocal Reflectance and Raman Spectroscopy

Another example of a multimodal instrument is a handheld device currently in development that combines confocal microscopy with Raman spectroscopy for imaging tissue (Arrasmith et al. 2009). While confocal microscopy images morphology at the nuclear and cellular level (Rajadhyaksha et al. 1999b), Raman spectroscopy detects biochemical signatures at the molecular level (Caspers et al. 2001, Gniadecka et al. 2004, Sigurdsson et al. 2004). The combination of confocal reflectance microscopy and Raman spectroscopy yields both morphological and functional information with sensitivity and specificity that may be higher than that with either mode alone (Caspers et al. 2003).

The compact design is driven by the use of a MEMS scanner (Microvision, Inc., Bothell WA) instead of the traditionally large rotating polygon and oscillating galvanometric scanners that are used within bench-top systems. The optical layout of the system is shown in Figure 25.15. The polarized 830 nm laser diode (Sacher Lasertechnik) fiber couples to the handheld component with polarization-maintaining (PM) optical fiber, and the confocal scanner follows the standard approach with a beam expander to fill the pupil of the objective lens, a telescope to relay the MEMS scanner to the pupil for telecentric scanning, the use of a quarter-wave plate to minimize instrumental back-reflected noise, and an APD (C5460, Hamamatsu) to detect the backscattered light. A dichroic mirror and long-pass filter separate the Stokes-shifted Raman signal from the reflectance signal that traverses to the fiber coupled Raman spectrograph (HoloSpec $f/1.8i$, Kaiser Optical Systems, Inc.). A second long-pass Raman filter removes any Raman light generated by the PM fiber from the illumination. The instrument also includes a BS behind the objective lens and a CCD camera that provides a macro-image of a large field of view to subsequently locate the confocal field of view and the site within the tissue for Raman spectroscopy. White light emitting diodes (LED) near the objective lens provide illumination for the macro-image. Objective lenses of NA from 0.55 to 1.1 can be interchanged to provide variable sectioning, resolution, and field of view. While the overall instrument is compact, the use of an objective lens with relative high NA provides sufficient optical sectioning and resolution. This differs from several existing miniaturized confocal endoscopes in which the numerical apertures are fairly low with corresponding decrease in image quality.

Imaging of human skin in vivo with 4 μm optical sections and the acquisition of Raman spectra with 10 μm resolution has been reported (Arrasmith et al. 2009). Figure 25.15b shows a confocal reflectance image of a hair pressed against the skin of a Caucasian male with nuclear and cellular architecture similar to that seen with earlier bench-top microscopes. The corresponding

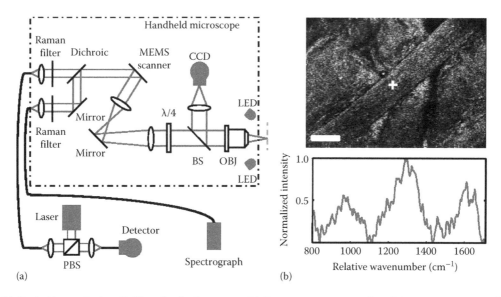

FIGURE 25.15 (a) Optical layout for handheld confocal microscope with Raman spectroscopy where λ/4 is a quarter-wave plate, BS is a beam-splitter, and OBJ is an objective. (Schematic adapted from Arrasmith, C.L. et al., *Proc. SPIE*, 7169, 7169N, 2009.) (b) Confocal reflectance image of a hair pressed against skin of Caucasian male and Raman spectra corresponding to white crosshair. Scale bar is 50 μm. (Courtesy of Chris Arrasmith and David Dickensheets, Montana State University, and Anita Mahadevan-Jansen, Vanderbilt University, Nashville, TN.)

Raman spectrum for the highlighted pixel (crosshair) is representative of the spectrum acquired with the device. The results to date show the promising possibility of integrating confocal imaging with Raman spectroscopy for interrogating tissue for multimodal morphologic and biochemical information. Work is in progress for development of a second generation miniaturized instrument followed by testing and use in clinical applications on skin and other tissues.

25.3.3 Line Scanning

Commercially available confocal microscopes are primarily based on point-scanning configurations, and research has continued to evolve point-scanning confocal microscopes into successful instruments with increasing utility among researchers and clinicians. However, point-scanners require scanning in two dimensions with relatively complex electronics, mechanical packaging, and manufacturing. Thus, they remain relatively large, expensive and not yet truly widespread in laboratory and clinical environments worldwide.

Line-scanning offers the opportunity to create a simpler scanning configuration for developing smaller, lower-cost confocal microscopes. A focused illumination line scans in only one dimension, thereby requiring a single scanner (Koester 1980, Koester et al. 1993, Brakenhoff and Visscher 1992, Gmitro and Aziz 1993), and the light collected by the objective descans and focuses through a slit (instead of a pinhole) to a linear detector, such as a CMOS or CCD linear array. Line-scanning is only confocal in the direction perpendicular to the line (slit) resulting in tails within the three-dimensional confocal line spread function (LSF) that extend far away from focus instead of dropping to zero in the Airy pattern for the confocal PSF (Wilson and Hewlett 1990).

The FWHM of the axial LSF is approximately 20% greater than that of the point, such that line scanning will provide optical sectioning that is only slightly worse than point scanning (Sheppard and Mao 1988). With sufficiently high NA (0.7–1.4), imaging of nuclear and cellular morphology may be expected similar to that with a point scanner, but the tails in the PSF produce multiply scattered out-of-focus background that reduces the contrast.

New developments in linear array detector technology allow the optics and electronics to be significantly simplified. The original pioneering designs of Koester, Brakenhoff, and Gmitro were relatively complex because the descanned detected beam required re-scanning onto a two-dimensional detector. The recent advent of high-quality linear CCD and CMOS array detectors allows simpler designs. Scanning only once (directly behind the pupil of the objective lens) and descanning once (directly onto a linear detector array) allows considerable simplification of the original designs.

Bench-top line-scanning confocal microscopes have been created that consist of 8–10 optical components, with total hardware costs of less than $15,000 (Dwyer et al. 2006, 2007, Gareau et al. 2009a). Two designs are currently being investigated for imaging in human skin and other tissues. One design is based on a divided pupil, with one half of the pupil for illumination and the other half for detection (i.e., separate illumination and detection paths that intersect at focus), resulting in the well-known theta-microscope configuration. The other design is based on a full pupil, with standard coaxial illumination and detection paths.

Figure 25.16a shows the design based on a divided pupil, also called a confocal theta-line scanner (Dwyer et al. 2006, 2007). The design is a combination of two concepts: use of a divided objective lens pupil (Koester 1980, Koester et al. 1993) and the theta microscope (Stelzer et al. 1995, Webb and Rogomentich

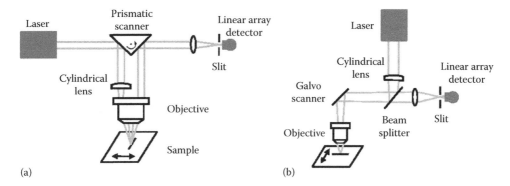

FIGURE 25.16 Optical layouts of line-scanning confocal microscopes, showing (a) divided-pupil and (b) full-pupil designs. Only a conceptual schematic is shown here in one dimension. Details of layout in two dimensions are included within references Dwyer et al. (2006, 2007) and Gareau et al. (2009).

1999). A collimated beam forms the illumination line with a cylindrical lens and an objective lens. The line scans through one-portion of the objective lens pupil. Physically, there is a strip that divides the objective lens pupil into an illumination portion and a detection portion. Scanning is, in principle, in the pupil of the objective lens such that no relay telescopes are necessary. In reality, the scanning is not exactly in, but as close to the pupil as possible. Thus, the beam slightly walks across the pupil. However, vignetting effects and loss of illumination at the edges of the field of view are minimized by sufficiently overfilling the pupil. Detection of backscattered light is through the symmetrically opposite portion of the pupil. The line is descanned and focused through a slit onto a linear array detector. The detector may be either a linear CMOS array (ELIS 1024, Panavision) or a linear CCD array (Reticon RL1024, Perkin Elmer).

Figure 25.16b shows the other design, based on a full pupil (Gareau et al. 2009a). A cylindrical lens produces a focused line in the back-focal plane of an objective lens, which is then Fourier-transformed to an orthogonally oriented line in the object plane within the tissue. A galvanometric mirror scans the focused line within the tissue to create the field of view. The light that is backscattered from the tissue is descanned and focused by a spherical lens through a slit onto a linear array detector. The optical configuration for both pupil designs is reasonably simple such that assembly is relatively straightforward.

The electronics are based on field programmable gate array (FPGA) logic. The pixel, line, and frame clocks and the sawtooth input signal to the galvanometric scanner are derived from a single pixel clock that is derived from the FPGA. The development of FPGA-based electronics offers the advantages of small footprint, rapid reconfigurability, and complete integration toward a stand-alone system.

With an illumination of 830 nm, a water-immersion lens of NA 0.9, and a slit width on the order of the diffraction-limited resolution, a divided-pupil confocal line-scanning microscope demonstrates nominal FWHM optical sectioning of 1.7 μm, which subsequently degrades to 9.2 μm through full-thickness human epidermis (Dwyer et al. 2006, 2007). The nominal lateral resolution is 1.0 μm, which degrades to 1.7 μm. As explained earlier in Section 25.2.5, the line spread function degrades due to scattering and aberration effects when imaging deeper in tissue. By comparison, the full-pupil design provides the nominal FWHM sectioning of 1.4 μm, which degrades through human epidermis to only 2.8 μm. The nominal lateral resolution is 0.8 μm which degrades to 1.6 μm. Thus, the full-pupil design provides significantly better optical sectioning with less tissue-induced variability through the scattering and aberrating conditions of human epidermis.

Figure 25.17 shows that nuclear and cellular detail is observed in human epidermis that confirms optical sectioning and

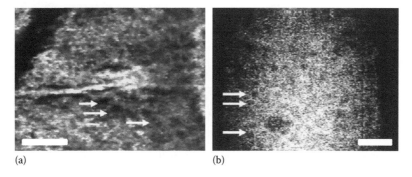

FIGURE 25.17 Images of human epidermis in vivo with reflectance contrast with (a) divided-pupil and (b) full-pupil line-scanning confocal microscopes. Dark nuclei (arrows) are seen with bright and grainy appearing cytoplasm. Scale bar is 50 μm.

imaging capability with both pupil configurations. The visually assessed contrast appears to be lower and the images somewhat more noisy with the full-pupil design than that seen with the divided pupil. Both out-of-focus multiply scattered background noise and speckle noise is evident in the images. However, the rejection of background light is superior in the divided-pupil design due to the angular separation of the detection path from the illumination path in the well-known theta-microscope geometry.

Both the full-pupil and divided-pupil (theta) confocal line-scanners achieve nuclear-level resolution in human skin. Line-scanning confocal microscopy may offer a competitive alternative to current point-scanning technology for imaging epithelial tissues. However, further improvements are needed to reach clinical-level capabilities. The choice of pupil configuration and the acceptable trade-off in optical sectioning and resolution *versus* contrast may depend on the scattering properties of tissue. In the long term, line-scanning with the use of FPGA-based electronics may enable simpler, smaller, lower-cost, stand-alone confocal microscopes for use at the bedside in diverse healthcare settings worldwide.

25.3.4 Spectral Encoding

A unique adaptation to line scanning spreads the wavelengths of a broadband light source spatially along a line such that each pixel along the illumination line corresponds to a separate wavelength (Tearney et al. 1998). The technique, termed spectral encoding, has provided the foundation for miniature endoscopy where a non-confocal fiber-optic probe has been created with a 350 μm maximum diameter that can be administered through a modified 23 gauge needle (Yelin et al. 2006).

Spectral encoding has been incorporated into confocal microscopy (SECM) using a broadband light source, such as a superluminescent diode, with a bandwidth of approximately 70 nm. The broadband light is incident at the Littrow angle,

such that the maximum amount of light transmits into the first-order at a diffraction angle equal to the incident angle, and continues along the optical path to the objective in Figure 25.18a. A relay telescope aligns the grating with the pupil of the objective, and a stage scans the illumination line in the orthogonal direction. Assuming a non-scanning system (Boudoux et al. 2005), the length of the line at the focus of the objective:

$$FOV_{line} = 2f \tan\left(\Delta\theta/2\right) \qquad (25.23)$$

determines one dimension of the field of view, where f is the effective focal length of the objective and $\Delta\theta$ is the angular deviation between the extremities of the spectral bandwidth ($\Delta\lambda$):

$$\Delta\theta = \frac{mG}{\cos\theta_i}\Delta\lambda, \qquad (25.24)$$

where

    m is the diffraction order
    G is the groove density (grooves/mm)
    θ_i is the incident angle of the light upon the grating.

The number of points (pixels) along the line can then be derived (Tearney et al. 1998, Boudoux et al. 2005) from the ratio of the spectral bandwidth of the source and the minimal resolvable spectral separation of the grating ($\delta\lambda$):

$$pixels = \frac{\Delta\lambda}{\delta\lambda} = \frac{mDG}{\lambda_0\cos\theta_i}\Delta\lambda, \qquad (25.25)$$

where

    D is the beam diameter incident upon the grating at angle θ_i
    λ_0 is the center wavelength of the broadband light source.

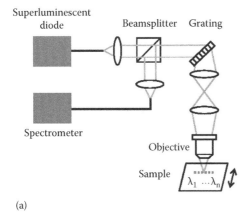

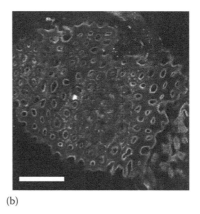

(a)   (b)

FIGURE 25.18 (a) Optical layout for spectral encoded confocal microscope that uses stage to scan illumination line in orthogonal direction. (Adapted from Kang, D.K. et al., *Gastrointest. Endosc.*, 71, 35, 2010). (b) SECM mosaic of human esophageal squamoglandular junctional mucosa biopsy. Scale bar is 500 μm. (Courtesy of DongKyun Kang, Brett Bouma, and Gary Tearney, Massachusetts General Hospital, Boston, MA.)

Various adaptations of SECM have been developed to decrease the complexity and reduce the size of the original device that include a spectrometer and a dual-prism GRISM (combination of a grating and a prism) (Pitris et al. 2003), a wavelength-swept source and point detector (Boudoux et al. 2005), and a helical scanner (Yelin et al. 2007a). However, all of these systems still require a scanner to develop a two-dimensional image. Two recent configurations have removed the need for any scanner. The first (Kim et al. 2006) couples light from a halogen light into a fiber-optic line guide to input a line of broadband light into the typical SECM configuration. The grating spreads the broadband line into a two-dimensional area within the sample and the backscattered light propagates back through the slit and a second grating to measure the signal with a two-dimensional CCD camera. The second, termed spectral-shower-encoded confocal microscopy and microsurgery (Goda et al. 2009, Tsia et al. 2009), uses a cylindrical lens to focus the light from an incoherent light source ($\Delta\lambda = 17$ nm) onto a tilted virtually imaged phased array (VIPA) that produces a line of collimated output beams with an angular propagation direction dependent on wavelength (Shirasaki 1996). The output of the VIPA reflects from a diffraction grating to create a "spectral shower" that illuminates a two-dimensional area with a separate wavelength at each point. The backscattered light returns through the VIPA and focuses into a single-mode fiber to an optical spectrum analyzer.

Figure 25.18b demonstrates the cellular resolution within a 2 mm × 2 mm SECM mosaic of a human esophageal squamoglandular junctional mucosa biopsy acquired on a bench-top system. Promising results have been shown for the translation of the technique into a clinical endoscopic tool for laryngeal investigation of the vocal fold (Boudoux et al. 2005), diagnosis of esophageal disease (Kang et al. 2009), and treatment with the combination of imaging and ablative therapy (Goda et al. 2009, Tsia et al. 2009).

25.3.5 Coherence Reflectance Confocal Microscopy

Another technique to improve optical sectioning is to incorporate coherence gating within the detection path. In coherence reflectance confocal microscopy, a portion of the undisturbed field from the light source interferes with the backscattered field from the sample before focusing through a pinhole to combine the coherence gating of optical coherence tomography with the axial focal gating of confocal microscopy. The axial response of such a system is the product of the axial focal gating of the objective lens, tube lens, and pinhole, and the coherence gating of the source bandwidth and interference detection. The combination of techniques provides the ability to reduce the NA of the objective while maintaining the optical sectioning of a typical high-NA objective confocal configuration. Reducing the NA provides a longer working distance and allows for deeper imaging, especially below the dermal–epidermal (DE) junction where the refractive-index difference between the dermis and epidermis and the tissue corrugation at the DE junction strongly aberrate the incident beam.

The coherence length of the light sources used in typical coherence gating systems generally range from 1 to 18 μm (measured in air) (Bouma and Tearney 2002). Generally, such broadband sources are either relatively inexpensive with lower power (<5 mW for superluminescent diodes) or expensive with high power (femtosecond Ti:Al₂O₃ laser). Kempe proposed the combination of multiple discrete lasers to create a single transverse-mode beam with the use of two or more single transverse-mode semiconductor lasers capable of producing approximately 150 mW (Kempe 2000, 2002). The optical layout of a coherence reflectance confocal microscope based on the Kempe laser design is shown in Figure 25.19 (Wang et al. 2006). The light source incorporates

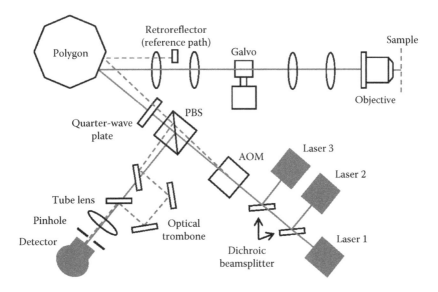

FIGURE 25.19 Optical layout for coherence reflectance confocal microscope. Three lasers create synthesized broadband light source and acousto-optic modulator (AOM) separates light into reference and object arms of an interferometer. (Adapted from Wang, Z. et al., Coherence reflectance confocal microscope using a synthesized broadband source, in *Opt. Soc. Am., Biomedical Topical Meeting (BIO)*, Fort Lauderdale, Fl, 2006.)

three co-aligned, linearly polarized, single transverse-mode laser diodes with wavelengths of 785, 810, and 850 nm (Microlaser, Garden Grove, California) that can be powered individually to provide a traditional confocal reflectance image, sequentially to provide a multi-wavelength confocal reflectance image, or simultaneously to provide a coherent confocal reflectance image. During coherent imaging, an acousto-optic (AO) modulator (Brimrose TEM-80-2) separates the sample and reference arms of the interferometer and produces an 80 MHz modulation on the reference beam. The sample and reference arms pass through a PBS and a quarter-wave plate before entering the 10 fps raster scanning system from a VivaScope 1000 confocal microscope (Lucid, Inc. Rochester, New York). The slightly translated reference arm reflects from a retroreflector positioned within the telescope between the spinning polygon and galvo mirror, and the sample arm illuminates the sample with a 30× objective lens (LOMO) that has an adjustable aperture stop to vary the NA. The reflected reference arm field and backscattered sample arm field propagate back through the quarter-wave plate and reflect from the PBS. The reference arm travels through an optical trombone to match the optical paths between the interferometer arms and mixes with the sample arm before focusing through a pinhole to the detector. The coherent optical sectioning of the synthesized broadband source was measured as 4.7 μm FWHM with a 0.5 NA objective, which compared to 2.9–3.9 μm FWHM with a 0.9 NA objective for single wavelength (noncoherent) 785, 810, and 850 nm illumination with Wang's tilted substrate method (Wang et al. 2007).

25.4 Summary

Confocal microscopy has provided insight into a variety of biomedical applications since its widespread acceptance within the biological community in the 1980s. Since that time, the term "confocal" has become synonymous with the concept of imaging individual planes within a sample. The ideal confocal instrument provides diffraction-limited resolution and optical sectioning, but the user must incorporate the tissue properties into the system to provide a true measure of the resolution, sectioning, and usefulness of the technique for the desired application. Numerous feasibility studies have shown that the confocal technique provides a relatively simple, lower-cost instrument that provides adequate resolution and fair penetration at video-rate acquisition speeds compared to other optical microscopy techniques. However, further technological development and understanding of tissue optics are needed to further reduce the size and cost of the instrumentation, improve the imaging capabilities, and transition confocal imaging from bench-top applications to the forefront of clinical diagnostics.

Acknowledgments

The authors greatly thank Jim Zavislan (University of Rochester) for providing the description of his work on coherence reflectance confocal imaging that has been supported by the National Institutes of Health (1R42CA110226); DongKyun Kang, Gary Tearney, and Brett Bouma (Massachusetts General Hospital) for providing information on SECM and Figure 25.18b; Chris Arrasmith and David Dickensheets (Montana State University) and Anita Mahadevan-Jansen (Vanderbilt University) for providing information on their handheld confocal microscope with Raman spectroscopy and Figure 25.15b; and Matthew Bouchard and Yogesh Patel for helping to prepare this chapter.

References

Amos, W.B. and White, J.G. 2003. How the confocal laser scanning microscope entered biological research. *Biol. Cell* 95: 335–342.

Arrasmith, C.L., Patil, C.A., Dickensheets, D.L. et al. 2009. A MEMS based handheld confocal microscope with Raman spectroscopy for *in-vivo* skin cancer diagnosis. *Proc. SPIE* 7169: 71690N.

Becker, V., Vercauteren, T., von Weyhern, C.H. et al. 2007. High-resolution miniprobe-base confocal microscopy in combination with video mosaicing (with video). *Gastrointest. Endosc.* 66: 1001–1007.

Bigelow, C.E. and Foster, T.H. 2006. Confocal fluorescence polarization microscopy in turbid media: Effects of scattering-induced depolarization. *J. Opt. Soc. Am. A* 23: 2932–2943.

Born, M. and Wolf, E. 1999. *Principles of Optics*, 7th edn. Cambridge: Cambridge University Press.

Botcherby, E.J., Booth, M.J., Juskaitis, R. et al. 2009. Real-time slit scanning microscopy in the meridional plane. *Opt. Lett.* 34: 1504–1506.

Botcherby, E.J., Juskaitis, R., Booth, M.J. et al. 2007. Aberration-free optical refocusing in high numerical aperture microscopy. *Opt. Lett.* 32: 2007–2009.

Botcherby, E.J., Juskaitis, R., Booth, M.J. et al. 2008. An optical technique for remote focusing in microscopy. *Opt. Comm.* 281: 880–887.

Boudoux, C., Yun, S.H., Oh, W.Y. et al. 2005. Rapid wavelength-swept spectrally encoded confocal microscopy. *Opt. Express* 13: 8214–8221.

Bouma, B. and Tearney, G. 2002. *Handbook of Optical Coherence Tomography*. New York: Marcel Dekker.

Brakenhoff, G.J., Blom, P., Barends, P. 1979. Confocal scanning light microscopy with high aperture immersion lenses. *J. Microsc.* 117: 219–232.

Brakenhoff, G.J. and Visscher, K. 1992. Confocal imaging with bilateral scanning and array detectors. *J. Microsc.* 165: 139–146.

Brock, R.S., Hu, X.-H., Weidner, D.A. et al. 2006. Effect of detailed cell structure on light scattering distribution: FDTD study of a B-cell with 3D structure constructed from confocal images. *J. Quant. Spectrosc. Radiat. Transfer* 102: 25–36.

Cagnet, M. 1962. *Atlas of Optical Phenomena*. Englewood Cliffs, NJ: Prentice Hall, p. 23.

Carlson, K., Chidley, M., Sung, K.-B. et al. 2005. In vivo fiber-optic confocal reflectance microscope with an injection-molded plastic miniature objective lens. *Appl. Opt.* 44: 1792–1797.

Caspers, P.J. Lucassen, G.W., Carter, E.A. et al. 2001. In vivo confocal Raman microspectroscopy of the skin: Noninvasive determination of molecular concentration profiles. *J. Invest. Dermatol.* 116: 434–442.

Caspers, P.J., Lucassen, G.W., and Puppels, G.J. 2003. Combined in vivo confocal Raman spectroscopy and confocal microscopy of human skin. *Biophys. J.* 85: 572–580.

Dabbs, T. and Glass, M. 1992. Fiber-optic confocal microscope: FOCON. *Appl. Opt.* 31: 3030–3035.

Denk, W., Strickler, J.H., and Webb, W.W. 1990. Two-photon laser scanning fluorescence microscopy. *Science* 248: 73–76.

Diaspro, A. 2002. *Confocal and Two-Photon Microscopy: Foundations, Applications, and Advances.* New York: Wiley-Liss.

Dickensheets, D.L. and Kino, G.S. 1996. Micromachined scanning confocal optical microscope. *Opt. Lett.* 21: 764–766.

Drezek, R., Dunn, A., and Richards-Kortum, R. 1999. Light scattering from cells: Finite-difference time-domain simulations and goniometric measurements. *Appl. Opt.* 38: 3651–3661.

Dunn, A.K. and Richards-Kortum, R. 1996. Three-dimensional computation of light scattering from cells. *IEEE J. Sel. Top. Quantum Electron.* 2: 898–905.

Dunn, A.K., Smithpeter, C., Welch, A.J. et al. 1996. Sources of contrast in confocal reflectance imaging. *Appl. Opt.* 35: 3441–3446.

Dwyer, P.J., DiMarzio, C.A., and Rajadhyaksha, M. 2007. Confocal theta line-scanning microscope for imaging human tissues. *Appl. Opt.* 46: 1843–1851.

Dwyer, P.J., DiMarzio, C.A., Zavislan, J.M. et al. 2006. Confocal reflectance theta line scanning microscope for imaging human skin *in vivo*. *Opt. Lett.* 31: 942–944.

Egger, M.D. and Petran, M., 1967. New reflected-light microscope for viewing unstained brain and ganglion cells. *Science* 157: 305–307.

Gareau, D.S., Abeytunge S, and Rajadhyaksha, M. 2009a. Line-scanning reflectance confocal microscopy: Comparison of full-pupil and divided-pupil configurations. *Opt. Lett.* 34: 3235–3237.

Gareau, D.S., Karen, J.K., Dusza, S.W. et al. 2009b. Sensitivity and specificity for detecting basal cell carcinomas in Mohs excisions for confocal fluorescence mosaicing microscopy. *J. Biomed. Opt.* 14: 034012.

Gareau, D.S., Li, Y., Huang, B. et al. 2008a. Confocal mosaicing microscopy in Mohs skin excisions: Feasibility of rapid surgical pathology. *J. Biomed. Opt.* 13: 054001.

Gareau, D.S., Patel, Y.G., Li, Y. et al. 2009c. Confocal mosaicing microscopy in skin excisions: A demonstration of rapid surgical pathology. *J. Microsc.* 233: 149–159.

Gareau, D.S., Patel, Y.G., and Rajadhyaksha, M. 2008b. Reflectance confocal microscopy of cutaneous tumors: An atlas with clinical, dermoscopic and histological correlations. In: eds. S. Gonzalez, M. Gill, and A.C. Halpern, *Basic Principles of Reflectance Confocal Microscopy.* London, U.K.: Informa Healthcare, pp. 1–7.

Gaskill, J. 1978. *Linear Systems, Fourier Transforms, and Optics.* New York: John Wiley & Sons.

Gmitro, A.F. and Aziz, D. 1993. Confocal microscopy through a fiber-optic imaging bundle. *Opt. Lett.* 18: 565–567.

Gniadecka, M., Philipsen, P., Sigurdsson, S. et al. 2004. Melanoma diagnosis by Raman spectroscopy and neural networks: Structure alterations in proteins and lipids in intact cancer tissue. *J. Invest. Dermatol.* 122: 443–449.

Goda, K., Tsia, K.K., and Jalali, B. 2009. Serial time-encoded amplified imaging for real-time observation of fast dynamic phenomena. *Nature* 458: 1145–1149.

Gong, W., Si, K., and Sheppard, C.J.R. 2009. Optimization of axial resolution in a confocal microscope with D-shaped apertures. *Appl. Opt.* 48: 3998–4002.

Goodman, J. 2004. *Introduction to Fourier Optics*, 3rd edn. Englewood Cliffs, NJ: Roberts & Company Publishers, pp. 103–108.

Göppert, M. 1929. Über die wahrscheinlichkeit des zusamenwirkens zweier licht quanten in einem elementarakt. *Die Naturwissenschaften*, 17: 932.

Göppert-Mayer, M. 1931. Über elementarakte mit zwei quantensprüngen. *Ann. Phys.* 9: 273–294.

Gu, M. 1996. *Principles of Three-Dimensional Imaging in Confocal Microscopes.* Singapore: World Scientific.

Gustafsson, M.G.L. 2000. Surpassing the lateral resolution limit by a factor of two using structured illumination microscopy. *J. Microsc.* 198: 82–87.

Im, K.-B., Han, S., Park, H. et al. 2005. Simple high-speed confocal line-scanning microscope. *Opt. Express* 13: 5151–5156.

Inoue, S. and Spring, K.R. 1997. *Video Microscopy.* New York: Plenum Press.

Jonkman, J.E.N. and Stelzer, E.H.K. 2002. Resolution and contrast in confocal and two-photon microscopy. In: *Confocal and Two-Photon Microscopy: Foundations, Applications, and Advances*, ed. A. Diaspro. New York: Wiley-Liss, pp. 101–125.

Kang, D.K., Suter, M.J., Boudoux, C. et al. 2010. Comprehensive imaging of gastroesophageal biopsy samples by spectrally encoded confocal microscopy. *Gastrointest. Endosc.* 71: 35–43.

Kempe, M. 2000. Confocal microscope, US Patent 6,151,127.

Kempe, M. 2002. Confocal heterodyne interference microscopy, US Patent 6,381,023.

Kim, J. Kang, D.K., and Gweon, D.G. 2006. Spectrally encoded slit confocal microscopy. *Opt. Lett.* 31: 1687–1689.

Koester, C.J. 1980. A scanning mirror microscope with optical sectioning characteristics: Applications in ophthalmology. *Appl. Opt.* 19: 1749–1757.

Koester, C.J., Auran, J.D., Rosskothen, H.D. et al. 1993. Clinical microscopy of the cornea utilizing optical sectioning and a high numerical aperture objective. *J. Opt. Soc. Am. A.* 10: 1670–1679.

Kumar, K., Hoshino, K., and Zhang, X. 2008. Handheld subcellular-resolution single-fiber confocal microscope using high-reflectivity two-axis vertical combdrive silicon microscanner. *Biomed. Microdevices* 10: 1387–2176.

Liang, C., Descour, M.R., Sung. K.-B. et al. 2001. Fiber confocal reflectance microscope (FCRM) for *in-vivo* imaging. *Opt. Express* 9: 821–830.

Lindek, S. and Stelzer, E.H.K. 1996. Optical transfer functions for confocal theta fluorescence microscopy. *J. Opt. Soc. Am. A* 13: 479–482.

Liu, J.T.C., Mandella, M.J., Ra, H. et al. 2007. Miniature near-infrared dual-axes confocal microscope utilizing a two-dimensional microelectromechanical systems scanner. *Opt. Lett.* 32: 256–258.

Maitland, K.C., Shin, H.J., Ra, H. et al. 2006. Single fiber confocal microscope with a two-axis gimbaled MEMS scanner for cellular imaging. *Opt. Express* 14: 8604–8612.

Makhlouf, H., Gmitro, A.F., Tanbakuchi, A.A. et al. 2008. Multispectral confocal microendoscope for *in vivo* and *in situ* imaging. *J. Biomed. Opt.* 13: 044016.

Masters, B.R. and So, P.T.C. 2008. *Handbook of Biomedical Nonlinear Optical Microscopy.* New York: Oxford University Press, pp. 42–59.

Maurice, D.M. 1974. A scanning slit optical microscope. *Invest. Ophthalmol.* 13: 1033–1037.

McNally, J.G., Karpova, T., Cooper, J. et al. 1999. Three-dimensional imaging by deconvolution microscopy. *Methods* 19: 373–385.

Minsky, M. 1955. Microscopy apparatus, U.S. Patent Number 3,013,467.

Minsky, M. 1988. Memoir on inventing the CSM. *Scanning* 10: 128–138.

Murphy, D.B. 2001. *Fundamentals of Light Microscopy and Electronic Imaging.* New York: Wiley-Liss.

Nakao, M., Yoshida, S., Tanaka, S. et al. 2008. Optical biopsy of early gastroesophageal cancer by catheter-based reflectance-type laser-scanning confocal microscopy. *J. Biomed. Opt.* 13: 054043.

Naora, H. 1951. Microspectrophotometry and cytochemical analysis of nucleic acids. *Science* 114: 279–280.

Neil, M.A.A., Juskaitis, R., and Wilson, T. 1997. Method of obtaining optical sectioning by using structured light in a conventional microscope. *Opt. Lett.* 22: 1905–1907.

Newmark, J.A., Warger, W.C. II, Chang, C. et al. 2007. Determination of the number of cells in preimplantation embryos by using noninvasive optical quadrature microscopy in conjunction with differential interference contrast microscopy. *Microsc. Microanal.* 13: 1–10.

Patel, Y.G., Nehal, K.S., Aranda, I. et al. 2007. Confocal reflectance mosaicing of basal cell carcinomas in Mohs surgical skin excisions. *J. Biomed. Opt.* 12: 034027.

Pawley, J.B. 2006. *Handbook of Biological Confocal Microscopy,* 3rd edn. New York: Springer.

Petran, M. and Hadravsky, M. 1970. Method and arrangement for improving the resolving power and contrast, U.S. Patent Number 3,517,980.

Pitris, C., Bouma, B.E., Shiskov, M. et al. 2003. A GRISM-based probe for spectrally encoded confocal microscopy. *Opt. Express* 11: 120–124.

Poher, V., Kennedy, G.T., Manning, H.B. et al. 2008. Improved sectioning in a slit scanning confocal microscope. *Opt. Lett.* 33: 1813–1815.

Poher, V., Zhang, H.X., Kennedy, G.T. et al. 2007. Optical sectioning microscopes with no moving parts using a micro-stripe array light emitting diode. *Opt. Express* 15: 11196–11206.

Poland, S.P., Girkin, J.M., Li, L. et al. 2009. External linearization of a MEMS electrothermal scanner with application in a confocal microscope. *Micro Nano Lett.* 4: 106–111.

Ra, H., Piyawattanametha, W., Mandella, M.J. et al. 2008. Three-dimensional *in vivo* imaging by a handheld dual-axes confocal microscope. *Opt. Express* 16: 7224–7232.

Ra, H., Piyawattanametha, W., Taguchi, Y. et al. 2007. Two-dimensional MEMS scanner for dual-axes confocal microscopy. *J. Microelectromech. Syst.* 16: 969–976.

Rajadhyaksha, M., Anderson, R., Webb, R. 1999a. Video-rate confocal scanning laser microscope for imaging human tissue in vivo. *Appl. Opt.* 38: 2105–2115.

Rajadhyaksha, M. and Gonzalez, S. 2003. Real-time in vivo confocal fluorescence microscopy. In: *Handbook of Biomedical Fluorescence,* eds. Mycek, M.A. and Pogue, B.W.. New York: Marcel Dekker, Inc., pp. 143–179.

Rajadhyaksha, M., Gonzalez, S., Zavislan, J. et al. 1999b. *In vivo* confocal scanning laser microscopy of human skin II: Advances in instrumentation and comparison with histology. *J. Invest. Dermatol.* 113: 293–303.

Rajadhyaksha, M., Gonzalez, S., and Zavislan, J.M. 2004. Detectability of contrast agents for confocal reflectance imaging of skin and microcirculation. *J. Biomed. Opt.* 9: 323–331.

Rajadhyaksha, M., Menaker, G., Flotte, T. et al. 2001. Confocal examination of nonmelanoma cancers in thick skin excisions to potentially guide Mohs micrographic surgery without frozen histopathology. *J. Invest. Dermatol.* 117: 1137–1143.

Rouse, A.R., Kano, A., Udovich, J.A. et al. 2004. Design and demonstration of a miniature catheter for a confocal microendoscope. *Appl. Opt.* 43: 5763–5771.

Saleh, B. and Teich, M. 1991. *Fundamentals of Photonics.* New York: Wiley-Interscience, p. 87.

Schmitt, J.M., Knuttel, A., and Yadlowsky, M. 1994. Confocal microscopy in turbid media. *J. Opt. Soc. Am. A* 11: 2226–2235.

Seibel, E.J. and Smithwick, Q.Y. 2002. Unique features of optical scanning, single fiber endoscopy. *Lasers Surg. Med.* 30: 177–183.

Sheppard, C.J.R. 2000. Confocal imaging through weakly aberrating media. *Appl. Opt.* 39: 6366–6368.

Sheppard, C.J.R. and Choudhury, A. 1977. Image formation in the scanning microscope. *Optica Acta* 24: 1051–1073.

Sheppard, C.J.R., Gan, X., Gu, M. et al. 2006. Signal-to-noise ratio in confocal microscopes. In: *Handbook of Biological Confocal Microscopy,* ed. J.B. Pawley, 3rd edn. New York: Springer, pp. 442–452.

Sheppard, C.J.R., Gong, W., and Si, W. 2008. The divided aperture technique for microscopy through scattering media. *Opt. Express* 16: 17031–17038.

Sheppard, C.J.R. and Gu, M. 1994. Imaging performance of confocal fluorescence microscopes with finite-sized source. *J. Mod. Opt.* 41: 1521–1530.

Sheppard, C.J.R. and Kompfner, R. 1978. Resonant scanning optical microscope. *Appl. Opt.* 17: 2879–2882.

Sheppard, C.J.R. and Mao, X.Q. 1988. Confocal microscopes with slit apertures. *J. Mod. Opt.* 35: 1169–1685.

Sheppard, C.J.R. and Matthews, H.J. 1987. Imaging in high-aperture optical systems. *J. Opt. Soc. Am. A* 4: 1354–1360.

Sheppard, C.J.R. and Shotton, D. 1997. *Confocal Laser Scanning Microscopy.* New York: BIOS Scientific Publishers.

Sheppard, C.J.R. and Wilson, T. 1978. Depth of field in the scanning microscope. *Opt. Lett.* 3: 115–117.

Sheppard, C.J.R. and Wilson, T. 1979. Effect of spherical aberration on the imaging properties of scanning optical microscopes. *Appl. Opt.* 18: 1058–1063.

Shin, H.-J., Pierce, M.C., Lee, D. et al. 2007. Fiber-optic confocal microscope using a MEMS scanner and miniature objective lens. *Opt. Express* 15: 9113–9211.

Shirasaki, M. 1996. Large angular dispersion by a virtually imaged phased array and its application to a wavelength demultiplexer. *Opt. Lett.* 21: 366–368.

Si, K., Gong, W., and Sheppard, C.J.R. 2009. Three-dimensional coherent transfer function for a confocal microscope with two D-shaped pupils. *Appl. Opt.* 48: 810–817.

Sigurdsson, S., Philipsen, P. Hansen, L. et al. 2004. Detection of skin cancer by classification of Raman spectra. *IEEE Trans. Biomed. Eng.* 51: 1784–1793.

Simon, B. and DiMarzio, C. 2007. Simulation of a theta line-scanning confocal microscope. *J. Biomed. Opt.* 12: 064020.

Smithpeter, C., Dunn, A., Drezek, R. et al. 1998a. Near real time confocal microscopy of cultured amelanotic cells: Sources of signal, contrast agents, and limits of contrast. *J. Biomed. Opt.* 3: 429–436.

Smithpeter, C.L., Dunn, A.K., Welch, A.J. et al. 1998b. Penetration depth limits of *in vivo* confocal reflectance imaging. *Appl. Opt.* 37: 2749–2754.

Stelzer, E. H. K., Lindek, S., Albrecht, S. et al. 1995. A new tool for the observation of embryos and other large specimens—Confocal theta-fluorescence microscopy. *J. Microsc.* 179: 1–10.

Tanaami, T., Otsuki, S., Tomosada, N. et al. 2002. High-speed 1-frame/ms scanning confocal microscope with a microlens and Nipkow disks. *Appl. Opt.* 41: 4704–4708.

Tearney, G.J., Webb, R.H., and Bouma, B.E. 1998. Spectrally encoded confocal microscopy. *Opt. Lett.* 23: 1152–1154.

Toomre, D. and Pawley, J.B. 2006. Disk-scanning confocal microscopy. In: *Handbook of Biological Confocal Microscopy*, ed. J.B. Pawley, 3rd edn. New York: Springer, pp. 221–238.

Torok, P., Hewlett, S.J., and Varga, P. 1997. The role of specimen-induced spherical aberration in confocal microscopy. *J. Microsc.* 188: 158–172.

Torok, P. and Wilson, T. 1997. Rigorous theory for axial resolution in confocal microscopes. *Opt. Comm.* 137: 127–135.

Tsia, K.K., Goda, K., Capewell, D. et al. 2009. Simultaneous mechanical-scan-free confocal microscopy and laser microsurgery. *Opt. Lett.* 34: 2099–2101.

Tsai, C.-L., Warger, W.C. II, Laevsky, G.S. et al. 2008. Alignment with sub-pixel accuracy for images of multi-modality microscopes using automatic calibration. *J. Microsc.* 232: 164–176.

Vercauteren, T., Perchant, A., Malandain, G. et al. 2006. Robust mosaicing with correction of motion distortions and tissue deformations for *in vivo* fibered microscopy. *Med. Image Anal.* 10: 673–692.

Vermot, J., Fraser, S.E., and Liebling, M. 2008. Fast fluorescence microscopy for imaging the dynamics of embryonic development. *HFSP J.* 2: 143–155.

Wallace, W., Schaefer, L.H., and Swedlow, J.R. 2001. A working-person's guide to deconvolution in light microscopy. *BioTechniques* 31: 1076–1097.

Wang, Z., Glazowski, C., and Zavislan, J.M. 2006. Coherence reflectance confocal microscope using a synthesized broad-band source. In *Opt. Soc. Am., Biomedical Topical Meeting (BIO)*, Fort Lauderdale, FL.

Wang, Z., Glazowski, C., and Zavislan, J.M. 2007. Modulation transfer function measurement of scanning reflectance microscopes. *J. Biomed. Opt.* 12: 051802.

Warger, W.C. II and DiMarzio, C.A. 2007. Dual-wedge scanning confocal reflectance microscope. *Opt. Lett.* 32: 40–42.

Warger, W.C. II and DiMarzio, C.A. 2009. Computational signal-to-noise ratio analysis for optical quadrature microscopy. *Opt. Express* 17: 2400–2422.

Warger, W.C. II, Laevsky, G.S., Townsend, D.J. et al. 2007. Multimodal optical microscope for detecting viability of mouse embryos *in vitro*. *J. Biomed. Opt.* 12: 044006.

Warger, W.C. II, Newmark, J.A., Warner, C.A. et al. 2008. Phase-subtraction cell-counting method for live mouse embryos beyond the eight-cell stage. *J. Biomed. Opt.* 13: 034005.

Webb, R.H. 1996. Confocal optical microscopy. *Rep. Prog. Phys.* 59: 427–471.

Webb, R.H. and Rogomentich, F. 1999. Confocal microscope with large field and working distance. *Appl. Opt.* 38: 4870–4875.

Wilson, T. 1990. *Confocal Microscopy.* London, U.K.: Academic Press.

Wilson, T. and Carlini, A.R. 1987. Size of the detector in confocal imaging systems. *Opt. Lett.* 12: 227–229.

Wilson, T. and Hewlett, S.J. 1990. Imaging in scanning microscopes with slit-shaped detectors. *J. Microsc.* 160: 115–139.

Wilson, T. and Juskaitis, R. 1995. The axial response of confocal microscopes with high numerical aperture objective lenses. *Bioimaging* 3: 35–38.

Wolleschensky, R., Zimmermann, B., and Kempe, M. 2006. High-speed confocal fluorescence imaging with a novel line scanning microscope. *J. Biomed. Opt.* 11: 064011.

Yelin, D., Rizvi, I., White, W.M. et al. 2006. Three-dimensional miniature endoscopy. *Nature.* 443: 765.

Yelin, D., Boudoux, C., Bouma, B.E. et al. 2007a. Large area confocal microscopy. *Opt. Lett.* 32: 1102–1104.

26

Fluorescence Microscopy with Structured Excitation Illumination

Alexander Brunner
Heidelberg University

Gerrit Best
Heidelberg University

Paul Lemmer
Heidelberg University

Roman Amberger
Heidelberg University

Thomas Ach
University Hospital Heidelberg

Stefan Dithmar
University Hospital Heidelberg

Rainer Heintzmann
*King's College London,
Institute of Photonic Technology,
and
University of Jena*

Christoph Cremer
*Heidelberg University,
The Jackson Laboratory
and
University of Maine*

26.1 Introduction 543
26.2 Abbe's Resolution Definition 544
26.3 Focusing SEI Methods 544
    Confocal Laser Scanning Microscopy • 4Pi-Microscopy • STED Microscopy
26.4 Wide-Field SEI Methods... 545
    Spatially Modulated Illumination Microscopy • Structured Illumination Microscopy Using
    One Objective Lens
26.5 The Conceptual Basis of the SIM Method 547
    Image Formation in Wide-Field Microscopy • Polarization Effects • Two-Beam
    Interference • Three-Beam Interference • Image Formation Using Structured
    Illumination • Image Reconstruction • Various Setups for SIM • Experimental Examples
    of the Resolution Improvement Using Structured Illumination • Perspective: Combination of
    SEI with Localization Microscopy
26.6 Discussion and Outlook 558
Acknowledgments.. 558
References.. 558

26.1 Introduction

Light microscopy has several advantages over non-light-optical microscopy methods in terms of sample preparation and cost-effectiveness, as well as in the variety of possibilities to extract biologically significant information. Especially the advances in fluorescent labeling and thus in fluorescence microscopy opened a broad diversity of utilizations.

However, the major drawback of standard fluorescence microscopy lies in the intrinsically limited resolving power of standard microscopes.

Already in the nineteenth century, this fundamental limit was recognized as depending on the wavelength of the light and on the numerical aperture (NA) of the objective lens. As the definition of the resolving power of a microscope is arbitrary, Abbe (1873) and Rayleigh (1896) both developed independent

resolution definitions on conditions they thought favorable. Over the years, other resolution definitions have been established delivering different absolute values for the resolution power of a microscope. Hence, it is always important to state the underlying definition if referring to the resolution of an optical system.

By the definition given by Abbe, which will be discussed below, the resolution limit of an objective lens with a high NA in a standard microscope is roughly 200 nm. Abbe's definition was originally developed for transmission light microscopes only but is also applicable for fluorescence microscopes with homogeneous illumination.

On the assumptions made (homogeneous illumination), this limit of optical resolution is still valid today.

However, Abbe already stated that the resolution limit of about half the wavelength used for imaging is valid only "... so lange nicht Momente geltend gemacht werden, die ganz ausserhalb der Tragweite der aufgestellten Theorie liegen …" ("… as long as no arguments are made which are completely beyond the theory stated here …") (Abbe 1873, p. 468).

Since the 1980s, this prophecy of Abbe has become true: Using approaches beyond the assumptions made by Abbe or by Rayleigh, techniques surpassing this limit have been developed. One experimental technique exploits the properties of the near field. One of the difficulties in the application of near field scanning optical microscopy (NSOM) (Lewis et al. 1984, Pohl et al. 1984) has been the required short working distance between the scanning tip and the object (fractions of a wavelength), thus being effectively limited to the imaging of surfaces only; since then, a variety of laser-optical far-field microscopy techniques (i.e., allowing a working distance in the $200\,\mu m$ range) based on fluorescence excitation has been developed to circumvent the Abbe limit of about 200 nm for visible light at least in one direction, either the lateral (x,y; object plane) or the axial direction (z; along the optical axis of the microscope system).

Essential to these techniques is illuminating and detecting light from fluorophores. Fluorescence maintains no phase relationship with its illuminating light, which allows illumination and detection to be treated independently both contributing to the achieved resolution. Finally, the techniques achieving the highest resolution make use of a distinctively nonlinear dependence of the emitted intensity on at least one of the illuminating beams. This is often achieved via utilizing reversible saturable optical fluorescence transitions (Hell 2007).

For these reasons, almost all super-resolution far-field imaging methods make use of a tailored spatial distribution of the excitation light. These structured excitation illumination (SEI) applications can be differentiated into focusing and wide-field (unfocused) methods.

The focusing methods evolved over the last decades from confocal laser scanning microscopy (CLSM) to more elaborate and more potent 4Pi- and stimulated emission depletion (STED) microscopy methods allowing a considerably improved optical resolution. However, to retrieve high-resolution information by SEI, the illumination intensity and the detected area do not necessarily have to each be a single diffraction-limited spot in the object plane. The wide-field methods of spatially modulated illumination (SMI) and structured illumination microscopy (SIM) apply a periodically modulated excitation intensity in the object.

In the following, the focus will be mainly on periodically modulated intensity by illumination and detection through one objective lens (SIM). Other SEI methods mentioned above will be presented briefly.

26.2 Abbe's Resolution Definition

In his famous contribution of 1873, Ernst Abbe defined the resolving power by the use of a fine grating with a small peak-to-peak distance.

The fine grating is located in the object plane of a transmitted light microscope. If the lattice spacing of the grating surpasses a certain distance, the grating becomes irresolvable and is thus not visible in the image plane. The resolution of the microscope is now defined by Abbe as the minimum lattice spacing (d) that can still be resolved by the microscope given by

$$d = \frac{\lambda}{2 \cdot NA} \qquad (26.1)$$

with the numerical aperature (NA)

$$NA = n \cdot \sin(\alpha) \qquad (26.2)$$

and with the refractive index n and the half opening angle of the objective α.

For a high NA oil objective (NA = 1.4) with blue-light illumination (420 nm), this formula leads to an Abbe resolution of 150 nm.

The Abbe definition of a minimum resolvable spacing that corresponds to a maximum transmittable frequency in Fourier Space will be used in the following to classify the resolving power of microscopes.

Ernst Abbe (Abbe 1873, p. 456) stated "Da nun auch beim Immersionssystem der Öffungswinkel durch kein Mittel erheblich über diejenige Größe, die 180° in Luft entsprechen würde, hinausgeführt werden kann, so folgt, dass, wie auch das Mikroskop in Bezug auf die förderliche Vergrößerung noch weiter vervollkommnet werden möchte, die Unterscheidungsgrenze…niemals über den der halben Wellenlänge des blauen Lichts um ein Nennenswertes hinausgehen wird." ("Since even in the case of an immersion system the aperture angle by no means can be significantly enlarged above the value corresponding to 180° in air, it follows that whatever microscopic perfection might be obtained in relation to the useful magnification, the limits of discrimination will never substantially surpass half the wavelength of blue light.") On the assumptions made, Abbe's limit of optical resolution is still valid today. This limit will be denoted as the Abbe limit in the following.

26.3 Focusing SEI Methods

26.3.1 Confocal Laser Scanning Microscopy

The first approach in principle already circumventing the Abbe limit by application of inhomogeneous illumination in combination with fluorescence (SEI) is the CLSM.

Already in 1957, Marvin Minsky filed a U.S. patent (Minsky 1961) for a "microscopy apparatus." This confocal scanning microscope already contained basic components of today's CLSM. Even though Minsky conceived the apparatus as a transmitted light microscope (not as fluorescence microscope) with the use of a second objective or an additional mirror, he intended to use illumination and detection pinholes in inter-mediate image planes to constrain the illuminated and detected area to a diffraction-limited spot. However, this revolutionary idea was ahead of its time, as computer capabilities at that time were insufficient to process the scanning data (Minsky intended a cathode ray tube to record the data), fluorescence microscopy was not popular, and first lasers were just being developed. The first spinning disk confocal using white light was presented in 1967 by Egger and Petrăn (1967). As the evolution of the tech-niques listed above progressed, first ideas to apply lasers in confocal microscopy to excite fluorescence arose (Cremer and Cremer 1978).

Today, CLSM has become an indispensable technique in biomedical studies. The main advantage over conventional fluorescence microscopy lies in the high contrast achieved in CLSM imaging by an effective suppression of out-of-focus light. The theoretical attainable resolution improvement of linear SEI using Abbe's definition is by a factor of two. To practically achieve this high value, however, the detection pinhole would have to be very small resulting in a low sig-nal to noise ratio. Therefore, in practice, a compromise has to be found and the resolution improvement over conventional wide-field microscopy is relatively small compared to newer, more elaborate SEI methods.

26.3.2 4Pi-Microscopy

The concept of 4Pi-microscopy originated from the fact that the resolving power of a standard wide-field microscope as well as of a standard confocal microscope is limited by the NA given by the objective lens.

Early ideas about a resolution improvement in confocal laser scanning fluorescence microscopy by the application of a "4π-geometry" focusing the illuminating laser beam from all sides (spatial angle 4π) were emerging already in the 1970s (Cremer and Cremer 1978).

The first experimental confocal laser scanning 4Pi-microscope was designed and realized in the early 1990s (Hell 1990, Hell and Stelzer 1992): Two opposing high NA lenses were used to concentrate two opposing laser beams constructively in a joint focus; although this did not allow to realize a full 4π-geometry for an isotropic resolution increase, it was sufficient to "point-spread-function-engi-neer" the focal diameter (for the excitation as well as for the detection) in the direction of the optical axis to values sub-stantially below 200 nm. Presently, confocal laser scanning 4Pi fluorescence microscopy applying either continuous wave visible laser light or femtosecond pulsed infrared laser wavelengths for excitation (Hell et al. 1994, Hänninen et al. 1995) has become an established light-optical "nanoscopy" method (Egner et al. 2002, Bewersdorf et al. 2006, Baddeley et al. 2006, Lang et al. 2010); an axial optical resolution better than 100 nm was experimentally realized, that is, about six to seven times better than in conventional confocal micros-copy. By combination of such a lens-based 4Pi-microscope

with micro-axial tomography techniques (Bradl et al. 1994, Heintzmann and Cremer 2002) even an isotropic optical res-olution (i.e., both laterally and axially) below the Abbe limit would be possible.

26.3.3 STED Microscopy

In experimental 4Pi-microscopy, so far a 3D optical resolution around $200 \times 200 \times 100$ nm^3 was realized. This means that cel-lular nanostructures with spatial features below about 100 nm would still remain unresolved. As lens-based 4Pi-microscopy, the STED microscopy conceived by Stefan Hell (Hell and Wichmann 1994) and realized in the following decade is based on scanning the object with a focused laser beam (Schrader et al. 1995). However, in contrast to 4Pi-microscopy approaches realized so far, STED microscopy surpassed the Abbe limit in the object plane (x,y) down to the 10 nm range and presently has found numerous applications in high-res-olution cell biology (Willing et al. 2007, Nagerl et al. 2008, Schmidt et al. 2008, Westphal et al. 2008). The basic idea of STED microscopy is to reduce the size of the region giving rise to fluorescence by using an excitation pulse that is followed by a "depletion" pulse ("STED pulse") acting in the vicinity of the center of the fluorescent region but not in its very center. By appropriate saturation intensities, the fluorophores in this annulus are induced to emit photons by stimulated emission of radiation at a red shifted wavelength compared to the fluo-rescence emission maximum. Remaining fluorescence, which can be spectrally separated from the very strong STED beam, can therefore only originate from the very center. As a conse-quence, fluorescence photons of a given energy are detected from a much smaller region. Since due to the scanning mech-anism, the position of this smaller fluorescent region can be known with an accuracy of few nanometers, the fluorescence signal obtained can now be assigned to this smaller region; hence, the optical resolution may be improved further. STED microscopy was the first implementation of utilizing nonlin-ear response in fluorescence microscopy. Other such concepts are "ground state depletion microscopy" (Willing et al. 2007, Rittweger et al. 2009) and nonlinear structured illumination (Heintzmann et al. 2002).

26.4 Wide-Field SEI Methods

26.4.1 Spatially Modulated Illumination Microscopy

SMI microscopy is a method of wide-field fluorescence micros-copy using structured illumination to obtain additional high-resolution information about sizes and relative positions of fluorescently marked target regions. To generate the illumina-tion pattern, two counter-propagating laser beams are brought to coherent interference, establishing a standing-wave field (Lanni et al. 1986, Bailey et al. 1993) (Figure 26.1). In contrast

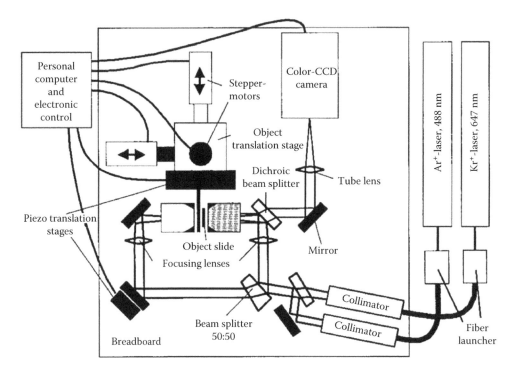

FIGURE 26.1 SMI microscope setup: Briefly, it consists of a rectangular interferometer with a 50:50 beam splitter. Two vertically polarized beams are focused into the back focal plane of a corresponding objective lens. A standing-wave field is produced between the two objective lenses. Objects mounted upon a conventional glass slide and covered by a conventional coverslip can be moved along the axial direction as the object translation stage is moved by stepper motors. For the detection, a cooled color CCD camera is used. (From Failla, A.V. et al., *Appl. Opt.*, 41(34), 7275, 2002.)

to focused laser light techniques (Hell et al. 1994, Klar et al. 2000, Dyba and Hell 2002, Egner et al. 2002) or other types of structured illumination with excitation intensity modulation along the object plane (Gustafsson 2000, Frohn et al. 2000, 2001, Heintzmann and Cremer 1999, Heintzmann et al. 2002), as described in the next paragraph, this method by itself is not suited for the generation of high-resolution images due to a range of missing intermediate spatial frequencies. However, in combination with high-precision axial positioning, this technique of far-field light microscopy allows the nondestructive analysis of complex spatial arrangements (Albrecht et al. 2002) inside relatively thick transparent specimens such as the cell nucleus and enables size measurements at molecular dimensions of a few tens of nanometers (Failla et al. 2002, 2003). SMI microscopy has become an established method for the analysis of topological arrangements of mammalian genome structure. In combination with novel approaches for fluorescence labeling, the SMI nanosizing technique has proved its applicability for a wide range of biological questions when using fixed cell preparations (Martin et al. 2004, Hildenbrand et al. 2005, Mathee et al. 2006, Baddeley et al. 2010). It has even been shown to allow nanosizing measurements in the living-cell nucleus (Reymann et al. 2008).

For example, the size of a specific chromatin region in a cell is regarded to have a decisive influence on its genetic activity; thus, size measurements may provide key information toward a better understanding of the regulation of gene activity. Another example is the identification of functional biomolecular machines by analyzing the co-localization of its constituents: It is obvious that a necessary requirement for the proper functioning of a biomolecular machine is that its constituents are sufficiently close together, that is, the diameter of the minimum volume enveloping these elements has to meet certain limits.

26.4.2 Structured Illumination Microscopy Using One Objective Lens

The use of only one objective lens for structured illumination has many obvious advantages. For example, inverse microscopy setups for living-cell observations are easy to implement.

In addition, the SMI approach described above so far did not allow a "true" increase in optical resolution; in particular, the lateral optical resolution is not improved in SMI.

In the following, we describe an approach of structured illumination to realize a substantial improvement of optical resolution both in the lateral and in the axial direction, using a single objective lens.

Since in this case the resolution improvement is based on two or three interfering excitation beams, the impact of the periodically distributed excitation intensity has to be considered in the process of image formation. The following sections of this contribution are focused on SIM beginning with the image-formation process.

26.5 The Conceptual Basis of the SIM Method

26.5.1 Image Formation in Wide-Field Microscopy

For linear shift invariant optical systems (Box 26.1), the relation between a fluorescence molecule distribution ρ and the formed image g is given by a convolution with the point spread function (PSF) (in the following, noise effects are not considered):

$$g = \text{PSF} \otimes \rho \qquad (26.3)$$

Transformation into Fourier space, abbreviated by FT, leads to (convolution theorem)

$$\text{FT}[g] = \text{OTF} \cdot \text{FT}[\rho] \quad \text{with} \quad \text{FT (PSF)} = \text{OTF} \qquad (26.4)$$

During image formation, the multiplication by the optical transfer function (OTF) (Box 26.2) acts like a low pass filter: only frequencies smaller than Abbe's cutoff frequency (derived from Equation 26.1) are transmitted by the optical system (e.g., the microscope, mainly the objective lens). Moreover, the OTF decreases in transfer strength from the origin to the cutoff frequency, which results in reduced contrast transmission. The reduced contrast can be compensated by an appropriate weighting, only limited by image quality (signal-to-noise ratio, abbreviated as SNR).

In contrast to a homogeneous wide-field illumination, the fluorescence molecule distribution has to be multiplied by the intensity distribution (I) of the illumination.

In case of structured illumination:

$$g = \text{PSF} \otimes (\rho \cdot I) \qquad (26.5)$$

This assumes that the fluorescence molecule emits photons at a rate proportional to the local excitation irradiance. In Fourier space, respectively:

$$\text{FT}[g] = \text{OTF} \cdot (\text{FT}[\rho] \otimes \text{FT}[I]) \qquad (26.6)$$

For further calculations, an expression for the intensity distribution (I) of the illumination is required. Typically, the intensity distribution (I) of the illumination is generated by the

interference of two or three beams coupled through a single objective.

In the following sections, we will discuss the case of two- and three-beam interference and the accordant outcome for image formation.

26.5.2 Polarization Effects

When regarding interference as it is used in SIM to generate the modulated excitation intensity, one must consider the polarization of the beams which affect the resulting interference pattern. Usually the interfering beams span one common plane, the interferometer plane. While the optional middle beam (three-beam interference) is parallel to the optical axis and enters the objective centrally, the respective angles of the two outer beams to the optical axis are typically $\alpha_l = \alpha$ and $\alpha_r = -\alpha$. To examine the polarization effects, one can analyze two extreme cases in two-beam interference: linear polarization perpendicular and parallel to the interferometer plane. The focusing by an lens causes a rotation of the propagation direction by an angle $\pm\alpha$ around the interferometer plane normal vector and thereby also a rotation of the polarization (Figure 26.2).

If the polarization is perpendicular (s) to the interferometer plane prior to entering the lens system, the refraction does not change the polarization. As the polarization of the two beams is within the interferometer plane (p), the intensity contrast can approach zero at 45° to the axis. The polarization vector stays parallel to the interferometer plane but is rotated by the angle $\pm\alpha$. Thereby, the interference decreases and even disappears at $\alpha = 45°$. At $\alpha > 45°$, the interference pattern is shifted by π compared to the pattern for perpendicular polarization (Figure 26.3).

To achieve maximum interference at different interferometer planes, the polarization of the incoming linear polarized beams has to be always azimuthal.

26.5.3 Two-Beam Interference

The intensity distribution (I) of the illumination generated by a two-beam interference can be calculated assuming the

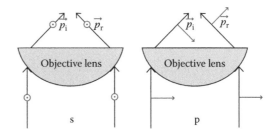

FIGURE 26.2 Effect of the lens on the polarization **p** of the beams in two-beam interference (s: for polarization perpendicular to the interferometer plane, p: for polarization parallel to the interferometer plane, $\vec{P}_l$: left polarization vector, $\vec{P}_r$: right polarization vector).

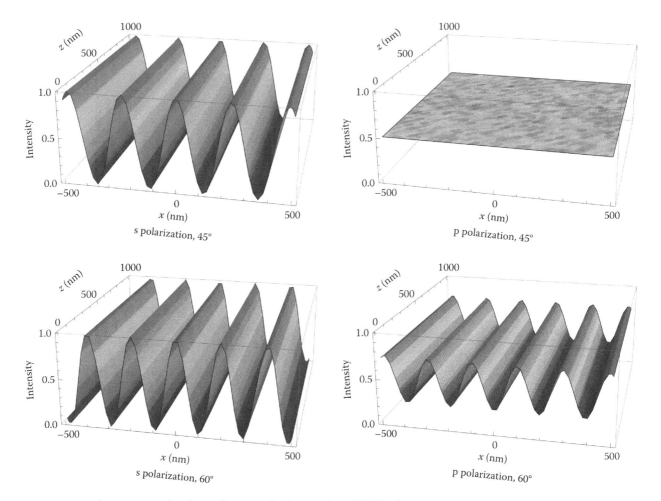

FIGURE 26.3   Simulated intensity distribution for perpendicular (s) and parallel (p) polarization in two-beam interference under different angles of incidence (NA = 1.4, λ_0 = 488 nm, n = 1.45).

interference of two planar coherent waves each described by a wave vector of size k at angle $\pm\alpha$ to the optic axis.

According to the linearity of the Maxwell equations, which implies that the sum of two amplitude distributions constitutes also a solution of the Maxwell equations (alternatively the d'Alembert equation), plane light waves can simply be added (Figure 26.4). Hence the y-component of the electric field is written as

$$E = E_{-1}e^{-i[k(-\sin(\alpha)\cdot x+\cos(\alpha)\cdot z)-\phi_1]} + E_{+1}e^{-i[k(\sin(\alpha)\cdot x+\cos(\alpha)\cdot z)-\phi_2]} \quad (26.7)$$

Here E_{-1} and E_{-2} are amplitudes of incident plane light waves, respectively, and ϕ is a phase shift as would be obtained by translating a grating that generates these beams. To achieve the highest signal to noise ratio, an optimal intensity contrast is required and achieved by interfering beams of equal intensity I_0:

$$E_{-1} = E_{+1} = \sqrt{I_0} \quad (26.8)$$

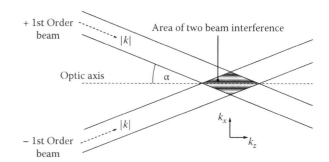

FIGURE 26.4   Schematic of the examined system for two-beam interference: two incident planar coherent waves, denoted as +1st order beam and −1st order beam each described by a wave vector of size k, interfere under an angle of 2α in the area of overlap.

For the squared absolute value (* indicates complex conjugation) of the electrical field, using the identity $\cos(x) = (1/2)(e^{ix} + e^{-ix})$, one obtains the intensity distribution:

$$I = E \cdot E^* = 2I_0[1 + \cos(2k \cdot \sin(\alpha) \cdot x + \phi)] \quad (26.9)$$

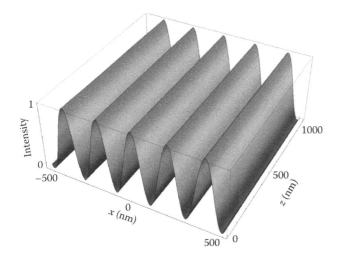

FIGURE 26.5 Intensity distribution for two-beam interference ($\alpha = 60°$, $\lambda_0 = 488\,\text{nm}$, $n = 1.45$, $\text{NA} = 1.4$, $E_{-1} = E_{+1}$).

with

$$\varphi = \frac{\varphi_1}{2} - \frac{\varphi_2}{2} \qquad (26.10)$$

Note that the intensity distribution has no dependency of the axial coordinate z (Figure 26.5).

26.5.4 Three-Beam Interference

Two-beam interference can be easily extended to three-beam interference by introducing a central plane coherent light wave. The angles between each incident light wave are fixed and denoted by α (Figure 26.6).

Similar to the approach in Section 26.5.3, one can specify the electrical field distribution as a sum of incident light waves:

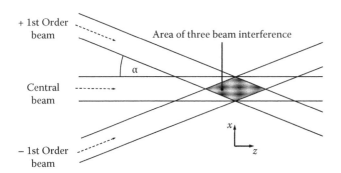

FIGURE 26.6 Schematic for a three-beam interference: in addition to a two-beam interference, a third plane wave that is coherent to the others, denoted as central beam incidences along the optic axis resulting in a three-beam interference in the area where all three beams overlap.

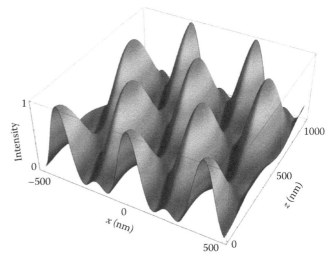

FIGURE 26.7 (See color insert.) Intensity distribution for a three-beam interference ($\alpha = 60°$, $\lambda_0 = 488\,\text{nm}$, $n = 1.45$, $\text{NA} = 1.4$, $E_0 = E_{-1} = E_{+1}$). In comparison to a two-beam interference, the intensity distribution is additionally modulated along the z-axis. (The intensity is in arbitrary units.)

$$E = E_0 e^{-ikz} + E_{-1} e^{-i[k(-\sin(\alpha)\cdot x + \cos(\alpha)\cdot z) - \phi_1]} + E_{+1} e^{-i[k(\sin(\alpha)\cdot x + \cos(\alpha)\cdot z) - \phi_2]} \qquad (26.11)$$

Again, the intensity distribution I is calculated by forming squared absolute value. The assumption $E_{+1} = E_{-1}$ yields

$$I = E_0^2 + 2E_{+1}^2 + 2E_{+1}^2 \cos(2(k \cdot \sin(\alpha) \cdot x - \varphi))$$
$$+ E_0 E_{+1} \cos(k[1 - \cos(\alpha)] \cdot z - \varphi) \cdot \cos(k \cdot \sin(\alpha) \cdot x - \varphi) \qquad (26.12)$$

with

$$\varphi = \frac{\varphi_1}{2} - \frac{\varphi_2}{2} \qquad (26.13)$$

An example is shown in Figure 26.7.

26.5.5 Image Formation Using Structured Illumination

In this chapter, the consequences of a periodically modulated intensity distribution (I), as it is used for SIM, on the image formation are further investigated (Frohn et al. 2001), starting from the equation as introduced earlier:

$$\text{FT}[g] = \text{OTF} \cdot \text{FT}[\rho \cdot I] \qquad (26.14)$$

In the following, we discuss the case of a three-beam interference more in depth. First, consider only the Fourier transform of the third term of Equation 26.12 multiplied by ρ for understanding the calculation steps:

$$\text{FT}[2 \cdot E_{+1}^2 \cos(2k\sin(\alpha)x - 2\varphi) \cdot \rho(x,y,z)] \qquad (26.15)$$

The identity $\cos(x) = (1/2)(e^{ix} + e^{-ix})$ is used to split the cosine function into two e-functions:

$$= \text{FT}[4 \cdot E_{+1}^2 (e^{2ik\sin(\alpha)x - 2i\varphi} + e^{-2ik\sin(\alpha)x - 2i\varphi})] \otimes \text{FT}[\rho(x,y,z)] \quad (26.16)$$

Executing the Fourier transformations leads to

$$= \left[4 \cdot E_{+1}^2 e^{2i\varphi} \left(\int_{-\infty}^{\infty} e^{2ik\sin(\alpha)x} e^{ik_x x} dx + \int_{-\infty}^{\infty} e^{-2ik\sin(\alpha)x} e^{ik_x x} dx \right) \right]$$

$$\otimes \left[p(k_x, k_y, k_z) \right] \qquad (26.17)$$

Now two δ-distributions emerge. This becomes clear if one rearranges the equation:

$$= \left[4 \cdot E_{+1}^2 e^{2i\varphi} \left(\int_{-\infty}^{\infty} e^{ix(k_x + 2k\sin(\alpha))} dx + \int_{-\infty}^{\infty} e^{ix(k_x - 2k\sin(\alpha))} dx \right) \right]$$

$$\otimes \left[p(k_x, k_y, k_z) \right]$$

$$= [8\pi \cdot E_{+1}^2 e^{2i\varphi} (\delta[k_x + 2k\sin(\alpha)] + \delta[k_x - 2k\sin(\alpha)])]$$

$$\otimes [p(k_x, k_y, k_z)] \qquad (26.18)$$

The distributivity property of the convolution allows splitting up into two convolutions:

$$= 8\pi \cdot E_{+1}^2 e^{2i\varphi} (\delta[k_x + 2k\sin(\alpha)] \otimes [p(k_x, k_y, k_z)]$$

$$+ 8\pi \cdot E_{+1}^2 e^{2i\varphi} \delta[k_x - 2k\sin(\alpha)] \otimes [p(k_x, k_y, k_z)] \qquad (26.19)$$

Executing the convolutions yields to the result:

$$= 8\pi \cdot E_{+1}^2 e^{2i\varphi} \int_{-\infty}^{\infty} \delta[\tau + 2k\sin(\alpha)] \cdot p(k_x - \tau, k_y, k_z) d\tau$$

$$+ 8\pi \cdot E_{+1}^2 e^{2i\varphi} \int_{-\infty}^{\infty} \delta[\tau - 2k\sin(\alpha)] \cdot p(k_x - \tau, k_y, k_z) d\tau$$

$$= 8\pi \cdot E_{+1}^2 p(k_x + 2k\sin(\alpha), k_y, k_z) \cdot e^{2i\varphi}$$

$$+ 8\pi \cdot E_{+1}^2 p(k_x - 2k\sin(\alpha), k_y, k_z) \cdot e^{2i\varphi} \qquad (26.20)$$

The remaining terms are calculated in a similar manner, leading to the Fourier transformation FT[ρ · I]:

$$\text{FT}[\rho(x,y,z) \cdot I] \propto (E_0^2 + 2E_{+1}^2) p(k_x, k_y, k_z)$$

$$+ E_{+1}^2 p(k_x - 2k\sin(\alpha), k_y, k_z) e^{-2i\varphi}$$

$$+ E_{+1}^2 p(k_x + 2k\sin(\alpha), k_y, k_z) e^{+2i\varphi}$$

$$+ E_0 E_{+1} p(k_x + k\sin(\alpha), k_y, k_z - k[1 - \cos(\alpha)]) e^{-i\varphi}$$

$$+ E_0 E_{+1} p(k_x - k\sin(\alpha), k_y, k_z - k[1 - \cos(\alpha)]) e^{+i\varphi}$$

$$+ E_0 E_{+1} p(k_x + k\sin(\alpha), k_y, k_z + k[1 - \cos(\alpha)]) e^{-i\varphi}$$

$$+ E_0 E_{+1} p(k_x - k\sin(\alpha), k_y, k_z + k[1 - \cos(\alpha)]) e^{+i\varphi} \qquad (26.21)$$

Interestingly, the final expression contains six additional copies of the Fourier-transformed molecule distribution p, which originate from higher frequency regions of the k_x–k_z-plain. How far the single additional copies are shifted in Fourier space is defined by the angle α and the wave vector k of the incident light waves, thus, by the resolution power of the objective lens. Remember the low pass filter property of the OTF (during image formation, the OTF is multiplied by FT[ρ…I]). Structured illumination leads to an increased flow of information through this filter, not in a form that directly leads to a high-resolution image, but in the mathematical structure of a sum.

During detection, the image plane at the camera plays a significant role. Considering a 3D fluorescent molecule distribution, only the focus plane is conjugate to the camera plane (in most configurations). The Fourier slice theorem states that a plane through the origin (the focus plane) of a 3D object in real space is equivalent to an integration along the direction parallel to the plane's normal vector in Fourier space. Therefore, the focus plane can be described by a projection along the k_z-direction, mathematically by an integration over k_z. First of all, the multiplication of the OTF by the Fourier-transformed fluorescence molecule distribution p is executed. Second, the detected image is calculated in Fourier space through integration over k_z accounting for projection. Due to projection, the number of additional copies is reduced from six to four, since the additional copies shifted in k_z-direction coincide and cannot be separated again.

After projection, five groups of summands can be separated (unmixed) and back-shifted to the correct position in the k_x–k_y-plain by means of image processing, denoted as reconstruction (see Section 26.5.6) to obtain a high-resolution image.

The process of image formation using structured illumination becomes more transparent if compared to a beat between two frequencies. Formally, a beat is generated by the interference of two waves. In case of structured illumination, the intensity distribution of the illumination is chosen in a

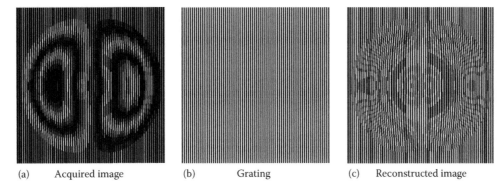

| (a)   Acquired image | (b)   Grating | (c)   Reconstructed image |

FIGURE 26.8 Display of the principle of structured illumination using a moiré pattern: the experimentally acquired image is a coarse beat of the fine structure of interest with the grating (a). Since the grating (b) is well known, the fine structure (c) can be reconstructed using a dedicated algorithm.

way that it creates a beat with the spatial frequencies of the fluorescence molecule distribution, which can reside beyond the cutoff frequency of the microscope. The beat frequency amounts to the difference between those two spatial frequencies. Thus, it can be smaller than the cutoff frequency, if proper experimental conditions are selected. In association with television and printing technology, a beat is known as moiré-effect.

26.5.6 Image Reconstruction

By application of structured illumination, the total information transferred by the objective lens increases. Additional information is transmitted in a sum and has to be separated and shifted to the correct position in Fourier space, before obtaining a high-resolution image. To separate the components, the acquisition of a few images of the same region of the object with different positions of the illumination pattern is required. Different positions mean different phases of the periodic modulation: using two-beam interference, three pictures with different phase positions, regarding three-beam interference five, are necessary to accomplish reconstruction (thus, the information theorem is not violated).

It is possible to determine the illumination pattern and the corresponding phases from the measured images via cross-correlations between the separated components, described in Figure 26.8. With this knowledge, a more detailed version of the fluorescence molecule distribution can be reconstructed, as compared to homogeneous wide-field illumination.

By introducing ϕ_i as different phase positions of the periodic illumination intensity distribution, a linear system of equations is obtained. If the values of ϕ_i are chosen such that the equation system is non-singular, it can be inverted to unmix the overlapping components in Fourier space. Focusing on three-beam interference, the Fourier-transformed images $B_{camera}(\phi_i)$ of a series with different phase positions ϕ_i can be written as

$$B_{camera}(\phi_i) = \int_{-\infty}^{\infty} dk_z FT[\rho(x,y,z) \cdot I(\phi_i)] \cdot OTF$$

$$= A + B_+ e^{+i \cdot \phi_i} + B_- e^{-i \cdot \phi_i} + C_+ e^{+2i \cdot \phi_i} + C_- e^{-2i \cdot \phi_i} \quad (26.22)$$

The five unknown parameters A, B_+, B_-, C_+, and C_- can be determined by measuring five images with different phase positions. Two methods are commonly used to solve the linear system of equation: Inverting the associated coefficient matrix or by a discrete Fourier transformation. The latter is the solution of the inversion problem for the special case of all phase steps being of equal size and sampling the coarsest nonzero spatial frequency once over the experiment.

Note also that the resolution is improved only along the direction of the *k*-vector of the intensity grating. An almost isotropic resolution improvement can be achieved, if the periodic illumination intensity distribution is rotated and image series are taken at different angles (typically at 0°, 60°, 120° or at 0°, 45°, 90°, 135°). The concept of SIM is summarized in Figure 26.9.

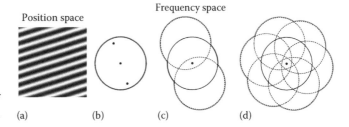

| (a) | (b) | (c) | (d) |

FIGURE 26.9 Extension of the accessible frequency region by SIM. The sinusoidal illumination pattern (a) contains three delta-peaks in frequency space as illustrated in the Fourier-transformed raw image (b). The continuous outline demonstrates the frequency limit. The raw data (b) consists of superposed original image information positioned at three different origins. After separation, the information can be shifted back to its respective position resulting in an expanded accessible frequency area (c). To expand the resolution in the object plane isotropically, the illumination is rotated (d).

Box 26.1 Introduction to Point Spread Function

Many optical configurations, such as lenses, wide-field microscopes, and confocal microscopes using incoherent fluorescent light, can be considered as linear shift invariant systems and a general concept for their resolving power, the PSF, is established. In the further discussion the z-dimension is excluded for clearness.

Input signals $f(x,y)$ are transformed into output signals $g(x,y)$ by linear shift invariant systems under the following conditions (for all constants a, b, α, and β):

- Linearity: If two input signals are multiplied with different constants and added subsequently, the overall output signal describes a sum of the individual output signals multiplied with constants, respectively.

$$a \cdot f_1(x,y) + b \cdot f_2(x,y) \Rightarrow a \cdot g_1(x,y) + b \cdot g_2(x,y) \qquad (26.23)$$

- Invariance in space: Shifting of input signal induces a shift of the output signal, but does not affect the shape of the output signal.

$$f(x - \alpha, y - \beta) \Rightarrow g(x - \alpha, y - \beta) \qquad (26.24)$$

Particularly in optics, the input signal corresponds to the fluorescent molecule distribution and the output signal corresponds to the image. If the input signal is a point-like object, like a single fluorescent molecule, then the output signal has a special denotation and is termed point spread function (PSF) (Figure 26.10). This fact can be expressed mathematically:

$$f(x,y) = \delta(x) \cdot \delta(y) \Rightarrow g(x,y) = \text{PSF}(x,y) \qquad (26.25)$$

In terms of resolution, the expansion of the PSF gives a quantitative parameter for the optical system resolving power. A highly extended PSF results in a poor resolution, since the minimum distance between optically distinguishable objects increases. Unfortunately, there is no way to define the expansion of the PSF and thus the resolving power unambiguously, but historically two criterions have been accomplished: Rayleigh's criterion and Sparrow's criterion. At this point it is crucial to underline that the transformation of the PSF into Fourier space permits a definition of the resolving power. (For further demonstration see Box 26.2).

Rayleigh's criterion: Two adjacent point-like objects are regarded as resolved when the principal PSF maximum of one image coincides with the first minimum of the other.

Sparrow's criterion: In case the combined signal generated by the PSFs of two adjacent point-like objects becomes a flat top, the objects can be resolved (Figure 26.11). It should be stressed that unlike Abbe criterion, both, the Rayleigh and the Sparrow criterion can be improved upon by blocking or phase-modifying areas of the back focal plane. In extreme cases, this allows constructing arbitrarily high resolution (according to the Rayleigh or Sparrow limit) at the expense of arbitrarily high side lobes (Di Francia 1955).

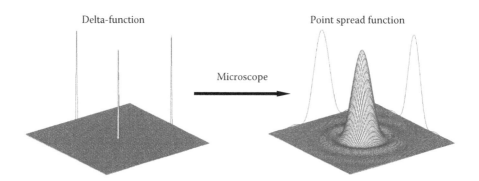

Delta-function · Microscope · Point spread function

FIGURE 26.10 The microscope blurs the (delta-distributed) molecule function into the point spread function. The *x*- and *y*-axis are in arbitrary units and the *z*-axis denotes the intensity distribution in arbitrary units. Line profiles through the center of the functions are shown at the edges of the 3D-plot.

Box 26.2 Introduction to Fourier Space and Optical Transfer Function

The process of image formation can be described in Fourier space. Thus, the Fourier-transformed fluorescence molecule distribution is multiplied by the Fourier-transformed PSF, which is denoted as optical transfer function (OTF). Applying a Fourier back-transformation leads to the resulting image in real space.

In Fourier transformation theory, it can be shown that any periodic function $f(x) = f(x+L)$ (L is the period length) is decomposable of sine and cosine functions (the so-called Fourier series, Figure 26.12) with different amplitudes (a_n, b_n) and frequencies ($2\pi n/L$):

$$f(x) = \frac{a_0}{2} + \sum_{n=1}^{\infty}\left[a_n\cos\left(\frac{2\pi n}{L}x\right) + b_n\sin\left(\frac{2\pi n}{L}x\right)\right] \quad (26.26)$$

Sine and cosine can be considered as the basis functions of the Fourier space and the associated scalar product is carried out by integration. Hence, the Fourier spectrum can be calculated:

$$a_n = \frac{2}{L}\int_{-2/L}^{2/L} f(x)\cos\left(\frac{2\pi n}{L}x\right)dx \quad \text{and}$$

$$b_n = \frac{2}{L}\int_{-2/L}^{2/L} f(x)\sin\left(\frac{2\pi n}{L}x\right)dx \quad (26.27)$$

This concept is equal to a vector space with infinitive base vectors. If one wants to express an arbitrary vector as a linear combination of a set of basis vectors, the factors are the scalar product of the basis vectors with the accordant arbitrary vector. Compared to $\Re^2$ with the canonical basis and the common scalar product, this means

$$\begin{pmatrix} x \\ y \end{pmatrix} = a \cdot \begin{pmatrix} 1 \\ 0 \end{pmatrix} + b \cdot \begin{pmatrix} 0 \\ 1 \end{pmatrix} \quad (26.28)$$

The factors are given by

$$a = \begin{pmatrix} x \\ y \end{pmatrix} \cdot \begin{pmatrix} 1 \\ 0 \end{pmatrix} = x \cdot 1 + y \cdot 0 \quad \text{and} \quad b = \begin{pmatrix} x \\ y \end{pmatrix} \cdot \begin{pmatrix} 0 \\ 1 \end{pmatrix} \quad (26.29)$$

Note the similarity to Fourier transformation.

The Fourier series is only applicable to periodic functions. However, the concept can be extended to nonperiodic functions (whose integral of their squared value yields a finite value). Therefore, a periodic function is considered in the border case $L \to \infty$, leading to a nonperiodic function. In this case, the Fourier series changes to a Fourier integral and the discrete Fourier spectrum becomes continuous (frequency spectrum $F(k)$). The continuous Fourier transformation and back transformation are defined by (scaling factor changes in literature):

$$f(x) = \int_{-\infty}^{\infty} F(k)e^{ikx}dk \quad \text{and} \quad F(x) = \frac{1}{2\pi}\int_{-\infty}^{\infty} f(k)e^{-ikx}dk \quad (26.30)$$

Image formation is expressed through a modification caused by multiplying the Fourier-transformed intensity distribution of the fluorescence molecules by the OTF. Since the OTF equals to zero, regarding the frequencies beyond the spatial frequency as given by Abbe, the cutoff frequency, it cannot transfer unlimited information. Hence, it acts like low pass filter (Figure 26.13).

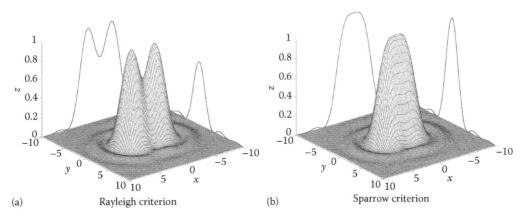

(a) Rayleigh criterion   (b) Sparrow criterion

FIGURE 26.11 Rayleigh (a) and Sparrow (b) criterion for two resolvable point objects. The x- and y-axis are in arbitrary units and the z-axis denotes the intensity distribution in arbitrary units. Line profiles through the center of the functions are shown at the edges of the 3D-plot.

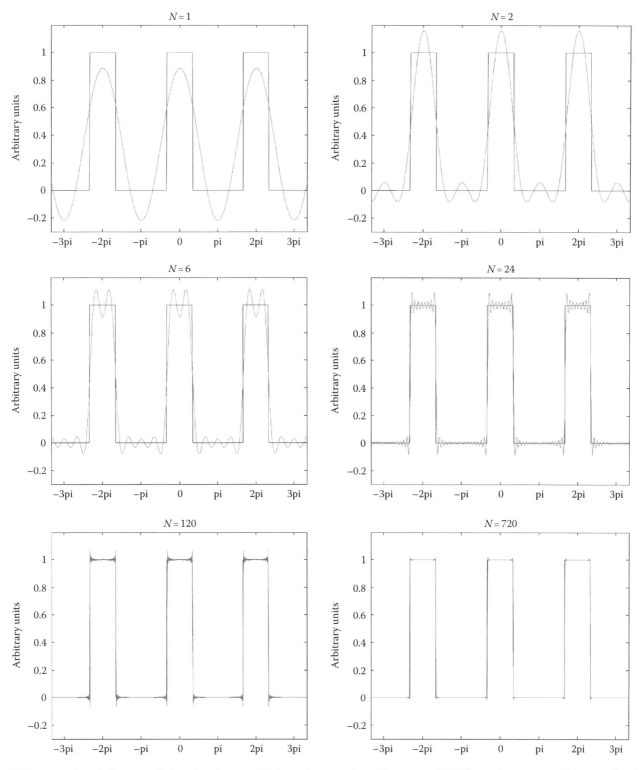

FIGURE 26.12 A periodic rectangle-function, shown in black, can be approximated by a sum of *N* different sin-functions, if their amplitudes, phases, and frequencies are calculated according to the Fourier series (Box 26.2). With increasing number *N* (here the cases of *N* = 1, 2, 6, 24, 120, 720 are depicted), the differences between the two curves vanish. In order to approximate sharp edges, sin-functions with very high frequencies are needed.

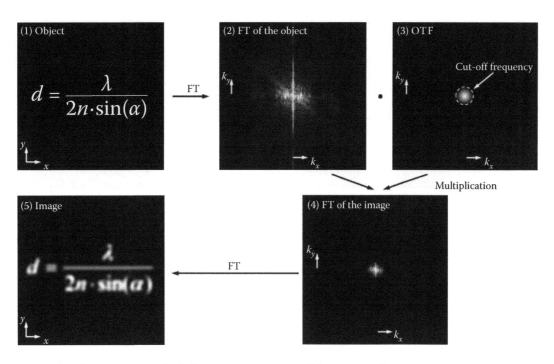

FIGURE 26.13 Considered in Fourier space, high frequencies are suppressed during image formation due to the multiplication of the Fourier transform of the object (2) by the OTF (3). Here a simplified model (a cone) of the OTF is used to describe the process. The OTF equals to zero for higher frequencies than the cutoff frequency (dotted circle). If the Fourier transform of the image (4) is inversely transformed to real space, the resulting image (5), is blurred compared to the object (1). This is expected, since high frequencies are needed to describe sharp edges. Compare with Figure 26.12.

26.5.7 Various Setups for SIM

There are different ways to generate the illumination pattern for SIM that provide different advantages and disadvantages.

These methods have in common that coherent excitation beams are guided through the objective into the object plane each collimated but under an angle, respectively. The resulting interference of the beams then leads to the illumination pattern used for the image enhancement.

An original approach is to couple previously separated beams with fibers into the objective.

A more convenient way is to use a fine solid grating diffracting the excitation beam into several orders. The central zero order and the two first diffraction orders are then coupled into the objective interfering in the object plane resulting in three-beam interference SIM. The zero-order beam can optionally be blocked in order to restrain the system to two-beam interference.

In this particular case, the period of the interference pattern is constantly related to the wavelength and the diffraction grating period. For a changed wavelength, a different grating has to be used in order to achieve an optimal resolution enhancement. To obtain the different illumination pattern positions relative to the object, either the sample or the diffraction grating is moved parallel to the modulation direction of the grating. To achieve an isotropic resolution enhancement in the focal plane, either the sample or the grating has to be rotated. Because these actions are time-consuming, the data

acquisition time of this method is usually long. Hence, more sophisticated ways have been developed. A further development of the diffraction grating method is to use a spatial light modulator (SLM), a pixelated matrix to generate the grating. Thus, the pattern can be shifted, rotated, and changed in its period in a much shorter time enabling time-resolved acquisitions of living cells (Gustafsson 2005, Hirvonen et al. 2009, Fiolka et al. 2008).

A different approach is to generate the interference pattern with an interferometer-like setup (Best et al. 2010) and to slide it by shifting the relative phase of the single beams with Piezo-Actuators. This method can be set up to offer two- or three-beam capability, has the advantage of a vast variability, and can be designed to reach high speeds of setting up the pattern reducing the acquisition time. Yet another way is the usage of acousto optical diffractive elements (AODs) to generate a diffracting phase grating. This is potentially a very fast way, but it is not easy to achieve gratings of the highest fringe quality over a reasonably large area of illumination.

26.5.8 Experimental Examples of the Resolution Improvement Using Structured Illumination

By structured illumination, the resolution can be increased by a factor of two. In addition, the axial separation of the optical information leads to a vastly improved confocality of the image compared to standard wide-field microscopy. Figures 26.14 and

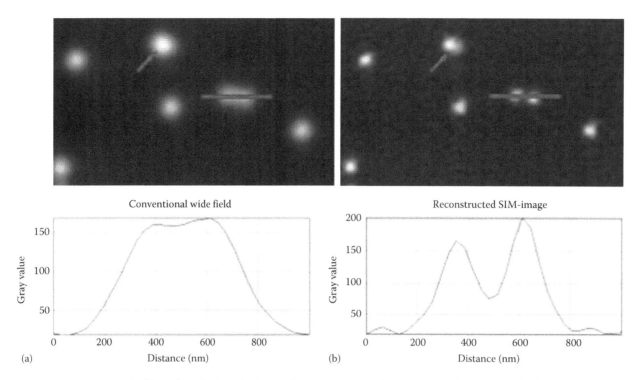

FIGURE 26.14 Improvement of lateral resolution: The image shows a specimen of 100-nm diameter fluorescent microspheres (G100 Duke Scientific Polymer Microspheres Green Fluorescent 0:1 μm). While the two spheres homogeneously illuminated (left) cannot be separated following Rayleigh's criterion, with structured illumination they can easily be identified solely. To further demonstration, the intensity is plotted over a length of 1 μm ($n = 1.4$, $\lambda_0 = 488$ nm, NA = 1.4).

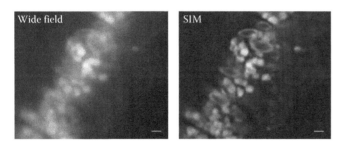

FIGURE 26.15 (See color insert.) Two-color image (green: 488 nm excitation; red: 568 nm excitation) of lipofuscin accretions in human retinal pigment epithelium. In addition to the lateral resolution enhancement, the improved suppression of out-of-focus light of SIM is beneficial. Scale bars are 1 μm.

26.15 are intended to demonstrate the effectiveness of the use of structured illumination.

Biological specimen: Structured illumination is applicable whenever wide-field fluorescent microscopy is applicable. In contrary to many other high resolving methods, it is not constrained to special fluorophores or other problematic preparations. Thus, a broad use in different biological specimen is possible.

In Figure 26.15, a SIM image of a human retinal pigment epithelium tissue section is displayed. The images show the combined autofluorescence distributions in the specimen excited by different excitation wavelengths (Ach et al. 2010, Best et al. 2010).

26.5.9 Perspective: Combination of SEI with Localization Microscopy

An additional important application of axially structured illumination is 3D localization of single molecules. The new approaches of single-molecule localization microscopy [SPDM,PALM,STORM] developed during the last decade (Cremer et al. 1996, 1999, Esa et al. 2000, Betzig et al. 2006) deliver impressive nanoscopic imaging far beyond the resolution limit. Whereas in conventional methods all fluorescent molecules are imaged simultaneously, the basic principle of the new methods is optical isolation in the time domain. The molecules are detected successively and their positions can be determined out of the optical isolated signals (no overlap with neighboring signals) with nanometer precision. Merged within a combined map, a high-resolution image is acquired. Thus, for the first time, investigations of small objects, before reserved to methods using ionizing and thus destructive radiation, can be performed applying visible light. In addition to a substantial increase of the lateral optical resolution, expansions like bi-plane detection (Juette et al. 2008) or utilized astigmatism (Huang et al. 2008) increase the axial optical resolution to about 100 nm, but they still are at least one order of magnitude away from achievable lateral resolution. Using structured illumination is an elegant way to eliminate this drawback. In the case of high photon numbers emitted from an individual, optically isolated molecule the conventional SPDM-localization concept described earlier for point-like objects like small beads can be used to achieve 3D localization with an accuracy in the single nanometer regime. Thus, 3D object resolution in the 10 nm range is achievable for light microscopy. Figure 26.16 shows how lateral super-resolution by SPDM can be combined with axial target positioning by SMI. SMI measurements and SPDM for object positions are performed to determine for every lateral object point x_0/y_0 the axial extension and the axial

position of the center of fluorescence (equivalent to center of mass). Combining this data results in the object volume in which a laterally localized single molecule must be located. The smaller the object's axial extension, the smaller is the accessible volume and the higher the achievable localization precision and therefore the resolution.

The application of 2D localization (Figure 26.17a) in combination with the additional axial information provided by SEI, as schematically shown in Figure 26.16, is expected to allow a reconstruction of the cell protrusion in Figure 26.17b with a 3D resolution of better than 50 nm. A detailed description of the method is given in Lemmer et al. (2008).

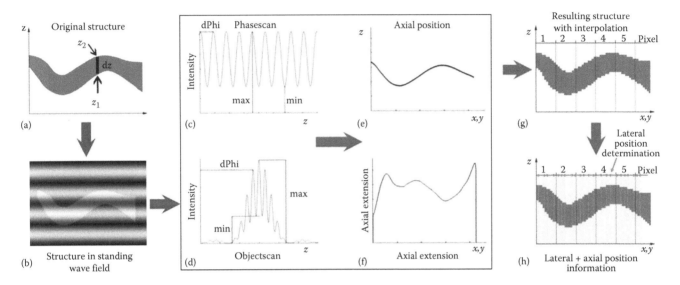

FIGURE 26.16 Scheme of the 3D SPDM/SMI process. The object of interest (a) is placed within a standing-wave field along the optical axis (b). By performing a phase scan (c) or an object scan (d), both the relative axial position z_0 (e) and the axial extension d_z can be obtained from the registered data. The interpolated axial information (g) can then be combined with the single-molecule positions of the 2D SPDM localization, resulting in a 3D image with an effective 3D optical resolution at the nanoscale (h) (for details see Lemmer et al.). (From Lemmer, P. et al., *Appl. Phys. B: Lasers Opt.*, 93(1), 1, 2008. With permission.)

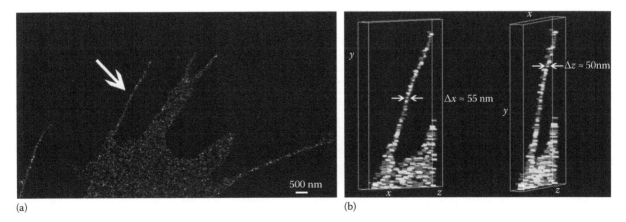

FIGURE 26.17 (a) 2D SPDM-localization step for SPDM/SMI. In total, the (x,y) positions of 4137 single molecules were determined and inserted into the pixel grid, which had an effective pixel size of 5 nm. The pixels with the localized molecules were blurred with a Gaussian kernel (standard deviation 10 nm). (b) 3D SPDM reconstruction of a human cancer cell protrusion. The volume rendering results are shown in two different orientations (for further details see Lemmer et al. 2008). (From Lemmer, P. et al., *Appl. Phys. B: Lasers Opt.*, 93(1), 1, 2008. With permission.)

26.6 Discussion and Outlook

The implementation of fluorescence microscopy in biomedical analysis has several explicit benefits over other imaging methods. It can be applied quickly and noninvasively and can thereby allow living-cell analysis. Also the costs for preparation and imaging stay relatively low. The use of fluorescent labeling additionally permits to highlight certain molecules of interest in the specimen and the concurrent application of different colors affords to distinguish the differently color-coded agents. For these reasons, fluorescence microscopy has become an important tool in biomedical research.

Unfortunately, the intracellular structures of interest for analysis with fluorescent labeling are often of a size significantly below 1 μm, and the resolving power of standard fluorescent microscopes turns out to be insufficient to render the objects' structural information adequately.

Therefore, the use of more complex non-light-optical imaging methods such as electron microscopy lacking the benefits stated above remains indispensable in today's research.

However, new light-optical super-resolution fluorescence techniques recently improved light-optical microscopy to reach into the previously unattainable realm beyond the Abbe limit.

A variety of super-resolution far-field microscopy SEI instrumentation in the visible range has been developed in the last decade. Most of them are able to achieve a true optical resolution or at least to deliver structural information beyond the Abbe Limit. Via these new techniques, an unprecedented application potential has been opened.

This potential spans throughout a wide range of scientific departments, but, of course, the results in life science remain to be of major interest.

Some possible topics could be as follows:

- Analysis of the spatial nano-organization of individual chromosome territories in cell nuclei during development and tumorigenesis, for example, of active/inactive genes, gene clusters and deserts, or chromosomal rearrangements
- Nanostructure of cancer-related recombination sites induced, for example, by ionizing radiation
- Nano-imaging of membrane structures, for example, the number and spatial distribution of ion channels, or of nuclear pore complexes
- Nano-imaging of the accumulation of beta-amyloid plaque pathology in transgenic mouse models of Alzheimer disease at the single-molecule level
- Spatial analysis of cellular and molecular mechanisms that regulate differentiation, self-renewal, and aging in hematopoietic stem cells, for example, expression and spatial distribution of candidate gene proteins on the single cell level and at molecular resolution
- Analysis of individual apoptosis signaling systems
- Analysis of aging processes on the single cell/single-molecule level

To summarize, while molecular cell biology so far focused on the type of molecules present in major cellular compartments (e.g., the nucleus, the nuclear membrane, the cytoplasm, and the cell membrane) and the biochemical reactions between them, the high-resolution techniques described here have the ability for cell imaging on a molecular level of resolution. This will allow merging molecular cell biology with molecular biophysics to an extent, which might have been thought impossible by many; consequently, it will be an essential step toward a complete "mechanistic" understanding of the machinery of life. This will open unprecedented perspectives for the therapy of diseases today still regarded to be resistant to conventional treatment.

Acknowledgments

For fruitful discussions, we thank Margund Bach, Sven Beichmanis, Heinz Eipel, Manuel Gunkel, Johann von Hase, Wei Jiang, Rainer Kaufmann, Patrick Mueller, Wladimir Schaufler, Alexander Urich, and Yanina Weiland. This work was supported by the Deutsche Forschungsgemeinschaft and the European Union.

References

Abbe, E. 1873. Beitraege zur Theorie des Mikroskops und der mikroskopischen Wahrnehmung. *Arch. Mikroskopisc. Anat.* 9: 413–420.

Ach, T., Best, G., Ruppenstein, M., Amberger, R., Cremer, C., and Dithmar, S. 2010. Hochauflösende Fluoreszenzmikroskopie des retinalen Pigmentepithels mittels strukturierter Beleuchtung–high resolution fluorescence microscopy of the retinal pigment epithelium using structured illumination. *Der Ophthalmologe* 107(11): 1037–1042.

Albrecht, B, Failla, A. V., Schweitzer, A., and Cremer, C. 2002. Spatially modulated illumination microscopy allows axial distance resolution in the nanometer range. *Appl. Opt.* 41: 80–87.

Baddeley, D., Carl, C., and Cremer, C. 2006. 4Pi microscopy deconvolution with a variable point-spread function. *Appl. Opt.* 45(27): 7056–7064.

Baddeley, D., Chagin, V. O., Schermelleh, L. et al. 2010. Measurement of replication structures at the nanometer scale using super-resolution light microscopy. *Nucleic Acids Res.* 38(2): e8.

Bailey, B., Farkas, D. L., Taylor, D. L., and Lanni, F. 1993. Enhancement of axial resolution in fluorescence microscopy by standing-wave excitation. *Nature* 366: 44–48.

Best, G., Amberger, R., Baddeley, D. et al. 2010. Structured illumination microscopy of autofluorescent aggregations in human tissue, *Micron* (in press).

Bradl, J., Hausmann, M., Schneider, B., Rinke, B., and Cremer, C. 1994. A versatile 2pi-tilting device for fluorescence microscopes. *J. Microsc.* 176: 211–221.

Betzig, E., Patterson, G. H., Sougrat, R. et al. 2006. Imaging intracellular fluorescent proteins at nanometer resolution. *Science* 313(5793): 1642.

Bewersdorf, J., Bennett, B., and Knight, K. 2006. H2AX chromatin structures and their response to DNA damage revealed by 4Pi microscopy. *Proc. Natl. Acad. Sci. U. S. A.* 103: 18137–18142.

Cremer, C. and Cremer, T. 1978. Considerations on a laser-scanning-microscope with high resolution and depth of field. *Microsc. Acta* 81(1): 31–44.

Cremer, C., Hausmann, M., Bradl, J., and Rinke, B. 1996. Verfahren zur multispektralen Praezisionsdistanzmessung in biologischen Mikroobjekten (Procedure for multispectral precision distance measurements in biological microobjects). German Patent application No. 196.54.824.1/DE, submitted Dec 23, 1996. European Patent EP 1997953660, 08.04.1999; Japanese Patent JP 1998528237, 23.06.1999; United States Patent US 09331644, 25.08.1999.

Cremer, C., Edelmann, P., Bornfleth, H. et al. 1999. Principles of spectral precision distance confocal microscopy for the analysis of molecular nuclear structure. In: *Handbook of Computer Vision and Applications*, eds. B. Jaehne, H. Haußecker, P. Geißler. San Diego, CA: Academic Press Vol. 3: 839–885.

Di Francia, G. T. 1955. Resolving power and information. *J. Opt. Soc. Am.* 45: 497–499.

Dyba, M. and Hell, S. 2002. Focal spots of size l/23 open up far-field fluorescence microscopy at 33 nm axial resolution. *Phys. Rev. Lett.* 88(16): 163901.

Egger, M. D. and Petrăn, M. 1967. New reflected-light microscope for viewing unstained brain and ganglion cells. *Science* 157(786): 305–307.

Egner, A., Jakobs, S., and Hell, S. 2002. Fast 100 nm resolution threedimensional microscope reveals structural plasticity of mitochondria in live yeast. *Proc. Natl. Acad. Sci. U. S. A.* 99(6): 3370–3375.

Esa, A., Edelmann, P., Kreth, G. et al. 2000. Three-dimensional spectral precision distance microscopy of chromatin nanostructures after triple-colour DNA labelling: A study of the BCR region on chromosome 22 and the Philadelphia chromosome. *J. Microsc.* 199(2): 96–105.

Failla, A. V., Spoeri, U., Albrecht, B., Kroll, A., and Cremer, C. 2002. Nanosizing of fluorescent objects by spatially modulated illumination microscopy. *Appl. Opt.* 41(34): 7275–7283.

Failla, A. V., Albrecht, B., Spoeri, U., Schweitzer, A., Kroll, A., Bach, M., and Cremer, C. 2003. Nanotechnology analysis using spatially modulated illumination (SMI) microscopy. *Complexus* 1: 29–40.

Fiolka, R., Beck, M., and Stemmer, S. 2008. Structured illumination in total internal reflection fluorescence microscopy using a spatial light modulator. *Opt. Lett.* 33(14): 1629–1631.

Frohn, J., Knapp, H., and Stemmer, A. 2000. True optical resolution beyond the Rayleigh limit achieved by standing wave illumination. *Proc. Natl. Acad. Sci. U. S. A.* 97(13): 7232–7236.

Frohn, J., Knapp, H., and Stemmer, A. 2001. Three-dimensional resolution enhancement in fluorescence microscopy by harmonic excitation. *Opt. Lett.* 26(11): 828–830.

Gustafsson, M. G. L. 2000. Surpassing the lateral resolution limit by a factor of two using structured illumination microscopy. *J. Microsc.* 198: 82–87.

Gustafsson, M. G. L. 2005. Nonlinear structured-illumination microscopy: wide-field fluorescence imaging with theoretically unlimited resolution. *Proc. Natl. Acad. Sci. U. S. A.* 102(37): 13081–13086.

Hänninen, P. E., Schrader, M., Soini, E., Hell, S.W. 1995. Two-photon excitation fluorescence microscopy using a semiconductor laser. *Bioimaging* 3(2): 70–75.

Hell, S. W. 1990. European Patent. OS 0491289.

Hell, S. W. 2007. Fluorescence nanoscopy: Breaking the diffraction barrier by the RESOLFT concept. *NanoBioTechnology*. 42: 296–297.

Hell, S. W. and Stelzer, E. H. K. 1992. Fundamental improvement of resolution with a 4Pi-confocal fluorescence microscope using two-photon excitation. *Opt. Commun.* 93: 277–282.

Hell, S. W., S. Lindek, S., C. Cremer, C., and Stelzer, E. H. K. 1994. Confocal microscopy with enhanced detection aperture: Type B 4Pi-confocal microscopy. *Opt. Lett.* 19: 222–224.

Hell, S. W. and Wichmann, J. 1994. Breaking the diffraction resolution limit by stimulated emission. *Opt. Lett.* 19(11): 780–782.

Heintzmann, R. and Cremer, C. 1999. Lateral modulated excitation microscopy: improvement of resolution by using a diffraction grating. In: *Optical Biopsies and Microscopic Techniques III*, eds. I. J. Bigio, H. Schneckenburger, J. Slavik, K. Svanberg, and P. M. Viallet, *Proc. SPIE* 3568: 185–196.

Heintzmann, R. and Cremer, C. 2002. Axial tomographic confocal fluorescence microscopy. *J. Microsc.* 206: 7–23.

Heintzmann, R., Jovin, T., and Cremer, C. 2002. Saturated patterned excitation microscopy—a concept for optical resolution improvement. *J. Opt. Soc. Am. A* 19(8): 1599–1609.

Hildenbrand, G., Rapp, A., Spoeri, U., Wagner, C., Cremer, C., and Hausmann, M. 2005. Nano-sizing of specific gene domains in intact human cell nuclei by spatially modulated illumination light microscopy. *Biophys. J.* 88(6): 4312–4318.

Hirvonen, L. M., Wicker, K., Mandula, O., and Heintzmann, R. 2009. Structured illumination microscopy of a living cell. *Eur. Biophys. J.* 38(6): 807–812.

Huang, B., Wang, W., Bates, M., and Zhuang, X. 2008. Three-dimension super-resolution imaging by stochastic optical reconstruction microscopy. *Science* 319(5864): 810–813.

Juette, M. F., Gould, T. J., Lessard, M. D. et al. 2008. Three-dimensional sub–100 nm resolution fluorescence microscopy of thick sample. *Nat. Methods* 5(6): 527–529.

Klar, T., Jakobs, S., Dyba, M., Egner, A., Hell, S. 2000. Fluorescence microscopy with diffraction resolution barrier broken by stimulated emission. *Proc. Natl. Acad. Sci. U. S. A.* 97: 8206–8210.

Lang, M., Jegou, T., Chung, I. et al. 2010. Three-dimensional organization of promyelocytic leukemia nuclear bodies. *J. Cell Sci.* 123: 392–400.

Lanni, F., Taylor, D. L., and Waggoner, A. S. 1986. Standing wave luminescence microscopy. US Patent 4621911.

Lemmer, P., Gunkel, M., Baddeley, D. et al. 2008. SPDM: Light microscopy with single-molecule resolution at the nanoscale. *Appl. Phys. B: Lasers and Opt.* 93(1): 1–12.

Lewis, A., Isaacson, M., Harootunian, A., and Murray, A. 1984. Development of a 500 Angstrom spatial resolution light microscope. *Ultramicroscopy* 13: 227–231.

Martin, S., Failla, A. V., Spoeri, U., Cremer, C., and Pombo, A. 2004. Measuring the size of biological nanostructures with spatially modulated illumination microscopy. *Mol. Biol. Cell* 15(5): 2449.

Mathee, H., Baddeley, D., Wotzlaw, C., Fandrey, J., Cremer, C., and Birk, U. 2006. Nanostructure of specific chromatin regions and nuclear complexes. *Histochem. Cell Biol.* 125(1): 75–82.

Minsky, M. 1961. Microscopy apparatus. US Patent 3013467.

Nagerl, U. V., Willig, K. I., Hein, B., Hell, S. W., and Bonhoeffer, T. 2008. Live-cell imaging of dendritic spines by STED microscopy. *Proc. Natl. Acad. Sci. U. S. A.* 105: 18982–18987.

Pohl, D. W., Denk, W., and Lanz, M. 1984. Optical stethoscopy: Image recording with resolution lambda/20. *Appl. Phys. Lett.* 44: 651–653.

Reymann, J., Baddeley, D., Gunkel, M. et al. 2008. High-precision structural analysis of subnuclear complexes in fixed and live cells via spatially modulated illumination (SMI) microscopy *Chromos. Res.* 16(3): 367–382.

Rittweger, E., Han, K. Y., Irvine, S. E., Eggeling, C., and Hell, S. W. 2009. STED microscopy reveals crystal colour centres with nanometric resolution. *Nat. Photonics* 3: 144–147.

Schmidt, R., Wurm, C. A., Jakobs, S., Engelhardt, J., and Egner, A., Hell, S. W. 2008. Spherical nanosized focal spot unravels the interior of cells. *Nat. Methods* 5(6): 539–544.

Schrader, A., Braune, M., and Engel, H. 1995. Dynamics of spiral waves in excitable media subjected to external periodic forcing. *Phys. Rev.* 52: 98–108.

Willig, K. I., Harke, B., Medda, R., and Hell, S. W., 2007. STED microscopy with continuous wave beams. *Nat. Methods* 4: 915–918.

Westphal, V., Rizzoli, S. O., Lauterbach, M. A., Kamin, D., Jahn, R., and Hell, S. W. 2008. Video-rate far-field optical nanoscopy dissects synaptic vesicle movement. *Science* 320(5873): 246–249.

27

Nonlinear Optical Microscopy for Biology and Medicine

Daekeun Kim
Dankook University

Heejin Choi
Massachusetts Institute of Technology

Jae Won Cha
Massachusetts Institute of Technology

Peter T. C. So
Massachusetts Institute of Technology

27.1 Introduction 561
    Overview and Historical Perspective • Role of Nonlinear Optical Microscopy among Other Biomedical Imaging Modalities
27.2 Principles and Methodology 563
    Theories of Nonlinear Excitation and 3D Imaging • Instruments for Nonlinear Optical Microscopy • Damage Mechanisms of Nonlinear Optical Microscopy
27.3 Frontiers of Nonlinear Optical Microscopy 567
    Improving Imaging Performance: Depth and Resolution • Novel Spectroscopic Contrast Mechanisms • Nonlinear Nanoprocessing • Nonlinear Optical Microscopy for Clinical Applications
27.4 Conclusion 580
Acknowledgments 580
References..580

27.1 Introduction

In 1990, the publication of "Two-photon laser scanning fluorescence microscopy" in *Science* by Denk, Strickler, and Webb has ushered in an era of unprecedented growth in applying nonlinear optical principles in biomedical microscopic imaging and manipulation (Denk et al. 1990). Two-photon microscopy, as described by Denk and coworkers, is based on fluorescence contrast and remains the most important and widely used form of nonlinear optical microscopy in biomedicine. Today, more advanced techniques based on other nonlinear processes, such as second (Gannaway and Sheppard 1978) and third harmonic generation (Barad et al. 1997) or Raman signals (Cheng et al. 2002), are also rapidly finding applications in biomedicine. With the initial development of nonlinear optical microscopy in the late 1970s, there were few, if any, important biomedical applications that resulted from this discovery for over two decades. After the seminal publication in *Science*, the growth of biomedical nonlinear optical microscopy is almost exponential as measured by the publication output. Today, this growth has shown no sign of abating. This chapter aims to provide an overview of the principles and practices of this important class of imaging techniques for biomedical systems and will also attempt to identify the most promising future directions.

27.1.1 Overview and Historical Perspective

While the field of biomedical nonlinear optical microscopy has become important in the life science during the last decade, the origin of this technology has its roots in the dawn of quantum mechanics in 1929. In Göttingen, Germany, Maria Göppert-Mayer (Göppert-Mayer 1931), under the direction of Max Born, chose to theoretically investigate the possibility for two photons to simultaneously interact with an atom to induce an electronic transition. Masters has published an English translation of her 1929 and 1931 papers together with introductory notes on her thesis research (Masters and So 2008). Most importantly, her PhD work provides theoretical foundation for multiphoton nonlinear interaction from the quantum mechanical view point. Her theoretical work covered the cases of two-photon absorption, two-photon emission, and the Raman effect. It is also interesting to note that in addition to stimulated absorption processes, her work also considered the possibility of stimulated emission processes. Clearly, Göppert-Mayer's work foreshadows much of the recent exciting developments in the field of nonlinear optical microscopy. It is quite fitting that Göppert-Mayer is recognized as the progenitor of the field and is honored by naming the unit of a fluorophores' two-photon cross section after her.

After the work of Göppert-Mayer and her contemporaries, the field of nonlinear optics became relatively dormant for about 30 years. After the completion of the theoretical work, it became clear that further progress in either fundamental research or in application required progress in the experimental front. However, this progress required a high intensity light source to induce efficient nonlinear optical interactions. The quest of high intensity light source, of course, results in the development of lasers in 1960s that opens the exciting field of nonlinear optics that has impact much beyond biomedical imaging.

In the field of nonlinear optical microscopy, the next relevant development probably was the first observation of second harmonic generation from ruby by Franken and coworkers (Franken et al. 1961). Shortly after, the observation of two-photon-induced fluorescence in CaF_2:Eu^{2+} was reported by Kaiser et al. (Kaiser and Garrett 1961) and three-photon-induced fluorescence was reported by Singh and coworkers (Singh and Bradley 1964). It may be interesting to note that most of these works were accomplished within a few short years after the invention of laser. The ability to make rapid progress in the experimental front can be attributed to the solid theoretical foundation being laid in the 1930s. It is also important to note that these research activities were in the field of basic physics and optics and far from the applications in biological sciences.

While most of the pioneers in nonlinear optics focused on nonbiological areas, a few researchers sought to utilize the fact that molecular excitation under nonlinear processes follow selection rules that differ from one-photon excitation selection rules; they thus allow different electronics states of biomolecules to be accessed. While these works were important for researchers in photophysics and photochemistry, the impact of these works in the broader community of biology and medicine were relatively limited. It should be noted that the pioneering work by researchers such as Birge, Fredrich, and McCain represented tour-de-force high precision measurements at that time since the intensities of available light sources were very weak compared with today's standard.

One of the most important developments in the decades between the invention of the laser and the publication of the Denk's paper was the work by the Oxford group in 1978 (Sheppard and Kompfner 1978). Under the leadership of Sheppard and Wilson, nonlinear optical properties of light were first utilized to improve the imaging performance of optical microscopes. Several important advantages of using nonlinear excitation processes for microscope imaging were realized at that time. Two of the most important discoveries are: (1) three-dimensionally resolved imaging is an inherent benefit of nonlinear excitation and (2) molecular states that are normally accessible by ultraviolet or deep infrared light may be studied using visible or near infrared radiation based on processes such as second- and higher-harmonic generation, nonlinear fluorescence excitation processes, and sum and difference frequency generation. Importantly, they have also constructed the first nonlinear microscope with three-dimensional (3D) resolution based on second harmonic generation as its contrast mechanism (Sheppard et al. 1977, Gannaway and Sheppard 1978). While the first nonlinear microscopes were built during this period, they

were relatively unnoticed by researchers in biology and medicine partly because of the papers focused on solid state specimens. More importantly, the available lasers at that time were difficult to operate and had low peak power resulting in a very low image frame rate.

Unlike previous publications that target optical specialists, the 1990 publication of Denk and coworkers appeared in the journal, *Science*, which is broadly read by life scientists; thus, ensuring the biomedical utility of this technique was recognized. One should also note that the popularization of confocal microscopy in life science a decade earlier prepared researchers well for this advance by amply demonstrating the need and the advantages of visualizing biological specimens in 3D. Further, the difficulty of using confocal microscopy for long-term in vivo imaging without photodamage of biological specimens ensures that life scientists are receptive of a less invasive 3D imaging technology. Therefore, the demonstration of noninvasive imaging of sea urchin embryo development by Denk and coworkers is appropriately noted for the advance that it represents. The subsequent applications of this technique by Denk, Svoboda, and Yuste in neurobiology showing important and previously unknown phenomena, such as dendritic spine remodeling, ensured that the biomedical importance of this new technique became well known (Denk 1994, Yuste and Denk 1995, Svoboda et al. 1996). Today, there are few cutting-edge neurobiology departments without at least several two-photon microscopes. It should also be noted that the initial research of Denk and coworkers used a difficult-to-tune colliding-pulse femtosecond dye laser for imaging. The broad acceptance of this new technique may not have happened without the almost concurrent invention of solid-state femtosecond Ti:Sapphire lasers (French et al. 1989, Spence et al. 1991). The availability of these stable, easy-to-use, and high-peak-power infrared lasers allows nonlinear microscopes to be placed in the laboratories of life scientists without the need of laser technicians on the research staffs. While there are now other femtosecond laser sources, well over 90% of nonlinear microscope systems today use the versatile Ti:Sapphire lasers.

During two decades after the publication by Denk and coworkers, the field of nonlinear optical microscopy has found applications in an ever-broader range of biomedical disciplines. At the same time, while the basic methodology of two-photon fluorescence microscopy is reaching maturity, the field of nonlinear optical microscopy is renewed by several technical innovations that can potentially greatly extend the biomedical utility of these new techniques.

27.1.2 Role of Nonlinear Optical Microscopy among Other Biomedical Imaging Modalities

Prior to an in depth discussion of nonlinear optical microscopy, it may be useful to examine this technology in the context of biomedical imaging needs. Unlike other biomedical imaging modalities such as MRI or CT, optical imaging inherently has relatively limited penetration depth. Compared with other optical imaging modalities in terms of tissue imaging depth, such as diffusive optical tomography (centimeter scale penetration depth, Chapters 19, 20, and 22) or optical coherence tomography (millimeter scale penetration depth, Chapters 13 through 15), optical microscope probably has the shallowest imaging depth of all that is typically on the order of less than 1 mm. While optical microscopy gives up penetration depth, it gains in terms of structural resolution. While optical coherence imaging has achieved resolution on the order of several microns, optical microscopy is an imaging technique that readily allows imaging of intracellular structures with resolution on the order of a couple hundred nanometers or better.

Potentially more important than resolution, optical microscopy is a technique that can readily integrate many spectroscopic measurement capabilities. Fluorescence spectroscopic measurement is probably the most powerful approach leveraging the rich library of fluorescent markers developed to demarcate specific cell and tissue structures, to monitor local biochemical environment and map the distribution of gene and protein expressions. In addition to fluorescence, many other contrast modalities can be, and have been, integrated into optical microscopy, For example, polarization-resolved imaging provides information of biomaterial nanoscale structure based on their birefringence (Tran et al. 1994, Katoh et al. 1997) (see Chapter 30). As another example, coherent anti-Stokes Raman imaging allows identification of the medically important tissue structures such as myelin sheets around neurons based on vibrational spectroscopic signature of myelin molecules (Wang et al. 2005).

Nonlinear optical microscopy has two distinctive features that guide their application of these techniques in biology and medicine. First, all nonlinear optical microscopy techniques have inherent 3D resolution. Therefore, unlike confocal microscopy that achieves 3D resolution based on rejecting out-of-focus signal using a conjugated pinhole, nonlinear optical microscopy can often map 3D structures of cells and tissues with better signal-to-noise ratio and reduced photobleaching and photodamage. As will be discussed later, this fact also contributes to significant deeper imaging for nonlinear optical microscopy as compared with confocal systems. Second, nonlinear optical microscopy often allows more versatile selection of excitation wavelength for a given contrast mechanism. As an example, many cell and tissue endogenous fluorophores are excited in the ultraviolet spectrum that is highly damaging to in vivo specimens. The use of two-photon fluorescence microscopy allows these fluorophores to be excited at twice the normal wavelength that shifts the excitation to the infrared range, by significantly reducing the potential for photodamage. In addition, the ability to separate excitation and signal spectra further enhances imaging signal to noise level.

27.2 Principles and Methodology

The basic principles and methodologies of nonlinear optical microscopy have been reviewed in many articles (Konig 2000, So et al. 2000), book chapters (Denk et al. 1995), and have recently

appeared in an edited book (Masters and So 2008). Instead of covering these basic materials in extensive length, a brief overview will be provided, and the focus of this chapter will be on the research "frontiers" of nonlinear optical microscopy.

27.2.1 Theories of Nonlinear Excitation and 3D Imaging

An understanding of the principles underlying nonlinear optical microscopy requires a familiarity with two classes of physical theories: nonlinear optical spectroscopy and microscopy image formation theory.

27.2.1.1 Nonlinear Optical Spectroscopy

Nonlinear optics and spectroscopy have been presented in a number of excellent text books (Shen 2002, Boyd 2008). Nonlinear optical processes can be divided into two classes: parametric and nonparametric processes. Nonparametric processes involve optical transitions that bring a molecule from an initial state to a different final state. The most important nonparametric processes for nonlinear optical microscopy are probably two-photon and three-photon fluorescence excitation. The typical residence time of the molecule in the final state is called the fluorescence lifetime that is on the order of nanoseconds. Parametric processes describe optical transitions after which the molecular initial and final states are the same. This class of processes includes many of the relevant nonlinear imaging processes described in this chapter including second and third harmonic generation (SHG, THG) and coherent anti-Stokes Raman scattering (CARS). These processes all involve the transition of a molecule to either one or several virtual intermediate states. This class of process can be considered to be almost instantaneous due the Heisenberg uncertainty principle. A schematic representation of some of these processes is presented in Figure 27.1.

It is beyond the scope of a short chapter to fully describe all the different nonlinear processes. Instead, we will focus on two-photon fluorescence excitation, the most common nonlinear optical microscopy imaging modality. Fluorescence excitation is an interaction between a fluorophore and an excitation electromagnetic field. This process can be described by the time-dependent Schrödinger equation where the Hamiltonian contains an electric dipole interaction term: $\vec{E}_\gamma \cdot \vec{r}$, where $\vec{E}_\gamma$ is the electric field vector of the photons and $\vec{r}$ is the position operator. This equation is typically solved by perturbation theory. The first-order solution corresponds to the one-photon excitation with transition probability P:

$$P \sim \left| \left\langle f \left| \vec{E}_\gamma \cdot \vec{r} \right| i \right\rangle \right|^2 \tag{27.1}$$

where

$|i\rangle$ denotes the ground electronic state
$|f\rangle$ denotes the excited electronic state

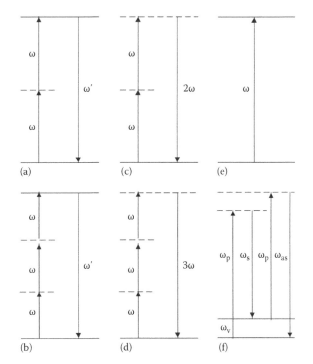

FIGURE 27.1 Energy diagrams for several common nonlinear optical microscopy contrast mechanisms: (a) two-photon fluorescence, (b) three-photon fluorescence, (c) second harmonic generation, (d) third harmonic generation, (e) absorption, and (f) coherent anti-Stokes Raman spectroscopy.

The *n*-photon transitions are represented by the *n*th order solutions. In the case of two-photon excitation, the transition probability between the molecular ground state $|i\rangle$ and the excited state $|f\rangle$ is represented by

$$P \sim \left| \sum_m \frac{\langle f | \vec{E}_\gamma \cdot \vec{r} | m \rangle \langle m | \vec{E}_\gamma \cdot \vec{r} | i \rangle}{\varepsilon_\gamma - \varepsilon_m} \right|^2 \tag{27.2}$$

where ε_γ is the photonic energy associated with the electric field vector $\vec{E}_\gamma$, and the summation is over all intermediate states *m* with energy ε_m.

Therefore, two-photon excitation of molecules is a nonlinear process involving the simultaneous absorption of two photons whose combined energy is sufficient to induce a molecular transition to an excited electronic state. Quantum mechanically, the first photon excites the molecule to a virtual intermediate state, and the molecule is eventually brought to the final excited state by the absorption of a second photon. In the absence of major vibrational perturbations, inductive effects, or solvent relaxation, the probability of the transition depends on the magnitude of the overlap integrals between the ground, intermediate, and excited states of a particular fluorophore. This transition probability is related to the two-photon excitation cross section of the fluorophore, as discussed earlier, measured in the unit of Göppert-Mayer (GM). Clearly, fluorophores with

higher two-photon cross sections allow more efficient excitation at lower excitation intensity. It should be noted that while the absorption cross sections for one-, two-, and multiphoton excitation of a given fluorophore are different, the molecule rapidly relaxes to the same vibrational level in the excited electronic state. The excited state residence time (fluorescence lifetime) and the fluorescence decay processes depend only on the molecular structure and its microenvironment. Therefore, the fluorescence quantum yield and emission spectra are independent of the initial excitation process.

27.2.1.2 Microscopy Image Formation Theory

One very important attribute of two-photon microscopy is its inherent 3D sectioning capability. The sectioning capability of this method originates from the quadratic and higher-order dependence of the fluorescence signal upon the excitation intensity distribution.

Consider the intensity distribution at the focal point of a high numerical aperture (NA) objective.

For an objective with numerical aperture, $NA = n \sin \alpha$, where *n* is the index of refraction of medium between the object and the specimen and α is the half angle subtended by the objective, the spatial profile of the diffraction limited focus is

$$I(u, v) = \left| 2 \int_0^1 J_0(v\rho) e^{-\frac{i}{2} u \rho^2} \rho \, d\rho \right|^2 \tag{27.3}$$

where J_0 is the zeroth order Bessel function, λ is wavelength of the excitation light, $u = 4k \sin^2(\alpha/2)z$, and $v = k \sin(\alpha)r$ are the respective dimensionless axial and radial coordinates normalized to wave number $k = 2\pi/\lambda$ (Sheppard and Gu 1990, Gu and Sheppard 1995). The point-spread-function (PSF), the fluorescent image of a point emitter, is a quadratic function of the excitation profile that is equal to $I^2(u/2, v/2)$. In contrast, one-photon fluorescence PSF has a functional form of $I(u, v)$. Through similar consideration, for *n*-photon process, the emission intensity profile is in general $I^n(u/n, v/n)$. For example, *n* equals two for SHG and *n* equals three for three-photon fluorescence and for THG. For CARS, it is effectively a third order process except that the point spread function needs to account for the different wavelengths of the pump and Stokes beams.

The most important difference between imaging modalities that depend linearly on excitation intensity distribution (one-photon) and those with higher-order dependence (such as two-photon fluorescence) is depth discrimination. The contribution of emission signal from each axial plane can be computed for one- and two-photon processes and are shown in Figure 27.2. Assuming negligible attenuation, the total signal generated is equal at each axial plane for one-photon microscopy. In contrast, the signal generated from two-photon processes falls off quickly from the focal plane resulting in axial localization. For higher-photon processes, the localization property is even more pronounced. Qualitatively, the localization of multiphoton

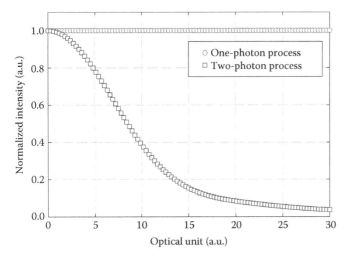

FIGURE 27.2 The contribution of emission signal from each axial plane can be computed for one- and two-photon processes. For one-photon process (circle), equal level of emission signal is observed independent of axial location. For two-photon process (square), emission signal is localized at the focal plane producing depth discrimination. The unit of X-axis is the dimensionless optical unit u as defined in Section 27.2.1. For high NA objectives and typical excitation wavelength, 10 optical units correspond to about 1 μm. The Y-axis is the normalized Intensity and is also dimensionless.

excitation can be understood by realizing that fluorescence generation is only appreciable at a region of high spatial photon density at the focal point of the microscope objective. The localization of the multiphoton excitation can be further appreciated by examining the dimensions of fluorescence PSF. Figure 27.3 illustrates the radial and axial PSFs for two-photon microscopy at 960 nm. The two-photon photointeraction volume is on the order of 0.1 fL. The axial depth discrimination of nonlinear optical microscopy greatly improves image contrast as compared with wide-field fluorescence microscopy, especially for thick specimens. This localization further reduces the photointeraction volume and greatly reduces photobleaching and photodamage in thick samples. Further, since photointeraction occurs in a femtoliter size volume, localized photochemical reactions can be initiated (Denk 1994).

27.2.2 Instruments for Nonlinear Optical Microscopy

Nonlinear optical excitation efficiency increases with the spatial and temporal density of the excitation photons. Spatial localization of photons is relatively straightforward by using high NA optics. Temporal localization of photons is comparatively more difficult and has not been efficiently achieved until the advent of femtosecond-pulsed lasers. Consider the two-photon fluorescence case, where the typical cross section for two-photon absorption is on the order of 1 GM which is equivalent to 10^{-50} cm^4 s. The need for temporal localization of photons using pulsed lasers can be observed by considering the absorption efficiency of a two-photon fluorophore with cross sections.

Two-photon absorption efficiency can be measured by n_a, the number of photons absorbed per fluorophore per pulse:

$$n_a \approx \frac{p_0^2 \delta}{\tau_p f_p^2} \left(\frac{(\mathrm{NA})^2}{2\hbar c\lambda} \right)^2 \qquad (27.4)$$

where
  τ_p is the pulse duration
  λ is the excitation wavelength
  p_0 is the average laser intensity
  f_p is the laser's repetition rate
  NA is the numerical aperture of the focusing objective
  $\hbar$ is Planck's constant divided by 2π
  c is the speed of light (Denk et al. 1990)

Equation 27.4 shows that for the same average laser power and repetition rate frequency, the excitation probability increases by using the higher NA focusing lens resulting in more spatially localizing the laser excitation. The absorption efficiency further improves linearly by reducing the pulse width of the laser indicating the importance of temporal localization.

For nonlinear optical microscopy, femtosecond, picosecond, and continuous-wave laser sources have been used. The most commonly used laser for multiphoton microscopy is femtosecond Ti:Sapphire lasers. These lasers characteristically produce 100 fs pulse train at 100 MHz repetition rate. The tuning range of Ti:Sapphire systems extends from 680–1050 nm. Cr:LiSAF and pulse compressed Nd:YLF lasers are some of the other femtosecond lasers used in multiphoton microscopy (Wokosin et al. 1996). Multiphoton excitation can also be generated using picosecond light sources although they have lower excitation efficiency. Two-photon excitation has been achieved using mode-locked Nd-YAG (~100 ps), picosecond Ti:Sapphire lasers, and pulsed dye lasers (~1 ps). Two-photon excitation using continuous-wave lasers has also been demonstrated using Ar/Kr laser and Nd:YAG laser (Hell et al. 1998).

Comparing these different laser sources, for fixed δ, p_0, and NA, Equation 27.4 implies that the difference in excitation efficiency per unit time for pulse and continuous-wave lasers for a two-photon fluorescence process is $\sqrt{f_p \tau_p}$. This factor is typically 300 for femtosecond Ti:Sapphire laser; two-photon excitation of a fluorophore that typically requires 1 mW of power using a femtosecond Ti:Sapphire laser will require 300 mW using a continuous-wave laser.

Since nonlinear excitation generates signal from a single point, the production of 3D images requires raster scanning of this excitation volume in 3D. As an example, a typical two-photon fluorescence microscope is shown in Figure 27.4. X-Y raster scanning is achieved by using a galvanometer-driven scanner. After beam power control and pulse width compensation optics, x–y scanner deflects the excitation light into a fluorescence microscope. Most nonlinear fluorescence microscopes use epi-luminescence geometry. The scan lens is positioned such that the x–y scanner is at its eye-point, while the

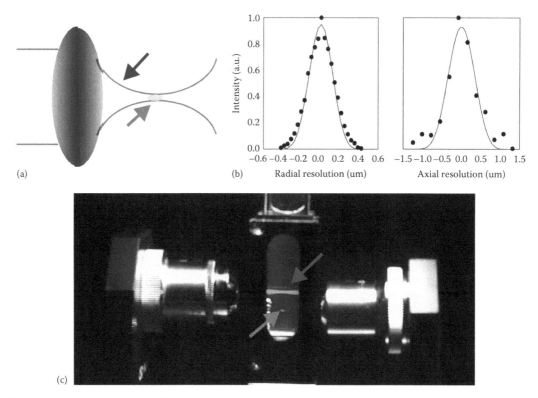

(a)

(b)   Radial resolution (um)   Axial resolution (um)

(c)

FIGURE 27.3 (See color insert.) Point spread function of two-photon fluorescence microscope. (a) In traditional one-photon excitation, fluorescence is generated throughout the double inverted cones (blue arrow). Two-photon excitation generates fluorescence only at the focal point (red arrow). (b) The submicron point spread function (PSF) of two-photon excitation at 960 nm. The full widths at half maximum are 0.3 μm radially and 0.9 μm axially. (c) An experimental visualization of the small excitation volume of two-photon fluorescence. One- and two-photon excitation light are focused by two objectives (equal numerical aperture) onto a fluorescein solution. Fluorescence is generated all along the path in the one-photon excitation case (blue arrow) whereas fluorescence is generated only in a 3D confined focal spot for two-photon excitation (red arrow). The reduced excitation volume is thought to lead to less photodamage.

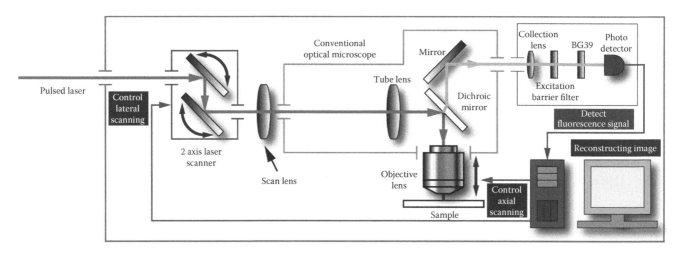

FIGURE 27.4 Schematics of a typical two-photon fluorescence microscope design.

field aperture plane is at its focal point. For a telecentric microscope, this arrangement ensures that angular scanning of the excitation light is converted to linear translation of the objective focal point. A tube lens is positioned to recollimate the excitation light directed towards the infinity-corrected objective

via a dichroic mirror. The scan lens and the tube lens function together as a beam expander that overfills the back aperture of the objective lens to ensure excitation light is diffraction-limited. Typically high NA objectives are used to maximize excitation efficiency. An objective positioner translates the focal point axially for 3D imaging. While nonlinear fluorescence microscope typically operates in the epi-geometry, harmonic generation microscopes are better operated in the transmission geometry because the phase matching condition dictates that harmonic-generated signal is forward scattered. On the other hand, for CARS microscopy, epi-detection improves the rejection of nonresonant background.

In the emission path, the signal emission is collected by the imaging objective and is transmitted through the dichroic mirror. Additional barrier filters are used to further attenuate the scattered excitation light and to select the emission band of interest. The signal is subsequently directed toward the detector system. It should be noted that no confocal aperture is required for 3D imaging. In fact, the use of a confocal aperture is not advised because it attenuates scattered emission photon. A number of photodetectors have been used in nonlinear optical microscopes including photomultiplier tubes (PMT), avalanche photodiodes, and charge-coupled device (CCD) cameras. PMT is the most common implementation because it is robust and low cost, has large active area, and has relatively good sensitivity. PMT and avalanche photodiode systems further allow the use of ultralow noise and high sensitivity single photon counting electronic circuitry. The electronic signal from the detector is converted to a digital signal and is recorded by the data acquisition computer. The control computer further permits the 3D rendering of these images.

27.2.3 Damage Mechanisms of Nonlinear Optical Microscopy

Although it is well known that nonlinear optical microscopy is particularly suitable for in vivo tissue imaging due to its minimization of specimen photodamage, it is important to realize that specimen photodamage still occurs. Most damage studies in nonlinear microscope have focused on two-photon fluorescence microscopy because of its early development and popularity. In a typical nonlinear optical microscopy experiment, the region of photodamage is restricted to within the axial depth of the excitation PSF. For specimen with thickness comparable to the width of the axial PSF, the use of nonlinear optical imaging does not significantly improve specimen viability. However, the decrease in photodamage is substantial for 3D imaging of thick tissues. Photodamage is approximately reduced by a factor equal to the ratio of the sample thickness to the width of axial PSF. For 3D imaging of a 100 μm thick specimen using multiphoton microscopy, the specimen is exposed to a light dosage about one hundred times less than other 3D microscopy technique such as confocal microscopy.

A number of embryology studies demonstrate well the minimal invasive nature of multiphoton microscopy. Long-term monitoring of *C. elegans* and hamster embryos using confocal microscopy have not been successful due to photodamage induced developmental arrest. However, these developing embryos can be repeatedly imaged with multiphoton microscopy for hours without appreciable damage (Mohler et al.

1998, Mohler and White 1998, Squirrell et al. 1999). Further, the hamster embryos that were reimplanted after the imaging experiments were successfully developed into normal adults.

Nonlinear optical microscopy photodamage typically proceeds through three main mechanisms. (1) Oxidative photodamage can be caused by two- or higher photon excitation of endogenous and exogenous fluorophores similar to ultraviolet exposure. Tissue fluorophores act as photosensitizers in oxidative processes (Keyse and Tyrrell 1990, Tyrrell and Keyse 1990). Photoactivation of these fluorophores results in the formation of reactive oxygen species, which triggers the subsequent biochemical damaging cascade in cells. Some studies found that the degree of photodamage follows a quadratic dependence on excitation power indicating that a two-photon process is the primary damage mechanism (Konig et al. 1996, 1999, Sako et al. 1997, Hockberger et al. 1999, Koester et al. 1999). However, higher-order damage has also been observed (Patterson and Piston 2000). Experiments have also been performed to measure the effect of laser pulse width on cell viability. Results indicate that the degree of photodamage is proportional to two-photon-excited fluorescence generation, independent of pulse width. Therefore, using shorter pulse width for more efficient two-photon excitation also produces greater photodamage (Koester et al. 1999, Konig et al. 1999). Flavin-containing oxidases have been identified as one of the primary endogenous targets for photodamage (Hockberger et al. 1999). (2) One- and multiphoton absorption of the high-power infrared radiation can also produce thermal damage. The thermal effect resulting from two-photon absorption by water has been estimated to be on the order of 1 mK for typical excitation power (Denk et al. 1995, Schonle and Hell 1998). Therefore, heating damage due to multiphoton water absorption is insignificant. However, there can be appreciable heating due to one-photon absorption in the presence of a strong infrared absorber such as melanin (Jacques and Mcauliffe 1991, Pustovalov 1995, Masters et al. 2004). Thermal damage has been observed in the basal layer of human skin (Buehler et al. 1999). (3) Photodamage may also be caused by mechanisms resulting from the high peak power of the femtosecond laser pulses causing plasma formation from multiphoton ionization process (Konig et al. 1996).

27.3 Frontiers of Nonlinear Optical Microscopy

A significant portion of this chapter is devoted to the "frontiers" in this field. Clearly, the choice of topics featured here is quite subjective. However, one may argue that a review chapter is the proper venue to point out some of the most exciting areas which new researchers work on. This section will focus mostly on technological improvements that either are greatly needed in this field or will potentially significantly extend the utility of this technology. Specifically, we will focus on (1) how to improve imaging depth and resolution in nonlinear optical microscopy, (2) how to study important biological structures that are not readily accessible by two-photon-induced fluorescence, and

(3) how to apply nonlinear optical microscopy, not for imaging but for modification of material properties. In addition to these topics, we will also examine the possibility of two-photon microscopy for clinical applications. We would like to clearly distinguish clinical applications from biological research. We will not focus on applications of nonlinear optical microscopy to study specific biological systems. Clearly, biological applications of nonlinear optical microscopy will grow rapidly in the foreseeable future; as an example, the applications of nonlinear optical microscopy in neurobiology probably deserve their own book. However, whether nonlinear optical microscopy will have significant impact on clinical practices, the future is less clear. In this section, we intend to provide a review of the many exciting progresses in clinical applications and also some major limitations that may hinder its ultimate adoption.

27.3.1 Improving Imaging Performance: Depth and Resolution

The value of nonlinear optical microscopy for biomedical imaging must be evaluated relative to its competing technologies. Specifically, one of the major strengths of nonlinear optical microscopy is its high resolution. The major weakness of nonlinear optical microscopy is its modest penetration depth. It is also useful to compare nonlinear optical microscopy with the two most important alternatives: confocal microscopy and optical coherent tomography (OCT). Comparing with confocal microscopy, clearly both techniques offer similar resolution. In fluorescent mode, confocal microscopy has significantly shorter tissue penetration depth due to significant absorption of the excitation and emission light. However, in reflected light mode, confocal microscopy has very comparable penetration depth to nonlinear optical microscopy. OCT offers significantly deeper penetration depth up to several millimeters. The drawback of OCT is its relatively modest axial resolution. However, while traditional OCT has resolution on the order of tens of microns, cutting-edge OCT has achieved axial resolution approaching one micron, which is no longer so different from nonlinear optical microscopy, and OCT can produce images much more rapidly (Fujimoto 2003, Vakoc et al. 2009). As OCT and confocal techniques are catching up on resolution and/or penetration depth, it is important to note that both techniques have substantially lower instrument cost and complexity. Therefore, the future relevance of nonlinear optical microscopy will depend on continuing improvement in imaging depth and resolution.

27.3.1.1 Image Penetration Depth

Significant effort has been devoted to enhance the imaging depth of nonlinear optical microscopy since its inception. One very significant publication was published by Oheim and coworkers (Oheim et al. 2001). This publication, and its erratum, clearly point out the relevant parameters that limit tissue imaging depth of nonlinear optical microscopy in two-photon fluorescence mode that is broadly applicable to many nonlinear

techniques (although it needs to be modified for higher order processes):

$$z_{max} = l_s^{(ex)} \ln\left(\alpha P_0 \sqrt{\Phi(z_{max}) \cdot \frac{1}{f_p \tau_p}} \right) \tag{27.5}$$

where

z_{max} is the maximum imaging depth
$l_s^{(ex)}$ is the tissue scattering length of the excitation light
$\Phi(z_{max})$ is the emission photon collection efficiency
P_0 is the average excitation power
τ_p is the pulse duration
f_p is the pulse repetition rate
α is a factor characterizing instrument performance such as fluorophore and detector quantum efficiencies

An important limitation of imaging depth is the scattering of the excitation light by tissues. The probability of tissue scattering is quantified by the tissue scattering length, $l_s^{(ex)}$, defined as the $1/e$ distance characterizing the exponential decay of excitation light intensity as a function of tissue imaging depth. The scattering lengths in tissues, such as the brain, are on the order of 200 microns. Since two-photon excitation efficiency is a quadratic function of excitation power, the maximum penetration is ultimately limited to a few scattering lengths due to the exponential attenuation. Although the scattering lengths are intrinsic properties of tissues, optical clearing is a viable method to greatly improve imaging depth by drastically increasing scattering length. Optical clearing of tissues has been employed in biomedical imaging for over a century. However, recent resurgence of this technique for nonlinear optical microscopy started several years ago and has found expanding areas of applications (Yeh et al. 2003, Plotnikov et al. 2006, Nadiarnykh and Campagnola 2009). Recently, whole mouse brain has been cleared and imaged optically (although not by nonlinear optical microscopy) (Dodt et al. 2007).

Clearly, it is fairly obvious that maximizing the laser average power, P_0 is advantageous for optimizing imaging depth. However, it should be noted that this factor has a clear upper bound. The average power is certainly bound by the specimen photodamage threshold. For the fluorescent contrast modalities, the maximum average power is further limited by fluorophore saturation. Specifically, as the excitation power increases, the finite lifetime of fluorophores results in electronic ground state depletion. As a consequence, the excitation PSF broadens when excitation probability at the peak of the PSF reaches 10%–20% (Nagy et al. 2005a,b). Parametric processes are not similarly bound since the molecule returns to the ground state virtually instantaneously.

The most significant contribution of Oheim and coworkers in this paper may be in pointing out the importance of emission photon collection efficiency, $\Phi(z_{max})$, as an important factor for nonlinear microscope design. Specifically, they point out that NA of the microscope objective plays an important role, but the objective field of view is equally important. The objective NA determines the collection efficiency of emission photons that have not been scattered by the intervening tissue. However, a majority of emission photons are scattered by the tissue and their collection efficiency is rather determined by the field of view of the objective. The publication of this paper has the effect of shifting the field from using high NA but small field of view objectives to modest NA, but large field of view objectives for tissue imaging. The Oheim paper further emphasized that the intermediate optics between the objective and the detector should be optimized to maximize the throughput of scattered photons. However, since individual microscope designs can be fairly different, this optimization is highly instrument-dependent. Finally, the active area of the detector is also important. PMTs are ideal detectors because they have large active areas and can efficiently collect scattered emission photons. In other detectors, such as single photon counting avalanche photodiodes with small active areas, a significant portion of emission photons are rejected due to its small solid angle. Similarly, the use of area detectors with small pixels such as CCD camera also degrade nonlinear optical microscopy signal-to-noise ratio and reduce tissue penetration depth (Kim et al. 2007).

The roles of pulse duration, τ_p, and pulse repetition rate, f_p also require careful examination. In fact, some publications appear to present contradictory views. It is clear that at constant average power and pulse repetition rate, two-photon excitation efficiency increases linearly with decreasing pulse duration and potentially resulting in greater imaging depth (Tang et al. 2006). However, it has been pointed out that as pulse width decreases, its equivalent spectral bandwidth becomes broader than the excitation spectra for most fluorophores in the fluorescent mode (Denk et al. 1995). In Raman scattering modes, it has long been recognized that picosecond pulse width is much better match to the Raman lines. In addition to matching spectral bandwidth, the use of ultrashort pulse also introduces very significant instrument complications. Laser pulse width below 50 fs can suffer significant nonlinear dispersive effects and may require complex dispersion compensation technology (Schelhas et al. 2006, Xi et al. 2009). While compensation for second-order group velocity dispersion is routinely implemented in many femtosecond laser systems, the compensation for higher-order dispersion will require the use of coherent control techniques that remain a topic of active research in optics.

The choice of laser pulse repetition rate is less straightforward. At the same average power, certain publications advocate increasing pulse repetition rate (Chu et al. 2001, Ji et al. 2008) while some other publications favor reducing the pulse repetition rate (Theer et al. 2003). The argument for increasing pulse repetition rate depends on the specific contrast mechanism for imaging. For fluorescent nonlinear optical microscopy, increasing the repetition rate will improve excitation efficiency and imaging depth. However, similar to increasing average power, eventually increasing the repetition rate will result in ground state depletion and a broadening of the imaging PSF. This situation is quite different for harmonic generation and Raman-based nonlinear optical microscopy. Since these processes are instantaneous, they are not limited by PSF broadening and can be benefited by increasing pulse repetition rate until other specimen damage mechanisms set in. However, it should be noted that increasing the pulse repetition rate is a benefit only when individual pulses have sufficient power to maximize excitation efficiency per pulse without photodamage or saturation because excitation efficiency has a nonlinear dependence on excitation power. For thin specimens, there is little concern with each pulse having sufficient power using typical Ti:Sapphire lasers. However, since excitation power in each pulse attenuates exponentially in tissues, nano-joule pulses available in typical Ti:Sapphire laser can become too weak for efficient excitation within about five scattering lengths (approximately one millimeter). This depth corresponds to approximately the maximum imaging depth achievable today with Ti:Sapphire lasers. A method to overcome this limitation is to use an excitation source with significantly higher peak power, such as an optical parametric amplifier. In this case, femtosecond pulses with micro-joule to milli-joule energy can be produced, although at the expense of reducing the pulse repetition rate to the kHz to 10 kHz range. This approach has pushed the imaging depth beyond one millimeter (Theer et al. 2003). While this is an interesting configuration, it is ultimately nonoptimal since lowering the repetition rate by almost five orders of magnitude reduced the image frame rate drastically and micro-joule level pulses cannot be effectively used at typical imaging depths without damage or saturation. A laser light source with intermediate pulse energy may be better suited for future studies.

Another important avenue for future imaging depth improvement may lie in shifting excitation wavelength further into the infrared spectrum (Balu et al. 2009, Kobat et al. 2009). While it has been widely recognized that nonlinear optical microscopy allows the use of infrared excitation that has significantly better penetration than ultraviolet or blue excitation wavelengths, it has also been recognized that if the emission wavelength is to remain in the visible, then there is a significant loss of emission photons due to scattering. Therefore, it is natural to shift the excitation wavelength further into the infrared spectrum. It should be noted that due to the higher power required for nonlinear excitation, it is impossible to shift excitation much beyond 1300 nm due to the logarithmic increase of infrared absorption by water and the damage of the specimen by heating. However, since scattering length is a highly nonlinear function with wavelength, shifting excitation wavelength to 1300 nm and the emission wavelength to the range of 600–800 nm can result in significant reduction of the scattering effect and greatly improvement the tissue imaging depth (Figure 27.5).

27.3.1.2 Image Resolution

Although image penetration depth is important, maintaining submicron image resolution with increasing imaging depth is critical. Given the importance of resolution, there are significant recent works that demonstrate an improvement of the resolution of nonlinear optical microscopy beyond the diffraction limit.

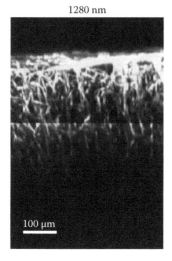

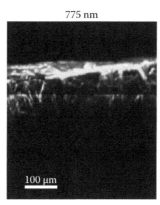

FIGURE 27.5 Ex vivo imaging of vasculature labeled with fluorescent agarose gel imaged at 1280 nm (left) vs 775 nm (right). (Imaged adapted from Balu, M. et al., *J. Biomed. Opt.*, 14, 010508, 2009; Kobat, D. et al., *Opt. Express*, 17, 13354, 2009.)

Improving image resolution has been investigated for over two decades. Several techniques, such as three-photon fluorescence excitation or combining two-photon fluorescence excitation with confocal detection, provide modest resolution improvement (Gu 2000). However, these techniques did not gain much traction, partly because the gain is modest but also because of their disadvantages; three-photon imaging is limited by a lack of appropriate contrast agents and confocal detection results in significant reduction of signal-to-noise ratio due to a loss of scattered emission photons.

The most significant advance in super-diffraction limited imaging is stimulated emission depletion (STED) microscopy that was first proposed by Hell and coworkers (Hell and Wichmann 1994). Fluorophores in a focal volume are excited by a pump laser. Super-resolution results from the addition of a probe laser that operates in the emission spectrum of the fluorophores. By the process of stimulated emission, the presence of the probe laser causes de-excitation of the fluorophores and quenching the fluorescence signal. The spatial distribution of the probe beam intensity is carefully engineered to have a 3D-donut geometry leaving the center of the excitation PSF minimally affected while quenching fluorescence emission from the boundaries of the PSF in all three dimensions. Hell and coworkers have shown that the quenching mechanism is "saturable," and quenching efficiency can be a highly nonlinear function of probe beam intensity. As a result, an extremely sharp boundary can be imposed between regions with and without quenching. This method results in an almost isotropic PSF function in 3D with 40 nm resolution (Figure 27.6). Fifteen years after the first proposal of this novel technique, STED microscopy is reaching maturity with an increasing number of exciting applications in biomedical sciences (Kittel et al. 2006, Willig et al. 2006, Eggeling et al. 2009). Recent development includes using two-photon fluorescence excitation for pump excitation and improving the spatial resolution by imposing a STED pulse (Moneron and Hell 2009).

While STED microscopy has achieved spectacular success in improving image resolution, the instrument complexity remains a limitation for this technique. A simpler approach (Vaziri et al. 2008) to achieve almost comparable resolution can be achieved by combining temporal focusing based wide-field two-photon imaging (Oron et al. 2005) with photoactivated localization microscopy (PALM) (Betzig et al. 2006) or stochastic optical reconstruction microscopy (STORM) (Rust et al. 2006) (Figure 27.6). Three-dimensional resolution is achieved by temporal focusing based wide-field two-photon fluorescence imaging. As discussed in Section 27.2.1, depth discrimination is achieved based on spatial focusing of the excitation light and the photon flux decreases away from the focal plane. For wide-field imaging, the image plane is illuminated uniformly by collimated light. Since the excitation light is collimated, the photon flux is uniform axially and there is no depth discrimination. Oron and coworkers elegantly overcome this difficulty by noting the excitation efficiency depends on photon flux that is a function of both space and time. By controlling the spectral bandwidth of the laser pulses, the excitation laser pulses can temporally broaden away from the focal plane, but regain their narrow femtosecond pulse width when they reach the focal plane. While wide-field two-photon imaging can provide very

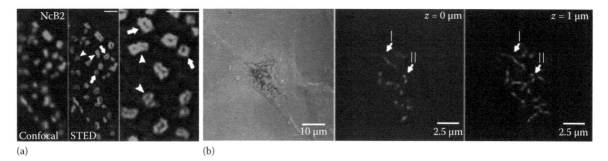

FIGURE 27.6 (See color insert.) (a) *Drosophila* coiled-coil domain protein Bruchpilot imaged using confocal (left) and STED microscopy (middle and right). Scale bars represent 1 μm. (Adapted from Kittel, R.J. et al., *Science*, 312, 1051, 2006.) (b) Mitochondria network imaged using wide field two-photon imaging with PALM. DIC and epi-fluorescence view (left) and two-photon wide field imaging combined with PALM views at two depths. (Adapted from Vaziri, A. et al., *Proc. Natl. Acad. Sci. U. S. A.*, 105: 20221, 2008.)

efficient 3D-resolved imaging, this method does not improve image resolution. To attain super-resolution, Vaziri and coworkers combine temporal focused two-photon wide-field imaging with the super-resolution approach of PALM/STORM. PALM and STORM are similar methods that utilize the fact that the centroid of the image of an isolated single fluorophore can be readily measured (localized) with resolution on the order of tens of nanometers. However, if two fluorophores are in close proximity of each other, their PSFs overlap and they can no longer be resolved with high precision. PALM and STORM overcome this limitation by enabling these fluorophores to be switched on and off stochastically such that only one of the two fluorophores is turned on for a majority of the time. By repeatedly turning these fluorophores on and off, the locations of both fluorophores can be determined again with tens of nanometer resolution by statistically analyzing the distribution of photons imaged. The combination of these two techniques has resulted in successful imaging of cellular membrane and mitochondrial structures with super-diffraction-limited resolution in 3D.

The enhancement of nonlinear optical microscopy resolution beyond the diffraction limit represents major progress. Since the major applications of nonlinear optical microscopy are in tissue imaging areas, the ability to maintain normal resolution at increasing imaging depth is also critical. It should be noted that tissue refractive index is inhomogeneous on all length scales (from nanometer scale upward). Light loss from interacting with small inhomogeneities (less than the wavelength of light) is what we typically consider as scattering. Scattering typically does not greatly affect resolution but limits image SNR and imaging depth as previously discussed. Inhomogeneities that are comparable or larger than the wavelength can cause the optical wavefront to be distorted as excitation light propagates through tissue. Sufficient wavefront distortion will clearly result in PSF broadening, resolution loss, and excitation efficiency loss. A number of interesting studies have proposed the use of adaptive optics system to compensate for wavefront distortions. In an adaptive optics system, the excitation light is reflected by a deformable mirror that is shaped to negate the wavefront distortion caused by the tissue. A major challenge is how to estimate the effect of the tissue on wavefront distortion. A number of methodologies have been proposed ranging from varying mirror shape based on genetic algorithm to maximizing signal intensity (Wright et al. 2005, Girkin et al. 2009), optimizing image feature quality (Debarre et al. 2009), and measuring the wavefront distortion of the reflected light (Rueckel et al. 2006). It should be noted that most of these two-photon fluorescence studies show some resolution improvement and about a factor of two improvement in imaging depth (Cha et al. 2010). This improvement appears to be consistent with the relatively weak wavefront distortion caused by tissue when the maximum imaging depth is limited by excitation loss from scattering. In contrast, another studying using adaptive optics in CARS microscopy shows almost six fold increase in CARS signal (Wright et al. 2007); this discrepancy may be due to the fact that CARS signal depends critically on the overlap of the PSFs of the pump and Stokes beams.

27.3.2 Novel Spectroscopic Contrast Mechanisms

Nonlinear optical microscopy is fortunate to have numerous spectroscopic contrast mechanisms. Spectroscopic measurements in the microscope not only allow the determination of tissue structures, quantify gene and protein expressions, and map the distribution of metabolites, but spectroscopic assays can also reveal the distribution of drug molecules, quantify stress and strain distributions in tissues, or map the presence of synapse junctions of neurons in the living brain. In fact, there are very few biomedical relevant parameters that cannot be measured using nonlinear optical microscopy. Of course, much of the power of nonlinear optical microscopy originates from the availability of an arsenal of fluorescent probes that have been developed by photochemists over the past century.

We do not intend to focus on fluorescence in this section as its importance is well recognized. Instead, this section will focus on contrast mechanisms beyond fluorescence. Clearly, it is impossible to review all the exciting spectroscopic analysis methodologies that have been implemented in nonlinear optical microscopy. Instead, we will feature two classes of spectroscopic-resolved nonlinear optical microscopy that have seen rapid growth and are finding important biomedical applications. Specifically, we will focus on nonlinear optical microscopy techniques that can image low quantum yield chromophores and molecular vibronic signatures.

27.3.2.1 Nonlinear Optical Microscopy of "Silent" Chromophores

Realizing that many biomedically important chromophores have distinct absorption spectra but negligible fluorescence, Warren and coworkers have pioneered nonlinear optical microscopy methods that target this class of "silent" chromophores (Tian and Warren 2002). Some of these more important "silent" chromophores include different tissue pigments, such as different melanins, myoglobin, and hemoglobin, and genetically expressible β-galactosidase cleaving chromatic substrate X-gal. Using traditional two-photon fluorescence microscopy, these chromophores either are nonfluorescent or have too low quantum yield for efficient imaging. The development of a spectrally-resolved imaging technique that can map the distribution of these fluorophores and tissues is clearly important.

Tian and Warren demonstrates that the small loss in the excitation beam absorbed by the chromophores can be measured using lock-in detection of intensity-modulated excitation light (Tian and Warren 2002). By further improving detection sensitivity and extending the technique to microscopic imaging, Tian and coworkers have demonstrated this technology in a number of biomedically interesting applications. In a measurement of two different types of melanin molecules in skin and hair, Fu and coworkers configured a microscope experiment similar to the classical pump-probe transient absorption experiment (Fu et al. 2008). In this case, the pump beam that depletes the ground state of the fluorophore was intensity-modulated. A constant intensity probe beam at a slightly different wavelength was incident upon the specimen. Because the ground state population was modulated in time by the pump beam, the absorption of the probe beam would also be modulated in time. The degree of the probe beam modulation provided a measure of chromophore concentration in the specimen, and the phase shift between the pump and the probe modulation was a measure of the excited state dynamics of the fluorophore. Furthermore, the modulated signal depends bilinearly on the pump and probe beam intensity and provides a 3D image PSF similar to standard two-photon fluorescence microscopy. In another study, a similar approach was taken to map oxyhemoglobin and deoxyhemoglobin distributions in blood vessels. The experimental arrangement was similar to the previously described melanin measurement except that the two-photon absorbance of the hemoglobins was measured instead of the one-photon absorbance of melanins (Fu et al. 2007a) (Figure 27.7).

Another very exciting microscopy study from Xie and coworkers recently offers an alternative method to measuring the presence of these "silent" fluorophores (Min et al. 2009). In previously described studies, chromophore absorption is measured by quantifying the depletion of the ground state. Min and coworkers instead quantify the excitation state population of these chromophores using stimulated emission. Some of these "silent" chromophores have very low quantum yield via spontaneous fluorescent emission because of the presence of very dominant nonradiative decay mechanisms. However, these chromophores can be induced to emit radiatively by stimulated emission provided that the probe beam has sufficient intensity and the temporal separation between pump and probe beams are sufficiently short such that the probability of nonradiative decay can be minimized. In this study, the authors demonstrated mapping melanin pigments as well as drug molecules in tissues. It is also important that this study demonstrate that chromoproteins and chromogenic reporters for gene expression can be imaged providing an important complementary technology to fluorescent protein technology that is widely used in the fluorescent mode.

27.3.2.2 Nonlinear Optical Microscopy Based on Raman Contrast

Typical contrast agents for both fluorescent imaging and absorption imaging are relatively large molecules that have electronic states that are optically active. However, there are many important molecules in biology and medicine that have no optically active electronic states. However, all molecules have vibronic states that may be probed optically. Traditionally, these vibronic states are studied using near-infrared absorption spectroscopy or spontaneous Raman scattering. However, neither of these techniques has been found to be ideal for microscopic imaging. Near-infrared absorption spectroscopy is limited by the strong water absorption in the relevant spectral range. Raman spectroscopy is limited by its signal level that has typically lower order of magnitude than that of fluorescence.

These limitations for quantifying the vibronic states of biomolecules have prompted researchers to develop higher sensitivity spectroscopic methods. Among the most promising methods, coherent anti-Stokes Raman spectroscopy (CARS) has demonstrated great promise. While the CARS technique was demonstrated long ago, the recent resurgence of this technique is mostly due to the work of Xie and coworkers. Initial implementations focused on purely spectroscopic measurement, subsequent work soon branched into nonlinear optical microscopy (Cheng et al. 2001). It should be pointed out that CARS provides a coherent signal that is significantly stronger than spontaneous Raman scattering. However, it also has the disadvantage; in addition to the resonant signal that represents the vibronic state of the molecule, there is also a strong nonresonant background. Therefore, much of the subsequent literature in CARS microscopy focused on how to recover the resonant signal from the nonresonant background. Several successful approaches have been developed such as polarization-resolved CARS (Cheng et al. 2001) and epi-detected CARS (Volkmer et al. 2001).

Today, CARS microscopy has found important applications in biomedicine. The main advantages of CARS for biomedical imaging are (1) the technique is probeless and (2) the technique has chemical specificity. It should be noted that CARS microscopy is still a relatively low sensitivity technique as compared with fluorescence microscopy. Therefore, CARS microscopy found applications in imaging molecules with relatively high abundance, at least several thousands in the excitation volume. For pharmaceutical development, mapping tissues containing drug molecules at sufficient concentration is a promising area. In terms of endogenous signals, the most important targets are

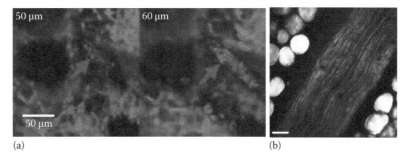

(a)   (b)

FIGURE 27.7 (a) Two-photon absorption imaging of blood vessels at two different depths. Scale bar represents 50 μm. (Adapted from Fu, D. et al., *J. Biomed. Opt.*, 13, 054036, 2007a.) (b) CARS microscope imaging of myelin sheaths on the sciatic nerves of a live mouse. Scale bar represents 25 μm. (Adapted from Huff, T.B. and Cheng, J.X., *J. Microsc.*, 225, 175, 2007.)

fatty acids and lipid molecules. Clearly, one of the most important applications of CARS is the study of lipid metabolisms in cells and tissues (Nan et al. 2006). An even more exciting application of CARS technique is in the imaging of myelin. Myelin sheath is an important lipid-rich insulation layer around neurons and is critical for proper neuronal communication. The degradation of myelin sheath is responsible for many important neuronal diseases including the very common multiple sclerosis. Recent research has successfully demonstrated that CARS is very adapted in quantifying the myelin state of neurons in brains (Fu et al. 2007b, Huff and Cheng 2007) (Figure 27.7). Subsequent work has shown that chemical imaging of whole mouse-brain slices can be performed efficiently using CARS (Evans et al. 2007).

While CARS has gained broad acceptance in the biomedical optics field, one recent advance promises to completely change the landscape of Raman imaging in nonlinear optical microscopy. This new approach is stimulated Raman spectroscopy. Similar to CARS, stimulated Raman spectroscopy (SRS) is also not a new technique and has been known for decades. However, with the advent of CARS microscopy, a number of research groups started to examine the advantage and disadvantage of stimulated Raman scattering approach relative to CARS (Ploetz et al. 2007, Freudiger et al. 2008, Nandakumar et al. 2009). The major advantage of the stimulated Raman scattering approach is that it has no nonresonant background. The major disadvantage is that stimulated Raman scattering signal represents a $1/10^6$ to $1/10^7$ change in the pump or the probe beam signal intensity. CARS has a signal spectrum very different from that of the pump or the probe beams, and the signal can be readily isolated spectrally. The weak stimulated Raman scattering has the same spectra and propagation direction as either the strong pump or the probe beam and cannot be readily isolated. To overcome this difficulty, stimulated Raman imaging utilizes high sensitivity lock-in detection circuitry. Improving signal to noise ratio of images remains an important research subject in this field (Ozeki et al. 2009). Recent publications have demonstrated that stimulated Raman scattering imaging can be successfully used to image cell and tissue lipid structures, including myelin sheath in brain and lipid droplets in cells very similar to CARS microscopy.

27.3.3 Nonlinear Nanoprocessing

Although a majority of nonlinear optical microscopy research focuses on imaging, a growing segment of applications utilize the ability of nonlinear excitation to modify material properties with 3D resolution on the nanoscale nonlinear processing. While imaging based on nonlinear excitation has been developed since late 1970s, the possibility of nanoprocessing was first demonstrated by 3D-localized photobleaching in Denk et al. (1990). Subsequently, Denk and coworkers first demonstrated that caged neural transmitters can be released locally in the vicinity of a neuron. The spatial distribution of neural transmitter receptors can be mapped by measuring neuronal electrical response as a function of the position of transmitter release. After these pioneering works, the field of nanoprocessing using nonlinear optical excitation progressed along many diverse directions. In order to facilitate our discussion, we will roughly divide these many areas of works into two subgroups: using nanoprocessing for material removal and using nanoprocessing for structure fabrication.

27.3.3.1 Nanoprocessing for Material Removal

Using nonlinear nanoprocessing to remove materials is the basis of many biomedical applications. The theoretical basis of how femtosecond laser pulses affect material removal is still a very active field of research (Vogel et al. 1999, Vogel and Venugopalan 2003). However, this process is typically considered to involve ionization of materials based on multiple photon absorption. Because multiple photon absorption is involved in this process, one would expect the photointeraction volume to be smaller than the typical quadratic process PSF. However, the actual volume where material is removed also depends on other parameters such as the thermal and mechanical properties of the material. In any case, nanoprocessing with resolution significantly better than diffraction-limited has been demonstrated.

Using nanoprocessing to remove material has found many applications in biomedicine. One particularly exciting area is to use this process to puncture cellular membranes to facilitate several important biotechnology applications such as transfecting a single selected cell with a specific gene (Soughayer et al. 2000, Tirlapur and Konig 2002a,b). This process also has been called optoporation. More recent work has shown that this technology is applicable to both animal and plant cells. Importantly, this process appears to be able to transfect cells with high efficiency while better maintaining cell viability as compared with other transfection techniques such as electroporation. One promising recent application is for genetic manipulation of stem cells that are resistant to transfection by other methods. Uchugonova and coworkers have demonstrated that this technology can be used to transfect stem cells efficiently while maintaining their viability (Uchugonova et al. 2008b).

Another promising application of nonlinear nanoprocessing is to knock out or destroy single cells in 3D tissue culture or inside tissues. The application of this technique to improve photodynamic therapy specificity has been proposed. However, the development of a clinical application based on this approach is relatively slow due to optical microscopy's relatively limited field of view and slow raster scanning speed. However, a number of recent works show that precise nanosurgery can be performed on animal models (Wang et al. 2007). Today, the ability to selectively knock out single cell may have found better applications both in the nanotechnology area and in cutting-edge biology research. In the area of biotechnology, the ability to isolate adult stem cells is important. To isolate adult stem cells, it is often necessary to "clean" its surrounding area by removing other cells. In this case, the destruction of selected cells based on nonlinear nanoprocessing inside tissue in situ can potentially improve stem cell recovery efficiency (Uchugonova et al. 2008a). Another very exciting application of this technique is in the area of developmental biology. Yanik and coworkers have demonstrated that single neuronal processes in living *C. elegans* can be selectively severed (Yanik et al. 2004, Guo et al. 2008). The change in the animal motility behavior after neuronal damage can be directly observed, and the processes of nerve regeneration can be studied. Since the whole neuronal circuits of model animals, like *C. elegans*, have already been mapped out at a single cell level, the ability to selectively disconnect one branch of a circuit is a particularly powerful research tool to decipher the function of this circuit (Figure 27.8). A similar approach has also found applications in neurobiology (Tsai et al. 2009). Instead of cutting a single dendrite of a neuron, two-photon ablation can also severe selected capillaries in the brain allowing the study of hemodynamics in vivo.

27.3.3.2 Nanoprocessing for Structure Fabrication

The potential of using a nonlinear optical microscope for micro- and nanofabrication was quickly recognized. Cumpston and coworkers (Cumpston et al. 1999) demonstrated the application of this technology for the fabrication of 3D photonic crystal structures shortly after the publication of Denk et al. (1990). Nonlinear optical microscopy based fabrication technique can be viewed as a subclass of lithographic fabrication.

Of course, lithographic fabrication is a cornerstone of the semiconductor industry, and the extension of the techniques to the micro- or nanoscale recently has found applications in many areas, including photonics (Campbell et al. 2000) and biomedicine (Vozzi et al. 2003, Hollister 2005). Photolithography, the most common microlithography technology, uses light to transfer the desired patterns on the substrate through several photochemical processes with its resolution limited by the diffraction limit of the light. To overcome this limitation, several nanoscale fabrication methods have been reported: electron beam lithography (Craighead 1985), which scans electron beam over the surface in a patterned fashion, nanoimprinting lithography (Chou et al. 1996), which stamps patterns on thin layers, dippen nanolithography (Piner et al. 1999), which transfers specific molecules on the surface, nanosphere lithography (Haynes and

Van Duyne 2001), which uses a nanosphere layer as a mask, and soft lithography (Xia and Whitesides 1998), which replicates structures with elastomer stamps or molds. Most nanoscale fabrications are limited to thin 2D applications, and there is a need to further develop 3D microfabrication techniques.

Three-dimensional fabrication techniques based on solid freeform fabrication have been used extensively for making prototypes during manufacturing. Stereolithography is one of the most commonly used rapid prototyping techniques in this class. It uses curable photopolymer resin. A laser traces the cross section of the parts at each layer curing the polymer, and the 3D structure is generated by stacking these patterned layers together. While this technique is very useful in generating the macroscale 3D prototypes, its fabrication resolution is limited to several hundred microns. One method to overcome this limitation is the development of microfabrication based on nonlinear excitation (Strickler and Webb 1991, Yang et al. 2005, Maruo and Fourkas 2008). Importantly, in addition to photopolymerization, this microfabrication technique can utilize a variety of photointeractions to alter material properties in 3D such as photochromism. Photochromism (Hirshberg 1956, Irie 2000, Mendonca et al. 2007) is a reversible process that optically transforms chemical species between two forms with different physicochemical properties such as absorption spectra, refractive index, or dielectric constant. This process may find applications in creating erasable memory media (Irie 2000) and display devices (Yao et al. 1992).

Photopolymerization (Odian 2004) is the most common approach to building structures on the micro- and nanoscale. Photopolymerization is a polymerization process in which the material is exposed to light, especially ultraviolet (UV) light. It is modeled as a free-radical chain polymerization. Either a photoinitiator or a photosensitizer is required to initiate polymerization, but these have different roles in the polymerization process. During photoinitiation, a photoinitiator absorbs energy by excitation and is decomposed into radicals, whereas a photosensitizer absorbs energy by excitation and its molecular interaction and transfers energy to other molecules, which become radicals. Regardless of their different roles, the polymerization rate is determined by their characteristics. Unlike other polymerization processes, photopolymerization can occur in the spatially confined specific region where the light is irradiated, but the process is affected by the material thickness due to the limited penetration depth of the light. Fabrication resolution by photopolymerization can be higher than diffraction-limited optical resolution since photopolymerization requires a threshold optical energy to initiate the polymerization process. After the pioneering works in this area that examine the theoretical and experimental consideration in nonlinear microscopic microfabrication (Campagnola et al. 2000, Tanaka et al. 2002, Sun et al. 2003b, Sun and Kawata 2004, Cunningham et al. 2006, Kaehr and Shear 2007), this approach has a broad range of applications in the field, such as 3D optical storage (Cumpston et al. 1999, Olson et al. 2002), tissue scaffold (Pins et al. 2006, Claeyssens et al. 2009), photonic crystal structure(Straub and Gu 2002,

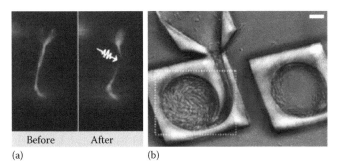

Before   After
(a)   (b)

FIGURE 27.8 (a) Severing axons of *C. elegans* by nanoprocessing that was later healed. (Adapted from Yanik, M.F. et al., *Nature*, 432, 822, 2004.) (b) Concentrating bacteria in a vortex microchamber. Scale bar is 5 μm. (Adapted from Kaehr, B. and Shear, J.B., *Lab Chip*, 9, 2632, 2009.)

Guo et al. 2005), and microfluidic devices (Wu et al. 2008a, Stoneman et al. 2009).

A particularly rich area for nonlinear microscopic fabrication for biomedical research is the use of femtosecond pulses to polymerize native biological materials (Pitts et al. 2000). The ability to polymerize native polymer materials is recognized to be useful for controlled release of drugs and proteins. Importantly, these fabrication processes have been shown to be biocompatible, allowing the trapping of proteins while maintaining their bioactivity (Basu and Campagnola 2004). Subsequently, this technology has found applications in a variety of biotechnology applications and bioscience research, including dynamically guiding neuronal growth on a 2D substrate (Kaehr et al. 2004), building 3D structures by polymerizing proteins that maintain their catalytic activity (Allen et al. 2005), directing cell migration in a 3D polymer scaffold (Tayalia et al. 2008), fashioning protein hydrogel for microactuation (Kaehr and Shear 2008), regulating the adhesion and migration of cancer cells (Chen et al. 2009), and directing bacteria motility in microfluidic devices (Kaehr and Shear 2009) (Figure 27.8). Clearly, the ability to fabricate 3D structures that are biologically compatible with potential biological activities may find many different uses in biology and medicine.

While 3D fabrication based on nonlinear nanoprocessing is a powerful technique allowing the formation of arbitrary structures with an excellent aspect ratio, there are several other competing 3D fabrication technologies. Three-dimensional fabrication with a dip pen (Gratson et al. 2004) using polymerizable ink for direct writing is also a promising technology. Finally, 3D nanoimprinting lithography (Li et al. 2001) has been reported, but the types of 3D structures that can be made are also quite limited. In this comparison, it is clear that most 3D fabrication techniques, including nonlinear nanoprocessing, are relatively slow compared to standard 2D photolithography, since most of these fabrication processes are carried out on a point-by-point basis. The throughput of fabrication is ultimately limited by the sequential nature of this method and the mechanical limit of the scanner. Although many of these 3D microfabrication techniques provide very attractive 3D structures at submicron resolution, the use of 3D fabrication today is limited to laboratory investigations and prototype fabrication. Therefore, there is a need to develop higher throughput methods to enable commercial mass-production. One approach that does not depend on nonlinear optics is based on generating a 3D interference pattern inside a solid specimen by illuminating the object with multiple coherent plane waves (Shoji and Kawata 2000, Kondo et al. 2001, Sun et al. 2003a). This rapid production method is restricted to producing 3D periodic structures and has found applications in the fabrication of photonic crystal structures. One approach, which can enhance the speed of nonlinear nanoprocessing beyond that of laser direct writing, is based on using multiple beamlets (Fittinghoff and Squier 2000, Fittinghoff et al. 2001) to generate multiple foci on the sample. This parallelization increases fabrication speed by a factor equal to the number of excitation foci. One limitation of this technique lies in the difficulty of individually controlling the scan pattern of an individual beamlet and registering the patterns produced by the different foci. Clearly, the development of simple, high-throughput 3D fabrication techniques based on nonlinear nanoprocessing is an exciting future research area.

27.3.4 Nonlinear Optical Microscopy for Clinical Applications

While nonlinear optical microscopy is routinely used in many fields of bioscience, its future in clinical medicine remains fairly unclear. On the one hand, the slow inroads made by this technology into medicine are not surprising. With significant safety concerns and administrative hurdles, the adaptation of any novel diagnostic technology into medical practice has always been slow. On the other hand, the clinical usage of nonlinear optical microscopy is confounded by the need for high intensity lasers with potential for tissue damage. Further, the presence of other lower-cost and safer competing technologies certainly clouds the clinical future of this field. Specifically, since reflected light confocal microscopy and optical coherence tomography are both in advanced stages of clinical trials, the breakthrough of a "comparable" technology becomes less probable. Of course, nonlinear optical microscopy has its advantages, including good resolution and reasonable penetration depth. However, it is unclear that these factors can overcome the significantly higher cost and the added danger of tissue damage. In competition with confocal microscopy and OCT, it is unlikely that nonlinear optical microscopy can gain acceptance in clinical medicine based solely on its ability to obtain tissue structure information. Instead, the possible clinical adaptation of nonlinear optical microscopy will depend on identifying specific diseases that can benefit from diagnoses that take advantage of a unique contrast mechanism afforded by this new technology.

27.3.4.1 Endogenous and Exogenous Contrast Agents for Clinical Imaging

The most important class of contrast agents in tissues is endogenous fluorescent molecules that can be imaged based on two-photon fluorescence excitation. Many proteins are fluorescent due to the presence of fluorescent amino acid residues: tryptophan and tyrosine. Two-photon-induced fluorescence from tryptophan and tyrosine in proteins is well studied (Lakowicz and Gryczynski 1992, Kierdaszuk et al. 1995). However, two-photon excitation wavelengths for these fluorophores fall well below the emission wavelength range of typical Ti:Sapphire lasers, and their usage requires the addition of laser modules, such as optical parametric oscillators, at additional cost and instrument complexity. The low transmittance of typical glass in microscope's optical components further limits signal detection. While point spectroscopy based on tryptophan has been used for clinical diagnosis, it is rarely used for imaging because the weak signal in the 300 nm range has very low tissue transmittance. Since tyrosine excites and emits even deeper into the UV, it is used even less frequently. Finally, imaging of these fairly ubiquitous proteins may not provide very informative structural or biochemical information.

The most important fluorescent cellular marker is probably endogenous β-nicotinamide-adenine dinucleotide phosphate (NAD(P)H) (Kierdaszuk et al. 1996). NAD(P)H is an important coenzyme associated with the cellular metabolism and the redox state (Masters and Chance 1999). The fluorescent emission spectra of NAD(P)H depends on whether it is protein bound; free NAD(P)H absorbs in the UV region at about 340 nm and fluoresces with a maximum at 460 nm with a fluorescence lifetime of about 400 ps, while bound NAD(P)H has a blue shifted emission spectrum and a lifetime of 2 ns. More importantly, the oxidized form of this compound NAD has over an order of magnitude lower quantum yield than NAD(P)H. Therefore, NAD(P)H has long been recognized as a redox intracellular sensor. NAD(P)H fluorescence has been used to monitor the redox state in cornea (Piston et al. 1995) and skin (Masters et al. 1997). For the cornea, a recent review further compared the utility of nonlinear imaging technique with OCT (Masters 1984, 2009). Another class of intracellular fluorophores that may be used for two-photon imaging is flavoproteins that are localized in cellular mitochondria. Flavoproteins have a one-photon excitation range of around 450 nm and emission in the range of 550 nm. Two-photon excitation spectrum of flavin mononucleotide (FMN) has been also quantified (Xu and Webb 1996). The combination of these two fluorophores has been applied to study redox mechanisms in the brain (Kasischke et al. 2004).

Nonlinear optical microscopy can also image extracellular matrix structures in tissues. Collagen and elastin are the two most important components of the extracellular matrix. Collagen can be imaged based on fluorescence. Either tropocollagen or collagen fibers has absorption in the range of 280–300 nm and emission ranges from 350 to 400 nm (LaBella and Gerald 1965, Dabbous 1966, Hoerman and Balekjian 1966). Elastin has been observed to have a redder emission spectrum at 400–450 nm with excitation wavelengths of 340–370 nm (LaBella 1961, LaBella and Lindsay 1963, Thomas et al. 1963). Elastin fibers can also be excited at 800 nm by two-photon fluorescence excitation. Importantly, collagen is also an efficient structure for second harmonic generation due to its chiral crystalline structure (Freund et al. 1986, Campagnola et al. 2001, Campagnola and Loew 2003, Theodossiou et al. 2001, 2002, Stoller et al. 2002). It should be noted that second harmonic generated light is forward-scattered due to phase matching condition and is best detected in the transmission geometry. However, epi-detection geometry is more practical in the clinical setting. Nonetheless, second harmonic generation in tissue can be readily detected since multiple scattering results in a significant fraction of photons returning toward the tissue surface.

Finally, lipid and lipid-rich structures are also important targets for clinical tissue imaging based on CARS and SRS contrast, as discussed in the previous section. It should be noted that lipid and fatty acid can also be imaged by three-photon fluorescence; however, the deep UV emission makes this imaging modality less desirable (Xu et al. 1996).

The use of NAD(P)H fluorescence has been successfully applied to monitor cancer development in animal models (Skala et al. 2005, 2007). A number of investigators, including Dong and coworkers, have pioneered using the ratio of NAD(P)H and collagen second harmonic signal for ex vivo diagnosis of many diseases (Lin et al. 2005). For many diseases, pathological progression results in a remodeling of the epithelial and stromal structures and a change in the relative cellular and extracellular matrix composition. This approach has been applied in many organs, including the skin, the bronchus (Zhuo et al. 2008), the lung (Wang et al. 2009), and the liver (Tai et al. 2009). It has even been applied to dentistry diagnosis for the teeth (Chen et al. 2007). Second harmonic imaging by itself has been very successful in monitoring collagen reorganization due to Lasik surgery in the eye (Hovhannisyan et al. 2008) and in the remodeling of myosin–actin structures in muscle diseases (LaComb et al. 2008, Plotnikov et al. 2008).

In addition to endogenous contrast agents, the application of exogenous contrast agents for clinical imaging is in its infancy. The development of exogenous agents for clinical applications faces a double jeopardy: the safety and regulatory hurdles associated with both introducing new drugs into patients and using a potentially photodamaging imaging technology. Nonetheless, the development of exogenous contrast agents holds great promises since exogenous agents can be tailored to target specific gene and protein products and utilize our growing understanding of molecular biology. Today, there are very few contrast agents that may be used in humans. The few exceptions are dye molecules without specific targeting moieties, such as indocyanine green and fluorescein. One of the very promising exogenous agents is 5-aminolevulinic acid. While it is not directly imaged, it induces fluorescent protoporphyrin IX production in some tissues, most notably in cancer cells (Madsen et al. 2000, Beck et al. 2007, Chen et al. 2008). Clinical evaluation of 5-aminolevulinic acid as contrast agent has been conducted for other imaging modalities, but a thorough clinical evaluation using nonlinear optical microscopy remains to be untaken. Another extremely exciting new work in this area is the development of targeted fluorescent heptapeptide (Hsiung et al. 2008). Wang and coworkers developed this new contrast agent, put it through all regulatory approval processes, and evaluated its efficacy clinically for the detection of colonic dysplasia in patients using a one-photon confocal endomicroscope. The further development of similar exogenous contrast agents will definitely impact the field of nonlinear microscopic imaging.

The use of fluorescent proteins for clinical imaging has also been considered. The class of fluorescent proteins is one of the most important recent innovations in molecular biology allowing the detection of protein expression (Chalfie et al. 1994). Today, fluorescence proteins with emission spectra covering the whole visible and near infrared spectral region have been developed (Tsien 1998, Tsien and Miyawaki 1998, Baird et al. 1999). The vectors of these fluorescence proteins can be strategically placed into specific loci in the genome of the host cells allowing the monitoring of the expression of specific genes. The vectors coding for a particular protein of interest also can be covalently linked to a fluorescent protein providing a method to monitor protein distribution and trafficking. Equally important, novel fluorescent proteins that are spectrally sensitive to their biochemical environment, such as calcium concentration, have been developed (Allen et al. 1999, Zaccolo et al. 2000). Two-photon imaging of these fluorescent proteins has clearly transformed biology research in cells and animal, as evidenced by the recent Nobel Prize in chemistry. Fluorescent protein technology also has potential in clinical medicine. One promising application may be in tumor margin detection by introducing viral vectors that express fluorescent protein and target cancer cells. However, the application of this technique in patients requires procedures equivalent to gene therapy and its associated precautions and hurdles. Recently, this type of technology has been applied to the detection of rare circulating tumor cells in patient blood, ex vivo (Fong et al. 2009).

Finally, one of the major problems in the applications of nonlinear optical microscopes for clinical imaging is the need for high intensity radiation with the associated danger of tissue photodamage or potential subsequent malignant transformation. Clearly, the development of contrast agents with high two-photon cross sections can potentially overcome this problem. The early development of high two-photon cross-section contrast agents has not been very successful. The problem lies not in creating high two-photon cross-section probes but in the cytotoxicity and the low solubility of these probes (Albota et al. 1998a,b). A number of recent probes hold greater promises, including two exciting new classes of nanoparticles such as quantum dots (Larson et al. 2003) and conjugated polymer nanoparticles (Moon et al. 2007, Rahim et al. 2009). Both classes of probes have two-photon cross section up beyond 10,000 GM as compared with rhodamine and fluorescein, which have cross sections of tens and hundreds of GM. However, the applications of these nanoparticles for clinical imaging still involve significant challenges. Most of these particles are fairly large (tens of nanometers) and delivering them to the relevant tissue sites is nontrivial. More importantly, for quantum dots containing heavy metal ions, their systematic clearance remains a major concern for the long-term health of patients. For conjugated polymer particles, their development is at an early stage and the synthesis of particles that can target relevant tissue structures, such as cancer cells, will be required before future clinical evaluation.

27.3.4.2 Development of Nonlinear Optical Endomicroscopy for Clinical Imaging

The application of nonlinear microscopes for clinical diagnosis is limited not only by the availability of contrast agents but also by the practical issue of bringing nonlinear optical microscopes to the patient. While some preliminary work had been done on human subjects using laboratory format microscopes (Masters et al. 1997), these instruments are ultimately unsuitable for clinical work. The most significant progress in clinical applications using laboratory-scale nonlinear optical microscopes is the

work of Konig, Kaatz, and coworkers on dermal lesion diagnosis (Konig et al. 2009); this group of investigators obtained, for the first time, regulatory approval to evaluate the clinical utility of nonlinear optical microscopy in the diagnosis of skin disorders with a substantial population of patients. While laboratory scale systems are useful for dermal applications, minimally invasive diagnosis using nonlinear microscopic imaging may have more impact on diseases of other organs such as the colorectal, esophageal, and cervical cancers. Many of these cancer types occur on the epithelial surface of interior body cavities, such as the gastro-enterological tract, which cannot be reached without miniaturizing and adapting a laboratory microscope into an endoscopic format—an endomicroscopy. The importance of this development can be seen from the incidence rate of some of these diseases. For example, the incidences of colorectal, esophageal, and cervical cancers are over 12,000, 150,000, and 15,000 new cases per year, respectively. To further illustrate the use of this new technology for cancer detection, we will examine the situation with cervical cancer in greater depth (Agnantis et al. 2003, Lambrou and Twiggs 2003, Apgar and Brotzman 2004, De Palo 2004, Dunleavey 2004, Gupta and Sodhani 2004). The onset of cervical cancers has been associated with sexually transmitted human papillomavirus (HPV) infection affecting over 40 million people. Approximately one million new cases of HPV are detected each year. Some forms of HPV have been identified as the primary cause of cervical intraepithelial neoplasia. HPV is often detected because of either the observation of genital warts or from an abnormal Pap Smear. The common Pap Smear cannot localize the lesions. Patients with an abnormal Pap Smear are often referred to colposcopy, a conventional endoscopic imaging method of the cervix. Colposcopy has had false positive rates as high as 40%–50%. Further, high-grade disease may be misclassified as low grade. Punch biopsy, another common technique, is subject to false negative results through incomplete sampling. This is an area where endomicroscopy may be applied to identify the most critical biopsy sites and provide more informative diagnosis since endomicroscopy can provide subcellular level images similar to histopathology.

In addition to the initial diagnosis, endomicroscopy may also benefit long-term, periodic monitoring of premalignant conditions. For example, Barrett's esophagus is very common and is the result of acid reflux. Barrett's esophagus has a high chance of developing into cancer and must be carefully monitored clinically (Falk 2002, Belo and Playford 2003, Pacifico et al. 2003, Ban et al. 2004, Faruqi et al. 2004, Guelrud and Ehrlich 2004, Kara et al. 2004, Lambert 2004, Paulson and Reid 2004, Peters and Wang 2004, Sharma 2004). Endomicroscopy may permit more accurate noninvasive monitoring of potential progress of the disease at an organ site where frequent and high-coverage excisional biopsy is highly undesirable. In the case of cervical cancer, genital warts often involve surgical intervention with the associated risks. Low-grade lesions may be topically treated with trichloroacetic acid or 20% podophyllin solution. More extensive cases are treated with cryosurgery, laser vaporization, or loop electrosurgical excision. Since early stage genital warts often regress and excision procedures may compromise patient fertility, a conservative management scheme has been often suggested. The success of conservative management schemes depends on the existence of reliable clinical markers for cancer progression and imaging techniques to monitor them. Endomicroscopy may allow a more accurate, noninvasive determination of the propensity for Barrett's esophagus and cervical lesions to progress into a more invasive form.

Clearly, these diseases demand the development of high-resolution endomicroscopes with 3D resolution that may provide minimally invasive biopsy as a complementary technology to traditional excisional biopsy and histopathology. Three classes of high-resolution endomicroscopes have been developed today based on OCT, confocal microscopy, and nonlinear optical microscopy. Of the three approaches, endoscopes based on OCT are the most advanced because the miniaturization of this technology is easier for two important reasons. First, OCT systems inherently obtain depth information based on the timing of the low-coherence optical pulses such that no mechanical scanning in the depth direction is required. 2D sagittal or coronal sections of the tissue can be generated by only mechanically scanning the light ray along a single direction either radially or axially. Since scanning along only one direction is required, this greatly simplifies micro-scanner design. Second, OCT automatically provides depth-resolved information when the imaging optics has a long depth of field, i.e., a relatively low numerical aperture. This longer depth of field, and hence longer working distances, significantly reduces the complexity involved in assembling the micro-optical components in the distal end of the endoscope at the expense of lower resolution. OCT endoscopes have entered into a number of clinical trials including the identification of Barrett's esophagus (Brand et al. 2000, Li et al. 2000, Pitris et al. 2000, Zuccaro et al. 2001, Jacobson and Van Dam 2002, Nishioka 2003, Poneros and Nishioka 2003, Faruqi et al. 2004, Poneros 2004, Sharma 2004) and colonic polyps (Brand et al. 2000, Asano and Mcleod 2002, Nishioka 2003). OCT endoscope development is advanced enough such that devices with outer diameter less than 2 mm have been fabricated and have been applied to the evaluation of intracoronary stenting and the study of atherosclerotic plagues in patients (Fujimoto et al. 1999, Yabushita et al. 2002, Bouma et al. 2003a,b, Tearney et al. 2003, Chau et al. 2004). While OCT endoscopes are finding important clinical applications, the combination of using low NA optics, relatively narrower bandwidth of light sources, and the inherently low contrast in OCT signal in tissues limits the lateral and axial resolutions of these systems to about 10 μm. Therefore, OCT does not provide images with cellular resolution to completely serve diagnostic needs.

Today, endomicroscopes with subcellular resolution are developed based on confocal detection or nonlinear optical excitation principles. Unlike OCT endoscopes, confocal and nonlinear optical endomicroscopes obtain information from a single point in the tissues. Scanning along two directions must be performed to generate a 2D image. The need to incorporate scanning along at least two directions presents a challenge in endomicroscope

design. Nonetheless, significant progress has been made with confocal-based systems. Today, some confocal endomicroscopes are becoming commercially available, and preliminary clinical tests using these devices have started (Sakashita et al. 2003, Kiesslich et al. 2004). Three classes of confocal endomicroscopes have been developed. The most popular approach for confocal endomicroscopes utilizes high density flexible fiber-optic bundles and performs scanning at the proximal end of the device (Liang et al. 2002, Watson et al. 2002, Sakashita et al. 2003, Rouse et al. 2004). The advantage of this class of endomicroscope is its simplicity. Since scanning is performed at the proximal end outside the patient, the scanning mechanism does not require miniaturization, but the resolution of these systems is limited by the fiber-optic bundle pitch size and the static pattern noise caused by imperfections in fiber-optic bundle fabrication. The second class of device incorporates into its distal end micro-scanning devices that drive a short length of fiber optics to scan in 2D (Kiesslich et al. 2004). This class of device is very promising as high resolution images have been shown. Significant progress has been also made in a third class of devices based on microelectromechanical systems (MEMS) mirrors or a MEMS mirror array integrated at the distal end of an endomicroscope (Dickensheets and Kino 1996, Himmer et al. 2001, Rector et al. 2003). While MEMS mirror-based confocal endomicroscopes have shown potentials for clinical imaging at an early date with good promise, there have been few significant clinical instrument breakthroughs along this direction recently due to the difficulty and cost of MEMS mirror fabrication.

Endomicroscopy based on nonlinear optical excitation provides complementary information to confocal systems. Nonlinear optical endomicroscopes can better utilize molecular level contrast as previously discussed. Nonlinear optical endomicroscopy has excitation penetration comparable to reflected light confocal endomicroscopes when operated in the infrared wavelengths. For fluorescence imaging, confocal endomicroscopes have a shallower penetration depth due to the use of shorter excitation wavelengths. Furthermore, the use of a detection aperture in confocal endomicroscopy results in significant rejection of scattered photons that can be retained in nonlinear optical endomicroscopes, allowing deeper imaging with higher signal-to-noise ratios in principle. Finally, nonlinear optical endomicroscopy produces less tissue photodamage compared to fluorescence confocal endomicroscopy due to the inherent localization of the excitation volume. Given these potentials, a number of nonlinear optical endomicroscopes have been developed and shown significant promise despite the greater technical challenges in constructing these systems.

There are three major technical challenges in designing a nonlinear optical endomicroscope for clinical application. First, the efficiency of second-order nonlinear optical excitation is a linear function of laser pulse width. Due to nonlinear optical effects, such as self-phase modulation, transmitting high-intensity light through a typical silica core optical fiber, femtosecond pulses can be significantly broadened in a fiber endomicroscope, greatly decreasing the efficiency of these systems. Second, due to pulse dispersion considerations, the successful confocal endomicroscope designs that utilize distal scanning through a fiber-optic bundle are not feasible for nonlinear optical excitation. The results from initial experiments are not optimal (Gobel et al. 2004). Therefore, micro-scanning mechanisms must be packaged in the distal end of the device while the instrument's outer diameter must be kept within a few millimeters. Third, fluorescence and second harmonic signals are weak, especially for endogenous contrast agents such as NAD(P)H. Therefore, unlike reflected light confocal systems, where strong optical signals can be obtained, nonlinear optical endomicroscopes must be designed to maximize light collection efficiency.

Significant progress has been made in overcoming the first challenge as the photonic bandgap crystal technology is coming of age, allowing the creation of ultra-broadband, omni-reflective mirrors and waveguides (Hart et al. 2002, Temelkuran et al. 2002). Photonic bandgap crystals technology is poised to produce novel optical switches and multiplexers (Soljacic et al. 2002, Attard 2003, Ibrahim et al. 2004). Most importantly, for the future of endomicroscopy, photonic bandgap crystal technology allows the creation of novel waveguides for femtosecond pulses with controllable dispersion characteristics (Myaing et al. 2003, McConnell et al. 2004).

Significant process has also been made in overcoming the need for multiple axis scanning mechanisms, based on our experiences in confocal endomicroscopes (Konig et al. 2007). Miniature nonlinear endomicroscopes started with miniature, handheld two-photon systems using Lissajous scanning of a short length single mode fibers first developed by Helmchen and coworkers (Helmchen et al. 2001). Some of these systems were used to image brain activity in freely moving rats. However, the clinical utility of these systems was limited by two factors. First, due to pulse dispersion, only centimeter lengths of fiber can be used. Second, the detection of fluorescence signal through the single mode fiber is very inefficient. Another class of nonlinear endomicroscope performs scanning on the proximal end using conventional scanning mechanisms and uses a short length of Graded Index Lens (GRIN) to conduct light. This technology allows the creation of a rigid laparoscope that is suitable for small animal imaging but has not yet been applied to patient imaging. These devices have been used to study green fluorescent protein

expressing neurons in living mice (Jung and Schnitzer 2003, Levene et al. 2003). While these devices allow cutting-edge study of neurobiology in small animals, they are not feasible for clinical human use due to their length limitation and their rigidity.

More clinically applicable endomicroscope started with a number of parallel developments, including further miniaturization of the Helmchen design by resonantly scanning a dual core photonic crystal fiber instead of single mode fiber to achieve imaging in 2D by the groups of Gu and Li (Myaing et al. 2006, Bao et al. 2008, Wu et al. 2009). The major advance here involves the use of a dual core fiber with single mode, low dispersion transmission of excitation light and multimode, higher efficiency collection of the emitted signal. The addition of an axial translation capability in a recent design enabled the first 3D scanning nonlinear optical endomicroscope (Figure 27.9). A parallel development by Chen and coworkers utilizing 2D-scanning MEMS mirrors has resulted in another class of nonlinear optical endomicroscopes based on 3D MEMS scanning mirrors (Jung et al. 2008, Tang et al. 2009) (Figure 27.9). In addition to actuation of the scanning algorithm, distal scanning endoscopes are limited by a number of other optical constraints, such as the NA of miniaturized lenses. Recent advances in coupling spherical lens to the tip of GRIN lens has first resulted in nonlinear optical endomicroscopes with objective numerical apertures approaching 0.8 (Konig et al. 2007, Le Harzic et al. 2008). These studies further evaluated the biosafety of some doping materials used in the fabrication of these high refractive index optical components. For compactness and simplicity, most nonlinear optical endomicroscopes use the same optical fiber for excitation and detection. Since the excitation and emission wavelengths in nonlinear microscope imaging are widely separated, the endomicroscope optics optimized for excitation are in general very suboptimal for emission detection due to chromatic aberration. This chromatic aberration results in significant light loss. By optimizing dual core photonic crystal fiber that allows single mode excitation light delivery and a large multimode core for light collection, this problem is alleviated. By further introduction an effective achromatic triplet as the objective, the chromatic misalignment problem is now largely overcome (Wu et al. 2008b). The main contrast mechanisms implemented in most nonlinear optical endomicroscopes are based on two-photon

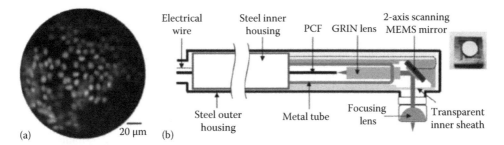

FIGURE 27.9 (a) Pig cornea stained with acridine orange imaged using three axis scanning endomicroscope. (Adapted from Wu, Y. et al., *Opt. Express*, 17, 7907, 2009.) (b) An endomicroscope design based on a 2D scanning MEMS mirror. (Adapted from Jung, W. et al., *Opt. Lett.*, 33, 1324, 2008.)

fluorescence and second harmonic generation. A system has been recently designed based on third harmonic generation and has been applied to study dermal fibrosis and hyperkeratosis (Lee et al. 2009). Finally, an exciting new system has been developed by Ben-Yakar and coworkers that combines imaging and microsurgery capability in the same compact system (Hoy et al. 2008).

Finally, in the absence of widely available exogenous probes, the signal from weak endogenous contrast agents remains a challenge for applying nonlinear optical endomicroscope for clinical imaging.

27.4 Conclusion

Nonlinear optical microscopy is one of the most important recent inventions in biological imaging. This technology enables noninvasive study of biological specimens in three dimensions with submicron resolution. The major advantages of this technique are its high resolution, its variety of contrast mechanisms, and its reasonable tissue penetration depth. It is also important that this technique induces less photodamage and photobleaching compared with some other techniques. As the technology of nonlinear optical microscopy reaches maturity, the range of applications in biology will surely increase. As discussed in this chapter, this exciting field is further sustained by a wide range of innovations in terms of both instrumental improvement and new biomedical applications.

Acknowledgments

The authors acknowledge funding support from NIH, Singapore-MIT Alliance, and Singapore-MIT Alliance Research & Technology Center. The authors also thank Barry R. Masters for help in the preparation of this chapter.

References

Agnantis, N. J., Sotiriadis, A., and Paraskevaidis, E. 2003. The current status of HPV DNA testing. *Eur. J. Gynaecol. Oncol.* 24: 351–356.

Albota, M., Beljonne, D., Bredas, J.-L. et al. 1998a. Design of organic molecules with large two-photon absorption cross sections. *Science* 281: 1653–1656.

Albota, M. A., Xu, C., and Webb, W. W. 1998b. Two-photon fluorescence excitation cross sections of biomolecular probes from 690 to 960 Nm. *Appl. Opt.* 37: 7352–7356.

Allen, G. J., Kwak, J. M., Chu, S. P. et al. 1999. Cameleon calcium indicator reports cytoplasmic calcium dynamics in *Arabidopsis* guard cells. *Plant J.* 19: 735–747.

Allen, R., Nielson, R., Wise, D. D., and Shear, J. B. 2005. Catalytic three-dimensional protein architectures. *Anal. Chem.* 77: 5089–5095.

Apgar, B. S. and Brotzman, G. 2004. Management of cervical cytologic abnormalities. *Am. Fam. Physician* 70: 1905–1916.

Asano, T. and Mcleod, R. S. 2002. Dietary fibre for the prevention of colorectal adenomas and carcinomas. *Cochrane Db. Syst. Rev.*, CD003430.

Attard, A. E. 2003. Optical bandwidth in coupling: The multicore photonic switch. *Appl. Opt.* 42: 2665–26673.

Baird, G. S., Zacharias, D. A., and Tsien, R. Y. 1999. Circular permutation and receptor insertion within green fluorescent proteins. *Proc. Natl. Acad. Sci. U. S. A.* 96: 11241–11246.

Balu, M., Baldacchini, T., Carter, J. et al. 2009. Effect of excitation wavelength on penetration depth in nonlinear optical microscopy of turbid media. *J. Biomed. Opt.* 14: 010508.

Ban, S., Mino, M., Nishioka, N. S. et al. 2004. Histopathologic aspects of photodynamic therapy for dysplasia and early adenocarcinoma arising in Barrett's esophagus. *Am. J. Surg. Pathol.* 28: 1466–1473.

Bao, H., Allen, J., Pattie, R., Vance, R., and Gu, M. 2008. Fast hand-held two-photon fluorescence microendoscope with a 475 microm X 475 microm field of view for in vivo imaging. *Opt. Lett.* 33: 1333–1335.

Barad, Y., Eisenberg, H., Horowitz, M., and Silberberg, Y. 1997. Nonlinear scanning laser microscopy by third harmonic generation. *Appl. Phys. Lett.* 70: 922–924.

Basu, S. and Campagnola, P. J. 2004. Enzymatic activity of alkaline phosphatase inside protein and polymer structures fabricated via multiphoton excitation. *Macromolecules* 5: 572–579.

Beck, T. J., Burkanas, M., Bagdonas, S. et al. 2007. Two-photon photodynamic therapy of C6 cells by means of 5-aminolevulinic acid induced protoporphyrin Ix. *J. Photochem. Photobiol. B* 87: 174–182.

Belo, A. C. and Playford, R. J. 2003. Surveillance for Barrett's oesophagus: Is there light the end of the metaplastic tunnel? *Surgeon* 1: 152–156.

Betzig, E., Patterson, G. H., Sougrat, R. et al. 2006. Imaging intracellular fluorescent proteins at nanometer resolution. *Science* 313: 1642–1645.

Bouma, B. E., Tearney, G. J., Yabushita, H. et al. 2003a. Evaluation of intracoronary stenting by intravascular optical coherence tomography. *Heart* 89: 317–320.

Bouma, J. L., Aronson, L. R., Keith, D. G., and Saunders, H. M. 2003b. Use of computed tomography renal angiography for screening feline renal transplant donors. *Vet. Radiol. Ultrasound* 44: 636–641.

Boyd, R. 2008. *Nonlinear Optics*. New York: Academic Press.

Brand, S., Poneros, J. M., Bouma, B. E. et al. 2000. Optical coherence tomography in the gastrointestinal tract. *Endoscopy* 32: 796–803.

Buehler, C., Kim, K. H., Dong, C. Y., Masters, B. R., and So, P. T. C. 1999. Innovations in two-photon deep tissue microscopy. *IEEE Eng. Med. Biol. Mag.* 18: 23–30.

Campagnola, P. J., Clark, H. A., Mohler, W. A., Lewis, A., and Loew, L. M. 2001. Second-harmonic imaging microscopy of living cells. *J. Biomed. Opt.* 6: 277–286.

Campagnola, P. J., Delguidice, D. M., Epling, G. A. et al. 2000. 3-Dimensional submicron polymerization of acrylamide by multiphoton excitation of xanthene dyes. *Macromolecules* 33: 1511–1513.

Campagnola, P. J. and Loew, L. M. 2003. Second-harmonic imaging microscopy for visualizing biomolecular arrays in cells, tissues and organisms. *Nat. Biotechnol.* 21: 1356–1360.

Campbell, M., Sharp, D. N., Harrison, M. T., Denning, R. G., and Turberfield, A. J. 2000. Fabrication of photonic crystals for the visible spectrum by holographic lithography. *Nature* 404: 53–56.

Cha, J. W., Ballesta, J., and So, P. T. C. 2010. A Shack-Hartmann wavefront sensor based adaptive optics system for multiphoton microscopy. *J. Biomed. Opt.* 15(4): 046022.

Chalfie, M., Tu, Y., Euskirchen, G., Ward, W. W., and Prasher, D. C. 1994. Green fluorescent protein as a marker for gene expression. *Science* 263: 802–805.

Chau, A. H., Chan, R. C., Shishkov, M. et al. 2004. Mechanical analysis of atherosclerotic plaques based on optical coherence tomography. *Ann. Biomed. Eng.* 32: 1494–1503.

Chen, M. H., Chen, W. L., Sun, Y., Fwu, P. T., and Dong, C. Y. 2007. Multiphoton autofluorescence and second-harmonic generation imaging of the tooth. *J. Biomed. Opt.* 12: 064018.

Chen, R., Huang, Z., Chen, G. et al. 2008. Kinetics and subcellular localization of 5-ala-induced PpIX in DHL cells via two-photon excitation fluorescence microscopy. *Int. J. Oncol.* 32: 861–867.

Chen, X., Brewer, M. A., Zou, C., and Campagnola, P. J. 2009. Adhesion and migration of ovarian cancer cells on crosslinked laminin fibers nanofabricated by multiphoton excited photochemistry. *Integrative Biology* 1: 469–476.

Cheng, J. X., Book, L. D., and Xie, X. S. 2001. Polarization coherent anti-stokes Raman scattering microscopy. *Opt. Lett.* 26: 1341–1343.

Cheng, J. X., Jia, Y. K., Zheng, G., and Xie, X. S. 2002. Laser-scanning coherent anti-stokes Raman scattering microscopy and applications to cell biology. *Biophys. J.* 83: 502–509.

Chou, S. Y., Krauss, P. R., and Renstrom, P. J. 1996. Imprint lithography with 25-nanometer resolution. *Science* 272: 85–87.

Chu, S. W., Chen, I. H., Liu, T. M. et al. 2001. Multimodal nonlinear spectral microscopy based on a femtosecond Cr:forsterite laser. *Opt. Lett.* 26: 1909–1911.

Claeyssens, F., Hasan, E. A., Gaidukeviciute, A. et al. 2009. Three-dimensional biodegradable structures fabricated by two-photon polymerization. *Langmuir* 25: 3219–3223.

Craighead, H. G. 1985. Ultra-high-resolution electron-beam lithography. *J. Electron Microsc. Tech.* 2: 147–155.

Cumpston, B. H., Ananthavel, S. P., Barlow, S. et al. 1999. Two-photon polymerization initiators for three-dimensional optical data storage and microfabrication. *Nature* 398: 51–54.

Cunningham, L. P., Veilleux, M. P., and Campagnola, P. J. 2006. Freeform multiphoton excited microfabrication for biological applications using a rapid prototyping Cad-based approach. *Opt. Express* 14: 8613–8621.

Dabbous, M. K. 1966. Inter- and intramolecula cross-linking in tyrosinase-treated tropocollagen. *J. Biol. Chem.* 241: 5307–5312.

De Palo, G. 2004. Cervical precancer and cancer, past, present and future. *Eur. J. Gynaecol. Oncol.* 25: 269–278.

Debarre, D., Botcherby, E. J., Watanabe, T. et al. 2009. Image-based adaptive optics for two-photon microscopy. *Opt. Lett.* 34: 2495–2497.

Denk, W. 1994. Two-photon scanning photochemical microscopy: Mapping ligand-gated ion channel distributions. *Proc. Natl. Acad. Sci. U. S. A.* 91: 6629–6633.

Denk, W., Strickler, J. H., and Webb, W. W. 1990. Two-photon laser scanning fluorescence microscopy. *Science* 248: 73–76.

Denk, W. J., Piston, D. W., and Webb, W. W. 1995. Two-photon molecular excitation laser-scanning microscopy. In: *Handbook of Biological Confocal Microscopy*, ed. Pawley, J. B., 2nd edn. New York: Plenum Press.

Dickensheets, D. L. and Kino, G. S. 1996. Micromachined scanning confocal optical microscope. *Opt. Lett.* 21: 764–766.

Dodt, H. U., Leischner, U., Schierloh, A. et al. 2007. Ultramicroscopy: Three-dimensional visualization of neuronal networks in the whole mouse brain. *Nat. Methods* 4: 331–336.

Dunleavey, R. 2004. Incidence, pathophysiology and treatment of cervical cancer. *Nurs. Times* 100: 38–41.

Eggeling, C., Ringemann, C., Medda, R. et al. 2009. Direct observation of the nanoscale dynamics of membrane lipids in a living cell. *Nature* 457: 1159–1162.

Evans, C. L., Xu, X., Kesari, S. et al. 2007. Chemically-selective imaging of brain structures with cars microscopy. *Opt. Express* 15: 12076–12087.

Falk, G. W. 2002. Barrett's esophagus. *Gastroenterology* 122: 1569–1591.

Faruqi, S. A., Arantes, V., and Bhutani, M. S. 2004. Barrett's esophagus: Current and future role of endosonography and optical coherence tomography. *Dis. Esophagus* 17: 118–123.

Fittinghoff, D. N., Schaffer, C. B., Mazur, E., and Squier, J. A. 2001. Time-decorrelated multifocal micromachining and trapping. *IEEE J. Sel. Top. Quant.* 7: 559–566.

Fittinghoff, D. N. and Squier, J. A. 2000. Time-decorrelated multifocal array for multiphoton microscopy and micromachining. *Opt. Lett.* 25: 1213–1215.

Fong, S. M., Lee, M. K., Adusumilli, P. S., and Kelly, K. J. 2009. Fluorescence-expressing viruses allow rapid identification and separation of rare tumor cells in spiked samples of human whole blood. *Surgery* 146: 498–505.

Franken, P. A., Hill, A. E., Peters, C. W., and Weinreich, G. 1961. Generation of optical harmonics. *Phys. Rev. Lett.* 7: 118.

French, P. M. W., Williams, J. a. R., and Taylor, J. R. 1989. Femtosecond pulse generation from a titanium-doped sapphire laser using nonlinear external cavity feedback. *Opt. Lett.* 14: 686–688.

Freudiger, C. W., Min, W., Saar, B. G. et al. 2008. Label-free biomedical imaging with high sensitivity by stimulated Raman scattering microscopy. *Science* 322: 1857–1861.

Freund, I., Deutsch, M., and Sprecher, A. 1986. Connective tissue polarity. Optical second-harmonic microscopy, crossed-beam summation, and small-angle scattering in rat-tail tendon. _Biophys. J._ 50: 693–712.

Fu, D., Ye, T., Matthews, T. E. et al. 2007a. High-resolution in vivo imaging of blood vessels without labeling. _Opt. Lett._ 32: 2641–2643.

Fu, D., Ye, T., Matthews, T. E. et al. 2008. Probing skin pigmentation changes with transient absorption imaging of eumelanin and pheomelanin. _J. Biomed. Opt._ 13: 054036.

Fu, Y., Wang, H., Huff, T. B., Shi, R., and Cheng, J. X. 2007b. Coherent anti-stokes Raman scattering imaging of myelin degradation reveals a calcium-dependent pathway in lyso-PtdCho-induced demyelination. _J. Neurosci. Res._ 85: 2870–2881.

Fujimoto, J. G. 2003. Optical coherence tomography for ultra-high resolution in vivo imaging. _Nat. Biotechnol._ 21: 1361–1367.

Fujimoto, J. G., Boppart, S. A., Tearney, G. J. et al. 1999. High resolution in vivo intra-arterial imaging with optical coherence tomography. _Heart_ 82: 128–133.

Gannaway, J. N. and Sheppard, C. J. R. 1978. Second harmonic imaging in the scanning optical microscope. _Opt. Quant. Electron._ 10: 435–439.

Girkin, J. M., Poland, S., and Wright, A. J. 2009. Adaptive optics for deeper imaging of biological samples. _Curr. Opin. Biotechnol._ 20: 106–110.

Gobel, W., Kerr, J. N., Nimmerjahn, A., and Helmchen, F. 2004. Miniaturized two-photon microscope based on a flexible coherent fiber bundle and a gradient-index lens objective. _Opt. Lett._ 29: 2521–2523.

Göppert-Mayer, M. 1931. Über elementarakte mit zwei quantensprüngen. _Ann. Phys. (Leipzig)_ 5: 273–294.

Gratson, G. M., Xu, M. J., and Lewis, J. A. 2004. Microperiodic structures—direct writing of three-dimensional webs. _Nature_ 428: 386.

Gu, M. 2000. _Advanced Optical Imaging Theory_. Berlin; New York: Springer.

Gu, M. and Sheppard, C. J. R. 1995. Comparison of three-dimensional imaging properties between two-photon and single-photon fluorescence microscopy. _J. Microsc._ 177: 128–137.

Guelrud, M. and Ehrlich, E. E. 2004. Enhanced magnification endoscopy in the upper gastrointestinal tract. _Gastrointest. Endosc. Clin. N. Am._ 14: 461–473.

Guo, R., Li, Z. Y., Jiang, Z. W. et al. 2005. Log-pile photonic crystal fabricated by two-photon photopolymerization. _J. Opt. A-Pure Appl. Op._ 7: 396–399.

Guo, S. X., Bourgeois, F., Chokshi, T. et al. 2008. Femtosecond laser nanoaxotomy lab-on-a-chip for in vivo nerve regeneration studies. _Nat. Methods_ 5: 531–533.

Gupta, S. and Sodhani, P. 2004. Why is high grade squamous intraepithelial neoplasia under-diagnosed on cytology in a quarter of cases? Analysis of smear characteristics in discrepant cases. _Indian J. Cancer._ 41: 104–108.

Hart, S. D., Maskaly, G. R., Temelkuran, B. et al. 2002. External reflection from omnidirectional dielectric mirror fibers. _Science_ 296: 510–513.

Haynes, C. L. and Van Duyne, R. P. 2001. Nanosphere lithography: A versatile nanofabrication tool for studies of size-dependent nanoparticle optics. _J. Phys. Chem. B_ 105: 5599–5611.

Hell, S. W., Booth, M., and Wilms, S. 1998. Two-photon near- and far-field fluorescence microscopy with continuous-wave excitation. _Opt. Lett._ 23: 1238–1240.

Hell, S. W. and Wichmann, J. 1994. Breaking the diffraction resolution limit by stimulated emission: Stimulated-emission-depletion fluorescence microscopy. _Opt. Lett._ 19: 780–782.

Helmchen, F., Fee, M. S., Tank, D. W., and Denk, W. 2001. A miniature head-mounted two-photon microscope. High-resolution brain imaging in freely moving animals. _Neuron_ 31: 903–912.

Himmer, P. A., Dickensheets, D. L., and Friholm, R. A. 2001. Micromachined silicon nitride deformable mirrors for focus control. _Opt. Lett._ 28: 1280–1282.

Hirshberg, Y. 1956. Reversible formation and eradication of colors by irradiation at low temperatures. A photochemical memory model. _J. Am. Chem. Soc._ 78: 2304–2312.

Hockberger, P. E., Skimina, T. A., Centonze, V. E. et al. 1999. Activation of flavin-containing oxidases underlies light-induced production of H_2O_2 in mammalian cells. _Proc. Natl. Acad. Sci. U. S. A._ 96: 6255–6260.

Hoerman, K. C. and Balekjian, A. Y. 1966. Some quantum aspects of collagen. _Federation Proc._ 25: 1016–1021.

Hollister, S. J. 2005. Porous scaffold design for tissue engineering. _Nat. Mater._ 4: 518–524.

Hovhannisyan, V., Lo, W., Hu, C., Chen, S. J., and Dong, C. Y. 2008. Dynamics of femtosecond laser photo-modification of collagen fibers. _Opt. Express_ 16: 7958–7968.

Hoy, C. L., Durr, N. J., Chen, P. et al. 2008. Miniaturized probe for femtosecond laser microsurgery and two-photon imaging. _Opt. Express_ 16: 9996–10005.

Hsiung, P. L., Hardy, J., Friedland, S. et al. 2008. Detection of colonic dysplasia in vivo using a targeted heptapeptide and confocal microendoscopy. _Nat. Med._ 14: 454–458.

Huff, T. B. and Cheng, J. X. 2007. In vivo coherent anti-stokes Raman scattering imaging of sciatic nerve tissue. _J. Microsc._ 225: 175–182.

Ibrahim, T. A., Amarnath, K., Kuo, L. C. et al. 2004. Photonic logic NOR gate based on two symmetric microring resonators. _Opt. Lett._ 29: 2779–2781.

Irie, M. 2000. Photochromism: Memories and switches—introduction. _Chem. Rev._ 100: 1683.

Jacobson, B. C. and Van Dam, J. 2002. Enhanced endoscopy in inflammatory bowel disease. _Gastrointest. Endosc. Clin. N. Am._ 12: 573–587.

Jacques, S. L. and Mcauliffe, D. J. 1991. The melanosome: Threshold temperature for explosive vaporization and internal absorption coefficient during pulsed laser irradiation. _Photochem. Photobiol._ 53: 769–775.

Ji, N., Magee, J. C., and Betzig, E. 2008. High-speed, low-photo-damage nonlinear imaging using passive pulse splitters. *Nat. Methods* 5: 197–202.

Jung, J. C. and Schnitzer, M. J. 2003. Multiphoton endoscopy. *Opt. Lett.* 28: 902–904.

Jung, W., Tang, S., Mccormic, D. T. et al. 2008. Miniaturized probe based on a microelectromechanical system mirror for multiphoton microscopy. *Opt. Lett.* 33: 1324–1326.

Kaehr, B., Allen, R., Javier, D. J., Currie, J., and Shear, J. B. 2004. Guiding neuronal development with in situ microfabrication. *Proc. Natl. Acad. Sci. U. S. A.* 101: 16104–16108.

Kaehr, B. and Shear, J. B. 2007. Mask-directed multiphoton lithography. *J. Am. Chem. Soc.* 129: 1904–1905.

Kaehr, B. and Shear, J. B. 2008. Multiphoton fabrication of chemically responsive protein hydrogels for microactuation. *Proc. Natl. Acad. Sci. U. S. A.* 105: 8850–8854.

Kaehr, B. and Shear, J. B. 2009. High-throughput design of microfluidics based on directed bacterial motility. *Lab Chip* 9: 2632–2637.

Kaiser, W. and Garrett, C. G. B. 1961. Two-photon excitation in $CaF_2:Eu^{2+}$. *Phys. Rev. Lett.* 7: 229–231.

Kara, M., Dacosta, R. S., Wilson, B. C., Marcon, N. E., and Bergman, J. 2004. Autofluorescence-based detection of early neoplasia in patients with Barrett's esophagus. *Digest. Dis.* 22: 134–141.

Kasischke, K. A., Vishwasrao, H. D., Fisher, P. J., Zipfel, W. R., and Webb, W. W. 2004. Neural activity triggers neuronal oxidative metabolism followed by astrocytic glycolysis. *Science* 305: 99–103.

Katoh, K., Langford, G., Hammar, K., Smith, P. J., and Oldenbourg, R. 1997. Actin bundles in neuronal growth cone observed with the pol-scope. *Biol. Bull.* 193: 219–220.

Keyse, S. M. and Tyrrell, R. M. 1990. Induction of the heme oxygenase gene in human skin fibroblasts by hydrogen peroxide and UVA (365 Nm) radiation: Evidence for the involvement of the hydroxyl radical. *Carcinogenesis* 11: 787–791.

Kierdaszuk, B., Gryczynski, I., Modrak-Wojcik, A. et al. 1995. Fluorescence of tyrosine and tryptophan in proteins using one- and two-photon excitation. *Photochem. Photobiol.* 61: 319–324.

Kierdaszuk, B., Malak, H., Gryczynski, I., Callis, P., and Lakowicz, J. R. 1996. Fluorescence of reduced nicotinamides using one- and two-photon excitation. *Biophys. Chem.* 62: 1–13.

Kiesslich, R., Burg, J., Vieth, M. et al. 2004. Confocal laser endoscopy for diagnosing intraepithelial neoplasias and colorectal cancer in vivo. *Gastroenterology* 127: 706–713.

Kim, K., Buehler, C., Bahlmann, K. et al. 2007. Multifocal multiphoton microscopy based on multianode photomultiplier tubes. *Opt. Express* 15: 11658–11678.

Kittel, R. J., Wichmann, C., Rasse, T. M. et al. 2006. Bruchpilot promotes active zone assembly, Ca^{2+} channel clustering, and vesicle release. *Science* 312: 1051–1054.

Kobat, D., Durst, M. E., Nishimura, N. et al. 2009. Deep tissue multiphoton microscopy using longer wavelength excitation. *Opt. Express* 17: 13354–13364.

Koester, H. J., Baur, D., Uhl, R., and Hell, S. W. 1999. Ca^{2+} fluorescence imaging with pico- and femtosecond two-photon excitation: Signal and photodamage. *Biophys. J.* 77: 2226–2236.

Kondo, T., Matsuo, S., Juodkazis, S., and Misawa, H. 2001. Femtosecond laser interference technique with diffractive beam splitter for fabrication of three-dimensional photonic crystals. *Appl. Phys. Lett.* 79: 725–727.

Konig, K. 2000. Multiphoton microscopy in life sciences. *J. Microsc.* 200: 83–104.

Konig, K., Becker, T. W., Fischer, P., Riemann, I., and Halbhuber, K.-J. 1999. Pulse-length dependence of cellular response to intense near-infrared laser pulses in multiphoton microscopes. *Opt. Lett.* 24: 113–115.

Konig, K., Ehlers, A., Riemann, I. et al. 2007. Clinical two-photon microendoscopy. *Microsc. Res. Tech.* 70: 398–402.

Konig, K., So, P. T. C., Mantulin, W. W., Tromberg, B. J., and Gratton, E. 1996. Two-photon excited lifetime imaging of autofluorescence in cells during UVA and NIR photostress. *J. Microsc.* 183: 197–204.

Konig, K., Speicher, M., Buckle, R. et al. 2009. Clinical optical coherence tomography combined with multiphoton tomography of patients with skin diseases. *J. Biophotonics* 2: 389–397.

LaBella, F. S. 1961. Studies on the soluble products released from purified elastic fibers by pancreatic elastase. *Arch. Biochm. Biophys.* 93: 72–79.

LaBella, F. S. and Gerald, P. 1965. Structure of collagen from human tendon as influence by age and sex. *J. Gerontol.* 20: 54–59.

LaBella, F. S. and Lindsay, W. G. 1963. The structure of human aortic elastin as influence by age. *J. Gerontol.* 18: 111–118.

LaComb, R., Nadiarnykh, O., Carey, S., and Campagnola, P. J. 2008. Quantitative second harmonic generation imaging and modeling of the optical clearing mechanism in striated muscle and tendon. *J. Biomed. Opt.* 13: 021109.

Lakowicz, J. R. and Gryczynski, I. 1992. Tryptophan fluorescence intensity and anisotropy decays of human serum albumin resulting from one-photon and two-photon excitation. *Biophys. Chem.* 45: 1–6.

Lambert, R. 2004. Diagnosis of esophagogastric tumors. *Endoscopy* 36: 110–119.

Lambrou, N. C. and Twiggs, L. B. 2003. High-grade squamous intraepithelial lesions. *Cancer J.* 9: 382–389.

Larson, D. R., Zipfel, W. R., Williams, R. M. et al. 2003. Water-soluble quantum dots for multiphoton fluorescence imaging in vivo. *Science* 300: 1434–1436.

Le Harzic, R., Weinigel, M., Riemann, I., Konig, K., and Messerschmidt, B. 2008. Nonlinear optical endoscope based on a compact two axes piezo scanner and a miniature objective lens. *Opt. Express* 16: 20588–20596.

Lee, J. H., Chen, S. Y., Yu, C. H. et al. 2009. Noninvasive in vitro and in vivo assessment of epidermal hyperkeratosis and dermal fibrosis in atopic dermatitis. *J. Biomed. Opt.* 14: 014008.

Levene, M. J., Dombeck, D. A., Kasischke, K. A., Molloy, R. P., and Webb, W. W. 2003. In vivo multiphoton microscopy of deep brain tissue. *J. Neurophysiol.* 1908–1912.

Li, M. T., Chen, L., and Chou, S. Y. 2001. Direct three-dimensional patterning using nanoimprint lithography. *Appl. Phys. Lett.* 78: 3322–3324.

Li, X. D., Boppart, S. A., Van Dam, J. et al. 2000. Optical coherence tomography: Advanced technology for the endoscopic imaging of Barrett's esophagus. *Endoscopy* 32: 921–930.

Liang, C., Sung, K. B., Richards-Kortum, R. R., and Descour, M. R. 2002. Design of a high-numerical-aperture miniature microscope objective for an endoscopic fiber confocal reflectance microscope. *Appl. Opt.* 41: 4603–4610.

Lin, S. J., Hsiao, C. Y., Sun, Y. et al. 2005. Monitoring the thermally induced structural transitions of collagen by use of second-harmonic generation microscopy. *Opt. Lett.* 30: 622–624.

Madsen, S. J., Sun, C. H., Tromberg, B. J., Wallace, V. P., and Hirschberg, H. 2000. Photodynamic therapy of human glioma spheroids using 5-aminolevulinic acid. *Photochem. Photobiol.* 72: 128–134.

Maruo, S. and Fourkas, J. T. 2008. Recent progress in multiphoton microfabrication. *Laser Photonics Rev.* 2: 100–111.

Masters, B. R. 1984. Noninvasive redox fluorometry: How light can be used to monitor alterations of corneal mitochondrial function. *Curr. Eye Res.* 3: 23–26.

Masters, B. R. 2009. Correlation of histology and linear and nonlinear microscopy of the living human cornea. *J. Biophotonics* 2: 127–139.

Masters, B. R. and Chance, B. 1999. Redox Confocal imaging: Intrinsic fluorescent probes of cellular metabolism. In: *Fluorescent and Luminescent Probes for Biological Activity*, ed. Mason, W. T. London: Academic Press.

Masters, B. R., So, P. T., Buehler, C. et al. 2004. Mitigating thermal mechanical damage potential during two-photon dermal imaging. *J. Biomed. Opt.* 9: 1265–1270.

Masters, B. R., So, P. T., and Gratton, E. 1997. Multiphoton excitation fluorescence microscopy and spectroscopy of in vivo human skin. *Biophys. J.* 72: 2405–2412.

Masters, B. R. and So, P. T. C. (eds.) 2008. *Handbook of Biomedical Non-Linear Optical Microscopy*. New York: Oxford University Press.

McConnell, G. and Riis, E. 2004. Two-photon laser scanning fluorescence microscopy using photonic crystal fiber. *J. Biomed. Opt.* 9: 922–927.

Mendonca, C. R., Baldacchini, T., Tayalia, P., and Mazur, E. 2007. Reversible birefringence in microstructures fabricated by two-photon absorption polymerization. *J. Appl. Phys.* 102: 013109.

Min, W., Lu, S., Chong, S. et al. 2009. Imaging chromophores with undetectable fluorescence by stimulated emission microscopy. *Nature* 461: 1105–1109.

Mohler, W. A., Simske, J. S., Williams-Masson, E. M., Hardin, J. D., and White, J. G. 1998. Dynamics and ultrastructure of developmental cell fusions in the *Caenorhabditis elegans* hypodermis. *Curr. Biol.* 8: 1087–1090.

Mohler, W. A. and White, J. G. 1998. Stereo-4-D reconstruction and animation from living fluorescent specimens. *Biotechniques* 24: 1006–1010, 1012.

Moneron, G. and Hell, S. W. 2009. Two-photon excitation STED microscopy. *Opt. Express* 17: 14567–14573.

Moon, J. H., Mcdaniel, W., Maclean, P., and Hancock, L. E. 2007. Live-cell-permeable poly (P-phenylene ethynylene). *Angew. Chem. Int. Ed.* 46: 8223–8225.

Myaing, M. T., Macdonald, D. J., and Li, X. 2006. Fiber-optic scanning two-photon fluorescence endoscope. *Opt. Lett.* 31: 1076–1078.

Myaing, M. T., Ye, J. Y., Norris, T. B. 2003. Enhanced two-photon biosensing with double-clad photonic crystal fibers. *Opt. Lett.* 28: 1224–1226.

Nadiarnykh, O. and Campagnola, P. J. 2009. Retention of polarization signatures in SHG microscopy of scattering tissues through optical clearing. *Opt. Express* 17: 5794–5806.

Nagy, A., Wu, J., and Berland, K. M. 2005a. Characterizing observation volumes and the role of excitation saturation in one-photon fluorescence fluctuation spectroscopy. *J. Biomed. Opt.* 10: 44015.

Nagy, A., Wu, J., and Berland, K. M. 2005b. Observation volumes and {gamma}-factors in two-photon fluorescence fluctuation spectroscopy. *Biophys. J.* 89: 2077–2090.

Nan, X., Potma, E. O., and Xie, X. S. 2006. Nonperturbative chemical imaging of organelle transport in living cells with coherent anti-stokes Raman scattering microscopy. *Biophys. J.* 91: 728–735.

Nandakumar, P., Kovalev, A., and Volkmer, A. 2009. Vibrational imaging based on stimulated Raman scattering microscopy. *New J. Phys.* 11: 033026.

Nishioka, N. S. 2003. Optical biopsy using tissue spectroscopy and optical coherence tomography. *Can. J. Gastroenterol.* 17: 376–380.

Odian, G. G. 2004. *Principles of Polymerization*. Hoboken, NJ: Wiley-Interscience.

Oheim, M., Beaurepaire, E., Chaigneau, E., Mertz, J., and Charpak, S. 2001. Two-photon microscopy in brain tissue: Parameters influencing the imaging depth. *J. Neurosci. Methods* 111: 29–37.

Olson, C. E., Previte, M. J. R., and Fourkas, J. T. 2002. Efficient and robust multiphoton data storage in molecular glasses and highly crosslinked polymers. *Nat. Mater.* 1: 225–228.

Oron, D., Tal, E., and Silberberg, Y. 2005. Scanningless depth-resolved microscopy. *Opt. Express* 13: 1468–1476.

Ozeki, Y., Dake, F., Kajiyama, S., Fukui, K., and Itoh, K. 2009. Analysis and experimental assessment of the sensitivity of stimulated Raman scattering microscopy. *Opt. Express* 17: 3651–3658.

Pacifico, R. J., Wang, K. K., Wongkeesong, L. M., Buttar, N. S., and Lutzke, L. S. 2003. Combined endoscopic mucosal resection and photodynamic therapy versus esophagectomy for management of early adenocarcinoma in Barrett's esophagus. *Clin. Gastroenterol. Hepatol.* 1: 252–257.

Patterson, G. H. and Piston, D. W. 2000. Photobleaching in two-photon excitation microscopy. *Biophys. J.* 78: 2159–2162.

Paulson, T. G. and Reid, B. J. 2004. Focus on Barrett's esophagus and esophageal adenocarcinoma. *Cancer Cell* 6: 11–16.

Peters, J. H. and Wang, K. K. 2004. How should Barrett's ulceration be treated? *Surg. Endosc.* 18: 338–344.

Piner, R. D., Zhu, J., Xu, F., Hong, S. H., and Mirkin, C. A. 1999. "Dip-Pen" nanolithography. *Science* 283: 661–663.

Pins, G. D., Bush, K. A., Cunningham, L. P., and Carnpagnola, P. J. 2006. Multiphoton excited fabricated nano and micro patterned extracellular matrix proteins direct cellular morphology. *J. Biomed. Mater. Res. A* 78A: 194–204.

Piston, D. W., Masters, B. R., and Webb, W. W. 1995. Three-dimensionally resolved Nad(P)H cellular metabolic redox imaging of the in situ cornea with two-photon excitation laser scanning microscopy. *J. Microsc.* 178: 20–27.

Pitris, C., Jesser, C., Boppart, S. A. et al. 2000. Feasibility of optical xoherence tomography for high-resolution imaging of human gastrointestinal tract malignancies. *J. Gastroenterol.* 35: 87–92.

Pitts, J. D., Campagnola, P. J., Epling, G. A., and Goodman, S. L. 2000. Submicron multiphoton free-form fabrication of proteins and polymers: Studies of reaction efficiencies and applications in sustained release. *Macromolecules* 33: 1514–1523.

Ploetz, E., Laimgruber, S., Berner, S., Zinth, W., and Gilch, P. 2007. Femtosecond stimulated Raman microscopy. *Appl. Phys. B.* 87: 389–393.

Plotnikov, S., Juneja, V., Isaacson, A. B., Mohler, W. A., and Campagnola, P. J. 2006. Optical clearing for improved contrast in second harmonic generation imaging of skeletal muscle. *Biophys. J.* 90: 328–339.

Plotnikov, S. V., Kenny, A. M., Walsh, S. J. et al. 2008. Measurement of muscle disease by quantitative second-harmonic generation imaging. *J. Biomed. Opt.* 13: 044018.

Poneros, J. M. 2004. Diagnosis of Barrett's esophagus using optical coherence tomography. *Gastrointest. Endosc. Clin. N. Am.* 14: 573–588.

Poneros, J. M., and Nishioka, N. S. 2003. Diagnosis of Barrett's esophagus using optical coherence tomography. *Gastrointest. Endosc. Clin. N. Am.* 13: 309–323.

Pustovalov, V. K. 1995. Initiation of explosive boiling and optical breakdown as a result of the action of laser pulses on melanosome in pigmented biotissues. *Kvantovaya Elecktron* 22: 1091–1094.

Rahim, N. A. A., Mcdaniel, W., Bardon, K. et al. 2009. Conjugated polymer nanoparticles for two-photon imaging of endothelial cells in a tissue model. *Adv. Mater.* 21: 3492–3496.

Rector, D. M., Ranken, D. M., and George, J. S. 2003. High-performance confocal system for microscopic or endoscopic applications. *Methods* 30: 16–27.

Rouse, A. R., Kano, A., Udovich, J. A., Kroto, S. M., and Gmitro, A. F. 2004. Design and demonstration of a miniature catheter for a confocal microendoscope. *Appl. Opt.* 43: 5763–5771.

Rueckel, M., Mack-Bucher, J. A., and Denk, W. 2006. Adaptive wavefront correction in two-photon microscopy using coherence-gated wavefront sensing. *Proc. Natl. Acad. Sci. U. S. A.* 103: 17137–17142.

Rust, M. J., Bates, M., and Zhuang, X. 2006. Sub-diffraction-limit imaging by stochastic optical reconstruction microscopy (Storm). *Nat. Methods* 3: 793–795.

Sakashita, M., Inoue, H., Kashida, H. et al. 2003. Virtual histology of colorectal lesions using laser-scanning confocal microscopy. *Endoscopy* 35: 1033–1038.

Sako, Y., Sekihata, A., Yanagisawa, Y. et al. 1997. Comparison of two-photon excitation laser scanning microscopy with UV-confocal laser scanning microscopy in three-dimensional calcium imaging using the fluorescence indicator indo-1. *J. Microsc.* 185: 9–20.

Schelhas, L. T., Shane, J. C., and Dantus, M. 2006. Advantages of ultrashort phase-shaped pulses for selective two-photon activation and biomedical imaging. *Nanomedicine* 2: 177–181.

Schonle, A. and Hell, S. W. 1998. Heating by absorption in the focus of an objective lens. *Opt. Lett.* 23: 325–327.

Sharma, P. 2004. Review article: Emerging techniques for screening and surveillance in Barrett's oesophagus. *Aliment. Pharmacol. Ther.* 20(Suppl 5): 63–70; discussion 95–96.

Shen, Y. R. 2002. *The Principles of Nonlinear Optics.* New York: Wiley Interscience.

Sheppard, C. J. R., Gannaway, J. N., Kompfner, R., and Walsh, D. 1977. Scanning harmonic optical microscope. *IEEE J. Sel. Top. Quant.* 13: D100.

Sheppard, C. J. R. and Gu, M. 1990. Image formation in two-photon fluorescence microscope. *Optik* 86: 104–106.

Sheppard, C. J. R. and Kompfner, R. 1978. Resonant scanning optical microscope. *Appl. Opt.* 17: 2879–2882.

Shoji, S. and Kawata, S. 2000. Photofabrication of three-dimensional photonic crystals by multibeam laser interference into a photopolymerizable resin. *Appl. Phys. Lett.* 76: 2668–2670.

Singh, S. and Bradley, L. T. 1964. Three-photon absorption in naphthalene crystals by laser excitation. *Phys. Rev. Lett.* 12: 162–164.

Skala, M. C., Riching, K. M., Gendron-Fitzpatrick, A. et al. 2007. In vivo multiphoton microscopy of NADH and FAD redox states, fluorescence lifetimes, and cellular morphology in precancerous epithelia. *Proc. Natl. Acad. Sci. U. S. A.* 104: 19494–19499.

Skala, M. C., Squirrell, J. M., Vrotsos, K. M. et al. 2005. Multiphoton microscopy of endogenous fluorescence differentiates normal, precancerous, and cancerous squamous epithelial tissues. *Cancer Res.* 65: 1180–1186.

So, P. T., Dong, C. Y., Masters, B. R., and Berland, K. M. 2000. Two-photon excitation fluorescence microscopy. *Annu. Rev. Biomed. Eng.* 2: 399–429.

Soljacic, M., Ibanescu, M., Johnson, S. G., Fink, Y., and Joannopoulos, J. D. 2002. Optimal bistable switching in nonlinear photonic crystals. *Phys. Rev. E* 66: 055601.

Soughayer, J. S., Krasieva, T., Jacobson, S. C. et al. 2000. Characterization of cellular optoporation with distance. *Anal. Chem.* 72: 1342–1347.

Spence, D. E., Kean, P. N., and Sibbett, W. 1991. 60-fsec pulse generation from a self-mode-locked Ti-sapphire laser. *Opt. Lett.* 16: 42–44.

Squirrell, J. M., Wokosin, D. L., White, J. G., and Bavister, B. D. 1999. Long-term two-photon fluorescence imaging of mammalian embryos without compromising viability. *Nat. Biotechnol.* 17: 763–767.

Stoller, P., Kim, B. M., Rubenchik, A. M., Reiser, K. M., and Da Silva, L. B. 2002. Polarization-dependent optical second-harmonic imaging of a rat-tail tendon. *J. Biomed. Opt.* 7: 205–214.

Stoneman, M., Fox, M., Zeng, C. Y., and Raicu, V. 2009. Real-time monitoring of two-photon photopolymerization for use in fabrication of microfluidic devices. *Lab Chip* 9: 819–827.

Straub, M. and Gu, M. 2002. Near-infrared photonic crystals with higher-order bandgaps generated by two-photon photopolymerization. *Opt. Lett.* 27: 1824–1826.

Strickler, J. H. and Webb, W. W. 1991. 3-Dimensional optical-data storage in refractive media by 2-photon point excitation. *Opt. Lett.* 16: 1780–1782.

Sun, H. B. and Kawata, S. 2004. Two-photon photopolymerization and 3d lithographic microfabrication. *Adv. Polym. Sci.* 170: 169–273.

Sun, H. B., Nakamura, A., Shoji, S., Duan, X. M., and Kawata, S. 2003a. Three-dimensional nanonetwork assembled in a photopolymerized rod array. *Adv. Mater.* 15: 2011–2014.

Sun, H. B., Takada, K., Kim, M. S., Lee, K. S., and Kawata, S. 2003b. Scaling laws of voxels in two-photon photopolymerization nanofabrication. *Appl. Phys. Lett.* 83: 1104–1106.

Svoboda, K., Tank, D. W., and Denk, W. 1996. Direct measurement of coupling between dendritic spines and shafts. *Science* 272: 716–719.

Tai, D. C., Tan, N., Xu, S. et al. 2009. Fibro-C-index: Comprehensive, morphology-based quantification of liver fibrosis using second harmonic generation and two-photon microscopy. *J. Biomed. Opt.* 14: 044013.

Tanaka, B., Sun, H. B., and Kawata, S. 2002. Rapid sub-diffraction-limit laser micro/nanoprocessing in a threshold material system. *Appl. Phys. Lett.* 80: 312–314.

Tang, S., Jung, W., Mccormick, D. et al. 2009. Design and implementation of fiber-based multiphoton endoscopy with microelectromechanical systems scanning. *J. Biomed. Opt.* 14: 034005.

Tang, S., Krasieva, T. B., Chen, Z., Tempea, G., and Tromberg, B. J. 2006. Effect of pulse duration on two-photon excited fluorescence and second harmonic generation in nonlinear optical microscopy. *J. Biomed. Opt.* 11: 020501.

Tayalia, P., Mendonca, C. R., Baldacchini, T., Mooney, D. J., and Mazur, E. 2008. 3d cell-migration studies using two-photon engineered polymer scaffolds. *Adv. Mater.* 20: 4494–4498.

Tearney, G. J., Yabushita, H., Houser, S. L. et al. 2003. Quantification of macrophage content in atherosclerotic plaques by optical coherence tomography. *Circulation* 107: 113–119.

Temelkuran, B., Hart, S. D., Benoit, G., Joannopoulos, J. D., and Fink, Y. 2002. Wavelength-scalable hollow optical fibres with large photonic nbandgaps for CO_2 laser transmission. *Nature* 420: 650–653.

Theer, P., Hasan, M. T., and Denk, W. 2003. Two-photon imaging to a depth of 1000 microm in living brains by use of a $Ti:Al_2O_3$ regenerative amplifier. *Opt. Lett.* 28: 1022–1024.

Theodossiou, T., Georgiou, E., Hovhannisyan, V., and Yova, D. 2001. Visual observation of infrared laser speckle patterns at half their fundamental wavelength. *Lasers Med Sci* 16: 34–39.

Theodossiou, T., Rapti, G. S., Hovhannisyan, V. et al. 2002. Thermally induced irreversible conformational changes in collagen probed by optical second harmonic generation and laser-induced fluorescence. *Lasers Med. Sci.* 17: 34–41.

Thomas, J., Elsden, D. F., and M., P. S. 1963. Degradation products from elastin. *Nature* 200: 651–652.

Tian, P. and Warren, W. S. 2002. Ultrafast measurement of two-photon absorption by loss modulation. *Opt. Lett.* 27: 1634–1636.

Tirlapur, U. K. and Konig, K. 2002a. Femtosecond near-infrared laser pulses as a versatile non-invasive tool for intra-tissue nanoprocessing in plants without compromising viability. *Plant J.* 31: 365–374.

Tirlapur, U. K. and Konig, K. 2002b. Targeted transfection by femtosecond laser. *Nature* 418: 290–291.

Tran, P. T., Inoue, S., Salmon, E. D., and Oldenbourg, R. 1994. Muscle fine structure and microtubule birefringence measured with a mew pol-scope. *Biol. Bull.* 187: 244–245.

Tsai, P. S., Blinder, P., Migliori, B. J. et al. 2009. Plasma-mediated ablation: An optical tool for submicrometer surgery on neuronal and vascular systems. *Curr. Opin. Biotechnol.* 20: 90–99.

Tsien, R. Y. 1998. The green fluorescent protein. *Annu. Rev. Biochem.* 67: 509–544.

Tsien, R. Y. and Miyawaki, A. 1998. Seeing the machinery of live cells. *Science* 280: 1954–1955.

Tyrrell, R. M. and Keyse, S. M. 1990. New trends in photobiology. The interaction of uva radiation with cultured cells. *J. Photochem. Photobiol. B* 4: 349–361.

Uchugonova, A., Isemann, A., Gorjup, E. et al. 2008a. Optical knock out of stem cells with extremely ultrashort femtosecond laser pulses. *J. Biophotonics* 1: 463–469.

Uchugonova, A., Konig, K., Bueckle, R., Isemann, A., and Tempea, G. 2008b. Targeted transfection of stem cells with sub-20 femtosecond laser pulses. *Opt. Express* 16: 9357–9364.

Vakoc, B. J., Lanning, R. M., Tyrrell, J. A. et al. 2009. Three-dimensional microscopy of the tumor microenvironment in vivo using optical frequency domain imaging. *Nat. Med.* 15: 1219–1223.

Vaziri, A., Tang, J., Shroff, H., and Shank, C. V. 2008. Multilayer three-dimensional super resolution imaging of thick biological samples. *Proc. Natl. Acad. Sci. U. S. A.* 105: 20221–20226.

Vogel, A., Noack, J., Nahen, K. et al. 1999. Energy balance of optical breakdown in water at nanosecond to femtosecond time scales. *Appl. Phys. B.* 68: 271–280.

Vogel, A. and Venugopalan, V. 2003. Mechanisms of pulsed laser ablation of biological tissues (Vol 103, Pg 577, 2003). *Chem. Rev.* 103: 2079.

Volkmer, A., Cheng, J. X., and Xie, X. S. 2001. Vibrational imaging with high sensitivity via epidetected coherent anti-stokes Raman scattering microscopy. *Phys. Rev. Lett.* 87: 023901.

Vozzi, G., Flaim, C., Ahluwalia, A., and Bhatia, S. 2003. Fabrication of PLGA scaffolds using soft lithography and microsyringe deposition. *Biomaterials* 24: 2533–2540.

Wang, B. G., Riemann, I., Schubert, H. et al. 2007. Multiphoton microscopy for monitoring intratissue femtosecond laser surgery effects. *Lasers Surg. Med.* 39: 527–533.

Wang, C. C., Li, F. C., Wu, R. J. et al. 2009. Differentiation of normal and cancerous lung tissues by multiphoton imaging. *J. Biomed. Opt.* 14: 044034.

Wang, H., Fu, Y., Zickmund, P., Shi, R., and Cheng, J. X. 2005. Coherent anti-stokes Raman scattering imaging of axonal myelin in live spinal tissues. *Biophys. J.* 89: 581–591.

Watson, T. F., Neil, M. A., Juskaitis, R., Cook, R. J., and Wilson, T. 2002. Video-rate confocal endoscopy. *J. Microsc.* 207: 37–42.

Willig, K. I., Rizzoli, S. O., Westphal, V., Jahn, R., and Hell, S. W. 2006. STED microscopy reveals that synaptotagmin remains clustered after synaptic vesicle exocytosis. *Nature* 440: 935–939.

Wokosin, D. L., Centonze, V. E., White, J. et al. 1996. All-solid-state ultrafast lasers facilitate multiphoton excitation fluorescence imaging. *IEEE J. Sel. Top. Quant.* 2: 1051–1065.

Wright, A. J., Burns, D., Patterson, B. A. et al. 2005. Exploration of the optimisation algorithms used in the implementation of adaptive optics in confocal and multiphoton microscopy. *Microsc. Res. Tech.* 67: 36–44.

Wright, A. J., Poland, S. P., Girkin, J. M. et al. 2007. Adaptive optics for enhanced signal in cars microscopy. *Opt. Express* 15: 18209–18219.

Wu, J., Day, D., and Gu, M. 2008a. A microfluidic refractive index sensor based on an integrated three-dimensional photonic crystal. *Appl. Phys. Lett.* 92: 071108.

Wu, Y., Leng, Y., Xi, J., and Li, X. 2009. Scanning all-fiber-optic endomicroscopy system for 3d nonlinear optical imaging of biological tissues. *Opt. Express* 17: 7907–7915.

Wu, Y., Xi, J., Cobb, M. J., and Li, X. 2008b. Fiber-optic endomicroscopy system for high-resolution nonlinear imaging of biological tissue. *Conf. Proc. IEEE Eng. Med. Biol. Soc.* 2008: 1851–1852.

Xi, P., Andegeko, Y., Pestov, D., Lovozoy, V. V., and Dantus, M. 2009. Two-photon imaging using adaptive phase compensated ultrashort laser pulses. *J. Biomed. Opt.* 14: 014002.

Xia, Y. N. and Whitesides, G. M. 1998. Soft lithography. *Annu. Rev. Mater. Sci.* 28: 153–184.

Xu, C. and Webb, W. W. 1996. Measurement of two-photon excitation cross sections of molecular fluorophores with data from 690 to 1050 Nm. *J. Opt. Soc. Am. B* 13: 481–491.

Xu, C., Zipfel, W., Shear, J. B., Williams, R. M., and Webb, W. W. 1996. Multiphoton fluorescence excitation: New spectral windows for biological nonlinear microscopy. *Proc. Natl. Acad. Sci. U. S. A.* 93: 10763–10768.

Yabushita, H., Bouma, B. E., Houser, S. L. et al. 2002. Characterization of human atherosclerosis by optical coherence tomography. *Circulation* 106: 1640–1645.

Yang, D., Jhaveri, S. J., and Ober, C. K. 2005. Three-dimensional microfabrication by two-photon lithography. *MRS Bull.* 30: 976–982.

Yanik, M. F., Cinar, H., Cinar, H. N. et al. 2004. Neurosurgery: Functional regeneration after laser axotomy. *Nature* 432: 822.

Yao, J. N., Hashimoto, K., and Fujishima, A. 1992. Photochromism induced in an electrolytically pretreated MoO_3 thin-film by visible-light. *Nature* 355: 624–626.

Yeh, A. T., Choi, B., Nelson, J. S., and Tromberg, B. J. 2003. Reversible dissociation of collagen in tissues. *J. Invest. Dermatol.* 121: 1332–1335.

Yuste, R. and Denk, W. 1995. Dendritic spines as basic functional units of neuronal integration. *Nature* 375: 682–684.

Zaccolo, M., De Giorgi, F., Cho, C. Y. et al. 2000. A genetically encoded, fluorescent indicator for cyclic AMP in living cells. *Nat. Cell Biol.* 2: 25–29.

Zhuo, S., Chen, J., Yu, B. et al. 2008. Nonlinear optical microscopy of the bronchus. *J. Biomed. Opt.* 13: 054024.

Zuccaro, G., Gladkova, N., Vargo, J. et al. 2001. Optical coherence tomography of the esophagus and proximal stomach in health and disease. *Am. J. Gastroenterol.* 96: 2633–2639.

28

Fluorescence Lifetime Imaging: Microscopy, Endoscopy, and Tomography

James McGinty
Imperial College London

Clifford Talbot
Imperial College London

Dylan Owen
Imperial College London

David Grant
Imperial College London

Sunil Kumar
Imperial College London

Neil Galletly
Imperial College London

Bebhinn Treanor
Imperial College London

Gordon Kennedy
Imperial College London

Peter M. P. Lanigan
Imperial College London

Ian Munro
Imperial College London

Daniel S. Elson
Imperial College London

Anthony Magee
Imperial College London

Dan Davis
Imperial College London

Gordon Stamp
Imperial College London

Mark Neil
Imperial College London

Christopher Dunsby
Imperial College London

Paul M. W. French
Massachusetts General Hospital
and
Harvard Medical School

28.1 Introduction 590
    Overview • Fluorescence
28.2 FLIM Techniques and Technology 592
    Introduction • Laser Scanning FLIM: Time-Domain Photon Counting/Binning Fluorescence
    Lifetime Measurements • Laser Scanning FLIM: Frequency-Domain Fluorescence Lifetime
    Measurement • High-Speed Laser Scanning FLIM • Wide-Field FLIM • Excitation Sources
    for FLIM
28.3 Applications of FLIM Microscopy 597
    FLIM Applied to Cell Biology: Fluorescence Lifetime Sensing • FLIM Applied to
    Cell Biology: FRET • FLIM Microscopy of Tissue Autofluorescence
28.4 FLIM Endoscopy 603
28.5 FLIM Tomography 604
28.6 Outlook 607
Acknowledgments 608
References 608

589

28.1 Introduction

28.1.1 Overview

This chapter aims to present the potential of fluorescence lifetime imaging (FLIM) as a widely applicable technique to obtain quantitative information from fluorescent samples ranging from single cells through small organisms and animals to clinical imaging. In general, fluorescence provides a powerful means of achieving optical molecular contrast (Lakowicz 2006) in spectrofluorometers, in cytometers and cell sorters, and in microscopes, endoscopes, and multiwell plate readers. For biological applications, fluorescent molecules (fluorophores) are typically used as "labels" to tag specific molecules of interest, particularly exploiting the growing family of genetically expressed fluorescent proteins (Tsien 2009) for molecular biology studies and live cell imaging. Fluorescence imaging enables, for example, the study of the spatiotemporal organization and interaction of proteins in cells, for example, using Förster resonant energy transfer (FRET) (Förster 1948, Clegg et al. 2003), thereby providing opportunities to study fundamental biological processes including those related to disease. For clinical applications, it is possible to utilize the fluorescence properties of endogenous target molecules themselves to provide *label-free molecular contrast*, which can be of immediate diagnostic utility, or to use exogenous labels, such as organic dyes or nanoparticles, to detect diseased tissue, for example, using photodynamic detection (Friesen et al. 2002). This chapter will briefly review the most common fluorescence imaging techniques used to analyze fluorescent signals, with particular reference to FLIM. Exemplar applications of FLIM to cell biology and tissue will be presented, illustrating implementations for microscopy, endoscopy, and tomography.

28.1.2 Fluorescence

The radiation emitted by a fluorophore can be described in terms of its intensity, excitation and emission spectra, quantum efficiency, polarization, and temporal decay after excitation (fluorescence lifetime). These parameters depend on the properties of the fluorophore itself and on its local environment, and so fluorescence imaging can map variations in chemical and physical properties of a sample as well as provide information on the distribution of the fluorophores and, therefore, of any molecules to which they are attached.

After a fluorophore has been excited to an upper electronic energy level, it may spontaneously decay back to the ground state either radiatively, with a rate constant, Γ, or non-radiatively, with a rate constant, k, as represented in Figure 28.1. The quantum efficiency, η, of the fluorescence process is defined as the ratio of the number of fluorescent photons emitted compared to the number of excitation photons absorbed and is equal to the radiative decay rate, Γ, divided by the total decay rate, $\Gamma + k$, as indicated. In general, the quantum efficiency is sensitive to the local fluorophore environment since it can be affected by any factors that

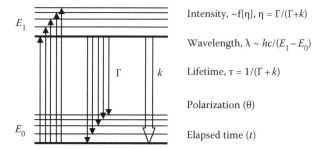

FIGURE 28.1 Overview of fluorescence parameters.

change the molecular electronic configuration or de-excitation pathways and, therefore, the radiative or non-radiative decay rates. Such environmental factors can include the local chemical environment, for example, pH, calcium, or oxygen concentration; the proximity of other fluorophores that result in, for example, quenching or FRET; or the local physical environment, for example, refractive index, temperature, viscosity, electric field, etc. Determining the quantum efficiency requires knowledge of the fluorophore concentration and the photon excitation and detection efficiencies. For many biological samples, however, quantitative measurements of intensity are compromised by optical scattering, internal reabsorption of fluorescence (inner filter effect), and background fluorescence from other (endogenous) fluorophores present in a sample. It can therefore be highly challenging to accurately map variations in quantum efficiency using fluorescence intensity imaging, particularly in biological tissue that is usually highly scattering and heterogeneous, often presenting multiple endogenous fluorophores.

More robust measurements can be made using *ratiometric* techniques to detect variations in the local fluorophore environment that impact the fluorescence process. If one can assume that unknown quantities such as excitation and detection efficiency, fluorophore concentration, and signal attenuation will be approximately the same in two or more spectral or time windows, they may be effectively "canceled out" in a ratiometric measurement. This approach is used in the spectral domain, for example, with ratiometric calcium sensing dyes (Grynkiewicz et al. 1985) and to provide intrinsic contrast between, for example, malignant and normal tissue (Andersson-Engels et al. 1990). Fluorescence lifetime measurement is also a ratiometric technique in which it is assumed that the various unknown quantities do not change significantly during the fluorescence decay time (typically ns) and the lifetime determination effectively compares the fluorescence signal at different delays after excitation. The fluorescence lifetime is the average time the molecule spends in an upper energy level before returning to the ground state. For a large ensemble of identical molecules (or a large number of excitations of the same molecule), the time-resolved fluorescence intensity profile (detected photon histogram) following instantaneous excitation will exhibit a monoexponential decay that may be described as

$$I(t) = I_0 e^{-\frac{t}{\tau}} + \text{const.} \qquad (28.1)$$

where I_0 is intensity of the fluorescence immediately after excitation and the constant term represents any background signal. In practice, the presence of multiple fluorophore species—or multiple states of a fluorophore species arising from interactions with the local environment—often result in more complex fluorescence decay profiles. This is often the case for autofluorescence of biological tissue. Such complex fluorescence decays are commonly modeled by an N-component multi-exponential decay model:

$$I(t) = \sum_{i=1}^{N} C_i e^{-\frac{t}{\tau_i}} + \text{const.} \qquad (28.2)$$

where each pre-exponential amplitude is represented by the value C_i. Alternative approaches to model complex fluorescence decay profiles include fitting to a stretched exponential model that corresponds to a continuous lifetime distribution (Lee et al. 2001), a power law decay (Wlodarczyk and Kierdaszuk 2003), or to Laguerre polynomials to generate empirical contrast (Jo et al. 2004).

The parameters describing the fluorescence decay profile can be determined by fitting the experimental fluorescence decay data to the appropriate model using, for example, an iterative nonlinear least squares Levenberg Marquardt algorithm, although in the case of a single exponential decay profile, Equation 28.1 can be simply linearized and τ can be calculated analytically. In general, as more information is used to describe a fluorescence signal, then more detected photons are required to be able to make a sufficiently accurate measurement of it (Grinvald and Steinberg 1974, James and Ware 1985). Thus fitting fluorescence decay profiles to more complex models requires increased data acquisition times, which is often undesirable in terms of temporal resolution, considerations of photobleaching or photodamage and the potential for motion artifacts. For many applications, it is therefore preferable to approximate complex fluorescence decay profiles to a single exponential decay model. The average fluorescence lifetime can still provide useful contrast since it will usually reflect changes in the decay times or relative contributions of different components but, of course, the interpretation of a change in the average lifetime of a complex fluorescence decay profile can be subject to ambiguity. Where multicomponent FLIM of fluorophores exhibiting complex fluorescence is required, it is possible to reduce the number of detected photons required by using a priori knowledge or assumptions, for example, about the magnitude of one or more component lifetimes, that is, τ_i, or about their relative contributions. It can also be useful to assume that some parameters, for example, τ_i, are global, that is, they take the same value in each pixel of the image. Such global analysis (Verveer et al. 2000a) is often used for the application of FLIM to FRET. In all cases, to extract the true fluorescence decay parameters, the effect that the instrument and detector have on the measured signal should to be taken into account. This can be addressed by measuring the instrument response function (IRF), that is, the response

of the whole system to a delta function input. This IRF can be convolved with the fluorescence decay model used to generate a new "model" to which the acquired data may be fitted in order to obtain the true decay parameters. In practice, it is not possible to determine the IRF using an ideal delta function, but ultrashort optical excitation pulses may be used in various ways to achieve a useful approximation, for example, by employing a scattering sample to simulate a very short lifetime sample or by using a reference lifetime standard (Boens et al. 2007).

A further ratiometric measurement approach is to resolve a fluorescence signal with respect to polarization, which can provide information concerning molecular orientation, and this can be combined with time-resolved fluorescence decay measurements to determine the molecular tumbling time (rotational correlation time) of a fluorophore, which can be used to report ligand binding or local solvent properties. Similarly, fluorescence lifetime measurements can be combined with spectrally resolved measurements to provide enhanced contrast or richer information about fluorophore environment. In recent years, there have been tremendous advances in fluorescence imaging tools and multidimensional fluorescence imaging (MDFI) instruments, partly driven by the progress in laser and photonics technology, including robust tunable and ultrafast excitation sources, relatively low-cost, high-speed detection electronics, high-performance imaging detectors, such as the EM-CCD camera, and low-cost computers. Such technological advances have facilitated the wide deployment of techniques such as multiphoton microscopy (Denk et al. 1990) and FLIM (Cubeddu et al. 2002), for which the availability of user-friendly femtosecond Ti:Sapphire lasers was a key development. In parallel, the development of genetically expressed fluorescent proteins (Tsien 1998, Zimmer 2002) created unprecedented opportunities to observe many biological processes in live cells and organisms with highly specific labeling. Such advances have driven the development and adoption of sophisticated fluorescence imaging techniques, of which one of the most significant for cell biology has been the widespread deployment of FRET for imaging protein interactions.

FRET is a fluorescence quenching process where the fluorescence of a particular molecular species (the donor) is affected by its proximity to a second chromophore (the acceptor). More precisely, the excitation energy of a donor fluorophore is transferred to a ground state acceptor molecule by a dipole–dipole coupling process (Förster 1948, Clegg et al. 2003, Stryer 1978). Although this process is non-radiative, the emission spectrum of the donor and the absorption spectrum of the acceptor must overlap for FRET to occur (Haugland et al. 1969) and the transition dipole moments of the donor and acceptor must not be perpendicular (Eisinger and Dale 1974). The FRET efficiency varies with the inverse sixth power of the distance between donor and acceptor and is usually negligible beyond 10 nm, which is of the order of the distance between proteins bound together in a complex. Measurements of FRET between appropriately labeled species can thus provide a gold standard for colocalization studies and facilitate mapping of protein interactions, such as ligand binding, in time and space. In some situations, it is possible to read out the distance between "FRETing" fluorophores—effectively realizing a so-called spectroscopic ruler (Stryer 1978, Dosremedios and Moens 1994) that can be used to analyze protein structure and monitor conformational changes of multi-labeled molecules.

FRET is thus a powerful tool to study molecular cell biology, particularly when combined with genetically expressed fluorophores, and is widely employed to study cell signaling mechanisms. However, detecting and quantifying resonant energy transfer is often not straightforward. Although FRET can be observed via fluorescence intensity imaging, for example, of the acceptor fluorescence when the donor is excited, this is susceptible to a number of artifacts, such as spectral "bleed-through" of the donor fluorescence into the acceptor detection channel. Such artifacts should be minimized through careful control measurements and correction calculations (Gordon et al. 1998); but this is not always practical and, increasingly, other fluorescence parameters are being utilized to improve the reliability of FRET experiments including fluorescence spectra, lifetime, and polarization anistropy. FLIM is emerging as one of the most robust approaches to read out FRET (e.g., Bastiaens and Squire 1999, Jares-Erijman and Jovin 2003, Suhling et al. 2005), because the donor fluorescence lifetime is largely independent of factors such as fluorophore concentration, excitation, and detection efficiency, inner filter, and multiple scattering effects, which can complicate or degrade absolute intensity measurements. In particular, the insensitivity of fluorescence lifetime measurements to fluorophore concentration make FLIM the preferred technique when imaging samples for which the precise stoichiometry of the donor and acceptor species is unknown—as is the case when monitoring interactions between separately labeled proteins (e.g., binding partners, enzyme/substrate pairs, etc.).

Besides providing a powerful tool to readout fluorescence experiments that take advantage of the ever increasing range of fluorescent labels, FLIM is also being increasingly applied to tissue autofluorescence for clinical studies. Spectrally resolved imaging of tissue autofluorescence is relatively well established (e.g., Richards-Kortum and Sevick-Muraca 1996), and advances in FLIM technology have resulted in autofluorescence lifetime now being actively investigated as a means of realizing or enhancing label-free diagnostic imaging in tissues (Das et al. 1997, Wagnieres et al. 1998, Dowling et al. 1998). Of course, there are applications of FLIM beyond cell biology and medical imaging. For example, there is a burgeoning interest in applying FLIM to biophotonic devices such as microfluidic systems for lab-on-a-chip applications (e.g., Benninger et al. 2006, Schaerli et al. 2009) and in utilizing FLIM and MDFI to study nanophotonics structures (e.g., Koeberg et al. 2003).

28.2 FLIM Techniques and Technology

28.2.1 Introduction

In general, FLIM techniques are categorized as time-domain or frequency-domain techniques, according to whether the instrumentation measures the fluorescence signal as a function of time delay following pulsed excitation or whether the lifetime information is derived from measurements of phase difference between a sinusoidally modulated excitation signal and the resulting sinusoidally modulated fluorescence signal. In principle, frequency- and time-domain approaches can provide equivalent information but specific implementations present different trade-offs with respect to cost and complexity, performance and acquisition time. In practice, the most appropriate FLIM methodology should be selected according to the target application. Historically, FLIM evolved from single channel fluorescence lifetime cuvette-measurements in time-resolved spectrofluorometers and initially frequency-domain methods could be developed with simpler electronic instrumentation and excitation source requirements. Following the tremendous advances in microelectronics and ultrafast laser technology, however, time-domain FLIM techniques today present a similar cost and complexity to most users. Further categorization of FLIM techniques can be made according to whether they are implemented in laser beam scanning or wide-field microscopes and whether they utilize sampling techniques, which use gated detection to determine the temporal properties of the fluorescence compared to the excitation signal, or photon counting/binning techniques, which assign detected photons to different time bins. For wide-field imaging, FLIM is usually implemented with gated detectors that sample the image pixels of the fluorescence signal in parallel while photon counting techniques have been implemented with sequential "single channel" pixel acquisition in laser scanning instruments. With respect to the latter, there is a further distinction between single-photon counting techniques, such as time-correlated single-photon counting (TCSPC), and techniques based on time-binning (of photoelectrons) or high-speed analogue-to-digital converters that can detect more than one photon per excitation pulse. References (Lakowicz 2006, Cubeddu et al. 2002, Gadella 2008, Esposito et al. 2007) provide extensive reviews of FLIM techniques and here we will briefly summarize the main time- and frequency-domain methods applied to laser scanning and wide-field FLIM.

28.2.2 Laser Scanning FLIM: Time-Domain Photon Counting/Binning Fluorescence Lifetime Measurements

Single-point time-domain measurement of fluorescence decay profiles has been implemented using a wide range of instrumentation such as fluorometers with ultrashort pulsed excitation typically being provided by mode-locked solid-state lasers or by gain-switched semiconductor diode lasers. For time-resolved detection, fast (GHz bandwidth) sampling oscilloscopes and streak cameras have been used since the 1970s but these have been increasingly supplanted by photon counting techniques that build up histograms of the decay profiles. For FLIM, the most widely used detection technique is probably TCSPC (O'Connor and Phillips 1984), which was demonstrated in the first FLIM microscope (Bugiel et al. 1989). The development of convenient and relatively low-cost TCSPC electronics (e.g., Zhang et al. 1999, Becker et al. 2004), for confocal and two-photon fluorescence scanning microscopes has greatly increased the uptake of FLIM generally, as well as impacting the development of single channel lifetime fluorometers.

The principle of TCSPC is illustrated in Figure 28.2 and requires that the excitation power is sufficiently low to ensure that less than one photon per excitation pulse is detected, such that a series of voltage signals that depend on the arrival time of individual detected photons relative to the train of excitation pulses can be recorded to build up a histogram of photon arrival times. By also recording spatial information from the scanning electronics of a confocal/multiphoton microscope, the photon arrival times can be allocated to their respective image pixels to form fluorescence lifetime images. This is straightforward to implement on a scanning microscope since it requires only additional electronic components (after the detector) and may be implemented on most scanning microscopes.

TCSPC is widely accepted as one of the most accurate methods of lifetime determination due to its shot-noise-limited detection, high photon economy, low temporal jitter, high temporal precision (offering large number of bins in the photon arrival time histogram) and high dynamic range (typically millions of photons can be recorded without saturation). Its main perceived drawback is a relatively low acquisition rate, owing to the requirement to operate at sufficiently low incident fluorescence intensity levels to ensure single-photon detection at a rate limited by the "dead-time" between measurement events, which is imposed by the time-to-amplitude (TAC) and constant fraction

discriminator (CFD) circuitry that determine the photon arrival times (Becker 2005). However, for modern TCSPC instrumentation, this limitation of the electronic circuitry is usually less significant than problems caused by "classical" photon pile up, which limits all single-photon counting techniques. Pulse pile up refers to the issue of more than one photon arriving in a single-photon detection period, which results in apparently shorter lifetimes being measured. This is avoided by decreasing the excitation power such that the excitation rate is much lower than the pulse repetition rate, which typically limits the maximum detection count rate to approximately 5% of the repetition rate of the laser (Becker 2005).

An alternative single-point photon-counting time-domain technique is based on temporal photon binning, for which the photoelectrons arising from the detected photons are accumulated in a number of different time bins and a histogram is built up accordingly (Buurman et al. 1992). This method does not use a TAC and so does not have the same dead-time and pulse-pile-up limitations as TCSPC, so it may be used with higher photon fluxes to provide higher imaging rates when implemented on a scanning fluorescence microscope. To date, however, it has not been commercially implemented with the same precision as TCSPC and so can provide FLIM at faster rates (approaching real time) but with lower lifetime precision (Gerritsen et al. 2002).

Of course the temporal resolution of fluorescence lifetime instrumentation is also limited by the temporal impulse response of the detector, as well as the electronic circuitry. Photon counting photomultipliers typically exhibit a response time of ~200 ps although faster multichannel plate (MCP) devices can have response times of a few tens of picoseconds. One significant exception to the above observation is the optical pump-probe approach where a second (probe) beam, which is delayed with respect to the excitation (pump) beam, interrogates the excited state population. A particularly elegant implementation of this technique for scanning fluorescence lifetime microscopy used a probe beam to induce stimulated emission from the excited state fluorophores. The net fluorescence signal is monitored as the probe beam pulse delay is swept with respect to the excitation pulse and this provides the fluorescence decay parameters (Dong et al. 1995). The temporal resolution of this and other pump-probe techniques is determined by the laser pulse durations and not the detection bandwidth. The requirement for temporal and spatial coincidence of the pump and probe pulses also provides optical sectioning in a manner analogous to two-photon microscopy.

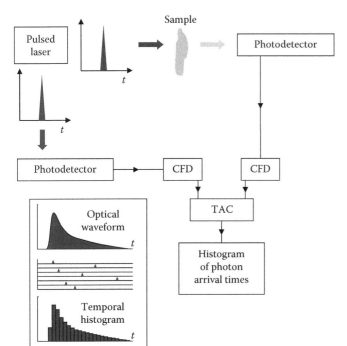

FIGURE 28.2 Schematic representation of TCSPC FLIM. The arrival times of photons are measured with respect to a reference signal from the excitation source and a temporal histogram is built up. CFD, constant fraction discriminator and TAC, time-to-amplitude converter.

28.2.3 Laser Scanning FLIM: Frequency-Domain Fluorescence Lifetime Measurement

Frequency-domain fluorescence lifetime measurements entail the demodulation of a fluorescence signal excited with modulated light. Initially this approach utilized sinusoidally modulated excitation sources that will produce a fluorescence signal that is also sinusoidally modulated but with a decreased

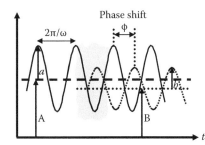

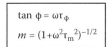

FIGURE 28.3 Schematic representation of frequency-domain FLIM signals.

modulation depth and a phase delay relative to the excitation signal, as illustrated in Figure 28.3.

The fluorescence lifetime can be determined from measurements of the relative modulation, m, and the phase delay, ϕ (Lakowicz 2006). For an ideal single exponential decay, the lifetime can be calculated from ϕ or m using the following equations:

$$\tau_\phi = \frac{1}{\omega}\tan\phi \tag{28.3}$$

$$\tau_m = \frac{1}{\omega}\left(\frac{1}{m^2}-1\right)^{\frac{1}{2}} \tag{28.4}$$

If the fluorescence excitation is not a pure sinusoid, the frequency-domain approach is still applicable since the lifetime can be calculated using the Fourier components of any modulated fluorescence signal, but the fluorescence signal must be sampled at more phase delays to permit the higher harmonic components to be resolved. It is sometimes desirable, for example, to use a pulsed excitation laser, for example, for multiphoton excitation (So et al. 1995) and, in fact, the implementation of frequency-domain FLIM with periodically pulsed excitation provides an improved signal-to-noise ratio (SNR) compared to a purely sinusoidal excitation (Philip and Carlsson 2003). If the fluorescence does not manifest a single exponential decay, however, the lifetime values calculated from Equations 28.3 and 28.4 will not be the same, and it is necessary to repeat the measurement at multiple harmonics of the excitation modulation frequency to build up a more complete (multi-exponential) description of the complex fluorescence profile. This is usually calculated by fitting the results to a set of dispersion relationships (e.g., Gratton and Limkeman 1983, Lakowicz and Maliwal 1985). The increased acquisition and data processing time associated with multiple frequency measurements of complex fluorescence signals can be undesirable. For applications requiring rapid imaging, it is sometimes useful to measure the fluorescence at only one modulation frequency and obtain an "average" lifetime by calculating the mean of the lifetime values from the modulation depth and phase delay measurements.

For frequency-domain laser scanning FLIM, one can use a sinusoidally modulated or pulsed excitation laser and measure the phase shift and demodulation of the fluorescence signal, for example, using a "lock-in" amplifier. Recently, advances in microelectronics have enabled frequency-domain laser scanning FLIM to be implemented with low-cost synchronous detection (Booth and Wilson 2004) or fast digitization circuitry (Colyer et al. 2008). This approach is not limited by the dead-time or maximum count rates associated with TCSPC and so can provide relatively high-speed scanning FLIM.

The ability to use sinusoidally modulated diode lasers or LED's as excitation sources makes frequency-domain FLIM attractive for applications demanding lower cost instrumentation, for which frequencies of less than 100 MHz are sufficient to provide nanosecond lifetime resolution. Achieving picosecond lifetime resolution, however, requires GHz modulation frequencies, and the necessary technology can introduce comparable complexity and expense to that associated with time-domain measurements.

28.2.4 High-Speed Laser Scanning FLIM

High-speed FLIM is desirable for studying dynamics, for example, of cell signaling networks using FRET, for high content analysis of sample arrays, and for clinical imaging of live subjects. For all laser scanning microscopes, the sequential pixel acquisition means that it is difficult to realize high-speed FLIM. For TCSPC, this is partly a function of the maximum photon detection rate imposed by the electronic circuitry. For all laser scanning microscopes, however, increasing the imaging speed requires a concomitant increase in excitation intensity, which can be undesirable due to photobleaching and phototoxicity considerations. This is a particular issue for FLIM, for which more photons need to be detected per pixel compared to intensity imaging (Gerritsen et al. 2002, Kollner and Wolfrum 1992). For example, it is estimated that detection of a few hundred photons is necessary to determine the lifetime (to ~10% accuracy) using a single exponential fluorescence decay model while a double exponential fit requires ~10^4 detected photons. For multiphoton FLIM microscopy, for which excitation is relatively inefficient and photobleaching scales nonlinearly with intensity (Patterson and Piston 2000), these considerations can result in FLIM acquisition times of many minutes for biological samples. For confocal FLIM microscopy with single-photon excitation, it is difficult to acquire useful FLIM images on timescales faster than a few seconds.

One way to significantly increase the imaging speed of multiphoton microscopy is to use multiple excitation beams in parallel (Bewersdorf et al. 1998, Fittinghoff et al. 2000, Kim et al. 2007), and this approach has been applied to TCSPC FLIM using 16 parallel excitation beams and a 16 channel PMT detector (Kumar et al. 2007). In general, parallel pixel excitation and detection is a useful approach to increase the practical imaging speed of all laser scanning microscopes. This can be extended to optically sectioned line-scanning microscopy using a rapidly scanned multiple beam array to produce a line of fluorescence emission that is relayed to the input slit of a streak camera (Krishnan et al. 2003). FLIM images have been acquired in less than 1 s using this approach, which is currently limited by the read-out rate of the streak camera system.

Alternatively, multiple scanning beam excitation can be applied with wide-field time-gated detection, as has been implemented with multibeam multiphoton microscopes (Straub and Hell 1998, Leveque-Fort et al. 2004, Benninger et al. 2005) and with single-photon excitation in spinning Nipkow disc microscopes (Grant et al. 2005, 2007, van Munster et al. 2007, Grant). A direct comparison of TSCPC and wide-field frequency-domain FLIM concluded that TCSPC provided a better SNR for weak fluorescence signals, the frequency-domain approach was faster and more accurate for bright samples (Gratton et al. 2003).

28.2.5 Wide-Field FLIM

The parallel nature of wide-field imaging techniques can support FLIM imaging rates of tens to hundreds of hertz (e.g., Agronskaia et al. 2003, Requejo-Isidro et al. 2004), although the maximum acquisition speed is still limited by the number of photons/pixel available from the sample. Wide-field FLIM is most commonly implemented using modulated image intensifiers with frequency or time-domain approaches, as represented in Figure 28.4. The frequency-domain approach was established first (e.g., Morgan et al. 1990, Gratton et al. 1990), utilizing frequency-modulated laser excitation and implementing synchronous detection by frequency modulating the gain of a microchannel plate (MCP) image intensifier to analyze the resulting fluorescence by acquiring a series of gated "intensified" images acquired at different relative phases between the MCP modulation and the excitation signal. Originally, the optical output image from the intensifier was read out using a linear photodiode array but this was rapidly superseded by CCD camera technology (Lakowicz and Berndt 1991, Gadella et al. 1993). It is possible to implement wide-field FLIM at a single modulation frequency using only three phase measurements (Lakowicz et al. 1992a) to calculate the fluorescence lifetime map, although it is common to use eight or more phase-resolved images to improve the accuracy. This can be necessary if the system exhibits unwanted nonlinear behavior that produces modulation signals at harmonic frequencies.

The development of wide-field frequency-domain FLIM was complemented by the demonstration of a streak camera–based approach (Minami and Hirayama 1990) and by the application of short pulse–gated MCP image intensifiers coupled to CCD cameras in an approach described as time-gated imaging for time-domain wide-field FLIM (e.g., Oida et al. 1993, Scully et al. 1996, Dowling et al. 1998). By gating the MCP image intensifier gain for short (ps–ns) periods after excitation, the fluorescence decay can be sampled such that the fluorescence lifetime images can be calculated. Initially, the shortest MCP gate widths achieved were over 5 ns but the technology quickly developed to provide subnanosecond gating times (Wang et al. 1991) and then the use of a wire mesh proximity-coupled to the photocathode led to devices with sub 100 ps resolution (Hares 1987).

While both frequency and time-domain FLIM offer parallel pixel acquisition compared to scanning microscopy techniques, they both suffer from inherently reduced photon economy owing to the time-varying gain applied to the image intensifier. This can be particularly significant in the time domain if very short (~100 ps) time gates are applied to sample the fluorescence decay profile. In many situations, however, such short time gates are not necessary since the fast rising and falling edges of the time gate provide the required time resolution and the gate can remain "open" for durations comparable to the fluorescence decay time (Munro et al. 2005). Thus the photon economy of time-gated imaging can approach that of wide-field frequency-domain FLIM. For both frequency-domain and time-domain measurements, it is necessary to acquire at least three time (phase)-gated images in order to obtain a mean fluorescence lifetime in the presence of an offset (background signal). If the background can be determined in a separate measurement and then subtracted from subsequent acquisitions, it is possible to implement high-speed FLIM using just two time (phase)-gated images for each fluorescence lifetime calculation. The required fluorescence lifetimes can be calculated analytically using rapid lifetime determination (RLD) such that real-time fluorescence lifetime images can be displayed for both frequency and time doman FLIM. A frequency-domain FLIM microscope achieved a FLIM rate of 0.7 Hz for a field of view of 300 × 220 pixels and also provided "lifetime-resolved" images based on the difference between only two phase resolved images at up to 55 Hz for

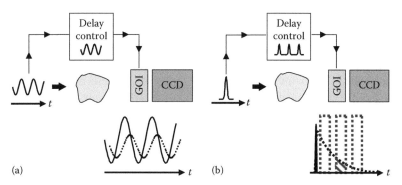

FIGURE 28.4 Schematic representation of general (a) frequency- and (b) time-domain wide-field approaches to FLIM.

164 × 123 pixels (Holub et al. 2000). Frequency-domain FLIM was extended to FLIM endoscopy, acquiring phase-resolved images at 25 Hz to achieve real-time FLIM for a field of view of 32 × 32 pixels (Mizeret et al. 1999).

In the time domain, RLD is often implemented using just two time gates (with the background level assumed to be zero or subtracted from the time-gated images) and employing an analytical approach to calculate the single exponential decay lifetime (Devries and Khan 1989). At Imperial we demonstrated video-rate wide-field RLD FLIM using two sequentially acquired time-gated fluorescence images (subtracting a previously acquired background image) using a gated optical imaging (GOI) intensifier–based system, which we also applied to FLIM endoscopy (Requejo-Isidro et al. 2004). The optimization of RLD has been studied (Chan et al. 2001) and, if both gates are of equal width, the minimum error in lifetime determination is obtained when the gate separation, Δt, is around twice the lifetime being investigated and the gate widths are equal to Δt. We generalized this analysis and showed analytically that the optimum gate separation for equal gate widths (of any duration) is 2.2τ (McGinty et al. 2009a). This optimized measurement is only accurate over a narrow range of sample fluorescence lifetimes and, to optimize RLD FLIM over a wider range of sample lifetimes, one can increase the width of the second time gate or use three or more time gates, albeit at the expense of imaging speed. There are also analytic expressions for RLD of a single exponential decay with an unknown background (Ballew and Demas 1991) or a biexponential decay using four time-gated images (Sharman et al. 1999). Further optimization of RLD FLIM can exploit the potential to adjust the CCD integration time for the acquisition of each time-gated image in order to collect more photons at the later stages of the decay profile (Munro et al. 2005, McGinty et al. 2009a).

High-speed FLIM is important for application to dynamic or moving samples. Time-gated (or phase-gated) wide-field imaging, however, is subject to severe artifacts if the sample moves between successive time-gated (or phase-gated) acquisitions (Elson et al. 2004a). This problem is less severe for TCSPC FLIM where sample motion during an image acquisition will degrade the spatial resolution but not drastically change the apparent fluorescence lifetimes. It is possible to avoid/address this issue of motion artifacts in RLD FLIM by acquiring the different time-gated (or phase-gated) images simultaneously. This was first implemented in a "single-shot" FLIM system that utilized an optical image splitter to produce two images incident on the same GOI but with the photons from one being delayed with respect to the other by an optical relay (Agronskaia et al. 2003). This elegant approach was demonstrated at up to 100 Hz imaging a rat neonatal myocyte stained with the calcium indicator, Oregon Green (BAPTA-1). Its main drawbacks are that the field of view is reduced with respect to the conventional sequential time-gated imaging acquisition approach and that the complexity of the optical imaging delay line makes it difficult to adjust the delay between the parallel images, in order to accommodate a range of sample fluorescence lifetimes. These issues can be addressed

using multiple independently gated GOI detectors to maintain the field of view (Young et al. 1988) at the cost of significantly increased system size and complexity, or by using a single GOI with a segmented photocathode, providing multiple image channels that can be independently gated through the introduction of resistive elements. The latter approach realized wide-field four-channel RLD FLIM at up to 20 Hz and was applied to label-free FLIM of ex vivo tissue (Elson et al. 2004a). This parallel acquisition of multiple time (phase)-gated images has recently been implemented for frequency-domain FLIM using a modulated CMOS detector that can accumulate photoelectrons in two parallel image stores according to a modulated voltage (Esposito et al. 2005). This permits fast FLIM (with two phase-resolved images) using solid-state camera technology that could potentially be cheaper and faster than current GOI technology, although to date it has only been realized with a 20 MHz modulation frequency and 124 × 160 pixels in each phase-gated image.

While high-speed FLIM is the compelling application of wide-field time/phase-gated imaging, it is not always applied using RLD. For applications such as FRET or tissue imaging, it is often desirable to extract information concerning different components of complex fluorescence decay profiles. In the time domain, this entails sampling the decay profiles with more time gates and (usually) iteratively fitting the data to a complex decay model using, for example, a nonlinear least squares Levenberg Marquardt algorithm, although it is possible to use an analytical modified RLD approach to analyze double exponential decay profiles (Sharman et al. 1999) or to use phasor analysis (Digman et al. 2008) to visualize complex FLIM data. In the frequency domain, it entails acquiring phase-gated fluorescence images for a number of excitation modulation frequencies and then determining multiple exponential decay components, for example, using Fourier analysis, or again visualizing the data using phasor analysis (Redford and Clegg 2005). A particularly elegant frequency-domain approach exploits a non-sinusoidal detector gain (e.g., a fast rectangular time-gated GOI detector function) to simultaneously determine phase delays at multiple harmonic frequencies in the excitation signal (Squire et al. 2000). This can reduce the sample exposure time, limiting photobleaching and photodamage compared to sequential acquisitions at different modulation frequencies.

Sampling and fitting complex decay profiles does, however, inevitably increase the FLIM acquisition time since significantly more detected photons are required for accurate lifetime determination. Where rapid FLIM of samples exhibiting complex fluorescence decays is required, it is usually necessary to use a priori information or assumptions such as fixing one or more component lifetimes or assuming that the unknown component lifetimes are the same in each pixel. This is particularly important for FLIM FRET where imaging speed is a consideration.

28.2.6 Excitation Sources for FLIM

A critical component for any FLIM instrument is the excitation source and the development of FLIM has been strongly influenced by developments in laser technology. The excitation wavelength is perhaps the most important parameter for fluorescence imaging since this will determine the extent to which different fluorophores are imaged. Ironically, this has been the most constrained parameter for many fluorescence imaging experiments, particularly for confocal laser scanning microscopy, which requires spatially coherent sources and so has been typically limited to a few discrete excitation wavelengths, and for FLIM, which requires ultrafast modulated radiation. While convenient tunable c.w. excitation has long been routinely available for wide-field microscopy using, for example, filtered lamp sources, the main ultrafast excitation sources available in the visible spectrum were c.w. mode-locked dye lasers before the advent of ultrafast Ti:Sapphire lasers. Today, however, there are many convenient all-solid-state ultrafast excitation sources available ranging from modulated LEDs and gain-switched laser diodes through diode-pumped mode-locked fiber lasers and solid-state lasers and, recently, supercontinuum sources. The most important of these for FLIM has probably been the mode-locked femtosecond Ti:Sapphire laser, partly because FLIM has so often been implemented with multiphoton microscopy. Multiphoton excitation uniquely facilitates fluorescence imaging at significantly increased depths in turbid media such as biological tissue compared to wide-field or confocal microscopy. Frequently, however, multiphoton excitation is employed with less challenging samples such as cell cultures, for which its disadvantages of reduced spatial resolution, lower excitation cross sections, and increased nonlinear photobleaching/photodamage in the focal plane compared to confocal microscopy are outweighed by the convenience and versatility of a (computer-controlled) tunable excitation laser that can also access important fluorophores normally requiring u.v. excitation. The ultrafast lasers used for multiphoton excitation are inherently applicable to FLIM, which then requires only a relatively modest additional investment, for example, to implement TCSPC. Nevertheless, there are situations—particularly for cell biology—where single-photon excitation is preferable, for example, to reduce nonlinear photodamage, to achieve the best possible spatial resolution, and to realize faster imaging, for example, of live (dynamic) samples or in a high-throughput context.

Tunable single-photon excitation may be provided by frequency-doubled Ti:Sapphire lasers over a range of ~350 to 520 nm but this will not excite many important fluorophores. An increasingly popular approach to provide tunable (and ultrafast) excitation over the visible and NIR spectrum is to use ultrafast laser-pumped supercontinuum generation in microstructured optical fibers (Ranka et al. 2000). Supercontinuum sources can be pumped using compact and relatively inexpensive ultrafast fiber lasers (Champert et al. 2002, Schreiber et al. 2003) to provide spectral coverage spanning from the ultraviolet to the near infrared (Kudlinski et al. 2006). Spectral selection from such a supercontinuum provides a relatively low-cost tunable source of ultrashort optical pulses applicable to single-photon and multiphoton microscopy and FLIM (Jureller et al. 2002, Deguil et al. 2004, Birk and Storz 2001, McConnell 2004, Dunsby et al. 2004).

FLIM can also be conveniently implemented at a range of discrete spectral wavelengths using relatively low-cost semiconductor diode laser and LED technology. Gain-switched picosecond diode lasers have been applied to time-domain FLIM in scanning (Bohmer et al. 2001) and wide-field (Elson et al. 2002) FLIM systems although their relatively low average output power (typically <1 mW) and sparse spectral coverage limit their range of applications. It is possible to reach higher average powers using sinusoidal modulation for frequency-domain FLIM (Dong et al. 2001, Booth and Wilson 2004) and, for wide-field FLIM where their low spatial coherence is an advantage, LEDs are increasingly interesting (van Geest and Stoop 2003), offering higher average powers and extended spectral coverage that extends to the deep ultraviolet (e.g., 280 nm (McGuinness et al. 2004)). An LED excitation source has also been used to realize frequency-domain FLIM in a scanning microscope (Herman et al. 2001).

The repetition rate of the excitation source used will define the upper limit on the fluorescence lifetimes that can be determined. Mode-locked solid-state lasers, such as Ti:Sapphire lasers, typically have repetition rates in the range of 80–100 MHz, which limits the sampling window to ~10 ns and prohibits the measurement of lifetimes much longer than this value. Fiber-based ultrafast excitation sources can provide repetition rates down to tens of MHz, which provides a more convenient sampling window (~50 ns) that is sufficient for most of the commonly encountered fluorophores (e.g., the family of fluorescent proteins and many endogenous fluorophores like collagen and NADH). Using gain-switched semiconductor lasers or pulse-picked and/or amplified laser systems, pulse repetition rates may be reduced arbitrarily. While this can facilitate measurements of fluorescence decays over ~μs and ms timescales, the reduced pulse rate leads to lower average signal powers and therefore longer acquisition times. In principle, it is possible to increase the average signal level by increasing the excitation pulse energies, but this is limited by the damage thresholds of typical biological samples.

28.3 Applications of FLIM Microscopy

28.3.1 FLIM Applied to Cell Biology: Fluorescence Lifetime Sensing

As discussed in Section 28.1, fluorescence lifetime measurements can report on the local fluorophore environment and so FLIM can provide molecular functional information concerning fluorescently labeled proteins as well as their localization. Fluorescence lifetime has been applied to read out many perturbations to the local molecular environments, including changes in temperature (Kitamura et al. 2003), viscosity (Siegel et al. 2003, Benninger et al. 2005), refractive index (Strickler and Berg 1962, Suhling et al. 2002), solvent polarity (Parasassi et al. 1998) and ionic concentrations, notably calcium (Lakowicz et al. 1992b, Herman et al. 1997, Sanders et al. 1994, Agronskaia et al. 2004), oxygen (Gerritsen et al. 1997) and pH (Sanders et al. 1995, Lin et al. 2003). Unlike intensity-based readouts, fluorescence lifetime measurements of such sensors do not require knowledge of fluorophore concentration or baseline fluorescence measurements (e.g., in the absence of analyte). Measuring FRET, which provides a robust read-out of molecular interactions and conformational changes (e.g., (Bastiaens and Squire 1999, Jares-Erijman and Jovin 2003, Miyawaki et al. 1997), can also be considered as sensing a change in the local fluorophore environment. The robust nature of lifetime measurements and the increasingly convenient implementation of FLIM have led to numerous applications in cell biology and particularly to the study of cell signaling processes. Although there is increasing interest in developing new probes for lifetime readouts, FLIM has often been applied to probes originally developed for intensity or spectral readouts, for which it offers the potential for enhanced contrast.

An example is illustrated in Figure 28.5, which presents the application of di-4-ANEPPDHQ (Obaid et al. 2004), a membrane staining dye originally designed as a membrane voltage probe and shown to also facilitate the imaging of the distribution of membrane lipid microdomains (MLM) or "lipid rafts," via a spectral shift in fluorescence emission with changes in the order of the lipid bilayer (Jin et al. 2005). Lipid rafts are thought to be associated with several important cell processes including signaling. Figure 28.5a shows the emission spectra of di-4-ANEPPDHQ labeling two samples of giant unilamellar vesicles (GUV), one formed from saturated phospholipids and cholesterol to mimic ordered MLMs and the other formed from unsaturated phospholipids to mimic the disordered bulk lipid bilayer. Although this spectral shift can be exploited to map the MLM distribution in cell membranes, our experiments suggest that FLIM can provide superior contrast of membrane lipid order (Owen et al. 2006). Figure 28.5b shows the fluorescence lifetime histograms for the same GUV samples, illustrating the clear lifetime contrast and Figure 28.5d illustrates the application to cell imaging. For comparison, Figure 28.5c shows a dual-channel (500–530 nm and 570 nm longpass) intensity merged image while Figure 28.5d shows the corresponding fluorescence lifetime image of the same sample of live HEK cells in full growth medium, supplemented with 5 mM of di-4-ANEPPDHQ dye 1 h before imaging. Both images show the increased lipid order of the plasma membrane compared to the intracellular membranes. FLIM, however, indicates regions in the plasma membrane enriched in liquid-ordered phase, which seem to be clustered around sites of membrane protrusion. Previous studies have shown that the actin cytoskeleton interacts with the membrane through cholesterol-enriched microdomains (Bodin et al. 2005). Other dyes have also been reported to show spectral (e.g., Laurdan (Gaus et al. 2003)) or fluorescence lifetime (e.g., PMI-COOH (Margineanu et al. 2007)) contrast as a function of

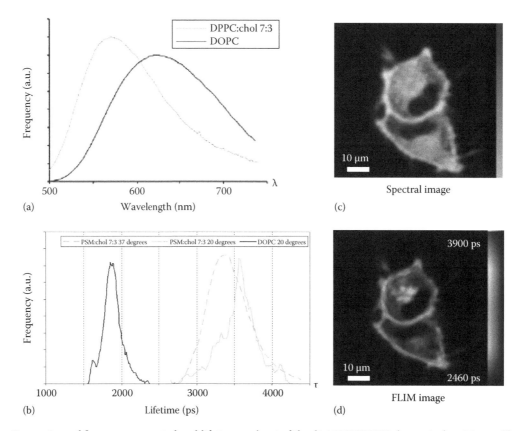

FIGURE 28.5 Comparison of fluorescence spectral and lifetime read-out of the di-4ANEPPDHQ dye excited at 473 nm. Change in emission spectra when staining giant unilamellar vesicles with and without cholesterol extraction by methyl-β-cyclodextrin (a) and the corresponding fluorescence lifetime histograms (b). The spectral ratiometric and fluorescence lifetime images, respectively, of live HEK cells labeled with di-4-ANEPPDHQ (c, d). (Adapted from figures 1 and 2, Owen, D.M. et al., *Biophys. J.*, 90, L80, 2006. With permission.)

membrane lipid order. Such membrane lipid order probes may be multiplexed with other fluorophores including genetically expressed fluorescence proteins, to permit studies correlating MLM accumulation with cell signaling pathways, which may be studied using FRET.

28.3.2 FLIM Applied to Cell Biology: FRET

The most widespread application of FLIM in cell biology is probably the mapping of FRET (Förster 1948) to report the interactions between suitably labeled specific proteins, lipids, enzymes, DNA, and RNA, as well as cleavage of a protein, or conformational changes within a protein (Clegg et al. 2003, Selvin 2000). A particularly vibrant area of FRET is the application to the study of cell signaling pathways, which is of crucial importance in understanding the means by which cells communicate with one another and respond to different stimuli. New insights into the spatiotemporal organization of protein–protein interactions can open up new possibilities for targeting disease at the molecular level.

FLIM FRET was initially implemented using small molecule dyes (e.g., Oida et al. 1993, Gadella and Jovin 1995, Bastiaens and Jovin 1996), and mainly limited to fixed cell imaging, but the advent and widespread deployment of probes based on genetically expressed fluorescent proteins (Tsien 1998, Lippincott-Schwartz and Patterson 2003, Giepmans et al. 2006) has made it possible to image a host of processes in live cells (e.g., Wouters et al. 2001, van Roessel and Brand 2002, Ng et al. 2001). A key example of the application of FLIM FRET to study cell signaling processes was the imaging of protein phosphorylation in live cells by FRET between GFP-tagged PKCa as the donor and a Cy3.5 labeled antibody as the acceptor that was specific to the phosphorylated protein (Ng et al. 1999). This technique was extended to be more generally applicable by imaging FRET between a phosphorylated protein, in this case, the ErB1 receptor tagged with GFP, and a Cy3-labeled antibody to phosphortyrosine (Wouters and Bastiaens 1999, Verveer et al. 2000b). Other cell signaling processes studied with FLIM FRET include dephosphorylation (Haj et al. 2002), caspase activity in individual cells during apoptosis (Harpur et al. 2001), NADH (Lakowicz et al. 1992c, Zhang et al. 2002) and the supramolecular organization of DNA (Murata et al. 2001).

At Imperial College London, we have adapted the approach of Wouters and Bastiaens (1999) to image receptor phosphorylation as a means to the study of signaling at the immunological synapse (IS). Specifically, we have applied FLIM to image FRET between the GFP-tagged KIR2DL1 (KIR) inhibitory receptor and a Cy3-tagged generic antiphosphotyrosine monoclonal antibody. This permitted us to visualize KIR phosphorylation in natural killer (NK) cells interacting with target cells expressing cognate major histocompatibility complex (MHC) class I proteins (Treanor et al. 2006). Once the live NK cell and target cells formed an IS, they were fixed and then treated with the Cy3-tagged antiphosphotyrosine antibody before imaging. This work revealed that inhibitory KIR signaling is spatially restricted to the IS, that it requires the presence of an Src family kinase and that KIR receptor phosphorylation occurs in microclusters on a spatial scale of the order of the spatial resolution of our confocal microscope, as shown in Figure 28.6. The fluorescence intensity and lifetime images were acquired in a confocal microscope with 470 nm excitation and TCSPC detection. The en face images of the IS were obtained by acquiring z-stacks of optical sectioned fluorescence lifetime images with 0.5 μm axial separation that were processed using 3-D rendering software to obtain the desired projections of the interface between the cells. Each optically sectioned FLIM image was acquired over 300 s and so each z-stack required many tens of minutes. This image acquisition rate was not a problem for imaging the fixed cells in this

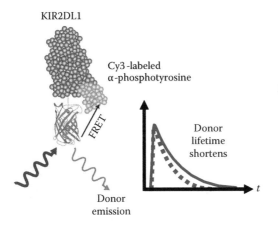

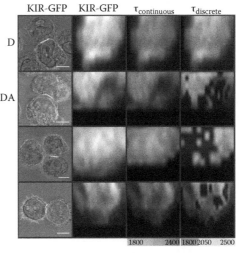

FIGURE 28.6 Schematic of FRET between EGFP-labeled KIR receptor and Cy3-labeled antiphosphotyosine with inset transmitted light images (where the IS are highlighted in white) and en face fluorescence intensity and lifetime images of FRET between NK cells and target cells at the IS with continuous and discrete lifetime scales. The scale bar is 8 μm. (Adapted from figure 5, Treanor, B. et al., *J. Cell Biol.*, 174, 153, 2006. With permission.)

experiment but would clearly preclude imaging dynamic processes in live cells.

In general, optically sectioned FLIM, as was required to obtain the FRET data represented in Figure 28.6, has been widely considered as a relatively slow imaging modality because it has mainly been implemented in laser scanning confocal or multiphoton microscopes, typically requiring FLIM acquisition times of minutes. Higher FLIM FRET imaging rates have been achieved with wide-field FLIM, for example, using the frequency-domain approach (Verveer et al. 2000b) but this did not provide optical sectioning. This issue can be addressed using optically sectioned Nipkow disc microscopes with wide-field FLIM detection implemented in the time domain (Grant et al. 2005) and frequency domain (van Munster et al. 2007). We have developed this approach combining a high-power supercontinuum excitation source with Nipkow disc microscopy and wide-field time-gated FLIM to acquire depth-resolved fluorescence lifetime images of live cells at rates of up to 10 time-gated images per second (Grant et al. 2007). Figure 28.7a shows a schematic of this high-speed optically sectioning Nipkow FLIM microscope, together with a FLIM image of live cells exhibiting donor (EGFP) only and FRET constructs (EGFP directly attached to mRFP with a short peptide linker) that was acquired at five frames/s. Also shown is a FLIM FRET image of live cells exhibiting FRET between EGFP-Raf-RBD and H-Ras-mRFP acquired in 5 s. This FRET experiment was designed to read out the activation of the Ras protein following stimulation of the EGF receptor signal pathway and illustrates the potential applications to studying cell signaling networks with high time resolution and/or with high throughout.

These advances permit us to capture fast time-lapse FLIM FRET sequences of live cell dynamics and to develop higher-throughput automated FLIM microscopes for high content analysis (HCA) and proteomics applications, for example, using FRET to screen protein interactions. Automated FLIM HCA has previously been demonstrated using frequency-domain FLIM in a wide-field (non-sectioning) microscope and we have recently demonstrated a Nipkow confocal FLIM and FRET multiwell plate reader that is able to automatically acquire optically sectioned FLIM images in less than 10 s per well, including the time for moving from well to well, autofocusing, and setting the FLIM acquisition parameters (Talbot et al. 2008).

28.3.3 FLIM Microscopy of Tissue Autofluorescence

In biological tissue, autofluorescence can provide a source of label-free optical molecular contrast and has the potential to discriminate between healthy and diseased tissue. The prospect of detecting molecular changes associated with the early manifestations of diseases such as cancer is particularly exciting. For example, accurate and early detection of cancer allows earlier treatment and significantly improves prognosis (Cancer facts and figures 2006). While exophytic tumors can often be directly identified by eye, the cellular and tissue perturbations of the peripheral components of neoplasia may not be apparent by direct inspection under visible light and are often beyond the discrimination of conventional noninvasive diagnostic imaging techniques.

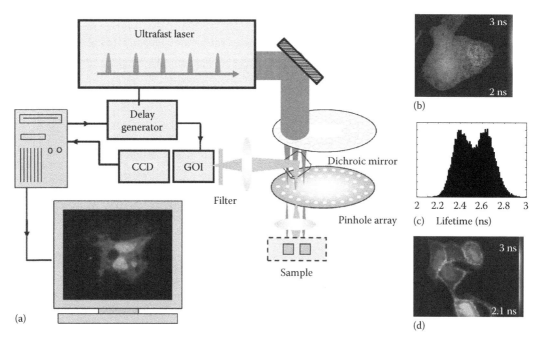

FIGURE 28.7 (a) schematic of a Nipkow disc FLIM microscope with (b) optically sectioned FLIM image of live cells expressing EGFP (left) and EGFP-mRFP FRET construct with (c) corresponding lifetime histogram and (d) FLIM FRET image recorded with 5 s acquisition of MDCK cells expressing EGFP-Raf-RBD and H-Ras-mRFP, stimulated with EGF for 10 min. (Adapted from figures 1 and 8, Grant, D.M. et al., *Opt. Express*, 15, 15656, 2007. With permission.)

In general, label-free imaging techniques are preferable for clinical imaging, particularly for diagnosis, as they avoid the need for administration of an exogenous agent, with associated considerations of toxicity and pharmacokinetics. A number of label-free modalities based on the interaction of light with tissue have been proposed to improve the detection of malignant change. These include fluorescence, elastic scattering, Raman, infrared absorption and diffuse reflectance spectroscopy's (Wagnieres et al. 1998, Sokolov et al. 2002, Bigio and Mourant 1997, Richards-Kortum and Sevick-Muraca 1996). To date, most of these techniques have been limited to point measurements, whereby only a very small area of tissue is interrogated at a time, for example, via a fiber-optic contact probe. This enables the acquisition of biochemical information, but provides no spatial or morphological information about the tissue. Autofluorescence can provide label-free "molecular" contrast that can be readily utilized as an imaging technique, allowing the rapid and relatively noninvasive collection of spatially resolved information from areas of tissue up to tens of centimeters in diameter.

The principal endogenous tissue fluorophores include collagen and elastin cross-links, reduced nicotinamide adenine dinucleotide (NADH), oxidized flavins (FAD and FMN), lipofuscin, keratin, and porphyrins. As shown in Figure 28.8 (Wagnieres et al. 1998), these fluorophores have excitation maxima in the UV-A or blue (325–450 nm) spectral regions and emit Stokes-shifted fluorescence in the near-UV to the visible (390–700 nm) region of the spectrum. The actual autofluorescence signal observed in biological tissue will depend on the concentration and the distribution of the fluorophores present, on the presence of chromophores (particularly hemoglobin) that absorb excitation and fluorescence light, and on the degree of light scattering that occurs within the tissue (Richards-Kortum and Sevick-Muraca 1996). Autofluorescence therefore reflects the biochemical and structural composition of the tissue and consequently is altered when tissue composition is changed by disease states such as atherosclerosis, cancer, and osteoarthritis.

While autofluorescence can be observed using conventional "steady-state" imaging techniques, it is challenging to make sufficiently quantitative measurements for diagnostic applications since the autofluorescence intensity signal may be affected by fluorophore concentration, variations in temporal and spatial properties of the excitation flux, the angle of the excitation light, the detection efficiency, attenuation by light absorption and scattering within the tissue, and spatial variations in the tissue microenvironment that alter local quenching of fluorescence. Two or more spectral emission windows can be used to improve quantitation through ratiometric imaging but the heterogeneity in the distribution of tissue fluorophores and their broad, heavily overlapping, emission spectra can limit the discrimination achievable (e.g., Butte et al. 2005). More sophisticated spectral unmixing approaches utilizing principal component analysis (PCA) and related techniques applied to large sample number training sets are being developed (e.g., de Veld et al. 2003) to identify signatures for diagnostic applications; but this is a highly challenging spectral unmixing problem. These considerations have so far restricted the widespread use of fluorescence imaging for the detection of malignancy, and current fluorescence imaging tools in clinical use are hampered by low specificity and a high rate of false positive findings (Bard et al. 2005, Beamis et al. 2004, Ohkawa et al. 2004).

Increasingly, there is interest in exploiting fluorescence lifetime contrast to analyze tissue autofluorescence signals since fluorescence decay profiles depend on relative (rather than absolute) intensity values, and FLIM is therefore largely unaffected by many factors that limit steady-state measurements (Chen et al. 2005). The principal tissue fluorophores exhibit

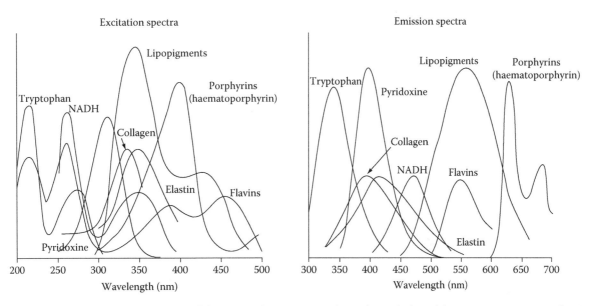

FIGURE 28.8 Excitation and emission spectra of the main endogenous tissue fluorophores. (Adapted from Wagnieres, G.A. et al., *Photochem. Photobiol.*, 68, 603, 1998. With permission.)

characteristic lifetimes (ranging from hundreds to thousands of picoseconds (Richards-Kortum and Sevick-Muraca 1996, Elson et al. 2004b) that enable spectrally overlapping fluorophores to be distinguished, and the sensitivity of fluorescence lifetime to changes in the local tissue microenvironment (e.g., pH, $[O_2]$, $[Ca^{2+}]$) (Lakowicz 2006) can provide a readout of biochemical changes indicating the onset or progression of disease.

Changes in kinetics of time-resolved fluorescence between normal versus malignant tissue were reported as early as 1986 (Tata et al. 1986) but the complexity and performance of instrumentation associated with picosecond-resolved measurements, and particularly imaging, hindered the development of FLIM as a clinical tool. Nevertheless, single-point measurements of autofluorescence lifetime have revealed differences between normal and neoplastic human tissue in tumors of the esophagus (Glanzmann et al. 1999, Dacosta et al. 2002, Pfeferr et al. 2003), colon (Mycek et al. 1998), brain (Marcu et al. 2004, Butte et al. 2005), oral cavity (Chen et al. 2005), lung (Glanzmann et al. 1999), breast (Pradhan et al. 1992, Jain et al. 1998), skin (De Beule et al. 2007), and bladder (Glanzmann et al. 1999). Of these results, (Glanzmann et al. 1999, Pfefer et al. 2003, Mycek et al. 1998, Chen et al. 2005) were obtained in vivo.

There have also been a number of FLIM studies of autofluorescence of cancerous tissues. To date, most of these FLIM studies have used laser scanning microscopes and particularly multiphoton microscopes, for example, to study cancer. These studies include ex vivo studies of brain (Kantelhardt et al. 2007), skin (Cicchi et al. 2007, Galletly et al. 2008), cervix (Elson et al. 2006), breast (Provenzano et al. 2008), and in vivo studies in animals (Skala et al. 2007) and human skin (Konig 2008). In general, multiphoton excitation provides a convenient approach to image tissue autofluorescence and, to date, the only FLIM microscope

licensed for in vivo application is a multiphoton microscope (JenLab GmbH, DermaInspect).

For many clinical applications such as diagnostic screening for disease, guided biopsy, or interoperative surgery, however, it is desirable to image a larger field of view than is usually possible with multiphoton microscopy. For such applications, wide-field FLIM can provide ~cm fields of view at high (hertz) frame rates and be applied to microscopes, endoscopes, or macroscopes. Ironically, since the first application of FLIM to autofluorescence for imaging cancer tissue (Mizeret et al. 1997), which was implemented using a wide-field frequency-domain FLIM system applied to ex vivo bladder and oral mucosa, there has been relatively little progress toward clinical wide-field FLIM instrumentation. At Imperial College London, we have worked to apply wide-field time-gated FLIM to tissue autofluorescence in microscopes (Tadrous et al. 2003), macroscopes (Elson et al. 2003, 2006), and endoscopes (Requejo-Isidro et al. 2004, Munro et al. 2005, Elson et al. 2006). Examples of potential of wide-field FLIM for label-free tissue imaging is illustrated in the macroscope images shown in Figure 28.9, which were obtained for excitation at 355 nm. Figure 28.9a shows a freshly resected colon displaying an advanced cancer for which the autofluorescence lifetime images and histograms clearly distinguish between normal mucosal tissue and the tumor. Figure 28.9d shows data from a section of liver displaying a metastatic colon tumor. Again FLIM can distinguish between the normal native liver tissue, vascular structures, the metastasis, and an area of tissue that was subject to radio-frequency ablation (RFA) during surgery. It is interesting to note that Figure 28.9e shows how the cancerous tissue can be contrasted by both longer and shorter autofluorescence lifetimes than the surrounding tissue. The origin for this lifetime contrast is not yet clear but contributions arising from

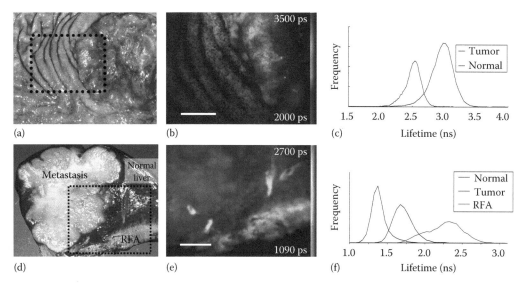

FIGURE 28.9 Macroscopic FLIM of freshly resected tissue. (a) Photograph and (b) FLIM image (indicated area) of colonic tumor with fluorescence lifetime histogram showing increased lifetime of tumor region. (d) Photograph and (e) FLIM image of a resected liver containing a metastatic colon cancer and a region of RFA damage. These regions can be clearly identified in the lifetime image as indicated by the lifetime histogram (f). Scale bar is 1 cm.

excitation at 355 nm can be expected from metabolites such as NADH and tissue matrix components such as collagen.

28.4 FLIM Endoscopy

While FLIM has demonstrated some diagnostic potential, it has not yet been translated into the clinic, although there have been some pioneering clinical applications, for example, in the eye (Schweitzer et al. 2004) and in skin (Konig 2008). These examples of in vivo FLIM were implemented using an ophthalmoscope and multiphoton microscope, respectively. Such instrumentation can facilitate FLIM of exterior surface tissues but cannot be used to image internal organs.

There have been a few reports of wide-field FLIM endoscopy using both rigid (Mizeret et al. 1997, 1999, Requejo-Isidro et al. 2004) and flexible (Munro et al. 2005, Elson et al. 2007, Sun et al. 2009) endoscopes. Figure 28.10a shows a coherent fiber-bundle-based flexible endoscope (10,000 fibers, Endoscan Ltd) coupled to the GOI photocathode of a wide-field time-gated FLIM system utilizing an ultrafast UV laser (Vanguard 350-HM355, Spectra-Physics) providing ~10 ps pulses at 355 nm. This was used to image a resected colon sample displaying an area of dysplasia (precancerous state), as shown in Figure 28.10b. Macroscopic white light and fluorescence lifetime images (b, c) are presented together with lifetime images acquired through the flexible endoscope. The figure indicates that the fluorescence lifetime contrast is maintained between the FLIM images

recorded using the macroscope (c, d) and the flexible endoscope (e, f), which entailed the image being relayed through ~1 m of glass fiber bundle. For compatibility with modern clinical video endoscopes, we developed a thin FLIM endoscope designed to pass through their biopsy port, which utilized a 1 mm diameter coherent fiber bundle with 30,000 fibers (IGN-08/30, Sumitomo Electric). Recently, time-gated wide-field FLIM endoscopes compatible with intravascular imaging (Elson et al. 2007) and in vivo animal studies (Sun et al. 2009) have been developed.

While wide-field FLIM endoscopy is suitable to image relatively large areas of tissue, for example, for diagnostic screening applications and optically guided biopsy, it is not able to provide depth-resolved imaging, which can be important, for example, to study the subsurface properties of lesions or to provide subcellular resolution, which is important for "optical biopsy" to correlate with histopathology. Optically sectioned images with subcellular resolution are routinely provided by laser scanning confocal and multiphoton microscopes and there has been considerable effort to translate these modalities to endoscopy, in order to assist in the early identification of cancerous and precancerous tissue. In general, there are two strategies to implement laser scanning confocal or multiphoton endomicroscopy. The first is to implement beam scanning at the proximal end of a coherent fiber-optic bundle (e.g., Gmitro and Aziz 1993), and the second is to employ miniature scanning devices at the distal end of an endoscope utilizing a single optical fiber (e.g., Dickensheets and Kino 1996). These concepts have been demonstrated using

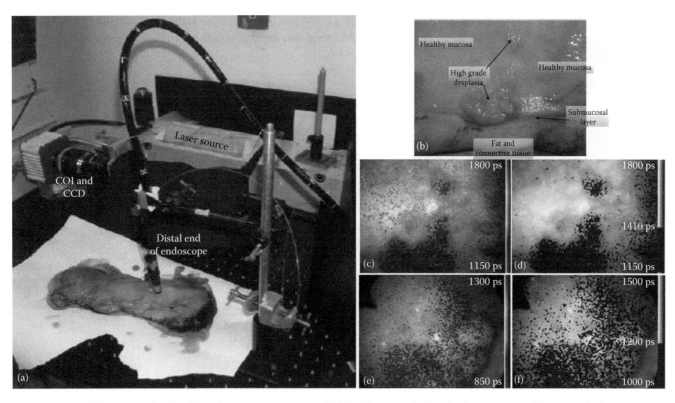

FIGURE 28.10 (a) Photograph of flexible endoscope FLIM setup and (b) freshly resected colon displaying an area of dysplasia. (c, d) Macroscopic and (e, f) endoscopic FLIM images of colon sample displayed on both continuous and discrete color scales.

a variety of imaging methods including reflected light (e.g., Dickensheets and Kino 1996, Liang et al. 2001, Tearney et al. 2002), confocal fluorescence (single-photon excitation) (e.g., Gmitro and Aziz 1993, Sabharwal et al. 1999, Laemmel et al. 2004), and multiphoton fluorescence (e.g., Helmchen et al. 2001, Lelek et al. 2007, Myaing et al. 2006). To access more spectral information, multispectral confocal endomicroscopy has been demonstrated (e.g., Jean et al. 2007, Rouse and Gmitro 2000), and we have recently reported confocal FLIM endomicroscopy (Kennedy et al. 2009), using the setup shown in Figure 28.11. We note that a single fiber-optic scanning FLIM microscope has been previously demonstrated (Ghiggino et al. 1992), although this did not permit endoscopic imaging.

Our confocal FLIM endomicroscope is based on a commercially available laser scanning single-photon confocal fluorescence endomicroscope (Cellvizio® GI, Mauna Kea Technologies) (Laemmel et al. 2004), which is licensed for clinical use in the GI tract and for bronchoscopy.

This instrument is capable of recording optically sectioned images with a variety of fiber-optic probes and normally uses an excitation wavelength of 488 nm. For this demonstration of confocal FLIM endomicroscopy, we used a coherent bundle consisting of 30,000 single mode fibers terminated with a miniature objective (Cellvizio® Mini O). The probe has a lateral resolution of ~1.4 μm, a working distance of 60 μm and a field of view of 240 μm. The diameter of the distal tip is 2.6 mm, which is compatible with any endoscope having a working channel of 2.8 mm diameter or greater. The excitation source was a fiber-delivered, mode-locked, frequency-doubled Ti:Sapphire laser (Mai-Tai, Newport Spectra Physics), providing femtosecond pulses with a center wavelength of 488 nm, which were

coupled into the internal beam path of the endoscope immediately before the scanning assembly. FLIM data was acquired using a commercial TCSPC system (SPC830, Becker and Hickl, GmbH). This configuration provided a fluorescence lifetime image size of 350 × 512 pixels, which was limited by the TCSPC card memory.

The confocal FLIM endomicroscope was applied to image pollen grains, unstained biological tissue, and fixed COS-7 cells expressing both EGFP and a tandem FRET construct comprising EGFP directly linked to mRFP (see Figure 28.12). We believe these results illustrate the potential of FLIM endomicroscopy for in vivo label-free clinical imaging exploiting autofluorescence, and small animal imaging for molecular cell biology and drug discovery.

28.5 FLIM Tomography

For drug discovery and disease-related research, there is an increasing appreciation that in vitro cell monolayers may exhibit non-physiological behavior that can impact the utility of cell-based assays and this has stimulated the development of techniques to image biological processes in 3-D cell cultures (e.g., Abbot 2003) in animal/embryo/engineered tissue samples (e.g., Weissleder and Ntziachristos 2003, Oldham et al. 2007, Keller et al. 2006), and in live organisms. While FLIM and FLIM FRET endomicroscopy provides a promising route to studying protein interactions in vivo, it is inevitably somewhat invasive, requires relatively large subjects, and is sometimes inconvenient to image time courses in live subjects. As discussed elsewhere in this volume, there are a number of approaches to acquire quantitative 3-D image data in subjects ranging from

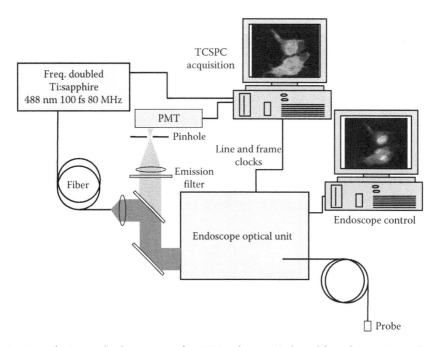

FIGURE 28.11 Schematic setup of microconfocal microscope for FLIM endoscopy. (Adapted from figure 1, Kennedy, G.T. et al., *J. Biophotonics*, 2, 103, 2009. With permission.)

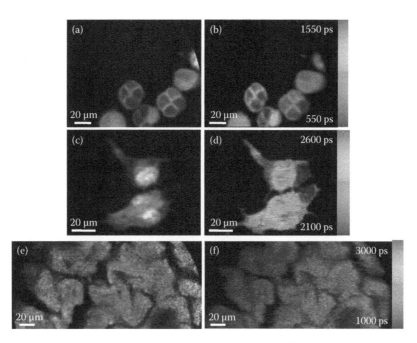

FIGURE 28.12 (a,c,e) Intensity and (b,d,f) FLIM microconfocal endoscopic images of pollen grains illustrating optical sectioning, live COS cells labeled with EGFP and EGFP-mRFP FRET construct and unstained rat tissue. (Adapted from figures 3 and 4, Kennedy, G.T. et al., *J. Biophotonics*, 2, 103, 2009. With permission.)

embryos to rodents, for which the main challenge is the impact of the strong scattering of optical radiation in biological tissue. For relatively transparent subjects, techniques such as optical projection tomography (OPT) (Sharpe et al. 2002), selective plane illumination microscopy (SPIM) (Huisken et al. 2004), and ultramicroscopy (Dodt et al. 2007) may be applied. These are essentially ballistic light imaging techniques for which it is often necessary to apply chemical clearing techniques to reduce the scattering of the optical radiation used. For live animal imaging, chemical clearing is not an option and so sophisticated inverse scattering techniques (e.g., Gibson et al. 2005) must be employed to reconstruct the 3-D image data from the diffuse light detected. Diffuse optical tomography (DOT) techniques are generally not able to achieve image resolution close to the diffraction limit and high resolution 3-D tomography is currently confined to the ballistic light imaging techniques. To date, ballistic and diffuse tomography techniques have been largely restricted to intensity-based imaging, providing mainly structural rather than functional information, and quantitative imaging can be degraded by intensity artifacts—particularly in scattering samples. As discussed above, ratiometric techniques can provide superior quantitative image data and so it is interesting to develop FLIM tomography systems.

We have recently extended OPT to FLIM OPT (McGinty et al. 2008), producing 3-D fluorescence lifetime images of samples in the range ~0.1 to 1 cm. OPT is the optical equivalent of x-ray computed tomography (x-ray CT), for which a stack of *X–Z* slices are reconstructed from a series of transverse (X-Y) projections acquired at a number of projection angles. Due to its similarity to x-ray CT, many of the image reconstruction techniques (e.g., Kak and Slaney 1988) can be directly applied providing

that the sample is treated (chemically cleared) to eliminate the optical scattering. The classic filtered back-projection technique assumes parallel ray paths and so optical beam propagation considerations limit its applicability samples that are smaller than the depth of focus of the imaging system (Sharpe et al. 2002, Sharpe 2004). Figure 28.13 shows a schematic of the experimental setup for FLIM OPT system. The sample is mounted beneath a rotation stage and suspended in a reservoir of index matching fluid. A fiber laser-pumped super-continuum-based light is used to excite the fluorescence, which is imaged onto the photocathode of a GOI and the resulting time-gated and intensified fluorescence signals are recorded by a CCD camera. The sample is rotated and typically a series of time-gated images are recorded at intervals of 1° for a full 360° rotation. For each time gate delay, the 3-D time-gated fluorescence intensity distribution is determined, and then the fluorescence lifetime distribution is calculated for each value of y from the intensity decay data at every pixel of the *X–Z–t* image data sets.

To reconstruct each *X–Z* plane of the sample, the series of projections acquired at each orientation was analyzed to determine the function, $P_\theta(r)$, for each y value, where *P* is the integrated signal along a projection, *r* is the radial coordinate in the *X–Z* plane, and θ is the angle at which each projection is acquired. A 1-D Fourier transform is then performed on each projection to give $S_\theta(w)$, where *w* is the spatial frequency coordinate. $S_\theta(w)$ is then multiplied by a filter function to account for unequal sampling of the spatial frequency components and the inverse Fourier transform is applied. The resulting filtered projections are then back-projected at their corresponding angles to reconstruct the *X–Z* fluorescence intensity distribution of the sample in that plane. The reconstructed image resolution may be improved

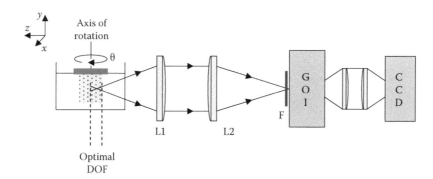

FIGURE 28.13 Schematic of FLIM OPT setup. (Adapted from figure 1, McGinty, J. et al., *J. Biophotonics*, 1, 390, 2008. With permission.)

further by including the point spread function in the filtering operation (Walls et al. 2007).

Figure 28.14a shows results from a FLIM OPT acquisition of a silicon elastomer phantom (Rhodorsil RTV 141, Bentley Chemicals Ltd) that was homogeneously stained with the dye Sulforhodamine 101 (1×10^{-6} mol dm^{-3}) and also contained a suspension of 30 μm diameter green fluorescent beads (35–6, Duke Scientific Corp). The resultant optically transparent phantom of 9 mm diameter and 25 mm length was suspended in liquid silicon as the index matching fluid. The fluorescent beads were excited at 480 ± 15 nm and 10 time-gated fluorescence OPT data sets were acquired at 0.5 ns intervals of time-gate delay after excitation (with 1 ns gate width). From this, the 3-D fluorescence lifetime distribution of the beads was determined. To reconstruct the surface of the phantom, the Sulforhodamine 101 was excited at 560 ± 20 nm and a single time-gated intensity image (immediately after excitation) was acquired. For both measurements, images were acquired at intervals of 1° for a full 360°

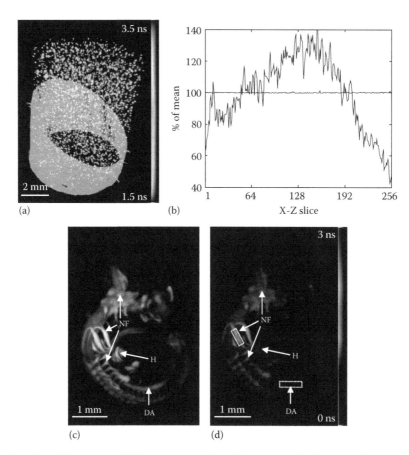

FIGURE 28.14 (a) Three-dimensional fluorescence lifetime reconstruction of fluorescence beads mounted in a transparent cylindrical phantom and (b) line profiles of mean fluorescence intensity (solid) and lifetime (dashed). (c) 3-D fluorescence intensity and (d) lifetime reconstructions of mouse embryo labeled with an Alexa-488 conjugated antibody. NF, neurofilament; H, heart; and DA, dorsal aorta. (Adapted from figures 3 and 4, McGinty, J. et al., *J. Biophotonics*, 1, 390, 2008. With permission.)

rotation. The intensity decay profiles in every voxel were subsequently fitted to a single exponential decay model and Figure 28.14a shows a 3-D rendered fluorescence lifetime reconstruction of the bead distribution along with a cut away of the phantom surface (Sulforhodamine 101 channel) in yellow. The surface was extracted from a 3-D rendering (performed using Volocity Visualisation, Improvision Inc). Figure 28.14b shows a line profile of the reconstructed intensity and lifetime distribution of the fluorescence beads through an *X–Z* plane. It should be noted that in the initial raw intensity images, there was a significant variation in the intensity profile (solid line), which was predominantly due to uneven illumination intensity although similar intensity artifacts could be caused by variations in fluorescence staining/expression levels and in collection/detection efficiency. The corresponding fluorescence lifetime profile (dashed line) does not vary significantly, as expected for an inherently ratiometric measurement. The mean reconstructed fluorescence lifetime of the beads, averaged over the whole phantom, was $2002 \pm 25\,ps$, which compares well to the value of $2060 \pm 100\,ps$ that was measured on a TSCPC-enabled confocal microscope.

Figure 28.14 shows 3-D time-integrated fluorescence intensity (c) and fluorescence lifetime (d) reconstructions of a mouse embryo imaged with FLIM OPT (McGinty et al. 2008). Prior to imaging, the embryo was fixed in formaldehyde and the neurofilament was labeled with an Alexa-488 conjugated antibody. The sample was then mounted in agarose, dehydrated in methanol, and finally chemically cleared in BABB (1:2 mixture of Benzyl Alcohol:Benzyl Benzoate). Fluorescence was excited at $485 \pm 10\,nm$ and the image acquisition involved time-gated measurements at five delay intervals spaced by 0.5 ns and 360 rotation steps of 1°. Reconstruction of the data and subsequent fitting to a single exponential decay yielded an average fluorescence lifetime of the Alexa-488 labeled neurofilament of $1360 \pm 180\,ps$ in this environment. Autofluorescence was also excited in, for example, the heart and dorsal aorta, which exhibited a shorter average lifetime of $1030 \pm 135\,ps$. These fluorescence lifetimes were calculated from the areas indicated in Figure 28.14d.

This experiment demonstrates the potential of FLIM OPT to image 3-D fluorescence lifetime distributions in anatomically relevant samples. It can distinguish between extrinsic and intrinsic fluorescence signals without prior anatomical knowledge of the labeling sites, which is not possible with steady-state fluorescence intensity imaging, and it could be applied to map protein–protein interactions using FLIM FRET. For drug discovery and biomedical research, however, it would be much more useful to be able to reconstruct 3-D fluorescence lifetime distributions in live embryos and other disease models including small animals. To realize this vision, it is necessary to address the issue of optical scattering in the diffuse imaging regime.

Serendipitously, the same time (phase)-gated imaging technology used for FLIM can be applied to DOT and diffuse fluorescence tomography (DFT), as discussed elsewhere in this volume, to aid tomographic reconstruction of fluorescence intensity images by inverse scattering techniques. Thus, it seems reasonable to apply the same technology for both DFT and FLIM

in order to reconstruct fluorescence lifetime distributions in turbid media (e.g., Paithankar et al. 1997). Tomographic FLIM has been implemented in experiments based on scattering phantoms using frequency-domain (e.g., Shives et al. 2002, Godavarty et al. 2005) and time-domain (e.g., Kumar et al. 2005, 2008) approaches. At Imperial College London, we have shown that we can reconstruct the fluorescence lifetime and other optical properties of strongly scattering samples with inclusions containing genetically expressed fluorophores exhibiting FRET (McGinty et al. 2009b). This was done using almost the same experimental setup, as shown in Figure 28.13 and employing reconstruction algorithm based on the minimization of a cost functional between the measured data and a model of light diffusion in the Fourier domain (Soloviev et al. 2009). We note that time-gated imaging has been applied to fluorescence molecular tomography (FMT) of live mice (Niedre et al. 2008) and that several groups are also making progress toward tomographic FLIM and FRET of live mice (e.g., Kumar et al. 2009, Nothdurft et al. 2009). It seems likely that in vivo tomographic FLIM and FRET of live disease models, including embryos and mice, will soon be practical for biomedical research and drug discovery.

28.6 Outlook

The range of applications of FLIM and MDFI are increasing rapidly and there is particular interest in the potential for FRET and other approaches to studying protein interactions. As the instrumentation becomes more robust and accessible, it is likely to be taken up in emerging areas such as in vivo imaging and high content analysis. The functionality of FLIM and MDFI instrumentation will continue to develop and the technological implementations will improve in terms of higher performance, lower cost, and more compact and ergonomic instrumentation. For high-throughput applications, automation will become increasingly important while for clinical applications, cost is likely to be a key issue that may be addressed through cheaper excitation sources, for example, based on LED or fiber laser technology, and cheaper, more sensitive detectors, for example, exploiting ultrafast CMOS technology.

In terms of functionality, FLIM will be increasingly combined with spectrally resolved and polarization-resolved imaging to provide multidimensional fluorescence images. Spectrally resolved TCSPC FLIM is commercially available for laser scanning microscopes, including a 16-channel "hyperspectral" FLIM system that incorporates a spectrometer and multi-anode photomultiplier with an electronic router and TCSPC electronics (Bird et al. 2004, Becker 2005). Faster imaging can be realized with parallel pixel acquisition using the so-called "push-broom" approach to hyperspectral imaging (Schultz et al. 2001). For hyperspectral FLIM, we have combined this with wide-field time-gated detection implemented in a line-scanning microscope configuration that also provides "semiconfocal" optical sectioning (De Beule et al. 2006). This MDFI approach can be further extended by incorporating a tunable excitation source, such as a supercontinuum source, to provide excitation-emission-lifetime-resolved fluorescence imaging (Owen et al. 2007). Combining spectral and fluorescence lifetime measurements can give enhanced information about FRET (e.g., Becker et al. 2006), and can enable unmixing of contributions from multiple fluorophores, which may be used to multiplex increased numbers of fluorescence labels or to disentangle complex autofluorescence signals from biological tissue. Of particular interest to biomedicine and drug discovery is the ability to simultaneously monitor multiple protein interactions by imaging multiplexed FRET constructs. We have demonstrated how FLIM may be combined with spectrally resolved imaging to simultaneous readout a FRET-based calcium sensor and the binding of ras to raf in an EGF receptor signal pathway via intermolecular FRET (Grant et al. 2008). As MDFI is implemented in automated microscopes and multiwell plate readers and applied to tissue-based assays and time course experiments, the volume of acquired data becomes huge. Today's MDFI experiments are already almost beyond the scope of ad hoc analysis by human investigators and sophisticated bioinformatics tools are now required to automatically analyze and present such data to identify trends and fluorescence "signatures."

Polarization-resolved FLIM can provide information about rotational mobility, clustering and homoFRET. While steady-state imaging of polarization-resolved fluorescence (i.e., fluorescence anisotropy) is a well-established technique (Lakowicz 2006), time-resolved imaging of fluorescence polarization anisotropy is not widely available although it has been realized for wide-field (Clayton et al. 2002, Siegel et al. 2003) and confocal scanning microscopy (Buehler et al. 2000, Bader et al. 2007). Such instruments can help address the ambiguities associated with steady-state fluorescence anisotropy measurements, for example, to provide information concerning protein clustering (Bader et al. 2007) and homoFRET (Gautier et al. 2001, Lidke et al. 2003), and can image rotational correlation time to map variations in viscosity (Siegel et al. 2003, Suhling et al. 2004, Benninger et al. 2005) and ligand binding (Benninger et al. 2007).

A further important trend in fluorescence microscopy is to image with resolution beyond the diffraction limit. Although FLIM has been applied to total internal reflection fluorescence (TIRF) microscopy (e.g., Schneckenburger et al. 2003) and to single-molecule imaging studies (e.g., Widengren et al. 2006), super-resolved microscopy is currently concerned mainly with intensity imaging although stimulated emission depletion microscopy (STED) (Hell and Wichmann 1994) has been adapted to FLIM in our laboratory (Auksorius et al. 2008). As yet, there has been no report of FLIM implemented with stochastically switched single-molecule localization techniques such as photoactivated localization microscopy (PALM) (Betzig et al. 2006) or STORM (Rust et al. 2006).

Perhaps the most exciting and challenging prospects for fluorescence imaging and FLIM are associated with the translation to in vivo imaging, where the ability to study protein function and interactions may provide tremendously valuable insights into the mechanisms underlying disease. Besides developing FLIM

endoscopes and tomographic imaging systems, we are working toward in vivo clinical FLIM using a multiphoton microscope that is approved for clinical use (JenLab GmbH, DermaInspect). In only a few years, we are likely to see the first live animal FLIM tomography systems and clinically deployable FLIM endoscopes. Together with vital advances in fluorescence probes and computational resource, these may permit molecular biology processes such as cell signal pathway activation to be directly imaged in vivo and applied to drug screening. We believe that FLIM and MDFI techniques will also find increasing uptake in fields outside biomedicine, for example, in plasmonics and nanophotonics.

Acknowledgments

The authors gratefully acknowledge funding from the Biotechnology and Biological Sciences Research Council (BBSRC), the UK Department of Trade and Industry (Technology Strategy Board Beacon award), the Engineering and Physical Sciences Research Council (EPSRC), an EU Framework VI Integrated Project (# LSHG-CT-2003-503259), a UK Joint Infrastructure Fund Award from the Higher Education Funding Council for England (HEFCE JIF), and a Wellcome Trust Showcase Award. Paul French and Dan Davis also acknowledge Royal Society Wolfson Research Merit Awards.

References

Abbott, A. 2003. Cell culture: Biology's new dimension. *Nature* 424: 870–872.

Agronskaia, A. V., Tertoolen, L., and Gerritsen, H. C. 2003. High frame rate fluorescence lifetime imaging. *J. Phys. D: Appl. Phys.* 36: 1655–1662.

Agronskaia, A. V., Tertoolen, L., and Gerritsen, H. C. 2004. Fast fluorescence lifetime imaging of calcium in living cells. *J. Biomed. Opt.* 9: 1230–1237.

Andersson-Engels, S., Johansson, J., Stenram, U., Svanberg, K., and Svanberg, S. 1990. Malignant-tumor and atherosclerotic plaque diagnosis using laser-induced fluorescence. *IEEE J. Quantum Electron.* 26: 2207–2217.

Auksorius, E., Boruah, B. R., Dunsby, C. et al. 2008. Stimulated emission depletion microscopy with a supercontinuum source and fluorescence lifetime imaging. *Opt. Lett.* 33: 113–115.

Bader, A. N., Hofman, E. G., Henegouwen, P., and Gerritsen, H. C. 2007. Imaging of protein cluster sizes by means of confocal time-gated fluorescence anisotropy microscopy. *Opt. Express* 15: 6934–6945.

Ballew, R. M. and Demas, J. N. 1991. Error analysis of the rapid lifetime determination method for single exponential decays with a nonzero base-line. *Anal. Chim. Acta* 245: 121–127.

Bard, M. P. L., Amelink, A., Skurichina, M. et al. 2005. Improving the specificity of fluorescence bronchoscopy for the analysis of neoptastic lesions of the bronchial tree by combination with optical spectroscopy: Preliminary communication. *Lung Cancer* 47: 41–47.

Bastiaens, P. I. H. and Jovin, T. M. 1996. Microspectroscopic imaging tracks the intracellular processing of a signal transduction protein: Fluorescent-labeled protein kinase C beta I. *Proc. Natl. Acad. Sci. U. S. A.* 93: 8407–8412.

Bastiaens, P. I. H. and Squire, A. 1999. Fluorescence lifetime imaging microscopy: Spatial resolution of biochemical processes in the cell. *Trends Cell Biol.* 9: 48–52.

Beamis, J. F., Ernst, A., Simoff, M., Yung, R., and Mathur, P. 2004. A multicenter study comparing autofluorescence bronchoscopy to white light bronchoscopy using a non-laser light stimulation system. In: *46th Annual Thomas L Petty Aspen Lung Conference*, Aspen, CO.

Becker, W. 2005. *Advanced Time-Correlated Single Photon Counting Techniques.* Berlin, Germany: Springer.

Becker, W., Bergmann, A., Haustein, E. et al. 2006. Fluorescence lifetime images and correlation spectra obtained by multidimensional time-correlated single photon counting. *Microsc. Res. Techn.* 69: 186–195.

Becker, W., Bergmann, A., Hink, M. A. et al. 2004. Fluorescence lifetime imaging by time-correlated single-photon counting. *Microsc. Res. Techn.* 63: 58–66.

Benninger, R. K. P., Hofmann, O., McGinty, J. et al. 2005. Time-resolved fluorescence imaging of solvent interactions in microfluidic devices. *Opt. Express* 13: 6275–6285.

Benninger, R. K. P., Hofmann, O., Onfelt, B. et al. 2007. Fluorescence-lifetime imaging of DNA-dye interactions within continuous-flow microfluidic systems (vol 46, pg 8525, 2007). *Angew. Chem. Int. Ed.* 46: 8536–8536.

Benninger, R. K. P., Koc, Y., Hofmann, O. et al. 2006. Quantitative 3D mapping of fluidic temperatures within microchannel networks using fluorescence lifetime imaging. *Anal. Chem.* 78: 2272–2278.

Betzig, E., Patterson, G. H., Sougrat, R. et al. 2006. Imaging intracellular fluorescent proteins at nanometer resolution. *Science* 313: 1642–1645.

Bewersdorf, J., Pick, R., and Hell, S. W. 1998. Multifocal multiphoton microscopy. *Opt. Lett.* 23: 655–657.

Bigio, I. J. and Mourant, J. R. 1997. Ultraviolet and visible spectroscopies for tissue diagnostics: Fluorescence spectroscopy and elastic-scattering spectroscopy. *Phys. Med. Biol.* 42: 803–814.

Bird, D. K., Eliceiri, K. W., Fan, C. H., and White, J. G. 2004. Simultaneous two-photon spectral and lifetime fluorescence microscopy. *Appl. Opt.* 43: 5173–5182.

Birk, H. and Storz, R. 2001. Illuminating device and microscope. US patent 6611643.

Bodin, S., Soulet, C., Tronchere, H. et al. 2005. Integrin-dependent interaction of lipid rafts with the actin cytoskeleton in activated human platelets. *J. Cell Sci.* 118: 759–769.

Boens, N., Qin, W. W., Basaric, N. et al. 2007. Fluorescence lifetime standards for time and frequency domain fluorescence spectroscopy. *Anal. Chem.* 79: 2137–2149.

Bohmer, M., Pampaloni, F., Wahl, M. et al. 2001. Time-resolved confocal scanning device for ultrasensitive fluorescence detection. *Rev. Sci. Instrum.* 72: 4145–4152.

Booth, M. J. and Wilson, T. 2004. Low-cost, frequency-domain, fluorescence lifetime confocal microscopy. *J. Microsc.(Oxf.)* 214: 36–42.

Buehler, C., Dong, C. Y., So, P. T. C., French, T., and Gratton, E. 2000. Time-resolved polarization imaging by pump-probe (stimulated emission) fluorescence microscopy. *Biophys. J.* 79: 536–549.

Bugiel, I., Konig, K., and Wabnitz, H. 1989. Investigation of cells by fluorescence laser scanning microscopy with subnanosecond time resolution. *Lasers Life Sci.* 3: 47–53.

Butte, P. V., Pikul, B. K., Hever, A. et al. 2005. Diagnosis of meningioma by time-resolved fluorescence spectroscopy. *J. Biomed. Opt.* 10(6): 064026.

Buurman, E. P., Sanders, R., Draaijer, A. et al. 1992. Fluorescence lifetime imaging using a confocal laser scanning microscope. *Scanning* 14: 155–159.

Cancer facts and figures. 2006. Atlanta, USA, American Cancer Society.

Champert, P. A., Popov, S. V., Solodyankin, M. A., and Taylor, J. R. 2002. Multiwatt average power continua generation in holey fibers pumped by kilowatt peak power seeded ytterbium fiber amplifier. *Appl. Phys. Lett.* 81: 2157–2159.

Chan, S. P., Fuller, Z. J., Demas, J. N., and DeGraff, B. A. 2001. Optimized gating scheme for rapid lifetime determinations of single-exponential luminescence lifetimes. *Anal. Chem.* 73: 4486–4490.

Chen, H. M., Chiang, C. P., You, C., Hsiao, T. C., and Wang, C. Y. 2005. Time-resolved autofluorescence spectroscopy for classifying normal and premalignant oral tissues. *Lasers Surg. Med.* 37: 37–45.

Cicchi, R., Massi, D., Sestini, S. et al. 2007. Multidimensional nonlinear laser imaging of basal cell carcinoma. *Opt. Express* 15: 10135–10148.

Clayton, A. H. A., Hanley, Q. S., Arndt-Jovin, D. J., Subramaniam, V., and Jovin, T. M. 2002. Dynamic fluorescence anisotropy imaging microscopy in the frequency domain (rFLIM). *Biophys. J.* 83: 1631–1649.

Clegg, R. M., Holub, O., and Gohlke, C. 2003. Fluorescence lifetime-resolved imaging: Measuring lifetimes in an image. *Biophotonics, Pt A.* San Diego, CA: Academic Press Inc., pp. 509–542.

Colyer, R. A., Lee, C., and Gratton, E. 2008. A novel fluorescence lifetime imaging system that optimizes photon efficiency. *Microsc. Res. Techn.* 71: 201–213.

Cubeddu, R., Comelli, D., D'Andrea, C., Taroni, P., and Valentini, G. 2002. Time-resolved fluorescence imaging in biology and medicine. *J. Phys. D: Appl. Phys.* 35: R61–R76.

Dacosta, R. S., Wilson, B. C., and Marcon, N. E. 2002. New optical technologies for earlier endoscopic diagnosis of premalignant gastrointestinal lesions. *World Congress of Gastroenterology*, Bangkok, Thailand.

Das, B. B., Liu, F., and Alfano, R. R. 1997. Time-resolved fluorescence and photon migration studies in biomedical and model random media. *Rep. Progr. Phys.* 60: 227–292.

De Beule, P., Owen, D. M., Manning, H. B. et al. 2006. Rapid hyperspectral fluorescence lifetime imaging. In: *Workshop on Advanced Multiphoton and Fluorescence Lifetime Imaging Techniques*, Ingbert, Germany.

De Beule, P. A. A., Dunsby, C., Galletly, N. P. et al. 2007. A hyperspectral fluorescence lifetime probe for skin cancer diagnosis. *Rev. Sci. Instrum.* 78(12): 123101.

de Veld, D. C. G., Skurichina, M., Witjes, M. J. H. et al. 2003. Autofluorescence characteristics of healthy oral mucosa at different anatomical sites. *Lasers Surg. Med.* 32: 367–376.

Deguil, N., Mottay, E., Salin, F., Legros, P., and Choquet, D. 2004. Novel diode-pumped infrared tunable laser system for multi-photon microscopy. *Microsc. Res. Techn.* 63: 23–26.

Denk, W., Strickler, J. H., and Webb, W. W. 1990. 2-photon laser scanning fluorescence microscopy. *Science* 248: 73–76.

JenLab GmbH. 1999, DermaInspect, Jena, Germany. Available at: http://www.jenlab.de/?id=29

Devries, P. D. and Khan, A. A. 1989. An efficient technique for analyzing deep level transient spectroscopy data. *J. Electron. Mater.* 18: 543–547.

Dickensheets, D. L. and Kino, G. S. 1996. Micromachined scanning confocal optical microscope. *Opt. Lett.* 21: 764–766.

Digman, M. A., Caiolfa, V. R., Zamai, M., and Gratton, E. 2008. The phasor approach to fluorescence lifetime imaging analysis. *Biophys. J.* 94: L14–L16.

Dodt, H. U., Leischner, U., Schierloh, A. et al. 2007. Ultramicroscopy: three-dimensional visualization of neuronal networks in the whole mouse brain. *Nat. Methods* 4: 331–336.

Dong, C. Y., So, P. T. C., French, T., and Gratton, E. 1995. Fluorescence lifetime imaging by asynchronous pump-probe microscopy. *Biophys. J.* 69: 2234–2242.

Dong, C. Y., Buehler, C., So, T. C., French, T., and Gratton, E. 2001. Implementation of intensity-modulated laser diodes in time-resolved, pump-probe fluorescence microscopy. *Appl. Opt.* 40: 1109–1115.

Dosremedios, C. G. and Moens, P. D. J. 1994. Fluorescence resonance energy-transfer spectroscopy is a reliable ruler for measuring structural-changes in proteins—dispelling the problem of the unknown orientation factor. In: *Workshop on Molecular Imaging of Cytoskeletal Protein Assembly*, Houston, TX, 1994.

Dowling, K., Dayel, M. J., Lever, M. J. et al. 1998. Fluorescence lifetime imaging with picosecond resolution for biomedical applications. *Opt. Lett.* 23: 810–812.

Dunsby, C., Lanigan, P. M. P., McGinty, J. et al. 2004. An electronically tunable ultrafast laser source applied to fluorescence imaging and fluorescence lifetime imaging microscopy. *J. Phys. D: Appl. Phys.* 37: 3296–3303.

Eisinger, J. and Dale, R. E. 1974. Interpretation of intramolecular energy-transfer experiments. *J. Mol. Biol.* 84: 643–647.

Elson, D., Requejo-Isidro, J., Munro, I. et al. 2003. Time-domain fluorescence lifetime imaging applied to biological tissue. In: *10th Congress of the European-Society-for-Photobiology*, Vienna, Austria.

Elson, D. S., Galletly, N., Talbot, C. et al. 2006. Multidimensional fluorescence imaging applied to biological tissue. In: *Reviews in Fluorescence 2006*, Eds. C. D. Geddes and J. R. Lakowicz. New York: Springer Science.

Elson, D. S., Jo, J. A., and Marcu, L. 2007. Miniaturized side-viewing imaging probe for fluorescence lifetime imaging (FLIM): Validation with fluorescence dyes, tissue structural proteins and tissue specimens. *New J. Phys.* 9: 127.

Elson, D. S., Munro, I., Requejo-Isidro, J. et al. 2004a. Real-time time-domain fluorescence lifetime imaging including single-shot acquisition with a segmented optical image intensifier. *New J. Phys.* 6: 1–13.

Elson, D. S., Requejo-Isidro, J., Munro, I. et al. 2004b. Time-domain fluorescence lifetime imaging applied to biological tissue. *Photochem. Photobiol. Sci.* 3: 795–801.

Elson, D. S., Siegel, J., Webb, S. E. D. et al. 2002. Fluorescence lifetime system for microscopy and multiwell plate imaging with a blue picosecond diode laser. *Opt. Lett.* 27: 1409–1411.

Esposito, A., Gerritsen, H. C., and Wouters, F. S. 2007. Optimizing frequency-domain fluorescence lifetime sensing for high-throughput applications: Photon economy and acquisition speed. *J. Opt. Soc. Am. A: Opt. Image Sci. Vis.* 24: 3261–3273.

Esposito, A., Oggier, T., Gerritsen, H. C., Lustenberger, F., and Wouters, F. S. 2005. All-solid-state lock-in imaging for wide-field fluorescence lifetime sensing. *Opt. Express* 13: 9812–9821.

Fittinghoff, D. N., Wiseman, P. W., and Squier, J. A. 2000. Widefield multiphoton and temporally decorrelated multifocal multiphoton microscopy. *Opt. Express* 7: 273–279.

Förster, T. 1948. Zwischenmolekulare energiewanderung und fluoreszenz. *Ann. Phys.* 437: 55–75.

Friesen, S. A., Hjortland, G. O., Madsen, S. J. et al. 2002. 5-Aminolevulinic acid-based photodynamic detection and therapy of brain tumors (review). *Int. J. Oncol.* 21: 577–582.

Gadella, T. W. J. 2008. *FRET and FLIM Techniques*. Amsterdam, the Netherlands: Elsevier.

Gadella, T. W. J. and Jovin, T. M. 1995. Oligomerization of epidermal growth-factor receptors on A431 cells studied by time-resolved fluorescence imaging microscopy—a stereochemical model for tyrosine kinase receptor activation. *J. Cell Biol.* 129: 1543–1558.

Gadella, T. W. J., Jovin, T. M., and Clegg, R. M. 1993. Fluorescence lifetime imaging microscopy (FLIM)—spatial-resolution of microstructures on the nanosecond time-scale. *Biophys. Chem.* 48: 221–239.

Galletly, N. P., McGinty, J., Dunsby, C. et al. 2008. Fluorescence lifetime imaging distinguishes basal cell carcinoma from surrounding uninvolved skin. *Br. J. Dermatol.* 159: 152–161.

Gaus, K., Gratton, E., Kable, E. P. W. et al. 2003. Visualizing lipid structure and raft domains in living cells with two-photon microscopy. *Proc. Natl. Acad. Sci. U. S. A.* 100: 15554–15559.

Gautier, I., Tramier, M., Durieux, C. et al. 2001. Homo-FRET microscopy in living cells to measure monomer-dimer transition of GFP-tagged proteins. *Biophys. J.* 80: 3000–3008.

Gerritsen, H. C., Asselbergs, M. A. H., Agronskaia, A. V., and Van Sark, W. 2002. Fluorescence lifetime imaging in scanning microscopes: Acquisition speed, photon economy and lifetime resolution. *J. Microsc. (Oxf.)* 206: 218–224.

Gerritsen, H. C., Sanders, R., Draaijer, A., Ince, C., and Levine, K. Y. 1997. Fluorescence lifetime imaging of oxygen in living cells. *J. Fluoresc.* 7: 11–16.

Ghiggino, K. P., Harris, M. R., and Spizzirri, P. G. 1992. Fluorescence lifetime measurements using a novel fiber-optic laser scanning confocal microscope. *Rev. Sci. Instrum.* 63: 2999–3002.

Gibson, A. P., Hebden, J. C., and Arridge, S. R. 2005. Recent advances in diffuse optical imaging. *Phys. Med. Biol.* 50: R1–R43.

Giepmans, B. N. G., Adams, S. R., Ellisman, M. H., and Tsien, R. Y. 2006. Review—the fluorescent toolbox for assessing protein location and function. *Science* 312: 217–224.

Glanzmann, T., Ballini, J. P., van den Bergh, H., and Wagnieres, G. 1999. Time-resolved spectrofluorometer for clinical tissue characterization during endoscopy. *Rev. Sci. Instrum.* 70: 4067–4077.

Gmitro, A. F. and Aziz, D. 1993. Confocal microscopy through a fiberoptic imaging bundle. *Opt. Lett.* 18: 565–567.

Godavarty, A., Sevick-Muraca, E. M., and Eppstein, M. J. 2005. Three-dimensional fluorescence lifetime tomography. *Med. Phys.* 32: 992–1000.

Gordon, G. W., Berry, G., Liang, X. H., Levine, B., and Herman, B. 1998. Quantitative fluorescence resonance energy transfer measurements using fluorescence microscopy. *Biophys. J.* 74: 2702–2713.

Grant, D. M., Elson, D. S., Schimpf, D. et al. 2005. Optically sectioned fluorescence lifetime imaging using a Nipkow disk microscope and a tunable ultrafast continuum excitation source. *Opt. Lett.* 30: 3353–3355.

Grant, D. M., McGinty, J., McGhee, E. J. et al. 2007. High speed optically sectioned fluorescence lifetime imaging permits study of live cell signaling events. *Opt. Express* 15: 15656–15673.

Grant, D. M., Zhang, W., McGhee, E. J. et al. 2008. Multiplexed FRET to image multiple signaling events in live cells. *Biophys. J.* 95, L69–L71.

Gratton, E., Breusegem, S., Sutin, J., and Ruan, Q. Q. 2003. Fluorescence lifetime imaging for the two-photon microscope: Time-domain and frequency-domain methods. *J. Biomed. Opt.* 8: 381–390.

Gratton, E., Feddersen, B., and Vandeven, M. 1990. Parallel acquisition of fluorescence decay using array detectors. In: *2nd Biennial Conference on Time-Resolved Laser Spectroscopy in Biochemistry*, Los Angeles, CA.

Gratton, E. and Limkeman, M. 1983. A continuously variable frequency cross-correlation phase fluorometer with picosecond resolution. *Biophys. J.* 44: 315–324.

Grinvald, A. and Steinberg, I. Z. 1974. On the analysis of fluorescence decay kinetics by the method of least-squares. *Anal. Biochem.* 59: 583–598.

Grynkiewicz, G., Poenie, M., and Tsien, R. Y. 1985. A new generation of Ca^{2+} indicators with greatly improved fluorescence properties. *J. Biol. Chem.* 260: 3440–3450.

Haj, F. G., Verveer, P. J., Squire, A., Neel, B. G., and Bastiaens, P. I. H. 2002. Imaging sites of receptor dephosphorylation by PTP1B on the surface of the endoplasmic reticulum. *Science* 295: 1708–1711.

Hares, J. D. 1987. Advances in sub-nanosecond shutter tube technology and applications in plasma physics. *X Rays from Laser Plasmas*. Proc. SPIE, 831: 165–170.

Harpur, A. G., Wouters, F. S., and Bastiaens, P. I. H. 2001. Imaging FRET between spectrally similar GFP molecules in single cells. *Nat. Biotechnol.* 19: 167–169.

Haugland, R. P., Yguerabide, J., and Stryer, L. 1969. Dependence of kinetics of singlet-singlet energy transfer on spectral overlap. *Proc. Natl. Acad. Sci. U. S. A.* 63: 23–30.

Hell, S. W. and Wichmann, J. 1994. Breaking the diffraction resolution limit by stimulated-emission-stimulated-emission-depletion fluorescence microscopy. *Opt. Lett.* 19: 780–782.

Helmchen, F., Fee, M. S., Tank, D. W., and Denk, W. 2001. A miniature head-mounted two-photon microscope: High-resolution brain imaging in freely moving animals. *Neuron* 31: 903–912.

Herman, B., Wodnicki, P., Kwon, S. et al. 1997. Recent developments in monitoring calcium and protein interactions in cells using fluorescence microscopy. *J. Fluoresc.* 7: 85–92.

Herman, P., Maliwal, B. P., Lin, H. J., and Lakowicz, J. R. 2001. Frequency-domain fluorescence microscopy with the LED as a light source. *J. Microsc. (Oxf.)* 203: 176–181.

Holub, O., Seufferheld, M. J., Gohlke, C., Govindjee, and Clegg, R. M. 2000. Fluorescence lifetime imaging (FLI) in real-time—a new technique in photosynthesis research. In: *3rd Regional Photosynthesis Workshop on the Chlorophyll Fluorescence Imaging and Its Application in Plant Science and Technology*, Lipno, Czech Republic.

Huisken, J., Swoger, J., Del Bene, F., Wittbrodt, J., and Stelzer, E. H. K. 2004. Optical sectioning deep inside live embryos by selective plane illumination microscopy. *Science* 305: 1007–1009.

Jain, B., Majumder, S. K., and Gupta, P. K. 1998. Time resolved and steady state autofluorescence spectroscopy of normal and malignant human breast tissue. *Lasers Life Sci.* 8: 163–173.

James, D. R. and Ware, W. R. 1985. A fallacy in the interpretation of fluorescence decay parameters. *Chem. Phys. Lett.* 120: 455–459.

Jares-Erijman, E. A. and Jovin, T. M. 2003. FRET imaging. *Nat. Biotechnol.* 21: 1387–1395.

Jean, F., Bourg-Heckly, G., and Viellerobe, B. 2007. Fibered confocal spectroscopy and multicolor imaging system for in vivo fluorescence analysis. *Opt. Express* 15: 4008–4017.

Jin, L., Millard, A. C., Wuskell, J. P., Clark, H. A., and Loew, L. M. 2005. Cholesterol-enriched lipid domains can be visualized by di-4-ANEPPDHQ with linear and nonlinear optics. *Biophys. J.* 89: L4–L6.

Jo, J. A., Fang, Q. Y., Papaioannou, T., and Marcu, L. 2004. Fast model-free deconvolution of fluorescence decay for analysis of biological systems. *J. Biomed. Opt.* 9: 743–752.

Jureller, J. E., Scherer, N. F., Birks, T. A., Wadsworth, W. J., and Russell, P. S. J. 2002. Widely tunable femtosecond pulses from a tapered fiber for ultrafast microscopy and multi-photon applications. In: *13th International Conference on Ultrafast Phenomena*, Vancouver, Canada.

Kak, A. C. and Slaney, M. 1988. *Principles of Computerized Tomographic Imaging*. New York: IEEE Press.

Kantelhardt, S. R., Leppert, J., Krajewski, J. et al. 2007. Imaging of brain and brain tumor specimens by time-resolved multiphoton excitation microscopy ex vivo. *Neuro-Oncology* 9: 103–112.

Keller, P. J., Pampaloni, F., and Stelzer, E. H. K. 2006. Life sciences require the third dimension. *Curr. Opin. Cell Biol.* 18: 117–124.

Kennedy, G. T., Manning, H. B., Elson, D. S. et al. 2009. A fluorescence lifetime imaging scanning confocal endomicroscope. *J. Biophotonics* 2: 103–107.

Kim, K. H., Buehler, C., Bahlmann, K. et al. 2007. Multifocal multiphoton microscopy based on multianode photomultiplier tubes. *Opt. Express* 15: 11658–11678.

Kitamura, N., Hosoda, Y., Iwasaki, C., Ueno, K., and Kim, H. B. 2003. Thermal phase transition of an aqueous poly(*N*-isopropylacrylamide) solution in a polymer microchannel-microheater chip. *Langmuir* 19: 8484–8489.

Koeberg, M., Elson, D. S., French, P. M. W., and Bradley, D. D. C. 2003. Spatially resolved electric fields in polymer light-emitting diodes using fluorescence lifetime imaging. In: *5th International Topical Conference on Optical Probes of Conjugated Polymers, Organic and Inorganic Nanostructure*, Venice, Italy.

Kollner, M. and Wolfrum, J. 1992. How many photons are necessary for fluorescence-lifetime measurements. *Chem. Phys. Lett.* 200: 199–204.

Konig, K. 2008. Clinical multiphoton tomography. *J. Biophotonics* 1: 13–23.

Krishnan, R. V., Saitoh, H., Terada, H., Centonze, V. E., and Herman, B. 2003. Development of a multiphoton fluorescence lifetime imaging microscopy system using a streak camera. *Rev. Sci. Instrum.* 74: 2714–2721.

Kudlinski, A., George, A. K., Knight, J. C. et al. 2006. Zero-dispersion wavelength decreasing photonic crystal fibers for ultraviolet-extended supercontinuum generation. *Opt. Express* 14: 5715–5722.

Kumar, A. T. N., Chung, E., Raymond, S. B. et al. 2009. Feasibility of in vivo imaging of fluorescent proteins using lifetime contrast. *Opt. Lett.* 34: 2066–2068.

Kumar, A. T. N., Raymond, S. B., Dunn, A. K., Bacskai, B. J., and Boas, D. A. 2008. A time domain fluorescence tomography system for small animal imaging. *IEEE Trans. Med. Imaging* 27: 1152–1163.

Kumar, A. T. N., Skoch, J., Bacskai, B. J., Boas, D. A., and Dunn, A. K. 2005. Fluorescence-lifetime-based tomography for turbid media. *Opt. Lett.* 30: 3347–3349.

Kumar, S., Dunsby, C., De Beule, P. A. A. et al. 2007. Multifocal multiphoton excitation and time correlated single photon counting detection for 3-D fluorescence lifetime imaging. *Opt. Express* 15: 12548–12561.

Laemmel, E., Genet, M., Le Goualher, G. et al. 2004. Fibered confocal fluorescence microscopy (Cell-viZio (TM)) facilitates extended imaging in the field of microcirculation—A comparison with intravital microscopy. *J. Vasc. Res.* 41: 400–411.

Lakowicz, J. R. 2006. *Principles of Fluorescence Spectroscopy*. New York: Springer Science + Business Media, LLC.

Lakowicz, J. R. and Berndt, K. W. 1991. Lifetime-selective fluorescence imaging using an RF phase-sensitive camera. *Rev. Sci. Instrum.* 62: 1727–1734.

Lakowicz, J. R. and Maliwal, B. P. 1985. Construction and performance of a variable-frequency phase-modulation fluorometer. *Biophys. Chem.* 21: 61–78.

Lakowicz, J. R., Szmacinski, H., Nowaczyk, K., Berndt, K. W., and Johnson, M. 1992a. Fluorescence lifetime imaging. *Anal. Biochem.* 202: 316–330.

Lakowicz, J. R., Szmacinski, H., Nowaczyk, K., and Johnson, M. L. 1992b. Fluorescence lifetime imaging of calcium using QUIN-2. *Cell Calcium* 13: 131–147.

Lakowicz, J. R., Szmacinski, H., Nowaczyk, K., and Johnson, M. L. 1992c. Fluorescence lifetime imaging of free and protein-bound NADH. *Proc. Natl. Acad. Sci. U. S. A.* 89: 1271–1275.

Lee, K. C. B., Siegel, J., Webb, S. E. D. et al. 2001. Application of the stretched exponential function to fluorescence lifetime imaging. *Biophys. J.* 81: 1265–1274.

Lelek, M., Suran, E., Louradour, F. et al. 2007. Coherent femtosecond pulse shaping for the optimization of a non-linear micro-endoscope. *Opt. Express* 15: 10154–10162.

Leveque-Fort, S., Fontaine-Aupart, M. P., Roger, G., and Georges, P. 2004. Fluorescence-lifetime imaging with a multifocal two-photon microscope. *Opt. Lett.* 29: 2884–2886.

Liang, C., Descour, M. R., Sung, K. B., and Richards-Kortum, R. 2001. Fiber confocal reflectance microscope (FCRM) for in-vivo imaging. *Opt. Express* 9: 821–830.

Lidke, D. S., Nagy, P., Barisas, B. G. et al. 2003. Imaging molecular interactions in cells by dynamic and static fluorescence anisotropy (rFLIM and emFRET). In: *Meeting on Intermolecular Associations in 2D and 3D*, Nottingham, England.

Lin, H. J., Herman, P., and Lakowicz, J. R. 2003. Fluorescence lifetime-resolved pH imaging of living cells. *Cytometry Part A* 52A, 77–89.

Lippincott-Schwartz, J. and Patterson, G. H. 2003. Development and use of fluorescent protein markers in living cells. *Science* 300: 87–91.

Marcu, L., Jo, J. A., Butte, P. V. et al. 2004. Fluorescence lifetime spectroscopy of glioblastoma multiforme. *Photochem. Photobiol.* 80: 98–103.

Margineanu, A., Hotta, J. I., Van der Auweraer, M. et al. 2007. Visualization of membrane rafts using a perylene monoimide derivative and fluorescence lifetime Imaging. *Biophys. J.* 93: 2877–2891.

McConnell, G. 2004. Confocal laser scanning fluorescence microscopy with a visible continuum source. *Opt. Express* 12: 2844–2850.

McGinty, J., Requejo-Isidro, J., Munro, I. et al. 2009a. Signal-to-noise characterization of time-gated intensifiers used for wide-field time-domain FLIM. *J. Phys. D: Appl. Phys.* 42: 135103.

McGinty, J., Soloviev, V. Y., Tahir, K. B. et al. 2009b. Three-dimensional imaging of Forster resonance energy transfer in heterogeneous turbid media by tomographic fluorescent lifetime imaging. *Opt. Lett.* 34: 2772–2774.

McGinty, J., Tahir, K. R., Laine, R. et al. 2008. Fluorescence lifetime optical projection tomography. *J. Biophotonics* 1: 390–394.

McGuinness, C. D., Sagoo, K., McLoskey, D., and Birch, D. J. S. 2004. A new sub-nanosecond LED at 280 nm: Application to protein fluorescence. *Meas. Sci. Technol.* 15: L19–L22.

Minami, T. and Hirayama, S. 1990. High-quality fluorescence decay curves and lifetime imaging using an elliptic scan streak camera. *J. Photochem. Photobiol. a-Chem.* 53: 11–21.

Miyawaki, A., Llopis, J., Heim, R. et al. 1997. Fluorescent indicators for Ca^{2+} based on green fluorescent proteins and calmodulin. *Nature* 388: 882–887.

Mizeret, J., Stepinac, T., Hansroul, M. et al. 1999. Instrumentation for real-time fluorescence lifetime imaging in endoscopy. *Rev. Sci. Instrum.* 70: 4689–4701.

Mizeret, J., Wagnieres, G., Stepinac, T., and VandenBergh, H. 1997. Endoscopic tissue characterization by frequency-domain fluorescence lifetime imaging (FD-FLIM). *Lasers Med. Sci.* 12: 209–217.

Morgan, C. G., Mitchell, A. C., and Murray, J. G. 1990. Nanosecond time-resolved fluorescence microscopy—Principles and practice. In: *Conference of the Royal Microscopical Society*, London, England.

Munro, I., McGinty, J., Galletly, N. et al. 2005. Toward the clinical application of time-domain fluorescence lifetime imaging. *J. Biomed. Opt.* 10(5): 051403.

Murata, S., Herman, P., and Lakowicz, J. R. 2001. Texture analysis of fluorescence lifetime images of nuclear DNA with effect of fluorescence resonance energy transfer. *Cytometry* 43: 94–100.

Myaing, M. T., MacDonald, D. J., and Li, X. D. 2006. Fiber-optic scanning two-photon fluorescence endoscope. *Opt. Lett.* 31: 1076–1078.

Mycek, M. A., Schomacker, K. T., and Nishioka, N. S. 1998. Colonic polyp differentiation using time-resolved autofluorescence spectroscopy. *Gastrointestin. Endosc.* 48: 390–394.

Ng, T., Parsons, M., Hughes, W. E. et al. 2001. Ezrin is a downstream effector of trafficking PKC-integrin complexes involved in the control of cell motility. *Embo J.* 20: 2723–2741.

Ng, T., Squire, A., Hansra, G. et al. 1999. Imaging protein kinase C alpha activation in cells. *Science* 283: 2085–2089.

Niedre, M. J., de Kleine, R. H., Aikawa, E. et al. 2008. Early photon tomography allows fluorescence detection of lung carcinomas and disease progression in mice in vivo. *Proc. Natl. Acad. Sci. U. S. A.* 105: 19126–19131.

Nothdurft, R. E., Patwardhan, S. V., Akers, W. et al. 2009. In vivo fluorescence lifetime tomography. *J. Biomed. Opt.* 14(2): 024004.

O'Connor, D. V. and Phillips, D. 1984. *Time-Correlated Single Photon Counting*. London: Academic Press.

Obaid, A. L., Loew, L. M., Wuskell, J. P., and Salzberg, B. M. 2004. Novel naphthylstyryl-pyridinium potentiometric dyes offer advantages for neural network analysis. *J. Neurosci. Methods* 134: 179–190.

Ohkawa, A., Miwa, H., Namihisa, A. et al. 2004. Diagnostic performance of light-induced fluorescence endoscopy for gastric neoplasms. *Endoscopy* 36: 515–521.

Oida, T., Sako, Y., and Kusumi, A. 1993. Fluorescence lifetime imaging microscopy (FLIMSCOPY)—Methodology development and application to studies of endosome fusion in single cells. *Biophys. J.* 64: 676–685.

Oldham, M., Sakhalkar, H., Wang, Y. M. et al. 2007. Three-dimensional imaging of whole rodent organs using optical computed and emission tomography. *J. Biomed. Opt.* 12(1): 014009.

Owen, D. M., Auksorius, E., Manning, H. B. et al. 2007. Excitation-resolved hyperspectral fluorescence lifetime imaging using a UV-extended supercontinuum source. *Opt. Lett.* 32: 3408–3410.

Owen, D. M., Lanigan, P. M. P., Dunsby, C. et al. 2006. Fluorescence lifetime imaging provides enhanced contrast when imaging the phase-sensitive dye di-4-ANEPPDHQ in model membranes and live cells. *Biophys. J.* 90: L80–L82.

Paithankar, D. Y., Chen, A. U., Pogue, B. W., Patterson, M. S., and SevickMuraca, E. M. 1997. Imaging of fluorescent yield and lifetime from multiply scattered light reemitted from random media. *Appl. Opt.* 36: 2260–2272.

Parasassi, T., Krasnowska, E. K., Bagatolli, L., and Gratton, E. 1998. Laurdan and prodan as polarity-sensitive fluorescent membrane probes. *J. Fluoresc.* 8: 365–373.

Patterson, G. H. and Piston, D. W. 2000. Photobleaching in two-photon excitation microscopy. *Biophys. J.* 78: 2159–2162.

Pfefer, T. J., Paithankar, D. Y., Poneros, J. M., Schomacker, K. T., and Nishioka, N. S. 2003. Temporally and spectrally resolved fluorescence spectroscopy for the detection of high grade dysplasia in Barrett's esophagus. *Lasers Surg. Med.* 32: 10–16.

Philip, J. and Carlsson, K. 2003. Theoretical investigation of the signal-to-noise ratio in fluorescence lifetime imaging. *J. Opt. Soc. Am. A: Opt. Image Sci. Vis.* 20: 368–379.

Pradhan, A., Das, B. B., Yoo, K. M. et al. 1992. Time-resolved UV photoexcited fluorescence kinetics from malignant and non-malignant human breast tissue. *Lasers Life Sci.* 4: 225–234.

Provenzano, P. P., Inman, D. R., Eliceiri, K. W. et al. 2008. Collagen density promotes mammary tumor initiation and progression. *BMC Med.* 6: 11.

Ranka, J. K., Windeler, R. S., and Stentz, A. J. 2000. Visible continuum generation in air-silica microstructure optical fibers with anomalous dispersion at 800 nm. *Opt. Lett.* 25: 25–27.

Redford, G. I. and Clegg, R. M. 2005. Polar plot representation for frequency-domain analysis of fluorescence lifetimes. *J. Fluoresc.* 15: 805–815.

Requejo-Isidro, J., McGinty, J., Munro, I. et al. 2004. High-speed wide-field time-gated endoscopic fluorescence-lifetime imaging. *Opt. Lett.* 29: 2249–2251.

Richards-Kortum, R. and Sevick-Muraca, E. 1996. Quantitative optical spectroscopy for tissue diagnosis. *Annu. Rev. Phys. Chem.* 47: 555–606.

Rouse, A. R. and Gmitro, A. F. 2000. Multispectral imaging with a confocal microendoscope. *Opt. Lett.* 25: 1708–1710.

Rust, M. J., Bates, M., and Zhuang, X. W. 2006. Sub-diffraction-limit imaging by stochastic optical reconstruction microscopy (STORM). *Nat. Methods* 3: 793–795.

Sabharwal, Y. S., Rouse, A. R., Donaldson, L., Hopkins, M. F., and Gmitro, A. F. 1999. Slit-scanning confocal microendoscope for high-resolution in vivo imaging. *Appl. Opt.* 38: 7133–7144.

Sanders, R., Draaijer, A., Gerritsen, H. C., Houpt, P. M., and Levine, Y. K. 1995. Quantitative pH imaging in cells using confocal fluorescence lifetime imaging microscopy. *Anal. Biochem.* 227: 302–308.

Sanders, R., Gerritsen, H. C., Draaijer, A., Houpt, P. M., and Levine, K. Y. 1994. Fluorescence lifetime imaging of free calcium in single cells. *Bioimaging* 2: 131–138.

Schaerli, Y., Wootton, R. C., Robinson, T. et al. 2009. Continuous-flow polymerase chain reaction of single-copy DNA in microfluidic microdroplets. *Anal. Chem.* 81: 302–306.

Schneckenburger, H., Stock, K., Strauss, W. S. L., Eickholz, J., and Sailer, R. 2003. Time-gated total internal reflection fluorescence spectroscopy (TG-TIRFS): Application to the membrane marker laurdan. *J. Microsc. (Oxf.)* 211: 30–36.

Schreiber, T., Limpert, J., Zellmer, H., Tunnermann, A., and Hansen, K. P. 2003. High average power supercontinuum generation in photonic crystal fibers. *Opt. Commun.* 228: 71–78.

Schultz, R. A., Nielsen, T., Zavaleta, J. R. et al. 2001. Hyperspectral imaging: A novel approach for microscopic analysis. *Cytometry* 43: 239–247.

Schweitzer, D., Hammer, M., Schweitzer, F. et al. 2004. In vivo measurement of time-resolved autofluorescence at the human fundus. *J. Biomed. Opt.* 9: 1214–1222.

Scully, D., MacRobert, A. J., Botchway, S. et al. 1996. Development of a laser-based fluorescence microscope with subnanosecond time resolution. *J. Fluoresc.* 6: 119–125.

Selvin, P. R. 2000. The renaissance of fluorescence resonance energy transfer. *Nat. Struct. Biol.* 7: 730–734.

Sharman, K. K., Periasamy, A., Ashworth, H., Demas, J. N., and Snow, N. H. 1999. Error analysis of the rapid lifetime determination method for double-exponential decays and new windowing schemes. *Anal. Chem.* 71: 947–952.

Sharpe, J. 2004. Optical projection tomography. *Annu. Rev. Biomed. Eng.* 6: 209–228.

Sharpe, J., Ahlgren, U., Perry, P. et al. 2002. Optical projection tomography as a tool for 3D microscopy and gene expression studies. *Science* 296: 541–545.

Shives, E., Xu, Y., and Jiang, H. B. 2002. Fluorescence lifetime tomography of turbid media based on an oxygen-sensitive dye. *Opt. Exp.* 10: 1557–1562.

Siegel, J., Suhling, K., Leveque-Fort, S. et al. 2003. Wide-field time-resolved fluorescence anisotropy imaging (TR-FAIM): Imaging the rotational mobility of a fluorophore. *Rev. Sci. Instrum.* 74: 182–192.

Skala, M. C., Riching, K. M., Gendron-Fitzpatrick, A. et al. 2007. In vivo multiphoton microscopy of NADH and FAD redox states, fluorescence lifetimes, and cellular morphology in precancerous epithelia. *Proc. Natl. Acad. Sci. U. S. A.* 104: 19494–19499.

So, P. T. C., French, T., Yu, W. M. et al. 1995. Time-resolved fluorescence microscopy using two-photon excitation. *Bioimaging* 3: 49–63.

Sokolov, K., Follen, M., and Richards-Kortum, R. 2002. Optical spectroscopy for detection of neoplasia. *Curr. Opin. Chem. Biol.* 6: 651–658.

Soloviev, V. Y., D'Andrea, C., Valentini, G., Cubeddu, R., and Arridge, S. R. 2009. Combined reconstruction of fluorescent and optical parameters using time-resolved data. *Appl. Opt.* 48: 28–36.

Squire, A., Verveer, P. J., and Bastiaens, P. I. H. 2000. Multiple frequency fluorescence lifetime imaging microscopy. *J. Microsc. (Oxf.)* 197: 136–149.

Straub, M. and Hell, S. W. 1998. Fluorescence lifetime three-dimensional microscopy with picosecond precision using a multifocal multiphoton microscope. *Appl. Phys. Lett.* 73: 1769–1771.

Strickler, S. J. and Berg, R. A. 1962. Relationship between absorption intensity and fluorescence lifetime of molecules. *J. Chem. Phys.* 37: 814–822.

Stryer, L. 1978. Fluorescence energy-transfer as a spectroscopic ruler. *Annu. Rev. Biochem.* 47: 819–846.

Suhling, K., French, P. M. W., and Phillips, D. 2005. Time-resolved fluorescence microscopy. *Photochem. Photobiol. Sci.* 4: 13–22.

Suhling, K., Siegel, J., Lanigan, P. M. P. et al. 2004. Time-resolved fluorescence anisotropy imaging applied to live cells. *Opt. Lett.* 29: 584–586.

Suhling, K., Siegel, J., Phillips, D. et al. 2002. Imaging the environment of green fluorescent protein. *Biophys. J.* 83: 3589–3595.

Sun, Y., Phipps, J., Elson, D. S. et al. 2009. Fluorescence lifetime imaging microscopy: In vivo application to diagnosis of oral carcinoma. *Opt. Lett.* 34: 2081–2083.

Tadrous, P. J., Siegel, J., French, P. M. W. et al. 2003. Fluorescence lifetime imaging of unstained tissues: Early results in human breast cancer. *J. Pathol.* 199: 309–317.

Talbot, C. B., McGinty, J., Grant, D. M. et al. 2008. High speed unsupervised fluorescence lifetime imaging confocal multiwell plate reader for high content analysis. *J. Biophotonics* 1: 514–521.

Tata, D. B., Foresti, M., Cordero, J. et al. 1986. Fluorescence polarization spectroscopy and time-resolved fluorescence kinetics of native cancerous and normal rat-kidney tissues. *Biophys. J.* 50: 463–469.

Tearney, G. J., Shishkov, M., and Bouma, B. E. 2002. Spectrally encoded miniature endoscopy. *Opt. Lett.* 27: 412–414.

Treanor, B., Lanigan, P. M. P., Kumar, S. et al. 2006. Microclusters of inhibitory killer immunoglobulin like receptor signaling at natural killer cell immunological synapses. *J. Cell Biol.* 174: 153–161.

Tsien, R. Y. 1998. The green fluorescent protein. *Annu. Rev. Biochem.* 67: 509–544.

Tsien, R. Y. 2009. Constructing and exploiting the fluorescent protein paintbox (Nobel Lecture). *Angew. Chem. Int. Ed.* 48: 5612–5626.

van Geest, L. K. and Stoop, K. W. J. 2003. FLIM on a wide field fluorescence microscope. In: *5th Australian Peptide Conference*, Daydream Isl, Australia.

van Munster, E. B., Goedhart, J., Kremers, G. J., Manders, E. M. M., and Gadella, T. W. J. 2007. Combination of a spinning disc confocal unit with frequency-domain fluorescence lifetime imaging microscopy. *Cytometry Part A* 71A: 207–214.

van Roessel, P. and Brand, A. H. 2002. Imaging into the future: Visualizing gene expression and protein interactions with fluorescent proteins. *Nat. Cell Biol.* 4: E15–E20.

Verveer, P. J., Squire, A., and Bastiaens, P. I. H. 2000a. Global analysis of fluorescence lifetime imaging microscopy data. *Biophys. J.* 78: 2127–2137.

Verveer, P. J., Wouters, F. S., Reynolds, A. R., and Bastiaens, P. I. H. 2000b. Quantitative imaging of lateral ErbB1 receptor signal propagation in the plasma membrane. *Science* 290: 1567–1570.

Wagnieres, G. A., Star, W. M., and Wilson, B. C. 1998. In vivo fluorescence spectroscopy and imaging for oncological applications. *Photochem. Photobiol.* 68: 603–632.

Walls, J. R., Sled, J. G., Sharpe, J., and Henkelman, R. M. 2007. Resolution improvement in emission optical projection tomography. *Phys. Med. Biol.* 52: 2775–2790.

Wang, X. F., Uchida, T., Coleman, D. M., and Minami, S. 1991. A 2-dimensional fluorescence lifetime imaging-system using a gated image intensifier. *Appl. Spectrosc.* 45: 360–366.

Weissleder, R. and Ntziachristos, V. 2003. Shedding light onto live molecular targets. *Nat. Med.* 9: 123–128.

Widengren, J., Kudryavtsev, V., Antonik, M. et al. 2006. Single-molecule detection and identification of multiple species by multiparameter fluorescence detection. *Anal. Chem.* 78: 2039–2050.

Wlodarczyk, J. and Kierdaszuk, B. 2003. Interpretation of fluorescence decays using a power-like model. *Biophys. J.* 85: 589–598.

Wouters, F. S. and Bastiaens, P. I. H. 1999. Fluorescence lifetime imaging of receptor tyrosine kinase activity in cells. *Curr. Biol.* 9: 1127–1130.

Wouters, F. S., Verveer, P. J., and Bastiaens, P. I. H. 2001. Imaging biochemistry inside cells. *Trends Cell Biol.* 11: 203–211.

Young, P. E., Hares, J. D., Kilkenny, J. D., Phillion, D. W., and Campbell, E. M. 1988. 4-frame gated optical imager with 120-ps resolution. *Rev. Sci. Instrum.* 59: 1457–1460.

Zhang, Q. H., Piston, D. W., and Goodman, R. H. 2002. Regulation of corepressor function by nuclear NADH. *Science* 295: 1895–1897.

Zhang, Y. L., Soper, S. A., Middendorf, L. R. et al. 1999. Simple near-infrared time-correlated single photon counting instrument with a pulsed diode laser and avalanche photodiode for time-resolved measurements in scanning applications. *Appl. Spectrosc.* 53: 497–504.

Zimmer, M. 2002. Green fluorescent protein (GFP): Applications, structure, and related photophysical behavior. *Chem. Rev.* 102: 759–781.

29

Application of Digital Holographic Microscopy in Biomedicine

29.1	Introduction	617
29.2	The Concept of Coherence	620
	Coherence Exploitation • Exploitation of the Mutual Coherence Properties of Crossing Beams	
29.3	Holography	624
	Square Terms • Cross Terms • Digital Holography by Analysis of Data in the Time Domain • OCT • Digital Holography by Analysis of Data in the Space Domain • Full Restoration of the Complex Wave Front	
29.4	Realization of Optical Setups for Holographic Microscopy	629
29.5	Space-Bandwidth Adaptation	631
	Optical Transfer Function	
29.6	"Electronic Focusing" and Extended Depth of Field	633
29.7	DHM Instrument	633
29.8	Further Developments	635
	Applications to Biology • Toward Full 3D Imaging of Specimens	
29.9	Conclusions and Outlook	639
	Acknowledgments	640
	References	640

Christian Depeursinge
*Federal Polytechnic
School of Lausanne*

Pierre Marquet
Lausanne University Hospital

Nicolas Pavillon
*Federal Polytechnic
School of Lausanne*

29.1 Introduction

To situate the interest of digital holographic microscopy (DHM) in biology, let us recall that, in the past, *phase contrast microscopy* (Zernike 1937, 1942, 1949, Françon 1949) has played a dominant role in the observation of living cells and has permitted the first precise determination of the refractive index of living cells (refractometry) (Barer and Joseph 1954, Barer 1957). At that time, the measuring procedure was quite complex and usually required the adjustment of the refractive index of an external medium in contact with the cell. These determinations have allowed establishing a precise relationship between the index of refraction and the "dry mass" (Davies and Wilkins 1952) or density of the cell constituents: protein, nucleic acids and nucleoproteins, lipids and lipoproteins, but also with carbohydrates and inorganic constituents, salts, etc. The relationship between microscopic observations and basal physiologic parameters such as metabolism, cell growth, or decline and death could be firmly established. Moreover, a first estimation of the optical dispersion laws could be established for proteins and opens the way to more precise characterization of cell constituents in the future. Technical adaptation of the design of the phase contrast microscope to a reflection arrangement (reflection contrast microscopy) permitted a precise and quantitative determination of refractive index of cellular cytoplasm (Zernike 1942). Interpretations of tiny variations of index are possible in terms of the cellular structure.

Interference microscopy was developed for applications requiring investigations of small objects and surfaces at a microscopic scale: e.g., surface profilometry and surface roughness measurements. Different arrangements in reflection were used, usually based on a Michelson geometry, Linnick configuration in particular, but may be also collinear (Mirau arrangement). The realization of the microscope objectives for interferometric measurements is delicate and often expensive. Gabor himself gave a first report on an interference microscope with total wave-front reconstruction (Gabor and Goss 1966). In some cases, interferometric microscopes have been developed to be used in transmittance: based on a Mach–Zehnder interferometer, they require the realization of two microscopes, placed side by side, with parallel axis. Such a system was manufactured by Leitz in the 1950s and has brought interesting perspectives on its applications in biology: determination of the dry mass of cells, measurement of the cell deformations and water membrane permeability. The prime interest of this technique was its capability to give quantitative phase measurements, but the excessive cost of the instrument limited its use in biology laboratories. A simpler and less expensive approach was developed by Nomarski (1955) by using the interference of two light beams slightly displaced by a Wollaston prism. The difficulty with the so-called differential interference contrast (DIC) microscope is that usually, no quantitative phase data is obtained. Several efforts have been made, however, to overcome this issue in the past (Lessor et al. 1979, Arnison et al. 2000, van Munster et al. 1997, Hartman et al. 1980), and investigation is pursued also recently (Arnison et al. 2004, Mehta and Sheppard 2009). Invaluable data on the refractive index of cell constituents could be first obtained by Beuthan et al. (1996) with an interferometric setup (Michelson type) coupled to a microscope. See also: (Eppich et al. 2000).

Confocal microscopy (Minsky 1988, Pawley 1995, Wilson and Sheppard 1984, Wilson 1990) has brought the capability of optical sectioning and given a first evaluation of 3D distribution of refractive indices (Schmidt et al. 2000).

At a larger scale, from the millimeter up to the micron scale, *optical coherence tomography* (OCT) *techniques* (Huang et al. 1991) appear the most promising candidates to give tomographic images of tissues possibly down to the cellular scale (see Chapters 13 through 15). However, intracellular imaging in mammal cells cannot be achieved yet.

Scanning techniques have been largely developed in a larger context: confocal scanning microscopy, optical scanning white light interferometry, or scanning OCT, etc. Scanning technique have also been implemented in holography to access the convolution of the fluorescence or scattering image of the objects with Fresnel zones (Schilling et al. 1997).

So-called *4π microscopy* brought a significant improvement in axial resolution by the coherent use of opposing lenses in fluorescence microscopy (Hell and Stelzer 1992, Hell 2003, 2004, Schmidt et al. 2000, Nagorni and Hell 2001a,b). A scanning version of the *4π fluorescence microscopy* and multichannel, parallel reading (Bewersdorf et al. 1998) have been proposed and more recently commercialized (Leica).

Nevertheless, a revival of *bright field* or "nonpoint" *scanning microscopy* techniques could be noted a few years ago with the proposal of a variant of 4π microscopy, called I^{5}M (Gustafsson 1999, Gustafsson et al. 1999). In this context, structured light illumination techniques have received increasing attention in microscopy during the last 10 years: (Juskaitis et al. 1996, Neil et al. 2000, Cole et al. 2000, Gustafsson 2000, 2005).

Structured illumination is an imaging technique that has a wide array of applications in 3D imaging: it can be mentioned that fringe projection for surface and volume measurement is an example of structured illumination that is common in metrology. This technique accommodates incoherent properties of light: white light illumination with incoherent sources, as well as interfering coherent sources (laser beams). It is an "intensity-based" imaging technique in the sense that the light intensity is the basic signal that will be finally processed. Practical instruments aiming at fast and low-cost sectioning have been brought to the market (Zeiss, Improvision). This approach is also very well suited to fluorescence imaging: fluorescence is a signal which is more or less linearly related to the intensity of the irradiating beam. The structuring of the irradiating field can be obtained by the interference of two coherent beams. The case of two counter-propagating plane waves, creating standing wave patterns, has been one of the first applications of structured illumination to fluorescence microscopy. It was soon acknowledged as a method to improve imaging resolution. The method is also referenced in microscopy as spatially modulated illumination (SMI). When SMI is used in the particular experimental configuration of two opposing microscope objectives, it was shown that axial resolution could be improved by a factor of more than four. The idea of creating standing waves patterns in the lateral direction was also exploited and showed a significant improvement of lateral resolution. This approach has proved fruitful in fluorescence microscopy in particular (Bailey et al. 1993, 1994, Albrecht et al. 2001, 2002, Failla et al. 2002a,b, 2003, Martin et al. 2004, Schweitzer et al. 2004, Spori et al. 2004, Hildenbrand et al. 2005, Frohn et al. 2001, 2000, Fedosseev et al. 2005). Impressive gain in resolution (down to 50 nm) could be obtained on a 2D preparation of fluorescent beads (Gustafsson 2005). It must be noted however that such a gain in resolution is extremely sensitive to the S/N ratio and apparently suited only for 2D arrangements of fluorophores. Structured illumination microscopy (SMI) appears henceforth as a technique capable of improving resolution. It must however still prove its immunity to perturbations met in practical biology context and in full 3D.

Structured illumination is a technique that has found a particularly attractive implementation in the form of the so-called scanning holographic microscopy. Introduced in 1997 (Schilling et al. 1997), the method provides the possibility of reconstructing the 3D distribution of fluorophores or more generally of scattering centers from the numerical processing of a 2D matrix formed by the intensities acquired sequentially on an integrating optical detector. In essence, this technique is similar to the reconstruction of an object from its hologram (2D → 3D), but, fundamentally, no coherent light is emitted from the collection of fluorophores. The structured illumination results from the interference of two waves originating from two-pupil functions: Lohmann highlighted the properties of such interfering pupils in the synthesis of the PSF a long time ago (Lohmann and Rhodes 1978). Appealing perspectives of the method have been given more recently (Indebetouw et al. 2005).

The idea of reconstructing a wave front from the hologram, yielding potentially the exact replica of the object optical properties and then, deductively, the object shape and dielectric composition, was a great breakthrough of the development of holography (Gabor 1949, 1951). Its practical implementation of digital processing methods brought fundamentally new perspectives in holography. The introduction of informatics was favored by the rapid development of digital processing techniques which brought computing facilities on a wider basis by the large spread of PCs performing well as digital signal and image processors. In digital holography (DH), the complex wave front is reconstructed numerically by using a computer. Data are provided by a hologram captured and digitalized with an electronic camera. The most recent informatics means have permitted the rapid calculation and simulation of wave propagation, so that this circumstance appeared as the second technological breakthrough, after the laser invention, which has promoted digital processing of holographic data. These developments, designated as "digital holography," have presented the great advantage of providing quantitative data as well as simplified optical manipulations and improved flexibility (Goodman and Lawrence 1967, Kronrod et al. 1972, Yaroslavsky and Merzlyakov 1980, Schnars and Jüptner 1994, Coquoz et al. 1995). For a review, one can see Schnars and Juptner (2002). At the same time, a major innovation of digital processing of holograms was also to provide straightforwardly a correction of lens defects, aberrations in particular, by pure digital means (Colomb et al. 2007). Through the perspective of postponing to a data processing step, many of the difficulties met in the practical implementation of sophisticated stigmatic optical systems appear, from now on, as a truly appealing perspective and prone to open new avenues in optical design.

The fact that the image acquisition in optics has become progressively dissociated from the step of forming a real image on a camera or the human eye is a major outcome of modern coherence imaging and will constitute a breakthrough in optical imaging technology. We present here the evidence of this evolution in the field of optical microscopy. The application of holography to microscopy has revealed a quite different field of research and is developed in an original context. First attempts to take holograms of microscopic objects have been proposed by Gabor in the first effort to develop holography in the 1960s (Gabor and Goss 1966). The realized interference microscope was operated with an optically filtered conventional high-pressure mercury lamp, and a total wave-front reconstruction was carried out by illumination of a pair of reflection holograms taken on the microscope. Later on, a holographic microscope was developed by Pluta (1987) in the 1980s, and by some others (unpublished results). The procedure was complex and did not appear very attractive. In the field of biology and medicine, holography was applied to cell and tissue investigations in 3D by using a photographic plate mounted in a Denisyuk configuration (de Haller et al. 1995). Holography was, however, rarely used in its early, nondigital phase. Later on, it appeared that the adaptation of digital holography to microscopy would soon revolutionize the domain. Several advantages favored the growth of DHM at the end of the 1990s.

First, it appeared that processing the hologram with the methods of digital holography would provide the means of reconstructing the complex wave front of the beam scattered by the specimen. This wave field has a particular significance in microscopy: the phase of the complex number can be interpreted without ambiguity as closely related to the true features of the object such as the morphology of the specimen and its dielectric properties: refractive index and birefringence. In particular, it provides true absolute phase contrast of the specimen, which will be interpreted in terms of biological content and structure such as protein dry mass, structural protein birefringence and cell volume changes.

Second, it appeared that the speckle—defined as the pattern generated by some randomness of the phase of the wave field after diffraction—if there is any, does not play the same role in holographic microscopy as it does in holography of macroscopic objects. The spot size of speckle is given by the diffraction limit of the microscope objective (MO), which, in microscopy, is most often of the same order of magnitude as the details of the specimen itself. Moreover, the details of the specimen morphology are searched at a scale commensurate with the wavelength, which has the consequence that the image features, which, in the context of a macroscopic object, would be considered as speckle noise, can be easily interpreted as specimen characteristics. Phase ambiguity can be most often avoided and unwrapping can be considered as straightforward and manageable by software in many situations, where smoothness and continuity of the specimen shape are verified. In other, more difficult cases where these conditions are not met, two or multiple wavelengths can yield unwrapped phase without ambiguity (Kühn et al. 2007, Montfort et al. 2006b) through so-called optical unwrapping. The phase of the reconstructed wave front therefore appears as absolute and meaningful for optical path length (OPL) measurements.

Third, wave-front reconstruction was deemed to provide true access to the 3D distribution of the wave field scattered by the specimen because the wave-front propagation can be simulated throughout the space surrounding the specimen and inside the specimen by the use of numerical models of wave propagation, fundamentally based on the Helmholtz equation. Provided that the aperture of the hologram could approach the maximum possible value just given by the refractive index (RI) of the medium, which appears as a reasonable achievement of synthetic aperture techniques, the promise of holography to provide true 3D data on the distribution of dielectric properties of the specimen is close to be fulfilled in microscopy.

Finally, a major asset of holography is that full 3D imaging in microscopy could be derived from the acquisition of a single hologram, which is acquired preferably in a very short period of time, avoiding thereby motion blur and vibration effects. This important advantage of holography paves the way to imaging at the nanoscale. This ideal situation can be approached in many favorable cases, which will be described in more details later on. It must be also emphasized that a major innovative aspect of DHM resides in the fact that there is no need of focusing before image acquisition. A drawback of the use of a high numerical aperture (NA) objective in microscopy is the reduced depth of field, which requires a permanent correction of the focus distance by displacing mechanically the MO in order to keep the specimen in focus. This tracking may become a heavy task when the specimen moves freely and rapidly. Resorting to holography provides a simple remedy to this problem. Taking a hologram rather than a focused image postpones focusing to an off-line step. This property of holography is promising for future applications where object identification must be performed by automated procedures. Indeed, the search for the right focus distance can be achieved by convenient algorithms or even off-line after the acquisition of the series of holograms.

29.2 The Concept of Coherence

A wave is a scalar or vector field that extends in the space-time manifold. Electric or magnetic vector fields are basically real data. However, mathematically, it was soon acknowledged that the electromagnetic waves, considered as a solution of d'Alembert equations, and as a direct consequence of Maxwell equations, could be described conveniently by an extension of the real function to the complex functions usually designated as complex analytical signals, spanning a full Hilbert space.

At this point, it must be stressed that the measurement of the energy or average power received by an appropriate detector or camera is insufficient to characterize a wave and to account for its imaging capabilities. It is appropriate to introduce a more general concept to describe the field. This concept is "coherence" which will describe accurately the information content carried by the wave and its potential to image a specimen. The concept of mutual coherence of the wave w_1 correlates its value at a point $\mathbf{r}_1$ and a time t_1 with the wave w_2 at a point $\mathbf{r}_2$ and a time t_2:

$$\Gamma_{1,2}\left(\mathbf{r}_1, t_1, \mathbf{r}_2, t_2\right) = \left\langle w\left(\mathbf{r}_1, t_1\right) w^*\left(\mathbf{r}_2, t_2\right) \right\rangle_e, \quad (29.1)$$

where the average $\langle\ \rangle_e$ is taken as an ensemble average over all realizations of the wave w considered as a random variable, for a given, stationary, optical arrangement.

A normalized version of the mutual coherence is called the partial degree of coherence: $\gamma_{1,2}(\mathbf{r}_1, t_1, \mathbf{r}_2, t_2) = \Gamma_{1,2}(\mathbf{r}_1, t_1, \mathbf{r}_2, t_2)/(\Gamma_{1,1}(\mathbf{r}_1, t_1)\Gamma_{2,2}(\mathbf{r}_2, t_2))$ and varies continuously from 0% to 100%.

The right-hand element describes the correlation both in time and space of the wave sampled at two different time and space points. Intuitively, it is easy to understand that nonzero correlations will help to reduce the number of degrees of freedom required to characterize the wave completely. Basically, this will contribute to evaluate the wave occupation of phase-space and contribute to the evaluation of the information content, usually called entropy. A photon gas can be described as a collection of quantum states generated by a compound wave or as a superposition of a plurality of elementary waves. In the extreme case, no correlation exists between any of the pair of elementary waves. This is the case where the entropy is maximal and the wave is considered as fully incoherent. This is also a qualifier that applies to the optical source: incoherent sources have been used for centuries in the early optical systems and microscopes in particular. This is still the dominant modality in contemporary microscopy: white light illumination, fluorescence imaging, confocal and further combinations of these modalities. On the opposite, for a wave emitted by a perfect laser (Gaussian or Bessel mode), the coherence is maximal for any pair of points.

Only quite recently, coherence properties have been considered in microscopy: coherence in the time domain has been firstly exploited in OCT (see Chapters 13 through 15). With holography, coherence is also exploited in the spatial domain; the interest is naturally an enlargement of the perspectives brought by the exploitation of coherence in the space-time domain.

In this theory, called second order coherence, the expression of the mutual coherence supposes no limitation or no drop of the correlations between distant points $\mathbf{r}_1$ and $\mathbf{r}_2$ in space and t_1 and t_2 in time. This assumption is unrealistic as any signal is supposed to be limited in space and time because their energy is supposed to be finite. The mathematical correlate is that the signals are described by square-integrable functions, as it is usually admitted that the space and time domains are homogeneous and do not contain any absolute origin. The calculation can therefore be carried out by referring the mutual coherence to the first $(\mathbf{r}_1, t_1)$ points and considering the differences: $\rho = \mathbf{r}_2 - \mathbf{r}_1$, and $\tau = t_1 - t_2$, we have: $\Gamma_{1,2}(\mathbf{r}_1, t_1, \mathbf{r}_2, t_2) = \Gamma(\mathbf{r}_1, t_1, \rho, \tau)$.

This suggests a so-called first-order coherence theory where the correlation is supposed to decrease to zero with the distance ρ and time τ. This reduction of the concept can be implemented with the concept of Wigner distribution.

Mutual coherence is a useful theoretical concept but does not end up with a practical measuring technology: measurement of the correlation between distant points is not straightforward. What is directly measured by a detector or a camera is the energy or average intensity in a 4D voxel-time interval: $\delta\mathbf{r} \times \delta t$. Therefore, the observable physical signal corresponds to the following conditions: $\mathbf{r}_1 = \mathbf{r}_2 = \mathbf{r} \pm \delta\mathbf{r}$ and $t_1 = t_2 = t \pm \delta t$, so that

$$\Gamma_{1,2}\left(\mathbf{r}_1, t_1, \mathbf{r}_2, t_2\right) \simeq \Gamma_{1,1}\left(\mathbf{r}, t\right) \simeq \Gamma_{2,2}\left(\mathbf{r}, t\right) = I\left(\mathbf{r}, t\right) = \left\langle i\left(\mathbf{r}, t\right) \right\rangle_e. \quad (29.2)$$

A different approach has been brought by holography: the mutual coherence is evaluated by decomposing the primary wave w into two or several waves with appropriate cross correlation properties: $w(\mathbf{r}, \mathbf{t}) = u(\mathbf{r}, \mathbf{t}) + v(\mathbf{r}, \mathbf{t})$ and where one of the waves is derived in a precise way and considered as a reference wave.

The energy collected by a physical detector, a recording medium, or a camera is the integrated intensity of the electromagnetic field over time and space. Considering the scalar waves, $u(\mathbf{r}, t)$ and $v(\mathbf{r}, t)$ not as mere sinusoidal waves, but as random waves in the space and time domain.

The statistical estimation of the wave intensity at the point $\mathbf{r}$ and time t is given by the expression:

$$\left\langle i(\mathbf{r}, t) \right\rangle_e = \left\langle \left(u(\mathbf{r}, t) + v(\mathbf{r}, t)\right)\left(u(\mathbf{r}, y) + v(\mathbf{r}, y)\right)^* \right\rangle_e$$

$$= \underbrace{\left\langle \left|u(\mathbf{r}, t)\right|^2 + \left|v(\mathbf{r}, t)\right|^2 \right\rangle_e}_{\text{diagonal terms}} + \underbrace{\left\langle u(\mathbf{r}, t) v^*(\mathbf{r}, t) \right\rangle_e}_{\text{cross term}}$$

$$+ \underbrace{\left\langle v(\mathbf{r}, t) u^*(\mathbf{r}, t) \right\rangle_e}_{\text{cross term (conjugate)}} \quad (29.3)$$

Considering the Fourier transform of u and v in the time domain: $U(\mathbf{r}, \omega)$ and $V(\mathbf{r}, \omega)$, the cross term in the Fourier domain becomes: $\left\langle U(\mathbf{r}, \omega)V^*(\mathbf{r}, \omega) \right\rangle_e$ which must be considered, according to the Wiener–Khintchine theorem, as the Fourier transform of the cross correlation of $u(\mathbf{r}, t)$ and $v(\mathbf{r}, t)$ denoted: $\left\langle (u \otimes v)(\tau) \right\rangle_e$, where τ is the delay introduced between both signals.

For an ergodic system or, practically, a stationary system (invariant in time), or a slowly varying nonstationary system, the ensemble average can be considered as an average over a time period which in principle goes to infinity, but practically can be taken as a finite interval, depending on the spectral characteristics of the signals: $\langle\ \rangle_e = \langle\ \rangle_{time}$. The average is taken over a time interval which goes to infinity.

Similarly, the same development holds in the spatial domain: considering the Fourier transform in the spatial domain: $\tilde{U}(\mathbf{k},\omega)$ and $\tilde{V}(\mathbf{k},\omega)$, the cross term in the Fourier domain becomes: $\langle\tilde{U}(\mathbf{k},\omega)\tilde{V}^*(\mathbf{k},\omega)\rangle_e$ which must be considered, according to the Wiener–Khinchine theorem, as the Fourier transform of the cross correlation, in the spatial domain of $U(\mathbf{r},\omega)$ and $V(\mathbf{r},\omega)$ and is denoted: $\langle(U \otimes V)(\rho,\tau)\rangle_e$, where ρ is the distance seperating both signals.

Deriving a value of the cross term $\langle u(\mathbf{r},t)v^*(\mathbf{r},t)\rangle_e$ from experimental measurements of the intensity by an appropriate algorithm (e.g., filtration) makes it possible the restoration of the full complex value of the wave $u(\mathbf{r},t)$ from the reference wave $v(\mathbf{r},t)$.

In this context, the cross correlations of two waves appear as an inner product of complex waves and therefore generate a metric space which confers to the set of complex waves the properties of a Hilbert space. Thanks to the linearity properties of those spaces, decomposing a complex wave in its elementary waves is made possible by their projections on a complete set of basis functions. These features are at the foundation of coherent imaging techniques. This projection can be achieved physically by the recourse to the measurement of the intensity of the interference resulting from the superposition of two optical waves, $u(\mathbf{r},t)$ and $v(\mathbf{r},t)$, as shown in Figure 29.1.

29.2.1 Coherence Exploitation

Several imaging approaches have been proposed to employ fully the mutual coherence (at any degree) of beams generated by optical sources used for their coherence properties. Coherence can be exploited to reconstruct the complex wave front emanating from the specimen. The requirements are indeed only slightly restrictive in microscopy. Only some degree of mutual coherence is needed to permit the evaluation of the mutual coherence between the object wave front (denoted O) scattered by the specimen, and a reference wave (denoted R) or even between O and itself, i.e., the space or time shifted O. The phase data can be derived from the intercorrelation or between O and R or possibly from the autocorrelation of the wave field O. In holographic microscopy, the coherence length either in the spatial or in the time domain has only to be comparable to the size of the specimen. Further on, it may be desirable to restrict the coherence length to delimitate the part of the specimen to be imaged: this "coherence gating" has been exploited soon in holography by introducing the concept of "time in flight holography," where the coherence length could be reduced by using pulsed laser sources, see Abramson (1983) or later Carlsson (1993). Further on, broadband or so-called ultra-broadband sources have been designed and used to reduce at will

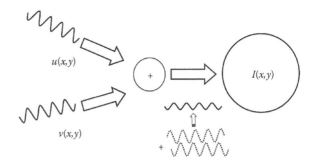

FIGURE 29.1 Use of the interference between the wave u and a well-characterized wave v for determination of the mutual coherence function through intensity measurement.

the coherence length in the time domain, whereas the spatial coherence is kept at high value by the use of pinholes or monomode optical fibers. The principle of coherence gating has been at the origin of OCT which is a widely used imaging technique of turbid media (Huang et al. 1991). For a review paper, refer to Fujimoto et al. (2000) and Chapters 13 through 15 of this book. Concerning coherence exploitation, see also Psaltis (2002).

It is worth to observe that a similar trend has been developed in the spatial domain: spatially distributed sources can be used to illuminate the object. They are selected for their reduced spatial coherence, whereas the temporal coherence has been kept large enough to keep noncanceling mutual coherence over the propagated field and therefore to preserve the capability of forming a hologram (Dubois et al. 1999). In a more general perspective, it is worth to bring out the fact that the mutual coherence of the random wave field and a reference wave field form a coherence wave which is propagated similarly to the optical wave field in holography. This property was coined as "coherence holography" by Takeda (Takeda et al. 2005) and can be obtained by computation. The source can therefore be made deliberately random in the spatial domain and provide an autocorrelation function approaching a delta function for infinite space-bandwidth product. The specimen hologram can therefore be obtained by computing the propagation of the mutual coherence function.

In an extreme case, time and spatial coherence can be practically reduced to zero by using spatially extended white light sources. This result can possibly be achieved using conventional thermal sources (Degroot and Deck 1995, Vabre et al. 2002, Roy et al. 2002, Dubois et al. 2004a). The possibility of recovering the wave field from mutual coherence data is however limited in this case and the reconstructed image is limited to the size of the field of view or spot where the autocorrelation function does not cancel. An in-depth scanning mechanism is therefore needed to restore the full 3D image, such as in white light interferometry.

29.2.2 Exploitation of the Mutual Coherence Properties of Crossing Beams

The considerations concerning the mutual coherence between the object wave O and the reference wave R can be generalized to a situation where the coherence properties of a beam *B* can be analyzed by the mutual coherence properties of beams B_1 and B_2 resulting from the division of *B*.

For beams presenting a high or, more generally, significant degree of mutual coherence, the superposition of beams, or wave fields, here called B_1 and B_2, having any, but some degree of coherence, i.e., nonvanishing mutual coherence terms, can be exploited to restore the wave front emitted by the object and therefore provide, after reconstruction, an indirect 3D image of the object. The sketches of Figure 29.2 illustrate schematically the principle of wave-front reconstruction from the detection of interfering beams.

The excitation beam *E* emitted from the sourced *S* is coherently divided into two beams: E_1 and E_2 by a conventional beamsplitter or by other optical means, like gratings and more generally diffracting optical elements or spatial light modulators (SLM). Both beams, or at least one beam can be transformed by an "optical operator" or what has been called coherent optical information processing system (COIPS), which, most generally, is an optical device which acts as a linear processor (e.g., see Goodman (2005), Butterweck (1977)). Its action appears as a convolution in the direct space or a product with a phase mask in the spatial frequency domain. Most often, this is a spatial filter, which cuts the high spatial frequency components and performs averaging of the wave field. The resulting phase appears as an averaged phase over the wave front.

Three principal scenarios can be developed to "structure" 3D space and derive full 3D imaging of object or specimen in microscopy:

1. According to Figure 29.2a, the source wave *E* goes up to or through the specimen, generating an object wave *O* which is analyzed by the instrument making interfere the beams O_1 and O_2 before detection by the detector or camera *D*.

    a. In one implementation, the object beam *O* is split by a beamsplitter into the beam O_1 and the beam O_2 which, after filtration of the high spatial frequency components (COIPS), make appear a filtered wave ("pseudo-reference") providing an average phase of the object wave which will interfere with O_1. This approach has been proposed by G. Indebetouw as "space-time digital holography" (Indebetouw and Klysubun 1999b, 2000, 2001, Indebetouw 2001). More recently, a similar approach, where the beamsplitter is composed of a transmission grating, has been proposed by G. Popescu and coined as "diffraction microscopy" (Popescu et al. 2006, 2008).

    b. Generalizing this beam analysis principle, one must consider the approach where a diffracting element is inserted in the object beam before impinging on the detector, generating thereby a plurality of diffracted beams which interfere with the primary beam, forming therefore interference patterns (lines) to be interpreted in term of local intensities and wave vectors. The idea is already ancient and has been applied first

to x-rays: the concept of Talbot self-imaging (Patorski 1989) has been developed to image objects and interpreted to provide x-ray phase contrast (Clauser and Reinsch 1992, Momose et al. 2003). More recently, different similar approaches have been applied to optics. A common-path interferometer can be realized by going through the object with the exciting or illuminating beam *E* which is then divided into two beams O_1 and O_2 by a grating situated on or just after the specimen plane, and then recombined by a second diffraction grating situated at some distance in front of the camera (Mico et al. 2006a,b). The high orders of diffraction bring, in combination with multiple off-axis incidences, an increase of the acceptance angle by aperture synthesis. The result is improved resolution or superresolution (Schwarz et al. 2003, Micó et al. 2008, Garcia-Sucerquia et al. 2008). More fundamentally, shearing interferometers based on the splitting of the common-path beam *B* by beamsplitters (Murty 1970), arrays of microlenses (Shack and Platt 1971) or gratings (Primot and Sogno 1995, Primot et al. 1997, Legarda-Sáenz et al. 2000, Nugumanov et al. 2000, Chanteloup 2005, Velghe et al. 2005) can be employed.

    c. In another implementation of the same principle, the source and the specimen are combined: the fluorescence of the object is used as an exciting beam to form an interferometric image. A two pupils system is used to split the object beam into two spherical waves interfering with each other: the results of the interference are also Fresnel patterns and can be applied to the reconstruction of the original wave by back-propagation. Moreover, the recipe can be applied to any punctual source and does not require particular coherence properties between the different sources. It can be applied to fluorescence emitters; the so-called Finch technique is based on this approach, where the two pupils are generated by a SLM at two orders of diffraction (Rosen and Brooker 2007a,b, 2008). This approach has sometimes been called incoherent holography.

2. Sketch in Figure 29.2b illustrates the arrangement where the object is inserted in one arm of the interferometer: the beam E_1 intercepts the object, generating thereby the object beam *O*. The other beam O_2 provides the reference beam *R*, which will interfere with *O*, forming the hologram that will be used for wave-front reconstruction.

    a. This situation corresponds to what can be called "coherent holography": the complex wave provides a "map" of the 3D space, pinpointing the scattering centers and is structured by the waves. Processing numerically the hologram permits the restoration of the complex wave, which can be then propagated back to the object volume. Algorithms have to be developed to retrieve true 3D data from one or several reconstructed wave fronts. This imaging modality

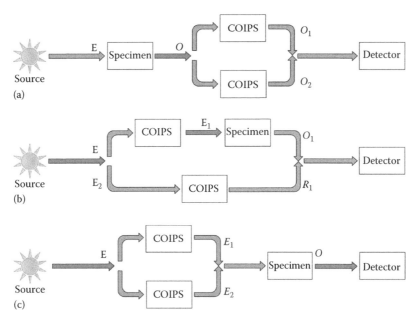

FIGURE 29.2 Reconstruction of a wave field from its interference with a reference wave: The beam E emitted from the optical source is divided into two beams: E_1 and E_2 by a conventional beamsplitter or another optical component. One beam (E_2 on the sketch) is just propagated in free space or transformed by an optical operator called coherent optical information processing system (COIPS), which can be a spatial filter and will serve as a reference wave. Both beams are then recombined to form an interference pattern recorded by a detector which is most often a camera. Three scenarios are presented: (a) the specimen is placed just after the source: the diffracted beam interferes with itself. (b) The object intercepts the beam E_1 and generates an object beam which interferes with the reference beam E_2: scenario of conventional holography. (c) E_1 and E_2 interfere to form an interference pattern, which illuminates the object: structured illumination. The object can be moved in the interference pattern: this is the scenario of scanning holography.

is certainly the most often used and the most developed. It has provided exploitable results in various application domains. The theoretical aspects as well as application domains will be described in details in the second part of the chapter.

b. "Reduced coherence holography": in many circumstances, it may be highly profitable to decrease significantly the coherence of the source. The speckle can be eliminated and the reconstructed image thus becomes much less noisy. The reduced coherence provides also a mean to delimitate in space the part of the object, which will deliver a wave with nonvanishing mutual coherence with the reference wave. This approach permits the exploitation of a particular tomographic imaging modality. Coherence can be reduced either in the time or in the spatial domain:

 i. *In the time domain*, by using broadband sources such as semiconductor diodes, superluminescent diodes, or filtered supercontinuum source obtained for instance with pulsed laser excited photonic crystal fibers. Coherence gated images have been obtained by different groups (Cuche et al. 1997, Indebetouw and Klysubun 1999a, Pedrini and Schedin 2001, Massatsch et al. 2005).

 ii. *In the space domain*, LED sources or rotating ground glass have been used by F. Dubois to achieve reduced spatial coherence and therefore a strong reduction of the coherent artifact noise (Dubois et al. 1999, 2002, 2004b, 2006a).

In fact, the requirement for an illuminating source having minimal temporal and spatial coherence can be completely abandoned, provided that mutual coherence properties can be fully exploited to establish and reconstruct the propagation, not of the wave field itself, but of the mutual coherence function:

c. "Incoherent holography" is an expression that has been used to describe the limiting case of weakly coherent or "quasi-coherent" light to generate holograms. Holograms can be generated if the optical path lengths of the object and reference paths are almost invariant, as it is the case in particular for conoscopy. It describes indeed the limit situation where each emitting point of the source interferes only with itself. More recently, other proposals have been, paradoxically, the exploitation of the coherence properties of a beam which is not restricted to the beams having full degree of coherence properties, as in auto-coherence or mutual coherence. Random wave fields having any degree of coherence, even if they have auto-coherence tending to zero, but mutual coherence with an arbitrary reference wave significantly different from zero. The properties of this mutual coherence field can be exploited to restore an image of the object: this approach has been coined by the designation of "coherence holography" (Takeda et al. 2005). In particular, the propagation of the wave through the specimen can be computed from the propagation of the mutual coherence function or wave. The exploitation of random fields can therefore be exploited to characterize the propagation of the wave field and could be used in the future to form an image of the object or specimen in biology.

3. The scenario evoked on Figure 29.2c is the following: the two exciting beams E_1 and E_2 are generated as two interfering beams which are superimposed in the object volume, the resulting pattern acts as illuminating or exciting beam of the object: this is a structured illumination providing an indirect map of 3D space. The resulting optical signal is captured by the detector and recorded during the translation of the object through the illuminating field: the signal appears as the result of the convolution of the object by the interference pattern. In a so-called two-pupils system, the excitation beams E_1 and E_2 are generated as spherical waves with different or even opposite curvature. The result of the interference appears as a set of fringes featuring paraboloid or hyperboloid, the section of which though planes perpendicular to the optical axis are Fresnel patterns. The restoration of the distributed optical signal emitted by the object appears therefore, in this context, as a convolution with Fresnel waves. Interestingly, the mathematical operation is similar to the reconstruction of the 3D distribution of wave field in digital holography. This similarity in the processing of the data justifies this designation of "scanning holography" (Poon 1985, Indebetouw et al. 2000, 2005, 2007, Schilling and Templeton 2001, Indebetouw and Zhong 2006). The nature of the optical signal collected by the detector is diverse: coherently scattered light (Indebetouw et al. 2006) or incoherently emitted fluorescent light (Indebetouw and Zhong 2006). Phase images as well as fluorescence images can be obtained in parallel. Phase shifts can be achieved by heterodyning E_1 or E_2 beams, providing a mean to capture and process in the temporal domain the hologram acquired in the spatial domain. Of course, the drawback is that a 2D scan is required to obtain the hologram of a 3D object, rendering the imaging process slow. However, it has been shown that scanning holographic microscopy offers a number of additional benefits in addition to being able to capture the 3D holographic information of incoherently scattering (fluorescent) specimens as well as providing a quantitative measure of the phase of the specimen. Perhaps the most significant benefit of scanning holographic microscopy is that it can operate in both an incoherent mode and a coherent mode simply by varying the detector geometry.

29.3 Holography

The development of holography is a direct consequence of the theory of diffraction. In its most general sense, holography covers the reconstruction of a full 3D wave field from one or several holograms taken on a 2D surface limiting or surrounding the volume of interest.

These properties of the wave fields are the basis of holography where holograms contain, in their cross terms, the result of the cross correlation of the wave field with basis functions formed by the reference waves: plane waves or more generally spherical waves. Embracing this point of view makes holography appear as a technique of reconstruction of the wave from its projections on a Riesz basis (Liebling et al. 2003).

The goal of holography is to restore exactly the wave field O from one or several holograms taken at different time or at different locations and orientations in space. A simple and robust way to achieve this goal is to use an external reference wave R having sufficient mutual coherence with the object wave O. A common technique is to illuminate the object and generate the reference beam with the same source designed to have convenient spatial and temporal coherence properties. Occasionally, R can be also derived directly from O itself. The result of the interference between O and R measured on a surface is traditionally the hologram. Its intensity distribution $I_H(x,y,t)$ on a 2D plane is given by the expression

$$I_H(x,y,t) = (R+O)^\star \cdot (R+O) = |R|^2 + |O|^2 + R^\star O + RO^\star \quad (29.4)$$

29.3.1 Square Terms

The two first terms, $|R|^2(x,y,t)$ and $|O|^2(x,y,t)$, of the right member of Equation 29.4 are the intensity distribution of the object and reference waves over the hologram plan. They are designated as the "zero-order terms" because their temporal and spatial variations are supposed to be weak and slow, and their bandwidths are limited both in time and space. It is therefore admitted that they do not give rise to high orders of diffraction. This is commonly the case of $|R|^2$, but not necessarily for $|O|^2$ which must be often artificially or naturally limited in its bandwidth to comply with this requirement.

Equation 29.4 has been worked out both in the time and spatial domain. In the past, spatial domain has been considered traditionally for holography because of the characteristics of the recording media: photographic plate and photopolymers. On the contrary, for the so-called coherence imaging, time domain was considered first because electronic detectors have been used as measuring device of the field intensity $I_H(t)$. Today, both approaches are converging and mixed solutions are often developed.

In the time domain, $|R|^2(t)$ appears as the sum of one, for monochromatic sources, or many complex exponentials for polychromatic or broadband sources. $|R|^2(\omega)$ can be calculated as the autocorrelation of the spectrum of $R(\omega)$ in the time domain, and therefore appears as a Dirac function for monochromatic sources and approximately as a bell shaped, often assimilated to a Gaussian envelope, surrounding the sinusoidal signal for broadband sources. The same is true for $|O|^2(t)$ and $|O|^2(\omega)$. These are permanent terms, present even if O and R waves have no mutual coherence. $|R|^2(t)$ is in most situation stationary or periodic, over time. $|O|^2(t)$ is more complex since it carries the data corresponding to the volume distribution of scattering matter and of its evolution in time. $|O|^2(\omega)$ can be interpreted as the autocorrelation of the complex spectrum of the complex object wave in the time domain. This autocorrelation is a complex signal that acts often as a perturbing term in the evaluation of I_H (Seelamantula et al. 2008) in the time domain as well as in the spatial domain.

Concerning the extension over space of the square terms, $|R|^2(x,y)$ is constant or slowly varying over space for most of usual reference wave: plane or weakly spherical. Its identification and elimination in $I_H(x,y,t)$ is in general easy: either in the space domain by subtracting the average intensity over space or in the spatial frequency domain (SFD) by eliminating the quasizero frequency component of the spatial spectrum. Delineating and eventually eliminating $|O|^2(x,y)$ is more complex: similarly to $|O|^2(\omega)$ in the time domain, The Fourier transform of $|O|^2(x,y)$ in the space domain $|O|^2(k_x,k_y)$ can be interpreted as the autocorrelation of the complex spatial spectrum of the complex object wave $O(x,y)$. Its extension in the space or in the spatial frequency domain depends on the spectrum of the object and of the scattered object wave, as well as of the transfer function of the optical setup. If this spectrum is limited, $|O|^2(k_x,k_y)$ can be delineated in the SFD and therefore removed in the SFD and therefore in the spatial domain. On the contrary, techniques have been developed to restore $O(x,y)$ from $|O|^2(x,y)$ in the space domain: the problem consists in making a "guess" on the complex wave field $O(x,y)$ and adjusting the propagated intensity $|O|^2(x,y,z)$ to the actual measured intensity. RMSE minimization scheme are developed to solve this task. In particular, iterative algorithms have been proposed for phase retrieval from intensity data (see Fienup 1982). In particular, the so-called Gerchberg and Saxton (Gerchberg and Saxton 1972), the error reduction approach (Fienup 1978, 1982) and Yang and Gu (Yang and Gu 1981) algorithms have been adapted to solve the problem of the determination of the complex wave field $O(x,y)$ in microscopy. In many situations, the problem appears however as ill-posed. It is computer intensive and the applications in optical microscopy appear quite limited. Another approach is based on the measurement of $|O|^2(x,y)$ on planes situated at a plurality of axial distances z (Teague 1983, 1985). Quantitative phase imaging can be derived from the so-called intensity transport equation (Gureyev et al. 1995). The method has been applied successfully to various domains in microscopy (Nugent et al. 2001).

29.3.2 Cross Terms

The last two terms in (29.1), R^*O (x,y,t) and $RO^*(x,y,t)$, are the "cross terms": $\langle R^*O(x,y,t)\rangle_{\text{time}}$ and $\langle R^*O(x,y,t)\rangle_{\text{time}}$ express the mutual coherence of the object wave and the reference wave. They do not vanish provided that some degree of coherence exists between both waves. Holography makes explicit use of these cross terms to provide a reconstructed complex wave front, but, more generally, most of the interferometric techniques, including white light interferometry and OCT have recourse to the quantization of cross terms in the time or spatial domains. In classical holography, the cross terms cause the appearance of the high-density fringes on the recording media. When illuminated by a light beam having characteristics similar to the reference, the object beam, diffracted to the first order, matches the original wave field, which is thereby fully restored in amplitude and phase. Similar approach has been developed in digital holography: the wave-field restoration is based on the digital evaluation of these cross terms: they provide a simple access to the true complex value of the wave front O, respectively O^*, just by multiplication of R, respectively R^*. The results are respectively $|R|^2O$ and $|R|^2O^*$ and therefore O, respectively O^*, because $|R|^2$ is approximately constant over the (x,y) detector, for a plane or weakly diverging reference wave. $|R|^2$ appear therefore only as a scale factor, and O matches the original wave field scattered by the object, which is often situated behind the detector: in this situation, it is called a virtual image of the object. O^* is to the complex conjugate of the object wave: when reconstructed in the plane of the detector, it appears similar to the original object wave, but with inverted value on the optical axis z after propagation. The wave appears therefore as a mirror image of the original wave, the mirror plane being given by the hologram plane. It will therefore converge on an image of the object situated most often in front of the detector plane or hologram plane. It is called the "real image" and can be restored in digital holography as a mirror image of the original object.

In digital holography, isolating one of the cross terms, generally R^*O which corresponds to the virtual image, is sufficient to fully reconstruct the complex object wave field, yielding indirectly the 3D object image. In a similar way, filtering the cross term RO^* yields the mirror image of the wave field, which therefore appears as the mirror image of the object. It is however mandatory to distinguish between them to avoid undesirable superposition of both. A first step is therefore to filter the hologram intensity $I_H(x, y, t)$, to keep only the contribution of R^*O. Practically, the exact value of the reference wave R that should be used for reconstruction of the virtual image is unknown. Therefore, an attempt of reference wave R_{attempt} is generated in the computer by using a mathematical model and must be further adjusted to match as closely as possible the original reference wave R:

$$O_{\text{reconstructed}}(x,y,t) = \frac{I_{\text{H filtered}}(x,y,t) \cdot R_{\text{attempt}}(x,y,t)}{\left| R_{\text{attempt}}(x,y,t) \right|^2} \quad (29.5)$$

The step of computing $I_{\text{H filtered}}$, i.e., separating, in the recorded intensity of the hologram $I_H(x,y,t)$, the contribution of the virtual image from the real (sometimes called twin image) and from the zero-order, is not a trivial task. This step can be considered as filtering $I_H(x,y,t)$ seen as a space and time function, so that filtering can be achieved in the time domain or in the space domain. Linear and nonlinear filtering techniques are appropriate for this task.

29.3.3 Digital Holography by Analysis of Data in the Time Domain

In the time domain, several holograms (at least three, but usually four) are taken successively in order to filter out the zero-order term and twin image: typically dephasing the reference wave by a multiple of $2\pi/3$ or $\pi/2$ permits both to eliminate the zero-order term as well as the twin image: subtracting the average taken of the three or four holograms from the hologram to be reconstructed permits to eliminate the zero-order term. Similarly, because advancing or retarding the phase of the reference beam modifies the cross terms with opposite signs, adding or subtracting selected terms of $I_H(x,y,t)$ in accordance to the phase shift permits also to eliminate the twin image. This method is directly derived from the so-called phase-shifting interferometry used in interferometry and has been proposed by I. Yamaguchi (Yamaguchi and Zhang 1997, Zhang and Yamaguchi 1998) for application in holography. The advantage is the preservation of the full spatial bandwidth, but a major inconvenient is that multiple holograms must be taken which renders the reconstruction sensitive to movement blur and vibrations. Instantaneous images are impossible to take. So-called heterodyne techniques have however been used to implement faster phase shifting techniques for numerical holography and interferometry (Le Clerc et al. 2000, Choi et al. 2007) and speckle interferometry (Aguanno et al. 2007, Chalut et al. 2007); the reference wave is frequency shifted in order to generate a beating wave as a result of the interference between object and reference wave. Modulating the reference wave (in general, with a pair of acousto-optic modulators to provide low frequency modulation, compatible with the imaging rate) permits to reach higher image acquisition speed. However, this rate is limited to the kilohertz range by the performance of the camera.

29.3.4 OCT

In the time domain, polychromaticity can be exploited to reduce the time window and therefore the spatial domain, where coherent imaging can be achieved. The time average of the cross term

$$\left\langle R^*\left(x,y,t_1\right)\cdot O\left(x,y,t_2\right)\right\rangle_{\text{time}} = R^* O\left(x,y,\tau\right), \quad (29.6)$$

where $\tau = \tau_2 - \tau_1$ is the time intercorrelation of R and O. Considering broadband sources, the autocorrelation of the illuminating and reference wave, given by $R^*R(x,y,\tau)$, is approximated by a sinusoid with a Gaussian envelope characterized by its mid-height width $\langle\tau\rangle$. The traveled distance during this time

interval is the coherence length of the beam. The intercorrelation $R^*O(x,y,\tau)$ divided by the autocorrelation $R^*R(x,y,\tau)$ provides the "echo" in amplitude and phase generated by a scattering center complex spectrum of the time intercorrelation of O with R. The amplitude of the provides a straightforward tomographic image "in depth" of the specimen, when the illuminating and object wave are traveling along the optical axis z. Scanning the τ value by moving a reflective mirror in the reference arm provides a scan of the specimen in depth. This tomographic technique is called, time domain OCT (TDOCT) (Huang et al. 1991). More recently, it has been taken into account that OCT data could be retrieved directly in the Fourier domain by considering the Wiener–Khinchine relation applied to the measurement of the spectral density of the cross correlation of the O and R waves. This Fourier transform can be carried out instrumentally just by dispersing the measured intensity with a spectrograph equipped, e.g., with a transmission or reflection grating. This is the so-called FDOCT (Fourier domain OCT) (Leitgeb et al. 2004). In this approach, the emphasis is given to the reconstruction of $O(x,y,\omega)$ from $R^*O(x,y,\omega)$. This can be achieved along the axis z and the step of computing the full 3D reconstruction by the computation of wave-field propagation is generally skipped because the depth of field is large enough to cover the specimen thickness. But this extended depth of field is usually achieved by limiting the NA of the optics or having recourse to Bessel beams, which has the consequence of limiting the lateral resolution to values higher than the micron. The reconstruction of $O(x,y,\omega)$ can be achieved punctually and the determination of the transverse component of the tomographic image is obtained by beam scanning. This approach has the advantage to combine the features of confocal microscopy in terms of out-of-focus light rejection and of interferometric sensitivity. Another approach consists in performing this operation in parallel and the result of the interference measured by an x–y camera matrix or a special CMOS device restoring the phase: this is the parallel OCT (Bourquin et al. 2001) or bright field OCT (Dubois et al. 2004a). These approaches require a vertical scanning if the depth of focus is limited. A tentative to restore the full 3D wave field by FDOCT and to propagate it in order to extend the depth of field by computation of the propagated wave field is however described in the paper of Yu et al. (Yu and Chen 2007).

29.3.5 Digital Holography by Analysis of Data in the Space Domain

Filtering the cross terms $R^*O(x,y,t)$ and $RO^*(x,y,t)$ is also performed in the spatial domain. Expression (29.1) can be also profitably analyzed in the spatial frequency domain (SFD): let us consider the spatial Fourier transform of $I_H(x,y,\omega)$: $I_H(k_x,k_y,\omega)$. Let us also consider first the case of a monochromatic source, corresponding to a fixed ω. In the spatial frequency spectrum, the zero-order term $|R|^2(k_x,k_y)$ appears usually as a localized contribution for a plane wave approximately described by a delta function at the origin $k_x,k_y = 0$, which can be easily removed from $I_H(k_x,k_y)$ in the SFD by subtracting the average intensity of the reference wave $|R|^2(x,y)$ or $|R|^2(k_x,k_y)$. The elimination of

$|O|^2(k_x,k_y)$ is more problematic because it depends on the spectrum of the object itself: $O(k_x,k_y)$ which is supposed to naturally or intentionally limited by the NA of the objective or more generally by the pupil of the optical setup. $|O|^2(k_x,k_y)$ is, mathematically, equal to the autocorrelation of the spatial spectrum of the object wave $O(k_x,k_y)$. It also provides the spectral density of the object wave in the (k_x,k_y) domain. In the case of a monochromatic illumination, the spatial spectrum is limited, usually by instrumental considerations, when a limiting aperture is involved in the hologram formation, but also, in the most favorable instrumental arrangement, where the aperture at its maximum, i.e., by the limited value of the beam wave vector in the observation medium, sometimes vacuum. Aperture is limited by the instrument when a lens or a beamsplitter is introduced in an optical setup used to generate the hologram, or simply by the limited size of the detector itself. These bandwidth limitations introduce naturally a band-stop or band rejection filter in the SFD. The presence of this band-stop can be built on to filter out the contribution of zero-order terms: The occupation of the SFD is therefore limited to an area or most often a disk of diameter twice as large as the bandwidth of the optical system bandwidth. This problem has been described in basic textbooks such as (Goodman 1968). Figure 29.3 illustrates the main steps in the wave-front reconstruction from a hologram taken in off-axis configuration: part a) shows the hologram formation by the interference of the original wave front propagated from the object in the $\xi-\eta$ plane: here a USAF test target to the hologram plane $x-y$ at a distance d. Due to the band-stop BS described earlier, k_x, k_y ?∞ BS, and accordingly, the angular spectrum, which is the complex amplitude as a function of the unit vector $\hat{s}\left(k_x/k, k_y/k, \sqrt{k^2-k_x^2-k_y^2}/k\right)$ of the beam diffracted by the object is included in a cone of half aperture ψ. A reference wave, usually a plane or a weakly diverging wave with curvature approximately matched to the object wave, impinges with an angle θ on the hologram plane and creates the interference pattern. High-density fringes corresponding to the cross terms

are formed on the hologram. The recording medium must be selected and positioned so that the information content carried by the scattered beam can be adequately captured. To perform well, the detector must provide both the sampling capacity permitting to satisfy the Shannon criterion for the considered spatial bandwidth (for an electronic camera, spacing between pixels shorter than half of the inverse of the largest wave vector), and the extent of the detector should match or exceed the diffracted image of the object in the Fresnel zone, which is approximately given by the size of the object (the USAF test target in plane $\xi-\eta$ on Figure 29.3) convolved with the interception of the diffraction cone of half aperture ψ with the detector plane $x-y$. In combination, these two requirements are not easily fulfilled by an electronic camera: the upper limit (for a 3.45 µm interpixel distance) corresponds to approximately 1,000 mm^{-1} wave-vector amplitude, which is far less than the wave-vector amplitude in vacuum: around 20,000 mm^{-1} or, in a $n=1.5$ RI medium, approximately 30,000 mm^{-1} for a wavelength around 600 nm. This mismatch must be compensated by limiting severely the angle of the reference beam with the object beam, so that $\theta < 10°$. This limitation could appear as a severe drawback in microscopy, where the need for high spatial bandwidth is crucial for high resolution imaging. However, it will be shown later on how the full bandwidth required for microscopy can be gained by space-spatial bandwidth adaptation: a perfect match of the space-spatial bandwidth product of the detector with the one of the beam scattered by the object can be performed by the recourse to optical components such as lenses or gratings.

Figure 29.3b illustrates the wave-front reconstruction technique classically developed for holography: illuminating the hologram with a beam matching the reference wave R provides a diffracted beam, which precisely reproduces the object wave field: intercepting the reconstruction plane $\xi-\eta$ at distance d, will restore precisely the real image of the object. The cones 0, +1, and −1 represented on part (b) encompass the angular spectrum corresponding, respectively, to the beams diffracted at order zero, plus and minus one. If the angular spectrum of the

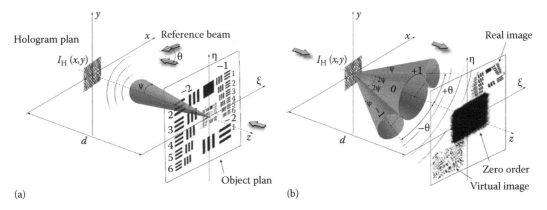

FIGURE 29.3 (a) Hologram formation: the illumination beam (to the right) is diffracted by the object in the plane $\xi-\eta$. The diffracted light inside the cone interferes with a reference wave incident with an angle θ to form the hologram $I_{H(x,y)}$ in the plane $x-y$. (b) The hologram $I_{H(x,y)}$, illuminated from the right, diffracts the illuminating beam at the zero-order and the ±1 order restoring the real image in the plane $\xi-\eta$, and the virtual image (unfocused) as well.

beam diffracted by the object, corresponding to the wave front $O(k_x,k_y)$, is included in a cone with aperture angle 2ψ (see part a)), the beam diffracted by the hologram at order zero $|O|^2(k_x,k_y)$ will be included in a cone with aperture angle 4ψ. The cross terms $R^*O\ (x,y,t)$ and $RO^*(x,y,t)$ diffract the light at the +1 and −1 order. Figure 29.3 shows that the angular spectra of these two terms are included in satellite cones denoted +1 and −1, with aperture angles of 2ψ. The axis of the cones diffracted by the hologram make an angle θ with the axis of the zero-order cone. These two orders of diffraction give rise to virtual and real images, which appear as twin images. If the angle θ is chosen carefully, i.e., $\geq 3\psi$, the cones of the ±1-order cones can be made completely distinct from the zero-order cone. On the contrary, if θ is chosen $<3\psi$, i.e., if the reference wave is taken close to "on-axis" geometry, i.e., propagating almost in parallel to the object wave O, real and virtual images would appear as superimposed in any plane where the reconstruction is performed. The specimen image would appear therefore as "blurred" by the presence of the unfocused twin image.

In summary, in classical holography, digital or nondigital, the optimal occupation of the angular spectrum or spatial frequency domain must therefore be based on the following considerations:

1. For free propagating O and R fields, the angle θ is limited to θ_{max}, a value that is given by the sampling capacity of the recording media, a camera for DH.
2. The band-stop of the object wave field should be defined by choosing the band-stop of the object field defined as half the aperture angle of the total angular spectrum: $\psi = \theta_{max}/6$.

Recently, however, a significant step has been passed toward getting free from the classical holography scheme imposed by optics considerations relevant to its physical nature: advantage can be taken of the full numerical processing of hologram data by introducing nonlinear treatment of the hologram intensity to intrinsically suppress the autocorrelation terms: in particular, considering the cepstrum of the hologram permits to get rid of the different order of diffraction composing the hologram and, therefore, to extend considerably the resolved bandwidth: $\psi = \theta_{max}$ (Pavillon et al. 2009b, Weng et al. 2008).

The off-axis geometry translates the object spatial frequency spectrum to high frequencies, beyond the contribution of zero-order terms. The off-axis reference wave (Cuche et al. 1999a,b, Carl et al. 2004) introduces a so-called spatial carrier frequency and after filtration, demodulation, i.e., multiplication by R, translates the R^*O spatial spectrum to the origin and restores the full O spatial frequency content of the wave front (Cuche et al. 2000a). This approach presents the main advantage of having all the information for reconstructing the complex wave field from a single hologram. This hologram can be acquired in a very short period of time: in the microsecond range or even much less by using pulsed lasers. The blur caused by the displacement or erratic movement of the sample or instrument can be therefore avoided. This advantage is of major concern when observing details in the nanorange. Very high acquisition rate is also

possible: many cameras permit acquisition rates in the kilohertz range and therefore reveals well adapted for exploring transient phenomena, in particular for cell dynamics studies in the millisecond time scale.

To conclude this section, the good news are that, as it will be further developed in the precise context of microscopy, it is possible, by a convenient instrumental control of the space-bandwidth product, to manage "at will" the occupation of the SFD by the different terms composing the hologram. In particular, the use of a microscope objective provides a simple mean to adjust the wave-vector scale in order to preserve the correct sampling of the full bandwidth of the MO, so that no loss of resolution will be conceded.

29.3.6 Full Restoration of the Complex Wave Front

Finally, the true 3D distribution of the wave field diffracted by the object can be in principle restored in full 3D from the exact knowledge of the wave-field distribution on a 2D surface delimiting the volume of interest. A result of the diffraction theory teaches us that the exact knowledge of the electromagnetic field on a surface limiting or surrounding a volume permits to derive the *E-M* value at each of the volume point, providing thereby the full restoration of the wave front O. Mathematical expression of the propagated field is given by the Rayleigh–Sommerfeld formula:

$$O(\mathbf{r}') = \iint_{S_A} \left(O(\mathbf{r}) \cdot \frac{\partial}{\partial n} G_k^+(\mathbf{r}-\mathbf{r}') \right) dS, \qquad (29.7)$$

where G_k^+ is the green function for the Helmholz equation, which coincides with the expression of the spherical wave:

$$G_k^+(\mathbf{r}-\mathbf{r}') = \frac{\exp(-ik|\mathbf{r}-\mathbf{r}'|)}{|\mathbf{r}-\mathbf{r}'|}. \qquad (29.8)$$

Easy computation of the derivative of G_k^+ yields the Huyghens–Fresnel expression at the basis of the computation of the propagation of the wave front:

$$O(\mathbf{r}') = -\frac{i}{\lambda} \iint_{S_A} \left(O(\mathbf{r}) \cdot G_k^+(\mathbf{r}-\mathbf{r}') \right) dS. \qquad (29.9)$$

In the context of digital holography, reconstructing the wave front in 3D will therefore be simply done by back-propagating the wave front generated in the hologram plane x–y:

$$\mathbf{r} = (x, y, 0) \qquad (29.10)$$

to the plane of the object ξ-η:

$$\mathbf{r}' = (\xi, \eta, d), \qquad (29.11)$$

situated at the distance d, called reconstruction distance of the object wave scattered by the object. Assuming that the paraxial approximation is valid for the considered geometries, the forward propagation of the wave field can be described by the Fresnel transform in the paraxial approximation, the back-propagation will be given by the inverse Fresnel transform:

$$O(\xi, \eta) = -i \cdot \exp(ikd) \cdot \mathbb{F}^{-1}_{\sigma \text{Fresnel}} \left[O(x, y) \right] \qquad (29.12)$$

where the Fresnel transform is given by

$$\mathbb{F}_{\sigma \text{ Fresnel}} \left[O(\xi, \eta) \right] =$$

$$\frac{1}{\sigma^2} \exp\left[\frac{-i\pi}{\sigma^2}(x^2 + y^2) \right] \cdot \mathbb{F}_{\text{Fourier}} \left\{ O(\xi, \eta) \cdot \exp\left[\frac{-i\pi}{\sigma^2}(\xi^2 + \eta^2) \right] \right\}$$

$$(29.13)$$

with

$$\sigma = \sqrt{\lambda d} = \sqrt{2\pi \frac{d}{k}}. \qquad (29.14)$$

When the reconstruction distance goes to infinity, the parameter σ will tend to infinity and the Fresnel transform becomes identical to the Fourier transform; such a representation, when given as a function of the unit vector, is called the angular spectrum of the object.

Convolution: another representation of the propagation can be obtained by considering the Huyghens–Fresnel expression (29.5) as a 2D convolution of the wave field $O(\mathbf{r}')$ in the plane of the object with $-(i/\lambda)G_k^+(\mathbf{r} - \mathbf{r}')$. The convolution is more complex to compute than Fresnel transform: it can be computed as the inverse Fourier transforms (FT) of the product of the FT of $O(\xi - \eta)$ and the FT of G_k^+. The propagated wave field is computed on a constant size matrix representing the image reconstructed from the hologram at a constant scale (Montfort et al. 2006a), which in many situations, like images reconstructed from synthesized aperture holograms, may appear more simple to handle because the scale and sampling of the wave field are constant. It is therefore most often preferred, even though the calculation is somewhat more complex and lengthy, because one or more Fourier transform have to be performed.

29.4 Realization of Optical Setups for Holographic Microscopy

As described extensively in the previous section, many ways have been opened to exploit coherence properties of beams.

Accordingly, many optical designs have been developed and proved feasible. A detailed review of the different realizations of microscopes is beyond the scope of this book.

A majority of optical setups have been realized according to Figure 29.2b, i.e., holography with coherent beam, or more precisely with some degree of coherence. In most of the proposed and realized optical setups, R is kept spatially separated from O. In other realizations, object and reference waves follow approximately the same optical path, in the so-called common path design. The reference beam is controllable both in intensity and polarization in order to optimize the contrast and signal. The different steps for hologram recording and reconstruction in this configuration are presented in Figure 29.4. The interference between the waves O and R is represented in Figure 29.4a; the physical propagation of the restored object wave O' when the hologram is illuminated by the field U is shown in Figure 29.4b. Finally, the principle of digital reconstruction in DHM is presented in Figure 29.4c.

Different optical setups have been proposed to perform microscopy with holography. In the past, lensless setups have been realized first. Strongly diverging wave fields such as spherical waves emitted by punctual sources have been proposed (Haddad et al. 1992). Holograms of microscopic objects are directly formed and collected on a CCD camera. The hologram therefore results from the interference of two spherical waves. The first achievements however revealed difficult to implement in practice due to the fact that positioning the pinhole emitting the reference beam very close to the specimen is difficult. From this point of view, digital in-line holographic microscopy (DIHM) arrangement with diverging beams has been proposed by Kreuzer et al. to give images of microscopic objects. The optical setup appears easy to realize (Kreuzer et al. 2001, Xu et al. 2001b). A bit more complex approach, the hybrid holographic microscopy, has been proposed by Takaki and Ohzu (Takaki et al. 1999, Takaki and Ohzu 1999, 2000) in order to eliminate in a more efficient way the zero-order and twin image: two beams are generated to illuminate the object. These two beams can be laterally shifted to shear the reference wave and/or be phase shifted to filter the cross terms in the time domain. DIHM has the advantage to be, by its realization, a common path interferometer, but presents the drawback to rely upon the hypothesis that the first Born approximation is respected for the transparent object, i.e., that the primary beam traverses the object without being strongly affected by the object, which is a requirement not met in general (Marquet 2003). Therefore, the technique appears as more appropriate to image 3D particles (Garcia-Sucerquia et al. 2006b) or when the object appears as relatively small on the field of view (Garcia-Sucerquia et al. 2008). The technique performs well also for lines and filaments (Kempkes et al. 2009). Variant approaches have been proposed (Repetto et al. 2004, 2005). In order to avoid the problem of the loss of primary beam in in-line holography, the introduction of a beamsplitter between the specimen and the camera appears as necessary to keep quantitative phase data for extended specimens (Xu et al. 2000), which presents the inconvenience to reduce further the numerical aperture

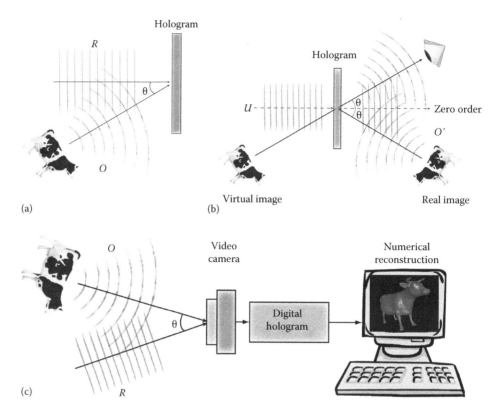

FIGURE 29.4 Representation of the different steps for hologram recording and reconstruction in off-axis digital holographic microscopy. (a) Hologram generation through the interference between the waves O and R. (b) Physical propagation of the restored object wave O' when the hologram is illuminated by the field U. (c) Principle of digital reconstruction in DHM, where the wave field behavior is modeled by digital means.

of the hologram and therefore the resolution of reconstructed images. Figure 29.5 illustrates the lensless optical arrangement for DHM: the source is a pinhole, a monomode fiber, or the focal point of a MO. The source illuminates the specimen at some distance, and the diffracted light is collected on the camera where it interferes with the primary or illuminating beam. For DIHM, only the cone illuminating the camera is considered. If a beam-splitter is added to the setup, the primary beam is maintained over the whole beam scattered by the object if the specimen is thick enough to modify the primary beam in a sizeable way. The reference beam is generated by a secondary source, which can be kept perfectly in line with the object or can be slightly off-axis to generate a carrier frequency on the hologram. The phase of the secondary source can be modulated in phase (phase-shifting holography) or heterodyned according to the description given earlier. Even though resolutions better than the micron could be achieved (Pedrini and Tiziani 2002, Garcia-Sucerquia et al. 2006a), lensless DHM has however fundamentally the disadvantage that the aperture, even if high optimization has been done, remains significantly less than the one achievable with the use of a high NA, oil immersion MO (NA > 1, typically 1.4). The reason is that, even if the cone capturing the light scattered by the object can be brought to a maximum, for instance by filling up the space separating the object from the detector (the detector itself excluded) (Garcia-Sucerquia et al. 2006a), the aperture of the cone cannot reach π, whereas this optimum is practically

obtained by some immersion MO. The resolution cannot therefore be as high as what can be obtained with MO. Hybrid techniques between lensless techniques and techniques with lenses or MO have been considered (Xu et al. 2001a), as well as in-line or off-axis techniques for the elimination of zero-order and twin image (Xu et al. 2003a).

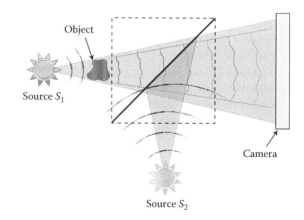

FIGURE 29.5 Optical setup for DIHM: the object is close to a punctual source. It diffracts the light from the source S_1, which interfere with the beam directly emitted by the source S_1. Optionally, a second source S_2, coherent with the source S_1, delivers a beam to serve as a reference beam and improve the contrast of the hologram in the case of a large and weakly transparent object.

So far, preference has been generally given to optical set-ups making use of a MO offering the largest numerical apertures (Cuche et al. 1999b, Dubois et al. 1999, Indebetouw and Klysubun 1999b, Yamaguchi et al. 2001, Ferraro et al. 2003a, Carl et al. 2004).

29.5 Space-Bandwidth Adaptation

Free-space propagation as well as the use of a microscope objective provides simple means to adapt the space-bandwidth product of the wave field radiated by the specimen to the characteristics of the detector, in our case, an electronic camera, which, at best and in the present state of the art, has a sampling capacity around a bit more than hundred line pairs per millimeter, which is still at least 20 times less than the sampling capacity provided by classical holographic plates. This discrepancy, which appeared insurmountable to many holographers in the first days of the introduction of electronic cameras, can be advantageously solved by the recourse to the general principle of space-bandwidth adaptation (Mendlovic et al. 1997, Zalevsky et al. 2000). The information content representing the scattered light by the object, both in space and wave-vector representation, and carried by free-space propagation and optical components like lenses, diffracting elements and other components, is traditionally given by the occupation of the so-called Wigner distribution function in the space-frequency domain (Mendlovic and Zalevsky 1997). The so-called space-frequency product is preserved by the use of optical components like lenses, MO in particular: a small field of view and a large bandwidth on the specimen side, limited by the input pupil of the MO, becomes a small bandwidth and a large field corresponding to the camera sensitive area on the image side. Practically, if the field of view is multiplied by the magnification factor M of the MO, the spectrum of the wave vector k is divided by M, which renders the sampling by the electronic camera optimal by a convenient choice of the magnification and positioning of the camera plane. Such concern will contribute to preserve the full bandwidth of the MO, so that no loss of resolution will be conceded.

The sketch of Figure 29.6 illustrates simply the space-bandwidth adaptation principle applied to DHM. It shows the microscope objective, together with the optional tube lens, which form at some distance S' a magnified image (by a factor M) of the specimen represented by a candle. The specimen is situated at a distance S of the primary principal plane PP, and the image position is formed at a distance S' of the secondary principal plane SP and is given by the well-known Gaussian lens formula. In modern microscopy, MO are infinity corrected, i.e., they are optimized for imaging the specimen at infinity. A tube lens is therefore inserted on the image side to obtain a focused image at its focal length. On the specimen side, the angular spectrum of the radiated field $O(x,y,z)$ or, similarly, the $k_x - k_y$ wave-vector spectrum fills up the entrance pupil of the objective. On the image side, the rays are bent according to the geometrical optics laws and form the beam $O'(x,y,z)$ converging on the image. The corresponding $k'_x - k'_y$ wave-vector spectrum is just scaled down by the magnification factor M. If we dispose the camera on a plane intercepting the beam O' converging toward the specimen image, $O'O'^*$ will contain twice the spectrum of O' and can be sampled according to the Shannon theorem, provided that a magnification M high enough is chosen.

Considering now the introduction of a reference beam incident on the camera plane at an angle θ, a hologram is formed. Depending on the value of the angle θ, the hologram space-bandwidth can increase considerably. The angular spectrum of the object will be turned by an angle equal to θ. The aperture angle ψ will be augmented by the value of θ. It may extend in

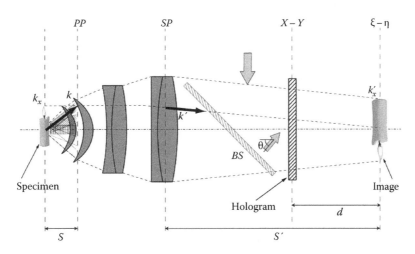

FIGURE 29.6 Illustration of the role of a microscope objective in the space-bandwidth adaptation of the beam scattered by the specimen and transmitted to a camera capturing the hologram with a limited spatial bandwidth. *PP*: primary principal plane, *SP*: secondary principal plane. *S*: distance of the specimen to the *PP* plane. *S'*: distance of the image to the *SP* plane. *d*: distance from the hologram plane to the image plane: reconstruction distance. BS: beam splitter. θ: incidence angle of the reference beam. *k*: wave vector of the wave emitted by the specimen. k_x: component along *x*. *k'*: wave vector of the wave converging on the image. k'_x: component along *x*.

principle up to the maximum achievable for a monochromatic beam, given, in principle, by the beam wave vector in vacuum, or in the traversed dielectric medium. Practically, it is limited by the sampling capacity of the camera, which is given by the inverse of the interpixel distance of the camera, multiplied by π. This upper bound is a basic parameter for a given camera with around 3 microns interpixel distance, which corresponds to wave vectors of amplitude more or less $1000\,\text{mm}^{-1}$.

This way of using the available space bandwidth offered by the detector or electronic camera may appear as extravagant and, at first sight, advice could be given against use of off-axis digital holography. These are the conclusions of some works comparing in-line and off-axis holography from the point of view of space-bandwidth product (SW) (Xu et al. 2005). However, one must keep in mind that, most often, four holograms must be acquired in succession to eliminate completely the zero-order and the twin image for time domain methods. Therefore, the total amount of data to reconstruct a wave front for in-line holography is around 2/3 of that required for full off-axis holography. This ratio, in favor of in-line holography could be inverted by the introduction of new algorithms based on nonlinear filtering of the holograms in the frequency domain (Seelamantula et al. 2009, Pavillon et al. 2009b, Weng et al. 2008), giving in this case an advantage to off-axis holography. Based on various arguments of present and future works, the debate is still open.

Anyway, off-axis arrangement presents the major advantage to provide an easy management of the space-bandwidth product and to provide a possibility to control "at will" the occupation of the SFD. The fact of being able to localize the cross terms or mutual coherence terms outside the SFD domain occupied by the zero-order terms permits to filter them out without losing any spatial frequency component delivered by the MO. This result can be achieved by choosing an incidence angle θ of the reference wave just high enough to keep the whole spectral contribution of the cross terms separated from the low frequencies zero-order terms but nevertheless low enough to preserve the full sampling capacity of the camera. This approach may imply an over-sampling by the camera, i.e., the density of pixels should be high enough to accommodate the fact that the total bandwidth must be higher by a factor 3/2 in the worst case, than the bandwidth of the object field intensity $|O|^2$ which is itself twice the bandwidth of the object field O. These requirements are nowadays not a critical issue because cameras can reach easily several megapixels sampling capacity. The first step is therefore to define a magnification factor M of the MO which permits the evaluation of the hologram bandwidth potentially larger (up to three times as large) than the object field bandwidth, which is given by the wave vector of the beam propagating in vacuum multiplied by the NA of the MO. Shannon sampling theorem must be respected: the number of camera pixels per unit length must be twice the inverse of the maximal k-vector defining the bandwidth. It must be pointed out that the recourse to infinity-corrected MO renders the determination of M more flexible because the focal length can be adjusted by the focal

length of the tube lens. The field of view is then defined accordingly by dividing the size of the camera chip by M. Following this recipe, the only parameter which finally depends on the characteristics of the camera chip is the field of view. The lateral resolution is anyway preserved and kept at the maximum achievable value for a given wavelength and NA. In many cases, the NA approaches nowadays the theoretical limit given by the refractive index of the propagation medium.

Further resolution improvement is hoped to be achieved by following different paths:

- Development of synthetic aperture methods: the gap between the actual NA and the upper limit can be filled up by collecting sequentially the partial bandwidth corresponding to the angular spectrum acquisition by the MO at different incidence angles and stitching up the different components of the bandwidth up to the theoretical limit. One can intuitively understand that the shorter the wavelength, the larger the space-bandwidth is.
- The knowledge of the optical transfer function (OTF) and its introduction in the image-forming model has allowed improving significantly fluorescence images by deconvolution in daily practice of microscopy. Similar expectancy lies in the improvement of DHM images. Even further, some hope can be placed in the particularities of the DHM context: whereas intensity images are obtained in fluorescence microscopy and the associated deconvolution is most often considered as an ill-posed problem, the question of deconvolution in DHM appears in a new light: amplitude and phase images are obtained, which are not positive definite value. The complex data of the wave field scattered by the object are accountable for the precise characteristics of the optical setup: in particular, the Amplitude Transfer Function can be precisely measured and introduced in a least square fit of experimental data (Marian et al. 2007, Charrière et al. 2007a).

29.5.1 Optical Transfer Function

The amplitude point spread function (APSF) of the optical imaging system (OIS) permitting the formation of the hologram is defined as the complex field distribution $O'(x',y',z')$, corresponding to the complex electromagnetic field (electric E or eventually magnetic H) of the beam delivered by the OIS and originating from a point source: a Dirac function of the space coordinates (x,y,z), designated by point object.

Provided that the response of the medium propagating the beam is linear, the field $O'(x',y',z')$ forms a distributed image of a 3D object generating the distributed field $O(x,y,z)$. It can be computed as

$$O'(x',y',z') = \iiint_{\text{ROI}} O(x,y,z)H(x,y,z,x',y',z')\,\mathrm{d}x\mathrm{d}y\mathrm{d}z, \quad (29.15)$$

where ROI is the region of interest for the specimen. The kernel of the convolution $H(x,y,z,x',y',z')$ is the APSF, which is a complex function. Its 3D Fourier transform is the amplitude optical transfer function, which is also a complex function, describing how the image is derived from the object wave field in the 3D spatial frequency domain. These functions have been precisely established by DHM (Marian et al. 2007). They characterize the OIS and can be introduced in the modeling of the image formation in DHM. These data provide also a mean to correct the aberrations and distortions introduced by the optical setup, and the MO more particularly (Stadelmaier and Massig 2000, Pedrini et al. 2001, De Nicola et al. 2002, Ferraro et al. 2003b, Montfort et al. 2006a, Colomb et al. 2006a,b,d, Liu et al. 2006, Marian et al. 2007, Charrière et al. 2007a, Miccio et al. 2007). The quantitative determination of the transfer characteristics of DHM, including aberrations and distortions, is hoped to improve the lateral as well as the axial resolution.

29.6 "Electronic Focusing" and Extended Depth of Field

In classical microscopy, the analysis of 3D images of extended, fixed, or, more particularly, moving targets requires focusing means in order to form a complete view of the specimen. These focusing means consist traditionally in mechanical and/or piezoelectric stages providing scanned images throughout the specimen. This is particularly important for confocal microscopy, where the depth of field is kept to a minimum value. This movement of the specimen or of the MO cannot be as fast as one could wish in cases where dynamical properties are essential aspects of the investigation. In confocal microscopy, these preoccupations have motivated the development of complex mechanical and electrooptical devices. From this point of view, digital holography brings a great asset in microscopy: the possibility to reconstruct the images of the specimen at different depths from a single hologram captured at a fixed distance from the MO. They are naturally obtained by reconstructing numerically, at different depths, the propagated wave front. The sketch of Figure 29.6 shows that the specimen is generally not fully contained in a single plane at a distance s of the primary principal plane, and a real image of the specimen is formed at finite distance s' of the secondary principal plane of the MO and therefore at a variable distance d of the fixed hologram plane x–y. Placing the hologram plane at a small distance from the specimen image gives rise to holograms in the Fresnel zone. Huyghens–Fresnel expression of diffraction will therefore be considered as valid for the calculation of the propagation of the reconstructed wave. On the contrary, placing the hologram plane close to the output pupil of the MO at a relatively large distance of the specimen image yields holograms in the Fraunhofer zone, which yields the Fourier decomposition of the specimen image.

The extended image of the specimen can be reconstructed by computing the wave field at variable reconstruction distances d.

In some cases, this reconstruction distance can be dynamically adjusted to keep the specimen in focus throughout a time-lapse experiment. This "dynamic focusing" is needed in situations where the specimen is moving and should be tracked (Langehanenberg et al. 2009).

The capability offered by DHM to reconstruct from a single hologram the wave field at a variable distance d can be exploited to form a multifocus image fusion. Some automatic algorithm can be developed or adapted from conventional microscopy to isolate the focused parts of the specimen, suppressing thereby the out-of-focus noise, and to stitch them in 3D to provide a synthetic image of the specimen called fusion image.

29.7 DHM Instrument

The realizations of the experimental setups implementing the different principles described in the first part of this chapter are too numerous to be reported in details here. Many systems however hinge on rather complex modulation techniques or sophisticated mechanical sliding or scanning techniques. We will concentrate on the implementation of coherent holography in microscopy, commonly called digital holographic microscopy (DHM) (Cuche et al. 1999b, Dubois et al. 1999, Indebetouw and Klysubun 1999b, Yamaguchi et al. 2001, Ferraro et al. 2003a, Coppola et al. 2004, Carl et al. 2004, Kemper et al. 2007b). In many aspects, the off-axis approach appears robust and simple and is convenient for current use. In the simplest DHM implementation, no time heterodyning or piezo-driven mirrors are required, and the extraction of the wave-front data is done in the SFD with simple algorithms. DHM brings quantitative data derived from the simultaneous evaluation of the amplitude and phase of the complex reconstructed wave front. With the help of a MO, microscopic objects can be imaged with the highest achievable resolution by optical means, both in the transmission and in the reflection geometry.

Typical arrangements are shown on Figure 29.7: the optical setups comprise the recording of a digital hologram of the specimen by means of a standard CCD or CMOS camera inserted at the exit of a Mach–Zehnder interferometer (see Figure 29.7a) or a modified Mach–Zehnder type, including a Michelson interferometer (see Figure 29.7b). The geometries have been chosen to image transparent specimens in transmission and reflective specimens in reflection. In transmission, the microscope measures the optical path length of optical rays crossing a transparent or translucent specimen: dielectrics, MOEMS, micro-optics, and biological preparations. In reflectance, the microscope collects the backscattered beam and permits imaging specimens reflecting light: MEMS, microsystems, wafers, mirrors, etc.

Other configurations are possible, according to the targeted application, but will not be reviewed here. An important issue is the need for a reference beam, which should be controllable both in intensity and polarization, since signal and therefore contrast is improved accordingly. Holographic microscopy permits also other invaluable concepts to be built on: in particular,

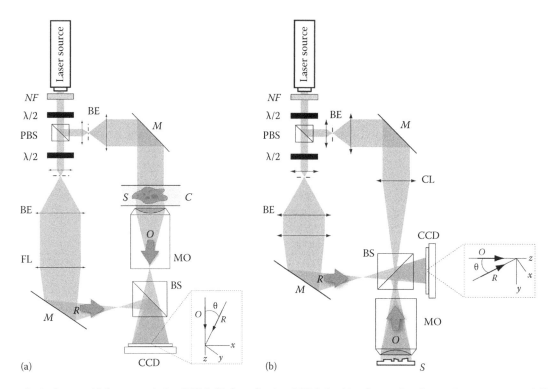

FIGURE 29.7 Optical setup: (a) for transmission DHM, (b) for reflection DHM. *O*: object beam, *R*: reference beam, *NF*: neural filter, BS: beam splitter, PBS: polarizing beam splitter, *M*: mirrors, BE: beam expander, FL: focusing length, CL: condenser lens, CCD camera, MO: microscope objective, *C*: cover slip, S: object or specimen.

the possibility of superimposing several holograms. Holograms with several reference waves corresponding to several polarization states can be generated in order to analyze the birefringence properties of specimens such as strained dielectrics or biological molecules (Colomb et al. 2002, 2005, 2004). Reference waves corresponding to different wavelengths can also be generated, permitting the generation of synthetic wavelengths, even simultaneously recorded on a single hologram (Kühn et al. 2007), through spatial multiplexing.

DHM provides an absolute phase image, considering that it is referenced to the phase measured, e.g., on the support of the specimen. This phase signal can be directly interpreted in terms of refractive index and/or profile of the object. Very high accuracies can be achieved, which are comparable to that provided by the highest quality interferometers, but DHM offers a better flexibility and the capability of adjusting the reference plane with the computer, i.e., without positioning the beam nor the object. This computerized procedure adds much flexibility so that all adjustment procedures can even be made transparent to the user.

The method requires the adjustment of several reconstruction parameters (Cuche et al. 1999b). These modifications can be performed easily with computer-aided methods developed in our group and others. Some image processing, such as hologram apodization, is also needed to improve the phase accuracy (Cuche et al. 2000b).

Concerning the lateral resolution, a detailed analysis (Kühn 2009) shows that the lateral resolution is, for high-enough magnifications, limited by the diffraction rather than by the

sampling capacity of the camera. This is even clearer for strongly scattering objects where the spatial spectrum is maximized. By using a high numerical aperture, submicron transverse resolutions are easily achieved, typically better than 600 nm of lateral resolution for high NA MO: 1.3, 100× oil-immersed.

Accuracies of approximately half a degree have been estimated for phase measurements. In reflection geometry, this corresponds to a vertical resolution currently less than 1 nanometer at a wavelength of 632 nm. Figure 29.8 shows the phase image of a test chromium strip of 8.9 ± 0.5 nm thickness (VLSI standard incorporated). Standard deviations down to 0.69 nm can be achieved, yielding accuracies in the picometer range (Kühn et al. 2008). These performances are illustrated in Figure 29.8. In the transmission geometry, the resolution is limited around a few nanometers for transparent specimens.

The achievement of such high accuracies raises the problem of the origin of noise and its effects on the formation of the hologram. Ultimately, its role in the reconstruction of the image makes appear some advantages of the holographic approach in microscopy: improvement of the S/N ratio can be expected in some conditions from the coherent detection of low level of scattered light (Charrière et al. 2006a, 2007b). In particular, the role of the so-called coherent detection gain resulting from the fact that the signal provided by the cross terms RO^* and R^*O is proportional to the amplitude of the reference signal R, and therefore pushed out of the background noise and Poisson noise is a known property of coherent detection and imaging systems. It has been documented and quantified in the case of DHM.

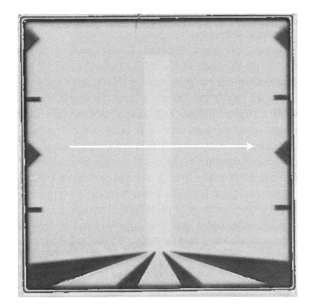

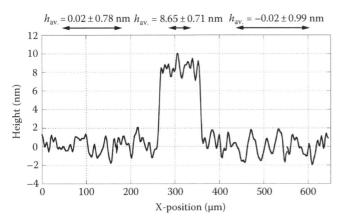

$h_{av.} = 0.02 \pm 0.78$ nm $h_{av.} = 8.65 \pm 0.71$ nm $h_{av.} = -0.02 \pm 0.99$ nm

FIGURE 29.8 Phase image of a test pattern formed of a test chromium strip of 8.9 ± 0.5 nm thickness (VLSI standard incorporated) illustrating the role of the residual noise compared to the phase signal corresponding to the 8.9 nm strip.

As it was explained in details, the use of an off-axis configuration enables capturing the whole image information by a single hologram acquisition (Cuche et al. 1999b). By using a time-gated camera or pulsed illumination sources, it is possible to acquire holograms in a few tens of microseconds and even much less with pulsed lasers. Such short acquisition times permit to avoid perturbations originating from parasitic movements, vibrations, or perturbing ambient light. On the other side, wave-front reconstruction rate may be as high as 15 frames per second with present standard computers, making DHM an ideal solution for performing real-time measurements on living specimens.

Acquisition rate can be as high as the camera can deliver and the computer can accept. Recently, Ethernet connections have provided very high transfer rate (Gbits/s) and large pixel matrices (larger than 2048×2048) that provide overall space-bandwidth product comparable to conventional recording media in holography (if we consider that acquisition time may last several

seconds in this last case). The phase can be derived from this instantaneous hologram which can be acquired at very high rate: many cameras permit acquisition rates in the kilohertz range and therefore, after off-line hologram reconstruction, appear as well adapted for exploring biological cell dynamics in the millisecond time scale.

29.8 Further Developments

The introduction of a precise method to evaluate simultaneously quantitative phase and amplitude contrast in microscopy (Cuche et al. 1997, 1999a) has been followed by many developments. It was made feasible to derive from the hologram taken in transmission or reflection precise and quantitative data (intensity and phase) of the wave front outgoing from a reflective and/or (semi-) transparent object (Cuche et al. 1999b, Grilli et al. 2001). It was also proved that the obtained phase distribution could be measured quantitatively, which gives access to the topography of well-characterized objects. In some cases, the surface and volume of homogeneous objects like single cells with well-resolved shape can be reconstructed in 3D and with nanometer accuracy on the optical axis. These developments have opened new perspectives in microscopy. By using this technique, one can reconstruct 3D objects very fast in comparison to more classical methods or scanning procedures. The transverse resolution was shown to approach closely the diffraction limit of the imaging system, and therefore to be practically as good as the lateral resolution achieved in classical optical microscopy (Carl et al. 2004, Mann et al. 2005, Colomb et al. 2006c, Marquet et al. 2005).

Digital holography, and especially DHM, took an increasing importance is several domains due in particular to the advantage of the DHM technique in terms of numerical aberration compensation (De Nicola et al. 2001, 2002, Ferraro et al. 2003b, 2008, Colomb et al. 2006a,b,d). The particular problem of the compensation of the aberrations introduced by the MO has been addressed by the precise measurement of the amplitude point spread function (APSF) (Marian et al. 2007, Charrière et al. 2007a). DHM was also extended to other spectral ranges such as near infrared (Repetto et al. 2005) or far infrared (Allaria et al. 2003, Mahon et al. 2006). Recent developments led to many different applications such as noise reduction (Liebling et al. 2003, Onural 2004, Palacios et al. 2004, 2005, Sotthivirat and Fessler 2004, Zhang et al. 2004b, Garcia-Sucerquia et al. 2005, Monnom et al. 2005), automatic focusing (Liebling et al. 2003, Ferraro et al. 2003a, Malkiel et al. 2004), numerical magnification and size control (Zhang et al. 2004a, Ferraro et al. 2004a,b), or extended focused image (de Nicola et al. 2005, Ferraro et al. 2005, Indebetouw et al. 2005). The capability to focus dynamically the image by wave-front reconstruction at various depths has been exploited for different applications (Dubois et al. 2006b, Langehanenberg et al. 2008). Furthermore, different experimental techniques have improved the applicability of DHM by the use of multiwavelength (Gass et al. 2003, Javidi et al. 2005), polarization (Colomb et al. 2004, 2002, 2005), infrared light sources (Repetto et al. 2005), or low-coherence sources

(Dubois et al. 2004b, Kozacki and Józwicki 2005, Martínez-León et al. 2005, Massatsch et al. 2005, Onodera et al. 2005).

DHM has the advantage of providing an "extended depth of focus" (Ferraro et al. 2005) because all focused images in the stack are obtained by the reconstruction from the data of a single hologram of the wave front at a variable reconstruction distance. No physical scan in depth is needed, so that at each depth, a robust autofocus method for detection of the focused parts of the image (Liebling and Unser 2004) can be employed. 3D appearance of the object appears as the result of a "fusion image."

29.8.1 Applications to Biology

One major field of application of DHM concerns cell biology. In this field, DHM has indeed the ability to visualize marker-free transparent living specimen thanks to its capability to accurately and quantitatively measure the phase of the transmitted light. Practically, DHM provides directly a full field quantitative measurement of the phase shift induced by the transparent specimen, including living cells (Cuche et al. 1999b, Klysubun and Indebetouw 2001, Carl et al. 2004, Mann et al. 2005, Marquet et al. 2005, Chalut et al. 2007, Kemmler et al. 2007, Mico et al. 2008). The phase shift, or the optical path difference (OPD), containing invaluable information about the cell morphology as well as various intracellular compartments can be regarded as a powerful endogenous contrast agent.

Practically, the DHM cell-induced phase shift has allowed the derivation of highly relevant cell parameters, including dry mass production, density, and spatial distribution. The evolution patterns of these parameters have permitted to noninvasively characterize cell cycle at the single cell level (Rappaz et al. 2009b). More generally, the relevant issue of the production of biological mass (proteins) and its transfer at short timescales (in the order of one second or less) can be efficiently investigated with the DHM technology. An absolute phase image, obtained for living neurons in culture, is presented in Figure 29.9. This image has

been obtained by representing in 3D the absolute phase or OPL expressed in degrees.

On the other hand, technical developments (Rappaz et al. 2005, 2008b, Kemper et al. 2007a) have allowed to separately measure, from the phase signal, the information about cell morphology and about the intracellular refractive index, related to the intracellular content, and particularly the protein concentration and water movements. As far as red blood cells (RBC) are concerned, such decomposition of the phase signal provides direct measurements of the mean corpuscular volume (MCV) and mean corpuscular hemoglobin concentration (MCHC), two parameters of great importance for diagnostic purposes (Rappaz et al. 2008a). Figure 29.10 presents the morphology of an RBC measured with DHM capabilities, where the typical shape of the cell can be identified in Figure 29.10a. Moreover, by using multiple-wavelength measurements in reflection, it was possible to determine the 3D conformation of the RBC membrane, as shown in Figure 29.10b.

In addition, the phase can be derived from a single hologram which can be acquired at a very high rate (camera limited) allowing to efficiently explore cell dynamics (Schnekenburger et al. 2007) or specifically cell membrane fluctuations within the millisecond time scale (Rappaz et al. 2009a).

Otherwise, the DHM numerical refocusing capability allows to address efficiently the question of the localization of biological objects: cells, organelles in living preparations, nanoparticles, etc. Concretely, DHM provides a direct method to study cell dynamics with a digitally based autofocus technique (Liebling and Unser 2004, Langehanenberg et al. 2008) or to perform 3D tracking of biological specimen migration (Xu et al. 2003b, DaneshPanah and Javidi 2007, Sheng et al. 2007, Sun et al. 2008). The approach performs well even when embedded in scattering media by using partially spatial coherence sources (Dubois et al. 2006a).

On the other hand, more elaborate techniques based on DHM including time domain digital holography or tomographic approaches enable performing true optical sectioning by eliminating out-of-focus disturbances. Practically, time domain digital holography has permitted visualizing the internal structure of scattering media, including constituted tissues (Massatsch et al. 2005, Tamano et al. 2006). More recently, tomographic approaches applied at a single cell level in a transmission configuration has provided the full 3D spatial distribution of the intracellular refractive index providing invaluable information concerning the intracellular content, including 3D distribution of organelle, nucleus vesicles, inclusions, etc. (Charrière et al. 2006c, Debailleul et al. 2009, Kim 2000).

Vesicles' trafficking is indeed of primary interest to understand biochemical exchanges and stimulations. More recently, it was demonstrated that tomographic images of semitransparent objects could be obtained with a submicrometer resolution (Sung et al. 2009). This "ultrahigh resolution" for a tomographic technique can be exploited to yield a full 3D visualization of living cells containing, in particular, tiny scattering structures, including structural molecules (actin, tubulin, etc.) whose

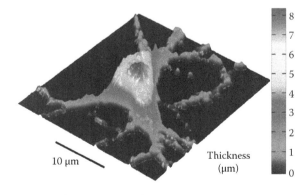

FIGURE 29.9 (See color insert.) Absolute phase image, obtained for a living neuron in culture, represented in 3D. Provided that the refractive index of the cytoplasm can be determined, this image is a very good representation of the topology of the cell surface.

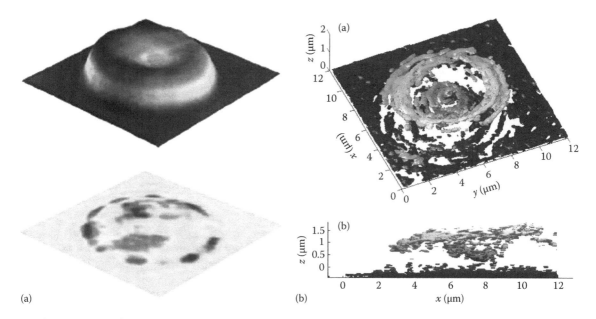

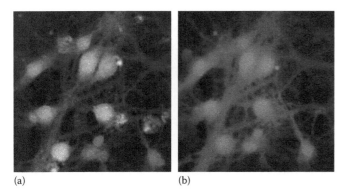

FIGURE 29.10 (See color insert.)   (a) Phase image of a living RBC on the surface of a cover slip. The lower part of the image shows the dynamical behavior of the living RBC. (b) Cell membrane fluctuations with amplitude of around 20 nm in a spectrum reaching a few tens of hertz. (From Kühn, J. et al., *Opt. Lett.*, 34, 653, 2009.)

activity plays a key role in the important issue of cell motility and plasticity.

Finally, a very attractive perspective of this research work is the development of multimodality imaging techniques combining, e.g., transmission DHM with an epifluorescence videomicroscopy, widely used in cell biology, in order to confront the DHM data with fluorescence data, in particular those concerning the intracellular ion dynamics measured with specific fluorescent probes (Pavillon et al. 2010). Figure 29.11 illustrates these perspectives: Figure 29.11a shows an absolute phase image of interconnected neurons. Figure 29.11b presents the corresponding fluorescence image of the same preparation labeled with Fura-2, a calcium sensitive marker. The images have been taken practically simultaneously on a microscope having

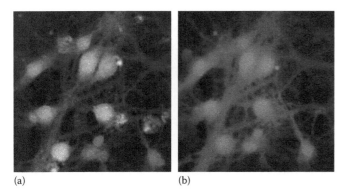

FIGURE 29.11   (a) Absolute phase contrast image of a preparation of living interconnected neurons. (b) Fluorescence image of the same preparation labeled with Fura-2, a calcium sensitive marker. The images have been taken practically simultaneously on a microscope having both DHM and fluorescence imaging capabilities.

both DHM and fluorescence imaging capabilities, where we observe a good match on neuronal bodies.

29.8.2 Toward Full 3D Imaging of Specimens

In general, a wave-front determination obtained from a single hologram does not suffice to obtain full 3D imaging of an object because the aperture of the hologram is practically limited by both the magnification optics, with the limited NA of the MO, and potentially also by the limited size of the camera. This is true for all imaging modalities, based on holography or not. The main benefit of holography is that it provides an easy way to combine the wave fields by simple addition. The resulting wave field simply corresponds to the wave field that would have provided an increased aperture, or more precisely a synthetic aperture. In order to be able to superimpose the different wave fields with the required precision, each wave field must be corrected to adjust the scale if necessary and to correct for the aberrations and eventual distortions. This task is facilitated by the capabilities of digital holography to give access to the complex wave field and to modify it. Data manipulation, image-processing techniques in particular, have contributed to found this new field of optics, coined as "digital optics" (DO) (Colomb et al. 2007).

The concept of synthetic aperture can be developed both in the time and in the spatial domain. One must admit that the true 3D shape of a given specimen cannot be derived from a one measurement obtained at a single wavelength. Broadening the spectrum of the illumination source can help in improving the resolution of the image.

A first variant of this approach consists in changing the wavelength (variable k-vector amplitude): such an approach is not really new and numerous papers report results obtained by changing the wavelength. The range of the wavelength scan is however very small and therefore the resolution along the wave propagation is very small. Arons et al. as "Fourier synthesis holography" (Arons et al. 1993) have also given reference to a similar approach. Multiple wavelengths have been used in digital holography to reconstruct 3D structures (Kim 1999, Dakoff et al. 2003, Montfort et al. 2006b, Potcoava and Kim 2008, Kühn et al. 2009). An alternative, but somehow similar approach is to use a wide bandwidth source and to form a hologram in the plane where the mutual coherence between object and reference wave is nonzero: this concept introduces coherence gating in the space domain, which proved to perform well (Cuche et al. 1997, Massatsch et al. 2005, Delacrétaz et al. 2009).

The second variant of this approach consists in varying the angle of the illumination waves (variable *k*-vector direction) and can be used in conjunction with the previous technique where the wavelength is changed (variable *k*-vector amplitude).

This second approach meets more exactly the concept found in the literature under the name of "diffraction tomography" for reconstruction of the scattering potential associated with the structure of diffracting objects (Wolf 1969). Diffracted waves can be collected and reconstructed from the holograms for various incidences. In microscopy, a tomographic approach, not

based on holography, was presented by Inoué and Lauer (Lauer 2002). Optical tomographic microscopy can be performed by applying two different techniques for varying the angle of the illumination waves (variable *k*-vector direction). On one hand, the specimen is rotated by 180° or 360°, as on the other hand, the incidence angle is varied by laterally scanning the illumination point in the back focal plane of the condenser lens: the incidence angle of the incident beam varies in the interval defined by the aperture of the condenser. These two variants imply different technical realization and different setups, which are shown on Figure 29.12. The approaches are however different and lead to different imaging performances as it was shown by Kou et al. (Kou and Sheppard 2008).

A simple way to reconstruct the scattered wave can be based on holography: the phase and amplitude of the diffracted wave is directly reconstructed from the hologram and can be used to compute the scattering potential at every point of the specimen. Tomography of cells based on DHM is new and original. In 2006, we proposed a first approach where we have rotated the specimen: a pollen grain (Charrière et al. 2006b) and an amoeba (Charrière et al. 2006c) (see Figure 29.13 (Pavillon et al. 2009a)). It was demonstrated that refractive index of the cell body could be measured in 3D with a resolution better than

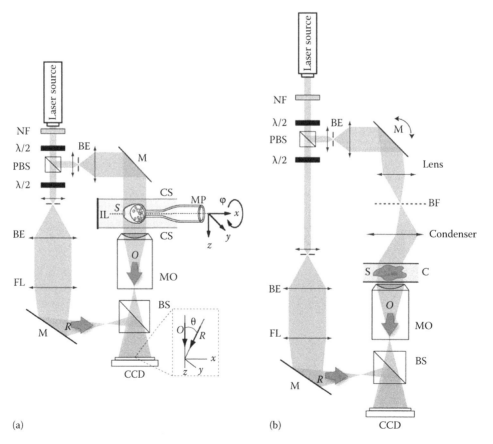

(a)   (b)

FIGURE 29.12 Optical tomographic microscopy: two different techniques for varying the angle of the illumination waves (variable *k*-vector direction) (a) the specimen is rotated by 360° (b) the incidence angle is varied by laterally scanning the illumination point in the back focal plane: the incidence angle of the incident beam varies in the interval defined by the aperture of the condenser.

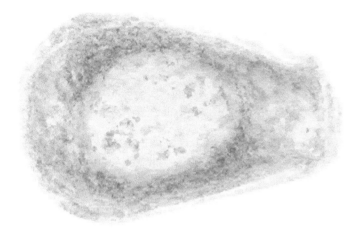

FIGURE 29.13 (See color insert.) Tomographic image of a testate amoebae *Hyalosphenia papilio*: tomographic image taken in the middle of the specimen, the refractive index is coded in color from around 1.4 to 1.6. The tomography has been obtained by rotating the specimen by 180°, one slice of the specimen is less than 3 μm. (From Pavillon, N. et al., Optical tomography by digital holographic microscopy. In: *Novel Optical Instrumentation for Biomedical Applications IV*, SPIE, 2009a.)

3 μm in all directions. Further works (Choi et al. 2007, 2008, Debailleul et al. 2008, 2009, Sung et al. 2009) have demonstrated the feasibility of the approach based on the rotation of the incident beam. The diffraction tomography approach is particularly efficient when the wavelength of the irradiating beam is comparable to the dimensions of the diffracting parts of the object. In the case of cells and of their constituents—nucleus and various organelles—the dimensions of the objects are of the order of one or a few microns and are therefore of the same order of magnitude as the irradiating wavelength. Much work is still required to achieve full tomography of single cells and tissues at the cellular scale.

Thanks to the capability of digital holography of giving access to the complex wave field, many of the modern digital data manipulation, image processing techniques in particular, can be extended to optics: this field could be coined as "Digital Optics" (DO) (Colomb et al. 2007). DO offers the very attractive prospect to emulate in all details the various functions of optical systems "just by a computer click." In optics, optical elements are used to transform, filter, and process optically wave fronts in order to magnify images, compensate for aberrations, or to suppress unwanted diffracted orders for example. Henceforth, the same can be achieved with numerical tools. It is expected that optical systems, with their imperfections, can be corrected with less efforts by digital processing, so that optical design and digital image processing are disciplines that will interpenetrate each other. The idea of digitally capturing a hologram rather than the focused image with a camera, connected or not to an optical instrument such as a microscope or an endoscope, and reconstructing the image from the wave front, usually called digital holography, has been proposed and demonstrated practically in different contexts. Developing numerical processes to mimic the optical parts of the system such

as lenses and microscope objectives is a realistic perspective in the context of DHM. It has been shown in particular how automatic or manual procedures have been developed to filter out the diffracted orders and/or parasitic interferences to compensate for aberrations and image distortions and finally control the position and magnification of reconstructed wave fronts (Montfort et al. 2006a, Colomb et al. 2006a,b,d). The aberrations generated by the MO can also be analysed in details and introduced in the numerical model of wave propagation (Marian et al. 2007). Deconvolution is a direct benefit of these developments. It is now feasible to apply the DO concept to compensate for chromatic aberrations in order to achieve perfect superposition of wave fronts reconstructed from digital holograms recorded at different wavelengths. Similarly, the adjustment of fields and spatial spectra is a worthy consequence of DO approach. This should have a great impact on synthetic aperture techniques, more and more often claimed as future tools to overcome the limitations of optical devices. Wave fronts, reconstructed from multiple holograms, can be combined to form a synthetic wave front converging on each voxel of the true 3D image of the specimen.

29.9 Conclusions and Outlook

Holography has found an attractive development field in microscopy. It brings many innovative aspects that are susceptible to transform progressively microscopy. Because of its flexibility, DHM could be already easily adapted to the need of microscopists, biologists in particular. Complex wave-front reconstruction not only yields absolute phase contrast, which deserves particular attention in biology, but also yields the unique possibility to capture in one shot most of the data permitting to reconstruct the image in 3D. Sometimes called electronic focusing, this feature also provides dynamic focusing and tracking capabilities.

In the reflection geometry, DHM permits the precise determination of the topology of the specimen. Accuracies in the nanometer or even in the sub-nanometer range with some statistical treatment are at reach. In the field of surface and interface sciences, and more generally in material sciences, DHM brings a detailed characterization of materials: refractive indices, dielectric properties, and birefringence. In the field of biology, it could be demonstrated that minute deformations and movements of living cells could be detected and monitored: vibrations associated with cell metabolism and minute migration of cytoplasm. Quantitative phase provides a measure of shapes, absolute volumes and refractive indices, thanks to decoupling procedures, which allow separating path lengths from RI values.

In biology, both morphology and functional studies can be achieved. Cell dynamics studies, in particular, can be carried out on living preparations. Because it is a nonscanning method, a very fast acquisition rate is possible, making feasible the observation of fast movements of membranes or cytoplasm. Polarization microscopy is also possible with DHM, thanks to the high sensitivity on the birefringence of large molecules such as structural molecules. The method has also the advantage to be completely noninvasive because the illumination levels are much lower than those of confocal microscopy and fluorescence, in general, as well as those of multiphoton microscopy. The absence of photodamage permits to perform measurements on living cell cultures for several hours and days. The hologram acquisition is performed at video frequency (25 Hz) or up to several thousands of hologram per second, and the reconstruction rate is of about 15 images per second. These characteristics make DHM a practical instrument in the hands of microscopists.

The future developments will help to realize instruments with full tomographic capabilities, yielding submicron resolution in all dimensions. These instruments will also be fast enough to get full functional data in biology, enlightening all the interactions in biology, taking place in full 3D: cell to cell interactions, cell to substrate or extracellular matrix, role of organelles, vesicles, interfaces, including the dynamical aspects, fluid exchanges, protein growth, movements, and vibrations. It will be a true help for the development of biomechanical models.

Fundamentally oriented toward the measurement of refractive indices and more generally dielectric properties of matter, DHM will incorporate novel approaches to image nonlinear properties of matter such as second, third, or fourth harmonic generation, by the so-called harmonic holography. Coherent anti-Stokes Raman scattering and, more generally, all coherent processes are candidates to coherent imaging by DHM. It must be also stressed out that incoherent scattering processes such as fluorescence or luminescence are also candidates to holographic imaging by a clever exploitation of their self-coherence properties.

Acknowledgments

The works presented in this chapter are the result of a close collaboration between the Advanced Photonics Laboratory of EPFL with Florian Charrière, Jonas Kühn, Etienne Shaffer, and Yann Cotte; the Laboratory of Neuroenergetics and Cellular Dynamics; Brain and Mind Institute of EPFL with Pierre Magistretti, Benjamin Rappaz, and Pascal Jourdain; the Neuropsychiatric Center; CHUV with Daniel Boss; and the company Lyncée Tec SA, Lausanne (http://www.lynceetec.com), with Etienne Cuche, Yves Emery, Tristan Colomb, Frédéric Montfort, Nicolas Aspert, and others.

References

Abramson, N. 1983. Light-in-flight recording—High-speed holographic motion-pictures of ultrafast phenomena. *Appl. Opt.* 22: 215–232.

Aguanno, M. V., Lakestani, F., Whelan, M. P., and Connelly, M. J. 2007. Heterodyne speckle interferometer for full-field velocity profile measurements of a vibrating membrane by electronic scanning. *Opt. Lasers Eng.* 45: 677–683.

Albrecht, B., Failla, A. V., Heintzmann, R., and Cremer, C. 2001. Spatially modulated illumination microscopy: Online visualization of intensity distribution and prediction of nanometer precision of axial distance measurements by computer simulations. *J. Biomed. Opt.*, 6: 292.

Albrecht, B., Failla, A. V., Schweitzer, A., and Cremer, C. 2002. Spatially modulated illumination microscopy allows axial distance resolution in the nanometer range. *Appl. Opt.* 41: 80.

Allaria, E., Brugioni, S., de Nicola, S., Ferraro, P., Grilli, S., and Meucci, R. 2003. Digital holography at 10.6 micron. *Opt. Commun.* 215: 257–262.

Arnison, M. R., Cogswell, C. J., Smith, N. I., Fekete, P. W., and Larkin, K. G. 2000. Using the Hilbert transform for 3D visualization of differential interference contrast microscope images. *J. Microsc.* 199: 79–84.

Arnison, M. R., Larkin, K. G., Sheppard, C. J. R., Smith, N. I., and Cogswell, C. J. 2004. Linear phase imaging using differential interference contrast microscopy. *J. Microsc.* 214: 7.

Arons, E., Dilworth, D., Shih, M., and Sun, P. C. 1993. Use of Fourier synthesis holography to image through inhomogeneities. *Opt. Lett.* 18: 1852–1854.

Bailey, B., Lanni, F., Farkas, D. L., and Taylor, D. L. 1994. High-resolution direct 3-D fluorescence microscopy of cells and organelles by standing-wave excitation-based optical subsectioning. *Biophys. J.* 66: A275.

Bailey, B., Farkas, D. L., Taylor, D. L., and Lanni, F. 1993. Enhancement of axial resolution in fluorescence microscopy by standing-wave excitation. *Nature* 366: 44.

Barer, R. 1957. Refractometry and interferometry of living cells. *J. Opt. Soc. Am.* 47: 545–556.

Barer, R. and Joseph, S. 1954. Refractometry of living cells part I. Basic principles. *Q. J. Microsc. Sci.* 95: 399–423.

Beuthan, J., Minet, O., Helfmann, J., Herrig, M., and Müller, G. 1996. The spatial variation of the refractive index in biological cells. *Phys. Med. Biol.* 41: 369–382.

Bewersdorf, J., Pick, R., and Hell, S. W. 1998. Multifocal multiphoton microscopy. *Opt. Lett.* 23: 655.

Bourquin, S., Seitz, P., and Salathé, R. P. 2001. Optical coherence topography based on a two-dimensional smart detector array. *Opt. Lett.* 26: 512–514.

Butterweck, H. J. 1977. General theory of linear, coherent optical data-processing systems. *J. Opt. Soc. Am.* 67: 60–70.

Carl, D., Kemper, B., Wernicke, G., and Bally, G. V. 2004. Parameter-optimized digital holographic microscope for high-resolution living-cell analysis. *Appl. Opt.* 43: 6536–6544.

Carlsson, T. E. 1993. Measurement of 3-dimensional shapes using light-in-flight recording by holography. *Opt. Eng.* 32: 2587–2592.

Chalut, K. J., Brown, W. J., and Wax, A. 2007. Quantitative phase microscopy with asynchronous digital holography. *Opt. Express* 15: 3047–3052.

Chanteloup, J. C. 2005. Multiple-wave lateral shearing interferometry for wave-front sensing. *Appl. Opt.* 44: 1559–1571.

Charrière, F., Colomb, T., Montfort, F., Cuche, E., Marquet, P., and Depeursinge, C. 2006a. Shot noise influence in reconstructed phase image SNR in digital holographic microscopy. *Appl. Opt.* 45: 7667–7673.

Charrière, F., Marian, A., Colomb, T., Marquet, P., and Depeursinge, C. 2007a. Amplitude point spread function measurement of high NA microscope objectives by digital holographic microscopy. *Opt. Lett.* 32: 2456–2458.

Charrière, F., Marian, A., Montfort, F., Kühn, J., Colomb, T., Cuche, E., Marquet, P., and Depeursinge, C. 2006b. Cell refractive index tomography by digital holographic microscopy. *Opt. Lett.* 31: 178–180.

Charrière, F., Pavillon, N., Colomb, T., Heger, T., Mitchell, E., Marquet, P., Rappaz, B., and Depeursinge, C. 2006c. Living specimen tomography by digital holographic microscopy: Morphometry of testate amoeba. *Opt. Express* 14: 7005–7013.

Charrière, F., Rappaz, B., Kühn, J., Colomb, T., Marquet, P., and Depeursinge, C. 2007b. Influence of shot noise on phase measurement accuracy in digital holographic microscopy. *Opt. Express* 15: 8818–8831.

Choi, W., Fang-Yen, C., Badizadegan, K., Dasari, R. R., and Feld, M. S. 2008. Extended depth of focus in tomographic phase microscopy using a propagation algorithm. *Opt. Lett.* 33: 171–173.

Choi, W., Fang-Yen, C., Badizadegan, K., Oh, S., Lue, N., Dasari, R. R., and Feld, M. S. 2007. Tomographic phase microscopy. *Nat. Methods* 4: 717–719.

Clauser, J. F. and Reinsch, M. W. 1992. New theoretical and experimental results in Fresnel optics with applications to matter-wave and x-ray interferometry. *Appl. Phys. B Photophys. Laser Chem.* 54: 380–395.

Cole, M. J., Siegel, J., Webb, S. E. D., Jones, R., Dowling, K., French, P. M. W., Lever, M. J., Sucharov, L. O. D., Neil, M. A. A., Juškaitis, R., and Wilson, T. 2000. Whole-field optically sectioned fluorescence lifetime imaging. *Opt. Lett.* 25: 1361.

Colomb, T., Cuche, E., Charrière, F., Kühn, J., Aspert, N., Montfort, F., Marquet, P., and Depeursinge, C. 2006a. Automatic procedure for aberration compensation in digital holographic microscopy and applications to specimen shape compensation. *Appl. Opt.* 45: 851–863.

Colomb, T., Cuche, E., Montfort, F., Marquet, P., and Depeursinge, C. 2004. Jones vector imaging by use of digital holography: Simulation and experimentation. *Opt. Commun.* 231: 137–147.

Colomb, T., Dahlgren, P., Beghuin, D., Cuche, E., Marquet, P., and Depeursinge, C. 2002. Polarization imaging by use of digital holography. *Appl. Opt.* 41: 27–37.

Colomb, T., Dürr, F., Cuche, E., Marquet, P., Limberger, H., Salathé, R.-P., and Depeursinge, C. 2005. Polarization microscopy by use of digital holography: Application to optical fiber bire-fringence measurements. *Appl. Opt.* 44: 4461–4469.

Colomb, T., Kühn, J., Charrière, F., Depeursinge, C., Marquet, P., and Aspert, N. 2006b. Total aberrations compensation in digital holographic microscopy with a reference conjugated hologram. *Opt. Express* 14: 4300–4306.

Colomb, T., Kuhn, J., Cuche, E., Charriere, F., Montfort, F., Marian, A., Aspert, N., Marquet, P., and Depeursinge, C. 2006c. Automatic procedure for aberrations compensation in digital holographic microscopy. In: Gorecki, C., Asundi, A. K. and Osten, W., eds. *Optical Micro- and Nanometrology in Microsystems Technology*, 2006c, Strasbourg, Proc. SPIE 6188, 618805.

Colomb, T., Marquet, P., Charrière, F., Kühn, J., Jourdain, P., Depeursinge, C., Rappaz, B., and Magistretti, P. 2007. Enhancing the performance of digital holographic microscopy. *SPIE, Newsroom* 10.1117/2.1200709.0872.

Colomb, T., Montfort, F., Kühn, J., Aspert, N., Cuche, E., Marian, A., Charrière, F., Bourquin, S., Marquet, P., and Depeursinge, C. 2006d. Numerical parametric lens for shifting, magnification and complete aberration compensation in digital holographic microscopy. *J. Opt. Soc. Am. A: Opt. Image Sci. Vis.* 23: 3177–3190.

Coppola, G., Ferraro, P., Iodice, M., de Nicola, S., Finizio, A., and Grilli, S. 2004. A digital holographic microscope for complete characterization of microelectromechanical systems. *Meas. Sci. Technol.* 15: 529–539.

Coquoz, O., Conde, R., Taleblou, F., and Depeursinge, C. 1995. Performances of endoscopic holography with a multicore optical fiber. *Appl. Opt.* 34: 7186–7193.

Cuche, E., Bevilacqua, F., and Depeursinge, C. 1999a. Digital holography for quantitative phase-contrast imaging. *Opt. Lett.* 24: 291–293.

Cuche, E., Depeursinge, C., and Marquet, P. 2000a. Spatial filtering and aperture apodization in digital holographic microscopy. In: *Biomedical Topical Meetings, Technical Digest*, 2000a, Washington, DC, 2000a, pp. 411–413. Optical Society of America.

Cuche, E., Marquet, P., and Depeursinge, C. 1999b. Simultaneous amplitude-contrast and quantitative phase-contrast micros-copy by numerical reconstruction of Fresnel off-axis holograms. *Appl. Opt.* 38: 6994–7001.

Cuche, E., Marquet, P., and Depeursinge, C. 2000b. Aperture apodization using cubic spline interpolation: Application in digital holographic microscopy. *Opt. Commun.* 182: 59–69.

Cuche, E., Poscio, P., and Depeursinge, C. 1997. Optical tomography by means of a numerical low-coherence holographic technique. *J. Opt.* 28: 260–264.

Dakoff, A., Gass, J., and Kim, M. K. 2003. Microscopic three-dimensional imaging by digital interference holography. *J. Electron. Imaging* 12: 643–647.

DaneshPanah, M. and Javidi, B. 2007. Tracking biological micro-organisms in sequence of 3D holographic microscopy images. *Opt. Express* 15: 10761–10766.

Davies, H. G. and Wilkins, M. H. F. 1952. Interference micros-copy and mass determination. *Nature* 169: 541.

de Haller, E. B., Bally, G. V., and Depeursinge, C. 1995. High-resolution holography and biopsy: Preliminary results. *Bioimaging* 3: 76–87.

de Nicola, S., Ferraro, P., Finizio, A., and Pierattini, G. 2001. Correct-image reconstruction in the presence of severe anamorphism by means of digital holography. *Opt. Lett.* 26: 974–976.

de Nicola, S., Ferraro, P., Finizio, A., and Pierattini, G. 2002. Wave front reconstruction of Fresnel off-axis holograms with compensation of aberrations by means of phase-shifting digital holography. *Opt. Lasers Eng.* 37: 331–340.

de Nicola, S., Finizio, A., Pierattini, G., Ferraro, P., and Alfieri, D. 2005. Angular spectrum method with correction of anamorphism for numerical reconstruction of digital holograms on tilted planes. *Opt. Express* 13: 9935–9940.

Debailleul, M., Georges, V., Simon, B., Morin, R., and Haeberlé, O. 2009. High-resolution three-dimensional tomographic diffractive microscopy of transparent inorganic and biological samples. *Opt. Lett.* 34: 79–81.

Debailleul, M., Simon, B., Georges, V., Haeberlé, O., and Lauer, V. 2008. Holographic microscopy and diffractive microtomography of transparent samples. *Meas. Sci. Technol.* 19(7), 1–8.

Degroot, P. and Deck, L. 1995. Surface profiling by analysis of white-light interferograms in the spatial-frequency domain. *J. Mod. Opt.* 42: 389–401.

Delacrétaz, Y., Pavillon, N., Lang, F., and Depeursinge, C. 2009. Off-axis low coherence interferometry contouring. *Opt. Commun.* 282: 4595.

Dubois, A., Grieve, K., Moneron, G., Lecaque, R., Vabre, L., and Boccara, C. 2004a. Ultrahigh-resolution full-field optical coherence tomography. *Appl. Opt.* 43: 2874–2883.

Dubois, F., Callens, N., Yourassowsky, C., Hoyos, M., Kurowski, P., and Monnom, O. 2006a. Digital holographic microscopy with reduced spatial coherence for three-dimensional particle flow analysis. *Appl. Opt.* 45: 864–871.

Dubois, F., Joannes, L., and Legros, J.-C. 1999. Improved three-dimensional imaging with a digital holography microscope with a source of partial spatial coherence. *Appl. Opt.* 38: 7085–7094.

Dubois, F., Minetti, C., Monnom, O., Yourassowsky, C., Legros, J. C., and Kischel, P. 2002. Pattern recognition with a digital holographic microscope working in partially coherent illumination. *Appl. Opt.* 41: 4108–4119.

Dubois, F., Requena, M. L. N., Minetti, C., Monnom, O., and Istasse, E. 2004b. Partial spatial coherence effects in digital holographic microscopy with a laser source. *Appl. Opt.* 43: 1131–1139.

Dubois, F., Yourassowsky, C., Monnom, O., Legros, J. C., Debeir, O., van Ham, P., Kiss, R., and Decaestecker, C. 2006b. Digital holographic microscopy for the three-dimensional dynamic analysis of in vitro cancer cell migration. *J. Biomed. Opt.* 11: 054032 (5 pages).

Eppich, B., Beuthan, J., Dressler, C., and Müller, G. 2000. Optical phase measurements on biological cells. *Laser Phys.* 10: 467.

Failla, A. V., Cavallo, A., and Cremer, C. 2002a. Subwavelength size determination by spatially modulated illumination virtual microscopy. *Appl. Opt.* 41: 6651.

Failla, A. V., Spoeri, U., Albrecht, B., Kroll, A., and Cremer, C. 2002b. Nanosizing of fluorescent objects by spatially modulated illumination microscopy. *Appl. Opt.* 41: 7275.

Failla, A. V., Spoeri, U., Albrecht, B., Kroll, A., and Cremer, C. 2003. Nanosizing of fluorescent objects by spatially modulated illuminated microscopy: Erratum. *Appl. Opt.* 42: 1308.

Fedosseev, R., Belyaev, Y., Frohn, J., and Stemmer, A. 2005. Structured light illumination for extended resolution in fluorescence microscopy. *Opt. Lasers Eng.* 43: 403.

Ferraro, P., Coppola, G., Alfieri, D., de Nicola, S., Finizio, A., and Pierattini, G. 2004a. Controlling images parameters in the reconstruction process of digital holograms. *IEEE J. Sel. Top. Quantum Electron.* 10: 829–839.

Ferraro, P., Coppola, G., de Nicola, S., Finizio, A., and Pierattini, G. 2003a. Digital holographic microscope with automatic focus tracking by detection sample displacement in real time. *Opt. Lett.* 28: 1257–1259.

Ferraro, P., de Nicola, S., Finizio, A., Coppola, G., Grilli, S., Magro, C., and Pierattini, G. 2003b. Compensation of the inherent wave front curvature in digital holographic coherent microscopy for quantitative phase-contrast imaging. *Appl. Opt.* 42: 1938–1946.

Ferraro, P., Grilli, S., Alfieri, D., Nicola, S. D., Finizio, A., Pierattini, G., Javidi, B., Coppola, G., and Striano, V. 2005. Extended focused image in microscopy by digital holography. *Opt. Express* 13: 6738–6749.

Ferraro, P., Grilli, S., Miccio, L., Alfieri, D., de Nicola, S., Finizio, A., and Javidi, B. 2008. Full color 3-D imaging by digital holography and removal of chromatic aberrations. *J. Display Technol.* 4: 97–100.

Ferraro, P., Nicola, S. D., Coppola, G., Finizio, A., Alfieri, D., and Pierattini, G. 2004b. Controlling image size as a function of distance and wavelength in Fresnel-transform reconstruction of digital holograms. *Opt. Lett.* 29: 854–856.

Fienup, J. R. 1978. Reconstruction of an object from modulus of its Fourier-transform. *Opt. Lett.* 3: 27–29.

Fienup, J. R. 1982. Phase retrieval algorithms—a comparison. *Appl. Opt.* 21: 2758–2769.

Françon, M. 1949. Nouveau dispositif a contraste de phase pour microscope. *C. R. Hebd. Acad. Sci.* 229: 183–185.

Frohn, J. T., Knapp, H. F. and Stemmer, A. 2000. True optical resolution beyond the Rayleigh limit achieved by standing wave illumination. *Proc. Natl. Acad. Sci. U. S. A.* 97: 7232.

Frohn, J. T., Knapp, H. F., and Stemmer, A. 2001. Three-dimensional resolution enhancement in fluorescence microscopy by harmonic excitation. *Opt. Lett.* 26: 828.

Fujimoto, J. G., Pitris, C., Boppart, S. A., and Brezinski, M. E. 2000. Optical coherence tomography: An emerging technology for biomedical imaging and optical biopsy. *Neoplasia* 2: 9–25.

Gabor, D. 1949. Microscopy by reconstructed wave-fronts. *Proc. R. Soc. Lond. Ser. A Math. Phys. Sci.* 197: 454–487.

Gabor, D. 1951. Microscopy by reconstructed wave fronts: II. *Proc. Phys. Soc.* 64: 449–470.

Gabor, D. and Goss, W. P. 1966. Interference microscope with total wavefront reconstruction. *J. Opt. Soc. Am.* 56: 849–858.

Garcia-Sucerquia, J., Ramirez, J. A. H., and Prieto, D. V. 2005. Reduction of speckle noise in digital holography by using digital image processing. *Optik* 116: 44–48.

Garcia-Sucerquia, J., Xu, W., Jericho, M., and Kreuzer, J. 2006a. Immersion digital in-line holographic microscopy. *Opt. Lett.* 31: 1211–1213.

Garcia-Sucerquia, J., Xu, W., Jericho, S. K., Jericho, M. H., and Kreuzer, H. J. 2008. 4-D imaging of fluid flow with digital in-line holographic microscopy. *Optik* 119: 419–423.

Garcia-Sucerquia, J., Xu, W., Jericho, S. K., Klages, P., Jericho, M. H., and Kreuzer, H. J. 2006b. Digital in-line holographic microscopy. *Appl. Opt.* 45: 836–850.

Gass, J., Dakoff, A., and Kim, M. K. 2003. Phase imaging without 2 pi ambiguity by multiwavelength digital holography. *Opt. Lett.* 28: 1141–1143.

Gerchberg, R. W. and Saxton, W. O. 1972. Practical algorithm for the determination of phase from image and diffraction plane pictures. *Optik* 35: 237–246.

Goodman, J. W. 1968. *Introduction to Fourier Optics*. San Francisco, CA: McGraw-Hill.

Goodman, J. W. 2005. *Introduction to Fourier Optics*. San Francisco, CA: Roberts & Company Publishers.

Goodman, J. W. and Lawrence, R. W. 1967. Digital image formation from electronically detected holograms. *Appl. Phys. Lett.* 11: 77–79.

Grilli, S., Ferraro, P., de Nicola, S., Finizio, A., Pierattini, G., and Meucci, R. 2001. Whole optical wavefields reconstruction by digital holography. *Opt. Express* 9: 294–302.

Gureyev, T. E., Roberts, A., and Nugent, K. A. 1995. Phase retrieval with the transport-of-intensity equation—Matrix solution with use of Zernike polynomials. *J. Opt. Soc. Am. a-Opt. Image Sci. Vis.* 12: 1932–1941.

Gustafsson, M. G. 1999. Extended resolution fluorescence microscopy. *Curr. Opin. Struct. Biol.* 9: 627.

Gustafsson, M. G. L. 2000. Surpassing the lateral resolution limit by a factor of two using structured illumination microscopy. *J. Microsc.* 198: 82.

Gustafsson, M. G. L. 2005. Nonlinear structured-illumination microscopy: Wide-field fluorescence imaging with theoretically unlimited resolution. *Proc. Natl. Acad. Sci. U. S. A.* 102: 13081–13086.

Gustafsson, M. G. L., Agard, D. A., and Sedat, J. W. 1999. I5M: 3D widefield light microscopy with better than 100 nm axial resolution. *J. Microsc.* 195: 10.

Haddad, W. S., Cullen, D., Solem, J. C., Longworth, J. W., McPherson, A., Boyer, K., and Rhodes, C. K. 1992. Fourier-transform holographic microscope. *Appl. Opt.* 31: 4973–4978.

Hartman, J. S., Gordon, R. L., and Lessor, D. L. 1980. Quantitative surface topography determination by Nomarski reflection microscopy—2. Microscope modification, calibration, and planar sample experiments. *Appl. Opt.* 19: 2998.

Hell, S. and Stelzer, E. H. K. 1992. Properties of a 4Pi confocal fluorescence microscope. *J. Opt. Soc. Am. A* 9: 2159.

Hell, S. W. 2003. Toward fluorescence nanoscopy. *Nat. Biotechnol.* 21: 1347.

Hell, S. W. 2004. Pushing fluorescence microscopy resolution to the nanoscale. *Cytometry Part A* 59: 29–29.

Hildenbrand, G., Rapp, A., Spöri, U., Wagner, C., Cremer, C., and Hausmann, M. 2005. Nano-sizing of specific gene domains in intact human cell nuclei by spatially modulated illumination light microscopy. *Biophys. J.* 88: 4312.

Huang, D., Swanson, E. A., Lin, C. P., Schuman, J. S., Stinson, W. G., Chang, W., Hee, M. R., Flotte, T., Gregory, K., Puliafito, C. A., and Fujimoto, J. G. 1991. Optical coherence tomography. *Science* 254: 1178–1181.

Indebetouw, G. 2001. Spatiotemporal digital microholography. *Holography* 12.1: 4.

Indebetouw, G. and Klysubun, P. 1999a. Optical sectioning with low coherence spatio-temporal holography. *Opt. Commun.* 172: 25–29.

Indebetouw, G. and Klysubun, P. 1999b. Space-time digital holography: A three-dimensional microscopic imaging scheme with an arbitrary degree of spatial coherence. *Appl. Phys. Lett.* 75: 2017–2019.

Indebetouw, G. and Klysubun, P. 2000. Imaging through scattering media with depth resolution by use of low-coherence gating in spatiotemporal digital holography. *Opt. Lett.* 25: 212–214.

Indebetouw, G. and Klysubun, P. 2001. Spatiotemporal digital microholography. *J. Opt. Soc. Am. a-Opt. Image Sci. Vis.* 18: 319–325.

Indebetouw, G., Klysubun, P., Kim, T., and Poon, T.-C. 2000. Imaging properties of scanning holographic microscopy. *J. Opt. Soc. Am. a-Opt. Image Sci. Vis.* 17: 380–390.

Indebetouw, G., Maghnouji, A. E., and Foster, R. 2005. Scanning holographic microscopy with transverse resolution exceeding the Rayleigh limit and extended depth of focus. *J. Opt. Soc. Am. A: Opt. Image Sci. Vis.* 22: 892–898.

Indebetouw, G., Tada, Y., and Leacock, J. 2006. Quantitative phase imaging with scanning holographic microscopy: an experimental assessment. *Biomed. Eng. Online* 5.

Indebetouw, G., Tada, Y., Rosen, J., and Brooker, G. 2007. Scanning holographic microscopy with resolution exceeding the Rayleigh limit of the objective by superposition of off-axis holograms. *Appl. Opt.* 46: 993–1000.

Indebetouw, G. and Zhong, W. 2006. Scanning holographic microscopy of three-dimensional fluorescent specimens. *J. Opt. Soc. Am. a-Opt. Image Sci. Vis.* 23: 1699–1707.

Javidi, B., Ferraro, P., Hong, S.-H., Nicola, S. D., Finizio, A., Alfieri, D., and Pierattini, G. 2005. Three-dimensional image fusion by use of multiwavelength digital holography. *Opt. Lett.* 30: 144–146.

Juskaitis, R., Wilson, T., Neil, M. A. A., and Kozubek, M. 1996. Efficient real-time confocal microscopy with white light sources. *Nature* 383: 804.

Kemmler, M., Fratz, M., Giel, D., Saum, N., Brandenburg, A., and Hoffmann, C. 2007. Noninvasive time-dependent cytometry monitoring by digital holography. *J. Biomed. Opt.* 12: 064002–10.

Kemper, B., Kosmeier, S., Langehanenberg, P., von Bally, G., Bredebusch, I., Domschke, W., and Schnekenburger, J. 2007a. Integral refractive index determination of living suspension cells by multifocus digital holographic phase contrast microscopy. *J. Biomed. Opt.* 12: 054009.

Kemper, B., Langehanenberg, P., and von Bally, G. 2007b. Digital holographic microscopy. *Opt. Photonik* 2: 41–44.

Kempkes, M., Darakis, E., Khanam, T., Rajendran, A., Kariwala, V., Mazzotti, M., Naughton, T. J., and Asundi, A. K. 2009. Three dimensional digital holographic profiling of microfibers. *Opt. Express* 17: 2938–2943.

Kim, M. K. 1999. Wavelength-scanning digital interference holography for optical section imaging. *Opt. Lett.* 24: 1693–1695.

Kim, M. K. 2000. Tomographic three-dimensional imaging of a biological specimen using wavelength-scanning digital interference holography. *Opt. Express* 7: 305–310.

Klysubun, P. and Indebetouw, G. 2001. A posteriori processing of spatiotemporal digital microholograms. *J. Opt. Soc. Am. a-Opt. Image Sci. Vis.* 18: 326–331.

Kou, S. S. and Sheppard, C. J. R. 2008. Image formation in holographic tomography. *Opt. Lett.* 33: 2362–2364.

Kozacki, T. and Józwicki, R. 2005. Digital reconstruction of a hologram recorded using partially coherent illumination. *Opt. Commun.* 252: 188–201.

Kreuzer, H. J., Jericho, M. J., Meinertzhagen, I. A., and Xu, W. 2001. Digital in-line holography with photons and electrons. *J. Phys.: Condensed Matter* 13: 10729–10741.

Kronrod, M. A., Yaroslavsky, L. P., and Merzlyakov, N. S. 1972. Computer synthesis of transparency holograms. *Sov. Phys.-Tech. Phys.* 17: 329–332.

Kühn, J. 2009. *Multiple Wavelength Digital Holgraphic Microscopy.* EPFL.

Kühn, J., Charrière, F., Colomb, T., Cuche, E., Montfort, F., Emery, Y., Marquet, P., and Depeursinge, C. 2008. Axial sub-nanometer accuracy in digital holographic microscopy. *Meas. Sci. Technol.* 19: 074007 (8 pages).

Kühn, J., Colomb, T., Montfort, F., Charrière, F., Emery, Y., Cuche, E., Marquet, P., and Depeursinge, C. 2007. Real-time dual-wavelength digital holographic microscopy with a single hologram acquisition. *Opt. Express* 15: 7231–7242.

Kühn, J., Montfort, F., Colomb, T., Rappaz, B., Moratal, C., Pavillon, N., Marquet, P., and Depeursinge, C. 2009. Submicrometer tomography of cells by multiple-wavelength digital holographic microscopy in reflection. *Opt. Lett.* 34: 653–655.

Langehanenberg, P., Ivanova, L., Bernhardt, I., Ketelhut, S., Vollmer, A., Dirksen, D., Georgiev, G., von Bally, G., and Kemper, B. 2009. Automated three-dimensional tracking of living cells by digital holographic microscopy. *J. Biomed. Opt.* 14: 014018–014027.

Langehanenberg, P., Kemper, B., Dirksen, D., and Bally, G. V. 2008. Autofocusing in digital holographic phase contrast microscopy on pure phase objects for live cell imaging. *Appl. Opt.* 47: D176–D182.

Lauer, V. 2002. New approach to optical diffraction tomography yielding a vector equation of diffraction tomography and a novel tomographic microscope. *J. Microsc.* 205: 165–176.

Le Clerc, F., Collot, L., and Gross, M. 2000. Numerical heterodyne holography with two-dimensional photodetector arrays. *Opt. Lett.* 25: 716–718.

Legarda-Sáenz, R., Rivera, M., Rodríguez-Vera, R., and Trujillo-Schiaffino, G. 2000. Robust wave-front estimation from multiple directional derivatives. *Opt. Lett.* 25: 1089–1091.

Leitgeb, R. A., Drexler, W., Unterhuber, A., Hermann, B., Bajraszewski, T., Le, T., Stingl, A., and Fercher, A. F. 2004. Ultrahigh resolution Fourier domain optical coherence tomography. *Opt. Express* 12: 2156–2165.

Lessor, D. L., Hartman, J. S., and Gordon, R. L. 1979. Quantitative surface topography determination by Normaski reflection microscopy I. Theory. *J. Opt. Soc. Am.* 69: 357–366.

Liebling, M., Blu, T., and Unser, M. 2003. Fresnelets: New multiresolution wavelet bases for digital holography. *IEEE Trans. Image Processing* 12: 29–43.

Liebling, M. and Unser, M. 2004. Autofocus for digital Fresnel holograms by use of a Fresnelet sparsity criterion. *J. Opt. Soc. Am. A-Opt. Image Sci. Vis.* 21: 2424–2430.

Liu, C., Liu, Y., Chen, H., and Yan, C. 2006. Aberration analysis of digital hologram reconstruction with a Fresnel integral. *Opt. Eng.* 45: 075802.

Lohmann, A. W. and Rhodes, W. T. 1978. Two-pupil synthesis of optical transfer functions. *Appl. Opt.* 17: 1141.

Mahon, R. J., Murphy, J. A., and Lanigan, W. 2006. Digital holography at millimetre wavelengths. *Opt. Commun.* 260: 469–473.

Malkiel, E., Abras, J. N., and Katz, J. 2004. Automated scanning and measurements of particle distributions within a holographic reconstructed volume. *Meas. Sci. Technol.* 15: 601–612.

Mann, C. J., Yu, L., Lo, C.-M., and Kim, M. K. 2005. High-resolution quantitative phase-contrast microscopy by digital holography. *Opt. Express* 13: 8693–8698.

Marian, A., Charrière, F., Colomb, T., Montfort, F., Kühn, J., Marquet, P., and Depeursinge, C. 2007. On the complex three-dimensional amplitude point spread function of lenses and microscope objectives: theoretical aspects, simulations and measurements by digital holography. *J. Microsc.* 225: 156–169.

Marquet, P. 2003. Développement d'une nouvelle technique de microscopie optique tridimensionnelle, la microscopie holographique digitale. Perspectives pour l'étude de la plasticite neuronale. Université de Lausanne, Lausanne, Switzerland.

Marquet, P., Rappaz, B., Magistretti, P. J., Cuche, E., Emery, Y., Colomb, T., and Depeursinge, C. 2005. Digital holographic microscopy: a noninvasive contrast imaging technique allowing quantitative visualization of living cells with sub-wavelength axial accuracy. *Opt. Lett.* 30: 468–470.

Martin, S., Failla, A. V., Spöri, U., Cremer, C., and Pombo, A. 2004. Measuring the size of biological nanostructures with spatially modulated illumination microscopy. *Mol. Biol. Cell* 15: 2449.

Martínez-León, L., Pedrini, G., and Osten, W. 2005. Applications of short-coherence digital holography in microscopy. *Appl. Opt.* 44: 3977–3984.

Massatsch, P., Charrière, F., Cuche, E., Marquet, P., and Depeursinge, C. 2005. Time-domain optical coherence tomography with digital holographic microscopy. *Appl. Opt.* 44: 1806–1812.

Mehta, S. B. and Sheppard, C. J. R. 2009. Quantitative phase-gradient imaging at high resolution with asymmetric illumination-based differential phase contrast. *Opt. Lett.* 34: 1924.

Mendlovic, D., Lohmann, A. W., and Zalevsky, Z. 1997. Space-bandwidth product adaptation and its application to super-resolution: examples. *J. Opt. Soc. Am. A* 14: 563–567.

Mendlovic, D. and Zalevsky, Z. 1997. Definition, properties and applications of the generalized temporal-spatial Wigner distribution function. *Optik (Jena)* 107: 49–56.

Miccio, L., Alfieri, D., Grilli, S., Ferraro, P., Finizio, A., de Petrocellis, L., and Nicola, S. D. 2007. Direct full compensation of the aberrations in quantitative phase microscopy of thin objects by a single digital hologram. *Appl. Phys. Lett.* 90: 041104 (3 pages).

Micó, V., Zalevsky, Z., Ferreira, C., and García, J. 2008. Superresolution digital holographic microscopy for three-dimensional samples. *Opt. Express* 16: 19260–19270.

Mico, V., Zalevsky, Z., and Garcia, J. 2006a. Superresolution optical system by common-path interferometry. *Opt. Express* 14: 5168–5177.

Mico, V., Zalevsky, Z., and García, J. 2008. Common-path phase-shifting digital holographic microscopy: A way to quantitative phase imaging and superresolution. *Opt. Commun.* 281: 4273–4281.

Mico, V., Zalevsky, Z., Garcia-Martinez, P., and Garcia, J. 2006b. Synthetic aperture superresolution with multiple off-axis holograms. *J. Opt. Soc. Am. a-Opt. Image Sci. Vis.* 23: 3162–3170.

Minsky, M. 1988. Memoir on inventing the confocal scanning microscope. *Scanning* 10: 128–138.

Momose, A., Kawamoto, S., Koyama, I., Hamaishi, Y., Takai, K., and Suzuki, Y. 2003. Demonstration of x-ray Talbot interferometry. *Jpn. J. Appl. Phys. Part 2: Lett.* 42.

Monnom, O., Dubois, F., Yourassowsky, C., and Legros, J. C. 2005. Improvement in visibility of an in-focus reconstructed image in digital holography by reduction of the influence of out-of-focus objects. *Appl. Opt.* 44: 3827–3832.

Montfort, F., Charrière, F., Colomb, T., Cuche, E., Marquet, P., and Depeursinge, C. 2006a. Purely numerical compensation for microscope objective phase curvature in digital holographic microscopy: influence of digital phase mask position. *J. Opt. Soc. Am. a-Opt. Image Sci. Vis.* 23: 2944–2953.

Montfort, F., Colomb, T., Charrière, F., Kühn, J., Marquet, P., Cuche, E., Herminjard, S., and Depeursinge, C. 2006b. Submicrometer optical tomography by multiple-wavelength digital holographic microscopy. *Appl. Opt.* 45: 8209–8217.

Murty, M. 1970. Compact lateral shearing Interferometer based on the Michelson interferometer. *Appl. Opt.* 9: 1146–1148.

Nagorni, M. and Hell, S. W. 2001a. Coherent use of opposing lenses for axial resolution increase in fluorescence microscopy. I. Comparative study of concepts. *J. Opt. Soc. Am. A: Opt. Image Sci. Vis.* 18: 36.

Nagorni, M. and Hell, S. W. 2001b. Coherent use of opposing lenses for axial resolution increase. II. Power and limitation of nonlinear image restoration. *J. Opt. Soc. Am. A: Opt. Image Sci. Vis.* 18: 49.

Neil, M. A. A., Squire, A., Juskaitis, R., Bastiaens, P. I. H., and Wilson, T. 2000. Wide-field optically sectioning fluorescence microscopy with laser illumination. *J. Microsc.* 197: 1.

Nomarski, G. 1955. Nouveau dispositif pour l'observation en contraste de phase differentiel. *J. Phys. Radium* 16, S88–S88.

Nugent, K. A., Paganin, D., and Gureyev, T. E. 2001. A phase odyssey. *Phys. Today* 54: 27–32.

Nugumanov, A. M., Smirnov, R. V., and Sokolov, V. I. 2000. A method for measuring the radiation wave front by using a three-wave lateral shearing interferometer. *Quantum Electron.* 30: 435–440.

Onodera, R., Wakaumi, H., and Ishii, Y. 2005. Measurement technique for surface profiling in low-coherence interferometry. *Opt. Commun.* 254: 52–57.

Onural, L. 2004. Some mathematical properties of the uniformly sampled quadratic phase function and associated issues in digital Fresnel diffraction simulations. *Opt. Eng.* 43: 2557–2563.

Palacios, F., Goncalves, E., Ricardo, J., and Valin, J. L. 2004. Adaptive filter to improve the performance of phase-unwrapping in digital holography. *Opt. Commun.* 238: 245–251.

Palacios, F., Ricardo, J., Palacios, D., Gonçalves, E., Valin, J. L., and Souza, R. D. 2005. 3D image reconstruction of transparent microscopic objects using digital holography. *Opt. Commun.* 248: 41–50.

Patorski, K. 1989. The self-imaging phenomenon and its applications. *Prog. Opt.* 27: 1–108.

Pavillon, N., Benke, A., Boss, D., Moratal, C., Kühn, J., Jourdain, P., Depeursinge, C., Magistretti, P. J., and Marquet, P. 2010. Cell morphology and intracellular ionic homeostasis explored with a multimodal approach combining epifluorescence and digital holographic microscopy. *J. Biophotonics*, 3(7), 432–436.

Pavillon, N., Kühn, J., Charrière, F., and Depeursinge, C. 2009a. Optical tomography by digital holographic microscopy. In: *Novel Optical Instrumentation for Biomedical Applications IV*, 2009a. SPIE, Munich, Germany.

Pavillon, N., Seelamantula, C. S., Kühn, J., Unser, M., and Depeursinge, C. 2009b. Suppression of the zero-order term in off-axis digital holography through nonlinear filtering. *Appl. Opt.* 48, H186–H195.

Pawley, J. B. 1995. *Handbook of Biological Confocal Microscopy.* New York: Plenum Press.

Pedrini, G. and Schedin, S. 2001. Short coherence digital holography for 3D microscopy. *Optik* 112: 427–432.

Pedrini, G., Schedin, S., and Tiziani, H. J. 2001. Aberration compensation in digital holographic reconstruction of microscopic objects. *J. Mod. Opt.* 48: 1035–1041.

Pedrini, G. and Tiziani, H. J. 2002. Short-coherence digital microscopy by use of a lensless holographic imaging system. *Appl. Opt.* 41: 4489–4496.

Pluta, M. 1987. *Holographic Microscopy*. London, U.K.: Academic Press.

Poon, T.-C. 1985. Scanning holography and two-dimensional image processing by acousto-optic two-pupil synthesis. *J. Opt. Soc. Am. A* 2: 521–527.

Popescu, G., Ikeda, T., Dasari, R., and Feld, M. 2006. Diffraction phase microscopy for quantifying cell structure and dynamics. *Opt. Lett.* 31: 775–777.

Popescu, G., Park, Y., Choi, W., Dasari, R. R., Feld, M. S., and Badizadegan, K. 2008. Imaging red blood cell dynamics by quantitative phase microscopy. *Blood Cells, Molecules, and Diseases* 41: 10–16.

Potcoava, M. C. and Kim, M. K. 2008. Optical tomography for biomedical applications by digital interference holography. *Meas. Sci. Technol.* 19: 074010 (8 pages).

Primot, J. and Sogno, L. 1995. Achromatic three-wave (or more) lateral shearing interferometer. *J. Opt. Soc. Am. A* 12: 2679–2685.

Primot, J., Sogno, L., Fracasso, B., and Heggarty, K. 1997. Wavefront sensor prototype for industrial applications based on a three-level phase grating. *Opt. Eng.* 36: 901–904.

Psaltis, D. 2002. Coherent optical information systems. *Science* 298: 1359–1363.

Rappaz, B., Barbul, A., Emery, Y., Korenstein, R., Depeursinge, C., Magistretti, P. J., and Marquet, P. 2008a. Comparative study of human erythrocytes by digital holographic microscopy, confocal microscopy, and impedance volume analyzer. *Cytometry Part A.*

Rappaz, B., Barbul, A., Hoffmann, A., Boss, D., Korenstein, R., Depeursinge, C., Magistretti, P. J., and Marquet, P. 2009a. Spatial analysis of erythrocyte membrane fluctuations by digital holographic microscopy. *Blood Cells, Molecules, and Diseases* 42: 228.

Rappaz, B., Cano, E., Colomb, T., Kuhn, J., Depeursinge, C., Simanis, V., Magistretti, P. J., and Marquet, P. 2009b. Noninvasive characterization of the fission yeast cell cycle by monitoring dry mass with digital holographic microscopy. *J. Biomed. Opt.* 14: 034049 (5 pages).

Rappaz, B., Charrière, F., Depeursinge, C., Magistretti, P. J., and Marquet, P. 2008b. Simultaneous cell morphometry and refractive index measurement with dual-wavelength digital holographic microscopy and dye-enhanced dispersion of perfusion medium. *Opt. Lett.* 33: 744–746.

Rappaz, B., Marquet, P., Cuche, E., Emery, Y., Depeursinge, C., and Magistretti, P. J. 2005. Measurement of the integral refractive index and dynamic cell morphometry of living cells with digital holographic microscopy. *Opt. Express* 13: 9361–9373.

Repetto, L., Chittofrati, R., Piano, E., and Pontiggia, C. 2005. Infrared lensless holographic microscope with a vidicon camera for inspection of metallic evaporations on silicon wafers. *Opt. Commun.* 251: 44–50.

Repetto, L., Piano, E., and Pontiggia, C. 2004. Lensless digital holographic microscope with light-emitting diode illumination. *Opt. Lett.* 29: 1132–1134.

Rosen, J. and Brooker, G. 2007a. Digital spatially incoherent Fresnel holography. *Opt. Lett.* 32: 912–914.

Rosen, J. and Brooker, G. 2007b. Fluorescence incoherent color holography. *Opt. Express* 15: 2244–2250.

Rosen, J. and Brooker, G. 2008. Non-scanning motionless fluorescence three-dimensional holographic microscopy. *Nat. Photon* 2: 190–195.

Roy, M., Svahn, P., Cherel, L., and Sheppard, C. J. R. 2002. Geometric phase-shifting for low-coherence interference microscopy. *Opt. Lasers Eng.* 37: 631–641.

Schilling, B. W., Poon, T.-C., Indebetouw, G., Storrie, B., Shinoda, K., Suzuki, Y. and Wu, M. H. 1997. Three-dimensional holographic fluorescence microscopy. *Opt. Lett.* 22: 1506.

Schilling, B. W. and Templeton, G. C. 2001. Three-dimensional remote sensing by optical scanning holography. *Appl. Opt.* 40: 5474–5481.

Schmidt, M., Nagorni, M., and Hell, S. W. 2000. Subresolution axial distance measurements in far-field fluorescence microscopy with precision of 1 nanometer. *Rev. Sci. Instrum.* 71: 2742.

Schnars, U. and Jüptner, W. 1994. Direct Recording of Holograms by a Ccd Target and Numerical Reconstruction. *Appl. Opt.* 33: 179–181.

Schnars, U. and Juptner, W. P. O. 2002. Digital recording and numerical reconstruction of holograms. *Meas. Sci. Technol.* 13, R85–R101.

Schnekenburger, J., Bredebusch, I., Domschke, W., Kemper, B., Langehanenberg, P., and von Bally, G. 2007. Digital holographic imaging of dynamic cytoskeleton changes. *Med. Laser Appl.* 22: 165–172.

Schwarz, C. J., Kuznetsova, Y., and Brueck, S. R. J. 2003. Imaging interferometric microscopy. *Opt. Lett.* 28: 1424–1426.

Schweitzer, A., Christian, W., and C. Cremer 2004. The nanosizing of fluorescent objects by 458 nm spatially modulated illumination microscopy using a simplified size evaluation algorithm. *J. Phys.-Condens. Matter* 16, S2393–S2404.

Seelamantula, C. S., Pavillon, N., Depeursinge, C., and Unser, M. 2009. Efficient suppression of the object-wave zero-order in digital holography. *J. Opt. Soc. Am. A* (submitted).

Seelamantula, C. S., Villiger, M. L., Leitgeb, R. A., and Unser, M. 2008. Exact and efficient signal reconstruction in frequency-domain optical-coherence tomography. *J. Opt. Soc. Am. A: Opt. Image Sci. Vis.* 25: 1762–1771.

Shack, R. V. and Platt, B. C. 1971. Production and use of a lenticular Hartmann screen. In: *Program of the 1971 Spring Meeting of the Optical Society of America*, Tucson, Arizona, 1971, p. 656. OSA.

Sheng, J., Malkiel, E., Katz, J., Adolf, J., Belas, R., and Place, A. R. 2007. Digital holographic microscopy reveals prey-induced changes in swimming behavior of predatory dinoflagellates. *Proc. Natl. Acad. Sci. U. S. A.* 104: 17512–17517.

Sotthivirat, S. and Fessler, J. A. 2004. Penalized-likelihood image reconstruction for digital holography. *J. Opt. Soc. Am. A: Opt. Image Sci. Vis.* 21: 737–750.

Spori, U., Failla, A. V., and Cremer, C. 2004. Superresolution size determination in fluorescence microscopy: A comparison between spatially modulated illumination and confocal laser scanning microscopy. *J. Appl. Phys.* 95: 8436–8443.

Stadelmaier, A. and Massig, J. H. 2000. Compensation of lens aberrations in digital holography. *Opt. Lett.* 25: 1630–1632.

Sun, H., Song, B., Dong, H., Reid, B., Player, M. A., Watson, J., and Zhao, M. 2008. Visualization of fast-moving cells in vivo using digital holographic video microscopy. *J. Biomed. Opt.* 13: 014007–014009.

Sung, Y., Choi, W., Fang-Yen, C., Badizadegan, K., Dasari, R. R., and Feld, M. S. 2009. Optical diffraction tomography for high resolution live cell imaging. *Opt. Express* 17: 266–277.

Takaki, Y., Kawai, H., and Ohzu, H. 1999. Hybrid holographic microscopy free of conjugate and zero-order images. *Appl. Opt.* 38: 4990–4996.

Takaki, Y. and Ohzu, H. 1999. Fast numerical reconstruction technique for high-resolution hybrid holographic microscopy. *Appl. Opt.* 38: 2204–2211.

Takaki, Y. and Ohzu, H. 2000. Hybrid holographic microscopy: visualization of three-dimensional object information by use of viewing angles. *Appl. Opt.* 39: 5302–5308.

Takeda, M., Wang, W., Duan, Z., and Miyamoto, Y. 2005. Coherence holography. *Opt. Express* 13: 9629–9635.

Tamano, S., Hayasaki, Y., and Nishida, N. 2006. Phase-shifting digital holography with a low-coherence light source for reconstruction of a digital relief object hidden behind a light-scattering medium. *Appl. Opt.* 45: 953–959.

Teague, M. R. 1983. Deterministic phase retrieval—a green-function solution. *J. Opt. Soc. Am.* 73: 1434–1441.

Teague, M. R. 1985. Image-Formation in Terms of the Transport-Equation. *J. Opt. Soc. Am. a-Opt. Image Sci. Vis.* 2: 2019–2026.

Vabre, L., Dubois, A., and Boccara, A. C. 2002. Thermal-light full-field optical coherence tomography. *Opt. Lett.* 27: 530–532.

van Munster, E. B., van Vliet, L. J., and Aten, J. A. 1997. Reconstruction of optical pathlength distributions from images obtained by a wide-field differential interference contrast microscope. *J. Microsc.* 188: 149.

Velghe, S., Primot, J., Guérineau, N., Cohen, M., and Wattellier, B. 2005. Wave-front reconstruction from multidirectional phase derivatives generated by multilateral shearing interferometers. *Opt. Lett.* 30: 245–247.

Weng, J., Zhong, J., and Hu, C. 2008. Digital reconstruction based on angular spectrum diffraction with the ridge of wavelettransform in holographic phase-contrast microscopy. *Opt. Express* 16: 21971–21981.

Wilson, T. 1990. *Confocal Microscopy*, New York: Academic Press.

Wilson, T. and Sheppard, C. 1984. *Theory and Practice of Scanning Optical Microscopy*. New York: Academic Press.

Wolf, E. 1969. Three-dimensional structure determination of semi-transparent object from holographic data. *Opt. Commun.* 1: 153–156.

Xu, L., Miao, J. M., and Asundi, A. 2000. Properties of digital holography based on in-line configuration. *Opt. Eng.* 39: 3214–3219.

Xu, L., Peng, X., Guo, Z., Miao, J., and Asundi, A. 2005. Imaging analysis of digital holography. *Opt. Express* 13: 2444–2452.

Xu, L., Peng, X. Y., Asundi, A. K., and Miao, J. M. 2001a. Hybrid holographic microscope for interferometric measurement of microstructures. *Opt. Eng.* 40: 2533–2539.

Xu, L., Peng, X. Y., Asundi, A. K., and Miao, J. M. 2003a. Digital microholointerferometer: development and validation. *Opt. Eng.* 42: 2218–2224.

Xu, W., Jericho, M. H., Kreuzer, H. J., and Meinertzhagen, I. A. 2003b. Tracking particles in four dimensions with in-line holographic microscopy. *Opt. Lett.* 28: 164–166.

Xu, W., Jericho, M. H., Meinertzhagendagger, I. A., and Kreuzer, H. J. 2001b. Digital in-line holography for biological applications. *PNAS* 98: 11301–11305.

Yamaguchi, I., Kato, J., Ohta, S., and Mizuno, J. 2001. Image formation in phase-shifting digital holography and applications to microscopy. *Appl. Opt.* 40: 6177–6186.

Yamaguchi, I. and Zhang, T. 1997. Phase-shifting digital holography. *Opt. Lett.* 22: 1268–1270.

Yang, G. and Gu, B. 1981. On the amplitude—Phase retrieval problem in the optical system. *Acta Phys. Sinica* 30: 410–413.

Yaroslavsky, L. P., and Merzlyakov, N. S. 1980. *Methods of Digital Holography*. New York: Consultant Bureau.

Yu, L. and Chen, Z. 2007. Digital holographic tomography based on spectral interferometry. *Opt. Lett.* 32: 3005–3007.

Zalevsky, Z., Mendlovic, D., and Lohmann, A. W. 2000. Understanding superresolution in Wigner space. *J. Opt. Soc. Am. A* 17: 2422–2430.

Zernike, F. 1937. Presentation on the luminous intensity of interference phenomenon and their interpretation. *Phys. Z.* 38: 994–994.

Zernike, F. 1942. Phase contrast, a new method for the microscopic observation of transparent objects. *Physica* 9: 686–698.

Zernike, F. 1949. A new method for the accurate measurement of phase differences. *J. Opt. Soc. Am.* 39: 1059–1059.

Zhang, F. C., Yamaguchi, I., and Yaroslavsky, L. P. 2004a. Algorithm for reconstruction of digital holograms with adjustable magnification. *Opt. Lett.* 29: 1668–1670.

Zhang, T. and Yamaguchi, I. 1998. Three-dimensional microscopy with phase-shifting digital holography. *Opt. Lett.* 23: 1221–1223.

Zhang, Y. M., Lu, Q. N., and Ge, B. Z. 2004b. Elimination of zero-order diffraction in digital off-axis holography. *Opt. Commun.* 240: 261–267.

30

Polarized Light Imaging of Biological Tissues

30.1 Introduction 649
  In This Chapter • A Brief History
30.2 Toolkit..651
  Nomenclature • Stokes Vector: The Description of Polarized Light • Poincaré
  Sphere • Mueller Matrix: Transport of Polarized Light • Data Matrix: Measurements of
  Polarized Light Transport in a Sample • Simulation Matrix: Monte Carlo Simulation of
  Polarized Light Transport • Testing Algorithms with an Identity
30.3 Depolarization Coefficients: μ_{LP} and μ_{CP} [cm^{-1}]... 658
  Optical Properties versus Transport Parameters • Relative Roles of Retardance and Scattering
  in Depolarization • Depolarization by Birefringence • Depolarization by Scattering
30.4 Experimental Measurements of Depolarization in Tissues... 662
  Experimental Evidence
30.5 A Polarized Light Camera for Clinical Use 665
  A Handheld Polarized Light Camera for Imaging Skin • Polarized Light Camera
  Images • Clinical Study: Finding Skin Cancer Margins
30.6 Future Directions... 668
Acknowledgments.. 668
Glossary.. 668
References... 668

Steven L. Jacques
*Oregon Health &
Science University*

30.1 Introduction

The more common imaging modalities rely on absorption of photons, scattering of photons, or regeneration of new photons at new wavelengths (fluorescence, Raman scattering). But a photon also has an orientation of the vibration of its electric field that characterizes the photon. When a population of photons share a common orientation of their electric fields, the light is described as polarized. Polarized light is commonly encountered in polaroid glasses with vertically oriented linear polarization filters to block the horizontally vibrating photons that skip off the surface of a road or water surface. Hence, the glasses block surface glare.

A common use of polarized light in medicine is the dermatologist using a dermatoscope that delivers linearly polarized light to the skin, while the dermatologist views through a linearly polarization filter that is cross-polarized, that is, oriented so it blocks the illumination light. Consequently, only light that penetrates the skin is randomized by propagation in the skin, such that half the light is now co-polarized with the viewing filter and will reach the eye of the dermatologist (Anderson 1991). This technique removes the glare from the skin surface. Alternatively, turning the viewing polarizer to be parallel with the orientation of the polarized illumination light will accent the surface glare and block half the multiply scattered light from the skin, allowing the dermatologist to better view the surface irregularities of the skin.

Perhaps the simplest demonstration of polarized light imaging is to deliver a narrow beam of light through a linear polarizer down onto a cup of milk (suspension of lipid droplets) and image the reflected light using a CCD camera that views through a second linear polarizer. The second polarizer is oriented either parallel to the polarization of the delivered light, or perpendicular. Hence, two images are created. Figure 30.1a shows an early experiment (Jacques et al. 1996) that used a HeNe laser and a milky suspension of lipid droplets (Intralipid™). The two images and the difference between the images are shown.

Figure 30.1b mimics the experiment using a Monte Carlo simulation (Ramella-Roman et al. 2005a,b). The electric field of the linearly polarized illumination light is oriented parallel to the horizontal axis, and this delivered light is called H. Viewing through a parallel polarization filter yields the image called HH. Viewing through a perpendicular polarization filter (V) yields an image called HV. The difference image HH − HV reveals the well-known Maltese cross or hourglass pattern, in which the vertical positive-value cross is dominated by HH photons and a

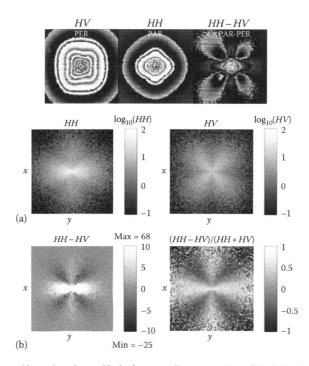

FIGURE 30.1 The reflectance of delivered linearly polarized light from a milky suspension of lipid droplets. (a) Experiment with Intralipid™ and HeNe laser (633 nm wavelength) (Jacques et al., 1996). Light is delivered at a slightly oblique angle, which explains why lower lobes are larger than upper lobes. (b) Monte Carlo simulation mimicking 500 nm dia. polystyrene microspheres (633 nm wavelength). In both a and b, linearly polarized light (*H*) is delivered with its electric field aligned parallel to the *x*-axis. Camera views through linear polarizer aligned parallel to delivered light to yield image called *HH*. Camera views through linear polarizer aligned perpendicular (*V*) to delivered light to yield image called *HV*. The difference image, *HH* − *HV*, and the degree of linear polarization image, (*HH* − *HV*)/(*HH* + *HV*), are shown.

second negative-value cross is oriented at 45° and is dominated by HV photons. This pattern is a simple illustration of polarized light imaging.

30.1.1 In This Chapter

In this chapter, we address the mechanism of contrast seen in polarized light images, which basically depends on the degree to which incident polarized light becomes randomized or depolarized.

First, we introduce the nomenclature and computational tools used to model polarized light transport. Second, we illustrate through simple computational simulations the attenuation of the degree of polarization (both linear and circular) due to two mechanisms: (1) loss due to tissue birefringence and (2) loss due to scattering. This section emphasizes the difference between transport, described as a Mueller matrix, and an optical property, described as the attenuation coefficient for depolarization (μ_{LP} [cm^{-1}] for linear polarization, μ_{CP} [cm^{-1}] for circular polarization). Third, we revisit early experimental measurements, which specify the values of μ_{LP} for various tissues (liver, muscle, skin, blood, myocardium, tendon). Fourth, we apply the lessons learned to the image contrast seen in imaging skin with polarized light.

30.1.2 A Brief History

A few reports pertinent to polarized light imaging in biological tissues are mentioned. This is not a full review of the literature, but

should provide an entrée for the reader who wishes to further study the history of polarized light imaging in medicine and biology.

Polarized light has long been used in ophthalmology to monitor retinal nerve fiber layer thickness, as in Weinreb et al. (1990). Schmitt et al. (1992) showed that polarized light could select photons that had only propagated a short path. Jacques et al. (1996) used a video camera and a polarization-maintaining optical fiber probe to record the point spread function (PSF) for backscatter of linearly polarized light from a microsphere solution, and found that the PSF was confined mainly within a radial distance 2.2/μ'_s, where μ'_s is the reduced scattering coefficient. Demos and Alfano (1997) reported on time-resolved reflectance of polarized light to illustrate polarized light gating, and demonstrated a polarized light image of skin. Hielscher et al. (1997a,b) reported on the PSF of polarized light in solutions of cells. De Boer et al. (1997) applied polarization-sensitive optical coherence tomography (psOCT) to the imaging of skin, and a great many psOCT papers have followed. Yao and Wang (1999) reported on backscattered polarized light imaging during optical coherence tomography in terms of the Mueller matrix. Jacques et al. (2000) described the rate of depolarization during transmission of linearly polarized light through various tissues and microsphere solutions, and built a polarized light camera for imaging skin with backscattered light. Sankaran et al. (2000a) reported on depolarization during transmission through several types of tissues, and Sankaran et al. (2000b) described how depolarization decreases when scatterers become too concentrated, which is an important caution when extrapolating the lessons of scattering by microspheres to the scattering by biological tissues. Studinski and Vitkin (2000) demonstrated that backscattered light retains some degree of polarization even in strongly scattering media, consistent with short pathlength photons escaping with only a few scatterings and returning some polarization. Mourant et al. (2001) developed a handheld polarized light optical fiber probe for characterizing tissues, and Mourant et al. (2002) studied the size distribution of scatterers in tissue responsible for polarized light scattering. Agafonov et al. (2002) demonstrated that increasing optical absorption in a tissue enabled backscattered light to retain its degree of polarization, consistent with the shorter photon pathlengths surviving and returning some polarization. Sankaran et al. (2002) reported that tissues depolarized linear polarization faster than circular polarization, suggesting that the scattering in tissues was dominated by Rayleigh scattering where scatterers are much smaller than the wavelength of light. Jacques et al. (2002) published a set of images of skin pathology using a polarized light camera, and imaged a razor blade edge onto skin to show that the lateral PSF of polarized light in skin using mid-visible light was ~390 µm (half-width half-max from edge of shadow). Baba et al. (2002) developed a Mueller matrix polarization imaging system using the backscattered light from an in vivo tissue site. Wang and Wang (2002) described a Monte Carlo simulation of polarized light transport in a tissue, including birefringence as a tissue optical property. Wang et al. (2003) discussed time-resolved propagation of polarized light, where depolarization versus time corresponds to depolarization versus photon path. Yaroslavsky et al. (2005) combined polarized light imaging with confocal fluorescence imaging to improve imaging of tissue pathology in biopsies. Jaillon and Saint-Jalmes (2005) studied the radial dependence of the PSF of linearly polarized light backscattered from turbid media, and reported that the radius of maximum PSF increased as the anisotropy of scattering (*g*) increased while keeping the reduced scattering coefficient, $\mu_s(1-g)$, constant. Wu and Walsh (2006) use polarimetry to measure the birefringence in rat tails. Ding et al. (2007) reported on the angular dependence of the Mueller matrix of light scattered by cells in suspension. Wood et al. (2007) described a Monte Carlo simulation for polarized light that was based on the Jones matrix formalism, and allowed for birefringence and chirality as tissue properties. Wood et al. (2009) described the process of decomposing a Mueller matrix description of a tissue into its components of diattenuation, retardance, and depolarization.

30.2 Toolkit

30.2.1 Nomenclature

The electric field (*E*) of a photon is described by a traveling wave as the photon propagates along the *z*-axis, which can be decomposed into two orthogonal electric field components, E_x and E_y:

$$E_x(z,t) = E_x \cos(\omega t - \beta z)$$

$$E_y(z,t) = E_y \cos(\omega t - \beta z + \phi) \quad (30.1)$$

with $\beta = 2\pi n/\lambda$ and $\omega = 2\pi f$, where

λ indicates the wavelength of light in the medium
n is the real refractive index of the medium
f [Hz] indicates the frequency of electric field oscillation

In Equation 30.1, as time progress, the factor ωt increases and must be balanced by the factor βz to keep a constant argument, $\omega t - \beta z$, within the cosine, so that a constant phase is maintained. Hence, a constant phase point moves in the +*z* direction as time increases. In other words, the function $E_x(z, t)$ moves along *z* as a function of time *t*, and the value of E_x at a particular position *z* oscillates between +1 and −1 in a sinusoidal manner. This is the classic traveling wave equation. The function $E_y(z, t)$ also propagates along *z*, but has an additional phase ϕ. Increasing ϕ is like increasing ωt, hence a positive ϕ causes the constant phase point of the traveling wave to advance along the *z*-axis by an increment $\Delta z = \phi/\beta = \phi\lambda/(2\pi)$. The total *E* is

$$E(z,t) = \sqrt{\left(E_x \cos(\omega t - \beta z)\right)^2 + \left(E_y \cos(\omega t - \beta z + \phi)\right)^2} \quad (30.2)$$

For convenience in this chapter, the magnitude of *E* is discussed as one unit of electric field, such that intensity $I = E^2$ is one unit of intensity.

Figure 30.2 illustrates the basic geometry of the electric field of a photon as it propagates along the *z*-axis. The *E* is decomposed into the two components, E_x aligned with the *x*-axis and E_y aligned with the *y*-axis. The phase difference ϕ between E_y and E_x in Figure 30.2 is $\pi/2$ or 90°, such that E_y has advanced along *z* by $\lambda/4$. E_y leads E_x by 90°, which corresponds to right circularly polarized light. Figure 30.2a shows the components E_y and E_x. Figure 30.2b shows the total *E*, which is the vector sum of E_y and E_x. The orientation of the field *E* rotates as a right-handed helix as it propagates.

How could the ϕ of E_y be made to equal $\pi/2$ such that E_y leads E_x? If the electric field *E* had passed through a thin slab of birefringent material, acting as a "retarder" (more discussion later), such that the fast axis was aligned with E_x and the slow axis was aligned with E_y, then E_y would have traveled more slowly than E_x, and would have undergone more cycles of oscillation within the retarder. When E_y exited the retarder, it would have resumed traveling at the same speed as E_x, but it would have a more advanced phase. If that phase were exactly $\pi/2$, then the polarization state of *E* would be right circularly polarized.

The relative magnitudes of E_x and E_y and the value of ϕ can vary and yield six distinct polarization states. Figure 30.3 shows these six states, which are specified as

Horizontal linear	H	$E_x = 1$	$E_y = 0$	$\phi = 0$
Vertical linear	V	$E_x = 0$	$E_y = 1$	$\phi = 0$
+45° linear	P+	$E_x = \sqrt{1/2}$	$E_y = \sqrt{1/2}$	$\phi = 0$
−45° linear	P−	$E_x = \sqrt{1/2}$	$E_y = -\sqrt{1/2}$	$\phi = 0$
Right circular	R	$E_x = 1$	$E_y = 1$	$\phi = \pi/2$
Left circular	L	$E_x = 1$	$E_y = 1$	$\phi = -\pi/2$

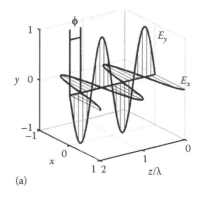
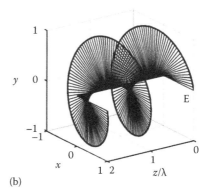

FIGURE 30.2 The electric field (*E*) of a photon can be decomposed into two components, E_x and E_y. (a) The E_x and E_y for right circular polarized light are shown. The phase ϕ is $\pi/2$. Note that at $z/\lambda = 2$, $E_x = 1$ and as time progresses and the wave advances along z, E_x will become zero then –1. In contrast, E_y is already zero at $z/\lambda = 2$, and E_y will immediately move toward –1. Therefore, E_y leads E_x by $\pi/2$ or 90°. (b) The total *E* is the vector sum of E_y and E_x, and *E* moves along *z* in a helical manner. The helix is a right-handed helix, and the *E* field propagates counterclockwise as it moves toward an observer, as observed by the observer. However, the *E* field at one detector plane, for example at $z/\lambda = 2$, will be seen to rotate in a clockwise manner as a function of time.

This chapter uses the symbols *H*, *V*, P^+, P^-, *R*, and *L* to denote these six polarization states. This chapter also uses the convention that horizontal refers to the *x*-axis that is parallel to the tabletop if one is doing an experiment. With this convention, *H* has $E_x = 1$, $E_y = 0$. These six polarization states can be experimentally observed as measurable intensities, which will be discussed later.

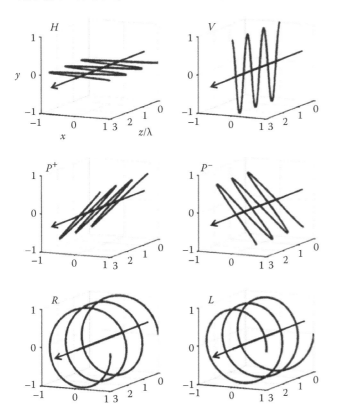

FIGURE 30.3 The *E* field moves along *z* in different patterns, depending on the phase difference between E_y and E_x. Six polarization states are shown, denoted as *H*, *V*, P^+, P^-, *R*, and *L*. The figure shows *E* at one moment in time as a function of *x*, *y*, and z/λ, where λ is wavelength. The *x* and *y* axes are in arbitrary units of amplitude (–1 to +1).

30.2.2 Stokes Vector: The Description of Polarized Light

The Stokes vector, *S*, describes the polarization state of light, and is specified by the intensities of the six polarization states. The *S* is a vector defined as

$$S = \begin{vmatrix} I \\ Q \\ U \\ V \end{vmatrix} = \begin{vmatrix} H+V \\ H-V \\ P^+ - P^- \\ R-L \end{vmatrix} \qquad (30.3)$$

where

   I indicates the total intensity of the light
   Q is the balance between *H* and *V*
   U is the balance between P^+ and P^-
   V is the balance between *R* and *L*

Note that $I = H + V$. Even if you have perfect right circular polarized light, the experimental determination of *H* will be ½ and the determination of *V* will be ½, hence the total intensity will be 1. If you have perfect P^+ light, the experimental determination of *H* will be $\left(\sqrt{1/2}\right)^2 = 1/2$, and the determination of *V* will be $\left(\sqrt{1/2}\right)^2 = 1/2$, hence the total intensity will be 1. While the total intensity, *I*, of the light may equal 1, the factors *Q*, *U*, and *V* can all be zero if all the light is equally balanced such that $H = L$, $P^+ = P^-$, and $R = L$. For totally unpolarized light, $S = [1\ 0\ 0\ 0]^T$. (Note: the superscript ^{T} transposes the row vector into a column vector, as in Equation 30.3.) The degree of polarization (*p*) is defined as

$$p = \frac{\sqrt{Q^2 + U^2 + V^2}}{I} \qquad (30.4)$$

The Stokes vector notations, $[I\ Q\ U\ V]^T$, for the six polarization states are summarized:

$$H = \begin{vmatrix} 1 \\ 1 \\ 0 \\ 0 \end{vmatrix}, \quad V = \begin{vmatrix} 1 \\ -1 \\ 0 \\ 0 \end{vmatrix}, \quad P^{+} = \begin{vmatrix} 1 \\ 0 \\ 1 \\ 0 \end{vmatrix}, \quad P^{-} = \begin{vmatrix} 1 \\ 0 \\ -1 \\ 0 \end{vmatrix}, \quad R = \begin{vmatrix} 1 \\ 0 \\ 0 \\ 1 \end{vmatrix}, \quad L = \begin{vmatrix} 1 \\ 0 \\ 0 \\ -1 \end{vmatrix}$$

(30.5)

30.2.3 Poincaré Sphere

A useful graphical presentation of S is the Poincaré sphere, shown in Figure 30.4. The sphere is drawn using Q, U, and V as the x, y, and z axes, respectively. The radius of the sphere is the degree of polarization (p). The figure shows the location of the six distinct polarization states. Later, the Poincaré sphere will be used to illustrate the depolarization of polarized light propagating through a model of birefringent tissue.

30.2.4 Mueller Matrix: Transport of Polarized Light

The Stokes vector S describes the polarization state of light. As light passes through some element or medium, changes in S can occur. For an input S_{in}, there is an output S_{out}, and the relationship between S_{in} and S_{out} defines the Mueller matrix M:

$$S_{out} = \begin{vmatrix} S(1) \\ S(2) \\ S(3) \\ S(4) \end{vmatrix}_{out} = \begin{vmatrix} I \\ Q \\ U \\ V \end{vmatrix}_{out} = \begin{vmatrix} M_{11} & M_{12} & M_{13} & M_{14} \\ M_{21} & M_{22} & M_{23} & M_{24} \\ M_{31} & M_{32} & M_{33} & M_{34} \\ M_{41} & M_{42} & M_{43} & M_{44} \end{vmatrix} \times \begin{vmatrix} I \\ Q \\ U \\ V \end{vmatrix}_{in} = M S_{in}$$

(30.6)

The first row, $[M_{11}\, M_{12}\, M_{13}\, M_{14}]$, is the diattenuation vector, which describes how the polarization states are attenuated differently. The first column $[M_{11}\, M_{21}\, M_{31}\, M_{41}]^{T}$ is the polarizance vector, which describes how unpolarized light is converted to polarized light. The inner matrix $[M_{22}\, M_{23}\, M_{24}; M_{32}\, M_{33}\, M_{34}; M_{42}\, M_{43}\, M_{44}]$

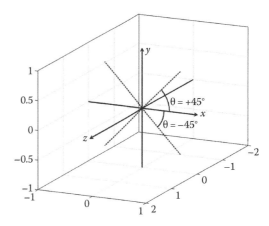

FIGURE 30.5 The convention for specifying the orientation of a linear polarizer or the orientation of a retarder fast axis. The θ is measured counterclockwise from the x-axis, as seen by the observer viewing upstream in the $-z$ direction toward the approaching light. A linear polarizer oriented at $\theta = 0°$ would pass horizontal linearly polarized light of type H. A linear polarizer oriented at $\theta = +45°$ would pass linearly polarized light of type P^{+}. A linear polarizer oriented at $\theta = 90°$ would pass vertical linearly polarized light of type V. A half-wave or quarter-wave retarder is often positioned with its fast axis at θ equal $+45°$ or $-45°$.

describes the retardance in which the phase of E_y relative to E_x is altered during transmission.

Note that often in the literature, one sees M_{00}, M_{01}, M_{02}, M_{03}, etc. rather than M_{11}, M_{12}, M_{13}, M_{14}, etc., for the Mueller matrix elements. Also, the literature often describes the Stokes vector elements as $S(0)$, $S(1)$, $S(2)$, and $S(3)$ rather than as $S(1)$, $S(2)$, $S(3)$, and $S(4)$ or I, Q, U, and V. Since this chapter uses the notation of the programming environment MATLAB® in the example programming code, and MATLAB arrays use indices starting with 1 rather than 0, we adopt the conventions of Equations 30.3 and 30.6.

To illustrate some examples of Mueller matrices, let us consider a retarder and a linear polarizer. Figure 30.5 shows the convention for defining the angle (θ) of orientation of the fast axis of a retarder or the orientation of a linear polarizer. As the wave approaches an observer, the observer viewing upstream in the $-z$ direction sees an angle θ that is counterclockwise from the x-axis.

One optical element we will need experimentally is a linear polarizer. The angle of the polarizer with respect to the x-axis is θ. The Mueller matrix is calculated:

```
function Mp = getMp(th)
% function Mp = getMp(th)
%    M = Mueller Matrix
%    th = angle of polarizer CCW from
       positive x-axis,
%    as viewed by the observer facing the
       approaching light.
c = cos(2*th);
s = sin(2*th);
```

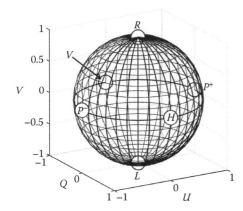

FIGURE 30.4 Poincaré sphere, which graphically depicts the Q, U, and V elements in the Stokes vector. The radius of the sphere is the degree of polarization p, and in this figure $p = 1$.

```
Mp=0.5*[
    1    c    s    0
    c    c^2  c*s  0
    s    c*s  s^2  0
    0    0    0    0];
```

The three types of linear polarizers that we use in this chapter are the horizontal polarizer, $M_p(\theta=0)$; the +45° polarizer, $M_p(\theta=\pi/4)$; and the vertical polarizer, $M_p(\theta=\pi/2)$:

$$\text{horizontal polarizer} = \texttt{getMp(0)} = \begin{vmatrix} 0.5 & 0.5 & 0 & 0 \\ 0.5 & 0.5 & 0 & 0 \\ 0 & 0 & 0 & 0 \\ 0 & 0 & 0 & 0 \end{vmatrix} \quad (30.7)$$

$$\text{+45° polarizer} = \texttt{getMp(}\pi/4\texttt{)} = \begin{vmatrix} 0.5 & 0 & 0.5 & 0 \\ 0 & 0 & 0 & 0 \\ 0.5 & 0 & 0.5 & 0 \\ 0 & 0 & 0 & 0 \end{vmatrix} \quad (30.8)$$

$$\text{vertical polarizer} = \texttt{getMp(}\pi/2\texttt{)} = \begin{vmatrix} 0.5 & -0.5 & 0 & 0 \\ -0.5 & 0.5 & 0 & 0 \\ 0 & 0 & 0 & 0 \\ 0 & 0 & 0 & 0 \end{vmatrix} \quad (30.9)$$

The following algorithm generates the Mueller matrix for a retarder whose fast axis is oriented at an angle θ relative to the x-axis, and has a retardation phase ϕ:

```
function Mr=getMr(th, ph)
% function Mr=getMr(th, ph)
%    th=fast axis angle CCW from positive
       x-axis(radians)
%    ph=retardance (radians)
c=cos(2*th);
s=sin(2*th);
cp=cos(ph);
sp=sin(ph);
Mr=[
    1    0          0          0
    0    c^2+cp*s^2 (1-cp)*s*c sp*s
    0    (1-cp)*s*c s^2+cp*c^2 -sp*c
    0    -sp*s      sp*c       cp];
```

The two types of retarders we use in this chapter are a half-wave plate (HWP) and a quarter-wave plate (QWP). The HWP has a retardation of half a wavelength ($\phi=\pi$ rad), and it is used with its fast axis oriented at 45° ($\theta=\pi/4$). The Mueller matrix is

$$\text{half-wave plate} = \texttt{getMr(}\pi/4,\pi\texttt{)} = \begin{vmatrix} 1 & 0 & 0 & 0 \\ 0 & -1 & 0 & 0 \\ 0 & 0 & 1 & 0 \\ 0 & 0 & 0 & -1 \end{vmatrix} \quad (30.10)$$

Passing V light through this half-wave plate will flip the linear polarization 90° and yield H light:

$$H = \begin{vmatrix} 1 \\ 1 \\ 0 \\ 0 \end{vmatrix} = \begin{vmatrix} 1 & 0 & 0 & 0 \\ 0 & -1 & 0 & 0 \\ 0 & 0 & 1 & 0 \\ 0 & 0 & 0 & -1 \end{vmatrix} \times \begin{vmatrix} 1 \\ -1 \\ 0 \\ 0 \end{vmatrix} = M_r(\pi/4,\pi)V \quad (30.11)$$

Incident H, R, or L light will flip to yield V, L, or R light, respectively, not shown.

The QWP is also used in our experiments. The QWP has a retardation of ¼ a wavelength ($\phi=\pi/2$ rad) and we will use it with its fast axis oriented at either +45° ($\theta=\pi/4$) or −45° ($\theta=-\pi/4$). The Mueller matrices are

$$\text{quarter-wave plate } (\theta=+45°)$$
$$= \texttt{getMr(}\pi/4,\pi/2\texttt{)} = \begin{vmatrix} 1 & 0 & 0 & 0 \\ 0 & 0 & 0 & 1 \\ 0 & 0 & 1 & 0 \\ 0 & -1 & 0 & 0 \end{vmatrix} \quad (30.12)$$

and

$$\text{quarter-wave plate } (\theta=-45°)$$
$$= \texttt{getMr(-}\pi/4,\pi/2\texttt{)} = \begin{vmatrix} 1 & 0 & 0 & 0 \\ 0 & 0 & 0 & -1 \\ 0 & 0 & 1 & 0 \\ 0 & 1 & 0 & 0 \end{vmatrix} \quad (30.13)$$

Incident H light is converted to right circular polarization (R) by a QWP ($\theta=-45°$):

$$R = \begin{vmatrix} 1 \\ 0 \\ 0 \\ 1 \end{vmatrix} = \begin{vmatrix} 1 & 0 & 0 & 0 \\ 0 & 0 & 0 & -1 \\ 0 & 0 & 1 & 0 \\ 0 & 1 & 0 & 0 \end{vmatrix} \times \begin{vmatrix} 1 \\ 1 \\ 0 \\ 0 \end{vmatrix} = M_r(-\pi/4,\pi/2)H \quad (30.14)$$

Incident V light is converted to left circular polarization (L), not shown.

Let H light pass through a horizontal linear polarizer, $M_p(0°)$, followed by a QWP oriented at −45°, $M_r(-\pi/4, \pi/2)$, to generate R light for delivery to an unknown sample. If this combination is now lifted off the table, rotated horizontally, and replaced on the table behind the sample, such that light transmitted through the sample now enters the QWP first, then passes through the horizontal polarizer before reaching an intensity detector, the combination can detect right circular polarized light R. This horizontal rotation causes our description of the QWP to become $\theta = +45°$. In this case,

$$H = \begin{vmatrix} 1 \\ 1 \\ 0 \\ 0 \end{vmatrix} = \begin{vmatrix} 1 & 0 & 0 & 0 \\ 0 & 0 & 0 & 1 \\ 0 & 0 & 1 & 0 \\ 0 & -1 & 0 & 0 \end{vmatrix} \times \begin{vmatrix} 1 \\ 0 \\ 0 \\ 1 \end{vmatrix} = M_r(\pi/4, \pi/2)R \quad (30.15)$$

In summary, this section has discussed the Stokes vector S, and its description of the six polarization states H, V, P^+, P^-, R, and L. The Mueller matrix M was introduced to describe the conversion of S_{in} to S_{out} when light passes through some medium or optical element. Particular examples of M for linear polarizers and retarders were shown, which are commonly used in experimental measurements.

30.2.5 Data Matrix: Measurements of Polarized Light Transport in a Sample

This section will consider how to use the tools to measure the polarization transport in some test sample of unknown optical properties. The term "transport" is used here to emphasize that both transmission and reflectance from a medium can be considered. In biological tissues, we often must make reflectance measurements. It is an issue of semantics whether to call reflected light "transmission" or "reflectance" when multiple scattering occurs.

Optical measurement of polarized light transport is shown in Figure 30.6, and involves generating a source of light with a particular polarization state, S_{in}, and detecting the transport of a particular polarization state, S_{out}. To characterize the Mueller matrix, which contains 16 elements, one must make at least 16 measurements. These are intensity measurements, using a simple detector.

In this chapter, we use four states of incident light $(H, V, P^+, R)_{in}$ and for each type of input, we detect the transported intensity associated with the same four states of light $(H, V, P^+, R)_{out}$. Hence, a four-by-four "data matrix" (DM) is acquired:

$$\text{data matrix DM} = \begin{vmatrix} HH & HV & HP^+ & HR \\ VH & VV & VP^+ & VR \\ P^+H & P^+V & P^+P^+ & P^+R \\ RH & RV & RP^+ & RR \end{vmatrix} \quad (30.16)$$

where

HH is the horizontal linear polarization intensity detected when H light is incident

HV is the vertical linear polarization intensity detected for incident H light

HP^+ is the +45° linear polarization intensity detected when H light is incident

HR is the right circular polarization intensity detected when H light is incident, etc.

So we need to describe how to create H, V, P^+, and R as the light source S_{in}, and how to measure H, V, P^+, and R to specify S_{out}. For the sake of completeness, the following lists how to generate all six polarization states for S_{in}, by passing unpolarized light through the following combinations of linear polarizer and retarder:

Unpolarized light	→	Horizontal linear polarizer $M_p(0)$		→	H
Unpolarized light	→	Vertical linear polarizer $M_p(\pi/2)$		→	V
Unpolarized light	→	+45° linear polarizer $M_p(\pi/4)$		→	P^+
Unpolarized light	→	−45° linear polarizer $M_p(-\pi/4)$		→	P^-
Unpolarized light	→	Horizontal linear polarizer $M_p(0)$	→ Quarter-wave plate $M_r(-\pi/4, \pi/2)$	→	R
Unpolarized light	→	Vertical linear polarizer $M_p(\pi/2)$	→ Quarter-wave plate $M_r(-\pi/4, \pi/2)$	→	L

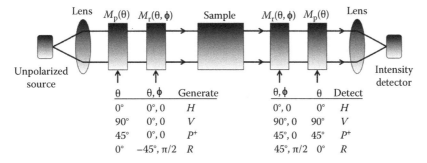

Lens	$M_p(\theta)$	$M_r(\theta, \phi)$	Sample	$M_r(\theta, \phi)$	$M_p(\theta)$	Lens

θ	θ, ϕ	Generate	θ, ϕ	θ	Detect
0°	0°, 0	H	0°, 0	0°	H
90°	0°, 0	V	90°, 0	90°	V
45°	0°, 0	P^+	45°, 0	45°	P^+
0°	−45°, π/2	R	45°, π/2	0°	R

FIGURE 30.6 An experimental setup for measuring polarization transport through an unknown sample. Light is collimated by a lens, passed through a polarizer $M_p(\theta)$ and retarder $M_r(\theta, \phi)$ to generate either H, V, P^+, or R light, passed through the sample, passed through a retarder $M_r(\theta, \phi)$ and a polarizer $M_p(\theta)$ to convert the transported light into either H, V, P^+ or R light, and focused on an intensity detector. Note that $M_r(0, 0)$ is equivalent to no retarder at all.

To detect the six intensities that specify the S_{out}, the transported light is passed through the following combination of retarder and linear polarizer before reaching the detector that measures intensity:

Transported light	$\rightarrow$	Horizontal linear polarizer $M_p(0)$		$\rightarrow$	H
Transported light	$\rightarrow$	Vertical linear polarizer $M_p(\pi/2)$		$\rightarrow$	V
Transported light	$\rightarrow$	+45° linear polarizer $M_p(\pi/4)$		$\rightarrow$	P^+
Transported light	$\rightarrow$	−45° linear polarizer $M_p(-\pi/4)$		$\rightarrow$	P^-
Transported light	$\rightarrow$	+45° quarter-wave plate $M_r(\pi/4, \pi/2)$	$\rightarrow$ Horizontal linear polarizer	$\rightarrow$	R
Transported light	$\rightarrow$	+45° quarter-wave plate $M_r(\pi/4, \pi/2)$	$\rightarrow$ Vertical linear polarizer	$\rightarrow$	L

Hence, a 6×6 data matrix can be generated, that is, 36 distinct measurements. This chapter uses a 4×4 data matrix, that is, 16 distinct measurements to specify the 4×4 Mueller matrix M. It is preferable to use more measurements to provide some redundancy and thereby overcome any noise in the measurements.

It should be mentioned that using the above set of combinations is experimentally a bit clumsy. One has to rotate a horizontal linear polarizer to a vertical linear polarizer or +45° linear polarizer, for generating H, V, and P. And one must add a QWP (−45°), placed after a horizontal linear polarizer, to generate R. This is a lot of manual manipulation. Similarly, one has to do a lot of manual manipulation on the detection side of an experimental setup. It is easier to control retarders and polarizers electronically. It is common to electronically switch the S_{in} to more than four states of polarization and detect the amount of intensity detected for a range of polarization states for S_{out}. A measurement matrix that relates S_{in} and S_{out} is experimentally calibrated by measuring known optical elements. This chapter uses only the minimal data matrix of Equation 30.16.

Now that we have the data matrix of Equation 30.16, we must convert these 16 values into the 16 elements of the Mueller matrix that characterizes the transport through the unknown sample. This step is accomplished by the following algorithm published by Yao and Wang (1999):

```
function M=D2M(DM)
% function M=D2M(DM)
%   Converts data matrix DM to Mueller
    matrix M
%   DM(in,out), where in=[H V P R]' and
    out=[H V P R]
%   from Yao G, LV Wang, Opt Lett 24:537-
    539, 1999
M(1,1)=D(1,1)+D(1,2)+D(2,1)+D(2,2);
M(1,2)=D(1,1)+D(1,2)-D(2,1)-D(2,2);
M(1,3)=2*D(3,1)+2*D(3,2)-M(1,1);
M(1,4)=2*D(4,1)+2*D(4,2)-M(1,1);
M(2,1)=D(1,1)-D(1,2)+D(2,1)-D(2,2);
```

```
M(2,2)=D(1,1)-D(1,2)-D(2,1)+D(2,2);
M(2,3)=2*D(3,1)-2*D(3,2)-M(2,1);
M(2,4)=2*D(4,1)-2*D(4,2)-M(2,1);
M(3,1)=2*D(1,3)+2*D(2,3)-M(1,1);
M(3,2)=2*D(1,3)-2*D(2,3)-M(1,2);
M(3,3)=4*D(3,3)-2*D(3,1)-2*D(3,2)-M(3,1);
M(3,4)=4*D(4,3)-2*D(4,1)-2*D(4,2)-M(3,1);
M(4,1)=2*D(1,4)+2*D(2,4)-M(1,1);
M(4,2)=2*D(1,4)-2*D(2,4)-M(1,2);
M(4,3)=4*D(3,4)-2*D(3,1)-2*D(3,2)-M(4,1);
M(4,4)=4*D(4,4)-2*D(4,1)-2*D(4,2)-M(4,1);
M=M/2;
```

The above algorithm D2M(DM) generates M when given the data matrix DM.

30.2.6 Simulation Matrix: Monte Carlo Simulation of Polarized Light Transport

A very useful tool is the Monte Carlo simulation of polarized light transport in a medium. In nonpolarized light propagation, Monte Carlo simulations propagate a weighted photon through a medium, with scattering events redirecting the trajectory of the photon with each scattering event and with absorption events causing a fraction of the photon's weight (energy) to be deposited in the local voxel bin. Hence, the Monte Carlo simulation propagates photon energy.

Polarized light Monte Carlo propagates the Stokes vector. The propagation of I would be identical to a normal Monte Carlo simulation. The additional information regarding the propagation of Q, U, and V provides a complete statement of the state of polarization of the photon at it propagates. The details of polarized light Monte Carlo are presented in Ramella-Roman et al. (2005a,b). For this chapter, we simply need to know that the simulation consists of four separate simulations, and in each simulation one of four types of polarized light (H, V, P^+, R) is launched. The output of each simulation is the Stokes vector $[I\ Q\ U\ V]^T$ that escapes the tissue for a given type of incident polarized light. Hence, a polarized light Monte Carlo simulation will yield a simulation matrix (SM):

$$\text{Simulation matrix } SM = \begin{vmatrix} HI & VI & P^+I & RI \\ HQ & VQ & P^+Q & RQ \\ HU & VU & P^+U & RU \\ HV & VV & P^+V & RV \end{vmatrix} \quad (30.17)$$

Figure 30.7 illustrates an example SM, for the case of transmission through and reflectance from a 200 μm-thick slab of aqueous medium containing 2 μm diameter polystyrene microspheres at a volume fraction of 0.01, using 0.632 nm wavelength light. Each element of the SM in the upper figures is an x–y map of the PSF for light escaping the slab. The lower figures show the summation value of all the pixels in each element, and display

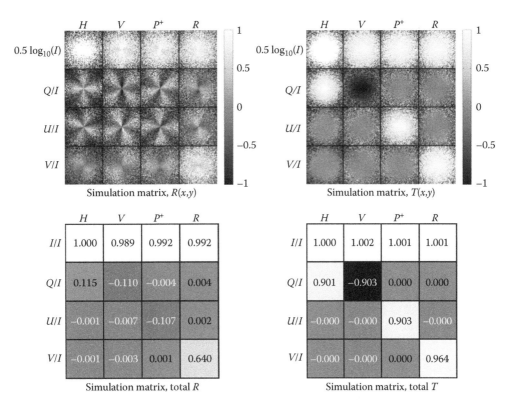

FIGURE 30.7 Polarized light Monte Carlo simulation. The SM is shown for the cases of (left) reflectance from and (right) transmission through a 200 μm-thick slab of aqueous medium containing 2 μm dia. polystyrene microspheres at a volume fraction of 0.01, using 0.632 nm wavelength light. (Top row) The SM contains 16 elements, and each element is an *x-y* map of the PSF of light escaping from the slab, over a square field of view of 0.834 cm × 0.834 cm. The columns are for each of the 4 types of input light (*H*, *V*, *P*⁺, *R*). The rows are for each output Stokes vector element. The 1st row shows $0.5 \log_{10}(I)$ so as to share the same color bar with other figures. The other rows show the ratios Q/I_{11}, U/I_{11} and V/I_{11}, where I_{11} is the intensity in upper-left element (*HI*). The color bar scale for all elements is −1 to +1 (black to white). (Bottom row) The summation of all reflected or transmitted light is shown as a numerical value and a color on the gray scale. The total reflectance is $R = 0.117$, and the total transmission is $T = 0.820$. The 1st row plots *HI/HI*, *VI/HI*, *PI/HI*, and *RI/HI*, so the upper-left element is 1.000 and the other values are very close to 1. (dia. = 2.0 mm, $v_f = 0.01$, $\lambda = 0.632$ mm, $\mu_s = 240$ cm⁻¹, $g = 0.913$, $\mu_a = 2$ cm⁻¹, $L = 0.0200$ cm, number of photons launched = 5.8×10^6, run time = 30 min for each of 4 simulations.)

this value as a color on the gray scale. For this thin slab, most of the light is transmitted. The reflectance image shows the Maltese cross pattern. Later in this chapter, we will use the total transmission value, as in the lower images, for an analysis of polarized light propagation.

To convert the simulation matrix SM into a Mueller matrix *M*, one first converts SM into a data matrix *D*, then converts *D* to *M*. This is accomplished by the expression:

$$M = D2M(IQUV2D(SM)) \tag{30.18}$$

which first uses the function `IQUV2D()`, listed below, then uses the function `D2M()`, listed previously.

The function `IQUV2D()` is listed:

```
function D = IQUV2D(IQUV)
%   function D = IQUV2D(IQUV)
%       IQUV = matrix of output Stokes vectors
```

```
%           1st column in response to H,
%           2nd column in response to V,
%           3rd column in response to P+,
%           4th column in response to R
%   DM = data matrix
%           1st row in response to H,
%           2nd row in response to V,
%           3rd row in response to P+,
%           4th row in response to R
for k=1:4 % selects input Stokes vector
H, V, P+ or R
    % detect horizontal linear
      polarization H
    H = (IQUV(1,k) + IQUV(2,k))/2; % <— H
    % detect vertical linear polariza-
      tion V
    V = (IQUV(1,k) - IQUV(2,k))/2; % <— V
    % detect +45deg polarization P+
    Mp = getMp(pi/4);
```

```
        X = Mp*IQUV(:,k); % +45deg linear
        polarizer
        P = X(1); % <—— P+
     % detect right circular polarization R
        Mr = getMr(pi/4, pi/2); % +45deg QWP
        Mp = getMp(0); % horizontal linear
        polarizer
        X = Mp*Mr*IQUV(:,k);
        R = X(1); % <—— R
     DM(k,:) = [H V P R]; % fill datacube
End
```

30.2.7 Testing Algorithms with an Identity

One can test the above algorithms by using the following routine called testIdentity, in which an arbitrary choice of M is generated, then the following identity is tested:

$$M = D2M(IQUV2D(M * IQUVin)) \qquad (30.19)$$

The matrix IQUVin is a 4×4 matrix that holds the set of input Stokes vectors for H, V, P+, and R. M is some arbitrary Mueller matrix. The matrix multiplication M*IQUVin yields a 4×4 matrix of output Stokes vectors called IQUV, one column $[I\ Q\ U\ V]^T$ of output for each of the 4 types of input S_{in}. This IQUV is converted to a data matrix DM by the algorithm IQUV2D(IQUV). Then, the DM is converted to a Mueller matrix M by D2M(DM). This procedure will return the original M regardless of the choice of the M. In other words, the identity of Equation 30.19 holds for any initial choice of M. The routine testIdentity is listed below.

```
% testIdentity.m
%   Checks the identity M ==
    D2M(IQUV2D(M*IQUVin))
% IQUVin matrix of input Stokes vectors
%   where columns indicate Stokes vector
    for H, V, P+, R
    IQUVin = [
    1   1   1   1
    1  -1   0   0
    0   0   1   0
    0   0   0   1]
% Mueller matrix M is filled with random
    numbers
    M = rand(4,4)
% IQUV matrix of output Stokes vectors is
    calculated
    IQUV = M*IQUVin
% Convert IQUVout to datacube
    DM = IQUVout2D(IQUV)
% Convert datacube to Mueller Matrix
    M = D2M(DM)
```

With these tools, one is now ready to make polarized light measurements and specify the light transport in terms of a Mueller matrix. One can deliver H, V, P+, and R light, and measure H, V, P+, and R light transport through the medium, to specify a data matrix DM. This DM specifies the Mueller matrix M = D2M(DM). Alternatively, one can conduct a Monte Carlo simulation of polarized light transport to yield an simulation matrix SM, then specify the Mueller matrix: M = D2M(IQUV2D(SM)).

30.3 Depolarization Coefficients: μ_{LP} and μ_{CP} [cm^{-1}]

30.3.1 Optical Properties versus Transport Parameters

The Mueller matrix describes transport parameters, not optical properties.

Consider the "transmission" (T) of unpolarized light through a 1 cm pathlength cuvette (L = 1 cm) filled with a nonscattering solution of absorbing dye. An input power P_{in} is delivered, and a transmitted power P_{out} is detected. The transmission is $T = P_{out}/P_{in}$. This T describes the transport through the cuvette, while the optical property of the dye solution is the absorption coefficient, μ_a [cm^{-1}], of the solution, such that $T = \exp(-\mu_a L)$. The T would vary if the cuvette pathlength, L, varied, despite the solution itself remaining constant. In other words, T is a measurement, not an optical property. It is the optical property μ_a that characterizes the solution of dye.

Similarly, a Mueller matrix that describes the transport of a Stokes vector is a measurement, not an optical property. If one knows the pathlength L of a photon through some medium, then it is possible to describe the tissue in terms of the change in the Stokes vector per incremental unit length. This is the approach that is followed in this chapter, describing the change in p per incremental L when either linearly or circularly polarized light is incident on the tissue.

In this section, the relative roles of birefringence and scattering on the depolarization coefficient are separately considered. The goal is to identify a depolarization coefficient for linear and circular polarization, μ_{LP} and μ_{CP} [cm^{-1}], respectively.

30.3.2 Relative Roles of Retardance and Scattering in Depolarization

As polarized light propagates through a tissue, it becomes depolarized. Each photon still has some polarization state, but the average polarization of the ensemble of photons becomes randomized. So far, we have discussed some basic tools for simulating polarization transport through a medium using the Mueller matrix. Let us introduce the depolarization coefficients μ_{LP} and μ_{CP} [cm^{-1}] that describe depolarization per unit length of photon path. This depolarization can occur due to three mechanisms: (1) depolarization due to tissue birefringence, (2) depolarization due to tissue scattering, and (3) depolarization due to the chirality of a tissue. This chapter discusses birefringence and scattering, but does not address chirality, which is a minor mechanism of depolarization in tissues.

In this section, the relative roles of birefringence and scattering in causing depolarization are discussed. The effect of birefringence occurs as light passes through oriented fibers. The effect of scattering occurs when the direction of the photon is redirected.

30.3.3 Depolarization by Birefringence

Many tissues, like skin or muscle, have birefringent fibers, such as collagen and actin-myosin fibers. If one observes such tissues through a polarizing microscope, there are microdomains of 10–100 μm size within which a group of fibers are oriented in the same direction (Jacques et al. 2001). Each microdomain is often oriented randomly relative to other microdomains. These microdomains act as incremental retarders. The birefringence is defined as the difference in refractive index of the fibers $\Delta n = n_e - n_o$, where n_e and n_o are the extraordinary and ordinary refractive indices where the electric field is perpendicular and parallel, respectively, to the fiber orientation. Typical values for Δn of biological birefringent fibers in the visible wavelength range are in the range of 10^{-3}–10^{-2} (Ramella-Roman and Jacques 2001). The incremental retardance due to such birefringence in a microdomain is $2\pi\Delta n\Delta L/\lambda$, where ΔL is the incremental thickness of the microdomain and λ [μm] is the wavelength. The incremental retardance per unit length is $\beta = 2\pi\Delta n/\lambda$ [rad/μm], and in this discussion we specify values for the retardance per microdomain, $\beta\Delta L$. For a 633 nm wavelength photon, a 10 μm thick microdomain with $\Delta n = 3 \times 10^{-3}$ will present $\beta\Delta L = 0.298 = \pi/10.5$.

To dissect the role of retardance from the complex behavior of transport through a tissue with both scattering and randomly oriented microdomains of birefringence, let us model the collimated transmission through a nonscattering medium composed of multiple incremental layers of randomly oriented birefringence. We will determine the contribution of such randomly oriented birefringent layers to the depolarization coefficients for polarized light.

Figure 30.8 shows the model, in which a slab of medium of thickness L [cm] consists of many incremental layers of thickness ΔL. Each incremental layer is a model for a birefringent microdomain. The incremental retardance of each microdomain is $\beta\Delta L$. The total thickness of the medium is L.

Figure 30.9 displays the status of transmitted photons on a Poincaré sphere when H photons have been delivered. In this example, the incremental retardance of each microdomain is $\beta\Delta L = \pi/10$. The thickness (L) of the medium is expressed as βL. Hence, the number of microdomains crossed by the photons is $N = (\beta L)/(\beta\Delta L) = L/\Delta L$. This figure shows the polarization status of each of 100 photons after propagating through N layers, for $N = 5$, 10, 20, and 40. For $N = 5$, the transmitted photons still have predominately H polarization and are closely distributed around the $Q = 1$, $U = 0$, $V = 0$ position on the Poincaré sphere, with a slight spread along the V axis. The average value of Q for the ensemble of photons, $\langle Q \rangle$, is 0.88. As N increases, the distribution of photons becomes increasingly more evenly distributed over the sphere, and the $\langle Q \rangle$ drops exponentially toward zero as N increases.

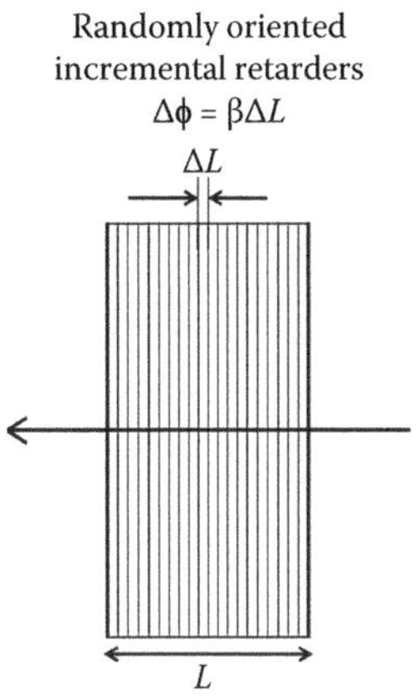

FIGURE 30.8 Slab of tissue of thickness L is divided into incremental layers of thickness ΔL that are randomly oriented retarders with an incremental retardance $\Delta\phi = \beta\Delta L$, where $\beta = 2\pi\Delta n/\lambda$, Δn is the birefringence, and λ is the photon wavelength.

Figure 30.10a shows the results for delivery of H light and observation of the transmitted Q element of the Stokes vector. Figure 30.10b shows the results for delivery of R light and observation of the transmitted V element of the Stokes vector.

In Figure 30.10a, the attenuation of $\langle Q \rangle$ occurs due to randomization of the value of Q for each photon. This randomization due to birefringence is described by a depolarization constant D_{LP} [microdomain^{-1}], which describes depolarization per microdomain of linearly polarized light by the mechanism of incremental retardance in randomly oriented microdomains. This D_{LP} applies to any incident linearly polarized light regardless of orientation, in other words for H, V, P^+, or P^- light, since this model is azimuthally symmetric. Figure 30.10c shows the value of D_{LP} for six values of $\beta\Delta L$: $\pi/5$, $\pi/10$, $\pi/20$, $\pi/50$, $\pi/100$, and $\pi/200$. As $\beta\Delta L$ increases, D_{LP} increases: $D_{LP} \approx 0.24(\beta\Delta L)^2$.

The attenuation of $\langle Q \rangle$ can be regarded as an attenuation of the degree of polarization (p) since the population of transmitted photons remains balanced with respect to P^+ vs. P^- and R vs. L, in other words $U = P^+ - P^- = 0$ and $V = R - L = 0$. In summary, the attenuation of $\langle Q \rangle$ due to depolarization by birefringence can be described by these alternative expressions:

$$\langle Q \rangle = e^{-D_{LP}N} = e^{-D_{LP}\left(L/\Delta L\right)} = e^{-\mu_{LP}L} \tag{30.20}$$

where the depolarization coefficient for linear polarization per unit length is $\mu_{LP} \approx D_{LP}/\Delta L$ [cm^{-1}]:

$$\mu_{LP} = \frac{D_{LP}}{\Delta L} = \frac{0.24(\beta\Delta L)^2}{\Delta L} = \frac{0.24(2\pi\Delta n\Delta L/\lambda)^2}{\Delta L} = 0.24(2\pi\Delta n)^2\frac{\Delta L}{\lambda^2} \tag{30.21}$$

For the example of $\Delta n = 3 \times 10^{-3}$, $\lambda = 0.633 \times 10^{-4}$ cm, and $\Delta L = 10 \times 10^{-4}$ cm, the value of μ_{LP} is 21 cm^{-1}. This μ_{LP} is an optical property that characterizes how fast a tissue depolarizes linearly

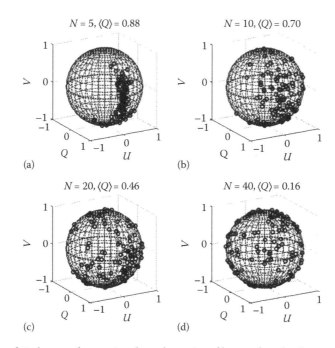

FIGURE 30.9 The depolarization of H photons after passing through a series of layers of randomly oriented birefringence. 100 photons were launched as H polarization (at $Q = 1$, $U = 0$, $V = 0$) through a slab of thickness L composed of N incremental layers (randomly oriented birefringent microdomains) of thickness DL with retardance $\beta\Delta L = \pi/10$. As the thickness increased, the final polarization state of the 100 photons became increasingly more evenly distributed over the Poincaré sphere. Hence, the average Q of the ensemble of photons, $\langle Q \rangle$, decays from 1 toward zero. (a) $N = 5$. (b) $N = 10$. (c) $N = 20$. (d) $N = 40$.

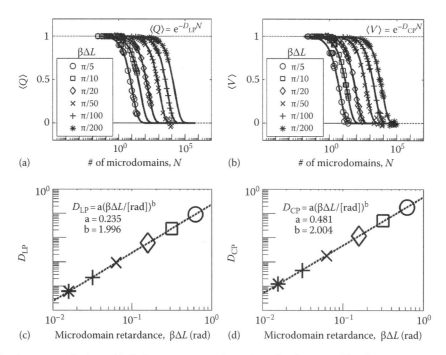

FIGURE 30.10 The depolarization of polarized light by transmission through randomly oriented birefringent microdomains with incremental retardances bDL [rad]. In the simulation, the polarization status (transmitted $\langle Q \rangle$ or $\langle V \rangle$ in response to 100 incident H or R photons) is plotted as the medium thickness is increased. The thickness of the nonscattering medium is expressed as the number of microdomains (N) crossed by the photons. (a) Depolarization of $\langle Q \rangle$ versus N. (b) Depolarization of $\langle V \rangle$ versus N. (c) The depolarization constant for linear polarization D_{LP} versus $\beta\Delta L$. (c) The depolarization constant for circular polarization D_{CP} versus $\beta\Delta L$. The value $\beta\Delta L$ is varied as $\pi/5$, $\pi/10$, $\pi/20$, $\pi/50$, $\pi/100$, and $\pi/200$. In (c), the plotted data behaves as $D_{LP} = (\text{slope})\beta\Delta L$, where slope $= 0.236 \pm 0.008$ ($n = 3$). In (d), the plotted data behaves as $D_{CP} = (\text{slope})\beta\Delta L$, where slope $= 0.468 \pm 0.026$ ($n = 3$). In summary, circularly polarized light, $\langle V \rangle$, depolarizes twice as fast as linearly polarized light, $\langle Q \rangle$.

polarized light per unit length of photon path. The mean survival length for linearly polarized light would be $1/\mu_{LPr}$ or 470 μm. The μ_{LPr} will increase proportionate to the average thickness of the microdomains of birefringence within a tissue, and fall with increasing wavelength as $1/\lambda^2$.

Figure 30.10b shows the attenuation of the average 4th element of the Stokes vector, $\langle V \rangle$, when right circularly polarized light (R) is delivered to the series of randomly oriented microdomains. The behavior is very similar to the behavior of $\langle Q \rangle$, but $\langle V \rangle$ drops faster than $\langle Q \rangle$. Figure 30.10d shows the factor D_{CP} that describes the depolarization of circularly polarized light. As $\beta \Delta L$ increases, D_{CP} increases proportionately, $D_{CP} \approx 0.48(\beta \Delta L)^2$, which is about twice as high a rate constant as the depolarization of linearly polarized light. The attenuation of $\langle V \rangle$ due to depolarization by birefringence can be described by these alternative expressions:

$$\langle V \rangle = e^{-D_{CP}N} = e^{-D_{CP}\left(L/\Delta L \right)} = e^{-\mu_{CP}L} \tag{30.22}$$

The coefficient of depolarization μ_{CP} for circularly polarized light is about twice the value of μ_{LP}.

30.3.4 Depolarization by Scattering

A scattering event redirects a photon. Simple redirection of a photon causes its orientation of polarization to appear different in the reference frame of an observer. Keep in mind that a Stokes vector is very dependent on the observer. If one observes linearly polarized light with its electric field oscillating horizontally with respect to an experimental tabletop, one might call this H polarization. But if one turned one's head sideways, one might call this light V polarization. Tilting one's head, one might call the light P^+ polarization. It is a matter of perspective. If one uses the tabletop as the frame of reference, but the photon scattering uses the scattering plane (specified by incident trajectory and scattered trajectory) as its frame of reference, and if the photon retains its polarization from its perspective but scatters into a new trajectory, from our perspective the photon may have

changed its polarization. Simple random redirection of photons by scattering can depolarize the transmitted light seen by an observer.

To dissect out the depolarization due to scattering, let us use Monte Carlo simulations (Ramella-Roman et al., 2005a,b) of polarized light transport in a scattering medium mentioned in Section 30.2, which uses the Mueller matrix for Mie scattering by spheres. In this section, polarized light, characterized by a Stokes vector, is transmitted through a slab of thickness L containing polystyrene microspheres of various sizes and concentrations in water. The transmitted photons are accumulated in a Stokes vector of transmission, $[I\ Q\ U\ V]^T$. The transmitted Q in response to incident H, and the transmitted V in response to incident R is reported here, specifying the coefficients of depolarization, μ_{LP} and μ_{CP}. It should be noted that all the transmitted light was used in this study. In general, the solid angle of collection will influence the degree of depolarization observed.

Figure 30.11a shows the range of sphere diameters (dia., x-axis) and volume fractions (v_f, y-axis) used in the simulations. For reference, the series of dia. are shown with different symbols. For each dia., a range of v_f values were used. Figure 30.11b shows the range of the scattering coefficient, μ_s [cm^{-1}], specified by Mie theory for these sphere parameters, plotted versus the anisotropy of scattering, g [dimensionless], which is specified by the sphere diameter. The choices of dia. and v_f were made to keep μ_s within a range such that attenuation of polarization was not too fast and not too slow.

Figure 30.12 shows the HQ and RV transmitted through slabs of varying thickness L (the x-axis), for the series of different sphere sizes and sphere concentrations. Each circle is one Monte Carlo simulation. Each line connects the simulations versus thickness L for one set of simulations for a given sphere size and concentration. The lines are exponentials that follow the forms:

$$HQ = \langle Q \rangle = e^{-\mu_{LP}L} = e^{-k_{LP}\mu_s L} \tag{30.23}$$

$$RV = \langle V \rangle = e^{-\mu_{CP}L} = e^{-k_{CP}\mu_s L} \tag{30.24}$$

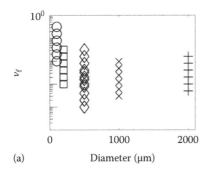

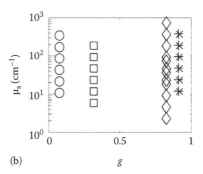

(a) Diameter (μm) (b) g

FIGURE 30.11 Experimental parameters of Monte Carlo simulations. (a) Choices of sphere diameter (dia. [μm]) and volume fraction (v_f). (b) Resulting anisotropy of scattering (g) and scattering coefficient (μ_s [cm^{-1}]). Symbols are the same in both figures, indicating the sphere diameter. The choices of dia and v_f yields similar ranges for μ_s.

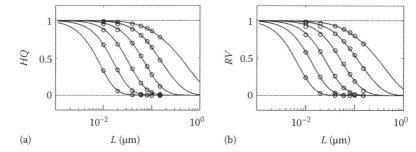

FIGURE 30.12 Depolarization of total transmitted light. (a) HQ vs. L, where H linearly polarized light is delivered and the Q element of the Stokes vector is measured after passing through increasing slab thicknesses L. (b) RV vs. L, where R circularly polarized is delivered and the V element of the Stokes vector is measured. For simplicity, the figures show only 42 of the simulations corresponding to 7 values of L and 6 choices of v_f using 100 μm dia. spheres. Since 6 sphere diameters were used, there were a total of 252 simulations. The solid lines show the fits by Equation 30.29, which behave very close to exponential decays.

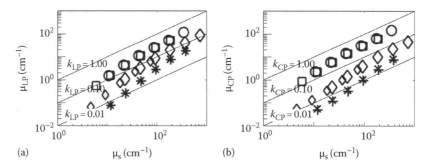

FIGURE 30.13 Relation between depolarization constants, k_{LP} and k_{CP}, for linearly and circularly polarized light, respectively, and the diffusivity χ. The straight solid line is $k_{LP} = \chi/2$ and $k_{CP} = \chi/2$.

Figure 30.13 shows the coefficient of depolarization μ_p versus the scattering coefficient μ_s for each of the simulations. The ratio $k_p = \mu_p/\mu_s$ is the efficiency of depolarization, that is, the amount of depolarization per scattering event. The lines in Figure 30.13 show the constant k_p lines for $k_p = 1, 0.1$, and 0.01. Figure 30.13a shows the μ_{LP} and k_{LP} for linearly polarized light. Figure 30.13b shows the μ_{CP} and k_{CP} for circularly polarized light. The efficiency of depolarization increases as the sphere size becomes smaller. There is an increase in the efficiency as the volume fraction of spheres increases, which yields higher values of μ_s.

Figure 30.14 plots the ratio k_{LP}/k_{CP} versus the volume fraction (v_f) and the sphere diameter (dia.). The efficiency of depolarization for small spheres is similar for both linearly and circularly polarized light. However, as the sphere size increases, the linear depolarization (k_{LP}) becomes two- to threefold greater than the circular depolarization (k_{CP}).

In summary, the Monte Carlo simulations show values of μ_{LP} and μ_{CP} that are always much smaller than the corresponding μ_s values. The efficiencies of depolarization, k_{LP} and k_{CP}, are greater for smaller spheres than for larger spheres. The efficiencies increase with increasing concentration of spheres, even when expressed as depolarization per scattering event.

In conclusion, we have compared depolarization due to randomly oriented birefringent layers and due to multiple scattering

in microsphere solutions. Depolarization based on birefringence yielded an attenuation of $\mu_{LP} \approx 21 \, \text{cm}^{-1}$ and $\mu_{CP} \approx 42 \, \text{cm}^{-1}$. Depolarization based on scattering by small-sized spheres relative to the wavelength yielded an attenuation of $\mu_{LP} \approx 10 \, \text{cm}^{-1}$ and $\mu_{CP} \approx 10 \, \text{cm}^{-1}$, when the concentration was sufficient to present $\mu_s \approx 100 \, \text{cm}^{-1}$, which is similar to the scattering coefficient encountered in tissues. For larger spheres, the efficiency of depolarization was less efficient, and k_{LP} was about two- to threefold greater than k_{CP}. Therefore, both mechanisms of depolarization, random birefringent microdomains and multiple scattering, play comparable roles in tissues.

30.4 Experimental Measurements of Depolarization in Tissues

30.4.1 Experimental Evidence

To illustrate the depolarization properties of biological tissues, data from three literature reports will be presented: Jarry et al. (1998), Jacques et al. (2000), and Sankaran et al. (2002). These studies measured the near-on-axis transport (T) of H and V light through tissue slices of varying thickness in response to delivered H light. Such measurements are called HH and HV measurements, indicating an incident H and transported H and V, respectively. In some cases, suspensions of polystyrene

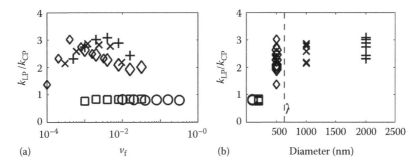

FIGURE 30.14 Ratio k_{LP}/k_{CP} versus volume fraction (v_f) and sphere diameter (dia.). Symbols indicate diameter. The efficiency of depolarization by small spheres is similar for linearly and circularly polarized light. For larger spheres comparable to or larger than the wavelength of light (see vertical dashed line in B at 633 nm), k_{LP} is two- to threefold greater than k_{CP}.

TABLE 30.1 Tissue Data on Depolarization of Linearly Polarized Light

Tissue	λ [nm]	x [rad²/mfp]	μ_s [cm⁻¹]	k_{LP} [—]	μ_{LP} [–]	Reference
Calf liver	633	0.0032	290	0.0016	0.46	Jarry et al. (1998)
Chicken liver	633	0.0030	290	0.0015	0.43	Jacques et al. (2000)
Chicken muscle	633	0.0617	250	0.0308	7.71	Jacques et al. (2000)
Pig muscle	543	0.0550	210	0.0275	5.78	Jacques et al. (2000)
Pig skin	543	0.2730	187	0.1365	25.53	Jacques et al. (2000)
Pig skin	543	0.2800	187	0.1400	26.18	Jacques et al. (2000)
Blood	633	(0.05)	(200)	0.10	(11.6)	Sankaran et al. (2002)
Myocardium	633	(0.23)	(200)	0.46	(52.3)	Sankaran et al. (2002)
Fat	633	(0.46)	(200)	0.92	(105.8)	Sankaran et al. (2002)
Tendon	633	(0.23)	(200)	0.45	(52.1)	Sankaran et al. (2002)

Note: Numbers in parentheses are based on assumed value of $\mu_s = 200\,\text{cm}^{-1}$.

microspheres of different sizes were measured. The details of the tissues and the measurements are summarized in Table 30.1.

The data analysis by Jacques et al. (1999) is used. The behavior of *HH* and *HV* was reported as a function of the optical depth (τ) of the tissue or microsphere solution. The τ was varied by varying the tissue sample thickness L or by varying the volume fraction v_f of microspheres in the cuvette. The light was initially launched as *H* linearly polarized light, and the orientation of the polarization was described as initially $\theta = 0°$. As the photons were scattered or retarded by the tissue, the orientation of the polarization began to diffuse into nonzero $\pm\theta$ values. This angular diffusion of the orientation was modeled as a random walk in θ space. The behavior was summarized:

$$p(\theta) = \frac{e^{-\theta^2/2\sigma^2}}{\sigma\sqrt{\pi/2}} \quad (30.25)$$

where σ [rad] is equal to the product of an angular diffusivity χ [rad²/mfp] and the optical depth of the tissues, τ [dimensionless], ($\tau = \mu_s L$, L = thickness of tissue [cm], μ_s = scattering coefficient [cm⁻¹], $1/\mu_s$ = mfp [cm] = mean free path between scatterers):

$$\sigma^2 = \chi\tau \quad (30.26)$$

Given incident *H* light, as the orientation of the electric (*E*) field of a photon broadened versus θ, the amount of the *E* field that coupled through a horizontal linear polarizer in front of the detector was proportional to $\cos\theta$. The amount of scattered radiant intensity that coupled to the detector was $\cos^2\theta$. Therefore, the total transmitted signal reaching the detector was

$$\langle\cos^2\theta\rangle = \int_{-\infty}^{\infty} p(\theta)\cos^2\theta\,d\theta = \frac{1}{2}\left(1 + e^{-\sigma^2/2}\right) = \frac{1}{2}\left(1 + e^{-\chi\tau/2}\right)$$

$$(30.27)$$

The transmitted *HH* and *HV* light were described as the combination of the primary collimated beam of light that was not scattered or absorbed, $T_c = e^{-\tau}$, and the scattered light components H_s and V_s, where the subscript s emphasizes this light has been scattered.

$$HH = T_c + H_s = T_c + C(1 - T_c)\frac{1}{2}\left(1 + e^{-\chi\tau/2}\right) \quad (30.28a)$$

$$HV = V_s = C(1 - T_c)\frac{1}{2}\left(1 - e^{-\chi\tau/2}\right) \quad (30.28b)$$

The factor C depends on the solid angle of collection (f) for the transmitted light, and depends on the attenuation of the scattered component by the tissue thickness, $C = f \exp(-\mu_{\text{atten}}L)$, where μ_{atten} is the attenuation coefficient for the scattered light and is comparable to μ_{eff} in diffusion theory although the light may not yet be fully diffuse and μ_{atten} is less than μ_{eff}. The degree of linear polarization (p) behaves as

$$p = \frac{HQ}{HI} = \frac{HH - HV}{HH + HV}$$

$$= \frac{T_c + C(1-T_c)\frac{1}{2}\left(\left(1+e^{-\chi\tau/2}\right)-\left(1-e^{-\chi\tau/2}\right)\right)}{T_c + C(1-T_c)\frac{1}{2}\left(\left(1+e^{-\chi\tau/2}\right)+\left(1-e^{-\chi\tau/2}\right)\right)} \approx e^{-\chi\tau/2} \quad \text{as } T_c \to 0$$

(30.29)

Hence, the decay of p is initially influenced by the values of T_c and C, but quickly becomes a simple exponential decay at large τ.

Since Equation 30.29 shows that the degree of linear polarization eventually starts dropping as a simple exponential, it is also possible to fit the experimental behavior of p versus L as

$$p = e^{-\mu_{\text{LP}}L}$$

(30.30)

where μ_{LP} [cm^{-1}] is the depolarization coefficient. Comparison of Equations 30.29 and 30.30 indicates that $\mu_{\text{LP}}L = k_{\text{LP}}\mu_s L = k_{\text{LP}}\tau = \tau\chi/2$. In other words, $k_{\text{LP}} = \chi/2$.

Figure 30.15 shows the generic behavior of the degree of linear polarization using both Equations 30.29 and 30.30, as a function of the tissue thickness for a generic tissue with varying values of k_p. For this example, C is set to $0.8 \exp(-0.5L)$. In Figure 30.15a, T_c decays quickly, HH slowly decays from 1 to $C/2$, and HV light builds up from zero to plateau at $C/2$. Figure 30.15B plots $p = (HH - HV)/(HH + HV)$, which falls exponentially to zero.

Figure 30.16 and Table 30.1 summarize the experimental data for tissues. The data for liver, muscle, skin, and microspheres were derived graphically from Jacques et al. (2000). Data for liver

from Jarry et al. (1998) is included. These data were analyzed in Jacques et al. (2000) using Equation 30.30 to specify the values of χ, and here are converted to the depolarization coefficient k_{PL}. The data for fat, myocardium, tendon, and blood from Sankaran et al. (2002) are also included, although the data did not specify the scattering coefficient and an average value of $\mu_s = 200\,\text{cm}^{-1}$ is assigned to this data. Values in parentheses in Table 30.1 indicate values that rely on this assignment. All the tissue data fall in the range of $\mu_s = 100\text{--}300\,\text{cm}^{-1}$, which is the expected range. The values of μ_{LP} for the data are in the range of $0.4\text{--}106\,\text{cm}^{-1}$, such that k_{LP} varies from $0.0015\text{--}0.92$.

The data of Jacques and Jarry showed the μ_{LP} values in the order of skin > muscle > liver, while the μ_s values were relatively similar. The values of anisotropy for these tissues are not so different ($g \approx 0.7\text{--}0.95$) for these tissues, so the strong inter-tissue differences in μ_{LP} and k_{LP} were thought to be due to differences in birefringence. However, the data of Sankaran show similar μ_{LP} and k_{LP} values even for fat and whole blood, which are not expected to be birefringent. Therefore, resolving the relative roles of birefringence and scattering as the mechanism of depolarization still awaits further experiments. But a key lesson is the relatively high values of k_{LP} seen in most tissues, which compare well with the depolarization by particles smaller than the wavelength of light. This suggests that cellular and tissue ultrastructures (scatterer sizes less than wavelength of light) are responsible for depolarization by the scattering mechanism.

The constant k_{LP} lines in Figure 30.16 indicate that tissues and polystyrene microspheres are comparable in their ability to depolarize linearly polarized light. The data for microspheres are located at lower μ_s values since the concentrations of solutions were usually <2% volume fraction to avoid aggregation of spheres. In the experimental sphere measurements, the two data at $k_{\text{LP}} > 0.1$ were due to spheres much smaller than the wavelength of light, while the data at $k_{\text{LP}} < 0.1$ were due to spheres comparable to or larger than the wavelength of light. The experimental sphere measurements are slightly lower in k_{LP} values than the Monte Carlo simulations, since the latter collected all transmitted light rather than a narrow solid angle of collection.

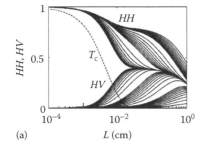 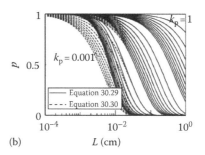

(a) (b)

FIGURE 30.15 A generic depolarization experiment. (a) The collimated transmission (T_c), the measured $HH = T_c + H_s$ where H_s is the H component of scattered light, and the measured $HV = V_s$ where V_s is the V component of scattered light, are plotted versus the tissue thickness L. The curves are for $k_{\text{LP}} = 0.001$ to 0.1. The model assumes $\mu_s = 200\,\text{cm}^{-1}$, $C = 0.7 \exp(-0.5L)$. (b) The degree of linear polarization, $p = (HH - HV)/(HH + HV)$, is plotted versus L. Solid lines use Equation 30.29, dashed lines use Equation 30.30. For $L > 10^{-2}$ cm (i.e., $\mu_s L > 2$), Equations 30.29 and 30.30 are indistinguishable.

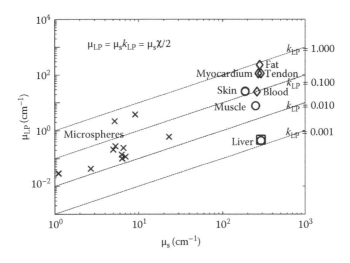

FIGURE 30.16 Experimental measurements of linear depolarization in tissues and microsphere solutions. The depolarization coefficient μ_{LP} is plotted versus the scattering coefficient μ_s. The lines show constant depolarization constant k_{LP} (0.001 to 1.0). (Circles and crosses are from Jacques et al. Square from Jarry et al. Diamonds show μ_{LP} from Sankaran et al., with x-axis value μ_s assigned a constant value of 200 cm^{-1} since the data did not specify μ_s.) The microspheres (crosses) with $k_{LP} > 0.10$ were about 300 nm in diameter, smaller than the 633 nm wavelengths, while the other microspheres with $k_{LP} < 0.10$ were in the 600 to 2000 nm range, comparable to or larger than the wavelengths (543 and 633 nm). All the experiments used a relatively narrow solid angle of collection. (Square from Jarry et al. (1998), circles and crosses from Jacques et al. (2002), and diamonds from Sankaran et al. (2002).)

In summary, experimental measurements of the transmitted degree of linear polarization demonstrated a randomization or depolarization of linearly polarized light. The angular diffusivity χ [rad^2/mfp] described the apparent random walk in angle space for this randomization of the orientation of the polarization. The value $\chi/2$ corresponds to the depolarization efficiency k_{LP}, such that $\mu_{LP} = k_{LP}\mu_s$. This μ_{LP} is an optical property of a tissue.

30.5 A Polarized Light Camera for Clinical Use

The previous sections have outlined the basics of polarized light transport. Our development of a practical polarized light camera for clinical use has concentrated on the simplest of measurements, reflectance mode HQ images ($HQ = HH - HV$), rather than pursuing a more complete study of Mueller matrix imaging. The rationale was to learn how useful polarized light might be in a clinical setting before pursuing more sophisticated measurements. Since linearly polarized light is so strongly depolarized by tissue, reflectance mode imaging using linearly polarized light was expected to only detect photons scattered by superficial layers. Since skin pathology usually arises in the superficial layers of the skin, the polarized light camera was designed to present an enhanced image of superficial tissue layers to the doctor.

30.5.1 A Handheld Polarized Light Camera for Imaging Skin

A polarized light camera is shown in Figure 30.17. Linearly polarized light is delivered obliquely to the skin, with a glass flat coupled to the skin by a film of water or gel. The surface glare from the air/glass, glass/water, and water/skin interfaces reflects obliquely and misses the overhead camera that viewed the skin. Only photons that enter the skin and backscatter toward the camera are collected. This technique of oblique illumination distinguishes this camera from common dermatoscopes and from the earlier image of Demos and Alfano (1997). The surface glare is stronger than the subsurface reflectance of interest. The oblique angle of illumination and the optical coupling enable the camera to image subsurface layers and avoid the surface glare. A linear polarization filter in front of the camera is oriented either parallel or perpendicular to the polarization of the illumination light to yield two images, called PAR (or HH) and PER (or HV), respectively, which collect both superficially scattered light that is still polarized and deeply multiply scattered light that is depolarized:

$$HH = \text{PAR} = I_o T_{mel} \, (\text{superficial} + 1/2 \, \text{deep}) \quad (30.31a)$$

$$HV = \text{PER} = I_o T_{mel} \, 1/2 \, \text{deep} \quad (30.31b)$$

The difference image $HH - HV$ images the superficial layer of the skin:

$$HQ = HH - HV = \text{PAR} - \text{PER} = I_o T_{mel} \, \text{superficial}$$

$$(30.32)$$

which includes the effect of any spatial variation in illumination, I_o, and any superficial filter such as the melanin of a freckle

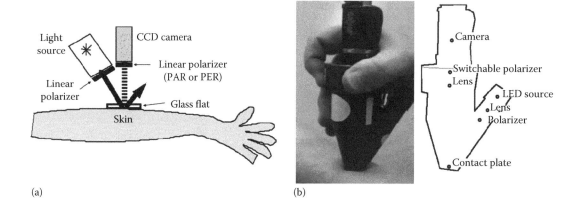

(a)　　　　　　　　　　　　　　　(b)

FIGURE 30.17 Polarized light camera. (a) Schematic diagram shows delivery of linearly polarized light obliquely (30° off the normal) onto the skin, through a glass plate optically coupled to the skin by a film of water or gel. Specular reflectance (glare) is reflected off the skin surface and misses the camera. The orientation of the electric field of the linearly polarized light is parallel to the scattering plane (laser–skin–camera triangle). The camera collects photons that enter the skin and backscatter upward toward the camera. The camera views through a second linear polarizer, which is oriented either parallel or perpendicular to the scattering plane to yield two images, *HH* and *HV*, respectively. (b) Picture and schematic for hand-held polarized light camera.

or age mark, T_{mel}. Alternatively, one can display the degree of polarization image:

$$p = \frac{HH - HV}{HH + HV} = \frac{I_0 T_{mel}\ \text{superficial}}{I_0 T_{mel}\ \text{total}} = \frac{\text{superficial}}{\text{total}} \quad (30.33)$$

where taking the ratio cancels the factor $I_0 T_{mel}$. This cancellation makes freckles disappear and corrects for nonuniform illumination. However, the denominator in Equation 30.33 is responsive to the deeper tissue layers, since the denominator is essentially the total diffuse reflectance. Hence, dermal variations contaminate the image of superficial layers. Therefore, we now routinely use just the difference image $HQ = HH - HV$.

30.5.2 Polarized Light Camera Images

Figure 30.18 shows two examples of polarized images compared with the associated normal light image, represented by an *HV* image that rejected surface glare but collected both superficially scattered light and multiply scattered light from deeper layers. The two examples are (1) the hair follicles of the forearm and (2) actinic keratoses (AK) lesions on the skin. In the images, both the region around the hair follicles and the AK lesions appear dark in comparison to the surrounding skin (~½ as bright). In the normal light images, both hair follicles and AK lesions appear as bright as the surrounding skin; in fact the AK lesions look a bit brighter. So there is plenty of light backscattered from these skin sites.

Why are these skin sites dark in the DOPL images? In the region around the hair follicle where the epidermis folds in around the hair shaft, and in the AK lesions where there is cellular growth displacing the superficial dermis, there is less collagen to backscatter light. So polarized light can penetrate deeper without backscatter, becoming increasingly depolarized before efficient backscatter can occur. There is a balance between the

rate of depolarization (μ_{LP}) and the rate of backscatter (μ_s) of the incident polarized light that affects the strength of the polarization signal. In other words, the depolarization efficiency k_{LP} is a key factor in the image contrast.

30.5.3 Clinical Study: Finding Skin Cancer Margins

In our clinical study, we are attempting to find the margins of skin cancer prior to Mohs surgery. The goal is to identify cancer margins and guide surgical excision. In this work, we do not find obvious differences in the magnitude of HQ between normal skin and skin cancer. Rather, we see roughly a 30% variation in the pixel values of HQ images. This variation is not random, but rather presents a "fabric pattern" that characterizes the skin. Cancer appears to randomize this fabric pattern, yielding a more uniform appearance. The surgeon visualizes the cancer margin as a disruption of this fabric pattern.

Figure 30.19 illustrates a clinical image of a basal cell carcinoma on the scalp. The Mohs surgeon marked a cancer margin with a pen based on visual appearance. Then a grid of 25 black dots was drawn on the skin through a template with holes to provide reference marks for co-registration of images. The handheld camera (Figure 30.17b) imaged each of the 4×4 or 16 regions, with each image containing four black dots (separated by 5 mm) to aid co-registration. These images were later stitched together into one big mosaic image. A standard handheld camera acquired images of the lesion area before and after the first excision, and after the final excision. The lines drawn in Figure 30.19c and d show the cancer margins predicted by the surgeon's eye and by viewing the HQ image. The HQ image indicated a wider lesion than the surgeon's eye. The pathology report (Figure 30.19e) agreed with the polarized light image, and a wider excision was required before the margins became cancer-free (Figure 30.19f).

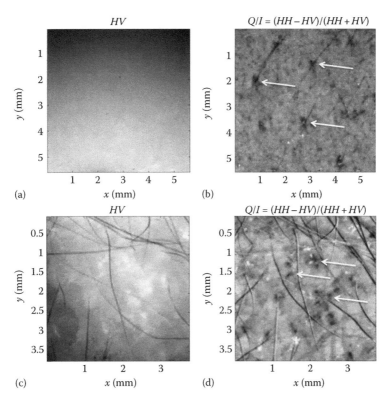

FIGURE 30.18 Polarized light images of hair follicles and actinic keratoses (AK) lesions. (a) Hair follicles using normal light (cross-polarized *HV* image). (b) Polarized light image of same hair follicle using degree of polarization, $p = (HH - HV)/(HH + HV)$. (c) Actinic keratoses lesions using normal light (*HV*). (d) Same as C using polarized light (*p*). In the polarized light images, both the region around the hair follicles and the AK lesions (arrows) are darker than the surrounding skin. The normal light images do not show these structures well.

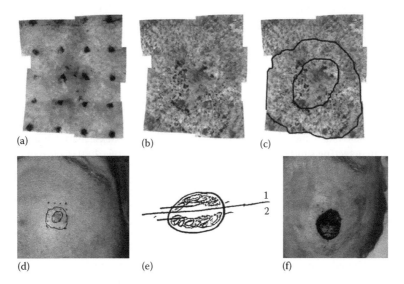

FIGURE 30.19 Clinical study to test polarized light camera guidance of surgical excision. (a) Normal light image (*HV*) of basal cell carcinoma with grid of dots for co-registering a set of 16 individual images into a mosaic image. Dots are ~5 mm apart. (b) Polarized light image (*HH – HV*). (c) Same as B, with inner circle showing Mohs surgeon's visual assessment of the cancer margin, and outer circle showing assessment of the cancer margin using the *HH – HV* image. The *HH – HV* image indicated that the cancer extended beyond the visible lesion. (d) View of patient's lesion on head, with cancer margin assessments graphically superimposed. (e) Pathology report showing that all the margins were still positive for cancer after the surgeon removed the visually assessed lesion. (f) After final excision when pathology finally confirmed all margins were cancer-free. (Collaboration by Ravikant Samatham, Scott Isenhath, Lindsay Severson, Ken Lee, Steven Jacques, Oregon Health & Science University, Portland, OR; sponsored by NIH RO1-CA CA113947.)

30.6 Future Directions

This chapter has discussed the depolarization of linearly polarized light (μ_{LP}) and the depolarization of circularly polarized light (μ_{CP}). But a 4×4 data matrix provides 16 elements that each describe the transport and transformation of polarized light. An optical property could be described as a coefficient of change, $\mu_{i,j}$ [cm^{-1}]:

$$\mu_{i,j} = \frac{-\partial M_{i,j}}{\partial L} \frac{1}{M_{i,j}} \quad (30.34)$$

where ∂L is an incremental change in photon path through the tissue. Future work might find that the values of $\mu_{i,j}$ corresponded to each of the 16 elements in the Mueller matrix are informative about a tissue. This ∂L could be calculated by polarized Monte Carlo simulations for different locations of escape from a tissue, or experimentally measured perhaps by using low coherence interferometry. The goal is to document the incremental change in transport per incremental change in photon pathlength. For randomly oriented tissues, this is probably not so useful. The simple linear depolarization coefficient μ_{LP} or circular depolarization coefficient μ_{CP} would be sufficient. But often epithelial tissues present structured architecture and are not random, and the coefficient $\mu_{i,j}$ [cm^{-1}] may prove very useful in characterizing a structured architecture and detecting deviation from native architecture. The $\mu_{i,j}$ may prove useful in characterizing constructs composed of nanomaterials that exhibit architecture.

In summary, this chapter has summarized some of the basic tools used to discuss and measure polarized light. The chapter has considered image contrast based on depolarization, a first step in such imaging. The depolarization of linearly and circularly polarized light was discussed, separating the effects of birefringence and scattering. The coefficient and efficiency of linear depolarization in some tissues were presented. Some examples of depolarization in clinical images were shown, and the efficiency of depolarization, k_{LP}, was suggested to be related to the intensity of polarized reflectance in such images. Finally, the future investigation of $\mu_{i,j}$ coefficients, describing the incremental change in the $M_{i,j}$ element of a Mueller matrix per incremental photon pathlength, may prove useful in detecting subtle changes in tissues or materials with a structural architecture. There is still much work to be done in the field of polarized light imaging.

Acknowledgments

This work has been supported over the past several years by the National Institutes of Health, United States, (RO1-CA80985, R24-CA84587, RO1-CA113947).

Glossary

$\beta 2\pi\Delta n/\lambda$	the phase shift per unit length [cm^{-1}] due to birefringence
ΔL	incremental thickness of birefringent layer [cm]
Δn	birefringence ($\Delta n = n_e - n_o$), n_e is extraordinary n, n_o is ordinary n
$\Delta\phi$	phase shift due to retardance [rad]
λ	wavelength of light [cm]
μ_a	absorption coefficient [cm^{-1}]
μ_{CP}	depolarization coefficient for circularly polarized light [cm^{-1}]
μ_{LP}	depolarization coefficient for linearly polarized light [cm^{-1}]
μ_s	scattering coefficient [cm^{-1}]
DM	Data matrix (see Equation 30.16)
g	anisotropy of scattering [dimensionless]
I	intensity of light, $I = H + V$
H	horizontal linearly polarized light intensity, electric field oriented horizontally
k_{LP}	efficiency of depolarization for linearly polarized light
k_{CP}	efficiency of depolarization for circularly polarized light
L	thickness of medium [cm]
M	Mueller matrix (see Equation 30.6)
n	refractive index [dimensionless]
P^+	linearly polarized light intensity oriented at $+45°$
Q	balance $H - V$
R	right circularly polarized light intensity
S	Stokes vector $[I\ Q\ U\ V]^T$, ^{T} indicates transpose vector to vertical column vector
SM	Simulation matrix (see Equation 30.17)
U	balance $P^+ - P^-$
V	vertical linearly polarized light intensity (distinguishable in context from $V = R - L$)
V	balance $R - L$ (distinguishable in context from $V =$ vertical linearly polarized light)

References

Agafonov, D. N., Zimnyakov, D. A., Sinichkin, Y. P., and Kiseleva, I. A. 2002. Influence of absorption on residual polarization of backscattered linearly polarized light. In: *Saratov Fall Meeting 2001: Coherent Optics of Ordered and Random Media II*, eds. D. A. Zimnyakov, Proc. of SPIE Vol. 4705, pp. 166–172.

Anderson, R. R. 1991. Polarized light examination and photography of the skin. *Arch. Dermatol.* 127(7): 1000–1005.

Baba, J. S., Chung, J.-R., DeLaughter, A. H., Cameron, B. D., and Cote, G. L. 2002. Development and calibration of an automated Mueller matrix polarization imaging system. *J. Biomed. Opt.* 7(3): 341–349.

De Boer, J. F., Milner, T. E., van Gemert, M. J. C., and Nelson, J. S. 1997. Two-dimensional birefringence imaging in biological tissue by polarization sensitive optical coherence tomography. *Opt. Lett.* 22: 934–936.

Demos, S. G. and Alfano, R. R. 1997. Optical polarization imaging. *Appl. Opt.* 36(1): 150–155.

Ding, H., Lu, J. Q., Brock, R. S., McConnell, T. J., Ojeda, J. F., Jacobs, K. M., and Hu, X.-H. 2007. Angle-resolved Mueller matrix study of light scattering by B-cells at three wavelengths of 442, 633, and 850 nm. *J. Biomed. Opt.* 12(3): 034032.

Hielscher, A., Eick, A., Mourant, J. et al. 1997a. Diffuse backscattering Mueller matrices of highly scattering media. *Opt. Express* 1(13): 441–453.

Hielscher, A. H., Mourant, J. R., and Bigio, I. J. 1997b. Influence of particle, size and concentration on the diffuse backscattering of polarized light from tissue phantoms and biological cell suspensions. *Appl. Opt.* 36: 125–135.

Jacques, S. L., Moody, A., and Ramella-Roman, J. C. 2001. Characterizing microscopic domains of birefringence in thin tissue sections. *Proc. SPIE* 4257: 464–468.

Jacques, S. L., Ostermeyer, M., Wang, L. et al. 1996. Polarized light transmission through skin using video reflectometry: Toward optical tomography of superficial tissue layers. In *Proc. SPIE*, Vol. 2671, pp. 199–210.

Jacques, S. L., Ramella-Roman, J. C., and Lee, K. 2002. Imaging skin pathology with polarized light. *J. Biomed. Opt.* 7: 329–340.

Jacques, S. L., Roman, J., and Lee, K. 2000. Imaging superficial tissues with polarized light. *Lasers Surg. Med.* 26: 119–129.

Jaillon, F. and Saint-Jalmes, H. 2005. Scattering coefficient determination in turbid media with backscattered polarized light. *J. Biomed. Opt.* 10(3): 034016.

Jarry, G., Steimer, E., Damaschini, V. et al. 1998. Coherence and polarization of light propagating through scattering media and biological tissues. *Appl. Opt.* 37: 7357–7367.

Mourant, J. R., Johnson, T. M., Carpenter, S. et al. 2002. Polarized angular dependent spectroscopy of epithelial cells and epithelial cell nuclei to determine the size scale of scattering structures. *J. Biomed. Opt.* 7(3): 378–387.

Mourant, J. R., Johnson, T. M., and Freyer, J. P. 2001. Characterizing mammalian cells and cell phantoms by polarized backscattering fiber-optic measurements. *Appl. Opt.* 40(28): 5114–5123.

Ramella-Roman, J. C. and Jacques, S. L. 2001. Mueller-matrix description of collimated light transmission through liver, muscle, and skin. *Proc. SPIE* 4257: 110–116.

Ramella-Roman, J. C., Prahl, S. A., and Jacques, S. L. 2005a. Three Monte Carlo programs of polarized light transport into scattering media: Part I. *Opt. Express* 13(12): 4420–4438.

Ramella-Roman, J. C., Prahl, S. A., and Jacques, S. L. 2005b. Three Monte Carlo programs of polarized light transport into scattering media: Part II. *Opt. Express* 13(25): 10392–10405.

Sankaran, V., Walsh, J. T. Jr., and Maitland, D. J. 2000a. Polarized light propagation through tissue and tissue phantoms. In *Coherence Domain Optical Methods in Biomedical Science and Clinical Applications IV*, eds. V. V. Tuchin, J. A. Lzatt, and J. G. Fujimoto, Proceedings of SPIE Vol. 3915, pp. 178–186.

Sankaran, V., Walsh, J. T. Jr., and Maitland, D. J. 2000b. Polarized light propagation through tissue phantoms containing densely packed scatterers. *Opt. Lett.* 25(4): 239–241.

Sankaran, V., Walsh, J. T. Jr., and Maitland, D. J. 2002. Comparative study of polarized light propagation in biologic tissues. *J. Biomed. Opt.* 7(3): 300–306.

Schmitt, J. M., Gandjbakhche, A. H., and Bonner, R. F. 1992. Use of polarized light to discriminate shortpath photons in a multiply scattering medium. *Appl. Opt.* 31: 6535–6546.

Studinski, R. C. N. and Vitkin, I. A. 2000. Methodology for examining polarized light interactions with tissues and tissuelike media in the exact backscattering direction. *J. Biomed. Opt.* 5(3): 330–337.

Wang, X. and Wang, L. V. 2002. Propagation of polarized light in birefringent turbid media: A Monte Carlo study. *J. Biomed. Opt.* 7(3): 279–290.

Wang, X., Wang, L. V., Sun, C.-H. et al. 2003. Polarized light propagation through scattering media: Time-resolved Monte Carlo simulations and experiments. *J. Biomed. Opt.* 8(4): 608–617.

Weinreb, R. N., Dreher, A. W., Coleman, A., Quigley, H., Shaw, B., and Reiter, K. 1990. Histopathologic validation of Fourier ellipsometry measurements of retinal nerve fiber layer thickness. *Arch. Ophthalmol.* 108: 557–560.

Wood, M. F. G., Ghosh, N., Moriyama, E. H., Wilson, B. C., and Vitkin, I. A. 2009. Proof-of-principle demonstration of a Mueller matrix decomposition method for polarized light tissue characterization *in vivo*. *J. Biomed. Opt.* 14(1): 014029, 1–5.

Wood, M. F. G., Guo, X., and Vitkin, I. A. 2007. Polarized light propagation in multiply scattering media exhibiting both linear birefringence and optical activity: Monte Carlo model and experimental methodology. *J. Biomed. Opt.* 12(1): 014029, 1–10.

Wu, P. J. and Walsh, J. T. Jr. 2006. Stokes polarimetry imaging of rat tail tissue in a turbid medium: Degree of linear polarization image maps using incident linearly polarized light. *Biomed. Opt.* 11(1): 014031, 1–10.

Yao, G. and Wang, L. V. 1999. Two-dimensional depth-resolved Mueller matrix characterization of biological tissue by optical coherence tomography. *Opt. Lett.* 24: 537–539.

Yaroslavsky, A. N., Barbosa, J., Neel, V. et al. 2005. Combining multispectral polarized light imaging and confocal microscopy for localization of nonmelanoma skin cancer. *J. Biomed. Opt.* 10(1): 014011.

V

Molecular Probe Development

31 Molecular Reporter Systems for Optical Imaging *Walter J. Akers and Samuel Achilefu* 673
Introduction • Reporter Systems • Functional Classifications of Molecular Imaging Agents • Discussion • List of Abbreviations • References

32 Nanoparticles for Targeted Therapeutics and Diagnostics *Timothy Larson, Kort Travis, Pratixa Joshi, and Konstantin Sokolov* 697
Introduction • Background • State of the Art: Nanoparticles in Biomedical Optics • Summary • References

33 Plasmonic Nanoprobes for Biomolecular Diagnostics of DNA Targets *Tuan Vo-Dinh and Hsin-Neng Wang* 723
Introduction • SERS Technique for Gene Diagnostics • Development of Molecular Sentinel Nanoprobes • Applications in Biomolecular Diagnostics • Conclusion • Acknowledgments • References

Molecular Reporter Systems for Optical Imaging

Walter J. Akers
Washington University School of Medicine

Samuel Achilefu
Washington University School of Medicine

31.1 Introduction 673
    Background
31.2 Reporter Systems 675
    Genetically Encoded Optical Reporters • Exogenous Molecular Probes and Nanomaterials
31.3 Functional Classifications of Molecular Imaging Agents............................. 679
    Nonspecific Molecular Imaging Agents • Affinity-Based Molecular Probes • Hybrid Molecular Imaging Agents • Multimodal Molecular Imaging
31.4 Discussion.. 688
List of Abbreviations 688
References.. 688

31.1 Introduction

The goal of molecular imaging is to identify or monitor specific molecular processes in target cells or tissue. In vivo, these events may be biological functions of normal physiology or signatures of pathological processes. The expression levels of a target molecule such as protein receptor or the activity of disease-related enzymes can be used diagnostically to give information about a disease state such as disease staging or predicting response to specific therapies.

Molecular reporter systems are molecular probes that enable specific detection of biochemical signatures in living cells or organisms with minimal interruption of native physiological processes. Unlike many areas of biomedical research where single specialties can drive new discoveries, biomedical imaging with molecular reporter systems, molecular imaging, incorporates diverse scientific and medical disciplines to solve biological problems. Currently, molecular imaging is accomplished by using various modalities that vary in spatial resolution and detection sensitivity.

X-ray computed tomography (CT), magnetic resonance imaging (MRI), and ultrasound imaging methods are the mainstay of clinical imaging because of their ability to provide exquisite resolution of anatomy and even physiology. CT contrast between tissues results from differences in the absorption of gamma radiation between various tissue types, with bone absorption being much greater than other tissues. Clinical MRI contrast originates from differential relaxivity of hydrogen ions in tissues (mainly from water), but may also come from halogens such as fluorine. However, molecular imaging with CT and MRI modalities has

been limited by poor sensitivity to externally administered, or exogenous, contrast agents. Contrast for ultrasound imaging can emanate from differences in tissue refractive index in response to ultrasound waves. In some cases, microbubbles have been administered to enhance contrast in human studies. The high resolution and real-time 3D display of ultrasound images are attractive in a variety of clinical settings. Major limitations of ultrasound imaging are its inability to penetrate bone tissue and the paucity of exogenous molecular contrast agents.

Nuclear imaging, including positron emission tomography (PET) and single photon emission computed tomography (SPECT), have significant advantages in molecular imaging due to the variety of available radiopharmaceuticals, excellent detection sensitivity of modern imaging systems, and surge in molecular imaging agent development. However, nuclear imaging uses ionizing radiation that has safety issues and it is limited to health centers that can generate or receive radionuclides. Relative to MRI and CT, nuclear imaging has poor anatomical resolution, which is a major reason for the recent increase in the use of multimodal PET/CT and SPECT/CT scanners.

The area of optical molecular imaging has also exploded in the last decade. Optical imaging uses low-energy light rather than potentially harmful ionizing radiation to report the location, functional status, or molecular events in cells and target tissue. Optical contrast may be obtained through intrinsic tissue properties such as absorption, scattering, and fluorescence, or through exogenous agents that can be administered for signal enhancement. Exogenous contrast agents for optical imaging include light-absorbing materials, also known as chromophores, or fluorophores, which are materials that reemit light after illumination at the excitation wavelength. Imaging signals from the materials can be detected using optical detectors and wavelength-limiting filters.

31.1.1 Background

31.1.1.1 Fluorescence

Fluorescence describes the phenomenon of photon emission from a molecule after absorption of higher energy photons or simultaneous absorption of multiple lower energy photons. The whole process occurs over a very short timescale (10^{-10}–10^{-8} s) during which the molecule is excited (photon absorption) to a higher energy state then relaxes via vibrational and rotational processes (nonradiative decay) or photon emission (radiative decay). The average time spent in the excited state is called the fluorescence lifetime (FLT). The fluorescence brightness of a fluorescent reporter is a product of the number of photons absorbed per dye molecule (absorption coefficient) and the probability of radiative emission per absorbed photon (quantum yield) (Lakowicz 1999).

Fluorescence intensity and fluorescence lifetime can be affected by reactions that occur while fluorophores are in the excited state. These factors include solvent polarity, viscosity, and pH as well as the presence of molecules that can accept energy or increase nonradiative relaxation. This sensitivity to environmental conditions provides sensing capabilities unique to optical imaging.

31.1.1.1.1 Fluorescence Resonance Energy Transfer

One unique feature of fluorescent molecules is the phenomenon of intermolecular energy transfer known as Förster or fluorescence resonance energy transfer (FRET). FRET is the process in which energy resulting from photon absorption by a fluorophore (donor) is directly transferred to a nearby chromophore (acceptor) resulting in either fluorescence from the acceptor molecule or nonradiative decay (Figure 31.1). FRET requires sufficient overlap of donor emission and acceptor excitation spectra and close proximity (20–90 Å) of the dye pairs (Lakowicz 1999). FRET between specific fluorescent protein pairs are often used as molecular calipers in fluorescence microscopy to measure protein–protein interactions by monitoring the change in donor–acceptor fluorescence intensity ratios or fluorescence lifetime changes (Kenworthy 2001). FRET is also the basis of many hybrid exogenous contrast agents for in vitro and in vivo imaging described in Section 31.3.3.

The detection sensitivity of optical methods is very high, with exquisite ability to detect single molecules through fluorescence microscopy. This high sensitivity allows molecular imaging using tracer levels of contrast agents, which are sufficiently low to induce no significant change in biological activity. Optical methods can also provide high spatial resolution down to nanoscopic level at small depths, but resolution quickly deteriorates with depth due to scattering. Scattering occurs due to interaction of light with biological components, dominating other effects on photon transport in biological tissues. Light penetration in tissue is also hindered by absorption. The dominant light-absorbing molecules, or chromophores, in mammalian tissues are hemoglobin in the visible and water in the infrared regions, respectively (Figure 31.2). A minima in optical absorption occurs in the 700–1000 nm near-infrared (NIR) region, often called the "Optical Window," in which biological tissue is relatively transparent. Scattering is, therefore, the dominant tissue–light interaction in this wavelength range.

Although some scholars consider molecular imaging a new field of study, tissue-specific delivery of molecular probes for imaging applications was successfully demonstrated by fluorescence microscopy many decades ago. For example, as far back as 1941, Coons and Kaplan labeled a monoclonal antibody with fluorescein to detect pneumococci in fixed tissue by fluorescence microscopy (Coons and Kaplan 1950). This achievement

Fluorescence resonance energy transfer

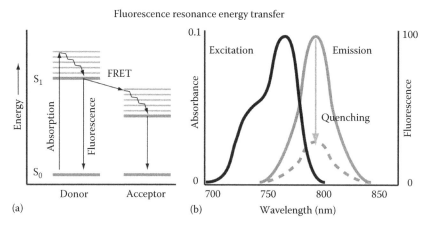

FIGURE 31.1 Depiction of fluorescence and FRET. (a) Upon photon absorption, a fluorophore is excited from the ground-state energy level (S_0) to an excited state. The fluorophore loses energy via vibrational and rotational mechanisms, reaching the first excited state energy level (S_1) from where fluorescence may occur as the molecule turns to the ground state. Energy may also be transferred to an acceptor species in a nonradiative manner, which may then fluoresce at a different wavelength or be expended in a nonradiative manner. The effect of FRET is lower emissivity and decreased quantum yield of the donor fluorophore, also called fluorescence quenching.

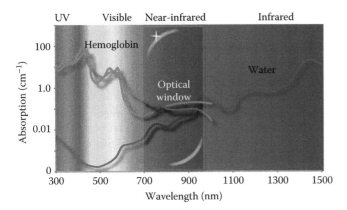

FIGURE 31.2 (See color insert.) Wavelength-dependent light absorption in biological tissues due to dominant chromophores: oxy- and deoxy-hemoglobin in the visible and water in the infrared regions, respectively. Absorption of light is 1–2 orders of magnitude less in the near-infrared region, resulting in relative transparency termed the "Optical Window."

targeting moieties labeled with contrast agents. Representative targeting molecules include antibodies, proteins, peptides, and synthetic molecules or materials. Hybrid molecular imaging strategies may employ nonspecific delivery mechanisms with target-specific signal amplification strategies, such as protease-activatable fluorescent probes (Figure 31.3). Alternatively, specific targeting may be coupled with specific signal generation. As the field of optical molecular imaging progresses, the excellent safety of optical methods and the unique capabilities of optical tracers will expand the use of optical imaging techniques in clinical applications.

In this chapter, we will review reporter systems and molecular targeting strategies for optical molecular imaging, focusing on recent progress in the field of optical molecular probes. Special emphasis will be given to NIR molecular probes that are widely used to image molecular processes in living organisms. We will conclude with a discussion of multimodality optical-based imaging probes and methods.

represents a multidisciplinary collaboration of biology, chemistry, and engineering that characterizes molecular imaging as we know it today (Coons 1961). More than 50 years later, Muguruma and others reported the development of an endoscopy system for cancer detection using NIR fluorescent-labeled antibody (Muguruma et al. 1998).

Optical molecular imaging approaches with exogenous molecular probes fall under three general methods: nonspecific, affinity-based, and hybrid techniques. Nonspecific molecular probes generate tissue-specific contrast by passive accumulation in tissues due to mechanisms such as the enhanced permeability and retention (EPR) effect that is predominant in cancerous tissue or inflammatory diseases. In addition, nonspecific uptake of molecular probes in tissues may also occur via affinity for specific environments such as low pH, hypoxia, high metabolic activity, changes in viscosity, or abnormal deposits such as neurofibrillary tangles in Alzheimer's disease (Hintersteiner et al. 2005). As demonstrated by the work of Coons and Kaplan (Coons and Kaplan 1950), affinity-based strategies employ active

31.2 Reporter Systems

31.2.1 Genetically Encoded Optical Reporters

Genetically engineered optical reporters have truly revolutionized our understanding of molecular events at cellular and tissue levels. This is evidenced by the award of the 2008 Nobel Prize in chemistry to the discoverer of green fluorescent protein (GFP) and two scientists who have contributed significantly to its applications and engineering many derivatives (Service 2008). In addition to fluorescent proteins, another form of genetically encoded genes produces bioluminescent proteins, which produce light through an energy-dependent photochemical reaction (bioluminescence). These reporter genes are significant because they can be inserted into the genome of eukaryotic cells, often with little or no effect on the cell growth and behavior. Cell transfection with reporter genes have many applications, including constitutive expression at high copy levels for cell tracking (Duda et al. 2007, Wilson et al. 2008) and quantifying cell numbers in infectious diseases (Doyle et al. 2004, Hutchens

Optical reporter systems		
Genetic optical reporters Bioluminescent reporters Fluorescent proteins	Intrinsic optical reporters Proteins Cofactors Intrinsic porphyrins	Exogenous optical reporters Organic dyes Inorganic nanomaterials

Molecular imaging agents		
Nonspecific pharmacologic modulation, e.g. Enhanced circulation time Vascular permeability Directed route of elimination	Hybrid activation mechanism, e.g. FRET Quenching Spectral change	Targeted molecular probe Molecular targeting agent e.g. Antibody Natural protein ligand High affinity peptide

FIGURE 31.3 Molecular reporter systems for optical imaging with fluorescent molecular probes.

and Luker 2007) and cancer (Gross and Piwnica-Worms 2005, Klerk et al. 2007). To improve detection sensitivity useful for enhancing our knowledge of molecular biology, the genes have been engineered to produce highly stable and bright luminescence at various wavelengths. These systems allow detection of protein–protein interactions and other events that cannot otherwise be readily observed. While the reporter systems are not directly applicable to clinical medicine at this time, further advances in gene therapy will undoubtedly benefit from using them as indicators of successful transfection and therapy. A general limitation of this approach is that gene transfection is a necessary part of the process. The insertion of extra DNA may change a cell's phenotype from that of the parent cells, thereby altering the growth rates, signaling pathways, and other factors. Therefore, these potential changes cannot be ignored when evaluating data acquired with genetically modified organisms. Many review articles have covered the advances in genetically encoded reporter systems (Ray and Gambhir 2007, Villalobos et al. 2007). Below, we briefly summarize the two major types of encoded reporters and techniques that further enhance the value of these molecular imaging tools.

31.2.1.1 Fluorescent Proteins

Most proteins generally reemit light at longer wavelengths upon absorption of photons and are the source of intrinsic fluorescence, also known as autofluorescence. The photophysical properties of fluorescent proteins are similar to organic fluorophores in several ways and can be viewed as macromolecular derivatives of small organic dyes. With the exception of a few proteins, light absorption and emission typically occur in the ultraviolet (UV) region of the electromagnetic spectrum. In this spectral window, the use of relatively high-energy photons could perturb biological systems and induce phototoxicity to cells. In addition, rapid attenuation of UV light precludes deep tissue imaging. For these reasons, proteins with fluorescence in the visible region of light are attractive. A notable example of this class of proteins is GFP, isolated from the jellyfish *Aequorea victoria* (Muller-Taubenberger and Anderson 2007). The structure of GFP has been engineered to produce fluorescence across the visible spectrum, loosely classified by the color range of fluorescence emission: cyan (CFP; 475 nm), green (GFP; 510 nm), yellow (YFP; 530 nm), red (RFP; 580–610 nm) (Shaner et al. 2005), and recent additions in the NIR region, infrared fluorescent protein (IFP; 700 nm) (Shu et al. 2009). Fluorescent proteins may exist as tetramers, dimers, or monomers and vary in brightness, maturation time, and stability (Shaner et al. 2005, Muller-Taubenberger and Anderson 2007). GFP and its derivatives have been incorporated into cells and extensively used in microscopy for studying protein expression and protein–protein interactions (Shaner et al. 2005, Muller-Taubenberger and Anderson 2007, Villalobos et al. 2007). Translation of fluorescent protein studies to live animals has allowed real-time imaging of molecular and physiological processes in natural tissue environments (Amoh et al. 2008). Mouse models have been developed that express fluorescent proteins in specific cells for use in the pharmaceutical industry

and basic research (Amoh et al. 2008). In one example, GFP expression was induced in endothelial cells permitting intravital microscopy of blood vessels and invading red fluorescent protein (RFP)-encoded cancer cells (Amoh et al. 2008). Other examples of transgenic mice developed with cell-specific fluorescent protein expression include cyan fluorescent protein in retinal ganglion cells for studying glaucoma progression (Tosi et al. 2009), identification of interleukin-7 producing cells by enhanced CFP (Mazzucchelli et al. 2009), and visualization of neuronal activation by *c-fos*-linked RFP expression (Fujihara et al. 2009).

Evidently, the prevalent use of fluorescent proteins in biological imaging continues to increase our knowledge of cell and molecular biology. Furthermore, these reporters provide valuable information regarding the response of various diseases to treatment in a longitudinal fashion, decreasing the number of animals needed for these studies (Hutchens and Luker 2007, Amoh et al. 2008, Tosi et al. 2009).

31.2.1.2 Luciferases

Similar to fluorescent proteins, luciferase genes have been isolated from insects (firefly and click beetle) and from Renilla and inserted into cellular genomes for stable expression (Ray and Gambhir 2007, Villalobos et al. 2007). Bioluminescent reporters have been genetically encoded in cells to quantify cell number in vitro and for estimation of tumor burden in vivo (Klerk et al. 2007).

For bioluminescence imaging (BLI), administration of the appropriate exogenous substrate, such as luciferin or coelenterazine, is necessary. The amount of light produced by this process is very small but with almost no natural light coming from other sources, the sensitivity of bioluminescence imaging is very high. Studies have shown that bioluminescence imaging can allow the detection of as few as 400–1000 viable cells beneath the skin of a mouse (Klerk et al. 2007).

Advantages of BLI include the relative simplicity of acquisition instrumentation and the high signal to background. Unlike fluorescent proteins, an external light source is not needed for BLI. Commercially available BLI systems consist of a light-tight chamber with supercooled and/or image intensifier charge-coupled device (CCD) camera. These systems are capable of rapid, even real-time bioluminescence detection. Moderately cooled CCD systems are also capable of bioluminescence detection but these devices require longer acquisition times.

Bioluminescent reporters are generally used in animal models for cell tracking (Helms et al. 2006) and high-throughput screening of interventional effects in cancer and infectious diseases (Doyle et al. 2004, Hutchens and Luker 2007). Currently, firefly (*Photinus pyralis*) luciferase is most commonly used in animal studies due to its rapid and sustained response and brightness relative to other luciferases (Villalobos et al. 2007). Applications in protein–protein interactions similar to FRET studies with fluorescent proteins have been reported (Villalobos et al. 2007). Co-expression of bioluminescent and fluorescent reporters is common in biomedical applications, enabling sensitive detection of molecular processes in living tissues and histologic validation by fluorescence microscopy of tissue sections (Villalobos et al. 2007). Similarly, a combination with genetic reporters detectable with other imaging modalities can enhance the molecular information gained through these systems (Deroose et al. 2007, Kang and Chung 2008, Ray et al. 2008).

31.2.2 Exogenous Molecular Probes and Nanomaterials

31.2.2.1 General Considerations

Although fluorescence from intrinsic proteins and other biomolecules, or autofluorescence signal, is widely used in optical imaging of diseases, exogenous fluorescent molecular probes have been shown to provide superior contrast between diseased and normal tissues in many cases (D'Hallewin et al. 2002). These imaging agents could be from organic or inorganic origin or a hybrid of both systems. Since they are not intrinsic to the organism and require administration of a foreign substance, many hurdles have to be overcome before their use in humans.

Pharmacological considerations are key issues in the development of optical molecular probes for biological imaging. Injected imaging agents must be nontoxic to the species. An interesting approach used by some researchers to overcome this requirement is to use known biocompatible targeting agent and signaling molecules. However, linkage of these units generates a completely new product that requires extensive and expensive safety studies. Because of the anticipated costs of clinical trials for imaging agents, translation of highly innovative imaging agents are confined to small animal imaging studies. When used at tracer level, as is the case with radiopharmaceuticals, most imaging agents do not have significant adverse pharmacological effects. However, at relatively high concentrations similar to those used in CT, MRI, and some optical imaging procedures, normal physiology and signaling pathways may be altered. Even at low concentrations, adverse effects could be observed. For example, a recent report indicates that low-level arginine-glycine-aspartic acid (RGD) peptide derivatives used in biological imaging could promote cancer growth (Reynolds et al. 2009). In addition, the metabolites of molecular probes may become toxic to healthy tissues, requiring detailed safety assessment of the imaging agents. Moreover, some organs are highly sensitive to the presence of exogenous materials and it is important to ascertain the biodistribution of the molecular probes in living organisms. Although this task is straightforward in small animals, the depth limitation of optical imaging methods precludes their use in whole-body human imaging studies. Consequently, it may be necessary to tag these molecular probes with radiotracers for accurate quantitation of the distribution of the probes in humans. This consideration falls within the purview of optical-nuclear multimodal imaging agents that are discussed in Section 31.3.4.

A variety of methods is available for modifying the bioavailability of molecular imaging agents. In all cases, there is an intricate balance between probe circulation time, tissue penetration, clearance, and signal amplification to enhance the contrast between a target tissue and its surrounding background. Enhanced imaging contrast may be achieved by rapid uptake in all tissues, followed by clearance from nontarget surrounding tissue. For many applications, the molecular probe should have a fast and long-lasting target interaction with the biomarker and fast clearance from nontarget tissues such that specific contrast is achieved rapidly and sustained sufficiently for imaging purposes. The primary consideration for activatable probes (see Section 31.3.3) is efficient delivery to the target tissue so that signal enhancement mechanisms can produce high fluorescence in target tissue relative to the circulating inactive fraction. From an imaging point of view, the low background signal in nontarget tissues facilitates the use of either long or short circulating imaging agents as long as sufficient materials are present at the target tissue.

Independent of the nature of optical molecular imaging agents, the first goal is to reach its target. A simple method to assess the viability of an imaging agent is through an initial in vitro evaluation of the selective cellular uptake, target binding affinity, cellular internalization, or fluorescence activation before proceeding to in vivo imaging. Therefore, it is important to consider what barriers exist between probe administration and delivery to the target for imaging probes that are deemed suitable for in vivo imaging. Some of these factors include protein binding in blood, blood–tissue barriers, and location of the molecular target. For example, the NIR dye indocyanine green (ICG) is highly protein-bound immediately after intravenous injection, and therefore shows low penetration into tissues distal from blood vessels. In fact, the clinical utility of ICG for measuring liver function in animals and humans arises from its rapid and complete clearance from the body by the liver. Thus, measurement of ICG clearance can be performed noninvasively by detecting the changes in tissue absorption or fluorescence over time rather than repeated blood collection. On the other hand, a small molecular probe that has less protein binding can quickly escape the vascular system and move into tissues, providing a mechanism to access nonvascular or poorly vascularized targets.

From the above discussion, it is logical to factor molecular weight and hydrodynamic size of the imaging agents into the design of molecular probes. As noted above, labeling of biomolecules with fluorescent dyes results in a new chemical entity with characteristic pharmacological properties. Physically, the effect of labeling also alters the molecular weight, hydrodynamic size, and surface properties of the imaging agent, which are reflected by the biodistribution of the molecule. The magnitude of this effect depends on the relative size of the fluorescent dye and biological carrier. As a simplistic model, the fluorescent reporter has a more significant effect on the biological behavior of molecular probes when the reporter size approaches that of the targeting agent. This is particularly the case when small biomolecules such as peptides are used as carriers. The reverse effect could occur when the size of the reporter system is much larger than the targeting moiety. For example, decorating nanoparticles such as quantum dots (Qdots) with a few copies of small biological carriers such as peptides could alter access of the targeting moiety to the molecular target and skew the biodistribution to that of the unlabeled nanoparticles. Large-sized reporter can be a hindrance when the target is intracellular. Therefore, it is imperative that the choice of targeting moiety and optical reporter be chosen with careful consideration of target location and biological properties.

Conversely, when large biomolecules such as antibodies are used, the size of the fluorescent dyes becomes insignificant, unless multiple copies of the dyes per protein are used. In this case, the inherent properties of the molecular carrier drive the binding and biodistribution of the imaging agent.

31.2.2.2 Synthetic Organic Dyes

Organic dyes have been and remain the mainstay of optical biological imaging agents today. The light-absorbing units for these compounds are formed from a careful arrangement of aromatic groups or unsaturated double bonds in the molecules. This arrangement increases the molecular orbital overlap, which in turn lowers the excited state energy level of the molecule. As the differences between the ground and excited states energy levels decrease, the absorption and fluorescence of these molecules shift from UV to visible (400–650 nm) and near-infrared (700–1000 nm) wavelengths.

A variety of visible dyes are available from commercial sources and they are widely used in fluorescence and multiphoton microscopy. These include fluoresceins, rhodamines, and cyanines (Figure 31.4). Without doubt, visible dyes have facilitated major discoveries in molecular cell biology and gene sequencing (Griffin and Griffin 1993, Nath and Johnson 2000, Pepperkok and Ellenberg 2006). For targeted imaging in microscopy, proteins and other biomolecules are typically tagged with visible dyes. For example, fluorescein-labeled Annexin *V* is used to detect apoptotic cells by in vitro assays and flow cytometry (van Engeland et al. 1998, Brumatti et al. 2008). A similar approach is used for noninvasive imaging of superficial tissues such as the skin or endoscope-accessible organs such as the gastrointestinal system and cervix (Pierce et al. 2008). The ability to visualize these dyes with the naked eye has been used in methylene blue-mediated sentinel lymph node mapping in breast cancer patients (Mathelin et al. 2009).

The heightened interest in translating molecular processes gleaned from cell microscopy to small animal models of human diseases have stimulated research in the use of NIR fluorescent reporters for in vivo optical imaging studies. Imaging in the NIR region improves depth profiling of molecular processes in tissues and minimizes autofluorescence, thereby reducing background fluorescence signal. Currently, ICG is the only NIR dye approved for human use. The dye is stable in serum and

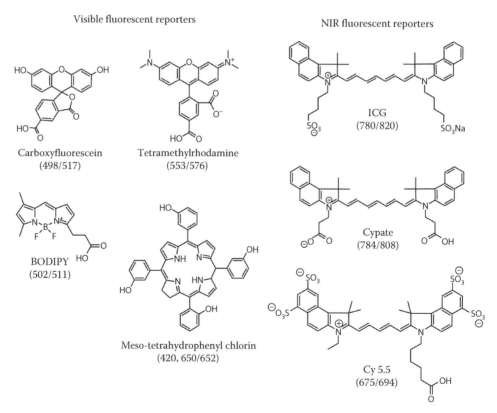

FIGURE 31.4 Representative fluorescent reporter dyes and their wavelengths (nm) of maximum excitation and emission wavelengths ($\lambda_{ex}/\lambda_{em}$) that are used in optical imaging studies and/or as labels for molecular imaging agents. (From Scherer, R.L. et al., *Mol. Imaging*, 7, 118, 2008; Zimmermann, A. et al., *Photochem. Photobiol.*, 74, 611, 2001; Perlitz, C. et al., *J. Fluoresc.*, 15, 443, 2005.)

has sufficient quantum yield for in vivo imaging but it does not have functional groups for conjugation with biological carriers. To overcome this limitation, several functional NIR fluorescent dyes have been developed, many of which are constructed on a carbocyanine platform (Figure 31.4).

In addition to the cyanine dyes shown above, many other organic NIR fluorescent probes have been developed, with improved photostability (Goncalves 2009). For example, chlorins, bacteriochlorins, napthalocyanines, and porphyrins targeted to specific disease biomarkers have been developed as imaging and therapeutic agents (Mew et al. 1983, Hasan et al. 1989). In some of these molecular probes, increased lipophilicity enhances nonspecific binding of the imaging agents, necessitating the synthesis of hydrophilic derivatives (Verdree et al. 2007). Sequestering part of the chromophore system may increase the photophysical properties of some dyes. This is typified by squaraine rotaxanes, which are new bright NIR dyes that have been developed for optical imaging applications (Johnson et al. 2007).

31.2.2.3 Inorganic Optical Contrast Agents

Inorganic materials have always played a major role in biological imaging. Examples include the use of radioactive metals for nuclear imaging and gadolinium chelates for MRI. Although some of these metal chelates are luminescent, their quantum efficiency is generally poor. However, recent advances in nanotechnology have led to the development of highly structured nanomaterials made of inorganic elements for use in disease diagnosis and treatment. For optical imaging applications, nanomaterials with high luminescence or absorption properties are highly desirable. Notable among these are Qdots and related semiconductor nanoparticles that have bright fluorescence characteristics useful for in vitro assays and in vivo imaging (Bentolila et al. 2009). Despite the numerous applications of Qdots in biological imaging, the perception of heavy metal toxicity lingers and could affect their use in humans at this time (Hauck et al. 2009). To alleviate this concern, development of materials with new surface coating consisting of biocompatible peptides and polymers (Schipper et al. 2009) and new luminescent materials using less toxic metals is in progress (Li et al. 2007).

31.3 Functional Classifications of Molecular Imaging Agents

31.3.1 Nonspecific Molecular Imaging Agents

Nonspecific imaging agents passively accumulate in diseased tissues without a specific targeting moiety. The extent of selective uptake in tissue determines their usefulness in biological imaging. In most cases, imaging contrast is produced by the rapid washout of the molecular probe from normal tissue or passive accumulation in target tissue. For example, protoporphyrin IX overproduction by cancer cells after administration of aminolevulinic acid (ALA) or hexaminolevulinate (HAL)

can improve detection and surgical resection of bladder cancer (Zaak et al. 2002, Witjes and Douglass 2007, Ray et al. 2009) and glioblastoma (Stepp et al. 2007, Eljamel et al. 2008, Hefti et al. 2008). Similarly, porphyrins and other fluorescent agents were observed to selectively stain diseased tissues such as cancer (Zimmermann et al. 2001, D'Hallewin et al. 2002). The tumor-specific uptake of these agents has been visualized via fluorescence endoscopy in gastrointestinal disease and bladder cancers (Moghissi et al. 2008), and by modified surgical microscopes for fluorescence-guided resection of glioblastoma (Zimmermann et al. 2001).

ICG is a typical example of a widely used NIR fluorescent dye for cancer detection by optical imaging via nonspecific accumulation in cancer tissues. The selective accumulation of this imaging probe is due to higher blood volume and the EPR effect of cancer tissue (Corlu et al. 2007, Alacam et al. 2008, Hagen et al. 2009). The EPR effect results from the highly tortuous and leaky vasculature present in many tumors and is responsible for most passive accumulation of drugs, molecular probes, and nanoparticles (Greish 2007). ICG is also known to be highly protein-bound after intravenous injection. This binding shortens the blood residence time of the compound, leading to rapid excretion by the liver. Noninvasive absorption spectroscopy has been used in humans after intravenous injection of ICG to assess liver function equivalent to serial blood samples (Hsieh et al. 2004). To overcome this problem, hydrophilic sugar-modified carbocyanine dyes were developed, which extended the circulation time and improved the renal clearance due to lower protein binding, as measured by noninvasive fluorescence spectroscopy (Licha et al. 2000). Many other low protein-binding ICG derivatives have been developed and widely used in optical imaging studies (Licha et al. 2000).

Nonspecific cyanine-based dyes have also been used in clinical trials of optical imaging of breast cancer based on the enhanced blood flow and vascular permeability within tumors. In these studies, the absorption (Alacam et al. 2008) and fluorescence (Corlu et al. 2007, van de Ven et al. 2009) properties of cyanine dyes were used to provide imaging contrast by diffuse optical tomography (DOT) imaging of human breasts. As with many nonspecific imaging agents using other imaging methods, it is important to develop robust pharmacokinetic models of ICG for quantitative imaging. Fortunately, these parameters are easily determined by noninvasive optical imaging, as demonstrated in animals (Gurfinkel et al. 2000) and in humans (Alacam et al. 2008). Due to the promising use of ICG in humans and preclinical subjects, many of the NIR fluorescent tags used for labeling targeting agents are based on the carbocyanine framework.

31.3.2 Affinity-Based Molecular Probes

Passive targeting is generally subject to low sensitivity and specificity because of low imaging contrast between disease and normal surrounding tissue. Therefore, other strategies have been developed to improve disease-specific imaging contrast. These methods include active targeting using targeted molecular probes with high affinity for disease-related targets and hybrid contrast enhancement that may combine passive delivery of the imaging agent with selective activation of the imaging signal at target tissue. A variety of approaches have been reported to achieve specific targeting or amplification of imaging signals.

31.3.2.1 Molecular Imaging Probe Strategies

31.3.2.1.1 Intrinsic Affinity Imaging Strategy

Some fluorescent molecular probes have natural affinity for biomarkers of diseases due to constituent chemical or structural factors. A good example is fluorescent molecular probes for imaging brain diseases and function. In general, optical molecular probes do not cross the blood–brain barrier but those that permeate the brain typically persist in brain tissue nonspecifically (Raymond et al. 2008). However, recent studies have shown that the properties of some fluorescent molecular probes increase their affinity to specific disease biomarkers. For example, several fluorescent compounds have been shown to have high affinity for amyloid beta plaque deposits in brain tissue of Alzheimer's disease (AD) patients and some of them have been radiolabeled for in vivo nuclear imaging (Okamura et al. 2004, Fodero-Tavoletti et al. 2009) and multiphoton microscopy (Klunk et al. 2002). These agents also demonstrate the potential for diagnosis of prion diseases such as Creutzfeldt–Jakob disease due to similar plaque formation (Sadowski et al. 2004). More recently, a NIR fluorescent oxazine dye was shown to be effective in detecting AD in transgenic mice using noninvasive planar (Hintersteiner et al. 2005) and tomographic (Hyde et al. 2009) optical imaging methods. Furthermore, fluorescent probes that change spectrally in response to *ex vivo* binding to amyloid beta plaques have been reported (Raymond et al. 2008). This contrast mechanism holds great promise for the use of intrinsic properties of optical molecular probes in molecular imaging of human diseases.

31.3.2.1.2 Multivalent and Polyvalent Imaging Strategy

Multivalency is the expression of multiple copies of a molecule, such as receptor or receptor ligand tethered on a central core to increase the probability of binding and/or binding affinity (Figure 31.5). Many natural ligands and targeting agents have multiple copies of binding moieties, resulting in higher affinity. These include antibodies, protein ligands, and virus particles (Mammen et al. 1998). Thus, multivalency increases binding avidity over monovalent equivalents, improving contrast (Ye et al. 2004, Rosca et al. 2007). For this reason, many researchers are developing synthetic multivalent imaging agents using small molecules such as peptides. An example is the development of multimeric cyclic RGD peptide-based molecular probes that showed higher uptake in tumors than their monomeric counterparts (Thumshirn et al. 2003). RGD multimers labeled with a NIR dye have been used for imaging $\alpha_v\beta_3$ integrin expression in glioblastoma xenografts. The results showed that the tetrameric RGD construct gave better tumor contrast than monomers and dimers (Wu et al. 2006). Using simple linear RGD peptides, significant improvement in the binding affinity and contrast was reported, demonstrating that biomolecules with low binding affinity would benefit more from multivalency than would naturally high affinity ligands (Ye et al. 2006). Similarly, NIR fluorescent dyes with multiple glucosamine groups showed excellent tumor to normal tissue contrast in vivo, although the tumor uptake did not necessarily correlate with the number of gluocosamine copies (Ye et al. 2005). Multivalency is a common

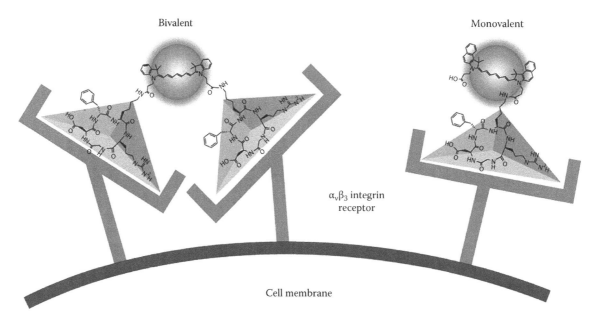

FIGURE 31.5 Illustration of multivalent strategy to improve targeted molecular probe binding affinity by increasing the number of ligands per molecule. In this example, bivalent cypate-[cyclic RGD]$_2$ can interact with multiple $\alpha_v\beta_3$ integrin receptors to enhance contrast relative to the monovalent cypate-cyclic RGD.

attribute of nanoparticles used as contrast agents since many copies of the targeting agent are required to increase the probability of binding. It has been shown that this multivalent effect may further enhance binding due to the separation between ligands, mimicking that of natural ligands such as immunoglobulins and virus particles.

Another strategy to improve biological targeting of molecular probes and enhance imaging specificity is to incorporate ligands for two or more molecular targets into a unifying core. Polyvalency is being exploited in the design of pharmaceuticals (Vance et al. 2008) and contrast agents (Ye et al. 2005). For example, a nuclear imaging with small peptide ligands for both $\alpha_v\beta_3$-integrin receptor and gastrin-releasing peptide receptor has demonstrated receptor-specific tumor uptake in xenograft models expressing both receptor systems (Liu et al. 2009a,b).

31.3.2.1.3 Agonist and Antagonist Molecular Strategy

While antibodies may bind with or without activating cellular signaling cascades related to the target, some ligand mimics bind specifically to the natural binding site and either inhibit or activate the cascade. Traditionally, molecular imaging agents have been selected by target binding affinity determined in vitro before proceeding to animal studies. The current dogma of using agonists for molecular imaging predicates on the assumption that internalization and intracellular retention of the molecular probe enhances imaging signal in cells expressing large amounts of the biomarker. It follows then that longevity of the molecular probe in the target tissue depends on receptor turnover rate and recycling time. Studies have shown that a somatostatin receptor agonist causes rapid (<2.5 min) internalization of the receptor in cells but the antagonist did not have a similar effect (Waser et al. 2009). Under this regime, antagonists would not generate strong imaging signal because of their typical inability to internalize in cells. While few direct comparisons have been made at this time, the paradigm of internalization has been challenged recently (Ginj et al. 2006, Edwards et al. 2008). Radiolabeled antagonist-based imaging agents have produced excellent tumor contrast in animal models, sometimes greater than that of traditional agonist molecules (Ginj et al. 2006, Han et al. 2007 Akgun et al. 2009). While most of this research has been investigated with radiolabeled imaging agents, optical molecular imaging agent development can benefit from this information.

The use of antagonists in molecular imaging is not limited to somatostatin receptor ligands. Improved tumor uptake of antagonist molecular probes relative to analogous agonist molecular probes has been demonstrated in tissues expressing cholecystokinin (Akgun et al. 2009), A3 adenosine (Kiesewetter et al. 2009), endothelin subtype-A (Mathews et al. 2008), urokinase plasminogen (Li et al. 2008c), and dopamine (Seneca et al. 2008) receptors. Therefore, a choice between agonist and antagonist imaging agent is important and depends on the intended application.

31.3.2.2 Molecular Imaging Agents

Early photodiagnostic work followed the "vital stain" methodology developed for histopathology in which certain compounds were found to be selective for abnormal tissues. Through advances in molecular biology, diverse molecular biomarkers that distinguish diseased cells and tissues from healthy ones have been discovered and have become the target for molecular imaging and targeted therapy. A variety of targeting molecules are currently used to deliver the reporter dyes to the tissue of interest. These biologically active molecules bind the biomarker with high affinity and the location of the target tissue could be visualized by optical imaging methods. Representative fluorescent dye-labeled targeting molecules and the variety of sizes available for optical imaging are shown in Figure 31.6. Depending on the intended usage, the targeting molecules are either labeled with visible or near-infrared fluorescent dyes or light-emitting nanomaterials. Commercial sources for these dyes and nanomaterials are available.

31.3.2.2.1 Monoclonal Antibodies

While antibodies have high cost of production, relatively large molecular weight, and the potential for elicit immune reactions, they continue to be popular in targeted therapy and molecular imaging research. Antibodies are relatively easy to raise against any antigen and are highly specific. A broad spectrum of monoclonal and polyclonal antibodies is available commercially and can be labeled with imaging tracers using commercially available kits.

Vascular endothelial growth factor receptor (VEGFR) is upregulated on the surface of nascent endothelial cells and is therefore a good target for therapy and detection of angiogenesis. The monoclonal antibody (mAb) to VEGF, Avastin® (Genentech, CA) is clinically approved for antiangiogenic therapy. Recently, Withrow et al. showed that labeling Avastin with a NIR fluorophore allowed detection of angiogenesis in animal models of head and neck cancer via intravital microscopy (Withrow et al. 2008). Another study used planar reflectance imaging to visualize Avastin labeled with Alexa Fluor 680 to measure angiogenesis after photodynamic therapy (Chang et al. 2008). Similarly, a mAb to VEGF type-2 receptor (VEGFR2) labeled with NIR dye was used to demonstrate decreased VEGF receptor expression in response to anti-VEGFR therapy (Virostko et al. 2009).

Labeling of antibodies with fluorescent dyes for imaging is not always a simple task and must be optimized for each case. Conjugation of fluorescent dyes to antibodies and proteins generally requires reactive derivatives of the dyes such as *N*-hydroxysulfosuccinimide (NHS) ester (Southwick et al. 1990, Ito et al. 1995) or isothiocyanates that react with amino groups (Southwick et al. 1990, Williams et al. 1993). Optimization of protein labeling is important to provide maximum signal while minimizing adverse effects on binding affinity and pharmacokinetics. Multiple labeling of antibodies is known to reduce antigen binding affinity, particularly if the labeling agent is in close proximity to the antibody binding domain (Tadatsu et al. 2006). Therefore, the labeling process must be tailored to the individual antibody to accommodate the wide distribution of reactive (amino or thiol) groups in the biomolecule (Schellenberger et al. 2004, Tadatsu et al. 2006, Qian et al. 2009).

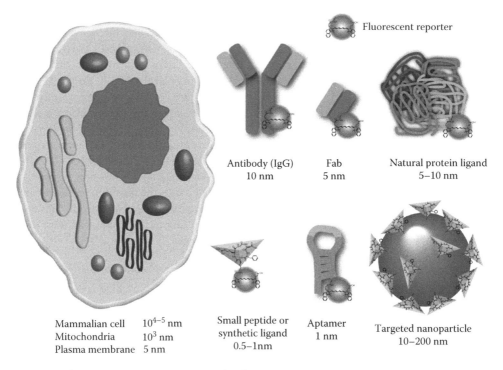

FIGURE 31.6 (See color insert.)   Representative fluorescent molecular imaging agents demonstrating the variety of targeting agents used in optical molecular imaging. The approximate sizes of a mammalian cell and organelles are given for comparison (right).

While kits are commercially available for antibody and protein labeling, success is dependent on many factors and may be difficult even for experienced researchers. This is because some of the dye molecules bind to antibodies and proteins non-covalently and could result in a heterogeneous distribution of unlabeled, single, multiple and non-covalently labeled proteins, which are reported as the average number or ratio of fluorophores per protein. If non-covalent labeling is predominant, the dye could be easily detached in vivo. This detachment could lead to misinterpretation of the data since the biodistribution of the dye and antibody may differ from each other. Unfortunately, there are no simple and unambiguous methods to distinguish covalent from non-covalent labeling of antibodies and proteins at this time.

The long circulating time of antibodies is not generally favorable in affinity-based optical imaging because of persistent background signal that takes several hours or days to clear. To overcome this obstacle and improve the pharmacokinetics of the molecular probes, smaller fragments of antibodies have been developed (Wu and Olafsen 2008).

Antibody fragments that retain target specificity and binding affinity are also candidates for molecular probe development (Wu and Olafsen 2008). For example, antibody fragments targeting carcinoembryonic antigen (CEA) (Arcitumomab) were labeled with a NIR dye for cancer detection in mice using a custom-built planar reflectance fluorescence imaging system (Lisy et al. 2008). Another fluorescent single-chain antibody (scFab), NovoMab-G2-scFv-Cy5.5, also demonstrated tumor-specific uptake in a shorter time than labeled antibodies (Ramjiawan

et al. 2000). Unfortunately, these fragments do not usually retain the stability and in vivo functionality of the parent antibodies (Colcher et al. 1998). Methods that improve in vivo stability, such as construction into nanoparticles, will improve their utility in biological imaging. Along this line, Hwang et al. decorated GMC-5-193-containing nanoparticles with anti-transferrin receptor scFabs and showed delivery in human cancer cell lines by GMC-5-193 fluorescence imaging (Hammer et al. 2002). Interesting newer antibody designs include "trimerbodies," which are constructed by attaching three single-chain antibody fragments to collagen domains for imaging (Cuesta et al. 2009). If these antibody derivatives could retain the stability and high specificity of the parent antibodies, they could supplant antibody-based molecular imaging in the future.

In addition to using smaller antibody fragments, various imaging strategies have been developed. One of them is the pre-targeting technique. Pre-targeting is a technique of administering the targeting molecule to achieve optimal specific uptake in target tissue and blood clearance prior to injecting the imaging reporter (Sharkey and Goldenberg 2006). For example, Hama et al. pre-targeted epidermal growth factor receptor (EGFR) in a peritoneal metastasis model using biotinylated monoclonal antibody against this receptor, followed with neutravidin-BODIPY-FL for fluorescence imaging (Hama et al. 2007).

31.3.2.2.2 *Protein Ligands*

Another targeted imaging strategy is to label the natural protein ligand of the receptor of interest with a fluorescent dye. Vascular endothelial growth factor (VEGF) is a dimeric protein that promotes angiogenesis through binding to its specific receptor, VEGFR. VEGF protein was conjugated to Qdots along with the metal chelator, 1,4,7,10-tetraazacyclododecane-1,4,7,10-tetraacetic acid (DOTA) for Cu-64 for PET imaging of VEGFR expression in glioblastoma xenografts (Chen et al. 2008). A single-chain VEGF with a cysteine tag was labeled with a NIR fluorescent for detection of angiogenesis in tumors in mice (Backer et al. 2007, Wang et al. 2009).

Other natural protein receptor ligands have been labeled with fluorescent tags for in vitro and in vivo imaging of receptor expression. These include transferring (Becker et al. 2000b), vasoactive intestinal peptide (Becker et al. 2000a), and epidermal growth factor (EGF) (Ke et al. 2003, Kovar et al. 2006), which have high affinity for their respective receptors that are highly expressed in many cancer tissues. For example, a Cy5.5-labeled EGF molecular probe selectively accumulated in EGF receptor-positive tumor xenografts with little uptake in EGF receptor-negative xenografts in mice (Ke et al. 2003). Similarly, a longer wavelength molecular probe, EGF-IRDye800CW, was used to image prostate cancer in small animals (Kovar et al. 2006).

Protein ligands have also been successfully used to report the functional status of intact tissue. A typical example is Annexin V (AnxV), a 36 kDa protein that binds trivalently to phosphatidylserine (PS). PS is a phospholipid component of cell membranes that is actively retained on the inner membrane but appears on the outer surface shortly after initiation of cell death (Ntziachristos et al. 2004). Radiolabeled AnxV has been the focus of clinical trials due to successful targeting of tumor tissues undergoing apoptosis or other forms of cell death. AnxV labeled with fluorescent reporters has been reported for in vitro and in vivo detection of apoptosis, and it is therefore useful for monitoring treatment response (Petrovsky et al. 2003, Schellenberger et al. 2003, 2004).

31.3.2.2.3 Small Peptide and Peptidomimetic Receptor Ligand Mimics

Despite advances in the engineering of antibodies and natural proteins, they remain relatively large and prone to allergic reactions. Development of small molecular probes can greatly enhance control over the biological activity of the imaging agents. Cell surface receptors are important targets for molecular imaging because of the ease of reaching the target relative to intracellular biomarkers. The peptide ligands for targeting these receptors are generally developed based on advances in protein sequencing and high-throughput screening for high affinity ligands (Aina et al. 2005, 2007).

These small molecules can provide better pharmacokinetic properties relative to larger proteins and antibodies. Natural and unnatural amino acids, synthetic organic molecules, peptide mimetics, and oligosaccharides, are examples of small molecules that have been used to develop optical molecular probes. Automated solid-phase peptide synthesis has revolutionized the development of peptide libraries and production of small molecular probes. Labeling of the small peptides with fluorescent dyes may be carried out on solid support or in solution phase reaction

(Goncalves 2009). Typically, high yield of homogenous product can be attained relative to protein labeling. Representative examples of the many peptide-based imaging agents include small peptide ligands of somatostatin (Becker et al. 2000a, 2001, Achilefu et al. 2002, Kostenich et al. 2005, 2008), integrin (Chen et al. 2004, Achilefu et al. 2005, Cheng et al. 2005, Kwon et al. 2005, Ye et al. 2006, Jin et al. 2007, Smith et al. 2008, Mulder et al. 2009), and bombesin (Bugaj et al. 2001, Achilefu et al. 2002, Ma et al. 2007) receptors, which are highly expressed in by many diseased tissues such as tumors.

A first demonstration using NIR dye-labeled small peptides targeted to a cell surface somatostatin receptor for in vivo whole-body optical imaging of tumors in small animals was reported about a decade ago (Achilefu et al. 2000). Somatostatin (sst-14) is a natural small peptide ligand of the somatostatin receptor family. This peptide was radiolabeled for nuclear imaging but the rapid metabolic degradation after injection mitigated its use as an effective imaging agent (Harris 1994). Shorter peptide analogues such as octreotide and octreotate have since been developed and labeled with fluorescent dyes for tumor detection in animals (Achilefu et al. 2000, Becker et al. 2000a, Kostenich et al. 2008).

Another widely targeted molecular system is the integrin family of cell-surface receptors. Particularly, $\alpha_v\beta_3$-integrin receptor and other integrin heterodimers are overexpressed on the surface of nascent blood vessels and some tumor cells. The RGD amino acid sequence has high affinity for $\alpha_v\beta_3$ and other integrin receptors. This sequence is recognized at similar binding sites as fibronectin, fibrinogen, vitronectin, von Willebrand factor, and osteopontin (Ruoslahti and Pierschbacher 1986, Heckmann and Kessler 2007). RGD-containing peptide sequences have highest affinity for the vitronectin receptor (Ruoslahti and Pierschbacher 1986), also known as $\alpha_v\beta_3$-integrin receptor. Cyclization of the linear RGD peptide improved the $\alpha_v\beta_3$-integrin receptor binding affinity (Aumailley et al. 1991) and reduced metabolic degradation of the peptide in vivo. This peptide sequence has since been the subject of numerous in vitro and in vivo studies of cancer, cardiovascular, and other diseases (Achilefu et al. 2005, Cai et al. 2008) Synthetic peptidomimetics have also shown selective accumulation in $\alpha_v\beta_3$-rich atherosclerotic plaque (Heroux et al. 2009) and cancer tissues (Burnett et al. 2005).

A complicating factor for small peptide ligands is that the fluorescent reporter may be equivalent in size or larger than the targeting small molecule. This situation could significantly alter the binding and pharmacokinetic properties of the molecular probe. Thus, for a small molecular probe, the contributions of both the reporter dye and delivery vehicle to effective molecular imaging contrast are important and must be considered in its development for in vivo imaging.

31.3.2.2.4 Non-Peptide Ligands

Folic acid is a non-peptide ligand in which uptake in cells is receptor mediated. The folate receptor is overexpressed in many cancer types (Sega and Low 2008). The binding of folic acid to folate receptor causes receptor internalization. This internalization occurs with a variety of imaging agents, including folic acid labeled with nuclear and optical imaging agents as well as nanoparticles (Reddy et al. 2005, Sega and Low 2008).

Another group of non-peptide ligands is glucose derivatives. The glucose analogue, radiolabeled fluorodexoxy glucose (^{18}FDG), which is taken up in cells by the GLUT1 transporter and retained after phosphorylation within cells, is successfully used in clinical cancer detection via PET (Pauwels et al. 2000). Similarly, the analogous fluorescent glucose, 2-(*N*-(7-nitrobenz-2-oxa-1,3-diazol-4-yl)amino)-2-deoxyglucose (2-NBDG), is presumably delivered in cells by glucose transporters (Roman et al. 2001). In a recent report, topical application of 2-NBDG showed a threefold increase in fluorescence intensity in neoplastic tissues relative to normal tissue using fluorescence microendoscopy (Nitin et al. 2009). NIR fluorescent molecular probes have also been developed for in vivo assessment of metabolism including cypate-mono-2-deoxy-glucose (Chen et al. 2003, Barbosa et al. 2004) and cypate-labeled glucosamine moieties with high uptake in tumor xenografts (Ye et al. 2005, Li et al. 2006a). Similarly, Pyro-2DG as imaging and PDT agent demonstrated efficient tumor destruction (Zhang et al. 2003). However, Cheng et al. reported that the uptake of glucose-based NIR probe was equivalent to NHS-functionalized control in tumor xenografts and suggested that the observed retention in tumor may not reflect metabolism (Cheng et al. 2006). Thus, it is still unclear as to whether uptake of these probes reflects metabolic activity and efforts to elucidate the mechanism of uptake is still needed.

31.3.2.2.5 Aptamers: DNA and RNA-Based Agents

Aptamers are short sequences of DNA or RNA that may or may not occur naturally and have high affinity for specific targets (Perkins and Missailidis 2007). These molecules are between single-chain antibodies and small peptide ligands in size. Systematic evolution of ligands by exponential enrichment (SELEX) is the method for identification of new aptamers, similar to phage display for peptides (Cerchia et al. 2009). Radiolabeled aptamers have been reported and one aptamer is approved for therapy in humans (Pegaptanib sodium, Pfizer, and Eyetech). Fluorescent-labeled aptamer probes may be targeted to extracellular receptors, resulting in rapid internalization (Li et al. 2008a). Fluorescent small molecule and nanoparticle aptamer probes have demonstrated in vitro detection of prostate-specific membrane antigen (PSMA) (Bagalkot et al. 2007) and tenascin-C (Hicke et al. 2006, Huang et al. 2008) expressed on the cell surface. FRET-based "Aptamer Switch Probes" for detection of angiogenin, a marker of angiogenesis, have been developed to measure target binding by fluorescence changes upon binding (Li et al. 2008b, Sheth et al. 2009) including a NIR fluorescent probe designed for in vivo imaging (Zhang et al. 2008). Recently, a quencher-labeled short DNA sequence complementary to the fluorophore-labeled aptamer has been developed (Sheth et al. 2009). Initially, this construct has very low fluorescence but in the presence of the target, ATP or human alpha-thrombin, the cDNA sequence is separated from the aptamer and fluorescence

is restored. Despite promising results in vitro, optical imaging with DNA and RNA constructs has not been very successful to date. The group led by Dr. Hnatowich has demonstrated tumor-specific fluorescence enhancement using quenched antisense DNA probe by planar reflectance fluorescence imaging in living mice, although nonspecific enhancement in other organs also occurred (Liang et al. 2009). Finally, aptamer binding can enhance fluorescence of some molecules (Constantin et al. 2008), a mechanism that has been proposed for genetically encodable units for fluorescent detection of specific RNAs (Babendure et al. 2003, Stojanovic and Kolpashchikov 2004). While aptamers are interesting molecular imaging agents, significant research is needed for adequate comparison with more established targeting agents.

31.3.3 Hybrid Molecular Imaging Agents

A major advantage of optical imaging over other modalities is the nature of fluorescence. Fluorescence from endogenous and exogenous reporters can be sensitive to physiological and molecular processes such as pH, protein binding, lipophilicity, and enzymatic cleavage of specific substrates contained in the molecular probe. Molecular fluorescent reporters for these applications produce significant changes in their fluorescence properties after interaction with the targeted tissue. These changes include alterations of fluorescence intensity, spectrum, and lifetime, and regional variation of these factors may exist due to physiologic differences between diseased and normal tissues. Therefore, optical imaging has the capability to not only detect the presence of molecular signatures but to also report molecular functional status within the tissue of interest.

One mechanism for detecting molecular function by optical imaging is the use of stealth or activatable probes utilizing FRET or other quenching mechanisms. The molecular probes are administered in a quenched or relatively nonfluorescent state and subsequent increase in the fluorescence yield when acted upon by specific biomolecules or medium. A classical example is the use of activatable probes to monitor the functional status of diagnostic proteases. These hydrolytic enzymes cleave specific amide bonds in proteins and peptides. Thus, enzyme-cleavable peptide sequences are incorporated into the molecular design, and dyes are arranged such that the cleavage process dequenches the fluorescence (Figure 31.7). Detection of the emitted light is then used to report the presence and functional status of a targeted protease.

The first protease-activatable probes for in vivo cancer imaging were reported by Weissleder et al. (1999). Capitalizing on the upregulated expression of specific enzymes in cancer tissues, the group designed activatable molecular probes to image the expression and activity of matrix metalloproteinases (MMPs). Various MMP enzymes are upregulated in aggressive cancers and associated with growth of new blood vessels and metastasis (Overall and Kleifeld 2006, Voorzanger-Rousselot et al. 2006).

A variety of approaches have been used to quench fluorescence of dyes for imaging enzyme activation. In one configuration,

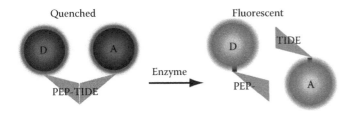

FIGURE 31.7 Conceptualized protease-activatable molecular probe. A small peptide is labeled on both ends with donor (D) and acceptor (A) fluorescent reporters held in close proximity resulting in quenching of fluorescence. Cleavage of the peptide by the target enzyme results in separation of the labeled fragments and restoration of fluorescence.

fluorescence quenching is achieved by stacking several copies of the fluorescent dye on a polymeric support (homoFRET). The close proximity of the fluorophores ensures that the energy absorbed by a fluorophore is transferred nonradiatively to the adjacent fluorophore, significantly reducing the probability of photon emission. HomoFRET provides high fluorescence amplification because cleavage of multiple fluorophores generates fluorescence from several dye molecules. The polymeric activatable probes also have larger molecular weight and concomitant longer plasma half-life than smaller molecules. This feature facilitates the nonspecific deposition of the probe in target tissue over extended period, enhancing the chances for the activation of multiple copies of the probe. This approach has been used to prepare activatable probes for imaging cathepsin (Weissleder et al. 1999) and MMP (Bremer et al. 2001a,b) activity in tumors with high levels of the targeted enzyme activity. Commercially available activatable probes are based on this structural template. While the detection sensitivity for the activatable probes is high, activation by nonspecific enzymes within the blood and healthy tissues may occur. In addition, complete quenching is also difficult to maintain in these molecular constructs. These difficulties must be overcome for developing highly sensitive and specific activatable molecular probes.

Another fluorescence quenching strategy for activatable probes is to combine one or more fluorescent molecules with an appropriate quencher. Recently, NIR quenchers have been developed that efficiently quench NIR fluorophores when placed in close proximity to the fluorescent moiety and have been used to develop activatable molecular probes (Pham et al. 2002, 2004, Stefflova et al. 2006, Bullok et al. 2007, Kozloff et al. 2009, Lee et al. 2009, Maxwell et al. 2009). A similar approach using donor–acceptor dyes in the conventional FRET mechanism has also been used to prepare activatable probes. Although only one fluorophore is released upon activation, quenching within the intact system is generally more efficient.

Most of the activatable probes are delivered to target tissue by nonspecific mechanisms, followed by enzyme activation. Combining activatable probes with molecular targeting moieties can result in enhanced contrast through specific delivery and activation. For example, a quenched NIR probe that included cell penetrating peptide has been developed that reports apoptosis in vivo by detecting intracellular activated caspase-3 enzyme

(Maxwell et al. 2009). Antibody-based targeted activatable molecular probes have also demonstrated tumor-to-normal fluorescence contrast (Ogawa et al. 2009a,b). Beyond enzyme activation, other types of optical imaging agents have been reported that are sensitive to environment polarity for detecting apoptosis after binding of AnxV to the cell membrane (Kim et al. 2009b).

Fluorescence intensity enhancement is the primary reporting strategy for activatable probes. However, errors arising from nonspecific activation or incomplete quenching complicate accurate quantification of fluorescence intensity measurements of activatable probes. One strategy to reduce these possible sources of errors is to include a reference tracer, detectable in a different channel after simultaneous administration (Baeten et al. 2009) or included within the activatable probe (Scherer et al. 2008, McIntyre and Matrisian 2009). In one study, a blood-pool agent was used as a reference control to correct for differences in tumor blood volume by co-administering it with protease-activatable probes. This combined agents accurately assessed the protease activity during tumor development in a murine spontaneous breast cancer model (Baeten et al. 2009). Importantly, the two contrast agents were similar in physical properties to minimize differences in pharmacokinetic properties that might have adversely affected the measurements (Baeten et al. 2009). Another study used a spectrally distinct, unquenched fluorophore as an internal standard within a polymeric activatable probe. Using ratiometric measurements, the authors were able to extract tumor-specific fluorescence activation (Scherer et al. 2008).

31.3.4 Multimodal Molecular Imaging

No single imaging modality currently available can provide complete diagnostic information. Therefore, the combination of different imaging modalities can provide additive or synergistic diagnostic information. For example, combination of PET or SPECT with CT, which all utilizes ionizing radiation for imaging purposes, enables accurate co-registration of molecular and anatomical information (Townsend 2008). In addition, CT data can be used to improve quantitation accuracy of PET and SPECT by correcting for signal attenuation in tissues (Patton and Turkington 2008, Wong et al. 2009). Similarly, the combination of optical imaging with other imaging modalities can improve the reconstruction of optical data.

One strength of optical imaging agents is the duration of their fluorescence signals. In contrast to the short-lived radioactivity of radiopharmaceuticals used in nuclear imaging, optical contrast agents, particularly endogenous and genetic reporters can be assessed for the life of the animal. Many exogenous optical molecular probes may also be detected for several weeks after administration. In addition, these agents can be safely administered multiple times. It is therefore logical that the combination of optical molecular imaging with other modalities promises synergistic improvements in biomedical imaging far beyond current techniques.

31.3.4.1 Multimodality Imaging Technologies and Methods

The first challenge to multimodal imaging is the combination of data acquired by each modality. In some cases, it is sufficient to visually compare the images to assess the location of signals when landmarks are evident. In other cases, precise location of signals is important, such as when searching for metastases using nuclear or optical imaging combined with CT or MRI. Co-registration of imaging data can be accomplished by image processing techniques using natural or fiducial landmarks to fit the different sets of data. Software co-registration is very useful when simultaneous acquisition is not possible. The most successful use of software co-registration has been in neuroimaging of the brain, where the skull provides rigid anatomical landmarks and brain tissues generally do not change significantly when repositioned. Disadvantages of this approach arise from the nonuniform movement of tissues that occurs with repositioning of some organs. For example, the abdominal contents can shift significantly in relation to each other and in overall shape when a patient is moved, making accurate co-registration difficult. This problem can be minimized by using simultaneous or sequential acquisition with minimal or no repositioning. In general, multimodality imaging is currently performed sequentially in different instruments with minimal change in animal position or post-imaging registration of data (Cherry 2006).

Strategies for combining optical imaging with MRI have been described, including combined bioluminescence tomography/MRI (Allard et al. 2007) and DOT/MRI (Ntziachristos et al. 2000). Another study used MRI-guided optical spectroscopy to adequately characterize human breast tumors (Ntziachristos et al. 2000). In addition, MRI-derived data have been used to constrain DOT reconstructions (Schweiger and Arridge 1999, Bamett et al. 2003, Guven et al. 2005, Yalavarthy et al. 2007). Similarly, contrast agents for both MRI (Gadolinium-DTPA) and DOT (ICG) have been used to differentiate viable and non-viable tumor tissues (Unlu et al. 2008). Other multimodality systems are under development to combine optical imaging with PET or SPECT (Alexandrakis et al. 2006, Li et al. 2009).

31.3.4.2 Multimodal Molecular Imaging Agents

The combination of multiple imaging reporters into a single molecule for multimodal molecular imaging can enhance the efficacy of detection beyond co-administration of two separate agents with disparate detection sensitivity. This approach also eliminates co-registration complications from differences in the distribution of the individual imaging agents. Additional synergism evolves from the quantitative capabilities of nuclear and MR imaging with the multifaceted capabilities of optical contrast agents described in the previous sections. Application of this strategy in gene-encoded luminescent reporters is widely used in small animal imaging studies. In particular, the expression of multiple reporter systems in single genetic constructs have been reported for multimodal cell tracking and tissue detection by bioluminescence, fluorescence, and nuclear imaging (Ponomarev et al. 2004, Ray et al. 2007, 2008).

For exogenous contrast agents, monomolecular multimodality imaging agents (MOMIAs) or polyvalent nanoparticles are commonly used. The MOMIA approach relies on the incorporation of different reporter moieties into a multifunctional molecule. In targeted multimodal imaging systems, the multifunctional molecule is typically a targeting protein such as monoclonal antibody or receptor-avid peptides. In both cases, the biological carrier is labeled with a fluorescent dye and a metal chelating group for MRI or nuclear imaging. It should be noted that optical-nuclear MOMIAs can also use radioactive halogens such as ^{18}F for PET or ^{125}I for SPECT, which do not require incorporation of chelating groups. Although recent reports demonstrate the feasibility of MRI-based imaging of molecular processes, many MOMIAs reported so far have focused on optical-nuclear multimodal imaging agents. This is partly due to the similarity in the detection sensitivities of both imaging methods.

The comparable detection sensitivity of both methods minimize the need to accommodate disparate reporting strategies, which would normally require selective amplification of reporters for imaging methods with relatively poor detection of contrast agents such as CT or MRI. Different optical-nuclear MOMIA constructs have been reported, including dual-labeled antibodies (Paudyal et al. 2009) and small molecules (Pandey et al. 2005, Li et al. 2006b, Edwards et al. 2008, 2009) for cancer detection. These MOMIAs play important roles in providing longitudinal follow-up after initial nuclear imaging with short-lived radionuclides (Edwards et al. 2009), validating quantitative optical imaging reconstructions, facilitating in situ histology using localized residual fluorescent signal in tissue, and enabling high-throughput optical screening of molecular processes with fluorescent dye-labeled non-radioactive MOMIAs.

The optical–nuclear MOMIAs are particularly useful for providing complementary information. Toward this goal, a small molecule multimodality molecular probe has been developed for quantitative whole-body imaging of MOMIAs biodistribution by SPECT and subsequent measurement of specific diagnostic molecular event such as enzyme activation by optical imaging (Figure 31.8) (Lee et al. 2009). A similar strategy was employed with a dual-labeled antibody probe that becomes fluorescent upon target binding and demonstrated differentiation of specific and nonspecific tumor uptake by optical imaging (Ogawa et al. 2009c).

Nanomaterials have also been used to construct multimodality imaging agents (MIAs). The relatively large size and high payload carrying capabilities of some nanomaterials enable the incorporation of different reporters for multiple modalities. For example, dendrimeric nanoparticles were labeled with ^{111}In for scintigraphy and different NIR fluorophores for optical imaging to demonstrate scintigraphy and multicolor fluorescence imaging in lymph node detection (Kobayashi et al. 2007). Highly fluorescent Qdot nanoparticles have also been labeled with radionuclides for combined optical and PET imaging of angiogenesis in cancer (Cai et al. 2007, Mulder et al. 2009) and cardiovascular

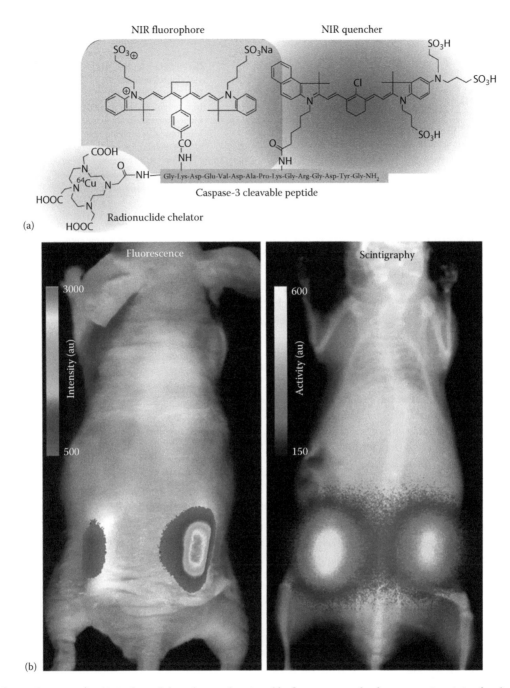

NIR fluorophore

NIR quencher

Caspase-3 cleavable peptide

Radionuclide chelator

Gly-Lys-Asp-Glu-Val-Asp-Ala-Pro-Lys-Gly-Arg-Gly-Asp-Tyr-Gly-NH₂

(a)

(b)

Fluorescence

Scintigraphy

Intensity (au)

Activity (au)

FIGURE 31.8 (See color insert.) (a) Multimodal nuclear and activatable fluorescent probe for caspase-3 activity for detecting apoptosis. Radiolabeling with Cu⁶⁴ via metal chelator (blue) enables PET imaging. Cleavage of peptide sequence by Caspase-3 separates the NIR quencher (gray) from the NIR fluorophore (orange), enhancing fluorescence yield. (b) Fluorescence (left) and scintigraphic (right) imaging of a mouse with subcutaneously implanted tubes containing equal concentrations of **A** in solution with either BSA (left side) or caspase-3 enzyme (right side) 2 h after implantation. The scintigraphy signal was used to quantitatively measure the probe concentration while the fluorescence signal indicated the enzyme activity with approximately sixfold increase in intensity.

disease models. For selective delivery to tissues of interest, targeted fluorescent-labeled MIA nanoparticles have been reported for multimodality MR and optical imaging of cardiovascular disease (Jaffer et al. 2009), cancer (Yang et al. 2009), and arthritis (Kim et al. 2009a). In this configuration, whole-body molecular imaging scans can be employed to locate suspicious lesions while fluorescent reporters provide molecular activity information, such as the levels of specific diagnostic enzyme activities (Jaffer et al. 2009). One unique advantage of MIA nanoparticles is the ability to selectively amplify the imaging signal of reporters with poor detection sensitivity for imaging agents. This provision allows seamless integration and optimization of optical and MRI reporters into MIA nanoparticles. In addition, these nanoparticles can be loaded with both imaging agents and drugs to monitor drug delivery and treatment response. It appears that this particular feature will galvanize the utility of MIA nanoparticles in future biomedical research and nanomedicine.

31.4 Discussion

Molecular optical imaging has evolved from cell microscopy and the use of vital stains and immunohistochemistry to real-time imaging of molecular processes in cells and tissues. Central to these advances is the development of a plethora of molecular probes for cellular and in vivo studies. Genetically encoded probes are particularly suitable for monitoring molecular interactions in cells and tissue and for longitudinal measurements, where signal from traditional dye-labeled cells could decrease with time. Extension of the fluorescent properties of exogenous dyes from the visible to the NIR region catalyzed the use of non-invasive optical methods in whole-body small animal imaging studies. Beginning with the use of fluorescent dye-labeled antibodies for receptor-mediated optical imaging (Ballou et al. 1995, Muguruma et al. 1999), targeted molecular imaging with therapeutic antibodies such as Herceptin (Hilger et al. 2004) and VEGF-mAb (Chang et al. 2008) has allowed the monitoring of treatment response. To overcome some of the limitations of antibodies, small but equally effective bio-functional analogues of antibodies have been developed for biological imaging and therapy. Additionally, small peptide and organic-based molecular probes add to the arsenal of reporters for molecular probes. The facile synthesis and ease of altering the biodistribution of these small molecular carriers are attractive for optimizing the structural features of targeted imaging agents. Moreover, these molecules can reach targets distal to the blood vessels through extravasation and subsequent diffusion controlled delivery to the target tissue. The advent of nanoparticles has expanded molecular imaging strategies to accommodate reporter molecules with different detection sensitivities for multimodal imaging applications. A bright future for these nanosystems lies in their ability to incorporate both imaging and therapeutic components for tissue-selective delivery, controlled drug release, and treatment monitoring.

List of Abbreviations

2-NBDG 2-(N-(7-nitrobenz-2-oxa-1,3-diazol-4-yl)amino)-2-deoxyglucose
AD Alzheimer's disease
ALA Aminolevulinic acid
AnxV Annexin V
BLI Bioluminescence imaging
CCD Charge-coupled device
cDNA Complementary DNA
CEA Carcinoembryonic antigen
CFP Cyan fluorescent protein
CT Computed tomography
DOTA 1,4,7,10-tetraazacyclododecane-1,4,7,10-tetraacetic acid
DOTA Diffuse optical tomography
EGF Epidermal growth factor
EGFR Epidermal growth factor receptor
EPR Enhanced permeability and retention

Fab Antibody fragment
FLT Fluorescence lifetime
FRET Förster or fluorescence resonance energy transfer
GFP Green fluorescent protein
HAL Hexaminolevulinate
ICG Indocyanine green
IFP Infrared fluorescent protein
mAb Monoclonal antibody
MIA Multimodality imaging agent
MMP Matrix metalloproteinase
MOMIA Monomolecular multimodality imaging agent
MRI Magnetic resonance imaging
NHS *N*-hydroxysulfosuccinimide
NIR Near-infrared
PET Positron emission tomography
PS Phosphatidylserine
PSMA Prostate-specific membrane antigen
Qdots Quantum dots
RFP Red fluorescent protein
RGD Arginine-glycine-aspartic acid
scFab Single-chain antibody fragment
SELEX Systematic evolution of ligands by exponential enrichment
SPECT Single photon emission counting tomography
UV Ultraviolet
VEGF Vascular endothelial growth factor
VEGFR Vascular endothelial growth factor receptor
YFP Yellow fluorescent protein

References

Achilefu, S., Bloch, S., Markiewicz, M. A., Zhong, T., Ye, Y., Dorshow, R. B., Chance, B., and Liang, K. 2005. Synergistic effects of light-emitting probes and peptides for targeting and monitoring integrin expression. *Proc. Natl. Acad. Sci. USA* 102: 7976–7981.

Achilefu, S., Dorshow, R. B., Bugaj, J. E., and Rajagopalan, R. 2000. Novel receptor-targeted fluorescent contrast agents for in vivo tumor imaging. *Invest. Radiol.* 35: 479–485.

Achilefu, S., Jimenez, H. N., Dorshow, R. B., Bugaj, J. E., Webb, E. G., Wilhelm, R. R., Rajagopalan, R., Johler, J., and Erion, J. L. 2002. Synthesis, in vitro receptor binding, and in vivo evaluation of fluorescein and carbocyanine peptide-based optical contrast agents. *J. Med. Chem.* 45: 2003–2015.

Aina, O. H., Liu, R., Sutcliffe, J. L., Marik, J., Pan, C. X., and Lam, K. S. 2007. From combinatorial chemistry to cancer-targeting peptides. *Mol. Pharm.* 4: 631–651.

Aina, O. H., Marik, J., Liu, R. W., Lau, D. H., and Lam, K. S. 2005. Identification of novel targeting peptides for human ovarian cancer cells using "one-bead one-compound" combinatorial libraries. *Mol. Cancer Ther.* 4: 806–813.

Akgun, E., Korner, M., Gao, F., Harikumar, K. G., Waser, B., Reubi, J. C., Portoghese, P. S., and Miller, L. J. 2009. Synthesis and in vitro characterization of radioiodinatable benzodiazepines selective for type 1 and type 2 cholecystokinin receptors. *J. Med. Chem.* 52: 2138–2147.

Alacam, B., Yazici, B., Intes, X., Nioka, S., and Chance, B. 2008. Pharmacokinetic-rate images of indocyanine green for breast tumors using near-infrared optical methods. *Phys. Med. Biol.* 53: 837–859.

Alexandrakis, G., Rannou, F. R., and Chatziioannou, A. F. 2006. Effect of optical property estimation accuracy on tomographic bioluminescence imaging: Simulation of a combined optical-PET (OPET) system. *Phys. Med. Biol.* 51: 2045–2053.

Allard, M., Cote, D., Davidson, L., Dazai, J., and Henkelman, R. M. 2007. Combined magnetic resonance and bioluminescence imaging of live mice. *J. Biomed. Opt.* 12: 034018.

Amoh, Y., Katsuoka, K., and Hoffman, R. M. 2008. Color-coded fluorescent protein imaging of angiogenesis: The AngioMouse models. *Curr. Pharm. Des.* 14: 3810–3819.

Aumailley, M., Gurrath, M., Muller, G., Calvete, J., Timpl, R., and Kessler, H. 1991. Arg-Gly-Asp constrained within cyclic pentapeptides. Strong and selective inhibitors of cell adhesion to vitronectin and laminin fragment P1. *FEBS Lett.* 291: 50–54.

Babendure, J. R., Adams, S. R., and Tsien, R. Y. 2003. Aptamers switch on fluorescence of triphenylmethane dyes. *J. Am. Chem. Soc.* 125: 14716–14717.

Backer, M. V., Levashova, Z., Patel, V., Jehning, B. T., Claffey, K., Blankenberg, F. G., and Backer, J. M. 2007. Molecular imaging of VEGF receptors in angiogenic vasculature with single-chain VEGF-based probes. *Nat. Med.* 13: 504–509.

Baeten, J., Haller, J., Shih, H., and Ntziachristos, V. 2009. In vivo investigation of breast cancer progression by use of an internal control. *Neoplasia* 11: 220–227.

Bagalkot, V., Zhang, L., Levy-Nissenbaum, E., Jon, S., Kantoff, P. W., Langer, R., and Farokhzad, O. C. 2007. Quantum dot-aptamer conjugates for synchronous cancer imaging, therapy, and sensing of drug delivery based on bi-fluorescence resonance energy transfer. *Nano. Lett.* 7: 3065–3070.

Ballou, B., Fisher, G. W., Waggoner, A. S., Farkas, D. L., Reiland, J. M., Jaffe, R., Mujumdar, R. B., Mujumdar, S. R., and Hakala, T. R. 1995. Tumor labeling in vivo using cyanine-conjugated monoclonal antibodies. *Cancer Immunol. Immunother.* 41: 257–263.

Bamett, A. H., Culver, J. P., Sorensen, A. G., Dale, A., and Boas, D. A. 2003. Robust inference of baseline optical properties of the human head with three-dimensional segmentation from magnetic resonance imaging. *Appl. Opt.* 42: 3095–3108.

Barbosa, E. M., Nonogaki, S., Katayama, M. L., Folgueira, M. A., Alves, V. F., and Brentani, M. M. 2004. Vitamin D3 modulation of plasminogen activator inhibitor type-1 in human breast carcinomas under organ culture. *Virchows. Arch.* 444: 175–182.

Becker, A., Hessenius, C., Bhargava, S., Grotzinger, C., Licha, K., Schneider-Mergener, J., Wiedenmann, B., and Semmler, W. 2000a. Cyanine dye labeled vasoactive intestinal peptide and somatostatin analog for optical detection of gastroenteropancreatic tumors. *Ann. N Y Acad. Sci.* 921: 275–278.

Becker, A., Hessenius, C., Licha, K., Ebert, B., Sukowski, U., Semmler, W., Wiedenmann, B., and Grotzinger, C. 2001. Receptor-targeted optical imaging of tumors with near-infrared fluorescent ligands. *Nat. Biotechnol.* 19: 327–331.

Becker, A., Riefke, B., Ebert, B., Sukowski, U., Rinneberg, H., Semmler, W., and Licha, K. 2000b. Macromolecular contrast agents for optical imaging of tumors: Comparison of indotricarbocyanine-labeled human serum albumin and transferrin. *Photochem. Photobiol.* 72: 234–241.

Bentolila, L. A., Ebenstein, Y., and Weiss, S. 2009. Quantum dots for in vivo small-animal imaging. *J. Nucl. Med.* 50: 493–496.

Bremer, C., Bredow, S., Mahmood, U., Weissleder, R., and Tung, C. H. 2001a. Optical imaging of matrix metalloproteinase-2 activity in tumors: Feasibility study in a mouse model. *Radiology* 221: 523–529.

Bremer, C., Tung, C. H., and Weissleder, R. 2001b. In vivo molecular target assessment of matrix metalloproteinase inhibition. *Nat. Med.* 7: 743–748.

Brumatti, G., Sheridan, C., and Martin, S. J. 2008. Expression and purification of recombinant annexin V for the detection of membrane alterations on apoptotic cells. *Methods* 44: 235–240.

Bugaj, J. E., Achilefu, S., Dorshow, R. B., and Rajagopalan, R. 2001. Novel fluorescent contrast agents for optical imaging of in vivo tumors based on a receptor-targeted dye-peptide conjugate platform. *J. Biomed. Opt.* 6: 122–133.

Bullok, K. E., Maxwell, D., Kesarwala, A. H., Gammon, S., Prior, J. L., Snow, M., Stanley, S., and Piwnica-Worms, D. 2007. Biochemical and in vivo characterization of a small, membrane-permeant, caspase-activatable far-red fluorescent peptide for imaging apoptosis. *Biochemistry* 46: 4055–4065.

Burnett, C. A., Xie, J., Quijano, J., Shen, Z., Hunter, F., Bur, M., LI, K. C., and Danthi, S. N. 2005. Synthesis, in vitro, and in vivo characterization of an integrin alpha(v)beta(3)-targeted molecular probe for optical imaging of tumor. *Bioorg. Med. Chem.* 13: 3763–3771.

Cai, W., Chen, K., Li, Z. B., Gambhir, S. S., and Chen, X. 2007. Dual-function probe for PET and near-infrared fluorescence imaging of tumor vasculature. *J. Nucl. Med.* 48: 1862–1870.

Cai, W., Niu, G., and Chen, X. 2008. Imaging of integrins as biomarkers for tumor angiogenesis. *Curr. Pharm. Des.* 14: 2943–2973.

Cerchia, L., Giangrande, P. H., Mcnamara, J. O., and de Franciscis, V. 2009. Cell-specific aptamers for targeted therapies. *Methods Mol. Biol.* 535: 59–78.

Chang, S. K., Rizvi, I., Solban, N., and Hasan, T. 2008. In vivo optical molecular imaging of vascular endothelial growth factor for monitoring cancer treatment. *Clin. Cancer Res.* 14: 4146–4153.

Chen, X., Conti, P. S., and Moats, R. A. 2004. In vivo near-infrared fluorescence imaging of integrin alphavbeta3 in brain tumor xenografts. *Cancer Res.* 64: 8009–8014.

Chen, K., Li, Z. B., Wang, H., Cai, W. B., and Chen, X. Y. 2008. Dual-modality optical and positron emission tomography imaging of vascular endothelial growth factor receptor on tumor vasculature using quantum dots. *Eur. J. Nucl. Med. Mol. Imaging* 35: 2235–2244.

Chen, Y., Zheng, G., Zhang, Z. H., Blessington, D., Zhang, M., Li, H., Liu, Q., Zhou, L., Intes, X., Achilefu, S., and Chance, B. 2003. Metabolism-enhanced tumor localization by fluorescence imaging: In vivo animal studies. *Opt. Lett.* 28: 2070–2072.

Cheng, Z., Levi, J., Xiong, Z., Gheysens, O., Keren, S., Chen, X., and Gambhir, S. S. 2006. Near-infrared fluorescent deoxyglucose analogue for tumor optical imaging in cell culture and living mice. *Bioconjug. Chem.* 17: 662–669.

Cheng, Z., Wu, Y., Xiong, Z., Gambhir, S. S., and Chen, X. 2005. Near-infrared fluorescent RGD peptides for optical imaging of integrin alphavbeta3 expression in living mice. *Bioconjug. Chem.* 16: 1433–1441.

Cherry, S. R. 2006. Multimodality in vivo imaging systems: Twice the power or double the trouble? *Annu. Rev. Biomed. Eng.* 8: 35–62.

Colcher, D., Pavlinkova, G., Beresford, G., Booth, B. J., Choudhury, A., and Batra, S. K. 1998. Pharmacokinetics and biodistribution of genetically-engineered antibodies. *Q. J. Nucl. Med.* 42: 225–241.

Constantin, T. P., Silva, G. L., Robertson, K. L., Hamilton, T. P., Fague, K., Waggoner, A. S., and Armitage, B. A. 2008. Synthesis of new fluorogenic cyanine dyes and incorporation into RNA fluoromodules. *Org. Lett.* 10: 1561–1564.

Coons, A. H. 1961. The beginnings of immunofluorescence. *J. Immunol.* 87: 499–503.

Coons, A. H. and Kaplan, M. H. 1950. Localization of antigen in tissue cells; improvements in a method for the detection of antigen by means of fluorescent antibody. *J. Exp. Med.* 91: 1–13.

Corlu, A., Choe, R., Durduran, T., Rosen, M. A., Schweiger, M., Arridge, S. R., Schnall, M. D., and Yodh, A. G. 2007. Three-dimensional in vivo fluorescence diffuse optical tomography of breast cancer in humans. *Opt. Express,* 15: 6696–6716.

Cuesta, A. M., Sanchez-Martin, D., Sanz, L., Bonet, J., Compte, M., Kremer, L., Blanco, F. J., Oliva, B., and Alvarez-Vallina, L. 2009. In vivo tumor targeting and imaging with engineered trivalent antibody fragments containing collagen-derived sequences. *PLoS One* 4: e5381.

Deroose, C. M., De, A., Loening, A. M., Chow, P. L., Ray, P., Chatziioannou, A. F., and Gambhir, S. S. 2007. Multimodality imaging of tumor xenografts and metastases in mice with combined small-animal PET, small-animal CT, and bioluminescence imaging. *J. Nucl. Med.* 48: 295–303.

D'Hallewin, M. A., Bezdetnaya, L., and Guillemin, F. 2002. Fluorescence detection of bladder cancer: A review. *Eur. Urol.* 42: 417–425.

Doyle, T. C., Burns, S. M., and Contag, C. H. 2004. In vivo bioluminescence imaging for integrated studies of infection. *Cell Microbiol.* 6: 303–317.

Duda, J., Karimi, M., Negrin, R. S., and Contag, C. H. 2007. Methods for imaging cell fates in hematopoiesis. *Methods Mol. Med.* 134: 17–34.

Edwards, W. B., Akers, W. J., Ye, Y., Cheney, P. P., Bloch, S., Xu, B., Laforest, R., and Achilefu, S. 2009. Multimodal imaging of integrin receptor-positive tumors by bioluminescence, fluorescence, gamma scintigraphy, and single-photon emission computed tomography using a cyclic RGD peptide labeled with a near-infrared fluorescent dye and a radionuclide. *Mol. Imaging* 8: 101–110.

Edwards, W. B., Xu, B., Akers, W., Cheney, P. P., Liang, K., Rogers, B. E., Anderson, C. J., and Achilefu, S. 2008. Agonist-antagonist dilemma in molecular imaging: Evaluation of a monomolecular multimodal imaging agent for the somatostatin receptor. *Bioconjug. Chem.* 19: 192–200.

Eljamel, M. S., Goodman, C., and Moseley, H. 2008. ALA and photofrin fluorescence-guided resection and repetitive PDT in glioblastoma multiforme: A single centre phase III randomised controlled trial. *Lasers Med. Sci.* 23: 361–367.

Fodero-Tavoletti, M. T., Rowe, C. C., Mclean, C. A., Leone, L., Li, Q. X., Masters, C. L., Cappai, R., and Villemagne, V. L. 2009. Characterization of PiB binding to white matter in Alzheimer disease and other dementias. *J. Nucl. Med.* 50: 198–204.

Fujihara, H., Ueta, Y., Suzuki, H., Katoh, A., Ohbuchi, T., Otsubo, H., Dayanithi, G., and Murphy, D. 2009. Robust up-regulation of nuclear red fluorescent-tagged fos marks neuronal activation in green fluorescent vasopressin neurons after osmotic stimulation in a double-transgenic rat. *Endocrinology* 150: 5633–5638.

Ginj, M., Zhang, H., Waser, B., Cescato, R., Wild, D., Wang, X., Erchegyi, J., Rivier, J., Macke, H. R., and Reubi, J. C. 2006. Radiolabeled somatostatin receptor antagonists are preferable to agonists for in vivo peptide receptor targeting of tumors. *Proc. Natl. Acad. Sci. USA* 103: 16436–16441.

Goncalves, M. S. 2009. Fluorescent labeling of biomolecules with organic probes. *Chem. Rev.* 109: 190–212.

Greish, K. 2007. Enhanced permeability and retention of macromolecular drugs in solid tumors: A royal gate for targeted anticancer nanomedicines. *J. Drug. Target.* 15: 457–464.

Griffin, H. G. and Griffin, A. M. 1993. DNA sequencing. Recent innovations and future trends. *Appl. Biochem. Biotechnol.* 38: 147–159.

Gross, S. and Piwnica-Worms, D. 2005. Spying on cancer: Molecular imaging in vivo with genetically encoded reporters. *Cancer Cell* 7: 5–15.

Gurfinkel, M., Thompson, A. B., Ralston, W., Troy, T. L., Moore, A. L., Moore, T. A., Gust, J. D., Tatman, D., Reynolds, J. S., Muggenburg, B., Nikula, K., Pandey, R., Mayer, R. H., Hawrysz, D. J., and Sevick-Muraca, E. M. 2000. Pharmacokinetics of ICG and HPPH-car for the detection of normal and tumor tissue using fluorescence, near-infrared reflectance imaging: A case study. *Photochem. Photobiol.* 72: 94–102.

Guven, M., Yazici, B., Intes, X., and Chance, B. 2005. Diffuse optical tomography with a priori anatomical information. *Phys. Med. Biol.* 50: 2837–2858.

Hagen, A., Grosenick, D., Macdonald, R., Rinneberg, H., Burock, S., Warnick, P., Poellinger, A., and Schlag, P. M. 2009. Late-fluorescence mammography assesses tumor capillary permeability and differentiates malignant from benign lesions. *Opt. Express,* 17: 17016–17033.

Hama, Y., Urano, Y., Koyama, Y., Choyke, P. L., and Kobayashi, H. 2007. Activatable fluorescent molecular imaging of peritoneal metastases following pretargeting with a biotinylated monoclonal antibody. *Cancer Res.* 67: 3809–3817.

Hammer, R. P., Owens, C. V., Hwang, S. H., Sayes, C. M., and Soper, S. A. 2002. Asymmetrical, water-soluble phthalocyanine dyes for covalent labeling of oligonucleotides. *Bioconjug. Chem.* 13: 1244–1252.

Han, L., Yang, D., and Kundra, V. 2007. Signaling can be uncoupled from imaging of the somatostatin receptor type 2. *Mol. Imaging* 6: 427–437.

Harris, A. G. 1994. Somatostatin and somatostatin analogues: Pharmacokinetics and pharmacodynamic effects. *Gut* 35: S1–S4.

Hasan, T., Lin, C. W., and Lin, A. 1989. Laser-induced selective cytotoxicity using monoclonal antibody-chromophore conjugates. *Prog. Clin. Biol. Res.* 288: 471–477.

Hauck, T. S., Anderson, R. E., Fischer, H. C., Newbigging, S., and Chan, W. C. 2009. In vivo quantum-dot toxicity assessment. *Small* 6: 138–144.

Heckmann, D. and Kessler, H. 2007. Design and chemical synthesis of integrin ligands. *Methods Enzymol.* 426: 463–503.

Hefti, M., von Campe, G., Moschopulos, M., Siegner, A., Looser, H., and Landolt, H. 2008. 5-Aminolevulinic acid induced protoporphyrin IX fluorescence in high-grade glioma surgery: A one-year experience at a single institution. *Swiss. Med. Wkly.* 138: 180–185.

Helms, M. W., Brandt, B. H., and Contag, C. H. 2006. Options for visualizing metastatic disease in the living body. *Contrib. Microbiol.* 13: 209–231.

Heroux, J., Gharib, A. M., Danthi, N. S., Cecchini, S., Ohayon, J., and Pettigrew, R. I. 2009. High-affinity alphavbeta3 integrin targeted optical probe as a new imaging biomarker for early atherosclerosis: Initial studies in Watanabe rabbits. *Mol. Imaging Biol.* 12: 2–8.

Hicke, B. J., Stephens, A. W., Gould, T., Chang, Y. F., Lynott, C. K., Heil, J., Borkowski, S., Hilger, C. S., Cook, G., Warren, S., and Schmidt, P. G. 2006. Tumor targeting by an aptamer. *J. Nucl. Med.* 47: 668–678.

Hilger, I., Leistner, Y., Berndt, A., Fritsche, C., Haas, K. M., Kosmehl, H., and Kaiser, W. A. 2004. Near-infrared fluorescence imaging of HER-2 protein over-expression in tumour cells. *Eur. Radiol.* 14: 1124–1129.

Hintersteiner, M., Enz, A., Frey, P., Jaton, A. L., Kinzy, W., Kneuer, R., Neumann, U., Rudin, M., Staufenbiel, M., Stoeckli, M., Wiederhold, K. H., and Gremlich, H. U. 2005. In vivo detection of amyloid-beta deposits by near-infrared imaging using an oxazine-derivative probe. *Nat. Biotechnol.* 23: 577–583.

Hsieh, C. B., Chen, C. J., Chen, T. W., Yu, J. C., Shen, K. L., Chang, T. M., and Liu, Y. C. 2004. Accuracy of indocyanine green pulse spectrophotometry clearance test for liver function prediction in transplanted patients. *World J. Gastroenterol.* 10: 2394–2396.

Huang, Y. F., Chang, H. T., and Tan, W. 2008. Cancer cell targeting using multiple aptamers conjugated on nanorods. *Anal. Chem.* 80: 567–572.

Hutchens, M. and Luker, G. D. 2007. Applications of bioluminescence imaging to the study of infectious diseases. *Cell Microbiol.* 9: 2315–2322.

Hyde, D., de Kleine, R., Maclaurin, S. A., Miller, E., Brooks, D. H., Krucker, T., and Ntziachristos, V. 2009. Hybrid FMT-CT imaging of amyloid-beta plaques in a murine Alzheimer's disease model. *Neuroimage* 44: 1304–1311.

Ito, S., Muguruma, N., Kakehashi, Y., Hayashi, S., Okamura, S., Shibata, H., Okahisa, T., Kanamori, M., Shibamura, S., Takesako, K., Nozawa, M., Ishida, K., and Shiga, M. 1995. Development of fluorescence-emitting antibody labeling substance by near-infrared ray excitation. *Bioorg. Med. Chem. Lett.* 5: 2689–2694.

Jaffer, F. A., Libby, P., and Weissleder, R. 2009. Optical and multimodality molecular imaging: Insights into atherosclerosis. *Arterioscler. Thromb. Vasc. Biol.* 29: 1017–1024.

Jin, Z. H., Razkin, J., Josserand, V., Boturyn, D., Grichine, A., Texier, I., Favrot, M. C., Dumy, P., and Coll, J. L. 2007. In vivo noninvasive optical imaging of receptor-mediated RGD internalization using self-quenched Cy5-labeled RAFT-c(-RGDfK-) (4). *Mol. Imaging* 6: 43–55.

Johnson, J. R., Fu, N., Arunkumar, E., Leevy, W. M., Gammon, S. T., Piwnica-Worms, D., and Smith, B. D. 2007. Squaraine rotaxanes: Superior substitutes for Cy-5 in molecular probes for near-infrared fluorescence cell imaging. *Angew. Chem. Int. Ed. Engl.* 46: 5528–5531.

Kang, J. H. and Chung, J. K. 2008. Molecular-genetic imaging based on reporter gene expression. *J. Nucl. Med.* 49(Suppl 2): 164S–179S.

Ke, S., Wen, X., Gurfinkel, M., Charnsangavej, C., Wallace, S., Sevick-Muraca, E. M., and Li, C. 2003. Near-infrared optical imaging of epidermal growth factor receptor in breast cancer xenografts. *Cancer Res.* 63: 7870–7875.

Kenworthy, A. K. 2001. Imaging protein-protein interactions using fluorescence resonance energy transfer microscopy. *Methods* 24: 289–296.

Kiesewetter, D. O., Lang, L., Ma, Y., Bhattacharjee, A. K., Gao, Z. G., Joshi, B. V., Melman, A., de Castro, S., and Jacobson, K. A. 2009. Synthesis and characterization of [76Br]-labeled high-affinity A3 adenosine receptor ligands for positron emission tomography. *Nucl. Med. Biol.* 36: 3–10.

Kim, J. S., An, H., Rieter, W. J., Esserman, D., Taylor-Pashow, K. M., Sartor, R. B., Lin, W., Lin, W., and Tarrant, T. K. 2009a. Multimodal optical and Gd-based nanoparticles for imaging in inflammatory arthritis. *Clin. Exp. Rheumatol.* 27: 580–586.

Kim, Y. E., Chen, J., Chan, J. R., and Langen, R. 2009b. Engineering a polarity-sensitive biosensor for time-lapse imaging of apoptotic processes and degeneration. *Nat. Methods*.

Klerk, C. P., Overmeer, R. M., Niers, T. M., Versteeg, H. H., Richel, D. J., Buckle, T., van Noorden, C. J., and van Tellingen, O. 2007. Validity of bioluminescence measurements for non-invasive in vivo imaging of tumor load in small animals. *Biotechniques* 43: 7–13, 30.

Klunk, W. E., Bacskai, B. J., Mathis, C. A., Kajdasz, S. T., Mclellan, M. E., Frosch, M. P., Debnath, M. L., Holt, D. P., Wang, Y., and Hyman, B. T. 2002. Imaging Abeta plaques in living transgenic mice with multiphoton microscopy and methoxy-X04, a systemically administered Congo red derivative. *J. Neuropathol. Exp. Neurol.* 61: 797–805.

Kobayashi, H., Koyama, Y., Barrett, T., Hama, Y., Regino, C. A., Shin, I. S., Jang, B. S., Le, N., Paik, C. H., Choyke, P. L., and Urano, Y. 2007. Multimodal nanoprobes for radionuclide and five-color near-infrared optical lymphatic imaging. *ACS Nano* 1: 258–264.

Kostenich, G., Livnah, N., Bonasera, T. A., Yechezkel, T., Salitra, Y., Litman, P., Kimel, S., and Orenstein, A. 2005. Targeting small-cell lung cancer with novel fluorescent analogs of somatostatin. *Lung Cancer* 50: 319–328.

Kostenich, G., Oron-Herman, M., Kimel, S., Livnah, N., Tsarfaty, I., and Orenstein, A. 2008. Diagnostic targeting of colon cancer using a novel fluorescent somatostatin conjugate in a mouse xenograft model. *Int. J. Cancer* 122: 2044–2049.

Kovar, J. L., Johnson, M. A., Volcheck, W. M., Chen, J., and Simpson, M. A. 2006. Hyaluronidase expression induces prostate tumor metastasis in an orthotopic mouse model. *Am. J. Pathol.* 169: 1415–1426.

Kozloff, K. M., Quinti, L., Patntirapong, S., Hauschka, P. V., Tung, C. H., Weissleder, R., and Mahmood, U. 2009. Non-invasive optical detection of cathepsin K-mediated fluorescence reveals osteoclast activity in vitro and in vivo. *Bone* 44: 190–198.

Kwon, S., Ke, S., Houston, J. P., Wang, W., Wu, Q., Li, C., and Sevick-Muraca, E. M. 2005. Imaging dose-dependent pharmacokinetics of an RGD-fluorescent dye conjugate targeted to alpha v beta 3 receptor expressed in Kaposi's sarcoma. *Mol. Imaging* 4: 75–87.

Lakowicz, J. 1999. *Principles of Fluorescence Spectroscopy*. New York: Kluwer Academic.

Lee, H., Akers, W. J., Cheney, P. P., Edwards, W. B., Liang, K., Culver, J. P., and Achilefu, S. 2009. Complementary optical and nuclear imaging of caspase-3 activity using combined activatable and radio-labeled multimodality molecular probe. *J. Biomed. Opt.* 14: 040507.

Li, Z. B., Cai, W., and Chen, X. 2007. Semiconductor quantum dots for in vivo imaging. *J. Nanosci. Nanotechnol.* 7: 2567–2581.

Li, C., Greenwood, T. R., Bhujwalla, Z. M., and Glunde, K. 2006a. Synthesis and characterization of glucosamine-bound near-infrared probes for optical imaging. *Organic. Lett.* 8: 3623–3626.

Li, Z. B., Niu, G., Wang, H., He, L., Yang, L., Ploug, M., and Chen, X. 2008c. Imaging of urokinase-type plasminogen activator receptor expression using a 64Cu-labeled linear peptide antagonist by microPET. *Clin. Cancer Res.* 14: 4758–4766.

Li, C., Wang, G., Qi, J., and Cherry, S. R. 2009. Three-dimensional fluorescence optical tomography in small-animal imaging using simultaneous positron-emission-tomography priors. *Opt. Lett.* 34: 2933–2935.

Li, C., Wang, W., Wu, Q., Ke, S., Houston, J., Sevick-Muraca, E., Dong, L., Chow, D., Charnsangavej, C., and Gelovani, J. G. 2006b. Dual optical and nuclear imaging in human melanoma xenografts using a single targeted imaging probe. *Nucl. Med. Biol.* 33: 349–358.

Li, W., Yang, X., Wang, K., Tan, W., He, Y., Guo, Q., Tang, H., and Liu, J. 2008a. Real-time imaging of protein internalization using aptamer conjugates. *Anal. Chem.* 80: 5002–5008.

Li, W., Yang, X., Wang, K., Tan, W., Li, H., and Ma, C. 2008b. FRET-based aptamer probe for rapid angiogenin detection. *Talanta* 75: 770–774.

Liang, M., Liu, X., Cheng, D., Nakamura, K., Wang, Y., Dou, S., Liu, G., Rusckowski, M., and Hnatowich, D. J. 2009. Optical antisense tumor targeting in vivo with an improved fluorescent DNA duplex probe. *Bioconjug. Chem.* 20: 1223–1227.

Licha, K., Riefke, B., Ntziachristos, V., Becker, A., Chance, B., and Semmler, W. 2000. Hydrophilic cyanine dyes as contrast agents for near-infrared tumor imaging: Synthesis, photophysical properties and spectroscopic in vivo characterization. *Photochem. Photobiol.* 72: 392–398.

Lisy, M. R., Goermar, A., Thomas, C., Pauli, J., Resch-Genger, U., Kaiser, W. A., and Hilger, I. 2008. In vivo near-infrared fluorescence imaging of carcinoembryonic antigen-expressing tumor cells in mice. *Radiology* 247: 779–787.

Liu, Z., Niu, G., Wang, F., and Chen, X. 2009a. (68)Ga-labeled NOTA-RGD-BBN peptide for dual integrin and GRPR-targeted tumor imaging. *Eur. J. Nucl. Med. Mol. Imaging* 36: 1483–1494.

Liu, Z., Yan, Y., Chin, F. T., Wang, F., and Chen, X. 2009b. Dual integrin and gastrin-releasing peptide receptor targeted tumor imaging using 18F-labeled PEGylated RGD-bombesin heterodimer 18F-FB-PEG3-Glu-RGD-BBN. *J. Med. Chem.* 52: 425–432.

Ma, L., Yu, P., Veerendra, B., Rold, T. L., Retzloff, L., Prasanphanich, A., Sieckman, G., Hoffman, T. J., Volkert, W. A., and Smith, C. J. 2007. In vitro and in vivo evaluation of Alexa Fluor 680-bombesin[7-14]NH2 peptide conjugate, a high-affinity fluorescent probe with high selectivity for the gastrin-releasing peptide receptor. *Mol. Imaging* 6: 171–180.

Mammen, M., Choi, S. K., and Whitesides, G. M. 1998. Polyvalent interactions in biological systems: Implications for design and use of multivalent ligands and inhibitors. *Angew. Chem. Int. Ed.* 37: 2755–2794.

Mathelin, C., Croce, S., Brasse, D., Gairard, B., Gharbi, M., Andriamisandratsoa, N., Bekaert, V., Francis, Z., Guyonnet, J. L., Huss, D., Salvador, S., Schaeffer, R., Grucker, D., Marin, C., and Bellocq, J. P. 2009. Methylene blue dye, an accurate dye for sentinel lymph node identification in early breast cancer. *Anticancer Res.* 29: 4119–4125.

Mathews, W. B., Murugesan, N., Xia, J., Scheffel, U., Hilton, J., Ravert, H. T., Dannals, R. F., and Szabo, Z. 2008. Synthesis and in vivo evaluation of novel PET radioligands for imaging the endothelin-A receptor. *J. Nucl. Med.* 49: 1529–1536.

Maxwell, D., Chang, Q., Zhang, X., Barnett, E. M., and Piwnica-Worms, D. 2009. An improved cell-penetrating, caspase-activatable, near-infrared fluorescent Peptide for apoptosis imaging. *Bioconjug. Chem.* 20: 702–709.

Mazzucchelli, R. I., Warming, S., Lawrence, S. M., Ishii, M., Abshari, M., Washington, A. V., Feigenbaum, L., Warner, A. C., Sims, D. J., Li, W. Q., Hixon, J. A., Gray, D. H., Rich, B. E., Morrow, M., Anver, M. R., Cherry, J., Naf, D., Sternberg, L. R., Mcvicar, D. W., Farr, A. G., Germain, R. N., Rogers, K., Jenkins, N. A., Copeland, N. G., and Durum, S. K. 2009. Visualization and identification of IL-7 producing cells in reporter mice. *PLoS One* 4: e7637.

McIntyre, J. O. and Matrisian, L. M. 2009. Optical proteolytic beacons for in vivo detection of matrix metalloproteinase activity. *Methods Mol. Biol.* 539: 1–20.

Mew, D., Wat, C. K., Towers, G. H., and Levy, J. G. 1983. Photoimmunotherapy: Treatment of animal tumors with tumor-specific monoclonal antibody-hematoporphyrin conjugates. *J. Immunol.* 130: 1473–1477.

Moghissi, K., Stringer, M. R., and Dixon, K. 2008. Fluorescence photodiagnosis in clinical practice. *Photodiagnosis Photodyn. Ther.* 5: 235–237.

Muguruma, N., Ito, S., Bando, T., Taoka, S., Kusaka, Y., Hayashi, S., Ichikawa, S., Matsunaga, Y., Tada, Y., Okamura, S. II, K., Imaizumi, K., Nakamura, K., Takesako, K., and Shibamura, S. 1999. Labeled carcinoembryonic antigen antibodies excitable by infrared rays: A novel diagnostic method for micro cancers in the digestive tract. *Intern. Med.* 38: 537–542.

Muguruma, N., Ito, S., Hayashi, S., Taoka, S., Kakehashi, H., II, K., Shibamura, S., and Takesako, K. 1998. Antibodies labeled with fluorescence-agent excitable by infrared rays. *J. Gastroenterol.* 33: 467–471.

Mulder, W. J., Castermans, K., van Beijnum, J. R., Oude Egbrink, M. G., Chin, P. T., Fayad, Z. A., Lowik, C. W., Kaijzel, E. L., Que, I., Storm, G., Strijkers, G. J., Griffioen, A. W., and Nicolay, K. 2009. Molecular imaging of tumor angiogenesis using alphavbeta3-integrin targeted multimodal quantum dots. *Angiogenesis* 12: 17–24.

Muller-Taubenberger, A. and Anderson, K. I. 2007. Recent advances using green and red fluorescent protein variants. *Appl. Microbiol. Biotechnol.* 77: 1–12.

Nath, J. and Johnson, K. L. 2000. A review of fluorescence in situ hybridization (FISH): Current status and future prospects. *Biotech. Histochem.* 75: 54–78.

Nitin, N., Carlson, A. L., Muldoon, T., El-Naggar, A. K., Gillenwater, A., and Richards-Kortum, R. 2009. Molecular imaging of glucose uptake in oral neoplasia following topical application of fluorescently labeled deoxy-glucose. *Int. J. Cancer.* 124: 2634–2642.

Ntziachristos, V., Schellenberger, E. A., Ripoll, J., Yessayan, D., Graves, E., Bogdanov, A., JR, Josephson, L., and Weissleder, R.

2004. Visualization of antitumor treatment by means of fluorescence molecular tomography with an annexin V-Cy5.5 conjugate. *Proc. Natl. Acad. Sci. USA* 101: 12294–12299.

Ntziachristos, V., Yodh, A. G., Schnall, M., and Chance, B. 2000. Concurrent MRI and diffuse optical tomography of breast after indocyanine green enhancement. *Proc. Natl. Acad. Sci. USA* 97: 2767–2772.

Ogawa, M., Kosaka, N., Choyke, P. L., and Kobayashi, H. 2009a. In vivo molecular imaging of cancer with a quenching near-infrared fluorescent probe using conjugates of monoclonal antibodies and indocyanine green. *Cancer Res.* 69: 1268–1272.

Ogawa, M., Regino, C. A., Choyke, P. L., and Kobayashi, H. 2009b. In vivo target-specific activatable near-infrared optical labeling of humanized monoclonal antibodies. *Mol. Cancer Ther.* 8: 232–239.

Ogawa, M., Regino, C. A., Seidel, J., Green, M. V., Xi, W., Williams, M., Kosaka, N., Choyke, P. L., and Kobayashi, H. 2009c. Dual-modality molecular imaging using antibodies labeled with activatable fluorescence and a radionuclide for specific and quantitative targeted cancer detection. *Bioconjug. Chem.* 20: 2177–2184.

Okamura, N., Suemoto, T., Shimadzu, H., Suzuki, M., Shiomitsu, T., Akatsu, H., Yamamoto, T., Staufenbiel, M., Yanai, K., Arai, H., Sasaki, H., Kudo, Y., and Sawada, T. 2004. Styrylbenzoxazole derivatives for in vivo imaging of amyloid plaques in the brain. *J. Neurosci.* 24: 2535–2541.

Overall, C. M. and Kleifeld, O. 2006. Tumour microenvironment—Opinion: Validating matrix metalloproteinases as drug targets and anti-targets for cancer therapy. *Nat. Rev. Cancer* 6: 227–239.

Pandey, S. K., Gryshuk, A. L., Sajjad, M., Zheng, X., Chen, Y., Abouzeid, M. M., Morgan, J., Charamisinau, I., Nabi, H. A., Oseroff, A., and Pandey, R. K. 2005. Multimodality agents for tumor imaging (PET, fluorescence) and photodynamic therapy. A possible "see and treat" approach. *J. Med. Chem.* 48: 6286–6295.

Patton, J. A. and Turkington, T. G. 2008. SPECT/CT physical principles and attenuation correction. *J. Nucl. Med. Technol.* 36: 1–10.

Paudyal, P., Paudyal, B., Iida, Y., Oriuchi, N., Hanaoka, H., Tominaga, H., Ishikita, T., Yoshioka, H., Higuchi, T., and Endo, K. 2009. Dual functional molecular imaging probe targeting CD20 with PET and optical imaging. *Oncol. Rep.* 22: 115–119.

Pauwels, E. K., Sturm, E. J., Bombardieri, E., Cleton, F. J., and Stokkel, M. P. 2000. Positron-emission tomography with [18F]fluorodeoxyglucose. Part I. Biochemical uptake mechanism and its implication for clinical studies. *J. Cancer Res. Clin. Oncol.* 126: 549–559.

Pepperkok, R. and Ellenberg, J. 2006. High-throughput fluorescence microscopy for systems biology. *Nat. Rev. Mol. Cell Biol.* 7: 690–696.

Perkins, A. C. and Missailidis, S. 2007. Radiolabelled aptamers for tumour imaging and therapy. *Q. J. Nucl. Med. Mol. Imaging* 51: 292–296.

Perlitz, C., Licha, K., Scholle, F. D., Ebert, B., Bahner, M., Hauff, P., Moesta, K. T., and Schirner, M. 2005. Comparison of two tricarbocyanine-based dyes for fluorescence optical imaging. *J. Fluoresc.* 15: 443–454.

Petrovsky, A., Schellenberger, E., Josephson, L., Weissleder, R., and Bogdanov, A., JR. 2003. Near-infrared fluorescent imaging of tumor apoptosis. *Cancer Res.* 63: 1936–1942.

Pham, W., Choi, Y., Weissleder, R., and Tung, C. H. 2004. Developing a peptide-based near-infrared molecular probe for protease sensing. *Bioconjug. Chem.* 15: 1403–1407.

Pham, W., Weissleder, R., and Tung, C. H. 2002. An azulene dimer as a near-infrared quencher. *Angew. Chem. Int. Ed. Engl.* 41: 3659–3662, 3519.

Pierce, M. C., Javier, D. J., and Richards-Kortum, R. 2008. Optical contrast agents and imaging systems for detection and diagnosis of cancer. *Int. J. Cancer* 123: 1979–1990.

Ponomarev, V., Doubrovin, M., Serganova, I., Vider, J., Shavrin, A., Beresten, T., Ivanova, A., Ageyeva, L., Tourkova, V., Balatoni, J., Bornmann, W., Blasberg, R., and Gelovani Tjuvajev, J. 2004. A novel triple-modality reporter gene for whole-body fluorescent, bioluminescent, and nuclear noninvasive imaging. *Eur. J. Nucl. Med. Mol. Imaging* 31: 740–751.

Qian, H., Gu, Y., Wang, M., and Achilefu, S. 2009. Optimization of the near-infrared fluorescence labeling for in vivo monitoring of a protein drug distribution in animal model. *J. Fluoresc.* 19: 277–284.

Ramjiawan, B., Maiti, P., Aftanas, A., Kaplan, H., Fast, D., Mantsch, H. H., and Jackson, M. 2000. Noninvasive localization of tumors by immunofluorescence imaging using a single chain Fv fragment of a human monoclonal antibody with broad cancer specificity. *Cancer* 89: 1134–1144.

Ray, E. R., Chatterton, K., Thomas, K., Khan, M. S., Chandra, A., and O'brien, T. S. 2009. Hexylaminolevulinate photodynamic diagnosis for multifocal recurrent nonmuscle invasive bladder cancer. *J. Endourol.* 23: 983–988.

Ray, P., De, A., Patel, M., and Gambhir, S. S. 2008. Monitoring caspase-3 activation with a multimodality imaging sensor in living subjects. *Clin. Cancer Res.* 14: 5801–5809.

Ray, P. and Gambhir, S. S. 2007. Noninvasive imaging of molecular events with bioluminescent reporter genes in living subjects. *Reporter Genes.*

Ray, P., Tsien, R., and Gambhir, S. S. 2007. Construction and validation of improved triple fusion reporter gene vectors for molecular imaging of living subjects. *Cancer Res.* 67: 3085–3093.

Raymond, S., Skoch, J., Hills, I., Nesterov, E., Swager, T., and Bacskai, B. 2008. Smart optical probes for near-infrared fluorescence imaging of Alzheimer's disease pathology. *Eur. J. Nucl. Med. Mol. Imaging* 35: 93–98.

Reddy, J. A., Allagadda, V. M., and Leamon, C. P. 2005. Targeting therapeutic and imaging agents to folate receptor positive tumors. *Curr. Pharm. Biotechnol.* 6: 131–150.

Reynolds, A. R., Hart, I. R., Watson, A. R., Welti, J. C., Silva, R. G., Robinson, S. D., da Violante, G., Gourlaouen, M., Salih, M., Jones, M. C., Jones, D. T., Saunders, G., Kostourou, V.,

Perron-Sierra, F., Norman, J. C., Tucker, G. C., and Hodivala-Dilke, K. M. 2009. Stimulation of tumor growth and angiogenesis by low concentrations of RGD-mimetic integrin inhibitors. *Nat. Med.* 15: 392–400.

Roman, Y., Alfonso, A., Louzao, M. C., Vieytes, M. R., and Botana, L. M. 2001. Confocal microscopy study of the different patterns of 2-NBDG uptake in rabbit enterocytes in the apical and basal zone. *Pflugers Arch.* 443: 234–239.

Rosca, E. V., Stukel, J. M., Gillies, R. J., Vagner, J., and Caplan, M. R. 2007. Specificity and mobility of biomacromolecular, multivalent constructs for cellular targeting. *Biomacromolecules* 8: 3830–3835.

Ruoslahti, E., and Pierschbacher, M. D. 1986. Arg-Gly-Asp: A versatile cell recognition signal. *Cell* 44: 517–518.

Sadowski, M., Pankiewicz, J., Scholtzova, H., Tsai, J., Li, Y., Carp, R. I., Meeker, H. C., Gambetti, P., Debnath, M., Mathis, C. A., Shao, L., Gan, W. B., Klunk, W. E., and Wisniewski, T. 2004. Targeting prion amyloid deposits in vivo. *J. Neuropathol. Exp. Neurol.* 63: 775–784.

Schellenberger, E. A., Bogdanov, A., JR., Petrovsky, A., Ntziachristos, V., Weissleder, R., and Josephson, L. 2003. Optical imaging of apoptosis as a biomarker of tumor response to chemotherapy. *Neoplasia* 5: 187–192.

Schellenberger, E. A., Weissleder, R., and Josephson, L. 2004. Optimal modification of annexin V with fluorescent dyes. *Chembiochem* 5: 271–274.

Scherer, R. L., Vansaun, M. N., Mcintyre, J. O., and Matrisian, L. M. 2008. Optical imaging of matrix metalloproteinase-7 activity in vivo using a proteolytic nanobeacon. *Mol. Imaging* 7: 118–131.

Schipper, M. L., Iyer, G., Koh, A. L., Cheng, Z., Ebenstein, Y., Aharoni, A., Keren, S., Bentolila, L. A., LI, J., Rao, J., Chen, X., Banin, U., Wu, A. M., Sinclair, R., Weiss, S., and Gambhir, S. S. 2009. Particle size, surface coating, and PEGylation influence the biodistribution of quantum dots in living mice. *Small* 5: 126–134.

Schweiger, M. and Arridge, S. R. 1999. Optical tomographic reconstruction in a complex head model using a priori region boundary information. *Phys. Med. Biol.* 44: 2703–2721.

Sega, E. I. and Low, P. S. 2008. Tumor detection using folate receptor-targeted imaging agents. *Cancer Metastasis. Rev.* 27: 655–664.

Seneca, N., Zoghbi, S. S., Skinbjerg, M., Liow, J. S., Hong, J., Sibley, D. R., Pike, V. W., Halldin, C., and Innis, R. B. 2008. Occupancy of dopamine D2/3 receptors in rat brain by endogenous dopamine measured with the agonist positron emission tomography radioligand [11C]MNPA. *Synapse* 62: 756–763.

Service, R. F. 2008. Nobel Prize in chemistry. Three scientists bask in prize's fluorescent glow. *Science* 322: 361.

Shaner, N. C., Steinbach, P. A., and Tsien, R. Y. 2005. A guide to choosing fluorescent proteins. *Nat. Methods,* 2: 905–909.

Sharkey, R. M. and Goldenberg, D. M. 2006. Advances in radioimmunotherapy in the age of molecular engineering and pretargeting. *Cancer Invest.* 24: 82–97.

Sheth, R. A., Upadhyay, R., Stangenberg, L., Sheth, R., Weissleder, R., and Mahmood, U. 2009. Improved detection of ovarian cancer metastases by intraoperative quantitative fluorescence protease imaging in a pre-clinical model. *Gynecol. Oncol.* 112: 616–622.

Shu, X., Royant, A., Lin, M. Z., Aguilera, T. A., Lev-Ram, V., Steinbach, P. A., and Tsien, R. Y. 2009. Mammalian expression of infrared fluorescent proteins engineered from a bacterial phytochrome. *Science* 324: 804–807.

Smith, B. R., Cheng, Z., De, A., Koh, A. L., Sinclair, R., and Gambhir, S. S. 2008. Real-time intravital imaging of RGD-quantum dot binding to luminal endothelium in mouse tumor neovasculature. *Nano Lett.* 8: 2599–2606.

Southwick, P. L., Ernst, L. A., Tauriello, E. W., Parker, S. R., Mujumdar, R. B., Mujumdar, S. R., Clever, H. A., and Waggoner, A. S. 1990. Cyanine dye labeling reagents—Carboxymethylindocyanine succinimidyl esters. *Cytometry* 11: 418–430.

Stefflova, K., Chen, J., Marotta, D., Li, H., and Zheng, G. 2006. Photodynamic therapy agent with a built-in apoptosis sensor for evaluating its own therapeutic outcome in situ. *J. Med. Chem.* 49: 3850–3856.

Stepp, H., Beck, T., Pongratz, T., Meinel, T., Kreth, F. W., Tonn, J., and Stummer, W. 2007. ALA and malignant glioma: Fluorescence-guided resection and photodynamic treatment. *J. Environ. Pathol. Toxicol. Oncol.* 26: 157–164.

Stojanovic, M. N. and Kolpashchikov, D. M. 2004. Modular aptameric sensors. *J. Am. Chem. Soc.* 126: 9266–9270.

Tadatsu, Y., Muguruma, N., Ito, S., Tadatsu, M., Kusaka, Y., Okamoto, K., Imoto, Y., Taue, H., Sano, S., and Nagao, Y. 2006. Optimal labeling condition of antibodies available for immunofluorescence endoscopy. *J. Med. Invest.* 53: 52–60.

Thumshirn, G., Hersel, U., Goodman, S. L., and Kessler, H. 2003. Multimeric cyclic RGD peptides as potential tools for tumor targeting: Solid-phase peptide synthesis and chemoselective oxime ligation. *Chem. Eur. J.* 9: 2717–2725.

Tosi, J., Wang, N. K., Zhao, J., Chou, C. L., Kasanuki, J. M., Tsang, S. H., and Nagasaki, T. 2009. Rapid and noninvasive imaging of retinal ganglion cells in live mouse models of glaucoma. *Mol. Imaging Biol.*

Townsend, D. W. 2008. Dual-modality imaging: Combining anatomy and function. *J. Nucl. Med.* 49: 938–955.

Unlu, M. B., Lin, Y., Birgul, O., Nalcioglu, O., and Gulsen, G. 2008. Simultaneous in vivo dynamic magnetic resonance-diffuse optical tomography for small animal imaging. *J. Biomed. Opt.* 13: 060501.

van de Ven, S., Wiethoff, A., Nielsen, T., Brendel, B., van der Voort, M., Nachabe, R., van der Mark, M., van Beek, M., Bakker, L., Fels, L., Elias, S., Luijten, P., and Mali, W. 2009. A novel fluorescent imaging agent for diffuse optical tomography of the breast: First clinical experience in patients. *Mol. Imaging Biol.*

van Engeland, M., Nieland, L. J., Ramaekers, F. C., Schutte, B., and Reutelingsperger, C. P. 1998. Annexin V-affinity assay: A review on an apoptosis detection system based on phosphatidylserine exposure. *Cytometry* 31: 1–9.

Vance, D., Shah, M., Joshi, A., and Kane, R. S. 2008. Polyvalency: A promising strategy for drug design. *Biotechnol. Bioeng.* 101: 429–434.

Verdree, V. T., Pakhomov, S., Su, G., Allen, M. W., Countryman, A. C., Hammer, R. P., and Soper, S. A. 2007. Water soluble metallo-phthalocyanines: The role of the functional groups on the spectral and photophysical properties. *J. Fluoresc.* 17: 547–563.

Villalobos, V., Naik, S., and Piwnica-Worms, D. 2007. Current state of imaging protein-protein interactions in vivo with genetically encoded reporters. *Annu. Rev. Biomed. Eng.* 9: 321–349.

Virostko, J., XIE, J., Hallahan, D. E., Arteaga, C. L., Gore, J. C., and Manning, H. C. 2009. A molecular imaging paradigm to rapidly profile response to angiogenesis-directed therapy in small animals. *Mol. Imaging Biol.*

Voorzanger-Rousselot, N., Juillet, F., Mareau, E., Zimmermann, J., Kalebic, T., and Garnero, P. 2006. Association of 12 serum biochemical markers of angiogenesis, tumour invasion and bone turnover with bone metastases from breast cancer: A crossectional and longitudinal evaluation. *Br. J. Cancer* 95: 506–514.

Wang, H., Chen, K., Niu, G., and Chen, X. 2009. Site-specifically biotinylated VEGF(121) for near-infrared fluorescence imaging of tumor angiogenesis. *Mol. Pharm.* 6: 285–294.

Waser, B., Tamma, M. L., Cescato, R., Maecke, H. R., and Reubi, J. C. 2009. Highly efficient in vivo agonist-induced internalization of sst2 receptors in somatostatin target tissues. *J. Nucl. Med.* 50: 936–941.

Weissleder, R., Tung, C. H., Mahmood, U., and Bogdanov, A., JR. 1999. In vivo imaging of tumors with protease-activated near-infrared fluorescent probes. *Nat. Biotechnol.* 17: 375–378.

Williams, R. J., Lipowska, M., Patonay, G., and Strekowski, L. 1993. Comparison of covalent and noncovalent labeling with near-infrared dyes for the high-performance liquid chromatographic determination of human serum albumin. *Anal. Chem.* 65: 601–605.

Wilson, K., Yu, J., Lee, A., and Wu, J. C. 2008. In vitro and in vivo bioluminescence reporter gene imaging of human embryonic stem cells. *J. Vis. Exp.*

Withrow, K. P., Newman, J. R., Skipper, J. B., Gleysteen, J. P., Magnuson, J. S., Zinn, K., and Rosenthal, E. L. 2008. Assessment of bevacizumab conjugated to Cy5.5 for detection of head and neck cancer xenografts. *Technol. Cancer Res. Treat.* 7: 61–66.

Witjes, J. A. and Douglass, J. 2007. The role of hexaminolevulinate fluorescence cystoscopy in bladder cancer. *Nat. Clin. Pract. Urol.* 4: 542–549.

Wong, K. K., Zarzhevsky, N., Cahill, J. M., Frey, K. A., and Avram, A. M. 2009. Hybrid SPECT-CT and PET-CT imaging of differentiated thyroid carcinoma. *Br. J. Radiol.* 82: 860–876.

Wu, Y., Cai, W., and Chen, X. 2006. Near-infrared fluorescence imaging of tumor integrin alpha v beta 3 expression with Cy7-labeled RGD multimers. *Mol. Imaging Biol.* 8: 226–236.

Wu, A. M. and Olafsen, T. 2008. Antibodies for molecular imaging of cancer. *Cancer J.* 14: 191–197.

Yalavarthy, P. K., Pogue, B. W., Dehghani, H., Carpenter, C. M., Jiang, S., and Paulsen, K. D. 2007. Structural information within regularization matrices improves near infrared diffuse optical tomography. *Opt. Express,* 15: 8043–8058.

Yang, L., Mao, H., Cao, Z., Wang, Y. A., Peng, X., Wang, X., Sajja, H. K., Wang, L., Duan, H., Ni, C., Staley, C. A., Wood, W. C., Gao, X., and Nie, S. 2009. Molecular imaging of pancreatic cancer in an animal model using targeted multifunctional nanoparticles. *Gastroenterology* 136: 1514–1525 e2.

Ye, Y., Bloch, S., and Achilefu, S. 2004. Polyvalent carbocyanine molecular beacons for molecular recognitions. *J. Am. Chem. Soc.* 126: 7740–7741.

Ye, Y., Bloch, S., Kao, J., and Achilefu, S. 2005. Multivalent carbocyanine molecular probes: Synthesis and applications. *Bioconjug. Chem.* 16: 51–61.

Ye, Y., Bloch, S., Xu, B., and Achilefu, S. 2006. Design, synthesis, and evaluation of near infrared fluorescent multimeric RGD peptides for targeting tumors. *J. Med. Chem.* 49: 2268–2275.

Zaak, D., Stepp, H., Baumgartner, R., Schneede, P., Waidelich, R., Frimberger, D., Hartmann, A., Kunchel, R., Hofstetter, A., and Hohla, A. 2002. Ultraviolet-excited (308 nm) autofluorescence for bladder cancer detection. *Urology* 60: 1029–1033.

Zhang, S., Metelev, V., Tabatadze, D., Zamecnik, P. C., and Bogdanov, A. 2008. Fluorescence resonance energy transfer in near-infrared fluorescent oligonucleoticle probes for detecting protein-DNA interactions. *Proc. Natl. Acad. Sci. USA* 105: 4156–4161.

Zhang, M., Zhang, Z., Blessington, D., Li, H., Busch, T. M., Madrak, V., Miles, J., Chance, B., Glickson, J. D., and Zheng, G. 2003. Pyropheophorbide 2-deoxyglucosamide: A new photosensitizer targeting glucose transporters. *Bioconjug. Chem.* 14: 709–714.

Zimmermann, A., Ritsch-Marte, M., and Kostron, H. 2001. mTHPC-mediated photodynamic diagnosis of malignant brain tumors. *Photochem. Photobiol.* 74: 611–616.

32

Nanoparticles for Targeted Therapeutics and Diagnostics

Timothy Larson
The University of Texas at Austin

Kort Travis
The University of Texas at Austin

Pratixa Joshi
The University of Texas at Austin

Konstantin Sokolov
The University of Texas at Austin

and

The University of Texas MD Anderson Cancer Center

32.1 Introduction 697
32.2 Background.. 698
32.3 State of the Art: Nanoparticles in Biomedical Optics.. 698
    Optical Properties of Nanoparticles • Specificity and Stealth of Nanoparticles as Influenced
    by Surface Properties • Nanoparticle Contrast in Imaging Modalities • Photothermal
    Therapy • Applications for Combined Diagnosis and Therapy • Research in Efficient
    Delivery, Biodistribution, Degradation, and Extraction from the Body
32.4 Summary...717
References..717

32.1 Introduction

Tremendous progress has been made in the understanding of molecular mechanisms of many devastating diseases such as cancer and atherosclerosis (Choudhury et al. 2004, Hanahan and Weinberg 2000, Packard and Libby 2008). It has been shown that molecular events associated with disease progression can be valuable diagnostic and prognostic biomarkers. Furthermore, this better understanding of molecular mechanisms of different pathologies has inspired novel molecular-oriented therapeutic strategies. These advances necessitate the development of technologies for molecular-specific imaging, therapy, and therapy monitoring in vivo.

Optical spectroscopic and imaging modalities can provide information about both tissue biochemistry and morphology; however, the number of natural chromophores that can be probed is limited. Nanotechnology has become an area of intense focus in the development of molecular-specific approaches in the field of biomedical optics due to unique optical properties provided by nanoparticles (Alivisatos 2004, Anker et al. 2008, Michalet et al. 2005). It has been shown that many inorganic nanoparticles inherently exhibit an ideal combination of unique size-dependant properties, which provide a bright signal for enhanced imaging contrast and sensitive detection of molecular-specific biomarkers of cancer and other pathologies. The most noticeable examples include plasmonic nanoparticles (Huang et al. 2006, Kumar et al. 2007, 2008, Loo et al. 2005, Skrabalak et al. 2007, Sokolov et al. 2003), luminescent quantum dots (Q-dots) (Choi et al. 2007a, Gao

et al. 2004), silica C-dots (Choi et al. 2007b), and carbon nanotubes (De La Zerda et al. 2008, Wang et al. 2009). Combining different types of nanoparticles and targeting molecules provides a common platform for multiple imaging applications with a high degree of flexibility (Alivisatos 2004, Larson et al. 2007, Michalet et al. 2005). For example, plasmonic nanoparticles have been combined with quantum dots (Jin and Gao 2009) and with iron oxide (Aaron et al. 2006, Ji et al. 2007, Larson et al. 2007) to form composite nanomaterials that provide contrast in more than one imaging modality. In addition, metal nanoparticles can have very strong optical absorption, scattering, and fluorescent properties that enable combined diagnostic imaging and therapeutic hyperthermia (Hirsch et al. 2003, Huang et al. 2006, Loo et al. 2004). Furthermore, molecular-targeted nanoparticles exhibit increased avidity, and they can be simultaneously decorated with different types of biomolecules that provide multiple functionalities such as molecular-targeting specificity, in vivo delivery, and therapeutic properties (Jiang et al. 2008, Kumar et al. 2008, Michalet et al. 2005). Therefore, nanoparticles readily provide solutions to one of the major challenges of modern medicine—minimally invasive molecular-specific diagnosis and therapy.

The scope of research in the field of biomedical application of nanoparticles is very broad, and therefore in this chapter we discuss only a selection of recent developments in the field of biomedical optics, with a specific focus on applications for in vivo imaging and therapy. This specific research is driven by the inherent optical properties of nanoparticles such as absorption, scattering, or luminescence. Recent reports have covered other exciting areas of study that we do not discuss further here, including the use of nanoparticles for live cell imaging (Lidke et al. 2004), for therapeutic delivery(Peer et al. 2007), and the application of the near-field enhancement around metal nanoparticles in the development of surface-enhanced Raman scattering (SERS) and surface- enhanced fluorescence contrast agents for sensitive multiplexed optical imaging (Keren et al. 2008, Kneipp et al. 2006, Qian et al. 2008, Souza et al. 2006).

First, we will introduce the fundamental optical properties of materials that are used in nanoparticle synthesis and discuss how these properties contribute to the optical characteristics of nanoparticles themselves. Then, we will discuss widely used conjugation chemistries for modification of the surface properties of nanoparticles that determine their in vivo behavior, with a focus on targeting specificity and so-called stealth properties. Next, we provide an outlook on application of nanoparticles in imaging and therapy. And finally, we discuss research activities of particular importance to the clinical translation of nanoparticles that include efficient delivery strategies, optimization of blood residence time and biodistribution, minimization of potential toxicity, and biodegradation and extraction from the body.

32.2 Background

Throughout written history, examples can be found of the use of nanoparticles for their optical properties. For example, one of the first known aesthetic applications dates to Roman times, when gold and silver nanoparticles were used in the coloring of glass (Wagner et al. 2000). The study of nanoparticles and their optical properties, with an associated recognition of the unique size-related aspects, began with Faraday's incidental exploration of gold films and colloids for the purpose of more fully understanding the properties of light (Faraday 1857). In his studies, he developed several methods for synthesizing colloidal gold, and despite the limited instrumentation available, was able to differentiate between the absorption and scattering properties of the colloidal dispersions and dried films that he produced. The electrodynamic theory to fully explain these optical properties was not available until the early 1900s, at which point a complete theory of the optical properties of metal nanoparticles was elaborated by Gustav Mie in a form that continues to be used today (Mie 1908).

Apart from application in photographic emulsion, the first widespread direct application of nanoparticles in biology was the use of colloidal gold conjugated to various proteins as labels for biomolecules in electron microscopy (Faulk and Taylor 1971). Contrast in electron microscopy is proportional to the square of the electron density of the material, making conjugated gold nanoparticles ideal molecular contrast agents, and they continue to be employed for this purpose. As part of the ongoing development of colloidal gold as a molecular contrast agent, non-covalent conjugation methods were refined for attaching a variety of proteins to gold nanoparticles, as were methods for synthesizing monodispersed gold spheres (Frens 1973, Geoghegan et al. 1977, Horisberger 1981). This work paved the

way for the use of gold nanoparticles as optical contrast agents, and in fact, the first papers published in this area used the same synthesis and conjugation methods that had been developed for electron microscopy (Sokolov et al. 2003, Yguerabide and Yguerabide 1998).

Much of the work associated with the development of fluorescent nanoparticles is grounded in the intense focus on the study of the properties of semiconductor materials associated with the technology transition from vacuum tube to transistor electronics beginning at Bell Labs in the early 1950s. In addition, specific optical studies were driven by the search for new laser-active materials, following the first successful demonstration of the LASER in 1960. More specifically, the development of fluorescent nanoparticles can be traced back to the study of the exciton structure in semiconductor materials. Detailed studies of CdS spectra were published in 1959 (Thomas and Hopfield 1959), and it was understood that size could impact the optical absorption of CdS crystals (Berry 1967). It was not until many years later that a detailed study of CdS nanocrystals within the context of quantum confinement led to the invention of quantum dots by Dr. Brus (Rossetti et al. 1983). Their unique optical properties have led to their use in many applications ranging from solar panels to confocal microscopy (Akerman et al. 2002, Chan and Nie 1998, Gao et al. 2007, Nozik 2002). This early work on the optical properties of nanoparticles and on conjugation methods suitable for biological applications has, over the past few decades, led to an explosion in research on the many applications of nanoparticles in biological and optical technology. Currently, many types of nanoparticles are commercially available in forms readily amenable to conjugation with biomolecules, lowering the barrier to entry into the field of nanoparticle applications for biology and medicine.

32.3 State of the Art: Nanoparticles in Biomedical Optics

Nanoparticles in biomedical optics applications are usually composite materials containing a core that acts as an optical contrast and/or therapeutic agent, and an organic surface coating that determines how the particle interacts with biological systems. Core materials include metals and metal oxides, semiconductors, dye-doped silica, carbon nanotubes, and polymers. Surface coatings generally contain passivating agents such as polyethylene glycol (PEG); targeting ligands including aptamers, peptides, antibodies and antibody fragments, and small molecules such as folic acid; and reporter molecules sensitive to enzymes, environmental factors such as pH, or other markers indicative of disease (Barbara et al. 2000, Lee et al. 2009, Michalet et al. 2005). In this section, we will detail the optical properties of these core materials and common methods for modifying their surface chemistry.

A dominant optical feature of biological systems is the so-called tissue optical window spanning a wavelength range from 600 to 1000 nm in the red to near-infrared (NIR) region, where the penetration depth of light is an order of magnitude higher than at other wavelengths (Richards-Kortum and Sevick-Muraca 1996, Welch and Gemert 1995). This wavelength range is delimited by the absorption of hemoglobin and melanin in the UV-visible wavelength range and the absorption of water at infrared wavelengths. For this reason, in the field of biomedical optics, nanoparticles have been optimized to operate at these wavelengths.

32.3.1 Optical Properties of Nanoparticles

32.3.1.1 Optical Properties of Bulk Materials

Optical properties of nanoparticles are best understood by qualifying the optical properties of the bulk materials that comprise the nanoparticles, and then applying empirical corrections that account for the very small size (Kreibig and Vollmer 1995).

Key to application of electrodynamics in human technology is the concept of the constitutive relation, which is an averaging relation allowing macroscopic electrodynamic properties to be expressed in terms of suitably averaged microscopic properties. This concept was initially advanced by Lorentz, through the development of relations for the macroscopic permittivity and permeability, which account for the specific electrodynamic properties of any material. At atomic length scales and at length scales of several times the wavelength of light in the medium, a simple formulation of the constitutive relations is usually sufficient for macroscopic electrodynamics. This leaves the intermediate nanometer length scale, where at present a unified formulation is problematic. At this intermediate length scale, the constitutive relations must be developed on a case-by-case basis. What this means for nanoparticles specifically is that the smallest nanoparticles, consisting of tens of atoms, are best understood quantum mechanically, and as such share properties with small macromolecules. For particles of intermediate size, useful empirical corrections can usually be obtained through consideration of the physical effects of the particle's surface in its interaction with the primary physical entity responsible for the key property concerned, such as electron–surface interactions in terms of their effect on absorption and refractive index properties of metals, surface-induced exciton confinement and recombination effects in semiconductors, and surface-induced crystal field perturbation effects on color center properties in dielectric materials.

In this discussion, bulk materials are divided into three primary groups: metals, semiconductors, and dielectrics. Boundaries between these groups are somewhat flexible and are determined by factors such as specific material preparation, temperature, or wavelength of application. We will focus on the material complex permittivity function to provide specific examples, although it should be understood that analogous considerations would also apply to the permeability function, which is relevant to materials with magnetic and magneto-optic properties. The material complex dielectric permittivity function $\varepsilon(\omega)$ relates directly to the complex refractive index n as $\varepsilon = n^2$, where the optical frequency $\omega = 2\pi c/\lambda$, with c and λ as the velocity and wavelength of light in the medium, respectively. For any material, the primary effect of increases in the real part of the refractive index leads to enhanced scattering, while

increases in the imaginary part of the refractive index leads to enhanced absorption. A useful form of the complex permittivity as a function of electromagnetic frequency, applicable to bulk materials in general, is also attributed to Lorentz (Jackson 1999):

$$\varepsilon(\omega) = 1 + \frac{4\pi N e^2}{m} \sum_j \frac{f_j}{\left(\omega_j^2 - \omega^2 - i\omega\gamma_j\right)} \quad (32.1)$$

This equation expresses the bulk material complex permittivity function in the form of a sum of resonances. Each of the resonance terms has an associated oscillator strength f_j, a resonance frequency ω_j, and peak-width γ_j, the latter corresponding to some degree of energy dissipation in the material; N, e, and m refer to the number of atoms per unit volume, the electron charge, and the mass, respectively. This expression is especially convenient in that each term in the series of resonance terms can be directly related to some physical resonance process in the bulk material. Further, these resonances are often directly observable through associated resonance-enhanced scattering and absorption.

At frequencies ω sufficiently higher than any of the ω_j, Equation 32.1 becomes simply

$$\varepsilon(\omega) = 1 - \frac{\omega_p^2}{\omega^2}, \quad (32.2)$$

where the plasma frequency $\omega_p = \frac{4\pi N Z e^2}{m}$ encapsulates the effect of all Z electrons per molecule, which now oscillate as if they are essentially free electrons (Jackson 1999). Equation 32.2 by itself is often used to describe the properties of metallic materials, and when combined with a phenomenological electron-collision damping factor, it is known as the Drude model (Aschroft and Mermon 1979). For bulk metals in common use, the optical frequency corresponding to the transition between Equations 32.1 and 32.2 occurs deep in the infrared, with the plasma frequency itself far in the ultraviolet.

For semiconductor materials, optical properties are described by Equation 32.1 when the energy corresponding to the optical frequency is below the bandgap energy, and by Equation 32.2 at higher frequencies. In contrast to common metals, in the semiconductor case this transition frequency often occurs in the visible range depending on the material concerned. The implication of this transition is that for wavelengths with energy insufficient to generate an electron–hole pair, optically the semiconductor acts as a pure dielectric, but at higher energies, the optical behavior is consistent with that of a metal.

Finally, for pure dielectric materials at visible wavelengths, the resonance wavelengths are often very far removed from the wavelength range of application, and thus only a very small number of terms is required in the sum of Equation 32.1 to accurately describe the bulk material.

32.3.1.2 Computational Electrodynamics

The ability to simulate the optical properties of nanoparticles or their ensembles is an important aspect in the development and optimization of new approaches in nanoparticle-mediated imaging and therapy. These simulations provide an opportunity to carry out virtual experiments and also allow better understanding of experimentally observed results. Therefore, much effort has been devoted to computational methods that can describe the optical properties of nanoparticles in a quantitative way, leading to significant advances in computational electrodynamics over the past few decades. The implication of this is that at present, with respect to computer resource utilization, the most efficacious techniques to be used for computational simulation of optical interactions with nanoparticles, and nanoparticle aggregates are the T-matrix techniques (Mishchenko et al. 1996), its extension to aggregates (Mackowski and Mishchenko 1996), and the discrete dipole approximation (Draine 1988). The primary advantages of these specific techniques are that in their native implementation they provide complete information about the optical cross sections, and also that well-validated numerical codes are available in the public domain for convenient download (Mishchenko and Travis 1998, Yurkin et al.). In contrast to these more modern techniques, the widely used finite difference time domain (FDTD) technique (Taflove and Hagness 2005) is more appropriate for simulation when the primary objective is to obtain the electrodynamic fields in and around the scattering objects, and for the simulation of structures with complicated spatial variation of the complex dielectric permittivity and magnetic permeability, such as for nanoparticles in interaction with guided wave structures. Although it is certainly possible to use FDTD to calculate optical cross sections, it is actually one of the least efficient techniques to use when this is the primary objective.

32.3.1.3 Plasmonic Nanoparticles

For metals, the frequency-dependent behavior of Equation 32.2 allows the real part of the permittivity to become negative, producing a corresponding resonance in the material polarizability function. For nanoparticles, this metallic resonance is readily seen in consideration of the denominator of the Claussius–Mossoti relation:

$$\alpha(\omega) = \frac{\varepsilon_s(\omega) - \varepsilon_m(\omega)}{\varepsilon_s(\omega) + 2\varepsilon_m(\omega)}, \qquad (32.3)$$

which expresses the volume polarizability α in terms of the complex permittivities of the particle ε_s, and the surrounding medium ε_m. The polarizability, and therefore also the scattering and absorption, are maximum when $\varepsilon_s \approx -2\varepsilon_m$. For small spherical particles, the location of this resonance is at an optical frequency $\omega \approx \omega_p/\sqrt{3}$ (Pitarke et al. 2007).

When analyzed using quantum mechanics, the resonance in the material polarizability corresponds to a quantized surface collective oscillation, and is technically known as a surface plasmon-polariton, colloquially referred to as a plasmon. Such plasmons can exist at the interface between any metallic and dielectric material. Plasmon-polaritons are the combination of a charge density oscillation of the conduction electrons with an

oscillation of the material electric polarization. The complete quantum mechanical treatment of plasmon-polaritons exists, which in general is significantly ahead of experimental observation; that is, there are physical predictions about the behavior of plasmons that have not yet been observed experimentally, yet the theoretical foundation has remained stable for many years (Pitarke et al. 2007). However, given that the basis of most material permittivity functions in use for electrodynamic modeling has been phenomenological, in many cases a full quantum mechanical treatment is not required. In fact, the ability to effectively calculate the electrodynamic interaction for small metal particles has been demonstrated since the early 1900s (Mie 1908), and these calculation results have not been significantly modified by the more recent quantum mechanics paradigm. For nanoparticles of silver, gold, and copper, the plasmon resonance is especially strong, and occurs in the visible region of the spectrum. For the smallest particles, some few nanometers in diameter, the absorption cross section significantly predominates; however, scattering exceeds absorption starting at a few tens of nanometer in diameter, and becomes highly efficient (Kreibig and Vollmer 1995). In general, the total scattering cross section of a single such particle will exceed the conversion cross section of any fluorescent dye molecule by many orders of magnitude. For gold nanoparticles specifically, the high degree of chemical inertness and immunity to photo-bleaching provide strong advantages for applications in biomedical optics.

Exciting recent developments make use of the high sensitivity of the plasmon resonance peak location to perturbations in the local optical field adjacent to the plasmonic nanoparticles (Reinhard et al. 2007, Rong et al. 2008, Sonnichsen et al. 2005, Storhoff et al. 2004). Using both the nonlinear enhancement of the intensity of the total scattering cross section, as well as the modification of the spectral properties of the cross section, imaging techniques have been developed, which have successfully demonstrated sensitivity to nanometer-scale morphology changes in observations of living cells (Figure 32.1) (Aaron et al. 2009).

In order to take advantage of the tissue optical window, metal nanoparticles have been synthesized with strong absorption and scattering in the NIR (Figure 32.2). These particles include gold-coated silica nanoshells (Lin et al. 2005), gold nanorods (Nikoobakht and El-Sayed 2003), hollow gold nanocages (Chen et al. 2005a,b, Sun et al. 2003), and silver prisms (Can and Chad 2007). It has also been demonstrated that molecular-specific clustering of targeted spherical gold nanoparticles produces the desired NIR shift in scattering and absorption properties of nanoparticles (Aaron et al. 2007, Larson et al. 2007, Zharov et al. 2005).

Gold-coated silica nanoparticles, known simply as gold nanoshells, have a broad extinction peak that is centered between approximately 700 and 1050 nm depending on the particular geometry, with most of the extinction due to scattering. As the shell-to-core diameter ratio increases, the optical properties of the particle shift toward those of a solid gold sphere of the same outer diameter, while thinner shells produce more redshifted

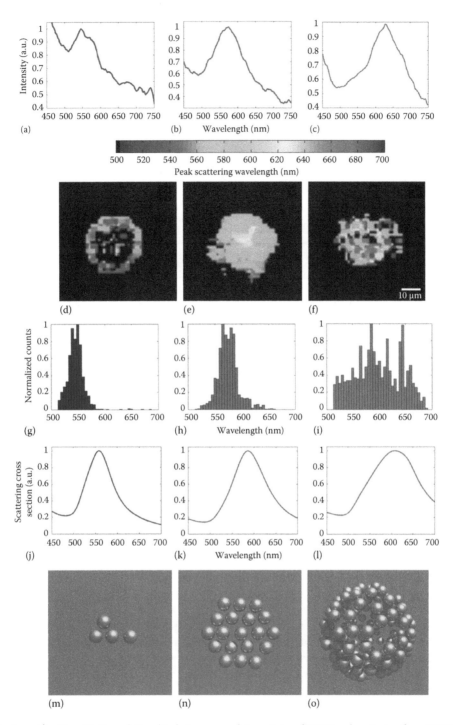

FIGURE 32.1 (See color insert.) Quantitative relationship between regulatory states of EGFR and scattering from EGFR-bound gold nanoparticles. Hyperspectral darkfield microscopy of cells labeled at 4°C (left column, a, d, g), 25°C (middle column, b, e, h), and 37°C (right column, c, f, i). As temperature increases, EGFR complexes with nanoparticles undergo a gradual transition from small clusters to a more 3D volume-filling morphology that is associated with the trafficking of EGF receptors from cell membrane to early and, finally, to late endosomes. Representative single-pixel spectra from a cell image display scattering peaks at ca. 546 (a), 576 (b), and 601 nm (c). Cell images in (d–f) are color-coded according to the scattering peak position at each pixel in the field of view. Distributions of the peak scattering wavelengths indicated in (d–f) are shown in (g–i). Electrodynamic simulations of scattering from nanoparticle aggregates are shown in (j–l) for the following three typical structures: a 4-particle chain-like structure (j, m), a 19-particle disklike structure (k, n), and a 130-particle volume-like structure (l, o). The total scattering cross section is plotted for each structure alongside a rendering of the corresponding detailed particle arrangement (m–o). Distinct redshifting and broadening of the scattering spectra is due to both the effect of the increased number of contributing particles and the effect of the transition from a 2D to a more 3D volume-filling morphology. (From Aaron, J. et al., *Nano Lett.*, 9, 3612, 2009. With permission.)

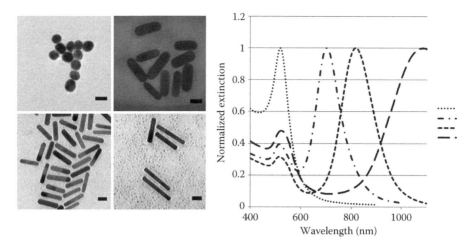

FIGURE 32.2 TEM images and extinction spectra of gold nanoparticles with increasing aspect ratios demonstrates a redshift as the aspect ratio increases. The scale bars are 20 nm, and the extinction spectra are normalized to 1.

extinction peaks. Gold nanocages, with their hollow interior, have properties similar to gold nanoshells (Chen et al. 2005). Gold nanorods have two extinction peaks, generally attributed to the transverse and longitudinal plasmon modes. The transverse peak is centered at 520 nm, near the peak for isolated gold spheres, while the longitudinal mode shifts toward the infrared with increasing aspect ratios. Aggregates of plasmonic nanoparticles have broad peak features that are highly dependent on the specific aggregate morphology, but in general the extinction will red-shift as the interparticle spacing decreases, and as more particles are added to the aggregate or the mean number of particle-to-particle neighbors increases(Aaron et al. 2009).

32.3.1.4 Semiconductors

For semiconductors, optical properties are markedly different depending on the wavelength position with respect to the wavelengths corresponding to the bandgap energy. For wavelengths with energy insufficient to generate an electron–hole pair, the semiconductor acts as a pure dielectric, but at higher energies, optical properties are dominated by broad absorption features. For bulk semiconductors, the kinetics relating to the formation and lifetime of the electron-hole bound state, or exciton, responsible for the most useful optical properties, are strongly temperature-limited, and in fact, usually excitons are difficult to observe as a separate spectral feature except during cryogenic measurements. Semiconductor nanoparticles, in contrast, possess strong size-induced quantum confinement effects that significantly perturb these kinetics, leading to excitons with strong optical interaction at room temperature.

Quantum dots (Q-dots) are semiconductor nanoparticles that have bright fluorescence due to the quantum confinement of the electrons and the associated presence of strongly interacting exciton states. Q-dot absorption is highest in the UV range and generally decreases with increasing wavelength for most types and sizes of Q-dots, while emission wavelength is highly dependent on particle size and composition and can be tuned from the UV to the infrared. Quantum dots that emit primarily in the visible and NIR are composed of cadmium and selenium, sulfur, or tellurium (Murray et al. 1993), while Q-dots with emissions farther into the infrared have been synthesized using indium and lead (Hines and Scholes 2003, Michalet et al. 2005). These nanoparticles range in size from a few nm in diameter to a few tens of nanometers. The wavelengths of emission maxima as a function of particle size are shown in Figure 32.3.

Bare quantum dots are hydrophobic and unstable under UV irradiation and thus require a capping shell of some form for both stability and aqueous solubility (Dabbousi et al. 1997). Suitable materials for the capping shell must have a high degree of chemical stability, a lattice-constant compatible with the core material, and a bandgap of sufficiently high energy so as to function essentially as a dielectric. For example, ZnS with a bandgap of >3.5 eV is compatible with Q-dot cores comprised of ZnSe/CdS/CdSe combinations, all of which have significantly lower bandgap energies. Q-dots are of particular interest in microscopy because their fluorescence emission peak is narrow and easily tuned throughout the visible spectrum by varying the particle size, while their excitation band is broad and does not vary greatly across different sizes of Q-dots (Yu et al. 2003). This allows a single excitation laser to be used with samples containing multiple Q-dot colors, greatly simplifying fluorescence microscopy experiments.

32.3.1.5 Dielectric and Organic Materials: Silica and Polymers

For dielectric materials at optical wavelengths, in many cases the resonances of Equation 32.1 occur far from the wavelength range of application. In this case, the formula for the complex permittivity given in Equation 32.1 reduces to the perhaps more familiar, few-parameter Sellmeier formula, which expresses the square of the material refractive index, as a function of wavelength:

$$n^2(\lambda) = 1 + \frac{B_1\lambda^2}{\lambda^2 - C_1} + \frac{B_2\lambda^2}{\lambda^2 - C_2} + \frac{B_3\lambda^2}{\lambda^2 - C_3} \text{(Sellmeier 1871)}$$

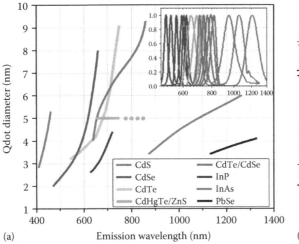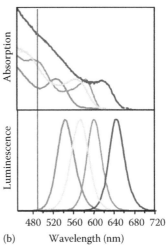

FIGURE 32.3 (See color insert.) (a) Emission maxima and sizes of quantum dots of different composition. The insert shows representative emission spectra for the many types of quantum dots. (b) Absorption and the corresponding luminescence spectra for different types of QDs. (From Michalet, X. et al., *Science*, 307, 538, 2005. With permission.)

Examples of dielectric materials are oxides such as SiO_2, or high bandgap energy chalcogenides, such as ZnS. These materials are useful largely as a matrix within which moieties possessing other desirable optical properties may be substituted, or to serve as a spacer in resonant structures formed of other types of material (such as capping in Q-dots, or the dielectric core of plasmonic core-shells). In notable cases, the matrix itself has properties such that its defects alone have strong optical resonances, as in the case of nitrogen-vacancy-induced color center formation in diamond nanoparticles.

Fluorophore-doped silica nanoparticles, also known as C-dots, provide an alternative to Q-dots that do not contain potentially cytotoxic elements, such as heavy metals (Figure 32.4) (Choi et al. 2007b). Further, organic fluorophores have enhanced photostability and quantum yield when incorporated into a silica matrix; these enhancements can be attributed to the rigidity of the oxide matrix surrounding the fluorophore and

to its relative chemical inertness, respectively. The presence of many fluorophores within each C-dot brings them to within an order of magnitude of the brightness of similarly sized Q-dots. The emission spectra generally match those of the incorporated fluorophore, and thus can be varied as required by simply choosing a suitable fluorophore. Examples in the literature include Ru(bpy) (Chang et al. 2006), TRITC, and Oregon Green 488 (Choi et al. 2007b).

32.3.1.6 Carbon Nanotubes

The general morphology of carbon nanotubes is quite complicated, and possibilities include multiwalled nanotubes (MWNT), extensive and distinct single-walled nanotubes (SWNT), and mixtures thereof, the latter of which will be designated generic carbon nanotubes (GNT). Optical properties of MWNT and GNT are related to those of SWNT, which have either semiconducting or metallic behavior depending on the specifics of the tube diameter and chiral angle, or twist (Odom et al. 1998, Wilder et al. 1998). Due to the broad range of possible nanotube properties, especially semiconducting properties, exciting engineering applications are expected as production and separation technologies improve.

Optical properties of carbon nanotubes are consistent with their classification as either a semiconductor or as a metallic conductor. These properties consist primarily of broad absorption throughout the UV and visible range, with strong ultraviolet and infrared absorption peaks. Due to the structural complexity of the various types of SWNT, and the possibility of mixtures, what is normally observed are three infrared absorption peaks: the first two centered at wavelengths of ~700 nm and ~1 μm due to transitions between different semiconductor bands, and the third, a broad absorption feature due to metallic conduction (Kataura et al. 1999). With respect to the complex permittivity function discussed previously, for a given batch of nanotubes, empirical measurements should be used to calculate

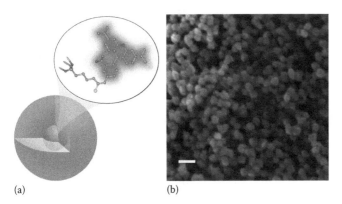

FIGURE 32.4 (a) Schematic of C dots showing TRITC fluorophore bound in the silica core. (b) SEM image of C dots. Scale bar of 50 nm. (From Choi, J. et al., *J. Biomed. Opt.*, 12, 064007, 2007. With permission.)

optimal resonance peak locations ω_j and peak widths γ_j to be used with the general permittivity Equation 32.1, or a single effective plasma frequency ω_p (and possibly a damping constant) to be used with Equation 32.2. With a certain degree of success, nanotubes may also be modeled as a tube comprised of a graphite sheet (Kataura et al. 1999), or as a simple dipole-antenna (De La Zerda et al. 2008), and the latter may be sufficient for applications depending primarily on absorption.

This broad absorption at all wavelengths makes nanotubes most suitable for use in photoacoustic and photothermal applications (Kataura et al. 1999, Levi-Polyachenko et al. 2009), which are discussed in Section 32.3.3.

32.3.2 Specificity and Stealth of Nanoparticles as Influenced by Surface Properties

The interaction between nanoparticles and biological systems is primarily governed by nanoparticle size, shape, and surface properties (Jiang et al. 2008). Surface properties can be modified through chemical conjugation with the goal to achieve certain general characteristics such as surface charge, hydrophilicity, and stability, as well as more specific functionality conferred by materials that include small ligands, peptides, nucleic acids, proteins, and other natural and synthetic polymers. More specific functionalities conferred by these surface moieties include enhanced binding affinity toward key molecular targets in cells and tissues, signaling in response to environmental cues, and efficient delivery to cellular compartments or to a specific disease site in vivo. More general functionalities include so-called stealth properties

such as non-reactivity with off-target molecules, avoiding of opsonization, and escaping quick clearance by the reticuloendothelial system (RES). Nanoparticle conjugation is critical for both in vitro and in vivo applications, as uncoated nanoparticles are colloidally unstable and often cytotoxic in biological solutions (Nastassja et al. 2008). Bare gold nanoparticles aggregate at physiological electrolyte concentrations, and uncoated quantum dots oxidize, decreasing quantum efficiency and photostability while releasing toxic elements into solution. Stable surface coatings can also reduce opsonization in vivo, leading to longer blood residence time and increased accumulation at the target site (Niidome et al. 2006: 253). Stability requirements are defined by the particular application and can range from hours to days and even longer in the case of in vivo applications. There are two main methods of conjugation: physical or non-covalent adsorption, and chemical or covalent attachment. Physical adsorption utilizes electrostatic and van der Waals interactions to promote attachment of larger proteins and polymers to the nanoparticle surface. In general, this form of attachment is not suitable for small molecules due to weak individual interactions. Covalent conjugation methods generally involve modifying either the molecule or the particle surface to add functional groups that are then used in standard bioconjugation techniques; these functional groups include amines, carboxylic acids, and thiols. Thiols are also widely used for their ability to bind strongly to both gold and to semiconductor nanoparticles. Schematics of these general conjugation approaches are shown in Figure 32.5.

Special care must be taken when developing conjugation protocols for biomolecules to avoid denaturation and loss of

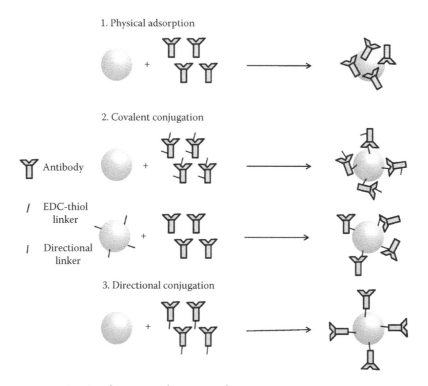

FIGURE 32.5 Common routes to conjugation chemistry with nanoparticles.

function. For example, many bioconjugation methods that utilize amine or carboxylic acid moieties may result in loss of control over the orientation of attached proteins, and may also reduce the activity of a protein due to the presence of amines and carboxylic acids in the protein-active site. Smaller peptides are often synthesized with a cysteine group at one end (Michalet et al. 2005), and nucleic acids can be synthesized with precise control over the placing of several functional groups for conjugation, allowing precise control over orientation (Javier et al. 2008, Rosi et al. 2006). Additionally, there are protocols available for conjugating specifically to the heavy chain of antibodies, allowing greater control over the antibody orientation (Kumar et al. 2008).

32.3.2.1 Physical Adsorption

Physical adsorption, or non-covalent attachment, is a process by which electrostatic and van der Waals interactions are used to attach proteins and polymers to nanoparticle surfaces. This process was the basis for the first conjugation methods developed for proteins and gold nanoparticles (Faulk and Taylor 1971), and has since been expanded to include many types of polymers and nanoparticles. The initial method involved adjusting the pH of a gold colloid solution to the isoelectric point of a protein, and then mixing the protein with gold colloid. More recent implementations involve using an organic buffering system such as HEPES or MPS to conjugate monoclonal antibodies to gold nanoparticles at room temperature (Sokolov et al. 2003).

A similar approach suffices for conjugating carbon nanotubes and can be accomplished by simply mixing cleaned nanotube suspensions with the protein of interest and relying on noncovalent interactions (Chin et al. 2007). Carbon nanotubes have also been conjugated to the RGD peptide by first coating them with an amphiphilic phospholipid-PEG polymer with an amino functional group at one end. The resulting conjugates were found to be stable in serum and showed specific accumulation in a mouse tumor xenograft model (De La Zerda et al. 2008, Liu et al. 2007). This implementation was also used to attach a chelating agent for radiological studies.

Physical adsorption was also applied for functionalization of CTAB-coated positively charged gold nanorods. The method used electrostatic interactions to conjugate antibodies to the nanorods and to make them biocompatible by coating the positive CTAB layer using the negatively charged polymer polystyrene sulfonate, and then coating the polystyrene sulfonate with antibodies through electrostatic interactions (Durr et al. 2007, Murphy et al. 2005). Although physical adsorption can in principle provide a very simple route to nanoparticle surface modification, the stability of the conjugates is suspect. Unfortunately, molecules with higher binding affinity to a nanoparticle surface can potentially replace physically adsorbed proteins. For example, it is energetically favorable for thiol-containing molecules to replace physically adsorbed compounds from gold or semiconductor nanoparticles leading to concomitant loss of functionality and destabilization. As discussed in the next section,

covalent attachment provides a more stable alternative to the physical adsorption methods.

32.3.2.2 Covalent Conjugation

Conjugation of metal and semiconductor nanoparticles often relies on the strong bond that forms between thiols and the nanoparticle surface (Kumar et al. 2007, Tkachenko et al. 2003). Specific implementations include adding thiol-containing linkers to proteins and peptides (Chanda et al. 2009, Kumar et al. 2007), synthesizing peptides with a terminal cysteine group, modifying the metal surface with thiolated linker molecules that contain a functional group suitable for a standard bioconjugation protocol (de la Fuente and Berry 2005), and using oligonucleotide handles with thiol groups at one end (Javier et al. 2008, Rosi et al. 2006). The oligonucleotide handle approach is detailed in Figure 32.6, which shows this method being used for attaching RNA aptamers to a nanoparticle surface. However, it should be noted that this method is not necessarily limited to conjugating nucleic acids; any molecule that can be tagged with a short oligonucleotide handle can potentially be attached to a nanoparticle surface using this technique. Below, we give specific examples for each type of nanoparticle that is generally used in biomedical optics.

Gold nanoparticles. Most synthesized spherical gold nanoparticles are formed in a citrate-containing solution and hence the nanoparticles are coated with negative citrate ions that electrostatically stabilize the gold colloid in salt-free solutions. Typical thiol-containing molecules can easily replace the citrate ions on gold surface to facilitate the stability of nanoparticles in biological environment and to aid in further bioconjugation of the nanoparticles. However, the synthesis procedure for gold nanorods involves the use of cationic surfactant molecules such as cetyl-trimethyl-ammonium-bromide (CTAB) (Jana et al. 2001, Nikoobakht et al. 2003). Unfortunately, these surfactants are cytotoxic and therefore must be replaced for in vivo

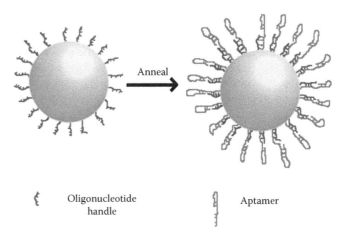

FIGURE 32.6 DNA conjugation schematic. Gold nanoparticles conjugated to oligonucleotide handles are shown on the left. After annealing, extensions on the aptamer facilitate aptamer binding to the oligonucleotide handles, shown on the right.

applications. Several approaches have been explored to replace the CTAB coating of gold nanorods to more biocompatible molecules that facilitate further bioconjugation of nanoparticles with targeting moieties (Yang and Cui 2008, Yu and Irudayaraj 2006). In particular, the surfactant coating has been replaced with the biocompatible molecules containing a thiol group, for example, mPEG-thiol, or the biological recognition molecules, such as antibodies, antibody fragments (Yu and Irudayaraj 2006), and peptides or aptamers with thiol groups (Chanda et al. 2009, Huang et al. 2008).

It is important to note that linkers with two adjacent sulfhydryl groups such as thioctic acid provide greater stability as compared to a single sulfhydryl group (Chanda et al. 2009). It has been shown that molecules conjugated using sulfhydryl-containing linkers can be replaced in biological environment by an excess of naturally present sulfhydryl-containing biomolecules such as glutathione (Oishi et al. 2006).

Quantum dots were synthesized with a tri-*n*-octylphosphine oxide (TOPO) coating and then replaced with thioacetic acid (Akerman et al. 2002). The thioacetic acid was then replaced by a mixture of thiolated PEG and thiolated peptides to enable molecular-specific targeting and greater in vivo stability. In another approach TOPO-capped Q-dots were coated with an amphiphilic block copolymer that contained carboxyl groups (Gao et al. 2004). The carboxyl groups were used for covalent attachment of antibodies and amino-PEG molecules using activation with 1-ethyl-3-[3-dimethylaminopropyl]carbodiimide hydrochloride (EDC). Care must be taken during Q-dot synthesis and functionalization, as exposure to air and ultraviolet light can cause surface oxidation, which then leads to increased cytotoxicity (Derfus et al. 2004).

Carbon nanotubes. At present, there are very limited data on covalent conjugation of carbon nanotubes as most conjugation approaches are based on physical adsorption. One recent example involves carboxylating carbon nanotubes by reflux in a mixture of sulfuric and nitric acid, followed by conjugation to amine-containing compounds using EDC chemistry (Wang et al. 2009).

Silica nanoparticles are typically modified by attaching functional groups, such as carboxyl or amine, on the nanoparticle surface using silane chemistry followed by the covalent attachment of polymers and biomolecules using standard bioconjugation protocols (Choi et al. 2007b). In one instance, C-dots were first coated with 3-aminopropyltriethoxysilane followed by the reaction of the amino groups with a maleimide-succinimidal ester. Maleimide, a thiol-reactive functional group, was used to attach a cysteine-containing oligopeptide (RGD) that targeted integrins.

32.3.2.3 Directional Conjugation

Directional conjugation involves the achievement of a high level of control over the orientation of the biomolecule on the nanoparticle surface. This is particularly important in the conjugation of antibodies to the particle surface as many potential orientations of the antibody lead to a loss of antibody binding

function, but in general orientation is also quite important for other proteins and molecules. Directional conjugation can be accomplished for molecules that have uniquely accessible functional groups. In the case of antibodies, one such functional group is the carbohydrate moiety located on the heavy chain, although it will be noted that this moiety is only available on glycosylated antibodies. This type of carbohydrate can be oxidized to an aldehyde in the presence of sodium periodate, and the resultant aldehyde will react specifically with hydrazide groups. An implementation of this approach is shown schematically in Figure 32.7, where the aldehydes were modified using a heterofunctional linker with hydrazide and dithiol groups on opposing sites of the molecule. The hydrazide moiety interacts with the aldehyde on the antibody and the dithiol then enables conjugation directly to the surface of gold nanoparticles (Kumar et al. 2008). Small peptides can be synthesized with a terminal cysteine group, and nucleic acids can also be synthesized with precise control over the location of a wide range of functional groups useful for conjugation.

32.3.3 Nanoparticle Contrast in Imaging Modalities

32.3.3.1 Confocal Reflectance Microscopy

The first thorough study of gold nanoparticles as scattering contrast agents for optical microscopy was published by Yguerabide and Yguerabide (1998). This idea was later extended for imaging cancer biomarkers using spherical gold nanoparticles conjugated to monoclonal antibodies (Sokolov et al. 2003). The antibodies targeted cell surface receptors that are over-expressed in cancer, and the nanoparticles were shown to greatly enhance backscattered light from labeled cancerous tissue, while showing very little nonspecific binding to healthy tissue (Aaron et al. 2007, Sokolov et al. 2003). Figure 32.8a compares confocal reflectance microscopy images obtained from cancerous tissue labeled with gold nanoparticles targeting epidermal growth factor receptor (EGFR), and from healthy tissue treated with the same nanoparticles. Standard histology was used to verify the tissue classification (Figure 32.8b). The signal intensity in the confocal reflectance showed a gradual increase with increasing progression of cervical dysplasia (Figure 32.8c). The difference in signal between healthy and cancerous tissues is associated with the increased expression of EGFR as cervical cancer progresses from benign to more advanced stages (Barnes and Kumar 2004). This work was the first demonstration of the phenomenon of plasmon resonance coupling for the molecular imaging of tissue ex vivo and in vivo. Activation of EGFR leads to dimerization and concurrent assembly of the receptors; this results in near-field coupling between spatially close gold nanoparticles that specifically interact with receptor molecules. The plasmon resonance coupling between spherical nanoparticles leads to coherent scattering, a significant redshift of more than 100 nm, and broadening of the absorbance of the nanoparticles. These effects allow optimization of imaging wavelengths

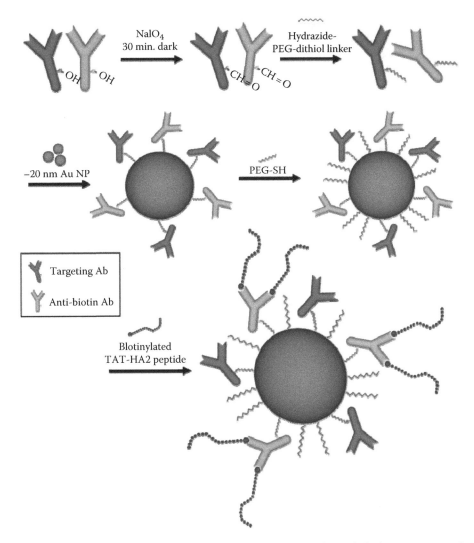

FIGURE 32.7 Schematic for antibody conjugation. First, sodium periodate is used to oxidize carbohydrate moieties on the antibody heavy chain into aldehydes. Next, a hydrazide-PEG-dithiol linker is used to convert the aldehydes to dithiol functional groups. These activated antibodies are then incubated with gold nanoparticles, leading to strong bonds forming between the dithiol linker and the gold surface. Finally, any remaining bare gold surface is passivated with methoxy-PEG-thiol. The figure illustrates simultaneous conjugation of two different antibodies with one of the antibodies (anti-biotin) used for attachment of biotinylated TAT-HA2 peptide that is used for cytosolic delivery of nanoparticles. (From Kumar, S. et al., *Nat. Protocols*, 3(2), 314, 2008. With permission.)

for highly sensitive detection of labeled cells, even in the presence of single unbound gold nanoparticles. The spectral shift associated with spheres was also used to follow EGFR trafficking in live cells following EGF stimulation (Figure 32.1) (Aaron et al. 2009). Aaron et al. used live cell dark-field microscopy to image dynamic behavior of EGF receptors labeled with spherical gold nanoparticles. After stimulation with a growth factor, the receptors dimerize and are then internalized into endosomes inside the cell, causing aggregation of the attached nanoparticles. These steps in receptor trafficking led to a progressive redshift in the scattering signal, enabling the determination of receptor regulatory states using a straightforward analysis of color images of labeled cells. It should be noted that more complex effects are associated with the coupling of plasmon resonances between nonsymmetric particles such as gold nanorods; in this

case, either a red or blue shift of the longitudinal peak can be observed depending on the relative orientation of the coupling nanoparticles (Funston et al. 2009, Thomas et al. 2004).

32.3.3.2 Optical Coherence Tomography

Initial studies have explored the use of the strong scattering properties of plasmonic nanoparticles for enhancement of molecular-specific contrast in both conventional and spectroscopic optical coherence tomography (OCT) (Cang et al. 2005a,b, Chen et al. 2005a,b, Gobin et al. 2007, Loo et al. 2005, Skrabalak et al. 2008). For example, contrast enhancement was observed in mouse xenograft tumors in vivo after systemic injection of nanoshells (Figure 32.9). However, there are certain limitations that are associated with this approach. Smaller gold particles, including spheres and nanorods below about 50 nm

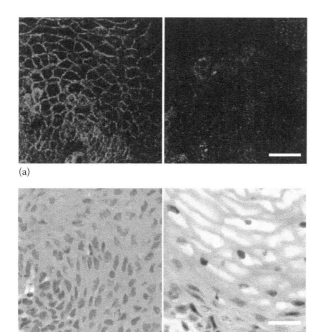

(a)

(b)

Histopathology of the abnormal sample	Reflectance intensity ratio of abnormal to normal samples after labeling with anti-EGFR gold nanoparticles
HPV associated changes	1.7
CIN 1	4.2
CIN 2	11.6
CIN 3	21

(c)

FIGURE 32.8 (a) Confocal reflectance images of transverse sections of abnormal (left) and normal (right) cervical biopsies incubated with gold nanoparticles conjugated to anti-EGFR antibodies. The abnormal biopsy was imaged directly above the basement membrane. (b) H&E staining of the same cervical biopsies as in (a). (c) The average signal intensity increase as a function of increasing dysplasia. As the tissue approaches a more cancerous state, reflectance signal from gold nanoparticle labeling increases. (From Aaron, J. et al., *J. Biomed. Opt.*, 12, 2007. With permission.)

in diameter or length, are primarily absorbers, with relatively little scattering contribution to their extinction cross sections. Because of this, their presence in tissue leads to a drop in the observed OCT signal, making them a form of negative contrast agent as has also been shown with gold nanocages (Chen et al. 2005a,b). Larger gold nanoshells have a much higher scattering to absorption ratio, but their scattering is mostly in the forward direction, away from the OCT detector, which observes backscattered light.

Because of this combination of factors, a more indirect method of detecting gold nanoparticles via OCT has been developed that takes advantage of the strong absorption of gold nanoparticles. This method uses phase-sensitive OCT in combination with an additional time-modulated laser beam to heat the local environment adjacent to the absorbing nanoparticles (Adler et al.

2008, Skala et al. 2008). The refractive index of the local environment shifts with the temperature, and this is detected by phase-sensitive OCT. This technique has been hypothesized to enable better discrimination between nanoparticles and tissue compared to the direct detection of the scattered light. In Figure 32.10, the presence of cells labeled with spherical gold nanoparticles in a tissue phantom led to the observation of detectable changes in phase due to heating induced by the time-modulated laser beam. The frequency and amplitude of modulation is detected using a Fourier transform of the phase versus time data.

32.3.3.3 Fluorescence Imaging

Fluorescence-based optical methods, with nanoparticles as bright and robust contrast agents, are used in a variety of applications. These include single-molecule detection and molecular imaging in cells (Giepmans et al. 2005, Matsuno et al. 2005), as well as in vivo applications such as sentinel lymph node mapping (Frangioni et al. 2007, Kim et al. 2004) and imaging molecular markers of cancer (Diagaradjane et al. 2008, Gao et al. 2004).

Sentinel lymph node mapping is an important diagnostic procedure that is used to determine tumor spread to lymph nodes, which is widely used for detection of metastasis in melanoma and breast cancer. The current technique requires multiple injections that include radionuclide tracers and an invasive surgical procedure. Therefore, Q-dots provide an attractive nonradioactive alternative that can significantly simplify the visualization of sentinel lymph nodes. In a recent study, Q-dots were injected intradermally in a mouse and a pig animal model and then a fluorescent imager was used to detect quantum dot accumulation in a sentinel lymph node (Figure 32.11) (Frangioni et al. 2007, Kim et al. 2004).

Because many different colors of Q-dots can be excited at the same wavelength, they are ideal for multiplex fluorescent imaging. Gao et al. used three different colors of Q-dots to visualize labeled-cell populations that were injected under the skin of a mouse (Figure 32.12) (Gao et al. 2004). Multiplexing with Q-dots in vivo was also demonstrated in more realistic settings by imaging two different molecular targets in a cancer tumor (Stroh et al. 2005), and by imaging multiple lymphatic basins in a mouse (Kobayashi et al. 2007).

It was also shown that Q-dots can be excited using the effect of bioluminescence resonance energy transfer (BRET). Q-dots were conjugated with luciferase that catalyzes oxidation of the substrate coelenterazine; this reaction results in the release of bioluminescence energy that is non-radiatively transferred to the Q-dots, which in turn emit red photons. These self-illuminating Q-dots can be used to obtain much higher signal-to-noise ratios by avoiding the need for external illumination, which generally causes unwanted background from tissue autofluorescence (So et al. 2006).

Multiphoton fluorescence microscopy has effectively utilized both plasmonic nanoparticles and quantum dots as contrast enhancement agents (Bateman et al. 2007, Durr et al. 2007). Multiphoton imaging provides advantages over traditional fluorescence microscopy in tissues due to the deeper penetration at

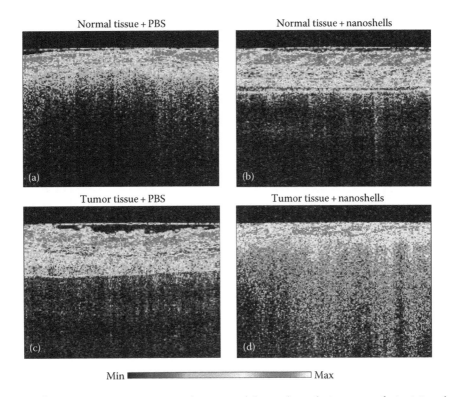

FIGURE 32.9 (See color insert.) Representative OCT images from normal skin and muscle tissue area of mice injected systemically with either PBS (a) or gold nanoshells (b). (c and d) show representative OCT images from tumor xenografts following systemic injection of either PBS (c) or gold nanoshells (d). Scale bar is 200 microns. (From Gobin, A.M. et al., *Nano Lett.*, 7, 1929, 2007. With permission.)

the excitation wavelength and to the inherent depth-sectioning accomplished without the use of a pinhole or other more complicated methods, thus allowing in vivo 3D imaging. In order to demonstrate the concept using gold nanoparticles, Durr et al. developed tissue phantoms consisting of cells overexpressing EGFR labeled with gold nanorods. The labeled cells were embedded in a collagen matrix and imaged using twophoton fluorescence. The increase in signal contrast of nanorodlabeled cells over unlabeled cells was three orders of magnitude, owing to the high signal of nanorods and low endogenous contrast of cells. Representative images acquired of tissue phantoms containing labeled and unlabeled cells are shown in Figure 32.13. Significantly higher power was used for imaging of unlabeled cells in comparison to that used for labeled cells. The spatial distribution of signal indicates that the nanorods were observed primarily on the surface membrane of the cells, and that they are consistent with the labeling of EGF receptors, whereas endogenous fluorophores are distributed throughout the cytosol.

Q-dots were used to image blood vessels in a transgenic mouse tumor model using multiphoton microscopy. In this case, Q-dots were adsorbed onto larger micron-sized micelles in order to ensure that they would remain in the blood vessels, as smaller constructs could extravasate into tissue causing blurring in images of the vessel. The Q-dot/micelle construct showed advantages over dye-conjugated dextran in the delineation of blood vessels (Figure 32.14). Figure 32.14c shows clear

separation between blood vessels labeled with blue QD470 Q-dots and GFP-labeled perivascular cells, and Figure 32.14d demonstrates simultaneous imaging of the blood vessels labeled with red-emitting QD660 Q-dots and second-harmonic signal from the surrounding tissue.

Another recent study showed that multiphoton confocal microscopy with quantum dot probes can improve resolution by 1.7-fold over conventional confocal microscopy (Hennig et al. 2009). This method relies on the recombination of three excitons in a single QD, leading to a blue-shift of the emitted photon and thus an increase in spatial resolution. While this method led to increased spatial resolution, its reliance on blue light emission makes it less suitable for in vivo applications.

32.3.3.4 Photoacoustic Imaging

Photoacoustic imaging is a combined optical and ultrasonic imaging modality that uses pulsed lasers to rapidly heat the local environment of strong absorbers, generating an acoustic response that is then detected using an ultrasonic transducer. Plasmonic nanoparticles with their strong optical absorption are ideal contrast agents for photoacoustic imaging (Copland et al. 2004). This was demonstrated using molecular-specific gold nanospheres undergoing receptor-mediated aggregation (Mallidi et al. 2007) and has also been demonstrated with gold nanorods (Song et al. 2009). Mallidi et al. used PEG-coated and antibody-coated gold nanoparticles targeting EGFR to demonstrate that photoacoustic imaging is able to distinguish between

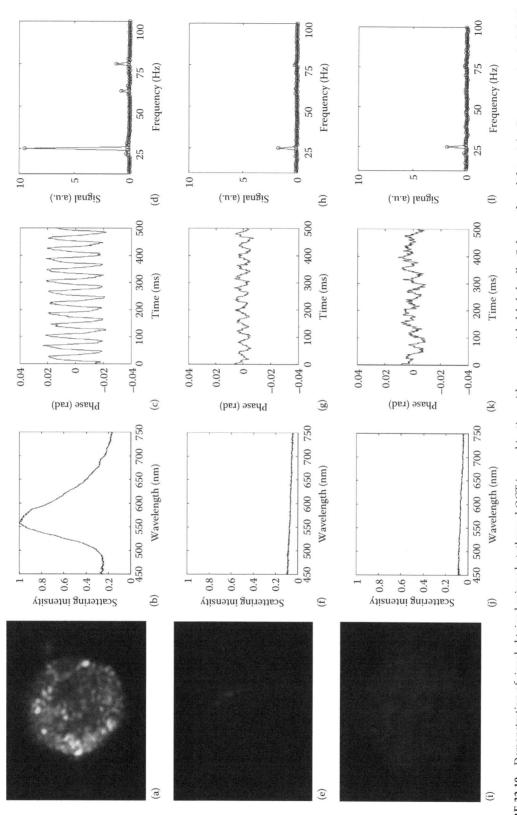

FIGURE 32.10 Demonstration of signal obtained using photothermal OCT in combination with nanoparticle-labeled cells. Columns from left to right: Representative dark field microscopic images, extinction spectra of tissue phantoms, phase shift plotted as a function of time, and frequency plots of the phase shift signal showing a peak at the modulation frequency. (a–d) EGFR(+) cells incubated with gold nanoparticles targeting EGFR. (e–h) EGFR(−) cells incubated with EGFR-targeted gold nanoparticles. (i–l) EGFR(+) cells alone. (From Skala, M.C. et al., *Nano Lett.*, 8, 3461, 2008. With permission.)

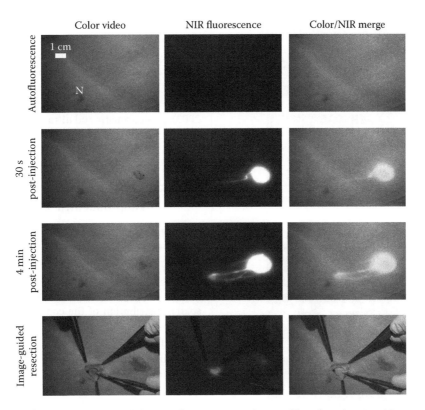

FIGURE 32.11 (See color insert.) Fluorescence-guided surgical resectioning of sentinel lymph node using QD contrast agents. The blue color is trypan blue, the current clinical standard for detecting sentinel lymph nodes. Quantum dots were observed to offer much greater contrast for detecting sentinel lymph nodes in combination with a fluorescence imaging system. (From Kim, S. et al., *Nat. Biotechnol.*, 22, 93, 2004. With permission.)

a mixture of cells and nontargeted gold nanoparticles and cells specifically labeled with anti-EGFR nanoparticles (Figure 32.15). The authors prepared three biologically relevant tissue models: unlabeled 3D tissue phantom containing just cells, cells in the presence of nontargeted PEG-coated nanoparticles, and specifically labeled cancer cells. It was demonstrated that photoacoustic imaging with 520 nm laser excitation is able to detect the presence of nanoparticles in tissue phantoms while 680 nm excitation selectively detects labeled cancer cells (Figure 32.15). This sensitivity of photoacoustic imaging with 680 nm excitation was attributed to the receptor-mediated aggregation of gold nanoparticles, which results in strong redshift and broadening of absorption and thereby enhances the detection using red and NIR excitation. Indeed, the peak absorption of isolated gold nanoparticles is near 520 nm, coincident with the 532 nm excitation, while gold nanoparticle aggregates have significantly redshifted absorption; this results in a significant increase in the signal at 680 nm excitation for gold nanoparticle aggregates. This concept has been expanded by the use of multiple illumination wavelengths to obtain spectroscopic data about the sample (Mallidi et al. 2009). This information can be used to distinguish between contrast agents and local tissue or to distinguish between contrast agents in different states, enabling spectroscopic photoacoustic imaging to obtain functional information about the biological environment.

Carbon nanotubes are another material showing promise as a photoacoustic contrast agent. De la Zerda et al. showed that nanotubes conjugated to an RGD peptide fragment accumulated in a mouse xenograft tumor to a greater extent than plain nanotubes at the 4 hour time point. This accumulation was detected with photoacoustic imaging (De La Zerda et al. 2008). In addition to their pharmacokinetic studies, the authors also quantified the photoacoustic signal as a function of carbon nanotube concentration injected intradermally into mice (Figure 32.16). It was shown that the signal is linear with increasing nanotube concentration in the range from 50 nM up to 600 nM.

32.3.4 Photothermal Therapy

There is much interest in the use of NIR-absorbing nanoparticles and nanoparticle clusters for photothermal therapy (Hirsch et al. 2003, Huang et al. 2006, Larson et al. 2007, Pitsillides et al. 2003, Zharov et al. 2005). In this application, nanoparticles are first delivered to the target tissue via either local injection or systemic delivery, and then NIR light is delivered to the tissue and selectively absorbed by the nanoparticles leading to the localized conversion of light to heat. This results in highly selective heating and photothermal destruction of the targeted tissue. This method has largely been proposed as a cancer therapy, although potentially it has application to other therapeutic objectives.

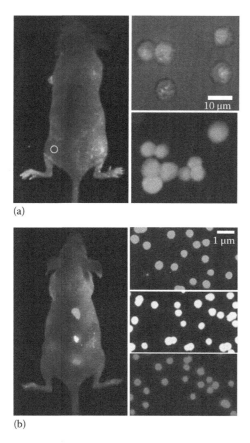

(a)

(b)

FIGURE 32.12 (See color insert.) Sensitivity and multicolor capability of QD imaging in live animals. (a) Comparison between QD-tagged (right flank) and GFP-transfected cancer cells (left flank) shows the difficulty of detecting implanted GFP-expressing cancer cells in vivo while QD-tagged cells are easily observed. (b) Simultaneous in vivo imaging of multicolor QD-encoded microbeads demonstrates the ability to image multiple colors of quantum dots simultaneously. (From Gao, X. et al., *Nat. Biotechnol.*, 22, 969, 2004. With permission.)

Photothermal therapy has been demonstrated using nanorods (Huang et al. 2006), nanoshells (Hirsch et al. 2003), nanocages (Au et al. 2008), and aggregates of nanospheres (Larson et al. 2007, Zharov et al. 2005). The first in vivo demonstration involved the use of gold nanoshells that were injected into a mouse xenograft tumor and CW NIR irradiation combined with the use of MRI for temperature monitoring (Hirsch et al. 2003). Tumor sites treated with nanoshells showed a maximum 34°C increase in temperature (Figure 32.17a), while the saline control showed a 13°C temperature increase as seen in Figure 32.17b. It has also been shown that nanoparticles could be targeted to specific cells, and even to specific proteins, and that these cells and proteins could be selectively destroyed using pulsed irradiation (Pitsillides et al. 2003). It is hypothesized that pulsed irradiation results in microcavitation, or small underwater explosions, which disrupts cellular membranes leading to cell necrosis, while CW irradiation leads to increased temperature over time and photothermal destruction of cells. Real-time monitoring of

temperature during treatment is important for two reasons: in order to ensure that the amount of heat is sufficient to kill all cancer cells within the tumor margins and also to ensure that collateral damage to the surrounding healthy tissue is minimized. In addition to MRI that was used for monitoring the treatment in the therapeutic study with nanoshells (Hirsch et al. 2003), photoacoustic and ultrasound imaging have been recently explored as alternative methods for monitoring photothermal therapy (Shah et al. 2008b).

32.3.5 Applications for Combined Diagnosis and Therapy

Nanoparticle systems have been proposed to provide feedback during therapy in order to individually tailor a course of treatment in real time (Bagalkot et al. 2007, Larson et al. 2007). Two diagnostic parameters that provide important information about the progress of the treatment are quantification of the spatial distribution of the concentration of delivered nanoparticles and quantification of therapeutic efficacy at the target site. Quantification of delivery is accomplished through the use of nanoparticles that are detectable using whole-body or deeply penetrating imaging modalities. For example, in the case of cancer, whole-body techniques such as x-ray, PET, and MRI allow the possibility for the detection of previously unknown micrometastases that express a sufficient concentration of the targeted cancer marker. As mentioned in previous sections, nanoparticles are detected in vivo using a variety of optical methods that are able to precisely delineate tumor boundaries, and significant fundamental research has been undertaken in order to calibrate signal strength to specific nanoparticle concentration. In the case where nanoparticle accumulation is associated with expression of disease biomarkers, nanoparticles can provide important information on disease progression while simultaneously serving as therapeutic contrast agents for photothermal therapy or as carriers for therapeutic compounds. This approach has been demonstrated in vitro using antibody-coated gold/iron oxide hybrid particles with MRI for detection, and pulsed NIR laser irradiation for photothermal destruction (Figure 32.18) (Larson et al. 2007), and with antibody-coated gold nanoshells and nanocages with OCT for detection and continuous wave NIR irradiation for photothermal destruction (Gobin et al. 2007). Recently, it was demonstrated in tissue mimicking phantoms that photoacoustic imaging can be applied for both the detection of plasmonic nanoparticles and for monitoring of temperature during nanopaticle-mediated photothermal therapy (Shah et al. 2008a).

In general, real-time feedback on therapeutic efficacy is difficult to obtain, but there are a few notable demonstrations. Bagalkot et al. designed a Q-dot-based system that targets tumor cells, using aptamers against cancer surface markers and simultaneously carrying doxorubicin, a highly toxic anticancer chemotherapeutic; a schematic of this system is shown in Figure 32.19. Doxorubicin is a fluorescent DNA-intercalating dye, and when incorporated into nucleic acids on a Q-dot surface a bi-FRET mechanism leads to the quenching of both the Q-dot and

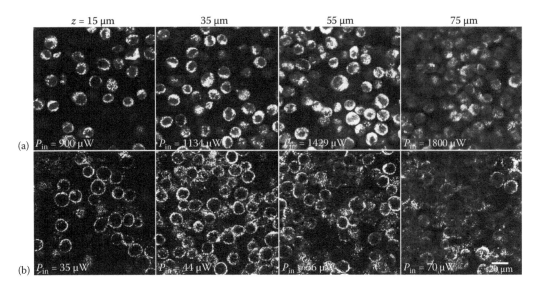

FIGURE 32.13 Two-photon imaging of cells embedded in collagen tissue phantoms composed of A431 cells. (a) Unlabeled A431 cells and (b) A431 cells labeled with gold nanorods targeting EGFR. Images are obtained at up to 75 microns depth in tissue phantoms. Note the difference in laser power required for imaging of labeled and unlabeled samples. (From Durr, N.J. et al., *Nano Lett.*, 7, 941, 2007.)

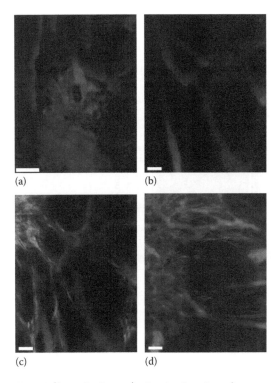

FIGURE 32.14 (See color insert.) In vivo imaging of a mouse xenograft tumor using multiphoton microscopy. (a) Depth projection of 20 images taken at 5 micron intervals demonstrates extravasation of fluorescent dextran causing blurring of tumor blood vessels. (b) Depth projection of 15 images at 5 micron intervals with QD470 contrast agent present in bloodstream. (c) Simultaneous imaging of GFP and QD470 used to visualize both perivascular cells and blood vessels, taken in the same region as (b). (d) Imaging of endogenous tissue second-harmonic generation in blue and QD660 contrast agents to visualize the blood vessels in red. (From Stroh, M. et al., *Nat. Med.*, 11, 678, 2005. With permission.)

the doxorubicin fluorescence. When the drug is released from the Q-dot inside target cells, both doxorubicin and the Q-dots fluoresce, providing feedback that the chemotherapeutic drug has been delivered and separated from its nanoparticle carrier.

32.3.6 Research in Efficient Delivery, Biodistribution, Degradation, and Extraction from the Body

Nanoparticle pharmacokinetics and long-term toxicity are important considerations for in vivo applications of nanoparticle-based diagnostic and therapeutic agents. Although preliminary work has been published on the in vitro stability of nanoparticles, it is essential to determine the biodistribution, biocompatibility, and clearance behavior of nanoparticles in order to ensure their safe use for in vivo biomedical applications.

There are two primary factors that need consideration while designing nanoparticle-based agents for clinical application. First, the agent should have sufficiently long blood residence time to accumulate at the target site after systemic delivery. Second, the agent should be cleared from the body after the intervention is complete in order to avoid potential long-term toxicity. Nanoparticles with hydrodynamic size less than about 6 nm undergo quick renal clearance that can significantly decrease their accumulation at target sites (Choi et al. 2007a). In contrast, larger nanoparticles do not clear quickly, but may have long-term retention in the body that potentially leads to adverse effects due to toxicity after degradation of the nanoparticles or chronic toxicity to the liver and spleen after prolonged presence. Intravenous administration is the most effective method for the delivery of imaging and therapeutic agents as the bloodstream very quickly distributes the administered agent throughout the body. The fate of any agent after injection depends on numerous

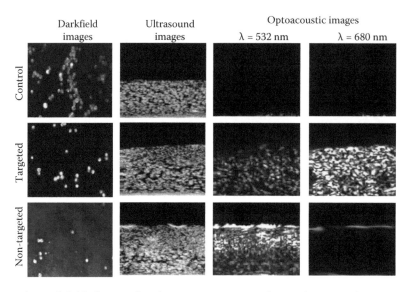

FIGURE 32.15 (See color insert.) Dark field, ultrasound, and optoacoustic images of tissue phantoms taken at 532 and 680 nm excitation wavelength. The control phantom contained cells and collagen, the nontargeted phantom contained cells and PEGylated spherical gold nanoparticles of 40 nm diameter, and the targeted tissue phantom contained cells specifically labeled with anti-EGFR gold nanoparticles, as seen in the dark field images. Photoacoustic imaging shows an increase in signal associated with receptor-mediated nanoparticle aggregation as seen by the difference in signal at 680 nm excitation between the PEGylated and EGFR-targeted nanoparticle tissue phantoms. (From Mallidi, S. et al., *Opt. Express*, 15, 6583, 2007.)

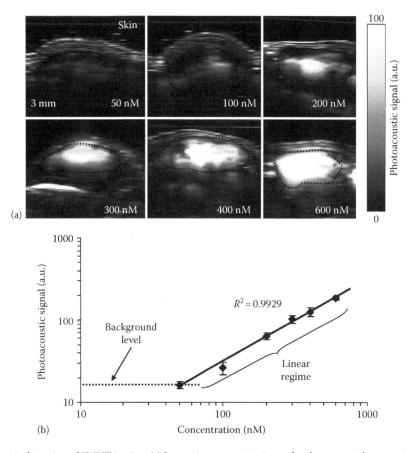

FIGURE 32.16 Photoacoustic detection of SWNT in vivo. (a) Increasing concentrations of carbon nanotubes were injected subdermally in mice and detected using photoacoustic imaging. (b) The photoacoustic signal as a function of nanotube concentration shows a linear behavior in the 50–600 nM range. (From De La Zerda, A. et al., *Nat. Nanotechnol.*, 3, 557, 2008. With permission.)

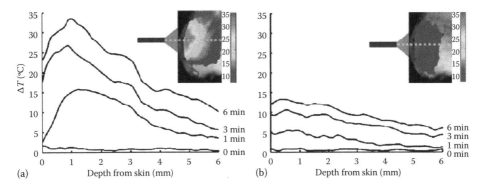

FIGURE 32.17 Photothermal therapy of tumors with nanoshells using MRI monitoring of temperature increase. The inserts show the temperature increase distribution throughout the tumor following laser irradiation. (a) A mouse xenograft tumor with nanoshells injected shows a maximal change in temperature of 34°C following NIR irradiation. (b) A xenograft tumor without nanoshells shows a much lower increase in temperature following the same NIR treatment. (From Hirsch, L.R. et al., *Proc. Natl. Acad. Sci. U. S. A.*, 100, 13549, 2003. With permission.)

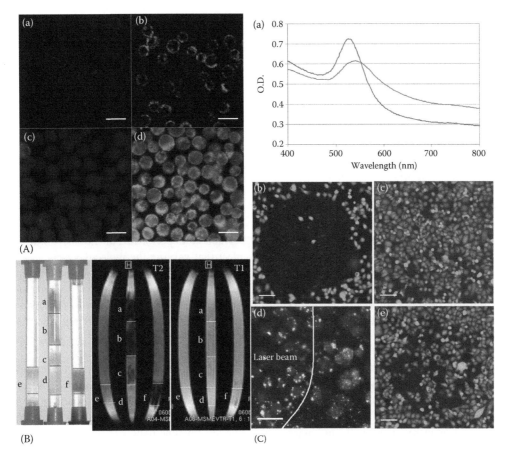

FIGURE 32.18 (See color insert.) Hybrid plasmonic magnetic nanoparticles (AuFeOx) with both optical and magnetic properties allow a combination of optical and magnetic resonance imaging and photothermal therapy. (A) Confocal (top row) and dark field (bottom row) reflectance microscopy demonstrate clear difference in signal between EGFR(+) (left column) and EGFR(−) cells (right column) after incubation with EGFR-targeted AuFeOx nanoparticles. (B) MRI images of tissue phantoms containing different concentrations of cells labeled with gold nanoparticles. Negative T2 contrast increases as the concentration of cells labeled with AuFeOx nanoparticles increases from (d) to (a) in the middle MRI tube. (C) (a) Absorbance spectra of unlabeled cells in the presence of PEGylated hybrid nanoparticles (green) and of labeled cells (red). The total amount of nanoparticles is the same in both cell suspensions. Fluorescence images of cells labeled with anti-EGFR hybrid nanoparticles (b) and cells pre-exposed to PEG-coated particles (c) after one 7 ns, 400 mJ/cm² laser pulse at 700 nm. Green fluorescence is due to calcein AM live stain that indicates cellular survival. (d) Dark field reflectance image of the irradiated labeled cells at the boundary of the laser beam spot shows that the cells do not detach after laser treatment. (e) Fluorescence image of cells pre-exposed to PEG-coated hybrid nanoparticles after 600 hundred 7 ns, 400 mJ/cm² pulses at 700 nm. The scale bars are ca. 100 μm in fluorescence images and ca. 50 μm in the dark field image. (From Larson, T.A. et al., *Nanotechnology*, 18, 325101, 2007. With permission.)

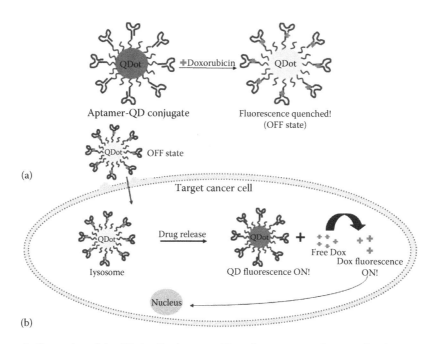

FIGURE 32.19 (a) Schematic illustration of the QD-Apt(Dox) system. First, the QD core is functionalized with an anti-PSMA aptamer, then doxorubicin (Dox) is added and intercalated into the aptamer on the QD. The presence of Dox leads to the quenching of both Dox and the QD. (b) Once the nanoparticle construct is internalized into cells, the Dox diffuses away from the QDs into the cytosol, allowing both the QDs and the Dox to be detected via fluorescence. (From Bagalkot, V. et al., *Nano Lett.*, 7, 3065, 2007. With permission.)

factors with size, shape, and surface coating of the nanoparticles playing crucial roles (Chithrani and Chan 2007, Jiang et al. 2008, Jong et al. 2008, Schipper et al. 2009). It has been shown that the clearance of circulating nanoparticles from the bloodstream and the associated uptake by the liver and spleen are significantly reduced by a neutral or hydrophilic coating of the particle surface such as is afforded by PEG (Niidome et al. 2006). In the case of applications in cancer imaging and therapy, administration of nanoparticles leads to specific accumulation in tumor sites due to enhanced permeability and retention (EPR) effect with the load in the tumor being directly proportional to the blood residence time. Eventually, nanoparticles are cleared from blood by the reticuloendothelial system (RES) or by the kidneys. Generally, particles larger than 200 nm are cleared by the spleen, while nanoparticles smaller than 100 nm are mainly cleared by the liver (Arruebo et al. 2007), and nanoparticles with hydrodynamic size smaller than about 5.5 nm undergo effective renal clearance (Sadauskas et al. 2007). The longer term fate of nanoparticles accumulated in the liver and spleen has not yet been fully studied, although more recent studies have indicated that small quantum dots (<4 nm) and gold nanoparticles (5 nm) can be excreted through feces (Balogh et al. 2007, Meike et al. 2009). The efficacy of this excreted pathway decreases by an order of magnitude when particle size increases to 11 nm.

Recently, several groups have recognized that the initial size of the nanoparticle agent does not necessarily limit longer term systemic clearance. For example, Park et al. reported luminescent silicon nanoparticles that biodegrade in vivo to sub 6 nm size to facilitate quick renal clearance (Park et al. 2009). Troutman et al. also demonstrated a similar approach through the reduction of

gold onto liposomes to result in a nanoshell-like geometry. These gold-liposome constructs were shown to degrade in vitro in the presence of surfactants to gold nanoparticles with sizes ranging from 4.5 to 8 nm (Troutman et al. 2008). In order to more fully explore the effects of surface coatings on pharmacokinetics, Perrault et al. performed an extensive study demonstrating the effect that varying the molecular weight of PEG molecules on the nanoparticle surface has on blood residence time (Perrault et al. 2009). In particular, they found that increasing the molecular weight of the PEG coating increases the blood residence time as much as 10-fold. Much work remains to determine the effects of mixing molecules such as PEG with targeting peptides and antibodies on pharmacokinetics.

Frangioni et al. demonstrated that the charge of the surface coating has a dramatic effect on the adsorption of serum proteins in vivo, actually leading to changes in the particle's hydrodynamic radius (Choi et al. 2007a). They demonstrated that the smallest possible nanoparticles with efficient clearance and an unaltered hydrodynamic size in vivo are fabricated using zwitterionic molecules. However, gold nanoparticles of 5 nm diameter coated with positive surface charge showed better excretion in urine and feces than negatively charged or neutral counterparts (Balogh et al. 2007). More work needs to be done to elaborate the details of surface coating-related effects on clearance patterns of different types of nanoparticle. It is possible that nanoparticle clearance from target organ sites such as tumor will proceed through reabsorption into blood capillaries for particles less than about 10 nm, while larger complexes between 10 and 100 nm will be filtered into the lymphatic system (Swartz 2001).

32.4 Summary

Initial research into the use of nanoparticles in biomedical applications has focused on the basics of nanoparticle synthesis, characterization of optical properties, and the conjugation of biomolecules to the particle surface. These results have enabled studies, primarily of an exploratory nature, of nanoparticle-enabled imaging and therapy in biologically relevant tissue models, ranging from tissue phantoms and cell cultures all the way to in vivo animal models. As synthesis and conjugation abilities have improved, more complex questions related to the behavior of nanoscale materials in vivo have been posed. These questions relate to the ability of nanoparticles to target specific tissues, the effect of different surface coatings, and the clearance or long-term accumulation of these nanoparticles in biological systems. Addressing these questions through further development of nanoparticle systems, in combination with imaging and therapeutic approaches, will determine the role that nanoparticles play in future clinical practice.

References

Aaron, J., Nitin, N., Travis, K. et al. 2007. Plasmon resonance coupling of metal nanoparticles for molecular imaging of carcinogenesis in vivo. *J. Biomed. Opt.* 12(3): 034007.

Aaron, J., Oh, J., Larson, T. A., Kumar, S., Milner, T. E., and Sokolov, K. V. 2006. Increased optical contrast in imaging of epidermal growth factor receptor using magnetically actuated hybrid gold/iron oxide nanoparticles. *Opt. Express* 14: 12930–12943.

Aaron, J., Travis, K., Harrison, N., and Sokolov, K. 2009. Dynamic imaging of molecular assemblies in live cells based on nanoparticle plasmon resonance coupling. *Nano Lett.* 9: 3612–3618.

Adler, D. C., Huang, S. W., Huber, R., and Fujimoto, J. G. 2008. Photothermal detection of gold nanoparticles using phase-sensitive optical coherence tomography. *Opt. Express* 16: 4376–4393.

Akerman, M. E., Chan, W. C. W., Laakkonen, P., Bhatia, S. N., and Ruoslahti, E. 2002. Nanocrystal targeting in vivo. *Proc. Natl. Acad. Sci.* 99: 12617–12621.

Alivisatos, P. 2004. The use of nanocrystals in biological detection. *Nat. Biotechnol.* 22: 47–52.

Anker, J. N., Hall, W. P., Lyandres, O., Shah, N. C., Zhao, J., and Van Duyne, R. P. 2008. Biosensing with plasmonic nanosensors. *Nat. Mater.* 7: 442–453.

Arruebo, M., Fernández-Pacheco, R., Ibarra, M. R., and Santamaría, J. 2007. Magnetic nanoparticles for drug delivery. *Nano Today* 2: 22–32.

Aschroft, N. and Mermon, N. 1979. *Solid State Physics.* Philadelphia, PA: W.B. Saunders.

Au, L., Zheng, D., Zhou, F., Li, Z.-Y., Li, X., and Xia, Y. 2008. A quantitative study on the photothermal effect of immuno gold nanocages targeted to breast cancer cells. *ACS Nano* 2: 1645–1652.

Bagalkot, V., Zhang, L., Levy-Nissenbaum, E. et al. 2007. Quantum dot-aptamer conjugates for synchronous cancer imaging, therapy, and sensing of drug delivery based on bi-fluorescence resonance energy transfer. *Nano Lett.* 7: 3065–3070.

Balogh, L., Nigavekar, S. S., Nair, B. M. et al. 2007. Significant effect of size on the in vivo biodistribution of gold composite nanodevices in mouse tumor models. *Nanomed. Nanotechnol. Biol. Med.* 3: 281–296.

Barbara, S., Silvia, A., Maria Teresa, P. et al. 2000. Design of folic acid-conjugated nanoparticles for drug targeting. *J. Pharm. Sci.* 89: 1452–1464.

Barnes, C. J. and Kumar, R. 2004. Biology of the epidermal growth factor receptor family. *Cancer Treat. Res.* 119: 1–13.

Bateman, R. M., Hodgson, K. C., Kohli, K., Knight, D., and Walley, K. R. 2007. Endotoxemia increases the clearance of mPEGylated 5000-MW quantum dots as revealed by multiphoton microvascular imaging. *J. Biomed. Opt.* 12: 064005–064008.

Berry, C. R. 1967. Structure and optical absorption of AgI microcrystals. *Phys. Rev.* 161: 848.

Can, X. and Chad, A. M. 2007. pH-switchable silver nanoprism growth pathways. *Angew. Chem. Int. Ed.* 46: 2036–2038.

Cang, H., Sun, T., Li, Z.-Y. et al. 2005. Gold nanocages as contrast agents for spectroscopic optical coherence tomography. *Opt. Lett.* 30: 3048–3050.

Chan, W. C. W. and Nie, S. 1998. Quantum dot bioconjugates for ultrasensitive nonisotopic detection. *Science* 281: 2016–2018.

Chanda, N., Shukla, R., Katti, K. V., and Kannan, R. 2009. Gastrin releasing protein receptor specific gold nanorods: Breast and prostate tumor avid nanovectors for molecular imaging. *Nano Lett.* 9: 1798–1805.

Chang, Z., Zhou, J., Zhao, K., Zhu, N., He, P., and Fang, Y. 2006. Ru(bpy)32 + -doped silica nanoparticle DNA probe for the electrogenerated chemiluminescence detection of DNA hybridization. *Electrochim. Acta* 52: 575–580.

Chen, J., Saeki, F., Wiley, B. J. et al. 2005a. Gold nanocages: Bioconjugation and their potential use as optical imaging contrast agents. *Nano Lett.* 5: 473–477.

Chen, J. Y., Wiley, B., Li, Z. Y. et al. 2005b. Gold nanocages: Engineering their structure for biomedical applications. *Adv. Mater.* 17: 2255–2261.

Chin, S.-F., Baughman, R. H., Dalton, A. B. et al. 2007. Amphiphilic helical peptide enhances the uptake of single-walled carbon nanotubes by living cells. *Exp. Biol. Med.* 232: 1236–1244.

Chithrani, B. D. and Chan, W. C. W. 2007. Elucidating the mechanism of cellular uptake and removal of protein-coated gold nanoparticles of different sizes and shapes. *Nano Lett.* 7: 1542–1550.

Choi, H. S., Liu, W., Misra, P. et al. 2007a. Renal clearance of quantum dots. *Nat. Biotechnol.* 25: 1165–1170.

Choi, J., Burns, A. A., Williams, R. M. et al. 2007b. Core-shell silica nanoparticles as fluorescent labels for nanomedicine. *J. Biomed. Opt.* 12: 064007–064011.

Choudhury, R. P., Fuster, V., and Fayad, Z. A. 2004. Molecular, cellular and functional imaging of atherothrombosis. *Nat. Rev. Drug Discov.* 3: 913–925.

Copland, J. A., Eghtedari, M., Popov, V. L. et al. 2004. Bioconjugated gold nanoparticles as a molecular based contrast agent: Implications for imaging of deep tumors using optoacoustic tomography. *Mol. Imaging Biol.* 6: 341–349.

Dabbousi, B. O., RodriguezViejo, J., Mikulec, F. V. et al. 1997. (CdSe)ZnS core-shell quantum dots: Synthesis and characterization of a size series of highly luminescent nanocrystallites. *J. Phys. Chem. B* 101: 9463–9475.

de la Fuente, J. M. and Berry, C. C. 2005. Tat peptide as an efficient molecule to translocate gold nanoparticles into the cell nucleus. *Bioconjug. Chem.* 16: 1176–1180.

De La Zerda, A., Zavaleta, C., Keren, S. et al. 2008. Carbon nanotubes as photoacoustic molecular imaging agents in living mice. *Nat. Nanotechnol.* 3: 557–562.

Derfus, A. M., Chan, W. C. W., and Bhatia, S. N. 2004. Probing the cytotoxicity of semiconductor quantum dots. *Nano Lett.* 4: 11–18.

Diagaradjane, P., Orenstein-Cardona, J. M., Colon-Casasnovas, N. E. et al. 2008. Imaging epidermal growth factor receptor expression in vivo: Pharmacokinetic and biodistribution characterization of a bioconjugated quantum dot nanoprobe. *Clin. Cancer Res.* 14: 731–741.

Draine, B. T. 1988. The discrete-dipole approximation and its application to interstellar graphite grains. *Astrophys. J.* 333: 848–872.

Durr, N. J., Larson, T., Smith, D. K., Korgel, B. A., Sokolov, K., and Ben-Yakar, A. 2007. Two-photon luminescence imaging of cancer cells using molecularly targeted gold nanorods. *Nano Lett.* 7: 941–945.

Faraday, M. 1857. Experimental relations of gold (and other metals) to light. *Philos. Trans. R. Soc. Lond.* 147: 145–181.

Faulk W. P. and Taylor, G. M. 1971. Immunocolloid method for the electron microscope. *Immunochemistry* 8: 1081–1083.

Frangioni, J. V., Kim, S.-W., Ohnishi, S., Kim, S., and Bawendi, M. G. 2007. Sentinel lymph node mapping with type-II quantum dots. *Methods Mol. Biol.* 374: 147–159.

Frens, G. 1973. Controlled nucleation for the regulation of the particle size in monodisperse gold suspensions. *Nat. Phys. Sci. (Lond.)* 241: 20–22.

Funston, A. M., Novo, C., Davis, T. J. and Mulvaney, P. 2009. Plasmon coupling of gold nanorods at short distances and in different geometries. *Nano Lett.* 9: 1651–1658.

Gao, X., Chung, L. W. K., and Nie, S. 2007. Quantum dots for in vivo molecular and cellular imaging. *Methods Mol. Biol.* 374: 135–145.

Gao, X., Cui, Y., Levenson, R. M., Chung, L. W. K., and Nie, S. 2004. In vivo cancer targeting and imaging with semiconductor quantum dots. *Nat. Biotechnol.* 22: 969–976.

Geoghegan, W. D. and Ackerman, G. A. 1977. Adsorption of horseradish peroxidase, ovomucoid and antiimmunoglobulin to colloidal gold for the indirect detection of concanavalin A, wheat germ agglutinin and goat antihuman immuno-

globulin G on cell surfaces at the electron microscopic level: A new method, theory and application. *J. Histochem. Cytochem.* 25: 1187–1200.

Giepmans, B. N. G., Deerinck, T. J., Smarr, B. L., Jones, Y. Z., and Ellisman, M. H., 2005. Correlated light and electron microscopic imaging of multiple endogenous proteins using quantum dots. *Nat. Methods* 2: 743–749.

Gobin, A. M., Lee, M. H., Halas, N. J., James, W. D., Drezek, R. A., and West, J. L. 2007. Near-infrared resonant nanoshells for combined optical imaging and photothermal cancer therapy. *Nano Lett.* 7: 1929–1934.

Hanahan, D. and Weinberg, R. A. 2000. The hallmarks of cancer. *Cell* 100: 57–70.

Hennig, S., van de Linde, S., Heilemann, M., and Sauer, M. 2009. Quantum dot triexciton imaging with three-dimensional subdiffraction resolution. *Nano Lett.* 9: 2466–2470.

Hines, M. A. and Scholes, G. D. 2003. Colloidal PbS nanocrystals with size-tunable near-infrared emission, observation of post-synthesis self-narrowing of the particle size distribution. *Adv. Mater.* 15: 1844–1849.

Hirsch, L. R., Stafford, R. J., Bankson, J. A. et al. 2003. Nanoshell-mediated near-infrared thermal therapy of tumors under magnetic resonance guidance. *Proc. Natl. Acad. Sci.* 100: 13549–13554.

Horisberger, M. 1981. Colloidal gold: A cytochemical marker for light and fluorescent microscopy and for transmission and scanning electron microscopy. *Scanning Electron Microsc.* 11: 9–31.

Huang, X., El-Sayed, I. H., Qian, W., and El-Sayed, M. A. 2006. Cancer cell imaging and photothermal therapy in the near-infrared region by using gold nanorods. *J. Am. Chem. Soc.* 128: 2115–2120.

Huang, Y.-F., Chang, H.-T., and Tan, W. 2008. Cancer cell targeting using multiple aptamers conjugated on nanorods. *Anal. Chem.* 80: 567–572.

Jackson, J. 1999. *Classical Electrodynamics.* New York: John Wiley & Sons.

Jana, N. R., Gearheart, L., and Murphy, C. J. 2001. Seed-mediated growth approach for shape-controlled synthesis of spheroidal and rod-like gold nanoparticles using a surfactant template. *Adv. Mater.* 13: 1389–1393.

Javier, D. J., Nitin, N., Levy, M., Ellington, A., and Richards-Kortum, R. 2008. Aptamer-targeted gold nanoparticles as molecular-specific contrast agents for reflectance imaging. *Bioconjug. Chem.* 19: 1309–1312.

Ji, X. J., Shao, R. P., Elliott, A. M. et al. 2007. Bifunctional gold nanoshells with a superparamagnetic iron oxide-silica core suitable for both MR imaging and photothermal therapy. *J. Phys. Chem. C* 111: 6245–6251.

Jiang, W., Kim, B. Y. S., Rutka, J. T., and Chan, W. C. W. 2008. Nanoparticle-mediated cellular response is size-dependent. *Nat. Nanotechnol.* 3: 145–150.

Jin, Y. and Gao, X. 2009. Plasmonic fluorescent quantum dots. *Nat. Nanotechnol.* 4(9): 571–576.

Jong, W. H. D., Hagens, W. I., Krystek, P., Burger, M. C., Sips, A. J. A. M., and Geertsma, R. E. 2008. Particle size-dependent organ distribution of gold nanoparticles after intravenous administration. *Biomaterials* 29: 1912–1919.

Kataura, H., Kumazawa, Y., Maniwa, Y. et al. 1999. Optical properties of single-wall carbon nanotubes. *Synth. Metals* 103: 2555–2558.

Keren, S., Zavaleta, C., Cheng, Z., de la Zerda, A., Gheysens, O., and Gambhir, S. S. 2008. Noninvasive molecular imaging of small living subjects using Raman spectroscopy. *Proc. Natl. Acad. Sci.* 105: 5844–5849.

Kim, S., Lim, Y. T., Soltesz, E. G. et al. 2004. Near-infrared fluorescent type II quantum dots for sentinel lymph node mapping. *Nat. Biotechnol.* 22: 93–97.

Kneipp, K., Kneipp, H., and Kneipp, J. 2006. Surface-enhanced Raman scattering in local optical fields of silver and gold nanoaggregates from single-molecule Raman spectroscopy to ultrasensitive probing in live cells. *Acc. Chem. Res.* 39: 443–450.

Kobayashi, H., Hama, Y., Koyama, Y. et al. 2007. Simultaneous multicolor imaging of five different lymphatic basins using quantum dots. *Nano Lett.* 7: 1711–1716.

Kreibig, U. and Vollmer, M. 1995. *Optical Properties of Metal Clusters*. Berlin, Germany: Springer-Verlag.

Kumar, S., Aaron, J., and Sokolov, K. V. 2008. Directional conjugation of antibodies to nanoparticles for synthesis of multiplexed optical contrast agents with both delivery and targeting moieties. *Nat. Protocols* 3: 314–320.

Kumar, S., Harrison, N., Richards-Kortum, R., and Sokolov, K. 2007. Plasmonic nanosensors for imaging intracellular biomarkers in live cells. *Nano Lett.* 7: 1338–1343.

Larson, T. A., Bankson, J., Aaron, J., and Sokolov, K. 2007. Hybrid plasmonic magnetic nanoparticles as molecular specific agents for MRI/optical imaging and photothermal therapy of cancer cells. *Nanotechnology* 18: 325101.

Lee, S., Ryu, J. H., Park, K. et al. 2009. Polymeric nanoparticle-based activatable near-infrared nanosensor for protease determination in vivo. *Nano Lett.* 9(12): 4412–4416.

Levi-Polyachenko, N. H., Merkel, E. J., Jones, B. T., Carroll, D. L., and Stewart, J. H. 2009. Rapid photothermal intracellular drug delivery using multiwalled carbon nanotubes. *Mol. Pharm.* 6(4): 1041–1051.

Lidke, D. S., Nagy, P., Heintzmann, R. et al. 2004. Quantum dot ligands provide new insights into erbB/HER receptor-mediated signal transduction. *Nat. Biotechnol.* 22: 198–203.

Lin, A. W. H., Lewinski, N. A., West, J. L., Halas, N. J., and Drezek, R. A. 2005. Optically tunable nanoparticle contrast agents for early cancer detection: Model-based analysis of gold nanoshells. *J. Biomed. Opt.* 10(6): 064035.

Liu, Z., Cai, W., He, L. et al. 2007. In vivo biodistribution and highly efficient tumour targeting of carbon nanotubes in mice. *Nat. Nanotechnol.* 2: 47–52.

Loo, C., Lin, A., Hirsch, L. et al. 2004. Nanoshell-enabled photonics-based imaging and therapy of cancer. *Technol. Cancer Res. Treat.* 3: 33–40.

Loo, C., Lowery, A., Halas, N., West, J., and Drezek, R. 2005. Immunotargeted nanoshells for integrated cancer imaging and therapy. *Nano Lett.* 5: 709–711.

Mackowski, D. W. and Mishchenko, M. I. 1996. Calculation of the T matrix and the scattering matrix for ensembles of spheres. *J. Opt. Soc. Am. A* 13: 2266–2278.

Mallidi, S., Larson, T., Aaron, J., Sokolov, K., and Emelianov, S. 2007. Molecular specific optoacoustic imaging with plasmonic nanoparticles. *Opt. Express* 15: 6583–6588.

Mallidi, S., Larson, T., Tam, J. et al. 2009. Multiwavelength photoacoustic imaging and plasmon resonance coupling of gold nanoparticles for selective detection of cancer. *Nano Lett.* 9: 2825–2831.

Matsuno, A., Itoh, J., Takekoshi, S., Nagashima, T., and Osamura, R. Y. 2005. Three-dimensional imaging of the intracellular localization of growth hormone and prolactin and their mRNA using nanocrystal (quantum dot) and confocal laser scanning microscopy techniques. *J. Histochem. Cytochem.* 53: 833–838.

Meike, L. S., Gopal, I., Ai Leen, K. et al. 2009. Particle size, surface coating, and PEGylation influence the biodistribution of quantum dots in living mice. *Small* 5: 126–134.

Michalet, X., Pinaud, F. F., Bentolila, L. A. et al. 2005. Quantum dots for live cells, in vivo imaging, and diagnostics. *Science* 307: 538–544.

Mie, G. 1908. Beiträge zur optik trüber Medien, speziell kolloidaler Metallösungen. *Annal. Phys.* 25: 377–445.

Mishchenko, M. I. and Travis, L. D. 1998. Capabilities and limitations of a current FORTRAN implementation of the T-matrix method for randomly oriented, rotationally symmetric scatterers. *J. Quant. Spectrosc. Radiat. Transfer* 60: 309–324.

Mishchenko, M. I., Travis, L. D., and Mackowski, D. W. 1996. T-matrix computations of light scattering by nonspherical particles: A review. *J. Quant. Spectrosc. Radiat. Transfer* 55: 535–575.

Murphy, C. J., Sau, T. K., Gole, A. M. et al. 2005. Anisotropic metal nanoparticles: Synthesis, assembly, and optical applications. *J. Phys. Chem. B* 109: 13857–13870.

Murray, C. B., Norris, D. J., and Bawendi, M. G. 1993. Synthesis and characterization of nearly monodisperse CdE (E = S, Se, Te) semiconductor nanocrystallites. *J. Am. Chem. Soc.* 115: 8706–8715.

Nastassja, L., Vicki, C., and Rebekah, D. 2008. Cytotoxicity of nanoparticles. *Small* 4: 26–49.

Niidome, T., Yamagata, M., Okamoto, Y. et al. 2006. PEG-modified gold nanorods with a stealth character for in vivo applications. *J. Control. Release* 114: 343–347.

Nikoobakht, B. and El-Sayed, M. A. 2003. Preparation and growth mechanism of gold nanorods (NRs) using seed-mediated growth method. *Chem. Mater.* 15: 1957–1962.

Nozik, A. J. 2002. Quantum dot solar cells. *Phys. E: Low Dimen. Syst. Nanostruct.* 14: 115–120.

Odom, T. W., Huang, J.-L., Kim, P., and Lieber, C. M. 1998. Atomic structure and electronic properties of single-walled carbon nanotubes. *Nature* 391: 62–64.

Oishi, M., Nakaogami, J., Ishii, T., and Nagasaki, Y. 2006. Smart PEGylated gold nanoparticles for the cytoplasmic delivery of siRNA to induce enhanced gene silencing. *Chem. Lett.* 35: 1046–1047.

Packard, R. R. S. and Libby, P. 2008. Inflammation in atherosclerosis: From vascular biology to biomarker discovery and risk prediction. *Clin. Chem.* 54: 24–38.

Park, J.-H., Gu, L., von Maltzahn, G., Ruoslahti, E., Bhatia, S. N., and Sailor, M. J. 2009. Biodegradable luminescent porous silicon nanoparticles for in vivo applications. *Nat. Mater.* 8: 331–336.

Peer, D., Karp, J. M., Homg, S., Farokhzad, O. C., Margalit, R., and Langer, R. 2007. Nanocarriers as an emerging platform for cancer therapy. *Nat. Nanotechnol.* 2: 751–760.

Perrault, S. D., Walkey, C., Jennings, T., Fischer, H. C., and Chan, W. C. W. 2009. Mediating tumor targeting efficiency of nanoparticles through design. *Nano Lett.* 9: 1909–1915.

Pitarke, J. M., Silkin, V. M., Chulkov, E. V., and Echenique, P. M. 2007. Theory of surface plasmons and surface-plasmon polaritons. *Rep. Prog. Phys.* 70: 1–87.

Pitsillides, C. M., Joe, E. K., Wei, X., Anderson, R. R., and Lin, C. P. 2003. Selective cell targeting with light-absorbing microparticles and nanoparticles. *Biophys. J.* 84(6): 4023–4032.

Qian, X., Peng, X.-H., Ansari, D. O. et al. 2008. In vivo tumor targeting and spectroscopic detection with surface-enhanced Raman nanoparticle tags. *Nat. Biotechnol.* 26: 83–90.

Reinhard, B. R. M., Sheikholeslami, S., Mastroianni, A., Alivisatos, A. P., and Liphardt, J. 2007. Use of plasmon coupling to reveal the dynamics of DNA bending and cleavage by single EcoRV restriction enzymes. *Proc. Natl. Acad. Sci.* 104: 2667–2672.

Richards-Kortum, R. and Sevick-Muraca, E. 1996. Quantitative optical spectroscopy for tissue diagnosis. *Annu. Rev. Phys. Chem.* 47: 555.

Rong, G., Wang, H., Skewis, L. R., and Reinhard, B. R. M. 2008. Resolving sub-diffraction limit encounters in nanoparticle tracking using live cell plasmon coupling microscopy. *Nano Lett.* 8: 3386–3393.

Rosi, N. L., Giljohann, D. A., Thaxton, C. S., Lytton-Jean, A. K. R., Han, M. S., and Mirkin, C. A. 2006. Oligonucleotide-modified gold nanoparticles for intracellular gene regulation. *Science* 312: 1027–1030.

Rossetti, R., Nakahara, S., and Brus, L. E. 1983. Quantum size effects in the redox potentials, resonance Raman spectra, and electronic spectra of CdS crystallites in aqueous solution. *J. Chem. Phys.* 79: 1086–1088.

Sadauskas, E., Wallin, H., Stoltenberg, M. et al. 2007. Kupffer cells are central in the removal of nanoparticles from the organism. *Particle Fibre Toxicol.* 4: 10.

Schipper, M. L., Iyer, G., Koh, A. L. et al. 2009. Particle size, surface coating, and PEGylation influence the biodistribution of quantum dots in living mice. *Small* 5: 126–134.

Sellmeier, W. 1871. Zur Erklärung der abnormen Farbenfolge im Spectrum einiger Substanzen. *Annal. Phys. Chem.* 219: 272–282.

Shah, J., Aglyamov, S. R., Sokolov, K., Milner, T. E., and Emelianov, S. Y. 2008a. Ultrasound imaging to monitor photothermal therapy: Feasibility study. *Opt. Express* 16: 3776–3785.

Shah, J., Park, S., Aglyamov, S. et al. 2008b. Photoacoustic imaging and temperature measurement for photothermal cancer therapy. *J. Biomed. Opt.* 13: 034024–034029.

Skala, M. C., Crow, M. J., Wax, A., and Izatt, J. A. 2008. Photothermal optical coherence tomography of epidermal growth factor receptor in live cells using immunotargeted gold nanospheres. *Nano Lett.* 8: 3461–3467.

Skrabalak, S. E., Chen, J., Au, L., Lu, X., Li, X., and Xia, Y. 2007. Gold nanocages for biomedical applications. *Adv. Mater.* 19: 3177–3184.

Skrabalak, S. E., Chen, J., Sun, Y. et al. 2008. Gold nanocages: Synthesis, properties, and applications. *Acc. Chem. Res.* 41: 1587–1595.

So, M.-K., Xu, C., Loening, A. M., Gambhir, S. S., and Rao, J. 2006. Self-illuminating quantum dot conjugates for in vivo imaging. *Nat. Biotech.* 24: 339–343.

Sokolov, K., Follen, M., Aaron, J. et al. 2003. Real-time vital optical imaging of precancer using anti-epidermal growth factor receptor antibodies conjugated to gold nanoparticles. *Cancer Res.* 63: 1999–2004.

Song, K. H., Kim, C., Maslov, K., and Wang, L. V. 2009. Noninvasive in vivo spectroscopic nanorod-contrast photoacoustic mapping of sentinel lymph nodes. *Eur. J. Radiol.* 70: 227–231.

Sonnichsen, C., Reinhard, B. M., Liphardt, J., and Alivisatos, A. P. 2005. A molecular ruler based on plasmon coupling of single gold and silver nanoparticles. *Nat. Biotechnol.* 23: 741–745.

Souza, G. R., Christianson, D. R., Staquicini, F. I. et al. 2006. Networks of gold nanoparticles and bacteriophage as biological sensors and cell-targeting agents. *Proc. Natl. Acad. Sci.* 103: 1215–1220.

Storhoff, J. J., Lucas, A. D., Garimella, V., Bao, Y. P., and Muller, U. R. 2004. Homogeneous detection of unamplified genomic DNA sequences based on colorimetric scatter of gold nanoparticle probes. *Nat. Biotechnol.* 22: 883–887.

Stroh, M., Zimmer, J. P., Duda, D. G. et al. 2005. Quantum dots spectrally distinguish multiple species within the tumor milieu in vivo. *Nat. Med.* 11: 678–682.

Sun, Y. G., Mayers, B., and Xia, Y. N. 2003. Metal nanostructures with hollow interiors. *Adv. Mater.* 15: 641–646.

Swartz, M. A. 2001. The physiology of the lymphatic system. *Adv. Drug Deliv. Rev.* 50: 3–20.

Taflove, A. and Hagness, S. 2005. *Computational Electrodynamics: The Finite-Difference Time-Domain Method.* Boston, MA: Artech.

Thomas, D. G. and Hopfield, J. J. 1959. Exciton spectrum of cadmium sulfide. *Phys. Rev.* 116: 573.

Thomas, K. G., Barazzouk, S., Ipe, B. I., Joseph, S. T. S., and Kamat, P. V. 2004. Uniaxial plasmon coupling through longitudinal self-assembly of gold nanorods. *J. Phys. Chem. B* 108: 13066–13068.

Tkachenko, A. G., Xie, H., Coleman, D. et al. 2003. Multifunctional gold nanoparticle-peptide complexes for nuclear targeting. *J. Am. Chem. Soc.* 125: 4700–4701.

Troutman, T. S., Barton, J. K., and Romanowski, M. 2008. Biodegradable plasmon resonant nanoshells. *Adv. Mater.* 20: 2604–2608.

Wagner, F. E., Haslbeck, S., Stievano, L., Calogero, S., Pankhurst, Q. A., and Martinek, K. P. 2000. Before striking gold in gold-ruby glass. *Nature* 407: 691–692.

Wang, C.-H., Huang, Y.-J., Chang, C.-W., Hsu, W.-M., and Peng, C.-A. 2009. In vitro photothermal destruction of neuroblastoma cells using carbon nanotubes conjugated with GD2 monoclonal antibody. *Nanotechnology* 20: 315101.

Welch, A. J. and Gemert, M. J. C. 1995. *Optical-Thermal Response of Laser-Irradiated Tissue.* New York: Plenum Publishing Corporation.

Wilder, J. W. G., Venema, L. C., Rinzler, A. G., Smalley, R. E., and Dekker, C. 1998. Electronic structure of atomically resolved carbon nanotubes. *Nature* 391: 59–62.

Yang, D.-P. and Cui, D.-X. 2008. Advanced and prospects of gold nanorods. *Chem. Asian J.* 3: 2010–2022.

Yguerabide, J. and Yguerabide, E. E. 1998. Light-scattering submicroscopic particles as highly fluorescent analogs and their use as tracer labels in clinical and biological applications: I. Theory. *Anal. Biochem.* 262: 137–156.

Yu, C. and Irudayaraj, J. 2006. Multiplex biosensor using gold nanorods. *Anal. Chem.* 79: 572–579.

Yu, W. W., Qu, L. H., Guo, W. Z., and Peng, X. G. 2003. Experimental determination of the extinction coefficient of CdTe, CdSe, and CdS nanocrystals. *Chem. Mater.* 15: 2854–2860.

Yurkin, M. A., Maltsev, V. P., and Hoekstra, A. G. 2007. The discrete dipole approximation for simulation of light scattering by particles much larger than the wavelength. *J. Quant. Spectrosc. Radiat. Transfer* 106: 546–557.

Zharov, V. P., Galitovskaya, E. N., Johnson, C., and Kelly, T. 2005. Synergistic enhancement of selective nanophotothermolysis with gold nanoclusters: Potential for cancer therapy. *Lasers Surg. Med.* 37: 219–226.

Plasmonic Nanoprobes for Biomolecular Diagnostics of DNA Targets

33.1 Introduction 723
33.2 SERS Technique for Gene Diagnostics...724
33.3 Development of Molecular Sentinel Nanoprobes...724
33.4 Applications in Biomolecular Diagnostics...726
    Detection of Infectious Diseases (HIV) • Multiplex Detection of Cancer Biomarkers
33.5 Conclusion 728
Acknowledgments..728
References...728

Tuan Vo-Dinh
Duke University

Hsin-Neng Wang
Duke University

33.1 Introduction

Techniques that use fluorescence probes based on fluorescence energy transfer (e.g., molecular beacons) have been widely applied with great success in biosensing assays. Nevertheless, the need for alternative, rapid, and selective assays has continued to encourage researchers to explore other technologies having comparable sensitivity as fluorescence but having additional unique and complementary advantages. Raman spectroscopy is an important analytical technique for chemical and biological analysis due to the wealth of information on molecular structures, surface processes, and interface reactions that can be extracted from experimental data. The spectral selectivity associated with the narrow emission lines and the molecular-specific vibrational bands of Raman labels make it an ideal tool for molecular genotyping. However, a limitation of Raman techniques for trace detection is the very weak Raman cross section. However, Raman spectroscopy has gained increasing interest as an analytical tool with the advent of the surface-enhanced Raman scattering (SERS) and surface-enhanced resonance Raman scattering (SERRS) effects, which can produce significant enhancement of the Raman signal. It is believed that the origin of the enormous Raman enhancement is produced by at least two main mechanisms that contribute to the SERS effect: (a) an electromagnetic effect occurring near metal surface structures associated with large local fields caused by electromagnetic resonances, often referred to as "surface plasmons" and (b) a chemical effect involving a scattering process associated with chemical interactions between the molecule and the metal surface. Plasmons are quanta associated with longitudinal waves propagating in matter through the collective motion of large numbers of electrons. According to classical electromagnetic theory, molecules on or near metal nanostructures experience enhanced fields relative to that of the incident radiation. When a metallic nanostructured surface is irradiated by an incident electromagnetic field (e.g., a laser beam), conduction electrons are displaced into frequency oscillations equal to those of the incident light. These oscillating electrons, called "surface plasmons," produce a secondary electric field, which adds to the incident field. These fields can be quite large (10^6–10^7-, even up to 10^{15}-fold enhancement at "hot spots"). When these oscillating electrons become spatially confined, as is the case for isolated metallic nanospheres or otherwise roughened metallic surfaces (nanostructures), there is a characteristic frequency (the plasmon frequency) at which there is a resonant response of the collective oscillations to the incident field. This condition yields intense localized fields that can interact with molecules in contact with or near the metal surface (Otto 1978, Gersten and Nitzan 1980, Schatz 1984, Zeman and Schatz 1987). In an effect analogous to a "lightning rod" effect, secondary fields can become concentrated at high curvature points on the roughened metal surface. In SERRS, the energy of the incoming laser is selected such that it coincides with an electronic transition of the molecule being monitored. An advantage of SERRS over SERS is the large increase in intensity of the Raman peaks.

Following the discovery of the SERS effect (Fleischmann et al. 1974, Albrecht and Creighton 1977, Jeanmaire and Vanduyne 1977), our laboratory has first demonstrated the general applicability of the SERS effect for trace analysis using solid substrates having silver-coated nanospheres (Vo-Dinh et al. 1984). In 1984, we exploited the spin-coating and electron beam evaporation techniques to fabricate close-packed arrays of nanospheres, onto which a thin silver shell was deposited, effectively forming a controlled, reproducible substrate of "hot spots" (Vo-Dinh et al. 1984). We developed and proposed the use of solid substrates consisting of nanosphere or nanoparticle arrays covered with nanolayer of metal (forming a "nanowave") as efficient and reproducible SERS-active media. During the following two decades our laboratory has extensively investigated the SERS technology and developed a wide variety of plasmonics-active SERS platforms for chemical sensing (Vo-Dinh et al. 1984, 1987, 1988, Enlow et al. 1986, Alak and Vo-Dinh 1987, Moody et al. 1987, Bello et al. 1989a,b) and for bioanalysis and biosensing (Vo-Dinh et al. 1994, 2002, 2005, 2010, Isola et al. 1998, Vo-Dinh 1998, Zeisel et al. 1998, Stokes et al. 2004, Wabuyele and Vo-Dinh 2005, Wabuyele et al. 2005, Vo-Dinh and Yan 2007, Dhawan et al. 2008, Khoury and Vo-Dinh 2008). These plasmonics substrate platforms have led to a wide variety of analytical applications including sensitive detection of a variety of chemicals of environmental, biological, and medical significance, such as DNA-adduct biomarkers (Vo-Dinh et al. 1987). The first application of SERS in DNA probe detection technology was reported (Vo-Dinh et al. 1994) and has led to subsequent development of this method for medical diagnostics (Vo-Dinh 1998, Isola et al. 1998; Vo-Dinh et al. 2002, 2005; Vo-Dinh and Yan 2007). This chapter discusses the development of a unique "molecular sentinel" (MS) nanoprobe technology designed to detect the presence of DNA/RNA biotargets, such as HIV (Wabuyele and Vo-Dinh 2005) and breast cancer genes (Wang and Vo-Dinh 2009) in homogeneous assays.

The SERS technology has now received increasing interest and contribution from many research groups worldwide and the reader is referred to a number of reviews and monographs for further details (Chang and Furtak 1982, Wokaun et al. 1982, Kerker 1984, Pockrand 1984, Schatz 1984, Moskovits 1985, 2005, Otto et al. 1992, Cao et al. 2002, Doering and Nie 2002, Kneipp et al. 2006, Scaffidi et al. 2009).

33.2 SERS Technique for Gene Diagnostics

Raman spectroscopy also offers some distinct features that are important for in situ monitoring of complex biological systems. Following laser irradiation of a sample, the observed Raman shifts are equivalent to the energy changes involved in molecular transitions of the scattering species and are therefore characteristic of it. These observed Raman shifts, which correspond to vibrational transitions of the scattering molecule, exhibit very narrow linewidths. For these reasons, Raman spectroscopy has a great potential for multiplexing detection, for example, in gene diagnostics. Therefore, many organic compounds with distinct Raman spectra may be used as dyes to label biological macromolecules and each labeled molecular species will be able to be distinguished on the basis of its unique Raman spectra. This is not the case with fluorescence, because the relatively broad spectral

characteristics of fluorescence excitation and emission spectra at room temperature result in large spectral overlaps if more than 3–4 fluorescent dyes are to be detected simultaneously.

Since we reported the first practical application of the SERS effect in analysis (Vo-Dinh et al. 1984), the development of SERS-active solid substrates has been an area of active research in our laboratory. We have extensively investigated the SERS technology in the development of nanostructure-based SERS substrates (Vo-Dinh 1998, Vo-Dinh and Yan 2007). These substrates consist of a plate having silver-coated dielectric nanoparticles or isolated dielectric nanospheres (30-nm diameter) coated with silver, producing a "nanowave" consisting an array of half-nanoshells. The fabrication process involves depositing nanoparticles on a substrate and then coating the nanoparticle base with a 50–150 nm layer of silver via vacuum deposition.

The development of practical and sensitive devices for screening multiple genes related to medical diseases and infectious pathogens is critical for early diagnosis and effective treatments of many illnesses as well as for high-throughput screening for drug discovery. To achieve the required level of sensitivity and specificity, it is often necessary to use a detection method that is capable of simultaneously identifying and differentiating a large number of biological constituents in complex samples. One of the most unambiguous and well-known molecular recognition events is the hybridization of a nucleic acid to its complementary target. Thus, the hybridization of a nucleic acid probe to its DNA (or RNA) target can provide a very high degree of accuracy for identifying complementary nucleic acid sequences.

We first reported the development and application of SERS plasmonics gene probe technology for DNA detection (Vo-Dinh et al. 1994). Our laboratory has further developed SERS-based gene probes for selective detection of HIV DNA and the breast cancer gene BRCA1 using plasmonics substrates (Isola et al. 1998, Allain and Vo-Dinh 2002, Vo-Dinh et al. 2002, Culha et al. 2003). To demonstrate the SERS gene detection scheme, we used pre-coated SERS-active solid substrates, on which DNA probes were bound and directly used for hybridization. Several factors affect the effectiveness of this scheme for DNA hybridization and SERS gene detection. It is desirable that the unlabeled DNA fragment does not exhibit any significant SERS signal that might interfere with the label signal. It is important to use a label that is SERS-active and compatible with the hybridization platform: an ideal label should exhibit a strong SERS signal when used with the SERS-active substrate of interest. Furthermore, the label should retain its strong SERS signal after being attached to a DNA probe. The use of SERS gene technology was demonstrated for the detection of the HIV gene sequence (Isola et al. 1998).

33.3 Development of Molecular Sentinel Nanoprobes

We have recently developed a unique "label-free" detection approach (i.e., the target does not need to be labeled) that incorporates the "SERS effect modulation" scheme associated with metallic nanoparticles and the DNA hairpin structure (Wabuyele and Vo-Dinh 2005). The SERS-based "Molecular Sentinel" (SERS-MS) technique uses the stem-loop structure similar to that of molecular beacons (MB) for DNA recognition. However, the detection scheme is fundamentally different from the MB detection scheme. Molecular beacons consist of a fluorescence molecule attached to the end of one arm and a quencher molecule attached to the end of the other arm of a hairpin structure. They are designed to report the presence of target DNA sequences (complementary to the hairpin DNA loop) using the principle of fluorescence resonance energy transfer (FRET) between the fluorescent molecule and the quenching molecule by generating a relatively strong fluorescent signal when complementary target sequences are hybridized, thus separating the quencher and the fluorophore. The fluorescence remains low (quenched) in the absence of a complementary sequence.

On the other hand, in the detection strategy of our SERS-MS probes, we exploit the change or modulation of the plasmonics enhancement of the SERS signal with the distance between the metallic nanoparticle and the Raman label. In the SERS-MS system (Figure 33.1), MS nanoprobes having a Raman label at one end are immobilized onto a metallic nanoparticle via a thiol group attached on the other end to form a SERS-MS nanoprobe. The metal nanoparticle is used as a signal-enhancing platform for the SERS signal associated with the label. The Raman enhancement is determined by the plasmonic effect at the metal surface. Theoretical studies of the plasmonic effect have shown that the SERS enhancement, G factor, falls off as $G = [r/(r + d)]^{12}$ for a single analyte molecule located a distance, d, from the surface of a metal particle of radius, r (Kneipp et al. 1997). Thus the electromagnetic SERS enhancement decreases significantly with increasing distance, due to the decay of a dipole over the distance $(1/d)^3$ to the fourth power, thus resulting in a total intensity decay of $(1/d)^{12}$ of the SERS signal. Since the Raman enhancement field decreases significantly away from the surface, a molecule (e.g., the Raman label) must be located within a very close range (0–10 nm) of the metal nanoparticle surface in order to experience the enhanced local field. Under normal conditions (i.e., in the absence of target genes), the hairpin configuration has the Raman label in contact or close proximity (<1 nm) to the nanoparticle, thus resulting in a strong SERS effect and indicating that no significant event has occurred (Figure 33.1b). However, when complementary target DNA is recognized and hybridized to the nanoprobes, the SERS signal of these molecular sentinels becomes significantly quenched, providing a warning sign of target recognition and capture (Figure 33.1c). In other terms, a SERS nanoprobe serves as molecular sentinel patrolling the sample solution with its SERS warning light "switched on" when no significant event occurs. Then, when a biotarget of interest is identified, the molecular sentinel binds to it and extinguishes its light, thus providing a warning sign.

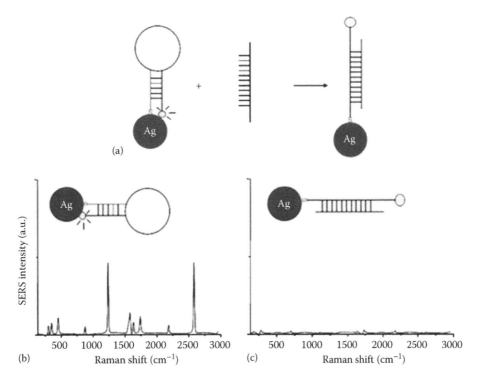

FIGURE 33.1 (a) Operating principle of a SERS molecular sentinel nanoprobe. (b) A SERS signal is observed when the MS probe is in the hairpin conformation in the absence of target sequences (closed state), (c) whereas in the open state, the signal is diminished in the presence of complementary target sequences. (Adapted from Wabuyele, M.B. and Vo-Dinh, T., *Anal. Chem.*, 77(23), 7810, 2005.)

33.4 Applications in Biomolecular Diagnostics

33.4.1 Detection of Infectious Diseases (HIV)

We have demonstrated the SERS-MS concept for detecting the presence of specific DNA sequences (HIV gene) in a homogeneous assay (Wabuyele and Vo-Dinh 2005). The HIV MS nanoprobes were designed to have a stem sequence that allowed the formation of stable hairpin structure at room temperature and incorporated a partial sequence for the HIV-1 isolate Fbr020, the reverse transcriptase (*pol*) gene. The HIV-1 MS nanoprobe (5′-HS-(CH$_2$)$_6$-CCTATCACAACAAAGAGCATACATAGGGATAGG-R6G) consisted of a 42-base DNA hairpin probe modified with rhodamine 6G on the 3′ end and a thiol substituent at the 5′ end; the 5′ thiol was used for covalent coupling to the surface of silver nanoparticles. The underlined portions of the sequence represent the complementary arms of the MS designed to form a stem-loop structure. The silver colloidal nanoparticles were prepared according to the citrate reduction method, yielding homogenously sized colloids. A 115-bp sequence in the *gag* region of the HIV-1 genome was amplified by PCR, using forward and reverse primers in the gag region of the genome. Following gel analysis, the PCR products were hybridized to the SERS-MS nanoprobes. The reaction volume was 40 μL and hybridization was performed at 55°C for 1 min. SERS measurements were performed using a Renishaw InVia Raman system equipped with a 50 mW, HeNe laser (Coherent, Model 106-1) emitting a 632.8 nm line used as the excitation source.

We have demonstrated the effectiveness of the MS technique by detecting the presence of a partial sequence of the HIV1 *gag* gene in a homogenous solution as depicted in Figure 33.2. In normal conditions, that is, in the absence of the target DNA (Figure 33.2, top curve), the SERS-MS nanoprobes have a stable hairpin conformation, which allows in a close proximity of the rhodamine 6G label with the surface of the silver nanoparticles (nanoenhancers). This situation produces an intense SERS signal from rhodamine 6G upon laser excitation due to the plasmonics effect. However, in the presence of a complementary HIV-1 target sequence (Figure 33.2, middle curve), the SERS HIV-1 MS nanoprobes bound to the target DNA, causes a physical separation of the rhodamine 6G label from the surface of the silver nanoparticles. As a consequence, the SERS signal was greatly diminished. On the other hand, the presence of noncomplementary target DNA template (Figure 33.2, bottom curve) did not significantly affect the SERS signal, indicating that the hairpin loop structure of the SERS HIV-1 MS nanoprobes remained unchanged. The results of this experiment show that the MS technology can effectively detect target DNA sequences in a homogenous solution. Using the most intense SERS intensity band at 1521 cm^{-1} as the marker band for rhodamine 6G, we estimated the SERS quenching efficiency to be ~75% upon hybridization of the SER HIV-1 MS nanoprobe to the HIV 1 DNA target. This result demonstrates the specificity and selectivity of SERS-MS nanoprobes and their potential application in selective diagnostics.

33.4.2 Multiplex Detection of Cancer Biomarkers

We have also demonstrated the feasibility of multiplex detection of cancer biomarkers using the SERS-MS technique (Wang

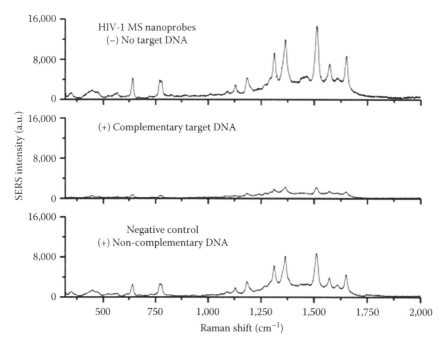

FIGURE 33.2 SERS spectra of HIV-1 SERS-MS nanoprobe with no target DNA sequence (top curve) and in the presence of a noncomplementary DNA target sequence (negative control: bottom curve) and a complementary HIV-1 DNA target (positive diagnostic: middle curve). (Adapted from Wabuyele, M.B. and Vo-Dinh, T., *Anal. Chem.*, 77(23), 7810, 2005.)

and Vo-Dinh 2009). The study was performed with two MS nanoprobes, ERBB2-MS and KI-67-MS, which were designed and separately prepared to detect two breast cancer biomarkers, *erbB-2* and *ki-67* genes, respectively. The ERBB2-MS (5′-SH-(CH$_2$)$_6$)-CGCCATCCACCCCCAAGACCACGACCAGCA GAATATGGCG-Cy3-3′) and KI-67-MS (5′-SH-(CH$_2$)$_6$)-GC GTATTCTGCACACC TCTTGACACTCCGATACGC-TAMR A-3′) nanoprobes were tagged with different Raman labels, Cy3 and 5-carboxytetramethylrhodamine (TAMRA), respectively. In this study, 6-Mercapto-1-hexanol (MCH) was used to displace the nonspecifically adsorbed DNA and passivate the silver surface. The multiplexing capability of the MS technique was demonstrated by mixing the ERBB2-MS and KI-67-MS nanoprobes in 20 mM Tris-HCl buffer (pH 8.0) containing 5 mM MgCl$_2$. Figure 33.3a shows the resulted SERS spectrum of the MS nanoprobe mixture. Note that all major Raman peaks of the two Raman labels tagged on the ERBB2-MS and KI-67-MS nanoprobes are clearly distinguishable from one another due to the narrow SERS peaks even though the fluorescence spectra of Cy3 and TAMRA strongly overlap. This feature underlines the advantage of the SERS-based MS nanoprobes over fluorescence-based assays for multiplex detection.

The specificity of the MS nanoprobe mixture was then demonstrated in the presence of target DNA sequences. Figure 33.3b shows that all SERS signals of the MS nanoprobe mixture were significantly quenched in the presence of both targets (0.5 μM for each target). The decreased SERS intensity indicates that both MS nanoprobes hybridized with their complementary DNA targets.

The specificity and selectivity of the MS nanoprobe mixture were then evaluated in the presence of individual complementary DNA target (i.e., only one of the two complementary targets). The results shown in Figure 33.3c and d indicate that only the SERS signal associated with the MS nanoprobes complementary to the present target was significantly quenched (indicated by arrows). These results demonstrate that the MS nanoprobe technique can provide a useful tool for multiplex DNA detection in a homogeneous solution for medical diagnostics and high-throughput bioassays. In addition, SERS measurements are performed immediately following the hybridization reactions without washing steps, which greatly simplifies the assay procedures.

The multiplex capability of Raman and SERS techniques to detect a large number of molecular processes simultaneously is also an important feature in molecular diagnostics as well as in systems biology research. Although luminescence techniques offer excellent detection sensitivity, the overlap of the relatively large bandwidth of fluorescence spectra limits the number of labels that can be used simultaneously. Therefore, alternative techniques such as SERS-based techniques with improved multiplex capability are desirable. Due to the narrow bandwidths of Raman bands, the multiplex capability of the SERS-MS probe is excellent in comparison to the other spectroscopic alternatives. For example, we can compare the detection of a commonly used spectroscopic label, crystal fast violet (CFV) dye, which is often used in fluorescence and SERS. The CFV label exhibits a fluorescence spectrum having a relatively broad bandwidth (approximately 50–60 nm halfwidth), whereas the bandwidth of

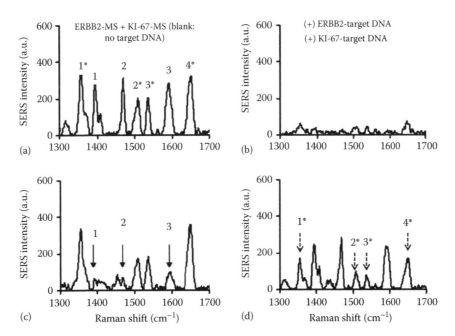

FIGURE 33.3 SERS spectra of the MS nanoprobe mixture (ERBB2-MS + KI-67-MS) in the presence or absence of target DNA. The major Raman bands from ERBB2-MS are marked with numbers, and the major Raman bands from KI-67-MS are marked with numbers with (*) sign: (a) blank (in the absence of any target DNA), (b) in the presence of two target DNA complementary to both MS nanoprobes, (c) in the presence of single target DNA complementary to the ERBB2-MS nanoprobes, and (d) in the presence of single target DNA complementary to the KI-67-MS nanoprobes. The solid and dashed arrow signs illustrate the decreased SERS intensity of the major Raman bands in the presence of corresponding ERBB2 and KI-67 target DNA, respectively. (Adapted from Wang, H.N. and Vo-Dinh, T., *Nanotechnology*, 20(6), 065101-1, 2009.)

its SERS spectrum is significantly narrower (<0.5 nm or 3 cm^{-1} halfwidth). In a typical Raman spectrum, a spectral interval of 3000 cm^{-1} can provide 3000/3 or 10^3 available individual spectral "intervals" at any given time. Even allowing a reduction factor of 10 due to possible spectral overlaps, it should be possible to find dozens to 100 labels that can be used for labeling multiple probes simultaneously. This multiplex advantage unique to Raman/SERS is extremely important for ultra-high-throughput analyses where multiple gene targets need to be screened in a highly parallel multiplex modality.

33.5 Conclusion

In conclusion, the SERS-MS gene probes could offer a unique combination of performance capabilities and analytical features of merit for use in biosensing. The SERS-MS gene probes are safer than radioactive labels and have excellent specificity due to the inherent specificity of Raman spectroscopy. With Raman scattering, multiple probes can be much more easily selected with minimum spectral overlap. This "label-multiplex" advantage can permit analysis of multiple probes simultaneously, resulting in much more rapid DNA detection, gene mapping, and improved ultra-high-throughput screening of small molecules for drug discovery. The SERS molecular sentinel probes could be used for a wide variety of applications in areas where nucleic acid identification is involved. The SERS-MS approach involving homogeneous assays greatly simplifies experimental procedures and could potentially be used in assays that require rapid, high-volume identification of genomic materials. Due to these unique properties, the SERS molecular sentinel approach could contribute to the development of the next generation of DNA/RNA diagnostic tools for molecular screening and molecular imaging.

Acknowledgments

This work was sponsored by the National Institutes of Health (Grants R01 EB006201 and R01 ES014774-01A1). The authors also acknowledge the contribution of M. Wabuyele.

References

Alak, A. M. and Vo-Dinh, T. 1987. Surface-enhanced Raman spectrometry of organophosphorus chemical agents. *Anal. Chem.* 59(17): 2149–2153.

Albrecht, M. G. and Creighton, J. A. 1977. Anomalously intense Raman spectra of pyridine at a silver electrode. *J. Am. Chem. Soc.* 99(15): 5215–5217.

Allain, L. R. and Vo-Dinh, T. 2002. Surface-enhanced Raman scattering detection of the breast cancer susceptibility gene BRCA1 using a silver-coated microarray platform. *Anal. Chim. Acta* 469: 149–154.

Bello, J. M., Stokes, D. L., and Vo-Dinh, T. 1989a. Silver-coated alumina as a new medium for surfaced-enhanced Raman scattering analysis. *Appl. Spectrosc.* 43(8): 1325–1330.

Bello, J. M., Stokes, D. L., and Vo-Dinh, T. 1989b. Titanium dioxide based substrate for optical monitors in surface-enhanced Raman scattering analysis. *Anal. Chem.* 61(15): 1779–1783.

Cao, Y. W. C., Jin, R. C., and Mirkin, C. A. 2002. Array-based electrical detection of DNA using nanoparticle probes. *Science* 297(5586): 1536–1540.

Chang, R. K. and Furtak T. E. 1982. *Surface Enhanced Raman Scattering.* New York: Plenum Press.

Culha, M., Stokes, D., Allain, L. R., and Vo-Dinh, T. 2003. Surface-enhanced Raman scattering substrate based on a self-assembled monolayer for use in gene diagnostics. *Anal. Chem.* 75: 6196–6201.

Dhawan, A., Muth, J. F., Leonard, D. N. et al. 2008. Focused in beam fabrication of metallic nanostructures on end faces of optical fibers for chemical sensing applications. *J. Vacuum Sci. Technol.* 26: 2168–2173.

Enlow, P. D., Buncick, M., Warmack, R. J., and Vo-Dinh, T, 1986. Detection of nitro polynuclear aromatic compounds by surface-enhanced Raman spectrometry. *Anal. Chem.* 58: 1119–1123.

Doering, W. E. and Nie, S. M. 2002. Single-molecule and single-nanoparticle SERS: Examining the roles of surface active sites and chemical enhancement. *J. Phys. Chem. B* 106(2): 311–317.

Fleischmann, M., Hendra, P. J., and McQuillan A. J. 1974. Raman spectra of pyridine adsorbed at a silver electrode. *Chem. Phys. Lett.* 26(2): 163–166.

Gersten, J. and Nitzan, A. 1980. Electromagnetic theory of enhanced Raman scattering by molecules adsorbed on rough surfaces. *J. Chem. Phys.* 73: 3023–3037.

Isola, N. R., Stokes, D. L., and Vo-Dinh, T. 1998. Surface-enhanced Raman gene probe for HIV detection. *Anal. Chem.* 70(7): 1352–1356.

Jeanmaire, D. L. and Vanduyne, R. P. 1977. Surface Raman spectroelectrochemistry: Part I. Heterocyclic, aromatic, and aliphatic amines adsorbed on the anodized silver electrode. *J. Electroanal. Chem.* 84(1): 1–20.

Kerker, M. 1984. Electromagnetic model for surface-enhanced Raman scattering (SERS) on metal colloids. *Acc. Chem. Res.* 17(8): 271–277.

Khoury, C. G. and Vo-Dinh, T. 2008. Gold nanostars for surface-enhanced Raman scattering: Synthesis, characterization and optimization. *J. Phys. Chem. C* 112(48): 18849–18859.

Kneipp, K., Moskovits, M., and Kneipp H. 2006. *Surface-Enhanced Raman Scattering: Physics and Applications.* New York: Springer.

Kneipp, K., Wang, Y, Kneipp, H. et al. 1997. Single molecule detection using surface-enhanced Raman scattering (SERS). *Phys. Rev. Lett.* 78: 1667–1670.

Moody, R. L., Vo-Dinh, T., and Fletcher, W. H. 1987. Investigation of experimental parameters for surface-enhanced Raman scattering (SERS) using silver-coated microsphere substrates. *Appl. Spectrosc.* 41(6): 966–970.

Moskovits, M. 1985. Surface-enhanced spectroscopy. *Rev. Mod. Phys.* 57(3): 783–826.

Moskovits, M. 2005. Surface-enhanced Raman spectroscopy: A brief retrospective. *J. Raman Spectrosc.* 36(6–7): 485–496.

Otto, A. 1978. Raman spectra of (CN)-adsorbed at a silver surface. *Surf. Sci.* 75: L392–L396.

Otto, A., Mrozek, I., Grabhorn, H., and Akemann, W. 1992. Surface-enhanced Raman scattering. *J. Phys. Condens. Matter* 4(5): 1143–1212.

Pockrand, I. 1984. *Surface Enhanced Raman Vibrational Studies at Solid/Gas Interfaces*. Berlin, Germany: Springer.

Scaffidi, J. P., Gregas, M. K., Seewaldt, V., and Vo-Dinh, T. 2009. <RF>SERS-based plasmonic nanobiosensing in single living cells. *Anal. Bioanal. Chem.* 393(4): 1135–1141.

Schatz, G. C. 1984. Theoretical studies of surface enhanced Raman scattering. *Acc. Chem. Res.* 17(10): 370–376.

Stokes, D. L., Chi, Z. H., and Vo-Dinh, T. 2004. Surface-enhanced-Raman-scattering-inducing nanoprobe for spectrochemical analysis. *Appl. Spectrosc.* 58(3): 292–298.

Vo-Dinh, T. 1998. Surface-enhanced Raman spectroscopy using metallic nanostructures. *TrAC-Trends Anal. Chem.* 17(8–9): 557–582.

Vo-Dinh, T., Allain, L. R., and Stokes, D. L. 2002. Cancer gene detection using surface-enhanced Raman scattering (SERS). *J. Raman Spectrosc.* 33(7): 511–516.

Vo-Dinh, T., Alak, A., and Moody, R. L. 1988. Recent advances in surface-enhanced Raman spectrometry for chemical analysis. *Spectrochim. Acta Part B: Atomic Spectrosc.* 43(4–5): 605–615.

Vo-Dinh, T., Hiromoto, M. Y. K., Begun, G. M., and Moody, R. L. 1984. Surface-enhanced Raman spectrometry for trace organic analysis. *Anal. Chem.* 56(9): 1667–1670.

Vo-Dinh, T., Houck, K., and Stokes, D. L. 1994. Surface-enhanced Raman gene probes. *Anal. Chem.* 66(20): 3379–3383.

Vo-Dinh, T., Uziel, M., and Morrison, A. L. 1987. Surface-enhanced Raman analysis of benzo [a] pyrene-DNA adducts on silver-coated cellulose substrates. *Appl. Spectrosc.* 41(4): 605–610.

Vo-Dinh, T., Yan, F., and Wabuyele, M. B. 2005. Surface-enhanced Raman scattering for medical diagnostics and biological imaging. *J. Raman Spectrosc.* 36(6–7): 640–647.

Vo-Dinh, T. and Yan, F. 2007. Gene detection and multi-spectral imaging using SERS nanoprobes and nanostructures. In: *Nanotechnology in Biology and Medicine*, eds. T. Vo-Dinh. New York: Taylor & Francis.

Vo-Dinh, T., Dhawan, A., Norton, S. J., Khoury, C. G., Wang, H.-N., Misra, V., and Gerhold, M. 2010. Plasmonic nanoparticles and nanowires: Design, fabrication and application in sensing. *J. Phys. Chem. C* 114(16): 7480–7488.

Wabuyele, M. B. and Vo-Dinh, T. 2005. Detection of human immunodeficiency virus type 1 DNA sequence using plasmonics nanoprobes. *Anal. Chem.* 77(23): 7810–7815.

Wabuyele, M. B., Yan, F., Griffin, G. D., and Vo-Dinh, T. 2005. Hyperspectral surface-enhanced Raman imaging of labeled silver nanoparticles in single cells. *Rev. Sci. Instrum.* 76(6): 063710-1–063710-7.

Wang, H. N. and Vo-Dinh, T. 2009. Multiplex detection of breast cancer biomarkers using plasmonic molecular sentinel nanoprobes. *Nanotechnology* 20(6): 065101-1–065101-6.

Wokaun, A., Gordon, J. P., and Liao, P. F. 1982. Radiation damping in surface-enhanced Raman scattering. *Phys. Rev. Lett.* 48(14): 957–960.

Zeisel, D., Deckert, V., Zenobi, R., and Vo-Dinh, T. 1998. Near-field surface-enhanced Raman spectroscopy of dye molecules adsorbed on silver island films. *Chem. Phys. Lett.* 283(5–6): 381–385.

Zeman, E. J. and Schatz, G. C. 1987. An accurate electromagnetic theory study of surface enhancement factors for Ag, Au, Cu, Li, Na, Al, Ga, In, Zn, and Cd. *J. Chem. Phys.* 91: 634–643.

VI

Phototherapy

34 Photodynamic Therapy *Jarod C. Finlay, Keith Cengel, Theresa M. Busch, and Timothy C. Zhu* 733
Introduction • PDT: Basic Science • PDT: Clinical Applications • Conclusions • References

35 Low Level Laser and Light Therapy *Ying-Ying Huang, Aaron C.-H. Chen, and Michael R. Hamblin* 751
Introduction • Biophysical and Biochemical Mechanisms • Biological Effects of LLLT in Cells • Medical Applications of LLLT • Summary • Future Perspective • Acknowledgment • References

Photodynamic Therapy

Jarod C. Finlay
*University of Pennsylvania
School of Medicine*

Keith Cengel
*University of Pennsylvania
School of Medicine*

Theresa M. Busch
*University of Pennsylvania
School of Medicine*

Timothy C. Zhu
*University of Pennsylvania
School of Medicine*

34.1 Introduction 733
    A Brief History of Photodynamic Therapy
34.2 PDT: Basic Science...733
    Basic Photochemistry and Photophysics • Optics of Sensitizers • Oxygen • Tumor Vascular
    and Microenvironmental Effects • Intracellular and Intercellular Signaling Effects
34.3 PDT: Clinical Applications...740
    Light Delivery Devices for Clinical PDT • PDT Dosimetry • Approved
    Indications • Clinical Trials
34.4 Conclusions...746
References...746

34.1 Introduction

In contrast to most of the chapters of this handbook, this chapter will address a therapeutic, rather than diagnostic, application of biomedical optics. The therapeutic effects of light have been appreciated since ancient times. Exposure to light has effects on health ranging from mood and psychological effects and production of vitamin D to UV damage and the induction of cancer. The majority of these effects can be reproduced or enhanced with artificial light. The term photodynamic therapy, however, refers specifically to photochemical reactions mediated by a pharmacological or endogenously generated photosensitizer. Because photosensitizers are involved, these effects are fundamentally different from the majority of naturally occurring effects expected from exposure to sunlight. This chapter will review the history of photodynamic therapy and the basic physics and optics underlying it, as well as the current clinical use of the modality.

34.1.1 A Brief History of Photodynamic Therapy

The history of photodynamic therapy (PDT) dates back to the observation by Raab and von Tappeiner in the late 1800s that acridine orange dye in combination with sunlight could cause cytotoxicity in paramecia. Von Tappeiner made immediate use of this effect, treating human skin tumors with the sensitizer eosin in 1903 (Ackroyd et al. 2001). The photodynamic effect received relatively little attention from the world of medicine for several decades. This changed in 1960 when researchers at the Mayo Clinic observed preferential fluorescence in the tumors of patients injected with hematoporphyrin (Hp)—a crude sensitizer derived from blood. Subsequent refinement led to the

development of hematoporphyrin derivative (HpD), which was later further purified into the FDA-approved drug Photofrin (Dougherty et al. 1998).

The development of these new photosensitizers, along with lasers capable of providing intense, collimated, monochromatic light, set the stage for a renaissance in the development of photodynamic therapy. The first large-scale clinical trials of the photodynamic therapy were performed at Roswell Park Cancer Center in the late 1960s and early 1970s. The success of these and other trials led to the regulatory approval of PDT for treatment of bladder and lung disease and macular degeneration in the 1990s and subsequent approval for skin and esophageal indications (Allison et al. 2004). The four decades since the first clinical trials have seen an explosion of optical technology, allowing much more sophisticated light sources, light delivery mechanisms, and light measurement capability. These developments have been paralleled by significant progress in the formulation of new sensitizers, leading to a new generation of sensitizers that have more favorable optical, chemical, and pharmacokinetic properties than their predecessors. These sensitizers have improved the repeatability, specificity, and side effect profile of PDT considerably and expanded its domain of treatment to a variety of cancerous and benign diseases. Photodynamic therapy has also benefited from the tremendous advances in the fields of biochemistry, molecular genetics, and tumor biology over the last several years.

34.2 PDT: Basic Science

34.2.1 Basic Photochemistry and Photophysics

There are a number of photochemical reactions that are capable of producing cytotoxic effects; however, the majority of the preclinical and clinical work in photodynamic therapy has focused on type II reactions, that is, those which rely on the generation of singlet oxygen and other reactive oxygen species (ROS). It is the reaction of these species with targets within the cell that causes the therapeutic effect.

The key to understanding the underlying photophysics of photodynamic therapy lies in understanding the chemistry of oxygen. Molecular oxygen in its ground state has two unpaired electrons in its outermost orbital. Because these electrons are unpaired, their combined spin can take on any of three quantum states, making the molecule a spectroscopic triplet. The excited state of oxygen responsible for the reactions involved in photodynamic therapy has two paired electrons in a single orbital. Because these electrons are paired, they are constrained to have only one spin quantum state, making the excited molecule a spectroscopic singlet. It should be noted that the excitation involved here is a change in electron spin quantum state, not a change in electron orbital energy level. The energy required for this transition is lower than that typically associated with electron energy level transitions seen in processes such as fluorescence. Despite the fact that visible light has sufficient energy to drive this transition, molecular quantum selection rules prevent the direct excitation of ground-state oxygen to its singlet excited state by optical means. It is therefore necessary to use a photosensitizing compound, or photosensitizer, to absorb energy in the form of light and transfer it to molecular oxygen.

The fundamental photophysics of photodynamic therapy can be summarized by the energy level diagram shown in Figure 34.1. The photodynamic process begins with the absorption of a photon of visible light by the sensitizer in its singlet ground state, denoted S_0. This absorption excites the sensitizer to its first excited state, which is also a spectroscopic singlet, S_1. Among the important attributes of a good photosensitizer is the ability to spontaneously decay from the singlet excited state S_1 to an excited triplet state T_1, a process known as intersystem crossing. Sensitizer molecules in the T_1 state may collide with ground

state molecular oxygen molecules ($^3O_2$), transferring energy to them. This interaction excites the oxygen molecule to its singlet state ($^1O_2$) and returns the sensitizer molecule to its ground state. In general, the process of generating singlet oxygen does not consume the sensitizer itself, allowing a single molecule of photosensitizer to generate many molecules of singlet oxygen. The singlet oxygen generated, however, generally reacts very quickly, so the process as a whole consumes oxygen.

Even photosensitizers with a high singlet oxygen yield generally possess non-singlet-oxygen-generating pathways for relaxation from the S_1 state. A direct relaxation from the S_1 state to the S_0 state results in the emission of a visible fluorescent photon. The majority of photosensitizers used clinically are fluorescent, and fluorescence spectroscopy has been used in a number of preclinical and clinical studies to evaluate the concentration and distribution of photosensitizers. Many photosensitizers exhibit irreversible photobleaching as a result of the reaction of singlet oxygen with the sensitizer molecule itself. It should be noted that while this photobleaching leads to a loss of fluorescence intensity, it is the result of a chemical reaction that destroys the sensitizer molecule, not a change in the photophysical parameters of the remaining sensitizer molecules.

34.2.2 Optics of Sensitizers

From the point of view of photophysics, the requirements of a good photosensitizer are relatively straightforward. First, the sensitizer must absorb light efficiently. The sensitizer's extinction coefficient (ε) must be high enough that it can absorb a significant fraction of the light present in the tissue. In terms of the fundamental photophysical quantities, the rate of excitation of the ground-state sensitizer is given by

$$I_a = \frac{\Phi \varepsilon c}{h\nu} \tag{34.1}$$

where

c and Φ are the concentrations of sensitizer and the fluence rate of light in the tissue, respectively

$h\nu$ is the energy per photon of light (Georgakoudi et al. 1997)

The higher the extinction coefficient, the less sensitizer must be administered, reducing the chances of dark toxicity. This is not necessarily a strict measure of goodness. It is possible to have a good sensitizer with a low extinction coefficient if its dark toxicity is low enough. Conversely, there are sensitizers with high extinction coefficients whose toxicity is limiting even at low concentrations. What is more important is the in vivo absorption coefficient of the sensitizer, given by the product of the extinction coefficient and the concentration achievable in the target tissue.

Regardless of the sensitizer's absorption coefficient, it is important to remember that the sensitizer competes with any other absorbers in the tissue for absorption of light. This is not explicitly accounted for in Equation 34.1; however, increased

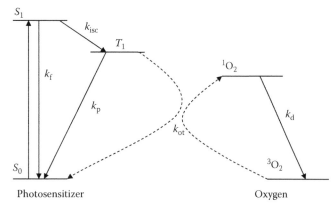

FIGURE 34.1 Energy level diagram for a typical photosensitizer (left) and molecular oxygen (right). The transitions among states and the rate constants for each transition are indicated by arrows. The notation is that of Foster et al. (From Georgakoudi, I. et al., *Photochem. Photobiol.*, 65, 135, 1997.)

absorption reduces the fluence rate in tissue (Φ). For a given incident irradiance, a sensitizer will be exposed to a lower fluence rate when the absorption coefficient in the tissue is high. In the extreme case where the sensitizer's extinction coefficient is very high, it is possible for the sensitizer itself to significantly change the absorption coefficient of the tissue. In this case, adding additional sensitizer may actually reduce the effectiveness of PDT by reducing the penetration of light into the tissue. Figure 34.2 shows the typical treatment wavelengths of several common photosensitizers against the background absorption of oxy- and deoxyhemoglobin, plotted at a concentration of 50 μM, and water, plotted at a concentration of 80% by volume. The hemoglobin spectrum is that of Finlay and Foster, corrected for the effects of confinement to erythrocytes (Finlay and Foster 2004) and the water spectrum is that of Kou et al. (1993). The background absorption in any given tissue will lie somewhere between the curves plotted here, because the blood in tissue will contain some combination of oxy- and deoxyhemoglobin. The advantage of a long-wavelength sensitizer such as Motexafin Lutetium (MLu), with an absorption peak at 732 nm, over Photofrin, typically excited at 630 nm, is clear, especially if the hemoglobin in the tissue of interest is partially deoxygenated. There is some additional advantage conferred by the fact that the scattering coefficient is also lower at longer wavelengths, however this is a much less important effect than that of absorption. It is important to note that the benefit of longer wavelengths is a function of the absorption spectrum of hemoglobin, not a fundamental principle of optics. In fact, for well-oxygenated tissue, the absorption coefficient increases with wavelength beyond approximately 700 nm, and the absorption due to water becomes increasingly important at wavelengths beyond 800 nm.

Tissue oxygenation also has an effect on the optics of light absorption. In the region of the optical spectrum where most PDT treatments are delivered (630–770 nm), the extinction coefficient of oxyhemoglobin is less than that of deoxyhemoglobin by as much as an order of magnitude. As a result, hypoxic tissue exhibits much higher background absorption than well-oxygenated tissue with the same blood content. For a given incident fluence rate and sensitizer concentration, the hypoxic tissue will therefore experience a lower rate of light absorption by the sensitizer. This effect has been observed quantitatively in animal models (Mitra and Foster 2004).

The second photophysical property a good photosensitizer requires is a high quantum yield of singlet oxygen. This, in turn, depends on a high triplet quantum yield, a high likelihood of collisional excitation with ground-state oxygen, and a ready supply of ground-state oxygen. The triplet yield (φ_t) is defined as the probability of generation of a triplet-state sensitizer molecule from one absorbed photon of light. This is determined by the balance between two competing processes, de-excitation to the ground state (including fluorescence and non-radiative relaxation) and intersystem crossing. The rate constants for these two processes are inherent characteristics of the molecule. In the notation of Foster et al. (Georgakoudi et al. 1997, Georgakoudi and Foster 1998), the triplet yield is given by

$$\varphi_t = \frac{k_{isc}}{k_f + k_{isc}}, \tag{34.2}$$

where k_f and k_{isc} are the rate constants for fluorescence and intersystem crossing, respectively. Because these two processes compete, a sensitizer with a high triplet yield necessarily has a low fluorescence yield. Because k_f includes radiative and non-radiative processes, it is possible for a molecule to be nonfluorescent and still have a nonzero k_f and a triplet yield less than 1.

Third, a good sensitizer requires the ability to convert the energy of its excited triplet state to excitation of singlet oxygen. As in the case of the triplet yield, the singlet oxygen yield from the triplet state is a balance between competing processes, in this case the spontaneous relaxation from the triplet state by either radiative or non-radiative pathways and the transfer of energy to oxygen. The radiative relaxation from the triplet state results in phosphorescence emission. The yield of singlet oxygen per molecule is given by

$$\phi_{ot} = S_\Delta \frac{k_{ot}\left[{}^3O_2\right]}{k_p + k_{ot}\left[{}^3O_2\right]}, \tag{34.3}$$

where

- k_{ot} and k_p are the rate constants for collision with ground-state oxygen and spontaneous relaxation of the triplet state, respectively
- S_Δ is the fraction of collisional interactions with oxygen that generate singlet oxygen, estimated to be approximately 0.5 (Nichols and Foster 1994)
- $[{}^3O_2]$ is the concentration of ground-state oxygen

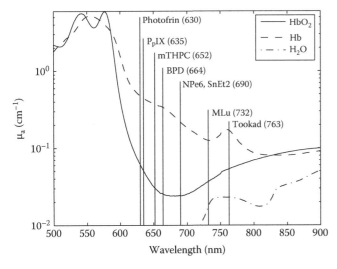

FIGURE 34.2 Absorption spectra of oxy- and deoxyhemoglobin and water, with the absorption maxima of several commonly used sensitizers. These absorption spectra assume a hemoglobin concentration of 50 μM and a water concentration of 80%.

The rate constant k_{ot} is a bimolecular rate constant because the rate of collisions between singlet oxygen and triplet sensitizer depend on the concentrations of both species. Equation 34.3 underscores a fundamental feature of the photodynamic process: the rate of singlet oxygen generation depends both on local concentration of oxygen, which is independent of the sensitizer chemistry, and on the inherent properties of the sensitizer (S_Δ, k_p, and k_{ot}). A good sensitizer should have a long triplet lifetime (and therefore a low k_p) so that triplet-state molecules have time to interact with oxygen, and a high S_Δ and k_{ot}, indicating a high likelihood of singlet oxygen generation when they do interact.

Finally, a factor that is often at least as important clinically as the absorption spectrum and quantum yields of a sensitizer is its pharmacokinetics. A sensitizer with excellent photophysical parameters is of little use if it is not taken up by the target tissue. Conversely, a sensitizer that produces prolonged normal tissue photosensitivity will require patients to avoid sunlight, which may have a significant impact on patient's quality of life. The ideal photosensitizer would be taken up or retained in tumors at higher concentrations than in normal tissues, and would clear from the skin and eyes very quickly, reducing photosensitivity. The first-generation sensitizers have an uptake in tumors that was only marginally greater than that in adjacent normal tissue (Hahn et al. 2006) and produce skin photosensitivity that lasts up to 6 weeks.

The photosensitizers that have thus far been most successful at meeting the criteria described above, come from several families of organic molecules related to the porphyrins (O'Connor et al. 2009). The essential optical characteristics of the porphyrins and chlorin sensitizers are a very strong absorption in the 400–450 nm region of the optical spectrum, known as the Soret peak, and a series of longer wavelength, often less intense, absorption bands throughout the visible spectrum. In most applications, it is the longest wavelength of these bands that is used for PDT treatment. These molecules exhibit some preferential uptake in tumors and are generally nontoxic when not exposed to light.

The first clinically viable sensitizer, hematoporphyrin derivative, (HpD) is a mixture of porphyrins derived from blood. It was the early work with HpD in humans that demonstrated the potential of PDT as a clinical modality (Dougherty et al. 1978). HpD was further refined to produce Photofrin, the first commercially successful photosensitizer. Photofrin has been remarkably successful in treating a large variety of diseases and has become the gold standard for PDT efficacy against which other sensitizers are measured. Researchers have sought to improve on Photofrin in subsequent sensitizer development. From the point of view of optics, the primary effort has been directed toward finding sensitizers with longer-wavelength absorption maxima. Equally important clinically are the pharmacokinetics and duration of sensitivity of normal tissue. Photofrin PDT requires patients to avoid direct sun exposure to the skin for as long as 6 weeks, which imposes a significant burden on the patient.

Since the introduction of Photofrin, several hundred sensitizers have been developed and tested in vitro. Of these, only a small fraction have demonstrated the low toxicity, favorable pharmacokinetics, and optical properties necessary for a good sensitizer. Rather than summarize this constantly changing field in detail, the following section will highlight a few of the sensitizers that are most unique and interesting from the point of view of optics. The interested reader can find more details in several recent reviews (O'Connor et al. 2009).

Among these successful second-generation sensitizers is meso-tetra hydroxyphenol chlorin (mTHPC), or Foscan. Early preclinical and clinical work with mTHPC showed it to be as much as 100 times more potent than Photofrin. A careful analysis of the combination of photophysical and optical properties of mTHPC reveals that this difference can be explained entirely in terms of optics, physics, and pharmacokinetics (Mitra and Foster 2005). mTHPC has a very high extinction coefficient and suffers far less competitive absorption of light by hemoglobin than Photofrin does. mTHPC does not, however, improve significantly on Photofrin's skin photosensitivity problem. mTHPC exhibits complex pharmacokinetics in vivo and even in suspension in blood (Hopkinson et al. 1999).

Subsequent development pushed the absorption wavelength even farther into the red. MLu absorbs at 732 nm, which is likely the longest wavelength at which increased wavelength is a significant advantage. MLu has faster clearance from the skin than Photofrin or mTHPC, potentially reducing side effects. MLu has been investigated for the treatment of prostate cancer (Stripp et al. 2004), but is not currently in clinical trials.

The sensitizer Tookad has a relatively long-wavelength absorption peak at 673 nm, and is unusual in that it is only minimally taken up by tissue. This confers an advantage in terms of normal tissue photosensitivity, as the drug clears from the body very quickly, however it complicates treatment planning. Tookad-mediated treatments are performed by delivering light while the sensitizer is still in circulation. The dynamics of the vascular supply of the drug therefore are much more important than they would be with a typical sensitizer. The effectiveness of Tookad has been demonstrated in preclinical (Lee et al. 1997, Huang et al. 2007) and clinical (Weersink et al. 2005) treatment of prostate cancer. The sensitizer Benzoporphrin derivative (BPD, trade name Verteporfin) has been investigated for treatment of a variety of tumors, and has been clinically approved for the treatment of macular degeneration, an overgrowth of vessels in the retina. Like Tookad, BPD can act primarily as a vascular sensitizer at short times after administration (e.g., 15 min); however, it exhibits more cellular damage at longer drug-light intervals (e.g., 3 h) (Chen et al. 2005a). When a drug is used during the period when it is primarily confined to the vasculature, as in the case of BPD or Tookad, the usual assumption that the drug is approximately homogeneously distributed on the microscopic scale may not be valid. There may be many cells with virtually no sensitizer, while the cells lining the vasculature may be exposed to a high concentration of sensitizer. In these cases, the sensitizer may suffer from the "pigment packaging effect," in which the high concentration of absorbers (both sensitizer and hemoglobin) within the vasculature reduces the effective absorption coefficient (Finlay et al. 2004b).

A third class of sensitizers are based on aminolevuinic acid (ALA). This compound takes advantage of the biosynthetic pathway responsible for the synthesis of heme. The penultimate step of this reaction pathway is the synthesis of protoporphyrin IX, which is an effective photosensitizer. Under normal conditions, PpIX generated by this pathway is quickly converted to heme by the incorporation of iron under the action of the enzyme ferrochelatase, so there is very little accumulation of PpIX. When sufficient exogenous ALA is administered, however, the rate of PpIX production exceeds the rate of conversion to heme, leading to a temporary accumulation of PpIX in the tissue. Because ALA requires an additional biochemical step to produce PpIX, its pharmacokinetics is complicated. This characteristic, though, has its advantages. For instance, ALA is water soluble, making it easier to dissolve in common solvents than PpIX, and the production and retention of PpIX produced by ALA sensitization shows some selectivity for tumor tissue. To increase the uptake in hydrophobic environments, several analogues of ALA have been made by attaching ester groups to the ALA molecule. These compounds have different pharmacokinetics than ALA, but the final product of the biosynthesis (PpIX) is the same. PpIX has an absorption maximum (at 635 nm) similar enough to that of Photofrin that the same light sources can often be used to excite both sensitizers. PpIX undergoes complicated photobleaching, producing two photoproducts. In addition, there is evidence that damage to mitochondria during PDT can lead to the accumulation of other porphyrin products produced in the same biosynthetic process. This feature makes the interpretation of fluorescence emission measurements from PpIX-treated tissue complicated but potentially very informative (Finlay et al. 2001).

The field of sensitizer development continues to evolve. A variety of approaches have been investigated to make sensitizers more tumor-selective, many of them taking advantages of recent advances in tumor-targeting in the larger cancer research field (Schrama et al. 2006, Sharkey and Goldenberg 2006). These include the conjugation of the photosensitizer to a tumor-selective molecule or particle (Solban et al. 2006), and photodynamic molecular beacons (PMB's), which combine a sensitizer with a quenching molecule to create a sensitizer that generates singlet oxygen only in the presence of a tumor-specific molecular target (Zheng et al. 2007). Related work has focused on developing sensitizers that exhibit enhanced imaging capability by incorporating contrast agents for optical, PET, or MR imaging (Chen et al. 2005b, Pandey et al. 2006).

34.2.3 Oxygen

34.2.3.1 Microscopic Oxygen Kinetics

The details of the sensitizer–oxygen interaction have been modeled using standard chemical reaction kinetics models. These models assume a single reaction rate constant for each reaction

involved in the process, from the absorption of light through the creation and reaction of singlet oxygen. These are used to generate a set of coupled differential equations describing the photodynamic process. There are three differential equations for the sensitizer, one describing the population of each state of the sensitizer (S_0, S_1, and T_1), and two equations for oxygen, one for ground-state oxygen and another for singlet oxygen. In the notation of Foster et al. (Georgakoudi et al. 1997) these equations are

$$\frac{d[S_0]}{dt} = -I_a + k_f[S_1] + k_P[T_1] + k_{ot}[T_1]\left[{}^3O_2\right] - k_{os}[S_0]\left[{}^1O_2\right]$$

$$\frac{d[S_1]}{dt} = I_a - k_f[S_1] - k_{isc}[S_1]$$

$$\frac{d[T_1]}{dt} = k_{isc}[S_1] - k_{ot}[T_1]\left[{}^3O_2\right] - k_P[T_1]$$

$$\frac{d\left[{}^3O_2\right]}{dt} = -S_\Delta k_{ot}[T_1]\left[{}^3O_2\right] + k_d\left[{}^1O_2\right]$$

$$\frac{d\left[{}^1O_2\right]}{dt} = +S_\Delta k_{ot}[T_1]\left[{}^3O_2\right] - k_d\left[{}^1O_2\right]$$

$$- k_{oa}[A]\left[{}^1O_2\right] - k_{os}[S_0]\left[{}^1O_2\right] \qquad (34.4)$$

It is possible to solve these equations numerically to obtain the population of each state; however, for the purposes of modeling the deposition of singlet oxygen dose in tissue, the details of the populations of the various excited states at very short timescales are irrelevant. A steady-state solution can be found by assuming that the excited states of the sensitizer and oxygen are always in equilibrium with their respective ground states. This assumption is justified by the fact that the lifetimes of these excited states are many orders of magnitude shorter than the timescales relevant to processes such as oxygen diffusion, cytotoxicity, and photobleaching. The result is a solution for the steady-state concentration of singlet oxygen:

$$\left[{}^1O_2\right] = S_\Delta \varphi_t I_a \left(\frac{k_{ot}\left[{}^3O_2\right]}{k_P + k_{ot}\left[{}^3O_2\right]}\right)\left(\frac{1}{k_d + k_{oa}[A]}\right). \qquad (34.5)$$

The dose of singlet oxygen deposited in tissue is proportional to the time integral of the singlet oxygen concentration.

Even the steady state model requires numerical solution if the effects of oxygen diffusion are considered. The equations listed here describe the minimal set of interactions required for type II photodynamic reactions. There are many other possible reactions that are not considered here, but that may be important. For instance, some photosensitizers bleach via reactions between the triplet-state sensitizer and other chemical species present in tissue. These effects can be modeled by adding another term to the kinetic equations, which can change the resulting dynamics substantially (Stratonnikov et al. 2003, Finlay et al. 2004a). Other reactions, such as singlet-state sensitizer reactions, can also be considered using a similar mechanism.

It is difficult or impossible to measure each of the photodynamic reaction rate constants independently, particularly in living tissue. However, by measuring the rate of oxygen consumption in multicellular tumor spheroids, Foster et al. have been able to measure quantities that relate directly to the ratios of several of these constants (Foster et al. 1993, Nichols and Foster 1994, Georgakoudi et al. 1997, 1998), or to place bounds on them based on observed photobleaching kinetics (Finlay et al. 2004a). Once the rate constants (or ratios among them) are determined, the kinetic model can make quantitative predictions about the dynamics of the photodynamic process. These predictions are often sensitive enough to differentiate between different mechanisms of photobleaching (Georgakoudi and Foster 1998). In the case where both the photodynamic effect and photobleaching are mediated primarily by singlet oxygen, it can be shown that the fraction of the initial sensitizer bleached is indicative of the absolute singlet oxygen dose delivered (Georgakoudi et al. 1997). While this relationship is quantitatively valid only on the microscopic scale, it provides a theoretical foundation for the use of photobleaching as a measure of photodynamic effect. This model also gives a quantitative prediction of the effects of oxygen depletion resulting from photochemical oxygen consumption. Recent work in the field of kinetic modeling includes extending the microscopic kinetics-with-diffusion model from the spherical geometry suitable for tumor spheroids to a Krogh cylinder model that approximates oxygen transport in vasculature (Wang et al. 2007). This model is still microscopic in scale, but can take into account the effects of blood flow in addition to oxygen diffusion, and can approximate the true in vivo geometry reasonably well.

34.2.3.2 Macroscopic Models and Effects

The microscopic kinetic model has proven useful in elucidating the underlying physics of PDT and providing qualitative explanations of the fundamental characteristics of clinical PDT response. However, it is strictly valid only at the microscopic level, and is therefore difficult to apply to clinical PDT, where only volume-averaged or bulk measurements of sensitizer concentration, fluorescence photobleaching, and tissue oxygenation are available. To bridge the gap between the detailed, microscopic model and the clinically feasible measurements, several researchers have developed bulk or "macroscopic" PDT kinetic models (Zhu et al. 2007, Zhou and Zhu 2008).

The key idea of the macroscopic modeling is to couple the rate equations describing oxygen consumption and singlet oxygen generation with the spatial distribution of the light transport and replacing the oxygen perfusion through the blood vessel with a macrosopic oxygen perfusion rate term (Hu et al. 2005, Zhu et al. 2007). One key discovery of the macroscopic modeling is the existence of a plateau area that corresponds to the region of oxygen depletion due to

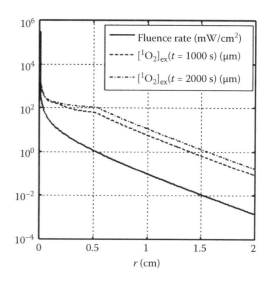

FIGURE 34.3 Light fluence rate (solid line) and reacted singlet oxygen concentration at different times: 1000 s (dashed) and 2000 s (dot-dashed) after the start of illumination with a point source. Notice a plateau area in the distribution of reacted singlet oxygen. (From Zhu, T.C. et al., *Proc. SPIE*, 6427, 642708, 2007. With permission.)

photodynamic consumption of oxygen. Figure 34.3 shows typical behavior from the macroscopic modeling. The plateau area could be used to correlate with the necrosis depth induced by PDT, which has exhibited well-defined boundaries of tissue necrosis.

Macroscopic kinetics modeling has the potential to include the effect from tissue oxygen consumption in the quantitative description of the PDT clinical outcome. In essence, the effective concentration of reacted singlet oxygen, $[^1O_2]_{rx}$, can be expressed as

$$[^1O_2]_{rx} = f \cdot \int_0^t \eta \cdot \rho \dot{D} dt, \tag{34.6}$$

where

$\dot{D}$ is the PDT dose rate, defined as the energy absorbed by the sensitizer per unit mass of tissue

ρ is the mass density of tissue

f is the fraction of singlet oxygen reactions with tumor tissue that contribute to the induction of necrosis

η is the singlet oxygen quantum yield,

$$\eta = S_\Delta \gamma \cdot \left(\frac{[^3O_2]}{[^3O_2] + \beta} \right) \cdot \alpha, \tag{34.7}$$

where α, β, γ, and S_Δ are photosensitizer-dependent constant parameters (Zhu et al. 2007). For a particular photosensitizer, the key is to obtain these parameters from a suitable in vivo model (Wang et al. 2009). These macroscopic parameters are related to the microscopic parameters k_f, k_{isc}, and so on, but cannot be assumed to be numerically identical.

34.2.4 Tumor Vascular and Microenvironmental Effects

The response of tumors to PDT is more complicated than even the most sophisticated singlet oxygen model would predict, because the relationship between tumor control and singlet oxygen dose is complicated and depends on the details of the mechanism of damage. Biological mechanisms of damage from PDT can include direct damage to the tumor cells themselves from the reactive oxygen species, as well as indirect effects, which include damage of the tumor-supporting blood vessels and the stimulation of a host immune response (Fingar et al. 2000, Chen et al. 2006, Gollnick et al. 2006, Kessel et al. 2006, Korbelik 2006). These are dynamic effects that change as a function of time elapsed during as well as after PDT, and their timing relative to each other can critically affect treatment efficacy (Busch 2006). For example, PDT-mediated insult to the vasculature can cause vasospasm and constriction during illumination that may or may not be reversible (Sitnik et al. 1998, Chen et al. 2005a). Vasoconstriction or vascular obstruction that occurs during (or after) PDT can favor therapeutic outcome by depriving the tumor cells of oxygen and nutrients essential for their survival and thus secondarily leading to their death. Yet, in some instances, vasoconstriction during PDT can reverse. Under these conditions, recovering vascular function in the hours after PDT will support the survival of tumor cells that escaped the direct cytotoxic effects of PDT. Making the situation even more complicated is the fact that direct cytotoxicity can be impaired when vasospasm occurs during illumination because it leads to limitations in the delivery of oxygen.

Both the direct and indirect components of tumor response to PDT can be affected by photophysical parameters, such as the local concentration of photosensitizer and light in the tissue (Henderson et al. 2006, Wang et al. 2007). Thus, the tumor microenvironment, and its effects on the distributions of light, photosensitizer, and oxygen in the target tissue are important considerations in PDT. In general, tumors are characterized by a poorly developed, heterogeneous vascular network that can limit the delivery of oxygen and drugs throughout its mass (Fukumura and Jain 2007). Preclinical investigations have found that individualization of light delivery in PDT to account for heterogeneity in photosensitizer uptake can reduce the variability in tumor response (Zhou et al. 2006). Moreover, many tumors, including those in patients who are to receive PDT, contain areas of hypoxia (Busch et al. 2004). PDT can greatly exacerbate this hypoxia through the limitations it imposes on oxygen delivery when vasospasm develops during illumination (discussed above), as well as through the process by which it consumes ground-state oxygen in the formation and destructive reaction of singlet oxygen (photochemical oxygen consumption) (Busch et al. 2002). The relevance of PDT-created hypoxia and

vascular effects to treatment outcome has been demonstrated by studies that find a correlation between the PDT-induced change in tumor hemoglobin oxygen saturation or blood flow in an individual animal and that animal's response to treatment (Pham et al. 2001, Wang et al. 2004, Yu et al. 2005, Standish et al. 2008).

Numerous approaches toward improving treatment outcomes to PDT through oxygen maintenance during illumination have been studied. Hyperbaric oxygen breathing during illumination has improved tumor responses in both clinical (Maier et al. 2000) and preclinical investigations (Chen et al. 2002). Alternatively, lower fluence rate of illumination can be used to slow the rate of photochemical oxygen consumption, thereby allowing better opportunity for oxygen delivery through the vasculature to keep pace with its consumption via PDT, and hyperfractionated light (e.g., cycles of 30 s "on", 30 s "off") can be used to permit recovery of tumor oxygenation during the "off" periods (Foster et al. 1991, Pogue and Hasan 1997). Both of these approaches have been effective in improving measures of PDT outcome, including photosensitizer photobleaching and tumor remission rate in clinical studies (Ericson et al. 2004, Pogue et al. 2008), and tumor cure rate, overall survival, cell death, and necrotic area in preclinical investigations (Henderson et al. 2004). Another advantage of methods that improve tumor oxygenation during PDT is the benefit provided to light penetration at PDT-relevant wavelengths due to differences in the absorption of this light by oxy- versus deoxyhemoglobin (Mitra and Foster 2004).

In addition to the protective effects of PDT-created hypoxia on direct and indirect tumor damage, this hypoxia, together with other microenvironmental effects of PDT, can lead to the production of survival factors that will limit treatment efficacy. For example, the growth of new blood vessels after PDT can be supported by PDT-induced increases in the pro-angiogenic molecule vascular endothelial growth factor (VEGF). This aspect of the interaction between PDT and tumor microenvironment has been recently reviewed (Gomer et al. 2006). To date, there has been success by several groups in the use of anti-angiogenic agents in combination with PDT to improve tumor responses (Ferrario et al. 2000, Ferrario and Gomer 2006, Kosharskyy et al. 2006, Bhuvaneswari et al. 2007).

34.2.5 Intracellular and Intercellular Signaling Effects

Newly emerging data suggests that signaling through growth factor receptors and post receptor signaling partners may be important in the survival of cancer cells following PDT (Solban et al. 2006). In this regard, the epidermal growth factor receptor (EGFR) pathway has been extensively studied. EGFR is a receptor tyrosine kinase that regulates important cellular functions including cell cycle progression and survival mediated through phosphatidylinositol 3'-kinase (PI3K)/AKT, proliferation through mitogen-activated protein kinases (MAPK), and protection from apoptosis through STAT3. Inhibitors of EGFR and post receptor signaling molecules have been shown to dramatically increase PDT-mediated cancer cell cytotoxicity, although the mechanisms for this effect remain incompletely understood. Some investigators have suggested that PDT-mediated EGFR activation is important for survival of cancer cells following PDT and that EGFR signaling is up-regulated by PDT. Others have found that PDT causes a temporary degradation/inactivation of cell surface receptors, including EGFR. Interestingly, in either of these cases, inhibition of EGFR signaling might be expected to augment PDT-mediated cancer cell killing. Similarly, PDT can lead to activation of the PI3K/AKT and MAPK pathways, although this effect is not universally observed. Some of these inconsistencies may be explained by incorporating the timing, schedule, and PDT dose dependence of signal transduction inhibition in combination with PDT. Alternatively, there may be cell type or photosensitizer-specific differences in EGFR pathway signaling following PDT. Nevertheless, it is clear that modulation of the molecular responses of cancer cells following PDT has the potential to significantly enhance the efficacy of PDT and will likely be the subject of intense preclinical and clinical investigation in the future.

PDT-mediated cytotoxicity occurs through multiple cellular mechanisms, including apoptosis, autophagy, and necrosis. The exact mechanism of cell death may be important in determining how PDT affects surrounding, untreated cells and tissues. For example, PDT-mediated necrotic cell death may stimulate a strong bystander effect, killing surrounding tumor cells that had been insufficiently damaged by PDT (Dahle et al. 2000). This effect is not observed when PDT kills cells by apoptosis and may be due to density-dependent alterations in intercellular signaling. In addition, necrotic cell death is predicted by some models to be more immunogenic than apoptotic cell death. These effects are also likely dependent on the gene expression profile of target cells, the dose of light and/or photosensitizer, and the identity of the specific photosensitizer being used.

34.3 PDT: Clinical Applications

The basic science described in the preceding section has been developed largely using in vitro models consisting of cells, cell colonies, or multicell tumor spheroids, or in small animal models. These models provide a useful platform for the development of new treatment techniques and new technologies, and allow controlled experiments to be performed to address specific scientific questions that relate to PDT. The remainder of this chapter will address the translation of the basic science developed in the laboratory environment into clinical practice. There are several fundamental ways in which the clinical practice of PDT differs from the basic science. First, most of the in vivo models used for basic science are small animals such as mice and rats with tumors that may be large in proportion to the animal's weight but are small (less than 1 cm) in absolute terms. While this is acceptable from the point of view of biology, it presents a challenge in that human tumors are often small in proportion to the body, but large (>1 cm) in absolute terms. While many aspects of the biology of these systems scale in proportion to body size,

the tissue optics scales only in absolute terms. A human 10 times the width of a mouse with a tumor four times the width of the mouse tumor will still have approximately the same absorption and scattering coefficients within the tumor tissue. The tumor size and body size of the human are therefore proportionately much larger in terms of the scale relevant to optics, namely, the transport mean free path length. This scaling requires that the light delivery devices and dosimetry calculations be more sophisticated for humans than for small animals, and further make testing of human-scale devices using small animal experiments difficult.

Second, in the treatment of humans, the first priority is the safety and effective treatment of the patient. This of necessity requires clinical treatment to be simpler, more robust, and more efficient than that used in animal experiments. Patient dosimetry often has to rely on fewer measurements and less complete data than the corresponding animal case. In research, it is the collection of data that drives the timing of treatment and the selection of procedures performed. In clinical treatment and even in clinical trials, it is the treatment of the patient that dictates the schedule. Complex optical measurements that require careful setup and calibration are often impossible in this context, especially in intraoperative cases, where additional time under anesthesia is detrimental to the patient's outcome. This distinction is key to understanding why the data collected in preclinical studies is often more sophisticated than that available to clinical trials and routine treatment.

34.3.1 Light Delivery Devices for Clinical PDT

Various systems have been designed to deliver light for PDT treatments. In general, these systems have been designed with the goal of delivering a uniform light distribution throughout the treatment field. Most PDT light delivery systems are based on lasers coupled into optical fibers terminated in delivery devices of some type. The details of the delivery system depend on the light source and treatment geometry, but are generally defined by the application: collimating lenses are used for surface applications, diffusers are used for interstitial and intracavitary applications, and focusing optics are used for intraocular applications. An alternative to optical fiber base delivery systems are the increasingly available light emitting diode (LED) systems.

34.3.1.1 Collimating Lenses for Surface Applications

The simplest system for PDT light delivery is the use of an optical fiber with some collimating optics for delivery of a uniform light field to a surface. The complexity of the collimating systems used clinically range from polished fibers, in which the fiber/air interface serves as the lens, to fibers with built-in or removable geometrical lenses or small gradient-index (GRIN) lenses. In each case, the resulting light field can be made uniform in intensity to within 10% over the area of the treatment field.

34.3.1.2 Diffusers for Interstitial or Interluminal Applications

There are many applications where light delivery with a collimated beam is not practical, for instance, in the interior of cavities or when the light must be delivered to the center of a bulky tumor or organ. In these cases, the applicator of choice is an optical fiber with a diffusing segment at the distal end. Cylindrical diffusers with lengths ranging from 2 mm to over 10 cm are commercially available. These fibers can tolerate very high powers, and deliver approximately uniform intensity along their length.

Alternatively, the fiber itself can be made to scatter light by etching or otherwise modifying the fiber itself. Fibers with diffusing tips created by various means are currently in routine clinical and preclinical use. More recently, the technology has been developed to allow the creation of diffusers with custom light emission distributions by optically etching grating structures into the cores of single-mode optical fibers (Vesselov et al. 2005). This method of fabrication has the advantage of potentially allowing fibers to be customized to deliver light distributions specific to each patient (Lilge et al. 2005). Cancers of the esophagus and bronchus are often treated using cylindrical diffusers built into transparent balloons, which center them in the cavity and may also reduce folding of the tissue.

34.3.1.3 Intercavity Applicators

Photodynamic therapy of large cavities requires the ability to deliver a uniform distribution of light over the interior surface of a cavity. This can be accomplished using a small, isotropic point source in the center of the cavity. This, however, is somewhat problematic because small light sources typically cannot tolerate the power required for PDT, and because maintaining a uniform fluence rate at the surface requires very accurate centering of the source in the center of the cavity. Star et al. (van Staveren et al. 1994) developed applicators for treatment of the bladder that address the centering problem using a "cage" of wire guides that maintain the shape of the bladder and the centering of the source at its center. Another, much more complicated, application for intercavity PDT is intraoperative PDT in which the irradiation is performed in a cavity opened by a surgical procedure. In this case, the light distribution must be kept uniform, but the cavity is often highly nonspherical. One solution, used in several ongoing clinical trials, is to construct an applicator consisting of an optical fiber terminated in a bulb of scattering liquid, which diffuses the incident light. This applicator is moved through the cavity, which is filled with light-scattering liquid as well, ensuring a uniform distribution of light (Cengel et al. 2007).

34.3.2 PDT Dosimetry

The most direct application of the field of biomedical optics in PDT has been in the development of measurements and theoretical models for dosimetry, or the quantification of dose deposited by the combination of sensitizer and light. Perhaps the most relevant field from which to draw guidance for the development of dosimetry is radiation oncology, which has developed very sophisticated methods for the calculation of dose deposited by various sources of ionizing radiation. In the case of ionizing radiation, the dose is defined as the amount of energy deposited per unit mass of tissue (Johns and Cunningham 1983). This definition of dose predicts biological response reasonably well for a variety of types of radiation over a large range of photon energies. The most computationally challenging calculations of dose involve corrections for the effects of inhomogeneities. These, however, can generally be quantified using volumetric imaging, and generally consist of variations in a single variable, the electron density of the medium. While tissue oxygenation can have an effect on ionizing radiation therapy, it is not typically considered explicitly in treatment planning.

In contrast, the cytotoxic effect of PDT depends on the distributions of light, sensitizer, and oxygen, all of which can vary independently. The light distribution depends on at least two variables, the absorption and scattering coefficients, which are not necessarily correlated with the distributions of oxygen and sensitizer. Throughout the history of PDT, several definitions of photodynamic dose have been developed, with varying degrees of sophistication and complexity. The ideal dosimetric quantity would be one that predicts the efficacy of treatment. This quantity would take into account the complex interplay among the distributions of light, oxygen, and sensitizer in the tissue. Thus far, however, no such dosimetric quantity has been developed to the point of clinical use, though several are under development. Wilson et al. have divided PDT dosimetry schemes into three broad categories, *explicit*, where the factors contributing to singlet oxygen production (light, sensitizer concentration, and tissue oxygenation) are measured directly, *implicit*, where a surrogate for singlet oxygen, such as fluorescence photobleaching, is used, and *direct*, where the production of singlet oxygen is measured directly (Wilson et al. 1997). Examples of each are given in the following section.

34.3.2.1 Delivered Irradiance

The simplest definition of dose is the product of the incident irradiance and the treatment time, delivered for a given sensitizer dose. The result is a measure of the energy delivered to the surface of the tissue. This quantity has the advantage of being simple to measure and easy to control, as the only variables the clinician needs to adjust are the treatment time and light intensity. The delivered irradiance is the standard of PDT dose for surface-applied PDT treatments such as those in dermatology. A distinct advantage of this method of dosimetry is the ease of measurement. In a typical application, where the delivery system produces a uniform, circular light field, the irradiance at the surface is given by the total power divided by the area of the field. The most critical measurement in this case becomes the power emitted by the delivery fiber, which can be measured by a calibrated power meter. In the case of interstitial light sources implanted in a lumen such as the esophagus, the prescription is typically written in terms of power per unit length delivered by the diffusing fiber.

The prescription of PDT dose in terms of incident irradiance has been used effectively; however, it is not sufficiently sophisticated to ensure uniform distribution of light within the tissue. This is due to three general effects. First, the geometry of tissue may vary from patient to patient. Oblique angles of incidents, variations in the distance between the delivery device and the surface, and surface irregularity may all lead to a fluence rate within the tissue that differs from the incident irradiance. Second, the optical properties of tissue have been observed to vary widely among tissues and patients, and even within a single tissue in a single patient. The transport of light is highly sensitive to changes in the absorption coefficient of tissue. Third, in treatments of cavities or lumens such as the esophagus, bronchus, or bladder, the combination of the concave cavity and the partially reflective surface creates the possibility of multiple reflections of light within the cavity known as the integrating sphere effect (van Staveren et al. 1994, 1995, 1996). When a significant portion of the incident light has been multiply reflected, the sensitivity to changes or variations in optical properties is enhanced. The integrating sphere effect can be partially compensated using *in situ* flat photodiode detectors to measure the incident irradiance directly (Hendren et al. 2001), however this strategy does not account for the effect of optical properties on the transport of light from the surface into the surface.

34.3.2.2 Measured Fluence Rate

One strategy to circumvent the shortcomings of delivered irradiance as a dosimetric quantity is to directly measure the fluence rate in the tissue. Some care is required in making such a measurement. Because tissue absorbs at least some light, there will always be a gradient in fluence rate within the tissue, implying that the fluence rate distribution is non-isotropic. Therefore, any measurement of fluence rate that depends on a detector with a limited acceptance angle, such as a planar photodiode is subject to error (Vulcan et al. 2000). This observation has led to the development of isotropic detectors of two distinct types.

Geometric isotropic detectors, described in detail by Marijnissen and Star (2002), achieve isotropic response *via* multiple scattering in a highly scattering, spherical bulb attached to the end of the fiber. Light entering the spherical detector is scattered sufficiently many times in the course of crossing the detector that its direction of incidence is lost. The efficiency of collection of photons in these detectors is limited by the fraction of light that escapes the detector after multiple scattering events. This fraction can be calculated based on multiple flux theory, assuming homogeneous, isotropically scattering media in both the detector and the surrounding tissue (Marijnissen and Star 1996). A significant factor in the sensitivity of any given detector is the refractive index mismatch between the two media. In cases where the mismatch is large (e.g., a glass, plastic, or epoxy detector in air), there is significant reflection of outgoing photons at the detector-air boundary, reducing the fraction of photons that escape the detector and increasing the detector's sensitivity. It is essential to calibrate the detectors in a medium whose index of refraction matches that where it will be used.

The second type of isotropic detector is based on a series of fluorophores embedded in an optical fiber (Pomerleau-Dalcourt and Lilge 2006). The fluorophores are chosen to absorb the treatment light efficiently. If spectrally resolved detectors are used to measure the emission, the contribution of multiple fluorophores can be separated, allowing a single fiber to measure the fluence rate at multiple locations. Because the fluorophore molecules are randomly oriented, their absorption and emission is, on average, isotropic.

Regardless of the technology used to measure fluence rate, a fundamental limitation arises from the fact that it is impractical to measure the fluence rate at every point in tissue, and the fluence rate often exhibits significant spatial variation. The fluence rate at any point in tissue, however, can be calculated using radiative transport theory if the light source distribution and tissue optical properties are known. Calculations of this type are particularly important in cases such as interstitial treatment of the prostate, where it is necessary to use multiple light sources, and the optical properties of the tissue may vary significantly (Zhu et al. 2005).

34.3.2.3 Measurements of Sensitizer Concentration

The measurement or modeling of fluence rate allows characterization of the light dose to tissue, which is one of the three essential components of photodynamic therapy. This parameter has been combined with the local drug concentration to give a quantity referred to as the "photodynamic dose" (Patterson et al. 1990), defined by

$$D = \int_0^t \varepsilon c \cdot \frac{\varphi(t')}{h\nu} \cdot \frac{1}{\rho} dt', \qquad (34.8)$$

where

ρ is the density of tissue

φ is the light fluence rate

$h\nu$ is the energy of a photon

c and ε are the concentration and extinction coefficients of the sensitizer, respectively

The photodynamic dose, therefore, represents the total number of photons absorbed by sensitizer molecules per unit mass of tissue. This definition is most similar in form to the definition of dose established for ionizing radiation. In conditions where oxygen is not limiting, this value is proportional to the energy transferred to oxygen per unit mass of tissue. As in the case of ionizing radiation dose, the exact conversion of this energy into cytotoxic effect is not taken into account, but is implicitly assumed to be uniform.

This definition of dose adds a layer of complexity, as it requires the quantification of the sensitizer concentration. This can be measured using absorption spectroscopy, which relies on the characteristic absorption peaks of the sensitizer to separate it from the background absorption present in the tissue. This method is particularly challenging for sensitizers such as HPD (absorption maximum ~630 nm), Photofrin (630 nm), and PpIX (635 nm), which have absorption maxima in the region where hemoglobin is highly absorbing. Because the concentration of hemoglobin is generally orders of magnitude greater than that of the sensitizer, its absorption easily overwhelms that of the sensitizer if the absorption peaks of the two coincide. For sensitizers with longer-wavelength absorption peaks, such as mTHPC (658 nm), MLu (732 nm), and BPD (690 nm), the long-wavelength absorption peak coincides with a relatively flat, featureless region of the hemoglobin absorption spectrum, making detection much easier.

In the extreme case of MLu, with an absorption peak at 732 nm, the absorption of the sensitizer is often greater than that of hemoglobin, making it possible to approximately quantify the sensitizer concentration based on the absorption at 732 nm alone, with no rigorous separation of the effects of absorption by hemoglobin (Zhu et al. 2005).

An alternative to absorption spectroscopy is fluorescence emission spectroscopy, applicable in cases where the sensitizer has sufficient fluorescence quantum yield (Finlay et al. 2006). Measurement of fluorescence is technically easier in that, unlike the case of absorption spectroscopy, there is no need to separate the effects of absorption and scattering. Fluorescence signal measured in vivo, however, is always conducted against a background of absorption and scattering, which must be corrected if a valid quantification of sensitizer concentration is to be achieved. A number of strategies have been developed to address this challenge, ranging from specially designed probes (Diamond et al. 2003) to analytic methods (Wu et al. 1993, Muller et al. 2001) to explicit models of the fluorescence excitation and emission wavelength effects of optical properties (Finlay and Foster 2005).

It has been demonstrated in animal models that the use of photodynamic dose leads to more uniform response to photodynamic therapy than prescription by fluence alone (Zhou et al. 2006).

34.3.2.4 Measurement of Local Tissue Oxygenation

The third parameter relevant to photodynamic therapy dosimetry is the local concentration of oxygen available in the tissue. In in vitro models, it is possible to measure the local oxygen concentration directly using oxygen-sensitive electrodes, and to rigorously model the rate of oxygen consumption during photodynamic therapy using the models described above (Foster et al. 1993, Nichols and Foster 1994). The most commonly used method of optical oxymetry is the measurement of absorption spectra and the extraction of the relative concentrations of oxy- and deoxyhemoglobin. The general theory behind these methods has been described in detail in Chapters 9 and 10 and will not be repeated here. However, because PDT is therapeutic, rather than a diagnostic modality, the details of the measurements differ somewhat. In the case of interstitial light delivery, for instance, the somewhat invasive step of placing catheters or needles in the tissue is required for treatment. Several researchers have developed systems to place optical detectors in the tissue as well, allowing interstitial measurements of optical properties (Zhu et al. 2005, Johansson et al. 2007). The ability to place light sources and detectors in the interior of solid tumors and organs allows reconstruction of optical properties throughout the volume—a task that would be difficult or impossible using exclusively noninvasive measurements. Because these measurements can be made immediately before the start of treatment, it is possible to develop an integrated dosimetry and control system capable of real-time measurement, dose calculation, and source optimization. Work in this area is ongoing (Johansson et al. 2007, Li et al. 2008).

In addition to optical-based oxymetry, approaches for noninvasive or minimally invasive imaging of either tissue oxygenation or hypoxia include techniques based on nuclear or magnetic resonance imaging (MRI), although not all of these are approved for clinical use. These techniques have recently been reviewed in Tatum et al. (2006). Nuclear imaging of tissue hypoxia has been performed using ^{18}F labeled versions of the hypoxia-labeling drugs (fluoro)misonidazole and EF5 [2-(2-nitroimidazol-1 [H]-yl)-*N*-(2,2,3,3,3-pentafluoropropyl)acetamide)], as well as other agents such as ^{62}Cu-ATSM [^{62}Cu-labeled diacetyl-bis (N^4-methylthiosemicarbazone)] and ^{124}I-β-D-IAZGP [^{124}I-labeled beta-D-iodinated azomycin galactopyranoside] (Kizaka-Kondoh and Konse-Nagasawa 2009). Magnetic resonance techniques to assess tissue oxygenation include, for example, a ^{19}F and perfluorcarbon-based approach, blood oxygen level-dependent (BOLD) MRI, and techniques based on electron paramagnetic resonance (Vikram et al. 2007).

Other, more invasive approaches used to study tissue oxygenation or hypoxia in human tissues have included needle probe electrodes and hypoxia-labeling markers. The Sigma/Eppendorf pO_2-histograph is a polarographic needle electrode that was used in key studies establishing the relevance of tumor hypoxia to therapy responses (Hockel et al. 1993), including PDT (Sitnik et al. 1998). Hypoxia-labeling drugs include derivatives of 2-nitroimidazole-based compounds, which were originally developed as sensitizers of hypoxic cells to radiation-induced cell death and bind to cells as an inverse function of oxygen tension (Evans and Koch 2003). Misonidazole and EF5, mentioned above for their use in noninvasive imaging, are two such drugs. In its nonradioactive form, EF5, as well as another 2-nitroimidazole derivative, pimonidazole, are among the most commonly used agents in in vivo labeling of tissue hypoxia in conjunction with ex vivo, antibody-based tissue analysis (Evans et al. 2003). Finally, intrinsic molecular markers, such as carbonic anhydrase 9 (CA9) and hypoxia inducible factor (HIF), are also studied as indicators of tissue hypoxia, albeit not without their own set of caveats (Williams et al. 2005, Kizaka-Kondoh and Konse-Nagasawa 2009).

34.3.2.5 Fluorescence-Based Implicit Dosimetry

The most viable method of implicit dosimetry developed thus far is based on the photobleaching of the sensitizer itself. If the photobleaching and cytotoxic dose are both mediated by singlet oxygen, the degree of sensitizer photobleaching can be shown to be indicative of the absolute dose of singlet oxygen delivered (Georgakoudi et al. 1997). While this relationship is relatively straightforward in the simplified environment of cell culture, it becomes more complicated in vivo. Despite this, it has been shown that for selected sensitizers, the degree of photobleaching is predictive of skin response (Robinson et al. 1998) and pain during PDT treatment (Cottrell et al. 2008). Care should be taken when using fluorescence bleaching as a dosimetric quantity; however, some sensitizers exhibit photobleaching and photodynamic dose deposition mediated by different mechanisms. This can lead to photobleaching that is difficult to interpret, as in the case of mTHPC (Finlay et al. 2002) or not predictive of outcome, as observed under some conditions for Photofrin (Finlay et al. 2004a).

34.3.2.6 Direct Measurement of Singlet Oxygen Production

Molecular oxygen in the excited singlet state spontaneously decays back to its ground state with the emission of a 1270 nm photon. This emission can be used for the direct detection of singlet oxygen, a technique known as singlet oxygen luminescence dosimetry (SOLD) (Jarvi et al. 2006). This conceptually simple strategy turns out to be remarkably complex in its practical implementation for several reasons. The intensity of the light emitted in a typical in vivo experiment is very low, and the efficiency of the detectors commonly used for light detection is low at 1270 nm. Recent developments in photodiode technology have addressed the latter problem (Jarvi et al. 2006). The low signal relative to the background, however, is a fundamental problem. This has been addressed by using pulsed lasers to generate singlet oxygen, allowing a very high peak fluence rate and, hence, a very high rate of singlet oxygen production. The problem of weak signal is compounded by the fact that the emission of light by singlet oxygen and the reaction with cellular substrates are competitive processes: efficient deposition of photodynamic damage reduces the observed signal from singlet oxygen. In microscope-based singlet oxygen detection, for example, it is common to observe lower emission intensity from cells (where singlet oxygen targets are plentiful) than from the surrounding medium (where they are sparse). The medium dependence of the signal makes interpretation of singlet oxygen measurements challenging. It is not simple to distinguish between a decrease in signal and an increase in singlet oxygen quenchers and targets. To resolve this ambiguity, time-resolved measurements of the singlet oxygen signal have been fit with a multi-exponential model to determine the rates of sensitizer triplet (T_1) decay and singlet oxygen decay. The T_1 decay rate is dependent on the likelihood of quenching by oxygen molecules in the immediate vicinity, allowing a determination of local oxygen concentration. The singlet oxygen decay rate is dependent on the local concentration of quenchers and targets. Combining these two pieces of information allows reconstruction of the photodynamic singlet oxygen dose, that is, the concentration of singlet oxygen that reacts with target substrates.

34.3.3 Approved Indications

PDT has been used to treat a vast array of malignant, premalignant, and benign conditions, only a few of which will be described in detail here. In many of these cases, PDT has been successful at treating patients. Its success in gaining widespread use has often been less remarkable, largely because of its technical complexity and the availability of competing modalities.

34.3.3.1 Intraluminal Applications

Among the first clinical applications of PDT were the treatment of lesions of the bronchus, lung, and esophagus. In these cases, light is typically delivered using a cylindrical fiber placed in the center of the lumen under endoscopic guidance. The fiber may be centered using a clear balloon. The light dose prescription has typically been specified in terms of energy per unit length of diffuser. Because of the confounding effects of optical properties and tissue geometry, this prescription can fail to predict outcome. In the case of esophageal PDT, it has been demonstrated that the fluence rate delivered to tissue can vary by more than a factor of two among patients treated with the same incident irradiance from the source (van Veen et al. 2006b). This observation has motivated the development of in-vivo dosimetry systems for endoscopic PDT. It is likely that the reproducibility of PDT treatment will improve as these systems are adopted in clinical practice.

34.3.3.2 Dermatological Applications

Photodynamic therapy's most prominent clinical applications have been in the field of dermatology. Skin is an ideal site for photodynamic treatment because it is accessible, generally relatively flat, and generally develops lesions that are, especially in the early stages, relatively thin. PDT treatment of superficial skin lesions is generally well tolerated by patients, produces durable destruction of the lesion, and has very favorable cosmetic outcome compared with surgery. PDT using topically applied ALA under the trade name Levulan is currently clinically available. This treatment has the advantage of sensitizing only the lesion and the immediately surrounding skin, reducing unwanted skin photosensitivity. In current practice, the prescription of PDT dose is typically given in terms of the incident irradiance, and a standard dose of light is often given in a single treatment.

Two recent areas of research may lead to changes in this standard. First, Foster et al. have shown that treatments at low incident irradiance can reduce pain (Cottrell et al. 2008). When the fluorescence emission of the PpIX produced in the skin is monitored, the extent of photobleaching can be used to determine when the optimal low-irradiance treatment has been accomplished. Thereafter, the irradiance can be increased with no additional pain.

Independently, it has been demonstrated that the treatment efficacy of PpIX-mediated skin PDT can be enhanced by delivering the treatment in two fractions separated by a period of 2 h, with the first fraction being much smaller than the second. This observation has been investigated extensively in animal models, and reproduced in a clinical trial in human skin cancers (de Haas et al. 2006). The explanation for the improved response in the fractionated irradiation scheme is complicated. The 2 h interval required to achieve the improvement in response is too long to be explained by replenishment of depleted oxygen. It cannot be explained solely in terms of additional synthesis of PpIX, and perhaps most counterintuitively, it is not observed in animal models treated with methyl-ALA, despite the fact that the PpIX accumulation is approximately the same (de Bruijn et al. 2007). This continues to be an area of active research.

34.3.3.3 Retinal PDT

Another area where PDT has joined the ranks of the standard treatments is in the treatment of age-related macular degeneration, or AMD. This condition results from an abnormal growth of blood vessels in the retina, which can obscure vision, leading to blindness. The typical photodynamic treatment for AMD targets the blood vessels themselves using the sensitizer Benzoporhphyrin derivative (BPD), marketed under the brand name Verteporfin. The treatment light is projected through the lens of the eye onto the retina a very short time after the intravenous administration of the drug, so that the bulk of the sensitizer is still in circulation. This leads to effective targeting of the vasculature, with minimal normal tissue toxicity. PDT was the first treatment found to be effective against the abnormal angiogenesis (vessel growth) that causes AMD (Wormald et al. 2007). However, the number of patients treated with PDT has declined in recent years, due mostly to the increased availability of anti-angiogenic drugs such as Ranibizumab (Kourlas and Abrams 2007), which have proven as effective as PDT and less technically challenging, although treatment approaches combining anti-angiogenics with PDT are under investigation. Because AMD is a disease of the superficial tissue, the usual concerns about the penetration of light into tissue do not apply here. In fact, there may be some advantage to confining the volume of active PDT only to specific vessels. This concept has inspired the development of two-photon PDT (Karotki et al. 2006). In this strategy, two simultaneously incident near-IR photons excite the sensitizer. The rate of excitation of the sensitizer is proportional to the square of the intensity of the incident light. This requires the use of focused, pulsed laser sources to reach sufficient intensity. Only the volume of tissue very close to the focus receives sufficient intensity to generate singlet oxygen, allowing the PDT effect to be confined to volumes on the order of femtoliters, analogous to the volume probed by two-photon microscopy (Chapter 27).

34.3.4 Clinical Trials

Over the past several decades, there have been dozens of clinical trials of photodynamic therapy for various indications. The following section will not cover them exhaustively, but rather point out the relevant characteristics of several different approaches from the point of view of optics.

34.3.4.1 Prostate

There has been significant investigation into the use of PDT for the treatment of prostate cancer over the past two decades. These studies have been summarized in a recent review (Moore et al. 2009). Prostate cancer is an attractive target for PDT for several reasons: It is a very common malignancy in men, and one that

has a significant number of cases that do not respond to conventional chemotherapy or radiation therapy. The disease is often disseminated microscopically throughout the prostate, and damage to the normal prostate tissue produces relatively little toxicity. Given this combination, treatment of the entire organ, without differentiation of cancerous and normal tissues within the organ, is acceptable and is often the treatment of choice. One of the standard treatments for prostate cancer is brachytherapy, the insertion of small radioactive seeds into the prostate *via* needles inserted through the perineum, typically under the guidance of transrectal ultrasound imaging. The prevalence of this procedure and the familiarity of clinicians with its use are advantageous for the implementation of PDT. Several groups have made modifications to standard brachytherapy equipment to allow the insertion of cylindrical diffusing fibers into the organ, typically *via* translucent catheters or needles (Nathan et al. 2002, Zaak et al. 2003, Weersink et al. 2005, Verigos et al. 2006). Among the sensitizers used in these studies, mTPHC and MLu are employed in the standard PDT methodology: administered intravenously followed by illumination after an interval of hours or days. In the studies involving Tookad, the light is delivered during continuous infusion of the drug, while it is actively circulating in the vasculature. ALA-induced PpIX is unique among these studies in that the ALA is administered orally, and that PpIX, unlike the other sensitizers used in prostate, has demonstrated selectivity for cancer cells, to the extent that its fluorescence can be used to guide surgical resection (Zaak et al. 2008).

The light fluence distribution inside prostate gland can be monitored by inserting optical fiber-based isotropic detectors during PDT treatment. A motorized system can be used to monitor light fluence rate at multiple locations from a point source to determine the optical properties (Dimofte et al. 2005, Zhu et al. 2005).

Photosensitizer concentration inside the prostate gland can be determined by either absorption spectroscopy (Finlay et al. 2004b) or fluorescence spectroscopy (Finlay et al. 2006). The fluorescence spectroscopy using a single fiber has the advantage of determining the local photosensitizer distribution because the excitation wavelength is shorter. However, it is necessary to apply an optical property correction to the measured data (Finlay et al. 2006). Absorption spectroscopy, on the other hand, measures the macroscopic average of photosensitizer distribution over a larger volume. This method has been used to obtain the drug spatial distribution by solving the inverse problem of the diffusion equation.

Work has been done to calculate the light fluence rate given a known set of source positions and intensities in the presence of heterogeneous optical properties (Li et al. 2008, Li and Zhu 2008). The overall goal is to optimize the PDT dose distribution to cover the entire prostate gland and to spare the critical structures, primarily the urethra and rectal wall. The light fluence rate can be optimized using the Cimmino algorithm in prostate gland with either homogeneous or heterogeneous optical properties as inputs (Altschuler et al. 2005, Li et al. 2008). Preliminary work has shown that it is feasible to optimize PDT dose directly using the Cimmino algorithm by adjusting the weight of light source intensities (Altschuler et al. 2009).

34.3.4.2 Head and Neck Cancers

Cancers of the head and neck have been treated extensively with PDT (Biel 2002). The majority of these patients have been treated with HPD, Photofrin, and mTHPC. Cancers in the oral cavity can be illuminated by a collimated beam of light, as in topical applications. Lesions in the larynx can be treated with cylindrical diffusing fibers. Bulky tumors can be treated using implanted cylindrical diffusers. As in the case of prostate PDT, the implantation of fibers in these tumors has taken advantage of the technology developed for radioactive seed implants. A specialized applicator has been developed for the purpose of treating the nasopharynx (Nyst et al. 2007). This device uses cylindrical diffusing fibers to provide the illumination of the target tissues, and a light-absorbing silicone shield custom made for each patient to shield the soft palate and other sensitive structures. This device has been implemented in a clinical trial using mTPHC (van Veen et al. 2006a). Early results for head and neck cancers have been very encouraging, and motivate the development of PDT as a primary modality as well as a salvage option for tumors that fail to respond to radiation therapy.

34.3.4.3 Intraoperative, Intraperitoneal, and Intrapleural PDT

A more technically challenging, but potentially highly beneficial use of PDT is in the intraoperative setting. In this case, the goal of PDT is not the ablation of the tumor, but the sterilization of microscopic disease remaining after surgical resection of the bulk disease. This requires coordination of sensitizer administration, surgery, and irradiation. Typically, the sensitizer is administered hours to days prior to the start of surgery. Following surgical resection of all grossly visible disease, PDT is performed with the goal of sterilizing microscopic disease from all serosal (peritoneal or pleural) surfaces. This introduces several significant challenges not normally encountered in clinical PDT. All devices used for these procedures must be compatible with the sterile surgical environment and the timing and duration of light delivery is much more critical because of the need to complete irradiation within the time allotted for surgery. In addition, the specific methods used to ensure relatively homogenous delivery light to a large, highly convoluted internal surface while maintaining adequate dosimetric constraints is extremely challenging. Currently, this is performed using a combination of incident light fields directed toward surgically externalized organs such as small bowel and internally distributed light using a spherically diffusing applicator and a light scattering medium such as dilute Intralipid (Cengel et al. 2007).

Intraoperative PDT for the treatment of brain tumors provides an additional challenge. Unlike the intraoperative treatments described in the previous section, in this case, the surrounding tissue is normal brain, and damage to this tissue has serious consequences for the patient. Fortunately, the physiology of

the brain confers an advantage in this case. It has been demonstrated that systemic administration of some sensitizers can lead to preferential accumulation of PpIX in brain tumors (Stylli and Kaye 2006). This is due at least in part to the disruption of the blood–brain barrier by the tumor. The difference in uptake between tumor and normal brain tissue is sufficient to allow the fluorescence of the sensitizer to be used to guide surgical resection. In addition, the preferential uptake allows PDT treatment of the microscopic disease at the surgical margin. One novel approach that has been developed in preclinical models is metronomic PDT (mPDT), in which the treatment light is delivered at low fluence rate over a period of days or weeks (Bisland et al. 2004). Preclinical studies have demonstrated the feasibility of mPDT using the sensitizer LS-11 and an implanted battery-powered LED light source.

34.4 Conclusions

The field of photodynamic therapy continues to evolve and develop. The fundamental challenges of light penetration in tissue and the confounding effects of heterogeneous and variable optical properties will continue to be active areas of research. Developments in the fields of nanotechnology, photochemistry, and biology will continue to drive new advances in photodynamic therapy. The future of PDT is likely to involve increased collaboration among these fields, and an increased role for an interdisciplinary understanding of the factors driving the biological response to therapy. PDT is already being investigated in combination with surgery, chemotherapy, and radiation therapy for treatment of cancer, and in combination with antiangiogenic therapy for vascular diseases. In addition, we are just beginning to understand the role that inflammatory and immune responses play in response to PDT to the point that treatments can be tailored to optimize these indirect effects as well as direct cytotoxicity.

Not surprisingly, the underlying biology, chemistry, and physics of PDT are in many ways more developed than its clinical application. Many applications of PDT have enjoyed remarkable success; however, it is not entirely clear where PDT will eventually fit in the medical landscape. It is currently practiced by dermatologists, surgeons, radiation oncologists, and ophthalmologists with a wide range of expertise and interest in biomedical optics. In many clinical trials, PDT has been used as a salvage therapy, but there is a good rationale for considering it as an adjunct to radiation therapy or chemotherapy. Whatever form PDT takes in its future clinical implementation, it will continue to be a modality that depends on progress in biomedical optics for its continued success, and generates new questions and new developments in the field as well.

References

Ackroyd, R., Kelty, C., Brown, N., and Reed, M. 2001. The history of photodetection and photodynamic therapy. *Photochem. Photobiol.* 74: 656–669.

Allison, R. R., Mota, H. C., and Sibata, C. H. 2004. Clinical PD/PDT in North America: An historical review. *Photodiagn. Photodyn. Ther.* 1: 263–277.

Altschuler, M. D., Zhu, T. C., Hu, Y. et al. 2009. A heterogeneous optimization algorithm for PDT dose optimization for prostate. *Proc. SPIE* 7164: 71640B.

Altschuler, M. D., Zhu, T. C., Li, J., and Hahn, S. M. 2005. Optimized interstitial PDT prostate treatment planning with the Cimmino feasibility algorithm. *Med. Phys.* 32: 3524–3536.

Bhuvaneswari, R., Yuen, G. Y., Chee, S. K. and Olivo, M. 2007. Hypericin-mediated photodynamic therapy in combination with Avastin (bevacizumab) improves tumor response by downregulating angiogenic proteins. *Photochem. Photobiol. Sci.* 6: 1275–1283.

Biel, M. A. 2002. Photodynamic therapy in head and neck cancer. *Curr. Oncol. Rep.* 4: 87–96.

Bisland, S. K., Lilge, L., Lin, A., Rusnov, R., and Wilson, B. C. 2004. Metronomic photodynamic therapy as a new paradigm for photodynamic therapy: Rationale and preclinical evaluation of technical feasibility for treating malignant brain tumors. *Photochem. Photobiol.* 80: 22–30.

Busch, T. M. 2006. Local physiological changes during photodynamic therapy. *Lasers Surg. Med.* 38: 494–499.

Busch, T. M., Hahn, S. M., Wileyto, E. P. et al. 2004. Hypoxia and photofrin uptake in the intraperitoneal carcinomatosis and sarcomatosis of photodynamic therapy patients. *Clin. Cancer Res.* 10: 4630–4638.

Busch, T. M., Wileyto, E. P., Emanuele, M. J. et al. 2002. Photodynamic therapy creates fluence rate-dependent gradients in the intratumoral spatial distribution of oxygen. *Cancer Res.* 62: 7273–7279.

Cengel, K. A., Glatstein, E., and Hahn, S. M. 2007. Intraperitoneal photodynamic therapy. *Cancer Treat. Res.* 134: 493–514.

Chen, B., Pogue, B. W., Hoopes, P. J., and Hasan, T. 2005a. Combining vascular and cellular targeting regimens enhances the efficacy of photodynamic therapy. *Int. J. Radiat. Oncol. Biol. Phys.* 61: 1216–1226.

Chen, B., Pogue, B. W., Hoopes, P. J., and Hasan, T. 2006. Vascular and cellular targeting for photodynamic therapy. *Critical Rev. Eukaryotic Gene Expr.* 16: 279–305.

Chen, Q., Huang, Z., Chen, H., Shapiro, H., Beckers, J., and Hetzel, F. W. 2002. Improvement of tumor response by manipulation of tumor oxygenation during photodynamic therapy. *Photochem. Photobiol.* 76: 197–203.

Chen, Y., Gryshuk, A., Achilefu, S. et al. 2005b. A novel approach to a bifunctional photosensitizer for tumor imaging and phototherapy. *Bioconjug. Chem.* 16: 1264–1274.

Cottrell, W. J., Paquette, A. D., Keymel, K. R., Foster, T. H., and Oseroff, A. R. 2008. Irradiance-dependent photobleaching and pain in delta-aminolevulinic acid-photodynamic therapy of superficial basal cell carcinomas. *Clin. Cancer Res.* 14: 4475–4483.

Dahle, J., Bagdonas, S., Kaalhus, O., Olsen, G., Steen, H. B., and Moan, J. 2000. The bystander effect in photodynamic inactivation of cells. *Biochim. Biophys. Acta* 1475: 273–280.

de Bruijn, H. S., de Haas, E. R., Hebeda, K. M. et al. 2007. Light fractionation does not enhance the efficacy of methyl 5-aminolevulinate mediated photodynamic therapy in normal mouse skin. *Photochem. Photobiol. Sci.* 6: 1325–1331.

de Haas, E. R., Kruijt, B., Sterenborg, H. J., Martino Neumann, H. A., and Robinson, D. J. 2006. Fractionated illumination significantly improves the response of superficial basal cell carcinoma to aminolevulinic acid photodynamic therapy. *J. Invest. Dermatol.* 126: 2679–2686.

Diamond, K. R., Patterson, M. S., and Farrell, T. J. 2003. Quantification of fluorophore concentration in tissue-simulating media by fluorescence measurements with a single optical fiber. *Appl. Opt.* 42: 2436–2442.

Dimofte, A., Finlay, J. C., and Zhu, T. C. 2005. A method for determination of the absorption and scattering properties interstitially in turbid media. *Phys. Med. Biol.* 50: 2291–2311.

Dougherty, T. J., Gomer, C. J., Henderson, B. W. et al. 1998. Photodynamic therapy. *J. Natl. Cancer Inst.* 90: 889–905.

Dougherty, T. J., Kaufman, J. E., Goldfarb, A., Weishaupt, K. R., Boyle, D., and Mittleman, A. 1978. Photoradiation therapy for the treatment of malignant tumors. *Cancer Res.* 38: 2628–2635.

Ericson, M. B., Sandberg, C., Stenquist, B. et al. 2004. Photodynamic therapy of actinic keratosis at varying fluence rates: Assessment of photobleaching, pain and primary clinical outcome. *Br. J. Dermatol.* 151: 1204–1212.

Evans, S. M. and Koch, C. J. 2003. Prognostic significance of tumor oxygenation in humans. *Cancer Lett.* 195: 1–16.

Ferrario, A. and Gomer, C. J. 2006. Avastin enhances photodynamic therapy treatment of Kaposi's sarcoma in a mouse tumor model. *J. Environ. Pathol. Toxicol. Oncol.* 25: 251–260.

Ferrario, A., von Tiehl, K. F., Rucker, N., Schwarz, M. A., Gill, P. S., and Gomer, C. J. 2000. Antiangiogenic treatment enhances photodynamic therapy responsiveness in a mouse mammary carcinoma. *Cancer Res.* 60: 4066–4069.

Fingar, V. H., Taber, S. W., Haydon, P. S., Harrison, L. T., Kempf, S. J., and Wieman, T. J. 2000. Vascular damage after photodynamic therapy of solid tumors: A view and comparison of effect in pre-clinical and clinical models at the University of Louisville. *In Vivo* 14: 93–100.

Finlay, J. C., Conover, D. L., Hull, E. L., and Foster, T. H. 2001. Porphyrin bleaching and PDT-induced spectral changes are irradiance dependent in ALA-sensitized normal rat skin *in vivo*. *Photochem. Photobiol.* 73: 54–63.

Finlay, J. C. and Foster, T. H. 2004. Effect of pigment packaging on diffuse reflectance spectroscopy of samples containing red blood cells. *Opt. Lett.* 29: 965–967.

Finlay, J. C. and Foster, T. H. 2005. Recovery of hemoglobin oxygen saturation and intrinsic fluorescence using a forward adjoint model of fluorescence. *Appl. Opt.* 44: 1917–1933.

Finlay, J. C., Mitra, S., and Foster, T. H. 2002. *In vivo* mTHPC photobleaching in normal rat skin exhibits unique irradiance-dependent features. *Photochem. Photobiol.* 75: 282–288.

Finlay, J. C., Mitra, S., and Foster, T. H. 2004a. Photobleaching kinetics of Photofrin *in vivo* and in multicell tumor spheroids indicate multiple simultaneous bleaching mechanisms. *Phys. Med. Biol.* 49: 4837–4860.

Finlay, J. C., Zhu, T. C., Dimofte, A. et al. 2006. Interstitial fluorescence spectroscopy in the human prostate during motexafin lutetium-mediated photodynamic therapy. *Photochem. Photobiol.* 82: 1270–1278.

Finlay, J. C., Zhu, T. C., Dimofte, A. et al. 2004b. In vivo determination of the absorption and scattering spectra of the human prostate during photodynamic therapy. *Proc. SPIE* 5315: 132–142.

Foster, T. H., Hartley, D. F., Nichols, M. G., and Hilf, R. 1993. Fluence rate effects in photodynamic therapy of multicell tumor spheroids. *Cancer Res.* 53: 1249–1254.

Foster, T. H., Murant, R. S., Bryant, R. G., Knox, R. S., Gibson, S. L., and Hilf, R. 1991. Oxygen consumption and diffusion effects in photodynamic therapy. *Radiat. Res.* 126: 296–303.

Fukumura, D. and Jain, R. K. 2007. Tumor microenvironment abnormalities: Causes, consequences, and strategies to normalize. *J. Cell Biochem.* 101: 937–949.

Georgakoudi, I. and Foster, T. H. 1998. Singlet oxygen- *versus* nonsinglet oxygen-mediated mechanisms of sensitizer photobleaching and their effects on photodynamic dosimetry. *Photochem. Photobiol.* 67: 612–625.

Georgakoudi, I., Nichols, M. G., and Foster, T. H. 1997. The mechanism of Photofrin photobleaching and its consequences for photodynamic dosimetry. *Photochem. Photobiol.* 65: 135–144.

Gollnick, S. O., Owczarczak, B., and Maier, P. 2006. Photodynamic therapy and anti-tumor immunity. *Lasers Surg. Med.* 38: 509–515.

Gomer, C. J., Ferrario, A., Luna, M., Rucker, N., and Wong, S. 2006. Photodynamic therapy: Combined modality approaches targeting the tumor microenvironment. *Lasers Surg. Med.* 38: 516–521.

Hahn, S. M., Putt, M. E., Metz, J. et al. 2006. Photofrin uptake in the tumor and normal tissues of patients receiving intraperitoneal photodynamic therapy. *Clin. Cancer Res.* 12: 5464–5470.

Henderson, B. W., Busch, T. M., and Snyder, J. W. 2006. Fluence rate as a modulator of PDT mechanisms. *Lasers Surg. Med.* 38: 489–493.

Henderson, B. W., Gollnick, S. O., Snyder, J. W. et al. 2004. Choice of oxygen-conserving treatment regimen determines the inflammatory response and outcome of photodynamic therapy of tumors. *Cancer Res.* 64: 2120–2126.

Hendren, S. K., Hahn, S. M., Spitz, F. R. et al. 2001. Phase II trial of debulking surgery and photodynamic therapy for disseminated intraperitoneal tumors. *Ann. Surg. Oncol.* 8: 65–71.

Hockel, M., Knoop, C., Schlenger, K. et al. 1993. Intratumoral pO_2 predicts survival in advanced cancer of the uterine cervix. *Radiother. Oncol.* 26: 45–50.

Hopkinson, H. J., Vernon, D. I., and Brown, S. B. 1999. Identification and partial characterization of an unusual distribution of the photosensitizer meta-tetrahydroxyphenyl chlorin (temoporfin) in human plasma. *Photochem. Photobiol.* 69: 482–488.

Hu, X. H., Feng, Y., Lu, J. Q. et al. 2005. Modeling of a type II photofrin-mediated photodynamic therapy process in a heterogeneous tissue phantom. *Photochem. Photobiol.* 81: 1460–1468.

Huang, Z., Chen, Q., Dole, K. C. et al. 2007. The effect of Tookad-mediated photodynamic ablation of the prostate gland on adjacent tissues—in vivo study in a canine model. *Photochem. Photobiol. Sci.* 6: 1318–1324.

Jarvi, M. T., Niedre, M. J., Patterson, M. S., and Wilson, B. C. 2006. Singlet oxygen luminescence dosimetry (SOLD) for photodynamic therapy: Current status, challenges and future prospects. *Photochem. Photobiol.* 82: 1198–1210.

Johansson, A., Axelsson, J., Andersson-Engels, S., and Swartling, J. 2007. Realtime light dosimetry software tools for interstitial photodynamic therapy of the human prostate. *Med. Phys.* 34: 4309–4321.

Johns, H. E. and Cunningham, J. R. 1983. *The Physics of Radiology.* Springfield, IL: Charles C. Thomas.

Karotki, A., Khurana, M., Lepock, J. R., and Wilson, B. C. 2006. Simultaneous two-photon excitation of photofrin in relation to photodynamic therapy. *Photochem. Photobiol.* 82: 443–452.

Kessel, D., Vicente, M. G., and Reiners, J. J. Jr. 2006. Initiation of apoptosis and autophagy by photodynamic therapy. *Lasers Surg. Med.* 38: 482–488.

Kizaka-Kondoh, S. and Konse-Nagasawa, H. 2009. Significance of nitroimidazole compounds and hypoxia-inducible factor-1 for imaging tumor hypoxia. *Cancer Sci.* 100: 1366–1373.

Korbelik, M. 2006. PDT-associated host response and its role in the therapy outcome. *Lasers Surg. Med.* 38: 500–508.

Kosharskyy, B., Solban, N., Chang, S. K., Rizvi, I., Chang, Y., and Hasan, T. 2006. A mechanism-based combination therapy reduces local tumor growth and metastasis in an orthotopic model of prostate cancer. *Cancer Res.* 66: 10953–10958.

Kou, L., Labrie, D., and Chylek, P. 1993. Refractive indices of water and ice in the 0.65 to 2.5 μm spectral range. *Appl. Opt.* 32: 3521–3540.

Kourlas, H. and Abrams, P. 2007. Ranibizumab for the treatment of neovascular age-related macular degeneration: A review. *Clin. Ther.* 29: 1850–1861.

Lee, L. K., Whitehurst, C., Chen, Q., Pantelides, M. L., Hetzel, F. W., and Moore, J. V. 1997. Interstitial photodynamic therapy in the canine prostate. *Br. J. Urol.* 80: 898–902.

Li, J., Altschuler, M. D., Hahn, S. M., and Zhu, T. C. 2008. Optimization of light source parameters in the photodynamic therapy of heterogeneous prostate. *Phys. Med. Biol.* 53: 4107–4121.

Li, J. and Zhu, T. C. 2008. Determination of in-vivo light fluence distribution in heterogeneous prostate during photodynamic therapy. *Phys. Med. Biol.* 53: 2103–2114.

Lilge, L., Vesselov, L., and Whittington, W. 2005. Thin cylindrical diffusers in multimode Ge-doped silica fibers. *Lasers Surg. Med.* 36: 245–251.

Maier, A., Tomaselli, F., Anegg, U. et al. 2000. Combined photo-dynamic therapy and hyperbaric oxygenation in carcinoma of the esophagus and the esophago-gastric junction. *Eur. J. Cardiothorac. Surg.* 18: 649–654; discussion 654–655.

Marijnissen, J. P. and Star, W. M. 1996. Calibration of isotropic light dosimetry probes based on scattering bulbs in clear media. *Phys. Med. Biol.* 41: 1191–1208.

Marijnissen, J. P. and Star, W. M. 2002. Performance of isotropic light dosimetry probes based on scattering bulbs in turbid media. *Phys. Med. Biol.* 47: 2049–2058.

Mitra, S. and Foster, T. H. 2004. Carbogen breathing significantly enhances the penetration of red light in murine tumours in vivo. *Phys. Med. Biol.* 49: 1891–1904.

Mitra, S. and Foster, T. H. 2005. Photophysical parameters, photosensitizer retention and tissue optical properties completely account for the higher photodynamic efficacy of meso-tetra-hydroxyphenyl-chlorin vs Photofrin. *Photochem. Photobiol.* 81: 849–859.

Moore, C. M., Pendse, D., and Emberton, M. 2009. Photodynamic therapy for prostate cancer—a review of current status and future promise. *Nat. Clin. Pract. Urol.* 6: 18–30.

Muller, M. G., Georgakoudi, I., Zhang, Q., Wu, J., and Feld, M. S. 2001. Intrinsic fluorescence spectroscopy in turbid media: Disentangling effects of scattering and absorption. *Appl. Opt.* 40: 4633–4646.

Nathan, T. R., Whitelaw, D. E., Chang, S. C. et al. 2002. Photodynamic therapy for prostate cancer recurrence after radiotherapy: A phase I study. *J. Urol.* 168: 1427–1432.

Nichols, M. G. and Foster, T. H. 1994. Oxygen diffusion and reaction kinetics in the photodynamic therapy of multicell tumour spheroids. *Phys. Med. Biol.* 39: 2161–2181.

Nyst, H. J., van Veen, R. L., Tan, I. B. et al. 2007. Performance of a dedicated light delivery and dosimetry device for photodynamic therapy of nasopharyngeal carcinoma: Phantom and volunteer experiments. *Lasers Surg. Med.* 39: 647–653.

O'Connor, A. E., Gallagher, W. M., and Byrne, A. T. 2009. Porphyrin and nonporphyrin photosensitizers in oncology: Preclinical and clinical advances in photodynamic therapy. *Photochem. Photobiol.* 85: 1053–1074.

Pandey, R. K., Goswami, L. N., Chen, Y. et al. 2006. Nature: A rich source for developing multifunctional agents. Tumor-imaging and photodynamic therapy. *Lasers Surg. Med.* 38: 445–467.

Patterson, M. S., Wilson, B. C., and Graff, R. 1990. In vivo tests of the concept of photodynamic threshold dose in normal rat liver photosensitized by aluminum chlorosulphonated phthalocyanine. *Photochem. Photobiol.* 51: 343–349.

Pham, T. H., Hornung, R., Berns, M. W., Tadir, Y., and Tromberg, B. J. 2001. Monitoring tumor response during photodynamic therapy using near-infrared photon-migration spectroscopy. *Photochem. Photobiol.* 73: 669–677.

Pogue, B. W. and Hasan, T. 1997. A theoretical study of light fractionation and dose-rate effects in photodynamic therapy. *Radiat. Res.* 147: 551–559.

Pogue, B. W., Sheng, C., Benevides, J. et al. 2008. Protoporphyrin IX fluorescence photobleaching increases with the use of fractionated irradiation in the esophagus. *J. Biomed. Opt.* 13: 034009.

Pomerleau-Dalcourt, N. and Lilge, L. 2006. Development and characterization of multi-sensory fluence rate probes. *Phys. Med. Biol.* 51: 1929–1940.

Robinson, D. J., de Bruijn, H. S., van der Veen, N., Stringer, M. R., Brown, S. B., and Star, W. M. 1998. Fluorescence photobleaching of ALA-induced protoporphyrin IX during photodynamic therapy of normal hairless mouse skin: The effect of light dose and irradiance and the resulting biological effect. *Photochem. Photobiol.* 67: 140–149.

Schrama, D., Reisfeld, R. A., and Becker, J. C. 2006. Antibody targeted drugs as cancer therapeutics. *Nat. Rev. Drug Discov.* 5: 147–159.

Sharkey, R. M. and Goldenberg, D. M. 2006. Targeted therapy of cancer: new prospects for antibodies and immunoconjugates. *CA Cancer J. Clin.* 56: 226–243.

Sitnik, T. M., Hampton, J. A., and Henderson, B. W. 1998. Reduction of tumour oxygenation during and after photodynamic therapy in vivo: Effects of fluence rate. *Br. J. Cancer* 77: 1386–1394.

Solban, N., Rizvi, I., and Hasan, T. 2006. Targeted photodynamic therapy. *Lasers Surg. Med.* 38: 522–531.

Standish, B. A., Lee, K. K., Jin, X. et al. 2008. Interstitial Doppler optical coherence tomography as a local tumor necrosis predictor in photodynamic therapy of prostatic carcinoma: An in vivo study. *Cancer Res.* 68: 9987–9995.

Stratonnikov, A. A., Douplik, A. Y., and Loschenov, V. B. 2003. Oxygen consumption and photobleaching in whole blood incubated with photosensitizer induced by laser irradiation. *Laser Phys.* 13: 1–21.

Stripp, D., Mick, R., Zhu, T. C. et al. 2004. Phase I trial of Motexfin Lutetium-mediated interstitial photodynamic therapy in patients with locally recurrent prostate cancer. *Proc. SPIE* 5315: 88–99.

Stylli, S. S. and Kaye, A. H. 2006. Photodynamic therapy of cerebral glioma—a review Part I—a biological basis. *J. Clin. Neurosci.* 13: 615–625.

Tatum, J. L., Kelloff, G. J., Gillies, R. J. et al. 2006. Hypoxia: importance in tumor biology, noninvasive measurement by imaging, and value of its measurement in the management of cancer therapy. *Int. J. Radiat. Biol.* 82: 699–757.

van Staveren, H. J., Beek, J. F., Keijzer, M., and Star, W. M. 1995. Integrating sphere effect in whole-bladder-wall photodynamic therapy: II. The influence of urine at 458, 488, 514 and 630 nm optical irradiation. *Phys. Med. Biol.* 40: 1307–1315.

van Staveren, H. J., Beek, J. F., Ramaekers, J. W., Keijzer, M., and Star, W. M. 1994. Integrating sphere effect in whole bladder wall photodynamic therapy: I. 532 nm versus 630 nm optical irradiation. *Phys. Med. Biol.* 39: 947–959.

van Staveren, H. J., Keijzer, M., Keesmaat, T. et al. 1996. Integrating sphere effect in whole-bladder-wall photodynamic therapy: III. Fluence multiplication, optical penetration and light distribution with an eccentric source for human bladder optical properties. *Phys. Med. Biol.* 41: 579–590.

van Veen, R. L., Nyst, H., Rai Indrasari, S. et al. 2006a. In vivo fluence rate measurements during Foscan-mediated photodynamic therapy of persistent and recurrent nasopharyngeal carcinomas using a dedicated light applicator. *J. Biomed. Opt.* 11: 041107.

van Veen, R. L., Robinson, D. J., Siersema, P. D., and Sterenborg, H. J. 2006b. The importance of in situ dosimetry during photodynamic therapy of Barrett's esophagus. *Gastrointest. Endosc.* 64: 786–788.

Verigos, K., Stripp, D. C., Mick, R. et al. 2006. Updated results of a phase I trial of motexafin lutetium-mediated interstitial photodynamic therapy in patients with locally recurrent prostate cancer. *J. Environ. Pathol. Toxicol. Oncol.* 25: 373–388.

Vesselov, L., Whittington, W., and Lilge, L. 2005. Design and performance of thin cylindrical diffusers created in Ge-doped multimode optical fibers. *Appl. Opt.* 44: 2754–2758.

Vikram, D. S., Zweier, J. L., and Kuppusamy, P. 2007. Methods for noninvasive imaging of tissue hypoxia. *Antioxid. Redox Signal* 9: 1745–1756.

Vulcan, T. G., Zhu, T. C., Rodriguez, C. E. et al. 2000. Comparison between isotropic and nonisotropic dosimetry systems during intraperitoneal photodynamic therapy. *Lasers Surg. Med.* 26: 292–301.

Wang, H. W., Putt, M. E., Emanuele, M. J. et al. 2004. Treatment-induced changes in tumor oxygenation predict photodynamic therapy outcome. *Cancer Res.* 64: 7553–7561.

Wang, K. K., Busch, T. M., Finlay, J. C., and Zhu, T. C. 2009. Optimization of physiological parameter for macroscopic modeling of reacted singlet oxygen concentration in an in-vivo model. *Proc. SPIE* 7164: 71640O.

Wang, K. K., Mitra, S., and Foster, T. H. 2007. A comprehensive mathematical model of microscopic dose deposition in photodynamic therapy. *Med. Phys.* 34: 282–293.

Weersink, R. A., Bogaards, A., Gertner, M. et al. 2005. Techniques for delivery and monitoring of TOOKAD (WST09)-mediated photodynamic therapy of the prostate: Clinical experience and practicalities. *J. Photochem. Photobiol. B* 79: 211–222.

Williams, K. J., Parker, C. A., and Stratford, I. J. 2005. Exogenous and endogenous markers of tumour oxygenation status: Definitive markers of tumour hypoxia? *Adv. Exp. Med. Biol.* 566: 285–294.

Wilson, B. C., Patterson, M. S., and Lilge, L. 1997. Implicit and explicit dosimetry in photodynamic therapy: A new paradigm. *Lasers Med. Sci.* 12: 182–199.

Wormald, R., Evans, J., Smeeth, L., and Henshaw, K. 2007. Photodynamic therapy for neovascular age-related macular degeneration. *Cochrane Database Syst. Rev.* CD002030.

Wu, J., Feld, M. S., and Rava, R. P. 1993. Analytical model for extracting intrinsic fluorescence in turbid media. *Appl. Opt.* 32: 3583–3595.

Yu, G., Durduran, T., Zhou, C. et al. 2005. Noninvasive monitoring of murine tumor blood flow during and after photodynamic therapy provides early assessment of therapeutic efficacy. *Clin. Cancer Res.* 11: 3543–3552.

Zaak, D., Sroka, R., Hoppner, M. et al. 2003. Photodynamic therapy by means of 5-ALA induced PPIX in human prostate cancer—Preliminary results. *Med. Laser Appl.* 18: 91–95.

Zaak, D., Sroka, R., Khoder, W. et al. 2008. Photodynamic diagnosis of prostate cancer using 5-aminolevulinic acid—First clinical experiences. *Urology* 72: 345–348.

Zheng, G., Chen, J., Stefflova, K., Jarvi, M., Li, H., and Wilson, B. C. 2007. Photodynamic molecular beacon as an activatable photosensitizer based on protease-controlled singlet oxygen quenching and activation. *Proc. Natl. Acad. Sci. U. S. A.* 104: 8989–8994.

Zhou, X., Pogue, B. W., Chen, B. et al. 2006. Pretreatment photosensitizer dosimetry reduces variation in treatment response. *Int. J. Radiat. Oncol. Biol. Phys.* 64: 1211–1220.

Zhou, X. and Zhu, T. C. 2008. Interstitial diffuse optical tomography using an adjoint model with linear sources. *Proc. SPIE* 6846: 68450C.

Zhu, T. C., Finlay, J. C., and Hahn, S. M. 2005. Determination of the distribution of light, optical properties, drug concentration, and tissue oxygenation in-vivo in human prostate during motexafin lutetium-mediated photodynamic therapy. *J. Photochem. Photobiol. B: Biol.* 79: 231–241.

Zhu, T. C., Finlay, J. C., Zhou, X., and Li, J. 2007. Macroscopic modeling of the singlet oxygen production during PDT. *Proc. SPIE* 6427: 642708.

35

Low Level Laser and Light Therapy

Ying-Ying Huang
Massachusetts General Hospital
and
Harvard Medical School
and
Aesthetic and Plastic Center
of Guangxi Medical University

Aaron C.-H. Chen
Massachusetts General Hospital
and
Boston University School
of Medicine

Michael R. Hamblin
Massachusetts General Hospital
and
Harvard Medical School
and
Harvard-MIT Division of Health
Sciences and Technology

35.1 Introduction 751
35.2 Biophysical and Biochemical Mechanisms..752
    Tissue Photobiology and Optics • Photoacceptors and Cellular
    Chromophores • Mitochondria and Cellular Respiratory Chain • Cytochrome c
    Oxidase • Nitric Oxide Release • Laser Speckle Effect on Mitochondria
35.3 Biological Effects of LLLT in Cells...755
    Initial Effects • Transcription Factors • Downstream Cellular Effects
35.4 Medical Applications of LLLT 758
    Dermatology • Wound Healing • Musculoskeletal Disorders • Nervous System • Pain
    Relief • Cardiovascular System • Dentistry • Miscellaneous Indications
35.5 Summary...764
35.6 Future Perspective 764
Acknowledgment..764
References..764

35.1 Introduction

In 1967, a few years after the first working laser was invented, Endre Mester in Hungary wanted to test if laser radiation might cause cancer in mice (Mester et al. 1968). He shaved the dorsal hair, divided them into two groups, and gave a laser treatment with a low powered ruby laser (694 nm) to one group. They did not get cancer and to his surprise the hair on the treated group grew back more quickly than the untreated group. This was the first demonstration of "laser biostimulation." Since then, medical treatment with coherent light sources (lasers) or noncoherent light sources consisting of filtered lamps or light-emitting diodes (LED) has spread throughout the world. Currently, low level laser (or light) therapy (LLLT), also known as "cold laser," "soft laser," "biostimulation," or "photobiomodulation," is routinely practiced as part of physical therapy and is under investigation for many serious conditions and even life-threatening diseases. In fact, light therapy is one of the oldest therapeutic methods used by humans (historically as solar therapy by Egyptians, later as UV therapy for which Nils Finsen won the Nobel prize in 1904 (Roelandts 2002)). The use of lasers and LEDs as light sources was the next step in the technological development of photomedicine and phototherapy. In LLLT, the question is no longer whether light has biological effects, but rather how energy from therapeutic lasers and LEDs works at the cellular and organism levels and what are the optimal light parameters for different uses of these light sources.

The optimum levels of energy density delivered are low when compared to other forms of laser therapy as practiced for ablation, cutting, and thermally coagulating tissue. In general, the power densities used for LLLT are lower than those needed to produce heating of tissue, less than 100 mW/cm^2 depending on wavelength and tissue type. One important point that has been demonstrated by multiple studies in cell culture (Pereira et al. 2002), in animal models (Kana et al. 1981), and in clinical studies is the concept of a biphasic dose response when the outcome is compared with the total delivered light energy density (fluence). It has been found that there exists an optimal dose of light for any particular application, and doses "lower" than this optimum value, or more significantly, "larger" than the optimum value will have a diminished therapeutic outcome, or for high doses of light a negative outcome may even result. Evidence suggests that both energy density and power density are key biological parameters for effectiveness of laser therapy and they may both operate with thresholds (i.e., a lower and an upper threshold for both parameters between which laser therapy is effective and outside of which laser therapy is too weak to have any effect or so intense that the tissue is inhibited) (Sommer et al. 2001).

35.2 Biophysical and Biochemical Mechanisms

35.2.1 Tissue Photobiology and Optics

An important consideration in LLLT involves the optical properties of tissue. Both the absorption and scattering of light in tissue are wavelength dependent (both much higher in the blue region of the spectrum than the red) and the principle tissue chromophores (hemoglobin and melanin) have high absorption bands at wavelengths shorter than 600 nm. Water begins to absorb significantly at wavelengths greater than 1150 nm. For these reasons there is a so-called optical window in tissue covering the red and near-infrared wavelengths, where the effective tissue penetration of light is maximized (Figure 35.1). Therefore, although blue, green, and yellow light may have significant effects on cells growing in optically transparent culture medium, the use of LLLT in animals and patients almost exclusively involves red and near-infrared light in the range of 600–1100 nm.

35.2.2 Photoacceptors and Cellular Chromophores

The first law of photobiology states that for low power visible light to have any effect on a living biological system, the photons must be absorbed by electronic absorption bands belonging to some molecular chromophore or photoacceptor (Sutherland 2002). There may be one (or more than one) chromophore that leads to the multitude of biological effects of LLLT, as shown in Figure 35.2. One approach to finding the identity of this chromophore(s) is to carry out action spectra. This is a graph representing biological photoresponse as a function of

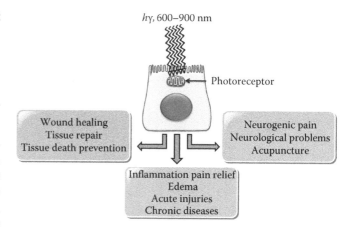

FIGURE 35.2 Schematic diagram showing how the absorption of red and NIR light by specific cellular chromophores or photoacceptors can lead to a wide range of beneficial medical effects.

wavelength and should resemble the absorption spectrum of the photoacceptor molecule. The fact that a structured action spectrum can be constructed supports the hypothesis of the existence of cellular photoacceptors and signaling pathways stimulated by light.

35.2.3 Mitochondria and Cellular Respiratory Chain

Mitochondria play an important role in energy generation and metabolism. Current research about the mechanism of LLLT effects inevitably involves mitochondria. Mitochondria are distinct organelles with two membranes and are usually rod shaped, ranging from 1 to 10 μm in size. Mitochondria are sometimes described as "cellular power plants," because they convert food molecules into energy in the form of ATP via the process of oxidative phosphorylation. Mitochondria continue the process of catabolism using metabolic pathways including the Krebs cycle, fatty acid oxidation, and amino acid oxidation.

The mitochondrial electron transport chain consists of a series of metalloproteins bound to the inner membrane of the mitochondria. The inner mitochondrial membrane contains five complexes of integral membrane proteins: NADH dehydrogenase (Complex I), succinate dehydrogenase (Complex II), cytochrome c reductase (Complex III), cytochrome c oxidase (Complex IV), and ATP synthase and two freely diffusible molecules, ubiquinone and cytochrome c, that shuttle electrons from one complex to the next (Figure 35.3). The respiratory chain accomplishes the stepwise transfer of electrons from NADH and FADH2 (produced in the citric acid or Krebs cycle) to oxygen molecules to form (with the aid of protons) water molecules harnessing the energy released by this transfer to the pumping of protons (H+) from the matrix to the intermembrane space. The gradient of protons formed across the inner membrane by this process of active transport forms a miniature battery. The protons can flow back down this gradient, reentering the matrix, only through another complex of integral

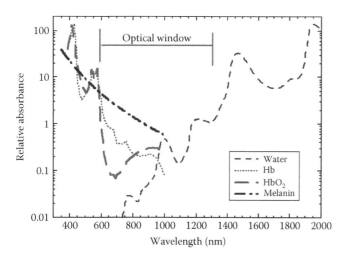

FIGURE 35.1 Absorption spectra of the main chromophores in living tissue on a log scale showing the optical window where visible and NIR light can penetrate deepest into tissue.

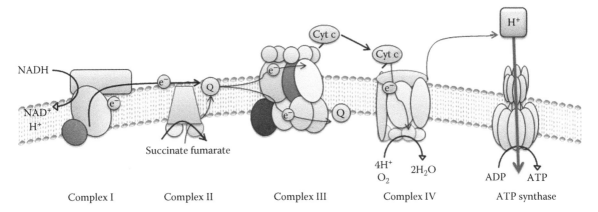

FIGURE 35.3 Mitochondrial respiratory chain consisting of five complexes of integral membrane proteins: NADH dehydrogenase (Complex I), succinate dehydrogenase (Complex II), cytochrome c reductase (Complex III), cytochrome c oxidase (Complex IV), and ATP synthase (Complex V).

proteins in the inner membrane, the ATP synthase complex (Turrens 2003).

35.2.4 Cytochrome c Oxidase

Absorption spectra obtained for cytochrome c oxidase in different oxidation states were recorded and found to be very similar to the action spectra for biological responses to light. Therefore, it was proposed that cytochrome c oxidase (Cox) is the primary photoacceptor for the red-NIR range in mammalian cells (Karu and Kolyakov 2005). Whelan's group has also suggested that Cox is the critical chromophore responsible for stimulatory effects of irradiation with infrared light (Eells et al. 2003). In vitro, Wong-Riley et al. demonstrated that infrared irradiation reversed the reduction in Cox activity produced by the blockade of voltage-dependent sodium channels with tetrodotoxin and upregulated Cox activity in primary neuronal cells (Wong-Riley et al. 2005). In vivo, Eells et al. demonstrated that rat retinal neurons are protected from damage induced by methanol intoxication. The actual toxic metabolite formed from methanol is formic acid, which inhibits Cox (Eells et al. 2003). Moreover, increased activity of cytochrome c oxidase and an increase in polarographically measured oxygen uptake were observed in illuminated mitochondria (Pastore et al. 1994).

Nitric oxide produced in the mitochondria can inhibit respiration by binding to cytochrome c oxidase and competitively displacing oxygen (Brown 2001). If light absorption displaced the nitric oxide and thus allowed the cytochrome c oxidase to recover and cellular respiration to resume, this would explain many of the observations made in LLLT (Lane 2006). Increases in ATP are one of the most often observed changes after LLLT is carried out in vitro, and increased cytochrome c oxidase activity would explain raised ATP levels due to increased cellular respiration and oxidative phosphorylation (Karu et al. 1995). Hypoxic, stressed, or damaged cells or tissues are likely to respond more to LLLT than normal cells and tissues, and their cytochrome c oxidase is more likely to be operating at suboptimal level due to NO inhibition. LLLT effects can keep working

for some time (hours or days) post-illumination because the displaced nitric oxide cannot easily return to inhibit cytochrome c oxidase. It has been proposed that LLLT might work by photodissociating the relatively weak bond that connects NO to Cox, thereby reversing the signaling consequences of excessive NO binding (Figure 35.4) (Karu et al. 2005). The removal of the inhibiting NO from cytochrome c oxidase will lead to a sharp drop in pO_2 in the cell as respiration resumes and if the cells have relatively low pO_2 levels to start with, this drop will lead to hypoxic signaling through stabilization of HIF-1alpha. Hypoxic signaling is one of the main factors leading to vascular endothelial growth factor (VEGF) synthesis and its consequent increase in angiogenesis as has been observed after LLLT for wound healing.

Light can indeed reverse the inhibition caused by NO binding to Cox, both in isolated mitochondria and in whole cells (Borutaite et al. 2000). Light can also protect cells against NO-induced cell death. These experiments used light in the visible spectrum, with wavelengths from 600 to 630 nm. NIR also seems to have effects on Cox in conditions where NO is unlikely to be present.

35.2.5 Nitric Oxide Release

Increased NO concentrations can sometimes be measured in cell culture or in animals after LLLT due to its photogenerated release from the mitochondria and cytochrome c oxidase. The release of NO from the mitochondria could explain the transient increases in blood flow measured in skin microcirculation after LLLT. Release of NO into the tissue could also explain the effect of LLLT in increasing lymphatic drainage in conditions like lymphedema and reducing swelling in trauma. A recent report (Zhang et al. 2009) suggests that biologically important amounts of nitric oxide can be released by light from other intracellular (or extracellular) NO stores (Shiva and Gladwin 2009). In addition to Cox these NO stores might consist of nitrosylated hemoglobin (Angelo et al. 2008) and nitrosylated myoglobin (Giuffre et al. 2005) (Figure 35.5).

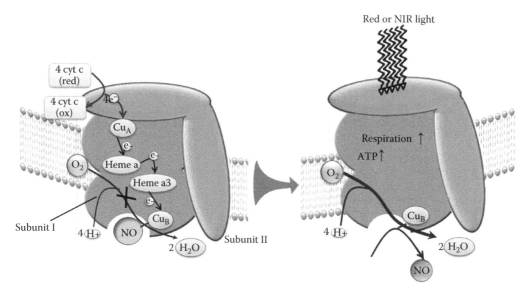

FIGURE 35.4 When NO is released from its binding to cytochrome c oxidase by light, respiration is restored to its former level leading to increased ATP synthesis.

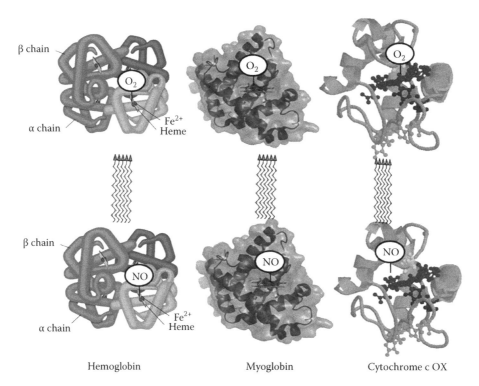

FIGURE 35.5 Light mediated release of nitric oxide from its tissue and cellular stores. Non-covalently bound NO is photodissociated from its binding sites within the molecules of hemoglobin, myoglobin, and cytochrome c oxidase.

35.2.6 Laser Speckle Effect on Mitochondria

It should be mentioned that there is another mechanism that has been proposed to account for low level laser effects on tissue. This explanation relies on the phenomenon of laser speckle, which is peculiar to laser light. The speckle effect is a result of the interference of many waves, having different phases, which add together to give a resultant wave whose amplitude, and therefore intensity, varies randomly. Each point on illuminated tissue acts as a source of secondary spherical waves. The light at any point in the scattered light field is made up of waves, which have been scattered from each point on the illuminated surface. If the surface is rough enough to create path-length differences exceeding one wavelength, giving rise to phase changes greater than 2π, the amplitude, and hence the intensity, of the resultant light varies randomly (Figure 35.6).

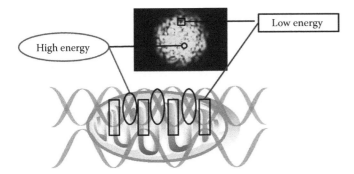

FIGURE 35.6 Diagram of how laser speckle may create a set of microscopic intensity gradients in mitochondria.

It is proposed that the variation in intensity between speckle spots that are about 1 μm apart can give rise to small but steep temperature gradients within subcellular organelles such as mitochondria without causing photochemistry (Rubinov and Afanas'ev 2005). These temperature gradients are proposed to cause some unspecified changes in mitochondrial metabolism. This hypothesis would explain reports that some LLLT effects in cells and tissues are more pronounced when coherent laser light is used than comparable noncoherent light from LED or filtered lamp sources that are of similar wavelength range although not monochromatic.

35.3 Biological Effects of LLLT in Cells

35.3.1 Initial Effects

Several pieces of evidence suggest that mitochondria are responsible for the cellular response to red visible and NIR light. The most popular system to study is the effects of He-Ne laser illumination on mitochondria isolated from rat liver, in which increased proton electrochemical potential and ATP synthesis was found (Passarella et al. 1984). Increased RNA and protein synthesis was demonstrated after $5 J/cm^2$ (Greco et al. 1989). Irradiation of mitochondria with light at other wavelengths such as 660 nm (Yu et al. 1997b), 650 nm, and 725 nm (Gordon 1960) also showed increases in oxygen consumption, membrane potential, and enhanced synthesis of NADH and ATP. Irradiation with light at 633 nm increased the mitochondrial membrane potential (MMP) and proton gradient, and increased the rate of ADP/ATP exchange (Passarella et al. 1988), as well as RNA and protein synthesis in the mitochondria. It is also believed that mitochondria are the primary targets when the whole cells are irradiated with light at 630, 632.8, or 820 nm [14,27,28].

Several classes of molecules such as reactive oxygen species (ROS) and reactive nitrogen species (RNS) are involved in the signaling pathways from mitochondria to nuclei. ROS first reported by Harman (Harman 1956) are ions or very small molecules and encompass a variety of partially reduced metabolites of oxygen (e.g., superoxide anions, hydrogen peroxide, and hydroxyl radicals) possessing higher reactivities than molecular oxygen due to the presence of unpaired valence shell electrons.

The combination of the products of the reduction potential and reducing capacity of the linked redox couples present in cells and tissues represent the redox environment (redox state) of the cell. Redox couples present in the cell include nicotinamide adenine dinucleotide (oxidized/reduced forms), nicotinamide adenine dinucleotide phosphate NADP/NADPH, glutathione/glutathione disulfide couple, and thioredoxin/thioredoxin disulfide couple (Schafer and Buettner 2001). The term redox signaling is widely used to describe a regulatory process in which the signal is delivered through redox chemistry, including bacteria, to induce protective responses against oxidative damage and to reset the original state of "redox homeostasis" after temporary exposure to ROS (Droge 2002). The primary ROS produced in mitochondria is superoxide anion ($O_2^{\bullet-}$), which is converted to H_2O_2 either by spontaneous dismutation or by the enzyme, superoxide dismutase (SOD). It has been reported that several molecules could be the sensors of ROS (Storz 2007). It is proposed that LLLT produces a shift in overall cell redox potential in the direction of greater oxidation (Karu 1999a). Lubart's group found that red light illumination increased ROS generation and cell redox activity (Lubart et al. 2005). They also reported that the ROS generation can be detected by electron spin resonance (ESR) techniques (Lavi et al. 2004). Furthermore several other laboratories (including ours (Chen et al. 2009a,b)) have demonstrated cellular generation of various ROS after red or NIR illumination of cells using fluorescent probes (Alexandratou et al. 2002, Zhang et al. 2008) or assays for hydrogen peroxide (Pal et al. 2007).

35.3.2 Transcription Factors

Nuclear factor kappa B (NF-κB) is a transcription factor regulating multiple gene expression (Wang et al. 2002), and has been shown to govern various cellular functions, including inflammatory and stress-induced responses and survival (Baichwal and Baeuerle 1997). NF-κB activation is governed by negative feedback by IκB, an inhibitor protein that binds to NF-κB, but can undergo ubiquitination and proteasomal degradation (Figure 35.7) (Henkel et al. 1993), thus freeing NF-κB to translocate to the nucleus and initiate transcription (Hoffmann et al. 2002). Understanding the activation mechanisms that govern NF-κB may be important in studying tissue repair or even cancer progression. NF-κB is a redox-sensitive transcription factor (D'Angio and Finkelstein 2000), that has been proposed to be the sensor for oxidative stress (Li and Karin 1999). Reactive oxygen species (ROS) can activate NF-κB (Schreck et al. 1991) and have been shown to be involved in NF-κB activation by other pro-inflammatory stimuli (Schreck et al. 1992). As discussed above the formation of ROS in cells in vitro after LLLT has been demonstrated, and it has been proposed that ROS are involved in the signaling pathways initiated after photons are absorbed by the mitochondria in cells (Tafur and Mills 2008). We have shown (Chen et al. 2009a) using a luciferase reporter assay that NF-κB is activated in mouse embryonic fibroblasts treated with 810 nm laser with a biphasic dose response pattern and a maximum activation 6 h after $0.3 J/cm^2$. NF-κB activation

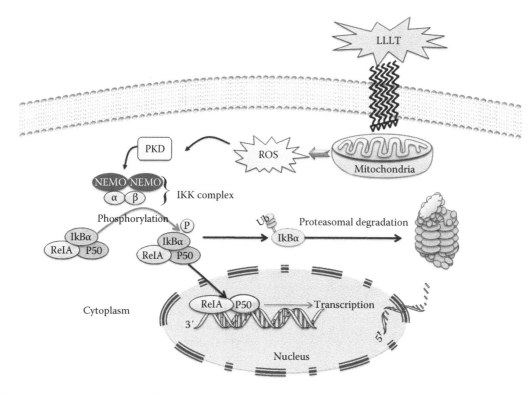

FIGURE 35.7 Reactive oxygen species (ROS) formed as a result of LLLT effects in mitochondria may activate the transcription factor NF-κB (relA-p50) via protein kinase D (PKD).

caused by either light or by hydrogen peroxide was blocked by the addition of antioxidants suggesting that light produces ROS that then activates NF-κB. The panoply of genes that are responsive to NF-κB include many involved in cell survival, proliferation, migration, and collagen synthesis that adds further weight to the hypothesis that some LLLT cell effects are mediated by ROS-induced activation of NF-κB (Chen et al. 2009b).

35.3.3 Downstream Cellular Effects

Although the underlying mechanisms of LLLT are still not clearly understood, in vitro studies and both animal experiments and clinical studies indicate LLLT can prevent cell apoptosis and improve cell proliferation, migration, and adhesion. All these effects can work together to support the wide clinical application of LLLT (Figure 35.8).

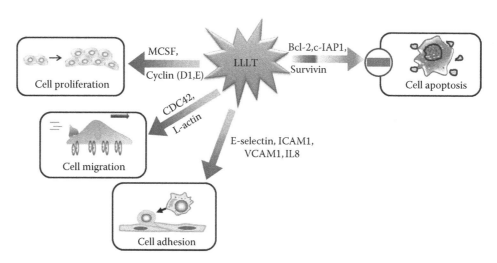

FIGURE 35.8 Some of the downstream cellular effects reported after LLLT treatment of cell cultures and some possible mediators of these effects. MCSF is a growth factor; cyclin D1 and E are responsible for cell proliferation; CDC42 is a chemokine and L-actin is used by cells to move; E-selectin, ICAM1, and VCAM1 are adhesion molecules for different types of cells; Bcl-2, c-IAR1 and survivin are proteins that inhibit apoptosis.

35.3.3.1 Proliferation

LLLT has been shown to enhance cell proliferation in vitro in several types of cells: fibroblasts [50–52], keratinocytes [53], endothelial cells (Moore et al. 2005), and lymphocytes [55, 56]. Almeida-Lopes et al. showed a shorter application time of laser radiation gave a higher increase in human gingival fibroblasts proliferation rate at the same fluence (2 J/cm^2) (Almeida-Lopes et al. 2001). The mechanism of proliferation was supposed that He-Ne irradiation elicited photostimulatory effects in mitochondria processes, which involved JNK/AP-1 activation and enhanced growth factor release, and ultimately led to cell proliferation [58].

35.3.3.2 Migration and Adhesion

Cell migration and adhesion can be enhanced by LLLT. Yu et al. observed that LLLT provided a microenvironment for inducing repigmentation in vitiligo (Yu et al. 2003). They proposed that He-Ne laser irradiation stimulated melanocyte migration and proliferation and mitogen release for melanocyte growth and may also rescue damaged melanocytes. It was also proposed that LLLT activates some signaling pathways in adhesive interactions between cells and extracellular matrices (Karu et al. 2001). Karu et al. showed that the attachment and proliferation of human gingival fibroblasts were enhanced by LLLT in a dose-dependent manner at doses of 1.5 and 3 J/cm^2 after 72 h and 7 days. Several reports (Hawkins and Abrahamse 2006, Houreld and Abrahamse 2007) have described the use of LLLT to "heal" a scratch wound made on a monolayer of cells growing on a plastic substrate in cell culture.

35.3.3.3 Anti-Apoptosis in Tissue Culture by Increasing Cell Migration

An important observation was reported that LLLT enabled cells to remain alive under the condition of nutritional deficiency. It was supposed that LLLT provided positive biomodulation, and acted to reestablish cellular homeostasis when the cells were maintained under the condition of nutritional stress, which may cause apoptosis (Carnevalli et al. 2003). An increased expression of the anti-apoptotic protein Bcl-2, and a reduced expression of pro-apoptotic protein BAX was observed for the first time. These changes were accompanied by a reduction in the expression of p53 and the cyclin-dependent kinase inhibitor p21, which implicated the regulation of these factors as part of the protective role of LLLT against apoptosis (Shefer et al. 2002).

35.3.3.4 Collagen Synthesis

Collagen is the main protein of connective tissue and the most abundant protein in mammals. LLLT has been shown to stimulate smooth muscle cell proliferation, stimulate collagen synthesis, and modulate the equilibrium between regulatory matrix remodeling enzymes using a 780 nm laser diode (1 and 2 J/cm^2) (Gavish et al. 2006). 625–635 nm LED maintained human fibroblast (HS68) viability and increased collagen synthesis. Stimulated cells showed increased basic fibroblast growth factor (bFGF) secretion, procollagen type I C-peptide production, and type I collagen synthesis (Huang et al. 2007). Transforming growth factor-beta (TGF-β) is a pleiotropic growth factor responsible (among others) for inducing collagen synthesis from fibroblasts and has been reported to be upregulated by LLLT (Khanna et al. 1999).

35.3.3.5 Myofibroblasts

Myofibroblasts are found subepithelially in many mucosal surfaces. In the wound tissue, they are implicated in wound strengthening by extracellular collagen fiber deposition and then helping wound contraction by intracellular contraction using smooth muscle fibers, and speeding wound repair. Laser therapy has been reported to reduce the inflammatory reaction, increase collagen deposition, and cause a greater proliferation of myofibroblasts in experimental cutaneous wounds (Medrado et al. 2003, 2008). It was reported that the number of myofibroblasts was increased when 685 nm polarized light is used (Pinheiro et al. 2005). It was found that in laser-irradiated lesions, smooth muscle alpha-actin positive cells predominated, which correspond to a higher number of myofibroblasts (de Araujo et al. 2007).

35.3.3.6 Growth Factors

The term growth factor refers to a naturally occurring substance capable of stimulating cellular growth, proliferation, and cellular differentiation, such as epidermal growth factor (EGF), transforming growth factor-beta (TGF-b), and vascular endothelial growth factor (VEGF) that play an essential role in wound healing and tissue repair. In an in vivo oral tooth extraction-healing model, an increased TGF-beta1 expression immediately following laser irradiation (Arany et al. 2007) was observed. The gene expression of PDGF and TGF-beta were significantly increased in rats' gingival and mucosal tissues with He-Ne laser of 7.5 J/cm^2 fluence (Safavi et al. 2008). Hou et al (Hou et al. 2008) found that 0.5 J/cm^2 of 635 nm laser stimulated secretion of VEGF and nerve growth factor from bone-marrow-derived stem cells.

35.3.3.7 Inflammatory Mediators

Inflammatory mediators are soluble, diffusible molecules that act locally at the site of tissue damage and infection, and at more distant sites, during infections and systemic inflammatory diseases. It has been found that 660 nm laser radiation induced an anti-inflammatory effect in rats characterized by inhibition of total or differential leukocyte influx, exudation, total protein, NO, interleukin (IL)-6, macrophage chemotactic protein-1, IL-10, and TNF-a, in a dose-dependent manner, in which 2.1 J was more effective than 0.9 and 4.2 J (Boschi et al. 2008). LLLT with 810 nm laser was highly effective in treating inflammatory arthritis in mice model (Lopes-Martins et al. 2006) and interestingly anti-inflammatory effect was abrogated by the steroid receptor antagonist mifepristone. LLLT also was shown to reduce inflammation by a mechanism in which IL-1beta seems to have an important role. There was reduced both IL-1beta in bronchial lymphoid tissue and IL-1beta mRNA expression in trachea obtained from animals subjected to LPS-induced inflammation (Aimbire et al. 2008).

35.3.3.8 Neuronal Modulation

Neuromodulation is the process in which several classes of neurotransmitters in the nervous system regulate diverse populations of neurons (one neuron may use different neurotransmitters to connect to several neurons). NIR lasers are widely used in LLLT to affect neurons both to stimulate and inhibit neuronal activity. It was shown that laser irradiation applied to intact skin produced a direct, localized effect upon conduction in underlying nerves (Baxter et al. 1994). It was first reported by Chow et al. (2007) that 830 nm (cw) laser blocked fast axonal flow and reduced mitochondrial membrane potential (MMP). In neurons, the decrease in MMP was significant, but the decrease of available ATP occurred more slowly that was required for nerve function, including maintenance of microtubules and molecular motors, dyneins and kinesins that was responsible for fast axonal flow. Laser-induced neural blockade is a consequence of such changes and provides a mechanism for a neural basis of laser-induced pain relief.

35.4 Medical Applications of LLLT

35.4.1 Dermatology

The ready accessibility of the skin to illumination and the highly visible results of therapy have led to several dermatological indications being investigated for LLLT.

35.4.1.1 LLLT for Hair Regrowth

Since the first pioneering publication of Mester (Mester et al. 1968) reported stimulation of hair growth in mice, there have been virtually no follow-up studies on LLLT stimulation for hair growth in animal models. There have been only a few literature reports containing some observations of LLLT-induced hair growth in patients, and amelioration or treatment of any type of alopecia (Yamazaki et al. 2003). A report from Finland (Pontinen et al. 1996) compared three different light sources used for male-pattern baldness (He-Ne laser, in GaAl diode laser at 670 nm and noncoherent 635 nm LED) and measured blood flow in the scalp. A recent report (Leavitt et al. 2009) detailed a randomized double-blind, sham device-controlled, multicenter clinical trial of a laser haircomb device designed to deliver 665 nm laser light to the scalp of men with androgenetic alopecia. Out of 110 subjects, 85% of the active group had a major increase in hair density, compared with only 5% of the sham group.

35.4.1.2 Acne Therapy

Acne vulgaris is one of the most common dermatologic disorders encountered in everyday practice. Light sources including blue lights and intense pulsed lights are becoming regular additions to routine medical management to enhance the therapeutic response. Recently, there have been some reports (Papageorgiou et al. 2000, Lee et al. 2007) that a combination of blue light and red light is effective in treating acne, because the porphyrin produced by *Propionibacterium acnes* absorbs blue light that kills the bacteria via PDT action. The red 633 nm light exerts an inflammatory action by photobiomodulation to lessen the severity of the acne lesions.

35.4.1.3 Wrinkle Reduction

Zelickson et al. (Zelickson et al. 1999) have shown that new collagen was formed 12 weeks after treatment with a 585 nm pulsed dye laser. Several reports showed improvement in skin collagen, and objectively measured data showed significant reductions of wrinkles (maximum: 36%) and increases of skin elasticity (maximum: 19%) compared to baseline on the treated face in treatment groups after low level laser irradiation (Chen et al. 2008). Omi reported the efficacy in collagen replenishment after irradiation of 3 J/cm², 585 nm laser (Chromogenex V3), with a spot size of 5 mm diameter in patients (Omi et al. 2005). Immunohistochemistry showed increases in several markers of healthy skin after LED phototherapy. They also found that 633 nm LED phototherapy is an effective approach for skin rejuvenation and can also activate the skin-homing immune system (Takezaki et al. 2006). A recent report (Alster and Wanitphakdeedecha 2009) describes the use of a 590 nm LED array to reduce facial erythema produced after undergoing full-face fractional laser skin resurfacing with a 1550 nm erbium-doped fiber laser.

35.4.2 Wound Healing

For many years LLLT has been tested as to whether it can stimulate wound healing in a variety of animal models, but the literature contains a plethora of both positive and negative studies. Nevertheless, a meta-analysis of publications on LLLT for wound healing (Woodruff et al. 2004) concluded that "laser therapy is an effective tool for promoting wound repair." The reasons for the conflicting reports, sometimes in very similar wound models, are probably diverse. The number of different possible combinations of variables such as wavelength, irradiance, fluence, pulse structure, and coherence that can be chosen in LLLT studies means that it is difficult to ensure that effective parameters are chosen in every case (Lowe et al. 1998). Many studies that have been published showing no significant benefit with the particular set of parameters chosen, may have shown different results with other sets of parameters. Moreover, it is probable that applications of LLLT in animal models will be more effective if carried out on models that have some intrinsic disease state that slows down or dysregulates wound healing (Karu 1999b). Although there have been several reports that processes such as wound healing are accelerated by LLLT in normal rodents (Demidova-Rice et al. 2007), an alternative approach is to inhibit healing by inducing some specific disease

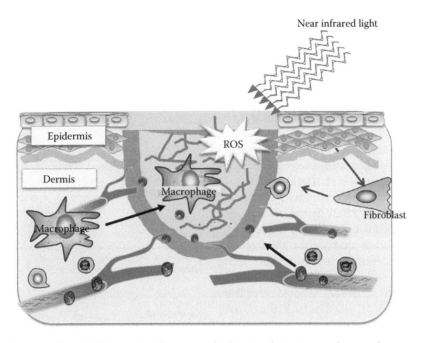

FIGURE 35.9 Schematic diagram of how LLLT may stimulate wound healing involving increased macrophage activity within the wound and fibroblast migration, transformation to myofibroblasts and more collagen synthesis.

state such as diabetes, a disease known to significantly depress wound healing in patients. LLLT significantly improves wound healing in both diabetic rats (Byrnes et al. 2004, Maiya et al. 2005) and diabetic mice (Yu et al. 1997a, Stadler et al. 2001). LLLT was also effective in x-radiation impaired wound healing in mice (Lowe et al. 1998). A study (Lyons et al. 1987) in hairless mice found improvement in the tensile strength of the He-Ne laser (632.7 nm)-irradiated wounds at 1 and 2 weeks. Furthermore, the total collagen content was significantly increased at 2 months when compared with controlled wounds. The beneficial effect of LLLT on wound healing can be explained by considering several basic biological mechanisms including the induction of expression cytokines and growth factors known to be responsible for the many phases of wound healing (Figure 35.9). First, there is a report (Yu et al. 1996) that He-Ne laser (632.7 nm) increased both protein and mRNA levels of IL-1α and IL-8 in keratinocytes. These are cytokines responsible for the initial inflammatory phase of wound healing. Second, there are reports (Poon et al. 2005) that LLLT can upregulate cytokines responsible for fibroblast proliferation and migration such as bFGF, HGF, and SCF. Third, it has been reported (Kipshidze et al. 2001) that LLLT can increase growth factors such as VEGF responsible for the neovascularization necessary for wound healing. Fourth, TGF-β is a growth factor responsible for inducing collagen synthesis from fibroblasts and has been reported to be upregulated by LLLT (Khanna et al. 1999). Fifth, there are reports (Neiburger 1999, Medrado et al. 2003, Demidova-Rice et al. 2007) that LLLT can induce fibroblasts to undergo the transformation into myofibroblasts, a cell type that expresses smooth muscle α-actin and desmin and has the phenotype of contractile cells that hasten wound contraction.

35.4.3 Musculoskeletal Disorders

These conditions are often chronic in nature, painful, and even disabling. There is a great demand among patients for relief from these disorders, and they may have failed several alternative treatments before trying LLLT. Joints, muscles, tendons, and cartilage are treated with LLLT, as shown in Figure 35.10.

The most common form of arthritis, osteoarthritis (degenerative joint disease) is a result of trauma to the joint, infection of the joint, or age. Our laboratory (Castano et al. 2007) reported that LLLT induced reduction of joint swelling in rats with inflammatory arthritis caused by intra-articular injection of zymosan, and this reduction in swelling correlated with reduction in the inflammatory marker serum prostaglandin E2 (PGE2) with illumination of 810 nm laser.

Brosseau (Brosseau et al. 2003) reported a meta-analysis of LLLT clinical trials for osteoarthritis. Thirteen trials were included, with 212 patients randomized to laser and 174 patients to placebo laser, and 68 patients received active laser on one hand and placebo on the opposite hand, relative to a separate control group. LLLT reduced pain by 70% relative to placebo and reduced morning stiffness. There are still more clinical

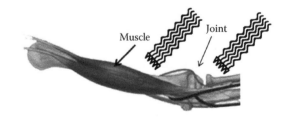

FIGURE 35.10 LLLT in the musculoskeletal system can improve dysfunction in joints, muscles, tendons, and cartilage.

trials required to establish the LLLT dosage and the treatment length. Recently, Bjordal et al. (Bjordal et al. 2003) conducted a systematic review of laser therapy with location-specific doses for chronic joint pain and found that this therapy significantly reduces pain and improves health status of patients with joint disorders. Rheumatoid arthritis is an autoimmune disease of joints that also responds to LLLT according to another systematic review (Brosseau et al. 2005).

Reports indicate that low level laser therapy is effective in relieving pain from several other musculoskeletal conditions including Achilles tendonitis (Bjordal et al. 2006), shoulder pain (Bingol et al. 2005) or frozen shoulder (Stergioulas 2008), muscle spasm and fibromyalgia (Gur et al. 2002), fibromyositis (Longo et al. 1997), and tennis elbow (medial and lateral epicondylitis) (Simunovic et al. 1998). While studying the effect of laser therapy on painful conditions, Baxter et al. reported that lasers achieved the premier overall ranking for pain relief compared with the other electro-physical modalities. A rapid relief from neck pain (Chow and Barnsley 2005) as well as back pain (Djavid et al. 2007) has been reported by several investigators following laser application. Carpal tunnel syndrome is a widespread affliction of the hand and wrist that responds well to LLLT (Chang et al. 2008) and, in fact, it was for this indication that the first FDA approval for a LLLT device was obtained (Weintraub 1997).

35.4.4 Nervous System

Diseases and injuries of the nervous system and especially in the central nervous system consisting of the brain and spinal cord are considered serious and even life-threatening. Although it might be thought that light would have difficulty penetrating into these deep organ structures, in fact, NIR light has been used in many studies to reduce damage or improve outcomes (Figure 35.11).

35.4.4.1 Traumatic Brain Injury

Traumatic brain injury (TBI) occurs when an outside force traumatically injures the brain. TBI can result in neurological impairment because of immediate CNS tissue disruption (primary injury), but, additionally, surviving cells may be secondarily damaged by complex mechanisms triggered by the primary event, leading to further damage and disability (Teasdale and Graham 1998). LLLT has been found to modulate various biological processes. Oron et al. (Oron et al. 2007) showed that TLT given 4 h following TBI provides a significant long-term functional neurological benefit. TBI was induced by a weight-drop device, and motor function was assessed 1 h post-trauma using a neurological severity score (NSS). Mice were then divided into three groups of eight mice each: one control group that received a sham LLLT procedure and was not irradiated, and two groups that received LLLT at two different doses (10 and 20 mW/cm²) transcranially. An 808 nm Ga-As diode laser was employed transcranially 4 h post-trauma to illuminate the entire cortex of the brain. Motor function was assessed up to 4 weeks, and lesion volume was measured. Further confirmatory trials are warranted.

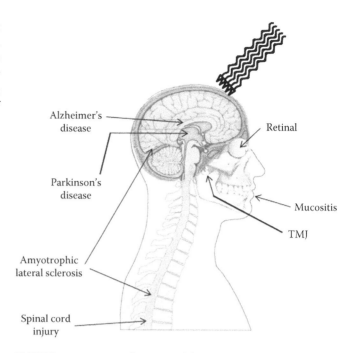

FIGURE 35.11 LLLT is being tested for many central nervous system applications including brain traumas and diseases, spinal cord repair, and retinal nerve preservation. Also illustrated are dental applications of LLLT for muscositis and TMJ disorder.

35.4.4.2 Stroke

The beneficial effects of NIR light on cells and neurons in vitro, together with the demonstrated ability of NIR light to penetrate into the brain, strongly suggested that transcranial LLLT should be studied as a therapy for stroke.

Transcranial laser therapy (TLT) at 808–810 nm can penetrate the brain and was shown to lead to enhanced production of ATP in the rat cerebral cortex (Streeter et al. 2004). Findings of increased neurogenesis in the subventricular zone (SVZ) were reported in an ischemic stroke animal model treated with TLT [119]. Based on these findings, it is thought that TLT may have multiple mechanisms of action and could be beneficial in acute ischemic stroke (Lampl 2007).

Recently, this second clinical trial of the effectiveness and safety trial of NIR laser treatment within 24 h from stroke onset (NEST-2) was finished (Zivin et al. 2009). Mortality rates and serious adverse events did not differ between groups with 17.5% and 17.4% mortality, 37.8% and 41.8% serious adverse events for TLT and sham, respectively. A third pivotal trial of TLT with refined baseline NIHSS exclusion criteria (<16) is now planned.

35.4.4.3 Neurodegenerative Diseases

A neurodegenerative disease is a disorder caused by the deterioration of certain nerve cells (neurons). Changes in these cells cause them to function abnormally, eventually bringing about their death. The diseases, Alzheimer's disease (AD), Parkinson's disease (PD), and amyotrophic lateral sclerosis (ALS), as well as multiple sclerosis (MS), are all due to neuronal degeneration in the central nervous system (Friedlander 2003). The chronic, unrelenting, progressive nature of these devastating degenerative diseases has motivated the search for therapies that could slow down or arrest the downward course experienced by most patients, and even more desirable would be a therapy that could actually reverse the neuronal damage. Transcranial light therapy is considered to have the potential to accomplish these goals.

35.4.4.4 Spinal Cord Repair

The spinal cord is a long, thin, tubular bundle of nerves that is an extension of the central nervous system from the brain and is enclosed in and protected by the bony vertebral column. Spinal cord injuries (SCI) can be caused by trauma to the spinal column. The vertebral bones or intervertebral disks can shatter, causing the spinal cord to be punctured by a sharp fragment of bone. Usually victims of spinal cord injuries will suffer loss of feeling in certain parts of their body.

As yet LLLT has only been studied in animal models of spinal cord injury. A pilot study by Rochkind (Rochkind et al. 2002) examined the effects of composite implants of cultured embryonal nerve cells and laser irradiation on the regeneration and repair of the completely transected spinal cord on rats. Three months after transection, implantation, and laser irradiation (with 780 nm, 250 mW, 30 min daily), intensive axonal sprouting occurred in the group with implantation and laser. The study suggests in vitro composite implants are a regenerative and reparative source for reconstructing the transected spinal cord. Postoperative low power laser irradiation enhances axonal sprouting and spinal cord repair. Rochkind's recent study showed postoperative 780 nm laser phototherapy enhanced the regenerative process of the peripheral nerve after reconnection of the nerve defect using a polyglycolic acid neurotube (Rochkind et al. 2007b). Byrnes et al. used 1600 J/cm² of 810 nm diode laser to improve healing and functionality in a T9 dorsal hemisection of the spinal cord in rats (Byrnes et al. 2005). Light (810 nm) was applied transcutaneously at the lesion site immediately after injury and daily for 14 consecutive days. A laser diode with an output power of 150 mW was used for the treatment. The daily dosage at the surface of the skin overlying the lesion site was 1589 J/cm². The average length of axonal regrowth in the rats in the light treatment groups with the hemisection and contusion injuries was significantly longer than the comparable untreated control groups (3.66 ± 0.26 mm, hemisection; 2.89 ± 0.84 mm, contusion). The total axon number in the light therapy groups was significantly higher compared to the untreated groups for both injury models ($p < 0.05$). Light therapy applied noninvasively promotes axonal regeneration and functional recovery in acute SCI caused by different types of trauma (Wu et al. 2009).

35.4.4.5 Peripheral Nerve Repair

The peripheral nervous system (PNS) consists of cranial and spinal nerves along with their associated ganglia. The nervous system is, on a small scale, primarily made up of neurons. Neurons exist in a number of different shapes and sizes and can be classified by their morphology and function. The anatomist Camillo Golgi grouped neurons into two types: type I with long axons used to move signals over long distances and type II without axons. Type I cells can be further divided by where the cell body or soma is located. The basic morphology of type I neurons, represented by spinal motor neurons, consists of a soma and a long thin axon, which is covered by the myelin sheath. Around the cell body is a branching dendritic tree that receives signals from other neurons. The end of the axon has branching terminals (axon terminal) that release transmitter substances into a gap called the synaptic cleft between the terminals and the dendrites of the next neuron (Squire et al. 2008).

Injury of a peripheral nerve frequently results in loss of sensory and motor functions, which leads to severe occupational and social consequences. Surgical repair is the preferred modality of treatment for the complete or severe peripheral nerve injury. In most cases, the repair can be successful if the surgery is performed in the first 6 months after injury, in comparison with long-term cases. Unfortunately, spontaneous recovery of long-term, severe, incomplete peripheral nerve injury is often unsatisfactory and involves degeneration of the axons and retrograde degeneration of the corresponding neurons of the spinal cord, followed by a very slow regeneration. Recovery may eventually occur, but it is slow and frequently incomplete. Laser phototherapy has been investigated for peripheral nerve repair (Gigo-Benato et al. 2005).

Animal studies showed that laser phototherapy had an immediate protective effect, maintained functional activity of the injured nerve, and decreased scar tissue formation at the injury site (Rochkind et al. 1987). Furthermore, there was decreased degeneration in corresponding motor neurons of the spinal cord, and significantly increased axonal growth and myelinization (Rochkind et al. 1990). A pilot clinical double-blind, placebo-controlled, randomized study showed was conducted in patients with incomplete long-term peripheral nerve injury (Rochkind et al. 2007a). Twenty-one consecutive daily sessions of laser (wavelength, 780 nm; power, 250 mW) or placebo irradiation were applied transcutaneously for 3 h to the injured peripheral nerve (energy density, 450 J/mm²) and for 2 h to the corresponding segments of the spinal cord (energy density, 300 J/mm²). The analysis of motor function ($p = 0.0001$) and electrophysiological analysis of recruitment of voluntary muscle activity ($p = 0.006$) showed statistically significant improvement in the laser-treated group compared to the placebo group.

35.4.5 Pain Relief

In recent years, there is growing interest in the use of laser biostimulation and phototherapy as a therapeutic modality for pain management. Alterations in neuronal activity have been suggested to play a role in pain relief by laser therapy. Many published reports documented the positive findings of the laser biostimulation in the pain management. This level of evidence relates to chronic neck pain (Chow and Barnsley 2005), tendonitis (Bjordal et al. 2006), chronic joint disorders (Bjordal et al.

2003), and chronic pain. Randomized controlled trials (RCTs) provided evidence for the efficacy of laser therapy in chronic low back pain. Pinheiro confirmed LLLT as an effective tool for trigeminal neuralgia, with treatment of 632.8, 670, and 830 nm diode lasers [132]. Reversal of diabetic neuropathy was associated with an immediate reduction in the absolute number of falls, a reduced fear of falling, and improved activities of daily living (Powell et al. 2006).

For clinical use, wavelength is generally recognized as one of the most important parameters. In the treatment of painful conditions, both visible (e.g., $\lambda = 632.8$, 670 nm) and infrared (e.g., $\lambda = 780$, 810–830, 904 nm) wavelengths have been used. Post-herpetic neuralgia is a particularly troublesome neurogenic pain that develops after shingles. Surgery in general has little to offer in this condition, so Yaksich et al. (Yaksich et al. 1993) employed low dose laser (He-Ne laser (Iijima et al. 1991)) therapy and showed that it was of value in a small series of cases. This has been both in the early phase soon after the vesicles have cleared and often late, many years after the onset of the pain. A good improvement rate was reported in approximately 60% of cases.

35.4.6 Cardiovascular System

Cardiovascular disease is the biggest killer of mankind (at least in the developed world) and the sheer volume of morbidity and mortality resulting from cardiovascular disorders has encouraged researchers to test LLLT in many indications in cardiology.

35.4.6.1 Mitigation of Damage after Heart Attack

It is known that much of the damage to the heart muscle or myocardium that is suffered by victims of myocardial infarctions does not occur during the period of acute ischemia when the coronary artery is blocked and blood supply to the heart is compromised, but rather occurs during the period of reperfusion when blood supply is restored to the heart muscle. This ischemia-reperfusion injury has been studied for many years but there have not been many advances in preventing it or mitigating the damage it causes (Tissier et al. 2008). Based on the known property of LLLT to prevent apoptosis and reduce tissue death and its anti-inflammatory properties it was plausible that delivering light to the heart could lessen the amount of heart damage after a heart attack.

Dr. Oron first found LLLT of the infarcted area in the surgically exposed myocardium of experimentally induced MI in rats, at the correct energy, duration, and timing, markedly reduces the loss of myocardial tissue (Ad and Oron 2001, Yaakobi et al. 2001). A similar beneficial decrease in amount of heart damage after induced MI was seen in a dog model (Oron et al. 2001a, b). NIR lasers around 803 nm were used and a biphasic dose response was seen with a maximum effect at 6 mW/cm² and lesser effects at both 2.5 mW/cm² and at 20 mW/cm² (Oron et al. 2001a). This phenomenon may have an important beneficial effect on patients after acute MI or ischemic heart disease.

35.4.6.2 Applications of LLLT in Atherosclerotic Disease

A common treatment for atherosclerosis in the coronary arteries is balloon angioplasty, a process in which the luminal narrowing that restricts blood flow to the heart is widened by insertion of a balloon catheter that is subsequently inflated. Unfortunately, an almost inevitable consequence of the initially successful restoration of blood flow is the process of restenosis, in which the injured portion of the artery undergoes an uncontrolled healing response and the smooth muscle cells proliferate to create another blockage that may be worse than the original one. The insertion of metal stents was employed in an attempt to reduce restenosis, and in recent years drug-releasing stents have become popular. Kipshidze and coworkers reasoned that the inability of the vascular endothelium to regenerate and cover the denuded area at the site of arterial injury may contribute to restenosis. Intravascular LLLT might stimulate rapid endothelial regeneration after balloon injury and this was demonstrated in rabbits (Kipshidze et al. 1998) and in pigs (De Scheerder et al. 1998). A He-Ne laser delivered doses of about 50 mW over 1 min (De Scheerder et al. 2001) and they showed induction of inducible nitric oxide synthase expression and increase of cyclic GMP in the arterial wall [141]. Clinical trials were carried [142–145] out that demonstrated safety and showed significant reductions in restenosis at both medium and long-term follow-ups.

35.4.6.3 Laser Irradiation of Blood

This procedure originated in Russia and has been used in various forms as a possible treatment for many diseases. It has been carried out principally by intravascular irradiation as described in Section 35.4.8.2, but also by a pheresis technique (Podil'chak and Nevzhoda 1994) in which blood is temporarily removed from the body, irradiated, and returned into the circulation and by intranasal (Gao et al. 2008) and transcutaneous light delivery (Schindl et al. 1997). The basic effects are supposed to be normalization of hemokinetic parameters, in other words allowing the circulation especially the microcirculation to improve and return to a more normal state. There is also some evidence of improvement in immunological parameters. The procedure has been used to treat patients suffering from heart attack (Kipshidze et al. 1990), peptic ulcer (Khamrabaeva and Aliavi 2003), tuberculosis (Ovsiankina et al. 2000), peripheral vascular disease (Losev et al. 1992), and even schizophrenia (Kut'ko et al. 1992).

35.4.7 Dentistry

Dentists have long been considered to be one of the medical specialties most likely to use LLLT techniques in their practice. This may be partly because dentists already use lasers for many other dental procedures and many offices are already equipped with one or more high power lasers, and all that is needed to turn down the power and defocus the spot for photobiomodulation procedures.

35.4.7.1 Mucositis

Mucositis is the painful inflammation and ulceration of the mucous membranes lining the digestive tract, usually as an adverse effect of chemotherapy and radiotherapy treatment for cancer. The potential for lesion and pain control using laser (He-Ne) treatment was initially studied in France (Ciais et al. 1992). Wong reported that an 830 nm laser with 0.7–0.8 J/cm² significantly reduced the incidence and the severity of mucositis in chemotherapy patients (Wong and Wilder-Smith 2002). Nes (Nes and Posso 2005) reported a significant reduction of mucositis pain, by using a GaAlAs laser, with a wavelength of 830 nm, a potency of 250 mW, and an energy given of 35 J/cm². Recently, Jaguar reported that patients received GaAlAs and diode laser therapy with 660 nm wavelength, power 10 mW, and the energy density delivered to the oral mucosa was 2.5 J/cm². LLLT significantly reduced the incidence and the severity of mucositis in chemotherapy patients (Jaguar et al. 2007).

35.4.7.2 Temporomandibular Joint Disorder

Temporomandibular joint (TMJ) disorder displays signs and symptoms that vary in their presentation and can involve muscles, nerves, tendons, ligaments, bones, connective tissue, and the teeth (Ficarra and Nassif 1991). Disorders of the muscles can present as trismus or limitation of jaw movement ranging from minor to severe. Muscle pain can sometimes be associated with trigger points and is termed myofascial pain syndrome. Pinheiro confirmed LLLT as an effective tool for disorders including TMJ pain, using 632.8, 670, and 830 nm diode lasers [132]. Mazzetto et al. (Mazzetto et al. 2007) demonstrated in a double-blind placebo-controlled clinical trial that infrared laser (780 nm, 70 mW, 10 s, 89.7 J/cm²), which applied in continuous mode on the affected temporomandibular region, inside the external auditive duct toward the retrodiskal region, twice a week, for 4 weeks, can decrease the pain level and lower sensitivity to palpation mainly for the active probe after eight treatments. Data showed a decrease in the pain level and lower sensitivity to palpation mainly for the active probe after eight treatments. Ficackova et al. (Fikackova et al. 2006) used a GaAlAs diode laser of 830 nm with an output power of 400 mW and an energy density of 15 J/cm² applied by contact mode on four targeted spots in 10 sessions. Reduction in pain and inflammation was observed and the effects of LLLT were confirmed by infrared thermography.

35.4.7.3 Dental Surgery

Many procedures carried out in dental surgery have been combined with LLLT to reduce pain and inflammation, and to hasten healing. Third molar extraction especially in the lower jaw leads to much pain and swelling and has been successfully treated with 4 J/cm² LLLT to the operative side intraorally 1 cm from the target tissue, and to the masseter muscle extraorally immediately after surgery (Aras and Gungormus 2009). Periodontal therapy (scaling and root planning) has been combined with LLLT (630 and 830 nm). In a published trial (Qadri et al. 2005), the clinical variables, that is, probing pocket depth, plaque and gingival indices were reduced more on the laser side than on the placebo. The total amount of matrix metalloproteinase (MMP-8, a marker of inflammation), increased on the placebo side but was lower on the laser side. LLLT has been studied as a method of encouraging the osseo-integration of dental implants in the bone of the jaw. An experimental study (Khadra et al. 2004) in rabbits used two coin-shaped titanium implants with a diameter of 6.25 mm and a height of 1.95 mm, which were implanted into cortical bone in the proximal tibia rabbits. GaAlAs laser was used immediately after surgery and carried out daily for 10 consecutive days. After 8 weeks, the mean tensile forces of the irradiated implants were significantly higher than controls and the histomorphometrical evaluation suggested that the irradiated group had more bone-to-implant contact than the controls.

35.4.7.4 Pain Relief

Adults undergoing endodontic surgery (*n* = 52) were randomized in a trial of LLLT pain reduction. Subsequently to suturing, 26 patients had the operation site treated with an 809 nm GaAlAs-laser at a power output of 50 mW and an irradiation time of 150 s. Laser treatment was simulated in further 26 patients. The results revealed that the pain level in the laser group was lower than in the placebo group throughout the 7 day follow-up period and was significant on the first postoperative day (Mann–Whitney *U*-test, *p* < 0.05) (Kreisler et al. 2004). Dentine hypersensitivity (the most frequently reported odontalgia) was treated with two types of lasers (660 and 830 nm) as dentine desensitizers (Ladalardo et al. 2004). A total of 40 teeth with cervical exposure were treated in four sessions. Dentine sensitivity to cold nociceptive stimulus was evaluated by means of a pain numeric scale from 0 to 10 before each treatment session, at 15 and 30 min after irradiation, and in a follow-up period of 15, 30, and 60 days after the end of treatment. Significant levels of dentinal desensitization were only found in patients ranging in age from 25 to 35 years. The 660 nm red diode laser was more effective than the 830 nm infrared laser and a higher level of desensitization was observed at the 15 and 30 min post-irradiation examinations.

35.4.8 Miscellaneous Indications

In this section, we will discuss some of the more controversial applications of LLLT. These applications tend to have more of the flavor of "alternative" or "complementary" medicine, and may partly explain why LLLT has had trouble in gaining mainstream acceptance among healthcare professionals.

35.4.8.1 Laser Acupuncture

Laser acupuncture is defined as the stimulation of traditional acupuncture points with low-intensity, nonthermal laser irradiation (Whittaker 2004). Broadly speaking laser acupuncture is used for similar indications as traditional needle acupuncture that was derived from Chinese medicine that described acupuncture meridians many centuries ago (Chen 1997). Although the use of laser acupuncture is rapidly gaining in popularity, objective evaluation of its efficacy in published studies is difficult because treatment parameters such as wavelength, irradiance, and beam profile are seldom fully described. The depth of laser energy transmission, likely an important determinant of efficacy, is governed not only by these parameters, but also by skin properties such as thickness, age, and pigmentation factors, which have also received little consideration in laser acupuncture. It has been pointed out that the usual method of LLLT application involves irradiating bulk tissues and there is a continuous spectrum of difference between that methodology and the tight-focused low power laser beams used in acupuncture (Chow 2006). A recent report showed visual cortex activation by laser acupuncture of foot points (Siedentopf et al. 2002). Moldovan (Moldovan 2007) could show measurable changes in thermal, electric, and optical properties of the LI4 (HEGU) acupoint, irradiated with a 685 nm, 30 mW laser not seen in adjacent non-acupoint areas in 27 healthy volunteers.

35.4.8.2 Tinnitus

LLLT targeting the inner ear has been discussed as a therapeutic procedure for cochlear dysfunction such as chronic cochlear tinnitus or sensorineural hearing loss. Often mastoidal and transmeatal irradiation has been performed without systematic light-dosimetric studies. A novel laser system, consisting of four diode lasers ($\lambda = 635$–830 nm) and a specific headset applicator, was developed. In a preliminary clinical study (Tauber et al. 2003), the TCL-system was applied to 35 patients with chronic tinnitus and sensorineural hearing loss, randomized to receive five single diode laser treatments ($\lambda = 635$ nm, 7.8 mW cw, $n = 17$ and $\lambda = 830$ nm, 20 mW cw, $n = 18$) with a space irradiation of 4 J/cm^2 site of maximal cochlear injury. After a follow-up period of 6 months, tinnitus loudness was attenuated in 13 of 35 irradiated patients, while 2 of 35 patients reported their tinnitus as totally absent. Hearing threshold levels and middle ear function remained unchanged. Another study (Gungor et al. 2008) used a 5 mW 650 nm laser or placebo laser, applied transmeatally for 15 min, once daily for a week. The loudness, duration, and degree of annoyance of tinnitus were improved, respectively, in up to 48.8%, 57.7%, and 55.5% of the patients in the active laser group. No significant improvement was observed in the placebo laser group.

35.5 Summary

Red and NIR light are absorbed in the mitochondria of mammalian cells. The energy deposited into mitochondrial chromophores is thought to release nitric oxide that is bound to members of the mitochondrial respiratory chain, in particular cytochrome c oxidase, and which inhibits respiration. The increased respiration has several effects inside the cell including an increase in the cell's energy supply (ATP) and an increase in ROS. The released nitric oxide may also have downstream signaling effects. These primary molecules (ATP, ROS, and NO) between them activate cellular transcription factors such as NF-κB and AP-1 that are responsible for the transcription of a host of gene products generally concerned with cell proliferation and survival, tissue repair, and reduction of pain and inflammation. The medical applications of LLLT are tremendously diverse. They include stimulation of healing in many situations of damage and trauma to multiple tissue types. LLLT is able to prevent cell apoptosis and tissue death in diverse diseases and situations where tissue is otherwise doomed to die. LLLT can reduce pain and inflammation in many traumatic, acute, and chronic injuries and diseases.

35.6 Future Perspective

It is only in relatively recent years that the basic molecular and cellular mechanisms of LLLT have begun to be understood. One of the reasons why LLLT has not found it easy to be accepted into mainstream medical practice is the lack of a clear mechanistic explanation for the beneficial effects that could be understood by biomedical scientists in these days of molecular biology and cell signaling pathways. We believe that further advances in mechanistic understanding will continue to be made in the near future. Moreover, there are a large number of optical parameters that can be carried in designing the light treatment for optimal therapeutic outcome. These include wavelength, fluence, power density, pulse duration and repetition rate, spot size, illumination time, coherence, polarization, and treatment timing. Moreover, it is well recognized that a biphasic dose response operates in many LLLT situations and a lower dose may be better for a particular patient than a higher dose. There will need to be a large number of clinical and experimental studies carried out before the effects of changing all these variables are completely understood. Nevertheless, as LLLT grows in acceptance in the future it is likely that more and more serious diseases will become amenable to phototherapy approaches including widespread killer diseases such as strokes and heart attacks.

Acknowledgment

Research in the Hamblin laboratory is supported by the U.S. NIH (grants R01CA/AI838801 and R01AI050875 to MRH).

References

Ad, N. and Oron, U. 2001. Impact of low level laser irradiation on infarct size in the rat following myocardial infarction. *Int. J. Cardiol.* 80: 109–116.

Aimbire, F., Ligeiro de Oliveira, A. P., Albertini, R. et al. 2008. Low level laser therapy (LLLT) decreases pulmonary microvascular leakage, neutrophil influx and IL-1beta levels in airway and lung from rat subjected to LPS-induced inflammation. *Inflammation* 31: 189–197.

Alexandratou, E., Yova, D., Handris, P., Kletsas, D., and Loukas, S. 2002. Human fibroblast alterations induced by low power laser irradiation at the single cell level using confocal microscopy. *Photochem. Photobiol. Sci.* 1: 547–552.

Almeida-Lopes, L., Rigau, J., Zangaro, R. A., Guidugli-Neto, J., and Jaeger, M. M. 2001. Comparison of the low level laser therapy effects on cultured human gingival fibroblasts proliferation using different irradiance and same fluence. *Laser Surg. Med.* 29: 179–184.

Alster, T. S. and Wanitphakdeedecha, R. 2009. Improvement of postfractional laser erythema with light-emitting diode photomodulation. *Dermatol. Surg.* 35: 813–815.

Angelo, M., Hausladen, A., Singel, D. J., and Stamler, J. S. 2008. Interactions of NO with hemoglobin: from microbes to man. *Methods Enzymol.* 436: 131–168.

Arany, P. R., Nayak, R. S., Hallikerimath, S. et al. 2007. Activation of latent TGF-beta1 by low-power laser in vitro correlates with increased TGF-beta1 levels in laser-enhanced oral wound healing. *Wound Repair Regen.* 15: 866–874.

Aras, M. H. and Gungormus, M. 2009. The effect of low-level laser therapy on trismus and facial swelling following surgical extraction of a lower third molar. *Photomed. Laser Surg.* 27: 21–24.

Baichwal, V. R. and Baeuerle, P. A. 1997. Activate NF-kappa B or die? *Curr. Biol.* 7: R94–R96.

Baxter, G. D., Walsh, D. M., Allen, J. M., Lowe, A. S., and Bell, A. J. 1994. Effects of low intensity infrared laser irradiation upon conduction in the human median nerve in vivo. *Exp. Physiol.* 79: 227–234.

Bingol, U., Altan, L., and Yurtkuran, M. 2005. Low-power laser treatment for shoulder pain. *Photomed. Laser Surg.* 23: 459–464.

Bjordal, J. M., Couppe, C., Chow, R. T., Tuner, J., and Ljunggren, E. A. 2003. A systematic review of low level laser therapy with location-specific doses for pain from chronic joint disorders. *Aust. J. Physiother.* 49: 107–116.

Bjordal, J. M., Lopes-Martins, R. A., and Iversen, V. V. 2006. A randomised, placebo controlled trial of low level laser therapy for activated Achilles tendinitis with microdialysis measurement of peritendinous prostaglandin E2 concentrations. *Br. J. Sports Med.* 40: 76–80; discussion 76–80.

Borutaite, V., Budriunaite, A., and Brown, G. C. 2000. Reversal of nitric oxide-, peroxynitrite- and S-nitrosothiol-induced inhibition of mitochondrial respiration or complex I activity by light and thiols. *Biochim. Biophys. Acta* 1459: 405–412.

Boschi, E. S., Leite, C. E., Saciura, V. C. et al. 2008. Anti-inflammatory effects of low-level laser therapy (660 nm) in the early phase in carrageenan-induced pleurisy in rat. *Lasers Surg. Med.* 40: 500–508.

Brosseau, L., Robinson, V., Wells, G. et al. 2005. Low level laser therapy (Classes I, II and III) for treating rheumatoid arthritis. *Cochrane Database Syst. Rev.* CD002049.

Brosseau, L., Welch, V., Wells, G. et al. 2003. Low level laser therapy (Classes I, II and III) for treating osteoarthritis. *Cochrane Database Syst. Rev.* CD002046.

Brown, G. C. 2001. Regulation of mitochondrial respiration by nitric oxide inhibition of cytochrome c oxidase. *Biochim. Biophys. Acta* 1504: 46–57.

Byrnes, K. R., Barna, L., Chenault, V. M. et al. 2004. Photobiomodulation improves cutaneous wound healing in an animal model of type II diabetes. *Photomed. Laser Surg.* 22: 281–290.

Byrnes, K. R., Waynant, R. W., Ilev, I. K. et al. 2005. Light promotes regeneration and functional recovery and alters the immune response after spinal cord injury. *Lasers Surg. Med.* 36: 171–185.

Carnevalli, C. M., Soares, C. P., Zangaro, R. A., Pinheiro, A. L., and Silva, N. S. 2003. Laser light prevents apoptosis in Cho K-1 cell line. *J. Clin. Laser Med. Surg.* 21: 193–196.

Castano, A. P., Dai, T., Yaroslavsky, I. et al. 2007. Low-level laser therapy for zymosan-induced arthritis in rats: Importance of illumination time. *Lasers Surg. Med.* 39: 543–550.

Chang, W. D., Wu, J. H., Jiang, J. A., Yeh, C. Y., and Tsai, C. T. 2008. Carpal tunnel syndrome treated with a diode laser: A controlled treatment of the transverse carpal ligament. *Photomed. Laser Surg.* 26: 551–557.

Chen, A. C. -H., Arany, P. R., Huang, Y. -Y. et al. 2009a. Low level laser therapy activates NF-kB via generation of reactive oxygen species in mouse embryonic fibroblasts. In: *Mechanisms for Low-Light Therapy IV*, eds. M. R. Hamblin, J. J. Anders, and R. W. Waynant, Vol. 7165. Bellingham, WA: The International Society for Optical Engineering, pp. doi: 10.1117/12.809605.

Chen, A. C. -H., Huang, Y. -Y., Arany, P. R., and Hamblin, M. R. 2009b. Role of reactive oxygen species in low level light therapy. In: *Mechanisms for Low-Light Therapy IV*, eds. M. R. Hamblin, J. J. Anders, and R. W. Waynant, Vol. 7165. Bellingham, WA: The International Society for Optical Engineering, pp. doi: 10.1117/12.814890.

Chen, C. H., Tsai, J. L., Wang, Y. H. et al. 2008. Low-level laser irradiation promotes cell proliferation and mRNA expression of type I collagen and decorin in porcine Achilles tendon fibroblasts in vitro. *J. Orthop. Res.* 27: 646–650.

Chen, Y. 1997. Silk scrolls: Earliest literature of meridian doctrine in ancient China. *Acupunct. Electrother. Res.* 22: 175–189.

Chow, R. 2006. Laser acupuncture studies should not be included in systematic reviews of phototherapy. *Photomed. Laser Surg.* 24: 69.

Chow, R. T. and Barnsley, L. 2005. Systematic review of the literature of low-level laser therapy (LLLT) in the management of neck pain. *Lasers Surg. Med.* 37: 46–52.

Chow, R. T., David, M. A., and Armati, P. J. 2007. 830 nm laser irradiation induces varicosity formation, reduces mitochondrial membrane potential and blocks fast axonal flow in small and medium diameter rat dorsal root ganglion neurons: Implications for the analgesic effects of 830 nm laser. *J. Peripher. Nerv. Syst.* 12: 28–39.

Ciais, G., Namer, M., Schneider, M. et al. 1992. [Laser therapy in the prevention and treatment of mucositis caused by anti-cancer chemotherapy]. *Bull. Cancer* 79: 183–191.

D'Angio, C. T. and Finkelstein, J. N. 2000. Oxygen regulation of gene expression: A study in opposites. *Mol. Genet. Metab.* 71: 371–380.

de Araujo, C. E., Ribeiro, M. S., Favaro, R., Zezell, D. M., and Zorn, T. M. 2007. Ultrastructural and autoradiographical analysis show a faster skin repair in He-Ne laser-treated wounds. *J. Photochem. Photobiol. B* 86: 87–96.

De Scheerder, I. K., Wang, K., Zhou, X. R. et al. 1998. Intravascular low power red laser light as an adjunct to coronary stent implantation evaluated in a porcine coronary model. *J. Invasive Cardiol.* 10: 263–268.

De Scheerder, I. K., Wang, K., Zhou, X. R. et al. 2001. Optimal dosing of intravascular low-power red laser light as an adjunct to coronary stent implantation: Insights from a porcine coronary stent model. *J. Clin. Laser Med. Surg.* 19: 261–265.

Demidova-Rice, T. N., Salomatina, E. V., Yaroslavsky, A. N., Herman, I. M., and Hamblin, M. R. 2007. Low-level light stimulates excisional wound healing in mice. *Lasers Surg. Med.* 39: 706–715.

Djavid, G. E., Mehrdad, R., Ghasemi, M. et al. 2007. In chronic low back pain, low level laser therapy combined with exercise is more beneficial than exercise alone in the long term: A randomised trial. *Aust. J. Physiother.* 53: 155–160.

Droge, W. 2002. Free radicals in the physiological control of cell function. *Physiol. Rev.* 82: 47–95.

Eells, J. T., Henry, M. M., Summerfelt, P. et al. 2003. Therapeutic photobiomodulation for methanol-induced retinal toxicity. *Proc. Natl. Acad. Sci. U S A* 100: 3439–3444.

Ficarra, B. J. and Nassif, N. J. 1991. Temporomandibular joint syndrome: Diagnostician's dilemma—A review. *J. Med.* 22: 97–121.

Fikackova, H., Dostalova, T., Vosicka, R. et al. 2006. Arthralgia of the temporomandibular joint and low-level laser therapy. *Photomed. Laser Surg.* 24: 522–527.

Friedlander, R. M. 2003. Apoptosis and caspases in neurodegenerative diseases. *N. Engl. J. Med.* 348: 1365–1375.

Gao, X., Zhi, P. K., and Wu, X. J. 2008. Low-energy semiconductor laser intranasal irradiation of the blood improves blood coagulation status in normal pregnancy at term. *Nan Fang Yi Ke Da Xue Xue Bao* 28: 1400–1401.

Gavish, L., Perez, L., and Gertz, S. D. 2006. Low-level laser irradiation modulates matrix metalloproteinase activity and gene expression in porcine aortic smooth muscle cells. *Lasers Surg. Med.* 38: 779–786.

Gigo-Benato, D., Geuna, S., and Rochkind, S. 2005. Phototherapy for enhancing peripheral nerve repair: A review of the literature. *Muscle Nerve* 31: 694–701.

Giuffre, A., Forte, E., Brunori, M., and Sarti, P. 2005. Nitric oxide, cytochrome c oxidase and myoglobin: Competition and reaction pathways. *FEBS Lett.* 579: 2528–2532.

Gordon, M. W. 1960. The correlation between in vivo mitochondrial changes and tryptophan pyrrolase activity. *Arch. Biochem. Biophys.* 91: 75–82.

Greco, M., Guida, G., Perlino, E., Marra, E., and Quagliariello, E. 1989. Increase in RNA and protein synthesis by mitochondria irradiated with helium-neon laser. *Biochem. Biophys. Res. Commun.* 163: 1428–1434.

Gungor, A., Dogru, S., Cincik, H., Erkul, E., and Poyrazoglu, E. 2008. Effectiveness of transmeatal low power laser irradiation for chronic tinnitus. *J. Laryngol. Otol.* 122: 447–451.

Gur, A., Karakoc, M., Nas, K. et al. 2002. Efficacy of low power laser therapy in fibromyalgia: A single-blind, placebo-controlled trial. *Lasers Med. Sci.* 17: 57–61.

Harman, D. 1956. Aging: A theory based on free radical and radiation chemistry. *J. Gerontol.* 11: 298–300.

Hawkins, D. and Abrahamse, H. 2006. Effect of multiple exposures of low-level laser therapy on the cellular responses of wounded human skin fibroblasts. *Photomed. Laser Surg.* 24: 705–714.

Henkel, T., Machleidt, T., Alkalay, I. et al. 1993. Rapid proteolysis of I kappa B-alpha is necessary for activation of transcription factor NF-kappa B. *Nature* 365: 182–185.

Hoffmann, A., Levchenko, A., Scott, M. L., and Baltimore, D. 2002. The IkappaB-NF-kappaB signaling module: Temporal control and selective gene activation. *Science* 298: 1241–1245.

Hou, J. F., Zhang, H., Yuan, X. et al. 2008. In vitro effects of low-level laser irradiation for bone marrow mesenchymal stem cells: proliferation, growth factors secretion and myogenic differentiation. *Lasers Surg. Med.* 40: 726–733.

Houreld, N. and Abrahamse, H. 2007. In vitro exposure of wounded diabetic fibroblast cells to a helium-neon laser at 5 and 16 J/cm². *Photomed. Laser Surg.* 25: 78–84.

Huang, P. J., Huang, Y. C., Su, M. F. et al. 2007. In vitro observations on the influence of copper peptide aids for the LED photo-irradiation of fibroblast collagen synthesis. *Photomed. Laser Surg.* 25: 183–190.

Iijima, K., Shimoyama, N., Shimoyama, M., and Mizuguchi, T. 1991. Evaluation of analgesic effect of low-power He:Ne laser on postherpetic neuralgia using VAS and modified McGill pain questionnaire. *J. Clin. Laser Med. Surg.* 9: 121–126.

Jaguar, G. C., Prado, J. D., Nishimoto, I. N. et al. 2007. Low-energy laser therapy for prevention of oral mucositis in hematopoietic stem cell transplantation. *Oral Dis.* 13: 538–543.

Kana, J. S., Hutschenreiter, G., Haina, D., and Waidelich, W. 1981. Effect of low-power density laser radiation on healing of open skin wounds in rats. *Arch. Surg.* 116: 293–296.

Karu, T. 1999a. Primary and secondary mechanisms of action of visible to near-IR radiation on cells. *J. Photochem. Photobiol. B* 49: 1–17.

Karu, T., Pyatibrat, L., and Kalendo, G. 1995. Irradiation with He-Ne laser increases ATP level in cells cultivated in vitro. *J. Photochem. Photobiol. B* 27: 219–223.

Karu, T. I. 1999b. A suitable model for wound healing: How many times are we to stumble over the same block? *Lasers Surg. Med.* 25: 283–284.

Karu, T. I. and Kolyakov, S. F. 2005. Exact action spectra for cellular responses relevant to phototherapy. *Photomed. Laser Surg.* 23: 355–361.

Karu, T. I., Pyatibrat, L. V., and Afanasyeva, N. I. 2005. Cellular effects of low power laser therapy can be mediated by nitric oxide. *Lasers Surg. Med.* 36: 307–314.

Karu, T. I., Pyatibrat, L. V., and Kalendo, G. S. 2001. Cell attachment to extracellular matrices is modulated by pulsed radiation at 820 nm and chemicals that modify the activity of enzymes in the plasma membrane. *Lasers Surg. Med.* 29: 274–281.

Khadra, M., Ronold, H. J., Lyngstadaas, S. P., Ellingsen, J. E., and Haanaes, H. R. 2004. Low-level laser therapy stimulates bone-implant interaction: An experimental study in rabbits. *Clin. Oral Implants Res.* 15: 325–332.

Khamrabaeva, F. I. and Aliavi, A. L. 2003. Laser infrared irradiation in the complex treatment of gastroduodenal ulcer. *Vopr Kurortol Fizioter Lech Fiz Kult* 33–35.

Khanna, A., Shankar, L. R., Keelan, M. H. et al. 1999. Augmentation of the expression of proangiogenic genes in cardiomyocytes with low dose laser irradiation in vitro. *Cardiovasc. Radiat. Med.* 1: 265–269.

Kipshidze, N., Nikolaychik, V., Keelan, M. H. et al. 2001. Low-power helium: Neon laser irradiation enhances production of vascular endothelial growth factor and promotes growth of endothelial cells in vitro. *Lasers Surg. Med.* 28: 355–364.

Kipshidze, N., Sahota, H., Komorowski, R., Nikolaychik, V., and Keelan, M. H. Jr. 1998. Photoremodeling of arterial wall reduces restenosis after balloon angioplasty in an atherosclerotic rabbit model. *J. Am. Coll. Cardiol.* 31: 1152–1157.

Kipshidze, N. N., Chapidze, G. E., Bokhua, M. R. et al. 1990. Intravascular laser therapy of acute myocardial infarction. *Angiology* 41: 801–808.

Kreisler, M. B., Haj, H. A., Noroozi, N., and Willershausen, B. 2004. Efficacy of low level laser therapy in reducing post-operative pain after endodontic surgery—A randomized double blind clinical study. *Int. J. Oral Maxillofac. Surg.* 33: 38–41.

Kut'ko, I. I., Pavlenko, V. V., and Voronkov, E. G. 1992. Use of intravascular laser irradiation of blood in the treatment of drug-resistant forms of schizophrenia. *Zh Nevropatol Psikhiatr Im S S Korsakova* 92: 53–56.

Ladalardo, T. C., Pinheiro, A., Campos, R. A. et al. 2004. Laser therapy in the treatment of dentine hypersensitivity. *Braz. Dent. J.* 15: 144–150.

Lampl, Y. 2007. Laser treatment for stroke. *Expert Rev. Neurother.* 7: 961–965.

Lane, N. 2006. Cell biology: Power games. *Nature* 443: 901–903.

Lavi, R., Sinyakov, M., Samuni, A. et al. 2004. ESR detection of 1O2 reveals enhanced redox activity in illuminated cell cultures. *Free Radic. Res.* 38: 893–902.

Leavitt, M., Charles, G., Heyman, E., and Michaels, D. 2009. HairMax LaserComb(R) laser phototherapy device in the treatment of male androgenetic alopecia: A randomized, double-blind, sham device-controlled, multicentre trial. *Clin. Drug Investig.* 29: 283–292.

Lee, S. Y., You, C. E., and Park, M. Y. 2007. Blue and red light combination LED phototherapy for acne vulgaris in patients with skin phototype IV. *Lasers Surg. Med.* 39: 180–188.

Li, N. and Karin, M. 1999. Is NF-kappaB the sensor of oxidative stress? *FASEB J.* 13: 1137–1143.

Longo, L., Simunovic, Z., Postiglione, M., and Postiglione, M. 1997. Laser therapy for fibromyositic rheumatisms. *J. Clin. Laser Med. Surg.* 15: 217–220.

Lopes-Martins, R. A., Albertini, R., Lopes-Martins, P. S. et al. 2006. Steroid receptor antagonist mifepristone inhibits the anti-inflammatory effects of photoradiation. *Photomed. Laser Surg.* 24: 197–201.

Losev, R. Z., Tsarev, O. A., and Gur'ianov, A. M. 1992. Intravascular laser irradiation of blood in multimodal treatment of patients with obliterating diseases of the lower limb vessels. *Grud Serdechnososudistaia Khir* 34–37.

Lowe, A. S., Walker, M. D., O'Byrne, M., Baxter, G. D., and Hirst, D. G. 1998. Effect of low intensity monochromatic light therapy (890 nm) on a radiation-impaired, wound-healing model in murine skin. *Lasers Surg. Med.* 23: 291–298.

Lubart, R., Eichler, M., Lavi, R., Friedman, H., and Shainberg, A. 2005. Low-energy laser irradiation promotes cellular redox activity. *Photomed. Laser Surg.* 23: 3–9.

Lyons, R. F., Abergel, R. P., White, R. A. et al. 1987. Biostimulation of wound healing in vivo by a helium-neon laser. *Ann. Plast. Surg.* 18: 47–50.

Maiya, G. A., Kumar, P., and Rao, L. 2005. Effect of low intensity helium-neon (He-Ne) laser irradiation on diabetic wound healing dynamics. *Photomed. Laser Surg.* 23: 187–190.

Mazzetto, M. O., Carrasco, T. G., Bidinelo, E. F., de Andrade Pizzo, R. C., and Mazzetto, R. G. 2007. Low intensity laser application in temporomandibular disorders: A phase I double-blind study. *Cranio* 25: 186–192.

Medrado, A. P., Soares, A. P., Santos, E. T., Reis, S. R., and Andrade, Z. A. 2008. Influence of laser photobiomodulation upon connective tissue remodeling during wound healing. *J. Photochem. Photobiol. B* 92: 144–152.

Medrado, A. R., Pugliese, L. S., Reis, S. R., and Andrade, Z. A. 2003. Influence of low level laser therapy on wound healing and its biological action upon myofibroblasts. *Lasers Surg. Med.* 32: 239–244.

Mester, E., Szende, B., and Gartner, P. 1968. The effect of laser beams on the growth of hair in mice. *Radiobiol. Radiother. (Berl.)* 9: 621–626.

Moldovan, C. 2007. Biophysics behavior of acupuncture points irradiated with low energy lasers. *Rom. J. Intern. Med.* 45: 281–285.

Moore, P., Ridgway, T. D., Higbee, R. G., Howard, E. W., and Lucroy, M. D. 2005. Effect of wavelength on low-intensity laser irradiation-stimulated cell proliferation in vitro. *Lasers Surg. Med.* 36: 8–12.

Neiburger, E. J. 1999. Rapid healing of gingival incisions by the helium-neon diode laser. *J. Mass Dent. Soc.* 48: 8–13, 40.

Nes, A. G. and Posso, M. B. 2005. Patients with moderate chemotherapy-induced mucositis: Pain therapy using low intensity lasers. *Int. Nurs. Rev.* 52: 68–72.

Omi, T., Kawana, S., Sato, S. et al. 2005. Cutaneous immunological activation elicited by a low-fluence pulsed dye laser. *Br. J. Dermatol.* 153 (Suppl 2): 57–62.

Oron, A., Oron, U., Streeter, J. et al. 2007. Low-level laser therapy applied transcranially to mice following traumatic brain injury significantly reduces long-term neurological deficits. *J. Neurotrauma* 24: 651–656.

Oron, U., Yaakobi, T., Oron, A. et al. 2001a. Attenuation of infarct size in rats and dogs after myocardial infarction by low-energy laser irradiation. *Lasers Surg. Med.* 28: 204–211.

Oron, U., Yaakobi, T., Oron, A. et al. 2001b. Low-energy laser irradiation reduces formation of scar tissue after myocardial infarction in rats and dogs. *Circulation* 103: 296–301.

Ovsiankina, E. S., Firsova, V. A., Dobkin, V. G., and Rusakova, L. I. 2000. Intravenous laser radiation treatment of acute and progressive forms of tuberculosis in teenagers. *Probl. Tuberk* 14–17.

Pal, G., Dutta, A., Mitra, K. et al. 2007. Effect of low intensity laser interaction with human skin fibroblast cells using fiber-optic nano-probes. *J. Photochem. Photobiol. B* 86: 252–261.

Papageorgiou, P., Katsambas, A., and Chu, A. 2000. Phototherapy with blue (415 nm) and red (660 nm) light in the treatment of acne vulgaris. *Br. J. Dermatol.* 142: 973–978.

Passarella, S., Casamassima, E., Molinari, S. et al. 1984. Increase of proton electrochemical potential and ATP synthesis in rat liver mitochondria irradiated in vitro by helium-neon laser. *FEBS Lett.* 175: 95–99.

Passarella, S., Ostuni, A., Atlante, A., and Quagliariello, E. 1988. Increase in the ADP/ATP exchange in rat liver mitochondria irradiated in vitro by helium-neon laser. *Biochem. Biophys. Res. Commun.* 156: 978–986.

Pastore, D., Greco, M., Petragallo, V. A., and Passarella, S. 1994. Increase in <−H+/e- ratio of the cytochrome c oxidase reaction in mitochondria irradiated with helium-neon laser. *Biochem. Mol. Biol. Int.* 34: 817–826.

Pereira, A. N., Eduardo Cde, P., Matson, E., and Marques, M. M. 2002. Effect of low-power laser irradiation on cell growth and procollagen synthesis of cultured fibroblasts. *Lasers Surg. Med.* 31: 263–267.

Pinheiro, A. L., Pozza, D. H., Oliveira, M. G., Weissmann, R., and Ramalho, L. M. 2005. Polarized light (400–2000 nm) and non-ablative laser (685 nm): A description of the wound healing process using immunohistochemical analysis. *Photomed. Laser Surg.* 23: 485–492.

Podil'chak, M. D. and Nevzhoda, O. A. 1994. The use of continuous plasmapheresis and extracorporeal laser irradiation of the blood in treating diabetic angiopathies of the lower extremities. *Klin Khir* 27–29.

Pontinen, P. J., Aaltokallio, T., and Kolari, P. J. 1996. Comparative effects of exposure to different light sources (He-Ne laser, InGaAl diode laser, a specific type of noncoherent LED) on skin blood flow for the head. *Acupunct. Electrother. Res.* 21: 105–118.

Poon, V. K., Huang, L., and Burd, A. 2005. Biostimulation of dermal fibroblast by sublethal Q-switched Nd:YAG 532 nm laser: Collagen remodeling and pigmentation. *J. Photochem. Photobiol. B* 81: 1–8.

Powell, M. W., Carnegie, D. H., and Burke, T. J. 2006. Reversal of diabetic peripheral neuropathy with phototherapy (MIRE) decreases falls and the fear of falling and improves activities of daily living in seniors. *Age Ageing* 35: 11–16.

Qadri, T., Miranda, L., Tuner, J., and Gustafsson, A. 2005. The short-term effects of low-level lasers as adjunct therapy in the treatment of periodontal inflammation. *J. Clin. Periodontol.* 32: 714–719.

Rochkind, S., Barrnea, L., Razon, N., Bartal, A., and Schwartz, M. 1987. Stimulatory effect of He-Ne low dose laser on injured sciatic nerves of rats. *Neurosurgery* 20: 843–847.

Rochkind, S., Drory, V., Alon, M., Nissan, M., and Ouaknine, G. E. 2007a. Laser phototherapy (780 nm), a new modality in treatment of long-term incomplete peripheral nerve injury: A randomized double-blind placebo-controlled study. *Photomed. Laser Surg.* 25: 436–442.

Rochkind, S., Leider-Trejo, L., Nissan, M. et al. 2007b. Efficacy of 780-nm laser phototherapy on peripheral nerve regeneration after neurotube reconstruction procedure (double-blind randomized study). *Photomed. Laser Surg.* 25: 137–143.

Rochkind, S., Shahar, A., Amon, M., and Nevo, Z. 2002. Transplantation of embryonal spinal cord nerve cells cultured on biodegradable microcarriers followed by low power laser irradiation for the treatment of traumatic paraplegia in rats. *Neurol. Res.* 24: 355–360.

Rochkind, S., Vogler, I., and Barr-Nea, L. 1990. Spinal cord response to laser treatment of injured peripheral nerve. *Spine* 15: 6–10.

Roelandts, R. 2002. The history of phototherapy: Something new under the sun? *J. Am. Acad. Dermatol.* 46: 926–930.

Rubinov, A. N. and Afanas'ev, A. A. 2005. Nonresonance mechanisms of biological effects of coherent and incoherent light. *Opt. Spectrosc.* 98: 943–948.

Safavi, S. M., Kazemi, B., Esmaeili, M. et al. 2008. Effects of low-level He-Ne laser irradiation on the gene expression of IL-1beta, TNF-alpha, IFN-gamma, TGF-beta, bFGF, and PDGF in rat's gingiva. *Lasers Med. Sci.* 23: 331–335.

Schafer, F. Q. and Buettner, G. R. 2001. Redox environment of the cell as viewed through the redox state of the glutathione disulfide/glutathione couple. *Free Radic. Biol. Med.* 30: 1191–1212.

Schindl, L., Schindl, M., Polo, L. et al. 1997. Effects of low power laser-irradiation on differential blood count and body temperature in endotoxin-preimmunized rabbits. *Life Sci.* 60: 1669–1677.

Schreck, R., Grassmann, R., Fleckenstein, B., and Baeuerle, P. A. 1992. Antioxidants selectively suppress activation of NF-kappa B by human T-cell leukemia virus type I Tax protein. *J. Virol.* 66: 6288–6293.

Schreck, R., Rieber, P., and Baeuerle, P. A. 1991. Reactive oxygen intermediates as apparently widely used messengers in the activation of the NF-kappa B transcription factor and HIV-1. *EMBO J.* 10: 2247–2258.

Shefer, G., Partridge, T. A., Heslop, L. et al. 2002. Low-energy laser irradiation promotes the survival and cell cycle entry of skeletal muscle satellite cells. *J. Cell Sci.* 115: 1461–1469.

Shiva, S. and Gladwin, M. T. 2009. Shining a light on tissue NO stores: Near infrared release of NO from nitrite and nitrosylated hemes. *J. Mol. Cell Cardiol.* 46: 1–3.

Siedentopf, C. M., Golaszewski, S. M., Mottaghy, F. M. et al. 2002. Functional magnetic resonance imaging detects activation of the visual association cortex during laser acupuncture of the foot in humans. *Neurosci. Lett.* 327: 53–56.

Simunovic, Z., Trobonjaca, T., and Trobonjaca, Z. 1998. Treatment of medial and lateral epicondylitis—Tennis and golfer's elbow—with low level laser therapy: A multicenter double blind, placebo-controlled clinical study on 324 patients. *J. Clin. Laser Med. Surg.* 16: 145–151.

Sommer, A. P., Pinheiro, A. L., Mester, A. R., Franke, R. P., and Whelan, H. T. 2001. Biostimulatory windows in low-intensity laser activation: Lasers, scanners, and NASA's light-emitting diode array system. *J. Clin. Laser Med. Surg.* 19: 29–33.

Squire, L. R., Bloom, F., and Spitzer, N. 2008. *Fundamental Neuroscience*. St Louis, MO: Academic Press.

Stadler, I., Lanzafame, R. J., Evans, R. et al. 2001. 830-nm irradiation increases the wound tensile strength in a diabetic murine model. *Lasers Surg. Med.* 28: 220–226.

Stergioulas, A. 2008. Low-power laser treatment in patients with frozen shoulder: Preliminary results. *Photomed. Laser Surg.* 26: 99–105.

Storz, P. 2007. Mitochondrial ROS—Radical detoxification, mediated by protein kinase D. *Trends Cell Biol.* 17: 13–18.

Streeter, J., De Taboada, L., and Oron, U. 2004. Mechanisms of action of light therapy for stroke and acute myocardial infarction. *Mitochondrion* 4: 569–576.

Sutherland, J. C. 2002. Biological effects of polychromatic light. *Photochem. Photobiol.* 76: 164–170.

Tafur, J. and Mills, P. J. 2008. Low-intensity light therapy: Exploring the role of redox mechanisms. *Photomed. Laser Surg.* 26: 323–328.

Takezaki, S., Omi, T., Sato, S., and Kawana, S. 2006. Light-emitting diode phototherapy at 630 +/− 3 nm increases local levels of skin-homing T-cells in human subjects. *J. Nippon Med. Sch.* 73: 75–81.

Tauber, S., Schorn, K., Beyer, W., and Baumgartner, R. 2003. Transmeatal cochlear laser (TCL) treatment of cochlear dysfunction: A feasibility study for chronic tinnitus. *Lasers Med. Sci.* 18: 154–161.

Teasdale, G. M. and Graham, D. I. 1998. Craniocerebral trauma: Protection and retrieval of the neuronal population after injury. *Neurosurgery* 43: 723–737; discussion 737–738.

Tissier, R., Berdeaux, A., Ghaleh, B. et al. 2008. Making the heart resistant to infarction: How can we further decrease infarct size? *Front. Biosci.* 13: 284–301.

Turrens, J. F. 2003. Mitochondrial formation of reactive oxygen species. *J. Physiol.* 552: 335–344.

Wang, T., Zhang, X., and Li, J. J. 2002. The role of NF-kappaB in the regulation of cell stress responses. *Int. Immunopharmacol.* 2: 1509–1520.

Weintraub, M. I. 1997. Noninvasive laser neurolysis in carpal tunnel syndrome. *Muscle Nerve* 20: 1029–1031.

Whittaker, P. 2004. Laser acupuncture: Past, present, and future. *Lasers Med. Sci.* 19: 69–80.

Wong, S. F. and Wilder-Smith, P. 2002. Pilot study of laser effects on oral mucositis in patients receiving chemotherapy. *Cancer J.* 8: 247–254.

Woodruff, L. D., Bounkeo, J. M., Brannon, W. M. et al. 2004. The efficacy of laser therapy in wound repair: a meta-analysis of the literature. *Photomed. Laser Surg.* 22: 241–247.

Wu, X., Dmitriev, A. E., Cardoso, M. J. et al. 2009. 810 nm wavelength light: An effective therapy for transected or contused rat spinal cord. *Lasers Surg. Med.* 41: 36–41.

Yaakobi, T., Shoshany, Y., Levkovitz, S. et al. 2001. Long-term effect of low energy laser irradiation on infarction and reperfusion injury in the rat heart. *J. Appl. Physiol.* 90: 2411–2419.

Yaksich, I., Tan, L. C., and Previn, V. 1993. Low energy laser therapy for treatment of post-herpetic neuralgia. *Ann. Acad. Med. Singapore* 22: 441–442.

Yamazaki, M., Miura, Y., Tsuboi, R., and Ogawa, H. 2003. Linear polarized infrared irradiation using Super Lizer is an effective treatment for multiple-type alopecia areata. *Int. J. Dermatol.* 42: 738–740.

Yu, H. S., Chang, K. L., Yu, C. L., Chen, J. W., and Chen, G. S. 1996. Low-energy helium-neon laser irradiation stimulates interleukin-1 alpha and interleukin-8 release from cultured human keratinocytes. *J. Invest. Dermatol.* 107: 593–596.

Yu, H. S., Wu, C. S., Yu, C. L., Kao, Y. H., and Chiou, M. H. 2003. Helium-neon laser irradiation stimulates migration and proliferation in melanocytes and induces repigmentation in segmental-type vitiligo. *J. Invest. Dermatol.* 120: 56–64.

Yu, W., Naim, J. O., and Lanzafame, R. J. 1997a. Effects of photostimulation on wound healing in diabetic mice. *Lasers Surg. Med.* 20: 56–63.

Yu, W., Naim, J. O., McGowan, M., Ippolito, K., and Lanzafame, R. J. 1997b. Photomodulation of oxidative metabolism and electron chain enzymes in rat liver mitochondria. *Photochem. Photobiol.* 66: 866–871.

Zelickson, B. D., Kilmer, S. L., Bernstein, E. et al. 1999. Pulsed dye laser therapy for sun damaged skin. *Lasers Surg. Med.* 25: 229–236.

Zhang, J., Xing, D., and Gao, X. 2008. Low-power laser irradiation activates Src tyrosine kinase through reactive oxygen species-mediated signaling pathway. *J. Cell Physiol.* 217: 518–528.

Zhang, R., Mio, Y., Pratt, P. F. et al. 2009. Near infrared light protects cardiomyocytes from hypoxia and reoxygenation injury by a nitric oxide dependent mechanism. *J. Mol. Cell Cardiol.* 46: 4–14.

Zivin, J. A., Albers, G. W., Bornstein, N. et al. 2009. Effectiveness and safety of transcranial laser therapy for acute ischemic stroke. *Stroke* 40: 1359–1364.

Index

A

Abbe theory of image formation, 26
Acousto-optic tunable filters (AOTF)
 advantage, 140
 center wavelength, applied acoustic
  frequency, 140
 diffracted wavelength, 139
 noncollinear, 139–140
Adriamycin/cytoxan (AC), DOSI, 188–189
Affinity-based molecular probes
 agonist and antagonist molecular strategy
  cholecystokinin, 681
  somatostatin receptor agonist, 681
 aptamers, 684
 intrinsic imaging strategy
  Alzheimer's disease (AD), 680
  brain diseases, 680
 monoclonal antibodies
  GMC-5-193-containing
   nanoparticles, 682
  N-hydroxysulfosuccinimide
   (NHS), 681
  non-covalent labeling, 682
  pre-targeting, 682
  VEGFR, 681
 multivalent and polyvalent imaging
  strategy
  ligands number, 680
  pharmaceuticals, 681
  RGD multimers, 680–681
 non-peptide ligands
  folic acid, 683–684
  glucose derivatives, 684
 passive and active, 679–680
 peptide and peptidomimetic receptor
  ligand mimics
  automated solid-phase peptide
   synthesis, 683
  cell surface receptors, 683
  somatostatin (sst-14), 683
  $\alpha_v\beta_3$-integrin receptor, 683
 protein ligands
  Annexin V (AnxV), 683
  epidermal growth factor (EGF), 683
  VEGF, 682–683
 "vital stain" methodology, 681

ALA, see Aminolevuinic acid (ALA)
Algebraic reconstruction technique
 (ART), 348
Aminolevuinic acid (ALA), 737
Amplitude point spread function, 22
Angular limit of resolution, see Resolution
Angular spacing, 39
Angular spectrum of plane waves, 19
Angular width, 39
Aorta, 93, 95
APD, see Avalanche photodiodes (APD)
Approximation error method, 351
Arrayed detectors, see Charge coupled
 device (CCD)
Arterial-spin-labeled perfusion MRI
 (ASL-MRI), 204
Artificial neural network (ANN), 117
Atherosclerotic plaque
 chronic total occlusions, 305–306
 fibrocalcific aortic plaque, 306
 fibrous coronary plaque, 306
 hypothesis, 304
 lipid-rich carotid plaque, 306
 macrophage detection, 305
 OCT vs. IVUS, 304–305
 qualitative image analysis, 306–307
 red and white thrombi, 306–307
 signal characteristics, 305
ATR, see Attenuated total internal reflectance
Attenuated total internal reflectance (ATR),
 248–249
Avalanche photodiodes (APD), 56–57, 60,
 182–183

B

Barrett's esophagus, 310–312
Beer–Lambert law, 108
Beer's law, 166
Benzoporphrin derivative (BPD)
 sensitizer, 736
Bessel beam, 23
Bessel function, 168
Bile ducts, see Pancreatobiliary ductal
 system
Biochemical chromophore, 166
Bio-optics

collimated light sources, 329
diffuse optical tomography (DOT),
 319–320
diffusion vs. transport solution
 diffusion-theory-based simulation, 329
 fluence distribution, 328
 numerical simulation, 327
 phantom, 328
 void-like region, 328
finger joints light transmission
 no-RS and RA case, 331
 rheumatoid arthritis, 330
 small-tissue fluid-filled case, 331
 synovial fluid, 331
fluorescence modeling, FD blocking-off
 method, 332
numerical methods
 Delta-Eddington method, 324–325
 finite-element even-parity method, 327
 finite-volume discrete-ordinate
  method, 325–327
 introduction, 322
 irregular geometries, finite-difference
  methods, 324
 regular geometries, finite-difference
  methods, 322–324
 SNR analysis
  frequency dependence, 330
  phase measurement, 329
  short-noise model, 329–330
  source-modulation frequency, 330
  time-dependent data, 329
 radiative transfer equation (RTE)-based
  models, 321–322
Boundary discretization methods, 342–343
Bound water index (BWI), 184, 186
BPD sensitizer, see Benzoporphrin derivative
 (BPD) sensitizer
Breast cancer
 chemotherapy effects, DOSI
  adriamycin/cytoxan (AC), 188–189
  deoxy-hemoglobin (ctHHb), serial
   images of, 188
  neoadjuvant chemotherapy (NAC),
   187–188
  physiological changes, 188
  TOI, 187, 189

detection, DOSI
    BWI, 186
    hormone replacement therapy, 185
    mammography, sensitivity of, 185
    signature regions, 185
    tissue optical index, 185–186
    water content, 186–187
diffuse optical tomography
    cancerous lesions, 409
    detection rates, 410
    hemoglobin concentration, 409–410
    multichannel TCSPC imaging, 409
    optical mammography, 408–409
    single wavelength and source-detector
        separation, 409
    time-domain imager, 409
    x-ray mammography, 408
optical coherence tomography, 314–315
Brewster's angle, 47
Broadband DOS, *see* Diffuse optical
    spectroscopic imaging (DOSI)
Broyden–Fletcher–Goldfarb–Shanno (BFGS)
    algorithm, 349

C

Camera obscura, 143
Cancer therapy monitoring, DCS, 204–206
Cardiology, 269
    atherosclerotic plaque (*see* Atherosclerotic
        plaque, OCT)
    coronary stents, 307–308
    noncoronary applications, 308
CARS, *see* Coherent anti-Stokes Raman
    scattering
Cartilage, 89–90, 94
CCD, *see* Charge coupled device (CCD)
CCD array, *see* Charge-coupled device
    (CCD) array
Cells and subcellular structures, light
    scattering, 167–168
Cerebral physiology and disease
    cerebrovascular hemodynamics, 210
    CMRO$_2$ estimates, NIRS-DCS, 207, 209
    cortical blood flow responses, 207
    3D rCBF tomography, CSD, 207–208
    noninvasive CBF measurements, 206–207
Cerebrovascular hemodynamics, 210
CFPI, *see* Confocal Fabry–Perot
    interferometer
Charge coupled device (CCD)
    APD and PMT arrays, 60
    characteristics and applications, 58–59
    complementary metal-oxide-
        semiconductor (CMOS)
        detectors, 60
    intensified CCD cameras, 60–61
    operational principle, 58
    spectrometers, 59–60
Charge-coupled device (CCD) array
        detectors, 239–240
Chemotherapy effects, in breast cancer,
    187–189

Cholangiocarcinoma, 313
Chronic total occlusion (CTO), 305–306
Circularly symmetric aperture, 14–15, 22
CLASS, *see* Confocal light absorption
        and scattering spectroscopic
        microscopy
CLSM, *see* Confocal laser scanning
        microscopy
CMOS detectors, *see* Complementary metal-
        oxide-semiconductor (CMOS)
        detectors
CMRO$_2$ estimates, NIRS-DCS, 207, 209
Coded apertures, 143–146
Coded aperture snapshot spectral imagers
        (CASSI), 144–145
Coded image capturing, SE-SI systems
    apertures
        CCD, 144
        cyclic S-matrix, 144
        Hadarmard/S-matrices, 143–144
        light detection scheme, 144
        non-linear reconstruction methods,
            144–145
        operation, 143
        phase-coded snapshot imager,
            145–146
        pinhole camera, 143
        schematic and code, CASSI imager,
            144–145
    CT, 142–143
CO$_2$ gas laser, 54
Coherence, digital holographic microscopy
    crossing beams, mutual
        beam analysis principle, 622
        coherent holography, 622–623
        incoherent holography, 623–624
        object and reference wave, 621–622
        reduced coherence holography, 623
        scanning holography, 624
        source and specimen, 622
        space-time digital holography, 622
        two pupils system, 624
    exploitation
        "gating," 621
        spatial domain, 621
        time and spatial, 621
        ultra-broadband sources, 621
    Fourier transform, 620–621
    mutual, 620
    partial degree, 620
    photon gas, 620
    properties, 620
Coherence length, 36
Coherence reflectance confocal microscopy
    acousto-optic (AO) modulator, 538
    gating systems, 537–538
    optical layout, 537
Coherent anti-Stokes Raman scattering
        (CARS)
    advantages, 572–573
    anti-Stokes signal, 236
    emission techniques, 237
    process, 247

Coherently stimulated Raman scattering, 247
Coherent transfer function
    4*f* system, 26
    grating object, 27–29
    imaging of line structure, 27
    pupil function, 27
    square wave object, 29
Cold laser, *see* Low level laser/light therapy
        (LLLT)
Colon, OCT, 312
Complementary metal-oxide-semiconductor
        (CMOS) detectors, 60
Compound microscope, 9–10
Computed tomography (CT), 142–143
Computed tomography imaging spectrometer
        (CTIS), 143
Computer-generated hologram (CGH),
        142–143
Concave lenses, 44
Confocal Fabry–Perot interferometer (CFPI)
    absorptive object, 432, 435
    central frequency transducers, 433
    experimental setup, 432, 434
    focal width and length, water, 435
    resolutions, 432
    typical A-line, 432, 434
    typical axial and lateral edge spread
        functions, 433, 435
Confocal laser scanning microscopy (CLSM)
    advantage, 545
    illumination and pinhole detection,
        544–545
Confocal light absorption and scattering
        spectroscopic (CLASS) microscopy
    in cell biology, applications of, 176–177
    principle
        carboxylate-modified microspheres,
            fluorescence image, 175–176
        depth sectioning, 175
        prototype/fluorescence
            microscope, 175
    single gold nanorods
        concentration of, 178
        normalized scattering spectrum
            for, 178
        optical properties of, 177
    time sequence of, 177
Confocal microscopy
    imaging
        beam walk, 528
        biomedical, 518
        concept, 518–520
        field of view, 528
        out-of-focus light, 517–518
        resolution, 522–528
        scanning, 520–522
        signal-to-noise ratio (SNR), 528–529
        turbid tissue, 529–530
    instrumentation
        acquisition rate, 530
        axial scanning, 531
        coherence reflectance confocal
            microscopy, 537–538

in-plane scanning, 531
line scanning, 534–536
miniaturization, 530
multimodal, 531–534
optical sectioning/working
distance, 531
spectral encoding, 536–537
Constructive interference, 38
Continuous-wave (CW) subsystems, DOSI,
182–183
Coronary stents, OCT imaging, 307–308
Crohn's disease, 312
CTO, *see* Chronic total occlusion (CTO)
Cylindrical waves, 34

D

Dark-field microscopy
vs. bright-field microscopy, 490–491
cardioid condenser lens, 485
central dark-field microscopy, 485
contrast generation, 484
direct illumination light, 485
Fourier analysis, 485
limitations, 485–486
numerical aperture (NA), condenser, 485
phase contrast microscopy, 486
Data preprocessing
color microscopy image, 150–151
covariance matrix, 150
dimensionality issue curse, 150
eigenvectors linear combination, 150
PCA, 150–151
Defocused pupil function, 25
Defocus effect, 22–23
Dentistry, 270
Deoxy-hemoglobin (ctHHb), DOSI, 188
Depolarization
birefringence
attenuation, 659–661
microdomains, 659
retardance, 659
Stokes vector, 659–660
transmitted photons, 659
optical properties *vs.* transport
parameters, 658
retardance and scattering
coefficients, 658
mechanisms, 658–659
scattering
efficiency, 662
Monte Carlo simulations, 661
Stokes vector, 661
total transmitted light, 661–662
tissues
angular diffusivity, 665
behavior, 663
data analysis, 663
linear polarization, 664
and microsphere solution, linear, 664
near-on-axis transport, 662–663
Depth of field, 8
Depth of focus, 7–8

Dermatological applications, 744–745
Dermis, 82
Destructive interference, 38
Dichroic/polychroic optics, 141–142
Diffraction
diffraction grating as monochromator, 39
N-slit interference, 38–39
and resolution, 40
single-slit interference, 39–40
two-slit interference, 37–38
Diffraction-free beam, *see* Bessel beam
Diffraction grating, 39
Diffraction optics
angular spectrum of plane waves, 19
circularly symmetric aperture, 14–15, 22
coherent transfer function
4*f* system, 26
grating object, 27–29
imaging of line structure, 27
pupil function, 27
square wave object, 29
defocus effect, 22–23
evanescent waves, 19–20
focus of lens, 21–22
Fraunhofer diffraction
annular aperture, 15
circular aperture, 13–14
geometry, 12–13
rectangular aperture, 13
regimes of diffraction, 11–12
single slit, 13
Fresnel diffraction
circular aperture, 15–16
circular obstruction, 18
by edge, 18
field in plane calculation, 15
half-plane, 17–18
rectangular aperture, 16–17
regimes of diffraction, 11–12
single slit, 17
Huygens–Fresnel diffraction formula, 12
Huygens' principle, 11–12
image formation
defocused case, 25
Fourier transformation by lens, 26
4*f* system, 26
imaging by lens, 23–24
satisfied lens law, 24
incoherent imaging
object intensity, 30
optical transfer function (OTP), 30–31
partially coherent, 29
two-point object, 30
Kirchhoff diffraction integral, 18–19
phase screen, 20
spatial filtering, 29
thin lens, 21
Diffuse correlation spectroscopy (DCS), *see*
Near-infrared-diffuse correlation
spectroscopy (NIRS-DCS)
Diffuse optical spectroscopic imaging (DOSI)
advantages of, 182
analysis tool, development of, 189–190

breast cancer detection
BWI, 186
hormone replacement therapy, 185
mammography, sensitivity of, 185
signature regions, 185
TOI, 185–186
water content range, 186–187
calculation of
tissue bound water content, 183–184
tissue chromophore
concentration, 183
chemotherapy effects, breast tumor
adriamycin/cytoxan (AC), 188–189
deoxy-hemoglobin (ctHHb), serial
images of, 188
neoadjuvant chemotherapy (NAC),
187–188
physiological changes, 188
TOI, 187, 189
conventional radiologic method,
co-registration with, 190
diffusion regime, 181
formation of, 184–185
handpiece, standard grid pattern,
184–185
hemoglobin parameters, 181–182
instrumentation and theoretical
framework
FDPM and CW subsystems, 182–183
theoretical reflectance, calculation
of, 183
molecular-targeted exogenous contrast
utilization, 190
quantitative tissue measurement, 181–182
spatial localization improvement, 190
Diffuse optical tomography (DOT), 319–320
continuous wave (CW) imaging, 395
contrast-detail analysis, 388–389
cutting-edge systems, 373
early-arrival photons, 396
frequency-domain measurements
advantages, 397–398
intensity-only measurement, 396–397
TPSFs, infant, 397
in vivo applications
brain, 411–412
breast characterization and imaging,
408–411
functional imaging, depth sensitivity,
412–413
tomographic brain imaging, 413
in vivo imaging
brain, 384–387
breast cancer, imaging, 382–384
Jacobian size, change, 387–.388
light scattering, 359
Moore–Penrose inverse formulation, 387
NIR light, 387
nonlinear inverse problem, 339–340
nonlinear reconstructions, 363
numerical modeling and reconstruction
algorithm, 387
photodetectors selection, 387

pulsed laser diodes, 396
reconstruction, optical properties
  absolute measurement, 406–407
  relative measurement, 407
  tomographic imaging, 407–408
sparsity, 388
temporal point spread function (TPSF)
  absorption, scatter and refractive
    index, 403
  characterization, 404–406
  correlated noise, 404
  coupling losses, 403–404
  deconvolution, 403
  definition, 395
  measure, tissue, 396
  modeling, 404
  noise, 404
  pulse width, 403
  sensitivity, noise, 406
theory and instrumentation
  amplitude and phase shift
    measurement, 373
  CCD arrays, 375
  extension, CW measurements, 377
  heterodyne detection, 374–375
  LEDs and LDs, 375
  light propagation and image
    reconstruction, FD systems,
    375–377
  measurement techniques, 374
  MSIR, 378–380
  multimodality imaging, 380–382
  principle, 374
  spectral fitting, 377–378
  wavelengths, 380
time-domain instrumentation
  divisions, 398
  TCSPC (*see* Time-correlated single-
    photon counting)
  time-gated systems, 399–403
  time-gating techniques, 396
  time-to-amplitude converter (TAC)
    systems, 396
x-ray computed tomography (CT), 359
Diffusion approximation (DA), 339
Diffusion theory, 226–227
Digital holographic microscopy (DHM)
advantage, 636
biology
  cell-induced phase shift, 636
  marker-free transparent living
    specimen, 636
  multimodality imaging
    techniques, 637
  phase signals, 636
  vesicles' trafficking, 636–637
coherence concept
  crossing beam, mutual properties,
    621–624
  exploitation, 621
  Fourier transform, 620–621
  interference, waves, 621
  mutual, 620

partial degree, 620
photon gas, 620
properties, 620
confocal microscopy, 618
dielectric properties, 640
domains, 635–636
electronic focusing and field extended
  depth
  DHM, 633
  3D images analysis, 633
  reconstruction, 633
holography
  complex wave front restoration,
    628–629
  cross terms, 625–626
  digital, time domain data analysis, 626
  OCT, 626
  space domain data analysis, digital,
    626–628
  square terms, 624–625
instrument
  accuracies, 634
  acquisition rate, 635
  lateral resolution, 634
  off-axis approach, 633
  optical setup, 634
  phase image, 635
  recording, 633
  reference beam, 633–634
interference microscopy, 617–618
optical coherence tomography (OCT), 618
optical setup realization
  DHM lensless, 630
  DIHM, 629–630
  immersion MO, 630–631
  lensless, 629
quantitative phase and amplitude, 635
reference beam, 633–634
reflection geometry, 639
scanning techniques, 618
space-bandwidth adaptation
  detector/electronic camera, 632
  free-space propagation, 631
  off-axis arrangement, 632
  optical transfer function, 632–633
  principle, DHM, 631
  reference beam incident, 631–632
  resolution improvement paths, 632
  Wigner distribution function, 631
specimen full 3D imaging
  benefit, 637
  diffraction tomography, 638
  digital holography, 639
  illumination wave angle, 638
  optical tomographic microscopy, 638
  scattered wave reconstruction,
    638–639
  wavelength change, 637–638
structured illumination, 618
wave front
  3D distribution, 619
  diffraction, 619
  reconstruction, 618–619

Direct methods, light transport, 343–344
Discrete filters/spatially variable filters,
  135–136
Doppler OCT
  flow velocity range, 283
  frequency shift (f_D), 282
  multifunctional imaging, 288–289
  validation, flow measurements, 283
  velocity profiles, 283
  volume rendering, angle measurement,
    282–283
  volumetric reconstruction, 283–284
Doppler ultrasound, premature infant brain,
  203–204
Dosimetry
  delivered irradiance, 741–742
  fluorescence-based implicit dosimetry,
    743–744
  local tissue oxygenation measurement, 743
  measured fluence rate, 742
  sensitizer concentration measurements,
    742–743
  singlet oxygen production direct
    measurement, 744
DOT, *see* Diffuse optical tomography (DOT)
Dual window (DW) method; *see also*
  Fourier domain low coherence
    interferometry (fLCI)
  processing method validation
    absorption phantom, 474–475
    chi-squared values, 476–477
    high-frequency modulation removal,
      476–477
    imaging fLCI system, 474
    interferometric data, 474–475
    local oscillations, 477
    spectral cross-sections and time
      marginals, 476
    spectral resolution, TFD, 475
Duodenum, OCT, 312
Dye lasers, 55
Dysplasia diagnosis, esophagus, 310–311
Dysplasia, LSS, 171–172

E

Electromagnetic spectrum, 36, 49–50
Electromagnetic waves and wave motion
  coherence, 36
  electromagnetic spectrum, 36
  energy and momentum, 35–36
  light, 33–34
  in matter, 35
  polarization, 35
  superposition, 35
  waves, notation, 33–34
Electromechanical tunable filters
  (EMTF), 135
Electronically tunable filters (ETF), 135
Empirical analysis methods, fluorescence
  spectroscopy
  brain, 224
  bronchus, 224

cancer, 224
limitations, 223
normal and cancerous human breast
tissue, 223
oral cavity, 223–224
uterine cervix, 224
Endoscopic polarized scanning spectroscopy
(EPSS)
advantages, 172
backscattered spectrum, 173–174
block diagram, 172–173
clinical instrument, 173
nondysplastic BE site, 174
nuclear size distribution, 174
parallel and perpendicular
representation, 174
pathologic examination, 174
Endoscopic retrograde
cholangiopancreatography (ERCP)
imaging, 313
Energy and momentum, 35–36
Enhanced Raman scattering
applications, 246
coherently stimulated, 247
resonance applications, 246–247
SERS, 247–248
Epithelial/mucous tissue, 86–87, 90, 94
EPSS, *see* Endoscopic polarized scanning
spectroscopy
Erbium-and neodymium-based lasers, 54
Esophagus
adenocarcinoma, 308
dysplasia identification, 310–311
normal squamous mucosa and gastric
cardia, 309
second generation OCT imaging, 310–312
SIM, identification of, 309–310
in vivo studies, 308–309
Euclidian distance (ED) algorithm, 153
Evanescent waves, 19–20
Excimer lasers, 54
Excitation mechanism, laser, 53
Extraordinary transmission, 146
Eye piece, 9

F

FDML, *see* Fourier domain mode locking
FDOT, *see* Fluorescence diffuse optical
tomography (FDOT)
Fermat's principle, 3, 41–42
Fiber-connected probes, 241–242
Fiber-optic catheter/endoscope, 268
Fiber-optic probes
arrangement and collection tip, 202–203
contact, non-contact and catheter, 202
source and detector, 202
Fingerprint region, 234
Finite-difference methods
irregular geometries
adaptive-mesh-refinement
method, 324
body-fitted-grid method, 324

regular geometries
cell surface grid point, 322
diamond-differencing scheme, 323
Euler-backward step, 323–324
octants yield, 322–323
quadrature rule, 322
SI technique, 323
FLIM, *see* Fluorescence lifetime imaging
Fluorescence diffuse optical tomography
(FDOT), 340
Fluorescence lifetime imaging (FLIM)
cell biology
di-4-ANEPPDHQ, 598
FRET, 599–600
liquid-ordered phase, plasma
membrane, 598–599
molecular functional information,
597–598
endoscopy
beam scanning, 603–604
coherent fiber-bundle-based
flexible, 603
diagnostic screening and optically
guided biopsy, 603
fiber-optic probes and excitation, 604
multispectral confocal, 604
excitation sources
all-solid-state ultrafast, 597
laser technology, 596–597
multiphoton, 597
repetition rate, 597
tunable single-photon, 597
fluorescence
decay profile, 591
fluorophore, 590
lifetime, 590–591
parameters, 590
ratiometric techniques, 590
FRET (*see* Förster resonant energy
transfer)
laser scanning (*see* Laser scanning FLIM)
and MDFI, 607–608
optical projection tomography (OPT)
3-D fluorescence lifetime images, 605
3-D time-integrated fluorescence
intensity and lifetime, 607
intensity decay, 607
scattering phantoms, 607
silicon elastomer phantom
acquisition, 606
time-gated fluorescence data, 606–607
X–Z plane, 605–606
polarization-resolved, 608
tissue autofluorescence
cancer, 600
decay profiles, 601–602
diagnostic screening, 602
label-free imaging techniques, 601
liver, 602–603
multiphoton excitation, 602
principal, endogenous, 601
"steady-state" imaging techniques, 601
time-resolved kinetics, 602

tomography
3-D cell cultures, 604–605
DOT and DFT, 607
filtered back-projection
technique, 605
intensity decay profiles, 607
OPT, 605–607
time-gated imaging, 607
wide-field
frequency-domain approach, 595
high-speed FLIM, 596
rapid lifetime determination (RLD),
595–596
time gates, 595
Fluorescence LOT, 366–367
Fluorescence microscopy, SEI
Abbe's resolution definition, 544
CLSM, 544–545
drawback, 543
4Pi-microscopy
4π-geometry, 545
point-spread-function-engineer, 545
SIM method, conceptual basis
image formation, structured
illumination, 549–551
image formation, wide-field
microscopy, 547
image reconstruction, 551–555
and localization microscopy, 556–557
polarization effects, 547
resolution improvement, 555–556
setups, 555
three-beam interference, 549
two-beam interference, 547–549
STED, 545
wide-field methods
spatially modulated illumination
(SMI) microscopy, 545–546
structured illumination microscopy
(SIM), 546–547
Fluorescence resonance energy transfer
(FRET), *see* Förster resonant
energy transfer (FRET)
Fluorescence spectroscopy (FS)
cancer and endogenous fluorescence,
218–219
empirical analysis methods
brain, 224
bronchus, 224
cancer, 224
normal and cancerous breast
tissue, 223
oral cavity, 223–224
uterine cervix, 224
fluorophore localization, 218–219
imaging spectroscopy, 228
instrumentation
brain tumor demarcation system,
220–221
bronchial cancer detection system,
220–221
depth-sensitivity, 220
phantom study, 222

probe designs and configuration,
220–222
probe with ball lens configuration,
221–222
model-based analysis methods
combined analytical methods, 227
diffusion theory, 226–227
Monte Carlo, 227–228
probabilistic photon migration,
225–226
vs. Raman spectroscopy, 237–238
Focal points, 43
Focusing of lens, 21–22
4*f* optical system, 26
Förster resonant energy transfer (FRET)
cell biology
FLIM, 600
fluorescence intensity and lifetime
images, 599–600
immunological synapse (IS), 599
Nipkow disc FLIM microscope, 600
small molecule dyes, 599
definition, 674
description, 591
efficiency, 591
hybrid molecular imaging, 675
measurements, 591–592
molecular cell biology, 592
monoclonal antibody, fluorescein,
674–675
nonspecific uptake, molecular probes, 675
sensitivity, 674
Foscan, *see* Meso-tetra hydroxyphenol
chlorin (mTHPC)
Fourier domain low coherence
interferometry (fLCI)
animal tissue studies
basal layer nuclear diameters, 479
depth-resolved spectroscopic plot,
478, 480
histopathology image and depth plots,
478, 480
left cheek pouch, 477
neoplastic transformation, 478
nuclear sizing algorithm, 481
raw data, 478–479
refractive index variation, 478
autocorrelation metric, 465
axial spatial cross correlation
function, 464
backscattered light, 463
cancer detection, 463
correlated scatterers, 479
depth-resolved spectroscopic
information, 464
dual window (DW) method
Cohen's class functions, 467
description, 466–467
processed signal, 467
properties, 467–469
standard deviation, 467
elastic scattering spectroscopy (ESS), 463
imaging

depth scan, 474
4f interferometer, 472–473
hamster cheek pouch epithelium,
scatter-mode image, 474
modes, number, 473
modified Michelson interferometer
geometry, 472
parallel frequency domain OCT
(pfdOCT) imaging, 473
raw interference spectra, 473–474
scatter-mode, 473
spectrometer, 472–473
instrumentation, 470
light, 464
OCT, 464
path scheme
depth resolved spectral data, 471
fluorescence microscopy, 471–472
nuclear size measurement, 471
T84 cells, nuclear diameter, 471–472
white light beam, 471
phase dispersion, 465
precancerous tissues detection, 479
scatterer structure determination, 464
spatial coherence, white light
interferometry
fringe visibility, 470
Gaussian functions, 470
OCT imaging, 470
Wigner distribution utility, 469
spectral analysis, tissue layers, 463–464
time-frequency distribution (TFD)
bilinear signal representations, 466
interferogram signal, 466
linear operations, 466
SOCT signals, 466
Wigner distribution, total electric
field, 466
transform analysis, 465
Fourier domain mode locking (FDML),
287–288
Fourier-domain optical coherence
tomography (FDOCT)
analytical theory, 263–265
axial range limitations, 265–266
development and history, 258
fiber-optic implementation of, 263
sensitivity advantage, 266
Fourier transform imaging (FTI), 135
Fourier transform imaging spectrometer
(FTIS)
applications, cytology, 137
common-path Sagnac interferometer
optical layout, 136
FTIR advantages, 137
Harvey Fletcher-Holmes interferometer,
Wollastone prisms, 136–137
Fourier transform infrared spectroscopy
(FTIR), 249
Fraunhofer diffraction
annular aperture, 15
circular aperture, 13–14
geometry, 12–13

normalized intensity, 15
rectangular aperture, 13
regimes of diffraction, 11–12
single slit, 13
Free space wavelength, 35
Frequency domain (FD) measurement systems
light propagation and image
reconstruction
angular distribution, light
intensity, 376
data difference, 376
diffusion approximation, 375–376
fitting procedure, 377
iterative procedure, 377
measured and model data, 376
sensitivity matrix, 376
Frequency-domain photon migration
(FDPM), 182–183
Fresnel diffraction
circular aperture, 15–16
circular obstruction, 18
by edge, 18
field in plane calculation, 15
half-plane, 17–18
Kirchhoff diffraction integral, 18–19
rectangular aperture, 16–17
regimes of diffraction, 11–12
single slit, 17
Fresnel half-period zone construction, 16
Fresnel's equation, *see* Reflection and
transmission
FRET, *see* Förster resonant energy transfer
FS, *see* Fluorescence spectroscopy
FTIR, *see* Fourier transform infrared
spectroscopy
Functional optical coherence tomography,
see Optical coherence tomography
(OCT)

G

Gamma rays, 50
Gas lasers, 54
Gastric cardia, 309
Gastroenterology
colon, 312
duodenum, 312–313
esophagus
adenocarcinoma, 308
dysplasia identification, 310–311
normal squamous mucosa and gastric
cardia, 309
second generation OCT imaging,
310–312
SIM, identification of, 309–310
in vivo studies, 308–309
pancreatobiliary ductal system, 313–314
Gegenbauer kernel phase function (GKPF),
73–74
Generalized Rayleigh criterion, 30
Geometrical optics
compound microscope, 9–10
definition, 3, 36

Fermat's principle, 3
magnifying lens, 8–9
matrix method in paraxial optics
    ray tracing through thin converging
        lens, 5
    ray transfer matrix, 3–5
thin converging lens
    depth of field, 8
    depth of focus, 7–8
    imaging, 5–7
    numerical aperture (NA), 7
    resolution, 7–8
GKPF, *see* Gegenbauer kernel phase function
    (GKPF)
Gold nanorods
    concentration of, 178
    normalized scattering spectrum for, 178
    optical properties of, 177
Green function, 19

H

Head and neck cancer, 746
Head/brain tissue, 83–85, 88, 90
Hematoporphyrin derivative (HpD), 736
Hemodynamic imaging
    Doppler OCT
        flow velocity range, 283
        frequency shift (f_D), 282
        validation, flow measurements, 283
        velocity profiles, 283
        volume rendering, angle measurement,
            282–283
        volumetric reconstruction, 283–284
    functional brain activation
        blood content changes, 285
        co-registered OISI and OCT images,
            286–287
    multifunctional Doppler OCT, 288
        maximum velocity *versus* hemoglobin
            saturation, 289–290
        4T1 tumor vasculature, 289–290
    speckle variance OCT
        advantage, 288
        Doppler phase shift, 287–288
        Dorsal skinfold window chamber
            model, 288
        FDML, 287
        intravital confocal microscopy, 288
        maximum intensity projection, 288
        microvascular information, 288
        phase-variance-based techniques, 288
        structural image intensity, 287
    SPFI, 282
    volumetric Doppler, bidirectional flow
        SPFI-SDOCT processing, 284–285
        two-dimensional projection, 285
HeNe laser, 54
Henyey–Greenstein phase function
    (HGPF), 73
HGPF, *see* Henyey–Greenstein phase function
    (HGPF)
Hierarchical cluster analysis (HCA), 117

High-grade dysplasia (HGD), 121
High-pressure arc lamp, 51
Holography
    complex wave front restoration
        convolution, 629
        2D and 3D, 628–629
        Fresnel transform, 629
    cross terms
        digital holography, 625–626
        object virtual image, 625
    digital, time domain data analysis, 626
    OCT, 626
    space domain data analysis, digital
        classical holography, 628
        off-axis geometry, 628
        spatial frequency domain (SFD),
            626–627
        wave-front reconstruction, 627–628
    square terms
        extension over space, 625
        spatial domain, 624
        time domain, 624–625
        "zero-order terms," 624
HpD, *see* Hematoporphyrin derivative (HpD)
Huygens–Fresnel diffraction formula, 12
Huygens' principle, 11–12
Hybrid fluorescence molecular tomography
    APP23 mice, Alzheimer's disease, 455
    fluorescent reporters, 453
    inverse problem, 454
    noncontact measurements, 453
    novel hybrid FMT-XCT system, 454
    nuclear imaging methods, 455
    photon propagation data, 454
    piecewise subdivision-specific
        fluorescence strength, 454
    sub-resolution imaging capacity, 453
Hyperspectral imaging (HSI) systems, 133

I

Image formation
    defocused case, 25
    Fourier transformation by lens, 26
    4*f* system, 26
    imaging by lens, 23–24
    satisfied lens law, 24
3D Image reconstruction, 362–364
Imaging spectroscopy, 228
Impulse response, 22
Incandescent lamp, 51
Incoherent imaging
    object intensity, 30
    optical transfer function (OTP), 30–31
    partially coherent, 29
    two-point object, 30
Infrared light, 50
Infrared spectroscopy, 238–239
Integrating sphere technique, 69–70
Intensified CCD cameras, 60–61
Intensity, 36
Inverse adding-doubling (IAD) method,
    71–72

Inverse Monte Carlo method, 72–74
Irradiance, 36
Irradiation guidelines
    at the cellular level
        photobleaching, 62–63
        photodamage, 63
        phototoxicity, 63–64
    safety practices, 64
    at tissue level
        photochemical damage, 61
        thermal damage, 62
        thermoacoustic damage, 62
        UV-induced risks, 61–62

K

Kaczmarz method, *see* Algebraic
    reconstruction technique (ART)
Keck 3D fusion microscope
    brightfield images, 531–532
    modalities, 533
    OQM, 532
    pinhole, 532–533
    two-photon images, 533
Keck 3D fusion microscope (OQM), 532
Kerr effect, 256
K-fold cross validation (CV), 158
Kirchhoff boundary conditions, 13, 19
Kirchhoff diffraction integral, 18–19
Krylov methods, 346–347
Kubelka–Munk (KM) and multi-flux
    approach, 70

L

Laminar optical tomography (LOT)
    combined OCT and line-scan LOT
        data acquisition, 369
        2D detector array, 369
        multimodality approach, 369
    contrast, body, 359–360
    depth-sensitive measurements, 370
    diffusion-approximation based
        models, 370
    3D image reconstruction, 362–364
    DOT-like imaging geometries, 369–370
    fiber-based systems, 370
    fluorescence
        detector arm configuration, 366–367
        electrical propagations, heart, 366
        image reconstruction, 366
        linear PMT array, 366
        phantom measurements, 366–367
        scanning point source, 367
    functional imaging, 370
    geometry determination, 364
        homogenous properties, 364
        least-square fitting, 364
        modeling types, 364–365
        simplified analysis, 364
    instrumentation, 361–362
    mesoscopic imaging techniques, 359
    modeling, 362

Monte Carlo modeling, 365–366
principles
    image resolution, 361
    light emission, 360
    source-detector (S-D) separations, 361
simultaneous multiwavelength
    absorption spectrum exploitation, 367
    diffraction grating, 368
    phantom imaging, 368
    polarization-maintaining optical fiber,
      368–369
Laser, 36
Laser Doppler flowmetry, 196–197
Laser scanning FLIM
frequency-domain fluorescence lifetime
    measurement
    calculation, 594
    "lock-in" amplifier, 594
    sinusoidally modulated excitation
      sources, 593–594
high-speed
    multiple excitation beams, 594–595
    multiple scanning beam
      excitation, 595
photon binning, 593
TCSPC (*see* Time-correlated single-
    photon counting)
Laser sources
components
    excitation mechanism, 53
    gain medium, 52–53
    resonant cavity, 53
properties of laser
    coherence, 52
    directionality, 52
    monochromaticity, 52
    short pulse duration, 52
pulsed laser operation
    modelocking, 54
    Q-switching, 53
types
    dye lasers, 55
    gas lasers, 54
    semiconductor lasers, 55
    solid-state lasers, 54–55
Laser speckle contrast analysis
    (LASCA), 197
LCI, *see* Low coherence interferometry
Leave one out cross validation
    (LOOCV), 158
LED, *see* Light-emitting diodes (LED)
Lenses
concave lenses, 44
focal points, 43
magnification, 44–45
paraxial regime, 43
planoconvex lens, 43
real and virtual images, 44
resolution, 45–46
spherical interface, 42–43
thin lenses, 44
Levenberg–Marquardt method, 348–349
Light, 33–34

Light delivery devices, 740–741
Light detectors
arrayed detectors, charge coupled device
    (CCD)
    APD and PMT arrays, 60
    characteristics and applications, 58–59
    complementary metal-oxide-
      semiconductor (CMOS)
      detectors, 60
    intensified CCD cameras, 60–61
    operational principle, 58
    spectrometers, 59–60
basic function, 55
photodiodes
    avalanche photodiodes (APDs), 56–57
    characteristics and applications, 56
    operational principle, 55–56
photomultiplier tubes
    characteristics and applications,
      57–58
    microchannel plates (MCPs), 58
    operational principle, 57
Light-emitting diodes (LED), 51
Light scattering spectroscopy (LSS)
absorption mechanism
    absorption coefficient, 166
    Beer's law, 166
    biochemical chromophore, 166
biomedical application, 171–172
cells and subcellular organelles, analytical
    approximations for
    Maxwell's wave equation, 168
    Rayleigh–Gans approximation, 169
    refractive index, 169
    scattering matrix, 168
cells and subcellular structures
    angular dependence, 168
    dimensional scales, hierarchy of, 168
    FDTD simulations, 168
    forward and near-forward scattering,
      167–168
CLASS microscopy (*see* Confocal
    light absorption and scattering
    spectroscopic microscopy)
coherence-gating method, 165
diffuse reflectance technique, 165
EPSS, cancer imaging and lesions (*see*
    Endoscopic polarized scanning
    spectroscopy (EPSS))
PLSS technique (*see* Polarized light
    scattering spectroscopy (PLSS))
polarization
    electric vector, 167
    scattering matrix, 167
    spherical scatterer, 167
principles
    backscattering component, 170
    discreet nuclear size distribution, 170
    epithelial nuclei, 169–170
    non-negativity constraints
      algorithm, 170
    reflectance spectrum, 170
    single scattering, 169–170

scattering mechanism
    elastic and inelastic, 166
    fluorescence, 166
    phosphorescence, 166
    Raman scattering, 166
Light sources
electromagnetic spectrum, 49–50
infrared, visible, and ultraviolet light, 50
laser sources (*see also* Laser sources)
    components, 52–53
    properties of laser, 51–52
    pulsed laser operation, 53–54
    types, 54–55
man-made light sources, 51
microwaves and radio waves, 50
nonlaser sources
    high-pressure arc lamp, 51
    incandescent lamp, 51
    light-emitting diodes (LED), 51
    low-pressure vapor lamps, 51
sun, 50–51
x-rays and gamma rays, 50
Light transport, inverse models
direct reconstruction method, 352
methods
    direct and semi-analytic inverse
      methods, 343–345
    forward models, 341–343
    numerical inversion, 345–350
    priors, 350–351
optical tomography, problems (*see* Optical
    tomography)
reconstruction methods, comparison of,
    351–352
structured light, 352–355
Linearly polarized wave, 35
Linearly variable filters (LVF), 135
Linear mixing model (LMM), 156
Line-scanning, confocal microscopy
bench-top, 534
confocal theta-line scanner, 534–535
field programmable gate array (FPGA)
    logic, 535
full-pupil and divided-pupil, 536
FWHM, 535
line spread function (LSF), 534
nuclear and cellular details, human
    epidermis, 535–536
optical layouts, 535
Liquid crystal tunable filters (LCTF)
drawbacks, 138–139
Fabry–Perot cavities, 139
Lyot filter, 137
"ordinary" and "extraordinary" rays, 137
spectral transmittances, two stage Lyot
    filter, 138
transmitted intensity/transfer
    function, 137
waveplate/polarizer assembly, 138
Liver, 91–92, 94–95
LLLT, *see* Low level laser/light therapy
Longitudinal magnification, 7
LOT, *see* Laminar optical tomography

Low coherence interferometry (LCI)
  biological application, 257
  schematic diagram, 256
  theory, 258–260
Lowgrade dysplasia (LGD), 121
Low level laser/light therapy (LLLT)
  biological effects, cells
    anti-apoptosis, 757
    collagen synthesis, 757
    growth factor, 757
    He-Ne laser illumination, 755
    inflammatory mediators, 757–758
    mediators, 756
    migration and adhesion, 757
    myofibroblasts, 757
    neuronal modulation, 758
    proliferation, 757
    redox couples, 755
    ROS and RNS, 755
    transcription factors, 755–756
  biphasic dose response, 751
  cardiovascular system
    atherosclerotic disease, 762
    laser irradiation, blood, 762
    mitigation, heart attack, 762
  cytochrome *c* oxidase (Cox)
    hypoxic signaling, 753
    infrared irradiation, 753
    nitric oxide, 753–754
  dentistry
    mucositis, 763
    pain relief, 763
    surgery, 763
    temporomandibular joint (TMJ)
      disorder, 763
    usage, 762
  dermatology
    acne therapy, 758
    hair regrowth, 758
    wrinkle reduction, 758
  energy and power density, 751–752
  laser acupuncture
    definition, 763
    visual cortex activation, 764
  laser biostimulation, 751
  laser speckle effect, mitochondria
    illuminated tissue, 754
    microscopic intensity gradients, 755
    temperature gradients, 755
  mitochondria and cellular respiratory
      chain
    cellular power plants, 752
    complexes, integral membrane
      proteins, 752–753
  musculoskeletal disorders
    arthritis, 759
    carpal tunnel syndrome, 760
    dysfunction, 759
    pain relief, 760
  nervous system
    neurodegenerative disease, 760–761
    peripheral nervous system (PNS), 761
    spinal cord repair, 761
      stroke, 760
      traumatic brain injury (TBI), 760
    nitric oxide release
      light mediated, 754
      lymphatic drainage, 753
    pain relief
      neuronal activity, alterations, 761
      randomized controlled trials
        (RCTs), 762
    photoacceptors and cellular
        chromophores
      absorption, red and NIR light, 752
      photobiology law, 752
    tinnitus, 764
    tissue photobiology and optics
      absorption spectra,
        chromophores, 752
      optical window, 752
    wound healing
      animal models, 758
      cytokines upregulation, 759
      increased macrophage activity, 759
Low-pressure vapor lamps, 51
LSS, *see* Light scattering spectroscopy
Lung tissue, 95

M

Magnification, 44–45
Magnifying lens, 8–9
Man-made light sources, 51
Maxima and minima, 39
Maximum likelihood expectation
      maximization (MLEM), 348
Maxwell's equations, 33
MCP, *see* Microchannel plates (MCP)
Meso-tetra hydroxyphenol chlorin
      (mTHPC), 736
Metal–insulator–metal (MIM) resonator, 147
Michelson interferometer, 258
Microchannel plates (MCP), 58
Microlens arrays, SE-SI systems, 142
Microvascular blood flow (BF) assessment, *see*
      Near-infrared diffuse-correlation
      spectroscopy (NIRS-DCS)
Microwaves, 50
Mid-infrared (MIR) spectroscopy, 248–249
Mie theory, 167–168
Modelocking, 54
Molecular imaging (MI), 281; *see also*
      Nanoparticle-based contrast
Molecular reporter systems, optical imaging
  affinity-based molecular probes
    agents, imaging, 681–684
    imaging strategies, 680–681
    passive and active targeting, 679–680
  exogenous molecular probes and
      nanomaterials
    enhanced imaging contrast, 677
    fluorescent reporter, 677
    indocyanine green (ICG), 677
    inorganic materials, 679
    pharmacological considerations, 677
      quantum dots (Qdots), 678
      synthetic organic dyes, 678–679
    fluorescence
      description, 674
      FRET, 674–675
    genetically encoded optical reporters
      bioluminescent proteins, 675
      fluorescent proteins, 676
      limitation, 676
      luciferases, 676–677
    hybrid molecular imaging agents
      fluorescent reporters, 684
      homoFRET, 685
      intensity enhancement, 685
      polymeric activatable probes, 685
      protease-activatable probes, 684
    multimodal
      additive/synergistic diagnostic
        information, 685
      agents, 686–688
      technologies and method, 686
    nonspecific molecular imaging agents
      cyanine-based dyes, 679
      ICG, 679
      porphyrins, 679
Momentum and energy, 35–36
MOMIAs, *see* Monomolecular multimodality
      imaging agents
Monochromator, 39
Monomolecular multimodality imaging
      agents (MOMIAs), 686
Monte Carlo modeling
  fluorescence spectroscopy, 227–228
  laminar optical tomography (LOT),
      365–366
MSIR, *see* Multispectral image reconstruction
mTHPC, *see* Meso-tetra hydroxyphenol
      chlorin (mTHPC)
Multi/hyper-spectral imaging
  camera hardware configurations and
      calibration, 133–135
  vs. color, 132–133
  data classification, unmixing and
      visualization
    benign and malignant, 148
    confusion matrix, 156
    cube, 148–149
    distribution strategies, 158
    individual class accuracies, 157
    K-means algorithm, 151–152
    overfitting, 157
    preprocessing, 150–151
    superviesd methods, 152–155
    taxonomy, 158–159
    thematic map, 149–150
    training set, 149
    unsuperviesd clustering methods,
      151–152
  scanning systems, electronically tunable
      filters
    AOTF, 139–140
    discrete/spatially variable filters,
      135–136

EMTF and LVF, 135
FTIS, 136–137
imaging wavelength tuning, 135
LCTF, 137–139
and single exposure (SE) systems
advantages, 141
coded image capturing, 142–146
image replication, filtering, and
projection optics, 141–142
surface plasmons and nanodevices,
146–148
tunable filter, 140
unmixing algorithms, 151, 155–156
Multi/hyper-spectral imaging devices, 107
Multimodal molecular imaging
additive/synergistic diagnostic
information, 685
agents
MOMIA approach, 686
nanomaterials, 686–688
technologies and method
co-registration, 686
MRI, 686
Multiple retarder/Wollastone prism
stages, 142
Multiple scattering limit (DWS), 198
Multispectral diffuse optical tomography
(MSDOT), 340–341
Multispectral image reconstruction (MSIR)
data difference, 380
iterative procedure, 379
minimization functional, 378
regularization parameter, 380
scatter power, 379
SDIR, 379
spectral fitting, 380
spectral Jacobian, 379
Multispectral imaging (MSI) devices, 133
Multispectral opto-acoustic tomography
(MSOT)
components, fluorochrome biodistribution, 447
fluorochromes
distribution, molecular probe, 448
fluorescent dyes, 447
high-resolution imaging, FP, 447
limits, penetration, 449
reporter molecules detection,
tissues, 447
three-dimensional *in vivo*
imaging, 448
hemoglobin, detection, 446
limitation, 455
multiwavelength illumination, 446
operation principle, 446
spectral processing, 447
technology components
back-projection algorithms, 450
bio-distribution, fluorochrome/
biomarker, 449
integrated pressure wave, 449
intense pulsed laser sources, 449
light attenuation correction, 450

model-based inverse methods, 450
optical absorption coefficient, 450
parameters, tissue, 449
reconstruction, 449
unprocessed single wavelength
images, 447
Multivariate calibration techniques, 242–243
Muscle, 92, 95
Myocardium tissue, 94–95

N

Nanoparticle-based contrast
absorption properties, 290
backscattering albedo, 295–296
backscattering-based
measured OCT intensity gain, 291
OCT images after PBS injection,
291–292
phosphate-buffered saline (PBS)
injection, 291–292
quantification of image intensity, 292
resonant character, enhancement,
292–293
coherent detection, 289
improved delivery, in vivo imaging,
296–297
magnetomotive OCT, 296
molecular OCT techniques, 289
photothermal OCT
antibody-conjugated
nanospheres, 295
nanoparticle concentration *vs.*
photothermal signal, 294
pump power *vs.* photothermal
signal, 294
schematic setup, 293
plasmon resonance, 289
Nanoparticles
cancer and atherosclerosis, 697
confocal reflectance microscopy
cervical biopsies, gold nanoparticles
incubation, 708
EGFR, 706
live cell dark-field microscopy, 707
core materials, 698
diagnosis and therapy, combined
doxorubicin, 712, 716
hybrid plasmonic magnetic
nanoparticles, 715
Q-dot-based system, 712–713
quantification, 712
fluorescence imaging
in vivo imaging, mouse xenograft
tumor, 713
multiphoton oeuorescence microscopy,
708–709
Q-dots, 708–709
sensitivity and multicolor capability,
QD imaging, 712
sentinel lymph node mapping, 708
surgical resectioning, 711
two-photon imaging, 713

gold films and colloids, 698
live cell imaging, 698
optical coherence tomography (OCT)
contrast enhancement, 707
gold nanoshells, 708
normal skin and muscle tissue area,
mice, 709
phase-sensitive, 708
photothermal, 710
optical properties
bulk materials, 699
carbon nanotubes, 703–704
computational electrodynamics,
699–700
dielectric and organic materials,
702–703
plasmonic nanoparticles, 700–702
semiconductors, 702
optical spectroscopic and imaging
modalities, 697
photoacoustic imaging
carbon nanotubes, 711
description, 709
detection, SWNT, 714
gold nanospheres, 709
tissue models, 711
tissue phantoms, 714
photothermal therapy
in vivo demonstration, 712
NIR light, 711
tumors, nanoshells, 715
plasmonic, 697
research, 713–716
semiconductor materials, 698
specificity and stealth, surface properties
amine/carboxylic acid moieties, 705
chemical conjugation, 704
covalent conjugation, 705–706
directional conjugation, 706
physical adsorption, 705
routes, conjugation, 704
stability requirements, 704
tissue optical window, 698–699
Narrow band imaging (NBI), 121
Nd:YAG laser, 54
Near-infrared absorption, 238
Near-infrared-diffuse correlation
spectroscopy (NIRS-DCS)
vs. ASL-MRI, human muscle, 204
correlation diffusion equation, DCS
normalized electric field
autocorrelation function, 200
semi-infinite geometry, 199
correlation diffusion equation theory, 197
vs. Doppler ultrasound, premature infant
brain, 203–204
eight-channel DCS instrument, 201–202
features, 196
fiber-optic probes
arrangement and collection tip,
202–203
contact, non-contact and catheter, 202
source and detector, 202

hybrid diffuse optical system, 201–202
in vivo applications
    cancer therapy monitoring, 204–206
    cerebral physiology and disease,
      206–210
    skeletal muscle hemodynamics,
      210–211
LASCA, 197
laser Doppler flowmetry, 196–197
multiple scattering limit, (DWS), 198
multi-tau correlator design, 201
single scattering, 197–198
static and dynamic method, 196
validation research, 203–204
Neoadjuvant chemotherapy (NAC), DOSI,
    187–188
Nervous system, LLLT
    neurodegenerative disease
      description, 760
      transcranial light therapy, 760
    peripheral nervous system (PNS)
      neuron types, 761
      surgical repair, 761
    spinal cord repair, 761
    stroke, 760
    traumatic brain injury (TBI), 760
Nipkow disc FLIM microscope, 600
Nonlaser light sources, *see* Light sources
Nonlaser sources
    high-pressure arc lamp, 51
    incandescent lamp, 51
    light-emitting diodes (LED), 51
    low-pressure vapor lamps, 51
Nonlinear conjugate gradient method
    (NCG), 349
Nonlinear Kaczmarz method, 349
Nonlinear nanoprocessing
    C. elegans, 574
    material removal
      biomedicine, 573
      biotechnology, 573–574
      3D tissue culture, 573
      femtosecond laser pulses, 573
      neurobiology, 574
    neural transmitters, 573
    structure fabrication
      biomedical research, 575
      lithographic, 574
      photopolymerization, 574–575
      three-dimensional (3D), 574–575
Nonlinear optical microscopy
    advantages, 562
    biomedical, 561–562
    biomedical imaging modalities
      features, 563
      fluorescence spectroscopic
        measurement, 563
      penetration depth, 562–563
    damage mechanisms
      femtosecond laser pulses, 567
      multiphoton microscopy, 567
      one-and multiphoton
        absorption, 567

oxidative photodamage, 567
    specimen photodamage, 567
endogenous and exogenous contrast
    agents
      extracellular matrix, 576
      fluorescent cellular marker, 576
      fluorescent proteins, 576–577
      high intensity radiation, 577
      lipid and lipid-rich structures, 576
      NAD(P)H fluorescence, 576
      two-photon fluorescence excitation,
        575–576
endomicroscopy
    Barrett's esophagus, 577–578
    cervical cancer, 577
    chromatic aberration, 579–580
    confocal systems, 578
    2D and 3D MEMS scanning
      mirrors, 579
    dermal lesion diagnosis, 577
    designing challenges, 578–579
    multiple axis scanning
      mechanisms, 579
    OCT, 578
    Pap smear, 577
    photonic bandgap crystal
      technology, 579
excitation and 3D imaging theories
    microscopy image formation theory,
      564–565
    nonlinear optical spectroscopy,
      563–564
imaging depth and resolution
    PALM and STORM, 570–571
    penetration, 568–569
    STED, 570
    three-photon fluorescence excitation,
      569–570
    tissue, 571
instruments
    excitation efficiency, 565
    laser sources, 565
    signal emission, 567
    telecentric microscope, 566–567
    two-photon absorption efficiency, 565
    two-photon fluorescence microscope,
      565–566
nanoprocessing
    material removal, 573–574
    structure fabrication, 574–575
nonbiology, 562
spectroscopic contrast mechanisms
    Raman contrast, 572–573
    "silent" chromophores, 571–572
    three-dimensional (3D) resolution, 562
Normal esophageal squamous mucosa, 309
N-slit interference, 38–39
Numerical aperture (NA), 7, 22
Numerical inversion methods, light transport
    data fit functionals, 345
    linear methods
      Krylov methods, 346–347
      Landweber method, 346

multiplicative ART and MLEM, 348
    simultaneous algebraic reconstruction
      technique, 347
    simultaneous iterative reconstruction
      technique, 347–348
    nonlinear methods
      Broyden–Fletcher–Goldfarb–Shanno
        algorithm, 349
      conjugate gradient method, 349
      damped Gauss–Newton method, 348
      Gauss–Newton method, 348
      ICD method, 350
      Kaczmarz method, 349
      Levenberg–Marquardt method,
        348–349
      PDE constrained method, 350
    regularized weighted least squares
      functional, 345
Numerical methods, RTE
    Delta-Eddington
      Legendre polynomials, 325
      scattering, 324
    finite-difference
      irregular geometries, 324
      regular geometries, 322–324
    finite-element even-parity, 327
    finite-volume discrete-ordinates
      linear equations, 326–327
      spatial differencing scheme, 326
      triangular elements and node-centered
        control volumes, 325

O

Objective of lens, 9
OCT, *see* Optical coherence tomography (OCT)
Ocular tissues, 80–82, 88
Oncology, 269
Ophthalmology, 268–269, 303–305
Optical and opto-acoustic molecular
    tomography
    animal imaging, MSOT
      components, fluorochrome bio-
        distribution, 447
      fluorochromes, 447–449
      hemoglobin, detection, 446
      multiwavelength illumination, 446
      operation principle, 446
      spectral processing, 447
      technology components, 449–450
      unprocessed single wavelength
        images, 447
    fluorescence, 443–444
    fluorescent protein (FPs), 443
    fluorochromes, NIR, 444
    probes, fluorescent, 443
    reporter technologies
      direct fluorescence imaging, 444–445
      indirect fluorescence imaging, 445
      photonic, 445
    small animal hybrid optical tomography
      Born/Rytov approximation, 451
      fluorescence, distribution, 451

forward problem, calculation, 452
hybrid fluorescence molecular
    tomography, 453–454
inversion, 452–453
measurement vector, 452
modulation frequency, 451
operation principle, 451
photon propagation, 451
priors, usage, 453
in vivo imaging, cell monolayer, 444
Optical axis, 3
Optical coherence tomography (OCT)
in biology, 270–271
breast cancer, 314–315
cardiology, 269
    atherosclerotic plaque, 304–307
    coronary stents, 307–308
    noncoronary applications, 308
commercial devices, 271
confocal microscopy, 303
delivery devices, 267–268
dentistry, 270
development and historical perspective,
    255–258
disease diagnosis, 268
endoscopic and catheter-based probes, 303
Fourier-domain, 258
gastroenterology
    colon, 312
    duodenum, 312–313
    esophagus, 308–312
    pancreatobiliary ductal system,
        313–314
hemodynamic imaging
    bidirectional flow, 284–285
    Doppler OCT, 282–284
    functional brain activation, 285–287
    multi Doppler OCT, 288–289
    speckle variance, 287–288
nanoparticle-based contrast
    backscattering albedo, 295–296
    backscattering-based nanoparticle
        contrast, 290–293
    magnetomotive OCT, 296
    photothermal OCT, 293–295
    in vivo imaging, 296–297
oncology, 269
ophthalmology, 268–269, 303–305
optical biopsy, 303
optical ranging, biological tissues
    Kerr shutter, 256
    LCI, 256–257
principle, 255
pulmonology, 315
quantitative imaging and 3D mapping,
    269–270
robust and reliable products, 315
spatial resolution in, 262
theory
    LCI, 258–260
    multiple scatterers signal, 260–261
    spatial resolution, 262
    spectral radar, 262–266

time domain OCT, 261–262
    wavelength tuning, 266–267
time-domain, 257
Optical detection system (ODS)
cervical tissue, 107
components, 107
uses, 113
Optical imaging
brain
    DOT, 384
    functional activation, 384
    functional MRI (fMRI), 384
    spatial correlation, 385
breast
    accuracy, 384
    clinical viability, 383
    disease contrast, 383–384
    edge-correction method, 383
    model-based approach, 383
    tissue scatter properties, 383
    transillumination light scanning,
        382–383
finger joints
    CW imaging system, 385
    DOT, 386
    reconstruction algorithm, 385–386
    rheumatoid arthritis (RA), 385
    single-wavelength systems, 386
    synovial fluid, 385
prostate
    dosimetry studies, 386–387
    endoscopy system, 387
    photodynamic therapy (PDT), 386
Optical mammography
detection rates, 410
limitation, 410
tumor detection, 408–409
Optical properties, nanoparticles
bulk materials
    constitutive relation, 699
    groups, 699
    semiconductor materials, 699
carbon nanotubes
    complex permittivity function,
        703–704
    infrared absorption peaks, 703
computational electrodynamics,
    699–700
dielectric and organic materials
    C-dots, 703
    Sellmeier formula, 702
plasmonic nanoparticles
    Claussius–Mossoti relation, 700
    gold-coated silica nanoparticles, 700
    gold nanocages and nanorods, 702
    plasmon–polaritons, 700
    sensitivity, nanometer-scale
        morphology, 700–701
semiconductors
    dielectric, 702
    quantum dots (Q-dots), 702
Optical tomographic imaging, *see* Diffuse
    optical tomography (DOT)

Optical tomography
inverse problem
    diffuse light, 339
    DOT, 339–340
    FDOT, 340
    forward mappings and derivatives,
        337–338
    multispectral DOT, 340–341
    radiative transport, 338–339
    structured illumination methods, 341
Optical window, 674–675
Optoacoustic tomography, *see* Photoacoustic
    tomography

P

Pancreatobiliary ductal system, 313–314
Paraxial optics
ray tracing through thin converging
    lens, 5
ray transfer matrix, 3–5
Paraxial rays, 3
Partial least squares (PLS), 243
PAT, *see* Photoacoustic tomography
PDT, *see* Photodynamic therapy (PDT)
Phase imaging microscopy
comparison
    dark-field microscopy *vs.* bright-field
        microscopy, 490–491
    microscopy modes, 494
    phase contrast microscopy *vs.* dark-
        field and bright-field microscopy,
        493–494
    phase contrast microscopy *vs.* DIC
        microscopy, 491–493
contrast generation, 484
dark-field microscopy
    cardioid condenser lens, 485
    central dark-field microscopy, 485
    contrast generation, 484
    direct illumination light, 485
    Fourier analysis, 485
    limitations, 485–486
    numerical aperture (NA),
        condenser, 485
    phase contrast microscopy, 486
DIC microscopy
    bias retardation, 489
    components, microscope, 489
    living cells, 490
    origin, contrast, 489
    qualitative technique, 488–489
    shear, 489
    Wollaston prism, 489
digital detection, 501
image formation models, 501
live-cell imaging, 501
microscope modes, 483–484
modes, comparison, 485
optical path-length (OPL) function, 484
phase contrast microscopy
    condenser annulus, 486
    configuration, 487

living cells, 488
optical train, 487
positive and negative phase contrast systems, 488
quarter wavelength plate, 486, 488
Zernike phase-contrast microscope, 486
phase function, 484
quantitative imaging
dark-field microscopy imaging model, 499
DIC imaging models, 499–500
2D phase, 501–506
image formation, 496–499
intensity, microscope images, 496
models, coherent imaging, 496
phase contrast microscopy imaging models, 499
system enhancements
apodized phase contrast microscopy, 495–496
configurations, 494
DIC microscopy, 496
phase contrast microscopy, high numerical aperture, 496
video-enhanced DIC, 495–496
three-dimensional phase imaging
beam-tilting phase-shifting interferometry, 507
Born approximation, 506–507
digital holographic microscopy (DHM), 508–509
Mach–Zehnder interferometric-based tomographic phase microscopy, 509
optical diffraction RI tomography, 508
optical tweezers–based rotation phase tomography, 509
phase contrast images, 506
quantitative phase amplitude tomography, 507–508
refractive index, 506
rotational oblique illumination computed tomography, 507
sample mounting, 507
synthetic aperture tomographic phase microscopy, 509–510
transparent specimens, 484
Phase screen diffraction, 20
Phosphate-buffered saline (PBS) injection, 291–292
Photoacoustic tomography (PAT)
acoustic lens system
advantages, 425–426
frequency-domain version, 426
acoustic-resolution and optical-resolution PA microscopy
bright-field confocal photoacoustic microscope, 425
dark-field PAM, 425
lateral resolution, 425
reciprocity principle, 424
reflection mode detection, 425

comparison, modalities
imaging contrast, 428
imaging depth and spatial resolution, 428
safety, 429
temporal resolution, 428–429
contrast agents
dyes, 427
endogenous EM absorbers, 426
injection, nanoshells, 427–428
nanoparticles, 427
description, 419, 429
groups, 422
mechanism, 420
molecular imaging
mechanisms, 427
single-wall carbon nanotubes (SWNT), 427–428
PA computed tomography (PACT)
acoustic detectors, 422
Fabry–Perot (FP) sensor head, 423
optical methods, 423
reconstruction algorithms, 423–424
setup, 422–423
ultrasonic transducer arrays, 423
PA signal generation, tissue
blood optical absorption coefficient, 421
EM pulses, 420
hemoglobin oxygen saturation, 420
molar extinction coefficients, 420
spectral dependence, absorption, 421
radiation safety
maximum permissible exposure (MPE), skin surface, 422
RF heating, 422
resolution, 429
spectroscopy and functional imaging
blood flow, 426
SO_2 distribution, 426
ultrasound-imaging, 420
waveequations
acoustic stress confinement, 421
Grueneisen parameter, 421
heating function, 421
illumination pulse, stress confinement, 422
PA pressure, 421
thermal confinement condition, 421
Photochemistry, 733–734
Photodiodes
avalanche photodiodes (APDs), 56–57
characteristics and applications, 56
operational principle, 55–56
Photodynamic therapy (PDT), 122
dermatological applications, 744–745
dosimetry
delivered irradiance, 741–742
fluorescence-based implicit dosimetry, 743–744
local tissue oxygenation measurement, 743
measured fluence rate, 742

sensitizer concentration measurements, 742–743
singlet oxygen production direct measurement, 744
head and neck cancer, 746
history, 733
human prostate *in situ* monitoring, 206
intraluminal applications, 744
intraoperative, intraperitoneal, and intrapleural PDT, 746
light delivery devices
collimating lenses for surface applications, 740–741
intercavity applicators, 741
murine tumors, 205
oxygen
intracellular and intercellular signaling effects, 739–740
macroscopic models and effects, 738–739
microscopic kinetics, 737–738
tumor vascular and microenvironmental effects, 739
photochemistry and photophysics, 733–734
photosensitizer, 734–737
prostate, 745–746
retinal PDT, 745
Photomultiplier tubes (PMT)
arrays, 60
characteristics and applications, 57–58
microchannel plates (MCPs), 58
operational principle, 57
Photophysics, 733–734
Photosensitizer, 734–737
Phototherapy, *see* Photodynamic therapy (PDT)
Physical optics, 36
Pigment packaging effect, 736–737
Pinhole camera, 143
Plane polarized wave, 35
Plane waves, 34
Planoconvex lens, 43
Plasmon-assisted light harvesting, 147
Plasmonic nanoprobes
detection, infectious diseases
noncomplementary target DNA template, 726
SERS spectra, HIV-1 SERS-MS nanoprobe, 726
silver colloidal nanoparticles, 726
DNA probe detection technology, 724
intense localized fields, 723
multiplex detection, cancer biomarkers
luminescence techniques, 727
MS nanoprobe mixture, 727
SERS-MS technique, 726–727
spectral interval, 728
sentinel, development
"label-free" detection, 724–725
metal nanoparticle, 725
molecular beacons, 725
operating principle, 725

SERS technique, gene diagnostics
  HIV DNA and BRCA1 gene, 724
  "nanowave," 724
  Raman spectroscopy, 724
  "surface plasmons," 723
  trace analysis, solid substrates, 723–724
Plasmon-resonant nanoparticles, 281–282
PLS, *see* Partial least squares
PLSS, *see* Polarized light scattering
    spectroscopy
Point-probe instruments
  ball lens-coupled optical fiber, 104
  endoscopic fiber-optic photograph and
      schematic diagram, 106
  fibers, 105
  illumination and collection fibers distance
      impact, 104, 106
  oblique polarized RS, 106
  optical and multifiber, 104
  tissue fiber-optic common end, 104, 106
Polariton, 146
Polarization, 35
Polarized light camera
  handheld, skin imaging
    image degree, 666
    PAR and PER, 665–666
    photons, 665
  images, 666
  skin cancer margins, 666–667
Polarized light imaging
  algorithms testing, identity, 658
  camera, clinical use
    images, 666
    skin, 665–666
    skin cancer margins, 666–667
  data matrix (DM)
    algorithm, 656
    optical measurement, 655
    "transport," 655
    transported intensity, 655–656
  depolarization, 650–651
  depolarization coefficients, μ^{LP} and μ^{CP}
      (cm^{-1})
    birefringence, 659–661
    optical properties *vs.* transport
        parameters, 658
    retardance and scattering roles,
        658–659
    scattering, 661–662
  dermatoscope, 649
  milky suspension, lipid droplets, 650
  Monte Carlo simulation, 649–650
  Monte Carlo simulation matrix
    calculation, 656–657
    conversion, Mueller matrix, 657
    function IQUV2D( ), 657–658
    Stokes vector, 656
  Mueller matrix
    calculation, 653–654
    definition, 653
    half-wave plate (HWP), 654
    quarter-wave plate (QWP), 654–655
    retarder and a linear polarizer, 653

nomenclature
  photon electric field, 651–652
  polarization states, 652
  retarder, 651
nomenclature and computational
    tools, 650
ophthalmology, 650
optical fiber probe, 651
photons, 649
Poincaré sphere, 653
Stokes vector
  definition, 652
  notations, 652–653
tissues, depolarization, 662–665
Polarized light scattering spectroscopy (PLSS)
  backscattered light, 170–171
  intestinal epithelial cells, 170–171
Poynting vector, 35
Principal components analysis (PCA), 111,
    150–151
Principle of least time, *see* Fermat's principle
Probabilistic photon migration, 225–226
Prostate, 745–746
Pulmonology, 315

Q

QE, *see* Quantum efficiency (QE)
Q-switching, 53
Quantitative 2D phase imaging
  computational DIC
    iterative model-based phase estimation
        algorithms, 504
    orthogonal shear directions, 503–504
    phase shifting, 502–503
    rotationally diverse images, 504
    spiral phase integration (SPI) method,
        502–503
  DIC-based phase imaging methods,
      501–502
  digital holographic phase microscopy, 505
  fast Fourier phase microscopy, 505
  Hilbert phase microscopy, 505
  Jones phase microscopy (JPM), 506
  limitations, 506
  OCT
    reflectivity measurement, 504
    SD-OCT, 505
  phase-dispersion microscopy (PDM), 505
  QSIP, 506
  quantitative phase amplitude (QPA)
      microscopy, 505
  scanning transmission microscopy, 505
  spiral phase contrast microscopy, 505
Quantitative spectroscopic imaging (QSI)
    system, 107
Quantum efficiency (QE), 134, 240

R

Radiative transfer equation (RTE)
  bioluminescence tomography (BLT), 322
  discretized control volume, 325

DOT, 320
gradient-search technique, 322
light propagation, 320
light transport, biological tissue, 321
numerical methods
  Delta-Eddington method, 324–325
  finite-element even-parity
      method, 327
  finite-volume discrete-ordinates
      method, 325–327
  introduction, 322
  irregular geometries, finite-difference
      methods, 324
  regular geometries, finite-difference
      methods, 322–324
time-independent, 320
Radiative transport equation (RTE), 110
Radiative transport theory, 338–339
Radio waves, 50
Raman scattering, 166
Raman spectroscopy
  of bacteria (*Streptococcus*), 234–235
  biofluids, 245
  classical model of Raman shift,
      236–237
  energetics
    energy level spacing, 235–236
    forbidden and virtual state, 236
    Jablonski energy level, 234–235
  equipments
    CCD array detectors, 239–240
    fiber-connected probes, 241–242
    folding mirror, 240
    imaging, 240–241
    lasers, 239
    microscopic volumes, 240
    oblique incidence, 240
  vs. fluorescence, 237–238
  hard tissue, 244–245
  initial states
    energy conservation rule, 235
    energy level diagrams, 235–236
  mathematical processing
    classification, 243
    quantification, 242–243
  microbes, 245
  multimodal systems, 246
  vs. near-infrared absorption, 238
  single cells, 245–246
  soft tissue, 244
  vibrational, 233
  vs. vibrational absorption, 238–239
  wave number units, 234
Random subsampling, 158
Rayleigh–Gans approximation, 169
Rayleigh–Sommerfeld I and II diffraction
    formulae, 19
Ray optics, *see* Geometrical optics
Rays, 3
Ray transfer matrix, 3–5
Real images, 44
Reflectance spectroscopy (RS)
  absorption

endogenous tissue absorbers, 103–104

extinction spectra plot, tissue absorbers, 104–105

applications

tissue diagnostics (*see* Tissue diagnostics, RS)

tissue therapeutics, 122

disease discrimination, normal tissue variations, 123

elastic scattering, 103

instrumentation and analysis procedures, 124

instruments

imaging, 107

point-probe, 104–106

modeling and data analysis

absorption coefficient, 108–110

diffusion theory, 110

light propagation and reflectance signal, tissue, 108

Monte Carlo simulations, 110–111

PCA, 111

RTE, 110

scattering coefficient, 108

tissue scattering and absorption, 108

prospective multisite studies, 124

tools development, 123–124

Reflection and transmission

Brewster's angle, 47

electric field parallel to plane of incidence, 47

electric field perpendicular to plane of incidence, 46–47

Refraction

Fermat's principle

description, 42

minimal time paths and Snell's law, 41–42

Snell's law, 41

total internal reflection, 42

Relative cerebral blood flow (rCBF) tomography, 207

Resolution, 7–8, 40, 45–46

conventional confocal microscopes

PSF, 523

Rayleigh criterion, 523–524

conventional full-field microscopes

coherent point spread function, 522

condenser lens, 522–523

optical coordinates, 522

point spread function (PSF), 522

Rayleigh criterion, 523

saddle-to-peak irradiance ratio, 523

high-NA objectives, 526

measurement

Airy function, 527

diffraction-limited lateral resolution, 527

Fourier optics theory, 527

FWHM, 527–528

FWHM and spot size, 526–527

optical sectioning

description, 524

field depth, 525

irradiance distribution, focused spot, 524–525

thickness measurement, 524

pinhole diameter

function, PSF, 525–526

FWHM, 526

pupil apodization

Gaussian distribution, 525

PSF, 525

Resonant cavity, laser, 53

Retinal nerve fiber layer (RNFL), 303–304

Retinal PDT, 745

RNFL, *see* Retinal nerve fiber layer

RTE, *see* Radiative transfer equation

S

Selective plane illumination microscopy (SPIM), 605

Semi-analytic methods, light transport, 345

Semiconductor lasers, 55

SERRS, *see* Surface-enhanced resonance Raman scattering

SERS, *see* Surface-enhanced Raman scattering

Shannon's theorem (Shannon–Nyquist sampling theorem), 144

Signal-to-noise ratio (SNR)

description, 528–529

light source output power, 529

shot noise, 529

SIM, *see* Specialized intestinal metaplasia (SIM); Structured illumination microscopy (SIM)

Simultaneous algebraic reconstruction technique (SART), 347

Simultaneous iterative reconstruction technique (SIRT), 347–348

Simultaneous multiwavelength LOT, 367–369

Single exposure-spectral imaging (SE-SI) systems, 140–148

Single-pass volumetric bidirectional flow imaging (SPFI), 292, 294–295

Single scattering, 197–198

Single-slit interference, 39–40

Skeletal muscle hemodynamics, 210–212

Skin and subcutaneous tissue, 75–82

Snell's law, 41

Snell's law of refraction, 3

Solid-state lasers, 54–55

Spatial filtering, 29

Spatially modulated illumination (SMI) microscopy, 545–546

Specialized intestinal metaplasia (SIM), 309–310

Specificity and stealth, nanoparticles

amine/carboxylic acid moieties, 705

chemical conjugation, 704

covalent conjugation

carbon nanotubes, 706

gold nanoparticles, 705–706

quantum dots, 706

silica nanoparticles, 706

directional conjugation

antibody, 706–707

carbohydrate moiety, 706

physical adsorption

carbon nanotubes, 705

description, 705

routes, conjugation, 704

stability requirements, 704

Speckle variance OCT

advantage, 288

Doppler phase shift, 287–288

Dorsal skinfold window chamber model, 288

FDML, 287

intravital confocal microscopy, 288

maximum intensity projection, 288

microvascular information, 288

phase-variance-based techniques, 288

structural image intensity, 287

Spectral angle mapper (SAM), 152–153

Spectral correlation angle (SCA), 153

Spectral correlation mapper (SCM), 153

Spectral cube data processing process, 158–159

Spectral encoding confocal microscopy (SECM)

broadband light, 536

cellular resolution, 536–537

non-scanning system, 536

VIPA, 537

Spectral imaging (SI) scanning systems

AOTF, 139–140

discrete/spatially variable filters, 135–136

EMTF and LVF, 135

FTIS, 136–137

imaging wavelength tuning, 135

LCTF, 137–139

Spectral information divergence (SID), 154

Spectral radar, *see* Fourier-domain optical coherence tomography (FDOCT)

Spectral unmixing algorithms

endmembers product and abundances, 156

image area, 155

invasive ductal carcinoma, 156–157

LMM, 156

processes, 156

Spectrometers, 59–60

Spectroscopic contrast mechanisms, nonlinear optical microscopy

Raman contrast

coherent anti-Stokes Raman spectroscopy (CARS), 572–573

molecules, vibronic states, 572

"silent" chromophores, 571–572

fluorophores, 572

pump-probe transient absorption experiment, 571–572

tissue pigments, 571

SPFI, *see* Single-pass volumetric bidirectional flow imaging

Spherical interface, 42–43
Spherical waves, 34
Stimulated emission depletion (STED)
  microscopy, 545
Structured illumination microscopy (SIM)
  image formation
    beat generation, 550–551
    3D fluorescent molecule
      distribution, 550
    Fourier space, 550
    Fourier transform, 549–550
  image reconstruction
    accessible frequency region, 551
    fluorescence molecule
      distribution, 551
    Fourier-transformed, 551
  one objective lens
    description, 546
    excitation intensity, 546–547
  polarization effects
    perpendicular, interferometer plane,
      547–548
    two-beam interference, 547
  resolution improvement
    biological specimen, 556
    human retinal pigment epithelium
      tissue, 556
    lateral, 555–556
  SEI and localization microscopy
    bi-plane detection, 556–557
    3D and 2D SPDM, 557
    optical isolation, time domain, 556
  setup, 546
  setups, 555
  three-beam interference, 549
  two-beam interference
    intensity distribution, 547–549
    Maxwell equations, 548
  wide-field microscopy, image formation
    Fourier space and optical transfer
      function, 553–555
    intensity distribution, 547
    point spread function (PSF), 547,
      552–553
Sun, 50–51
Superposition principle, 35
Superviesed classification methods
  distance similarity measures
    Euclidian distance measure, 153–154
    SAM, 152–153
    SCM/SCA, 153
    SID, 154
  nongeometric similarity measures
    class membership, 154
    discriminant function, 154–155
    Lagrangian, 155
    maximum-margin hyperplane and
      margins, SVM, 154–155
Surface-enhanced Raman scattering (SERS),
  247–248
  DNA probe detection technology, 724
  mechanisms, 723
  nanoprobe, 725

operating principle, molecular sentinel
  nanoprobe, 725
  vs. SERRS, 723
  SERS-based "Molecular Sentinel"
    (SERS-MS), 725
  SERS HIV-1 MS nanoprobes, 726
  trace analysis, solid substrates, 723–724
Surface-enhanced resonance Raman
  scattering (SERRS)
  enhancement, Raman signal, 723
  incoming laser, energy, 723
  vs. SERS, 723
Surface plasma polaritons (SPP), 146
Surface plasmon resonance, 282
Surface plasmons and nanodevices
  antenna effect, 147
  EM resonances, 146
  MIM waveguide, insulator, 147–148
  multiple MIMs hexagonal packing,
    147–148
  periodic structures, 147
  spatial filtering and detection process, 147
  SPP, 146
Swept source OCT, *see* Wavelength tuning
  OCT
Synthetic organic dyes
  in vivo optical imaging studies, 678
  imaging and therapeutic agents, 679
  light-absorbing units, 678
  visible, 678
System matrix, 6

T

TCSPC, *see* Time-correlated single-photon
  counting
Temporal point spread function (TPSF)
  data types, characterization
    absorption and scatter, 405–406
    central moments, 404
    intensity and mean photon flight
      advantages, 404–405
    Laplace transform, 405
    Mellin–Laplace transform, 405
    multiple, 405
    temporal moments, 404
    time-gating, 405
  modeling
    Green's function, 404
    image reconstruction, 404
Therapeutic window, 104
Thin converging lens
  depth of field, 8
  depth of focus, 7–8
  imaging, 5–7
  numerical aperture (NA), 7
  resolution, 7–8
Thin lenses, 21, 44
Thin-lens matrix, 5
Time-correlated single-photon counting
  (TCSPC)
  approaches, 399
  description, 592–593

detectors, 399
drawback, 593
light sources
  pulsed laser diodes, 399
  wavelengths, 398
MONSTIR, 399
operation principle
  advantage, 398
  detection system, 398
  "reversed start-stop" operation, 398
principle, 593
source fiber and detector fiber
  bundles, 399
Time-domain optical coherence tomography
  development and history, 257
  theory, 261–262
Time-gated systems
  devices
    brain imaging, 402
    data analysis, 401
    simultaneous detection, multiple
      temporal gates, 402
    single-photon avalanche diode,
      402–403
    transmission geometry, 401
  gated detectors
    image intensifier, 400–401
    intensified charge-coupled device
      (ICCD) camera, 400–401
    "null" source-detector separation, 401
    photocathode response, 401
  light sources
    signal-to-noise ratio (SNR), 400
    wavelength, 400
  operation principle
    laser pulse emission, 400
    pulsed source and time-resolved
      detection, 399
Ti:sapphire laser, 54–55
Tissue diagnostics, RS
  breast
    ANN and HCA, 117
    boxplots, 114–116
    diffuse reflectance spectra
      representative spectra, pathologies,
        114–115
    hemoglobin and β-carotene, 117
    lumpectomy surgery, 114
    normal and metastatic nodes,
      116–117
    point-probe reflectance studies
      results, 116
    results, multistep algorithms, 114–115
  cervix
    algorithms scheme, 112
    cHb_T, 114
    colposcopy, 111
    dysplasia detection, 113
    point-probe reflectance studies
      results, 112
    representative I1/I3 and I1/I4 spectra,
      112–113
    unbiopsied and biopsied sites, 113

esophagus and gastrointestinal tract,
121–122
lungs, 120–121
oral cavity
detection scheme limitations, 117
keratinized and nonkeratinized
sites, 119
leukoplakia, 118
nonkeratinized tissue average spectra,
diagnosis, 118–119
point-probe reflectance studies
results, 118
spectral features, 119
Tissue optical index (TOI), 185–187, 189
Tissue optical properties
aorta, 93, 95
breast tissue, 88, 94
cartilage, 89–90, 94
direct measurement of the scattering
phase function, 74–75
epithelial/mucous tissue, 86–87, 90, 94
head/brain tissue, 83–85, 88, 90
integrating sphere technique, 69–70
inverse adding-doubling (IAD) method,
71–72
inverse Monte Carlo method, 72–74
Kubelka–Munk (KM) and multi-flux
approach, 70
liver, 91–92, 94–95
lung tissue, 95
measurement, 67–69
muscle, 92, 95
myocardium tissue, 94–95
ocular tissues, 80–82, 88
skin and subcutaneous tissue, 75–82
Tomographic brain imaging
absorption and scatter coefficients, 413
neonatal brain, 413
Tookad sensitizer, 736
Total internal reflection, 42
TPSF, *see* Temporal point spread function
Translation matrix, 4–6

Transscleral cyclophotocoagulation
(TSCPC), 82
Turbid tissue imaging
FWHM, 529–530
intralipid, 529
refractive-index mismatches, 529
Two-slit interference, 37–38

U

Ultrasound-modulated optical tomography
(UOT)
CFPI (*see* Confocal Fabry–Perot
interferometer)
coherent pulsed laser
B-scan image, 438
double-pass configuration, 438
experimental setup, 437
master oscillator power amplifier
(MOPA), 437
description, 439
mechanisms, 431
autocorrelation function, 430
incoherent intensity modulation,
light, 430
modulation depth, 431
optical phase changes, 430
power spectral density, 430–431
power spectrum, 431
spectral intensity, frequency, 431
ultrasound-induced displacements, 430
motivation
diffuse optical tomography
(DOT), 429
low SNR, 429
principle, 429
photorefractive interferometer-based
detection
avalanche photodiode (APD), 431
experimental setup, 432
local oscillator (LO), 431
typical A-lines, 431, 433

spectral-hole-burning-based detection
advantages, 433–434
B-scan images, 436
experimental setup, 434, 436
medium, 434
normalized spectral absorption
profile, 436
transmitted diffusive light, 436
Ultraspectral imaging (USI), 133
Ultraviolet light, 50
Unsupervised clustering methods, *K*-means
algorithm, 151–152
UOT, *see* Ultrasound-modulated optical
tomography

V

Verteporfin, *see* Benzoporphrin derivative
(BPD) sensitizer
Virtual images, 44
Visible light, 50
Volume discretization methods, 341–342
Volumetric Doppler OCT
SPFI-SDOCT processing, 284–285
two-dimensional projection, 285

W

Wavelength tuning OCT, 266–267
Wiener–Khinchin theorem, 259–260

X

X-ray computed tomography (CT)
optical image reconstructions, 363
sensitivity distribution, 359
X-rays, 50

Z

Zernike phase contrast, 29
Zero accommodation, 8

T - #0619 - 071024 - C0 - 276/219/36 - PB - 9780367576943 - Gloss Lamination